BIOLOGY

The Unity and Diversity of Life

15TH EDITION

STARR
TAGGART
EVERS
STARR

CENGAGE

Australia • Brazil • Canada • Mexico • Singapore • United Kingdom • United States

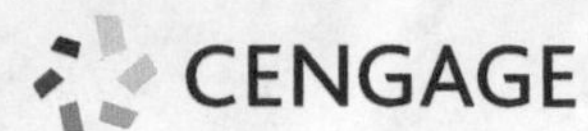

***Biology: The Unity and Diversity of Life*, Fifteenth Edition**

Cecie Starr, Ralph Taggart, Christine Evers, Lisa Starr

Product Director: Dawn Giovanniello

Product Team Manager: Kelsey Churchman

Senior Content Developer: Jake Warde

Product Team Assistant: Vanessa Desiato

Executive Marketing Manager: Tom Ziolkowski

Senior Designer: Helen Bruno

Manufacturing Planner: Karen Hunt

Content Project Manager: Hal Humphrey

Content Digitization Project Manager: Maya Whelan

Project Managers: Michael McGranaghan, Matthew Fox & Phil Scott, SPi Global

Intellectual Property Project Manager: Erika Mugavin

Intellectual Property Analyst: Christine Myaskovsky

Photo Researcher: Cheryl DuBois, Lumina Datamatics

Text Researcher: Rameshkumar P.M., Lumina Datamatics

Illustrators: Lisa Starr, Gary Head, ScEYEnce Studios, SPi Global

Compositor: SPi Global

Text Designer: Liz Harasymczuk

Cover Designer: Helen Bruno

Cover and Title Page Image:
TIM LAMAN/National Geographic Creative

A male satin bowerbird decorates his "bower" by painting its component twigs with berry juice. Building and decorating the bower is an essential aspect of his courtship behavior. Female satin bowerbirds preferentially mate with males who have the best built and decorated bowers.

Library of Congress Control Number: 2017955998

Student Edition ISBN: 978-1-337-40833-2
Loose-leaf Edition ISBN: 978-1-337-40841-7

Cengage
200 Pier 4 Boulevard
Boston, MA 02210
USA

Cengage is a leading provider of customized learning solutions with employees residing in nearly 40 different countries and sales in more than 125 countries around the world. Find your local representative at **www.cengage.com**.

To learn more about Cengage platforms and services, register or access your online learning solution, or purchase materials for your course, visit **www.cengage.com.**

Printed at CLDPC, USA, 08-22

Contents in Brief

Detailed Contents

INTRODUCTION

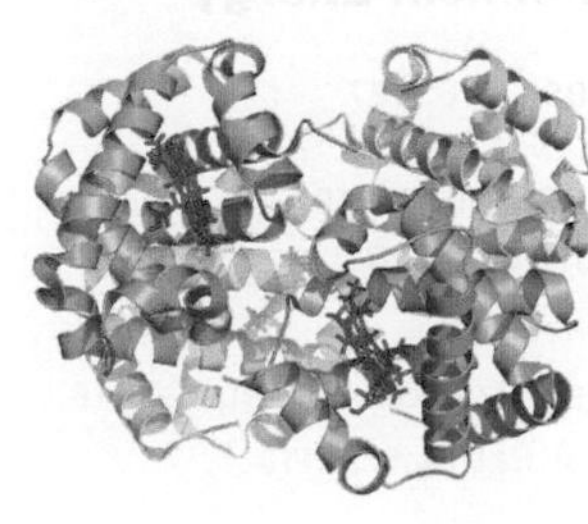

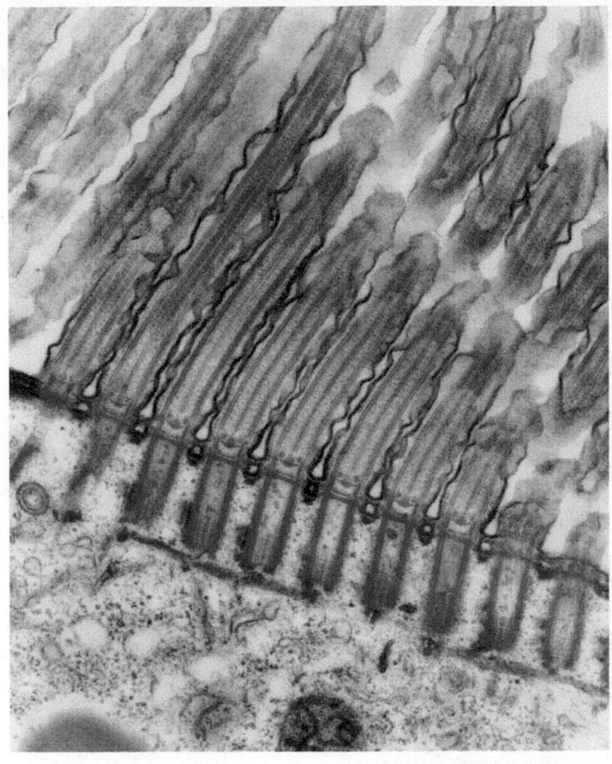

Detailed Contents (continued)

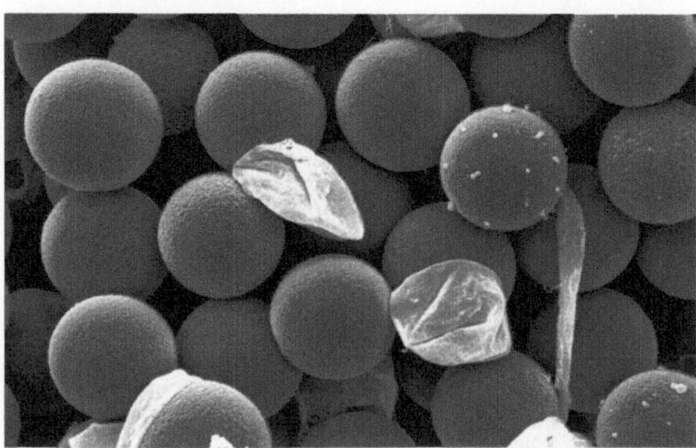

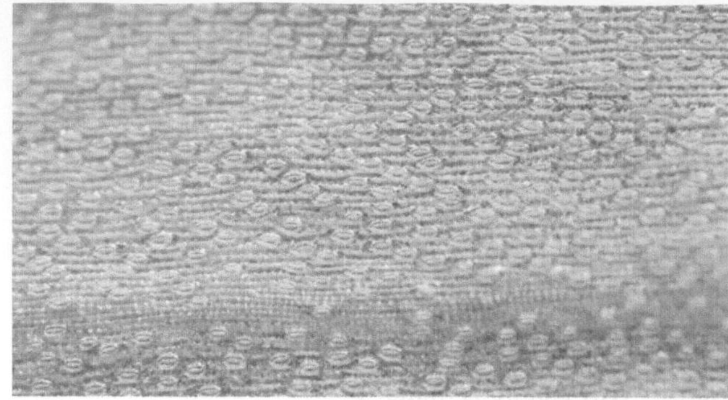

UNIT II GENETICS

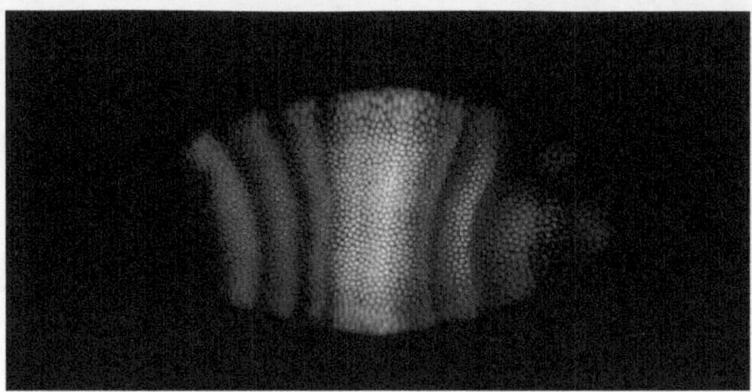

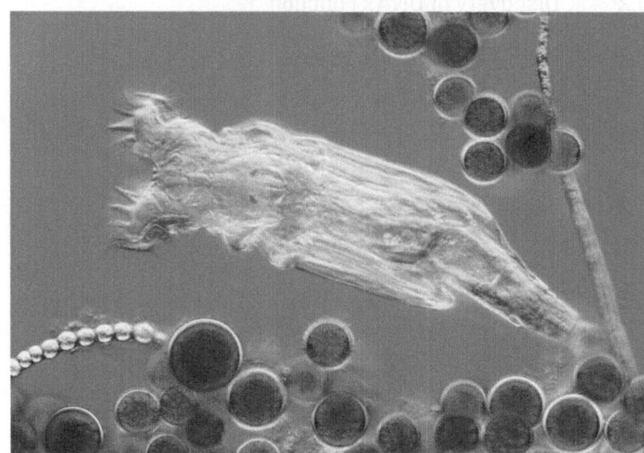

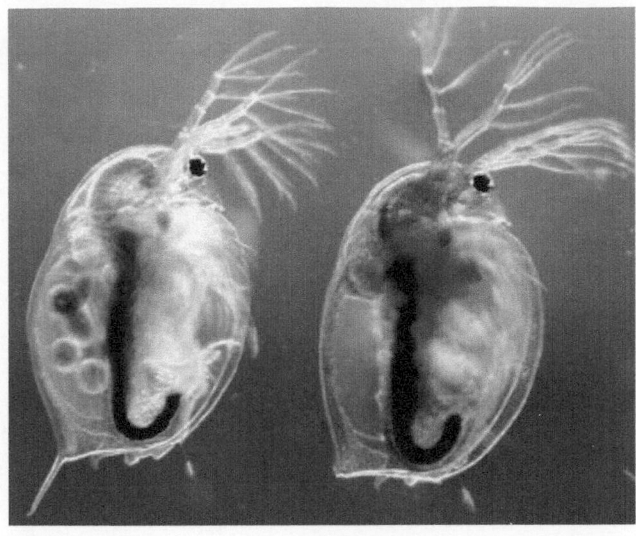

Detailed Contents (continued)

UNIT III PRINCIPLES OF EVOLUTION

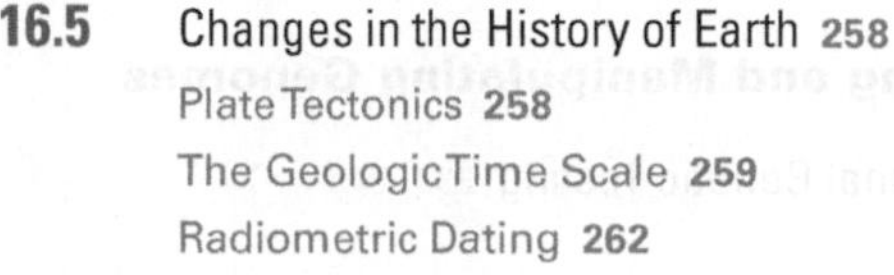

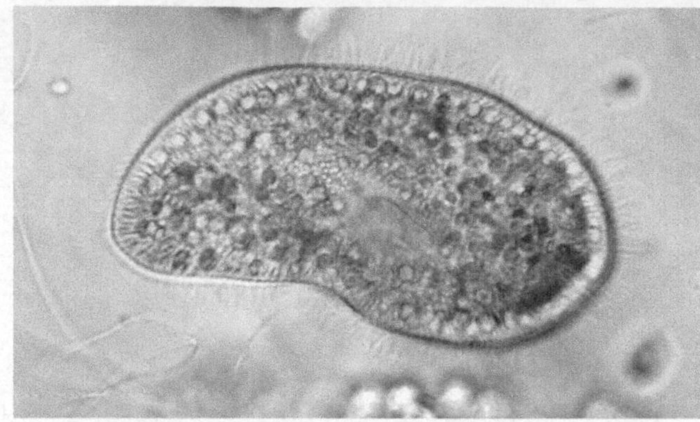

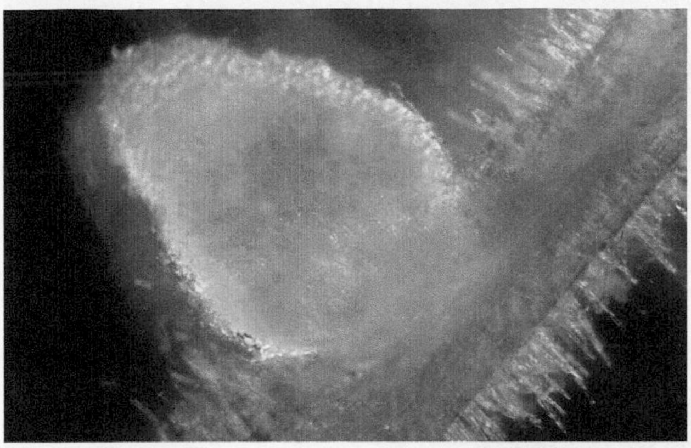

Detailed Contents (continued)

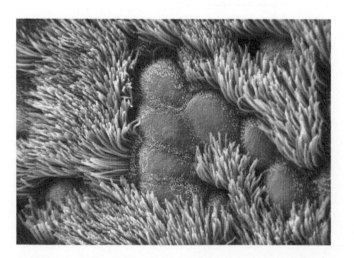

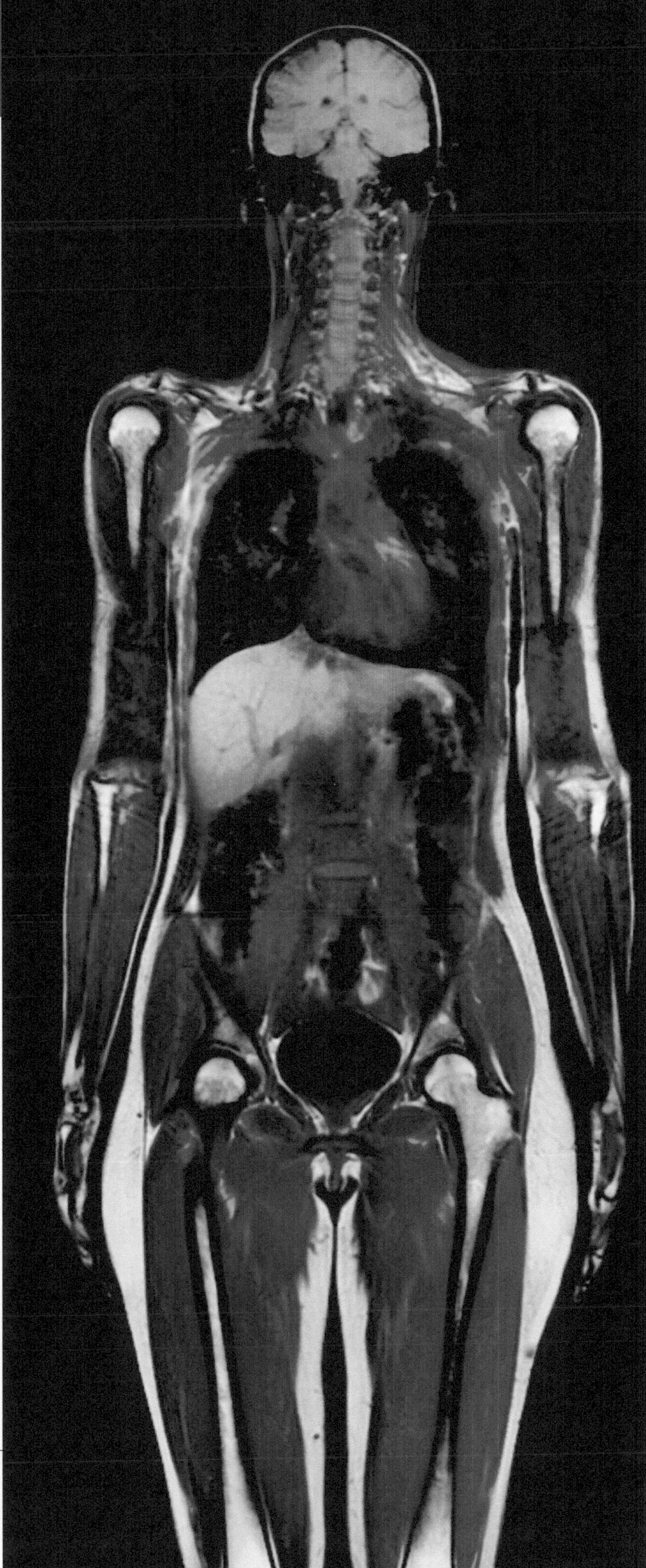

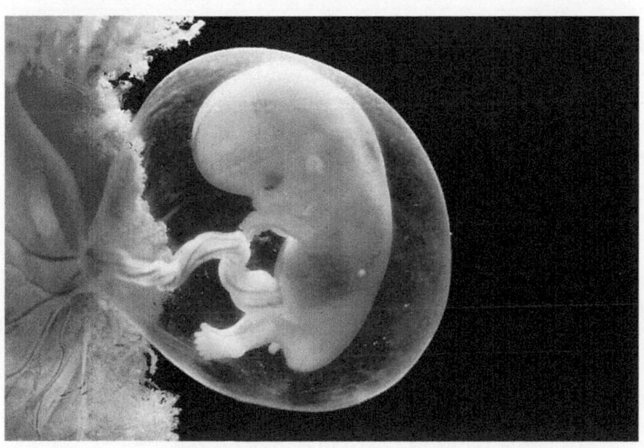

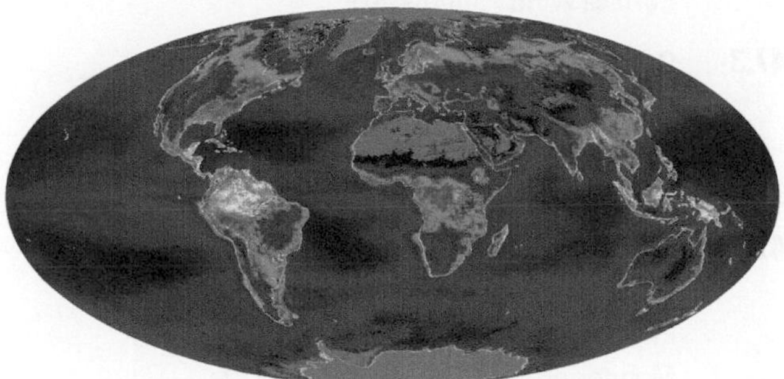

Appendices

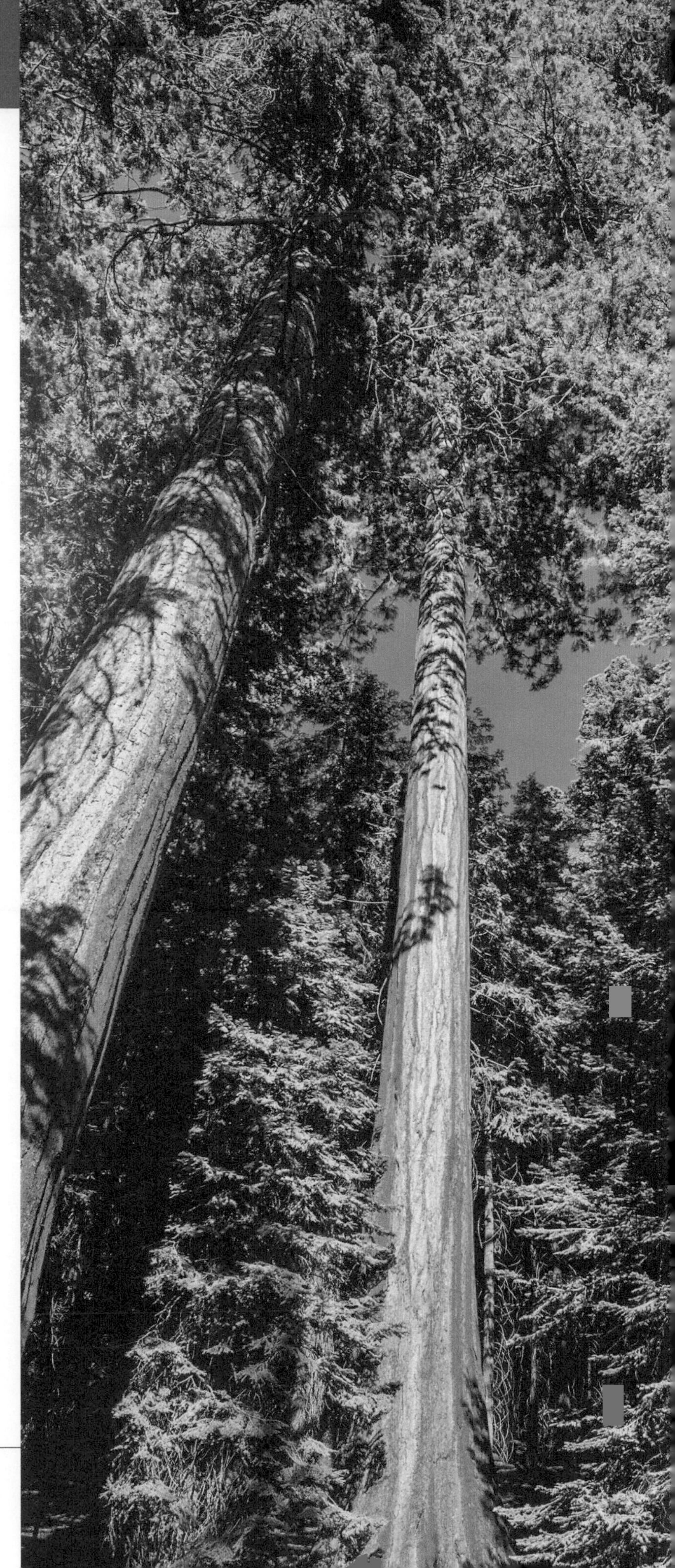

Preface

A revolution in the way information is shared has fundamentally changed the nature of biological inquiry. Interdisciplinary collaborations facilitated by instant, global access to data and ideas have fostered entirely new areas of research, both theoretical and practical. New discoveries and new technologies emerging from these collaborations are altering the way biologists think about their work—and the field in general.

Realizing that a traditional life science education would not adequately prepare students for the changing field, the American Association for the Advancement of Science and the National Science Foundation initiated a series of national conversations among leading life scientists, policymakers, educators, and students. The result was a document, "Vision and Change in Undergraduate Biology Education," that calls for a fundamental change in the way life sciences are taught to undergraduate students. A broad consensus recommends that science education become much more active, because personal experience with the process and limits of science better prepares students to evaluate scientific content and differentiate it from other information. A more concept-oriented approach that uses fundamental biological principles as a context for information (rather than the reverse) better prepares students to understand the rapidly changing field. Our future citizens and leaders will need this understanding to confront urgent societal problems such as climate change, threats to biodiversity, and the global spread of disease.

This book has been revised in alignment with "Vision and Change" recommendations. As always, recent discoveries are integrated in an accessible and appealing introduction to the study of life. This edition also includes tools to explore core biological concepts from a variety of perspectives (molecular, cellular, organismal, ecological, and so on).

Features of This Edition

Setting the Stage

Each chapter opens with a dramatic photo. A brief Links to Earlier Concepts paragraph reminds students of relevant information in previous chapters. A summary of chapter content is organized and presented in terms of Core Concepts: evolution; information flow; systems; pathways of transformation; structure/function; or the process of science.

Section-Based Learning Objectives

The content of every chapter is organized as a series of sections. Learning Objectives associated with each section are phrased as activities that the student should be able to carry out after reading the text.

On-Page Glossary

An On-Page Glossary includes boldface key terms introduced in each section. This section-by-section glossary offers pronunciations, definitions in alternate wording, and it can be used as a quick study aid. All glossary terms also appear in boldface in the Chapter Summary.

Emphasis on Relevance

We continue to focus on real-world applications, including social issues arising from new research and developments—particularly the many ways in which human activities continue to alter the environment and threaten both human health and Earth's biodiversity. Each chapter begins and ends with a section that explains a current topic in light of the chapter content.

Self-Assessment Tools

Many figure captions include a Figure-It-Out question designed to engage students in an active learning process; an upside-down answer allows a quick check of understanding. At the end of each chapter, Self-Quiz and Critical Thinking Questions provide additional self-assessment material. Another active-learning feature, the in-text Data Analysis Activity, sharpens analytical skills by asking the student to interpret data presented in graphic or tabular form. The data is presented with relevant chapter content, and is from a published scientific study in most cases.

Some Updates in This Revision

1 Invitation to Biology Expanded section "The Nature of Science" includes new, detailed coverage of pseudoscience and how it differs from science.

2 Life's Chemical Basis New table compares elemental composition of the human body with Earth's crust, seawater, and the universe. Updated art more clearly demonstrates vacancies. New art illustrates and compares bond polarity.

3 Molecules of Life Revised text further emphasizes levels of protein structure as related to protein function. New art illustrates patterns of secondary structure; new content reflects current research elucidating misfolded prion structure and pathogenesis of amyloid diseases.

4 Cell Structure New, major art depicts interactions among components of the endomembrane system. New art reveals ultrastructural details of cell junctions per recent discoveries. New photos illustrate a beneficial biofilm, cuticle, and basement membrane. Expanded section on the nature of life now includes theory of living systems.

5 Ground Rules of Metabolism New photos and art use firefly luciferase to illustrate energy flow in metabolism. Updated art clarifies selective permeability of cell membranes and tonicity, and directional orientation of membrane proteins during exocytosis. Expanded coverage of ATP as a coenzyme.

6 Where It Starts—Photosynthesis Chapter has been reorganized for a better introductory sequence. New art illustrates a special pair, light-harvesting complex, and cyclic photophosphorylation. Expanded discussion of the cyclic pathway emphasizes the interplay between both versions of light reactions, and the evolutionary significance of dual pathways. Expanded discussion of photorespiration incorporates new research on its adaptive value.

7 Releasing Chemical Energy Art updated throughout. New introductory table and art in each section detail inputs and outputs linking steps in aerobic respiration. Revised diagram better illustrates dual PGAL breakdown in glycolysis. New Data Analysis Activity concerns reprogramming of mitochondria in brown fat by dietary fat overload.

8 DNA Structure and Function Nucleotide structure art revised to better illustrate composition and two-dimensional accuracy. Updated art including model of a pyrimidine dimer clarifies how replication errors become mutations.

9 From DNA to Protein New art illustrates the overall structure of a gene and its relationship to RNA translated from it. New table compares DNA and RNA. Revised art better illustrates post-transcriptional modification; major translation figure updated to clarify elongation and polysomes. Expanded section on mutations includes material on a beneficial hemoglobin mutation (E6K, HbC) that offers resistance to malaria without the health consequences of HbS; and how a mutation in a regulatory site (an intron) can affect gene expression (resulting in hairlessness in cats).

10 Control of Gene Expression Updated art better illustrates points of control in eukaryotic gene expression. New flow chart shows gene expression cascade in *Drosophila* development; new art illustrates random nature of X chromosome inactivation. Added coverage of circadian cycles of gene expression; expanded discussion reflects current understanding of epigenetic mechanisms.

11 How Cells Reproduce New ultra-high resolution confocal live-cell images better illustrate mitosis and the spindle. Cell cycle illustration now correlated with illustration of ploidy changes in mitosis. Revised text and art showing cytokinesis include ultrastructural details/processes per current research and paradigms. Expanded material on telomeres now includes telomere-associated triggering and consequences of senescence.

12 Meiosis and Sexual Reproduction New figure illustrates how fertilization restores the chromosome number. Newly discovered mechanism of gene acquisition by individual rotifers added to revisited section.

13 Observing Patterns in Inherited Traits Marfan syndrome discussion updated to reflect change in life expectancy due to increased awareness, accompanied by new photo of former Baylor University basketball star Isaiah Austin.

14 Chromosomes and Human Inheritance Molecular pathogenesis of Huntington's and DMD updated to reflect current research. Table of genetic abnormalities now broken by section. Updated art better depicts chromosome structural changes.

15 Studying and Manipulating Genomes Art depicting recombinant DNA production and reverse transcription revised for clarity. Art and text revised to include structure and utility of eukaryotic expression vectors. New figure illustrates exponential amplification of DNA by PCR. New table lists human genome statistics. Gene therapy section revised to include mechanism, application, and social implications of CRISPR-Cas9 gene editing.

16 Evidence of Evolution Cetacean evolutionary sequence updated to reflect currently accepted narrative. Current research informed revisions of plate tectonics art. Paleogeography art revised to show Mercatur projections.

17 Processes of Evolution Updated material on antibiotic resistance and overuse of antibiotics in livestock. Illustration of HbS allele frequency vs. incidence of malaria updated to reflect recent data in Gabon. New photo of bumblebee on white sage flower added to mechanical isolation figure. Discussion of sympatric speciation in wheat revised to reflect current research.

18 Organizing Information About Species Phylogeny section and parsimony analysis figure revised for clarity. Convergent and divergent evolution terminology introduced. New photo of stem reptile fossil added to divergent evolution figure. New close-up of saguaro cactus spines juxtaposed with Euphorbia spines for better illustration of convergent evolution.

19 Life's Origin and Early Evolution Expanded coverage of the Precambrian, including timeline. Updated information about the possibility of an RNA world, the earliest proposed fossil life, and the archaeal and bacterial ancestors of eukaryotes. New Data Analysis Activity about the effect of some antibiotics on mitochondria.

20 Viruses, Bacteria, and Archaea New opening essay about the human microbiota. Improved art comparing viral structures. Updated information about Ebola, AIDS, and Zika virus. Updated figure showing binary fission. New figure illustrating mechanisms of horizontal gene transfer in prokaryotes. New section about bacteria as pathogens. New Data Analysis Activity about antibiotics inspired by the study of bacteriophages.

21 The Protists Chapter reorganized to reflect our current understanding of the major eukaryotic supergroups. New overview of protist cell structure.

22 Plant Evolution Revised life cycle graphics throughout the chapter; improved figure illustrating generalized process of seed production.

23 Fungi Added photos of athlete's foot and ringworm. Updated information about white nose syndrome. Increased coverage of the use of fungi in food production, research, biotechnology, and as sources of medicine.

24 Animals I: Major Invertebrate Groups Added information about sponge regeneration. New figure showing schistosomes. Coverage of placozoans, rotifers, and tardigrades deleted.

25 Animals II: The Chordates New information about bone loss in the evolution of cartilaginous fishes.

26 Human Evolution Updated information about fossil hominids, including discussion of Denisovans and *Homo naledi*.

27 Plant Tissues New art includes illustration of stem structure and location of the vascular cylinder in a root.

28 Plant Nutrition and Transport Added information about use of phytoremediation at Fukushima. New micrographs and associated art detail the flow of water through xylem cells. Updated translocation art coordinates with new art illustrating sieve tube structure.

29 Life Cycles of Flowering Plants Detail added to plant life cycle art for accuracy. New photo illustrates root suckers.

30 Communication Strategies in Plants Current research informed updates to mechanisms of hormonal action. Table summarizing plant hormones has been broken by section. Added material includes the role of ABA in stress-related stomata closure; nastic movements and accompanying new photos; and explanation of *Phylloxera* resistance in American grapevines based on enhanced hypersensitive response involving resveratrol.

31 Animal Tissues and Organ Systems New summary table describing tissue types. Improved graphic illustrating relative volumes of the fluid components of a human body. New information about brown fat versus white fat and white matter and gray matter. Updated information about research on and clinical use of induced pluripotent stem cells (IPSCs) and about transdifferentiation as an alternative source of replacement cells.

32 Neural Control Updated information about brain damage among professional football players. New opening overview of intracellular signaling mechanisms. Revised/reorganized coverage of the peripheral nervous system. New subsection covers tissues and fluid of the CNS; information about neuroglia moved here.

33 Sensory Perception Updated illustration/discussion of retina anatomy to include light-channeling neuroglia.

34 Endocrine Control Added information about sites of human steroid hormone production (gonads/adrenals). Consolidated information about hormones in a single table. Moved discussion of pineal gland to follow discussion of pituitary/hypothalamus. Deleted coverage of the thymus. New Data Analysis Activity covers the disruptive effect of BPA on insulin secretion.

35 Structural Support and Movement Improved figures depicting locomotion of fly and earthworm. Revised figure showing the structure of skeletal muscle.

36 Circulation Updated photo depicting measurement of blood pressure. Expanded coverage of venule function. New art depicting atherosclerosis. Added discussion of heart attack symptoms and of causes and symptoms of stroke. New Data Analysis Activity on how hypertension affects the risk of stroke and heart attack.

37 Immunity New application section that details vaccination and benefits of herd immunity features a narrative about an unvaccinated child with permanent health consequences of contracting measles. New photos illustrate microbial sectors of a dental plaque biofilm, agglutination, and anaphylaxis. Revised art better illustrates specificity of antigen binding sites in antibodies. New content includes role of keratinocytes as immune cells in contact allergies.

38 Respiration Improved photo of insect tracheal system. Increased emphasis on evolutionary trends in vertebrate lung structure. Added information about the risks of vaping. New Data Analysis Activity addresses effects of tobacco and marijuana smoke on the lungs.

39 Digestion and Human Nutrition New opening essay about the role of human digestive enzymes and genetic variations in these enzymes. Added coverage of sponge digestion with new graphic. New figure compares length of digestive tract regions in a mammalian carnivore and herbivore. New graphics depict peristalsis and segmentation. Coverage of beneficial gut microflora moved to the section about the large intestine.

40 Maintaining the Internal Environment Material reorganized with separate sections describing formation and types of wastes and types of excretory organs. Variations on kidney structure in fish and desert rat kidneys now covered in a separate section after discussion of human kidney structure. Added information about human body hair as a temperature-related adaptation. Deleted coverage of human variation in sodium reabsorption.

41 Animal Reproduction Updated information about STDs, including new discussion of *Mycoplasma genitalium* infection.

42 Animal Development Morphogens now discussed in the context of embryonic induction. Added information about the teratogenic effects of the Zika virus.

43 Animal Behavior New information about epigenetic effects in a variety of contexts. Improved honeybee dance language figure. New information about tent caterpillars, a pre-social species. Added information about reciprocal altruism.

44 Population Ecology New information about research indicating that human hunting altered the life-history traits of wooly mammoths. Updated human population statistics.

45 Community Ecology Added coverage and photos of sundews and spiders that complete for insect prey. Updated photo of Surtsey.

46 Ecosystems Updated information about the increasing level of atmospheric carbon dioxide and the evidence that human activities are responsible for this increase.

47 The Biosphere Updated information about the movement of radioactive compounds released at Fukushima. Figure depicting seasonal overturn in a lake revised. Increased coverage of the importance of dissolved oxygen in aquatic habitats. New coverage of ocean acidification as a threat to reefs.

48 Human Effects on the Biosphere New opening section about the proposal to recognize human effects by naming a new geological epoch—the Anthropocene. New Data Analysis Activity about bioaccumulation of radioactive materials in tuna. Revised, updated coverage of biodiversity hotspots. New closing section about how citizen science can help document the distribution and decline of biodiversity.

Student and Instructor Resources

MindTap

MindTap is an outcome-driven application that propels students from memorization to mastery. MindTap is the only platform that gives you complete ownership of your course—to provide engaging content, to challenge every individual, and to build students' confidence.

MindTap brings ultimate convenience for students and instructors by allowing you to access everything in one place. In addition, the MindTap Mobile App gives students complete flexibility to read, listen, and study anytime, anywhere on their phones—and learn on their terms.

MindTap for *Biology: The Unity and Diversity of Life 15e*

The MindTap Learning Path includes these engaging learning opportunities in every chapter, and more!

Make It Relevant

Based on the chapter core applications, the popular *How Would You Vote?* feature provides scenarios for further research and critical thinking on topics that show the relevance of biology to student lives. After reading the chapter, students can revisit their original vote in a follow-up activity called *How Would You Vote Now?*

Assign and Grade Content that Matters

The following activities are included and can be set to auto-graded using a simple button-click.

- **Data Analysis Activities** are fully assignable in MindTap! Ensure your students are sharpening their analytical skills by assigning these engaging activities. The data is related to the chapter material, and is taken from a published scientific study in most cases. Other assignable and gradable activities in every chapter include:
- **Conceptual Learning Assignments**
- **Critical Thinking Questions**
- **Chapter Test**

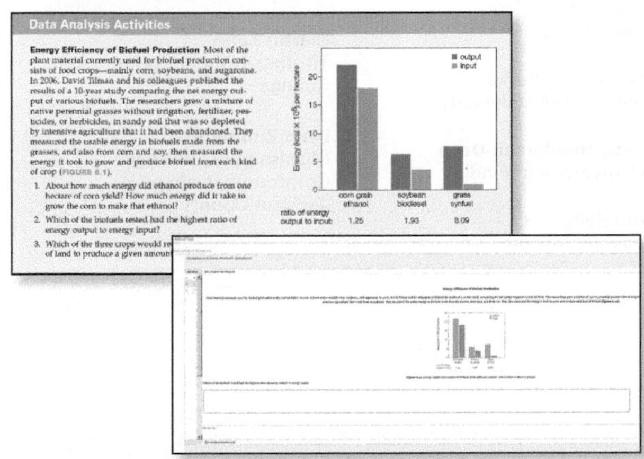

Data Analysis Activities
Energy Efficiency of Biofuel Production Most of the plant material currently used for biofuel production consists of food crops—mainly corn, soybeans, and sugarcane. In 2006, David Tilman and his colleagues published the results of a 10-year study comparing the net energy output of various biofuels. The researchers grew a mixture of native perennial grasses without irrigation, fertilizer, pesticides, or herbicides, in sandy soil that was so depleted by intensive agriculture that it had been abandoned. They measured the usable energy in biofuels made from the grasses, and also from corn and soy, then measured the energy it took to grow and produce biofuel from each kind of crop (FIGURE 6.1).
1. About how much energy did ethanol produce from one hectare of corn yield? How much energy did it take to grow the corn to make that ethanol?
2. Which of the biofuels tested had the highest ratio of energy output to energy input?
output
input
Energy (kcal × 10⁶) per hectare
corn grain ethanol
soybean biodiesel
grass synfuel
ratio of energy output to input:
1.25
1.93
8.09

Data Analysis activities in every chapter...assigned in MindTap.

Just for Students: Study and Practice

The new MindTap features two sets of activities to ensure students are not only doing the reading but are also engaging and learning the topics.

- **Student Study Card** For each section in every chapter, students read a brief summary, complete answers to learning objective "quick-checks" and get immediate feedback on their responses. Direct links to media related to the section content are included for many sections.
- **Study Guide** For student practice only; this set of detailed questions provides students further practice on the topics in each section.

MindTap is fully customizable to meet your course goals. Easily assign students the content you want them to learn, in the order you want them to learn it.

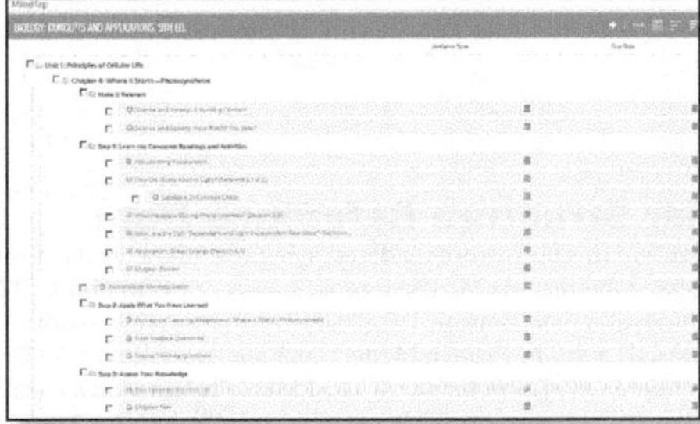

Make it your own. Insert your own materials—slides, videos, and lecture notes—wherever you want your students to see them.

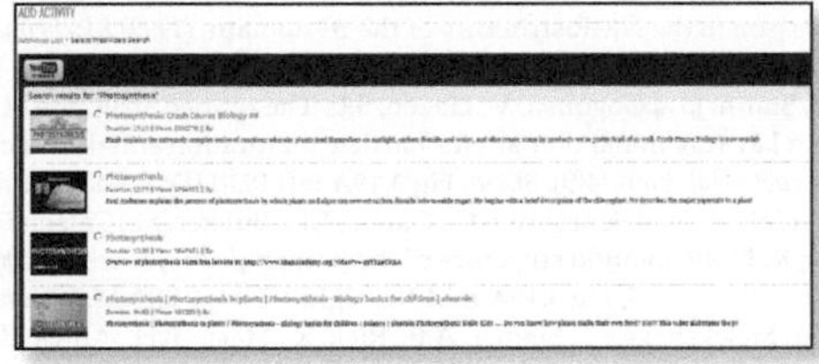

MindTap Course Development

Effectively introducing digital solutions into your classroom—online or on-ground—is now easier than ever. We're with you every step of the way.

Our Digital Course Support Team

When you adopt from Cengage Learning, you have a dedicated team of Digital Course Support professionals, who will provide hands-on start to-finish support, making digital course delivery a success for you and your students.

"The technical support provided by Cengage staff in using MindTap was superb. The Digital Support Team was proactive in training and prompt in answering follow-up questions." —K. Sata Sathasivan, The University of Texas at Austin

Instructor Companion Site

Everything you need for your course in one place! This collection of book-specific lecture and class tools is available online via www.cengagebrain.com/login. Access and download PowerPoint presentations, images, instructor's manuals, videos, test banks, and more.

Acknowledgments

Writing, revising, and illustrating a biology textbook is a major undertaking for two full-time authors, but our efforts constitute only a small part of what is required to produce and distribute this one. We are truly fortunate to be part of a huge team of very talented people who are as committed as we are to creating and disseminating an exceptional science education product.

Biology is not dogma; paradigm shifts are a common outcome of the fantastic amount of research in the field. Ideas about what material should be taught and how best to present that material to students changes even from one year to the next. It is only with the ongoing input of our many academic reviewers and advisors (see the list of Class Testers and Reviewers for this edition, right) that we can continue to tailor this book to the needs of instructors and students while integrating new information and models. We continue to learn from and be inspired by these dedicated educators.

Thanks to Hal Humphrey, Production Manager; Jake Warde, Content Developer; Tom Ziolkowski, Marketing Manager; and Marina Starkey, Product Assistant. Thank you also to Cheryl DuBois and Christine Myaskovsky for help with photo research.

Lisa Starr and Christine Evers, August 2017

Cengage acknowledges and appreciates Lisa Starr's contribution of more than 300 pieces of art to this edition.

Molecular Structure Reference Data

Structural models in this book were rendered using the following data from RCSB PDB (www.rcsb.org, Berman, H.M., Westbrook, J., Feng, Z., Gilliland, G., Bhat, T.N., Weissig, H., Shindyalov, I.N., Bourne, P.E. (2000) The Protein Data Bank. *Nucleic Acids Research*, 28: 235-242) and The Protein Model DataBase (Castrignanò, T., De Meo, P.D., Cozzetto, D., Talamo, I.G., Tramontano, A. The PMDB Protein Model Database. (2006) *Nucleic Acids Res.* 34 Database issue: D306-9). **Fig 3.6 & Fig 3.16(3,4,5) & Fig 4.3 & Fig 5.18 left & Fig 9.14 & Fig 38.15** PDB ID: 1BBB. Silva, M.M., Rogers, P.H., Arnone, A. A third quaternary structure of human hemoglobin A at 1.7-Å resolution. (1992) *J.Biol.Chem.* 267: 17248-17256. **Fig 3. 17 middle** PDB ID: 1PGX. Achari, A., Hale, S.P., Howard, A.J., Clore, G.M., Gronenborn, A.M., Hardman, K.D., Whitlow, M. 1.67 Å X-ray structure of the B2 immunoglobulin-binding domain of streptococcal protein G and comparison to the NMR structure of the B1 domain. (1992) *Biochemistry* 31: 10449-10457. **Fig 3.18** PMDB ID: PM0074956. Wu, Z., Wagner, M.A., Zheng, L., Parks, J.S., Shy, J.M., Smith, J.D., Gogonea, V., Hazen, S.L. The refined structure of nascent HDL reveals a key functional domain for particle maturation and dysfunction. (2007) *Nat. Struct. Mol. Biol.* 14(9): 861-8. **Fig 3.19A left** PDB ID: 1QM2. Zahn, R., Liu, A., Luhrs, T., Riek, R., Von Schroetter, C., Garcia, F.L., Billeter, M., Calzolai, L., Wider, G., Wuthrich, K. NMR solution structure of the human prion protein. (2000) Proc. *Natl. Acad. Sci.* USA 97: 145. **Fig 3.19A middle, right** PDB ID: 2RNM. Wasmer, C., Lange, A., Van Melckebeke, H., Siemer, A.B., Riek, R., Meier, B.H., Amyloid fibrils of the HET-s(218-289) prion form a beta solenoid with a triangular hydrophobic core. (2008) *Science* 319: 1523-1526. **Page 47** PDB ID: 2W5J. Vollmar, M., Shlieper, D., Winn M., Buechner, C., Groth, G. Structure of the C14 rotor ring of the proton translocating chloroplast ATP synthase. (2009) *J.Biol.Chem.* 284: 18228. **Fig 5.10B** PDB ID: 1HKB. Aleshin, A.E., Zeng, C., Bourenkov, G.P., Bartunik, H.D., Fromm, H.J., Honzatko, R.B. The mechanism of regulation of hexokinase: new insights from the crystal structure of recombinant human brain hexokinase complexed with glucose and glucose-6-phosphate. (1998) *Structure* 6: 39-50. **Fig 5.10A** PDB ID: 1HKC. Aleshin, A.E., Zeng, C., Bartunik, H.D., Fromm, H.J., Honzatko, R.B. Regulation of hexokinase I: crystal structure of recombinant human brain hexokinase complexed with glucose and phosphate. (1998) *J.Mol.Biol.* 282: 345-357. **Fig 6.10** PDB ID: 2BHW. Standfuss, J., Terwisscha Van Scheltinga, A.C., Lamborghini, M., Kuehlbrandt, W. Mechanisms of Photoprotection and Nonphotochemical Quenching in Pea Light-Harvesting Complex at 2.5Å Resolution. (2005) *Embo J.* 24: 919. **Page 141** PDB ID: 1TTD. McAteer, K., Jing, Y., Kao, J., Taylor, J.S., Kennedy, M.A. Solution-state structure of a DNA dodecamer duplex containing a Cis-syn thymine cyclobutane dimer, the major UV photoproduct of DNA. (1998) *J.Mol.Biol.* 282: 1013-1032. **Fig 9.2 right** PDB ID: 2AAI. Rutenber, E., Katzin, B.J., Ernst, S., Collins, E.J., Mlsna, D., Ready, M.P., Robertus, J.D. Crystallographic refinement of ricin to 2.5Å. (1991) *Proteins* 10: 240-250. **Fig 9.10** PDB ID: 3O30. Ben-Shem, A., Jenner, L., Yusupova, G., Yusupov, M. Crystal structure of the eukaryotic ribosome. (2010) *Science* 330: 1203-1209. **Fig 9.11 top left** PDB ID: 1EVV. Jovine, L., Djordjevic, S., Rhodes, D. The crystal structure of yeast phenylalanine tRNA at 2.0 Å resolution: cleavage by Mg(2+) in 15-year old crystals. (2000) *J.Mol.Biol.* 301: 401-414. **Fig 9.2 right** PDB ID: 2AAI. Rutenber, E., Katzin, B.J., Ernst, S., Collins, E.J., Mlsna, D., Ready, M.P., Robertus, J.D. Crystallographic refinement of ricin to 2.5Å. (1991) *Proteins* 10: 240-250. **Fig 10.10** PDB ID: 1IG4. Ohki, I., Shimotake, N., Fujita, N., Jee, J., Ikegami, T., Nakao, M., Shirakawa, M. Solution structure of the methyl-CpG binding domain of human MBD1 in complex with methylated DNA. (2001) *Cell* (Cambridge, Mass.) 105: 487-497. **Fig 37.9** PDB ID: 1IGT. Harris, L.J., Larson, S.B., Hasel, K.W., McPherson, A. Refined structure of an intact IgG2a monoclonal antibody. (1997) *Biochemistry* 36: 1581-1597. **Fig 37.16 left** PDB ID: 1MI5. Kjer-Nielsen, L., Clements, C.S., Purcell, A.W., Brooks, A.G., Whisstock, J.C., Burrows, S.R., McCluskey, J., Rossjohn, J. A structural basis for the selection of dominant alphabeta T Cell receptors in antiviral immunity. (2003) *IMMUNITY* 18: 53-64.

Class Testers and Reviewers

Laura Almstead
University of Vermont

Dan Ardia
Franklin & Marshall College

Erin Baumgartner
Western Oregon University

Joel Benington
St. Bonaventure University

Amanda Brammer
Delgado Community College

Sarah Cooper
Arcadia University

Leigh Delaney-Tucker
University of South Alabama

Daron Goodloe
Northwest-Shoals Community College

Lisa Boggs
Southwestern Oklahoma State University

Debra Chapman
Wilkes University

Teresa Cowan
Baker College

Victor Fet
Marshall University

Steven Fields
Winthrop University

Ted Gregorian
St. Bonaventure University

Adam Hrincevich
Louisiana State University

Joseph Daniel Husband
Florida State College at Jacksonville

Petura McCaa-Burke
Alabama A&M University

Bethany Henderson-Dean
The University of Findlay

Martin Kelly
D'Youville College

Michael Loui
Mississippi Gulf Coast Community College

Andrew Petzold
University of Minnesota Rochester

Kumkum Prabhakar
Nassau Community College

Paul Ramp
Pellissippi State Community College

Christina Russin
Northwestern University

Leslie Saucedo
University of Puget Sound

Amanda Schaetzel
University of Colorado

Takrima Sadikot
Washburn University

Kim Sadler
Middle Tennessee State University

Jack Shurley
Idaho State University

Lakhbir Singh
Chabot College

Sharon Thoma
University of Wisconsin

Sylvia Fromherz Sharp
Saginaw Valley State University

Roy Wilson
Mississippi Gulf Coast Community College

Martin Zahn
Thomas Nelson Community College

Lynn Zimmerman
Mississippi Gulf Coast Community College—Jackson County

BIOLOGY

The Unity and Diversity of Life

15TH EDITION

STARR
TAGGART
EVERS
STARR

1 INVITATION TO BIOLOGY

CORE CONCEPTS

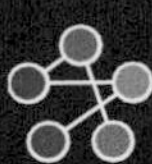

Systems

Complex properties arise from interactions among components of a biological system.

We can understand life by studying it at increasingly inclusive levels, starting with atoms that compose matter, and extending to the biosphere. Each level is a biological system composed of interacting parts. Interactions among the components of a system give rise to complex properties not found in any of the components. The movement of matter and energy through ecosystems influences interactions among organisms and their environment.

Evolution

Evolution underlies the unity and diversity of life.

Shared core processes and features that are widely distributed among organisms provide evidence that all living things are linked by lines of descent from common ancestors. All biological systems are sustained by the exchange of matter and energy; all store, retrieve, transmit, and respond to information essential for life.

Process of Science

The field of biology consists of and relies upon experimentation and the collection and analysis of scientific evidence.

Science addresses only testable ideas about observable events and processes. Observation, experimentation, quantitative analysis, and critical thinking are key aspects of scientific research. Carefully designed experiments that yield objective data help researchers unravel cause-and-effect relationships in complex biological systems.

Links to Earlier Concepts

Whether or not you have studied biology, you already have an intuitive understanding of life on Earth because you are part of it. Every one of your experiences with the natural world—from the warmth of the sun on your skin to the love of your pet—contributes to that understanding.

1.1 Application: Secret Life of Earth

In this era of cell phone GPS, could there possibly be any places left on Earth that humans have not yet explored? Actually, there are plenty. Consider a 2-million-acre cloud forest in the Foja Mountains of New Guinea that was not penetrated by humans until 2005. Since then, about forty new **species**—unique types of organisms—have been discovered there, including a rhododendron plant with flowers the size of dinner plates, a rat the size of a cat, and a frog the size of a pea. Also discovered among the forest's inhabitants were hundreds of species on the brink of extinction in other parts of the world, and some that supposedly had been extinct for decades. A few new or rare species were discovered accidentally, as animals that had never learned to be afraid of humans wandered casually through campsites (FIGURE 1.1).

How do we know what species a particular organism belongs to? What is a species, anyway, and why should discovering a new one matter to anyone other than a biologist? You will find the answers to such questions in this book. They are part of the scientific study of life, **biology**, which is one of many ways we humans try to make sense of the world around us. Ironically, the more we learn about the natural world, the more we realize we have yet to learn. But don't take our word for it. Find out what biologists know, and what they do not, and you will have a solid foundation upon which to base your own opinions about how humans fit into this world. By reading this book, you are choosing to learn about the human connection—your connection—with all life on Earth.

FIGURE 1.1 The Pinocchio frog. Biologist Paul Oliver discovered this tiny tree frog perched on a sack of rice during the first survey of a cloud forest in New Guinea. It was named after the Disney character because the male frog's long nose inflates and points upward during times of excitement.

biology The scientific study of life.
species (SPEE-sheez) A unique type of organism.

1.2 Life Is More Than the Sum of Its Parts

LEARNING OBJECTIVES

- Describe the successive levels of life's organization.
- Explain the idea of emergent properties and give an example.

Biologists study life. What, exactly, is "life?" We may never actually come up with a concise definition, because living things are too diverse, and they consist of the same basic components as nonliving things. When we try to define life, we end up with a long list of complex properties that differentiate living from nonliving things. These properties often emerge from the interactions or arrangements of basic components (FIGURE 1.2).

Consider a complex behavior called swarming that is characteristic of honeybees. When bees swarm, they fly en masse to establish a hive in a new location. Each bee is autonomous, but the new hive's location is decided collectively based on an integration of signals from hivemates. A characteristic of a system (a swarm's collective intelligence, for example) that does not appear in any of the system's components (individual bees) is called an **emergent property**.

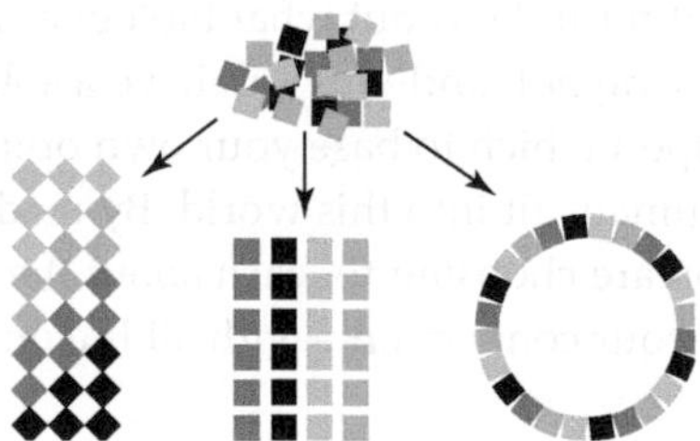

FIGURE 1.2 **The same materials, assembled in different ways, form objects with different properties.** The property of "roundness" emerges when these squares are assembled in a certain way.

atom The smallest unit of a substance; composes matter.
biosphere (BY-oh-sfeer) All regions of Earth where organisms live.
cell Smallest unit of life.
community All populations of all species in a defined area.
ecosystem A community interacting with its environment.
emergent property (ee-MERGE-ent) A characteristic of a system that does not appear in any of the system's components.
molecule (MAUL-ick-yule) Two or more atoms bonded together.
organ In multicelled organisms, a structure that consists of tissues engaged in a collective task.
organism (ORG-uh-niz-um) An individual that consists of one or more cells.
organ system In multicelled organisms, a set of interacting organs that carry out a particular body function.
population A group of interbreeding individuals of the same species living in a defined area.
tissue In multicelled organisms, specialized cells organized in a pattern that allows them to perform a collective function.

Life's Organization

Biologists view life in increasingly inclusive levels of organization (FIGURE 1.3). This organization begins with the **atom** ❶, the smallest unit of a substance. Atoms and the fundamental particles that compose them are the building blocks of all matter. Atoms bond together to form **molecules** ❷. There are no atoms unique to living things, but there are unique molecules. A **cell** ❸, which is the smallest unit of life, consists of many of these "molecules of life."

Some cells live and reproduce independently. Others do so as part of a multicelled organism. An **organism** is an individual that consists of one or more cells ❼. In most multicelled organisms, cells are organized as tissues ❹. A **tissue** consists of specific types of cells organized in a particular pattern. The arrangement allows the cells to collectively perform a special function such as protection from injury (dermal tissue) or movement (muscle tissue).

An **organ** is a structure composed of tissues that collectively carry out a particular task or set of tasks ❺. An **organ system** is a set of interacting organs and tissues that fulfill one or more body functions ❻. Examples of organ systems include the aboveground parts of a plant (the shoot system), and the heart and blood vessels of an animal (the circulatory system).

A **population** is a group of interbreeding individuals of the same species living in a given area. For example, all of the California poppies growing in California's Antelope Valley Poppy Reserve form a population ❽. A **community** consists of all populations of all species in a given area. The Antelope Valley Reserve community includes the California poppy population, as well as populations of other plants, animals, microorganisms, and so on ❾. Communities may be large or small, depending on the area defined.

The next level of organization is the **ecosystem**, which is a community interacting with its physical and chemical environment ❿. Earth's largest ecosystem is the **biosphere,** and it encompasses all regions of the planet's crust, waters, and atmosphere in which organisms live ⓫.

TAKE-HOME MESSAGE 1.2

✔ Biologists study life by thinking about it at successive levels of organization. Emergent properties occur at each level.

✔ All matter consists of atoms. Atoms join as molecules, and molecules make up cells. The cell is the smallest unit of life.

✔ Organisms, populations, communities, ecosystems, and the biosphere are successively higher levels of life's organization.

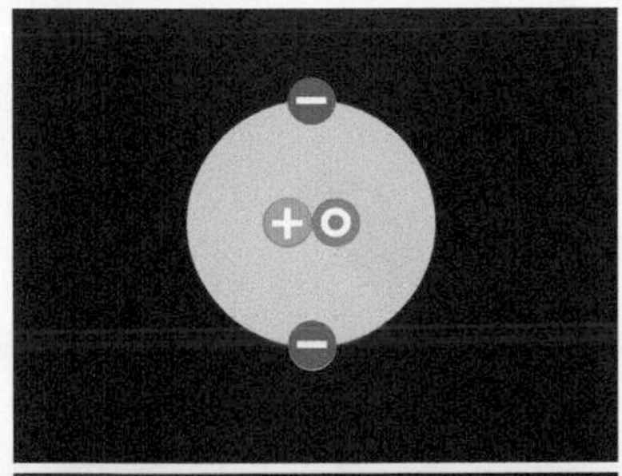

❶ **Atom**

Atoms and the particles that compose them make up all matter.

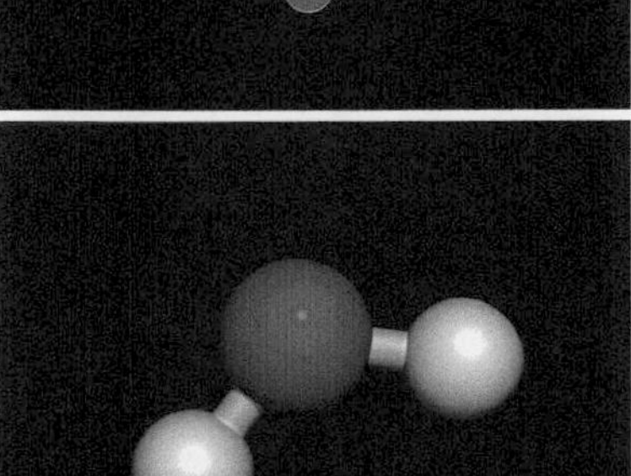

❷ **Molecule**

Atoms join other atoms in molecules. This is a model of a water molecule. The molecules special to life are much larger and more complex than water.

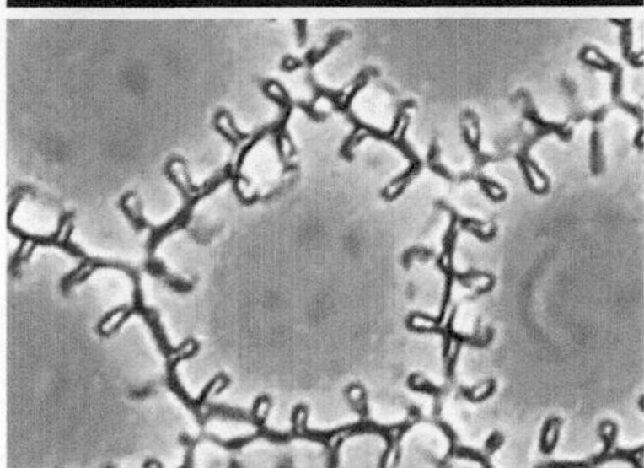

❸ **Cell**

The cell is the smallest unit of life. Some, like these plant cells, live and reproduce as part of a multicelled organism; others do so on their own.

❹ **Tissue**

An organized array of cells that interact in a collective task. This is dermal tissue on the outer surface of a flower petal.

❺ **Organ**

A structural unit of interacting tissues. Flowers are the reproductive organs of some plants.

❻ **Organ system**

A set of interacting organs. The shoot system of this poppy plant includes its aboveground parts: leaves, flowers, and stems.

❼ **Multicelled organism**

An individual that consists of more than one cell. Cells of this California poppy plant make up its shoot system and root system.

❽ **Population**

A group of single-celled or multicelled individuals of a species in a given area. This population of California poppy plants is in California's Antelope Valley Poppy Reserve.

❾ **Community**

All populations of all species in a specified area. These plants are part of the community in the Antelope Valley Poppy Reserve.

❿ **Ecosystem**

A community interacting with its physical environment through the transfer of energy and materials. Sunlight and water sustain the community in the Antelope Valley.

⓫ **Biosphere**

The sum of all ecosystems: every region of Earth's waters, crust, and atmosphere in which organisms live.

FIGURE 1.3 Levels of life's organization. Emergent properties appear at each successive level.

FIGURE IT OUT At which level does the emergent property of "life" appear?

Answer: The cell

CREDITS: (3) 3, 4: © Umberto Salvagnin; 5: California Poppy, © 2009, Christine M. Welter; 6: Lady Bird Johnson Wildflower Center; 7: Michael Szoenyi/Science Source; 8: James Randklev / Exactostock-1672/SuperStock; 9: © Sergei Krupnov, www.flickr.com/photos/7969319@N03; 10: © Mark Koberg Photography; 11: NASA.

1.3 How Living Things Are Alike

LEARNING OBJECTIVES

- Distinguish producers from consumers.
- Explain why homeostasis is important for sustaining life.
- Explain how DNA is the basis of similarities and differences among organisms.

All living things share a particular set of key features. You already know one of these features: Because the cell is the smallest unit of life, all organisms consist of at least one cell. For now, we introduce three more: All living things require ongoing inputs of energy and raw materials; all sense and respond to change; and all use DNA as the carrier of genetic information (TABLE 1.1).

TABLE 1.1

Some Key Features of Life

Cellular basis	All living things consist of one or more cells.
Requirement for energy and nutrients	Life is sustained by ongoing inputs of energy and nutrients.
Homeostasis	Living things sense and respond to change.
DNA is hereditary material	Genetic information in the form of DNA is passed to offspring.

Organisms Require Energy and Nutrients

Not all living things eat, but all require energy and nutrients on an ongoing basis. A **nutrient** is a substance that an organism needs for growth and survival but cannot make for itself.

Both nutrients and energy are essential to maintain the organization of life, so organisms spend a lot of time acquiring them. However, the source of energy and the type of nutrients required differ among organisms. These differences allow us to classify all living things into two categories: producers and consumers (FIGURE 1.4). A **producer** makes its own food using energy and simple raw materials it obtains from nonbiological sources ❶. Plants are producers. By a process called **photosynthesis**, plants can use the energy of sunlight to make sugars from carbon dioxide (a gas in air) and water. Consumers, by contrast, cannot make their own food. A **consumer** obtains energy and nutrients by feeding on other organisms ❷. Animals are consumers. So are decomposers, which feed on the wastes or remains of other organisms. Leftovers from consumers' meals end up in the environment, where they serve as nutrients for producers. Said another way, nutrients cycle between producers and consumers ❸.

Unlike nutrients, energy is not cycled. It flows through the world of life in one direction: from the environment ❹, through organisms, and then back to the environment ❺. This flow maintains the organization of every living cell and body, and it also influences how individuals interact with one another and their environment. The energy flow is one way, because with each transfer, some energy escapes as heat, and cells

❶ producer acquiring energy and nutrients from the environment
❷ consumer acquiring energy and nutrients by eating a producer

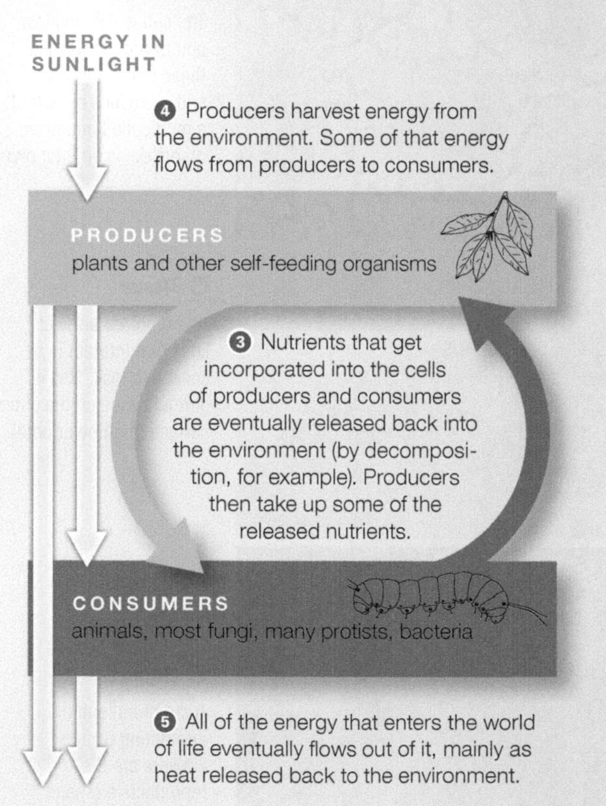
ENERGY IN SUNLIGHT
❹ Producers harvest energy from the environment. Some of that energy flows from producers to consumers.
PRODUCERS
plants and other self-feeding organisms
❸ Nutrients that get incorporated into the cells of producers and consumers are eventually released back into the environment (by decomposition, for example). Producers then take up some of the released nutrients.
CONSUMERS
animals, most fungi, many protists, bacteria
❺ All of the energy that enters the world of life eventually flows out of it, mainly as heat released back to the environment.

FIGURE 1.4 **The one-way flow of energy and cycling of materials through the world of life.**

cannot use heat as an energy source. Thus, energy that enters the world of life eventually leaves it (we return to this topic in Chapter 5).

Homeostasis

An organism cannot survive for very long unless it responds appropriately to specific stimuli inside and outside of itself. For example, humans and some other animals normally perspire (sweat) when the body's internal temperature rises above a certain set point (**FIGURE 1.5**). The moisture cools the skin, which in turn helps cool the body.

All of the internal fluids that bathe the cells in your body are collectively called your internal environment. Temperature and many other conditions in that environment must be kept within certain ranges, or your cells will die (and so will you). By sensing and adjusting to change, all organisms keep conditions in their internal environment within ranges that favor cell survival. **Homeostasis** is the name for this process, and it is one of the defining features of life.

DNA Is Hereditary Material

With little variation, the same types of molecules perform the same basic functions in every organism. For example, information in an organism's **DNA** (deoxyribonucleic acid) guides ongoing cellular activities that sustain the individual through its lifetime. Such functions include **development**: the process by which the first cell of a new individual gives rise to a multicelled adult; **growth**: increases in cell number, size, and volume; and **reproduction**: processes by which organisms produce offspring. **Inheritance**, the transmission of DNA to offspring, occurs during reproduction. All organisms inherit their DNA from one or more parents.

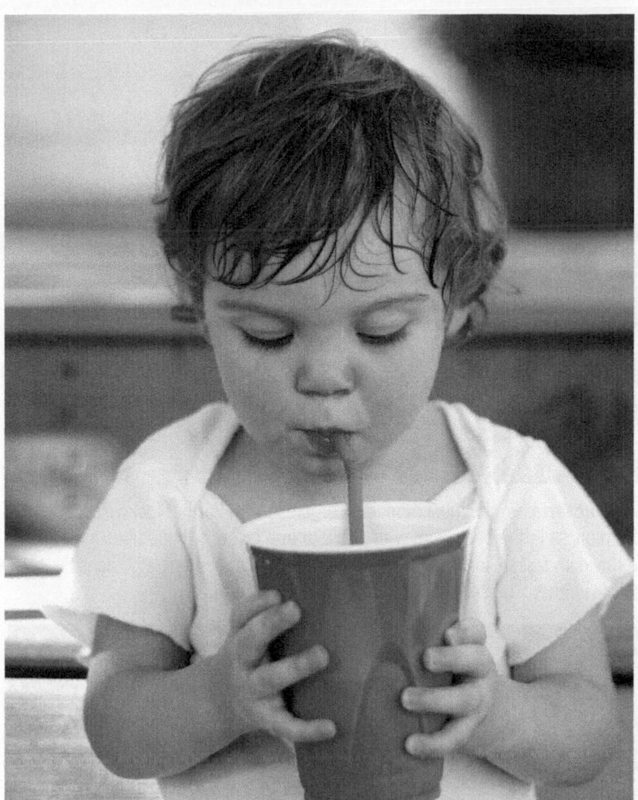

FIGURE 1.5 Living things sense and respond to their environment. Sweating is a physiological response to an internal body temperature that exceeds the normal set point. The response cools the skin, which in turn helps return the internal temperature to the set point.

Individuals of every natural population are alike in most aspects of body form and behavior because their DNA is very similar: Humans look and act like humans and not like poppy plants because they inherited human DNA, which differs from poppy plant DNA in the information it carries. Individuals of almost every natural population also vary—just a bit—from one another: One human has blue eyes, the next has brown eyes, and so on. Such variation arises from small differences in the details of DNA molecules, and herein lies the source of life's diversity. As you will see in later chapters, differences among individuals of a species are the raw material of evolutionary processes.

TAKE-HOME MESSAGE 1.3

✔ A one-way flow of energy and a cycling of materials maintain life's complex organization.

✔ Organisms sense and respond to conditions inside and outside themselves. They make adjustments that keep conditions in their internal environment within a range that favors cell survival, a process called homeostasis.

✔ All organisms use information in the DNA they inherited from their parent or parents to develop, grow, and reproduce. DNA is the basis of similarities and differences among organisms.

consumer (kun-SUE-murr) An organism that gets energy and nutrients by feeding on tissues, wastes, or remains of other organisms.
development (dih-VELL-up-ment) Processes by which the first cell of a multicelled organism gives rise to an adult.
DNA Deoxyribonucleic (dee-ox-ee-ribe-oh-new-CLAY-ick) acid; molecule that carries hereditary information; guides development and other activities.
growth Increase in the number, size, and volume of cells.
homeostasis (home-ee-oh-STAY-sis) Process in which organisms keep their internal conditions within tolerable ranges by sensing and responding appropriately to change.
inheritance (in-HAIR-ih-tunce) Transmission of DNA to offspring.
nutrient (NEW-tree-unt) A substance that an organism acquires from the environment to support growth and survival.
photosynthesis (foe-toe-SIN-thuh-sis) Process by which producers use light energy to make sugars from carbon dioxide and water.
producer An organism that makes its own food using energy and nonbiological raw materials from the environment.
reproduction (ree-pruh-DUCK-shun) Processes by which organisms produce offspring.

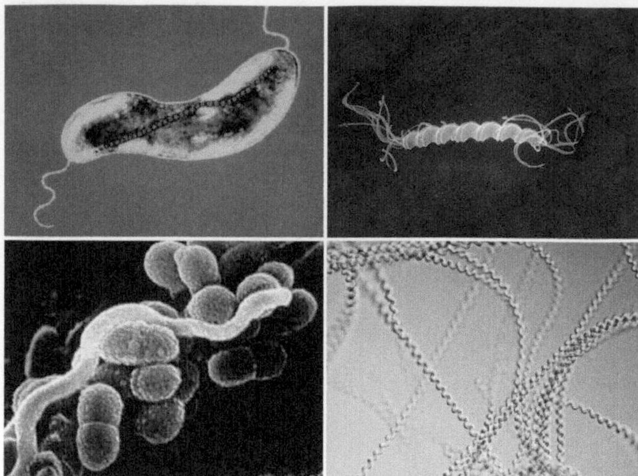

A Bacteria are the most numerous organisms on Earth. Clockwise from upper left, a bacterium with a row of iron crystals that serves as a tiny compass; a common bacterial resident of cat and dog stomachs; photosynthetic bacteria; bacteria found in dental plaque.

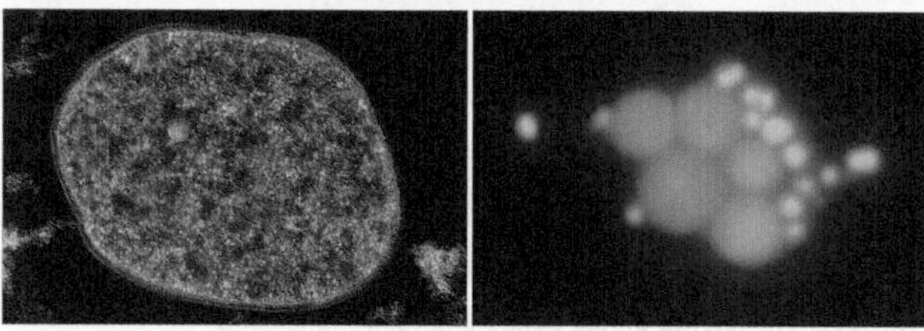

B Archaea resemble bacteria, but are more closely related to eukaryotes. Left, an archaeon that grows in sulfur hot springs. Right, two types of archaea from a seafloor hydrothermal vent.

FIGURE 1.6 A few representative prokaryotes.

1.4 How Living Things Differ

LEARNING OBJECTIVES

- List two characteristics of prokaryotes.
- Name the four main groups of eukaryotes.

You will see in later chapters how differences in the details of DNA molecules are the basis of a tremendous range of differences among types of organisms. Various classification schemes help us organize what we understand about this variation, which is an important aspect of Earth's biodiversity. For example, organisms can be grouped on the basis of whether they have a nucleus, which is a sac-like structure that contains a cell's DNA. **Bacteria** (singular, bacterium) and **archaea** (singular, archaeon) are the organisms whose DNA is *not* contained within a nucleus (**FIGURE 1.6**). All bacteria and archaea are single-celled, which means each individual consists of one cell. Collectively, these organisms are the most diverse representatives of life. Different kinds are producers or consumers in nearly all regions of Earth. Some inhabit such extreme environments as frozen desert rocks, boiling sulfurous lakes, and nuclear reactor waste. The first cells on Earth may have faced similarly hostile conditions.

Traditionally, organisms without a nucleus have been classified as **prokaryotes**, but the designation is now used only informally. This is because bacteria and archaea are less related to one another than we once thought, despite their similar appearance. Archaea turned out to be more closely related to **eukaryotes**, which are organisms whose DNA is contained within a nucleus. Some eukaryotes live as individual cells; others are multicelled. Eukaryotic cells are typically larger and more complex than prokaryotes.

There are four main groups of eukaryotes: protists, fungi, plants, and animals (**FIGURE 1.7**).

Protist is the common term for a collection of eukaryote groups that are not plants, animals, or fungi. Collectively, they vary dramatically, from single-celled consumers to giant, multicelled producers.

Fungi (singular, fungus) are eukaryotic consumers that secrete substances to break down food externally, then absorb nutrients released by this process. Many fungi are decomposers. Most fungi, including those that form mushrooms, are multicellular. Fungi that live as single cells are called yeasts.

Plants are multicelled eukaryotes, and the vast majority of them are photosynthetic producers that live on land. Besides feeding themselves, plants also serve as food for most other land-based organisms.

Animals are multicelled consumers that ingest other organisms or components of them. Unlike fungi, animals break down food inside their body. They also develop through a series of stages that lead to the adult form. All animals actively move about during at least part of their lives.

animal A multicelled consumer that develops through a series of stages and moves about during part or all of its life.
archaea (are-KEY-uh) Singular, archaeon. Group of single-celled organisms that lack a nucleus but are more closely related to eukaryotes than to bacteria.
bacteria Singular, bacterium. The most diverse and well-known group of single-celled organisms that lack a nucleus.
eukaryote (you-CARE-ee-oat) An organism whose cells characteristically have a nucleus.
fungus Plural, fungi. A single-celled or multicelled eukaryotic consumer that breaks down material outside itself, then absorbs nutrients released from the breakdown.
genus (JEE-nuss) Plural, genera. A group of species that share a unique set of traits.
plant A multicelled eukaryotic producer; typically photosynthetic.
prokaryote (pro-CARE-ee-oat) A single-celled organism without a nucleus.
protist Common term for a eukaryote that is not a plant, animal, or fungus.
specific epithet Second part of a species name.
taxonomy (tax-ON-oh-me) The practice of naming and classifying species.
trait An inherited characteristic of an organism or species.

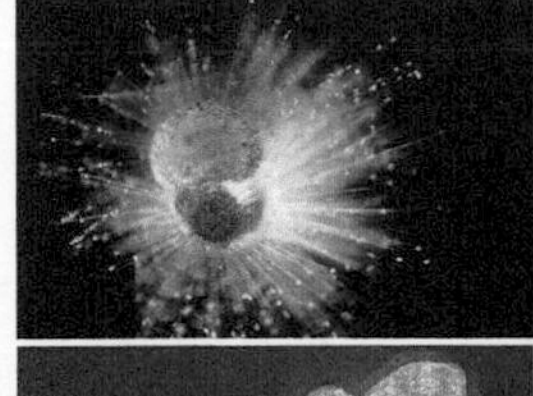

Protists are a group of extremely diverse eukaryotes that range from microscopic free-living cells (left) to giant multicelled seaweeds (right).

Fungi are eukaryotic consumers that secrete substances to break down food outside their body. Some are single-celled (left); most are multicelled (right).

Plants are multicelled eukaryotes, most of which are photosynthetic. Nearly all have roots, stems, and leaves.

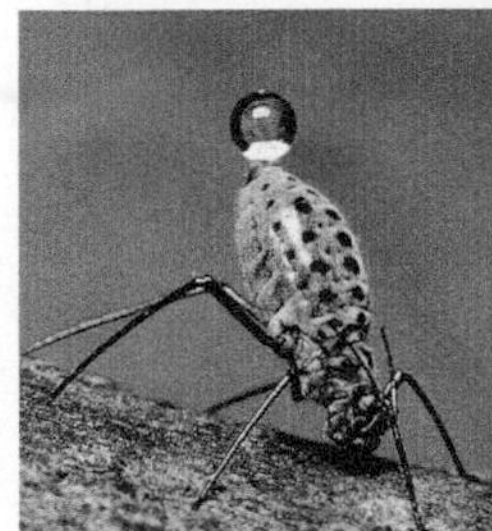

Animals are multicelled eukaryotes that ingest other organisms or their parts, and they actively move about during part or all of their life cycle.

FIGURE 1.7 A few representative eukaryotes.

TAKE-HOME MESSAGE 1.4

✔ Details of appearance and other characteristics vary greatly among living things.

✔ Bacteria and archaea are organisms whose DNA is not contained in a nucleus. All are single-celled.

✔ Protists, fungi, plants, and animals are eukaryotes: organisms whose DNA is contained in a nucleus. Most are multicelled.

1.5 Organizing Information About Species

LEARNING OBJECTIVES

- Explain how organisms are named in the Linnaean system.
- Describe the way species are classified in taxa.
- List the taxa from species to domain.
- Describe the "biological species concept" and explain its limitations.

A Rose by Any Other Name . . .

Each time we discover a new species, we name it, a practice called **taxonomy**. Taxonomy began thousands of years ago, but naming species in a consistent way did not become a priority until the eighteenth century. At the time, European explorers who were just discovering the scope of life's diversity started having more and more trouble communicating with one another because species often had multiple names. For example, the dog rose (a plant native to Europe, Africa, and Asia) was alternately known as briar rose, witch's briar, herb patience, sweet briar, wild briar, dog briar, dog berry, briar hip, eglantine gall, hep tree, hip fruit, hip rose, hip tree, hop fruit, and hogseed—and those are only the English names! Species often had multiple scientific names too, in Latin that was descriptive but often cumbersome. The scientific name of the dog rose was *Rosa sylvestris inodora seu canina* (odorless woodland dog rose), and also *Rosa sylvestris alba cum rubore, folio glabro* (pinkish white woodland rose with smooth leaves).

An eighteenth-century naturalist, Carolus Linnaeus, standardized a naming system that we still use. By the Linnaean system, each species is given a unique two-part scientific name. The first part of a scientific name is the **genus** (plural, genera), which is defined as a group of species that share a unique set of features. The second part of the name is the **specific epithet**. Together, the genus name and the specific epithet designate one species. Thus, the dog rose now has one official name, *Rosa canina*, that is recognized worldwide.

Genus and species names are always italicized. For example, *Panthera* is a genus of big cats. Lions belong to the species *Panthera leo*. Tigers belong to a different species in the same genus (*Panthera tigris*), and so do leopards (*P. pardus*). Note how the genus name may be abbreviated after it has been spelled out.

Distinguishing Species

The individuals of a species share a unique set of inherited characteristics, or **traits**. For example, giraffes normally have very long necks, brown spots on white

CREDIT: (7) Protists: top left, Courtesy of Allen W.H. Bé and David A. Caron; bottom left, MI Walker/Science Source; right, © worldswildlifewonders/Shutterstock.com; Plants: left, © Jag_cz/Shutterstock.com; right, Martin Ruegner/Radius Images/Getty Images; Fungi: left, London Scientific Films / Exactostock-1598/SuperStock; right, Kichigin/Shutterstock.com; Animals: left, Martin Zimmerman, *Science*, 1961, 133:73-79, © AAAS; middle, Thomas Eisner, Cornell University; right, iStockphoto.com/keeweeboy.

	wild carrot	marijuana	apple	prickly rose	dog rose
domain	Eukarya	Eukarya	Eukarya	Eukarya	Eukarya
kingdom	Plantae	Plantae	Plantae	Plantae	Plantae
phylum	Magnoliophyta	Magnoliophyta	Magnoliophyta	Magnoliophyta	Magnoliophyta
class	Magnoliopsida	Magnoliopsida	Magnoliopsida	Magnoliopsida	Magnoliopsida
order	Apiales	Rosales	Rosales	Rosales	Rosales
family	Apiaceae	Cannabaceae	Rosaceae	Rosaceae	Rosaceae
genus	*Daucus*	*Cannabis*	*Malus*	*Rosa*	*Rosa*
species	*carota*	*sativa*	*domestica*	*acicularis*	*canina*

FIGURE 1.8 **Taxonomic classification of five species that are related at different levels.** Each species has been assigned to ever more inclusive groups, or taxa: in this case, from genus to domain.

FIGURE IT OUT Which of the plants shown here are in the same order? Answer: Marijuana, apple, prickly rose, and dog rose

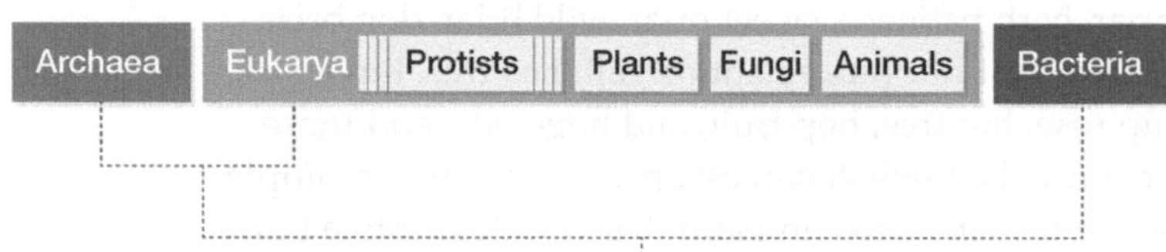
Archaea
Eukarya
Protists
Plants
Fungi
Animals
Bacteria

FIGURE 1.9 **The big picture of life.** This diagram summarizes one hypothesis about how all life is connected by shared ancestry. Lines indicate proposed evolutionary connections between the domains.

TABLE 1.2

All of Life in Three Domains

Archaea	Single cells, no nucleus. Evolutionarily closer to eukaryotes than to bacteria.
Eukarya	Eukaryotic cells (with a nucleus). Includes single-celled and multicelled species of protists, plants, fungi, and animals.
Bacteria	Single cells, no nucleus.

coats, and so on. These are morphological (structural) traits. Individuals of a species also share biochemical traits (they make and use the same molecules) and behavioral traits (they respond the same way to certain stimuli, as when hungry giraffes feed on tree leaves). We can rank a species into ever more inclusive categories based on some subset of traits it shares with other species. Each rank, or **taxon** (plural, taxa), is a group of organisms that share a unique set of inherited traits. Each category above species—genus, family, order, class, phylum (plural, phyla), kingdom, and domain—consists of a group of the next lower taxon (FIGURE 1.8). Using this system, we can sort all life into a few categories (FIGURE 1.9 and TABLE 1.2). Later chapters return to details of these and other classification systems.

It is easy to tell that humans and dog roses are different species because they appear very different. Distinguishing other species that are more closely related may be much more challenging (FIGURE 1.10). In addition, traits shared by members of a species often vary a bit among individuals.

How do biologists decide whether similar-looking organisms belong to different species? The short answer to that question is that they rely on whatever information is available. Early naturalists studied anatomy and distribution—essentially the only methods available at the time—so species were named and classified according to what they looked like and where they lived. Today's biologists are able to compare traits that the early naturalists did not even know about, including biochemical ones such as DNA molecules.

Consider that the information in a molecule of DNA changes a bit each time it passes from parents to offspring, and it has done so since life began. Over long periods of time, these tiny changes have added up to big differences between species such as humans and dog roses. Thus, differences in DNA are one way to measure relative relatedness: The fewer differences between species, the closer the relationship. For example, we know that the DNA of humans is more similar to chimpanzee DNA than it is to cat DNA, so we can assume that chimpanzees and humans are closer relatives than cats and humans. Every living species known

FIGURE 1.10 Four butterflies, two species: Which are which?

The top row shows two forms of the butterfly species *Heliconius melpomene*; the bottom row, two forms of *H. erato*.

DNA comparisons confirmed that *H. melpomene* and *H. erato* are different species; however, they never cross-breed. The alternate but similar patterns of coloration evolved as a shared warning signal to predatory birds that these butterflies taste terrible.

has DNA in common with every other species, so every living species is related to one extent or another. Unraveling these relationships has become a major focus of biology (later chapters return to this topic).

The discovery of new information sometimes changes the way we distinguish a particular species or how we group it with others. For example, Linnaeus grouped plants by the number and arrangement of reproductive parts, a scheme that resulted in odd pairings such as castor-oil plants with pine trees. Having more information today, we place these plants in separate phyla.

Evolutionary biologist Ernst Mayr defined a species as one or more groups of individuals that potentially can interbreed, produce fertile offspring, and do not interbreed with other groups. This "biological species concept" is useful but not universally applicable. For example, we may never know whether populations of some species could interbreed because impassable geographical barriers keep them separated. We return to speciation and how it occurs in Chapter 17, but for now it is important to remember that a "species" is a convenient but artificial construct of the human mind.

TAKE-HOME MESSAGE 1.5

✔ Each type of organism, or species, is given a unique name that consists of the genus and specific epithet.

✔ We define and group species based on shared, inherited traits.

critical thinking Act of evaluating information before accepting it.
science Systematic study of the physical universe.
taxon (TAX-on) Plural, taxa. A rank of organisms that share a unique set of traits.

1.6 The Science of Nature

LEARNING OBJECTIVES

- Describe critical thinking and give some examples of how to do it.
- Distinguish between inductive and deductive reasoning.
- Use suitable examples to explain dependent and independent variables.
- Explain how a control group is used in an experiment.
- List the tasks that are part of the scientific method.
- Give an example, real or hypothetical, of an experiment in which a dependent variable is affected by an independent variable.

Thinking About Thinking

Most of us assume that we do our own thinking, but do we, really? You might be surprised to find out how often we let others think for us. Consider how a school's job (which is to impart as much information to students as quickly as possible) meshes perfectly with a student's job (which is to acquire as much knowledge as quickly as possible). In this rapid-fire transfer of information, it can be very easy to forget about the merit of what is being transferred. Any time you accept information without evaluating it, you let someone else think for you.

Critical thinking is the deliberate process of judging the quality of information before accepting it. "Critical" comes from the Greek *kriticos* (discerning judgment). When you use critical thinking, you move beyond the content of new information to consider supporting evidence, bias, and alternative interpretations. This may sound complicated, but it just involves a bit of extra awareness. There are many ways to do it. For example, you might ask yourself some of the following questions while learning something new:

› What message am I being asked to accept?
› Is the message based on fact or opinion?
› Is there a different way to interpret the facts?
› What biases might the presenter have?
› Do I have biases that might influence the way I hear this message?

Questions like these are a way of being conscious about learning. They can help you decide whether to allow new information to guide your beliefs and actions.

The Scientific Method

Critical thinking is a crucial part of **science**, the systematic study of the observable world and how it works. A line of inquiry in biology typically begins with a researcher's curiosity about something observable in nature, such as, say, an unusual decrease

CREDIT: (10) From Meyer A., Repeating Patterns of Mimicry. *PLoS Biology* Vol. 4, No. 10, e341 doi:10.1371/journal.pbio.0040341.

TABLE 1.3

The Scientific Method

1. **Observe** some aspect of nature.
2. **Think of an explanation** for your observation (in other words, make a hypothesis).
3. **Evaluate the hypothesis.**
    a. Make a prediction based on the hypothesis (if . . . then).
    b. Test the prediction using experiments or surveys.
    c. Analyze the results (data).
4. **Decide** whether the results support your hypothesis or not (form a conclusion).
5. **Report** your experiment, data, and conclusion to the scientific community.

in the number of birds occupying a particular area. The researcher reads about what others have discovered before making a **hypothesis**, a testable explanation for a natural phenomenon. An example of a hypothesis would be: The number of birds is decreasing because the number of cats is increasing. Making a hypothesis this way is an example of **inductive reasoning**, which means arriving at a conclusion based on one's observations. Inductive reasoning is the way we come up with new ideas about groups of objects or events. Note that a scientific hypothesis must be testable in the real world. Consider how, for example, there is no way to test the following hypothesis: The number of birds is decreasing because undetectable aliens are taking them to another planet. (Section 1.8 returns to the topic of testability.)

A **prediction**, or statement of some condition that should occur if the hypothesis is correct, comes next. Making predictions is called the if–then process, in which the "if" part is the hypothesis, and the "then" part is the prediction: *If* the number of birds is decreasing because the number of cats is increasing, *then* removing cats from the area should stop the decline in the bird population. Using a hypothesis to make a prediction is a form of **deductive reasoning**, the logical process of using a general premise to draw a conclusion about a specific case.

Next, a researcher tests the "then" part of the prediction. Some predictions are tested by systematic observation; others require experimentation. **Experiments** are tests designed to determine whether a prediction is valid. In an investigation of our hypothetical bird–cat relationship, the researcher may remove all cats from the area. If working with a subject or event directly is not possible, experiments may be performed on a **model**, or analogous system (for example, animal diseases are often used as models of human diseases).

A typical experiment explores a cause and effect relationship using variables. A **variable** is an experimental factor that varies, for example a characteristic or that differs among individuals, or an event that differs over time. An **independent variable** is defined or controlled by the person doing the experiment (in the bird–cat experiment, the independent variable is the presence or absence of cats). A **dependent variable** is presumed to be influenced by the independent variable (the dependent variable in the bird–cat experiment is the number of birds).

Biological systems are complex because they typically involve many interdependent variables. It can be difficult or even impossible to study one variable separately from the rest. Thus, biology researchers often test two groups of individuals simultaneously. An **experimental group** is a set of individuals that have a certain characteristic or receive a certain treatment. This group is tested side by side with a **control group**, which is identical to the experimental group except for one independent variable: the characteristic or the treatment being tested. Any difference in experimental results between the two groups is likely to be an effect of changing the variable.

Test results—**data**—offer a quantifiable way of evaluating the hypothesis. Data that validate the prediction are evidence in support of the hypothesis. Data that show the prediction is invalid are evidence that the hypothesis is flawed and should be revised.

A necessary part of science is reporting one's data and conclusions in a standard way, such as in a peer-reviewed journal article. The communication gives other scientists an opportunity to evaluate the tested hypotheses, both by repeating the experiments and by checking conclusions drawn from them.

Forming a hypothesis based on observation, and then systematically testing and evaluating the hypothesis, are collectively called the **scientific method** (**TABLE 1.3**). However, scientific research—particularly in biology—rarely proceeds in a direct, start-to-finish fashion as Table 1.3 might suggest. Researchers often describe their work as a nonlinear process of exploration, asking questions, testing hypotheses, and changing directions. Hypotheses are often wrong and experimental results are often unpredictable. Research usually raises more questions than it answers, so there may be no endpoint. The unpredictability might be frustrating at times, but researchers typically say they enjoy their work because of the surprising twists and turns it takes.

Research in the Real World

Biology is the branch of science concerning past and present life, and it comprises hundreds of research specializations (**FIGURE 1.11** and **TABLE 1.4**). To give

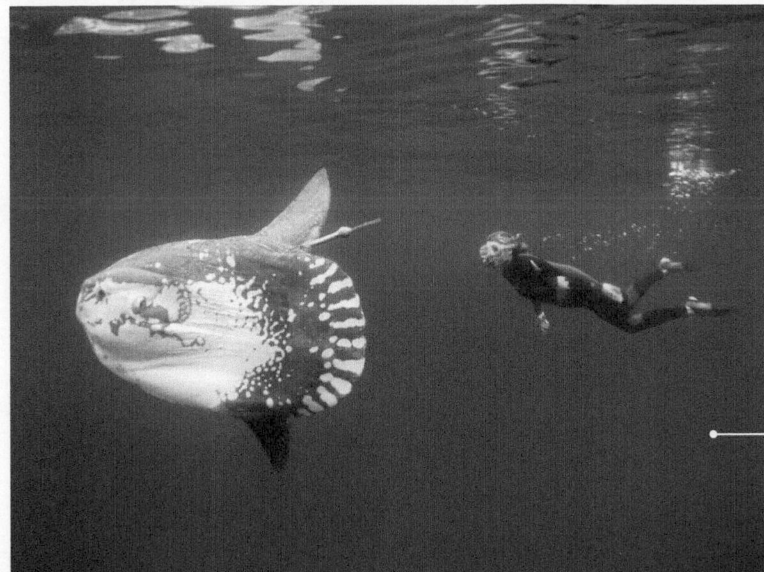

TABLE 1.4

A Few Research Specializations in Biology

Field	Focus
Astrobiology	Potential life elsewhere in the universe
Biogeography	Distribution of life on Earth
Bioinformatics	Development of tools to analyze data
Botany	Plant structure and processes
Cell biology	Cell structure and processes
Ecology	Interactions among organisms, and among organisms and their environment
Ethology	Animal behavior
Genetics	Inheritance
Marine biology	Life in saltwater habitats
Medicine	Human health
Paleontology	Life in the ancient past
Structural biology	Architecture-dependent function of large biological molecules

FIGURE 1.11 Examples of scientific research in biology. Left, National Geographic Explorer Tierney Thys travels the world's oceans to study the giant sunfish (mola). "When it comes to fishes, the mola really pushes the boundary of fish form," she says. "It seems a somewhat counterintuitive design for plying the waters of the open seas—a rather goofy design—and yet the more I learn about it, the more respect and admiration I have for it."

you a sense of how biology research works, we summarize two published studies here.

Potato Chips and Stomachaches In 1996, the U.S. Food and Drug Administration (the FDA) approved Olestra®, a fat replacement manufactured from sugar and vegetable oil, as a food additive. Potato chips were the first Olestra-containing food product to be sold in the United States. Controversy about the chip additive soon raged. Many people complained of intestinal problems after eating the chips, and thought that the Olestra was at fault. Two years later, researchers at the Johns Hopkins University School of Medicine designed an experiment to test whether Olestra causes cramps.

The researchers made the following prediction: If Olestra causes cramps, then people who eat Olestra should be more likely to get cramps than people who do not eat it. To evaluate their prediction, they used a Chicago theater as a "laboratory." They asked 1,100 people between the ages of thirteen and thirty-eight to watch a movie and eat their fill of potato chips. Each person received an unmarked bag that contained 13 ounces of chips.

In this experiment, the individuals who received Olestra-containing potato chips constituted the experimental group, and individuals who received regular chips were the control group. The independent variable was the presence or absence of Olestra in the chips.

A few days later, the researchers contacted everyone and collected reports of any post-movie gastrointestinal problems (the dependent variable). Of 563 people making up the experimental group, 89 (15.8 percent) complained about cramps. However, so did 93 of the 529 people (17.6 percent) making up the control group—who had eaten the regular chips. People were about as likely to get cramps whether or not they ate chips made with Olestra. These results showed that the prediction was invalid, so the researchers concluded that eating Olestra does not cause cramps.

Butterflies and Birds The peacock butterfly is a winged insect named for the large, colorful spots on its wings. In 2005, researchers reported the results

control group A group of individuals identical to an experimental group except for the independent variable under investigation.
data (DAY-tuh) Experimental result(s).
deductive reasoning Using a general idea to make a conclusion about a specific case.
dependent variable In an experiment, a variable that is presumably affected by an independent variable being tested.
experimental group In an experiment, group of individuals who have a certain characteristic or receive a certain treatment as compared with a control group.
experiment A test designed to support or falsify a prediction.
hypothesis (high-POTH-uh-sis) A testable explanation of a natural phenomenon.
independent variable A variable that is controlled by an experimenter in order to explore its relationship to a dependent variable.
inductive reasoning Drawing a conclusion based on observation.
model An analogous system used for testing hypotheses.
prediction A statement, based on a hypothesis, about a condition that should occur if the hypothesis is correct.
scientific method Making, testing, and evaluating hypotheses.
variable (VAIR-ee-uh-bull) In an experiment, a characteristic or event that differs among individuals or over time.

A With wings folded, a peacock butterfly resembles a dead leaf, so it is appropriately camouflaged from predatory birds.

B When a predatory bird approaches, a butterfly flicks its wings open and closed, revealing brilliant spots and producing hissing and clicking sounds.

C Researchers tested whether the wing-flicking and sound-making behaviors of peacock butterflies affected predation by blue tits (a type of songbird).

Experimental Treatment	Number of Butterflies Eaten
Wing spots concealed	5 of 10 (50%)
Wings silenced	0 of 8 (0%)
Wing spots painted out and wings silenced	8 of 10 (80%)
No treatment	0 of 9 (0%)

**Proceedings of the Royal Society of London, Series B (2005) 272: 1203–1207.*

D The researchers painted out the spots of some butterflies, cut the sound-making part of the wings on others, and did both to a third group; then exposed each butterfly to a hungry blue tit for 30 minutes. Results support only the hypothesis that peacock butterfly spots deter predatory birds.

FIGURE 1.12 Testing the defensive value of two peacock butterfly behaviors.

FIGURE IT OUT What was the dependent variable in this series of experiments?

Answer: Getting eaten

of experiments investigating whether certain behaviors help these butterflies defend themselves against insect-eating birds. The study began with the observation that a resting peacock butterfly sits motionless, wings folded (**FIGURE 1.12A**). The dark underside of the wings provides appropriate camouflage. However, when a predator approaches, the butterfly exposes its brilliant spots by repeatedly flicking its wings open and closed (**FIGURE 1.12B**). At the same time, it moves the hindwings in a way that produces a hissing sound and a series of clicks. A colorful, moving, noisy insect is usually very attractive to insect-eating birds, so the researchers were curious about why the peacock butterfly moves and makes noises only in the *presence* of predators. After they reviewed earlier studies, the scientists made two hypotheses that might explain the wing-flicking behavior:

- Hypothesis 1: Peacock butterflies flick their wings in the presence of a predator because it exposes brilliant wing spots, thereby reducing predation. Peacock butterfly wing spots resemble owl eyes, and anything that looks like owl eyes is known to startle insect-eating birds.
- Hypothesis 2: Peacock butterflies hiss and click when they flick their wings because these sounds reduce predation by birds. The sounds may be an additional defense that startles insect-eating birds.

The researchers used these hypotheses to make the following predictions:

- Prediction 1: If exposing brilliant wing spots by wing-flicking reduces predation by insect-eating birds, then removing the wingspots from peacock butterflies should make them more likely to get eaten.
- Prediction 2: If the hissing and clicking sounds produced during wing-flicking reduce predation by insect-eating birds, then silencing peacock butterflies should make them more likely to get eaten.

Next came the experiments. The researchers used a black marker to conceal the wing spots of some butterflies. They cut the wings of other butterflies to disable the sound-making parts. A third group had both treatments: Their spots were concealed and their wings were silenced. Each butterfly was then put into a large cage with a hungry blue tit (**FIGURE 1.12C**), and the pair was watched for 30 minutes.

FIGURE 1.12D lists the results. All of the butterflies with unmodified wing spots survived, regardless of whether they made sounds. These results were consistent with the first hypothesis: By exposing brilliant spots, peacock butterfly wing-flicking behavior decreases predation by blue tits.

In contrast, a large proportion of butterflies with concealed spots got eaten, whether or not they could make sounds. Experimental results were not consistent

CREDITS: (12A) © Matt Rowlings, www.eurobutterflies.com; (12B) © Adrian Vallin/Stockholm University; (12C) © Antje Schulte.

Data Analysis Activities

Peacock Butterfly Predator Defenses The photographs below represent the experimental and control groups used in the peacock butterfly experiment discussed in Section 1.6. See if you can identify the experimental groups and match them up with the relevant control group(s). *Hint:* Identify which variable is being tested in each group (each variable has a control).

A Wing spots painted out

B Wing spots visible; wings silenced

C Wing spots painted out; wings silenced

D Wings painted but spots visible

E Wings cut but not silenced

F Wings painted, spots visible; wings cut, not silenced

with the second hypothesis that peacock butterfly sounds reduce predation by birds. Other questions raised by these results offer an example of how research often leads to more research: Do other predatory birds respond differently to the sounds than blue tits? If not, do the sounds reduce predation by other organisms (such as mice) that eat peacock butterflies? If sound-making is unrelated to predation, what is its function? Several additional experiments would be necessary to answer these questions.

TAKE-HOME MESSAGE 1.6

- ✔ Judging the quality of information before accepting it is an active process called critical thinking. Critical thinking is central to science.
- ✔ The scientific method consists of making, testing, and evaluating hypotheses about natural phenomena.
- ✔ Experiments determine how changing an independent variable affects a dependent variable.
- ✔ Natural systems are typically complex and influenced by many interacting variables. Researchers can unravel cause-and-effect relationships by changing one independent variable at a time.

1.7 Analyzing Experimental Results

LEARNING OBJECTIVES

- Explain why it is risky to draw generalizations from a subset.
- Name some ways that researchers minimize sampling error and bias.
- Describe statistical significance.
- Discuss some areas of inquiry that science does not address.

Sampling Error

In a natural setting, researchers can rarely observe all individuals of a population or all instances of an event. For example, the biologists who explored the cloud forest you read about in Section 1.1 did not—and could not—survey every uninhabited part of the Foja Mountains. The cloud forest itself cloaks more than 2 million acres, so surveying all of it would take unrealistic amounts of time and effort.

When researchers cannot directly observe an entire set of subjects or events, they may test or survey a subset. Results from the subset are then used to make generalizations about the whole. However, generalizing from a subset is risky because subsets are not necessarily representative of the whole. Consider the golden-mantled tree kangaroo, which was first discovered in 1993 on a single forested mountaintop in New Guinea. For more than a decade, the species was never seen outside of that habitat, which is getting smaller every year because of human activities. Thus, the golden-mantled tree kangaroo was considered to be one of the most endangered animals on the planet. Then, in 2005, explorers discovered that this kangaroo species is fairly common in a cloud forest of the Foja Mountains (**FIGURE 1.13**). As a result, biologists now believe its future is secure, at least for the moment.

FIGURE 1.13 Kris Helgen holds a golden-mantled tree kangaroo he found during a 2005 survey of a cloud forest in the Foja Mountains.

This kangaroo species is extremely rare in other areas, so it was thought to be critically endangered prior to the expedition.

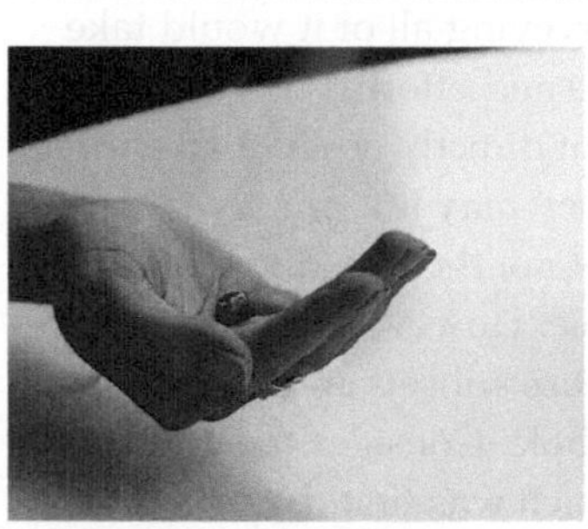

A Natalie chooses a random jelly bean from a jar.

She is blindfolded, so she does not know that the jar contains 120 green and 280 black jelly beans.

The jar is hidden from Natalie's view before she removes her blindfold.

She sees one green jelly bean in her hand and assumes that the jar must hold only green jelly beans (100 percent are green).

This assumption is incorrect: 30 percent of the jelly beans in the jar are green, and 70 percent are black. The deviation is sampling error.

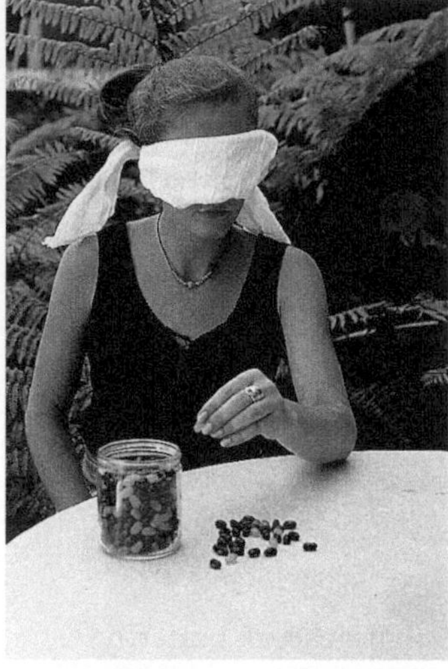

B Still blindfolded, Natalie randomly picks out 50 jelly beans from the jar. She chooses 10 green and 40 black ones.

The larger sample leads Natalie to estimate that one-fifth of the jar's jelly beans are green (20 percent) and four-fifths are black (80 percent). The larger sample more closely approximates the jar's actual green-to-black ratio of 30 percent to 70 percent.

The more jelly beans that Natalie chooses, the closer her estimates will be to the actual ratio.

FIGURE 1.14 Demonstration of sampling error.

Sampling error is a difference between results obtained from a subset, and results from the whole (**FIGURE 1.14A**). Sampling error may be unavoidable, but knowing how it can occur helps researchers design their experiments to minimize it. For example, sampling error can be a substantial problem with a small subset, so experimenters try to start with a relatively large sample (**FIGURE 1.14B**), and they typically repeat their tests.

To understand why these practices reduce the risk of sampling error, think about flipping a coin. There are two possible outcomes of each flip: The coin lands heads up, or it lands tails up. Thus, the chance that the coin will land heads up upon being flipped is one in two (1/2), which is a 50 percent chance. However, when you flip a coin repeatedly, it often lands heads up, or tails up, several times in a row. With just 4 flips, the proportion of times that heads actually land up may not even be close to 50 percent. With 1,000 flips, however, the overall proportion of times the coin lands heads up is much more likely to approach 50 percent.

In cases such as flipping a coin, it is possible to calculate **probability**, which is the measure, expressed as a percentage, of the chance that a particular outcome will occur. That chance depends on the total number of possible outcomes. For instance, if 10 million people enter a drawing, each has the same probability of winning: 1 in 10 million, or (an extremely improbable) 0.00001 percent.

A result that is very unlikely to have occurred by chance alone is **statistically significant**. In this context, the word "significant" does not refer only to the result's importance. Rather, it means that a rigorous statistical analysis has shown a very low probability (usually 5 percent or less) of the result being incorrect because of sampling error.

Variation in data is often shown as error bars on a graph (**FIGURE 1.15**). Depending on the graph, error bars may indicate (for example) variation around an average, the difference between two sample sets.

Bias in Interpreting Results

Experimenting with a single variable apart from all others is not often possible, particularly when studying humans. For example, remember that the people who participated in the Olestra experiment were chosen randomly, which means the study was not controlled for gender, age, weight, medications taken, and so on. These variables may well have influenced the experiment's results.

Scientists, like all other humans, are subjective by nature. Researchers risk interpreting their results

CREDIT: (14) Gary Head.

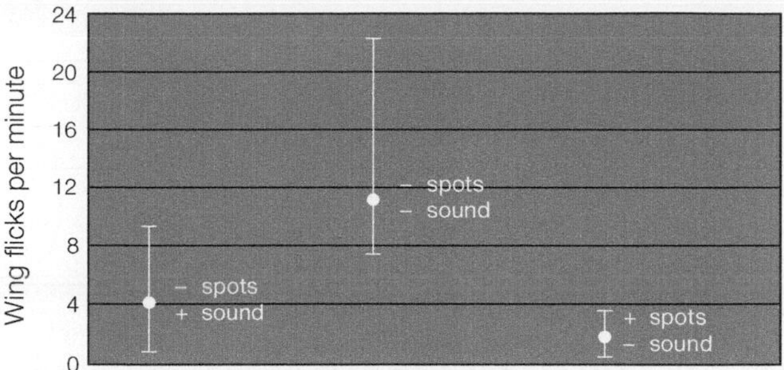
Wing flicks per minute
24
20
16
12
8
4
0
– spots
+ sound
– spots
– sound
+ spots
– sound

FIGURE 1.15 Example of error bars in a graph.
This graph was adapted from the peacock butterfly research described in Section 1.6. The researchers recorded the number of times each butterfly flicked its wings in response to an attack by a bird.

The dots represent average frequency of wing flicking for each sample set of butterflies. The error bars that extend above and below the dots indicate the range of values—the sampling error.

FIGURE IT OUT What was the fastest rate at which a butterfly with no spots or sound flicked its wings?

Answer: 22 times per minute

in terms of what they want to find out. That is why they typically design experiments that will yield quantitative results, which are measurements or some other data that can be gathered objectively. Such results minimize the potential for bias.

This last point gets us back to the role of critical thinking in science. Scientists expect one another to recognize and put aside bias in order to test their hypotheses in ways that may prove them wrong. If a researcher does not, then others will, because exposing errors is just as useful as applauding insights. The scientific community consists of critically thinking people trying to poke holes in one another's ideas. Their collective efforts make science a self-correcting endeavor.

TAKE-HOME MESSAGE 1.7

- ✔ Checks and balances inherent in the scientific process help researchers to be objective about their observations.
- ✔ Sampling error is minimized by using large sample sizes and by repeating experiments.
- ✔ Probability calculations can show whether a result is statistically significant: highly unlikely to have occurred by chance alone.
- ✔ Science is, ideally, a self-correcting process because the scientific community uses critical thinking to challenge results and conclusions.

probability The chance that a particular outcome of an event will occur; depends on the total number of outcomes possible.
sampling error A difference between results derived from testing an entire group of events or individuals, and results derived from testing a subset of the group.
scientific theory A hypothesis that has not been falsified after many years of rigorous testing, and is useful for making predictions about a wide range of phenomena.
statistically significant Refers to a result that is statistically unlikely to have occurred by chance alone.

1.8 The Nature of Science

LEARNING OBJECTIVES

- Name the criteria that qualify a hypothesis for status as a scientific theory.
- Explain what happens when data arise that are inconsistent with a theory.
- Distinguish science from pseudoscience.

What Science Is

A scientific hypothesis, remember, is a testable explanation for a natural phenomenon, and testing a hypothesis may prove it false. Suppose a hypothesis survives after serious attempts to falsify it by years of rigorous testing. It is consistent with existing evidence, and scientists use it to make successful predictions about a wide range of other phenomena. A hypothesis that meets these criteria is a **scientific theory** (**TABLE 1.5**).

Consider the hypothesis that matter consists of atoms and their tiny components, which are called subatomic particles. Researchers no longer spend time testing this hypothesis for the compelling reason that, since we started looking 200 years ago, no one has discovered matter that consists of anything else. Thus, the hypothesis—now part of atomic theory—has been incorporated into our general understanding of matter. This understanding underpins research in many fields, including biology.

Scientific theories are our best objective descriptions of the natural world. However, they can never be proven under every possible circumstance. For example, proving that atomic theory holds under all circumstances would mean the composition of all matter in the universe would have to be checked—an impossible task even if someone wanted to try. Thus, scientists avoid using the word "proven" to describe a theory. Instead, a theory is "accepted," along with

TABLE 1.5

Examples of Scientific Theories

Atomic theory	All matter consists of atoms and their smaller subatomic parts; properties of matter arise from these components.
Big bang	The universe originated with an explosion of high-density matter.
Cell theory	All organisms consist of one or more cells, the cell is the basic unit of life, and all cells arise from preexisting cells.
Climate change	Earth's average temperature is rising; human activities are part of the cause.
Evolution by natural selection	Environmental pressures drive change in the inherited traits of a population.
Plate tectonics	Earth's lithosphere (crust and upper mantle) is cracked into pieces that move in relation to one another.

the possibility—however remote—that new data inconsistent with it might be found.

What happens if someone discovers an exception to a theory? By definition, a scientific theory has been tested rigorously and repeatedly. New results do not invalidate previous results, but the interpretation of what the results mean can change. Thus, new data inconsistent with a theory may trigger its revision. Atomic theory has been modified many times since it was proposed thousands of years ago. Ancient Greeks defined atoms as indivisible units of matter (*atomo* is the Greek word for "uncuttable"), but they had no way of knowing what atoms actually are. Their definition prevailed until 1897, when the newly invented cathode ray tube allowed Joseph J. Thomson to discover the existence of subatomic particles. Atoms were not indivisible after all! This new understanding did not falsify atomic theory, but it did prompt a revision that removed "indivisible" from the definition of atoms. Even today's version of atomic theory is revised periodically as improved equipment reveals new information about subatomic particles. If someone ever discovers matter that does not consist of atoms and subatomic particles, atomic theory would be revised to include the exception (all matter consists of atoms and subatomic particles except . . .). If many exceptions accumulate, the theory will be rewritten so it better accounts for the discrepancies.

The theory of evolution by natural selection, which holds that environmental pressures can drive change in the inherited traits of a population, still stands after more than a century of concerted observations, testing, and challenges. Natural selection is not the only mechanism by which evolution occurs, but it is by far the most studied. Few other scientific theories have withstood as much scrutiny.

You may hear people apply the word "theory" to a speculative idea, as in the phrase "It's just a theory." This everyday usage of the word differs from the specific way it is used in science. A scientific theory also differs from a **law of nature**, which describes a phenomenon that always occurs under certain circumstances. Laws do not include mechanisms, so they do not necessarily have a complete scientific explanation. The laws of thermodynamics, which describe energy, are examples. We understand how energy behaves, but not exactly why it behaves the way it does (Chapter 5 returns to energy).

law of nature A generalization that describes a consistent natural phenomenon that has an incomplete scientific explanation.

pseudoscience A claim, argument, or method that is presented as if it were scientific, but is not.

TABLE 1.6

Distinguishing Science from Pseudoscience

Science	Pseudoscience
Is the claim about an observation of a natural phenomenon?	
Addresses only the observable, natural world.	Often involves supernatural or mysterious phenomena.
Is the explanation for the observation testable?	
Hypotheses are testable in ways that might falsify them. Tests are repeatable and yield data.	Claims are not testable or falsifiable by valid experimentation or observation.
Are the findings supported by valid scientific evidence?	
Based on systematic observations and experiments that have been evaluated by members of the scientific community.	Based on invented or unverified "facts," anecdotes, rhetoric, or the absence of scientific evidence.
Is the conclusion consistent with other scientific evidence?	
Consistent with most or all existing data and observations.	Typically indifferent to scientific evidence and often denies it.
Is the process self-correcting?	
Rigorous standards for ensuring accuracy and objectivity.	No requirement for accuracy, truthfulness, or verification.

What Is Not Science

Not everything that uses scientific vocabulary is actually science. Claims, arguments, or methods that are presented as scientific but do not follow scientific principles are **pseudoscience** (*pseudo* means false).

Distinguishing pseudoscience from the real thing can be tricky, but your critical-thinking skills will help (**TABLE 1.6**). Consider how the scientific method begins with observations (**FIGURE 1.16**) and ends with conclusions. By contrast, pseudoscience often begins with a conclusion ("eating bananas results in weight loss," for example), followed by a search for supporting information. That information may be invented, unverified, anecdotal ("my aunt said she lost weight when she ate bananas"), or taken from a resource that presents such information. Conflicting scientific evidence is typically ignored, dismissed, or denied.

Pseudoscience can involve mysterious phenomena: "Earth appears older than it is because it came into existence that way," for example. Such claims are impossible to test or falsify—in this case, because no measurement or observation could possibly reveal Earth's true age. A statement that cannot be potentially proven wrong by testing is not part of science. Remember, testing a scientific hypothesis is technically a deliberate attempt to falsify it.

Accuracy and objectivity are primary goals of science. Tests are repeatable, and results are communicated to and evaluated by a community of skeptics. Pseudoscience has no requirement for accuracy, truthfulness, or objectivity; communication usually consists of rhetoric intended to persuade the general public.

FIGURE 1.16 Observing an aspect of nature. Near a tent serving as a makeshift laboratory, herpetologist Paul Oliver records the call of a frog on the first expedition to New Guinea's Foja Mountains cloud forest in 2005.

What Science Is Not

Science helps us to be objective because it is limited to testable ideas about observable aspects of nature. For example, science does not address philosophical questions such as "Why do I exist?" Answers to questions like this one can only come from within, as an integration of all the personal experiences and mental connections that shape our consciousness. This is not to say subjective answers have no value, because no human society can function for long unless its individuals share standards for making judgments, even if they are subjective. Moral, aesthetic, and philosophical standards vary from one society to the next, but all help people decide what is important and good. All give meaning to our lives.

Neither does science address the supernatural (anything that is "beyond nature"). Science neither assumes nor denies that supernatural phenomena occur, but scientists may cause controversy when they discover a natural explanation for something that was thought to have none. Such controversy often arises when a society's moral standards are interwoven with its understanding of nature. Consider what happened after Nicolaus Copernicus proposed in the early 1500s that Earth orbits the sun. Today that idea is generally accepted, but the prevailing belief system had Earth as the immovable center of the universe. In 1610, astronomer Galileo Galilei published evidence for the Copernican model of the solar system, an act that resulted in his imprisonment. He was publicly forced to recant his work, spent the rest of his life under house arrest, and was never allowed to publish again.

As Galileo's story illustrates, exploring a traditional view of the natural world from a scientific perspective may be misinterpreted as a violation of morality. As a group, scientists are no less moral than anyone else, but their work occurs by a particular process—the scientific method—that other professions do not require.

Science helps us objectively communicate our experiences of the natural world, and as such, it may be as close as we can get to a universal language. We are fairly sure, for example, that the laws of gravity apply everywhere in the universe. If we discover intelligent beings on a distant planet, they would likely understand the concept of gravity. Thus, we might well use gravity or another scientific concept to communicate with them, or anyone, anywhere. The point of science, however, is not to communicate with aliens. It is to find common ground here on Earth.

TAKE-HOME MESSAGE 1.8

- ✔ Science is concerned only with testable ideas about observable aspects of the observable world.
- ✔ Because a scientific theory is thoroughly tested and revised until no one can prove it wrong, it is our best way of describing reality.
- ✔ Pseudoscience is a claim, argument, or method that is presented as science but does not follow scientific practices or principles of accuracy, honesty, repeatability, and objectivity.

1.1 Application: Secret Life of Earth (revisited)

New species are discovered all the time, even in places much less exotic than a cloud forest in New Guinea. A survey in 2015 revealed 30 new insect species living in urban Los Angeles, for example. Each discovery is a reminder that we do not yet know all of the organisms that share our planet. We don't even know how many to look for. Why does that matter? Trying to understand the immense scope of life on Earth gives us some perspective on where we fit into it. For example, hundreds of new species are discovered every year, but about 20 species become extinct every minute in rain forests alone—and those are only the ones we know about. The current rate of extinctions is about 1,000 times faster than normal, and human activities are responsible for the acceleration. At this rate, we will never know about most of the species that are alive on Earth today. Does that matter? Biologists think so. Whether or not we are aware of it, humans are intimately connected with the world around us. Our activities are profoundly changing the entire fabric of life on Earth. These changes are, in turn, affecting us in ways we are only beginning to understand. ●

CREDIT: (16) Tim Laman/National Geographic Creative.

STUDY GUIDE

Section 1.1 **Biology** is the scientific study of life. We know of only a fraction of the **species** that live on Earth, in part because we have explored only a fraction of its inhabited regions. Understanding the scope of Earth's biodiversity gives us perspective on where we fit into it.

Section 1.2 Biologists think about life at different levels of organization, with **emergent properties** appearing at successive levels.

All matter, living or not, consists of **atoms** and their subatomic components. Atoms combine as **molecules**. The unique properties of life emerge as certain kinds of molecules become organized into a **cell**. **Organisms** are individuals that consist of one or more cells. In larger multicelled organisms, cells are organized as **tissues**, **organs**, and **organ systems**.

A **population** is a group of interbreeding individuals of a **species** in a given area; a **community** is all populations of all species in a given area. An **ecosystem** is a community interacting with its environment. The **biosphere** includes all regions of Earth that hold life.

Section 1.3 Life has underlying unity in that all living things have similar characteristics. For example, all organisms require energy and **nutrients** to sustain themselves. **Producers** harvest energy from the environment to make their own food by processes such as **photosynthesis**; **consumers** ingest other organisms, their wastes, or remains. All organisms sense and respond to change, making adjustments that keep conditions in their internal environment within tolerable ranges—a process called **homeostasis**. Information in an organism's **DNA** guides its **growth**, **development**, and **reproduction**. The passage of DNA from parents to offspring is called **inheritance**. DNA is the basis of similarities and differences among organisms.

Section 1.4 The many types of organisms that currently exist on Earth differ greatly in form and function. **Bacteria** and **archaea** are **prokaryotes**: single-celled organisms whose DNA is not contained within a nucleus. The DNA of single-celled or multicelled **eukaryotes** (**protists**, **plants**, **fungi**, and **animals**) is contained within a nucleus.

Section 1.5 Each species is given a two-part name. The first part is the **genus** name. When combined with the **specific epithet**, it designates the particular species. With **taxonomy**, species are ranked into ever more inclusive **taxa** (genus, family, order, class, phylum, kingdom, domain) on the basis of shared inherited **traits**.

Section 1.6 **Critical thinking**, the act of judging the quality of information as one learns, is an important part of **science**. Generally, a researcher observes something in nature, uses **inductive reasoning** to form a **hypothesis** (testable explanation) for it, then uses **deductive reasoning** to make a testable **prediction** about what might occur if the hypothesis is correct. Predictions are evaluated with observations, experiments, or both. **Experiments** involve **variables**: A researcher typically changes an **independent variable**, then observes the effects of the change on a **dependent variable**. A **model** may be used if working directly with a subject or event is not possible. Results from testing an **experimental group** are compared with results from a **control group**. Conclusions are drawn from experimental **data**.

The **scientific method** consists of making, testing, and evaluating hypotheses, and sharing results with the scientific community. Research in the real world rarely proceeds in a linear manner; rather, it tends to be a nonlinear process involving exploration, asking questions, testing hypotheses, and changing directions. Biological systems in particular are complex and typically influenced by many interacting variables, so it can be difficult to study a single cause and effect relationship in biology research.

Section 1.7 Checks and balances inherent in the scientific process help researchers to be objective about their observations. **Sampling error** is minimized by using large sample sizes and by repeating experiments. **Probability** calculations can show whether a result has **statistical significance** (it is very unlikely to have occurred by chance alone). Science is ideally self-correcting because it is carried out by a large community of people systematically checking one another's data and conclusions.

Section 1.8 Opinion and belief have value in human culture, but they are not part of science. Science addresses only testable ideas about observable aspects of the natural world. Testing a hypothesis is a deliberate attempt to falsify it. A **scientific theory** is a hypothesis that stands after years of rigorous testing, and is useful for making predictions about a wide range of other phenomena. Scientific theories may be revised upon the discovery of new data. They are our most objective way of describing the natural world.

A **law of nature** describes a consistent natural phenomenon but not an explanation for it.

Pseudoscience is a claim, argument, or method that is not scientific but presented as if it were; there is no requirement for following scientific practices, or for accuracy, honesty, repeatability, or objectivity.

SELF-ASSESSMENT

Answers in Appendix VII

1. The smallest unit of any substance is the ____ .
   a. atom b. molecule c. cell
2. The smallest unit of life is the ______ .
   a. atom c. cell
   b. molecule d. organism

3. Organisms require ______ and ______ to maintain themselves, grow, and reproduce.
   a. DNA; energy
   b. food; sunlight
   c. nutrients; energy
   d. DNA; cells
4. By sensing and responding to change, organisms keep conditions in the internal environment within ranges that cells can tolerate. This process is called ______ .
   a. reproduction
   b. development
   c. homeostasis
   d. inheritance
5. DNA ______ .
   a. guides form and function
   b. is the basis of traits
   c. is transmitted from parents to offspring
   d. all of the above
6. ______ is the transmission of DNA to offspring.
   a. Reproduction
   b. Development
   c. Homeostasis
   d. Inheritance
7. A butterfly is a(n) ______ (choose all that apply).
   a. organism
   b. domain
   c. species
   d. eukaryote
   e. consumer
   f. producer
   g. prokaryote
   h. trait
8. ______ move around for at least part of their life.
   a. Living organisms
   b. Animals
   c. Species
   d. Eukaryotes
9. A bacterium is ______ (choose all that apply).
   a. an organism
   b. single-celled
   c. an animal
   d. a eukaryote
10. Bacteria, Archaea, and Eukarya are three ______ .
11. A control group is ______ .
   a. a set of individuals that have a certain characteristic or receive a certain treatment
   b. the standard against which an experimental group is compared
   c. the experiment that gives conclusive results
12. Fifteen randomly selected students are found to be taller than 6 feet. The researchers concluded that the average height of a student is greater than 6 feet. This is an example of ______ .
   a. experimental error
   b. sampling error
   c. a subjective opinion
   d. experimental bias
13. Science only addresses that which is ______ .
   a. alive
   b. observable
   c. variable
   d. indisputable
14. Which of the following statements cannot be falsified?
   a. Some of the fish in Lake Michigan are brown.
   b. None of the fish in Lake Michigan are brown.
   c. All of the fish in Lake Michigan are brown.

15. Match the terms with the most suitable description.

___ data	a. if–then statement
___ probability	b. unique type of organism
___ species	c. experimental results
___ hypothesis	d. testable explanation
___ prediction	e. measure of chance
___ producer	f. makes its own food

CRITICAL THINKING

1. A person is declared dead upon the irreversible ceasing of spontaneous body functions: brain activity, blood circulation, and respiration. Only about 1 percent of a body's cells have to die in order for all of these things to happen. How can a person be dead when 99 percent of his or her cells are alive?

2. What is the difference between a one-celled organism and a single cell of a multicelled organism?

3. Why would you think twice about ordering from a restaurant menu that lists the specific epithet but not the genus name of its offerings? *Hint:* Look up *Homarus americanus, Ursus americanus, Ceanothus americanus, Bufo americanus, Lepus americanus,* and *Nicrophorus americanus.*

4. Once there was a highly intelligent turkey that had nothing to do but reflect on the world's regularities. Morning always started out with the sky turning light, followed by the master's footsteps, which were always followed by the appearance of food. Other things varied, but food always followed footsteps. The sequence of events was so predictable that it eventually became the basis of the turkey's theory that footsteps bring food. One morning, after more than 100 confirmations of this theory, the turkey listened for the master's footsteps, heard them, and had its head chopped off.

   Scientific theories can be revised upon the discovery of inconsistent evidence. Suggest how the turkey's theory might be modified so the remaining members of the flock would find it more useful for making predictions.

5. In 2005, researcher Woo-suk Hwang reported that he had made immortal stem cells from human patients. His research was hailed as a breakthrough for people affected by degenerative diseases, because stem cells may be used to repair a person's own damaged tissues. Hwang published his results in a peer-reviewed journal. In 2006, the journal retracted his paper after other scientists discovered that Hwang's group had faked their data. Does the incident show that results of scientific studies cannot be trusted? Or does it confirm the usefulness of a scientific approach, because other scientists discovered and exposed the fraud?

2 LIFE'S CHEMICAL BASIS

CORE CONCEPTS

Systems

Complex properties arise from interactions among components of a biological system. When components of a biological system interact, they enable properties not found in the individual components alone. The behavior of atoms arises from the number and type of subatomic particles that compose them. Similarly, the properties of molecules arise from the types and arrangement of their component atoms, and the properties of substances arise from the kinds and amounts of their component molecules. Unique properties that make water essential to life arise from the polarity of individual molecules.

Pathways of Transformation

Organisms exchange matter and energy with the environment in order to grow, maintain themselves, and reproduce. All organisms harvest atomic and molecular building blocks from their environment. Electrons have energy that can be stored in and retrieved from atoms and molecules they compose. The movement of atoms and molecules is influenced by external factors such as temperature.

Information Flow

Living systems store, retrieve, transmit, and respond to information essential for life. Homeostatic mechanisms allow living things to maintain themselves by responding dynamically to internal and external conditions, including shifts in pH and temperature. Organisms keep their internal environment stable by returning a changed condition back to its target set point, but extreme variation can overwhelm these mechanisms.

Links to Earlier Concepts

In this chapter, you will explore the first level of life's organization—atoms—as you encounter an example of how the same building blocks, arranged different ways, form different products (Section 1.2). You will also see one aspect of homeostasis, the process by which organisms keep themselves in a state that favors cell survival (Section 1.3).

2.1 Mercury Rising

Mercury is a naturally occurring toxic element. Most of it is locked away in rocky minerals, but volcanic activity and other geologic processes release it into the atmosphere. So do human activities, especially burning coal (**FIGURE 2.1**). Airborne mercury can drift long distances before settling to Earth's surface, where microorganisms combine it with carbon to form a substance called methylmercury. Methylmercury is even more toxic than mercury because it easily crosses skin and mucous membranes. In water, it ends up in the tissues of aquatic organisms. All fish and shellfish contain it. Humans contain it too, mainly as a result of eating seafood.

Any form of mercury that enters body tissues can damage the nervous system, brain, kidneys, and other organs. An average-sized adult who ingests as little as 200 micrograms of methylmercury may experience blurred vision, tremors, itching or burning sensations, and loss of coordination. Exposure to larger amounts can result in thought and memory impairment, coma, and death. Methylmercury in a pregnant woman's blood passes to her unborn child, along with a legacy of permanent developmental problems.

It takes months or even years for mercury to be cleared from the body, so the toxin can build up to high levels if even small amounts are ingested on a regular basis. That is why large predatory fish have a lot of mercury in their tissues. It is also why the U.S. Environmental Protection Agency recommends that adult humans ingest less than 0.1 microgram of mercury per kilogram of body weight per day. That is about 7 micrograms per day—not a big amount if you eat seafood. A typical 6-ounce can of albacore tuna contains about 60 micrograms of mercury, and

FIGURE 2.1 The Navajo Generating Station, a coal-fired power plant in Arizona. Emissions from such plants are by far the largest source of mercury pollution worldwide.

CREDITS: (opposite) Francois Gohier/Science Source; (1) Ed Darack/Science Faction/SuperStock.

the occasional can has many times that amount. It does not matter if the fish is canned, grilled, or raw, because methylmercury is unaffected by cooking. Eat a medium-sized tuna steak, and you could be getting more than 700 micrograms of mercury along with it.

With this chapter, we turn to the first of life's levels of organization: atoms. Interactions between atoms make the molecules that can sustain life, and also some that destroy it. ●

2.2 Building Blocks of Matter

LEARNING OBJECTIVES

- Use an example to explain why we say that an atom is the smallest unit of a substance.
- Explain the difference between an atom and an element.
- Describe the properties of a radioisotope.

Atoms and Elements

You learned in Chapter 1 that an atom is the smallest unit of a substance. To understand what that means, you need to know about subatomic particles—the particles that make up atoms: protons, neutrons, and electrons (FIGURE 2.2). **Protons** (p^+) are positively charged (**charge** is an electrical property in which opposite charges attract, and like charges repel). One or more protons occupy the central core, or **nucleus**, of every atom. Most atoms also have uncharged **neutrons** in their nucleus. Negatively charged **electrons** (e^-) move at high speed around the nucleus.

Different atoms can have different numbers of subatomic particles, but most have about the same number of electrons as protons. The negative charge of an electron is the same magnitude as the positive charge of a proton, so the two charges cancel one another. Thus, an atom with exactly the same number of electrons and protons carries no charge.

All atoms have protons. The number of protons in the nucleus is called the **atomic number**, and it determines the element. **Elements** are pure substances, each consisting only of atoms with the same number of

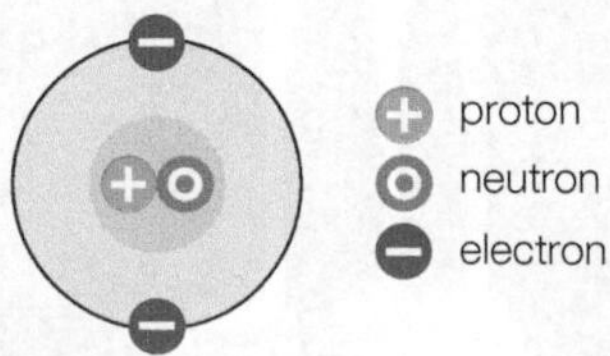

proton
neutron
electron

FIGURE 2.2 Atoms. Atoms consist of electrons moving around a nucleus of protons and neutrons. Models such as this one do not show what atoms look like. Electrons move in defined, three-dimensional spaces about 10,000 times bigger than the nucleus.

TABLE 2.1

Comparing Elemental Compositions by Mass

	Oxygen (O)	**Carbon** (C)	**Hydrogen** (H)	**Other**
Human body	61.00%	23.00%	10.00%	6.00%
Seawater	85.70%	<0.01%	10.78%	3.52%
Earth's crust	46.00%	0.18%	0.15%	53.67%
Universe	1.00%	0.50%	75.00%	23.50%

Source: www.webelements.com/oxygen/, /carbon/, /hydrogen/

protons in their nucleus. For example, the atomic number of carbon is 6; all atoms with six protons are carbon atoms, no matter how many electrons or neutrons they have. Elemental carbon (the substance) consists only of carbon atoms, and all of those atoms have six protons.

Each of the 118 known elements is represented by a symbol that is an abbreviation of its name. For example, the symbol for lead, Pb, is an abbreviation of its Latin name: *plumbum*. The word "plumbing" is related (ancient Romans made their water pipes with lead). Carbon's symbol, C, is from *carbo*, the Latin word for coal, which is mostly carbon. Carbon is a major component of living things, including humans (TABLE 2.1).

In 1869, chemist Dmitry Mendeleyev arranged the elements known at the time by their chemical properties. The arrangement, which he called the **periodic table** (FIGURE 2.3), turned out to be by atomic number, even though subatomic particles would not be discovered until the early 1900s.

Isotopes and Radioisotopes

All atoms of an element have the same number of protons, but some differ in the number of other subatomic particles. For example, one carbon atom may have six neutrons, and another may have seven. We call

atomic number Number of protons in the atomic nucleus; determines the element.
charge Electrical property in which opposite charges attract, and like charges repel.
electron Negatively charged subatomic particle.
element A pure substance that consists only of atoms with the same number of protons.
isotopes (ICE-oh-topes) Forms of an element that differ in the number of neutrons their atoms carry.
mass number Total number of protons and neutrons in the atomic nucleus.
neutron (NOO-tron) Uncharged subatomic particle in the atomic nucleus.
nucleus (NOO-klee-us) Core of an atom; occupied by protons and neutrons.
periodic table Tabular arrangement of elements by their atomic number.
proton (PRO-tawn) Positively charged subatomic particle that occurs in the nucleus of all atoms.
radioactive decay Process by which atoms of a radioisotope emit energy and/or subatomic particles when their nucleus spontaneously breaks up.
radioisotope An isotope with an unstable nucleus.
tracer A substance that can be traced via its detectable component.

these two carbon atoms isotopes. **Isotopes** are atoms of the same element that have different numbers of neutrons. The total number of neutrons and protons in the nucleus of an isotope is its **mass number**. Mass number is written as a superscript to the left of the element's symbol. For example, the most common isotope of hydrogen has one proton and no neutrons, so it is designated ^{1}H. Other isotopes include deuterium (^{2}H, one proton and one neutron), and tritium (^{3}H, one proton and two neutrons).

The most common isotope of carbon has six neutrons, so its mass number is 12. This isotope is called carbon 12, and it is abbreviated ^{12}C. The other naturally occurring isotopes of carbon are ^{13}C and ^{14}C (**FIGURE 2.4**). Carbon 14 is an example of a **radioisotope**, or radioactive isotope. Atoms of a radioisotope have an unstable nucleus that breaks up spontaneously. As a nucleus breaks up, it emits radiation (subatomic particles, energy, or both), a process called **radioactive decay**. The atomic nucleus cannot be altered by heat or any other ordinary means, so radioactive decay is unaffected by external factors such as temperature, pressure, or whether the atoms are part of molecules.

Each radioisotope decays at a predictable rate into predictable products. For example, when carbon 14 decays, one of its neutrons splits into a proton and an electron. The electron is emitted as radiation. Thus, a carbon atom with eight neutrons and six protons (^{14}C) becomes a nitrogen atom, with seven neutrons and seven protons (^{14}N, **FIGURE 2.4C**). This decay occurs at a predictable rate: About half of the atoms in any sample of ^{14}C will be ^{14}N atoms after 5,730 years. The rate of radioisotope decay is so predictable that scientists can estimate the age of a rock or fossil by measuring its isotope content (Section 16.5 returns to this topic).

You will see in the next section why the chemical behavior of an atom arises from the number of protons and electrons it has. Neutrons have little effect on chemistry, so all isotopes of an element generally have the same chemical properties—and all are interchangeable in a biological system. Researchers take advantage of the interchangeability when they use radioactive tracers to study biological processes. A **tracer** is any substance with a detectable component; a radioactive tracer is a molecule in which an atom (such as ^{12}C) has been replaced with a radioisotope (such as ^{14}C). When delivered into a biological system such as a cell or a body, a radioactive tracer may be followed by detecting the radiation emitted during decay.

Radioactive tracers are widely used in research. A famous example is a series of experiments carried out by Melvin Calvin and Andrew Benson. These researchers synthesized carbon dioxide with ^{14}C, then let green algae take up the radioactive gas. Using instruments that detect electrons emitted by the radioactive decay of ^{14}C, they tracked carbon through steps by which the algae—and all plants—make sugars.

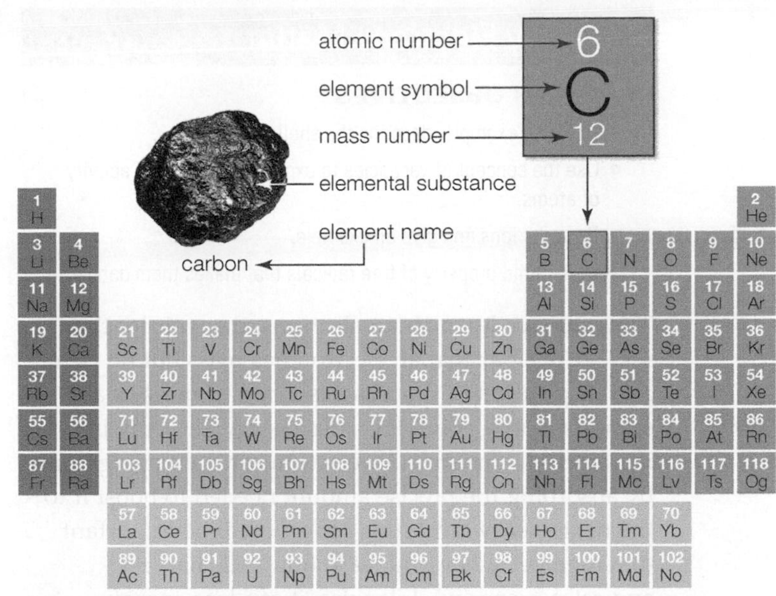
atomic number
element symbol
mass number
elemental substance
element name
carbon
6
C
12
1 H
2 He
3 Li 4 Be 5 B 6 C 7 N 8 O 9 F 10 Ne
11 Na 12 Mg 13 Al 14 Si 15 P 16 S 17 Cl 18 Ar
19 K 20 Ca 21 Sc 22 Ti 23 V 24 Cr 25 Mn 26 Fe 27 Co 28 Ni 29 Cu 30 Zn 31 Ga 32 Ge 33 As 34 Se 35 Br 36 Kr
37 Rb 38 Sr 39 Y 40 Zr 41 Nb 42 Mo 43 Tc 44 Ru 45 Rh 46 Pd 47 Ag 48 Cd 49 In 50 Sn 51 Sb 52 Te 53 I 54 Xe
55 Cs 56 Ba 71 Lu 72 Hf 73 Ta 74 W 75 Re 76 Os 77 Ir 78 Pt 79 Au 80 Hg 81 Tl 82 Pb 83 Bi 84 Po 85 At 86 Rn
87 Fr 88 Ra 103 Lr 104 Rf 105 Db 106 Sg 107 Bh 108 Hs 109 Mt 110 Ds 111 Rg 112 Cn 113 Nh 114 Fl 115 Mc 116 Lv 117 Ts 118 Og
57 La 58 Ce 59 Pr 60 Nd 61 Pm 62 Sm 63 Eu 64 Gd 65 Tb 66 Dy 67 Ho 68 Er 69 Tm 70 Yb
89 Ac 90 Th 91 Pa 92 U 93 Np 94 Pu 95 Am 96 Cm 97 Bk 98 Cf 99 Es 100 Fm 101 Md 102 No

FIGURE 2.3 The periodic table of the elements. Atomic number and element symbols are shown; Appendix I has a table with element names.

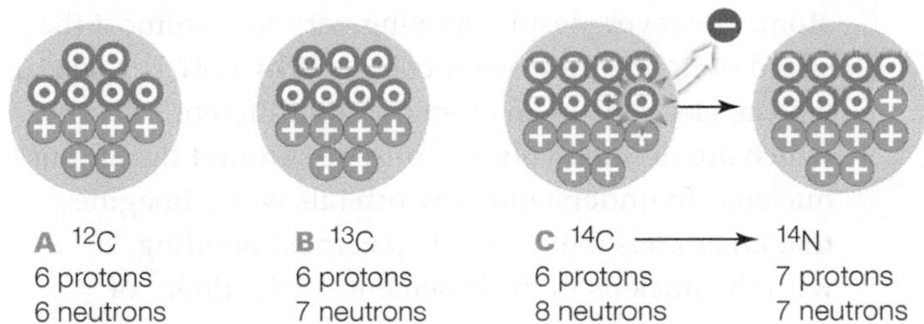
A ^{12}C
6 protons
6 neutrons
B ^{13}C
6 protons
7 neutrons
C ^{14}C
6 protons
8 neutrons
^{14}N
7 protons
7 neutrons

FIGURE 2.4 Isotopes of carbon.
These drawings show nuclei of the three naturally occurring isotopes of carbon. Carbon 14 (^{14}C) is a radioisotope, and it decays into nitrogen 14 when one of its neutrons spontaneously splits into a proton and an electron. The electron is emitted as radiation.

TAKE-HOME MESSAGE 2.2

✔ All matter consists of atoms, tiny particles that in turn consist of electrons moving around a nucleus of protons and neutrons.

✔ The number of protons in an atom determines the element. An element is a pure substance that consists only of atoms with the same number of protons.

✔ Isotopes are atoms of an element that have different numbers of neutrons. Isotopes are referenced by mass number.

✔ Unstable nuclei of radioisotopes emit radiation as they spontaneously break down (decay). Radioisotopes decay at a predictable rate to form predictable products.

✔ Radioisotopes can be used as tracers in biological processes.

2.3 Why Electrons Matter

LEARNING OBJECTIVES

- Using an example, describe the shell model.
- Use the concept of vacancies to explain the chemical activity of atoms.
- Describe ions and give an example.
- Explain the property of free radicals that makes them dangerous.

The more we learn about electrons, the weirder they seem. Consider that an electron has mass but no size, and its position in space is described as more of a smudge than a point. It carries energy, but only in incremental amounts. An electron can gain energy only by absorbing the precise amount needed to boost it to the next energy level (this concept will be important to remember when you learn how cells harvest and release energy). Likewise, it can lose energy only by emitting the exact difference between two energy levels.

A lot of electrons may be occupying the same atom. However, despite moving very fast—almost the speed of light—they never collide. Why not? For one reason, electrons in an atom occupy different orbitals, which are defined volumes of space around the atomic nucleus. To understand how orbitals work, imagine that an atom is a multilevel apartment building, with the nucleus in the basement. Each "floor" of the building corresponds to a certain energy level, and each has a certain number of "rooms" (orbitals) available for rent. Up to two electrons can occupy each room. Pairs of electrons populate rooms from the ground floor up; in other words, they fill orbitals from lower to higher energy levels. The farther an electron is from the nucleus in the basement, the more energy it has. An electron can move to a room on a higher floor if an energy input gives it a boost, but it quickly emits the extra energy and moves back down.

Shell models help us visualize how electrons populate atoms (**FIGURE 2.5**). In these models, successive "shells" correspond to successively higher energy levels. Thus, each shell includes all of the rooms (orbitals) on one floor (energy level) of our atomic apartment building.

We draw a shell model of an atom by filling it with electrons (represented as balls or dots), from the innermost shell out, until there are as many electrons as the atom has protons. There is only one room on the first floor, one orbital at the lowest energy level. It fills up first. In hydrogen, the simplest atom, a single electron occupies that room (**FIGURE 2.5A**). Helium, with two protons, has two electrons that fill the room—and the first shell. In larger atoms, more electrons rent the second-floor rooms (**FIGURE 2.5B**). When the second floor fills, more electrons rent third-floor rooms (**FIGURE 2.5C**), and so on.

FIGURE 2.5 Shell models. Each circle (shell) represents one energy level. To make these models, we fill the shells with electrons from the innermost shell out, until there are as many electrons as the atom has protons. The number of protons in each model is indicated.

FIGURE IT OUT Which of these models show unpaired electrons in the outer shell?

Answer: Hydrogen, carbon, oxygen, sodium, and chlorine

first shell

A The first shell corresponds to the first energy level, and it can hold up to 2 electrons. Hydrogen has 1 proton, so it has 1 electron and one vacancy. A helium atom has 2 protons, 2 electrons, and no vacancies.

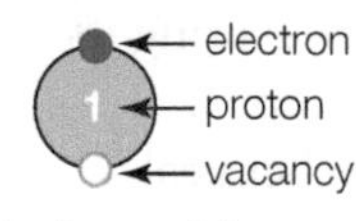
electron
proton
vacancy
1
hydrogen (H)

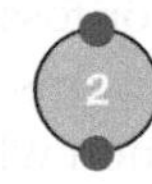
2
helium (He)

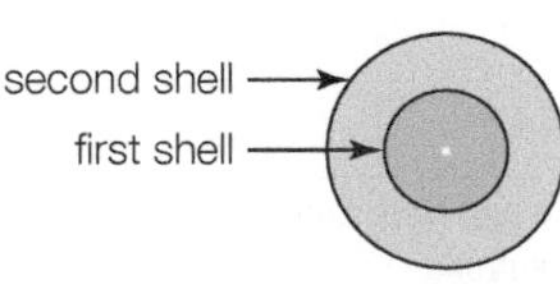
second shell
first shell

B The second shell corresponds to the second energy level, and it can hold up to 8 electrons. Carbon has 6 electrons, so its first shell is full. Its second shell has 4 electrons and four vacancies. Oxygen has 8 electrons and two vacancies. Neon has 10 electrons and no vacancies.

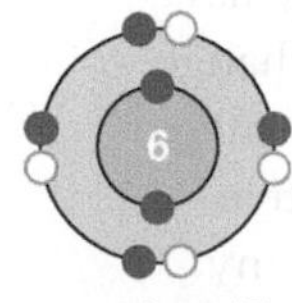
6
carbon (C)

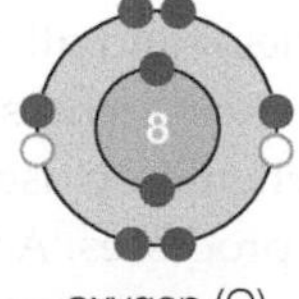
8
oxygen (O)

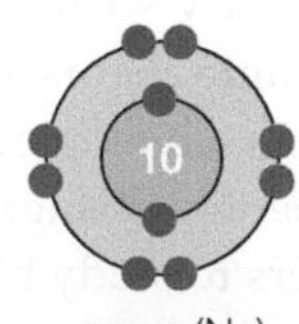
10
neon (Ne)

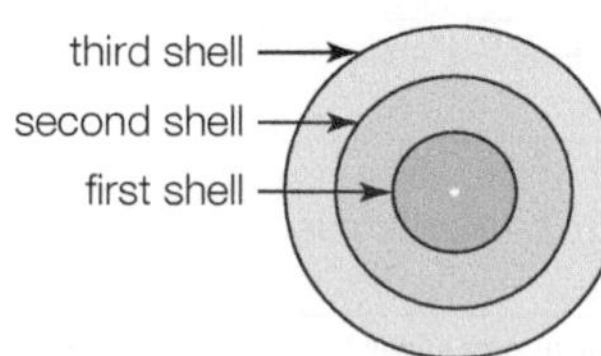
third shell
second shell
first shell

C The third shell corresponds to the third energy level, and it can hold up to 8 electrons. A sodium atom has 11 electrons, so its first two shells are full and the third shell has 1 electron. Thus, sodium has seven vacancies. Chlorine has 17 electrons and one vacancy. Argon has 18 electrons and no vacancies.

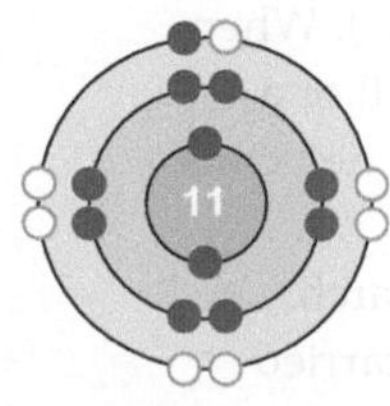
11
sodium (Na)

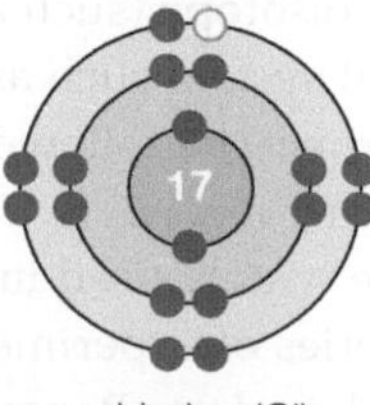
17
chlorine (Cl)

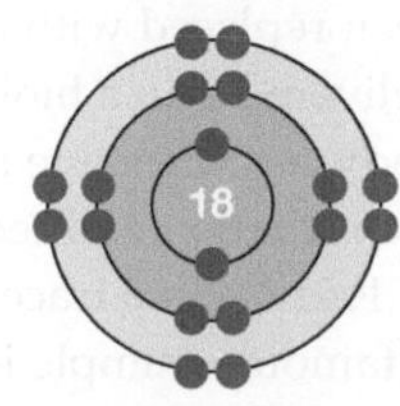
18
argon (Ar)

When an atom's outermost shell is filled with electrons, we say that it has no vacancies. Any atom is in its most stable state when it has no vacancies. Helium, neon, and argon are examples of elements with no vacancies. Atoms of these elements are chemically stable, which means they have no tendency to interact with other atoms. Thus, these elements occur most frequently in nature as solitary atoms.

vacancy
no vacancy

When an atom's outermost shell has room for an electron, we say that it has a vacancy. Atoms with vacancies tend to get rid of them by interacting with other atoms; in other words, they are chemically active. For example, the sodium atom (Na) depicted in **FIGURE 2.5C** has one electron in its outer (third) shell, which can hold eight. With seven vacancies, we can predict that this atom is chemically active.

In fact, this particular sodium atom is not just active, it is extremely active. Why? The shell model shows that a sodium atom has an unpaired electron; in the real world, electrons really like to be in pairs when they occupy orbitals. Atoms that have unpaired electrons are called **free radicals**. With a few exceptions, free radicals are very unstable, easily forcing electrons upon other atoms or ripping electrons away from them. This property makes free radicals dangerous because it damages DNA and other molecules of life (we return to this topic in Section 5.6).

A free-radical sodium atom tends to force its one unpaired electron upon another atom. When that happens, the sodium atom's second shell—which is full of electrons—becomes its outermost, and no vacancies remain. This is the most stable state of a sodium atom. The vast majority of sodium atoms on Earth are like this one, with 11 protons and 10 electrons.

Atoms with an unequal number of protons and electrons are called ions. **Ions** are atoms or molecules that carry a net (overall) charge because they have gained or lost electrons. Sodium atoms that lose an electron become positively charged sodium ions (**FIGURE 2.6**). Note how an ion's charge is indicated by a superscript to the right of the element symbol; Na^+, for example, is the designation for a sodium ion.

chemical bond Attractive force that arises between two atoms when their electrons interact. Links atoms in molecules.
free radical An atom with an unpaired electron.
ion (EYE-on) An atom or molecule that carries a net charge.
shell model Model of electron distribution in an atom.

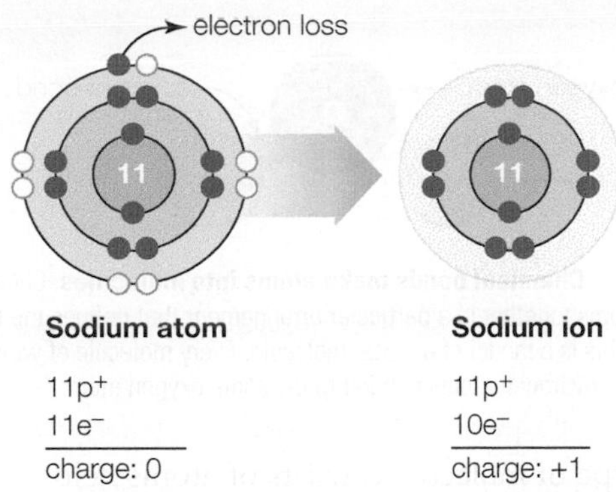
electron loss
11
11
Sodium atom
11p+
11e−
charge: 0
Sodium ion
11p+
10e−
charge: +1

FIGURE 2.6 Ion formation. Here, a sodium atom (Na) becomes a positively charged sodium ion (Na^+) when it loses the electron in its third shell. The atom's full second shell is now its outermost, so it has no vacancies.

Atoms of other elements accept electrons and become negatively charged. Chlorine is an example. An uncharged chlorine atom has 17 protons and 17 electrons. Its outermost shell can hold eight electrons, but has only seven. With one vacancy and one unpaired electron, we can predict—correctly—that this atom is chemically very active. An uncharged chlorine atom easily fills its third shell by pulling an electron off of another atom. When that happens, the chlorine atom becomes chloride: a negatively charged chlorine ion (Cl^-) with 17 protons and 18 electrons.

TAKE-HOME MESSAGE 2.3

- ✔ An atom's electrons are the basis of its chemical behavior.
- ✔ Shell models are a way of visualizing how electrons populate atoms. Each shell corresponds to an electron energy level.
- ✔ An atom has a vacancy when its outermost shell is not full of electrons. Atoms with vacancies tend to interact with other atoms.
- ✔ Atoms with unpaired electrons are free radicals. Most free radicals are extremely reactive; they are dangerous to life because they can destroy biological molecules.

2.4 Chemical Bonds

LEARNING OBJECTIVES

- Describe a chemical bond.
- Distinguish between ionic and covalent bonds.
- Explain polarity in terms of ionic bonds.
- Write the structural formula for a molecule of water.
- Use appropriate examples to distinguish between polar and nonpolar covalent bonds.

An atom can get rid of vacancies by participating in a **chemical bond**, which is an attractive force that arises between two atoms when their electrons interact. Chemical bonds make molecules out of atoms.

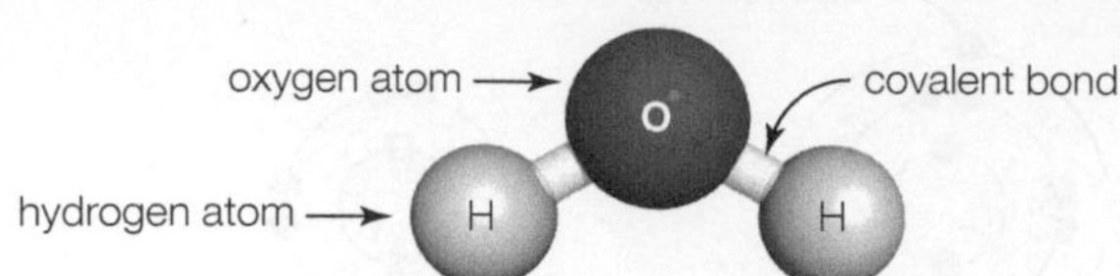
oxygen atom
covalent bond
hydrogen atom
O
H
H

FIGURE 2.7 Chemical bonds make atoms into molecules. Chemical bonds hold atoms together in a particular arrangement that defines the type of molecule. This is a model of a water molecule. Every molecule of water consists of two hydrogen atoms bonded to the same oxygen atom.

Each type of molecule consists of atoms held together in a particular number and arrangement by chemical bonds. Consider the molecules that make up water. A water molecule has three atoms: two hydrogen atoms bonded to the same oxygen atom (**FIGURE 2.7**). Every water molecule has the identical configuration, whether it is part of an ocean, floating in space, making up vapor in a cloud, and so on.

Because a water molecule consists of two or more elements, it is called a **compound**. Other molecules have atoms of one element only; molecular oxygen (a gas in air) is an example.

The term "bond" applies to a continuous range of atomic interactions. However, we can categorize most bonds into distinct types based on their properties. In this book, we discuss two kinds of chemical bonds: ionic and covalent. An **ionic bond** is a strong mutual attraction between ions of opposite charge. For example, a molecule of sodium chloride (NaCl) consists of a sodium ion and a chloride ion held together by an ionic bond (**FIGURE 2.8A**). Molecules of sodium chloride make up the substance we know as table salt. Each crystal of salt is a tiny lattice of sodium and chloride ions interacting in ionic bonds (**FIGURE 2.8B**).

Ions retain their full respective charges when participating in an ionic bond (**FIGURE 2.8C**). Thus, one "end" of an ionic bond has a positive charge, and the other "end" has a negative charge. This and any other separation of charge into distinct positive and negative regions is called **polarity**.

The ionic bond in sodium chloride is completely polar because the chloride ion keeps a very strong hold on its extra electron. In other words, chlorine is strongly electronegative. **Electronegativity** is a measure of an atom's ability to pull electrons away from another atom. Electronegativity is not the same as charge. Rather, an atom's electronegativity depends on its size, how many vacancies it has, and its interactions with other atoms in a molecule.

Atoms that can form an ionic bond with each other have a large difference in electronegativity. When

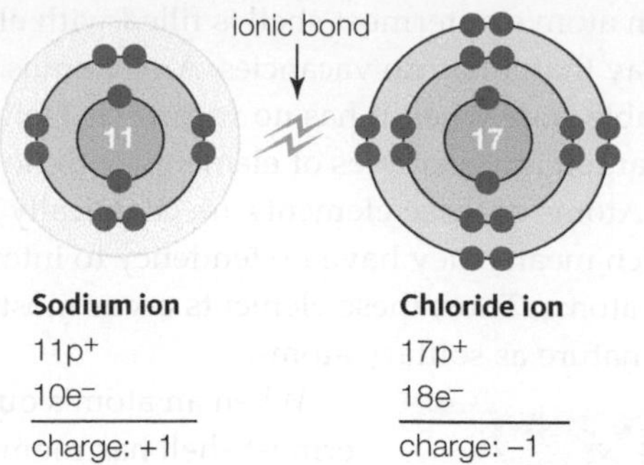
ionic bond
11
17
Sodium ion
11p+
10e−
charge: +1
Chloride ion
17p+
18e−
charge: −1

A The strong mutual attraction of opposite charges holds a sodium ion and a chloride ion together in an ionic bond.

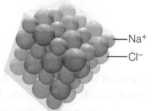

Na+
Cl−

B Tiny crystals of sodium chloride (left) compose table salt. Each crystal consists of many sodium and chloride ions locked together in a cubic lattice by ionic bonds (right).

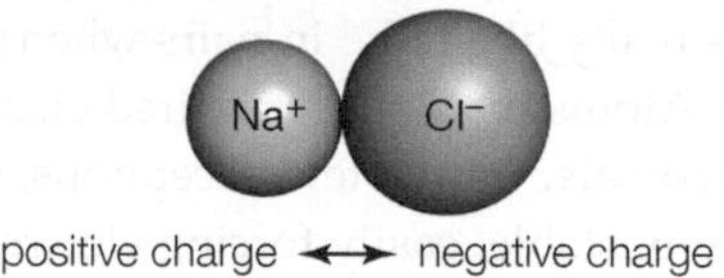
Na+
Cl−
positive charge ←→ negative charge

C Ions taking part in an ionic bond retain their respective charges, so the bond is completely polar. The sodium ion is positively charged (+), and the chloride ion is negatively charged (–).

FIGURE 2.8 Ionic bonds in table salt (NaCl).

atoms with a small difference in electronegativity interact, they tend to form covalent bonds. In a **covalent bond**, two atoms share a pair of electrons, so each rids itself of vacancies (**FIGURE 2.9**). Sharing electrons links the two atoms, just as sharing a pair of earphones links two friends (left). Covalent bonds can be stronger than ionic bonds, but they are not always so.

TABLE 2.2 shows some different ways of representing covalent bonds. In structural formulas, a line between two atoms represents a single covalent bond. For example, a covalent bond links two atoms in molecular hydrogen, so the structural formula of this molecule is H—H. Multiple covalent bonds may form between two atoms when they share multiple pairs of electrons. For example, two atoms sharing two pairs of electrons are connected by two covalent

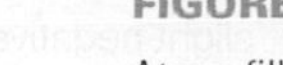
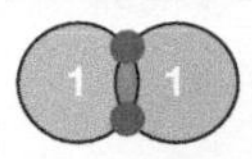

FIGURE 2.9 Covalent bonds. Atoms fill vacancies by sharing electrons. Two electrons are shared in each covalent bond. This is a model of hydrogen, a gas in air. Two hydrogen atoms, each with one proton, share two electrons in a covalent bond.

bonds, which are represented by a double line between the atoms. A double bond links the two oxygen atoms in molecular oxygen (O=O). Three lines indicate a triple bond, in which two atoms share three pairs of electrons. A triple covalent bond links the two nitrogen atoms in molecular nitrogen (N≡N). Comparing bonds between the same two atoms: A triple bond is stronger than a double bond, which is stronger than a single bond.

A structural formula consists of letters and sticks. By contrast, a structural model is a three-dimensional representation of atoms and bonds. Double and triple bonds are not distinguished from single bonds in structural models. All covalent bonds are shown as one stick connecting two balls, which represent atoms. We use a common color-coding scheme to distinguish elements in structural models:

carbon hydrogen oxygen nitrogen phosphorus

Bond Polarity Atoms participating in an ionic bond do not share electrons, so these bonds are completely polar (**FIGURE 2.10A**). Atoms taking part in a covalent bonds share electrons, so these bonds are less polar than ionic bonds. If one atom is more electronegative than the other, then the sharing is unequal and the bond is polar (**FIGURE 2.10B**). The polar covalent bond that links a hydrogen and chlorine atom in hydrochloric acid (HCl) is an example. The chlorine is more electronegative than the hydrogen, so it pulls the electrons a bit more toward its side of the bond. Thus, it has a slight negative charge, and the hydrogen is left with a slight positive charge. If the atoms taking part in a covalent bond have equal electronegativity, they share electrons equally and the bond is nonpolar (**FIGURE 2.10C**). A nonpolar covalent bond links two hydrogen atoms in molecular hydrogen. Neither of the atoms in this molecule carries a charge.

compound Molecule that has atoms of more than one element.
covalent bond (co-VAY-lent) Type of chemical bond in which two atoms share a pair of electrons.
electronegativity Measure of the ability of an atom to pull electrons away from other atoms.
ionic bond (eye-ON-ick) Type of chemical bond in which a strong mutual attraction links ions of opposite charge.
polarity Any separation of charge into positive and negative regions.

TABLE 2.2
Six Ways of Representing the Same Molecule

Common name	Water	Familiar term.
Chemical name	Dihydrogen oxide	Describes elemental composition.
Chemical formula	H_2O	Indicates unvarying proportions of elements. Subscripts show number of atoms of an element per molecule. The absence of a subscript means one atom.
Structural formula	H—O—H	Represents each covalent bond as a single line between atoms.
Structural model		Shows relative sizes and positions of atoms in three dimensions.
Shell model	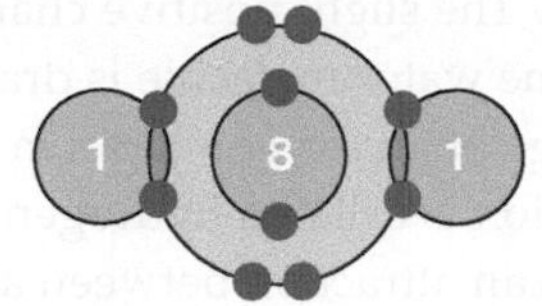	Shows how pairs of electrons are shared in covalent bonds.

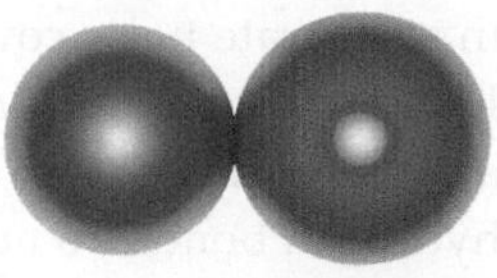

A An ionic bond
In an ionic bond, each atom retains its full respective charge, so the bond is completely polar.

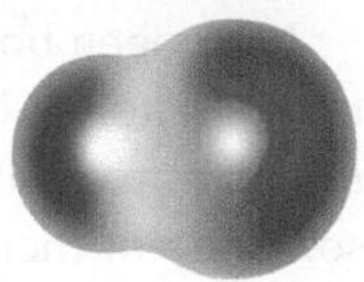

B A polar covalent bond
In a polar covalent bond, one atom pulls the shared electrons more than the other, so it has a slight negative charge. The other atom ends up with a slight positive charge. The bond is polar, but less so than an ionic bond.

C A nonpolar covalent bond
The atoms participating in a nonpolar share electrons equally, so neither has a charge. The bond is nonpolar.

FIGURE 2.10 Comparing bond polarity. Bonding refers to a range of atomic interactions. Here, polarity is represented in color; red indicates negative charge; blue, positive charge. Uncharged regions are white.

TAKE-HOME MESSAGE 2.4

- ✔ A chemical bond forms between atoms when their electrons interact. Chemical bonds make molecules out of atoms.
- ✔ A chemical bond may be ionic or covalent depending on the electronegativity of the atoms taking part in it.
- ✔ An ionic bond is a strong mutual attraction between ions of opposite charge. An ionic compound is highly polar because the ions retain their full respective charges.
- ✔ Atoms share a pair of electrons in a covalent bond. When the atoms share electrons equally, the bond is nonpolar. When the sharing is unequal, the bond is polar.

2.5 Hydrogen Bonding and Water

LEARNING OBJECTIVES

- Draw a hydrogen bond between two water molecules.
- Describe the way an ionic solid dissolves in water.
- Distinguish between hydrophilic and hydrophobic substances.
- Using appropriate examples, explain how hydrogen bonding gives rise to three properties of water that are essential to life.

Hydrogen Bonds

Consider the two polar covalent bonds in a water molecule. Overall, the molecule has no charge, but the oxygen atom carries a slight negative charge, and each of the hydrogen atoms carries a slight positive charge. Thus, the molecule itself is polar (**FIGURE 2.11A**).

The polarity of individual water molecules attracts them to one another. The slight positive charge of a hydrogen atom in one water molecule is drawn to the slight negative charge of an oxygen atom in another. This type of interaction is called a hydrogen bond. A **hydrogen bond** is an attraction between a covalently bonded hydrogen atom and another atom taking part in a separate polar covalent bond (**FIGURE 2.11B**). Like ionic bonds, hydrogen bonds form by the mutual attraction of opposite charges. Unlike ionic bonds, hydrogen bonds do not make molecules out of atoms, so they are not chemical bonds.

Hydrogen bonds are on the weaker end of the spectrum of atomic interactions, and they form and break much more easily than covalent or ionic bonds. Even so, many of them form, and collectively they can be very strong. As you will see in later chapters, hydrogen bonds stabilize the characteristic structures of biological molecules such as DNA and proteins. They also form in tremendous numbers among water molecules (**FIGURE 2.11C**). Extensive hydrogen bonding among water molecules gives liquid water several special properties that make life possible.

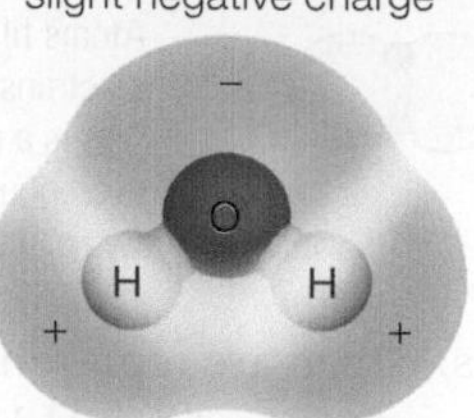
slight negative charge
O
H
H
slight positive charge

A Polarity of a water molecule. Each of the hydrogen atoms bears a slight positive charge (represented by a blue overlay). The oxygen atom carries a slight negative charge (red overlay).

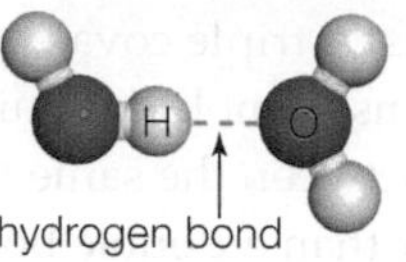
H
O
hydrogen bond

B A hydrogen bond is an attraction between a hydrogen atom and another atom taking part in a separate polar covalent bond.

C The many hydrogen bonds that form among water molecules impart special properties to liquid water.

FIGURE 2.11 Hydrogen bonds.

Water's Special Properties

Water Is an Excellent Solvent The ability of water molecules to form hydrogen bonds make water an excellent **solvent**, which means that many other substances can dissolve in it. Substances that dissolve easily in water are **hydrophilic** (water-loving).

Ionic solids such as sodium chloride (NaCl) are hydrophilic. They dissolve in water because the slight positive charge on each hydrogen atom in a water molecule attracts negatively charged ions (Cl^-), and the slight negative charge on the oxygen atom attracts positively charged ions (Na^+). Hydrogen bonds among many water molecules are collectively stronger than an ionic bond, so the solid dissolves as hydrogen-bonded water molecules tug the ions apart and surround each one (right).

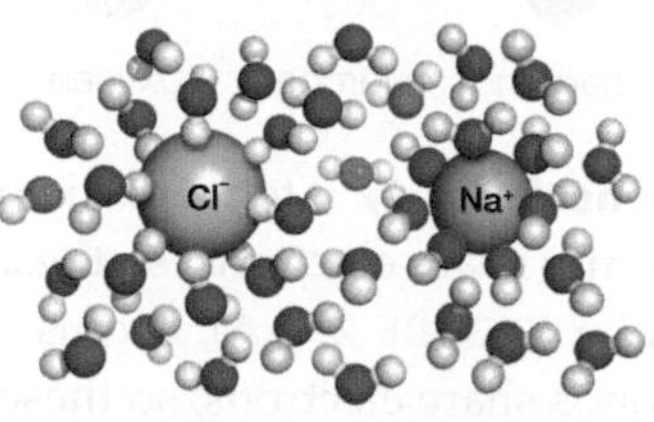
Cl⁻
Na⁺

When a substance such as NaCl dissolves, its component ions disperse uniformly among molecules of the solvent, and it becomes a **solute**. Sodium chloride is called a **salt** because it releases ions other than H^+ and OH^- when it dissolves in water (more about these ions in the next section). A uniform mixture such as salt dissolved in water is called a **solution**. Chemical bonds do not form between molecules of solute and solvent, so the proportions of the two substances in a solution can vary. The amount of a solute that is dissolved in a given volume of fluid is its **concentration**.

Many compounds other than salts are hydrophilic and dissolve easily in water. Sugars are examples. Molecules of these substances have one or more polar covalent bonds, and atoms participating in a polar covalent bond can form hydrogen bonds with water molecules. When one of these solids is added to water, it dissolves because hydrogen bonding with water pulls its individual molecules away from one another and keeps them apart. Unlike ionic compounds, molecules of these compounds do not dissociate into atoms when they dissolve.

Water does not interact with **hydrophobic** (water-dreading) substances such as oils. Oils consist of nonpolar molecules, and hydrogen bonds do not form between nonpolar molecules and water. Vigorous shaking can disperse water in oil, but hydrogen bonding causes the water to coalesce. As this happens, the water excludes molecules of oil and pushes them together into drops that rise to the surface of the mixture. The same interactions occur at the thin, oily membrane that separates the watery fluid inside cells from the watery fluid outside of them (Chapter 5 returns to this topic).

Water Has Cohesion Molecules of some substances resist separating from one another, and this resistance gives rise to a property called **cohesion**. Water has cohesion because hydrogen bonds collectively exert a continuous pull on its individual molecules. You can see cohesion in water as surface tension, which means that the surface of liquid water behaves a bit like a sheet of elastic (left).

Cohesion is a part of many processes that sustain multicelled bodies. Consider how sweating helps keep your body cool during hot, dry weather. Sweat, which is about 99 percent water, cools the skin as it evaporates. Why? **Evaporation** is the process in which molecules escape from the surface of a liquid and become vapor. The evaporation of water is resisted by hydrogen bonding among individual water molecules. In other words, overcoming water's cohesion takes energy. Thus, evaporation sucks energy (in the form of heat) from liquid water, and this lowers the water's surface temperature.

Cohesion works inside organisms, too. Consider how plants absorb water from soil as they grow. Water molecules evaporate from leaves, and replacements are pulled upward from roots. Cohesion makes it possible for columns of liquid water to rise from roots to leaves inside narrow pipelines of vascular tissue. In some trees, these pipelines extend hundreds of feet above the soil (Section 28.4 returns to this topic).

Water Stabilizes Temperature All atoms jiggle nonstop, so the molecules they make up jiggle too. We measure the energy of this motion as **temperature**. Adding energy (in the form of heat, for example) speeds up the jiggling, so the temperature rises. Extensive hydrogen bonding keeps water molecules from moving as much as they would otherwise, so it takes more heat to raise the temperature of water compared with other liquids. Water's resistance to temperature change is an important part of homeostasis, because most of the molecules of life function properly only within a certain range of temperature.

Below 0°C (32°F), water molecules do not jiggle enough to break hydrogen bonds between them, and they become locked in the rigid, lattice-like bonding pattern of ice crystals. Individual water molecules pack less densely in ice than they do in water, which is why ice floats on water—another unusual property. Sheets of ice that form on the surface of ponds, lakes, and streams can insulate the water under them from subfreezing air temperatures. Such "ice blankets" protect aquatic organisms during long, cold winters.

TAKE-HOME MESSAGE 2.5

✔ In a hydrogen bond, a negatively charged region of a polar molecule attracts a hydrogen atom in another molecule.

✔ Extensive hydrogen bonding among water molecules arises from the polarity of the individual molecules.

✔ Hydrogen bonding among water molecules imparts cohesion to liquid water, and also gives it the ability to stabilize temperature and dissolve many substances.

2.6 Acids and Bases

LEARNING OBJECTIVES

- Define pH and give some examples.
- Differentiate between acids and bases.
- Explain how buffers maintain the pH of solutions.
- Use an example to explain why pH stability is important in biological systems.

A hydrogen atom, remember, is just a proton and an electron. When a hydrogen atom participates in a polar covalent bond, its electron is pulled away from its proton, just a bit. Hydrogen bonding in water can pull that proton right off of the molecule. The detached proton is called a hydrogen ion (H^+). The electron stays with

cohesion (ko-HE-zhun) Property of a substance that arises from the tendency of its molecules to resist separating from one another.
concentration Amount of solute per unit volume of solution.
evaporation Transition of a liquid to a vapor.
hydrogen bond Attraction between a covalently bonded hydrogen atom and another atom taking part in a separate covalent bond.
hydrophilic Describes a substance that dissolves easily in water.
hydrophobic Describes a substance that resists dissolving in water.
salt Ionic compound that releases ions other than H^+ and OH^- when it dissolves in water.
solute (SAWL-yute) A dissolved substance.
solution Uniform mixture of solute completely dissolved in solvent.
solvent Liquid in which other substances dissolve.
temperature Measure of molecular motion.

the rest of the molecule and makes it negatively charged (ionic). For example, a water molecule that loses a proton becomes a hydroxyl ion (OH^-). The loss is more or less temporary, because these two ions easily get back together to form a water molecule:

$$\underset{\text{water molecule}}{H_2O} \longrightarrow \underset{\text{hydroxide ion}}{OH^-} + \underset{\text{hydrogen ion}}{H^+} \longrightarrow \underset{\text{water molecule}}{H_2O}$$

With other molecules, the loss of a hydrogen ion in water is essentially permanent. An **acid** is a substance that gives up hydrogen ions in water. By contrast, a **base** accepts hydrogen ions in water.

We use a value called **pH** as a measure of the number of hydrogen ions in a water-based fluid. The higher the number of these ions, the lower the pH. If a fluid has an equal number of H^+ and OH^- ions, its pH is 7 (or neutral). Pure water is like this. If there are more H^+ ions than OH^- ions, the pH is below 7 (or acidic). Conversely, if there are fewer H^+ ions than OH^- ions, the pH is above 7 (or basic). A one-unit difference in pH corresponds to a tenfold difference in hydrogen ion concentration (**FIGURE 2.12**). One way to get a sense of this pH scale is to taste dissolved baking soda (pH 9), pure water (pH 7), and lemon juice (pH 2).

Acids can be strong or weak. Strong acids ionize completely in water to give up all of their H^+ ions. Hydrochloric acid (HCl) is an example. When HCl dissolves in water, all of the molecules give up H^+ ions, leaving Cl^- ions behind. The H^+ released from HCl makes gastric fluid in your stomach very acidic (pH 1–2). By contrast, weak acids do not ionize completely in water. Carbonic acid is an example. It forms when carbon dioxide gas dissolves in plasma, the fluid portion of human blood:

$$\underset{\text{carbon dioxide}}{CO_2} + \underset{\text{water molecule in plasma}}{H_2O} \longrightarrow \underset{\text{carbonic acid}}{H_2CO_3}$$

Carbonic acid is a weak acid, so only some of its molecules give up a hydrogen ion in water. When carbonic acid loses a hydrogen ion, it becomes an ionic molecule called bicarbonate:

$$\underset{\text{carbonic acid}}{H_2CO_3} \longrightarrow \underset{\text{hydrogen ion}}{H^+} + \underset{\text{bicarbonate}}{HCO_3^-}$$

Bicarbonate can act like a base by accepting a hydrogen ion. When it does, carbonic acid forms again:

$$\underset{\text{hydrogen ion}}{H^+} + \underset{\text{bicarbonate}}{HCO_3^-} \longrightarrow \underset{\text{carbonic acid}}{H_2CO_3}$$

Together, carbonic acid and bicarbonate constitute a buffer. A **buffer** is a set of chemicals that can keep the pH of a solution stable by alternately donating and accepting ions that contribute to pH.

more basic
more acidic
14 drain cleaner
13 oven cleaner
hair remover
12 household ammonia
11 bleach
milk of magnesia
bicarbonate
10 bar soap
Tums
9 detergents
baking soda
8 seawater
egg white
blood, tears
7 pure water
toothpaste
milk, urine
butter
6 corn
typical rain
5 black coffee
tea
acid rain
4 bananas, beer
carbonic acid
tomatoes, wine
3 orange juice
cola
vinegar
2 lemon juice
gastric fluid
1 hydrochloric acid
battery acid
0
COLA

FIGURE 2.12 A pH scale. This pH scale ranges from 0 (most acidic) to 14 (most basic). A change of one unit on the scale corresponds to a tenfold change in the amount of H^+ ions. Red dots signify hydrogen ions (H^+) and grey dots signify hydroxyl ions (OH^-). Also shown are the approximate pH values for some common solutions.

Consider how the pH of pure water declines when even a tiny amount of acid is added to it. This is because the acid gives up hydrogen ions in water, and these ions contribute to pH. If the same amount of acid is added to blood plasma, the carbonic acid–bicarbonate buffer system keeps the pH from declining. Hydrogen ions released by the acid combine with bicarbonate, so they do not contribute to pH. The same

Data Analysis Activities

Mercury Emissions by Continent By weight, coal does not contain much mercury, but we burn a lot of it. Several industries besides coal-fired power plants contribute substantially to atmospheric mercury pollution. FIGURE 2.13 shows mercury emissions by industry from different regions of the world in 2010.

1. About how many metric tons of mercury were released in total from these regions?
2. Which industry tops the list of mercury emitters? Which industry is next on the list?
3. Which region emitted the most mercury from burning fossil fuels?
4. About how many metric tons of mercury were released from metal production in Asia?

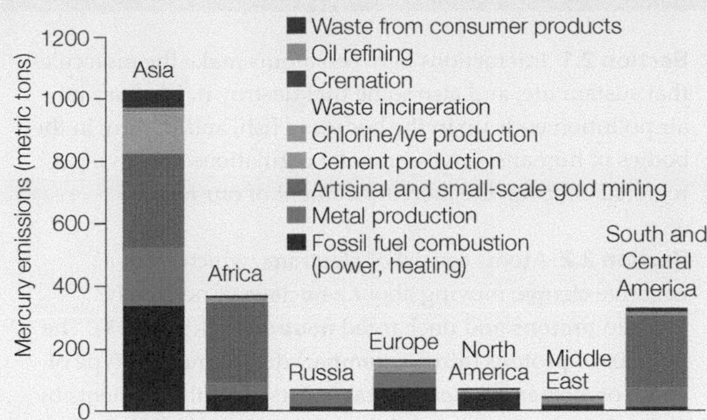

Mercury emissions (metric tons)
1200
1000
800
600
400
200
0
Asia
Africa
Russia
Europe
North America
Middle East
South and Central America
Waste from consumer products
Oil refining
Cremation
Waste incineration
Chlorine/lye production
Cement production
Artisinal and small-scale gold mining
Metal production
Fossil fuel combustion (power, heating)

FIGURE 2.13 **Global mercury emissions by sector, 2010.**

buffer system also prevents the pH of blood plasma from rising when a basic substance is added. Bases remove hydrogen ions from a solution; in plasma, carbonic acid replaces these ions as it becomes bicarbonate. In both cases, the proportion of carbonic acid and bicarbonate molecules in plasma shifts, but the pH stays stable (typically between 7.35 and 7.45).

The addition of too much acid or base can overwhelm a buffer's capacity to stabilize pH. This outcome can be catastrophic in a cell or body because most biological molecules function properly only within a narrow range of pH. Even a slight deviation from that range can halt cellular processes. Consider what happens when breathing is impaired suddenly. Carbon dioxide accumulates in tissues, so too much carbonic acid forms in plasma. The excess acid reduces the pH of the plasma. If it falls below 7.3, a dangerous level of unconsciousness called coma can be the outcome. By contrast, hyperventilation (sustained rapid breathing) causes the body to lose too much CO_2. The loss results in a rise in blood pH. If blood pH rises too much, prolonged muscle spasm (tetany) or coma may occur.

Burning fossil fuels such as coal releases a lot of carbon dioxide: about 36 billion metric tons in 2015 alone. More than a third of this CO_2 ends up in the oceans, where it combines with water molecules to form carbonic acid. The carbonic acid releases hydrogen ions into the water. Ocean water is not buffered, so these ions lower its pH. Ocean acidification is already harming sea-dwelling life worldwide, for example by dissolving the shells of marine animals.

acid Substance that releases hydrogen ions in water.
base Substance that accepts hydrogen ions in water.
buffer Set of chemicals that can keep the pH of a solution stable by alternately donating and accepting ions that contribute to pH.
pH Measure of the number of hydrogen ions in a fluid.

TAKE-HOME MESSAGE 2.6

✔ The number of hydrogen ions in a fluid is measured as pH.

✔ Acids release hydrogen ions in water; bases accept them.

✔ Most biological systems function properly only within a narrow range of pH. Buffers help keep pH stable. Inside organisms, they play a role in homeostasis.

2.1 Mercury Rising (revisited)

All human bodies now have detectable amounts of mercury; the average adult living in the United States has about 4 micrograms of it circulating in his or her blood. Some comes from dental fillings, imported skin-bleaching cosmetics, pharmaceuticals, and broken thermostats and light bulbs. However, most comes from dietary seafood: The more fish and shellfish you eat, the more mercury your body has. A diet that consists of a high proportion of seafood can result in a blood mercury content 30 times the average.

Humans release thousands of tons of mercury into the atmosphere every year, and that number is rising as coal-burning increases globally. However, environmental regulations in the United States have been reducing mercury emissions by U.S. industries (mainly coal-fired power plants), and the mercury content of North Atlantic bluefin tuna is mirroring this trend, declining at a rate of about 4 percent annually. The research shows that reducing emissions can have a direct and rapid effect on our environment—and our health. ●

Section 2.1 Interactions between atoms make the molecules that sustain life, and also some that destroy it. Mercury in air pollution ends up in the bodies of fish, and in turn, in the bodies of humans. Environmental regulations are having a regional effect on the mercury content of our food.

Section 2.2 Atoms consist of **electrons**, which carry a negative **charge**, moving about a **nucleus** of positively charged **protons** and uncharged **neutrons** (TABLE 2.3). The number of protons (**atomic number**) determines the type of atom, or **element**. A **periodic table** lists all of the elements by atomic number. **Isotopes** of an element differ in the number of neutrons. The total number of protons and neutrons is the **mass number**. **Tracers** can be made with **radioisotopes**, which, by a process called **radioactive decay**, emit particles and energy when their nucleus spontaneously breaks up.

Section 2.3 An atom's electrons are the basis of its chemical behavior. In a **shell model**, the energy levels of an atom's electrons are represented as concentric circles (shells). Atoms are in their most stable state when their outermost shell is full of electrons.

When an atom's outermost shell is not full of electrons, it has a vacancy. Atoms with vacancies tend to get rid of them by interacting with other atoms, for example by gaining or losing electrons and becoming charged **ions**.

Electrons in atoms like to be in pairs. An atom with an unpaired electron is a **free radical**. Extreme chemical reactivity makes most free radicals dangerous to life.

Section 2.4 A **chemical bond** is an attractive force that unites two atoms as a molecule. A molecule that has atoms of two or more elements is a **compound**. The difference in **electronegativity** between two atoms influences the type of bond that can form between them.

An **ionic bond** is a strong mutual attraction between two ions with opposite charges. Ionic bonds are completely polar (**polarity** is a separation of charge). Atoms share a pair of electrons in a **covalent bond**, which is nonpolar if the sharing is equal, and polar if it is not.

Section 2.5 Two polar covalent bonds give each water molecule an overall polarity. **Hydrogen bonds** that form among water molecules in tremendous numbers are the basis of water's unique life-sustaining properties. Water has **cohesion** and a capacity to act as a **solvent** that dissolves **salts** and other polar **solutes**; it also resists **temperature** changes. **Hydrophilic** substances dissolve easily in water to form **solutions**; **hydrophobic** substances do not. The amount of solute in a given volume of fluid is the solute's **concentration**. **Evaporation** is the transition of a liquid to vapor.

Section 2.6 The number of hydrogen ions (H^+) in a fluid determines its **pH**. At neutral pH (7), there are an equal number of H^+ and OH^- ions. **Acids** release hydrogen ions in water, thus lowering pH; **bases** accept hydrogen ions, thus raising pH. A **buffer** can stabilize the pH of a solution. Most cell and body fluids are buffered because most molecules of life work only within a narrow range of pH.

TABLE 2.3

Players in the Chemistry of Life

Atoms	Particles that are building blocks of all matter
Proton (p^+)	Positively charged subatomic particle of the nucleus
Electron (e^-)	Negatively charged subatomic particle that can occupy a defined volume of space (orbital) around the nucleus
Neutron	Uncharged subatomic particle of the nucleus
Element	Pure substance that consists entirely of atoms with the same, characteristic number of protons
Isotopes	Atoms of an element that differ in the number of neutrons
Radioisotope	Isotope with an unstable nucleus that emits radiation when it decays (breaks up)
Tracer	Substance with a detectable component (such as a radioisotope) that can be followed as it moves through a biological system
Ion	Atom or molecule that carries a charge after it has gained or lost one or more electrons
Molecule	Two or more atoms joined in a chemical bond
Compound	Molecule of two or more different elements
Solute	Substance dissolved in a solvent
Hydrophilic	Refers to a substance that dissolves easily in water
Hydrophobic	Refers to a substance that resists dissolving in water
Acid	Compound that releases H^+ when dissolved in water
Base	Compound that accepts H^+ when dissolved in water
Salt	Ionic compound that releases ions other than H^+ or OH^- when dissolved in water
Solvent	Substance that can dissolve other substances
Buffer	Set of chemicals that can stabilize pH

SELF-QUIZ

Answers in Appendix VII

1. What atom has only one proton?
   a. hydrogen
   b. an isotope
   c. helium
   d. a free radical
   e. a radioisotope
   f. oxygen

2. A molecule into which a radioisotope has been incorporated can be used as a(n) ________ .
   a. compound   c. salt
   b. tracer   d. acid

3. Which of the following statements is *in*correct?
   a. Isotopes have the same atomic number and different mass numbers.
   b. Atoms have about the same number of electrons as protons.
   c. All molecules consist of atoms.
   d. Free radicals are dangerous because they emit energy.

4. In the periodic table, the elements are arranged according to ________ .
   a. size   c. mass number
   b. charge   d. atomic number

5. All ions have at least one ________ (select all that are correct).
   a. proton   c. electron
   b. neutron   d. all of the above

6. Rank the following chemical bonds in order of increasing polarity.
   a. nonpolar covalent   b. ionic   c. polar covalent

7. The measure of an atom's ability to pull electrons away from another atom is called ________ .
   a. electronegativity
   b. charge
   c. polarity

8. The mutual attraction of opposite charges holds atoms together as molecules in a(n) ________ bond.
   a. ionic   c. polar covalent
   b. hydrogen   d. nonpolar covalent

9. Atoms share electrons unequally in a(n) ________ bond.
   a. ionic   c. polar covalent
   b. hydrogen   d. nonpolar covalent

10. A(n) ________ substance repels water.
    a. acidic   c. hydrophobic
    b. basic   d. polar

11. A salt does not release ________ in water.
    a. ions   b. H^+

12. Hydrogen ions (H^+) are ________ .
    a. in blood   c. indicated by a pH scale
    b. protons   d. all of the above

13. When dissolved in water, a(n) ________ donates H^+; a(n) ________ accepts H^+.
    a. acid; base   c. buffer; solute
    b. base; acid   d. base; buffer

14. A ________ can help keep the pH of a solution stable.
    a. covalent bond   c. buffer
    b. hydrogen bond   d. pH

15. A ________ is dissolved in a solvent.
    a. molecule   c. solute
    b. salt   d. chemical bond

16. Match the terms with their most suitable description.

____ hydrophilic	a. protons > electrons
____ atomic number	b. number of protons in nucleus
____ hydrogen bonds	c. polar; dissolves easily in water
____ positive charge	d. collectively strong
____ negative charge	e. protons < electrons
____ temperature	f. measure of molecular motion
____ pH	g. water-dreading
____ covalent bond	h. electron sharing
____ hydrophobic	i. reflects H^+ concentration

CRITICAL THINKING

1. Alchemists were medieval scholars and philosophers who were the forerunners of modern-day chemists. Many spent their lives trying to transform lead (atomic number 82) into gold (atomic number 79). Explain why they never succeeded.

2. Draw a shell model of a lithium atom (Li), which has 3 protons. Predict whether the majority of lithium atoms on Earth are uncharged, positively charged, or negatively charged.

3. Polonium is a rare element with 33 radioisotopes. The most common one, ^{210}Po, has 82 protons and 128 neutrons. When ^{210}Po decays, it emits an alpha particle, which is a helium nucleus (2 protons and 2 neutrons). ^{210}Po decay is tricky to detect because alpha particles do not carry very much energy compared to other forms of radiation. They can be stopped by, for example, a sheet of paper or a few inches of air. This property is one reason why authorities failed to discover toxic amounts of ^{210}Po in the body of former KGB agent Alexander Litvinenko until after he died suddenly and mysteriously in 2006.

   What element does an atom of ^{210}Po decay into after it emits an alpha particle?

4. Some undiluted acids are not as corrosive as when they are diluted with water. That is why lab workers are told to wipe off acid splashes with a towel before washing skin. Explain.

3 MOLECULES OF LIFE

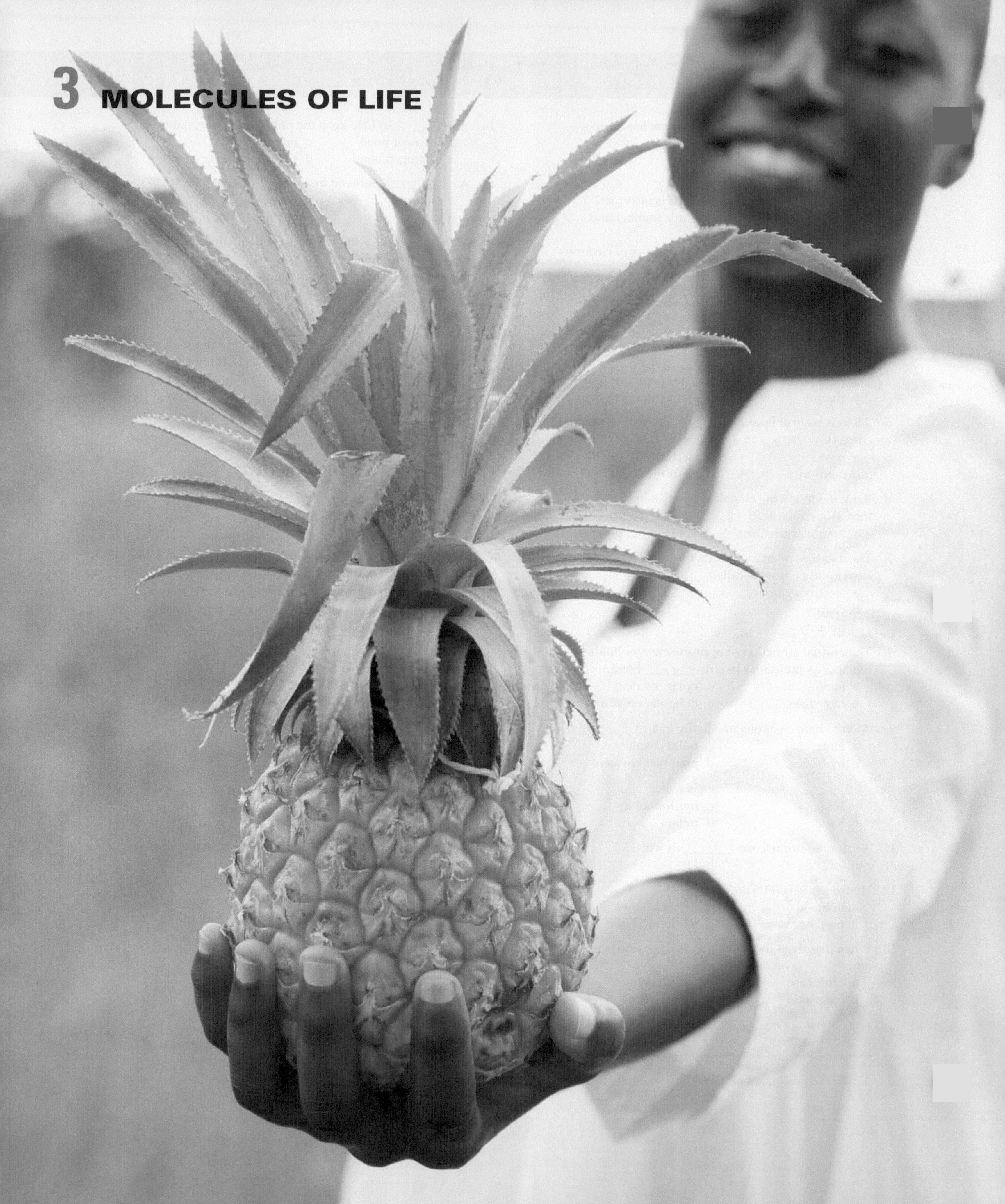

CORE CONCEPTS

Pathways of Transformation

Organisms exchange matter and energy with the environment in order to grow, maintain themselves, and reproduce.

All organisms take up carbon-containing compounds from the environment and use them to build the molecules of life: complex carbohydrates and lipids, proteins, and nucleic acids. Nitrogen is essential for building proteins and nucleic acids; phosphorus is incorporated into lipids and nucleic acids. All biological molecules are eventually broken down, and their components are cycled back to the environment in by-products, wastes, and remains.

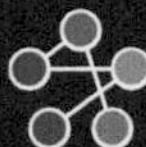

Systems

Complex properties arise from interactions among components of a biological system.

The molecules of life are assembled from simpler organic subunits. The structure and function of a biological molecule arises from (and depends on) the order, orientation, and interactions of its component subunits. The dual hydrophilic and hydrophobic properties of individual phospholipids give rise to the lipid bilayer structure of cell membranes.

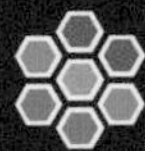

Structure and Function

The three-dimensional form and arrangement of biological structures give rise to their function and interactions.

The same sugar molecules, bonded together in slightly different ways, form carbohydrates with very different properties. A protein's function depends on its structure, which arises from interactions among its amino acid components. Biological information encoded in a nucleic acid consists of the sequence of nucleotide monomers that compose it.

Links to Earlier Concepts

Having learned about atoms and their interactions (Section 2.3), you are now in a position to understand the structure of the molecules of life. Keep the big picture in mind by reviewing Section 1.2. You will be building on your knowledge of covalent bonding (2.4), acids and bases (2.6), and the effects of hydrogen bonds (2.5).

3.1 Fear of Frying

The human body requires only about a tablespoon of fat each day to stay healthy, but most people in developed countries eat far more than that. The average American eats about 70 pounds of fat per year, which may be part of the reason why the average American is overweight. Being overweight increases one's risk for many chronic illnesses. However, the total quantity of fat in the diet may have less of an impact on health than the types of fats eaten.

Molecules that make up oils and other fats have three fatty acid tails, each a long chain of carbon atoms that can vary a bit in structure. Fats that have a certain arrangement of hydrogen atoms around those carbon chains are called *trans* fats (**FIGURE 3.1**). Small amounts of *trans* fats occur naturally in red meat and dairy products. However, the main source of these fats in the Western diet is an artificial food product called partially hydrogenated vegetable oil. Hydrogenation is a manufacturing process that adds hydrogen atoms to oils in order to change them into solid fats, and it creates abundant *trans* bonds in fatty acid tails.

In 1911, Procter & Gamble Co. introduced partially hydrogenated cottonseed oil as a substitute for the

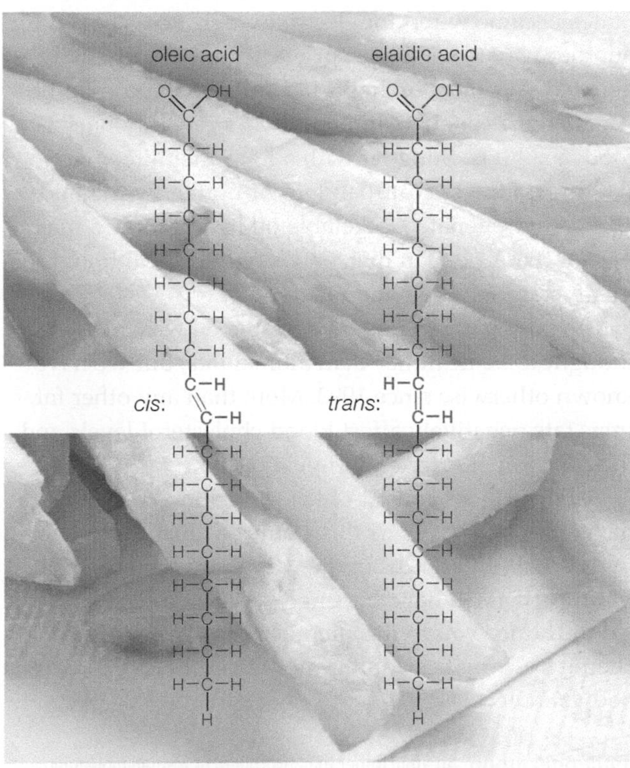
oleic acid
elaidic acid
cis:
trans:

FIGURE 3.1 ***Trans* fats.** Fatty acids are components of molecules that make up oils and other fats. Double bonds in the tails of fatty acids are *cis* or *trans*, depending on their architecture. Fats that have *trans* bonds in their fatty acid tails are called *trans* fats, and they are particularly unhealthy ingredients in food. Partially hydrogenated oils have a high proportion of these fats.

CREDITS: (opposite) Blend/Corbis; (1) photo, Kentoh/Shutterstock.com; art, © Cengage Learning.

Data Analysis Activities

Effects of Dietary Fats on Lipoprotein Levels
Cholesterol that is made by the liver or that enters the body from food cannot dissolve in blood, so it is carried through the bloodstream in clumps called lipoprotein particles. Low-density lipoprotein (LDL) particles carry cholesterol to body tissues such as artery walls, where they can form deposits associated with cardiovascular disease. Thus, LDL is often called "bad" cholesterol. High-density lipoprotein (HDL) particles carry cholesterol away from tissues to the liver for disposal, so HDL is often called "good" cholesterol. In 1990, Ronald Mensink and Martijn Katan published a study that tested the effects of different dietary fats on blood lipoprotein levels. Their results are shown in **FIGURE 3.2**.

FIGURE 3.2 Effect of diet on lipoprotein levels. Researchers placed 59 men and women on a diet in which 10 percent of their daily energy intake consisted of *cis* fatty acids, *trans* fatty acids, or saturated fats.

The amounts of LDL and HDL in the blood were measured after three weeks on the diet; averaged results are shown in mg/dL (milligrams per deciliter of blood). All subjects were tested on each of the diets. The ratio of LDL to HDL is also shown.

1. In which group was the level of LDL ("bad" cholesterol) highest?
2. In which group was the level of HDL ("good" cholesterol) lowest?
3. An elevated risk of heart disease has been correlated with increasing LDL-to-HDL ratios. Which group had the highest LDL-to-HDL ratio?
4. Rank the three diets from best to worst according to their potential effect on heart disease.

	Main Dietary Fats			
	cis fatty acids	*trans* fatty acids	saturated fats	optimal level
LDL	103	117	121	<100
HDL	55	48	55	>40
ratio	1.87	2.44	2.2	<2

more expensive solid animal fats they had been using to make candles and soaps. The demand for candles began to wane as more households in the United States became wired for electricity, and P&G looked for another way to sell its proprietary fat. Partially hydrogenated vegetable oil looks like lard, so the company began aggressively marketing it as a revolutionary new food: a solid cooking fat with a long shelf life, mild flavor, and lower cost than lard or butter. By the mid-1950s, hydrogenated vegetable oil had become a major part of the American diet, and today it is still found in many manufactured and fast foods.

Partially hydrogenated vegetable oil was once thought to be healthier than animal fats, but we have known otherwise since 1993. More than any other fat, *trans* fats negatively affect blood cholesterol levels and the function of arteries and veins.

All organisms consist of the same kinds of biological molecules. However, as this example illustrates, seemingly small differences in the way those molecules are put together can have big effects. With this concept, we introduce you to the chemistry of life. This is your chemistry. It makes you far more than the sum of your body's molecules. ●

functional group An atom (other than hydrogen) or a small molecular group bonded to a carbon of an organic compound; imparts a specific chemical property.
hydrocarbon Compound that consists only of carbon and hydrogen atoms.
organic Describes a compound that consists mainly of carbon and hydrogen atoms.

3.2 The Chemistry of Biology

LEARNING OBJECTIVES

- Use appropriate examples to describe functional groups.
- Explain why biological molecules are modeled in different ways.
- Explain how the molecules of life are polymers.
- Give an example of a metabolic reaction.

The Carbon Backbone

The same elements that make up a living body also occur in nonliving things, but their proportions differ. For example, compared to sand or seawater, a human body has a much larger proportion of carbon atoms (Section 2.2). Why? Unlike sand or seawater, a body has a lot of the molecules of life—complex carbohydrates and lipids, proteins, and nucleic acids—which, in turn, consist of a high proportion of carbon atoms. Compounds that consist mainly of carbon and hydrogen are said to be **organic**. The term is a holdover from a time when these molecules were thought to be made only by living things, as opposed to the "inorganic" molecules that form by nonliving processes. We now know that organic compounds existed on Earth long before life arose. They even form in deep space.

A carbon atom is unusual among elements because it can bond stably with many other elements. It also has four vacancies (Section 2.3), so it can form four covalent bonds with other atoms—including other carbon atoms. Most organic molecules have a chain of carbon atoms, and this backbone often forms rings (**FIGURE 3.3**).

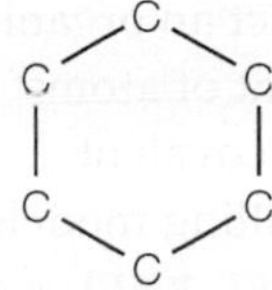
C
C C
C C
C

A Carbon's versatile bonding behavior allows it to form a variety of structures, including rings.

B Carbon rings form the framework of many sugars, starches, and fats (including those that make up doughnuts).

FIGURE 3.3 Carbon rings.

glucose (open-chain form)

glucose (ring form)

FIGURE 3.4 Glucose ($C_6H_{12}O_6$).
This simple sugar converts from a straight-chain form into a ring when the aldehyde group combines with a hydroxyl group. The cyclic form is the main one in water.

The versatility of carbon atoms means that they can be assembled into a wide variety of organic compounds. A molecule that consists only of carbon and hydrogen atoms is called a **hydrocarbon**, and it is completely nonpolar. The molecules of life have other elements in addition to carbon and hydrogen.

Functional Groups

Methane, the simplest hydrocarbon, is one carbon atom bonded to four hydrogen atoms. Other organic molecules, including the molecules of life, have at least one functional group. A **functional group** is an atom (other than hydrogen) or small molecular group covalently bonded to a carbon atom of an organic compound.

TABLE 3.1 lists some common functional groups in biological molecules. Each imparts chemical properties such as acidity or polarity. For example, a hydroxyl group (—OH) adds polar character to an organic compound, thus increasing its ability to dissolve in water. Sugars have a lot of hydroxyl groups, so they are very soluble. A methyl group adds nonpolar character, and may dampen the effect of a polar functional group. Methyl groups added to DNA can act like an "off" switch for this molecule; acetyl groups can act like an "on" switch (we return to this topic in Chapter 10). Acetyl groups also carry two carbons from one molecule to another during some processes of metabolism.

Carboxyl groups make amino acids and fatty acids acidic; amine and amide groups make nucleotide bases basic. Aldehyde and ketone groups are part of simple sugars. Some of these sugars convert to a ring form when the highly reactive aldehyde group on one carbon of the backbone combines with a hydroxyl group on another (**FIGURE 3.4**).

Bonds between sulfhydryl groups stabilize the structure of many proteins, including those that make up human hair. Heat and some kinds of chemicals temporarily break sulfhydryl bonds, which is why we can curl straight hair and straighten curly hair.

TABLE 3.1

Some Functional Groups in Biological Molecules

Functional Group	Structural Formula	General Character	Chemical Formula	Examples of Occurrence
methyl	—C—CH_3	nonpolar	—CH_3	diverse organic compounds
hydroxyl	—C—OH	polar	—OH	alcohols, sugars
sulfhydryl	—C—SH	polar, reactive	—SH	proteins, cofactors
carbonyl	—C—C(=O)—(?)	polar, reactive	—CO	sugars
carboxyl	—C—C(=O)—OH	polar, acidic	—COOH	fatty acids, amino acids
ketone	—C—C(=O)—C—	polar	—CO—	simple sugars, nucleic acids
aldehyde	—C—C(=O)—H	polar, reactive	—CHO	simple sugars
acetyl	—C—C(=O)—CH_3	polar, acidic	—$COCH_3$	proteins, coenzymes
amide	—C—C(=O)—N—	polar	—C(O)N—	proteins, nucleic acids
amine	—C—NH_2	basic	—NH_2	nucleic acids, proteins
phosphate	—C—O—P(=O)(—O^-)—O^-	polar, acidic	—PO_4	nucleic acids, phospholipids

Modeling Organic Compounds

As you will see in the next few sections, the function of an organic molecule arises from and depends on its structure. Researchers make models of organic compounds in order to study different aspects of this relationship. Models can reveal surface properties, changes during synthesis or other biochemical processes, sites of molecular recognition, and so on. Different models allow us to visualize different characteristics.

Structural formulas of organic molecules can be quite complex, even when the molecules are relatively small (**FIGURE 3.5A**). For clarity, some of the features may be implied but not represented: element symbols, for example, or hydrogen atoms bonded to a carbon backbone (**FIGURE 3.5B**). Carbon rings may be simplified as polygons (**FIGURE 3.5C**).

A A structural formula that shows all the bonds and atoms can be very complicated, even for a simple organic molecule. The overall structure is obscured by detail.

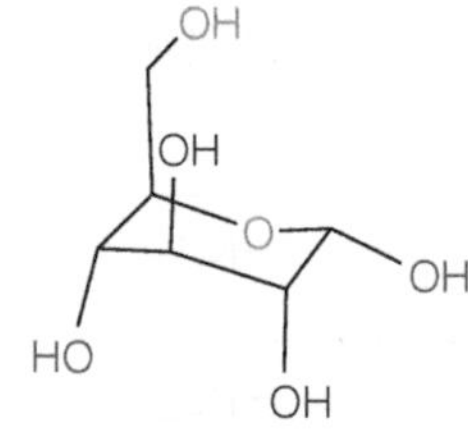

OH
OH
O
OH
HO
OH

B For clarity, some features of structural formulas may be implied but not drawn. In this three-dimensional representation, carbon atoms forming the backbone are not labeled. Neither are the hydrogen atoms bonded to the carbons.

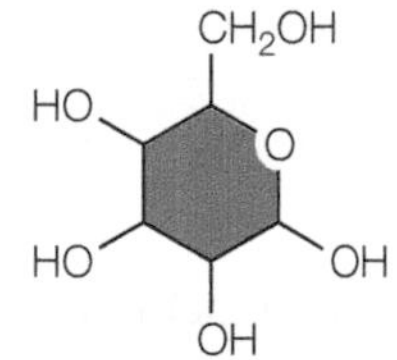

CH_2OH
HO
O
HO
OH
OH

C Using colored polygons as symbols for rings can simplify a structural formula. Some bonds and element labels are omitted.

D A ball-and-stick model is often used to show the arrangement of atoms and bonds in three dimensions.

E A space-filling model can be used to show a molecule's overall shape. Individual atoms are visible in this model.

FIGURE 3.5 Modeling a small organic compound. All of these models represent the same molecule: glucose, a simple sugar.

Ball-and-stick models are used to depict an organic molecule's three-dimensional arrangement of atoms (**FIGURE 3.5D**). Single, double, and triple covalent bonds are all shown as one stick. Space-filling models reveal the molecule's overall shape (**FIGURE 3.5E**).

Many organic molecules are so large that ball-and-stick or space-filling models of them may be incomprehensible. **FIGURE 3.6** shows three different ways to represent hemoglobin, a large molecule that functions as the main oxygen carrier in your blood. Some interesting features of this molecule are not visible in a space-filling model (**FIGURE 3.6A**). Consider that a properly functioning hemoglobin molecule has embedded hemes, which are small carbon-ring structures with an iron atom at their center (Section 5.6 returns to hemes). The hemes are impossible to distinguish in a space-filling model of hemoglobin, but become visible when depicted in surface models such as the one shown in **FIGURE 3.6B**.

Other types of models further reduce visual complexity. Proteins and nucleic acids are often represented as ribbon structures, which, as you will see in Sections 3.5 and 3.6, show how the backbone of these molecules folds and twists. In a ribbon model of hemoglobin (**FIGURE 3.6C**), you can see that the molecule consists of four coiled protein components, each folded around a heme. Such structural details are clues about function: Oxygen binds at the hemes, so each hemoglobin molecule can carry up to four molecules of oxygen.

Metabolic Reactions

All biological systems are based on the same organic molecules, a similarity that is one of many legacies of life's common origin. However, the details of those molecules differ. Just as atoms bonded in different numbers and arrangements form different molecules, simple organic building blocks bonded in different numbers and arrangements form different versions of the molecules of life. The building blocks—sugars, fatty acids, amino acids, and nucleotides—are **monomers** when used as subunits of larger molecules. A molecule that consists of repeated monomers is called a **polymer**.

condensation Chemical reaction in which an enzyme builds a large molecule from smaller subunits; water also forms.
enzyme Organic molecule that speeds a reaction without being changed by it.
hydrolysis (hy-DRAWL-uh-sis) Water-requiring chemical reaction in which an enzyme breaks a molecule into smaller subunits.
metabolism All of the enzyme-mediated reactions in a cell.
monomer Molecule that is a subunit of a polymer.
polymer Molecule that consists of repeated monomers.
reaction Process of molecular change.

Cells link polymers to form monomers, and break apart polymers to release monomers. These and other processes of molecular change are called **reactions**. Cells constantly run reactions as they acquire and use energy to stay alive, grow, reproduce, and so on. Collectively, these reactions are called **metabolism**. Metabolism requires **enzymes**, which are organic molecules (usually proteins) that speed up reactions without being changed by them. Enzymes remove monomers from polymers in a common metabolic reaction called **hydrolysis** (**FIGURE 3.7A**). Hydrolysis requires water (hence the name). The reverse of hydrolysis is a reaction called **condensation**, in which an enzyme joins one monomer to another (**FIGURE 3.7B**). Water forms during condensation, so the reaction is also called dehydration.

TAKE-HOME MESSAGE 3.2

✔ All of the molecules of life are organic, which means they consist mainly of carbon and hydrogen atoms.

✔ The structure of an organic molecule starts with a chain of carbon atoms (the backbone) that may form a ring. Functional groups attached to the backbone impart chemical characteristics to the molecule.

✔ We use different types of molecular models to visualize different structural characteristics. Considering a molecule's structural features gives us insight into its function.

✔ By processes of metabolism, cells assemble the molecules of life from monomers, and break apart polymers into component monomers.

A The complexity of a space-filling model of hemoglobin obscures many interesting features of the molecule.

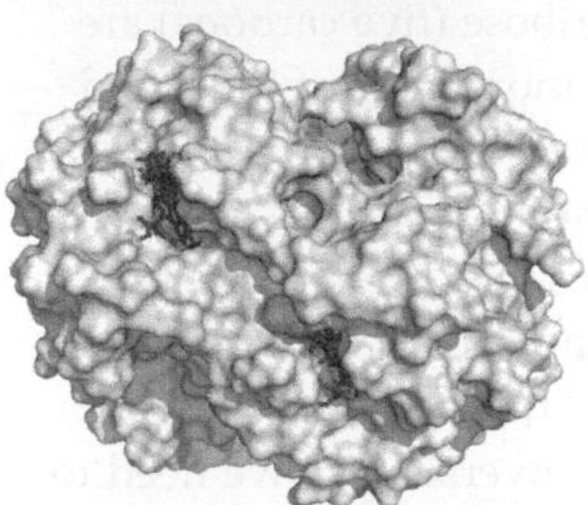

B A surface model of the same molecule reveals crevices and folds that are important for its function. Hemes, in red, are cradled in pockets of the molecule.

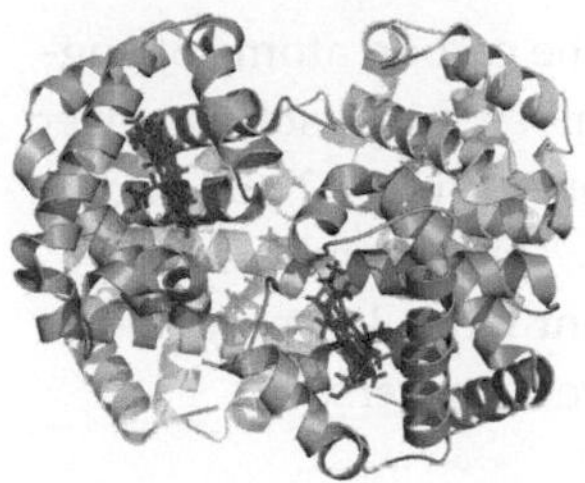

C A ribbon model of hemoglobin reveals all four hemes, also in red. The hemes are held in place by the coiled backbones of the molecule's four protein components.

FIGURE 3.6 Modeling a large organic compound. All of these models represent the same molecule: hemoglobin, the oxygen-transporting molecule in human blood. **A** Models that show individual atoms usually depict them color-coded by element (Section 2.4). **B, C** Other types of models use other colors, depending on which features are being highlighted.

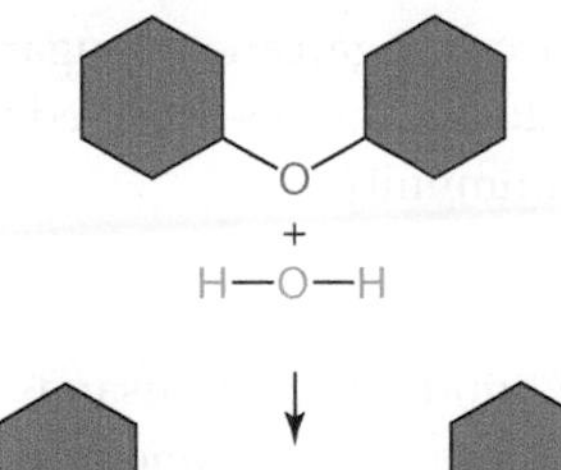
O
+
H—O—H
OH + HO

A Hydrolysis. Cells use this water-requiring reaction to split polymers into monomers. An enzyme attaches a hydroxyl group and a hydrogen atom (both from water) at the site of the split.

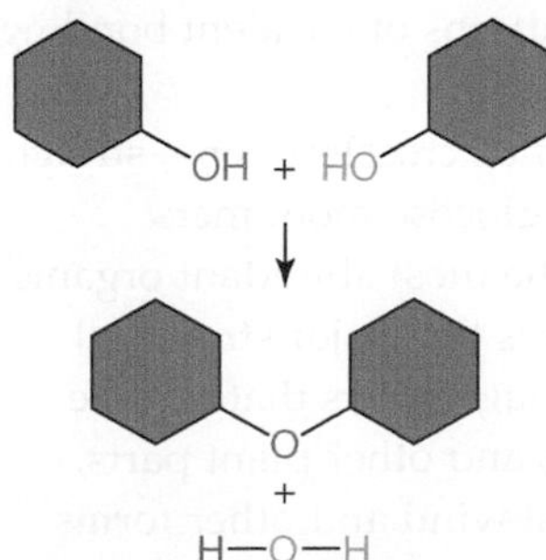
OH + HO
O
+
H—O—H

B Condensation. Cells use this reaction to build polymers from monomers. An enzyme removes a hydroxyl group from one molecule and a hydrogen atom from another. A covalent bond forms between the two molecules. Water forms, so this reaction is also called dehydration.

FIGURE 3.7 Examples of metabolic reactions. Common reactions by which cells build and break down organic molecules are shown.

3.3 Carbohydrates

LEARNING OBJECTIVES

- Describe the structure of carbohydrates and explain their roles in cells.
- Using an example, explain how the structure of a polysaccharide gives rise to its function.
- Name the function that glycogen serves in the human body.

Carbohydrates are organic compounds that consist of carbon, hydrogen, and oxygen in a 1:2:1 ratio. Cells use different kinds as structural materials, for fuel, and for storing and transporting energy. All carbohydrates are either a sugar or a polymer made from sugar monomers, so they are also called saccharides (from Latin *saccharum*, which means sugar).

Simple Sugars

Monosaccharides (one sugar) are often called simple sugars because they are the simplest carbohydrate, and many of them have a sweet taste. Common monosaccharides have a backbone of five or six carbon atoms, one carbonyl group (—C=O), and two or more hydroxyl groups (—OH). The polar functional groups impart solubility to a sugar molecule, which means that monosaccharides move easily through the water-based internal environments of all organisms.

Monosaccharides have extremely important biological roles. Cells break the bonds of glucose (six carbons) to release energy that can be harnessed to power other reactions (we return to this metabolic process in Chapter 7). Ribose and deoxyribose (five carbons) are components of the nucleotide monomers of RNA and DNA, respectively. Cells also use monosaccharides as structural materials to build larger molecules, and as precursors, or parent molecules, that are remodeled into other molecules. For example, cells of plants and many animals make vitamin C from glucose. Human cells are unable to make this conversion, so we need to get vitamin C from our food.

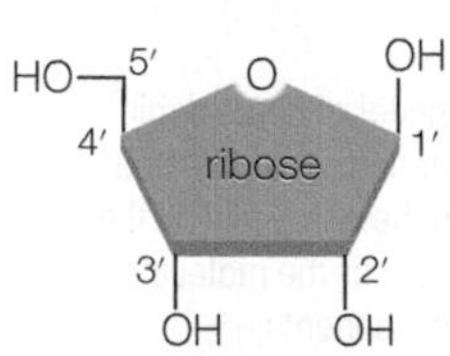
HO
5′
O
OH
4′
ribose
1′
3′
2′
OH
OH

Note that the carbon atoms of sugars are numbered in a standardized way: 1′ ("one prime") 2′, 3′, and so on (left). These numbers will be important when learning about DNA in later chapters.

Oligosaccharides

Short chains of covalently bonded monosaccharides are called oligosaccharides (*oligo–* means a few). **Disaccharides** consist of two sugar monomers. The lactose in milk, with one glucose and one galactose, is a disaccharide. So is sucrose, with one glucose and one fructose (**FIGURE 3.8**). Sucrose is the most plentiful sugar in nature; extracted from sugarcane or sugar beets, it is our table sugar. Oligosaccharides attached to lipids or proteins function in immunity.

Polysaccharides

Polysaccharides are chains of hundreds or thousands of monosaccharide monomers. The most common polysaccharides—cellulose, starch, and glycogen—all consist only of glucose monomers, but as substances their properties are very different. Why? The answer begins with differences in patterns of covalent bonding that link their monomers.

In **cellulose**, hydrogen bonds crosslink long, straight chains of covalently bonded glucose monomers (**FIGURE 3.9A**). Cellulose is the most abundant organic molecule on Earth because it is the major structural material of plants. It forms tough fibers that act like reinforcing rods inside stems and other plant parts, helping these structures resist wind and other forms of mechanical stress. Cellulose is insoluble (it does not dissolve) in water, and it is not easily broken down. Some bacteria and fungi make enzymes that can break it apart into its component sugars, but humans and other mammals do not. Dietary fiber, or "roughage," usually refers to the indigestible cellulose in our vegetable foods. Bacteria that live in the guts of termites and grazers such as cattle and sheep help these animals digest the cellulose in plants.

In **starch**, a different covalent bonding pattern between glucose monomers makes a chain that coils up into a spiral (**FIGURE 3.9B**). Starch does not dissolve easily in water, but it is easier to break down than cellulose. These properties make the molecule ideal for storing sugars in the watery, enzyme-filled interior of plant cells. For example, sugars are moved into developing seeds and then converted to starch for storage. A mature seed may rest for a long time before it sprouts. Then, hydrolysis enzymes break the bonds between the starch's glucose monomers. The released glucose is used to fuel growth of the young plant until it can make its own food (Section 29.7 returns to development in plants). Humans also have enzymes that

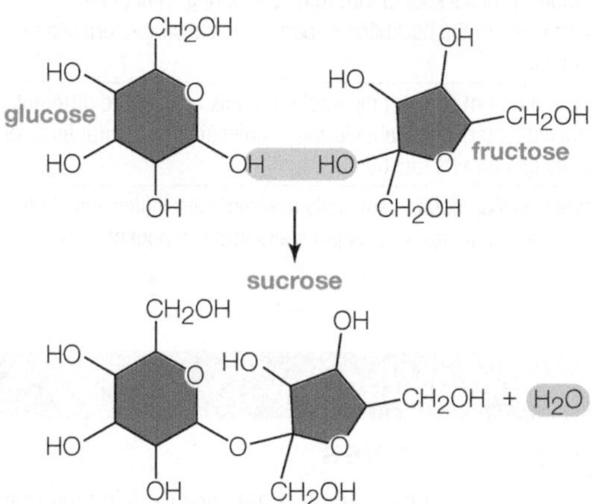
CH2OH
HO
glucose
O
HO
OH
OH
OH
HO
CH2OH
HO
O
fructose
CH2OH
sucrose
CH2OH
OH
HO
O
HO
CH2OH + H2O
HO
O
O
OH
CH2OH

FIGURE 3.8 Formation of sucrose, a disaccharide.

FIGURE IT OUT What type of metabolic reaction is this?

Answer: Condensation

carbohydrate (car-bow-HI-drait) Molecule that consists primarily of carbon, hydrogen, and oxygen atoms in a 1:2:1 ratio.

cellulose (SELL-you-low-ss) Crosslinked polysaccharide of glucose monomers; the major structural material in plants.

chitin (KIE-tin) Nitrogen-containing polysaccharide that composes fungal cell walls and arthropod exoskeletons.

disaccharide (die-SACK-uh-ride) Carbohydrate that consists of two monosaccharide monomers.

fatty acid Lipid that consists of an acidic carboxyl group "head" and a long hydrocarbon "tail."

glycogen (GLIE-ko-jen) Highly branched polysaccharide of glucose monomers. Principal form of stored sugars in animals.

lipid Fatty, oily, or waxy organic compound.

monosaccharide (mon-oh-SACK-uh-ride) Simple sugar; consists of one sugar unit so it cannot be broken apart into monomers.

polysaccharide (paul-ee-SACK-uh-ride) Carbohydrate that consists of hundreds or thousands of monosaccharide monomers.

starch Polysaccharide; energy reservoir in plant cells.

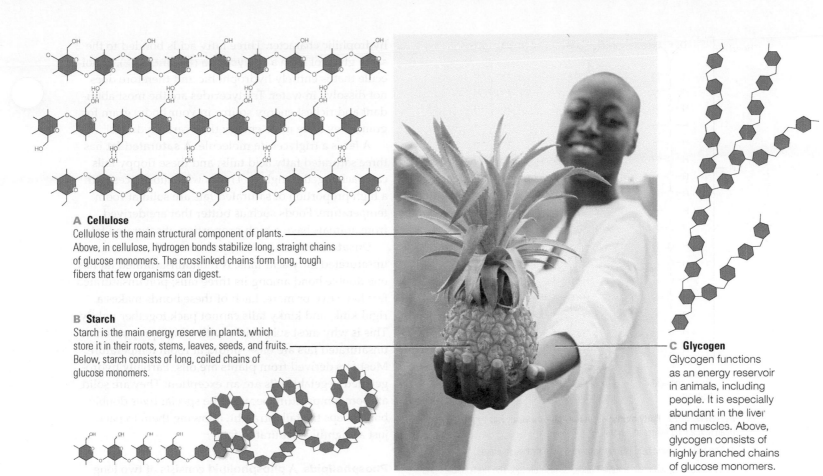

A Cellulose
Cellulose is the main structural component of plants. Above, in cellulose, hydrogen bonds stabilize long, straight chains of glucose monomers. The crosslinked chains form long, tough fibers that few organisms can digest.

B Starch
Starch is the main energy reserve in plants, which store it in their roots, stems, leaves, seeds, and fruits. Below, starch consists of long, coiled chains of glucose monomers.

C Glycogen
Glycogen functions as an energy reservoir in animals, including people. It is especially abundant in the liver and muscles. Above, glycogen consists of highly branched chains of glucose monomers.

FIGURE 3.9 Three of the most common complex carbohydrates and their locations in a few organisms. Each polysaccharide consists only of glucose subunits, but different bonding patterns result in substances with very different properties.

break down starch, so this carbohydrate is an important component of our food.

Animals store sugars in the form of **glycogen**, a polysaccharide that consists of highly branched chains of glucose monomers (**FIGURE 3.9C**). Muscle and liver cells contain most of the body's glycogen. When the blood sugar level falls, liver cells break down the glycogen, and the released glucose subunits enter the blood.

In **chitin**, a polysaccharide similar to cellulose, long, unbranching chains of nitrogen-containing sugar monomers are linked by hydrogen bonds (**FIGURE 3.10**). As a structural material, chitin is durable, translucent, and flexible. It reinforces the cell wall of fungi, and strengthens the outer body covering of animals such as insects, spiders, and shrimp.

TAKE-HOME MESSAGE 3.3

✔ Some monosaccharides are used as structural materials or precursors; others are broken down for energy.

✔ Cells assemble monosaccharide monomers into polysaccharides. Cellulose, starch, and glycogen consist of glucose monomers.

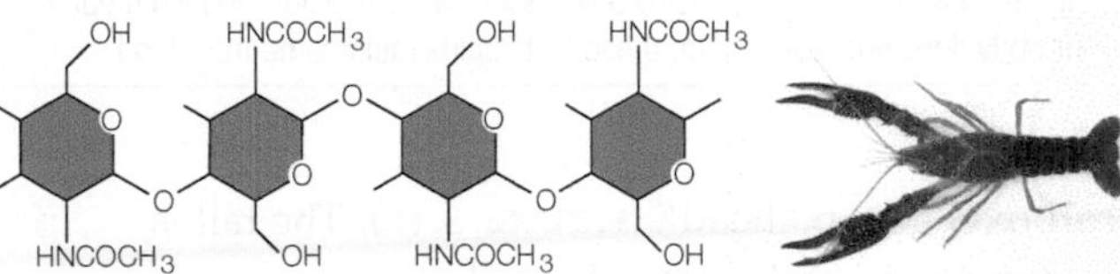
OH
HNCOCH3
OH
HNCOCH3
HNCOCH3
OH
HNCOCH3
OH

FIGURE 3.10 Chitin. This nitrogen-containing polysaccharide is a polymer of modified glucose monomers. It strengthens the exoskeleton (external skeleton) of small animals such as shrimp.

3.4 Lipids

LEARNING OBJECTIVES

- Describe a fat, and identify the difference between saturated and unsaturated fats.
- Explain why a phospholipid is both hydrophilic and hydrophobic.
- Describe the lipid bilayer.
- Give one example of a molecule that is made from cholesterol.

Lipids in Biological Systems

Lipids are fatty, oily, or waxy organic compounds. They vary in structure, but all are hydrophobic (Section 2.5). One type of lipid, a **fatty acid**, is a small organic molecule that consists of a long hydrocarbon "tail" with a

CREDITS: (9 photo) Blend/Corbis; (10) Valerio Pardi/Shutterstock.com.

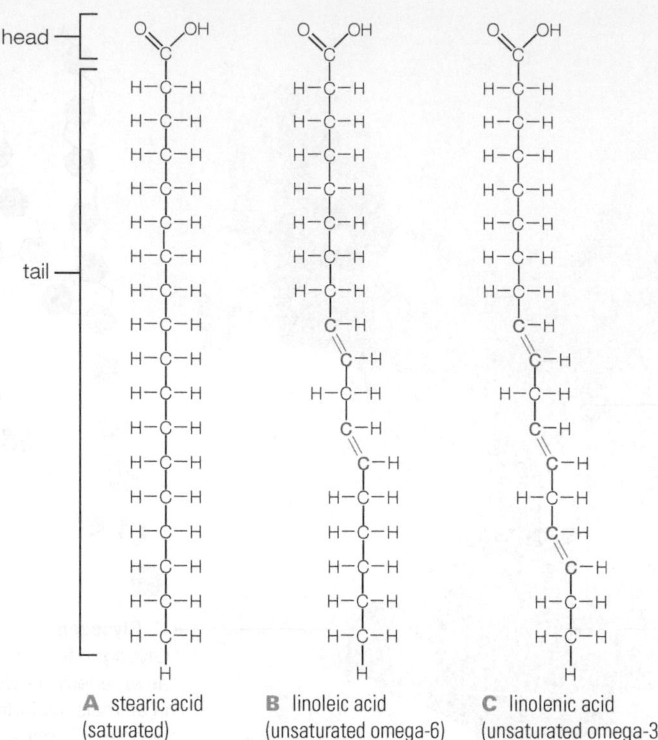
head
tail
A stearic acid (saturated)
B linoleic acid (unsaturated omega-6)
C linolenic acid (unsaturated omega-3)

FIGURE 3.11 Fatty acids. Each fatty acid molecule has a carboxyl group head and a long hydrocarbon tail.

A The tail of stearic acid is fully saturated with hydrogen atoms.

B Linoleic acid, with two double bonds, is unsaturated. The first double bond occurs at the sixth carbon from the end, so linoleic acid is called an omega-6 fatty acid.

C Linolenic acid is unsaturated. The first double bond occurs at the third carbon from the end, so linolenic acid is called an omega-3 fatty acid. Omega-3 and omega-6 fatty acids are "essential fatty acids," which means your body does not make them; these fatty acids must come from food.

carboxyl group "head" (**FIGURE 3.11**). The tail is hydrophobic (fatty), and the carboxyl group head is hydrophilic (and acidic). You are already familiar with the dual properties of fatty acids, because these molecules are the main component of soap: The hydrophobic tails attract oily dirt, and the hydrophilic heads dissolve the dirt in water.

Saturated fatty acids have only single bonds linking the carbons in their tails. In other words, their carbon chains are fully saturated with hydrogen atoms (**FIGURE 3.11A**). Saturated fatty acid tails are flexible and they wiggle freely. Double bonds between carbons kink the tail of an unsaturated fatty acid (**FIGURE 3.11B,C**) and limit its flexibility. These bonds are *cis* or *trans*, depending on the way the hydrogens are arranged around them (as shown in Figure 3.1).

Fats The carboxyl group head of a fatty acid can easily react with another molecule. When it reacts with a glycerol (a type of alcohol), the fatty acid loses its hydrophilic character. Three fatty acids bonded to the same glycerol form a **triglyceride** (**FIGURE 3.12**), a molecule that is entirely hydrophobic and therefore does not dissolve in water. Triglycerides are the most abundant and richest energy source in your body; gram for gram, they store more energy than carbohydrates.

A **fat** is a triglyceride molecule. A **saturated fat** has three saturated fatty acid tails, and these floppy tails can pack together tightly. This is why substances with a high proportion of saturated fats are solid at room temperature. Foods such as butter that are derived from animals have a high proportion of saturated fats.

Unsaturated fats are triglycerides with one or more unsaturated fatty acid tails. A monounsaturated fat has one double bond among its three tails; polyunsaturated fats have two or more. Each of these bonds makes a rigid kink, and kinky tails cannot pack together tightly. This is why most substances with a high proportion of unsaturated fats are oils—liquid at room temperature. Most fats derived from plants are oils. Partially hydrogenated vegetable oils are an exception: They are solid at room temperature because the special *trans* double bond keeps the tails straight, allowing them to pack just as tightly as saturated fats.

Phospholipids A **phospholipid** consists of two long hydrocarbon tails (which, in most organisms, are derived from fatty acids) and a head with a phosphate

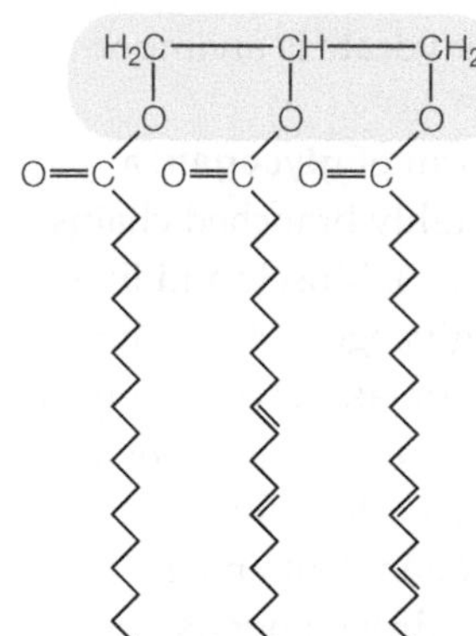
H_2C — CH — CH_2
O O O
O=C O=C O=C

FIGURE 3.12 A triglyceride. A triglyceride (fat) has three fatty acid tails attached to a glycerol head (shaded blue). This molecule is entirely hydrophobic.

FIGURE IT OUT Is this fat saturated or unsaturated?

Answer: Unsaturated

fat A triglyceride.
lipid bilayer Double layer of lipids arranged tail-to-tail; structural foundation of cell membranes.
phospholipid (foss-foe-LIP-id) A lipid with two (hydrophobic) fatty acid tails and a (hydrophilic) phosphate group in its head.
saturated fat Triglyceride molecule with three saturated fatty acid tails.
steroid (STARE-oyd) Lipid with four carbon rings, no fatty acid tails.
triglyceride (tri-GLISS-a-ride) A molecule with three fatty acids bonded to a glycerol; a fat.
unsaturated fat Triglyceride molecule with one or more unsaturated fatty acid tails.
wax Water-repellent substance that is a complex, varying mixture of lipids with long fatty acid tails bonded to long-chain alcohols.

group (FIGURE 3.13A). The tails are hydrophobic, and the polar phosphate group makes the head very hydrophilic. As you will see in Chapter 4, these opposing properties give rise to the **lipid bilayer**, a two-layered sheet of phospholipids that is the basic structure of all cell membranes (FIGURE 3.13B). The heads of one layer face the cell's fluid interior, and the heads of the other layer face the cell's watery surroundings. Tails of all the phospholipids are sandwiched between the heads, so the interior of a lipid bilayer is highly hydrophobic. Some of the lipids in a cell membrane have oligosaccharides attached to their head; these glycolipids have important functions in immunity.

Steroids **Steroids** are lipids with no fatty acid tails; they have a rigid backbone that consists of 20 carbon atoms arranged in a characteristic pattern of four rings. Functional groups attached to the rings define the type of steroid. These molecules serve varied and important physiological functions in plants and animals. Cholesterol, the most common steroid in animal tissues, is remodeled into other molecules such as vitamin D (required to keep teeth and bones strong) and steroid hormones (FIGURE 3.14).

Waxes A **wax** is a water-repellent substance that consists of a complex, varying mixture of lipids with long fatty acid tails bonded to long-chain alcohols. Waxes are firm and water-repellent because these molecules can pack together very tightly. Plants secrete waxes onto their exposed surfaces to restrict water loss, and to repel parasites and other pests. Other types of waxes protect, lubricate, and soften skin and hair. Waxes, together with fats and fatty acids, make feathers waterproof. Bees store honey and raise new generations of bees inside a honeycomb of secreted beeswax, which consists mainly of two lipids (right).

beeswax

TAKE-HOME MESSAGE 3.4

- ✔ Fatty acids are lipids with dual chemical character: a hydrophilic head, and hydrophobic tails.
- ✔ A triglyceride is a fat. Saturated fats have three fatty acid tails; unsaturated fats have one or more unsaturated tails.
- ✔ Phospholipids are the main component of lipid bilayers.
- ✔ Steroids are lipids with four carbon rings. They serve varied and important physiological roles in plants, fungi, and animals.
- ✔ Waxes are mixtures of lipids with complex, varying structures.

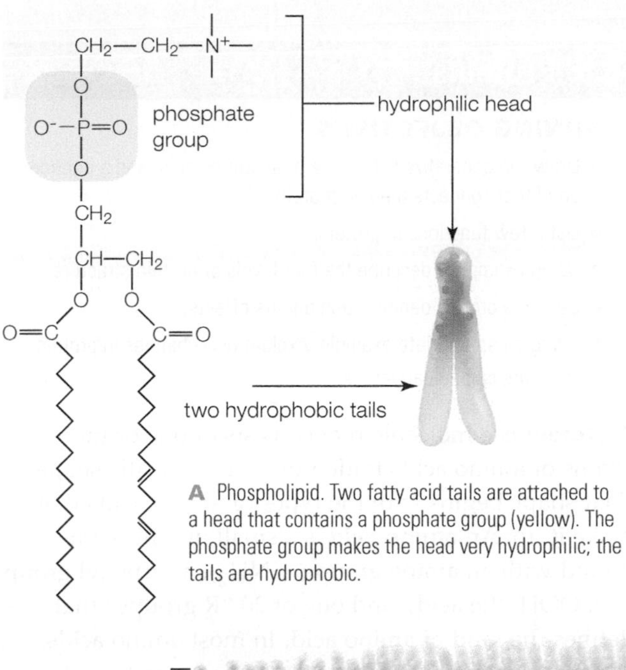
phosphate group
hydrophilic head
two hydrophobic tails

A Phospholipid. Two fatty acid tails are attached to a head that contains a phosphate group (yellow). The phosphate group makes the head very hydrophilic; the tails are hydrophobic.

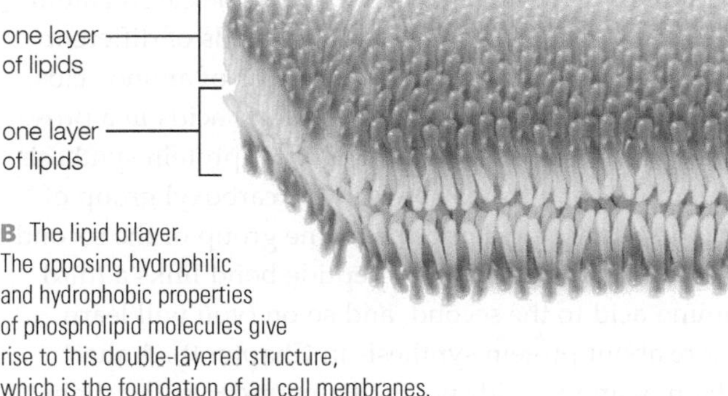
one layer of lipids
one layer of lipids

B The lipid bilayer. The opposing hydrophilic and hydrophobic properties of phospholipid molecules give rise to this double-layered structure, which is the foundation of all cell membranes.

FIGURE 3.13 **Phospholipids make up cell membranes.**

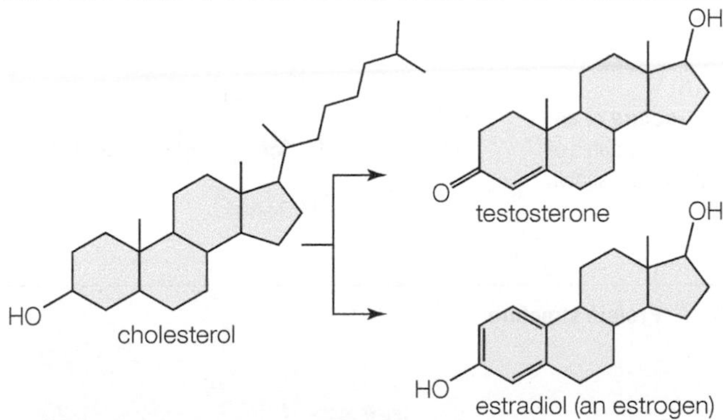
cholesterol
testosterone
estradiol (an estrogen)

FIGURE 3.14 **Steroids.** Above, cells remodel cholesterol (a steroid) into many other compounds, including estradiol and testosterone: two steroid hormones that govern reproduction and secondary sexual traits. Despite the small differences in structure, the two hormones have very different effects in the body. They are the source of gender-specific traits in many species, including wood ducks (left).

♂
♀

3.5 Proteins

LEARNING OBJECTIVES

- Draw the generalized structure of an amino acid, and a peptide bond that connects them in proteins.
- List a few functions of proteins.
- Using examples, describe the four levels of protein structure.
- Describe protein denaturation and its effects.
- Using an appropriate example, explain why changes in protein structure can be dangerous.

A **protein** is a molecule that consists of one or more chains of amino acids folded up into a specific shape. That shape begins with the types of amino acids that make it up. An **amino acid** is a small organic compound with an amine group ($—NH_2$), a carboxyl group (—COOH, the acid), and one of 20 "R groups" that defines the kind of amino acid. In most amino acids, all three groups are attached to the same carbon atom (**FIGURE 3.15**). Cells make the thousands of different proteins they need from only 20 kinds of amino acids.

The covalent bond that links amino acids in a protein is called a **peptide bond**. During protein synthesis, a peptide bond forms between the carboxyl group of the first amino acid and the amine group of the second (**FIGURE 3.16** ❶). Another peptide bond links a third amino acid to the second, and so on (you will learn more about protein synthesis in Chapter 9). A short chain of amino acids is called a **peptide**; as the chain lengthens, it becomes a **polypeptide**.

From Structure to Function

Proteins participate in all processes that sustain life. Structural proteins support cell parts and, as part of tissues, multicelled bodies. Most enzymes that carry out metabolic reactions are proteins. Proteins move substances, help cells communicate, and defend the body. The idea that structure dictates function is particularly appropriate as applied to proteins, because the diversity in biological activity among these molecules arises from differences in their three-dimensional shape.

Primary and Secondary Structure Protein structure starts with a series of amino acids that become joined into a polypeptide during protein synthesis ❷. The order of the amino acids, which is called primary structure, defines the type of protein.

Primary structure also determines the higher orders of structure that make up a protein's final shape. This shape begins to arise even before protein synthesis has finished, as hydrogen bonds that form between amino acids cause the lengthening polypeptide to twist and turn in three dimensions. The hydrogen bonds pull sections of the polypeptide into characteristic patterns such as coils (helices) and sheets connected by flexible loops and tight turns (**FIGURE 3.17**). These are patterns of secondary structure ❸. Each type of protein has a unique primary structure, but almost all proteins have similar patterns of secondary structure.

amino acid (uh-ME-no) Small organic compound that is a monomer of proteins. Consists of a carboxyl group, an amine group, and one of 20 side groups (R), all typically bonded to the same carbon atom.
peptide Short chain of amino acids linked by peptide bonds.
peptide bond A bond between the amine group of one amino acid and the carboxyl group of another. Joins amino acids in proteins.
polypeptide Long chain of amino acids linked by peptide bonds.
protein (PRO-teen) Organic molecule that consists of one or more polypeptides folded into a specific shape.

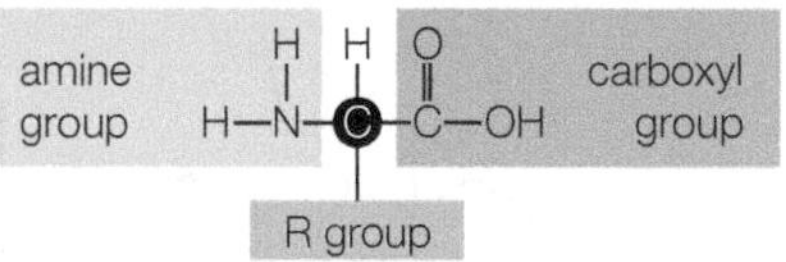
amine group
carboxyl group
R group

FIGURE 3.15 **Generalized structure of an amino acid.** The structures of the twenty R groups are shown in Appendix II.

FIGURE 3.16 **Protein structure.**

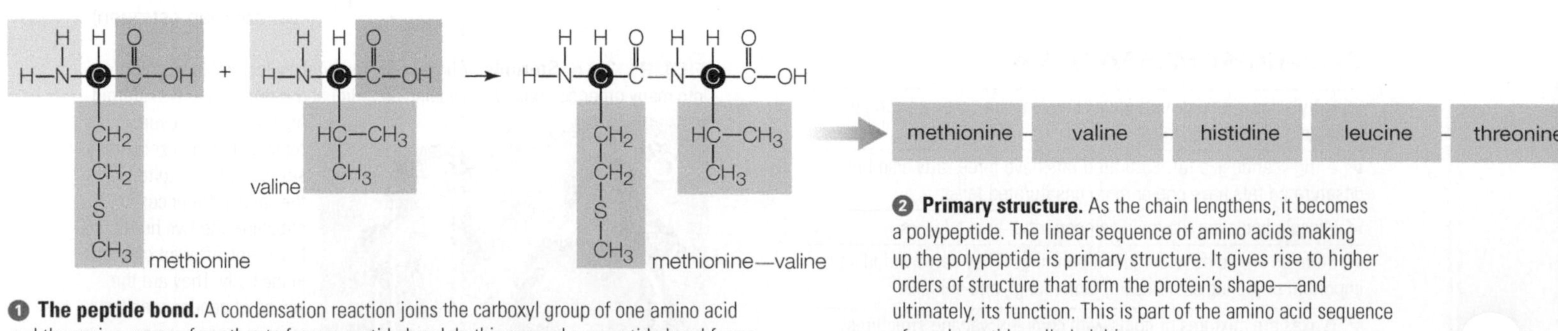
methionine
valine
methionine—valine
methionine - valine - histidine - leucine - threonine

❶ **The peptide bond.** A condensation reaction joins the carboxyl group of one amino acid and the amine group of another to form a peptide bond. In this example, a peptide bond forms between the amino acids methionine and valine.

❷ **Primary structure.** As the chain lengthens, it becomes a polypeptide. The linear sequence of amino acids making up the polypeptide is primary structure. It gives rise to higher orders of structure that form the protein's shape—and ultimately, its function. This is part of the amino acid sequence of a polypeptide called globin.

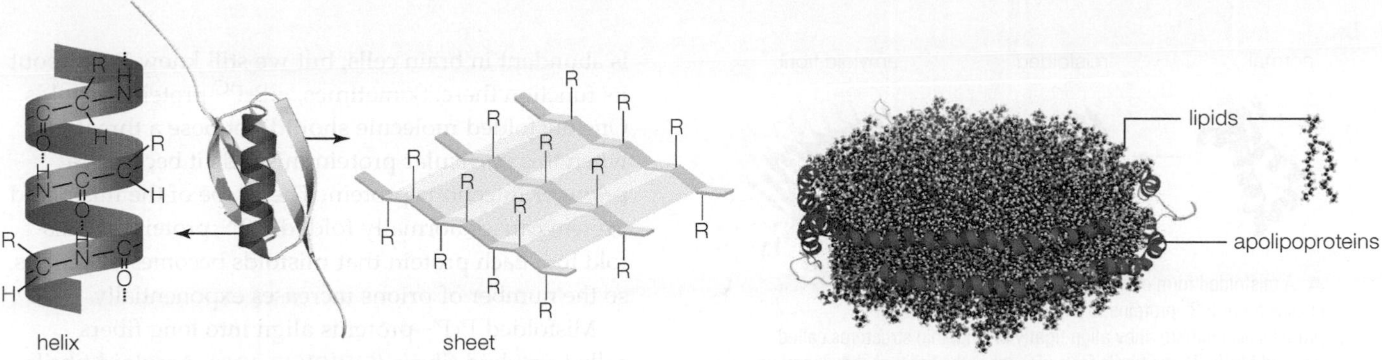
helix
sheet
lipids
apolipoproteins

FIGURE 3.17 Examples of secondary structure in proteins. The small protein in the middle has a helix (in red) and a sheet (in yellow). Tight turns reverse the direction of the polypeptide to form the sheet; flexible loops connect the sheet to the helix.

FIGURE 3.18 A lipoprotein particle. The one depicted here (HDL, which is often called "good" cholesterol) consists of thousands of lipids lassoed into a clump by apolipoproteins (a type of lipoprotein).

Tertiary and Quaternary Structure Much as an overly twisted rubber band coils back upon itself, hydrogen bonding and other interactions between helices and sheets can make them fold up together into compact, functional domains ❹. These domains are the protein's tertiary structure, and they make it a working molecule. Some domains are simple; others are complex structures that resemble tiny barrels (left), propellers, sandwiches, and so on. Each of these domains has a particular function that is more or less separate from the rest of the protein. For example, some barrel domains rotate like motors in small molecular machines. Others form tunnels through cell membranes that allow ions to cross. A large protein may have several domains, each contributing a particular structural or functional property to the molecule.

Domains are modular, and, like evolutionary building blocks, many have been reused and reassembled into proteins that have different functions. A common domain called a zinc finger, for example, holds a zinc ion that helps proteins bind to DNA. Small changes in primary structure have resulted in slightly different versions of the domain, each recognizing a different region of DNA. A protein that incorporates one of these domains can bind to a specific area in a DNA molecule (Chapter 10 returns to the function of these DNA-binding proteins).

Many proteins also have quaternary structure, which means they consist of two or more polypeptide chains that are closely associated or covalently bonded together ❺. Most enzymes are like this, with multiple polypeptide chains that collectively form a roughly spherical shape. So are fibrous proteins, which aggregate by many thousands into much larger structures.

Finishing Touches Carbohydrates, lipids, or both may get attached to a protein after synthesis. A protein with one or more oligosaccharides attached to it is called a glycoprotein. Molecules that allow a tissue or a body to recognize its own cells are glycoproteins, as are other molecules that help cells interact in immunity. A protein that can bind to lipids is called a lipoprotein. Lipoproteins form particles that allow fats and other hydrophobic molecules to move through watery fluid inside cells and bodies (**FIGURE 3.18**).

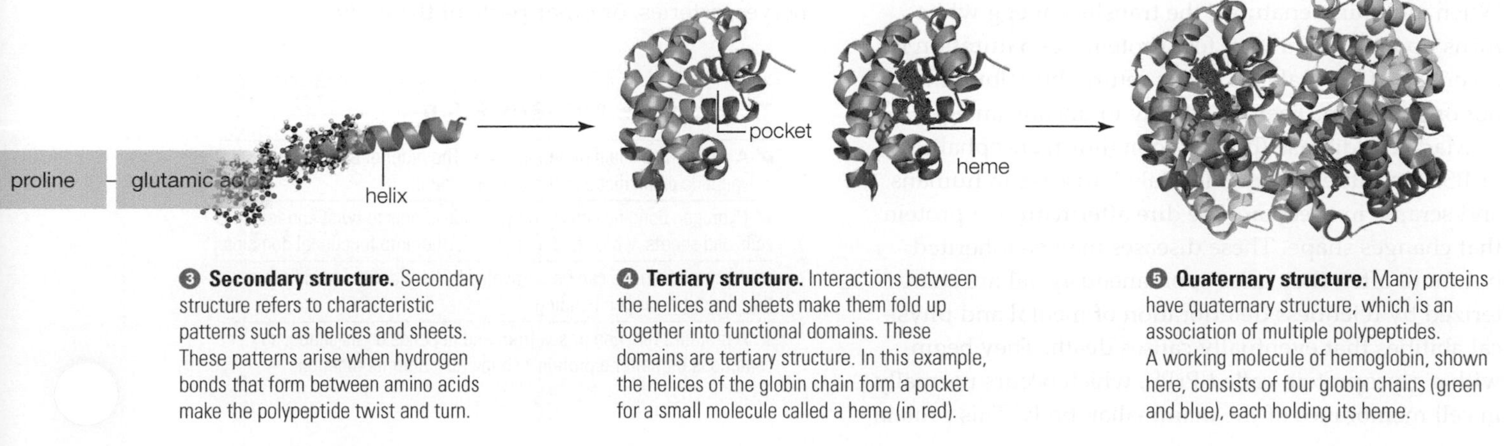
proline
helix
pocket
heme

❸ **Secondary structure.** Secondary structure refers to characteristic patterns such as helices and sheets. These patterns arise when hydrogen bonds that form between amino acids make the polypeptide twist and turn.

❹ **Tertiary structure.** Interactions between the helices and sheets make them fold up together into functional domains. These domains are tertiary structure. In this example, the helices of the globin chain form a pocket for a small molecule called a heme (in red).

❺ **Quaternary structure.** Many proteins have quaternary structure, which is an association of multiple polypeptides. A working molecule of hemoglobin, shown here, consists of four globin chains (green and blue), each holding its heme.

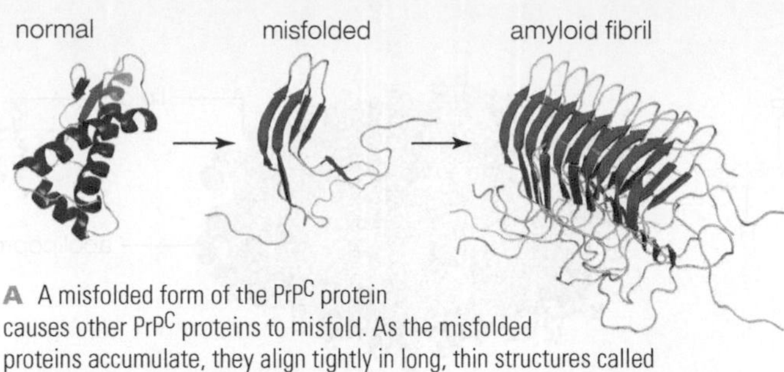
normal
misfolded
amyloid fibril

A A misfolded form of the PrP^C protein causes other PrP^C proteins to misfold. As the misfolded proteins accumulate, they align tightly in long, thin structures called amyloid fibrils. These fibrils form plaques in the brain as they lengthen.

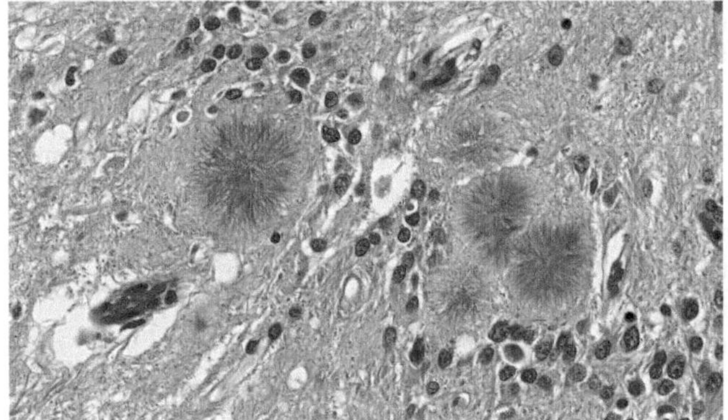

B Amyloid fibrils radiating from several plaques are visible in this slice of brain tissue from a person with vCJD.

FIGURE 3.19 Variant Creutzfeldt–Jakob disease (vCJD).

FIGURE IT OUT How does PrP^C secondary structure change when it misfolds?

Answer: It loses helices and gains sheets.

The Importance of Protein Structure

Hydrogen bonds that maintain the shape of a protein can be disrupted by shifts in pH or temperature, or by exposure to detergents or salts. Such disruption causes a protein to undergo **denaturation**, which means it loses its secondary, tertiary, and quaternary structure. As a protein's shape unravels, so does its function.

Consider what happens when you cook an egg. The white of an egg consists mainly of a protein called albumin. Cooking does not disrupt the covalent bonds of albumin's primary structure, but it does destroy the hydrogen bonds that maintain the protein's shape. When albumin denatures, the translucent egg white turns opaque. For a very few proteins, denaturation is reversible if normal conditions return, but albumin is not one of them. There is no way to uncook an egg.

Mad cow disease (bovine spongiform encephalitis, or BSE) in cattle, Creutzfeldt–Jakob disease in humans, and scrapie in sheep are the dire aftermath of a protein that changes shape. These diseases may be inherited, but more often they arise spontaneously; all are characterized by relentless deterioration of mental and physical abilities that eventually causes death. They begin with a glycoprotein called PrP^C, which occurs normally in cell membranes of the mammalian body. This protein is abundant in brain cells, but we still know little about its function there. Sometimes, a PrP^C protein misfolds. One misfolded molecule should not pose a threat, but when this particular protein misfolds it becomes a **prion**, or infectious protein. The shape of the misfolded protein causes normally folded PrP^C proteins to misfold too. Each protein that misfolds becomes infectious, so the number of prions increases exponentially.

Misfolded PrP^C proteins align into long fibers called amyloid fibrils (FIGURE 3.19A). Amyloid fibrils resist cellular mechanisms that destroy misfolded proteins. They grow from their ends as newly misfolded proteins join the aggregation, forming characteristic patches in the brain called plaques (FIGURE 3.19B). The misfolded PrP^C protein is toxic to brain cells, and holes form in the brain as its cells die. Progressive symptoms such as confusion, memory loss, and lack of coordination precede death.

In the mid-1980s, an epidemic of mad cow disease in Britain was followed by an outbreak of a new variant of Creutzfeldt–Jakob disease (vCJD) in humans. A prion similar to the one in scrapie-infected sheep was found in cows with BSE, and also in humans affected by vCJD. How did the prion get from sheep to cattle to people? Prions resist denaturation, so treatments such as cooking that inactivate other infectious agents have little effect on them. The cattle became infected by the prion after eating feed prepared from the remains of scrapie-infected sheep, and people became infected by eating beef from the infected cattle.

PrP^C is not the only protein that is harmful when misfolded. In the last decade, researchers discovered similar mechanisms that underlie several neurodegenerative disorders, including Alzheimer's, Parkinson's, and Huntington's diseases, atherosclerosis, and amyotrophic lateral sclerosis (ALS). Like vCJD, all involve proteins that are normally found in the body. When these proteins misfold, they cause normal versions of the protein to misfold too. The misfolded proteins are toxic, and they disrupt normal function of the brain, nerves, arteries, or other parts of the body.

TAKE-HOME MESSAGE 3.5

- ✔ A protein is a chain of amino acids. The order of amino acids in a polypeptide chain dictates the type of protein.
- ✔ Hydrogen bonding causes polypeptide chains to twist and fold into coils and sheets, which fold and pack further into functional domains.
- ✔ A protein's shape can be unraveled by heat or other conditions that disrupt hydrogen bonding.
- ✔ A protein's function arises from and depends on its shape, so conditions that alter a protein's shape also alter its function.

CREDIT: (19B) Sherif Zaki, MD PhD, Wun-Ju Shieh, MD PhD; MPH/ CDC.

3.6 Nucleic Acids

LEARNING OBJECTIVES

- Describe the structure of a nucleotide and the general structure of a nucleic acid.

Nucleotides are small organic molecules that function as energy carriers, enzyme helpers, chemical messengers, and subunits of DNA and RNA. Each consists of a monosaccharide ring bonded to a nitrogen-containing "base" and one, two, or three phosphate groups (FIGURE 3.20). The monosaccharide is a five-carbon sugar, either ribose or deoxyribose; and the base is one of five small compounds with a flat ring structure (we return to the structure of nucleotide bases in Sections 8.3 and 9.2). When the third phosphate group of a nucleotide is transferred to another molecule, energy is transferred along with it. As you will see in Section 5.6, the nucleotide **ATP** (adenosine triphosphate) serves an especially important role as a type of energy currency in cells.

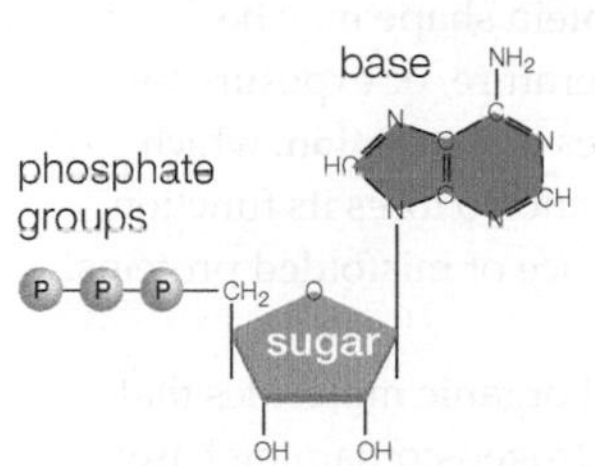

base
NH_2
phosphate groups
P
P
P
CH_2
sugar
OH
OH

FIGURE 3.20 A nucleotide. This one is ATP.

Nucleic acids are polymers, chains of nucleotides in which the sugar of one nucleotide is joined to the phosphate group of the next (FIGURE 3.21A). **RNA**, or ribonucleic acid, is named after the ribose sugar of its component nucleotides. An RNA is a single chain of nucleotide monomers (FIGURE 3.21B), one of which is ATP. Different types of RNA interact with DNA during protein synthesis.

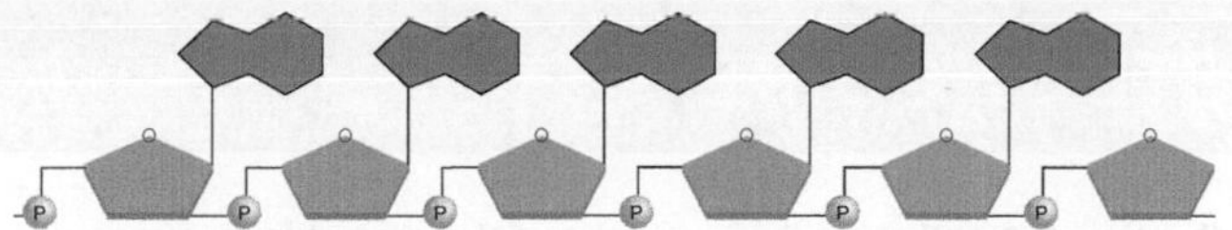

P
P
P
P
P
P

A A chain of nucleotides is a nucleic acid. The sugar of one nucleotide is covalently bonded to the phosphate group of the next, forming a sugar–phosphate backbone.

B The nucleic acid RNA.

C The nucleic acid DNA.

FIGURE 3.21 Nucleic acid structure.

DNA, or deoxyribonucleic acid, is a nucleic acid named after the deoxyribose sugar of its component nucleotides. DNA has two chains of nucleotides twisted into a double helix (FIGURE 3.21C). The order of nucleotides composing a DNA molecule is biological information: an idea so important that we devote an entire unit to it (Unit 2, Genetics). Each cell starts life with DNA inherited from a parent cell. That DNA contains all of the information necessary to build a new cell and, in the case of multicelled organisms, an entire individual. The cell uses the order of nucleotide bases in DNA—the DNA sequence—to guide production of RNA and proteins.

ATP Adenosine triphosphate. (uh-DEN-uh-seen) Nucleotide monomer of RNA; also is an important energy carrier in cells.

denaturation (dee-nay-turr-AY-shun) Loss of a protein's three-dimensional shape, as by a shift in temperature.

DNA Deoxyribonucleic acid. (dee-ox-ee-rye-bo-new-CLAY-ick) Nucleic acid that carries hereditary information.

nucleic acid (new-CLAY-ick) Polymer of nucleotides; DNA or RNA.

nucleotide (NEW-klee-uh-tide) Small molecule with a five-carbon sugar, a nitrogen-containing base, and phosphate groups. Monomer of nucleic acids; some have additional metabolic roles.

prion (PREE-on) Infectious protein.

RNA Ribonucleic acid. (rye-bo-new-CLAY-ick) Nucleic acid that carries out protein synthesis.

TAKE-HOME MESSAGE 3.6

✔ Nucleotides are monomers of nucleic acids. The nucleotide ATP also has an important metabolic role as an energy carrier.

✔ The nucleic acid DNA holds information necessary to build cells and multicelled individuals.

✔ RNAs are nucleic acids that carry out protein synthesis.

3.1 Fear of Frying (revisited)

In the United States, partially hydrogenated oils will be banned from manufactured foods in 2018. Until then, nutritional labels on packaged foods are required to include *trans* fat content. These products may be marked "zero grams of *trans* fat" even when a single serving contains up to half a gram.

Eating as little as half a gram of *trans* fat per day (about 0.4 teaspoon of hydrogenated vegetable oil) measurably increases one's risk of atherosclerosis (hardening of the arteries), heart attack, and diabetes. A small serving of French fries made with hydrogenated vegetable oil contains about 5 grams of *trans* fat.

You can minimize your intake of *trans* fat by avoiding foods fried in and prepared with partially hydrogenated oils. Check ingredient labels: Many packaged foods—including margarine, non-dairy creamer, microwave popcorn, cakes, biscuits, cookies, crackers, and puddings—are still made with these oils.

STUDY GUIDE

Section 3.1 All organisms consist of the same kinds of molecules. Seemingly small differences in the way those molecules are put together can have big effects inside a living organism. A minor architectural difference between *cis* and *trans* bonds in fatty acid tails makes a major difference in the human body. Fats with *trans* bonds in their fatty acid tails (*trans* fats) are particularly unhealthy foods; only a tiny amount increases the risk of serious disease. *Trans* fats are abundant in partially hydrogenated vegetable oils.

Section 3.2 Molecules that consist mainly of carbon and hydrogen atoms are **organic**. **Hydrocarbons** have only carbon and hydrogen atoms. The structure of the molecules of life—complex carbohydrates and lipids, proteins, and nucleic acids—starts with a chain of carbon atoms (the backbone) that may form rings. **Functional groups** attached to the backbone influence the molecule's chemical character, and thus its function. Different molecular models reveal different aspects of structure.

Metabolism includes all enzyme-mediated **reactions** in a cell. In **condensation** reactions, **enzymes** build **polymers** from smaller **monomers**. **Hydrolysis** releases monomers by breaking apart polymers.

Section 3.3 Cells use simple carbohydrates (sugars) for energy and to build other molecules. **Monosaccharides** (simple sugars) are bonded together to form **disaccharides** (two sugars), oligosaccharides (a few sugars), and **polysaccharides** (many sugars). **Cellulose, starch,** and **glycogen** are polysaccharides that consist of the same glucose monomers, bonded different ways. **Chitin** is a polysaccharide of nitrogen-containing sugar monomers.

Section 3.4 **Lipids** in biological systems are partially or entirely nonpolar. A **fatty acid** is a lipid with a carboxyl group head and a long hydrocarbon tail. Fatty acids have a dual chemical character: the carboxyl group is hydrophilic, and the hydrocarbon tail is hydrophobic. Only single bonds link the carbons in the tail of a saturated fatty acid; the tail of an unsaturated fatty acid has one or more double bonds.

Fats are **triglycerides**, which have three fatty acid tails bonded to a glycerol head. Triglycerides are entirely hydrophobic. A **saturated fat** has no double bonds in its fatty acid tails (all three are saturated). By contrast, an **unsaturated fat** has one or more unsaturated fatty acid tails. A monounsaturated fat has one double bond among its three fatty acid tails; a polyunsaturated fat has two or more.

The basic structure of cell membranes is the **lipid bilayer**, which consists mainly of **phospholipids**.

Steroids, with four carbon rings and no fatty acid tails, serve important physiological roles such as starting materials for sex hormone synthesis.

Waxes are water-repellent substances that consist of complex, varying mixtures of lipids.

Section 3.5 **Peptides** and **polypeptides** are (short and long) chains of **amino acids** linked by **peptide bonds.** A **protein** consists of one or more polypeptides. The order of amino acids making up a polypeptide (primary structure) dictates the type of protein and its shape.

A protein's shape is the source of its function. Each type of protein has a unique primary structure, but almost all proteins have similar patterns of secondary structure—helices, sheets, loops, and turns—that form as the polypeptide lengthens and hydrogen bonds form between its amino acids.

Helices, sheets, loops, and turns of a lengthening polypeptide fold into functional domains (tertiary structure). Many proteins, including most enzymes, consist of two or more polypeptides (quaternary structure). Fibrous proteins aggregate into much larger structures.

A protein that can bind to lipids is a lipoprotein; a protein with attached oligosaccharides is a glycoprotein.

Changes in a protein's structure may alter its function. Hydrogen bonds that stabilize protein shape may be disrupted by shifts in pH or temperature, or exposure to detergent or some salts. This causes **denaturation**, which means the protein loses its shape, and so loses its function. **Prion** diseases are a fatal consequence of misfolded proteins.

Section 3.6 **Nucleotides** are small organic molecules that consist of a five-carbon sugar; a nitrogen-containing base; and one, two, or three phosphate groups. Nucleotides are monomers of **DNA** and **RNA**, which are **nucleic acids**. Some nucleotides have additional roles. **ATP**, for example, is an important energy carrier in cells. DNA encodes heritable information; different types of RNAs interact with DNA in protein synthesis.

SELF-QUIZ

Answers in Appendix VII

1. Organic molecules consist mainly of _______ atoms.
   a. carbon c. carbon and hydrogen
   b. carbon and oxygen d. carbon and nitrogen

2. Each carbon atom can bond with as many as _______ other atom(s).
   a. one c. three
   b. two d. four

3. _______ groups are the "acid" part of amino acids and fatty acids.
   a. Hydroxyl (—OH) c. Methyl (—CH_3)
   b. Carboxyl (—COOH) d. Phosphate (—PO_4)

4. _______ is a simple sugar (a monosaccharide).
   a. Ribose c. Starch
   b. Sucrose d. all are monosaccharides

5. Name three carbohydrates that can be built using only glucose monomers.

6. Unlike saturated fats, the fatty acid tails of unsaturated fats incorporate one or more ______ .

7. Is this statement true or false? Unlike saturated fats, all unsaturated fats are beneficial to health because their fatty acid tails kink and do not pack together.

8. Steroids are among the lipids with no ______ .
   a. double bonds c. hydrogens
   b. fatty acid tails d. carbons

9. Which of the following is a class of molecules that encompasses all of the other molecules listed?
   a. triglycerides c. waxes e. lipids
   b. fatty acids d. steroids f. phospholipids

10. ______ are to proteins as ______ are to nucleic acids.
    a. Sugars; lipids c. Amino acids; hydrogen bonds
    b. Sugars; proteins d. Amino acids; nucleotides

11. A denatured protein has lost its ______ .
    a. hydrogen bonds c. function
    b. shape d. all of the above

12. A ______ is an example of protein secondary structure.
    a. barrel c. domain
    b. polypeptide d. helix

13. In the following list, identify the carbohydrate, the fatty acid, the amino acid, and the polypeptide:
    a. methionine-valine-proline-leucine-serine
    b. $C_6H_{12}O_6$
    c. $NH_2—CHR—COOH$
    d. $CH_3(CH_2)_{16}COOH$

14. Match the molecules with the best description.

___ wax	a. protein primary structure
___ starch	b. an energy carrier
___ triglyceride	c. water-repellent secretions
___ DNA	d. richest energy source
___ polypeptide	e. sugar storage in plants
___ ATP	f. sugar storage in animal muscle
___ glycogen	g. carries heritable information

15. Match each polymer with its component monomers.

___ protein	a. phosphate, fatty acids
___ phospholipid	b. amino acids, sugars
___ glycoprotein	c. glycerol, fatty acids
___ fat	d. nucleotides
___ nucleic acid	e. glucose only
___ cellulose	f. sugar, phosphate, base
___ nucleotide	g. amino acids
___ lipoprotein	h. glucose, fructose
___ sucrose	i. lipids, amino acids

CRITICAL THINKING

1. Abundant *trans* bonds make partially hydrogenated vegetable oil a very unhealthy food choice. Vegetable oil can also be hydrogenated until it becomes fully saturated with hydrogen atoms. Would the physical properties of the hydrogenated and partially hydrogenated oils differ? If so, how and why would the differences occur? Do you think that full hydrogenation makes vegetable oil more or less healthy to eat, or does it have no effect?

2. Lipoprotein particles are relatively large, spherical clumps of protein and lipid molecules (see Figure 3.18) that circulate in the blood of mammals. They are like suitcases that move cholesterol, fatty acid remnants, triglycerides, and phospholipids from one place to another in the body. Given what you know about the solubility of lipids in water, which types of lipids would you predict to be on the outside of a lipoprotein clump, bathed in the water-based fluid portion of blood?

3. In 1976, a team of chemists in the United Kingdom was developing new insecticides by modifying sugars with chlorine (Cl_2), phosgene (Cl_2CO), and other toxic gases. One young member of the team misunderstood his verbal instructions to "test" a new molecule. He thought he had been told to "taste" it. Luckily for him, the molecule was not toxic, but it was very sweet. It became the food additive sucralose.

   Sucralose has three chlorine atoms substituted for three hydroxyl groups of sucrose (table sugar):

   sucrose sucralose

   The altered sugar binds so strongly to the sweet-taste receptors on the tongue that the human brain perceives it as 600 times sweeter than sucrose. Sucralose was originally marketed as an artificial sweetener called Splenda®, but it is now available under several other brand names.

   Researchers investigated whether the body recognizes sucralose as a carbohydrate by feeding sucralose labeled with ^{14}C to volunteers. Analysis of the radioactive molecules in the volunteers' urine and feces showed that 92.8 percent of the sucralose passed through the body without being altered.

   Some people are worried that the chlorine atoms impart toxicity to sucralose. How would you respond to that concern?

4 CELL STRUCTURE

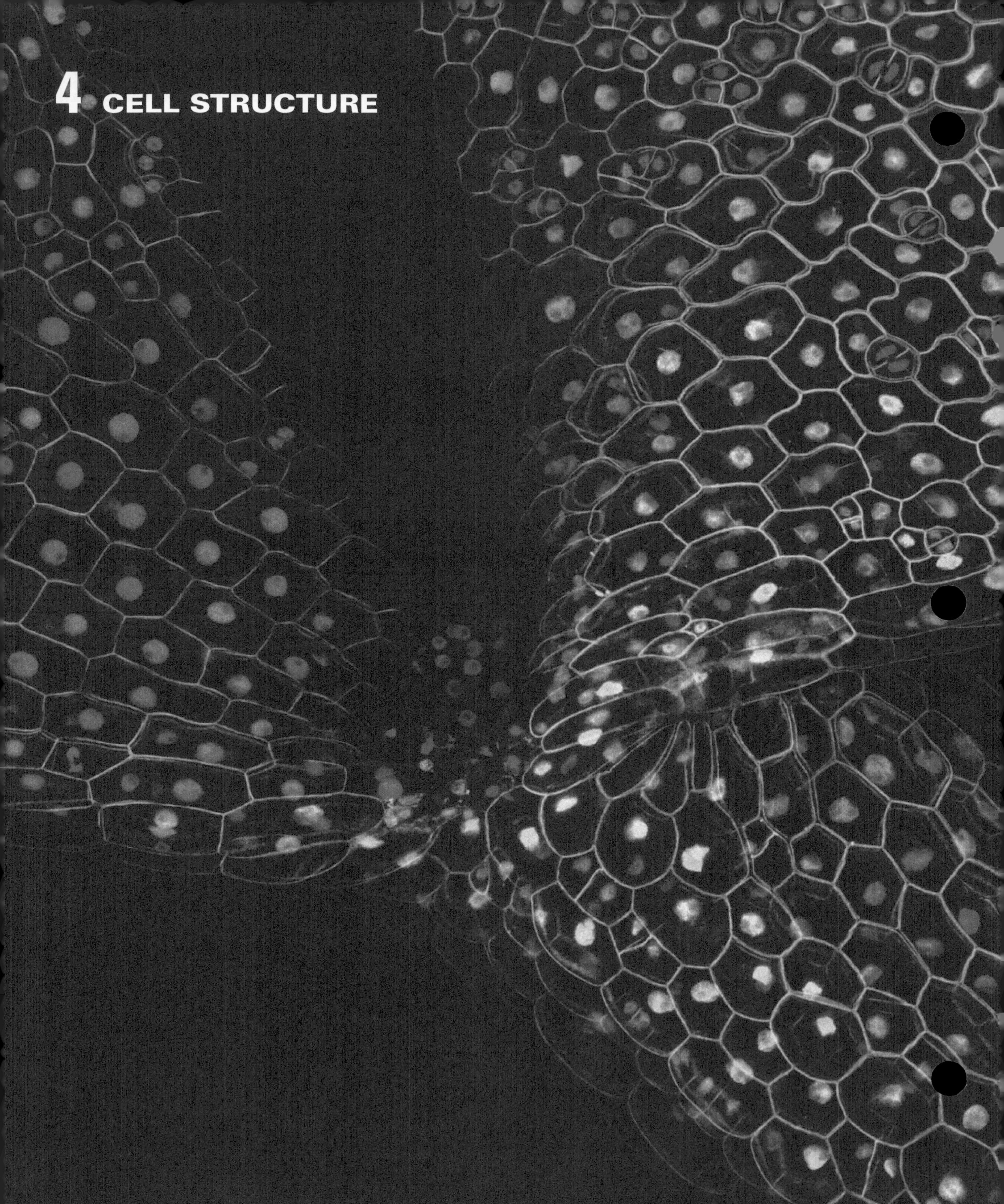

CORE CONCEPTS

 Pathways of Transformation

Organisms exchange matter and energy with the environment in order to grow, maintain themselves, and reproduce.

All cells must acquire energy and nutrients, and all must eliminate metabolic wastes. These requirements have shaped the way metabolic tasks are carried out. Physical constraints involving the exchange of matter across the plasma membrane limit cell size and influence cell shape. All cells have areas or compartments in which functions related to energy and matter occur.

 Systems

Complex properties arise from interactions among components of a biological system.

When components of a biological system interact, they enable properties not found in the individual components alone. Cellular functioning arises from and depends on interactions among cell parts specialized for different tasks. Some cells have structures that allow them to carry out special processes such as photosynthesis. Life, which is impossible to define, has a unique set of properties that arise from cellular components.

 Structure and Function

The three-dimensional form and arrangement of biological structures give rise to their function and interactions.

A plasma membrane separates a cell from the external environment and allows it to maintain internal conditions that support life. Internal membranes that partition cytosol into specialized regions maximize cellular efficiency. Proteins assemble into structures that structurally and functionally link cells to one another in tissues.

Links to Earlier Concepts

Reflect on life's levels of organization in Section 1.2. In this chapter, you will consider the location of DNA and the function of ATP (3.6), and expand your understanding of the roles of proteins (3.5). You will also revisit the nature of science (1.8), the major categories of organisms (1.4), the use of tracers (2.2), and cellular uses of carbohydrates (3.3) and lipids (3.4).

4.1 Food for Thought

There are about as many microorganisms living in and on a human body as there are human cells. Most are bacteria in the digestive tract, where they help with digestion, make vitamins that mammals cannot, prevent the growth of dangerous germs, and shape the immune system.

One of the most common intestinal bacteria is *Escherichia coli*. Most of the hundreds of types, or strains, of *E. coli* are helpful, but some make a toxic protein that can severely damage the lining of the intestine. After ingesting as few as ten cells of a toxic strain, a person may become ill with severe cramps and bloody diarrhea that lasts up to ten days. In some people, complications of infection result in kidney failure, blindness, paralysis, and death. Each year, about 265,000 people in the United States become infected with toxin-producing *E. coli*.

Strains of *E. coli* that are toxic to people live in the intestines of other animals without sickening them. Humans are exposed to the bacteria when they come into contact with feces of animals that harbor them, for example, by eating contaminated ground meat. An animal's feces can contaminate its meat during slaughter. Bacteria in the feces stick to the meat, then get thoroughly mixed into it during the grinding process. Unless the contaminated meat is cooked to at least 71°C (160°F), live bacteria will enter the digestive tract of anyone who eats it.

People also become infected with toxic *E. coli* by eating fresh fruits and vegetables that have contacted animal feces. Washing produce with water does not remove all of the bacteria because they are sticky. In 2011, more than 4,000 people in Europe were sickened after eating sprouts, and 49 of them died. The outbreak was traced to a single shipment of contaminated sprout seeds from Egypt.

The impact of such outbreaks, which occur with unfortunate regularity, extends beyond casualties. The contaminated sprouts cost growers in the European Union at least $600 million in lost sales. In 2011 alone, the United States Department of Agriculture (USDA) recalled 36.7 million pounds of ground meat products contaminated with toxic bacteria, at a cost in the billions of dollars. Such costs are eventually passed to taxpayers and consumers.

Food growers and processors are implementing new procedures intended to reduce the number and scope of these outbreaks. Meat and produce are being tested for some bacteria before sale, and improved documentation should allow a source of contamination to be pinpointed more quickly.

CREDIT: (opposite) Fernan Federici & Jim Haseloff/Wellcome Images.

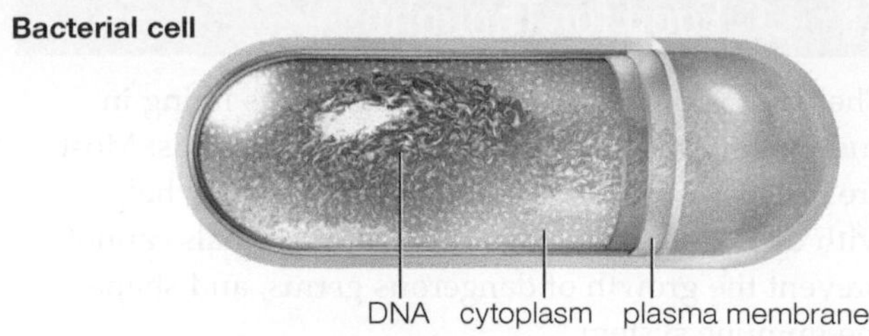
Bacterial cell
DNA
cytoplasm
plasma membrane

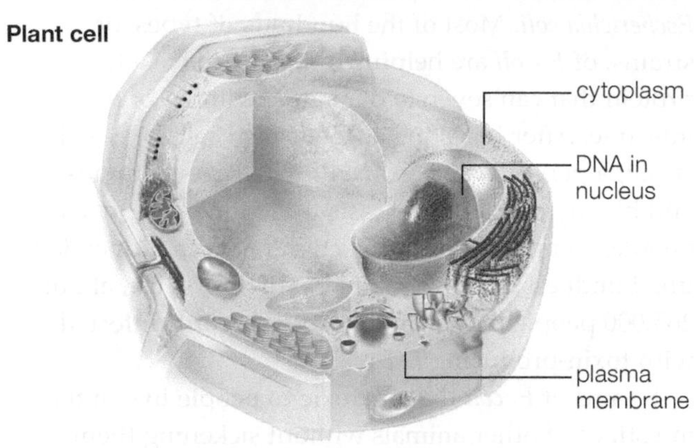
Plant cell
cytoplasm
DNA in nucleus
plasma membrane

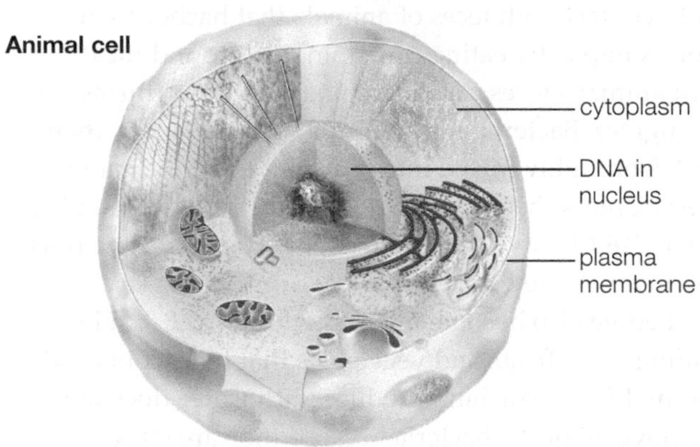
Animal cell
cytoplasm
DNA in nucleus
plasma membrane

FIGURE 4.1 General organization of cells.

All cells start out life with a plasma membrane, cytoplasm, and DNA. Archaea are similar to bacteria in overall structure; both are typically much smaller than eukaryotic cells. If the cells depicted above had been drawn to the same scale, the bacterium would be about this big:

4.2 What Is a Cell?

LEARNING OBJECTIVES

- List the four generalizations that constitute cell theory.
- Describe the three components that all cells have.
- Explain how the surface-to-volume ratio limits cell size.
- Explain how a light microscope works.
- Describe the use of fluorescent dyes in microscopy.
- Explain the difference between a transmission electron micrograph (TEM) and a scanning electron micrograph (SEM).

Components of All Cells

Cells vary in shape and in what they do, but all share certain organizational and functional features: a plasma membrane, DNA, and cytoplasm (**FIGURE 4.1**).

A cell's **plasma membrane** separates its contents from the external environment (**FIGURE 4.2**). The basic structure of a plasma membrane is a lipid bilayer, as it is for all cell membranes. Many different proteins embedded in the lipid bilayer or attached to one of its surfaces carry out particular membrane functions (Chapter 5 returns to this topic). Only certain materials can cross cell membranes. Thus, a plasma membrane controls the exchange of materials between the cell's internal and external environments.

The plasma membrane encloses a jellylike mixture of water, sugars, ions, and proteins called **cytosol**. A major part of a cell's metabolism occurs in cytosol, and the cell's other internal components, including organelles, are suspended in it. **Organelles** are structures that carry out special functions inside a cell. Those with membranes compartmentalize substances and activities.

Every cell starts out life with DNA. A eukaryotic cell's DNA is contained in an organelle called the **nucleus** (plural, nuclei). In eukaryotic cells, the cytosol, organelles, and all other cellular components between the nucleus and plasma membrane are collectively

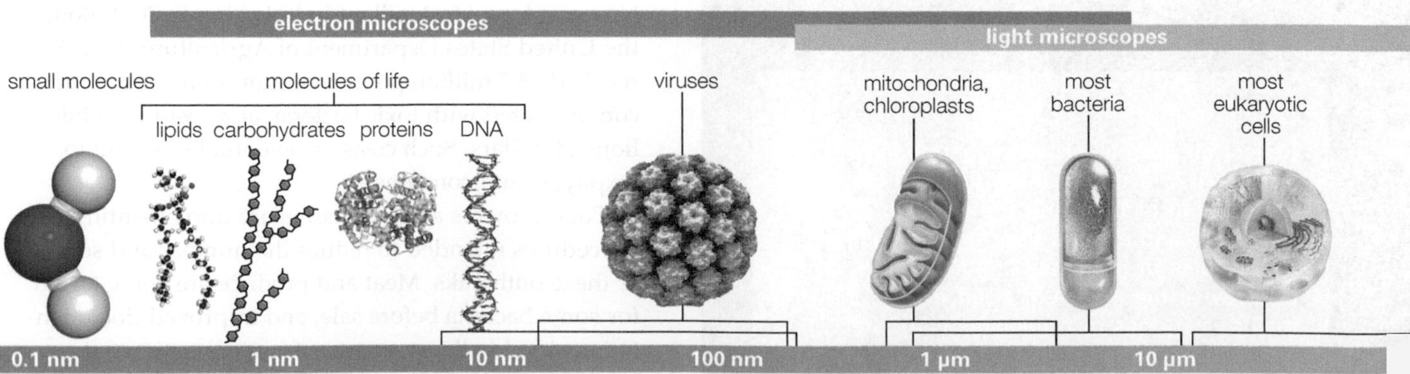
electron microscopes
light microscopes
small molecules
molecules of life
lipids
carbohydrates
proteins
DNA
viruses
mitochondria, chloroplasts
most bacteria
most eukaryotic cells
0.1 nm
1 nm
10 nm
100 nm
1 µm
10 µm

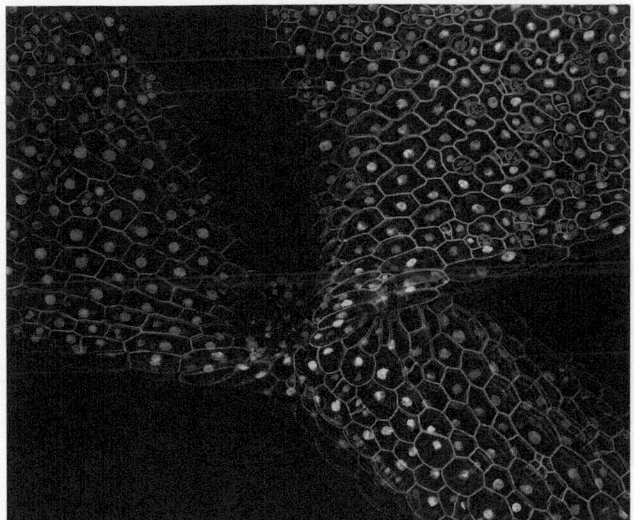

FIGURE 4.2 A plasma membrane separates a cell from its environment. Each cell making up this plant seedling is surrounded by a plasma membrane (blue-green). Each also contains a nucleus (orange spots), which is the defining characteristic of eukaryotes.

called **cytoplasm**. In nearly all prokaryotes (bacteria and archaea), DNA is suspended directly in cytosol, so the definition of cytoplasm differs slightly between these cells. In prokaryotes, cytoplasm includes everything inside the plasma membrane, including the DNA.

Visualizing Cells

Before microscopes were invented, no one knew that cells existed because nearly all are invisible to the naked eye. By 1665, Antoni van Leeuwenhoek had constructed an early microscope that revealed tiny organisms in rainwater, insects, fabric, semen, feces, and other samples. In scrapings of tartar from his teeth, Leeuwenhoek saw "many very small animalcules, the motions of which were very pleasing to behold." He (incorrectly) assumed that movement defined life, and (correctly) concluded that the moving "beasties" he saw were alive. Robert Hooke, a contemporary of Leeuwenhoek, observed cork under a microscope and discovered it to consist of "a great many little Boxes." He named the tiny compartments cells, after the small chambers that monks lived in.

TABLE 4.1

Equivalent Units of Length

Unit	Equivalent Meter(s)	Inch(es)
kilometer	1,000	39,370
meter (m)	1	39.37
centimeter (cm)	1/100	0.4
millimeter (mm)	1/1,000	0.04
micrometer (μm)	1/1,000,000	0.00004
nanometer (nm)	1/1,000,000,000	0.00000004

Modern microscopes allow us to observe objects in the micrometer range of size. One micrometer (μm) is one-thousandth of a millimeter, which is one-thousandth of a meter (**TABLE 4.1**). Most cells are 10–20 micrometers in diameter, about 50 times smaller than the unaided human eye can perceive (**FIGURE 4.3**).

Visible light illuminates a sample in a light microscope. As you will learn in Chapter 6, all light travels in waves. This property of light causes it to bend when passing through a curved glass lens. Inside

cytoplasm (SITE-uh-plaz-um) In a prokaryote, collective term for everything inside the cell's plasma membrane. In a eukaryote, everything between the plasma membrane and the nucleus.

cytosol (SITE-uh-sall) Jellylike mixture of water and solutes enclosed by a cell's plasma membrane.

nucleus (NEW-klee-us) Plural, nuclei. Of a eukaryotic cell, organelle with a double membrane that holds the cell's DNA.

organelle (or-guh-NEL) Structure that carries out a specialized metabolic function inside a cell.

plasma membrane (PLAZ-muh) Membrane that encloses a cell and separates it from the external environment.

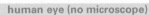

FIGURE 4.3 Relative sizes. Below, the diameter of most cells is in the range of 1 to 100 micrometers. **TABLE 4.1** shows conversions among units of length; also see Units of Measure, Appendix VI.

FIGURE IT OUT Which one is smallest: a protein, a lipid, or a water molecule?

Answer: A water molecule

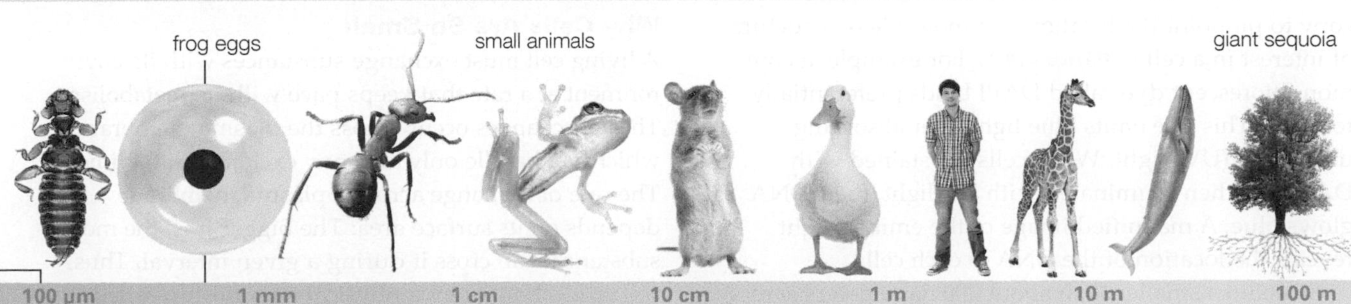
human eye (no microscope)
frog eggs
small animals
giant sequoia
100 μm
1 mm
1 cm
10 cm
1 m
10 m
100 m

CREDITS: (2) Courtesy of © Johannes Kästner, Universität Stuttgart; (3) from left 1–5, © Cengage Learning; Virus, CDC; Mitochondria, Chloroplast, Bacteria, and Eukaryotic cells, © Cengage Learning; Louse, Edward S. Ross; Frog egg, © Cengage Learning; Ant, VladimirDavydov/Getty Images; Frog, © A Cotton Photo/Shutterstock; Rat, © Pakhnyushcha/Shutterstock; Goose, © photomaster/Shutterstock; Boy, PhotoMediaGroup/Shutterstock; Giraffe, © Valerie Kalyuznnyy/Photos.com; Whale, Dorling Kindersley/Getty Images; Tree, © Cengage Learning; (Table 4.1) © Cengage Learning.

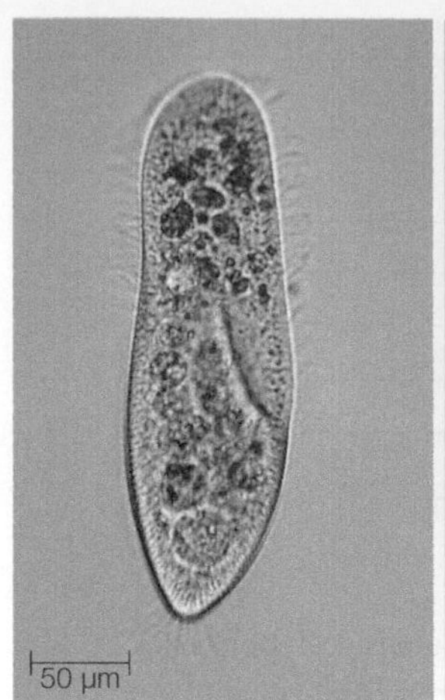
50 µm

A The green blobs visible in this light micrograph (LM) of a living cell are ingested algae. Hairlike structures on the cell's surface are waving cilia that propel this motile organism through fluids.

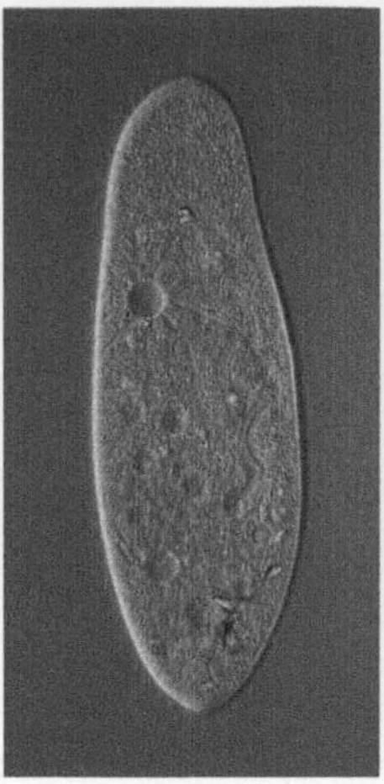

B A light micrograph taken with polarized light shows edges in relief. This technique reveals ingested algae and some internal structures not visible in A.

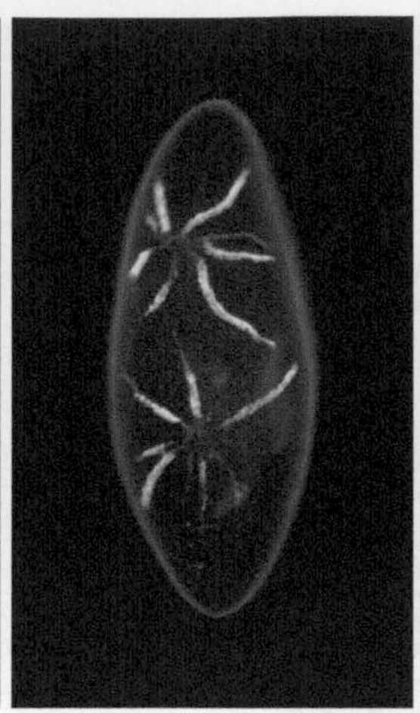

C In this fluorescence micrograph, yellow pinpoints the location of a particular protein in the membrane of organelles called contractile vacuoles. These organelles are also visible in B.

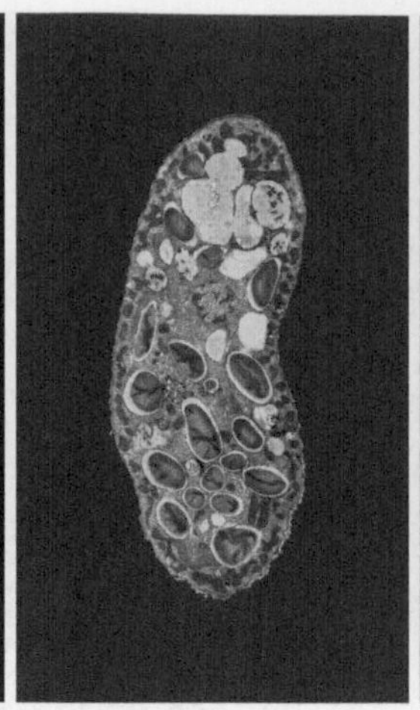

D A colorized transmission electron micrograph (TEM) reveals several types of internal structures in a plane (slice). Ingested algae are being broken down inside food vacuoles.

E A scanning electron micrograph (SEM) shows details of the cell's surface, including its thick coat of cilia. The cell ingests its food via the indentation (also visible in A).

FIGURE 4.4 **Different microscopy techniques reveal different characteristics.** All of these micrographs show the same organism, a protist called *Paramecium*, that is about 250 µm long.

FIGURE IT OUT What is the width of a *Paramecium*?

Answer: Approximately 50 µm

a light microscope, such lenses focus light that passes through a specimen, or bounces off of one, into a magnified image (FIGURE 4.4A). Microscopes that use polarized light can yield images in which the edges of some structures appear in three-dimensional relief (FIGURE 4.4B). Photographs of images enlarged with a microscope are called micrographs; those taken with visible light are called light micrographs (LM).

Most cells are nearly transparent, so their internal details may not be visible unless they are first stained. Staining a cell or other sample means exposing it to dyes or other substances that only some of its parts soak up. Staining results in an increase in contrast (the difference between light and dark) that allows us to see a greater range of detail.

Fluorescent dyes consist of molecules that absorb light of a particular color, then emit light of a different color. These dyes are often used as tracers in microscopy to pinpoint the location of a molecule or structure of interest in a cell (FIGURE 4.4C). For example, a common fluorescent dye called DAPI binds preferentially to DNA. This dye emits blue light after absorbing ultraviolet (UV) light. When cells are stained with DAPI and then illuminated with UV light, their DNA glows blue. A magnified image of the emitted light reveals the location of the DNA in each cell.

Structures smaller than about 200 nanometers across appear blurry under light microscopes. To observe objects of this size range clearly, we would have to switch to an electron microscope. There are two types of electron microscope; both use magnetic fields as lenses to focus a beam of electrons onto a sample. A transmission electron microscope directs electrons through a thin specimen. The specimen's internal details appear as shadows in the resulting image, which is called a transmission electron micrograph, or TEM (FIGURE 4.4D). A scanning electron microscope directs a beam of electrons back and forth across the surface of a specimen that has been coated with a thin layer of gold or other metal. The irradiated metal emits electrons and X-rays, which are converted into an image (a scanning electron micrograph, or SEM) of the surface (FIGURE 4.4E). SEMs and TEMs are always black and white; colored versions have been digitally altered to highlight specific details.

Why Cells Are So Small

A living cell must exchange substances with its environment at a rate that keeps pace with its metabolism. These exchanges occur across the plasma membrane, which can handle only so many exchanges at a time. The rate of exchange across a plasma membrane depends on its surface area: The bigger it is, the more substances can cross it during a given interval. Thus, cell size is limited by a physical relationship called the **surface-to-volume ratio**. By this ratio, an object's

CREDITS: (4A) © iStockphoto.com/Nancy Nehring; (4B) Michael Abbey/Science Source; (4C) © Dennis Kunkel Microscopy, Inc./PhototakeUSA.com; (4D) © Microworks/PhototakeUSA.com; (4E) Steve Gschmeissner/Science Source.

Diameter (cm)	2	3	6
Surface area (cm^2)	12.6	28.2	113
Volume (cm^3)	4.2	14.1	113
Surface-to-volume ratio	3:1	2:1	1:1

FIGURE 4.5 Surface-to-volume ratio.
The physical relationship between surface area (πd^2) and volume ($\pi d^3/6$) constrains cell size and shape. Three examples are listed.

volume increases with the cube of its diameter, but its surface area increases only with the square.

Let's apply the surface-to-volume ratio to a cell. As **FIGURE 4.5** shows, when a round cell expands in diameter, its volume increases faster than its surface area does. Imagine that the cell expands until it is four times its original diameter. The cell's volume has increased 64 times (4^3), but its surface area has increased only 16 times (4^2). Each unit of plasma membrane must now handle exchanges with four times as much cytoplasm ($64 \div 16 = 4$). If the cell gets too big, the inward flow of nutrients and the outward flow of wastes across that membrane will not be fast enough to keep the cell alive.

The surface-to-volume ratio also constrains cell form in colonial and multicelled organisms. For example, cells of some colonial algae attach end to end in long strands. This arrangement allows each cell to interact directly with the environment. As another example, consider that some muscle cells run the length of your thigh. Although surprisingly long, each of these cells is very thin, so it exchanges substances efficiently with fluids in the surrounding tissue.

Properties of All Cells

In the early 1800s, Matthias Schleiden, a botanist, hypothesized that a plant cell is an independent living unit even when it is part of a plant. Schleiden compared notes with zoologist Theodor Schwann, and both concluded that the tissues of animals as well as plants are composed of cells and their products. Together, the two scientists recognized that cells have a life of their own even as part of a multicelled body. Schleiden and Schwann's observations became part of a larger picture of cellular life.

cell theory Set of principles that constitute the foundation of modern biology: every living organism consists of one or more cells; the cell is the basic structural and functional unit of life; all cells come from division of preexisting cells; and all cells pass hereditary material (DNA) to offspring.
surface-to-volume ratio Relationship in which the volume of an object increases with the cube of the diameter, and the surface area increases with the square.

TABLE 4.2

Cell Theory
1. Every living organism consists of one or more cells.
2. The cell is the basic structural and functional unit of life. Cells are individually alive even as part of a multicelled organism.
3. All living cells arise by division of preexisting cells.
4. Cells contain hereditary material (DNA), which they pass to their offspring during processes of reproduction.

Today, we know that every living cell carries out metabolism and homeostasis, and reproduces either on its own or as part of a larger organism. By this definition, each cell is alive even if it is part of a multicelled body, and all living organisms consist of one or more cells. Cells reproduce by dividing, so it follows that all existing cells must have arisen by division of other cells. As a cell divides, it passes its hereditary material—DNA—to offspring. Taken together, these principles constitute what is known as **cell theory,** which is a foundation of modern biology (**TABLE 4.2**).

TAKE-HOME MESSAGE 4.2

✔ All cells start life with a plasma membrane, cytoplasm, and DNA. In eukaryotic cells only, the DNA is contained in a nucleus.

✔ Most cells are visible only with the help of microscopes. Different microscopes and techniques reveal different aspects of cell structure.

✔ The surface-to-volume ratio limits cell size and influences cell shape.

✔ Early observations of cells led to the cell theory: All organisms consist of one or more cells; the cell is the smallest unit of life; each new cell arises from another cell; and a cell passes hereditary material to its offspring.

4.3 Introducing the Prokaryotes

LEARNING OBJECTIVES

- Identify the two characteristics that originally defined prokaryotes.
- Describe some of the structures shared by bacteria and archaea.
- Explain the difference between bacterial and archaeal cell walls.
- Distinguish between a prokaryotic flagellum and a pilus.
- Describe a biofilm and explain how it benefits its inhabitants.

All bacteria and archaea are single-celled, but individual cells of many species cluster in filaments or colonies. Outwardly, cells of the two groups appear so similar that archaea were once thought to be an unusual group of bacteria. Neither have a nucleus,

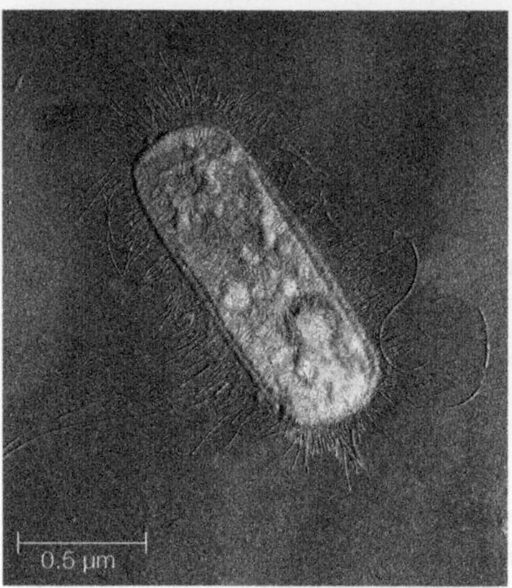
0.5 μm

A Bacteria *Escherichia coli* is a common bacterial inhabitant of human intestines. Short, hairlike structures are pili; longer ones are flagella.

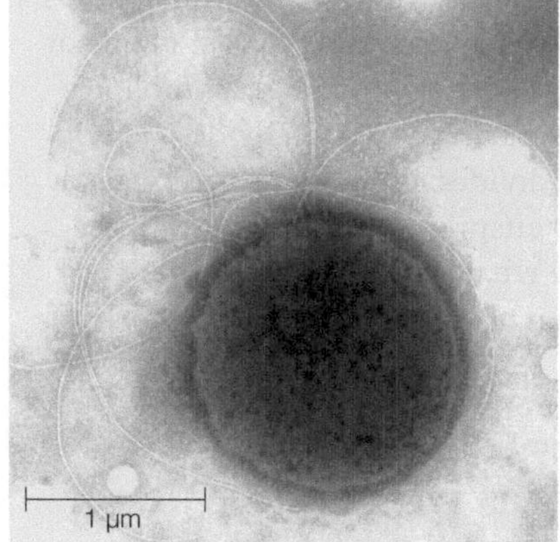
1 μm

B Archaea *Thermococcus gammatolerans* is a species of archaea discovered at a deep-sea hydrothermal vent, under extreme conditions of salt, temperature, and pressure. *T. gammatolerans* is the most radiation-resistant organism ever discovered, capable of withstanding thousands of times more radiation than humans can.

FIGURE 4.6 Representative prokaryotes.

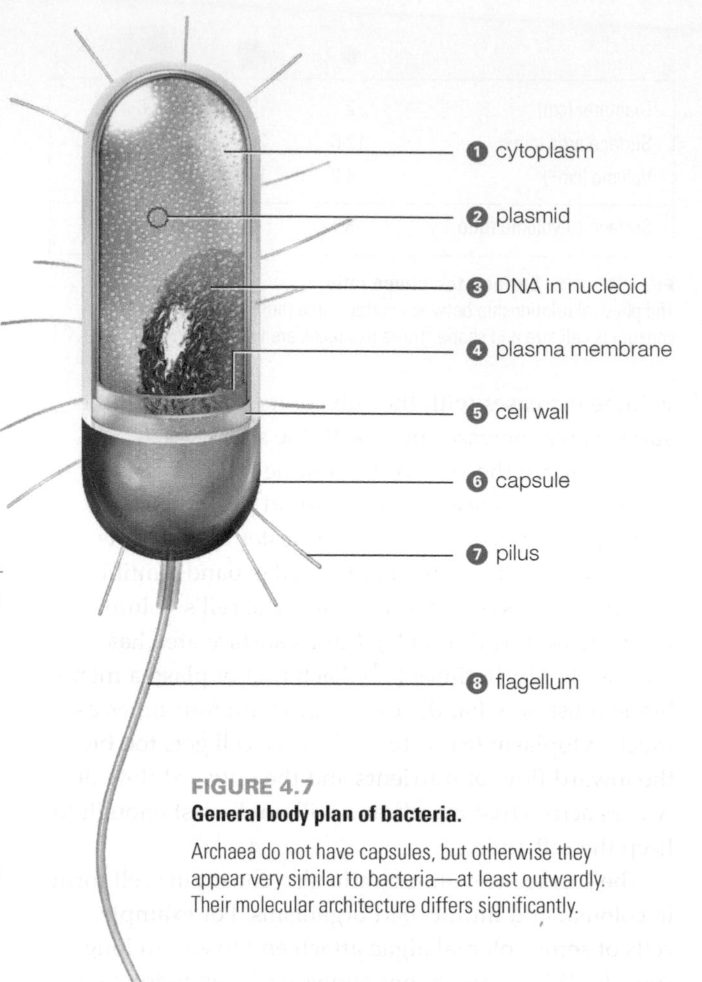
1 cytoplasm
2 plasmid
3 DNA in nucleoid
4 plasma membrane
5 cell wall
6 capsule
7 pilus
8 flagellum

FIGURE 4.7
General body plan of bacteria.

Archaea do not have capsules, but otherwise they appear very similar to bacteria—at least outwardly. Their molecular architecture differs significantly.

so they were named prokaryotes: a word that means "before the nucleus." By 1977, it had become clear that archaea are more closely related to eukaryotes than to bacteria, so they were given a separate domain. The term "prokaryote" is now an informal designation only. Chapter 20 revisits them in more detail; here we present an overview of structures common to both groups.

Structural Features

Bacteria and archaea are the smallest and most metabolically diverse forms of life that we know about (FIGURE 4.6). These cells have a less elaborate internal framework than eukaryotic cells (FIGURE 4.7). Protein filaments under the plasma membrane reinforce the cell's shape and provide a scaffolding for internal structures. The cytoplasm ❶ contains many **ribosomes**—organelles upon which polypeptides are assembled—and in some species, additional organelles. Cytoplasm of many species also contains **plasmids** ❷, which are small circles of DNA that carry a few genes (units of inheritance). Plasmid genes can provide advantages such as resistance to antibiotics; the cell's essential genetic information occurs on one or two circular DNA molecules located in an irregularly shaped region of cytosol called the **nucleoid** ❸. Membrane-enclosed organelles are generally uncommon among prokaryotes, but a few bacterial groups have them.

Bacteria and archaea, like all cells, have a plasma membrane ❹. Almost all have a rigid layer of secreted material called a **cell wall** ❺ that encloses the plasma membrane. A wall protects the cell and supports its shape. Cell walls are not membranes and they do not

biofilm Community of microorganisms living within a shared mass of secreted slime.
cell wall Rigid but permeable structure that surrounds the plasma membrane of some cells.
flagellum (fluh-JEL-um) Plural, flagella. Long, slender cellular structure used for motility.
nucleoid (NEW-klee-oyd) Of a bacterium or archaeon, region of cytoplasm where the DNA is concentrated.
pilus (PIE-luss) Plural, pili. Protein filament that projects from the surface of some prokaryotic cells.
plasmid (PLAZ-mid) Small circle of DNA in some bacteria and archaea.
ribosome (RYE-buh-sohm) Organelle of protein synthesis.

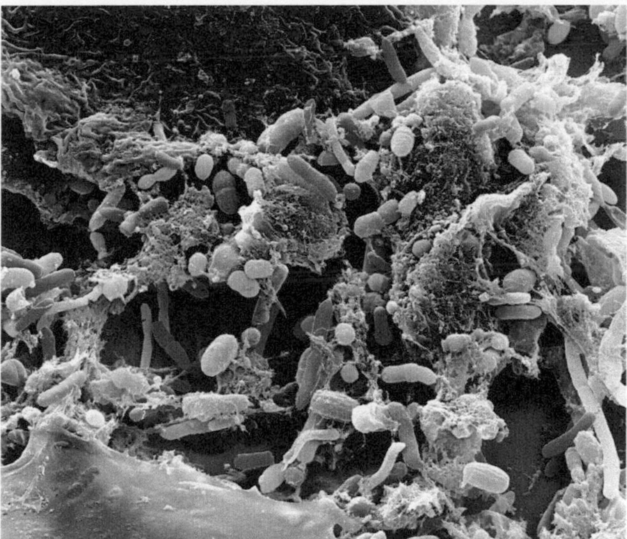

FIGURE 4.8 A beneficial biofilm.
This SEM of human fecal matter reveals a few of the more than 30,000 species of bacteria that normally inhabit our large intestine. These organisms form a biofilm that protects intestinal surfaces from colonization by harmful species, while producing needed vitamins and helping us digest our food.

consist of lipids; rather, they are crystalline arrays of other molecules. Unlike lipid bilayers, they are permeable to ions and other solutes. Bacterial cell walls consist of polysaccharides, proteins, and peptidoglycan (a polymer of peptides and sugars unique to bacteria). Archaeal cell walls contain no peptidoglycan. In most species of archaea, the cell wall consists of a single layer of tightly arrayed proteins or glycoproteins.

A second membrane surrounds the cell wall of many bacteria (and a few archaea). Like the plasma membrane, this outer membrane consists of a lipid bilayer with embedded proteins. An outer membrane prevents the peptidoglycan wall from taking up a common stain called Gram, so bacteria that have it are said to be Gram-negative. Gram-positive bacteria have no second membrane, so their cell wall takes up the Gram stain.

In many bacteria, a thick capsule ❻ that consists of proteins and/or polysaccharides encloses the cell wall (or the second membrane in species that have one). This sticky capsule helps the cells adhere to many types of surfaces, and also offers protection against some predators and toxins. Some archaea have a polysaccharide coating that is analogous to the bacterial capsule.

Protein filaments called **pili** (singular, pilus) ❼ project from the surface of some prokaryotes. Pili help these cells move across or cling to surfaces. One kind, a "sex" pilus, attaches to another bacterium and then shortens, reeling in the attached cell. When the cells make contact, DNA is transferred from one to the other. Many prokaryotes also have one or more flagella projecting from their surface. **Flagella** (singular, flagellum) are long, slender cellular structures used for motion ❽. A prokaryotic flagellum rotates like a propeller that drives the cell through fluid habitats.

Biofilms

Bacteria often live so close together that an entire community shares a loose capsule called a slime layer. A communal living arrangement in which single-celled organisms occupy a shared mass of slime is called a **biofilm**. A biofilm is often attached to a solid surface, and may include bacteria, algae, fungi, protists, and/or archaea. Participating in a biofilm allows the cells to linger in a favorable spot rather than be swept away by fluid currents, and to reap the benefits of living communally. For example, rigid or netlike secretions of some species serve as permanent scaffolding for others; species that break down toxic chemicals allow more sensitive ones to thrive in habitats that they could not withstand on their own; and waste products of some serve as raw materials for others. The human gastrointestinal tract hosts a large and beneficial biofilm (**FIGURE 4.8**). Other biofilms, including the dental plaque that forms on our teeth, can be harmful to human health.

TAKE-HOME MESSAGE 4.3

✔ Prokaryotic cells (bacteria and archaea) do not have a nucleus. Most or all of their DNA occurs in a nucleoid. Plasmids carry additional genetic information in cells that have them.

✔ Almost all prokaryotes have a cell wall that surrounds, reinforces, and protects the plasma membrane. Some have a second membrane around the cell wall, and many have a sticky outer capsule.

✔ A community of single-celled organisms living within a shared mass of slime is called a biofilm.

4.4 Introducing the Eukaryotic Cell

LEARNING OBJECTIVES

- Identify some cellular components unique to eukaryotes.
- Explain the function of membranes that enclose organelles.
- List the functions of the cell nucleus and explain how these functions arise from its structure.

Every cell on Earth is either prokaryotic or eukaryotic. The presence or absence of a nucleus is the main characteristic that distinguishes the two cell types, but there are many other structural differences. For example, all prokaryotes are single-celled, but not all eukaryotes are (Section 1.4). A eukaryotic cell is larger

CREDIT: (8) Eye of Science/Science Source.

than a prokaryote, and a typical one has many more organelles (TABLE 4.3 and FIGURE 4.9).

Organelles enclosed by lipid bilayer membranes, including the nucleus, endoplasmic reticulum, Golgi bodies, chloroplasts, and mitochondria, are characteristic of eukaryotic cells. An enclosing membrane allows an organelle to regulate the types and amounts of substances that enter and exit. Through this control, the organelle maintains a special internal environment that allows it to carry out a particular function—for example, isolating toxic or sensitive substances from the rest of the cell, storing large amounts of water, transporting substances through cytoplasm, maintaining fluid balance, or providing a favorable environment for a special process. Compartmentalizing such processes maximizes the cell's metabolic efficiency.

The rest of this section details the nucleus. Later sections return to other organelles.

The Nucleus

A nucleus serves two important functions. First, it keeps the cell's genetic material—its one and only copy of DNA—safe from metabolic processes that might damage it. Isolated in its own compartment, DNA stays separated from the bustling activity of the cytoplasm. Second, a nucleus controls the passage of certain molecules across its membrane. FIGURE 4.10 shows the components of the nucleus. TABLE 4.4 lists their functions. Let's zoom in on the individual components.

A nucleus, as you know, contains the cell's DNA. Proteins associate with and organize each DNA molecule so it can pack tightly—and precisely—into the nucleus (Chapter 8 returns to this topic). All of the DNA in a cell's nucleus, together with associated

TABLE 4.3

Some Components of Eukaryotic Cells

Organelles with membranes	
Nucleus	Protects and controls access to DNA
Endoplasmic reticulum (ER)	Makes and modifies new polypeptides and lipids, among other tasks
Golgi body	Modifies and sorts polypeptides and lipids
Vesicle	Transports, stores, or breaks down substances
Mitochondrion	Makes ATP by glucose breakdown
Chloroplast	Makes sugars (in plants and some protists)
Lysosome	Intracellular digestion
Peroxisome	Breaks down fatty acids, amino acids, toxins
Vacuole	Stores, breaks down substances
Organelles without membranes	
Ribosome	Assembles polypeptides
Centriole	Anchors cytoskeleton
Other components	
Cytoskeleton	Contributes to cell shape, internal organization, and movement

FIGURE 4.9 **Some components of typical eukaryotic cells.**

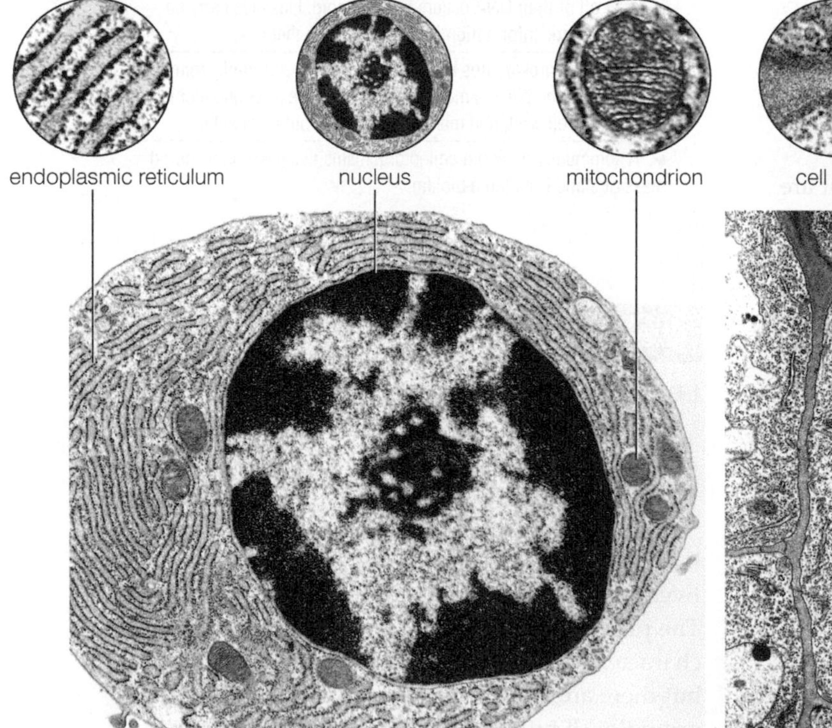
endoplasmic reticulum
nucleus
mitochondrion

An animal cell. This is a white blood cell from a guinea pig.

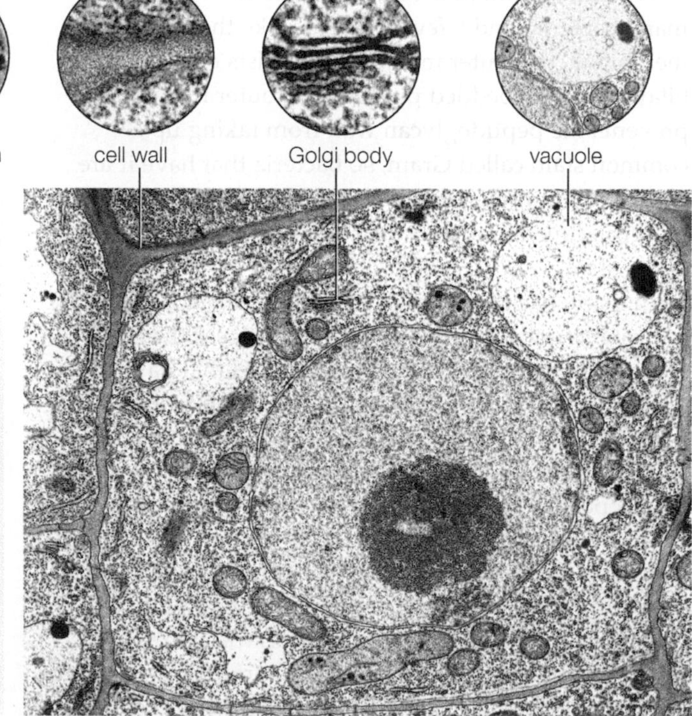
cell wall
Golgi body
vacuole

A plant cell. This is a cell from the root of a plant called thale cress.

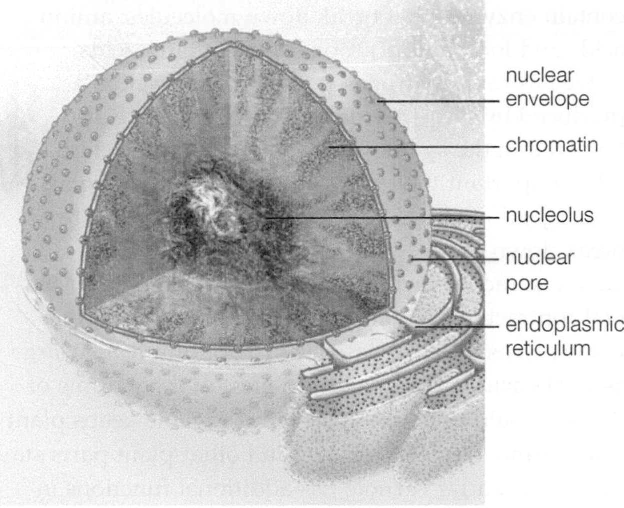
nuclear envelope
chromatin
nucleolus
nuclear pore
endoplasmic reticulum

FIGURE 4.10 The cell nucleus.

TABLE 4.4

Components of the Nucleus

Chromatin	DNA and associated proteins in a cell nucleus
Nucleoplasm	Semifluid interior portion of the nucleus
Nuclear envelope	Double membrane with nuclear pores that control which substances enter and exit the nucleus
Nucleolus	Dense region of proteins and nucleic acid where ribosomal subunits are being produced

proteins, is collectively called **chromatin**. Chromatin is suspended in **nucleoplasm**, a viscous fluid similar to cytosol, that fills the nucleus.

Depending on a cell's metabolic state, its nucleus contains one or more nucleoli. A **nucleolus** (plural, nucleoli) is a dense, irregularly shaped region rich in proteins and nucleic acids. Subunits of ribosomes are being produced in these regions. The subunits pass through nuclear pores into the cytoplasm, where they join and become active in protein synthesis (Section 9.4 returns to ribosome structure and function). New research is revealing additional roles for nucleoli, for example in cell division, cell death, and responses to cellular stress.

A special membrane called the **nuclear envelope** encloses the nucleus. This membrane consists of two lipid bilayers folded together around an intermembrane space (FIGURE 4.11A). The outer lipid bilayer is continuous with the lipid bilayer of the endoplasmic reticulum, and like rough endoplasmic reticulum it is covered with ribosomes (more about endoplasmic reticulum in the next section). The inner lipid bilayer is covered and supported by the nuclear lamina, a dense mesh of fibrous proteins.

Thousands of structures called nuclear pores span the two bilayers of the nuclear membrane (FIGURE 4.11B). Each is an assembly of hundreds of proteins, and changes in the shape of these proteins open and close a tiny hole in the nuclear envelope (FIGURE 4.11C). Nuclear pores are important because large molecules, including RNA and proteins, cannot cross lipid bilayers on their own (Chapter 5 returns to this topic). Nuclear pores function as gateways for

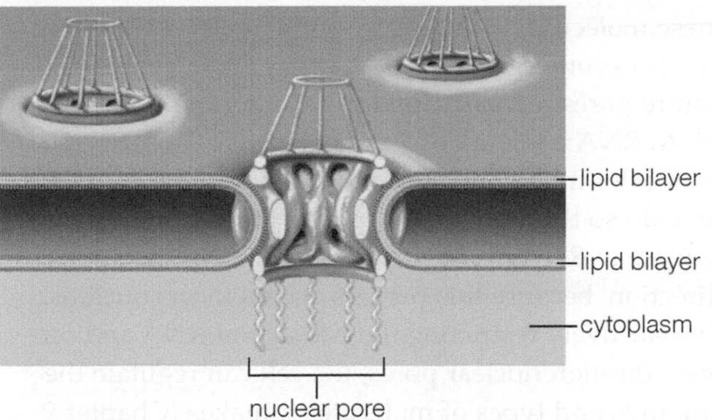
lipid bilayer
lipid bilayer
cytoplasm
nuclear pore

A The nuclear envelope consists of two lipid bilayers that connect at nuclear pores, each a complex structure formed by hundreds of proteins. Part of a pore is a basket-shaped, multifunctional scaffold that can, for example, bind to chromatin. Pores form closable holes through the envelope; when they are open, cytoplasmic fibrils guide molecules through them.

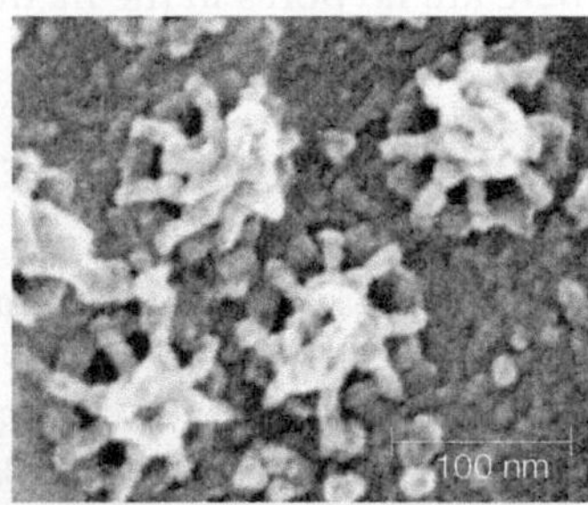
100 nm

B The proteins that compose a nuclear pore allow only certain molecules to cross the nuclear membrane. The "baskets" on the inside of the nuclear envelope are visible in this SEM. All of these pores are closed.

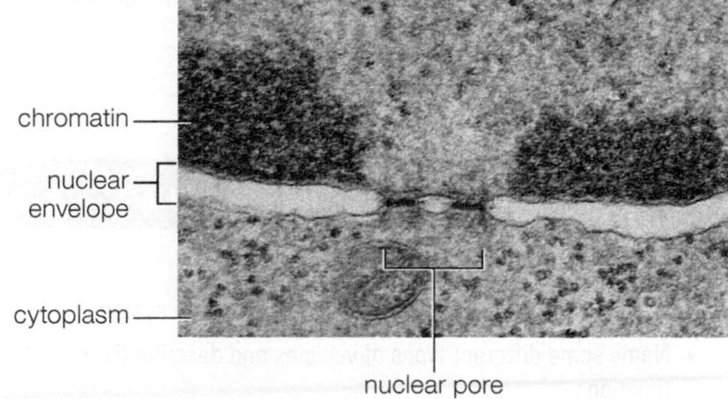
chromatin
nuclear envelope
cytoplasm
nuclear pore

C This TEM shows a pore in the nuclear envelope of a bone marrow cell nucleus. The pore structure is around 100 nm wide; the hole in the middle, about 40 nm across.

FIGURE 4.11 Structure of the nuclear envelope.

chromatin Collective term for all of the DNA and associated proteins in a cell nucleus.
nuclear envelope Double membrane that constitutes the outer boundary of the nucleus. Pores in the membrane control which molecules can cross it.
nucleolus (new-KLEE-oh-luss) In a cell nucleus, a dense, irregularly shaped region rich in proteins and nucleic acids; site of ribosome subunit assembly.
nucleoplasm (NEW-klee-oh-plaz-um) Viscous fluid inside the nucleus.

these molecules to enter and exit a nucleus. Consider protein synthesis, a process that occurs in cytoplasm and requires the participation of many molecules of RNA. RNA is produced in the nucleus. Thus, RNA molecules must move from nucleus to cytoplasm, and they do so through nuclear pores. Proteins that participate in RNA production must move in the opposite direction, because this process occurs in the nucleus. By selectively restricting the passage of RNA and proteins through nuclear pores, the cell can regulate the amounts and types of molecules it makes (Chapter 9 returns to protein synthesis; and Chapter 10, to cellular controls over RNA and protein production).

Note that some bacteria have a membrane enclosing their DNA, but this structure is not a nucleus because there are no pores in the membrane.

TAKE-HOME MESSAGE 4.4

✔ All eukaryotic cells start life with a nucleus and other membrane-enclosed organelles. Organelles enclosed by membranes maximize cellular efficiency by compartmentalizing metabolic processes.

✔ A nucleus protects and controls access to a eukaryotic cell's DNA.

✔ The nuclear envelope is a double lipid bilayer. Proteins embedded in the bilayer form pores that control the passage of molecules between the nucleus and cytoplasm.

4.5 The Endomembrane System

LEARNING OBJECTIVES

- List the main components of the endomembrane system.
- Name some different types of vesicles and describe their functions.
- Explain the differences between rough and smooth endoplasmic reticulum.
- Describe the function of Golgi bodies.

The **endomembrane system** is a multifunctional network of membrane-enclosed organelles that occur throughout a cell's cytoplasm (**FIGURE 4.12**). Here we introduce the functions of its main components: vesicles, endoplasmic reticulum, and Golgi bodies.

Vesicles are sacs that form by budding from other organelles or when a patch of plasma membrane sinks into the cytoplasm. Many types carry substances from one organelle to another, or to and from the plasma membrane. Some are a bit like trash cans that collect or dispose of waste, debris, or toxins. Enzymes in vesicles called **lysosomes** break down particles such as cellular debris and bits of waste. Bacteria, cell parts, and other particles taken in by a cell are delivered to lysosomes for digestion ❶. Small vesicles called **peroxisomes** contain enzymes that break down molecules: amino acids and long hydrocarbon chains of fatty acids, as well as toxic hydrogen peroxide and ammonia produced by these reactions. Peroxisomes are also involved in the synthesis of cholesterol, bile acids, and other important molecules in various animal organs.

Large, fluid-filled vesicles called **vacuoles** store or break down waste, debris, toxins, or food. Some cells have contractile vacuoles that expel excess water (several contractile vacuoles are visible in Figure 4.4). In plants, lysosome-like vesicles fuse to form a very large **central vacuole** that makes up most of the volume of the cell. Fluid pressure in a central vacuole keeps plant cells plump, so stems, leaves, and other plant parts stay firm. The central vacuole has additional functions in some cells.

The nuclear envelope is often considered to be part of the endomembrane system. This is because the system of sacs and tubes that make up **endoplasmic reticulum (ER)** extends from the outer lipid bilayer of the nuclear envelope. The space between the nuclear envelope's two lipid bilayers is continuous with the space enclosed by the ER membrane.

Two kinds of ER, rough and smooth, are named for their appearance in micrographs. The membrane of rough ER is typically folded into flattened sacs, and has thousands of attached ribosomes that give it a "rough" appearance. These ribosomes make polypeptides that thread into the ER's interior as they are assembled ❷. The polypeptides take on their tertiary structure in the interior of rough ER, and many assemble with other polypeptides (Section 3.5). Cells that make, store, and secrete proteins have a lot of rough ER. For example, ER-rich cells in the pancreas make digestive enzymes that they secrete into the small intestine.

Some proteins made in rough ER become part of its membrane. Others migrate through it to smooth ER. Smooth ER has no ribosomes, so it does not make its

central vacuole Very large fluid-filled vesicle of plant cells.
endomembrane system Multifunctional network of membrane-enclosed organelles (endoplasmic reticulum, Golgi bodies, and vesicles).
endoplasmic reticulum (ER) (en-doh-PLAZ-mick ruh-TICK-you-lum) System of sacs and tubes that is a continuous extension of the nuclear envelope. Smooth ER makes phospholipids, stores calcium, and has additional functions in some cells; ribosomes on the surface of rough ER make proteins.
Golgi body (GOAL-jee) Organelle that modifies proteins and lipids, then packages the finished products into vesicles.
lysosome (LICE-uh-sohm) Enzyme-filled vesicle that breaks down cellular wastes and debris.
peroxisome (purr-OX-uh-sohm) Enzyme-filled vesicle that breaks down molecules and has other specialized functions in animal organs.
vacuole (VAK-you-ole) Large, fluid-filled vesicle that isolates or breaks down waste, debris, toxins, or food.
vesicle (VESS-ih-cull) Saclike, membrane-enclosed organelle; different kinds store, transport, or break down their contents.

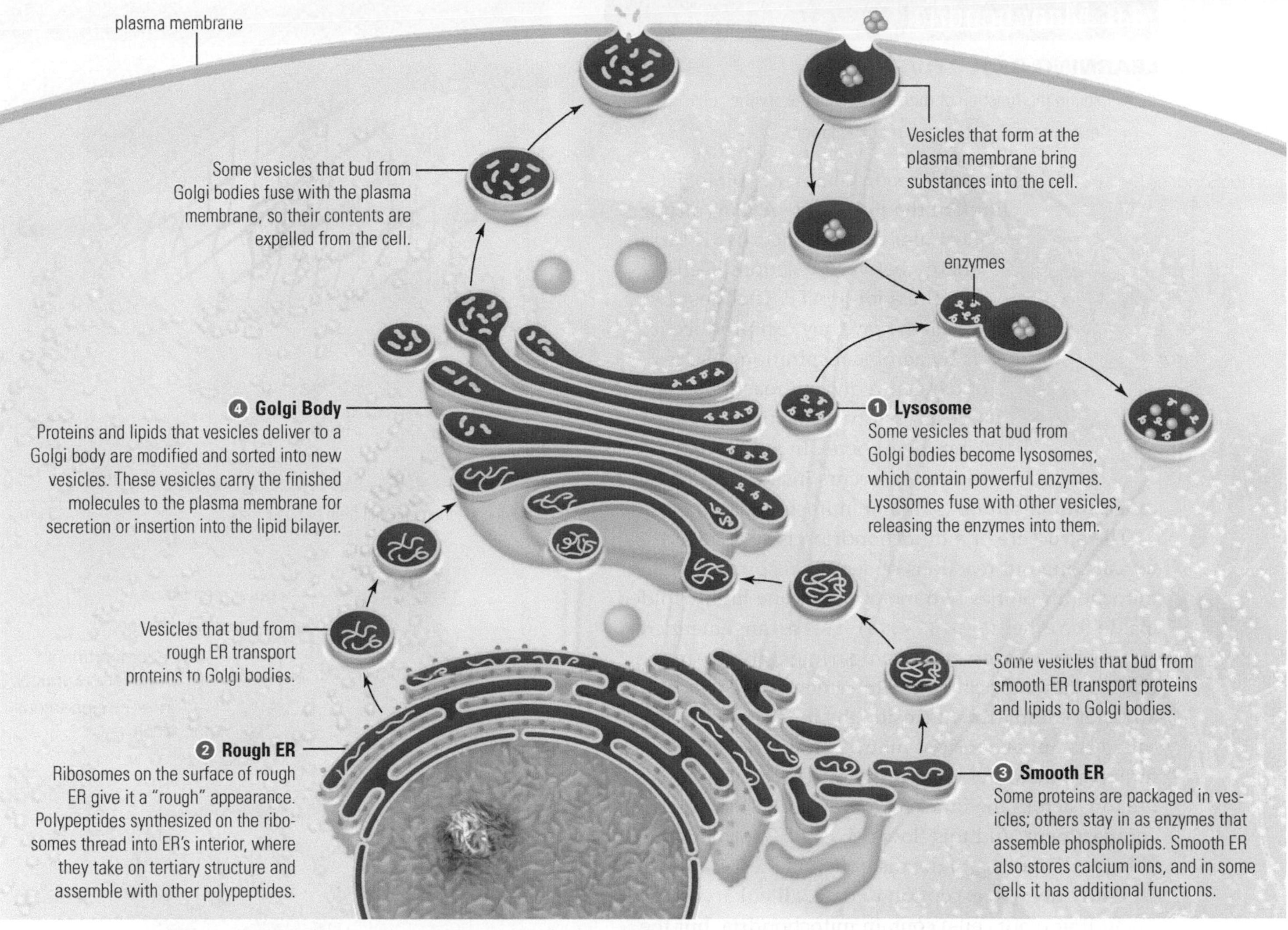
plasma membrane
Some vesicles that bud from Golgi bodies fuse with the plasma membrane, so their contents are expelled from the cell.
Vesicles that form at the plasma membrane bring substances into the cell.
enzymes
4 Golgi Body
Proteins and lipids that vesicles deliver to a Golgi body are modified and sorted into new vesicles. These vesicles carry the finished molecules to the plasma membrane for secretion or insertion into the lipid bilayer.
1 Lysosome
Some vesicles that bud from Golgi bodies become lysosomes, which contain powerful enzymes. Lysosomes fuse with other vesicles, releasing the enzymes into them.
Vesicles that bud from rough ER transport proteins to Golgi bodies.
Some vesicles that bud from smooth ER transport proteins and lipids to Golgi bodies.
2 Rough ER
Ribosomes on the surface of rough ER give it a "rough" appearance. Polypeptides synthesized on the ribosomes thread into ER's interior, where they take on tertiary structure and assemble with other polypeptides.
3 Smooth ER
Some proteins are packaged in vesicles; others stay in as enzymes that assemble phospholipids. Smooth ER also stores calcium ions, and in some cells it has additional functions.

FIGURE 4.12 Some interactions among components of the endomembrane system.

own proteins ❸. Some of the proteins delivered from rough ER are packaged into vesicles for delivery elsewhere. Others become enzymes that stay and carry out various functions of the smooth ER, including synthesis of phospholipids for the cell's membranes. Smooth ER also stores calcium ions in its interior (the concentration of these ions in cytosol is normally kept very low). Some cells have a lot of smooth ER, and in these cells it has special functions. For example, the abundant smooth ER in liver cells releases glucose from glycogen, and also breaks down hydrophobic toxins. Peroxisomes form as vesicles that bud from smooth ER; these acquire proteins directly from ribosomes.

The folded membrane of a **Golgi body** (also called a Golgi apparatus) often looks a bit like a stack of pancakes ❹. Enzymes inside Golgi bodies put finishing touches on proteins and lipids that have been delivered from ER, for example by attaching sugars, oligosaccharides, or phosphate groups. The finished products—membrane proteins and lipids, proteins for secretion, and enzymes—are sorted and packaged in new vesicles. Some of the new vesicles deliver their cargo to the plasma membrane; others become lysosomes. In plant cells, Golgi bodies have an additional, major function: They make pectin and other complex, branched polysaccharides for the cell wall.

TAKE-HOME MESSAGE 4.5

- ✔ The endomembrane system is a set of membrane-enclosed organelles: endoplasmic reticulum (ER), Golgi bodies, and vesicles.
- ✔ Rough ER produces enzymes and other proteins. Smooth ER produces lipids and breaks down carbohydrates, fatty acids, and toxins.
- ✔ Golgi bodies modify proteins and lipids, then package the finished molecules into vesicles for delivery to the plasma membrane.
- ✔ Various vesicles transport, store, and break down substances.

4.6 Mitochondria

LEARNING OBJECTIVES

- Explain the function of mitochondria in eukaryotic cells.
- Describe the structure of a mitochondrion.

mitochondrion

As you will see in Chapter 5, biologists think of the nucleotide ATP as a type of cellular currency because it carries energy between reactions. Cells require a lot of ATP. The most efficient way they can produce it is by aerobic respiration, a series of oxygen-requiring reactions that harvests energy from sugars by breaking their bonds. In eukaryotes, aerobic respiration occurs inside organelles called **mitochondria** (singular, mitochondrion).

The structure of a mitochondrion is specialized for carrying out reactions of aerobic respiration. Each mitochondrion has two membranes, one highly folded inside the other (**FIGURE 4.13**). This arrangement creates two spaces: an outer compartment (between the two membranes) called the intermembrane space, and an inner compartment (inside the inner membrane) called the mitochondrial matrix. Hydrogen ions accumulate in the outer compartment. The buildup pushes the ions across the inner membrane, into the inner compartment, and this flow drives ATP formation (Chapter 7 returns to details of aerobic respiration).

With only one exception to date, all eukaryotic cells (including plant cells) contain mitochondria, but the number varies by the type of cell and by the organism. For example, a single-celled organism such as a yeast cell may have only one mitochondrion, but a human skeletal muscle cell normally has a thousand or more. In general, cells that have the highest demand for energy tend to have the most mitochondria.

Mitochondria vary in size depending on the cell, but most are between 1 and 4 micrometers in length. Each is able to change its shape, for example by enlarging, shrinking, and branching. It may also split or fuse with another mitochondrion.

Mitochondria resemble bacteria. For example, they have their own DNA, which is circular and otherwise similar to bacterial DNA. They also divide independently of the cell, and have their own ribosomes. These features led to the theory that mitochondria evolved from bacteria that took up permanent residence inside a host cell (we return to this topic in Section 19.7).

Some eukaryotes that live in oxygen-free environments have modified mitochondria called hydrogenosomes that produce hydrogen in addition to ATP. Like mitochondria, hydrogenosomes have two membranes. Unlike mitochondria, they have no DNA, so they cannot divide independently of the cell.

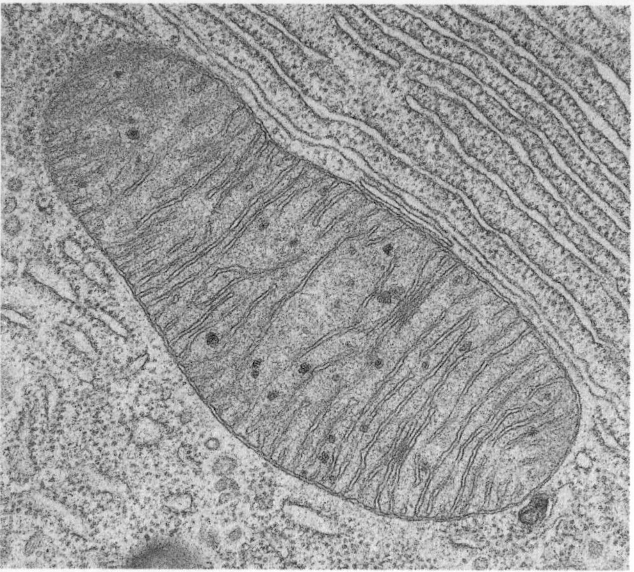
0.5 μm

A Mitochondrion in a cell from bat pancreas.

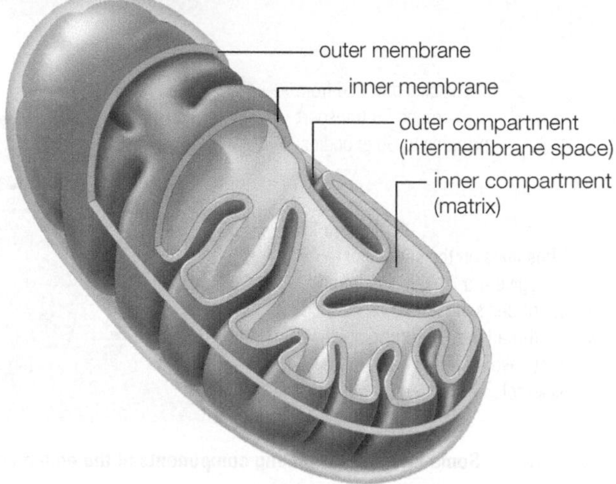
outer membrane
inner membrane
outer compartment (intermembrane space)
inner compartment (matrix)

B A mitochondrion has two membranes, one highly folded inside the other. The outer compartment formed by these membranes is called the intermembrane space; the inner compartment is called mitochondrial matrix.

FIGURE 4.13 The mitochondrion. This eukaryotic organelle specializes in producing ATP.

FIGURE IT OUT What organelle is visible in the upper right-hand corner of the TEM?

Answer: Rough ER

TAKE-HOME MESSAGE 4.6

- ✔ Mitochondria specialize in producing ATP. Each has two membranes, one highly folded inside the other.
- ✔ The number, size, and shape of mitochondria varies by cell type.
- ✔ Mitochondria have their own DNA and ribosomes, and they divide independently of the cell.

4.7 Chloroplasts and Other Plastids

LEARNING OBJECTIVES

- Describe the structure of chloroplasts.
- Explain the function of chloroplasts and amyloplasts.

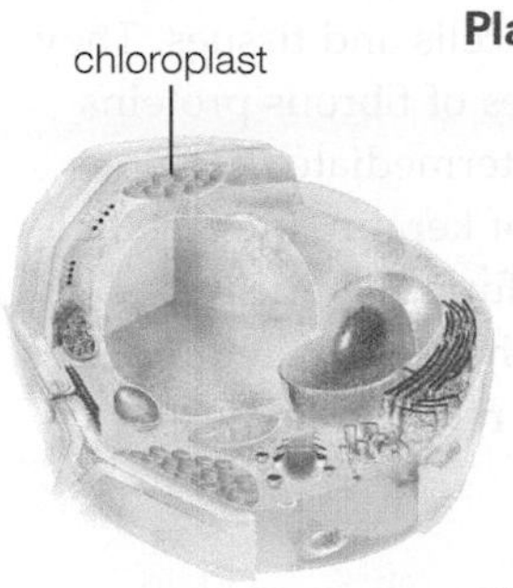
chloroplast

Plastids are double-membraned organelles that function in photosynthesis, storage, and/or pigmentation in plant and algal cells. Photosynthetic cells of plants and many protists contain **chloroplasts**, which are plastids specialized for photosynthesis (**FIGURE 4.14**).

Plant chloroplasts are oval or disk-shaped. Each has two outer membranes enclosing a semifluid interior, the stroma, that contains enzymes and the chloroplast's own DNA. In the stroma, a third, highly folded membrane forms a single, continuous compartment. Photosynthesis occurs at this inner membrane, which is called the thylakoid membrane.

A thylakoid membrane incorporates many pigments, including a green one called chlorophyll (the abundance of chlorophyll in plant cell chloroplasts is the reason why most plants are green). During photosynthesis, these pigments capture energy from sunlight, and pass it to other molecules that use the energy to make ATP. The ATP produced by this process is used to power reactions in the stroma that build sugars from carbon dioxide and water. (Chapter 6 details these processes.) In many ways, chloroplasts resemble the photosynthetic bacteria from which they evolved.

Chromoplasts are plastids that make and store pigments other than chlorophylls. They often contain red or orange pigments called carotenoids that color flowers, leaves, roots, and fruits (**FIGURE 4.15**). Chromoplasts are related to chloroplasts, and the two types of plastids are interconvertible. For example, as fruits such as tomatoes ripen, green chloroplasts in their cells are converted to red chromoplasts, so the color of the fruit changes (Section 30.5 returns to the process of ripening in fruits).

Amyloplasts are unpigmented plastids that make and store starch grains. They are notably abundant in cells of stems, fruits, and seeds. Like chromoplasts, amyloplasts are related to chloroplasts, and one type can change into the other. Starch-packed amyloplasts are dense and heavy compared to cytosol; in some plant cells, they function as gravity-sensing organelles (we return to this topic in Section 30.6).

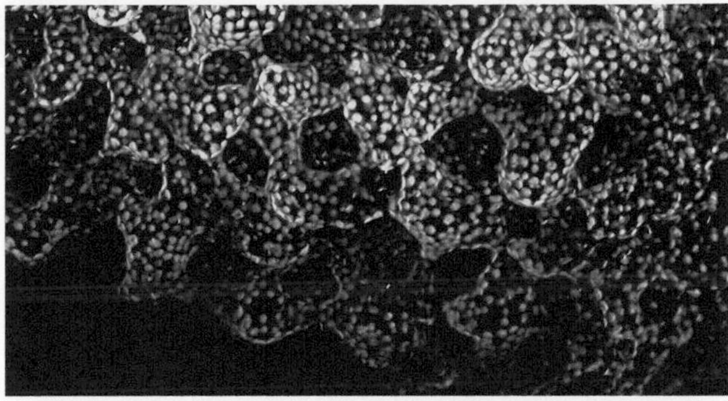

A Chloroplast-packed cells make up a leaf of a flowering plant.

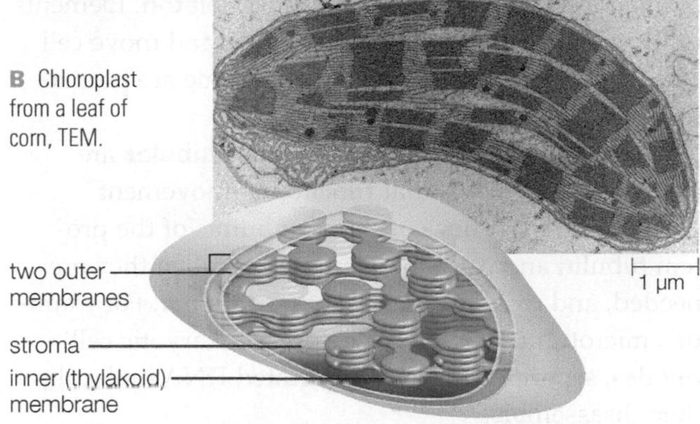
two outer membranes
stroma
inner (thylakoid) membrane
1 µm

B Chloroplast from a leaf of corn, TEM.

C Each chloroplast has two outer membranes. Photosynthesis occurs at a third, much-folded inner membrane that is called the thylakoid membrane.

FIGURE 4.14 The chloroplast.

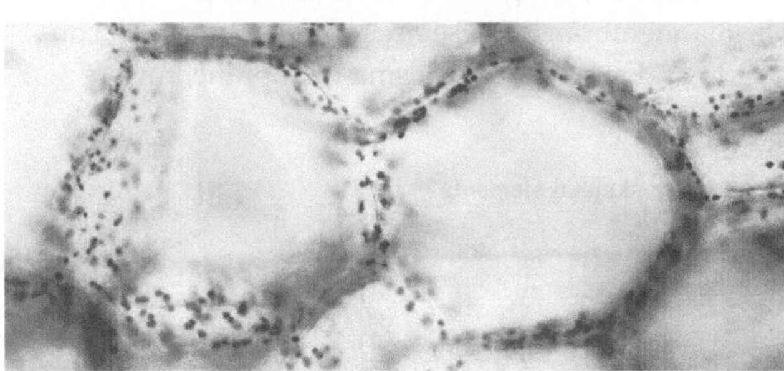

FIGURE 4.15 Chromoplasts. Pigments inside these plastids impart color to cells and the structures they compose. These are chromoplasts in cells of a red bell pepper.

TAKE-HOME MESSAGE 4.7

✔ Plastids occur in plants and some protists; different types function in photosynthesis, storage, and/or pigmentation.

✔ Chloroplasts are plastids that carry out photosynthesis. Each has two outer membranes and a highly folded inner (thylakoid) membrane.

chloroplast (KLOR-uh-plast) Organelle of photosynthesis in the cells of plants and photosynthetic protists.
mitochondrion (my-tuh-CON-dree-un) Double-membraned organelle that produces ATP by aerobic respiration in eukaryotes.
plastid One of several types of double-membraned organelles in plants and algal cells; for example, a chloroplast or amyloplast.

4.8 The Cytoskeleton

LEARNING OBJECTIVES

- Using appropriate examples, name three types of cytoskeletal elements and explain their function.
- With examples, describe how motor proteins move cell parts.
- Describe the movement of eukaryotic cilia and flagella and how this movement arises.

Cytoskeletal Elements

Between the nucleus and plasma membrane of all eukaryotic cells is a system of interconnected protein filaments collectively called the **cytoskeleton**. Elements of the cytoskeleton reinforce, organize, and move cell structures, and often the whole cell. Some are permanent; others form only at certain times.

Long, hollow cylinders called **microtubules** are cytoskeletal elements that function in movement (**FIGURE 4.16A**). They consist of subunits of the protein tubulin and can rapidly assemble when they are needed, and disassemble when they are not. For example, microtubules assemble before a eukaryotic cell divides, separate the cell's duplicated DNA molecules, then disassemble.

Fine fibers called **microfilaments** consist primarily of subunits of the protein actin (**FIGURE 4.16B**). In many cells, a mesh of microfilaments is part of the **cell cortex,** a region of cytoplasm just inside the plasma membrane. These microfilaments, which connect to and support the plasma membrane, form a scaffolding for proteins that function in cellular movement, contraction, shape changes, and migration (**FIGURE 4.17**).

Animal cells and some protists have an additional type of cytoskeletal element: the intermediate filament. **Intermediate filaments** form a stable framework that lends structure and resilience to cells and tissues. They are assembled from various types of fibrous proteins (**FIGURE 4.16C**). For example, intermediate filaments that make up your hair consist of keratin. Intermediate filaments that consist of lamins form the nuclear lamina of animal cells, and also help regulate processes inside the nucleus such as DNA replication.

Cellular Movement

Motor proteins that associate with cytoskeletal elements move cell parts when energized by a phosphate-group transfer from ATP (Section 3.6). A cell is like a bustling train station, with molecules and structures moving continuously throughout its interior. Motor proteins are like freight trains, dragging cellular cargo along tracks of microtubules and microfilaments (**FIGURE 4.18**).

A motor protein called myosin interacts with microfilaments to bring about muscle cell contraction. Another motor protein, dynein, interacts with microtubules to bring about movement of eukaryotic flagella and cilia. **Cilia** (singular, cilium) are similar to flagella except they are shorter, and they often occur in clumps that beat in unison.

FIGURE 4.16 Cytoskeletal elements.

tubulin subunit
25 nm

A Microtubule Involved in moving cell parts or the whole cell.

actin subunit
6–7 nm

B Microfilament Reinforces cell membranes; functions in muscle contraction.

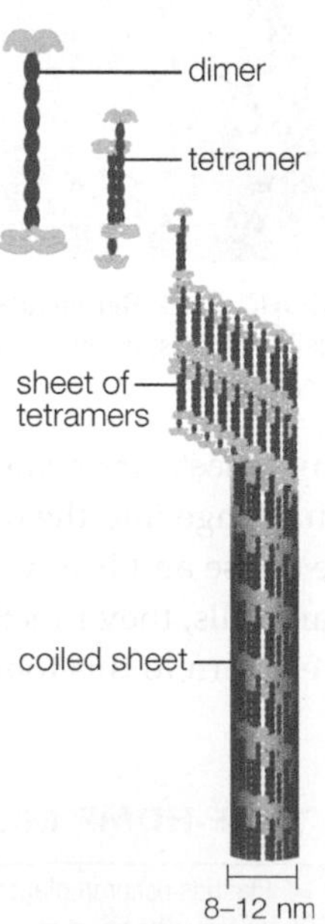
dimer
tetramer
sheet of tetramers
coiled sheet
8–12 nm

C Intermediate filament Structurally supports cell membranes and tissues. Most stable element.

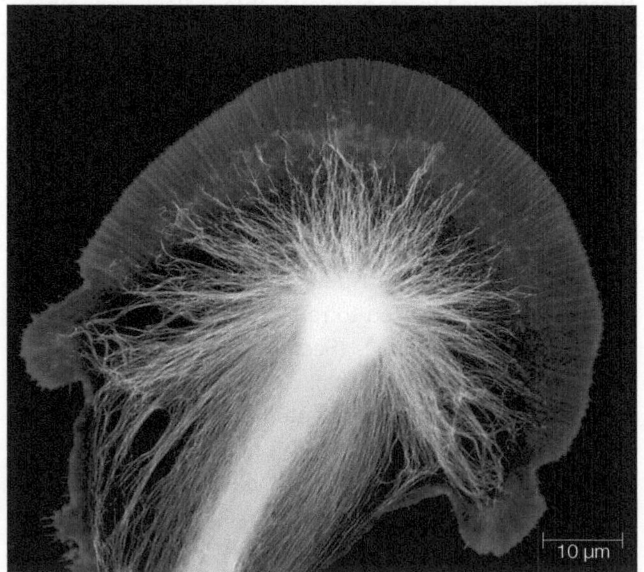
10 µm

FIGURE 4.17 Cytoskeletal elements in a nerve cell. This fluorescence micrograph shows microtubules (yellow) and microfilaments (blue) in the growing end of a nerve cell. These cytoskeletal elements support and guide the cell's lengthening in a particular direction.

Running lengthwise through each cilium or flagellum is a ring of nine microtubule pairs surrounding a central pair, in what is called a 9+2 array (**FIGURE 4.19A**). The microtubules grow from a barrel-shaped organelle called a **centriole**, which remains below the finished array as a **basal body** (**FIGURE 4.19B**). ATP-energized dynein causes the outer microtubule pairs to slide past one another, so the whole structure bends (**FIGURE 4.19C**). The whiplike motion that arises from this interaction—a power stroke that begins at the base and propagates to the tip—differs from the propeller-like motion of a prokaryotic flagellum, but both types of flagella can propel motile cells through fluid (**FIGURE 4.19D**).

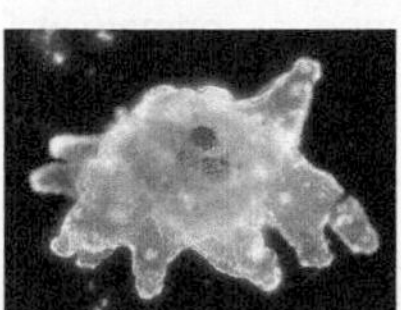

Some eukaryotic cells, including the amoeba at left, form **pseudopods**, or "false feet." As these temporary, irregular lobes bulge outward, they move the cell and engulf a target such as prey. Elongating microfilaments force the lobe to advance in a steady direction. Motor proteins attached to the microfilaments drag the plasma membrane along with them.

TAKE-HOME MESSAGE 4.8

- ✔ A cytoskeleton of protein filaments is the basis of eukaryotic cell shape, internal structure, and movement.
- ✔ Microtubules reinforce and help move cell parts.
- ✔ Networks of microfilaments reinforce cell shape and function in movement, for example as part of the cell cortex.
- ✔ Intermediate filaments strengthen and maintain the shape of cell membranes and tissues, and form external structures such as hair.
- ✔ ATP-energized motor proteins associated with microtubules and microfilaments can move cell parts or the whole cell.

basal body (BASE-ull) Organelle that develops from a centriole.
cell cortex Region of cytoplasm just inside the plasma membrane; often contains a mesh of microfilaments and associated proteins.
centriole (SEN-tree-ole) Barrel-shaped organelle from which microtubules lengthen.
cilium (SILL-ee-uh) Plural, cilia. Short, hairlike structures that project from the plasma membrane of some eukaryotic cells.
cytoskeleton (sigh-toe-SKEL-uh-ton) Network of protein filaments that support, organize, and move eukaryotic cells and their internal structures.
intermediate filament Stable cytoskeletal element of animals and some protists; different types are assembled from different fibrous proteins.
microfilament Reinforcing cytoskeletal element involved in cell movement; fiber of actin subunits.
microtubule (my-crow-TUBE-yule) Cytoskeletal element that can form a dynamic scaffolding for many cellular processes involved in movement. Hollow filament of tubulin subunits.
motor protein Type of energy-using protein that interacts with cytoskeletal elements to move the cell's parts or the whole cell.
pseudopod (SUE-doh-pod) A temporary protrusion that helps some eukaryotic cells move and engulf prey.

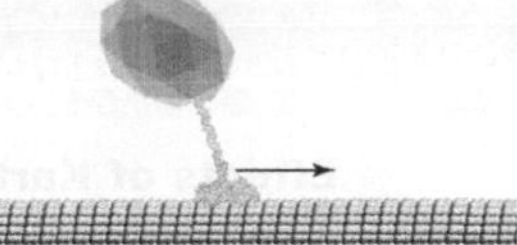

FIGURE 4.18 A motor protein. Here, kinesin (tan) drags a vesicle (pink) along a microtubule.

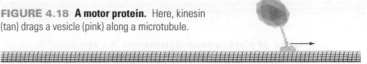

FIGURE 4.19 How eukaryotic flagella and cilia move.

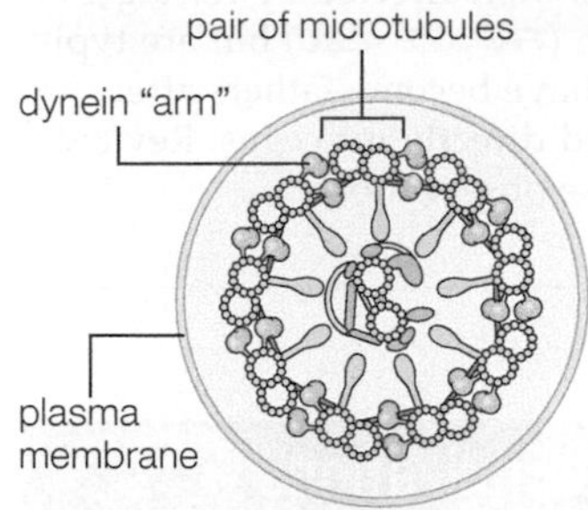
pair of microtubules
dynein "arm"
plasma membrane

A A 9+2 array consists of a ring of nine pairs of microtubules plus one central pair. Stabilizing spokes and linking elements connect the microtubules and keep them aligned in this pattern. Projecting from each pair of microtubules are "arms" of the motor protein dynein.

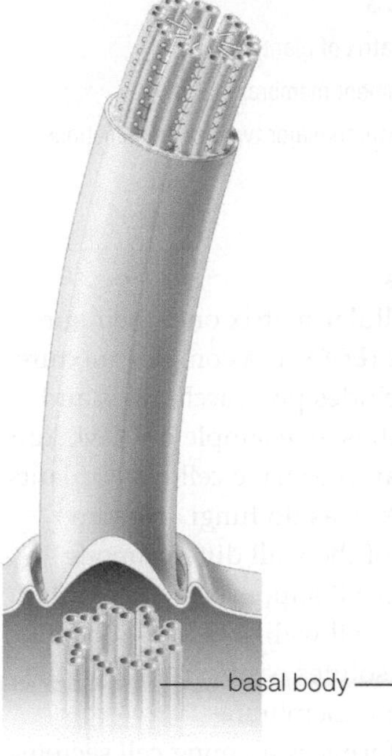
basal body

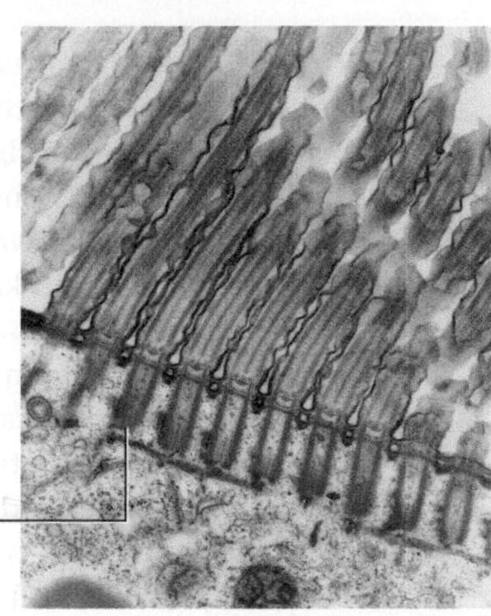

B Microtubules of a developing 9+2 array lengthen from a centriole, which remains under the finished array as a basal body. The micrograph below shows basal bodies underlying cilia of *Paramecium*, the protist pictured in Figure 4.4.

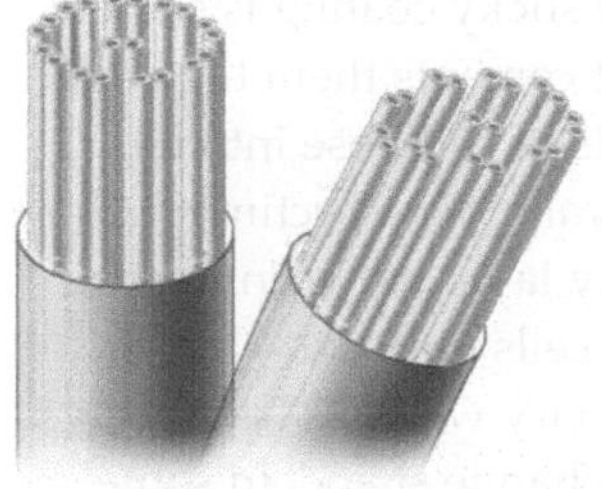

C Phosphate-group transfers from ATP cause the dynein arms in a 9+2 array to repeatedly bind the adjacent pair of microtubules, bend, and then disengage.

The dynein arms "walk" along the microtubules, so adjacent microtubule pairs slide past one another. The short, sliding strokes of the dynein arms occur in a coordinated sequence around the ring, down the length of the microtubules. The movement causes the entire structure to bend.

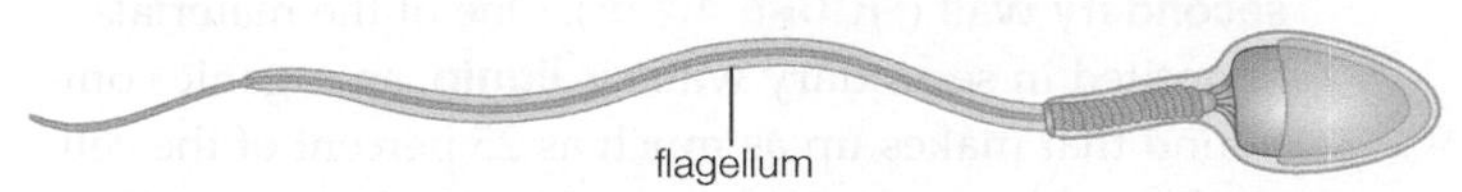
flagellum

D A 9+2 array gives rise to the whiplike motion of a flagellum that can propel mobile eukaryotic cells such as sperm through fluid surroundings.

Data Analysis Activities

Effects of Kartagener Syndrome An abnormal form of the motor protein dynein causes Kartagener syndrome, a genetic disorder characterized by chronic sinus and lung infections. Biofilms form in the thick mucus that collects in the airways, and the resulting bacterial activities and inflammation damage tissues. Men affected by Kartagener syndrome can produce sperm (**FIGURE 4.20**) but are typically infertile; some of them have become fathers after their sperm cells were injected directly into eggs. Review Figure 4.19, then explain these observations.

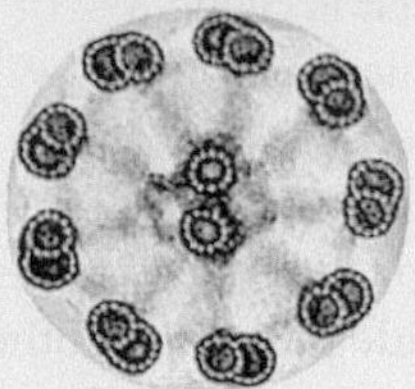

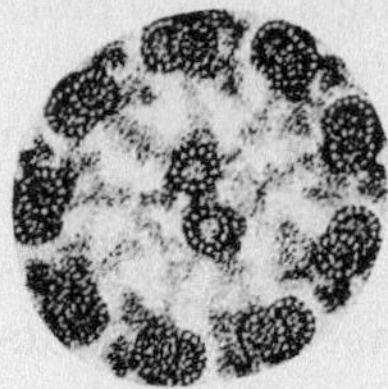

FIGURE 4.20 An effect of Kartagener syndrome. Cross-section of a sperm flagellum from an affected man (left) and an unaffected man (right).

4.9 Cell Surface Specializations

LEARNING OBJECTIVES

- Describe an extracellular matrix of plants.
- Explain the function of basement membrane.
- Explain the functions of the three major types of cell junctions in animals.
- Describe a plasmodesma.

Extracellular Matrices

Many cells secrete extracellular matrix onto their surfaces. **Extracellular matrix (ECM)** is a complex mixture of molecules that often includes polysaccharides and fibrous proteins. A cell wall is an example of ECM. You already learned that prokaryotes have cell walls; plants have them too (**FIGURE 4.21**), as do fungi and some protists. The composition of the wall differs among these groups, but in all cases it supports and protects the cell. Like a prokaryotic cell wall, a eukaryotic cell wall is porous: Water and solutes easily cross it on the way to and from the plasma membrane.

In plants, the cell wall forms as a young cell secretes polysaccharides such as pectin onto the outer surface of its plasma membrane. The sticky coating is shared between adjacent cells, and it cements them together. Each cell then secretes strands of cellulose into the coating, forming a **primary wall**. Some pectin remains as the middle lamella, a sticky layer between the primary walls of abutting plant cells (**FIGURE 4.21A**). Being thin and pliable, a primary wall allows a growing plant cell to enlarge and change shape. In some plants, mature cells secrete material onto the primary wall's inner surface. These deposits form a firm **secondary wall** (**FIGURE 4.21B**). One of the materials deposited in secondary walls is **lignin**, an organic compound that makes up as much as 25 percent of the cell walls in older stems and roots. Lignified plant parts are stronger, more waterproof, and less susceptible to plant-attacking organisms than younger tissues.

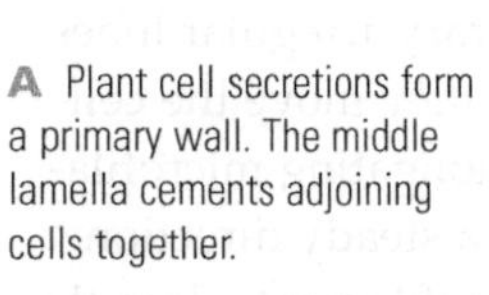

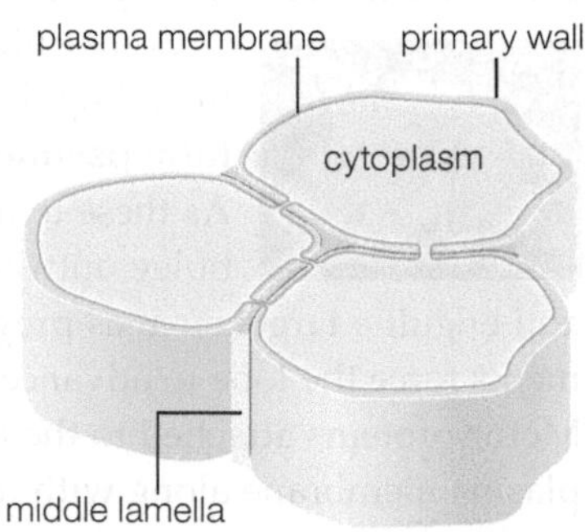
plasma membrane
primary wall
cytoplasm
middle lamella

A Plant cell secretions form a primary wall. The middle lamella cements adjoining cells together.

B Cells in older plant tissues secrete materials in layers on the inner surface of their primary wall. These layers form layers of a sturdy secondary wall. In some tissues, the wall remains after the cells die, becoming part of pipelines that carry water through the plant

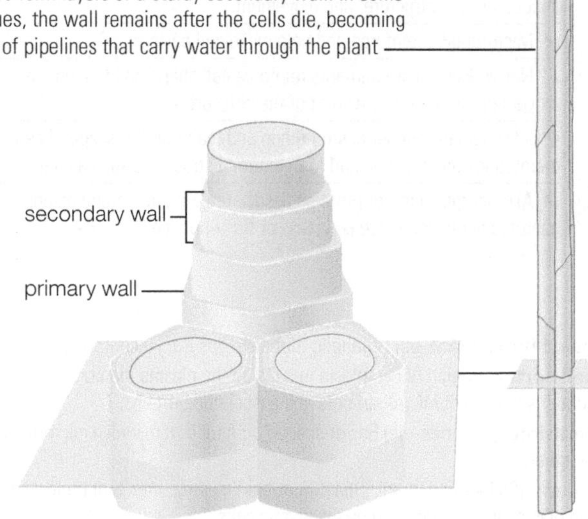
secondary wall
primary wall

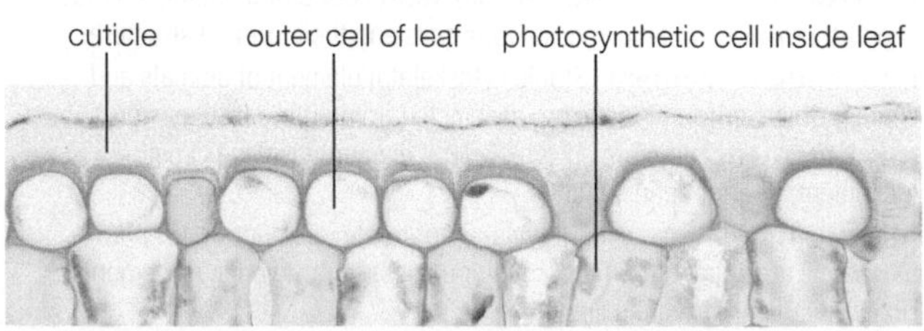
cuticle
outer cell of leaf
photosynthetic cell inside leaf

C Section through a plant leaf showing cuticle, a protective covering secreted by living cells.

FIGURE 4.21 Extracellular matrices of plants.

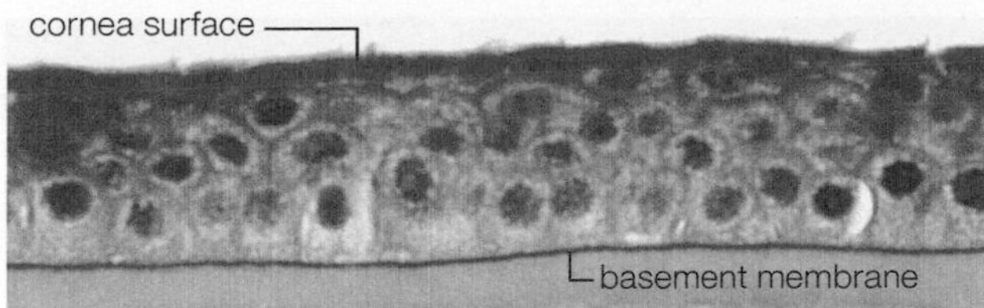
cornea surface
basement membrane

FIGURE 4.22 Basement membrane. In this section of human cornea (a part of the eye), basement membrane is visible as a dark line under the cells.

Animal cells have no walls, but some types secrete an extracellular matrix called **basement membrane** (FIGURE 4.22). Despite the name, basement membrane is not a cell membrane because it does not consist of a lipid bilayer. Rather, it is a sheet of fibrous material that structurally supports and organizes tissues, and it has roles in cell signaling. Bone is an ECM of the fibrous protein collagen hardened into a mineral by bonding with calcium, magnesium, and phosphate ions.

A covering called **cuticle** is a type of ECM secreted by cells at a body surface. In plants, a cuticle of waxes and proteins helps stems and leaves retain water and fend off insects (FIGURE 4.21C). The cuticle of crabs, spiders, and other arthropods consists mainly of chitin (Section 3.3).

Cell Junctions

In multicelled species, cells can interact with one another and their surroundings by way of cell junctions. **Cell junctions** are structures that connect a cell directly to other cells or to its environment. Cells send and receive substances and signals through some junctions. Other junctions help cells recognize and stick to each other in tissues.

Three types of cell junctions are common in animal tissues (FIGURE 4.23). In tissues that line body surfaces and internal cavities, rows of **tight junctions** fasten the plasma membranes of adjacent cells together and prevent body fluids from seeping between them (FIGURE 4.23A). For example, the lining of the stomach is leak-proof because the cells that make it up are linked by tight junctions. These junctions keep gastric fluid, which contains acid and destructive enzymes, safely inside the stomach. If a bacterial infection damages the stomach lining, gastric fluid leaks into and damages the underlying layers. A painful peptic ulcer results.

Adhesion proteins assemble into various types of **adhering junctions**, which fasten cells to one another and to basement membrane (FIGURE 4.23B–D). These junctions make a tissue quite strong because they connect to cytoskeletal elements inside the cells. Contractile tissues (such as heart muscle) have a lot of adhering junctions, as do tissues subject to abrasion or stretching (such as skin).

Gap junctions are closable channels that connect the cytoplasm of adjoining animal cells (FIGURE 4.23E).

adhering junction Cell junction that fastens an animal cell to another cell, or to basement membrane. Connects to cytoskeletal elements inside the cell; composed of adhesion proteins.
basement membrane Extracellular matrix that structurally supports and organizes animal tissues.
cell junction Structure that connects a cell to another cell or to ECM.
cuticle (KEW-tih-cull) Secreted covering at a body surface.
extracellular matrix (ECM) Complex mixture of substances secreted by a cell onto its surface; composition and function vary by cell type.
gap junction Cell junction that forms a closable channel across the plasma membranes of adjoining animal cells.
lignin (LIG-nin) Material that strengthens cell walls of some plants.
primary wall The first cell wall of young plant cells.
secondary wall Lignin-reinforced wall that forms inside the primary wall of a plant cell.
tight junction Cell junction that fastens together the plasma membrane of adjacent animal cells. Collectively prevent fluids from leaking between cells.

FIGURE 4.23 Cell junctions in animal tissues. Tight junctions (A), three types of adhering junctions (B, C, and D), and gap junctions (E) are shown.

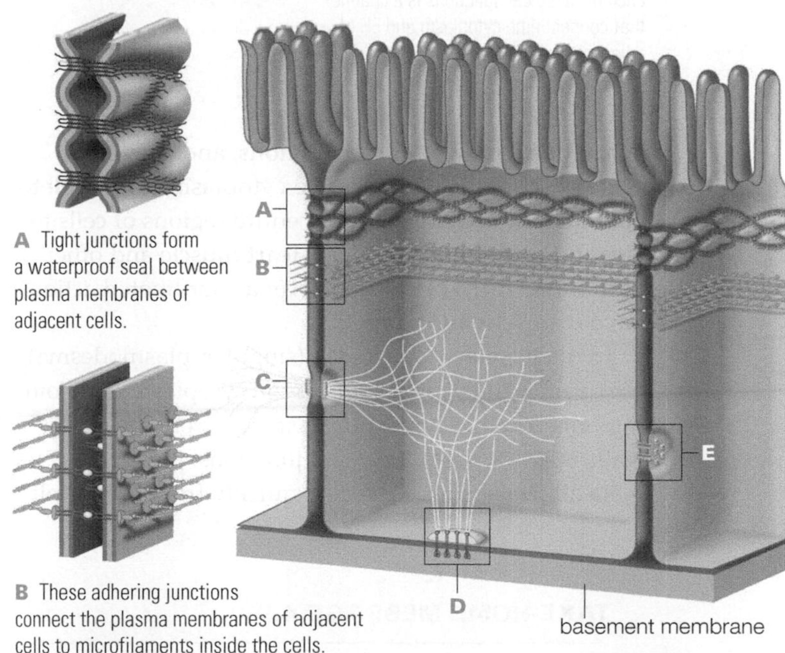
A
B
C
D
E
basement membrane

A Tight junctions form a waterproof seal between plasma membranes of adjacent cells.

B These adhering junctions connect the plasma membranes of adjacent cells to microfilaments inside the cells.

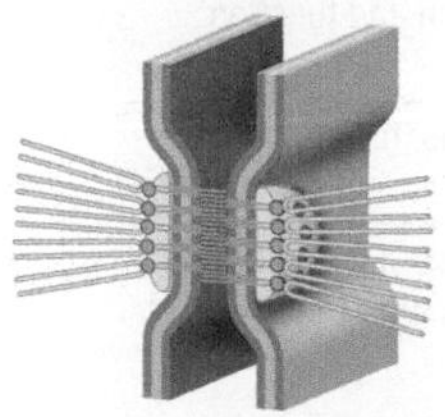

C These adhering junctions connect the plasma membranes of adjacent cells to intermediate filaments inside the cells.

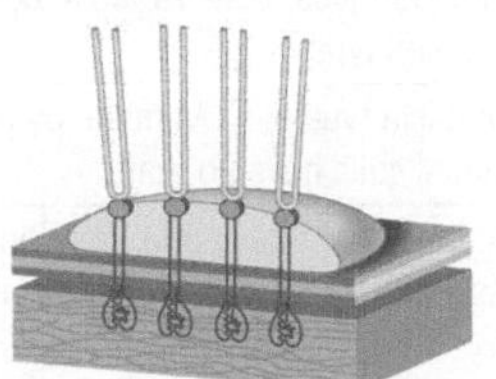

D These adhering junctions attach the plasma membrane of a cell to basement membrane and to intermediate filaments inside the cell.

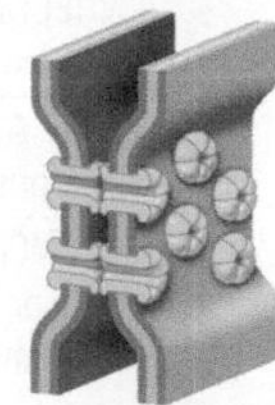

E Gap junctions are closable channels that connect the cytoplasm of adjacent cells.

CREDITS: (22) Ralph C. Eagle, Jr./Science Source; (23) © Cengage Learning.

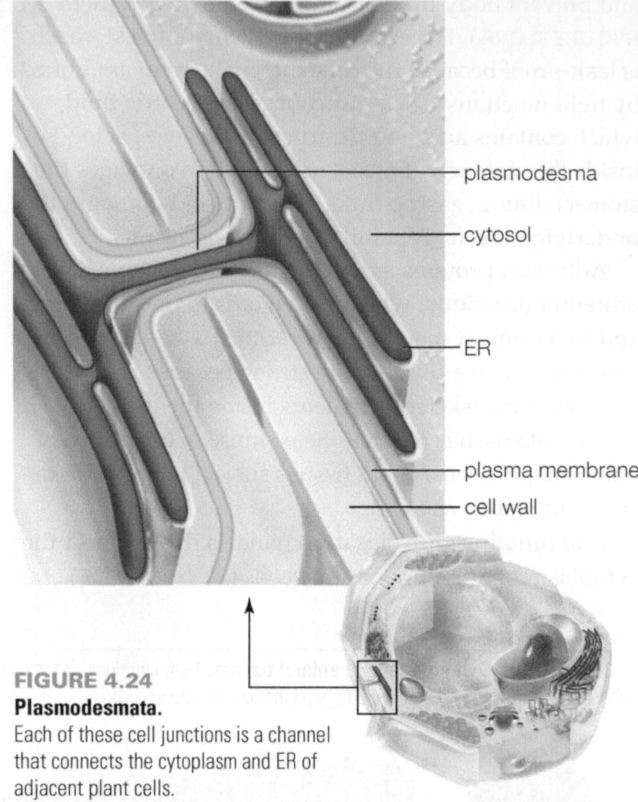
plasmodesma
cytosol
ER
plasma membrane
cell wall

FIGURE 4.24
Plasmodesmata.
Each of these cell junctions is a channel that connects the cytoplasm and ER of adjacent plant cells.

When open, they permit water, ions, and small molecules to pass directly from the cytoplasm of one cell to another. These channels allow entire regions of cells to respond to a single stimulus. Heart muscle and other tissues in which the cells perform a coordinated action have many gap junctions.

In plants, **plasmodesmata** (singular, plasmodesma) are open channels that connect the cytoplasm of adjoining cells. These cell junctions extend across the cell walls (**FIGURE 4.24**). Like gap junctions, plasmodesmata allow substances to flow quickly from cell to cell.

TAKE-HOME MESSAGE 4.9

✔ Many cells secrete material that forms an extracellular matrix (ECM) on their surfaces. ECM varies in composition and function depending on cell type.

✔ A cell wall is a type of ECM made by plant cells, fungi, and some protists. Animal cells have no wall.

✔ Cell junctions structurally and functionally connect cells in tissues. In animal tissues, cell junctions also connect cells with basement membrane.

plasmodesmata (plaz-mow-dez-MA-tah) Singular, plasmodesma. Type of cell junction that forms an open channel between the cytoplasm of adjacent plant cells.

4.10 The Nature of Life

LEARNING OBJECTIVES

- Explain why it is difficult to define "life."
- List some of the properties that are associated with living systems.

Carbon, hydrogen, oxygen, and other atoms of organic molecules are the stuff of you, and us, and all of life. Yet it takes more than organic molecules to complete the picture. In this chapter, you learned about the structure of cells, which have at minimum a plasma membrane, cytoplasm, and DNA. Differences in other cellular components—the presence or absence of a particular organelle, for example—are often used to categorize life's diversity. What about life's commonality? The cell is the smallest unit of life, but a cell does not spring to life from cellular components mixed in the proper amounts and proportions. What is it, exactly, that makes a cell, or an organism that consists of them, alive?

Many brilliant people have thought about that question, but we still don't have a satisfactory answer. When we try to define life, we end up with a very long list of properties that collectively set the living apart from the nonliving (Section 1.3). However, even that list can be tricky. For example, living things have a high proportion of the organic molecules of life, but so do the remains of dead organisms in seams of coal. Living things use energy to reproduce themselves, but computer viruses, which are arguably not alive, can do that too. Living things can pass hereditary material to offspring, but a single living rabbit (for example) cannot. No list of properties reliably and unambiguously unites all things that we would consider alive, and excludes all things we would consider not alive.

So how do biologists, who study life as a profession, define it? According to evolutionary biologist Gerald Joyce, the simplest definition of life might well be "that which is squishy." He says, "Life, after all, is protoplasmic and cellular. It is made up of cells and organic stuff and is undeniably squishy." The point is that defining life may be impossible—and pointless from a scientific standpoint. The property that we call "life" is not the same as properties (such as homeostasis, reproduction, and metabolism) associated with the state of being alive; a description of an organism that is alive is not a description of the life of the organism.

However, that very long list of properties is far from useless: It constitutes a working theory of living systems, a way of looking at a biological system as part of a network of relationships rather than an assembly

TABLE 4.5

Collective Properties of Living Systems

A living system:
Consists of one or more cells
Harvests matter and energy from the environment
Requires water
Makes and uses complex carbon-containing molecules
Engages in self-sustaining biological processes such as homeostasis and metabolism
Has the capacity to respond to internal and external stimuli
Has the capacity for growth
Changes over its lifetime, for example by maturing and aging
Passes hereditary material (DNA) to offspring
Has the capacity to adapt to environmental pressures over successive generations

of individual components. You have already been introduced to many of the properties associated with biological systems (**TABLE 4.5**). The remaining units of this book explore biologists' current understanding of these properties.

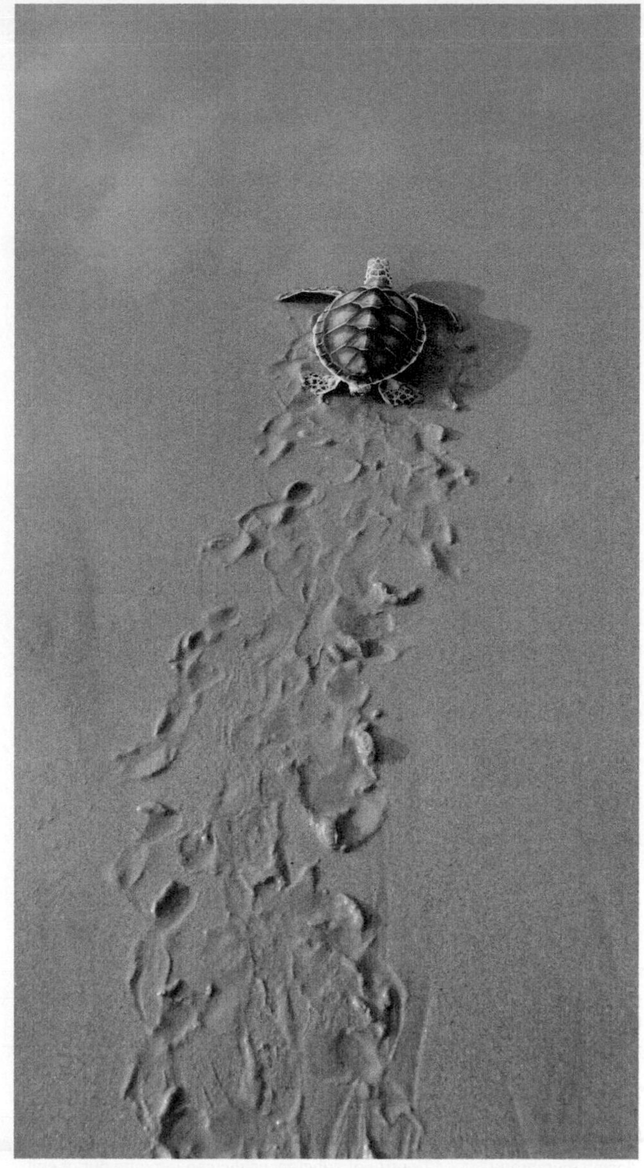

TAKE-HOME MESSAGE 4.10

✔ We describe the characteristic of "life" in terms of properties that are collectively unique to living things.

✔ Organisms make and use the organic molecules of life. DNA is their hereditary material.

✔ In living things, the molecules of life are organized as one or more cells that engage in self-sustaining biological processes.

✔ Living things change over lifetimes, and also over successive generations.

4.1 Food for Thought (revisited)

The United States Department of Agriculture (USDA) recalls food products in which toxic bacteria are discovered. Recalled meat is not necessarily discarded; often, it is cooked and processed into ready-to-eat products such as canned chili and frozen dinners. Sterilization by cooking or other means kills bacteria, and it is one way to ensure food safety.

Raw beef trimmings, which have a high risk of contact with fecal matter during the butchering process, are effectively sterilized when sprayed with ammonia. Ground to a paste and formed into pellets or blocks, the resulting product is termed "lean finely textured beef" or "boneless lean beef trimmings." Although this product cannot be sold directly to consumers, lean finely textured beef has been and continues to be routinely used as a filler in prepared food products such as hamburger patties, fresh ground beef, hot dogs, lunch meats, sausages, frozen entrees, canned foods, and other items sold to quick service restaurants, hotel and restaurant chains, institutions, and school lunch programs.

A series of news reports in 2012 provoked public outrage at the widespread, unlabeled use of lean finely textured beef, pejoratively nicknamed pink slime. Meat industry organizations and the USDA agree that lean finely textured beef, appetizing or not, is perfectly safe to eat because it has been sterilized—any live bacteria in it have been killed. However, the public controversy prompted some companies to discontinue use of ground beef containing the filler product.

Section 4.1 A huge number of bacteria live in and on the human body. Most of them are helpful; a few types can cause disease. Contamination of food with disease-causing bacteria can result in illness that is sometimes fatal.

Section 4.2 Early observations of cells led to **cell theory**, a set of principles that, taken together, constitute the foundation of modern biology:

1. Every living organism consists of one or more cells.
2. The cell is the basic structural and functional unit of life. Cells are individually alive even as part of a multicelled organism.
3. All living cells arise by division of preexisting cells.
4. All cells pass hereditary material (DNA) to offspring when they divide.

Cells share certain structural and functional features; all have a plasma membrane, DNA, and cytoplasm. The **plasma membrane** separates the cell from its environment and, like all membranes, its basic structure is a lipid bilayer. Proteins embedded in the bilayer carry out particular membrane functions. Only certain materials can cross a cell membrane, so a plasma membrane controls the exchange of materials between the cell and the external environment. The plasma membrane encloses jellylike **cytosol**.

All cells start out life with DNA. A eukaryotic cell's DNA is contained within a **nucleus**, which is one type of membrane-enclosed **organelle**. Most have many additional components (**TABLE 4.6** and **FIGURE 4.25**). In eukaryotic cells, cytosol and everything else between the plasma membrane and the nucleus is collectively called **cytoplasm**. In prokaryotic cells, DNA is suspended in cytosol, so cytoplasm is cytosol and everything else enclosed by the plasma membrane.

A cell's surface area increases with the square of its diameter, while its volume increases with the cube. This **surface-to-volume ratio** limits cell size and influences cell shape. Almost all cells are far too small to see with the naked eye, so we use microscopes to observe them. Different types of microscopes and staining techniques reveal different internal and external details of cells.

Section 4.3 Bacteria and archaea, informally grouped as prokaryotes, are single-celled organisms with no nucleus. All have **ribosomes** and one or two circular molecules of DNA in the **nucleoid**. Many have **plasmids** that carry some additional genetic information, and some have motile structures (**flagella**) and other projections (**pili**).

Almost all prokaryotic species have a protective, rigid **cell wall** that surrounds the plasma membrane. Some have a second membrane around the wall. A sticky capsule helps some bacteria fend off predators and stick to surfaces.

Bacteria and other microbial organisms may live together in a shared mass of slime as a **biofilm**.

TABLE 4.6

Comparing Components of Prokaryotic and Eukaryotic Cells

			Eukaryotes			
Cell Component	**Example(s) of Function**	**Prokaryotes**	**Protists**	**Fungi**	**Plants**	**Animals**
Cell wall	Protection, structural support	+	+	+	+	–
Plasma membrane	Control of substances moving into and out of cell	+	+	+	+	+
Nucleus	Physical separation of DNA from cytoplasm	–	+	+	+	+
Nucleolus	Assembly of ribosome subunits	–	+	+	+	+
DNA	Encoding of hereditary information	+	+	+	+	+
RNA	Protein synthesis	+	+	+	+	+
Ribosome	Protein synthesis	+	+	+	+	+
Endoplasmic reticulum	Protein, lipid synthesis; carbohydrate, fatty acid breakdown	–	+	+	+	+
Golgi body	Final modification of proteins, lipids	–	+	+	+	+
Lysosome	Intracellular digestion	–	+	+	+	+
Peroxisome	Breakdown of fatty acids, amino acids, and toxins	–	+	+	+	+
Mitochondrion	Production of ATP by aerobic respiration	–	+	+	+	+
Hydrogenosome	Anaerobic production of ATP	–	+	+	+	+
Chloroplast	Photosynthesis; starch storage	–	+	–	+	–
Central vacuole	Increasing cell surface area; storage	–	–	+	+	–
Flagellum	Locomotion through fluid surroundings	+	+	+	+	+
Cilium	Locomotion through fluid surroundings; movement of surrounding fluid	+	+	–	+	+
Cytoskeleton	Physical reinforcement; internal organization; movement	+	+	+	+	+

+ found in at least some species; – not found in any species to date.

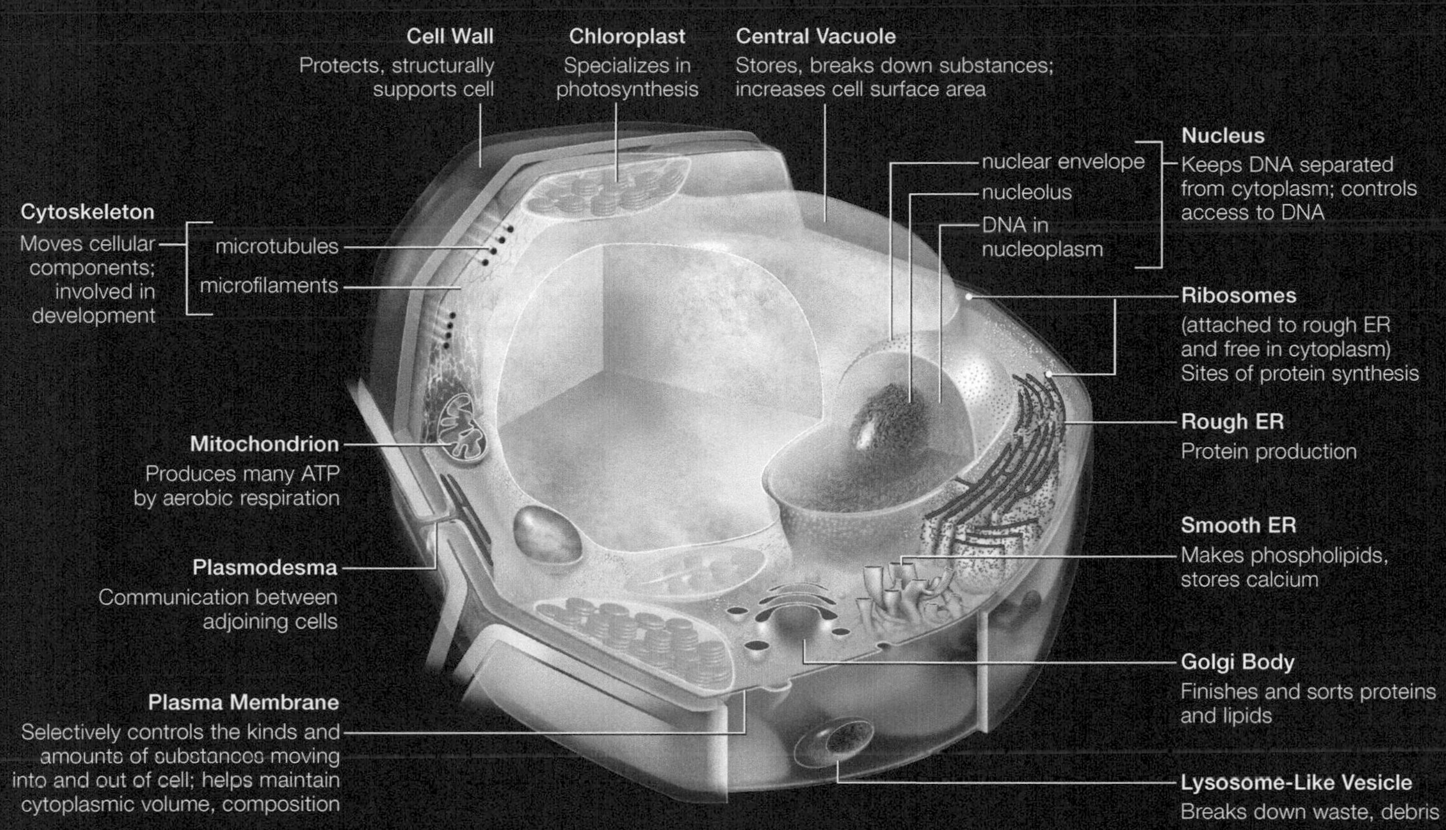

Cell Wall
Protects, structurally supports cell
Chloroplast
Specializes in photosynthesis
Central Vacuole
Stores, breaks down substances; increases cell surface area
Nucleus
Keeps DNA separated from cytoplasm; controls access to DNA
nuclear envelope
nucleolus
DNA in nucleoplasm
Cytoskeleton
Moves cellular components; involved in development
microtubules
microfilaments
Ribosomes
(attached to rough ER and free in cytoplasm) Sites of protein synthesis
Rough ER
Protein production
Mitochondrion
Produces many ATP by aerobic respiration
Smooth ER
Makes phospholipids, stores calcium
Plasmodesma
Communication between adjoining cells
Golgi Body
Finishes and sorts proteins and lipids
Plasma Membrane
Selectively controls the kinds and amounts of substances moving into and out of cell; helps maintain cytoplasmic volume, composition
Lysosome-Like Vesicle
Breaks down waste, debris

A Typical plant cell components.

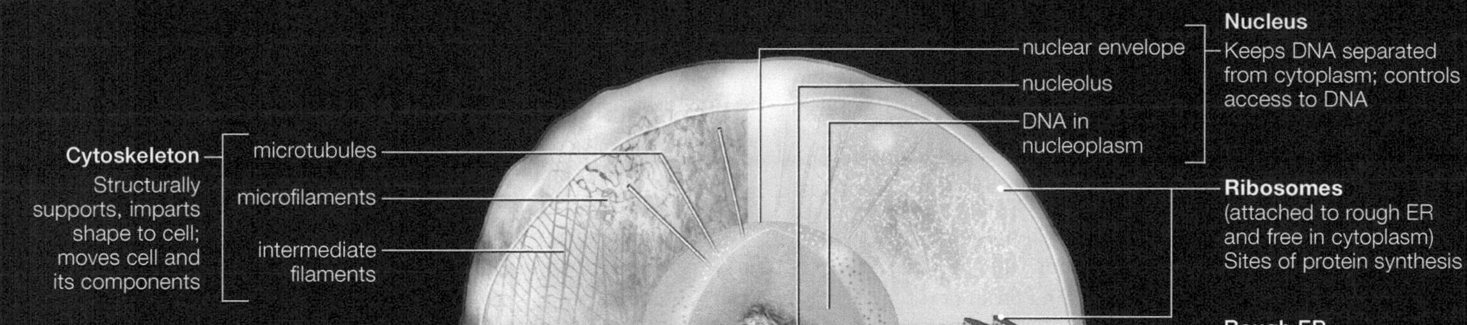

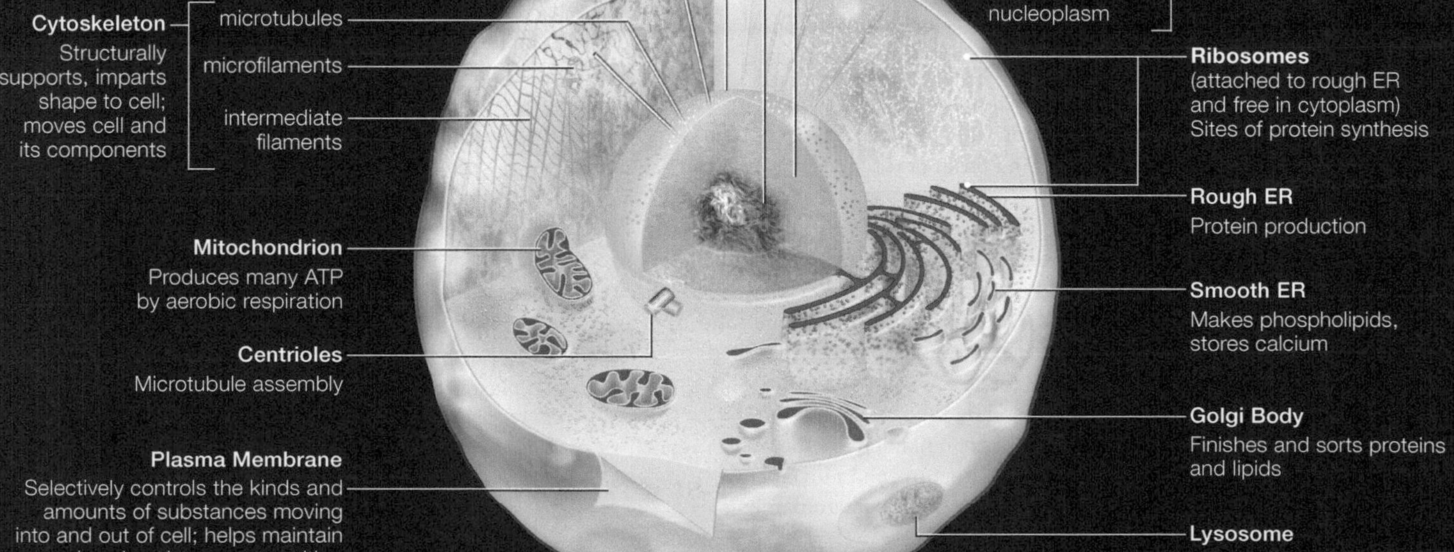

Nucleus
Keeps DNA separated from cytoplasm; controls access to DNA
nuclear envelope
nucleolus
DNA in nucleoplasm
Cytoskeleton
Structurally supports, imparts shape to cell; moves cell and its components
microtubules
microfilaments
intermediate filaments
Ribosomes
(attached to rough ER and free in cytoplasm) Sites of protein synthesis
Rough ER
Protein production
Mitochondrion
Produces many ATP by aerobic respiration
Smooth ER
Makes phospholipids, stores calcium
Centrioles
Microtubule assembly
Golgi Body
Finishes and sorts proteins and lipids
Plasma Membrane
Selectively controls the kinds and amounts of substances moving into and out of cell; helps maintain cytoplasmic volume, composition
Lysosome
Breaks down waste, debris

B Typical animal cell components.

FIGURE 4.25 Organelles and structures typical of eukaryotes, illustrated in a plant cell (**A**) and an animal cell (**B**).

STUDY GUIDE

Section 4.4 All eukaryotic cells start out life with a nucleus and other organelles. Membrane-enclosed organelles maximize cellular efficiency by compartmentalizing tasks and substances that may affect or be affected by other parts of the cell.

A nucleus protects and controls access to a eukaryotic cell's DNA. The outer boundary of the nucleus is a **nuclear envelope,** a double lipid bilayer studded with special pores that allow some molecules, but not others, to pass into and out of the nucleus.

Inside the nuclear envelope, **chromatin** is suspended in viscous **nucleoplasm**. Ribosome subunits are produced in dense, irregularly shaped regions called **nucleoli**.

Section 4.5 The **endomembrane system** is a series of organelles (endoplasmic reticulum, Golgi bodies, and vesicles) that interact mainly to make lipids, enzymes, and proteins for insertion into membranes or secretion.

Different types of **vesicles** store, break down, or transport substances through the cell. Enzymes in **peroxisomes** break down molecules such as amino acids, fatty acids, and toxins. **Lysosomes** contain enzymes that break down particles such as cellular debris. Fluid-filled **vacuoles** store or break down waste, food, and toxins; contractile vacuoles expel excess water. Fluid pressure inside a **central vacuole** keeps plant cells plump, which in turn keeps plant parts firm.

Endoplasmic reticulum (**ER**) is a continuous system of sacs and tubes extending from the nuclear envelope. Polypeptides made by ribosomes on rough ER take on tertiary and quaternary structure in the ER compartment. Smooth ER makes phospholipids and stores calcium ions. **Golgi bodies** modify proteins and lipids. Some vesicles that bud from Golgi bodies deliver the finished molecules to the plasma membrane; others become lysosomes.

Section 4.6 **Mitochondria** are organelles with two membranes, one folded inside the other. This structural specialization allows mitochondria to produce ATP by aerobic respiration, an oxygen-requiring series of reactions that breaks down carbohydrates.

Section 4.7 **Plastids** of plants and some protists function in photosynthesis, storage, and pigmentation. In eukaryotes, photosynthesis takes place at the inner (thylakoid) membrane of **chloroplasts**. Pigment-filled chromoplasts and starch-filled amyloplasts are plastids used for storage and other purposes.

Section 4.8 A **cytoskeleton** of protein filaments is the basis of eukaryotic cell shape, internal structure, and movement. Cytoskeletal elements include microtubules, microfilaments, and intermediate filaments.

ATP-driven **motor proteins** bring about the movement of cell parts and/or the whole cell. For example, motor proteins interact with a special 9+2 array of microtubules inside **cilia** and eukaryotic flagella to move these structures. A 9+2 array arises from a **centriole**, which remains at the base of a cilium or flagellum as a **basal body**.

Networks of **microfilaments** in the **cell cortex** form a mesh just under the plasma membrane. The mesh connects to the plasma membrane and forms a scaffold for motor proteins that bring about movement, for example, of **pseudopods**.

Stable **intermediate filaments** composed of various fibrous proteins reinforce the shape of animal cell membranes and tissues. They also form external structures such as hair. The nuclear lamina consists of intermediate filaments.

Section 4.9 Many cells secrete an **extracellular matrix**, or **ECM**, onto their surfaces The composition and function of ECM varies depending on cell type. In animals, an ECM called **basement membrane** supports and organizes cells in tissues. A cell wall is another example of ECM. Cells of plants, fungi, and some protists have walls, but animal cells do not. Older plant cells secrete a rigid, **lignin**-containing **secondary wall** inside their pliable **primary wall**. Many eukaryotic cell types secrete protective **cuticle**.

Cell junctions structurally and functionally connect cells in tissues. **Plasmodesmata** connect the cytoplasm of adjacent plant cells. In animals, **gap junctions** are closable channels that connect adjacent cells. **Adhering junctions** that connect to cytoskeletal elements fasten animal cells to one another and to basement membrane. **Tight junctions** form a waterproof seal between cells.

Section 4.10 Differences among cell components allow us to categorize life, but not to define it. We can describe the quality of "life" as a set of properties that are collectively unique to living things.

Living systems consist of cells that harvest energy and matter from the environment. They also make and use complex carbon-based molecules; require water; grow; engage in self-sustaining biological processes; respond to stimuli; change over their lifetime, and over generations; and pass hereditary material (DNA) to offspring.

SELF-QUIZ

Answers in Appendix VII

1. All cells have these three things in common: ________ .
   a. cytoplasm, DNA, and organelles with membranes
   b. a plasma membrane, DNA, and a nucleus
   c. cytoplasm, DNA, and a plasma membrane
   d. a cell wall, cytoplasm, and DNA

2. Which of the following is *not* a principle of the cell theory?
   a. Every cell arises from another cell.
   b. A cell is alive even as part of a multicelled body.
   c. Eukaryotic cells have a nucleus, and prokaryotic cells do not.
   d. The cell is the smallest unit of life.

3. The surface-to-volume ratio ________ .
   a. does not apply to prokaryotic cells
   b. constrains cell size
   c. is part of the cell theory

4. A plasma membrane _______ .
   a. controls exchanges between the cell and the external environment
   b. is a mesh of intermediate filaments
   c. is a type of extracellular matrix

5. True or false? Ribosomes are only found in eukaryotes.

6. Unlike eukaryotic cells, prokaryotic cells ________ .
   a. have no nucleus
   b. have RNA but not DNA
   c. have no ribosomes
   d. a and c

7. What controls the passage of molecules into and out of the nucleus?
   a. endoplasmic reticulum, an extension of the nucleus
   b. nuclear pores, which consist of membrane proteins
   c. nucleoli, in which ribosome subunits are made
   d. dynamically assembled microtubules
   e. tight junctions

8. The main function of the endomembrane system is:
   a. building and modifying proteins and lipids
   b. isolating DNA from toxic substances
   c. secreting extracellular matrix onto the cell surface
   d. producing ATP by aerobic respiration

9. Enzymes contained in ________ break down worn-out organelles, bacteria, and other particles.
   a. lysosomes   c. endoplasmic reticulum
   b. amyloplasts   d. peroxisomes

10. Put the following structures in order according to the pathway of a secreted protein:
    a. plasma membrane   c. endoplasmic reticulum
    b. Golgi bodies   d. post-Golgi vesicles

11. Which of the following statements is false?
    a. The plasma membrane is the outermost component of all cells.
    b. The cell is the basic unit of life.
    c. All living cells contain cytoplasm.
    d. Some eukaryotes are single-celled and others are multicelled.

12. Cell junctions called ________ connect the cytoplasm of adjacent plant cells.
    a. plasmodesmata   d. adhesion proteins
    b. adhering junctions   e. chloroplasts
    c. tight junctions   f. gap junctions

13. Which of the following organelles contains no DNA?
    a. nucleus   c. mitochondrion
    b. Golgi body   d. chloroplast

14. No animal cell has a ________ .
    a. plasma membrane   c. lysosome
    b. flagellum   d. cell wall

15. Match each cell component with a function.

___	mitochondrion	a. protein synthesis
___	chloroplast	b. breaks down debris
___	cell junction	c. stores starch
___	smooth ER	d. breaks down toxins
___	Golgi body	e. finishes proteins and lipids
___	rough ER	f. assembles phospholipids
___	peroxisome	g. photosynthesis
___	amyloplast	h. ATP production
___	flagellum	i. protection
___	cuticle	j. movement
___	lysosome	k. connection

CRITICAL THINKING

1. Explain the difference between cytosol and cytoplasm in both prokaryotes and eukaryotes.

2. In a classic episode of *Star Trek*, a gigantic amoeba engulfs an entire starship. Spock blows the cell to bits before it can reproduce. Think of at least one inaccuracy that a biologist would identify in this scenario.

3. In plants, the cell wall forms as a young plant cell secretes polysaccharides onto the outer surface of its plasma membrane. Being thin and pliable, this primary wall allows the cell to enlarge and change shape. In mature woody plants, cells in some tissues deposit a sturdy secondary wall onto the primary wall's inner surface. Why doesn't a secondary wall form on the outer surface of the primary wall?

4. What type of micrograph is shown below? Is the organism pictured prokaryotic or eukaryotic? How can you tell? Which structures can you identify?

CREDIT: (in text CT) P. L. Walne and J. H. Arnott, *Planta*, 77:325-354, 1967.

5 GROUND RULES OF METABOLISM

CORE CONCEPTS

 Pathways of Transformation

Organisms exchange matter and energy with the environment in order to grow, maintain themselves, and reproduce.

Energy cannot be created or destroyed, but it can be transferred and converted to other forms. With each transfer or conversion, some energy disperses. Maintaining the organization of living systems requires ongoing inputs of energy to offset this loss.

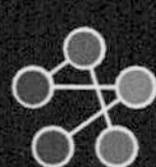 Systems

Complex properties arise from interactions among components of a biological system.

Cells store energy by building organic molecules, and retrieve energy by breaking them down. Many metabolic pathways couple energy-releasing reactions with energy-requiring reactions. The coupled reactions of electron transfer chains allow cells to harvest energy from electrons in small, manageable increments.

 Information Flow

Living systems store, retrieve, transmit, and respond to information essential for life.

Homeostasis depends on constant movement of substances across cell membranes, which are not permeable to ions or most polar molecules. Membrane proteins that allow cells to sense and respond appropriately to external conditions are a critical part of homeostatic control. Regulating the steps in metabolic pathways can quickly shift a cell's activities in response to changes in its internal and external conditions.

Links to Earlier Concepts

In this chapter, you will gain insight into the one-way flow of energy (Section 1.3) through the world of life (1.2) as you learn more about specific types of energy and the laws of nature (1.8) that describe it. This chapter also revisits atoms (2.2); electron energy (2.3); hydrogen bonding and water (2.5); pH (2.6); organic compounds, enzymes, and metabolism (3.2); carbohydrates (3.3); lipids (3.4); protein structure (3.5); and cell components (4.3, 4.4, 4.5, 4.6).

5.1 A Toast to Alcohol Dehydrogenase

Most college students are under the legal drinking age, but alcohol abuse continues to be the most serious drug problem on college campuses throughout the United States. Almost half of university undergraduates admit to regularly consuming five or more alcoholic beverages within a two-hour period—a self-destructive behavior called binge drinking. Drinking large amounts of alcohol in a brief period of time is an extremely risky behavior, both for the drinkers and people around them. Every year, around 600,000 students injure themselves while under the influence of alcohol; intoxicated students physically assault 690,000 people and sexually assault 97,000 others. About 1,800 students die from alcohol-related injuries.

Before you drink, consider what you are consuming. All alcoholic drinks—beer, wine, hard liquor, and so on—contain the same psychoactive ingredient: ethanol. Ethanol molecules move quickly from the stomach and small intestine into the bloodstream. Almost all of the ethanol ends up in the liver. Liver cells make an enzyme, alcohol dehydrogenase (ADH), that is part of a metabolic pathway that detoxifies alcohols.

If you drink more alcohol than your enzymes can deal with, then you will damage your body. Ethanol breakdown harms liver cells, so the more a person drinks, the fewer liver cells are left to do the breaking down. Ethanol also interferes with normal processes of metabolism. For example, oxygen that would ordinarily take part in breaking down fatty acids is diverted to breaking down the ethanol. As a result, fats tend to accumulate in large globules in the tissues of heavy drinkers.

Long-term heavy drinking causes alcoholic hepatitis, a disease characterized by inflammation and destruction of liver tissue. It also causes cirrhosis, a condition in which the liver becomes so scarred, hardened, and filled with fat that it loses its function. (The term cirrhosis is from the Greek *kirros*, meaning "orange-colored," after the abnormal skin color of people with the disease.) The liver has many important functions besides making ADH. In addition to breaking down fats and toxins, it helps regulate the body's blood sugar level, and it makes proteins essential for blood clotting, immune function, and maintaining the solute balance of body fluids. Loss of these functions can be deadly.

With this sobering example, we invite you to learn about the mechanisms that allow your cells to break down organic compounds, including toxic ones such as ethanol.

CREDIT: (opposite) Hristo Svinarov/Shutterstock.com.

5.2 Energy in the World of Life

LEARNING OBJECTIVES

- Describe entropy in terms of chemical bonding.
- Compare kinetic energy with potential energy.
- Use the first and second laws of thermodynamics to explain why the total amount of energy available for doing work in the universe always decreases.

Energy is formally defined as the capacity to do work, but this definition is not very satisfying. We do have an intuitive understanding of energy just by thinking about familiar forms of it, such as light, heat, electricity, and motion. We can also understand intuitively that one form of energy can be converted to another. Think about how a lightbulb changes electricity into light, or how an automobile changes gasoline into the energy of motion, which is called **kinetic energy**.

The formal study of heat and other forms of energy is thermodynamics (*therm* is a Greek word for heat; *dynam* means energy). By making careful measurements, thermodynamics researchers discovered that the total amount of energy before and after every conversion is always the same. In other words, energy cannot be created or destroyed—a phenomenon that is the **first law of thermodynamics**.

Energy also tends to spread out, or disperse, until no part of a system holds more than another part. In a kitchen, for example, heat always flows from a hot pan to cool air until the temperature of both is the same. We never see cool air raising the temperature of a hot pan. **Entropy** is a measure of how much the energy of a particular system has become dispersed. We can use the hot pan in a cool kitchen as an example of a system. As heat flows from the pan into the air, the entropy of the system increases (**FIGURE 5.1**). Entropy continues to increase until the heat is evenly distributed throughout the kitchen, and there is no longer a net (or overall) flow of heat from one area to another. Our system has now reached its maximum entropy with respect to heat. The tendency of entropy to increase is the **second law of thermodynamics.** This is the formal way of saying that energy tends to spread out spontaneously.

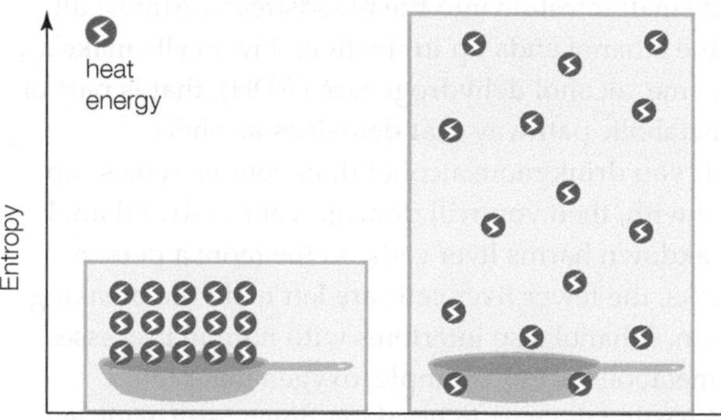
heat
energy
Entropy
Time

FIGURE 5.1 Entropy.

Entropy tends to increase, which means that energy tends to spread out spontaneously. The total amount of energy in any system stays the same.

FIGURE 5.2 Feed conversion ratio.

It takes more than 10,000 pounds of feed to raise a 1,000-pound steer. Where do the other 9,000 pounds go? About half of the steer's food is indigestible and passes right through it. The animal's body breaks down molecules in the remaining half to access energy stored in chemical bonds. Only about 15% of that energy goes toward building body mass. The rest is lost during energy conversions, as heat.

Biologists use the concept of entropy as it applies to chemical bonding, because energy flow in living things occurs mainly by the making and breaking of chemical bonds. How is entropy related to chemical bonding? Think about it just in terms of motion. Two unbound atoms can vibrate, spin, and rotate in every direction, so they are at high entropy with respect to motion. A covalent bond between the atoms restricts their movement, so they are able to move in fewer ways than they did before bonding. Thus, the entropy of two atoms decreases when a bond forms between them. Such entropy changes are part of the reason why some reactions occur spontaneously and others require an energy input, as you will see in the next section.

Work occurs as a result of energy transfers. Consider how it takes work to push a box across a floor. In this case, a body (you) transfers energy to another body (the box) to make it move. Similarly, a plant cell works to make sugars. Inside the cell, one set of molecules harvests energy from light, then transfers it to another set of molecules. The second set of molecules uses the energy to build the sugars from carbon dioxide

energy The capacity to do work.
entropy (EN-truh-pee) Measure of how much the energy of a system is dispersed.
first law of thermodynamics Energy cannot be created or destroyed.
kinetic energy (kih-NEH-tick) The energy of motion.
potential energy Stored energy.
second law of thermodynamics Energy tends to disperse spontaneously.

CREDIT: (2) Martin Metsemakers/Shutterstock.

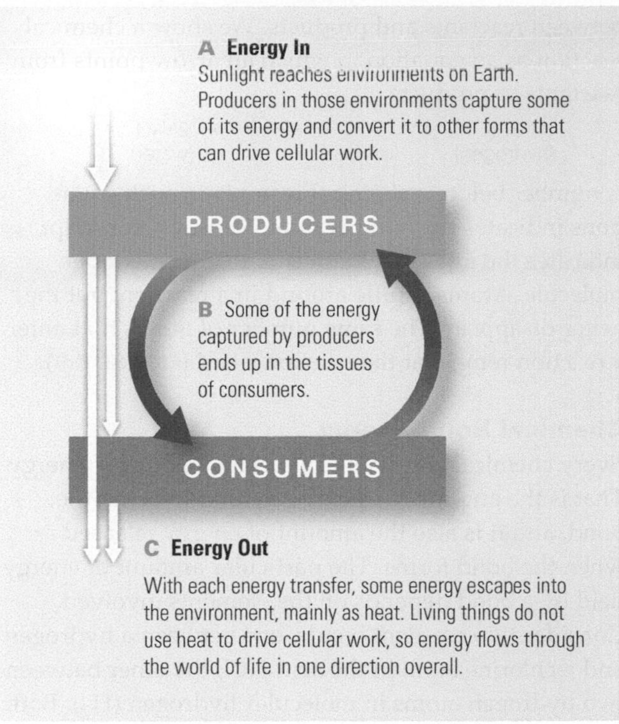
A Energy In
Sunlight reaches environments on Earth. Producers in those environments capture some of its energy and convert it to other forms that can drive cellular work.
PRODUCERS
B Some of the energy captured by producers ends up in the tissues of consumers.
CONSUMERS
C Energy Out
With each energy transfer, some energy escapes into the environment, mainly as heat. Living things do not use heat to drive cellular work, so energy flows through the world of life in one direction overall.

FIGURE 5.3 Energy movement through the world of life.

Energy (yellow arrows) flows from the environment into living organisms, then back to the environment. The flow drives a cycling of materials (green arrows) among producers and consumers.

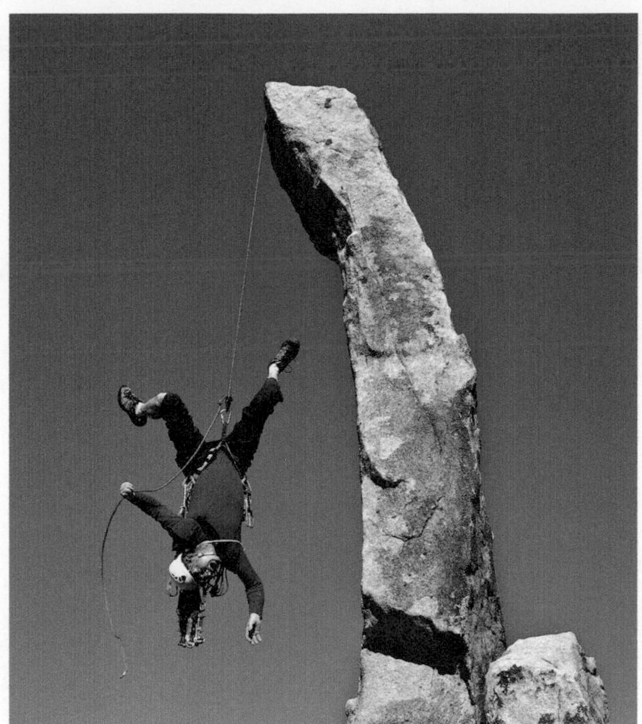

FIGURE 5.4 Illustration of potential energy.

By opposing gravity's downward pull, the rope attached to the rock keeps the man from falling. Similarly, a chemical bond attaches two atoms and keeps them from flying apart.

and water. This particular energy transfer involves the conversion of light energy to chemical energy. Most other types of cellular work occur by the transfer of chemical energy from one molecule to another.

Every time energy is transferred, a bit of it disperses—usually in the form of heat. As a simple example, an incandescent lightbulb converts only about 5 percent of the energy of electricity into light. The remaining 95 percent ends up as heat that disperses from the bulb.

Dispersed heat is not useful for doing work, and it is not easily converted to a more useful form of energy (such as electricity). Because some of the energy in every transfer disperses as heat, and dispersed heat is not useful for doing work, we can say that the total amount of energy available for doing work in the universe is always decreasing.

Is life an exception to this inevitable flow? An organized body is hardly dispersed. Energy becomes concentrated in each new organism as the molecules of life assemble and organize into cells. Even so, living things constantly use energy to grow, to move, to acquire nutrients, to reproduce, and so on. Some energy is lost in every one of these processes (FIGURE 5.2). Unless the losses are replenished with energy from another source, life's complex organization will end.

The energy that fuels most life on Earth comes from the sun. That energy flows through producers, then consumers (FIGURE 5.3). During its journey, the energy is transferred many times. With each transfer, some energy escapes as heat until, eventually, all of it is permanently dispersed. However, the second law of thermodynamics does not specify how quickly entropy occurs. Energy's spontaneous dispersal is resisted by chemical bonds. The energy in chemical bonds is a type of **potential energy**, which is energy stored in the position or arrangement of objects in a system (FIGURE 5.4). Think of all the bonds in the countless molecules that make up your skin, heart, liver, and other body parts. Those bonds hold the molecules, and you, together—at least for the time being.

TAKE-HOME MESSAGE 5.2

- ✔ Energy, which is the capacity to do work, cannot be created or destroyed.
- ✔ Energy disperses spontaneously.
- ✔ Energy can be transferred between systems or converted into another form, but some is lost (usually as heat) during each exchange.
- ✔ Sustaining life's organization requires ongoing energy inputs to counter energy loss. Organisms stay alive by replenishing themselves with energy they harvest from someplace else.

5.3 Energy in the Molecules of Life

LEARNING OBJECTIVES

- Explain chemical bond energy.
- Distinguish between endergonic and exergonic reactions.
- Describe activation energy.
- Explain how cells store and harvest energy using chemical reactions.

Remember from Section 3.2 that reactions change molecules into other molecules. All cells store and retrieve energy in chemical bonds of the molecules of life, and these activities occur by way of reactions.

During a reaction, one or more **reactants** (molecules that enter a reaction and become changed by it) become one or more **products** (molecules that are produced by the reaction). Intermediate molecules may form between reactants and products. We show a chemical reaction as an equation in which an arrow points from reactants to products:

$$\underset{\text{(hydrogen)}}{2H_2} + \underset{\text{(oxygen)}}{O_2} \longrightarrow \underset{\text{(water)}}{2H_2O}$$

A number before a chemical formula in such equations indicates the number of molecules; a subscript indicates the number of atoms of that element per molecule. Atoms shuffle around in a reaction, but they never disappear: The same number of atoms that enter a reaction remain at the reaction's end (**FIGURE 5.5**).

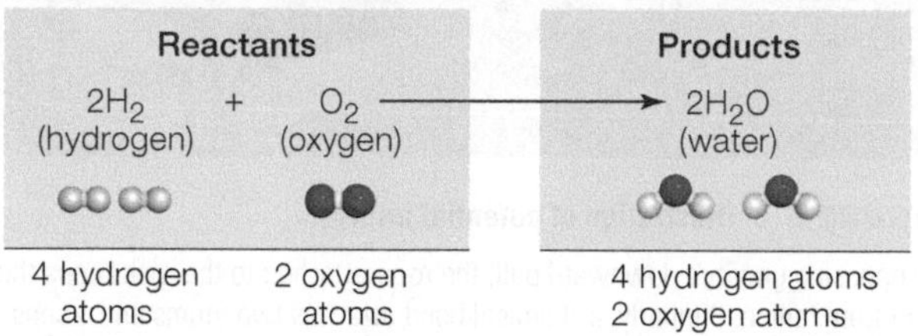
Reactants
Products
$2H_2$ + O_2 → $2H_2O$
(hydrogen)
(oxygen)
(water)
4 hydrogen atoms
2 oxygen atoms
4 hydrogen atoms
2 oxygen atoms

FIGURE 5.5 Chemical bookkeeping. In equations that represent chemical reactions, reactants are written to the left of an arrow that points to the products. A number before a formula indicates the number of molecules. Atoms may shuffle around, but the same number of atoms that enter a reaction remain at its end.

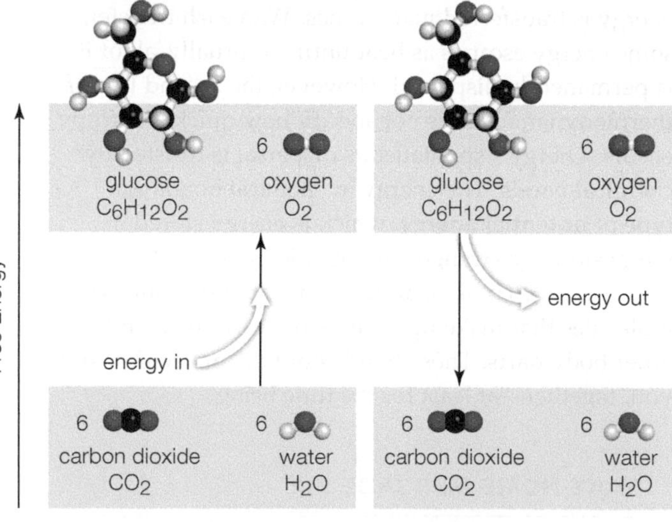
Free Energy
6
glucose
$C_6H_{12}O_2$
oxygen
O_2
energy in
energy out
6
carbon dioxide
CO_2
6
water
H_2O

A Endergonic reactions convert molecules with lower free energy to molecules with higher free energy, so they require a net energy input to proceed.

B Exergonic reactions convert molecules with higher free energy to molecules with lower free energy, so they end with a net energy release.

FIGURE 5.6 The ins and outs of energy in chemical reactions.

FIGURE IT OUT Which law of thermodynamics explains energy inputs and outputs in chemical reactions?

Answer: The first law

Chemical Bond Energy

Every chemical bond holds a certain amount of energy. That is the amount of energy required to break the bond, and it is also the amount of energy released when the bond forms. The particular amount of energy held by a bond depends on the elements involved. Consider two covalent bonds, one between a hydrogen and a chlorine atom in the acid HCl, the other between two hydrogen atoms in molecular hydrogen (H_2). Both of these bonds hold energy, but different amounts of it.

Bond energy and entropy both contribute to a molecule's free energy, which is the amount of energy that is available ("free") to do work. In most reactions, the free energy of reactants differs from the free energy of products. If the reactants have less free energy than the products, the reaction will not proceed without a net energy input. Such reactions are **endergonic**, which means "energy in" (**FIGURE 5.6A**). If the reactants have more free energy than the products, the reaction will end with a net release of energy. Such reactions are **exergonic**, which means "energy out" (**FIGURE 5.6B**).

Why Earth Does Not Go Up in Flames

The molecules of life release energy when they combine with oxygen. Think of how a spark ignites wood in a campfire. Wood is mostly cellulose, which consists of long chains of repeating glucose monomers (Section 3.3). A spark starts a reaction that converts cellulose (in wood) and oxygen (in air) to water and carbon dioxide. The reaction is highly exergonic, which means it releases a lot of energy—enough to initiate the same reaction with other cellulose and oxygen molecules. That is why wood keeps burning after it has been lit (**FIGURE 5.7A**).

Earth is rich in oxygen—and in potential exergonic reactions. Why doesn't it burst into flames? Luckily for us, chemical bonds do not break without at least a small input of energy, even in an energy-releasing reaction. We call this input activation energy. **Activation energy**, the minimum amount of energy required to get a chemical reaction started, is a bit like a hill that reactants must climb before they can coast down the other side to become products (**FIGURE 5.7B**).

Both endergonic and exergonic reactions have activation energy, but the amount varies with the reaction. Consider guncotton (nitrocellulose), a highly explosive derivative of cellulose. Christian Schönbein accidentally discovered a way to manufacture it when he used his wife's cotton apron to wipe up a nitric acid spill on his kitchen table, then hung it up to dry next to the oven. The apron exploded, and being a chemist in the 1800s, Schönbein was thrilled. He immediately tried marketing guncotton as a firearm explosive, but it proved too unstable to manufacture. So little activation energy is needed to make guncotton react with oxygen that it tends to explode unexpectedly. Several manufacturing plants burned to the ground before guncotton was abandoned for use as a firearm explosive. The substitute? Gunpowder, which has a higher activation energy for a reaction with oxygen.

Energy In, Energy Out

Cells store energy by running endergonic reactions that build organic compounds (**FIGURE 5.8A**). For example, light energy drives the overall reactions of photosynthesis, which produce sugars such as glucose from carbon dioxide and water. Unlike light, glucose (and the energy in its bonds) can be stored in a cell. Cells release stored energy by running exergonic reactions that break the bonds of organic compounds (**FIGURE 5.8B**). Most cells do this when they carry out the overall reactions of aerobic respiration, which releases the energy of glucose by breaking the bonds between its carbon atoms. You will see in the next sections how cells use energy released from some reactions to drive others.

TAKE-HOME MESSAGE 5.3

- ✔ Endergonic reactions will not run without a net input of energy. Exergonic reactions end with a net release of energy.
- ✔ Both endergonic and exergonic reactions require an input of activation energy to begin.
- ✔ Cells store energy in chemical bonds by running endergonic reactions that build organic compounds. To release this stored energy, they run exergonic reactions that break the bonds.

activation energy Minimum amount of energy required to start a reaction.
endergonic (end-er-GON-ick) Describes a reaction that requires a net input of free energy to proceed.
exergonic (ex-er-GON-ick) Describes a reaction that ends with a net release of free energy.
product A molecule that is produced by a reaction.
reactant (ree-ACK-tunt) A molecule that enters a reaction and is changed by participating in it.

A Wood continues to burn after it has been lit. The reaction between cellulose and oxygen releases enough energy to trigger the reaction again with other molecules. Activation energy keeps this and other exergonic reactions from starting without an energy input.

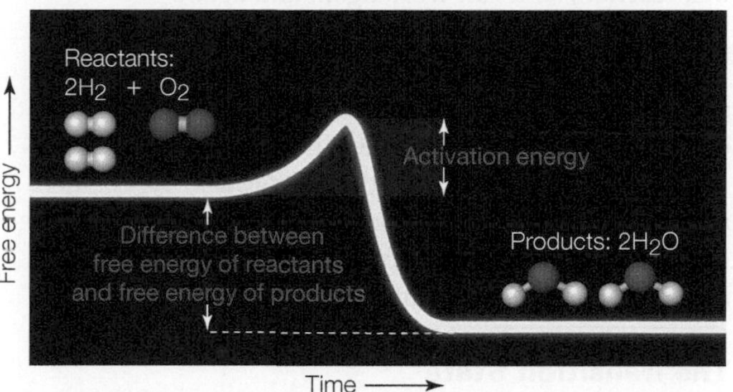
Reactants:
$2H_2$ + O_2
Activation energy
Difference between free energy of reactants and free energy of products
Products: $2H_2O$
Free energy
Time

B Most reactions will not begin without at least a small input of energy. The minimum energy required to start a reaction is called activation energy, and it is shown on a graph as an energy hill. This graph shows an energy-releasing reaction; energy-requiring reactions also have activation energy.

FIGURE 5.7 Activation energy.

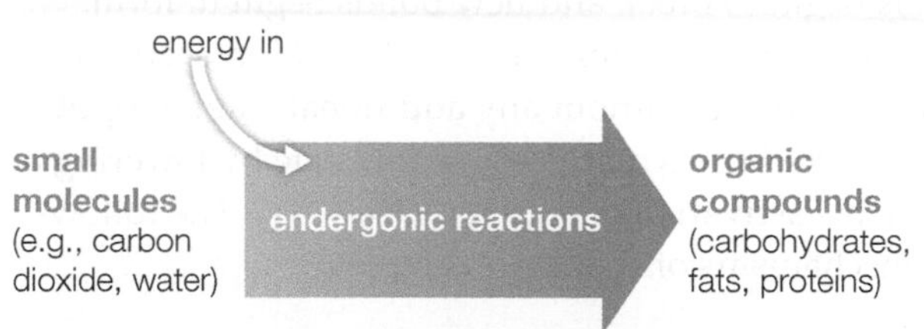
energy in
small molecules (e.g., carbon dioxide, water)
endergonic reactions
organic compounds (carbohydrates, fats, proteins)

A Cells run endergonic reactions to store energy in the bonds of organic compounds.

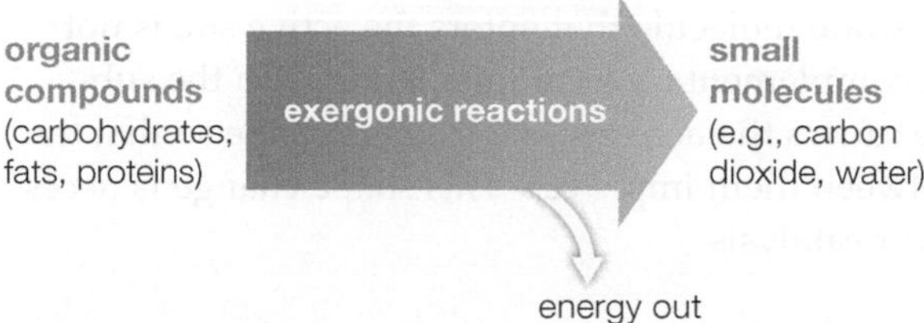
organic compounds (carbohydrates, fats, proteins)
exergonic reactions
small molecules (e.g., carbon dioxide, water)
energy out

B Cells run exergonic reactions to retrieve energy stored in the bonds of organic compounds.

FIGURE 5.8 Cells store and retrieve energy in bonds.

5.4 How Enzymes Work

LEARNING OBJECTIVES

- Explain enzyme specificity.
- Describe four ways that enzymes speed reactions.
- Using appropriate examples, explain how environmental factors affect enzyme activity.

Metabolism requires enzymes. Why? Consider the breakdown of sugar to carbon dioxide and water. This reaction can occur on its own, but it might take decades. The same conversion takes just seconds inside your cells. Enzymes make the difference. In a process called **catalysis**, an enzyme makes a reaction run much faster than it would on its own (in other words, it catalyzes the reaction). An enzyme is unchanged by participating in a reaction, so it can work again and again.

Some enzymes are RNAs, but most are proteins. Each has an **active site**, which is a pocket where catalysis occurs (FIGURE 5.9). An enzyme acts only on molecules that "fit" its active site—those complementary in shape, size, polarity, and charge (FIGURE 5.10). Such molecules are called the enzyme's **substrates**. Some enzymes have multiple substrates, but each catalyzes a specific reaction or set of reactions.

The Transition State

When we talk about activation energy, we are really talking about the energy required to bring reactant bonds to their breaking point. Before a reaction, there are molecules of reactants. Afterward, there are molecules of products. What happens between? At a particular moment between reactants and products, existing bonds begin to break and new bonds begin to form. At that in-between moment—the **transition state**—the reaction can run without any additional energy input. An enzyme brings on the transition state by lowering a reaction's activation energy (FIGURE 5.11). The following mechanisms offer some examples.

Inducing a Fit Between Enzyme and Substrate

By the **induced-fit model**, the "perfect" fit between active site and substrate occurs at the transition state. A substrate molecule that enters the active site is not quite complementary to it; interacting with the substrate causes the active site to change shape so that the fit between them improves. This shape change is necessary for catalysis.

Forcing Substrates Together Entering an active site brings two substrates in close physical proximity. The closer the substrates are to one another, the more likely they are to react.

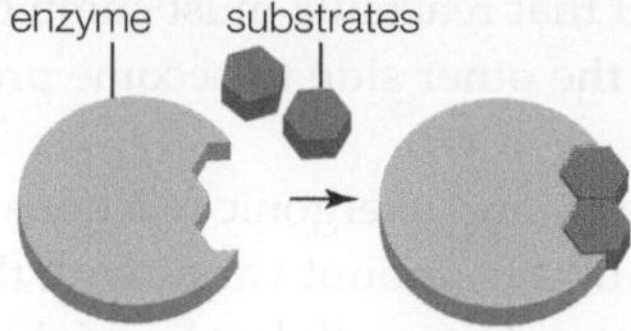

enzyme
substrates

A An enzyme can act only on molecules that "fit" its active site. Such molecules are called the enzyme's substrates.

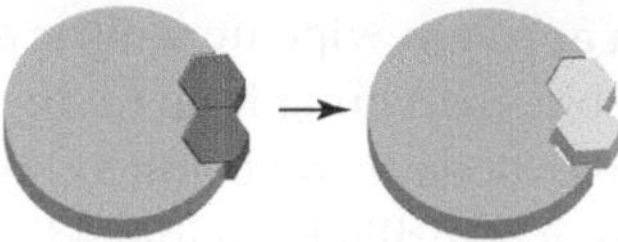

B An active site squeezes substrates together, influences their charge, or causes some change that lowers activation energy, so the reaction proceeds.

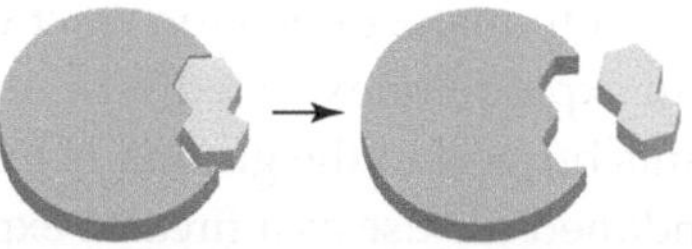

C The product leaves the active site after the reaction is finished. The enzyme is unchanged, so it can work repeatedly.

FIGURE 5.9 **How an active site works.** Each enzyme acts only on molecules (substrates) that "fit" its active site. For simplicity, active sites are often depicted as blobs or geometric shapes.

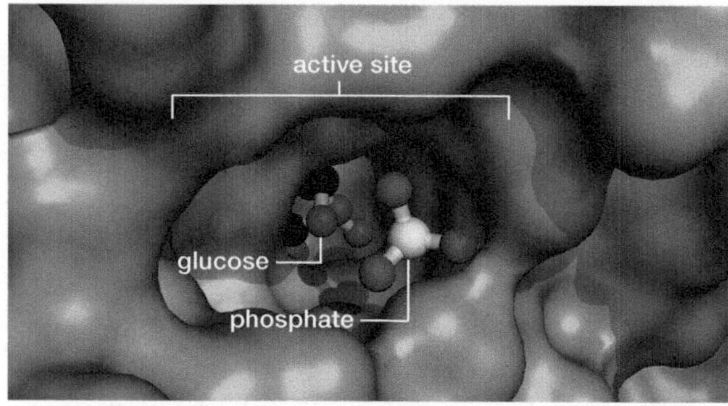

active site
glucose
phosphate

A A glucose molecule meets up with a phosphate in the active site of a hexokinase molecule (tan).

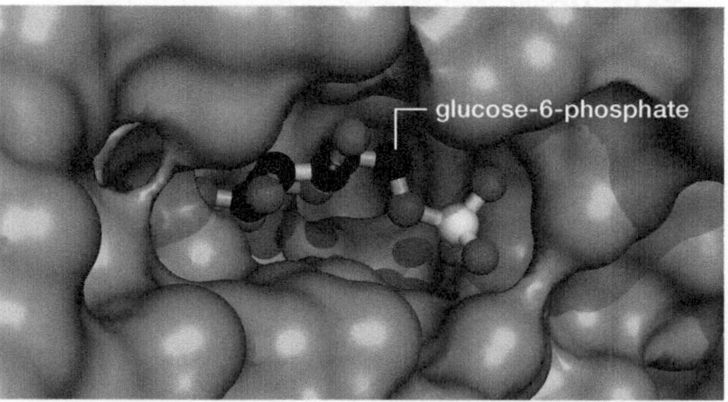

glucose-6-phosphate

B The enzyme has catalyzed the reaction between glucose and phosphate. The product of the reaction, glucose-6-phosphate, is shown leaving the active site. Notice how the shape of the active site has changed.

FIGURE 5.10 **Example of an active site.** The models show the actual contours of an active site in hexokinase, an enzyme that adds a phosphate group to glucose and other six-carbon sugars.

Orienting Substrates in Positions That Favor Reaction Substrate molecules in a solution collide from random directions. By contrast, an active site orients substrates in positions that favor reaction.

Excluding Water The active site of some enzymes repels water when a substrate binds. Intermediates of some reactions are unstable in water; minimizing the presence of water molecules allows these intermediates to persist a bit longer, thus increasing the rate of the final reaction.

Environmental Influences

Environmental factors such as pH, temperature, and salt concentration influence an enzyme's shape, which in turn influences its function (Section 3.5). Each enzyme works best in a particular range of conditions that reflect the environment in which it evolved.

Consider two enzymes that play a role in human digestion (**FIGURE 5.12A**). The enzyme pepsin, which works best at low pH, begins the process of protein breakdown in the very acidic environment of the stomach (pH 2). During digestion, the stomach's contents pass into the small intestine, where the pH rises to about 7.5. Pepsin denatures (unfolds) above pH 5.5, so this enzyme becomes inactivated in the small intestine. Protein breakdown continues with the assistance of trypsin, an enzyme that functions well at the higher pH.

Adding heat boosts free energy, which is why the jiggling motion of atoms and molecules increases with temperature (Section 2.5). The greater the free energy of reactants, the closer they are to reaching activation energy. Thus, the rate of an enzymatic reaction typically increases with temperature—but only up to a point. An enzyme denatures above a characteristic temperature. Then, the reaction rate falls sharply as the shape of the enzyme changes and it stops working (**FIGURE 5.12B**). Body temperatures above 42°C (107.6°F) adversely affect the function of many of your enzymes, which is why severe fevers are dangerous.

The activity of enzymes is also influenced by the amount of salt in the solution. Too little salt, and polar

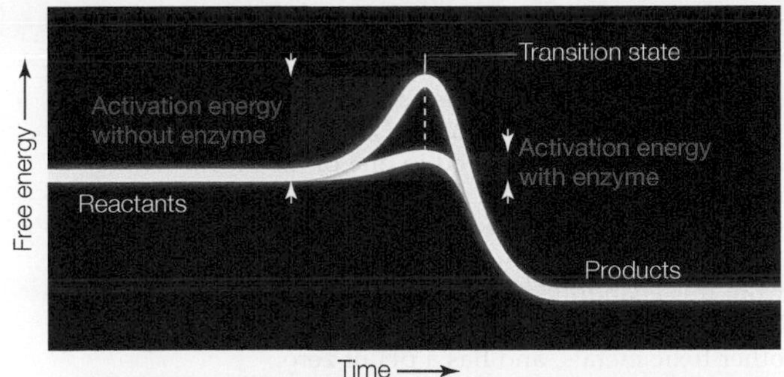
Transition state
Activation energy without enzyme
Activation energy with enzyme
Reactants
Products
Free energy
Time

FIGURE 5.11 The transition state. An enzyme increases the rate of a reaction by lowering activation energy, thus bringing on the transition state.

FIGURE IT OUT Is this reaction endergonic or exergonic?

Answer: It is exergonic.

A The pH-dependent activity of two digestive enzymes. Pepsin acts in the stomach, where the normal pH is 2. Trypsin acts in the small intestine, where the pH is normally around 7.5.

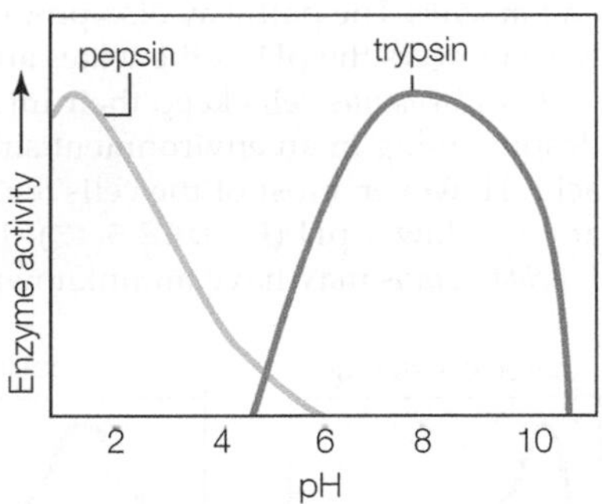
pepsin
trypsin
Enzyme activity
2
4
6
8
10
pH

B The temperature-dependent activity of a DNA synthesis enzyme from two species of bacteria: *E. coli*, which inhabits the human gut (normally 37°C); and *Thermus aquaticus*, which lives in hot springs around 70°C.

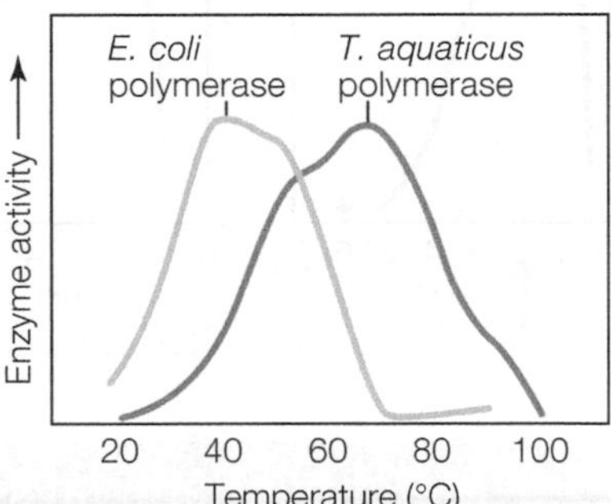
E. coli polymerase
T. aquaticus polymerase
Enzyme activity
20
40
60
80
100
Temperature (°C)

FIGURE 5.12 Enzymes, temperature, and pH. Each enzyme functions best within a characteristic range of conditions—generally, the same conditions that occur in the environment where the enzyme is normally found.

parts of the enzyme attract one another so strongly that the enzyme's shape changes. Too much salt interferes with the hydrogen bonds that hold the enzyme in its characteristic shape, and the enzyme denatures.

active site Pocket in an enzyme where substrates react and are converted to products.

catalysis (cut-AL-ih-sis) The acceleration of a reaction rate by a molecule that is unchanged by participating in the reaction.

induced-fit model Substrate binding to an active site improves the fit between the two; this fit is necessary for catalysis.

substrate Molecule that an enzyme acts upon and converts to a product; reactant in an enzyme-catalyzed reaction.

transition state During a reaction, the moment at which substrate bonds are at the breaking point and the reaction will run spontaneously.

TAKE-HOME MESSAGE 5.4

- ✔ Enzymes catalyze (greatly enhance the rate of) specific reactions.
- ✔ Binding at an enzyme's active site causes a substrate to reach its transition state. In this state, the substrate's bonds are at the breaking point, and the reaction can run spontaneously to completion.
- ✔ Each enzyme works best within a certain range of conditions that include temperature, pH, pressure, and salt concentration.

Data Analysis Activities

One Tough Bug The genus *Ferroplasma* consists of a few species of acid-loving archaea. One species, *Ferroplasma acidarmanus*, was discovered in one of the most contaminated sites in the United States: Iron Mountain Mine in California. *F. acidarmanus* is the main constituent of slime streamers (a type of biofilm) growing in water draining from this abandoned copper mine (right). The water is hot (about 40°C, or 104°F), heavily laden with arsenic and other toxic metals, and has a pH of zero.

F. acidarmanus cells have an ancient energy-harvesting pathway that uses electrons pulled from iron–sulfur compounds in minerals such as pyrite. Removing electrons from these compounds dissolves the minerals, so groundwater in the mine ends up with extremely high concentrations of solutes, including metal ions such as copper, zinc, cadmium, and arsenic. The pathway also produces sulfuric acid, which lowers the pH of the water around the cells to zero.

F. acidarmanus cells keep their internal pH at a cozy 5.0 despite living in an environment similar to hot battery acid. However, most of the cells' enzymes function best at much lower pH (**FIGURE 5.13**). Thus, researchers think *F. acidarmanus* may have an unknown type of internal compartment that keeps their enzymes in a highly acidic environment.

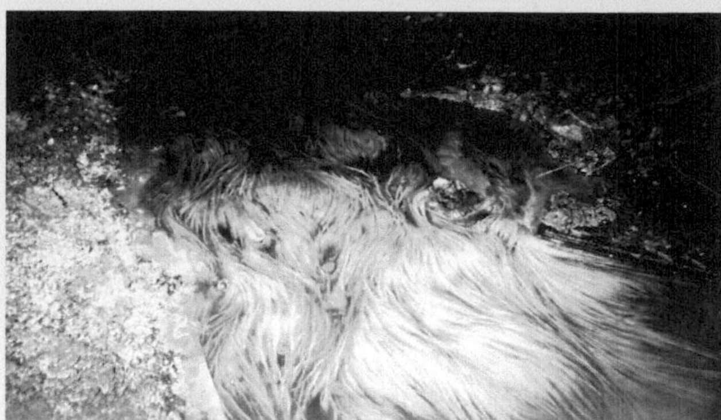

1. What do the dashed lines in the graphs signify?
2. Of the four enzymes profiled in the graphs, how many function optimally at a pH lower than 5? How many retain significant function at pH 5?
3. What is the optimal pH for the carboxylesterase?

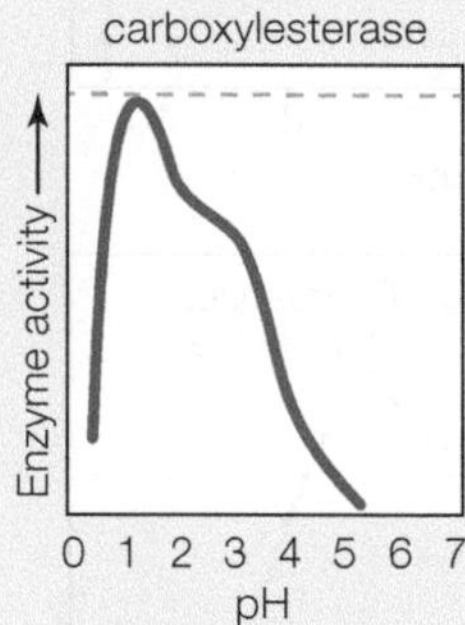

carboxylesterase
Enzyme activity
0 1 2 3 4 5 6 7
pH

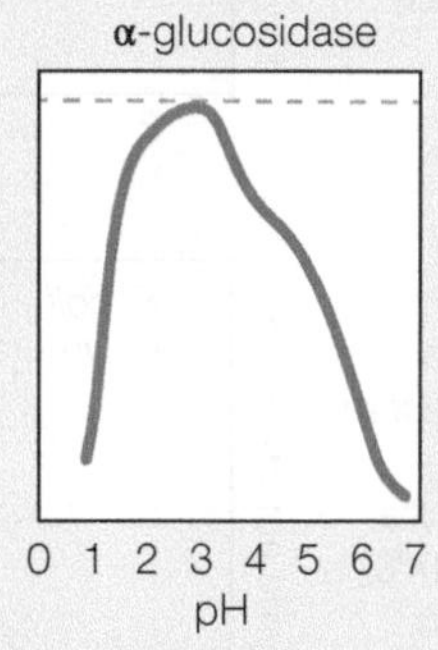

α-glucosidase
0 1 2 3 4 5 6 7
pH

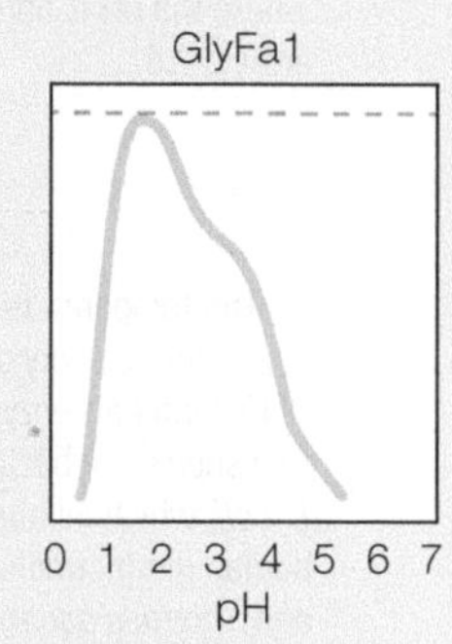

GlyFa1
0 1 2 3 4 5 6 7
pH

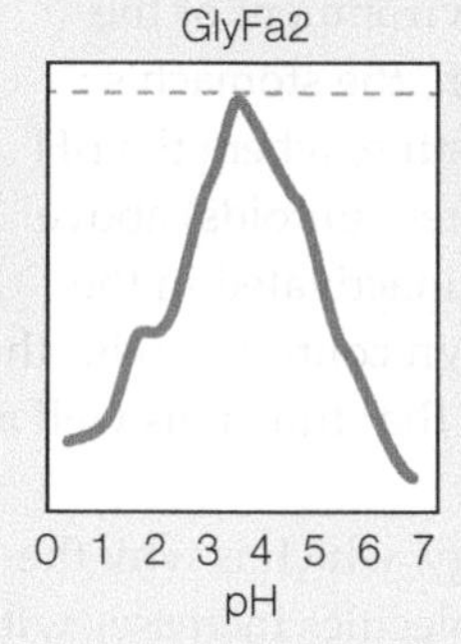

GlyFa2
0 1 2 3 4 5 6 7
pH

FIGURE 5.13 pH anomaly of *Ferroplasma acidarmanus*.

Left, graphs showing pH activity profiles of four enzymes isolated from *Ferroplasma acidarmanus*. Researchers had expected all of these enzymes to function best at the cells' cytoplasmic pH (5.0).

5.5 Metabolic Pathways

LEARNING OBJECTIVES

- Describe metabolic pathways.
- Use an example to explain how cells can control enzyme activity.
- Describe how feedback inhibition affects a metabolic pathway.
- Explain the role of redox reactions in electron transfer chains.

Building, rearranging, or breaking down an organic molecule often occurs stepwise, in a series of enzymatic reactions called a **metabolic pathway**. Many pathways are quite complex, involving thousands of molecules, and they interconnect in an even more complex network that constitutes the cell's metabolism. Some pathways are linear, meaning that the reactions run straight from reactant to product (**FIGURE 5.14A**). Others are cyclic. In a cyclic pathway, the last step regenerates a reactant of the first step (**FIGURE 5.14B**). Both linear and cyclic pathways are common in cells; later chapters detail the steps in some important pathways.

Controls Over Metabolism

A cell conserves energy and resources by making only what it needs at any given moment—no more, no less. Several mechanisms help it maintain, raise, or lower the production of thousands of different substances. Consider that reactions do not only run from reactants to products. Many also run in reverse at the same time, with some products being converted back to reactants:

$$\text{reactant} \rightleftharpoons \text{product}$$

The rates of the forward and reverse reactions often depend on the relative concentrations of reactants and products. A high concentration of reactants favors the forward reaction, and a high concentration of products favors the reverse reaction:

reactant ⟶ product

reactant ⟵ **product**

CREDIT: (in text) Katrina J. Edwards.

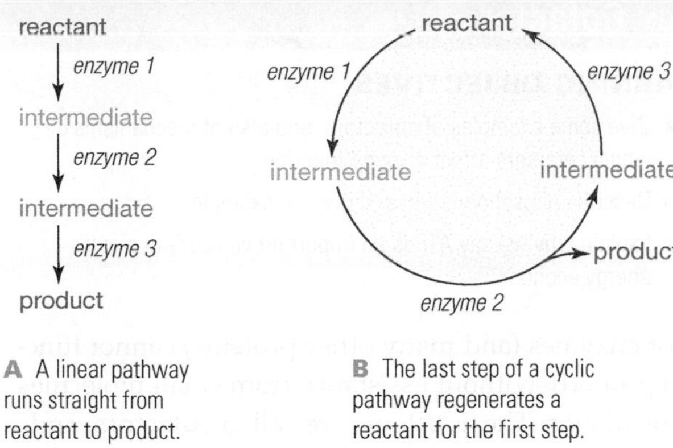
reactant
enzyme 1
intermediate
enzyme 2
intermediate
enzyme 3
product
reactant
enzyme 1
enzyme 3
intermediate
intermediate
product
enzyme 2

A A linear pathway runs straight from reactant to product.

B The last step of a cyclic pathway regenerates a reactant for the first step.

FIGURE 5.14 Linear and cyclic metabolic pathways.

Other mechanisms more actively regulate metabolic pathways. The catalytic activity of many enzymes changes when specific ions or molecules bind to them, for example. These substances act as regulators that enhance or inhibit an enzyme's activity. Consider how some enzymes are active only when they have an attached phosphate group; such enzymes can be "switched on" by adding a phosphate group, and "switched off" by removing it. The location of a regulatory binding site varies among enzymes; if it occurs outside of the active site (**FIGURE 5.15**), the mechanism of control is called **allosteric regulation** (*allo–* means other; *–steric* means structure).

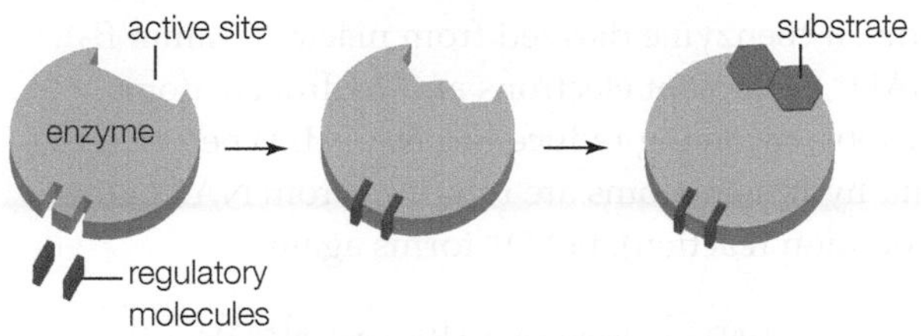
active site
substrate
enzyme
regulatory
molecules

FIGURE 5.15 Example of allosteric regulation.

Specific ions or molecules may regulate the activity of an enzyme by binding to it. If they bind outside of the active site, the regulation is allosteric.

FIGURE IT OUT Is this enzyme's activity enhanced or inhibited by regulatory molecule binding? Answer: Enhanced

allosteric regulation (al-oh-STARE-ick) Regulation of enzyme activity by the binding of a specific ion or molecule outside the active site.
electron transfer chain In a cell membrane, a series of enzymes and other molecules that accept and give up electrons in turn, thus releasing the energy of the electrons in steps.
feedback inhibition Regulatory mechanism in which a change that results from some activity decreases or stops the activity.
metabolic pathway A series of enzyme-mediated reactions by which cells build, remodel, or break down an organic molecule.
redox reaction (REE-docks) Oxidation–reduction reaction. In a typical redox reaction, one molecule accepts electrons from another molecule. The transfer oxidizes the electron donor and reduces the electron acceptor.

Changes in the activity of a single enzyme can affect an entire metabolic pathway. In some cases, the end product of a series of enzymatic reactions inhibits the activity of one of the enzymes in the series (**FIGURE 5.16**). This type of regulatory mechanism, in which a change that results from an activity decreases or stops the activity, is called **feedback inhibition.**

A pathway that synthesizes the amino acid serine is regulated by feedback inhibition. The first enzyme in the pathway has four binding sites for serine. When all four of these sites are occupied by serine molecules, the enzyme becomes inactive. When serine is being used faster than it is being produced (which can occur during times of rapid protein synthesis), its concentration in cytoplasm falls. Then, the enzyme releases the bound serines and becomes active again.

reactant
enzyme 1
intermediate
enzyme 2
product

FIGURE 5.16 Feedback inhibition. In this example, three different enzymes act in sequence to convert a substrate to a product. The product inhibits the activity of the first enzyme.

FIGURE IT OUT Is this pathway cyclic or linear? Answer: Linear

Electron Transfers

Capturing and harvesting energy can be dangerous for cells. For example, the bonds of organic molecules such as glucose hold enough energy to harm a cell if released all at once, as occurs during combustion (burning) reactions with oxygen. Most cells use oxygen to break the bonds of organic molecules, but they have no way to harvest the uncontrolled burst of energy released by combustion. Instead, they break the bonds of an organic molecule one by one, in steps that release energy in small, manageable amounts. These steps are reactions in metabolic pathways, and most are redox (oxidation–reduction) reactions. A typical **redox reaction** is an electron transfer, in which an electron is transferred from one molecule to another. The transfer oxidizes the electron donor, and reduces the electron acceptor. To remember what reduced means, think of how the negative charge of an electron "reduces" the charge of the acceptor.

When an electron is transferred, it loses energy. In Chapters 6 and 7, you will learn why this energy loss is important in electron transfer chains. An **electron transfer chain** is a series of membrane-bound enzymes and other molecules that give up and accept electrons in turn. Electrons are at a higher energy level (Section 2.3) when they enter a chain than when they leave. Energy given off by an electron

CREDITS: (14, 15, 16) © Cengage Learning.

A Right, combustion of glucose, a highly exothermic redox reaction in which electrons are transferred directly from oxygen to glucose. The transfer breaks the bonds of both molecules and releases energy as light and heat.

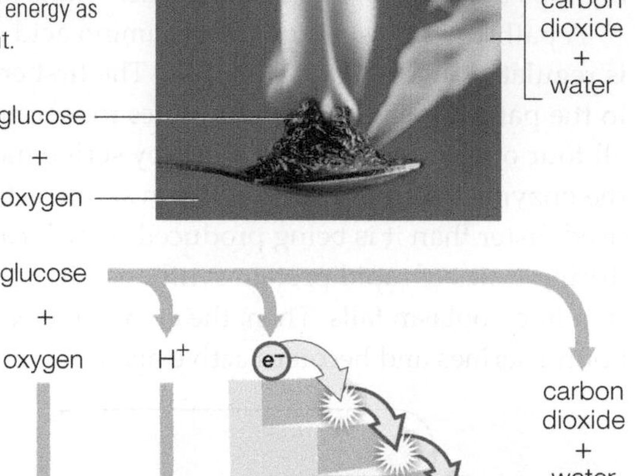
energy
carbon dioxide + water
glucose + oxygen
glucose + oxygen
H+
e−
carbon dioxide + water
e−

B Above, the bond energy of glucose is released stepwise, in small amounts that cells are able to harness.

Glucose reacts with oxygen in a series of redox reactions carried out by an electron transfer chain (represented here as a staircase). Electrons released at each step lose energy ( ) as they move through an electron transfer chain. Energy released by the electrons is harnessed for cellular work.

FIGURE 5.17 Comparing uncontrolled (A) and controlled (B) energy release. The overall reaction is the same in both cases:

$$C_6H_{12}O_6 + O_2 \longrightarrow CO_2 + H_2O + \text{energy}$$
glucose oxygen carbon dioxide water energy

as it drops to a lower energy level is harvested by molecules of the electron transfer chain to do cellular work (FIGURE 5.17). You will see in later chapters the importance of electron transfer chains that help power ATP synthesis.

TAKE-HOME MESSAGE 5.5

✔ A metabolic pathway is a stepwise series of enzyme-mediated reactions.

✔ Cells conserve energy and resources by producing only what they need at a given time. Such metabolic control can arise from mechanisms (such as regulatory molecule binding to an enzyme) that start, stop, or alter the rate of a single reaction. Other mechanisms (such as feedback inhibition) can influence an entire pathway.

✔ Many metabolic pathways involve electron transfers. Electron transfer chains are sites of energy exchange.

5.6 Cofactors

LEARNING OBJECTIVES

- Give some examples of cofactors, and also of mechanisms by which cofactors affect enzyme function.
- Describe phosphorylation and give an example.
- Explain why we say ATP is an important currency in a cell's energy economy.

Most enzymes (and many other proteins) cannot function properly without assistance from small molecules or metal ions. The assistants are called **cofactors,** and they may bind to an enzyme in its active site or in an allosteric site. Some essential dietary vitamins and minerals are cofactors; many others are converted into cofactors when they enter the body.

Metal ions can act as cofactors. These ions can stabilize enzyme structure, which means an enzyme denatures if the ions are removed. Metals also play a functional role in many reactions by interacting with electrons in nearby atoms. A metal cofactor can help bring on the transition state by donating or accepting electrons, or simply by tugging on them.

Organic cofactors are called **coenzymes** (TABLE 5.1). Coenzymes carry chemical groups, atoms, or electrons from one reaction to another, and often into or out of organelles. Unlike enzymes, many coenzymes are modified by taking part in a reaction. They are regenerated in separate reactions.

Consider NAD^+ (nicotinamide adenine dinucleotide), a coenzyme derived from niacin (vitamin B_3). NAD^+ can accept electrons and hydrogen atoms, thereby becoming reduced to NADH. When electrons and hydrogen atoms are removed from NADH (an oxidation reaction), NAD^+ forms again:

$$NAD^+ + \text{electrons} + H^+ \longrightarrow \mathbf{NADH}$$

$$\mathbf{NADH} \longrightarrow NAD^+ + \text{electrons} + H^+$$

In some reactions, cofactors participate by temporarily associating with the enzyme. In others, they stay tightly bound. Catalase, an enzyme of peroxisomes, has four tightly bound cofactors called hemes. A heme is a small organic compound with an iron atom at its center (FIGURE 5.18). Catalase's substrate, hydrogen peroxide (H_2O_2), is a highly reactive molecule that forms during

antioxidant Substance that prevents oxidation of other molecules.
ATP/ADP cycle Process by which cells use and regenerate ATP. ADP forms when ATP loses a phosphate group. ATP forms when ADP is phosphorylated.
coenzyme An organic cofactor.
cofactor (KO-fack-ter) A coenzyme or metal ion that associates with an enzyme and is necessary for its function.
phosphorylation Addition of a phosphate group to an organic molecule.

CREDIT: (17) left, Martyn F. Chillmaid/Science Source; right, © Cengage Learning.

TABLE 5.1

Some Common Coenzymes

Coenzyme	Example of Function
Ascorbic acid (vitamin C)	Collagen fiber formation; carries electrons during peroxide breakdown (in lysosomes)
ATP	Transfers energy with a phosphate group
NAD, NAD^+	Carries electrons during glycolysis
NADP, NADPH	Carries electrons, hydrogen atoms during photosynthesis
FAD, FADH, $FADH_2$	Carries electrons during aerobic respiration
CoA	Carries acetyl group ($COCH_3$) during glycolysis
Coenzyme Q10	Carries electrons in electron transfer chains of aerobic respiration
Heme	Accepts and donates electrons

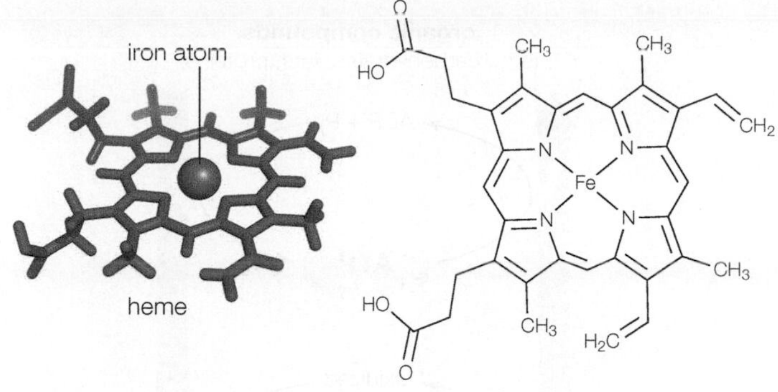
iron atom
heme
Fe
N
HO
O
CH_3
CH_2
H_2C

FIGURE 5.18 Heme. This small molecule is part of the active site in many enzymes. In proteins such as hemoglobin, it carries oxygen.

FIGURE IT OUT Is heme a cofactor or a coenzyme? Answer: It is both.

some normal metabolic reactions. Hydrogen peroxide is dangerous because it can easily oxidize and damage the organic molecules of life, or form free radicals that do (Section 2.3). Catalase neutralizes this threat. The enzyme's active site holds a hydrogen peroxide molecule close to a heme. Two hydrogen peroxide molecules alternately oxidize and then reduce the heme's iron atom, an interaction that causes the molecules to break down. Water and oxygen are products.

Substances such as catalase that interfere with the oxidation of other molecules are called **antioxidants**. Antioxidants minimize the amount of damage that cells sustain as a result of oxidation by free radicals or other molecules, so they are essential to health. Oxidative damage is associated with many diseases, including cancer, diabetes, atherosclerosis, stroke, and neurodegenerative problems such as Alzheimer's disease.

ATP: A Special Coenzyme

Many coenzymes are multifunctional molecules. The nucleotide ATP (adenosine triphosphate, Section 3.6) is a component of RNA molecules, and it also functions as a coenzyme in many reactions by donating and accepting phosphate groups. Bonds between phosphate groups hold a lot of energy compared to other bonds, and ATP has two of them holding its three phosphate groups together (FIGURE 5.19A). When a phosphate group is transferred to or from a nucleotide, this bond energy is transferred along with it. Thus, the nucleotide can receive energy from an exergonic reaction, and it can donate energy to an endergonic one. ATP in particular is such an important currency in a cell's energy economy that we use a cartoon coin to symbolize it.

A reaction in which an enzyme attaches a phosphate to an organic molecule is called a **phosphorylation**. ADP (adenosine diphosphate) forms when an enzyme transfers a phosphate group from ATP to another

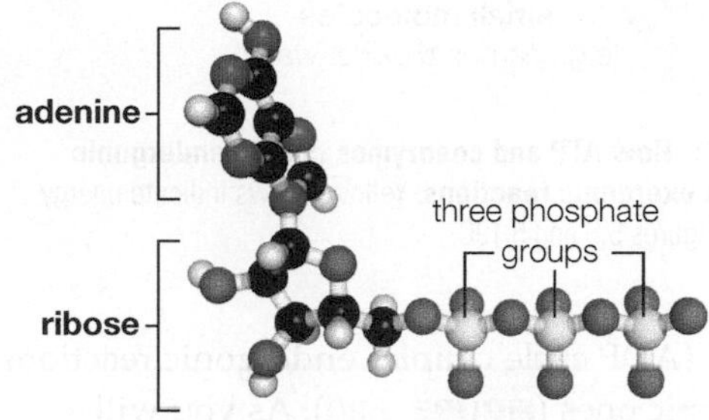
adenine
ribose
three phosphate groups

A Bonds between the phosphate groups of ATP hold a lot of energy.

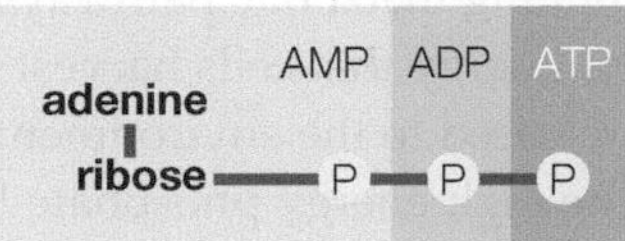
AMP
ADP
ATP
adenine
ribose
P
P
P

B After ATP loses one phosphate group, the nucleotide is ADP (adenosine diphosphate); after losing two, it is AMP (adenosine monophosphate).

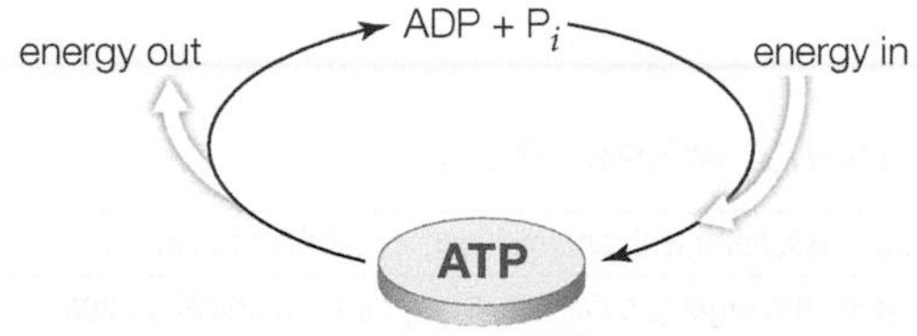
energy out
ADP + P_i
energy in
ATP

C The ATP/ADP cycle. ADP forms in a reaction that removes a phosphate group from ATP (P_i is an abbreviation for phosphate group). Energy released in this exergonic reaction drives other reactions that are the stuff of cellular work. ATP forms again in endergonic reactions that phosphorylate ADP.

FIGURE 5.19 The nucleotide ATP, an important energy currency in a cell's metabolism.

molecule during a phosphorylation (FIGURE 5.19B). Cells constantly run this reaction in order to drive a variety of endergonic reactions. Thus, they must constantly replenish their stockpile of ATP—by running exergonic reactions that phosphorylate ADP. The cycle of using and then replenishing ATP is called the **ATP/ADP cycle** (FIGURE 5.19C).

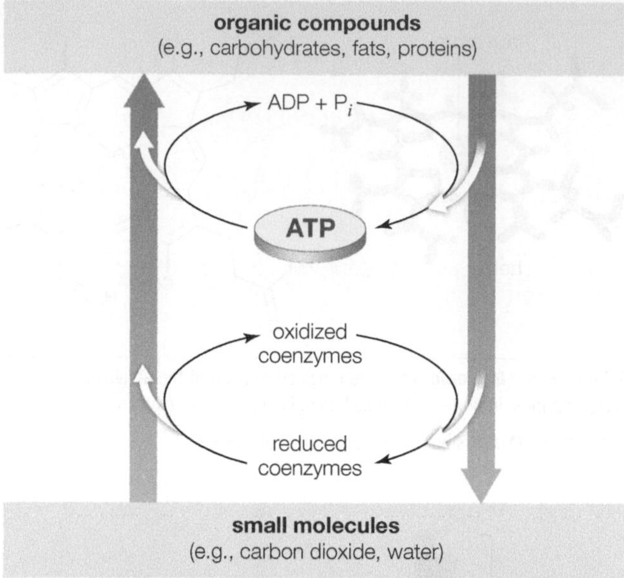
organic compounds
(e.g., carbohydrates, fats, proteins)
ADP + P_i
ATP
oxidized coenzymes
reduced coenzymes
small molecules
(e.g., carbon dioxide, water)

FIGURE 5.20 How ATP and coenzymes couple endergonic reactions with exergonic reactions. Yellow arrows indicate energy flow. Compare Figures 5.8 and 5.19C.

The ATP/ADP cycle couples endergonic reactions with exergonic ones (**FIGURE 5.20**). As you will see in Chapter 7, cells harvest energy from organic compounds by running metabolic pathways that break them down. Energy that cells harvest in these pathways is not released to the environment, but rather stored in the high-energy phosphate bonds of ATP molecules and in electrons carried by reduced coenzymes. Both the ATP and the reduced coenzymes that form in these pathways can be used to drive many types of endergonic reactions (**FIGURE 5.21**).

TAKE-HOME MESSAGE 5.6

- ✔ Cofactors associate with enzymes and assist their function.
- ✔ Coenzymes are organic cofactors. Many carry chemical groups, atoms, or electrons from one reaction to another.
- ✔ The formation of ATP from ADP is an endergonic reaction. ADP forms again when a phosphate group is transferred from ATP to another molecule.
- ✔ When a phosphate group is transferred from ATP to another molecule, energy is transferred along with it. This energy can be used to drive cellular work.

adhesion protein Plasma membrane protein that helps cells stick together in animal tissues. Some types form adhering junctions and tight junctions.
fluid mosaic Model of a cell membrane as a two-dimensional fluid of mixed composition.
receptor protein Membrane protein that triggers a change in cell activity in response to a stimulus such as binding a hormone.
transport protein Protein that moves specific ions or molecules across a membrane.

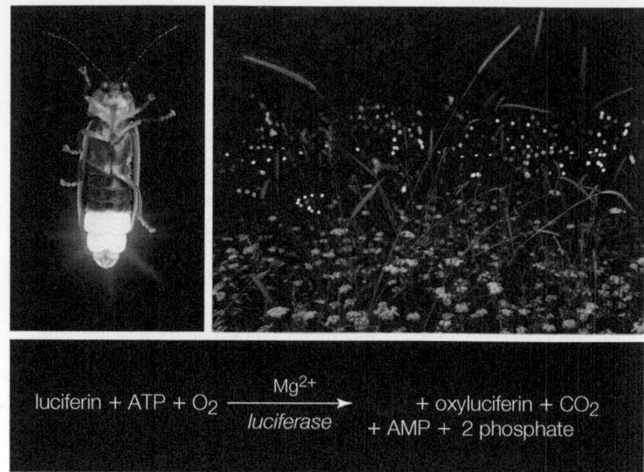
luciferin + ATP + O_2 $\xrightarrow[\textit{luciferase}]{Mg^{2+}}$ + oxyluciferin + CO_2 + AMP + 2 phosphate

FIGURE 5.21 ATP energy drives a metabolic pathway: bioluminescence in fireflies. Bioluminescence is light emitted by a living organism. Like many other animals, the firefly (top left) emits light to attract mates and prey (top right). Bottom, the light-producing metabolic pathway in these beetles (summarized) runs on energy provided by ATP.

5.7 A Closer Look at Cell Membranes

LEARNING OBJECTIVES

- Describe the structure of a lipid bilayer.
- Explain the fluid mosaic model of cell membranes.
- List the functions of four common types of membrane proteins.

The Fluid Mosaic Model

The foundation of cell membranes is a lipid bilayer that consists mainly of phospholipids. Remember from Section 3.4 that a phospholipid has a phosphate-containing head and two fatty acid tails. The head is polar and hydrophilic, so it interacts with water molecules. The two long hydrocarbon tails are nonpolar and hydrophobic, so they do not interact with water molecules. As a result of these opposing properties, phospholipids swirled into water will spontaneously organize themselves into lipid bilayer sheets or bubbles, with hydrophobic tails together, hydrophilic heads facing the watery surroundings (right). A cell's basic structure is essentially a lipid bilayer bubble filled with fluid.

fluid

Other molecules, including cholesterol, proteins, glycoproteins, and glycolipids, are embedded in or attached to the lipid bilayer of a cell membrane. Many of these molecules move around the membrane more or less freely. The **fluid mosaic** model describes a membrane as a two-dimensional liquid of mixed composition. The "mosaic" part of the name comes from the many molecules, and many different types of molecules, in the membrane. Membrane fluidity occurs

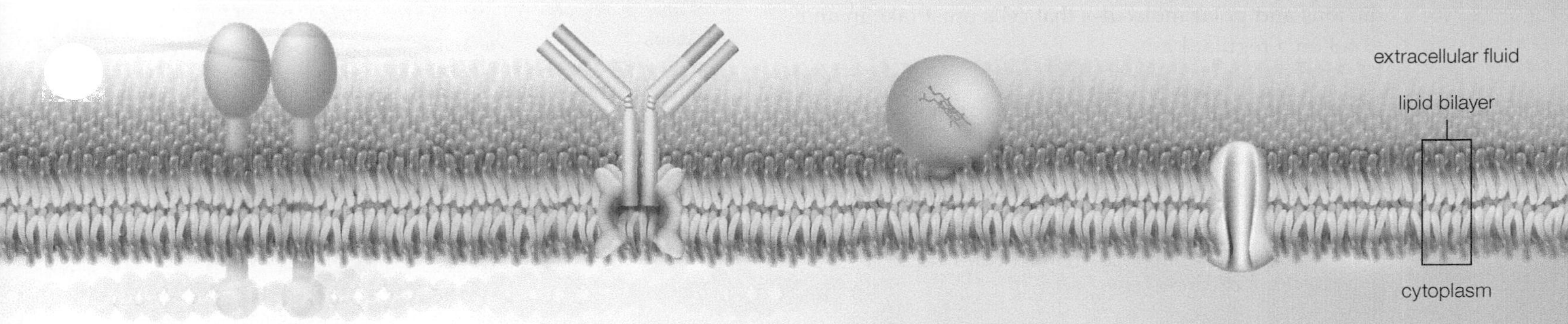
extracellular fluid
lipid bilayer
cytoplasm

A Adhesion proteins fasten cells together or to external proteins. This one connects microfilaments inside the cell to extracellular matrix proteins.

B Receptor proteins trigger a change in cellular activity in response to a stimulus such as binding to a particular substance. This one occurs on cells of the immune system.

C Enzymes catalyze reactions at membranes. This one is part of electron transfer chains that break down drugs and other organic toxins.

D Transport proteins bind to molecules on one side of the membrane, and release them on the other side. This one transports glucose.

FIGURE 5.22 Examples of common membrane proteins.
Many membrane proteins have an extremely complex structure; for clarity, they are often modeled as blobs or geometric shapes.

because phospholipids in the bilayer are not chemically bonded to one another; they stay organized as a result of collective hydrophobic and hydrophilic attractions. These interactions are, on an individual basis, relatively weak. Thus, individual phospholipids in the bilayer drift sideways and spin around their long axis, and their tails wiggle.

A cell membrane's properties vary depending on the types and proportions of molecules composing it. For example, membrane fluidity decreases with increasing cholesterol content. A membrane's fluidity also depends on the length and saturation of its phospholipids' fatty acid tails. Archaea do not even use fatty acids to build their phospholipids. Instead, they use molecules with reactive side chains, so the tails of archaeal phospholipids form covalent bonds with one another. As a result of this rigid crosslinking, archaeal phospholipids do not drift, spin, or wiggle in a bilayer. This makes the membranes of archaea stiffer than those of bacteria or eukaryotes, a characteristic that may help these cells survive in extreme habitats.

Proteins Add Function

Many different proteins are associated with a cell membrane, and the type of association varies by the type of protein. With integral membrane proteins, the association is more or less permanent. Domains sunk deeply into the lipid bilayer's hydrophobic core anchor these proteins in the membrane. Integral proteins that span the entire bilayer are called transmembrane proteins. Peripheral membrane proteins are those that temporarily attach to one of the lipid bilayer's surfaces by interacting with lipid heads or integral proteins.

A cell membrane physically separates an external environment from an internal one, but that is not its only task. Each kind of protein in a membrane imparts a specific function to it. Thus, different cell membranes can carry out different tasks depending on which proteins they include. A plasma membrane incorporates certain proteins that no internal cell membrane has, so it has functions that no other membrane does. For example, cells stay organized in animal tissues because **adhesion proteins** in their plasma membranes fasten them together and hold them in place (FIGURE 5.22A). This arrangement strengthens a tissue, and can constrain certain membrane proteins to an upper or lower surface of the cell. Adhesion proteins are the sticky components of adhering and tight junctions in animal tissues (Section 4.9). They also provide a cell with information about its position relative to other cells or structures.

Plasma membranes and some internal membranes have **receptor proteins**, which trigger a change in the cell's activities in response to a stimulus (FIGURE 5.22B). Each type of receptor protein receives a particular stimulus such as a hormone binding to it. The response may involve (for example) metabolism, movement, division, or even cell death.

Enzymes are associated with all cell membranes (FIGURE 5.22C). Some of these enzymes act on other proteins or lipids that are part of the membrane; others such as those that compose electron transfer chains use the membrane as a scaffold. All membranes also have **transport proteins**, which move specific substances across the bilayer (FIGURE 5.22D). Transport proteins are important because, as you will see shortly, lipid bilayers are impermeable to most substances, including the ions and polar molecules that cells must take in and expel on a regular basis.

TAKE-HOME MESSAGE 5.7

✔ The foundation of almost all cell membranes is the lipid bilayer: two layers of lipids (mainly phospholipids), with tails sandwiched between heads.

✔ Proteins embedded in or attached to a lipid bilayer add specific functions to each cell membrane.

5.8 Diffusion Across Membranes

LEARNING OBJECTIVES

- Name five factors that influence the rate and direction of diffusion.
- Describe tonicity and how it determines the direction of osmosis.
- Explain the relationship between turgor and osmotic pressure.

Factors That Affect Diffusion

Metabolic pathways require the participation of molecules that must move across membranes and through cells. **Diffusion**, the spontaneous spreading of atoms or molecules through a fluid or gas, is an essential way in which this movement occurs. An atom or molecule is always jiggling, and this internal movement causes it to randomly bounce off of nearby objects, including other atoms or molecules. Rebounds from such collisions propel solutes through a liquid, with the result being a gradual and complete mixing (you can see this mixing when tea diffuses through hot water, left). Five factors influence diffusion:

Concentration A difference in solute concentration (Section 2.5) between adjacent regions of a solution is called a concentration gradient. Solutes tend to diffuse "down" their concentration gradient, from a region of higher concentration to one of lower concentration. The greater the difference in concentration, the faster diffusion occurs. Why? Consider that moving objects (such as molecules) collide more often as they get more crowded. Thus, during a given interval, more molecules get bumped out of a region of higher concentration than get bumped into it.

Temperature Atoms and molecules jiggle faster at higher temperature, so they move more quickly. Thus, diffusion occurs more quickly at higher temperatures.

Charge Each ion or charged molecule in a fluid contributes to the fluid's overall electric charge.

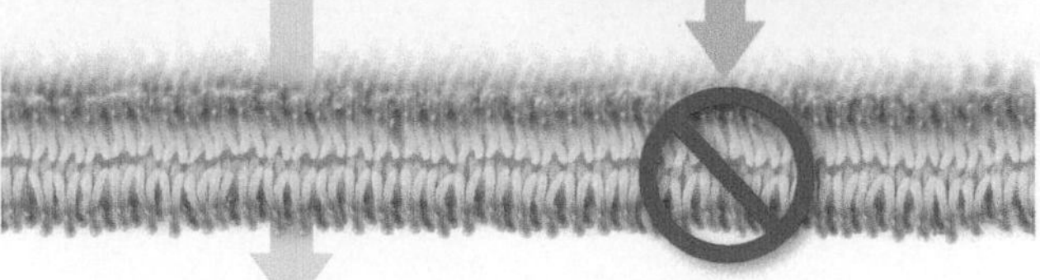
Hydrophobic molecules
steroids, fats, oils
Gases
O_2, CO_2, N_2
Small polar molecules
water, urea
Ions
Na^+, H^+, HCO_3^-
Large polar molecules
sugars, proteins, nucleic acids, amino acids

FIGURE 5.23 Selective permeability of lipid bilayers.
Hydrophobic molecules, gases, and small polar molecules can cross a lipid bilayer on their own. Ions and large polar molecules cannot.

A difference in charge between two regions of fluid can affect the rate and direction of diffusion between them. For example, positively charged ions will tend to diffuse toward a region with an overall negative charge.

Molecular Size It takes more energy to move a large object than it does to move a small one, so ions and small molecules diffuse more quickly than large molecules.

Pressure Pressure squeezes atoms and molecules closer together. Atoms and molecules that are more crowded collide and rebound more frequently. Thus, diffusion occurs faster at higher pressures.

Osmosis

Lipid bilayers are selectively permeable, which means that only some substances can diffuse across them (**FIGURE 5.23**). The long, nonpolar phospholipid tails make the core of a bilayer quite hydrophobic (Section 3.4). Hydrophobic molecules (such as steroids), as well as gases and small polar molecules (such as water) can easily move through this core. However, ions and large polar molecules such as glucose cannot.

When a lipid bilayer separates two fluids with differing solute concentrations, water will diffuse across

diffusion (dif-YOU-zhun) Spontaneous spreading of molecules or atoms through a fluid or gas.
hypertonic Describes a fluid that has a high overall solute concentration relative to another fluid.
hypotonic Describes a fluid that has a low overall solute concentration relative to another fluid.
isotonic Describes two fluids with identical solute concentrations.
osmosis (oz-MOE-sis) Diffusion of water across a selectively permeable membrane; occurs in response to a difference in solute concentration between the fluids on either side of the membrane.
osmotic pressure (oz-MAH-tick) Amount of turgor that prevents osmosis into cytoplasm or other hypertonic fluid.
turgor (TR-gr) Pressure that a fluid exerts against a structure that contains it.

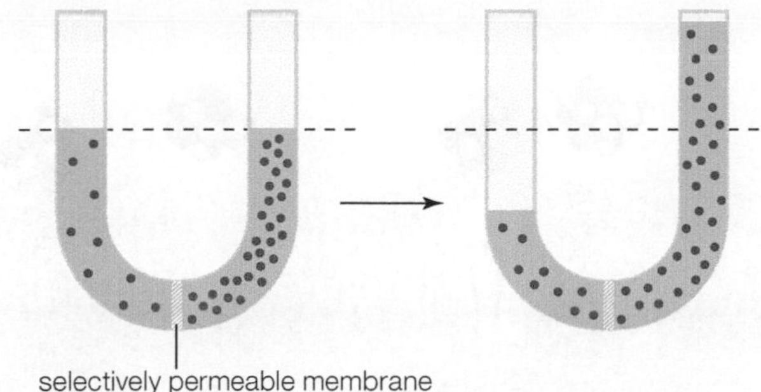
selectively permeable membrane

FIGURE 5.24 Osmosis. Water moves across a selectively permeable membrane that separates two fluids of differing solute concentration. The fluid volume changes in the two compartments as water diffuses across the membrane from the hypotonic solution to the hypertonic one.

it. The direction and rate of this diffusion depend on the relative solute concentration of the two fluids, which we describe in terms of tonicity. If the solute concentration differs between the two fluids, the fluid with the lower solute concentration is said to be **hypotonic** (*hypo–*, under). The other one, with the higher solute concentration, is **hypertonic** (*hyper–*, over). Water diffuses from a hypotonic fluid into a hypertonic one. The diffusion will continue until the two fluids are **isotonic**, which means they have the same overall solute concentration. The movement of water across membranes is so important in biology that it is given a special name: **osmosis** (**FIGURE 5.24**).

If a cell's cytoplasm becomes hypertonic with respect to the fluid outside of its plasma membrane, water will diffuse into the cell. If the cytoplasm becomes hypotonic, water will diffuse out. In either case, the solute concentration of the cytoplasm will change. If it changes enough, the cell's enzymes will stop working, with lethal results. Most cells have homeostatic mechanisms that compensate for osmosis when the solute concentration of cytoplasm differs from extracellular (external) fluid. In cells with no such mechanism, the volume—and solute concentration—of cytoplasm changes when water diffuses into or out of the cell (**FIGURE 5.25**).

Turgor

Even in a hypotonic environment, a cell wall can resist an increase in the volume of cytoplasm. Plant cells usually contain more solutes than soil water. Thus, water usually diffuses from soil into a plant—but only up to a point. Stiff walls keep plant cells from expanding very much, so an influx of water causes pressure to build up inside them. Pressure that a fluid exerts against a structure that contains it is called **turgor**. When enough pressure builds up inside a plant cell, water stops diffusing into it. The amount of turgor that is enough to stop osmosis is called **osmotic pressure**.

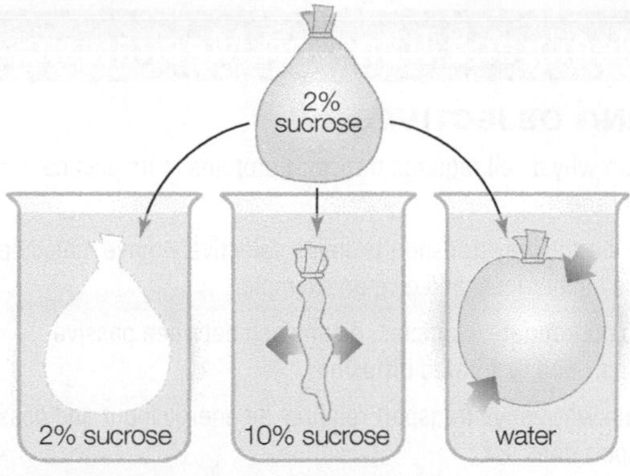
2%
sucrose
2% sucrose
10% sucrose
water

A What happens when a semipermeable bag containing a solution of water and sucrose is immersed in solutions of different tonicities.

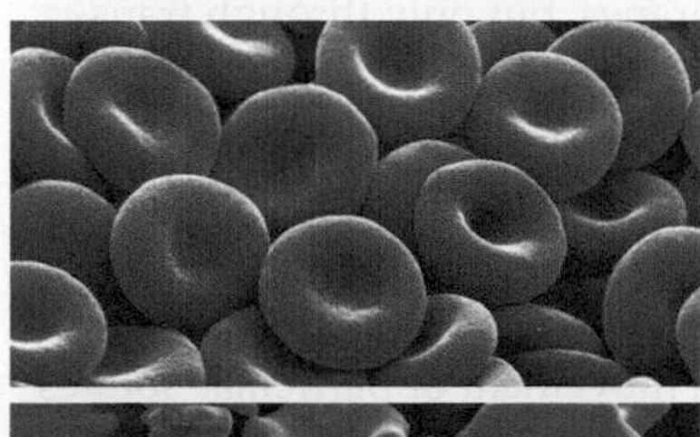

B Red blood cells in an isotonic solution (such as the fluid portion of blood) have an indented disk shape.

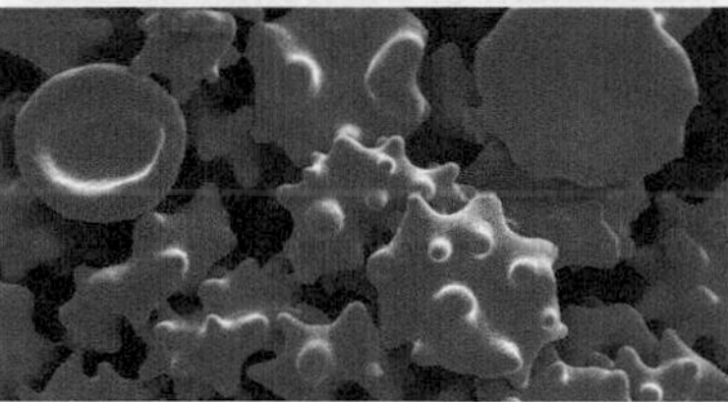

C Red blood cells immersed in a hypertonic solution shrivel up because water diffuses out of them.

2 µm

D Red blood cells immersed in a hypotonic solution swell up because water diffuses into them. Some of these have burst.

FIGURE 5.25 Effects of tonicity.

Like a semipermeable bag filled with a solution of sucrose and water (**A**), red blood cells have no mechanism to compensate for differences in solute concentration across their membrane (**B–D**).

Osmotic pressure keeps walled cells plump, just as air pressure inside a tire keeps it inflated. A young plant can resist gravity to stay erect because turgor in its cells is high. If the plant does not get enough water to replace what it uses, the cytoplasm of its cells shrinks. As turgor inside the cells decreases, the plant wilts.

TAKE-HOME MESSAGE 5.8

✔ Solutes tend to diffuse from regions of higher to lower concentration. The rate of diffusion is affected by differences in concentration, temperature, molecular size, charge, and pressure.

✔ When two fluids of different solute concentration are separated by a selectively permeable membrane, water diffuses from the hypotonic to the hypertonic fluid until the two fluids become isotonic. This movement, osmosis, is opposed by turgor.

CREDITS: (25B) Annie Cavanagh/Wellcome Images; (25C-D) CMSP/Getty Images.

5.9 Membrane Transport Mechanisms

LEARNING OBJECTIVES

- Explain why a cell requires transport proteins in its plasma membrane.
- Describe the way transport proteins selectively move molecules across membranes, and give an example.
- Using appropriate examples, distinguish between passive transport and facilitated diffusion.
- Explain why active transport requires an energy input and passive transport does not.

Solutes that cannot diffuse directly through lipid bilayers do cross cell membranes, but only through transport proteins. Each type of transport protein allows a specific substance to cross: Calcium pumps pump only calcium ions; glucose transporters transport only glucose; and so on. This specificity is an important part of homeostasis. Consider how the composition of cytoplasm depends on movement of particular solutes across the plasma membrane, which in turn depends on transport proteins in the lipid bilayer. Glucose is an important source of energy for most cells, so they normally take up as much as they can from extracellular fluid. This uptake occurs via glucose transporters in the plasma membrane. As soon as a molecule of glucose enters cytoplasm, an enzyme called hexokinase phosphorylates it (as shown in Figure 5.9D). Phosphorylation traps the molecule inside the cell because the transporters are specific for glucose, not phosphorylated glucose. Thus, phosphorylation prevents the molecule from moving back through the transporter and leaving the cell.

Passive Transport

Osmosis is one example of **passive transport**, a membrane-crossing mechanism that requires no energy input. Another example is **facilitated diffusion**, in which a solute follows its concentration gradient across a membrane by diffusing through a transport protein. The movement of a solute through a passive transport protein is driven entirely by the solute's concentration gradient, so it requires no input of energy.

Some transport proteins form pores (permanently open channels through a membrane). Others are gated, which means they open and close in response to a stimulus such as a shift in electric charge or binding to a signaling molecule. Still others change shape when they bind and release a solute. Typically, binding of a solute triggers a change in the protein's shape, and the shape change releases the solute to the opposite side of the membrane. A glucose transporter works this way (**FIGURE 5.26**).

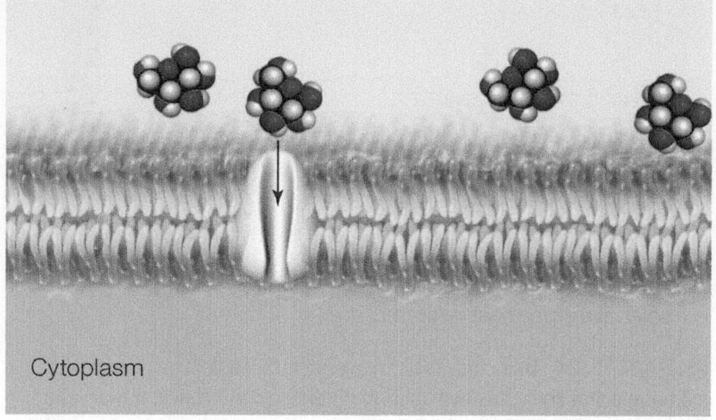
Cytoplasm

A A glucose molecule (here, in extracellular fluid) binds to a glucose transporter (gray) in the plasma membrane.

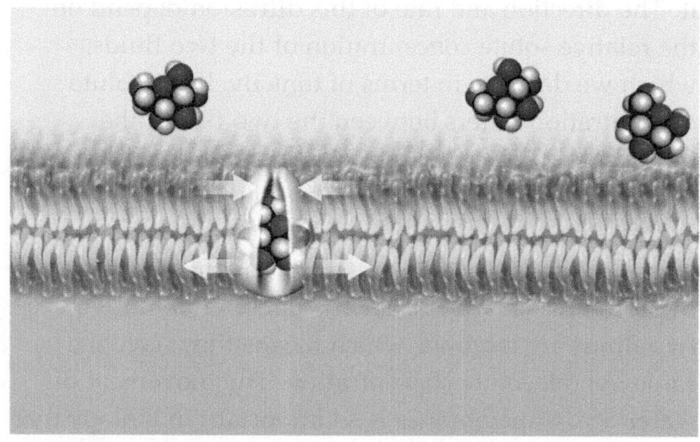

B Binding causes the transport protein to change shape.

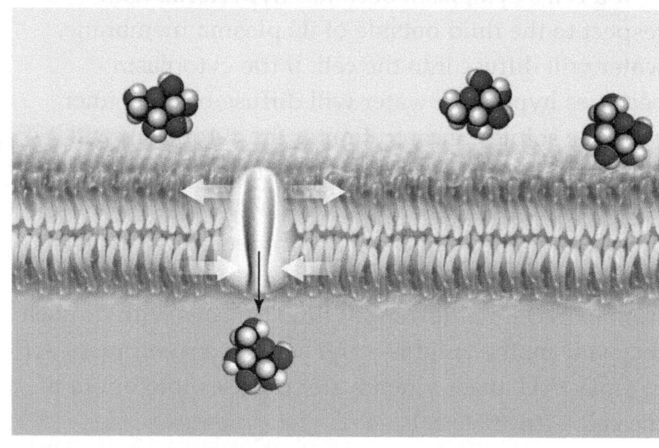

C The transport protein releases the glucose on the other side of the membrane (in cytoplasm) and resumes its original shape.

FIGURE 5.26 An example of facilitated diffusion.

FIGURE IT OUT In this example, which fluid is hypotonic with respect to glucose: extracellular fluid or the cytoplasm?

Answer: Cytoplasm

active transport Energy-requiring mechanism in which a transport protein pumps a solute across a cell membrane against the solute's concentration gradient.

facilitated diffusion Passive transport mechanism in which a solute follows its concentration gradient across a membrane by moving through a transport protein.

passive transport Membrane-crossing mechanism that requires no energy input.

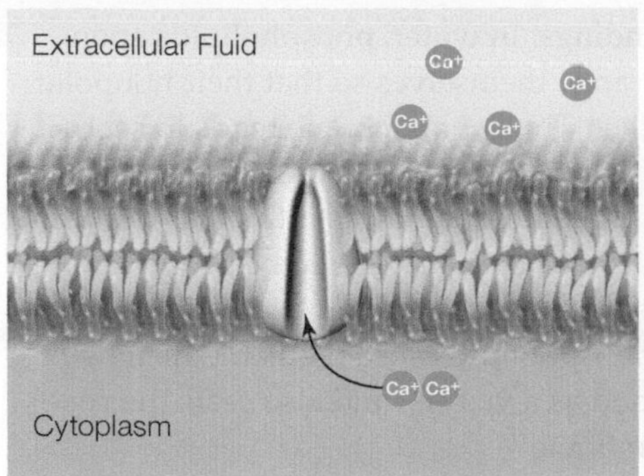
Extracellular Fluid
Ca+
Ca+
Ca+
Ca+
Ca+ Ca+
Cytoplasm

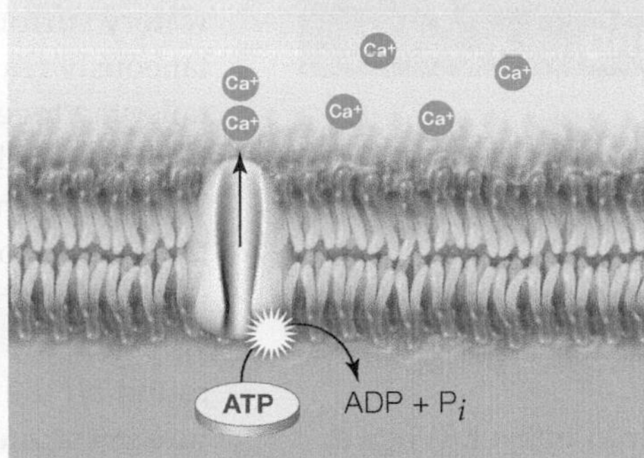
Ca+
Ca+
Ca+
Ca+
Ca+
Ca+
ATP
ADP + P_i

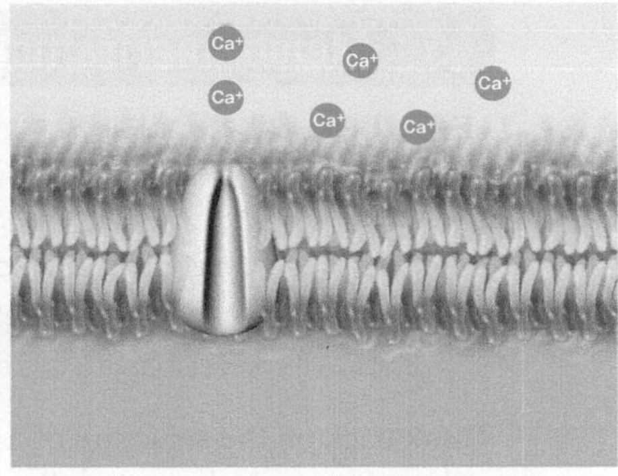
Ca+
Ca+
Ca+
Ca+
Ca+
Ca+

A Two calcium ions (blue) bind to the transport protein (a calcium pump, gray).

B A phosphate group transfer from ATP causes the protein to change shape so that the calcium ions are ejected to the opposite side of the membrane.

C After it loses the calcium ions, the transport protein resumes its original shape.

FIGURE 5.27 Active transport of calcium ions.

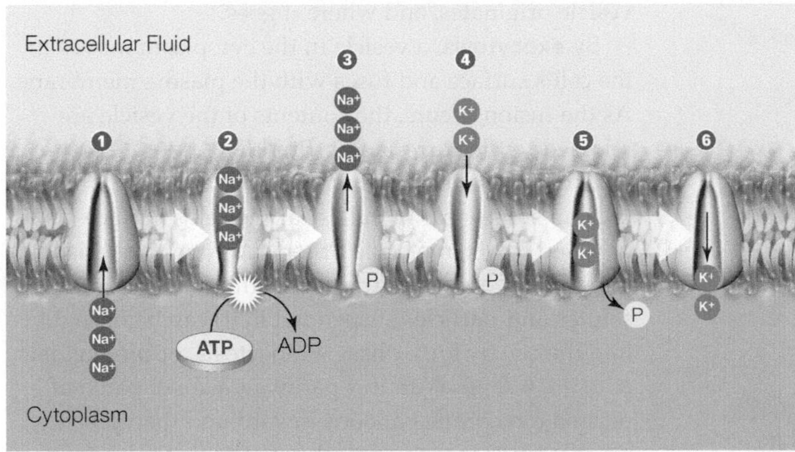
Extracellular Fluid
1
2
3
4
5
6
Na+
Na+
Na+
K+
K+
P
P
P
ATP
ADP
Cytoplasm

FIGURE 5.28 The sodium–potassium pump. This protein (gray) actively transports sodium ions (Na^+) from cytoplasm to extracellular fluid, and potassium ions (K^+) in the other direction. ATP provides energy required for transporting both ions against their concentration gradient.

1. Sodium ions in cytoplasm diffuse into the pump's open channel and bind to its interior.
2. The transfer of a phosphate group (P) from ATP causes the pump to change shape so that its channel opens to extracellular fluid, where it releases the sodium ions 3.
4. Potassium ions from extracellular fluid diffuse into the channel and bind to its interior.
5. The transport protein releases the phosphate group and reverts to its original shape.
6. The channel opens to the cytoplasm, where it releases the potassium ions.

Active Transport

Solutes required for many cellular processes must be moved across a membrane against their concentration gradient, and this requires energy. With **active transport**, a transport protein uses energy to pump a solute against its gradient across a cell membrane. Typically, an energy input (often in the form of a phosphate-group transfer from ATP) changes the shape of an active transport protein. The shape change causes the protein to release a bound solute to the other side of the membrane.

Consider calcium ions, which act as potent messengers that trigger various processes inside cells. Thus, the concentration of these ions in cytoplasm must be kept thousands of times lower than in extracellular fluid. This gradient is maintained by calcium pumps, which export calcium ions from a cell by active transport (**FIGURE 5.27**).

Another example of active transport involves sodium–potassium pumps. Nearly all of the cells in your body have these transport proteins, which pump two substances in opposite directions across the membrane: sodium ions from cytoplasm to extracellular fluid, and potassium ions from extracellular fluid to cytoplasm (**FIGURE 5.28**).

Bear in mind that the membranes of all cells, not just those of animals, have proteins that carry out active transport. In plants, for example, transport proteins in the plasma membranes of photosynthetic cells pump sugar molecules from cytoplasm into tubes that thread throughout the plant body.

TAKE-HOME MESSAGE 5.9

- ✔ Many types of molecules and ions can cross a lipid bilayer only with the help of transport proteins.
- ✔ Each type of transport protein moves a specific solute across a cell membrane. The types and amounts of solutes that cross a membrane depend on the transport proteins embedded in it.
- ✔ In facilitated diffusion (a type of passive transport), a solute follows its concentration gradient through a transport protein. The movement requires no energy input.
- ✔ In active transport, a transport protein pumps a solute across a membrane against its concentration gradient. The movement requires an energy input, as from ATP.

5.10 Membrane Trafficking

LEARNING OBJECTIVES

- Identify and explain the process by which cells expel materials in bulk.
- Explain how a vesicle forms at the plasma membrane.
- Explain the mechanism by which cells sample extracellular fluid.
- Describe the way cells take in and digest large particles.

Think back on the structure of a lipid bilayer. When a cell membrane is disrupted, the fatty acid tails of the phospholipids in the bilayer become exposed to their watery surroundings. In water, phospholipids spontaneously rearrange themselves so that their nonpolar tails stay together. Thus, a membrane tends to seal itself after a disruption. Vesicles form the same way. When a patch of membrane bulges into the cytoplasm, the hydrophobic tails of the lipids in the bilayer are repelled by the watery fluid on both sides. The fluid "pushes" the phospholipid tails together, which helps round off the bud as a vesicle, and also seals the rupture in the membrane.

Vesicles are constantly carrying materials to and from a cell's plasma membrane. This movement requires energy because it involves motor proteins that drag the vesicles along cytoskeletal elements. We describe the movement based on where and how the vesicle originates, and where it goes.

By **exocytosis**, a vesicle in the cytoplasm moves to the cell's surface and fuses with the plasma membrane. As the fusion occurs, the contents of the vesicle are released to the surrounding fluid (**FIGURE 5.29A**).

There are several pathways of **endocytosis**, but all take up substances in bulk near the cell's surface (as opposed to one molecule or ion at a time via transport proteins). A drop of extracellular fluid (along with solutes and particles suspended in it) can be brought into the cell by bulk-phase endocytosis, or **pinocytosis** (**FIGURE 5.29B**). With this pathway, a small patch of plasma membrane balloons inward and then pinches off as it sinks into the cytoplasm. The balloon becomes a vesicle that encloses extracellular fluid (along with whatever solutes and particles are suspended in it). Bulk-phase endocytosis is nonspecific about what it brings into the cell, so it is used for sampling the composition of extracellular fluid.

Receptor-mediated endocytosis is more selective than pinocytosis (**FIGURE 5.29C**). With this pathway, molecules of a targeted substance trigger endocytosis. Receptor proteins in the plasma membrane bind to these molecules. The binding triggers a shallow pit to form in the membrane, just under the receptors. The pit sinks into the cytoplasm and traps the targeted substance in a vesicle as it closes back on itself. LDL and other lipoprotein particles (Section 3.5) enter cells this way (**FIGURE 5.29D**).

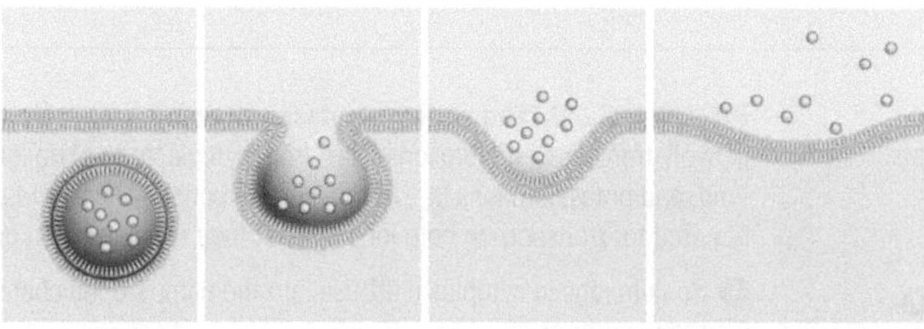

A Exocytosis. A vesicle in cytoplasm fuses with the plasma membrane. Lipids and proteins of the vesicle's membrane become part of the plasma membrane as its contents are expelled to the environment.

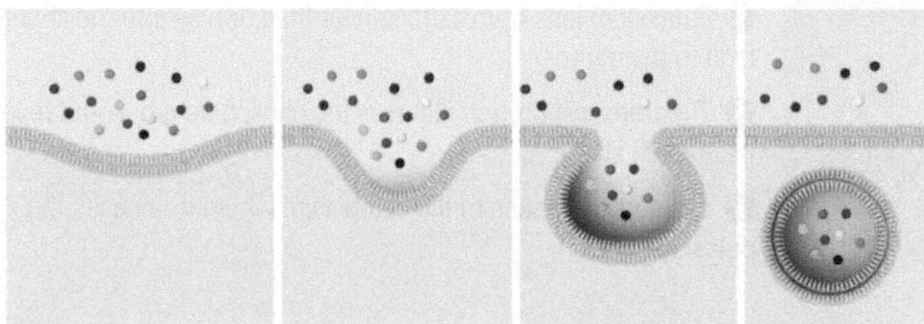

B Pinocytosis. A pit in the plasma membrane traps any fluid, solutes, and particles near the cell's surface in a vesicle as it sinks into the cytoplasm.

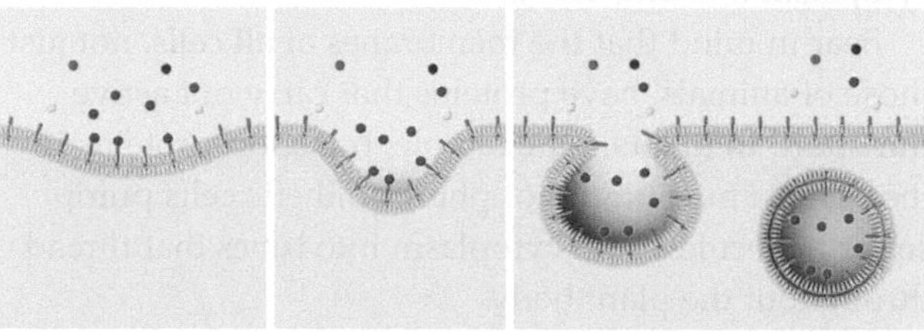

C Receptor-mediated endocytosis. Cell surface receptors (green) bind a target molecule and trigger a pit to form in the plasma membrane. The target molecules are trapped in a vesicle as the pit sinks into the cytoplasm.

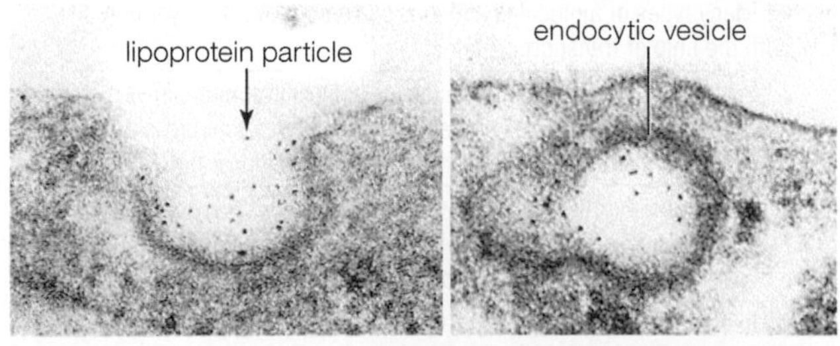
lipoprotein particle
endocytic vesicle

D Example of receptor-mediated endocytosis: intake of lipoprotein particles.

FIGURE 5.29 Exocytosis and endocytosis.

endocytosis (en-doe-sigh-TOE-sis) Process by which a cell takes in a small amount of extracellular fluid (and its contents) by the ballooning inward of the plasma membrane.

exocytosis (ex-oh-sigh-TOE-sis) Process by which a cell expels a vesicle's contents to extracellular fluid.

phagocytosis (fag-oh-sigh-TOE-sis) "Cell eating"; type of receptor-mediated endocytosis in which a cell engulfs a large solid particle such as another cell.

pinocytosis (pin-oh-sigh-TOE-sis) Endocytic pathway by which fluid and materials in bulk are brought into the cell.

Phagocytosis (which means "cell eating") is a type of receptor-mediated endocytosis in which mobile cells engulf microorganisms, cellular debris, or other large particles. Many single-celled protists such as amoebas feed by this pathway. As you will see in Chapter 34, your white blood cells use phagocytosis to engulf microorganisms and cellular debris.

Phagocytosis begins when receptor proteins bind to a particular extracellular target. The binding causes microfilaments to assemble in a mesh under the plasma membrane. The microfilaments then contract, forcing a lobe of membrane-enclosed cytoplasm to bulge outward as a pseudopod (Section 4.8). Pseudopods that merge around a target trap it inside a vesicle that sinks into the cytoplasm. Typically, the vesicle merges with a lysosome, and the object taken in by phagocytosis is digested by lysosomal enzymes (Section 4.5).

Recycling Membrane

The composition of a plasma membrane begins in the endoplasmic reticulum (ER). There, membrane proteins and lipids are made and modified, and both become part of vesicles that transport them to Golgi bodies for final modification (Section 4.5). New plasma membrane forms when the finished proteins and lipids are repackaged as vesicles that travel to the plasma membrane and fuse with it. Membrane proteins automatically end up in the correct orientation in the plasma membrane (**FIGURE 5.30**).

As long as a cell is alive, exocytosis and endocytosis continually replace and withdraw patches of its plasma membrane. If the cell is not enlarging, membrane lost as a result of endocytosis is replaced by membrane gained during exocytosis. Thus, the total area of the plasma membrane remains more or less constant.

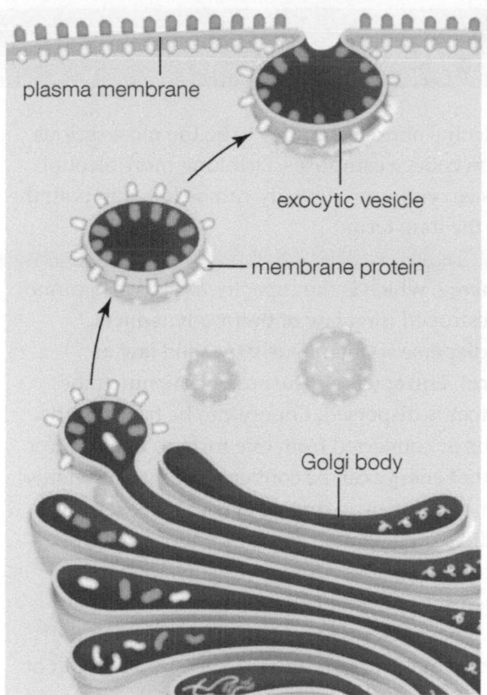
plasma membrane
exocytic vesicle
membrane protein
Golgi body

FIGURE 5.30 **How membrane proteins become oriented to the inside or the outside of a cell.**

Proteins of the plasma membrane are assembled in the ER, and finished inside Golgi bodies.

The proteins become part of vesicle membranes that bud from the Golgi. When the vesicles fuse with the plasma membrane (exocytosis), these proteins automatically become oriented in the proper direction.

TAKE-HOME MESSAGE 5.10

- ✔ By processes of exocytosis and endocytosis, cells take in and expel bulk substances and particles too big for transport proteins.
- ✔ In exocytosis, a cytoplasmic vesicle fuses with the plasma membrane and releases its contents to the cell's exterior.
- ✔ In endocytosis, a patch of plasma membrane sinks inward and forms a vesicle in the cytoplasm.
- ✔ Some cells can engulf large particles by phagocytosis.

5.1 A Toast to Alcohol Dehydrogenase (revisited)

The main function of the enzyme alcohol dehydrogenase (ADH) is to detoxify tiny amounts of alcohols that form in some metabolic pathways. ADH converts ethanol to acetaldehyde, which is even more toxic than ethanol and the most likely source of various hangover symptoms. A second enzyme, ALDH, converts acetaldehyde to acetate, which is a nontoxic salt. Both ADH and ALDH use the coenzyme NAD^+. Thus, the overall pathway of ethanol metabolism in humans is:

$$\text{ethanol} \xrightarrow[NAD^+ \;\; NADH]{ADH} \text{acetaldehyde} \xrightarrow[NAD^+ \;\; NADH]{ALDH} \text{acetate}$$

In the average healthy adult, this pathway can detoxify between 7 and 14 grams of ethanol per hour. The average alcoholic beverage contains between 10 and 20 grams of ethanol, which is why having more than one drink in any two-hour interval may result in a hangover.

In the United States, alcohol abuse is the leading cause of cirrhosis of the liver. A cirrhotic liver stops making albumin, a protein that maintains tonicity of blood and tissue fluids. Without enough albumin, body tissues swell with watery fluid, especially in the legs and abdomen. Drugs and other toxins can no longer be removed from the blood, so they accumulate in the brain—which impairs mental functioning and alters personality. Restricted blood flow through the liver causes veins to enlarge and rupture, so internal bleeding is a risk. The damage to the body results in a heightened susceptibility to diabetes and liver cancer. Once cirrhosis has been diagnosed, a person has about a 50 percent chance of dying within 10 years.

CREDIT: (30) © Cengage Learning.

STUDY GUIDE

Section 5.1 Alcohol abuse continues to be the most serious drug problem on college campuses. Drinking more alcohol than the body's enzymes can detoxify can be lethal in both the short term and the long term.

Section 5.2 **Energy**, which is the capacity to do work, cannot be created or destroyed (**first law of thermodynamics**), and it tends to disperse spontaneously (**second law of thermodynamics**). **Entropy** is a measure of how much the energy of a system is dispersed. Energy can be transferred between systems or converted from one form to another (for example, **potential energy** can be converted to **kinetic energy**), but some is lost, often as heat, during every such exchange.

Sustaining life's organization requires ongoing energy inputs to counter energy loss. Organisms stay alive by replenishing themselves with energy they harvest from someplace else. Energy flows in one direction through the biosphere, starting mainly from the sun, then into and out of ecosystems. Producers and then consumers use the captured energy to assemble, rearrange, and break down organic molecules that cycle among organisms in an ecosystem.

Section 5.3 In chemical reactions, **reactants** are converted to **products**. Cells store energy in chemical bonds by running **endergonic** reactions that build organic compounds. To release this stored energy, they run **exergonic** reactions that break the bonds. Both endergonic and exergonic reactions require an input of **activation energy** to begin.

Section 5.4 Enzymes greatly enhance the rate of reactions without being changed by them, a process called **catalysis**. Interacting with an enzyme's active site causes a **substrate** to reach its **transition state**. Enzymes lower a reaction's activation energy, for example, by improving the fit between substrate and **active site** (**induced-fit model**), or by forcing substrates together. Each enzyme works best within a certain range of environmental conditions that include temperature, pH, and salt concentration.

Section 5.5 A **metabolic pathway** is a stepwise series of enzyme-mediated reactions that collectively build, remodel, or break down an organic molecule. Cells conserve energy and resources by producing only what they need at a given time. Such control can arise from mechanisms that start, stop, or alter the rate of a single reaction in a metabolic pathway.

The relative concentration of reactants and products affects the rate of a reaction. Also, the binding of specific ions or molecules to an enzyme can enhance or inhibit its activity. With **allosteric regulation**, this binding occurs in a region other than the active site. The products of some metabolic pathways inhibit their own production, a regulatory mechanism called **feedback inhibition**. **Redox** (oxidation–reduction) **reactions** in **electron transfer chains** allow cells to harvest energy stepwise, in small, manageable amounts.

Section 5.6 **Cofactors** associate with enzymes and assist their function. Cofactors help some **antioxidant** enzymes prevent dangerous oxidation reactions by interfering with the oxidation of other molecules. Some cofactors are metal ions; organic cofactors are **coenzymes**.

Many coenzymes carry chemical groups, atoms, or electrons from one reaction to another. ATP is often used as a coenzyme that carries energy between reactions. Energy harvested in exergonic reactions can be stored in ATP's high-energy phosphate bonds. When a phosphate group is transferred from ATP to another molecule, energy transferred with it can drive an exergonic reaction. Thus, phosphate-group transfers (**phosphorylations**) to and from ATP couple exergonic with endergonic reactions. Cells regenerate ATP in the **ATP/ADP cycle**.

Section 5.7 The foundation of a cell membrane is the lipid bilayer—two layers of lipids (mainly phospholipids), with tails sandwiched between heads. A bacterial or eukaryotic cell membrane can be described as a **fluid mosaic** of lipids and proteins; archaeal membranes are not fluid. Proteins embedded in or attached to a lipid bilayer add specific functions to each type of cell membrane. All cell membranes have enzymes, and all have **transport proteins** that help substances cross the lipid bilayer. Plasma membranes also incorporate **adhesion proteins** that lock cells together in tissues. Plasma membranes and some internal membranes have **receptor proteins** that trigger a change in cell activities in response to a specific stimulus.

Section 5.8 The rate of **diffusion** is influenced by concentration gradients, temperature, molecular size, charge, and pressure. Nonpolar molecules as well as gases and small polar molecules such as water can diffuse across a lipid bilayer. Large polar molecules and ions cannot.

Solutes tend to diffuse into an adjoining region of fluid in which they are not as concentrated. When two fluids of different solute concentrations are separated by a selectively permeable membrane such as a lipid bilayer, water diffuses across the membrane from the **hypotonic** to the **hypertonic** fluid (there is no net movement of water between **isotonic** solutions). This movement, **osmosis**, is opposed by **turgor** (fluid pressure against a cell membrane or wall). **Osmotic pressure** is the amount of turgor sufficient to halt osmosis.

Section 5.9 Ions and large polar molecules can cross cell membranes only with the help of transport proteins. In **facilitated diffusion**, a solute binds to a transport protein that releases it to the opposite side of the membrane. Because the movement is driven by the solute's concentration gradient, it is a type of **passive transport** (no energy input is required). With **active transport**, a transport protein uses energy (often in the form of a phosphate-group transfer from ATP) to pump a solute across a membrane against its concentration gradient.

Section 5.10 Exocytosis and endocytosis move particles and substances in bulk across plasma membranes. With **exocytosis**, a cytoplasmic vesicle fuses with the plasma membrane, and its contents are released to the outside of the cell. With **endocytosis**, a patch of plasma membrane balloons into the cell, taking with it a drop of extracellular fluid. The balloon forms a vesicle that sinks into the cytoplasm. **Pinocytosis** is a type of endocytosis that is not specific about what it takes in; receptor-mediated endocytosis targets specific molecules. Some cells engulf large particles by the endocytic pathway of **phagocytosis**.

SELF-QUIZ

Answers in Appendix VII

1. Which of the following statements is *not* correct?
   a. Energy cannot be created or destroyed.
   b. Energy cannot change from one form to another.
   c. Energy tends to disperse spontaneously.

2. _______ is life's primary source of energy.
   a. Food b. Water c. Sunlight d. ATP

3. Entropy _______ .
   a. tends to disperse c. tends to decrease, overall
   b. is free energy d. is a measure of disorder

4. If we liken a chemical reaction to an energy hill, then a(n) _______ reaction is, overall, a downhill run.
   a. exergonic c. catalytic
   b. endergonic d. both a and c

5. In an endergonic reaction, activation energy is a bit like _______ .
   a. a burst of speed
   b. coasting downhill
   c. an energy hill
   d. putting on the brakes

6. _____ are always changed by participating in a reaction.
   a. Enzymes c. Reactants
   b. Cofactors d. Coenzymes

7. Name one environmental factor that typically influences enzyme function.

8. A metabolic pathway _______ .
   a. may build or break down molecules
   b. generates heat
   c. can include redox reactions
   d. all of the above

9. A molecule that donates electrons becomes _______ , and the one that accepts the electrons becomes _______ .
   a. reduced; oxidized c. oxidized; reduced
   b. ionic; electrified d. electrified; ionic

10. All antioxidants _______ .
    a. prevent other molecules from being oxidized
    b. are coenzymes
    c. balance charge
    d. deoxidize free radicals

11. Solutes tend to diffuse from a region where they are _______ concentrated to an adjacent region where they are _______ concentrated.
    a. more, less c. movement is independent of concentration
    b. less, more

12. _______ cannot diffuse across a lipid bilayer.
    a. Water molecules c. Ions
    b. Gases d. Nonpolar molecules

13. A transport protein requires ATP to pump sodium ions across a membrane. This is a case of _______ .
    a. passive transport c. osmosis
    b. active transport d. facilitated diffusion

14. Immerse a human red blood cell in a hypotonic solution, and water _______ .
    a. diffuses into the cell c. shows no net movement
    b. diffuses out of the cell d. moves in by endocytosis

15. Match each term with its most suitable description.

___ reactant	a. assists enzymes
___ phagocytosis	b. forms at reaction's end
___ lipid bilayer	c. enters a reaction
___ cyclic pathway	d. enzyme action
___ product	e. one cell engulfs another
___ cofactor	f. electron exchange
___ passive transport	g. no energy input required
___ catalysis	h. phospholipids + water
___ redox reaction	i. goes in circles
___ first law of thermodynamics	j. basis of diffusion
___ concentration gradient	k. energy cannot be created or destroyed

CENGAGE brain .com To access course materials, please visit www.cengagebrain.com.

CRITICAL THINKING

1. Beginning physics students are often taught the basic concepts of thermodynamics with two phrases: First, you can never win. Second, you can never break even. Explain.

2. Describe diffusion in terms of entropy.

3. What is a redox reaction?

4. The enzyme trypsin is sold as a dietary supplement. What happens to trypsin taken with food?

5. Catalase combines two hydrogen peroxide molecules ($H_2O_2 + H_2O_2$) to make two molecules of water. A gas also forms. What is the gas?

6 WHERE IT STARTS—PHOTOSYNTHESIS

CORE CONCEPTS

Pathways of Transformation

Organisms exchange matter and energy with the environment in order to grow, maintain themselves, and reproduce.

The main flow of energy through the biosphere starts with photosynthesis. Photosynthesis captures light energy from the environment and converts it to chemical energy that supports photosynthetic organisms as well as most consumers. During photosynthesis, carbon (CO_2) moves from the environment to organisms that incorporate it into sugars. The main pathway of photosynthesis uses water and produces O_2.

Systems

Complex properties arise from interactions among components of a biological system.

Photosynthesis begins with pigments, which have a particular pattern of molecular structure that allows them to absorb light energy. In a series of coordinated reaction pathways, photosynthetic pigments and other molecules embedded in special membranes capture the energy of light, then store it in ATP and reduced coenzymes. Energy stored in these molecules powers the synthesis of sugars from carbon dioxide. In eukaryotes, these processes occur in chloroplasts.

Evolution

Evolution underlies the unity and diversity of life.

Variations in photosynthetic pathways are evolutionary adaptations that allow photosynthetic organisms to thrive in a variety of environments. Like building blocks, molecules and processes in complex metabolic pathways are often evolutionarily reused in pathways that offer enhanced benefits such as improved photosynthetic efficiency.

Links to Earlier Concepts

This chapter explores the main metabolic pathways (Sections 5.5) by which organisms harvest energy from the sun (5.2, 5.3). We revisit experimental design (1.6), electrons and energy levels (2.3), bonds (2.4), carbohydrates (3.3), plastids and cell walls of plants (4.7, 4.9), free energy (5.3), ATP as an energy carrier (5.6), membrane proteins (5.7), active transport (5.9), and concentration gradients (5.8).

6.1 Biofuels

Today, the expression "food is fuel" is not just about eating. Consider how the human population continues to expand exponentially, but Earth has a finite amount of resources available to sustain us. For hundreds of years, we have been using up one resource—fossil fuels—to provide the energy that we consume for transportation, lighting, cooking, heating, and so on. Coal, crude oil, natural gas, and other fossil fuels are the remains of ancient swamp forests that decayed and compacted over millions of years. Earth has a limited supply of these materials, and we are well on our way to exhausting them.

Plants and other organic materials not derived from fossilized matter can also be used as a source of energy. Oils, gases, and alcohols made from these materials are called biofuels. Like fossil fuels, biofuels can power our cars. Unlike fossil fuels, they are a renewable energy resource: We can always make more simply by growing more plants.

Corn and other food crops are rich in oils, starches, and sugars that can be easily converted to biofuels. The starch in corn kernels, for example, can be enzymatically broken down to glucose, which is converted into a biofuel by bacteria or yeast. Making biofuels from other types of plant matter requires additional steps, because these materials contain a higher proportion of cellulose. Breaking down this tough, insoluble carbohydrate to its glucose monomers adds a lot of cost to the biofuel product. Thus, almost all biofuels are produced from food crops. Growing food crops for our expanding population is using up another one of Earth's limited resources: physical space suitable for agriculture. Thus, making biofuels is removing food from our tables.

How did we end up competing with our vehicles for food? We both run on the same fuel: energy that plants have stored in the chemical bonds of organic molecules. Fossil fuels consist mainly of molecules originally assembled by ancient plants. Biofuels—and foods—consist mainly of molecules originally assembled by recently living plants.

The energy that plants store in organic molecules comes from sunlight. Photosynthesis is the process in which plants capture energy from the sun and store it in organic molecules. That energy fuels us and other consumers, as when an animal cell powers ATP synthesis by breaking the bonds of sugars (a topic detailed in Chapter 7). It also fuels our cars, which run on energy released by burning biofuels or fossil fuels. Both processes are fundamentally the same: They release energy by breaking the bonds of organic molecules.

Data Analysis Activities

Energy Efficiency of Biofuel Production Most of the plant material currently used for biofuel production consists of food crops—mainly corn, soybeans, and sugarcane. In 2006, David Tilman and his colleagues published the results of a 10-year study comparing the net energy output of various biofuels. The researchers grew a mixture of native perennial grasses without irrigation, fertilizer, pesticides, or herbicides, in sandy soil that was so depleted by intensive agriculture that it had been abandoned. They measured the usable energy in biofuels made from the grasses, and also from corn and soy, then measured the energy it took to grow and produce biofuel from each kind of crop (**FIGURE 6.1**).

1. About how much energy did ethanol produced from one hectare of corn yield? How much energy did it take to grow the corn to make that ethanol?
2. Which of the biofuels tested had the highest ratio of energy output to energy input?
3. Which of the three crops would require the least amount of land to produce a given amount of biofuel energy?

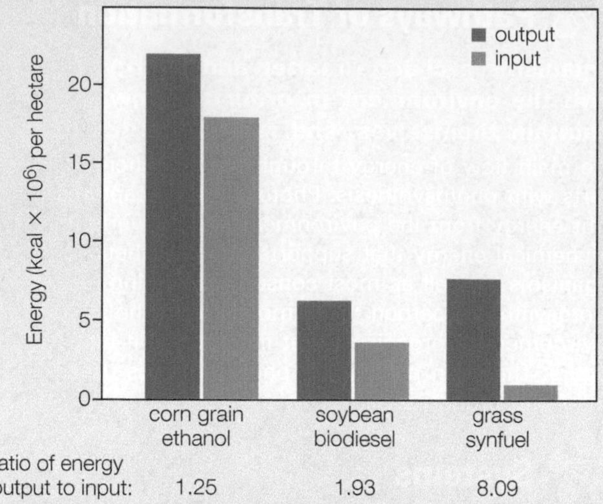
output
input
Energy (kcal × 10^6) per hectare
20
15
10
5
0
corn grain ethanol
soybean biodiesel
grass synfuel
ratio of energy output to input:
1.25
1.93
8.09

FIGURE 6.1 Energy inputs and outputs of biofuels made from three different crops. One hectare is about 2.5 acres.

6.2 Overview of Photosynthesis

LEARNING OBJECTIVES

- Explain why we say that photosynthesis feeds most life on Earth.
- Describe the structure of the thylakoid membrane.
- Write an equation that summarizes the overall pathway of photosynthesis.

All life is sustained by inputs of energy, but not all forms of energy can sustain life. Sunlight, for example, is abundant here on Earth, but it cannot directly power energy-requiring metabolic reactions. For this purpose, the energy of sunlight must first be converted to the energy of chemical bonds (Section 5.3). Unlike light, chemical energy can power the reactions of life, and it can be stored for later use.

Energy flow through most ecosystems begins with **autotrophs**, which are organisms that make their own food by harvesting energy directly from the environment (*auto-* means self; *-troph* refers to nourishment). All organisms need carbon to build the molecules of life; autotrophs obtain it from inorganic molecules such as carbon dioxide (CO_2). Plants and most other autotrophs harvest energy from the environment by photosynthesis. **Photosynthesis** is a metabolic pathway that uses light to drive the assembly of carbohydrates—sugars—from carbon dioxide and water. The sugars can be stored as polysaccharides for later use, remodeled into other compounds, or broken down to release energy held in their bonds (a topic of Chapter 7).

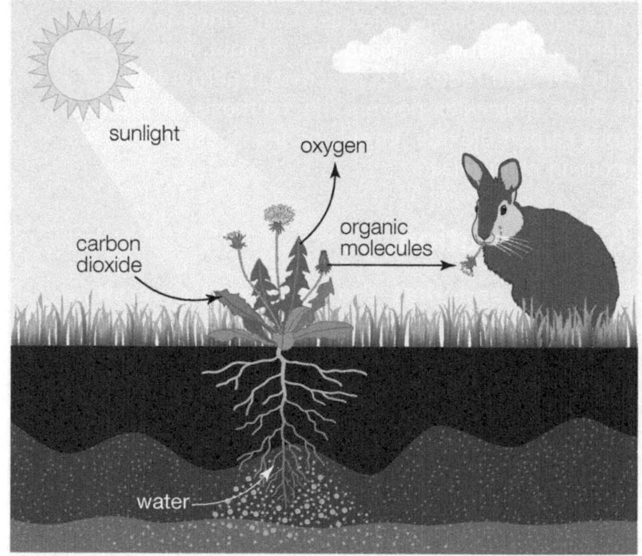
sunlight
oxygen
organic molecules
carbon dioxide
water

FIGURE 6.2 How photosynthesis sustains life. Photosynthetic autotrophs get energy from sunlight, and carbon from carbon dioxide. Most heterotrophs get their energy and carbon from organic molecules assembled by photosynthetic autotrophs.

Autotrophs are an ecosystem's producers (Section 1.3). **Heterotrophs** are consumers; they get their carbon by breaking down organic molecules that were assembled by other organisms (*hetero-* means other). Humans and almost all other heterotrophs obtain carbon and energy from organic molecules originally assembled by photosynthesizers. Thus, directly or indirectly, photosynthesis feeds most life on Earth (**FIGURE 6.2**).

Two Stages of Reactions

The main pathway of photosynthesis is often summarized by this equation:

$$CO_2 + H_2O \xrightarrow{\text{light energy}} \text{sugars} + O_2$$

CO_2 is carbon dioxide, and O_2 is oxygen; both gases are abundant in the atmosphere. Photosynthesis is not a single reaction, however; it is a metabolic pathway with many reactions that occur in two stages. The reactions of the first stage require light, so they are collectively called the **light-dependent reactions**. The "photo" in photosynthesis means light, and it refers to the conversion of light energy to the bond energy of ATP during this stage. In addition to making ATP, the main light-dependent pathway splits water molecules and releases oxygen. Hydrogen ions and electrons from the water molecules are loaded onto $NADP^+$, so NADPH forms (**FIGURE 6.3A**).

The "synthesis" part of photosynthesis refers to the reactions of the second stage, which build sugars from CO_2 and water. These are collectively called the **light-independent reactions** because they are not powered by light. Light-independent reactions run on energy delivered by NADPH and ATP that formed during the light-dependent reactions (**FIGURE 6.3B**). At the end of the second stage, $NADP^+$ and ADP are recycled to work again in the light-dependent reactions (**FIGURE 6.3C**).

In plants, photosynthetic protists, and cyanobacteria, the light-dependent reactions are carried out by molecules embedded in a **thylakoid membrane**. This membrane encloses a continuous internal compartment, and, in eukaryotes, it occurs in chloroplasts (**FIGURE 6.4**). A chloroplast's two outer membranes enclose a single thylakoid membrane that is folded into stacks of interconnected disks. The thylakoid membrane is suspended in a cytosol-like fluid called **stroma**, as are the chloroplast's own DNA and ribosomes. The light-independent reactions take place in stroma.

autotroph (AH-toe-trof) Organism that makes its own food using energy from the environment and carbon from inorganic molecules such as CO_2.
heterotroph (HET-er-oh-trof) Organism that obtains carbon from organic compounds assembled by other organisms.
light-dependent reactions First stage of photosynthesis; collectively convert light energy to chemical energy of ATP. The main pathway also splits water and produces oxygen and NADPH.
light-independent reactions Second stage of photosynthesis; collectively use ATP and NADPH to assemble sugars from CO_2 and water.
photosynthesis Metabolic pathway by which most autotrophs use the energy of light to make sugars from carbon dioxide and water.
stroma (STROH-muh) Cytosol-like fluid between the thylakoid membrane and the two outer membranes of a chloroplast.
thylakoid membrane (THIGH-la-koyd) Inner membrane system that carries out light-dependent reactions in chloroplasts and cyanobacteria.

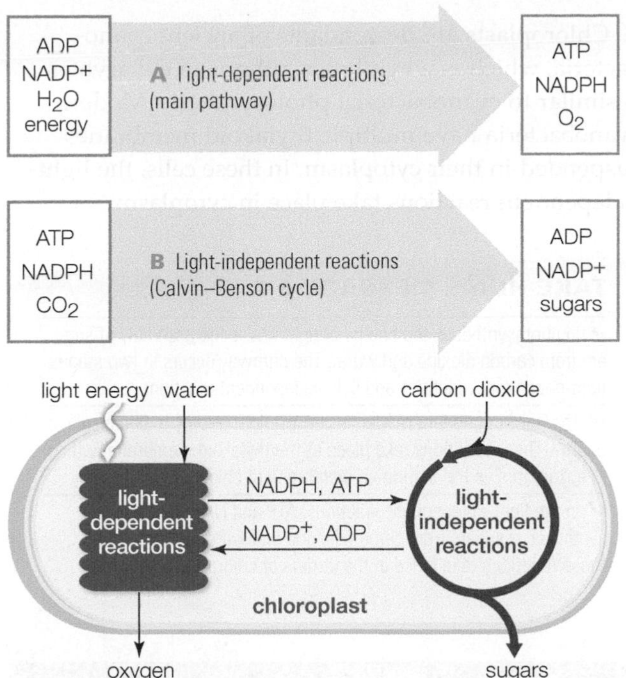

ADP
NADP+
H2O
energy
A Light-dependent reactions (main pathway)
ATP
NADPH
O2
ATP
NADPH
CO2
B Light-independent reactions (Calvin–Benson cycle)
ADP
NADP+
sugars
light energy
water
carbon dioxide
light-dependent reactions
NADPH, ATP
NADP+, ADP
light-independent reactions
chloroplast
oxygen
sugars

C Light energy drives the production of ATP in the light-dependent reactions; the main pathway also produces NADPH and O_2. NADPH and ATP drive sugar production in the light-independent reactions.

FIGURE 6.3 How coenzymes connect the first- and second-stage reactions of photosynthesis. Substrates and products of the main pathways are shown as they occur in chloroplasts. With some variation, cyanobacteria use the same pathways.

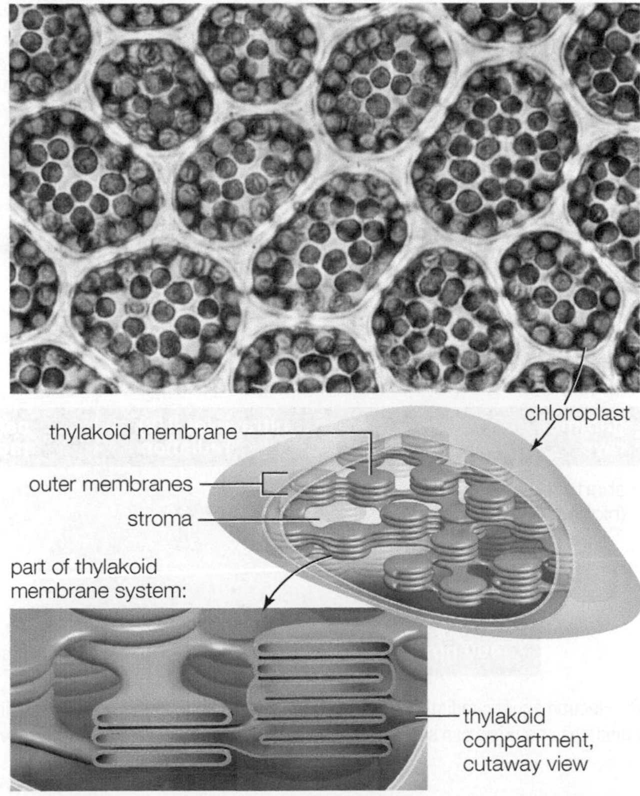

chloroplast
thylakoid membrane
outer membranes
stroma
part of thylakoid membrane system:
thylakoid compartment, cutaway view

FIGURE 6.4 Zooming in on chloroplast structure. The micrograph shows chloroplast-stuffed cells of a moss (*Rhizomnium*).

Chloroplasts are descendants of ancient cyanobacteria, which is why photosynthesis in eukaryotes is similar to cyanobacterial photosynthesis. Modern cyanobacteria have multiple thylakoid membranes suspended in their cytoplasm. In these cells, the light-independent reactions take place in cytoplasm.

TAKE-HOME MESSAGE 6.2

- ✔ In photosynthesis, the energy of light drives the assembly of sugars from carbon dioxide and water. The pathway occurs in two stages: light-dependent reactions and light-independent reactions.
- ✔ The light-dependent reactions convert light energy to chemical energy. These reactions take place at the thylakoid membrane, which, in eukaryotes, is the innermost membrane of chloroplasts.
- ✔ In the light-independent reactions, ATP and NADPH drive the synthesis of sugars from carbon dioxide and water. In eukaryotes, these reactions take place in the stroma of chloroplasts.

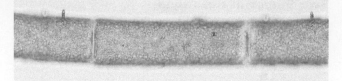

A Each cell in a strand of *Cladophora* is filled with a single chloroplast. Theodor Engelmann used this and other species of green algae in a series of experiments to determine if some colors of light are better for photosynthesis than others.

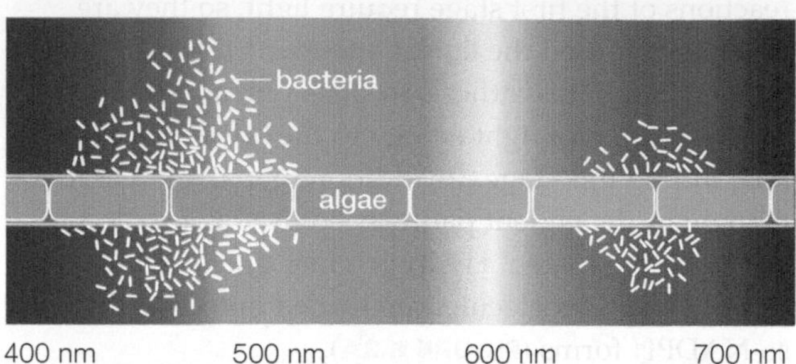
bacteria
algae
400 nm
500 nm
600 nm
700 nm
Wavelength

B Engelmann directed light through a prism so that bands of colors crossed a water droplet on a microscope slide. The water held a strand of photosynthetic algae, and also oxygen-requiring bacteria.

The bacteria swarmed around the algal cells that were releasing the most oxygen—the ones most actively engaged in photosynthesis. Those cells were under blue and red light.

FIGURE 6.5 Discovery that photosynthesis is driven by particular wavelengths of light.

6.3 Sunlight as an Energy Source

LEARNING OBJECTIVES

- Summarize Theodor Engelmann's photosynthesis experiment.
- Describe the relationship between a photon's wavelength and its energy.
- Explain how a pigment absorbs light.

In 1882, botanist Theodor Engelmann designed a series of experiments to test his hypothesis that the color of light affects the rate of photosynthesis. It had long been known that photosynthesis releases oxygen, so Engelmann used oxygen emission as an indirect measure of photosynthetic activity. He directed a spectrum of light across individual strands of green algae suspended in water (**FIGURE 6.5A**). Oxygen-sensing equipment had not yet been invented, so Engelmann used motile, oxygen-requiring bacteria to show him where the oxygen concentration in the water was highest. The bacteria moved through the water and gathered mainly where blue and red light fell across the algal cells (**FIGURE 6.5B**). Engelmann concluded that photosynthetic cells illuminated by light of these colors were releasing the most oxygen—a sign that blue and red light are the best for driving photosynthesis.

FIGURE 6.6 Properties of light.

visible light

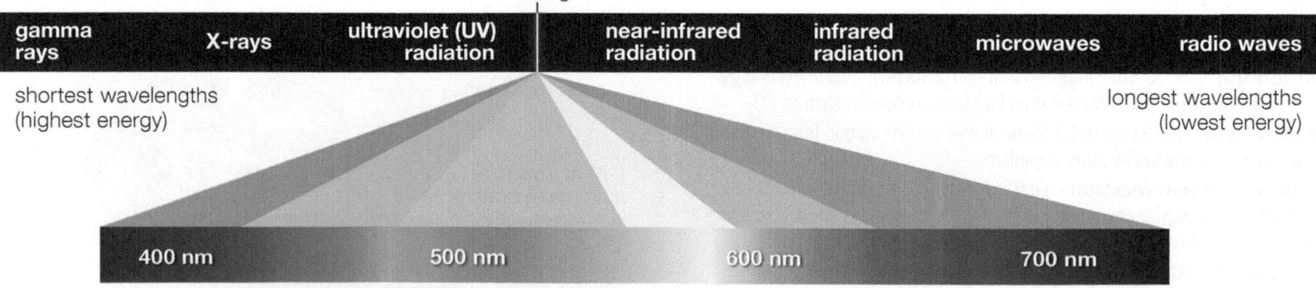
gamma rays
X-rays
ultraviolet (UV) radiation
near-infrared radiation
infrared radiation
microwaves
radio waves
shortest wavelengths (highest energy)
longest wavelengths (lowest energy)
400 nm
500 nm
600 nm
700 nm

A Electromagnetic radiation moves through space in waves that we measure in nanometers (nm). Visible light makes up a very small part of this energy. Raindrops or a prism can separate visible light's different wavelengths, which we see as different colors. About 25 million nanometers are equal to 1 inch.

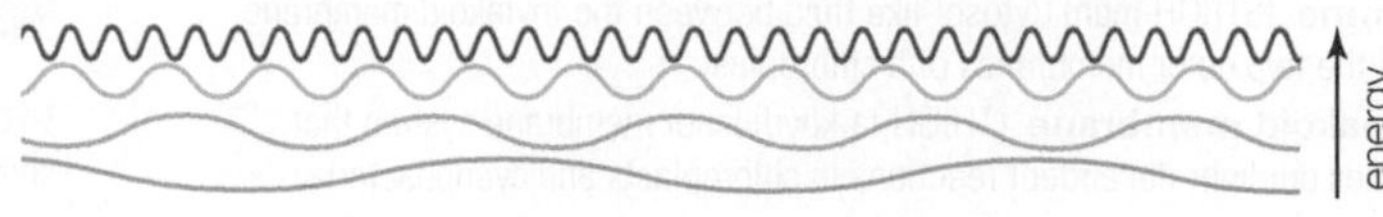
energy

B Light is organized as packets of energy called photons. The shorter a photon's wavelength, the greater its energy. Thus, photons of violet light carry more energy than photons of blue light, and so on.

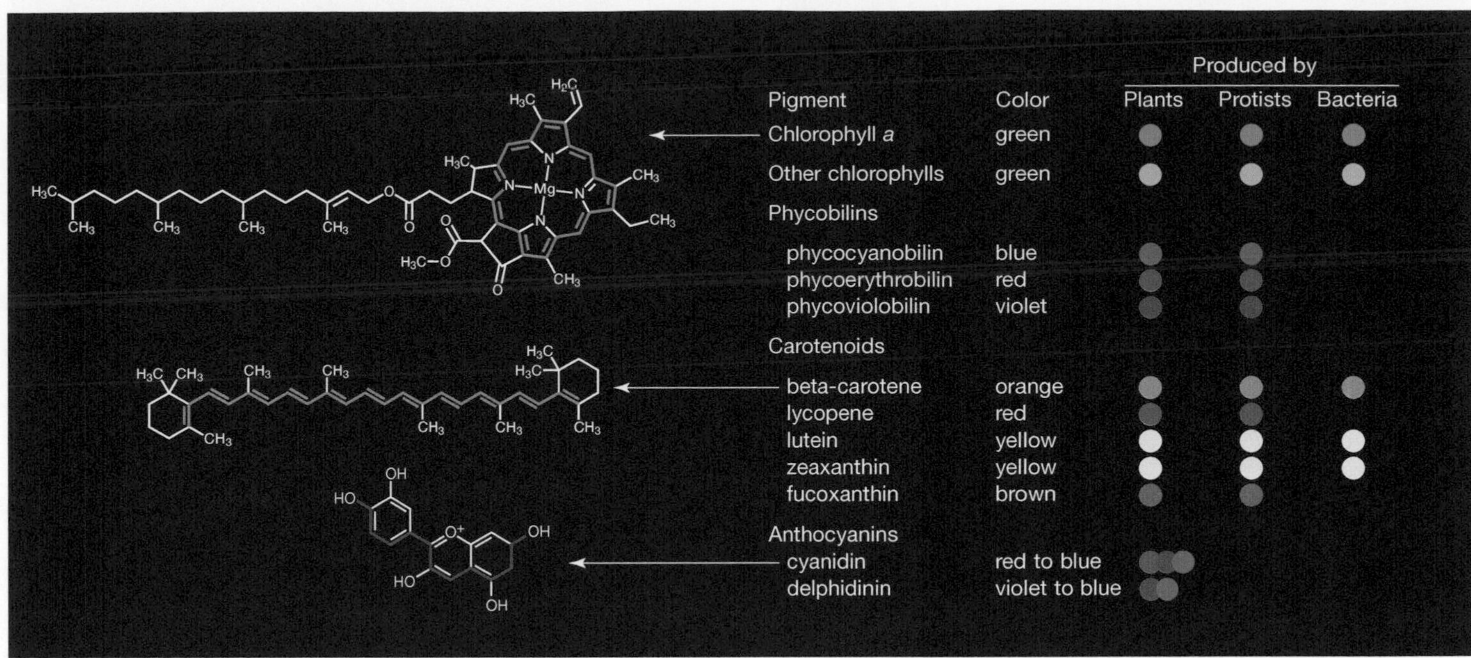

Produced by
Pigment
Color
Plants
Protists
Bacteria
Chlorophyll *a*
green
Other chlorophylls
green
Phycobilins
phycocyanobilin
blue
phycoerythrobilin
red
phycoviolobilin
violet
Carotenoids
beta-carotene
orange
lycopene
red
lutein
yellow
zeaxanthin
yellow
fucoxanthin
brown
Anthocyanins
cyanidin
red to blue
delphidinin
violet to blue

FIGURE 6.7 Examples of photosynthetic pigments. Left, the light-catching part of a pigment (shown in color) is the region in which single bonds alternate with double bonds. These and many other pigments are derived from evolutionary remodeling of the same compound (compare the structure of chlorophyll *a* with heme, Section 5.6, which is the functional part of the pigment hemoglobin). Animals convert dietary beta-carotene into a similar pigment (retinal) that is the basis of vision.

Engelmann's conclusion was correct: Blue and red light are indeed best for driving photosynthesis in many types of cells. Why? To answer that question, you need to know a bit about the nature of light.

Light is electromagnetic radiation, a type of energy that moves through space in waves, a bit like waves move across an ocean. The distance between the crests of two successive waves is called **wavelength**, and it is measured in nanometers (nm).

Light that humans can see is a small part of the spectrum of electromagnetic radiation emitted by the sun. Visible light travels in wavelengths between 380 and 750 nm (**FIGURE 6.6A**). Our eyes perceive particular wavelengths in this range as different colors; light with all of these wavelengths combined appears white. White light separates into its component colors when it passes through a prism, or through raindrops that act as tiny prisms (left). A prism bends longer wavelengths more than it bends shorter ones, so a rainbow of colors forms.

Light travels in waves, but it is also organized in packets of energy called photons. A photon's wavelength and its energy are related, so all photons traveling at the same wavelength carry the same amount of energy. Photons with the most energy travel in shorter wavelengths; those with the least energy travel in longer wavelengths (**FIGURE 6.5B**).

To Catch a Rainbow

Photosynthesizers use pigments to capture photons of visible light. A **pigment** is an organic molecule that selectively absorbs light of specific wavelengths. This molecule is a bit like an antenna specialized for receiving light. It has a light-trapping region: a carbon chain or ring in which single bonds alternate with double bonds. Electrons (Section 2.2) can move freely among all the atoms of this region. These electrons can easily absorb a photon—but not just any photon. Only those photons with precisely enough energy to boost the electrons to a higher energy level (shell) are absorbed. This is why a pigment absorbs light of only certain wavelengths.

Wavelengths of light that are not absorbed give each pigment its characteristic color (**FIGURE 6.7**). The most common photosynthetic pigment in plants, cyanobacteria, and photosynthetic protists is **chlorophyll *a***. Chlorophyll *a* absorbs violet, red, and orange light, and it reflects green light, so it appears green to us. Most photosynthetic cells also have accessory pigments, including other chlorophylls, that work together with chlorophyll *a*. Accessory pigments maximize the range

chlorophyll *a* (KLOR-uh-fil) Main photosynthetic pigment in plants.
pigment An organic molecule that can absorb light of certain wavelengths.
wavelength Distance between the crests of two successive waves.

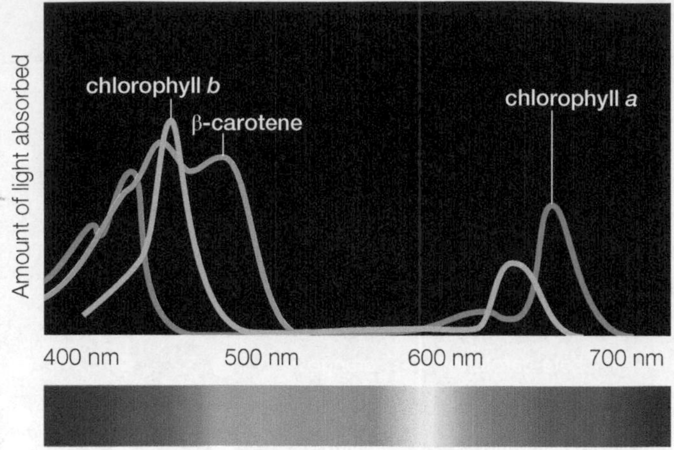
Amount of light absorbed
chlorophyll *b*
β-carotene
chlorophyll *a*
400 nm
500 nm
600 nm
700 nm

FIGURE 6.8 Efficiency with which a few photosynthetic pigments absorb wavelengths of visible light. In this absorption spectrum, line color indicates the characteristic color of each pigment. Compare Figure 6.5B.

FIGURE 6.9 An organism's photosynthetic pigments are an evolutionary adaptation to life in a certain habitat. Algae that can live far below the sea surface such as this *Polysiphonia* tend to be rich in phycobilins and other pigments that absorb green and blue-green light—wavelengths that penetrate water most efficiently.

of wavelengths that an organism can use for photosynthesis (**FIGURE 6.8**).

In plants, accessory pigments have roles in addition to photosynthesis. Appealing colors, for example, attract pollinators to flowers and animals to fruits. You are already familiar with accessory pigments: Carrots, for example, are orange because they contain beta-carotene (β-carotene), which functions as an antioxidant and also in signaling pathways between chloroplast and nucleus. Roses are red and violets are blue because their cells make anthocyanin, another antioxidant and also a natural sunscreen.

Different photosynthetic species use different combinations of accessory pigments during photosynthesis. Why? Like all organisms, photosynthesizers are adapted to the environment in which they evolved, and light that reaches different environments varies in its proportions of wavelengths. Consider that seawater absorbs green and blue-green light less efficiently than other colors. Thus, more green and blue-green light penetrates deeper ocean water. Algae that evolved in deep water tend to be rich in accessory pigments that absorb green and blue-green light (**FIGURE 6.9**).

In green plants, chlorophylls are usually so abundant that they mask the colors of the other pigments. Plants that change color during autumn are preparing for a period of dormancy; they conserve resources by moving nutrients from tender parts that would be damaged by winter cold (such as leaves) to protected parts (such as roots). Chlorophylls are not needed during dormancy, so they are disassembled and their components recycled. Yellow and orange accessory pigments are also recycled, but not as quickly as chlorophylls. Their colors begin to show as the chlorophyll content declines in leaves. Anthocyanin synthesis also increases in some plants, adding red and purple tones to turning leaf colors.

TAKE-HOME MESSAGE 6.3

- ✔ The sun emits electromagnetic radiation (light). Visible light is the main form of energy that drives photosynthesis.
- ✔ Light travels in waves and is organized as photons. We see different wavelengths of visible light as different colors.
- ✔ Pigments absorb light at specific wavelengths. Photosynthetic species use pigments such as chlorophyll *a* to harvest the energy of light for photosynthesis.
- ✔ A combination of pigments allows a photosynthetic cell to efficiently capture the wavelengths of light most abundant in the habitat in which it evolved.

6.4 The Light-Dependent Reactions

LEARNING OBJECTIVES

- Describe what happens when a pigment that is part of a light-harvesting complex absorbs light.
- Explain the process of electron transfer phosphorylation.
- Compare the cyclic and noncyclic light-dependent reactions.
- Explain the evolutionary advantages of the noncyclic pathway.

Photosynthesis begins when energy is absorbed by a photosystem. A **photosystem** is a very large complex of molecules—pigments, proteins, and cofactors—embedded in the thylakoid membrane. The core of a photosystem is called a reaction center, and it contains a "special pair" of closely associated chlorophyll *a*

electron transfer phosphorylation Process in which electron flow through electron transfer chains sets up a hydrogen ion gradient that drives ATP formation.

photosystem Large protein complex in the thylakoid membrane; each consists of pigments and other molecules that collectively convert light energy to chemical energy in photosynthesis.

molecules (FIGURE 6.10A) that jointly absorb energy. When a special pair absorbs energy, it emits electrons.

There are two kinds of photosystems, type I and type II (named in order of their discovery). The special pair in a photosystem I is called p700 because the photons it absorbs best are 700 nm in wavelength; the special pair in a photosystem II best absorbs photons that are 680 nm in wavelength, so it is called p680. However, most of the energy absorbed by photosystems comes from nearby light-harvesting complexes.

A light-harvesting complex is a circular array of chlorophylls, various accessory pigments, lipids, and proteins (FIGURE 6.10B). When a pigment in a light-harvesting complex absorbs a photon, one of its electrons jumps to a higher energy level (shell, Section 2.3). The electron quickly drops back down to a lower shell by emitting its extra energy. Light-harvesting complexes hold on to that emitted energy by passing it back and forth, a bit like volleyball players pass a ball among team members. A light-harvesting complex can pass this energy to the reaction center of a nearby photosystem.

Photosystems operate in two versions of light-dependent reactions, a noncyclic pathway and a cyclic pathway. Both produce ATP. The noncyclic pathway, which is the primary one in plants, photosynthetic protists (such as algae), and cyanobacteria, also produces oxygen (O_2) and NADPH.

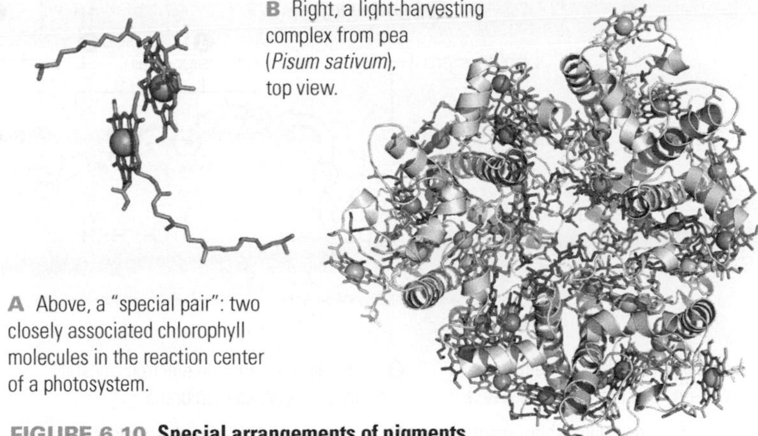

A Above, a "special pair": two closely associated chlorophyll molecules in the reaction center of a photosystem.

B Right, a light-harvesting complex from pea (*Pisum sativum*), top view.

FIGURE 6.10 Special arrangements of pigments in the thylakoid membrane. Chlorophylls are shown in green; beta-carotene and other pigments, in orange and yellow. Proteins are shown in tan, and lipids are in pink.

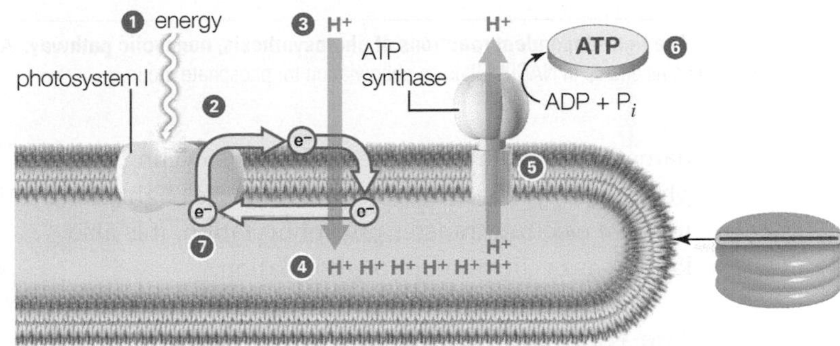

1 energy
photosystem I
2
3 H+
ATP synthase
H+
ATP
6
ADP + P_i
e−
e−
e−
5
7
H+
4 H+ H+ H+ H+ H+ H+

FIGURE 6.11 Light-dependent reactions, cyclic pathway.

1 A special pair in photosystem I absorbs light energy and emits electrons (e−).

2 The electrons enter an electron transfer chain in the thylakoid membrane.

3 Energy released by the electrons as they move through the electron transfer chain is used to actively transport hydrogen ions (H^+) from the stroma into the thylakoid compartment.

4 The movement of electrons through the electron transfer chain sets up and maintains a hydrogen ion gradient across the thylakoid membrane.

5 Hydrogen ions in the thylakoid compartment follow their gradient across the thylakoid membrane by flowing through ATP synthases.

6 Hydrogen ion flow through ATP synthases causes these proteins to phosphorylate ADP, so ATP forms in the stroma (electron transfer phosphorylation).

7 Electrons that reach the end of the electron transfer chain cycle back to photosystem I, thus replacing its lost electrons.

The Cyclic Pathway

Light-dependent reactions in the cyclic pathway use only photosystem I (FIGURE 6.11). The pathway begins when the special pair in a photosystem I absorbs energy 1. Absorbing energy causes the special pair to release electrons, and these enter an electron transfer chain in the thylakoid membrane 2. Remember that electron transfer chains can harvest the energy of electrons in a series of redox reactions, releasing a bit of energy at each step (Section 5.5). In this case, molecules of the electron transfer chain use the released energy to actively transport hydrogen ions (H^+) across the membrane, from the stroma to the thylakoid compartment 3. Thus, the movement of electrons through the electron transfer chain sets up and maintains a hydrogen ion gradient across the thylakoid membrane 4.

The hydrogen ion gradient set up by the movement of electrons through the electron transfer chain is a type of potential energy that can be tapped to make ATP. The ions want to follow their gradient by moving back into the stroma, but ions cannot diffuse through a lipid bilayer (Section 5.8). Hydrogen ions leave the thylakoid compartment only by moving through ATP synthases in the thylakoid membrane 5. An ATP synthase is a small molecular motor that is both a transport protein and an enzyme. When hydrogen ions in the thylakoid compartment flow through its interior, an ATP synthase phosphorylates ADP, so ATP forms in the stroma 6. Any process in which the flow of electrons through electron transfer chains drives ATP formation is called **electron transfer phosphorylation.**

A photosystem that loses electrons must replace them immediately. Photosystem I does this by accepting electrons that have reached the end of the electron transfer chain 7. The cyclic pathway is

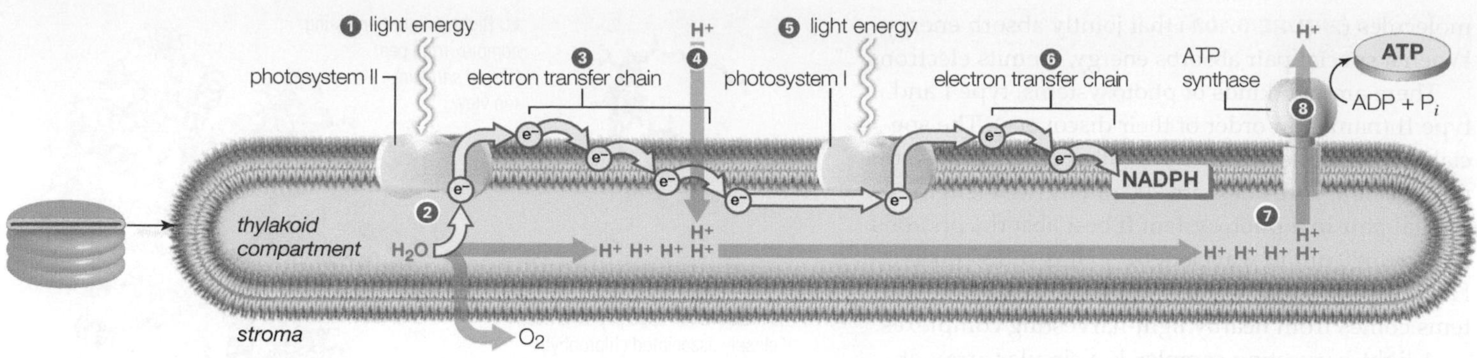

1 light energy
photosystem II
3 electron transfer chain
H+
4
5 light energy
photosystem I
6 electron transfer chain
ATP synthase
H+
ATP
ADP + P_i
8
e−
NADPH
2
thylakoid compartment
H_2O
H+ H+ H+ H+
7
H+ H+ H+ H+
stroma
O_2

1 A special pair in photosystem II absorbs energy and emits electrons (e−).

2 The photosystem pulls replacement electrons from water molecules, which then break apart into hydrogen ions and oxygen atoms. The oxygen leaves the cell in O_2 gas.

3 The electrons enter an electron transfer chain in the thylakoid membrane.

4 Energy released by the electrons as they move through the chain is used to actively transport hydrogen ions (H^+) from the stroma into the thylakoid compartment. A hydrogen ion gradient forms across the thylakoid membrane.

5 A photosystem I absorbs energy and its special pair emits electrons. Replacement electrons come from photosystem II via an electron transfer chain.

6 Electrons from photosystem I enter an electron transfer chain, then combine with $NADP^+$ and H^+ to form NADPH.

7 Hydrogen ions in the thylakoid compartment follow their gradient across the thylakoid membrane by flowing through ATP synthases.

8 Hydrogen ion flow causes ATP synthases to phosphorylate ADP, so ATP forms in the stroma.

FIGURE 6.12 The light-dependent reactions of photosynthesis, noncyclic pathway. ATP and oxygen gas are produced in this pathway. Electrons that travel through two electron transfer chains end up in NADPH. P_i is an abbreviation for phosphate group.

named after the movement of electrons cycling back to photosystem I. Because this pathway uses light energy to drive electron transfer phosphorylation, it is also known as cyclic photophosphorylation.

The Noncyclic Pathway

Light-dependent reactions that occur in the noncyclic pathway use both photosystem I and photosystem II (**FIGURE 6.12**). The pathway begins when the special pair in a photosystem II absorbs energy 1 and emits electrons. The electrons immediately enter an electron transfer chain in the thylakoid membrane.

A photosystem II that has lost electrons pulls replacements off of water molecules in the thylakoid compartment. A water molecule does not give up electrons easily; doing so breaks it apart into hydrogen ions and oxygen atoms 2. This and any other process by which a molecule is broken apart by light energy is called **photolysis**. The hydrogen ions stay in the thylakoid compartment; oxygen atoms combine and diffuse out of the cell as oxygen gas (O_2).

Meanwhile, the electrons from photosystem II move through the electron transfer chain in the thylakoid membrane 3. As in the cyclic pathway, molecules of the electron transfer chain use energy released by the electrons to actively transport hydrogen ions (H^+) across the membrane, from the stroma to the thylakoid compartment 4. The movement of electrons through the electron transfer chain sets up and maintains a hydrogen ion gradient across the thylakoid membrane.

Electrons that have reached the end of the electron transfer chain are accepted by a photosystem I. When the reaction center of this photosystem absorbs energy, its special pair releases electrons 5. Photosystem I can release these electrons to one of two electron transfer chains. The first chain operates in the cyclic pathway, and it shuttles the electrons back to photosystem I. The second chain operates in the noncyclic pathway 6. At the end of this second electron transfer chain, the coenzyme $NADP^+$ accepts the electrons along with H^+, so NADPH forms:

$$NADP^+ + \text{electrons} + H^+ \longrightarrow \text{NADPH}$$

NADPH is a powerful reducing agent (electron donor) that will be used later, in the light-dependent reactions.

As in the cyclic pathway, the hydrogen ion gradient that forms across the thylakoid membrane forces hydrogen ions through ATP synthases 7. The flow causes these proteins to phosphorylate ADP, so ATP forms in the stroma 8 (electron transfer phosphorylation). Because the noncyclic pathway uses light energy to drive electron transfer phosphorylation, it is also known as noncyclic photophosphorylation.

Evolution of the Two Pathways

Photosynthesis offers an example of how metabolism runs on energy harvested from the environment: Just as water flowing downhill can drive a turbine, electrons flowing down a free energy gradient can drive an ATP generator (**FIGURE 6.13**).

Why are there two pathways of light-dependent reactions? The cyclic pathway evolved first. It allowed early organisms to produce ATP photosynthetically, but not the NADPH required for sugar-producing

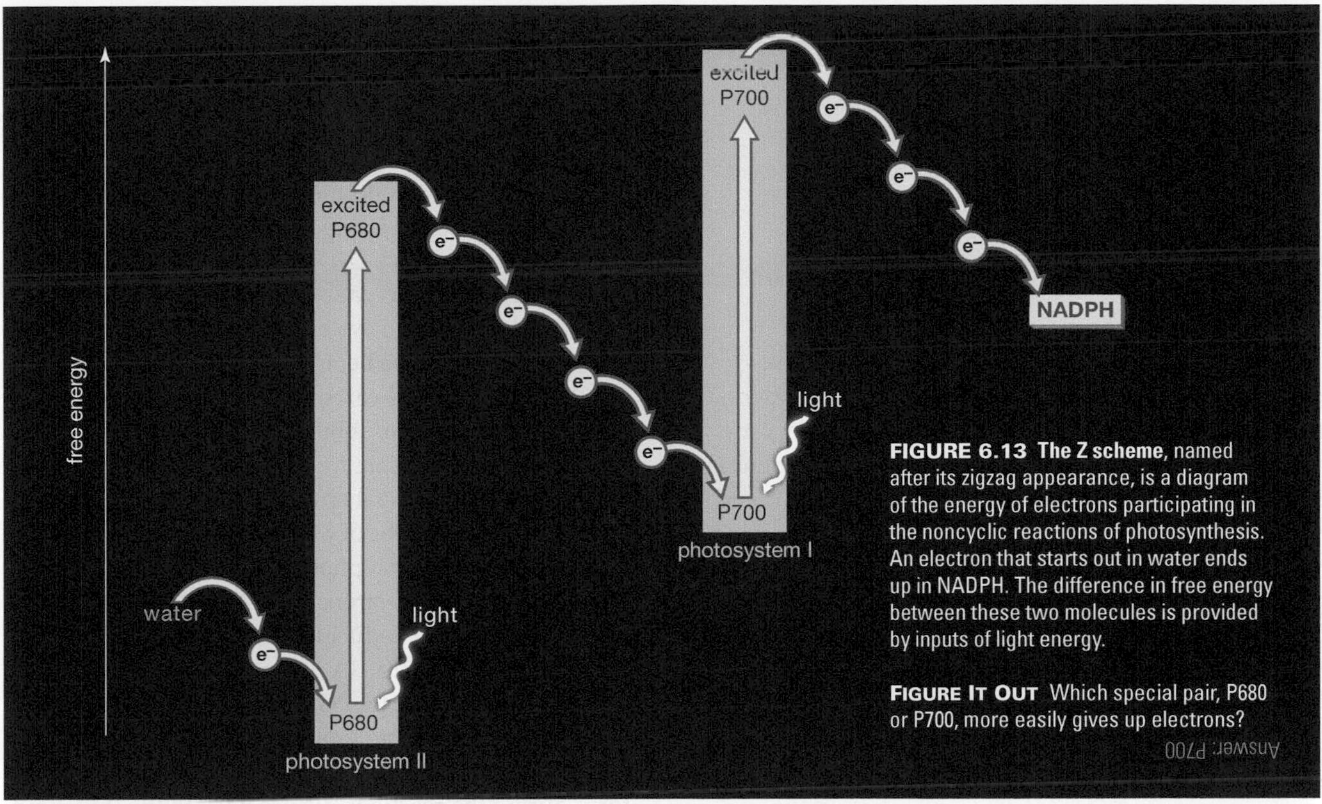

free energy
water
e⁻
light
P680
photosystem II
excited
P680
e⁻
e⁻
e⁻
e⁻
excited
P700
P700
photosystem I
light
e⁻
e⁻
e⁻
NADPH

FIGURE 6.13 The Z scheme, named after its zigzag appearance, is a diagram of the energy of electrons participating in the noncyclic reactions of photosynthesis. An electron that starts out in water ends up in NADPH. The difference in free energy between these two molecules is provided by inputs of light energy.

FIGURE IT OUT Which special pair, P680 or P700, more easily gives up electrons?

Answer: P700

reactions. Both processes run on electron energy, so they require a continuous supply of electrons. The early photosynthesizers harvested electrons from a limited resource: reduced inorganic compounds that formed by geologic processes.

As photosynthetic organisms evolved, their molecular machinery got an upgrade. Photosystem I became remodeled into photosystem II, and the new photosystem was incorporated into the existing reactions. The resulting noncyclic pathway was advantageous for cells that used it. First, it freed these organisms from the requirement for scarce inorganic compounds: Cells can live only where there is water, so their supply of electrons became convenient and essentially unlimited.

Second, the new pathway made sugar production much more efficient. Photosystem II is a very strong oxidizer—it is the only biological system strong enough to pull electrons from water molecules. Thus, it can acquire more electrons from the environment than can photosystem I. More electrons power the production of more ATP, and more ATP powers the production of more sugars. Electrons reduce NAD^+ at the end of the noncyclic pathway, so extra electron-harvesting reactions became unnecessary for making the NADPH required for sugar production.

photolysis (foe-TALL-ih-sis) Process by which light energy breaks down a molecule.

Today, all plants and cyanobacteria use the noncyclic pathway. However, on its own, this pathway does not yield enough ATP to balance NADPH use in sugar production. The cyclic pathway provides extra ATP for this purpose. It also allows light-dependent reactions to continue when the noncyclic pathway stalls, for example as occurs under intense illumination. Light energy in excess of what can be used for photosynthesis causes dangerous free radicals to form (Section 2.3). Photosystem II minimizes this effect. In strong light, it stops releasing electrons to electron transfer chains; instead, it emits absorbed energy as heat. The cyclic pathway predominates in this circumstance.

TAKE-HOME MESSAGE 6.4

- ✔ The light-dependent reactions convert light energy to chemical energy held in ATP and NADPH.
- ✔ Photosynthetic pigments in the thylakoid membrane transfer the energy of light to photosystems. Absorbing energy causes a photosystem to emit electrons that enter electron transfer chains.
- ✔ In both noncyclic and cyclic pathways, the flow of electrons through the transfer chains sets up hydrogen ion gradients that drive ATP formation, a process called electron transfer phosphorylation.
- ✔ The noncyclic pathway uses two photosystems. Water molecules are split, oxygen is released, and electrons end up in NADPH.
- ✔ The cyclic pathway uses only photosystem I. No NADPH forms, and no oxygen is released.

6.5 The Light-Independent Reactions

LEARNING OBJECTIVES

- State the role of carbon fixation in photosynthesis.
- Explain how coenzymes connect the light-dependent reactions with the Calvin–Benson cycle.
- Describe stomata and their function.
- Explain how C4 and CAM plants minimize photorespiration.

The Calvin–Benson Cycle

You learned in Section 6.2 that the reactions of the second stage of photosynthesis are light-independent because light energy does not power them: They can run night and day. Energy that drives these reactions is provided by phosphate-group transfers from ATP, and electrons from NADPH. Both molecules are products of the light-dependent reactions.

The light-independent reactions produce sugars, and they are collectively called the **Calvin–Benson cycle** (FIGURE 6.14). This cyclic pathway uses carbon atoms from CO_2 to build the carbon backbones of sugar molecules. Extracting carbon atoms from an inorganic source (such as CO_2) and incorporating them into an organic molecule is called **carbon fixation**.

The Calvin–Benson cycle runs in the stroma of chloroplasts. The reactions begin when the enzyme **rubisco** attaches CO_2 to a five-carbon organic compound called ribulose bisphosphate, abbreviated as RuBP ❶. The six-carbon intermediate that forms by this carbon-fixation reaction is unstable, so it splits right away into two three-carbon molecules of phosphoglycerate (PGA).

Each PGA receives a phosphate group from ATP, and hydrogen and electrons from NADPH ❷. Thus, ATP energy and the reducing power of NADPH convert each molecule of PGA into a molecule of phosphoglyceraldehyde (PGAL). PGAL is a three-carbon simple sugar with a phosphate group, and it is the product of the Calvin–Benson cycle ❸. (The $NADP^+$ and ADP that form during the reactions diffuse back to the thylakoid for reuse in the light-dependent reactions.)

Most of the PGAL produced in the Calvin–Benson cycle is used to regenerate RuBP, the starting compound of the cycle ❹. In plants, the remaining PGAL is usually exported from the chloroplast to the cell's cytoplasm ❺. PGAL is an intermediate in several metabolic pathways, and it can be assembled into a variety of carbohydrates. Most of the PGAL molecules that enter plant cell cytoplasm are combined to make sucrose ❻, the main sugar in plants (Section 3.3). Sucrose produced by photosynthetic cells is loaded into vascular tissues for transport to other parts of the plant. (Section 28.5 returns to the transport of organic molecules through the plant body.)

When sucrose production exceeds demand (for example, during a sunny day when the light-dependent reactions are running at top speed), some PGAL molecules are not exported from chloroplasts, and these are assembled into starch. Starch is disassembled at night, and its monosaccharide monomers are used to produce sucrose. An uninterrupted supply of sucrose can sustain the plant's metabolism and growth even in the dark.

FIGURE 6.14 The Calvin–Benson cycle.

This sketch shows a cross-section of a chloroplast with Calvin–Benson reactions cycling in the stroma. The steps are a summary of three cycles of reactions. Water is also a substrate, but not shown for clarity (Appendix III details the reactions). Black balls signify carbon atoms.

❶ Three CO_2 molecules diffuse into a photosynthetic cell, and then into a chloroplast. Rubisco attaches each to a five-carbon RuBP molecule. The resulting intermediate splits, so six molecules of three-carbon PGA form.

❷ Each PGA gets a phosphate group from ATP, plus hydrogen and electrons from NADPH, so six molecules of phosphoglyceraldehyde (PGAL) form ❸. This phosphorylated sugar is the product of the Calvin–Benson cycle.

❹ Five PGAL continue in reactions that regenerate 3 RuBP.

❺ The remaining PGAL is usually exported from the chloroplast. In plant cell cytoplasm, PGAL molecules are combined to form sucrose ❻.

FIGURE IT OUT For every molecule of carbon dioxide that enters the Calvin–Benson cycle, how many NADPH are reduced?

Answer: Two

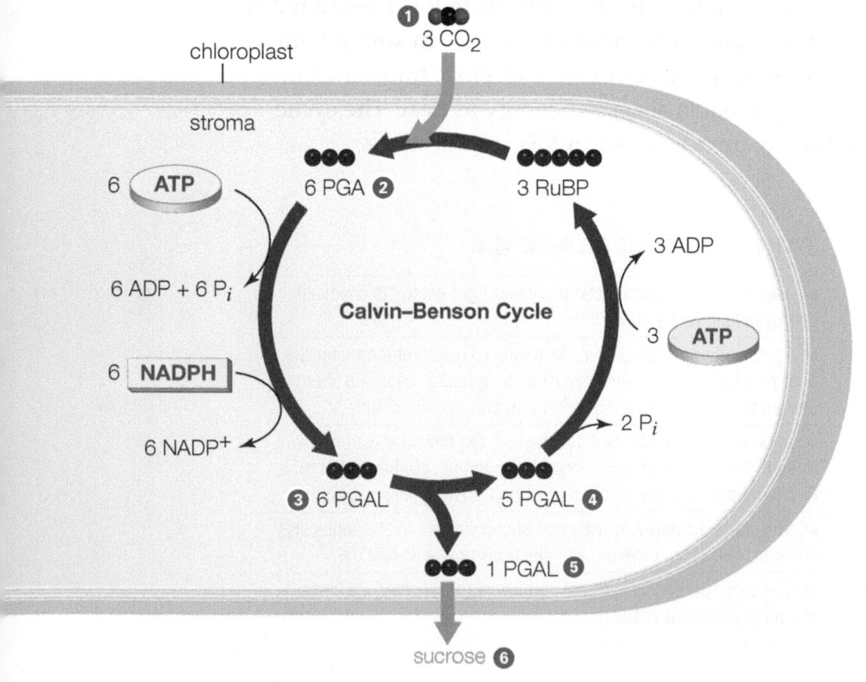

❶ 3 CO_2
chloroplast
stroma
6 ATP
6 PGA ❷
3 RuBP
3 ADP
6 ADP + 6 P_i
Calvin–Benson Cycle
3 ATP
6 NADPH
2 P_i
6 $NADP^+$
❸ 6 PGAL
5 PGAL ❹
1 PGAL ❺
sucrose ❻

Photorespiration

Most plants have a thin, waterproof cuticle that limits evaporative water loss from their aboveground parts. Gases cannot diffuse across the cuticle, but carbon dioxide needed for the Calvin–Benson cycle must enter the plant, and oxygen produced by the light-dependent reactions must escape it. Thus, the surfaces of leaves and stems are studded with tiny, closable pores called **stomata** (FIGURE 6.15A). When stomata are open, CO_2 diffuses from the air into photosynthetic tissues, and O_2 diffuses out of the tissues into the air. Stomata close to conserve water on hot, dry days. When that happens, gas exchange comes to a halt.

Plants that fix carbon only by the Calvin–Benson cycle are called **C3 plants** because a three-carbon molecule (PGA) is the first stable intermediate to form in their light-independent reactions. In C3 plants, both stages of photosynthesis run during the day. With stomata closed, the O_2 level in the plant's tissues rises, and the CO_2 level declines. This outcome can reduce the efficiency of sugar production because both gases are substrates of rubisco, and they compete for its active site.

Rubisco initiates the Calvin–Benson cycle by attaching CO_2 to RuBP. It also initiates a pathway called **photorespiration** by attaching O_2 to RuBP. The remainder of the photorespiration pathway converts the product of this reaction to a substrate of the Calvin–Benson cycle. ATP is required, and intermediates must be transported among three organelles (FIGURE 6.15B).

Photorespiration produces CO_2 (so carbon is lost instead of being fixed), and also ammonia that must be detoxified (requiring additional ATP). These extra steps and energy requirements make photorespiration an extremely inefficient way to produce sugars. C3 plants compensate for the inefficiency by making a lot of rubisco: It is the most abundant protein on Earth.

Despite being wasteful in terms of sugar production, however, photorespiration is an essential part of a wide range of other processes in plants. For example, it is a major source of hydrogen peroxide in photosynthetic cells, and as such it is required for several signaling pathways that govern growth and defense responses. (Chapter 27 returns to this topic.)

Alternative Pathways in Plants

In some plant lineages, structural and metabolic adaptations have evolved that minimize photorespiration. These plants also close stomata on hot, dry days, but additional steps in their light-independent reactions keep sugar production high.

C4 plants Corn is an example of a **C4 plant**, so named because the first stable intermediate to form in its light-independent reactions is a four-carbon compound called oxaloacetate. Like C3 plants, C4 plants close stomata on dry days, but their sugar production does not decline. C4 plants compensate for rubisco's inefficiency by fixing carbon twice, in two kinds of

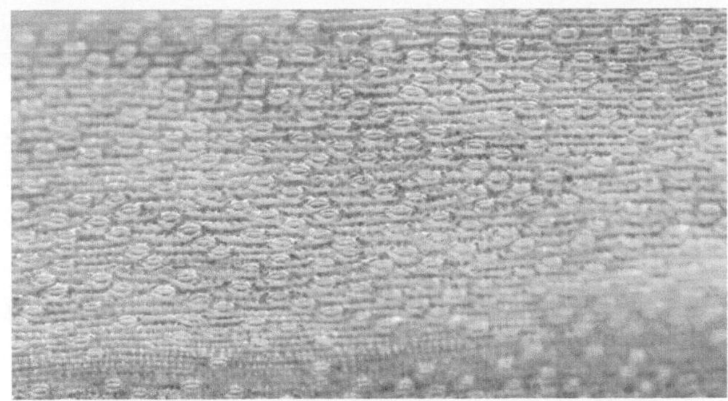

A Stomata on the surface of a leaf. When these tiny pores are open, they allow gas exchange between the plant's internal tissues and air.

Stomata close to conserve water on hot, dry days, and then gas exchange stops. Oxygen produced by the light-dependent reactions cannot exit the plant, and CO_2 required for the light-independent reactions cannot enter it. Thus, the amount of oxygen rises and the amount of CO_2 declines in photosynthetic tissues.

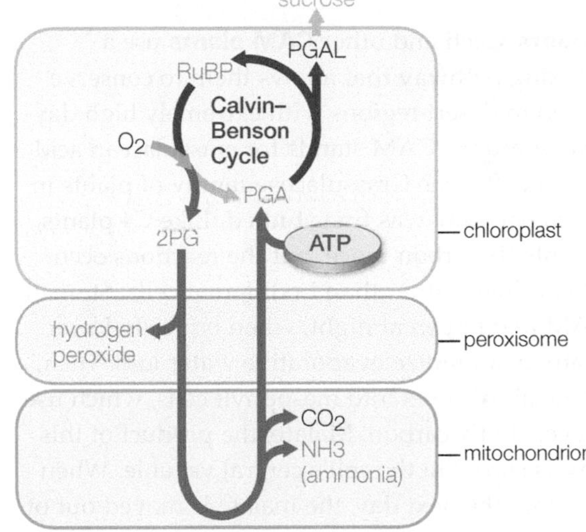

sucrose
PGAL
RuBP
Calvin–Benson Cycle
O2
PGA
2PG
ATP
chloroplast
hydrogen peroxide
peroxisome
CO2
NH3 (ammonia)
mitochondrion

B Photorespiration occurs when the oxygen level rises in photosynthetic tissues. A high O_2/CO_2 ratio causes rubisco to use oxygen as a substrate.

Several reactions are necessary to convert the product of this reaction to a molecule that can reenter the Calvin–Benson cycle. Intermediates are transported from the chloroplast to a peroxisome, then a mitochondrion, a peroxisome again, and back to a chloroplast. The energy required to carry out these extra steps makes photorespiration an inefficient way to produce sugars.

FIGURE 6.15 Photorespiration.

C3 plant Type of plant that uses only the Calvin–Benson cycle to fix carbon.
C4 plant Type of plant that minimizes photorespiration by fixing carbon twice, in two cell types.
Calvin–Benson cycle Cyclic carbon-fixing pathway that builds sugars from CO_2; light-independent reactions (second stage) of photosynthesis.
carbon fixation Process in which carbon from an inorganic source such as carbon dioxide gets incorporated into an organic molecule.
photorespiration Pathway initiated when rubisco attaches oxygen instead of carbon dioxide to RuBP (ribulose bisphosphate).
rubisco (roo-BIS-co) Ribulose bisphosphate carboxylase. Carbon-fixing enzyme of the Calvin–Benson cycle.
stomata (stow-MA-tuh) Singular, stoma. Gaps that open on plant surfaces; allow water vapor and gases to diffuse into and out of plant tissues.

cells (**FIGURE 6.16A**). The C4 reactions begin in mesophyll cells, where carbon is fixed by an enzyme that does not use oxygen even when the CO_2 level is low (**FIGURE 6.16B**). The product of this pathway, a four-carbon molecule called malate, is moved into bundle-sheath cells, where it is converted back to CO_2.

Rubisco then fixes carbon for the second time as the CO_2 enters the Calvin–Benson cycle. Bundle-sheath cells in C4 plants have chloroplasts that carry out light-dependent reactions, but only in the cyclic pathway. No oxygen is released, so the oxygen level near rubisco stays low. This, along with the high CO_2 level provided by the C4 reactions, minimizes photorespiration. The pathway uses an extra ATP to produce each molecule of PGAL, but the increased efficiency of sugar production in hot, dry weather more than compensates for the loss (**FIGURE 6.16C**).

CAM plants Cacti and other **CAM plants** use a carbon-fixing pathway that allows them to conserve water even in desert regions with extremely high daytime temperatures. CAM stands for crassulacean acid metabolism, after the Crassulaceae family of plants in which this pathway was first studied. Like C4 plants, CAM plants fix carbon twice, but the reactions occur at different times rather than in different cells. Stomata on a CAM plant open at night, when typically lower temperatures minimize evaporative water loss. Then, CO_2 in the air diffuses into mesophyll cells, which use the C4 cycle to fix carbon. Malate, the product of this pathway, is stored in the cell's central vacuole. When stomata close the next day, the malate is moved out of the vacuole and converted back to CO_2. Rubisco then fixes carbon for the second time as the CO_2 enters the Calvin–Benson cycle. As in C4 plants, providing a high level of CO_2 near rubisco minimizes photorespiration.

TAKE-HOME MESSAGE 6.5

- ✔ NADPH and ATP produced by the light-dependent reactions power the light-independent reactions of the Calvin–Benson cycle.
- ✔ The Calvin–Benson reactions use carbon atoms from CO_2 to build sugar molecules.
- ✔ Incorporating carbon atoms from an inorganic source (such as CO_2) into an organic molecule is called carbon fixation.
- ✔ When stomata close on hot, dry days, they also prevent the exchange of gases between plant tissues and the air.
- ✔ Rubisco can initiate photorespiration by attaching oxygen (instead of carbon dioxide) to RuBP. Photorespiration reduces the efficiency of sugar production, especially in C3 plants.
- ✔ Plants adapted to hot, dry conditions limit photorespiration by fixing carbon twice. C4 plants separate the two sets of reactions in space; CAM plants separate them in time.

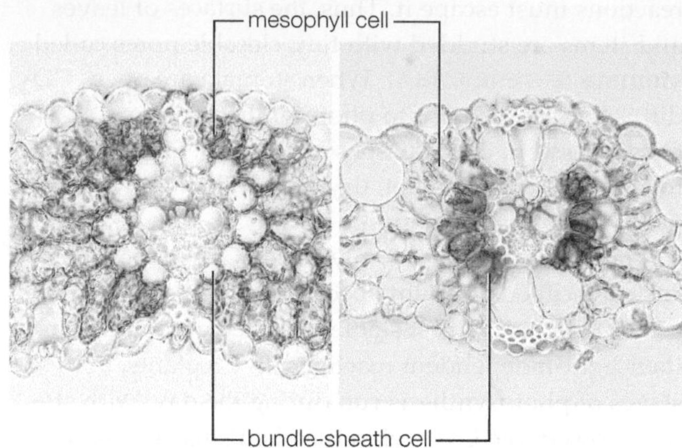
mesophyll cell
bundle-sheath cell

A In a C3 plant (barley, left), chloroplasts—the sites of carbon fixation—occur mainly in mesophyll cells. A C4 plant (millet, right) has chloroplasts in mesophyll cells and in bundle-sheath cells, so it can carry out photosynthesis in both types of cells. C4 plants minimize photorespiration by fixing carbon twice, in the two cell types.

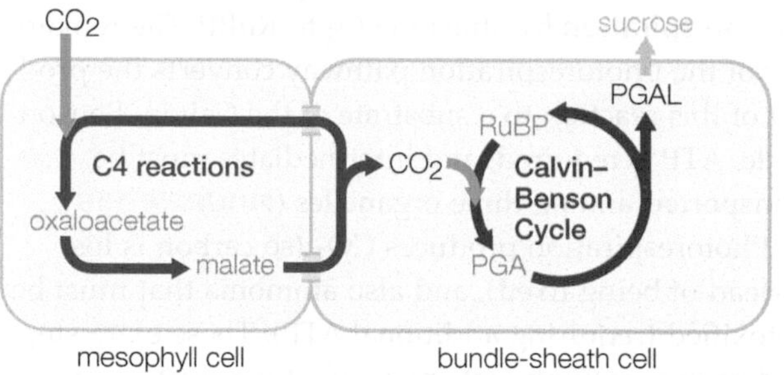
CO2
sucrose
PGAL
RuBP
C4 reactions
CO2
Calvin–
Benson
Cycle
oxaloacetate
malate
PGA
mesophyll cell
bundle-sheath cell

B In C4 plants, as in C3 plants, oxygen builds up when stomata close during photosynthesis. However, C4 plants fix carbon twice, first in mesophyll cells and then in bundle-sheath cells. This strategy minimizes photorespiration because it maintains a high CO_2 concentration near rubisco.

C Crabgrass "weeds" overgrowing a lawn. Crabgrasses, which are C4 plants, thrive in hot, dry summers, when they easily outcompete Kentucky bluegrass and other fine-leaved C3 grasses commonly planted in residential lawns.

FIGURE 6.16 C4 plant adaptations minimize photorespiration.

CAM plant Type of C4 plant that minimizes photorespiration by fixing carbon twice, at different times of day.

CREDITS: (16A) (*left*) Masahiro Yamada, Michio Kawasaki, Tatsuo Sugiyama, Hiroshi Miyake, Mitsutaka Taniguchi; "Differential Positioning of C4 Mesophyll and Bundle Sheath Chloroplasts: Aggregative Movement of C4 Mesophyll Chloroplasts in Response to Environmental Stresses": *Plant and Cell Physiology*, (2009) 50(10: 1736–1749). (*right*) Eri MAai, Shouu Shimada, Masahiro Yamada, Tatsuo Sugiyama, Hiroshi Miyake, Mitsutaka Taniguchi; The avoidance and aggregative movements of mesophyll chloroplasts in C4 monocots in response to blue light and abscisic acid; *Journal of Experimental Botany*, May 2011, Volume 62, issue number 9, 3213–3221, by permission of Oxford University Press; (16C) Image courtesy msuturfweeds.net.

You and other heterotrophs, remember, ingest tissues of other organisms to get carbon. Thus, your body is built from organic compounds obtained from other organisms. The carbon atoms in those compounds may have passed through other heterotrophs before you ate them, but at some point they were part of photosynthetic organisms. Plants and other photosynthesizers in the human food chain obtain their carbon from carbon dioxide. Your carbon atoms—and those of most other organisms that live on land—were recently part of Earth's atmosphere, in molecules of CO_2.

Photosynthesis removes carbon dioxide from the atmosphere, and fixes its carbon atoms in organic compounds that make up living things. When you and other organisms break down organic compounds for energy, carbon atoms are released in the form of CO_2, which then reenters the atmosphere. Since photosynthesis evolved, these two processes have constituted a more or less balanced cycle of the biosphere: The amount of carbon dioxide that photosynthesis removes from the atmosphere is roughly the same amount that organisms release back into it. At least it was, until humans came along.

As early as 8,000 years ago, we began burning forests to clear land for agriculture. When trees and other plants burn, most of the carbon locked in their tissues is released into the atmosphere as carbon dioxide. Fires that occur naturally release CO_2 the same way. However, we burn a lot more today than our ancestors ever did. In addition to wood, we burn fossil fuels to satisfy our growing population's increasing demand for energy. When fossil fuels burn, carbon that has been locked in organic molecules for hundreds of millions of years is released into the atmosphere as CO_2.

Human activities are now adding far more carbon dioxide to the atmosphere than photosynthetic organisms are removing from it, and the resulting imbalance is the major cause of global climate change (Chapter 46 returns to this topic). We have been releasing more CO_2 every year since the industrial revolution in the mid-1800s: 36 billion tons in 2014 alone—more than three times the amount released in 2010, and six times the amount in 1990. By far, most of it comes from burning fossil fuels (**FIGURE 6.17**). How do we know? By measuring the ratio of carbon isotopes in a sample of atmospheric CO_2, researchers can determine how long ago its carbon atoms were part of living organisms (you will read more about radioisotope dating in Section 16.5). These calculations are confirmed by global statistics on fossil fuel extraction, refining, and trade.

FIGURE 6.17 Air pollution—smog—from fossil fuel use blankets Lianyungang, China. The brown color of smog comes from nitric oxide, a gas that is toxic in large amounts. Carbon dioxide is an invisible component of smog.

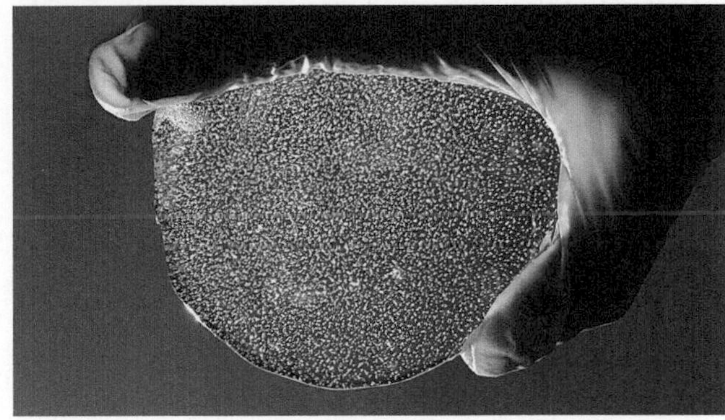

FIGURE 6.18 Antarctic ice core sample. Bubbles in ice cores are tiny pockets of air that became trapped in the ice when it formed: the deeper the ice, the older the air in the bubbles. Researchers study ice cores to determine the composition of Earth's ancient atmosphere.

Tiny pockets of Earth's ancient atmosphere remain in Antarctica, preserved in snow and ice that have been accumulating in layers, year after year, for millions of years (**FIGURE 6.18**). Air and dust trapped in each layer reveal the composition of the atmosphere during the time the layer formed. The layers have provided us with data for the last 900,000 years: In all that time, the CO_2 level has fluctuated but never risen above 300 ppm (parts per million)—until the industrial revolution. Since then it has been rising, and today it has surpassed 400 ppm. Measurements that are less direct than ice layer data indicate that the last time the atmosphere held 400 ppm of CO_2 was almost *3 million years* ago.

Such alarming statistics are why developing cost-effective biofuels is a priority. Burning biofuels also releases CO_2 into the atmosphere, but it contributes less to global climate change. Most biofuels are made from plant matter, and growing plants for fuel recycles carbon that is already in the atmosphere. ●

STUDY GUIDE

Section 6.1 Photosynthesis removes CO_2 from the atmosphere, and the metabolic activity of most organisms puts it back. Humans have been disrupting this cycle by burning fossil fuels, an activity that has been adding far more CO_2 to the atmosphere than photosynthesis can remove. The resulting imbalance is the major cause of climate change. Growing plant matter for biofuels contributes less to climate change because it recycles carbon already in the atmosphere.

Section 6.2 Plants and other **autotrophs** make their own food using energy from the environment and carbon from inorganic sources such as carbon dioxide. By metabolic pathways of **photosynthesis**, plants and most other autotrophs capture the energy of light and use it to build sugars from water and CO_2. Humans and almost all other **heterotrophs** obtain carbon and energy from organic molecules originally assembled by photosynthetic organisms.

Photosynthesis is a metabolic pathway that occurs in two stages: the **light-dependent reactions**, which are driven by light; and the **light-independent reactions**, which are not. The light-dependent reactions collectively produce ATP, and the main pathway also produces NADPH and O_2. ATP and NADPH drive the synthesis of sugars from water and carbon dioxide in the light-independent reactions.

In eukaryotes and cyanobacteria, the light-dependent reactions are carried out by molecules in the **thylakoid membrane**. This membrane encloses a single, continuous compartment. Eukaryotic photosynthesis occurs inside chloroplasts, which have a highly folded thylakoid membrane suspended in **stroma**. In chloroplasts, the light-independent reactions take place in the stroma. In cyanobacteria, these reactions occur in cytoplasm.

Section 6.3 Visible light is a very small part of the spectrum of electromagnetic energy radiating from the sun. That energy travels in waves, and it is organized as photons. A photon's **wavelength** is related to its energy: the shorter the wavelength, the higher the energy.

Wavelengths of light that we can see—visible light—drive photosynthesis, which begins when photons are absorbed by photosynthetic pigments. A **pigment** absorbs light of particular wavelengths only; wavelengths not captured are reflected as its characteristic color.

The main photosynthetic pigment in eukaryotes and cyanobacteria is **chlorophyll *a***, which absorbs violet and red light so it appears green. Accessory pigments absorb additional wavelengths, thus maximizing the amount of energy that can be used for photosynthesis. Many accessory pigments have additional functions.

Section 6.4 Photosynthetic pigments are part of light-harvesting complexes in the thylakoid membrane. The light-dependent reactions begin when light-harvesting complexes absorb photons and pass the energy to **photosystems**. There are two types of photosystems: photosystems I and II. Both types have a special pair of chlorophylls in their reaction center. Absorbing energy causes a photosystem's special pair to emit electrons.

In the cyclic pathway, electrons released from photosystem I move through an electron transfer chain, then cycle back to photosystem I.

In the noncyclic pathway, electrons released from photosystem II flow through an electron transfer chain, then to photosystem I. An input of energy causes photosystem I to release electrons that move through a second electron transfer chain. $NADP^+$ accepts the electrons at the end of this chain, so NADPH forms. Photosystem II replaces lost electrons by pulling them from water, which then splits into H^+ and O_2 (an example of **photolysis**).

In both pathways, electron flow through electron transfer chains sets up a hydrogen ion gradient that drives ATP formation, a process called **electron transfer phosphorylation**. Energy lost by electrons moving through the chains drives active transport of hydrogen ions into the thylakoid compartment. The ions follow their gradient back across the membrane through ATP synthases, and the flow causes these transport proteins to phosphorylate ADP.

The cyclic pathway was evolutionarily remodeled into the noncyclic pathway. The new pathway offered an unlimited supply of electrons (in water) and increased efficiency of sugar production.

Section 6.5 NADPH and ATP produced by the light-dependent reactions power the light-independent reactions of the **Calvin–Benson cycle**, which builds sugars from CO_2. The reactions begin when the enzyme **rubisco** carries out **carbon fixation** by attaching CO_2 to an organic molecule. The product of the Calvin–Benson cycle is PGAL, a phosphorylated three-carbon sugar that plant cells usually convert to sucrose.

On hot, dry days, a plant conserves water by closing the **stomata** on its aboveground surfaces. Then, carbon dioxide for the light-independent reactions cannot enter the plant's tissues, and oxygen produced by the light-dependent reactions cannot leave. The resulting high O_2/CO_2 ratio near photosynthetic cells can shift sugar production toward **photorespiration**. This inefficient pathway limits the growth rate of **C3 plants** in hot, dry climates. Other types of plants minimize photorespiration by fixing carbon twice, thus keeping the CO_2 level high near rubisco. **C4 plants** carry out the two sets of reactions in different cell types; **CAM plants** carry them out at different times.

SELF-QUIZ

Answers in Appendix VII

1. A cat eats a bird, which ate a caterpillar that chewed on a weed. Which organisms are autotrophs? Which ones are heterotrophs?

2. Photosynthesis runs on the energy of _______ .
   a. light c. O_2
   b. hydrogen ions d. CO_2
3. Most of the carbon dioxide that plants use for photosynthesis comes from _______ .
   a. glucose c. rainwater
   b. the atmosphere d. photolysis
4. Which of the following statements is incorrect?
   a. Pigments absorb light of certain wavelengths only.
   b. Some accessory pigments are antioxidants.
   c. Chlorophyll is green because it absorbs green light.
5. In cyanobacteria and photosynthetic eukaryotes, the light-dependent reactions proceed in/at the _______ .
   a. thylakoid membrane c. stroma
   b. plasma membrane d. cytoplasm
6. When a photosystem absorbs light, _______ .
   a. sugar phosphates are produced
   b. electrons are transferred to ATP
   c. RuBP accepts electrons
   d. its special pair emits electrons
7. In the light-dependent reactions, _______ .
   a. carbon dioxide is fixed c. CO_2 accepts electrons
   b. ATP forms d. sugars form
8. In the light-dependent reactions, what accumulates in the thylakoid compartment of chloroplasts?
   a. sugars c. O_2
   b. hydrogen ions d. CO_2
9. The atoms in oxygen molecules released during photosynthesis come from _______ .
   a. sugars b. water c. O_2 d. CO_2
10. In chloroplasts, the light-independent reactions proceed at/in the _______ .
   a. thylakoid membrane c. stroma
   b. plasma membrane d. cytoplasm
11. The Calvin–Benson cycle starts with _______ .
   a. the absorption of photon energy
   b. carbon fixation
   c. the release of electrons from photosystem II
   d. $NADP^+$ formation
12. Which of the following substances does *not* participate in the Calvin–Benson cycle?
   a. ATP d. PGAL
   b. NADPH e. O_2
   c. RuBP f. CO_2
13. Closed stomata _______ .
   a. limit gas exchange c. prevent photosynthesis
   b. permit water loss d. absorb light

14. In C3 plants, _______ makes sugar production inefficient when stomata close during the day.
   a. photosynthesis c. photorespiration
   b. photolysis d. carbon fixation
15. Match each term with the best description.

___ PGAL	a. absorbs light
___ CO_2 fixation	b. converts light to chemical energy
___ photolysis	c. self-feeder
___ ATP forms; NADPH does not	d. electrons cycle back to photosystem I
___ photorespiration	e. problem in C3 plants
___ photosynthesis	f. Calvin–Benson cycle product
___ pigment	g. water molecules split
___ autotroph	h. rubisco function

CRITICAL THINKING

1. About 200 years ago, Jan Baptista van Helmont wanted to know where growing plants get the materials necessary for increases in size. He planted a tree seedling weighing 2.2 kilograms (5 pounds) in a barrel filled with 90 kilograms (200 pounds) of soil and then watered the tree regularly. After five years, the tree had gained almost 75 kilograms (164 pounds), and the soil's weight was unchanged. He incorrectly concluded that the tree had gained all of its additional weight by absorbing water. How did the tree really gain most of its weight?

2. While gazing into an aquarium, you see bubbles coming from an aquatic plant (left). What are the bubbles?

3. A C3 plant absorbs a carbon radioisotope (as part of $^{14}CO_2$). In which compound does the labeled carbon appear first? Which compound forms first if a C4 plant absorbs the same radioisotope?

4. In 2005, a new species of green sulfur bacteria was discovered near a geothermal vent at the bottom of the ocean. An exceptionally efficient light-harvesting mechanism allows these bacteria to carry out photosynthesis where the only illumination is a dim volcanic glow emanating from the vent. Cell membranes are required for electron transfer chains in the light-dependent reactions, but these bacteria have no internal membranes. How do you think they carry out the light-dependent reactions?

7 RELEASING CHEMICAL ENERGY

SPAR
UKR
BARCELONA 20
SPAR
POL
BARCELONA 2010

CORE CONCEPTS

Evolution

Evolution underlies the unity and diversity of life.

The evolution of photosynthesis in bacteria permanently changed Earth's atmosphere by adding oxygen gas, which wiped out most ancient life. Aerobic respiration evolved in some of the survivors; molecules and processes that already existed were reused in this new pathway. Today, photosynthesis and respiration are linked by reactants, products, and energy flow. Cells of most modern organisms can switch between aerobic respiration and fermentation as needed.

Pathways of Transformation

Organisms exchange matter and energy with the environment in order to grow, maintain themselves, and reproduce.

All organisms require ongoing inputs of energy. Cells store acquired energy in the bonds of organic molecules, and retrieve energy stored in organic molecules by breaking apart the carbon backbones. Redox reactions that capture electrons in coenzymes are typically part of energy-harvesting pathways.

Structure and Function

The three-dimensional form and arrangement of biological structures give rise to their function and interactions.

Aerobic respiration and anaerobic fermentation are pathways that release energy from carbohydrates. Both store the released energy in ATP, and both begin with the same reactions in cytoplasm, but their concluding reactions use different electron acceptors. Aerobic respiration, the more efficient way of making ATP, uses oxygen to accept electrons.

Links to Earlier Concepts

This chapter focuses on metabolic pathways (Section 5.5) that harvest energy (5.2) stored in the chemical bonds of sugars (3.3). Some reactions (3.2, 5.3) of these pathways occur in mitochondria (4.6). You will revisit free radicals (2.3), lipids (3.4), proteins (3.5), electron transfer phosphorylation (6.4), coenzymes (5.6), membrane transport (5.7, 5.8), and photosynthesis (6.2).

7.1 Risky Business

The first cells we know of appeared on Earth about 3.4 billion years ago, and they lived in a world we would not recognize. The atmosphere contained no oxygen, and it would stay that way for another billion years. Organisms at the time were necessarily **anaerobic**, which means they lived in the absence of oxygen. Current research suggests these cells released energy from sugars by cellular respiration. **Cellular respiration** refers to any pathway that uses an electron transfer chain (Section 5.5) to harvest energy from organic molecules and make ATP. You already learned about one pathway that couples ATP synthesis with electron transfer chains: electron transfer phosphorylation, which is part of the light-dependent reactions of photosynthesis (Section 6.4). Electron transfer phosphorylation is also part of cellular respiration, but the electrons come from organic molecules, not from water as occurs in noncyclic photophosphorylation.

When the noncyclic pathway of photosynthesis evolved, oxygen gas (O_2) released from water molecules began seeping out of photosynthetic cells. In what may have been the earliest case of catastrophic air pollution, the new abundance of O_2 exerted tremendous pressure on all life at the time. Why? Molecules that carry electrons are a critical part of metabolism. Oxygen gas easily removes the electrons from (oxidizes) these molecules, destroying their function and also producing dangerous free radicals in the process (Section 2.3). Most cells had no way to counter these effects, and were wiped out everywhere except in deep water, muddy sediments, and other anaerobic habitats

By lucky circumstance, a few types of cells were already making antioxidants (Section 5.6) that could prevent damage caused by O_2 gas. They were the first **aerobic** organisms—they could live in the presence of oxygen. As these organisms evolved in environments with abundant oxygen, so did their metabolic pathways. Oxygen, the molecule that had poisoned most life on Earth, turned out to be a useful addition to cellular respiration, and many lineages incorporated it into their pathways.

One of the new pathways, aerobic respiration, put the oxidative properties of O_2 to use. In modern eukaryotic cells, most of the aerobic respiration pathway takes place in mitochondria (Section 4.6). An internal folded membrane system allows these organelles

aerobic (air-OH-bick) Involving or occurring in the presence of oxygen.
anaerobic (an-air-OH-bick) Occurring in or requiring the absence of oxygen.
cellular respiration Pathway that breaks down an organic molecule to form ATP and includes an electron transfer chain.

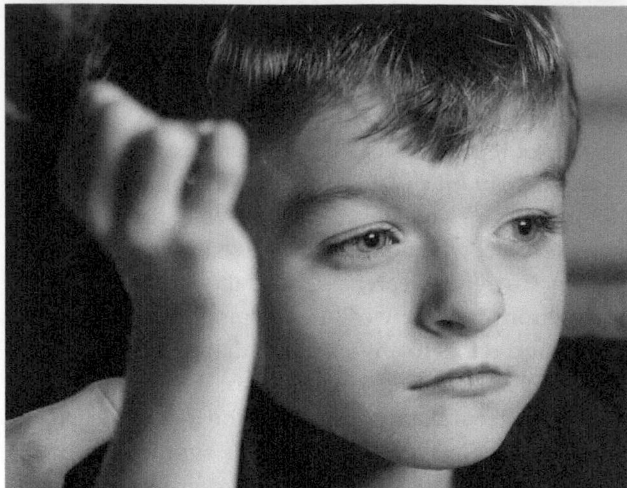

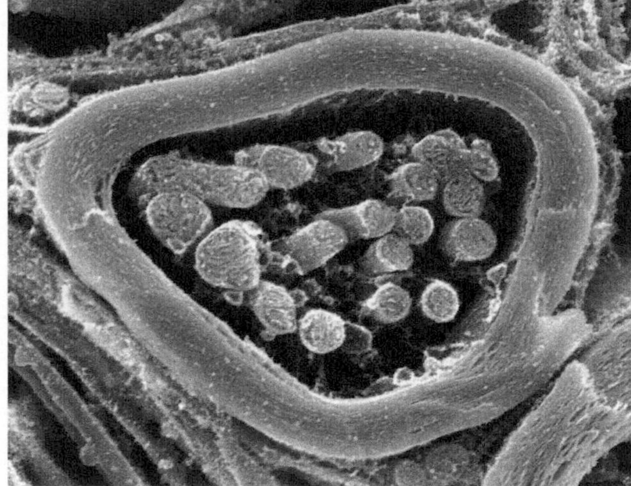

FIGURE 7.1 Mitochondrial disorders. Top, Carter, one of thousands of children born every year with a devastating mitochondrial disorder, died when he was 7 years old. Bottom, nerve cells, which are particularly affected by mitochondrial malfunction, are packed with mitochondria (gold).

to make ATP very efficiently. Electron transfer chains in this membrane set up hydrogen ion gradients that power ATP synthesis. Oxygen molecules accept electrons at the end of the chains.

ATP participates in almost all cellular reactions, so a cell benefits from making a lot of it. However, aerobic respiration is a dangerous occupation. When an oxygen molecule accepts electrons from an electron transfer chain, it dissociates into oxygen atoms. Most of the time, the atoms simultaneously combine with hydrogen ions and end up in water molecules. Occasionally, however, an oxygen atom escapes. The atom has an unpaired electron, so it is a free radical.

Mitochondria cannot detoxify free radicals, so they rely on antioxidant enzymes and vitamins in the cell's cytoplasm to do it for them. The system easily neutralizes the occasional free radical that forms during aerobic respiration. However, this normal cellular balance sometimes goes awry. Free radicals accumulate and destroy first the function of mitochondria, then the cell. The resulting tissue damage is called oxidative stress.

At least 83 proteins are directly involved in mitochondrial electron transfer chains. A malfunction of any one of them—or in any of the thousands of other proteins made by mitochondria—can wreak havoc in the body. Oxidative stress caused by mitochondrial malfunction is involved in aging, and also in the progression of many conditions such as cancer, hypertension, Alzheimer's and Parkinson's diseases, and autism.

Mitochondria, remember, contain their own DNA and replicate independently of the cell (Section 4.6). Defects in their function can arise during a person's lifetime, for example if the DNA gets damaged in most of a cell's mitochondria. Mitochondrial defects can also be inherited: Hundreds of genetic (heritable) disorders are known to be associated with them, and more are being discovered all the time. Nerve cells, which require a lot of ATP, are particularly affected. Symptoms of mitochondrial disorders range from mild to major progressive loss of neurological and muscular function, blindness, deafness, diabetes, strokes, seizures, gastrointestinal malfunction, and disabling weakness. These disorders can be devastating and incurable (**FIGURE 7.1**).●

7.2 Introduction to Carbohydrate Breakdown Pathways

LEARNING OBJECTIVES

- Compare the processes of cellular respiration and fermentation.
- Write an equation that summarizes aerobic respiration.
- Explain the difference between substrate-level phosphorylation and electron transfer phosphorylation.
- Describe the transfer of energy in glycolysis.

Autotrophs harvest energy directly from the environment and store it in the form of sugars and other carbohydrates. They and all other organisms use energy stored in sugars to power various endergonic reactions that sustain life (Section 5.3). However, in order to use the energy stored in sugars, cells must first transfer it to molecules—especially ATP—that can participate directly in these reactions.

Cells harvest energy from an organic molecule by breaking its carbon backbone, one bond at a time. This

aerobic respiration Oxygen-requiring cellular respiration; breaks down organic molecules (particularly glucose) and produces ATP, carbon dioxide, and water.

fermentation Anaerobic glucose-breakdown pathway that produces ATP without use of an electron transfer chain.

CREDITS: (1 top) AP Images/J. Scott Park; (1 bottom) Dr David Furness, Wellcome Images.

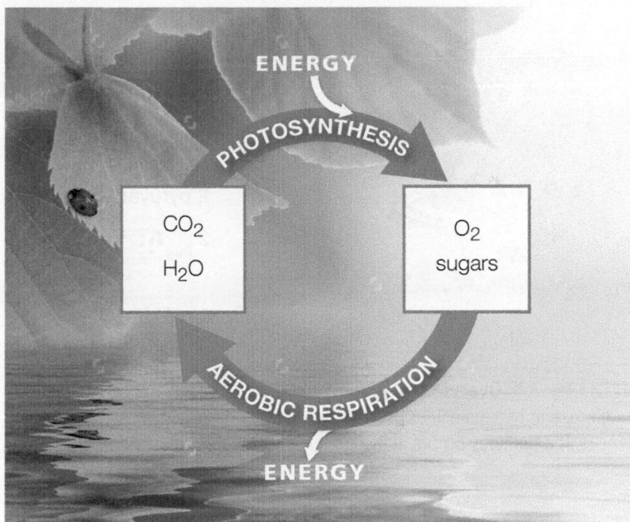
ENERGY
PHOTOSYNTHESIS
CO_2
H_2O
O_2
sugars
AEROBIC RESPIRATION
ENERGY

FIGURE 7.2 Substrates and products link photosynthesis with aerobic respiration. Photosynthesis produces sugars and, in most organisms, it also releases oxygen. The breakdown of sugars in aerobic respiration requires oxygen, and it produces carbon dioxide and water—the raw materials from which photosynthetic organisms make the sugars in the first place.

releases the energy of the molecule stepwise, in small increments that can be captured for cellular work. A number of different pathways drive ATP synthesis by breaking the bonds of sugar molecules this way. All of them are ancient.

Today, most cellular respiration uses oxygen. Oxygen-requiring cellular respiration is called **aerobic respiration**. Aerobic respiration means "breathing air," and the name is a reference to this pathway's requirement for oxygen. With each breath, you take in oxygen for your trillions of aerobically respiring cells. You exhale the pathway's products: carbon dioxide and water. CO_2 and water are the raw materials of photosynthesis, which produces sugars and, in most organisms, releases oxygen (Section 6.2). Thus, raw materials and products link the two pathways in most of the biosphere (**FIGURE 7.2**).

Reaction Pathways

Aerobic respiration harvests energy from an organic molecule—glucose, in particular—by completely breaking apart its carbon backbone, bond by bond. The following equation summarizes the overall pathway:

$$C_6H_{12}O_6 + O_2 \longrightarrow CO_2 + H_2O + \text{ATP}$$

The equation means that glucose and oxygen are converted to carbon dioxide and water, for a yield of ATP. However, aerobic respiration is not a single reaction. Rather, it consists of many reactions that occur in four stages: glycolysis, acetyl–CoA formation, the citric acid cycle, and electron transfer phosphorylation (**FIGURE 7.3**). These pathways are linked by products and substrates (**TABLE 7.1**), and all involve electron transfers (redox reactions, Section 5.5).

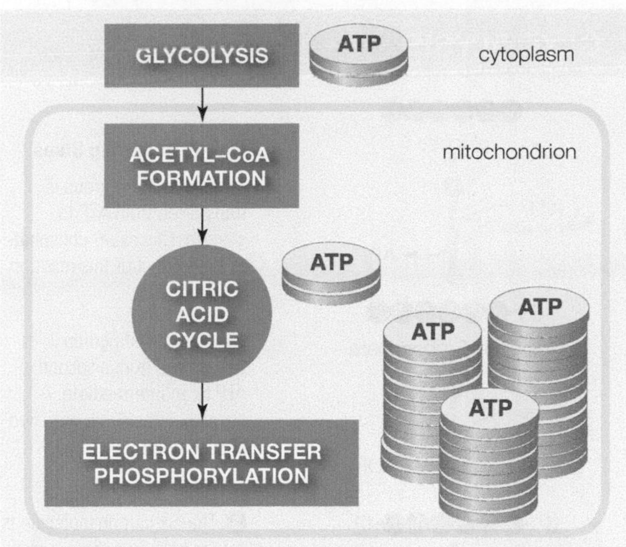
GLYCOLYSIS
ATP
cytoplasm
ACETYL–CoA FORMATION
mitochondrion
CITRIC ACID CYCLE
ATP
ATP
ATP
ATP
ELECTRON TRANSFER PHOSPHORYLATION

FIGURE 7.3 Overview of aerobic respiration. Aerobic respiration consists of four pathways: glycolysis, acetyl–CoA formation, the citric acid cycle, and electron transfer phosphorylation. In eukaryotes only, the last three steps occur in mitochondria. Compare Table 7.1.

TABLE 7.1

Inputs and Outputs Linking Steps in Aerobic Respiration

Pathway	Main Reactants	Net Yield
Glycolysis	1 six-carbon sugar (such as glucose)	2 pyruvate 2 NADH 2 ATP
Acetyl–CoA formation	2 pyruvate	2 acetyl–CoA 2 CO_2 2 NADH
Citric acid cycle	2 acetyl–CoA	4 CO_2 6 NADH 2 $FADH_2$ 2 ATP
Electron transfer phosphorylation	10 NADH 2 $FADH_2$ O_{2v}	32 ATP (approx.) H_2O
Overall	1 six-carbon sugar O_2	6 CO_2 H_2O 36 ATP (approx.)

Fermentation refers to glucose-breakdown pathways that make ATP without the use of oxygen or electron transfer chains. Many organisms supplement cellular respiration with fermentation; a few species of bacteria that have lost the ability to carry out respiration use fermentation exclusively. Fermentation pathways do not break all of the carbon–carbon bonds in glucose, so their products include two-carbon or three-carbon organic molecules.

Like aerobic respiration, fermentation begins with glycolysis. Unlike aerobic respiration, fermentation

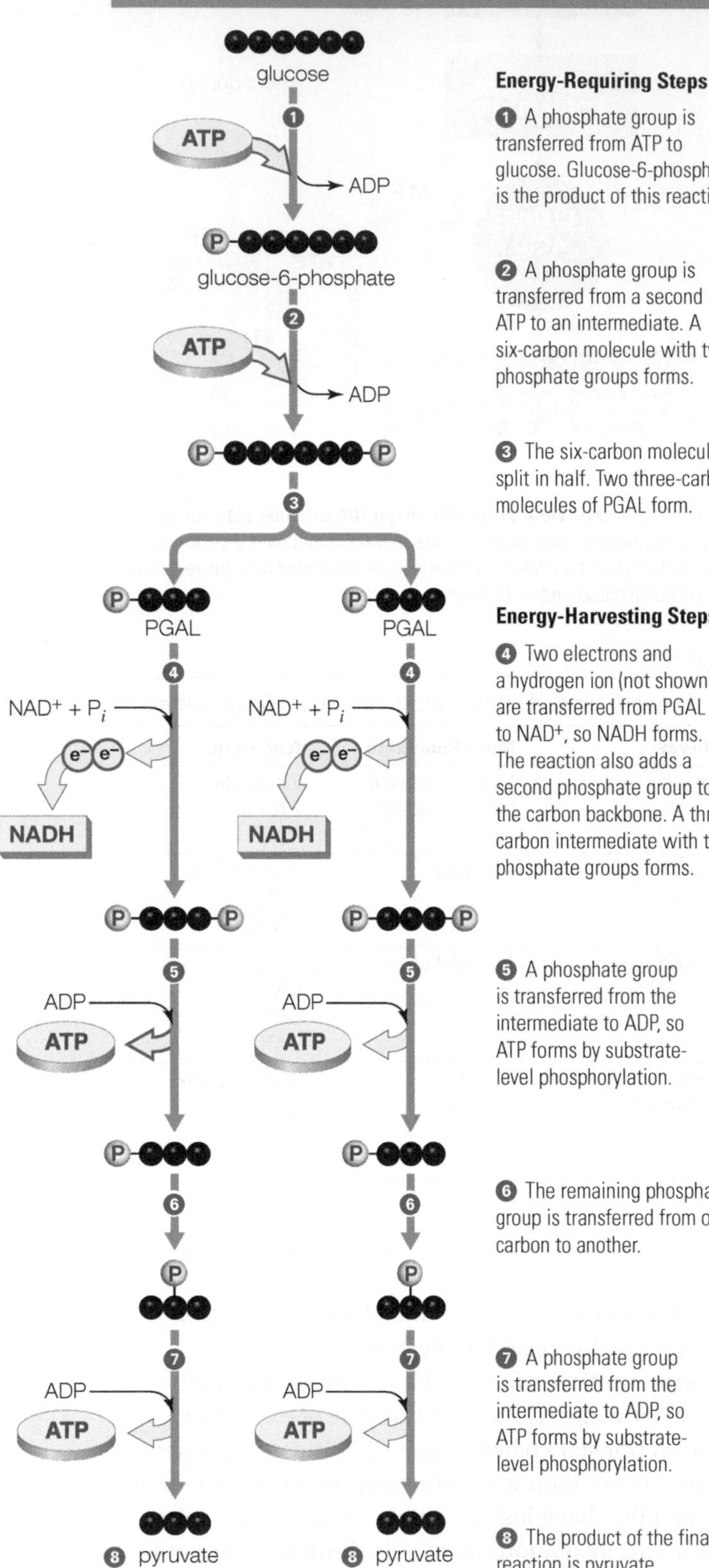
GLYCOLYSIS
glucose
ATP
ADP
glucose-6-phosphate
ATP
ADP
PGAL
PGAL
NAD+ + P$_i$
NAD+ + P$_i$
NADH
NADH
ADP
ATP
ADP
ATP
ADP
ATP
ADP
ATP
pyruvate
pyruvate
Energy-Requiring Steps
1 A phosphate group is transferred from ATP to glucose. Glucose-6-phosphate is the product of this reaction.
2 A phosphate group is transferred from a second ATP to an intermediate. A six-carbon molecule with two phosphate groups forms.
3 The six-carbon molecule is split in half. Two three-carbon molecules of PGAL form.
Energy-Harvesting Steps
4 Two electrons and a hydrogen ion (not shown) are transferred from PGAL to NAD+, so NADH forms. The reaction also adds a second phosphate group to the carbon backbone. A three-carbon intermediate with two phosphate groups forms.
5 A phosphate group is transferred from the intermediate to ADP, so ATP forms by substrate-level phosphorylation.
6 The remaining phosphate group is transferred from one carbon to another.
7 A phosphate group is transferred from the intermediate to ADP, so ATP forms by substrate-level phosphorylation.
8 The product of the final reaction is pyruvate.

FIGURE 7.5 Reactions of glycolysis.
This first stage of sugar breakdown starts and ends in the cytoplasm of all cells.

For clarity, we track only the six carbon atoms (black balls) of the glucose molecule that enters the pathway. All reactions are catalyzed by enzymes. Appendix III has more details for interested students.

glucose
GLYCOLYSIS
2 pyruvate
2 ATP
2 NADH

FIGURE 7.4 Overview of glycolysis. Glycolysis converts one molecule of glucose to two molecules of pyruvate, for a net yield of two ATP and two NADH. The pathway occurs in the cytoplasm of all cells.

does not include electron transfer phosphorylation, and it does not produce many ATP. Fermentation can sustain some single-celled species. It also helps cells of multicelled species produce ATP, especially under anaerobic conditions. However, aerobic respiration is a much more efficient way of harvesting energy from glucose. You and other large, multicelled organisms could not live without its higher yield.

Glycolysis: Sugar Breakdown Begins

Glycolysis is a series of reactions that produce ATP by converting glucose to **pyruvate**, an organic compound with a three-carbon backbone. The pathway occurs in the cytoplasm of all cells (FIGURE 7.4), and it is the first step in both aerobic respiration and fermentation. Its name (which is derived from the Greek words *glyk-*, sweet, and *-lysis*, loosening) refers to the release of chemical energy from sugars. Glycolysis breaks one carbon–carbon bond of a glucose molecule. The energy released when that bond breaks is captured in electrons carried by NADH, and in high-energy phosphate bonds of ATP. The reactions use two ATP and produce four, so we say the net yield of glycolysis is two ATP. (Other six-carbon sugars such as galactose and fructose can enter glycolysis, but we focus here on glucose, for clarity.)

Glycolysis begins when a molecule of glucose enters a cell through a glucose transporter, a passive transport protein that you encountered in Section 5.9. The cell invests two ATP in the endergonic reactions that begin the pathway (FIGURE 7.5). In the first reaction, a phosphate group is transferred from ATP to the glucose, thus forming glucose-6-phosphate ❶ (hexokinase, the enzyme that catalyzes this reaction, is shown in Figure 5.10). A phosphate group from a second ATP is transferred to an intermediate ❷, so a six-carbon molecule with two phosphate groups forms. This molecule is split in half to form two molecules of PGAL ❸, a phosphorylated three-carbon sugar. Both PGAL molecules continue in glycolysis, so the remaining reactions are carried out in duplicate.

At this point in glycolysis, two ATP have been invested to break one carbon–carbon bond, but no energy has been harvested. The energy-harvesting steps begin with a redox reaction that transfers electrons and a hydrogen ion from PGAL to the coenzyme NAD^+, which is thereby reduced to NADH ❹. Aerobic respiration's final stage requires this NADH, as does fermentation (Sections 7.5 and 7.6 detail the final stages of these pathways).

The next reaction transfers one of the phosphate groups to ADP, so ATP forms ❺. The remaining phosphate group is transferred from one carbon to another ❻ and then to ADP, so ATP forms again ❼. Pyruvate is the product of the final reaction ❽. As you will see in the next sections, pyruvate is a substrate for the second-stage reactions of aerobic respiration, and also for fermentation reactions.

Comparing Other Pathways

Aerobic respiration has several molecules and processes in common with photosynthesis, an outcome of evolutionary repurposing (later chapters return to this topic). For example, glycolysis and the Calvin–Benson cycle use some of the same enzymes (Section 6.5), and some of the same intermediates form (such as PGAL). Also, ATP forms by substrate-level phosphorylation in both pathways. In **substrate-level phosphorylation**, a phosphate group is transferred directly from a phosphorylated molecule to ADP, so ATP forms. Note that substrate-level phosphorylation differs from electron transfer phosphorylation, which is the way ATP forms during the light-dependent reactions of photosynthesis (Section 6.4). In electron transfer phosphorylation, the movement of electrons through electron transfer chains sets up a hydrogen ion gradient, and the resulting flow of hydrogen ions through ATP synthases causes these proteins to attach a free phosphate group to ADP.

A pathway that builds glucose from pyruvate is almost the reverse of glycolysis, and the two pathways must be tightly regulated or the cell would make glucose and break it down at the same time. Consider how the two-ATP investment during the first reactions of glycolysis prevents the pathway from running in reverse. Remember from Section 5.5 that most reactions can run backward as well as forward. After two inputs of ATP energy, the reverse pathway would require too much energy to run spontaneously.

glycolysis (gly-COLL-ih-sis) First stage of aerobic respiration and fermentation; set of reactions that convert glucose to two pyruvate for a net yield of two ATP and two NADH.
pyruvate (pie-ROO-vait) Three-carbon product of glycolysis.
substrate-level phosphorylation ATP formation by the transfer of a phosphate group from a phosphorylated molecule to ADP.

TAKE-HOME MESSAGE 7.2

✔ Most cells can make ATP by breaking down glucose in aerobic respiration, fermentation, or both.

✔ Fermentation does not require oxygen.

✔ Aerobic respiration is cellular respiration that requires oxygen. It yields much more ATP per glucose molecule than fermentation.

✔ Glycolysis is the first stage of sugar breakdown in aerobic respiration and in fermentation. The reactions of glycolysis occur in the cytoplasm of all cells.

✔ Glycolysis converts one molecule of glucose to two molecules of pyruvate, for a net energy yield of two ATP. Two NADH also form.

7.3 Aerobic Respiration Continues

LEARNING OBJECTIVES

- Explain why aerobic respiration releases CO_2.
- Describe the movement of energy during acetyl–CoA formation and the citric acid cycle.

Each of the two pyruvate molecules that formed in glycolysis has two carbon–carbon bonds. The next two steps of aerobic respiration, acetyl–CoA formation and the citric acid cycle, break both of these bonds. Energy released when the bonds break is captured in electrons carried by NADH, and in high-energy phosphate bonds of ATP (**FIGURE 7.6**). All of the carbon atoms that were once part of glucose end up in CO_2, which diffuses out of the cell (remember, gases freely cross cell membranes, Section 5.8). In prokaryotes, acetyl–CoA formation and the citric acid cycle occur in cytoplasm; in eukaryotes, they occur in mitochondria.

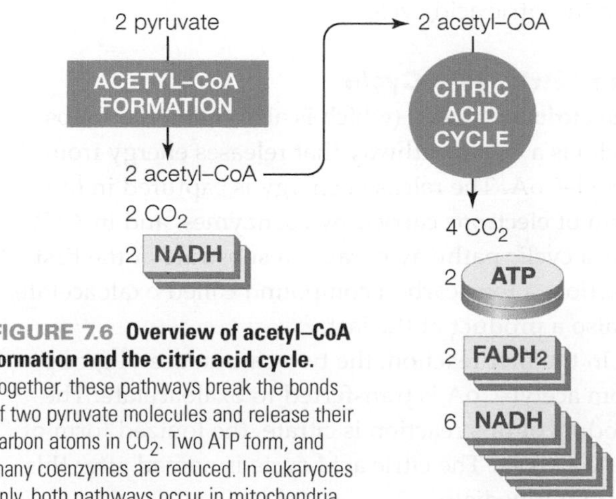
2 pyruvate
ACETYL–CoA FORMATION
2 acetyl–CoA
$2\ CO_2$
2 NADH
2 acetyl–CoA
CITRIC ACID CYCLE
$4\ CO_2$
2 ATP
2 $FADH_2$
6 NADH

FIGURE 7.6 Overview of acetyl–CoA formation and the citric acid cycle. Together, these pathways break the bonds of two pyruvate molecules and release their carbon atoms in CO_2. Two ATP form, and many coenzymes are reduced. In eukaryotes only, both pathways occur in mitochondria.

A All eukaryotic cells have mitochondria. An inner membrane divides the interior of a mitochondrion into two compartments. The outer compartment is called the intermembrane space. The inner compartment is called the mitochondrial matrix.

B Pyruvate produced by glycolysis is transported from cytoplasm into a mitochondrion, across both membranes, and into the matrix. Aerobic respiration continues here, in the matrix, with acetyl–CoA formation and the citric acid cycle.

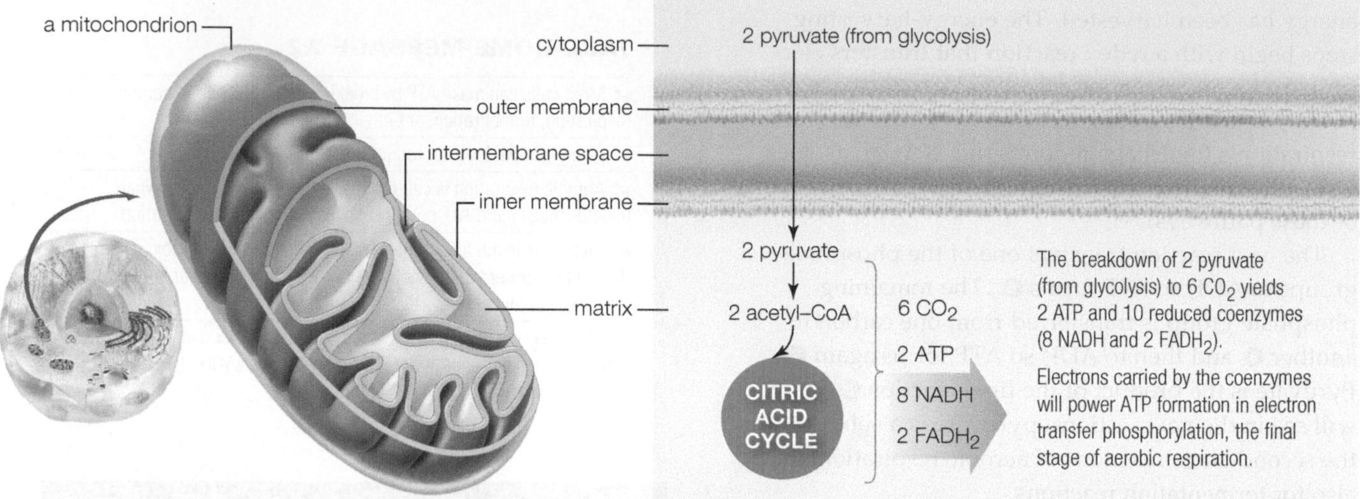
a mitochondrion
cytoplasm
outer membrane
intermembrane space
inner membrane
matrix
2 pyruvate (from glycolysis)
2 pyruvate
2 acetyl–CoA
CITRIC ACID CYCLE
6 CO_2
2 ATP
8 NADH
2 $FADH_2$
The breakdown of 2 pyruvate (from glycolysis) to 6 CO_2 yields 2 ATP and 10 reduced coenzymes (8 NADH and 2 $FADH_2$).
Electrons carried by the coenzymes will power ATP formation in electron transfer phosphorylation, the final stage of aerobic respiration.

FIGURE 7.7 Acetyl–CoA formation and the citric acid cycle in mitochondria.

Acetyl–CoA Formation

In eukaryotes, aerobic respiration continues when the two pyruvate molecules that formed during glycolysis enter a mitochondrion. Pyruvate is transported across the mitochondrion's two membranes and into the inner compartment, or matrix (**FIGURE 7.7A**). There, a redox reaction splits a carbon from the pyruvate, and this carbon diffuses out of the cell in CO_2 (**FIGURE 7.8** ❶). The reaction transfers electrons and a hydrogen ion to NAD^+, thus reducing it to NADH. The remaining two-carbon fragment of pyruvate ends up as an acetyl group (—$COCH_3$) attached to a coenzyme called coenzyme A (abbreviated CoA). The product of the reaction is called acetyl–CoA. Both pyruvates from glycolysis undergo these reactions, so two acetyl–CoA molecules form. Each now carries two carbon atoms into the citric acid cycle.

The Citric Acid Cycle

The **citric acid cycle** (which is also called the Krebs cycle) is a cyclic pathway that releases energy from acetyl–CoA. The released energy is captured in the form of electrons carried by coenzymes, and in ATP. It is a cyclic pathway because a substrate of the first reaction, a four-carbon compound called oxaloacetate, is also a product of the last.

In the first reaction, the two-carbon acetyl group from acetyl–CoA is transferred to oxaloacetate. The product of this reaction is citrate, the ionized form of citric acid ❷. The citric acid cycle is named after this first intermediate.

A series of redox reactions follows. Electrons and hydrogen ions are transferred to NAD^+, so NADH forms (❸ and ❹). The reactions remove two carbon atoms from intermediates, and these depart the cell in CO_2. No more carbons are split from the backbone of the four-carbon intermediate that forms; the remaining reactions of the citric acid cycle harvest energy from this molecule by rearranging its bonds.

In the next reaction, ATP forms by substrate-level phosphorylation ❺. Then, electrons and hydrogen atoms are transferred from an intermediate to the coenzyme FAD (flavin adenine dinucleotide), which becomes reduced to $FADH_2$ ❻.

A final redox reaction transfers electrons and a hydrogen to NAD^+, so another NADH forms ❼. The product of this reaction is oxaloacetate ❽.

It takes two rounds of citric acid cycle reactions to harvest the energy from two acetyl–CoA molecules. At this point in aerobic respiration, the carbon backbone of one glucose molecule has been broken apart completely, its six carbon atoms having exited the cell in six molecules of CO_2:

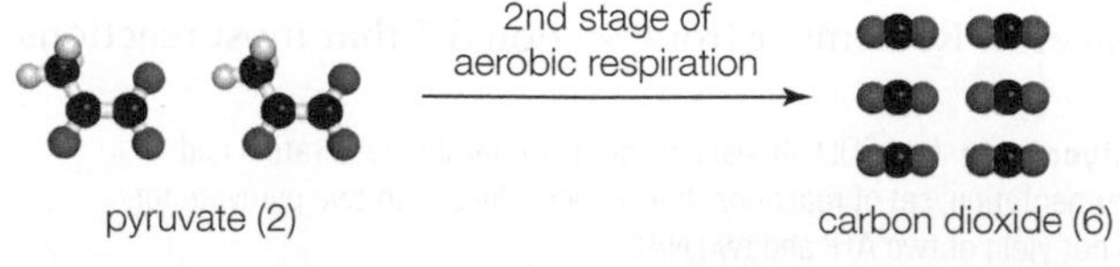
2nd stage of aerobic respiration
pyruvate (2)
carbon dioxide (6)

citric acid cycle Also called the Krebs cycle. Cyclic pathway that reduces many coenzymes by breaking down acetyl–CoA; part of aerobic respiration.

ACETYL–CoA FORMATION AND THE CITRIC ACID CYCLE

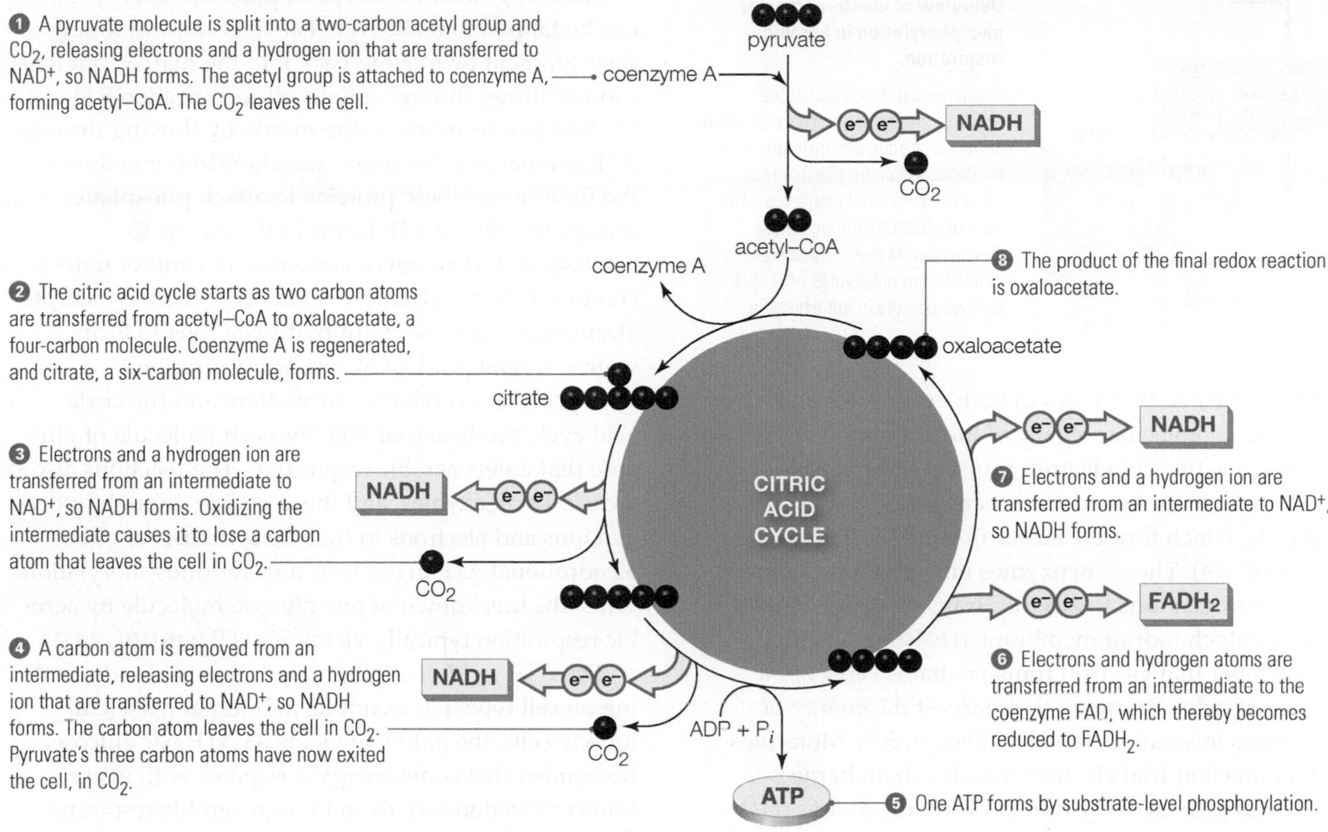

1 A pyruvate molecule is split into a two-carbon acetyl group and CO_2, releasing electrons and a hydrogen ion that are transferred to NAD^+, so NADH forms. The acetyl group is attached to coenzyme A, forming acetyl–CoA. The CO_2 leaves the cell.
pyruvate
coenzyme A
NADH
CO_2
acetyl–CoA
coenzyme A
2 The citric acid cycle starts as two carbon atoms are transferred from acetyl–CoA to oxaloacetate, a four-carbon molecule. Coenzyme A is regenerated, and citrate, a six-carbon molecule, forms.
citrate
8 The product of the final redox reaction is oxaloacetate.
oxaloacetate
3 Electrons and a hydrogen ion are transferred from an intermediate to NAD^+, so NADH forms. Oxidizing the intermediate causes it to lose a carbon atom that leaves the cell in CO_2.
NADH
CO_2
CITRIC ACID CYCLE
NADH
7 Electrons and a hydrogen ion are transferred from an intermediate to NAD^+, so NADH forms.
$FADH_2$
6 Electrons and hydrogen atoms are transferred from an intermediate to the coenzyme FAD, which thereby becomes reduced to $FADH_2$.
4 A carbon atom is removed from an intermediate, releasing electrons and a hydrogen ion that are transferred to NAD^+, so NADH forms. The carbon atom leaves the cell in CO_2. Pyruvate's three carbon atoms have now exited the cell, in CO_2.
NADH
CO_2
ADP + P_i
ATP
5 One ATP forms by substrate-level phosphorylation.

FIGURE 7.8 Acetyl–CoA formation and the citric acid cycle.

Follow the carbons ● to understand the movement of matter; follow the electrons (e^-) to see the movement of energy. Each time these reactions run, they break down one pyruvate molecule. Thus, it takes two sets of reactions to break down the two pyruvate molecules that formed during glycolysis. Then, all six carbons that entered glycolysis in one glucose molecule have exited the cell, in six CO_2. Electrons and hydrogen ions are released as each carbon is removed from the backbone of intermediate molecules; these combine with ten coenzymes (thus reducing them). Not all reactions are shown; see Appendix III for details.

The two ATP that form during the citric acid cycle add to the small net yield of two ATP from glycolysis. However, acetyl–CoA formation reduced two coenzymes, and the citric acid cycle reduced eight more (FIGURE 7.7B). Add in the two coenzymes reduced in glycolysis, and the full breakdown of each glucose molecule has a big potential payoff: Twelve coenzymes will deliver electrons—and the energy they carry—to the final stage of aerobic respiration: electron transfer phosphorylation.

TAKE-HOME MESSAGE 7.3

✔ In eukaryotes, the second and third stages of aerobic respiration, (acetyl–CoA formation and the citric acid cycle) occur in the inner compartment (matrix) of mitochondria.

✔ Acetyl-CoA formation and the citric acid cycle convert the two pyruvate that formed in glycolysis to six CO_2. Two ATP form, and ten coenzymes (eight NAD^+ and two FAD) are reduced.

7.4 Aerobic Respiration Ends

LEARNING OBJECTIVES

- Describe ATP formation in the last stage of aerobic respiration.
- Explain why aerobic respiration requires O_2.

The last stage of aerobic respiration is electron transfer phosphorylation. This pathway occurs at the inner membrane of mitochondria, and at the plasma membrane of aerobic bacteria. (Aerobic archaea also carry out this pathway, but it is not as well understood in these organisms.)

Electron transfer phosphorylation in aerobic respiration is generally the same as in the light-dependent reactions of photosynthesis: The flow of electrons through electron transfer chains sets up a hydrogen ion (H^+) gradient that drives ATP synthesis. In fact, several of the molecules that carry out electron transfer phosphorylation are highly conserved (similar or identical)

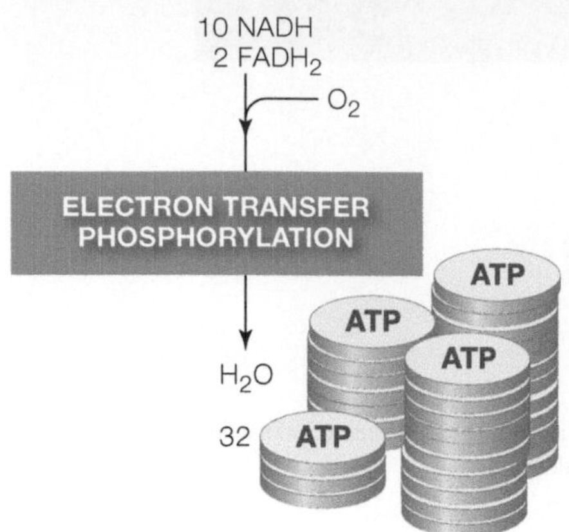
10 NADH
2 FADH2
O2
ELECTRON TRANSFER PHOSPHORYLATION
ATP
ATP
ATP
H2O
32
ATP

FIGURE 7.9 **Overview of electron transfer phosphorylation in aerobic respiration.**

Coenzymes reduced during the previous stages of aerobic respiration deliver electrons and hydrogen ions to electron transfer chains in the inner mitochondrial membrane. The flow of electrons through these chains powers the formation of many ATP. In eukaryotes only, this pathway occurs in mitochondria.

among the two pathways in both eukaryotes and bacteria, a molecular legacy of life's common origin.

The reactions of electron transfer phosphorylation begin with the reduced coenzymes NADH and $FADH_2$, which formed earlier in aerobic respiration (**FIGURE 7.9**). These coenzymes now give up electrons and hydrogen ions to electron transfer chains in the inner mitochondrial membrane (**FIGURE 7.10** ❶). Remember that electron transfer chains carry out a series of redox reactions that harvest the energy of electrons in small increments (Section 5.5). Molecules of a mitochondrial electron transfer chain harness that energy to actively transport hydrogen ions (H^+) across the inner membrane, from the matrix to the intermembrane space ❷. Thus, the movement of electrons through the electron transfer chain sets up and maintains a hydrogen ion gradient across the inner mitochondrial membrane.

This H^+ gradient is a type of potential energy that can be tapped to make ATP. The ions want to follow their gradient by moving back into the matrix, but ions cannot diffuse through a lipid bilayer (Section 5.8). Hydrogen ions return to the matrix by flowing through ATP synthases in the inner mitochondrial membrane; the flow causes these proteins to attach phosphate groups to ADP, so ATP forms in the matrix ❸.

Oxygen (O_2) accepts electrons at the end of mitochondrial electron transfer chains ❹. When O_2 accepts electrons, it combines with hydrogen ions to form water—a final product of aerobic respiration.

Glycolysis, acetyl–CoA formation, and the citric acid cycle produce four ATP for each molecule of glucose that enters aerobic respiration. The reactions also reduce 12 coenzymes, and these deliver enough hydrogen ions and electrons to fuel the synthesis of about 32 additional ATP in electron transfer phosphorylation. Thus, the breakdown of one glucose molecule by aerobic respiration typically yields 36 ATP (**FIGURE 7.11**).

The ATP yield of aerobic respiration varies depending on cell type. For example, in brain and skeletal muscle cells, the pathway yields 38 ATP per glucose. Remember that some energy dissipates with every transfer (Section 5.2). Even though aerobic respiration is a very efficient way of retrieving energy from carbohydrates, about 60 percent of the energy harvested in this pathway disperses as metabolic heat.

FIGURE 7.10 **The final step of aerobic respiration**, electron transfer phosphorylation, in a mitochondrion.

❶ NADH and $FADH_2$ deliver electrons and hydrogen ions to electron transfer chains in the inner membrane.

❷ Energy lost by electrons moving through electron transfer chains fuels the active transport of hydrogen ions (H^+) from the matrix to the intermembrane space. A hydrogen ion gradient forms across the inner membrane.

❸ Hydrogen ion flow back to the matrix through ATP synthases drives the formation of ATP from ADP and phosphate (P_i).

❹ Oxygen accepts electrons and hydrogen ions at the end of the electron transfer chains, so water forms.

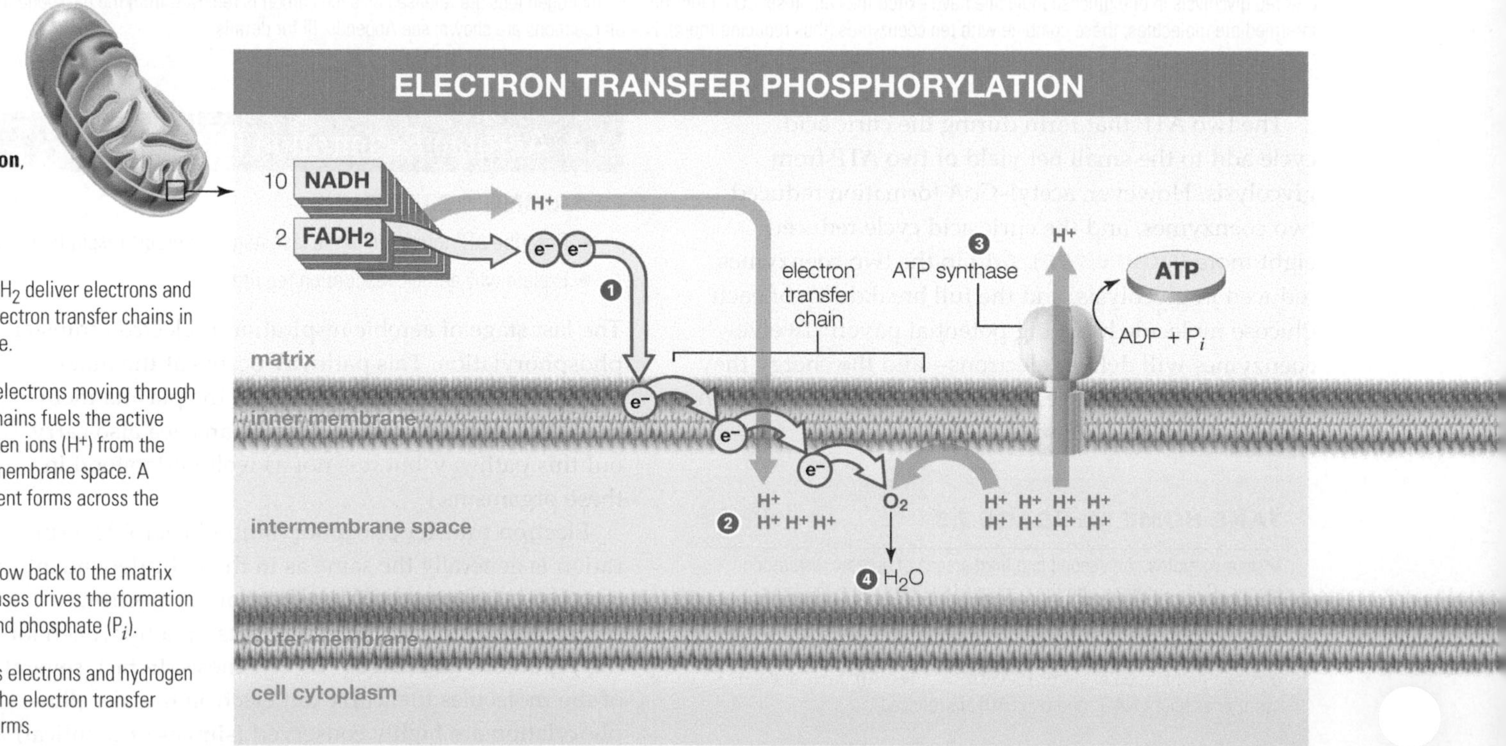

ELECTRON TRANSFER PHOSPHORYLATION

TAKE-HOME MESSAGE 7.4

✔ Aerobic respiration's final stage is electron transfer phosphorylation. Coenzymes reduced in earlier stages of aerobic respiration deliver electrons and hydrogen ions to electron transfer chains in the inner mitochondrial membrane.

✔ Energy released by electrons as they pass through electron transfer chains is used to pump H^+ from the mitochondrial matrix to the intermembrane space. The H^+ gradient that forms across the inner mitochondrial membrane drives the flow of hydrogen ions through ATP synthases, which results in the attachment of phosphate to ADP.

✔ The operation of mitochondrial electron transfer chains is ultimately captured in high energy phosophate bonds of about 32 ATP.

✔ A typical net yield of all stages of aerobic respiration is 36 ATP per glucose.

7.5 Fermentation

LEARNING OBJECTIVES

- Describe ATP formation in fermentation pathways.
- List some commercial uses of alcoholic and lactate fermentation.
- Explain why fermentation, unlike aerobic respiration, requires no oxygen.

Almost all cells can carry out fermentation, and many can switch between aerobic respiration and fermentation as needed. Fermentation reactions vary greatly, but all occur in cytoplasm. They require no oxygen because the final acceptor of electrons is an organic molecule (not oxygen).

FIGURE 7.11 Summary of aerobic respiration in eukaryotes.

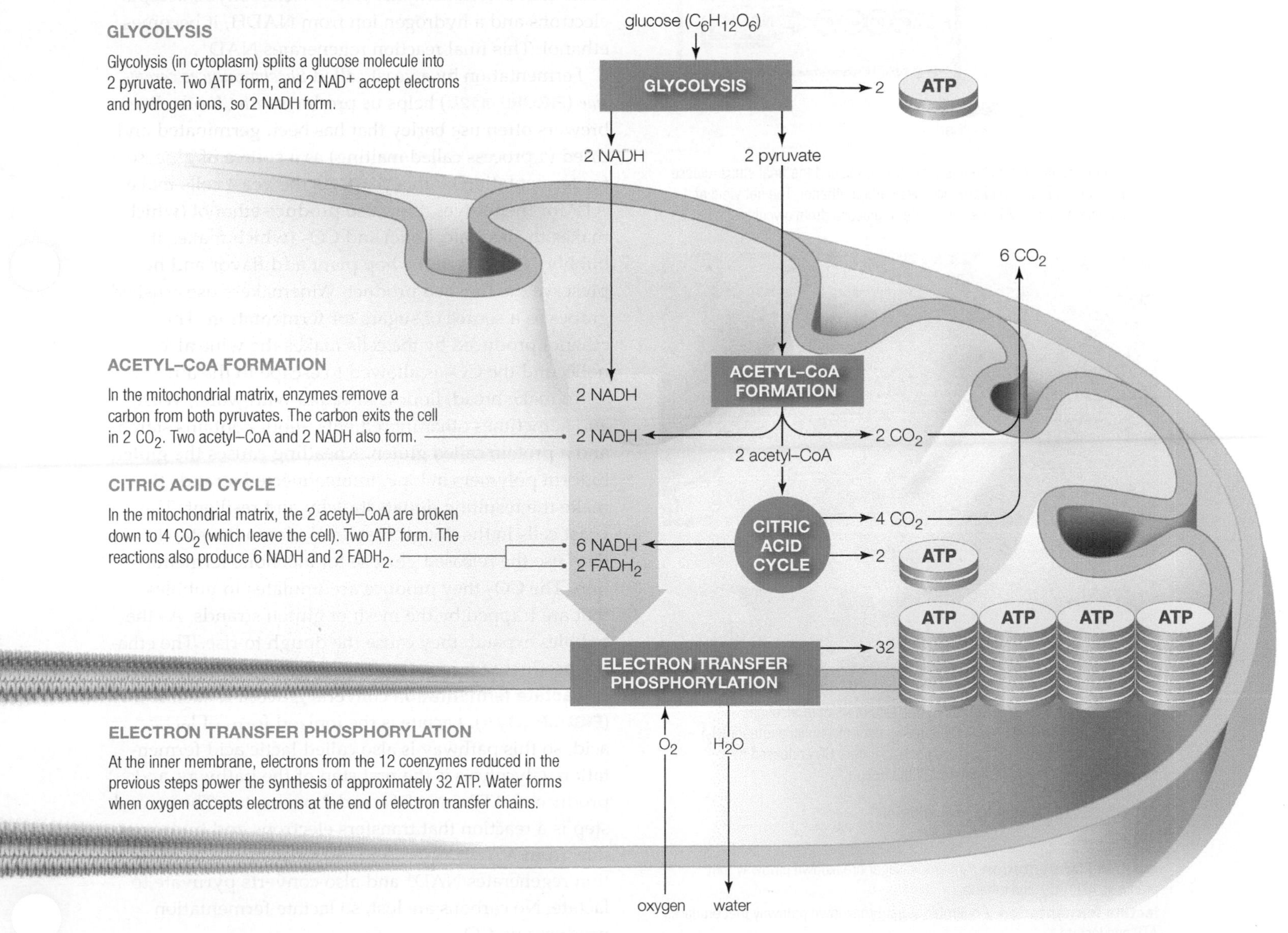

GLYCOLYSIS
Glycolysis (in cytoplasm) splits a glucose molecule into 2 pyruvate. Two ATP form, and 2 NAD+ accept electrons and hydrogen ions, so 2 NADH form.
glucose ($C_6H_{12}O_6$)
GLYCOLYSIS
2
ATP
2 NADH
2 pyruvate
6 CO_2
ACETYL–CoA FORMATION
In the mitochondrial matrix, enzymes remove a carbon from both pyruvates. The carbon exits the cell in 2 CO_2. Two acetyl–CoA and 2 NADH also form.
2 NADH
2 NADH
ACETYL–CoA FORMATION
2 CO_2
2 acetyl–CoA
CITRIC ACID CYCLE
In the mitochondrial matrix, the 2 acetyl–CoA are broken down to 4 CO_2 (which leave the cell). Two ATP form. The reactions also produce 6 NADH and 2 $FADH_2$.
6 NADH
2 $FADH_2$
CITRIC ACID CYCLE
4 CO_2
2
ATP
ATP
ATP
ATP
ATP
32
ELECTRON TRANSFER PHOSPHORYLATION
ELECTRON TRANSFER PHOSPHORYLATION
At the inner membrane, electrons from the 12 coenzymes reduced in the previous steps power the synthesis of approximately 32 ATP. Water forms when oxygen accepts electrons at the end of electron transfer chains.
O_2
H_2O
oxygen
water

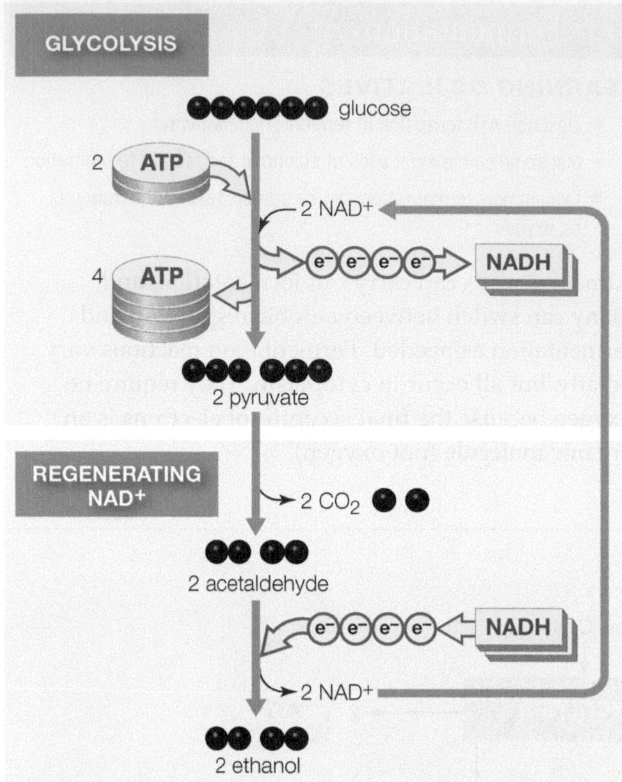
GLYCOLYSIS
glucose
2 ATP
2 NAD+
NADH
4 ATP
2 pyruvate
REGENERATING NAD+
2 CO2
2 acetaldehyde
NADH
2 NAD+
2 ethanol

A Alcoholic fermentation begins with glycolysis, and the final steps release CO_2, regenerate NAD^+, and produce two-carbon ethanol. The net yield of these reactions is two ATP per molecule of glucose (from glycolysis).

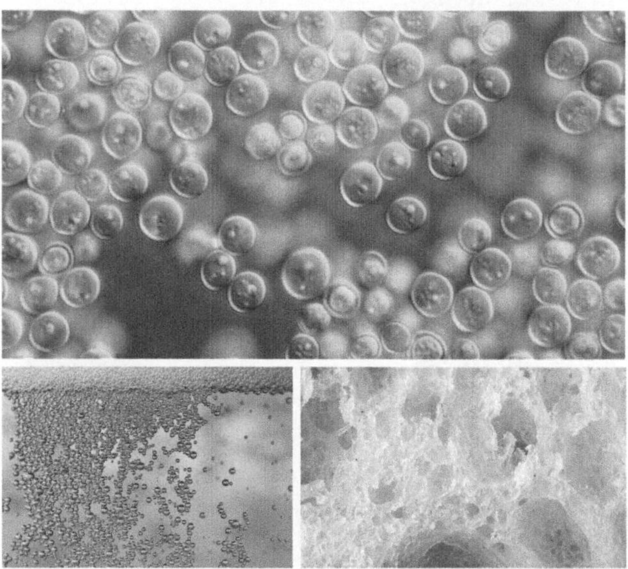

B *Saccharomyces cerevisiae* cells (top). One product of alcoholic fermentation by this yeast (ethanol) makes beer alcoholic; another (CO_2) makes it bubbly. Holes in bread are pockets where CO_2 released by fermenting yeast cells accumulated in the dough.

FIGURE 7.12 Alcoholic fermentation.

alcoholic fermentation Anaerobic sugar breakdown pathway that produces ATP, CO_2, and ethanol.

lactate fermentation Anaerobic sugar breakdown pathway that produces ATP and lactate.

Fermentation pathways, like aerobic respiration, begin with glycolysis. Unlike aerobic respiration, neither pathway fully breaks down the carbon backbone of glucose. The reactions after glycolysis simply serve to remove electrons and hydrogen ions from NADH, so NAD^+ forms. Regenerating this coenzyme allows glycolysis—and the ATP yield it offers—to continue. Thus, the net yield of fermentation consists of the two ATP that form in glycolysis.

Fermentation pathways are named after their end product. **Alcoholic fermentation**, for example, converts glucose to ethyl alcohol—ethanol (**FIGURE 7.12A**). Glycolysis is the first part of the pathway, and as you know it produces 2 ATP, 2 NADH, and 2 pyruvates. One carbon is removed from each of the pyruvate molecules, and this carbon leaves the cell in CO_2. The remaining fragment of pyruvate is an organic molecule called acetaldehyde. When acetaldehyde accepts electrons and a hydrogen ion from NADH, it becomes ethanol. This final reaction regenerates NAD^+.

Fermentation by a yeast called *Saccharomyces cerevisiae* (**FIGURE 7.12B**) helps us produce some foods. Beer brewers often use barley that has been germinated and dried (a process called malting) as a source of glucose for fermentation by this yeast. As the yeast cells make ATP for themselves, they also produce ethanol (which makes the beer alcoholic) and CO_2 (which makes it bubbly). Flowers of the hop plant add flavor and help preserve the finished product. Winemakers use crushed grapes as a source of sugars for fermentation. The ethanol produced by the cells makes the wine alcoholic, and the CO_2 is allowed to escape to the air.

To make bread, flour is kneaded with water, yeast, and sometimes other ingredients. Flour contains starch and a protein called gluten. Kneading causes the gluten to form polymers in long, interconnected strands that make the resulting dough stretchy and resilient. The yeast cells in the dough first break down the starch, then use the released glucose for alcoholic fermentation. The CO_2 they produce accumulates in bubbles that are trapped by the mesh of gluten strands. As the bubbles expand, they cause the dough to rise. The ethanol product of fermentation evaporates during baking.

Lactate fermentation converts glucose to lactate (**FIGURE 7.13A**). Lactate is the ionized form of lactic acid, so this pathway is also called lactic acid fermentation. Glycolysis is the first step of the pathway, and produces 2 ATP, 2 NADH, and 2 pyruvates. The second step is a reaction that transfers electrons and hydrogen ions from NADH directly to the pyruvates. The reaction regenerates NAD^+ and also converts pyruvate to lactate. No carbons are lost, so lactate fermentation produces no CO_2.

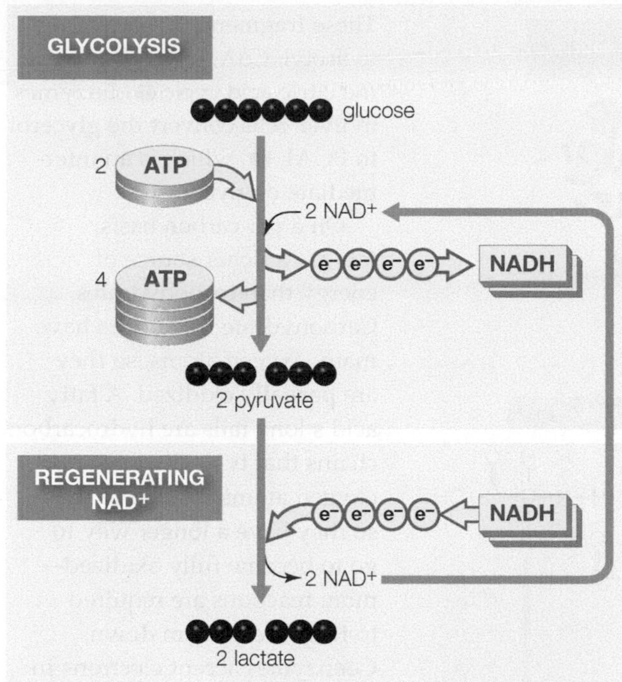
GLYCOLYSIS
glucose
2 ATP
2 NAD+
NADH
4 ATP
2 pyruvate
REGENERATING NAD+
NADH
2 NAD+
2 lactate

A Lactate fermentation begins with glycolysis, and the final steps regenerate NAD^+ and produce three-carbon lactate. The net yield of these reactions is two ATP per molecule of glucose (from glycolysis).

FRA
UKR
POL

B Intense activity such as sprinting quickly depletes oxygen in muscles. Under anaerobic conditions, ATP is produced mainly by lactate fermentation. Fermentation does not make enough ATP to sustain strenuous activity for long.

FIGURE 7.13 Lactate fermentation.

We use lactate fermentation by beneficial bacteria to prepare many foods. Yogurt, for example, is made by allowing bacteria such as *Lactobacillus bulgaricus* and *Streptococcus thermophilus* to grow in milk. Milk contains a disaccharide (lactose) and a protein (casein). The cells first break down the lactose into its monosaccharide subunits, then use the sugars for lactate fermentation. The lactate they produce reduces the pH of the milk, which imparts tartness and causes the casein to form a gel.

Cells in animal skeletal muscles are fused as long fibers that carry out aerobic respiration, lactate fermentation, or both. Aerobic respiration predominates under most circumstances. However, there are times when fermentation is required, for example when intense exercise depletes oxygen in muscles faster than it can be replenished. Under the resulting anaerobic conditions, muscle cells produce ATP mainly by lactate fermentation. This pathway makes ATP quickly, so it is useful for strenuous bursts of activity, but the low ATP yield does not support prolonged exertion (**FIGURE 7.13B**).

Contrary to popular opinion, the buildup of lactate in muscles after exercise does not cause muscle soreness the following day. Lactate very quickly leaves muscles and enters the bloodstream, where it is taken up by cells that are not oxygen-depleted. These cells convert the lactate back to pyruvate for use in aerobic respiration.

Lactate fermentation sustains a few animals that hibernate without oxygen for long periods of time. Consider how some freshwater turtles spend winter months unable to breathe because they are buried in mud or trapped under ice. The normal metabolic rate of a turtle is much lower than a warm-blooded animal in the first place, but during these times it drops even lower, to about 1/10,000 the rate of a resting mammal. Very little ATP is required to maintain such a low metabolism, so the individual can meet its energy requirement by using only lactate fermentation. Excess lactate produced by the pathway is taken up by the animal's bones and shell.

TAKE-HOME MESSAGE 7.5

✔ Prokaryotes and eukaryotes use fermentation pathways, which are anaerobic, to produce ATP by breaking down glucose.

✔ Fermentation's small ATP yield (two per molecule of glucose) occurs by glycolysis. The final steps in these pathways regenerate NAD^+ for glycolysis, but do not produce ATP.

7.6 Alternative Energy Sources in Food

LEARNING OBJECTIVES

- Explain how oxidizing an organic compound can fuel ATP production.
- Describe how organic molecules other than sugars can be broken down in aerobic respiration.

During the first two stages of aerobic respiration, electrons released by the breakdown of glucose are transferred to coenzymes. In other words, glucose becomes oxidized (it gives up electrons) and coenzymes become reduced (they accept electrons). Oxidizing an organic molecule breaks the covalent bonds of its carbon backbone. Aerobic respiration generates a lot of ATP by fully oxidizing glucose, completely dismantling it carbon by carbon. Coenzymes that are reduced during the

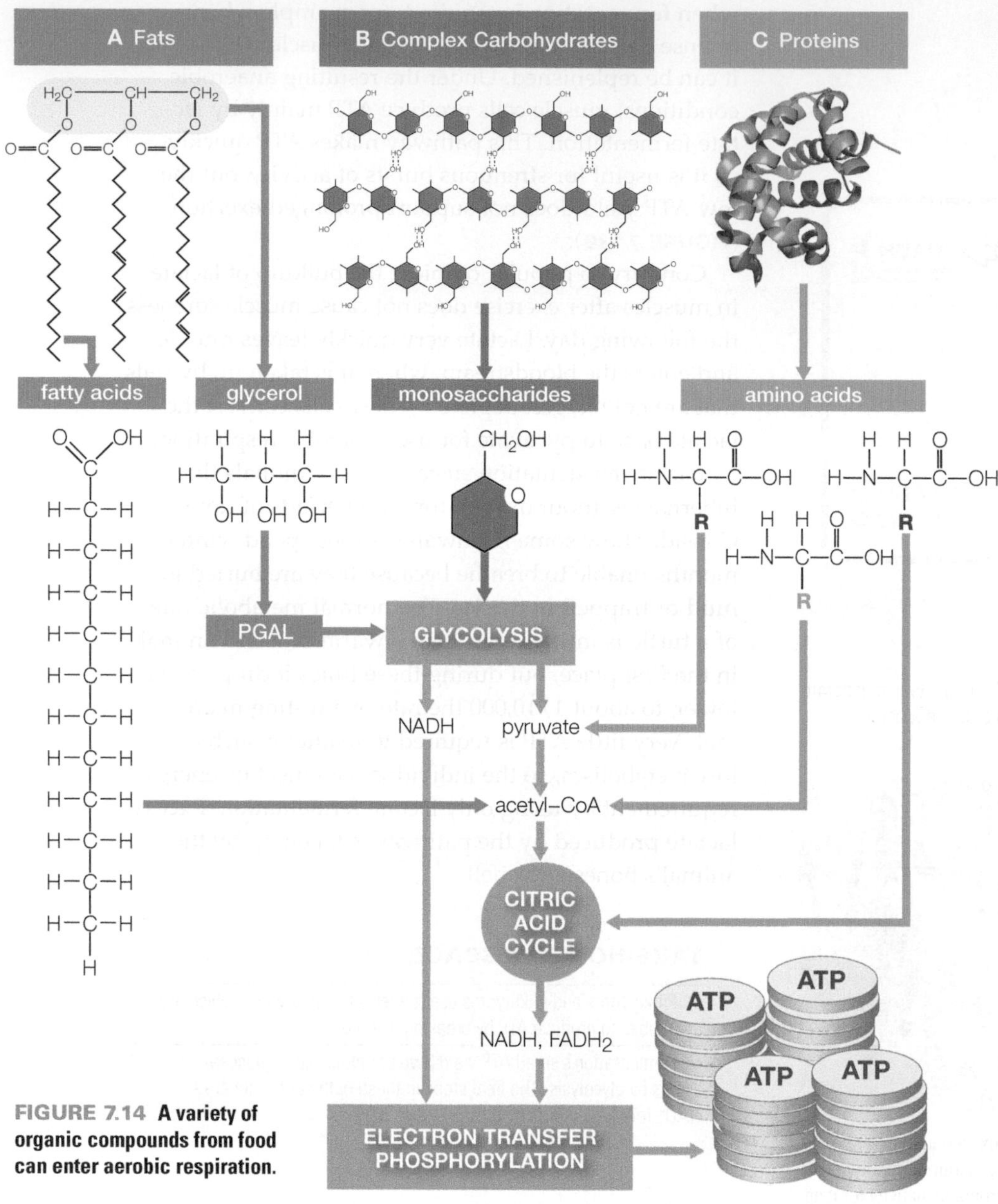

A Fats
B Complex Carbohydrates
C Proteins
fatty acids
glycerol
monosaccharides
amino acids
PGAL
GLYCOLYSIS
NADH
pyruvate
acetyl–CoA
CITRIC ACID CYCLE
NADH, $FADH_2$
ELECTRON TRANSFER PHOSPHORYLATION
ATP
ATP
ATP
ATP

FIGURE 7.14 A variety of organic compounds from food can enter aerobic respiration.

process deliver electrons to electron transfer chains, and the energy of these electrons powers ATP synthesis.

Cells also harvest energy from other organic molecules by oxidizing them. Fats, complex carbohydrates, and proteins in food can be converted to molecules that enter aerobic respiration at various stages. As in glucose breakdown, many coenzymes are reduced, and the energy of the electrons they carry ultimately drives the synthesis of ATP in electron transfer phosphorylation.

A triglyceride molecule has three fatty acid tails attached to a glycerol (Section 3.4). Cells dismantle triglycerides by first breaking the bonds that connect fatty acid tails to the glycerol (**FIGURE 7.14A**). Nearly all cells in the body can oxidize the released fatty acids by splitting their long backbones into two-carbon fragments. These fragments are converted to acetyl–CoA, which can enter the citric acid cycle ❶. Enzymes in liver cells convert the glycerol to PGAL ❷, which is an intermediate of glycolysis.

On a per carbon basis, fats are a richer source of energy than carbohydrates. Carbohydrate backbones have many oxygen atoms, so they are partially oxidized. A fatty acid's long tails are hydrocarbon chains that typically have no oxygen atoms bonded to them, so they have a longer way to go to become fully oxidized—more reactions are required to fully break them down. Coenzymes accept electrons in these oxidation reactions. The more coenzymes that become reduced, the more electrons can be delivered to the ATP-forming machinery of electron transfer phosphorylation.

In humans and other mammals, the digestive system breaks down starch and other complex carbohydrates to monosaccharides (**FIGURE 7.14B**). Cells quickly take up these simple sugars. Glucose is immediately converted to glucose-6-phosphate, which can continue in glycolysis ❸. Six-carbon sugars other than glucose also enter glycolysis. Fructose, for example, can be phosphorylated by hexokinase (the enzyme that phosphorylates glucose). The product of this reaction, fructose-6-phosphate, is an intermediate of glycolysis.

When a cell produces more ATP than it uses, the concentration of ATP rises in the cytoplasm. A high ATP concentration causes glucose-6-phosphate to be diverted away from glycolysis and into a pathway that builds glycogen (Section 3.3). Liver and muscle cells especially favor the conversion of glucose to glycogen, and these cells contain the body's largest stores of it. Between meals, the liver maintains the glucose level in blood by converting the stored glycogen back to glucose. What happens if you eat too many carbohydrates? When the blood level of glucose gets too high,

Data Analysis Activities

Dietary Fat Overload Reprograms Mitochondria

The bodies of humans and other mammals store triglycerides in two kinds of tissue: white fat and brown fat. The main function of white fat is to store triglycerides for energy. Brown fat has more mitochondria than white fat, and these are used to generate body heat.

Mitochondria in brown fat make thermogenin, a transport protein that allows hydrogen ions to move across the inner mitochondrial membrane. Hydrogen ions that move through this protein instead of ATP synthase power no ATP formation. Thus, thermogenin uncouples the operation of mitochondrial electron transfer chains from ATP synthesis.

In cells of brown fat, organic compounds are rapidly broken down by aerobic respiration. Electrons pass through mitochondrial electron transfer chains, which actively transport hydrogen ions from the matrix to the intermembrane space. However, hydrogen ion gradient formation is inefficient across the "leaky" inner membrane, so not much ATP forms. Thus, the major product of aerobic respiration in these cells is heat lost from the ongoing electron transfers.

In 2015, Daniele Barbato and his colleagues investigated the effect of a high-fat diet on brown fat in mice. They maintained an experimental group of mice on a high-fat diet (60 percent fat), and a control group of mice on a normal diet (12 percent fat). Results are shown in **FIGURE 7.15**.

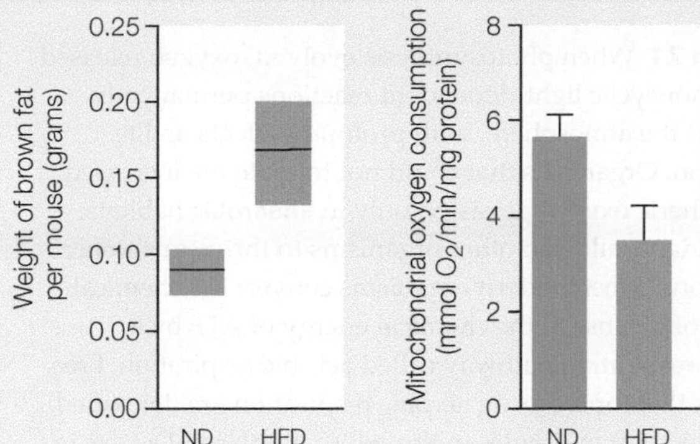

Weight of brown fat per mouse (grams)
0.25
0.20
0.15
0.10
0.05
0.00
ND
HFD
Mitochondrial oxygen consumption (mmol O2/min/mg protein)
8
6
4
2
0
ND
HFD

FIGURE 7.15 Effects of a high-fat diet on brown fat. ND, normal diet group. HFD, high-fat diet group. Oxygen consumption was normalized for protein concentration.

1. The number of mitochondria per gram of brown fat tissue was measured and found to be identical in both groups. Which group had the highest total number of brown fat mitochondria per mouse?
2. In which group was the oxygen consumption per mitochondrion greatest?
3. How did a high-fat diet affect brown fat?

acetyl–CoA is diverted away from the citric acid cycle and into a pathway that builds fatty acids. This is why excess dietary carbohydrates end up as fat.

Enzymes in the digestive system split dietary proteins into their amino acid subunits (**FIGURE 7.14C**), which are absorbed into the bloodstream. Cells use these free amino acids to build proteins or other molecules. When you eat more protein than your body needs for this purpose, the amino acids are broken down. The amino (—NH_2) group is removed, and the carbon backbone is split. The removed amino group is converted to ammonia (NH_3), a waste product that is eliminated in urine. The carbon-containing fragments are converted to pyruvate, acetyl–CoA, or an intermediate of the citric acid cycle, depending on the amino acid ❹. These molecules enter aerobic respiration at the acetyl–CoA formation stage or the citric acid cycle.

TAKE-HOME MESSAGE 7.6

✔ Oxidizing organic molecules breaks their carbon backbones, releasing electrons. The energy of the released electrons can be harnessed to drive ATP formation in aerobic respiration.

✔ First the digestive system and then individual cells convert molecules in food (fats, complex carbohydrates, and proteins) into substrates of glycolysis or aerobic respiration's second-stage reactions.

7.1 Risky Business (revisited)

A child inherits mitochondria from one parent only: the mother. For a variety of reasons, some women's eggs contain a lot of defective mitochondria. Children born to these women have a higher than normal risk of mitochondrial disorders.

A reproductive technology called mitochondrial donation, or three-person IVF (*in vitro* fertilization), can help women with defective mitochondria in their eggs have healthy babies. For example, nuclear DNA from the mother and father can be inserted into a donor egg from a woman with healthy mitochondria. The embryo that forms from the hybrid egg is implanted into the mother, much as normal IVF works.

Babies born from three-person IVF have three parents: Their cells have nuclear DNA inherited from the mother and father, and mitochondria inherited from the donor. Opponents point out that donated mitochondria are passed to future generations, so using it puts us on a slippery slope toward genetically modifying the human race. ●

Section 7.1 When photosynthesis evolved, oxygen released by the noncyclic light-dependent reactions permanently changed the atmosphere, with profound effects on life's evolution. Organisms that could not tolerate the increased atmospheric oxygen persisted only in **anaerobic** habitats. Antioxidants allowed other organisms to thrive in **aerobic** conditions. Most modern organisms convert the chemical energy of glucose to the chemical energy of ATP by a **cellular respiration** pathway called aerobic respiration. Free radicals that form during aerobic respiration are detoxified by antioxidant molecules in the cell's cytoplasm. Defects in mitochondrial electron transfer chain components can cause a buildup of free radicals that damage the cell—and ultimately, the individual. Symptoms can be lethal. Oxidative stress due to mitochondrial malfunction plays a role in many illnesses and genetic disorders.

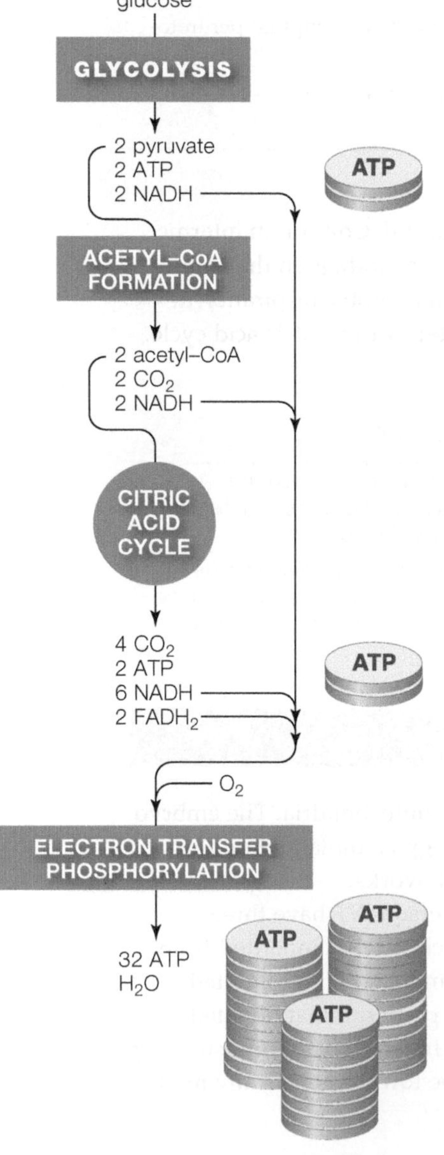
glucose
GLYCOLYSIS
2 pyruvate
2 ATP
2 NADH
ATP
ACETYL–CoA FORMATION
2 acetyl–CoA
2 CO_2
2 NADH
CITRIC ACID CYCLE
4 CO_2
2 ATP
6 NADH
2 $FADH_2$
ATP
O_2
ELECTRON TRANSFER PHOSPHORYLATION
32 ATP
H_2O
ATP
ATP
ATP

FIGURE 7.16
Reviewing the steps of aerobic respiration.

Section 7.2 Most modern organisms convert the chemical energy of glucose to the chemical energy of ATP by the pathway **aerobic respiration**, a cellular respiration pathway that includes glycolysis, acetyl–CoA formation, the citric acid cycle, and electron transfer phosphorylation (FIGURE 7.16). Pathways of **fermentation** begin with glycolysis, do not require oxygen, and yield fewer ATP than aerobic respiration.

Glycolysis, the first stage of aerobic respiration and fermentation, occurs in the cytoplasm of all cells. The reactions convert one six-carbon molecule of glucose to two three-carbon molecules of **pyruvate**. Electrons and hydrogen ions released by the reactions are transferred to two NAD, yielding two NADH. Two ATP also form by **substrate-level phosphorylation**.

Section 7.3 In eukaryotes, aerobic respiration continues in mitochondria, when the two pyruvate molecules that formed in glycolysis are transported into the mitochondrial matrix. The first reaction, acetyl–CoA formation, converts pyruvate to acetyl–CoA and CO_2, and also reduces NAD^+ to NADH. The acetyl–CoA then enters the **citric acid cycle**. Reactions of the citric acid cycle release CO_2 and transfer electrons and hydrogen ions to NAD^+ and FAD (which are thereby reduced to NADH and $FADH_2$). ATP forms by substrate-level phosphorylation.

After two sets of citric acid cycle reactions, the glucose molecule that entered glycolysis has been fully dismantled, all of its six carbons having exited the cell in CO_2. Four ATP have formed, and electrons and hydrogen ions released during the reactions are now carried by 12 reduced coenzymes.

Section 7.4 The final stage of aerobic respiration is electron transfer phosphorylation. Coenzymes reduced in earlier reactions deliver electrons and hydrogen ions to electron transfer chains in the inner mitochondrial membrane. Energy lost by electrons moving through the chains is harnessed to move hydrogen ions from the matrix to the intermembrane space. The resulting hydrogen ion gradient across the inner membrane drives the ions through ATP synthases. The flow of hydrogen ions through these proteins drives ATP synthesis. Oxygen accepts electrons at the end of the electron transfer chains and combines with hydrogen ions, so water forms. The ATP yield of aerobic respiration varies, but typically it is about 36 ATP per glucose.

Section 7.5 Almost all cells carry out fermentation, and many can switch back and forth between aerobic respiration and fermentation. Fermentation pathways produce ATP by breaking down glucose in cytoplasm. They require no oxygen to run: An organic molecule (instead of oxygen) accepts electrons at the end of these pathways.

The end product of **alcoholic fermentation** is ethyl alcohol, or ethanol. The end product of **lactate fermentation** is lactate. Both pathways begin with glycolysis. Their final reactions regenerate the NAD^+ required for glycolysis to continue, but produce no ATP. Thus, fermentation's overall yield of 2 ATP per glucose yield comes entirely from glycolysis.

We use fermentation in microorganisms to produce some foods. Fermentation also supplements metabolism of multicelled eukaryotes.

Section 7.6 Oxidizing an organic molecule breaks its carbon backbone. Aerobic respiration fully oxidizes glucose, dismantling its backbone carbon by carbon. Each carbon removed releases electrons that drive ATP formation in electron transfer phosphorylation. Organic molecules other than sugars can also be broken down (oxidized) for energy. In humans and other mammals, first the digestive system and then individual cells convert fats, proteins, and complex carbohydrates in food to molecules that are intermediates in glycolysis or other pathways of aerobic respiration.

SELF-QUIZ

Answers in Appendix VII

1. Is the following statement true or false? Unlike animals, which make many ATP by aerobic respiration, plants make all of their ATP by photosynthesis.

2. Glycolysis starts and ends in the ________ .
   a. nucleus c. plasma membrane
   b. mitochondrion d. cytoplasm

3. Which of the following pathways requires molecular oxygen (O_2)?
   a. glycolysis
   b. electron transfer phosphorylation in mitochondria
   c. the citric acid cycle
   d. alcoholic fermentation

4. Which molecule does not form during glycolysis?
   a. NADH c. $FADH_2$
   b. pyruvate d. ATP

5. In eukaryotes, the final reactions of aerobic respiration are completed in the ________ .
   a. nucleus c. plasma membrane
   b. mitochondrion d. cytoplasm

6. After the citric acid cycle reactions run ________ , one six-carbon glucose molecule has been completely broken down to CO_2.
   a. once b. twice c. six times d. twelve times

7. In the final stage of aerobic respiration, ________ is the final acceptor of electrons.
   a. water c. oxygen (O_2)
   b. hydrogen d. NADH

8. Most of the energy that aerobic respiration releases from glucose ends up in ________ .
   a. NADH c. heat
   b. ATP d. electrons

9. Put the following pathways in the order they occur during aerobic respiration.
   a. electron transfer phosphorylation
   b. acetyl–CoA formation
   c. citric acid cycle
   d. glycolysis

10. Which of the following is *not* produced by an animal muscle cell operating under anaerobic conditions?
    a. heat d. ATP
    b. pyruvate e. lactate
    c. PGAL f. oxygen

11. ________ accepts electrons in alcoholic fermentation.
    a. Oxygen c. Acetaldehyde
    b. Pyruvate d. Ethanol

12. One of the main differences between aerobic respiration and fermentation is ________ .
    a. fermentation occurs only in prokaryotic cells
    b. ATP forms only in aerobic respiration
    c. fermentation uses no electron transfer chains
    d. aerobic respiration requires no oxygen

13. Which of the following molecules can be oxidized to produce ATP?
    a. amino acids d. complex carbohydrates
    b. six-carbon sugars e. pyruvate
    c. fatty acids f. all of the above

14. Match the reactions with the events.
    ___ glycolysis a. ATP, NADH, $FADH_2$, and CO_2 form
    ___ fermentation b. glucose to two pyruvate
    ___ citric acid cycle c. NAD^+ regenerated, little ATP
    ___ electron transfer phosphorylation d. H^+ flow via ATP synthases

15. Match the term with the best description.
    ___ mitochondrial matrix a. needed for glycolysis
    ___ pyruvate b. inner space
    ___ NAD^+ c. makes many ATP
    ___ mitochondrion d. product of glycolysis
    ___ NADH e. reduced coenzyme
    ___ anaerobic f. no oxygen required

CRITICAL THINKING

1. The higher the altitude, the lower the oxygen level in air. Climbers of very tall mountains risk altitude sickness, which is an illness characterized by shortness of breath, weakness, dizziness, and confusion. The early symptoms of cyanide poisoning are the same as those for altitude sickness. Cyanide binds tightly to cytochrome *c* oxidase, the protein that reduces oxygen molecules in the final step of mitochondrial electron transfer chains. Cytochrome *c* oxidase with bound cyanide can no longer transfer electrons. Explain why cyanide poisoning starts with the same symptoms as altitude sickness.

2. The bar-tailed godwit is a type of shorebird that makes an annual migration from Alaska to New Zealand and back. The birds make each 11,500-kilometer (7,145-mile) trip by flying over the Pacific Ocean in about nine days, depending on weather, wind speed, and direction of travel. One bird was observed to make the entire journey uninterrupted, a feat that is comparable to a human running a nonstop seven-day marathon at 70 kilometers per hour (43.5 miles per hour). Would you expect the flight (breast) muscles of bar-tailed godwits to use mainly aerobic respiration or fermentation? Explain your answer.

8 DNA STRUCTURE AND FUNCTION

CORE CONCEPTS

Information Flow

Living systems store, retrieve, transmit, and respond to information essential for life.

DNA offers a simple, efficient way of storing a gigantic amount of information that can be copied with high fidelity for transmission to offspring. The DNA in each cell encodes all of the information necessary for growth, survival, and reproduction of the cell and (in multicelled organisms) the individual. This information passes from parent to offspring during processes of reproduction. DNA replication ensures the continuity of hereditary information, so it maintains the continuity of life.

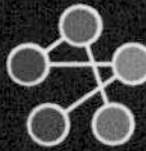

Systems

Complex properties arise from interactions among components of a biological system.

Random changes in DNA that are not repaired become mutations, which can alter the structure and function of cells, and also of multicelled organisms. DNA replication is a source of mutations because it is an imperfect process. Exposure to some environmental factors can also result in mutations.

Process of Science

The field of biology consists of and relies on experimentation and the collection and analysis of scientific evidence.

A series of historical experiments led to the discovery of DNA's structure and function. Early experiments revealed that DNA is the primary source of heritable information for all life. Decades of further investigation elucidated DNA's double-stranded structure, and how that structure allows the molecule to encode heritable information.

Links to Earlier Concepts

Radioisotope tracers (Section 2.2) were used in research that led to the discovery that DNA (3.6), not protein (3.5), is the hereditary material of all organisms (1.3). This chapter revisits experimental design (1.6, 1.7), free radicals (2.3, 5.6), the cell nucleus (4.4), enzymes (5.4), and metabolic pathways (5.5). Your knowledge of carbohydrate ring numbering (3.3) will help you understand DNA replication.

8.1 A Hero Dog's Golden Clones

On September 11, 2001, Constable James Symington drove his search dog Trakr from Nova Scotia to Manhattan. Within hours of arriving, the dog led rescuers to the area where the final survivor of the World Trade Center attacks was buried. She had been clinging to life, pinned under rubble from the building where she had worked. Symington and Trakr helped with the search and rescue efforts for three days nonstop, until Trakr collapsed from smoke and chemical inhalation, burns, and exhaustion (**FIGURE 8.1**).

Trakr survived the ordeal, but later lost the use of his limbs from a degenerative neurological disease probably linked to toxic smoke exposure at Ground Zero. The hero dog died in April 2009, but his DNA lives on in his genetic copies—his **clones**. Symington's essay about Trakr's superior nature and abilities as a search and rescue dog won the Golden Clone Giveaway, a contest to find the world's most clone-worthy dog. Trakr's DNA was shipped to Korea, where it was inserted into donor dog eggs, which were then implanted into surrogate mother dogs. Five puppies, all clones of Trakr, were delivered to Symington in July 2009. Symington trained the clones to be search and rescue dogs for his humanitarian organization, Team Trakr Foundation.

Cloning works because each cell inherits DNA from a parent cell. That DNA is like a master blueprint: It contains all of the information necessary to rebuild the cell and, in the case of animals and other multicelled organisms, the entire individual.

With this unit, we turn to genetics: the information that DNA holds, how cells use it, and how it passes from one generation to the next.

POLICE
POLICE

FIGURE 8.1 **James Symington and his dog Trakr assisting in the search for victims at Ground Zero, September 2001.**

clone Genetic copy of an organism.

CREDITS: (opposite) zagorodnaya/Shutterstock; (1) Splash News/Newscom.

8.2 Discovery of DNA's Function

LEARNING OBJECTIVES

- Describe the four properties required of any genetic material.
- Summarize the classic experiments of Griffith, Avery, and Hershey and Chase that demonstrated DNA is genetic material.

Deoxyribonucleic acid, or DNA (**FIGURE 8.2**), was first described in 1869 by Johannes Miescher, a chemist who extracted it from cell nuclei. Miescher determined that DNA is not a protein, and that it is rich in nitrogen and phosphorus, but he never learned its function.

Sixty years later, Frederick Griffith unexpectedly found a clue about DNA's function. Griffith was studying pneumonia-causing bacteria in the hope of making a vaccine. He isolated two types, or strains, of the bacteria, one harmless (R), the other lethal (S). Griffith used R and S cells in a series of experiments testing their ability to cause pneumonia in mice (**FIGURE 8.3**). He discovered that heat destroyed the ability of lethal S bacteria to cause pneumonia, but it did not destroy their hereditary material, including whatever specified "kill mice." That material could be transferred from dead S cells to live R cells, which put it to use. The transformation was permanent and heritable: Even after hundreds of generations, descendants of transformed R cells retained the ability to kill mice.

What was the substance that transformed harmless R cells into killers? Oswald Avery, Colin MacLeod, and Maclyn McCarty decided to identify this "transforming principle." The team extracted the lipids, proteins, and nucleic acids from S cells, then used a process of elimination to determine which transformed bacteria. Treating the extract with enzymes that destroy lipids and proteins did not destroy the transforming principle. Avery and McCarty realized that the substance they were seeking must be nucleic acid—DNA or RNA. DNA-degrading enzymes destroyed the extract's ability to transform cells, but RNA-degrading enzymes did not. Thus, DNA had to be the transforming principle.

Like most other scientists, the researchers had assumed that proteins were the material of heredity. After all, traits are diverse, and proteins are the most diverse biological molecules. DNA was widely assumed to be too simple (experts of the era called it "a stupid molecule") because it has only four nucleotide components (Section 8.3 returns to the details of DNA structure). Avery and his team were so skeptical that they published their results only after they had convinced themselves, by years of painstaking experimentation, that DNA was indeed hereditary material. They were also careful to point out that they had not proven DNA was the *only* hereditary material.

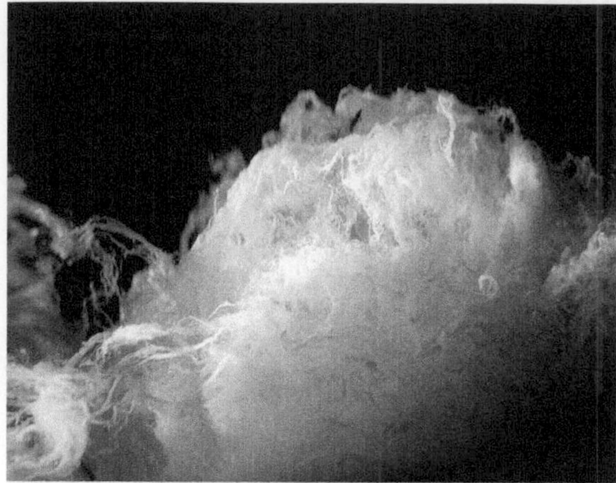

FIGURE 8.2 **DNA, the substance**, extracted from human cells.

A Griffith's first experiment showed that R cells were harmless. When injected into mice, the bacteria multiplied, but the mice remained healthy.

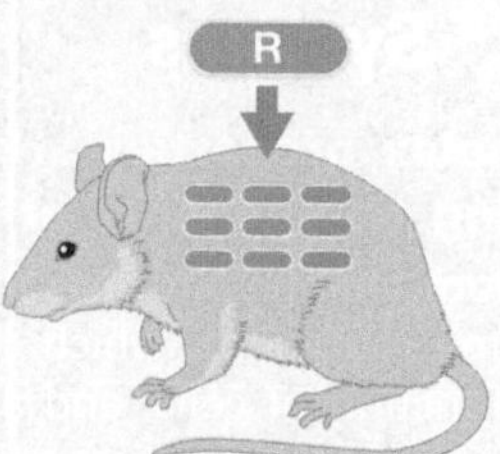
R

B The second experiment showed that an injection of S cells caused mice to develop fatal pneumonia. Their blood contained live S cells.

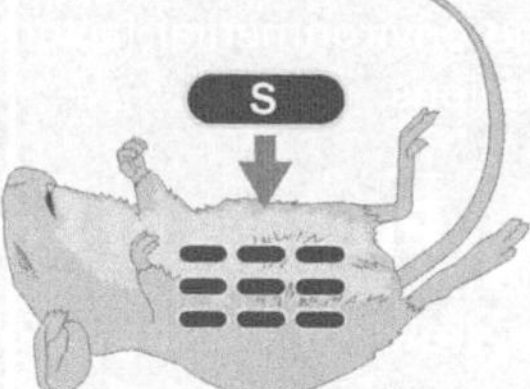
S

C For a third experiment, Griffith killed S cells with heat before injecting them into mice. The mice remained healthy, indicating that the heat-killed S cells were harmless.

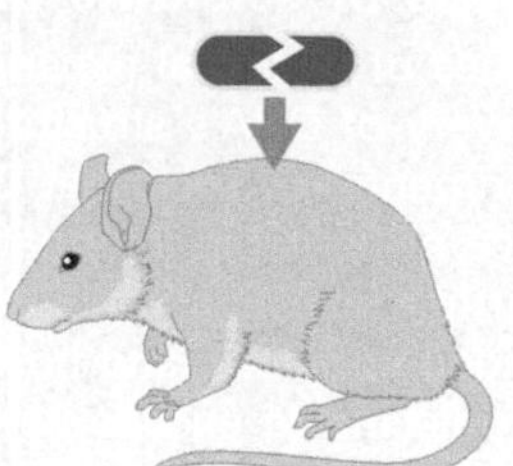

D In his fourth experiment, Griffith injected a mixture of heat-killed S cells and live R cells. To his surprise, the mice became fatally ill, and their blood contained live S cells.

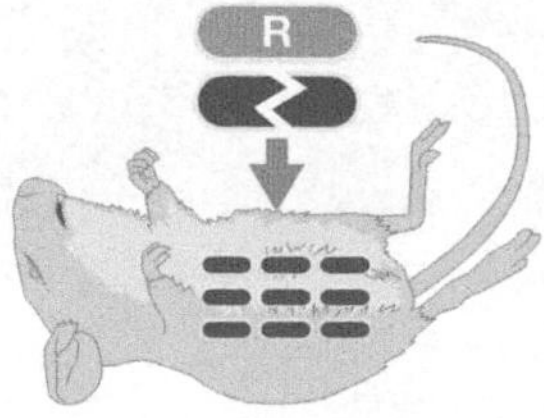
R

FIGURE 8.3 **Fred Griffith's experiments.** Griffith discovered that hereditary information passes between two strains (R and S) of *Streptococcus pneumoniae* bacteria.

FIGURE IT OUT In the fourth experiment, did hereditary material pass from R cells to S cells, or from S cells to R cells? Answer: From S cells to R cells

bacteriophage (bac-TEER-ee-oh-fayj) Virus that infects bacteria.

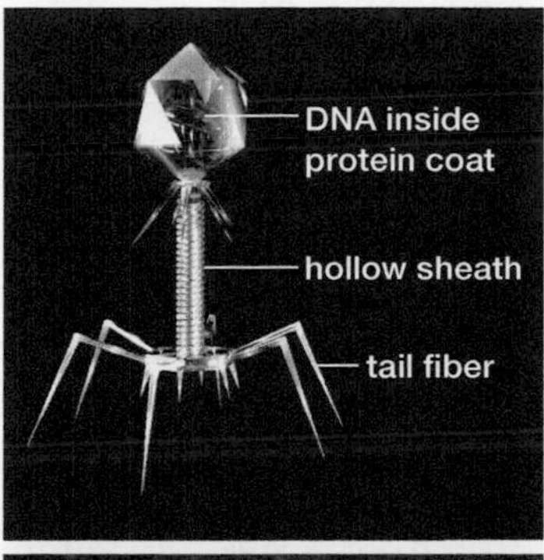
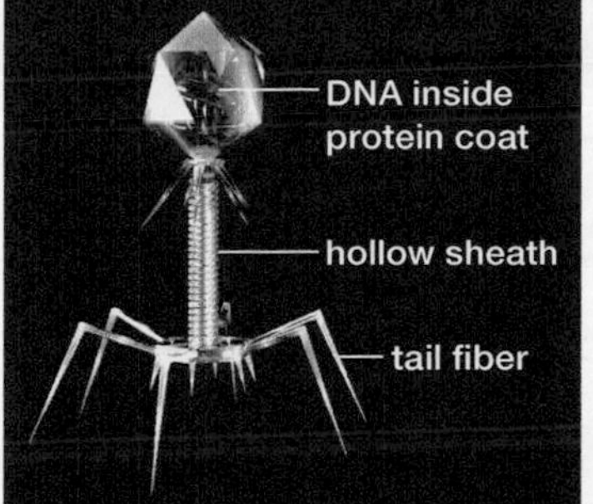
DNA inside
protein coat
hollow sheath
tail fiber

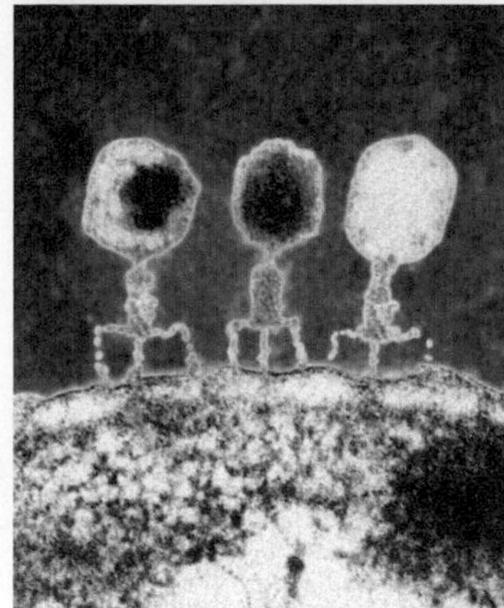

A Top left, model of a bacteriophage. Bottom left, micrograph of bacteriophage injecting DNA into a bacterium.

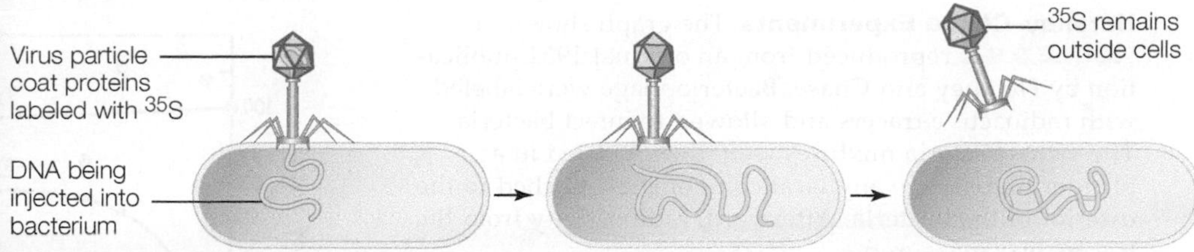
Virus particle
coat proteins
labeled with 35S
DNA being
injected into
bacterium
35S remains
outside cells

B In one experiment, bacteriophage were labeled with a radioisotope of sulfur (^{35}S), a process that makes their protein components radioactive. The labeled viruses were mixed with bacteria long enough for infection to occur, and then the mixture was whirled in a kitchen blender. Blending dislodged viral parts that remained on the outside of the bacteria. Afterward, most of the radioactive sulfur was detected outside the bacterial cells. The viruses had not injected protein into the bacteria.

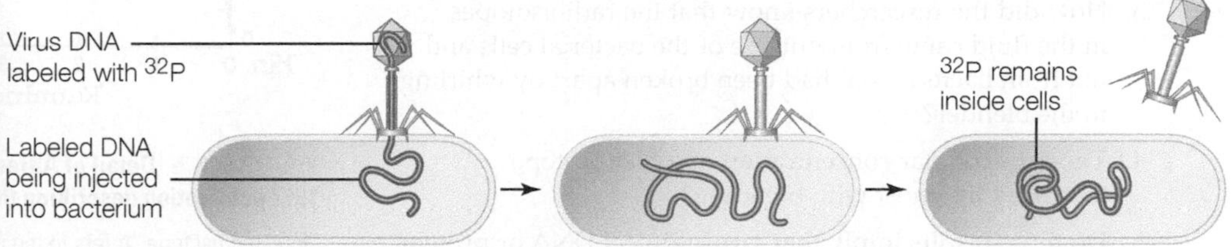
Virus DNA
labeled with 32P
Labeled DNA
being injected
into bacterium
32P remains
inside cells

C In another experiment, bacteriophage were labeled with a radioisotope of phosphorus (^{32}P), a process that makes their DNA radioactive. The labeled viruses were allowed to infect bacteria. After the external viral parts were dislodged from the bacteria, the radioactive phosphorus was detected mainly inside the bacterial cells. The viruses had injected DNA into the cells—evidence that DNA is the genetic material of this virus.

FIGURE 8.4 The Hershey–Chase experiments.

Alfred Hershey and Martha Chase carried out experiments to determine the composition of the hereditary material that bacteriophage inject into bacteria. The experiments were based on the knowledge that proteins contain more sulfur (S) than phosphorus (P), and DNA contains more phosphorus than sulfur.

Avery, MacLeod, and McCarty's surprising result prompted a stampede of other scientists into the field of DNA research, and the resulting explosion of discovery confirmed DNA's role as the molecule of heredity. Key in this advance was the realization that any molecule, DNA or otherwise, had to have a certain set of properties in order to function as the sole hereditary (genetic) material. First, a full complement of genetic information must be transmitted along with the molecule from one generation to the next; second, cells of a given species should contain the same amount of it; third, it must be exempt from major change in order to function as a genetic bridge between generations; and fourth, it must be capable of encoding the unimaginably huge amount of information required to build a new individual.

In the late 1940s, Alfred Hershey and Martha Chase proved that DNA, and not protein, satisfies the first property of a hereditary molecule: It transmits a full complement of genetic information. Hershey and Chase worked with **bacteriophage**, a type of virus that infects bacteria (**FIGURE 8.4A**). Like all viruses, these infectious particles carry hereditary material that specifies how to make new viruses. After a virus injects a cell with this material, the cell starts making new virus particles. Hershey and Chase carried out a series of experiments proving a bacteriophage injects DNA into bacteria, not protein (**FIGURE 8.4B,C**).

In 1948, the second property expected of a hereditary molecule was pinned on DNA. By meticulously measuring the amount of DNA in cell nuclei from a number of species, André Boivin and Roger Vendrely proved that body cells of any individual of a species contain precisely the same amount of DNA. Proof that DNA has the third property expected of a hereditary molecule came from a demonstration that DNA is not involved in metabolism: Daniel Mazia's laboratory discovered that the protein and RNA content of cells varies over time, but the DNA content does not. The fourth property—that a hereditary molecule must somehow encode a huge amount of information—would be proven along with the discovery of DNA's structure, a topic we continue in the next section.

TAKE-HOME MESSAGE 8.2

✔ Many decades of experiments with bacteria and bacteriophage proved that DNA, and not protein, is the molecule that carries hereditary information.

Data Analysis Activities

Hershey–Chase Experiments The graph shown in FIGURE 8.5 is reproduced from an original 1952 publication by Hershey and Chase. Bacteriophage were labeled with radioactive tracers and allowed to infect bacteria. The virus–bacteria mixtures were then whirled in a blender to dislodge any viral components attached to the exterior of the bacteria. Afterward, radioactivity from the tracers was measured.

1. Before blending, what percentage of each isotope, ^{35}S and ^{32}P, was extracellular (outside the bacteria)?
2. After 4 minutes in the blender, what percentage of each isotope was extracellular?
3. How did the researchers know that the radioisotopes in the fluid came from outside of the bacterial cells and not from bacteria that had been broken apart by whirling in the blender?
4. The extracellular concentration of which isotope increased the most with blending?
5. Do these results imply that viruses inject DNA or protein into bacteria? Why or why not?

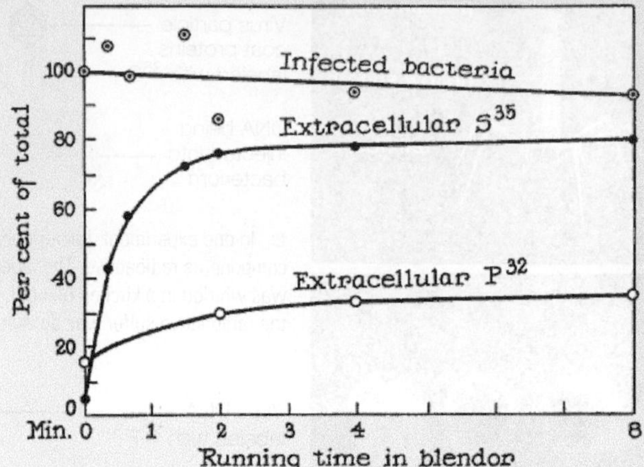
Infected bacteria
Extracellular S 35
Extracellular P 32
Per cent of total
100
80
60
40
20
0
Min. 0
1
2
3
4
8
Running time in blendor

FIGURE 8.5 **Detail of Alfred Hershey and Martha Chase's 1952 publication describing their experiments with bacteriophage.**

"Infected bacteria" refers to the percentage of bacteria that survived the blender.

8.3 Discovery of DNA's Structure

LEARNING OBJECTIVES

- Summarize the events that led to the discovery of DNA's structure.
- Identify the subunits of DNA and how they differ.
- Describe the structure of a DNA molecule.
- Describe base pairing.
- Explain how DNA holds information.

Building Blocks of DNA

Section 3.6 introduced nucleotides as monomers of nucleic acids (DNA and RNA). Each nucleotide is built from three smaller molecules: a nitrogen-containing base, a five-carbon sugar, and a chain of three phosphate groups (FIGURE 8.6). The base is attached to the 1′ (one prime) carbon of the sugar, and the phosphate groups are attached to the 5′ carbon (Section 3.3).

DNA takes its name from deoxyribose, the sugar component of its nucleotides. Only four types of nucleotides make up DNA, and all have a deoxyribose. The four types differ in their component base (FIGURE 8.7): adenine (abbreviated A), guanine (G), thymine (T), or cytosine (C). Adenine and guanine are called purines; their bases have double carbon rings. Thymine and cytosine are pyrimidines; their bases have single carbon rings. Nucleotide structure had been worked out in the early 1900s, but just how the four kinds are arranged in a DNA molecule was a puzzle that took 50 more years to solve.

Fame and Glory

Clues about DNA's structure started coming together around 1950, when Erwin Chargaff (one of many researchers investigating DNA's function) made two important discoveries about the molecule. First, the amounts of thymine and adenine are identical in any DNA molecule, as are the amounts of cytosine and guanine (A = T and G = C). We call this discovery Chargaff's first rule. Chargaff's second discovery, or rule, is that the DNA of different species differs in its proportions of adenine and guanine.

Around the same time, James Watson and Francis Crick were sharing ideas about the structure of DNA. The helical (coiled) pattern of secondary structure that occurs in many proteins (Section 3.5) had just been discovered, and Watson and Crick suspected that the DNA molecule was also a helix. The two spent many hours arguing about the size, shape, and bonding requirements of the four DNA nucleotides. They pestered chemists to help them identify bonds they might have overlooked, fiddled with cardboard cutouts, and made models from scraps of metal connected by suitably angled "bonds" of wire.

Meanwhile, Rosalind Franklin had been investigating the structure of DNA. Franklin specialized in x-ray crystallography, a technique in which x-rays are directed through a purified and crystallized substance. Atoms in the substance's molecules scatter the x-rays in a pattern that can be captured as an image. Researchers use the pattern to calculate the size, shape, and spacing

CREDIT: (5) © From A.D. & M. Chase, Independent Functions of Viral Protein and Nucleic Acid. Acid in Growth of Bacteriophage. *J. Gen. Physiol.* (1952) 36:39-56.

FIGURE 8.6 Components of a nucleotide.

Each nucleotide has three components: a nitrogenous base, a deoxyribose sugar, and phosphate groups. The base is attached to the sugar's 1′ carbon; the phosphate group(s), to its 5′ carbon. Numbering the carbons in the sugars allows us to keep track of the orientation of nucleotide chains.

between any repeating elements of the molecules—all of which are details of molecular structure.

Franklin had been told she would be the only one in her department working on the structure of DNA, so she did not know that Maurice Wilkins was already doing the same thing just down the hall. No one had told Wilkins about Franklin's assignment; he assumed she was a technician hired to do his x-ray crystallography work. And so a clash began. Wilkins thought Franklin displayed an appalling lack of deference that technicians of the era usually accorded researchers. To Franklin, Wilkins seemed prickly and oddly overinterested in her work.

Wilkins and Franklin had been given identical samples of DNA for crystallography. As molecules go, DNA is gigantic, and it was difficult to crystallize given the techniques of the time. Franklin's meticulous work with her sample yielded the first clear x-ray diffraction image of DNA as it occurs in cells (left). She gave a presentation on this work in 1952. She used the image to calculate that a DNA molecule is very long compared to its diameter of 2 nanometers. She also identified a repeating pattern every 0.34 nanometer along its length, and another every 3.4 nanometers. She concluded that DNA had two chains twisted into a double helix, with a backbone of phosphate groups on the outside, and bases arranged in an unknown way on the inside. She had calculated DNA's diameter, the distance between its chains and between its bases, the pitch (angle) of the helix, and the number of bases in each coil. Crick, with his crystallography background, would have recognized the significance of the work—if he had been there. Watson was in the audience but he was not a crystallographer, and he did not understand the implications of Franklin's x-ray diffraction image or her calculations.

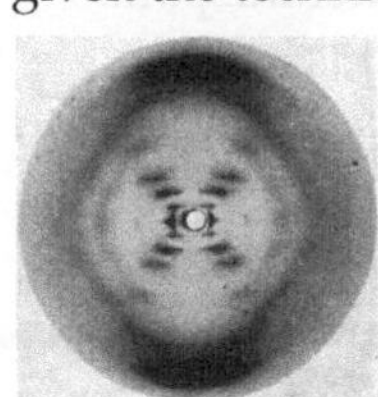

Franklin started to write a research paper on her findings. Meanwhile, and perhaps without her knowledge, Watson and Wilkins reviewed Franklin's x-ray diffraction image, and Watson and Crick read a report detailing Franklin's unpublished data. Crick, who had more experience with molecular modeling than Franklin, immediately understood what the image and the data meant. Watson and Crick now had all

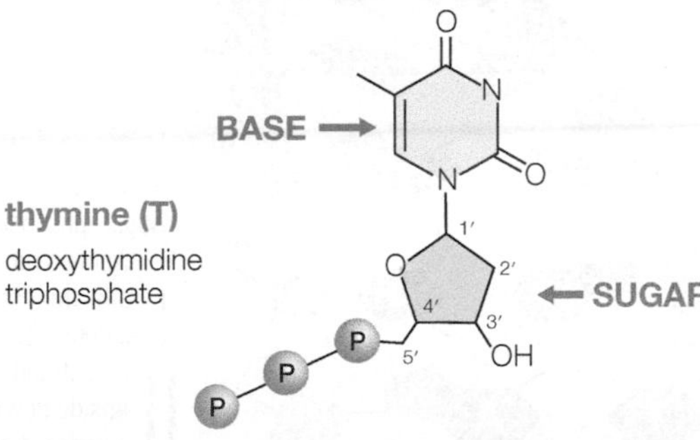
BASE
thymine (T)
deoxythymidine
triphosphate
SUGAR

FIGURE 8.7 The four nucleotides that make up DNA.

Nucleotides that become assembled into a DNA molecule have three phosphate groups. They differ in their component bases (blue): adenine (A), guanine (G), cytosine (C), and thymine (T). Adenine and guanine are called purines; thymine and cytosine are called pyrimidines.

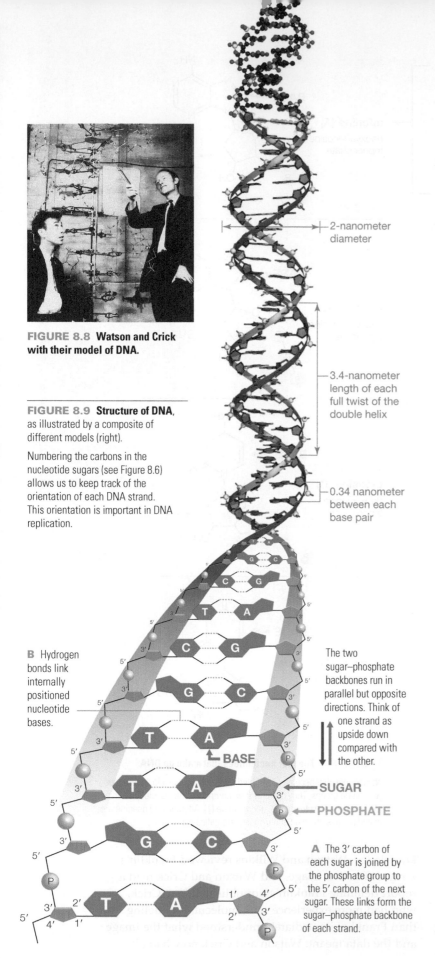

FIGURE 8.8 **Watson and Crick with their model of DNA.**

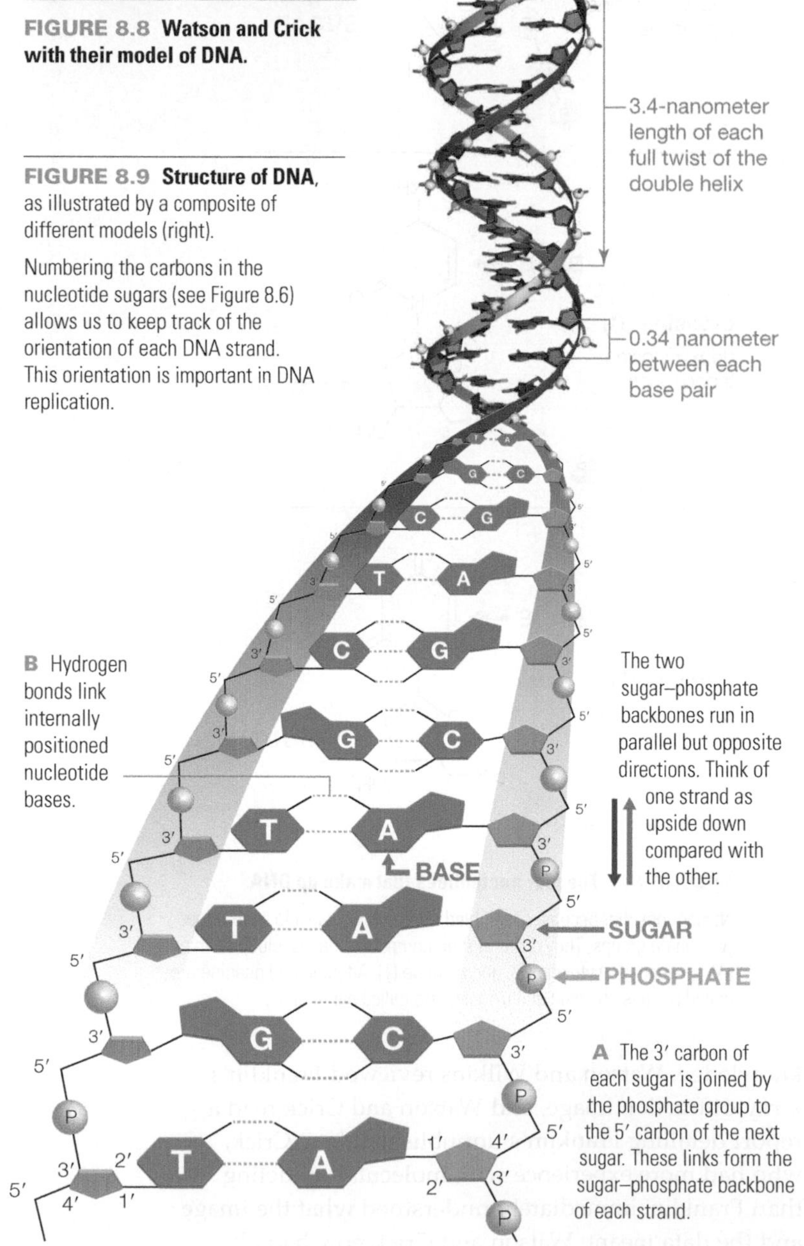

FIGURE 8.9 **Structure of DNA**, as illustrated by a composite of different models (right).

Numbering the carbons in the nucleotide sugars (see Figure 8.6) allows us to keep track of the orientation of each DNA strand. This orientation is important in DNA replication.

2-nanometer diameter

3.4-nanometer length of each full twist of the double helix

0.34 nanometer between each base pair

B Hydrogen bonds link internally positioned nucleotide bases.

The two sugar–phosphate backbones run in parallel but opposite directions. Think of one strand as upside down compared with the other.

BASE

SUGAR

PHOSPHATE

A The 3′ carbon of each sugar is joined by the phosphate group to the 5′ carbon of the next sugar. These links form the sugar–phosphate backbone of each strand.

the information they needed to build the first accurate model of DNA (FIGURE 8.8).

In 1962, Watson, Crick, and Wilkins were awarded a Nobel prize for the discovery of the structure of DNA. Rosalind Franklin did not share in the honor because the Prize is not given posthumously. She died in 1958, at the age of 37, of ovarian cancer probably caused by extensive exposure to x-rays during her work.

The Anatomy of DNA

Each DNA molecule has two chains (strands) of nucleotides coiled into a double helix (FIGURE 8.9). Covalent bonds link the deoxyribose of one nucleotide and a phosphate group of the next, forming a sugar–phosphate backbone. The backbone of the two strands runs in opposite directions, with the nucleotide bases positioned internally. Most scientists had assumed that the bases had to be positioned on the outside of the helix, so they would be accessible to DNA-copying enzymes (you will see shortly how these enzymes access internally positioned bases). Hydrogen bonds between paired bases hold the strands together:

T G A G G A C T C C T C
A C T C C T G A G G A G

one base pair

Notice how the strands "fit." They are complementary, which means the base of each nucleotide on one strand pairs with a suitable partner base on the other. Only two types of base pairings form: adenine pairs with thymine; guanine, with cytosine. This pattern (A–T, G–C) is the same in all molecules of DNA (which explains Chargaff's first rule).

How can just two kinds of base pairings give rise to the incredible diversity of traits we see among living things? Even though DNA is composed of only four nucleotides, the *order* in which one base follows the next along a strand—the **DNA sequence**—varies tremendously among species (which explains Chargaff's second rule). DNA molecules can be hundreds of millions of nucleotides long, so their sequence can encode a massive amount of information (we return

centromere (SEN-truh-meer) Of a duplicated eukaryotic chromosome, constricted region where sister chromatids attach to each other.
chromosome A structure that consists of DNA together with associated proteins; carries part or all of a cell's genetic information.
DNA sequence Order of nucleotide bases in a strand of DNA.
histone (HISS-tone) Type of protein that associates with DNA and structurally organizes eukaryotic chromosomes.
nucleosome (NEW-klee-uh-sohm) A length of DNA wound twice around histone proteins.
sister chromatids (KROME-uh-tid) The two attached DNA molecules of a duplicated eukaryotic chromosome.

to the nature of that information in Chapter 9). DNA sequence variation is the basis of traits that define species and distinguish individuals. Thus DNA, the molecule of inheritance in every cell, is the basis of life's unity. Variations in its sequence are the foundation of life's diversity.

TAKE-HOME MESSAGE 8.3

✔ A DNA molecule has two nucleotide chains (strands) coiled into a double helix and held together by hydrogen bonds between internally positioned bases. A pairs with T, and G with C.

✔ The sequence of bases along a DNA strand varies among species.

8.4 Eukaryotic Chromosomes

LEARNING OBJECTIVES

- Describe the way DNA is organized in a chromosome.
- Explain the meaning of diploid.
- Distinguish between autosomes and sex chromosomes.

The DNA in a human cell has about 3 billion base pairs, which would be about 2 meters (6.5 feet) long if stretched out. How can that much DNA cram into a nucleus less than 10 micrometers in diameter? Proteins that associate with DNA pack it very tightly by organizing the molecule into structure called **chromosomes** (**FIGURE 8.10**). In eukaryotes, a DNA molecule ❶ wraps twice at regular intervals around proteins called **histones**. These DNA–histone spools, which are called **nucleosomes** ❷, look like beads on a string in micrographs. Interactions among histones coil the DNA into a tight fiber ❸. This fiber coils and packs into a hollow cylinder a bit like an old-style landline phone cord ❹.

Most of the time, a eukaryotic chromosome consists of one DNA molecule (and its associated proteins). A cell preparing to divide will duplicate its DNA. After duplication, a eukaryotic chromosome consists of two identical DNA molecules attached to one another at a constricted region called the **centromere**. The two identical halves of a duplicated eukaryotic chromosome are called **sister chromatids**:

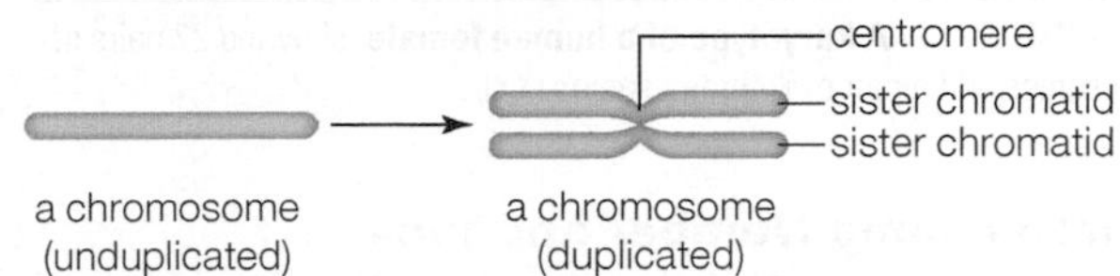

centromere
sister chromatid
sister chromatid
a chromosome (unduplicated)
a chromosome (duplicated)

The duplicated chromosomes condense into their familiar "X" shapes ❺ just before the cell divides.

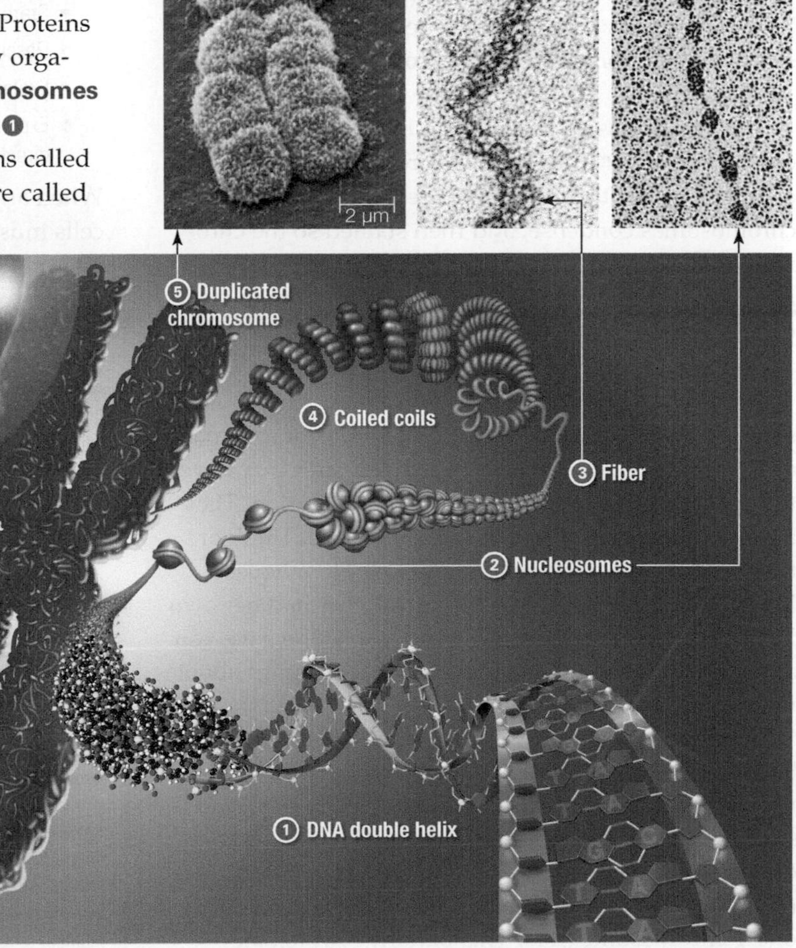

2 μm
⑤ Duplicated chromosome
④ Coiled coils
③ Fiber
② Nucleosomes
① DNA double helix

FIGURE 8.10
The organization of DNA in a eukaryotic chromosome.

❶ The two strands of a DNA molecule form a double helix.

❷ At regular intervals, the DNA molecule (blue) wraps around a core of histone proteins (purple), forming a nucleosome.

❸ The DNA and proteins associated with it twist into a fiber.

❹ The fiber packs further into a hollow cylinder.

❺ At its most condensed, a duplicated eukaryotic chromosome has an X shape. This structure consists of two identical DNA molecules (sister chromatids) attached at a centromere.

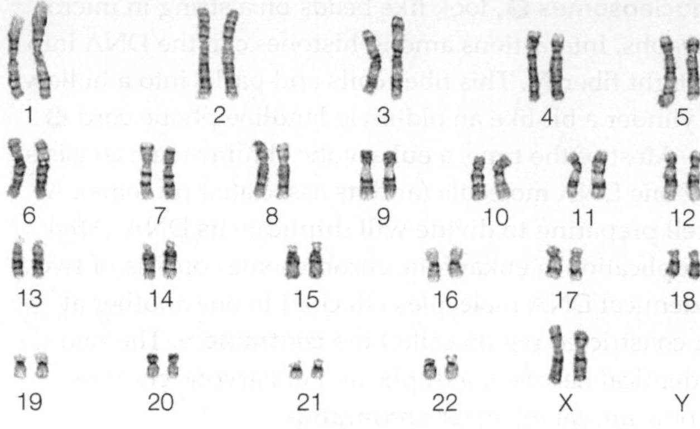
1
2
3
4
5
6
7
8
9
10
11
12
13
14
15
16
17
18
19
20
21
22
X
Y

FIGURE 8.11 A karyotype of a human female, showing 22 pairs of autosomes and a pair of X chromosomes (XX).

Chromosome Number and Type

Most prokaryotes have a single circular chromosome. By contrast, the DNA of a eukaryotic cell is divided among some number of linear chromosomes that differ in length and centromere location. The number of chromosomes in a cell of a given species is called the **chromosome number**, and it is a characteristic of the species. For example, the chromosome number of humans is 46, so our cells have 46 chromosomes.

Actually, human body cells have two sets of 23 chromosomes: two of each type. Cells with two sets of chromosomes are **diploid**, or $2n$ (n stands for "number").

A **karyotype** is an image of an individual cell's chromosomes (**FIGURE 8.11**). To create a karyotype, cells taken from the individual are treated to make the chromosomes condense, and then stained so the chromosomes can be distinguished under a microscope. A micrograph of a single cell is digitally rearranged so the chromosomes are lined up by centromere location and arranged according to size, shape, and length.

In a human body cell, 22 of the 23 pairs of chromosomes are autosomes. The two **autosomes** of each pair are the same in both females and males, and they have the same length, shape, and centromere location. They also hold information about the same traits. Think of them as two sets of books on how to build a house. Your father gave you one set. Your mother had her own ideas about wiring, plumbing, and so on. She gave you an alternate set that says slightly different things about many of those tasks.

Members of a pair of **sex chromosomes** differ between females and males, and the differences determine an individual's anatomical sex. The sex chromosomes of humans are called X and Y. The body cells of typical human females have two X chromosomes (XX); those of typical human males have one X and one Y chromosome (XY). This pattern—XX females and XY males—is the rule among fruit flies, mammals, and many other animals, but there are other patterns. Female butterflies, moths, birds, and certain fishes have two nonidentical sex chromosomes; the two sex chromosomes of males are identical. Environmental factors (not chromosomes) determine sex in some species of invertebrates, frogs, and turtles. As an example, the temperature of the sand in which sea turtle eggs are buried determines the sex of the hatchlings.

TAKE-HOME MESSAGE 8.4

- ✔ In eukaryotic cells, proteins that associate with a DNA molecule organize it as coiled coils that further pack into a chromosome.
- ✔ A eukaryotic cell's DNA is divided among a characteristic number of linear chromosomes that differ in length and shape.
- ✔ Diploid cells have two copies of each chromosome.
- ✔ Members of a pair of sex chromosomes differ between males and females. Chromosomes that are the same in both sexes are called autosomes.

8.5 DNA Replication

LEARNING OBJECTIVES

- State the purpose of DNA replication and describe the process.
- Describe nucleic acid hybridization.
- Explain semiconservative replication.
- Explain why DNA replication proceeds only in the 5′ to 3′ direction.

When any cell reproduces, it divides. Both descendant cells must inherit a complete and accurate copy of their parent's genetic information, or they will not be like

autosome Chromosome of a pair that is the same in males and females; a chromosome that is not a sex chromosome.
chromosome number The total number of chromosomes in a cell of a given species.
diploid (DIP-loyd) Having two of each type of chromosome characteristic of the species ($2n$).
DNA ligase (LIE-gayce) Enzyme that seals gaps or breaks in double-stranded DNA.
DNA polymerase Enzyme that carries out DNA replication. Uses a DNA template to assemble a complementary strand of DNA.
DNA replication Process by which a cell duplicates its DNA before it divides.
karyotype (CARE-ee-uh-type) Image of a cell's chromosomes arranged by size, length, shape, and centromere location.
hybridization (hi-brih-die-ZAY-shun) Spontaneous establishment of base pairing between two nucleic acid strands.
primer Short, single strand of DNA or RNA that serves as an attachment point for DNA polymerase.
semiconservative replication Describes the process of DNA replication in which one strand of each copy of a DNA molecule is new, and the other is a strand of the original DNA.
sex chromosome Chromosome involved in determining anatomical sex; member of a chromosome pair that differs between males and females.

the parent cell. Thus, in preparation for division, the cell copies its chromosomes so that it contains two sets: one for each of its future offspring.

Semiconservative Replication

A cell copies its DNA by way of an energy-intensive pathway called **DNA replication** (**FIGURE 8.12**). Before DNA replication, a chromosome consists of one molecule of DNA—one double helix. As replication begins, proteins called initiators bind to certain sequences of nucleotides in the DNA ❶. Initiator proteins allow other molecules to bind, including enzymes that pry apart the two DNA strands: one (topoisomerase) that untwists the double helix, and another (helicase) that breaks the hydrogen bonds holding the double helix together. The two DNA strands begin to unwind from one another ❷.

Separating the DNA strands exposes their internally positioned bases. Another enzyme (primase) then starts making primers. A **primer** is a short, single strand of DNA or RNA that serves as an attachment point for the enzyme that carries out DNA synthesis. Exposed nucleotide bases on a single strand of DNA form hydrogen bonds with complementary bases of a primer. In other words, a primer base-pairs with a complementary strand of DNA ❸. The establishment of base pairing between two strands of DNA (or DNA and RNA) is called **hybridization** (left). Hybridization is spontaneous and is driven entirely by hydrogen bonding.

DNA
+
primer
hybridization

The enzyme that carries out DNA synthesis is called **DNA polymerase**. There are several types of DNA polymerase, but all can attach to a hybridized primer and begin DNA synthesis. A DNA polymerase moves along a DNA strand, using the sequence of exposed nucleotide bases as a template (guide) to assemble a new strand of DNA from free nucleotides ❹. Each nucleotide provides energy for its own attachment. Remember from Section 5.6 that the bonds between phosphate groups hold a lot of energy. Two of a nucleotide's three phosphate groups are removed when a nucleotide is added to a DNA strand. Breaking those bonds releases enough energy to drive the attachment.

DNA polymerase follows base-pairing rules: When it reaches an A in the template strand, it adds a T to the new DNA strand; when it reaches a G, it adds a C; and so on. Thus, the DNA sequence of each new strand is complementary to the template (parental) strand. The enzyme **DNA ligase** seals any gaps in the sugar–phosphate backbones of the new strands ❺.

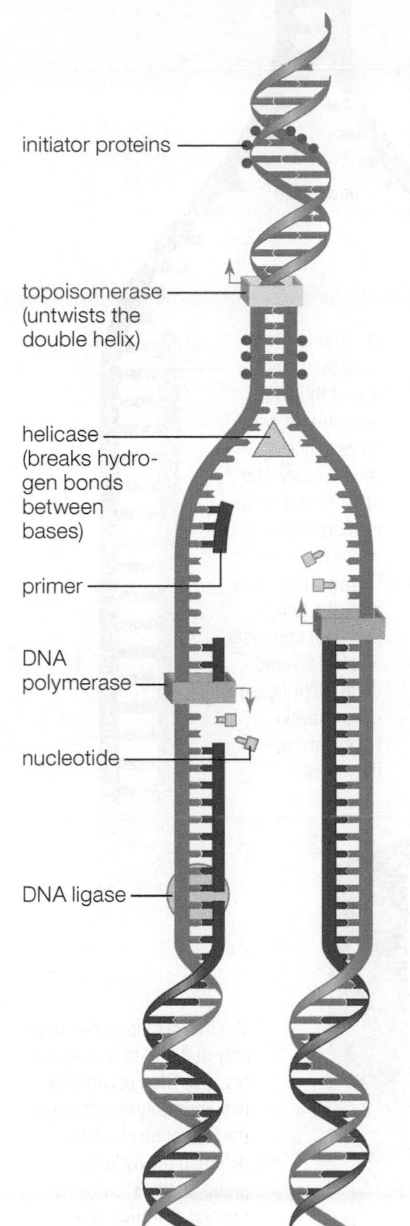
initiator proteins
topoisomerase (untwists the double helix)
helicase (breaks hydrogen bonds between bases)
primer
DNA polymerase
nucleotide
DNA ligase
❶ As replication begins, many initiator proteins attach to the DNA at certain sites in the chromosome. Eukaryotic chromosomes have many of these origins of replication; DNA replication proceeds more or less simultaneously at all of them.
❷ Enzymes that bind to the initiator proteins begin unwinding the two strands of DNA from one another.
❸ Primers base-pair with the exposed single DNA strands.
❹ Starting at primers, DNA polymerases (green boxes) assemble new strands of DNA from nucleotides, using the parent strands as templates.
❺ DNA ligase seals any gaps in the sugar–phosphate backbone of each strand.
❻ Each parental DNA strand (blue) serves as a template for assembly of a new strand of DNA (magenta). The new DNA strand winds up with its template, so two double-stranded DNA molecules result. One strand of each is parental (conserved), and the other is new, so DNA replication is said to be semiconservative.

FIGURE 8.12 DNA replication.

Green arrows show the direction of synthesis for each strand. The Y-shaped structure where the DNA molecule is being unwound is called a replication fork.

Both strands of the parent molecule are copied at the same time. As each new DNA strand lengthens, it winds up with the template strand into a double helix. So, after replication, two double-stranded molecules of DNA have formed ❻. One strand of each molecule is conserved (parental), and the other is new; hence the name of the process, **semiconservative replication**. Both double-stranded molecules produced by DNA replication are replicas of the parent molecule. In a eukaryotic cell, these molecules are sister chromatids that remain attached at the centromere until cell division occurs.

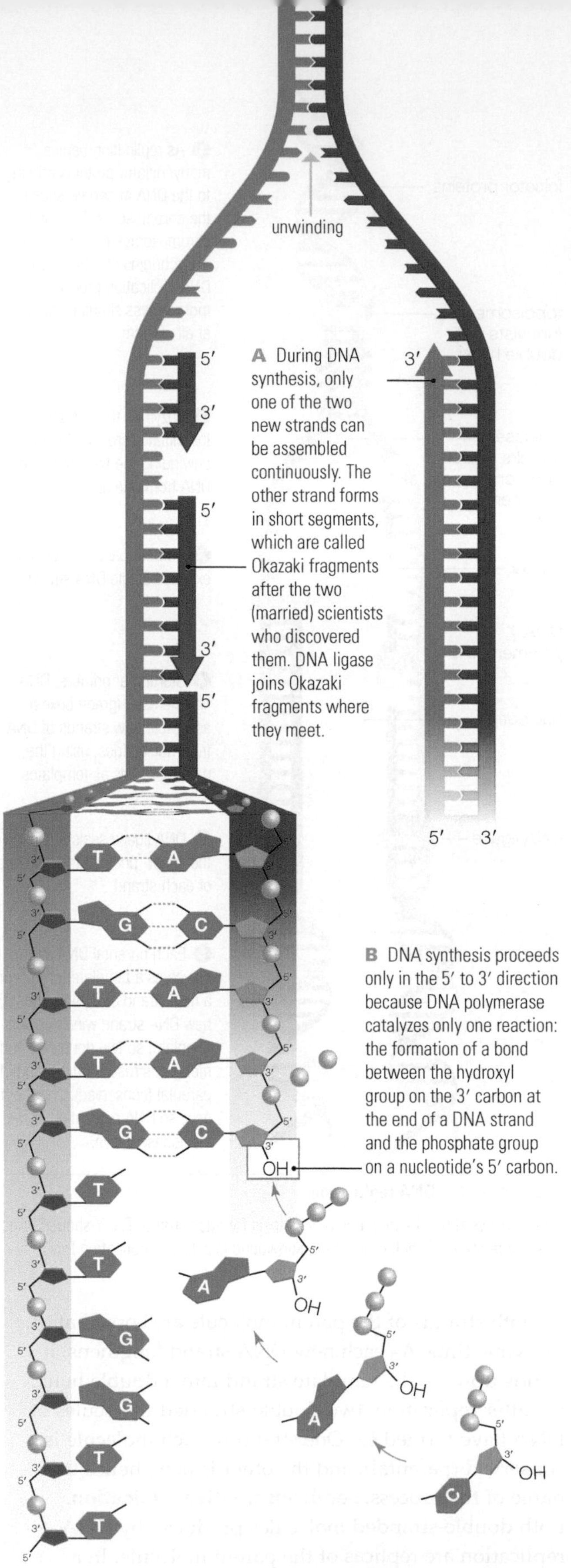

A During DNA synthesis, only one of the two new strands can be assembled continuously. The other strand forms in short segments, which are called Okazaki fragments after the two (married) scientists who discovered them. DNA ligase joins Okazaki fragments where they meet.

B DNA synthesis proceeds only in the 5′ to 3′ direction because DNA polymerase catalyzes only one reaction: the formation of a bond between the hydroxyl group on the 3′ carbon at the end of a DNA strand and the phosphate group on a nucleotide's 5′ carbon.

FIGURE 8.13 Discontinuous synthesis of DNA.
This close-up of a replication fork shows that only one of the two new DNA strands is assembled continuously. The other is assembled in short segments.

Directional Synthesis

Numbering the carbons of the deoxyribose sugars in nucleotides allows us to keep track of the orientation of strands in a DNA double helix (see Figures 8.7 and 8.9). Each strand has two ends. The last carbon atom on one end of the strand is a 5′ carbon of a sugar; the last carbon atom on the other end is a 3′ carbon of a sugar:

DNA polymerase can attach a nucleotide only to a 3′ end. Thus, during DNA replication, only one of two new strands of DNA can be constructed in long pieces (**FIGURE 8.13**). Synthesis of the other strand occurs in short segments that must be joined by DNA ligase where they meet up. This is why we say that DNA synthesis proceeds only in the 5′ to 3′ direction.

TAKE-HOME MESSAGE 8.5

- ✔ Before dividing, a cell copies its chromosomes by DNA replication.
- ✔ DNA replication is an energy-intensive pathway that requires the participation of DNA polymerase and many other enzymes.
- ✔ During replication of a molecule of DNA, each strand of its double helix serves as a template for synthesis of a new, complementary strand of DNA from free nucleotides.
- ✔ Replication of a molecule of DNA produces two double helices that are duplicates of the parent molecule. One strand of each helix is parental; the other is new.

8.6 Mutations: Cause and Effect

LEARNING OBJECTIVES

- Using examples, explain how mutations can arise.
- Describe two cellular mechanisms that can prevent mutations from occurring.

Sometimes, a newly replicated DNA strand is not exactly complementary to its parent strand. A nucleotide may get deleted during DNA replication, or an extra one inserted. Occasionally, the wrong nucleotide is added. Most of these replication errors occur simply because DNA polymerases work very fast. A eukaryotic polymerase can add about 50 nucleotides per second to a strand of DNA; a prokaryotic polymerase can add up to 1,000 per second. Mistakes are inevitable, and some DNA polymerases make a lot of them. Luckily, most DNA polymerases also proofread

mutation Permanent change in the DNA sequence of a chromosome.

their work, reversing the synthesis reaction to remove an error, then resuming synthesis in the forward direction. Even if a polymerase does not correct an error during DNA replication, other proteins can recognize and repair it afterward.

When both proofreading and repair mechanisms fail, a replication error becomes a **mutation**—a permanent change in the DNA sequence of a chromosome. Repair proteins cannot detect an error after a second round of replication, because both strands will then base-pair properly (**FIGURE 8.14**). Thus, a mutation passes to the cell's descendants, their descendants, and so on.

Exposure to some chemicals causes replication errors that can lead to mutations. Several of the cancer-causing chemicals in tobacco smoke, for example, bind directly and irreversibly to nucleotide bases in DNA. The body converts other chemicals in tobacco smoke to DNA-binding compounds that bind to DNA. Both outcomes cause replication errors.

Replication errors also occur after DNA gets broken or otherwise damaged, because DNA polymerases do not copy damaged DNA very well. Consider how free radicals easily open the carbon rings in deoxyribose and nucleotide bases. The resulting oxidative damage (Section 5.6) includes DNA strand breaks and lost or mispaired bases.

Exposure to ultraviolet (UV) light also damages DNA. UV light in the range of 320 to 380 nanometers causes the ring of a pyrimidine (cytosine or thymine) base to form a covalent bond with the ring of an adjacent pyrimidine. The resulting dimer (left) kinks the DNA strand. A specific set of proteins can recognize, remove, and replace pyrimidine dimers before replication begins. If this repair fails, mutations can occur because DNA polymerases tend to copy the kinked DNA incorrectly. Mutations that arise as a result of pyrimidine dimers are the cause of most skin cancers. Exposing unprotected skin to sunlight increases the risk of cancer mainly because UV wavelengths in the light cause a lot of dimers to form. For every second a skin cell spends in the sun, 50 to 100 dimers form in its DNA.

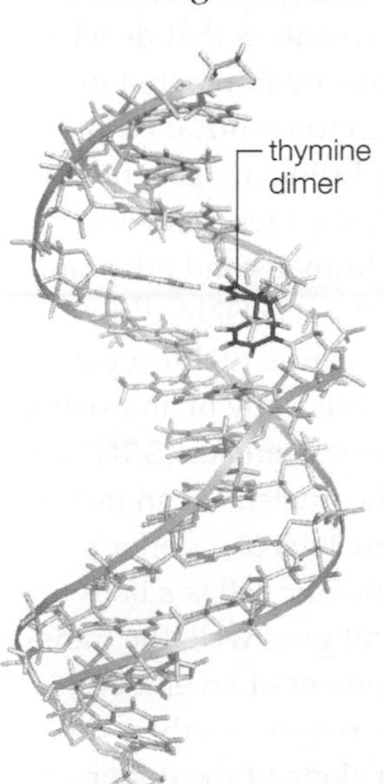
thymine dimer

Electromagnetic energy with a wavelength shorter than 320 nanometers (gamma rays, X-rays, and short-wave UV light) has enough energy to slam electrons out of atoms, thus ionizing them. Exposure to ionizing radiation breaks DNA into pieces (**FIGURE 8.15**). (Chapter 14 considers the effects of chromosomal breaks.) Ionizing radiation also causes covalent bonds to form between bases on opposite strands of the double helix, an outcome that permanently blocks DNA replication. The nucleotide bases themselves can be irreparably damaged by ionizing radiation. Repair enzymes can remove bases damaged in this way, but

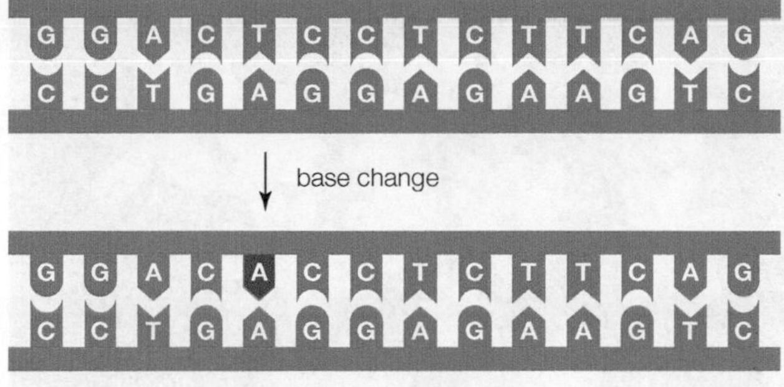
G G A C T C C T C T T C A G
C C T G A G G A G A A G T C
base change
G G A C A C C T C T T C A G
C C T G A G G A G A A G T C

A Repair enzymes can recognize a mismatched base (magenta), but sometimes fail to correct it before DNA replication.

DNA replication
G G A C A C C T C T T C A G
C C T G T G G A G A A G T C
G G A C T C C T C T T C A G
C C T G A G G A G A A G T C

B After replication, both strands base-pair properly. Repair enzymes can no longer recognize the error, which has now become a mutation that will be passed on to the cell's descendants.

FIGURE 8.14 How a replication error becomes a mutation.

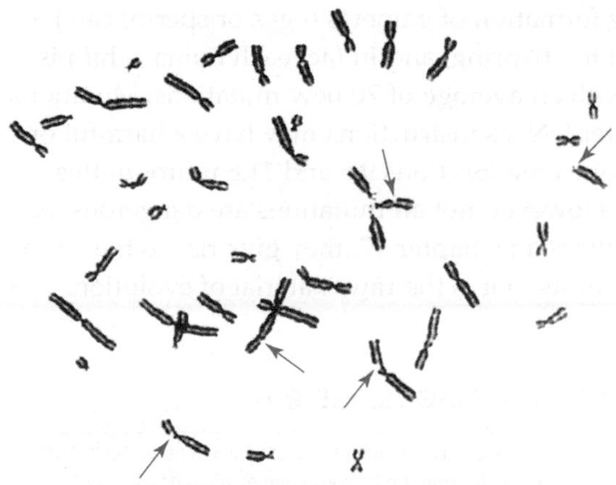

FIGURE 8.15 DNA damaged by ionizing radiation. These are the chromosomes of a human white blood cell that has been exposed to ionizing radiation. Red arrows indicate major breaks. Pieces of chromosomes broken like this often become lost during DNA replication.

CREDIT: (15) Olga Shovman, Andrew C. Riches, Douglas Adamson, and Peter E. Bryant. An improved assay for radiation-induced chromatid breaks using a colcemid block and calyculin-induced PCC combination. *Mutagenesis* (2008) 23(4): 267-270 first published online March 6, 2008 doi:10.1093/mutage/gen009, by permission of Oxford University Press.

normal
flower

FIGURE 8.16 Mutated flowers from Chernobyl.

In 1986, a catastrophic accident at the Chernobyl nuclear power plant in Ukraine released a massive amount of radiation. These *Ranunculus* flowers were recently collected in the Exclusion Zone, an area of about 1,000 square miles that is still dangerously contaminated. All but one are abnormal—an effect of mutations caused by radiation exposure.

they leave an empty space in the double helix, or a strand break that requires further repair. Any of these events can result in mutations (**FIGURE 8.16**).

Mutations can occur in any type of cell, and they can alter any part of the cell's DNA. Those that arise during formation of gametes (eggs or sperm) can be passed to offspring, and in fact each human child is born with an average of 70 new mutations. Mutations that alter DNA's instructions may have a harmful or lethal outcome (Sections 9.6 and 11.6 return to this topic). However, not all mutations are dangerous: As you will see in Chapter 17, they give rise to the variation in traits that is the raw material of evolution.

TAKE-HOME MESSAGE 8.6

✔ Proofreading and repair mechanisms usually maintain the integrity of a cell's genetic information by correcting mispaired bases and fixing damaged DNA before replication occurs.

✔ Uncorrected nucleotide mismatches and unrepaired DNA damage can lead to mutations—permanent changes in the DNA sequence of a chromosome. Mutations are passed to a cell's descendants.

✔ DNA is damaged by exposure to environmental agents such as UV light, free radicals, and some types of chemicals.

✔ Mutations in gametes can be passed to offspring.

differentiation Process by which cells become specialized during development; occurs as different cells in an embryo begin to use different subsets of their DNA.

reproductive cloning Laboratory procedure such as SCNT that produces animal clones from a single cell.

somatic cell nuclear transfer (SCNT) Reproductive cloning method in which the DNA of a donor's body cell is transferred into an enucleated egg.

8.7 Cloning Adult Animals

LEARNING OBJECTIVES

- Discuss the role of DNA in the continuity of life.
- Explain how and why animal clones can be produced from a single body cell.
- Describe differentiation.

The continuity of life is based on DNA. When a cell reproduces, it passes a copy of its chromosome(s) to each of its descendant cells. Mutations aside, the DNA of the offspring is identical with the DNA of the parent cell. Information in that DNA is the basis of cellular form and function, and, in the case of multicelled animals, it also directs the development and functioning of the entire body. Thus, the inheritance of DNA is the reason why offspring are similar to their parent(s) in both form and function, generation after generation.

Consider how identical twins have identical DNA, so their bodies develop the same way (which is why they appear identical, as in the chapter opening photo). Identical twins are the product of a natural process called embryo splitting. The first few divisions of a fertilized egg form a tiny ball of cells that sometimes splits spontaneously. If both halves develop independently, identical twins result. Animal breeders have long exploited this process with a technique called artificial embryo splitting. A ball of cells grown from a fertilized egg is teased apart into two halves that develop as separate embryos. The embryos are implanted in surrogate mothers, who give birth to identical twins.

Twins produced by embryo splitting are genetically identical, but they are not genetically identical with their parents. This is because, in humans and other animals, offspring have two parents whose DNA differs slightly in sequence (Chapter 12 returns to this topic). Animal breeders who want a genetic copy of an existing individual use **somatic cell nuclear transfer (SCNT)**, a laboratory procedure in which the nucleus of an unfertilized egg is replaced with the nucleus of a donor's somatic cell (**FIGURE 8.17A**). A somatic cell is a body cell (as opposed to a gamete). If all goes well, the transplanted DNA directs the development of an embryo, which is then implanted into a surrogate mother. The animal born to the surrogate is a clone of the donor.

"Cloning" means making an identical copy of something, and it can refer to deliberate interventions intended to produce a genetic copy of an organism. **Reproductive cloning** refers to SCNT and any other technology that produces clones of an animal from a single cell. Animal breeders use these technologies because cloned animals have the same desirable features as their DNA donors (**FIGURE 8.17B**). Among

CREDIT: (16) Taavi Tuulik.

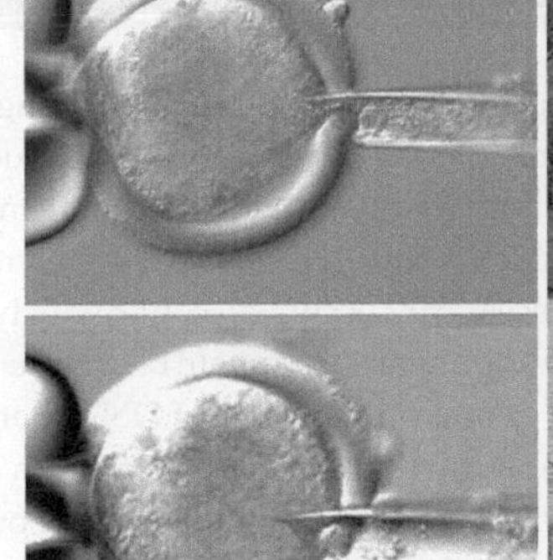

A SCNT with cells from cattle. These micrographs were taken by scientists at Cyagra, a company that specializes in cloning livestock.

Top, a micropipette punctures the egg and sucks out the DNA. All that remains inside the egg's plasma membrane is cytoplasm.

Bottom, another micropipette then delivers a cell grown from the skin of a donor animal into the egg. An electric current applied to the egg will cause the foreign cell to release its nucleus into the cytoplasm. If the egg begins to divide, an embryo forms.

B Champion dairy cow Nelson's Estimate Liz (right) and her clone, Nelson's Estimate Liz II (left), who was produced by SCNT.

Liz II started winning championships before she was one year old.

FIGURE 8.17 Cloning cattle by SCNT (somatic cell nucler transfer).

other benefits, offspring can be produced from a donor animal that is castrated or even dead.

SCNT has been used since 1997, when a lamb named Dolly was cloned from a mammary cell of an adult sheep (the clone was named after voluptuous performer Dolly Parton). Dolly looked and acted like a normal sheep, but she died early. Signs of premature aging suggested that her early demise may have been an effect of the cloning procedure (Section 11.4 returns to this topic).

Twenty years later, the outcome of SCNT can still be unpredictable. Depending on the species, few implanted embryos may survive until birth, and many that do survive have serious health issues. Why the problems? SCNT works because the DNA in a somatic cell contains all the information necessary to build the individual from scratch. However, during early development, an embryo's cells start using different subsets of their DNA. As they do, the cells become different in form and function, a process called **differentiation**. Differentiation in animals is usually a one-way path: Once a cell specializes, all of its descendant cells will be specialized the same way. By the time a liver cell, muscle cell, or other differentiated cell forms, most of its DNA has been turned off and is no longer used (Chapter 10 returns to this topic). In order to produce a clone from a somatic cell, the cell's DNA must be reprogrammed so the parts that trigger embryonic development are turned back on. Researchers still have a lot to learn about that, but SCNT techniques have improved so much that health problems are much less common in animals cloned today. For example, four clones have been produced from Dolly's cells, and these animals have reached old age in good health.

TAKE-HOME MESSAGE 8.7

✔ DNA is the basis of the continuity of life. Information in DNA, the source of form and function, passes from parent to offspring.

✔ A cell's DNA has all the information necessary to build a new cell, and in the case of multicelled organisms, a new body.

✔ Clones appear identical because they have identical DNA. Reproductive cloning technologies such as SCNT can be used to produce animal clones from a single cell.

8.1 A Hero Dog's Golden Clones (revisited)

Many other adult mammals have been cloned since Trakr. As SCNT techniques improve, cloning humans is no longer only within the realm of science fiction. SCNT is already being used to produce human embryos for research purposes. For example, cells of cloned embryos are helping researchers unravel the molecular mechanisms of human genetic disorders. Replacement tissues and organs for people with incurable diseases are also being generated from cloned human cells. Human cloning is not the intent of such research, but if it were, SCNT would indeed be the first step toward that end.

Cloning animals brings us closer to the possibility of cloning humans, both technically and ethically. The idea raises uncomfortable questions. For example, if cloning a lost animal for a grieving owner is acceptable, why would it not be acceptable to clone a lost child for a grieving parent? Different people have very different answers to such questions, so controversy over cloning continues to rage even as the technique improves.

CREDIT: (17) Courtesy of Cyagra, Inc., www.cyagra.com.

Section 8.1 The practice of making **clones** (exact genetic copies) of adult animals is now common. The technique is useful for research into human diseases and genetic disorders. Cloning animals continues to raise ethical questions, particularly because the improving technology brings us closer to the possibility of cloning humans.

Section 8.2 Identifying deoxyribonucleic acid (DNA) as the hereditary material of life took decades of research involving many scientists. Experiments with bacteria and **bacteriophage** were key to the discovery.

Section 8.3 A nucleotide has three components: a five-carbon sugar, a nitrogen-containing base, and phosphate groups. Bonds between the sugar of one nucleotide and a phosphate group of the next form the sugar–phosphate backbone of nucleic acid chains (strands). Four types of nucleotides compose DNA strands. Each has a deoxyribose sugar, a chain of three phosphate groups, and one of four bases: adenine (A), guanine (G), cytosine (C), or thymine (T).

A molecule of DNA consists of two nucleotide strands coiled into a double helix, with the sugar–phosphate backbones running in parallel but opposite directions. Hydrogen bonds between internally positioned bases of the nucleotides hold the two strands together. The bases pair in a consistent way: A with T, and G with C. The order of bases along a strand of DNA—the **DNA sequence**—varies among species, and this variation is the basis of life's diversity.

Section 8.4 Proteins that associate with DNA organize and pack it tightly into a structure called a **chromosome**. In eukaryotic chromosomes, the DNA wraps around **histones** to form **nucleosomes**.

The DNA of a eukaryotic cell is divided among a number of chromosomes that differ in length and **centromere** location. When duplicated, a eukaryotic chromosome consists of two double helices attached at the centromere as **sister chromatids**.

Diploid (2*n*) cells have two sets of chromosomes (two of each type of chromosome). **Chromosome number** is the total number of chromosomes in a cell of a given species, and it is a characteristic of the species. For example, in humans, a normal body cell has 23 pairs of chromosomes. A micrograph showing the complete set of chromosomes in an individual's cells is called a **karyotype**.

Members of a pair of **sex chromosomes** differ among males and females. Chromosomes of a pair that are the same in males and females are **autosomes**.

Section 8.5 Before a cell divides, it copies its chromosomes by the energy-intensive process of **DNA replication**. During DNA replication, enzymes unwind and separate the two strands of the double helix, and assemble **primers**. The primers base-pair with complementary nucleotides exposed on the single DNA strands, a spontaneous process called nucleic acid **hybridization**. Starting at the primers, **DNA polymerase** enzymes use the sequence of bases on each strand as a template to assemble new, complementary strands of DNA from free nucleotides. Synthesis of one strand necessarily occurs discontinuously. **DNA ligase** seals any gaps in the sugar–phosphate backbone of each new strand.

For each molecule of DNA that is copied, two DNA molecules are produced; each is a duplicate of the parent. One strand of each molecule is new, and the other is parental; hence the name **semiconservative replication**.

Section 8.6 DNA replication is not a perfect process, so errors such as incorrect, missing, or extra nucleotides are inevitable. Proofreading by DNA polymerases corrects most replication errors as they occur. Uncorrected errors become **mutations**, which are permanent changes in the DNA sequence of a chromosome. Mutations are passed to descendant cells.

Exposure to environmental agents such as ultraviolet light and some chemicals can damage DNA. DNA polymerase does not copy damaged DNA very well, so these agents lead to mutations. Cancer begins with mutations, but not all mutations are harmful.

Section 8.7 **Somatic cell nuclear transfer** (**SCNT**) and other types of **reproductive cloning** technologies can produce genetically identical individuals (clones) from a body cell of an adult animal. These technologies work because the DNA in each body cell contains all the information necessary to build a new individual. The outcome of SCNT can be unpredictable because **differentiation** is usually a one-way path in animals. During development, cells of an embryo become specialized as they begin to use different subsets of their DNA. Reprogramming the DNA of a differentiated cell to trigger development of an embryo can be unpredictable.

SELF-QUIZ

Answers in Appendix VII

1. Which is *not* a nucleotide base in DNA?
   a. adenine d. thymine
   b. glutamine e. cytosine
   c. guanine f. all are in DNA

2. What are the base-pairing rules for DNA?
   a. A–G, T–C c. A–C, T–G
   b. A–T, G–C d. A–A, G–G, C–C, T–T

3. Similarities in _______ are the basis of similarities in traits.
   a. karyotype c. the double helix
   b. DNA sequence d. chromosome number

4. One species' DNA differs from others in its _______ .
   a. nucleotides c. double helix
   b. DNA sequence d. sugar–phosphate backbone

5. In eukaryotic chromosomes, DNA wraps around _____ .
   a. histone proteins
   b. sister chromatids
   c. the double helix
   d. centromeres
   e. nucleosomes
   f. mutations

6. The chromosome number _______ .
   a. refers to a particular chromosome in a cell
   b. is a characteristic feature of a species
   c. is the number of autosomes in cells of a given type
   d. is the same in all species

7. Human body cells are diploid, which mean they _______ .
   a. divide to form two cells
   b. have two full sets of chromosomes
   c. contain two chromosomes

8. When DNA replication begins, _______ .
   a. the two DNA strands unwind from each other
   b. the two DNA strands condense for base transfers
   c. old strands move to find new strands

9. Which of the following is/are not required for DNA replication to occur?
   a. DNA polymerase
   b. nucleotides
   c. template DNA
   d. primers
   e. helicase
   f. all are required

10. Energy that drives the attachment of a nucleotide to the end of a growing strand of DNA comes from _______ .
    a. the nucleotide
    b. phosphate-group transfers from ATP
    c. DNA polymerase

11. The phrase "5 prime to 3 prime" refers to the _______ .
    a. timing of DNA replication
    b. directionality of DNA synthesis
    c. number of phosphate groups

12. After DNA replication, a eukaryotic chromosome _____ .
    a. consists of two sister chromatids
    b has a characteristic X shape
    c. is constricted at the centromere
    d. all of the above

13. Match the terms appropriately.

___ bacteriophage	a. nitrogen-containing base, sugar, phosphate group(s)
___ clone	b. copy of an organism
___ nucleotide	c. usually one-way
___ diploid	d. injects DNA
___ DNA ligase	e. seals gaps
___ mutation	f. a pyrimidine
___ differentiation	g. can cause cancer
___ cytosine	h. something old, something new
___ semiconservative replication	i. two of each chromosome

14. All mutations _______ .
    a. result from radiation
    b. lead to evolution
    c. are caused by DNA damage
    d. change the DNA sequence

15. _______ is an example of reproductive cloning.
    a. Somatic cell nuclear transfer (SCNT)
    b. Multiple offspring from the same pregnancy
    c. Artificial embryo splitting
    d. *In vitro* fertilization

CRITICAL THINKING

1. Determine the complementary strand of DNA that forms on this template DNA fragment during replication:

   5′—GGTTTCTTCAAGAGA—3′

2. Review Figures 8.12 and 8.13. In cells, the primers for DNA synthesis are short strands of RNA, so each newly-synthesized strand of DNA has a segment of RNA at its 5′ end. As replication proceeds, DNA polymerases remove these RNA segments and fill in the resulting gaps with DNA. However, the gaps at the very 5′ ends of the new strands cannot be filled in with DNA. Why not? DNA replication leaves exposed about 100 nucleotides at the 5′ end of each template strand, and these single-stranded ends are removed. What are the effects of this "end problem" on a cell's DNA as it continues to divide?

3. Woolly mammoths have been extinct for about 4,000 years, but we often find their well-preserved remains in Siberian permafrost. Research groups are now planning to use SCNT to resurrect these huge elephant-like mammals. No mammoth eggs have been recovered yet, so elephant eggs would be used instead. An elephant would also be the surrogate mother for the resulting embryo. The researchers may try a modified SCNT technique used to clone a mouse that had been dead and frozen for 16 years. Ice crystals that form during freezing break up cell membranes, so cells from the frozen mouse were in bad shape. Their DNA was transferred into donor mouse eggs, and cells from the resulting embryos were fused with undifferentiated mouse cells. Four healthy clones were born from the hybrid embryos. What are some of the pros and cons of cloning an extinct animal?

4. Xeroderma pigmentosum is an inherited disorder characterized by rapid formation of many skin sores that develop into cancers. All forms of radiation trigger these symptoms, including fluorescent light, which contains UV light in the range of 320 to 400 nm. In most affected individuals, at least one of nine particular proteins is missing or defective. What is the collective function of these proteins?

9 FROM DNA TO PROTEIN

CORE CONCEPTS

Information Flow

Living systems store, retrieve, transmit, and respond to information essential for life. An organism's growth, survival, and reproduction depend on the genetic information carried by the DNA in its cells. In order for cells to use that information, it must be expressed: genes are transcribed into RNA (DNA→RNA), and in many cases translated into protein (RNA→protein). RNA and protein products of gene expression make other structural and functional components of cells.

Structure and Function

The three-dimensional form and arrangement of biological structures give rise to their function and interactions. The chemical structures of DNA and RNA allow a cell to preserve its genetic information even while using it. Both nucleic acids are polymers of very similar nucleotides, but important structural differences between the two types of monomers give rise to major differences in molecular stability and function. Three types of RNA interact during protein synthesis.

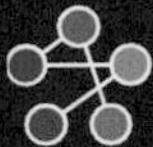

Systems

Complex properties arise from interactions among components of a biological system. The organizational basis of cells and multicelled organisms is heritable information. The way genes encode that information is highly conserved among all organisms, indicating a common origin of life. A mutation that changes a gene's expression or its product can result in observable changes in anatomy and physiology. Depending on the mutation and its location in DNA, these effects may be harmful, neutral, or even advantageous.

Links to Earlier Concepts

Your knowledge of base pairing (Section 8.3) and chromosomes (8.4) will help you understand the role of nucleic acids (3.6) in protein synthesis (3.5). This chapter also revisits cell membranes (5.7) and membrane-crossing mechanisms (5.8, 5.10), the nucleus (4.4) and endomembrane system (4.5); hydrophobicity (2.5); cofactors (5.6); enzymes (5.4) and metabolic pathways (5.5); DNA replication (8.5); and mutations (8.6).

9.1 Ricin, RIP

Castor-oil plants (*Ricinus communis*) grow wild in tropical regions, and they are widely cultivated for their seeds. The seeds of the plant are rich in castor oil, which is used as an ingredient in plastics, cosmetics, paints, polishes, soaps, and many other items. They are also rich in ricin, a toxic protein that effectively deters beetles, birds, mammals, and other animals from eating them. After castor oil is extracted from the seeds, the leftover ricin-containing seed pulp is usually discarded.

Inhaled, ingested, or injected ricin is extremely toxic to humans. Most cases of ricin poisoning occur as a result of eating castor-oil seeds, and just a few of these are enough to kill an adult. Purified, a dose of ricin as small as a few grains of salt is lethal, and there is no known antidote.

Ricin's toxicity was known as long ago as 1888, and since then several countries have tried (unsuccessfully) to weaponize it. Most countries have outlawed production, stockpiling, possession, and use of toxic chemicals as weapons. Ricin falls into this category. However, no special skills or equipment are required to manufacture the toxin from easily obtained raw materials, so controlling its production is impossible. Thus, ricin appears periodically in the news, mostly in reports of amateur criminals getting caught extracting it or trying to poison someone with the purified material.

Perhaps the most famous ricin poisoning occurred in 1978 at the height of the Cold War, when defectors from countries under Russian control were targets for assassination. Bulgarian journalist Georgi Markov had defected to England and was working for the BBC. As he made his way to a bus stop on a London street, an assassin used a modified umbrella (**FIGURE 9.1**) to fire a tiny, ricin-laced ball into Markov's leg. Markov died in agony three days later.

Ricin is called a ribosome-inactivating protein (RIP) because it inactivates ribosomes, the organelles that assemble amino acids into proteins. These proteins

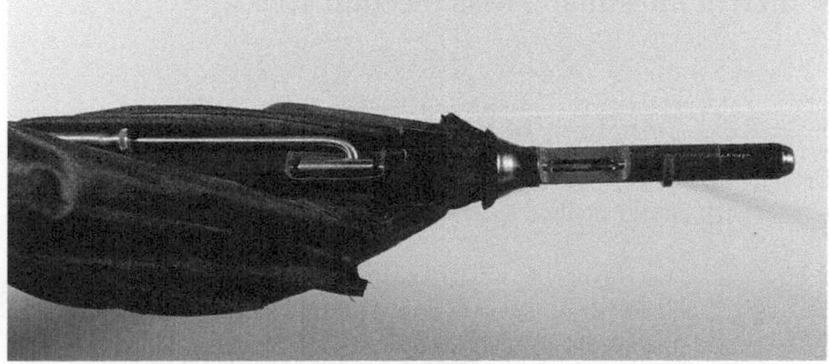

FIGURE 9.1 Bulgarian spy's weapon: an umbrella modified to fire a tiny pellet of ricin into a victim. An umbrella like this one was used to assassinate Georgi Markov on the streets of London in 1978.

Castor-oil seeds
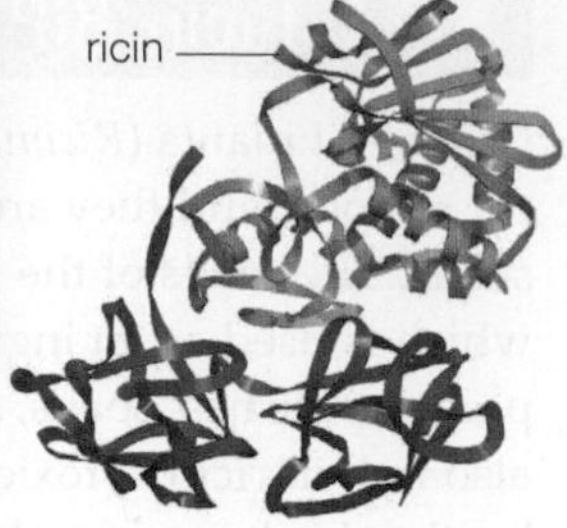
ricin

FIGURE 9.2 Ricin: source and structure.

Left, ricin is a toxic component of castor-oil seeds. Right, structure of ricin. Like other toxic RIPs, ricin has two polypeptide chains. One (red) helps the molecule cross a cell's plasma membrane via endocytosis. The other (gold) is an enzyme that inactivates ribosomes.

interfere with ribosome function, so they prevent the assembly of amino acids into proteins. Other RIPs are made by some bacteria, mushrooms, algae, and many plants (including food crops such as tomatoes, barley, and spinach), but most of these proteins are not particularly toxic to humans because they do not cross intact cell membranes very well. By contrast, ricin and other toxic RIPs can enter cells. These proteins have a domain that binds tightly to plasma membrane glycolipids or glycoproteins (**FIGURE 9.2**). Binding of an RIP to these molecules triggers endocytosis (Section 5.10), so the cell takes in the RIP. Once inside the cell, the second domain of the RIP—an enzyme—begins to inactivate ribosomes. One molecule of ricin can inactivate more than 1,000 ribosomes per minute. If enough ribosomes are affected, protein synthesis grinds to a halt. Proteins are critical to all life processes, so cells that cannot make them die quickly.

Fortunately, few people actually encounter ricin. Contact with other toxic RIPs is much more common. Bracelets made from beautiful seeds were recalled from stores in 2011 after a botanist recognized them as jequirity beans. These beans contain abrin, an RIP even more toxic than ricin. Shiga toxin, an RIP made by *Shigella dysenteriae* bacteria, causes a severe bloody diarrhea (dysentery) that can be lethal. Some strains of *E. coli* bacteria make Shiga-like toxin, an RIP that is the source of intestinal illness (Section 4.1).

9.2 DNA, RNA, and Gene Expression

LEARNING OBJECTIVES

- Compare the components and structure of DNA and RNA molecules.
- Explain how the function of RNA differs from the function of DNA.
- Describe the basic structure of a gene.
- Explain the flow of information during gene expression.

You learned in Chapter 8 that chromosomes are like a set of books that provide instructions for building and operating an individual. You already know the alphabet used to write those books: the four letters A, T, G, and C, for the four nucleotide bases in DNA: adenine, thymine, guanine, and cytosine. The "instructions" in a chromosome are encoded in the sequence of bases in its DNA, and they occur in units called **genes**. Cells can use the sequence of bases in a gene—its coding sequence—to make an RNA or protein product. Converting information in a gene to the gene's product is a multistep process called **gene expression**.

Gene expression begins with **transcription**, an energy-intensive process that copies a gene's coding sequence into RNA form. The RNA is a copy of the gene, just as a paper transcript of a conversation carries the same information in a different format. Regulatory sites that function in transcription flank a gene's coding sequence (**FIGURE 9.3**). Molecules that carry out transcription bind to the gene's **promoter**, a short series of bases upstream (in the 5′ direction) of the coding sequence. Transcription ends at another short series of bases called a terminator. (The next section returns to details of transcription).

Most DNA is double-stranded and most RNA is single-stranded, but otherwise the two molecules are very similar in structure. Like a strand of DNA, a strand of RNA consists of four kinds of nucleotides, each with phosphate groups, a sugar, and one of four bases. The nucleotides differ a bit between DNA and RNA (**TABLE 9.1**). Three bases—adenine, cytosine, and guanine—occur in both RNA and DNA nucleotides, but the fourth base differs slightly (**FIGURE 9.4A**). In DNA, the fourth base is thymine; in RNA, it is uracil (U). The sugar component of nucleotides differs slightly between RNA and DNA too. RNA (ribonucleic acid) is named

FIGURE 9.3 Anatomy of a gene.

During transcription, a gene is copied into RNA form. Regulatory sites that function in transcription flank the gene's coding sequence.

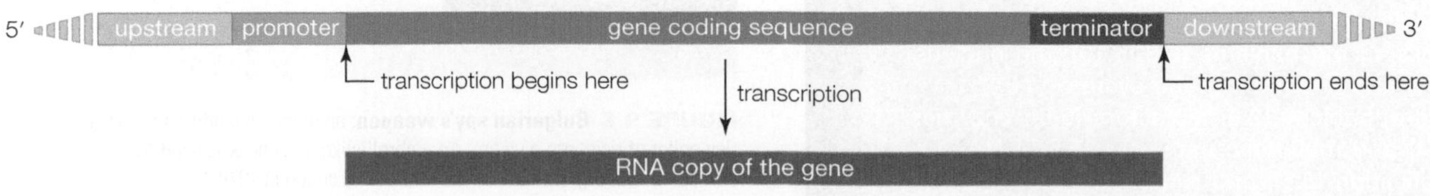
5′
upstream
promoter
gene coding sequence
terminator
downstream
3′
transcription begins here
transcription
transcription ends here
RNA copy of the gene

CREDITS: (2) Amawasri Pakdara/Shutterstock; (3) © Cengage Learning.

after ribose, the sugar in its nucleotides. Ribose has a hydroxyl group on its 2′ carbon; deoxyribose, the sugar in DNA nucleotides, does not (**FIGURE 9.4B**).

These small differences in nucleotide structure (**FIGURE 9.4C**) give rise to the very different functions of DNA and RNA. DNA's important but only role is to store a cell's genetic information. By contrast, a cell makes several kinds of RNA, each with a different function. MicroRNA and long noncoding RNA are important in gene control, which is the subject of Chapter 10. Three other types of RNA have roles in protein synthesis. **Ribosomal RNA** (abbreviated **rRNA**) is the main component of ribosomes (Section 4.4), which are the organelles that assemble amino acids into polypeptides (Section 3.5). **Transfer RNA (tRNA)** delivers amino acids to ribosomes, one by one, in the order specified by a **messenger RNA (mRNA)**.

Messenger RNA was named for its function as the "messenger" between DNA and protein. As you will see in Section 9.3, an mRNA carries a protein-building message in the sequence of its nucleotide bases. In an energy-intensive process called **translation**, that message guides the assembly of a polypeptide.

During gene expression, information flows from DNA to RNA to protein:

DNA → TRANSCRIPTION → RNA → TRANSLATION → PROTEIN

Expression of genes that encode RNA products (such as tRNA and rRNA) involves transcription only. Expression of genes that encode protein products involves both transcription and translation.

The DNA sequence of a cell's chromosome(s) contains all of the information required to make the molecules of life. Transcription produces RNAs that interact in translation. Some of the resulting proteins have structural roles in the cell; others (enzymes in particular) assemble lipids and carbohydrates, replicate DNA, make RNA, and so on.

TABLE 9.1

Comparing DNA and RNA

	DNA	RNA
Form	double helix	most are single-stranded
Monomers	deoxyribonucleotides	ribonucleotides
Sugar	deoxyribose	ribose
Bases	adenine, guanine, cytosine, thymine	adenine, guanine, cytosine, uracil
Base pairing	A–T, G–C	A–U, G–C
Function	stores genetic information	protein synthesis, other roles depending on type

A DNA and RNA are polymers of four types of nucleotides. One base differs between DNA and RNA: Uracil occurs only in RNA; thymine, in DNA.

uracil (U) thymine (T)

B The sugar in an RNA nucleotide is a ribose (as opposed to a deoxyribose in a DNA nucleotide). Ribose has a hydroxyl group on its 2′ carbon.

ribose deoxyribose

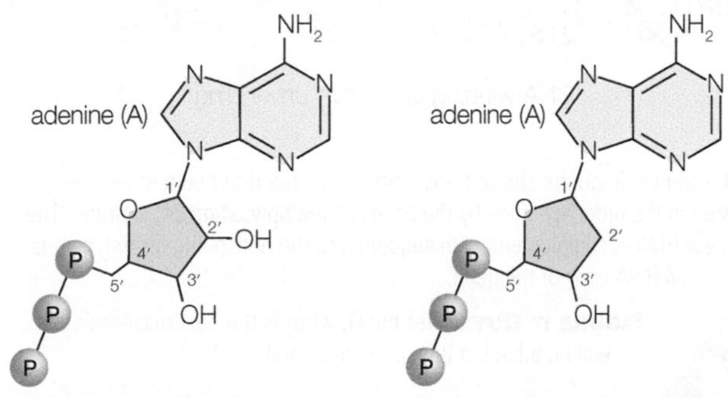

adenine (A)
adenine (A)

adenosine triphosphate (an RNA nucleotide)

deoxyadenosine triphosphate (a DNA nucleotide)

C Three of four bases (adenine, guanine, and cytosine) are identical between DNA and RNA nucleotides. Only the hydroxyl group on the sugar differs.

FIGURE 9.4 Comparing nucleotides of DNA and RNA. The very different functions of DNA and RNA arise from very small differences in the structure of their nucleotides.

TAKE-HOME MESSAGE 9.2

- ✔ Information encoded in the sequence of nucleotide bases in DNA occurs in units called genes.
- ✔ In gene expression, information in DNA flows to RNA and then to protein. Gene expression includes transcription and translation.
- ✔ In transcription, a gene is copied into RNA form; translation converts protein-building information in an mRNA into a polypeptide.

gene (JEEN) Unit of information encoded in the sequence of nucleotide bases in DNA.

gene expression Multistep process of converting information in a gene to an RNA or protein product. Includes transcription and translation.

messenger RNA (mRNA) RNA that carries a protein-building message.

promoter Short sequence of bases that functions as a binding site in DNA for molecules that carry out transcription.

ribosomal RNA (rRNA) RNA component of ribosomes.

transcription Process in which a gene is copied (transcribed) into RNA form.

transfer RNA (tRNA) RNA that delivers amino acids to a ribosome during translation.

translation Process in which a polypeptide is assembled according to information in an mRNA.

9.3 Transcription: DNA to RNA

LEARNING OBJECTIVES

- Compare DNA replication with transcription.
- Explain three types of post-transcriptional modifications to RNA.

Transcription, the process of copying a gene into RNA form, is similar in many ways to DNA replication (Section 8.5). Base-pairing rules are followed, for example. In DNA replication, cytosine (C) pairs with guanine (G), and adenine (A) pairs with thymine (T):

DNA
C C T C T T C A G
DNA
G G A G A A G T C

The very same base-pairing rules also govern RNA synthesis in transcription, except that RNA, remember, has uracil instead of thymine. Uracil (U) base-pairs with adenine (A):

RNA
C C U C U U C A G
DNA
G G A G A A G T C

Note that the drawing above shows hybridization between strands of DNA and RNA (Section 8.5). During transcription, a strand of DNA acts as a template upon which a strand of RNA is assembled from nucleotides. A nucleotide can be added to the end of a growing RNA only if it base-pairs with the corresponding nucleotide of the template DNA. Thus, a new RNA is complementary in base sequence to the DNA template. As in DNA replication, two of a nucleotide's three phosphate groups are removed when it is added to an RNA strand; breaking those bonds releases enough energy to drive the attachment.

Transcription is also similar to DNA replication in that it proceeds in the 5′ to 3′ direction, and one strand of a nucleic acid serves as a template for synthesis of another. However, in contrast with DNA replication, transcription produces a single strand of RNA (not two DNA double helices). Also, only part of one DNA strand is used as a template for transcription.

Genes and their promoters occur on both strands of DNA. However, the coding sequence of each gene occurs on only one strand—the coding strand. The base sequence of the other (noncoding) strand is complementary to the gene's coding sequence.

RNA polymerase is the enzyme that carries out transcription. In eukaryotic cells, the process occurs in the nucleus; in prokaryotes, it occurs in cytoplasm. Transcription begins when an RNA polymerase binds to a gene's promoter (**FIGURE 9.5** ❶). After binding, the polymerase starts moving over the gene, unwinding

FIGURE 9.5 Transcription, in which RNA polymerase assembles a strand of RNA from nucleotides. A gene's noncoding strand serves as the template for RNA synthesis.

❶ The enzyme RNA polymerase binds to a gene's promoter.

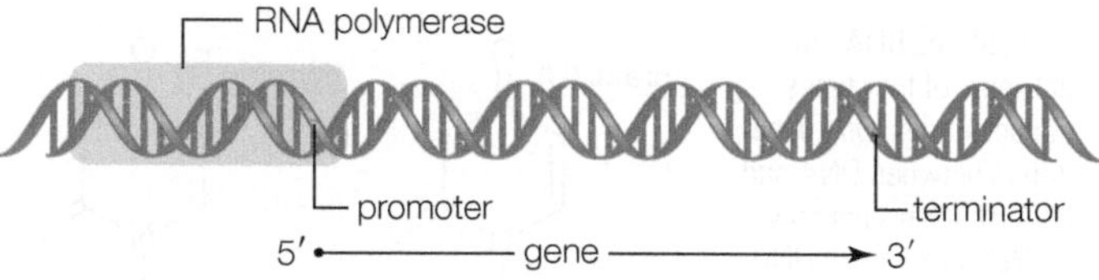

RNA polymerase
promoter
terminator
5′
gene
3′

❷ The RNA polymerase moves over the gene, unzipping the DNA's double helix to form a "transcription bubble." As it moves, the polymerase assembles a strand of RNA. The DNA winds up again into a double helix after the polymerase passes.

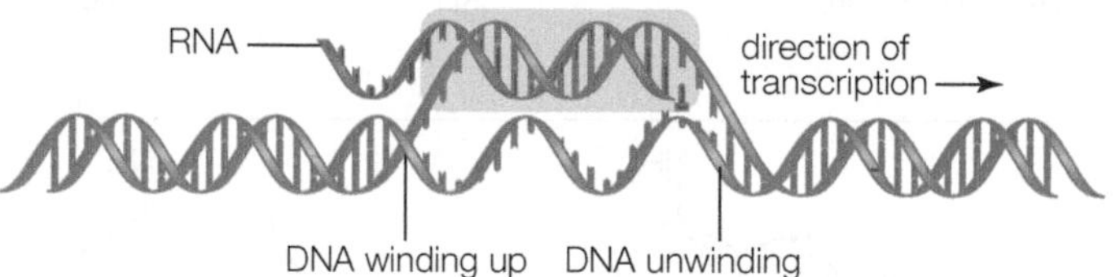

RNA
direction of transcription
DNA winding up
DNA unwinding

❸ Zooming in on the site of transcription, we see that nucleotides are linked in the order specified by the bases of the noncoding DNA strand. The new RNA is complementary in sequence to the noncoding strand, so it is an RNA copy of the gene.

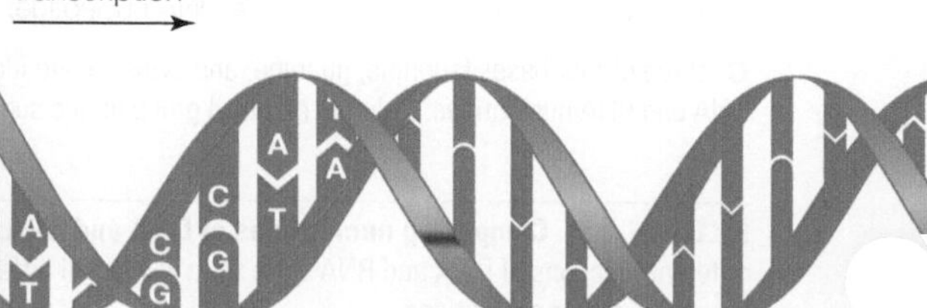

5′
RNA
noncoding strand
coding strand
transcription

FIGURE IT OUT After the G, what is the next nucleotide that will be added to this growing strand of RNA?

Answer: Another G

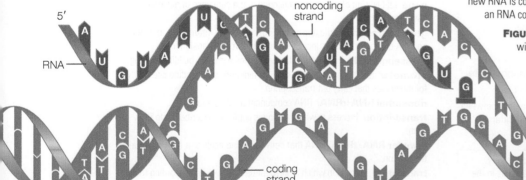

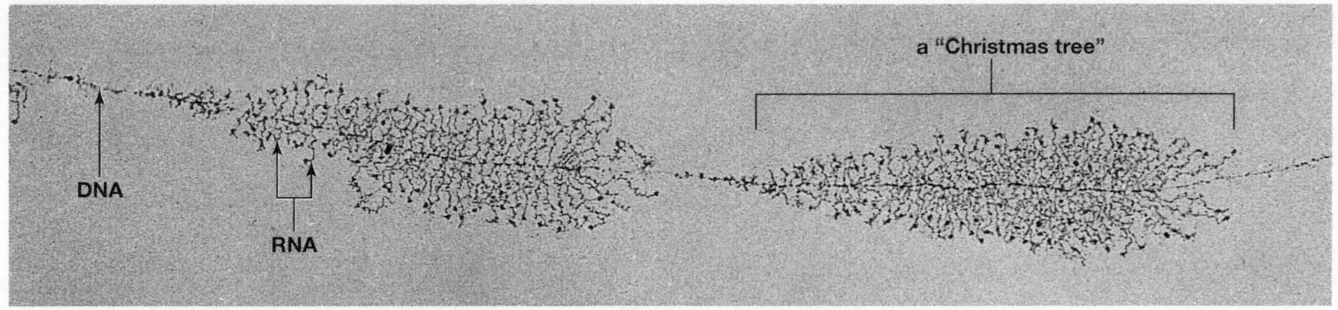
a "Christmas tree"
DNA
RNA

FIGURE 9.6 RNA polymerases in action. Many RNA polymerases may transcribe the same gene simultaneously, producing a structure called a "Christmas tree" after its characteristic shape. Here, multiple polymerases are transcribing each of two genes next to one another on a chromosome.

FIGURE IT OUT Are the polymerases transcribing this DNA molecule moving from left to right or from right to left?

Answer: Left to right

the double helix to form an opening called a transcription bubble ❷. As the enzyme moves, it "reads" the base sequence of the noncoding strand. That sequence serves as a template for the polymerase to join free nucleotides into a new strand of RNA ❸. The double helix winds back up as the polymerase passes.

When the polymerase reaches the gene's terminator, it releases the DNA and the new RNA. Because RNA polymerase follows base-pairing rules, the new RNA is complementary to the noncoding DNA strand: It is an RNA copy of the gene. Many polymerases may transcribe a gene at the same time (**FIGURE 9.6**).

Post-Transcriptional Modifications

Just as a dressmaker may snip off loose threads or add bows to a dress before it leaves the shop, so do eukaryotic cells tailor their RNA before it leaves the nucleus. Consider that most eukaryotic genes contain regions called **introns** that are removed in chunks from a newly transcribed RNA. Introns intervene between **exons**, which are regions that remain in the finished RNA.

Short base sequences at the ends of introns and exons mark their boundaries. Molecules that remove introns from RNA recognize these boundaries as sites where exons are to be joined. These molecules also carry out **alternative splicing**, which means they can rearrange exons and splice them together in different combinations (**FIGURE 9.7**). Alternative splicing allows one gene to encode multiple versions of a protein.

An RNA that will become an mRNA is further tailored after splicing. For example, 50 to 300 adenines are added to the 3′ end of a new mRNA. This poly-A tail helps regulate the timing and duration of the mRNA's translation, and in eukaryotes it is also a signal that allows the molecule to be exported from the nucleus. Both eukaryotes and prokaryotes add a modified nucleotide "cap" to the 5′ end of a newly transcribed mRNA. Molecules that bind to the cap splice the RNA and add a poly-A tail to it. A cap also helps ribosomes bind to the finished mRNA, and it helps protect the molecule from being broken down prematurely.

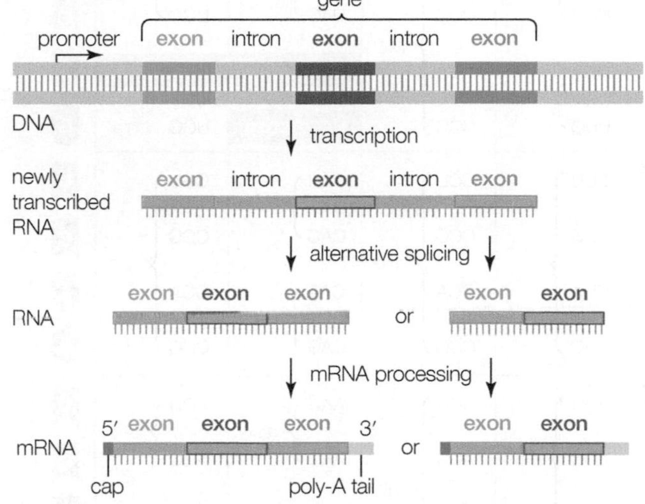
gene
promoter
exon
intron
exon
intron
exon
DNA
transcription
newly transcribed RNA
exon
intron
exon
intron
exon
alternative splicing
RNA
exon
exon
exon
or
exon
exon
mRNA processing
mRNA
5′
exon
exon
exon
3′
or
exon
exon
cap
poly-A tail

FIGURE 9.7 Post-transcriptional modification of RNA. Introns are removed from a new RNA, and the remaining exons are spliced together in combinations that can vary (alternative splicing). Messenger RNAs also get a poly-A tail and modified guanine "cap."

TAKE-HOME MESSAGE 9.3

- ✔ In transcription, RNA polymerase assembles an RNA copy of a gene. The enzyme uses the noncoding strand of DNA as a template for RNA synthesis.
- ✔ Eukaryotic genes include segments called introns, which intervene between exons.
- ✔ Introns are removed from a new RNA. Other post-transcriptional modifications include the addition of a poly-A tail and 5′ cap.

alternative splicing Post-transcriptional RNA modification process in which exons are rearranged and/or joined in different combinations.

exon Gene segment that remains in an RNA after post-transcriptional modification.

intron Gene segment that intervenes between exons and is removed during post-transcriptional modification.

RNA polymerase Enzyme that carries out transcription.

CREDIT: (6) O. L. Miller.

FIGURE 9.8 The genetic code. Top, a codon table lists all 64 mRNA codons. The first base of each codon is given to the left of the table; the second, in the top row; and the third, to the right of the table. Together, the three bases specify one amino acid (bottom).

Sixty-one codons encode amino acids; one of those, AUG, both codes for methionine and serves as a signal to start translation. Three codons are signals that stop translation.

FIGURE IT OUT Which codon(s) specify the amino acid lysine?

Answer: AAA and AAG

The Genetic Code

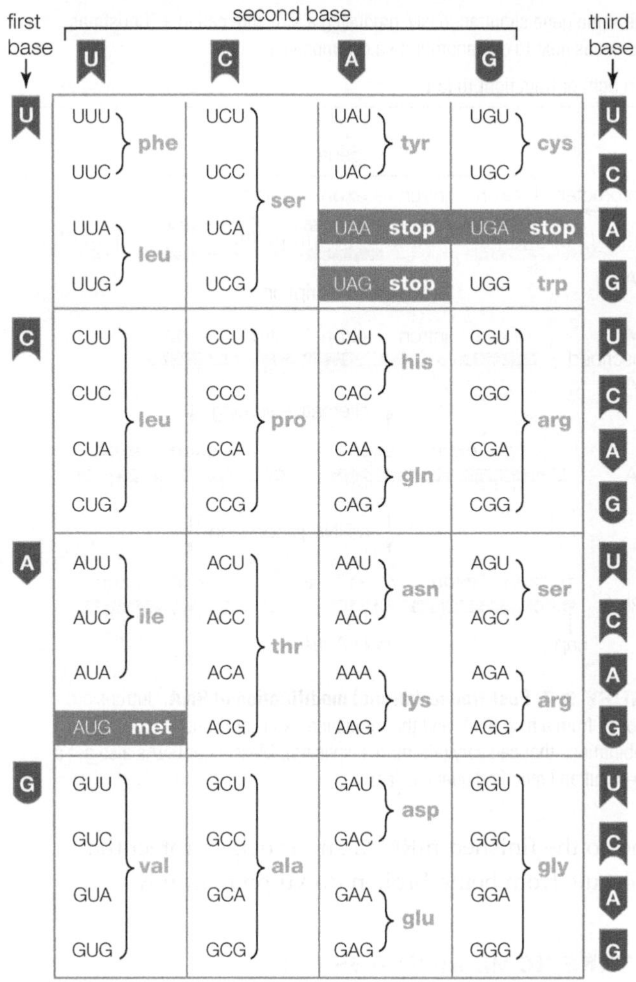

first base	second base U	second base C	second base A	second base G	third base
U	UUU phe	UCU ser	UAU tyr	UGU cys	U
U	UUC phe	UCC ser	UAC tyr	UGC cys	C
U	UUA leu	UCA ser	UAA stop	UGA stop	A
U	UUG leu	UCG ser	UAG stop	UGG trp	G
C	CUU leu	CCU pro	CAU his	CGU arg	U
C	CUC leu	CCC pro	CAC his	CGC arg	C
C	CUA leu	CCA pro	CAA gln	CGA arg	A
C	CUG leu	CCG pro	CAG gln	CGG arg	G
A	AUU ile	ACU thr	AAU asn	AGU ser	U
A	AUC ile	ACC thr	AAC asn	AGC ser	C
A	AUA ile	ACA thr	AAA lys	AGA arg	A
A	AUG met	ACG thr	AAG lys	AGG arg	G
G	GUU val	GCU ala	GAU asp	GGU gly	U
G	GUC val	GCC ala	GAC asp	GGC gly	C
G	GUA val	GCA ala	GAA glu	GGA gly	A
G	GUG val	GCG ala	GAG glu	GGG gly	G

The Amino Acids

ala alanine (A)	gly glycine (G)	pro proline (P)
arg arginine (R)	his histidine (H)	ser serine (S)
asn asparagine (N)	ile isoleucine (I)	thr threonine (T)
asp aspartic acid (D)	leu leucine (L)	trp tryptophan (W)
cys cysteine (C)	lys lysine (K)	tyr tyrosine (Y)
glu glutamic acid (E)	met methionine (M)	val valine (V)
gln glutamine (Q)	phe phenylalanine (F)	

anticodon In a tRNA, set of three nucleotides that base-pairs with an mRNA codon.
codon (CO-don) In an mRNA, a nucleotide base triplet that codes for an amino acid or stop signal during translation.
genetic code Complete set of 64 mRNA codons.

9.4 RNA and the Genetic Code

LEARNING OBJECTIVES

- Describe codons and their function.
- Explain the signals that start and stop translation.
- Explain how an mRNA specifies the order of amino acids in a polypeptide.
- Summarize the role of rRNA and tRNA in translation.

Three types of RNA interact in protein synthesis: messenger RNA, transfer RNA, and ribosomal RNA. An mRNA is essentially a disposable copy of a gene: its job, to carry the gene's protein-building information into translation. That protein-building information is a sequence of genetic "words" that occur one after another along its length. Like the words of a sentence, a series of these genetic words can form a meaningful parcel of information—in this case, a sequence of amino acids that constitutes the primary structure of a protein.

Each genetic "word" in an mRNA is three bases long, and each is a code—a **codon**—for a particular amino acid. The sequence of bases in a triplet determines which amino acid the codon specifies (FIGURE 9.8). For instance, the codon UUU specifies the amino acid phenylalanine, and UUA specifies leucine. With four possible bases (G, A, U, or C) in each of the three positions of a codon, there are a total of sixty-four (or 4^3) mRNA codons. Collectively, the sixty-four codons constitute the **genetic code**. These codons specify a total of twenty naturally occurring amino

FIGURE 9.9 Example of the correspondence between DNA, RNA, and protein. A gene region in a strand of chromosomal DNA is transcribed into an mRNA, and the codons of the mRNA specify the primary structure of a protein.

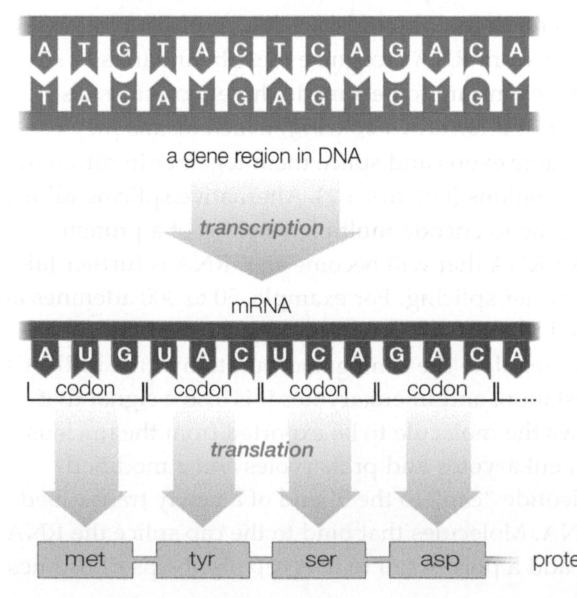
A T G T A C T C A G A C A
T A C A T G A G T C T G T
a gene region in DNA
transcription
mRNA
A U G U A C U C A G A C A
codon
codon
codon
codon
translation
met
tyr
ser
asp
protein

acids, so some amino acids are specified by more than one codon. For instance, the amino acid tyrosine (tyr) is specified by two codons: UAU and UAC.

Other codons signal the beginning and end of a protein-coding sequence. In almost all cases, the first AUG in an mRNA serves as the signal to start translation. AUG is also the codon for methionine, so methionine is always the first amino acid in new polypeptides (in prokaryotes, it is a modified methionine). The codons UAA, UAG, and UGA do not specify an amino acid. These are signals that stop translation, so they are called stop codons. A stop codon marks the end of the protein-coding sequence in an mRNA.

The genetic code is highly conserved, which means all organisms use essentially the same code and probably always have. Bacteria, archaea, and a few eukaryotes use a few codons that differ from the standard genetic code. So do mitochondria and chloroplasts—a clue that led to a theory of how these two organelles evolved (we return to this topic in Section 19.7).

Codons occur one after the next in an mRNA. When the mRNA is translated, the order of its codons determines the order of amino acids in the resulting polypeptide. Thus, the DNA sequence of a gene is transcribed into the nucleotide sequence of an mRNA, which is in turn translated into an amino acid sequence (**FIGURE 9.9**).

rRNA and tRNA

Ribosomes are the organelles that build polypeptides (Section 3.5 and 4.4). A ribosome has two subunits, one large and one small (**FIGURE 9.10**). Each subunit consists of rRNA and associated structural proteins. As translation begins, a large and a small ribosomal subunit converge as an intact ribosome on an mRNA. Ribosomal RNA is one example of RNA with enzymatic activity: The rRNA components of a ribosome (not the protein components) catalyze the formation of peptide bonds between amino acids.

Each tRNA has an **anticodon**, which is a triplet of nucleotides that base-pairs with an mRNA codon (**FIGURE 9.11A**). The tRNA also has an attachment site that binds to an amino acid—the one specified by the codon. Transfer RNAs with different anticodons carry different amino acids (**FIGURE 9.11B**).

During translation, tRNAs deliver amino acids to a ribosome, one after the next in the order specified by the codons in an mRNA. As the amino acids are delivered, the ribosome joins them via peptide bonds into a new polypeptide. Thus, the order of codons in an mRNA—DNA's protein-building message—becomes translated into a protein. The next section details this process.

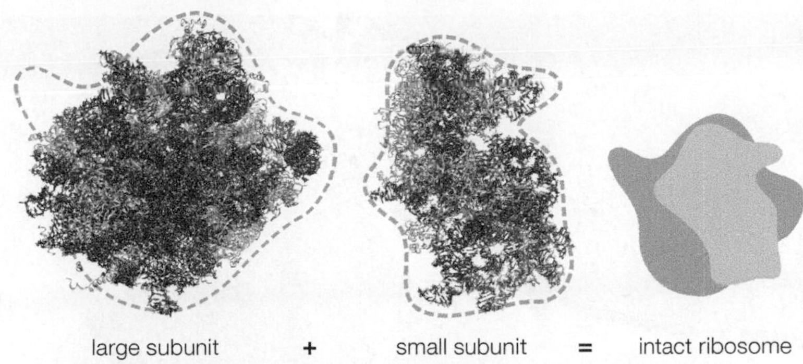
large subunit
+
small subunit
=
intact ribosome

FIGURE 9.10 Ribosome structure. An intact ribosome consists of a large and a small subunit. Protein components of both subunits are shown in the ribbon models in green; rRNA components, in brown.

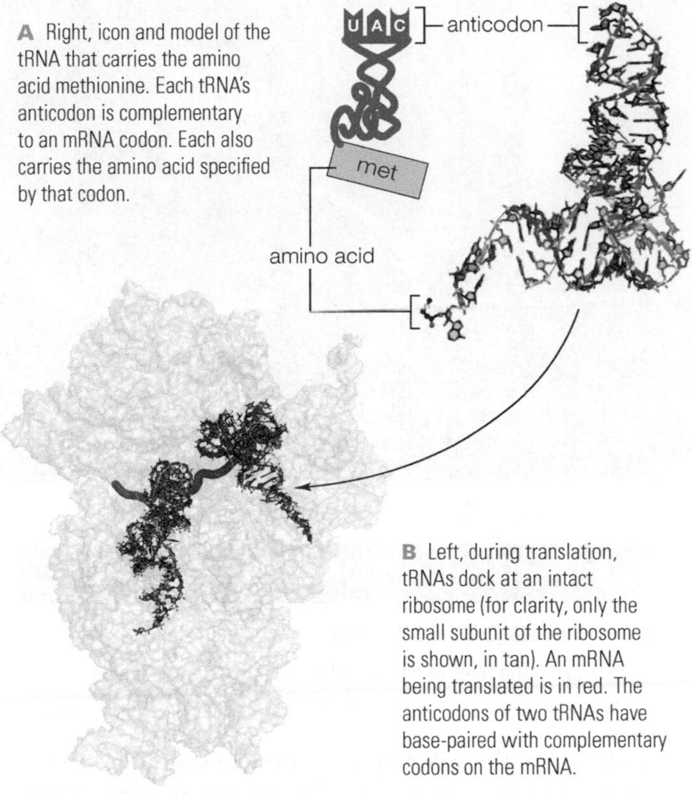
UAC
anticodon
met
amino acid

A Right, icon and model of the tRNA that carries the amino acid methionine. Each tRNA's anticodon is complementary to an mRNA codon. Each also carries the amino acid specified by that codon.

B Left, during translation, tRNAs dock at an intact ribosome (for clarity, only the small subunit of the ribosome is shown, in tan). An mRNA being translated is in red. The anticodons of two tRNAs have base-paired with complementary codons on the mRNA.

FIGURE 9.11 tRNA structure.

TAKE-HOME MESSAGE 9.4

- ✔ An mRNA carries a gene's protein-building message into translation. A series of mRNA codons (base triplets) encode that message.
- ✔ A codon specifies an amino acid or a stop signal that ends translation. The genetic code consists of all 64 codons.
- ✔ The order of codons in an mRNA determines the order of amino acids in the polypeptide chain translated from it.
- ✔ The two subunits of a ribosome consist of proteins and ribosomal RNA. The rRNA (not the protein) catalyzes the formation of peptide bonds during translation.
- ✔ Each tRNA has an anticodon that base-pairs with a codon in mRNA, and a binding site for the amino acid specified by that codon.

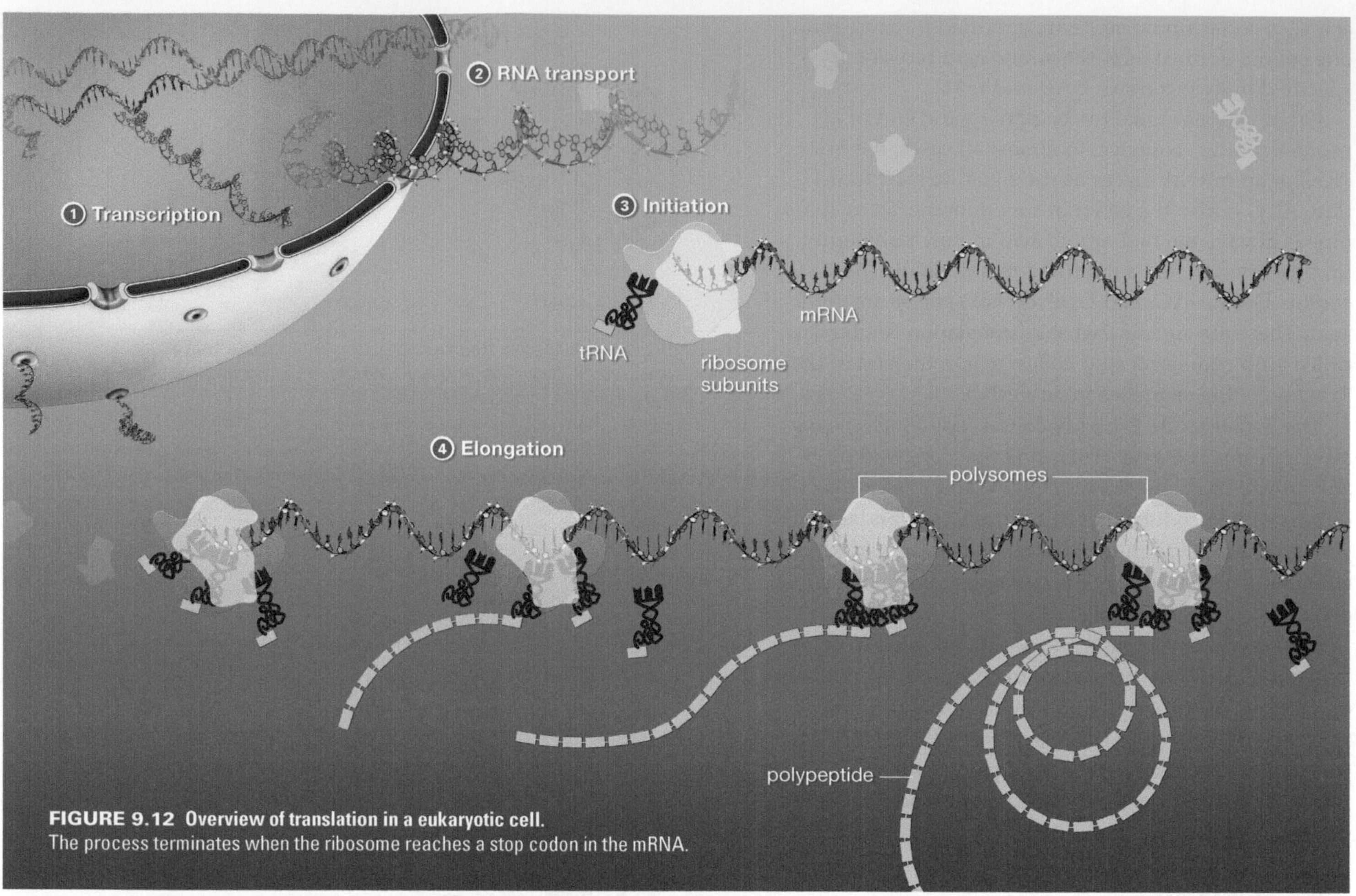

① Transcription
② RNA transport
③ Initiation
tRNA
ribosome subunits
mRNA
④ Elongation
polysomes
polypeptide

FIGURE 9.12 Overview of translation in a eukaryotic cell.
The process terminates when the ribosome reaches a stop codon in the mRNA.

9.5 Translation: RNA to Protein

LEARNING OBJECTIVES

- Explain the roles of mRNA, tRNA, and rRNA in translation.
- Describe the way a polypeptide is synthesized during translation.

Translation, the second part of protein synthesis, occurs in the cytoplasm of all cells. The process uses molecules abundant in cytoplasm: free amino acids, ribosomes, and tRNAs.

In a eukaryotic cell, RNAs that participate in translation (like all other RNAs) are produced by transcription in the nucleus (**FIGURE 9.12** ①), then transported through nuclear pores into the cytoplasm ②.

Translation proceeds in three stages: initiation, elongation, and termination. Initiation occurs when ribosomal subunits and tRNAs converge on an mRNA ③. First, a small ribosomal subunit binds to the mRNA, and the anticodon of a special tRNA called an initiator base-pairs with the mRNA's first AUG codon. Then, a large ribosomal subunit joins the small subunit.

The complex of molecules is now ready to carry out protein synthesis. In the elongation stage, the intact ribosome moves along the mRNA and assembles a polypeptide ④. **FIGURE 9.13** zooms in on this process. Initiator tRNAs carry methionine, so the first amino acid of new polypeptides is methionine. Another tRNA joins the complex when its anticodon base-pairs with the second codon in the mRNA ❶. This tRNA brings with it the second amino acid. The ribosome then catalyzes formation of a peptide bond (Section 3.5) between the first two amino acids ❷.

As the ribosome moves to the next codon, it releases the first tRNA. Another tRNA brings the third amino acid to the complex as its anticodon base-pairs with the third codon of the mRNA ❸. The ribosome catalyzes the formation of a peptide bond between the second and third amino acids, so a peptide forms.

The amino acid chain continues to elongate as amino acids delivered by successive tRNAs are added. In eukaryotes, many ribosomes associate with rough endoplasmic reticulum (Section 4.5). Polypeptides made by these ribosomes thread into the interior of the ER as they lengthen.

The new polypeptide chain continues to elongate as amino acids are delivered by successive tRNAs.

Translation terminates when the ribosome reaches a stop codon in the mRNA ❹. The mRNA and the polypeptide detach from the ribosome, and the ribosomal subunits separate from each other.

Many ribosomes may simultaneously translate the same mRNA, in which case they are called polysomes. In bacteria and archaea, transcription and translation both occur in cytoplasm, and these processes are closely linked in time and space.

Translation is energy intensive. Most of that energy is provided in the form of phosphate-group transfers from the RNA nucleotide GTP to molecules that help the ribosome move from one codon to the next.

TAKE-HOME MESSAGE 9.5

- ✔ Translation is an energy-requiring process that converts a series of codons in an mRNA into a polypeptide.
- ✔ In initiation, ribosomal subunits and tRNAs converge on an mRNA.
- ✔ In elongation, amino acids are delivered to the ribosome by tRNAs in the order dictated by successive mRNA codons. As amino acids arrive, the ribosome joins each to the end of the polypeptide.
- ✔ Termination occurs when the ribosome encounters a stop codon in the mRNA.

9.6 Consequences of Mutations

LEARNING OBJECTIVES

- Describe three types of mutations.
- Explain how mutations can affect protein structure.
- Using an example, explain why some mutations are not harmful.

Mutations, remember, are permanent changes in the DNA sequence of a chromosome (Section 8.6). Mutations are relatively uncommon events in a normal cell. Consider that the chromosomes in the nucleus of a human body cell collectively consist of about 6.5 billion nucleotides, any of which may be copied incorrectly each time that cell divides. The mutation rate in human somatic cells has been measured: About five nucleotides change every time DNA replication occurs. Less than 2 percent of human DNA encodes gene products, however, so there is a low probability that any mutation will occur in a coding region. The redundancy of the genetic code offers an additional margin of safety for protein-coding genes. For example, a mutation that changes a codon from GUA to GUG may have no further effect, because both of these codons specify valine.

Rarely, a mutation changes a gene's product or interferes with its expression. Such mutations can have drastic effects in an organism, particularly if they occur during gamete formation. Consider hemoglobin, an oxygen-transporting protein in your red blood cells.

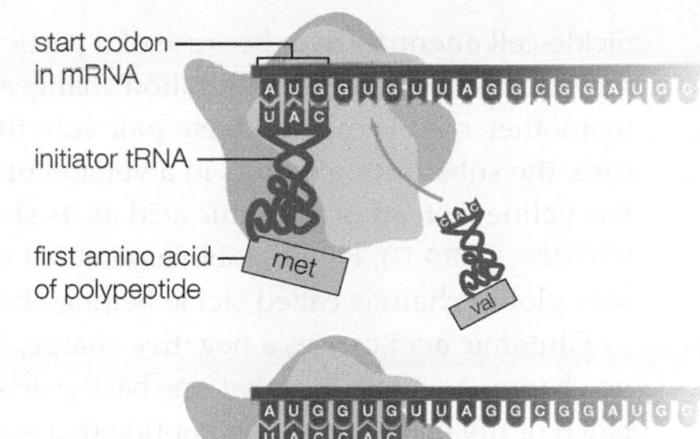
start codon in mRNA
❶ Ribosomal subunits and an initiator tRNA converge on an mRNA. A second tRNA binds to the second codon.
initiator tRNA
first amino acid of polypeptide
met
val

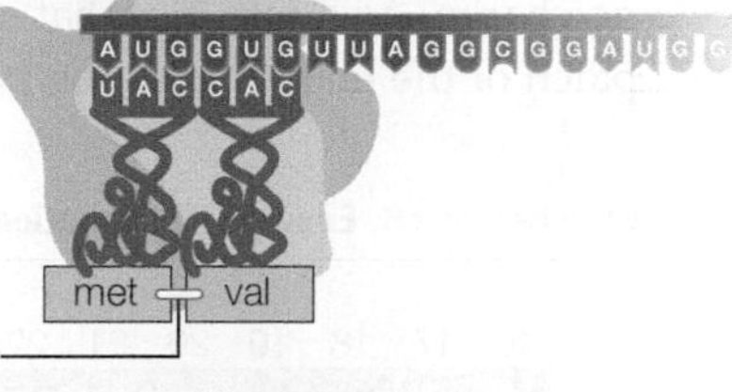
❷ A peptide bond forms between the first two amino acids.
met
val
peptide bond

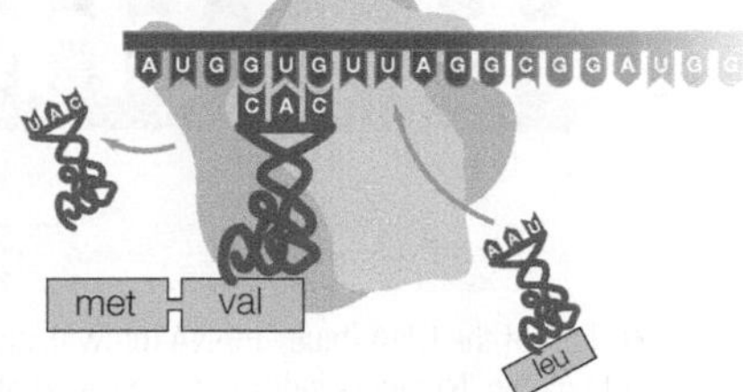
❸ The first tRNA is released and the ribosome moves to the next codon. A third tRNA binds to the third codon.
met
val
leu

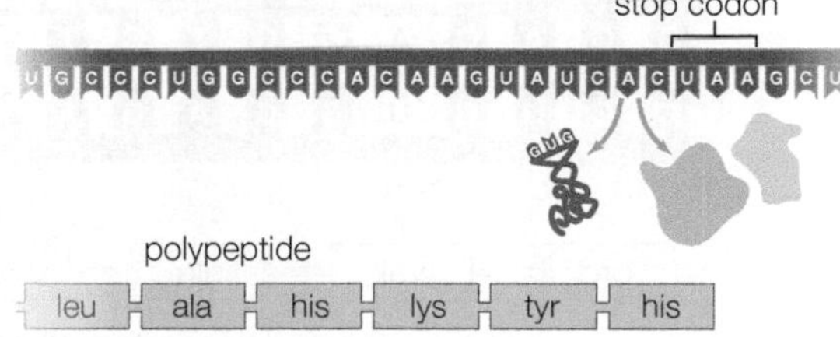
❹ The process repeats until the ribosome encounters a stop codon. Then, the new polypeptide is released and the ribosome subunits separate.
stop codon
polypeptide
leu
ala
his
lys
tyr
his

FIGURE 9.13 Zooming in on translation.

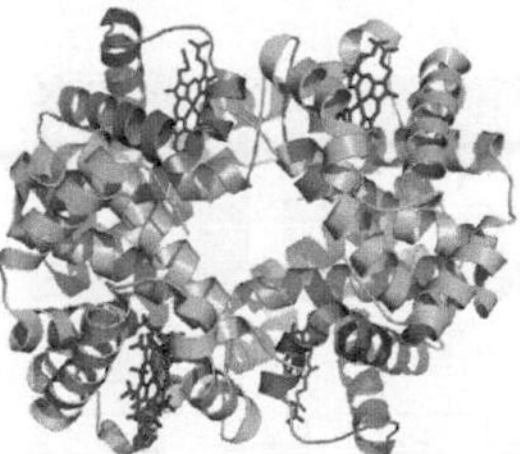

FIGURE 9.14 Hemoglobin, an oxygen-binding protein in red blood cells. A working molecule of hemoglobin has four polypeptides: two alpha globins (blue) and two beta globins (green). Each globin has a pocket that cradles a heme (red). Oxygen molecules bind to the iron atom at the center of each heme.

Hemoglobin's structure allows it to bind and release oxygen. In adult humans, a hemoglobin molecule consists of four polypeptides called globin chains or globins: two alpha globins and two beta globins (**FIGURE 9.14**), each folded around a heme (Section 5.6). Oxygen molecules bind at those hemes. Mutations that affect either globin chain also affect hemoglobin function, so people born with them often have health issues.

Mutations in globin genes cause a condition called anemia, in which a person's blood is deficient in hemoglobin or in red blood cells. Both outcomes limit the blood's ability to carry oxygen, and the resulting symptoms can range from mild to life-threatening.

Sickle-cell anemia arises because of a particular mutation in the beta globin gene. The mutation changes one base pair to another, so it is called a **base-pair substitution**. In this case, the substitution results in a version of beta globin that has valine instead of glutamic acid as its sixth amino acid (**FIGURE 9.15A,B**). Hemoglobin assembled with this altered beta globin chain is called sickle hemoglobin, or HbS.

Glutamic acid carries a negative charge, but valine carries no charge. As a result of that one base-pair substitution, a tiny patch of the beta globin polypeptide that is normally hydrophilic becomes hydrophobic. This change alters the behavior of hemoglobin. Under certain conditions, HbS molecules stick together and form large, rodlike clumps. Red blood cells that contain the clumps are distorted into a crescent, or sickle shape (right). Sickled cells clog tiny blood vessels, disrupting blood circulation throughout the body. Over time, repeated episodes of sickling damage organs and eventually cause death.

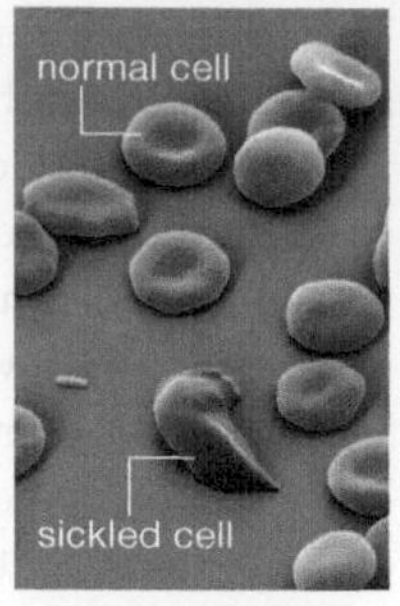
normal cell
sickled cell

FIGURE 9.15 **Examples of mutations.**

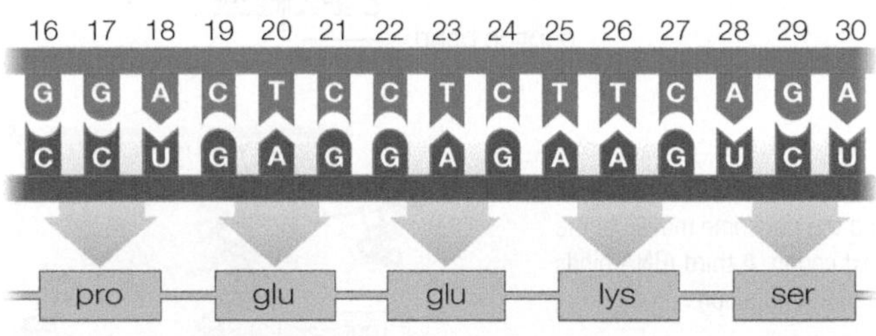
16 17 18 19 20 21 22 23 24 25 26 27 28 29 30
G G A C T C C T C T T C A G A
C C U G A G G A G A A G U C U
pro
glu
glu
lys
ser

A Part of the DNA (blue), mRNA (brown), and amino acid sequence (green) of human beta globin. Numbers indicate the position of the nucleotide in the mRNA.

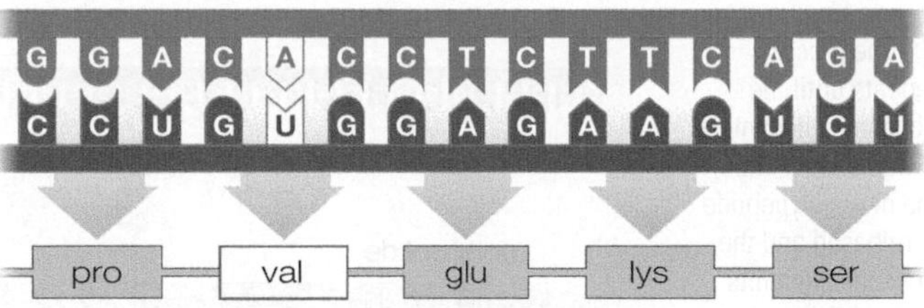
G G A C A C C T C T T C A G A
C C U G U G G A G A A G U C U
pro
val
glu
lys
ser

B A base-pair substitution replaces a thymine with an adenine in the beta globin gene. When the altered mRNA is translated, valine replaces glutamic acid as the sixth amino acid. Hemoglobin with this form of beta globin is called sickle hemoglobin, or HbS.

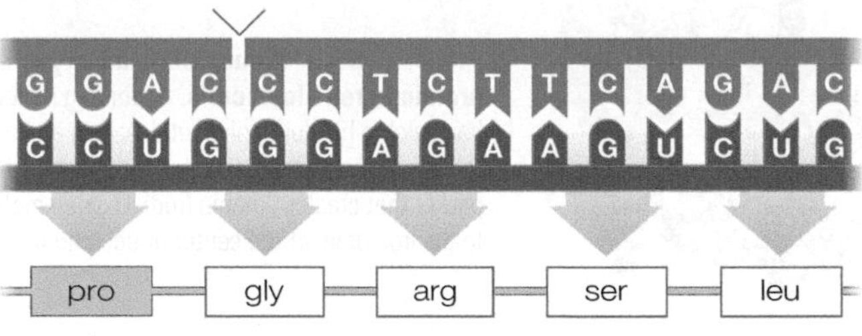
G G A C C C T C T T C A G A C
C C U G G G A G A A G U C U G
pro
gly
arg
ser
leu

C A deletion of one nucleotide shifts the reading frame for the rest of the mRNA, so a completely different protein product forms. The mutation shown results in a defective beta globin. The outcome is beta thalassemia, a genetic disorder in which a person has an abnormally low amount of hemoglobin.

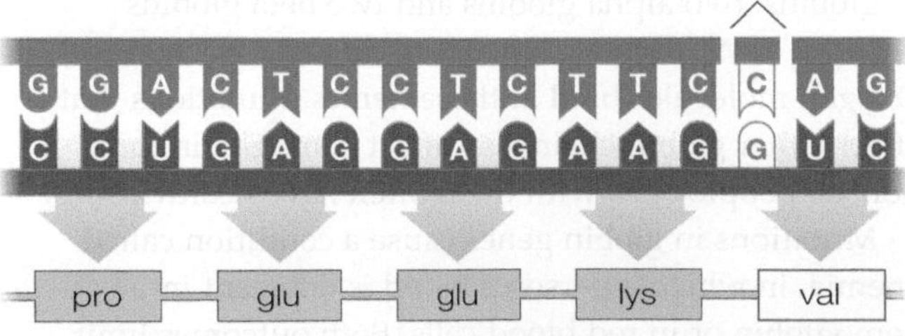
G G A C T C C T C T T C C A G
C C U G A G G A G A A G G U C
pro
glu
glu
lys
val

D An insertion of one nucleotide causes the reading frame for the rest of the mRNA to shift. The protein translated from this mRNA is too short and does not assemble correctly into hemoglobin molecules. As in **C**, the outcome is beta thalassemia.

A different type of anemia called beta thalassemia is caused by a **deletion**, which is a mutation in which one or more nucleotides is lost from the DNA. In this case, the twentieth nucleotide in the coding region of the beta globin gene is lost (**FIGURE 9.15C**). Like most other deletions, this one causes the reading frame of the mRNA codons to shift. A frameshift usually has drastic consequences because it garbles the genetic message, just as incorrectly grouping a series of letters garbles the meaning of a sentence:

The fat cat ate the sad rat.

Th efa tca tat eth esa dra t.

The frameshift caused by the beta globin deletion results in a polypeptide that is very different from normal beta globin in amino acid sequence and in length. This outcome is the source of the anemia. Beta thalassemia can also be caused by an **insertion**, which is a mutation in which nucleotides are added to the DNA (**FIGURE 9.15D**). Insertions, like deletions, often cause frameshifts.

Not all mutations that change proteins are harmful. Consider the sickle-cell anemia mutation shown in Figure 9.15B. A different mutation in the same codon, a base-pair substitution that changes the GAG to an AAG, results in a beta globin with lysine as its sixth amino acid. Hemoglobin assembled from this globin is called HbC. Unlike HbS, HbC does not clump or distort red blood cells. It can cause a mild anemia, but most people who carry the mutation have no symptoms at all. These people are particularly resistant to infection by the parasite that causes malaria, a trait that is quite helpful in malaria-ridden regions of the world.

A mutation can affect a gene's expression without changing any codons, for example if it occurs in a promoter or other regulatory site. Consider a mutation that causes the hairless appearance of the sphynx cat (left). In this case, a

base-pair substitution Type of mutation in which a single base pair changes.

deletion Mutation in which one or more nucleotides are lost.

insertion Mutation in which one or more nucleotides are inserted.

CREDIT: (in text) Eye of Science/Science Source; Glennis Siverson.

Data Analysis Activities

RIPs as Cancer Drugs Researchers are taking a page from the structure–function relationship of RIPs in their quest for cancer treatments. The most toxic RIPs, remember, have one domain that interferes with ribosomes, and another that carries them into cells. Melissa Cheung and her colleagues incorporated a peptide that binds to skin cancer cells into the enzymatic part of an RIP, the *E. coli* Shiga-like toxin. The researchers created a new RIP that specifically kills skin cancer cells, which are notoriously resistant to established therapies. Some of their results are shown in **FIGURE 9.17**.

1. Which cells had the greatest response to an increase in concentration of the engineered RIP?
2. At what concentration of RIP did all of the different kinds of cells survive?
3. Which cells survived best at 1 microgram per liter RIP?
4. Why are some of the data points linked by curved lines?

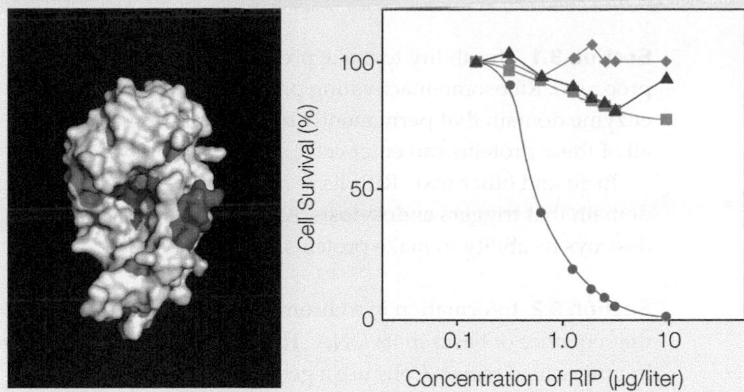

Cell Survival (%)
100
50
0
0.1
1.0
10
Concentration of RIP (μg/liter)

FIGURE 9.17 **Effect of an engineered RIP on cancer cells.** The model on the left shows the enzyme portion of *E. coli* Shiga-like toxin engineered to carry a small sequence of amino acids (in blue) that targets skin cancer cells. (Red indicates the active site.) The graph on the right shows the effect of this engineered RIP on human cancer cells of the skin (●); breast (◆); liver (▲); and prostate (■).

base-pair substitution disrupts an intron–exon splice site in a gene for keratin, a fibrous protein. The intron is not removed during post-transcriptional processing, and it introduces a premature stop codon in the finished mRNA (**FIGURE 9.16**). The altered protein translated from the resulting mRNA is too short and cannot properly assemble into filaments that make up hair. Cats that have this mutation still make hair, but it falls out before it gets very long.

FIGURE 9.16 **How a mutation in an intron affects gene expression.** A mutation that causes hairlessness in sphynx cats (chapter opening photo) is a base-pair substitution that changes a G to A in an intron–exon splice site. The altered site is no longer recognized during RNA processing, so the finished mRNA ends up with an intron in it. A stop codon in the intron sequence cuts short the protein translated from the mRNA.

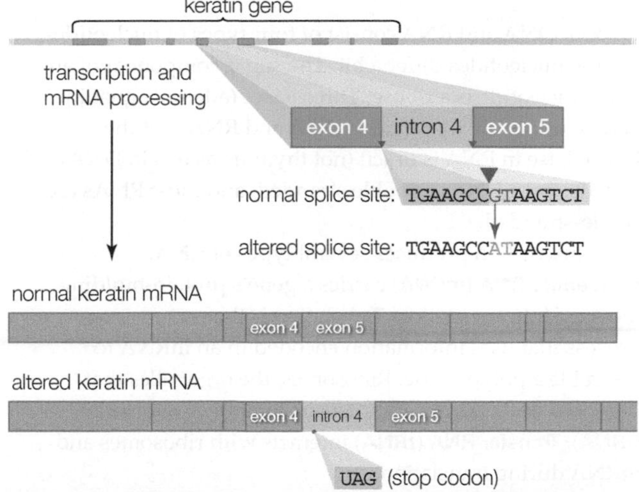

keratin gene
transcription and
mRNA processing
exon 4
intron 4
exon 5
normal splice site: TGAAGCCGTAAGTCT
altered splice site: TGAAGCCATAAGTCT
normal keratin mRNA
exon 4
exon 5
altered keratin mRNA
exon 4
intron 4
exon 5
UAG (stop codon)

TAKE-HOME MESSAGE 9.6

- ✔ A mutation may alter a gene's product or interfere with its expression. The effects may be harmful, but they are not always so.
- ✔ A base-pair substitution can change an amino acid in a protein.
- ✔ Insertions and deletions can alter an mRNA's codon reading frame. This frameshift garbles the mRNA's protein-building instructions.
- ✔ Any type of mutation can change a regulatory site in the DNA.

9.1 Ricin, RIP (revisited)

RIPs remove a particular adenine base from one of the rRNAs in the ribosome's large subunit. The adenine is part of a binding site for proteins involved in GTP-requiring steps of elongation. After the base has been removed, the ribosome can no longer bind to these proteins, and elongation stops.

Despite their toxicity, the main function of RIPs may not be destroying ribosomes. Many are part of plant immune systems, but it is their antiviral and anticancer activity that has researchers abuzz. Plants that make RIPs have been used as traditional medicines for many centuries; now, Western scientists are investigating RIPs as drugs to combat HIV and cancer. For example, researchers who design drugs for cancer therapy have modified ricin's glycolipid-binding domain to recognize plasma membrane proteins (Section 5.7) especially abundant in cancer cells. The modified ricin preferentially enters—and kills—cancer cells. Ricin's toxic enzyme has also been attached to an antibody that can find cancer cells in a person's body. The intent of both strategies: to assassinate the cancer cells without harming normal ones. ●

Section 9.1 The ability to make proteins is critical to all life processes. Ribosome-inactivating proteins (RIPs) have an enzyme domain that permanently disables ribosomes, but not all of these proteins can enter cells, so not all are toxic.

Ricin and other toxic RIPs have an additional protein domain that triggers endocytosis. An RIP that enters a cell destroys its ability to make proteins.

Section 9.2 Information in a chromosome is encoded in the sequence of bases in its DNA. That information occurs in units called **genes**. Cells use a gene's coding sequence to produce an RNA or protein product.

During **gene expression**, information flows from DNA to RNA to protein:

DNA → TRANSCRIPTION → RNA → TRANSLATION → PROTEIN

Transcription is the energy-intensive process that copies a gene into RNA form. Transcription of a gene requires a **promoter** upstream of the coding sequence and a terminator at its end:

promoter | gene coding sequence | terminator

Both DNA and RNA consist of four types of nucleotides, but the nucleotides differ a bit. The sugar component of an RNA nucleotide is a ribose. Three bases (adenine, guanine, and cytosine) are the same in DNA and RNA, but the fourth base in RNA is uracil (not thymine as it is in DNA). In cells, most DNA is double stranded, and most RNAs are single-stranded.

Transcription produces several types of RNA. **Messenger RNA (mRNA)** carries a gene's protein-building message into translation. **Translation** is the energy-intensive process that uses information encoded in an mRNA to assemble a polypeptide. Ribosomes, the organelles that carry out protein synthesis, consist mainly of **ribosomal RNA (rRNA)**. **Transfer RNA (tRNA)** interacts with ribosomes and mRNA during translation.

Section 9.3 Transcription occurs in the nucleus of eukaryotes, and in the cytoplasm of prokaryotes. The enzyme **RNA polymerase** carries out the process. This enzyme binds to a gene's promoter, then unwinds the DNA as it moves along the gene region. The polymerase uses the base sequence of the (noncoding) strand as a template to assemble a strand RNA from nucleotides. The new RNA strand is a copy of the gene in RNA form.

In eukaryotes, newly transcribed RNA is typically modified before leaving the nucleus. **Intron** sequences are removed, and the remaining **exon** sequences may be rearranged and spliced in different combinations (a process called **alternative splicing**). Messenger RNAs are further modified, receiving a cap and poly-A tail.

Section 9.4 The protein-building information in an mRNA consists of a series of **codons**. Most specify a particular amino acid during translation; some amino acids are specified by multiple codons. One codon is a signal to begin translation, and three terminate translation. All 64 codons constitute the **genetic code**.

Each tRNA has an **anticodon** that base-pairs with a codon. Each tRNA also binds to the amino acid specified by that codon. During translation, tRNAs bring amino acids to ribosomes. Proteins and rRNAs make up the two subunits of a ribosome. The rRNA components of a ribosome catalyze formation of a peptide bond between amino acids during translation.

Section 9.5 During translation, a polypeptide is assembled according to codons in an mRNA. The order of codons in the mRNA determines the order of amino acids in the resulting polypeptide.

Translation begins when two ribosomal subunits and an initiator tRNA converge on an mRNA. Other tRNAs deliver amino acids to the ribosome in the order dictated by successive mRNA codons. As the amino acids arrive, the ribosome joins them via peptide bonds. Translation ends when the ribosome encounters a stop codon in the mRNA and releases the new polypeptide.

Section 9.6 **Deletions**, **insertions**, and **base-pair substitutions** change the sequence of bases in DNA, and these mutations may affect gene products. Deletions and insertions often result in frameshifts that garble the information in a gene, thus affecting the base sequence of RNA translated from it. Mutations that occur in regulatory sequences such as intron–exon splice sites can also affect a gene's product.

A mutation that changes a gene's product or interferes with its expression may have harmful effects, but this is not always the case. Sickle-cell anemia, which is caused by a base-pair substitution in the gene for the beta globin chain of hemoglobin, is one example of a harmful outcome of a mutation. A different base-pair substitution in the same codon results in HbC, a form of hemoglobin that provides protection from malaria and has no negative effects on health. This is an example of a helpful outcome of a mutation.

SELF-QUIZ

Answers in Appendix VII

1. A chromosome contains many different genes that are transcribed into different ________.
   a. proteins
   b. polypeptides
   c. RNAs
   d. a and b

2. A binding site for RNA polymerase is called a ________.
   a. gene
   b. promoter
   c. codon
   d. protein

3. In cells, most RNA molecules are _______ ; most DNA molecules are _______ .
   a. single-stranded; double-stranded
   b. double-stranded; single-stranded

4. Match each process with its product.
   ___ transcription   a. DNA
   ___ replication   b. RNA
   ___ translation   c. protein

5. The main function of a DNA molecule is to _______ .
   a. store heritable information
   b. carry a translatable message
   c. form peptide bonds between amino acids

6. The main function of an mRNA molecule is to _______ .
   a. store heritable information
   b. carry a translatable message
   c. form peptide bonds between amino acids

7. Transcription is similar to DNA replication because both processes _______ .
   a. use the same enzyme
   b. copy both strands
   c. require the same nucleotides
   d. proceed in the 5′ to 3′ direction

8. Most codons specify a(n) _______ .
   a. protein   c. amino acid
   b. polypeptide   d. mRNA

9. Energy that drives transcription is provided mainly by _______ .
   a. ATP   c. GTP
   b. RNA nucleotides   d. RNA polymerase

10. Anticodons pair with _______ .
    a. mRNA codons   c. RNA anticodons
    b. DNA codons   d. amino acids

11. What is the maximum number of amino acids that can be encoded by a gene with 45 bases plus a stop codon?
    a. 15   c. 90
    b. 45   d. 135

12. _______ is/are removed from a new mRNA.
    a. Introns   c. A poly-A tail
    b. Exons   d. Amino acids

13. Transcription take place in the _______ of prokaryotic cells.
    a. nucleus   c. cytoplasm
    b. plasma membrane   d. all of the above

14. Energy that drives translation is provided mainly by _______ .
    a. ATP   c. GTP
    b. amino acids   d. all of the above

15. Match the terms with the best description.

___ genetic message	a. protein-coding segment
___ promoter	b. transcription begins here
___ polysome	c. read as base triplets
___ exon	d. removed before translation
___ genetic code	e. occurs only in groups
___ intron	f. 64 codons
___ anticodon	g. destroys ribosomes
___ RIP	h. often causes a frameshift
___ deletion	i. enzymatic RNA
___ rRNA	j. binds to a codon

CRITICAL THINKING

1. Researchers are designing and testing antisense drugs as therapies for a variety of diseases, including cancer, AIDS, diabetes, and muscular dystrophy. The drugs are also being tested to fight infection by deadly viruses such as Ebola. Antisense drugs consist of short RNA strands complementary in sequence to mRNAs that form during the progression of a disease. How do you think these drugs work?

2. An anticodon has the sequence GCG. What amino acid does this tRNA carry? What would be the effect of a mutation that changed the C of the anticodon to a G?

3. Each position of a codon can be occupied by one of four nucleotides. What is the minimum number of nucleotides per codon necessary to specify all 20 of the amino acids that are found in proteins?

4. Refer to Figure 9.7, then translate the following mRNA nucleotide sequence into an amino acid sequence, starting at the first base:

   5′—UGUCAUGCUCGUCUUGAAUCUUGUGAU
   GCUCGUUGGAUUAAUUGU—3′

5. Translate the sequence of bases in the previous question, starting at the second base.

6. Can you spell your name (or someone else's) using the one-letter amino acid abbreviations shown in Figure 9.7? If so, construct an mRNA sequence that encodes your "protein" name.

7. Bacteria use the same stop codons as eukaryotes. However, bacterial transcription is also terminated in places where the mRNA folds back on itself to form a hairpin-looped structure like the one shown at right. How do you think that this structure stops transcription?

10 CONTROL OF GENE EXPRESSION

CORE CONCEPTS

Information Flow

Living systems store, retrieve, transmit, and respond to information essential for life.

All cells respond to internal and external change, and these responses involve adjustments to gene expression. In multicelled eukaryotes, normal embryonic development depends on appropriate cellular responses to molecules that regulate gene expression. Environmental factors that influence gene expression during an individual's lifetime can have multigenerational effects.

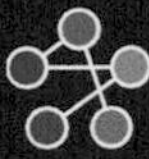

Systems

Complex properties arise from interactions among components of a biological system.

The timing and coordination of specific molecular and cellular events are regulated by mechanisms that govern gene expression. All cells of a multicelled organism have the same DNA, but different genes are active in the different cell types. The genes that a cell expresses determine the products it makes, and these in turn determine the type of cell it is. Dividing cells of an early embryo use the same genes, then differentiate as they start using different subsets of genes. Development is an outcome of gene expression cascades during embryonic development.

Evolution

Evolution underlies the unity and diversity of life.

Cascades of regulatory gene expression during embryonic development give rise to complex multicelled bodies. Many of these genes are evolutionarily conserved even among distantly related species.

Links to Earlier Concepts

This chapter explores metabolism (Sections 5.5–5.7) in the context of gene expression (9.2). You will be applying what you know about DNA structure (8.2, 8.3) and replication (8.5); mutations (8.6, 9.6), and differentiation (8.7), as well as transcription (9.3) and translation (9.5). You will also revisit functional groups (3.2), carbohydrates (3.3), nuclear pores (4.4), and fermentation (7.5).

10.1 Between You and Eternity

You are in college, your whole life ahead of you. Your risk of developing cancer is as remote as old age, an abstract statistic that is easy to forget. "There is a moment when everything changes—when the width of two fingers can suddenly be the total distance between you and eternity." Robin Shoulla wrote those words after being diagnosed with breast cancer. She was seventeen years old.

At an age when most young women are thinking about school, friends, parties, and potential careers, Robin was dealing with radical mastectomy: the removal of a breast, all lymph nodes under the arm, and skeletal muscles in the chest wall under the breast. She was pleading with her oncologist not to use her jugular vein for chemotherapy and wondering if she would survive to see the next year.

Robin's ordeal became part of a statistic, one of more than 200,000 new cases of breast cancer diagnosed in the United States each year. About 5,700 of those cases occur in women and men under thirty-four years of age.

Cancer occurs when abnormally dividing cells disrupt body tissues. Every second, millions of your cells—in skin, bone marrow, gut lining, liver, and so on—are dividing to replace their worn-out, dead, and dying predecessors. A normal, healthy cell does not divide at random; it has a finely tuned system of molecular mechanisms that trigger division at appropriate times, and prevent division at other times. Anything that interferes with these mechanisms can cause uncontrolled cell divisions characteristic of cancer.

Most cancer-causing mutations disrupt expression of genes involved in cell division control. These genes are called tumor suppressors because tumors are more likely to form when their products are missing, defective, or underproduced. Two tumor suppressors, *BRCA1* and *BRCA2*, were named because their products are compromised in most breast cancer cells.

In most cases of breast cancer, exposure to environmental factors has altered the expression of an otherwise functional *BRCA* gene. The remaining breast cancers have a *BRCA* gene mutation. Robin Shoulla is one of the unlucky people born with mutations in both *BRCA1* and *BRCA2*, so she had a very high risk of developing breast cancer.

Radical mastectomy is rarely performed today, but at the time it was Robin's only option. The treatment helped her to survive. Now, she has what she calls a normal life: career, husband, children. Her goal as a cancer survivor: "To grow very old with gray hair and spreading hips, smiling." ●

CREDIT: (opposite) 7th Son Studio/Shutterstock.com.

10.2 Regulating Gene Expression

LEARNING OBJECTIVES

- Explain gene expression control and why it is necessary.
- Use examples to describe transcription factors.

In all cells, homeostasis depends on tight regulation of molecules being produced at a given time, and this regulation involves adjustments to gene expression. Individual cells regulate gene expression to adjust metabolism, for example in response to the availability of a nutrient. Cells of multicelled organisms also adjust gene expression to alter body form and function. During development, an embryo's cells adjust gene expression in response to signals from other cells, an interplay that helps sculpt a developing body.

Consider how the genes a cell uses determine the molecules it produces, which in turn determine the type of cell it is. A multicelled body starts out as a tiny cluster of identical cells, all expressing the same genes in the same DNA. These cells divide repeatedly, and their descendants begin to differentiate as they start using different subsets of genes. As the cells diverge in form and function, they give rise to different parts of a body. A typical differentiated cell uses only about 10 percent of its genes at any given time. Some of the expressed genes affect structural features and metabolic functions common to all cells; others are used only by certain subsets of cells. For example, all of your body cells express genes for glycolysis enzymes, but only immature red blood cells express globin genes.

Switching Genes On and Off

Regulating gene expression does not involve changes to DNA sequence. Cells use "switches" to turn genes on or off. These switches consist of molecules that can promote or inhibit individual steps of gene expression. Every one of those steps is regulated, from transcription to delivery of a gene's RNA or protein product to its final destination in the cell (**FIGURE 10.1**).

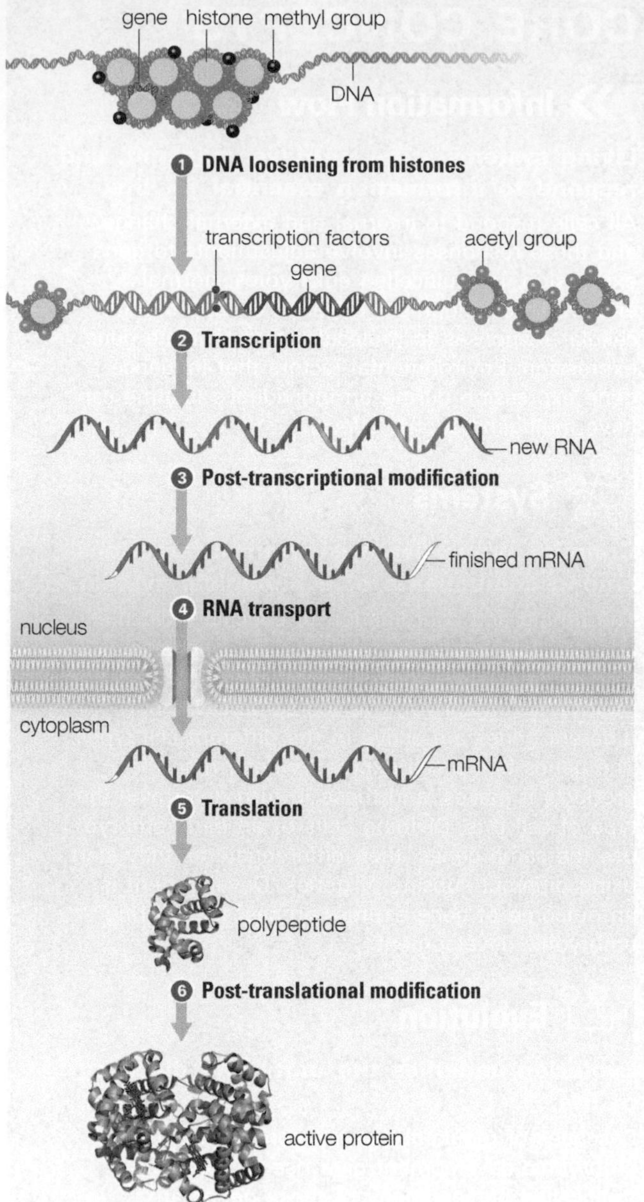
gene
histone
methyl group
DNA
1 DNA loosening from histones
transcription factors
acetyl group
gene
2 Transcription
new RNA
3 Post-transcriptional modification
finished mRNA
4 RNA transport
nucleus
cytoplasm
mRNA
5 Translation
polypeptide
6 Post-translational modification
active protein

FIGURE 10.1 Points of control over gene expression. Expression of a eukaryotic gene with a protein product is illustrated. Models are not to scale.

1 DNA–Histone Interactions Remember from Section 8.4 that a DNA molecule in a eukaryotic chromosome is wrapped around histones. Reversible modifications to these proteins make them loosen or tighten their grip on DNA wrapped around them. Tightly wrapped DNA is inaccessible to RNA polymerase, so it cannot be transcribed. Adding acetyl groups (—$COCH_3$) to a histone loosens the DNA around it, so enzymes that attach these functional groups to histones promote transcription. Conversely, removing acetyl groups tightens the DNA and represses transcription. Methyl groups (—CH_3) on histones also affect transcription of DNA wrapped around them. Adding methyl groups to (methylating) certain amino acids of a histone tightens the DNA and represses transcription; methylating other amino acids loosens the DNA and enhances transcription. Typically, multiple functional groups are attached to the same histone, and these have a collective effect on local transcription.

2 Transcription A cell can change the rate of transcription of particular genes by changing the transcription factors it makes. A **transcription factor** is a protein that binds directly to DNA and affects whether and how fast a gene is transcribed. Transcription of most genes

◄gene
promoter
insulator
silencer
enhancer
promoter
gene►

FIGURE 10.2 Sites in DNA where transcription factors bind.
Repressors slow or stop transcription by binding to a promotor or silencer; activators encourage transcription of nearby genes by binding to an enhancer. Insulators prevent activators or repressors from inappropriately affecting transcription of a neighboring gene.

is governed by multiple transcription factors that have overlapping effects. The complexity allows RNA production to be adjusted in a precise and nuanced way.

Repressors are transcription factors that slow or stop transcription, either by preventing RNA polymerase from accessing a promoter, or by blocking the enzyme's progress. Some repressors work by binding directly to the promoter (**FIGURE 10.2**). Others bind to a DNA sequence called a silencer that may be thousands of base pairs away from the gene.

Activators are transcription factors that speed up transcription by helping RNA polymerase access a promoter. Many activators work by binding to the promoter; others bind to DNA sequences called **enhancers**, which, like silencers, may be very far away from the gene they affect. A silencer or enhancer operating on a distant gene can inappropriately affect a nearby gene. Insulators prevent this from occurring. An insulator is a region of DNA that, upon binding a transcription factor, can block the effect of a silencer or enhancer on a neighboring gene.

❸ RNA Processing In eukaryotic cells, RNAs that are used in cytoplasm must first exit the nucleus. This movement occurs through nuclear pores, which are selective about the molecules that can pass through them (Section 4.4). Certain proteins must be attached to an RNA before it is allowed out of the nucleus, and these proteins attach to an RNA only after it has been processed—spliced, capped, and finished with a poly-A tail, in the case of mRNA. Mechanisms that delay post-transcriptional modification of an RNA also delay its export from the nucleus.

Other mechanisms that regulate processing of an mRNA affect the form of a protein. Consider fibronectin, a protein made by all vertebrate animals. Cells called fibroblasts produce a fibrous form of the protein that is a major component of extracellular matrix (Section 4.9). Liver cells make a soluble fibronectin that circulates in blood plasma. Both forms of fibronectin arise from expression of the same gene. The difference in the gene's product begins with tissue-specific expression of molecules that splice RNA: Fibroblasts and liver cells produce different mRNAs from the same fibronectin gene by alternative splicing.

❹ RNA Transport In many cases, an mRNA is transported to a particular region of the cell where its protein product is required. In an egg, for example, cytoplasmic localization of mRNA is crucial for proper development of the future embryo. How does localization occur? In the case of eukaryotic mRNA, a short nucleotide sequence called a "zip code" near the poly-A tail specifies a particular destination. Protein chaperones that attach to the zip code help deliver the mRNA to its destination. Other proteins can influence the timing or location of delivery by binding to these chaperones.

❺ Translation Translation of an mRNA depends on many proteins that associate with it. For example, some zip code-binding proteins prevent translation until an mRNA reaches its destination. Other proteins produced in response to environmental signals can attach to and stabilize (or destabilize) specific mRNAs. Still other proteins bind to a poly-A tail and help the mRNA attach to a ribosome for translation. The length of the tail influences this interaction, with shorter tails being less efficient at promoting transcription. As soon as an mRNA shows up in cytoplasm, enzymes start removing its tail. When the tail gets too short, transcription ends. Cells can recycle these used-up mRNAs by producing enzymes that rebuild their poly-A tails.

Translation of particular mRNAs can be prevented by other RNAs. For example, in a process called **RNA interference**, small segments of double-stranded RNA are taken up by protein complexes that remove one of their two strands. The remaining strand—the guide strand—is complementary in sequence to a particular mRNA. When that mRNA shows up in cytoplasm, it base-pairs with the guide strand. The protein complex then cuts the mRNA into pieces, or in some cases it prevents ribosomes from accessing the mRNA.

❻ Post-Translational Modification Many newly synthesized polypeptide chains must be modified before they become functional. For example, some enzymes become active only after they have been phosphorylated (Section 5.6). Such modifications inhibit, activate,

activator Transcription factor that increases the rate of transcription when it binds to a promoter or enhancer.
enhancer Binding site for an activator.
repressor Transcription factor that slows or stops transcription.
RNA interference Mechanism in which translation of a particular mRNA is suppressed by other RNA molecules.
transcription factor Regulatory protein that binds to DNA and influences transcription; e.g, an activator or repressor.

or stabilize many molecules, including the enzymes that participate in transcription and translation.

TAKE-HOME MESSAGE 10.2

- ✔ Mechanisms that govern gene expression affect the kinds and amounts of substances that are present in a cell at any given time.
- ✔ Regulating gene expression allows individual cells to respond appropriately to internal and external change. It is also the basis of differentiation during development of multicelled bodies.
- ✔ Many molecules work together or in opposition to enhance or inhibit each step of gene expression.
- ✔ Transcription factors regulate transcription by binding to DNA at specific sequences such as promoters.

10.3 Regulating Gene Expression in Development

LEARNING OBJECTIVES

- Describe the role of master regulators in embryonic development.
- Give an example of a homeotic gene and explain its function.
- Using an appropriate example, explain why homeotic genes offer evidence of shared ancestry.
- Explain how adjustments to gene expression affect the form and function of male and female mammals.

How Genes Direct Embryonic Development

As an embryo develops, its differentiating cells organize into body parts. Each tissue, organ, and limb forms in a predictable area of the embryo, at a predictable stage of development. The whole process is orchestrated by transcription factors. Hundreds of transcription factors are produced in cascades of gene expression: The transcription factor product of one gene affects expression of other genes, whose products in turn affect expression of others, and so on. Some of these genes are master regulators. Expression of a **master regulator** begins a gene expression cascade that ultimately changes cells in a lineage from one type to other, more differentiated types—a bit like a master switch turns on a whole system with a single flip.

Embryonic development begins with master regulators called maternal effect genes. Different maternal effect genes are transcribed in an egg as it forms. The resulting mRNAs are delivered to different regions of the egg's cytoplasm, but they are not translated until the egg is fertilized.

What happens next has been studied most in the common fruit fly, *Drosophila melanogaster* (**FIGURE 10.3**). Development in flies differs from development in plants or animals, but corresponding genes govern the process in all organisms with bodies. After a fly egg is fertilized, its nucleus begins to divide repeatedly.

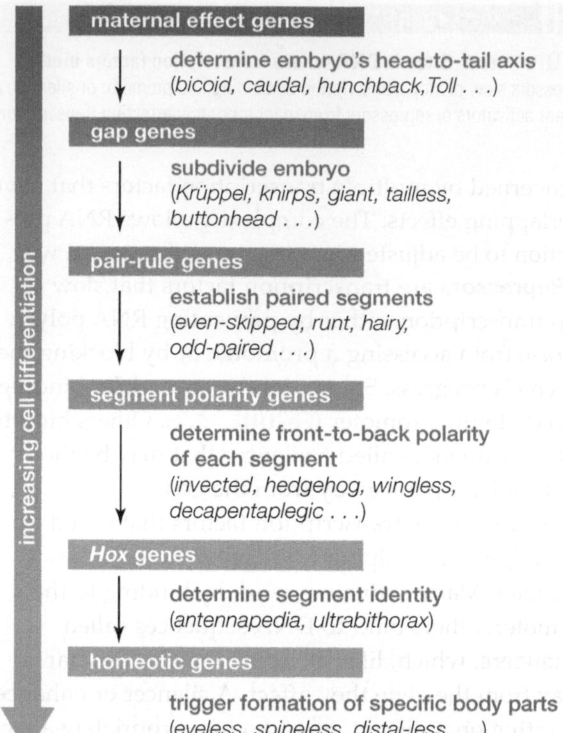
increasing cell differentiation
maternal effect genes
determine embryo's head-to-tail axis
(bicoid, caudal, hunchback, Toll . . .)
gap genes
subdivide embryo
(Krüppel, knirps, giant, tailless, buttonhead . . .)
pair-rule genes
establish paired segments
(even-skipped, runt, hairy, odd-paired . . .)
segment polarity genes
determine front-to-back polarity of each segment
(invected, hedgehog, wingless, decapentaplegic . . .)
Hox genes
determine segment identity
(antennapedia, ultrabithorax)
homeotic genes
trigger formation of specific body parts
(eyeless, spineless, distal-less . . .)

FIGURE 10.3 **The cascade of regulatory gene expression that results in formation of a fruit fly.**

Corresponding genes expressed in similar cascades direct development in all other animals.

Transcription of maternal effect mRNAs also begins. These mRNAs are localized to specific regions, but their protein products—transcription factors—diffuse away in gradients that span the whole embryo. The position of each nucleus in the embryo determines which and how much of these transcription factors it is exposed to. This in turn determines which of its own genes are turned on. The products of those genes also form gradients. Still other genes are transcribed depending on where a nucleus falls within these gradients, and so on. The genes expressed in this cascade first establish the front-to-back axis of the embryo, then divide the embryo into segments (**FIGURE 10.4**), then establish the front-to-back polarity and identity of each segment. Eventually, the cascade activates **homeotic genes**: master regulators whose expression results in the formation of specific body parts such as an eye or a flower.

Most genes that operate during development have been discovered via mutations that cause abnormal body form. The abnormalities are clues to the function of the gene's product. Researchers use a technique called a **knockout** to introduce an inactivating

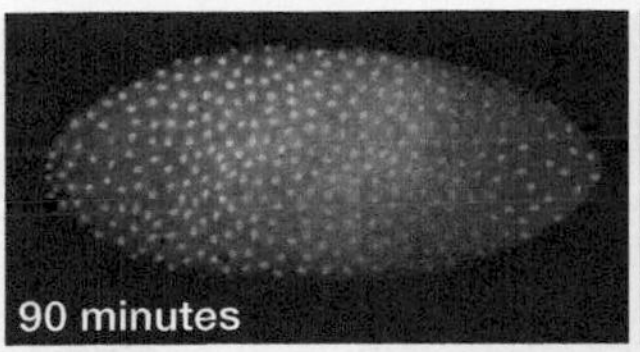
90 minutes
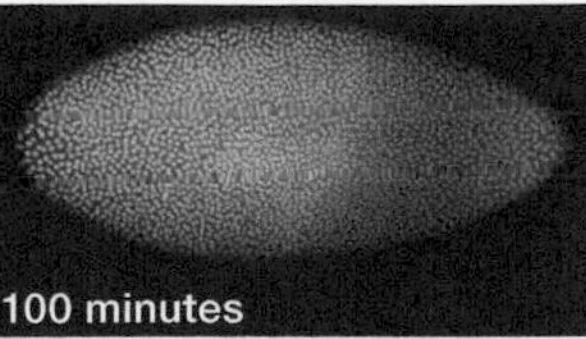
100 minutes

A Two maternal effect genes, *caudal* and *bicoid*, are expressed in opposite ends of the embryo. The products of these genes are visible in green (*caudal*) and blue (*bicoid*). Each diffuses away from its site of translation to form a gradient that spans the embryo. Another gene, *even-skipped*, is expressed only where *caudal* and *bicoid* proteins overlap. Its product is visible in red.

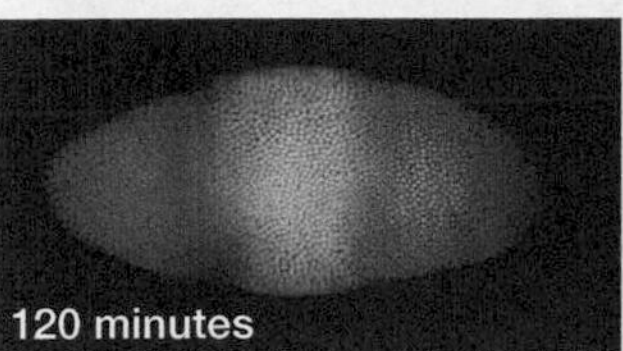
120 minutes
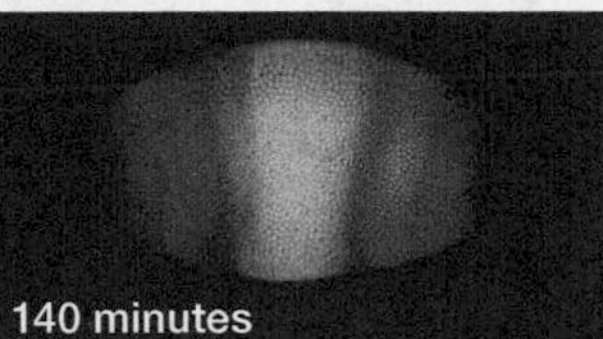
140 minutes

B The products of several genes, including *Krüppel* (in green) and *giant* (blue) further confine the expression of *even-skipped* (red). Pink and yellow areas are regions in which red fluorescence overlaps with blue or green.

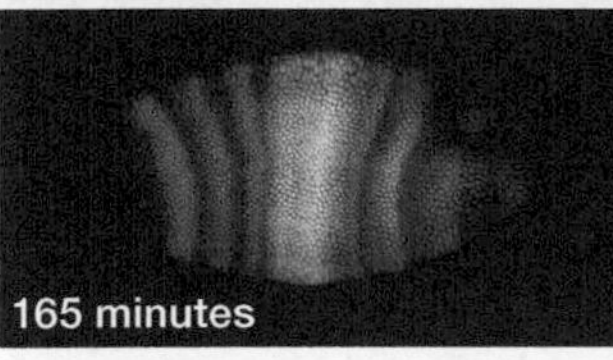
165 minutes
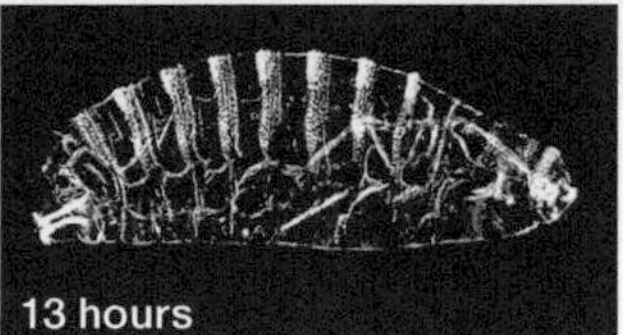
13 hours

C 165 minutes after fertilization, expression of *even-skipped* has been confined to seven stripes (red). A day later, seven segments have formed in corresponding positions.

FIGURE 10.4 How gene expression control makes a fly, as illuminated by segment formation. Bright dots pinpoint nuclei in these micrographs of developing *Drosophila* embryos (time after fertilization is indicated). Many genes are expressed, but only five are tracked here.

mutation into a gene (Chapter 15 returns to genetic engineering). Gene expression cascades that sculpt development have been deciphered by knocking out genes one by one in flies and other organisms.

Mutations that disrupt the earliest genes in expression cascades have the greatest effect on development. For example, a mutation that disables a maternal effect gene called *bicoid* causes a fly embryo to form with two back ends and no head end (embryos with this degree of disorganization cannot survive). Mutations that disrupt genes further downstream in expression cascades cause body parts to form in the wrong places. Flies with a mutation that disables their *antennapedia* gene, for example, have legs in place of antennae on the head (**FIGURE 10.5A**). Homeotic gene mutations disorganize individual organs. For example, flies with an inactive *eyeless* gene lack eyes (**FIGURE 10.5B**).

homeotic gene (home-ee-OTT-ic) Type of master regulator; its expression results in the formation of a specific body part during development.

knockout Technique of introducing a mutation that inactivates an organism's gene.

master regulator Gene whose expression triggers a gene expression cascade that ultimately changes cells in a lineage from one type to other, more differentiated types.

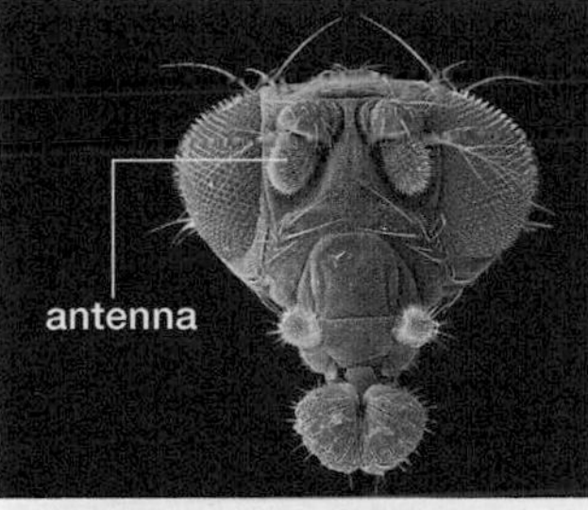
antenna
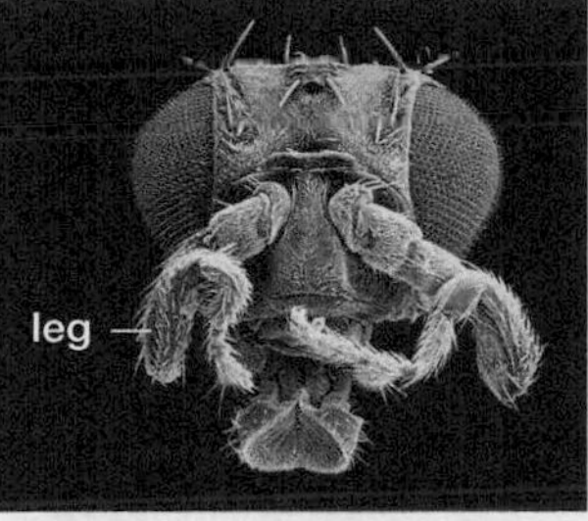
leg

A The head of a normal fruit fly (left) has two antennae. Right, a mutation in the *antennapedia* gene causes legs to form instead of antennae.

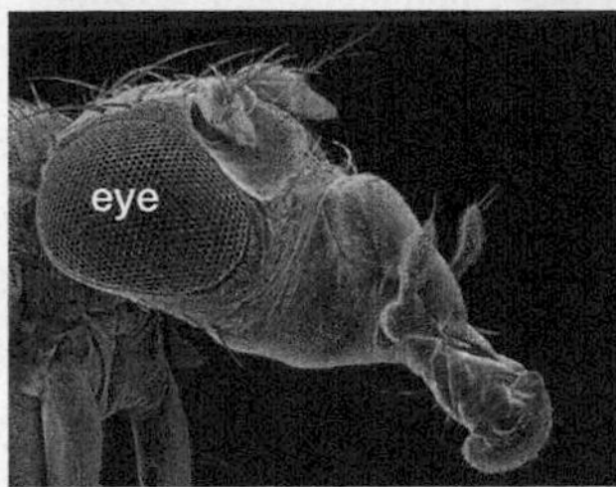
eye
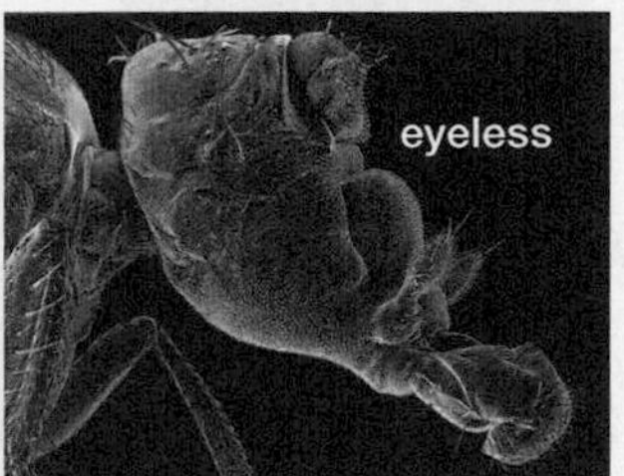
eyeless

B A normal fruit fly (left) has large, round eyes. A fruit fly with a mutation in its *eyeless* gene (right) develops without eyes.

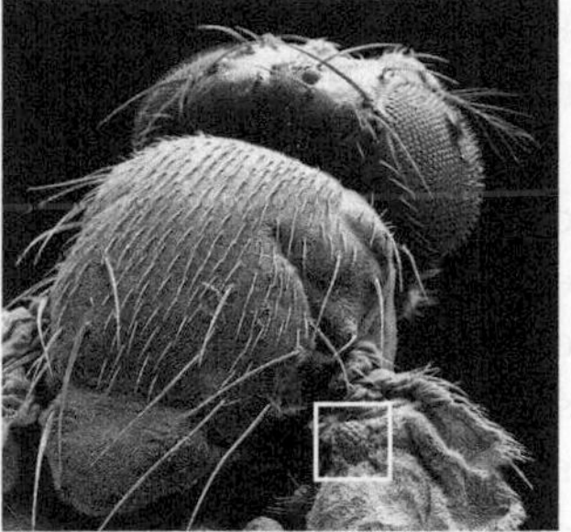

C Eyes form wherever the *eyeless* gene is expressed in fly embryos—here, on the head and also on the wing.

The *PAX6* gene of humans, mice, squids, and some other animals is so similar to *eyeless* that it similarly triggers eye development in fruit flies.

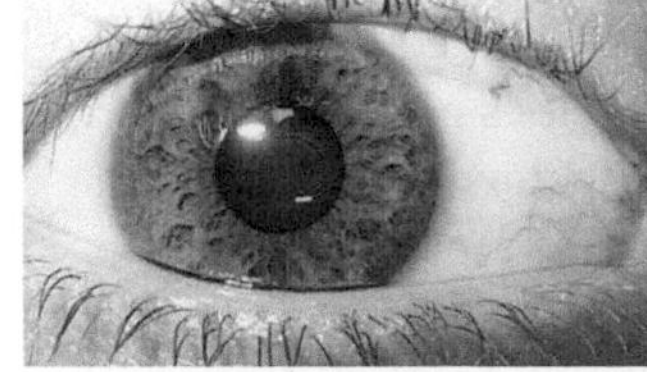
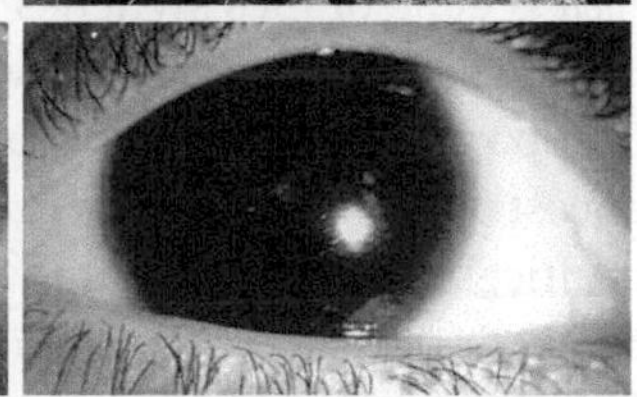

D A normal human eye has a colored iris surrounding the pupil (dark area where light enters). Mutations in *PAX6* cause eyes to develop without an iris, a condition called aniridia.

FIGURE 10.5 Some effects of mutations in developmental genes.

Many homeotic genes are interchangeable among species. Consider that eyes form in fly embryos wherever the *eyeless* gene is expressed, which is normally in tissues of the head. If the *eyeless* gene is expressed in another part of the developing embryo, eyes form there too (**FIGURE 10.5C**). Humans, squids, mice, fishes, and many other animals have a gene called *PAX6*, which is very similar to the *eyeless* gene of flies. In humans, mutations in *PAX6* cause eye disorders such as aniridia, in which the irises are underdeveloped or missing (**FIGURE 10.5D**). If a *PAX6* gene from a human

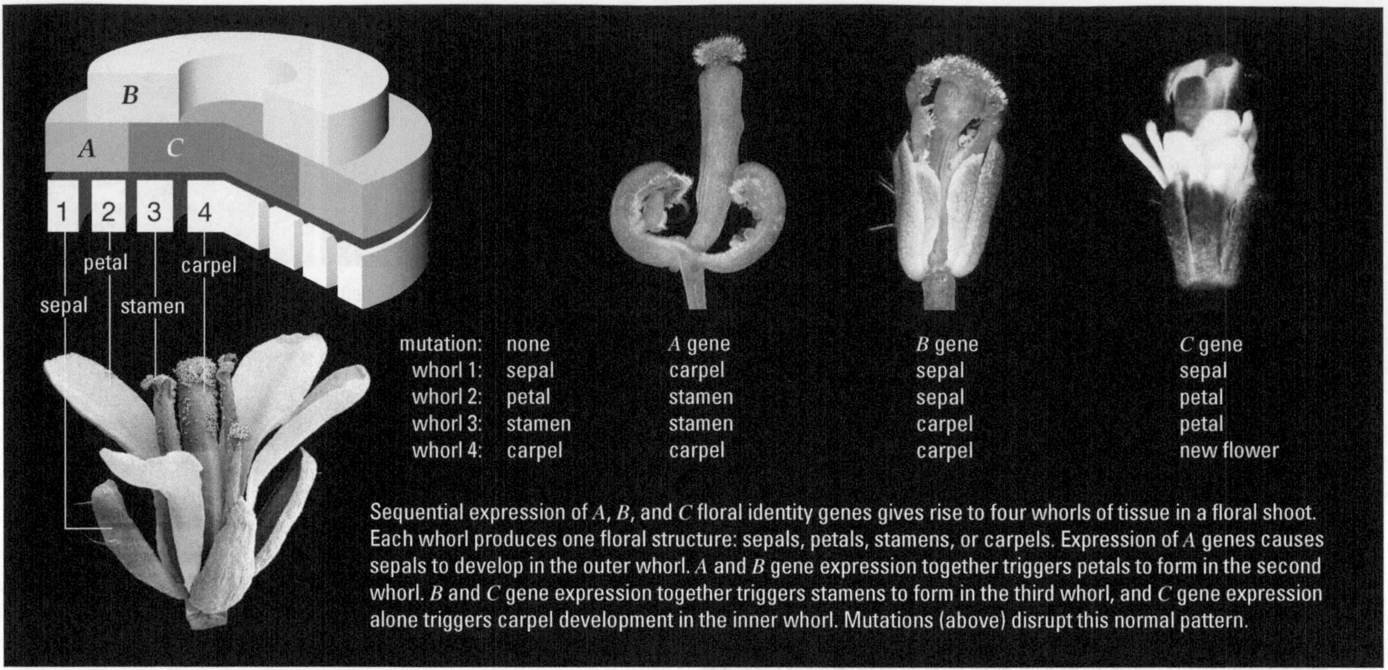
B
A
C
1
2
3
4
petal
carpel
sepal
stamen
mutation: none
whorl 1: sepal
whorl 2: petal
whorl 3: stamen
whorl 4: carpel
A gene
carpel
stamen
stamen
carpel
B gene
sepal
sepal
carpel
carpel
C gene
sepal
petal
petal
new flower
Sequential expression of A, B, and C floral identity genes gives rise to four whorls of tissue in a floral shoot. Each whorl produces one floral structure: sepals, petals, stamens, or carpels. Expression of A genes causes sepals to develop in the outer whorl. A and B gene expression together triggers petals to form in the second whorl. B and C gene expression together triggers stamens to form in the third whorl, and C gene expression alone triggers carpel development in the inner whorl. Mutations (above) disrupt this normal pattern.

FIGURE 10.6 Control of flower formation, as revealed by mutations in homeotic genes of thale cress (*Arabidopsis thaliana*) plants.

FIGURE IT OUT Mutations in which floral identity gene give(s) rise to flowers with no petals?

Answer: A or B

or mouse is inserted into a fly embryo, it has the same effect as the *eyeless* gene: An eye forms wherever it is expressed. (Because the *PAX6* gene product is just a switch, the eye that forms is a fly eye, not a human or mouse eye.) The same principle applies in reverse: The *eyeless* gene of flies switches on eye formation in frogs.

The correspondence between homeotic genes of evolutionarily distant animals provides evidence of shared ancestry. Homeotic genes guide development in all multicelled eukaryotes, and some are even similar to transcription factor genes in yeast. Thus, we can infer that they evolved before multicellularity did.

Examples of Developmental Outcomes

Formation of Flowers in Plants Plant development differs from animal development in that it continues after maturity, but it similarly occurs via regulation of gene expression. Studies of mutations in thale cress (*Arabidopsis thaliana*) revealed genes that govern flower formation. Transcription factors produced by three sets of floral identity genes—*A*, *B*, and *C*—establish the identity of sections of a floral shoot. Sequential and overlapping expression of *ABC* genes in cells of a shoot tip cause the cells to give rise to whorls (rings) of tissue, one over the other like layers of an onion. Each whorl develops into one type of floral structure—sepals, petals, stamens, or carpels (**FIGURE 10.6**).

The *A* genes are switched on first. Expression of these master regulators triggers a gene expression cascade that causes the outer whorl to form and give rise to sepals. *B* genes switch on before *A* genes turn off. Together, *A* and *B* gene products cause the second whorl to form and produce petals. Next, *A* genes turn off and the *C* gene switches on (there is only one *C* gene). Together, the products of the *B* and *C* genes trigger formation of the third whorl, which makes stamens. Finally, the *B* gene turns off, and the *C* gene product on its own gives rise to the fourth, inner whorl, which produces the carpel.

Dosage Compensation A female mammal inherits two X chromosomes (XX), one from her mother, the other from her father. In each of her body cells, one X chromosome is always tightly condensed (**FIGURE 10.7A**). We call the condensed X chromosomes **Barr bodies**, after Murray Barr, who discovered them. Most of the genes on a Barr body are not expressed. The mechanism that condenses an X chromosome is called **X chromosome inactivation**, and it ensures that most genes on one X chromosome are not expressed. This and any other mechanism of equalizing gene expression between the sexes is called **dosage compensation**. Body cells of male mammals (XY) have one set of X chromosome genes. Body cells of females have two, but female embryos do not develop properly when both sets are expressed.

How does just one of two X chromosomes get inactivated? An X chromosome gene called *XIST* is transcribed on only one X chromosome. The gene's product, a long noncoding RNA, sticks to the chromosome that expresses the gene. The RNA also causes the

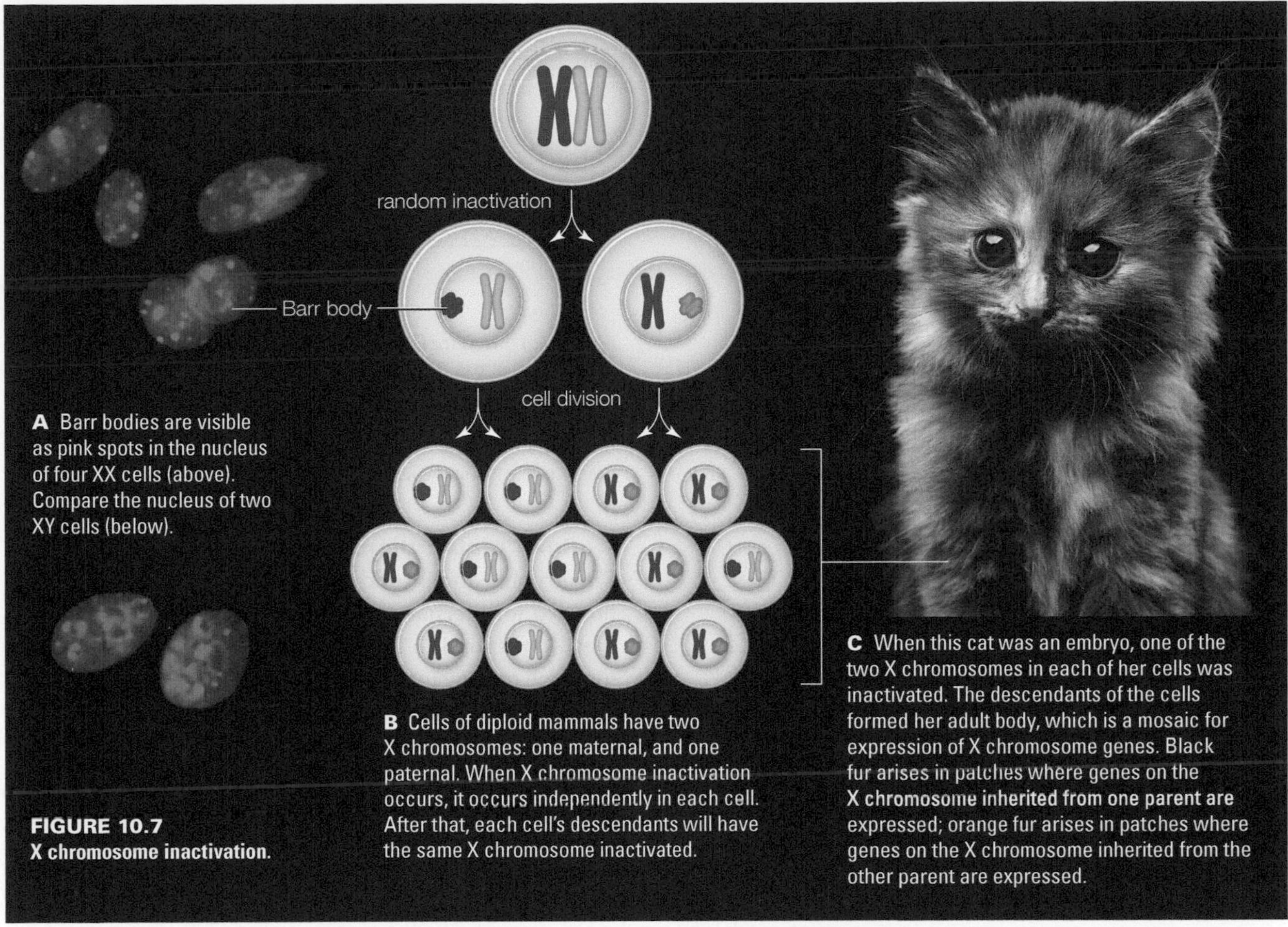
random inactivation
Barr body
cell division

FIGURE 10.7
X chromosome inactivation.

A Barr bodies are visible as pink spots in the nucleus of four XX cells (above). Compare the nucleus of two XY cells (below).

B Cells of diploid mammals have two X chromosomes: one maternal, and one paternal. When X chromosome inactivation occurs, it occurs independently in each cell. After that, each cell's descendants will have the same X chromosome inactivated.

C When this cat was an embryo, one of the two X chromosomes in each of her cells was inactivated. The descendants of the cells formed her adult body, which is a mosaic for expression of X chromosome genes. Black fur arises in patches where genes on the X chromosome inherited from one parent are expressed; orange fur arises in patches where genes on the X chromosome inherited from the other parent are expressed.

chromosome to migrate toward the nuclear lamina, the mesh of intermediate filaments that supports the nuclear envelope (Section 4.8). The nuclear lamina also serves as a warehouse for proteins involved in gene expression. Many of the proteins are repressors, and these shut down transcription on the X chromosome coated with *XIST* RNA. Other interactions with the nuclear lamina condense the chromosome into a Barr body. The homologous X chromosome does not express *XIST*, so it does not come into contact with repressors in the nuclear lamina, and it does not condense into a Barr body.

X chromosome inactivation occurs when an embryo is a ball of about 200 cells. In humans and most other mammals, it occurs independently in every cell of a female embryo. Which X chromosome condenses is random: The maternal X chromosome may get inactivated in one cell, and the paternal or maternal X chromosome may get inactivated in a cell next to it

Barr body Condensed, inactivated X chromosome in a cell of a female mammal. The other X chromosome is active.
dosage compensation Any mechanism of equalizing gene expression between males and females.
X chromosome inactivation Developmental shutdown of one of the two X chromosomes in the cells of female mammals.

(**FIGURE 10.7B**). Once a cell makes a selection, all of its descendants make the same selection as they continue dividing and forming tissues. As a result of random inactivation, an adult female mammal is a "mosaic" for expression of X chromosome genes. She has patches of tissue in which genes of the maternal X chromosome are expressed, and patches in which genes of the paternal X chromosome are expressed (**FIGURE 10.7C**).

Male Sex Determination in Humans More than 800 protein-coding genes have been identified on the human X chromosome, and only a few are associated with traits such as body fat distribution that differ between males and females. Most of the other X chromosome genes govern nonsexual traits such as blood clotting and color perception. Such genes are expressed in both males and females. Males, remember, also inherit one X chromosome.

The human Y chromosome has only 63 protein-coding genes, but one of them is *SRY*—the master regulator that determines male sex in mammals. Its expression in XY embryos triggers the formation of male reproductive organs (testes). Some of the cells in testes make testosterone, a sex hormone that triggers the emergence of male secondary sexual traits such as facial hair, increased musculature, and deepened voice. We know that *SRY* controls the emergence of male sexual traits because mutations in this gene cause XY individuals to develop external genitalia that appear female. An XX embryo has no Y chromosome, no *SRY* gene, and much less testosterone, so primary female reproductive organs (ovaries) form instead of testes.

TAKE-HOME MESSAGE 10.3

- ✔ Animal development is orchestrated by cascades of transcription factor gene expression in cells of embryos. The expression of master regulators triggers these cascades.
- ✔ A homeotic gene is a master regulator whose expression results in the formation of a specific body part. Many homeotic genes function in similar ways in evolutionarily distant animals.
- ✔ In plants, expression of *ABC* floral identity genes governs development of the specialized parts of a flower.
- ✔ X chromosome inactivation balances expression of X chromosome genes between female (XX) and male (XY) mammals.
- ✔ *SRY* expression triggers development of male traits in mammals.

10.4 Regulating Gene Expression to Adjust Metabolism

LEARNING OBJECTIVES

- Explain circadian rhythms in terms of gene expression.
- Use the *lac* operon as an example to describe one way that bacteria regulate gene expression.

Circadian Rhythms

A **circadian rhythm** is a cycle of biological activity that repeats every 24 hours or so (circadian means "about a day"). Circadian rhythms are the outcome of cycles of gene expression. These cycles arise from regulatory loops: networks of transcription factors affecting expression of the genes that encode them. Daily environmental cues set the loops in motion, but an established circadian rhythm will continue for a while even after the cues stop.

Our sleep–wake cycle is one example of a circadian rhythm, and it is the reason we get jet lag after traveling to a different time zone. Plants have circadian rhythms too (FIGURE 10.8). Cyclic shifts in gene expression underlie cyclic shifts in plant metabolism, for example between daytime starch-building reactions and nighttime starch-consuming reactions. Components of rubisco, photosystem II, ATP synthase, and other proteins used in photosynthesis are among the many gene products produced during the day and broken down at night. Similarly, many molecules used at night are broken down during the day.

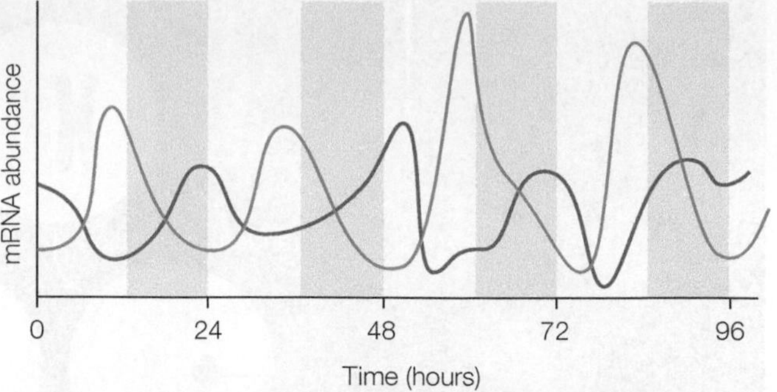

mRNA abundance
0
24
48
72
96
Time (hours)

FIGURE 10.8 Circadian cycles of gene expression in plant temperature stress responses. Molecules that help the plants tolerate cold nights or hot days are produced in circadian cycles, just before they are needed. The blue line shows averaged expression of 46 genes involved in cold tolerance; the red line, averaged expression of 30 genes involved in heat tolerance. Plants were kept in constant light; gray bars show nighttime hours.

Lactose Metabolism in Bacteria

Prokaryotes do not undergo development, so they have no need for mechanisms that orchestrate it. However, they do respond to environmental changes by adjusting gene expression—mainly at transcription. For example, when a preferred nutrient becomes available, a prokaryotic cell begins transcribing genes whose products allow the cell to use the nutrient. When the nutrient is no longer available, transcription of those genes stops. Thus, the cell does not waste energy and resources producing gene products that are not needed.

In bacteria, genes that are used together often occur in a group, one after the next on the chromosome. One promoter precedes the group, so all of the genes are transcribed together into a single RNA strand. Thus, their transcription can be controlled in a single step. This step often involves a repressor binding to an **operator**, which is a type of silencer sequence in DNA. A group of genes together with a promoter and one or more operators that control their transcription are collectively called an **operon**. Operons were discovered in bacteria, but they also occur in archaea and eukaryotes.

Escherichia coli bacteria that live in the gut of mammals dine on nutrients traveling past. Their carbohydrate of choice is glucose, but they can make use of other sugars such as lactose, a disaccharide in milk. An operon called *lac* allows *E. coli* cells to metabolize lactose (FIGURE 10.9). The *lac* operon includes three genes and a promoter that is flanked by two operators ❶.

One gene in the *lac* operon encodes an active transport protein (Section 5.9) that brings lactose across the plasma membrane into the cell. Another encodes an enzyme that breaks the bond between lactose's two monosaccharide monomers: glucose and galactose. (The third gene encodes an enzyme whose function in this context is still being investigated.) Bacteria make these three proteins only when lactose is present.

When lactose is not present, a repressor binds to the two operators and twists the region of DNA with the promoter into a loop ❷. RNA polymerase cannot access the twisted-up promoter, so the *lac* operon's genes cannot be transcribed.

When lactose is present, some of it is converted to another sugar that binds to the repressor and changes its shape (an example of allosteric regulation, Section 5.5). The altered repressor releases the operators, and the looped DNA unwinds ❸. The promoter is now accessible to RNA polymerase, and transcription of lactose-metabolizing genes begins ❹.

In bacteria, glucose metabolism is a shorter pathway than lactose metabolism, so it requires less energy and fewer resources. Accordingly, when both lactose and glucose are present, the cells will use up all of the available glucose before switching to lactose metabolism.

Like *E. coli*, individual cells composing the bodies of multicelled organisms also regulate expression of genes involved in ongoing metabolic processes. These cells adjust gene expression for their own needs, and also for the needs of the whole organism. Thus, they respond to nutritional cues as well as to input from other parts of the body (via hormones, for example).

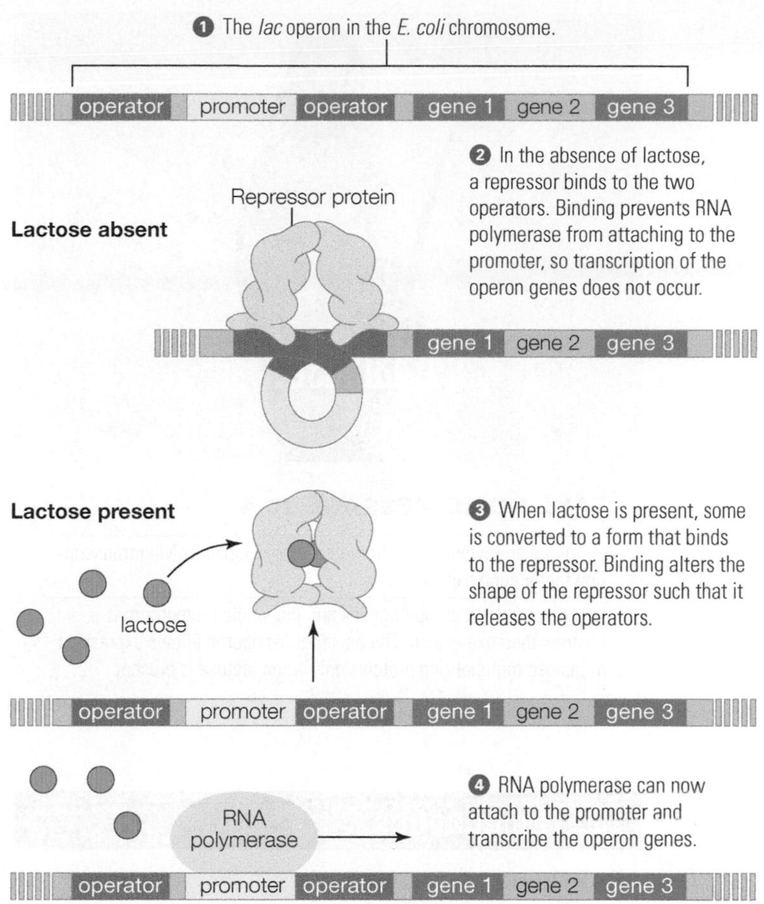
❶ The *lac* operon in the *E. coli* chromosome.
operator
promoter
operator
gene 1
gene 2
gene 3
Lactose absent
Repressor protein
❷ In the absence of lactose, a repressor binds to the two operators. Binding prevents RNA polymerase from attaching to the promoter, so transcription of the operon genes does not occur.
Lactose present
lactose
❸ When lactose is present, some is converted to a form that binds to the repressor. Binding alters the shape of the repressor such that it releases the operators.
RNA polymerase
❹ RNA polymerase can now attach to the promoter and transcribe the operon genes.

FIGURE 10.9 Example of metabolic control in bacteria: the lactose (*lac*) operon on a bacterial chromosome. The operon consists of a promoter flanked by two operators, and three genes for lactose-metabolizing proteins.

Lactose Metabolism in Humans

Humans and other mammals can break down lactose in milk, but most do so only when young. An individual's ability to digest lactose ends at a certain age that depends on the species. In the majority of humans worldwide, this metabolic change occurs at about age five, when transcription of the gene for lactase slows. Lactase is an enzyme that breaks apart lactose into glucose and galactose; cells in the intestinal lining secrete it into the small intestine during digestion. These monosaccharides are absorbed directly by the small intestine, but lactose and other disaccharides are not. Thus, when lactase production slows, undigested lactose passes through the small intestine into the large intestine. Huge numbers of bacteria that inhabit the large intestine normally subsist on cellulose and other complex carbohydrates passing through, but they much prefer lactose. The cells switch their *lac* operon genes, and the resulting abundance of glucose fuels rapid fermentation. Gaseous products of this pathway accumulate quickly in the large intestine, distending its wall and causing pain. Lactose also disrupts the large intestine's solute–water balance, and diarrhea results. These symptoms characterize a common condition known as lactose intolerance.

About one-third of humans can digest milk into adulthood, but this was not always the case. Analyses of DNA from well-preserved skeletons show that the vast majority of adult humans living about 8,000 years ago in Europe were lactose intolerant. Around that time, a mutation appeared in the DNA of prehistoric people inhabiting a region between what is now central Europe and the Balkans. This mutation allowed its bearers to continue digesting milk as adults, and it spread rapidly to the rest of the continent along with the practice of dairy farming. Today, most adults of northern and central European ancestry are able to digest milk because they carry this mutation, a single base-pair substitution that doubles the effectiveness of a lactase gene enhancer. Other mutations in the same enhancer arose independently in North Africa, southern Asia, and the Middle East. Some people descended from these populations can continue to digest milk as adults.

circadian rhythm (sir-KAY-dee-un) A biological activity that is repeated about every 24 hours.

operator Part of an operon; a DNA binding site for a repressor.

operon (OPP-er-on) Group of genes together with a promoter–operator DNA sequence that controls their transcription.

got lactase?

TAKE-HOME MESSAGE 10.4

✔ Circadian rhythms arise from regulatory loops involving transcription factor gene expression.

✔ An operon is a group of genes and the single promoter that controls their expression. The bacterial *lac* operon allows expression of lactose-metabolizing proteins only when lactose is present.

10.5 Epigenetics

LEARNING OBJECTIVES

- Explain why a DNA methylation is passed to all of a cell's descendants.
- List some environmental factors that can affect an individual's DNA methylation patterns.

You learned in Section 10.2 that transcription can be inhibited by histone methylation. Methylation of nucleotide bases in a promoter also inhibits transcription, often more permanently than histone modifications.

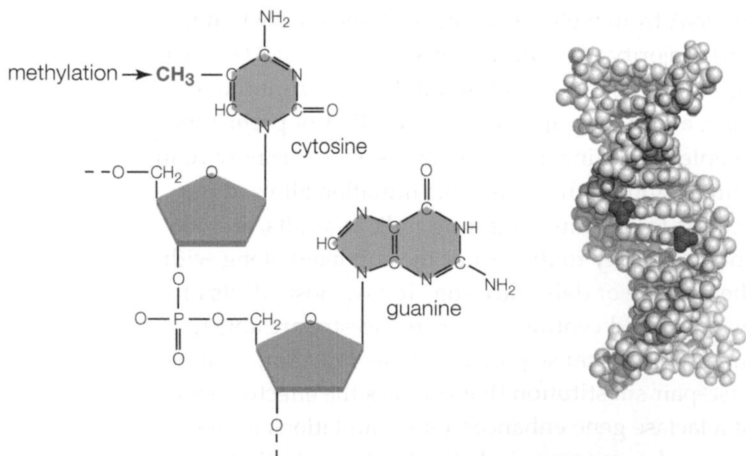
methylation → CH_3
NH_2
cytosine
guanine
CH_2
NH_2
NH
O

FIGURE 10.10 DNA methylation. Left, a methyl group (red) is most often attached to a cytosine that is followed by a guanine. Right, a model of DNA shows methylations (red) on both strands. When the cytosine on one strand is methylated, enzymes methylate the cytosine on the other strand. This is why a methylation tends to persist in a cell's descendants.

In eukaryotes, methyl groups are usually added to a cytosine that is followed by a guanine (**FIGURE 10.10**). When a cytosine on one DNA strand becomes methylated, enzymes methylate the cytosine on the other strand. Once a particular nucleotide has become methylated in a cell's DNA, it will usually stay methylated in all of the DNA of the cell's descendants. This is an outcome of semiconservative replication (Section 8.5): If the parental strand is methylated, then enzymes methylate the new strand too.

DNA methylation is necessarily a part of differentiation, so it begins very early in embryonic development. Genes actively expressed in cells of an embryo become silenced as their promoters get methylated, and this silencing part of the selective gene expression that drives differentiation. Each cell's DNA continues to acquire methylations during development.

Between 3 and 6 percent of the DNA is methylated in normal, differentiated body cells, but which sites are methylated varies by the individual. This is because methylation is influenced by environmental factors experienced during an individual's lifetime. Consider how patterns of DNA methylation in an individual's cells change after exposure to environmental contaminants. For example, cigarette smoke substantially alters methylation of certain promoters in a pattern that also occurs in cancer cells. Synthetic compounds such as BPA and phthalates (found in plastics) alter methylation patterns of a number of genes, and this is thought to contribute to the adverse health effects associated with exposure to these chemicals.

Methylation patterns also change depending on nutritional intake, starting at conception. Offspring conceived during a period of highly restricted or excessive nourishment end up with dramatically different patterns of methyl groups attached to certain genes; the resulting differences in gene expression give rise to substantial differences in glucose and fatty acid metabolism. Methylation patterns associated with cancer, obesity, and other health conditions can be reversed by certain bioactive molecules particularly abundant in foods such as soybeans, kale, blueberries, turmeric, garlic, green tea, and even chocolate (Chapter 30 returns to the health benefits of compounds made by plants).

These and other factors that influence DNA methylation patterns can have multigenerational effects. When an organism reproduces, it passes its DNA to offspring. Methylation of parental DNA is normally "reset" in gametes (eggs and sperm), with new methyl groups being added and old ones being removed. After

epigenetic Refers to potentially heritable modifications to DNA that affect gene expression without changing the DNA sequence.

Data Analysis Activities

Effect of Paternal Grandmother's Food Supply on Infant Mortality Widely available historical data on periods of famine show that before the industrial revolution, a failed harvest in one autumn often led to severe food shortages the following winter. Retrospective studies have correlated infant mortality with the abundance of food during a grandparent's childhood. FIGURE 10.11 shows results from one of these studies.

1. Compare the mortality risk of girls whose paternal grandmothers ate well at age 2 with girls whose grandmothers experienced famine at the same age. Which girl was more likely to die early? How much more likely was she to die?
2. Children have a period of slow growth around age 9. What trend in this data can you see around that age?
3. There was no correlation between early death of a male child and eating habits of his paternal grandmother, but there was a strong correlation with the eating habits of his paternal grandfather. What does this tell you about the location of epigenetic changes that gave rise to these data?

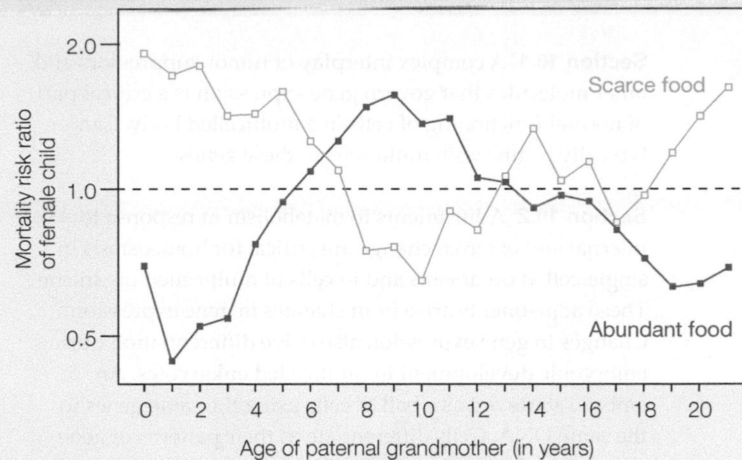
2.0
1.0
0.5
Mortality risk ratio of female child
Scarce food
Abundant food
0 2 4 6 8 10 12 14 16 18 20
Age of paternal grandmother (in years)

FIGURE 10.11 **Relative risk of early death of a female child**, correlated with the age at which her paternal grandmother experienced a winter with a food supply that was scarce (blue) or abundant (red) during childhood. The dotted line represents no difference in risk of mortality. A value above the line means increased risk; one below the line indicates reduced risk.

fertilization, DNA is demethylated again in the zygote. However, this reprogramming does not remove all of the parental methyl groups, so some methylations acquired during an individual's lifetime are passed to future offspring—as are their effects. Transient exposure of pregnant rats to vinclozolin (a common fungicide) reduced fertility in four generations of male offspring. The rats' DNA sequence did not change, but their chromosomes had altered methylation patterns in dozens of different regions. In humans, methylation patterns in hundreds of chromosomal regions change after exposure to lead (a toxic metal). These changes are passed to children—and grandchildren.

Potentially heritable modifications to DNA that affect gene expression without changing the DNA sequence are said to be **epigenetic**. DNA methylations are epigenetic, as are histone modifications. Epigenetic modifications are not considered to be evolutionary because they do not involve DNA sequence changes (we return to evolutionary processes in Chapter 17). However, epigenetic inheritance can adapt offspring to an environmental challenge much more quickly than evolution, and the changes can be reversed much more quickly after an environmental challenge has faded.

TAKE-HOME MESSAGE 10.5

- ✔ Potentially heritable modifications to DNA that change gene expression without changing DNA sequence are said to be epigenetic.
- ✔ DNA methylations accumulate throughout an individual's lifetime.
- ✔ Exposure to environmental factors can change an individual's DNA methylation patterns. Methylations are epigenetic, so these patterns can be passed to offspring.

10.1 Between You and Eternity (revisited)

Inherited mutations in *BRCA* genes increase the risk of breast cancer, but they cause a minority of cases. More than 90 percent of breast cancers have no *BRCA* mutations. However, in half of these cancers, expression of the *BRCA1* gene is suppressed because its promoter has become methylated. The protein product of *BRCA1* is part of regulatory mechanisms that maintain the structure and number of chromosomes in a dividing cell. It participates directly in DNA repair, so any reduction in *BRCA1* gene expression also reduces a cell's capacity to repair damaged DNA. Mutations accumulate quickly and cause the cells to malfunction—a hallmark of cancer cells. The other hallmark is uncontrolled division. Normal cells in breast and ovarian tissue divide mainly during cyclic renewals. Abundant estrogen receptors trigger events that cause these cells to divide, but only when the receptors are activated. Interactions with many other molecules, including the *BRCA1* gene product, regulate estrogen receptor activity. When a cell's DNA is damaged, the *BRCA1* gene product inhibits the activity of estrogen receptors until the damage can be repaired. In cells that underproduce the *BRCA1* gene product, estrogen receptors are not inhibited, and cell divisions are uncontrolled. ●

Section 10.1 A complex interplay of tumor suppressors and other molecules that govern gene expression is a critical part of normal functioning of cells in a multicelled body. Cancer typically begins with mutations in these genes.

Section 10.2 Adjustments to metabolism in response to internal and external change are critical for homeostasis in single-celled organisms and in cells of multicelled organisms. These adjustments arise from changes in gene expression. Changes in gene expression also drive differentiation during embryonic development in multicelled eukaryotes. An embryo starts out as a ball of cells using the same genes in the same DNA. Cells differentiate as their patterns of gene expression change during development. As the cells start using different subsets of their genes, they become different in form and function.

Gene expression can be switched on or off, or speeded or slowed, at each step from transcription to delivery of the gene's product to its final destination. Many molecules interact together or in opposition to bring about this control.

Transcription factors such as **activators** and **repressors** influence transcription by binding directly to chromosomal DNA. Sequences in the DNA such as promoters, **enhancers**, and silencers are binding sites for these regulatory proteins. Small segments of RNA can prevent transcription of a particular mRNA, a mechanism called **RNA interference**.

Section 10.3 During development, cells of an embryo organize into tissues and body parts that form at particular times and in particular regions. The process is orchestrated by transcription factors produced in cascades of gene expression.

Gene expression cascades begin with **master regulators** expressed in different regions of a fertilized egg. The products of these genes diffuse away in gradients. The gradients determine where in the embryo other transcription factor genes are expressed. The products of these genes diffuse away in gradients, and so on. The embryo's cells begin to differentiate as they begin to express different genes. Gene expression patterns establish the polarity of the embryo and the identity of its different regions.

Eventually, the cascade of gene expression activates **homeotic genes**, each a master regulator whose expression triggers the development of a specific body part. Homeotic genes function in similar ways among evolutionarily distant organisms, so they provide evidence of shared ancestry.

Mutations in any of the genes in a developmental cascade of expression can result in abnormal body form or function. The cascades were deciphered using **knockouts** that disable one gene at a time. Mutations in the earliest genes in a cascade result in the greatest disruption of embryonic development.

Examples of developmental outcomes of gene expression control include the formation of flowers in plants. Overlapping expression of floral identity genes governs development of the specialized parts of a flower. Differentiating cells in a shoot tip form sepals, petals, stamens, or carpels depending on the floral identity genes they express.

In cells of female mammals, one of the two X chromosomes is condensed as a **Barr body**. Most of the genes on a Barr body are not expressed. This **X chromosome inactivation** balances the expression of X chromosome genes between female (XX) and male (XY) mammals, a mechanism of **dosage compensation**. X chromosome inactivation occurs because interactions with the noncoding RNA product of the *XIST* gene shut down the one X chromosome that transcribes it.

Expression of the *SRY* gene on the Y chromosome triggers development of male traits in mammals.

Section 10.4 **Circadian rhythms** result from cyclic changes in gene expression brought about by regulatory loops. In these loops, transcription factors affect expression of the genes that encode them.

Prokaryotes, being single-celled, do not undergo development. Most of their control over gene expression involves reversible adjustments to transcription. The adjustments occur in response to environmental conditions, especially nutrient availability. For example, bacteria have a ***lac* operon** that includes three genes whose products allow the cell to use lactose when it is present. Two **operators** that flank the promoter are binding sites for a repressor that blocks transcription. The presence of lactose removes the repressor so the genes can be transcribed.

Most adult humans do not produce the enzyme that breaks down lactose. When undigested lactose enters the large intestine, resident bacteria switch on their *lac* operons and begin rapid fermentation. Gaseous products of their fermentation accumulate quickly, causing the bloating and pain associated with lactose intolerance.

Section 10.5 Methylation of nucleotide bases in a promoter suppresses gene expression. DNA methylation patterns change during development, and also during an individual's lifetime as a result of environmental factors. These patterns can be passed to offspring. Methylations and other potentially heritable modifications to DNA that affect gene expression without changing the DNA sequence are **epigenetic**.

SELF-QUIZ

Answers in Appendix VII

1. Gene expression does not vary by _______ .
   a. cell type
   b. extracellular conditions
   c. stage of development
   d. the genetic code

2. Binding of _______ to _______ in DNA can increase the rate of transcription of specific genes.
   a. activators; repressors
   b. activators; enhancers
   c. repressors; operators
   d. repressors; enhancers

3. Muscle cells differ from bone cells because they _______ .
   a. carry different genes   c. are eukaryotic
   b. express different genes   d. are different ages
4. Proteins that influence RNA synthesis by binding directly to DNA are called _______ .
   a. promoters   c. operators
   b. transcription factors   d. enhancers
5. Mechanisms that govern gene expression do not operate during _______ .
   a. transcription   c. translation
   b. RNA processing   d. knockouts
6. Control over eukaryotic gene expression drives _______ .
   a. transcription factors   c. embryonic development
   b. nutrient availability   d. all of the above
7. Homeotic gene expression _______ .
   a. occurs via bacterial operons
   b. maps out the overall body plan in embryos
   c. triggers formation of a specific body part
8. A gene that is knocked out is _______ .
   a. duplicated   c. expressed
   b. inactivated   d. a master regulator
9. Which of the following includes all of the others?
   a. homeotic genes   c. *SRY* gene
   b. master regulators   d. *PAX6*
10. The expression of *ABC* genes _______ .
   a. occurs in layers of an onion
   b. controls flower formation
   c. causes mutations in flowers
11. During X chromosome inactivation, _______ .
   a. female cells shut down
   b. RNA sticks to a chromosome
   c. pigments form
12. A cell with a Barr body is _______ .
   a. a bacterium   c. from a female mammal
   b. a sex cell   d. infected by the Barr virus
13. Operons _______ .
   a. only occur in bacteria
   b. include multiple genes
   c. involve selective gene expression
14. Which of the following statements is *in*correct?
   a. Some gene expression patterns can be passed to an individual's offspring.
   b. Expression of a master regulator triggers a gene expression cascade.
   c. X chromosome inactivation is necessary for normal development of male mammals.
15. Match the terms with the most suitable description.

___ operon	a. makes a man out of you
___ Circadian rhythm	b. binding site for repressor
___ Barr body	c. can be epigenetic
___ differentiation	d. inactivated X chromosome
___ mRNA zip code	e. controls multiple genes
___ DNA methylation	f. localization mechanism
___ *eyeless*	g. speeds transcription
___ activator	h. required for eye formation
___ *SRY* gene	i. effect of regulatory loops
___ operator	j. cells become specialized

CRITICAL THINKING

1. Explain why some—but not all—of an organism's genes are expressed.
2. Do the same mechanisms that govern gene expression operate in bacterial cells and eukaryotic cells? Explain your answer.

3. A "calico" cat (left) is mainly white with patches of black and orange fur.
   a. Almost all calico cats are female. Why?
   b. Why do you think that, of the few male calico cats that are born, almost all are sterile.
   c. Each calico cat has a unique pattern of white, black, and orange fur. Review Figure 10.7, then propose a mechanism that would give rise to the white fur.
4. The photos below show flowers from two *Arabidopsis* plants. The plant on the left is wild-type (unmutated); the other carries a mutation that causes its flowers to have sepals and petals instead of stamens and carpels. The mutation inactivated one of the plant's *ABC* floral identity genes. Refer to Figure 10.8 and decide which gene (*A*, *B*, or *C*) has been inactivated.

wild-type
mutated

CREDITS: CT #3, Thy Le/Shutterstock; #4, left, Peggy Greb/USDA/Science Source; right, © Jose Luis Riechmann.

11 HOW CELLS REPRODUCE

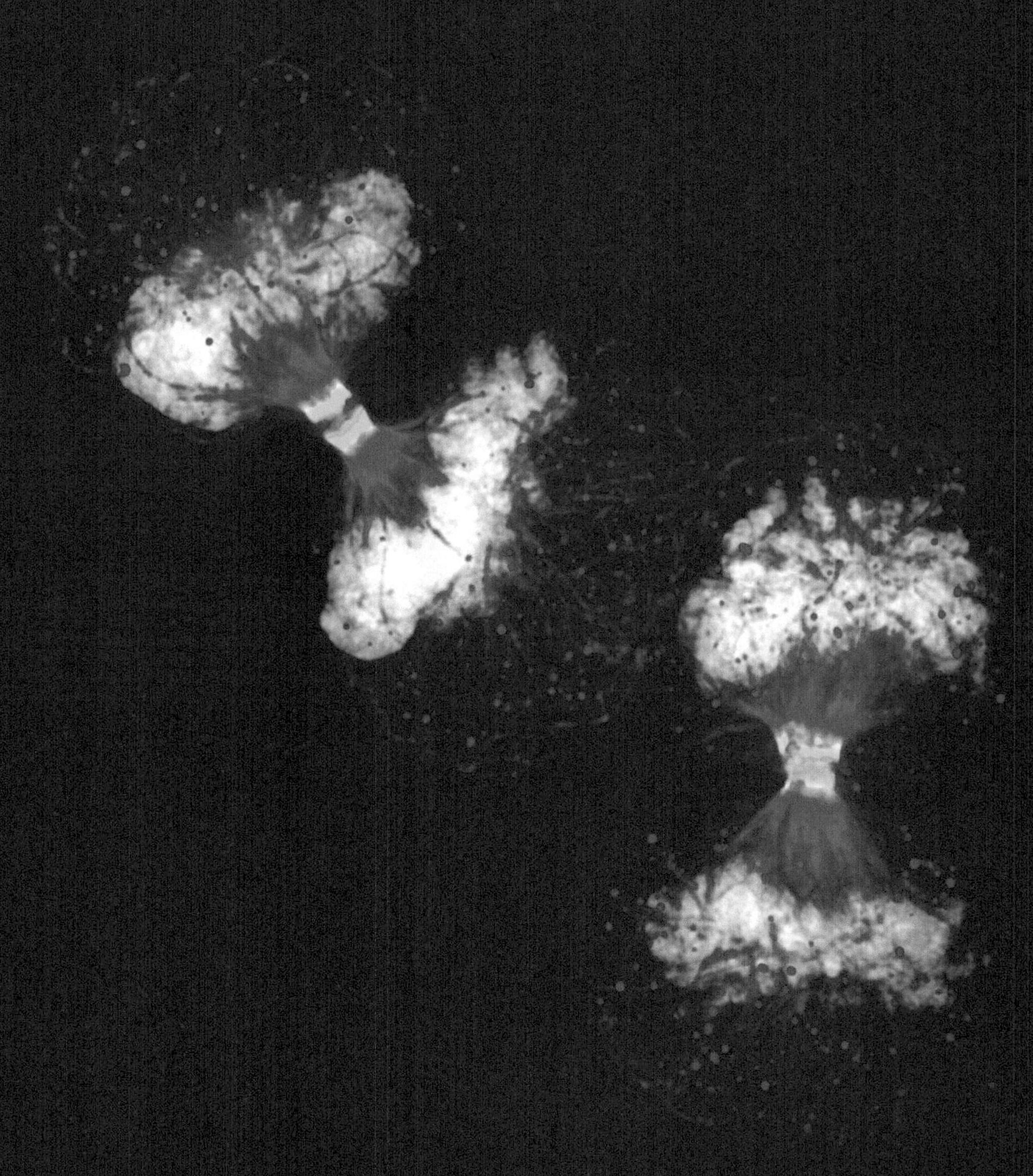

CORE CONCEPTS

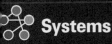

Evolution

Evolution underlies the unity and diversity of life.

Cells of all multicelled eukaryotes reproduce by mitosis and cytoplasmic division. Together, these processes are the basis of growth, tissue repair, and in some species asexual reproduction. Because mitosis links one generation of cells to the next, it is a mechanism by which the continuity of life occurs. All multicelled organisms use similar molecules to regulate their cell cycle.

Information Flow

Living systems store, retrieve, transmit, and respond to information essential for life.

Cells transmit essential genetic information to their offspring in the form of DNA. When a cell divides by mitosis, it passes a complete set of chromosomes to each of its two descendant cells. The descendant cells are identical to one another and to the parent cell.

Systems

Complex properties arise from interactions among components of a biological system.

Mechanisms that control gene expression orchestrate the timing and coordination of molecular events required for proper cellular function. Mutations that alter the molecules involved in these mechanisms may have complex effects on the cell, tissue, or whole organism. Cancer arises from disruptions in molecular interactions governing the cell cycle.

Links to Earlier Concepts

Before beginning this chapter, be sure you understand cell theory (Section 4.2) and structures (4.4, 4.5, 4.8, 4.9); chromosome structure (8.4); and DNA replication (8.5) and repair (8.6). What you know about membrane proteins (5.7), the consequences of mutations (9.6), and gene expression control (10.2) will help you understand how cancer develops.

11.1 Henrietta's Immortal Cells

Each human starts out as a fertilized egg. By the time of birth, that single cell has given rise to about a trillion other cells, all organized as a human body. Even in the adult, billions of cells divide every day as new cells replace worn-out ones. However, human cells grown in the laboratory tend to divide a limited number of times and die within weeks; a cell division limit caps the number of times that body cells can divide. As early as the mid-1800s, researchers were trying to coax human cells to keep dividing outside of the body because they realized immortal cell lineages—cell lines—would allow them to study human diseases (and potential cures) without experimenting on people.

The quest to create a human cell line continued unsuccessfully for over one hundred years. George and Margaret Gey had been trying for nearly thirty years when, in 1951, their assistant Mary Kubicek prepared a new sample of human cancer cells. Mary named the cells HeLa, after the first and last names of the patient from whom the cells had been taken.

The HeLa cells began to divide, again and again. The cells were astonishingly vigorous, quickly coating the inside of their bottles and consuming their nutrient broth. Four days later, there were so many cells that the researchers had to transfer them to more bottles. The cells were dividing every twenty-four hours and coating the inside of the bottles within days. Sadly, cancer cells in the patient were dividing just as fast. Only six months after she had been diagnosed with cervical cancer, malignant cells had invaded tissues throughout her body. Two months after that, Henrietta Lacks, a young woman from Baltimore, was dead.

Even after Henrietta had passed away, her cells lived on in the Geys' laboratory. The Geys discovered how to grow poliovirus in HeLa cells, a practice that enabled them to determine which strains of the virus cause polio. That work was a critical step in the development of polio vaccines, which have since saved millions of lives.

Henrietta Lacks was just thirty-one, a wife and mother of five, when runaway cell divisions of cancer killed her. Her cells, however, are still dividing, again and again, more than sixty years after she died. Frozen away in tiny tubes, HeLa cells continue to be shipped among laboratories all over the world. They are still widely used to investigate cancer, viral growth, protein synthesis, the effects of radiation, and countless other processes important in medical research. HeLa cells helped several researchers win Nobel Prizes, and they even traveled into space for experiments on satellites.

CREDIT: (opposite) Dr. Paul D. Andrews/University of Dundee.

11.2 Multiplication by Division

LEARNING OBJECTIVES

- Describe the main events in the eukaryotic cell cycle.
- Explain why we say mitosis maintains the chromosome number.
- Distinguish sister chromatids from homologous chromosomes.
- Identify two body processes that occur by mitosis.
- Describe the purpose of cell cycle checkpoints.

A life cycle is a sequence of recognizable stages that occur during an organism's lifetime, from the first cell of the new individual until its death. Multicelled organisms and free-living cells have life cycles, but what about cells that make up a multicelled body? Biologists consider such cells to be individually alive, each with its own lifetime. A cell's life passes through a series of recognizable intervals and events collectively called the **cell cycle** (FIGURE 11.1).

A typical body cell spends most of its life in interphase ❶. During **interphase**, the cell increases in size, replicates its DNA, and produces components required for division. These activities proceed in three major stages of interphase: G_1, S, and G_2. G_1 and G_2 were named "Gap" phases because outwardly they seem to be periods of inactivity, but they are not. A cell starts out life in G_1 ❷. During this phase, the cell roughly doubles the number of its cytoplasmic components, and it produces the molecules it will need for DNA replication (Section 8.5). Cells in G_1 may temporarily or permanently exit the cell cycle by entering a reversible, non-dividing state called G_0 ❸. Differentiated cells going about their metabolic business are in G_0. DNA replication occurs after G_1, in the S phase ❹. During the next phase, G_2 ❺, cells make proteins and other cellular components needed for division. The remainder of the cell cycle consists of the division process. Because each cell arises by division of a preexisting cell (Section 4.2), the cycle begins and ends with division.

A eukaryotic cell reproduces by dividing, and each of its two cellular offspring receives DNA packaged in a nucleus. Thus, eukaryotic cell division necessarily includes a mechanism by which the nucleus divides. **Mitosis** ❻ is the mechanism of nuclear division that maintains the chromosome number. Consider how your body cells are diploid, which means their nucleus contains two sets of chromosomes (Section 8.4). One set came from your mother; the other, from your father. Thus, your cells have *two of each type* of chromosome. Except for a pairing of nonidentical sex chromosomes (XY) in males, the chromosomes of each pair are homologous. **Homologous chromosomes** have the same length, shape, and genes (*hom–* means alike). Note that homologous chromosomes differ from sister chromatids, which are identical molecules of DNA produced

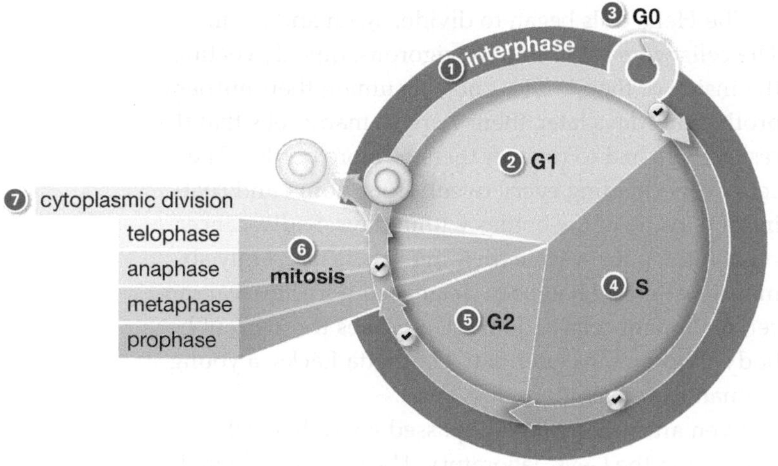
③ G0
① interphase
② G1
④ S
⑤ G2
⑥ mitosis
⑦ cytoplasmic division
telophase
anaphase
metaphase
prophase

FIGURE 11.1 A eukaryotic cell cycle. G_1, G_0, S, and G_2 are part of interphase. The length of these intervals differs among cells.

❶ A cell spends most of its life in interphase.

❷ Each cell starts out life in G_1, during which it increases in size and produces molecules required for DNA replication.

❸ A cell in G_1 may exit the cycle to temporarily or permanently stay in a state called G_0. A cell in G_0 is going about its metabolic business and not preparing to divide.

❹ During S, a cell copies its chromosomes by DNA replication.

❺ During G_2, the cell produces proteins and other components that are necessary for division.

❻ The nucleus divides during mitosis.

❼ After mitosis, the cytoplasm may divide. Each descendant cell begins the cycle anew, in G_1.

✔ Built-in checkpoints stop the cycle from proceeding until certain conditions are met.

FIGURE 11.2 How mitosis maintains chromosome number. For clarity, only one homologous pair of chromosomes in a diploid cell is shown.

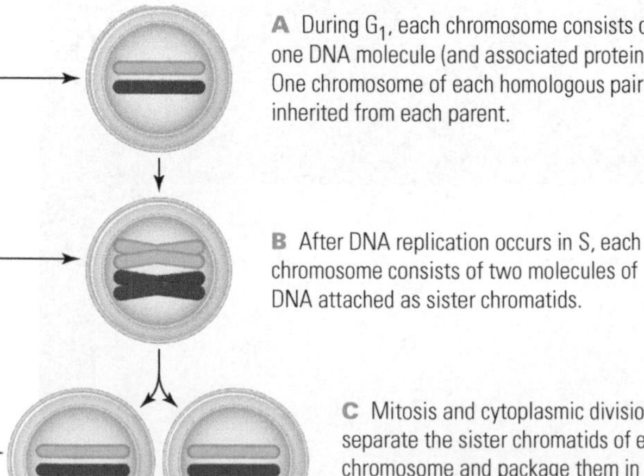

A During G_1, each chromosome consists of one DNA molecule (and associated proteins). One chromosome of each homologous pair was inherited from each parent.

B After DNA replication occurs in S, each chromosome consists of two molecules of DNA attached as sister chromatids.

C Mitosis and cytoplasmic division separate the sister chromatids of each chromosome and package them in the nuclei of two new cells. Each cell has the chromosome number of the parent.

by DNA replication. Sister chromatids are attached at the centromere to form one X-shaped chromosome:

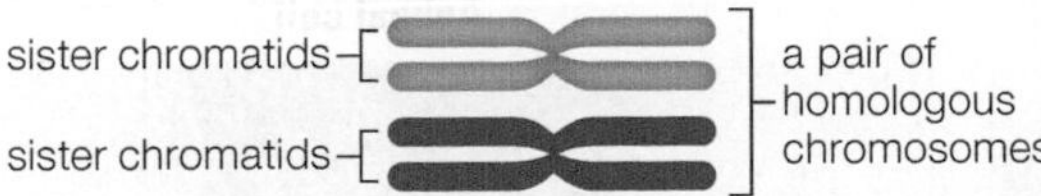
sister chromatids
sister chromatids
a pair of homologous chromosomes

By contrast, the two homologous chromosomes of a pair are not identical. Each was inherited from two parents that differ genetically, so their DNA sequence differs slightly (Section 12.2 returns to this topic).

Mitoses Maintains the Chromosome Number

A diploid human body cell has a chromosome number of 46; when it divides by mitosis, both cellular offspring also have a chromosome number of 46. However, if only the total number of chromosomes mattered, then one cell might inherit, say, two pairs of chromosome 22 and no chromosome 9. A cell cannot function normally unless it has a proper complement of chromosomes.

When a cell is in G_1, each of its chromosomes consists of one double-stranded DNA molecule (**FIGURE 11.2A**). DNA replication occurs during S, so by G_2 each chromosome consists of two double-stranded DNA molecules attached as sister chromatids (**FIGURE 11.2B**). The sister chromatids stay attached until mitosis is almost over, and then they are pulled apart and packaged into two separate nuclei.

When sister chromatids are pulled apart, each becomes an individual chromosome with one double-stranded DNA molecule. Thus, the two new nuclei that form by mitosis contain the same number *and types* of chromosomes as the parent cell. When the cytoplasm divides ❼, these nuclei are packaged into separate cells, each with the chromosome number of the parent (**FIGURE 11.2C**).

Controlling the Cell Cycle

When a cell divides is determined by molecules that regulate gene expression (Section 10.2). Like the accelerator of a car, some of these molecules cause the cell cycle to advance. Others are like brakes that prevent the cycle from proceeding. This control occurs at checkpoints built into the cell cycle. Protein products of "checkpoint genes" interact to ensure (for example) that the cell's DNA has been copied completely, that it is not damaged, and even that enough nutrients are available to support division. For example, at a checkpoint that operates in S, the cell's DNA is monitored for damage during replication. Checkpoint gene products that function as sensors recognize damaged DNA and bind to it. Upon binding, they trigger other events that slow DNA replication and enhance expression of genes involved in DNA repair. After the problem has been corrected, the brakes are lifted and the cell cycle proceeds normally. If the problem remains uncorrected, other checkpoint gene products may initiate a series of events that eventually cause the cell to self-destruct (you will read more about cell suicide, or apoptosis, in Section 42.4).

Why Cells Divide by Mitosis

In multicelled organisms, mitosis and cytoplasmic division are the basis of body formation and increases in size during development: Continuing mitotic divisions of embryonic cells as they differentiate produce tissues and body parts of the embryo. Juvenile development into the adult form also occurs by mitosis, as do ongoing replacements of damaged or dead cells. Scrape your knee, and mitotically dividing cells in your skin repair the wound. Skin cells and many other types of body cells are continually replaced by mitosis.

Mitosis is part of **asexual reproduction**, a reproductive mode in which offspring are produced by one parent only. Many plants can reproduce this way, for example when a stem breaks off and takes root in soil. Many single-celled eukaryotes also reproduce asexually by mitosis. Prokaryotes do not have a nucleus and do not undergo mitosis (we discuss their reproduction in Section 20.5).

asexual reproduction Reproductive mode by which offspring arise from a single parent only.
cell cycle The collective series of intervals and events of a cell's life, from the time it forms until it divides.
homologous chromosomes In a nucleus, chromosomes with the same length, shape, and genes.
interphase In a eukaryotic cell cycle, the interval during which a cell grows, roughly doubles the number of its cytoplasmic components, and replicates its DNA in preparation for division.
mitosis (my-TOE-sis) Mechanism of nuclear division that maintains the chromosome number.

TAKE-HOME MESSAGE 11.2

- ✔ A cell's life occurs in a recognizable series of intervals and events collectively called the cell cycle. The cycle begins and ends with division by mitosis. An interval called interphase separates each division.
- ✔ Division of a eukaryotic cell typically occurs in two steps: nuclear division followed by cytoplasmic division.
- ✔ Homologous chromosomes are inherited from each of two parents. A diploid body cell has a full set of homologous chromosomes.
- ✔ Each cell produced by mitosis has the same number and types of chromosomes as the parent.
- ✔ The timing of cell division is determined by molecules that regulate gene expression.
- ✔ Built-in checkpoints that delay or stop the cell cycle allow problems to be corrected before the cycle proceeds.

FIGURE 11.3 Mitosis. Micrographs show plant cells (onion root, left) and animal cells (fertilized eggs of a roundworm, right). A diploid ($2n$) animal cell with two chromosome pairs is illustrated.

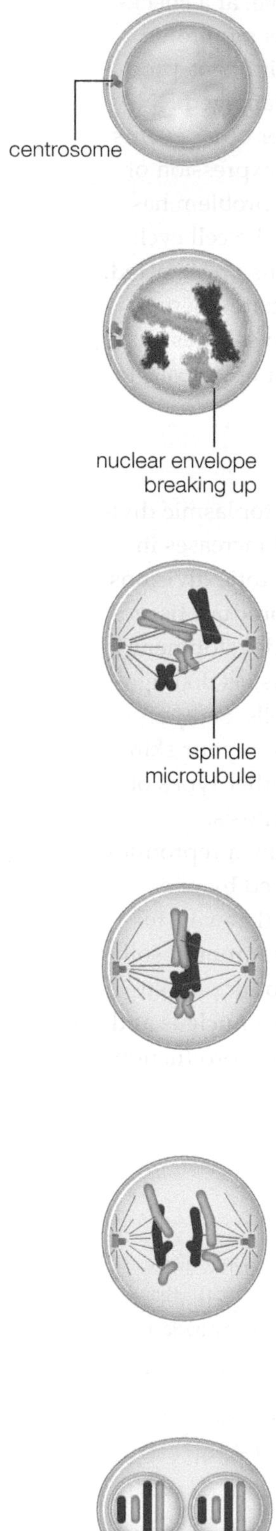

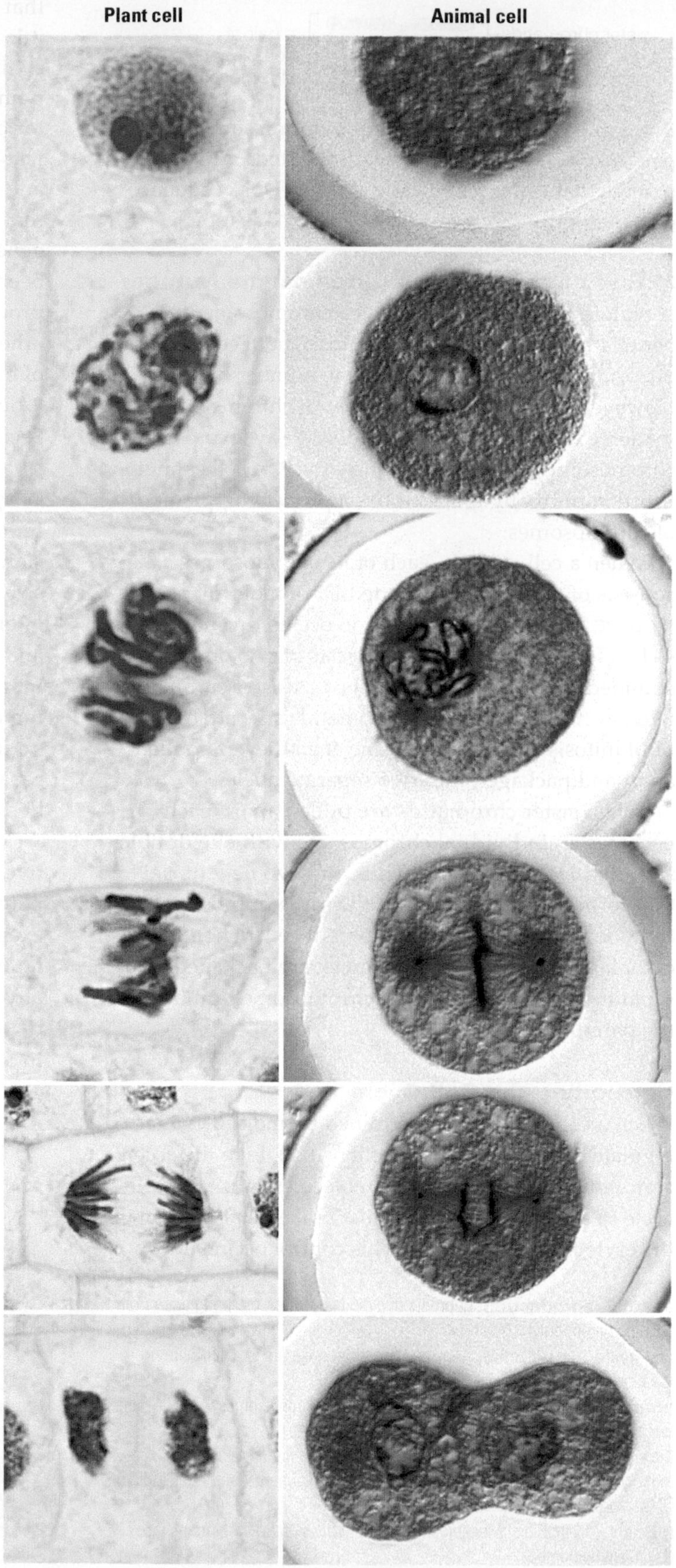

Plant cell | Animal cell

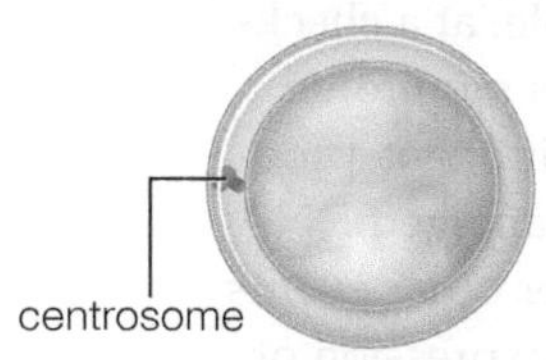
centrosome

1 Interphase

Interphase cells are shown for comparison, but interphase is not part of mitosis. The nuclear envelope is intact, and the loosened chromosomes are not easily visible under a light microscope.

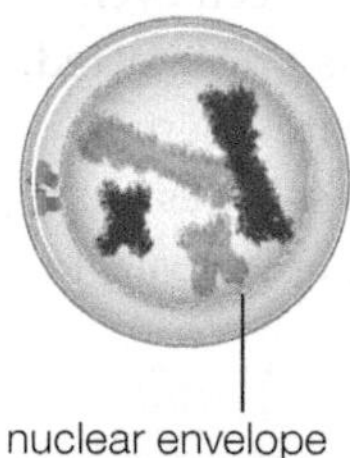
nuclear envelope breaking up

2 Early Prophase

Mitosis begins as DNA replication stops. The chromosomes begin to appear grainy as they start to condense. The nuclear envelope begins to break up and the centrosome gets duplicated.

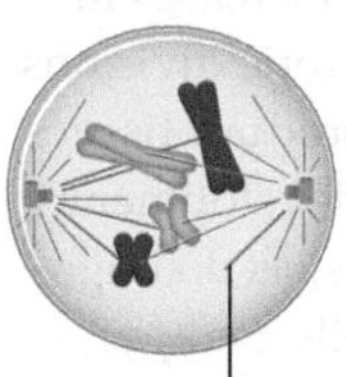
spindle microtubule

3 Prophase

The duplicated chromosomes become visible as they condense. One of the two centrosomes moves to the opposite side of the cell as the nuclear envelope breaks up completely. Spindle microtubules assemble and bind to chromosomes at the centromere, attaching sister chromatids to opposite centrosomes.

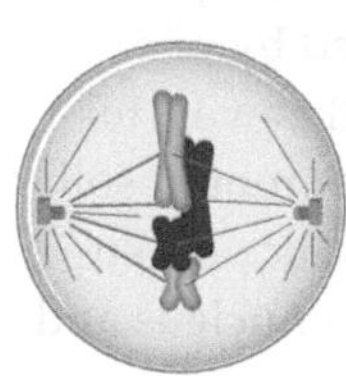

4 Metaphase

All of the chromosomes are aligned midway between the spindle poles.

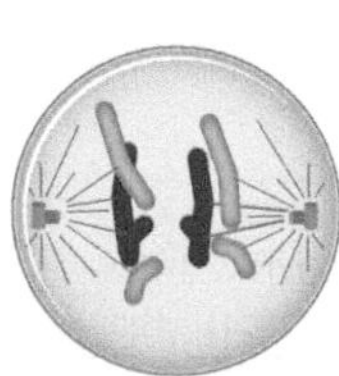

5 Anaphase

Spindle microtubules separate the sister chromatids and move them toward opposite spindle poles. Each sister chromatid has now become an individual, unduplicated chromosome.

6 Telophase

The chromosomes reach opposite sides of the cell and loosen up. Mitosis ends when a new nuclear envelope forms around each cluster of chromosomes.

11.3 A Closer Look at Mitosis

LEARNING OBJECTIVES

- List the main events that occur during each stage of mitosis.
- Describe the role of microtubules in nuclear division.
- Explain how mitosis packages two complete sets of chromosomes into two new nuclei.

When a cell is in interphase, its chromosomes are loosened to allow transcription and DNA replication (**FIGURE 11.3**, opposite). Loosened chromosomes are spread out, so they are not easily visible under a light microscope ❶. In preparation for nuclear division, the chromosomes begin to pack tightly ❷. Transcription and DNA replication end as the chromosomes condense into their most compact "X" forms. This condensation keeps the chromosomes from getting tangled and breaking during mitosis.

Mitosis proceeds in four main stages: prophase, metaphase, anaphase, and telophase. During **prophase**, the chromosomes condense so much that they become visible under a light microscope ❸. "Mitosis" is from *mitos*, the Greek word for thread, after the threadlike appearance of chromosomes during the process.

Most animal cells have a centrosome, which is a region of protein-rich cytoplasm that contains a pair of centrioles (Section 4.8) and many tubulin subunits. A centrosome becomes duplicated during prophase, and the copy moves to the opposite side of the cell. Microtubules then begin to assemble and lengthen from the two centrosomes (or from other structures in cells with no centrosomes). The lengthening microtubules form a **spindle**, a temporary structure that moves chromosomes during nuclear division (**FIGURE 11.4**). The areas where the spindle originates on both sides of the cell are called spindle poles.

As prophase continues, the nuclear envelope breaks up. Lengthening spindle microtubules penetrate the nuclear region and attach to the chromosomes at their centromeres. By the end of prophase, one sister chromatid of each chromosome has become attached to microtubules extending from one spindle pole; the other sister chromatid has become attached to microtubules extending from the other spindle pole. The opposing sets of microtubules then begin a tug-of-war. By adding or losing subunits, the microtubules drag the chromosome toward the middle of the cell. When all the microtubules are the same length, the chromosomes are aligned midway between spindle poles ❹. The alignment marks **metaphase** (from *meta*, the ancient Greek word for between).

spindle pole

FIGURE 11.4 **The spindle.**

A spindle is visible in this cell, a freshly fertilized egg (of a worm, *Cerebratulus*) undergoing mitosis. Microtubules are visible in yellow; chromosomes, in blue. The bright spot on the edge of the egg is an unlucky (extra) sperm.

FIGURE IT OUT
In which stage of mitosis is this cell?

Answer: Prophase

During **anaphase**, the sister chromatids detach and move toward opposite sides of the cell ❺. When sister chromatids separate, each becomes an individual, unduplicated chromosome.

Telophase begins when one set of chromosomes arrives at each spindle pole and forms a cluster ❻. Each cluster has the same number and kinds of chromosomes as the parent cell nucleus had: two of each type of chromosome, if the parent cell was diploid. A new nuclear envelope forms around the clusters as the chromosomes loosen up. At this point, mitosis is over.

TAKE-HOME MESSAGE 11.3

✔ DNA replication occurs before mitosis. As mitosis begins, each chromosome consists of two DNA molecules attached at the centromere as sister chromatids.

✔ The four main stages of mitosis are prophase, metaphase, anaphase, and telophase.

✔ In prophase, the chromosomes condense. A spindle forms and attaches the sister chromatids of each chromosome to opposite poles.

✔ At metaphase, all chromosomes are aligned midway between the two spindle poles.

✔ In anaphase, sister chromatids separate and move toward opposite spindle poles. At the moment of separation, each sister chromatid becomes an individual chromosome.

✔ In telophase, two clusters of chromosomes reach opposite spindle poles. A new nuclear envelope forms around each cluster.

anaphase (ANN-uh-faze) Stage of mitosis during which sister chromatids separate and move toward opposite spindle poles.
metaphase (MEH-tuh-faze) Stage of mitosis at which all chromosomes are aligned midway between spindle poles.
prophase (PRO-faze) Stage of mitosis during which chromosomes condense and become attached to a newly forming spindle.
spindle (SPIN-dull) Temporary structure that moves chromosomes during nuclear division; consists of microtubules.
telophase (TELL-uh-faze) Stage of mitosis during which chromosomes arrive at opposite spindle poles, decondense, and become enclosed by a new nuclear envelope.

11.4 Cytoplasmic Division

LEARNING OBJECTIVES

- Explain why cytokinesis is a necessary part of the cell cycle.
- Describe and compare cytokinesis in animal and plant cells.

After mitosis, a cell's cytoplasm divides in most cases. The mechanism of cytoplasmic division, which is called **cytokinesis**, requires many molecules and structures that are in place well before mitosis ends. For example, during mitosis, spindle microtubules deliver vesicles from Golgi bodies and the plasma membrane to the future plane of division. New membranes that forms from the fusion of these vesicles will partition the two new cells at the end of division.

Animal cells have a mesh of cytoskeletal elements under the plasma membrane (Section 4.8). This cell cortex reorganizes during anaphase to form a band of microfilaments and motor proteins that wraps around the cell (**FIGURE 11.5** ①). The band is called a contractile ring because it contracts after telophase, pulling the membrane inward as it does. The sinking plasma membrane becomes visible on the outside of the cell as an indentation called a **cleavage furrow** ②. Vesicles guided to the plane of division by spindle microtubules add membrane to the furrow region, and this prevents the plasma membrane from stretching at the furrow. When the contractile ring shrinks to its smallest diameter, spindle microtubules inside the ring are cut, and more vesicles fuse to form new membrane that partitions the two new cells ③.

A contractile ring minimizes the amount of new plasma membrane required to partition descendant cells, but this strategy works only in animal cells—which have no walls. Plant cells, by contrast, have a stiff cell wall around their plasma membrane (Section 4.9), so their strategy of cytokinesis differs (**FIGURE 11.6**). Before mitosis, a ring of parallel microtubules forms around the future plane of division ①. These microtubules reorganize during mitosis to form the spindle, and by late anaphase have guided vesicles to the plane of division. During telophase, the vesicles start to fuse into a flat **cell plate** ②. The membranes of the merged vesicles will become the plasma membranes separating the two descendant cells, and the vesicles' contents

cell plate Disk-shaped structure that forms and partitions descendant cells during cytokinesis in a plant cell.
cleavage furrow In a dividing animal cell, the indentation where cytoplasmic division will occur.
cytokinesis (site-oh-kye-KNEE-sis) Cytoplasmic division.
senescent (suh-NESS-ent) Refers to a metabolically active but malfunctioning cell stuck in an irreversibly arrested cell cycle.
telomere (TELL-a-meer) Region of noncoding DNA at the end of a chromosome. Proteins that bind to it protect the chromosome from end damage.

① A ring of microfilaments and motor proteins (yellow) forms under the plasma membrane during anaphase. This contractile ring wraps around the cell.

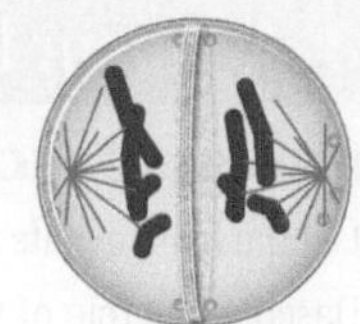

② After telophase, the contractile ring pulls the cell surface inward, forming a cleavage furrow as it constricts. Spindle microtubules guide vesicles to the plane of division.

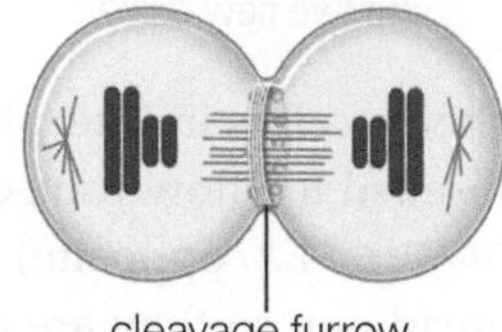
cleavage furrow

③ When the contractile ring is at its smallest diameter, spindle microtubules inside of it are cut. Vesicles fuse to form new membrane that partitions the two new cells.

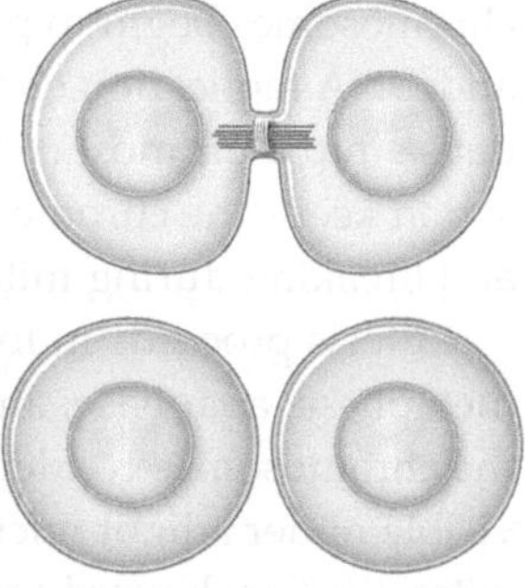

FIGURE 11.5 Cytoplasmic division of an animal cell.
The small blue dots represent vesicles.

① A network of parallel microtubules (orange) determines the future plane of division in a plant cell. During G_2, this network develops into a dense ring of microtubules around the nucleus at the future plane of division.

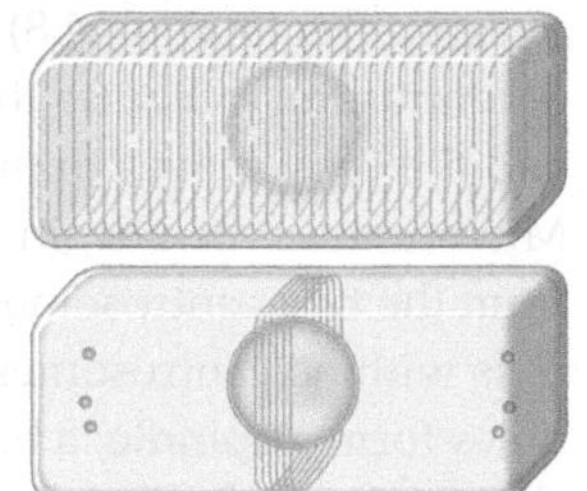

② During mitosis, microtubules of the ring reorganize to form a spindle. The spindle is perpendicular to the former ring. In late anaphase, spindle remnants begin to guide vesicles to the future plane of division. By telophase, the vesicles have begun to fuse and form a cell plate.

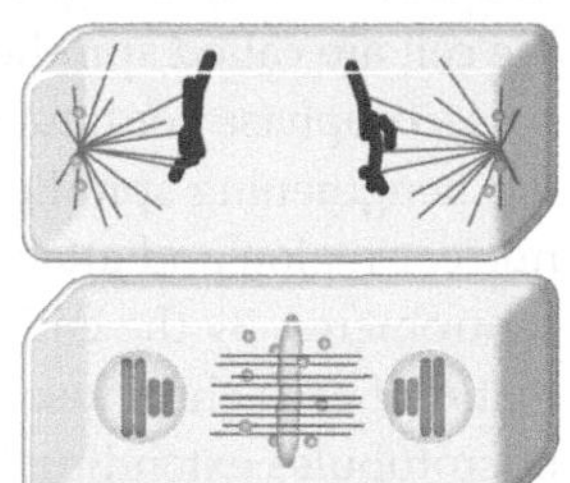

③ The cell plate expands along the plane of division as more vesicles merge with it. When the cell plate fuses with the plasma membrane, the cytoplasm is partitioned so two new cells form.

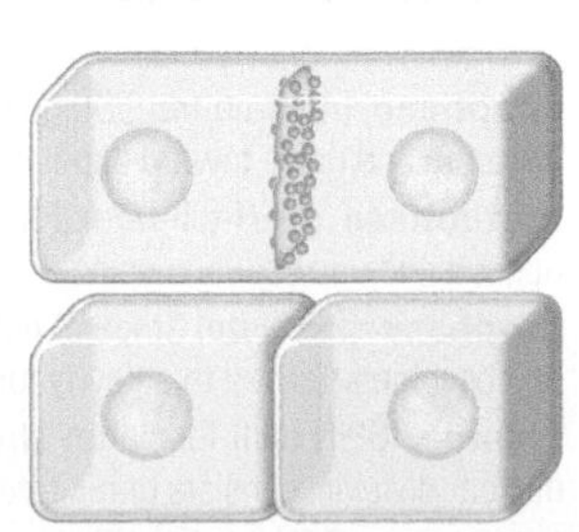

FIGURE 11.6 Cytoplasmic division of a plant cell.
The small blue dots represent vesicles.

will be converted to walls. The cell plate expands at its edges as more vesicles merge, until it reaches and fuses with the cell's plasma membrane ❸.

TAKE-HOME MESSAGE 11.4

✔ In eukaryotes, the cell cytoplasm typically divides after mitosis. Two descendant cells form, each with its own nucleus.

✔ In animal cell cytokinesis, a contractile ring pinches the cytoplasm in two. In plant cell cytokinesis, a cell plate that forms in the middle of the cell partitions the cytoplasm.

11.5 Marking Time with Telomeres

LEARNING OBJECTIVES

- Describe the relationship between telomeres and senescence.
- Explain the cell division limit and its fail-safe function.

The short life span of Dolly, the first clone produced by SCNT (Section 8.7), may have been the result of short telomeres. **Telomeres** are regions of noncoding DNA at the ends of eukaryotic chromosomes (FIGURE 11.7). Vertebrate telomeres consist of a short DNA sequence, 5′-TTAGGG-3′, repeated perhaps thousands of times. Proteins that bind to these repeats shield the ends of a chromosome from potential damage. A chromosome's exposed ends can be damaged by normal cellular processes, including repair mechanisms that inappropriately recognize them as double-stranded DNA breaks.

Telomeres shorten every time a cell divides. When they get too short, the protective proteins can no longer bind to them, and DNA at the ends of the chromosomes becomes exposed. Checkpoint gene products then bind to the DNA and target it for repair. One of these gene products is called p53, and it operates at the checkpoint in G_1. When a cell's DNA is damaged, this checkpoint moves the cell cycle into G_0 until repair enzymes fix the damage. These enzymes can repair minor DNA damage, but they cannot fix short telomeres, so the cell stays in G_0.

Cell cycle arrest due to telomere shortening tends to be irreversible. A cell in this state eventually dies or becomes **senescent**: permanently stuck in G_0, metabolically active but not functioning properly. The result can be dangerous, in the same way it is dangerous for the accelerator of a car to get stuck even when the brakes are working. A senescent cell becomes progressively abnormal, for example enlarging and releasing too much of a protein that promotes inflammation. It will continue interacting dysfunctionally with other cells until the immune system detects and eliminates it.

Most body cells can divide only a certain number of times before they senesce, and that number—the cell division limit—varies by species. Human cells, for example, can divide about 50 times before senescing. The cell division limit is a fail-safe mechanism in case a cell loses control over the cell cycle and begins to divide again and again. Limiting the number of divisions prevents these rogue cells from overrunning the body.

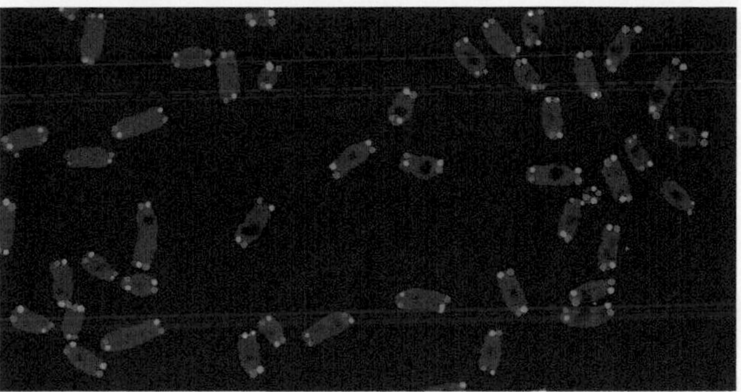

FIGURE 11.7 **Telomeres.** The bright dots at the end of each DNA strand in these duplicated chromosomes show telomere sequences.

Telomere shortening may be the only physiological barrier to immortality, and it is a likely part of the mechanism that sets an organism's life span. This is because, as an animal ages, the number of senescent cells increases in its tissues. When enough of these cells accumulate, they disrupt tissue and organ function: a common problem with advancing age. Dolly's DNA came from a cell taken from an adult sheep. When Dolly was two years old, her telomeres were as short as those of a six-year-old sheep—the exact age of the animal that had been her genetic donor.

A few normal cells in an adult body retain the ability to divide indefinitely. Their descendants replace cell lineages that eventually die out when they reach their division limit. These cells are called stem cells, and they are immortal because they continue to make telomerase—an enzyme that lengthens telomeres by adding TTAGGG repeats to them. In an adult body, only stem cells make telomerase; differentiated cells normally do not. Like stem cells, cancer cells characteristically have high levels of telomerase, which is why, also like stem cells, they can divide indefinitely.

TAKE-HOME MESSAGE 11.5

✔ Telomeres are noncoding sequences at the ends of chromosomes. Protective proteins that bind to them shield the ends from damage.

✔ Telomeres shorten with every division of a normal, differentiated body cell. When they become too short, the cell stops dividing and becomes senescent.

✔ Telomere shortening imposes a limit on cell division—a fail-safe mechanism in case a cell loses control over its cell cycle.

✔ Stem cells can divide indefinitely because they continue to make telomerase that rebuilds their telomeres.

CREDIT: (7) ISM/Phototake.

11.6 When Mitosis Is Dangerous

LEARNING OBJECTIVES

- Explain why mutations can give rise to neoplasms.
- Use an example to describe tumor suppressors.
- Describe some hallmarks of malignant cells.

Some mutations can affect a checkpoint gene so that its protein product no longer works properly. Other mutations can disrupt the gene's expression, so a cell makes too much or too little of its product. In both cases, the cell begins dividing when it should not. If the checkpoint malfunction allows the cell cycle to continue even when the DNA is damaged, more mutations acccumulate. Signaling mechanisms that cause abnormal cells to die may stop working. The mutations are inherited by the cell's descendants, which form a **neoplasm**—an accumulation of abnormally dividing cells. A neoplasm that forms a lump in the body is called a **tumor**, but the two terms are used interchangeably. Almost all mutations that affect cell cycle regulation occur in body cells. Those that occur in reproductive cells can be passed to offspring, which is why some types of tumors run in families.

Some tumor-causing mutations increase the activity or number of molecules that stimulate cell division. Genes in which such mutations can occur are called **proto-oncogenes**. A proto-oncogene that acquires a tumor-causing mutation becomes an **oncogene**. Consider a proto-oncogene called *EGFR*, which encodes a plasma membrane receptor for epidermal growth factor (EGF). EGF is a protein, and like other **growth factors** it stimulates cell division. When the EGF receptor binds to EGF, it becomes activated and triggers mitosis. When EGF is not present, the receptor is in an inactive form and does not trigger mitosis. Tumor-causing mutations in *EGFR* change the receptor so that it triggers mitosis even in the absence of EGF. Most neoplasms carry mutations resulting in an overactivity or overabundance of this particular receptor.

Other tumor-causing mutations decrease the activity or number of tumor suppressors (Section 10.1). p53 is a tumor suppressor that is missing or malfunctioning in more than half of human cancers. The protein product of the *BRCA1* gene is also a tumor suppressor. Among other functions, this protein dampens expression of *EGFR*. Thus, mutations that result in underproduction of the BRCA1 protein also result in overproduction of the EGF receptor. A cell with too many EGF receptors is overly responsive to EGF, so it divides too often.

Viruses such as HPV (human papillomavirus) cause a cell to produce proteins that bind to tumor suppressors and prevent them from working. Benign

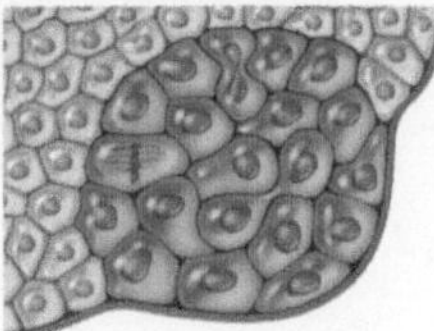

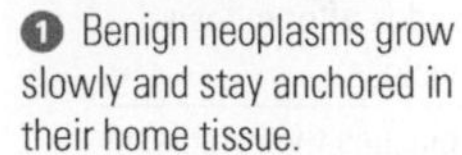

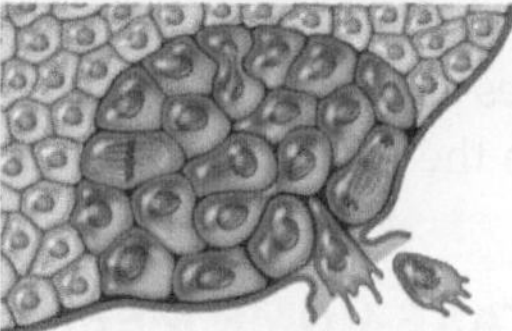

❶ Benign neoplasms grow slowly and stay anchored in their home tissue.

❷ Cells of malignant neoplasms grow quickly and can break away from their home tissue.

❸ Malignant cells become attached to the wall of a lymph vessel or blood vessel (as shown here). They release digestive enzymes that create an opening in the wall, then enter the vessel.

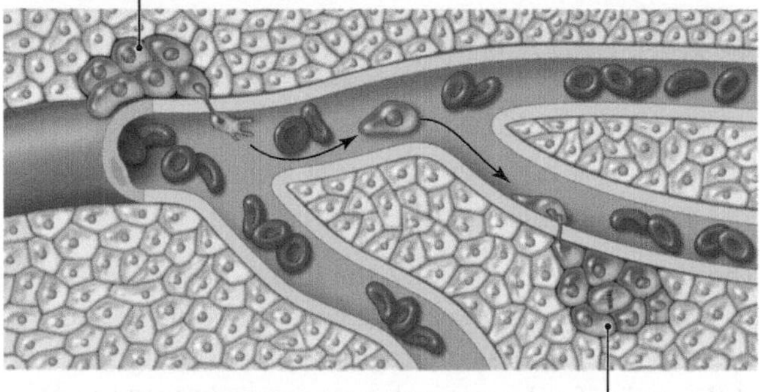

❹ The cells move through the vessel, then exit the same way they got in. Migrating cells may start growing in other tissues, a process called metastasis.

FIGURE 11.8 Neoplasms and metastasis.

neoplasms called warts are an outcome of HPV infection. A benign neoplasm grows very slowly and stays anchored in its home tissue (**FIGURE 11.8** ❶). By contrast, a malignant neoplasm progressively worsens, and is dangerous to health. Malignant cells typically display the following three characteristics.

First, like cells of all neoplasms, malignant cells divide abnormally. Controls that usually keep cells from getting overcrowded in tissues are lost in malignant cells, so their populations may reach extremely high densities with cell division occurring very rapidly.

Second, the cytoplasm and plasma membrane of malignant cells are altered. Both are indications of cellular malfunction. The cytoskeleton may be shrunken, disorganized, or both. Malignant cells typically have an abnormal chromosome number, with some chromosomes present in multiple copies, and others missing or damaged. The balance of metabolism is often shifted,

cancer Disease that occurs when a malignant neoplasm physically and metabolically disrupts body tissues.
growth factor Molecule that stimulates mitosis and differentiation.
metastasis (meh-TASS-tuh-sis) The process in which malignant cells spread from one part of the body to another.
neoplasm (KNEE-oh-plaz-um) An accumulation of abnormally dividing cells.
oncogene (ON-ko-jean) Gene with a mutation that results in the overabundance or overactivity of a molecule that stimulates cell division.
proto-oncogene Gene that, by mutation, can become an oncogene.
tumor (TOO-murr) A neoplasm that forms a lump.

Data Analysis Activities

HeLa Cells Are a Genetic Mess HeLa cells can vary in chromosome number. Defects in proteins that orchestrate cell division result in descendant cells with too many or too few chromosomes, an outcome that is one of the hallmarks of cancer cells. The panel of chromosomes in **FIGURE 11.9**, originally published in 1989, shows all of the chromosomes in a single metaphase HeLa cell.

1. What is the chromosome number of this HeLa cell?
2. How many extra chromosomes does this cell have, compared to a normal human body cell?
3. Can you tell that this cell came from a female? How?

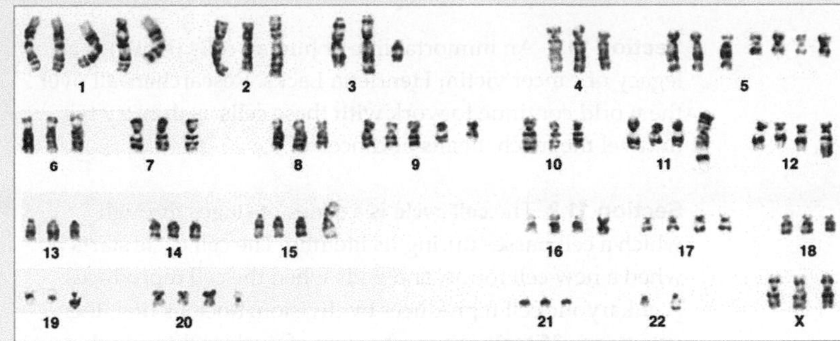
1
2
3
4
5
6
7
8
9
10
11
12
13
14
15
16
17
18
19
20
21
22
X

FIGURE 11.9 Karyotype of HeLa showing chromosomes in one cell.

such as ATP synthesis occurring mainly by fermentation rather than aerobic respiration.

Altered or missing proteins impair the function of the plasma membrane of malignant cells. For example, these cells do not stay anchored properly in tissues ❷ because their plasma membrane adhesion proteins are defective or missing. Malignant cells can slip easily into and out of vessels of the circulatory and lymphatic systems ❸. By migrating through these vessels, the cells can establish neoplasms elsewhere in the body ❹. The process in which malignant cells break loose from their home tissue and invade other parts of the body is called **metastasis**. Metastasis is the third hallmark of malignant cells.

The disease called **cancer** occurs when the abnormally dividing cells of a malignant neoplasm disrupt body tissues, both physically and metabolically. Unless chemotherapy, surgery, or another procedure eliminates malignant cells from the body, they can put an individual on a painful road to death. Each year, cancer causes 15 to 20 percent of all human deaths in developed countries. The good news is that mutations in multiple checkpoint genes are required to transform a normal cell into a malignant one, and such mutations may take a lifetime to accumulate. Lifestyle choices such as not smoking and avoiding exposure of skin to sunlight can reduce one's risk of acquiring mutations in the first place. Some neoplasms can be detected with periodic screening such as gynecology or dermatology exams. Detected early, many types of malignant neoplasms can be removed before metastasis occurs.

TAKE-HOME MESSAGE 11.6

- ✔ Neoplasms and tumors form when checkpoint mechanisms fail and cells begin dividing when they should not.
- ✔ Mutations in multiple checkpoint genes can give rise to a malignant neoplasm that gets progressively worse.
- ✔ Cancer is a disease that occurs when the abnormally dividing cells of a malignant neoplasm disrupt body tissues.
- ✔ Although some mutations are inherited, lifestyle choices and early intervention can reduce one's risk of cancer.

11.1 Henrietta's Immortal Cells (revisited)

HeLa cells were used in early tests of Paclitaxel, a drug that keeps microtubules from disassembling. Spindle microtubules that cannot shrink also cannot properly position the cell's chromosomes during metaphase, and this triggers a checkpoint that stops the cell cycle. Shortly thereafter, the cell either exits mitosis or dies. Frequent divisions make cancer cells more vulnerable to this microtubule poison than normal cells.

A more recent example of cancer research is shown in the chapter opening photo, a fluorescence micrograph that shows some of the internal components of two HeLa cells in the process of mitosis. Chromosomes appear white, and spindle microtubules are red. Blue fluorescence shows the location of an enzyme, Aurora B, that is the product of a checkpoint gene called *AURKB*. Aurora B associates with centromeres and helps mediate several key events in mitosis, including attachment of sister chromatids to microtubules of opposite spindle poles. *AURKB* is overexpressed in HeLa cells, as it is in cells of many other aggressive cancers. When a cell with an abnormally high level of Aurora B divides, its offspring have an abnormally high risk of inheriting the wrong number of chromosomes—a common feature of cancer cells.

CREDIT: (9) Courtesy of © Dr. Thomas Ried, NIH and the American Association for Cancer Research.

STUDY GUIDE

Section 11.1 An immortal line of human cells (HeLa) is a legacy of cancer victim Henrietta Lacks. Researchers all over the world continue to work with these cells as they try to unravel the mechanisms of cancer.

Section 11.2 The **cell cycle** is a series of stages through which a cell passes during its lifetime. The cell cycle starts when a new cell forms, and ends when the cell reproduces. A eukaryotic cell reproduces by division: nucleus first, then cytoplasm. **Mitosis** is a mechanism of nuclear division that maintains the chromosome number. For example, mitosis in a diploid cell produces diploid offspring, each with a full set of **homologous chromosomes**. Mitosis is the basis of **asexual reproduction** in many species, and development, growth, and tissue repair in multicelled organisms.

The interval of the cell cycle between cell divisions is **interphase**, and it occurs in three major stages: G_1, S, and G_2. Cells enlarge and make the components required for DNA replication during G_1. Most differentiated body cells exit the cell cycle in G_1, and remain temporarily or permanently in the non-dividing G_0 state. DNA replication occurs during S. Components for division are made during G_2.

Section 11.3 DNA replication occurs before mitosis, so each chromosome consists of two DNA molecules attached as sister chromatids. Mitosis proceeds in four stages: prophase, metaphase, anaphase, and telophase. During **prophase**, the chromosomes condense, the nuclear envelope breaks up, and a **spindle** forms. Microtubules that extend from one spindle pole attach to one chromatid of each chromosome; microtubules that extend from the opposite spindle pole attach to its sister chromatid. These microtubules move each chromosome toward the center of the cell. At **metaphase**, all of the chromosomes are aligned between spindle poles.

During **anaphase**, the sister chromatids of each chromosome separate and move toward opposite spindle poles. Each DNA molecule is now an individual chromosome.

During **telophase**, a complete set of chromosomes reaches each spindle pole and forms a cluster. A nuclear envelope forms around each cluster. Two new nuclei, each with the parental chromosome number, are the result.

Section 11.4 In most cases, **cytokinesis** follows nuclear division. Vesicles guided by microtubules to the future plane of division merge to separate the two new cells produced by cytokinesis. In animal cells, a contractile ring of microfilaments pulls the plasma membrane inward (forming a **cleavage furrow**). When the contractile ring is at its smallest, merging vesicles form new plasma membrane that partitions the two new cells. In plant cells, vesicles merge as a **cell plate** that expands and fuses with the parent cell wall. The cell plate becomes a cross-wall that partitions the two new cells.

Section 11.5 **Telomeres** (regions of noncoding DNA at the end of eukaryotic chromosomes) shorten every time DNA replication occurs. Normal body cells can divide only a certain number of times before their telomeres get too short. Cells that have too-short telomeres become **senescent**. A cell division limit is a fail-safe mechanism in case the cell loses control over its cell cycle.

Section 11.6 The products of checkpoint genes work together to control the cell cycle. These molecules monitor the integrity of the cell's DNA, and can pause the cycle until breaks or other problems are fixed. When checkpoint mechanisms fail, a cell loses control over its cell cycle, and its abnormally dividing descendants form a **neoplasm**. Neoplasms may form lumps called **tumors**.

Genes encoding **growth factor** receptors are examples of **proto-oncogenes**, which means mutations can turn them into tumor-causing **oncogenes**. Mutations in multiple checkpoint genes can give rise to a malignant neoplasm that gets progressively worse. Cells of malignant neoplasms can break loose from their home tissues and colonize other parts of the body, a process called **metastasis**. **Cancer** occurs when malignant neoplasms physically and metabolically disrupt normal body tissues.

SELF-QUIZ

Answers in Appendix VII

1. Mitosis and cytoplasmic division function in ________ .
   a. asexual reproduction of single-celled prokaryotes
   b. development and tissue repair in multicelled species
   c. sexual reproduction in plants and animals
2. A duplicated chromosome has how many chromatids?
3. In the diagrams of the nucleus shown below, fill in the blanks with the name of each interval.

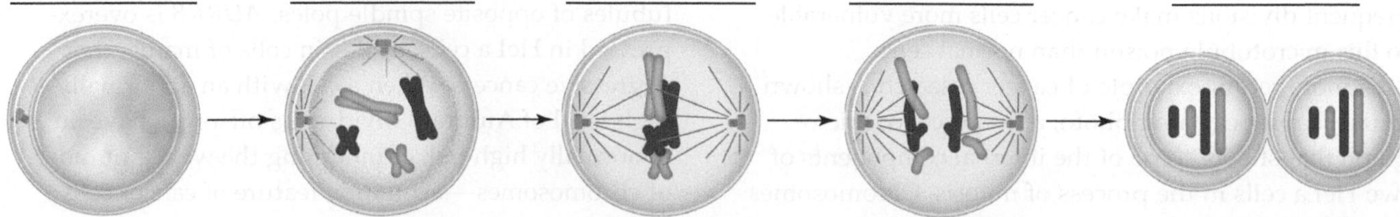

4. Homologous chromosomes _______ .
   a. are inherited from two parents
   b. are sister chromatids
   c. are different in size and in length
5. Most healthy body cells spend the majority of their lives in _______ .
   a. prophase
   b. metaphase
   c. anaphase
   d. telophase
   e. interphase
   f. senescence
6. The spindle attaches to chromosomes at the _______ .
   a. centriole
   b. contractile ring
   c. centromere
   d. centrosome
7. Which is not a stage of mitosis?
   a. prophase
   b. metaphase
   c. interphase
   d. anaphase
8. In intervals of interphase, G stands for _______ .
   a. gap
   b. growth
   c. Gey
   d. gene
9. Interphase is the part of the cell cycle when _______ .
   a. a cell ceases to function
   b. a cell forms its spindle apparatus
   c. a cell grows and replicates its DNA
   d. mitosis proceeds
10. After mitosis, the chromosome number of a descendant cell is _______ the parent cell's.
    a. the same as
    b. one-half of
    c. rearranged compared to
    d. double that of
11. Cytoplasm of a plant cell divides by the process of ____ .
    a. telekinesis
    b. nuclear division
    c. fission
    d. cytokinesis
12. Match each term with its best description.

___ cell plate	a. lump of abnormal cells
___ spindle	b. made of microfilaments
___ tumor	c. divides plant cells
___ cleavage furrow	d. spindle originates here
___ contractile ring	e. dangerous metastatic cells
___ cancer	f. made of microtubules
___ centrosomes	g. indentation
___ telomere	h. shortens with age

13. Match each stage with the events listed.

___ metaphase	a. sister chromatids move apart
___ prophase	b. chromosomes condense
___ telophase	c. new nuclei form
___ interphase	d. chromosomes aligned midway between spindle poles
___ anaphase	e. G_1, S, G_2
___ cytokinesis	f. cytoplasmic division

14. The *BRCA1* gene _______ .
    a. is a checkpoint gene
    b. is a proto-oncogene
    c. encodes a tumor suppressor
    d. all of the above
15. _______ are characteristic of cancer.
    a. Malignant cells b. Neoplasms c. Tumors

CRITICAL THINKING

1. When a cell reproduces by mitosis and cytoplasmic division, does its life end?
2. How is the human life cycle related to the cell cycle of the cells that make up our bodies?

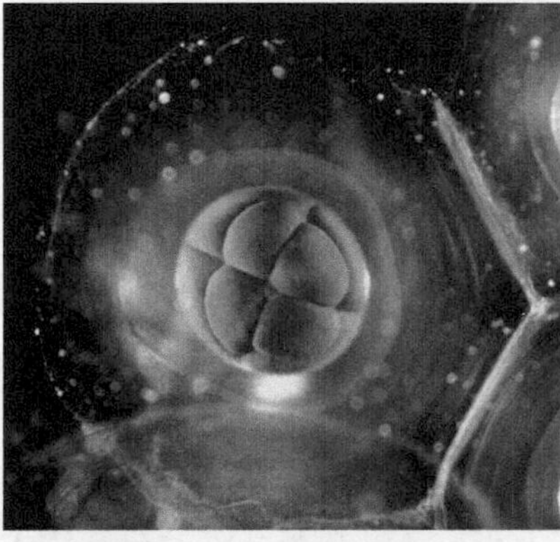

3. The photo on the left shows an early frog embryo after three mitotic divisions of a fertilized egg. This embryo has how many cells?

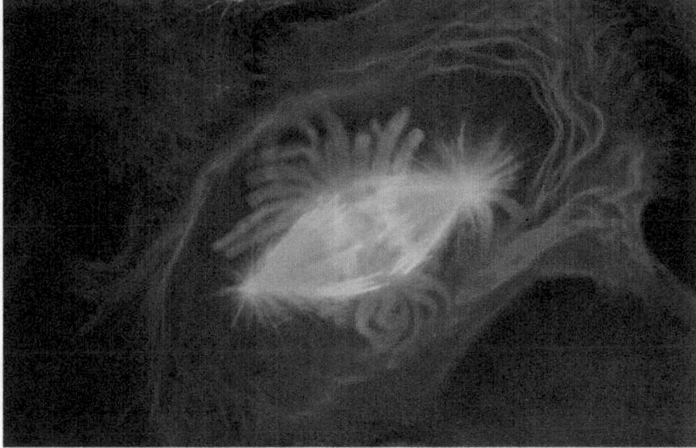

4. The cell above is a lung cell of a salamander. In which stage of mitosis is it? What are the structures visible in green and blue fluorescence?

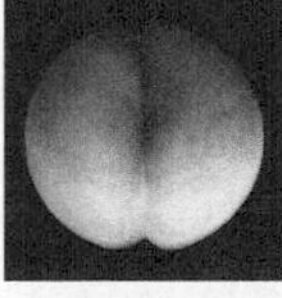

5. The eukaryotic cell at left is in the process of cytoplasmic division. Is this cell from a plant or an animal? How do you know?
6. Exposure to radioisotopes or other sources of radiation can damage DNA. Humans exposed to high levels of radiation face a condition called radiation poisoning. Why do you think that exposure to radiation is used as a therapy to treat some kinds of cancers?

CREDITS: (in text CT) #3, Carolina Biological Supply Company/Phototake; Dr. Conly Rieder; #5, ISM/Phototake.

12 MEIOSIS AND SEXUAL REPRODUCTION

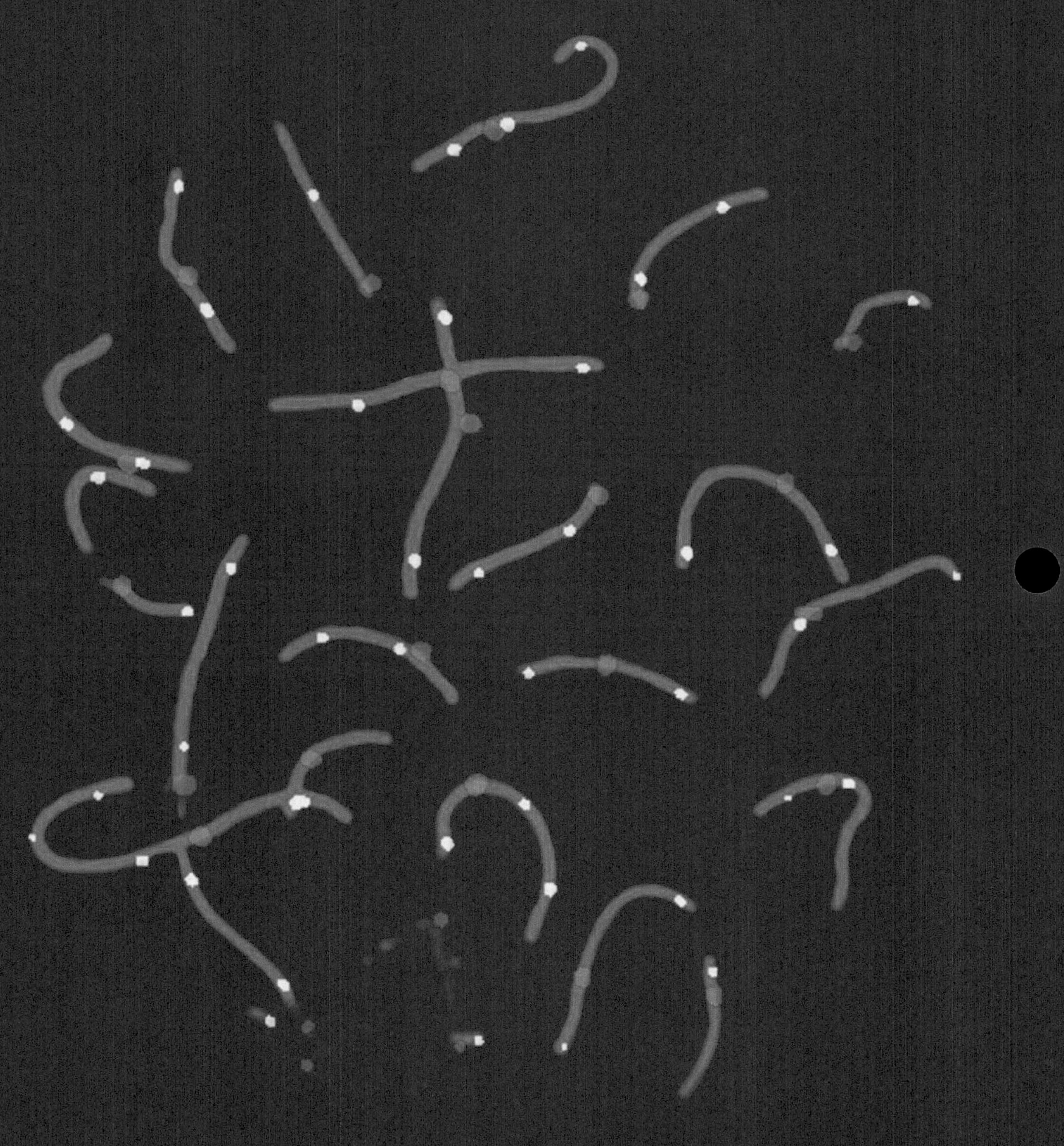

CORE CONCEPTS

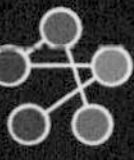

Systems

Complex properties arise from interactions among components of a biological system.

In general, the most resilient biological systems have many different components. Individuals of sexually reproducing species vary in the details of their DNA. Such variation is almost always advantageous, because genetically diverse populations are more resilient to environmental change than populations with low diversity. Asexual reproduction produces genetically identical individuals, so populations of asexual reproducers typically have low genetic diversity.

Information Flow

Living systems store, retrieve, transmit, and respond to information essential for life.

All organisms transmit genetic information to their offspring in the form of DNA. Sexual reproduction is one way in which this occurs, and it involves meiosis (which reduces the chromosome number for forthcoming gametes) and fertilization. Both processes mix genetic information from two parents who differ in the details of their DNA, so sexual reproduction produces offspring genetically different from one another and from the parents.

Evolution

Evolution underlies the unity and diversity of life.

Like building blocks, molecules and processes are often evolutionarily reused in improved pathways that offer enhanced benefit. The outcomes of mitosis and meiosis differ, but the processes are similar. Both use some of the same molecules that are also involved in DNA repair.

Links to Earlier Concepts

This chapter will draw on your knowledge of genes (Section 9.2) and the effects of mutation (8.6), as well as mitosis (11.2, 11.3) and cytoplasmic division (11.4). You will revisit DNA replication (8.5), controls over the cell cycle (11.2), and clones (8.1, 8.7).

12.1 Why Sex?

You learned in Chapter 11 that mitosis and cytoplasmic division are part of asexual reproduction in eukaryotes, but not all species can use this reproductive mode. Many eukaryotes reproduce sexually, either exclusively or most of the time. With **sexual reproduction**, offspring arise from two parents and inherit genes from both.

If the function of reproduction is the perpetuation of one's genes, then an asexual reproducer would seem to win the evolutionary race. When it reproduces, it passes all of its genes to every one of its offspring. Only about half of a sexual reproducer's genes are passed to each offspring. So why is sex so widespread?

Consider how all offspring produced by asexual reproduction are clones: They have the same traits as one another (and their one parent). This consistency is a good strategy in a favorable, unchanging environment; traits that help an organism survive and reproduce do the same for its descendants. However, most environments are constantly changing, and change is not always favorable. Individuals that are identical are equally vulnerable to environmental challenges.

Variation in traits is characteristic of sexual reproducers. Sexual reproduction randomly mixes up the genetic information of two parents with different traits, so it produces offspring that differ from one another and from parents. In a changing environment, diversity in traits gives sexual reproducers the evolutionary edge. An environmental challenge may be unfavorable for some individuals, but others may be perfectly suited to the change. Thus, as a group, they have a better chance of surviving environmental change than clones.

To understand why environments constantly change, think about one example: interactions between a predatory species and its prey. To the prey species, the predators are an environmental challenge. Prey individuals that are best at escaping predation tend to leave more offspring. Thus, traits that enhance an individual's ability to escape a predator tend to become more common in a prey species over generations. At the same time, prey species that become better at escaping predation are an environmental challenge to a predator species. Predator individuals that are best at capturing prey tend to leave more offspring. Thus, traits that enhance an individual's ability to capture prey tend to become more common in the predator species over generations (Chapter 17 returns to this

sexual reproduction Reproductive mode by which offspring arise from two parents and inherit genes from both.

CREDITS: (opposite) Reprinted from *Fertility and Sterility*, Vol 87 /edition 3, Fei Sun,Paul Turek,Calvin Greene,Evelyn Ko,Alfred Rademaker,Renée H. Martin, Abnormal progression through meiosis in men with nonobstructive azoospermia, 565-571, Copyright 2007, with permission from Elsevier. Image supplied by Renee H. Martin, Ph.D.

and other evolutionary processes). The two species are locked in a constant race, with each genetic improvement in one countered by a genetic improvement in the other. This idea is called the Red Queen hypothesis, a reference to Lewis Carroll's book *Through the Looking Glass*. In the book, the Queen of Hearts tells Alice, "It takes all the running you can do, to keep in the same place." Environmental challenges that constantly change favor sexual reproduction because the variation in traits it fosters is evolutionarily advantagous in a changing environment.

Another advantage of sexual reproduction involves the inevitable occurrence of harmful mutations that can be passed to offspring. A population of sexual reproducers has a better chance of weathering the evolutionary effects of such mutations. With asexual reproduction, individuals bearing a moderately harmful mutation necessarily pass it to all of their offspring. This outcome is relatively rare with sexual reproduction, because each offspring of a sexual union has a 50 percent chance of inheriting a parent's mutation. Thus, a moderately harmful mutation tends to spread more slowly through a population of sexual reproducers. ●

12.2 Meiosis in Sexual Reproduction

LEARNING OBJECTIVES

- Describe alleles and why they occur on homologous chromosomes of sexual reproducers.
- Explain why sexual reproduction requires meiosis.

Introducing Alleles

Genes are the basis of an organism's traits—its form and function. Variation in DNA sequence is the basis of variation in traits among species, and it is also the basis of variation in forms of traits among individuals of sexually reproducing species. Remember from Section 11.2 that your body cells are diploid ($2n$), which means they contain pairs of homologous chromosomes. The two chromosomes of each pair have the same genes, but their DNA sequence is not identical. This is because you inherited your chromosomes from two parents who differ genetically, and unique mutations accumulated in your parents' separate lines of descent over time. Thus, the DNA sequence of any of your genes may differ a bit from a corresponding gene on a homologous chromosome (**FIGURE 12.1**). Different forms of the same gene are called **alleles**.

New alleles arise by mutation. They may encode slightly different forms of a gene's product, and such differences influence the details of shared, inherited traits. Members of a species have the same traits because they have the same genes, but almost every

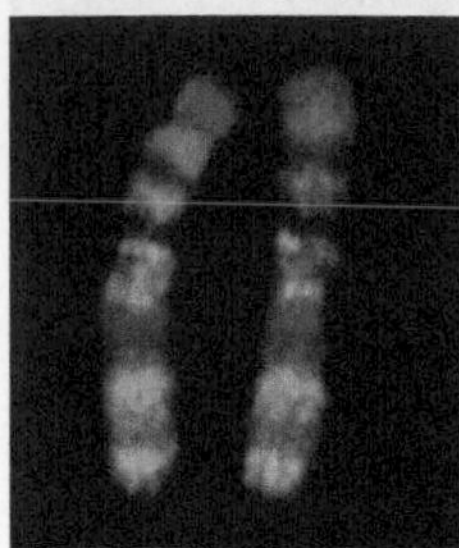

A Corresponding colored patches in this fluorescence micrograph indicate corresponding DNA sequences in a homologous chromosome pair.

Homologous chromosomes carry the same set of genes.

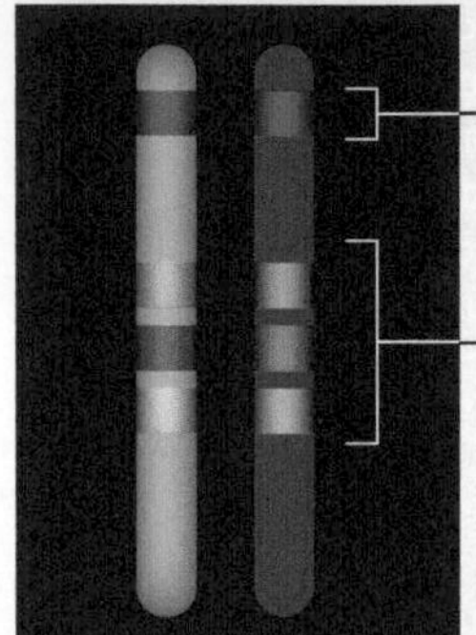

B Genes occur in pairs on homologous chromosomes.

The members of each pair of genes may be identical in DNA sequence, or they may differ slightly, as alleles (color variations represent DNA sequence differences).

FIGURE 12.1
Genes on homologous chromosomes. Different forms of a gene are called alleles.

shared trait varies a bit among individuals. Alleles of the shared genes are the basis of this variation. Consider just one human gene, *HBB*, which encodes beta globin (Section 9.6). *HBB* has more than 700 alleles; a few cause sickle-cell anemia, several cause beta thalassemia, and so on. There are more than 20,000 human genes, and most of them have multiple alleles. (Chapter 14 returns to the genetic basis of human traits.)

Meiosis Halves the Chromosome Number

Mitosis is the nuclear division mechanism that maintains the chromosome number (Section 11.2). When a diploid ($2n$) cell undergoes mitosis, it produces diploid offspring, each with two copies of every chromosome. Sexual reproduction requires a different nuclear division mechanism, **meiosis**, that halves the chromosome number. When a diploid cell undergoes meiosis, it produces haploid cells. A cell that is **haploid** (n) has one copy of each chromosome.

Meiosis is part of the process by which mature, haploid reproductive cells called **gametes** form. In multicelled eukaryotes, gametes arise by division

alleles (uh-LEELZ) Forms of a gene with slightly different DNA sequences; may encode different versions of the gene's product.
fertilization Fusion of two gametes to form a zygote.
gamete (GAM-eat) Mature, haploid reproductive cell; e.g., a sperm.
germ cell Immature reproductive cell that gives rise to haploid gametes when it divides.
haploid (HAP-loyd) Having one copy of each type of chromosome (n).
meiosis (my-OH-sis) Nuclear division process that halves the chromosome number. Required for sexual reproduction.
zygote (ZYE-goat) Diploid cell that forms when two gametes fuse; the first cell of a new individual.

of immature reproductive cells called **germ cells**. In plants and animals, germ cells form in organs set aside for reproduction, but the two groups make gametes somewhat differently. In plants, haploid germ cells (spores) form by meiosis. These cells divide by mitosis to form structures that produce or contain gametes. In animals, germ cells are part of the germline, which is a lineage of cells dedicated to producing gametes. Meiosis in diploid germ cells gives rise to eggs (female gametes) or sperm (male gametes).

The first part of meiosis is similar to mitosis. A cell replicates its DNA before either nuclear division process begins, so each chromosome consists of two sister chromatids. As in mitosis, a spindle forms, and its microtubules move the chromosomes. However, meiosis sorts the chromosomes into new nuclei not once but two times, so it results in the formation of four nuclei. The two consecutive nuclear divisions are called meiosis I and meiosis II:

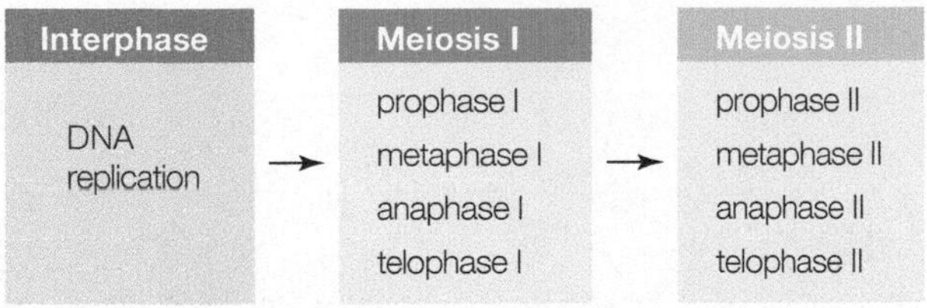

Interphase
DNA replication
Meiosis I
prophase I
metaphase I
anaphase I
telophase I
Meiosis II
prophase II
metaphase II
anaphase II
telophase II

During meiosis I, every chromosome aligns with its homologous partner (FIGURE 12.2 ❶). Then the homologous chromosomes are pulled away from one another and packaged into separate nuclei ❷. Each of the two new nuclei is haploid (*n*), so its chromosome number is half that of the diploid parent cell.

After meiosis I, each chromosomes still consists of two sister chromatids. During meiosis II, the chromatids are pulled apart, and each becomes an individual, unduplicated chromosome ❸. The chromosomes are sorted into four new nuclei. Each new nucleus still has one copy of each chromosome, so it is haploid (*n*).

In summary, meiosis partitions the chromosomes of one diploid nucleus (2*n*) into four haploid (*n*) nuclei. The next section zooms in on the details of this process.

Fertilization Restores the Chromosome Number

We leave details of sexual reproduction for later chapters, but you will need to understand a few concepts before getting there. You already know that meiosis is required for the formation of haploid gametes. A male gamete fuses with a female gamete at **fertilization**. The fusion of haploid gametes at fertilization produces a diploid cell called a **zygote**, which is the first cell of a new individual. Thus, meiosis halves the chromosome number, and fertilization restores it (FIGURE 12.3).

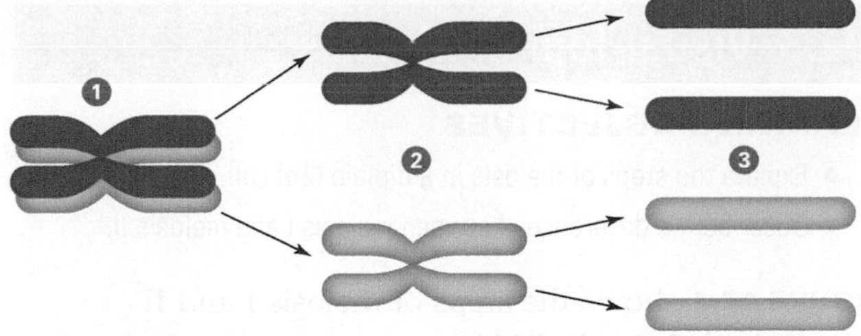

❶
❷
❸

❶ Chromosomes are duplicated before meiosis begins, so each consists of two sister chromatids. During meiosis I, each chromosome in the nucleus aligns with its homologous partner. The nucleus contains two of each chromosome, so it is diploid (2*n*).

❷ Homologous partners separate and are packaged into two new nuclei. Each new nucleus contains one of each chromosome, so it is haploid (*n*). The chromosomes still have two sister chromatids.

❸ Sister chromatids separate in meiosis II and are packaged into four new nuclei. Each new nucleus has one of each chromosome, so it is haploid (*n*).

FIGURE 12.2 **How meiosis halves the chromosome number.**

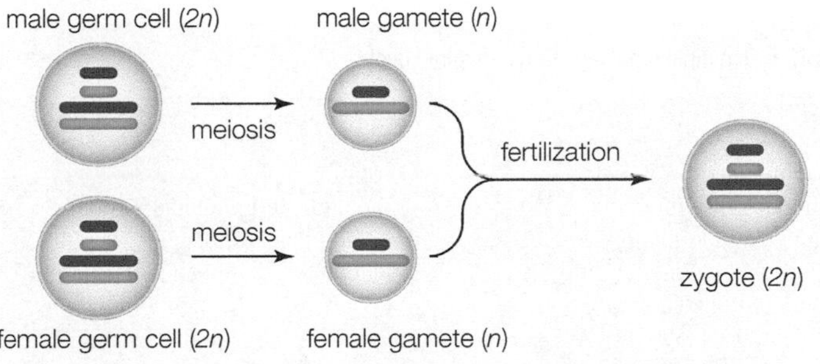

male germ cell (2n)
male gamete (n)
meiosis
fertilization
meiosis
zygote (2n)
female germ cell (2n)
female gamete (n)

FIGURE 12.3 **Meiosis halves the chromosome number, and fertilization restores it.** Animal cells are illustrated. In animal cells, meiosis in diploid germ cells produces haploid gametes. At fertilization, two haploid gametes meet and form a diploid zygote.

If meiosis did not precede fertilization, the chromosome number would double with every generation. An individual's chromosomes are like a fine-tuned blueprint that must be followed exactly, in order to build a body that functions normally. If the chromosome number changes, so do the individual's genetic instructions. As you will see in Chapter 14, such changes can have drastic consequences for health, particularly in animals.

TAKE-HOME MESSAGE 12.2

- ✔ Paired genes on homologous chromosomes may vary in DNA sequence as alleles. Alleles arise by mutation.
- ✔ Alleles give rise to differences in shared traits among individuals of a sexually reproducing species.
- ✔ Meiosis occurs in cells set aside for reproduction.
- ✔ Meiosis halves the diploid (2*n*) chromosome number, to the haploid number (*n*), for forthcoming gametes.
- ✔ When two gametes fuse at fertilization, the diploid chromosome number is restored in the resulting zygote.

12.3 Visual Tour of Meiosis

LEARNING OBJECTIVES

- Explain the steps of meiosis in a diploid ($2n$) cell.
- Describe the differences between meiosis I and meiosis II.

FIGURE 12.4 shows the steps of meiosis I and II in a diploid ($2n$) cell. DNA replication occurs before meiosis I, so each chromosome starts out with two sister chromatids.

The first stage of meiosis I is prophase I ❶. During this phase, the chromosomes condense, and homologous chromosomes align tightly and swap segments (more about this segment-swapping in the next section). The nuclear envelope breaks up. A spindle forms, and by the end of prophase I, microtubules attach one chromosome of each homologous pair to one spindle pole, and the other to the opposite spindle pole. These microtubules grow and shrink, pushing and pulling the chromosomes as they do. At metaphase I ❷, all of the microtubules are the same length, and the chromosomes are aligned midway between the spindle poles.

During anaphase I ❸, the homologous chromosomes of each pair move away from one another and toward opposite spindle poles. During telophase I, the two sets of chromosomes reach the spindle poles, and a new nuclear envelope forms around each set as the DNA loosens up ❹. The two new nuclei are haploid (n); each contains one complete set of chromosomes. Each chromosome still consists of two sister chromatids. The cytoplasm often divides at this point.

FIGURE 12.4 Meiosis. Illustrations show two pairs of chromosomes in a diploid ($2n$) animal cell; homologous chromosomes are indicated in blue and pink. Micrographs show meiosis in a cell from lily (*Lilium regale*).

FIGURE IT OUT Which phase of meiosis reduces the chromosome number?

Answer: Anaphase I

MEIOSIS I One diploid nucleus to two haploid nuclei

① **Prophase I**
Homologous chromosomes condense, pair up, and swap segments. Spindle microtubules attach to them as the nuclear envelope breaks up.

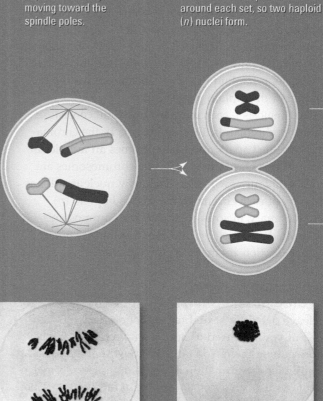

plasma membrane
spindle
nuclear envelope breaking up

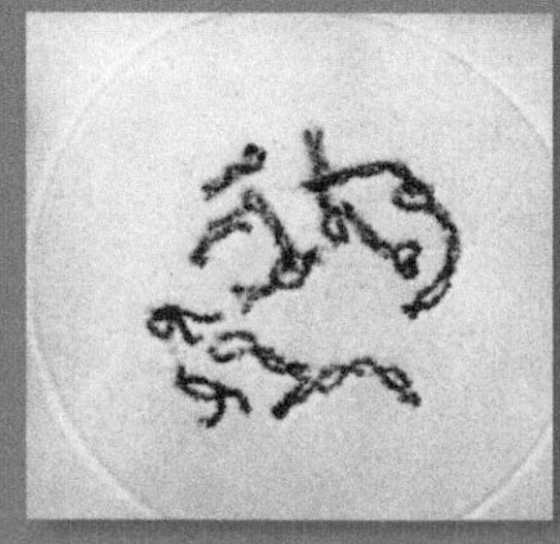

② **Metaphase I**
Homologous chromosome pairs are aligned between spindle poles. Spindle microtubules have attached the two chromosomes of each pair to opposite spindle poles.

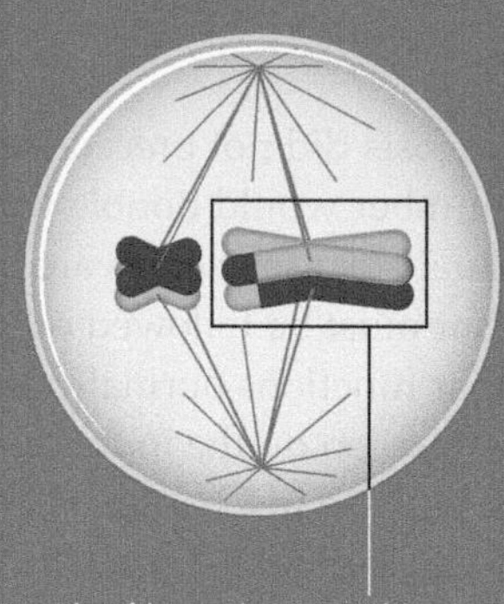
pair of homologous chromosomes

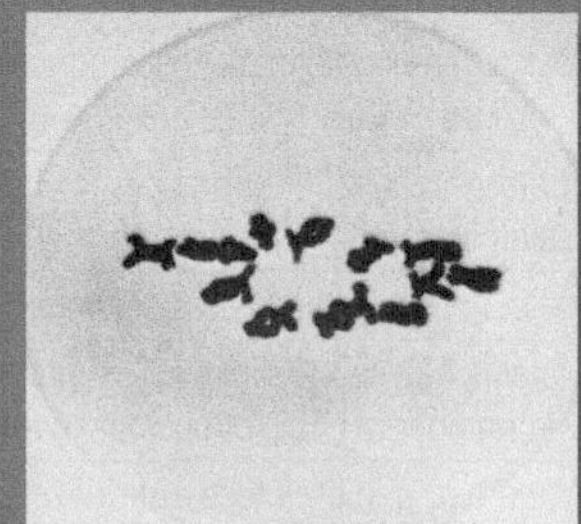

③ **Anaphase I**
Homologous chromosomes separate and begin moving toward the spindle poles.

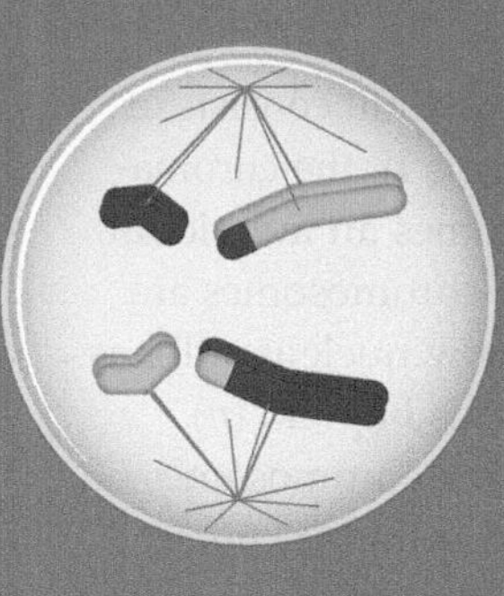

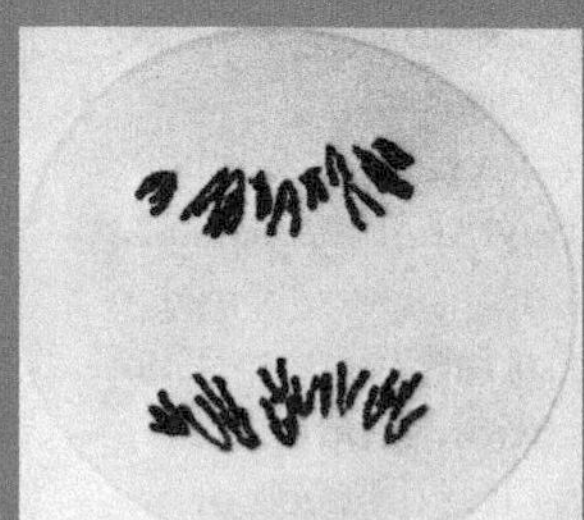

④ **Telophase I**
A complete set of chromosomes clusters at each spindle pole. A nuclear envelope forms around each set, so two haploid (n) nuclei form.

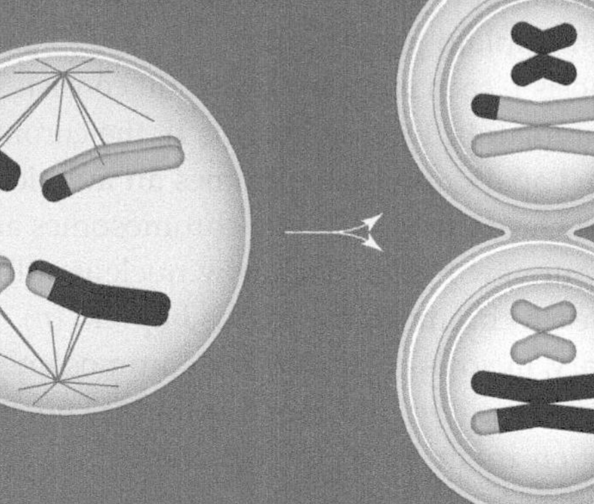

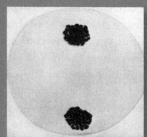

CREDITS: (4) bottom 1-4, With thanks to the John Innes Foundation Trustees, computer enhanced by Gary Head.

In some cells, meiosis II occurs immediately after meiosis I. In other cells, a period of protein synthesis—but no DNA replication—intervenes between the divisions.

Meiosis II proceeds simultaneously in both nuclei that formed in meiosis I. During prophase II ❺, the chromosomes condense and the nuclear envelope breaks up. A new spindle forms. By the end of prophase II, spindle microtubules attach each chromatid to one spindle pole, and its sister chromatid to the opposite spindle pole. These microtubules push and pull the chromosomes, midway between spindle poles at metaphase II ❻.

During anaphase II ❼, the sister chromatids separate and move toward opposite spindle poles, thus becoming individual chromosomes. During telophase II ❽, the chromosomes reach the spindle poles. New nuclear envelopes form around the four clusters of chromosomes as the DNA loosens up. The four nuclei that form are haploid (n); each contains one complete set of chromosomes. The cytoplasm often divides at this point.

TAKE-HOME MESSAGE 12.3

✔ Meiosis distributes the chromosomes of a diploid ($2n$) nucleus into four haploid (n) nuclei. Each new nucleus that forms during the process has a full set of chromosomes—one of each type.

✔ DNA replication occurs before meiosis, so each chromosome consists of two sister chromatids as the process begins.

✔ Meiosis I separates homologous chromosomes and packages them into two new haploid (n) nuclei.

✔ Meiosis II separates sister chromatids and packages them into two new haploid nuclei.

MEIOSIS II Two haploid nuclei to four haploid nuclei

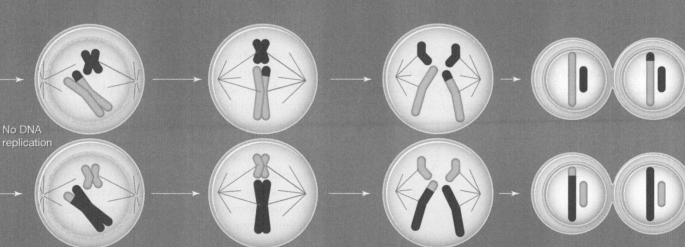

⑤ **Prophase II**
The chromosomes condense. Spindle microtubules attach to each sister chromatid as the nuclear envelope breaks up.

⑥ **Metaphase II**
The chromosomes are aligned midway between spindle poles. Spindle microtubules have attached their sister chromatids to opposite spindle poles.

⑦ **Anaphase II**
Sister chromatids separate and move toward opposite spindle poles. When sister chromosomes separate, each becomes an individual chromosome.

⑧ **Telophase II**
A complete set of chromosomes clusters at both ends of the cell. A new nuclear envelope forms around each set, so four haploid (n) nuclei form.

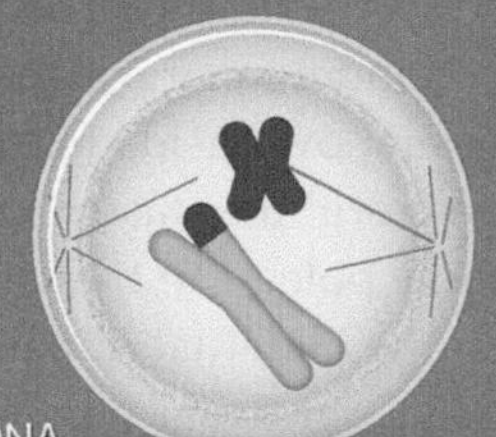

No DNA replication

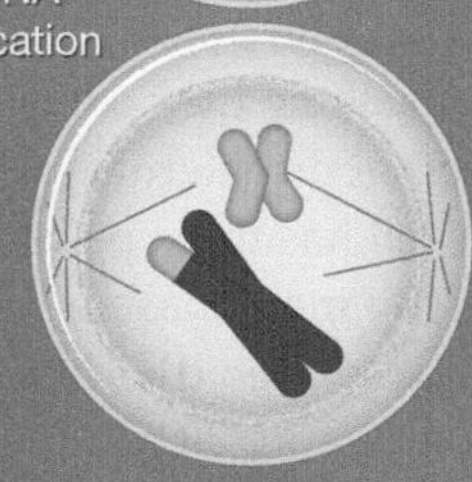

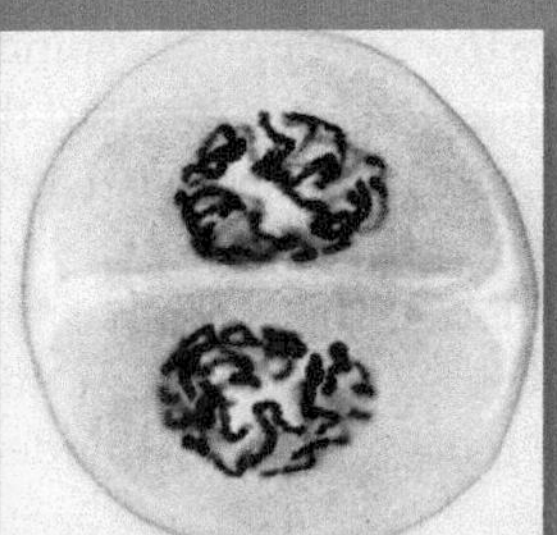

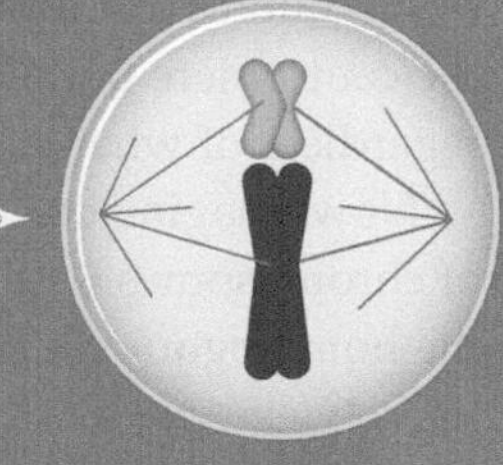

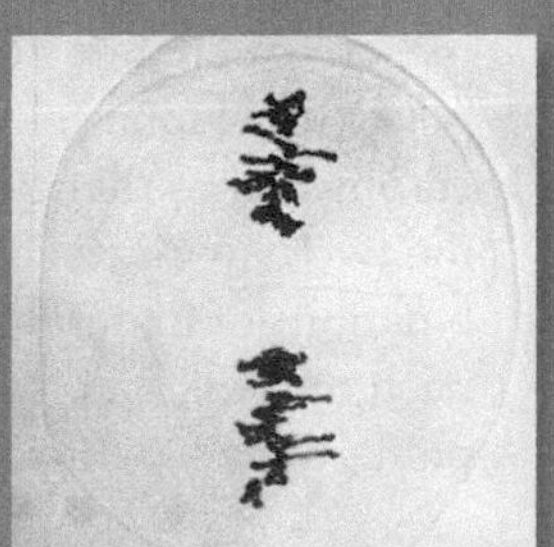

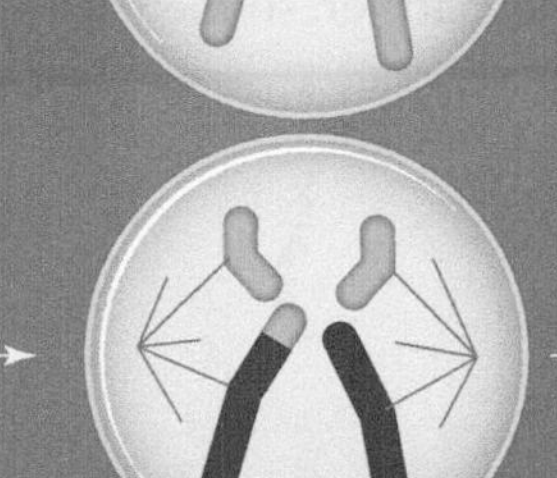

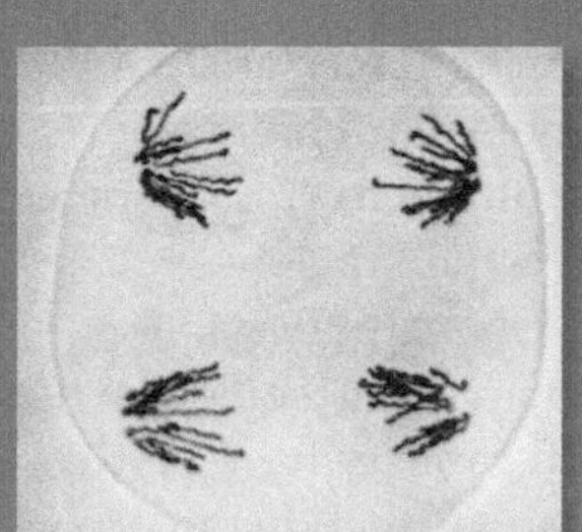

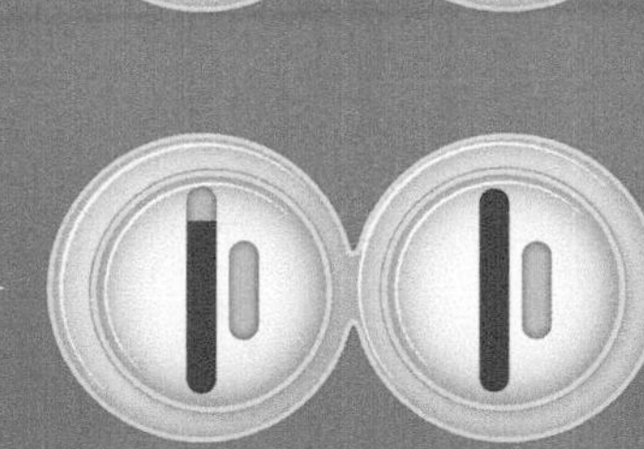

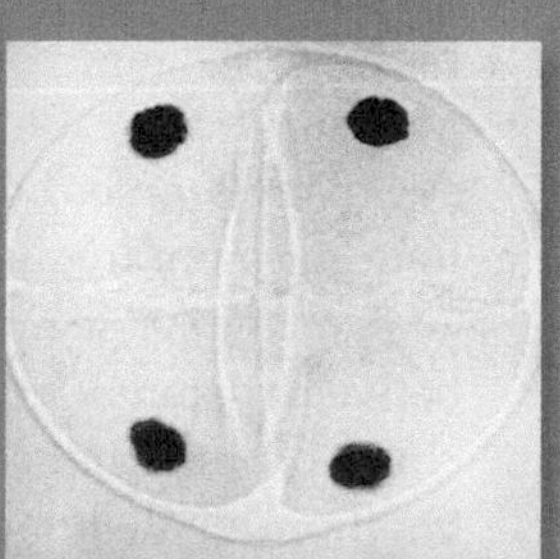

CREDITS: (4) bottom 5-8, With thanks to the John Innes Foundation Trustees, computer enhanced by Gary Head.

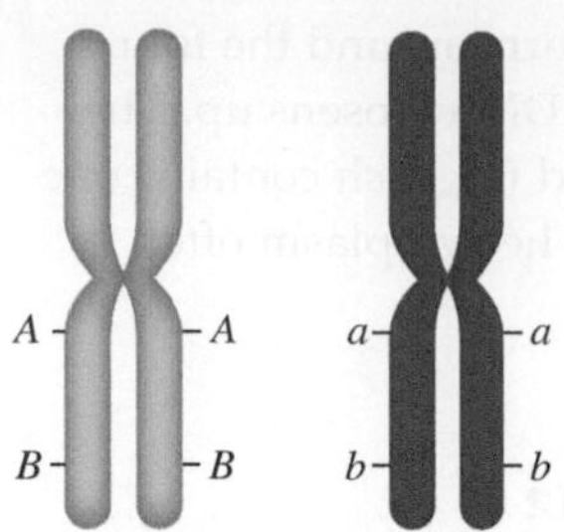
A
A
a
a
B
B
b
b

A In this example, one gene has alleles A and a; the other gene has alleles B and b.

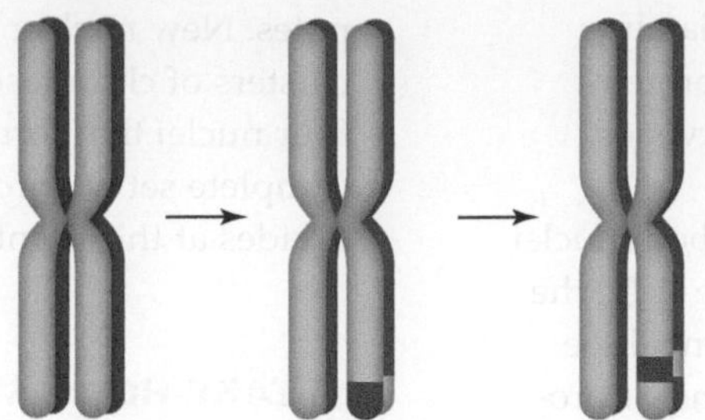

B Close contact between homologous chromosomes promotes crossing over, in which nonsister chromatids exchange corresponding pieces. Multiple crossovers are not uncommon.

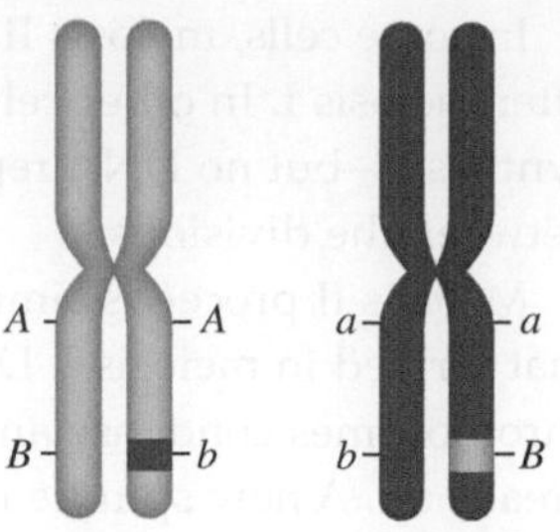
A
A
a
a
B
b
b
B

C After crossing over, paternal and maternal alleles have been mixed up on homologous chromosomes.

FIGURE 12.5 Crossing over. For clarity, we track only two genes on one pair of homologous chromosomes. Blue signifies the paternal chromosome, and pink, the maternal chromosome.

12.4 Meiosis Fosters Genetic Diversity

LEARNING OBJECTIVES

- Describe crossing over and how it introduces variation in traits among the offspring of sexual reproducers.
- Explain the random nature of chromosome segregation in gamete formation, and its significance in terms of variation in traits.

During prophase I of meiosis, homologous chromosomes align and swap segments. Then, they separate and become packaged in different nuclei. Together with fertilization, these processes give rise to offspring with different combinations of alleles—genetic variation that is the basis of variation in shared traits among individuals of a sexually reproducing species.

Crossing Over

Early in prophase I of meiosis, all chromosomes in the cell condense. When they do, each chromosome is drawn close to its homologous partner, so their corresponding chromatids align:

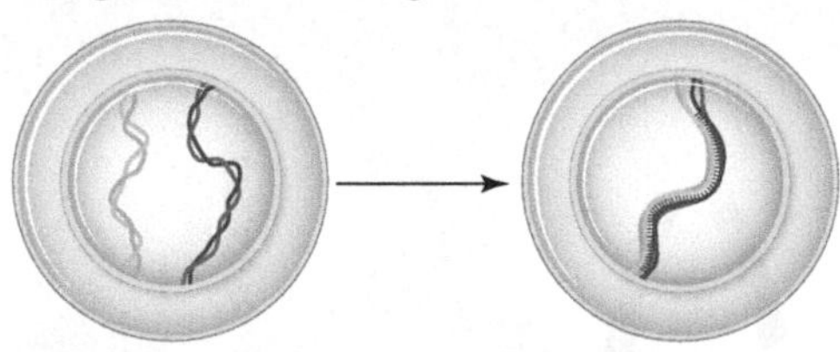

This tight, parallel orientation favors **crossing over**, a process by which a chromosome and its homologous partner exchange corresponding pieces of DNA during meiosis (FIGURE 12.5). Homologous chromosomes may swap any segment of DNA, although crossovers tend to occur more frequently in certain regions.

Exchanging segments of DNA shuffles alleles between parental homologous chromosomes, so the chromosomes that end up in gametes have new combinations of alleles: combinations that do not occur in either parent. Thus, crossing over creates new combinations of traits among offspring. It is a required process—meiosis will not finish unless it happens—but the rate of crossing over varies among species and among chromosomes. In humans, between 46 and 95 crossovers occur per meiosis, so each chromosome crosses over at least once, on average.

Chromosome Segregation

In meiosis I, one diploid nucleus gives rise to two haploid nuclei. Each new nucleus receives one chromosome of each homologous pair, but which of the two chromosomes ends up in a particular nucleus is entirely random. Why? The answer has to do with the way meiosis I distributes homologous chromosomes (segregates them) into new nuclei.

The process of chromosome segregation begins in prophase I. Imagine a germ cell undergoing meiosis. For simplicity, we will put crossing over aside for a moment. During prophase I, microtubules fasten the cell's chromosomes to the spindle poles, but there is no pattern to the attachment of the maternal or paternal homologous chromosome to a particular pole. Microtubules extending from a spindle pole attach to the centromere of the first chromosome they contact, regardless of whether it is maternal or paternal.

Now imagine that the germ cell has three pairs of chromosomes (FIGURE 12.7). By metaphase I, the homologous chromosomes have been divided up between the two spindle poles in one of four ways ❶. In anaphase I, the maternal and paternal chromosomes separate and move toward opposite spindle poles. In telophase I, a new nucleus forms around the chromosomes that cluster at each spindle pole. Each of the two new nuclei contains one of eight possible combinations of maternal and paternal chromosomes ❷. Each nucleus divides again during meiosis II. At this point, three (3) chromosome pairs have been divided

crossing over Process by which homologous chromosomes exchange corresponding segments of DNA during prophase I of meiosis.

Data Analysis Activities

BPA and Abnormal Meiosis In 1998, researchers at Case Western University were studying meiosis in mouse oocytes (germ cells) when they saw an unexpected and dramatic increase of abnormal events (**FIGURE 12.6**). Improper segregation of chromosomes during meiosis is one of the main causes of human genetic disorders.

The spike in abnormal meiosis began right after the mouse facility started washing the animals' plastic cages and water bottles in a new, alkaline detergent. The detergent had damaged the plastic, which as a result was leaching bisphenol A (BPA). BPA is a synthetic chemical that mimics estrogen, the main female sex hormone in animals. Though it has since been banned for use in baby bottles, BPA is still widely used to manufacture other plastic items and epoxies (such as the coating on the inside of metal cans of food). BPA-free plastics are often manufactured with a related compound, bisphenol S (BPS), that has effects similar to BPA.

1. What percentage of mouse oocytes displayed abnormalities of meiosis with no exposure to damaged caging?
2. Which group of mice had the most meiotic abnormalities in their oocytes?
3. What is abnormal about metaphase I as it is occurring in the oocytes shown in **FIGURE 12.6B**, **C**, and **D**?

Caging materials	Total number of oocytes	Abnormalities
Control: New cages with glass bottles	271	5 (1.8%)
Damaged cages with glass bottles		
Mild damage	401	35 (8.7%)
Severe damage	149	30 (20.1%)
Damaged plastic bottles	197	53 (26.9%)
Damaged cages with damaged plastic bottles	58	24 (41.4%)

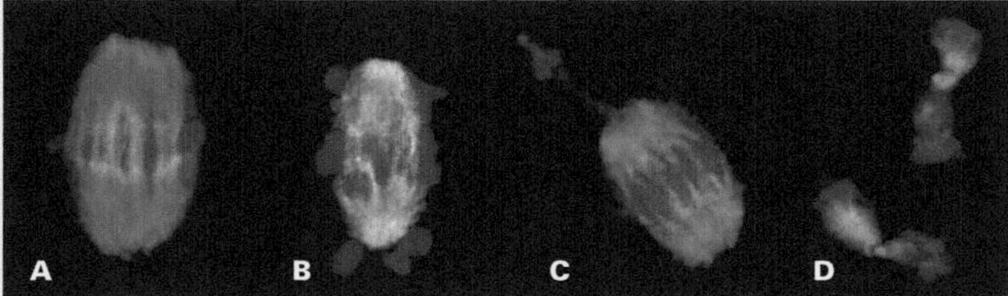
A
B
C
D

FIGURE 12.6 Abnormalities in meiosis that occurred after exposure to BPA.

Top, the most abnormal meiosis events occurred in mice that were housed in damaged plastic caging with damaged plastic bottles. Damaged plastic releases BPA.

Bottom, fluorescent micrographs show the chromosomes (red) and spindle (green) in nuclei of mouse germ cells in metaphase I. **A** Normal metaphase; **B–D** abnormal metaphase.

up between two (2) spindle poles. The resulting four nuclei have one of eight (2^3) possible combinations of maternal and paternal chromosomes ❸.

Cells that give rise to human gametes have twenty-three pairs of homologous chromosomes, not three. Each time a human germ cell undergoes meiosis, the four gametes that form end up with one of 8,388,608 (or 2^{23}) possible combinations of homologous chromosomes. In addition, any number of genes may occur as different alleles on the maternal and paternal chromosomes, and crossing over makes mosaics of that genetic information. Then, out of all the male and female gametes that form, which two actually get together at fertilization is a matter of chance. Are you getting an idea

FIGURE 12.7 Segregation of three chromosome pairs during meiosis. Maternal chromosomes are pink in this hypothetical example; paternal chromosomes are blue. For simplicity, we show no crossing over, so all sister chromatids are identical.

❶ The homologous chromosomes of three pairs can be divided between two spindle poles in four possible ways.

❷ There are eight possible combinations of maternal and paternal chromosomes in the two nuclei that form at telophase I.

❸ There are eight possible combinations of maternal and paternal chromosomes in the four nuclei that form at telophase II.

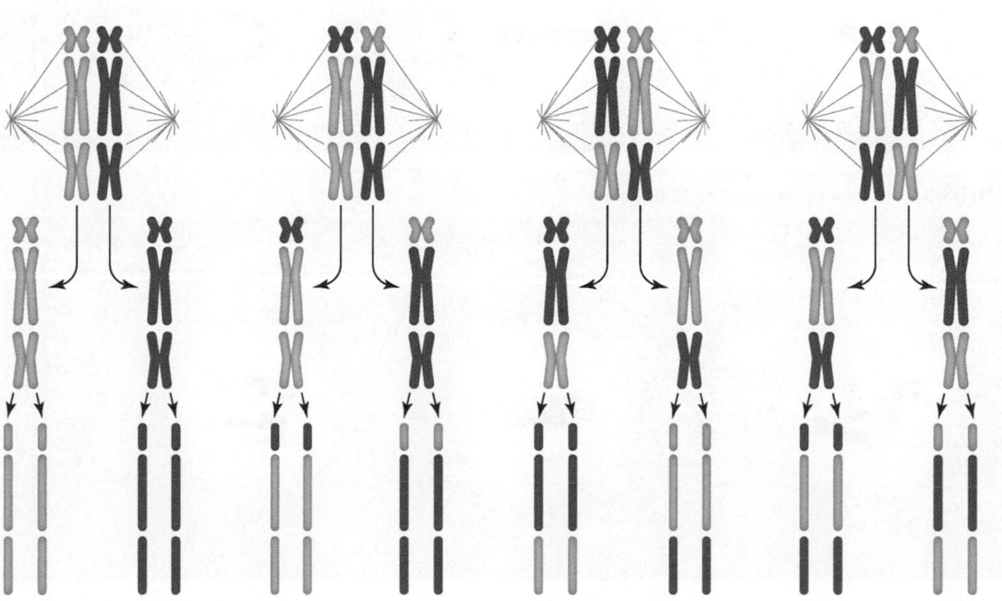

of why such fascinating combinations of traits show up among the generations of your own family tree?

TAKE-HOME MESSAGE 12.4

✔ Homologous chromosomes exchange corresponding segments during prophase I of meiosis, a process called crossing over.

✔ Each chromosome of a homologous pair can end up in either of the two new nuclei that form in meiosis I.

✔ Crossing over and random segregation of chromosomes into gametes give rise to new combinations of alleles among offspring.

12.5 An Ancestral Connection

LEARNING OBJECTIVES

- Describe the similarities and differences between mitosis and meiosis II.
- Offer evidence in support of the hypothesis that meiosis evolved from mutations in the process of mitosis.

The earliest eukaryotes were almost certainly single-celled and haploid, and they reproduced by mitosis. Meiosis evolved later, from mitosis. It seems like a giant evolutionary step from producing clones to producing genetically varied offspring, but was it really?

The outcome of mitosis and meiosis differ. By mitosis and cytoplasmic division, one cell becomes two new cells that have copies of the parental chromosomes. Mitotic (asexual) reproduction produces clones of the parent. By contrast, meiotic (sexual) reproduction produces offspring that differ from one another, and also from parents. Crossing over during meiosis gives rise to haploid gametes that vary in genetic makeup. Gametes of two parents fuse to form a zygote, which is a cell of mixed parentage. Though the end results differ, there are striking parallels. For example, the four stages of mitosis and meiosis II are quite similar (**FIGURE 12.8**).

Long ago, the molecular machinery of mitosis became remodeled into meiosis. Evidence for this hypothesis includes a host of shared molecules, including the products of the *BRCA* genes (Sections 10.1 and 11.6) that all modern eukaryotes have. By monitoring and fixing problems with the DNA—such as

FIGURE 12.8 A visual comparison between meiosis II and mitosis.

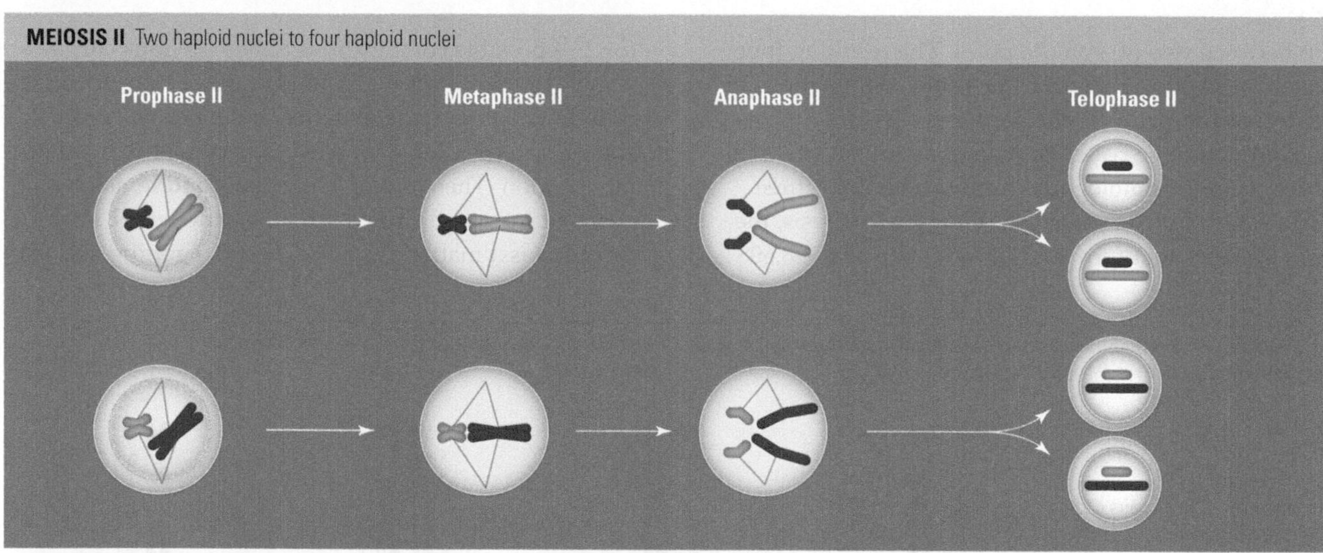
MEIOSIS II Two haploid nuclei to four haploid nuclei
Prophase II
Metaphase II
Anaphase II
Telophase II

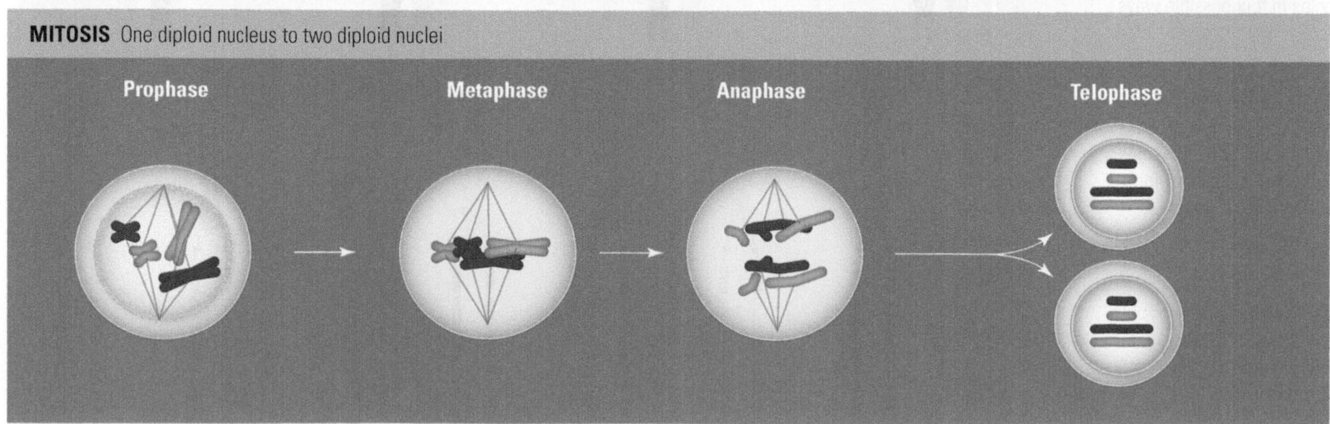
MITOSIS One diploid nucleus to two diploid nuclei
Prophase
Metaphase
Anaphase
Telophase

damaged or mismatched bases (Section 8.6)—these molecules actively maintain the integrity of a cell's chromosomes, particularly during DNA replication and mitosis. Many of the same molecules also help homologous chromosomes cross over in prophase I of meiosis (**FIGURE 12.9**). As another example of shared molecules, the same regulatory molecules involved in checkpoints of mitosis are also involved in checkpoints of meiosis.

During anaphase of mitosis, sister chromatids are pulled apart. What would happen if the connections between the chromatids did not break? Both sister chromatids would be pulled to one or the other spindle pole—which is exactly what happens during anaphase I of meiosis. The shared molecules and mechanisms imply a shared evolutionary history; sexual reproduction probably originated with mutations that affected processes of mitosis. As you will see in later chapters, the remodeling of existing processes into new ones is a common evolutionary theme.

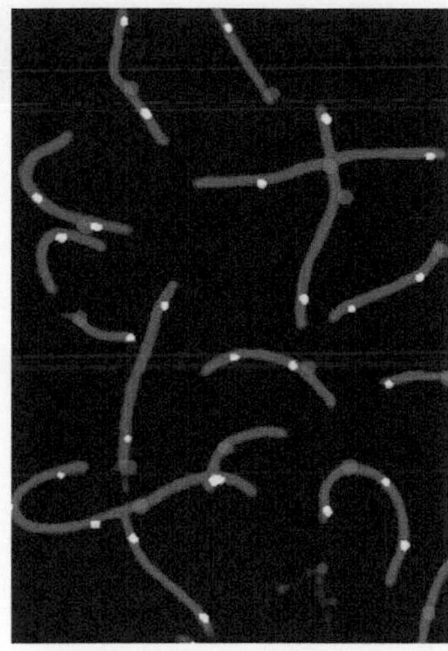

FIGURE 12.9 Mitosis and meiosis use the same molecular machinery.

This fluorescence micrograph shows paired homologous chromosomes (red) crossing over in prophase I of meiosis. Centromeres are visible in blue. Yellow pinpoints the location of a protein called MLH1.

MHL1 has several functions in meiosis, including chromosomal repair after crossovers and preventing inappropriate crossovers.

MLH1 also has several functions in mitosis. For example, it helps repair DNA replication errors, and it prevents homologous chromosomes from crossing over.

TAKE-HOME MESSAGE 12.5

✔ Molecular similarities offer evidence that meiosis evolved by the remodeling of existing mechanisms of mitosis.

12.1 Why Sex? (revisited)

Why do males exist? No male has ever been found among the tiny freshwater creatures called bdelloid rotifers (**FIGURE 12.10**). Females have been reproducing for 80 million years solely through cloning themselves. Bdelloids are one of the few groups of animals to have completely abandoned sex.

How have these organisms survived for 80 million years without sex? Compared to sexual reproduction, asexual reproduction is often seen as a poor long-term strategy because it does not include crossing over or random segregation of chromosomes into gametes—the chromosomal shufflings that bring about genetic diversity thought to give species an adaptive edge in the face of new challenges. Bdelloids have contradicted this theory by being very successful: Over 400 species are thriving today.

An unusual ability may help to explain the success of the bdelloids despite their rejection of sex. These rotifers can apparently import genes from bacteria, fungi, protists, and even plants. The direct swapping of genetic material was thought to be incredibly rare in complex organisms, but bdelloids are bringing in external genes to an extent unheard of in animals. Each rotifer is a genetic mosaic whose DNA spans almost all the major kingdoms of life: About 10 percent of its active genes have been pilfered from other organisms. If the main advantage of sex is that it promotes genetic diversity, why worry about it when you have the genes of entire kingdoms available to you?

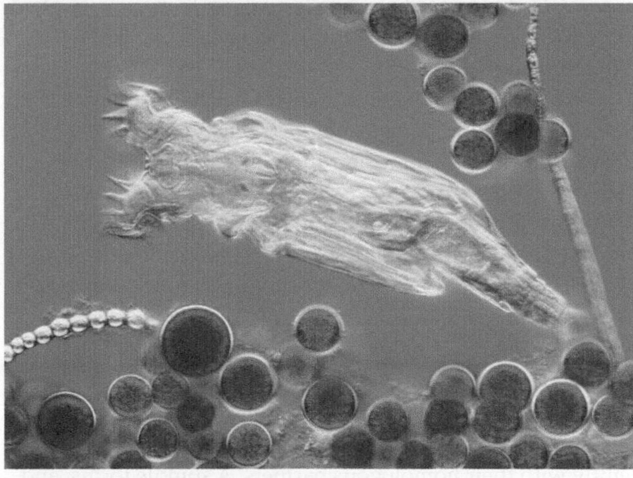

FIGURE 12.10 A bdelloid rotifer. All of these tiny animals are female. Red balls are algal cells.

Bdelloid rotifers can dry out and survive for decades in a dormant state, reanimating when they encounter liquid again. Researchers suspect that foreign genes slip into the animals during the reanimation process. When a rotifer dries out, its chromosomes break into pieces and its cell membranes become leaky. Fragments of foreign DNA in the environment easily cross the leaky membranes. When the individual reanimates, DNA repair enzymes reassemble its shattered chromosomes. As the reassembly occurs, any foreign DNA fragments that entered the rotifer when it was dried out can become incorporated into its chromosomes. ●

STUDY GUIDE

Section 12.1 **Sexual reproduction** mixes up the genetic information of two parents who differ in the details of shared, inherited traits. Thus, offspring produced by sexual reproduction typically differ from one another and from the parents. Particularly in a changing environment, genetic variation in a population can offer an evolutionary advantage over identical offspring produced by asexual reproduction.

Section 12.2 In cells of sexual reproducers, one chromosome of each homologous pair was inherited from the mother; the other, from the father. The two chromosomes of each pair carry the same genes, but each gene may vary slightly in DNA sequence. Different forms of the same gene are called **alleles**. Alleles are the basis of differences in shared, heritable traits among individuals of a sexually reproducing population. They arise by mutation.

Meiosis is a nuclear division mechanism that halves the chromosome number. Cells produced by meiosis are **haploid** (n): They have a complete set of chromosomes, but only one of each type.

Meiosis is the basis of sexual reproduction in eukaryotes. It occurs only in special cells set aside for sexual reproduction. In animals, divisions of **germ cells** give rise to mature, haploid reproductive cells called **gametes**. Gametes form somewhat differently in plants.

The fusion of two haploid gametes during **fertilization** restores the diploid parental chromosome number in the **zygote**, the first cell of the new individual.

Section 12.3 Meiosis of a diploid nucleus produces four haploid nuclei. DNA replication occurs before meiosis, so each chromosome consists of two molecules of DNA attached at the centromere as sister chromatids. Two consecutive nuclear divisions occur during meiosis: meiosis I and meiosis II. The first nuclear division (meiosis I) includes prophase I, metaphase I, anaphase I, and telophase I.

During prophase I, the chromosomes condense and align tightly with their homologous partners. A spindle forms, and its lengthening microtubules penetrate the nuclear region as the nuclear envelope breaks up. The microtubules tether one chromosome of each homologous pair to opposite spindle poles, then move all of the chromosomes to the middle of the cell. The chromosomes are lined up midway between the spindle poles at metaphase I.

During anaphase I, the chromosome number is reduced as the homologous chromosomes separate and move toward opposite spindle poles. During telophase I, a complete set of chromosomes reaches each spindle pole. A nuclear envelope forms around each set, so two haploid nuclei form. The cytoplasm may divide at this point, but DNA replication does not occur.

The second nuclear division (meiosis II) includes prophase II, metaphase II, anaphase II, and telophase II; and it proceeds simultaneously in both nuclei that formed during meiosis I. The chromosomes condense during prophase II, and spindle microtubules move them toward the middle of the cell. At metaphase II, the chromosomes are aligned between spindle poles. During anaphase II, the sister chromatids of each chromosome separate (becoming individual chromosomes), then move to opposite spindle poles. During telophase II, a complete set of chromosomes reaches each spindle pole. A new nuclear envelope forms around each set, so four haploid nuclei form.

Section 12.4 Two processes that occur during meiosis—**crossing over** and random segregation of chromosomes into gametes—are the basis of variation in traits among offspring of sexual reproducing organisms. During prophase I, homologous chromosomes cross over by exchanging corresponding segments, thus mixing up alleles between maternal and paternal chromosomes. Crossing over gives rise to combinations of alleles not present in either parental chromosome, so it gives rise to combinations of traits not present in either parent. Meiosis also gives rise to variation in traits among offspring by randomly segregating homologous chromosomes into gametes. Microtubules can attach the maternal or the paternal chromosome of each pair to one or the other spindle pole. Either chromosome may end up in any new nucleus, and in any gamete.

Section 12.5 The same mechanisms that operate during meiosis also operate during mitosis, and many of the same molecules function the same way in both processes. The similarities are evidence that meiosis evolved from mitosis by the repurposing of an existing molecular system.

SELF-QUIZ

Answers in Appendix VII

1. One evolutionary advantage of sexual over asexual reproduction is that it produces ________ .
   a. more offspring per individual
   b. more variation among offspring
   c. healthier offspring

2. Meiosis is a necessary part of sexual reproduction because it ________ .
   a. divides two nuclei into four new nuclei
   b. reduces the chromosome number for gametes
   c. gives rise to new alleles

3. Meiosis ________ .
   a. occurs only in animals
   b. supports growth and tissue repair in multicelled species
   c. gives rise to genetic diversity among offspring
   d. is part of the life cycle of all cells

4. Sexual reproduction in animals requires ________ .
   a. meiosis
   b. fertilization
   c. gametes
   d. all of the above

5. Meiosis _______ the parental chromosome number.
   a. doubles c. maintains
   b. halves d. mixes up

6. Dogs have a diploid chromosome number of 78. How many chromosomes do their gametes have?
   a. 39 c. 156
   b. 78 d. 234

7. The cell in the diagram to the right is in anaphase I, not anaphase II. I know this because _______ .

8. Which of the following cells can undergo meiosis?
   a. the diploid body cells of an animal
   b. haploid gametes
   c. germ cells

9. The cell pictured to the right is in which stage of nuclear division?
   a. metaphase II c. anaphase II
   b. anaphase I d. interphase

10. Crossing over mixes up _______ .
    a. chromosomes c. zygotes
    b. alleles d. gametes

11. Crossing over happens during which phase of meiosis?
    a. prophase I c. anaphase I
    b. prophase II d. anaphase II

12. Which phase of meiosis reduces the chromosome number?
    a. prophase I c. anaphase I
    b. prophase II d. anaphase II

13. Which of the following is one of the very important differences between mitosis and meiosis?
    a. Chromosomes align midway between spindle poles only in meiosis.
    b. Homologous chromosomes separate only in meiosis.
    c. DNA is replicated before mitosis only.
    d. Sister chromatids separate only in meiosis.
    e. Interphase occurs only in mitosis.

14. Match each term with the best description.

___ DNA replication	a. basis of variation in traits
___ metaphase I	b. useful for varied offspring
___ alleles	c. none between meiosis I and meiosis II
___ prophase I	d. similar to mitosis
___ zygotes	e. haploid
___ gametes	f. form at fertilization
___ males	g. mash-up time
___ meiosis II	h. chromosome lineup

15. Which of the following does *not* directly contribute to variation in traits among the offspring of sexual reproducers?
    a. Crossing over during meiosis
    b. The union of gametes at fertilization
    c. Equal distribution of chromosomes into gametes
    d. Alleles on homologous chromosomes
    e. Mutation

CRITICAL THINKING

1. The diploid chromosome number for the body cells of a frog is 26. What would that number be after three generations if meiosis did not occur before gamete formation?

2. In your own words, describe how genetic diversity makes a population resilient to environmental change.

3. Make a simple sketch of meiosis in a cell with a diploid chromosome number of 4. Now try it when the chromosome number is 3.

4. Asexually reproducing animal populations are rare, but they typically consist of females who produce female offspring, so all offspring can produce more offspring. By contrast, half of the offspring of the sexual reproducers are male and cannot produce offspring. Thus, a clonal population of females can expand much more quickly than a population of sexual reproducers.

   Potamopyrgus antipodarum is a species of freshwater snail that has spread far beyond its native New Zealand. Some populations of these tiny snails reproduce sexually, and other populations reproduce asexually. Huge asexual populations are now disrupting ecosystems all over the world, particularly where the water is contaminated with phosphorus. Fertilizers and detergents contain phosphorus, so agricultural runoff and other forms of water pollution are common in these regions.

   Like most animals that cannot reproduce sexually, the asexual snails have at least three sets of chromosomes; like most sexual organisms, the sexual snails have two. Explain why sexually reproducing snail populations predominate in regions where the water is unpolluted.

5. Figure 12.7 shows how meiosis in a cell with three pairs of homologous chromosomes produces eight unique gametes (crossovers aside). Use the same technique to determine how many unique gametes can be produced by a cell that has four pairs of homologous chromosomes.

   A human female can release about 350 eggs during her reproductive years. What is the chance that she would generate the same gamete twice in her lifetime?

CREDIT: (in text S-Q #9) Michael Clayton/ University of Wisconsin, Department of Botany.

13 OBSERVING PATTERNS IN INHERITED TRAITS

CORE CONCEPTS

 **Information Flow**

Living systems store, retrieve, transmit, and respond to information essential for life.
The chromosomal basis of inheritance explains patterns in which traits appear among generations of offspring. For example, alleles of some genes are usually inherited together; alleles of others are inherited independently. Meiosis and fertilization produce variation in shared traits among offspring of sexual reproducers.

 **Process of Science**

The field of biology consists of and relies on experimentation and the collection and analysis of scientific evidence
By applying mathematical reasoning to a biological phenomenon, Gregor Mendel discovered some simple patterns of inheritance in garden pea plants. Quantitative analysis of the distribution of traits among generations of offspring can reveal Mendelian patterns of inheritance, and can also confirm non-Mendelian patterns.

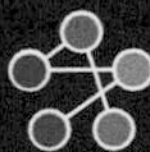 **Systems**

Complex properties arise from interactions among components of a biological system.
Most traits are inherited in complex patterns. For example, one trait may be affected by multiple genes, or one gene may influence multiple traits. Traits of an individual may also be influenced by the environment. Changes in genetic information, such as occur by mutation, may result in observable changes in traits. Outcomes can include abnormal development and/or function.

Links to Earlier Concepts

You may want to review what you know about traits (Section 1.5), chromosomes (8.4), genes (9.2) and gene expression (10.2), mutations and hemoglobin (9.6), sexual reproduction (12.1), alleles (12.2), and meiosis (12.3, 12.4). This chapter revisits probability and sampling error (1.7), laws of nature (1.8), transport proteins (5.7), pigments (6.3), gene expression control (10.2, 10.3), epigenetics (10.5), and homologous chromosomes (11.2).

13.1 Menacing Mucus

Cystic fibrosis (CF), the most common fatal genetic disorder in the United States, occurs in people homozygous for a mutated allele of the *CFTR* gene. The gene encodes an active transport protein that moves chloride ions out of epithelial cells. Sheets of these cells line passageways and ducts of the lungs, liver, pancreas, intestines, and reproductive system. When the CFTR protein pumps chloride ions out of these cells, water follows the solute by osmosis, a two-step process that maintains a thin, watery film on the surface of epithelial cell sheets. Mucus slides easily over the wet sheets of cells.

The allele most commonly associated with CF has a deletion of three base-pairs. It is called *ΔF508* because it encodes a CFTR protein missing the phenylalanine (F) that is normally the 508th amino acid (Δ means deleted). A CFTR protein missing this amino acid misfolds in a tiny region. This small defect interferes with cellular processes that would otherwise finish the protein and install it in the plasma membrane. Normally, a newly translated CFTR polypeptide is modified by the endoplasmic reticulum (ER) and exported to a Golgi body, which attaches carbohydrates to it. The finished protein is then packaged in vesicles routed to the plasma membrane. CFTR polypeptides with the missing amino acid are produced properly, but a cellular quality control mechanism recognizes the misfolded region and destroys most of them before they leave the ER. A few misfolded proteins that reach the plasma membrane are quickly taken back into the cell by endocytosis and destroyed.

Epithelial cell membranes that lack the CFTR protein cannot transport chloride ions. Too few chloride ions leave these cells. Not enough water leaves them either, so the surfaces of epithelial cell sheets are too dry. Mucus that normally slips through the body's tubes sticks to the walls of the tubes instead. Thick globs of mucus accumulate and clog passageways and ducts throughout the body. Breathing becomes difficult as the mucus obstructs the smaller airways of the lungs. Digestive problems arise as ducts that lead to the gut become clogged. Males are typically infertile because the flow of sperm is hampered through tubes of the reproductive system.

In addition to its role in chloride ion transport, the CFTR protein also helps alert the immune system to the presence of disease-causing bacteria in the lungs. It functions as a receptor by binding directly to these bacteria and causing them to be taken into the cell by endocytosis. In epithelial cells lining the respiratory tract, endocytosis of bacteria triggers an immune response. Bacteria-fighting molecules that are produced

CREDITS: (opposite) Photograph from left: first row, Anemone/Shutterstock; Jason Salmon/Shutterstock; Anemone/Shutterstock; Villedieu Christophe/Shutterstock; second row, szefei/Shutterstock; Aaron Amat/Shutterstock; photoJS/Shutterstock; Radius / Superstock; third row, Tatiana Makotra/Shutterstock; J. Helgason/Shutterstock; Tischenko Irina/Shutterstock; photoJS/Shutterstock; fourth row, vaaka/Shutterstock; Anemone/Shutterstock; rawcaptured photography/Shutterstock; evantravels/Shutterstock; fifth row, Andrey Armyagov/Shutterstock; photoJS/Shutterstock; lightpoet/Shutterstock; Jason Salmon/Shutterstock.

in the response keep microbial populations at bay. When the cells lack CFTR, this early alert system fails, so bacteria have time to multiply before being detected by the immune system. Thus, chronic bacterial infections of the lungs are a hallmark of cystic fibrosis. Antibiotics help control infections, but there is no cure for the disorder. Most affected people die before age thirty, when their tormented lungs fail. ●

13.2 Mendel, Pea Plants, and Inheritance Patterns

LEARNING OBJECTIVES

- Explain Gregor Mendel's contribution to the study of inheritance.
- Describe the difference between a homozygous and heterozygous genotype, and represent each symbolically with an example.
- Use an example to describe dominant and recessive alleles.

In the nineteenth century, people thought that hereditary material must be some type of fluid, with fluids from both parents blending at fertilization like milk into coffee. However, the idea of "blending inheritance" failed to explain what people could see with their own eyes: Children sometimes have traits such as freckles that do not appear in either parent, for example. A cross between a black horse and a white one does not produce gray offspring.

The naturalist Charles Darwin had no hypothesis to explain such phenomena, even though inheritance was central to his theory of natural selection (we return to Darwin and natural selection in Chapter 16). At the time, no one knew that hereditary information (DNA) is divided into discrete units (genes), an insight that is critical to understanding how traits are inherited. However, even before Darwin presented his theory, someone had been gathering evidence that would support it. Gregor Mendel (left), an Austrian monk, had been carefully breeding thousands of pea plants. By keeping detailed records of traits that passed from one generation to the next, Mendel had been collecting evidence of how inheritance works.

Mendel's Experiments

Mendel cultivated the garden pea plant (FIGURE 13.1). This species is naturally self-fertilizing, which means each plant's flowers produce male and female gametes ❶ that form viable embryos when they meet up. In order to study inheritance, Mendel had to carry out controlled matings between individuals with specific traits. First, he prevented pea plants from self-fertilizing by removing the pollen-bearing parts (anthers) from their flowers. Second, he cross-fertilized the plants by brushing the egg-bearing parts (carpels) of their flowers with pollen from other plants ❷. Third, he collected seeds ❸ from the cross-fertilized individuals, and recorded the traits of the resulting pea plant offspring ❹.

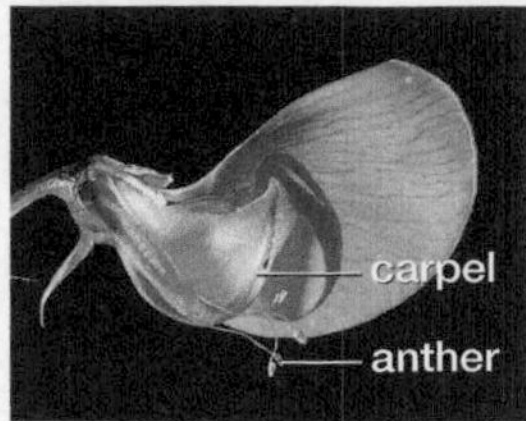
carpel
anther

❶ In garden pea plants (left), pollen grains that form in anthers of the flowers (right) produce male gametes. Female gametes form in carpels.

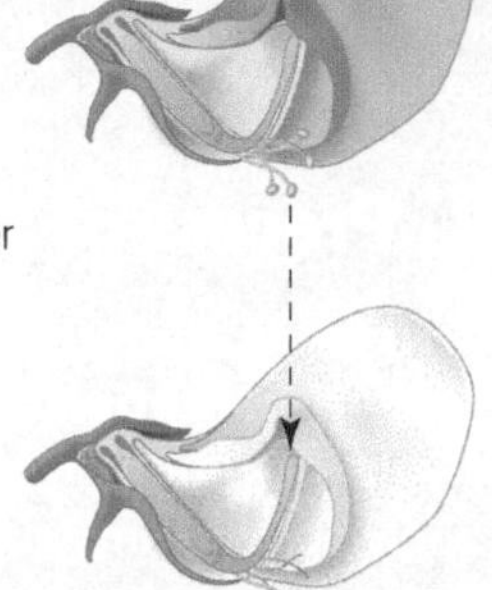

❷ Experimenters control the transfer of hereditary material from one pea plant to another by cutting off a flower's pollen-producing anthers (to prevent it from self-fertilizing), then brushing pollen from another flower onto its egg-producing carpel.

In this example, pollen from a plant with purple flowers is brushed onto the carpel of a white-flowered plant.

❸ Later, seeds develop inside pods of the cross-fertilized plant. When the seeds are planted, the embryo in each develops into a pea plant.

❹ In this example, every plant that arises from the cross has purple flowers. Predictable patterns such as this are evidence of how inheritance works.

FIGURE 13.1 **Breeding experiments with the garden pea.**

dominant Refers to an allele that masks the effect of a recessive allele paired with it in heterozygous individuals.
genotype (JEEN-oh-type) The particular set of alleles that is carried by an individual's chromosomes.
heterozygous (het-er-uh-ZYE-guss) Genotype in which homologous chromosomes have different alleles at the same locus.
homozygous (ho-mo-ZYE-guss) Genotype in which homologous chromosomes have the same allele at the same locus.
hybrid (HI-brid) A heterozygous individual.
locus (LOW-cuss) A gene's location on a chromosome.
phenotype (FEEN-oh-type) An individual's observable traits.
recessive Refers to an allele with an effect that is masked by a dominant allele on the homologous chromosome.

CREDITS: (in text) Moravian Museum; (1-1) left, Yasonya/Shutterstock; right, Jean M. Labat/ Ardea London.

Many of Mendel's experiments, which are called crosses, started with plants that "breed true" for particular traits such as white flowers or purple flowers. Breeding true for a trait means that, new mutations aside, all offspring have the same form of the trait as the parent(s), generation after generation. For example, all offspring of pea plants that breed true for white flowers also have white flowers. As you will see in Section 13.3, Mendel cross-fertilized pea plants that breed true for different forms of a trait, and discovered that the traits of the offspring often appear in predictable patterns. Mendel's meticulous work tracking pea plant traits led him to conclude (correctly) that hereditary information passes from one generation to the next in discrete units.

Inheritance in Modern Terms

Mendel discovered hereditary units, which we now call genes, almost a century before the discovery of DNA (Section 8.2). Today, we know that individuals of a species share certain traits because their chromosomes carry the same genes.

The location of a gene on a chromosome is called its **locus** (plural, loci, **FIGURE 13.2**). Diploid cells have pairs of homologous chromosomes (Section 11.2), so they have two copies of each gene; in most cases, both copies are expressed at the same level. The two copies of any gene may be identical, or they may vary as alleles (Section 12.2).

An individual with the same allele on both homologous chromosomes is **homozygous** for the allele (*homo-* means the same). Organisms breed true for a trait because they are homozygous for alleles governing the trait. By contrast, an individual with different alleles at a gene locus is **heterozygous** (*hetero-* means mixed). A **hybrid** is a heterozygous individual produced by a cross or mating between parents that breed true for different forms of a trait.

Homozygous and heterozygous are examples of **genotype**, the particular set of alleles an individual carries. Genotype is the basis of **phenotype**, which refers to an individual's observable traits. "White-flowered" and "purple-flowered" are examples of pea plant phenotypes that arise from differences in genotype.

The phenotype of a heterozygous individual depends on how the products of its two different alleles interact. In many cases, the effect of one allele influences the effect of the other, and the outcome of this interaction is reflected in the individual's phenotype. An allele is **dominant** when its effect masks that of a **recessive** allele paired with it. A dominant allele is often represented by an uppercase italic letter such as *A*; a recessive allele, by a lowercase italic letter such

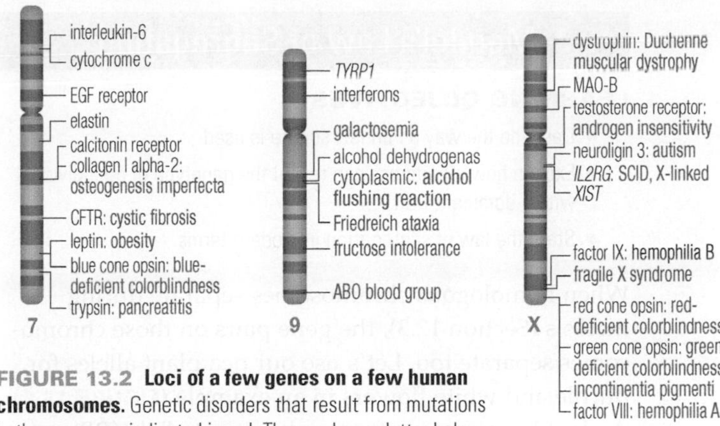

interleukin-6
cytochrome c
EGF receptor
elastin
calcitonin receptor
collagen I alpha-2: osteogenesis imperfecta
CFTR: cystic fibrosis
leptin: obesity
blue cone opsin: blue-deficient colorblindness
trypsin: pancreatitis
7
TYRP1
interferons
galactosemia
alcohol dehydrogenas
cytoplasmic: alcohol flushing reaction
Friedreich ataxia
fructose intolerance
ABO blood group
9
dystrophin: Duchenne muscular dystrophy
MAO-B
testosterone receptor: androgen insensitivity
neuroligin 3: autism
IL2RG: SCID, X-linked
XIST
factor IX: hemophilia B
fragile X syndrome
red cone opsin: red-deficient colorblindness
green cone opsin: green-deficient colorblindness
incontinentia pigmenti
factor VIII: hemophilia A
X

FIGURE 13.2 Loci of a few genes on a few human chromosomes. Genetic disorders that result from mutations in the genes are indicated in red. The number or letter below a chromosome is its name. Characteristic banding patterns appear after staining. A similar map of all 23 chromosomes is in Appendix IV.

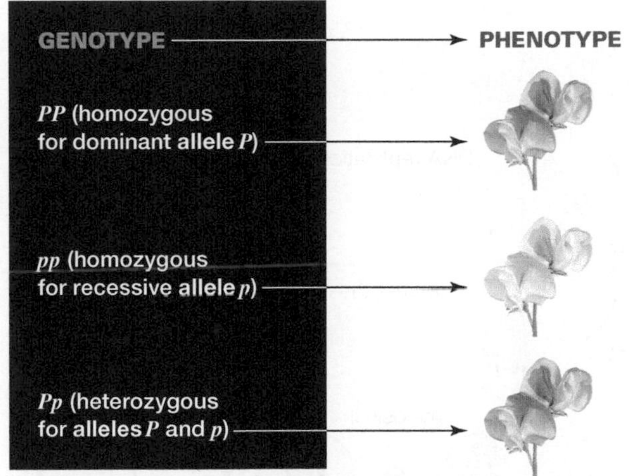

GENOTYPE
PHENOTYPE
PP (homozygous for dominant allele *P*)
pp (homozygous for recessive allele *p*)
Pp (heterozygous for alleles *P* and *p*)

FIGURE 13.3 Genotype gives rise to phenotype. In this example, the dominant allele *P* specifies purple flowers; the recessive allele *p*, white flowers.

FIGURE IT OUT Which individual is a hybrid?

Answer: The heterozygous one

as *a*. Consider the purple- and white-flowered pea plants that Mendel studied. In these plants, the allele that specifies purple flowers (let's call it *P*) is dominant over the allele that specifies white flowers (*p*). Thus, a pea plant homozygous for the dominant allele (*PP*) has purple flowers; one homozygous for the recessive allele (*pp*) has white flowers (**FIGURE 13.3**). A heterozygous plant (*Pp*) has purple flowers.

TAKE-HOME MESSAGE 13.2

- ✔ Genotype refers to the particular set of alleles that an individual carries. Genotype is the basis of phenotype, which refers to the individual's observable traits.
- ✔ A homozygous individual has two identical alleles of a gene. A heterozygous individual has two nonidentical alleles.
- ✔ A dominant allele masks the effect of a recessive allele paired with it in a heterozygous individual.

13.3 Mendel's Law of Segregation

LEARNING OBJECTIVES

- Describe the way a Punnett square is used.
- Explain how a testcross can reveal the genotype of an individual with a dominant trait.
- State the law of segregation in modern terms.

When homologous chromosomes separate during meiosis (Section 12.3), the gene pairs on those chromosomes separate too. Let's use our pea plant alleles for purple and white flowers in an example (**FIGURE 13.4**). A plant homozygous for the dominant allele (*PP*) can only make gametes that carry the dominant allele *P* ❶. A plant homozygous for the recessive allele (*pp*) can only make gametes that carry the recessive allele *p* ❷. If the two plants are crossed (*PP* × *pp*), only one outcome is possible: A gamete carrying allele *P* meets up with a gamete carrying allele *p* ❸. All offspring of this cross will have both alleles—they will be heterozygous (*Pp*). A grid called a **Punnett square** is helpful for predicting the outcomes of such crosses (**FIGURE 13.5**).

In a **testcross**, an individual that has a dominant trait (but unknown genotype) is crossed with an individual known to be homozygous for the recessive allele. The pattern of traits among the offspring of the cross can reveal whether the tested individual is heterozygous or homozygous. If all of the offspring of the testcross have the dominant trait (as occurred in our example above), then the parent with the unknown genotype is likely to be homozygous for the dominant allele. If some of the offspring have the recessive trait, then the parent is heterozygous.

Dominance relationships between alleles determine the phenotypic outcome of a **monohybrid cross**, in which individuals that are identically heterozygous at one gene locus are crossed (*Pp* × *Pp*). The frequency at which traits associated with the alleles appear among the offspring depends on whether one of the alleles is dominant over the other.

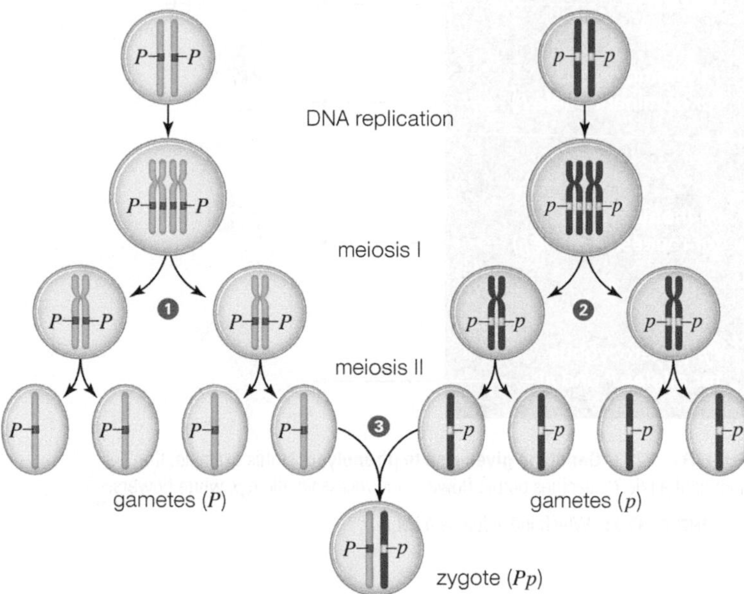

DNA replication
meiosis I
meiosis II
❶
❷
❸
gametes (*P*)
gametes (*p*)
zygote (*Pp*)

FIGURE 13.4 Segregation of genes on homologous chromosomes into gametes. Homologous chromosomes separate during meiosis, so the genes on them separate too. Each of the resulting gametes carries one of the two members of every gene pair. For clarity, only one set of chromosomes is illustrated.

❶ All gametes made by a parent homozygous for a dominant allele carry that allele.

❷ All gametes made by a parent homozygous for a recessive allele carry that allele.

❸ If these two parents are crossed, the union of any of their gametes at fertilization produces a zygote (Section 12.2) with both alleles. All offspring will be heterozygous.

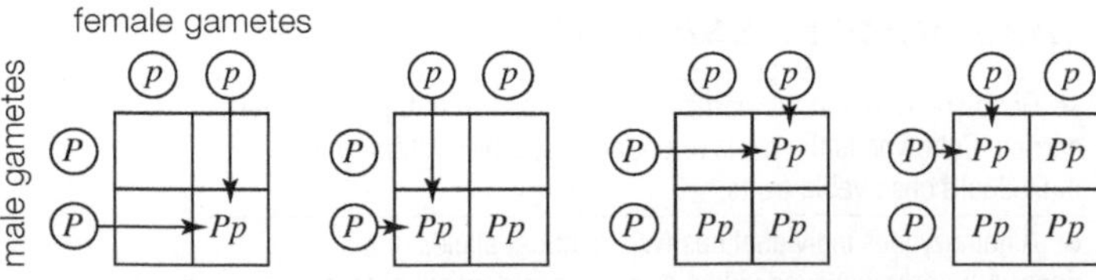

female gametes
male gametes

FIGURE 13.5 Making a Punnett square. Parental gametes are circled on the top and left sides of a grid. Each square is filled with the combination of alleles that would result if the gametes in the corresponding row and column met up.

TABLE 13.1

Mendel's Seven Pea Plant Traits

Trait	Dominant Form	Recessive Form
Seed Shape	Round	Wrinkled
Seed Color	Yellow	Green
Pod Texture	Smooth	Wrinkled
Pod Color	Green	Yellow
Flower Color	Purple	White
Flower Position	Along Stem	At Tip
Stem Length	Tall	Short

Making a monohybrid cross starts with two individuals that breed true for different forms of a trait. In pea plants, flower color (purple and white) is one example of a trait with two distinct forms, but there are many others. Mendel investigated seven of them: stem length (tall and short), seed color (yellow and green), pod texture (smooth and wrinkled), and so on (**TABLE 13.1**). A cross between individuals that breed true for different forms of a trait yields hybrid offspring, all with the same set of alleles governing the trait (**FIGURE 13.6A**). A cross between two of these F_1 (first generation) hybrids is the monohybrid cross. The frequency at which the two traits appear in the F_2 (second generation) offspring offers information about a dominance relationship between the two alleles. (F is an abbreviation for filial, which means offspring.)

A cross between two purple-flowered heterozygous individuals (*Pp*) is an example of a monohybrid cross. Each of these plants can make two types of gametes: ones that carry a *P* allele, and ones that carry a *p* allele (**FIGURE 13.6B**). So, in a monohybrid cross between two *Pp* plants (*Pp* × *Pp*), the two types of gametes can meet up in four possible ways at fertilization:

Possible Event	Probable Outcome
Sperm *P* meets egg *P* →	zygote genotype is *PP* → purple flowers
Sperm *P* meets egg *p* →	zygote genotype is *Pp* → purple flowers
Sperm *p* meets egg *P* →	zygote genotype is *Pp* → purple flowers
Sperm *p* meets egg *p* →	zygote genotype is *pp* → white flowers

Three of four possible outcomes of this cross include at least one copy of the dominant allele *P*. In other words, each time fertilization occurs, there are 3 chances in 4 that the resulting offspring will have a *P* allele (and the individual will make purple flowers). There is 1 chance in 4 that the zygote will have two *p* alleles (and the individual will make white flowers). Thus, the probability that a particular offspring of this cross will have purple or white flowers is 3 purple to 1 white—a ratio of 3:1 (**FIGURE 13.6C**). If the probability of one individual inheriting a particular genotype is difficult to imagine, think about it in terms of many offspring. In this example, there will be roughly three purple-flowered plants for every white-flowered plant.

law of segregation A diploid cell has two copies of every gene that occurs on its homologous chromosomes. Two alleles at any locus are distributed into separate gametes during meiosis.
monohybrid cross Cross between two individuals identically heterozygous for alleles of one gene; for example, *Aa* × *Aa*.
Punnett square (PUN-it) Diagram used to predict the genotypic and phenotypic outcomes of a cross.
testcross Method of determining the genotype of an individual with a dominant phenotype: a cross between the individual and another individual known to be homozygous recessive.

A A monohybrid cross starts with two individuals that breed true for different forms of a trait. Each is homozygous for one allele and only makes gametes with that allele.

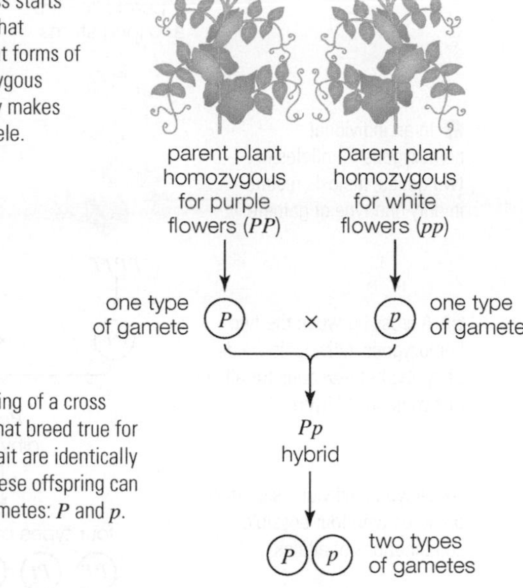
parent plant homozygous for purple flowers (PP)
parent plant homozygous for white flowers (pp)
one type of gamete
P
×
p
one type of gamete
Pp
hybrid
P
p
two types of gametes

B All of the F_1 offspring of a cross between two plants that breed true for different forms of a trait are identically heterozygous (*Pp*). These offspring can make two types of gametes: *P* and *p*.

C A cross between the F_1 offspring (*Pp* × *Pp*) is a monohybrid cross. In this example, the chance that a particular F_2 offspring has purple flowers is 3:1 (3 purple to 1 white). Among many offspring, there are about three purple-flowered individuals for every white-flowered individual.

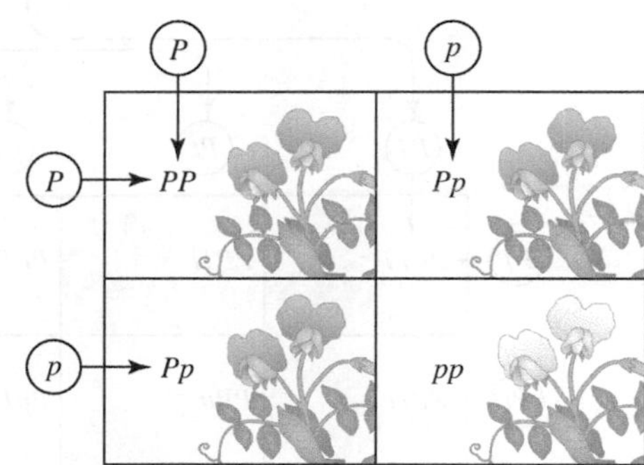
P
p
P
PP
Pp
p
Pp
pp

FIGURE 13.6 A monohybrid cross. In this example, *P* and *p* are dominant and recessive alleles for purple and white flower color.

FIGURE IT OUT How many genotypes are possible in the F_2 generation?

Answer: Three: *PP*, *Pp*, and *pp*

Our example illustrated a pattern so predictable that it can be used as evidence of a dominance relationship between alleles. The phenotype ratios in the F_2 offspring of Mendel's monohybrid crosses were all close to 3:1. These results became the basis of his **law of segregation**, which we state here in modern terms: A diploid cell has two copies of every gene that occurs on its homologous chromosomes, and the two copies may vary as alleles. Two alleles at any locus are distributed into separate gametes during meiosis.

TAKE-HOME MESSAGE 13.3

✔ Recurring patterns of inheritance offer observable evidence of how heredity works.

✔ Homologous chromosomes carry pairs of genes. The two genes of each pair are separated from each other during meiosis, so they end up in different gametes.

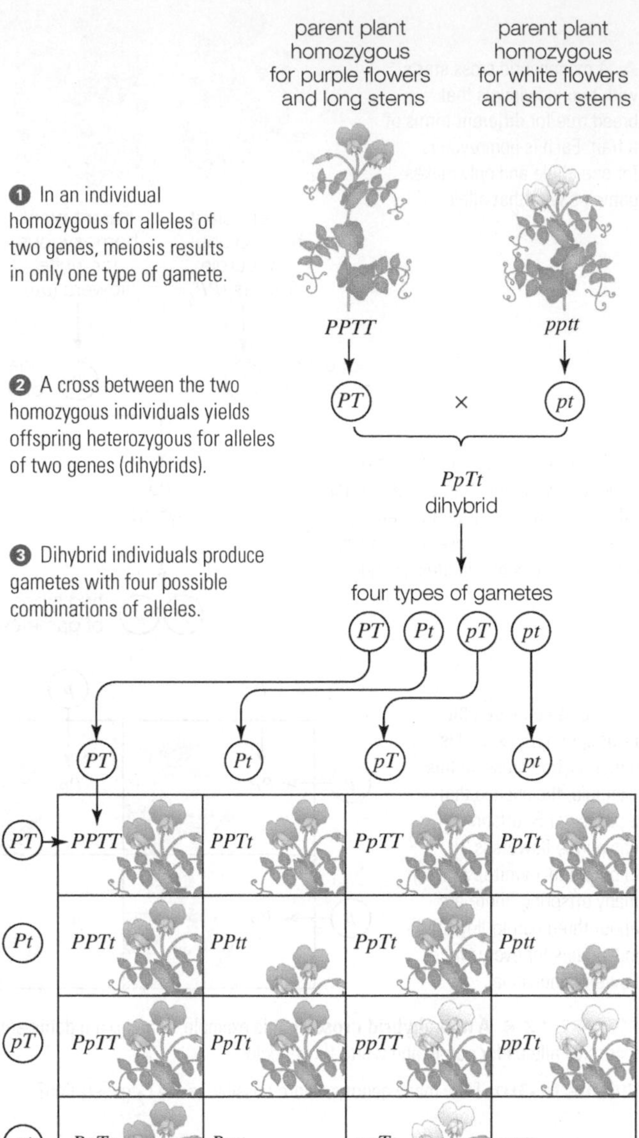
parent plant homozygous for purple flowers and long stems
parent plant homozygous for white flowers and short stems
1 In an individual homozygous for alleles of two genes, meiosis results in only one type of gamete.
PPTT
pptt
PT
×
pt
2 A cross between the two homozygous individuals yields offspring heterozygous for alleles of two genes (dihybrids).
PpTt
dihybrid
3 Dihybrid individuals produce gametes with four possible combinations of alleles.
four types of gametes
PT
Pt
pT
pt
PPTT
PPTt
PpTT
PpTt
PPTt
PPtt
PpTt
Pptt
PpTT
PpTt
ppTT
ppTt
PpTt
Pptt
ppTt
pptt

4 If two of the dihybrid individuals are crossed, the four types of gametes can meet up in 16 possible ways. Of 16 possible offspring genotypes, 9 will result in plants that are purple-flowered and tall; 3, purple-flowered and short; 3, white-flowered and tall; and 1, white-flowered and short. Thus, the ratio of phenotypes is 9:3:3:1.

FIGURE 13.7 A dihybrid cross.

In this example, *P* and *p* are dominant and recessive alleles for purple and white flower color; *T* and *t* are dominant and recessive alleles for tall and short plant height.

FIGURE IT OUT What do the flowers inside the boxes represent?

Answer: Phenotypes of the F_2 offspring

dihybrid cross Cross between two individuals identically heterozygous for alleles of two genes; for example *AaBb* × *AaBb*.

law of independent assortment During meiosis, alleles at one gene locus on homologous chromosomes tend to be distributed into gametes independently of alleles at other loci.

linkage group A set of genes whose alleles do not assort independently into gametes.

13.4 Mendel's Law of Independent Assortment

LEARNING OBJECTIVES

- State the law of independent assortment in modern terms.
- Explain why the relative location of two genes on a chromosome can affect the way their alleles are distributed into gametes.

A monohybrid cross allows us to study a dominance relationship between alleles of one gene. What about alleles of two genes? An individual heterozygous for two alleles at two loci (*AaBb*, for example) is called a dihybrid, and a cross between two such individuals is a **dihybrid cross**. As with a monohybrid cross, the frequency of traits appearing among the offspring of a dihybrid cross depends on the dominance relationships between the alleles.

To make a dihybrid cross, we would start with individuals that breed true for two different traits. Let's use a gene for flower color (*P*, purple; *p*, white) and one for plant height (*T*, tall; *t*, short) in an example (FIGURE 13.7). The dihybrid cross begins with one parent plant that breeds true for purple flowers and tall stems (*PPTT*), and one that breeds true for white flowers and short stems (*pptt*). The *PPTT* plant only makes gametes with the dominant alleles (*PT*); the *pptt* plant only makes gametes with the recessive alleles (*pt*) 1. So, all offspring from a cross between these parent plants (*PPTT* × *pptt*) will be dihybrids (*PpTt*) with purple flowers and tall stems 2.

Four combinations of alleles are possible in gametes made by *PpTt* dihybrids 3. If two *PpTt* plants are crossed (a dihybrid cross, *PpTt* × *PpTt*), the four types of gametes can combine in sixteen possible ways at fertilization 4. Nine of the sixteen genotypes would give rise to tall plants with purple flowers; three, to short plants with purple flowers; three, to tall plants with white flowers; and one, to short plants with white flowers. Thus, the ratio of phenotypes among the offspring of this dihybrid cross would be 9:3:3:1.

Mendel discovered this 9:3:3:1 ratio, but he had no idea what it meant. He could only say that "units" specifying one trait (such as flower color) are inherited independently of "units" specifying other traits (such as plant height). In time, Mendel's hypothesis became known as the **law of independent assortment**, which we state here in modern terms: During meiosis, two alleles at one locus tend to be sorted into gametes independently of alleles at other loci.

Mendel published his results in 1866, but apparently his work was read by few and understood by no one at the time. In 1871 he was promoted, and his pioneering experiments ended. When he died in 1884, he did not

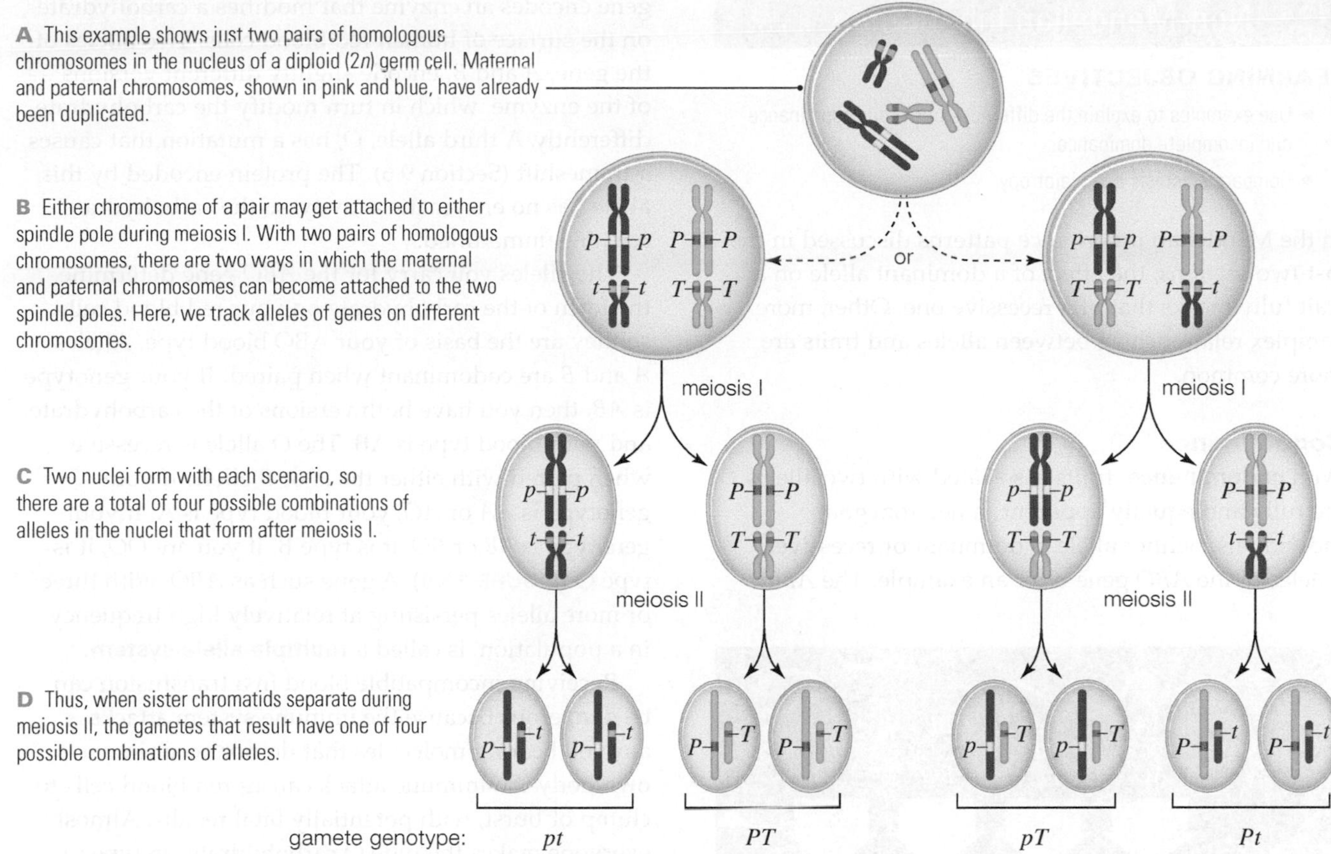
A This example shows just two pairs of homologous chromosomes in the nucleus of a diploid (2n) germ cell. Maternal and paternal chromosomes, shown in pink and blue, have already been duplicated.
B Either chromosome of a pair may get attached to either spindle pole during meiosis I. With two pairs of homologous chromosomes, there are two ways in which the maternal and paternal chromosomes can become attached to the two spindle poles. Here, we track alleles of genes on different chromosomes.
C Two nuclei form with each scenario, so there are a total of four possible combinations of alleles in the nuclei that form after meiosis I.
D Thus, when sister chromatids separate during meiosis II, the gametes that result have one of four possible combinations of alleles.
or
meiosis I
meiosis I
meiosis II
meiosis II
gamete genotype:
pt
PT
pT
Pt

FIGURE 13.8 Independent assortment. Alleles of genes on different chromosomes assort independently into gametes. Alleles of genes that are far apart on the same chromosome usually assort independently too, because crossovers typically separate them.

know that his work with pea plants would be the starting point for modern genetics.

The Contribution of Crossovers

How alleles at two gene loci get sorted into gametes depends partly on whether the two genes are on the same chromosome. When the members of a pair of homologous chromosomes separate during meiosis, either chromosome can end up in either of the two new nuclei that form (Section 12.4). This random assortment happens independently for each pair of homologous chromosomes in the cell. Thus, alleles of genes on one chromosome pair tend to assort into gametes independently of alleles of genes on other chromosome pairs (**FIGURE 13.8**).

What about genes on the same chromosome? Pea plants have seven pairs of homologous chromosomes. Mendel studied seven pea genes, and alleles of all of them assorted into gametes independently of one another. Was he lucky enough to choose one gene on each of those chromosomes? As it turns out, some of the genes Mendel studied *are* on the same chromosome. These genes are far enough apart that crossing over occurs between them very frequently—so frequently that their alleles tend to assort into gametes independently, just as if the genes were on different chromosomes. By contrast, alleles of genes that are very close together on a chromosome usually do not assort independently into gametes, because crossing over does not happen between them very often. Thus, gametes usually end up with parental combinations of alleles of these genes. Genes whose alleles do not assort independently into gametes are said to be linked. A **linkage group** is a set of linked genes.

TAKE-HOME MESSAGE 13.4

✔ During meiosis, gene pairs on homologous chromosomes tend to be distributed into gametes independently of how other gene pairs are distributed.

✔ Independent assortment of genes on the same chromosome depends on proximity. Genes that are closer together get separated less frequently by crossovers, so gametes often receive parental combinations of alleles of these genes.

✔ Genes whose alleles do not assort independently into gametes are said to be linked.

13.5 Non-Mendelian Inheritance

LEARNING OBJECTIVES

- Use examples to explain the difference between codominance and incomplete dominance.
- Compare epistasis and pleiotropy.

In the Mendelian inheritance patterns discussed in the last two sections, the effect of a dominant allele on a trait fully masks that of a recessive one. Other, more complex relationships between alleles and traits are more common.

Codominance

With **codominance**, traits associated with two alleles are fully and equally apparent in heterozygous individuals; neither allele is dominant or recessive. Alleles of the *ABO* gene offer an example. The *ABO* gene encodes an enzyme that modifies a carbohydrate on the surface of human red blood cells. Two alleles of the gene, *A* and *B*, encode slightly different versions of the enzyme, which in turn modify the carbohydrate differently. A third allele, *O*, has a mutation that causes a frameshift (Section 9.6). The protein encoded by this allele has no enzymatic activity, so the carbohydrate remains unmodified.

The alleles you carry for the *ABO* gene determine the form of the carbohydrate on your red blood cells, so they are the basis of your ABO blood type. Alleles *A* and *B* are codominant when paired. If your genotype is *AB*, then you have both versions of the carbohydrate, and your blood type is AB. The *O* allele is recessive when paired with either the *A* or *B* allele. If your genotype is *AA* or *AO*, your blood type is A. If your genotype is *BB* or *BO*, it is type B. If you are *OO*, it is type O (FIGURE 13.9). A gene such as *ABO*, with three or more alleles persisting at relatively high frequency in a population, is called a **multiple allele system**.

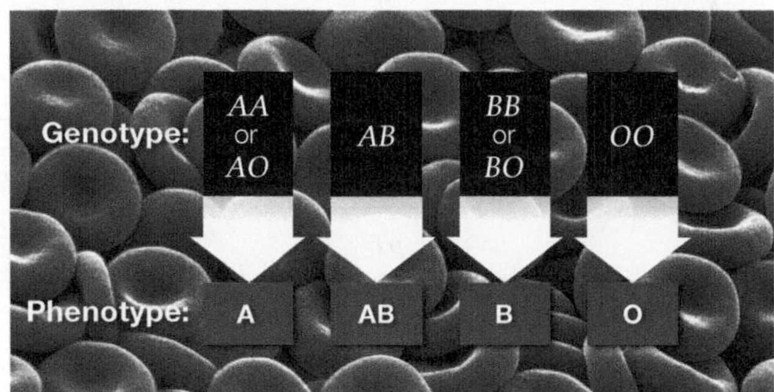
Genotype:
AA or AO
AB
BB or BO
OO
Phenotype:
A
AB
B
O

FIGURE 13.9 **Combinations of alleles** (genotype) **that are the basis of human blood type** (phenotype).

Receiving incompatible blood in a transfusion can be dangerous because the immune system attacks any cell bearing molecules that do not occur in one's own body. An immune attack causes red blood cells to clump or burst, with potentially fatal results. Almost everyone makes the type O carbohydrate, so type O blood does not trigger an immune response in most transfusion recipients. People with type O blood are called universal donors because they can donate blood to anyone. However, because their body is unfamiliar with the carbohydrates made by people with type A or B blood, they can receive type O blood only. People with type AB blood can receive a transfusion of any blood type, so they are called universal recipients.

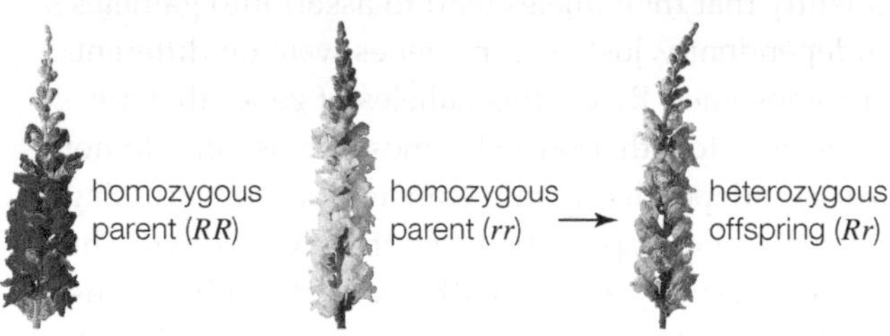
homozygous parent (RR)
homozygous parent (rr)
heterozygous offspring (Rr)

A Cross a red-flowered snapdragon plant with a white-flowered plant, and all of the offspring will have pink flowers.

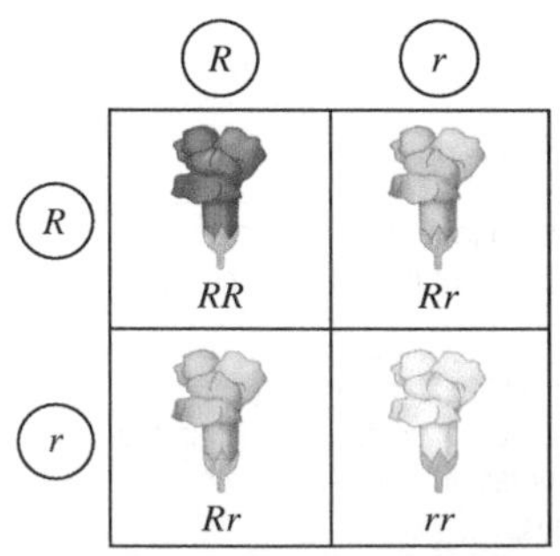
R
r
R
r
RR
Rr
Rr
rr

B If two of the pink-flowered snapdragons are crossed, the phenotypes of their offspring will occur in a 1:2:1 ratio.

FIGURE 13.10 **Incomplete dominance in heterozygous (pink) snapdragons.** One allele (*R*) results in the production of a red pigment; the other (*r*) results in no pigment.

Incomplete Dominance

With **incomplete dominance**, one allele is not fully dominant over the other, so the heterozygous phenotype is an intermediate blend of the two homozygous phenotypes. Alleles of a gene that affects flower color in snapdragons offer an example. One allele (*R*) encodes an enzyme that makes a red pigment. The enzyme

codominance The full and separate phenotypic effects of two alleles are apparent in heterozygous individuals.
epistasis (epp-ih-STAY-sis) Form of polygenic inheritance in which the effect of an allele on a trait masks the effect of a different gene.
incomplete dominance One allele is not fully dominant over another, so the heterozygous phenotype is an intermediate blend between the two homozygous phenotypes.
multiple allele system Gene for which three or more alleles persist in a population at relatively high frequency.
pleiotropic (ply-uh-TROH-pick) Refers to a gene that affects multiple traits.
polygenic inheritance (pall-ee-JENN-ick) Pattern of inheritance in which the form of a single trait is collectively determined by alleles of several genes.

encoded by a mutated allele (*r*) cannot make any pigment. Plants homozygous for the *R* allele (*RR*) make a lot of red pigment, so they have red flowers. Plants homozygous for the *r* allele (*rr*) make no pigment, so their flowers are white. Heterozygous plants (*Rr*) make only enough red pigment to tint their flowers pink (**FIGURE 13.10**). A cross between two heterozygous plants yields red-, pink-, and white-flowered offspring in a 1:2:1 ratio.

Polygenic Inheritance

In a pattern called **polygenic inheritance**, alleles of two or more genes collectively determine the form of a single trait. Hundreds of genes may be involved, with each gene making a small contribution to the phenotype. Consider how fur color in dogs and other animals arises from pigments called melanins. The color of brown or black fur arises from a dark brown form of melanin; a reddish melanin colors fur yellow or tan. The products of several genes interact to make these melanins and deposit them in fur. In Labrador retriever dogs, alleles of two genes determine whether the individual has black, brown, or yellow fur (**FIGURE 13.11**). The product of the *TYRP1* gene helps make the brown melanin. A dominant allele (*B*) of this gene results in a higher production of brown melanin than the recessive allele (*b*). The product of the *MC1R* gene affects the type of melanin produced. A dominant allele (*E*) of this gene triggers production of the brown melanin; its recessive partner (*e*) has a mutation that results in production of the reddish form. Dogs homozygous for the *e* allele are yellow because they make only the reddish melanin. This pattern of polygenic inheritance, in which an allele of one gene masks the effect of a different gene, is called **epistasis**. Note that epistasis is not the same as dominance, which describes a relationship between alleles of the same gene.

Pleiotropy

A **pleiotropic** gene influences multiple traits, so mutations that affect its expression or its product affect all of the traits. Many complex genetic disorders, including sickle-cell anemia and cystic fibrosis, are caused by mutations in single genes (Section 9.6). Marfan syndrome, another example, is a result of mutations in the gene for fibrillin. Long fibers of this protein are part of elastic tissues that make up the heart, skin, blood vessels, tendons, and other body parts. When the fibrillin gene carries a mutation that alters its product, body tissues form with defective fibrillin or none at all. The largest blood vessel leading from the heart, the aorta, is particularly affected. Without a proper scaffold of fibrillin, the aorta's thick wall is not as elastic as it should be, and it eventually stretches and becomes leaky. Thinned and weakened, the aorta can rupture during exercise—an abruptly fatal outcome. About 1 in 5,000 people have Marfan syndrome, and there is no cure. However, increased awareness of the symptoms of Marfan has saved many lives (**FIGURE 13.12**).

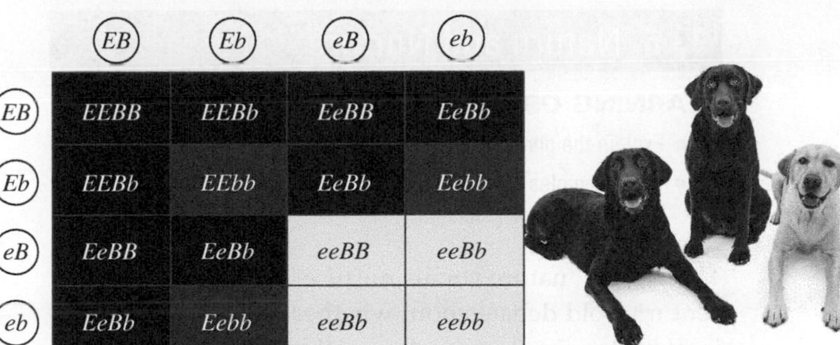

	EB	*Eb*	*eB*	*eb*
EB	*EEBB*	*EEBb*	*EeBB*	*EeBb*
Eb	*EEBb*	*EEbb*	*EeBb*	*Eebb*
eB	*EeBB*	*EeBb*	*eeBB*	*eeBb*
eb	*EeBb*	*Eebb*	*eeBb*	*eebb*

FIGURE 13.11 An example of epistasis. Interactions among products of two genes affect fur color in Labrador retrievers. Dogs with alleles *E* and *B* have black fur. Those with an *E* and two recessive *b* alleles have brown fur. Dogs homozygous for the recessive *e* allele have yellow fur.

BAYLOR
21

FIGURE 13.12 A heartbreaker: Marfan syndrome. Isaiah Austin was diagnosed with Marfan syndrome just days before the 2014 NBA draft. He was considered a first-round prospect, but learning that his heart could rupture unexpectedly during a game ended his dream of becoming a professional basketball player. The diagnosis probably saved Isaiah's life.

TAKE-HOME MESSAGE 13.5

- ✔ Mendel studied traits with distinct forms arising from alleles that have a clear dominant–recessive relationship. Other, more complex relationships between alleles and traits are more common.
- ✔ With incomplete dominance, one allele is not fully dominant over another, so the heterozygous phenotype is an intermediate blend of the two homozygous phenotypes.
- ✔ Codominant alleles have full and separate effects, so the heterozygous phenotype comprises both homozygous phenotypes.
- ✔ With polygenic inheritance, multiple genes collectively influence the form of one trait. Epistasis is an example.
- ✔ A pleiotropic gene affects multiple traits.

13.6 Nature and Nurture

LEARNING OBJECTIVES

- Explain the phrase "nature vs. nurture."
- Give examples of environmental factors that affect phenotype by altering gene expression.

The phrase "nature versus nurture" refers to a centuries-old debate about whether human behavioral traits arise from one's genetics (nature) or from environmental factors (nurture). Today, we know that both play a substantial role. The environment affects the expression of many genes, which in turn affects phenotype—including behavioral traits. We can summarize this thinking with an equation:

$$\text{genotype} + \text{environment} \longrightarrow \text{phenotype}$$

Epigenetics research is revealing that the environment makes an even greater contribution to this equation than biologists had suspected (Section 10.5). Environmental cues initiate cell-signaling pathways that in turn trigger changes in gene expression (you will learn more about such pathways in later chapters). Some of these cell-signaling pathways methylate particular regions of DNA, so they suppress gene expression in those regions. In humans and other animals, DNA methylation patterns can be permanently and heritably affected by diet, stress, and exercise, and also by exposure to drugs and toxins such as tobacco, alcohol, arsenic, and asbestos.

Examples of Environmental Effects

Mechanisms that adjust phenotype in response to external cues are part of an individual's normal ability to adapt to its environment, as the following examples illustrate.

Alternative Phenotypes in Water Fleas Water fleas (*Daphnia*) are tiny aquatic animals that inhabit seasonal ponds and other standing pools of fresh water. Conditions such as temperature, oxygen content, and salinity vary dramatically over time and between different areas of these pools. For example, water at the top of a still summer pond is typically warmer than water at the bottom, and it contains more dissolved oxygen. Individual water fleas that move from one region of a pond to another can acclimate to environmental differences by adjusting their gene expression. The adjustment provides an appropriate set of proteins to maintain cellular function in the different conditions.

Daphnia have a lot of genes, and the abundance offers a striking flexibility of phenotype. Environmental cues trigger adjustments in gene expression that change an individual's form and function to suit its current environment. For example, a water flea that swims to the bottom of a pond can survive the low oxygen conditions there by turning on expression of genes involved in the production of hemoglobin—and turning red (**FIGURE 13.13A**). Hemoglobin is a protein that carries oxygen (Section 9.6), and it enhances the individual's ability to absorb oxygen from the water.

Other environmental factors also affect water flea phenotype. The presence of insect predators causes water fleas to form a protective pointy helmet and lengthened tail spine, for example. Individual water fleas also switch between asexual and sexual modes of reproduction depending on the season. During early spring, food and space are typically abundant, and competition for these resources is minimal. Under these conditions, water fleas reproduce rapidly by asexual means, giving birth to large numbers of female offspring that quickly fill the ponds. Later in the season, competition intensifies as the pond water becomes warmer, saltier, and more crowded. Then, some of the water fleas start giving birth to males, and the population begins to reproduce sexually. The increased genetic diversity of sexually produced offspring offers an advantage in the more challenging environment.

Seasonal Changes in Coat Color In many mammals, seasonal changes in temperature and length of day affect production of melanin and other pigments that color skin and fur. These species have different color phases in different seasons (**FIGURE 13.13B**). Hormonal signals triggered by the seasonal changes cause fur to be shed, and new fur grows back with different types and amounts of pigments deposited in it. The resulting shift in phenotype provides the animals with seasonally appropriate camouflage from predators.

Effect of Altitude on Yarrow In plants, a flexible phenotype gives immobile individuals an ability to thrive in diverse habitats. For example, genetically identical yarrow plants grow to different heights at different altitudes (**FIGURE 13.13C**). More challenging temperature, soil, and water conditions are typically encountered at higher altitudes. Differences in altitude are also correlated with changes in the reproductive mode of yarrow: Plants at higher altitude tend to reproduce asexually, and those at lower altitude tend to reproduce sexually.

Psychiatric Disorders Researchers recently discovered several mutations associated with five human psychiatric disorders: autism, depression, schizophrenia, bipolar disorder, and attention deficit/hyperactivity

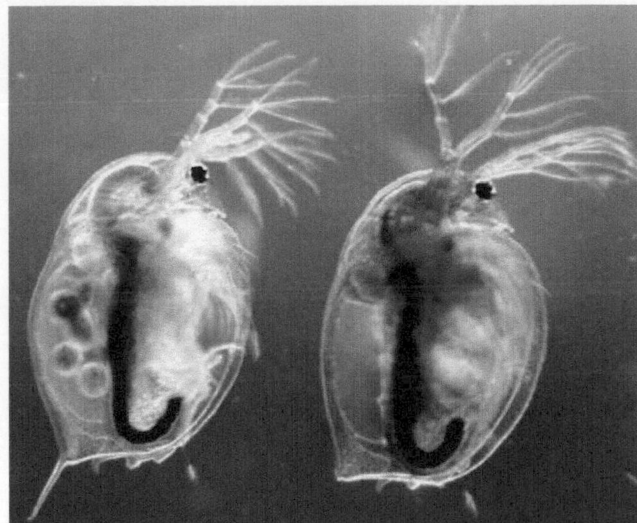

A Under low-oxygen conditions, a water flea (*Daphnia*) switches on genes involved in producing hemoglobin. Making this red protein enhances the individual's ability to take up oxygen from water. The water flea on the left has been living in water with a normal oxygen content; the one on the right, in water with a low oxygen content.

B The color of the snowshoe hare's fur varies by season. A hare's summer fur is brown (left); its winter fur is white (right). Both color forms offer seasonally appropriate camouflage from predators.

60
Height (cm)
0
3060
1400
30

C The height of a mature yarrow plant depends on the elevation at which it grows.

FIGURE 13.13 Some environmental effects on individual phenotype.

disorder (ADHD). However, there must be environmental components to these disorders too, because the majority of people who have the mutations never end up with a psychiatric disorder. Moreover, one person with the mutations might get one type of disorder, while a relative with the same mutations might get another: two different phenotypic results from the same genetic underpinnings.

Animal models are helping us unravel some of the mechanisms by which environment can influence mental state. For example, we have discovered that learning and memory are associated with dynamic and rapid DNA modifications in animal brain cells. Mood is, too. Stress-induced depression causes methylation-based silencing of a particular nerve growth factor gene; some antidepressants work by reversing this methylation. As another example, rats whose mothers are not very nurturing end up anxious and having a reduced resilience for stress as adults. The difference between these rats and ones who had nurturing maternal care is traceable to epigenetic DNA modifications that result in a lower-than-normal level of another nerve growth factor. Drugs can reverse these modifications—and their effects. We do not yet know all of the genes that influence human mental state, but the implication of such research is that future treatments for many psychiatric disorders will involve deliberate modification of methylation patterns in an individual's DNA.

TAKE-HOME MESSAGE 13.6

✔ Genotype, together with environmental influences, determine many phenotypes.

✔ An individual's normal ability to adapt to its environment includes mechanisms that adjust phenotype in response to external cues.

✔ Cell-signaling pathways link environmental cues with changes in gene expression. These changes, in turn, can alter phenotype.

13.7 Complex Variation in Traits

LEARNING OBJECTIVE

- Use examples to explain continuous variation and its causes.

The pea plant phenotypes that Mendel studied appeared in two distinct forms, which made them easy to track through generations. However, many other traits do not appear in distinct forms. Such traits are often the result of complex genetic interactions—multiple genes, multiple alleles, or both—with added environmental influences (we return to this topic in Chapter 17, as we consider some evolutionary consequences of variation in shared traits). The complexity often makes these traits difficult to study, which is why

63 64 65 66 67 68 69 70 71 72 73 74 75 76 77

59 60 61 62 63 64 65 66 67 68 69 70 71

A Continuous variation in height. Photos show University of Florida biology students (men, top; women, bottom) lined up according to height in inches.

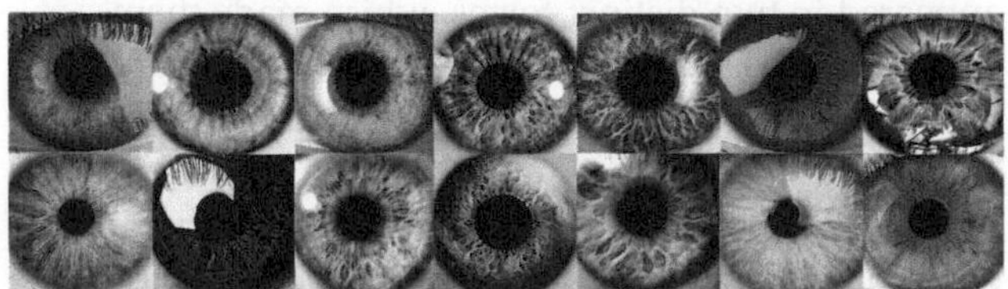

B Continuous variation in iris color.

FIGURE 13.14 Examples of continuous variation in humans.

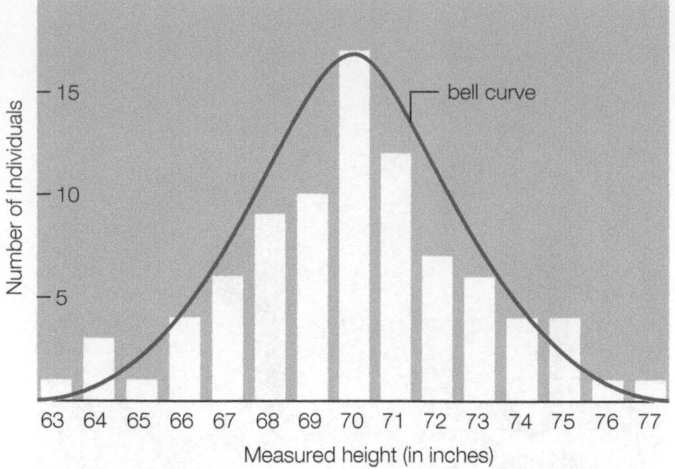
Number of Individuals
15
10
5
bell curve
63 64 65 66 67 68 69 70 71 72 73 74 75 76 77
Measured height (in inches)

FIGURE 13.15 Characteristic bell curve that results from graphing the distribution of values for a trait that varies continuously. Height of male biology students in Figure 13.14A provided the data for this graph.

the genetic basis of many of them has not yet been completely unraveled.

Continuous Variation

Some traits occur in a range of small differences that is called **continuous variation**. Continuous variation can be an outcome of epistasis, in which multiple genes affect a single trait. The more genes and environmental factors that influence a trait, the more continuous is its variation.

Human height varies continuously (**FIGURE 13.14A**), as does human eye color (**FIGURE 13.14B**). The colored part of the eye is a doughnut-shaped structure called the iris. Iris color, like skin color, is the result of interactions among gene products that make and distribute melanins. The more melanin deposited in the iris, the less light is reflected from it and the darker it appears.

How do we know that a particular trait varies continuously? First, we divide the total range of phenotypes into measurable categories. The number of individuals in each category reveals the frequencies of phenotypes across the range of values. When the data are plotted as a bar chart, a graph line around the top of the bars shows the distribution of values for the trait (**FIGURE 13.15**). If the line is a bell-shaped curve, or **bell curve**, then the trait varies continuously.

Traits that arise from genes with a lot of alleles may vary continuously. Consider that some genes have regions in which a series of 2 to 6 nucleotides is repeated many times in a row. The number of these

bell curve Bell-shaped curve; typically results from graphing frequency versus distribution for a trait that varies continuously.

continuous variation Range of small differences in a trait.

short tandem repeat In chromosomal DNA, sequences of a few nucleotides repeated multiple times in a row.

Data Analysis Activities

Carrying the Cystic Fibrosis Allele Offers Protection from Typhoid Fever Epithelial cells that lack the CFTR protein cannot take up bacteria by endocytosis. Endocytosis is an important part of the respiratory tract's immune defenses against common *Pseudomonas* bacteria, which is why *Pseudomonas* infections of the lungs are a chronic problem in cystic fibrosis patients. Endocytosis is also the way that *Salmonella typhi* bacteria (shown at right) enter cells of the gastrointestinal tract, where internalization of this bacteria can result in typhoid fever.

Typhoid fever is a common worldwide disease. Its symptoms include extreme fever and diarrhea, and the resulting dehydration causes delirium that may last several weeks. If untreated, it kills up to 30 percent of those infected. Around 600,000 people, most of whom are children, die annually from typhoid fever.

Gerald Pier and his colleagues compared the uptake of *S. typhi* by different types of epithelial cells: those homozygous for the normal allele, and those heterozygous for the *ΔF508* allele associated with CF. (Cells that are homozygous for the mutation do not take up any *S. typhi* bacteria.) Some of their results are shown in **FIGURE 13.16**.

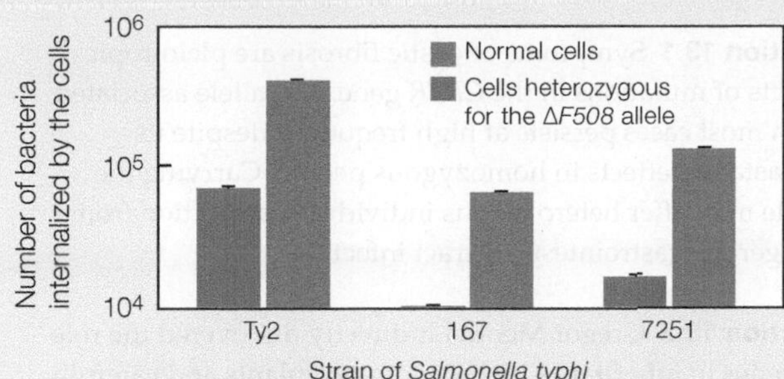
Number of bacteria internalized by the cells
10^6
10^5
10^4
Normal cells
Cells heterozygous for the ΔF508 allele
Ty2
167
7251
Strain of Salmonella typhi

FIGURE 13.16 Effect of the *ΔF508* mutation on the uptake of three different strains of *Salmonella typhi* bacteria by epithelial cells.

1. Regarding the Ty2 strain of *S. typhi*, about how many more bacteria were able to enter normal cells than cells heterozygous for the *ΔF508* allele?
2. Which strain of bacteria entered normal epithelial cells most easily?
3. Entry of all three *S. typhi* strains into the heterozygous epithelial cells was inhibited. Is it possible to tell from this graph which strain was most inhibited?

short tandem repeats can spontaneously increase or decrease during DNA replication and repair, at a rate that is much faster than other mutations. The resulting expansion or contraction of a repeat region may be preserved as an allele. For example, changes in the number of short tandem repeats have given rise to 12 alleles of a homeotic gene that influences face length in dogs. Alleles with more repeats are associated with longer faces (**FIGURE 13.17**).

TAKE-HOME MESSAGE 13.7

✔ The more genes and other factors that influence a trait, the more continuous is its range of variation.

FIGURE 13.17 Face length varies continuously in dogs.

A gene with 12 alleles influences this trait; all arose by spontaneous expansion or contraction of short tandem repeat regions. Variation in the number of repeats correlates with variation in face length: the more repeats, the longer the face.

13.1 Menacing Mucus (revisited)

The *ΔF508* allele that causes cystic fibrosis is at least 50,000 years old and very common: In some populations, 1 in 25 people are heterozygous for it. Why does the allele persist if it is so harmful? The *ΔF508* allele is eventually lethal in homozygous individuals, but not in those who are heterozygous. It is codominant with the normal allele. Heterozygous individuals typically have no symptoms of cystic fibrosis because their cells have plasma membranes with enough CFTR to transport chloride ions normally. The *ΔF508* allele may offer these individuals an advantage in surviving certain deadly infectious diseases. CFTR's receptor function is an essential part of the immune response to bacteria in the respiratory tract. However, the same function allows bacteria to enter cells of the gastrointestinal (GI) tract, where they can be deadly. Thus, people who carry *ΔF508* are probably less susceptible to dangerous bacterial diseases that begin in the GI tract. ●

Section 13.1 Symptoms of cystic fibrosis are pleiotropic effects of mutations in the *CFTR* gene. The allele associated with most cases persists at high frequency despite its devastating effects in homozygous people. Carrying the allele may offer heterozygous individuals protection from dangerous gastrointestinal tract infections.

Section 13.2 Gregor Mendel indirectly discovered the role of genes in inheritance by breeding pea plants and carefully tracking traits of the offspring over many generations. **Genotype** (an individual's alleles) is the basis of **phenotype** (the individual's observable traits). Each gene occurs at a **locus**, or location, on a chromosome. A **homozygous** individual has the same allele of a gene on both homologous chromosomes. A **heterozygous** individual, or **hybrid**, has two different alleles. A **dominant** allele masks the effect of a **recessive** allele in a heterozygous individual.

Section 13.3 Crossing two individuals that breed true for different forms of a trait yields identically heterozygous offspring. A cross between such offspring is called a **monohybrid cross**. The frequency at which traits appear in offspring of a **testcross** can reveal the genotype of an individual with a dominant phenotype. **Punnett squares** are useful for determining the probability of offspring genotype and phenotype. Mendel's monohybrid cross data led to his **law of segregation**, stated here in modern terms: A diploid cell has two copies of every gene that occurs on its homologous chromosomes. Two alleles at any locus separate from each other during meiosis, so they end up in different gametes.

Section 13.4 Crossing individuals that breed true for two forms of two traits yields F_1 offspring identically heterozygous for alleles governing those traits. A cross between these F_1 offspring is a **dihybrid cross**. The frequency at which the two traits appear in F_2 offspring can reveal dominance relationships between the alleles for each trait.

Mendel's dihybrid cross data led to his **law of independent assortment**, stated here in modern terms: Alleles at one locus tend to assort into gametes independently of alleles at other loci. **Linkage groups** are an exception. Crossovers do not often separate genes that are close together on a chromosome, so alleles of these genes tend to be inherited together.

Section 13.5 With **incomplete dominance**, the phenotype of heterozygous individuals is an intermediate blend of the two homozygous phenotypes. With **codominant** alleles, heterozygous individuals have both homozygous phenotypes. Codominance may occur in **multiple allele systems** such as the one underlying ABO blood type. With **polygenic inheritance**, two or more genes affect the same trait. **Epistasis** is a form of polygenic inheritance in which an allele of one gene masks the effect of a different gene. A **pleiotropic** gene affects two or more traits.

Section 13.6 Changes in phenotype are part of an individual's ability to adapt to its environment. Environmental cues alter gene expression by way of cell-signaling pathways that change gene expression (for example, by methylating DNA).

Section 13.7 A trait that is influenced by multiple genes often occurs in a range of small increments of phenotype called **continuous variation**. A **bell curve** in the range of values is typical of a trait that varies continuously. Multiple alleles such as those that arise in regions of **short tandem repeats** can give rise to continuous variation.

SELF-QUIZ

Answers in Appendix VII

1. A heterozygous individual has ______ for a trait being studied.
   a. the same allele on both homologous chromosomes
   b. two different alleles of a gene
   c. a haploid condition, in genetic terms

2. An organism's observable traits constitute its ______ .
   a. phenotype c. genotype
   b. variation d. pedigree

3. Independent assortment means ______ .
   a. alleles at one locus assort into different gametes
   b. alleles at different loci assort into gametes independently of each other
   c. the assortment of genes on homologous chromosomes varies independently

4. The second-generation offspring of a cross between individuals who are homozygous for different alleles of a gene are called the ______ .
   a. F_1 generation c. F_2 generation
   b. hybrid generation d. daughters

5. The offspring of the cross *AA* × *aa* are ______ .
   a. all *AA* c. all *Aa*
   b. all *aa* d. half are *AA* and half are *aa*

6. Refer to question 5. Assuming complete dominance, the F_2 generation will show a phenotypic ratio of ______ .
   a. 3:1 b. 9:1 c. 1:2:1 d. 9:3:3:1

7. A testcross is a way to determine ______ .
   a. phenotype b. genotype c. dominance

8. Assuming complete dominance, a cross between dihybrid F_1 pea plants produces F_2 phenotype ratios of ______ .
   a. 1:2:1 b. 3:1 c. 1:1:1:1 d. 9:3:3:1

9. The probability of a crossover occurring between two genes on the same chromosome ______ .
   a. is unrelated to the distance between them
   b. decreases with increasing distance between them
   c. increases with the distance between them

10. True or false? All traits are inherited in a Mendelian pattern.

11. One gene that affects three traits is an example of ____.
    a. dominance
    b. codominance
    c. pleiotropy
    d. epistasis

12. The phenotype of individuals heterozygous for ________ alleles comprises both homozygous phenotypes.
    a. epistatic
    b. codominant
    c. pleiotropic
    d. hybrid

13. ________ in a trait is indicated by a bell curve.
    a. An epigenetic effect
    b. Pleiotropy
    c. Incomplete dominance
    d. Continuous variation

14. Match the terms with the best description.
    ___ dihybrid cross — a. *bb*
    ___ monohybrid cross — b. *AaBb* × *AaBb*
    ___ homozygous condition — c. *Aa*
    ___ heterozygous condition — d. *Aa* × *Aa*

GENETICS PROBLEMS

Answers in Appendix VII

1. Mendel crossed a true-breeding pea plant with green pods and a true-breeding pea plant with yellow pods. All the F_1 plants had green pods. Which color is recessive?

2. Assuming that independent assortment occurs during meiosis, what type(s) of gametes will form in individuals with the following genotypes?
   a. *AABB* b. *AaBB* c. *Aabb* d. *AaBb*

3. Refer to problem 2. Determine the predicted genotype frequencies among the offspring of the following crosses:
   a. *AABB* × *aaBB*
   b. *AaBB* × *AABb*
   c. *AaBb* × *aabb*
   d. *AaBb* × *AaBb*

4. For each genotype listed, what allele combinations will occur in gametes?
   a. *AABBCC*
   b. *AaBBcc*
   c. *AaBBCc*
   d. *AaBbCc*

5. Heterozygous individuals perpetuate some alleles that have lethal effects in homozygous individuals. A mutated allele (M^L) associated with taillessness in Manx cats (left) is an example. Cats homozygous for this allele (M^LM^L) typically die before birth due to severe spinal cord defects. In a case of incomplete dominance, cats heterozygous for the M^L allele and the normal, unmutated allele (M) have a short, stumpy tail or none at all. Two M^LM cats mate. What is the probability that any one of their surviving kittens will be heterozygous?

6. Suppose you identify a new gene in mice. One of its alleles specifies white fur, another specifies brown. You want to see if these alleles are inherited in a Mendelian pattern, or with incomplete dominance. What crosses would give you the answer?

7. Mutations in the *TYR* gene may render its enzyme product—tyrosinase—nonfunctional. Individuals homozygous for such mutations cannot make the pigment melanin. Albinism, the absence of melanin, results. Humans and many other organisms can have this phenotype (right). Mutated tyrosinase alleles are recessive when paired with the normal allele in heterozygous individuals. In the following situations, what are the probable genotypes of the father, the mother, and their children?

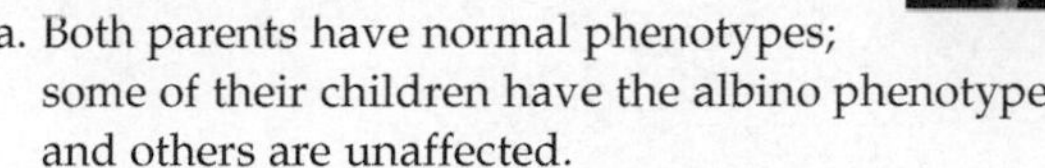

   a. Both parents have normal phenotypes; some of their children have the albino phenotype and others are unaffected.
   b. Both parents and children have the albino phenotype.
   c. The mother and three children are unaffected; the father and one child have the albino phenotype.

8. In sweet pea plants, an allele for purple flowers (P) is dominant when paired with a recessive allele for red flowers (p). An allele for long pollen grains (L) is dominant when paired with a recessive allele for round pollen grains (l). Bateson and Punnett crossed a plant having purple flowers and long pollen grains with one having white flowers and round pollen grains. All F_1 offspring have purple flowers and long pollen grains. Among the F_2 generation, the researchers observed the following phenotypes:
   296 purple flowers/long pollen grains
   19 purple flowers/round pollen grains
   27 red flowers/long pollen grains
   85 red flowers/round pollen grains
   What is the best explanation for these results?

9. Red-flowering snapdragons are homozygous for allele R^1. White-flowering snapdragons are homozygous for a different allele (R^2). Heterozygous plants (R^1R^2) bear pink flowers. What phenotypes should appear among first-generation offspring of the crosses listed? What are the expected proportions for each phenotype?
   a. $R^1R^1 \times R^1R^2$
   b. $R^1R^1 \times R^2R^2$
   c. $R^1R^2 \times R^1R^2$
   d. $R^1R^2 \times R^2R^2$

CREDITS: (in text Genetics Problems #5) Leslie Falteisek/Clacritter Manx; (in text Genetics Problems #8) © Rick Guidotti, Positive Exposure.

14 CHROMOSOMES AND HUMAN INHERITANCE

CORE CONCEPTS

 Information Flow

Living systems store, retrieve, transmit, and respond to information essential for life.

A cell's DNA encodes all of the instructions necessary for growth, survival, and reproduction of the cell and (in multicelled organisms) the individual. The chromosomal basis of inheritance explains patterns in which this genetic information is transmitted to offspring. Some human traits arise from single genes. The inheritance pattern of a single-gene trait can be revealed by tracking the appearance of the trait's alternative forms in individuals through generations of family trees.

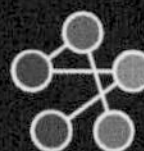 Systems

Complex properties arise from interactions among components of a biological system.

Changes in genotype can affect human phenotype. Mutations in single genes give rise to many genetic disorders, and errors in mitosis or meiosis that change chromosome number or structure often result in developmental disorders. Growth, reproduction, and homeostasis depend on proper timing and coordination of specific molecular events. Mutations that alter expression of genes involved in these events may have complex effects on the cell, tissue, or whole organism.

 Evolution

Evolution underlies the unity and diversity of life.

Individuals of sexually reproducing species vary in the details of their DNA. Such variation is almost always advantageous, because genetically diverse populations are more resilient than populations with low diversity. Genetic diversity is fostered by processes inherent in sexual reproduction.

Links to Earlier Concepts

Be sure you understand inheritance patterns (Sections 13.2, 13.5–13.7); chromosomes (8.4); mutations (8.6, 9.6); gene expression (9.2, 9.3, 10.3); and meiosis (12.2, 12.3). This chapter revisits sampling error (1.7), amyloid fibrils (3.5), eukaryotic cell components (4.4, 4.5, 4.8, 4.9), transport proteins (5.7), mitochondrial malfunction (7.1), melanin genes (13.5), telomeres and senescence (11.5), proto-oncogenes (11.6), cystic fibrosis (13.1), and short tandem repeats (13.7).

14.1 Shades of Skin

The color of human skin begins with melanosomes, which are organelles that make melanin pigments. Most people have about the same number of melanosomes in their skin cells. Variations in skin color arise from differences in the size, shape, and cellular distribution of melanosomes in the skin, as well as in the kinds and amounts of melanins they make.

Variation in human skin color may have evolved as a balance between vitamin production and protection against harmful ultraviolet (UV) radiation in the sun's rays. Dark skin would have been (and still is) beneficial under the intense sunlight of African savannas where humans first evolved. Melanin is a natural sunscreen: It prevents UV radiation from breaking down folate, a vitamin essential for normal sperm formation and embryonic development.

Early human groups that migrated to regions with cold climates were exposed to less sunlight. In these regions, lighter skin color is beneficial. Why? Exposure to UV radiation stimulates skin cells to make a molecule the body converts to vitamin D. Where sunlight exposure is minimal, UV radiation is less of a risk than vitamin D deficiency, which has serious health consequences for children and developing fetuses. In regions with long, dark winters, people with dark, UV-shielding skin have a high risk of this deficiency.

Skin color, like most other human traits, has a genetic basis; at least 100 gene products are involved in pigmentation. The evolution of regional variations in human skin color began with mutations in these genes. Consider a gene on chromosome 15, *SLC24A5*, that encodes a transport protein in melanosome membranes. Nearly all people of African, Native American, or east Asian descent carry the same allele of this gene. Between 6,000 and 10,000 years ago, a mutation gave rise to a different allele. The mutation, a single base-pair substitution (Section 9.6), changed the 111th amino acid of the transport protein from alanine to threonine. The change results in less melanin—and lighter skin color—than the original African allele does. Today, nearly all people of European descent are homozygous for the mutated allele.

A person of mixed ethnicity may make gametes that contain different combinations of alleles for dark and light skin. It is fairly rare that one of those gametes contains all of the alleles for dark skin, or all of the alleles for light skin, but it happens. Skin color is only one of many human traits that vary as a result of single nucleotide mutations. The small scale of such changes offers a reminder that all of us share the genetic legacy of common ancestry. ●

CREDIT: (opposite) Ciarra, photo by © Michelle Harmon.

Data Analysis Activities

Skin Color Survey of Native Peoples In 2000, researchers measured the average amount of UV radiation received in more than fifty regions of the world, and correlated it with the average skin reflectance of people native to those regions (reflectance is a way to measure the amount of melanin pigment in skin: Skin with more melanin is darker, so it reflects less light). Some of the results of this study are shown in **FIGURE 14.1**.

1. Which country receives the most UV radiation?
2. Which country receives the least UV radiation?
3. People native to which country have the darkest skin?
4. People native to which country have the lightest skin?
5. According to these data, how does skin color of human populations correlate with the amount of UV radiation incident in their native regions?

Country	Skin Reflectance	UVMED
Australia	19.30	335.55
Kenya	32.40	354.21
India	44.60	219.65
Cambodia	54.00	310.28
Japan	55.42	130.87
Afghanistan	55.70	249.98
China	59.17	204.57
Ireland	65.00	52.92
Germany	66.90	69.29
Netherlands	67.37	62.58

FIGURE 14.1 Average skin reflectance of people native to ten countries correlated with regional incident UV radiation. Skin reflectance measures how much light of 685-nanometer wavelength is reflected from skin; UVMED is the annual average UV radiation received at Earth's surface.

14.2 Human Chromosomes

LEARNING OBJECTIVES

- Explain why pedigrees are used to study human inheritance patterns.
- Differentiate between a genetic disorder and a genetic abnormality.

Few human traits are inherited in Mendelian patterns. Like the flower color of Mendel's pea plants, such traits arise from a single gene with alleles that have a clear dominance relationship. Consider the *MC1R* gene that you learned about in Section 13.5. In humans, dogs, and other animals, *MC1R* encodes a protein that triggers production of the brownish melanin. Mutations can result in a defective protein; an allele with one of these mutations is recessive when paired with an unmutated allele. A person who has two mutated *MC1R* alleles does not make the brownish melanin—only the reddish type—so this individual has red hair.

Red hair is fairly common in human populations, but other single-gene traits (one gene $\rightarrow$ one trait) are not. Most human traits are inherited in patterns much more complex than the Mendelian model. Human skin color, for example, varies continuously (above) because it is an outcome of interactions among the products of about 350 genes.

Our understanding of inheritance patterns in humans comes mainly from research involving genetic disorders, because this information helps us develop treatments for affected people. Note that a rare or uncommon version of a trait, such as having six fingers on a hand, or a web between two toes, is called a genetic abnormality. Genetic abnormalities are not inherently life-threatening, and how you view them is a matter of opinion. By contrast, a genetic disorder sooner or later causes medical problems that may be quite severe.

Studying Human Genetics

Even though single-gene traits are the least common kind in humans, we actually know most about inheritance of single-gene disorders. This is partly an outcome of relative complexity. Genetic disorders usually have multiple symptoms (a syndrome) that may vary among affected individuals. Environmental effects on phenotype (Section 13.6) further complicate inheritance patterns. Diabetes, asthma, obesity, cancers, heart disease, multiple sclerosis, and many other disorders can be inherited, but in patterns so complex that their genetic underpinnings remain unclear despite decades of research. For example, mutations associated with an increased risk of autism (a developmental disorder) have been found on almost every chromosome, but most people who carry these mutations do not have autism.

Researching human inheritance patterns is intrinsically problematic. Consider how pea plants and fruit flies are ideal for genetics research. Breeding them in a controlled manner poses few ethical problems. They reproduce quickly, so it does not take long to follow a trait through many generations. Humans, however, live under variable conditions, in different places, and we live as long as the geneticists who study us. Most of us choose our own mates and reproduce if and when we want to. Most human families are small, so sampling error (Section 1.7) is unavoidable.

Because of these challenges, geneticists often use historical records to track traits through many generations of a family. They make and study **pedigrees**, which are charts that marks the appearance of a trait among generations of family members. A pedigree is constructed in a standard way, with individuals represented as polygons and relationships between them as lines (**FIGURE 14.2**). Males are represented as squares; females, as circles. A filled-in polygon indicates an individual with the trait being studied. A marriage (or mating) is signified by a line connecting parents. Any offspring of the match are connected to the marriage by a perpendicular line, and every generation appears in a row designated by a roman numeral (I, II, III, and so on).

A pedigree allows geneticists to estimate the probability that a phenotype will reappear in future generations. It can also reveal whether a single-gene trait is associated with a dominant or recessive allele, and whether the allele is on an autosome or a sex chromosome (Section 8.4).

Alleles that give rise to severe genetic disorders are generally rare in populations because they compromise the health and reproductive ability of their bearers. Why do they persist? Inevitable mutations periodically reintroduce them. In some cases, a codominant allele persists because it offers a survival advantage in a particular environment. You learned about one example in Section 13.1, the $\Delta F508$ allele that causes cystic fibrosis: People who are

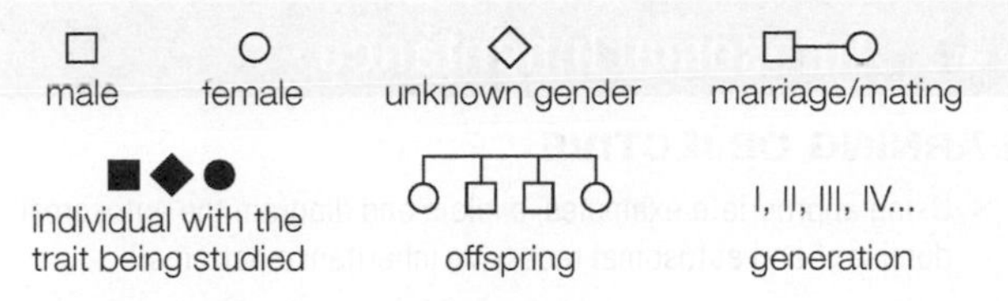
male
female
unknown gender
marriage/mating
individual with the trait being studied
offspring
I, II, III, IV...
generation

A Standard symbols used in pedigrees.

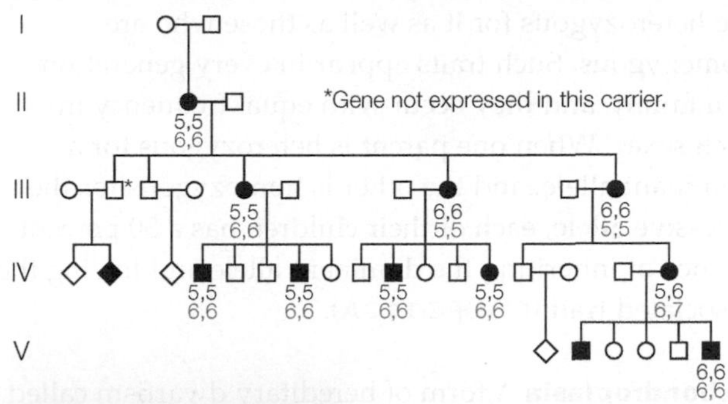
I
II
III
IV
V
*Gene not expressed in this carrier.

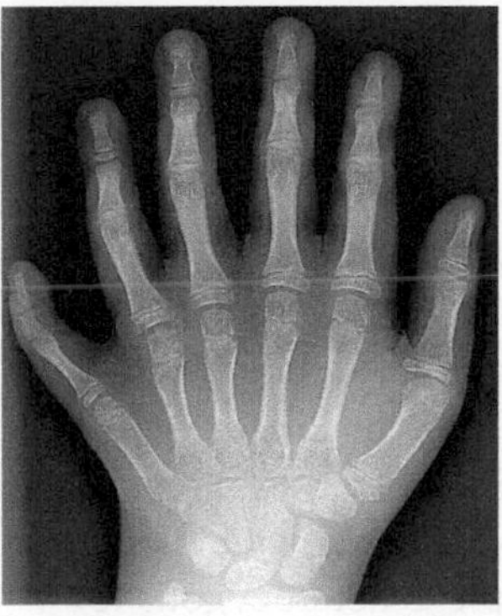

B Above, a pedigree that tracks a genetic abnormality through five generations of a family tree. The trait, polydactyly, is characterized by extra fingers (left), toes, or both. The pedigree shows the number of fingers on each hand in black; toes on each foot, in red. Dominant alleles give rise to polydactyly on its own. This abnormality also occurs as part of some syndromes (such as Ellis–van Creveld syndrome), and in these cases it arises from recessive alleles.

FIGURE 14.2 Pedigrees.

FIGURE IT OUT In this family, does polydactyly arise from dominant or recessive allele?

Answer: A recessive allele

heterozygous for this allele are protected from infection by bacteria that cause typhoid fever. You will learn about other examples in later chapters.

Collectively, single-gene disorders affect about 1 in 100 people. Appendix IV shows the loci of several genes known to be associated with disorders and abnormalities.

TAKE-HOME MESSAGE 14.2

✔ Geneticists study inheritance patterns in humans by tracking genetic disorders and abnormalities through generations of families. Pedigrees are charts that mark these traits among family members.

✔ A few genetic disorders are governed by single genes inherited in a Mendelian fashion. Most human traits are polygenic, and some have environmental contributions.

✔ A genetic disorder causes medical problems; a genetic abnormality is a rare but harmless trait.

pedigree Chart that marks the appearance of a phenotype through generations of a family tree.

CREDITS: (2A) © Cengage Learning; (2B) Courtesy of Irving Buchbinder, DPM, DABPS, Community Health Services, Hartford CT.

14.3 Autosomal Inheritance

LEARNING OBJECTIVE

- Using appropriate examples, explain and diagram the autosomal dominant and autosomal recessive inheritance patterns.

The Autosomal Dominant Pattern

A trait associated with a dominant allele on an autosome (TABLE 14.1) appears in people who are heterozygous for it as well as those who are homozygous. Such traits appear in every generation of a family, and they occur with equal frequency in both sexes. When one parent is heterozygous for a dominant allele, and the other is homozygous for the recessive allele, each of their children has a 50 percent chance of inheriting the dominant allele and having the associated trait (FIGURE 14.3A).

Achondroplasia A form of hereditary dwarfism called achondroplasia offers an example of an autosomal dominant disorder (one caused by a dominant allele on an autosome). Mutations associated with achondroplasia occur in a gene for a growth factor receptor (Section 11.6), the activity of which inhibits growth and differentiation of cells that give rise to bone. The mutations interfere with the normal cellular process that recycles the receptor after it is no longer needed. The cells end up with receptors when they should not have them, so bone growth is inhibited when it should not be. About 1 in 10,000 people is heterozygous for one of these mutations. As adults, affected people are, on average, about 4 feet 4 inches (1.3 meters) tall, with arms and legs that are short relative to torso size (FIGURE 14.3B). An allele that causes achondroplasia can be passed to children because its expression does not interfere with reproduction, at least in heterozygous people. The homozygous condition results in severe skeletal malformations that are lethal before birth or shortly after.

TABLE 14.1

Some Autosomal Abnormalities and Disorders

Disorder/Abnormality	Main Symptoms
Autosomal dominant inheritance pattern	
Achondroplasia	One form of dwarfism
Aniridia	Defects of the eyes
Huntington's disease	Degeneration of the nervous system
Marfan syndrome	Cardiovascular system malfunction
Progeria	Drastic premature aging
Autosomal recessive inheritance pattern	
Albinism	Absence of pigmentation
Cystic fibrosis	Difficulty breathing; lung infections
Ellis–van Creveld syndrome	Dwarfism, heart defects, polydactyly
Phenylketonuria (PKU)	Mental impairment
Sickle-cell anemia	Anemia, swelling, frequent infections
Tay–Sachs disease	Deterioration of mental and physical abilities; early death

Huntington's Disease Huntington's disease is an autosomal dominant disorder. The disease is caused by expansions of a short tandem repeat (Section 13.7) in a

FIGURE 14.3 Autosomal dominant inheritance.

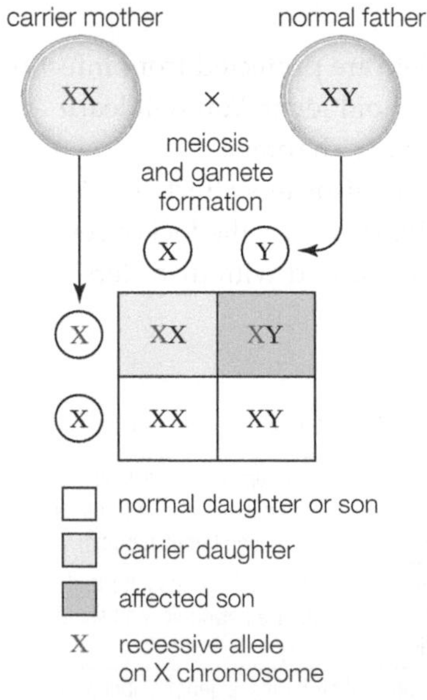
carrier mother
normal father
XX
×
XY
meiosis
and gamete
formation
X
Y
X
X
XX
XY
XX
XY
normal daughter or son
carrier daughter
affected son
X recessive allele
on X chromosome

A A dominant allele (red) on an autosome affects all heterozygous people.

B Achondroplasia affects Ivy (left), as well as her brother, father, and grandfather.

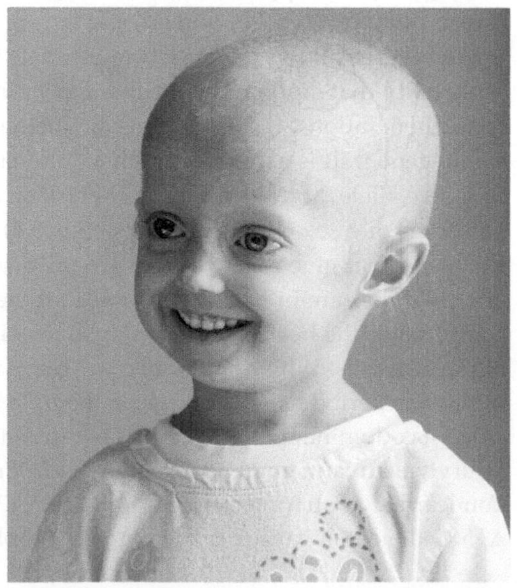

C Symptoms of Hutchinson–Gilford progeria are already apparent in Megan, age 5.

gene for a cytoplasmic protein, the many functions of which are still being discovered. The protein encoded by the altered gene has a section in which the amino acid glutamine is repeated 30 or more times. These repeats cause the protein to misfold, and also to resist cellular processes that normally dismantle and recycle defective proteins. The misfolded protein accumulates to high levels, particularly in brain cells involved in movement, thinking, and emotion. Cellular functioning is hampered, and a stress response is triggered that eventually causes the cell to die. Glutamine-rich protein fragments accumulate in amyloid fibrils, so amyloid plaques form in the brain (Section 3.5). In the most common form of Huntington's, symptoms do not start until after the age of thirty. Voluntary muscle control gradually gives way to involuntary jerking, twitching, and writhing movements. Eventually, serious problems with swallowing cause many patients to die from choking or malnutrition in their forties or fifties. With this and other late-onset disorders, people tend to reproduce before symptoms appear, so an allele is often passed unknowingly to children.

Hutchinson–Gilford Progeria Drastically accelerated aging characterizes an autosomal dominant disorder called Hutchinson–Gilford progeria. A mutation in the gene for lamin A is the cause. Lamins are protein subunits of intermediate filaments that support the nuclear envelope (Section 4.4). They also act as a bridge between the inner nuclear membrane and chromosomes, with roles in mitosis, DNA synthesis and repair, and transcriptional regulation. The progeria mutation is a single base-pair substitution that adds a signal for an intron–exon splice site (Section 9.3), resulting in a polypeptide that is too short and cannot be processed correctly after translation. The defective protein binds abnormally to other molecules that usually interact with lamin A, and interferes with their function. Cells that carry the mutation have a nucleus that is grossly abnormal, with nuclear pore complexes that do not assemble properly and membrane proteins localized to the wrong side of the nuclear envelope. The function of the nucleus as protector of chromosomes is severely impaired, and DNA damage accumulates quickly. Abnormal DNA methylation patterns and unusually short telomeres result in early senescence of the individual's cells (Section 11.5).

The gene for lamin A is pleiotropic (Section 13.5), so the progeria mutation affects multiple traits. Outward symptoms begin to appear before age two, as skin that should be plump and resilient starts to thin, muscles weaken, and bones soften. Premature baldness is inevitable (**FIGURE 14.3C**). Most people with the disorder die in their early teens as a result of a stroke or heart attack brought on by hardened arteries, a condition typical of advanced age. Progeria does not run in families because affected people do not live long enough to reproduce.

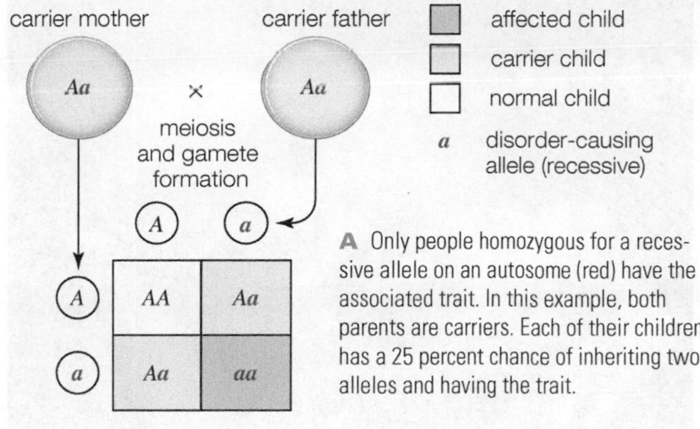
carrier mother
carrier father
Aa
×
Aa
meiosis and gamete formation
A
a
A
a
AA
Aa
Aa
aa
affected child
carrier child
normal child
a disorder-causing allele (recessive)

A Only people homozygous for a recessive allele on an autosome (red) have the associated trait. In this example, both parents are carriers. Each of their children has a 25 percent chance of inheriting two alleles and having the trait.

FIGURE 14.4 Autosomal recessive inheritance.

The Autosomal Recessive Pattern

Traits that arise from a recessive allele on an autosome appear only in homozygous individuals; heterozygous individuals are called carriers because they have the allele but not the associated trait. These traits appear in both sexes at equal frequency, and they tend to skip generations. Any child of two carriers has a 25 percent chance of inheriting the allele from both parents and developing the trait (**FIGURE 14.4**).

Albinism Albinism, a phenotype characterized by an abnormally low level of the pigment melanin, is inherited in an autosomal recessive pattern. Mutations associated with albinism affect proteins involved in melanin synthesis. Skin, hair, or eye pigmentation may be reduced or missing. In the most dramatic form, the skin is very white and does not tan, and the hair is white. The irises of the eyes appear red because the lack of pigment allows underlying blood vessels to show through. Melanin also plays a role in the retina, so vision problems are typical. In skin, melanin acts as a sunscreen; without it, the skin is defenseless against UV radiation. Thus, people with the albino phenotype have a very high risk of skin cancer.

Tay–Sachs Disease Tay–Sachs disease is an autosomal recessive disorder. In the general population, about 1 in 300 people is a carrier, but the incidence is ten times higher in Jews of eastern European descent and some other groups. The gene altered in Tay–Sachs encodes a lysosomal enzyme responsible for breaking down a particular type of lipid. Mutations cause this enzyme to misfold and become destroyed, so cells make the lipid

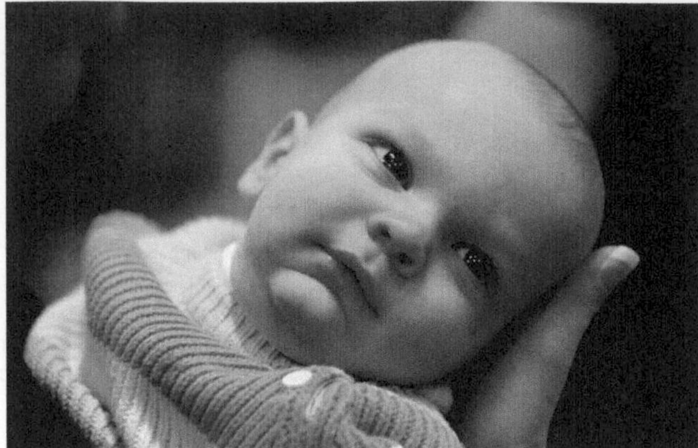

FIGURE 14.5 Tay–Sachs disease, an autosomal recessive disorder. This is a photo of Conner, who was 7 months old when he was diagnosed with Tay–Sachs. He died before his second birthday.

but cannot break it down. Typically, newborns homozygous for a Tay–Sachs allele are healthy, but within three to six months they become irritable, listless, and may have seizures as the lipid accumulates in their nerve cells. Blindness, deafness, and paralysis follow. Affected children usually die by age five (**FIGURE 14.5**).

TAKE-HOME MESSAGE 14.3

✔ A trait associated with a dominant allele on an autosome appears in every generation. Everyone with the allele, homozygous or heterozygous, has the trait.

✔ A trait associated with a recessive allele on an autosome skips generations. Only persons homozygous for the allele have the trait.

14.4 X-Linked Inheritance

LEARNING OBJECTIVE

- Explain why X-linked recessive disorders are more common in males than in females.

Traits inherited in an X-linked pattern (**TABLE 14.2**) arise from genes on the X chromosome. In most cases, X chromosome alleles that cause genetic disorders are recessive, and these leave two inheritance clues. First, an affected father never passes one of these alleles to a son, because all children who inherit their father's X chromosome are female (**FIGURE 14.6A**). Second, the disorder appears more often in males than females. Having only one X chromosome, a male must inherit only one allele to be affected by the disorder; a female must inherit two, and inheriting two disorder-causing alleles is statistically less common than inheriting one.

An X-linked recessive disorder can have uneven effects in heterozygous females because of X chromosome inactivation (Section 10.3). About half of the cells making up the body of a heterozygous female express the recessive allele on one X chromosome; the other half express the dominant allele on the other X chromosome. Mild symptoms may appear if the effect of the recessive allele on her body is not fully masked by that of the dominant allele.

Red–Green Color Blindness Color blindness refers to a range of genetic abnormalities in which an individual

FIGURE 14.6 X-linked recessive inheritance.

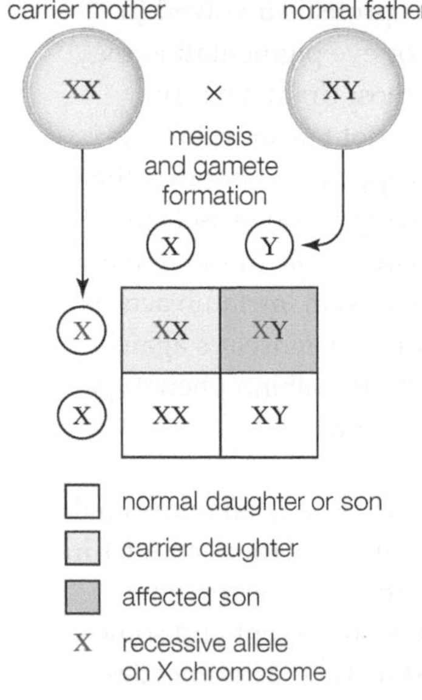
carrier mother
normal father
XX
×
XY
meiosis
and gamete
formation
X
Y
X
X
XX
XY
XX
XY
normal daughter or son
carrier daughter
affected son
X recessive allele
on X chromosome

A In this example, the mother carries a recessive allele on one of her two X chromosomes (red).

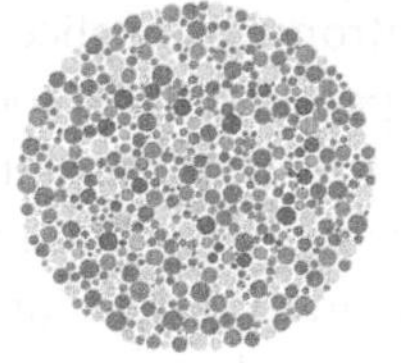

You may have one form of red–green color blindness if you see a 7 in this circle instead of a 29.

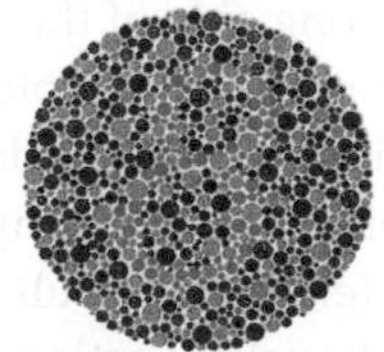

You may have another form of red–green color blindness if you see a 3 instead of an 8 in this circle.

B A view of color blindness, a trait inherited in an X-linked recessive pattern. The photo on the left shows how a person with red–green color blindness sees the photo on the right. The perception of blues and yellows is normal; red and green appear similar. The circle diagrams are part of a standardized test for color blindness. A set of 38 of these diagrams is commonly used to diagnose deficiencies in color perception.

TABLE 14.2

Some X-Linked Traits

Disorder/Abnormality	Main Symptoms
X-linked recessive inheritance pattern	
Androgen insensitivity syndrome	XY individual but having some female traits; sterility
Red–green color blindness	Inability to distinguish red from green
Hemophilia	Impaired blood clotting ability
Muscular dystrophies	Progressive loss of muscle function
X-linked anhidrotic dysplasia	Mosaic skin (patches with or without sweat glands); other ill effects
X-linked dominant inheritance pattern	
Fragile X syndrome	Intellectual, emotional disability
Incontinentia pigmenti	Abnormalities of skin, hair, teeth, nails, eyes; neurological problems

cannot distinguish among some or all colors of visible light. These conditions are typically inherited in an X-linked recessive pattern, because most of the genes involved in color vision are on the X chromosome.

Human eyes normally sense differences between 150 colors, and this perception depends on receptors that respond to red, blue, or green light. Thus, mutations that result in abnormal or defective receptors affect color vision. The brain distinguishes colors only by comparing signals between receptors, so color vision depends on having at least two types of working receptors. Thus, people with mutations that affect two (or three) types of receptors see no color at all, but this is the rarest form of color blindness. The more common red–green color blindness is caused by mutations that alter receptors for red or green light (**FIGURE 14.6B**). Blue–yellow color blindness is caused by mutations that alter receptors for blue light.

Hemophilia Hemophilias are genetic disorders in which the blood does not clot properly. Most people have a blood clotting mechanism that quickly stops bleeding from minor injuries. That mechanism involves two proteins, clotting factors VIII and IX, both products of X chromosome genes. Mutations in these two genes cause two types of hemophilia (A and B, respectively). Males who carry one of these mutations have prolonged bleeding, as do homozygous females (heterozygous females make about half the normal amount of the clotting factor, but this is generally enough to sustain normal clotting). Affected people bruise easily, but internal bleeding is their most serious problem. Repeated bleeding inside the joints disfigures them and causes chronic arthritis.

In the nineteenth century, the incidence of hemophilia was relatively high among European and Russian royals (**FIGURE 14.7**), in part because a centuries-old practice of inbreeding among close relatives kept the allele circulating in royal families. Today, about 1 in 7,500 people in the general population is affected. That number may be rising because the disorder is now treatable, so more affected people are living long enough to transmit a mutated allele to children.

Duchenne Muscular Dystrophy Progressive muscle degeneration characterizes a severe genetic disorder called Duchenne muscular dystrophy (DMD). The disorder is caused by mutations in the X chromosome gene for dystrophin, a rod-shaped, flexible protein found mainly in skeletal and heart (cardiac) muscle. Dystrophin connects contractile units in the cytoplasm of muscle cells to proteins that anchor the cell in extracellular matrix (Section 4.9). The connection is flexible,

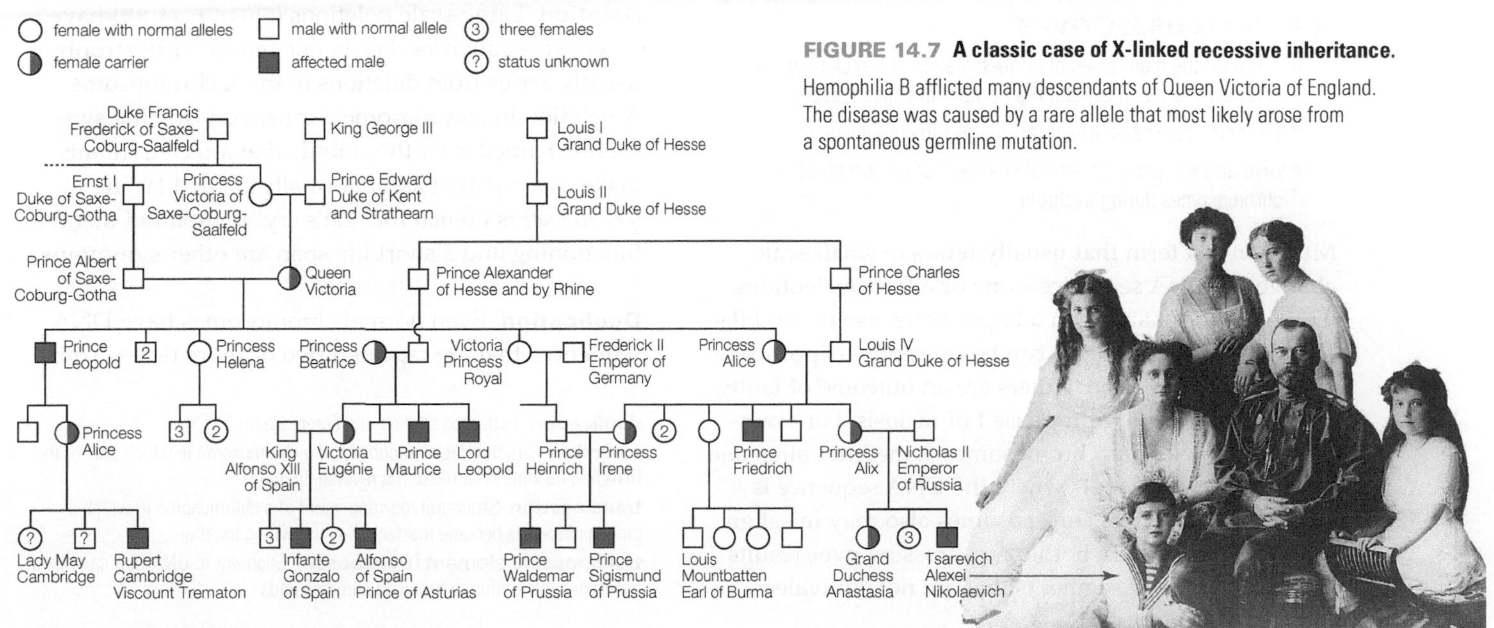
female with normal alleles
male with normal allele
three females
female carrier
affected male
status unknown
Duke Francis Frederick of Saxe-Coburg-Saalfeld
King George III
Louis I Grand Duke of Hesse
Ernst I Duke of Saxe-Coburg-Gotha
Princess Victoria of Saxe-Coburg-Saalfeld
Prince Edward Duke of Kent and Strathearn
Louis II Grand Duke of Hesse
Prince Albert of Saxe-Coburg-Gotha
Queen Victoria
Prince Alexander of Hesse and by Rhine
Prince Charles of Hesse
Prince Leopold
Princess Helena
Princess Beatrice
Victoria Princess Royal
Frederick II Emperor of Germany
Princess Alice
Louis IV Grand Duke of Hesse
Princess Alice
King Alfonso XIII of Spain
Victoria Eugénie
Prince Maurice
Lord Leopold
Prince Heinrich
Princess Irene
Prince Friedrich
Princess Alix
Nicholas II Emperor of Russia
Lady May Cambridge
Rupert Cambridge Viscount Trematon
Infante Gonzalo of Spain
Alfonso of Spain Prince of Asturias
Prince Waldemar of Prussia
Prince Sigismund of Prussia
Louis Mountbatten Earl of Burma
Grand Duchess Anastasia
Tsarevich Alexei Nikolaevich

FIGURE 14.7 A classic case of X-linked recessive inheritance. Hemophilia B afflicted many descendants of Queen Victoria of England. The disease was caused by a rare allele that most likely arose from a spontaneous germline mutation.

and it imparts strength to muscle tissue. It also protects muscle cell membranes from mechanical damage during contraction.

Mutations associated with DMD result in defective or missing dystrophin. Without dystrophin, muscle cell plasma membranes are easily damaged during contraction, and the cells become flooded with calcium ions. Calcium ions act as potent messengers in cells, so their concentration in cytoplasm is normally kept very low (Section 5.9). Among other negative effects, a chronic calcium ion overload causes mitochondrial malfunction (Section 7.1). Muscle cell mitochondria produce too little ATP to support normal cellular function, so muscles are abnormally weak. Free radicals accumulate and kill the cells. Eventually, the tissue's capacity to regenerate is overwhelmed by rapid cell turnover, and it becomes replaced by fat and connective tissue.

DMD affects about 1 in 3,500 people, almost all of them boys. Symptoms begin around age four and progress very quickly. Anti-inflammatory drugs can slow the progression of DMD, but there is no cure. When an affected boy is about ten years old, he will begin to need a wheelchair and his heart will start to fail. Even with the best care, he will probably die before the age of thirty, from a heart disorder or respiratory failure.

TAKE-HOME MESSAGE 14.4

✔ Traits associated with recessive alleles on the X chromosome appear more frequently in men than in women.

✔ A man can inherit an X chromosome allele from his mother only.

14.5 Changes in Chromosome Structure

LEARNING OBJECTIVES

- Describe the main types of large-scale structural changes in chromosomes, and explain their potential consequences.
- List some causes of structural changes in chromosomes.
- Give an example of structural change that has occurred in chromosomes during evolution.

Mutation is a term that usually refers to small-scale changes in DNA sequence—one or a few nucleotides. Chromosome changes on a larger scale also occur. Like mutations, these changes can be caused by exposure to chemicals or radiation. Others are an outcome of faulty crossing over during prophase I of meiosis. For example, nonhomologous chromosomes sometimes align and swap segments at spots where the DNA sequence is similar. Homologous chromosomes also may misalign along their length. In both cases, crossing over results in the exchange of segments that are not equivalent.

TABLE 14.3

A Few Effects of Human Chromosome Structure Change

Disorder	Structural Change	Chromosome(s) Affected
Fragile X syndrome	Duplication	X
Huntington's disease	Duplication	4
Cri-du-chat syndrome	Deletion	5
Prader–Willi syndrome	Deletion	15
Hemophilia A	Inversion	X
Burkitt lymphoma	Translocation	8⟷14
Chronic myelogenous leukemia	Translocation	9⟷22

Types of Chromosomal Change

Large-scale changes in chromosome structure are categorized by the type of change. Any type can have a drastic effect on health (TABLE 14.3); about half of all miscarriages are due to chromosome abnormalities of the developing embryo.

Insertions Large-scale insertions (FIGURE 14.8A) often result from the activity of **transposable elements** (**transposons**), which are segments of DNA hundreds or thousands of nucleotides long that can move spontaneously ("jump") within or between chromosomes. Transposons are very common in the chromosomes of all species; about 45 percent of human DNA consists of them and their evolutionary remnants. The jumping mechanism involves enzymes encoded by the DNA sequence of the transposon itself, or by another transposon. A transposon that jumps into a gene becomes a major insertion that destroys the function of the gene product. In most cases the location of insertion is random; the effect depends on the gene interrupted.

Deletion Large-scale deletions (FIGURE 14.8B) have severe consequences. Duchenne muscular dystrophy usually arises from deletions in the X chromosome. A deletion in chromosome 5 causes cri-du-chat syndrome, named after the sounds that affected infants make as a result of an abnormally shaped larynx (*cri du chat* is French for "cat's cry"). Impaired mental functioning and a short life span are other symptoms.

Duplication Even normal chromosomes have DNA sequences that are repeated two or more times. These

duplication Repeated section of a chromosome.
inversion Structural rearrangement of a chromosome in which part of the DNA has become oriented in the reverse direction.
translocation Structural rearrangement of a chromosome in which a broken piece has become reattached in the wrong location.
transposable element (transposon) Segment of DNA that can move spontaneously within or between chromosomes.

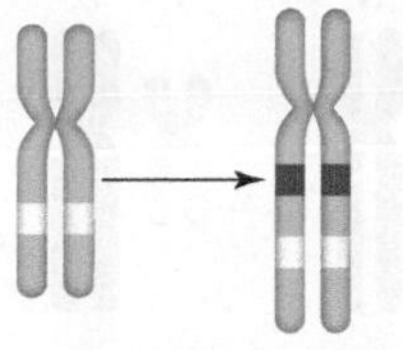

A Insertion A section of DNA is inserted into a chromosome.

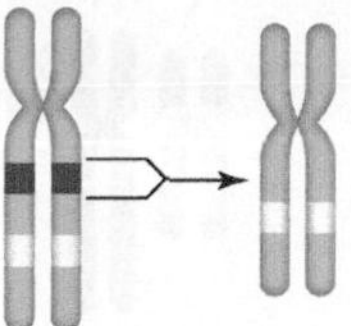

B Deletion A section of DNA is lost from a chromosome.

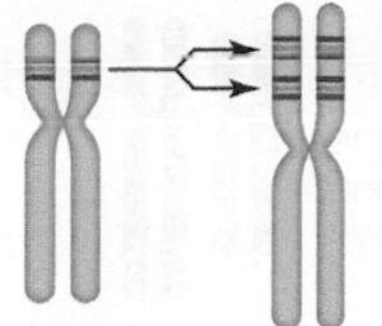

C Duplication A section of DNA in a chromosome is repeated.

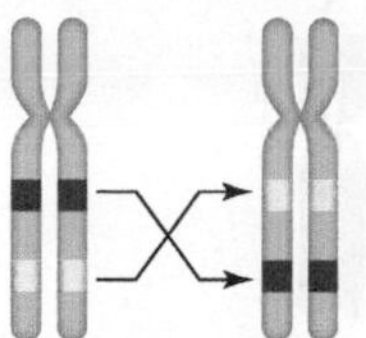

D Inversion A section of a chromosome is flipped so it runs in the opposite orientation.

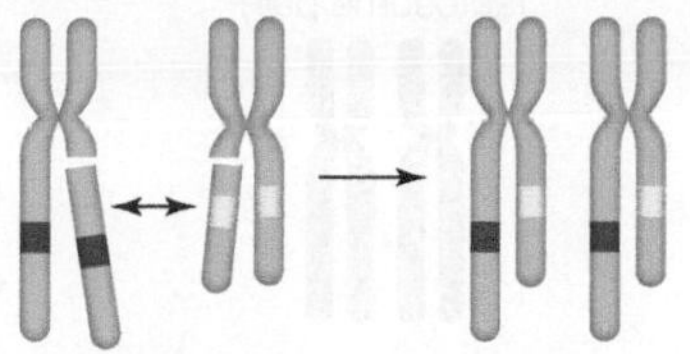

E Translocation A piece of a broken chromosome is reattached in the wrong place. This example shows a reciprocal translocation, in which two nonhomologous chromosomes exchange chunks.

FIGURE 14.8 **Major structural changes in chromosomes.**

repetitions are called **duplications** (FIGURE 14.8D). Some newly occurring duplications, such as the expansion mutations that cause Huntington's disease, cause genetic abnormalities or disorders. Others, as you will soon see, have been evolutionarily important.

Inversion Sometimes a segment of a chromosome breaks off and gets reattached in the reverse direction, with no loss of DNA (FIGURE 14.8D). This type of structural change is called an **inversion**. An inversion may not affect a carrier's health if the break point does not occur in a gene or control region, because the individual's cells still contain their full complement of genetic material. However, fertility may be compromised because a chromosome with an inversion does not pair properly with its homologous partner during meiosis. Crossovers may occur between the mispaired chromosomes, producing other chromosome abnormalities that reduce the viability of forthcoming embryos. Some people learn that they carry an inversion when their karyotype is checked because of fertility problems.

Translocation If a chromosome breaks, the broken part may get attached to a different chromosome, or to a different part of the same chromosome. This type of structural change is called a **translocation**. Most translocations are reciprocal, meaning that two nonhomologous chromosomes exchange broken parts (FIGURE 14.8E). A reciprocal translocation between chromosomes 8 and 14 is the usual cause of Burkitt's lymphoma, an aggressive cancer of the immune system. This translocation moves a proto-oncogene to a region that is vigorously transcribed in white blood cells, resulting in uncontrolled cell divisions that are characteristic of cancer (Section 11.6).

Many other reciprocal translocations have no adverse effects on health, but like inversion they can compromise fertility. During meiosis, translocated chromosomes pair abnormally and segregate improperly; about half of the resulting gametes carry major duplications or deletions. When one of these gametes unites with a normal gamete at fertilization, the resulting embryo almost always dies. As with inversions, people who carry a translocation may not find out about it until they have difficulty with fertility.

Chromosome Changes in Evolution

There is evidence of major structural alterations in the chromosomes of all known species. For example, duplications have often allowed a copy of a gene to mutate while the original carried out its unaltered function. The multiple and strikingly similar globin chain genes of mammals apparently evolved by this process. Globin chains, remember, associate to form molecules of hemoglobin (Section 9.6). Two identical genes for the alpha chain—and five other slightly different versions of it—form a cluster on chromosome 16. The gene for the beta chain clusters with four other slightly different versions on chromosome 11.

Some chromosome structure changes contributed to differences among closely related organisms such as apes and humans. Human somatic cells have twenty-three pairs of chromosomes; cells of chimpanzees, gorillas, and orangutans have twenty-four. Thirteen human chromosomes are nearly identical with chimpanzee chromosomes. Nine more are similar, except for some inversions. One human chromosome matches up with two in chimpanzees and the other great apes (FIGURE 14.9). During human evolution, two chromosomes evidently fused end to end and formed our chromosome 2. How do we know? The region where the fusion occurred contains remnants of a telomere (Section 11.5).

telomere sequence

human chimpanzee

FIGURE 14.9 **Human chromosome 2 compared with chimpanzee chromosomes 2A and 2B.**

Consider also the human X and Y chromosomes. Long ago, these two chromosomes were homologous

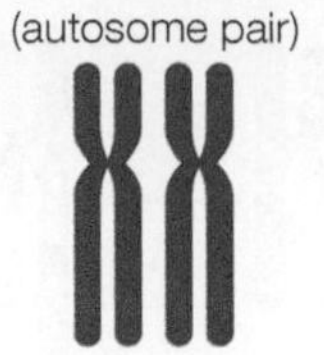
(autosome pair)

Ancestral reptiles
>350 mya

A Before 350 mya, sex was determined by temperature, not by chromosome differences.

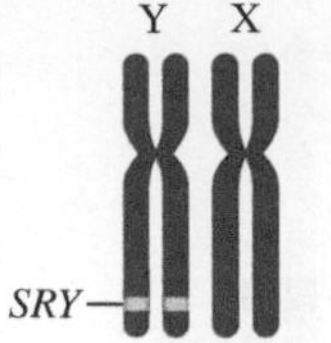
Y X
SRY

Ancestral reptiles
350 mya

B 350 mya, the *SRY* gene began as a mutation that interfered with crossing over. The homologous chromosomes diverged as other mutations accumulated separately in each one.

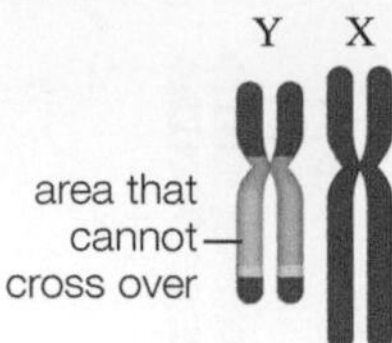
Y X
area that
cannot
cross over

Monotremes
320–240 mya

C By 320–240 mya, the DNA sequences of the chromosomes are so different that the pair can no longer cross over in one region. The Y chromosome begins to shorten.

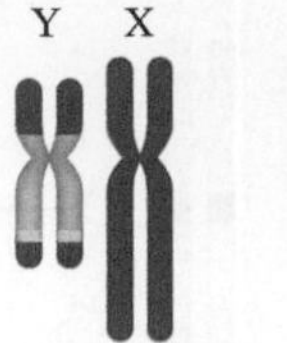
Y X

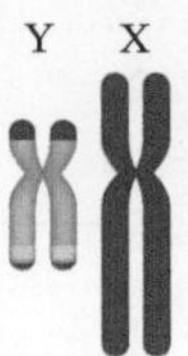
Y X

Marsupials 170–130 mya | Monkeys 130–80 mya | Humans 50–30 mya

D Three more times, the pair stops crossing over in yet another region. Each time, the DNA sequences of the chromosomes diverge, and the Y chromosome shortens. Today, the X and Y can cross over only at a small region near the ends.

FIGURE 14.10 Evolution of the Y chromosome. Today, the *SRY* gene on the Y chromosome determines male sex, but this was not always the case. Homologous regions are pink; mya, million years ago. Monotremes are egg-laying mammals; marsupials are pouched mammals.

autosomes in the ancient, reptile-like ancestors of mammals (**FIGURE 14.10**). Ambient temperature probably determined gender in those organisms, as it still does in turtles and some other modern reptiles. About 350 million years ago, a gene on one of the two homologous chromosomes mutated. The mutation interfered with crossing over during meiosis, and it was the beginning of the male sex determination gene *SRY* (Section 10.3). A reduced frequency of crossing over allowed mutations to accumulate separately in the two chromosomes, so they began to diverge in sequence around the changed region. Over evolutionary time, the chromosomes became so different that they no longer crossed over at all in these regions, so they diverged even more. Today, the Y chromosome is much smaller than the X, and is homologous with it only in a tiny part. The Y completes meiosis by crossing over with itself, translocating duplicated regions of its own DNA.

TAKE-HOME MESSAGE 14.5

✔ A segment of DNA in a chromosome may become inserted, duplicated, deleted, inverted, or translocated.

✔ In most cases, structural changes in chromosomes are harmful. A change may be conserved in the rare circumstance that it has a neutral or beneficial effect.

14.6 Changes in Chromosome Number

LEARNING OBJECTIVES

- Distinguish between polyploidy and aneuploidy.
- Explain nondisjunction and its potential effects.

Individuals of some species have three or more complete sets of chromosomes, a condition called **polyploidy**. About 70 percent of flowering plant species are polyploid, as are some insects, fishes, and other animals—but not humans. In our species, inheriting more than two full sets of chromosomes is invariably fatal, although in normal adult tissues some somatic cells are polyploid.

TABLE 14.4

A Few Effects of Aneuploidy in Humans

Cause	Syndrome	Main Symptoms
Trisomy 21	Down	Mental impairment; heart defects
Trisomy 18	Edwards	Severe disability; low survival rate
Trisomy 13	Patau	Severe disability; low survival rate
XO	Turner	Abnormal ovaries and sexual traits
XXY	Klinefelter	Sterility; mild mental impairment
XXX	Trisomy X	Minimal abnormalities
XYY	Jacob's	Mild mental impairment or no effect

A few babies survive with too many or too few copies of a chromosome, and this condition is called **aneuploidy** (**TABLE 14.4**). Aneuploidy is usually an outcome of **nondisjunction**, the failure of chromosomes to separate properly during nuclear division. If the failure occurs in meiosis (**FIGURE 14.11**), the chromosome number at fertilization may be affected. Suppose a normal gamete (n) fuses with a gamete that has an extra chromosome ($n+1$). The resulting zygote will have three copies of one type of chromosome and two of every other type ($2n+1$), an aneuploid condition called trisomy. If a normal gamete (n) fuses with a gamete missing a chromosome ($n-1$), the new individual will have one copy of one chromosome and two of every other type ($2n-1$), a condition called monosomy.

Down Syndrome

In many cases, autosomal aneuploidy in humans is fatal before birth or shortly thereafter. An important exception is trisomy 21. A person born with three chromosomes 21 has Down syndrome and a high likelihood of surviving infancy. Mild to moderate mental impairment and health problems such as heart disease

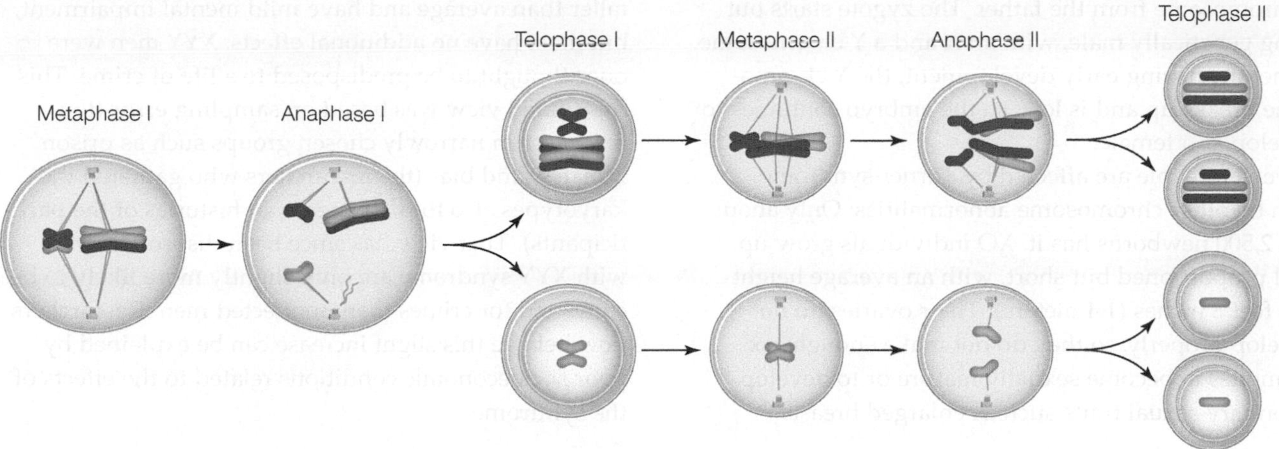
Metaphase I
Anaphase I
Telophase I
Metaphase II
Anaphase II
Telophase II

FIGURE 14.11 An example of nondisjunction. Of the two pairs of homologous chromosomes shown here, one fails to separate properly during anaphase I of meiosis. The chromosome number is altered in the resulting gametes.

are hallmarks of this disorder. Other phenotypic effects may include a somewhat flattened facial profile, a fold of skin that starts at the inner corner of each eyelid, white spots on the iris (FIGURE 14.12), and one deep crease (instead of two shallow creases) across each palm. The skeleton develops abnormally, so older children have short body parts, loose joints, and misaligned bones of the fingers, toes, and hips. Muscles and reflexes are weak, and motor skills such as speech develop slowly. With medical care, affected individuals live about fifty-five years. Early training can help these individuals learn to care for themselves and to take part in normal activities.

Alzheimer's disease, a condition of progressive mental deterioration, is associated with Down syndrome. A gene on chromosome 21 is the culprit. Having three of these chromosomes, a person with Down syndrome makes an excess of the gene's product, a protein that forms the main component of amyloid fibrils (Section 3.5) characteristic of Alzheimer's. All persons with Down syndrome will have these fibrils in their brain by the age of forty. Down syndrome occurs in about 1 of 700 live births, and the risk increases with maternal age.

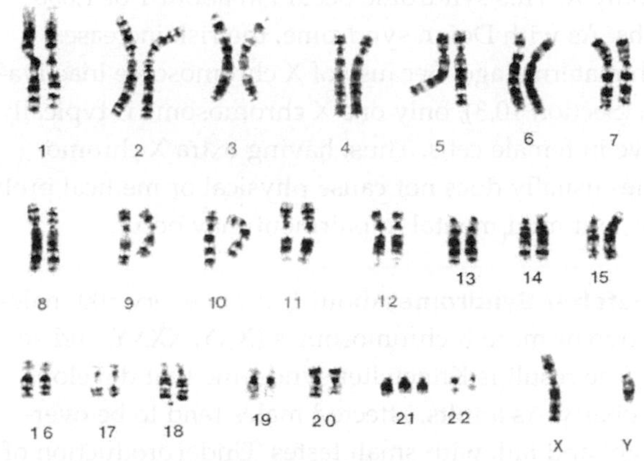
1
2
3
4
5
6
7
8
9
10
11
12
13
14
15
16
17
18
19
20
21
22
X
Y

FIGURE 14.12 Down syndrome: genotype and phenotype.

Above, karyotype of an individual with three copies of chromosome 21 and Down syndrome. Below, a person with Down syndrome typically has a somewhat flattened facial profile, and a fold of skin that starts at the inner corner of each eyelid. Excess tissue deposits on the iris also give rise to a ring of starlike white speckles, a lovely effect of the chromosome number change.

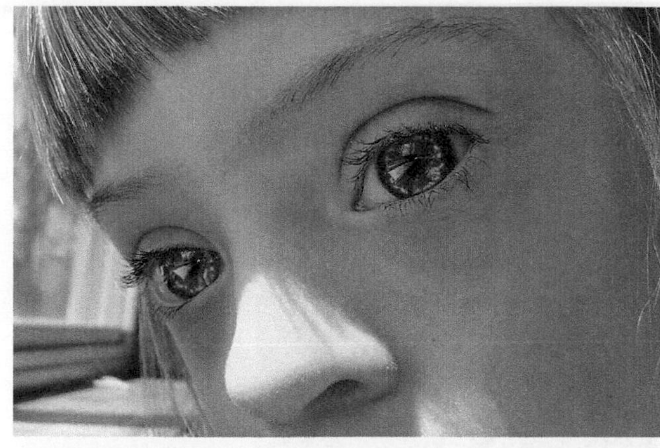

Sex Chromosome Aneuploidy

About 1 in 400 human babies is born with an atypical number of sex chromosomes. Most often, such alterations lead to mild difficulties in learning and impairment in motor skills such as speech, but these problems may be very subtle.

Turner Syndrome Individuals with Turner syndrome have an X chromosome and no corresponding X or Y chromosome (XO). The syndrome is thought to arise most often as an outcome of inheriting an unstable

aneuploidy (AN-you-ploy-dee) Condition of having too many or too few copies of a particular chromosome.
nondisjunction Failure of chromosomes to separate properly during mitosis or meiosis.
polyploidy (PALL-ee-ploy-dee) Condition of having three or more of each type of chromosome characteristic of the species.

Y chromosome from the father. The zygote starts out being genetically male, with an X and a Y chromosome. Sometime during early development, the Y chromosome breaks up and is lost, so the embryo continues to develop as a female.

Fewer people are affected by Turner syndrome than by other chromosome abnormalities: Only about 1 in 2,500 newborns has it. XO individuals grow up well proportioned but short, with an average height of 4 feet 8 inches (1.4 meters). Their ovaries do not develop properly, so they do not make enough sex hormones to become sexually mature or to develop secondary sexual traits such as enlarged breasts.

XXX Syndrome A female may inherit multiple X chromosomes, a condition called XXX syndrome or trisomy X. This syndrome occurs in about 1 of 1,000 births. As with Down syndrome, the risk increases with maternal age. Because of X chromosome inactivation (Section 10.3), only one X chromosome is typically active in female cells. Thus, having extra X chromosomes usually does not cause physical or medical problems, but mild mental impairment may occur.

Klinefelter Syndrome About 1 out of every 500 males has two or more X chromosomes (XXY, XXXY, and so on). The result is Klinefelter syndrome that develops at puberty. As adults, affected males tend to be overweight and tall, with small testes. Underproduction of the hormone testosterone interferes with sexual development and can result in sparse facial and body hair, a high-pitched voice, enlarged breasts, and infertility. Testosterone injections during puberty can minimize this feminization.

XYY Syndrome About 1 in 1,000 males is born with an extra Y chromosome (XYY), a result of nondisjunction of the Y chromosome during sperm formation. Adults with the resulting Jacob's syndrome tend to be taller than average and have mild mental impairment, but most have no additional effects. XYY men were once thought to be predisposed to a life of crime. This misguided view was based on sampling error (too few cases in narrowly chosen groups such as prison inmates) and bias (the researchers who gathered the karyotypes also took the personal histories of the participants). That view has since been disproven: Men with XYY syndrome are only slightly more likely to be convicted for crimes than unaffected men. Researchers now believe this slight increase can be explained by poor socioeconomic conditions related to the effects of the syndrome.

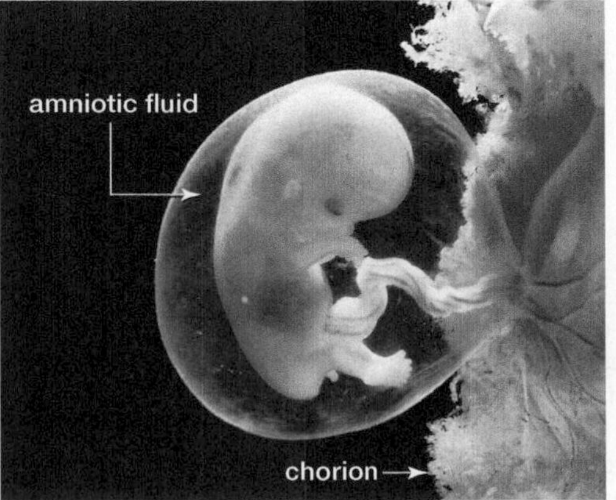

amniotic fluid
chorion

FIGURE 14.13 Tissues tested in prenatal diagnosis. With amniocentesis, fetal cells that have been shed into amniotic fluid are tested for genetic disorders. Chorionic villus sampling tests cells of the chorion, which is part of the placenta.

TAKE-HOME MESSAGE 14.6

- ✔ Polyploidy is the condition of having multiple complete sets of chromosomes. Many types of flowering plants and some other organisms are polyploid, but the condition is fatal in humans.
- ✔ Aneuploidy is the condition of having too many or too few copies of a chromosome. It can arise from nondisjunction during meiosis.
- ✔ Sex chromosome aneuploidy typically has milder effects on health than autosomal aneuploidy.
- ✔ Most types of autosomal aneuploidy are fatal in humans. Trisomy 21, which causes Down syndrome, is an exception.

14.7 Genetic Screening

LEARNING OBJECTIVES

- Explain how early genetic screening can help a baby.
- Discuss the benefits and risks of three prenatal diagnosis methods.

Studying human inheritance patterns has given us many insights into how genetic disorders arise and progress, and how to treat them. Surgery, prescription drugs, hormone replacement therapy, and dietary controls can minimize and in some cases eliminate the symptoms of a genetic disorder. Some disorders can be detected early enough to start countermeasures before symptoms develop; thus, most hospitals in the United States now screen newborns for mutations such as those that cause PKU (phenylketonuria). PKU mutations affect an enzyme that converts one amino acid (phenylalanine) to another (tyrosine). Without this enzyme, the body becomes deficient in tyrosine, and phenylalanine accumulates to high levels. The imbalance inhibits protein synthesis in the brain, which in turn results in permanent intellectual disability. Restricting all intake of phenylalanine can slow the progression of PKU, so routine early screening has resulted in fewer individuals suffering from the symptoms of the disorder.

Prospective parents may opt for genetic screening. The probability that a future child will inherit a genetic disorder can be estimated by testing parents for certain alleles. Karyotypes and pedigrees are also useful in this type of screening, which can help people make informed decisions about family planning.

Genetic screening after conception but before birth is called prenatal diagnosis (prenatal means before birth). Prenatal diagnosis checks an embryo or fetus for physical abnormalities and genetic disorders. Dozens of disorders are detectable prenatally, including aneuploidy, hemophilia, Tay–Sachs disease, sickle-cell anemia, muscular dystrophy, and cystic fibrosis. Early diagnosis gives parents time to prepare for the birth of an affected child, and an opportunity to decide whether to continue with the pregnancy. If a disorder is treatable, early detection may allow the newborn to receive prompt and appropriate treatment. A few conditions are surgically correctable before birth.

As an example of how prenatal diagnosis works, consider a woman who becomes pregnant at age thirty-five. Her doctor will probably perform a procedure called obstetric sonography, in which ultrasound waves directed across the woman's abdomen form images of the fetus's limbs and internal organs. If the images reveal a physical defect that may be the result of a genetic disorder, a more invasive technique would be recommended. With fetoscopy, sound waves pulsed from inside the mother's uterus yield images much higher in resolution than ultrasound. Samples of tissue or blood are often taken at the same time.

Human genetics studies show that our thirty-five-year-old woman has about a 1 in 80 chance that her baby will be born with a chromosomal abnormality, a risk more than six times greater than when she was twenty years old. Thus, even if no physical abnormalities are detected by ultrasound, she probably will be offered an additional diagnostic procedure in which the fetus' cells are karyotyped or otherwise tested for genetic disorders. In amniocentesis, a small sample of fluid is drawn from the amniotic sac enclosing the fetus (**FIGURE 14.13**). The fluid contains cells shed by the fetus, and those cells are tested. Chorionic villus sampling (CVS) can be performed earlier than amniocentesis. With this technique, a few cells from the chorion are removed and tested (the chorion is a membrane that surrounds the amniotic sac).

An invasive procedure often carries a risk to the fetus. The risks vary by procedure. Amniocentesis has improved so much that, in the hands of a skilled physician, it no longer increases the risk of miscarriage. CVS occasionally disrupts the placenta's development, causing underdeveloped or missing fingers and toes in 0.3 percent of newborns. Fetoscopy raises the miscarriage risk by a whopping 2 to 10 percent, so it is rarely performed unless surgery or another medical procedure is required before the baby is born.

Couples who discover they are at high risk of having a child with a genetic disorder may opt for reproductive interventions such as *in vitro* fertilization. With this procedure, sperm and eggs taken from the prospective parents are mixed in a test tube. If an egg becomes fertilized, the resulting zygote will begin to divide. In about forty-eight hours, it will have become an embryo that consists of a ball of eight undifferentiated cells. One cell can be removed and its genes analyzed, a procedure called preimplantation diagnosis. The withdrawn cell will not be missed. If the embryo has no detectable genetic defects, it is inserted into the woman's uterus to develop. Most of the resulting "test-tube babies" are born in good health.

TAKE-HOME MESSAGE 14.7

- ✔ Studying inheritance patterns for genetic disorders has helped researchers develop treatments for some of them.
- ✔ Symptoms of some genetic disorders can be minimized or eliminated with early detection and treatment.
- ✔ Genetic screening can provide prospective parents with information about the health of their future children.

14.1 Shades of Skin (revisited)

Individuals of European descent and those of east Asian descent have several alleles in common that influence skin pigmentation. However, most people of east Asian descent carry a particular mutation in their *OCA2* gene—a single base-pair substitution in which an adenine changes to a cytosine—that results in lightened skin color. The product of the *OCA2* gene is named after the condition that occurs when the protein is missing: oculocutaneous albinism type II.

The *OCA2* mutation that lightens east Asian skin is uncommon in people of European ancestry. The *SLC24A5* allele that lightens European skin is uncommon in people of east Asian ancestry. Taken together, the distribution of these alleles suggests that (1) an African population with dark skin was ancestral to both east Asians and Europeans, and (2) east Asian and European populations separated before their pigmentation genes mutated and their skin color changed. ●

STUDY GUIDE

Section 14.1 Like most other human traits, skin color has a genetic basis. Minor differences in the alleles that govern melanin production and the deposition of melanosomes affect skin color. Skin color differences probably evolved as a balance between vitamin production and protection against harmful UV radiation.

Section 14.2 Relatively few human traits are governed by single genes inherited in a Mendelian pattern; most are inherited in a polygenic pattern with environmental contributions. Standardized charts called **pedigrees** can reveal inheritance patterns for alleles that are associated with specific phenotypes.

Geneticists study human inheritance by tracking disorders and abnormalities through generations of families. Most of what we know about human inheritance involves genetic disorders, because this information helps us develop treatments for affected people.

A genetic abnormality is an uncommon but relatively harmless version of a heritable trait. A genetic disorder is a heritable condition that sooner or later results in mild or severe medical problems. The symptoms of a genetic disorder usually occur in a syndrome.

Section 14.3 In an autosomal dominant inheritance pattern, a dominant allele associated with a trait occurs on an autosome, so the trait appears in everyone who has the allele (homozygous or heterozygous). The trait occurs in every generation of a family, and both sexes are affected with equal frequency. In an autosomal recessive pattern, a recessive allele associated with a trait occurs on an autosome, and the trait appears only in homozygous people. Both sexes are affected with equal frequency, but the trait can skip generations.

Section 14.4 In an X-linked inheritance pattern, an allele associated with a trait occurs on the X chromosome. Alleles that cause most X-linked disorders are recessive, and these tend to appear in men more often than in women. Heterozygous women have a dominant, normal allele that can mask the effects of the recessive one; men do not. Men cannot transmit an allele on an X chromosome to sons; a male child can inherit an X chromosome allele only from his mother.

Section 14.5 Faulty crossovers and the activity of **transposable elements** (**transposons**) can give rise to large-scale changes in chromosome structure, including insertions, deletions, **duplications** (in which a section of a chromosome is repeated), **inversions** (rearrangements in which part of a chromosome has become oriented in reverse), and **translocations** (rearrangement in which a broken piece of a chromosome has become reattached in the wrong place). Many of these changes are harmful or lethal; others affect fertility. Even so, many have accumulated in the chromosomes of all species over evolutionary time.

Section 14.6 Occasionally, abnormal events occur before or during meiosis, and new individuals end up with the wrong chromosome number. Consequences of such changes range from minor to lethal alterations in form and function.

Chromosome number changes are usually an outcome of **nondisjunction**, in which chromosomes fail to separate properly during nuclear division. Nondisjunction in meiosis may lead to **polyploidy**, the condition of having three or more complete sets of chromosomes (as compared to the two full sets of chromosomes in diploid cells). Polyploidy is lethal in humans, but not in flowering plants and some insects, fishes, and other animals. Nondisjunction may also lead to **aneuploidy**, the condition of having too many or too few copies of a particular chromosome. In humans, most types of autosomal aneuploidy are lethal. Trisomy 21, which causes Down syndrome, is an exception. Sex chromosome aneuploidy often results in minor impairment in learning and motor skills.

Section 14.7 Prospective parents can have genetic screening to estimate their risk of transmitting a harmful allele to offspring. The procedure involves analysis of parental pedigrees and genotype by a genetic counselor. Amniocentesis and other methods of prenatal genetic testing can reveal a genetic disorder before birth.

SELF-QUIZ

Answers in Appendix VII

1. Constructing a pedigree is particularly useful when studying inheritance patterns in organisms that ______ .
   a. produce many offspring per generation
   b. produce few offspring per generation
   c. have a very large chromosome number
   d. have a fast life cycle

2. Pedigree analysis is necessary when studying human inheritance patterns because _______ .
   a. humans have more than 20,000 genes
   b. of ethical problems with experimenting on humans
   c. inheritance in humans is more complicated than it is in other organisms
   d. genetic disorders occur only in humans

3. A recognized set of symptoms that characterize a genetic disorder is a(n) _______ .
   a. syndrome
   b. disease
   c. abnormality
   d. inheritance pattern

4. Tay–Sachs disease is caused by recessive alleles on an autosome. In which case(s) could two parents with a normal phenotype have a child with Tay–Sachs?
   a. Both parents are homozygous for a Tay–Sachs allele.
   b. Both parents are heterozygous for a Tay–Sachs allele.
   c. One parent is homozygous for a Tay–Sachs allele, and the other is heterozygous.

5. A trait that is present in a male child but not in either of his parents is characteristic of _______ inheritance.
   a. autosomal dominant
   b. autosomal recessive
   c. X-linked recessive
   d. It is impossible to answer this question without more information.

6. Choose the statement that is *in*correct.
   a. A son can inherit a recessive allele on an X chromosome from either parent.
   b. An individual may inherit three or more of each type of chromosome characteristic of the species, a condition called polyploidy.
   c. A female child inherits one X chromosome from her mother and one from her father.
   d. Pedigree analysis can be used to determine a future child's chance of being born with achondroplasia.

7. Color blindness is inherited in a(n) _______ pattern.
   a. autosomal dominant c. X-linked dominant
   b. autosomal recessive d. X-linked recessive

8. Nondisjunction in meiosis can result in _______ .
   a. base-pair substitutions c. crossing over
   b. aneuploidy d. pleiotropy

9. Klinefelter syndrome (XXY) can most be easily diagnosed by _______ .
   a. pedigree analysis c. karyotyping
   b. aneuploidy d. phenotypic treatment

10. Match the chromosome terms appropriately.

___ polyploid	a. symptoms of a genetic disorder
___ deletion	b. chromosomal mashup
___ aneuploidy	c. extra sets of chromosomes
___ translocation	d. gets around
___ syndrome	e. chromosome loses a chunk of DNA
___ transposable element	f. one extra chromosome

GENETICS PROBLEMS

Answers in Appendix VII

1. Does the phenotype indicated by the red circles and squares in this pedigree show an inheritance pattern that is autosomal dominant, autosomal recessive, or X-linked?

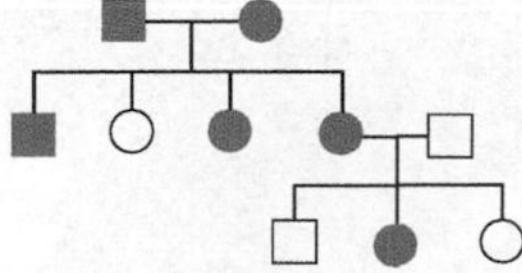

CENGAGE brain.com To access course materials, please visit www.cengagebrain.com.

2. G6PD deficiency is an X-linked recessive disorder. When people who have this disorder eat fava beans, they have a dangerous acute reaction in which a large number of their red blood cells rupture. A mother has the disorder; the father does not. What is the chance that any child of this union will have the trait?

3. Marfan syndrome (Section 13.5) is inherited in an autosomal dominant pattern. What is the chance that a child will inherit the associated allele if one parent does not carry it and the other is heterozygous?

4. Duchenne muscular dystrophy, which is inherited in an X-linked recessive pattern, nearly always occurs in males. Explain why.

5. Human females have two X chromosomes (XX); males have one X and one Y chromosome (XY).
   a. With respect to X chromosome alleles, how many different types of gametes can a male produce?
   b. If a female is homozygous for an allele on an X chromosome, how many types of gametes can she produce with respect to that allele?
   c. If a female is heterozygous for an X chromosome allele, how many types of gametes can she produce with respect to that allele?

6. A mutation on an autosome causes a particular protein to be overproduced, and the excess protein accumulates in the liver and damages it. Would the resulting disorder most likely be inherited in an autosomal dominant or recessive pattern?

7. Expression of the *SRY* gene on the Y chromosome gives rise to the male phenotype in humans. What do you think the inheritance pattern of *SRY* alleles is called?

8. The somatic cells of most individuals with Down syndrome contain an extra chromosome 21, for a total of forty-seven chromosomes.
   a. At which stage(s) of meiosis could nondisjunction alter the chromosome number?
   b. A few individuals with Down syndrome have forty-six chromosomes: two normal-appearing chromosomes 21, and a longer-than-normal chromosome 14. Speculate on how this chromosome abnormality may arise.

9. Mutations in the genes for clotting factor VIII and IX cause hemophilia A and B, respectively. A woman may be heterozygous for mutations in both genes, with a mutated factor VIII allele on one X chromosome, and a mutated factor IX allele on the other. All of her sons should have either hemophilia A or B. However, on rare occasions, one of these women gives birth to a son who does not have hemophilia, and his one X chromosome does not have either mutated allele. Explain.

15 STUDYING AND MANIPULATING GENOMES

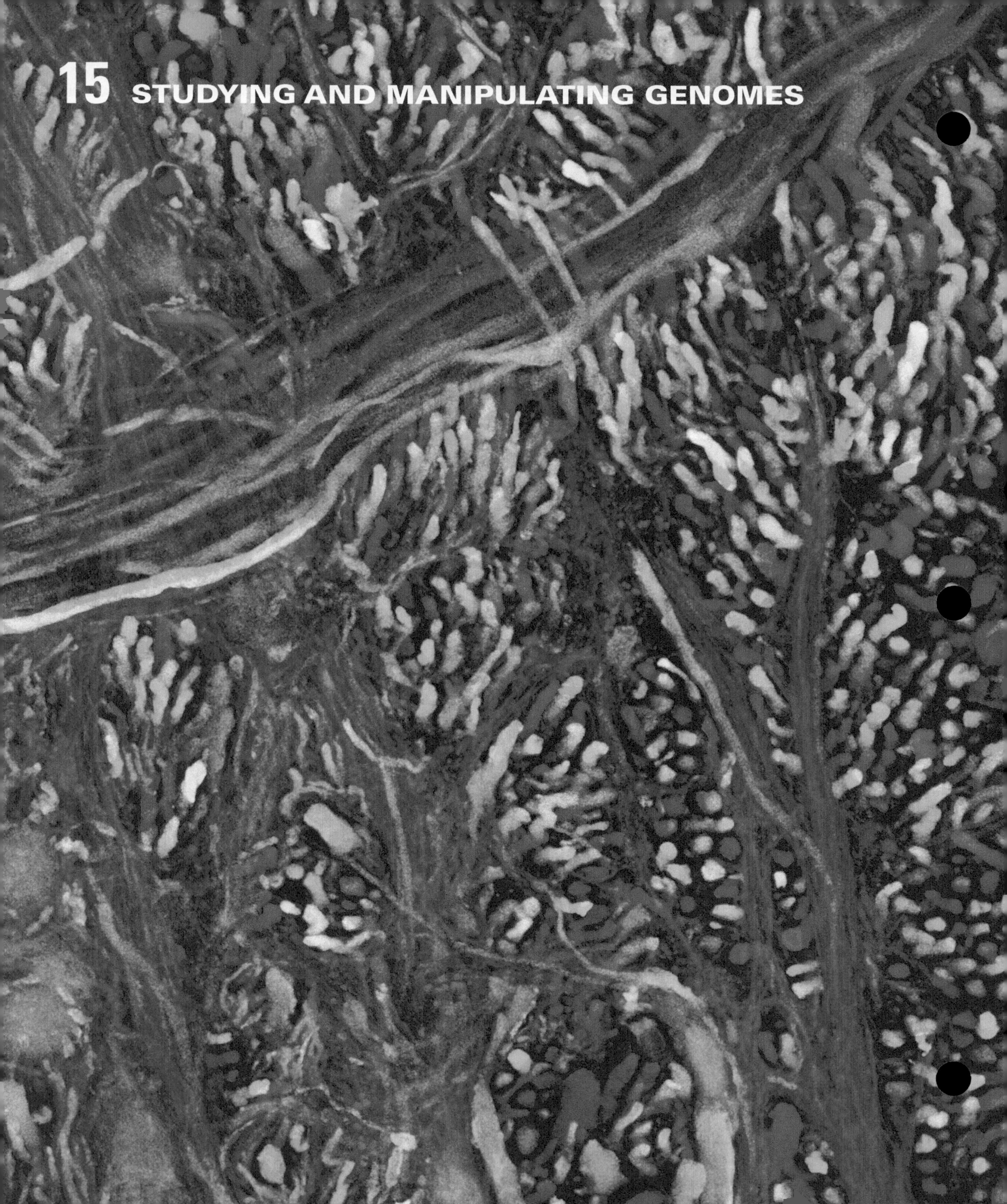

CORE CONCEPTS

Information Flow

Living systems store, retrieve, transmit, and respond to information essential for life.

An organism's form and function arise from and depend on expression of genetic information in its DNA. Differences in genes are the basis of differences between species; differences in alleles are the basis of variation in shared traits among individuals of a species. Researchers can change an organism's phenotype in a specific way by altering its genes, or by transferring genes from another species into it.

Evolution

Evolution underlies the unity and diversity of life.

Shared core processes and features widely distributed among organisms alive today provide evidence that all living things are descended from a common ancestor. Genetic and biochemical similarities among organisms of different taxa help us understand our own genes, and also allow us to manipulate heritable information by techniques of genetic engineering.

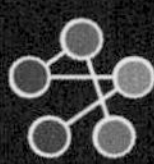

Systems

Complex properties arise from interactions among components of a biological system.

Genetic modification raises social and ethical issues. Biological systems are too complex to accurately predict long-term effects of genetic manipulation on humans, or on the environment.

Links to Earlier Concepts

In this chapter, we build on the relationship between genotype and phenotype (Section 13.2). Your knowledge of DNA structure (8.3, 8.4) and replication (8.5); clones (8.1); hybridization (8.5); gene expression (9.2); introns (9.3); and alleles (12.2) will help you understand genetic engineering. The chapter also revisits tracers (2.2), triglycerides and lipoproteins (3.5), β-carotene (6.3), bacteriophage (8.2), mutations (9.6), gene expression control (10.2, 10.5), knockouts (10.3), proto-oncogenes (11.6), the *MC1R* gene (13.5), environmental effects on gene expression (13.6), short tandem repeats (13.7), and evolutionary changes in chromosome structure (14.5).

15.1 Personal Genetic Testing

If you compared your DNA with your neighbor's, about 2.97 billion nucleotide bases of the two sequences would be identical; the remaining 30 million nonidentical bases are sprinkled throughout your chromosomes. The sprinkling is not entirely random because some regions of DNA vary less than others; these conserved regions are of particular interest to researchers because they are most likely to have an essential function. If a conserved sequence does vary among people, the variation tends to be in single bases at a particular locus. A base-pair substitution that is carried by a measurable percentage of a population, usually above 1 percent, is called a **single-nucleotide polymorphism**, or **SNP** (pronounced "snip").

Alleles of most genes differ by single bases, and differences in alleles are the basis of the variation in human traits that makes each person unique (Section 12.2). SNPs account for many of differences in appearance, and they also have a lot to do with differences in the way our bodies work—how we age, respond to drugs, weather assaults by pathogens and toxins, and so on.

Consider the lipoprotein particles that carry fats and cholesterol through our bloodstreams (Section 3.5). These particles consist of variable amounts and types of lipids and proteins. One of these proteins is called apolipoprotein E, and it is encoded by the *APOE* gene. About one in four people carries an allele of this gene, *E4*, with a SNP: a cytosine instead of the more common thymine at a particular location in its sequence. How this substitution affects the function of the protein is not yet clear, but we do know that having the *E4* allele increases one's risk of developing Alzheimer's disease later in life, particularly in people homozygous for it.

If you want to know your own genotype, finding out has never been easier. Genetic testing companies can extract DNA from a sample such as a drop of spit, then analyze it for SNPs. Personalized genetic testing is revolutionizing medicine, for example by allowing physicians to determine a patient's ability to respond to certain drugs before treatment begins. Cancer treatments are being tailored for individual patients and their tumor cells. People who discover they carry SNPs associated with a heightened risk of a medical condition are being encouraged to make lifestyle changes that could delay the condition's onset or prevent it entirely; preventive medical treatments based on these SNPs are becoming more common.

single-nucleotide polymorphism (SNP) One-nucleotide DNA sequence variation carried by a measurable percentage of a population.

15.2 DNA Cloning

LEARNING OBJECTIVES

- Describe restriction enzymes and explain why their discovery was important for DNA research.
- List the steps involved in making recombinant DNA.
- Explain the process of DNA cloning, and why researchers clone cDNA.

In the 1950s, excitement over the discovery of DNA's structure (Section 8.3) gave way to frustration: No one could determine the base sequence of DNA in a chromosome. Identifying single bases among thousands or millions of others turned out to be a huge technical challenge. Research in a seemingly unrelated field yielded a solution when Werner Arber, Hamilton Smith, and their coworkers made a discovery: Bacteria can resist infection by bacteriophage (Section 8.2) because they have enzymes that chop up any injected viral DNA. The enzymes restrict viral replication; hence their name, restriction enzymes. A **restriction enzyme** cuts DNA wherever a specific sequence of bases occurs (**FIGURE 15.1**). For example, the enzyme *Eco*RI (named after *E. coli*, the bacteria from which it was isolated) cuts DNA at the base sequence G-A-A-T-T-C ❶. Other restriction enzymes cut at different sequences.

The discovery of restriction enzymes allowed researchers to cut chromosomal DNA into manageable chunks. It also allowed them to combine DNA fragments from different organisms. How? Many restriction enzymes leave single-stranded tails on DNA fragments. Because the chemical structure of DNA is the same in all organisms, complementary tails will base-pair regardless of the source of DNA ❷. The tails are called "sticky ends" because two DNA fragments stick together when their matching tails base-pair (hybridization, Section 8.5). The enzyme DNA ligase can be used to seal the gaps between hybridized sticky ends, so continuous DNA strands form ❸. Thus, using restriction enzymes and DNA ligase, DNA can be cut from different sources and pasted together. The result, a hybrid molecule that consists of genetic material from two or more organisms, is called **recombinant DNA**.

Making recombinant DNA is the first step in **DNA cloning**, a set of laboratory methods that uses living cells to mass-produce specific DNA fragments. Researchers clone a fragment of DNA by inserting it into a **vector**, which in this context is a molecule that can carry foreign DNA into host cells. Plasmids may be used as cloning vectors in bacteria, yeast, and other cells (**FIGURE 15.2**). When a cell reproduces, its offspring inherit a full complement of genetic information—one chromosome plus plasmids. A recombinant plasmid gets replicated and distributed to descendant cells just like other plasmids (**FIGURE 15.3**).

A host cell into which a cloning vector has been inserted can be grown in the laboratory (cultured) to yield a huge population of genetically identical descendants. Each of these clones (Section 8.1) contains a copy of the recombinant plasmid. The plasmid can be harvested in quantity from the clones, and the hosted

FIGURE 15.1 Restriction enzymes and recombinant DNA.

FIGURE IT OUT Why does the enzyme cut both strands of DNA?

Answer: Because the recognition sequence occurs on both strands.

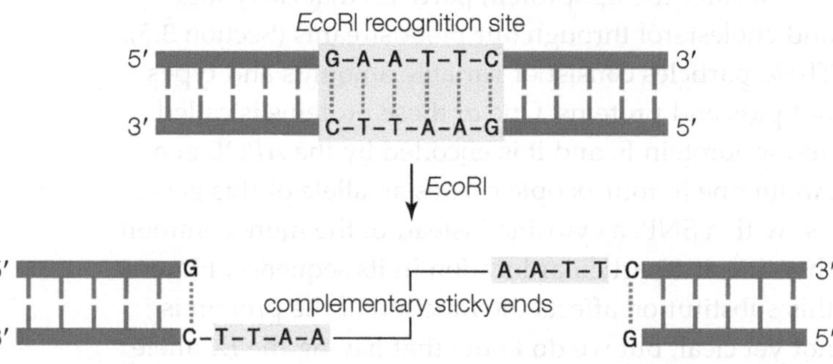

*Eco*RI recognition site
5′
G-A-A-T-T-C
3′
3′
C-T-T-A-A-G
5′
*Eco*RI
5′
G
A-A-T-T-C
3′
complementary sticky ends
3′
C-T-T-A-A
G
5′

❶ The restriction enzyme *Eco* RI recognizes a specific base sequence in DNA and cuts it. *Eco*RI leaves single-stranded tails ("sticky ends").

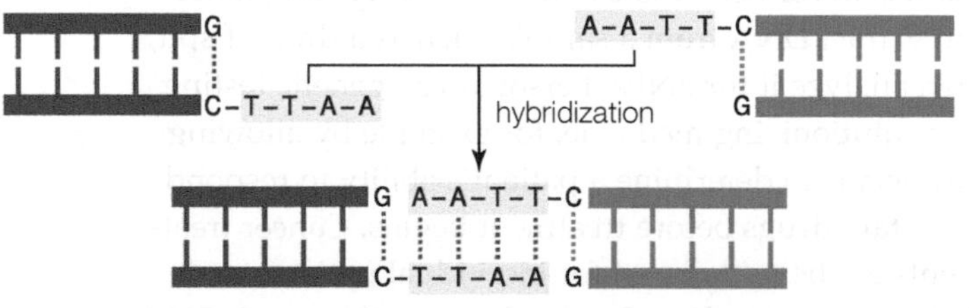

G
A-A-T-T-C
C-T-T-A-A
hybridization
G
G A-A-T-T-C
C-T-T-A-A G

❷ When DNA fragments from two sources are cut with *Eco* RI and mixed together, matching sticky ends base-pair.

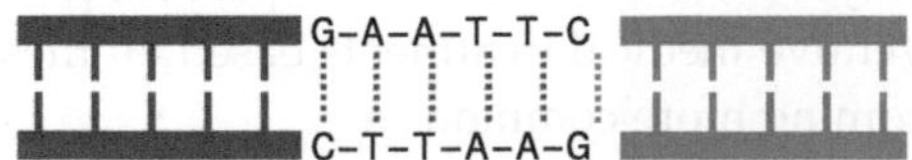

G-A-A-T-T-C
C-T-T-A-A-G

❸ DNA ligase seals the gaps between hybridized DNA fragments. The resulting hybrid molecules are called recombinant DNA.

cDNA Complementary strand of DNA synthesized from an RNA template by the enzyme reverse transcriptase.

DNA cloning Set of methods that uses living cells to mass-produce targeted DNA fragments.

recombinant DNA (ree-COM-bih-nent) A DNA molecule that contains genetic material from more than one organism.

restriction enzyme Type of enzyme that cuts DNA at a specific base sequence.

reverse transcriptase (trance-CRYPT-ace) An enzyme that uses mRNA as a template to make a strand of cDNA.

vector A DNA molecule that can accept foreign DNA and be replicated inside a host cell.

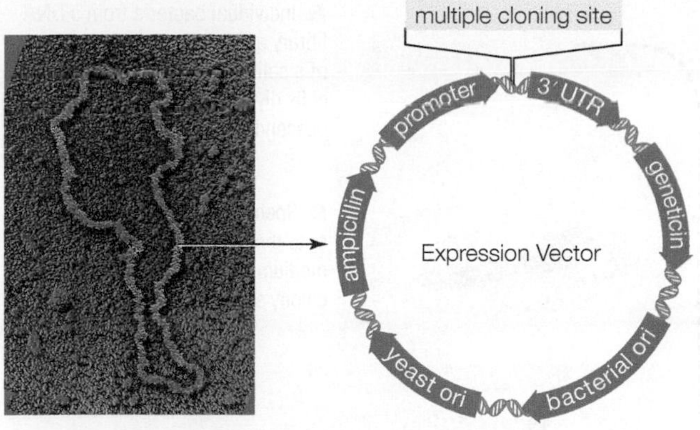
multiple cloning site
promoter
3' UTR
geneticin
ampicillin
Expression Vector
yeast ori
bacterial ori

FIGURE 15.2 A plasmid cloning vector.

A fragment of DNA can be inserted into a vector's multiple cloning site, which has several restriction enzyme recognition sequences useful for this purpose. Bacteria or yeast cells are then induced to take up the recombinant plasmid. Antibiotic resistance genes (here, ampicillin for bacteria, and geneticin for yeast) help researchers to identify cells that take up the plasmid, because only those cells survive treatment with the antibiotic.

This vector is also useful for *in vitro* expression of cloned DNA fragments. Yeast and bacterial origins of replication (ori) are sites where DNA replication of the plasmid begins in these cells. A eukaryotic promoter is necessary for yeast cells to transcribe RNA from the inserted DNA, as is a 3′ untranslated region (UTR) that includes required regulatory signals.

DNA fragment can be excised from the vector using the same restriction enzyme used for cloning.

Why Clone DNA?

Learning how to clone DNA was a key step in the quest to understand the information it encodes, which is why Arber and Smith won a 1978 Nobel Prize for the discovery of restriction enzymes. The sequence of bases in a cloned fragment of DNA could now be determined, and sequence information from many organisms started accumulating (Section 15.4 returns to this topic). The next step, finding the genes, became a major topic of research. This work was difficult because eukaryotic genes occur discontinuously on a chromosome—their coding sequences are interrupted by very long introns. Researchers realized that post-transcriptional processing removes introns (Section 9.3), so they developed a method for cloning mRNA.

An mRNA cannot be cut with restriction enzymes or pasted with DNA ligase, because these enzymes work only on double-stranded DNA. Thus, cloning mRNA requires reverse transcriptase, a replication enzyme made by some viruses. **Reverse transcriptase** uses an RNA template to assemble a strand of complementary DNA, or **cDNA**:

mRNA
Reverse transcriptase
cDNA
mRNA

DNA polymerase can be used to replace the mRNA with a new strand of DNA. The resulting double-stranded cDNA may be cut with restriction enzymes:

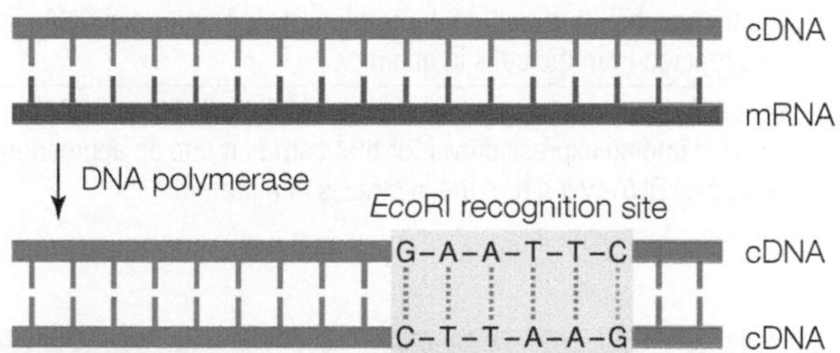
cDNA
mRNA
DNA polymerase
*Eco*RI recognition site
G-A-A-T-T-C
C-T-T-A-A-G
cDNA
cDNA

The outcome is a double-stranded cDNA version of the original mRNA. Like any other double-stranded DNA, it can be used for cloning or for other purposes.

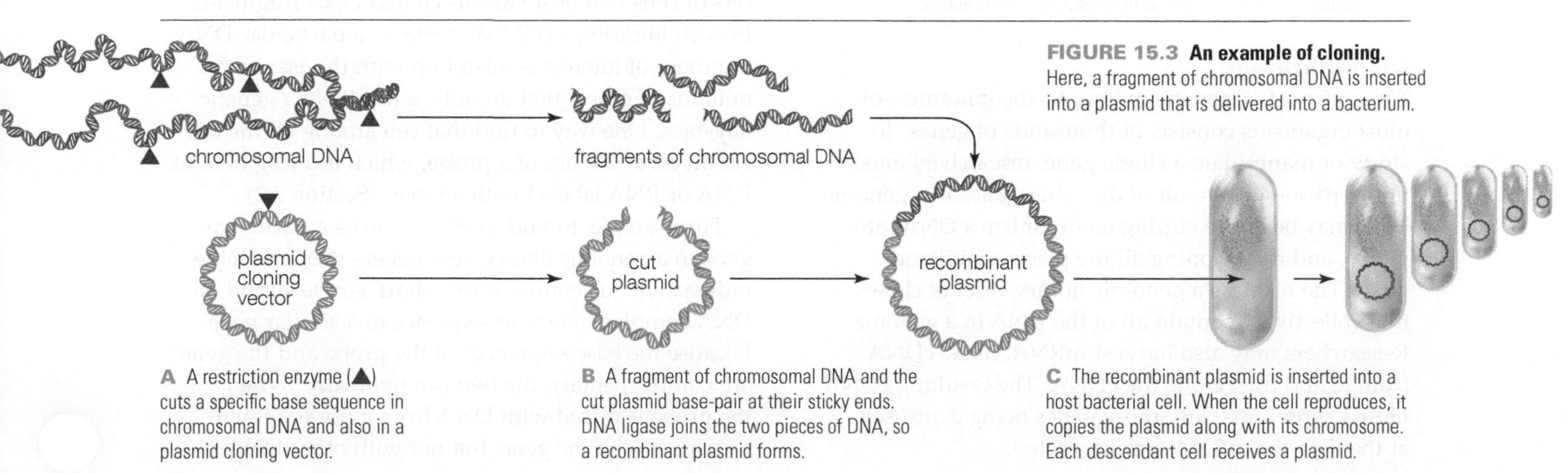
chromosomal DNA
fragments of chromosomal DNA
plasmid cloning vector
cut plasmid
recombinant plasmid

FIGURE 15.3 An example of cloning. Here, a fragment of chromosomal DNA is inserted into a plasmid that is delivered into a bacterium.

A A restriction enzyme (▲) cuts a specific base sequence in chromosomal DNA and also in a plasmid cloning vector.

B A fragment of chromosomal DNA and the cut plasmid base-pair at their sticky ends. DNA ligase joins the two pieces of DNA, so a recombinant plasmid forms.

C The recombinant plasmid is inserted into a host bacterial cell. When the cell reproduces, it copies the plasmid along with its chromosome. Each descendant cell receives a plasmid.

A cDNA contains all of the protein-building information encoded in a eukaryotic mRNA, but in DNA form. Inserted into an appropriate vector and host, a cDNA will be transcribed into mRNA, which in turn will be translated into a protein product (the plasmid shown in Figure 15.2 is useful for this purpose). Cloning cDNA allows researchers to manipulate eukaryotic genes, for example by changing them, knocking them out (Section 10.3), or inserting them into other species. The method also allows genes from complex eukaryotes to be expressed in bacteria or yeast, thus bypassing normal gene expression controls that would otherwise limit a protein's production in its cell of origin. Particularly in bacteria, foreign genes can be expressed at extremely high levels. Finally, bacteria and yeast can be easily grown on an industrial scale, for example to produce medically relevant proteins (Section 15.5 returns to this topic).

TAKE-HOME MESSAGE 15.2

✔ DNA cloning uses living cells to mass-produce targeted DNA fragments. Restriction enzymes cut DNA into fragments, then DNA ligase seals the fragments into cloning vectors. Recombinant DNA molecules result.

✔ Vectors can carry foreign DNA into host cells. When the host cell divides, it gives rise to huge populations of genetically identical cells (clones), each with a copy of the foreign DNA. The foreign DNA can be extracted from the cells in quantity.

✔ Researchers studying gene expression make and clone cDNA. Inserted into an expression vector that carries it into an appropriate host, the cDNA will direct the synthesis of a protein.

15.3 Isolating Genes

LEARNING OBJECTIVES

- Describe two ways of mass-producing a targeted section of DNA.
- Differentiate between a genomic library and a cDNA library.
- Explain the use of a probe in screening a DNA library.

DNA Libraries

The entire set of genetic material—the **genome**—of most organisms consists of thousands of genes. To study or manipulate a single gene, researchers must first separate it from all of the other genes in a genome. They may begin by cutting an organism's DNA into pieces, and then cloning all the pieces simultaneously. The result is a genomic library, a set of clones that collectively contain all of the DNA in a genome. Researchers may also harvest mRNA, make cDNA from it, and then clone the cDNA. The resulting cDNA library represents only those genes being expressed at the time the mRNA was harvested.

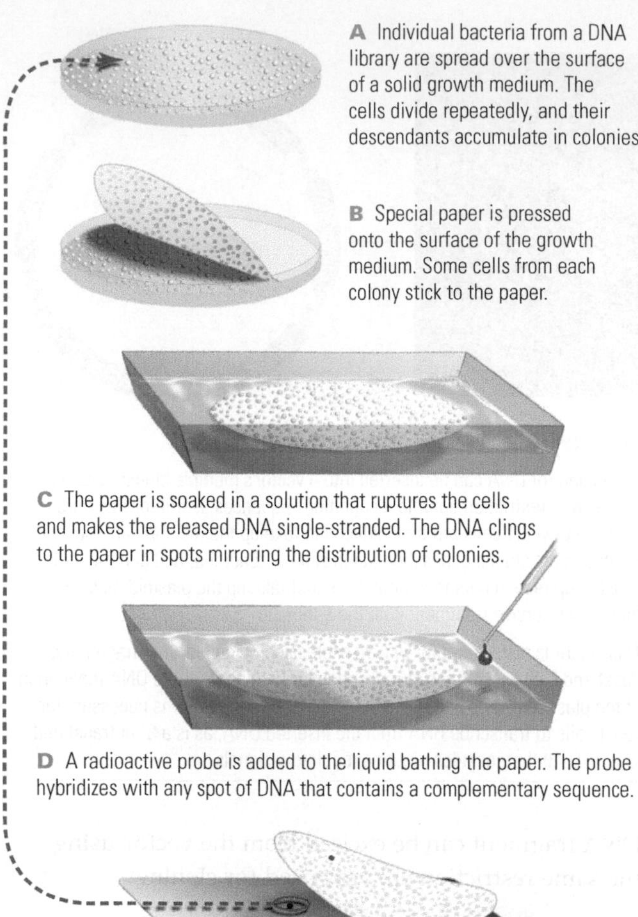
A Individual bacteria from a DNA library are spread over the surface of a solid growth medium. The cells divide repeatedly, and their descendants accumulate in colonies.
B Special paper is pressed onto the surface of the growth medium. Some cells from each colony stick to the paper.
C The paper is soaked in a solution that ruptures the cells and makes the released DNA single-stranded. The DNA clings to the paper in spots mirroring the distribution of colonies.
D A radioactive probe is added to the liquid bathing the paper. The probe hybridizes with any spot of DNA that contains a complementary sequence.
E The paper is pressed against X-ray film. The radioactive probe darkens the film in a spot where it has hybridized. The spot's position is compared to the positions of the original bacterial colonies. Cells from the colony that corresponds to the spot are cultured, and their DNA is harvested.

FIGURE 15.4 Using a probe to screen a DNA library.
In this example, a radioactive probe helps identify a colony of bacterial cells that host a targeted fragment of DNA.

Genomic and cDNA libraries are **DNA libraries**, sets of cells that host various cloned DNA fragments. In such libraries, a cell that contains a particular DNA fragment of interest is mixed up with thousands or millions of others that do not—a needle in a genetic haystack. One way to find that cell among all the others involves the use of a **probe**, which is a fragment of DNA or RNA labeled with a tracer (Section 2.2).

For example, to find a cell that hosts a particular gene in a genomic library, researchers may assemble radioactive nucleotides into a short, single strand of DNA complementary in sequence to a similar gene. Because the base sequences of the probe and the gene are complementary, the two can hybridize. When the probe is mixed with DNA from a library, it will hybridize with the gene, but not with other DNA

(**FIGURE 15.4**). Researchers pinpoint a cell that hosts the gene by detecting the label on the probe. That cell is isolated and cultured. The gene can then be extracted in bulk from the cultured cells.

PCR

The **polymerase chain reaction** (**PCR**) is a technique used to mass-produce copies of (amplify) a particular section of DNA. PCR is often used in place of cloning because it quickly and easily transforms a needle in a haystack—that one-in-a-million fragment of DNA—into a huge stack of needles with a little hay in it.

The starting material for PCR is a sample that contains at least one molecule of DNA. Essentially anything with DNA in it can be used as the source of the sample: a sperm, a hair from a crime scene, ancient remains, and so on. The technique is based on DNA replication (**FIGURE 15.5**), and it requires two synthetic primers designed to base-pair at opposite ends of a targeted section of DNA. Researchers mix these primers with the starting (template) DNA, nucleotides, and DNA polymerase ➊. Then, they expose the mixture to repeated cycles of high and low temperatures. A few seconds at high temperature disrupts the hydrogen bonds that hold the two strands of a DNA double helix together (Section 8.3), so every molecule of DNA unwinds and becomes single-stranded. As the temperature of the reaction mixture is lowered, the primers hybridize with any DNA strands that have the targeted sequence ➋.

The DNA polymerases of most organisms denature at the high temperature required to separate DNA strands. PCR makes use of *Taq* polymerase, an enzyme isolated from *Thermus aquaticus*. These bacteria live in hot springs and hydrothermal vents, so their polymerase necessarily tolerates heat. Like other DNA polymerases, *Taq* polymerase recognizes hybridized primers as places to start DNA synthesis ➌. Synthesis proceeds along the template strand until the temperature is raised again and the DNA separates into single strands ➍. The newly synthesized DNA is a copy of the targeted section. When the mixture is cooled, the primers rehybridize, and DNA synthesis begins again. Each cycle of heating and cooling takes only a few minutes, but it can double the number of copies of the targeted section of DNA ➎. PCR is so efficient that thirty cycles

DNA library Collection of cells that host different fragments of foreign DNA, often representing an organism's entire genome.

genome An organism's complete set of genetic material.

polymerase chain reaction (**PCR**) Method that rapidly generates many copies of a specific section of DNA.

probe Short fragment of DNA labeled with a tracer; designed to hybridize with a base sequence of interest.

FIGURE 15.5 Two rounds of PCR.

➊ DNA (blue) that has a targeted sequence (gray) is mixed with primers (pink), nucleotides, and heat-tolerant *Taq* DNA polymerase.

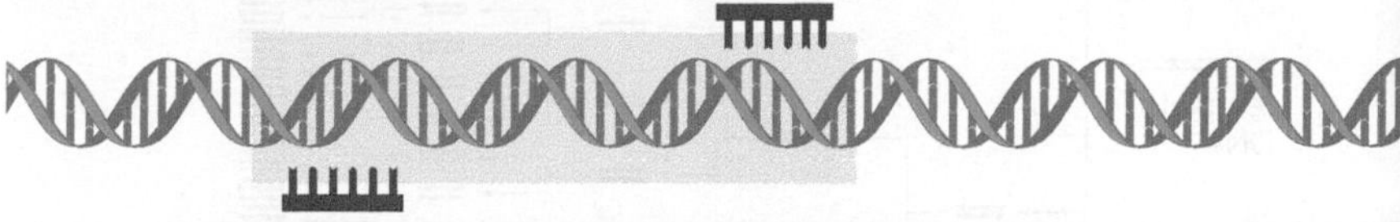

➋ When the mixture is heated, the double-stranded DNA separates into single strands. When the mixture is cooled, some of the primers base-pair with the DNA at opposite ends of the targeted sequence.

➌ *Taq* polymerase begins DNA synthesis at the primers, so it produces complementary strands of the targeted DNA sequence.

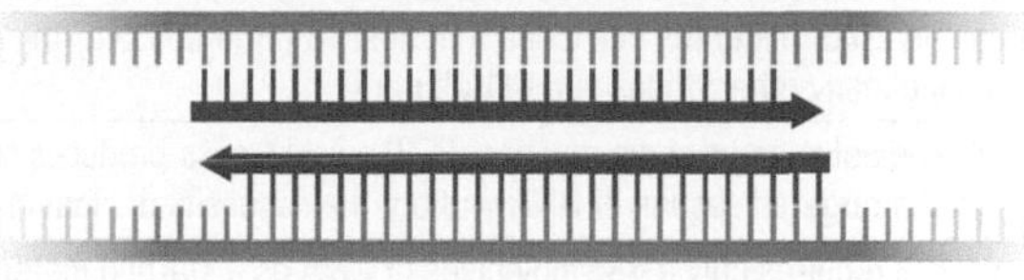

➍ The mixture is heated again, so all double-stranded DNA separates into single strands. When it is cooled, primers base-pair with the targeted sequence in the original template DNA and in the new DNA strands.

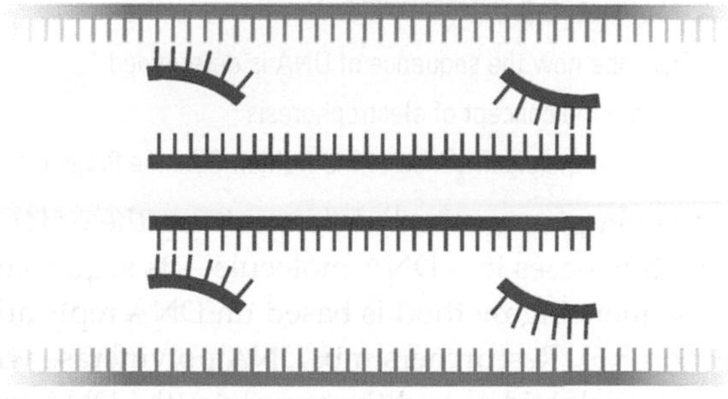

➎ Each cycle of heating and cooling can double the number of copies of the targeted DNA section.

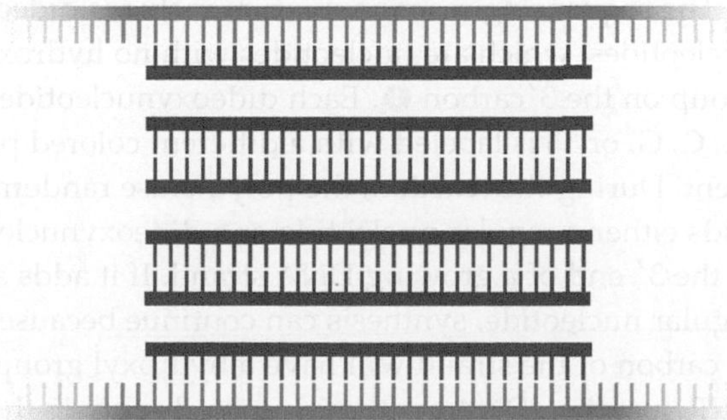

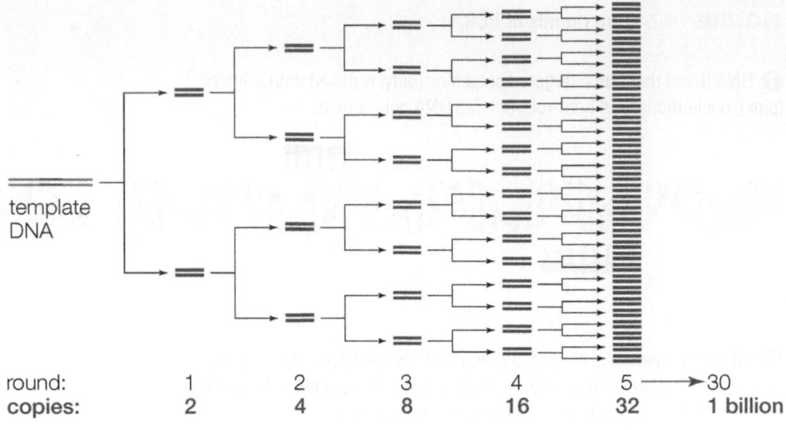
template DNA
round: 1 2 3 4 5 → 30
copies: 2 4 8 16 32 1 billion

FIGURE 15.6 Exponential amplification of DNA by PCR.

of reactions may amplify by a billionfold the number of copies of DNA in a sample (**FIGURE 15.6**).

TAKE-HOME MESSAGE 15.3

✔ A DNA library can be made from cDNA or genomic DNA. Probes can be used to identify one clone that hosts a targeted DNA fragment among many other clones in a DNA library.

✔ The polymerase chain reaction (PCR) quickly mass-produces copies of a targeted section of DNA without the necessity of cloning.

✔ PCR requires only a few molecules of DNA as a starting material.

15.4 DNA Sequencing

LEARNING OBJECTIVES

- Describe how the sequence of DNA is determined.
- Explain the concept of electrophoresis.
- Interpret the significance of the Human Genome Project.

Sequencing is a method of determining the order of nucleotide bases in a DNA molecule—its sequence. One sequencing method is based on DNA replication (Section 8.5). Researchers mix DNA polymerase with a primer, nucleotides, and the template (the DNA to be sequenced). Starting at the primer, the polymerase joins the nucleotides into a new strand of DNA, in the order dictated by the sequence of the template (**FIGURE 15.7**).

The sequencing reaction mixture includes dideoxynucleotides, which are nucleotides with no hydroxyl group on the 3′carbon ❶. Each dideoxynucleotide base (A, C, G, or T) is labeled with a different colored pigment. During the reaction, the polymerase randomly adds either a regular nucleotide or a dideoxynucleotide to the 3′ end of a growing DNA strand. If it adds a regular nucleotide, synthesis can continue because the 3′ carbon of the strand will have a hydroxyl group on it. (Remember, DNA synthesis proceeds only in the 5′ to 3′ direction.) If the polymerase adds a dideoxynucleotide, the 3′ carbon of the strand will not have a hydroxyl group, so synthesis ends there ❷.

Millions of DNA strands of different lengths are produced, each a partial, complementary copy of the template DNA ❸. Every strand of a given length ends with the same dideoxynucleotide. For example, if the template strand's tenth base was guanine, then any newly synthesized strand that is 10 nucleotides long ends with a cytosine.

The DNA strands are then separated by length. With a technique called **electrophoresis**, an electric field pulls the strands through a semisolid gel. Strands of different sizes move through the gel at different rates: The shorter the strand, the faster it moves, because shorter strands slip through the tangled molecules of the gel faster than longer strands do. All strands of the same length move through the gel at the same speed, so they gather into bands. The strands in a band have the same modified nucleotide at their ends, and that pigment now imparts a distinct color to the band ❹. Each color designates one of the four modified nucleotides, so the order of colored bands in the gel reflects the sequence of the template DNA ❺.

The Human Genome Project

Ten years after DNA sequencing had been invented, it had become so routine that scientists began to consider sequencing the entire human genome. Proponents of the idea said it could provide huge payoffs for medicine and research. Opponents said this daunting task would divert attention and funding from more urgent research. Given the existing methods, sequencing 3 billion bases would take at least 50 years. However, the methods continued to improve, and with each improvement more bases could be sequenced in less time. Automated (robotic) sequencing and PCR had just been invented. Both were still too cumbersome and expensive to be useful in routine applications, but they would not be so for long. Waiting for faster, cheaper technologies seemed the most efficient way to sequence the genome, but just how fast and cheap did they need to be before the project should begin?

A few privately owned companies decided not to wait, and started sequencing. One of them intended to determine the genome sequence in order to patent it. The idea of patenting the human genome provoked widespread outrage, but it also spurred commitments in the public sector. In 1988, the National Institutes of

electrophoresis (eh-lek-troh-fur-EE-sis) Technique that separates DNA fragments by size.
sequencing Method of determining DNA sequence.

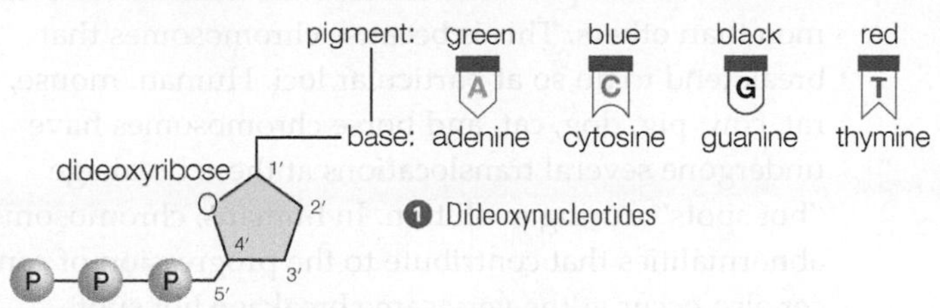
template DNA
pigment:
green
blue
black
red
A
C
G
T
base: adenine
cytosine
guanine
thymine
dideoxyribose
1′
2′
3′
4′
5′
P
P
P
1 Dideoxynucleotides

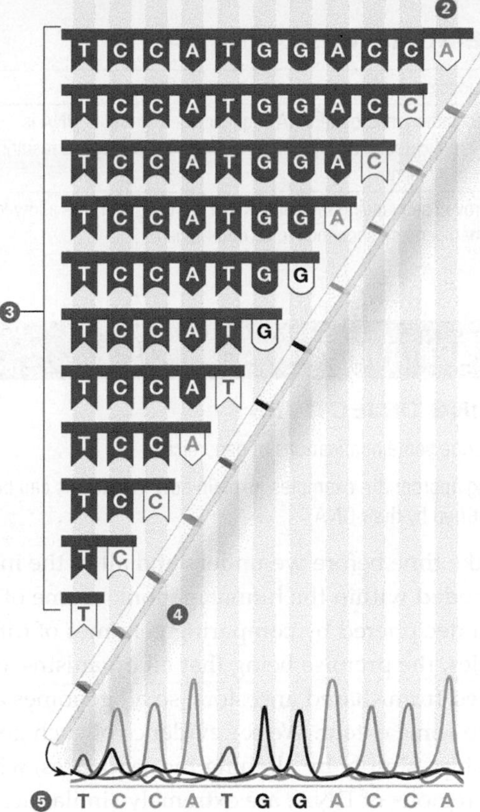
T C C A T G G A C C A
DNA polymerase movement
A G G T A C C T G G T
2
3
4
5 T C C A T G G A C C A

FIGURE 15.7 Sequencing. In this method of sequencing, DNA polymerase incompletely replicates a section of DNA.

1 The sequencing reaction requires dideoxynucleotides: nucleotides, which have a hydrogen atom instead of a hydroxyl group on their 3′ carbon (compare the structure with those in Figure 8.7). Each dideoxynucleotide base (G, A, T, or C) is labeled with a different colored pigment.

2 DNA polymerase uses a section of DNA as a template to synthesize new strands of DNA. Synthesis of each new strand stops when a dideoxynucleotide is added.

3 At the end of the reaction, there are many incomplete copies of the template DNA in the mixture.

4 Electrophoresis separates the copied DNA strands into bands according to their length. All DNA strands in each band end with the same base, so each band is the color of the base's tracer pigment.

5 A computer detects and records the color of successive bands on the gel. The order of colors of the bands reflects the sequence of the DNA.

Health (NIH) essentially took over the project by hiring James Watson (of DNA structure fame) to head an official Human Genome Project, and providing $200 million per year to fund it. A partnership formed between the NIH and international institutions that were sequencing different parts of the genome. Watson set aside 3 percent of the funding for studies of ethical and social issues arising from the work. He later resigned over a patent disagreement, and geneticist Francis Collins took his place.

Amid ongoing squabbles over patent issues, Celera Genomics formed in 1998. With biologist Craig Venter at its helm, the company intended to commercialize the human genome sequence. Celera invented faster sequencing technologies because the first to complete the genome sequence had a legal basis for patenting it. The competition motivated the international partnership to accelerate its efforts. Then, in 2000, U.S. President Bill Clinton and British Prime Minister Tony Blair jointly declared that the sequence of the genome could not be patented. Celera kept sequencing anyway, and, in 2001, the competing governmental and corporate teams published about 90 percent of the sequence. In 2003, fifty years after the discovery of the structure of DNA, the sequence of the human genome was completed. The sequence is freely accessible worldwide (see www.ncbi.nlm.nih.gov/genome). Anyone can search it and see current statistics (**TABLE 15.1**).

TABLE 15.1

Some Human Genome Statistics

Chromosomes (number of)*	24
Nucleotides (number of)*	3,547,121,844
Genes	
Protein-coding genes (number of)*	20,313
Non-protein-coding genes (number of)*	25,180
Pseudogenes (gene duplications, number of)*	14,453
Exons (number of)	381,122
Average exon length (base pairs)	163
Introns (number of)	378,089
Average intron length (base pairs)	5,849
Variation	
Short tandem repeats (number of)	741,587
Single-nucleotide polymorphisms (number of)‡	87,339,846

**Ensembl release 84.38.*
‡ NCBI dbSNP build 146.

From start to finish, the human genome project required about 16 years and $2.7 billion to complete, but it spurred development of much more efficient technologies for sequencing and data analysis. Today, an entire genome can be sequenced in a few days, for less than $1,000.

TAKE-HOME MESSAGE 15.4

- ✔ Sequencing reveals the order of bases in DNA.
- ✔ In a common method of DNA sequencing, a strand of DNA is partially replicated. Electrophoresis is used to separate the resulting fragments by length.
- ✔ Improved sequencing techniques and worldwide efforts allowed the human genome sequence to be determined.

15.5 Genomics

LEARNING OBJECTIVES

- Describe some applications of genomics.
- Using appropriate examples, explain how individuals can be identified by their DNA.

It will take time before we understand all of the information coded within the human genome. Some of it has been deciphered by comparing genomes of different species, the premise being that all organisms are descended from shared ancestors, so all genomes are related to some extent. We see evidence of such genetic relationships simply by aligning the sequences, which, in some regions of DNA, are extremely similar across many species (**FIGURE 15.8**).

Similarities among genomes allow us to discover the function of human genes by studying their counterparts in other species. For example, researchers comparing human and mouse genomes discovered a human version of a mouse gene, *APOA5*, that encodes a lipoprotein (Section 3.5). Mice with an *APOA5* knockout have four times the normal level of triglycerides in their blood. The researchers then looked for—and found—a correlation between APOA5 mutations and high triglyceride levels in humans. High triglycerides are a risk factor for coronary artery disease.

The study of whole-genome structure and function is called **genomics**, and it has given us powerful insights into evolution. For example, comparing primate genomes revealed a change in chromosome structure that occurred during the evolution of our species (Section 14.5). Comparing genomes also showed us that structural changes in chromosomes are not entirely random: Some specific alterations are much more common than others. This is because chromosomes that break tend to do so at particular loci. Human, mouse, rat, cow, pig, dog, cat, and horse chromosomes have undergone several translocations at these breakage "hot spots" during evolution. In humans, chromosome abnormalities that contribute to the progression of cancer also occur at the very same breakage hot spots.

A genome is relatively constant, but gene expression is not. The amounts and types of proteins in a cell differ dramatically by cell type and over time (Section 10.2). Proteomics, an offshoot of genomics, involves the study of proteins on a large scale: the comprehensive set of proteins produced by an organism or cellular system. Studying patterns and shifts in protein production helps us understand how genes are used in cellular functioning.

Genomics and proteomics require computational tools for analyzing and interpreting biological data. Developing these tools is called bioinformatics, and this field is profoundly shifting the focus of biology research. Instead of trying to understand a complex biological system by studying its parts, researchers can now study the system as a whole. Until recently, for example, the role of genetics in medicine was limited to diagnosis of single-gene disorders (Section 14.2). Bioinformatics is bringing us closer to unraveling conditions that have complex inheritance patterns (such as cancer, diabetes, and cardiovascular disease).

DNA Profiling

About 99 percent of your DNA is exactly the same as everyone else's. The shared part is what makes you human; the differences make you a unique member of the species. In fact, those differences are so unique that

DNA profiling Identifying an individual by analyzing the unique parts of his or her DNA.

genomics The study of whole-genome structure and function.

FIGURE 15.8 Genomic DNA alignment. This is a region of the gene for a DNA polymerase. Bases that differ from the human sequence (top) are highlighted. The chance that any two of these sequences would randomly match is about 1 in 10^{46}.

```
758 GATAATCCTGTTTTGAACAAAAGGTCAAATTGCTGAATAGAAA-GTCTTGATTAACTAAAAGATGTACAAAGTGGAATTA 836 Human
752 GATAATCCTGTTTTGAACAAAAGGTCAAATTGCTGAATAGAAA-GTCTTGATTAACTAAAAGATGTACAAAGTGGAATTA 830 Mouse
751 GATAATCCTGTTTTGAACAAAAGGTCAAATTGCTGAATAGAAA-GTCTTGATTAACTAAAAGATGTACAAAGTGGAATTA 829 Rat
754 GATAATCCTGTTTTGAACAAAAGGTCAAATTGCTGAATAGAAA-GTCTTGATTAACTAAAAGATGTACAAAGTGGAATTA 832 Dog
782 GATAATCCTGTTTTGAACAAAAGGTCAAATTGCTGAATAGAAA-GTCTTGATTAACTAAAAGATGTACAAAGTGGAATTA 860 Chicken
758 GATAATCCTGTTTTGAACAAAAGGTCAAATTGCTGAATAGAAA-GTCTTGATTAAGTAAAAGATGTACAAAGTGGAATTA 836 Frog
823 GATAATCCTGTTTTGAACAAAAGGTCAGATTGCTGAATAGAAAAGGCTTGATTAAAGCAGAGATGTACAAAGTGGACGCA 902 Zebrafish
763 GATAATCCTGTTTTGAACAAAAGGTCAAATTGTTGAATAGAGACGCTTTGATAAAGCGGAGGAGGTACAAAGTGGGACC- 841 Pufferfish
```

they can be used to identify you. Identifying an individual by his or her DNA is called **DNA profiling**.

One DNA profiling method uses SNP-chips (**FIGURE 15.9**). A SNP-chip is a glass plate with microscopic spots of DNA stamped on it. The DNA in each spot is a short, synthetic single strand with a unique SNP. When an individual's DNA is washed over a SNP-chip, it hybridizes only to spots with matching SNPs. Probes reveal where the DNA has hybridized—and which SNPs are in it.

Another method of DNA profiling involves short tandem repeats (Section 13.7). These repeated sequences of nucleotide bases usually occur at predictable loci, but the number of times a sequence is repeated differs among individuals. For example, one person's DNA may have fifteen repeats of the bases TTTTC at a certain locus. Another person's DNA may have four repeats of this sequence in the same locus. DNA polymerases stutter or slip over these regions during DNA replication, so the number of repeats can increase or decrease from one generation to the next. Unless two people are identical twins, the chance that they have identical short tandem repeats at even three loci is 1 in a quintillion (10^{18}), which is far more than the number of people who have ever lived. Thus, an individual's array of short tandem repeats is, for all practical purposes, unique.

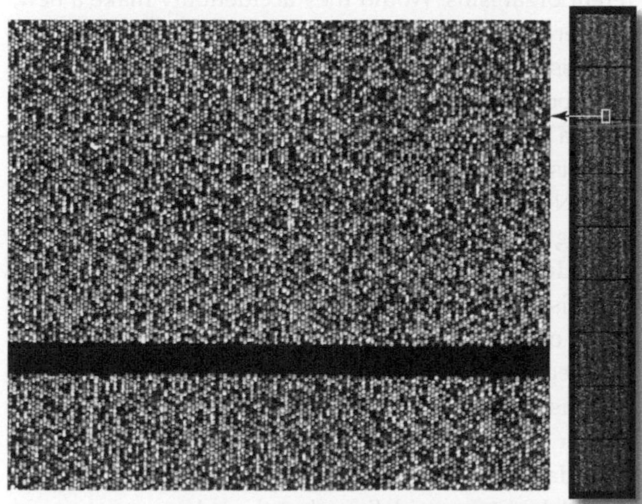

FIGURE 15.9 A SNP-chip analysis. Each spot is a region where an individual's DNA has hybridized with a SNP sequence. A red or green dot means that the individual has two copies of a SNP (homozygosity); a combined signal (yellow dot) indicates heterozygosity. The entire chip, shown right, tests for 550,000 SNPs; magnified portion is indicated by the small white box.

Analyzing short tandem repeats in a person's DNA begins with PCR, which is used to amplify ten to thirteen loci known to have repeats. The lengths of the PCR-amplified DNA fragments differ among most individuals because the number of repeats at those loci differs. Thus, electrophoresis can be used to reveal an individual's unique array of short tandem repeats (**FIGURE 15.10**).

Geneticists compare short tandem repeats on the Y chromosome to determine relationships among male relatives, and to trace an individual's ethnic heritage. They also track mutations that accumulate in populations over time by comparing DNA profiles of living humans with those of people who lived long ago. Such studies are allowing us to study population dispersals that happened in the ancient past.

Short tandem repeat profiles are routinely used to resolve kinship disputes and to provide evidence in criminal cases. Within the context of a criminal or forensic investigation, DNA profiling is called DNA fingerprinting. In 2017, the database of DNA fingerprints maintained by the Federal Bureau of Investigation (the FBI) contained the short tandem repeat profiles of almost 12 million convicted offenders, and had been used in more than 300,000 criminal investigations.

TAKE-HOME MESSAGE 15.5

✔ Comparing the genomes of humans and other species helps us understand how the human body works.

✔ Unique sequences of genomic DNA can be used to distinguish an individual from all others.

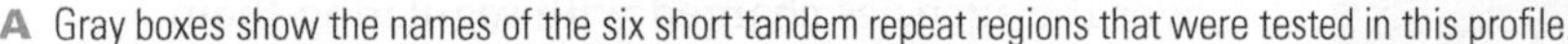

FIGURE 15.10 A partial short tandem repeat profile. Remember, human body cells are diploid, with pairs of homologous chromosomes. Double peaks appear on a profile when homologous chromosomes of a pair carry a different number of repeats.

A Gray boxes show the names of the six short tandem repeat regions that were tested in this profile.

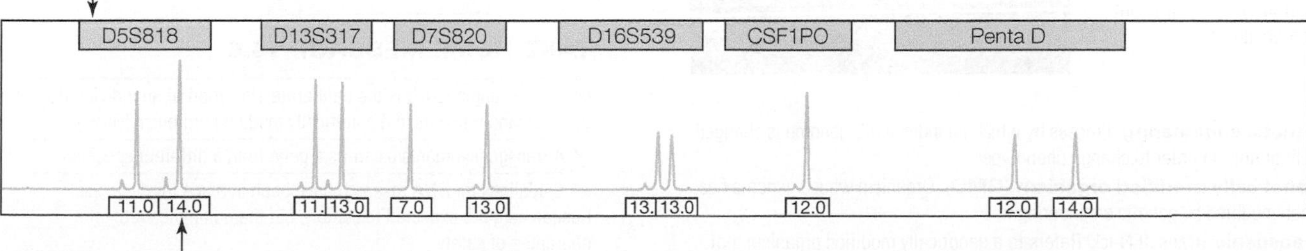
D5S818
D13S317
D7S820
D16S539
CSF1PO
Penta D
11.0
14.0
11.
13.0
7.0
13.0
13.
13.0
12.0
12.0
14.0

B The number of repeats is shown in a box below each peak. Peak size reflects the amount of DNA produced by the PCR reaction.

15.6 Genetic Engineering

LEARNING OBJECTIVES

- Differentiate between a GMO and a transgenic organism.
- Using appropriate examples, describe some of the ways that genetically engineered microorganisms are used.
- Describe safety measures put in place for genetic modification.

Genetically Modified Organisms

Traditional cross-breeding methods can alter genomes, but only if individuals with the desired traits will interbreed. Genetic engineering takes gene-swapping to an entirely different level. **Genetic engineering** is a process by which a genome is deliberately modified with the intent of changing its phenotype. An individual whose genome has been engineered is a **genetically modified organism** (**GMO**). Some GMOs are **transgenic**, which means a gene from a different species has been inserted into their DNA.

Most genetic engineering involves yeast and bacteria (**FIGURE 15.11**). Both types of cells have the metabolic machinery to make complex organic molecules, and they are easily engineered to produce, for example, medically important proteins. People with diabetes were among the first beneficiaries of such organisms. Insulin for their injections was once extracted from animals, but some people had allergic reactions to this insulin. Human insulin, which does not provoke allergic reactions, has been produced by transgenic *E. coli* since 1982. Slight modifications of the gene have yielded fast-acting and slow-release forms of insulin.

Genetically engineered microorganisms also make proteins used in food production. For example, we use enzymes produced by modified microorganisms to improve the taste and clarity of beer and fruit juice, to slow bread staling, and to modify fats. Cheese is traditionally made with an enzyme (chymosin) extracted from calf stomachs. Today, almost all cheese is made with calf chymosin produced by transgenic yeast.

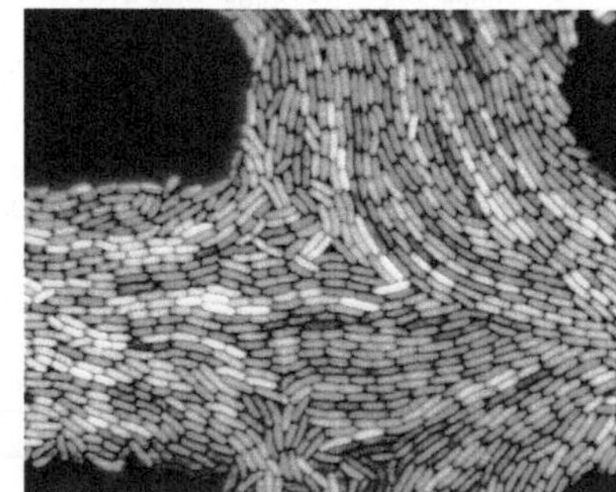

FIGURE 15.11
Bacteria transgenic for a fluorescent jellyfish protein. Variation in fluorescence among these genetically identical *E. coli* cells reveals differences in gene expression that may help us understand why some bacteria become dangerously resistant to antibiotics, and others do not.

Safety Issues

When James Watson and Francis Crick presented their model of DNA in 1953, they ignited a global blaze of optimism. The very book of life seemed to be open for scrutiny, but in reality, no one could read it. New techniques would have to be invented before that book would be readable. Twenty years later, Paul Berg and his coworkers discovered how to make recombinant organisms by fusing DNA from two species of bacteria, thus providing a tool for studying DNA sequences. They began to clone DNA from many different organisms. The technique of genetic engineering was born, and suddenly everyone was worried about it. Researchers knew that DNA itself was not toxic, but they could not predict with certainty what would happen each time they fused genetic material from different organisms. Would they accidentally make a new, dangerous form of life by fusing DNA of two harmless organisms? What if an engineered organism escaped the laboratory and transformed other organisms?

In a remarkably quick act of self-regulation, scientists reached a consensus on new safety guidelines for DNA research. Adopted at once by the NIH, these guidelines included precautions for laboratory procedures. They covered the design and use of host organisms that could survive only under a narrow range of conditions inside the laboratory. Researchers stopped using DNA from pathogenic or toxic organisms for recombinant DNA experiments until proper containment facilities were developed.

Today, all genetic engineering should be done under these laboratory guidelines, but the rules are not a guarantee of safety. We are still learning about escaped GMOs and their effects, and regulation is a problem. For example, the expense of removing government restrictions on the use of a GMO is prohibitive for organizations in the public sector. Thus, most commercial GMOs are produced by large private companies—the same ones that wield tremendous political influence over the agencies charged with regulating them.

TAKE-HOME MESSAGE 15.6

✔ Genetic engineering is the deliberate alteration of an individual's genome, and it results in a genetically modified organism (GMO).

✔ A transgenic organism carries a gene from a different species.

✔ Guidelines for DNA research have been in place for decades. Researchers are expected to comply, but the guidelines are not a guarantee of safety.

genetic engineering Process by which an individual's genome is changed deliberately, in order to change phenotype.

genetically modified organism (GMO) Organism whose genome has been modified by genetic engineering.

transgenic (trans-JEN-ick) Refers to a genetically modified organism that carries a gene from a different species.

CREDIT: (11) Photo Courtesy of Systems Biodynamics Lab, P. I. Jeff Hasty, UCSD Department of Bioengineering, and Scott Cookson.

15.7 Designer Plants

LEARNING OBJECTIVES

- Describe a method of making transgenic plants.
- Describe some applications of genetically engineered plants.
- Explain an unintended environmental effect of GMO crop use.

As crop production expands to keep pace with human population growth, farmers have begun to rely on transgenic crops. Foreign genes can be introduced into plant cells by way of electric or chemical shocks, by blasting them with microscopic DNA-coated pellets, or by using *Agrobacterium tumefaciens*, a type of bacteria that infects plants. *A. tumefaciens* carries a plasmid with genes that cause tumors to form on a host plant; hence the name Ti plasmid (for Tumor-inducing). Tumor-inducing genes are replaced with foreign or modified genes, and then the recombinant plasmid is used as a vector to deliver the genes into plant cells. Whole plants can be grown from cells that integrate a recombinant plasmid into their chromosomes (**FIGURE 15.12A,B**).

Some genetically modified plants carry genes that impart resistance to diseases or pests. Consider that farmers often spray organic crops with spores of Bt (*Bacillus thuringiensis*), a species of bacteria that makes a protein toxic only to some insect larvae. The gene for the Bt protein has been transferred into crop plants such as soy and corn (**FIGURE 15.12C**). Cells of the engineered plants produce the Bt protein, and larvae die shortly after eating their first and only GMO meal.

Transgenic crop plants are also being developed for impoverished regions of the world. Genes that confer drought tolerance, insect resistance, and enhanced nutritional value are being introduced into plants such as corn, rice, beans, sugarcane, cassava, cowpeas, banana, and wheat. The resulting GMO crops may help people in these regions who rely mainly on agriculture for food and income.

Genetic modifications can make food plants more nutritious. For example, rice plants have been engineered to make β-carotene (Section 6.3), an orange photosynthetic pigment that is remodeled into vitamin A by cells of the small intestine. These rice plants carry two genes in the β-carotene synthesis pathway: one from corn, the other from bacteria. One cup of the engineered rice seeds—grains of Golden Rice—has enough β-carotene to satisfy a child's daily need for vitamin A.

Worldwide, more than 330 million acres are currently planted in GMO crops, the majority of which are corn, sorghum, cotton, soy, canola, and alfalfa genetically engineered for resistance to the herbicide glyphosate. Rather than tilling the soil to control weeds, farmers can spray their fields with glyphosate, which kills the weeds but not the GMO crops.

Genes that confer glyphosate resistance are now appearing in weeds and other wild plants, as well as in unmodified crops—which means that recombinant DNA can (and does) escape into the environment. Glyphosate resistance genes are probably being transferred from transgenic plants to nontransgenic ones via pollen carried by wind or insects.

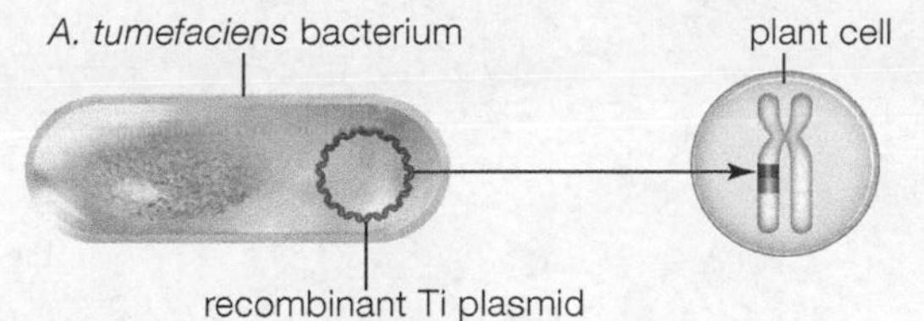
A. tumefaciens bacterium
plant cell
recombinant Ti plasmid

A A Ti plasmid carrying a foreign gene is inserted into an *Agrobacterium tumefaciens* bacterium. The bacterium infects a plant cell and transfers the Ti plasmid into it. The recombinant plasmid becomes integrated into one of the cell's chromosomes.

B The infected plant cell divides, and its descendants are induced to form embryos that develop into plants. All of the cells of the new plants carry the foreign gene and may express it.

C Cells of transgenic plants carry a gene from a different species. The genetically modified plants that produced the corn on the left express a gene from the bacterium *Bacillus thuringiensis* (Bt) that confers insect resistance. Compare the corn from unmodified plants on the right. No pesticides were used on either crop.

FIGURE 15.12 Transgenic plants.

TAKE-HOME MESSAGE 15.7

✔ Plants can be genetically modified for resistance to pests and pathogens, or to enhance nutritional value and hardiness.

✔ The widespread use of GMO crops has had unintended environmental effects. Herbicide resistant weeds are now common, and recombinant genes have spread to wild plants and non-GMO crops.

CREDITS: (12A) © Cengage Learning; (12B) Pascal Goetgheluck/Science Source; (12C left and right) The Bt and Non-Bt corn photos were taken as part of field trial conducted on the main campus of Tennessee State University at the Institute of Agricultural and Environmental Research. The work was supported by a competitive grant from the CSREES, USDA titled Southern Agricultural Biotechnology Consortium for Underserved Communities, (2000–2005). Dr. Fisseha Tegegne and Dr. Ahmad Aziz served as Principal and Co-principal Investigators respectively to conduct the portion of the study in the State of Tennessee.

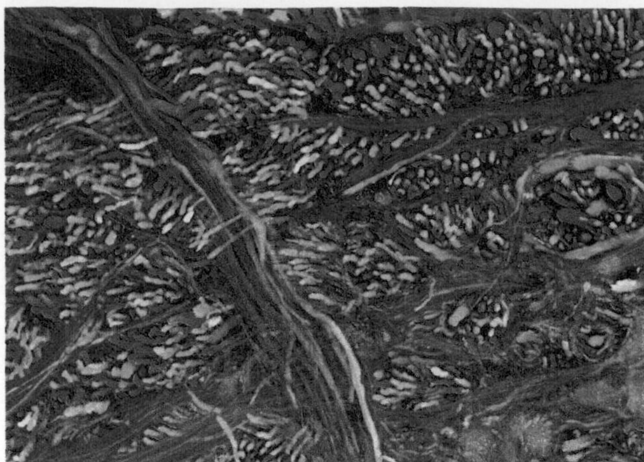

A Mice transgenic for multiple pigments ("brainbow mice") are allowing researchers to map the complex neural circuitry of the brain. Individual nerve cells in the brain stem of a brainbow mouse are visible in this fluorescence micrograph.

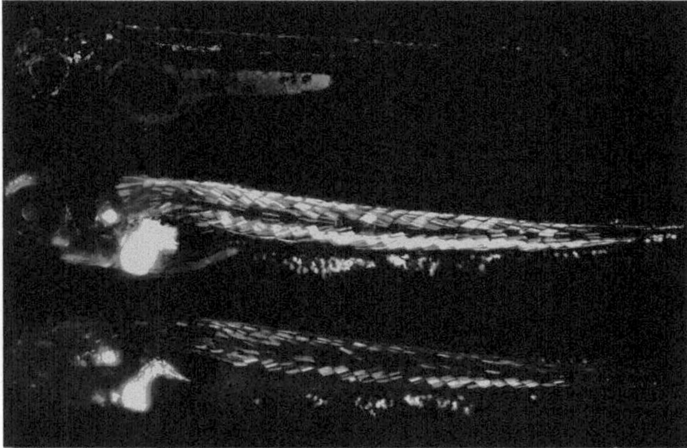

B Zebrafish engineered to glow in places where BPA, an endocrine-disrupting chemical, is present. The fish are literally illuminating where this pollutant acts in the body—and helping researchers discover what it does when it gets there.

C Transgenic goats produce human antithrombin, an anticlotting protein. Antithrombin harvested from their milk is used as a drug during surgery or childbirth to prevent blood clotting in people with hereditary antithrombin deficiency. This genetic disorder carries a high risk of life-threatening clots.

FIGURE 15.13 Examples of genetically modified animals.

15.8 Biotech Barnyards

LEARNING OBJECTIVE

- Describe some transgenic animals and their uses.

Genetically modified animals can be produced by injecting recombinant DNA into a fertilized egg, and then implanting the resulting embryo in a surrogate to complete its development.

We have discovered the function of many human genes (including the *APOA5* gene discussed in Section 15.5) by inactivating their counterparts in mice. Genetically modified mice are also used as models of human anatomy (**FIGURE 15.13A**) and diseases. For example, researchers inactivated the molecules involved in the control of glucose metabolism, one by one, in mice. Studying the effects of these knockouts resulted in much of our current understanding of how diabetes works in humans.

Genetically modified animals other than mice are also useful in research (**FIGURE 15.13B**), and some make molecules that have medical and industrial applications (**FIGURE 15.13C**). Various transgenic goats produce proteins used to treat blood clotting disorders, cystic fibrosis, heart attacks, and even nerve gas exposure. Milk from goats transgenic for lysozyme, an antibacterial protein in human milk, may protect infants and children in developing countries from acute diarrheal disease. Goats transgenic for a spider silk gene produce the silk protein in their milk; researchers can spin this protein into nanofibers that have applications in medicine and electronics. Rabbits make human interleukin-2, a protein that can be used as a cancer drug because it triggers immune cells to divide.

Genetic engineering has also given us allergen-free cats, pigs with heart-healthy fat and environmentally friendly low-phosphate feces, muscle-bound trout, chickens that do not transmit bird flu, and cows that do not get mad cow disease. In 2015, the FDA approved the first GMO animal for use as human food: a transgenic Atlantic salmon called AquAdvantage Salmon. These fish have been engineered to carry a promoter from ocean pout (a type of fish) governing a growth hormone from a Chinook salmon. The transgenic fish grow to full size about twice as fast as their unmodified counterparts. AquAdvantage Salmon are all female and triploid (so they are sterile), and have been approved only for aquaculture in warehouse-type facilities.

Manipulating genes in animals raises a host of ethical dilemmas. Consider animals genetically engineered to carry mutations associated with human disorders such as multiple sclerosis, cystic fibrosis, diabetes, cancer, or Huntington's disease. Researchers produce these

Data Analysis Activities

Enhanced Spatial Learning Ability in Mice Engineered to Carry an Autism Mutation Autism is a neurobiological disorder with symptoms that include impaired social interactions and repetitive, stereotyped patterns of behavior. Around 10 percent of autistic people also have an extraordinary skill or talent such as greatly enhanced memory.

Mutations in the gene for neuroligin 3, an adhesion protein that connects brain cells, have been associated with autism. One of these mutations is called *R451C* because the altered gene encodes a protein with an amino acid substitution: a cysteine (C) instead of an arginine (R) in position 451.

In 2007, Katsuhiko Tabuchi and his colleagues introduced the *R451C* mutation into the neuroligin 3 gene of mice. The researchers discovered that the genetically modified mice had impaired social behavior and superior spatial learning ability.

Spatial learning in mice is tested with a water maze, which consists of a small platform submerged a bit below the surface of a pool of water so it is invisible to a swimming mouse. Mice do not particularly enjoy swimming, so they try to locate the hidden platform as quickly as they can. When tested again later, they remember the platform's location by checking visual cues around the edge of the pool. How quickly they remember is a measure of their spatial learning ability. **FIGURE 15.14** shows some of Tabuchi's results.

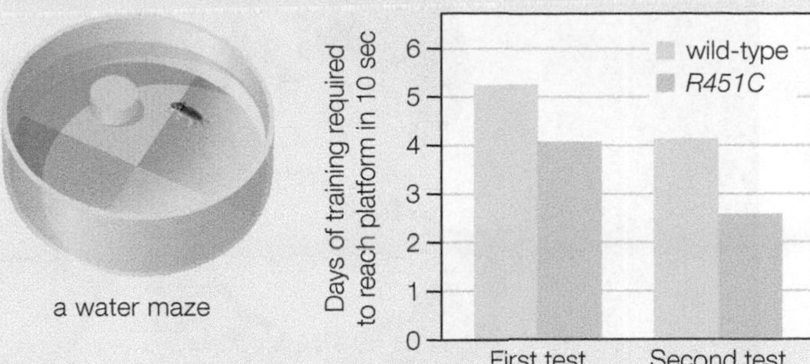
a water maze
Days of training required to reach platform in 10 sec
0
1
2
3
4
5
6
wild-type
R451C
First test
Second test

FIGURE 15.14 Spatial learning ability in mice. Mice with a mutation in neuroligin 3 (*R451C*) were tested for learning performance as compared with unmodified (wild-type) mice.

1. In the first test, how many days did it take unmodified mice to learn to find the location of a hidden platform in 10 seconds?
2. Did the modified or the unmodified mice learn the location of the platform faster in the first test?
3. Which mice learned faster the second time around?
4. Which mice had the greatest improvement in memory?

animals in order to study the disorders (and potential treatments) without experimenting on humans. However, the modified animals often suffer the same terrible symptoms of the condition as humans do.

Many people worry that our ability to tinker with genetics has surpassed our ability to understand the impact of the tinkering. Controversy raised by GMO production and use invites you to read the research and form your own opinions. The alternative is to be swayed by media hype (the term "Frankenfood," for instance) or by reports from potentially biased sources (such as herbicide manufacturers).

TAKE-HOME MESSAGE 15.8

✔ Genetic engineering of animals has substantial benefits for medical research and in industrial applications.

✔ Production and use of engineered animals raises ethical concerns, for example regarding animal welfare.

15.9 Editing Genomes

LEARNING OBJECTIVE

- Describe some methods of gene therapy explain their benefits and risks.

We know of more than 15,000 serious genetic disorders. Collectively, they cause 20 to 30 percent of infant deaths each year, and account for half of all mentally impaired patients and a fourth of all hospital admissions. Drugs and other treatments can minimize the symptoms of some genetic disorders, but gene therapy is the only cure. **Gene therapy** is the transfer of recombinant DNA into an individual's body cells, with the intent to correct a genetic defect or treat a disease.

DNA can be introduced into human cells in many ways, for example by direct injection, electrical pulses, lipid clusters, nanoparticles, or genetically engineered viruses. Viruses have molecular machinery that delivers their genomes into cells they infect. Those used as vectors have DNA that splices itself into the infected cells' chromosomes, along with foreign DNA that has been inserted into it.

Gene therapy is a compelling reason to embrace the idea of genetically engineering people. It is being tested as a treatment for AIDS, muscular dystrophy, heart attack, sickle-cell anemia, cystic fibrosis, hemophilia A, Parkinson's disease, Alzheimer's disease, inherited diseases of the eye, the ear, and the immune system, and several types of cancer. For example, a viral vector has been used to insert a gene into immune cells extracted from leukemia patients (leukemia is a potentially fatal cancer of bone marrow cells). When the modified cells are reintroduced into the patients' bodies, the inserted gene directs the destruction of cancer cells

gene therapy Treating a disease or genetic defect by repairing or replacing a gene in an affected individual.

FIGURE 15.15 To save a child. Six-year-old Emily was days from death when she underwent an experimental gene therapy for acute lymphocytic leukemia. Five years later, she remains cancer-free.

(**FIGURE 15.15**). The therapy seems to work astonishingly well: In one patient, all traces of the leukemia vanished in eight days.

Despite the successes, manipulating a gene within the context of a living individual can be unpredictable. Consider SCID-X1, a severe X-linked genetic disorder that stems from a mutated allele of the *IL2RG* gene. The gene encodes a receptor for an immune signaling molecule. Without treatment, people affected by this disorder can survive only in germ-free isolation tents because they cannot fight infections; a diminished life in a sterile isolation tent was the source of the term "bubble boy." Researchers used a genetically engineered virus to insert unmutated copies of *IL2RG* into cells taken from the bone marrow of twenty boys with SCID-X1. Each child's modified cells were infused back into his bone marrow. Within months of their treatment, eighteen of the boys left their isolation tents. Gene therapy had repaired their immune systems. However, five of the boys later developed leukemia and died. Developers of the gene therapy could not have known that the very gene targeted for repair, especially when combined with the virus that delivered it, would cause cancer. The recombinant viral vector preferentially inserted the gene into chromosomes at a site near a proto-oncogene (Section 11.6). The insertion triggered inappropriate transcription of the gene, and that is how the leukemia began.

CRISPR

As the previous example illustrates, gene therapy with viral vectors is risky. A recombinant viral vector sometimes inserts a gene into an unexpected site in chromosomes, for example, and DNA remnants of the vector that remain in the genome may have unknown consequences. A powerful new method of editing chromosomal DNA does not have these risks, and it is profoundly changing conversations about editing genomes. The technique, named CRISPR-Cas9 after its components, is based on a system of RNA interference (Section 10.2) that prokaryotes use to defend themselves against viral infection.

CRISPR includes a type of restriction enzyme that is guided to DNA by a piece of RNA. The enzyme cuts the DNA wherever its "guide" RNA hybridizes (**FIGURE 15.16**). Then, a piece of DNA that includes a template sequence is used to repair the break. Here is the relevant part: The RNA and DNA can be designed to precisely target and change (respectively) essentially any part of a genome in a living cell. Researchers deliver all three molecules into cells, for example in the form of genes carried by a plasmid expression vector. A cell that expresses the genes will have both of its chromosomes edited, and the change will be passed to all of the cell's descendants.

CRISPR is more accurate and efficient than viral vectors or any other method of altering genomes. Accordingly, its use in research has been spreading like wildfire. The system has not yet been perfected—for example, it sometimes cuts DNA in unexpected places—but given the pace of research, the kinks will probably be worked out quickly.

The CRISPR method holds phenomenal promise in a variety of applications. For example, crop plants can be enhanced without leaving traces of vector. HIV can be edited out of an infected patient's cells to prevent relapse after remission. Removing mutations from somatic cells can permanently cure many genetic disorders; already, the technique has been used to repair mutations causing muscular dystrophy, achondroplasia, and cancer in cells taken from human patients.

CRISPR can also be used to remove mutations or enhance genetic traits in gametes and embryos, in which case the alterations would become part of the human germline and pass to subsequent generations. A premature and unsuccessful attempt to correct a β-thalassemia mutation (Section 9.6) in human embryos prompted an international summit in 2015; the meeting concluded with the recommendation that no germline editing should occur until all potential risks associated with it are understood and resolved.

Consider how CRISPR is the first successful method of making gene drives, which are genes that can "drive" themselves into other chromosomes and spread through a population. The eradication of malaria is a worthwhile application of the technology. Already, mosquitoes have been engineered to carry

CREDIT: (15) Courtesy of the Emily Whitehead Foundation

genes preventing growth of the parasite that causes the disease (mosquitoes in which the parasite cannot grow are unable to pass it to humans). With a typical genetic manipulation, the engineered genes would quickly be diluted out of a population because only half of a mosquito's offspring would inherit them. However, the outcome is different in mosquitoes engineered with CRISPR. When one of them mates with an unmodified mosquito, 100 percent of their offspring inherit the engineered genes. This is because the CRISPR system that delivered the modification is part of the mosquitoes' chromosomes, so it also edits homologous chromosomes in zygotes. If the engineered mosquitoes were to be released in the wild, the worldwide population of mosquitoes would be modified within a few years. No more malaria.

Some people are concerned that altering genomes in any way has us on a slippery slope that may result in irreversible damage to ourselves and to the biosphere. For example, if ending malaria by engineering mosquitoes seems like a good idea, what about the gene drive that would eliminate mosquitoes entirely? Who needs mosquitoes anyway? However, given the complexity of biological systems, such interventions could have effects that are impossible to predict. Even more unpredictable are humans. For example, the CRISPR system could be designed—accidentally or deliberately—to deliver unwanted genetic modifications into humans.

In this brave new world, the questions before you are these: What do we stand to lose if risks are not taken? And, does anyone have the right to impose unknown consequences on people who would choose not to take those risks?

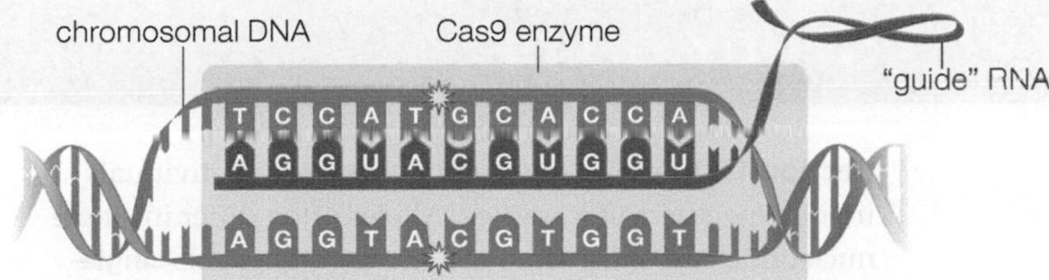

chromosomal DNA
Cas9 enzyme
"guide" RNA
T C C A T G C A C C A
A G G U A C G U G G U
A G G T A C G T G G T

A The Cas9 enzyme cuts chromosomal DNA wherever a "guide" RNA hybridizes. The guide RNA can be customized to target any sequence in the DNA.

template DNA used for repair

B Repairing a double-stranded break in a chromosome requires a DNA template. Normally, a cell uses the homologous chromosome for this purpose. With the CRISPR system, researchers design a short fragment of DNA that serves as the template for repair. The sequence of the fragment can be customized to include, for example, an insertion or deletion. This template includes an insertion of a stop codon (TAA).

C After the chromosome is repaired, its sequence includes the change introduced by the researcher. When the cell divides, the altered chromosome is replicated and passed to descendant cells.

FIGURE 15.16 An example of CRISPR-Cas9 use.

TAKE-HOME MESSAGE 15.9

✔ Genetic engineering technologies can be used to correct genetic defects or treat diseases in humans.

✔ The CRISPR technique targets and alters specific sequences of DNA in an individual's cells. It can be used to make gene drives.

15.1 Personal Genetic Testing (revisited)

Results of a personal genetic test may include estimated risks of developing conditions associated with your particular set of SNPs. For example, the test may determine whether you are homozygous for certain alleles of the *MC1R* gene (Section 13.5). If so, then your results will say that you have red hair. Few SNPs have such a clear effect, however; most human traits arise from a complex interplay of genes and environmental factors that are still being unraveled (Section 13.6). A DNA test can accurately determine an individual's SNPs, but it cannot reliably predict the effect of most of those SNPs on the individual. For example, if you are heterozygous for the *E4* allele of the *APOE* gene, a genetic testing company cannot tell you whether you will develop Alzheimer's. However, it may report your lifetime risk of developing the disease, which is about 47 percent, as compared with about 20 percent for someone with no *E4* allele. What, exactly, does a 47 percent lifetime risk of Alzheimer's mean? The number is a probability statistic: It means, on average, 47 of every 100 people who have one *E4* allele eventually get the disease. However, a risk is just that. Not everyone who has the allele develops Alzheimer's, and not everyone who develops the disease has the allele. Other unknown factors, including epigenetic modifications of DNA, contribute to the disease.

We still have a limited understanding of how genes contribute to many health conditions, particularly age-related ones such as Alzheimer's. Geneticists believe that it will be many years before genetic testing can be used to accurately predict an individual's future health problems. ●

STUDY GUIDE

Section 15.1 Alleles of shared genes make individuals unique members of a species. Alleles often differ in single nucleotide bases. Personal genetic testing reveals **single-nucleotide polymorphisms (SNPs)** in an individual's DNA. This type of testing is revolutionizing medicine. Personal treatment plans and lifestyle recommendations can be based on SNPs that have been linked to drug responsiveness and health risks. However, it is not yet possible to accurately predict future health problems based on genotype.

Section 15.2 In **DNA cloning**, researchers use **restriction enzymes** to cut a sample of DNA into pieces and insert the fragments into plasmids or other cloning **vectors**. The resulting molecules of **recombinant DNA** are delivered into host cells such as bacteria or yeast. Division of host cells produces huge populations of genetically identical descendant cells, each with a copy of the recombinant DNA. The cloned DNA can be harvested from the host cells.

Researchers can study eukaryotic gene expression by using **reverse transcriptase** to transcribe mRNA into **cDNA**. cDNA can be cloned like any other DNA. Inserted into an expression vector and delivered into appropriate host cells, a cDNA will direct the synthesis of an mRNA, which in turn will be used by the host cell to produce a protein.

Section 15.3 A **DNA library** is a collection of cells that host different fragments of DNA, often representing an organism's entire **genome**. A DNA library made from cDNA includes only those genes being expressed at the time the mRNA was harvested. Researchers can use **probes** to identify cells in a library that carry a specific fragment of DNA.

The **polymerase chain reaction (PCR)** uses primers and the heat-resistant *Taq* DNA polymerase to rapidly amplify (increase the number of copies of) a targeted section of DNA.

Section 15.4 Advances in **sequencing**, which reveals the order of nucleotide bases in DNA, allowed the human genome to be sequenced.

A common sequencing technique uses DNA polymerase to partially replicate a DNA template. The reaction mixture includes four types of deoxynucleotides, each labeled with a different colored pigment. A dideoxynucleotide randomly added during DNA synthesis terminates the new strand, so millions of DNA fragments of different lengths are produced. Each fragment ends with a deoxynucleotide; fragments of the same length end with the same deoxynucleotide base. **Electrophoresis** separates the fragments by length into bands. The dideoxynucleotides impart color to each band. The order of band colors reflects the sequence of the DNA template.

Section 15.5 **Genomics** and proteomics give us insights into genome function. Similarities between genomes of different organisms are evidence of evolutionary relationships, and can be used as a predictive tool in research.

DNA profiling identifies a person by the unique parts of his or her DNA. Aside from identical twins, each individual has a unique array of **short tandem repeats** and SNPs. Within the context of a criminal investigation, a DNA profile is called a DNA fingerprint.

Sections 15.6–15.8 Recombinant DNA technology is the basis of **genetic engineering**, the directed modification of a genome with the intent to modify phenotype. An organism whose genome has been modified by genetic engineering is called a **genetically modified organism (GMO)**. A genetically modified organism that carries a gene from a different species is **transgenic**.

Bacteria and yeast, the most common genetically engineered organisms, produce proteins that have medical and industrial value. Transgenic crops are in widespread, worldwide use; most have been modified for resistance to pests and pathogens. Animals being produced by genetic engineering are used for medical applications, research, and food production.

Section 15.9 With **gene therapy**, a gene is repaired or replaced in order to correct a genetic defect or treat a disease. Potential benefits of gene therapy must be weighed against potential risks. CRISPR is a type of gene therapy that can be used to edit a gene in the context of a living organism. CRISPR can also be used to create gene drives that affect an entire species. Recent advances in CRISPR techniques are accelerating gene therapy research and applications.

SELF-QUIZ

Answers in Appendix VII

1. _______ cut(s) DNA molecules at specific sites.
   a. DNA polymerase
   b. DNA probes
   c. Restriction enzymes
   d. Reverse transcriptase

2. A _______ is a molecule that can be used to carry a fragment of DNA into a host organism.
   a. vector
   b. chromosome
   c. GMO
   d. cDNA

3. Reverse transcriptase assembles a(n) _______ on a(n) _______ template.
   a. mRNA; DNA
   b. cDNA; mRNA
   c. DNA; ribosome
   d. protein; mRNA

4. For each species, all _______ in the complete set of chromosomes is the _______ .
   a. genomes; library
   b. DNA; genome
   c. mRNA; start of cDNA
   d. cDNA; start of mRNA

5. A set of cells that host various DNA fragments collectively representing an organism's entire set of genetic information is a _______ .
   a. genome
   b. clone
   c. genomic library
   d. GMO

6. _______ is the technique of determining the order of nucleotide bases in a sample of DNA.
   a. PCR c. Electrophoresis
   b. Sequencing d. Nucleic acid hybridization

7. Electrophoresis separates fragments of DNA according to _______ .
   a. sequence b. length c. species

8. PCR can be used _______ .
   a. as a cloning vector
   b. in DNA profiling
   c. to modify a human genome

9. Put the following tasks in the order they would occur during a DNA cloning experiment.
   a. using DNA ligase to seal DNA fragments into vectors
   b. using a probe to identify a clone in the library
   c. sequencing the DNA of the clone
   d. making a DNA library of clones
   e. cutting genomic DNA with restriction enzymes

10. An individual's unique set of _______ can be used in DNA profiling.
   a. DNA sequences c. SNPs
   b. short tandem repeats d. all of the above

11. A transgenic organism _______ .
   a. carries a gene from another species
   b. has been genetically modified
   c. both a and b

12. True or false? Some transgenic organisms can pass their foreign genes to offspring.

13. *Taq* polymerase is used for PCR because it _______ .
   a. tolerates the high temperature needed to separate DNA strands
   b. is an enzyme from a bacterium
   c. does not require primers
   d. is genetically modified

14. _______ can correct a genetic defect in an individual.
   a. Cloning vectors c. Sequencing
   b. Gene therapy d. Electrophoresis

15. Match each term with the most suitable description.
   ___ DNA profile a. GMO with a foreign gene
   ___ Ti plasmid b. alleles commonly have them
   ___ probes c. unique array of short tandem repeats
   ___ SNPs d. used to find clones
   ___ transgenic e. mouse vs. human
   ___ genomics f. used in plant gene transfers
   ___ CRISPR g. based on RNA interference

16. Match the method with the appropriate enzyme.
   ___ PCR a. *Taq* polymerase
   ___ cutting DNA b. DNA ligase
   ___ cDNA synthesis c. reverse transcriptase
   ___ DNA sequencing d. restriction enzyme
   ___ pasting DNA e. DNA polymerase (not *Taq*)

CRITICAL THINKING

1. In 1918, an influenza pandemic that originated with avian flu killed 50 million people. Researchers isolated samples of that virus from bodies of infected people preserved in Alaskan permafrost since 1918. From the samples, they sequenced the viral genome, then reconstructed the virus. The reconstructed virus is 39,000 times more infectious than modern influenza strains, and 100 percent lethal in mice. Understanding how this virus works can help us defend ourselves against other strains that may arise. For example, discovering what makes it so infectious and deadly would help us design more effective vaccines. Critics of the research are concerned: If the virus escapes the containment facilities (even though it has not done so yet), it might cause another pandemic. Worse, the published DNA sequence and methods to make the virus could be used for criminal purposes. Do you think this research makes us more or less safe?

2. Restriction enzymes in bacterial cytoplasm cut injected bacteriophage DNA wherever certain sequences occur. A bacteria's chromosome also contains these sequences, and, being suspended in cytoplasm, it is also exposed to the enzymes. Why do you think the enzymes do not cut the chromosome?

3. The results of a paternity test using short tandem repeats are listed in the table below. Who's the daddy? How sure are you?

Locus	Mother	Baby	Alleged Father #1	Alleged Father #2
CSF1PO	15, 17	17, 23	23, 27	17, 15
FGA	9, 9	9, 9	9, 12	9, 12
THO1	29, 29	29, 27	27, 28	29, 28
TPOX	14, 18	18, 20	15, 20	17, 22
VWA	14, 14	14, 14	14, 14	14, 16
D3S1358	11, 14	14, 16	12, 16	14, 20
D5S818	11, 13	10, 13	8, 10	18, 18
D7S820	7, 13	13, 13	13, 19	13, 13
D8S1179	13, 13	13, 15	12, 15	10, 12
D13S317	12, 12	10, 12	8, 10	12, 17
D16S539	12, 14	14, 12	14, 14	18, 25
D18S51	5, 6	6, 22	22, 6	5, 22

16 EVIDENCE OF EVOLUTION

CORE CONCEPTS

Process of Science

The field of biology consists of and relies on experimentation and the collection and analysis of scientific evidence.

We often reconstruct history by studying physical evidence of events that took place in the past. Reconstructing the history of life relies on a premise that is supported by a vast amount of data from many scientific disciplines: Natural phenomena that occurred in the distant past can be explained by the same physical, chemical, and biological processes operating today.

Evolution

Evolution underlies the unity and diversity of life.

A record of ancient life that persists in fossils provides information about the evolution of ancient organisms and the modern species descended from them. The composition of sedimentary rock layers provide an environmental and chronological context for fossillized organisms found in the layers. Morphological and biogeographic patterns among species offer clues about evolutionary relationships.

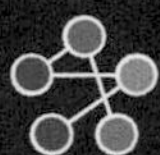

Systems

Complex properties arise from interactions among components of a biological system.

Members of a population must compete for resources that are limited. Individuals vary in the details of their shared traits, and those with forms of a trait that confer some advantage in this competition tend to leave more offspring (a process called natural selection). Thus, advantageous traits tend to become more common among members of a population over time. Change in a line of descent is called evolution.

Links to Earlier Concepts

You may wish to review critical thinking (Section 1.6) before reading this chapter, which discusses a clash between belief and science (1.8). What you know about genes (9.2) and alleles (12.2) will help you understand natural selection. The chapter also revisits sampling error (1.7), radioisotopes (2.2), redox reactions (5.5), and the effect of photosynthesis on Earth's early atmosphere (7.1).

16.1 Reflections of a Distant Past

Asteroids are space rocks—chunks of rock or metal hurtling through space—that range in size from a few meters to hundreds of kilometers in diameter. They are cold, stony leftovers from the formation of our solar system. Millions of them orbit the sun.

Space rocks enter Earth's atmosphere by the ton, but most of them burn up before they hit the ground. Those that do reach Earth's surface are called meteorites, and they can cause significant damage at the site of impact. Consider Meteor Crater, a mile-wide hole in the desert sandstone near Flagstaff, Arizona (**FIGURE 16.1**). This crater formed 50,000 years ago, when an asteroid 45 meters in diameter slammed into Earth. The impact had so much energy that most of the meteorite and part of the ground vaporized instantly.

No humans were around 50,000 years ago to witness anything, so how could anyone know what happened at Meteor Crater? We often reconstruct history by studying physical evidence of past events. Geologists were able to infer the most probable cause of the crater by analyzing tons of meteorite fragments, melted sand, and other rocky clues at the site.

Similar evidence points to impacts of even larger asteroids. For example, fossil hunters have long known about a mass extinction that occurred 66 million years ago. The event is marked by an unusual, worldwide layer of a sedimentary rock called the K–Pg boundary. There are plenty of dinosaur fossils embedded in rock below this layer. Above it, in layers of rock that were deposited more recently, there are no dinosaur fossils, anywhere. A gigantic impact crater encompassing the town of Chicxulub on the Yucatán Peninsula is about

FIGURE 16.1 Evidence to inference: Meteor crater.
This photo was taken from an airplane; the buildings to the left of the crater give an idea of its size. Geologists used rocky evidence to at the site infer that the impact of a 300,000-ton asteroid made the crater 50,000 years ago.

CREDIT: (opposite) Jonathan Blair/Getty Images; © Brad Snowder.

66 million years old. Coincidence? Many scientists say no. The Chicxulub crater is 180 kilometers wide and 20 kilometers deep; to make it, a meteorite 15 kilometers in diameter would have hit Earth with a force *a billion times* more powerful than the Hiroshima bomb. An impact with this much energy would have caused a worldwide ecological catastrophe, one of sufficient scale to wipe out the dinosaurs along with most other life on Earth.

You are about to make an intellectual leap through time to consider events that were unknown even a few centuries ago. We invite you to launch yourself from this premise: Natural phenomena that occurred in the ancient past can be explained by the same physical, chemical, and biological processes that operate today. That premise is the foundation for scientific research into the history of life. The research represents a shift from experience to inference—from the known to what can only be surmised—and it gives us astonishing glimpses into the distant past. ●

16.2 Old Beliefs and New Discoveries

LEARNING OBJECTIVE

- Using appropriate examples, explain how three types of observations of the natural world challenged traditional belief systems in the 1800s.

About 2,300 years ago, the Greek philosopher Aristotle described nature as a continuum of organization, from lifeless matter through complex plants and animals. Aristotle's work greatly influenced later European thinkers, who adopted his view of nature and modified it in light of their own beliefs. By the fourteenth century, Europeans generally believed that a "great chain of being" extended from the lowest form of life (plants), up through animals, humans, and spiritual beings. Each link in the chain was a species, and each was said to have been forged at the same time, in one place, and in a perfect state. The chain was complete. Because everything that needed to exist already did, there was no room for change.

In the 1800s, European naturalists embarked on globe-spanning survey expeditions and brought back tens of thousands of plants and animals from Asia, Africa, North and South America, and the Pacific Islands. Each newly discovered species was carefully cataloged as another link in the chain of being. The explorers began to see patterns in where species live and similarities in body plans, and they started to think about natural forces that shape life. These explorers were pioneers in **biogeography**, the study of patterns in the geographic distribution of species and communities.

Some of these patterns raised questions that could not be answered within the framework of existing belief systems. For example, the explorers had discovered plants and animals living in extremely isolated places. The isolated species looked suspiciously similar to species living on the other side of impassable mountain ranges, or across vast expanses of open ocean. Consider the emu, rhea, and ostrich, three species native to three different continents. These birds share a set of unusual traits (**FIGURE 16.2**). Alfred Wallace, an explorer interested in the geographic distribution of animals, thought that the shared traits might mean that the birds descended from a common ancestor (and he was correct), but how could they have ended up on widely separated continents?

Naturalists of the time also had trouble classifying organisms that are very similar in some features, but different in others. For example, both plants shown in **FIGURE 16.3** live in hot deserts where water is seasonally scarce. Both have rows of sharp spines that deter

A Emu, native to Australia

B Rhea, native to South America

C Ostrich, native to Africa

FIGURE 16.2 Similar-looking, related animals native to distant geographic realms.

These birds are unlike most others in several unusual features, including their large size; long, muscular legs; and inability to fly. All are native to open grassland regions about the same distance from the equator.

CREDITS: (2A) Earl & Nazima Kowall/Getty Images; (2B) Novarc Images / Alamy; (2C) Rebecca Yale/Getty Images.

herbivores, and both store water in their thick, fleshy stems. However, their reproductive parts are very different, so these plants cannot be (and are not) as closely related as their outward appearance might suggest.

Observations such as these are examples of **comparative morphology**, the study of anatomical patterns: similarities and differences among the body plans of organisms. Today, comparative morphology is only one of several aspects of taxonomy (Section 1.5), but in the nineteenth century it was the only way to distinguish species. In some cases, comparative morphology revealed anatomical details (such as body parts with no apparent function) that added to the mounting confusion. If every species had been created in a perfect state, then why were there useless parts such as wings in birds that do not fly, eyes in moles that are blind, or remnants of a tail in humans (**FIGURE 16.4**)?

Given the new discoveries in the nineteenth century, naturalists began trying to wrap prevailing belief systems around increasing evidence that life on Earth had changed over time.

Among them was Jean-Baptiste Lamarck (left). Lamarck proposed the idea that a species gradually improved over generations because of an inherent drive toward perfection, up the chain of being. The drive directed an unknown "fluida" into body parts needing change. By Lamarck's hypothesis, environmental pressures cause an internal requirement for change in an individual's body, and the resulting change is inherited by offspring. Lamarck's understanding of how inheritance works was incomplete, but he was the first to propose a mechanism of driving **evolution**, or change in a line of descent. A line of descent is also called a **lineage**.

Georges Cuvier, an expert in the new field of comparative morphology, vehemently rejected Lamarck's hypothesis. Like most others of his time, Cuvier (left) was a proponent of **catastrophism**, the idea that Earth's current landscape had been shaped by periodic, violent geologic events unlike any that had ever been experienced. Cuvier had been studying newly unearthed fossil skeletons of animals (such as gigantic dinosaurs) that were unlike any living ones. If these animals were perfect at the time of creation, then what had happened to them? Cuvier dismissed a popular idea that the missing animals were somehow hidden in large forests or unexplored regions of the world. Instead, he proposed a different idea that was startling for the time: Many species had been killed off during the violent geologic events. Cuvier thought that new species appeared after each event, and that there was not enough time between catastrophes for these species to change. He argued that evidence for change did not exist: No one had yet discovered any fossils representing intermediate forms between the gigantic animals and living species.

Charles Lyell, a geologist, thought that global catastrophes were not necessary to explain Earth's current

FIGURE 16.3 Similar-looking, unrelated species.
Left, saguaro cactus (*Carnegiea gigantea*), native to the Sonoran Desert of Arizona. Right, an African milk barrel plant (*Euphorbia horrida*), native to the Great Karoo desert of South Africa.

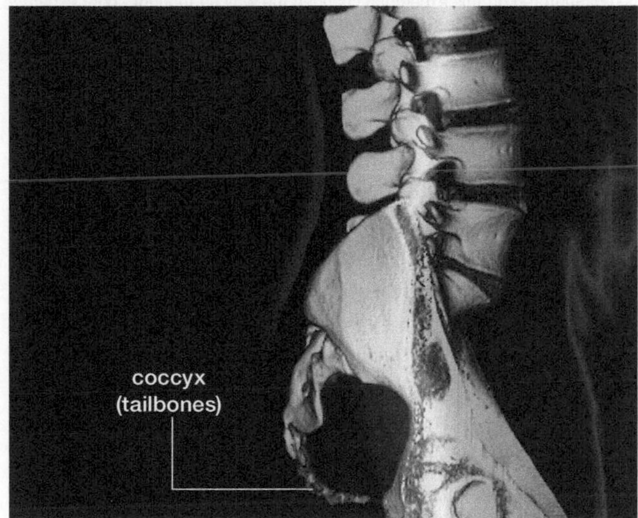
coccyx
(tailbones)

FIGURE 16.4 Human tailbones. Nineteenth-century naturalists were well aware of—but had trouble explaining—body structures such as human tailbones that had apparently lost most or all function.

biogeography Study of patterns in the geographic distribution of species and communities.
catastrophism Idea that catastrophic geologic events have periodically shaped Earth's surface.
comparative morphology (more-FALL-uh-jee) Scientific study of similarities and differences in body plans.
evolution Change in a line of descent.
lineage (LINN-edge) Line of descent.

CREDITS: (3) left, © Richard J. Hodgkiss, www.succulent-plant.com; right, Marka / Marka / Superstock; (4) Zephyr/Science Photo Library/Science Source.

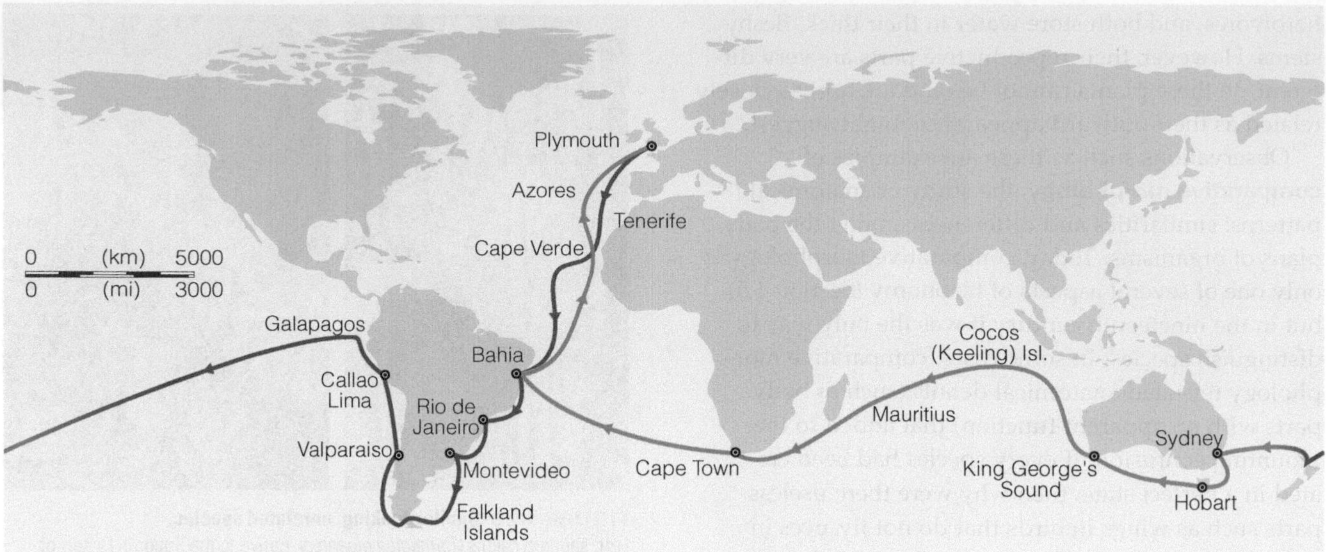
Plymouth
Azores
Tenerife
Cape Verde
0 (km) 5000
0 (mi) 3000
Galapagos
Bahia
Callao
Lima
Rio de Janeiro
Valparaiso
Montevideo
Falkland Islands
Cape Town
Cocos (Keeling) Isl.
Mauritius
King George's Sound
Sydney
Hobart

FIGURE 16.5 Voyage of the HMS *Beagle*. With Darwin aboard as ship's naturalist, the vessel originally set sail to map the coast of South America, but ended up circumnavigating the globe (bottom). The path of the voyage is shown from red to blue. Darwin's detailed observations of the geology, fossils, plants, and animals he encountered on this expedition changed the way he thought about evolution.

landscape. He was a proponent of **uniformitarianism**, the idea that gradual, everyday geologic processes such

as erosion shaped Earth's surface. Lyell (left) had seen plenty of evidence to support uniformitarianism. For many years, geologists had been chipping away at sandstones, limestones, and other types of rocks that form from accumulated sediments. These rocks provided evidence that gradual processes of geologic change operating in the present were the same ones that operated in the distant past. Uniformitarianism challenged the prevailing belief that Earth was thousands of years old. By Lyell's calculations, gradual processes that took place over millions of years could have sculpted Earth's surface into its current form.

Taken as a whole, accumulating discoveries from comparative morphology, biogeography, and geology began to challenge traditional beliefs that life had been created in a perfect state a few thousand years ago. Perhaps species did indeed change over time.

TAKE-HOME MESSAGE 16.2

✔ Belief systems that are inconsistent with observations of the physical world tend to change over time.

✔ In the nineteenth century, increasingly extensive observations of nature did not fit into the framework of prevailing beliefs. Cumulative findings led naturalists to question traditional ways of interpreting the natural world.

✔ In the 1800s, many naturalists realized that Earth—and life—had changed over time. New thoughts emerged about forces that could drive such change.

16.3 Evolution by Natural Selection

LEARNING OBJECTIVES

- Explain Darwin's hypothesis of evolution by natural selection.
- Discuss the relationship between evolutionary adaptations and fitness.

Lamarck, Cuvier, and Lyell influenced the thinking of another naturalist, Charles Darwin. Darwin (left) had earned a theology degree from Cambridge after an attempt to study medicine. All through school, however, he had spent most of his time with faculty members and other students who embraced natural history. In 1831, when he was 22, Darwin joined a 5-year survey expedition on a ship named *Beagle* (**FIGURE 16.5**). During the expedition, Darwin found many unusual fossils, and he saw diverse species living in environments that ranged from the sandy shores of remote islands to plains high in the Andes.

Among the thousands of specimens Darwin collected during his voyage were fossil glyptodons. These armored mammals are extinct, but they have many unusual traits in common with modern armadillos (**FIGURE 16.6**). Armadillos also live only in places where glyptodons once lived. Could the odd shared traits and restricted distribution mean that glyptodons were ancient relatives of armadillos? If so, perhaps traits of their common ancestor had changed in the lineage that led to armadillos. But why would such changes have occurred?

After Darwin returned to England, he pondered his notes and fossils, and read an essay by economist

CREDITS: (5) Cengage Learning; (in text) Painting by George Richmond.

A Fossil of a glyptodon, a huge mammal that existed from about 2 million to 10,000 years ago. This one is about 11 feet long.

B A modern armadillo. This one is about 11 inches long.

FIGURE 16.6 Ancient relatives: glyptodon and armadillo. These animals are widely separated in time, but they share a restricted distribution and unusual traits, including a shell and helmet of keratin-covered bony plates—a material similar to crocodile and lizard skin. (The fossil in **A** is missing its helmet.) The similarities were a clue that helped Darwin develop the theory of evolution by natural selection.

Thomas Malthus (left). Malthus had proposed the idea that famine, disease, and war limit the size of human populations. When people reproduce beyond the capacity of their environment to sustain them, they run out of food and start competing for resources that become limited. Some survive this struggle for existence, and some do not. Darwin realized that Malthus's ideas had wider application: Individuals of all populations, not just human ones, struggle for existence by competing for limited resources.

Reflecting on his journey, Darwin started thinking about how individuals of a species often vary a bit in the details of shared traits such as size and coloration.

adaptive trait A form of a heritable trait that enhances an individual's fitness. An evolutionary adaptation.
adaptation Adaptive trait.
fitness Degree of adaptation to an environment, as measured by the individual's relative genetic contribution to future generations.
natural selection Differential survival and reproduction of individuals of a population based on differences in shared, heritable traits.
uniformitarianism Idea that gradual repetitive processes occurring over long time spans shaped Earth's surface.

TABLE 16.1

Natural Selection

Observations About Populations

- ✔ Natural populations have an inherent capacity to increase in size over time.
- ✔ As a population expands, resources that are used by its members (such as food and living space) eventually become limited.
- ✔ Members of a population compete for resources that are limited.

Observations About Genetics

- ✔ Individuals of a species share certain traits.
- ✔ Individuals of a natural population vary in the details of those shared traits.
- ✔ Shared traits have a heritable basis, in genes. Slightly different versions of those genes (alleles) give rise to variation in shared traits.

Inferences

- ✔ A certain form of a shared trait may make its bearer better able to survive.
- ✔ Individuals of a population that are better able to survive tend to leave more offspring.
- ✔ Thus, an allele associated with an adaptive trait tends to become more common in a population over time.

He saw this variation among finch species on isolated islands of the Galápagos archipelago. This island chain is separated from South America by 900 kilometers (550 miles) of open ocean, so most species living on the islands had no opportunity to interbreed with mainland populations. The Galápagos island birds resembled finches in South America, but had unique traits suited to their particular island habitats.

Darwin was familiar with dramatic variations in traits that selective breeding could produce in pigeons, dogs, and horses, and reasoned that natural environments could similarly "select" certain traits. Having a particular form of a shared trait might give an individual an advantage over competing members of its species. In any population, some individuals have forms of shared traits that make them better suited to their environment than others. In other words, individuals of a natural population vary in fitness. Today, we define **fitness** as the degree of adaptation to a specific environment, and measure it by the individual's relative genetic contribution to future generations. An evolutionary **adaptation**, or **adaptive trait**, is a form of a heritable trait that enhances fitness.

Darwin realized that individuals with adaptive traits would tend to leave more offspring than their less fit rivals, a process he named **natural selection**. He understood that natural selection causes a population to change. **TABLE 16.1** summarizes this reasoning: If an individual has an adaptive trait that makes it better suited to an environment, then it is better able to

CREDITS: (6) top, 2004 Arent, commons.wikimedia.org/wiki/File:Glyptodon-1.jpg; bottom, ©John White; (in text) De Agostini Picture Library/Getty Images.

survive. If an individual is better able to survive, then it has a better chance of living long enough to produce offspring. If individuals with an adaptive trait produce more offspring than other individuals, then the frequency of the adaptive trait in the population will increase over successive generations (the population will evolve). If Earth was millions of years old (as Lyell had proposed), then there had been ample time for natural selection to drive evolution.

Darwin wrote out his hypothesis of evolution by natural selection in the late 1830s, but was so conflicted about its implications that he collected

evidence for twenty years without sharing it. Meanwhile, Alfred Wallace (left), who was studying wildlife in the Amazon basin and the Malay Archipelago, wrote to Darwin about patterns in the geographic distribution of species. When he was ready to publish his own ideas about evolution, he sent them to Darwin for advice. To Darwin's shock, Wallace had come up with the same hypothesis: that evolution can be driven by natural selection.

In 1858, the hypothesis of evolution by natural selection was presented at a scientific meeting. Darwin and Wallace were credited as authors, but neither attended the meeting. The next year, Darwin published *On the Origin of Species*, which laid out detailed evidence in support of the hypothesis. By that time, many people had accepted the idea of descent with modification (evolution). However, there was a fierce debate over the idea that natural selection drives the process. Decades would pass before experimental evidence from the field of genetics led to its widespread acceptance by the scientific community. As with all scientific theories, the theory of evolution by natural selection has not been falsified after years of rigorous testing, and it has proved useful for making predictions about a wide range of other phenomena.

As you will see in the remainder of this unit, the theory of evolution by natural selection is supported by and helps explain the fossil record as well as similarities and differences in the form, function, and biochemistry of living things.

TAKE-HOME MESSAGE 16.3

✔ Darwin's observations of species in different parts of the world led him to propose natural selection as a driving force of evolution.

✔ With natural selection, individuals of a population survive and reproduce with differing success depending on the details of their shared, heritable traits.

✔ A form of a heritable trait that enhances fitness is an evolutionary adaptation, or adaptive trait.

A Fossil skeleton of an ichthyosaur that lived about 200 million years ago. These marine reptiles were about the same size as modern porpoises, breathed air like them, and probably swam as fast, but the two groups are not closely related.

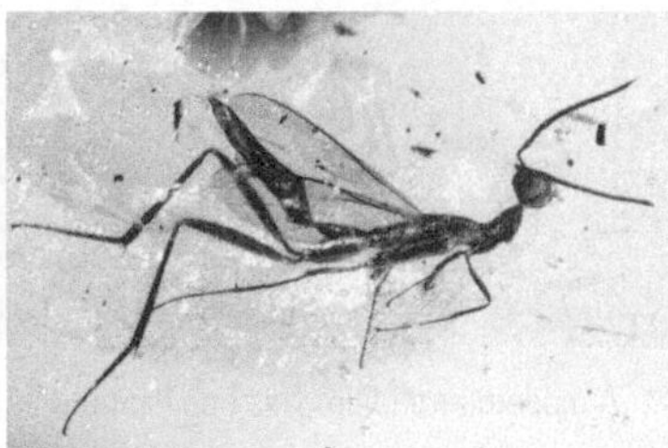

B Extinct wasp encased in amber, which is ancient tree sap. This 9-mm-long insect lived about 20 million years ago.

C Fossilized leaf from a 260-million-year-old *Glossopteris*, a type of plant called a seed fern.

D Fossilized footprints of a theropod, a name that means "beast foot." This group of carnivorous dinosaurs, which includes the familiar *Tyrannosaurus rex*, arose about 250 million years ago. Each of these footprints is about 18 inches (45 cm) long.

E Coprolite (fossilized feces). Fossilized food remains and parasitic worms inside coprolites offer clues about the diet and health of extinct species. A foxlike animal excreted this one.

FIGURE 16.7 Examples of fossils.

16.4 Fossils: Evidence of Ancient Life

LEARNING OBJECTIVES

- Explain why fossils are relatively rare, and give examples of information that they provide about past life.
- Describe the kinds of evidence about the history of life that can be found in sedimentary rock formations.

Fossils are the remains or traces of organisms that lived long ago—stone-hard evidence of ancient life (FIGURE 16.7). Most fossils consist of mineralized bones, teeth, shells, seeds, spores, or other durable parts of an ancient organism. Trace fossils such as footprints and other impressions, nests, burrows, trails, eggshells, or feces are evidence of activities.

The process of fossilization typically begins when an organism or its traces become covered by sediments, mud, or ash. Groundwater then seeps into the remains, filling spaces around and inside of them. Minerals dissolved in the water gradually replace minerals in bones and other hard tissues, and they can crystallize inside cavities and impressions to form detailed imprints of internal and external structures. Sediments that slowly accumulate on top of the site exert increasing pressure, and, after a very long time, extreme pressure transforms the mineralized remains into rock.

Most fossils are found in layers of sedimentary rock (FIGURE 16.8). The deepest layers in a formation were the first to form, and those closest to the surface formed most recently. Thus, in general, the deeper the layer, the older the fossils it contains (FIGURE 16.9). Sedimentary rock forms as a river washes silt, sand, volcanic ash, or other materials from land to sea. Mineral particles in the materials settle on seafloors in layers that vary in thickness and composition. After many millions of years, the layers of sediments become buried and compacted into layers of rock. Geologic processes can move an entire rock formation so its layers become exposed by erosion (FIGURE 16.10).

Biologists study sedimentary rock formations in order to understand the historical context of ancient life. Each layer contains information about environmental conditions and events that occurred as it formed. Consider banded iron, a unique sedimentary rock named after its striped appearance (left). Huge deposits of it are the source of most iron we mine for steel today, but they also provide a record of how the evolution of the noncyclic pathway of photosynthesis changed the chemistry of Earth. Banded iron became abundant

fossil Physical evidence of an organism that lived in the ancient past.

FIGURE 16.8 **Examples of sedimentary rock.** Sandstones (left) form from grains of sand or minerals; shales (right) form from clay or mud.

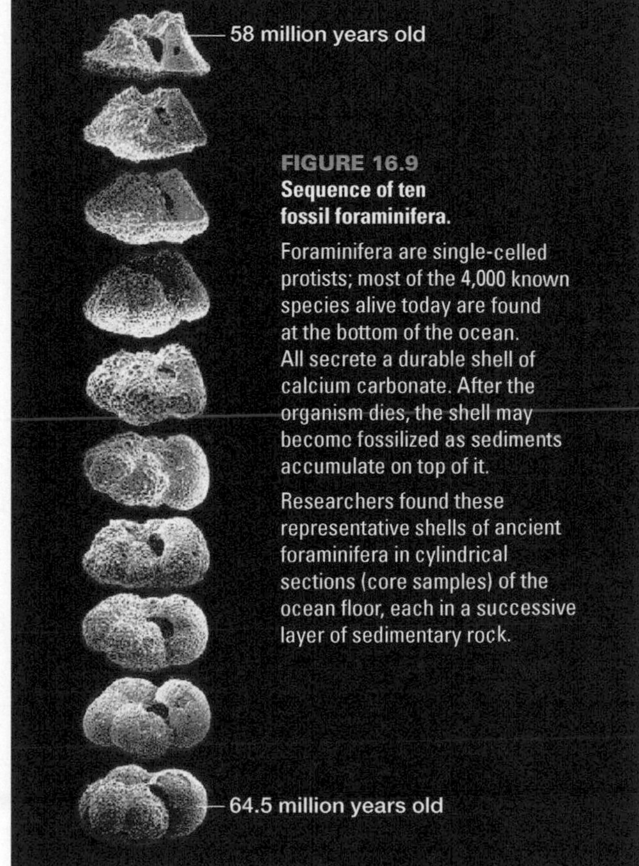
58 million years old
64.5 million years old

FIGURE 16.9 **Sequence of ten fossil foraminifera.**

Foraminifera are single-celled protists; most of the 4,000 known species alive today are found at the bottom of the ocean. All secrete a durable shell of calcium carbonate. After the organism dies, the shell may become fossilized as sediments accumulate on top of it.

Researchers found these representative shells of ancient foraminifera in cylindrical sections (core samples) of the ocean floor, each in a successive layer of sedimentary rock.

FIGURE 16.10 **Sedimentary rock displaced by geologic processes.** High in the Andes, a scientist infers the stride of an extinct dinosaur by measuring the distance between its fossilized footprints on an ancient shore. Geologic processes have tilted this sedimentary rock and lifted it far above sea level, where the layers—and fossils in them—have become exposed.

CREDITS: (in text) Natural History Museum, London/Science Photo Library/Science Source; (8) left, right, Jupiter Images/Getty Images; (9) Courtesy of Daniel C. Kelley, Anthony J. Arnold, and William C. Parker, Florida State University Department of Geological Science; (10) Jonathan Blair/Getty Images.

about 2.4 billion years ago, after photosynthetic organisms evolved. Before then, the atmosphere and ocean contained very little oxygen, so almost all iron was in a reduced form (Section 5.5). Reduced iron dissolves in water, so runoff from surface rocks carried a lot of it into ocean water. Oxygen released into the ocean by photosynthesis combined with the dissolved iron. The resulting oxidized iron compounds are insoluble in water, and they began to rain down on the ocean floor in massive quantities. These compounds accumulated in sediments that would eventually become compacted into banded iron formations. This process continued for about 600 million years. After that, ocean water no longer contained very much dissolved iron, and oxygen gas bubbling out of it had oxidized the iron in rocks exposed to the atmosphere.

The Fossil Record

We have fossils for more than 250,000 known species. Considering the current range of biodiversity, there must have been many millions more, but we will never know all of them because fossils are relatively rare. Few individuals of any species become fossilized. Typically, when an organism dies, its remains are quickly obliterated by scavengers. Organic materials decompose in the presence of moisture and oxygen, so remains that escape scavenging endure only if they dry out, freeze, or become encased in an air-excluding material such as mud. If remains do become fossilized, geologic assaults usually crush, scatter, or dissolve them.

In order for us to know about an extinct species that existed long ago, we have to find a fossil of it. At least one specimen had to be buried before it decomposed or something ate it. The burial site had to escape destruction and end up in an accessible place. Most ancient species had no hard parts to fossilize, so we do not find much evidence of them. For example, there are many fossils of bony fishes and mollusks with hard shells, but few fossils of the jellyfishes and soft worms that were probably much more common. Also think about relative numbers of organisms. Fungal spores and pollen grains are typically released by the billions. By contrast, the earliest humans lived in small bands and few of their offspring survived. The odds of finding even one fossilized human bone are much smaller than the odds of finding a fossilized fungal spore. Finally, imagine two species, one that existed only briefly and the other for a hundred million years. Which one is more likely to be represented in the fossil record?

Missing Links in the Fossil Record

The discovery of intermediate forms of cetaceans (an order of animals that includes whales, dolphins, and porpoises) offers an example of how fossil finds can be used to reconstruct evolutionary history of current lineages. Skeletons of modern cetaceans have remnants of a pelvis and hindlimbs (**FIGURE 16.11A**), so evolutionary biologists had long thought that the ancestors of this group walked on land. However, no one had found fossils demonstrating a clear link between cetaceans and ancient land animals, so the rest of the story remained speculative. Intact fossil skeletons of extinct whale-like *Dorudon atrox* had been discovered, but these animals were clearly cetaceans, with little resemblance to land animals (**FIGURE 16.11B**). *Dorudon* lived about 37 million years ago, and like modern whales it was a tail-swimmer and fully aquatic: The tiny hindlimbs could not have supported the animal's huge body out of water.

In the early 1990s, DNA sequence comparisons suggested that modern cetaceans are more related to artiodactyls than to other groups. Artiodactyls are hooved mammals with an even number of toes (two or four) on each foot; modern members of the lineage include hippopotamuses, antelopes, sheep, and so on. The DNA findings were controversial. Some thought the ancestors of cetaceans were mesonychids, a lineage of carnivorous mammals, but the ancestors of artiodactyls were herbivores resembling tiny deer with long tails. Either way, a seemingly unimaginable number of skeletal and physiological changes would have been required for small, land-based animals to evolve into whales with gigantic bodies uniquely adapted to deep ocean swimming.

Then, in 2000, Philip Gingerich and his colleagues unearthed complete fossilized skeletons of two cetaceans in a 47-million-year-old rock formation in Pakistan. Later named *Artiocetus clavis* and *Rodhocetus balochistanensis*, both animals had whalelike skulls and robust hindlimbs (**FIGURE 16.11C**). Their bodies were built to swim with their feet (not their tails as whales do), and their ankle bones had clearly distinctive features of artiodactyls—settling the debate about cetacean ancestry (**FIGURE 16.11D**). *Rodhocetus* and *Artiocetus* were not direct ancestors of modern whales, but their telltale ankle bones mean they were long-lost relatives. Both were offshoots of the artiodactyl-to-modern-whale lineage as it transitioned from land to water.

TAKE-HOME MESSAGE 16.4

✔ Fossils are evidence of organisms that lived in the remote past, a stone-hard historical record of life.

✔ The fossil record will never be complete because fossils are relatively rare. It is slanted toward hard-bodied species that lived in large populations and persisted for a long time.

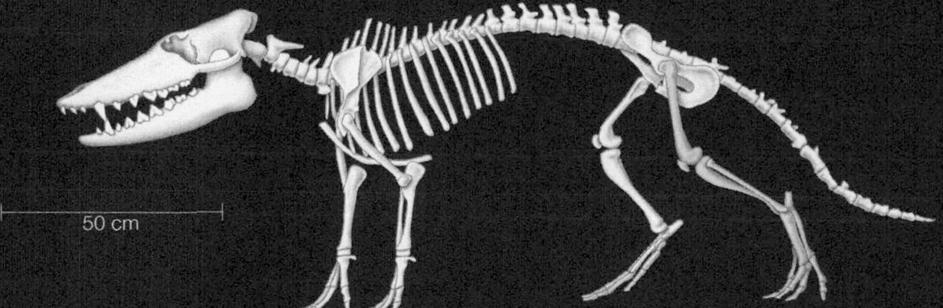
50 cm

D *Pakicetus attocki*, a small animal that lived about 50 million years ago, was semi-aquatic, but its body was specialized for running. *Pakicetus* belonged to the artiodactyl lineage, but it also had some traits unique to cetaceans. It is considered to be one of the earliest cetaceans.

FIGURE 16.11 Comparison of cetacean skeletons. The ancestor of whales was an artiodactyl that walked on land. Over millions of years, the lineage transitioned from life on land to life in water, and as it did, bones of the hindlimb (highlighted in blue) became smaller.

CREDITS: (11A, B, C left, D) © Cengage Learning; (11B middle and right) © Philip Gingerich, University of Michigan.

16.5 Changes in the History of Earth

LEARNING OBJECTIVES

- Describe some geologic features that form as a result of plate tectonics, and explain how these features arise.
- Explain how researchers determine the age of ancient materials.

Plate Tectonics

Wind, water, and other forces continuously sculpt Earth's surface, but they are only part of a much bigger picture of geologic change. For example, all continents on Earth today were once part of a bigger supercontinent called **Pangea** that split into fragments and drifted apart about 200 million years ago. The idea that continents move around, originally called continental drift, was proposed in the early 1900s to explain why the Atlantic coasts of South America and Africa seem to "fit" like jigsaw puzzle pieces. The concept of continental drift also explained why the magnetic poles of gigantic rock formations point in different directions on different continents. As magma solidifies into rock, some iron-rich minerals in it become magnetic, and their magnetic poles align with Earth's poles when they do. If the continents never moved, then all of these ancient rocky magnets should be aligned north-to-south, like compass needles. Indeed, the magnetic poles of rocks in each formation are aligned with one another, but the alignment is not always north-to-south. Either Earth's magnetic poles have veered dramatically from their north–south axis, or the continents have wandered.

Continental drift was first greeted with skepticism because there was no known mechanism for continents to move. Then, in the late 1950s, deep-sea explorers found huge ridges and trenches stretching thousands of kilometers across the seafloor. The discovery led to an explanation of continental drift (**FIGURE 16.12**). By the **plate tectonics theory**, Earth's outer layer of rock is cracked into huge plates, like a gigantic cracked eggshell. Magma welling up at an undersea ridge ❶ or continental rift at one edge of a plate pushes old rock at the opposite edge into a trench ❷. The movement is like that of a conveyor belt that slowly transports continents on top of it to new locations. The plates move no more than 10 centimeters (4 inches) a year, but that is enough to carry a continent halfway around the world after about 200 million years.

Evidence of tectonic plate movement is all around us, in geologic features such as faults. Faults are cracks in Earth's crust, and they often occur where plates

The San Andreas Fault, which extends 800 miles through California, marks the boundary between two tectonic plates.

FIGURE 16.12 Plate tectonics. Slowly moving pieces of Earth's outer layer of rock convey continents around the globe.

❶ At oceanic ridges, magma (red) welling up from Earth's interior drives the movement of tectonic plates. New crust spreads outward as it forms on the surface, forcing adjacent tectonic plates away from the ridge and into trenches elsewhere.

❷ At trenches, the advancing edge of one plate plows under an adjacent plate and buckles it.

❸ Faults (ruptures in Earth's crust) often occur where plates meet.

❹ Magma ruptures Earth's crust at what are called "hot spots."

❺ An archipelago forms as a tectonic plate moves over a hot spot. The Hawaiian Islands have been forming from magma that continues to erupt from a hot spot under the Pacific Plate. This and other tectonic plates are shown in Appendix V.

FIGURE IT OUT Is the tectonic plate over the hot spot moving left to right, or right to left?

Answer: Left to right

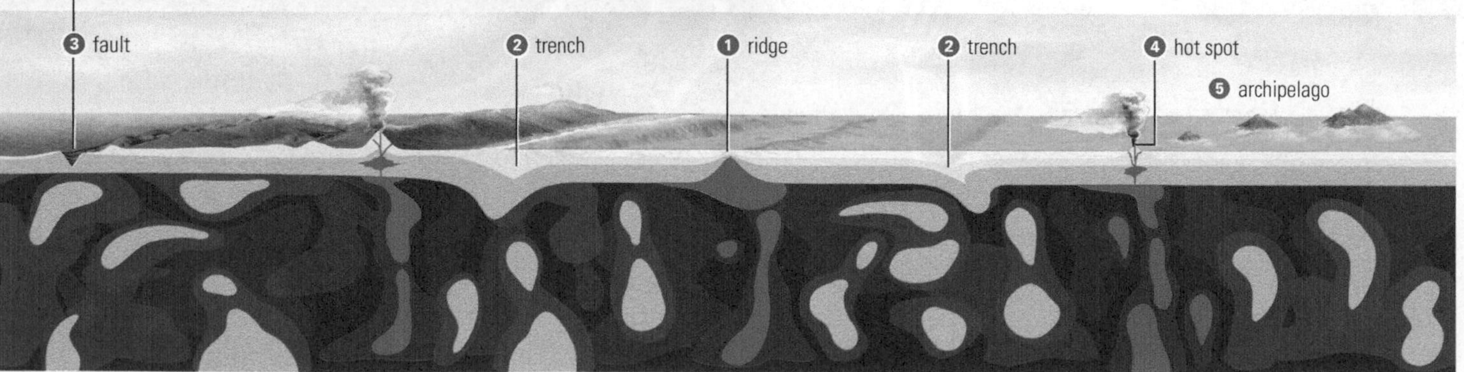
❸ fault
❷ trench
❶ ridge
❷ trench
❹ hot spot
❺ archipelago

CREDITS: (12) top, Kevin Schafer/Getty Images; bottom, © Cengage Learning.

meet ❸. As another example, consider volcanic hot spots, which are places where plumes of molten rock well up from deep inside Earth and rupture its crust ❹. Volcanic island chains (archipelagos) form as a plate moves across an undersea hot spot ❺.

The fossil record also provides evidence in support of plate tectonics. Consider an unusual geologic formation that occurs in a belt across Africa. The formation consists of a complex sequence of rock layers that is unlikely to have formed more than once, but identical sequences occur in huge belts spanning India, South America, Madagascar, Australia, and Antarctica. Across all of these continents, the layers are the same ages. They also hold fossils found nowhere else, including remains of the seed fern *Glossopteris* (shown in Figure 16.7C), which lived 299–252 million years ago, and an early reptile called *Lystrosaurus* that existed 270–225 million years ago. The rock formation must have formed on a single continent that later broke up.

At least five supercontinents formed and split up again since Earth's outer layer of rock solidified 4.55 billion years ago. One of them, a supercontinent called **Gondwana**, existed around 540 million years ago. Over the next 260 million years, Gondwana wandered across the South Pole, then drifted north until it merged with another supercontinent about 300 million years ago to form Pangea (**FIGURE 16.13**). Most of the landmasses currently in the Southern Hemisphere as well as India and Arabia were once part of Gondwana. Some modern species, including the ratite birds pictured in Figure 16.2, live only in these places.

Continental collisions brought together populations of land-based organisms that had been living apart, and split up ocean-dwelling populations. Continental breakups split up populations on land, and brought together populations that had been living in different parts of the ocean. As you will see in the next chapter, events like these have been a major driving force of evolution.

The Geologic Time Scale

Charles Lyell, the proponent of uniformitarianism, could not have imagined how plate tectonics has changed Earth's landscape over millions of years. Georges Cuvier, the proponent of catastrophism, never knew about asteroid impacts that permanently altered the course of life in a geologic instant. Today, we

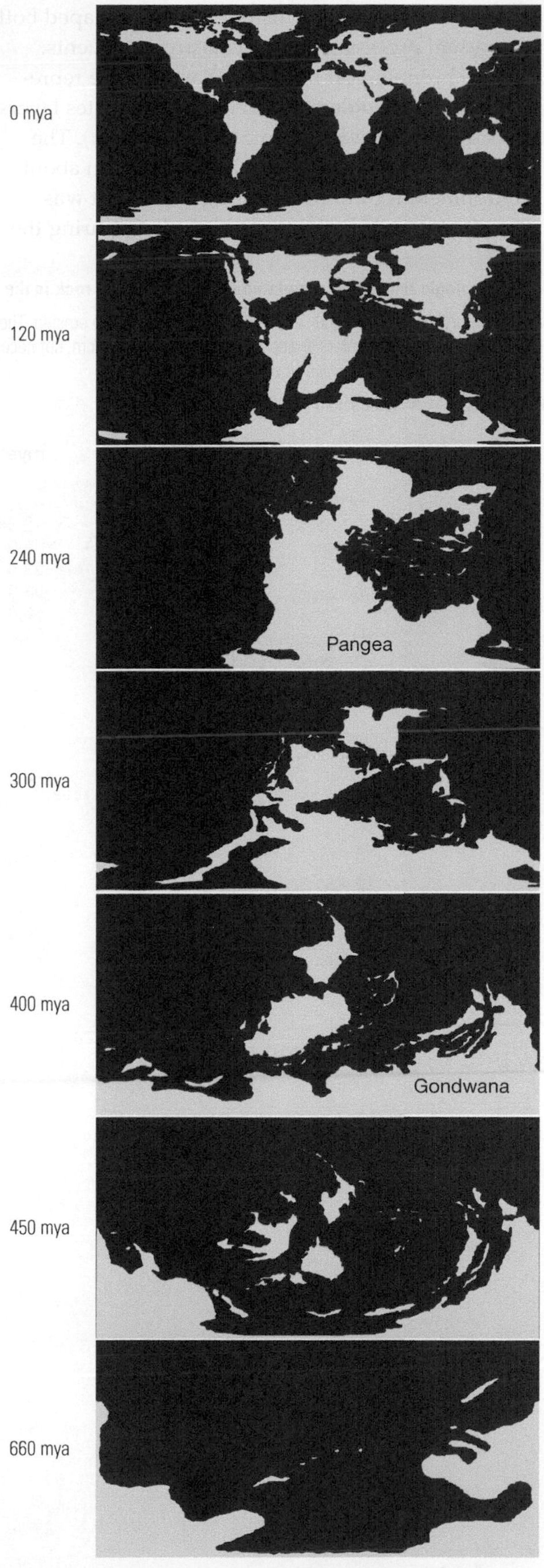
0 mya
120 mya
240 mya
Pangea
300 mya
400 mya
Gondwana
450 mya
660 mya

FIGURE 16.13 A series of reconstructions of drifting continents. mya: million years ago.

Gondwana (gond-WAN-uh) Supercontinent that existed before Pangea, more than 500 million years ago.

Pangea (pan-JEE-uh) Supercontinent that began to form about 300 million years ago; broke up 100 million years later.

plate tectonics theory Theory that Earth's outermost layer of rock is cracked into plates, the slow movement of which conveys continents to new locations over geologic time.

understand that Earth's history has been shaped both by gradual processes and by catastrophic events.

The chronological history of Earth can be represented as a **geologic time scale** that correlates layers of rock with intervals of time (**FIGURE 16.14**). The composition of each layer holds information about environmental conditions during the time it was deposited; fossils in it are a record of life during the same period. Because the layers differ in composition and fossil content, a transition between them implies an environmental change that affected life. These transitions mark boundaries between intervals of the geologic time scale. Consider Hermit Shale, a sedimentary rock formation that stretches across Arizona,

geologic time scale Chronology of Earth's history.

FIGURE 16.14 The geologic time scale correlated with sedimentary rock in the Grand Canyon.

Opposite, layers of sedimentary rock in the Grand Canyon have been exposed by erosion. The layers are correlated with the geologic time scale below. Red triangles mark great mass extinctions; "first appearance" refers to appearance in the fossil record, not necessarily first appearance on Earth. mya: million years ago.

Dates are from the International Commission on Stratigraphy, 2014.

FIGURE IT OUT Which formation in the photo is marked with the ?

Answer: Tapeats Sandstone

Eon	Era	Period	Epoch	mya*	Major Geologic and Biological Events
Phanerozoic	Cenozoic	Quaternary	Holocene	0.01	Modern humans evolve. Major extinction event is now under way.
			Pleistocene	2.6	
		Neogene	Pliocene	5.3	Tropics, subtropics extend poleward. Climate cools; dry woodlands and grasslands emerge. Adaptive radiations of mammals, insects, birds.
			Miocene	23.0	
		Paleogene	Oligocene	33.9	
			Eocene	56.0	
			Paleocene	66.0 ◀	**Major extinction event**
	Mesozoic	Cretaceous	Upper		Flowering plants diversify; sharks evolve. All dinosaurs and many marine organisms disappear at the end of this epoch.
				100.5	
			Lower		Climate very warm. Dinosaurs continue to dominate. Important modern insect groups appear (bees, butterflies, termites, ants, and herbivorous insects including aphids and grasshoppers). Flowering plants originate and become dominant land plants.
				145.0	
		Jurassic			Age of dinosaurs. Lush vegetation; abundant gymnosperms and ferns. Birds appear. Pangea breaks up.
				201.3 ◀	**Major extinction event**
		Triassic			Recovery from the major extinction at end of Permian. Many new groups appear, including turtles, dinosaurs, pterosaurs, and mammals.
				252 ◀	**Major extinction event**
	Paleozoic	Permian			Supercontinent Pangea and world ocean form. Adaptive radiation of conifers. Cycads and ginkgos appear. Relatively dry climate leads to drought-adapted gymnosperms and insects such as beetles and flies.
				299	
		Carboniferous			High atmospheric oxygen level fosters giant arthropods. Spore-releasing plants dominate. Age of great lycophyte trees; vast coal forests form. Ears evolve in amphibians; penises evolve in early reptiles (vaginas evolve later, in mammals only).
				359 ◀	**Major extinction event**
		Devonian			Land tetrapods appear. Explosion of plant diversity leads to tree forms, forests, and many new plant groups including lycophytes, ferns with complex leaves, seed plants.
				419	
		Silurian			Radiations of marine invertebrates. First appearances of land fungi, vascular plants, bony fishes, and perhaps terrestrial animals (millipedes, spiders).
				443 ◀	**Major extinction event**
		Ordovician			Major period for first appearances. The first land plants, fishes, and reef-forming corals appear. Gondwana moves toward the South Pole and becomes frigid.
				485	
		Cambrian			Earth thaws. Explosion of animal diversity. Most major groups of animals appear (in the oceans). Trilobites and shelled organisms evolve.
				541	
Precambrian	Proterozoic				Oxygen accumulates in atmosphere. Origin of aerobic metabolism. Origin of eukaryotic cells, then protists, fungi, plants, animals. Evidence that Earth mostly freezes over in a series of global ice ages between 750 and 600 mya.
				2,500	
	Archean and earlier				3,800–2,500 mya. Origin of bacteria and archaea. 4,600–3,800 mya. Origin of Earth's crust, first atmosphere, first seas. Chemical, molecular interactions lead to origin of life (from protocells to anaerobic single cells).

Ka Kaibab Limestone
To Toroweap Formation
Permian
Co Coconino Sandstone
He Hermit Shale
Es Esplanade Sandstone
We Wescogame Formation
Carboniferous
Ma Manakacha Formation
Wa Watahomigi Formation
Re Redwall Limestone
Te Temple Butte Formation
Mu Muav Limestone
Cambrian
Br Bright Angel Shale
Ta Tapeats Sandstone
*Chuar Group
Nankoweap Formation
*Unkar Group
Proterozoic
Vi Vishnu Basement Rocks
Ka
To
Co
He
Es
We
Ma
Wa
Re
Te
Mu
Br
Ta
Vi
?

Layers not visible in this view of the Grand Canyon

California, Nevada, and Utah. It consists mainly of mud and silt with abundant fossils of land plants and insects, so it was probably laid down by a river system that stretched across a broad, wet coastal plain. Above Hermit Shale is another sedimentary rock formation, Coconino Sandstone, that consists mainly of weathered sand. Ripple marks and reptile tracks are the only fossils in it. These and other characteristics indicate that Coconino Sandstone accumulated as a vast sand desert similar to the modern Sahara. Thus, a major climactic shift must have occurred between the deposition of Hermit Shale and Coconino Sandstone.

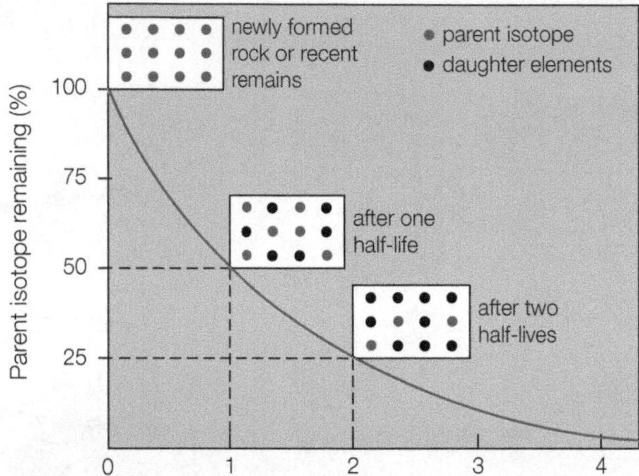

newly formed rock or recent remains
parent isotope
daughter elements
after one half-life
after two half-lives
Parent isotope remaining (%)
100
75
50
25
0
1
2
3
4

FIGURE 16.15 Half-life.

FIGURE IT OUT How much of a radioisotope remains after two of its half-lives have passed?

Answer: 25 percent

A Long ago, ^{14}C and ^{12}C were incorporated into the tissues of a nautilus. The carbon atoms were part of organic molecules in the animal's food. ^{12}C is stable and ^{14}C decays, but the proportion of the two isotopes in the nautilus's tissues remained the same. Why? The nautilus continued to gain both carbon isotopes in the same proportions from its food.

B The nautilus stopped eating when it died, so its body stopped gaining carbon. The ^{12}C atoms in its tissues were stable, but the ^{14}C atoms (represented as red dots) were decaying into nitrogen atoms. Thus, over time, the amount of ^{14}C decreased relative to the amount of ^{12}C. After 5,730 years, half of the ^{14}C had decayed; after another 5,730 years, half of what was left had decayed, and so on.

C Fossil hunters discover the fossil and measure its content of ^{14}C and ^{12}C. They use the ratio of these isotopes to calculate how many half-lives passed since the organism died. For example, if its ^{14}C to ^{12}C ratio is one-eighth of the ratio in living organisms, then three half-lives $(½)^3$ must have passed since it died. Three half-lives of ^{14}C is 17,190 years.

FIGURE 16.16 Example of how radiometric dating is used to find the age of a carbon-containing fossil.

Carbon 14 (^{14}C) is a radioisotope of carbon that decays into nitrogen. It forms in the atmosphere and combines with oxygen to become CO_2, which enters food chains by way of photosynthesis.

Radiometric Dating

How do we know the age of events that occurred in the ancient past? The answer involves radioisotopes (Section 2.2). Radioactive decay is not influenced by temperature, pressure, chemical bonding state, or moisture; it is influenced only by time. Thus, like the ticking of a perfect clock, each type of radioisotope decays at a constant rate into predictable products (daughter elements). The time it takes for half of the atoms in a sample of radioisotope to decay is called a **half-life** (FIGURE 16.15).

Almost all of the carbon on Earth is in the form of ^{12}C; a tiny but constant amount of ^{14}C forms in the upper atmosphere and becomes incorporated into molecules of carbon dioxide. The ratio of ^{14}C to ^{12}C in atmospheric CO_2 is relatively stable; because carbon dioxide is life's main source of carbon, the same ratio occurs in the body of a living organism. As long as the organism lives, it continues to assimilate both isotopes in the same proportions. After it dies, no more carbon is assimilated, and the ratio of ^{14}C to ^{12}C in its remains declines over time as the ^{14}C decays. The half-life of ^{14}C is known (5,730 years) so the ratio of ^{14}C to ^{12}C in the organism's remains can be used to calculate how long ago it died (FIGURE 16.16).

We have just described **radiometric dating**, a method that can reveal the age of a material by measuring its radioisotope content. Carbon isotope dating can be used to find the age of a biological material less than 60,000 years old (essentially no ^{14}C remains in older material). The age of older fossils can be estimated by dating volcanic rocks in lava flows above and below the fossil-containing layer of sedimentary rock.

How is rock dated? The original source of most rock on Earth is magma, a hot, molten material deep under Earth's surface. Atoms swirl and mix in it. When magma cools, for example after reaching the surface as lava, it hardens and becomes rock. As this occurs, the atoms in it join and crystallize as different kinds of minerals, each with a characteristic structure

half-life Characteristic time it takes for half of a quantity of a radioisotope to decay.

radiometric dating Method of estimating the age of a rock or a fossil by measuring the content and proportions of a radioisotope and its daughter elements.

Data Analysis Activities

Discovery of Iridium in the K–Pg Boundary Layer
In the late 1970s, geologist Walter Alvarez was investigating the composition of the K–Pg boundary layer in different parts of the world. He asked his father, Nobel Prize–winning physicist Luis Alvarez, to help him analyze the elemental composition of the layer. The Alvarezes and their colleagues tested the K–Pg boundary layer in Italy and Denmark, and discovered that it contains a much higher iridium content than the surrounding rock layers (**FIGURE 16.17**). Iridium belongs to a group of elements that are much more abundant in asteroids and other solar system materials than they are in Earth's crust. The Alvarez group concluded that the K–Pg boundary layer must have originated with extraterrestrial material.

1. What was the iridium content of the K–Pg boundary layer?
2. How much higher was the iridium content of the boundary layer than the sample taken 0.7 meter above it?

Sample Depth	Average Abundance of Iridium (ppb)
+ 2.7 m	< 0.3
+ 1.2 m	< 0.3
+ 0.7 m	0.36
boundary layer	41.6
– 0.5 m	0.25
– 5.4 m	0.30

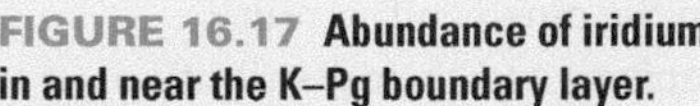

FIGURE 16.17 Abundance of iridium in and near the K–Pg boundary layer.

Iridium content of rock samples above, below, and at the K–Pg boundary layer in Stevns Klint, Denmark. Sample depths are given as meters above or below the layer. ppb, parts per billion. An average Earth rock contains 0.4 ppb iridium; the average meteorite, 550 ppb. The photo shows Luis and Walter Alverez next to the K–Pg boundary layer in Stevns Klint.

zircon

and composition. Consider zircon (left), a mineral that consists mainly of zirconium silicate molecules ($ZrSiO_4$). Some of the molecules in a newly formed zircon crystal have uranium atoms substituted for zirconium atoms, but never lead atoms. Uranium is a radioactive element with a half-life of 4.5 billion years. Its daughter element, thorium, is another radioactive element, and it decays at a predictable rate into another radioactive element, and so on until the atom becomes lead, a stable element. Over time, uranium atoms disappear from a zircon crystal, and lead atoms accumulate in it. The ratio of uranium atoms to lead atoms in a zircon crystal can be measured and used to calculate how long ago the crystal formed (its age). Using this technique, we know that the oldest known terrestrial rock, a tiny zircon crystal from the Jack Hills in Western Australia, is 4.404 billion years old.

TAKE-HOME MESSAGE 16.5

✔ Over geologic time, movements of Earth's crust have caused dramatic changes in continents and oceans. These changes profoundly influenced the course of evolution.

✔ The geologic time scale is a chronology of life's history that correlates layers of rock with intervals of time.

✔ The predictability of radioisotope decay can be used to determine the age of fossils and ancient rocks.

16.1 Reflections of a Distant Past (revisited)

The K–Pg boundary layer consists of an unusual white clay (**FIGURE 16.18**) that is rich in iridium, an element rare on Earth's surface but common in asteroids. After researchers discovered the iridium, they looked for evidence of a meteorite impact massive enough to cover the entire Earth with extraterrestrial debris. They found the Chicxulub crater, which is so big that no one had realized it was a crater before then. The K–Pg boundary layer also contains shocked quartz (left) and small glass spheres called tektites—rocks that form when quartz or sand (respectively) undergoes a sudden, violent application of extreme pressure. The only known processes on Earth that produce shocked quartz and tektites are atomic bomb explosions and asteroid impacts. ●

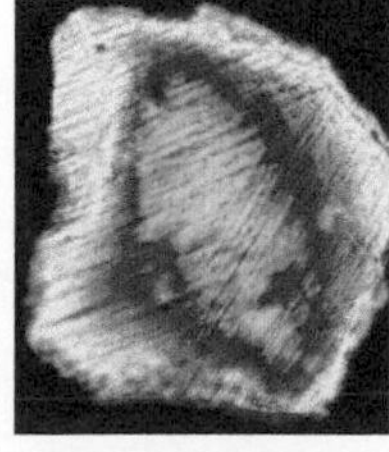
shocked quartz

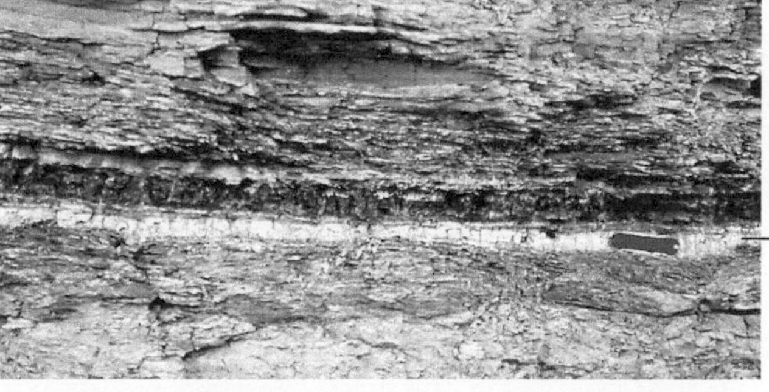
K–Pg boundary layer

FIGURE 16.18 The K–Pg boundary layer. The red pocketknife is shown for scale.

CREDITS: (in text) Courtesy of Stan Celestian/Glendale Community College Earth Science Image Archive; 16.1 revisited, U.S. Geological Survey. (17) left, © Cengage Learning; right, Lawrence Berkeley National Laboratory; (18) © David A. Kring, NASA/Univ. Arizona Space Imagery Center.

STUDY GUIDE

Section 16.1 Events of the ancient past can be explained by the same physical, chemical, and biological processes that operate today. The dinosaurs disappeared 66 million years ago in a mass extinction that was probably caused by a catastrophic asteroid impact. The impact left physical traces in a worldwide sedimentary rock formation called the K–Pg boundary layer.

Section 16.2 Expeditions in the nineteenth century yielded observations of organisms and their distribution (**biogeography**). **Comparative morphology** yielded new observations of similarities and differences between organisms and fossils. Taken together, the new, increasingly detailed evidence implied that Earth and life on it had changed over time, an idea that could not be explained within the framework of existing belief systems.

New ways of thinking about the natural world were an outcome of attempts to reconcile the new evidence with traditional beliefs. Evidence of **evolution**, or change in a line of descent (a **lineage**), led Lamarck to propose a mechanism that might drive it. Discoveries of fossil animals with no living counterparts led Cuvier to propose that some species had become extinct. **Catastrophism**, the idea that periodic, violent geologic events shaped Earth's current landscape, offered an explanation for extinction. Evidence that geologic processes operating in the present are the same as those in the past led Lyell to propose that Earth was much older than previously believed. Gradual, familiar geologic processes could have shaped Earth's landscape (an idea called **uniformitarianism**) if they had occurred over very long spans of time.

Section 16.3 The new proposals about nature set the stage for Darwin's ideas about evolution. His voyage on the *Beagle* gave him the opportunity to encounter species in diverse habitats, and to collect thousands of specimens to study when he returned. These, together with exposure to Malthus' ideas about human competition for resources, led him to the idea of **natural selection**, explained here in modern terms: A natural population tends to grow until it exhausts resources in its environment. As that happens, competition for those resources intensifies among the population's members. Individuals with forms of shared, heritable traits that make them more competitive for the resources tend to produce more offspring. These evolutionary **adaptations** (**adaptive traits**) impart greater **fitness** to individuals, so they end to become more common in a population over generations. Both Darwin and Wallace realized that natural selection drives evolution.

Section 16.4 **Fossils** are mineralized remains or traces of ancient organisms. They are typically found in layers of sedimentary rock, which forms from sand, silt, and other materials deposited on the bottom of seas and other bodies of water. Geologic processes can lift and tilt sedimentary rock formations far above sea level.

Typically, the most recent fossils in a sedimentary rock formation occur in layers deposited most recently, on top of older fossils in older layers.

A fossil is a stone-hard record of ancient life, and its historical context may be inferred from the layer of sedimentary rock in which it formed. The composition of the rock layer holds information about environmental conditions that prevailed when it was deposited. Differences between layers of a sedimentary rock formation reflect some type of change in the environment.

Fossils are relatively rare, so the fossil record will never be complete. Even so, we have discovered enough fossils to reconstruct the evolutionary history of many ancient and modern species.

Section 16.5 By the **plate tectonics theory**, Earth's crust is cracked into giant plates that carry landmasses to new positions as they move. The course of life's evolution has been profoundly influenced by this movement, for example as populations were brought together or split up during the formation and breakup of supercontinents such as **Gondwana** and **Pangea**.

The **geologic time scale** is a chronology of Earth's history in which layers of rock in worldwide formations are correlated with intervals of time. Biologists use the scale to integrate geologic and evolutionary events of the ancient past.

The predictability of radioactive decay allows us to determine the age of rocks and fossils. Each radioisotope has a characteristic **half-life**. Half-life calculations can reveal the age of a rock or fossil, a technique called **radiometric dating**. With this technique, a sample's content of radioisotope and daughter elements is measured. The ratio of these two numbers can be used to calculate the age of the sample.

SELF-QUIZ

Answers in Appendix VII

1. The number of species on an island depends on the size of the island and its distance from a mainland. This statement would most likely be made by ________ .
   a. an explorer
   b. a biogeographer
   c. a geologist
   d. a philosopher

2. Evolution ________ .
   a. is change in a line of descent
   b. is the same as natural selection
   c. is the goal of natural selection
   d. explains the origin of life

3. The process in which environmental pressures result in the differential survival and reproduction of individuals of a population is called ________ .
   a. catastrophism
   b. evolution
   c. natural selection
   d. genetics

4. The dinosaurs died out ________ million years ago.

5. The bones of a bird's wing are similar to the bones in a bat's wing. This observation is an example of ________.
   a. uniformitarianism
   b. evolution
   c. comparative morphology
   d. a lineage

6. A trait is adaptive if it ________.
   a. arises by mutation
   b. increases fitness
   c. is passed to offspring
   d. occurs in fossils

7. Darwin and Wallace proposed the hypothesis that ________.
   a. natural selection drives evolution
   b. change occurs in a line of descent
   c. new species arise after geologic events
   d. dinosaurs perished in the aftermath of an asteroid impact

8. In which type of rock are you most likely to find a fossil?
   a. basalt, a dark, fine-grained volcanic rock
   b. limestone, composed of sedimented calcium carbonate
   c. slate, a volcanically melted and cooled shale
   d. granite, which forms by crystallization of magma cooling below Earth's surface

9. Which of the following is *not* a fossil?
   a. Dried-out animal bones in a layer of desert sand
   b. A patch of fur encased in amber
   c. A woolly mammoth frozen in Arctic permafrost
   d. Remains of an extinct whalelike animal
   e. The impression of a plant leaf in sandstone
   f. The wrinkly texture of shale imprinted with biofilms in sediments

10. The geologic time scale ________.
    a. explains plate tectonics
    b. is a type of biological clock
    c. correlates intervals of time with layers of sedimentary rock
    d. is an example of uniformitarianism

11. If the half-life of a radioisotope is 20,000 years, then a sample in which three-quarters of that radioisotope has decayed is ________ years old.
    a. 15,000 b. 26,667 c. 30,000 d. 40,000

12. Did Pangea or Gondwana form first?

13. On the geologic time scale, life originated in the ______.
    a. Archean
    b. Proterozoic
    c. Phanerozoic
    d. Cambrian

14. Forces that cause geologic change include ________ (select all that are correct).
    a. erosion
    b. natural selection
    c. volcanic activity
    d. tectonic plate movement
    e. wind
    f. asteroid impacts

15. Match the terms with the most suitable description.

___ lineage	a. measured in terms of reproductive success
___ half-life	b. geologic change occurs continuously
___ fossils	c. geologic change occurs in unusual major events
___ natural selection	d. evidence of life in the distant past
___ fitness	e. survival of the fittest
___ uniformitarianism	f. characteristic of a radioisotope
___ catastrophism	g. a line of descent

CRITICAL THINKING

1. Natural selection makes an adaptive trait more common in a population. After many generations, will all individuals in the population have the same adaptive trait?

2. Radiometric dating does not measure the age of an individual atom. It is a measure of the age of a quantity of atoms—a statistic. As with any statistical measure, its values may deviate around an average (see sampling error, Section 1.7). Imagine that one sample of rock is dated ten different ways. Nine of the tests yield an age close to 225,000 years. One test yields an age of 3.2 million years. Do the nine consistent results imply that the one that deviates is incorrect, or does the one odd result invalidate the nine that are consistent?

3. If you think of geologic time spans as minutes, life's history might be plotted on a clock such as the one shown below. According to this clock, the most recent epoch started in the last 0.1 second before noon. Where does that put you?

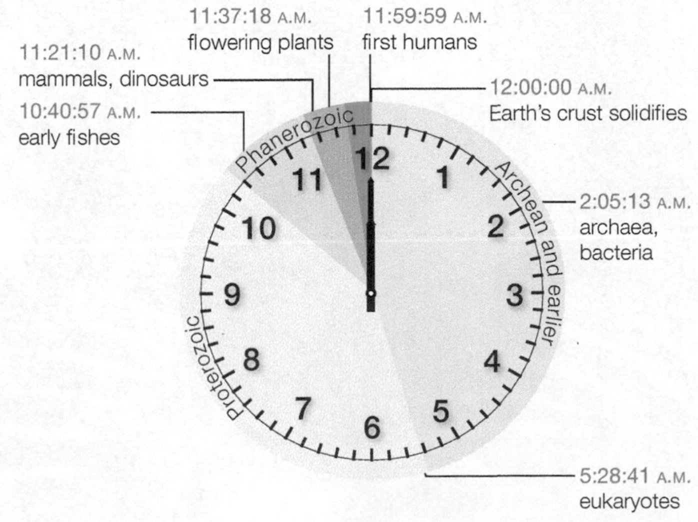

11:37:18 A.M.
flowering plants
11:59:59 A.M.
first humans
11:21:10 A.M.
mammals, dinosaurs
12:00:00 A.M.
Earth's crust solidifies
10:40:57 A.M.
early fishes
Phanerozoic
Archean and earlier
Proterozoic
2:05:13 A.M.
archaea, bacteria
5:28:41 A.M.
eukaryotes

17 PROCESSES OF EVOLUTION

CORE CONCEPTS

Evolution

Evolution underlies the unity and diversity of life.

Natural selection operates on variation in shared, heritable traits, which have their basis in a population's gene pool. The frequency of alleles in a gene pool changes constantly in natural populations because natural selection and other mechanisms that drive microevolution are always in play. These forces of evolutionary change are the source of life's diversity.

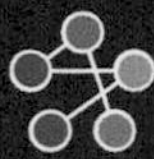

Systems

Complex properties arise from interactions among components of a biological system.

New species arise in the context of geographic, genetic, and ecological factors that influence the timing and direction of evolutionary change, from the population level up to entire kingdoms of life. Diversity at all levels of life offers resilience to environmental challenges.

Process of Science

The field of biology consists of and relies upon experimentation and the collection and analysis of scientific evidence.

Mathematical approaches that reveal changes in allele frequency provide evidence that microevolution occurs continually in natural populations. Evolution is dynamic and often messy, but we can study its processes (such as natural selection, speciation, and macroevolution) by modeling them in patterns.

Links to Earlier Concepts

This chapter builds on your knowledge of species (Section 1.5), mutations (8.6, 9.6), variation in traits and sexual reproduction (12.1–12.3), the chromosomal basis of inheritance (13.2–13.4), complex variation in traits (13.5–13.7), and natural selection (16.3). You will revisit experimental design (1.6), globin mutations (9.6), *BRCA* genes (10.1, 11.6), polyploidy (14.6), transgenic plants (15.7), plate tectonics (16.5), and the geologic time scale (16.5).

17.1 Superbug Farms

Scarlet fever, tuberculosis, and pneumonia once caused one-fourth of the annual deaths in the United States. Since the 1940s, we have been relying on antibiotics to fight these and other dangerous bacterial diseases. We have also been using them in other, less dire circumstances. For an unknown reason, antibiotics promote growth in cattle, pigs, poultry, and even fish. The agricultural industry uses a lot of antibiotics, mainly for this purpose. In the United States, 13.7 million kilograms (about 30 million pounds) of antibiotics were used for animal production in 2011 alone—four times the amount used for human medical purposes in the same year. Despite recommendations to stop the practice, agricultural use of antibiotics is still rising.

Farms where antibiotics are used to promote growth (**FIGURE 17.1**) are hot spots for the evolution of antibiotic-resistant bacteria and their spread to humans. People who work with animals on these farms tend to have more antibiotic-resistant bacteria in their bodies, as do neighbors living within a mile radius. The bacteria spread much farther than that, however. Bacteria on an animal's skin or in its digestive tract easily contaminate its meat during slaughter, and contaminated meat ends up in restaurant and home kitchens. A 2013 investigation found "worrisome" amounts of antibiotic-resistant bacteria in 97 percent of the chicken meat in stores across the United States. About half of the samples were contaminated with superbugs (bacteria resistant to multiple antibiotics), and one in ten contained multiple types of superbugs. An earlier study found antibiotic-resistant bacteria in more than half of supermarket pork chops and ground beef, and in over 80 percent of ground turkey. Bacteria can be killed by the heat of cooking, but it is almost impossible to prevent them from spreading from contaminated meat to kitchen surfaces—and to people—during the process.

FIGURE 17.1 A hot spot for the evolution of antibiotic-resistant bacteria. The vast majority of chickens raised for meat in the United States spend their lives in gigantic flocks that crowd huge buildings like this one. Growth-promoting antibiotics are given to the entire flock in food, a practice that pressures normal bacterial populations to become antibiotic resistant.

CREDITS: (opposite) David McIntyre; (1) Bob Nichols/USDA photo.

We have only a limited number of antibiotic drugs, and developing new ones is much slower than bacterial evolution. As resistant bacteria become more common, the number of antibiotics that can be used to effectively treat infections in humans dwindles. Using a particular antibiotic only in animals, or only in humans, is not a solution to this problem because there are just a few mechanisms by which these drugs kill bacteria; resistance to one antibiotic often confers resistance to others. For example, bacteria that become resistant to Flavomycin® (an antibiotic used only in animals) also resist vancomycin (an antibiotic used only in humans). Superbugs resistant to most antibiotics are turning up at a very alarming rate; recently, bacteria resistant to all 26 antibiotics currently available in the United States have been found—in deadly circumstances.

All of this amounts to bad news. We are now paying the price for overuse of antibiotics: An infection with antibiotic-resistant bacteria tends to be longer, more severe, and more likely to be deadly than one more easily treatable with antibiotics. Superbugs cause more than 2 million cases of serious illness each year in the United States alone, and they outright kill 23,000 of these people. Many, many more die because the infection complicates another, preexisting illness. ●

17.2 Alleles in Populations

LEARNING OBJECTIVES

- Using appropriate examples, explain variation in shared traits among individuals of a species.
- Explain allele frequency.
- Describe microevolution.

Variation in Shared Traits

Remember from Section 1.2 that a **population** is a group of interbreeding individuals of the same species in some specified area. Individuals of a population (and a species) have the same genes, so they share certain features. Giraffes, for example, normally have long necks, brown spots on white coats, and so on. These are examples of morphological traits (*morpho–* means form). Individuals of a species also share physiological traits, such as details of metabolism. They also respond the same way to certain stimuli, as when hungry giraffes feed on tree leaves. These are behavioral traits.

Sexual reproduction produces offspring with different combinations of alleles (Section 12.2), so almost every shared trait varies a bit among members of a sexually reproducing population (FIGURE 17.2). Many traits have two or more distinct forms, or morphs. A trait with two forms is called a dimorphism (*di–* means two). Flower color in the pea plants that Gregor Mendel studied is an example of a dimorphic trait (Section 13.3). In these plants, two alleles with a clear dominance relationship give rise to the dimorphism: purple or white flowers.

TABLE 17.1

Some Sources of Variation in Shared, Heritable Traits

Genetic Event	Effect
Mutation	Original source of new alleles
Crossing over at meiosis I	Mixes up maternal and paternal alleles on homologous chromosomes for forthcoming gametes
Independent assortment at meiosis I	Randomly distributes homologous chromosomes into gametes
Fertilization	Combines alleles from two individuals

A trait with three or more distinct forms is called a polymorphism (*poly–* means many). ABO blood type in humans, which is determined by the codominant alleles of the *ABO* gene, is an example (Section 13.5). Most other traits are complex, as is their genetic basis. Any or all of the genes that influence such traits may have multiple alleles.

In earlier chapters, you learned that alleles arise by mutation; other events shuffle alleles among individuals of a population (TABLE 17.1). To understand the potential scope of variation that results from these events, consider our own species. Humans have more than 20,000 genes, all with multiple alleles. For all practical purposes, there is an infinite number of potential combinations of human alleles. Thus, unless you have an identical twin, it is essentially impossible that another person with your particular complement of alleles has ever lived, or ever will.

An Evolutionary View of Mutations

Mutations—the original source of new alleles—are the raw material of evolution. We cannot predict when or in which individual a particular gene will mutate. We can, however, predict the average mutation rate of a species, which is the probability that a mutation will

gene pool All alleles of all genes in a population; a pool of genetic resources.

lethal mutation Mutation that alters phenotype so drastically that it causes death.

neutral mutation A mutation that has no effect on survival or reproduction.

population A group of organisms of the same species who live in a specific location and breed with one another more often than they breed with members of other populations.

FIGURE 17.2 Sampling morphological variation among zigzag Nerite snails and humans. Variation in shared traits among individuals is mainly an outcome of variations in alleles that influence those traits.

occur in a given interval. In humans, that rate has been measured at about 1.2×10^{-8} mutations per nucleotide per generation. In other words, each child is born with an average of 64 new mutations.

Most mutations are neutral. All mutations change the DNA sequence of a chromosome, but a **neutral mutation** has no effect on survival or reproduction—it neither helps nor hurts the individual. Consider a mutation that keeps your earlobes attached to your head instead of swinging freely. In itself, having attached earlobes should not stop you from surviving and reproducing as well as anybody else, so natural selection would not affect the frequency of this mutation in the human population.

Other mutations are not benign. Consider what happens if a gene for collagen mutates. Collagen is a protein component of the skin, bones, tendons, lungs, blood vessels, and other vertebrate organs, so a mutation that alters its function affects the entire body. A mutation in this or any other pleiotropic gene (Section 13.5) can change phenotype so drastically that it results in death, in which case it is called a **lethal mutation.**

Mutations that are beneficial tend to become more common in a population over time, even if they bestow only a slight advantage. This is because natural selection operates on traits with a genetic basis. With natural selection, remember, environmental pressures result in an increase in the frequency of an adaptive form of a trait in a population over generations (Section 16.3). Mutations have been altering genomes for billions of years, and they continue to do so. Cumulatively, they have given rise to Earth's staggering biodiversity. Think about it: The reason you do not look like an earthworm or even your neighbor began with mutations that occurred in different lines of descent.

Allele Frequency

Together, all the alleles of all the genes of a population comprise a pool of genetic resources—a **gene pool.**

Members of a population breed with one another more often than they breed with members of other populations, so their gene pool is more or less isolated.

Allele frequency is the proportion of one allele relative to all copies of the gene in a population—the fraction of chromosomes that have the allele. For example, if half of the members of a population are homozygous for a particular allele, then the allele's frequency is 50 percent, or 0.5. If half of the population is heterozygous, then the allele's frequency is 25 percent, or 0.25.

Allele frequency can change over time, and this change is called **microevolution**. Microevolution occurs constantly in natural populations because natural selection and other processes that drive it are always operating. As you learn about these processes, remember an important point: Evolution is not purposeful; it simply fills the nooks and crannies of opportunity.

TAKE-HOME MESSAGE 17.2

✔ Individuals of a natural population share morphological, physiological, and behavioral traits characteristic of the species.

✔ Details of shared traits tend to differ among members of a population. Alleles, the main basis of these differences, arise by mutation.

✔ All alleles of all genes make up a population's gene pool.

✔ The frequency of a particular allele is its proportion relative to the total number of copies of the gene in a gene pool. Allele frequency is expressed as a fraction.

✔ A change in an allele's frequency in a population is called microevolution.

✔ Microevolution is always occurring in natural populations because processes that drive it are always operating.

17.3 Genetic Equilibrium

LEARNING OBJECTIVES

- Explain genetic equilibrium, and list the conditions necessary for it to occur.
- Use an example to describe how genetic equilibrium can be used as a benchmark.

Early in the twentieth century, Godfrey Hardy (a mathematician) and Wilhelm Weinberg (a physician) independently applied the rules of probability to population genetics. Both realized that, under certain theoretical conditions, allele frequencies in a sexually reproducing population's gene pool would remain stable from one generation to the next. The population would stay in this stable state, which they called **genetic equilibrium**, as long as all of the following five conditions are met:

1. Mutations never occur.
2. The population is infinitely large.
3. The population is isolated from all other populations of the species—no individual enters or leaves.
4. Mating is random.
5. Natural selection does not occur.

As you can imagine, meeting all five of these conditions is impossible for a natural population. If at least one condition is not being met, we can expect to see an allele's frequency change in a shared gene pool.

The concept that genetic equilibrium can occur only under ideal conditions is called the Hardy–Weinberg principle. To understand how we use this principle, start with an idea that you already know: All members of a population have the same genes. If all members of the population carry a particular allele of a gene on both chromosomes (they are identically homozygous, Section 13.2), then the frequency of that allele in the population is 100 percent, or 1.0. If two alleles of the gene occur among the population's members, then the frequency of both alleles together must also be 100 percent (1.0). We can express this concept as an equation:

$$p + q = 100 \text{ percent } (1.0)$$

where p is the frequency of one allele in the population, and q is the frequency of the other.

Consider a hypothetical gene that encodes a blue pigment in daisies. A plant homozygous for one allele (*BB*) has dark blue flowers. A plant homozygous for the other allele (*bb*) has white flowers. If these two alleles are inherited in a pattern of incomplete dominance, a heterozygous plant (*Bb*) will have medium-blue flowers (**FIGURE 17.3A**). Assuming these alleles assort independently into gametes during meiosis (Section 13.4), the Punnett square in **FIGURE 17.3B** shows the possible ways in which those gametes can meet up at fertilization: The predicted fraction of offspring that inherit two *B* alleles (*BB*) is $p \times p$, or p^2; the fraction that inherit two *b* alleles (*bb*) is q^2; and the fraction that inherit one *B* allele and one *b* allele (*Bb*) is $2pq$. Note that the frequencies of the three genotypes, whatever they may be, add up to 100 percent (1.0):

$$p^2 + 2pq + q^2 = 1.0$$

Imagine that a population of daisies consists of 1,000 plants: 490 homozygous (*BB*), 420 heterozygous (*Bb*), and 90 homozygous (*bb*). These plants are diploid, so

allele frequency Abundance of a particular allele among all copies of the gene in a population's gene pool.

genetic equilibrium Theoretical state in which an allele's frequency never changes.

microevolution Change in allele frequency.

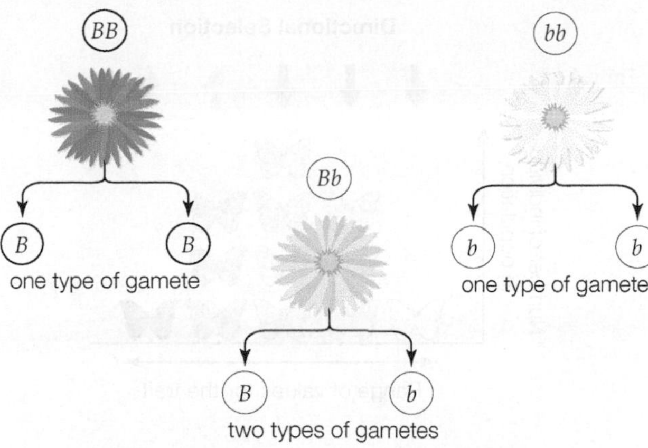
BB
bb
Bb
B
B
one type of gamete
b
b
one type of gamete
B
b
two types of gametes

A In this two-allele system, *B* specifies dark blue flowers; *b*, white. Plants that are homozygous (*BB* or *bb*) make one kind of gamete (*B* or *b*, respectively). Heterozygous plants (*Bb*) have light blue flowers and make two kinds of gametes (*B* and *b*).

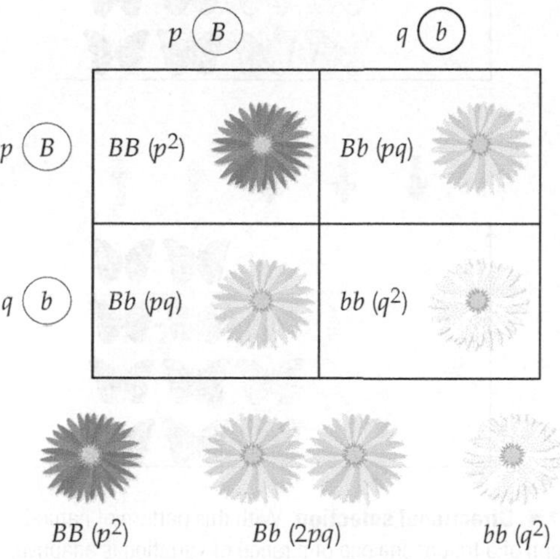
p B
q b
p B
q b
BB (p²)
Bb (pq)
Bb (pq)
bb (q²)
BB (p²)
Bb (2pq)
bb (q²)

B Say p is the proportion of *B* alleles in the gene pool, and q is the proportion of *b* alleles. This Punnett square shows that in each generation of a randomly mating population, the predicted proportion of offspring that will inherit two *B* alleles is $p \times p = p^2$. Likewise, the proportion that will inherit both alleles is $2pq$, and the proportion that will inherit two *b* alleles is q^2.

FIGURE 17.3 Hardy–Weinberg calculations. In this example, two alleles show incomplete dominance over flower color.

FIGURE IT OUT If 1/4 of this population has dark blue flowers and 1/4 has white flowers, what proportion of the next generation will have light blue flowers (assuming genetic equilibrium)?

Answer: Half of the gametes have a *B* allele; the other half have a *b* allele. Both p and $q = 0.5$, so $2pq$ = 50 percent.

the population's gene pool has a total of 2,000 copies of the gene. If each plant makes just two gametes, then the *BB* individuals make a total of 980 gametes, all with the *B* allele (490 × 2). The *Bb* individuals make a total of 840 gametes (420 × 2), and half have the *B* allele. Thus, the frequency of the *B* allele among the gametes is:

$$B\ (p) = \frac{980 + 420}{2{,}000 \text{ alleles}} = \frac{1{,}400}{2{,}000} = 0.7$$

The *bb* individuals make a total of 180 gametes, each with the *b* allele. The other half of the 840 gametes made by the *Bb* individuals also have the *b* allele. Thus, the frequency of the *b* allele among the gametes is:

$$b\ (q) = \frac{180 + 420}{2{,}000 \text{ alleles}} = \frac{600}{2{,}000} = 0.3$$

(Notice that $p + q = 0.7 + 0.3 = 1.0$, as expected.) Using our Punnett square, we can calculate the predicted proportion of individuals in the next generation:

$$BB\ (p^2) = (0.7)^2 = 0.49$$

$$bb\ (q^2) = (0.3)^2 = 0.09$$

$$Bb\ (2pq) = 2\ (0.7 \times 0.3) = 0.42$$

These proportions are the same as the ones in the parent population. As long as the five conditions required for genetic equilibrium are met, traits specified by the alleles should show up in the same proportions in each generation. If they do not, the population is not at genetic equilibrium, and microevolution is occurring.

Genetic equilibrium is a hypothetical state, but it is often used as a benchmark. Consider how the Hardy–Weinberg equations were used in early studies of an allele that causes hereditary hemochromatosis. Individuals affected by this disorder absorb too much iron from their food, and the excess accumulates in tissues and organs. Untreated, the iron overload results in fatigue and pain; in serious cases, cirrhosis of the liver, heart disease, diabetes, and other problems can occur. The allele is inherited in an autosomal recessive pattern, so carriers (people heterozygous for the allele) have no symptoms. People of northern European descent have a very high risk of hereditary hemochromatosis compared with other populations; about 1 in 83 people of Irish descent have the disorder. Researchers discovered that about 28 percent of normal (unaffected) Irish individuals were carriers, which means the frequency of the allele among members of the Irish population (q) is 0.14. Thus, the frequency of Irish people who are homozygous for the allele (q^2) should be 0.0196, or 1 in 51 individuals. This calculated frequency is higher than the observed prevalence of the disorder, so the researchers concluded (correctly) that other factors probably contribute to the progression of symptoms.

The Hardy–Weinberg equations were also used in a study of *BRCA* gene frequency. Mutations in these genes are linked to breast cancer (Sections 10.1 and 11.6). A deviation from predicted allele frequencies suggested that *BRCA* mutations have effects even before birth. Researchers investigating the frequency of mutated *BRCA* alleles among newborn girls found fewer individuals homozygous for these alleles than expected, based on the number of heterozygous individuals. The most likely explanation for the discrepancy is that, in homozygous form, *BRCA* mutations impair the survival of female embryos.

TAKE-HOME MESSAGE 17.3

✔ A theoretical baseline of genetic equilibrium can be used as a benchmark to track microevolution in a population.

17.4 Patterns of Natural Selection

LEARNING OBJECTIVES

- Describe three patterns of natural selection.
- Explain how natural selection drives microevolution.
- Use an example to explain how a change in the environment can favor a mutation that had previously been harmful.

Natural selection is one of several mechanisms that can drive microevolution. By operating on forms of a trait that differ among members of a population, natural selection affects the frequency of alleles that influence the trait. Depending on a population's environment, this process may occur in recognizable patterns. In some cases, natural selection causes a directional shift in a range of phenotypes. In other cases, environmental pressures favor or eliminate a midrange form of a trait.

Directional Selection

With **directional selection**, a form of a trait at one end of a range of variation is adaptive (**FIGURE 17.4**). The following examples illustrate how directional selection can drive microevolution.

The Peppered Moth A well-documented case of directional selection involves coloration changes in the peppered moth. In preindustrial England, the vast majority of these insects were light with black speckles, and a small number were black. At the time, the air was clean, and light-gray lichens grew on the trunks and branches of most trees. Light moths that rested on lichen-covered trees were well camouflaged, but black moths were not (**FIGURE 17.5A**).

By the 1850s, black moths had become much more common than light moths. Peppered moths feed and mate at night, then rest on trees during the day. Scientists suspected that predation by birds was the selective pressure that shaped moth coloration in local populations. The industrial revolution had begun, and smoke emitted by coal-burning factories was killing lichens. The black moths were better camouflaged from

Directional Selection

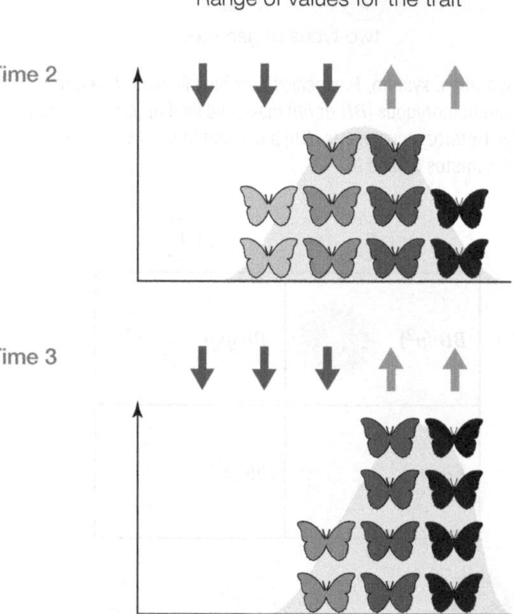

Time 1
Number of individuals in population
Range of values for the trait
Time 2
Time 3

FIGURE 17.4 Directional selection. With this pattern of natural selection, a form of a trait at one end of a range of variation is adaptive. Red arrows indicate which forms are being selected against; green, forms that are adaptive.

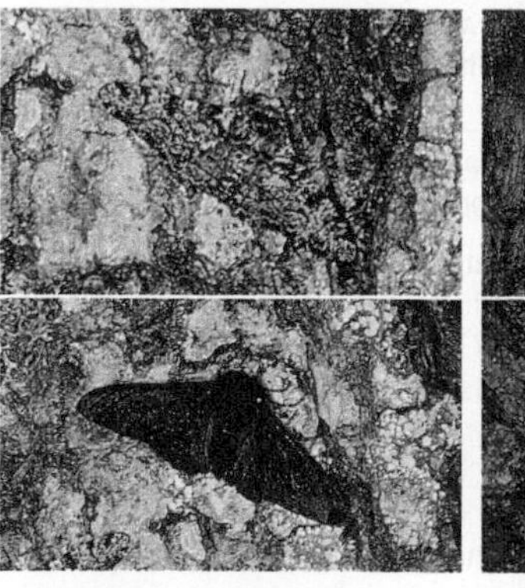

A Light-colored moths on a nonsooty tree trunk (top) are hidden from predators; black moths (bottom) stand out.

B In places where soot darkens tree trunks, the black color (bottom) provides more camouflage than the light color (top).

FIGURE 17.5 Adaptive value of color forms of the peppered moth.

directional selection Mode of natural selection in which a form of a trait at one end of a range of variation is adaptive.

Data Analysis Activities

Resistance to Rodenticides in Wild Rat Populations Beginning in 1990, rat infestations in northwestern Germany started to intensify despite continuing use of rat poisons. In 2000, Michael Kohn and his colleagues tested wild rat populations around Münster. In five towns, they trapped and tested wild rats for resistance to warfarin and the more recently developed poison bromadiolone. The results are shown in **FIGURE 17.6**.

1. In which of the five towns were most of the rats susceptible to warfarin?
2. Which town had the highest percentage of poison-resistant wild rats?
3. What percentage of rats in Olfen were warfarin resistant?
4. In which town do you think the application of bromadiolone was most intensive?

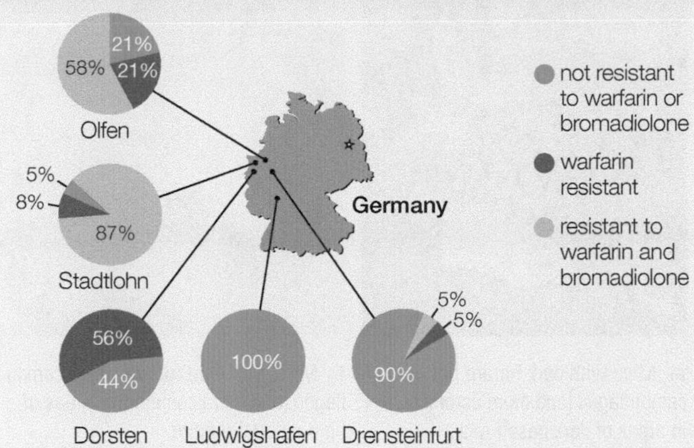

58%
21%
21%
Olfen
5%
8%
87%
Stadtlohn
56%
44%
Dorsten
100%
Ludwigshafen
5%
5%
90%
Drensteinfurt
Germany
not resistant to warfarin or bromadiolone
warfarin resistant
resistant to warfarin and bromadiolone

FIGURE 17.6 Poison resistance in wild rats in Germany, 2000.

predatory birds on lichen-free, soot-darkened trees (**FIGURE 17.5B**).

In the 1950s, H. B. Kettlewell set out to test this hypothesis. He bred both color morphs in captivity, marked them for easy identification, then released them in several areas. His team recaptured more of the black moths in the polluted areas, and more of the light moths in the less polluted areas. The researchers also observed predatory birds eating more light moths in soot-darkened forests, and more black moths in cleaner, lichen-rich forests. Black moths were clearly at a selective advantage in industrialized areas.

Pollution controls went into effect in 1952. As a result, tree trunks gradually became free of soot, and lichens made a comeback. Kettlewell observed that moth phenotypes shifted too: Wherever pollution decreased, the frequency of black moths in local populations decreased as well.

Later research confirmed Kettlewell's results. In peppered moths, having a color form that matches tree color is an adaptive trait, and directional selection for this trait drives microevolution in local populations. Moth color is determined by a single gene. Individuals with a dominant allele of the gene are black, and those homozygous for a recessive allele are light. Populations of moths living in sooty forests have a higher frequency of the dominant allele; populations in unpolluted forests have a higher frequency of the recessive allele.

Warfarin-Resistant Rats Human activities that affect the environment can result in directional selection. Consider that the average large city sustains about one rat for every ten people. Rats thrive in urban centers where garbage is plentiful and natural predators are not. Part of their success stems from an ability to reproduce very quickly: Rat populations can expand within weeks to match the amount of garbage available for them to eat.

For decades, people have been using poisons to fight rat infestations. Baits laced with warfarin, an organic compound that interferes with blood clotting, became popular in the 1950s. Warfarin inhibits the function of an enzyme called VKORC1. This enzyme regenerates vitamin K, which functions as a coenzyme in the production of blood clotting factors. When vitamin K is not regenerated, clotting factors are not properly produced, and clotting cannot occur. Rats that eat warfarin baits die within days after bleeding internally or losing blood through cuts or scrapes.

Warfarin quickly became popular because it was extremely effective, and it had less impact on harmless species than other poisons. By 1980, however, about 10 percent of rats living in urban areas were resistant to warfarin. Exposure to the poison had exerted directional selection favoring resistance, and this selection pressure had driven microevolution in rat populations. Rats resistant to warfarin have a mutated version of the *VKORC1* gene; the enzyme encoded by this allele is insensitive to warfarin. Rats with the normal *VKORC1* allele die after eating warfarin; the lucky ones with a mutated allele survive and pass it to offspring. With each onslaught of warfarin, the frequency of the mutated *VKORC1* allele increases in rat populations.

After warfarin treatment ends, the number of resistant rats declines. Why? Mutations that confer resistance also reduce the activity of the VKORC1 enzyme, so resistant rats require a lot of extra vitamin K. These individuals cannot easily obtain enough vitamin K from their diet to sustain normal blood clotting and bone formation, but being deficient in vitamin K is not

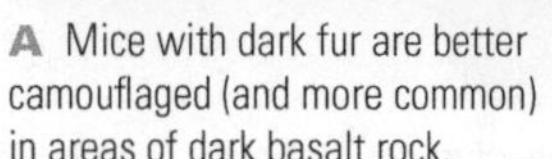

A Mice with dark fur are better camouflaged (and more common) in areas of dark basalt rock.

B Mice with light fur are better camouflaged (and more common) in areas of light-colored granite.

FIGURE 17.7 Directional selection in the rock pocket mouse (*Chaetodipus intermedius*). Predators preferentially eliminate individuals with coat colors that do not match their surroundings.

so bad when compared with being dead from rat poison. However, in the absence of warfarin, rats with a mutated *VKORC1* allele are at a serious disadvantage. Thus, when warfarin exposure ends, the frequency of the allele declines in rat populations—another example of microevolution driven by directional selection.

Rock Pocket Mice Directional selection also affects the coat color of rock pocket mice, which are small mammals that inhabit deserts in Arizona and New Mexico. These environments are dominated by light brown granite that is punctuated by widely separated patches of dark basalt. Rock pocket mice inhabit both areas. Individual mice have no preference for the basalt or the granite, but their coat color differs depending on the rock they live in. Almost all of the mice inhabiting the dark basalt have black coats (FIGURE 17.7A). Almost all of the mice inhabiting the light brown granite have light brown coats (FIGURE 17.7B). The difference arises because mice that match the rock color in each habitat are camouflaged from their natural predators. Rock pocket mice forage for seeds mainly at night, when they are visible to night-flying owls. Owls use their keen sense of vision to find prey, so they preferentially eliminate easily seen mice. In this environment, coat color that matches rock color is an adaptive trait, and predation by owls exerts directional selection on populations of mice living in each type of rock. This natural selection is driving microevolution. Remember from Section 13.5 that the products of several genes interact to determine fur color in animals. Populations of black mice have a high frequency of alleles that carry mutations affecting melanin production; populations of brown mice do not.

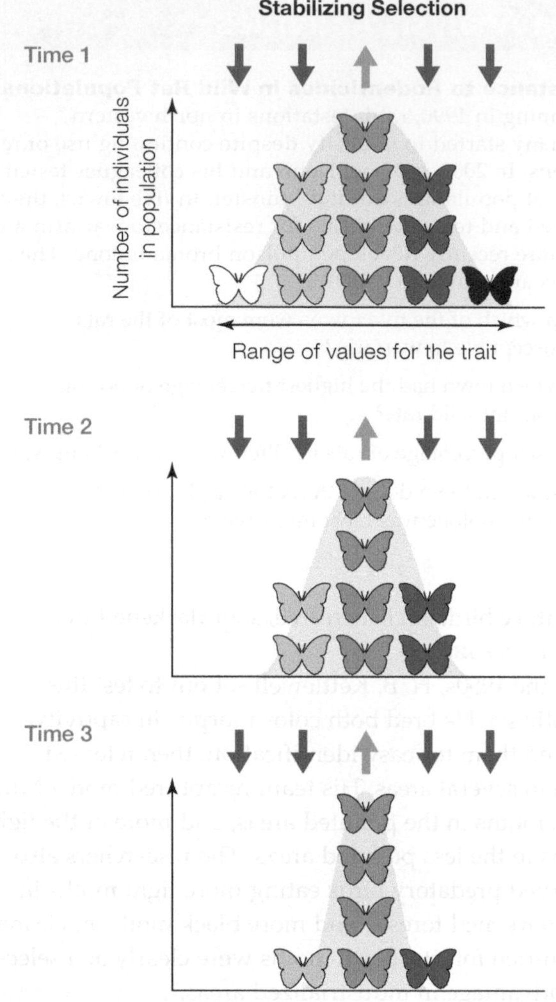
Stabilizing Selection
Time 1
Number of individuals in population
Range of values for the trait
Time 2
Time 3

FIGURE 17.8 Stabilizing selection eliminates extreme forms of a trait and maintains an intermediate form. Red arrows indicate which forms are being selected against; green, the form that is adaptive. Compare the data set from a field experiment, Figure 17.9.

Stabilizing Selection

With **stabilizing selection**, an intermediate form of a trait is adaptive, and extreme forms are selected against (FIGURE 17.8). Consider how environmental pressures maintain an intermediate body mass in populations of sociable weavers, which are birds that build large communal nests in the African savanna. Between 1993 and 2000, Rita Covas and her colleagues investigated selection pressures that operate on sociable weaver body mass. The results of this study indicated that optimal body mass in sociable weavers is a trade-off between the risks of starvation and predation. Fatter birds are less likely to starve, but they also spend more time eating, which in this species means foraging in open areas where they are easily accessible to predators. Plump birds are also more attractive to predators and not as agile when escaping. Thus, predators are agents of selection that eliminate the fattest

CREDITS: (7A–B) Courtesy of Hopi Hoekstra, University of California, San Diego; (8) © Cengage Learning.

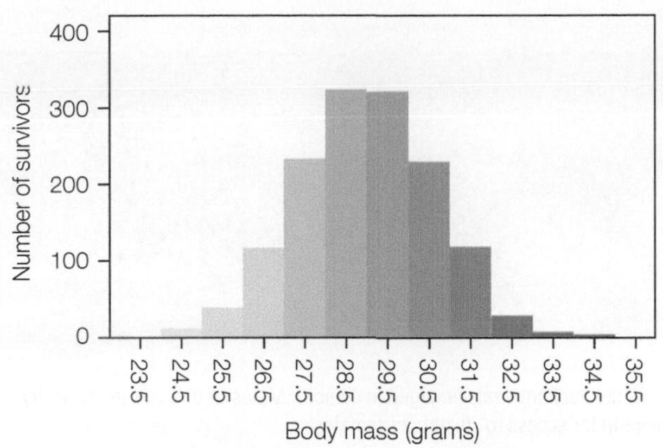
Number of survivors
400
300
200
100
0
23.5
24.5
25.5
26.5
27.5
28.5
29.5
30.5
31.5
32.5
33.5
34.5
35.5
Body mass (grams)

FIGURE 17.9 Stabilizing selection in sociable weavers. Graph shows the number of birds (out of 977) that survived a breeding season.

FIGURE IT OUT What is the optimal weight of a sociable weaver?

Answer: About 29 grams

individuals. Birds of intermediate weight have the selective advantage, and as a result they make up the bulk of sociable weaver populations (**FIGURE 17.9**).

Most populations are at least fairly well adapted to their environment, so stabilizing selection is thought to be the most common form of natural selection. However, stabilizing selection typically operates on complex traits that vary continuously, and the genetic underpinnings of these traits are the most difficult to study (Section 13.7). Thus, we have little information about allele frequencies that are affected by this form of natural selection.

Disruptive Selection

With **disruptive selection**, forms of a trait at both ends of a range of variation are favored, and intermediate forms are selected against (**FIGURE 17.10**).

Disruptive selection maintains a dimorphism in black-bellied seedcrackers. These colorful birds are native to Cameroon, Africa, and the size of their beaks (bills) has a genetic basis. The bill of a typical black-bellied seedcracker, male or female, is either 12 millimeters wide, or wider than 15 millimeters (**FIGURE 17.11**). Birds with a bill size between 12 and 15 millimeters are uncommon. It is as if every human adult were 4 feet or 6 feet tall, with no one of intermediate height.

Large-billed and small-billed African seedcrackers inhabit the same geographic range, and they breed randomly with respect to bill size. The dimorphism is maintained by environmental factors that affect feeding performance. The birds feed mainly on the seeds of two types of sedge, a grasslike plant. One sedge produces hard seeds; the other produces soft seeds. Small-billed birds are better at opening the soft seeds, but large-billed birds are better at cracking the hard ones.

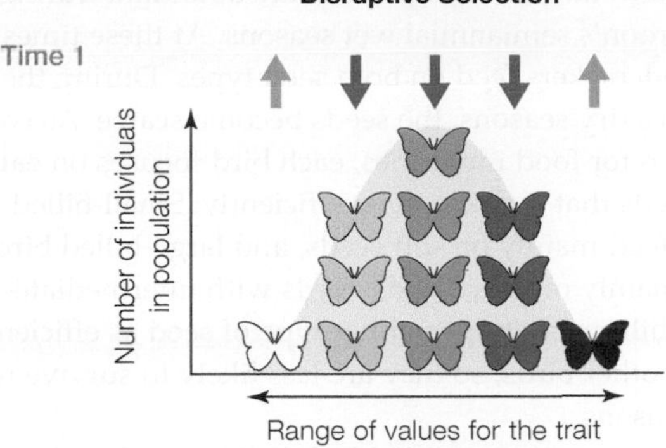
Disruptive Selection
Time 1
Number of individuals in population
Range of values for the trait

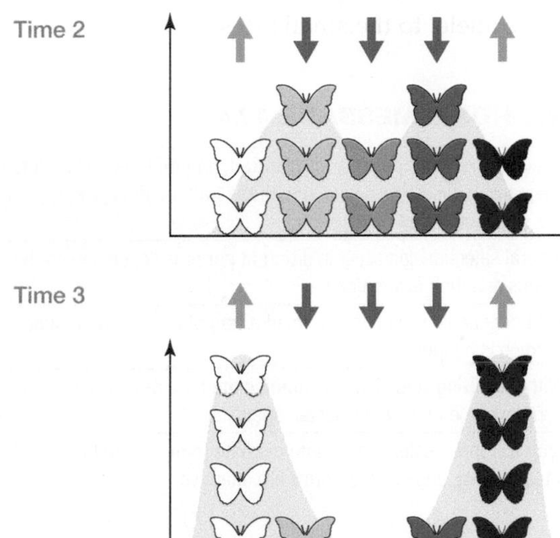
Time 2
Time 3

FIGURE 17.10 Disruptive selection eliminates midrange forms of a trait, and maintains extreme forms. Red arrows indicate which form is being selected against; green, forms that are adaptive.

lower bill 12 mm wide

lower bill 15 mm wide

FIGURE 17.11 Disruptive selection in African seedcracker populations maintains a dimorphism in bill size. Competition for scarce food during dry seasons favors birds with bills that are either 12 millimeters wide (left) or 15 to 20 millimeters wide (right). Birds with bills of intermediate size are selected against.

disruptive selection Mode of natural selection in which extreme forms of a trait are adaptive, and intermediate forms are not.

stabilizing selection Mode of natural selection in which an intermediate form of a trait is adaptive, and extreme forms are not.

Both hard and soft sedge seeds are abundant during Cameroon's semiannual wet seasons. At these times, all seedcrackers feed on both seed types. During the region's dry seasons, the seeds become scarce. As competition for food intensifies, each bird focuses on eating the seeds that it opens most efficiently: Small-billed birds feed mainly on soft seeds, and large-billed birds feed mainly on hard seeds. Birds with intermediate-sized bills cannot open either type of seed as efficiently as the other birds, so they are less likely to survive the dry seasons.

Inheritance patterns in black-bellied seedcrackers indicate that bill size arises from alleles of a single gene: A dominant allele gives rise to the large bill size; a recessive allele, to the small bill size.

TAKE-HOME MESSAGE 17.4

- ✔ By operating on different forms of a trait, natural selection affects the frequency of alleles associated with the trait. In other words, natural selection drives evolution.
- ✔ Natural selection can occur in different patterns depending on the organisms and their environment.
- ✔ With directional selection, a form of a trait at one end of a range of variation is adaptive.
- ✔ With stabilizing selection, an intermediate form of a trait is adaptive, and extreme forms are selected against.
- ✔ With disruptive selection, an intermediate form of a trait is selected against, and extreme forms are adaptive.

17.5 Natural Selection and Diversity

LEARNING OBJECTIVES

- Use examples to explain sexual selection and its outcomes.
- Describe two ways that a balanced polymorphism can be maintained.
- Explain why a harmful allele can persist at high frequency in a population.

Survival of the Sexiest

Not all evolution is driven by selection for traits that enhance survival. Competition for mates is another selective pressure that can shape form and behavior. Consider how individuals of many sexually reproducing species have a distinct male or female phenotype. A trait that differs between males and females is called a **sexual dimorphism**. Individuals of one sex (often males) are more colorful, larger, or more aggressive than individuals of the other sex. These traits can seem puzzling because they take energy and time away from activities that enhance survival, and some actually hinder an individual's ability to survive. Why, then, do they persist?

A Male elephant seals engaged in combat. Males of this species typically compete for access to clusters of females.

B A male peacock engaged in a flashy courtship display has caught the eye (and, perhaps, the sexual interest) of a female. Females are choosy; a male mates with any female that accepts him.

C Mating stalk-eyed flies. Female stalk-eyed flies prefer to mate with males that have the longest eyestalks, a trait that provides no obvious selective advantage other than sexual attractiveness.

FIGURE 17.12 Sexual selection in action.

balanced polymorphism Maintenance of two or more alleles of a gene at high frequency in a population.

frequency-dependent selection Natural selection in which a trait's adaptive value depends on its frequency in a population.

sexual dimorphism A trait that differs between males and females of a species.

sexual selection Mode of natural selection in which some individuals outreproduce others of a population because they are better at securing mates.

The answer is **sexual selection**, in which the evolutionary winners outreproduce others of a population because they are better at securing mates. In this type of natural selection, adaptive traits are those that help individuals defeat rivals for mates, or are most attractive to the opposite sex. For example, the females of some species cluster in defensible groups when they are sexually receptive, and males compete for sole access to the groups. Competition for the ready-made harems favors brawny, combative males (**FIGURE 17.12A**).

Males or females that are choosy about mates act as selective agents on their own species. The females of some species shop for a mate among males that display species-specific cues such as a highly specialized appearance or courtship behavior (**FIGURE 17.12B**). The cues often include flashy body parts or movements, traits that tend to attract predators and in some cases are a physical hindrance. However, in terms of reproductive success, the ability to command the sexual attention of females can offset a survival handicap imposed by flashiness. Selected males pass alleles for their attractive traits to the next generation of males, and females pass alleles that influence mate preference to the next generation of females. Highly exaggerated traits can be an evolutionary outcome (**FIGURE 17.12C**).

Maintaining Multiple Alleles

Natural selection may maintain two or more alleles of a gene at relatively high frequency in a population's gene pool, a state called **balanced polymorphism**. For example, sexual selection maintains multiple alleles that govern eye color in *Drosophila* fruit flies. Female flies prefer to mate with rare white-eyed males, until the white-eyed males become more common than red-eyed males, at which point the red-eyed flies are again preferred. This is an example of **frequency-dependent selection**, in which the adaptive value of a particular form of a trait depends on its frequency in a population.

Balanced polymorphism can also arise in environments that favor heterozygous individuals. Consider the gene that encodes the beta globin chain of hemoglobin (Section 9.6). *HbA* is the normal allele; the *HbS* allele has a mutation that causes sickle-cell anemia in homozygous people. Without treatment, the vast majority of these individuals die in early childhood, an outcome of body damage caused by the abnormal sickle shape of their red red blood cells.

Despite being so harmful, the *HbS* allele persists at very high frequency among the human populations in tropical and subtropical regions of Asia, Africa, and the Middle East (**FIGURE 17.13A**). Why? Populations with the highest frequency of the *HbS* allele also have the highest incidence of malaria (**FIGURE 17.13B**).

Gabon

A Distribution of people who carry the sickle-cell allele in Gabon.

Yellow indicates regions where more than 10 percent of the population are carriers of the *HbS* allele. In the white regions, less than 10 percent of the population are carriers.

Gabon

B Distribution of malaria cases in Gabon.

Yellow indicates regions where more than 30 percent of the population have malaria. In the white regions, less than 30 percent of the population have the disease.

FIGURE 17.13 Frequency of the *HbS* allele and incidence of malaria in Gabon, Africa in 2014.

Mosquitoes transmit *Plasmodium*, the parasitic protist that causes malaria, to human hosts. *Plasmodium* multiplies in the liver and then in red blood cells, which rupture and release new parasites during recurring bouts of severe illness. The *HbA* and *HbS* alleles are codominant, so heterozygous people make both normal and sickle hemoglobin. Red blood cells of these individuals can sickle under some circumstances, but not enough to cause severe symptoms. One of the circumstances under which sickling occurs is infection with *Plasmodium*. The abnormal shape brings the cells to the attention of the immune system, which destroys them along with the parasites they harbor. The action of the immune system can prevent the infection from spreading to other red blood cells. Heterozygous individuals are more likely to survive malaria than individuals homozygous for the normal *HbA* allele. *Plasmodium*-infected red blood cells of people who make only normal hemoglobin do not sickle, so the parasite may remain hidden from the immune system.

In areas where malaria is common, the persistence of the *HbS* allele is a matter of relative evils. Malaria and sickle-cell anemia are both potentially deadly. Heterozygous individuals may not be completely healthy, but they do have a better chance of surviving malaria than people homozygous for the normal allele (*HbA/HbA*). With or without malaria, people who have both alleles (*HbA/HbS*) are more likely to live long enough to reproduce than individuals homozygous for the sickle allele (*HbS/HbS*). The result is that nearly one-third of people living in the most malaria-ridden regions of the world carry the *HbS* allele.

TAKE-HOME MESSAGE 17.5

✔ Sexual selection is a form of natural selection in which adaptive traits give an individual an advantage in securing mates. It can result in exaggerated form or behavior.

✔ Any mode of natural selection may maintain multiple alleles in a population at relatively high frequency (a balanced polymorphism).

✔ Balanced polymorphism can be an outcome of frequency-dependent selection, or of environmental pressures that favor heterozygous individuals.

17.6 Nonselective Evolution

LEARNING OBJECTIVES

- With suitable examples, explain how mechanisms that do not involve adaptive traits can change allele frequency.
- Explain why smaller populations are more vulnerable to the loss of genetic diversity.
- Describe the way gene flow stabilizes allele frequency.

Natural selection is a major driver of evolution, but it is not the only one. Evolution can be influenced by external mechanisms that do not involve adaptive traits, as the following examples illustrate.

Factors That Reduce Genetic Diversity

Genetic Drift Members of a natural population survive and reproduce with differing success, and the differences are not always an outcome of natural selection. By chance, a perfectly fit and healthy individual may not pass its alleles to offspring, for example by dying in a random event before the opportunity to reproduce arises. Such events can change a population's allele frequencies. Change in allele frequency brought about by chance alone is called **genetic drift**.

The effects of genetic drift are more pronounced in smaller populations (**FIGURE 17.14**). To understand why, imagine a hypothetical gene with two alleles, neither of which confers a selective advantage. These alleles (let's call them *A* and *a*) occur at a frequency of 95 percent and 5 percent, respectively. In a population with 10 members, one individual would be heterozygous (*Aa*) and the remaining nine would be homozygous (*AA*). A random event that eliminates the heterozygous individual from the population before it reproduces also eliminates allele *a* from the population's gene pool. If this occurs, the other allele (*A*) becomes **fixed**—all individuals of the population are homozygous for it. The frequency of a fixed allele will not change unless a new mutation occurs, or an individual bearing another allele enters the population.

Now imagine that larger population has the same alleles (*A* and *a*) at the same frequency (95 percent). This hypothetical population consists of 100 members, and ten individuals are heterozygous for allele *a*. In order for the allele to be lost from the population's gene pool, all ten individuals would have to be eliminated before reproducing. The chance of this outcome is much smaller in the larger population: In general, larger populations are less vulnerable to the loss of genetic diversity.

Bottlenecks Even a large population can lose genetic diversity in the case of a **bottleneck**, which is a drastic reduction in population size. Consider northern elephant seals (shown in Figure 17.12A). In the late 1890s, overhunting left only about ten individuals of this species alive. Since then, hunting restrictions have allowed the population to recover, but genetic diversity among its members has been greatly reduced. The bottleneck and subsequent genetic drift eliminated many alleles at all loci tested.

The Founder Effect A loss of genetic diversity can also occur when a small group of individuals establishes a new population. If the founding group is not representative of the original population in terms of allele frequencies, then the new population will not be representative of it either. This outcome is called the **founder effect** (**FIGURE 17.15**). Consider that all three *ABO* alleles for blood type are common in most human populations. Native Americans are an exception, with the majority of individuals being homozygous for the *O* allele. Native Americans are descendants of early humans who migrated from Asia between 14,000 and 21,000 years ago, across a narrow land bridge that

bottleneck Reduction in population size so severe that it reduces genetic diversity.
fixed Refers to an allele for which all members of a population are homozygous.
founder effect After a small group of individuals found a new population, allele frequencies in the new population differ from those in the original population.
gene flow The movement of alleles into and out of a population.
genetic drift Change in allele frequency due to chance alone.
inbreeding Mating among close relatives.

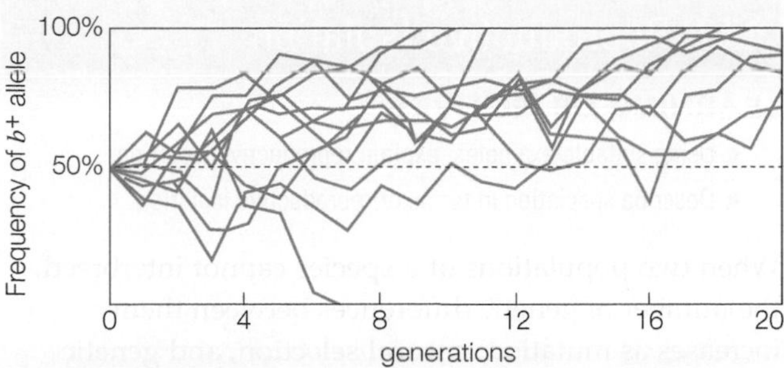

100%
50%
Frequency of b^+ allele
0
4
8
12
16
20
generations

A In these experiments, population size was maintained at 10 beetles. Several populations were tested; notice that allele b^+ was lost in one population (one graph line ends at 0).

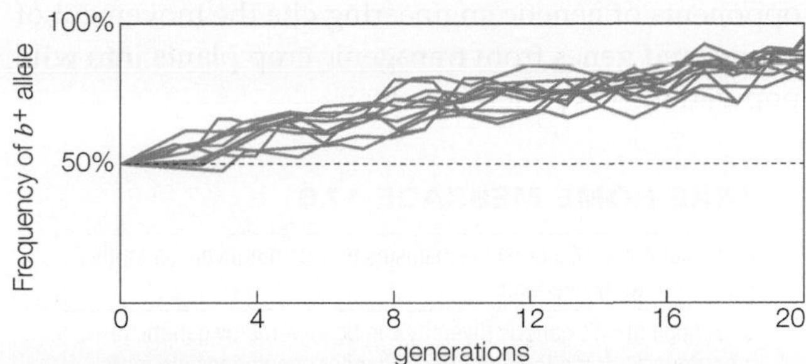

100%
50%
Frequency of b^+ allele
0
4
8
12
16
20
generations

B In these experiments, population size was maintained at 100 beetles. Genetic drift was less pronounced in these populations than in the 10-beetle populations in **A**.

FIGURE 17.14 Genetic drift experiment in the flour beetle (*Tribolium castaneum*), shown left on a flake of cereal. In two sets of experiments, beetles heterozygous for alleles b^+ and b were maintained for 20 generations in populations of 10 individuals (**A**) or 100 individuals (**B**). Graph lines in **B** are smoother than in **A**, which means that less genetic drift occurred in the larger populations. Notice that the average frequency of allele b^+ rose at the same rate in both groups, an indication that natural selection was at work too: Allele b^+ was weakly favored.

FIGURE IT OUT In how many populations did allele b^+ become fixed during these experiments?

Answer: Six

FIGURE 17.15 The founder effect. A group that founds a new population is not genetically representative of the original population, so allele frequencies differ between the two populations.

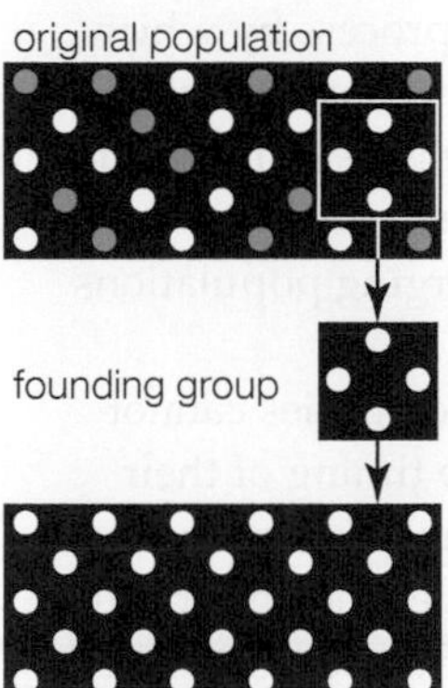

original population
founding group
new population

once connected Siberia and Alaska. Analysis of DNA from ancient skeletal remains reveals that most early Americans were also homozygous for the *O* allele. Modern Siberian populations have all three alleles. Thus, the first humans in the Americas were probably members of a small group that had reduced genetic diversity compared with the general population.

Inbreeding Small founding populations and those that have undergone a bottleneck are necessarily inbred. **Inbreeding** is mating between close relatives, and it can have negative effects on a population with low genetic diversity. Closely related individuals tend to share more harmful recessive alleles than nonrelatives do, and inbreeding keeps these alleles circulating in a gene pool. Thus, inbred populations often have an unusually high incidence of genetic disorders. This outcome is minimized in human populations that discourage inbreeding and forbid incest (mating between parents and children or between siblings).

The Old Order Amish in Lancaster County, Pennsylvania, offer an example of the effects of inbreeding. Amish people marry only within their community. Intermarriage with other groups is not permitted, and no "outsiders" are allowed to join the community. As a result, Amish populations are moderately inbred, and many of their members are homozygous for harmful recessive alleles. The Lancaster community has an unusually high frequency of a recessive allele that causes Ellis–van Creveld syndrome, a genetic disorder characterized by dwarfism, heart defects, and polydactyly (extra fingers or toes), among other symptoms. This allele has been traced to a man and his wife, two of a group of 400 Amish people who immigrated to the United States in the mid-1700s. As a result of the founder effect and inbreeding since then, about 1 of 8 people in the Lancaster community is now heterozygous for the allele, and 1 in 200 is homozygous.

Gene Flow

Individuals tend to mate or breed most frequently with other members of their own population. However, not all populations of a species are completely isolated from one another, and nearby populations may occasionally interbreed. Also, individuals sometimes leave one population and join another. **Gene flow**, the movement of alleles between populations, occurs in both cases, and it can stabilize allele frequencies.

Gene flow is common among populations of animals, but it also occurs in less mobile organisms. Consider the acorns that jays disperse when they gather nuts for the winter. Every fall, these birds visit acorn-bearing oak trees repeatedly, then store the acorns in the soil of territories as much as a mile away. The jays transfer acorns (and the alleles carried by these seeds) among populations of oak trees that may otherwise be genetically isolated. Gene flow also occurs when wind or an animal transfers pollen from one plant to another, often over great distances. Many opponents of genetic engineering cite the movement of engineered genes from transgenic crop plants into wild populations via pollen transfer.

TAKE-HOME MESSAGE 17.6

✔ Evolution can occur by mechanisms that do not involve adaptive traits or natural selection.

✔ A population's genetic diversity can be lowered by genetic drift, a bottleneck, or the founder effect. Smaller populations are more vulnerable to this outcome.

✔ Genetic drift is change in allele frequency due to chance alone.

✔ A bottleneck drastically reduces population size.

✔ A new population founded by a small group may not be genetically representative of the original population (the founder effect).

✔ Inbreeding keeps harmful alleles circulating in a population that has reduced genetic diversity.

✔ Gene flow between populations can change or stabilize allele frequencies.

FIGURE 17.16 How reproductive isolation prevents interbreeding between species.

Different species form and . . .
Reproduction occurs at different times in the two species (temporal isolation).
The species inhabit different environments, so their members never meet up for sex (ecological isolation).
Cues required for sex differ between the species (behavioral isolation).
Physical incompatibilities prevent sex between individuals of the two species (mechanical isolation).
Mating occurs and . . .
Fertilization does not occur (gamete incompatibility).
Zygotes form and . . .
Hybrid individuals or their offspring have reduced fitness (hybrid inviability).
Hybrid individuals cannot produce offspring (hybrid sterility).
Interbreeding is successful

17.7 Reproductive Isolation

LEARNING OBJECTIVES

- Using suitable examples, explain reproductive isolation.
- Describe speciation in terms of reproductive isolation.

When two populations of a species cannot interbreed, the number of genetic differences between them increases as mutation, natural selection, and genetic drift occur independently in each. Over time, the populations may become so different that we consider them to be different species. The emergence of a new species—the splitting of one lineage into two—is called **speciation**. Evolution is a dynamic, extravagant, messy, and ongoing process that can be challenging for people who like clear categories. Speciation offers a perfect example, because it rarely occurs at a precise moment: Individuals often continue to interbreed even as populations are diverging, and populations that have already diverged may come together and interbreed again.

Every time speciation happens, it happens in a unique way, which means that each species is a product of its own unique evolutionary history. However, there are recurring patterns. For example, **reproductive isolation**, the end of gene flow between populations, is always part of the process by which sexually reproducing species achieve and maintain separate identities. Several mechanisms of reproductive isolation prevent successful interbreeding, and thus reinforce differences between diverging populations (**FIGURE 17.16**).

Some closely related species cannot interbreed because the timing of their reproduction differs, an effect called temporal isolation. Consider the periodical cicada (left). Larvae of these insects feed on roots as they mature underground, and adults emerge to reproduce. Three cicada species reproduce every 17 years. Each has a related sibling species with nearly identical form and behavior that emerges on a 13-year cycle instead of a 17-year cycle. Sibling species have the potential to interbreed, but they can get together only once every 221 years!

Adaptation to different environmental conditions may prevent closely related species from interbreeding, a mechanism called ecological isolation. For example, two species of manzanita, a plant native to the Sierra Nevada mountain range, rarely hybridize. One species that lives on dry, rocky hillsides is better adapted for conserving water. The other species, which requires

reproductive isolation The end of gene flow between populations.
speciation (spee-see-A-shun) Emergence of a new species.

FIGURE 17.17 Behavioral isolation. A male peacock spider (*Maratus volans*) approaches a female, signaling his intent to mate with her by raising and waving a colorful flap, and gesturing his legs in time with abdominal vibrations. If his courtship display fails to impress her, she may eat him.

more water, lives on lower slopes where water stress is not as intense. The physical separation makes interbreeding unlikely.

Behavioral isolation occurs when differences in behavior prevent mating between related animal species. For example, males and females of many species engage in courtship displays before sex (**FIGURE 17.17**). In a typical pattern, the female recognizes the sounds and movements of a male of her species as an overture to sex; females of different species do not.

Mechanical isolation occurs when the size or shape of an individual's reproductive parts prevent it from mating with members of related species. For example, closely related species of sage plants grow in the same areas, but their flowers are specialized for different pollinators so cross-pollination rarely occurs and hybrids rarely form. Small-flowered black sage is pollinated most effectively by small bees and other insects; large-flowered white sage, by larger insects (**FIGURE 17.18**).

Even if gametes of different species do meet up, they often have molecular incompatibilities that prevent a zygote from forming. For example, the molecular signals that trigger pollen germination in flowering

FIGURE 17.18 Mechanical isolation in sage.

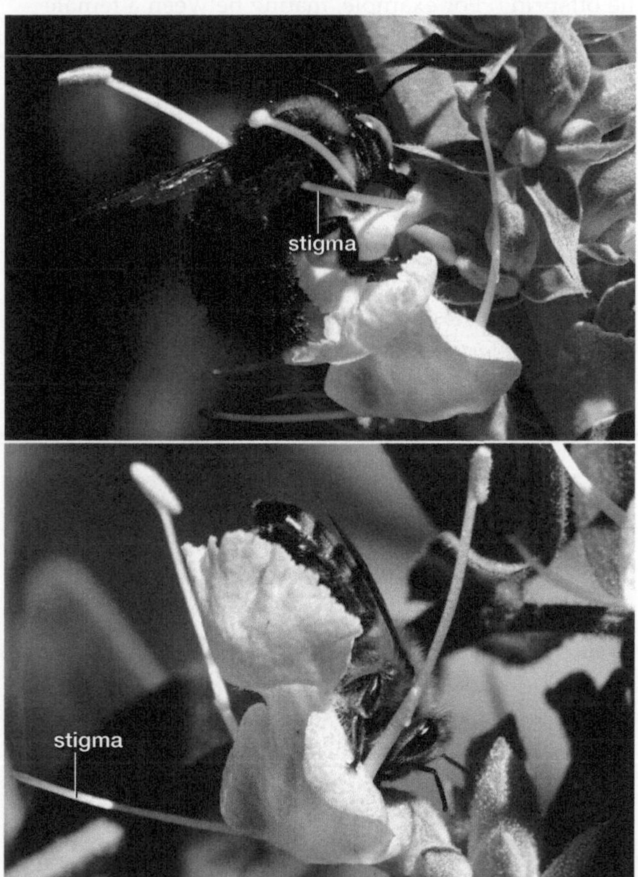
stigma
stigma

A Black sage is pollinated mainly by small bees and other insects that touch the reproductive parts of the flowers as they sip nectar (top). Larger insects cannot perch on small sage flowers; they access nectar by piercing the petals (bottom), so they do not touch the flower's reproductive parts (stigma and pollen-bearing stamens).

B Flowers of white sage are pollinated mainly by insects heavy enough to activate a tripping mechanism. When a big insect lands on the flower to sip nectar, it pushes down the large top petal (top). This causes stamens and stigma to move inward and touch the pollinator's body. Insects that do not weigh enough to trigger this mechanism do not touch the flower's reproductive parts (bottom).

plants are species-specific; thus, pollen grains typically will not germinate if they land on a flower of a different species (Section 29.4). Gamete incompatibility may be the primary speciation route among animals that release eggs and free-swimming sperm into water.

Genetic changes are the basis of divergences in form, function, and behavior. Even chromosomes of species that diverged relatively recently may be different enough that a hybrid zygote ends up with extra or missing genes, or genes with incompatible products. Such outcomes typically disrupt embryonic development. Hybrid individuals that do survive embryonic development often have reduced fitness. For example, hybrid offspring of lions and tigers have more health problems and a shorter life expectancy than individuals of either parent species. If hybrids live long enough to reproduce, their offspring often have lower and lower fitness with each successive generation. Incompatible nuclear and mitochondrial DNA may be the cause (mitochondrial DNA is inherited from the mother only).

Some interspecies crosses produce robust but sterile offspring. For example, mating between a female horse (64 chromosomes) and a male donkey (62 chromosomes) produces a mule (63 chromosomes: 32 from the horse, and 31 from the donkey). Mules are healthy but their chromosomes cannot pair up properly during meiosis, so this animal makes few viable gametes.

TAKE-HOME MESSAGE 17.7

✔ Speciation is an evolutionary process in which new species form.

✔ The details of speciation differ every time it occurs, but reproductive isolation mechanisms are always part of the process.

17.8 Models of Speciation

LEARNING OBJECTIVE

- Describe some models of speciation and give examples of each.

In most cases, speciation is associated with some type of geographic separation, for example after a physical barrier arises and separates populations. It can also occur with no apparent barrier to gene flow. These categories are useful as models, but the relationship between pattern and process is rarely simple. As you learn about speciation, remember that it typically occurs in a continuum of geographic, genetic, and ecological contexts.

Allopatric Speciation

A physical barrier that arises between two populations can interrupt gene flow between them. Whether the barrier slows gene flow or prevents it entirely depends on the barrier and the species—how it travels (such as by swimming, walking, or flying), and how it reproduces (for example, by internal fertilization or by pollen dispersal). If gene flow becomes sufficiently hampered, genetic changes accumulate independently in the two populations. The divergences cause the populations to become separate, reproductively isolated lineages. This pattern, in which speciation occurs after a physical barrier interrupts gene flow between populations, is called **allopatric speciation** (*allo–* means different; *patria* means fatherland).

A barrier that fosters allopatric speciation can arise in an instant, or over an eon. The Great Wall of China is an example of a barrier that arose abruptly. When it was built, the wall interrupted gene flow among nearby populations of insect-pollinated plants. Today, genetic divergences are occurring between populations of trees, shrubs, and herbs on either side of the wall.

Other barriers to gene flow arose over much longer periods of time. For example, it took millions of years of tectonic plate movements (Section 16.5) to bring the two continents of North and South America close enough to collide. The land bridge where the two continents now connect is called the Isthmus of Panama. When this isthmus formed about 3 million years ago, it cut off the flow of water—and gene flow among populations of aquatic organisms—as it separated the Pacific Ocean from the Atlantic. Today, populations of many species on opposite sides of the isthmus are closely related but reproductively isolated. For example, Atlantic Ocean populations of snapping shrimp are so similar to Pacific populations that they might interbreed, but when brought together they snap their claws aggressively at one another instead of mating.

Allopatric speciation is common on isolated archipelagos. Archipelagos are island chains that arise from hot spots on the ocean floor (Section 16.5). Because the islands start out as the tops of volcanoes, we can assume that their fiery surfaces are initially barren and inhospitable to life. Sooner or later, individuals migrate to the islands from a mainland or from other islands.

Islands in close proximity to a mainland are quickly colonized. Ongoing gene flow keeps island and mainland populations similar, so speciation is rare in these cases. Islands that are very far from a mainland are populated in a different pattern. Few species have the

allopatric speciation Speciation pattern in which a physical barrier that interrupts gene flow between populations fosters genetic divergences.
sympatric speciation Speciation pattern in which genetic changes within a population lead to reproductive isolation.

'Akepa
(Loxops coccineus)
'Akeke'e
(Loxops caeruleirostris)
'Akikiki
(Oreomystis bairdi)
Palila
(Loxioides bailleui)
'I'iwi
(Drepanis coccinea)
'Akohekohe
(Palmeria dolei)
'Apapane
(Himatione sanguinea)
'Akiapola'au
(Hemignathus wilsoni)
Kiwikiu
(Pseudonestor xanthophrys)
Maui 'Alauahio
(Paroreomyza montana)
Kaua'i 'Amakihi
(Chlorodrepanis stejnegeri)
Hawai'i 'Amakihi
(Chlorodrepanis virens)

Rosefinch
(Carpodacus)

FIGURE 17.19 A few Hawaiian honeycreepers (above). All are descended from a few Eurasian rosefinches (left) that migrated 2,000 miles across the open ocean to the Hawaiian archipelago (right).

Unique traits adapt each honeycreeper species to a particular habitat. For example, a special bill shape allows exploitation of a specific food source: a particular type of insect, seed, fruit, nectar in a floral cup, and so on.

Hawaiian
archipelago

ability to cross large expanses of ocean, so the arrival of colonizing individuals from a distant mainland can be very sporadic.

Consider the Hawaiian archipelago, which includes 19 islands separated from other landmasses by more than 2,000 miles of open ocean. As the islands formed, a few individuals of mainland species were delivered to them by winds or currents. Descendants of these individuals reproduced, and the lack of gene flow with mainland populations allowed the island populations to diverge. Today, thousands of species are unique to this island chain. Among them are Hawaiian honeycreepers, descendants of Eurasian rosefinches that arrived on the islands millions of years ago (**FIGURE 17.19**). Isolated from gene flow with mainland bird populations, the island colonizers diverged. Habitats on the Hawaiian islands vary dramatically: from lava beds, rain forests, and grasslands to dry woodlands and snow-capped peaks. Selection pressures differ within and between these habitats, and, over many generations, populations of birds living in the different habitats became hundreds of separate species. The cumulative result of these divergences is a spectacular array of honeycreepers. Unique forms and behaviors allow each species to exploit special opportunities presented by its particular island habitat.

Sympatric Speciation

Genetic changes that lead to reproductive isolation can occur within a single population, a pattern called **sympatric speciation** (*sym–* means together). Sympatric speciation occurs in the absence of a physical barrier. The process may occur in a single generation, for example when the chromosome number increases.

This is a common way for new plant species to arise: Individuals of different taxa sometimes hybridize, and the offspring of the union inherit all of the chromosomes of both parents. Our bread wheat, *Triticum aestivum*, arose this way. *T. aestivum* is hexaploid (6*n*), having inherited its three distinct sets of chromosomes by way of multiple hybridizations between grasses of different genera (**FIGURE 17.20**).

ancestral species
Triticum urartu
AA
Aegilops speltoides
BB
1
2
Triticum turgidum
AABB
Aegilops tauschii
DD
3
Triticum aestivum
AABBDD

FIGURE 17.20 Sympatric speciation in wheat.

The wheat genome occurs in several slightly different forms: A, B, D, and so on.

① About 5.5 million years ago, a diploid (2*n*) einkorn grass, *Triticum urartu* (AA), hybridized with another diploid grass, *Aegilops speltoides* (BB). The union produced another diploid grass, *A. tauschii* (DD).

② A few million years later, the same two parental species hybridized again, this time producing the tetraploid (4*n*) emmer wheat *T. turgidum* (AABB).

③ Our hexaploid (6*n*) bread wheat, *T. aestivum* (AABBDD), is the offspring of a hybridization between *T. turgidum* (AABB) and *A. taushii* (DD). The match is shown for scale.

New polyploid plant species also arise in a single generation as a result of nondisjunction during mitosis. If the nucleus of a somatic cell fails to divide during mitosis, for example, the resulting cell—which is polyploid—may proliferate and give rise to shoots and flowers. If the flowers can self-fertilize, a new polyploid species may be the result.

Sympatric speciation can occur with no change in chromosome number. The mechanically isolated sage plants you learned about in the last section speciated like this. As another example, more than 500 species of cichlid fishes arose by sympatric speciation in the shallow waters of Lake Victoria. This large freshwater lake sits isolated from river inflow on an elevated plain in Africa's Great Rift Valley. Lake Victoria has dried up completely several times during its 400,000-year history. DNA sequence comparisons indicate that almost all of the cichlid species in this lake arose since it was last dry, about 12,400 years ago.

How could hundreds of cichlid species arise so quickly in the same body of water? The answer involves sexual selection. Consider how the color of ambient light differs in different parts of a lake. The light in Lake Victoria's shallower, clear water is mainly blue; light that penetrates the deeper, muddier water is redder. Cichlid species also vary in color (**FIGURE 17.21**). Female cichlids prefer to mate with brightly colored males of their own species. This preference has a genetic basis, in alleles of genes for light-sensitive pigments of the retina (part of the eye). Retinal pigments made by species that live mainly in shallow areas of the lake are more sensitive to blue light. The males of these species are also the bluest.

Pundamilia nyererei, a Lake Victoria cichlid that breeds in deep water

Pundamilia pundamilia, a Lake Victoria cichlid that breeds in shallow water

FIGURE 17.21 Red fish, blue fish: sympatric speciation of cichlids native to Lake Victoria, Africa.

Mutations that affect female cichlids' perception of the color of ambient light in deeper (red) or shallower (blue) regions of the lake also affect their choice of mates. Female cichlids prefer to mate with males that they perceive to be most brightly colored. The outcome of this sexual selection: Species that inhabit different regions of the lake match the color of ambient light.

macroevolution Evolutionary change in taxa above the species level.
parapatric speciation Two populations become different species while in contact along a common border.

Retinal pigments made by species that prefer deeper areas of the lake are more sensitive to red light. Males of these species are redder. In other words, the color that a female cichlid sees best is also the color best displayed by males of her species. Thus, mutations that affect color perception are likely to affect a female's choice of mates. Mutations like these are probably the way sympatric speciation begins in these fishes.

Sympatric speciation has also occurred in greenish warblers of central Asia. A chain of populations of this bird encircles the Tibetan plateau (**FIGURE 17.22**). Adjacent populations of greenish warblers interbreed easily, except for the two populations at the ends of the chain. These two populations overlap in northern Siberia, but their individuals do not interbreed because they do not recognize one another's songs (an example of behavioral isolation). Small genetic differences between adjacent populations have added up to major differences between the two populations at the ends of the chain. Greenish warbler populations that make up the chain are collectively called a ring species. Ring species present one of those paradoxes for people who like neat categories: Gene flow occurs continuously all around the chain, but the two populations at the ends of the chain are clearly different species. Where should we draw the line that divides those two species?

Parapatric Speciation

Parapatric speciation is an uncommon pattern in which populations in contact across a shared border become different species. Divergences spurred by local selection pressures are reinforced because hybrids that form in the contact zone are less fit than individuals on either side of it. Consider velvet walking worms, which resemble caterpillars but are only distantly related to insects. The worms are predatory, shooting streams of gluey slime from their head to entangle insect prey (**FIGURE 17.23**). Two rare species are native to the island of Tasmania: the giant velvet walking worm and the blind velvet walking worm. These species can interbreed, but they do so only in a tiny area where their habitats overlap. Hybrid offspring are sterile, which may be the main reason the two species are maintaining separate identities in the absence of a physical barrier between their adjacent populations.

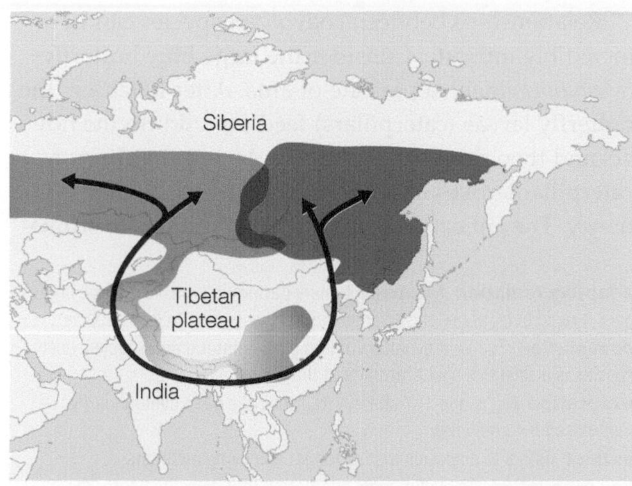
Siberia
Tibetan plateau
India

FIGURE 17.22 A ring species. The range of the greenish warbler *Phylloscopus trochiloides* extends in a ring around the Tibetan plateau. Adjacent populations interbreed—except for the two populations that join the ring in Siberia (blue and red). These two populations are behaviorally isolated.

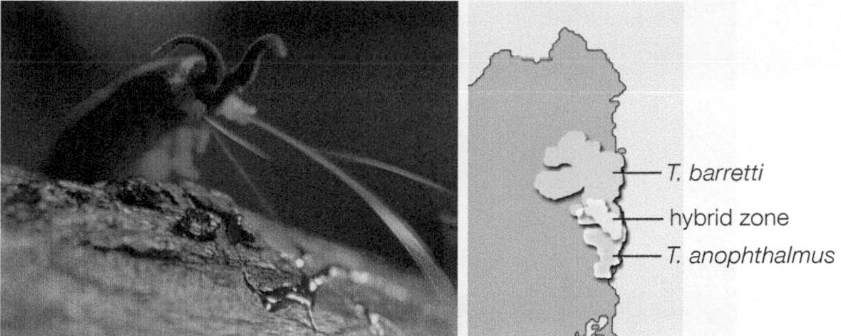
T. barretti
hybrid zone
T. anophthalmus

FIGURE 17.23 Parapatric speciation in the Tasmanian velvet walking worm. Populations of *Tasmanipatus barretti* and *T. anophthalmus* remain in contact across a common border. Sterile hybrids form where the two populations meet.

TAKE-HOME MESSAGE 17.8

- ✔ Speciation occurs within a range of geographic, ecological, and genetic contexts. We can model the process in patterns.
- ✔ A physical barrier can interrupt gene flow between two populations. The populations then diverge genetically and become different species. This pattern is called allopatric speciation.
- ✔ Genetic divergence within a population can lead to reproductive isolation, a pattern called sympatric speciation.
- ✔ With parapatric speciation, populations speciate while in contact along a common border.

17.9 Macroevolution

LEARNING OBJECTIVES

- Distinguish between microevolution and macroevolution.
- Use examples to describe some patterns of macroevolution.

Microevolution is change in allele frequency, and it occurs within a population. **Macroevolution** is evolutionary change in higher taxa. Patterns of macroevolution include large-scale trends such as land plants evolving from green algae, the disappearance of the dinosaurs, and so on.

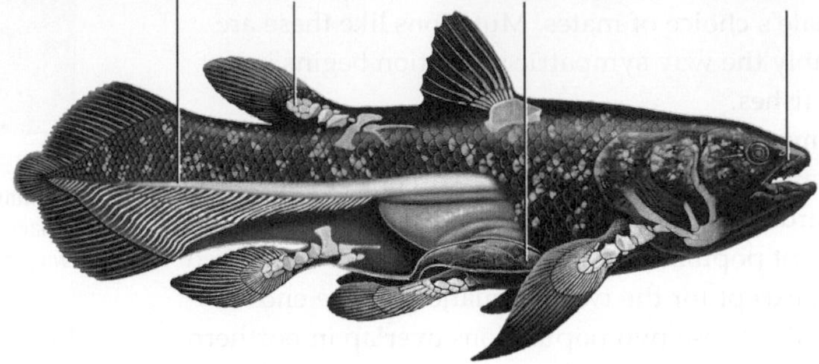
Notochord
This tough, elastic tube, which is partially hollow and filled with fluid, is ancestral to the spinal cord.
Lobed fins
These fleshy fins retain a few of the ancestral bones that gave rise to legs and arms in other lineages.
Long gestation
Coelacanths give birth to litters of up to 26 fully developed "pups" after gestation of more than a year.
Rostral organ
A sensory organ that responds to electrical impulses in water, it probably helps the fish locate prey in dark ocean depths.

FIGURE 17.24 Stasis in the coelacanth. Until a fisherman caught one in 1938, coelacanths were thought to have become extinct 70 million years ago. Left, compare a 320-million-year-old fossil found in Montana with a live fish. Right, a few of the coelacanth's unusual ancestral features that have been lost in almost all other fish lineages over evolutionary time.

In a macroevolutionary pattern called **stasis**, very little change occurs in a lineage over a very long period of time. Consider coelacanths, an ancient order of lobe-finned fish. In its unique form and other traits, modern coelacanth species are similar to fossil specimens hundreds of millions of years old (**FIGURE 17.24**).

Major evolutionary novelties often stem from the adaptation of an existing structure for a completely new purpose. A trait that has been evolutionarily repurposed is called an **exaptation**. The feathers that allow modern birds to fly, for example, are an exaptation for flight. These structures are derived from dinosaur feathers, which could not have sustained flight when they first evolved. The earliest feathers probably served as insulation and may have been useful for courtship displays or camouflage.

By current estimates, more than 99 percent of all species that ever lived are now **extinct**, which means they no longer have living members. In addition to continuing extinctions, the fossil record indicates that there have been more than twenty mass extinctions, which are simultaneous losses of many lineages. These include five catastrophic events in which the majority of species on Earth disappeared (Section 16.5).

With **adaptive radiation**, one lineage rapidly diversifies into many. Adaptive radiation can occur after a population colonizes a new environment that has a variety of habitats and few competitors. Speciation occurs as adaptations to the different habitats evolve. The Hawaiian honeycreepers arose this way.

A geologic or climatic event that eliminates some species from a habitat can spur adaptive radiation; species that survive the event then have access to resources from which they had previously been excluded. This is the way mammals were able to undergo an adaptive radiation after the dinosaurs disappeared 66 million years ago.

Adaptive radiation may also occur after a key innovation evolves. A **key innovation** is an adaptive trait that allows its bearer to exploit a habitat more efficiently or in a novel way. The evolution of lungs offers an example, because lungs were a key innovation that opened the way for an adaptive radiation of vertebrates on land.

Two species that have close ecological interactions may evolve jointly, a pattern called **coevolution**. One species acts as an agent of selection on the other, and each adapts to changes in the other. Over evolutionary time, the two species may become so interdependent that they can no longer survive without one another.

Relationships between coevolved species can be incredibly intricate. Consider the large blue butterfly (*Maculinea arion*), a parasite of ants. After hatching, the butterfly larvae (caterpillars) feed on wild thyme flowers and then drop to the ground. An ant that finds a caterpillar strokes it, which makes the caterpillar exude honey. The ant eats the honey and continues to stroke

adaptive radiation Macroevolutionary pattern in which a lineage undergoes a rapid burst of genetic divergences that gives rise to many species.
coevolution The joint evolution of two closely interacting species; each species is a selective agent for traits of the other.
exaptation (eggs-app-TAY-shun) A trait that has been evolutionarily repurposed for a new use.
extinct Refers to a species that no longer has living members.
key innovation An evolutionary adaptation that gives its bearer the opportunity to exploit a particular environment much more efficiently or in a new way.
stasis (STAY-sis) Macroevolutionary pattern in which a lineage persists with little or no change over evolutionary time.

the caterpillar, which exudes more honey. This interaction continues for hours, until the caterpillar suddenly hunches itself up into a shape that appears (to an ant) very much like an ant larva (**FIGURE 17.25**). The deceived ant then carries the caterpillar back to the ant nest, where, in most cases, other ants kill it—unless the ants are of the species *Myrmica sabuleti*. The caterpillar secretes the same chemicals as *M. sabuleti* larvae, and makes the same sounds as their queen—behaviors that trick the ants into adopting the caterpillar and treating it better than their own larvae. The adopted caterpillar feeds on ant larvae for about 10 months, then undergoes metamorphosis, changing into a butterfly that emerges from the ground to mate. Eggs are deposited on wild thyme near another *M. sabuleti* nest, and the cycle starts anew. This relationship between ant and butterfly is typical of coevolved relationships in that it is extremely specific. Any increase in the ants' ability to identify a caterpillar in their nest selects for caterpillars that better deceive the ants, which in turn select for ants that can better identify the caterpillars. Each species exerts directional selection on the other.

A This *Maculinea arion* caterpillar is interacting with a *Myrmica sabuleti* ant. The beguiled ant is preparing to carry the honey-exuding, hunched-up caterpillar back to its nest, where the caterpillar will feed on ant larvae for the next 10 months until it becomes a pupa.

B *Maculinea arion* butterflies emerge from pupae to feed, mate, and lay eggs on wild thyme flowers. Larvae that emerge from the eggs will survive only if a colony of *M. sabuleti* ants adopts them.

FIGURE 17.25 Coevolved species: butterfly and ant.

Evolutionary Theory

Biologists do not doubt that macroevolution occurs, but many disagree about how it occurs. However we choose to categorize evolutionary processes, the very same genetic change may be at the root of all evolution—fast or slow, large-scale or small-scale. Dramatic jumps in form, if they are not artifacts of gaps in the fossil record, may be the result of mutations in homeotic or other regulatory genes. Macroevolution may include more processes than microevolution, or it may not. It may be an accumulation of many microevolutionary events, or it may be an entirely different process. Evolutionary biologists may disagree about these and other hypotheses, but all of them are trying to explain the same thing: how all species are related by descent from common ancestors.

TAKE-HOME MESSAGE 17.9

- ✔ Macroevolution comprises evolutionary trends and patterns that occur in taxa above the species level.
- ✔ Very little change may occur in a lineage over long periods of time, a pattern called stasis.
- ✔ A species that no longer has living members is extinct.
- ✔ A key innovation can trigger adaptive radiation, a pattern in which one lineage rapidly diversifies into many.
- ✔ Two species with close ecological interactions may evolve jointly, a pattern called coevolution.

17.1 Superbug Farms (revisited)

A natural population of bacteria is diverse, and it can evolve astonishingly fast. Consider how each cell division is an opportunity for mutation. The intestinal bacteria *E. coli* can divide every 17 minutes, so even if a population starts out as clones, its cells diversify quickly. Bacteria can swap genes even among distantly related species, and this adds even more genetic diversity to their gene pools. When a natural bacterial population is exposed to an antibiotic, some cells in it are likely to survive because they have an allele that offers resistance. As susceptible cells die and the survivors reproduce, the frequency of antibiotic-resistance alleles increases in the population (an example of directional selection). A typical two-week course of antibiotics can exert selection pressure on over a thousand generations of bacteria. The pressure drives genetic change in bacterial populations so they become composed mainly of antibiotic-resistant cells. Thus, using antibiotics on an ongoing basis effectively guarantees the production of antibiotic-resistant bacterial populations. ●

CREDITS: (25A) Natural Visions; (25B) Roger Meerts/Shutterstock.

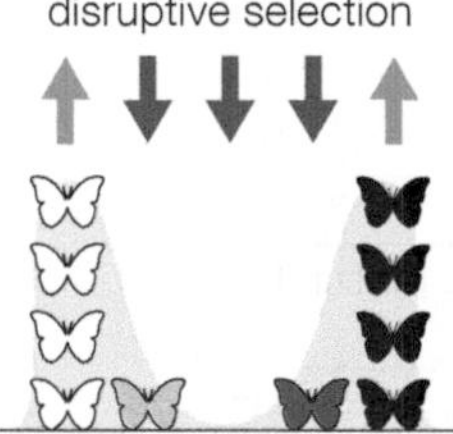
population before selection
directional selection
stabilizing selection
disruptive selection

FIGURE 17.26 Comparing three modes of natural selection.

Section 17.1 Our overuse of antibiotics exerts directional selection favoring resistant bacterial populations, which are now common in the environment. We are running out of effective antibiotics to use as human drugs.

Section 17.2 Mutations are the source of new alleles, and they can be **neutral**, **lethal**, or adaptive. All alleles of all genes in a **population** constitute a **gene pool**. **Allele frequency** is the abundance of an allele among all copies of the gene in a gene pool (the proportion of chromosomes with the allele). **Microevolution**, or change in allele frequency, is always occurring in natural populations because processes that drive it are always operating.

Section 17.3 **Genetic equilibrium** is a hypothetical state in which a population is not evolving. Researchers use genetic equilibrium as a benchmark to track microevolution in a population.

Section 17.4 Natural selection is one mechanism that drives evolution, and it can occur in patterns (**FIGURE 17.26**) that depend on selection pressures in a population's environment. **Directional selection** is a mode of natural selection in which a form of a trait at one end of a range of variation is adaptive. An intermediate form of a trait is adaptive in **stabilizing selection**; extreme forms are adaptive in **disruptive selection**.

TABLE 17.2

Comparing Speciation Models

	Allopatric	Sympatric	Parapatric
Original population(s)			
Initiating event:	physical barrier arises	genetic change	localized selection pressures
Reproductive isolation occurs			
New species arise:	in isolation	within existing population	in contact along common border

Section 17.5 **Sexual dimorphism** can be an outcome of **sexual selection**, a mode of natural selection in which adaptive traits are those that make their bearers better at securing mates. **Frequency-dependent selection** or any other type of natural selection can maintain a **balanced polymorphism**.

Section 17.6 Evolution can occur by external mechanisms that do not involve adaptive traits or natural selection. **Genetic drift**, a change in allele frequency due to chance alone, is most pronounced in small populations, and it can cause alleles to become **fixed**. The **founder effect** may be an outcome of an evolutionary **bottleneck**. **Inbreeding** can keep harmful recessive alleles circulating in a small population. **Gene flow** can stabilize allele frequencies between populations.

Section 17.7 **Reproductive isolation**, the end of gene flow between populations, is always a part of **speciation**: the splitting of one lineage into two. The moment of speciation is often impossible to pinpoint.

Section 17.8 Speciation patterns are useful as models (**TABLE 17.2**), but the process is rarely as simple as these models suggest. Each speciation occurs within a continuum of geographic, genetic, and ecological factors. With **allopatric speciation**, a geographic barrier arises and interrupts gene flow between populations. The interruption allows genetic divergences to occur independently in each population, and this can result in separate species. Speciation can also occur in the absence of a physical barrier to gene flow. **Sympatric speciation** occurs by genetic divergence within a population. Polyploid species of many plants have originated this way. With **parapatric speciation**, populations become separate species while in contact along a common border.

Section 17.9 **Macroevolution** refers to large-scale patterns and trends in evolution. With **stasis**, a lineage changes very little over long spans of time. An **exaptation** is a trait that has been evolutionarily repurposed for a new use. A **key innovation** can result in an **adaptive radiation**, the rapid splitting of one lineage into many. **Coevolution** occurs when two species act as agents of selection upon one another. A lineage with no more living members is **extinct**; in a mass extinction, many lineages become extinct simultaneously.

SELF-QUIZ

Answers in Appendix VII

1. ________ is the original source of new alleles.
   a. Mutation
   b. Natural selection
   c. Gene flow
   d. Genetic drift

2. A neutral mutation ________ .
   a. has a pH of 7.0
   b. has no effect on survival
   c. does not alter the DNA sequence

3. Change in allele frequency of a population is called ________ .
   a. macroevolution c. inbreeding
   b. adaptive radiation d. microevolution
4. A wild population of pea plants has two alleles for flower color in its gene pool. A dominant allele (*P*) specifies purple flowers; a recessive allele (*p*) specifies white flowers. If the frequency of allele *P* is 0.75, what is the expected frequency of allele *p*?
5. Which of the following is *not* part of how we define a species?
   a. Its individuals appear different from other species.
   b. It is reproductively isolated from other species.
   c. Its populations can interbreed.
   d. Fertile offspring are produced.
6. Sexual selection frequently influences aspects of body form and can lead to ________ .
   a. a sexual dimorphism c. exaggerated traits
   b. male aggression d. all of the above
7. The persistence of sickle-cell anemia in a population with a high incidence of malaria is a case of ________ .
   a. bottlenecking c. the founder effect
   b. inbreeding d. a balanced polymorphism
8. ________ tends to keep populations of a species similar to one another.
   a. Genetic drift c. Mutation
   b. Gene flow d. Natural selection
9. Which is required for a population to evolve?
   a. genetic diversity c. gene flow
   b. selection pressure d. none of the above
10. In many bird species, sex is preceded by a courtship dance. If a male's dance is unrecognized by the female, she will not mate with him. This is an example of _____ .
    a. sexual selection c. reproductive isolation
    b. behavioral isolation d. all of the above
11. After fire devastates all of the trees in a wide swath of forest, populations of a species of tree-dwelling frog on either side of the burned area diverge and become separate species. This is an example of ________ .
12. True or false? Inbreeding can increase the frequency of a harmful allele in a population's gene pool.
13. The difference between sympatric and parapatric speciation is ________ .
    a. parapatric speciation occurs only in velvet walking worms
    b. sympatric speciation requires a barrier to gene flow
    c. the extent of overlap in range
    d. reproductive isolation does not occur

14. Natural selection does not explain ________ .
    a. genetic drift d. mutations
    b. the founder effect e. inheritance
    c. gene flow f. any of the above
15. Match the evolution concepts.
    ___ gene flow a. outcome can be interdependence
    ___ sexual selection b. changes in a population's allele frequencies due to chance alone
    ___ extinct c. alleles enter or leave a population
    ___ genetic drift d. adaptive traits make their bearers better at securing mates
    ___ coevolution e. no more living members
    ___ adaptive radiation f. burst of divergences from one lineage into many

CRITICAL THINKING

1. Species have traditionally been characterized as "primitive" and "advanced." For example, mosses were considered to be primitive, and flowering plants advanced; crocodiles were primitive and mammals were advanced. Why do most biologists of today think it is incorrect to refer to any modern species as primitive?
2. Rama the cama, a llama–camel hybrid, was born in 1998. The idea was to breed an animal that has the camel's strength and endurance, and the llama's gentle disposition. However, instead of being large, strong, and sweet, Rama is smaller than expected and has a camel's short temper. The breeders plan to mate him with Kamilah, a female cama. What potential problems with this mating should the breeders anticipate?
3. Two species of antelope, one from Africa, the other from Asia, are put into the same enclosure in a zoo. To the zookeeper's surprise, individuals of the different species begin to mate and produce healthy, hybrid baby antelopes. Explain why a biologist might not view these offspring as evidence that the two species of antelope are in fact one.
4. Some human traits may have arisen by sexual selection. Over thousands of years, women attracted to charming, witty men perhaps prompted the development of human intellect beyond what was necessary for mere survival. Men attracted to women with juvenile features may have shifted the species as a whole to be less hairy and softer featured than any of our simian relatives. Can you think of a way to test this hypothesis?
5. About 70 percent of flowering plants are polyploid, so having extra sets of chromosomes must be adaptive. What advantage do you think polyploidy offers?

18 ORGANIZING INFORMATION ABOUT SPECIES

CORE CONCEPTS

Evolution

Evolution underlies the unity and diversity of life.

A record of ancient life persists in the form and function of every modern species. Shared core processes, developmental programs, and structures provide evidence that evolution has occurred throughout the history of life, and also that all living things are linked by lines of descent from a common ancestor. In general, lineages that diverged more recently have more structures and processes in common.

Structure and Function

The three-dimensional form and arrangement of biological structures give rise to their function and interactions.

The structural and biochemical underpinnings of traits are commonly remodeled and repurposed during evolution. Molecular programs that direct core processes such as embryonic development have been highly conserved among evolutionarily distant species. Some structures that are outwardly similar in different lineages have been shaped by similar environmental constraints.

Process of Science

The field of biology consists of and relies upon experimentation and the collection and analysis of scientific evidence.

Biologists base their hypotheses about evolutionary relationships on the premise that all life is interconnected by shared ancestry. Comparing physical, biochemical, and developmental traits among living and extinct organisms yields evidence for these hypotheses.

Links to Earlier Concepts

This chapter adds concepts of evolution (Sections 16.1–16.4) to taxonomy (1.5). Before starting, you should review what you learned about alleles and mutations (17.2), reproductive isolation and speciation (17.7), and macroevolution (17.9). The chapter also revisits the genetic code (9.4), master regulators (10.3), gene duplications (14.5), PCR (15.3), DNA sequencing (15.4), genomics (15.5), gene flow (17.6), and honeycreeper speciation (17.8).

18.1 Bye Bye Birdie

Some finch species migrate far outside of their normal range when food becomes scarce in a preferred overwintering spot, traveling in flocks of thousands or even tens of thousands of individuals. About seven million years ago, one of these flocks was migrating through southern Asia when it became caught in the winds of a huge storm. The birds were blown at least 6,000 miles (9,600 kilometers) across the open ocean to the Hawaiian archipelago. Enough individuals survived what must have been a very difficult journey to found a new population.

The arrival of the finches on the Hawaiian islands had been preceded by insects and plants, but no bird-eating predators, so their descendants thrived. Isolation from mainland finch populations fostered adaptive radiation that gave rise to at least 62 species of Hawaiian honeycreeper (Section 17.8).

The first Polynesians arrived on the islands sometime before 1000 A.D.; Europeans followed in 1778. Hawaii's rich ecosystem was hospitable to the newcomers as well as their livestock, pets, and crops. Escaped livestock began to eat and trample rain forest plants that had provided honeycreepers with food and shelter. Entire forests were cleared to grow imported crops, and plants that escaped cultivation began to crowd out native plants. Mosquitoes accidentally introduced in 1826 spread diseases such as avian malaria from imported songbirds to native birds. Stowaway rats ate their way through populations of native birds and their eggs. Mongooses deliberately imported to eat the rats preferred to eat birds and bird eggs.

The isolation that had allowed honeycreepers to arise by adaptive radiation also made them vulnerable to these challenges. Divergence from the ancestral finch species had led to the loss of unnecessary traits such as defenses against mainland predators and diseases. Traits that had previously been adaptive for honeycreepers—such as a long, curved beak matching the flower of a particular plant—became hindrances as the plant disappeared.

By 1778, only 44 Hawaiian honeycreeper species remained, and another 26 have disappeared since then. The surviving species are still being pressured by established populations of nonnative plants and animals. Rising global temperatures are also allowing mosquitoes to invade high-altitude habitats that had previously been too cold for the insects, so honeycreepers in these habitats are now succumbing to mosquito-borne diseases. Today, sixteen of the eighteen remaining Hawaiian honeycreeper species are in immediate danger of extinction.

CREDIT: (opposite) John Steiner/Smithsonian Institution.

18.2 Phylogeny

LEARNING OBJECTIVES

- Distinguish between a character and a derived character.
- Describe cladistics and compare it to Linnaean taxonomy.

Today's biologists work from the premise that every living thing is related if you just look back far enough in time. Grouping species according to evolutionary relationships is a way to fill in the details of this bigger picture of evolution. Thus, reconstructing **phylogeny**, the evolutionary history of a species or a group of species, is a priority. Phylogeny is a kind of genealogy that follows evolutionary relationships through time.

Humans were not around to witness the evolution of most species, but there is plenty of evidence to help us understand ancient evolutionary events (Chapter 16). Consider how each species bears traces of its own unique evolutionary history in its characters. A **character** is a quantifiable, heritable trait such as the number of segments in a backbone, the nucleotide sequence of ribosomal RNA, or the presence of hair (TABLE 18.1). A **derived character** is one that is present in a group under consideration, but not in any of the group's ancestors. A group whose members share one or more defining derived characters is called a **clade**. By definition, a clade is a **monophyletic group**, which consists of an ancestor (in which a derived character evolved) together with all of its descendants.

It is the relative newness of a derived character that defines a clade. Consider how alligators look a lot more like lizards than birds. In this case, the similarity in appearance does indicate shared ancestry, but it is a more distant relationship than alligators have with birds. A unique set of traits that include a gizzard and a four-chambered heart evolved in the lineage that gave rise to alligators and birds, but not in the lineage that gave rise to lizards. As another example, consider that humans, mice, and bacteria use some of the same proteins to repair DNA, for example, but so do almost all other organisms on Earth. Both humans and mice, however, make proteins such as keratin (a component of hair) that bacteria do not. Thus, we can predict that humans and mice share a more recent ancestor than they share with bacteria.

Making hypotheses such as these about clades and their evolutionary relationships is called **cladistics**. Cladistics differs from Linnaean taxonomy even though both are based on shared traits. Taxonomy is a system of naming species and categorizing them into taxa based on similarities (Section 1.5). Cladistics focuses on the ancestral relationships that gave rise to the similarities in the first place. Many taxa are equivalent to clades—flowering plants, for example, constitute both a phylum and a clade—but this is not always the case. For example, the traditional Linnaean class Reptilia ("reptiles") includes crocodiles, alligators, tuataras, snakes, lizards, turtles, and tortoises, but these groups would not constitute a clade unless birds are included, as you will see in Chapter 25. Taxa that do not include all descendants of the last shared ancestor are not clades.

Clades are defined by derived traits, and connections between them are organized as hierarchical branches. The branchings can be represented as **evolutionary trees**, which are diagrams that show ancestral connections. A **cladogram** is an evolutionary tree that visually summarizes a hypothesis about how a group of clades are related (FIGURE 18.1). Data from an

TABLE 18.1

Examples of Characters

	Bird	Bat	Cat
Warm-blooded	Y	Y	Y
Hair	N	Y	Y
Echolocation	N	Y	N
Retractable claws	N	N	Y
Feathers	Y	N	N

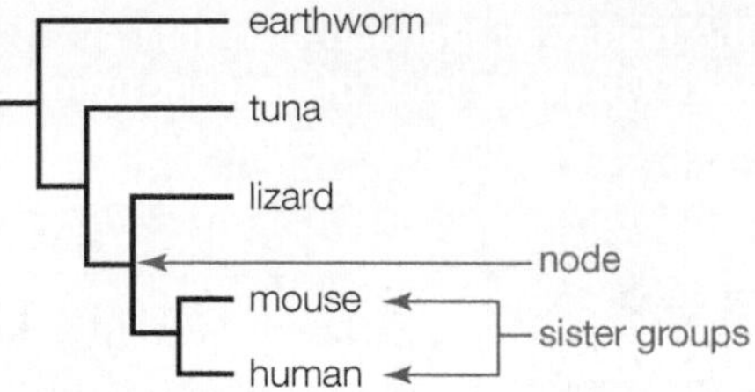

earthworm
tuna
lizard
node
mouse
sister groups
human

A Evolutionary connections among clades are represented as lines on a cladogram. Sister groups emerge from a node, which represents a common ancestor.

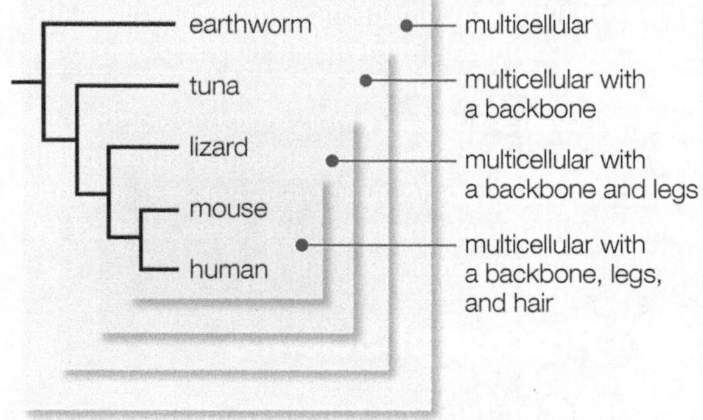

earthworm
tuna
lizard
mouse
human
multicellular
multicellular with a backbone
multicellular with a backbone and legs
multicellular with a backbone, legs, and hair

B A cladogram can be viewed as "sets within sets" of derived characters. Each set (an ancestor together with all of its descendants) is a clade.

FIGURE 18.1 Cladograms.

FIGURE IT OUT Which groups in this cladogram diverged most recently?

Answer: Human and mouse

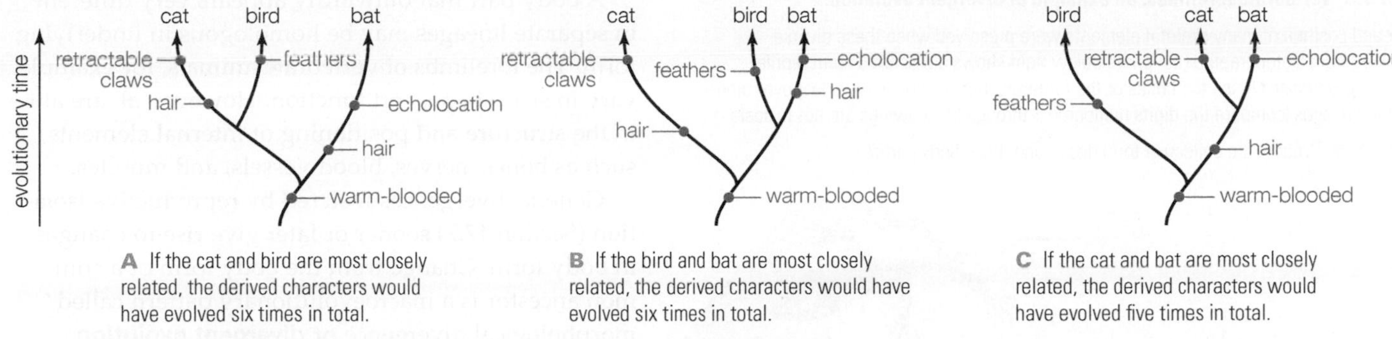
evolutionary time
cat
bird
bat
retractable claws
feathers
hair
echolocation
hair
warm-blooded
A If the cat and bird are most closely related, the derived characters would have evolved six times in total.
cat
bird
bat
retractable claws
feathers
echolocation
hair
hair
warm-blooded
B If the bird and bat are most closely related, the derived characters would have evolved six times in total.
bird
cat
bat
retractable claws
echolocation
feathers
hair
warm-blooded
C If the cat and bat are most closely related, the derived characters would have evolved five times in total.

FIGURE 18.2 An example of cladistics, using parsimony analysis with the characters listed in Table 18.1. **A**, **B**, and **C** show the three possible evolutionary pathways that could connect cats, birds, and bats; red indicates the appearance of a derived character in a lineage. The pathway most likely to be correct (**C**) is the simplest—the one in which the derived characters would have had to evolve the fewest number of times in total.

outgroup (a species not closely related to any member of the group under study) may be included in order to "root" the tree. Each line is a lineage, which may split into two lineages at a node. A node represents a common ancestor, and the two lineages that emerge from it are called **sister groups**. Any complete branch that can be cut from a cladogram—including the smallest branch possible (a species)—is a clade.

One way of constructing cladograms involves the logical rule of simplicity: When there are multiple ways that a group of clades can be connected, the simplest evolutionary pathway is probably the correct one. By comparing all of the possible connections among the clades, we can identify the simplest—the one in which the defining derived characters evolved the fewest number of times (**FIGURE 18.2**). The process of finding the simplest pathway is called parsimony analysis.

All species are interconnected by shared ancestry; an evolutionary biologist's job is to figure out the connections. As you learn about these relationships, remember that evolutionary history does not change because of events in the present: A species' ancestry remains the same no matter how it evolves. However, we can make mistakes when we group organisms based on incomplete information. Thus, a taxon or clade may change when new discoveries are made. As with all hypotheses, the more data in support of an evolutionary grouping, the less likely it is to require revision.

character Quantifiable, heritable trait.
clade (CLAYD) A group whose members share a defining derived character.
cladistics (cluh-DISS-ticks) Making hypotheses about evolutionary relationships among clades.
cladogram (CLAD-oh-gram) Evolutionary tree diagram that shows how a group of clades are related.
derived character A trait shared by all members of a clade but not in any of the clade's ancestors.
evolutionary tree Branching diagram of ancestral connections.
homologous structures Body structures that are similar in different lineages because they evolved in a common ancestor.
monophyletic group (ma-no-fill-EH-tick) A group including an ancestor in which a derived character evolved, together with all of its descendants.
phylogeny (fie-LA-juh-knee) Evolutionary history of a species or group of species.
sister groups The two lineages that emerge from a node on a cladogram.

TAKE-HOME MESSAGE 18.2

✔ Evolutionary biologists reconstruct phylogeny in order to understand how all species are connected by shared ancestry.

✔ A clade is a monophyletic group whose members share one or more derived characters. Cladistics is a method of making hypotheses about evolutionary relationships among clades.

18.3 Comparing Form and Function

LEARNING OBJECTIVES

- Describe how body form and function can offer clues about evolutionary relationships.
- Distinguish between divergent and convergent evolution.
- Use examples to explain the difference between analogous structures and homologous structures.

To biologists, remember, evolution means change in a line of descent. How do they reconstruct evolutionary events that occurred in the ancient past? Evolutionary biologists are a bit like detectives, using clues to piece together history that no human witnessed. Fossils provide some clues. The body form and function of modern organisms provide others.

Divergent Evolution

Body parts that appear similar in separate lineages because they evolved in a common ancestor are called **homologous structures** (*hom–* means "the same"). Homologous structures may be used for different purposes in different groups, but the very same genes direct their development.

FIGURE 18.3 Vertebrate forelimbs: an example of divergent evolution.

The number and position of many skeletal elements were preserved when these diverse animals evolved from a stem reptile (the photo below right shows a fossilized stem reptile of the genus *Captorhinus*). Notice the bones of the forearms. Certain bones were lost over time in some of the lineages (compare the digits numbered 1 through 5). Drawings are not to scale.

FIGURE IT OUT What is the collective term describing these body parts?

Answer: Homologous structures

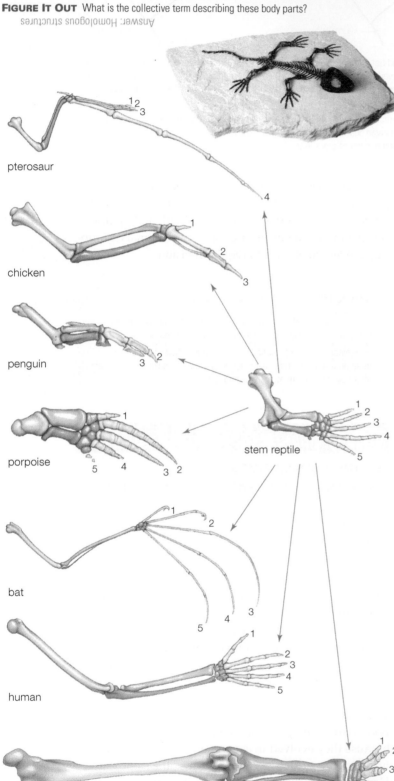

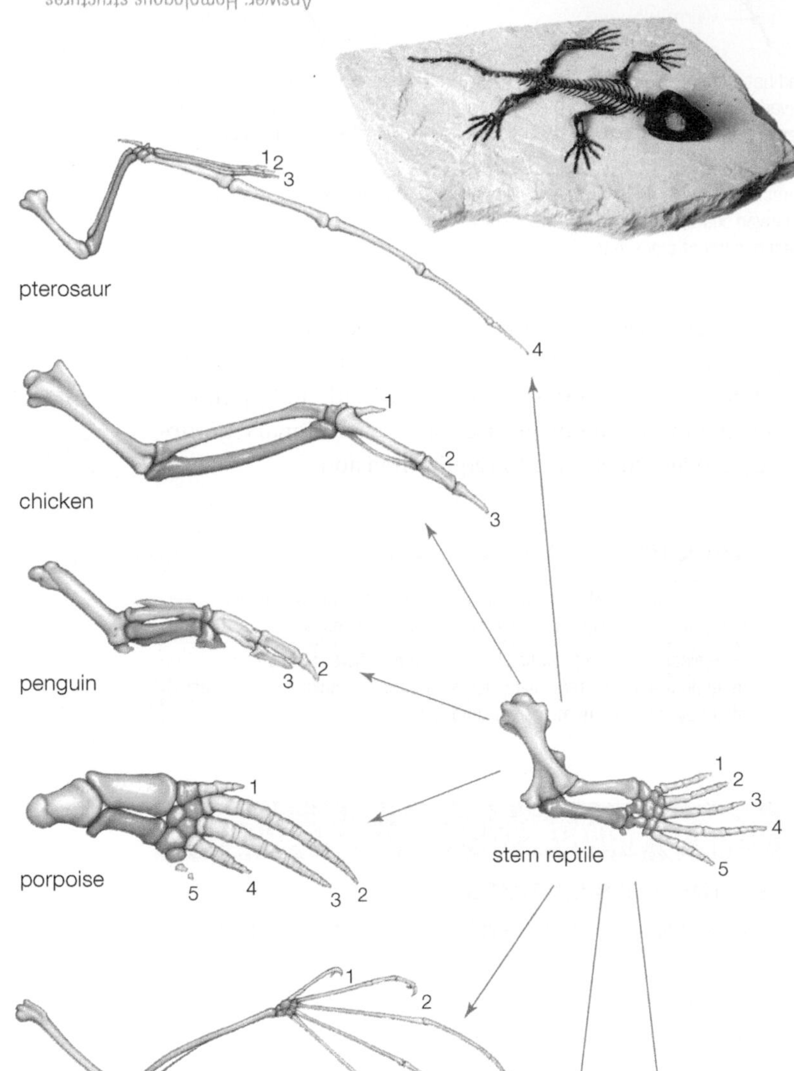

pterosaur
1 2 3
4
chicken
1
2
3
penguin
3 2
porpoise
1
5 4 3 2
stem reptile
1 2 3 4 5
bat
1 2
5 4 3
human
1 2 3 4 5
elephant
1 2 3 4 5

A body part that outwardly appears very different in separate lineages may be homologous in underlying form. The forelimbs of vertebrate animals, for example, vary in size, shape, and function. However, all are alike in the structure and positioning of internal elements such as bones, nerves, blood vessels, and muscles.

Genetic divergences fostered by reproductive isolation (Section 17.7) sooner or later give rise to changes in body form. Change from the body form of a common ancestor is a macroevolutionary pattern called morphological divergence or **divergent evolution**. Consider the homologous limbs of modern vertebrate animals. Fossil evidence suggests that a family of ancient "stem reptiles" with five-toed limbs was ancestral to many vertebrate lineages. Descendants of this ancestral group diversified over millions of years, and eventually gave rise to modern reptiles, birds, and mammals. A few lineages that had become adapted to walking on land even returned to aquatic living (Section 16.4). As the diversifications occurred, five-toed limbs became adapted for appropriate purposes (**FIGURE 18.3**). Forelimbs became modified for flight in extinct reptiles called pterosaurs and in bats and most birds. In penguins and cetaceans, they are now flippers useful for swimming. Human forelimbs are arms and hands with four fingers and an opposable thumb. Elephant limbs are strong and pillarlike, capable of supporting a great deal of weight. The five-toed limb was reduced to a nub in pythons and boa constrictors, and it disappeared entirely in other snakes.

Convergent Evolution

Body parts that appear similar in different species are not always homologous; they sometimes evolve independently in lineages subjected to the same environmental pressures. The independent evolution of similar body parts in different lineages is an evolutionary pattern called morphological convergence or **convergent evolution**. Body parts that look alike but evolved independently after lineages diverged are **analogous structures**.

Consider how bird, bat, and insect wings all perform the same function, which is flight (**FIGURE 18.4A**). However, several clues tell us that the wing surfaces are not homologous. All of the wings are adapted to the same physical constraints that govern flight, but each is adapted in a different way. In the case of birds and bats, the limbs themselves are homologous, but the adaptations that make those limbs useful for flight differ. The surface of a bat wing is a thin, membranous extension of the animal's skin. By contrast, the surface of a bird wing is a sweep of feathers, which are specialized structures derived from skin. Insect wings differ even more. An insect wing forms as a saclike extension of the body wall. Except at forked veins, the sac flattens and fuses into a thin membrane. The sturdy, chitin-reinforced veins structurally support the wing. Unique adaptations for flight are evidence that wing surfaces of birds, bats, and insects are analogous structures that evolved after the ancestors of these modern groups diverged.

As another example of analogous structures, consider the similar appearance of the saguaro cactus and the African milk barrel plant (see Figure 16.3). Both species are adapted to similarly harsh desert environments where rain is scarce. Accordion-like pleats allow the plant body to swell with water when rain does come; water stored in the plants' tissues allows them to survive long dry periods. As the stored water is used, the plant body shrinks, and the folded pleats provide some shade in an environment that typically has none. Despite these similarities, a closer look reveals many differences that indicate the two plants are not closely related (**FIGURE 18.4B**). For example, cactus spines have a simple fibrous structure; they are modified leaves that arise from dimples on the plant's surface. *Euphorbia* spines project smoothly from the plant surface, and they are not modified leaves: In many species the spines are dried flower stalks.

TAKE-HOME MESSAGE 18.3

✔ Physical similarities are often evidence of shared ancestry.

✔ Body parts often become modified differently as lineages diverge from a common ancestor (divergent evolution). Such body parts are homologous structures.

✔ Similar body parts may evolve in separate lineages (convergent evolution). Such body parts are called analogous structures.

18.4 Comparing Molecules

LEARNING OBJECTIVE

- Explain why similarities between protein sequences (or between DNA sequences) can be used as a measure of relative relatedness.

Over time, inevitable mutations change a genome's DNA sequence. Most of these mutations are neutral. Neutral mutations have no effect on an individual's survival or reproduction, so we can assume they accumulate at a constant rate. For example, a nucleotide substitution that changes one codon from AAA to AAG in a gene's protein-coding region would probably not affect the protein product, because both codons specify lysine (Section 9.4). In other cases, a neutral mutation

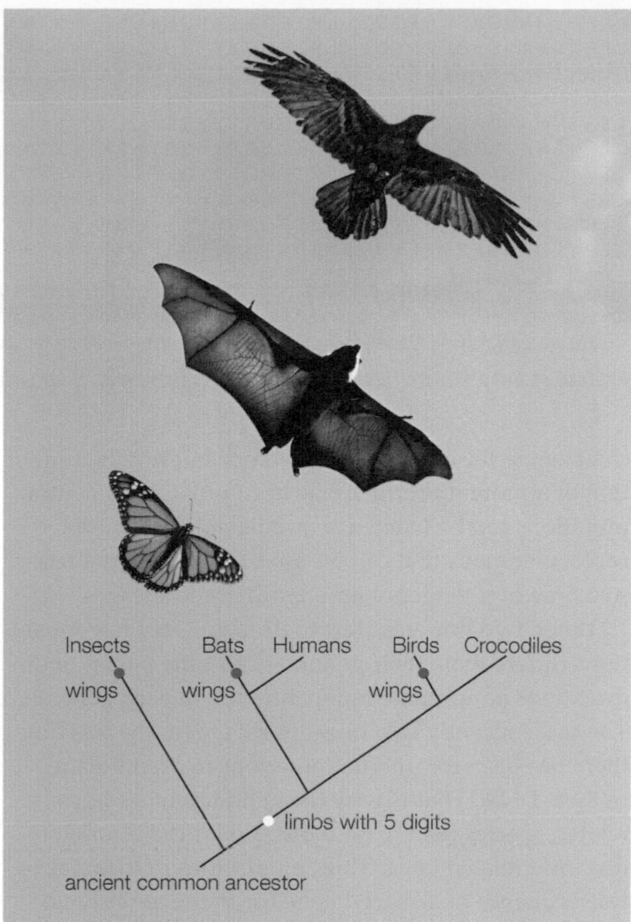
Insects
Bats
Humans
Birds
Crocodiles
wings
wings
wings
limbs with 5 digits
ancient common ancestor

A The flight surfaces of an insect wing, a bat wing, and a bird wing are analogous structures. The cladogram shows how the evolution of wings (red dots) occurred independently in the three lineages.

B Spines of a saguaro cactus (left) differ from spines of an African milk barrel plant (*Euphorbia*, right). This and some other differences indicate the two plants are not closely related despite their similar appearances (compare Figure 16.3). Many similarities between these species are an outcome of convergent evolution.

FIGURE 18.4 Examples of convergent evolution.

analogous structures Similar body structures that evolved separately in different lineages (by convergent evolution).

convergent evolution Morphological convergence. Macroevolutionary pattern in which similar body parts evolve separately in different lineages.

divergent evolution Morphological divergence. Evolutionary pattern in which a body part of an ancestor changes in its descendants.

CREDITS: (4A) top, © iStockphoto.com/DanCardiff; middle, © Taro Taylor, www.flickr.com/photos/tjt195; bottom, Alberto J. Espiñeira Francés - Alesfra/Getty Images; (4B) left, George Burba/Shutterstock; right, James C. Gaither, www.flickr.com/people/jim-sf/.

```
LIRNLHANGASFFFICIYLHIGRGIYYGSYLNK--ETWNIGVILLLTLMATAFVGYVLPWGQMSFWG...  honeycreepers (10)
LIRNLHANGASFFFICIYLHIGRGIYYGSYLNK--ETWNVGIILLLALMATAFVGYVLPWGQMSFWG...  song sparrow
LIRNIHANGASFFFICIYLHIGRGLYYGSYLYK--ETWNVGVILLLTLMATAFVGYVLPWGQMSFWG...  Gough Island finch
LIRYMHANGASMFFICLFLHVGRGMYYGSYTFT--ETWNIGIVLLFAVMATAFMGYVLPWGQMSFWG...  deer mouse
IIRYMHANGASMFFICLFMHVGRGLYYGSYLLS--ETWNIGIILLFTVMATAFMGYVLPWGQMSFWG...  Asiatic black bear
LIRNLHANGASFFFICIYLHIGRGLYYGSYLYK--ETWNIGVVLLLLVMGTAFVGYVLPWGQMSFWG...  bogue (a fish)
IIRYLHANGASMFFICLFLHIGRGLYYGSFLYS--ETWNIGIILLLATMATAFMGYVLPWGQMSFWG...  human
LLRYMHANGASMFLIVVYLHIFRGLYHASYSSPREFVWCLGVVIFLLMIVTAFIGYVLPWGQMSFWG...  thale cress (a plant)
MVRSIHANGASWFFIMLYSHIFRGLWVSSFTQP--LVWLSGVIILFLSMATAFLGYVLPWGQMSFWG...  baboon louse
ILRYLHANGASFFFMVMFMHMAKGLYYGSYRSPRVTLWNVGVIIFTLTIATAFLGYCCVYGQMSHWG...  baker's yeast
```

FIGURE 18.5 Comparing proteins. Here, partial amino acid sequences of mitochondrial cytochrome *b* from 19 species are aligned. This protein is a crucial component of mitochondrial electron transfer chains. The honeycreeper sequence is identical in ten species of honeycreeper; amino acids that differ in the other species are shown in red. Dashes reflect insertion or deletion mutations. Single-letter amino acid codes provided in Figure 9.8.

FIGURE IT OUT Based on this comparison, which species is the closest relative of honeycreepers?

Answer: The song sparrow

can change the amino acid sequence but not the function of a protein product. The accumulation of neutral mutations can be likened to predictable ticks of a **molecular clock** that can be used to estimate the relative time of divergence among different lineages.

When two lineages start to diverge, very few mutations differentiate their genomes. As time passes, more mutations accumulate independently in each genome. The more recently two lineages diverged, the less time there has been for unique mutations to accumulate in their DNA. That is why the genomes of closely related species tend to be more similar than those of distantly related ones. Thus, similarities in the nucleotide sequence of a shared gene (or in the amino acid sequence of a shared protein) can be used to study evolutionary relationships. Molecular comparisons like these may be combined with morphological comparisons, in order to provide data for hypotheses about shared ancestry.

DNA and Protein Sequence Comparisons

Two species with a distant evolutionary relationship have few similarities between their proteins. Evolutionary biologists often compare a protein's amino acid sequence among several species, and use the number of differences as a measure of relative relatedness (**FIGURE 18.5**).

The amino acids that differ are also clues. For example, a leucine to isoleucine change may not affect the function of a protein very much, because both amino acids are nonpolar, and both are about the same size. Such changes are called conservative amino acid substitutions. By contrast, the substitution of a lysine (which is basic) for an aspartic acid (which is acidic) may dramatically change the character of a protein. Nonconservative substitutions can affect phenotype. Most mutations that affect phenotype are selected against, but occasionally one proves adaptive. Thus, the longer it has been since two lineages diverged, the more nonconservative amino acid substitutions we are likely to see when comparing their proteins.

Among lineages that diverged relatively recently, many proteins have identical amino acid sequences. Nucleotide sequence differences may be instructive in such cases. Even if the amino acid sequence of a protein is identical among species, the nucleotide sequence of the gene that encodes the protein may differ because of redundancies in the genetic code. The DNA from

FIGURE 18.6 DNA barcoding. This example allows a quick visual comparison of 591 nucleotides of a mitochondrial gene called *ND2* (NADH dehydrogenase subunit 2) from three Hawaiian honeycreeper species and a rosefinch.

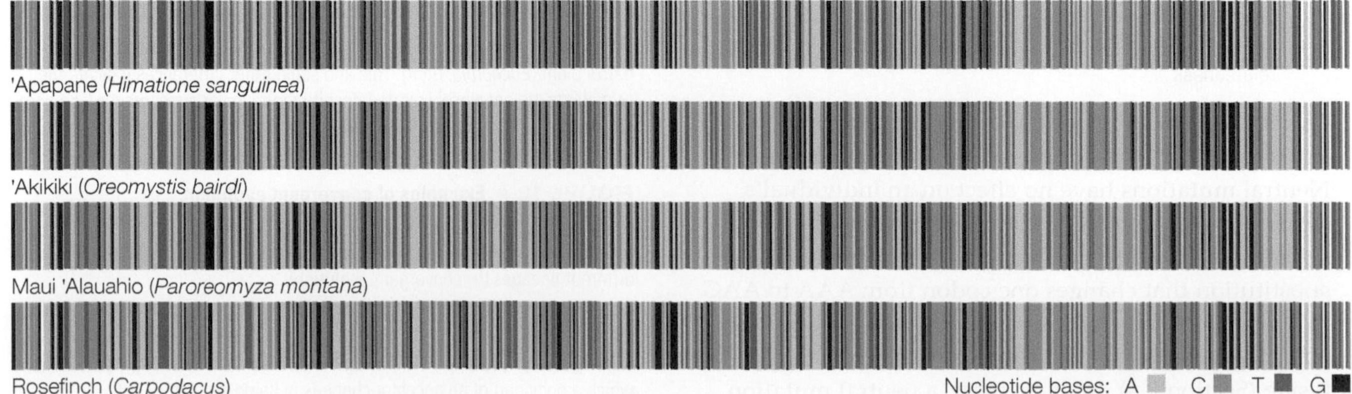
'Apapane (Himatione sanguinea)
'Akikiki (Oreomystis bairdi)
Maui 'Alauahio (Paroreomyza montana)
Rosefinch (Carpodacus)
Nucleotide bases: A C T G

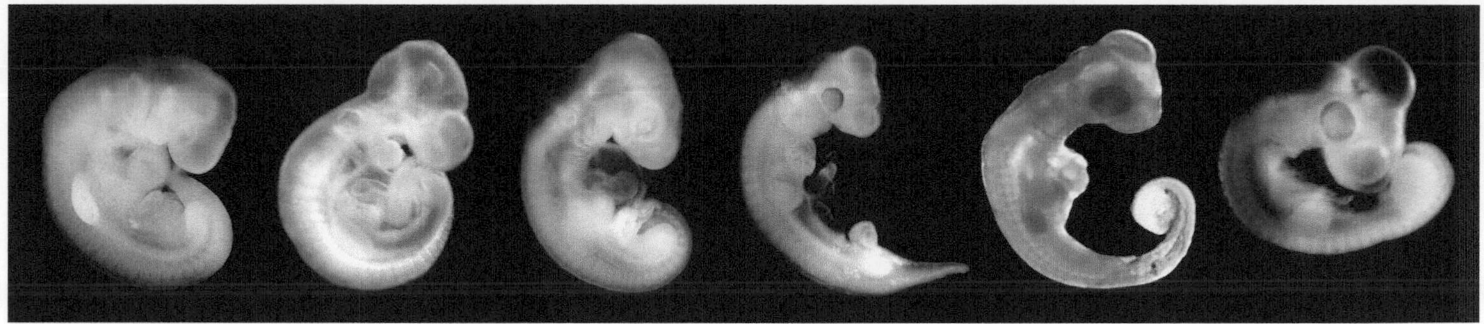

FIGURE 18.7 Comparing vertebrate embryos. All vertebrates go through an embryonic stage in which they have four limb buds, a tail, and divisions called somites along their back. From left to right: human, mouse, lizard, turtle, alligator, chicken.

nuclei, mitochondria, or chloroplasts can be used in nucleotide comparisons.

Mitochondrial DNA acquires mutations faster than nuclear DNA. Thus, comparing mitochondrial DNA sequences can reveal relationships among closely related individuals—even members of the same family. In most animals, mitochondria are inherited intact from only one of the parents (the mother, in humans). Mitochondrial chromosomes do not undergo crossing over, so any changes in their DNA sequence occur by mutation. Thus, in most cases, differences in mitochondrial DNA sequences between maternally related individuals are due to new mutations.

As you will see in the next section, some essential genes are highly conserved (their DNA sequences have changed very little or not at all over evolutionary time). Other genes are not conserved, and these underlie differences that distinguish species. DNA sequence differences are the basis of a method called DNA barcoding, which is used to identify an individual as belonging to a particular species. A standard region of the individual's DNA is amplified using PCR (Section 15.3) and then sequenced. The sequence is illustrated as a barcode in which each base is represented by a different color (**FIGURE 18.6**). This technique can be used when other methods of identification are difficult or impossible. For example, researchers can use DNA barcoding of feces to identify components of an endangered animal's diet.

Getting useful information from DNA comparisons requires a lot more data than protein comparisons. This is because coincidental homologies are statistically more likely to occur with DNA comparisons—there are only four nucleotides in DNA versus twenty amino acids in proteins. However, DNA sequencing has become so fast that there is a lot of data available to compare. Genomics studies with such data have shown us (for example) that about 86 percent of the mouse genome sequence is identical with the human genome, as is 51 percent of the bee genome, 19 percent of the thale cress genome, and 9 percent of the bacterial genome.

molecular clock Method of comparing the number of neutral mutations in the genomes of two lineages to estimate how long ago they diverged.

TAKE-HOME MESSAGE 18.4

✔ In general, more closely related species share more similarities in their DNA (and their proteins).

18.5 Comparing Development

LEARNING OBJECTIVE

- Explain why similarities in patterns of embryonic development can be used as a measure of relative relatedness.

In general, the more closely related animals are, the more similar is their development. For example, all vertebrates go through a stage during which a developing embryo has four limb buds, a tail, and a series of somites: divisions of the body that give rise to the backbone and associated skin and muscle (**FIGURE 18.7**).

Animals have similar patterns of embryonic development because the very same master regulators direct the process. These genes are expressed in cascades that orchestrate development, so the failure of any one of them can result in a drastically altered body plan (Section 10.3). Because a mutation in a master regulator typically unravels development completely, these genes tend to be highly conserved. Even among lineages that diverged a very long time ago, many master regulators retain similar sequences and functions.

If the same genes direct development in all vertebrate lineages, how do the adult forms end up so different? Part of the answer is that there are differences in the timing of early steps in development. The differences arise from variations in expression patterns of master regulators. Such variation can arise (for example) after a gene becomes duplicated, and mutations alter the duplicates (Section 14.5). Consider

CREDIT: (7) from left, Lennart Nilsson/ Bonnierforlagen AB; Courtesy of Anna Bigas, IDIBELL-Institut de Recerca Oncologica, Spain; Catherine May; USGS; Courtesy of Prof. Dr. G. Elisabeth Pollerberg, Institut für Zoologie, Universität Heidelberg, Germany.

homeotic genes called *Hox*, which, like other homeotic genes, trigger formation of specific body parts during embryonic development. Insects have ten *Hox* genes, and vertebrates have multiple versions of all of them. You learned in Section 10.3 about one insect *Hox* gene called *antennapedia*. A leg forms wherever this gene is expressed, which, in a normal embryo, occurs in the midsection of the body (the thorax). Humans and other vertebrates have a version of antennapedia called *Hoxc6*. Expression of *Hoxc6* in a vertebrate embryo causes ribs to develop on a vertebra. Vertebrae of the neck and tail normally develop with no *Hoxc6* expression, and no ribs (FIGURE 18.8).

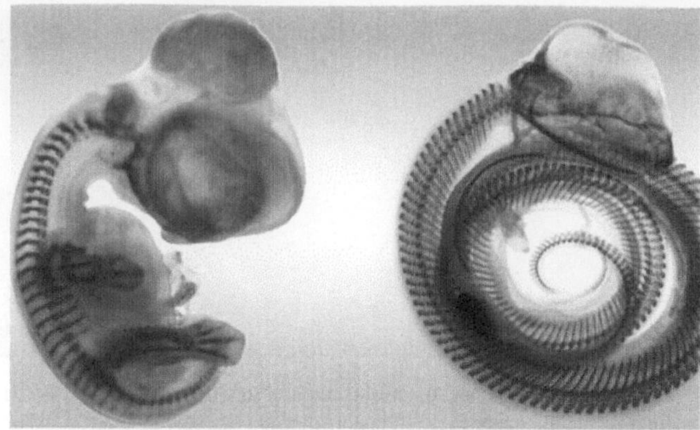

FIGURE 18.8 **Differences in body form arise from differences in master regulator expression.** Expression of the *Hoxc6* gene is indicated by purple stain in two vertebrate embryos: chicken (left) and garter snake (right). Expression of this gene causes a vertebra to develop ribs as part of the back. Chickens have 7 vertebrae in their back and 14 to 17 vertebrae in their neck; snakes have upwards of 450 back vertebrae and essentially no neck.

Hox genes also regulate limb formation. Body appendages as diverse as crab legs, beetle legs, sea star arms, butterfly wings, fish fins, and mouse feet start out as clusters of cells that bud from the surface of the embryo. The buds form wherever a homeotic gene called *Dlx* is expressed. *Dlx* encodes a transcription factor that signals clusters of embryonic cells to "stick out from the body" and give rise to an appendage. *Hox* genes suppress *Dlx* expression in all parts of an embryo that will not have appendages.

TAKE-HOME MESSAGE 18.5

✔ Similarities in patterns of development are the result of master regulators that have been conserved over evolutionary time.

18.6 Phylogeny Research

LEARNING OBJECTIVE

- Use examples to describe some applications of phylogeny.

Deciphering ancestral connections can help species that are still living. Consider how, as more and more honeycreeper species become extinct (FIGURE 18.9), the group's reservoir of genetic diversity is dwindling. The lowered diversity means the group as a whole is less resilient to change, and more likely to suffer catastrophic losses. As you will see in later chapters, the stability of a natural ecosystem depends on its biodiversity—the diversity of organisms that make it up. The disappearance of an entire group directly or indirectly affects all of the other organisms in a community. In some cases, the loss of a single species can unravel an entire ecosystem. Funding for conservation efforts continues to decrease, so understanding phylogeny allows us to focus available resources on species whose extinction would have the greatest impact on an ecosystem.

A Palila (*Loxioides bailleui*). Palila feed on seeds of a native Hawaiian plant that are toxic to most other birds. The one remaining population is declining because these plants are being trampled and eaten by domestic livestock. Between 1997 and 2016, the number of Palila dropped from 4,396 to 1,934.

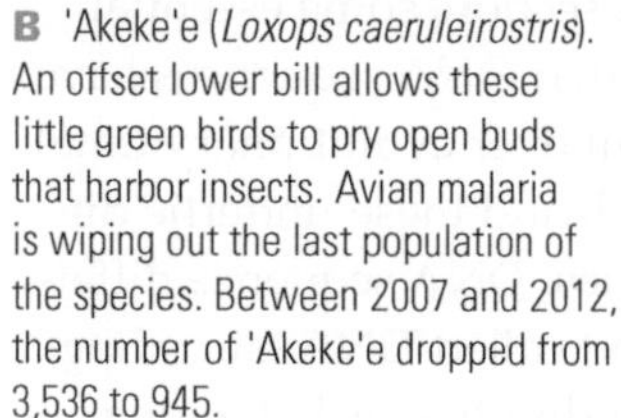

B 'Akeke'e (*Loxops caeruleirostris*). An offset lower bill allows these little green birds to pry open buds that harbor insects. Avian malaria is wiping out the last population of the species. Between 2007 and 2012, the number of 'Akeke'e dropped from 3,536 to 945.

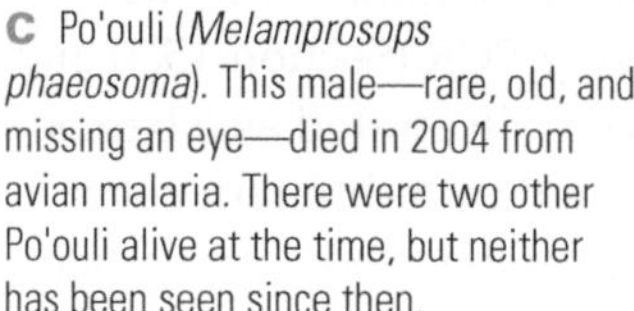

C Po'ouli (*Melamprosops phaeosoma*). This male—rare, old, and missing an eye—died in 2004 from avian malaria. There were two other Po'ouli alive at the time, but neither has been seen since then.

FIGURE 18.9 **Three honeycreeper species:** going, going, gone.

Cladistics also helps us understand ongoing evolutionary processes. For example, evolutionary biologists discovered that a decline in antelope populations in African savannas is at least partly due to competition with domestic cattle. Comparisons of mitochondrial DNA sequences suggested that current populations of blue wildebeest are genetically less similar than they should be, based on other antelope groups of similar age. Combined with behavioral and geographic data, the analysis helped conservation biologists realize that a patchy distribution of preferred food plants is

CREDITS: (in text) John Steiner/Smithsonian Institution; (8) Courtesy of Ann C. Burke/Wesleyan University; (9A) Jack Jeffrey/Minden Pictures; (9B) Courtesy of © Lucas Behnke; (9C) Bill Sparklin/Ashley Dayer.

Data Analysis Activities

Hawaiian Honeycreeper Phylogeny The po'ouli (*Melamprosops phaeosoma*) was discovered in 1973 by a group of students from the University of Hawaii. Its membership in the Hawaiian honeycreeper clade had been controversial, mainly because its appearance and behavior are so different from other living honeycreepers. It particularly lacked the "old tent" odor characteristic of other honeycreepers.

In 2011, Heather Lerner and her colleagues deciphered phylogeny of the 19 Hawaiian honeycreepers that were not yet officially declared to be extinct at the time, including the po'ouli. The researchers sequenced mitochondrial and nuclear DNA samples taken from the honeycreepers, and also from 28 other birds (outgroups). Phylogenetic analysis of these data firmly establishes the po'ouli as a member of the clade, and also reveals the Eurasian rosefinch as the clade's closest relative (FIGURE 18.10).

1. Which species on the cladogram represents an outgroup?
2. Which species is most closely related to the 'Apapane (*Himatione sanguinea*)?
3. What is the sister group of the 'Akikiki (*Oreomystis bairdi*)?
4. Which species is more closely related to the Palila (*Loxioides bailleui*): the 'I'iwi (*Vestiaria coccinea*) or the Maui 'Alauahio (*Paroreomyza montana*)?

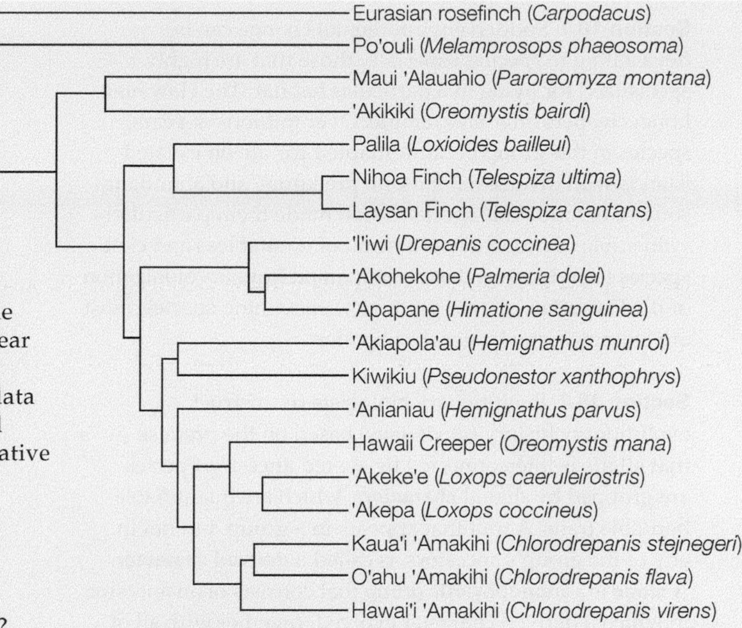
Eurasian rosefinch (Carpodacus)
Po'ouli (Melamprosops phaeosoma)
Maui 'Alauahio (Paroreomyza montana)
'Akikiki (Oreomystis bairdi)
Palila (Loxioides bailleui)
Nihoa Finch (Telespiza ultima)
Laysan Finch (Telespiza cantans)
'I'iwi (Drepanis coccinea)
'Akohekohe (Palmeria dolei)
'Apapane (Himatione sanguinea)
'Akiapola'au (Hemignathus munroi)
Kiwikiu (Pseudonestor xanthophrys)
'Anianiau (Hemignathus parvus)
Hawaii Creeper (Oreomystis mana)
'Akeke'e (Loxops caeruleirostris)
'Akepa (Loxops coccineus)
Kaua'i 'Amakihi (Chlorodrepanis stejnegeri)
O'ahu 'Amakihi (Chlorodrepanis flava)
Hawai'i 'Amakihi (Chlorodrepanis virens)

FIGURE 18.10 Phylogeny of Hawaiian honeycreepers. This cladogram was constructed using sequence comparisons of mitochondrial DNA (whole genome), and 13 nuclear DNA loci of 19 Hawaiian honeycreepers and 28 other finch species.

preventing gene flow among blue wildebeest populations. The absence of gene flow can lead to the loss of genetic diversity, an outcome that can devastate populations under pressure. Restoring appropriate grasses in intervening, unoccupied areas of savanna would allow isolated wildebeest populations to reconnect.

Researchers also use cladistics to study the evolution of infectious agents such as viruses. Viruses are not alive, but they can mutate every time they infect a host, so their genetic material changes over time. Thus, they can be grouped into clades based on genetic changes. Consider the H5N1 strain of influenza (flu) virus, which infects birds and other animals. H5N1 has a very high mortality rate in humans, but human-to-human transmission has been rare to date. However, the virus replicates in pigs without causing symptoms. Pigs transmit the virus to other pigs—and apparently to humans too. A phylogenetic analysis of H5N1 isolated from pigs showed that the virus "jumped" from birds to pigs at least three times since 2005, and that one of the isolates had acquired the potential to be transmitted among humans. Understanding how and when this virus adapts to new hosts is a necessary part of efforts to prevent a possible pandemic.

TAKE-HOME MESSAGE 18.6

✔ Among other applications, phylogeny research can help us to prioritize efforts to preserve endangered species, and to understand the spread of infectious diseases.

18.1 Bye Bye Birdie (revisited)

We now know the Po'ouli to be the most distant relative in the Hawaiian honeycreeper family, but the knowledge came too late because this species is probably extinct. Researchers captured one of the three Po'oulis that remained in 2004, with the intent of starting a captive breeding program before the inevitable extinction. They were unable to capture a female to mate with this male before he died in captivity. Cells from this last bird were frozen, and may be used for cloning. However, with no parents left to demonstrate the species' natural behavior to chicks, cloned birds would probably never be able to establish themselves as a natural population. The extinction of the Po'ouli means the loss of a large part of evolutionary history of the Hawaiian honeycreepers: One of the longest branches of the group's family tree is gone forever.

Section 18.1 Sudden environmental change can be devastating to species, especially those that are highly specialized for living in a particular habitat. The Hawaiian honeycreepers offer an example. Over millions of years, species in this group became adapted for life on isolated islands with diverse habitats, no predators, and abundant sources of food. This specialization made them particularly vulnerable to extinction as a result of habitat loss and exotic species introductions that accompanied human colonization of the Hawaiian Islands. Of the few remaining species, most are in immediate danger of extinction.

Section 18.2 Evolutionary biologists reconstruct evolutionary history (**phylogeny**) based on the premise that all life is interconnected by shared ancestry. Species are grouped by shared **characters**, which are quantifiable, heritable traits. A trait that appears in a group, but not in any of the group's ancestors, is called a **derived character**. A **clade** is a **monophyletic group** that consists of an ancestor in which a derived character evolved, together with all of its descendants.

Making hypotheses about the evolutionary history of a group of clades is called **cladistics**. These hypotheses are often represented as **cladograms**, which are a type of **evolutionary tree** with branching connections between clades. In a cladogram, each line represents a lineage. One lineage branches into two **sister groups** at a node, which represents a shared ancestor.

Section 18.3 Comparative morphology can reveal evidence of evolutionary connections among lineages. **Homologous structures** are similar body parts derived from a common ancestor and modified differently in different lineages (a pattern called **divergent evolution**). However, not all similar body parts reflect common ancestry. **Analogous structures** are body parts that look alike in different lineages but did not evolve in a common ancestor. Rather, they evolved separately after the lineages diverged, a pattern called **convergent evolution**. Analogous structures reflect adaptations to similar environmental constraints.

Section 18.4 Neutral mutations accumulate independently in the genomes of separate lineages, and they occur at the same rate in every genome. Thus, the number of differences among the DNA of different lineages can be used as a **molecular clock** to estimate relative times of divergence. Researchers can discover and clarify evolutionary relationships by comparing amino acid sequences of proteins, or nucleotide sequences of DNA, from different organisms. In general, these sequences are more similar among lineages that diverged more recently.

Section 18.5 Similarities in patterns of embryonic development reflect shared ancestry that can be evolutionarily ancient. Master regulators tend to be highly conserved even among lineages that are only distantly related. Mutations that alter the expression patterns of master regulators can alter the timing of early steps in development, and thus result in different adult body forms.

Section 18.6 Phylogeny research informs conservation efforts by revealing a clade's diversity. We use it to prioritize allocation of limited funding on species whose extinction will have the greatest impact. The loss of an entire group or even a single species from a community can unravel the connections that stabilize an ecosystem. Cladistics also helps us study the spread of viruses and other agents of infectious diseases.

SELF-QUIZ

Answers in Appendix VII

1. _______ is a way of reconstructing evolutionary history based on derived characters.
   a. Cladistics
   b. Linnaean taxonomy
   c. Convergent evolution
   d. Divergent evolution
2. Which of the following assumptions is *not* part of cladistics?
   a. A lineage's characters change over time.
   b. All species share common ancestry.
   c. Each lineage branches from another lineage.
   d. Similar body parts are evidence of a recent divergence.
3. The only taxon that is always the same as a clade is the _______ .
   a. genus
   b. family
   c. species
   d. kingdom
4. A clade is _______ .
   a. defined by a derived character
   b. a monophyletic group
   c. a hypothesis
   d. all of the above
5. In a cladogram, each node represents a(n) _______ .
   a. single lineage
   b. extinction
   c. common ancestor
   d. adaptive radiation
6. Constructing a cladogram _______ .
   a. may involve parsimony analysis
   b. helps us rank species into taxa
   c. reveals convergent structures
   d. is a necessary part of DNA barcoding
7. In cladograms, sister groups are _______ .
   a. inbred
   b. the same age
   c. represented by nodes
   d. in the same family
8. Through _______ , a body part of an ancestor is modified differently in different lines of descent.
   a. homologous evolution
   b. convergent evolution
   c. analogous evolution
   d. divergent evolution
9. Homologous structures among major groups of organisms may differ in _______ .
   a. size
   b. shape
   c. function
   d. all of the above

10. A diagram that shows evolutionary connections between clades is called a _______ .
    a. karyotype c. cladogram
    b. molecular clock d. homologous structure

11. Molecular clocks are based on comparisons of the number of _______ mutations between species.
    a. lethal c. conservative
    b. neutral d. nonconservative

12. Mitochondrial DNA sequences are often used to study a relationship between _______ .
    a. different species
    b. individuals of the same species
    c. different taxa
    d. evolutionary structures
    e. homologous structures

13. All of the following data types can be used as evidence of shared ancestry except similarities in _______ .
    a. amino acid sequences
    b. embryonic development
    c. DNA sequences
    d. fossil morphologies
    e. form due to convergent evolution

14. Choose the statement that is *in*correct.
    a. Phylogeny helps us study the spread of viruses through human populations.
    b. DNA barcoding can identify an individual as belonging to a particular species.
    c. Lineages that are more distantly related have more similarities in the DNA sequence of their genomes.
    d. Nonconservative amino acid substitutions can cause drastic changes in phenotype.

15. A mutation that alters the expression pattern of a _____ may lead to major differences in the adult form.
    a. derived character d. homologous structure
    b. master regulator e. lineage
    c. molecular clock f. all of the above

16. Match the terms with the most suitable description.

___ phylogeny	a. similar across many taxa
___ cladogram	b. evolutionary history
___ master regulators	c. human arm and bird wing
___ convergent evolution	d. present in a group but not in its ancestors
___ derived character	e. based on neutral mutations
___ molecular clock	f. insect wing and bird wing
___ analogous structures	g. diagram of sets within sets
___ divergent evolution	h. different lineages evolve in a similar way
___ homologous structures	i. closely related lineages evolve in different ways

CRITICAL THINKING

1. In the late 1800s, a biologist studying animal embryos coined the phrase "ontogeny recapitulates phylogeny," meaning that the physical development of an animal embryo (ontogeny) seemed to retrace the changing form of the species during its evolutionary history (phylogeny). Why would embryonic development retrace evolutionary steps?

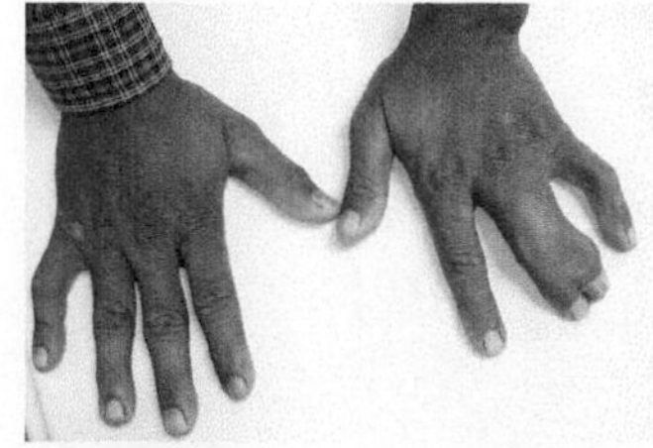
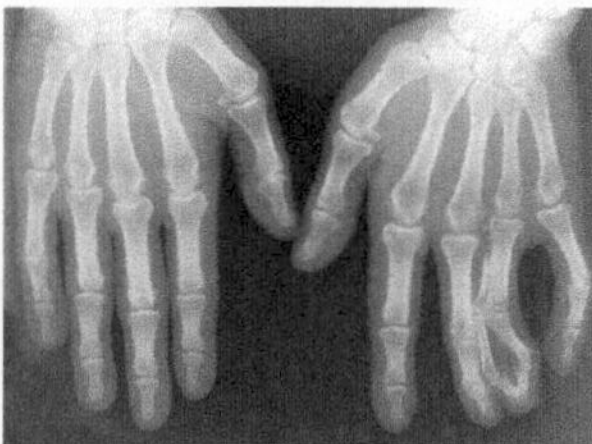

2. The photos shown above illustrate a case of synpolydactyly, a genetic abnormality characterized by two phenotypes: partially or completely duplicated fingers or toes, and webbing between fingers or toes. The same mutations that give rise to the human phenotype also give rise to a similar phenotype in mice. In which family of genes do you think these mutations occur?

3. Bonobos and chimpanzees are sister groups. Humans, bonobos, and chimpanzees share a more recent ancestor than do humans and gorillas. Orangutans share an ancestor with all of these groups. Draw a cladogram that represents these relationships.

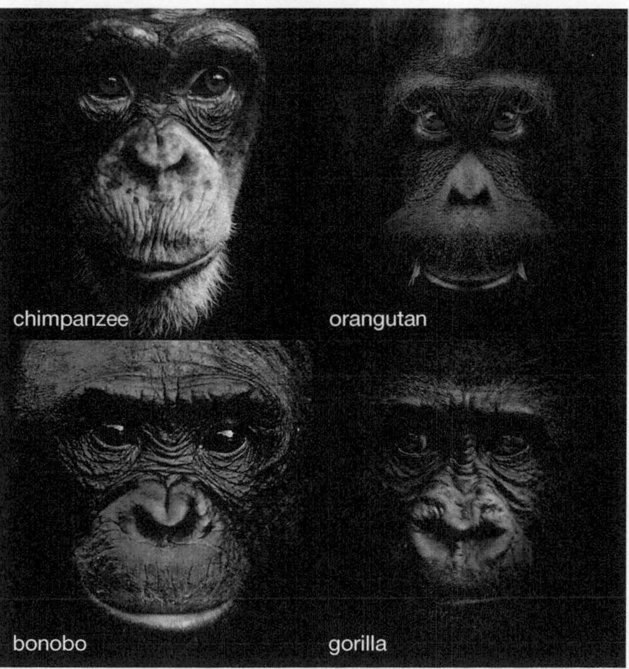
chimpanzee
orangutan
bonobo
gorilla

CREDITS: (in text) #1, Courtesy of Dr. Sajid Malik, from www.biomedcentral.com/1471-2350/8/78; #3 from left: top, Dennis Jacobsen/Shutterstock; bottom, Sergey Uryadnikov/Shutterstock; from right: top and bottom, Ron Meijer/Shutterstock.

19 LIFE'S ORIGIN AND EARLY EVOLUTION

CORE CONCEPTS

Process of Science

The field of biology consists of and relies upon experimentation and the collection and analysis of scientific evidence.

Hypotheses supported by experimental evidence provide a natural explanation for the origin of life. Simulations show that chemical and physical processes can produce the building blocks of life and the assembly of these simple subunits into more complex molecules.

Evolution

Evolution underlies the unity and diversity of life.

Shared core processes and features observed among modern and fossil organisms provide evidence of life's common origin. Divergences from shared ancestors and adaptation to different environments produced the diversity of life.

Pathways of Transformation

Organisms exchange matter and energy with the environment in order to grow, maintain themselves, and reproduce.

Life first evolved in an oxygen-free environment and relied upon anaerobic energy-releasing pathways. The evolution of oxygen-producing photosynthesis created a new selection pressure that favored organisms capable of aerobic respiration. The evolutionary transformation of aerobic bacteria into mitochondria provided eukaryotic cells with an energy advantage over prokaryotes.

Links to Earlier Concepts

This chapter explains how the molecular subunits of life introduced in Section 3.3 could have originated and assembled to form the first cells. It considers the evolution of eukaryotic traits (Sections 4.4–4.9) and draws on your understanding of fossils and the geologic time scale (16.4 and 16.5), aerobic respiration (7.1, 7.2), photosynthesis (6.2), and protein synthesis (9.2–9.5).

19.1 Looking for Life

We live in a vast universe that we have only begun to explore. So far, the only planet known to have life is Earth. In addition, the biochemical, genetic, and metabolic similarities among Earth's species imply that all evolved from a common ancestor that lived billions of years ago. What properties of the ancient Earth allowed life to arise, survive, and diversify? Could similar processes occur on other planets? These are some of the questions posed by **astrobiology**, the study of life's origins and distribution in the universe.

Astrobiologists study Earth's extreme habitats to determine the range of conditions that can support life. They have found species that withstand extraordinary levels of temperature, pH, salinity, and pressure. For example, some bacteria live deep in the soil of Chile's Atacama Desert, the driest place on Earth. Other bacteria thrive at high pressure and temperature 3 kilometers (almost 2 miles) beneath the soil surface in Virginia. There is even a biofilm (Section 4.3) under Antarctica's western ice sheet.

Understanding how these species withstand such extreme conditions on Earth informs the search for life elsewhere. For example, the details of metabolism vary, but in all known species metabolic reactions occur in or at the boundary of water-based solutions. Thus, liquid water is considered an essential requirement for life as we know it. Excitement over the discovery of water on Mars, our closest planetary neighbor, stems from this assumption. A robotic lander (**FIGURE 19.1**) found water frozen in the soil of Mars, and geologic formations suggest that, billions of years ago, water flowed across the planet's surface and pooled in giant lakes.

astrobiology Scientific study of life's origin and distribution in the universe.

FIGURE 19.1 Exploring Mars. This composite selfie was taken by NASA's Curiosity Rover, which is looking for evidence that Mars supports or previously supported microbial life.

CREDITS: (opposite) roboriginal/iStockphoto.com; (1) NASA/JPL-Caltech/MSSS.

Other features of Mars suggest a small amount of water may still flow seasonally in some areas.

Suppose scientists do find evidence that microbial life exists or existed on Mars, or on some other planet. Why would it matter? The discovery of extraterrestrial microbes would lend credence to the idea that nonhuman intelligent life exists somewhere else in the universe. The more places microbial life exists, the more likely it is that complex, intelligent life evolved on planets other than Earth.

This chapter is your introduction to a slice through time. We begin with Earth's formation and move on to life's chemical origins and the evolution of traits present in modern eukaryotes. The picture we paint here sets the stage for the next unit, which takes you along lines of descent to the present range of biodiversity. ●

19.2 The Early Earth

LEARNING OBJECTIVE

- Explain what study of ancient rocks has revealed about conditions on the early Earth.

Origin of the Universe and Earth

According to the **big bang theory**, the universe began in a single instant, about 13.7 billion years ago. In that instant, all existing matter and energy exploded outward from a single point. Then, over millions of years, gravity drew gases together to form the stars of the first galaxies.

Explosions of the early stars scattered heavier elements from which today's galaxies formed. About 5 billion years ago, a cloud of dust and rocks (asteroids) orbited the star we now call our sun. The asteroids collided and merged into bigger asteroids. The heavier these preplanetary objects became, the more gravitational pull they exerted, and the more material they gathered. By about 4.6 billion years ago, this gradual buildup of materials had formed Earth and the other planets of our solar system.

FIGURE 19.2 The Hadean eon. Artist's depiction of the early Earth.

Conditions on the Early Earth

On the geologic time scale, Earth's formation marks the onset of the **Precambrian era**, an interval that encompasses almost 90 percent of Earth's history (**FIGURE 19.2**). The first part of the Precambrian, the **Hadean eon**, runs from 4.6 billion to 4.0 billion years ago. *Hadean* means "hell-like," and early Earth lived up to this name (**FIGURE 19.3**). Planet formation did not remove all the material orbiting the Sun. As a result, interplanetary objects such as asteroids, meteoroids (smaller rocks), and comets (balls of ice and dust) constantly bombarded Earth. These extraterrestrial materials, together with gases and rocky material released by

big bang theory Well-supported hypothesis that the universe originated by a nearly instant distribution of matter through space.

Hadean eon (HAY-dee-un) Precambrian interval that runs from Earth's formation to 4 billion years ago.

Precambrian era Interval from Earth's formation to 541 million years ago.

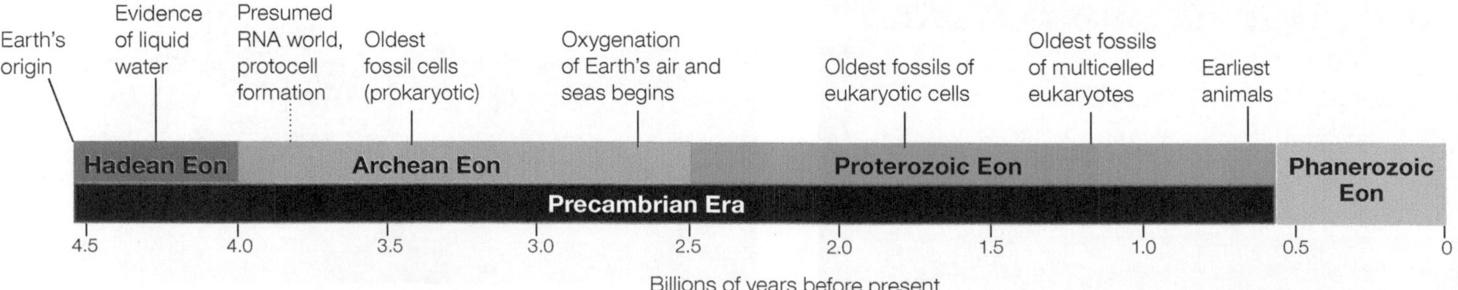
Earth's origin
Evidence of liquid water
Presumed RNA world, protocell formation
Oldest fossil cells (prokaryotic)
Oxygenation of Earth's air and seas begins
Oldest fossils of eukaryotic cells
Oldest fossils of multicelled eukaryotes
Earliest animals
Hadean Eon
Archean Eon
Proterozoic Eon
Phanerozoic Eon
Precambrian Era
4.5
4.0
3.5
3.0
2.5
2.0
1.5
1.0
0.5
0
Billions of years before present

FIGURE 19.3 Milestones of the Precambrian era. Approximate dates are based on current data.

CREDITS: (2) NASA; (3) © Cengage Learning.

frequent volcanic eruptions, provided the components of Earth's land, waters, and atmosphere.

At first, Earth's surface was molten rock, so any water present was in the form of vapor. Then, as Earth cooled, water began to pool on its surface. Determining when this cooling took place is important because, as previously noted, life as we know it requires liquid water. Zircon crystals that date back 4.4 billion years provide the earliest evidence of water. Composition of the crystals indicates they formed when water-saturated rocks melted, then cooled. The age of the crystals was determined by radiometric dating (Section 16.5).

Ancient rocks also provide information about Earth's early atmosphere. When iron-containing rocks are exposed to free oxygen (O_2), their iron becomes oxidized (it rusts). However, the iron in Earth's oldest rocks shows no sign of oxidation, indicating that the atmosphere was initially oxygen free. A lack of oxygen gas probably made life's origin possible. Had oxygen been present, it would have oxidized any small organic molecules as soon as they formed. Such molecules serve as the building blocks of life.

TAKE-HOME MESSAGE 19.2

- ✔ During Earth's early years, meteorites pummeled the planet's surface and volcanic eruptions were common.
- ✔ As Earth cooled, liquid water began to pool on its surface.
- ✔ Earth's early atmosphere had no oxygen gas.

19.3 Organic Monomers Form

LEARNING OBJECTIVES

- Describe how scientists have tested the hypothesis that simple organic molecules could have formed on the early Earth.
- List three locations where the building blocks of Earth's first life may have formed.

Organic Molecules from Inorganic Precursors

All living things are made from the same organic subunits: amino acids, fatty acids, nucleotides, and simple sugars. Until the early 1800s, chemists thought that such organic molecules possessed a special "vital force" and that only living organisms could make them. This idea gradually fell into disfavor as chemists discovered ways to produce organic compounds by artificial methods. In 1828, a German chemist synthesized urea, an organic molecule abundant in urine. Synthetic production of acetic acid (the acid in vinegar) and alanine (an amino acid) soon followed.

Sources of Life's First Building Blocks

Once scientists understood that organic molecules can form in the absence of life, they began to think about how simple organic compounds could have formed on the early Earth. We consider three possible mechanisms for this process below. Keep in mind that these mechanisms are not mutually exclusive. All three may have operated simultaneously and contributed to an accumulation of simple organic compounds in Earth's early seas. Note also that once such compounds formed, they would have persisted longer than they do today because there was no oxygen to break them down, and organisms capable of metabolizing them had not yet evolved.

Lightning-Fueled Atmospheric Reactions In 1953, Stanley Miller and Harold Urey tested the hypothesis that lightning could have powered synthesis reactions in Earth's early atmosphere. To simulate this process, they filled a reaction chamber with methane, ammonia, and hydrogen gas, and zapped it with sparks from electrodes (**FIGURE 19.4**). Within a week, a variety of organic molecules formed, including amino acids that are common in living things.

We now know that the mix of gases in this initial experiment probably did not accurately represent those present on early Earth. However, this work remains notable because it was the first experimental research into the origin of life. Later experimental variations used the same sort of apparatus with other mixtures of gases. For example, one experiment included hydrogen sulfide gas and mimicked atmospheric conditions

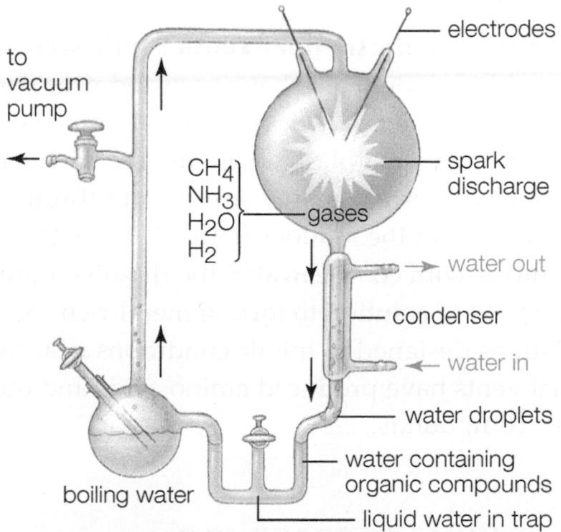

electrodes
to vacuum pump
spark discharge
CH4
NH3
H2O
H2
gases
water out
condenser
water in
water droplets
water containing organic compounds
boiling water
liquid water in trap

FIGURE 19.4 The Miller–Urey Experiment. This apparatus was used to test whether organic compounds could form spontaneously on the early Earth. Water vapor, hydrogen gas (H_2), methane (CH_4), and ammonia (NH_3) simulated Earth's early atmosphere. Sparks from an electrode simulated lightning.

FIGURE IT OUT What was the source of the nitrogen for the amino acids formed in this apparatus?

Answer: Ammonia

FIGURE 19.5 Hydrothermal vent on the seafloor. Mineral-rich water heated by geothermal energy streams out into cold ocean water. As the water cools, dissolved minerals come out of solution and form a chimney-like structure around the vent.

around ancient volcanoes. This variation produced even more amino acids than the initial test.

Delivery from Space Modern-day meteorites that fall to Earth sometimes contain amino acids, sugars, and nucleotide bases. These compounds (or precursors of them) have also been discovered in gas clouds surrounding nearby stars. Thus it is possible that some of the many meteorites that fell on the early Earth carried organic monomers that had formed in outer space.

Reactions at Hydrothermal Vents Synthesis of life's building blocks could also have occurred near deep-sea hydrothermal vents. A **hydrothermal vent** is like an underwater geyser, a place where mineral-rich water heated by geothermal energy streams out through a rocky opening in the seafloor (FIGURE 19.5). When this water mixes with cool seawater, the dissolved minerals come out of solution to form mineral-rich rock. Simulations designed to mimic conditions near hydrothermal vents have produced amino acids and other organic compounds.

TAKE-HOME MESSAGE 19.3

✔ Small organic molecules that serve as the building blocks for living things can be formed by nonliving mechanisms.

✔ These molecules form in space, near hydrothermal vents, and may have formed in the atmosphere of the early Earth.

19.4 From Polymers to Protocells

LEARNING OBJECTIVES

- List the proposed steps that led to the first cellular life.
- Explain the RNA world hypothesis.
- Describe the properties of laboratory-produced protocells.

Properties of Cells

In addition to sharing the same molecular components, all cells have a plasma membrane with a lipid bilayer. They have a genome of DNA that enzymes transcribe into RNA, and ribosomes that translate RNA into proteins. All cells replicate, and they pass on copies of their genetic material to their descendants. The many similarities in structure, metabolism, and replication processes among all life provide evidence of descent from a common cellular ancestor.

No one was around to witness the origin of life on Earth, and time has erased all traces of the earliest cells. However, scientists can still investigate this first chapter in life's history by making hypotheses about how life began, and testing those hypotheses experimentally. Their work has shown that cells could have arisen in a stepwise process beginning with inorganic materials (FIGURE 19.6). Each step on this hypothetical road to life can be explained by familiar chemical and physical mechanisms that still operate today.

Origin of Metabolism

Modern cells take up organic monomers, concentrate them, and assemble them into organic polymers. Before there were cells, a nonbiological process that concentrated organic subunits would have increased the chance of polymer formation. By one hypothesis, this process occurred on clay-rich tidal flats. Clay particles have a slight negative charge, so positively charged molecules (such as organic subunits dissolved in seawater) stick to them. Such binding would have concentrated the subunits. Evaporation at low tide would have continued this process, and energy from the Sun might have triggered polymerization. Amino acids do form short chains under simulated tidal flat conditions.

The **iron–sulfur world hypothesis**, proposed by Günter Wächtershäuser, holds that life originated at deep-sea hydrothermal vents. The porous rocks that form around these vents are rich in iron sulfides, com-

hydrothermal vent Underwater opening from which mineral-rich water heated by geothermal energy streams out.

iron–sulfur world hypothesis Hypothesis that the metabolic reactions that led to the first cells took place on the porous surface of iron–sulfide-rich rocks at hydrothermal vents.

pounds that easily donate electrons to inorganic gases dissolved in the hot seawater spewing from the vents. Accepting electrons causes the gases (such as carbon dioxide and hydrogen cyanide) to react and form organic molecules (such as pyruvate). Keep in mind spontaneous chemical reactions of this type proceed more quickly in a warm environment (such as the area near a hydrothermal vent) than in a cooler one. Organic molecules that formed would have adhered to iron–sulfur compounds and accumulated inside tiny, cell-sized chambers of the rocks. Later, such molecules became catalytic: early versions of enzymes that carry out modern metabolic reactions. The iron–sulfur cofactors required by all modern organisms may be a legacy of life's rocky beginnings at hydrothermal vents.

Origin of the Genome

All modern cells have a genome of DNA. They pass copies of their DNA to descendant cells, which use instructions encoded in the DNA to build proteins. Some of these proteins are enzymes that synthesize new DNA, which is passed along to descendant cells, and so on. Thus, protein synthesis depends on DNA, which is built by proteins. How did this cycle begin?

In the 1960s, Francis Crick and Leslie Orgel addressed this dilemma by suggesting that RNA may have been the first molecule to encode genetic information, a concept known as the **RNA world hypothesis**. Evidence that RNA can both store genetic information and function like an enzyme in protein synthesis supports this hypothesis. **Ribozymes**, or RNAs that function as enzymes, are common in living cells. For example, the rRNA in ribosomes speeds formation of peptide bonds during protein synthesis (Section 9.4). Other ribozymes cut noncoding bits (introns) out of newly formed mRNAs.

To investigate the likelihood that an RNA world existed, some researchers try to create synthetic RNAs with the properties that such a world would require. Such research recently yielded a synthetic RNA that can copy most other RNAs, thus mimicking the function of the enzyme RNA polymerase. However, no one has yet produced a synthetic RNA that can copy both other RNAs and itself.

Another drawback to the RNA world hypothesis is its inability to explain how a system based entirely on RNA gave rise to the current system that uses both RNA and DNA. Recent experiments have shown that molecules containing both RNA and DNA are less stable than either RNA or DNA alone, so it is unlikely that such hybrid molecules served as an intermediate.

protocell Membranous sac that contains interacting organic molecules; hypothesized to have formed prior to the earliest life-forms.
RNA world hypothesis Hypothesis that RNA served as the genetic information of early life.
ribozyme (RYE-boh-zime) RNA that functions as an enzyme.

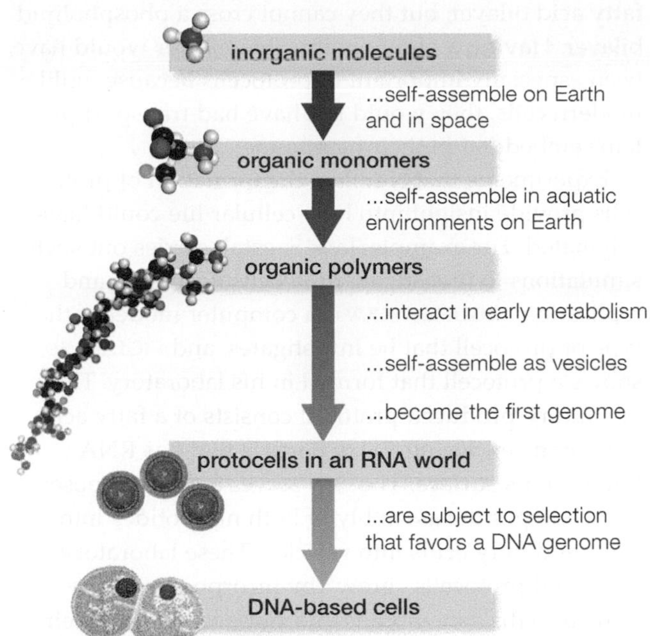
inorganic molecules
...self-assemble on Earth and in space
organic monomers
...self-assemble in aquatic environments on Earth
organic polymers
...interact in early metabolism
...self-assemble as vesicles
...become the first genome
protocells in an RNA world
...are subject to selection that favors a DNA genome
DNA-based cells

FIGURE 19.6 Proposed steps on the path to cellular life. Scientists carry out experiments and simulations to test the feasibility of each step.

Origin of the Plasma Membrane

Self-replicating molecules and products of other early synthetic reactions would have floated away from one another unless something enclosed them. In modern cells, a plasma membrane serves this function. If the first reactions took place in tiny rock chambers, the rock would have acted as a boundary. Over time, lipids produced by reactions inside a chamber could have accumulated and lined the chamber wall, forming a protocell. A **protocell** is a membrane-enclosed collection of interacting molecules that can take up material and replicate. Protocells are hypothesized to be the ancestors of cellular life.

Membranes of early protocells probably consisted of a variety of fatty acids, rather than the phospholipids that make up the bulk of modern cell membranes (Section 3.4). Like phospholipids, fatty acids have water-loving (hydrophilic) and water-hating (hydrophobic) parts, so they form a lipid bilayer when mixed with water and can self-assemble as vesicles (membranous, fluid-filled chambers). However, fatty acids are structurally simpler than phospholipids and would have been more abundant on early Earth. A fatty acid membrane also provides an advantage in terms of permeability. Nucleotides and ions move through a fatty acid bilayer, but they cannot cross a phospholipid bilayer. Having a highly permeable bilayer would have been especially important to protocells because, unlike modern cells, they would not have had transport proteins embedded in their membrane.

Experiments that simulate the formation of protocells provide insight into how cellular life could have originated. For example, Jack Szostak carries out such simulations to investigate protocells formation and replication. **FIGURE 19.7A** is a computer model of the type of protocell that he investigates, and **FIGURE 19.7B** shows a protocell that formed in his laboratory. The laboratory-produced protocell consists of a fatty acid membrane enclosing a clay particle that has RNA bound to its surface. The clay serves a dual purpose; it encourages the assembly of both nucleotides into RNA and fatty acids into vesicles. These laboratory-produced protocells "grow" by incorporating fatty acids into their membrane and nucleotides into their RNA. New protocells are produced when large ones break into smaller units as a result of mechanical force. However, the resulting "daughter cells" do not have the same RNA sequence as their parent.

TAKE-HOME MESSAGE 19.4

✔ Metabolic reactions may have begun when molecules became concentrated on clay particles or inside tiny rock chambers near hydrothermal vents.

✔ RNA can serve as an enzyme, as well as a genome. An RNA world may have preceded evolution of DNA-based genomes.

✔ Vesicle-like structures with outer membranes form spontaneously when certain organic molecules are mixed with water.

✔ Protocells, vesicles that contain interacting molecules, may have been an intermediate step on the road to cellular life.

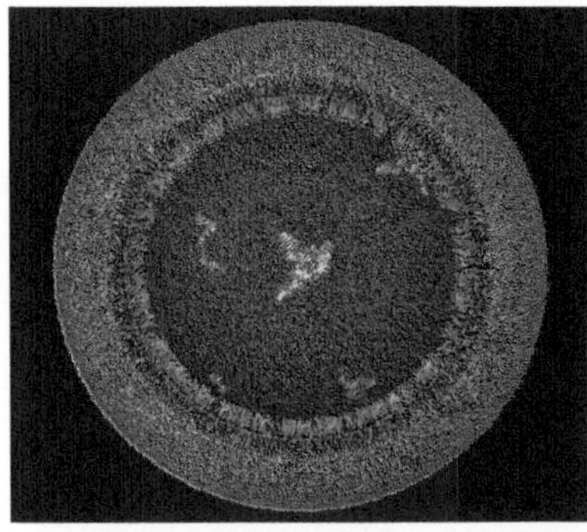

A Illustration of a laboratory-produced protocell with a bilayer membrane of fatty acids and strands of RNA inside.

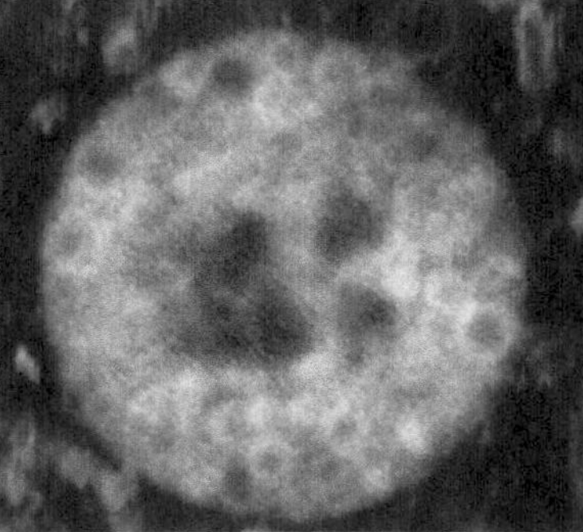

B Laboratory-formed protocell consisting of RNA-coated clay (red) surrounded by fatty acids and alcohols.

FIGURE 19.7 Protocells. Scientists test hypotheses about protocell formation through laboratory simulations and field experiments.

19.5 The Age of Prokaryotes

LEARNING OBJECTIVES

- List the probable traits of LUCA, the last common ancestor of all life.
- Describe the oldest fossils of early life and the environments in which these fossil cells lived.

The Last Common Ancestor of All Life

The processes described in Section 19.4 may have produced cellular life more than once. If so, all but one of those early cell lineages became extinct. Similarities among modern genomes indicates that all living organisms descended from the same ancestral species, which scientists refer to as LUCA (last universal common ancestor). This ancestor probably lived early in the **Archean eon**, which extends from 4 billion years ago to 2.5 billion years ago. "Archean" means beginning.

What do we know about LUCA? Given the relationships among modern species, most scientists agree that LUCA was prokaryotic, meaning it did not have a nucleus (Section 1.4). Oxygen was scarce on the early Earth, so LUCA must have been anaerobic (capable of living without oxygen).

To learn more about where and how LUCA lived, researchers recently looked at the functions of genes that modern bacteria and archaea share. These are the genes that were most likely passed down from LUCA. Analysis of these genes suggests that LUCA lived in a high-temperature environment rich in metal ions, where it obtained energy by metabolizing hydrogen. This result can be taken as support for the hypothesis that life began near hydrothermal vents. Keep in mind, however, that LUCA was not the first cellular life, but rather the last shared ancestor of all modern life. It remains possible that life arose in shallow waters before colonizing the deep hydrothermal vent habitat where LUCA lived.

Fossil Evidence of Early Life

Finding and identifying signs of early cells poses a challenge. Cells are microscopic and most have no hard parts to fossilize. In addition, few ancient rocks that could hold early fossils still exist. Tectonic plate movements have destroyed nearly all rocks older than about 4 billion years, and most slightly younger rocks have been heated or undergone other processes that destroy traces of biological material. To add to the difficulty, structures formed by nonbiological mechanisms

Archean eon (ahr-KEY-an) Precambrian interval extending from 4 billion years ago to 2.5 billion years ago. First eon for which there is evidence of life.

FIGURE 19.8 Modern stromatolites. Each of these rocky structures in Australia's Shark Bay consists of living cells atop the layered remains of earlier generations of cells and the sediment that they trapped.

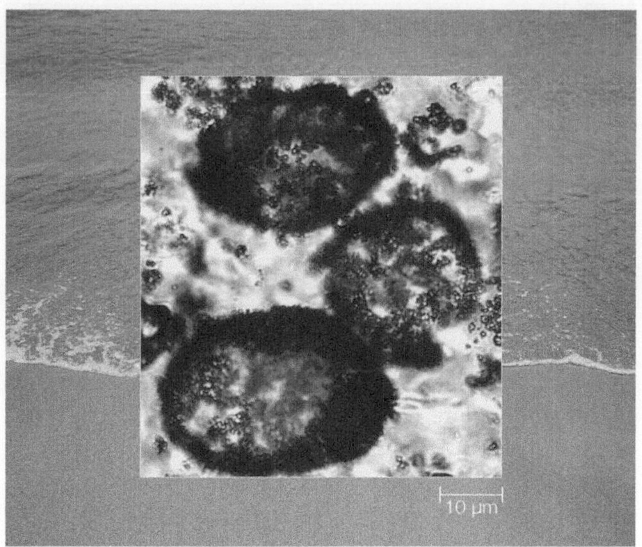
10 μm

FIGURE 19.9 Fossil cells from an ancient beach. Microscopic 3.4-billion-year-old fossil cells nestle among sand grains in a sandstone. They resemble bacteria that live in modern mud flats.

sometimes resemble fossils. To avoid mistakenly accepting such material as a genuine fossil, scientists constantly reanalyze purported fossil finds and they often question one another's conclusions. For example, 3.5-billion-year-old filaments discovered in 1993 by William Schopf were long considered the earliest fossil cells. Recently, however, scientists used new methods to show that these filaments are crystals rather than biological material.

Fossils of Early Cells

At present, the oldest proposed fossil cells come from Canadian rocks that date back at least 3.77 billion years. The rocks contain tiny tubes and filaments that resemble those formed by cells near modern deep sea hydrothermal vents. Australian fossils that date back 3.2 billion years are also thought to be the remains of cells that lived near a hydrothermal vent.

The most widespread evidence of early cellular life comes from fossil stromatolites. A **stromatolite** is a rocky layered structure that forms in shallow sunlit water when a mat of prokaryotic cells traps minerals and sediment. When sediment covers the living cells atop the stromatolite, new cells grow over that sediment and then trap more sediment to form a new layer. Scientists can observe this process in modern stromatolites such as those in Australia's Shark Bay (**FIGURE 19.8**). Here, photosynthetic bacteria that live on the stromatolites' upper, sunlit surface grow as a mat that traps sediment. As new sediment is added, bacterial cells grow up through it and trap more sediment. Archaea and heterotrophic bacteria live in the stromatolite's deeper layers. Ancient stromatolites probably contained both bacteria and archaea too.

Some proposed stromatolite fossils from Greenland date back 3.7 billion years. However, the oldest undisputed stromatolite fossils date to 2.8 billion years ago. Stromatolites reached their peak distribution 1.25 billion years ago, when they were common worldwide.

Australian sandstone rocks that date back 3.4 billion years provide another glimpse of early life. These rocks include spherical fossils nestled among sand grains from an ancient beach (**FIGURE 19.9**). The presence of the mineral pyrite (iron sulfide) in and around the fossil cells suggests the cells were metabolically similar to modern sulfur bacteria. These bacteria, which are common in mud flats, use sulfur as the final electron acceptor in their energy-producing reactions. Pyrite forms and accumulates as a by-product of these reactions.

Taken together, the variety of early fossils indicates that by about 3 billion years ago, prokaryotic life was widespread in Earth's waters, from sandy shorelines to the deep ocean floor.

TAKE-HOME MESSAGE 19.5

- ✔ The ancestor of all modern life arose by about 4 billion years ago.
- ✔ The first cells were most likely anaerobic and prokaryotic.
- ✔ Fossils indicate that by about 3 billion years ago prokaryotic cells were widespread in the ocean.

stromatolite (stroh-MAT-oh-lite) Rocky, layered structure that forms over time as a mat of aquatic prokaryotic cells traps sediments.

19.6 A Rise in Oxygen

LEARNING OBJECTIVES

- Explain what caused the increase in oxygen in Earth's air and waters.
- Describe how the increase in oxygen affected the existing organisms and how it set the stage for the later movement of life onto land.

The Cause of Oxygenation

As previously noted, life began when Earth had little or no free oxygen. Today, oxygen makes up 21 percent of the atmosphere, and most of Earth's waters contain some dissolved oxygen. The increase in oxygen is a consequence of the evolution of the oxygen-releasing pathway of photosynthesis in cyanobacteria. Many types of bacteria carry out photosynthesis, but only the cyanobacteria do so by a pathway that produces oxygen as a by-product.

Cyanobacterial production of oxygen may have begun as early as 2.9 billion years ago. At first, most oxygen these aquatic cells released reacted with materials such as iron-containing rocks and dissolved organic compounds. Then, about 2.5 billion years ago, the amount of free oxygen in the atmosphere and surface waters began to rise significantly.

Effects of Oxygenation

The increasing concentration of oxygen had dramatic effects on the diversity and distribution of life. Oxygen was toxic to many species that had evolved in its absence. Inside cells, oxygen reacts with metal ions, forming free radicals (Section 2.3) that can damage DNA and other essential cell components. Some species had a limited capacity to detoxify free radicals and, as the level of oxygen increased, selection favored an expansion of this capacity. Eventually, descendants of these oxygen-tolerant species evolved the ability to carry out aerobic respiration, the energy-releasing pathway in which oxygen serves as the final electron acceptor (Section 7.2). Aerobic respiration is the main energy-releasing pathway in eukaryotes, whose evolution we consider in the next section. Species that were unable to detoxify the free radicals either went extinct or became restricted to low-oxygen environments.

The increasing oxygen concentration also led to formation of the **ozone layer**, a region of the upper atmosphere that has a high concentration of ozone gas (O_3). Ozone serves as a sort of sunblock by absorbing ultraviolet (UV) wavelengths that can cause mutations (Section 8.6). Before the ozone layer formed, life was restricted to aquatic habitats, where water shielded organisms from incoming UV radiation. Only after the ozone layer formed was it possible for life to move onto land.

TAKE-HOME MESSAGE 19.6

✔ Evolution of the oxygen-releasing pathway of photosynthesis in cyanobacteria resulted in the oxygenation of Earth's atmosphere and waters.

✔ The rise in oxygen favored organisms that could detoxify oxygen or use it in aerobic respiration.

✔ Addition of oxygen to the atmosphere led to formation of the ozone layer, which allowed life to move onto land.

19.7 Origin and Evolution of Eukaryotes

LEARNING OBJECTIVE

- Describe the likely origins of the eukaryotic nucleus, endomembrane system, mitochondria, and chloroplasts.

Eukaryotic Traits and Traces

Let's take a moment to review how eukaryotes differ from both bacteria and archaea. The genome of a typical bacterial or archaeal cell consists of a circle of DNA that, in almost all species, is in contact with the cytoplasm, rather than enclosed within a membrane. By contrast, the DNA of a eukaryotic cell is distributed among multiple linear chromosomes and is always inside a nucleus. The nucleus is associated with a uniquely eukaryotic membrane system that includes the endoplasmic reticulum (ER) and Golgi bodies. All eukaryotes have mitochondria or similar energy-releasing organelles derived from mitochondria, and many have chloroplasts. No bacteria or archaea have either mitochondria or chloroplasts.

Nuclei and other organelles are almost never preserved during fossilization, but other traits can indicate that a fossilized cell was eukaryotic. Eukaryotic cells are generally larger than prokaryotic ones. A cell wall with complex patterns, spines, or spikes also indicates that a cell is most likely eukaryotic.

To date, the oldest fossils that most scientists agree are eukaryotic date to about 1.8 billion years ago during the Proterozoic eon. The **Proterozoic eon** is the final portion of the Precambrian, extending from 2.5 billion to 541 million years ago. These oldest fossils resemble the resting stage cells (cysts) of some modern single-celled algae. The age of these fossils is consis-

ozone layer (OH-zone) Upper atmospheric region rich in ozone (O_3) that screens out incoming ultraviolet (UV) radiation.

Proterozoic eon (pro-ter-oh-ZOE-ik) Final eon of the Precambrian era; 2.5 billion to 541 million years ago.

tent with molecular clock studies that date the origin of eukaryotes to between 1.9 and 1.7 billion years ago. Such studies use the degree of genetic differences among existing lineages to estimate the time since these lineages shared a common ancestor.

A Mixed Heritage

The first eukaryote must have evolved from a prokaryotic ancestor, so we can ask whether that ancestor was bacterial or archaeal. The answer is complicated and still not fully understood. Eukaryotes are similar to archaea with regard to how they store their DNA (wrapped around histone proteins) and in some details of how they transcribe, replicate, and repair their DNA. On the other hand, many eukaryotic genes, especially those involved in energy metabolism, are more similar to bacterial genes than archaeal ones.

The mix of archaeal and bacterial traits in eukaryotes indicates that this group has a composite heritage.

Origin of the Nucleus

The signature feature of eukaryotes is their nucleus. Formation of a nucleus probably began in the archaeal ancestor of eukaryotes when portions of the plasma membrane folded inward (**FIGURE 19.10**). Over time, some of this now internal membrane came to surround the DNA, forming the start of a nuclear membrane. Other portions of internalized membrane gave rise to additional components of the endomembrane system.

Study of the few modern prokaryotes known to have internal membranes illustrates that membrane infolding does occur and how it can be advantageous. Membrane-bound enzymes carry out many essential reactions in prokaryotes. Thus, membrane infoldings can provide a selective advantage by increasing the surface area available to hold such enzymes. An internal membrane around the nucleus can also protect a genome from physical threats such as radiation or biological threats such as viruses.

What was the archaeal ancestor of eukaryotes like? It may have resembled *Lokiarchaeum*, a recently discovered group of archaea that is referred to as "Loki." Scientists first detected Loki when they analyzed DNA samples taken near a hydrothermal vent deep in the Arctic Sea. No one has been able to grow Loki in the laboratory yet, so we do not know much about its structure. However, we do know that of all archaeal genomes examined so far, Loki's has the greatest number of eukaryote-like genes. For example, Loki is the only archaeal group with genes homologous to the genes that regulate membrane trafficking in eukaryotes.

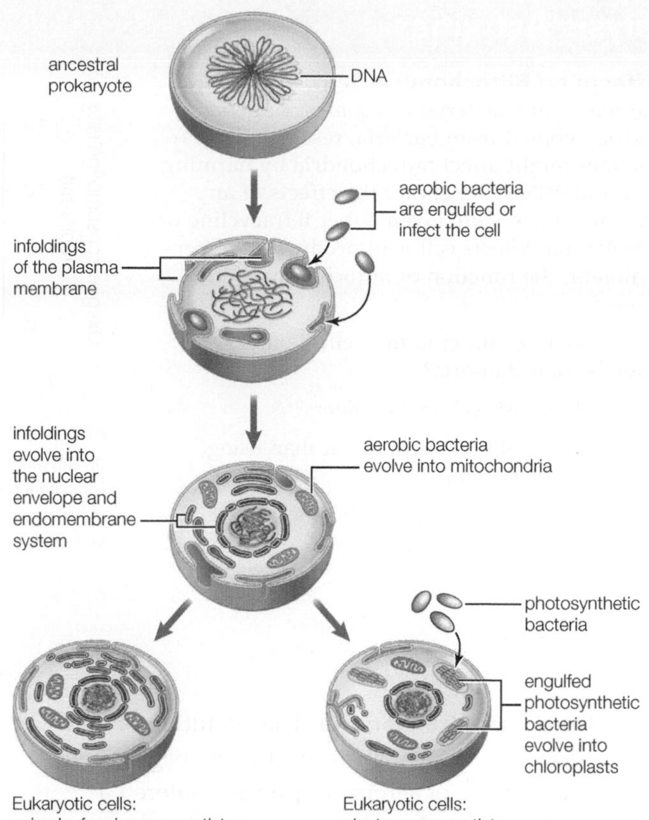

ancestral prokaryote
DNA
aerobic bacteria are engulfed or infect the cell
infoldings of the plasma membrane
infoldings evolve into the nuclear envelope and endomembrane system
aerobic bacteria evolve into mitochondria
photosynthetic bacteria
engulfed photosynthetic bacteria evolve into chloroplasts
Eukaryotic cells: animals, fungi, some protists
Eukaryotic cells: plants, some protists

FIGURE 19.10 Proposed steps in the evolution of eukaryotes.

The Endosymbiont Hypothesis

Mitochondria and chloroplasts resemble bacteria in their size and shape, and these organelles replicate independently of the eukaryotic cell that holds them. These organelles have their own genome which, as in most prokaryotes, is a single circular chromosome. They have ribosomes that resemble bacterial ribosomes. They also have at least two outer membranes, and their innermost membrane is similar to a bacterial plasma membrane.

Recognition of these similarities led to the formulation of the **endosymbiont hypothesis**, which states that mitochondria and chloroplasts evolved from bacteria. (Endosymbiosis means living inside and refers to a relationship in which one organism lives inside another.) Nearly all eukaryotic lineages have mitochondria or mitochondria-like organelles, but only some have chloroplasts. Thus, biologists postulate that the two types of organelles were acquired independently in the sequence illustrated in **FIGURE 19.10**.

---

endosymbiont hypothesis (en-doh-SIM-by-ont) Hypothesis that mitochondria and chloroplasts evolved from bacteria.

Data Analysis Activities

Antibiotic Effects on Mitochondria Tetracyclines are antibiotics that inactivate bacterial ribosomes. Knowing that mitochondria evolved from bacteria, researchers suspected tetracyclines might affect mitochondria by harming their ribosomes. **FIGURE 19.11** shows the effects of an experiment in which mice were given either tetracycline or a control antibiotic that affects cell wall production. After 10 days of treatment, the function of mitochondria in the mice was assessed.

1. How did the two drugs differ in their effect on oxygen consumption by mitochondria?
2. Which rats had the least ATP in their livers?
3. Why was a control antibiotic used, rather than using untreated mice as the control?
4. Do these results support the hypothesis that tetracyclines impair mitochondrial function?

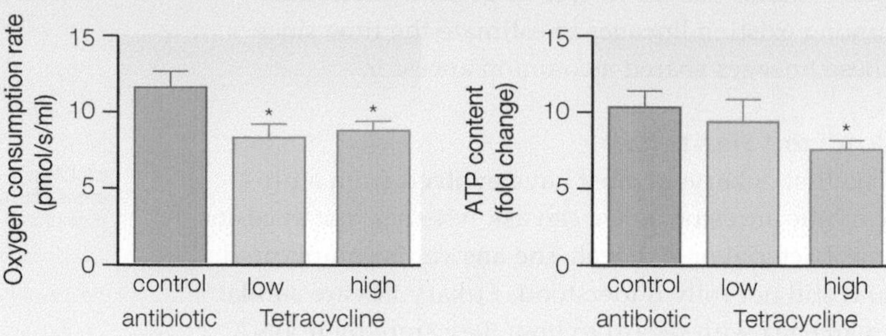

Oxygen consumption rate (pmol/s/ml)
15
10
5
0
*
*
control antibiotic
low
high
Tetracycline
ATP content (fold change)
15
10
5
0
*
control antibiotic
low
high
Tetracycline

FIGURE 19.11 Effects of tetracycline on mitochondria. Mice received low-dose tetracycline, high-dose tetracycline, or a control antibiotic that targets bacterial cell walls. Mitochondrial function was assessed by looking at oxygen consumption rate and ATP content in homogenized mouse livers. Bar graphs show means with standard error. An * indicates a statistically significant difference between the experimental and the control treatments.

Evolution of Mitochondria The evolution of mitochondria by endosymbiosis most likely began when bacteria capable of aerobic respiration entered a archaeal cell or early eukaryote descended from archaea. When the host cell divided, it passed some "guest" bacteria, referred to as endosymbionts, along to its offspring. As the host and its bacterial endosymbionts lived together over many generations, both evolved. The endosymbionts lost the capacity to build structures they no longer needed, such as a cell wall. Some metabolism-related genes from the endosymbiont moved into the host's genome. Redundant genes were eliminated; a gene could lose its function in either partner if a similar gene carried by the other partner still worked. Eventually, the host and endosymbiont became incapable of living independently. At that point, the endosymbionts had become mitochondria.

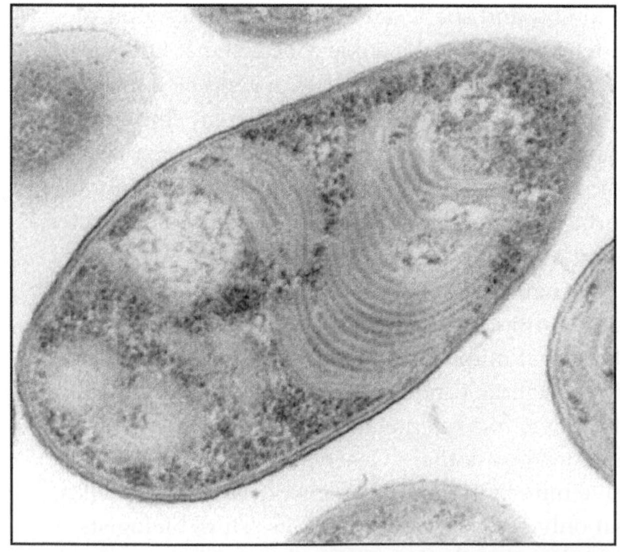

FIGURE 19.12 One proposed modern relative of mitochondria. *Methylobacter*, a methane-feeding species of bacteria that has stacked internal membranes.

The endosymbiotic hypothesis can explain why the information-processing traits of eukaryotes resemble those of archaea, while traits related to energy metabolism resemble those of bacteria. Information-processing traits were passed down from the lineage that served as the endosymbiotic host. Metabolism-related traits were passed down from the bacterial endosymbionts that gave rise to mitochondria.

What do we know about the bacteria that gave rise to mitochondria? The high degree of similarity among all modern mitochondrial genomes indicates that this organelle arose only once. In other words, all mitochondria are descended from the same bacterial ancestor. To determine what that ancestor could have been like, researchers look for the closest relatives of mitochondria among modern bacteria. Such analysis places methanotropic (methane-feeding) bacteria among the closest relatives of mitochondria. In addition to being metabolically similar to mitochondria, some members of this aerobic group have stacks of internal membranes (**FIGURE 19.12**). These stacks resemble the internal membranes (cristae) of mitochondria.

Recently some scientists have proposed that the evolution of mitochondria was an essential step toward the greater complexity that characterizes eukaryotes. They point out that because gene expression is an energy-requiring process, a cell with mitochondria can express a larger genome and produce a greater variety of proteins than one that lacks these organelles. Keep in mind that the genome of the average eukaryotic cell is hundreds of times larger than that of the typical prokaryote. The energy boost that mitochondria provided would have also opened the way for the evolution of some new energy-intensive processes. For example, membrane trafficking, which is typical of eukaryotic cells (Section 5.10), does not occur in prokaryotes.

Evolution of Chloroplasts Chloroplasts first arose through an endosymbiosis between a host eukaryote and its guest cyanobacteria. As stated earlier, cyanobacteria are the only bacteria that carry out photosynthesis by the oxygen-reducing pathway as chloroplasts do. The result of this endosymbiotic event was a photosynthetic alga, a type of protist. As explained in Chapter 20, some photosynthetic protists would later be taken up by other protists and evolve into chloroplasts inside of them. Thus, although chloroplasts evolved repeatedly, all still ultimately trace their ancestry to the same cyanobacterial ancestor.

Diversification of Eukaryotes

Early eukaryotic cells had a nucleus, endomembrane system, mitochondria, and—in certain lineages—chloroplasts. These single-celled organisms were the first members of the collection of lineages that we now refer to as protists. As you will learn in Chapter 20, branches from the various protist lineages would later give rise to the more familiar eukaryotic groups: plants, fungi, and animals.

The oldest animal fossils date to about 650 million years ago, during the final period of the Proterozoic (the Ediacaran period). Animal lineages that arose during the Ediacaran period and survived to the present include sponges, cnidarians (such as jellies and sea anemones), and annelids (such as sandworms and earthworms). However, no known fossils of snails, crabs, sea stars, or fish date to the Precambrian. The groups to which these animals belong first appear in the fossil record during an adaptive radiation in the Cambrian. We discuss animal evolution in detail in Chapters 23 and 24.

Fungi were also present in the seas of the late Proterozoic. However, they were not the sort of mushroom-making fungi with which you are probably familiar. The fungi that evolved in the late Proterozoic lived as aquatic, flagellated cells. As you will learn in Chapter 22, some fungi still produce flagellated spores, a legacy of this group's aquatic origins.

The green algal ancestors of plants were present in the Proterozoic, but plants themselves did not appear until about 450 million years ago. We discuss plant evolution in Chapter 21.

Note that Precambrian history is the story of life in the seas. For most of this interval, the ozone layer remained so thin that the amount of incoming UV radiation made life on land impossible. Life survived only in the seas, where seawater screened out some of this damaging radiation.

TAKE-HOME MESSAGE 19.7

- ✔ A nucleus and other membrane-bound organelles are defining features of eukaryotic cells.
- ✔ The nucleus and endomembrane system could have arisen through modification of infoldings of the plasma membrane.
- ✔ Mitochondria and chloroplasts evolved through endosymbiosis and are descended from bacteria.

19.1 Looking for Life (revisited)

When it comes to sustaining life, Earth is just the right size. If the planet were much smaller, it would not exert enough gravitational pull to keep atmospheric gases from drifting off into space. The photo to the right shows the sizes of Earth and Mars. As you can see, Mars is only about half the size of Earth. Mars has less gravity than Earth and is less able to hold on to atmospheric gases. The relatively small amount of atmosphere that remains consists mainly of carbon dioxide, some nitrogen, and only traces of oxygen. Thus, if life exists on Mars, it is almost certainly anaerobic. ●

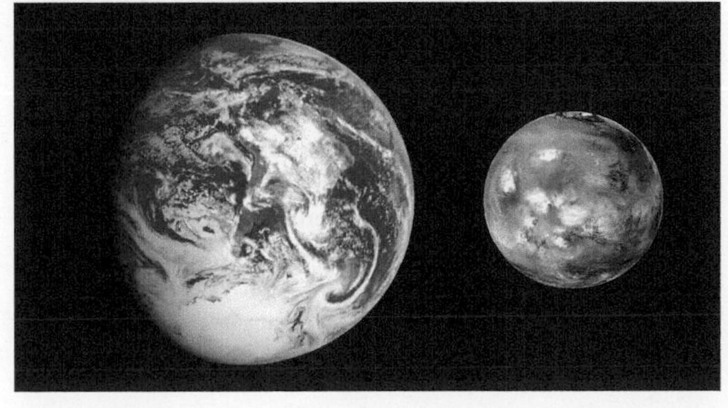

CREDIT: (in text) NASA/JPL.

STUDY GUIDE

Section 19.1 Astrobiology is the study of life's origin and distribution in the universe. The presence of cells in deserts and deep below Earth's surface suggests life may exist in similar settings on other planets. Mars, our closest planetary neighbor, has some water, but a much thinner atmosphere. Robotic rovers are currently searching Mars for signs of life.

Section 19.2 According to the **big bang theory**, the universe formed in an instant 13 to 15 billion years ago. Earth and other planets formed about 4.6 billion years ago from material released by explosions of giant stars. The first portion of the **Precambrian era** is known as the **Hadean eon**. During the Hadean, the planet had little or no free oxygen, and it received a constant hail of meteorites. Oxygen is highly reactive, so a lack of oxygen would have facilitated assembly of simple organic compounds. Earth's surface was initially molten, but by 4.3 billion years ago it had cooled enough for life-sustaining water to pool on its surface.

Section 19.3 Laboratory simulations provide indirect evidence that organic monomers could have formed by lightning-fueled reactions in Earth's early atmosphere or in the hot, mineral-rich water around **hydrothermal vents**. Examination of meteorites shows that such compounds form in deep space and are transported to Earth by meteorites.

Section 19.4 Proteins that function in metabolic pathways might have first formed when amino acids that became stuck to clay on tidal flats bonded under the heat of the Sun. Alternatively, the **iron–sulfur world hypothesis** postulates that metabolism began on the surface of rocks at hydrothermal vents.

Membrane-like structures and vesicles form when lipids are mixed with water. They serve as a model for **protocells**, which may have preceded cells. Laboratory-produced protocells are used to test hypotheses about how and where cells could have formed.

The **RNA world hypothesis** proposes that the evolution of DNA-based systems was preceded by an interval during which RNA was the genetic material. Discovery of **ribozymes**, RNAs that act as enzymes, lends support to the RNA world hypothesis, as does production of synthetic ribozymes capable of copying RNAs. However, the RNA world hypothesis does not account for the switch from RNA-based to DNA-based genomes.

Section 19.5 Genetic comparisons among modern organisms indicate that all modern life descended from the same cellular ancestor. This last universal common ancestor (LUCA) probably lived during the **Archean eon**. It was most likely anaerobic and prokaryotic. Studies of shared genes in modern prokaryotes (genes that could have been passed down from LUCA) suggest LUCA lived near a hydrothermal vent.

Searching for fossils of early cells is a difficult task. As a result of geologic processes, there are few rocks that date back more than 4 million years. In addition, mineral materials can sometimes appear to be fossils.

The most prevalent fossil evidence of early cells comes from **stromatolites**, which are rocky structures that form in shallow water when mats of prokaryotic cells trap sediments. The oldest proposed stromatolites date to 3.7 billion years ago. Other ancient fossils of cells indicate that by about 3 billion years ago, prokaryotic cells lived among sand grains on ancient beaches and near deep-sea hydrothermal vents.

Section 19.6 By 2.5 billion years ago, oxygen released as a by-product of photosynthesis by cyanobacteria had begun to enter Earth's waters and atmosphere. The presence of free oxygen created a new selective pressure that favored cells capable of withstanding it. Some such cells evolved the ability to use oxygen in aerobic respiration. Addition of oxygen to the atmosphere also led to formation of the **ozone layer**. By screening out some dangerous UV radiation, the ozone layer made it possible for life to move onto land.

Sections 19.7 Eukaryotes arose during the **Proterozoic eon**. They have a composite ancestry. Genes governing information processing are most like those of archaea, and genes related to metabolism are most like those of bacteria.

Internal membranes of eukaryotic cells such as a nuclear membrane and endoplasmic reticulum probably evolved from infoldings of the plasma membrane.

According to the **endosymbiont hypothesis**, mitochondria and chloroplasts evolved from bacteria. Mitochondria evolved from aerobic bacteria that entered and lived inside a host archaeon. Scientists have identified modern relatives of the bacteria and the archaeon involved in this partnership. The evolution of mitochondria increased the amount of energy available in cells, thus opening the way to a larger genome and energy-intensive processes such as membrane trafficking.

Chloroplasts first evolved from cyanobacteria. Later, protists with such chloroplasts were taken up and evolved into chloroplasts inside other protists.

Protists were the first eukaryotes. Over time, various protist lineages gave rise to plants, fungi, and animals.

SELF-QUIZ

Answers in Appendix VII

1. Which of the following is NOT true of conditions during the Hadean eon?
   a. The atmosphere held no oxygen.
   b. Rocks began to rust.
   c. Volcanic eruptions were common
   d. Meteorites struck the Earth continually.

2. An abundance of ________ in Earth's early atmosphere would have interfered with assembly of organic compounds.
   a. carbon dioxide　　c. water
   b. ammonia　　d. oxygen

3. Stanley Miller's experiment demonstrated that ________ .
   a. Earth is more than 4 billion years old
   b. under some conditions, amino acids can assemble spontaneously from inorganic materials
   c. oxygen is necessary for all life
   d. RNA was the first genome

4. According to one hypothesis, negatively charged clay particles played a role in early ________ .
   a. protein formation　　c. photosynthesis
   b. DNA replication　　d. oxygen declines

5. The prevalence of ________ in living organisms is taken as support for the idea that life arose near deep-sea vents.
   a. mitochondria　　c. DNA
   b. iron–sulfide cofactors d. a plasma membrane

6. An RNA that functions as an enzyme is a ________ .
   a. protein　　c. ribosome
   b. protocell　　d. ribozyme

7. Among prokaryotes, only the cyanobacteria ________ .
   a. live near hydrothermal vents
   b. produce oxygen during photosynthesis
   c. cannot tolerate oxygen
   d. have a nucleus-like structure

8. The evolution of ________ resulted in the eventual formation of the ozone layer.
   a. DNA-based genomes
   b. aerobic respiration
   c. sexual reproduction
   d. oxygen-producing photosynthesis

9. Evolution of ________ increased the energy available to cells, making a larger genome and processes such as membrane trafficking possible.
   a. the nucleus　　c. chloroplasts
   b. endoplasmic reticulum　d. mitochondria

10. Infoldings of the plasma membrane into the cytoplasm of some ancestral cells may have evolved into the ________ .
   a. nuclear envelope　　c. primary cell wall
   b. ER membranes　　d. both a and b

11. The host cell in the endosymbiosis that produced the first chloroplasts was a(n) ________.
   a. early eukaryote　　c. aerobic bacterium
   b. archaeon　　d. alga

12. A stromatolite is a structure ________ .
   a. produced by endosymbiosis
   b. that forms when a meteorite hits the Earth
   c. consisting of layered bacteria and sediment
   d. that expels hot water from deep in the Earth

13. The first eukaryotes were ________ .
   a. fungi　　c. protists
   b. plants　　d. animals

14. Which of the following events did NOT occur during the Precambrian?
   a. Formation of the ozone layer
   b. RNA world
   c. Evolution of the first animals
   d. Evolution of the first plants

15. Chronologically arrange the evolutionary events, with 1 being the earliest and 6 the most recent.

___ 1	a. onset of oxygen-releasing pathway of photosynthesis
___ 2	b. liquid water appears
___ 3	c. origin of protocells
___ 4	d. emergence of the first eukaryotes
___ 5	e. origin of chloroplasts
___ 6	f. the big bang

CRITICAL THINKING

1. Researchers looking for fossils of the earliest life-forms face many hurdles. For example, few sedimentary rocks date back more than 3 billion years. Review what you learned about plate tectonics in Section 16.5. Then, explain why so few remaining samples of these early rocks remain.

2. The astronomer Carl Sagan once said, "We are made of star stuff." Explain why this is true of all life on Earth.

3. How did the evolution of oxygen-releasing photosynthesis in cyanobacteria increase the likelihood that mitochondria would one day evolve?

4. Just as all life shares a last universal common ancestor, all eukaryotes share a last eukaryotic common ancestor (LECA). In considering what LECA was like, scientists look for features and processes common to all or nearly all eukaryotic groups. They assume that these structures emerged before the groups diverged and thus were present in LECA. Make a list of the features and processes that you think might have been passed down from LECA to modern eukaryotes.

20 VIRUSES, BACTERIA, AND ARCHAEA

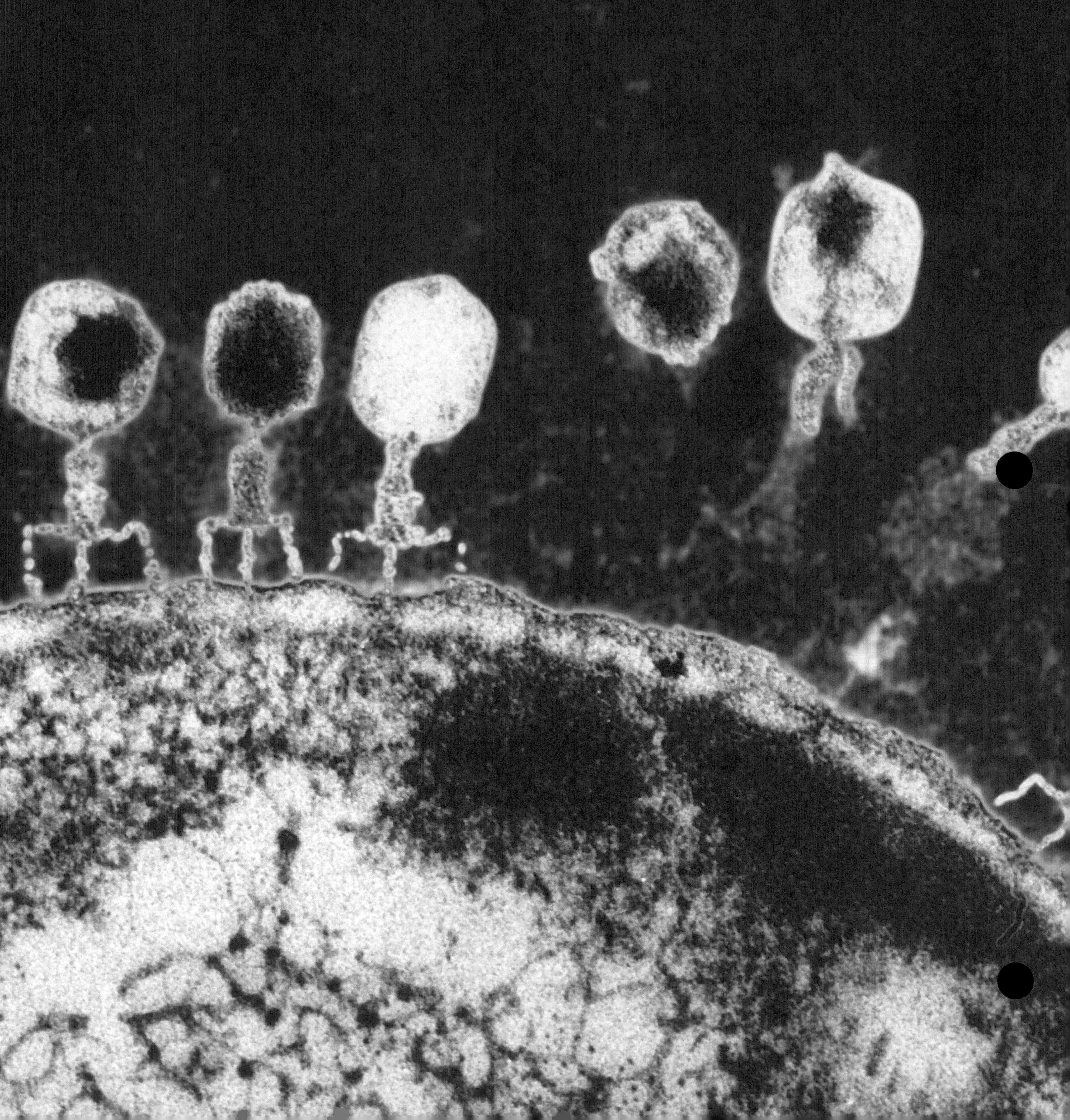

CORE CONCEPTS

Pathways of Transformation

Organisms exchange matter and energy with the environment in order to grow, maintain themselves, and reproduce.

As a group, bacteria and archaea have more ways of meeting their energy and nutrient needs than eukaryotes, so they can live in a wider range of environments. Viruses, which are noncellular, have no metabolic machinery and must use energy and materials obtained by their host.

Evolution

Evolution underlies the unity and diversity of life.

Bacteria, archaea, and viruses have a capacity for genetic change through mutation. They also can exchange genes among existing individuals. In populations of these microbes, traits that increase fitness in a particular environment tend to become more common over time.

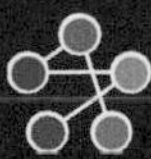

Systems

Complex properties arise from interactions among components of a biological system.

Disruptions at the molecular and cellular levels can negatively affect the health of organisms. Viruses disrupt cell function in their host, causing the host to put aside essential activities in favor of virus production. Toxins made by disease-causing bacteria either harm cells directly or elicit a whole body immune response that causes symptoms.

Links to Earlier Concepts

Section 1.4 gave you an early glimpse of the bacteria and archaea. Section 4.3 began our discussion of their structure and Section 19.5 put these groups into an evolutionary time frame. Section 17.1 focused on antibiotic resistance in bacteria. This chapter also discusses bacteriophages, a type of virus that will be familiar from Section 8.2.

20.1 The Human Microbiota

Your body is home to vast numbers of microbes. By the most recent estimates, a healthy adult hosts somewhere between 30 trillion and 50 trillion bacteria. This means that your body contains at least as many bacteria as it does human cells. In addition, you are home to archaea, fungi, protists, and viruses. Taken together, your many microbial components account for about 3 percent of your body weight.

The number and diversity of microbes is greatest in the gastrointestinal tract, especially the colon. Microbes also are abundant on the skin and in the mouth, respiratory tract, and urogenital tract. Collectively, all of the types of microbes capable of living in or on humans constitute the **human microbiota**.

The first glimpse of the human microbiota occurred late in the seventeenth century, when the pioneering microscopist Antonie van Leewenhoek (Section 4.2) viewed bacteria in samples of dental plaque and feces. In the 1860s, scientists came to the realization that some microbes can cause disease. A disease-causing microbe is called a **pathogen**, and identifying pathogens was a major focus of twentieth century microbiology. We now know of about 1,400 types of human pathogens. This number includes bacteria, fungi, viruses, and protists.

Keep in mind that pathogens constitute only a tiny fraction of our microbes. Most members of the human microbiota are either harmless or helpful. For example, components of our normal gut microbiota produce essential vitamins and aid in digestion.

Today, studies of the human microbiota benefit from advanced gene sequencing technology and computerized analyses of sequence data. Use of these tools has revealed a previously unsuspected abundance of microbial diversity. For example, a recent study estimates that perhaps as many as 40,000 bacterial species can live in the human intestines. Each species of bacteria in turn hosts a different array of viruses.

Sequencing studies have also revealed that microbiota diversity varies among human populations. To date, the most diverse microbiota known is that of the Yanomami people. The Yanomami are hunter-gatherers who live in the Amazon rainforest, where they have almost no contact with outsiders. A study of rural, subsistence farmers in New Guinea also revealed a comparatively diverse microbiota. The intestinal microbiota of this group included about 50 types of bacteria not

human microbiota (my-crow-by-O-tuh) All the microbes (cellular organisms and viruses) that can live on or in humans.
pathogen Cellular organism or virus that causes disease in its host.

CREDIT: (opposite), Eye of Science/Science Source.

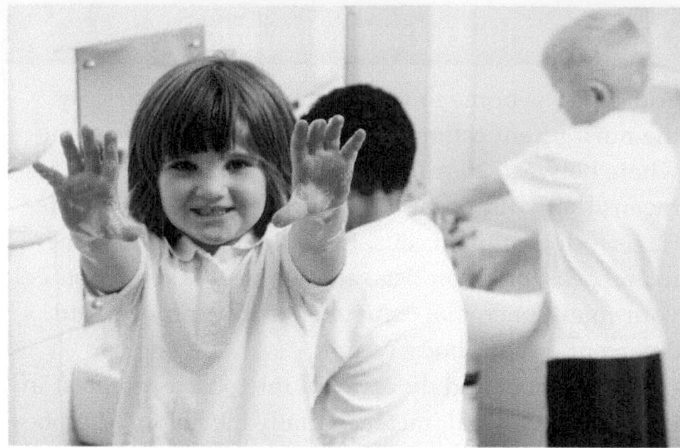

FIGURE 20.1 Sanitation and the microbiota. Hand washing and other hygiene practices that help prevent the spread of infectious disease may also reduce the diversity of our microbiota.

present in residents of the United States. In industrialized countries, the use of antibiotics, limited time outdoors, a diet low in fiber, and improved hygiene and sanitation (**FIGURE 20.1**) have combined to reduce the diversity of the human microbiota.

The reduced microbiota diversity observed in industrialized nations may have negative health effects. By one hypothesis, the human immune system, which evolved in the presence of a diverse microbiota, becomes disordered when microbial diversity declines. The result is a rise in immune-related disorders such as asthma, food allergies, and inflammatory bowel disease. Such disorders are more common in developed countries than in less-developed ones.

20.2 Virus Structure and Function

LEARNING OBJECTIVES

- List the components shared by all free viral particles (virions).
- Explain the difference between enveloped viruses and nonenveloped (naked) viruses.
- Describe the structure of a bacteriophage.

We begin our survey of the microbes with viruses, the most abundant infectious agents on the planet.

In the late 1800s, scientists studying stunted tobacco plants discovered a new kind of pathogen. It was so small that it passed through screens that filtered out bacteria, and it could not be seen with a light microscope. The scientists called this unseen infectious entity a virus, a term that means "poison" in Latin.

A **virus** is a noncellular, infectious particle that can replicate only inside a living cell. It is an obligate intracellular parasite. It does not have ribosomes or other metabolic machinery, and it cannot make ATP. To replicate, the virus must insert its genetic material into a cell of a specific host organism.

We do not know how viruses are related to cellular life. The fact that they can replicate only inside cells suggests they might have evolved from cells, perhaps being derived from bits of DNA or RNA that escaped a plasma membrane. Alternatively, viruses may be remnants of a time before cells. This could explain why some viral genomes consist largely of genes that have no counterparts in modern cells.

A free viral particle, or **virion**, is typically so small that it can only be seen with an electron microscope (about 25 to 300 nanometers in diameter). However, some recently discovered viruses, appropriately referred to as giant viruses, can be larger than many bacteria. A virion always includes a viral genome enclosed within a protein coat called the capsid. The viral genome may be RNA or DNA, and it may be a single-stranded or double-stranded. The capsid may also enclose one or more viral enzymes.

The capsid consists of protein subunits that typically assemble in a repeating pattern to produce a helical or many-sided (polyhedral) shape. A capsid protects the viral genetic material and facilitates its delivery into a host cell. In all viruses, some components of the viral coat bind to proteins at the surface of a host cell.

Viruses infect and replicate in all cellular organisms, no matter how simple or complex. A viral infection often decreases a host's ability to survive and reproduce, so viruses affect ecological interactions among species throughout the biosphere.

The tobacco mosaic virus, which infects tobacco plants, has a virion with a helical structure (**FIGURE 20.2**). Its coat proteins assemble in a tight helix around the genetic material, a single strand of RNA. Plant viruses usually enter their host through a wound

bacteriophage (bac-TEER-ee-oh-fayj) Virus that infects bacteria.
virion (VEER-ee-on) Free viral particle.
virus Noncellular, infectious particle of protein and nucleic acid; replicates only in a host cell.

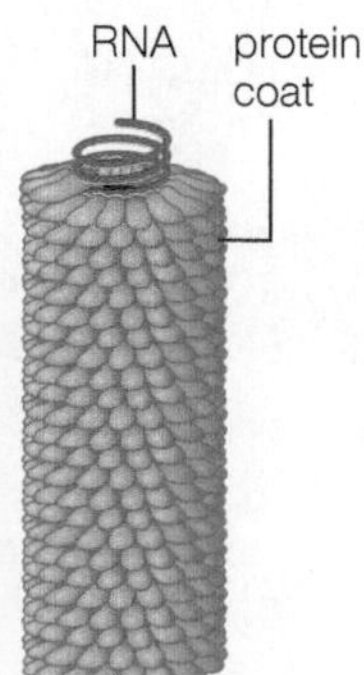

RNA
protein coat

FIGURE 20.2 Tobacco mosaic virus. This virus infects tobacco (left) and related plants. The helical arrangement of the capsid subunits (right) gives the virus a rodlike structure. The genome is single-stranded RNA.

made by an insect, pruning, or another mechanical injury. Symptoms of viral infection typically include stunting and curling, yellowing, or spotting of leaves.

Bacteriophages, sometimes called phages, are viruses that infect bacteria. You learned in Section 8.2 how Hershey and Chase used bacteriophages in experiments that identified DNA as the genetic material. Bacteriophages have a complex structure (FIGURE 20.3). The "head" of the virus is a polyhedral capsid that encloses the viral DNA. A hollow helical "tail" extends from the head. During infection, the DNA will pass through this tail and into the host cell. At the base of the tail, six leglike tail fibers attach to a baseplate. During infection, the tail fibers bind to the surface of the host cell.

The tobacco mosaic virus and a bacteriophage are referred to as nonenveloped or "naked" viruses because their outermost layer is the protein coat. By contrast, viruses that infect animals usually have an outer viral envelope. The viral envelope consists of membrane from the cell in which the virus formed.

Human immunodeficiency virus (HIV), the virus that causes AIDS, is an example of an enveloped RNA virus (FIGURE 20.4). The conical HIV capsid is surrounded by a layer of matrix proteins and by the viral envelope. The capsid encloses the viral genome and some viral enzymes that are essential for replication. There are two copies of the viral genome, each a single strand of RNA.

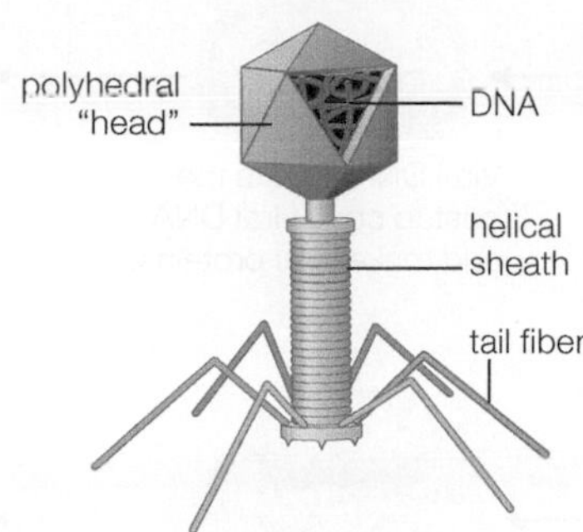
polyhedral "head"
DNA
helical sheath
tail fiber

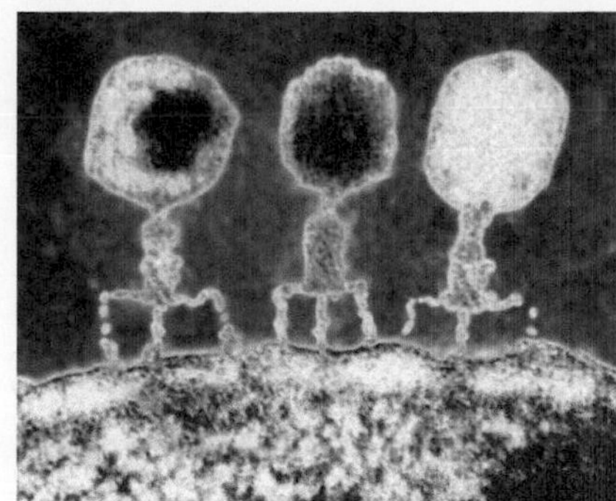

FIGURE 20.3 Bacteriophage. The micrograph at the right shows three phages infecting a bacterial cell.

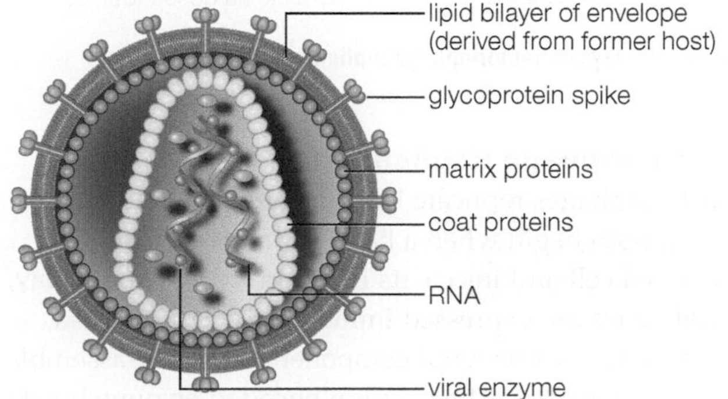
lipid bilayer of envelope (derived from former host)
glycoprotein spike
matrix proteins
coat proteins
RNA
viral enzyme

FIGURE 20.4 HIV (Human Immunodeficiency Virus). This enveloped retrovirus causes AIDS.

TAKE-HOME MESSAGE 20.2

- ✔ Viruses are noncellular infectious particles that consist of nucleic acid surrounded by capsid proteins and, in some cases, an envelope derived from membrane of its host.
- ✔ A virus is an obligate intracellular parasite. It has no metabolic machinery of its own and can multiply only inside living cells.
- ✔ Viruses infect all types of organisms, from bacteria to animals.

TABLE 20.1

Steps in Most Viral Replication Cycles

1. **Attachment** Proteins on viral particle chemically recognize and lock onto specific receptors at the host cell surface.
2. **Penetration** Either the viral particle or its genetic material crosses the plasma membrane of a host cell and enters the cytoplasm.
3. **Replication and synthesis** Viral DNA or RNA directs host to make viral nucleic acids and viral proteins.
4. **Assembly** Viral components self-assemble as new viral particles.
5. **Release** The new viral particles are released from the cell.

20.3 Viral Replication

LEARNING OBJECTIVES

- List the step common to all viral replication pathways.
- Explain the difference between the two bacteriophage replication pathways.
- Describe how HIV replicates.

Steps in Viral Replication

Details of viral replication processes vary, but all involve the steps outlined in TABLE 20.1.

A virus cannot move itself toward a host, so infection begins with a chance encounter. During attachment, viral proteins bind receptor proteins on the surface of a host cell.

After the virus has attached, its genetic material enters the host cell and takes over the host's metabolic and genetic machinery. Under the virus's influence, the cell puts aside its normal tasks and turns to replicating and expressing viral genes. The result is production of viral proteins and nucleic acid.

When viral proteins and nucleic acid come into contact, they self-assemble spontaneously as new virions. The virus either buds from the host cell or escapes when the host cell lyses.

CREDITS: (3) left, © Cengage Learning; right, Eye of Science/Science Source; (4) © Cengage Learning.

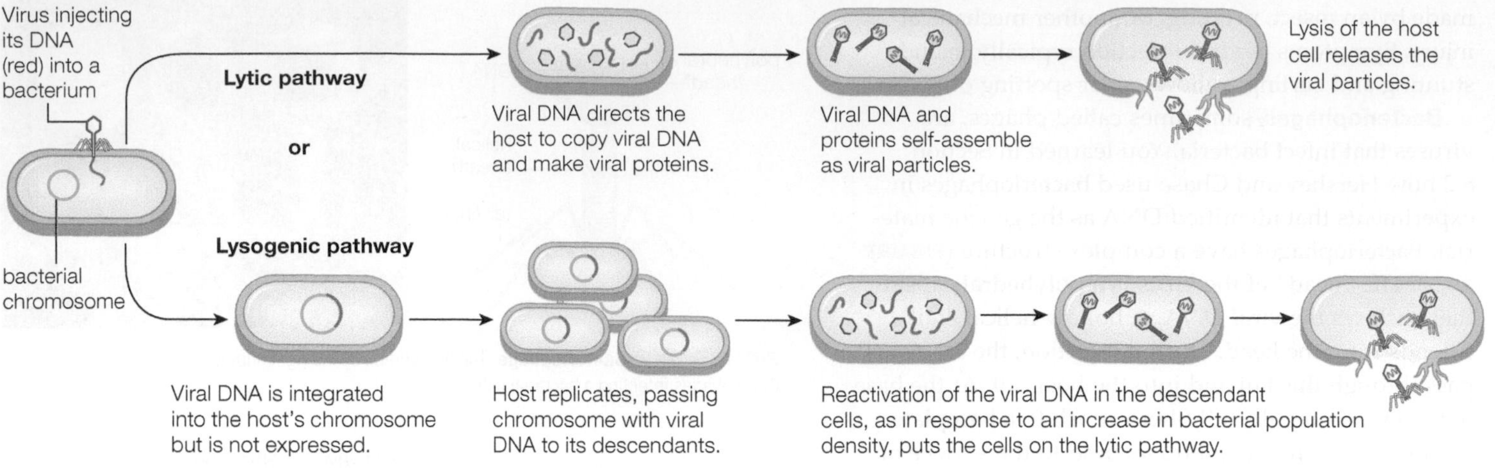
Virus injecting its DNA (red) into a bacterium
Lytic pathway
or
bacterial chromosome
Lysogenic pathway
Viral DNA directs the host to copy viral DNA and make viral proteins.
Viral DNA and proteins self-assemble as viral particles.
Lysis of the host cell releases the viral particles.
Viral DNA is integrated into the host's chromosome but is not expressed.
Host replicates, passing chromosome with viral DNA to its descendants.
Reactivation of the viral DNA in the descendant cells, as in response to an increase in bacterial population density, puts the cells on the lytic pathway.

FIGURE 20.5 Two pathways of bacteriophage replication.

Bacteriophage Replication

Bacteriophages replicate by two pathways (**FIGURE 20.5**). Both begin when a bacteriophage attaches to a bacterial cell and injects its DNA. In the **lytic pathway**, viral genes are expressed immediately. The infected host first produces viral components that self-assemble as virus particles. Then, a viral-encoded enzyme breaks down the bacterial cell wall, triggering lysis—the cell disintegrates and dies.

In the **lysogenic pathway**, viral DNA is injected into a bacterial cell and integrated into the host's genome. Viral genes are not expressed, so this cell stays healthy. When the host cell reproduces, viral DNA is copied and passed on along with the host's genome. Like miniature time bombs, viral DNA inside these descendant cells awaits a cue to enter the lytic pathway.

Some bacteriophages always replicate by the lytic pathway. They kill their host cell quickly and are not passed from one bacterial generation to the next. Others embark upon either the lytic or lysogenic pathway, depending on conditions in the host cell.

Replication of HIV

HIV is an enveloped virus that replicates inside human white blood cells (**FIGURE 20.6**). It attaches to a cell via a glycoprotein that extends through its envelope ❶. The glycoprotein binds two different proteins on the host cell. After attachment, the viral envelope fuses with the blood cell's plasma membrane, releasing viral enzymes and RNA into the cell ❷.

HIV is a **retrovirus**, a virus whose RNA genome is used as a template to produce double-stranded DNA

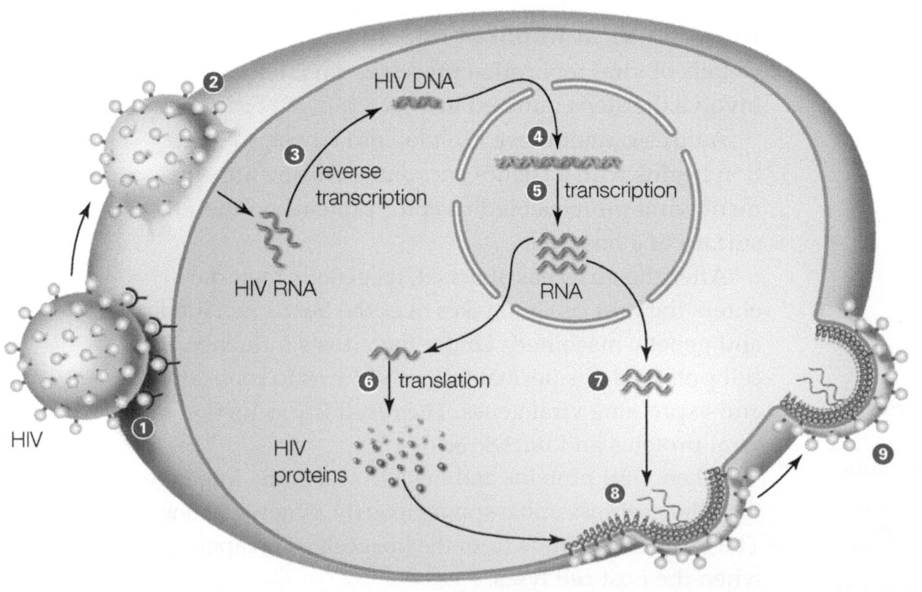
HIV DNA
reverse transcription
transcription
HIV RNA
RNA
translation
HIV proteins
HIV

FIGURE 20.6 Replication of HIV, a retrovirus.

❶ Viral glycoprotein spikes bind to proteins at the surface of a white blood cell.

❷ The viral envelope fuses with the host cell plasma membrane, allowing viral RNA and enzymes to enter the cell.

❸ Viral reverse transcriptase uses viral RNA to make double-stranded viral DNA.

❹ Viral DNA enters the nucleus and becomes integrated into the host genome.

❺ Transcription produces viral RNA.

❻ Some viral RNA is translated to produce viral proteins.

❼ Other viral RNA forms the new viral genome.

❽ Viral proteins and viral RNA self-assemble at the host plasma membrane.

❾ New virus buds from the host cell, with an envelope of host plasma membrane.

inside a host cell. This conversion is carried out by a viral enzyme called reverse transcriptase ❸. DNA produced by reverse transcriptase enters the nucleus together with a viral enzyme called integrase. Integrase inserts the DNA into one of the host's chromosomes ❹. Once integrated, viral DNA is replicated and transcribed along with the host genome ❺.

Some of the resulting viral RNA is translated into viral proteins ❻ and some becomes the genetic material of new viruses ❼. HIV particles self-assemble at the plasma membrane ❽. As the virus buds from the host cell, some of the host's plasma membrane becomes the viral envelope ❾. Each new virus can then infect another white blood cell. New HIV-infected cells are also produced when an infected cell replicates.

Drugs designed to fight HIV take aim at steps in viral replication. Some interfere with the way HIV binds to a host cell. Others impair the viral reverse transcriptase. Integrase inhibitors prevent viral DNA from integrating into a human chromosome. Protease inhibitors prevent the processing of newly translated polypeptides into mature viral proteins.

These antiviral drugs lower the number of HIV particles, so a person stays healthier. With such treatment, an HIV-infected person can expect a near-to-normal life span. Less HIV in body fluids also reduces the risk of passing the virus to others. At present, no drug or vaccine has yet been found that can eliminate the virus entirely, so treatment must continue for life.

TAKE-HOME MESSAGE 20.3

✔ A virus binds to a specific type of host cell. Then viral genetic material enters the cell. Viral genes direct the production of viral components that self-assemble as new viral particles.

20.4 Viruses and Human Health

LEARNING OBJECTIVES

- List some common diseases caused by viruses.
- Describe the ways that viruses can spread between hosts.
- Explain what an emerging disease is and give an example.
- Explain how new strains of flu arise.

The Threat of Infectious Disease

An infection occurs when one organism enters another and replicates inside it. An infectious disease arises when activities of the "guests" interfere with a host's normal functions. Communicable infectious diseases spread from person to person through contact with mucus, blood, or other pathogen-containing body fluid. Washing your hands regularly is the best defense against such diseases. Other infectious diseases require a **disease vector**, an animal that carries the pathogen from host to host. Biting insects and ticks are the most common vectors for human pathogens.

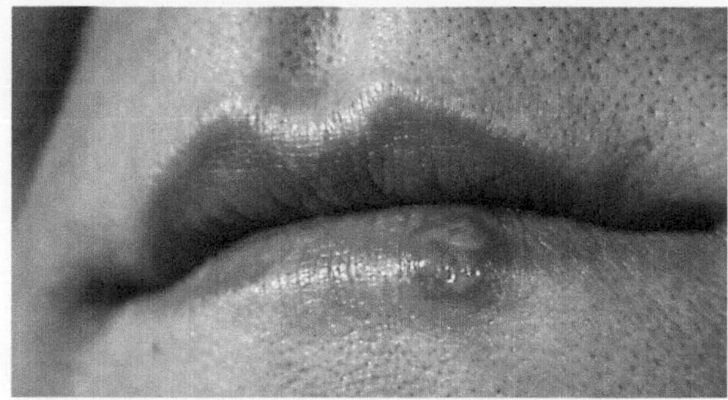

FIGURE 20.7 Sign of an active herpes simplex virus 1 infection. Fluid rich in viral particles leaks from the open sore.

Common Viral Diseases

Most viral diseases cause mild symptoms and trouble us only briefly. For example, some adenoviruses infect the membranes of our upper respiratory system and cause common colds. Others colonize the lining of our gut and cause a brief bout of vomiting and diarrhea.

Viral diseases can also be more persistent. Various types of herpesviruses cause cold sores, genital herpes, mononucleosis, or chicken pox. Typically the initial infection causes symptoms that subside quickly. However, the virus remains in the body in a latent state, and can reawaken later on. After a person has been infected by herpes simplex virus 1 (HSV-1), the virus can lie latent in nerve cells for years. When activated, the virus replicates and causes painful "cold sores" on the edge of the lips (**FIGURE 20.7**). Similarly, the virus responsible for a childhood case of chicken pox can later cause the disease shingles.

Measles, mumps, rubella (German measles), and chicken pox are viral diseases of childhood that, until recently, were common worldwide. Today, most children in developed countries have been vaccinated against these illnesses. Administering a vaccine primes the body to fight off a specific pathogen, a process explained in detail in Section 37.7.

disease vector Animal that transmits a pathogen between hosts.
lysogenic pathway (lice-oh-JEN-ik) Bacteriophage replication mechanism in which viral DNA becomes integrated into the host's chromosome and is passed to the host's descendants.
lytic pathway (LIH-tik) Bacteriophage replication mechanism in which a virus replicates in its host and kills it quickly.
retrovirus Virus having an RNA genome that is used as a template to produce double-stranded viral DNA within a host cell.

CREDIT: (7) CDC/Dr. Hermann.

A few types of viral infections increase the risk of cancer. Infection by certain strains of sexually transmitted human papillomaviruses (HPV) can cause cancer of the cervix, anus, mouth, or throat. Infection by some hepatitis viruses raises the risk of liver cancer. Similarly, Epstein-Barr virus, the agent of infectious mononucleosis, also raises the risk of lymphoma (a blood cell cancer).

Emerging Viral Diseases

Viruses cause a number of emerging diseases. An emerging disease is a disease that has only recently been detected in humans or is now expanding its range. Below we consider three examples.

HIV/AIDS AIDS (acquired immunodeficiency syndrome) is an emerging viral disease that arises as a result of infection by HIV. The virus is spread by sex or exposure to infected blood. Once inside a human, HIV disarms the immune system, allowing the body to be overrun by other pathogens. We discuss the effects of HIV on immunity in detail in Section 37.9.

Gene sequence comparisons have shown that the most common strain of HIV (HIV-1) evolved from a simian immunodeficiency virus (SIV) that infects chimpanzees in west central Africa. HIV arose after SIV entered and survived inside a human, then mutated. The person may have become infected while butchering an infected animal. (Chimpanzees and other primates are hunted and butchered as "bushmeat" in many parts of Africa.) Alternatively, the person could have been bitten by an SIV-infected animal.

HIV-1 probably arose by the 1900s, and it spread from Africa to Haiti in the mid-1960s. The virus diversified in Haiti and acquired distinctive mutations not seen in Africa. By 1969, HIV-1 with these mutations had been introduced to the United States. Once there, it spread quietly until AIDS was identified as a threat in 1981. Since then, HIV has caused more than 35 million deaths. An estimated 37 million people are currently infected by HIV.

Zika virus The Zika virus is an RNA virus that was discovered in Africa during the 1950s. The virus is spread primarily by mosquitoes, but can also be transmitted sexually. From Africa, the virus traveled to Asia, and from there to South America, where it spread north. By 2016, small outbreaks of Zika had occurred in Florida and Texas.

Most people infected by the Zika virus have only mild symptoms, but some develop a temporary paralysis. An infection during pregnancy raises the risk of miscarriage. It can also cause microcephaly (an unusually small brain and head) or other neurological defects in the developing child.

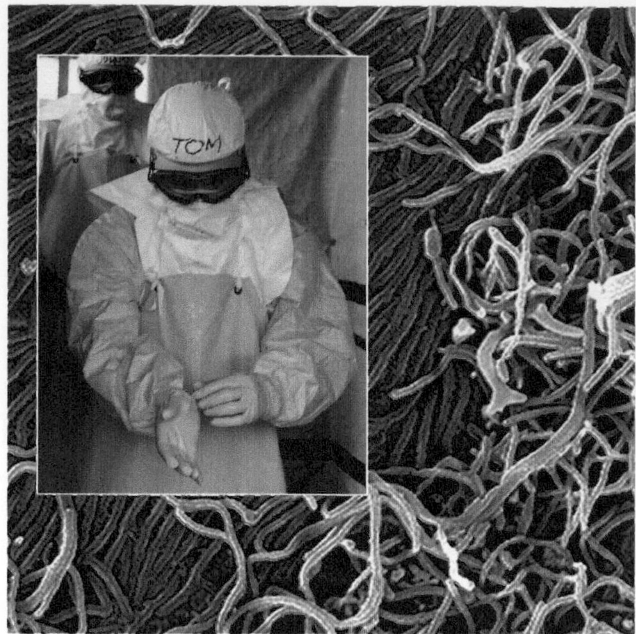
TOM

FIGURE 20.8 Fighting Ebola. Dr. Tom Frieden, head of the U.S. Centers for Disease Control and Prevention, at an Ebola treatment center in Liberia during the 2014 outbreak. The micrograph shows the Ebola virus, which has a threadlike structure.

Ebola virus Ebola hemorrhagic fever is caused by an enveloped RNA virus that first discovered in Africa in 1976. Until recently, Ebola outbreaks had occurred only in limited regions of Africa and had affected fewer than 500 people. However, an outbreak that began in Guinea in 2013 killed more than 11,000 people before ending in 2016. Sporadic African outbreaks have occurred since and are likely to continue.

Within three weeks of infection by the Ebola virus, a person develops flulike symptoms, followed by a rash, vomiting, diarrhea, and bleeding from all body openings. About half of those infected die. The virus is transmitted by contact with body fluids, so caregivers must wear protective gear (**FIGURE 20.8**).

Viral Mutation and Reassortment

Like cellular organisms, viruses have genomes that can be altered by mutation. RNA viruses such as HIV and influenza virus mutate especially quickly. The viral reverse transcriptase makes frequent replication errors. These errors remain uncorrected because the host's proofreading and repair mechanisms evolved to fix errors that arise during DNA replication and do not operate during reverse transcription.

To keep up with ongoing mutations in influenza viruses, scientists create a new flu shot every year.

CREDITS: (8 background) CDC/NIAID; (8 inset) CDC/Sally Ezra.

The flu shot is a vaccine designed to thwart the newly mutated influenza viruses that scientists predict are most likely to pose a threat during the upcoming flu season. Unfortunately, determining which flu strains will be circulating is not an exact science. Even after a flu shot, a person remains susceptible to a virus that differs from the virus types targeted by the vaccine.

Influenza subtypes are named for the structure of two proteins at the viral surface (**FIGURE 20.9**). One protein, hemagglutinin (H), is a glycoprotein that allows the virus to bind to a host cell. The other is an enzyme, neuraminidase (N), that helps new viral particles exit an infected cell.

New influenza subtypes sometimes arise as a result of **viral reassortment**, the swapping of genes between viruses that infect a host at the same time (**FIGURE 20.10**). For example, in April 2009, a new version of the H1N1 subtype of influenza appeared. It had a composite genome, with genes from a human flu virus, bird flu virus, and two different swine flu viruses. The 2009 H1N1 virus was discovered when it caused an epidemic in Mexico. An **epidemic** is an outbreak of disease in a limited region. Within months, there was a **pandemic**, an outbreak of disease that simultaneously affects people throughout the world. Fortunately, governments quickly released reserves of antiviral drugs to treat those infected. A vaccine was created to prevent new infections, and the World Health Organization declared the H1N1 pandemic over in August 2010.

Some strains of influenza that usually infect birds also occasionally infect people who have direct contact with birds. In humans, the death rate from these viruses can be disturbingly high. From 2003 through 2016, the World Health Organization received reports of 856 human cases of avian influenza H5N1, mainly in Asia. About half of these cases were fatal. Fortunately, person-to-person transmission of the H5N1 virus remains exceedingly rare.

Health officials continue to monitor influenza viruses. These viruses mutate, and the coexistence of many viral strains raises the possibility of a potentially disastrous gene exchange. If an easily transmissible strain were to pick up genes from deadly on, the result could be a virus that is easily transmissible and deadly.

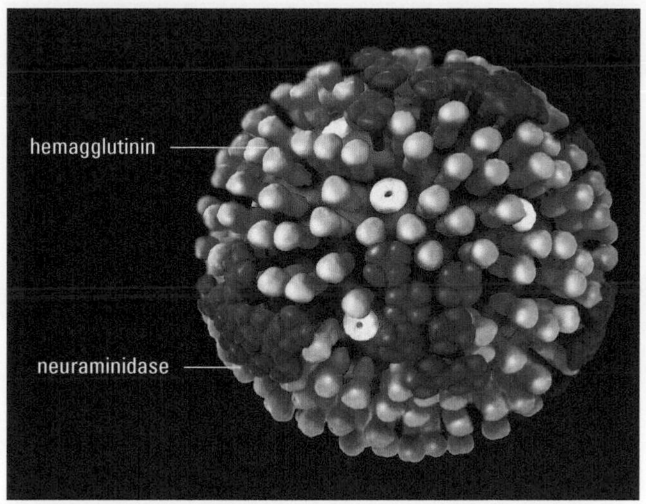
hemagglutinin
neuraminidase

FIGURE 20.9 Influenza virus. Subtypes such as H1N1 or H5N1 are defined by the structure of viral proteins—hemagglutinin (H) and neuraminidase (N)—that extend through the outer envelope.

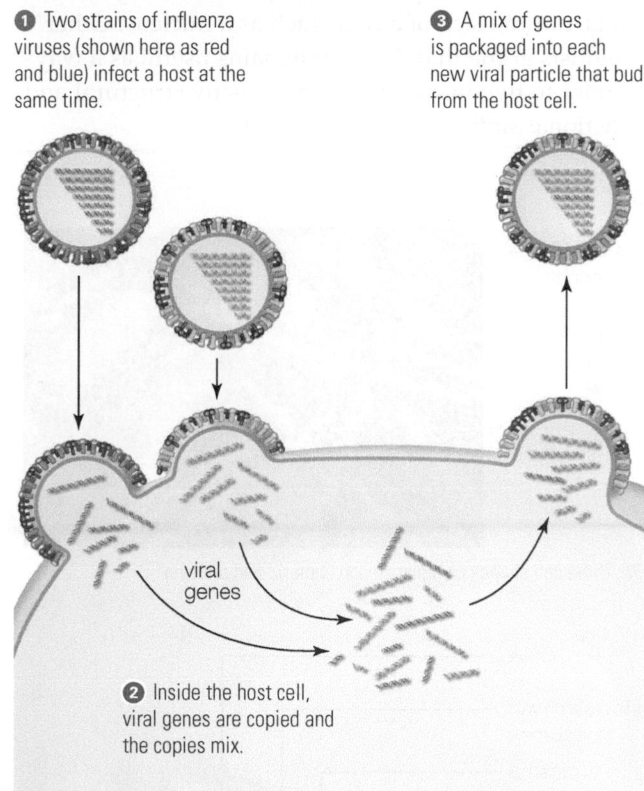
1 Two strains of influenza viruses (shown here as red and blue) infect a host at the same time.
3 A mix of genes is packaged into each new viral particle that buds from the host cell.
viral genes
2 Inside the host cell, viral genes are copied and the copies mix.

FIGURE 20.10 Viral reassortment. When a host cell is infected by two viruses of the same type, copies of viral genes reassort to form new combinations.

TAKE-HOME MESSAGE 20.4

✔ Viruses cause many diseases, most of brief duration and relatively mild, but some can persist or be deadly.

✔ Viruses can be transmitted from animals to humans.

✔ Viral pathogens can change by mutation or reassortment.

epidemic Disease outbreak that occurs in a limited region.
pandemic Disease outbreak with cases worldwide.
viral reassortment Two related viruses infect the same individual simultaneously and swap genes.

CREDITS: (9) CDC/Dan Higgins; (10) © Cengage Learning.

20.5 Prokaryotic Structure and Function

LEARNING OBJECTIVES

- Describe some of the structural traits shared by bacteria and archaea.
- Explain how prokaryotic cells reproduce.
- Describe the three processes of horizontal gene transfer.

Biologists have historically divided all life into two groups. Cells without a nucleus were **prokaryotes** and those with a nucleus were eukaryotes. More recently it was discovered that prokaryotes constitute two lineages: the domains Bacteria and Archaea. **Bacteria** are the more well-known and widespread group of cells that do not have a nucleus. **Archaea** are less well studied, and many live in extreme habitats.

In light of the realization that bacteria and archaea are not a monophyletic group, some microbiologists have advocated abandoning the term prokaryote. They point out that biological groups are defined by shared traits, not the lack of a trait, such as a nucleus. Other scientists argue that the term remains useful as a way to refer to two lineages that share many structural and functional similarities (TABLE 20.2).

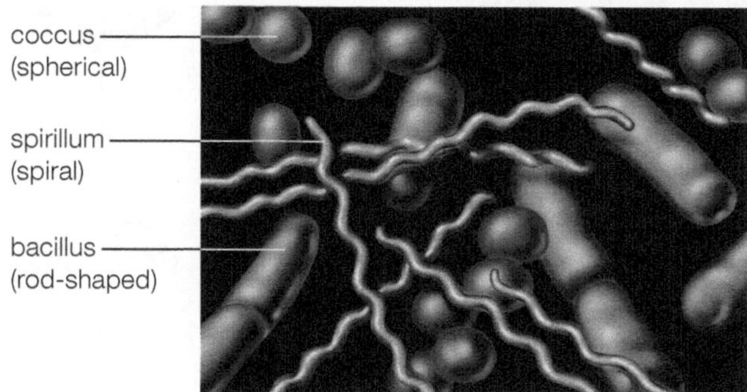
coccus
(spherical)
spirillum
(spiral)
bacillus
(rod-shaped)

A Three cell shapes common among bacteria and archaea.

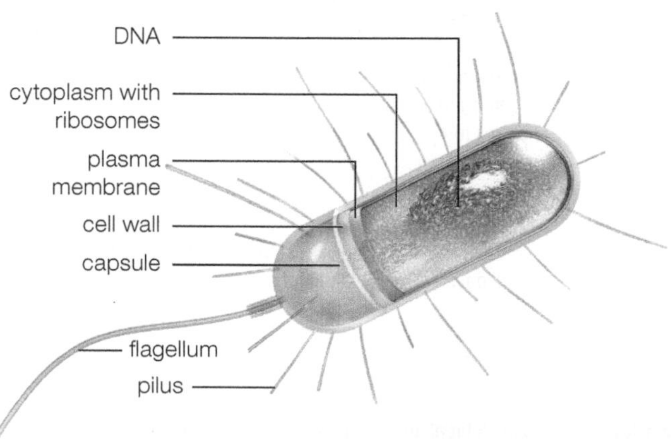
DNA
cytoplasm with ribosomes
plasma membrane
cell wall
capsule
flagellum
pilus

B Features of a typical bacterial cell.

FIGURE 20.11 **Structure of prokaryotic cells.**

TABLE 20.2

Traits Common to Bacteria and Archaea
1. No nuclear envelope; chromosome in nucleoid
2. Typically a single chromosome (a circular DNA molecule); many species also contain plasmids
3. Cell wall (in most species)
4. Ribosomes distributed in the cytoplasm
5. Asexual reproduction by binary fission
6. Capacity for gene exchange among cells by way of conjugation, transduction, and transformation

Structural Traits

A typical bacterial or archaeal cell cannot be seen without a light microscope. Three cell shapes are common in both groups (FIGURE 20.11A). Rod-shaped cells are referred to as bacilli (singular, bacillus), spherical cells are called cocci (singular, coccus), and spiral cells are called spirilla (singular, spirillum).

Most bacteria and archaea secrete a semirigid, porous cell wall around their plasma membrane (FIGURE 20.11B). Bacteria may also have a slime layer or capsule around the cell wall. Slime helps a cell stick to surfaces. A capsule helps some pathogenic bacteria evade the immune defenses of their vertebrate hosts.

Bacteria and archaea typically have a single circular chromosome that consists of double-stranded DNA. The chromosome resides in a cytoplasmic region called the **nucleoid**. There is no nuclear envelope as in eukaryotes, although at least one bacterial species has a membrane surrounding its DNA. Membranes of some bacteria fold inward, but no bacteria or archaea have an endoplasmic reticulum or Golgi bodies. Both

Archaea (ar-KEY-uh) One of the two domains of prokaryotes; the most recently discovered and less well-known domain.

Bacteria One of the two domains of prokaryotes; the most diverse and well-studied domain of prokaryotes.

binary fission Cell reproduction process of bacteria and archaea.

conjugation (con-juh-GAY-shun) Mechanism of horizontal gene transfer in which one bacterial or archaeal cell passes a plasmid to another.

horizontal gene transfer Transfer of genetic material by a mechanism other than inheritance from a parent or parents.

nucleoid (NEW-klee-oid) Of a bacterium or archaeon, region of the cytoplasm where the DNA is concentrated.

pilus (PIE-lus) Protein filament that projects from the surface of some bacterial and archaeal cells.

plasmid Of many bacteria and archaea, a small ring of DNA replicated independently of the chromosome.

prokaryote Member of one of two single-celled domains (Bacteria and Archaea) that do not have a nucleus; a bacterium or archaeon.

transduction In bacteria and archaea, a mechanism of horizontal gene transfer in which a bacteriophage transfers DNA between cells.

transformation In bacteria and archaea, a type of horizontal gene transfer in which DNA is taken up from the environment.

bacteria and archaea have ribosomes distributed through their cytoplasm.

Many bacteria and archaea have a flagellum. However, a bacterial or archaeal flagellum does not contain microtubules and it does not bend side to side. Rather, the flagellum rotates like a propeller. The exact structure of the flagellum, as well as its mechanism of operation differs between bacteria and archaea.

Hairlike filaments called **pili** (singular, pilus) may also extend from the cell surface. Some cells use pili to stick to surfaces. Others glide along by using their pili as grappling hooks. A pilus extends out to a surface, sticks to it, then shortens, drawing the cell forward. Another type of retractable pilus draws cells together for gene exchanges.

Reproduction

Bacteria and archaea replicate by **binary fission**, a type of asexual reproduction (**FIGURE 20.12**). The process begins when the cell replicates its single chromosome, which is suspended in the cytoplasm ❶. DNA Replication typically begins at a single point. It is carried out by enzymes at the center of the cell ❷. Once the chromosome is replicated, plasma membrane and cell wall material are deposited across the cell's midsection ❸. This material partitions the cell, eventually yielding two genetically identical descendant cells ❹.

Gene Transfers

In addition to inheriting DNA "vertically" from a parent cell, bacteria and archaea occasionally take part in **horizontal gene transfers**, in which an individual acquires genes by a mechanism other than inheritance. The gene donor can be a cell of the same species or a different species. Three mechanisms allow horizontal gene transfers in prokaryotes: transformation, transduction, and conjugation (**FIGURE 20.13**).

❶ A typical bacterium has one circular chromosome suspended in the cytoplasm.

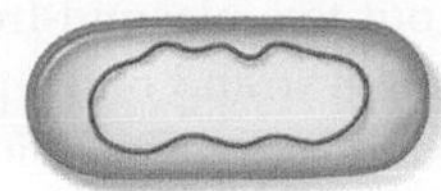

❷ Enzymes in the center of the cell replicate the bacterial chromosome. Replication usually begins at a single point.

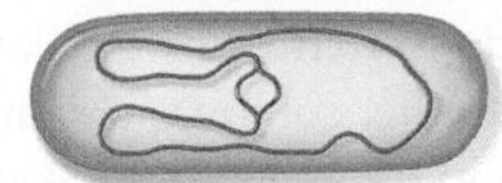

❸ After the chromosome is duplicated, membrane and wall are deposited between the two daughter chromosomes.

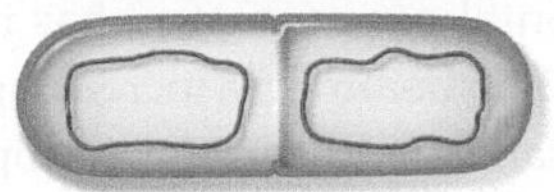

❹ The result is two identical daughter cells.

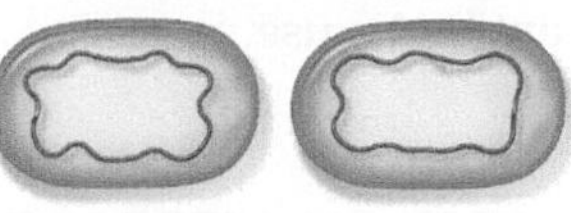

FIGURE 20.12 Binary fission. The reproductive mechanism of bacteria and archaea.

Transformation occurs when a prokaryotic cell takes up DNA from its environment and integrates the DNA into its genome (**FIGURE 20.13A**). This process was discovered by Fredrick Griffith in 1928. As described in Section 8.2, Griffith observed that mixing a harmless strain of bacteria with heat-killed pneumonia-causing bacteria transformed the harmless cells into lethal ones. The harmless bacteria took up DNA from the dead lethal cells and integrated it into their own genome.

With **transduction**, bacteriophages move genes between cells (**FIGURE 20.13B**). A virus picks up some host DNA while in one cell, then introduces that DNA into its next host.

Conjugation involves movement of genes on a plasmid (**FIGURE 20.13C**). A **plasmid** is a small circle of double-stranded DNA separate from the chromosome. During conjugation, a plasmid-bearing cell extends a

FIGURE 20.13 Horizontal gene transfers. There are three mechanisms of gene exchange among prokaryotic cells.

FIGURE IT OUT Which method of gene transfer requires two live prokaryotic cells?

Answer: Conjugation.

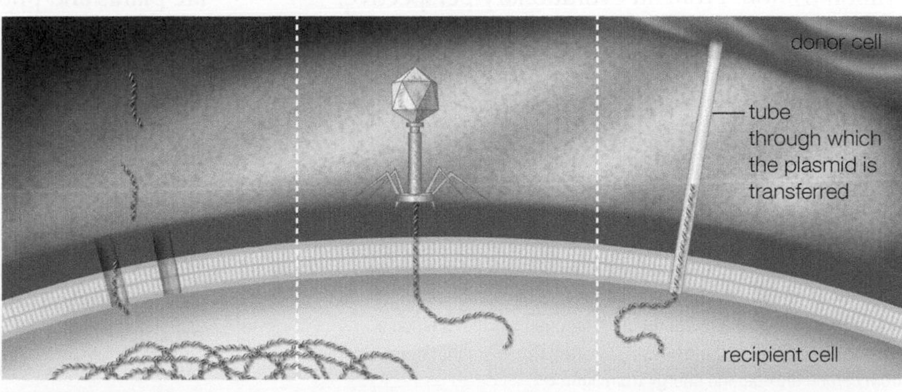

donor cell
tube through which the plasmid is transferred
recipient cell

A Transformation: taking up DNA from the environment. **B** Transduction: transfer of DNA by means of a virus. **C** Conjugation: direct transfer of a plasmid between cells.

sex pilus out to a plasmid-free cell and draws it close. Then, a single strand of the plasmid DNA is moved from the donor to the recipient cell. Both cells then create a complementary strand of plasmid DNA, so each ends up with a complete, functional plasmid. Because plasmids sometimes are incorporated into a bacterial chromosome, conjugation can transfer both plasmid genes and genes derived from the chromosome.

The ability of bacteria to acquire new genes through horizontal gene transfers has important implications for public health. By increasing the speed with which a gene can spread through a population of bacteria, horizontal gene transfers enhance the population's ability to respond to a selection pressure such as that arising from antibiotic use.

TAKE-HOME MESSAGE 20.5

- ✔ Both bacteria and archaea cells are typically walled and have a single chromosome that is not enclosed within a nucleus.
- ✔ Prokaryotic cells reproduce asexually by binary fission.
- ✔ Horizontal gene transfers allow prokaryotes to acquire new genes.

20.6 Metabolic Diversity in Prokaryotes

LEARNING OBJECTIVES

- List the four modes of nutrition used by bacteria.
- Explain the function of endospores.
- Describe the process of nitrogen fixation and explain its ecologic importance.

Bacteria and archaea are usually smaller and structurally simpler than eukaryotes, but simplicity does not imply inferiority. Bacteria and archaea existed before eukaryotes and have coexisted with them for more than 2 billion years. The number of bacterial cells currently living on Earth has been estimated at 5 million trillion trillion. From an evolutionary perspective,

chemoautotroph (keem-oh-AWE-toe-trof) Organism that uses carbon dioxide as its carbon source and obtains energy by oxidizing inorganic molecules.

chemoheterotroph (keem-oh-HET-ur-o-trof) Organism that obtains both energy and carbon by breaking down organic compounds.

decomposer Organism that breaks organic wastes and remains down into their inorganic subunits.

endospore Spore (resting structure) formed by some soil bacteria; contains a dormant cell and is highly resistant to adverse conditions.

nitrogen fixation Incorporation of nitrogen gas into ammonia.

photoautotroph (foe-toe-AWE-toe-trof) Organism that obtains carbon from carbon dioxide and energy from light.

photoheterotroph (foe-toe-HET-ur-o-trof) Organism that obtains carbon from organic compounds and its energy from light.

CARBON SOURCE	ENERGY SOURCE: Light	ENERGY SOURCE: Chemicals
Inorganic source such as CO_2	**Photoautotrophs** bacteria, archaea, photosynthetic protists, plants	**Chemoautotrophs** bacteria, archaea
Organic source such as glucose	**Photoheterotrophs** bacteria, archaea	**Chemoheterotrophs** bacteria, archaea, fungi, animals, nonphotosynthetic protists

FIGURE 20.14 Modes of nutrition in bacteria and archaea.
FIGURE IT OUT Which group of organisms can build their own food from CO_2 in the dark?

Answer: Chemoautotrophs

bacteria and archaea are highly successful. Their extraordinary metabolic diversity contributes to their success.

Diverse Modes of Nutrition

Organisms harvest energy and nutrients from the environment in a wide variety of ways, but we can categorize four modes of nutrition. All of these nutritional modes occur among bacteria, archaea, or both (**FIGURE 20.14**). In addition, some bacteria and archaea can switch from one metabolic mode to another.

As you learned in Section 6.2, autotrophs build their own food using carbon dioxide (CO_2) as their carbon source. There are two subgroups of autotrophs: those that obtain energy from light, and those that obtain energy from chemicals.

Photoautotrophs are photosynthetic. They use the energy of light to assemble organic compounds from CO_2 and water. Many bacteria are photoautotrophs, as are plants and photosynthetic protists. As explained in Section 19.6, the ancestors of eukaryotic chloroplasts are cyanobacteria, a type of photosynthetic bacteria. Like plants, cyanobacteria capture light using chlorophylls and they release oxygen as a by-product.

Chemoautotrophs obtain energy by oxidizing (removing electrons from) inorganic molecules such as hydrogen sulfide or methane. They use energy released by this process to build organic compounds from CO_2. Chemoautotrophic bacteria and archaea are the main producers in dark environments such as the seafloor. We know of no eukaryotic chemoautotroph.

Heterotrophs cannot tap into inorganic sources of carbon, so they obtain carbon from organic molecules in their environment. **Photoheterotrophs** harvest energy from light, and carbon from alcohols, fatty acids, or other small organic molecules. Heliobacteria that live in the soils of rice paddies are an example. No photoheterotrophic eukaryotes are known.

Chemoheterotrophs obtain both energy and carbon by breaking down carbohydrates, lipids, and proteins. Most bacteria and some archaea are chemoheterotrophs, as are fungi, animals, and nonphotosynthetic protists. All pathogenic bacteria are chemoheterotrophs that extract organic compounds from their host.

Other chemoheterotrophic bacteria live in or on a host without causing any harm, and some even provide benefits. For example, chemoheterotrophic bacteria are part of the digestive tract microbiota of many animals. They assist in digestion and produce vitamins and other nutrients that the host requires.

Still other bacterial chemoheterotrophs serve as decomposers. **Decomposers** break down the complex organic molecules in wastes and remains into inorganic components that plants can take up and use. Decomposers can also break down pesticides and pollutants, thus improving the environment.

Aerobes and Anaerobes

With rare exceptions, eukaryotic organisms rely on aerobic respiration (Section 7.2) and thus require oxygen. By contrast, many bacteria and most archaea are anaerobes. Some can tolerate an oxygen-free environment but are not harmed by oxygen. Others are obligate anaerobes, meaning oxygen either slows their growth or kills them outright. Such cells are harmed by oxygen because oxidation reactions damage their biological molecules and, unlike aerobic cells, they have no enzymes that repair such damage. We find obligate anaerobes in aquatic sediments and the animal gut. They also can infect deep wounds. *Clostridium tetani*, a species of bacteria that causes the disease tetanus when it infects wounds, is an example.

Nitrogen Fixation

Some bacteria and archaea provide an important ecological service by converting atmospheric nitrogen to a form that other organisms can use. By the process of **nitrogen fixation**, these prokaryotes incorporate nitrogen from the air into ammonia (NH_3).

Photosynthetic eukaryotes need nitrogen, but they cannot use the gaseous form ($N{\equiv}N$) because they do not have an enzyme that can break the molecule's triple bond. They can, however, take up ammonium, which forms when the ammonia produced by nitrogen fixation dissolves in water. Some nitrogen-fixing bacteria in the genus *Rhizobium* form a mutually beneficial partnership with legumes, which are plants such as peas. The bacteria live inside a root nodule, which is a small growth that forms on a root (**FIGURE 20.15**). The plant provides the bacteria with photosynthetically produced sugars, and in return the bacteria supply the plant with ammonium.

FIGURE 20.15 Root nodules. The round growths on these soybean roots contain *Rhizobium* bacteria that fix nitrogen.

Dormant Resting Structures

When conditions do not favor growth, some bacteria can enter a state of suspended animation. Depending on the group and how the structure forms, it may be called a spore or a cyst.

Some soil bacteria such as *Clostridium* can produce an especially resilient resting structure called an **endospore**. An endospore consists of a stripped-down, dehydrated bacterial cell within a thick protective covering. Unlike metabolically active cells, endospores withstand heating, freezing, drying out, and exposure to ultraviolet radiation.

Endospores can survive in a dormant state for many years. Consider that scientists have extracted bacterial endospores from the gut of a bee that had been encased in amber (fossilized tree sap) for at least 20 million years. When given nutrients and moisture, the endospores germinated; the cells within the spores became active once again.

TAKE-HOME MESSAGE 20.6

- ✔ Both bacteria and archaea have survived for billions of years and continue to coexist beside the eukaryotes.
- ✔ Bacteria and archaea are Earth's most abundant organisms.
- ✔ Prokaryotes are metabolically diverse; they utilize all four modes of nutrition and may be aerobic or anaerobic.
- ✔ Only prokaryotes can carry out nitrogen fixation.

CREDIT: (15) iStockphoto.com/DS70.

20.7 Major Bacterial Lineages

LEARNING OBJECTIVES

- Compare the structure of Gram-positive and Gram-negative bacteria.
- Describe some members of the proteobacteria.
- Explain the ecological importance of cyanobacteria.

There are many bacterial lineages, and new ones are constantly being discovered. Here we consider a few major groups to provide insight into bacterial diversity.

Gram-Positive Bacteria

In **Gram-positive bacteria**, the outermost layer of the cell is a thick wall of peptidoglycan. Peptidoglycan is a polymer of amino acids and sugars that is unique to bacteria. The cell wall of Gram-positive bacteria is colored purple when a technique called Gram staining is used to prepare samples for microscopy. In other bacterial lineages, the cell wall has a thinner peptidoglycan layer and is covered by an outer membrane. Such cells are colored pink by Gram staining and are described as "Gram-negative."

Most Gram-positive bacteria are chemoheterotrophs, and many serve as decomposers. Actinomycetes are Gram-positive decomposers that grow through soil as long, branching chains of cells. Their presence gives freshly exposed soil its distinctive "earthy" smell. Many antibiotics, including streptomycin and vancomycin, were first isolated from actinomycetes. The actinomycetes make these bacteria-killing compounds to eliminate their competitors. They are not themselves harmed by the antibiotic they release.

Soil also contains members of the Gram-positive genus *Lactobacillus*. These cells ferment sugars and produce lactate. Lactobacilli sometimes spoil milk, but they are also used to produce sour foods such as sauerkraut and yogurt (**FIGURE 20.16**). Streptococci, which are close relatives of lactobacilli, include species that cause strep throat and the skin disease impetigo.

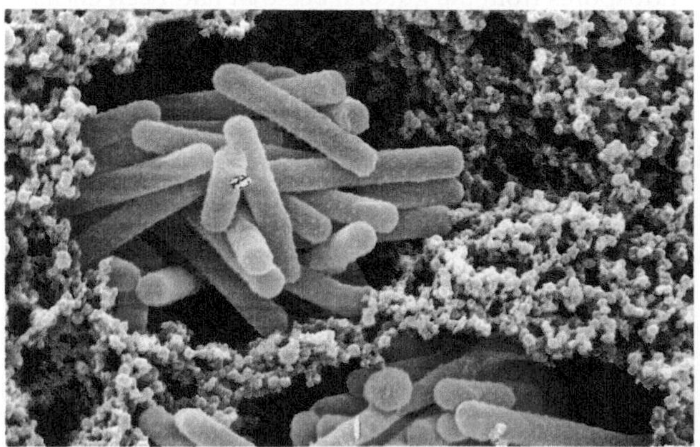

FIGURE 20.16 Gram-positive bacteria. The *Lactobacillus* cells shown here were used to turn milk into yogurt.

The endospore-forming soil bacteria belong to the Gram-positive lineage as well. Various members of this group cause the muscle-paralyzing disease tetanus, the deadly food poisoning known as botulism, and the respiratory disorder anthrax.

Cyanobacteria

Photosynthesis evolved in many bacterial lineages, but only **cyanobacteria** utilize the noncyclic pathway and release free oxygen. Given that chloroplasts evolved from ancient cyanobacteria, we have cyanobacteria and their chloroplast descendants to thank for nearly all the oxygen in Earth's atmosphere.

Some cyanobacteria partner with fungi to form lichens (which we will discuss in Section 23.7) and others grow on the surface of soils, but most are aquatic. Aquatic cyanobacteria grow as single cells or as filaments of cells arranged end to end within a secreted mucus sheath (**FIGURE 20.17**). When conditions become unfavorable for growth, some species of filamentous cyanobacteria can produce thick-walled resting spores that remain dormant until environmental conditions improve.

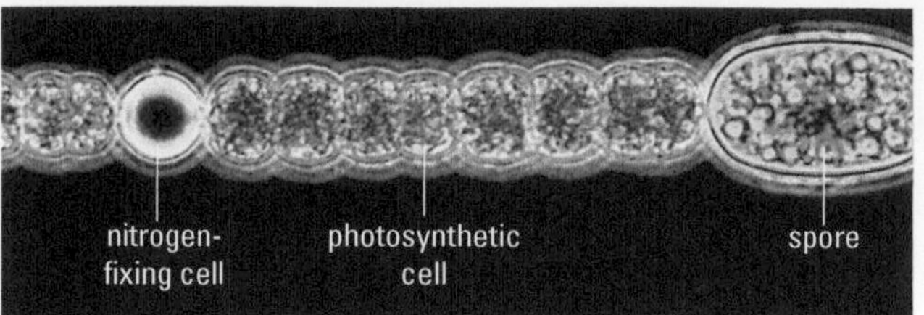
nitrogen-
fixing cell
photosynthetic
cell
spore

FIGURE 20.17 Cyanobacteria. *Anabaena*, an filamentous aquatic species that can fix nitrogen and form dormant spores.

Proteobacteria

The largest and most diverse bacterial group is the **proteobacteria**. They include about a third of all known bacteria. The group is named for Proteus, a Greek ocean god who could take many forms. The bacterial cell ancestral to mitochondria (Section 19.7) is thought to have been an ancient proteobacteria.

The proteobacteria include *Thiomargarita namibiensis*, the largest bacterial cell known. It is so large it can be seen without a microscope. It gets energy by stripping electrons from sulfur that it stores in a giant vacuole.

Many chemoheterotrophic proteobacteria live inside other organisms. The nitrogen-fixing genus *Rhizobium* lives inside plant roots. Rickettsias are tiny, nonmotile proteobacteria that infect vertebrate cells. They are transmitted by ticks, fleas, and other arthropods.

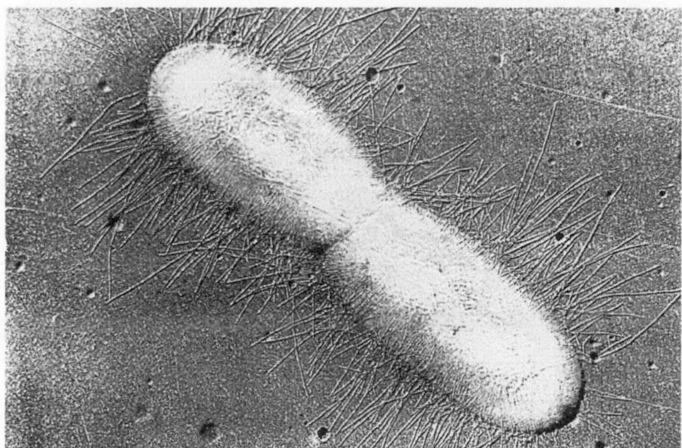

A *Escherichia coli*, lives in the mammalian gut. It is shown here dividing by binary fission.

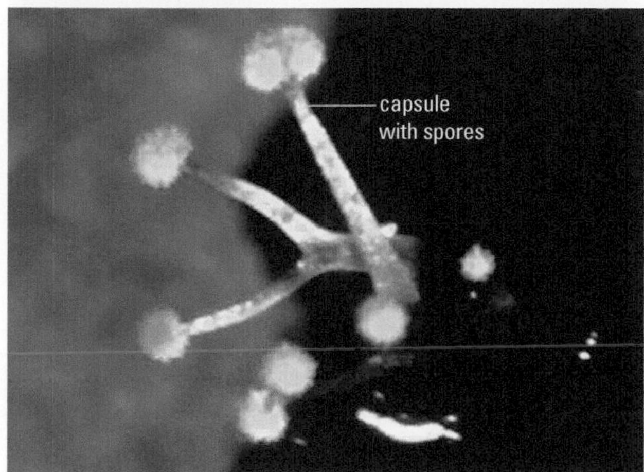
capsule
with spores

B *Chondromyces crocatus*, a myxobacterium, hunts other soil bacteria. When food runs out, thousands of cells form a fruiting body with spores at its tip.

FIGURE 20.18 Proteobacteria.

The most studied prokaryote is the proteobacterial species *Escherichia coli*, which normally lives in the mammalian gut (**FIGURE 20.18A**). Researchers often study genetic and metabolic processes in *E. coli* because it is easily grown in laboratories. *E. coli* is also used in industrial biotechnology. Recombinant *E. coli* now make hormones and other proteins for medical use.

When biotechnologists want to alter a plant's genome, they may turn to *Agrobacterium*. These soil proteobacteria have a plasmid that gives them the capacity to infect plants and cause a tumor. As explained in Section 15.7, scientists produce recombinant plants by inserting genes into the tumor-inducing plasmid, then infecting a plant with recombinant bacteria.

The proteobacteria known as slime bacteria, or myxobacteria, are notable for their collective behavior. They glide about as a swarm and eat other bacteria. When food dwindles, hundreds of thousands of cells form a multicelled fruiting body with spores at its tip (**FIGURE 20.18B**). Spores released into the environment remain dormant until conditions favor growth.

chlamydias (klah-MID-ee-ahs) Tiny bacteria that live as intracellular parasites of vertebrates.

cyanobacteria (sigh-AN-no-BAK-tee-ree-ah) Oxygen-producing photosynthetic bacteria.

Gram-positive bacteria Lineage of bacteria with a thick cell wall that are colored purple by Gram staining.

proteobacteria (PRO-tee-o-bak-TEAR-ee-ah) Most diverse bacterial lineage; includes species that carry out photosynthesis and that, fix nitrogen. Some cause disease.

spirochetes (spy-row-KEY-teez) Lineage of bacteria shaped like a stretched-out spring.

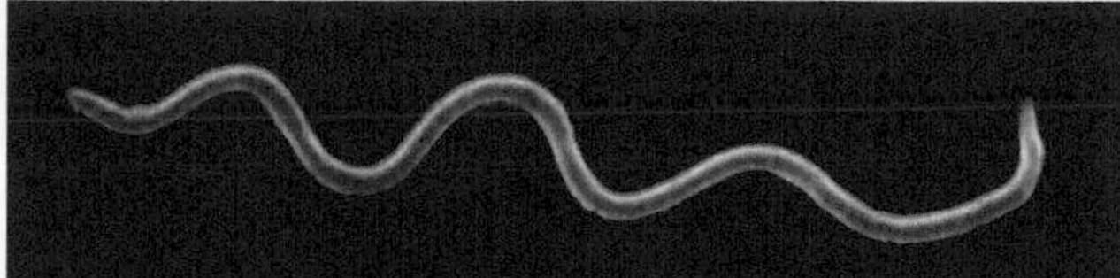

FIGURE 20.19 Spirochete. This species causes Lyme disease.

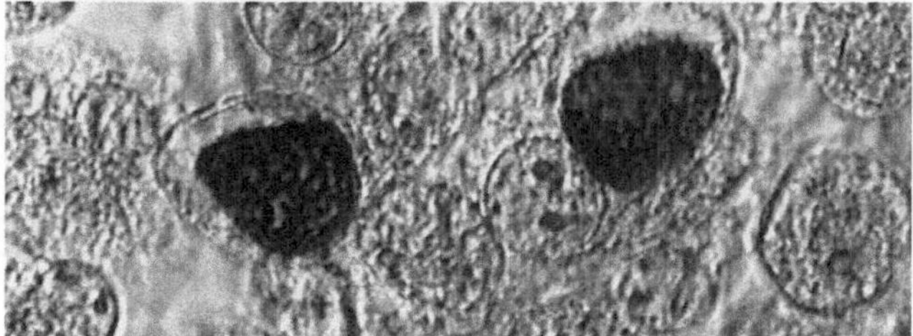

FIGURE 20.20 Chlamydias. Culture of human cells with two cells colored dark by an abundance of chlamydia living as parasites inside them.

Spirochetes and Chlamydias

Spirochetes and chlamydias are two unrelated lineages of extremely small bacteria. They can barely be seen with a light microscope.

Spirochetes are shaped like a stretched-out spring (**FIGURE 20.19**). Some live in the stomach of cattle and sheep, where they help their host break down cellulose. Others are aquatic decomposers and some fix nitrogen. Still others are parasites and some, such as the species that causes syphilis, are human pathogens.

All **chlamydias** live inside cells of vertebrates (**FIGURE 20.20**). The species, *Chlamydia trachomatis*, causes the common sexually transmitted disease known as chlamydia.

TAKE-HOME MESSAGE 20.7

- ✔ Bacteria vary widely in their size, structure, and metabolism.
- ✔ Bacteria benefit other organisms by putting oxygen into the air, fixing nitrogen, and aiding their host's digestion.
- ✔ We use bacteria in scientific research, biotechnology, and food production.

CREDITS: (18A) Courtesy James Evarts; (18B) Universität des Saarlandes; (19) Stem Jems/Science Source; (20) CDC/Dr. E. Arum and Dr. N. Jacobs.

20.8 Bacteria as Pathogens

LEARNING OBJECTIVES

- Distinguish between endotoxins and exotoxins.
- Describe some mechanisms of antibiotic action and antibiotic resistance.

Bacterial Toxins

Bacteria cause many common diseases (TABLE 20.3). Most pathogenic bacteria harm us by way of toxins that disrupt our health. The toxin may be a substance that bacteria release into their environment (an exotoxin), or a molecule integral to the cell's wall (an endotoxin).

Exotoxins are secreted proteins that bind to and directly harm our body cells. For example, botulinum exotoxin released by the soil bacteria *Clostridium botulinum* is one of the most poisonous substances known. *C. botulinum* produces endospores which can enter the body though a wound or in improperly canned food. Once inside the body, the spores germinate and the bacteria releases the deadly toxin. The result is the disease known as botulism. Botulinum endotoxin binds to nerve cells and prevents them from releasing a chemical signal they use to communicate with other cells. The initial effects are impaired vision, drooping eyelids, slurred speech, difficulty swallowing, and muscle weakness. If untreated, botulism poisoning paralyzes muscles involved in breathing and can be fatal.

Purified botulinum exotoxin is now available for medical use under the brand names Botox and Myobloc. The toxin can be injected into neck muscles to relieve chronic spasms or into the wall of the bladder to treat an overactive bladder. It is also used for cosmetic purposes. Injection of Botox into facial muscles weakens these muscles and can reduce the appearance of wrinkles.

Shigella species are Gram-negative bacteria that produce Shiga exotoxin. Like the poison ricin, Shiga exotoxin is a ribosome-inactivating protein (Section 9.1). It acts on cells lining small blood vessels, especially those of the digestive system and kidney. Ingesting food or water containing *Shigella* bacteria results in abdominal cramps and bloody diarrhea. Feces from an infected person contains live bacteria that can, if ingested, start a new infection. One strain of *E. coli* (O157:H7) also produces Shiga exotoxin and has been involved in several large outbreaks of food poisoning.

Endotoxins are lipopolysaccharides integral to the outer membrane of Gram-negative bacteria. Endotoxins are not secreted. They are released when a bacterial cell dies and fragments. The toxin does not directly harm host cells, but its presence elicits a host immune response. This response results in symptoms such as fever and aches. Entry of bacterial endotoxins into the bloodstream, can elicit a runaway immune response called shock.

All Gram-negative bacteria have endotoxins, but exotoxins are secreted by some Gram-negative bacteria and some Gram-positives.

endotoxin Substance that is integral to the membrane of a Gram-negative bacterial pathogen and causes symptoms of disease in the bacteria's host.
exotoxin Substance that is secreted by a bacterial pathogen and causes symptoms of disease in the bacteria's host.

TABLE 20.3

Examples of Bacterial Diseases

Disease	Description
Whooping cough	Childhood respiratory disease
Tuberculosis	Respiratory disease
Impetigo, boils	Blisters, sores on skin
Strep throat	Sore throat, can damage heart
Cholera	Diarrheal illness
Syphilis	Sexually transmitted disease
Gonorrhea	Sexually transmitted disease
Chlamydia	Sexually transmitted disease
Lyme disease	Rash, flulike symptoms, spread by ticks
Botulism, tetanus	Muscle paralysis by bacterial toxin

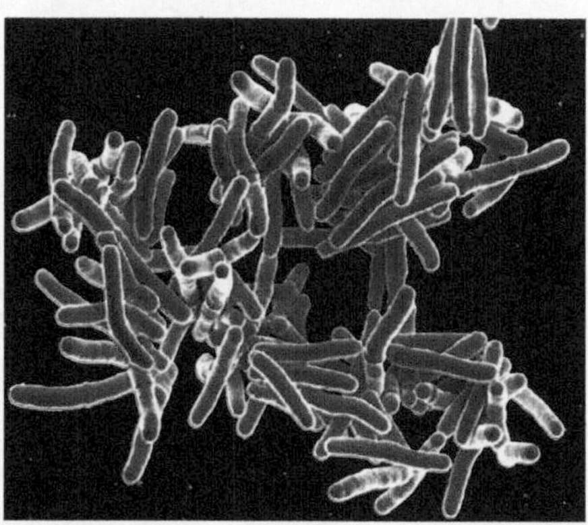

FIGURE 20.21 **Bacteria that cause tuberculosis.** *Mycobacterium tuberculosis* infects about a third of the human population.

Antibiotics

We are fortunate to live at a time when nearly all bacterial infections can be cured by an antibiotic. Antibiotics first came into use in the 1940s. Prior to that time, there was no effective treatment for bacterial diseases.

Penicillin, one of the earliest antibiotics, was initially isolated from a fungus (*Penicillium*) that lives in the soil. It was the first of many antibiotics to be isolated from soil microbes. Penicillin acts by inhibiting synthesis of bacterial cell walls, the most common mechanism of antibiotic action.

Tetracyclines and streptomycins are antibiotics that were discovered in actinomycetes (also soil bacteria).

CREDIT: (21) CDC/National Institute of Allergy and Infectious Diseases (NIAID).

Data Analysis Activities

Bacteriophage-Inspired Antibiotics Although bacteriophages have been infecting bacteria for billions of years, no mechanism has evolved in bacteria to prevent the viruses from lysing the cell walls of their hosts. Now, scientists are targeting the same bacterial wall components that bacteriophages do. The goal is to develop antibiotics that bacteria will be less likely to develop resistance to.

FIGURE 20.22 shows the results of a study to test Epimerox, a new bacteriophage-inspired antibiotic, against Bacillus anthracis, the bacterial species that causes the disease anthrax.

1. How long did it take for all the mice that received the drug-free buffer alone to die? What function did this group play in the experiment?
2. What do these data indicate regarding the optimal time to begin Epimerox treatment?
3. In studies with *Bacillus anthracis* cells grown in culture, no Epimerox-resistant cells were observed. Explain why this result is consistent with the scientists' goal for developing this drug.

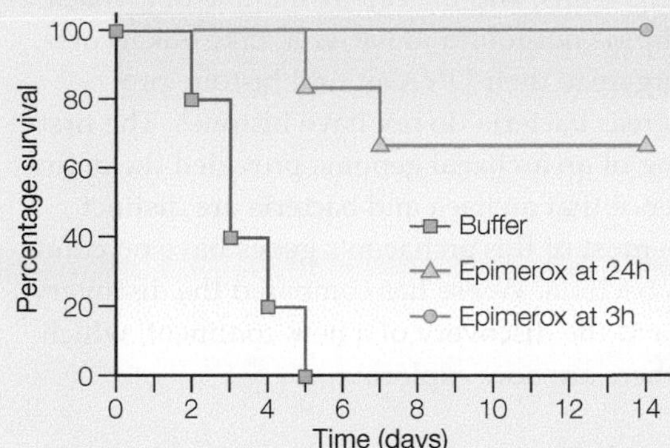

Percentage survival
100
80
60
40
20
0
0
2
4
6
8
10
12
14
Time (days)
Buffer
Epimerox at 24h
Epimerox at 3h

FIGURE 20.22 Effect of Epimerox on the survival of mice with anthrax. Mice were infected with the bacteria *B. anthracis.* One group of 15 then began receiving a drug-free buffer solution 3 hours later. Another 15 were treated with Epimerox beginning 3 hours after infection. A third group of 15 was treated with Epimerox beginning 24 hours after infection.

These compounds disable bacterial ribosomes. Eukaryotic ribosomes, which have a different structure are unaffected by these compounds. Soil bacteria were also the source of bacitracin, an antibiotic that is applied topically to cuts and wounds. It acts by disrupting bacterial cell membranes.

Antibiotic Resistance

Gram-negative bacteria are naturally more resistant to antibiotics than Gram-positive bacteria, because many compounds cannot cross the envelope that surrounds their cell wall. Mycobacteria, such as those that cause tuberculosis (**FIGURE 20.21**) have a waxy coating that makes them similarly difficult to treat.

An arms race, in which scientists search for new antibiotics and bacteria evolve resistance to those replacements, is ongoing. As noted in Section 17.1, our widespread use of antibiotics favors bacterial strains that can survive in the presence of these drugs. Genes that confer resistance can arise through mutation or be acquired through horizontal gene transfer.

A variety of resistance mechanisms are known. Some bacteria use an enzyme to break down the antibiotic. In others, the structure of the drug's target has been modified so the drug cannot bind to the target or is no longer effective when it does bind. Some bacteria actively transport the antibiotic out of their cytoplasm. In others, the drug cannot enter because the structure of proteins in the cell membrane has been altered.

TAKE-HOME MESSAGE 20.8

✔ Bacterial diseases arise from the effects of bacterial toxins.

✔ Antibiotics target the metabolic processes of pathogenic bacteria. Changes in bacterial structure or metabolism can lead to antibiotic resistance.

20.9 Archaea

LEARNING OBJECTIVES

- Explain how the domain Archaea was discovered.
- List some ways that archaea differ from bacteria.
- Describe two extreme habitats where archaea are found.

Discovery of the Third Domain

The distinctive features of archaea first came to light in the 1970s. Carl Woese was comparing the ribosomal RNAs (rRNAs) among what he thought were bacterial species to find out how they were related. All organisms have rRNAs, so they are often used for these sorts of genetic comparisons. Woese discovered that some species fell into a distinct group based on their rRNA genes. They were more like eukaryotes than bacteria. On the basis of this evidence, Woese proposed a classification system with three domains: Bacteria, Archaea, and Eukarya (Section 1.5).

Woese's ideas were initially greeted with skepticism, but as years went by, evidence in support of them

CREDIT: (22) Based on Schuch R, Pelzek AJ, Raz A, Euler CW, Ryan PA, Winer BY, et al. (2013) Use of a Bacteriophage Lysin to Identify a Novel Target for Antimicrobial Development. PLoS ONE 8(4): e60754. doi:10.1371/journal.pone.0060754.

mounted. Archaea differ from bacteria in many ways. Unlike bacteria, archaea do not have peptidoglycan in their cells walls, and the cell membrane of archaea contains lipids not found in bacteria. Like eukaryotes, archaea organize their DNA around histone proteins, whereas bacteria do not have histones. The first sequencing of an archaeal genome provided the definitive evidence that archaea and bacteria are distinct lineages—most of this archaeon's genes have no counterpart in bacteria. Woese has compared the discovery of archaea to the discovery of a new continent, which he and others are now exploring.

Here, There, Everywhere

Many archaea thrive in seemingly hostile habitats. The first archaeon to have its genome sequenced was discovered near a hydrothermal vent on the seafloor. It is an **extreme thermophile**, an organism that grows only at a very high temperature. Some archaea that live near hydrothermal vents can grow even at 110°C (230°F). Heat-loving archaea also live in volcanically heated geysers and hot springs (**FIGURE 20.23A**).

Other archaea are among the **extreme halophiles**, organisms that live in highly salty water. Salt-loving archaea live in the Dead Sea, the Great Salt Lake, and many smaller brine-filled lakes (**FIGURE 20.23B**). One extreme halophile, *Halobacterium*, is a photoheterotroph with gas-filled vesicles that keep it afloat in well-lit surface waters. Its plasma membrane has a purple pigment called bacteriorhodopsin. When excited by light, this protein pumps protons (H^+) out of the cell, against their gradient. The H^+ flows back into the cell through ATP synthases, thus driving the formation of ATP.

Many archaea, including some extreme thermophiles and extreme halophiles, are **methanogens**, or methane makers. These chemoautotrophs form ATP by pulling electrons from hydrogen gas or acetate. The reactions produce methane (CH_4), an odorless gas. Methanogenic archaea are strict anaerobes, meaning they cannot live in the presence of oxygen. They abound in sewage, marsh sediments, and the animal gut (**FIGURE 20.23C**). Cattle have methanogens in their stomach and they release methane gas primarily by belching.

Methane is a greenhouse gas; it contributes to global warming. Thus, the release of methane by cattle is a

extreme halophile Organism adapted to life in a highly salty environment.

extreme thermophile Organism adapted to life in a very high-temperature environment.

methanogen Organism that produces methane gas as a metabolic by-product.

A Thermally heated waters. Pigmented archaea color the rocks in waters of this hot spring in Nevada.

B Highly salty waters. Pigmented extreme halophiles such as *Halobacterium* color the brine in this California lake.

C The gut of many animals. Cows belch frequently to release the methane produced by archaea in their stomach.

FIGURE 20.23 Examples of archaeal habitats.

matter of concern. A cow can release up to 600 liters of methane a day, all of it produced by the methanogens in its gut.

Some people release methane too. About a third of the human population has significant numbers of methanogens in their intestine, so their flatulence (farts) contains methane. Archaea have also been discovered in the human mouth and vagina. Although no archaea are known to cause disease on their own, some may contribute to ill health. For example, some methanogens that can live in the human mouth encourage gum disease. By taking up hydrogen, the archaea make the mouth more hospitable to the pathogenic bacteria that cause this disease.

Similarly, some methanogenic archaea live in the human large intestine, and their abundance may affect our weight. Hydrogen uptake by methanogens improves the environment for bacteria that break down carbohydrates. As a result, the more methanogens in your gut, the more calories you can extract from food. Some studies have found a correlation between an abundance of gut methanogens and obesity.

Researchers continue to investigate the distribution and diversity of archaea, and to delve into the evolutionary relationships among them. They are finding that archaea live alongside bacteria nearly everywhere. In deep, dark ocean waters, archaea are the most abundant cells.

Phylogeny of the archaea remains a work in progress. A few major lineages have been identified. Most of the extreme thermophiles belong to the lineage Crenarchaeota, and most methanogenic and salt-loving archaea belong to the lineage Euryarchaeota.

Lokiarchaeota, a group discovered near a deep-sea hydrothermal vent includes the most eukaryote-like archaea discovered so far (Section 19.7).

Many scientists now think that eukaryotes arose when an archaeon something like Loki partnered with one of the proteobacteria. The proteobacteria evolved into what are now mitochondria. By this hypothesis, we are descendants of both archaea and bacteria.

TAKE-HOME MESSAGE 20.9

✔ Archaea are single cells without a nucleus that are closer to eukaryotes than to bacteria. Many live in very hot or very salty habitats, but there are archaea nearly everywhere.

✔ No archaea are known to be human pathogens, although some may foster the presence of pathogenic bacteria.

20.1 The Human Microbiota (revisited)

When we eat, we feed not only ourselves, but also the trillions of bacteria in our digestive tract. Different types of bacteria are adapted to utilize different types of nutrients, so one would expect variations in the diet to be reflected in our microbiota.

Experimental studies have confirmed that different diets select for different intestinal microbes. In one study, volunteers were put on a six-day diet of either all animal products (meat, eggs, and cheese) or all plant products (grains, legumes, fruits, and vegetables). Those who ate only animal products had an increase in the proportion of bile-tolerant bacteria in their feces. (Bile is a fluid that is produced by the liver and plays a role in the digestion of fats.) The all-animal diet also decreased the proportion of bacteria that ferment plant polysaccharides. Eating the all-plant diet had the opposite effect; bile-tolerant bacteria decreased and fermenters of plant polysaccharides increased.

If sustained, a shift toward bile-tolerant bacteria could have a negative effect on health. These bacteria release metabolic by-products that promote liver cancer and encourage inflammatory bowel disease. The compounds also discourage the growth of bacteria that ferment plant material. Compounds released by those bacteria promote health by suppressing the growth of pathogens, serving as an energy sources for cells in the gut lining, and dampening inflammation.

The plants-only diet was rich in prebiotics, which are nondigestible polysaccharides that encourage the growth of beneficial intestinal bacteria. Inulin is an example of a prebiotic. It is abundant in garlic, onions, artichokes, asparagus, and barley. Like starch, inulin is a polysaccharide consisting of many simple sugars (Section 3.3). Unlike starch, inulin cannot be broken down by human digestive enzymes. It remains intact until it reaches the large intestine. Here, beneficial bacteria with the appropriate enzymes break inulin down and ferment it as their source of energy.

Eating plenty of plant foods containing inulin and other prebiotic compounds will keep the beneficial bacteria in your gut well fed. Think of it like providing food to some particularly useful pets.

CREDIT: (in text) Palokha Tetiana/Shutterstock.

Section 20.1 The **human microbiota** includes all the microbes that can live on or in the human body. Some of these are **pathogens**, but most are harmless or beneficial. The microbiota differs among cultures and among individuals. People in industrialized cultures tend to have a less diverse microbiota than hunter-gathers or subsistence farmers. A diet rich in prebiotics helps maintain helpful bacteria in the gut.

Section 20.2 A **virus** is an intracellular parasite that replicates only inside a cell of a specific host type. Viruses infect all types of organisms. They may have evolved before cells, or they may be descended from cells or cell components.

When outside a host cell, the virus is a **virion**. A virion consists of nucleic acid (DNA or RNA) inside a protein coat (a capsid). Plant viruses enter plants through wounds. **Bacteriophages** inject their genetic material into bacteria. Most viruses that infect animals are enclosed within a viral envelope (a bit of membrane from the host cell in which the virus formed).

Section 20.3 To replicate, a virus attaches to a host cell and its genetic material enters the cell. Viral genes and enzymes direct host machinery to replicate the viral genome and make viral proteins. These components self-assemble as viral particles, which are then released.

Bacteriophages may multiply by a **lytic pathway**, in which the new viral particles are made fast and released by lysis, or by a **lysogenic pathway**, in which viral DNA becomes part of the host chromosome and is passed on to descendant cells.

HIV (human immunodeficiency virus) is an enveloped **retrovirus** that infects human white blood cells. It has a genome of RNA. The viral enzyme reverse transcriptase uses this RNA as its template to make DNA that becomes integrated into the genome of the host cell.

Section 20.4 Viral diseases may be spread by contact with a viral particle or delivered into the body by a **disease vector** such as a tick. Most viral diseases such as common colds cause symptoms only briefly. Some viruses persist in the body and reawaken after a latent period. Virus genomes change when viral genes mutate or viruses swap genes in a host, a process called **viral reassortment**. An **epidemic** is an outbreak of disease in only a limited region. A **pandemic** is a worldwide outbreak. AIDS is an emerging disease, as are diseases caused by the Zika virus and the Ebola virus.

Section 20.5 The organisms known as **prokaryotes** actually constitute two distinct lineages: **bacteria** and **archaea**. Unlike eukaryotes, these lineages do not typically have a nucleus or endomembrane system, and they do not reproduce sexually. Members of both groups typically have cell walls. Many have surface projections such as flagella and **pili**. The chromosome is a circular molecule of DNA that resides in the cytoplasm, in a region called the **nucleoid**. There may also be one or more **plasmids**, circles of DNA that are separate from the chromosome and carry a few genes. Reproduction occurs by **binary fission**, a type of asexual reproduction.

Three types of **horizontal gene transfer** move genes between existing prokaryotic cells. **Conjugation** transfers a plasmid from one cell to another. Virus-assisted transfer of genes is **transduction**. With **transformation**, cells take up DNA from the environment.

Section 20.6 Bacteria and archaea are abundant, and, as a group, metabolically diverse. Some are aerobic and others cannot tolerate oxygen. **Photoautotrophs** carry out photosynthesis. **Photoheterotrophs** capture light, but get carbon from organic molecules rather than CO_2. **Chemoautotrophs** build food from CO_2 using energy from inorganic substances. **Chemoheterotrophs** such as **decomposers** get energy and carbon from organic molecules. Only prokaryotes can carry out **nitrogen fixation**, the incorporation of nitrogen from nitrogen gas into ammonia. Some bacteria and archaea escape adverse conditions by becoming inactive, as when soil bacteria form **endospores**.

Section 20.7 Bacteria are the more known and diverse lineage of prokaryotic organisms. Many are ecologically important. **Cyanobacteria** produce oxygen as a by-product of photosynthesis. **Proteobacteria** is the most diverse bacterial lineage. It includes the ancestors of mitochondria. The **Gram-positive bacteria** have a thick wall of peptidoglycan. Spiral-shaped cells called **spirochetes** and intracellular parasites called **chlamydias** are also major lineages.

Section 20.8 Both Gram-negative and Gram-positive bacteria can cause disease by releasing an **exotoxin** that impairs cell function. An **endotoxin** in the cell membrane of Gram-negative bacteria triggers an immune reaction that results in symptoms such as fever. Antibiotics act by interfering with bacterial metabolism or replication. Antibiotic resistance arises through changes in structure or metabolism.

Section 20.9 Archaea are the more recently discovered lineage of prokaryotes. Comparisons of structure, function, and genetic sequences position them in a separate domain, between eukaryotes and bacteria. Many live in extreme environments. **Extreme halophiles** live in very salty waters and **extreme thermophiles** live at very high temperatures. Some archaea are photoheterotrophs with a unique purple protein that captures light energy. Most, including the **methanogens**, are chemoautotrophs. Some archaea live in our bodies, but none are considered pathogens.

SELF-QUIZ

Answers in Appendix VII

1. The genome of _______ can be either RNA or DNA.
   a. a bacterium c. a virus
   b. a eukaryote d. an archaeon

2. The capsid of a virion consists of _______ .
   a. DNA c. protein
   b. RNA d. lipids

3. Bacteriophages kill their host quickly by _______ .
   a. binary fission c. a lysogenic pathway
   b. a lytic pathway d. transformation

4. The genetic material of HIV (a retrovirus) is _______ .
   a. DNA c. protein
   b. RNA d. lipids

5. Bacteria and archaea reproduce by _______ .
   a. binary fission c. conjugation
   b. transformation d. the lytic pathway

6. One cell transfers a plasmid to another by _______ .
   a. binary fission c. conjugation
   b. transformation d. the lytic pathway

7. _______ carry out oxygen-releasing photosynthesis.
   a. Spirochetes c. Cyanobacteria
   b. Chlamydias d. Proteobacteria

8. Bacteria that serve as decomposers are _______ .
   a. photoautotrophs c. chemoautotrophs
   b. photoheterotrophs d. chemoheterotrophs

9. New flu strains can arise by _______ .
   a. meiosis c. viral reassortment
   b. mitosis d. binary fission

10. Formation of a(n) _______ allows some soil bacteria to survive adverse conditions.
    a. pilus c. endospore
    b. nucleoid d. plasmid

11. _______ in the stomach of a cow release methane.
    a. Bacteria c. Aerobes
    b. Archaea d. Viruses

12. A plasmid is a circle of _______ .
    a. RNA c. either RNA or DNA
    b. DNA d histone proteins

13. A bacterial exotoxin is a _______ .
    a. membrane lipid c. secreted protein
    b. membrane protein d. cell wall component

14. Both archaea and eukaryotes have _______ .
    a. peptidoglycan c. a nucleus
    b. histone proteins d. a capsid

15. Match the terms with their most suitable description.
    ___ methanogen a. component of bacterial cell wall
    ___ nucleoid b. has RNA genome, protein coat
    ___ retrovirus c. halts cell wall formation
    ___ plasmid d. releases methane gas
    ___ extreme halophile e. region with DNA
    ___ peptidoglycan f. circle of nonchromosomal DNA
    ___ penicillin g. salt lover

CRITICAL THINKING

1. Antibiotics enter the environment in wastewater from pharmaceutical manufacturing plants, in wastes from farms where they are used in animals, and in sewage from human populations. (Many antibiotics are excreted in their active form.) The influx of antibiotics into the environment is a concern because the presence of these compounds in the environment selects for antibiotic resistance. As pollutants, antibiotics can also have ecological effects. Describe some of the ecological roles of bacteria, and explain why disruption of these roles by antibiotic pollution would have a negative impact.

2. Adenoviruses that cause colds do not have a lipid envelope. Such naked viruses tend to remain infectious outside the body for longer than enveloped viruses. Naked viruses are also less likely to be destroyed by soap and water. Why are naked viruses hardier than enveloped viruses?

3. The antibiotic penicillin acts by interfering with the production of peptidoglycan. Cells treated with penicillin do not die immediately, but they cannot reproduce. Explain how penicillin halts binary fission and why it is most effective against Gram-positive bacteria.

4. Raw red alga of the genus *Porphyra* is part of a traditional Japanese diet. This alga is rich in porphyran, a polysaccharide that human digestive enzymes cannot break down. Marine bacteria that live on the alga produce a porphyran-digesting enzyme. In Japan, a homologous enzyme is common in one species of intestinal bacteria. Outside of Japan, people have the same species of intestinal bacteria, but they do not have the gene for the porphyran-digesting enzyme. The marine bacteria and the intestinal bacteria are not close relatives, so they most likely did not inherit the enzyme from a shared ancestor. Propose a scenario to explain the observed distribution of the enzyme.

21 PROTISTS—THE SIMPLEST EUKARYOTES

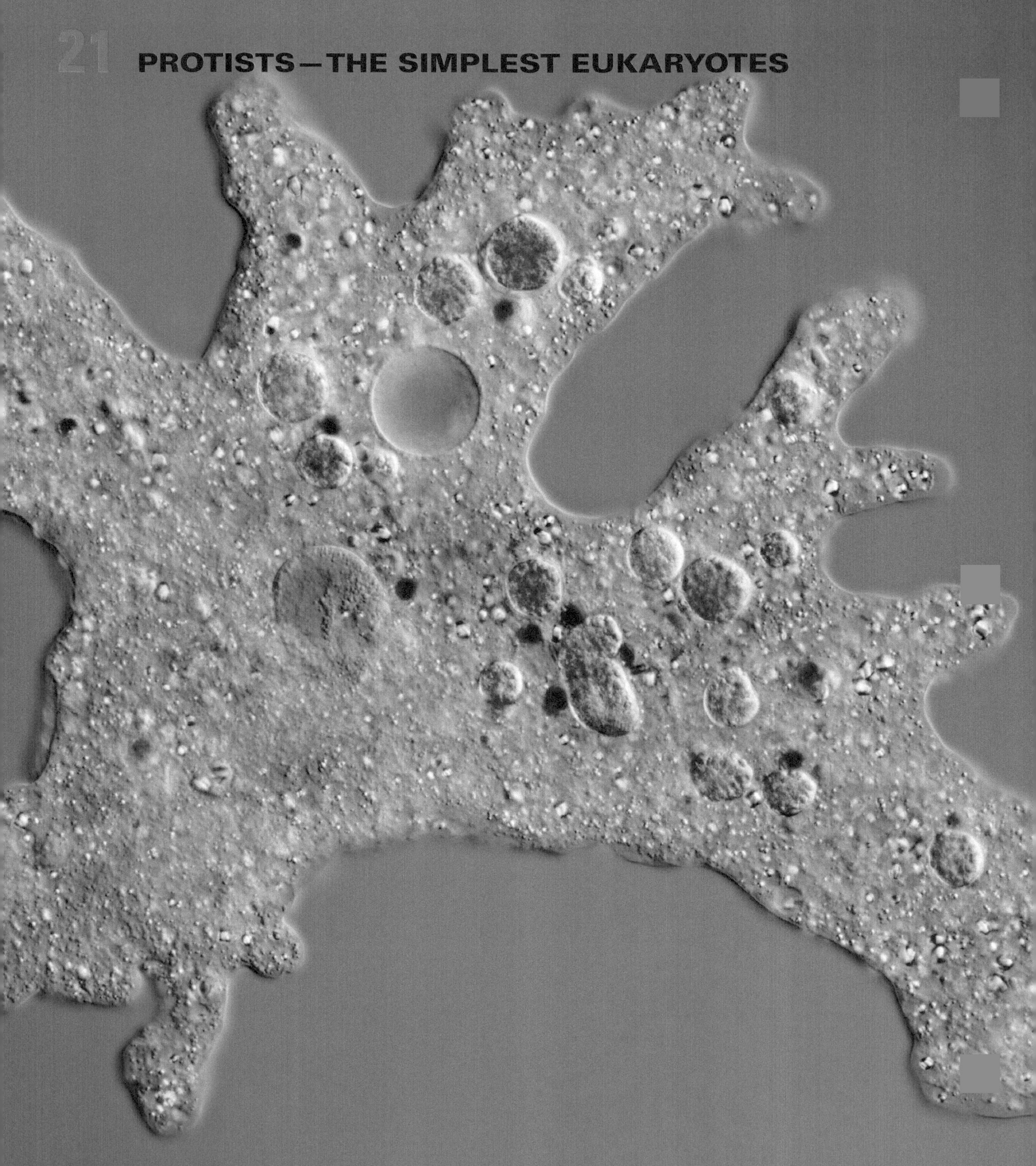

CORE CONCEPTS

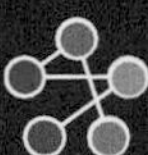

Evolution

Evolution underlies the unity and diversity of life.

All protists are eukaryotes. Their diverse structure and metabolic processes are adaptations to different environments. Protists live in freshwater, saltwater, damp places on land, and inside other organisms. Most are aerobic, but some are anaerobic. Some protists are heterotrophs; others have chloroplasts that evolved from cyanobacteria or an alga. Fungi, plants, and animals all have protist ancestors.

Systems

Complex properties arise from interactions among components of a biological system.

Protists interact with one another and with other lineages. They can be predators, prey, beneficial partners, or harmful pathogens. These interactions influence the structure and properties of the diverse communities that include protists.

Process of Science

The field of biology consists of and relies upon experimentation and the collection and analysis of scientific evidence.

Determining the relationships among the various protist groups remains a work in progress. Use of gene comparisons now makes it possible to trace the ancestry of chloroplasts and to determine how protist groups are related to each other and to the plants, animals, and fungi.

Links to Earlier Concepts

This chapter is devoted to the protists, a eukaryotic group introduced in Section 1.4. Section 4.4 described traits of eukaryotic cells and Section 19.7 explored eukaryotic origins. The chapter's discussion of relationships among protist groups will draw on your understanding of cladistics, a topic presented in detail in Section 18.2.

21.1 Malaria: A Protistan Disease

Eukaryotes that are not fungi, plants, or animals are collectively referred to as the **protists**. The term is derived from the Greek *protistos*, meaning "the very first." Protists include the oldest eukaryotic lineages.

The vast majority of protists are free-living single cells. Such protists play important ecological roles as producers in aquatic habitats and as tiny predators on microorganisms. A few single-celled protist groups include important human pathogens. Most notably, the protist *Plasmodium* causes malaria, a disease that kills more than 500,000 people each year.

Plasmodium is transmitted from person to person by mosquitoes. The parasite invades and multiplies inside human liver cells and oxygen-carrying red blood cells. Infection destroys red blood cells (**FIGURE 21.1**), resulting in fatigue and weakness. When blood cells infected by *Plasmodium* get into the brain, they can cause blindness, seizures, coma, and death.

Like AIDS (Section 20.4), malaria has an evolutionary connection to a primate disease. The *Plasmodium* species responsible for most cases of human malaria is the descendant of a parasite that infects gorillas in western Africa. Unlike AIDS, malaria has been affecting human populations for millennia. For example, researchers reported finding bits of *Plasmodium* DNA in the 3,300-year-old mummified remains of the Egyptian pharaoh Tutankhamun, also known as "King Tut."

FIGURE 21.1 Malaria. The graphic depicts an infected human red blood cell rupturing and releasing new *Plasmodium* cells (blue).

protist General term for member of one of the eukaryotic lineages that is not a fungus, animal, or plant.

As explained in Section 17.5, mortality from malaria has affected human evolution. Selection resulting from malaria has driven up the frequency of the hemoglobin allele associated with sickle-cell anemia (HbS) in some human populations. Having the HbS allele reduces the risk of dying from malaria.

Malaria was common in the southern United States until the 1940s, when crews drained swamps and ponds where mosquitoes breed, and sprayed the insecticide DDT inside millions of homes. Today, nearly all cases of malaria in the United States occur in people who contracted the disease elsewhere.

Malaria remains a threat in many tropical regions. It is endemic (constantly present) in parts of Mexico, Central America, and South America, as well as Asia and the Pacific Islands. However, the disease takes its greatest toll in Africa. More than 90 percent of malaria deaths occur in sub-Saharan Africa. ●

21.2 A Diverse Collection of Lineages

LEARNING OBJECTIVES

- Explain why protists are not considered a clade.
- Compare primary endosymbiosis and secondary endosymbiosis.
- Describe the two modes of nutrition in protists.

Classification and Phylogeny

Protists do not constitute a valid biological group (a clade) and no shared trait defines them. Protists include all members of some clades, plus some members of other clades. As FIGURE 21.2 illustrates, some protists are more closely related to plants, fungi, or animals than to other protists.

Exactly how all the various eukaryotes are related remains under investigation, but a classification system that distributes the eukaryotes among six supergroups is now in wide use. Four of these supergroups include only protists: excavates, the SAR supergroup, amoebozoa, and a small group that we will not discuss. A fifth group (archaeplastids) includes plants and their protist relatives. Animals, fungi, and their protist relatives are members of the sixth group (opisthokonts). We discuss the features that distinguish each of these groups later in this chapter.

Level of Organization

Most protists live as single cells, but some lineages also include colonial or multicellular members (FIGURE 21.3). A **colonial organism** consists of cells that live together but remain self-sufficient. Each cell retains the traits required to survive and reproduce on its own. By contrast, cells of a **multicellular organism** have a division of labor and are interdependent. In a multicellular organism, only specialized cells produce gametes. Multicellularity has evolved independently in many different lineages.

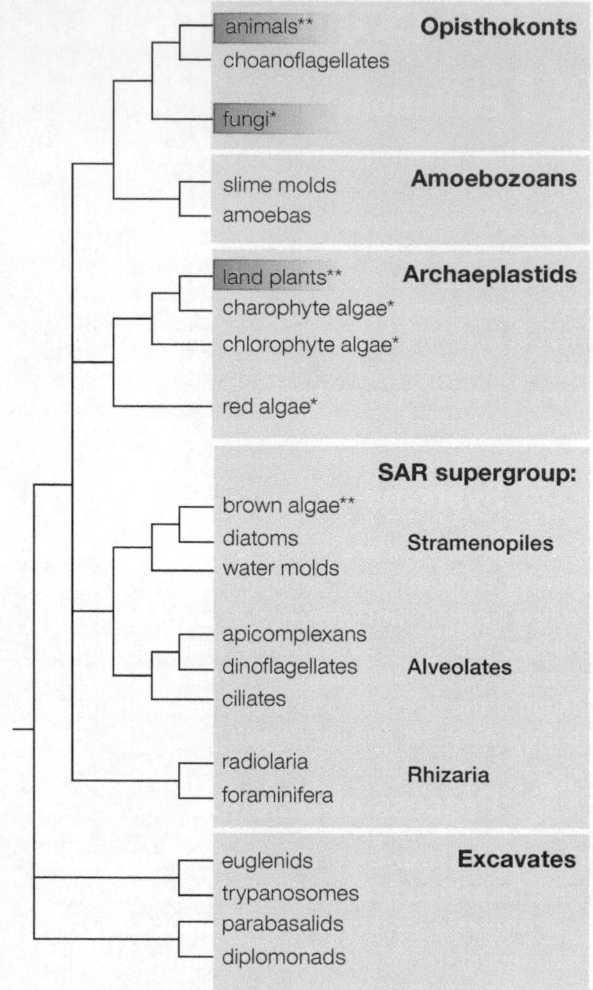
animals**
choanoflagellates
fungi*
Opisthokonts
slime molds
amoebas
Amoebozoans
land plants**
charophyte algae*
chlorophyte algae*
red algae*
Archaeplastids
SAR supergroup:
brown algae**
diatoms
water molds
Stramenopiles
apicomplexans
dinoflagellates
ciliates
Alveolates
radiolaria
foraminifera
Rhizaria
euglenids
trypanosomes
parabasalids
diplomonads
Excavates

*Some multicelled species **All species multicelled

FIGURE 21.2 Proposed phylogenetic tree for eukaryotic groups discussed in this book. Protist groups are indicated by black boxes. There are many additional protists, and new groups are continually being named.

FIGURE IT OUT Are land plants more closely related to the red algae or the brown algae?

Answer: Red algae

colonial organism Organism composed of many integrated cells, each capable of living and reproducing on its own.
contractile vacuole In freshwater protists, an organelle that collects and expels excess water.
eyespot In some protists, a pigmented organelle that detects light.
multicellular organism Organism composed of interdependent cells that vary in their structure and function.
pellicle Outer layer of plasma membrane and elastic proteins that protects and gives shape to many unwalled, single-celled protists.
primary endosymbiosis Evolution of an organelle from bacteria that entered a host cell and lived inside it.
secondary endosymbiosis Evolution of a chloroplast from a protist that itself contains chloroplasts that arose by primary endosymbiosis.

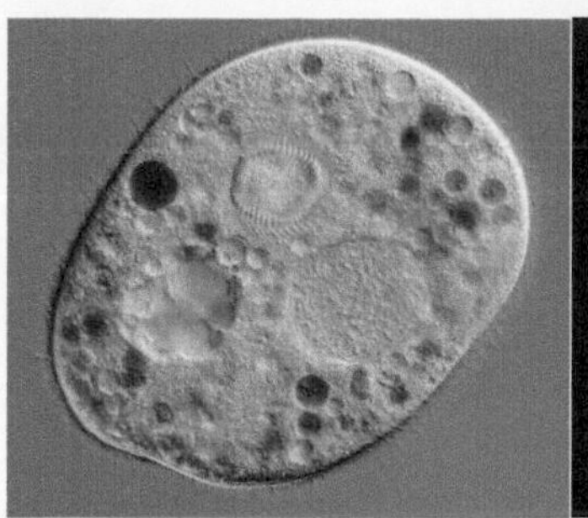

A Single-celled ciliate (*Nassula*), about 0.25 millimeters across.

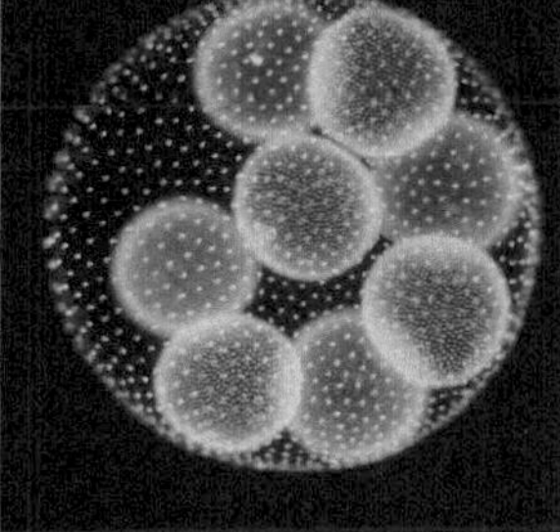

B Colonial green alga (*Volvox*), about 1 millimeter across.

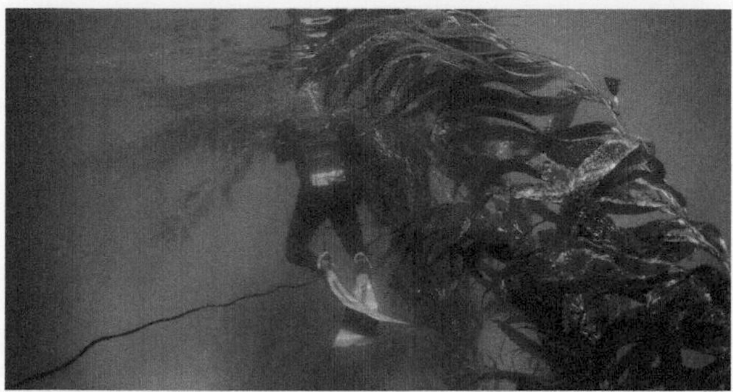

C Diver with a multicelled giant kelp (*Macrocystis*). These brown algae can be 45 meters (150 feet) long.

FIGURE 21.3 Diversity of protist size and organization.

Cell Structure

Protists are eukaryotes, so they all have a nucleus. Most have all other standard eukaryotic organelles such as endoplasmic reticulum (ER) and Golgi bodies. Almost all protists have at least one mitochondrion, or a modified mitochondrion that functions in anaerobic energy production. The eukaryotic organelles can be seen in FIGURE 21.4, which illustrates the body plan of the single-celled, freshwater protist *Euglena*.

Most single-celled protists have a protective layer at their cell surface. Green algae and dinoflagellates secrete a cellulose-containing cell wall. Radiolaria and diatoms make a glassy silica shell. Foraminifera build a calcium carbonate shell. By contrast, the outer layer of *Euglena* is a flexible **pellicle** that consists of a plasma membrane and a thin layer of elastic proteins just under it. A pellicle constrains the cell's shape somewhat, but flexes enough to allow the organism to bend and to squeeze through narrow openings.

Protists have a variety of structures that allow them to move from place to place. Some, such as *Euglena*, propel themselves by moving one or more flagella. Others move by means of cilia. Still others crawl along by extending lobes of cytoplasm called pseudopods.

The interior of a freshwater protist such as *Euglena* is saltier than its freshwater habitat, so water tends to diffuse into the cell. To keep from bursting, a euglena operates one or more **contractile vacuoles**, organelles that collect excess water from the cytoplasm, then contract and expel it from the cell through a pore.

Many protists have chloroplasts. By the process of **primary endosymbiosis** cyanobacteria evolved into chloroplasts in the common ancestor of red algae and green algae (FIGURE 21.5A). These chloroplasts have two membranes. Later, green algae and red algae evolved into chloroplasts in other protists by **secondary endosymbiosis** (FIGURE 21.5B). These chloroplasts have four membranes. *Euglena's* chloroplasts have four membranes and evolved from a green algae by secondary endosymbiosis.

Some motile photosynthetic protists, including *Euglena*, have an **eyespot**. This pigmented organelle allows them to detect light so they can move toward areas that are best for photosynthesis.

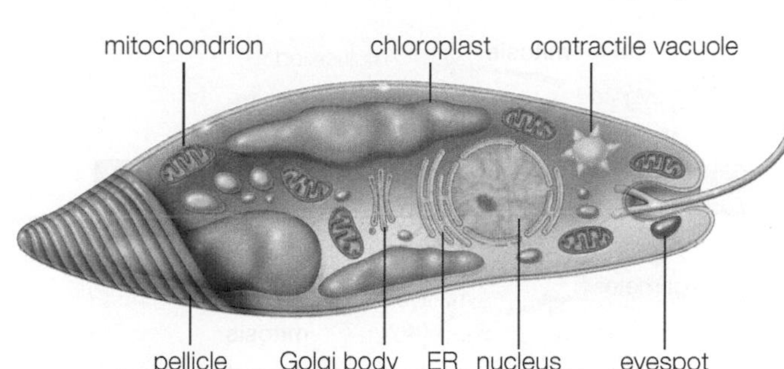
long flagellum
mitochondrion
chloroplast
contractile vacuole
pellicle
Golgi body
ER
nucleus
eyespot

FIGURE 21.4 Structure of a single-celled protist. *Euglena* is a freshwater excavate.

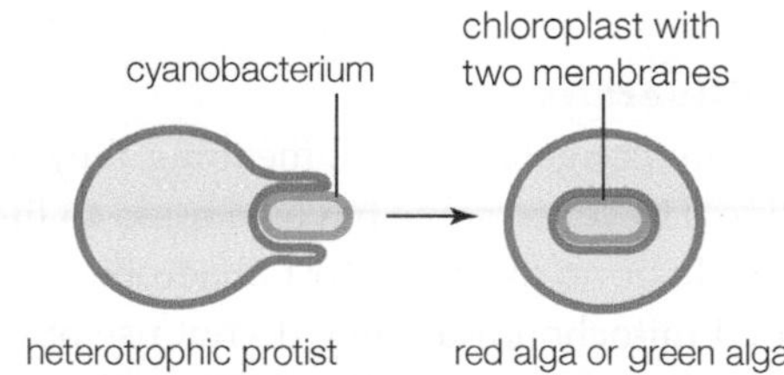
cyanobacterium
chloroplast with two membranes
heterotrophic protist
red alga or green alga

A Primary endosymbiosis. A heterotrophic protist engulfs a cyanobacterium that evolves into a chloroplast.

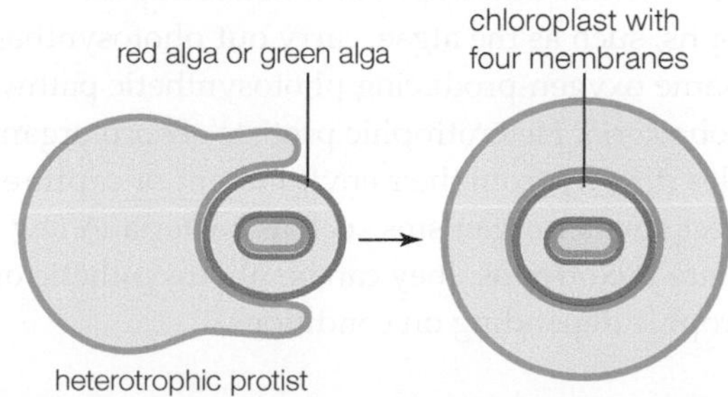
red alga or green alga
chloroplast with four membranes
heterotrophic protist

B Secondary endosymbiosis. A heterotrophic protist engulfs a red alga or green alga, which then evolves into a chloroplast.

FIGURE 21.5 Endosymbiotic origin of protist chloroplasts. Primary endosymbiosis is illustrated in more detail in Figure 19.10.

CREDITS: (3A) Lebendkulturen.de/Shutterstock; (3B) Lebendkulturen.de/Shutterstock; (3C) iStock/vernonwiley. (4) © Cengage Learning; (5) © Cengage Learning.

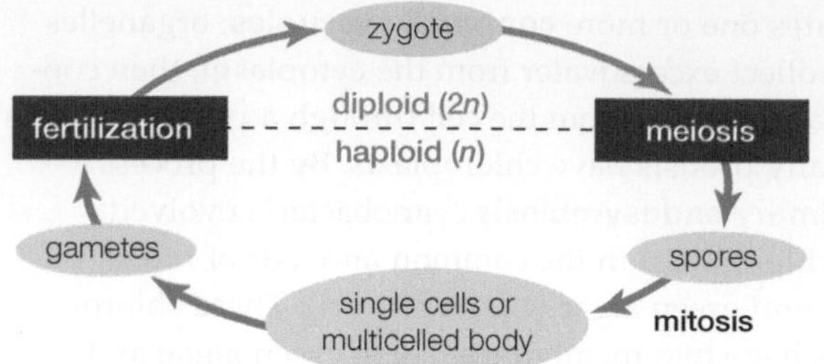
zygote
fertilization
diploid (2n)
haploid (n)
meiosis
gametes
spores
single cells or multicelled body
mitosis
A Haploid-dominant cycle; the zygote is the only diploid cell.

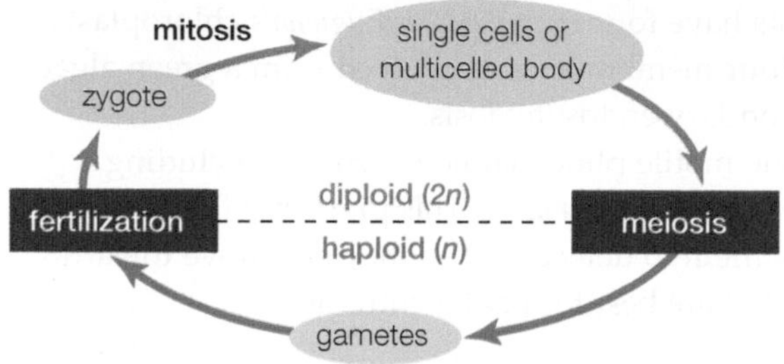
mitosis
single cells or multicelled body
zygote
fertilization
diploid (2n)
haploid (n)
meiosis
gametes
B Diploid-dominant cycle; gametes are the only haploid cells.

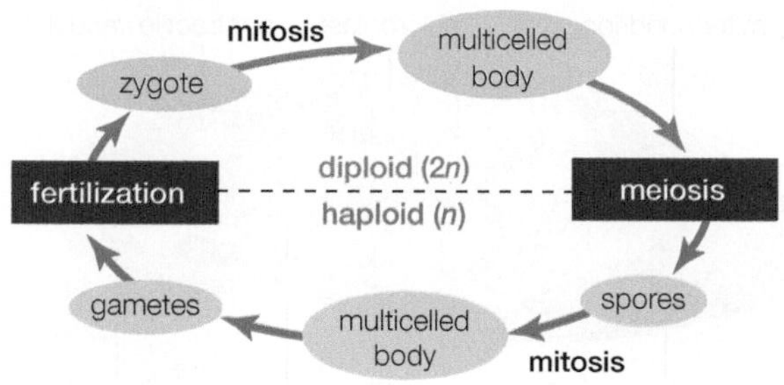
mitosis
multicelled body
zygote
fertilization
diploid (2n)
haploid (n)
meiosis
gametes
multicelled body
spores
mitosis
C Alternation of generations; multicelled haploid and diploid stages.

FIGURE 21.6 Life cycles in sexually reproducing eukaryotes. Collectively, protists utilize all three types of life cycles.

Metabolic Diversity

Most protists are obligate aerobes, meaning they need oxygen to live. However, some have adapted to life in places with little or no oxygen. Most anaerobic protists have modified mitochondria. They do not use oxygen as the final electron acceptor in their ATP-making pathway as aerobic organisms do (Section 7.5).

Protists include autotrophs and heterotrophs. Autotrophs, such as the algae, carry out photosynthesis by the same oxygen-producing photosynthetic pathway as cyanobacteria. Heterotrophic protists absorb organic molecules directly from their environment or capture and ingest smaller organisms such as bacteria. A few protists are mixotrophs; they can be photosynthetic or heterotrophic depending on conditions.

Habitats

The overwhelming majority of protists are free-living and aquatic, living in seawater or freshwater habitats. Many are components of the **plankton**, the collective term for the tiny drifting and swimming organisms in lakes and seas. There are two type of plankton. **Phytoplankton** includes all the photosynthetic organisms. **Zooplankton** refers to the heterotrophic ones.

Some protists live in moist soils. Others live on or inside other organisms. Some such protists benefit their hosts. For example, photosynthetic protists called dinoflagellates live in the tissues of corals, where they supply their animal host with sugars. Protists can also be pathogens, including the species that causes the deadly tropical disease malaria.

Life Cycles

Most protists reproduce asexually when conditions favor growth, but switch to sexual reproduction when they do not. Some protists have a cycle dominated by haploid forms (**FIGURE 21.6A**). Other protists, like all animals, have a cycle in which diploid forms dominate (**FIGURE 21.6B**). Some multicellular algae (and all plants) have an **alternation of generations**, in which both haploid and diploid bodies form (**FIGURE 21.6C**).

TAKE-HOME MESSAGE 21.2

✔ Protists do not have a specific defining trait; they are a collection of many eukaryotic lineages rather than a clade.

✔ Most protists are single-celled, but there are multicelled and colonial species.

✔ Protists are metabolically diverse. They include autotrophs, heterotrophs, and mixotrophs. Most are aerobic, but some can live without oxygen.

✔ Protist live in seas, freshwater, damp land environments, and in the bodies of other organisms.

21.3 Excavates

LEARNING OBJECTIVES

- Explain the metabolic trait that differentiates diplomonads and parabasalids from other excavate protists.
- Give three examples of medically important excavate protists.
- Compare the structure of a trypanosome and a euglenid.

With this section, we begin a survey of the protists in each eukaryote supergroup. All **excavates** (members of the supergroup Excavata) are unwalled cells with one or more flagella. The group name refers to an "excavation," a groove that functions in feeding. There are two major lineages: metamonads and euglenids.

Metamonads

Protists that lack typical mitochondria and have multiple flagella are grouped as **metamonads**. They live

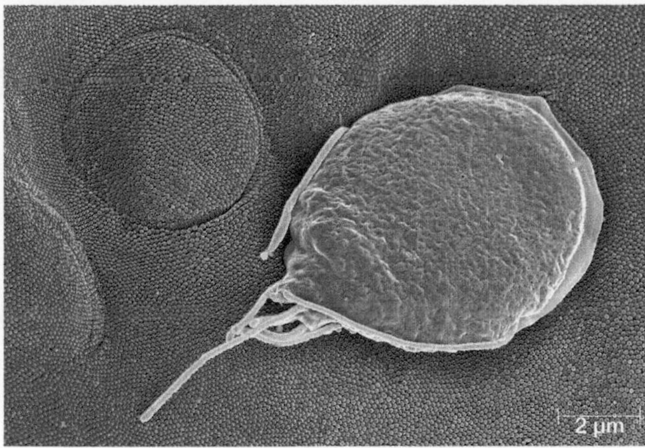
2 µm

FIGURE 21.7 Diplomonad. *Giardia lamblia* in the small intestine. (Colorized electron micrograph.) This parasite attaches via a suction cup-like disk that leaves behind a circular imprint where intestinal villi (red), tiny projections that absorb nutrients, have been damaged.

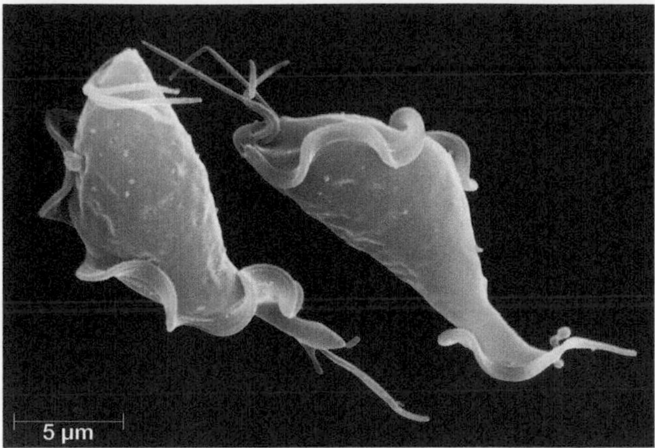
5 µm

FIGURE 21.8 Parabasalids. Colorized electron micrograph of *Trichomonas vaginalis*, which causes trichomoniasis. Cells in vaginal secretions can be seen under a light microscope, and can be identified by their twitching movements.

in oxygen-free of oxygen-poor habitats and rely on anaerobic mechanisms of ATP production.

Diplomonads have two identical nuclei and several freely beating flagella. They have only nonfunctional remnants of mitochondria, so they make their ATP by way of reactions in their cytoplasm. Most diplomonads live in animals. For example, *Giardia lamblia* attaches to the intestinal lining of a mammal and sucks up nutrients (**FIGURE 21.7**). In humans, *G. lamblia* causes the disease giardiasis. Symptoms can include cramps, nausea, and severe diarrhea. Infected people or animals excrete cysts of *G. lamblia* in their feces. (A protistan cyst is a dormant cell enclosed within a thick protective wall.) Ingesting even a few *G. lamblia* cysts in contaminated water can result in an infection.

A parabasalid has several freely beating flagella as well as an undulating membrane, which is a finlike, membrane-enclosed flagellum. Modified mitochondria called hydrogenosomes produce ATP and hydrogen. The parabasalid *Trichomonas vaginalis* (**FIGURE 21.8**) infects human reproductive and urinary tracts, causing the disease trichomoniasis. *T. vaginalis* does not make cysts, so it cannot survive very long outside the human body. Fortunately for the parasite, sexual intercourse delivers it directly into hosts. In women, symptoms include vaginal soreness, itching, and a yellowish discharge. Infected males typically show no symptoms. Untreated infections damage the urinary tract, cause infertility, and increase risk of HIV infection. An antiprotozoal drug will provide a quick cure.

Euglenozoans

Euglenozoans are flagellated protists that have mitochondria. Euglenids and kinetoplastids are the two major lineages.

Euglenids are free-living flagellated protists that have multiple mitochondria. A few are marine, but most, including *Euglena* (**FIGURE 21.4**) live in fresh water. Many photosynthetic euglenids, including *Euglena,* are mixtrophs. They can function as heterotrophs when conditions prevent photosynthesis. Some euglenids lack chloroplasts and are always heterotrophs.

Kinetoplastids have a single large mitochondrion. Inside the mitochondrion, near the base of the flagellum, is a clump of mitochondrial DNA. This clump is the kinetoplast for which the group is named.

Some free-living kinetoplastids prey on bacteria in fresh water and seas. However, the most diverse kinetoplast subgroup, the **trypanosomes**, parasitizes plants or animals. Trypanosomes are long, tapered cells with an undulating membrane.

Biting insects act as vectors for many trypanosomes that parasitize humans. Bloodsucking bugs transmit *Trypanosoma cruzi*, the cause of Chagas disease.

alternation of generations Life cycle in which both haploid and diploid multicelled bodies form; occurs in some algae and all land plants.

euglenid (you-GLEEN-id) Free-living, flagellated, excavate protist with mitochondria; most live in fresh water; e.g., *Euglena*.

excavates Protist in a supergroup defined by their feeding groove; all are single, unwalled flagellated cells; includes diplomonads, parabasalids, trypanosomes, and euglenids.

metamonads Anaerobic excavate protists, including diplomonads and parabasalids; have multiple flagella and no mitochondria.

phytoplankton (FIGHT-o-plank-tun) Photosynthetic members of the plankton; tiny, aquatic autotrophs.

plankton Collection of tiny organisms that drift and swim in lakes and seas.

trypanosome (trih-PAN-o-SOHM) Parasitic excavate protist with a single mitochondrion and an undulating membrane.

zooplankton (zoh-ah-PLANK-ton) Heterotrophic members of the plankton; tiny, aquatic heterotrophs.

CREDITS: (7) CDC/ Dr. Stan Erlandsen; (8) David M. Phillips/Science Source.

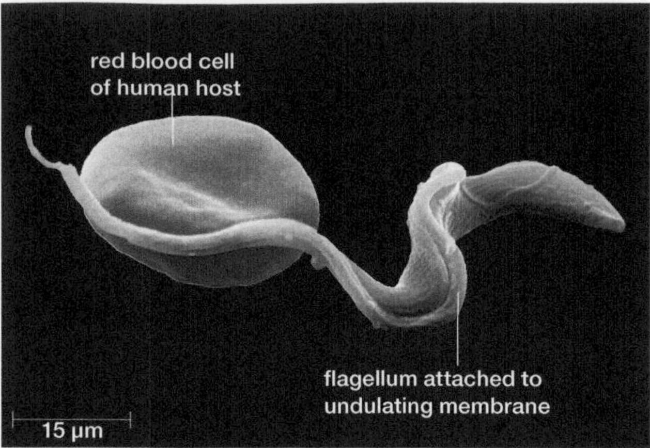
red blood cell
of human host
flagellum attached to
undulating membrane
15 µm

FIGURE 21.9 Trypanosome with red blood cell. Colorized electron micrograph of the *Trypanosoma brucei.* A bite of an infected tsetse fly delivers this parasite into the blood. It lives in the fluid portion of the blood (the plasma) and absorbs nutrients across its body wall.

Untreated, an infection can harm the heart and digestive organs. Chagas disease is prevalent in parts of Central and South America, and occurs at low frequency in the southern United States.

Tsetse flies spread *Trypanosoma brucei* (**FIGURE 21.9**), which causes African trypanosomiasis, or African sleeping sickness. The disease affects the brain and impairs the sleep cycle. If untreated, it is typically fatal.

TAKE-HOME MESSAGE 21.3

✔ Excavates are single-celled, unwalled protists with flagella.

✔ Diplomonads and parabasalids are heterotrophs with multiple flagella. They live in oxygen-poor habitats and produce ATP by anaerobic processes. Some cause disease in humans.

✔ Trypanosomes are heterotrophs that have a single large mitochondrion and an undulating membrane. Some are human pathogens.

✔ Euglenids are free-living, with multiple mitochondria. Most live in fresh water. They can be autotrophs, heterotrophs, or mixotrophs.

21.4 Stramenopiles

LEARNING OBJECTIVES

- Compare the structure of diatoms and kelps.
- List some commercial products derived from stramenopiles.
- Explain how water molds differ from true fungal molds.
- Describe the importance of water molds as pathogens.

The **SAR supergroup** is the most diverse of the exclusively protistan eukaryotic supergroups. It is named for the three related lineages it includes: stramenopiles, alveolates, and rhizarians. The affinity between members of this group is not readily apparent, but gene sequencing studies indicate that they are relatives.

The group name Stramenopila means "straw-haired" and refers to a short anterior flagellum with shaggy hairlike extensions. Many stramenopiles have two flagella at some point during their life cycle, the short, shaggy one denoted by the group name, and a longer smooth one.

Most **stramenopiles** are aquatic and photosynthetic, but the group also includes some free-living and parasitic heterotrophs. Chloroplasts of photosynthetic stramenopiles contain a brown accessory pigment (fucoxanthin), along with chlorophylls *a* and *c*.

Gene comparisons indicate that chloroplasts in all SAR supergroup lineages descend from the same red alga. That alga partnered with a common ancestor of all these lineages and, over time, evolved into a chloroplast—an example of secondary endosymbiosis. Later, some members of the various lineages lost their chloroplasts and became heterotrophs.

Diatoms

The **diatoms** (Bacillariophyceae) are diploid, photosynthetic, protists that secrete a protective silica shell (**FIGURE 21.10**). Some live as a single cell and others are colonial. The diatom shell consists of two overlapping parts that fit one inside the other, like a shoe box and its lid. Many diatoms have a cylindrical shape, but others are triangular, square, or needlelike.

Diatoms live in lakes, seas, and damp soils. They are among the most abundant members of the phytoplankton in temperate and polar waters. When marine diatoms die, their shells accumulate on the seafloor. In places where accumulations of diatom shells have been lifted onto land by geologic processes, people mine the resulting "diatomaceous earth." This silica-rich material is used in filters and cleaners, and as an insecticide that is nontoxic to vertebrates. Diatomaceous earth kills crawling insects by nicking their outer covering, causing them to dry out and die.

Diatom cells contain a large amount of oil. Oil is less dense than water, and its presence helps these

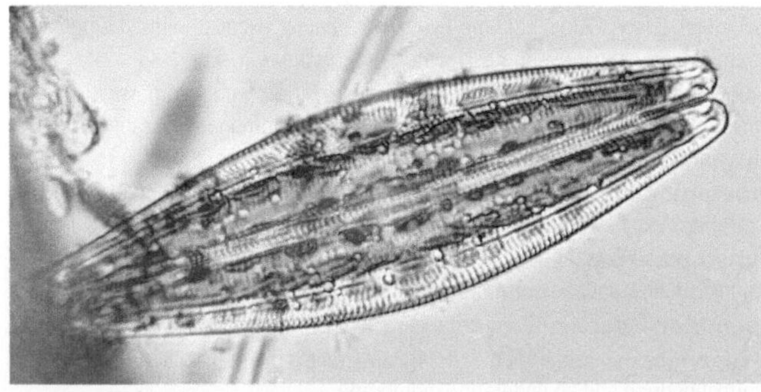

FIGURE 21.10 Living diatom. Chloroplasts are visible through the cell's glassy, silica shell.

photosynthetic cells stay afloat in sunlit waters. The oil also serves as a store of energy. In some places, geologic processes have transformed deposits of ancient diatom oil into petroleum, which we extract to produce gasoline. Researchers are currently investigating the use of diatoms to produce biofuels.

Brown Algae

All **brown algae** (Phaeophyceae) are multicelled and most live in cool coastal waters. Although some brown algae appear plantlike, this similarity is a result of morphological convergence. Brown algae evolved from a different single-celled ancestor than the lineage that includes the red algae, green algae, and land plants.

About 1,500 species of brown algae range in size from microscopic filaments to kelps that stand 30 meters (100 feet) tall. Giant kelps form forestlike stands in coastal waters of the Pacific Northwest (**FIGURES 21.3C** and **21.11**). Like trees in a forest, these kelps shelter a wide variety of organisms.

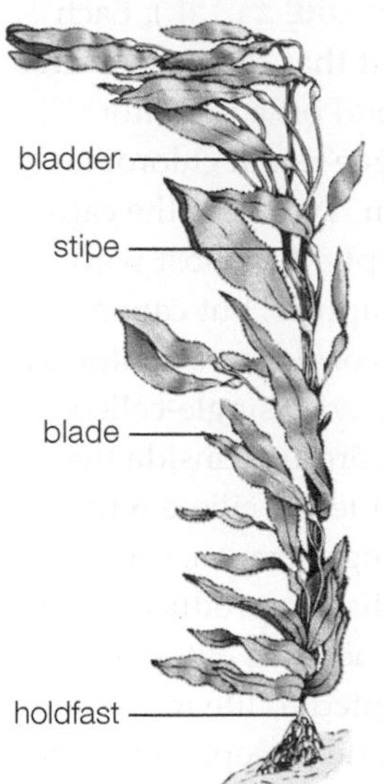
bladder
stipe
blade
holdfast

FIGURE 21.11 Structure of a giant kelp. A holdfast anchors the kelp. The stemlike stipe has leaflike blades. Many gas-filled bladders make the stipes and blades buoyant.

The kelp life cycle is an alternation of generations. The large, spore-bearing seaweed is a long-lived diploid body. Gametes form on a microscopic shorter-lived, haploid body.

Brown algae have several commercial uses. Some species are harvested for use as food or as nutritional supplements. Kelps are the source of alginates. These polysaccharides are extracted from kelp cell walls and used to thicken foods, beverages, cosmetics, and body lotions.

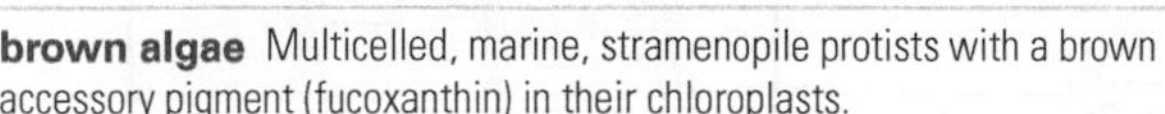

brown algae Multicelled, marine, stramenopile protists with a brown accessory pigment (fucoxanthin) in their chloroplasts.

diatoms (DIE-ah-tom) Single-celled photosynthetic stramenopile protists with a brown accessory pigment (fucoxanthin) and a two-part silica shell.

SAR supergroup Major lineage of eukaryotes; includes three protist lineages, the stramenopiles, alveolates, and rhizarians.

stramenopiles (struh-men-O-piles) Protist lineage that includes the photosynthetic diatoms and brown algae, as well as the heterotrophic water molds.

water molds Heterotrophic stramenopiles that grow as a mesh of nutrient-absorbing filaments.

FIGURE 21.12 Water mold. Filaments of *Saprolegnia* on a fish.

Water Molds

Water molds (Oomycota) are heterotrophs that form a mesh of nutrient-absorbing filaments. Water molds were once classified with the fungi, because both groups have a similar growth habit. However, the filaments of a water mold consist of diploid cells that have cell walls made of cellulose. By contrast, fungal filaments are haploid and have cell walls made of chitin.

Most water molds decompose organic matter in aquatic habitats, but some are parasites. *Saprolegnia* frequently infects fish in aquariums, fish farms, and hatcheries (**FIGURE 21.12**). Another water mold, *Pythium insidiosum*, infects mammals. *P. insidiosum* spores enter the body through a wound or in drinking water. The resulting disease (pythiosis) can be fatal. Most reported cases are in domestic animals such as dogs and horses, but rare human infections do occur. Pythiosis is primarily a tropical disease. In the United States, most cases are in Florida, Louisiana, and Texas.

Water molds are important plant pathogens. *Phytophthora* means "plant destroyer," and this genus of water molds lives up to its name. In the mid-1800s, one species caused an outbreak of late blight disease that destroyed Ireland's potato crop. The result was a widespread famine that caused at least one million deaths and led to the emigration of millions more. Today, *Phytophthora* species cause an estimated 5 billion dollars in crop losses every year. Wild species are susceptible as well. In California and Oregon, an ongoing outbreak of "sudden oak death" caused by the water mold *P. ramorum* is devastating native forests.

TAKE-HOME MESSAGE 21.4

✔ The single-celled, silica-shelled diatoms and the multicelled brown algae are photosynthetic stramenopiles. Diatoms are mainly aquatic and all brown algae are marine.

✔ The water molds are heterotrophic stramenopiles that grow as filaments. Water molds parasitize both plants and animals.

21.5 Alveolates

LEARNING OBJECTIVES

- Compare the structure and means of locomotion of ciliates and dinoflagellates.
- Describe the life cycle of the apicomplexan that causes malaria.

Alveolates are the second major lineage in the SAR supergroup. **Alveolates** are defined by sacs beneath their plasma membrane. (Alveolus means sac.) Depending on the group, the sacs can be empty or contain cellulose. There are three major alveolate lineages: ciliates, dinoflagellates, and apicomplexans.

Ciliates

The presence of short motile structures called cilia (Section 4.8) defines the **ciliates** (Ciliophora). This diverse group of heterotrophs lives just about anywhere there is water. Most eat other microorganisms. Some are colonial, but the majority live as single cells.

Paramecium is a common freshwater ciliate (**FIGURE 21.13A**). The cilia that cover its surface function in feeding and locomotion. They sweep water laden with food particles into an oral groove at the cell surface, and then push food to the gullet. Enzyme-filled vesicles in the gullet digest food, then digestive wastes are expelled by exocytosis. Two contractile vacuoles rid the cell of the excess water that enters by osmosis.

A ciliate has two types of nuclei. A large polyploid macronucleus controls daily function. During asexual reproduction this nucleus divides by mitosis. A smaller micronucleus functions in sexual reproduction. During sexual reproduction, two ciliates get together and their micronuclei undergo meiosis. The cells swap one of their now haploid micronuclei, then use their new combination of micronuclei to form a macronucleus with a mixed genome. After the cells separate, each divides and passes a macronucleus with a mix of genes to its daughter cells.

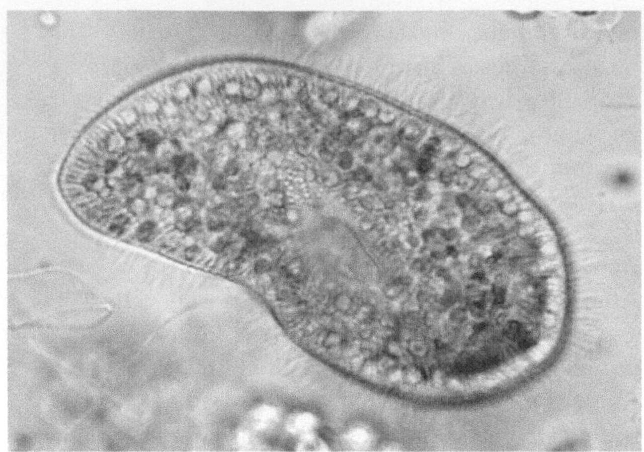

FIGURE 21.14 A ciliate-alga partnership. *Paramecium bursaria* with green algal symbionts living in its cytoplasm.

Like other ciliates, *Paramecium* has organelles called trichocysts beneath its pellicle (**FIGURE 21.13B**). Each trichocyst holds a protein filament that can be extruded to help the cell capture prey or fend off a predator.

Many ciliates that graze on algae retain chloroplasts from their prey in their cytoplasm. Typically, the captive chloroplasts survive and supply their host with sugars for some time after their capture, but cannot replicate. By contrast, one species of *Paramecium* forms a mutually beneficial relationship with single-celled green algae. The algae live and reproduce inside the ciliate (**FIGURE 21.14**). They provide the ciliate with oxygen and sugars, while receiving nitrogen, carbon dioxide, and shelter. When the ciliate reproduces, each descendant cell receives some algae in its cytoplasm.

A few ciliate species have adapted to life in the animal digestive tract. Some help cattle, sheep, and related grazers break down the plant material. Others help termites digest wood. Only one ciliate, *Balantidium coli*, is a known human pathogen. It can live in the digestive tract of both humans and pigs. Human infections usually arise when dormant *B. coli* cysts that were shed in feces get into food or water. Gut-dwelling ciliates are

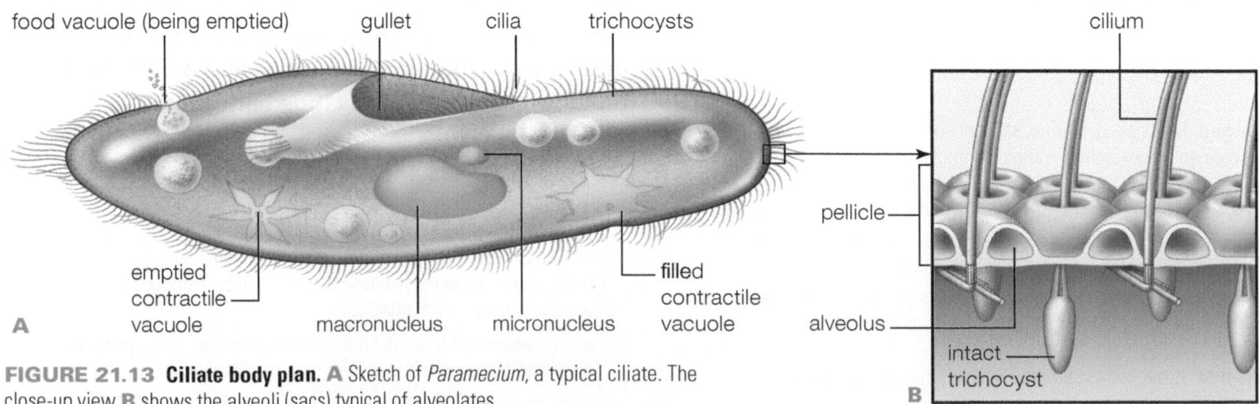

food vacuole (being emptied)
gullet
cilia
trichocysts
emptied contractile vacuole
macronucleus
micronucleus
filled contractile vacuole
A
cilium
pellicle
alveolus
intact trichocyst
B

FIGURE 21.13 Ciliate body plan. **A** Sketch of *Paramecium*, a typical ciliate. The close-up view **B** shows the alveoli (sacs) typical of alveolates.

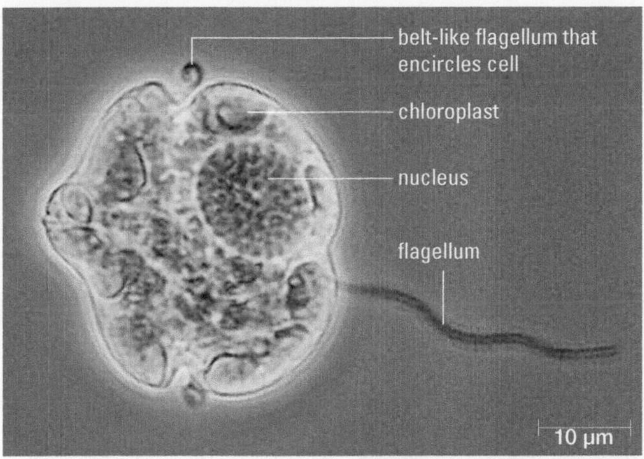
belt-like flagellum that encircles cell
chloroplast
nucleus
flagellum
10 μm

FIGURE 21.15 Dinoflagellate. This species, *Karenia brevis*, is a mixotroph. It has chloroplasts, but can also ingest microorganisms.

FIGURE 21.16 Algal bloom. A population explosion of dinoflagellates tints waters off Florida's Gulf Coast red.

typically anaerobic and, like parabasalids, they have hydrogenosomes derived from mitochondria.

Dinoflagellates

The **dinoflagellates** are solitary cells that typically have two flagella, one extending out from the base of the cell, and the other around the cell's middle like a belt (**FIGURE 21.15**). The group name means "whirling flagellate." Combined action of the two flagella causes the cell to rotate as it moves. Many dinoflagellates deposit cellulose in the alveoli beneath their plasma membrane. The cellulose accumulates as thick but porous plates.

Some dinoflagellates prey on bacteria and some are parasitic. However, most are photosynthetic, with chloroplasts descended from the same red algal ancestor as those of stramenopiles. A few photosynthetic dinoflagellate species are endosymbionts inside animals, such as reef-building corals. The dinoflagellates supply the coral with oxygen needed for aerobic respiration and with photosynthetically produced sugar. The coral provides the protists with nutrients, shelter, and the carbon dioxide necessary for photosynthesis. Without its dinoflagellate partners, a reef-building coral will die.

The vast majority of dinoflagellates are part of the marine plankton, and they are especially abundant in warm climates. In tropical seas, dinoflagellates are the most common source of **bioluminescence**, light produced by a living organism. When disturbed, dinoflagellates produce a blue or blue-green glow by an oxidation–reduction reaction (Section 5.5). The light may protect the protist by startling a predator that was about to eat it. Or, the flash of light may function like a car alarm, attracting the attention of predators that pursue the would-be dinoflagellate eaters.

Free-living photosynthetic dinoflagellates or other photosynthetic cells sometimes undergo great increases in population size, a phenomenon known as an **algal bloom**. In habitats enriched with nutrients, as from agricultural runoff, a liter of water may hold millions of cells. Blooms of some dinoflagellates cause "red tides," so named because the abundance of these protists tints the water red (**FIGURE 21.16**).

Algal blooms can pose a threat to other organisms. Aerobic bacteria that feed on algal remains can use up the oxygen in the water, so that aquatic animals suffocate. Some dinoflagellates also produce toxins. For example, *Karenia brevis* produces a toxin that binds to transport proteins in the plasma membrane of nerve cells. Eat shellfish contaminated with this toxin and you might end up dizzy and nauseated from neurotoxic shellfish poisoning. Symptoms usually develop hours after the meal and persist for a few days.

Apicomplexans

The largest lineage of parasitic protists is the apicomplexans, with more than 5,000 species. **Apicomplexans** (Apicomplexa) are single-celled protists that spend part of their life cycle inside cells of their animal hosts. The group name refers to a complex of microtubules and secretory vesicles at the apex (upper tip) of the cell. Components of this complex allow the parasite to invade a host cell. The life cycle is complicated, involving both sexual and asexual reproduction. Consider the

algal bloom Population explosion of photosynthetic cells in an aquatic habitat.

alveolate (al-VEE-o-late) Member of a protist lineage having small sacs beneath the plasma membrane; dinoflagellate, ciliate, or apicomplexan.

apicomplexan (ay-PEE-com-PLEKS-an) Single-celled alveolate protist that lives as a parasite inside animal cells; some cause malaria or toxoplasmosis.

bioluminescence (BY-oh-loom-ih-NESS-sense) Light produced by an organism.

ciliate (Sil-ee-ets) Single-celled or colonial protist with multiple cilia.

dinoflagellate (di-no-FLAJ-o-let) Single-celled, aquatic protist with cellulose plates and two flagella; may be heterotrophic or photosynthetic.

CREDITS: (15) Bob Andersen and D. J. Patterson; (16) Peter Johnson.

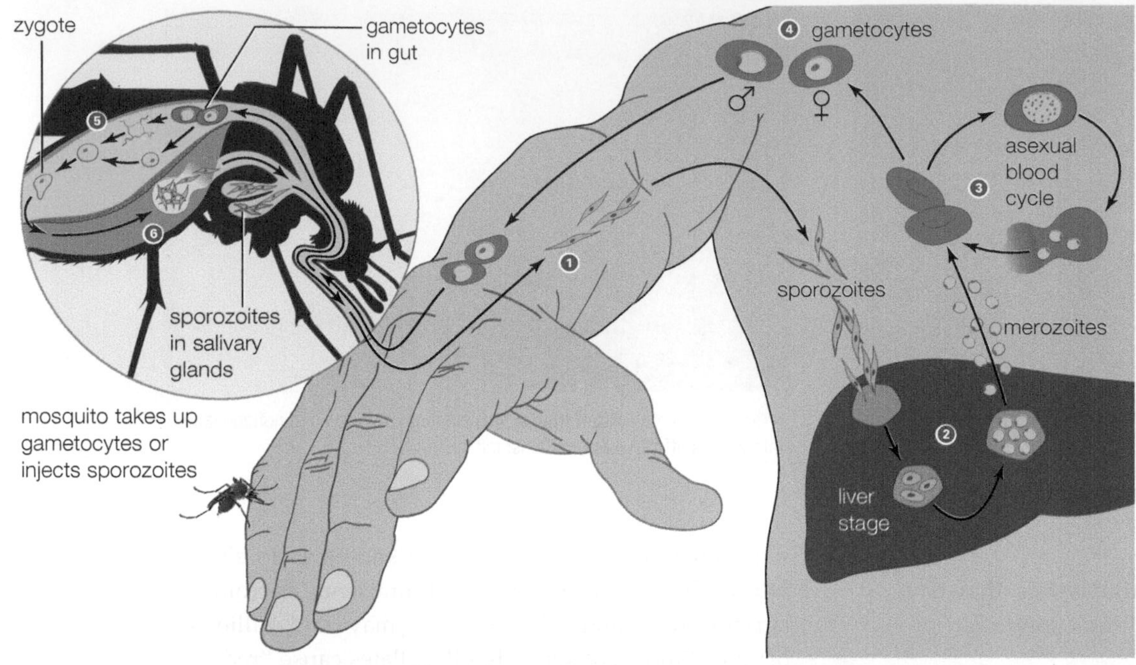
zygote
gametocytes in gut
④ gametocytes
♂
♀
⑤
⑥
asexual blood cycle
③
①
sporozoites in salivary glands
sporozoites
merozoites
mosquito takes up gametocytes or injects sporozoites
②
liver stage

FIGURE 21.17 Life cycle of the apicomplexan that causes malaria (*Plasmodium*).

❶ A mosquito bites a human. Haploid sporozoites enter blood, which carries them to the liver.

❷ Sporozoites reproduce asexually in liver cells, then mature into merozoites. Merozoites leave the liver and infect red blood cells.

❸ Merozoites reproduce asexually in some red blood cells.

❹ In other red blood cells, merozoites become gametocytes.

❺ A mosquito bites and sucks blood from the infected person. Gametocytes in blood enter its gut and mature into gametes, which fuse to form diploid zygotes.

❻ Meiosis of zygotes produces cells that develop into sporozoites. The sporozoites migrate to the mosquito's salivary glands.

life cycle of *Plasmodium*, the apicomplexan that causes malaria (FIGURE 21.17). A female mosquito transmits the motile infective form of the parasite (the sporozoite) to a vertebrate, such as a human ❶. The sporozoite travels to liver cells, then reproduces asexually ❷. Some of the resulting cells (merozoites) enter red blood cells and liver cells, where they divide asexually ❸. Others develop into immature gametes (gametocytes) ❹. A mosquito sucks up gametocytes along with blood, and they mature into gametes in its gut ❺. When two gametes fuse, the resulting zygote develops into a new sporozoite ❻. Sporozoites migrate to the insect's salivary glands, where they await transfer to a new host.

Fever and weakness usually start a week or two after the initial mosquito bite, as infected liver cells rupture, releasing merozoites and cellular debris into

Data Analysis Activities

Summoning Mosquitoes Parasites sometimes alter their host's behavior in a way that increases their chances of transmission to another host.

Dr. Jacob Koella and his associates hypothesized that *Plasmodium* might benefit by making its human host more attractive to hungry mosquitoes when gametocytes are available in the host's blood. Gametocytes taken up by the mosquito will mature into gametes and mate inside its gut.

To test their hypothesis, the researchers recorded the response of mosquitoes to the odor of *Plasmodium*-infected children and uninfected children over the course of 12 trials on 12 separate days. FIGURE 21.18 shows their results.

1. On average, which group of children was most attractive to mosquitoes?
2. Which group of children averaged the fewest mosquitoes attracted?
3. What percentage of the total number of mosquitoes were attracted to the most attractive group?
4. Did the data support Dr. Koella's hypothesis?

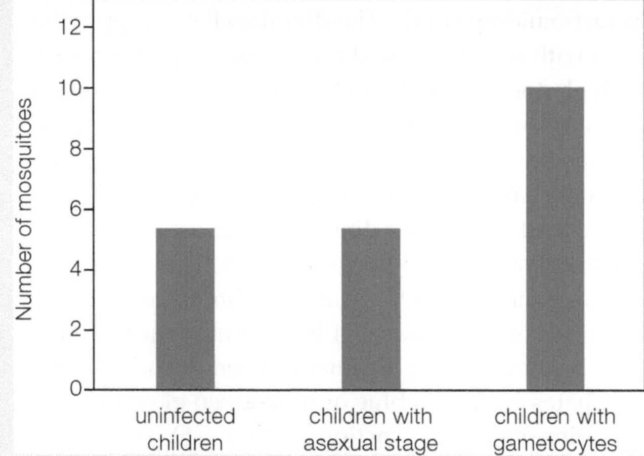
Number of mosquitoes
0
2
4
6
8
10
12
uninfected children
children with asexual stage
children with gametocytes

FIGURE 21.18 Attracting mosquitoes. The graph shows the number of mosquitoes (out of 100) attracted to uninfected children, children harboring the asexual stage of *Plasmodium*, and children with gametocytes in their blood. The bars show the average number of mosquitoes attracted to that category of child over the course of 12 separate trials.

the bloodstream. After the initial episode, symptoms may subside. However, continued infection harms organs throughout the body and—if the infection remains untreated—eventually kills the host.

Another apicomplexan disease, toxoplasmosis, is caused by *Toxoplasma gondii*. Infected people typically have no obvious symptoms. However, an infection can be fatal in immune-suppressed people and an infection during pregnancy can cause miscarriage or result in neurological birth defects. Toxoplasmosis most often arises after ingestion of cysts in undercooked meat. Cats that spend time hunting outdoors can shed *T. gondii* cysts in their feces, so pregnant women should avoid contact with cat feces.

TAKE-HOME MESSAGE 21.5

✔ Alveolates are characterized by sacs beneath their membrane. They include ciliates, dinoflagellates, and apicomplexans. All live primarily as single cells.

✔ Ciliates use cilia to move about and feed. Most are free-living predators, but some live inside animals and some get a nutritional boost from chloroplasts or algae that survive ingestion.

✔ Dinoflagellates move with a whirling motion caused by the action of their two flagella. Some are heterotrophs. Others are photosynthetic members of plankton or endosymbionts in reef-building corals.

✔ Apicomplexans such as those that cause malaria are single-celled parasites with a complex life cycle. During part of this cycle, the parasite lives inside host cells.

21.6 Rhizarians

LEARNING OBJECTIVES

- Compare the structure of foraminifera and radiolarians.
- Explain how rhizarians feed.
- List the types of rocks that form from rhizarian remains.

Rhizarians (Rhizaria) are the third lineage in the SAR supergroup. Nearly all **rhizarians** are single-celled, marine heterotrophs with a protective shell. They capture their prey using cytoplasmic extensions that protrude through small holes in their shell. Foraminifera and radiolarians are the two main rhizarian lineages.

Foraminifera have a hard shell that, in most species, consists of multiple secreted chambers of calcium carbonate ($CaCO_3$). Including its shell, an individual foraminiferan can be as big as a grain of sand. Most species live on the seafloor, where they probe the water and sediments for prey. A few are members of the marine plankton (FIGURE 21.19). Planktonic foraminifera often have smaller photosynthetic protists such as diatoms or algae living on or in them.

200 μm

FIGURE 21.19 **Foraminiferan.** A planktonic foraminiferan with algal cells (yellow dots) living on its cytoplasmic extensions.

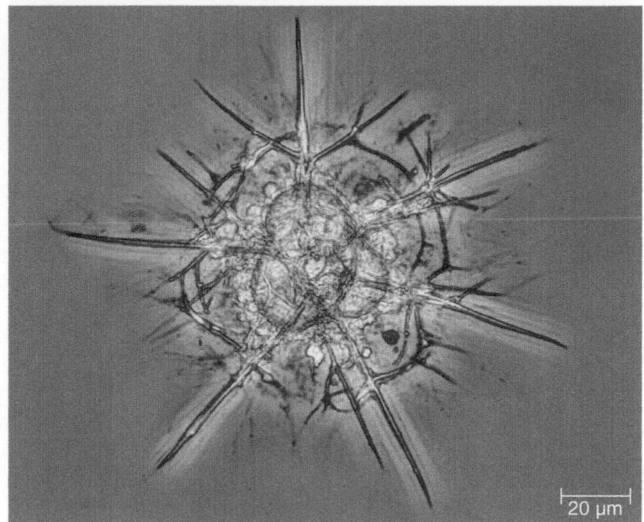
20 μm

FIGURE 21.20 **Radiolarian.**

Radiolaria secrete a glassy silica (SiO_2) shell (FIGURE 21.20) and have two cytoplasmic layers. The inner layer holds all the typical eukaryotic organelles and the outer one has numerous gas-filled vacuoles that keep the cell afloat. Radiolaria are planktonic, and they are most abundant in nutrient-rich tropical waters.

Foraminifera and radiolaria have lived and died in the oceans for more than 500 million years, so remains of countless cells have fallen to the seafloor. Over time, geologic processes transformed some accumulations of

foraminifera (for-am-in-IF-er-ah) Heterotrophic single-celled protists with a porous calcium carbonate shell and long cytoplasmic extensions.
radiolaria (ray-dee-oh-LAIR-ee-ah) Heterotrophic single-celled protists with a porous silica shell and long cytoplasmic extensions.
rhizarians (riz-ahr-EE-ans) Members of the eukaryote supergroup Rhizaria; single-celled protists with cytoplasmic extensions protruding through a secreted sievelike shell; e.g., foraminifera and radiolaria.

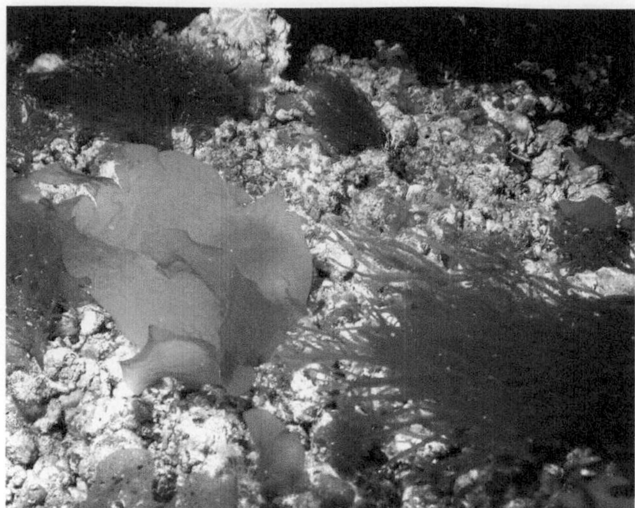

FIGURE 21.21 Red algae. Branching and sheetlike red algae growing 75 meters (250 ft) beneath the sea surface in the Gulf of Mexico.

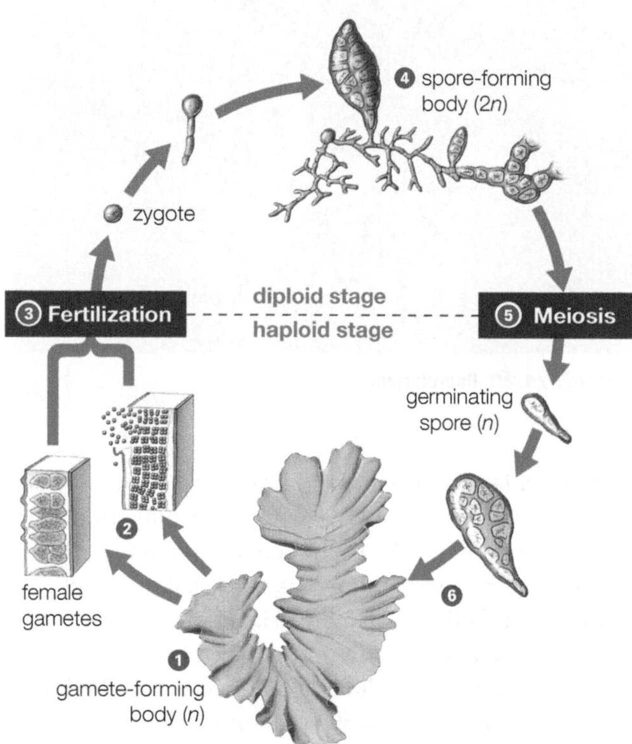

4 spore-forming body (2n)
zygote
3 Fertilization
diploid stage
haploid stage
5 Meiosis
germinating spore (n)
2
female gametes
6
1
gamete-forming body (n)

FIGURE 21.22 Life cycle of a red alga. *Porphyra* has an alternation of generations, with both haploid and diploid multicelled bodies.

1. The haploid gamete-forming body (gametophyte) is sheetlike.
2. Gametes form at its edges.
3. Fertilization produces a diploid zygote.
4. The zygote develops into a diploid spore-forming body (a sporophyte).
5. Haploid spores form by meiosis and are released.
6. Spores germinate and develop into a new gamete-forming body.

foraminiferal shells into chalk and limestone, two types of calcium carbonate-rich sedimentary rock. The giant blocks of limestone used to build the great pyramids of Egypt consist largely of the shells of foraminifera that fell to the seafloor about 50 million years ago. Similarly, geologic processes transformed silica-rich deposits of radiolarian shells into a rock called chert.

TAKE-HOME MESSAGE 21.6

✔ Rhizarians are single cells encased in a mineralized shell. Most are heterotrophs that use the cytoplasmic extensions that protrude through openings in their shell to capture prey.

✔ Most foraminiferans have a calcium carbonate shell and live on the seafloor. Radiolarians have a silica shell and are planktonic.

21.7 Archaeplastids

LEARNING OBJECTIVES

- Explain the relationship among red algae, green algae, and land plants.
- Describe the life cycle of the red alga *Porphyra*.
- List some human uses of archaeplastids.

Archaeplastids are members of the eukaryotic supergroup Archaeplastida. All archaeplastids have cells with cellulose walls, and their chloroplasts have two membranes. These chloroplasts evolved from cyanobacteria by primary endosymbiosis in the ancestor of this group. Archaeplastids include red algae, two lineages of green algae, and the land plants. The land plants evolved from one lineage of the green algae after the red algae and green algae diverged from their common ancestor.

Red Algae

The **red algae** (Rhodophyta) are photosynthetic, typically multicelled protists that live in clear, warm seas. Most of the 6,000 or so species grow as thin sheets or in a branching pattern (**FIGURE 21.21**). Chloroplasts of red algae have chlorophyll *a*, as well as red accessory pigments called phycobilins. Phycobilins absorb blue-green light, which penetrates deeper into water than light of other wavelengths. Thus, red algae can live in deeper water than other algae.

Agar and carrageenan are two gelatinous polysaccharides extracted from red algae. They are used as thickeners or stabilizers in foods, cosmetics, and personal care products. Agar is also widely used in microbiology. When mixed with appropriate nutrients, agar can be used to form a semisolid culture medium for growing bacteria or fungi.

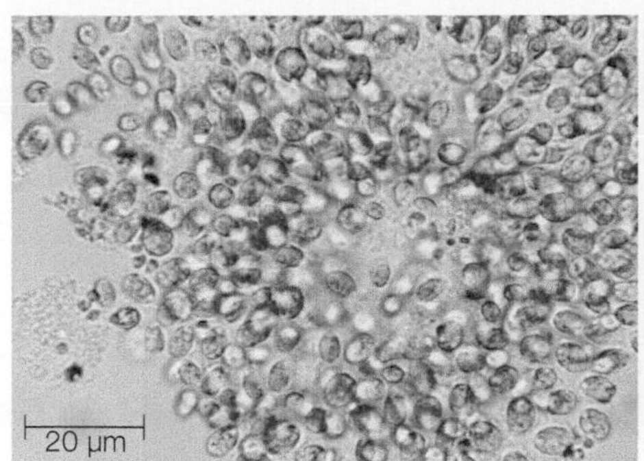
20 µm

A *Chlorella*, a nonmotile, single-celled freshwater chlorophyte.

B *Ulva*, a marine chlorophyte that is commonly called sea lettuce.

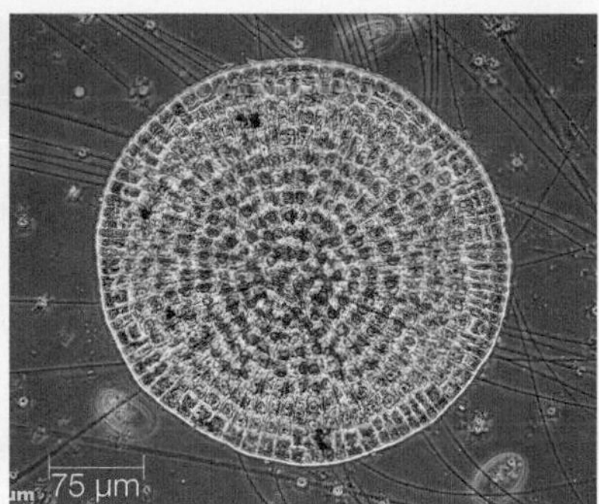
75 µm

C The charophyte alga *Coleochaete orbicularis.* It grows as a multicellular disk attached to a surface.

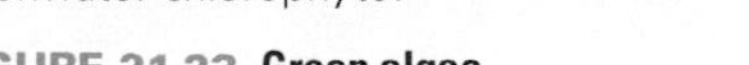

FIGURE 21.23 Green algae.

Nori, the sheets of seaweed used to wrap sushi, are derived from a red alga (*Porphyra*). Like many multicelled algae and all plants, *Porphyra* has an alternation of generations (**FIGURE 21.22**). The haploid, gamete-forming body (the gametophyte) is a sheetlike seaweed ❶ that produces gametes by mitosis ❷. When two gametes join at fertilization, they form a diploid zygote ❸ that grows and develops into the diploid spore-forming body (the sporophyte) ❹. The sporophyte is a tiny, branching filament that produces haploid spores by meiosis ❺. After its release, a spore germinates, then grows and develops into a new gamete-forming body ❻.

Green Algae

Green algae is an informal group of about 7,000 photosynthetic species belonging two lineages. Green algae range in size from microscopic cells to multicelled filamentous or branching forms more than a meter long. Like land plants, and unlike other protists, green algae have chloroplasts that contain both chlorophyll *a* and chlorophyll *b*. Also as in plants, these chloroplasts store excess carbohydrates as starch.

Chlorophyte algae (clade Chlorophyta) are the most diverse lineage of green algae. They include both freshwater and marine species. Some single-celled chlorophytes live in the soil or grow as a film on surfaces, including ice. Others partner with a fungus and form a composite organism known as a lichen. However, most single-celled chlorophytes live in fresh water.

Chlorella is a single-celled chlorophyte (**FIGURE 21.23A**). Melvin Calvin used *Chlorella* to study the light-independent reactions of photosynthesis (Section 6.5). Today, it is grown as a health food and is used in experimental production of biofuel. *Chlorella* is nonmotile, but many single-celled chlorophytes have flagella.

Other freshwater chlorophytes form long filaments. Theodor Engelmann used one of these, *Cladophora*, in studies of photosynthesis (Section 6.3).

Volvox is a colonial freshwater chlorophyte. Hundreds to thousands of flagellated cells are joined together by thin cytoplasmic strands to form a whirling, spherical colony (shown earlier in **FIGURE 21.3B**). Daughter colonies form inside the parental sphere, which eventually ruptures and releases them.

Many "seaweeds" are chlorophytes. Wispy sheets of *Ulva* cling to coastal rocks worldwide (**FIGURE 21.23B**). In some species, sheets grow longer than your arm, but are less than 40 microns thick. *Ulva* is commonly known as sea lettuce and people harvest it as food.

Gene sequence similarities indicate that land plants are descended from a second lineage of green algae, the Charophyta (**FIGURE 21.23C**). Like plants, and unlike other green algae, **charophyte algae** divide their cytoplasm by cell plate formation (Section 11.4) and have plasmodesmata (cytoplasmic connections between neighboring cells, Section 4.9).

TAKE-HOME MESSAGE 21.7

- ✔ Red algae and green algae are descendants of a common ancestor in which bacteria evolved into chloroplasts.
- ✔ Red algae are mostly multicellular and marine. They have red accessory pigments that allow them to live in deep water.
- ✔ Green algae live in many habitats and may be single-celled, colonial, or multicelled. One subgroup, the charophyte algae, includes the closest living relatives of land plants.

archaeplastids (ark-ee-PLAST-ids) Members of the eukaryotic supergroup Archaeplastida; red algae, two lineages of green algae, and the land plants.
charophyte algae (kar-UH-fight) Green algal lineage that includes the ancestors of land plants.
chlorophyte algae (KLOR-uh-fight) Most diverse lineage of green algae.
red algae Photosynthetic protists; typically multicelled, with chloroplasts containing red accessory pigments (phycobilins).

CREDITS: (23A) iStockphoto.com/Nancy Nehring; (23B) Andrew J. Martinez/Science Source; (23C) Dr. Ralf Wagner, www.dr-ralf-wagner.de.

21.8 Amoebozoans and Opisthokonts

LEARNING OBJECTIVES

- Describe the structure and feeding of free-living amoebas.
- Explain why slime molds are described as social amoebas.
- Describe the protist group most closely related to animals.

Amoebozoans are members of the eukaryotic supergroup Amoebozoa. They are shape-shifting heterotrophic cells that typically lack a wall or pellicle. Amoebozoans move about and capture smaller cells by extending thick lobes of cytoplasm called pseudopods (Section 4.8). There are two major groups of amoebozoan protists: amoebas and slime molds.

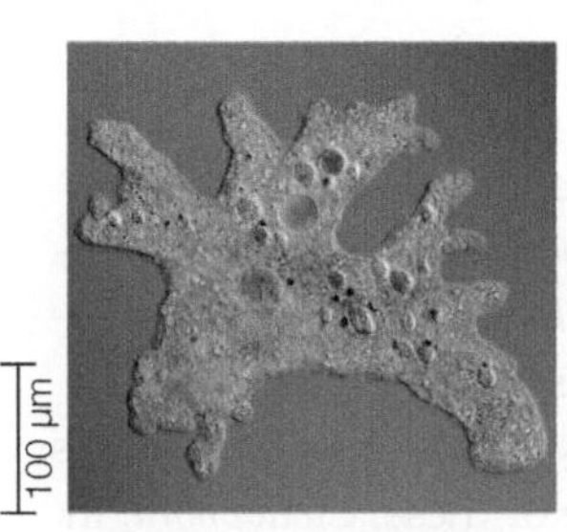
100 μm

FIGURE 21.24 **Amoeba.**

Amoebas are solitary cells (FIGURE 21.24). Most are predators in soil or freshwater, but some are marine and a few are parasites.

Entamoeba infects the human gut, causing cramps, aches, and bloody diarrhea. Infections are most common in developing nations where amoebic cysts contaminate drinking water. Exposing your sinuses to the freshwater amoeba *Naegleria fowleri* can cause a rare but deadly brain infection. Most people become infected while swimming in a warm pond or lake. Inadequate sterilization of contact lenses or swimming with such lenses on can result in an eye infection by *Acanthamoeba*.

Slime molds are sometimes described as "social amoebas." They are especially common on the floor of temperate forests.

Cellular slime molds spend the bulk of their existence as individual haploid amoeboid (amoeba-like) cells. Consider the life cycle of *Dictyostelium discoideum* (FIGURE 21.25), a species that is widely studied. Each cell eats bacteria and reproduces by mitosis ❶. When food runs out, thousands of cells aggregate to form a multicelled mass ❷. Environmental gradients in light and moisture induce the mass to crawl along as a cohesive unit often referred to as a "slug" ❸. When the slug reaches a suitable spot, its component cells differentiate to form a spore-bearing structure called a fruiting body. Some cells become a stalk, and others become spores atop the stalk ❹. When a spore germinates, it releases a cell that starts the life cycle anew ❺.

Plasmodial slime molds are usually encountered as a multinucleated mass called a plasmodium. The plasmodium forms when a diploid cell divides its nucleus repeatedly by mitosis, but does not undergo cytoplasmic division. The resulting mass can be as big as a dinner plate. It streams out along the forest floor feeding on microbes and organic matter. When food supplies dwindle, a plasmodium develops into spore-bearing fruiting bodies (FIGURE 21.26). Later, after the spores disperse, a cell will emerge from each spore.

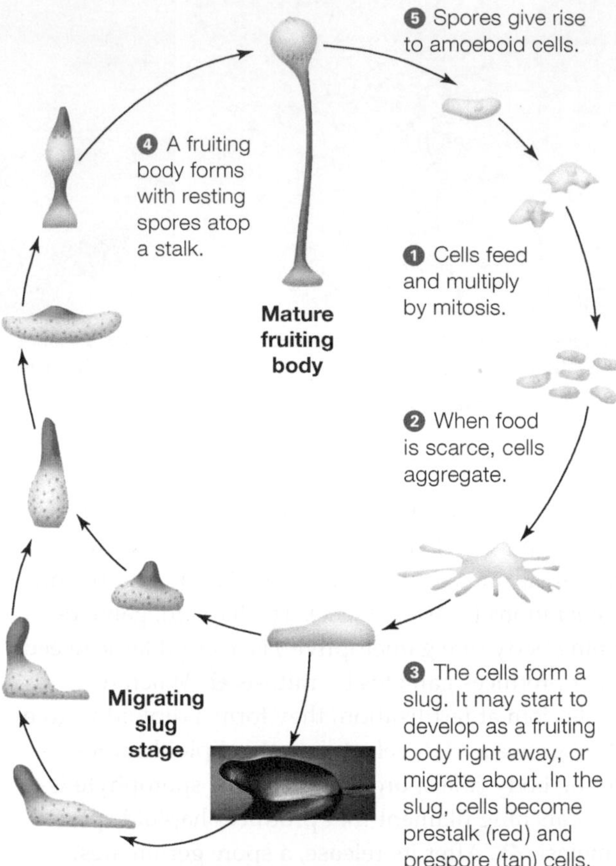
❺ Spores give rise to amoeboid cells.
❹ A fruiting body forms with resting spores atop a stalk.
Mature fruiting body
❶ Cells feed and multiply by mitosis.
❷ When food is scarce, cells aggregate.
Migrating slug stage
❸ The cells form a slug. It may start to develop as a fruiting body right away, or migrate about. In the slug, cells become prestalk (red) and prespore (tan) cells.

FIGURE 21.25 **Life cycle of a cellular slime mold** (*Dictyostelium discoideum*).

FIGURE 21.26 **Plasmodial slime mold.** Dog vomit slime mold (*Fuligo septica*) developing into a fruiting body on a log.

Opisthokonts

The sister group to the Amoebozoans is the supergroup Opisthokonta. **Opisthokonts** include fungi, animals,

CREDITS: (24) iStockphoto.com/micro_photo; (25) Carolina Biological Supply Company; (26) iStock/Mantonature.

and the protistan relatives of both groups. The closest living relatives of fungi are a group of amoeba-like cells (nucleariids) that live in soil and freshwater. We discuss them along with the fungi in Chapter 22.

The **choanoflagellates,** a group of aquatic, heterotrophic protists are the closest known protistan relatives of animals. A choanoflagellate cell has a flagellum surrounded by a "collar" of threadlike projections. Movement of the flagellum sets up a current that draws food-laden water through the collar.

Most choanoflagellates live as single cells, but some form colonies (**FIGURE 21.27**). Colonies arise by mitosis, when descendant cells do not separate after division. Instead, cells stick together with the help of adhesion proteins. Choanoflagellate adhesion proteins are similar to those of animals, and even solitary choanoflagellates have these proteins. By one hypothesis, these proteins helped a solitary common ancestor of animals and choanoflagellates to catch prey. Later, these proteins helped choanoflagellates stick together to form multicelled colonies. Later still, the proteins allowed animal cells to adhere to one another in multicelled bodies. This is an example of exaptation, an evolutionary process by which a trait that evolves with one function later takes on another (Section 17.9).

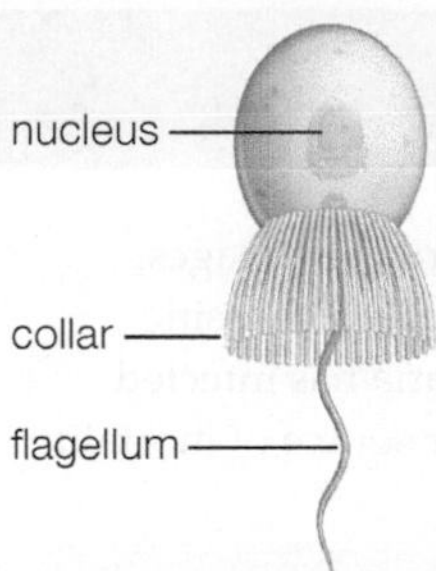
nucleus
collar
flagellum

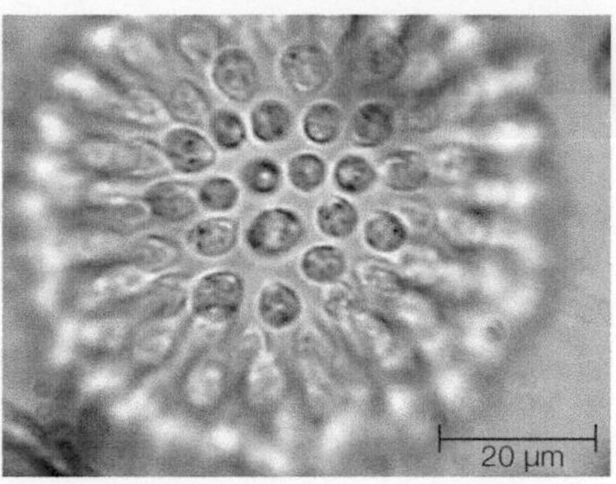
20 μm

FIGURE 21.27 Choanoflagellates. The colony at the right is made up of single cells of the type illustrated at the left.

amoeba (uh-ME-buh) Single-celled, unwalled protist that extends pseudopods to move and to capture prey.

amoebozoans (uh-me-buh-ZOE-ans) Members of the eukaryotic super group Amoeboza; includes amoebas and slime molds.

cellular slime mold Soil-dwelling protist that feeds as solitary cells, but congregates under adverse conditions to form a cohesive unit that develops into a fruiting body.

choanoflagellates (ko-an-oh-FLAJ-el-ets) Heterotrophic protists closely related to animals; collared flagellates that strain food from water.

opisthokonts (oh-PIS-thuh-konts) Members of the eukaryotic super group Opisthokonta; includes fungi, animals, and the protistan relatives of these groups.

plasmodial slime mold (plaz-MOH-dee-ul) Soil-dwelling protist that feeds as a multinucleated mass and develops into a fruiting body under adverse conditions.

TAKE-HOME MESSAGE 21.8

- ✔ Amoebozoans, such as amoebas and slime molds, are unwalled heterotrophs that feed and move by extending cytoplasmic extensions (pseudopods).
- ✔ Opisthokonts include fungi, animals, and the closest protistan relatives of these groups.
- ✔ Choanoflagellates are flagellated, heterotrophic cells that are considered close relatives of the animals.

21.1 Malaria: A Protistan Disease (revisited)

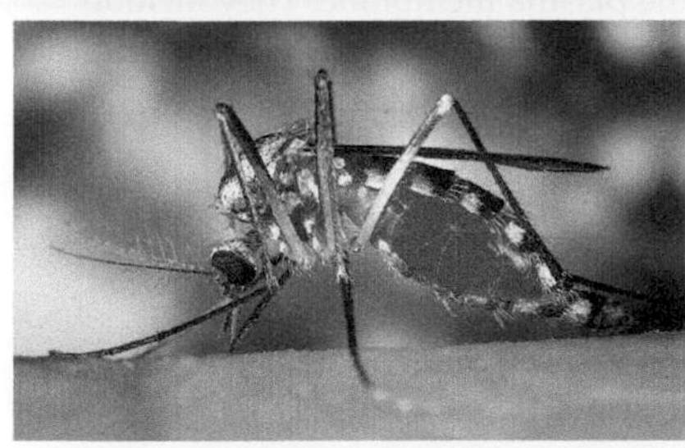

Biting people is a risky business. Each time a mosquito bites, it runs the risk of being detected and swatted. To facilitate speedy feedings and quick getaways, mosquitos inject their host with apyrase, an enzyme that impairs blood clotting. An injection of apyrase before a meal helps ensure that blood will flow smoothly up the insect's proboscis, rather than turning chunky and clogging it.

After *Plasmodium* (the apicomplexan that causes malaria) has replicated in a mosquito's gut, it moves to the insect's salivary glands. Here, it inhibits production of the anticlotting enzyme. A mosquito with infectious *Plasmodium* in its saliva has to bite more often because blood clots as it feeds. This is bad for the mosquito, but good for *Plasmodium*. A mosquito that is forced to eat many small meals, rather than a single large one, is more likely to deliver sporozoites to many new hosts.

Other apicomplexan parasites also alter host behavior to their own benefit. Consider the parasite that causes toxoplasmosis, *Toxoplasma gondii*. It reproduces asexually in rodents, but must enter a cat to complete its life cycle. Parasite-induced changes in the rat's behavior facilitate the parasite's movement from rat to cat. Unlike healthy rats, those infected by *T. gondii* do not avoid cat-scented areas but rather show a preference for such areas. Exactly how the parasite alters the rodent's behavior is not clear, but we do know that *T. gondii* infects the amygdala, the region of the brain that governs fear and anxiety.

CREDITS: (in text) Photo by James Gathany, Centers for Disease Control; (27) Courtesy of Damian Zanette.

STUDY GUIDE

Section 21.1 **Protists** include many eukaryotic lineages. *Plasmodium*, the protist that causes malaria, is a parasitic protist with important health effects. Malaria has infected humans for centuries and remains a major source of mortality in many less-developed regions.

TABLE 21.1

Organization and Nutrition of Major Protist Groups

Group	Organization	Nutritional Mode
Excavates		
Diplomonads	Single cell	Heterotrophs; free-living or parasites of animals
Parabasalids	Single cell	Heterotrophs; free-living or parasites of animals
Kinetoplastids (e.g., trypanosomes)	Single cell	Heterotrophs; mostly parasites, some free-living
Euglenids	Single cell	Free-living heterotrophs; autotrophs and mixotrophs, most free-living
Stramenopiles		
Diatoms	Single cell	Mostly autotrophs; a few heterotrophs or mixotrophs
Brown algae	Multicelled	Autotrophs
Water molds	Single cell or multicelled	Heterotrophs; free-living or parasites
Alveolates		
Ciliates	Single cell or colony	Heterotrophs; most free-living, some parasites of animals
Dinoflagellates	Single cell	Autotrophs; mixotrophs; heterotrophs; free-living, parasitic, or endosymbiotic
Apicomplexans	Single cell	Heterotrophs; all are parasites of animals
Rhizarians		
Radiolaria	Single cell	Free-living heterotrophs
Foraminifera	Single cell	Free-living heterotrophs
Archaeplastids		
Red algae	Most multicelled	Autotrophs
Green algae	Single cell, colony, or multicelled	Autotrophs
Amoebozoans		
Amoebas	Single cell	Heterotrophs; most free-living, a few parasites of animals
Cellular slime molds	Single cells that aggregate and show collective behavior	Heterotrophs
Plasmodial slime molds	Multinucleated mass	Heterotrophs
Opisthokonts		
Choanoflagellates	Single cell or colony	Heterotrophs

Section 21.2 Most protists live as single cells, but some are **colonial organisms** or **multicellular organisms** (**TABLE 21.1**). Autotrophic protists have chloroplasts derived from bacteria (**primary endosymbiosis**) or other protists (**secondary endosymbiosis**). Some protists have a cell wall, and others have a flexible **pellicle**. Single-celled freshwater protists have a **contractile vacuole** to expel water. An **eyespot** allows some motile photosynthetic protists to seek well-lit areas. Some protists are part of the **plankton** (either the photosynthetic **phytoplankton** or the heterotrophic **zooplankton**).

Life cycles vary: In some, diploid cells dominate, in others only the zygote is diploid, and in still others there is an **alternation of generations**.

Section 21.3 **Excavates** are a eukaryotic supergroup of single-celled, unwalled flagellates. **Metamonads** such as diplomonads and parabasalids are anaerobic excavates that lack typical mitochondria. **Trypanosomes** are parasitic excavates with a giant mitochondrion and an undulating membrane. Some diplomonads, parabasilids, and trypanosomes are human pathogens. **Euglenids** live mainly in fresh water. Most have chloroplasts derived from a green alga.

Section 21.4 The **SAR supergroup** includes three major protist lineages: stramenopiles, alveolates, and rhizarians. **Stramenopiles** are named for their flagellum, which has hairlike filaments. Diatoms and brown algae are photosynthetic stramenopiles. **Diatoms** are photosynthetic single cells with a two-part silica shell. Like diatoms, **brown algae** such as giant kelps contain the pigment fucoxanthin. **Water molds** are heterotrophic stramenopiles that grow as filaments. Some are animal or plant pathogens.

Sections 21.5 **Alveolates** are single cells characterized by tiny sacs (alveoli) beneath the plasma membrane. They include ciliates, dinoflagellates, and apicomplexans. **Ciliates** are aquatic heterotrophs with many cilia. Most are freshwater predators. Some live in the gut of animals. **Dinoflagellates** are aquatic heterotrophs and autotrophs that move with a whirling motion. Some are capable of **bioluminescence**. In nutrient-enriched water, dinoflagellates and other photosynthetic protists undergo population explosions known as **algal blooms**. The **apicomplexans** are parasites that spend part of their time inside host cells. They cause the diseases malaria and toxoplasmosis.

Section 21.6 **Rhizarians** are a eukaryotic supergroup of single-celled heterotrophs with secreted shells. They include foraminifera and radiolaria. **Foraminifera** typically live on the seafloor and secrete a calcium carbonate shell. **Radiolaria** drift as plankton and have a silica shell. Remains of both lineages accumulate in abundance on the seafloor and can be transformed over time into sedimentary rock.

Section 21.7 The red algae, green algae, and plants are **archaeplastids**. They share a common ancestor in which chloroplasts evolved from cyanobacteria. **Red algae** are usually multicelled. Their phycobilin pigments allow them to survive in deeper water than most photoautotrophs.

Green algae can be single-celled, colonial, or multicelled. **Chlorophyte algae** are the most diverse group of green algae. The closest relatives of land plants are **charophyte algae.**

Section 21.8 **Amoebozoans** are a eukaryotic supergroup of heterotrophic protists that move and feed by forming pseudopods. They include amoebas and slime molds. **Amoebas** live as solitary cells in aquatic habitats and the animal gut. **Plasmodial slime molds** feed as a multinucleated mass. **Cellular slime molds** feed as individual amoeba-like cells. When food becomes scarce, both types of slime molds form fruiting bodies that produce resting spores.

Animals, fungi, and the protist relatives of these groups are **opisthokonts**. The closest protist relatives of animals are **choanoflagellates**. They have a flagellum surrounded by threadlike projections that filter food from the water. Choanoflagellates may live on their own or in colonies of cells held together by adhesion proteins.

SELF-QUIZ

Answers in Appendix VII

1. A contractile vacuole ______ .
   a. captures prey c. expels water
   b. stores food d. detects light
2. Diplomonads and parabasalids lack ______ .
   a. a nucleus c. a plasma membrane
   b. mitochondria d. flagella
3. Radiolaria and diatoms have a shell of ______ .
   a. cellulose c. calcium carbonate
   b. silica d. chitin
4. Some ______ enter and replicate inside human cells.
   a. apicomplexans c. trypanosomes
   b. water molds d. ciliates
5. Diatoms produce ______ that help(s) them float.
   a. phycobilins c. diatomaceous earth
   b. oils d. alginates
6. Chloroplasts of green algae evolved from ______ .
   a. cyanobacteria b. archaea c. red algae
7. ______ are members of the phytoplankton .
   a. Water molds c. Diatoms
   b. Giant kelps d. Cellular slime molds
8. To reproduce sexually, two ciliates join and ______ .
   a. infect cells c. emit light
   b. form spores d. exchange micronuclei
9. Which group does not include human pathogens?
   a. Apicomplexans c. Euglenids
   b. Ciliates d. Parabasalids
10. The ______ include animals, fungi, and some protists.
    a. amoebozoans c. opisthokonts
    b. stramenopiles d. alveolates
11. Like plants, ______ have chlorophylls *a* and *b*.
    a. diatoms c. green algae
    b. red algae d. giant kelps
12. Choanoflagellates are considered the sister group to, or closest living relatives of, ______ .
    a. plants c. animals
    b. fungi d. dinoflagellates
13. Bioluminescent dinoflagellates ______ .
    a. have a clear silica shell
    b. emit light when disturbed
    c. live inside most corals
    d. are multicelled heterotrophs
14. Match each material with its most suitable description.
    ___ limestone a. extracted from kelps
    ___ chert b. a dried red alga
    ___ alginates c. rocky radiolarian remains
    ___ nori d. extracted from red algae
    ___ carrageenan e. remains of foraminifera
15. Match each protist with its most suitable description.
    ___ diplomonad a. whirling cell
    ___ water mold b. silica-shelled producer
    ___ dinoflagellate c. grows as filaments
    ___ diatom d. anaerobic parasite
    ___ amoeba e. closest relative of plants
    ___ red alga f. closest relative of animals
    ___ choanoflagellate g. contains phycobilins
    ___ charophyte alga h. feeds using pseudopods

CRITICAL THINKING

1. Which groups of protists would you be most likely to find as fossils? Why?
2. Apicomplexans evolved from a photosynthetic ancestor and have the remnant of a chloroplast. This organelle no longer acts in photosynthesis, but remains essential to the protist. Why might targeting this organelle yield an antimalarial drug with minimal side effects in humans?
3. Diplomonads have neither mitochondria nor hydrogenosomes. All ATP is produced in their cytoplasm. What energy-releasing pathways might they use to make ATP?

22 THE LAND PLANTS

CORE CONCEPTS

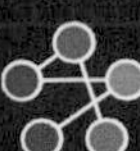

Systems

Complex properties arise from interactions among components of a biological system.

Plants take part in diverse interactions with other organisms. They provide animals with food and shelter. Plants in turn depend on animals for assistance in their reproduction and dispersal. These types of interactions influence properties of biological communities. Plants are also the source of human foods, fabrics, medicines, and lumber.

Evolution

Evolution underlies the unity and diversity of life.

Plants evolved from an aquatic green alga. Over time, changes in structure, life cycle, and reproductive processes allowed plants to live in increasingly drier climates. The oldest plant lineages, such as mosses and ferns, have adaptations to moist habitats, where their flagellated sperm can swim to eggs through films of water. Seed plants, which evolved later, are better suited to drier environments.

Pathways of Transformation

Organisms exchange matter and energy with the environment in order to grow, maintain themselves, and reproduce.

All organisms harvest atomic and molecular building blocks from their environment. Plants are autotrophs that capture energy from sunlight. The simplest plants absorb essential water and nutrients across their body surface. Most plants have roots that take water and nutrients up from the soil. Internal pipelines (vascular tissues) distribute water and nutrients through the body.

Links to Earlier Concepts

This chapter continues the story of how plants evolved from green algae (Section 21.7). Like some algae, plants have an alternation of generations (21.4, 21.7). Recalling how scientists reconstruct phylogeny (18.2) will help you comprehend how plants are now grouped. We discuss variations in carbon-fixing pathways (6.5) and the importance of lignin (4.9) as a structural material.

22.1 Saving Seeds

Plant diversity is declining. We know of threats to about 12,000 wild plant species, and the actual number of threatened plants is no doubt much higher. As a result, many valuable sources of food, medicine, and other products could disappear before we ever learn their value.

A decline in the diversity of cultivated plants raises additional concerns. In the past, food crops were locale specific. People developed and planted crop varieties that did well in their region, and they saved seeds from their crops to plant the following year. Now, many traditional varieties of crop plants are disappearing as farmers worldwide turn to the same few large companies for seeds. Different varieties of plants are resistant to different diseases, so widespread planting of one variety increases the risk that a disease could decimate the global supply of a crop. The more varieties of a crop we plant, the more likely it is that some will resist a particular pathogen. Sustaining the wild relatives of crop plants provides a form of insurance. Such plants represent a reservoir of genetic diversity that plant breeders can draw upon to meet future challenges.

One way to ensure the survival of potentially useful plants is by storing their seeds in seed banks. There are now more than 1,500 seed banks around the world. The most ambitious of these, the Svalbard Global Seed Vault, was built in 2008 on a Norwegian island 700 miles from the North Pole. This so-called doomsday vault serves as a backup for other seed banks. Here, deep inside a mountain, seeds are stored in a permanently chilled, earthquake-free zone 400 feet above sea level (**FIGURE 22.1**). The location was chosen to ensure

FIGURE 22.1 Inside the Svalbard Global Seed Vault. To learn more about the vault and the seeds stored inside it, visit www.croptrust.org.

CREDITS: (opposite) Dr. Annkatrin Rose, Appalachian State University; (1) Jim Richardson/National Geographic Creative.

that the seeds will remain high and dry even if global climate change causes the polar ice caps to melt. The vault also has an advanced security system and has been engineered to withstand any nearby explosions.

The Svalbard vault now holds the world's most diverse collection of seeds, with more than 750,000 samples and material from nearly every nation. The United States has stored seeds of its native crops such as chile peppers from New Mexico, as well as seeds that American scientists collected elsewhere in the world. Under the deep-freeze conditions in Svalbard, the seeds are expected to survive about 100 years. ●

22.2 Plant Ancestry and Diversity

LEARNING OBJECTIVES

- Explain when land plants evolved and the earliest evidence for these plants.
- List the traits that land plants share with green algae in general and with charophyte algae in particular.
- Explain the trait that defines land plants as a distinct lineage.
- Describe the alternation of generations in land plants.

From Algal Ancestors to Embryophytes

The **plants** are multicelled, typically photosynthetic eukaryotes adapted to life on land. They are close relatives of red algae and green algae, and the charophyte algae are considered their sister group:

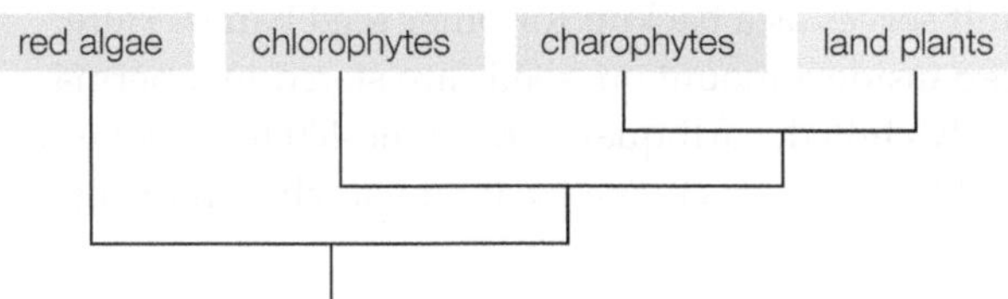

red algae
chlorophytes
charophytes
land plants

Like all green algae, plants have cell walls made of cellulose, chloroplasts with chlorophylls *a* and *b*, and they store sugars as starch. Shared traits that unite the plants and the charophyte algae include the mechanism by which a new cell wall forms after cell division, the presence of plasmodesmata, and the arrangement of flagella on sperm (among those species that have flagellated sperm).

A developmental trait defines the plant clade. They are called **embryophytes** (meaning "embryo-bearing plants"), because their embryos form and develop within a chamber composed of maternal tissues. In some multicelled charophytes the zygote forms in a chamber on the parental body, but that zygote is released into the environment. The charophyte embryo then grows and develops without any parental protection or assistance.

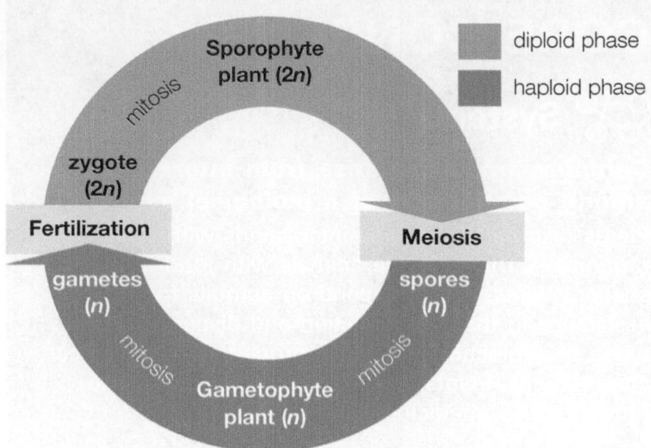

Sporophyte plant (2n)
diploid phase
haploid phase
mitosis
zygote (2n)
Fertilization
Meiosis
gametes (n)
spores (n)
mitosis
mitosis
Gametophyte plant (n)

FIGURE 22.2 Generalized life cycle for land plants. Like some algae, all plants have an alternation of generations in which a multicelled haploid generation alternates with a multicelled diploid generation.

An Adaptive Radiation on Land

Scientists estimate that the first land plants evolved by about 500 million years ago. By that time, enormous numbers of photosynthetic cells had come and gone. Oxygen released by photosynthetic cyanobacteria and protists had altered the composition of the atmosphere. High above Earth, the sun's energy had converted some of this oxygen into a dense ozone layer, which screened out much incoming ultraviolet radiation. UV radiation is a mutagen, so the decrease in the amount reaching Earth's surface made land much more hospitable to life.

The earliest evidence of land plants comes from fossil spores that date to about 473 million years ago. (In plants, a spore is a haploid cell with a thick wall.) The fossil spores were discovered in Argentina. However, at the time the spores were released, this area was in the eastern part of the supercontinent Gondwana (Section 16.5).

Like some algae, plants have an alternation of generations, a life cycle in which a diploid generation alternates with a haploid one (FIGURE 22.2). During the diploid generation, a multicellular **sporophyte** produces spores by meiosis. When spores germinate (resume their activity), they undergo mitosis and grow into the next generation. During this generation, a multicelled haploid **gametophyte** produces gametes by mitosis. When male and female gametes meet up, fertilization occurs and a diploid zygote forms. The zygote grows and develops into a new sporophyte.

In their structure, the oldest fossil spores resemble spores of modern liverworts. Liverworts, mosses, and hornworts are three early plant lineages informally

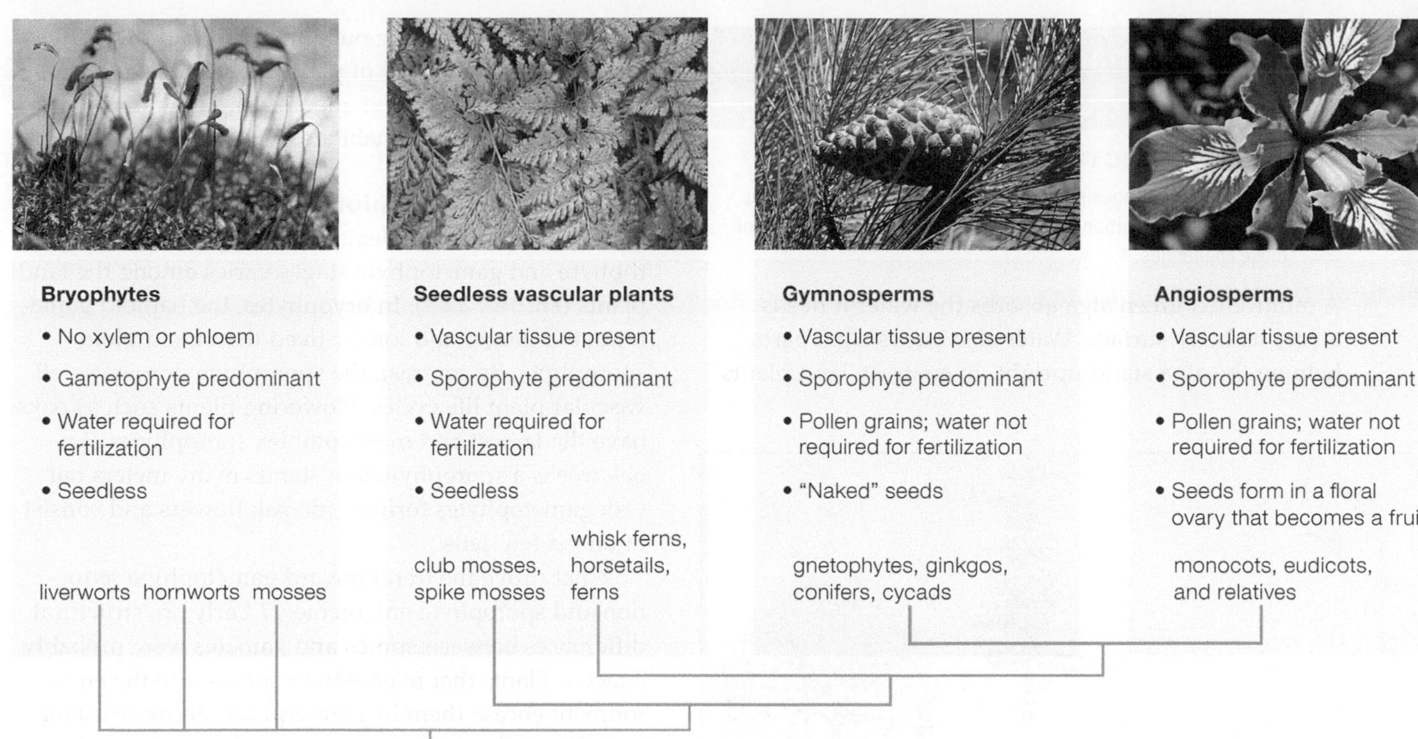
Bryophytes
• No xylem or phloem
• Gametophyte predominant
• Water required for fertilization
• Seedless
Seedless vascular plants
• Vascular tissue present
• Sporophyte predominant
• Water required for fertilization
• Seedless
Gymnosperms
• Vascular tissue present
• Sporophyte predominant
• Pollen grains; water not required for fertilization
• "Naked" seeds
Angiosperms
• Vascular tissue present
• Sporophyte predominant
• Pollen grains; water not required for fertilization
• Seeds form in a floral ovary that becomes a fruit
liverworts
hornworts
mosses
club mosses, spike mosses
whisk ferns, horsetails, ferns
gnetophytes, ginkgos, conifers, cycads
monocots, eudicots, and relatives
ancestral alga

FIGURE 22.3 Traits of major plant groups and relationships among them.

referred to as **bryophytes** (**FIGURE 22.3**). In all bryophytes, the gametophyte is larger and longer-lived than the sporophyte.

The first vascular plants evolved about 430 million years ago. In **vascular plants**, the sporophyte is larger and longer-lived than the gametophyte, and it has specialized internal pipelines that transport water and sugars. Because bryophytes do not have these specialized tissues, they are sometimes described as nonvascular plants.

Seed plants evolved about 385 million years ago, about the same time the first insects evolved. **Seed plants** are vascular plants that hold on to their spores and disperse by releasing seeds. Seed plants are also the only plants that produce pollen. As Section 22.3 explains, producing seeds and pollen allowed seed plants to expand into dry habitats that seedless plants could not tolerate.

Two lineages of seed plants survived to the present: gymnosperms and angiosperms. The gymnosperms are the more ancient of the two groups. Modern conifers such as pines, firs, and spruces are gymnosperms.

The flowering plants, or angiosperms, arose from a gymnosperm ancestor in the late Jurassic or early Cretaceous. In less than 40 million years, they would become the dominant plants in most habitats.

bryophyte (BRY-oh-fite) Member of an early-evolving plant lineage with a gametophyte-dominant life cycle; a moss, liverwort, or hornwort.
embryophytes (EM-bree-oh-fite) Land plants; clade of multicelled, photosynthetic species that protect and nourish the embryo on the parental body.
gametophyte (gam-EET-o-fite) Haploid, multicelled body that produces gametes; forms during the life cycle of all land plants and some algae.
plants Lineage of multicelled, typically photosynthetic eukaryotes adapted to life on land.
seed plant Plant that produces seeds and pollen; an angiosperm or gymnosperm.
sporophyte (SPOR-o-fite) Diploid, multicelled body that produces spores; forms during the life cycle of all land plants and some algae.
vascular plant Plant having specialized tissues (xylem and phloem) that transport water and sugar within the plant body.

TAKE-HOME MESSAGE 22.2

✔ Land plants are embryophytes; their embryo forms in a chamber on the parental body and is nourished by the parent.

✔ Three early-evolving plant lineages are known collectively as bryophytes. These nonvascular plants disperse by releasing spores.

✔ Vascular plants have internal pipelines that transport water and sugars. Some produce spores; others produce seeds.

✔ Gymnosperms and angiosperms are two modern lineages of seed plants. Angiosperms (flowering plants) are the most diverse plant lineage, and the most recently evolved.

22.3 Evolutionary Trends Among Plants

LEARNING OBJECTIVE

- Explain the challenges plants faced as they moved from water onto land and evolutionary changes that occurred as a result of those challenges.

A multicelled green alga absorbs the water it needs across its body surface. Water also buoys algal parts, helping the alga stand upright. In contrast, land plants face the threat of drying out and must hold themselves upright. Thus, the story of plant evolution is a tale of adaptation to life on land and an adaptive radiation into increasingly drier habitats.

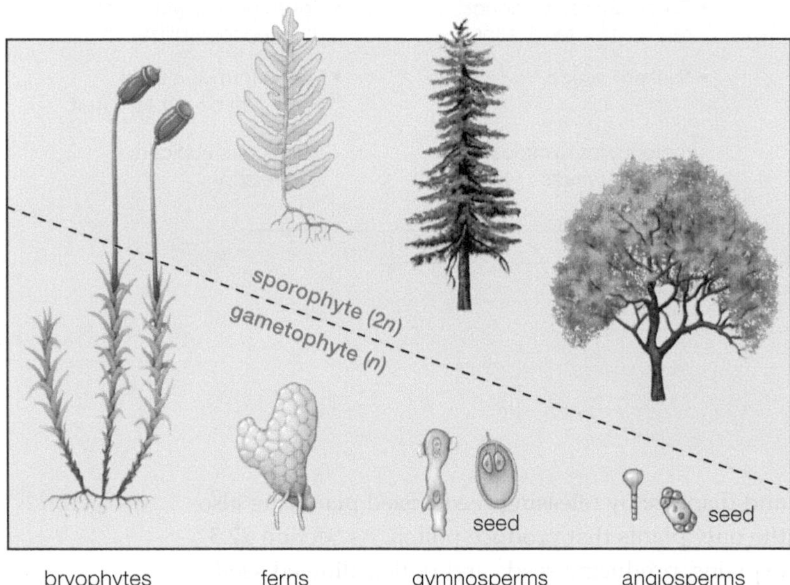

sporophyte (2n)
gametophyte (n)
seed
seed
bryophytes
ferns
gymnosperms
angiosperms

FIGURE 22.4 Evolutionary trend in plant life cycles. Mosses and other bryophytes put the most energy into making gametophytes. Later-evolving groups invested increasingly in making sporophytes.

FIGURE IT OUT A pine tree is a gymnosperm. Which is the larger, more prominent phase in its life cycle, the sporophyte or gametophyte?

Answer: Sporophyte

FIGURE 22.5 Diagram of a vascular plant leaf in cross section. Structural traits that contribute to the success of this group are included in this illustration.

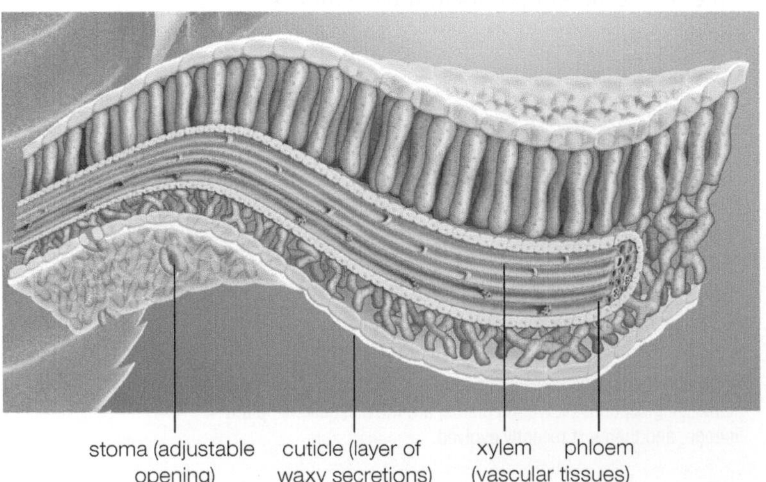

stoma (adjustable opening)
cuticle (layer of waxy secretions)
xylem
phloem
(vascular tissues)

From Haploid to Diploid Dominance

The relative size, complexity, and longevity of the sporophyte and gametophyte stages varies among the land plants (FIGURE 22.4). In bryophytes, the haploid gametophyte is larger and longer-lived than the diploid sporophyte. By contrast, the sporophyte dominates all vascular plant life cycles. Flowering plants such as oaks have the largest and most complex sporophytes. An oak tree is a sporophyte that stands many meters tall. Oak gametophytes form inside oak flowers and consist of only a few cells.

What drove the trend toward gametophyte reduction and sporophyte enlargement? Early on, structural differences between spores and gametes were probably a factor. Plants that release their spores into the environment encase them in a waterproof, decay-resistant wall. By contrast, gametes are unwalled cells. Under dry conditions, spores are more likely to survive than gametes, so increased spore production provides a greater advantage than increased gamete production.

Genetic factors may also have encouraged a trend toward sporophyte dominance. In a haploid body, any mutant allele that arises is expressed and exposed to selection. In a diploid body, a dominant allele (Section 13.2) can mask the effect of a mutant allele, allowing it to persist even if it is somewhat deleterious. Thus, expanding the diploid stage of the life cycle allowed an increase in genetic diversity. Remember that genetic diversity within a population is the raw material upon which selection acts. A high degree of genetic diversity in sporophyte-dominated lineages may have facilitated their adaptive radiation into diverse habitats. We discuss some of their relevant adaptations below.

Structural Adaptations

Like algae, bryophytes absorb all the water and dissolved minerals they require across their body surface. They also lose water by evaporation across this surface. In some bryophytes and all vascular plants, the sporophyte secretes a waxy covering, or **cuticle**, that helps reduce evaporative water loss (FIGURE 22.5). Closable pores called **stomata** (singular, stoma) extend across the cuticle. Depending on conditions, stomata open to allow gas exchange or close to conserve water.

Early plants had threadlike structures to hold them in place, but these structures did not deliver water and dissolved minerals to other parts of the plant body, as roots do. Moving substances from roots to other body regions requires **vascular tissues**, a system of internal pipelines. **Xylem** is a vascular tissue that distributes water and mineral ions. **Phloem** is a vascular tissue that distributes sugars made in photosynthetic cells. Most modern plants have xylem and phloem, and thus are vascular plants (tracheophytes).

Vascular tissues not only distribute material, they also provide structural support. An organic compound called **lignin** (Section 4.9) stiffens the walls of xylem cells. Evolution of lignin-stiffened tissue allowed vascular plants to stand taller than bryophytes and to branch, giving them an advantage in spore dispersal.

Most vascular plants also have leaves. Leaves are flattened, aboveground organs that increase the surface area available for intercepting sunlight and for gas exchange. They contain veins of vascular tissue.

Pollen and Seeds

Novel reproductive traits gave seed-bearing vascular plants a competitive edge. All bryophytes, and some vascular plants such as the ferns, release sperm that must swim through the environment to eggs. Only seed-bearing vascular plants release pollen grains. A **pollen grain** is a walled, immature gametophyte that will give rise to the male gametes. After pollen grains are released, they travel to female gametophytes on the wind or on the bodies of animals, most often insects. An ability to produce pollen gave seed plants an ability to reproduce even in dry environments.

Bryophytes and seedless vascular plants release spores (**FIGURE 22.6A**), but seed plants protect spores within their tissues and release seeds. A **seed** consists of an embryo sporophyte and some nutritive tissue enclosed within a waterproof seed coat. Many seeds have features that facilitate their dispersal away from the parent plant. Angiosperms, or flowering plants, disperse the seeds inside a fruit that forms from floral tissue (**FIGURE 22.6B**). The overwhelming majority of modern plants are angiosperms.

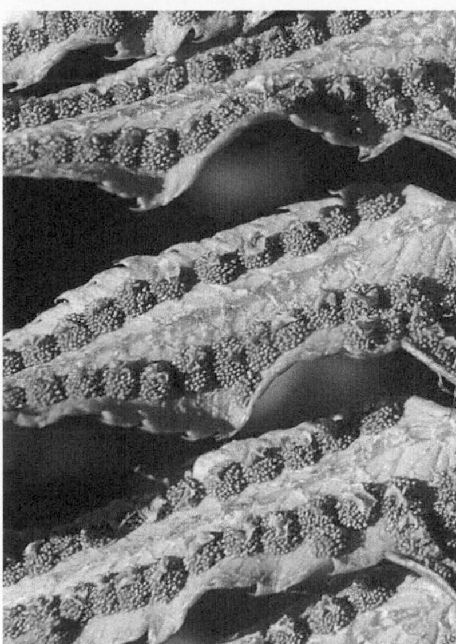

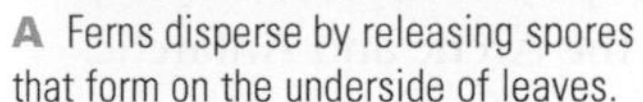

A Ferns disperse by releasing spores that form on the underside of leaves.

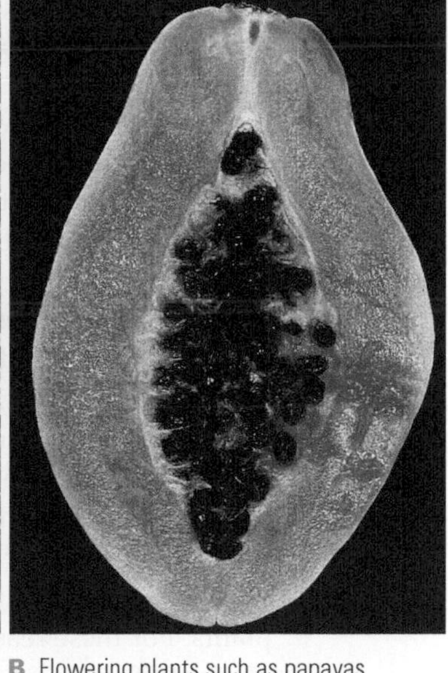

B Flowering plants such as papayas hold on to their spores and disperse by releasing seeds enclosed in fruit.

FIGURE 22.6 Mechanisms of dispersal.

TAKE-HOME MESSAGE 22.3

- ✔ Plant life cycles shifted from a gametophyte-dominated cycle in bryophytes to a sporophyte-dominated cycle in vascular plants.
- ✔ Life on land favored water-conserving features such as a cuticle. In vascular plants, a system of vascular tissue—xylem and phloem—distributes material through the leaves, stems, and roots of sporophytes.
- ✔ Bryophytes and seedless vascular plants release spores. Only seed plants release embryos inside protective seeds. Only in the flowering plants do seeds form inside floral tissue that later develops into a fruit.

22.4 Bryophytes

LEARNING OBJECTIVES

- List the three lineages of bryophytes and the traits that they share.
- Explain why there are no tree-sized bryophytes.
- Describe the life cycle of a moss.

All nonvascular plants (plants lacking xylem and phloem) were traditionally grouped together as bryophytes. Today, the term bryophyte is still used as an informal designation for a nonvascular plant. However, biologist now recognize three distinct nonvascular lineages: the liverworts (Marchantiophyta), mosses (Bryophyta), and hornworts (Anthocerotophyta).

Bryophytes are the only plants in which the gametophyte is larger and longer-lived than the sporophyte. However, even the gametophytes tend to be small and

---

cuticle Secreted covering at a body surface; in land plants, a waxy layer.
lignin Material that stiffens cell walls of vascular plants.
phloem (FLOW-um) In vascular plants, tissue that distributes photosynthetically produced sugars through the plant body.
pollen grain Male gametophyte of a seed plant.
seed Embryo sporophyte of a seed plant packaged with nutritive tissue inside a protective coat.
stomata (STOH-ma-tuh) Singular, stoma. Adjustable opening across a plant's cuticle and epidermis; can be opened for gas exchange or closed to prevent water loss.
vascular tissue In vascular plants, tissue (xylem and phloem) that distributes water and nutrients through the plant body.
xylem (ZY-lum) In vascular plants, the tissue that distributes water and dissolved minerals through the plant body.

low-growing. A lack of vascular tissue constrains bryophyte stature. Without some sort of pipelines, it is difficult to move materials through a tall plant. The few mosses that do grow more than a few inches high have simple water-conducting tubes.

Reproductive factors constrain bryophyte size too. The gametophyte (the dominant body form in bryophytes) produces eggs and flagellated sperm. Small gametophyte size results in a matlike growth pattern that minimizes the distance that sperm must swim to reach eggs. Growing as a mat also helps retain the film of moisture that sperm require to make their journey.

Despite their small size, bryophytes can have a large effect on their community. They often colonize rocky sites where vascular plants cannot take root. They also withstand drought and cold better than vascular plants. For these reasons, bryophytes are the only types of plant life in some parts of the Arctic and Antarctic.

Mosses

Mosses are the most diverse and familiar bryophytes. The most economically important bryophytes are the 350 or so species of peat moss (*Sphagnum*). Peat mosses are the dominant plants in **peat bogs** that cover hundreds of millions of acres in high-latitude regions of Europe, Asia, and North America. Many peat bogs have persisted for thousands of years, and layer upon layer of plant remains have become compressed as a carbon-rich material called peat. Blocks of peat are cut, dried, and burned as fuel, especially in Ireland. Freshly harvested peat moss is also an important commercial product. The moss is dried and added to planting mixes to help soil retain moisture.

Like other nonvascular plants, mosses do not have true leaves or roots with vascular tissue. However, their gametophytes have leaflike photosynthetic parts arranged around a central stalk (**FIGURE 22.7**). Threadlike **rhizoids** anchor the gametophyte and wick up water. The moss gametophyte can also absorb water and nutrients across its leaflike parts.

A moss sporophyte consists of a **sporangium** (a spore-producing structure) atop a stalk. A sporophyte is not photosynthetic, so it is attached to and dependent on the gametophyte even when mature. Meiosis of diploid cells inside the sporangium produces haploid spores ❶. After dispersal by the wind,

FIGURE 22.7 Life cycle of a moss. In the species depicted here (*Polytrichum*), each gametophyte produces either eggs or sperm.

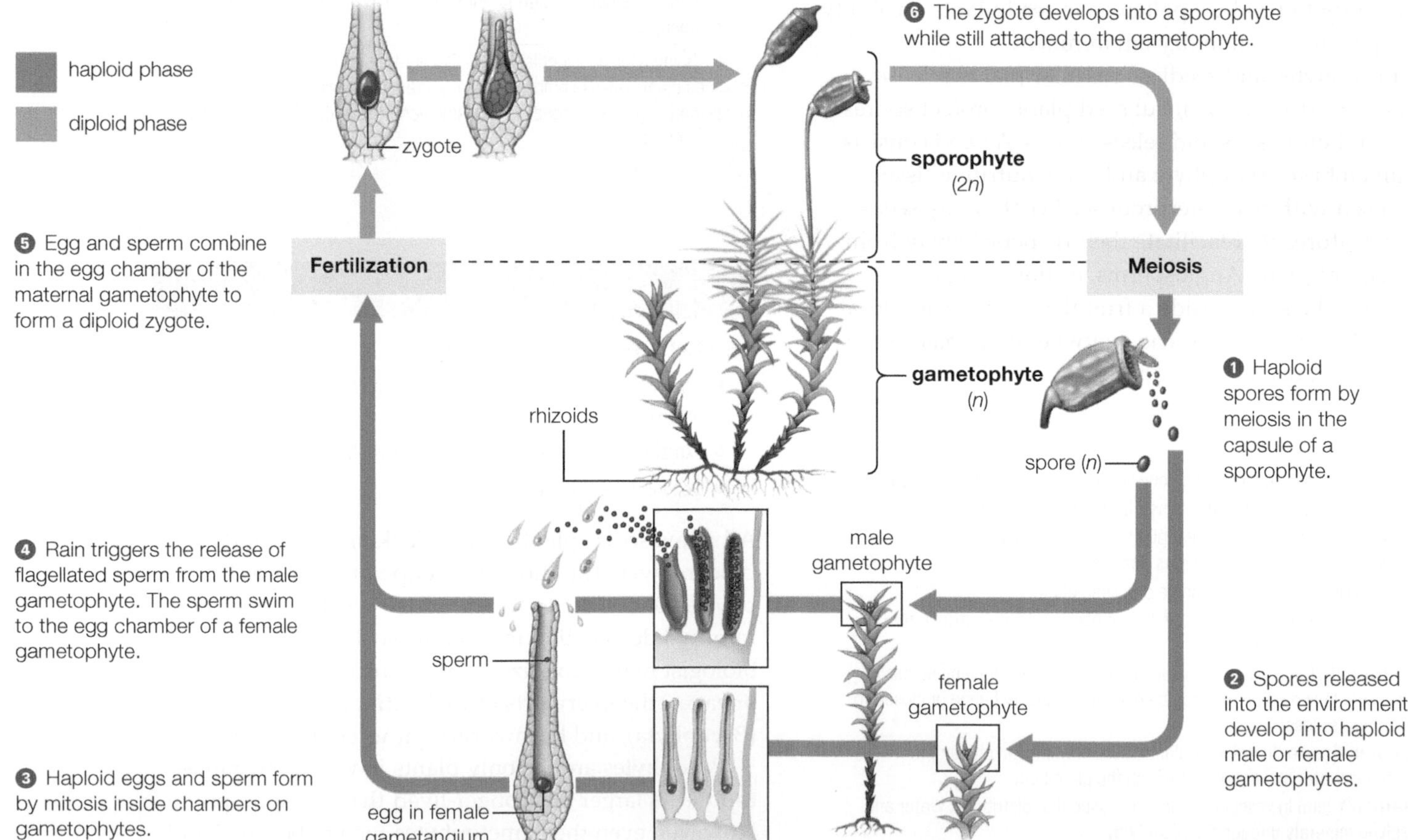
haploid phase
diploid phase
zygote
❻ The zygote develops into a sporophyte while still attached to the gametophyte.
sporophyte
(2n)
❺ Egg and sperm combine in the egg chamber of the maternal gametophyte to form a diploid zygote.
Fertilization
Meiosis
gametophyte
(n)
rhizoids
❶ Haploid spores form by meiosis in the capsule of a sporophyte.
spore (n)
male gametophyte
❹ Rain triggers the release of flagellated sperm from the male gametophyte. The sperm swim to the egg chamber of a female gametophyte.
sperm
female gametophyte
❷ Spores released into the environment develop into haploid male or female gametophytes.
❸ Haploid eggs and sperm form by mitosis inside chambers on gametophytes.
egg in female gametangium

Data Analysis Activities

Insect-Assisted Fertilization in Moss Moss sperm can swim, but plant ecologist Nils Cronberg suspected that they sometimes hitch a ride on crawling insects or mites (tiny animals related to spiders). To test this hypothesis he carried out an experiment. He placed patches of male and female moss gametophytes in dishes, either next to one another or with water-absorbing plaster between them. The plaster prevented sperm from swimming between plants. He then looked at how the presence or absence of insects affected the number of sporophytes formed. FIGURE 22.8 shows his results.

1. Why is sporophyte formation a good way to determine if fertilization occurred?
2. How close did the male and female patches have to be for sporophytes to form in the absence of insects?
3. Does this study support the hypothesis that insects and mites aid moss fertilization?

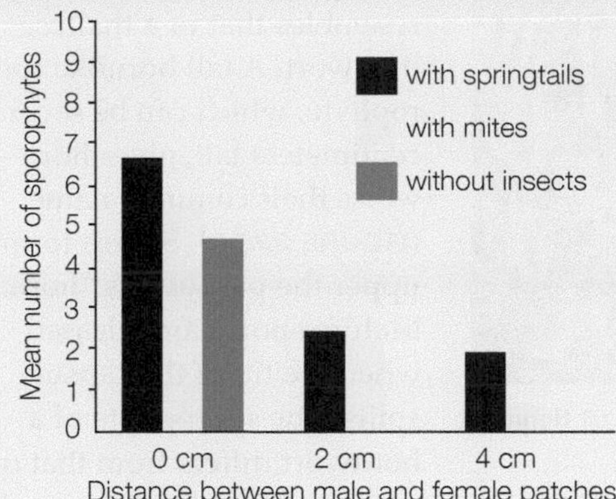

Mean number of sporophytes
10
9
8
7
6
5
4
3
2
1
0
with springtails
with mites
without insects
0 cm
2 cm
4 cm
Distance between male and female patches

FIGURE 22.8 **Sporophyte production in female moss patches with and without either crawling insects (springtails) or mites.** No sporophytes formed in the animal-free dishes when moss patches were 2 or 4 centimeters apart.

a spore germinates and grows into a filament of cells from which one or more gametophytes develop ❷.

Multicellular **gametangia** (gamete-producing structures) develop in or on the gametophyte. Eggs and sperm form by mitosis inside a gametangium ❸. The moss we are using as our example has separate sexes, but in some species each gametophyte produces both eggs and sperm.

Rain stimulates the release of sperm that swim to eggs ❹. Fertilization occurs inside the maternal egg chamber and produces a zygote ❺. The zygote then grows and develops into a new sporophyte ❻.

In addition to reproducing sexually, many mosses can reproduce asexually by fragmentation. A bit of the gametophyte breaks off and develops into a new plant.

Liverworts

The oldest evidence of plants comes from fossil spores that resemble the spores of liverworts. These fossils, together with genetic evidence, suggest that liverworts may have been the earliest land plants.

Two body forms are common among modern liverworts. The gametophyte of a leafy liverwort resembles a moss gametophyte. It consists of a central stalk that bears leaflike structures. By contrast, the gametophyte of a thallose liverwort looks like a multilobed green pancake or ribbon. (In botany, a thallus is a plant body that is not differentiated into distinct parts such as leaves and stems.) A resemblance between a thallose liverwort and a human liver (both are multilobed and flattened) is the source of the group's common name.

Marchantia is a common genus of thallose liverwort. It reproduces asexually by producing gemmae (little clusters of cells) in cups on the surface of the gametophyte thallus (FIGURE 22.9A). Raindrops disperse the gemmae, which grow into new gametophytes. During sexual reproduction, stalked gametangia-bearing structures form on the *Marchantia* thallus (FIGURE 22.9B). After fertilization, the sporophyte develops while still attached to the gametangium. As in mosses, the liverwort sporophyte is not photosynthetic.

A Gametophyte thallus with cups that contain mitotically produced cell clusters (gemmae).

B Umbrella-like structures on which female gametangia (pink) is formed. Sporophytes with yellow capsules remain attached.

FIGURE 22.9 **Liverworts.** *Marchantia* is a thallose liverwort.

gametangium (gam-et-TAN-gee-um) Plural, gametangia. Gamete-producing organ of a plant.
peat bog High-latitude community dominated by *Sphagnum* moss.
rhizoid Threadlike structure that anchors a bryophyte.
sporangium (spor-AN-gee-um) Plural, sporangia. A structure in which haploid spores form by meiosis.

FIGURE 22.10 **Hornwort.**

Hornworts

A hornwort gametophyte resembles that of a thallose liverwort. A tall hornlike sporophyte, which can be several centimeters tall, gives hornworts their common name (FIGURE 22.10). Spores form in upper the part of this "horn." Mature spores are released when the tip of the capsule splits. The sporophyte of a hornwort differs from that of other bryophytes in several ways. It grows continually from its base, so it can make and release spores over an extended period. It also has chloroplasts and in some species can survive after the death of the gametophyte.

TAKE-HOME MESSAGE 22.4

✔ Bryophytes include three lineages of plants: mosses, liverworts, and hornworts. All are low-growing with no lignin-reinforced vascular tissues. All have flagellated sperm that require water to swim to eggs, and all disperse by releasing spores.

✔ Bryophytes have a life cycle in which the gametophyte is the dominant generation. The sporophyte remains attached to the gametophyte even when mature.

22.5 Seedless Vascular Plants

LEARNING OBJECTIVES

- List and describe the major groups of seedless vascular plants.
- Describe the fern life cycle.

Like bryophytes, seedless vascular plants disperse by releasing spores. Also as in bryophytes, sperm are flagellated and swim to the eggs. However, sporophytes of seedless vascular plants are large, long-lived, and have roots, stems, and leaves serviced by vascular tissues. The sporophytes release spores that develop into tiny, short-lived gametophytes that are photosynthetic.

Two lineages of seedless vascular plants survived to the present. The monilophytes includes ferns, horsetails, and whisk ferns. The lycophytes include plants that are commonly known as club mosses and spike mosses, but are not actually mosses.

Ferns

With 12,000 or so species, ferns are the most diverse and familiar seedless vascular plants. Most live in the tropics. Fern sporophytes vary in size and form. Some floating ferns have fronds only 1 millimeter long. Tree ferns can be 25 meters (80 feet) high. Many tropical ferns are epiphytes; they attach to and grow on a trunk or branch of another plant without harming that plant.

FIGURE 22.11 **Life cycle of a fern.**

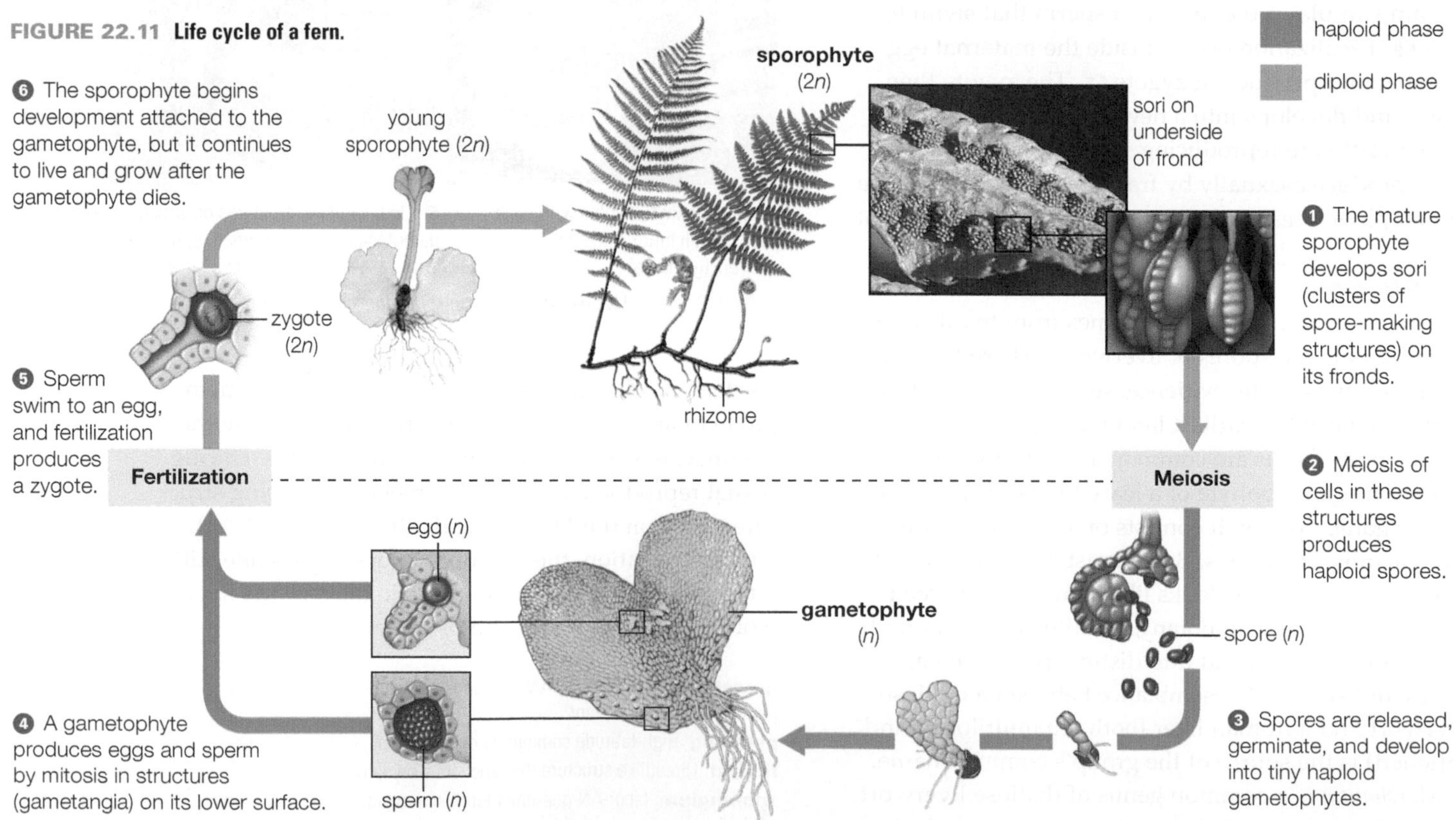

haploid phase
diploid phase
sporophyte (2n)
young sporophyte (2n)
sori on underside of frond
rhizome
zygote (2n)
Fertilization
Meiosis
egg (n)
gametophyte (n)
spore (n)
sperm (n)
❶ The mature sporophyte develops sori (clusters of spore-making structures) on its fronds.
❷ Meiosis of cells in these structures produces haploid spores.
❸ Spores are released, germinate, and develop into tiny haploid gametophytes.
❹ A gametophyte produces eggs and sperm by mitosis in structures (gametangia) on its lower surface.
❺ Sperm swim to an egg, and fertilization produces a zygote.
❻ The sporophyte begins development attached to the gametophyte, but it continues to live and grow after the gametophyte dies.

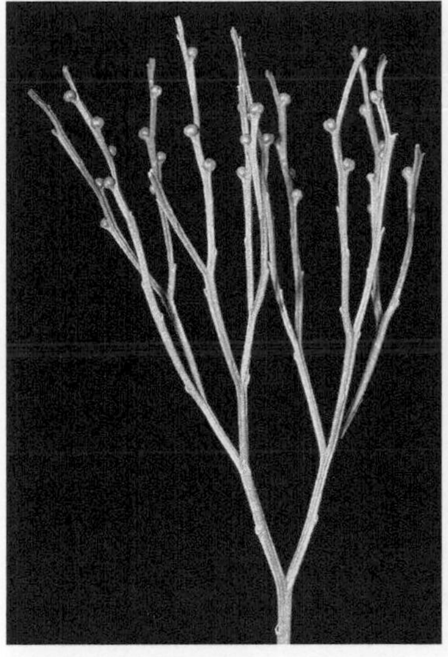

A Whisk fern (*Psilotum*).

B Horsetail (*Equisetum*) vegetation.

C Horsetail strobilus.

D Club moss (*Lycopodium*) with strobilus.

FIGURE 22.12 Diversity of seedless vascular plants.

A typical fern sporophyte has fronds (leaves) and roots that grow out from a **rhizome**, a stem that grows horizontally along or under the ground (FIGURE 22.11). Fiddleheads, the young, tightly coiled fronds of some ferns, are harvested from the wild as food.

Fern spores form by meiosis in a **sorus** (plural, sori), a cluster of capsules that develop on the lower surface of fronds ❶, ❷. The capsules are sporangia, and they pop open to release spores that disperse in the wind. After a spore germinates, it grows into a photosynthetic, heart-shaped gametophyte just a few centimeters wide ❸. Eggs and flagellated sperm form in chambers on the underside of the tiny gametophyte ❹. The chambers are gametangia that release sperm when rainwater wets them. The sperm then swim through a film of water to reach and fertilize an egg ❺. The resulting zygote develops into a new sporophyte ❻, and its parental gametophyte dies.

Whisk Ferns and Horsetails

Whisk ferns (*Psilotum*) are native to the southeastern United States. Their sporophytes have underground rhizomes, but no roots. The photosynthetic stems appear leafless (FIGURE 22.12A). Spores form in fused sporangia at the tips of short branches. Florists often use branches of whisk ferns in mixed bouquets.

The 25 *Equisetum* species include plants commonly known as horsetails. Their sporophytes have rhizomes and hollow stems with reduced leaves at the joints. (FIGURE 22.12B). Deposits of silica in the stem support the plant and help fend off herbivores. Before scouring powders and pads were widely available, people used stems of *Equisetum* species as pot scrubbers.

Equisetum produces spores in a **strobilus** (plural, strobili) a cone-shaped, spore-producing structure. Depending on the species, *Equisetum* strobili form either at tips of photosynthetic stems or on specialized reproductive stems that do not make chlorophyll (FIGURE 22.12C).

Club Mosses

Most of the 1,200 modern lycophytes are club mosses. Members of the club moss genus *Lycopodium* often grow on the floor of temperate forests (FIGURE 22.12D). Their sporophytes resemble miniature pine trees and are sometimes called ground pines. Roots and upright stems with tiny leaves grow from a rhizome. When a plant is several years old, it produces a strobilus seasonally. *Lycopodium* spores have a waxy coating and ignite easily when suspended in the air. In the past, they were used as a flash powder for photography and for staged magic tricks.

TAKE-HOME MESSAGE 22.5

✔ Club mosses and relatives belong to one seedless vascular lineage (lycophytes). Ferns, horsetails, and whisk ferns belong to the other (monilophytes).

✔ A sporophyte with vascular tissues dominates the seedless vascular plant life cycle; dispersal occurs by release of spores.

rhizome Horizontal stem that grows along or under the ground.
sorus (SORE-us) Plural, sori. Cluster of spore-producing capsules of a fern.
strobilus (STROH-bill-us) Plural, strobili. Of some nonflowering plants such as horsetails and cycads, a cluster of spore-producing structures.

Ordovician	Silurian	Devonian	Carboniferous	Permian	Triassic	Jurassic	Cretaceous	Tertiary
Origin of first land plants (bryophytes) by 475 mya.	Origin of seedless vascular plants.	Bryophytes and seedless vascular plants diversify. Seed plants arise by 385 mya.	Lycophytes and horsetails diversify. Cycads and conifers arise late in this period.	Ginkgos appear. Horsetails and lycophytes decline in diversity.	Diversification of conifers; by the end of the Jurassic they are the dominant trees. Cycad, ginkgos, tree ferns remain common.		Flowering plants appear in the early Cretaceous, undergo adaptive radiation, and become dominant.	

488 443 416 359 299 251 200 146 66

Millions of years ago (mya)

FIGURE 22.13 A time line for major events in plant evolution.

22.6 History of the Vascular Plants

LEARNING OBJECTIVES

- Describe the types of plants that lived in a Carboniferous coal forest, and explain how coal forms.
- Explain how the reproductive traits of seed plants give them an advantage over seedless plants in dry habitats.
- Compare the function of megaspores and microspores in the seed plant life cycle, and explain where each type of spore forms.
- Discuss how the evolution of wood provided an advantage to seed plants.

From Tiny Branches to Coal Forests

The earliest fossils of vascular plants are spores that date to about 450 million years ago, during the late Ordovician period (FIGURE 22.13). Later fossils of plant bodies show that the earliest lineages stood only a few centimeters high and had a simple branching pattern, with no leaves or roots. Spores formed at branch tips. Over time, more complex branching patterns and greater size evolved in many lineages.

The oldest forest that we know about existed about 385 million years ago, during the middle Devonian. It was at a site that is now Gilboa, New York. Fossil stumps and fronds from this area indicate that giant monilophytes (relatives of horsetails and ferns) dominated the forest. Some of these seedless vascular plants were as large as 2 meters in diameter and stood up to 10 meters tall.

Later, during the Carboniferous period (359–299 million years ago), vast areas of swampy forests were dominated by tall seedless vascular plants (FIGURE 22.14). After these forests first formed, climates changed, and the sea level rose and fell many times. When the waters receded, the forests flourished. After the sea moved back in, submerged trees became buried in sediments that protected them from decomposition. Layers of sediments accumulated one on top of the other. Their weight squeezed the water out of the saturated, undecayed remains, and the compaction generated heat. Over time, pressure and heat transformed the compacted organic remains into **coal**.

FIGURE 22.14 Carboniferous coal forest. Tree-sized relatives of modern horsetails and club mosses shaded an understory of ferns.

FIGURE 22.15 Early angiosperms. Flowering plants (foreground) evolved while dinosaurs still walked the Earth.

It took millions of years of photosynthesis, burial, and compaction to form coal. When you hear about annual production of coal or other fossil fuels, keep in mind that we do not really "produce" these materials, we only extract them. This is why fossil fuels are said to be nonrenewable sources of energy.

Rise of the Seed Plants

The first gymnosperms evolved from a seedless ancestor late in the Devonian. Representatives of some modern gymnosperm lineages (cycads and conifers) first appeared in the late Carboniferous. Angiosperms, or flowering plants, arose by about 120 million years ago, during the Cretaceous, when dinosaurs reigned (**FIGURE 22.15**).

Reproductive traits of seed-bearing plants gave them a selective advantage over earlier lineages. Gametophytes of seedless vascular plants develop in the environment. By contrast, gametophytes of seed plants develop within the protection of a sporophyte body (**FIGURE 22.16**). Sporangia called **pollen sacs** produce microspores by meiosis ❶. **Microspores** develop into sperm-making gametophytes (pollen grains) ❷. Sporangia called **ovules** produce megaspores ❶. Egg-producing gametophytes develop from the **megaspores** ❹. A sporophyte releases pollen grains, but holds on to its eggs. Wind or animals can deliver pollen from one seed plant to the ovule of another, a process known as **pollination** ❺. Because the sperm of seed plants do not need to swim through a film of water to reach eggs, these plants can reproduce even during dry times. After pollination, a pollen tube grows through the ovule ❻ and delivers the sperm to the egg. After fertilization, an ovule develops into a seed ❼.

Releasing seeds puts seed plants at an advantage over spore-bearing plants. A seed contains a multicelled embryo sporophyte and stored food that the embryo can draw on during early development. By contrast, seedless plants release single-celled spores without any stored food.

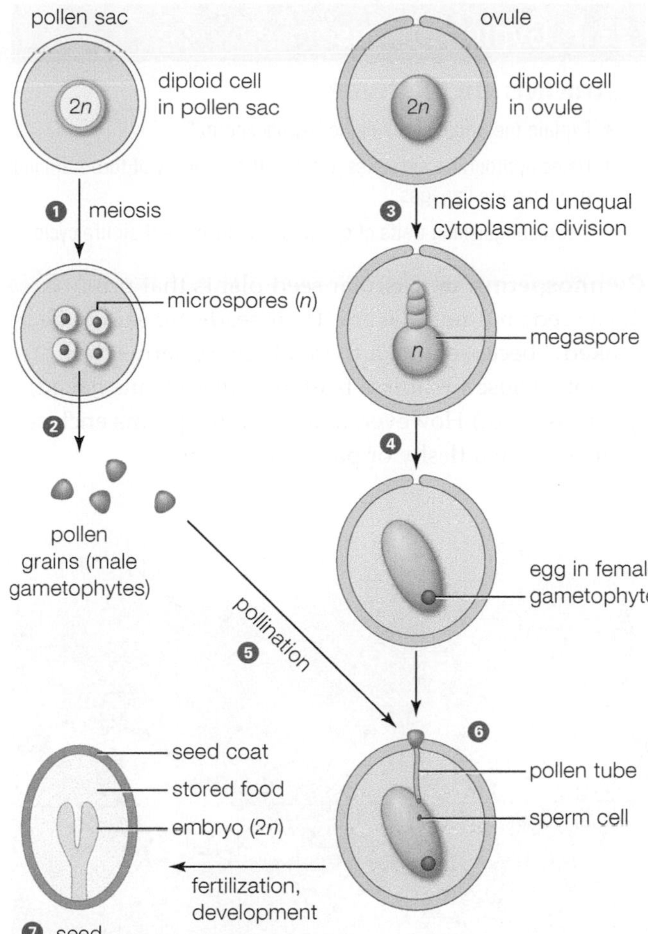

pollen sac
ovule
diploid cell in pollen sac
2n
diploid cell in ovule
2n
❶ meiosis
❸ meiosis and unequal cytoplasmic division
microspores (n)
megaspore
n
❷
❹
pollen grains (male gametophytes)
egg in female gametophyte
❺ pollination
❻
seed coat
stored food
embryo (2n)
pollen tube
sperm cell
fertilization, development
❼ seed

FIGURE 22.16 How a seed forms. Pollen sacs and ovules are sporangia that form on the sporophytes of seed plants. Spores are not released, but rather develop into gametophytes inside these structures. Fertilization takes place in an ovule that then develops into a seed.

Structural traits also gave seed plants an advantage over seedless lineages. Some seed plants undergo secondary growth (growth in diameter) and produce wood. Wood is lignin-stiffened tissue that strengthens and protects older stems and roots. The giant nonvascular plants that lived in Carboniferous forests did undergo secondary growth. However, because their trunks were softer and more flexible than those of woody seed plants, they could not grow as tall.

TAKE-HOME MESSAGE 22.6

✔ Seedless vascular plants became widespread during the Carboniferous. Coal is the remains of some of these plants.

✔ The first seed plants evolved in the late Devonian. Because they produced pollen and seeds, they were able to live in drier places than other plants. An ability to produce wood allowed some seed plants to grow taller than any previous plants.

coal Fossil fuel formed over millions of years by compaction and heating of plant remains.

megaspore Haploid spore formed in ovule of seed plants; develops into an egg-producing gametophyte.

microspore Haploid spore formed in pollen sacs of seed plants; develops into a sperm-producing gametophyte (a pollen grain).

ovule Of seed plants, sporangium that produces megaspores that develop into egg-producing gametophytes.

pollen sac Of seed plants, sporangium where microspores form and develop into pollen grains.

pollination Arrival of a pollen grain on the egg-bearing part of a seed plant.

22.7 Gymnosperms

LEARNING OBJECTIVES

- Explain the origin of the name "gymnosperm."
- Using appropriate examples, discuss the diversity of forms among gymnosperm lineages.
- List distinguishing traits of conifers and describe their life cycle.

Gymnosperms are vascular seed plants that produce seeds but not flowers. Their seeds are said to be "naked," because unlike those of angiosperms they are not enclosed within a fruit. (*Gymnos* means naked; *sperma* is seed.) However, many gymnosperms enclose their seeds in a fleshy or papery covering.

A Cycad (*Cycas*) growing in the understory of a Cambodian forest. Seeds are near its base.

B Fan-shaped leaves and fleshy seeds of *Ginkgo biloba*, the maidenhair tree.

C Pollen cones of *Ephedra*, a shrubby gnetophyte.

D *Welwitschia*, a gnetophyte, with seed cones and two long, wide leaves.

FIGURE 22.17 Gymnosperm diversity.

Gymnosperm diversity

Cycads and ginkgos are gymnosperm lineages whose diversity peaked while dinosaurs were abundant. Today, they are the only seed plants with flagellated sperm. Their ovules produce a fluid that the sperm delivered by a pollen grain can swim through.

The 130 modern **cycads** are native to dry tropics and subtropics. Many resemble palms (**FIGURE 22.17A**), but palms and cycads are not close relatives.

The single living species of **ginkgo**, *Ginkgo biloba*, is native to China (**FIGURE 22.17B**). It is deciduous, meaning it loses all its leaves at once seasonally. The ginkgo's pretty fan-shaped leaves and resistance to insects, disease, and air pollution make it a popular tree along city streets. Female trees produce fleshy plum-sized seeds that emit a strong unpleasant odor, so male trees are preferred for urban landscaping.

Gnetophytes include tropical trees, desert shrubs, and leathery vines. *Ephedra*, a twiggy shrub (**FIGURE 22.17C**), produces ephedrine, a chemical sold as an herbal stimulant and weight loss aid. *Welwitschia* lives in Africa's Namib desert (**FIGURE 22.17D**). It has a taproot, woody stem, and two long, straplike leaves that split lengthwise repeatedly, giving the plant a shaggy appearance.

The most diverse gymnosperm lineage is **conifers**, which includes about 600 species of trees and shrubs with woody cones. Conifers typically have needlelike or scalelike leaves with a thick cuticle. They tend to be more resistant to drought and cold than hardwood trees (angiosperms) and are the dominant plants in high-latitude forests of the Northern Hemisphere. Most conifers are evergreen: They shed some leaves year round but always retain enough to remain green. Conifers include the tallest trees in the Northern Hemisphere (redwoods) and the most abundant (pines). They also include the longest-lived plants; some bristlecone pines are more than 4,000 years old.

Conifers are of great economic importance. We mulch our gardens with fir bark, use oils from cedar in cleaning products, and eat the seeds, or "pine nuts," of some pines. Pines also provide lumber for building homes and furniture. Some pines make a sticky resin

conifer Gymnosperm with nonmotile sperm and woody cones; for example, a pine.

cycad (SIGH-cad) Tropical or subtropical gymnosperm with flagellated sperm, palmlike leaves, and fleshy seeds.

ginkgo Deciduous gymnosperm with flagellated sperm, fan-shaped leaves, and fleshy seeds.

gnetophyte (NEAT-o-fite) Shrubby or vinelike gymnosperm, with nonmotile sperm; for example, *Ephedra*.

gymnosperm (JIM-no-sperm) Seed plant that does not make flowers or fruits; for example, a conifer.

that deters insects from boring into them. We use this resin to make turpentine, a paint solvent.

Gymnosperm Life Cycle

As an example of a gymnosperm life cycle, consider the life of a pine. Pines are conifers that belong to the genus *Pinus*. A pine tree is a diploid sporophyte that produces spores on specialized strobili called cones. In most pine species, each tree makes only one type of cone: either small, soft pollen cones or large, woody, ovulate cones. Both types of cones have scales arranged around a central axis.

Each scale of a pollen cone contains pollen sacs (**FIGURE 22.18** ➊). Microspores form by meiosis in these sacs ➋, then develop into pollen grains ➌. Pollen cones release these tiny grains to drift with the winds.

Each scale of an ovulate cone holds a pair of ovules ➍. Meiosis of diploid cells in an ovule produces megaspores ➎ that develop into egg-producing female gametophytes ➏.

An unpollinated ovule exudes a sugary "pollination droplet" that can capture windborne pollen. The pollen grain then germinates, and a pollen tube grows through the ovule tissue to deliver sperm to the egg ➐. The pollen tubes of pines grow very slowly, so it takes more than a year for the tube to reach an ovule. Fertilization produces a zygote ➑. The zygote develops into an embryo sporophyte that, along with tissues of the ovule, becomes a seed. The seed is released, germinates, then develops into a new sporophyte ➒.

TAKE-HOME MESSAGE 22.7

✔ Gymnosperms include conifers, cycads, ginkgos, and gnetophytes.

✔ These seed-bearing vascular plants produce seeds on strobili. In the case of conifers, the seed-bearing strobilus is a woody cone.

FIGURE 22.18 **Life cycle of a pine.**

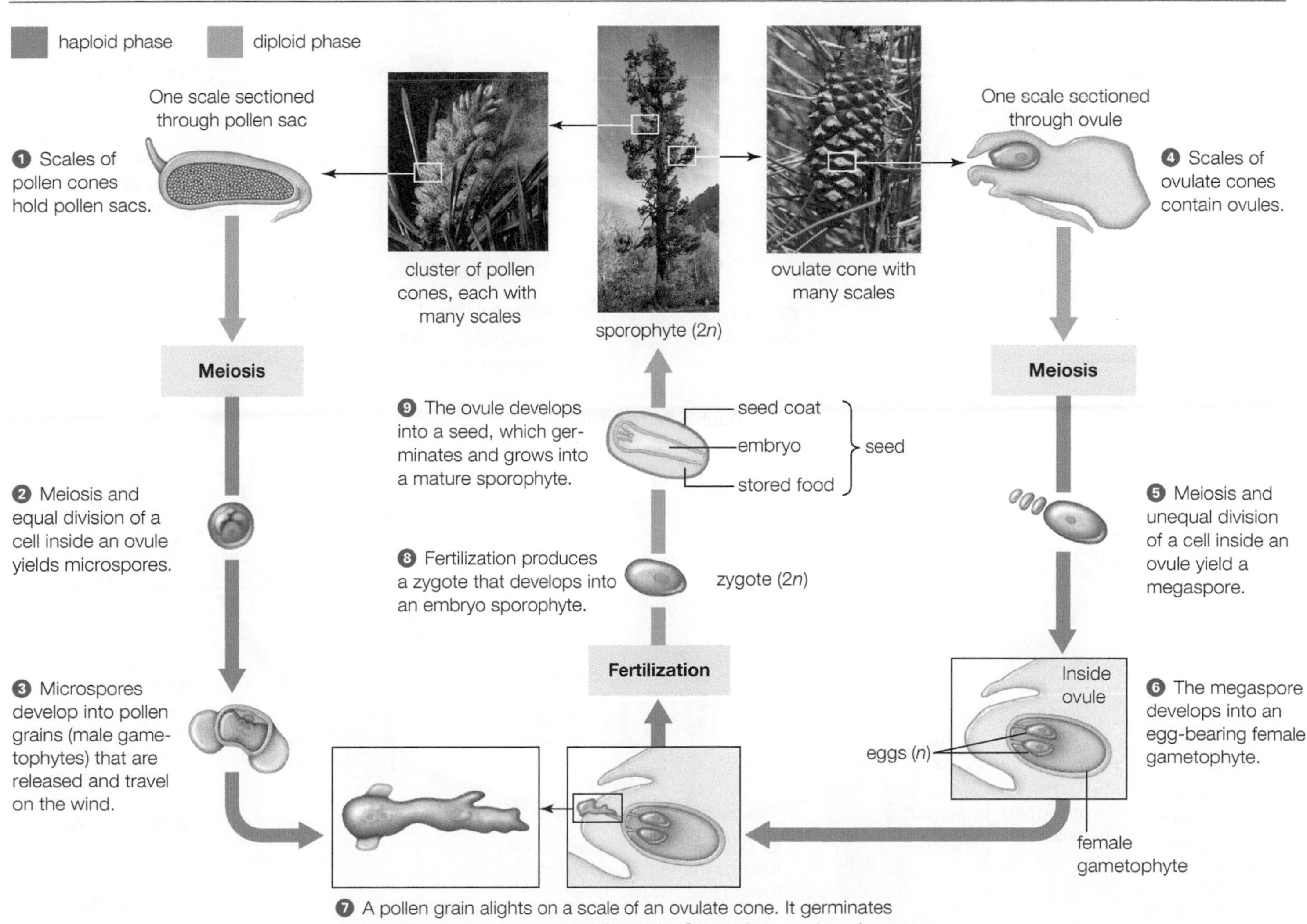
haploid phase
diploid phase
One scale sectioned through pollen sac
➊ Scales of pollen cones hold pollen sacs.
cluster of pollen cones, each with many scales
sporophyte (2n)
ovulate cone with many scales
One scale sectioned through ovule
➍ Scales of ovulate cones contain ovules.
Meiosis
Meiosis
➒ The ovule develops into a seed, which germinates and grows into a mature sporophyte.
seed coat
embryo
stored food
seed
➋ Meiosis and equal division of a cell inside an ovule yields microspores.
➎ Meiosis and unequal division of a cell inside an ovule yield a megaspore.
➑ Fertilization produces a zygote that develops into an embryo sporophyte.
zygote (2n)
Fertilization
Inside ovule
➏ The megaspore develops into an egg-bearing female gametophyte.
eggs (n)
➌ Microspores develop into pollen grains (male gametophytes) that are released and travel on the wind.
female gametophyte
➐ A pollen grain alights on a scale of an ovulate cone. It germinates and a pollen tube grows toward the ovule. Sperm form as the tube grows.

CREDIT: (18 photos) left, R. J. Erwin/Science Source; middle, Robert Potts, California Academy of Sciences; right, AndrÃ© Pascal/Science Source.

22.8 Angiosperm Traits

LEARNING OBJECTIVES

- Describe the components of a flower and their functions.
- Explain how endosperm forms and its function.
- Discuss how components of a flower give rise to fruits and seeds.

Angiosperms are vascular seed plants that make flowers and fruits. A **flower** is a specialized reproductive shoot that consists of modified leaves arranged in concentric whorls (**FIGURE 22.19**). Sepals form the outermost whorl; petals, the next whorl; pollen-bearing **stamens**, the next whorl. The innermost part of the flower is the **carpel**. The **ovary**, a chamber at the base of the carpel, contains one or more ovules where eggs form. After fertilization, an ovule matures into a seed, and the ovary becomes the **fruit**. The name "angiosperm" refers to the development of seeds inside an ovary. (*Angio–* means enclosed chamber; *sperma*, seed.)

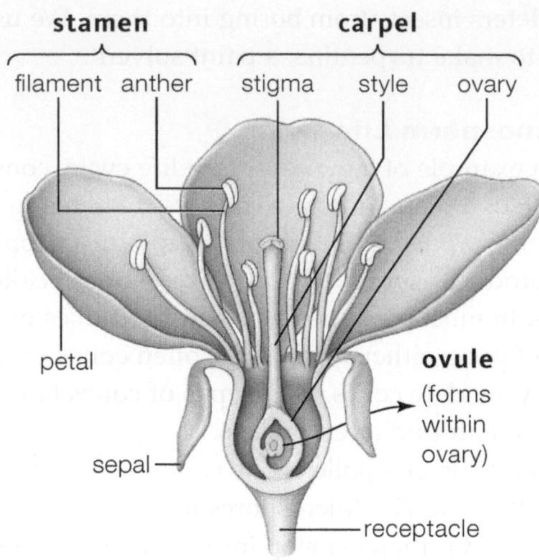

stamen
carpel
filament
anther
stigma
style
ovary
petal
ovule
(forms within ovary)
sepal
receptacle

FIGURE 22.19 Components of a flower. The example shown has both an egg-producing carpel and pollen-producing stamens. In some species, carpels and stamens form on different flowers. We discuss the diversity of flower types in detail in Section 29.2.

FIGURE 22.20 Flowering plant life cycle.

haploid phase

diploid phase

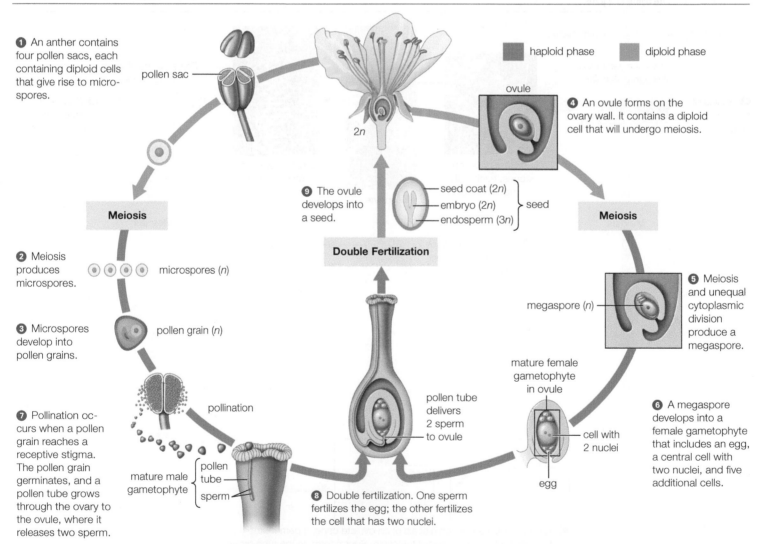

❶ An anther contains four pollen sacs, each containing diploid cells that give rise to microspores.

pollen sac

2n

ovule

❹ An ovule forms on the ovary wall. It contains a diploid cell that will undergo meiosis.

Meiosis

❾ The ovule develops into a seed.

seed coat (2n)

embryo (2n)

endosperm (3n)

seed

Meiosis

Double Fertilization

❷ Meiosis produces microspores.

microspores (n)

megaspore (n)

❺ Meiosis and unequal cytoplasmic division produce a megaspore.

❸ Microspores develop into pollen grains.

pollen grain (n)

mature female gametophyte in ovule

pollination

pollen tube delivers 2 sperm to ovule

❻ A megaspore develops into a female gametophyte that includes an egg, a central cell with two nuclei, and five additional cells.

❼ Pollination occurs when a pollen grain reaches a receptive stigma. The pollen grain germinates, and a pollen tube grows through the ovary to the ovule, where it releases two sperm.

cell with 2 nuclei

mature male gametophyte

pollen tube

sperm

egg

❽ Double fertilization. One sperm fertilizes the egg; the other fertilizes the cell that has two nuclei.

The Angiosperm Life Cycle

FIGURE 22.20 shows a generalized life cycle for a flowering plant. The upper portion of a stamen is a bilobed anther that holds four pollen sacs, two in each lobe. Inside those sacs are diploid cells ❶ that give rise to microspores by meiosis ❷. The microspores develop into pollen grains (immature male gametophytes) ❸.

A flowering plant's ovules form on the wall of an ovary at the base of a carpel ❹. Meiosis of a cell in the ovule yields four haploid megaspores ❺. Only one of these megaspores develops into a female gametophyte, which consists of a haploid egg, a cell with two nuclei, and a few additional cells ❻.

Pollination occurs when a pollen grain arrives on a receptive stigma, the uppermost part of the carpel ❼. The pollen grain germinates, and a pollen tube grows through the style (the structure that elevates the stigma) to the ovary at the base of the carpel. Two nonmotile sperm form inside the tube as it grows.

Double fertilization occurs when a pollen tube delivers the two sperm into the ovule ❽. One sperm fertilizes the egg to create a zygote. The other sperm fuses with a cell that has two nuclei, forming a triploid ($3n$) cell. After double fertilization, the ovule matures into a seed ❾. The zygote develops into an embryo sporophytes and the triploid cell divides to become the **endosperm**, a nutrient-storing tissue that will serve as a source of food for the developing embryo.

TAKE-HOME MESSAGE 22.8

- ✔ A flower is a reproductive shoot composed of modified leaves.
- ✔ Angiosperm ovaries are enclosed within an ovule. After pollination, the ovule becomes a seed and the ovary becomes a fruit.
- ✔ As a result of double fertilization, angiosperm seeds contain endosperm. This triploid tissue serves as food for the embryo.

angiosperms (AN-gee-o-sperms) Most diverse seed plant lineage. Only group that makes flowers and fruits.
carpel Of flowering plants, a reproductive structure that produces female gametophytes; consists of a stigma, a style, and an ovary.
coevolution The joint evolution of two closely interacting species; each species is a selective agent for traits of the other.
double fertilization In flowering plants only, one sperm fertilizes an egg to produce a zygote and another fertilizes a diploid cell to create a triploid cell that will develop into endosperm.
endosperm Triploid ($3n$) nutritive tissue in an angiosperm seed.
flower Specialized reproductive shoot of a flowering plant.
fruit Mature ovary of a flowering plant; encloses a seed or seeds.
ovary In flowering plants, the enlarged base of a carpel, inside which one or more ovules form and eggs are fertilized.
pollinator Animal that moves pollen, thus facilitating pollination.
stamen (STAY-men) Pollen-producing structure of a flowering plant.

A Bee transferring pollen.

B Bird dispersing seeds.

FIGURE 22.21 **Animal assistants to flowering plants.**

22.9 Angiosperm Diversity

LEARNING OBJECTIVES

- List the factors that contributed to the success of angiosperms.
- Compare the traits of monocots and eudicots.
- Describe the ways in which humans benefit from angiosperms.

Factors Contributing to Angiosperm Success

Today, nearly 90 percent of all plant species are flowering plants. Several factors gave angiosperms a selective advantage over gymnosperms.

Accelerated Life Cycle Compared to gymnosperms, most angiosperms have a shorter life cycle. A dandelion or grass can grow from a seed, mature, and produce seeds of its own within a month or so. In contrast, gymnosperms typically tend to take years to mature. Producing and dispersing seeds quickly helps angiosperms expand their range faster than gymnosperms.

Animal-Pollinated Flowers Evolution of flowers gave angiosperms an advantage by facilitating animal-assisted pollination. After seed plants evolved, some insects began feeding on protein-rich pollen. The plants lost a bit of pollen, but benefited when the insects inadvertently transferred pollen from one plant to another of the same species. An animal that aids in pollination by transferring pollen between plants is a **pollinator**. Most pollinators are insects (FIGURE 22.21A), but birds, bats, and other animals also fulfill this role.

Over time, many flowering plants coevolved with pollinators. **Coevolution** refers to the joint evolution of two or more species as a result of their close ecological interactions. Producing sugary nectar encouraged more pollinator visits, which improved pollination rates and enhanced seed production. Conspicuous petals and distinctive scents that attracted pollinators provided a selective advantage. At the same time, selection favored pollinators that searched out nectar-rich flowers and accessed the nectar reward.

Enhanced Seed Dispersal Fruit production also contributed to angiosperm success. Fleshy, sugary fruits attract animals that carry the fruits (and their enclosed seeds) away from a parent plant (**FIGURE 22.21B**). Other fruits have shapes that help them catch the wind or stick to fur. As a group, gymnosperms have fewer structural adaptations for seed dispersal.

Angiosperm Lineages

Gene comparisons have identified the oldest angiosperm lineages that include modern representatives. Among living groups, *Amborella*, a small shrub native to New Caledonia, has the most ancient ancestry. Like gymnosperms, and unlike all other angiosperms, it has only one type of xylem. Water lilies (**FIGURE 22.22A**) and star anise are also modern representatives of ancient plant lineages.

Genetic divergences gave rise to other groups that became dominant: magnoliids, eudicots (true dicots), and monocots. Among the 9,200 or so magnoliids are magnolias (**FIGURE 22.22B**) and avocados. The 80,000 or so **monocots** include orchids, palms, lilies, grasses, and irises (**FIGURE 22.22C**). The approximately 170,000 **eudicots** (**FIGURE 22.22D**) include familiar broadleaf plants such as tomatoes, cabbages, roses, and poppies, as well as flowering shrubs, trees, and cacti.

Monocots and eudicots derive their group names from the number of seed leaves, or cotyledons, in the embryo. Monocots have one cotyledon and parallel leaf veins; eudicots have two cotyledons and branching veins. The two groups also differ in the arrangement of their vascular tissues, number of flower petals, and other traits. Some eudicots undergo secondary growth and become woody, but no monocots produce true wood. We discuss the differences between monocots and eudicots in greater detail in Section 27.2.

A Water lily (basal angiosperm).

B Magnolia (magnoliid).

C Iris (monocot).

D Chickweed (eudicot).

FIGURE 22.22 Representatives of four angiosperm lineages. The vast majority of angiosperms are monocots or eudicots.

Ecological and Economic Importance

It would be nearly impossible to overestimate the importance of angiosperms. As the most abundant plants in most habitats, they provide food, shelter, and nesting materials for a variety of animals. Animals feed on angiosperm roots, shoots, and fruits, and they sip floral nectar.

Angiosperms provide nearly all human food, either directly or as food for livestock. Cereal grains are the most widely planted crops. All are grasses. Like other monocots, cereals store nutrients in their endosperm mainly as starch. In the United States, corn is the top cereal crop. Worldwide, rice feeds the greatest number of people. Other widely grown cereal crops include rye, oats, barley, sorghum, and wheat (**FIGURE 22.23A**).

The widespread use of cereal crops is partially a consequence of their varied carbon-fixing pathways (Section 6.5). C3 grasses such as rice, wheat, oats, and barley grow well under cool conditions. C4 grasses such as corn and sorghum thrive in hot, dry climates. Thus, people throughout the world can grow a cereal grain that is well suited to their climate.

Soybeans, lentils, peas, and peanuts are among the legumes, the second most important source of human food. Legumes are eudicots, and their endosperm stores nutrients as proteins and oils, in addition to carbohydrates. Legumes can be paired with grains to provide all the amino acids that a human body needs to synthesize proteins.

Humans enliven their diet with a variety of plant parts. We dine on leaves of lettuce and spinach, roots of carrots, immature floral shoots of broccoli and cauliflower, and fleshy fruits of tomatoes and blueberries. Stamens of crocus flowers provide the spice saffron, and grating the bark of a tropical tree gives us cinnamon. Maple syrup is fluid tapped from a tree's xylem and boiled down to a syrupy consistency.

Angiosperms are the source of fabrics such as linen, ramie, hemp, burlap, and cotton. The fruit of a cotton plant, the cotton boll, is nearly pure cellulose

CREDITS: (22A) Eakachai Leesin/Shutterstock; (22B) Donald Johansson/iStockphoto.com; (22C) iStockphoto.com/Wendy Townrow; (22D) Mikhail Melnikov/Shutterstock.

A Mechanized harvesting of wheat, a eudicot.

B Cotton (a eudicot) with bolls ready for harvest.

FIGURE 22.23 Angiosperms as crops.

(**FIGURE 22.23B**). Fibers from the leaves of agave are used to make sisal rugs.

Oils extracted from seeds of eudicots such as rapeseed and hemp are used in detergents, skin care products, and as industrial lubricants and fuel.

Eudicot woods are "hardwoods," as opposed to gymnosperm "softwoods." Furniture and flooring are often made from oak or other hardwoods. Hardwoods are also preferred as firewood.

We use secondary metabolites of plants as medicines and mood-altering drugs. A **secondary metabolite** is a compound with no known metabolic role in the organism that makes it. Many plant secondary metabolites are defenses against grazing animals. Aspirin is derived from a compound made by willows, and digitalis used to strengthen heartbeats is from foxglove plants. Caffeine from coffee beans and nicotine from tobacco leaves are widely used stimulants. Marijuana is one of the United States' most valuable cash crops. Worldwide, cultivation of opium poppies (the source of heroin) and coca (the source of cocaine) have wide-reaching health, economic, and political effects.

TAKE-HOME MESSAGE 22.9

✔ Most flowering plants are monocots or eudicots.

✔ The grains that serve as staples of the human diet are monocots. Many other crop plants are eudicots. We use the secondary metabolites of some flowering plants as medicines or mood-altering drugs.

eudicots (you-DIE-cots) Most diverse lineage of angiosperms; members have two seed leaves, branching leaf veins.

monocots Highly diverse angiosperm lineage; includes plants such as grasses that have one seed leaf and parallel veins.

secondary metabolite Molecule that is produced by an organism but does not play any known role in its metabolism. Some serve as defense against predation.

22.1 Saving Seeds (revisited)

The seed bank at Svalbard is maintained at a temperature just below freezing. Most flowering plants make seeds that remain viable for long periods after being dried and frozen at this temperature. However, seeds of some plants, most often tropical species, are usually short-lived when frozen under standard conditions. To preserve such species, botanists turn to cryopreservation, which is the storage of biological material at an ultralow temperature. Typically, the seeds are stored in tanks of liquid nitrogen at –190°C (–320°F), as shown in **FIGURE 22.24**.

Cryopreservation can also be used to preserve the spores of seedless plants and vegetative tissues of plants that reproduce mainly by asexual mechanisms such as fragmentation. ●

LIQUID NITROGEN

FIGURE 22.24 Freezing seeds. A botanist lowers a container of seeds into a vat of liquid nitrogen for long-term storage. Cryopreservation has the advantage of killing most plant pathogens in the plant material being preserved.

CREDITS: (23A) Photo USDA; (23B) Photo by Scott Bauer, USDA/ARS; (24) Photo by Scott Bauer/USDA Agricultural Research Service.

STUDY GUIDE

Section 22.1 Sustaining many varieties of crop plants and their wild relatives ensures that plant breeders will have a reservoir of genetic diversity to tap into if widely planted varieties fail. Seed banks can help us maintain a wide variety of potentially valuable plant species. Seeds and other plant parts can be frozen for long-term storage.

Section 22.2 **Plants** are multicelled, typically photosynthetic eukaryotes. They are closely related to the charophyte algae. They are classified as **embryophytes**, because they form a multicelled embryo on the parental body. The earliest evidence of plants comes from fossil spores that date back about 450 million years.

All plants have an alternation of generations, in which a haploid **gametophyte** alternates with a diploid **sporophyte**. The gametophyte generation produces haploid gametes by mitosis. These gametes unite at fertilization, and the resulting zygote develops into a diploid sporophyte. The sporophyte produces haploid spores by meiosis. Fossil spores are the oldest evidence of plant life.

The oldest plant lineages are the **bryophytes**. The **vascular plants** were the next to evolve. Vascular plants include both seedless lineages and the **seed plants**. There are two lineages of seed plants: gymnosperms and angiosperms.

Section 22.3 The oldest plant lineages, the bryophytes, have a life cycle dominated by the gametophyte. Over time, there has been an evolutionary trend toward a sporophyte-dominated life cycle. A **cuticle** and **stomata** that minimize water loss are adaptations to life on land.

Evolution of **xylem** and **phloem** (**vascular tissues**) allowed vascular plants to grow larger than most bryophytes. The **lignin** in cell walls of xylem provides support to the bodies of vascular plants. Evolution of **pollen grains** allowed seed plants to reproduce even in dry habitats. Seed plants protect their embryo sporophytes inside **seeds**.

Section 22.4 Three lineages (mosses, liverworts, and hornworts) are collectively referred to as bryophytes. Bryophytes are nonvascular plants and most are low-growing. They are the only plants in which the gametophyte generation is dominant.

Mosses are the most diverse bryophytes and the main plants in **peat bogs**. A moss gametophyte has green leaflike parts arranged around a central axis. **Rhizoids** attach the gametophyte to soil or another surface. **Gametangia** at the tips of the gametophyte produce gametes. Sperm are flagellated and swim to eggs through a film of water. After fertilization, a nonphotosynthetic sporophyte develops while attached to and dependent upon the gametophyte. The sporophyte consists of a stalk and a capsule that contains the **sporangia**. Spores released from the capsule develop into new gametophytes. Mosses also reproduce by fragmentation.

Liverworts may be the oldest surviving plant lineage. Their gametophyte may be mosslike or a flattened thallus. Some reproduce asexually by forming clumps of cells. Hornworts have an elongated hornlike sporophyte. They are the only bryophytes with a photosynthetic sporophyte.

Section 22.5 Seedless vascular plants have a long-lived sporophyte supported by vascular tissues. The gametophyte is tiny and short-lived. Both generations are free-living and photosynthetic. Sperm are flagellated and swim to eggs.

Club mosses belong to one lineage of seedless vascular plants; horsetails, whisk ferns, and true ferns belong to another. Typically the roots and shoots of the sporophyte grow from a horizontal stem, or **rhizome**. Fern sporophytes produce spores in **sori** on fronds. Horsetails and club mosses make spores in a **strobilus**.

Section 22.6 The earliest vascular plants were tiny and had a simple branching pattern. By the Carboniferous, swamp forests were dominated by giant lycophytes. Geologic processes transformed remains of these forests into **coal**.

Seed plants, which began to diversify during the Carboniferous, do not release spores. Their sporophytes have **pollen sacs**, where **microspores** form and develop into pollen grains (male gametophytes). **Megaspores** form inside **ovules** and develop into female gametophytes. **Pollination** unites the egg and sperm of a seed plant. A seed is a mature ovule. It includes nutritive tissue and a tough seed coat that protects the embryo sporophyte inside the seed from harsh conditions.

Section 22.7 **Gymnosperms** include the **conifers** and the lesser-known **cycads**, **ginkgos**, and **gnetophytes**. Most conifers are evergreen trees, and the group is an important source of lumber. Conifers produce soft, pollen-bearing cones as well as woody, ovulate cones. They are wind pollinated.

Sections 22.8, 22.9 **Angiosperms** are the most diverse plants. They alone produce **flowers**, which are modified shoots. The **stamens** of a flower produce pollen. An **ovary** at the base of a **carpel** holds one or more egg-producing ovules. After pollination and **double fertilization**, a floral ovary becomes a **fruit** containing one or more seeds. A flowering plant seed includes an embryo sporophyte and nutritious **endosperm**.

Factors that contributed to angiosperm success include a short life cycle, coevolution with **pollinators**, and a variety of mechanisms for dispersing fruits.

As the dominant plants in most land habitats, flowering plants are ecologically important, as well as essential to human existence. The two major lineages are **monocots** and **eudicots**. The starch-rich endosperm of monocot seeds such as wheat and rice makes them staples of human diets throughout the world. Angiosperms also supply us with vegetables, spices, fiber, furniture, oils, medicines, and mood-altering drugs. Many useful plant compounds are **secondary metabolites**, compounds that probably help defend the plant against predation.

SELF-QUIZ

Answers in Appendix VII

1. The first land plants were _______ .
   a. gnetophytes c. bryophytes
   b. gymnosperms d. lycophytes

2. Lignin supports the stems of _______ .
   a. mosses c. ferns
   b. hornworts d. liverworts

3. A waxy cuticle helps land plants _______ .
   a. conserve water c. reproduce
   b. take up carbon dioxide d. stand upright

4. True or false? Ferns produce seeds inside sori.

5. _______ attach mosses to soil.
   a. Rhizoids c. Roots
   b. Rhizomes d. Strobili

6. Bryophytes alone have a relatively large _______ and an attached, dependent _______ .
   a. sporophyte; gametophyte
   b. gametophyte; sporophyte

7. Club mosses, horsetails, and ferns are _______ plants.
   a. multicelled aquatic c. seedless vascular
   b. nonvascular seed d. seed-bearing vascular

8. Coal consists primarily of compressed remains of the _______ that dominated Carboniferous swamp forests.
   a. seedless vascular plants c. flowering plants
   b. conifers d. hornworts

9. The sperm of _______ swim to eggs.
   a. horsetails c. conifers
   b. eudicots d. monocots

10. A seed is a(n) _______ .
   a. female gametophyte c. mature pollen tube
   b. mature ovule d. immature microspore

11. The gametophytes of ferns _______ .
   a. are photosynthetic c. are diploid
   b. produce spores in sori d. are dependent on the sporophyte for nutrition

12. Which angiosperm lineage includes the most species?
   a. magnoliids c. monocots
   b. eudicots d. water lilies

13. Match the terms appropriately.
   ___ bryophyte a. has seeds, but no fruits
   ___ seedless vascular plant b. has flowers and fruits
   ___ gymnosperm c. has xylem and phloem, but no pollen
   ___ angiosperm d. no xylem or phloem

14. Match the terms appropriately.
   ___ ovule a. gamete-producing body
   ___ cuticle b. spore-producing body
   ___ gametophyte c. where eggs form
   ___ sporophyte d. underground stem
   ___ fruit e. mature ovary
   ___ endosperm f. nutritive tissue in seed
   ___ rhizome g. where fern spores form
   ___ sorus h. secreted waxy layer

15. Place these groups in order of their appearance with the oldest lineage first and the most recently evolved last.
   ___ 1 a. ferns
   ___ 2 b. cycads
   ___ 3 c. eudicots
   ___ 4 d. mosses

CRITICAL THINKING

1. Early botanists admired ferns but found their life cycle perplexing. In the 1700s, they learned to propagate ferns by sowing what appeared to be tiny dustlike "seeds" from the undersides of fronds. Despite many attempts, the scientists could not find the pollen source, which they assumed must stimulate these "seeds" to develop. Imagine you could write to one of these botanists. Compose a note that would clear up the confusion.

2. In animals, meiosis of germ cells produces gametes. Some plant cells also undergo meiosis, but the resulting cells are not gametes. Explain what the products of plant meiosis are, what they develop into, and how plants produce gametes.

3. Consider a cherry pit, which is a seed with a cherry embryo inside it. Trace the paternal ancestry of that embryo. Explain how the sperm that fathered the embryo came to unite with the egg, and the process by which that sperm originated.

4. Suppose you wanted to make a science fiction film in which human time travelers went back to the Carboniferous. What types of modern plants would you use and which would you avoid if you wanted to depict the Carboniferous setting as accurately as possible?

5. Lignin and vascular tissue first evolved in relatives of club moss, and some extinct species stood 40 meters (130 feet) high. Explain how the evolution of vascular tissues and lignin would have allowed a dramatic increase in plant height. How might being tall give one plant species a competitive advantage over another?

23 FUNGI

CORE CONCEPTS

Systems

Complex properties arise from interactions among components of a biological system.

Populations of different organisms in a biological community interact in complex ways. Nutrients released by fungal decomposers can be taken up by other organisms in the environment. Some fungi partner with algae to form lichens, and others partner with plant roots. Still other fungi are parasites or pathogens that affect the health of their plant or animal hosts.

Evolution

Evolution underlies the unity and diversity of life.

Fungi are eukaryotes that share a protist ancestor with animals. Fungi range in size from single cells to multicelled networks that extend through the soil for miles. They produce spores asexually and sexually. Fungal groups are defined by their mechanism of sexual reproduction. Elaborate spore-producing structures evolved in some groups.

Pathways of Transformation

Organisms exchange matter and energy with the environment in order to grow, maintain themselves, and reproduce.

Fungi are heterotrophs and most are decomposers. All fungi meet their nutritional needs by secreting digestive enzymes onto organic matter and absorbing the resulting breakdown products. The porous structure of fungal cell walls ensures that food and water taken up by one part of a fungus can be shared with cells in other regions of the fungal body.

Links to Earlier Concepts

Fungi are eukaryotes, so their cells have the features you learned about in Section 4.4. Fungi have cell walls of chitin (Section 3.3), and some carry out fermentation (7.5). We discussed the photosynthetic components of lichens earlier (20.7 and 21.7). Here we look at the fungal component.

23.1 High-Flying Fungi

Fungi are not known for their mobility. You probably don't think of mushrooms and their relatives as world travelers, but some do get around. Fungi produce microscopic spores that can lodge in crevices on tiny dust particles. When winds carry these particles aloft, spores go along for the ride. Dustborne fungal spores can travel long distances riding the winds that swirl high above Earth's surface.

Dust storms in North African deserts sometimes lift fungus-laden particles more than 4.5 kilometers (3 miles) above the desert floor. Winds carry this dust out over the Atlantic Ocean, and sometimes completely across it. Most fungal spores that hitchhike on African dust are harmless, but occasionally spores of a fungus that causes plant disease make the journey. For example, winds of a 1978 cyclone introduced sugar cane rust (a fungal disease) from Cameroon to the Dominican Republic. Similarly, winds probably introduced coffee rust fungus from Angola to Brazil in 1980.

Today, an African outbreak of an old fungal foe has agricultural officials around the world worried. The fungus that arouses their concern, *Puccinia graminis*, causes wheat stem rust disease. Like other rust fungi, *P. graminis* is an obligate plant parasite, meaning it can grow and reproduce only in living plant tissue. An infection begins when a spore lands on the leaf of a wheat plant. The spore germinates, and a fungal filament enters the plant through a stoma. As the fungus grows through the plant's tissues, it sucks up photosynthetic sugars that the plant would normally use to meet its own needs. As a result, an infected plant is stunted and produces little or no wheat. About a week after a plant becomes infected, tens of thousands of rust-colored fungal spores appear on its stem (FIGURE 23.1). Each spore can disperse and infect a new plant.

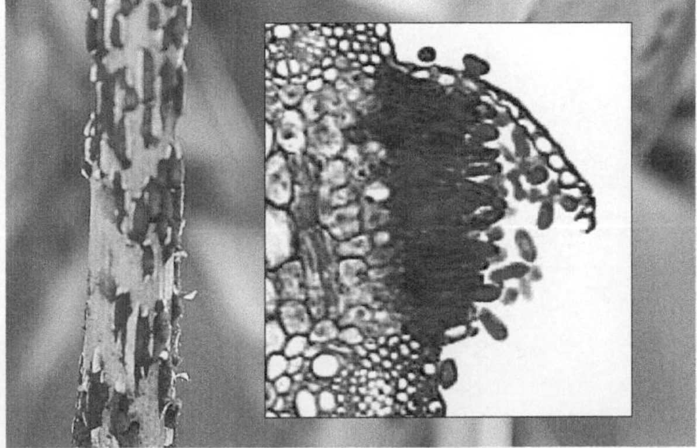

FIGURE 23.1 **Wheat stem, with rust-colored fungal sporangia (spore-producing structures) on its surface.** The inset micrograph shows spores (red) escaping from the sporangia.

CREDITS: (opposite) Courtesy of Christine Evers; (1) background, Photo by Yue Jin/USDA; inset, Courtesy of Charles Good, Ohio State University at Lima.

Wheat stem rust disease routinely decimated wheat crops worldwide until the 1960s, when strains resistant to *P. graminis* became available. Worldwide use of rust-resistant wheats provided a respite from outbreaks of wheat stem rust for decades. Then, in 1999, a new strain of wheat stem rust called Ug99 was discovered in Uganda. Ug99 has mutations that allowed it to infect most wheat varieties that were previously considered resistant to wheat stem rust.

Since then, windblown spores of Ug99 have dispersed within Africa, crossed the Red Sea to Yemen, and from there crossed the Persian Gulf to Iran. Ug99 is now present in thirteen countries and winds could potentially spread it worldwide. Fungicides can help minimize the damage but are too expensive for farmers in many developing nations.

With one of the world's most important crops at risk, plant scientists are working to develop new varieties of wheat with genes that will provide resistance against Ug99. Some Ug99-resistant wheats have been developed, but multiplying enough seed and deploying it to farmers who currently grow susceptible varieties is a challenge. Farmers save their own wheat seed and resow it, so the resistant wheat is not of commercial interest to seed companies that have the capacity to distribute it. ●

23.2 Fungal Traits and Diversity

LEARNING OBJECTIVES

- Explain how fungi are similar to and different from plants and animals.
- Describe the structure of multicelled fungi.
- Explain how fungi meet their nutritional needs.

A **fungus** (plural, fungi) is a spore-forming, eukaryotic heterotroph that digests its food externally. It secretes enzymes onto food, and then absorbs nutrients released by the action of the enzymes.

Like plants, fungi have cell walls and a life cycle that involves producing spores. However, fungi are more closely related to animals than plants. Both animals and fungi belong to the eukaryotic supergroup Opisthokonts. Like animals, fungi are heterotrophs that store excess carbohydrates as glycogen.

Fungus Structure

Unlike plants and animals, the fungi include both single-celled and multicellular species. Fungi that live as single cells are commonly called **yeasts**. Molds, mildews, and mushrooms are familiar examples of multicelled fungi (**FIGURE 23.2**). These fungi grow as a **mycelium** (plural, mycelia), a network of microscopic interwoven filaments. Each filament, or **hypha** (plural, hyphae), is a strand of walled cells arranged end to end (**FIGURE 23.3**). The walls of a fungal cell consist primarily of chitin, a polysaccharide also found in the body covering of crabs and insects (Section 3.3).

The structure of fungal hyphae varies among groups. The oldest fungal lineages include chytrids, zygote fungi (zygomycetes), and glomeromycetes (**FIGURE 23.4A–C**). In these groups, cells of a hypha do not have cross-walls between them. Each hypha is a long tube full of cytoplasm and nuclei. Hyphae divided into compartments by cross-walls, or septae (singular,

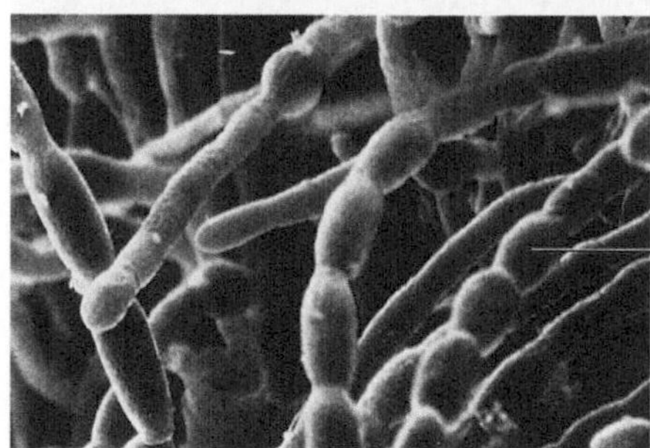
one cell (part of one hypha of the mycelium)

FIGURE 23.3 A mycelium. A mass of threadlike hyphae. Each hypha is a strand of cells attached one to the other.

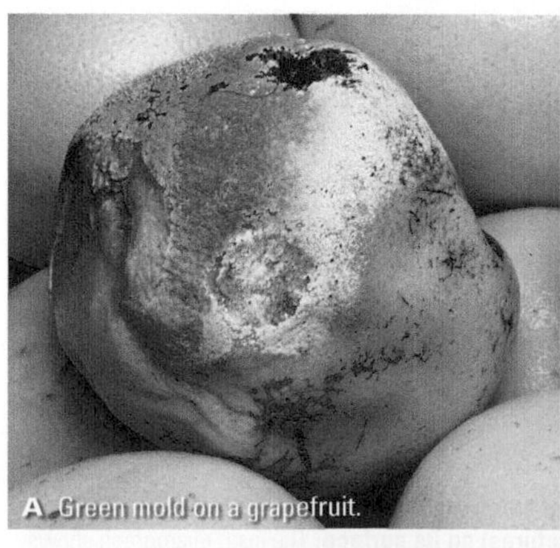

A Green mold on a grapefruit.

B Powdery mildew on leaves.

C Mushrooms on a forest floor.

FIGURE 23.2 Growth forms of multicelled fungi.

CREDITS: (2A) Photo by Scott Bauer/USDA; (2B) Nigel Cattlin/Science Source; (2C) Courtesy of Christine Evers; (3) Micrograph Garry T. Cole, University of Texas, Austin/BPS.

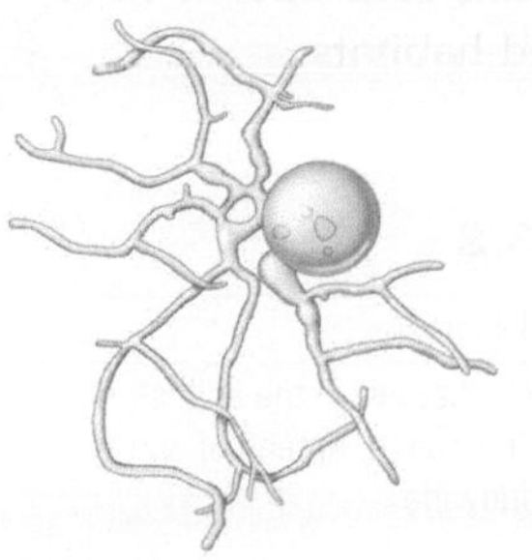

A Chytrids. Mostly aquatic, produce flagellated spores.

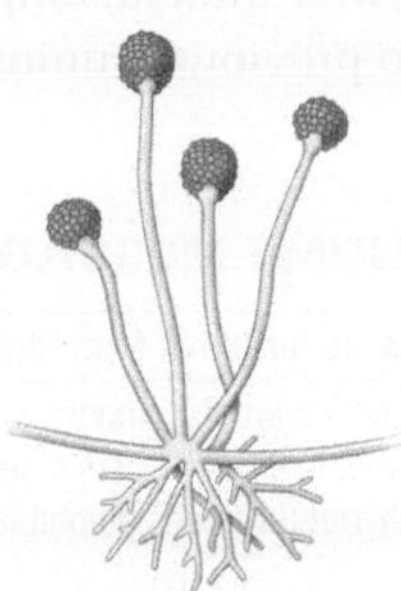

B Zygote fungi. Mostly molds, some parasites.

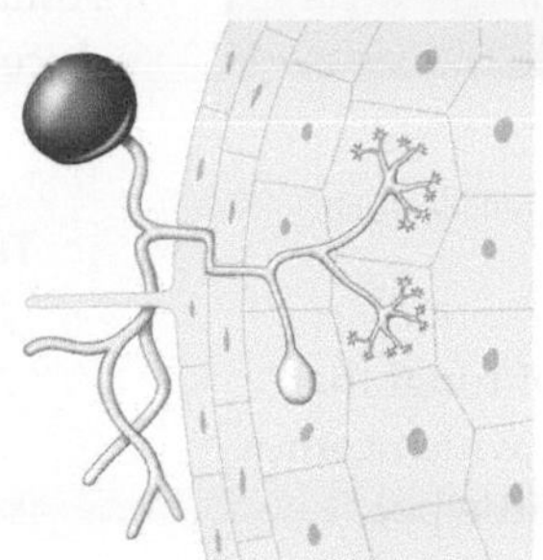

C Glomeromycetes. Soil fungi with hyphae that branch inside root cells.

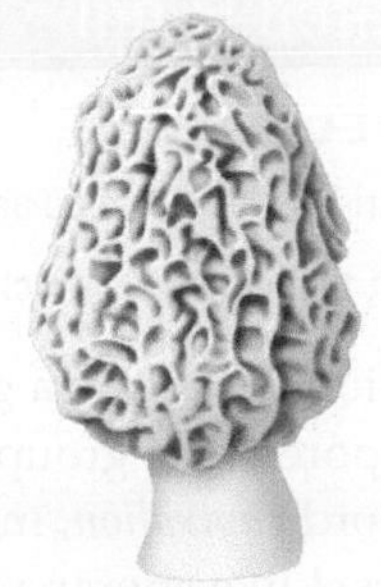

D Sac fungi. Most diverse group. Yeasts, molds, parasites, species that produce fruiting bodies.

E Club fungi. Mushrooms, other species that produce fruiting bodies, some plant pathogens.

FIGURE 23.4 Major fungal groups. Graphics show examples of spore-bearing structures.

septa), evolved in the common ancestor of sac fungi and club fungi. These are the most diverse fungal lineages, and septate hyphae contributed to their success.

When cross-walls are present, they are porous, so materials can flow between adjacent cells:

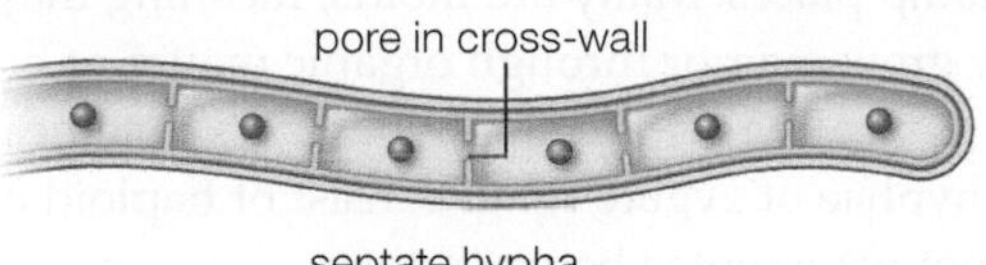
pore in cross-wall
septate hypha

Thus, all multicelled fungi can share nutrients or water taken up in one part of a mycelium with cells in other parts of the fungal body. The presence of cross-walls makes hyphae sturdier and allowed the evolution of larger, more elaborate spore-producing bodies, also known as fruiting bodies (**FIGURE 23.4C,D**). Fungi with septate hyphae also are more resistant to drying out than those with nonseptate hyphae.

Fungus Life Cycles

Fungi produce spores both asexually (by mitosis) and sexually (by meiosis). Major fungal groups are defined by their sexual production of spores.

The life cycle of a sexually reproducing fungus differs from that of a plant or animal in that it includes a dikaryotic stage (**FIGURE 23.5**). A **dikaryotic** structure contains two genetically distinct types of nuclei (n+n). In plants and animals, cytoplasmic fusion of an egg and sperm is followed immediately by fusion of their nuclei. In fungi, cytoplasmic fusion creates a dikaryotic structure. The length of the dikaryotic phase differs among fungal groups, as you will see below. When the nuclei do fuse, the result is a diploid zygote that undergoes meiosis to produce haploid spores. Those spores germinate, releasing cells that divide by mitosis to form a new haploid mycelium.

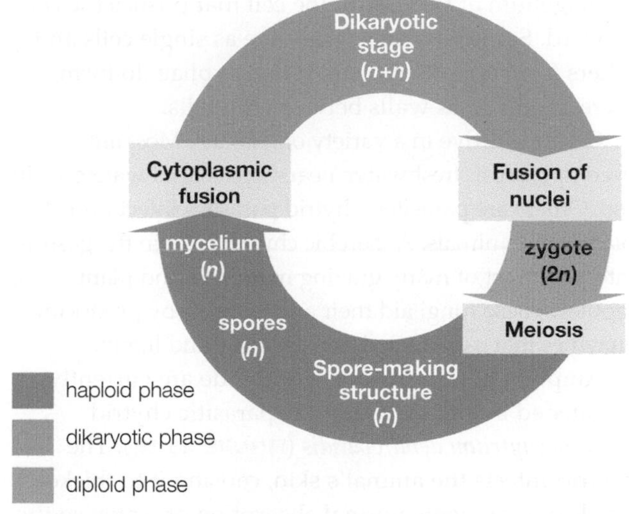
Dikaryotic stage (n+n)
Cytoplasmic fusion
Fusion of nuclei
mycelium (n)
zygote (2n)
spores (n)
Meiosis
Spore-making structure (n)
haploid phase
dikaryotic phase
diploid phase

FIGURE 23.5 Sexual reproduction cycle for a multicellular fungus.

dikaryotic (die-kary-OTT-ik) Having two different nuclei in a cell (n+n).
fungus Eukaryotic heterotroph with cell walls of chitin; obtains nutrients by extracellular digestion and absorption.
hypha (HI-fah) Component of a fungal mycelium; a filament made up of cells arranged end to end.
mycelium (my-SEAL-ee-um) Mass of threadlike filaments (hyphae) that make up the body of a multicelled fungus.
yeast Single-celled fungus.

TAKE-HOME MESSAGE 23.2

✔ Fungi are heterotrophs that secrete digestive enzymes to break down organic material, then absorb the released nutrients.

✔ Some fungi are single cells; others grow as a multicelled mycelium. All have cell walls with chitin.

✔ Fungi disperse by producing spores, which may be produced either by mitosis or meiosis.

23.3 Flagellated Fungi

LEARNING OBJECTIVES

- Explain why chytrids are considered an ancient fungal group.
- Describe the trait that distinguishes chytrids.

Chytrids (Chytridiomycota) are a group of fungi that form flagellated spores. The group name, which comes from the Greek word *chytridion*, means "little pot." It refers to the vessel-like sporangium in which the spores form (FIGURE 23.6A). The oldest fossil fungi, which date to 408–360 million years ago, resemble the sporangia of modern chytrids. Genetic comparisons among modern fungi also indicate that chytrids are one of the oldest fungal lineages.

Like animal sperm, motile chytrid spores have a single flagellum at the rear of the cell that pushes the cell forward. Some chytrid species live as single cells and others form simple hyphae. When hyphae do form, there are no cross-walls between the cells.

Chytrids thrive in a variety of habitats. Most are decomposers in freshwater, nearshore coastal waters, or in soil. Others are parasites; chytrid parasites infect protists, plants, and animals. Anaerobic chytrids live in the gastrointestinal tract of many grazing mammals and plant-eating reptiles. These fungi aid their animal hosts by producing enzymes that can break down cellulose and lignin.

Amphibian populations worldwide are currently threatened by the spread of the parasitic chytrid *Batrachochytrium dendrobatidis* (FIGURE 23.6B). The chytrid infects the animal's skin, causing it to thicken. Thick skin prevents normal absorption of water, so the amphibian eventually dies of dehydration. *B. dendrobatidis* is native to Africa, where it infects African clawed frogs without sickening them. From the 1930s to the 1950s, these frogs were exported worldwide for use in pregnancy tests. *B. dendrobatidis* traveled with the frogs to regions where native frogs have no resistance to it. The pathogen now occurs on all continents except Antarctica, and trade in amphibians continues to introduce it into previously uninfected habitats.

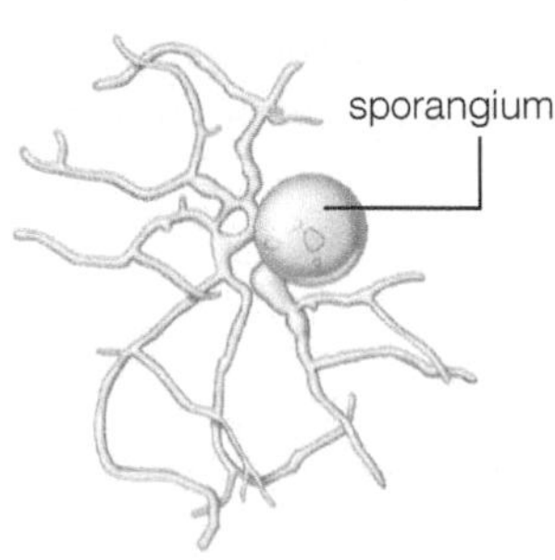
sporangium

A Structure of a hypha-forming chytrid. Flagellated zoospores form inside the sporangium.

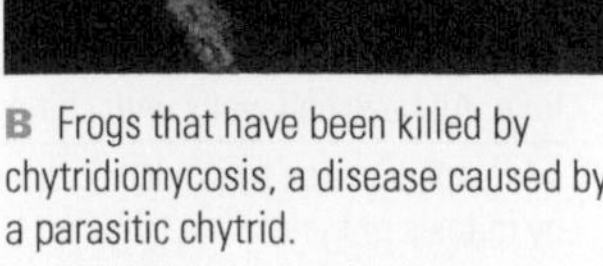

B Frogs that have been killed by chytridiomycosis, a disease caused by a parasitic chytrid.

FIGURE 23.6 **Chytrids.**

TAKE-HOME MESSAGE 23.3

✔ Chytrids are fungi that form flagellated spores.

✔ Most chytrids are decomposers. Chytrids that live in the animal digestive tract help their hosts digest plant material. Spread of one parasitic chytrid threatens amphibians worldwide.

23.4 Zygote Fungi and Relatives

LEARNING OBJECTIVES

- Describe the growth and reproduction of a zygote fungus.
- Explain how a microsporidian infects a host cell.
- Describe the structure of a glomeromycete fungus.

Zygote Fungi

The 1,100 or so species of **zygote fungi** (zygomycetes) live in damp places. Many are molds, meaning they typically grow over or through organic matter as a mass of asexually reproducing hyphae. As noted earlier, the hyphae of zygote fungi consist of haploid cells that do not have septae between them.

Black bread mold (*Rhizopus stolinifera*) has a life cycle typical of zygote fungi (FIGURE 23.7). As long as food is plentiful, it grows as a haploid mycelium and produces spores by mitosis ❶. If the food supply dwindles and a compatible sexual partner is nearby, hyphae develop special gamete-forming side branches (gametangia) ❷. Zygote fungi do not have walls between their cells, so many haploid nuclei from each hypha flow into each gametangium. When gametangia of two individuals come into contact, the walls where they touch break down, and their cytoplasm fuses. The result is an immature dikaryotic zygospore (a zygospore containing multiple nuclei from each parent) ❸.

As the zygospore matures, haploid nuclei from both parents pair up and form diploid nuclei. A mature zygospore contains multiple diploid nuclei and has a thick, protective wall ❹. The zygote is the only diploid stage in a zygote fungus life cycle.

When a zygospore germinates, a hypha emerges and cells at its tip undergo meiosis to produce haploid spores ❺. Each spore is capable of giving rise to a new, haploid mycelium.

Another zygote fungus, *Pilobolus*, is notable for its interesting mechanism of spore dispersal. Grazing animals ingest *Pilobolus* spores along with grass. The spores pass through the animal's gut and are deposited in feces. They germinate and grow into a mycelium

CREDIT: (6A) © Cengage Learning; (6B) JOEL SARTORE/National Geographic Creative

FIGURE 23.7 Life cycle of black bread mold, a zygote fungus.

❶ A haploid mycelium grows in size and produces spores by mitosis.

❷ When nutrients are limited, two individuals produce special hyphae that grow toward one another. As these hyphae develop, a wall forms behind their tips, isolating multiple nuclei inside each of the resulting gametangia.

❸ Cytoplasmic fusion of the gametangia produces a zygospore that contains many haploid nuclei from each parent.

❹ Nuclei within the zygospore pair up and fuse to produce a mature zygospore with many diploid nuclei. A thick protective wall develops around the zygospore.

❺ The zygospore undergoes meiosis and germinates. A hypha emerges and produces spores, each capable of giving rise to a new haploid mycelium.

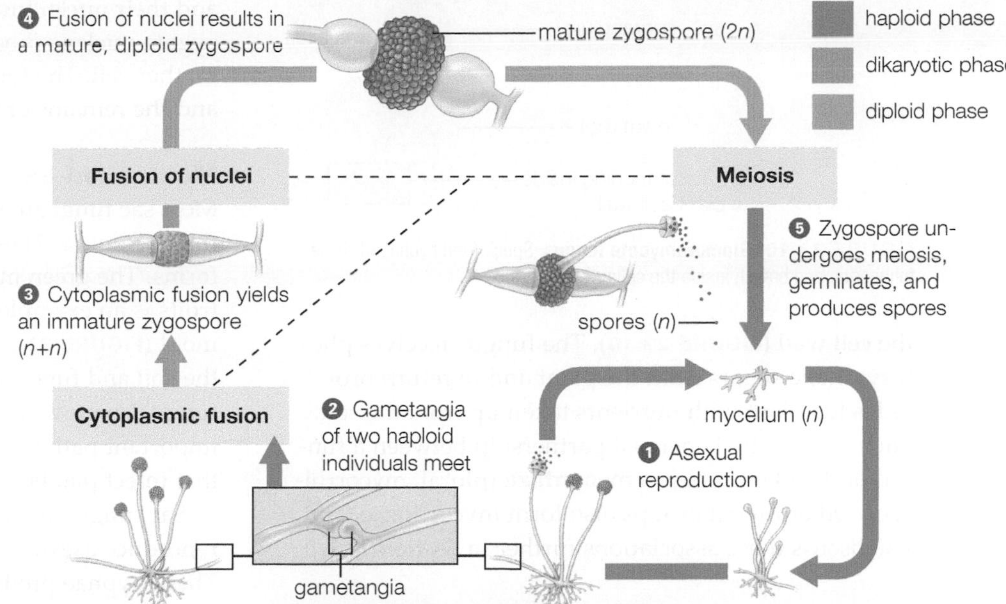
❹ Fusion of nuclei results in a mature, diploid zygospore
mature zygospore (2n)
haploid phase
dikaryotic phase
diploid phase
Fusion of nuclei
Meiosis
❺ Zygospore undergoes meiosis, germinates, and produces spores
❸ Cytoplasmic fusion yields an immature zygospore (n+n)
spores (n)
mycelium (n)
Cytoplasmic fusion
❷ Gametangia of two haploid individuals meet
❶ Asexual reproduction
gametangia

that produces spore-bearing hyphae (**FIGURE 23.8**). At the tip of each hypha, a tiny dark sporangium perches atop a large fluid-filled vesicle. When fluid pressure builds up in the vesicle, it bursts, hurling the sporangium as far as 2 meters (about 6.5 feet) away. Animals ingest the sporangium and the process begins again.

1 mm

FIGURE 23.8 Spore-bearing structures of *Pilobolus*. The name means "hat-thrower." The dark "hats" are sporangia (spore sacs).

Microsporidia—Intracellular Parasites

Microsporidia (singular, microsporidium) include about 1,300 named species of single-celled intracellular parasites. Most infect animals, with fish and insects being the most common hosts. The relationship of microsporidians to other fungi remains unclear, although genetic similarities suggest that they are closer to zygote fungi than to other fungal groups.

A microsporidium spore has a tough coat that allows it to survive adverse conditions for years. Beneath the coat, a long tube coils in the cytoplasm (**FIGURE 23.9**). During infection, the tube uncoils and perforates a host cell. The microsporidium's nucleus and cytoplasm flow through the tube into the host.

Some microsporidia can infect the human gut, but they generally cause symptoms only in people with a weakened immune system. Microsporidia of the genus *Nosema* infect the digestive tract of bees causing a fatal disease. Nosema disease is the most common disease in honeybee colonies, and its spread is contributing to an ongoing loss of these essential pollinators.

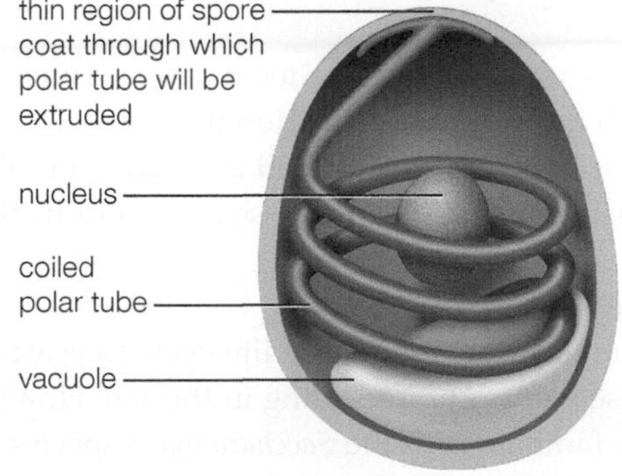
thin region of spore coat through which polar tube will be extruded
nucleus
coiled polar tube
vacuole

FIGURE 23.9 Microsporidium spore.

Glomeromycetes

All 160 or so species of **glomeromycetes** live only in close association with plant roots. Hyphae of these fungi penetrate the wall of a root cell and branch inside

chytrids (KIH-trids) Fungi that produce flagellated spores.
glomeromycete (GLOW-mer-o-MY-seat) Fungus with hyphae that grow inside the wall of a plant root cell.
microsporidia Single-celled, spore-forming fungi that are intracellular animal parasites.
zygote fungus Fungus that forms a zygospore during sexual reproduction; hyphae do not have septa.

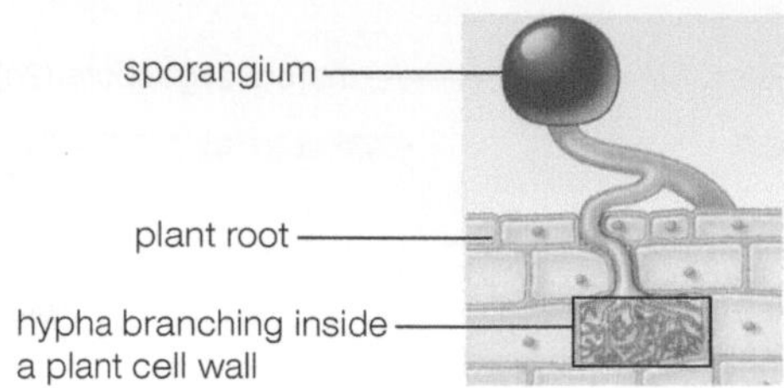
sporangium
plant root
hypha branching inside
a plant cell wall

FIGURE 23.10 Glomeromycete fungus. Specialized hyphae of these fungi enter and branch inside the cells of a plant root.

the cell wall (FIGURE 23.10). The fungus receives photosynthetic sugars from the plant and in return provides the plant with nutrients taken up from the soil. Such a mutually beneficial partnership between a fungus and a plant root is a **mycorrhiza** (plural, mycorrhizae). Other fungal groups also form mycorrhizae, and we discuss these associations further in Section 23.7.

TAKE-HOME MESSAGE 23.4

- ✔ Zygote fungi form a thick-walled, diploid zygospore during sexual reproduction. Many are molds.
- ✔ Microsporidia are single-celled parasites that invade an animal cell by way of a polar tube.
- ✔ Glomeromycetes are mycorrhizal fungi; their hyphae grow inside the cell wall of plant root cells.

23.5 Sac Fungi

LEARNING OBJECTIVES

- Describe the variety of growth forms in sac fungi.
- Explain the function of a sac fungus ascocarp and how this structure forms.

Sac fungi (Ascomycota) are the most diverse fungal lineage, with more than 64,000 known species. Their defining trait is a saclike cell called an **ascus** (plural, asci) in which spores form during sexual reproduction.

Sac Fungal Yeasts

Some sac fungi spend their entire life cycle as yeasts. Most of these yeasts are free-living in the soil. However, you may be familiar with the *Saccharomyces* species that are used in baking and brewing (Section 7.5).

Sac fungal yeasts most commonly reproduce asexually by **budding**, a process in which mitosis is followed by unequal cytoplasmic division. During budding, a descendant cell with a small amount of cytoplasm buds from its parent (FIGURE 23.11A).

When conditions do not favor growth, yeasts produce spores through sexual reproduction. Two haploid yeast cells of different mating types come together and their nuclei fuse to form a zygote. Meiosis of the zygote nucleus produces four haploid nuclei within the mother cell. The four nuclei give rise to four spores, and the remains of the mother cell becomes the ascus.

Multicelled Sac Fungi

Most sac fungi are multicelled during part or all of their life cycle. These species have a variety of growth forms. The green mold that frequently spoils citrus fruits is an example of a sac fungus that grows as a mold (FIGURE 23.2A). Most sac fungal molds grow in the soil and function as decomposers. Multicelled sac fungi also grow on or in other organisms, and some are important pathogens. For example, powdery mildews that infect plants are sac fungi (FIGURE 23.2B).

Sac fungi that grow as molds and mildews typically reproduce asexually by means of specialized hyphae. These hyphae produce spores by mitosis in cells at their tips (FIGURE 23.11B). Mitotically produced spores of sac fungi are called conidiospores or conidia. *Conidia* is the Greek word for dust.

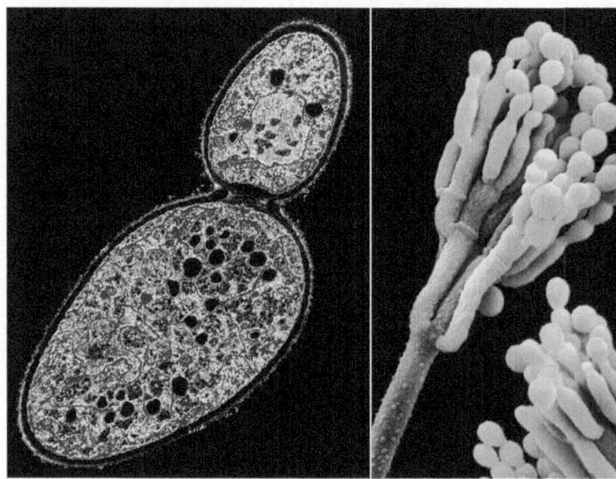

A Asexual reproduction by budding in a yeast (*Saccharomyces*). **B** Conidia (asexually produced spores) of the mold *Eupenicillium*.

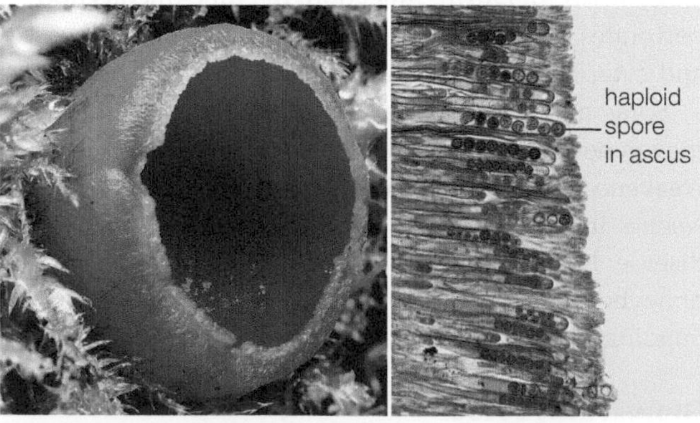
haploid
spore
in ascus

C Cup fungi reproduce sexually by producing an ascocarp (left). Spores (right) form by meiosis of cells on the cup's concave surface.

FIGURE 23.11 Modes of reproduction in sac fungi.

In multicelled sac fungi, sexual reproduction begins when the haploid hyphae of two compatible individuals meet and cells at their tips undergo cytoplasmic fusion. The resulting dikaryotic cell divides by mitosis to produce dikaryotic hyphae. These hyphae intertwine with haploid hyphae to form a fruiting body called an ascocarp. Ascocarps vary widely in their form. They may be spherical, globular, flask shaped, or cup shaped (**FIGURE 23.11C**). Asci form at the tips of dikaryotic hyphae on or inside the ascocarp. Fusion of nuclei in the ascus forms a zygote that undergoes meiosis to produce four haploid cells. Typically, these cells divide by mitosis to yield eight haploid spores.

TAKE-HOME MESSAGE 23.5

✔ Sac fungi are the most diverse group of fungi. Some are yeasts, but most are multicelled.

✔ Sac fungal yeasts reproduce asexually by budding, and multicelled sac fungi reproduce asexually by forming conidia. Sac fungi that reproduce sexually form spores inside an ascus.

23.6 Club Fungi

LEARNING OBJECTIVES

- Describe the club fungus life cycle.
- Compare the length of the dikaryotic phase in club fungi, sac fungi, and zygote fungi.

Life Cycle

Club fungi (Basidiomycota) form spores inside club-shaped cells during sexual reproduction. Most often the spores form on a spore-bearing structure (a basidiocarp) composed of dikaryotic hyphae. Basidocarp-producing club fungi do not produce spores asexually.

Of all fungi, club fungi have the most prolonged dikaryotic phase in their life cycle. Consider the life cycle of the button mushroom (**FIGURE 23.12**). After a haploid spore of this club fungus germinates, mitotic divisions produce a haploid mycelium ❶. If hyphae of two compatible individuals meet, they undergo cytoplasmic fusion, giving rise to a dikaryotic mycelium ❷.

Embryonic mushrooms composed of dikaryotic hyphae form on the mycelium. When it rains, these tiny mushrooms expand and break through the soil surface ❸. Producing fruiting bodies immediately after a rain ensures that spores will disperse when conditions favor their germination and survival.

The underside of a mushroom's cap has thin tissue sheets (gills) fringed with club-shaped, dikaryotic cells ❸. Fusion of the nuclei in one of these cells yields a diploid zygote ❹. The zygote undergoes meiosis, forming four haploid spores ❺. After dispersal, these spores germinate and the life cycle begins again ❻.

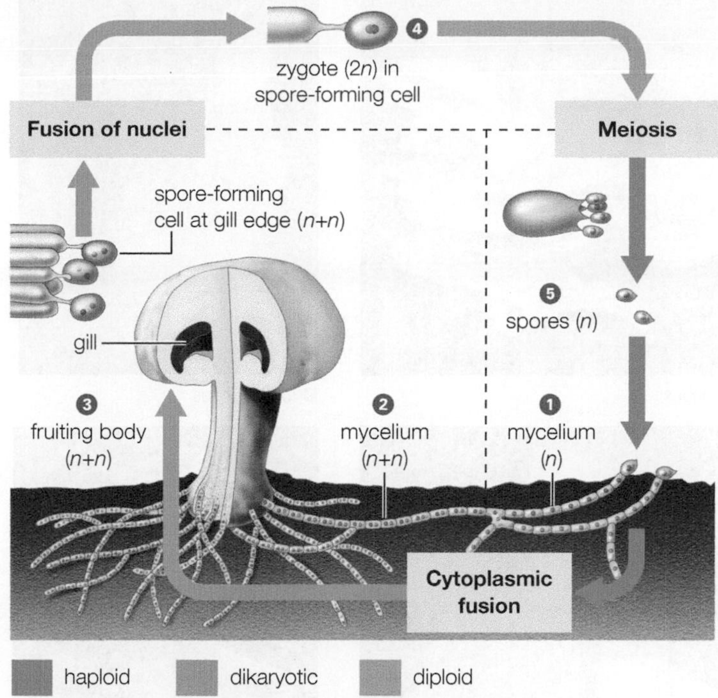

❹
zygote (2n) in spore-forming cell
Fusion of nuclei
Meiosis
spore-forming cell at gill edge (n+n)
gill
❺
spores (n)
❸
fruiting body (n+n)
❷
mycelium (n+n)
❶
mycelium (n)
Cytoplasmic fusion
haploid
dikaryotic
diploid

FIGURE 23.12 Life cycle of a button mushroom, a club fungus.

❶ A haploid spore gives rise to a haploid mycelium. (Red dots and blue dots represent genetically distinct nuclei.)

❷ Cytoplasmic fusion of two haploid mycelia produces a dikaryotic cell that gives rise to a dikaryotic mycelium.

❸ A fruiting body (basidiocarp) composed of dikaryotic hyphae develops.

❹ Fusion of nuclei in cells at the edge of the mushroom's gills produces zygotes.

❺ Meiosis of a zygote produces haploid spores.

FIGURE IT OUT Is there a diploid multicelled body in this life cycle?

Answer: No. Only the zygote is diploid.

The fruiting body is the most conspicuous part of the club fungus life cycle, but it is relatively small and exists only briefly. A dikaryotic mycelium, on the other hand, can be enormous and very long lived. In one Oregon forest, the mycelium of a single honey mushroom extends across nearly 4 square miles (10 km^2). It has been growing for an estimated 2,400 years.

Club Fungus Diversity

There are more than 32,000 club fungus species. Many club fungi are capable of breaking down lignin, thus

ascus (Plural, asci) Of sac fungi, saclike cell in which spores form by meiosis.
budding In yeasts, a mechanism of asexual reproduction by which a small cell forms on a parent, then is released.
club fungi Fungi that have septate hyphae and during sexual reproduction spores are produced in club-shaped cells.
mycorrhiza (my-kuh-RYE-zah) Mutually beneficial partnership between a fungus and a plant root.
sac fungi Most diverse group of fungi; hyphae are septate and sexual reproduction produces spores inside a saclike cell (an ascus).

A Shelf fungus

B Jelly fungus

C Giant puffball

D Stinkhorn

FIGURE 23.13 Diverse club fungus fruiting bodies (basidiocarps).

making this group important decomposers and parasites of woody plants.

All commercially grown mushrooms are club fungi. However, not all club fungal fruiting bodies are mushrooms with gills. Depending on the subgroup, club fungi can produce spores on their surface, deep inside a fruiting body, or in specialized pores. **FIGURE 23.13** provides a sample of the fruiting body diversity within club fungi.

Shelf or bracket fungi (**FIGURE 23.13A**) feed on dead wood and on living trees. They make spores within pores on their underside.

Jelly fungi (**FIGURE 23.13B**) form a gelatinous fruiting body on logs and tree trunks. They do not feed on wood. Rather, they steal nutrients from the mycelia of other fungi that do feed on wood.

Puffballs produce some of the largest fruiting bodies (**FIGURE 23.13C**). These decomposers can be as large as a meter across. They produce spores internally. When mature, the outer covering of the puffball turns dark and splits, allowing spores to escape.

Stinkhorns have a phallic shape and an aroma like rotting meat or feces (**FIGURE 23.13D**). Flies and beetles attracted by the scent land on the spore-laden tip of the fungus. They inadvertently pick up spores, which they then distribute to new locations.

TAKE-HOME MESSAGE 23.6

✔ Club fungi are a group of mostly multicelled fungi. Multicelled species spend most of their life cycle as a dikaryotic mycelium and have the largest, most complex spore-bearing structures of all fungi.

✔ Club fungi are important in forests both as decomposers and as pathogens.

✔ Mushrooms are one type of club fungus fruiting body. The fruiting bodies of club fungi vary widely in their size and shape.

23.7 Biological Roles of Fungi

LEARNING OBJECTIVES

- Using appropriate examples, describe the ways that fungi can benefit and harm plants and animals.
- Describe the effects of two types of fungal infections common in humans.
- List some of the ways that humans use fungi.

Nature's Recyclers

Most fungi provide an important ecological service by breaking down complex compounds in organic wastes and remains. Some soluble nutrients released by this process enter soil or water. Plants and other producers can then take up the nutrients to meet their own needs. Bacteria also serve as decomposers, but they tend to grow mainly on surfaces. By contrast, fungal hyphae can extend deep into a decaying log or other bulky food source.

Fungal Partnerships

Many fungi take part in a **mutualism**, which is a mutually beneficial partnership between different species.

Mycorrhizae A mycorrhiza (plural, mycorrhizae) is a type of mutualism in which a soil fungus forms a partnership with the root of a vascular plant (**FIGURE 23.14**). An estimated 80 percent of the vascular plants form mycorrhizae with a glomeromycete fungus. Hyphae of these fungi enter root cells and branch in the space between the cell wall and the plasma membrane. Some sac fungi and club fungi form mycorrhizae in which hyphae surround a root and grow between its cells. Most forest mushrooms are fruiting bodies of mycorrhizal club fungi.

Hyphae of all mycorrhizal fungi functionally increase the absorptive surface area of their plant

CREDITS: (13A) iStockphoto.com/busik1; (13B) Christine Evers; (13C) Dave Shemanske, Fermilab; (13D) Courtesy of Dr. Darrell Hensley/University of Tennessee

partner. Hyphae are thinner than even the smallest roots and can grow through tiny spaces between soil particles. The fungus shares water and nutrients taken up by its hyphae with root cells. In return, the plant supplies the fungus with sugars.

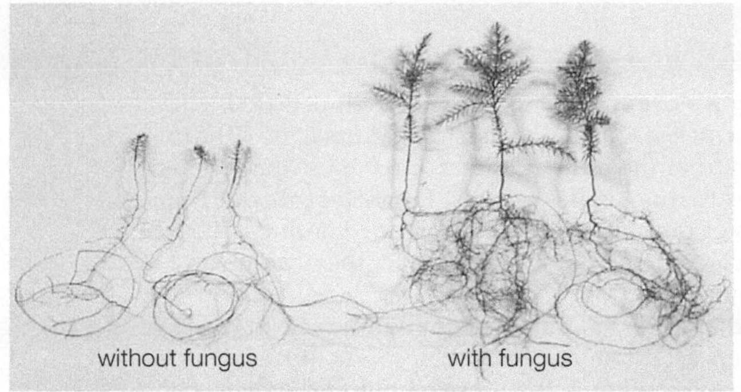
without fungus
with fungus

FIGURE 23.14 Benefit of mycorrhizal fungi. Juniper seedlings were grown with or without mycorrhizal fungi in sterilized, phosphorus-poor soil. The photo shows the results after 6 months.

Lichens A **lichen** is a composite organism consisting of a fungus or fungi and a photosynthetic partner. The fungus makes up most of a lichen's mass (**FIGURE 23.15**) Cyanobacteria or green algae provide the fungus with sugars and are held in place by fungal hyphae. If these cells are cyanobacteria, they can also provide fixed nitrogen.

Lichens are generally described as a mutualism. However, in some cases, the fungus may be parasitically exploiting captive photosynthetic cells that would do better on their own.

Lichens disperse by fragmentation or by releasing special structures that contain cells of both partners. In addition, the fungus can release spores. The fungus that germinates from a spore can form a new lichen only if it alights near an appropriate photosynthetic cell. This is not as unlikely as it may seem; the required algae and bacteria are common as free-living cells. Lichens colonize places too hostile for most organisms, such as newly exposed bedrock. They break down rock by releasing acids and by holding water that freezes and thaws. When soil conditions improve, plants move in and take root. Long ago, lichens may have preceded plants onto land.

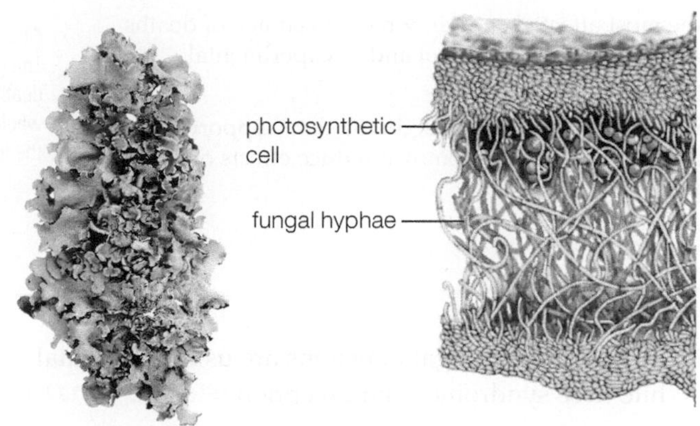
photosynthetic cell
fungal hyphae

FIGURE 23.15 A leafy (foliose) lichen. The graphic shows a cross section through the body of such a lichen.

Digestive Assistants Fungal partners also enhance the nutrition of some animals. Chytrid fungi that live in the stomachs of grazing hoofed mammals such as cattle, deer, and moose aid their hosts by breaking down otherwise indigestible cellulose. Similarly, fungal partners of some ants and termites serve as an external digestive system. Leaf-cutter ants gather bits of leaves to feed a fungus that lives in their nest. The ants cannot digest leaves, but they do eat some of the fungus.

Parasites and Pathogens

Many sac fungi and club fungi are plant parasites. Powdery mildews (sac fungi) and rusts and smuts (club fungi) are parasites that grow in living plants. Their hyphae extend into plant cells, where they suck up photosynthetically produced sugars. The resulting loss of nutrients stunts the plant, minimizes seed production, and may eventually kill it. Typically, a plant does not die before the fungus has produced spores on the plant's infected parts. The wheat rust spores shown in **FIGURE 23.1** are an example.

Other fungi are pathogens that kill plant tissue with their toxins, then feed on the remains. For example, the club fungus *Armillaria* causes root rot by infecting trees and woody shrubs. Once an infected tree dies, the fungus decomposes the stumps and logs left behind.

Many more fungi infect plants than animals. Among animals, those that do not maintain a high body temperature are most vulnerable to fungal infections. You learned earlier about a chytrid that infects amphibians.

Hundreds of fungal species infect insects, and some turn their hosts into seeming zombies, the better to disperse their spores. A fly infected by the zygote fungus *Entomophthora muscae* ceases normal activities, climbs to the top of a stem, and clings there with outstretched wings. It soon dies, and spore-bearing hyphae erupt through its abdominal wall to shed spores that can cause new infections.

lichen (LIE-kin) Composite organism consisting of a fungus and a single-celled alga or a cyanobacterium.

mutualism Mutually beneficial relationship between two species.

Data Analysis Activities

Fighting a Forest Fungus The honey mushroom, *Armillaria ostoyae*, is a parasite of living trees; it withdraws nutrients from them. If an infected tree dies, the fungus continues to dine on its remains. Hyphae grow out from the roots of infected trees and from dead stumps. If these hyphae contact roots of a healthy tree, they can invade and cause a new infection.

Canadian forest pathologists hypothesized that removing stumps after logging could help prevent tree deaths. To test this hypothesis, they carried out an experiment. In half of a forest, they removed stumps after logging. In a control area, they left stumps behind. For more than 20 years, they recorded tree deaths and whether *A. ostoyae* caused them. **FIGURE 23.16** shows the results.

1. Which tree species was most often killed by *A. ostoyae* in control forests? Which was least affected by the fungus?
2. For the most affected species, what percentage of deaths did *A. ostoyae* cause in control and in experimental forests?
3. Looking at the overall results, do the data support the hypothesis? Does stump removal reduce effects of *A. ostoyae*?

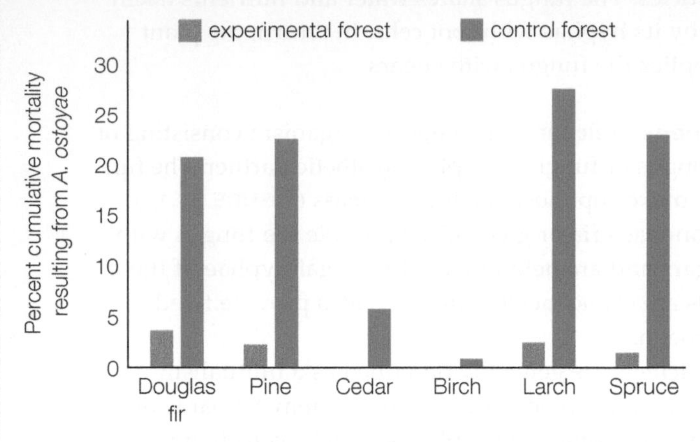
experimental forest
control forest
Percent cumulative mortality resulting from A. ostoyae
30
25
20
15
10
5
0
Douglas fir
Pine
Cedar
Birch
Larch
Spruce

FIGURE 23.16 Effect of stump removal on the spread of a fungal pathogen. The graph shows results of a long-term study of how logging practices affect tree deaths caused by the fungus *A. ostoyae*. In the experimental portion of the forest, whole trees—including stumps—were removed (brown bars). The control portion of the forest was logged conventionally, with stumps left behind (blue bars).

In mammals, fungal infections are usually not fatal. White nose syndrome is an exception (**FIGURE 23.17A**).

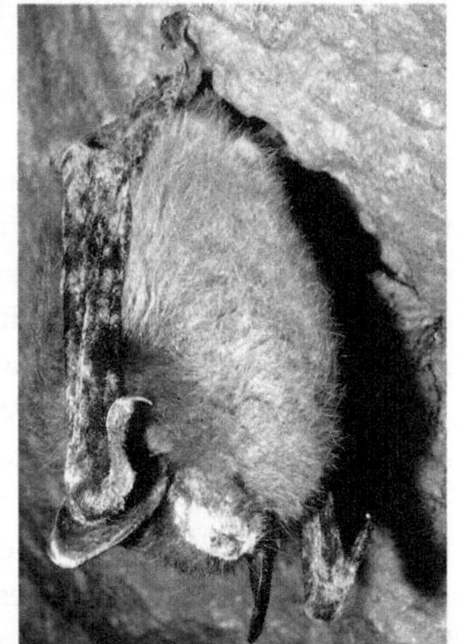

A Bat with white nose syndrome. Such bats have fuzzy white filaments of the sac fungus *Pseudogymnoascus destructans* on their wings, ears, and muzzle. They fly when they should be hibernating, lose weight, and eventually die.

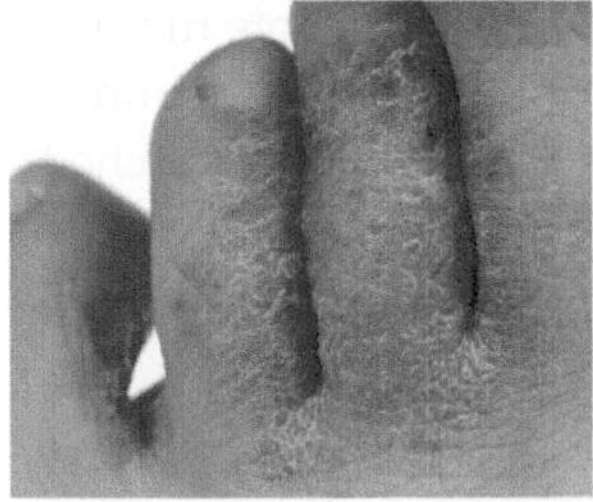

B Athlete's foot.

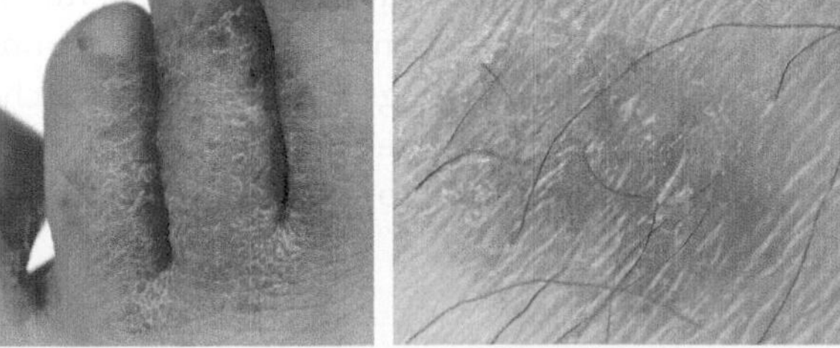

C Ringworm.

FIGURE 23.17 Fungi as mammalian parasites.

This fungal disease of bats was unknown in North America until it was detected in upstate New York in 2006. By early 2017, it had spread to 26 states and five Canadian provinces. The disease has killed millions of North American bats. The sac fungus that causes the disease was introduced to North America from Europe, where the bats have evolved resistance to the fungus and are not sickened by it.

Most human fungal infections involve body surfaces. Infected areas become raised, red, and itchy, as when a fungus infects skin between the toes and on the sole of the foot, causing "athlete's foot" (**FIGURE 23.17B**). Fungal vaginitis (a vaginal yeast infection) occurs when a yeast (*Candida*) that is normally present in the vagina in low numbers undergoes a population explosion. Fungi also cause skin infections misleadingly known as "ringworm." No worm is involved. Rather, a ring-shaped lesion forms as fungal hyphae grow outward from the initial infection site (**FIGURE 23.17C**). Fungal infections in humans usually are life threatening only in people whose immune response is weak as a result of other factors.

Human Uses of Fungi

Many fungal fruiting bodies serve as human food. Button mushrooms, shiitake mushrooms, and oyster mushrooms are club fungi that decompose organic matter, so these species are easily cultivated. By contrast, edible mycorrhizal fungi such as chanterelles, porcini mushrooms, morels, and truffles need a living plant partner, so they are typically gathered from the wild. Picking wild mushrooms is best done with the aid of someone experienced in identifying species. Each year thousands of people become ill after eating poisonous mushrooms they mistook for edible ones.

An interesting dispersal strategy has evolved in truffle-producing fungi, which are highly prized by gourmets. Truffles are the fruiting bodies of some mycorrhizal sac fungi that partner with tree roots. Truffles form underground near their host tree and, when mature, produce an odor similar to that of an amorous male wild pig. Female wild pigs detect the scent and root through the soil in search of their seemingly subterranean suitor. When they unearth the truffles, they eat them, then disperse fungal spores in their feces. Human truffle hunters usually rely on trained dogs to sniff out the fungi.

Fungal fermentation reactions help us make a variety of products. A *Rhizopus* species is used to produce tempeh, a high-protein fermented soy product with a chewy, meatlike consistency. Hyphae of the mold bind the partially fermented soybeans together into a dense, sliceable form. Fermentation by another mold (*Aspergillus*) helps make soy sauce. Still another mold produces the tangy blue veins in cheeses such as Roquefort (**FIGURE 23.18**). A packet of baker's yeast holds spores of the sac fungus *Saccharomyces cerevisiae*. When these spores germinate in bread dough, they ferment sugar and produce carbon dioxide that causes the dough to rise. Other strains of *S. cerevisiae* are used to make wine, beer, and ethanol that is used as a biofuel.

Geneticists and biotechnologists also use yeasts. Like *E. coli* bacteria, yeasts grow readily in laboratories, and they have the added advantage of being eukaryotes, like us. Checkpoint genes that regulate the eukaryotic cell cycle (Section 11.2) were discovered in the yeast *S. cerevisiae*. This discovery was the first step toward our current understanding of how mutations in these genes cause human cancers. Genetically engineered *S. cerevisiae* and other yeasts are used to produce proteins for use as vaccines and medicines.

FIGURE 23.18 Foods produced with the aid of fungi.

Some naturally occurring fungal compounds have medicinal or psychoactive properties. The initial source of the antibiotic penicillin was a sac fungus that lives in the soil. Another soil fungus gave us cyclosporin, an immune suppressant used to prevent rejection of transplanted organs. Ergotamine, a compound used to relieve migraines, was first isolated from ergot (*Claviceps purpurea*), a club fungus that infects rye plants. Ergotamine is also used in synthesis of LSD, a hallucinogen.

The hallucinogen psilocybin is the active ingredient in so-called magic mushrooms. Psilocybin affects perception by increasing communication among regions of the brain. Researchers are investigating whether medically supervised use of psilocybin can treat some psychological disorders. Preliminary studies indicate that doctor-supervised use of psilocybin can lessen symptoms of depression and post-traumatic stress disorder (PTSD), and relieve the anxiety of people who have been diagnosed with fatal cancers.

TAKE-HOME MESSAGE 23.7

- ✔ Fungi decompose materials, thus releasing nutrients that producers can take up and use.
- ✔ Fungi form mutually beneficial partnerships with a variety of organisms.
- ✔ Some fungi are parasites that harm other organisms.
- ✔ Humans use fungi as food, in biotechnology, in research, and as a source of medicines and psychoactive drugs.

23.1 High-Flying Fungi (revisited)

Plant breeders have produced some new bread wheat varieties resistant to the pathogen Ug99 by inducing mutations through radiation, then selecting for Ug99 resistance among offspring of the radiated plants. They have also investigated the source of Ug99 resistance in some relatives of bread wheat. Recently, they isolated Ug99 resistance genes from two such relatives, einkorn wheat (the first wheat that was cultivated) and wild goatgrass. The newly isolated genes do not occur in any of the cultivated bread wheats. However, it may be possible to introduce one or both of these genes into cultivated strains of bread wheat and thus produce transgenic Ug99-resistant wheats that are suitable for cultivation. ●

Section 23.1 Fungi disperse by releasing spores. Winds can distribute spores far from their point of origin. Windblown spores of a new strain of wheat stem rust fungus threatens global food supplies.

Section 23.2 **Fungi** are heterotrophs that secrete digestive enzymes onto organic matter and absorb released nutrients. Most feed on organic wastes or remains (they are decomposers), but some live in or on other organisms. Fungi are more closely related to animals than to plants. They include single-celled **yeasts** and multicelled species. In the multicelled species, spores germinate and give rise to filaments called **hyphae**. Hyphae typically grow as an extensive mesh called a **mycelium**. The hyphae of sac fungi and club fungi have septae (walls) between cells. Other fungal hyphae do not have walls between cells.

Fungi reproduce by producing spores either asexually or sexually. During sexual reproduction, cytoplasmic fusion of haploid cells produces a **dikaryotic** (n+n) cell. Fusion of nuclei in a dikaryotic cell produces a zygote that undergoes meiosis to produce haploid spores.

Section 23.3 **Chytrids** are an ancient group of fungi and the only modern fungi that produce flagellated spores. One parasitic species is the cause of a disease that threatens many amphibian populations.

Section 23.4 Hyphae of **zygote fungi**, which include molds that spoil bread, are continuous tubes with no cross-walls (septae). A thick-walled, diploid zygospore forms during sexual reproduction. Meiosis of cells inside the zygospore produces haploid spores that germinate and produce a haploid mycelium. Mycelia also produce asexual spores.

Microsporidia are possible relatives of zygote fungi and live inside animal cells, sometimes causing disease. **Glomeromycetes**, also relatives of zygote fungi, partner with plant roots in a relationship called a **mycorrhiza**.

Section 23.5 **Sac fungi** are the most diverse group of fungi. They include single-celled yeasts and multicelled species that have hyphae with cross-walls. Yeasts reproduce asexually by **budding**. Other sac fungi grow as haploid hyphae and produce asexual spores called conidia. Sexual spores are produced an **ascus**. In multicelled species, these saclike structures form on an ascocarp consisting of intertwined haploid and dikaryotic hyphae.

Section 23.6 The mostly multicelled **club fungi** have hyphae with cross-walls. This group produces the largest and most complex fruiting bodies. Typically, a dikaryotic mycelium dominates the club fungus life cycle. This mycelium grows by mitosis and, in some species, extends through a vast volume of soil. When conditions favor reproduction, a basidiocarp, also made up of dikaryotic hyphae, develops. Haploid spores form by meiosis at the tips of club-shaped cells in the basidiocarp. A mushroom is a familiar basidiocarp, but the size and shape of basidiocarps varies widely.

Section 23.7 Fungi play an important ecological role as decomposers, and many take part in a **mutualism** with another species. A mycorrhiza is a mutualism involving a fungus and a plant. The fungal hyphae supplement the plant's surface area for absorbing water and nutrients. The fungus shares some absorbed nutrients with the plant and receives sugars in return.

A **lichen** is a composite organism that usually consists of a sac fungus and one or more photoautotrophs, such as green algae or cyanobacteria. The fungus, which makes up the bulk of the lichen, obtains nutrients from its photosynthetic partner. Lichens fix nitrogen, and they contribute to the breakdown of rocks to soil.

Gut-dwelling fungi help some animals digest plant materials. In other cases, insects grow fungus on plant material in their nest and eat the fungus.

Fungal infections are more common in plants than in animals. In humans, most fungal infections occur at a body surface such as the skin, mouth, or vaginal lining. Such infections seldom pose a severe threat unless the immune system is impaired.

Many fungal fruiting bodies are edible, although some produce dangerous toxins. Fungi that carry out fermentation reactions help us produce food products, alcoholic beverages, and ethanol for use as a biofuel.

Fungi also are used in scientific studies and to produce medicines. The study of yeasts can provide insights into eukaryotic genetics. Recombinant yeasts produce vaccines and other desired proteins. Some compounds extracted from fungi are useful as medicines or psychoactive drugs.

SELF-QUIZ

Answers in Appendix VII

1. All fungi ______ .
   a. are multicelled
   b. form flagellated spores
   c. are heterotrophs
   d. live on land

2. Most fungi obtain nutrients from ______ .
   a. nonliving organic matter
   b. living plants
   c. living animals
   d. photosynthesis

3. The hyphae of ______ have no cross-walls.
   a. zygote fungi
   b. sac fungi
   c. club fungi
   d. all of the above

4. The yeasts whose fermentation activity produces the carbon dioxide that makes bread rise are a ______ .
   a. chytrid
   b. zygote fungus
   c. sac fungus
   d. club fungus

5. In most _______ , an extensive dikaryotic mycelium is the longest-lived phase of the life cycle.
   a. chytrids c. sac fungi
   b. zygote fungi d. club fungi

6. The mycelium of a multicelled fungus is a mesh of filaments, each called a _______ .
   a. septa b. hypha c. spore

7. A mushroom is _______ .
   a. a fungal digestive organ
   b. the only part of the fungal body made of hyphae
   c. a reproductive structure that releases sexual spores
   d. made of haploid hyphae

8. Spores released from a mushroom's gills are _______ .
   a. flagellated c. dikaryotic
   b. produced by mitosis d. haploid

9. _______ are fungi that produce flagellated spores.
   a. Chytrids c. Zygote fungi
   b. Sac fungi d. Club fungi

10. Some nitrogen-fixing cyanobacteria can partner with a fungus to form a _______ .
    a. mycelium c. mycorrhiza
    b. lichen d. fruiting body

11. _______ are fungi that live as intracellular parasites.
    a. Glomeromycetes c. Microsporidia
    b. Chytrids d. Club fungi

12. Fungal infections are most common in _______ .
    a. plants c. mammals
    b. insects d. birds

13. All _______ form partnerships with plant roots.
    a. Glomeromycetes c. Microsporidia
    b. Chytrids d. Club fungi

14. Budding is a mechanism of _______ in yeasts.
    a. extracellular digestion c. defense
    b. asexual reproduction d. toxin production

15. Match the terms appropriately.

___ club fungus	a. first discovered in a soil-dwelling sac fungus
___ chitin	b. component of fungal cell walls
___ penicillin	c. partnership between a fungus and one or more photoautotrophs
___ sac fungus	d. basidocarp producer
___ zygote fungus	e. fungus–plant partnership
___ lichen	f. forms sexual spores in an ascus
___ mycorrhiza	g. bread mold is an example
___ hypha	h. fungal skin disease
___ ringworm	i. strand of many cells arranged end to end

CRITICAL THINKING

1. Researchers working in a Brazilian rain forest recently discovered eight species of bioluminescent mushrooms at a single site. The mushrooms continually emit a faint glow that, although undetectable in daylight, makes them visible at night. Suggest a mechanism by which glowing in the dark could benefit a mushroom. Why do you think so many species with this unusual trait live in the same region?

2. Fungi play an important role in the breakdown of the cellulose in plant cell walls. Review Chapter 3 and decide what type of chemical reaction this breakdown entails. Which simple sugar is the released by this reaction?

3. Health professionals refer to fungal skin diseases as "tineas" and name them according to the region affected (**TABLE 23.1**). Fungal skin diseases are persistent, in part because fungi can penetrate deeper layers of skin than can ointments and creams. There are fewer antifungal drugs than antibacterial ones, and antifungals often have more severe side effects. Reflect on the evolutionary relationships among bacteria, fungi, and humans. Why it is harder to fight fungi than bacteria?

4. A giant puffball (**FIGURE 23.13C**) produces trillions of spores that disperse on wind. A stinkhorn (**FIGURE 23.13D**) makes fewer spores and they are dispersed by insects. How are these differences analgous to differences in pollen production and dispersal among plants?

TABLE 23.1

Fungal Diseases of Skin

Disease	Infected Body Parts
Tinea corporis (ringworm)	Trunk, limbs
Tinea pedis (athlete's foot)	Feet, toes
Tinea capitis	Scalp, eyebrows, eyelashes
Tinea cruris (jock itch)	Groin, perianal area
Tinea barbae	Bearded areas
Tinea unguium	Toenails, fingernails

24 ANIMAL EVOLUTION—THE INVERTEBRATES

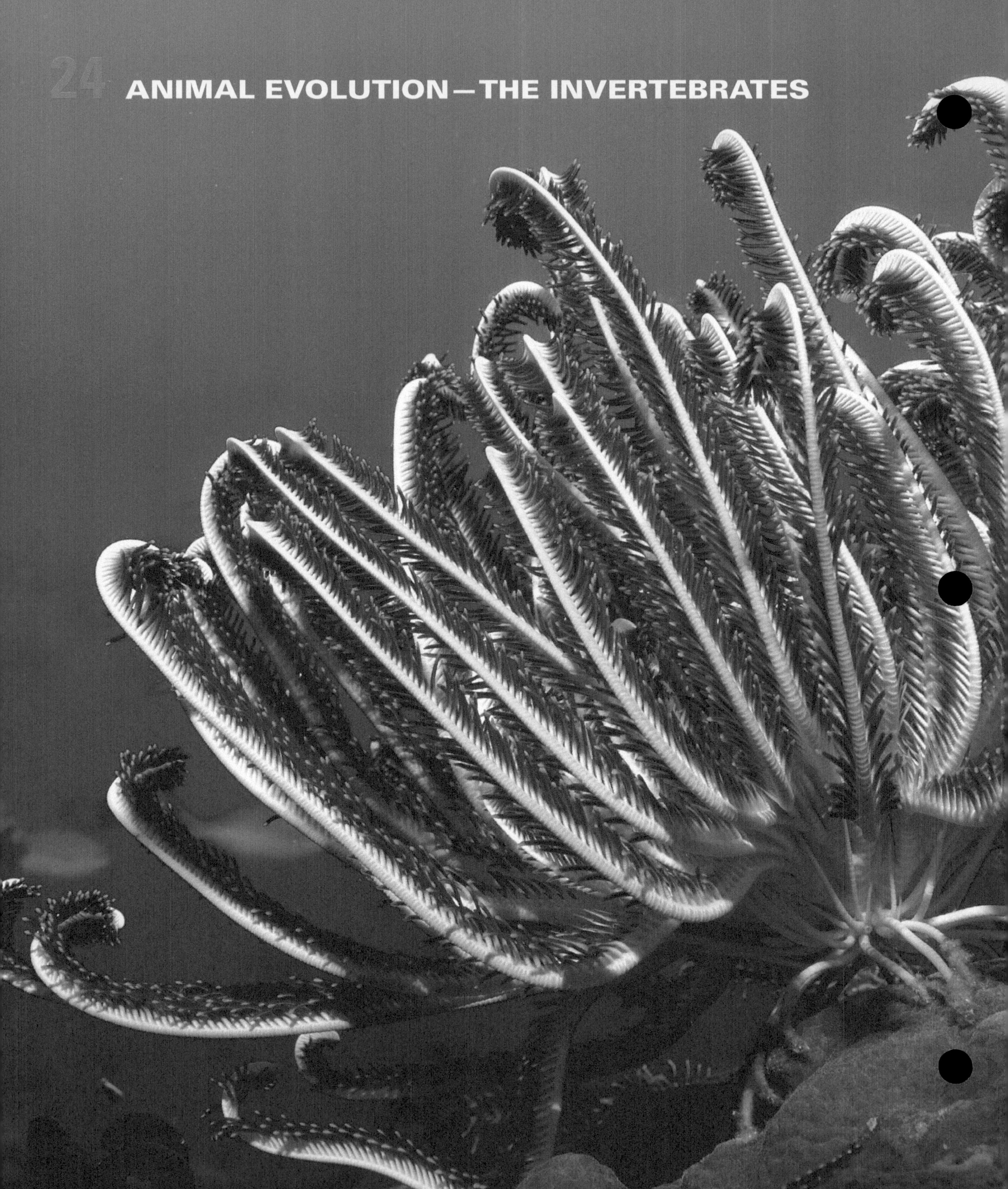

CORE CONCEPTS

Evolution

Evolution underlies the unity and diversity of life.

Shared core processes and features provide evidence that all living things are descended from a common ancestor. All animals have unwalled cells connected by cell junctions and a collagen-containing extracellular matrix. Animals evolved from a colonial protist ancestor by about 600 million years ago.

Pathways of Transformation

Organisms exchange matter and energy with the environment in order to grow, maintain themselves, and reproduce.

Animals are heterotrophs that ingest and digest their food. In the simplest animals, the gut is saclike and nutrients diffuse through tissues. Animals with more complex body plans have a tubular gut, and some have a circulatory system. All animals undergo development and move from place to place during part or all of the life cycle.

Information Flow

Living systems store, retrieve, transmit, and respond to information essential for life.

Homeostatic mechanisms that involve feedback control allow living things to maintain themselves by responding dynamically to internal and external conditions. Many animals have organ systems that allow them to control the solute level of their body fluids. Most have sensory organs that alert them to changes in their environment.

Links to Earlier Concepts

This chapter discusses how a protist ancestor (Section 21.8) could have given rise to animals. Understanding the geologic time line (16.5) will help you put the events we discuss in perspective. We draw on earlier discussions of homeotic genes (10.3), comparative genomics (15.5), and patterns of development (18.5).

24.1 Medicines from the Sea

Animal life began in the sea, and the oceans remain the greatest reservoir of animal diversity. In the oceans, as on land, the majority of animals are **invertebrates**, which means they do not have a backbone. Only about 5 percent of animals have a backbone and thus are considered vertebrates. Invertebrates evolved long before vertebrates, and they remain the most diverse and numerous animals. The longevity and diversity of invertebrate lineages attests to how well they have adapted to their environment.

Many marine invertebrates produce secondary metabolites (Section 22.9) that help them survive. These compounds may protect an animal from predators, help it fend off pathogens, or assist in the capture of prey. Some such compounds also have effects in the human body, and so can be useful as medicines.

Consider the venom that some fish-eating cone snails inject into their prey to subdue it (**FIGURE 24.1**). The snail's venom anesthetizes and paralyzes a fish, thus preventing it from struggling with and possibly harming the snail as it is captured and consumed. Human nerves and fish nerves use the same chemical communication signals, so compounds used by a cone snail to drug fish can also affect the function of the human nervous system. A person stung by a fish-eating cone snail may become numb at the site of injection, suffer from temporary paralysis, or even die.

A synthetic version of one peptide in cone snail venom is now used as a pain reliever. The drug, ziconotide (Prialt), is injected to suppress severe pain that cannot be controlled by other means. Other peptides isolated from cone snail venom are being tested as treatments for epilepsy, diabetes, or cancer.

FIGURE 24.1 Cone snail with a fish in its mouth.

invertebrate Animal that does not have a backbone.

24.2 Animal Traits and Body Plans

LEARNING OBJECTIVES

- List the features common to all animals.
- Describe some of the ways in which animal body plans can vary.

What Is an Animal?

Animals are multicelled heterotrophs that take food into their body, where they digest it and absorb the released nutrients. An animal body consists of a few to hundreds of types of unwalled cells. These cells become specialized as an animal develops from an embryo (an individual in the earliest stage of development) to an adult. Most animals reproduce sexually, some reproduce asexually, and some do both. All are motile (move from place to place) during part or all of their life.

Variation in Animal Body Plans

FIGURE 24.2 shows an evolutionary tree for the animal groups covered in this book, and we will use it to discuss evolutionary trends. All animals are multicellular ❶ and constitute the clade Metazoa (also called Animalia). The earliest animals were aggregations of cells, and this level of organization persists in sponges. However, most modern animals have cells organized as tissues ❷. Tissue organization begins in animal embryos. Embryos of jellies and other cnidarians have two tissue layers: an outer ectoderm and an inner endoderm. In other modern animals, embryonic cells typically rearrange themselves to form a middle tissue layer called mesoderm (FIGURE 24.3). Evolution of a three-layer embryo allowed an important increase in structural complexity. Most internal organs in animals develop from embryonic mesoderm.

Animals with the simplest structural organization, such as sponges, are asymmetrical; their body cannot be divided into two halves that are mirror images. Jellies, sea anemones, and other cnidarians have **radial symmetry**: Body parts are repeated around a central axis, like spokes of a wheel ❸. Radial animals typically attach to an underwater surface or drift along. A radial body plan allows them to capture food that can arrive from any direction. By contrast, animals with a three-layer body plan typically have **bilateral symmetry**: The body's left and right halves are mirror images ❹. Such lineages typically show **cephalization**, an evolutionary trend whereby many nerve cells and sensory structures become concentrated at the front of the body. These structures help the animal find food or avoid threats as it moves headfirst through its environment.

The two lineages of bilateral animals are defined in part by developmental differences. In **protostomes**, the first opening that appears on an embryo becomes a mouth ❺. *Proto–* means first and *stoma* means opening. Most bilateral invertebrates are protostomes. In **deuterostomes**, the mouth develops from the second embryonic opening ❻. *Deutero-* means second.

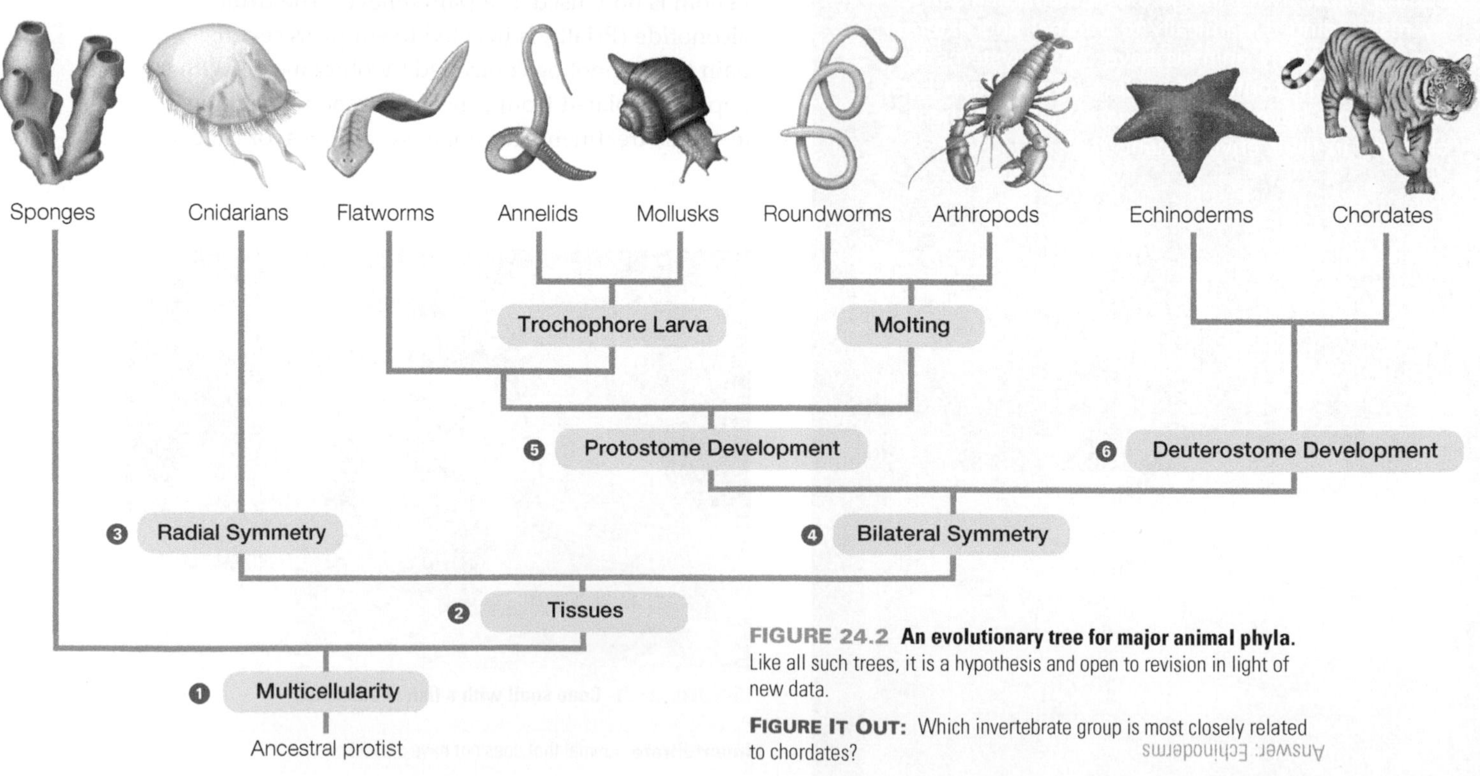
Sponges
Cnidarians
Flatworms
Annelids
Mollusks
Roundworms
Arthropods
Echinoderms
Chordates
Trochophore Larva
Molting
❺ Protostome Development
❻ Deuterostome Development
❸ Radial Symmetry
❹ Bilateral Symmetry
❷ Tissues
❶ Multicellularity
Ancestral protist

FIGURE 24.2 An evolutionary tree for major animal phyla. Like all such trees, it is a hypothesis and open to revision in light of new data.

FIGURE IT OUT: Which invertebrate group is most closely related to chordates?

Answer: Echinoderms

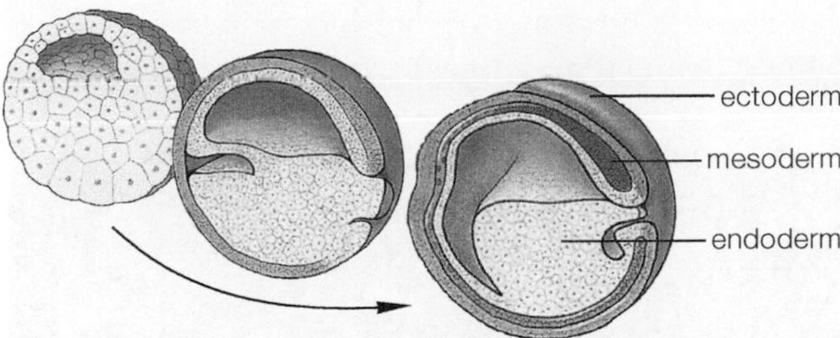
ectoderm
mesoderm
endoderm

FIGURE 24.3 How a three-layer animal embryo forms through cell movements. Most animals have a three-layer embryo.

Deuterostomes include some invertebrate animals and all vertebrates.

Animals also differ in how they digest food. In sponges, digestion is intracellular. Cnidarians and flatworms digest food inside a saclike gut called a **gastrovascular cavity**. Food enters this cavity through the same opening that expels wastes. The cavity also functions in gas exchange. Most bilateral animals have a **complete digestive tract**, which is a tubular gut with an opening at either end. Parts of the tube specialize in taking in food, digesting it, absorbing nutrients, or compacting waste. A tubular gut carries out all of these tasks simultaneously, whereas a gastrovascular cavity cannot. With a gastrovascular cavity, one load of food must be fully digested and wastes expelled before the next load of food can be ingested and processed.

In flatworms, a mass of tissues and organs surrounds the gut (FIGURE 24.4A). Most other animals have a "tube within a tube" body plan. Their gut runs through a fluid-filled body cavity. In roundworms, this cavity is partially lined with tissue derived from mesoderm and is called a **pseudocoelom** (FIGURE 24.4B). More typically, bilateral animals have a **coelom**, a body cavity fully lined with tissue derived from mesoderm (FIGURE 24.4C). Sheets of tissue called mesentery suspend the gut in the center of a coelom. Coelomic fluid cushions the gut and keeps it from being distorted by body movements. The fluid also helps distribute material through the body, and in some animals it plays a role in locomotion.

TAKE-HOME MESSAGE 24.2

- ✔ Animals are multicelled heterotrophs with bodies made of unwalled cells. Animals are motile for part or all of their life.
- ✔ Sponges have an asymmetrical body plan. Later, radially symmetrical and then bilaterally symmetrical bodies evolved.
- ✔ Most animals with bilateral symmetry have a fluid-filled coelom.
- ✔ The two lineages of bilaterally symmetrical animals, protostomes and deuterostomes, differ in the details of their development, but both develop from an embryo that has three tissue layers.

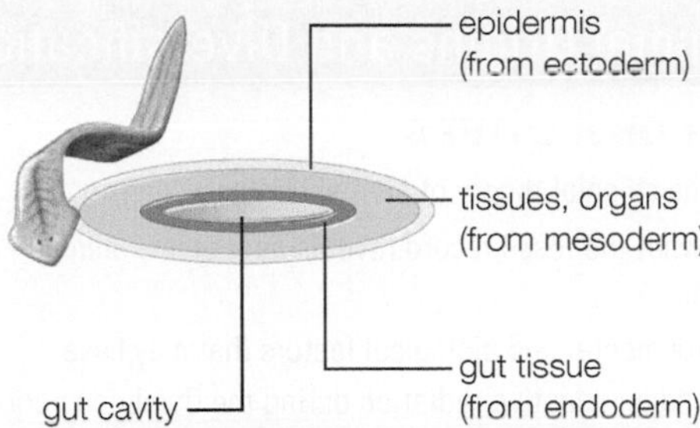
epidermis
(from ectoderm)
tissues, organs
(from mesoderm)
gut tissue
(from endoderm)
gut cavity

A Flatworms have no body cavity except their gut.

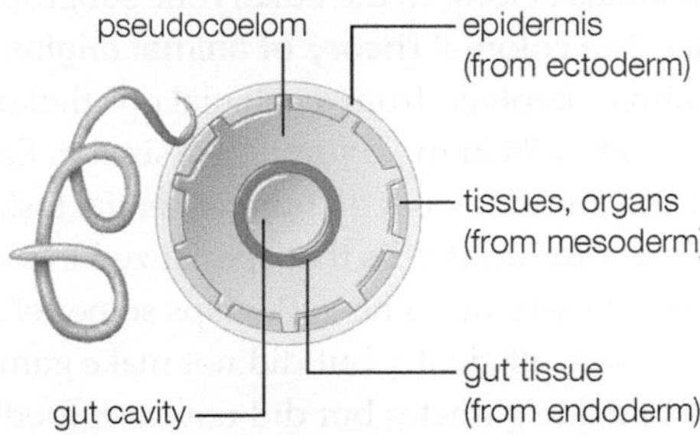
pseudocoelom
epidermis
(from ectoderm)
tissues, organs
(from mesoderm)
gut tissue
(from endoderm)
gut cavity

B Roundworms have a fluid-filled pseudocoelom.

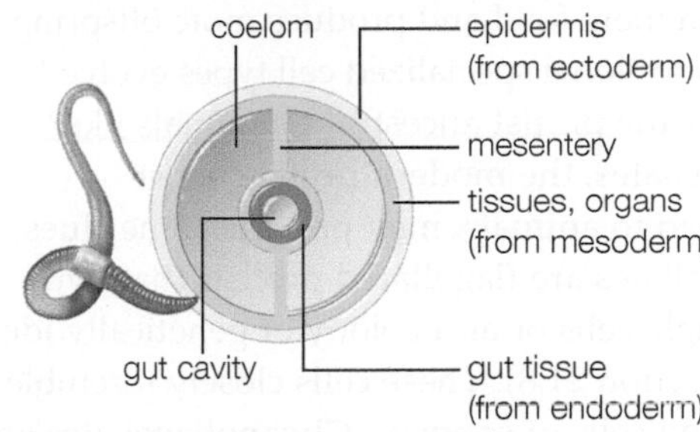
coelom
epidermis
(from ectoderm)
mesentery
tissues, organs
(from mesoderm)
gut cavity
gut tissue
(from endoderm)

C Annelids have a fluid-filled coelom with sheets of tissue (mesentery) that hold the gut in place.

FIGURE 24.4 Variations in body plans among three bilateral invertebrates. Graphics are schematics of a cross section through the body. Tissue layers are shown, but their relative width is not to scale.

animals Multicelled heterotrophs with unwalled cells; are motile during part or all of the life cycle.

bilateral symmetry Having paired structures so the right and left halves are mirror images.

cephalization (sef-uhl-iz-AY-shun) Evolutionary trend whereby nerve and sensory cells become concentrated in an animal's front end.

coelom (SEE-lum) In many animals, a body cavity that surrounds the gut and is lined with tissue derived from mesoderm.

complete digestive tract Tubular gut.

gastrovascular cavity In some invertebrates, a saclike cavity that functions in digestion and in gas exchange.

deuterostomes (DUE-ter-oh-stomes) Lineage of bilateral animals in which the second opening on the embryo surface develops into a mouth.

protostomes (PRO-toe-stomes) Lineage of bilateral animals in which the first opening on the embryo surface develops into a mouth.

pseudocoelom (SUE-doe-see-lum) Animal body cavity that is only partially lined with tissue derived from mesoderm.

radial symmetry Having parts arranged around a central axis, like spokes around a wheel.

24.3 Animal Origins and Diversification

LEARNING OBJECTIVES

- Explain the colonial theory of animal origins.
- Discuss what the fossil record reveals about early animal evolution.
- List environmental and biological factors that may have encouraged an adaptive radiation during the Cambrian period.

Colonial Origins

Like fungi, animals belong to the eukaryotic subgroup Opisthokonta. The **colonial theory of animal origins** states that animals evolved from a colonial opisthokont protist. At first, all cells in the colony were similar. Each could reproduce and carry out all other essential tasks. Later, mutations produced cells that specialized in some tasks and did not carry out others. Perhaps some cells captured food more efficiently but did not make gametes, whereas others made gametes but did not catch food. The division of labor among interdependent cells made the colony as a whole more efficient, allowing such colonies to obtain more food and produce more offspring. Over time, additional specialized cell types evolved.

What was the protist ancestral to animals like? Choanoflagellates, the modern protists most closely related to animals, may provide some clues. Choanoflagellates are flagellated protists that live either as single cells or as a colony of genetically identical cells (Section 21.8). These cells closely resemble the flagellated cells of sponges. Choanoflagellates and animals are the only living organisms that make both cadherins (a type of adhesion molecule that holds cells together) and collagen (a structural protein that is the main component of animal extracellular matrices). Evolution of these proteins in the protists may have opened the way to the later evolution of animals.

Early Animals

The oldest widely accepted evidence of animals dates to the final period of the Precambrian: the Ediacaran (635–541 million years ago). There are many Ediacaran trace fossils—rocks with trails or burrows created when unidentified early animals moved across or tunneled through sediments on the ancient seafloor. There are also a variety of body fossils (**FIGURE 24.5A**).

All known Ediacaran animals were soft bodied and aquatic. Some were unlike any modern animal. Others resemble and may be related to some modern invertebrates (**FIGURE 24.5B**). Various late Ediacaran fossils are considered possible relatives of sponges, cnidarians, annelids, mollusks, or arthropods. Thus, the origin of bilateral body plans and the divergence between protostomes and deuterostomes are thought to have occurred during the Ediacaran.

colonial theory of animal origins Hypothesis that the first animals evolved from a colonial protist.

A Reconstruction of an Ediacaran sea based on body fossils.

B Fossil of *Spriggina*, about 3 centimeters (1 inch) long. It may have been a soft-bodied ancestor of arthropods, a group that includes crabs and insects.

FIGURE 24.5 Ediacaran animals.

The Cambrian Explosion

A dramatic adaptive radiation occurred during the Cambrian (542–488 million years ago). By the end of this period, all major animal lineages were present in the seas. Environmental factors probably encouraged diversification. During this period, global climate warmed and the oxygen concentration rose in the seas. Both factors made the environment more hospitable to animal life. With rare exceptions, animals require oxygen. In modern seas, highly oxygenated waters support a greater species diversity than lower-oxygen ones. Also during the Cambrian, the supercontinent Gondwana (Section 16.5) underwent a dramatic rotation. Movement of this landmass could have isolated populations, thus increasing the likelihood that allopatric speciation events would occur.

Biological factors could also have encouraged diversification. After predatory animals arose, evolution of novel prey defenses such as protective hard parts would have been favored. Duplications and divergence of homeotic genes (Section 10.3) may have facilitated diversification. Changes in these genes have dramatic effects on body plans. Some mutations produced adaptive traits that better allowed animals to more easily escape predators or to survive in novel habitats.

TAKE-HOME MESSAGE 24.3

✔ It is thought that animals evolved from a colonial protist that resembled modern choanoflagellates. Proteins that connect animal cells together or make up the extracellar matrix between them first evolved in the protist ancestor.

✔ Some representatives of modern animal lineages are found among the Ediacarans. A great adaptive radiation of animals took place during the Cambrian period.

24.4 Sponges

LEARNING OBJECTIVES

- Describe the types of cells in a sponge body, and explain how they work together to keep the animal alive.
- Explain how sponges reproduce sexually.
- Describe how sponges can reproduce asexually and their capacity to regenerate.

Sponges (phylum Porifera) are aquatic animals that do not have tissues or organs. Most of the more than 5,000 species live in tropical seas. Some are as small as a fingertip, whereas others stand meters tall. An asymmetrical vaselike or columnar shape is most common, but some grow as a thin crust.

An adult sponge is a **sessile animal**, meaning it lives attached to a surface. The body has many pores (FIGURE 24.6). Flat, nonflagellated cells cover the body's outer surface, flagellated collar cells line the inner surface, and a jellylike extracellular matrix fills the space in between. In many sponges, cells in this matrix secrete fibrous proteins, glassy silica spikes, or both. These materials structurally support the body and discourage predation by making the sponge difficult to eat and digest. Some protein-rich sponges are harvested from the sea, dried, cleaned, and bleached. Their rubbery protein remains are used for bathing and cleaning.

The typical sponge is a **suspension feeder**, meaning it filters its food from the surrounding water. Whiplike motion of flagella on collar cells draws food-laden water through the pores in the sponge body wall. The collar cells engulf food by phagocytosis, then digest it intracellularly. Amoeba-like cells in the matrix receive breakdown products of digestion from collar cells. They distribute these nutrients to other cells in the sponge body.

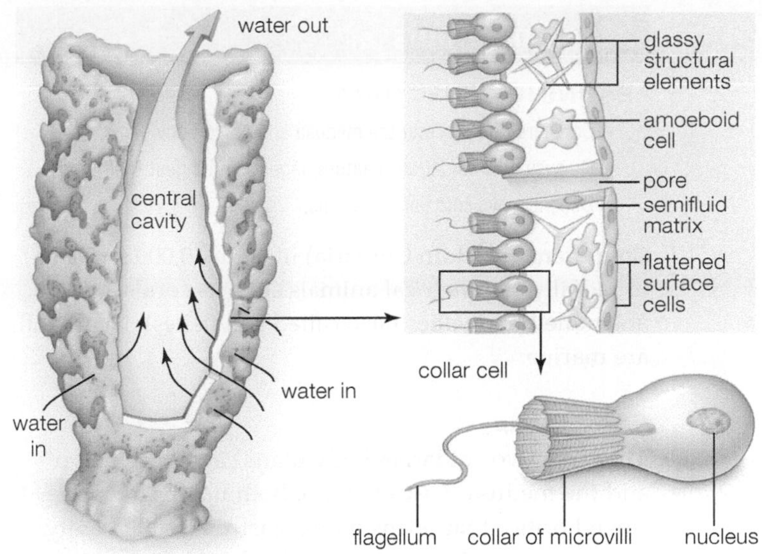
water out
central cavity
water in
water in
glassy structural elements
amoeboid cell
pore
semifluid matrix
flattened surface cells
collar cell
flagellum
collar of microvilli
nucleus

FIGURE 24.6 **Adult sponge, with a porous, asymmetrical body.**

Most sponges are **hermaphrodites**, meaning each individual can produce both eggs and sperm. Typically, a sponge releases sperm into the water but holds on to its eggs. Fertilization produces a zygote that develops into a ciliated larva. A **larva** is a young, sexually immature animal with a body form that differs from that of an adult. Sponge larvae swim briefly, then settle and develop into adults.

Of all animals, sponges have the greatest ability to regenerate. Many sponges reproduce asexually when small buds or fragments break away and grow into new sponges. Some can even recover after being broken into individual cells. When thus isolated, amoeba-like sponge cells reaggregate. Some then differentiate to become missing cell types. If cells of two different sponge species are mixed, cells of each species will seek out and reaggregate with their own kind alone.

Some freshwater sponges survive dry conditions by producing gemmules, which are tiny clumps of resting cells encased in a hardened coat. Gemmules are dispersed by the wind. Those that land in a hospitable habitat become active and grow into new sponges.

hermaphrodite (her-MAFF-roe-dite) An individual animal that produces both eggs and sperm during its life cycle.
larva Plural, larvae. In some animal life cycles, a sexually immature form that differs from the adult in its body plan.
sessile animal Animal that lives fixed in place on some surface.
sponge Aquatic invertebrate that has no body symmetry, tissues, or organs; feeds by filtering food from the water.
suspension feeder Animal that filters food from water around it.

TAKE-HOME MESSAGE 24.4

✔ Sponges are mostly marine animals that have an asymmetrical body without any tissues or organs.

✔ Ciliated larvae swim briefly, but adults are sessile suspension feeders that strain bits of food from the water that flows through their pores.

24.5 Cnidarians

LEARNING OBJECTIVES

- Compare and contrast the medusa and polyp body plans.
- Discuss how cnidarians capture, ingest, and digest their food.
- Describe two colonial cnidarians.

Cnidarians (phylum Cnidaria) include 10,000 species of radially symmetrical animals such as corals, sea anemones, and jellies (also called jellyfishes). Nearly all are marine.

Body Plans

There are two cnidarian body plans called the polyp and the medusa (FIGURE 24.7). Both have a tentacle-ringed orifice that opens onto a gastrovascular cavity. This is the only body cavity, and it functions in digestion and gas exchange. Nutrients and oxygen diffuse from the gastrovascular cavity to body cells. The **polyp** is tubular. Typically the tentacle-fringed mouth faces upward and the lower end of the body attaches to some surface. A **medusa** is shaped like a bell or umbrella, with a mouth on the lower surface. Most medusae swim or drift from place to place.

Cnidarians have two tissue layers. Their outer epidermis develops from embryonic ectoderm, and the inner gastrodermis from endoderm. Mesoglea, a jelly-like secreted matrix, lies between the tissue layers.

Cnidarians are predators that use their tentacles to sting and capture prey. The name Cnidaria is from *cnidos*, the Greek word for nettle, a kind of stinging plant. The epidermis of the tentacles has cells called **cnidocytes** that contain specialized organelles (nematocysts). These organelles function like a jack-in-the-box (FIGURE 24.8). When something brushes against a trigger, a coiled thread pops out and captures prey. Tentacles push the captured prey through the mouth, into the saclike gastrovascular cavity. Gland cells of the gastrodermis secrete enzymes that digest the prey, then digestive remains are expelled through the mouth.

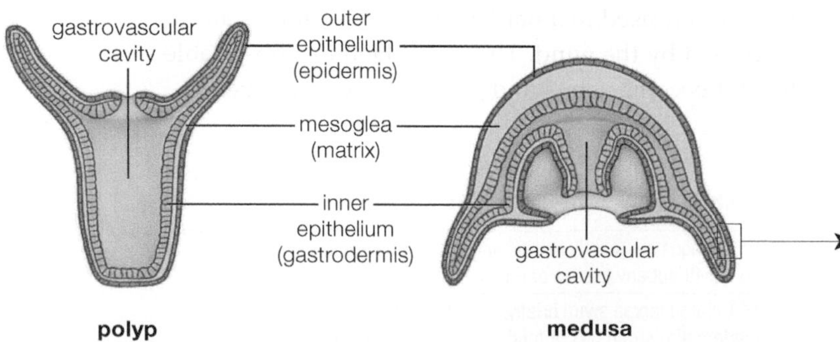
gastrovascular cavity
outer epithelium (epidermis)
mesoglea (matrix)
inner epithelium (gastrodermis)
gastrovascular cavity
polyp
medusa

FIGURE 24.7 **The two cnidarian body plans.**

Cnidarians are brainless, but interconnecting nerve cells extend through their tissues, forming a **nerve net**. Body parts move when cells of the nerve net signal contractile cells. In a manner analogous to squeezing a water-filled balloon, the contractions redistribute mesoglea or water trapped in the gastrovascular cavity. A fluid-filled cavity or cellular mass on which contractile cells exert force is a **hydrostatic skeleton**.

Diversity and Life Cycles

There are four cnidarian classes: hydrozoans, anthozoans, cubozoans, and scyphozoans.

Some marine hydrozoans such as *Obelia* have a life cycle that includes polyp, medusa, and larval stages (FIGURE 24.9). A cnidarian larva, called a planula, is ciliated and has bilateral symmetry. *Hydra*, another hydrozoan, is a freshwater predatory polyp about 20 millimeters (3/4 inch) tall (FIGURE 24.10A). There is no medusa stage, and reproduction usually occurs asexually when new polyps bud from existing ones.

Anthozoans such as corals and sea anemones also lack a medusa stage (FIGURE 24.10B). Gametes form on polyps. Coral reefs consist of colonies of polyps that enclose themselves in a framework of secreted calcium carbonate. Photosynthetic dinoflagellates (Section 21.5) in the polyp's tissues provide it with sugars.

Cubozoans and scyphozoans have a medusa-dominated life cycle. Cubozoans, commonly called box jellies, are active swimmers with structurally complex eyes, complete with a lens. Some box jellies that live in the Indian and Pacific oceans make a powerful

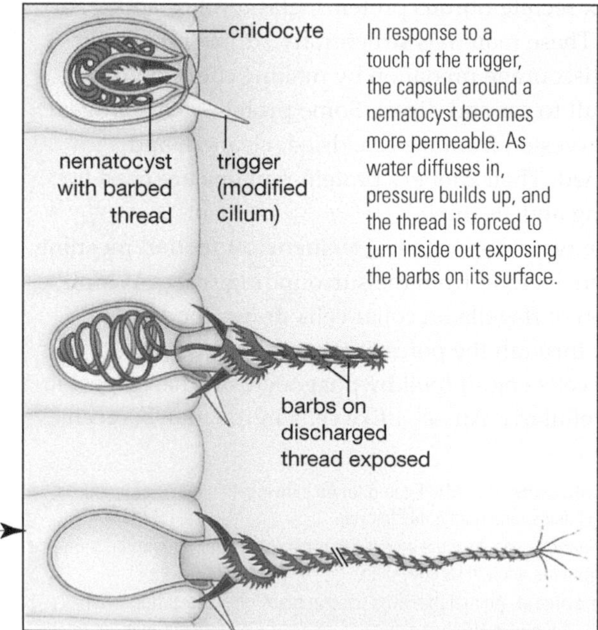
cnidocyte
In response to a touch of the trigger, the capsule around a nematocyst becomes more permeable. As water diffuses in, pressure builds up, and the thread is forced to turn inside out exposing the barbs on its surface.
nematocyst with barbed thread
trigger (modified cilium)
barbs on discharged thread exposed

FIGURE 24.8 **Example of cnidocyte action.** Cnidocytes are nematocyst-containing cells in the epidermis of cnidarian tentacles.

FIGURE 24.9 Life cycle of a hydrozoan (*Obelia*).

❶ A medusa makes and releases eggs or sperm into the water. Gametes combine and a zygote forms.

❷ The zygote develops into a ciliated bilateral larva.

❸ The larva settles and develops into a polyp.

❹ The polyp grows and reproduces asexually, eventually producing a branching colony.

❺ Some branches of the colony capture and consume prey.

❻ Other branches produce and release medusae.

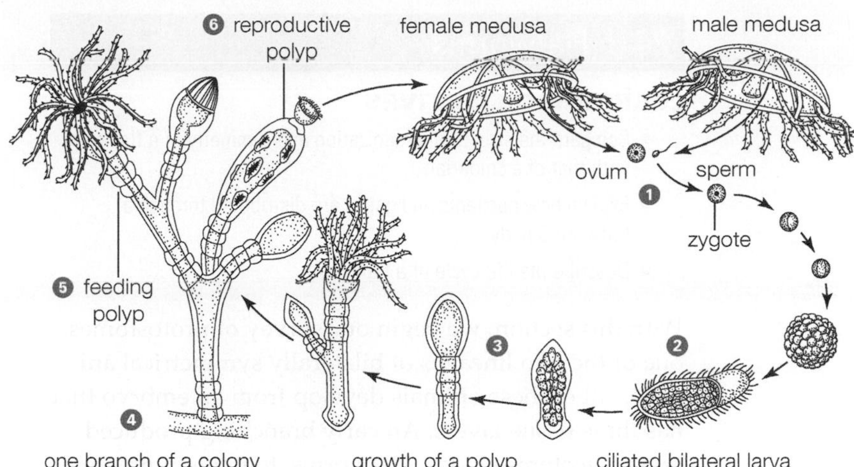

6 reproductive polyp
female medusa
male medusa
ovum
sperm
1
zygote
5 feeding polyp
2
3
4
one branch of a colony
growth of a polyp
ciliated bilateral larva

venom, and their sting can be deadly to humans (FIGURE 24.10C).

Scyphozoans include most jellies that commonly wash up on beaches. Some are harvested and dried for use as food, especially in Asia.

---

cnidarian (nigh-DARE-ee-en) Radially symmetrical invertebrate that has tentacles with stinging cells (cnidocytes).
cnidocyte (NIGH-duh-site) Stinging cell unique to cnidarians.
hydrostatic skeleton Of soft-bodied invertebrates, a fluid-filled chamber that muscles exert force against, redistributing the fluid.
medusa (meh-DUE-suh) Plural, medusae. Dome-shaped cnidarian body; fringed by tentacles.
nerve net Mesh of nerve cells with no central control organ.
polyp (PAH-lup) A pillarlike cnidarian body topped with tentacles.

TAKE-HOME MESSAGE 24.5

✔ Cnidarians are radial, typically marine predators that have two tissue layers.

✔ There are two body plans: medusa and polyp. In both, tentacles with cnidocytes surround a mouth that opens onto a gastrovascular cavity. A nerve net interacts with the hydrostatic skeleton to allow movement.

✔ Cnidarians include sea anemones, jellies, and corals. The reef-building corals form reefs by secreting calcium carbonate.

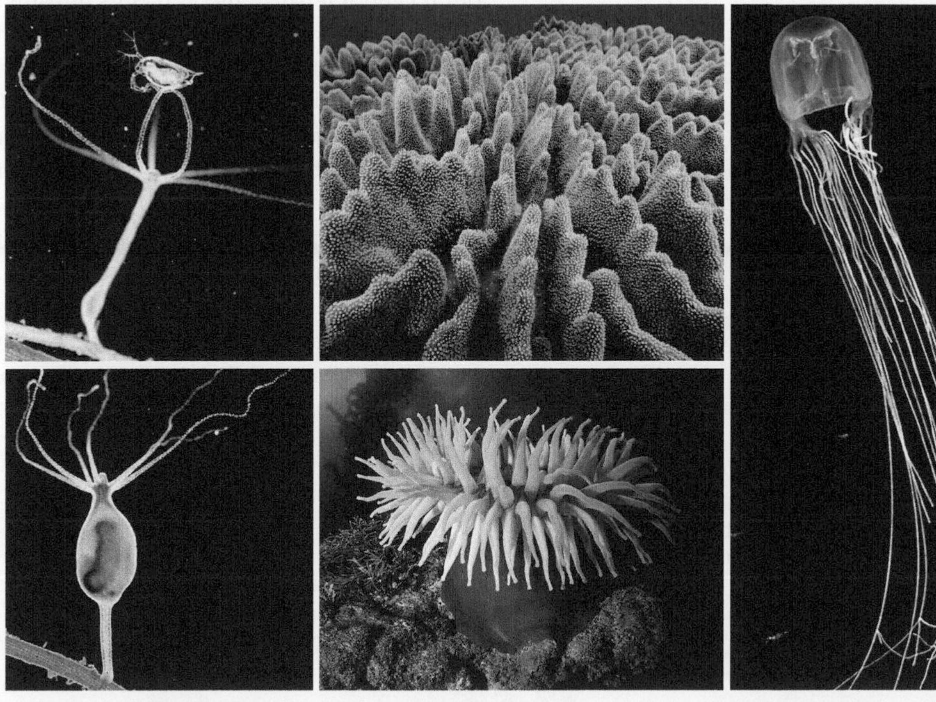

A Hydrozoan. *Hydra*, a freshwater species, capturing and digesting a water flea (a tiny crustacean).

B Anthozoans. A coral with polyps extended for feeding (top), and a sea anemone (bottom).

C Cubozoan. The box jelly *Chironex* makes a toxin that can kill a person.

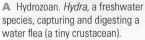

FIGURE 24.10 Representatives of the cnidarian classes.

24.6 Flatworms

LEARNING OBJECTIVES

- Compare the structural organization and symmetry of a flatworm with that of a cnidarian.
- Explain how nutrients and gases are distributed through a flatworm's body.
- Describe the life cycle of a tapeworm.

With this section, we begin our survey of protostomes, one of the two lineages of bilaterally symmetrical animals. All of these animals develop from an embryo that has three tissue layers. An early branching produced two protostome clades. Flatworms, together with annelids and mollusks, belong to the clade Lophotrochozoa. The second protostome clade, the Ecdysozoa, includes the roundworms and arthropods.

Flatworm Traits

The simplest protostomes are **flatworms** (phylum Platyhelminthes; pronounced pla-tee-hell-MIN-theez). Flatworms have a flat body with an array of organ systems, but no body cavity other than the gastrovascular cavity. Like cnidarians they rely on diffusion alone to move nutrients and gases through their body. Some flatworms are free-living and many others are parasites. Nearly all are hermaphrodites.

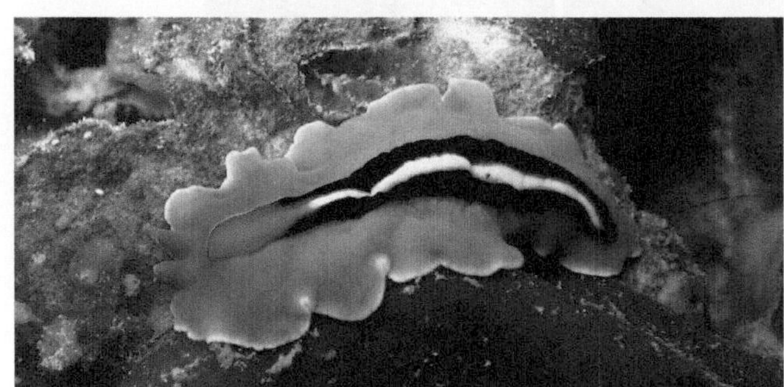

FIGURE 24.11 Marine flatworm. Brightly colored flatworms are common inhabitants of coral reef ecosystems.

Free-Living Flatworms

Most free-living flatworms live in tropical seas, and many of these are brilliantly colored (**FIGURE 24.11**). A lesser number live in fresh water, and a few live in damp places on land. Free-living flatworms can glide along, propelled by the movement of cilia that cover the body surface.

FIGURE 24.12 shows the structure of a planarian, a free-living flatworm common in ponds. A planarian takes in food and expels wastes through a muscular, tubular **pharynx**. The pharynx connects to the branched gastrovascular cavity (**FIGURE 24.12A**). There is no circulatory system. Nutrients and oxygen diffuse from the gastrovascular cavity to all body cells.

A planarian's head has chemical receptors and eyespots that detect light. These sensory structures send messages to a simple brain consisting of paired ganglia. A **ganglion** is a cluster of nerve cell bodies. Bundles of nerve fibers called **nerve cords** extend from the head and run the length of the body (**FIGURE 24.12B**).

A planarian's body fluid has a higher solute concentration than the fresh water around it, so water tends to move into the body by osmosis. A system of tubes regulates internal water and solute levels by driving excess fluids out through a pore at the body surface (**FIGURE 24.12C**).

A planarian is a hermaphrodite (**FIGURE 24.12D**), but it cannot fertilize its own eggs. Freshwater planarians typically swap sperm. By contrast, some marine flatworms battle over who will assume the male role. In a behavior described as "penis fencing," each flatworm attempts to stab its penis into a partner's body and squirt in some sperm, while fending off its partner's attempts to do the same.

Many planarians reproduce asexually, and some have an amazing capacity for regeneration. During asexual reproduction, the body splits in two near the middle, then each piece regrows the missing parts. This capacity for regrowth is also put to use replacing lost parts when a planarian is injured.

FIGURE 24.12 Organ systems of a planarian. Planarians are freshwater flatworms.

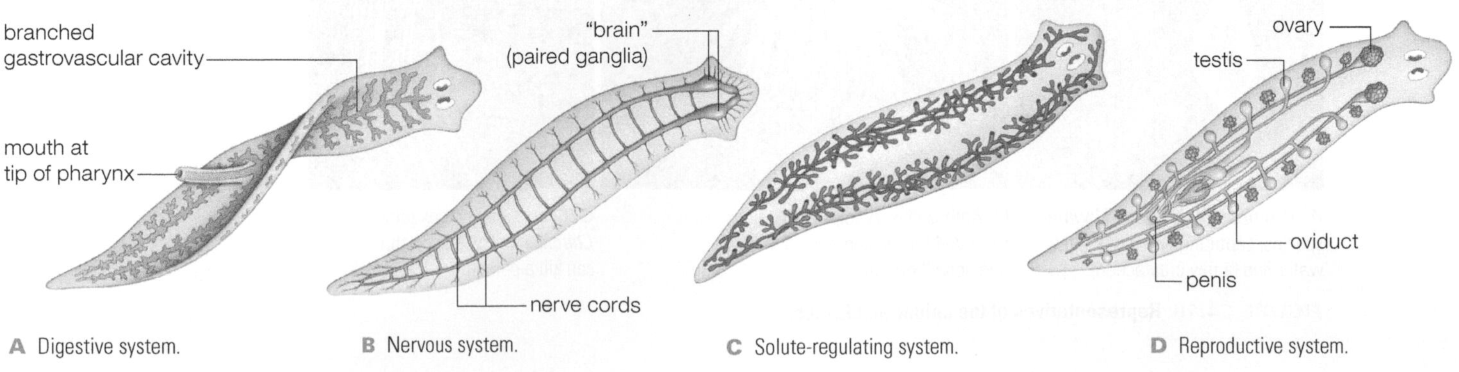
branched gastrovascular cavity
mouth at tip of pharynx
"brain" (paired ganglia)
nerve cords
ovary
testis
oviduct
penis

A Digestive system. **B** Nervous system. **C** Solute-regulating system. **D** Reproductive system.

CREDIT: (11) Kim Briers/Shutterstock.

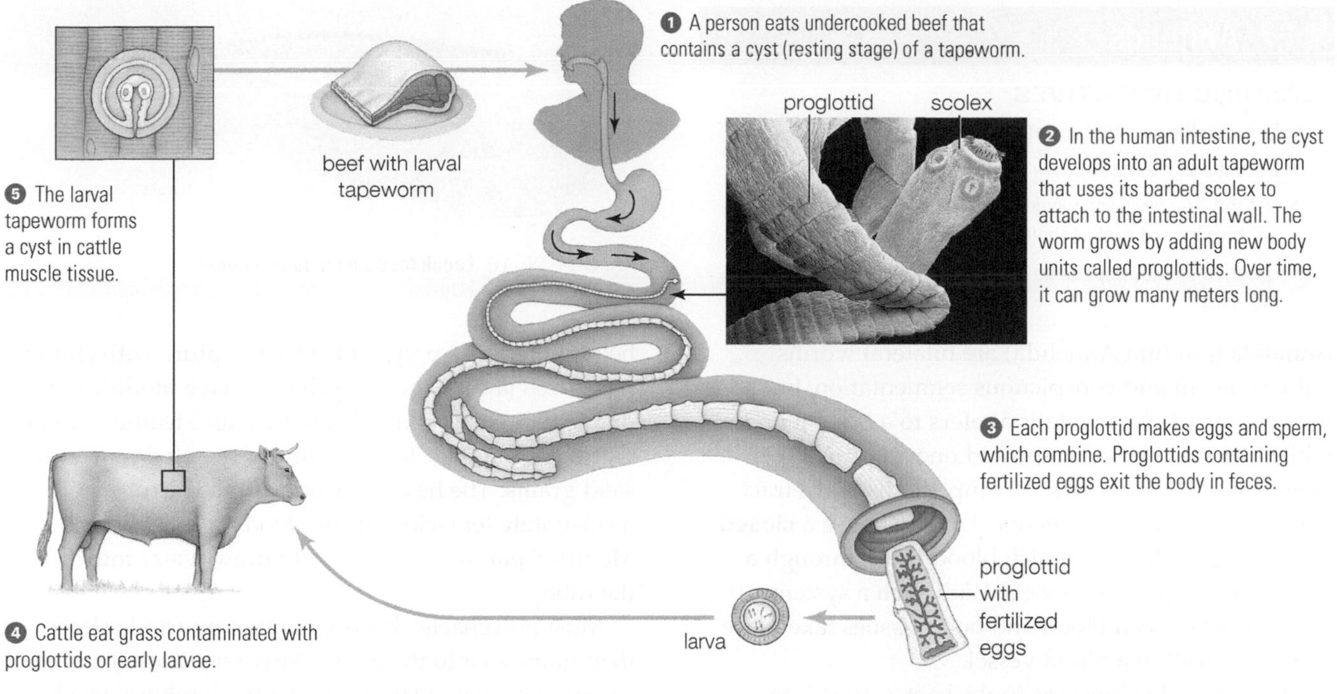
❶ A person eats undercooked beef that contains a cyst (resting stage) of a tapeworm.
proglottid
scolex
❷ In the human intestine, the cyst develops into an adult tapeworm that uses its barbed scolex to attach to the intestinal wall. The worm grows by adding new body units called proglottids. Over time, it can grow many meters long.
beef with larval tapeworm
❺ The larval tapeworm forms a cyst in cattle muscle tissue.
❸ Each proglottid makes eggs and sperm, which combine. Proglottids containing fertilized eggs exit the body in feces.
proglottid with fertilized eggs
larva
❹ Cattle eat grass contaminated with proglottids or early larvae.

FIGURE 24.13 Life cycle of a beef tapeworm.

Parasitic Flatworms

Tapeworms and flukes are parasitic flatworms. Typically, their larvae reproduce asexually in one or more intermediate hosts before developing into adults. Adults reproduce sexually in a definitive (final) host.

FIGURE 24.13 shows the life cycle of a beef tapeworm. Cattle are its intermediate host, and humans are its definitive host. The worm attaches to the wall of the human intestine by means of hooks or suckers on its scolex (head). Behind the scolex are body units called proglottids. Unlike the planarians and flukes, tapeworms do not have a gastrovascular cavity. They absorb nutrients directly across their body wall.

The blood fluke (*Schistosoma*) causes the tropical disease schistosomiasis. Free-swimming fluke larvae develop inside aquatic snails (**FIGURE 24.14A**). The larvae can cross human skin and enter the bloodstream. They travel to veins where they mature into sexually reproducing adults (**FIGURE 24.14B**). The presence of actively reproducing adult flukes in the human body results in weakness, fever, and damage to internal organs. More than 200 million people currently suffer from schistosomiasis.

flatworm Acoelomate, unsegmented worm; a planarian, fluke, or tapeworm.
ganglion (GANG-lee-on) Plural, ganglia. Cluster of nerve cell bodies.
nerve cord Bundle of nerve fibers that runs the length of the body in many invertebrates.
pharynx (FAHR-inks) Tube connecting the mouth to the digestive tract.

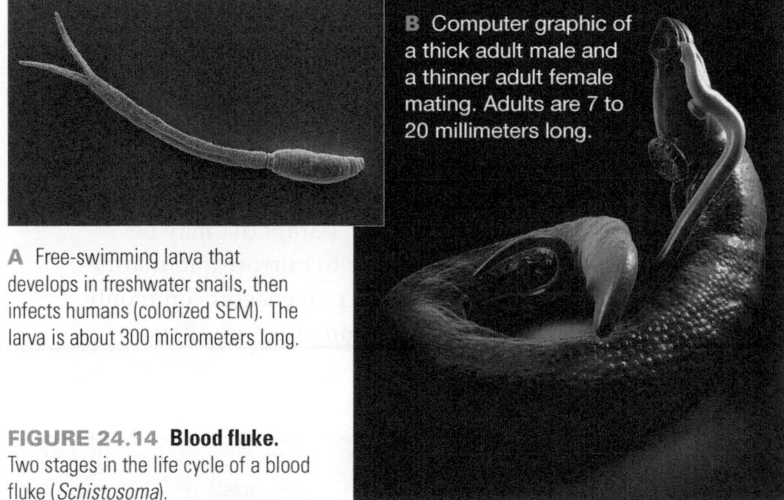

A Free-swimming larva that develops in freshwater snails, then infects humans (colorized SEM). The larva is about 300 micrometers long.

B Computer graphic of a thick adult male and a thinner adult female mating. Adults are 7 to 20 millimeters long.

FIGURE 24.14 Blood fluke. Two stages in the life cycle of a blood fluke (*Schistosoma*).

TAKE-HOME MESSAGE 24.6

✔ Flatworms are bilateral, acoelomate animals with a gastrovascular cavity, a cephalized nervous system, and a system for regulating water and solutes.

✔ Planarians live in fresh water, and some flatworms live in moist places on land. Other free-living flatworms are marine.

✔ Flukes and tapeworms are parasitic flatworms. Both groups include species that infect humans.

CREDITS: (13) Andrew Syred/Science Source; (14A) David Scharf/Science Source; (14B) Science Picture Co/Science Source.

24.7 Annelids

LEARNING OBJECTIVES

- Describe the annelid circulatory system.
- Describe the features of the three subgroups of annelids.
- List the structures you would see in a cross section through an earthworm, and explain their functions.
- Describe the developmental trait that unites annelids and mollusks.

Annelids (phylum Annelida) are bilateral worms with a coelom and conspicuous segmentation, both inside and out. Segmentation refers to a body plan in which similar units are repeated one after the other along a body's main axis. A complete digestive tract extends through all segments. Annelids have a **closed circulatory system**, in which blood flows through a continuous system of vessels. With such a system, all exchanges between blood and body tissues take place across the wall of a blood vessel.

Two annelid subgroups (polychaetes and oligochaetes) are named for the bristles called chaetae on their segments. Polychaetes have many bristles and oligochaetes have few. (*Poly–* means many; *oligo–* means few.) Polychaete bristles are composed of chitin, a nitrogen-containing polysaccharide (Section 3.3). Members of a third subgroup, the leeches, lack bristles.

Polychaetes

A typical polychaete has a pair of bristle-tipped lobed appendages called parapodia on each segment. Depending on the species, parapodia may be used to swim, to crawl, or to burrow into marine sediments. The predatory polychaetes commonly called sandworms are common in mudflats (**FIGURE 24.15A**). Their eyes and sensory tentacles help them detect prey, which they capture with chitin-reinforced jaws. Other polychaetes have modifications of this basic body plan. Fan worms and feather duster worms spend their life in a tube of secreted mucus and sand grains. The head protrudes from the tube, and its elaborate tentacles capture food (**FIGURE 24.15B**). Modified parapodia are used to draw water into the tube.

Most polychaetes have separate sexes and release their gametes into the water. After fertilization, a free-swimming ciliated larva called a trochophore develops. Some mollusks have the same type of larva, and this developmental similarity is taken as evidence of a close relationship between the annelid and mollusk lineages. Genetic studies unite the flatworms, annelids and mollusks in the clade Lophotrochozoa.

FIGURE 24.16 Leech feeding on human blood.

Leeches

Some leeches live in the ocean and damp habitats on land, but they are most common in fresh water. The leech body lacks conspicuous bristles and has a sucker at either end.

Most leeches are scavengers or predators of small invertebrates. An infamous few attach to a vertebrate, pierce its skin, and suck blood (**FIGURE 24.16**). Their saliva has a protein that keeps blood from clotting

A Sandworm (*Nereis*), an active predator.

B Feather duster worm, a suspension feeder.

FIGURE 24.15 Polychaetes.

CREDITS: (15A) Robert DeGoursey/Visuals Unlimited, Inc.; (15B) Bruce Coleman, Inc.; (16) sydeen/Shutterstock.

while the leech feeds. For this reason, doctors who reattach a severed finger or ear often apply leeches to the reattached body part. The presence of the feeding leeches prevents clots from forming inside the newly reconnected blood vessels.

Oligochaetes

Oligochaetes include species that live in marine or freshwater sediments, but the soil-dwelling earthworms are most familiar (**FIGURE 24.17**). In the United States, most earthworms observed in backyards, bait shops, and compost heaps are introduced species native to Europe. However, some undisturbed habitats retain native earthworms, including a species in Oregon that can be more than a meter in length.

An earthworm's skin is protected by a cuticle of secreted proteins and a layer of mucus. Visible grooves at the body surface correspond to internal partitions between coelomic chambers.

Gas exchange occurs across the body surface. The surface must be moist for such exchanges to occur; if an earthworm dries out, it will smother. A closed circulatory system speeds the distribution of gases. Five paired, heavily muscled blood vessels in the front of the worm function like hearts. Their contractions provide the pumping power that moves the worm's blood. Large blood vessels run the length of the worm along its the upper and lower surfaces.

A simple brain at the worm's front end connects to a nerve cord that extends along the lower length of the body. The brain coordinates locomotion and receives sensory information. Earthworms can sense light (which they avoid), detect touch and vibration, and distinguish specific odors emitted by their food.

The complete digestive tract extends the length of the body inside the coelom. Earthworms eat their way through soil or decaying material. They digest both organic debris and any microorganisms that cling to it. The worms improve soil by loosening its particles and by excreting tiny bits of organic matter that decomposers can easily break down. Excreted earthworm "castings" are sold as a natural fertilizer.

Most body segments have a pair of excretory organs. These organs, called nephridia, regulate the solute composition and volume of coelomic fluid. They take in coelomic fluid in one coelomic compartment, adjust its composition, and expel wastes and excess solutes through a pore in the body wall of the adjacent compartment.

Two sets of muscles in the body wall function in locomotion. Longitudinal muscles parallel the body's long axis, and circular muscles ring the body. The worm's coelomic fluid is a hydrostatic skeleton.

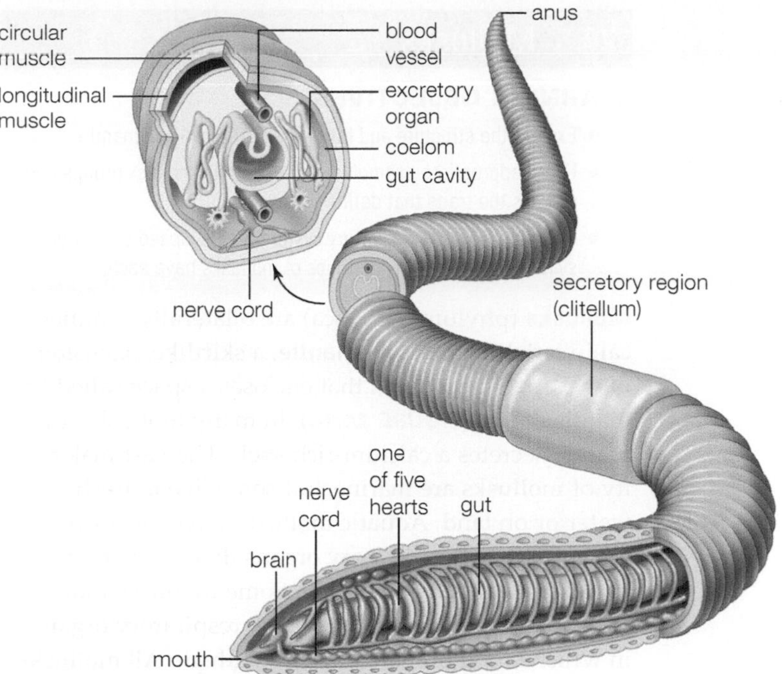
circular muscle
longitudinal muscle
blood vessel
anus
excretory organ
coelom
gut cavity
nerve cord
secretory region (clitellum)
one of five hearts
nerve cord
gut
brain
mouth

FIGURE 24.17 Earthworm body plan. Each segment has a coelomic chamber full of organs. A gut, ventral nerve cord, and dorsal and ventral blood vessels extend through all the chambers.

FIGURE IT OUT Which muscles contract to shorten a segment?

Answer: Longitudinal muscles.

Contraction of muscles puts pressure on fluid trapped inside body segments, causing them to change shape. When a segment's longitudinal muscles contract, the segment gets shorter and fatter. When circular muscles contract, a segment gets longer and thinner. Coordinated waves of contraction that run along the body propel the worm through soil.

Earthworms are hermaphrodites. During mating, a secretory organ called the clitellum produces mucus that glues two worms together while they swap sperm. Later, the same organ secretes a silky case that protects fertilized eggs as they develop in the soil.

TAKE-HOME MESSAGE 24.7

✔ Annelids are bilateral, segmented worms with a coelom. They have a complete digestive tract and a closed circulatory system.

✔ Larval similarities imply a close relationship between annelids and mollusks.

✔ Polychaetes have parapodia and most are marine. Leeches have a sucker at either end of their body and are mainly a freshwater group. A few suck blood from vertebrates. Most oligochaetes are aquatic, but the earthworms live in soil.

annelid Segmented worm with a coelom, complete digestive system, and closed circulatory system. A polychaete, leech, or oligochaete.

closed circulatory system System in which blood flows to and from a heart or hearts through a continuous series of vessels.

CREDIT: (17) art, After Solomon, 8th edition, p624, figure 29-4.

24.8 Mollusks

LEARNING OBJECTIVES

- Explain the structure and functions of the mollusk mantle.
- Using appropriate examples, name the four mollusk groups and discuss the traits that define them.
- Compare an open circulatory system with a closed circulatory system, and note which groups of mollusks have each.

Mollusks (phylum Mollusca) are bilaterally symmetrical invertebrates with a **mantle**, a skirtlike extension of the upper body wall that encloses a space called the mantle cavity (FIGURE 24.18). In many mollusks, the mantle secretes a calcium-rich shell. The vast majority of mollusks are marine, but some live in fresh water or on land. Aquatic mollusks have one or more **gills**, which are respiratory organs that facilitate the exchange of gases with water. Some mollusks that live on land have a **lung**, a saclike respiratory organ in which blood exchanges gases with air. All mollusks have a complete digestive tract and a reduced coelom.

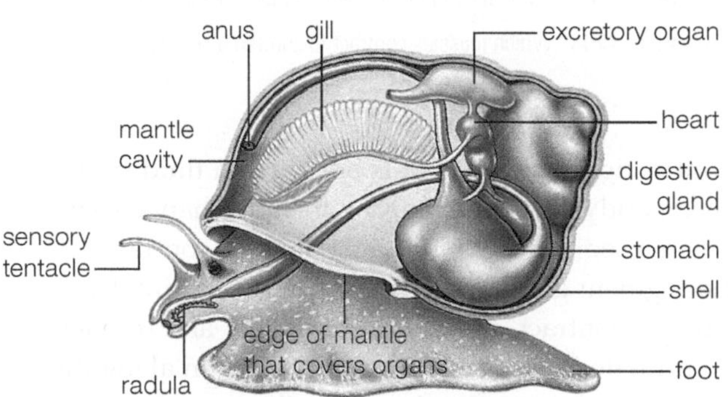
anus
gill
excretory organ
heart
mantle cavity
digestive gland
sensory tentacle
stomach
shell
edge of mantle that covers organs
radula
foot
FIGURE 24.18 **Body plan of an aquatic snail.**

A **Chitons** (overlapping plates).

B **Gastropods** (belly footed).

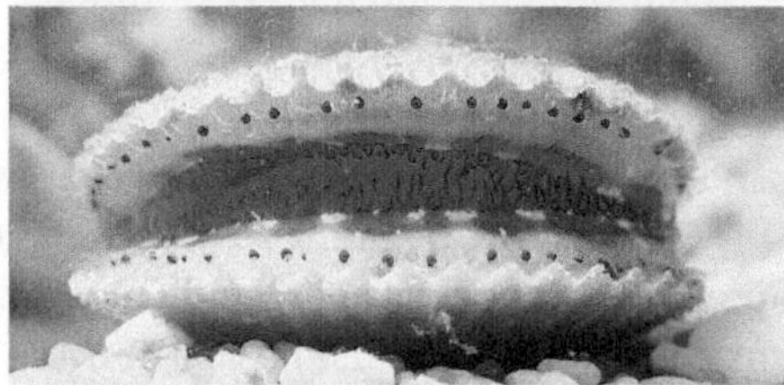
C **Bivalves** (two-part shell).

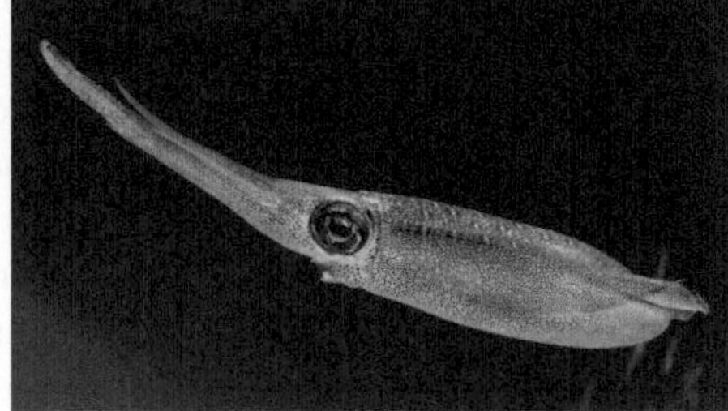
D **Cephalopods** (jet-propelled).

FIGURE 24.19 **Representatives of the four mollusk classes.**

Mollusk Diversity

With more than 100,000 living species, mollusks are second only to arthropods in diversity. There are four major classes: chitons, gastropods, bivalves, and cephalopods (FIGURE 24.19).

Chitons are probably the most similar to ancestral mollusks. All are marine and have a dorsal shell that consists of eight plates (FIGURE 24.19A). Chitons cling to rocks and scrape up algae using their **radula**, which is a tonguelike organ hardened with chitin. Many types of mollusks have a radula.

Like gastropods and bivalves, chitons have an **open circulatory system**, in which fluid leaves vessels and seeps among tissues before returning to the heart. Such a system requires less energy to operate than a closed circulatory system, in which blood is always confined within vessels or a heart. However, an open circulatory system moves materials more slowly than a closed one.

With more than 65,000 species of snails and slugs, the gastropods are the most diverse mollusks. Their name means "belly foot." Most species glide about on a broad muscular foot that makes up most of the lower body mass (FIGURES 24.18 and 24.19B). A gastropod shell, when present, is one-piece and often coiled.

Gastropods have a distinct head that usually has eyes and sensory tentacles. In many aquatic species, a part of the mantle forms an inhalant siphon, a tube

gill Respiratory organ that facilitates gas exchanges with water.
lung Saclike respiratory organ in which blood exchanges gases with the air.
mantle In mollusks, extension of the body wall.
mollusk Invertebrate with a reduced coelom and a mantle, includes chitons, bivalves, gastropods, and cephalopods.
open circulatory system Circulatory system in which fluid leaves vessels and mingles with tissue fluid before returning to the heart.
radula In many mollusks, a tonguelike organ hardened with chitin.

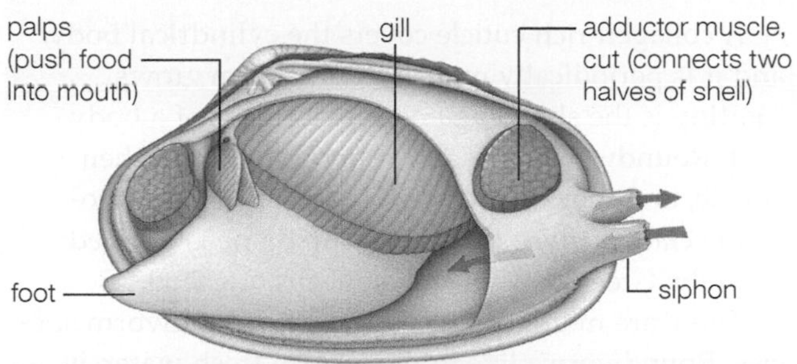
palps
(push food
into mouth)
gill
adductor muscle,
cut (connects two
halves of shell)
foot
siphon

FIGURE 24.20 Body plan of a clam, a bivalve.

through which water is drawn into the mantle cavity. The cone snails discussed in Section 24.1 use their siphon to sniff out prey. Cone snails are predatory, harpooning prey with a modified radula, but most of the gastropods are herbivores.

Nudibranchs, commonly called sea slugs, are marine gastropods with a greatly reduced shell. Nudibranch means "naked gills." Many nudibranchs defend themselves from predators using defensive weapons obtained from their prey. Sponge-eating nudibranchs store sponge toxins and secrete them when disturbed. Cnidarian-eating nudibranchs store the cnidarians' stinging cells in outpouchings on their body. Any predator that bites such a nudibranch is stung when the stored cnidocytes discharge.

Gastropods include the only terrestrial mollusks. Glands on the foot continually secrete mucus that protects the animal as it moves across dry, abrasive surfaces. Most mollusks have separate sexes, but land dwellers are typically hermaphrodites. Unlike other mollusks, which produce a swimming trochophore larva, these groups develop directly into adults.

Bivalves include many of the mollusks that end up on our dinner plates, including mussels, oysters, clams, and scallops (**FIGURE 24.19C**). All bivalves have a hinged, two-part shell. Powerful adductor muscles connect the two parts (**FIGURE 24.20**). Contraction of these muscles pulls the shell shut, enclosing the body and protecting it from predation or drying out. A bivalve has a reduced head, but eyes arrayed around the edge of its mantle alert it to danger.

Bivalves have a large triangular foot that most use to burrow. A clam burrows beneath sand and extends its siphons into the water. Like other bivalves, it lacks a radula. It feeds by drawing water into its mantle cavity and trapping bits of food in mucus on its gills. Waving cilia on the gills direct particle-laden mucus to a pair of fleshy extensions called palps that sort out particles and sweep food into the mouth.

Squids (**FIGURE 24.19D**), nautiluses (**FIGURE 24.21A**), octopuses (**FIGURE 24.21B**), and cuttlefish are cephalopods. The name of this group means "head-footed," and their foot has been modified into arms and/or tentacles that extend from the head. All cephalopods are predators and most have beaklike, biting mouthparts in addition to a radula.

A Chambered nautilus, one of six living nautilus species. The shell consists of many gas-filled chambers. The numerous tentacles do not have suckers.

B Octopus. Its eight arms are covered with many individually controlled suckers that allow it to hold on to objects, and to adhere to or walk along surfaces.

FIGURE 24.21 Cephalopods.

Cephalopods move by jet propulsion. They draw water into the mantle cavity, then force it out through a funnel-shaped siphon. A closed circulatory system supports their high activity level.

Cephalopods include the fastest (squids), biggest (giant squid), and smartest (octopuses) invertebrates. Of all invertebrates, octopuses have the largest brain relative to body size, and the most complex behavior.

TAKE-HOME MESSAGE 24.8

✔ Mollusks are invertebrates with a bilateral body plan, a reduced coelom, and a mantle that drapes over their internal organs. In most species, the mantle secretes a protective hardened shell.

✔ Most mollusks are aquatic, but some gastropods have adapted to life on land. In addition to gastropods, mollusks include chitons, bivalves, and cephalopods.

24.9 Roundworms

LEARNING OBJECTIVES

- Describe the roundworm body plan.
- Explain why roundworms are used as models in scientific studies.
- Describe the ecological roles played by roundworms.

Roundworms, or nematodes (phylum Nematoda), are unsegmented worms that have a pseudocoelom, a complete digestive tract, and a nervous system. There are no circulatory or respiratory organs (**FIGURE 24.22**). Sexes are separate.

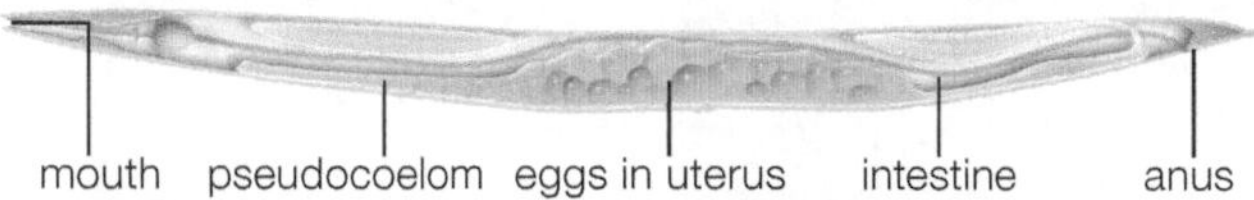
mouth
pseudocoelom
eggs in uterus
intestine
anus

FIGURE 24.22 Body plan of a free-living roundworm.

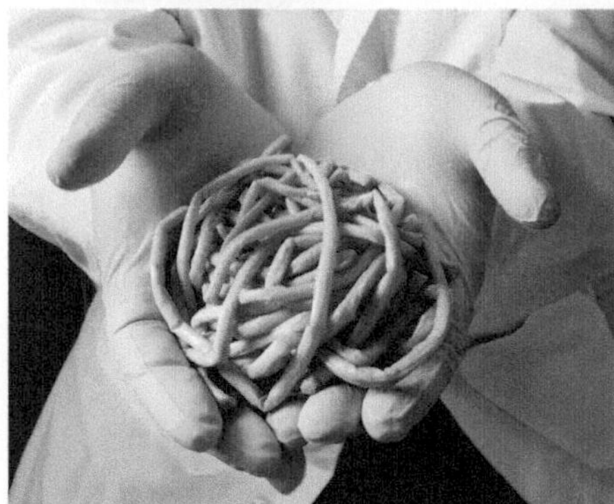

A Intestinal parasite, *Ascaris* passed by a child.

B The grossly enlarged leg of the man on the left is a sign of lymphatic filariasis.

C Plant-infecting roundworm entering a root.

FIGURE 24.23 Parasitic roundworms.

A collagen-rich cuticle covers the cylindrical body, and it is periodically molted as the worm grows. **Molting** is the shedding and replacement of a body part. Roundworms and arthropods both molt their cuticle, and they are grouped together in the protostome clade Ecdysozoa. The group name is derived from the Greek *ekdusis*, which means "to shed."

There are more than 25,000 named roundworm species. Roundworms live in the seas, in fresh water, in damp soil, and inside other animals. Most roundworms are free-living decomposers less than a millimeter long. Parasitic species tend to be larger. One species that parasitizes whales can be up to 8 meters long.

The soil roundworm *Caenorhabditis elegans* is frequently used in scientific studies. It serves as a model for developmental processes because it has the same tissue types as more complex organisms, but it is transparent, has less than 1,000 body cells, reproduces fast, and has a small genome. It was the first multicelled organism to have its genome sequenced. Studies involving *C. elegans* have revealed how death of specific cells shapes its organs during development, and how mutations in some genes can extend life span.

Several kinds of parasitic roundworms infect humans. The intestinal parasite *Ascaris lumbricoides* (**FIGURE 24.23A**) currently infects more than one billion people. Most of those affected live in developing tropical nations, but occasional infections occur in rural parts of the American Southeast. Pinworms (*Enterobius vermicularis*) infect children worldwide. The small worms, about the size of a staple, live in the rectum. Mosquito-transmitted roundworms cause a disfiguring tropical disease called lymphatic filariasis. The worms travel in the body's lymph vessels and destroy the vessels' valves so lymph pools in the lower limbs (**FIGURE 24.23B**). Elephantiasis, the common name for this disease, refers to fluid-filled, "elephant-like" legs.

Parasitic roundworms also infect our livestock and pets. A roundworm that lives in pigs can also infect humans who eat undercooked pork, causing the disease trichinosis. Dogs are susceptible to infection by heartworms, which are roundworms transmitted by mosquitoes. In cats, roundworm infections usually occur after the cat feeds on an infected rodent that serves as an intermediate host.

Many crop plants are susceptible to nematode parasites. In some cases, the worms suck on plant roots and in others they actually enter the plant (**FIGURE 24.23C**).

molting Shedding and replacement of an animal body part.
roundworm Nematode. An unsegmented pseudocoelomate worm with a cuticle that is molted periodically as the animal grows.

Either way, the infection stunts the plant's growth and lowers crop yields.

TAKE-HOME MESSAGE 24.9

✔ Roundworms are unsegmented, pseudocoelomate worms with a cuticle that they molt as they grow.

✔ Most roundworms are decomposers, but some are parasites. One free-living species is often used by scientists in studies development.

24.10 Arthropods

LEARNING OBJECTIVES

- Describe the features that have contributed to the success of the arthropods.
- List the living arthropod subgroups and the features of each group.
- Using appropriate examples, discuss the ecological roles of arthropods and their economic and health effects on humans.

Arthropods (phylum Arthropoda) are the most diverse animal phylum. Modern representatives include spiders, lobsters, crabs, centipedes, and insects.

All arthropods are bilaterally symmetrical animals that have a reduced coelom, a complete digestive tract, and an open circulatory system, as well as respiratory and excretory organs. Sexes are usually separate, although hermaphrodites occur in some groups.

Key Arthropod Adaptations

A variety of traits have contributed to the evolutionary success of arthropods.

Hardened Exoskeleton with Jointed Appendages A tough cuticle stiffened by chitin (Section 3.3) protects and supports the arthropod body. An arthropod cuticle is an **exoskeleton**, an external skeleton to which muscles attach. The exoskeleton does not inhibit movement because it is thin at the joints where two body parts connect. *Arthropod* means jointed leg. Because the exoskeleton consists of secreted materials and does not grow, arthropods must periodically molt their exoskeleton and replace it (FIGURE 24.24).

The arthropod exoskeleton took on additional functions among arthropod groups that live on land. In these groups, it supports the body against the force of gravity and helps conserve water.

Specialized Segments In early arthropods, the body segments were distinct, and all appendages were similar. In many of their descendants, the segments became fused into structural units such as a head, a thorax (midsection), and an abdomen (hind section). Appendages on segments also became modified for special tasks. In some species, including American lobsters, the appendages on one segment evolved into large, muscular claws that can crush food (FIGURE 24.25). In flying insects, one or two body segments have wings.

Well-developed sensory structures on some segments allow arthropods to monitor their environment. Most arthropods have one or more pairs of eyes. Insects and crustaceans have **compound eyes**, which consist of many individual units, each with a lens. Such eyes excel at detecting movement. Most arthropods also have paired **antennae** that can detect touch and waterborne or airborne chemicals.

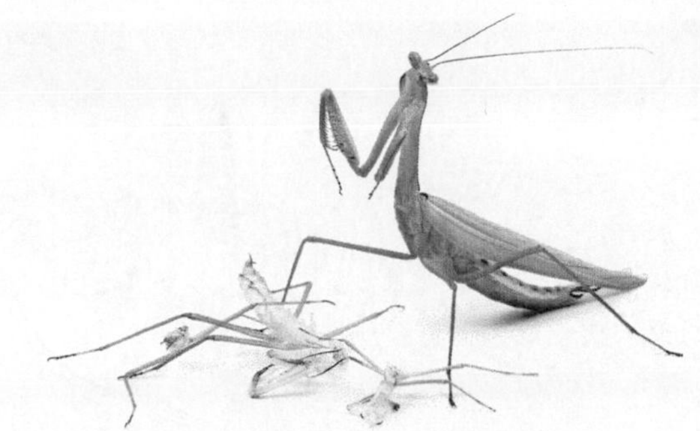

FIGURE 24.24 **Arthropod exoskeleton.** A praying mantis beside the chitin-rich exoskeleton that it recently molted.

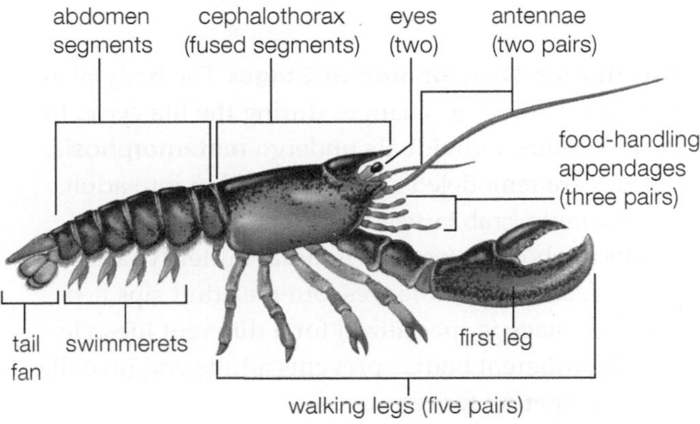
abdomen segments
cephalothorax (fused segments)
eyes (two)
antennae (two pairs)
food-handling appendages (three pairs)
tail fan
swimmerets
first leg
walking legs (five pairs)

FIGURE 24.25 **Body plan of an American lobster.**

antenna Plural antennae. Of some arthropods, sensory structure on the head that detects touch and odors.
arthropod Invertebrate with jointed legs and a hard exoskeleton that is periodically molted; for example, an insect or crustacean.
compound eye Of some arthropods, an eye that consists of many individual units; each with a lens; excels at detecting movement.
exoskeleton Of some invertebrates, hard external parts that muscles attach to and move.

FIGURE 24.26 **Fossil trilobite.** Over 20,000 species of this now-extinct group have been identified from fossils.

FIGURE 24.27 **Horseshoe crabs.** These marine chelicerates come ashore once a year to mate and lay their eggs.

Specialized Developmental Stages The body plan of many arthropods changes during the life cycle. In many groups, individuals undergo **metamorphosis**: Tissues are remodeled as larvae develop into adults. For example, crab larvae float in the plankton, but adults are bottom-feeders. A larval butterfly (a caterpillar) feeds on plant leaves, but the adult sips nectar. Each stage is specialized for a different lifestyle. Having different bodies prevents adults and juveniles from competing for resources.

Arthropod Diversity

Trilobites, a now-extinct lineage, were the most abundant and diverse arthropods in Cambrian seas, and they left many fossils (FIGURE 23.26). Trilobites became extinct about 250 million years ago. Despite this loss, arthropods remain the most diverse animal phylum. There are four major subphyla: chelicerates, myriapods, crustaceans, and insects.

Chelicerates The **chelicerates** (subphylum Chelicerata) have a body with two regions, a cephalothorax (fused head and thorax) and an abdomen. Walking legs attach to the cephalothorax. The head has eyes, but no antennae. Paired feeding appendages near the mouth, called chelicerae, give the group its name.

Four species of horseshoe crabs are the only marine chelicerates and members of the oldest surviving arthropod lineage. Horseshoe crabs are bottom feeders that eat clams and worms. A horseshoe-shaped shield covers the cephalothorax, and the last segment has evolved into a long spine. Each spring, large num-

Data Analysis Activities

Sustainable Use of Horseshoe Crabs Horseshoe crab blood clots immediately upon exposure to bacterial toxins, so it can be used to test injectable drugs for the presence of dangerous bacteria. To keep horseshoe crab populations stable, blood is extracted from captured animals, which are then returned to the wild. Concerns about the survival of animals after bleeding led researchers to do an experiment. They compared survival of animals captured and maintained in a tank with that of animals captured, bled, and kept in a similar tank. FIGURE 24.28 shows the results.

1. In which trial did the most control crabs die? In which did the most bled crabs die?
2. Looking at the overall results, how did the mortality of the two groups differ?
3. Based on these results, would you conclude that bleeding harms horseshoe crabs more than capture alone does?

	Control Animals		Bled Animals	
Trial	Number of crabs	Number that died	Number of crabs	Number that died
1	10	0	10	0
2	10	0	10	3
3	30	0	30	0
4	30	0	30	0
5	30	1	30	6
6	30	0	30	0
7	30	0	30	2
8	30	0	30	5
Total	200	1	200	16

FIGURE 24.28 **Mortality of young male horseshoe crabs kept in tanks during the 2 weeks after their capture.** Half the animals were bled on the day of their capture. Control animals were handled, but not bled. This procedure was repeated 8 times with different sets of horseshoe crabs.

abdomen
cephalothorax
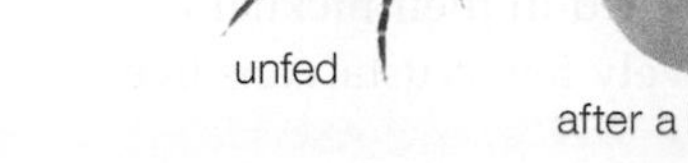
unfed
after a blood meal

A Spider, a predator. B Scorpion, a predator. C Tick, a parasite of vertebrates. D Mite, a predator as an adult; larvae suck fluid from insects.

FIGURE 24.29 Arachnids. All have four pairs of walking legs.

bers of Atlantic horseshoe crabs come ashore on sandy beaches to lay their eggs (**FIGURE 24.27**). The eggs feed millions of migratory shorebirds.

The **arachnids**, a group of terrestrial chelicerates, include spiders, scorpions, ticks, and mites (**FIGURE 24.29**). Nearly all live on land. All have four pairs of walking legs, and a pair of appendages (pedipalps) between the chelicerae and the front legs.

Scorpions and spiders are venomous predators. Spider chelicerae are fanglike and have venom glands. Of 46,600 spider species, about 30 produce venom harmful to humans. Most spiders benefit us by eating insect pests. Some weave prey-catching webs made of silk that is exuded from glands on their abdomen. Others such as tarantulas (**FIGURE 24.29A**) are active hunters. Scorpions hunt prey and dispense venom through a stinger on their last segment (**FIGURE 24.29B**). Their pedipalps have evolved into claws.

Ticks are parasites of vertebrates (**FIGURE 24.29C**). They pierce a host's skin with their chelicerae, and then suck blood. Some are vectors for the pathogens that cause Lyme disease or other human diseases.

Mites are the smallest arachnids (**FIGURE 24.29D**); most are less than a millimeter long. Dust mites are scavengers that can cause problems for people who are allergic to their feces. Other mites are predators or parasites. Some larval mites, commonly called chiggers, attach to human skin at hair follicles and sip our tissue fluids. Mites that burrow beneath the skin cause scabies in humans and mange in dogs.

Myriapods Centipedes and millipedes are **myriapods** (subphylum Myriapoda). These nocturnal ground dwellers have an elongated body composed of many similar segments (**FIGURE 24.30**). The head has a pair of antennae and two simple eyes. Myriapod means "many feet," and aptly describes these animals.

Centipedes have a low-slung, flattened body with a single pair of legs per segment, for a total of 30 to 350 (**FIGURE 24.30A**). They are fast-moving predators. Their first pair of legs has become modified as fangs that inject paralyzing venom. Most centipedes prey on insects, but some large tropical species capture and eat small vertebrates, including lizards, rodents, and birds.

Millipedes are slower-moving animals that feed on decaying vegetation. They have a cylindrical body with a cuticle hardened by calcium carbonate (**FIGURE 24.30B**). As a result of segment fusion during development, adults have two pairs of legs per segment. Most have a few hundred pairs of legs. Because millipedes are not predators, they do not produce venom. However, they do make and release toxic compounds as a defense. When threatened, a millipede curls up and secretes a foul-smelling, unpalatable fluid.

Myriapods have a long history. The earliest known fossil of a land-dwelling, air-breathing animal is of a myriapod that lived 428 million years ago. Another

A Centipede, a speedy predator.

B Millipede, a scavenger of decaying plant material.

FIGURE 24.30 Myriapods. All have two antennae and many legs.

arachnids (uh-RAK-nidz) Land-dwelling chelicerate arthropods with four pairs of walking legs; spiders, scorpions, mites, and ticks.

chelicerates (kell-ISS-er-ates) Arthropod group that has specialized feeding structures (chelicerae) and no antennae. Includes the horseshoe crabs and arachnids.

metamorphosis Remodeling of body form during the transition from larva to adult.

myriapods (MEER-ee-uh-pods) Land-dwelling arthropod group with two antennae and an elongated body with many segments; the millipedes and centipedes.

CREDITS: (29A) Eric Isselee/Shutterstock; (29B) wacpan/Shutterstock; (29C) Sarah2/Shutterstock; (29D) iStock.com/KozyrevAnton; (30) Eric Isselee/Shutterstock.

extinct myriapod is one of the largest known arthropods. It lived during the Carboniferous and reached more than 2.5 meters (8 feet) in length.

Crustaceans Most aquatic arthropods are **crustaceans** (subphylum Crustacea). Their remarkable diversity and abundance of this group is reflected in their nickname "the insects of the seas." Relatively few crustaceans live on land.

Like chelicerates, crustaceans have two distinct body regions, a cephalothorax and an abdomen. In many crustaceans, the exoskeleton is stiffened in some regions by incorporation of calcium carbonate. The head has two pairs of antennae. Some species have simple eyes, and others compound eyes.

You are probably familiar with some of the decapod crustaceans. This group of bottom-feeding scavengers includes the lobsters, crayfish, crabs, and shrimps. *Decapod* means "ten-footed," and decapods typically have five pairs of walking legs. In some lobsters, crayfish, and crabs, the first pair of legs has become modified into claws.

An aquatic crustacean starts life as a microscopic, planktonic larva that typically bears little resemblance to the adult animal. The transition from larval form to adult form takes place over the course of several molts. For example, larvae of Dungeness crabs harvested along the Pacific coast of North America live as members of the plankton for months and molt repeatedly before taking on their adult form (FIGURE 24.31A).

Krill and copepods are small swimmers that feed on plankton. They in turn serve as a major food source for larger animals. Krill (euphausiids) are relatives of the decapods and have a shrimplike body up to 6 centimeters long (FIGURE 24.31B). They swim through cool marine waters worldwide. Krill have historically been so abundant that a 30-ton blue whale could sustain itself largely on the krill that it filtered from the water. Recent overharvesting of krill from North Pacific and Antarctic waters has caused steep population declines and may endanger the animals that depend on this resource. Humans harvest krill to feed farm-raised salmon and as a source of omega-3 fatty acids.

Free-swimming copepods abound in seas and freshwater habitats (FIGURE 24.31C). They are probably the most numerous of all crustaceans. Their name means "oar feet" and refers to the oarlike motion of two swimming legs. Adults are typically about 1 to 2 millimeters long, although parasitic species can be larger. Some copepod species have a negative effect on human

FIGURE 24.31 Crustaceans. See also Figure 24.25.

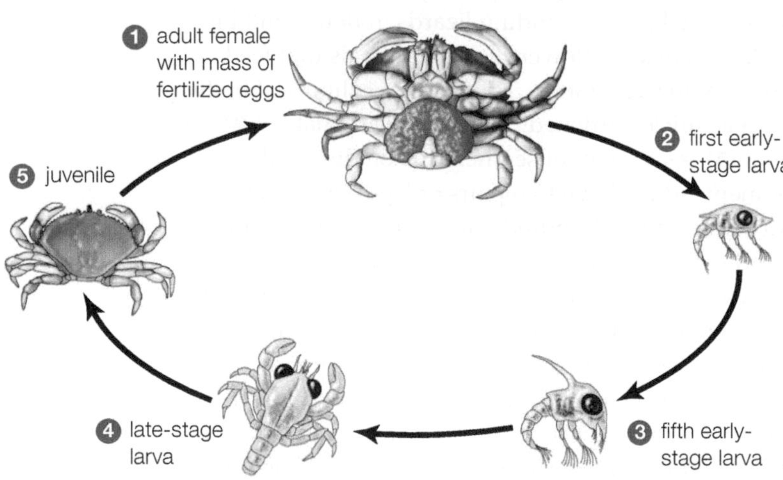

1 adult female with mass of fertilized eggs
2 first early-stage larva
3 fifth early-stage larva
4 late-stage larva
5 juvenile

A Development of the Dungeness crab (*Cancer magister*). The crabs live for 8–10 years.

1 After a female crab has mated, she fertilizes eggs with stored sperm, then holds them under her abdomen for 2–3 months until they hatch.

2 Within an hour of hatching, an early-stage planktonic larva (called a zoea) develops.

3 The early-stage larva grows and molts repeatedly, altering its form only slightly.

4 The fifth molt of the early-stage larva produces a late-stage planktonic larva with an altered body form (a megalops).

5 The late-stage larva molts to become a bottom-dwelling juvenile crab. Growth and repeated molting will produce a sexually mature adult in about two years.

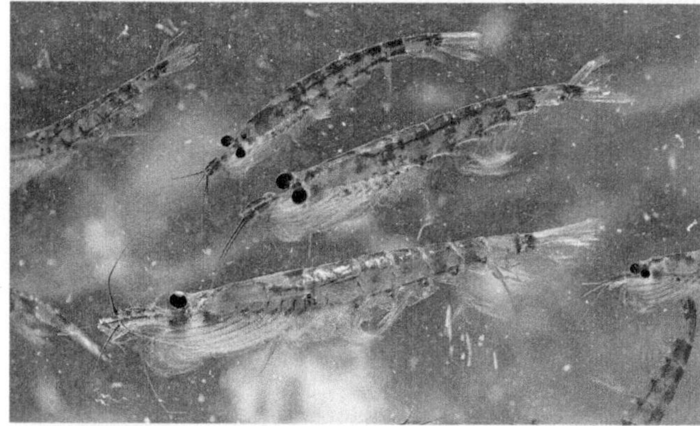

B Antarctic krill.

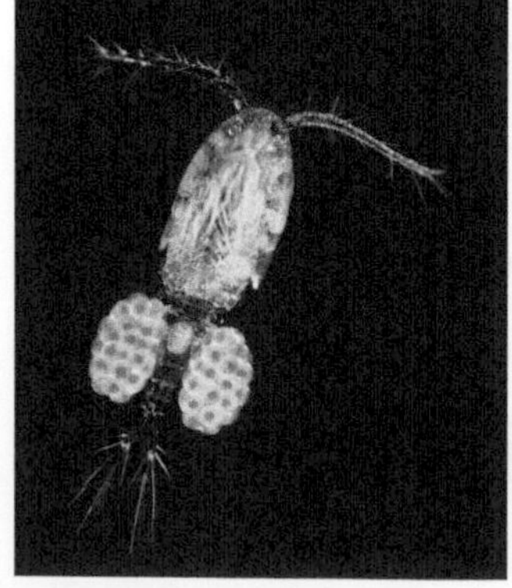

C Copepod.

D Barnacle.

health by serving as the intermediate host for bacteria that cause the disease cholera.

Larval barnacles are free-swimming plankton feeders. An adult lives fixed in place within a thick calcified shell that it secreted and filters food from the water with its feathery legs (FIGURE 24.31D). Some barnacles are notable for the length of their penis, which can be eight times that of their body.

Isopods are a mostly marine group, but you are probably familiar with the species that have adapted to life on land. Sow bugs and pill bugs are often found beneath flowerpots or rotting logs. They require a damp habitat where they can feed on organic debris or on soft, young plant parts. Some can be pests of agricultural crops such as soybeans. The species commonly known as pill bugs defend themselves from threats by rolling into a ball (left).

Insect Traits and Diversity

Insects (subphylum Hexapoda) are the most diverse arthropod lineage. Scientists recently estimated that there may be as many as 5.5 million insect species.

An insect such as a grasshopper has a three-part body plan, with a head, thorax, and abdomen (FIGURE 24.32). The head has one pair of antennae and two compound eyes. Near the mouth are jawlike mandibles and other feeding appendages. Three pairs of walking legs attach to the thorax.

In some insect groups, one or two pairs of wings also attach to the thorax. Insects are the only winged invertebrates, and the ability to fly gives them dispersal abilities unrivaled among other land invertebrates. Unlike the wings of birds, insect wings are not modified limbs. Insect wings evolved from outgrowths of the exoskeleton and do not contain muscle. or bone.

Insects are overwhelmingly terrestrial. The respiratory system consists of branching tracheal tubes that convey air from openings across the exoskeleton to tissues deep inside the body. An insect abdomen contains digestive organs, sex organs, and water-conserving excretory organs called **Malpighian tubules**. These tubules take up coelomic fluid and deliver wastes from that fluid to the terminal portion of the digestive tract for excretion.

Until recently, insects were thought to be close relatives of myriapods. Both of these arthropod groups have a single pair of antennae and unbranched legs. However, new gene comparisons made scientists rethink the connections. The currently favored hypothesis holds that insects are most closely related to crustaceans. Specifically, insects are thought to be descended

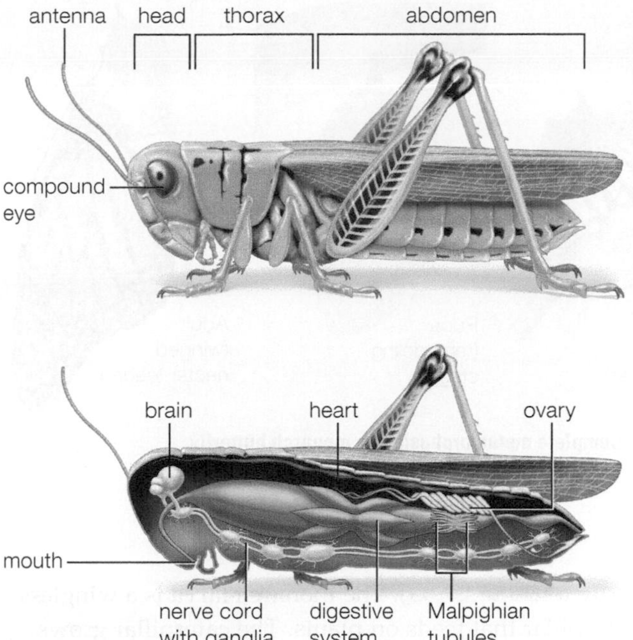
antenna
head
thorax
abdomen
compound eye
brain
heart
ovary
mouth
nerve cord with ganglia
digestive system
Malpighian tubules

FIGURE 24.32 **Body plan of a female grasshopper.**

from freshwater crustaceans. If this is correct, then insects are the crustaceans of the land.

The earliest insects were wingless, ground-dwelling scavengers that did not undergo metamorphosis. A few modern insects such as bristletails and silverfish retain this type of body form and development. When they hatch from an egg, they look like a tiny adult, and they simply grow larger with each molt.

Most groups of modern insects have wings and undergo metamorphosis. In groups that undergo incomplete metamorphosis, an egg hatches into a nymph that differs somewhat in form from the adult. It gradually achieves adult form over the course of several molts. Cockroaches, grasshoppers, and dragonflies develop in this manner. A dragonfly lives as an aquatic nymph for one to three years, molting several times, before emerging from the water and undergoing a final molt to the winged adult form.

With complete metamorphosis, a larva grows and molts without altering its form, then undergoes pupation; it becomes a pupa. A pupa is a nonfeeding body in which larval tissues are remodeled into the adult form. Consider the development of a monarch but-

---

crustaceans (krus-TAY-shuns) Mostly marine arthropod group with two pairs of antennae and a calcium-stiffened exoskeleton.
insects Six-legged arthropods with two antennae and two compound eyes. The most diverse class of animals.
Malpighian tubule Water-conserving excretory organ of insects.

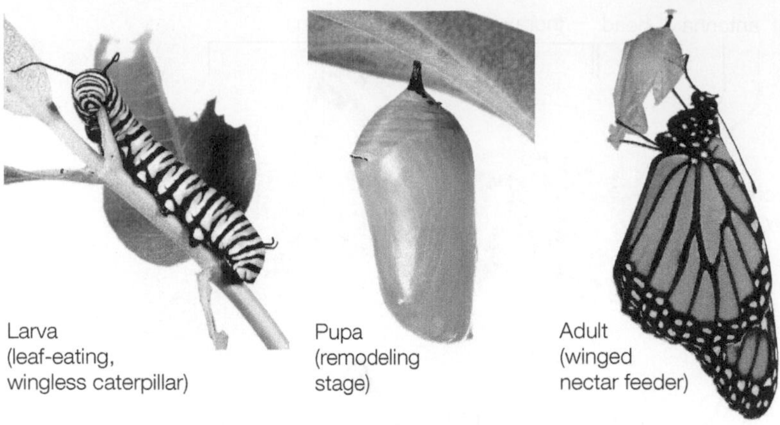
Larva
(leaf-eating,
wingless caterpillar)
Pupa
(remodeling
stage)
Adult
(winged
nectar feeder)

FIGURE 24.33 Complete metamorphosis in a monarch butterfly.

A Bumblebee pollinating flowers.

B Dung beetles disposing of manure.

FIGURE 24.34 Helpful insects.

terfly (FIGURE 24.33). The monarch larva is a wingless caterpillar that feeds on plants. The caterpillar grows and molts its skin five times. After the final molt, the caterpillar attaches its tail end to a stem and becomes a pupa. During pupation, it is transformed into an adult butterfly with six legs and two wings.

The four most diverse insect orders all have wings and undergo complete metamorphosis. The beetles (order Coleoptera) include an estimated 1.5 million species, and the flies (Diptera) more than a million. The order Hymenoptera includes more than 180,000 species of wasps, ants, and bees. Moths and butterflies are Lepidoptera with more than 120,000 species. As a comparison, consider that there are about 5,500 species of mammals.

Importance of Insects

Given the diversity and numbers of insects, it would be difficult to overestimate their importance.

Insects have many beneficial effects. They are important pollinators (FIGURE 24.34A). Members of the four most diverse insect orders (Coleoptera, Diptera, Hymenoptera, and Lepidoptera) are especially likely to be pollinators. Other insect orders contain few or no pollinators. By one hypothesis, the interactions between insects that serve as pollinators and the flowering plants that they pollinate contributed to an increased rate of speciation in both groups.

Insects are also important as food for wildlife. Most songbirds nourish their nestlings on a diet consisting largely of insects. Migratory songbirds often travel long distances to nest and raise young in areas where insect abundance is seasonally high. Aquatic larvae of insects such as dragonflies, mayflies, and mosquitoes serve as food for trout and other freshwater fish. Amphibians and reptiles feed mainly on insects. Even humans eat insects. In many cultures, they are considered a source of protein.

Insects dispose of wastes and remains. Flies and beetles are quick to discover an animal corpse or a pile of feces. They lay their eggs in or on this organic material, and the larvae that hatch devour it. By their actions, these insects keep organic wastes and remains from accumulating and help distribute nutrients through the ecosystem. Dung beetles (FIGURE 24.34B) are especially helpful in this regard. As manure dries out, it releases nitrogen into the atmosphere. By burying manure while it is still fresh, dung beetles maximize its nutritional value to plants.

Many insects eat or parasitize other insects. Some of these insects benefit us by keeping pest species in check. Use of such insects can serve as an alternative to chemical pest control methods that can sometimes harm non-target species, including humans.

A variety of economically important products are made by insects. Honey is floral nectar that has been concentrated and stored by honeybees. Larval silk moths produce silk as they spin a cocoon in which to pupate. Asian bark-feeding bugs excrete shellac, a resin

CREDITS: (33) left and middle, Jacob Hamblin/Shutterstock; right, Laurie Barr/Shutterstock; (34A) Oleksii Zelivianskyi/Shutterstock; (34B) john michael evan potter/Shutterstock.

we use to give a shiny coating to wood products and to candies such as jelly beans. Cochineal and carmine, two red colorings used in foods and personal care products, are derived from scale insects.

We also use insects in scientific studies. The fruit fly *Drosophila melanogaster* has been used in genetic studies for over a century. Among other discoveries, research on fruit flies revealed that genes are carried on chromosomes and that x-rays can damage DNA and cause deleterious mutations. Fruit flies can also help us understand the basis of human genetic disorders. Many such disorders are caused by mutation of genes that we share in common with fruit flies.

On the other hand, insects are our main competitors for food and other plant products. About a quarter to a third of all crops grown in the United States are lost to insects. In this age of global trade and travel, we have more than just homegrown pests to worry about. Consider the Mediterranean fruit fly, or medfly (**FIGURE 24.35A**). Medflies lay eggs in citrus and other fruits, as well as many vegetables. Damage done to plants and fruits by larvae of the Medfly can cut crop yield in half. Medflies are not native to the United States, and there is an ongoing inspection program for imported produce, but some Medflies still slip in. So far, eradication efforts have been successful, but they have cost hundreds of millions of dollars.

Termites, beetles, and ants that feed on wood or bore into wood to make their nest can also be a problem. The activities of these insects can weaken beams and otherwise cause damage to wooden structures.

Some insects spread human diseases. Mosquitoes are probably the most important vectors. They transmit malaria, which causes more than 500,000 deaths each year (Section 21.1). Other mosquito-borne diseases include Zika virus (Section 20.4) and lymphatic filariasis (Section 24.9). Biting flies transmit trypanosomes that cause African sleeping sickness (Section 21.3). True bugs spread Chagas disease. Fleas that bite rats and then bite humans can transmit deadly bubonic plague. Body lice transmit typhus and other diseases.

Head lice and pubic lice do not cause disease, but their bites can elicit an immune reactions that causes itching. Pubic lice are commonly referred to as "crabs." They live in human pubic hair and can be transferred through sexual intercourse. Head lice live on the human scalp and eyebrows. They are especially common among children.

Bites from bedbugs (**FIGURE 24.35B**) can also cause itching. The bugs hide in bedding, carpets or funiture, during the day, and then emerge at night to feed on human blood. Bedbugs do not transmit disease, but an infestation can have a negative economic impact.

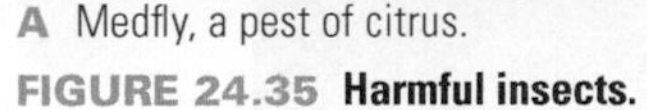

A Medfly, a pest of citrus.

B Bedbug, parasite of humans.

FIGURE 24.35 Harmful insects.

TAKE-HOME MESSAGE 24.10

- ✔ Arthropods are the most diverse animal phylum. They have a jointed exoskeleton that must be periodically molted as they grow.
- ✔ Chelicerates are named for their distinctive mouthparts. They include the marine horseshoe crabs and the land-dwelling arachnids such as spiders, scorpions, ticks, and mites.
- ✔ Myriapods have elongated bodies and live on land. They include the millipedes and centipedes.
- ✔ Crustaceans are primarily aquatic. They include decapods such as crabs and lobsters, as well as shrimp, krill, copepods, and barnacles.
- ✔ Insects are the most diverse arthropods. They have six legs and most species have wings. Insect adaptations to life on land include a system of tracheal tubes that function in respiration and Malpighian tubules that excrete wastes while conserving water. The most diverse insect groups undergo complete metamorphosis.
- ✔ Insects play important ecological roles and are useful in scientific studies, but some destroy crops or spread human diseases.

24.11 The Spiny-Skinned Echinoderms

LEARNING OBJECTIVES

- Explain the structure and function of the echinoderm water–vascular system.
- Describe how a sea star feeds.
- Using appropriate examples, discuss the variety of echinoderm body plans.

The Protostome–Deuterostome Split

In Section 24.2, we introduced the two major lineages of bilateral animals, protostomes and deuterostomes. The flatworms, annelids, mollusks, roundworms, and insects are all protostomes. With this section we begin our survey of deuterostome lineages. Echinoderms are the largest group of invertebrate deuterostomes, and we consider them here. Vertebrates and their closest invertebrate relatives are deuterostomes too. We will discuss these groups in Chapter 25.

CREDITS: (35A) Scott Bauer/ USDA; (35B) CDC/Piotr Naskrecki.

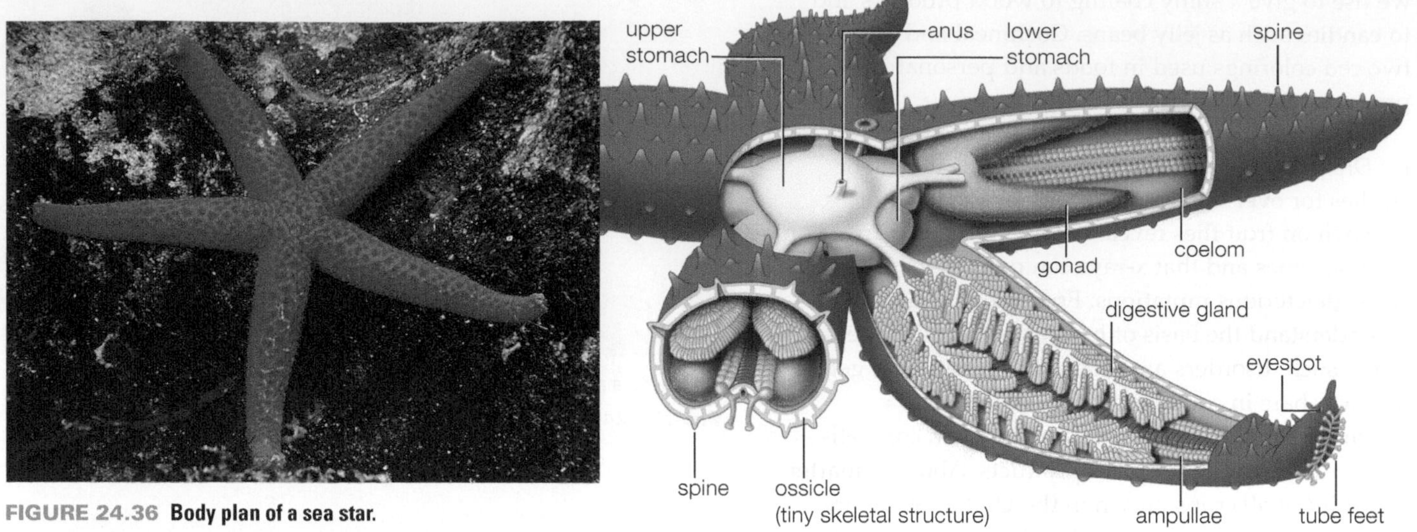
upper
stomach
anus
lower
stomach
spine
gonad
coelom
digestive gland
eyespot
spine
ossicle
(tiny skeletal structure)
ampullae
tube feet

FIGURE 24.36 **Body plan of a sea star.**

Echinoderm Characteristics and Body Plan

The **echinoderms** (phylum Echinodermata) include about 7,000 marine invertebrates. Their name means "spiny-skinned" and refers to interlocking spines and plates of calcium carbonate embedded in their skin. These plates constitute an internal skeleton referred to as an **endoskeleton**.

Adult echinoderms are coelomate animals and most have a radial body, with five parts (or multiples of five) around a central axis. The larvae are bilateral, which suggests that the ancestor of echinoderms was a bilateral animal.

Sea stars (also called starfish) are the most familiar echinoderms, and we will use them to illustrate the general body plan (FIGURE 24.36). Sea stars do not have a brain, but they do have a nerve net. A nerve ring surrounds the mouth, and nerves that branch from this ring extend into the arms. Eyespots at the tips of arms detect light and movement.

A typical sea star is an active predator that moves about on tiny, fluid-filled tube feet. Tube feet are part of a **water–vascular system** unique to echinoderms. The system includes a central ring and fluid-filled canals that extend into each arm. Side canals deliver coelomic fluid into muscular ampullae, which are tiny bulbs that function like the bulb on a medicine dropper. Contraction of an ampulla forces fluid into the attached tube foot, extending the foot. A sea star glides along as coordinated contraction and relaxation of the ampullae redistribute fluid among its tube feet.

Sea stars typically feed on bivalve mollusks. A sea star's mouth is on its lower surface. To feed, the animal slides its stomach out through its mouth and into a bivalve's shell. The stomach secretes acid and enzymes that kill the mollusk and begin the process of digestion. Partially digested food is taken into the stomach, then digestion is completed with the aid of digestive glands in the arms.

Echinoderms do not have specialized respiratory or circulatory organs. Gas exchange occurs by diffusion across the tube feet and tiny skin projections at the body surface.

Sexes are separate. A sea star's arms contain sexual organs that release eggs or sperm into the water that surrounds them. Fertilization produces an embryo that develops into a ciliated, bilateral larva. The larva swims about briefly, then undergoes metamorphosis into the adult form.

Sea stars and other echinoderms have a remarkable ability to regenerate lost body parts. If a sea star is cut into pieces, any portion with some of the central disk can regrow the missing body parts.

Echinoderm Diversity

Brittle stars are echinoderms that have a central disk and highly flexible arms that move in a snakelike way (FIGURE 24.37A). They are the most diverse and abundant echinoderms, but are less familiar than sea stars because they generally live in deeper water. Most brittle stars are scavengers on the seafloor.

Crinoids are suspension feeders that capture food with the tube feet on their multiple arms. Those that attach to the sea floor by a stalk are commonly known as sea lilies for their flower-like appearance (FIGURE 24.37B). Most modern sea lilies live in deep water. Feather stars are crinoids that lack a stalk and can swim with a graceful motion by moving their arms. They live in both deep and shallow water.

The calcium carbonate plates of sea urchins form a stiff, rounded covering from which spines protrude (**FIGURE 24.37C**). The spines provide protection and are used in movement. Some urchins graze on algae. Others act as scavengers or prey on invertebrates.

In sea cucumbers, hardened parts have been reduced to microscopic plates embedded in a soft body. Most sea cucumbers have a wormlike body (**FIGURE 24.37D**). Like earthworms, they eat their way through sediments and digest any organic material that they take in. Others are suspension feeders that look somewhat like sea anemones. Some deep-sea species can swim with the aid of capelike body extensions.

Lacking spines or sharp plates, sea cucumbers have alternative defenses. When threatened, they expel a sticky mass of specialized threads or internal organs out through their anus. If this maneuver successfully distracts the predator, the sea cucumber escapes, and its missing parts grow back.

Sea urchin roe (eggs) and sea cucumbers are popular foods in Asia, and overharvesting for this market threatens species in both groups. Farming these animals may provide a way to meet growing consumer demand while protecting natural ecosystems.

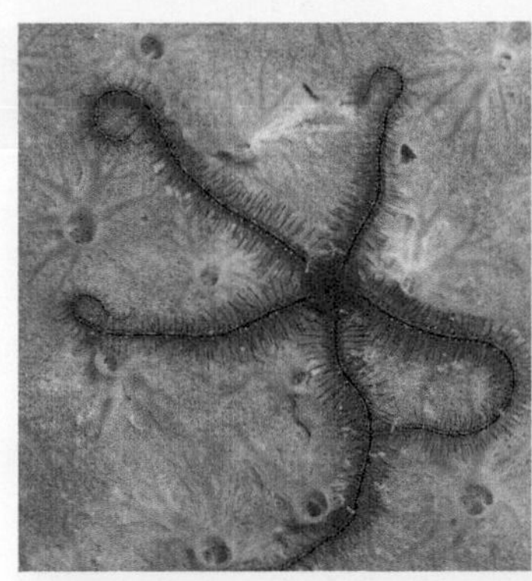

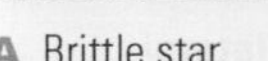

A Brittle star.

B Sea lily (a crinoid).

C Sea urchin.

D Sea cucumber.

FIGURE 24.37 Echinoderm diversity.

TAKE-HOME MESSAGE 24.11

✔ Echinoderms are deuterostome invertebrates that have a radial body as adults. They are brainless and have a unique water–vascular system that functions in locomotion.

echinoderms (eh-KYE-nuh-derms) Invertebrates with hardened plates and spines embedded in the skin or body, and a water–vascular system.
endoskeleton An internal skeleton.
water–vascular system In echinoderms, a system of fluid-filled tubes and tube feet that function in locomotion.

 26.1 Medicines from the Sea (revisited)

A variety of compounds derived from marine invertebrates are also already in use. AZT (azidothymidine), the first drug developed for treating AIDS, is a synthetic derivative of a molecule that was initially discovered in sponges. Other compounds from sponges can be used to treat infections caused by herpesviruses. Gorgonians, which are relatives of sea anemones, are the source of anti-inflammatory compounds, one of which is used in an "anti-aging" face cream.

Many invertebrates make chemicals that inhibit the growth of bacteria or protists. To find new drugs with these properties, researchers extract compounds from invertebrates, then test the ability of the compounds to kill pathogens cultured in the laboratory. The effect of each compound is also tested on cultured human cells, because an ideal candidate drug does no harm to these cells. Chemicals are also tested for their ability to kill cancer cells, while sparing normal ones.

Finding a compound that may have medicinal value is only the first step in drug development. For a compound to be used in clinical tests, researchers must obtain a sufficient amount of it. This can be difficult because many compounds of interest occur only at very low concentrations in animals. Consider Eribulin (Halaven), a drug used to fight some metastatic breast cancers. Sponges make the compound on which it is modeled, but only in miniscule amounts. Obtaining enough of the compound to test its efficiency as a cancer drug (300 milligrams) required processing more than one metric ton of sponges.

Generally, once the structure of a useful compound has been determined, chemists can find a way to produce it or a compound that has a similar structure and properties. Eribulin is a variant of the sponge compound, and it is now synthesized, rather than extracted from sponges. ●

CREDITS: (37A) Stubblefield Photography/Shutterstock; (37B) kaschibo/Shutterstock; (37C) Derek Holzapfel/photos.com; (37D) Andrew David, NOAA/NMFS/SEFSC Panama City, Lance Horn, UNCW/NURC-Phantom II ROV operator.

STUDY GUIDE

Section 24.1 Most animals are **invertebrates** (animals with no backbone). Some compounds that invertebrates produce to fight off infections or to help them capture prey can be used as medicines in humans.

Section 24.2 **Animals** are multicelled heterotrophs with unwalled cells. TABLE 24.1 summarizes the traits of groups covered in this chapter. Some have no body symmetry; others have **radial symmetry**, like a wheel. Most have **bilateral symmetry** and **cephalization**, a concentration of nerves and sensory structures at the head end. The two lineages of bilateral animals, **protostomes** and **deuterostomes**, develop from a three-layered embryo. Some animals digest food in a saclike **gastrovascular cavity**, others in a tubular **complete digestive tract**. A digestive tract is usually inside a fluid-filled cavity (a **coelom** or a **pseudocoelom**).

Sections 24.3 According to the **colonial theory of animal origins**, animals evolved from a colonial protists similar to choanoflagellates, a type of protist. The oldest animal body fossils are of Ediacarans and date back about 600 million years. A great adaptive radiation during the Cambrian gave rise to most modern lineages.

Section 24.4 **Sponges** are asymmetrical and do not have tissues. They are **suspension feeders** with flagellated collar cells that cause water to move through pores in the body wall. Sponges are **hermaphrodites**: Each makes eggs and sperm. Adult sponges are **sessile animals**, but the **larvae** swim.

Section 24.5 **Cnidarians** have two radially symmetrical body forms: **medusa** and **polyp**. A coral or sea anemone is a polyp. A jelly is a medusa. Both medusae and polyps have tentacles with stinging cells called **cnidocytes** that help them catch prey. Both also have two tissue layers with a jellylike layer of mesoglea between them. Cnidarians have a **hydrostatic skeleton**. Their **nerve net** controls movements of contractile cells. Food is digested in a gastrovascular cavity that also functions in gas exchange.

Section 24.6 **Flatworms** are bilateral acoelomate worms with organ systems. They include free-living species such as planarians as well as the parasitic tapeworms and flukes. In the planarians, a **pharynx** on the lower surface sucks up food and delivers it to the gastrovascular cavity. **Nerve cords** connect to **ganglia** in the head that serve as a control center.

Section 24.7 **Annelids** include polychaetes, oligochaetes, and leeches. They have a **closed circulatory system**, in which blood is always within blood vessels or a heart. Nephridia regulate body fluid composition. Like mollusks, some annelids have a trochophore larva.

Section 24.8 **Mollusks** include chitons, gastropods, bivalves, and cephalopods. Their **mantle** is an extension of the body wall. Most have a food-scraping **radula** and use **gills** or (in land dwellers) a **lung** to breathe. All except the cephalopods have an **open circulatory system**.

Section 24.9 The **roundworms** (nematodes) have an unsegmented body covered by a cuticle. Periodic **molting** of the cuticle allows the animal to grow. There is a complete gut, and a false coelom. Some roundworms are human parasites, but most are free-living in soil or water. One free-living soil roundworm is frequently used in scientific studies.

TABLE 24.1

Comparison of Invertebrate Groups Surveyed in This Chapter

Animal Phylum	Representative Groups	Living Species	Organization	Symmetry	Digestion	Body Circulation
Porifera	Sponges	8,000	Connected cells	None	Intracellular	Diffusion
Cnidaria	Sea anemones, jellies, corals	9,000	2 tissue layers	Radial	Gastrovascular cavity	Diffusion
Platyhelminthes (Flatworms)	Planarians, tapeworms, flukes	20,000	3 tissue layers	Bilateral	Gastrovascular cavity	Diffusion
Annelida	Polychaetes, oligochaetes, leeches	17,000	3 tissue layers	Bilateral	Complete gut	Closed system
Mollusca	Snails, slugs, clams, octopuses, squids	110,000	3 tissue layers	Bilateral	Complete gut	Open in most, closed in cephalopods
Nematoda (Roundworms)	Pinworms, hookworms	25,000	3 tissue layers	Bilateral	Complete gut	Diffusion
Arthropoda	Horseshoe crabs, spiders, scorpions, ticks, crabs, lobsters, shrimp, krill, copepods, millipedes, centipedes, insects	>5,000,000	3 tissue layers	Bilateral	Complete gut	Open system
Echinodermata	Sea stars, brittle stars, sea lilies, sea urchins, sea cucumbers	7,000	3 tissue layers	Larvae bilateral; adults radial	Complete gut	Open system

Section 24.10 **Arthropods**, the largest phylum of animals, have a jointed **exoskeleton** that is periodically molted. Most have sensory **antennae**; insects and crustaceans have **compound eyes**. Development often includes **metamorphosis**, a change in body form.

Chelicerates include marine horseshoe crabs and the eight-legged, land-dwelling **arachnids**. The **crustaceans** have an exoskeleton hardened with calcium, and most are marine. **Myriapods** include centipedes and millipedes.

Insects, the most diverse arthropods, include the only winged invertebrates. Tracheal tubes and **Malpighian tubules** adapt insects to life on land.

Insects pollinate plants, dispose of wastes, serve as food, and are used in scientific research. Some insects eat crops, destroy wooden structures, or transmit disease.

Section 24.11 **Echinoderms** are deuterostomes. Their skin is hardened with bits of calcium carbonate to form an **endoskeleton**. A **water–vascular system** with tube feet helps most echinoderms glide about. Adults are radial with some bilateral features, and the larvae are bilateral.

SELF-QUIZ

Answers in Appendix VII

1. True or false? Animal cells do not have walls.
2. A body cavity fully lined with tissue derived from mesoderm is called a _______ .
   a. coelom c. complete digestive tract
   b. pseudocoelom d. gastrovascular cavity
3. Flatworms, annelids, and roundworms are all _______ .
   a. arthropods c. protostomes
   b. aceolomate d. radially symmetrical
4. The oldest body fossils of a land animal are of _______ .
   a. insects c. cnidarians
   b. roundworms d. myriapods
5. _______ function(s) in the movement of cnidarians.
   a. A hydrostatic skeleton c. Cnidocytes
   b. Tube feet d. Malpighian tubules
6. Earthworms are most closely related to _______ .
   a. tapeworms c. planarians
   b. roundworms d. leeches
7. Annelids and _______ have a closed circulatory system.
   a. insects c. flatworms
   b. cephalopods d. sea stars
8. Which invertebrate phylum includes the most species?
   a. mollusks c. arthropods
   b. roundworms d. flatworms

CENGAGE brain .com To access course materials, please visit www.cengagebrain.com.

9. A _______ can reassemble if broken into individual cells.
   a. sea anemone c. sea star
   b. roundworm d. sponge
10. A slug is a land-dwelling _______ .
    a. arthropod c. cephalopod
    b. gastropod d. crustacean
11. A barnacle is a shelled _______ .
    a. arthropod c. cephalopod
    b. gastropod d. crustacean
12. _______ include the only winged invertebrates.
    a. Cnidarians c. Insects
    b. Echinoderms d. Crustaceans
13. The _______ and _______ are grouped as Ecdysozoa because they both molt.
    a. cnidarians/arthropods c. annelids/mollusks
    b. echinoderms/flatworms d. insects/roundworms
14. Match each structure with its descriptions.
    ___ radula a. internal skeleton
    ___ exoskeleton b. external skeleton
    ___ endoskeleton c. stinging cell of cnidarians
    ___ cnidocyte d. excretory organ of insects
    ___ compound eye e. mollusk feeding device
    ___ tube foot f. larva of annelid or mollusk
    ___ trochophore g. has many lenses
    ___ Malpighian tubule h. moves an echinoderm
    ___ gastrovascular cavity i. digests a planarian's food
15. Match the organisms with their descriptions.
    ___ echinoderms a. stinging cells in tentacles
    ___ sponges b. simplest organ systems
    ___ cnidarians c. no tissues, filters out food
    ___ flatworms d. jointed exoskeleton
    ___ roundworms e. mantle over body mass
    ___ annelids f. segmented worms
    ___ arthropods g. spiny skinned deuterostomes
    ___ mollusks h. complete gut, pseudocoelom

CRITICAL THINKING

1. A massive die-off of lobsters in the Long Island Sound was blamed on pesticides sprayed to control mosquitoes that carry West Nile virus. Why might a chemical designed to kill insects also harm lobsters but have no effect on sea stars?

2. A rise in atmospheric carbon dioxide increases ocean acidity, making it harder for animals to build calcium carbonate parts. List some invertebrate animals that increased ocean acidity could harm in this way.

25 ANIMAL EVOLUTION—THE VERTEBRATES

CORE CONCEPTS

Evolution

Evolution underlies the unity and diversity of life.
Shared core processes and features that are widely distributed among organisms provide evidence that all living things are linked by lines of descent from common ancestors. The four shared traits that link all chordates (invertebrate and vertebrate) can be observed in their embryos: a supporting rod (notochord), a dorsal nerve cord, a pharynx with gill slits, and a tail that extends past the anus.

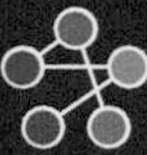

Systems

Complex properties arise from interactions among components of a biological system.
Vertebrates evolved in water and their transition to land involved many changes in their component body parts. The gills and fins of fishes are adaptations to life in water. Lungs, a more efficient circulatory system, and bony limbs allow amphibians to spend time on land. Waterproof skin and eggs that can develop away from water allow reptiles, birds, and mammals to spend their entire life on land.

Process of Science

The field of biology consists of and relies upon experimentation and the collection and analysis of scientific evidence.
Science addresses only testable ideas about observable events and processes. Hypotheses about animal evolution are tested by examining fossils and analyzing the similarities and differences among living species. Hypotheses are often revised when new data come to light.

Links to Earlier Concepts

This chapter introduces the remaining lineages of deuterostomes, a group defined in Section 24.2. We refer to discussions of fossils (16.4), plate tectonics and the geologic time scale (16.5), and adaptive radiation (17.9). Understanding cladistics (18.2) and how comparative studies help determine relationships (18.4–18.6) will be extremely useful.

25.1 Very Early Birds

In Darwin's time, acceptance of his theory of evolution by natural selection was hindered, in part, by an apparent absence of transitional fossils. Skeptics wondered, if new species evolve from existing ones, where are the fossils that represent these transitions? In fact, one of these "missing links" was unearthed by workers at a limestone quarry in Germany just one year after Darwin's *On the Origin of Species* was published.

That fossil, about the size of a large crow, looked like a small dinosaur. It had a long, bony tail, three clawed fingers on each forelimb, and a heavy jaw with short, spiky teeth, but it also had feathers (**FIGURE 25.1**). The fossil species was named *Archaeopteryx* (meaning "ancient winged one"). To date, eight fossilized members of this species have been unearthed. Radiometric dating indicates that they lived about 150 million years ago.

Piecing together the evolutionary history of animals, like that of other groups, relies on evidence of past events that no one witnessed. Fossils and biochemical comparisons of living organisms provide clues to the past. However, interpretations of any historical event can differ when evidence of it is incomplete. For this reason, evolutionary biologists often disagree about the relationships among lineages, both modern and extinct. These disagreements do not call into question the fundamental idea that all animal lineages arose by descent with modification from a common ancestor. Rather, the disputes—which involve details such as the timing of a divergence, or a taxonomic grouping—are part of the process of science. If data that are inconsistent with a hypothesis are discovered, the hypothesis is revised. The fossil record will always be incomplete, so it is not surprising that new discoveries prompt major revisions in our understanding of evolutionary history.

FIGURE 25.1 Early bird fossil. This *Archaeopteryx* lived about 150 million years ago. It had feathers, teeth, and a long bony tail.

CREDITS: (opposite) Soren Egeberg Photography/Shutterstock; (1) Aredea London.

25.2 Chordate Traits and Evolutionary Trends

LEARNING OBJECTIVES

- List the four traits that define chordates.
- Describe the two groups of invertebrate chordates, and draw a tree showing how they relate to one another and to vertebrates.
- Explain how an endoskeleton differs from an exoskeleton.

Chordate Characteristics

Chapter 24 introduced echinoderms, a deuterostome lineage. The other major deuterostome lineage is the **chordates** (phylum Chordata), a group of bilaterally symmetrical, coelomate animals, that have a complete digestive system and a closed circulatory system. Four traits unique to chordate embryos define the lineage:

1. A **notochord**, a rod of flexible connective tissue, extends the length of the embryonic body and supports it.
2. A hollow nerve cord parallels the notochord and runs along the dorsal (upper) surface.
3. Gill slits (narrow openings) extend across the wall of the pharynx (the throat region).
4. A muscular tail extends beyond the anus.

Depending on the group, some, none, or all of the defining chordate traits persist in the adult.

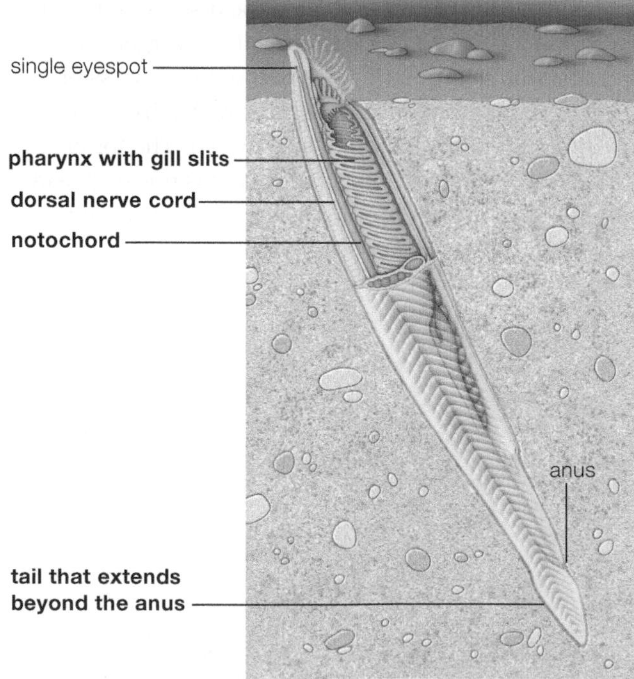
single eyespot
pharynx with gill slits
dorsal nerve cord
notochord
anus
tail that extends beyond the anus

FIGURE 25.2 Lancelet. Lancelets can swim, but spend most of their time with their lower body buried in sediment. Both larvae and adults have all four chordate traits (bold labels).

Most of the 65,000 or so chordates are **vertebrates,** which are animals that have a backbone. The group also includes two lineages of marine invertebrates.

Invertebrate Chordates

Lancelets (subphylum Cephalochordata) have a fish-shaped body (**FIGURE 25.2**). The adult is about 5 centimeters (2 inches) long, and it retains all four characteristic chordate traits. The nerve cord extends into the head, where there is a simple brain. An eyespot at the end of the nerve cord detects light, but there are no paired sensory organs as in fishes. A lancelet spends most of its time buried up to its mouth in sediments. Waving of cilia inside the pharynx moves water into the pharynx and out through gill slits. The gill slits filter food particles out of the water.

In **tunicates** (subphylum Urochordata), only larvae have all the typical chordate traits (**FIGURE 25.3A**). The larva does not feed. Almost immediately after it hatches, metamorphosis transforms it into a barrel-shaped adult that secretes a "tunic" of polysaccharides (**FIGURE 25.3B**). Adults feed by sucking in water through a tube, capturing food on their pharynx, then expelling the filtered water through another tube.

Overview of Chordate Evolution

Until recently, lancelets were considered the closest invertebrate relatives of vertebrates. An adult lancelet

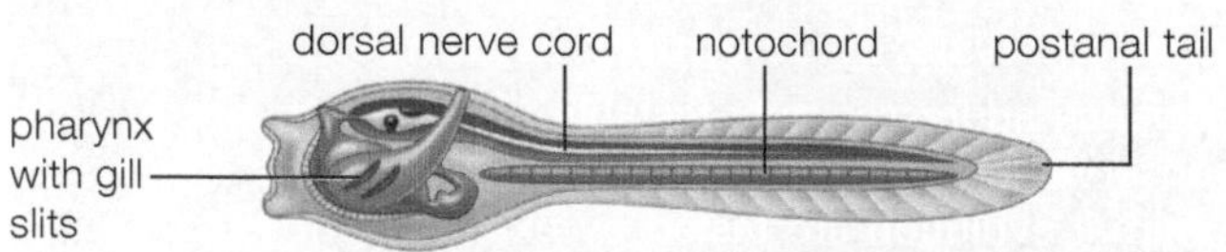
dorsal nerve cord
notochord
postanal tail
pharynx with gill slits

A Free-swimming tunicate larva. It has all the defining chordate traits.

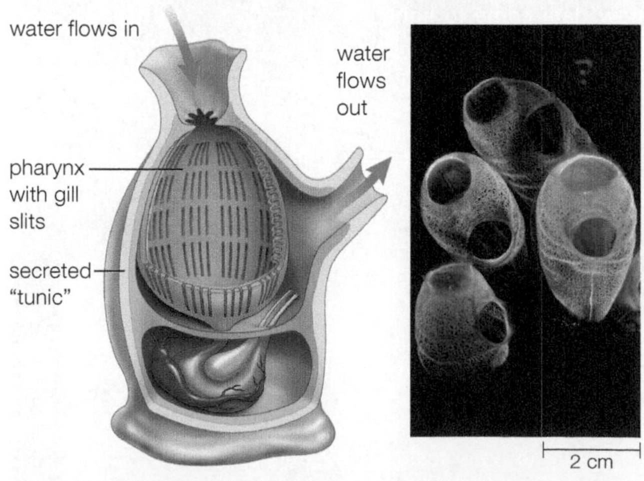
water flows in
water flows out
pharynx with gill slits
secreted "tunic"
2 cm

B Adult tunicate. The only defining chordate trait it retains is the pharynx with gill slits. The species in the photo is sessile as an adult.

FIGURE 25.3 Tunicates.

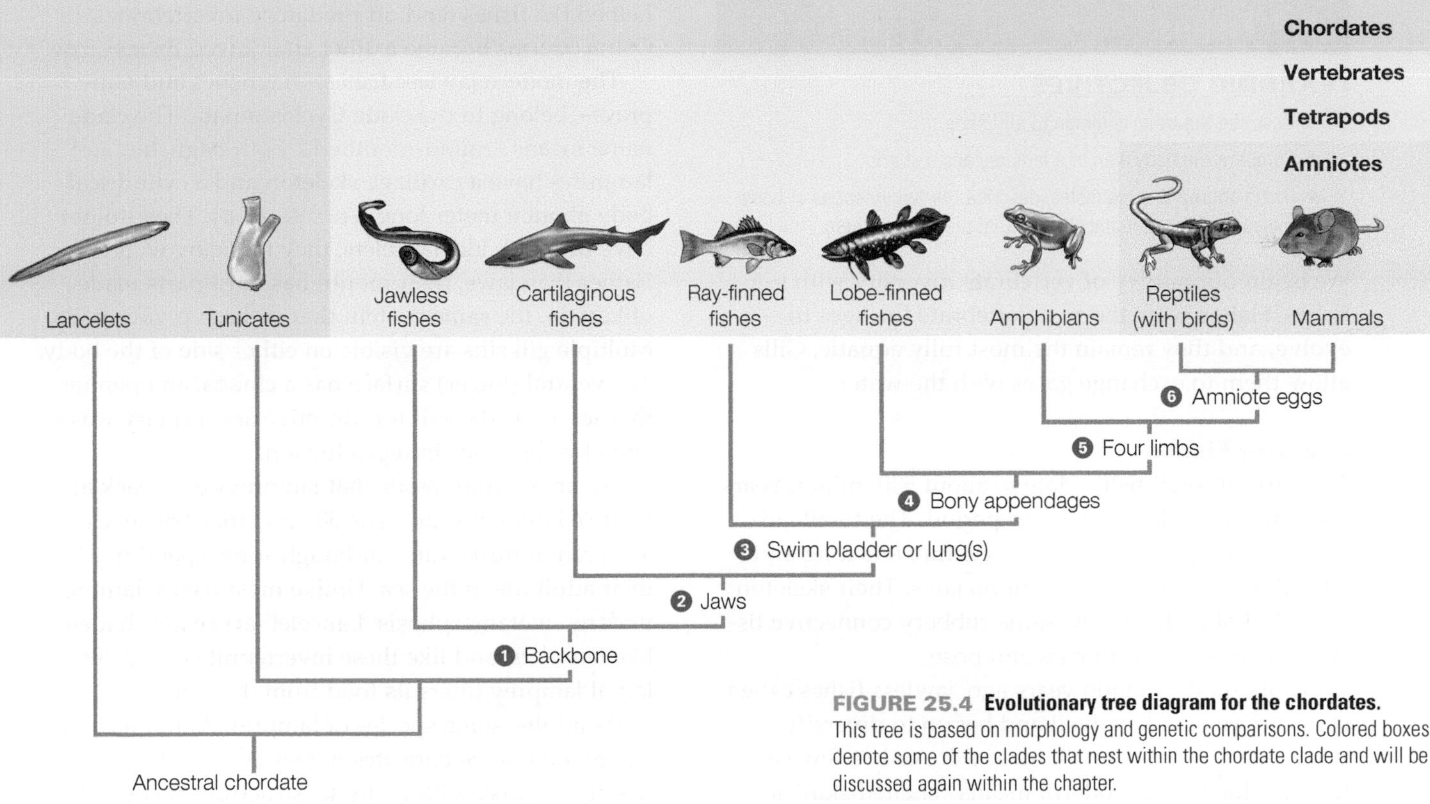

Chordates
Vertebrates
Tetrapods
Amniotes
Lancelets
Tunicates
Jawless fishes
Cartilaginous fishes
Ray-finned fishes
Lobe-finned fishes
Amphibians
Reptiles (with birds)
Mammals
6 Amniote eggs
5 Four limbs
4 Bony appendages
3 Swim bladder or lung(s)
2 Jaws
1 Backbone
Ancestral chordate

FIGURE 25.4 Evolutionary tree diagram for the chordates. This tree is based on morphology and genetic comparisons. Colored boxes denote some of the clades that nest within the chordate clade and will be discussed again within the chapter.

looks more like a fish than an adult tunicate does, but morphological traits can be deceiving. Studies of developmental processes and gene sequences revealed that tunicates are the invertebrate lineage most closely related to the vertebrates (FIGURE 25.4).

Most chordates have a backbone and thus are vertebrates ❶. The backbone and other skeletal elements are components of the vertebrate **endoskeleton**, or internal skeleton. A vertebrate endoskeleton consists of living cells, so it grows with the animal and does not have to be molted as an exoskeleton does.

The first vertebrates were jawless fishes that sucked up or scraped up food. Later, hinged skeletal elements called jaws evolved ❷. Jaws allowed their bearers to exploit new strategies for feeding. The vast majority of fishes and other modern vertebrates have jaws.

Later evolutionary modifications allowed animals to move onto land. In one group of fishes, two outpouchings on the side of the gut wall evolved into lungs: moist, internal sacs that enhance gas exchange with the air ❸. Fins with bones inside them evolved in a subgroup of these fishes ❹. These fins would later evolve into limbs of the four-legged walkers, or **tetrapods** ❺.

Early tetrapods spent time on land, but laid their eggs in water. Later, eggs that enclosed an embryo within a series of waterproof membranes evolved in one lineage. These specialized eggs allowed animals known as **amniotes** to become the most diverse group of vertebrates on land ❻.

TAKE-HOME MESSAGE 25.2

✔ All chordate embryos have a notochord, a dorsal tubular nerve cord, a pharynx with gill slits in its wall, and a tail that extends past the anus. There are two groups of invertebrate chordates: lancelets and tunicates.

✔ Most chordates have a backbone and so are vertebrates. Limbs evolved in one lineage that later colonized the land. Amniotes, a tetrapod subgroup with specialized eggs, are the predominant vertebrates on land.

amniote (AM-nee-oat) Vertebrate with eggs that enclose the embryo within waterproof membranes.
chordate (CORE-date) Animal with an embryo that has a notochord, dorsal nerve cord, pharyngeal gill slits, and a tail that extends beyond the anus. For example, a lancelet or a vertebrate.
endoskeleton Internal skeleton.
lancelet (LANCE-uh-lit) Invertebrate chordate with a fishlike shape; retains all the defining embryonic chordate traits into adulthood.
notochord (NO-toe-cord) Stiff rod of connective tissue that runs the length of the body in chordate larvae or embryos.
tetrapod (TEH-trah-pod) Vertebrate that has four bony limbs or is a descendant of a four-limbed ancestor.
tunicate (TOON-ih-kit) Marine invertebrate chordate; a fish-shaped, swimming larva undergoes metamorphosis into a barrel-shaped, sessile adult.
vertebrate Animal with a backbone.

25.3 Fishes

LEARNING OBJECTIVES

- Describe the traits common to all fishes.
- Compare the body plan of a lamprey and a shark.
- Using appropriate examples, describe the two lineages of bony fishes and the anatomical traits that distinguish them.

We begin our survey of vertebrate diversity with the fishes. **Fishes** were the first vertebrate lineages to evolve, and they remain the most fully aquatic. Gills allow them to exchange gases with the water.

Jawless Fishes

The earliest fossil fishes date to about 530 million years ago, during the late Cambrian period. The fossil animals had a tapered body a few centimeters long and a head with a pair of eyes, but no jaws. Their skeleton consisted of cartilage, the same rubbery connective tissue that supports your ears and nose.

By about 480 million years ago, jawless fishes called ostracoderms had evolved and begun to diversify. Body size remained small; most were only a few centimeters long. Ostracoderm means "shelled skin" and refers to bony external plates that covered the head or, in some cases, the entire body. The plates probably helped the fishes fend off predatory invertebrates. Ostracoderms became extinct after jawed fishes arose.

A Parasitic lamprey. It attaches to another fish with its oral disk and scrapes off bits of flesh.

B Hagfish. It feeds on worms and scavenges on the seafloor.

FIGURE 25.5 Jawless fishes.

The modern jawless fishes—hagfishes and lampreys—belong to the clade Cyclostomata. The clade name means "round-mouthed." Both hagfishes and lampreys have a cartilage skeleton and a cylindrical body about a meter long (**FIGURE 25.5**). They do not have fins and, like lancelets, they move by wiggling. Rather than jaws, their mouth has hard parts made of keratin, the same protein that makes up your nails. Multiple gill slits are visible on either side of the body. The ventral (lower) surface has a **cloaca,** an opening that serves as the exit for digestive and urinary waste, and also functions in reproduction.

We know from fossils that lampreys date back at least 360 million years. The 50 or so modern species all breed in fresh water, although some spend most of their adult life in the sea. Unlike most fishes, lampreys undergo metamorphosis. Lancelet larvae are shaped like lancelets, and like these invertebrate chordates a larval lamprey filters its food from the water.

As adults, some species of lamprey do not feed at all and others are parasites of fish. A parasitic lamprey has an oral disk with toothlike structures made of keratin (**FIGURE 25.5A**). The lamprey used its oral disk to attach to a fish, then scrapes off bits of flesh. Fish rarely survive this attack. In the early 1900s, newly built canals allowed parasitic lampreys from the Atlantic ocean to enter the Great Lakes. The lampreys decimated native fish populations. Fishery managers now lower lamprey numbers with dams, nets, and poisons.

The 60 or so species of flexible-bodied hagfishes are marine bottom-feeders (**FIGURE 25.5B**). Hagfishes have poor eyesight and use sensory tentacles near their mouth to locate worms and carcasses. Their mouth has dental plates covered with sharp barbs of keratin that are used to grab and pull apart food. When frightened, a hagfish secretes a compound that combines with water to form a gelatinous slime. Slime deters most predators but has not kept humans from harvesting hagfish. Most belts, wallets, and other products labeled as "eelskin" are actually made of hagfish skin.

Evolution of Jawed Fishes

Jawed vertebrates (subphylum Gnathostomata) evolved during the late Silurian period, by about 420 million years ago. Jaws evolved from gill arches, which are skeletal elements that support fish gills (**FIGURE 25.7**). Jaws help a fish catch and kill prey, and to tear large prey into easier-to-swallow chunks.

Jawed fishes were the first animals with scales and paired fins. **Scales** are hard, flattened structures that grow from and often cover the skin. Fins are flattened

CREDITS: (5A) Heather Angel/Natural Visions; (5B) NHPA / NHPA / Superstock.

Data Analysis Activities

Deathly Lamprey Repellent Predation by sea lampreys on native fishes in the Great Lakes is an ongoing problem. To help solve it, Michael Wagner and his team test methods of repelling lampreys. They carried out an experiment to investigate reports that sea lampreys detect the scent of lamprey carcasses and tend to avoid them. The researchers made alcohol-based lamprey carcass extracts, then observed what happened when lampreys were put in tanks and exposed to either this extract or to alcohol alone. FIGURE 25.6 shows their results.

1. Why was it necessary to test the response of lampreys to the scent of alcohol alone?
2. What was the lowest proportion on lampreys on the scented side of the tank when the scent was alcohol? When the scent was alcohol-based carcass extract?
3. Do the results indicate that lampreys detect and avoid the scent of dead lampreys?
4. Do the results show that lampreys cannot detect alcohol?

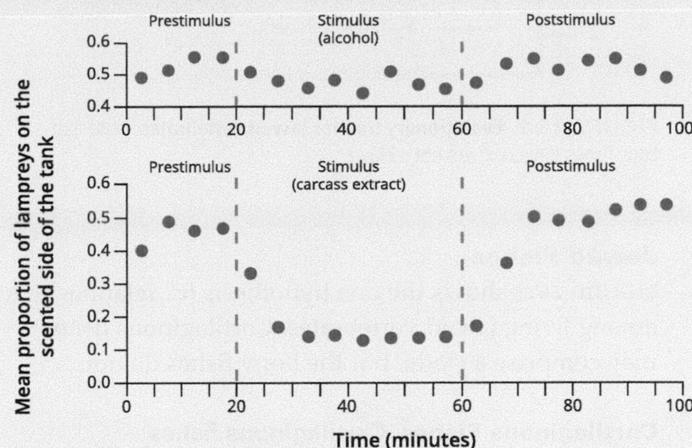
Prestimulus
Stimulus (alcohol)
Poststimulus
Prestimulus
Stimulus (carcass extract)
Poststimulus
Mean proportion of lampreys on the scented side of the tank
0.6
0.5
0.4
0.3
0.2
0.1
0.0
0
20
40
60
80
100
Time (minutes)

FIGURE 25.6 Effectiveness of carcass extracts as a lamprey repellent. Mean proportion of lampreys on the scented side of the test tank during 8 trials with 10 lampreys. Lampreys were placed in the tank for 20 minutes before exposure to alcohol or carcass extract and remained there for 40 minutes after exposure. The upper graph shows results with alcohol as the stimulus; the lower shows the results with carcass extracts. Bars indicate standard error.

appendages used to propel and steer a body while swimming. Most jawed fishes have paired fins.

The Devonian period (416–359 million years ago) is called the "Age of Fishes," and jawed fishes called placoderms were the most numerous and diverse vertebrates in Devonian seas. About 200 different genera of placoderms are known from fossils. Placoderm means "tablet skin" and refers to bony armor that covered the animal's head and neck. Sharp bony plates functioned like teeth. Placoderms grew larger than the jawless fishes that preceded them, and some were enormous. *Dunkleosteus*, which once inhabited a shallow sea in what is now Ohio, grew as long as a bus (FIGURE 25.8). Placoderms are the earliest vertebrates for which we have fossil evidence of internal fertilization and development. Scientists recently discovered a fossilized female that apparently died while giving birth. Placoderms became extinct at the end of the Devonian.

Another group of early jawed fish lineages is collectively referred to as acanthodians (spiny fins). Acanthodians arose at about the same time as placoderms, but were small (only centimeters long), less diverse, and lacked bony armor. They left fewer fossils than placoderms, so we know less about them. Acanthodians became extinct at the end of the Permian.

cloaca (klo-AY-kuh) In some vertebrates, a body opening through which both wastes and gametes exit.
fishes Aquatic vertebrates that have gills and lack limbs; jawless fishes, cartilaginous fishes, and bony fishes.
scales In some vertebrates, flattened structures that grow from and sometimes cover the skin.

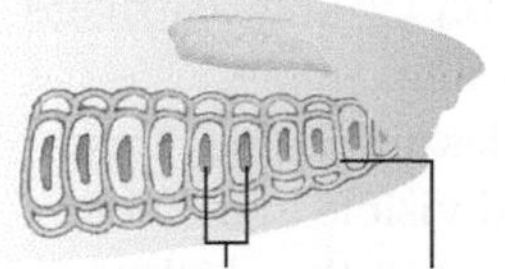
A Ancestral jawless fish
gill slits
gill support

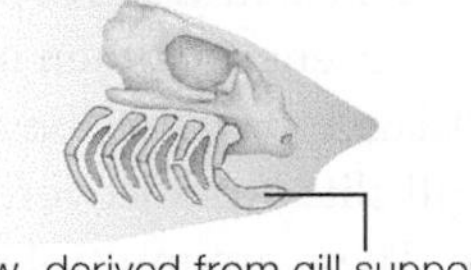
B Jawed fish
jaw, derived from gill supports

FIGURE 25.7 Proposed mechanism for the evolution of jaws. Modification of gill supports of a jawless fish **A** produced a hinged, two-part jaw in early jawed fishes **B**.

FIGURE 25.8 Placoderm. Artist's depiction of *Dunkleosteus*, a placoderm that lived during the Devonian. These early jawed fishes had bony coverings on their head and bony plates that functioned like teeth. Like most jawed fishes, they had paired fins.

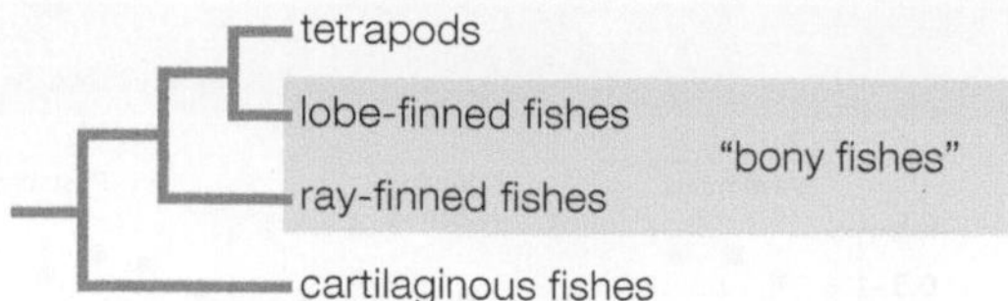
tetrapods
lobe-finned fishes
"bony fishes"
ray-finned fishes
cartilaginous fishes

FIGURE 25.9 Evolutionary tree for jawed vertebrates. Note that bony fishes (blue box) are not a clade.

Jawed Fishes

FIGURE 25.9 shows the one hypothesis for relationships among living jawed vertebrates. Cartilaginous fishes may comprise a clade, but the bony fishes do not.

Cartilaginous Fishes **Cartilaginous fishes** (Chondrichthyes) are jawed fishes whose skeleton is composed primarily or entirely of cartilage. Acanthodians are considered the likely ancestors of cartilaginous fishes and, like this group, early cartilaginous fishes had some bone. The skeleton of modern cartilaginous fishes, which consists entirely of cartilage, may be a product of natural selection that favored a lightweight body. Cartilage weighs less than bone, so fishes in which bone is replaced by cartilage expend less energy when swimming.

Cartilaginous fishes have scales and paired fins. Their gill slits are uncovered and visible at the body surface. There are multiple rows of teeth, which are continually shed and replaced. Most cartilaginous fishes have a cloaca. Sexes are separate and fixed for life. Eggs typically develop in an egg case inside the mother's body. When the young are ready to hatch, the egg case ruptures and they are released into the environment. Less commonly, females release egg cases containing developing embryos.

There are more than 900 species of cartilaginous fishes. Most are sharks or rays. Some sharks are speedy predators that chase down and tear apart prey (**FIGURE 25.10A**). Others are bottom-feeders that suck up invertebrates and act as scavengers. Still others strain plankton from seawater. The largest living fish, the whale shark, feeds in this manner. It can weigh several tons.

Rays have a flattened body with large pectoral fins. Manta rays glide through warm seas and feed by filtering out plankton (**FIGURE 25.10B**). Stingrays are bottom-feeders. Their barbed tail has a venom gland that serves as a defense against predators.

A Galápagos shark, a streamlined, fast-moving predator.

B Manta ray, a plankton feeder.

FIGURE 25.10 Cartilaginous fishes. Note the visible gill slits.

Bony Fishes **Bony fishes** are fishes in which the skeleton consists largely of bone. Most bony fishes have moveable pelvic and pectoral fins and a movable cover over their gill slits. Bony fishes also have a gas-filled organ or organs derived from outpouchings of the pharynx (throat). In most bony fishes, urinary waste, digestive waste, and gametes exit through three separate body openings. There are two modern lineages of bony fishes: ray-fined fishes and lobe-finned fishes.

The **ray-finned fishes** (Actinopterygii) have membranous fins with thin, flexible fin supports derived from skin (**FIGURE 25.11A**). Most members of this lineage have a **swim bladder**, a sac whose volume of gas can be varied to adjust buoyancy (**FIGURE 25.11B**).

With about 30,000 living species, ray-finned fishes are the most diverse group of modern fishes. Sturgeons are modern representatives of one ancient ray-finned lineage. Humans harvest eggs from some sturgeons for use as caviar. Gars, predatory fish with an elongated body, are members of a second ray-finned lineage.

A third ray-finned lineage, the teleosts, includes 99 percent of ray-finned fishes, and about half of all vertebrates. Comparisons of the teleost genome with that of other ray-finned fishes indicate that the teleost lineage underwent whole-genome duplication early in its history. Gene duplications can speed evolution by allowing one copy of a gene to mutate and take on a new function while the other copy retains its original role. The diversification of copied genes is thought to have facilitated an adaptive radiation of this group.

Teleosts have a vast array of body forms, and the group includes most fishes that humans harvest as food, including anchovies, salmon, sardines, bass, swordfish, trout, tuna, halibut, carp, and cod. Teleosts also have a wider variety of reproductive patterns than other fishes. In some species, individuals are hermaphrodites that simultaneously produce eggs and sperm. In others, individuals change sex during the course of their lifetime. In still others, sexes are separate and fixed for life.

Thick, fleshy fins with supporting bones inside them characterize **lobe-finned fishes** (Sarcopterygii). There are two lineages, the marine coelacanths and the freshwater lungfishes. In both, sexes are separate and fixed. Until 1938, coelacanths were known only from fossils, which showed these fish were abundant from the Devonian through the Cretaceous period. In 1938, a living coelacanth was found in the Indian Ocean. We now know that there are several modern species (**FIGURE 25.12A**).

As their common name implies, the lungfishes (**FIGURE 25.12B**) have lungs—air-filled sacs with an associated network of tiny blood vessels. A lungfish gulps air to fill its lungs, then oxygen diffuses from the lungs into its blood.

As Section 25.4 explains, bony fins and simple lungs facilitated the evolutionary move to land. Tetrapods descended from an ancestral lobe-finned fish. Genome comparisons indicate that, of the two modern lobe-finned groups, tetrapods are closest to lungfishes.

TAKE-HOME MESSAGE 25.3

- ✔ Like the earliest fishes, modern lampreys and hagfishes do not have jaws, scales, or paired fins.
- ✔ Jaws evolved through modification of gill supports.
- ✔ Jawed fishes include cartilaginous fishes and bony fishes. They have paired fins and scales. Modern cartilaginous fishes have a skeleton of cartilage. Bony fishes have a skeleton made primarily of bone.
- ✔ The ray-finned lineage of bony fishes is the most diverse group of vertebrates. Their fins are reinforced by rays derived from skin.
- ✔ Lobe-finned fishes have thick, fleshy fins supported by internal bones. They include coelocanths and the lungfishes, which are the closest living relatives of the tetrapods.

bony fish Fish that has a skeleton consisting primarily of bone; a ray-finned fish or lobe-finned fish.
cartilaginous fish Jawed fish with a skeleton that consists mainly or entirely of cartilage; for example, a shark or ray.
lobe-finned fish Bony fish that has fleshy fins supported by bones.
ray-finned fish Bony fish that has fin supports derived from skin.
swim bladder In ray-finned fishes, a gas-filled sac that allows the fish to regulate its buoyancy.

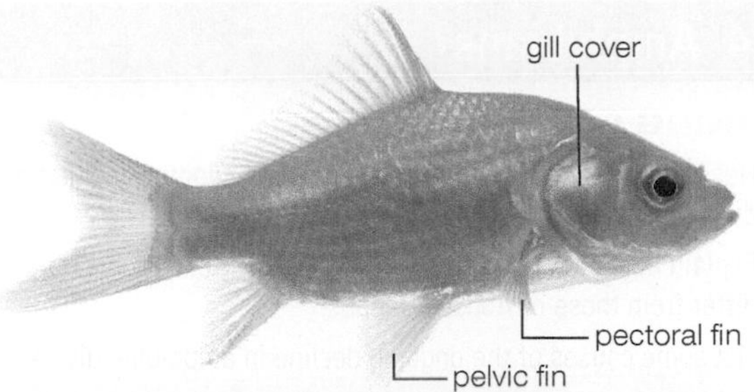
gill cover
pectoral fin
pelvic fin

A Goldfish (a type of carp). Note the flexible fins supported by thin rays. As in most bony fishes, a bony cover hides the gills.

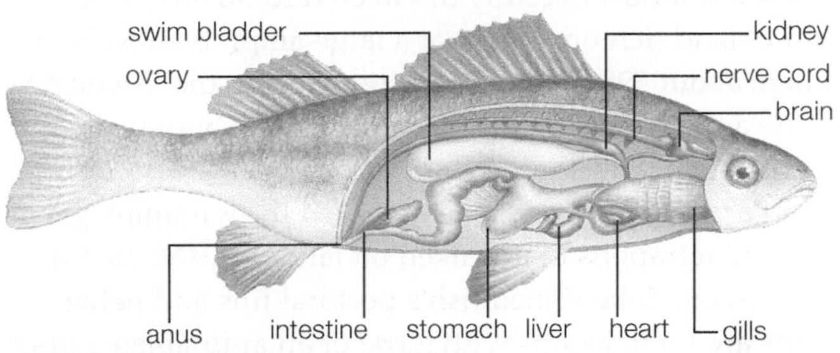
swim bladder
ovary
kidney
nerve cord
brain
anus
intestine
stomach
liver
heart
gills

B Anatomy of a perch. The swim bladder allows the fish to adjust its buoyancy (its tendency to float).

FIGURE 25.11 Ray-finned fishes. Both carps and perches are members of the teleost lineage, the most diverse lineage of fishes.

A Coelacanth.

B Lungfish.
pelvic fin
pectoral fin

FIGURE 25.12 Lobe-finned fishes. These fishes have fleshy pelvic and pectoral fins supported by sturdy bones. Lungfishes have been observed to "walk" underwater. They thrust their pelvic fins against the bottom to propel their body forward.

25.4 Amphibians

LEARNING OBJECTIVES

- Describe the anatomical and physiological adaptations that arose during the transition to life on land.
- Explain how the body form and development of salamanders differ from those of frogs and toads.
- List some causes of the ongoing decline in amphibian diversity.

Adapting to Life on Land

Amphibians are scaleless, land-dwelling vertebrates that typically breed in water. Amphibians were the first tetrapods. Recently discovered fossil footprints in Poland demonstrate that a large amphibian walked here about 395 million years ago, during the Devonian. The animal that left these footprints was about 2.5 meters (8 feet) long.

Fossils show how fishes adapted to swimming gave rise to tetrapods that walked on land (**FIGURE 25.13**). Bones of a lobe-finned fish's pectoral fins and pelvic fins are homologous with those of an amphibian's front and hind limbs. During the transition to land, these bones became larger and better able to bear weight. Ribs enlarged and a distinct neck emerged, allowing the head to move independently of the rest of the body.

The transition to land was not simply a matter of skeletal changes. Lungs became larger and more complex. Division of the previously two-chambered heart into three chambers allowed blood to flow in two circuits, one to the body and one to those increasingly important lungs. Changes to the inner ear improved detection of airborne sounds. Evolution of eyelids prevented delicate eye tissues from drying out.

FIGURE 25.14 Salamander. Like ancestral amphibians, salamanders have equal-sized forelimbs and hind limbs.

What drove the move onto land? An ability to spend time out of water would have been favored in seasonally dry places. In addition, it would have allowed escape from aquatic predators and access to a new food source—insects—which also arose in the Devonian.

Modern Amphibians

Modern amphibians include salamanders, caecilians, frogs, and toads. All are carnivores as adults, preying mainly on insects and worms.

The 655 species of salamanders and related newts live mainly in North America, Europe, and Asia. In body form, they are the modern group most like early tetrapods. Forelimbs and back limbs are of similar size, and there is a long tail (**FIGURE 25.14**). As salamanders walk, their body bends from side to side, like the body of a swimming fish.

Larval salamanders look like small versions of adults, except for the presence of gills. Typically, they lose their gills and develop lungs as they mature. However, some salamanders (axolotls) retain gills even

3
2
1

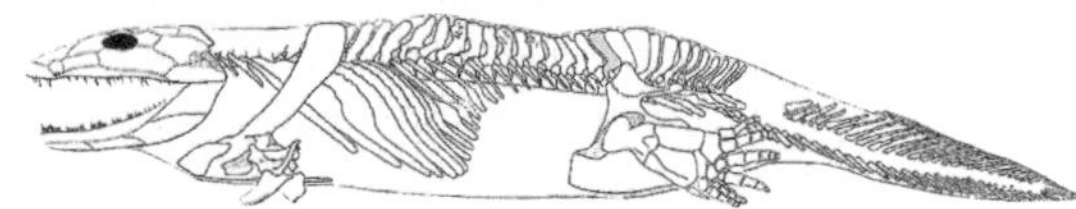

❸ Early amphibian (*Icthyostega*) with well-developed ribs and thick limbs with distinct digits.

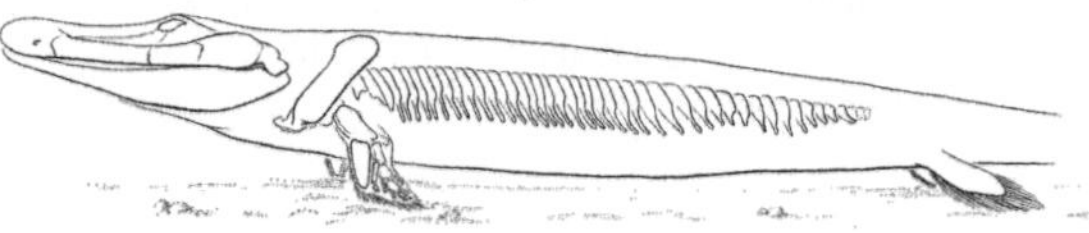

❷ Fish (*Tiktaalik*) with sturdier weight-bearing pectoral fins, wristlike bones, and enlarged ribs.

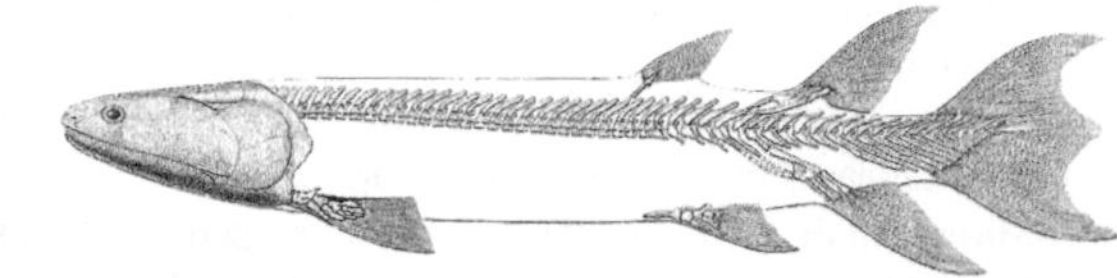

❶ Fish (*Eusthenopteron*) with bony fins.

FIGURE 25.13 Transition to land. Fossil species from the late Devonian illustrate how a fish body became adapted for life on land.

A Long, muscular hind limbs allow an adult frog to make spectacular leaps. Frogs typically have smooth skin and spend much of their time in water.

as adults. Other salamander species lose the gills but do not develop lungs. In these species, gas exchange occurs across the skin.

Caecilians are amphibians closely related to salamanders. They live in the tropics and have a wormlike form that adapts them to their burrowing life. There are about 200 species, all limbless and blind.

Frogs and toads belong to the most diverse amphibian order, with more than 5,000 species. The order's name, Anura, means "without a tail" and adults are always tailless.

The long, muscular hind limbs of an adult frog allow it to swim, hop, and make spectacular leaps (**FIGURE 25.15A**). The much smaller forelimbs help to absorb the impact of landings. Frogs usually have a smooth, thin skin and live in a moist environment.

Toads are better adapted to dry conditions. They have thick, bumpy skin (**FIGURE 25.15B**). Compared with frogs, toads have a stubbier body and their hind legs are proportionately shorter. Toads can hop, but more often they walk.

Both frogs and toads undergo metamorphosis, during which a gilled, tailed larva (**FIGURE 25.15C**) transforms itself into an adult with lungs and no tail.

Declining Diversity

We are in the midst of an alarming decline in amphibian numbers. Population reductions are best documented in North America and Europe, but similar declines are occurring worldwide. In the United States, more than thirty amphibian species are now listed as threatened or endangered.

A variety of factors contribute to the threat. An amphibian's thin, scaleless skin makes it relatively easy for parasites, pathogens, and pollutants to enter the body. Section 23.3 described the effects of an introduced chytrid fungus on frogs. Habitat loss is another important threat. In many places, people have filled in low-lying areas that once collected water from seasonal rains. Such seasonal pools are important breeding sites for amphibians.

B Toads have rougher, thicker skin than frogs and typically spend less time in water. The flattened disk visible behind this cane toad's eye is its eardrum.

C Aquatic larva (a tadpole), with gills and a tail.

FIGURE 25.15 Anurans: frogs and toads.

TAKE-HOME MESSAGE 25.4

✔ Amphibians are carnivorous vertebrates that typically live on land but breed in water.

✔ The transition to life on land involved changes to the skeleton, lungs, circulatory system, and sensory systems.

✔ The most diverse amphibian group includes the frogs and toads, which undergo metamorphosis from gilled larvae. Salamanders and the closely related caecilians are less diverse lineages.

✔ An amphibian's scaleless, permeable skin makes it vulnerable to pollutants, and its requirement for water in which to breed makes it sensitive to habitat alteration.

amphibian Tetrapod with a three-chambered heart and scaleless skin; typically develops in water, then lives on land as an air-breathing carnivore.

CREDITS: (15A) Mc Donald Wildlife Photog./Animals Animals; (15B) Chris Ison/Shutterstock; (15C) iStockphoto.com/Tommounsey.

25.5 Amniote Evolution

LEARNING OBJECTIVES

- Describe amniote adaptations to life on land.
- List the two amniote lineages and the modern groups in each.
- Compare the physiology of endotherms and ectotherms, and provide examples of each.

Amniotes branched off from an amphibian ancestor during the Carboniferous. A variety of traits adapt them to life in dry places. They have lungs throughout their life. Their skin is rich in keratin, a protein that makes it waterproof. A pair of well-developed kidneys help conserve water, and fertilization takes place in the female's body. Sexes are typically separate and fixed for life. Amniotes produce eggs in which an embryo develops bathed in fluid, so amniotes can develop on dry land (FIGURE 25.16). Membranes within the egg function in gas exchange, nutrition, and waste removal.

A branching of the amniote lineage during the Carboniferous produced two clades: Synapsida and Reptilia. Mammals belong to the clade Synapsida. The clade Reptilia includes all modern lizards, snakes, turtles, crocodilians, and birds:

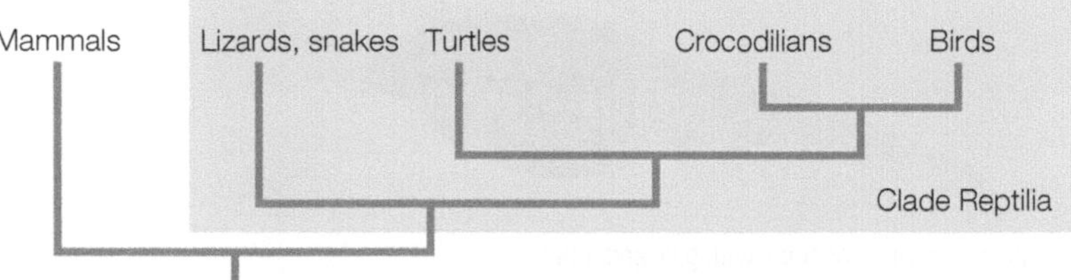
Mammals
Lizards, snakes
Turtles
Crocodilians
Birds
Clade Reptilia

Reptilia also includes the dinosaurs. **Dinosaurs** are an amniote group defined by skeletal features such as the anatomy of the pelvis and hips. One dinosaur lineage, the theropods, included many feathered species. Like modern birds and mammals, these may have been **endotherms**, which maintain their body temperature by adjusting metabolic heat production. All modern snakes, lizards, turtles and crocodilians are **ectotherms**, animals whose temperature varies with that of their environment.

FIGURE 25.16 **Amniote eggs.** Snakes hatching from leathery eggs.

Birds branched off from a theropod dinosaur lineage during the Jurassic (FIGURE 25.17) and are the only surviving descendants of dinosaurs. All dinosaurs became extinct by the end of the Cretaceous, most likely as a result of an asteroid impact (Section 16.1). Mammals also arose during the Jurassic, so some lived alongside dinosaurs. However, the group to which most modern mammals belong—the placental mammals—probably did not evolve until after the dinosaurs' demise.

TAKE-HOME MESSAGE 25.5

✔ Amniotes are vertebrates that produce eggs in which the young can develop away from water. They have waterproof skin and highly efficient kidneys. Fertilization is internal.

✔ An early divergence created two amniote lineages. Synapsida includes the mammals. Reptilia include turtles, lizards, snakes, crocodilians, and birds, as well as the now extinct dinosaurs.

FIGURE 25.17 **A Jurassic scene.** In the foreground, the early bird *Archaeopteryx* glides along. Behind the birds, a meat-eating dinosaur sizes up a larger plant-eating one. At the far right, an early mammal surveys the scene from its perch on a tree.

FIGURE IT OUT Which of the animals mentioned above do biologists consider amniotes? Which do they group as Reptilia?

Answer: All are amniotes. The dinosaurs and birds are members of Reptilia.

25.6 Reptiles

LEARNING OBJECTIVES

- Explain how snakes are related to lizards.
- Describe how turtle anatomy has changed over time.
- Describe some ways that crocodilians are similar to birds.

The amniote clade Reptilia includes the animals informally referred to as reptiles (lizards, snakes, turtles, and crocodilians), as well as the birds. We discuss reptiles here, and birds in Section 25.7.

Lizards and Snakes

Lizards and snakes constitute the Squamata, the most diverse group of modern reptiles. There are more than 10,000 species. Members of this group have overlapping scales, and they periodically shed their skin as they grow. All have teeth. Iguanas are herbivores, but most lizards eat insects. The largest lizard, the Komodo dragon (**FIGURE 25.18A**), grows up to 3 meters (10 feet) long. After inflicting a venomous bite, it trails its prey for hours or days until the poisoned animal collapses.

Snakes first evolved during the Cretaceous, from short-legged, long-bodied lizards. Some modern snakes have bony remnants of hind limbs, but most lack limb bones entirely. All snakes are predators, but only some have fangs. Rattlesnakes and other fanged snakes subdue prey with a venom they produce in modified salivary (saliva-producing) glands. Other snakes are constrictors that suffocate a prey animal by wrapping it so tightly that it cannot expand its chest to inhale.

Turtles

The 300 or so species of turtles (Testudines) have a bony, keratin-covered shell attached to their skeleton (**FIGURE 25.18B**). Most turtles that live in the sea feed on invertebrates such as sponges or jellies; others feed mainly on seagrass. Freshwater turtles eat fishes and invertebrates. Land-dwelling turtles, which are commonly called tortoises, feed on plants.

Fossils that date back more than 200 million years reveal how turtles have evolved. The chest and belly of the fossil turtles is covered by a shell derived from fused, expanded ribs. However, there is no shell on the back, only widened ribs. The fossil turtles also have teeth. By contrast, modern turtles are toothless. Like birds, they have jaws that are covered with keratin to form a horny beak.

Crocodilians

Crocodiles, alligators, and caimans (Crocodilia) are stealthy predators with a long snout and many sharp peglike teeth (**FIGURE 25.18C**). They spend much of their time in water, where a long, powerful tail propels them when they swim. A marine crocodile that lived during the Cretaceous was 10 meters long.

Crocodilians are the closest living relatives of birds. Like many birds, they are highly vocal and engage in complex parental behavior. During courtship, males and females grunt and bellow. After a female mates, she digs a nest, lays eggs, then buries and guards them. Young that are ready to hatch call to summon their mother, who helps them dig their way to the surface.

A Komodo dragon, the largest lizard.

B Turtle swimming in pond.

C Crocodile with its fish prey.

FIGURE 25.18 **Reptile diversity.**

dinosaur Group of reptiles that includes the ancestors of birds; became extinct at the end of the Cretaceous.
ectotherm Animal whose body temperature varies with that of its environment.
endotherm Animal that maintains its temperature by adjusting its production of metabolic heat; for example, a bird or mammal.

CREDITS: (18A) Soren Egeberg Photography/Shutterstock; (18B) Michelson, Robert S/Animals Animals; (18C) Johan Swanepoel/Shutterstock.

Crocodilians and birds also share another trait: a four-chambered heart. In lizards, snakes, and turtles (as in amphibians) the heart has three chambers. In one of these chambers, oxygen-rich and oxygen-poor blood mix a bit. In a four-chambered heart, oxygen-poor blood from body tissues never mixes with oxygen-rich blood from the lungs.

TAKE-HOME MESSAGE 25.6

✔ Lizards and snakes have skin covered with overlapping scales. Most are predators.

✔ Turtles have a bony, keratin-covered shell and are toothless, with a horny keratin beak.

✔ Crocodilians are aquatic predators with a four-chambered heart and a long snout with peglike teeth.

25.7 Birds

LEARNING OBJECTIVES

- Give examples of structural and physiological traits that are adaptations to flight.
- List some functions of bird feathers.
- Describe some ways in which bird body form varies.

Birds (Aves) are only the living animals with feathers. They arose during the Jurassic and are descended from feathered therapod dinosaurs. **Feathers** are filamentous structures derived from scales and made mainly of keratin.

FIGURE 25.19 **Bird in flight.**

Bird feathers have a variety of functions. Birds are endotherms, and downy feathers help them maintain their temperature. Feathers slow the loss of metabolic heat in cool environments and prevent heat gain in hot ones. Feathers also shed water and thus help to keep a bird's skin dry. In many birds, colorful plumage plays a role in courtship. Feather color can result from the way that keratin reflects light or from deposition of pigments derived from the diet. For example, dietary carotenoids deposited in feathers make flamingos pink.

Adaptations to Flight

Feathers also play a role in flight. Like humans, birds stand upright on their hind limbs, and their wings are homologous with our arms. Each wing is covered with long flight feathers that extend outward and increase the wing's surface area (FIGURE 25.19). Flight feathers give the wing a shape that helps lift the bird as air passes over the wing.

Birds typically have a large sternum (breastbone) with a bladelike vertical protrusion called a keel at its center (FIGURE 25.20). Flight muscles extend from the keel to bones of the upper limb. Contraction of one set of flight muscles produces a powerful downstroke that lifts the bird. A less powerful set of muscles contracts to raise the wing.

Most birds are surprisingly lightweight, and this feature helps them become and remain airborne. Bird bones are stronger and stiffer than mammal bones, but internal air cavities make bird bones lighter than their mammalian equivalents. The beak, which consists of keratin, weighs much less than a jaw with bony teeth. Birds also do not have a bladder, an organ that stores urinary waste in many other vertebrates.

Supplying the ATP that flight muscles need to contract requires a steady supply of oxygen for aerobic respiration (Section 7.2). A system of air sacs keeps air flowing continually through a bird's lungs and thus maximizes oxygen intake. A four-chambered heart moves blood to and from the tissues quickly.

Flying requires good eyesight and a great deal of coordination. Compared to a lizard of similar body mass, a bird has a larger brain and much larger eyes.

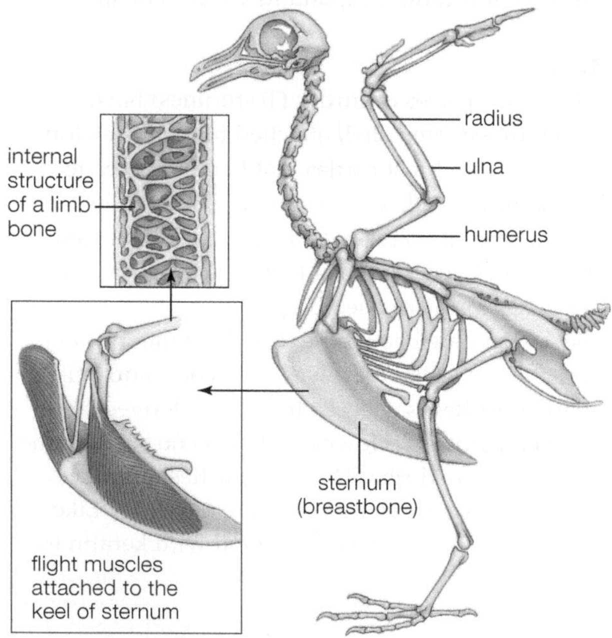
radius
ulna
humerus
internal structure of a limb bone
sternum (breastbone)
flight muscles attached to the keel of sternum

FIGURE 25.20 **Bird skeleton.** A bird's skeleton is made up of lightweight bones with internal air pockets. A wing is a modified forelimb (see FIGURE 18.5). Powerful flight muscles attach to the keel of the large breastbone, or sternum.

bird Feathered, endothermic amniote with a beak and wings.
feather Filamentous keratin stucture derived from ancestral scales.

CREDITS: (19) Eric Isselee/Shutterstock; (20) © Cengage Learning.

Each eye is protected within a circle of bone that holds it fixed in place. Thus, a bird cannot move its eyes to look to the side, as you can. Instead, the bird must turn its highly flexible neck.

Reproduction and Development

As in other amniotes, fertilization is internal. However, most male birds do not have a penis. To inseminate a female, a male presses his cloaca against hers in a maneuver poetically described as a cloacal kiss (**FIGURE 25.21A**). After fertilization occurs, the female lays a fertilized egg that has the characteristic amniote membranes (**FIGURE 25.21B**). Nutrients from the egg's yolk and water from the egg white (albumen) sustain the developing embryo. Like some turtles and all crocodilians, birds encase their eggs within a rigid shell hardened by calcium carbonate.

The egg must be kept warm in order for the embryo within it to develop. In nearly all birds, one or both parents incubate the eggs until they hatch. Parental responsibilities typically continue even after eggs hatch (**FIGURE 25.21C**).

Avian Diversity

More than half of the approximately 10,000 living bird species belong to the order of perching birds (Passeriformes). This group includes familiar backyard birds such as sparrows, crows, jays, starlings, swallows, finches, robins, warblers, orioles, and cardinals. All have four toes on each foot, three facing forward and one facing backward. Together, the toes can encircle a branch when the bird perches. Most birds can vocalize, but a subgroup of the perching birds is notable for their elaborate songs. Birds do not have a larynx with vocal cords like we do. Their vocal organ is a bony syrinx. Paired structures in the syrinx allow birds to produce two different notes simultaneously.

The second most diverse order of birds includes 450 species, most of them hummingbirds. Hummingbirds are agile fliers and the only birds capable of flying backward. They have an elongated beak and a lengthy tongue that allow them to take up nectar from deep inside flowers.

The structure of a bird's beak generally reflects the bird's diet. Seed-eating birds have stout beak for breaking open their food. Flesh-eating birds such as eagles, owls, and vultures have a sharp beak for cutting open prey. Herbivorous ducks have a flattened bill for straining vegetation from the water.

About 20 percent of birds make a seasonal migration. These birds fly from one region to another in response to a seasonal change, such as a shift in day length. Migration typically involves a spring flight to a breeding site where insects are abundant in the summer, then an autumn flight to an overwintering site. One shorebird monitored by researchers flew from Alaska to New Zealand, a distance of 11,500 kilometers (7,145 miles), without stopping to feed or rest.

At the other extreme, penguins and ratite birds such as ostriches cannot fly. Penguins live along coasts of the Southern Hemisphere. They flap their wings to propel themselves through water. The ostrich, the largest living bird, is native to Africa. It can weigh up to

A Mating. A male bird does not have a penis. To inseminate his partner, he must balance on her back and bend his body so his cloaca meets hers.

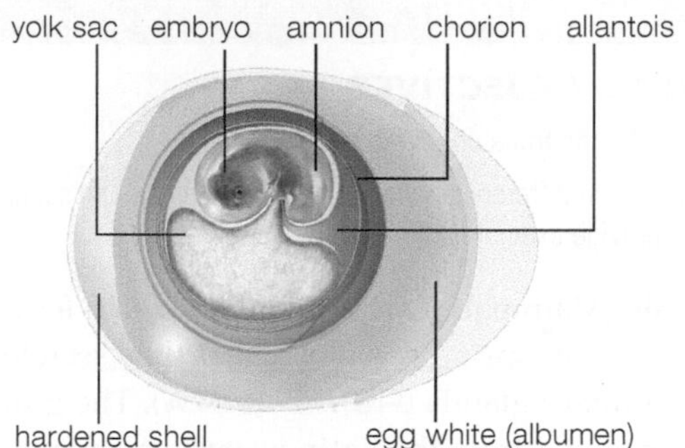
yolk sac
embryo
amnion
chorion
allantois
hardened shell
egg white (albumen)

B Egg structure. The egg has a calcium-hardened shell that encloses the embryo and four characteristic amniote membranes (yolk sac, amnion, chorion, and allantois).

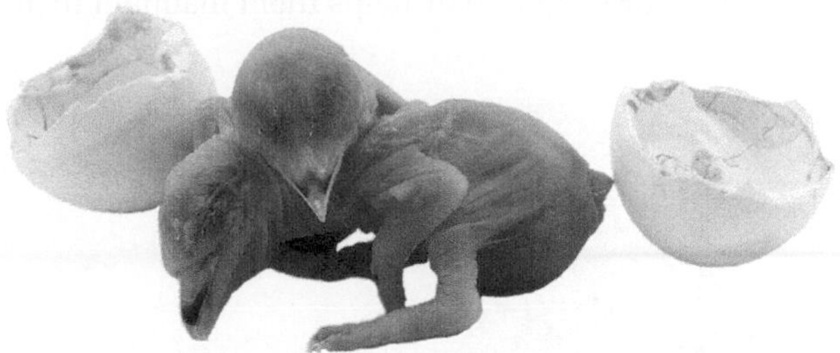

C Hatchlings. Many birds, including these parrots, hatch in a relatively undeveloped state and require extensive parental care before they can live on their own.

FIGURE 25.21 Bird reproduction and development.

CREDITS: (21A) en.wikipedia.org/wiki/User:Jmgarg1 'Creating awareness of Indian Flora & Fauna' Image Resource of thousands of my images of Birds, Butterflies, Flora etc. (arranged alphabetically & place-wise): commons.wikimedia.org/wiki/Category:J.M.Garg. For learning about Indian Flora, visit/ join Google e-group- Indiantreepix:http://groups.google.co.in/group/indiantreepix hl=en; (21C) Jane Burton/ Nature Picture Library.

150 kilograms (330 pounds). Other ratite birds include Australia's emus and cassowaries, New Zealand's kiwis, and South America's rheas. Unlike other birds, ratites have a flat sternum; there is no keel.

TAKE-HOME MESSAGE 25.7

✔ Birds evolved from therapod dinosaurs. They are the only living animals with feathers, which are derived from ancestral scales.

✔ Adaptations for flight include lightweight bones, air sacs that increase the efficiency of respiration, and a four-chambered heart.

✔ Birds have a toothless beak. The structure of the beak reflects the bird's diet.

25.8 Mammals

LEARNING OBJECTIVES

- Describe the traits unique to mammals.
- Explain the differences among the three mammalian subgroups and provide examples of each.

Mammals (Mammalia) are animals in which females nourish their offspring with milk that they secrete from mammary glands (**FIGURE 25.22A**). The group name is derived from the Latin *mamma*, meaning breast. Mammary glands are modified sweat glands. Mammals have hair or fur that, like feathers, is made of the protein keratin. Mammals are endotherms, and a coat of fur or head of hair helps them maintain their body temperature. Mammals also include the only animals that sweat, although not all mammals do so. The mammalian heart is four-chambered, and gas exchange occurs in a pair of well-developed lungs.

Mammals have distinctive skeletal and dental traits. Compared to other vertebrates, they have a larger skull and brain for their body size. Their lower jaw consists of a single bone, whereas other jawed vertebrates have multiple bones in the lower jaw. Mammals are also the only vertebrates that have three bones in their middle ear. In the ancestor of mammals, there was a single middle ear bone and multiple lower jaw bones, as in reptiles. Over time, two of the jaw bones moved and became modified to serve as bones of the middle ear.

As a group, mammals have four different types of teeth, each with a distinctive shape (**FIGURE 25.22B**). Incisors are used to gnaw, canines tear and rip flesh, and premolars and molars grind and crush hard foods. Not all mammals have teeth of all four types, but most have some combination. In other vertebrates, an individual's teeth may differ in size, but they are all the same shape. Having a variety of tooth shapes gives mammals an ability to eat a wider variety of foods than other vertebrates.

A Mammary glands and hair or fur. A young seal sucks milk from its mother's nipples. Seal nipples retract inward when a pup is not nursing.

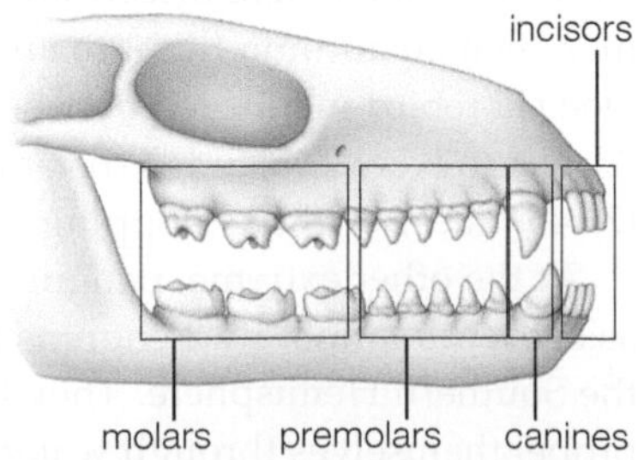
incisors
molars
premolars
canines

B Multiple types of teeth and a single bone in the lower jaw.

FIGURE 25.22 Distinctive mammalian traits.

Mammalian Origins and Diversification

As previously noted, a divergence during the Carboniferous produced two amniote lineages, Reptilia and Synapsida. Our focus here is the Synapsida.

Mammals belong to a synapsid clade called the therapsids. Therapsids arose and underwent an adaptive radiation during the late Permian. By the end of this period, they had become the dominant animals on land. Then, disaster struck. The end of the Permian period (250 million years ago) marks the most extensive extinction event known; 70 percent of species that lived on land disappeared. Most synapsids perished at this time, but a group of therapsids called cynodonts was among the survivors. Cynodont means "dog toothed," and these animals had canines, incisors, and molars. Most likely they were endotherms and had insulating fur or hair. Mammals evolved from a cynodont ancestor during the Jurassic.

Three mammal lineages survived to the present: monotremes, marsupials, and placental mammals. An early divergence separated the lineage leading to monotremes from that leading to marsupial and placental mammals:

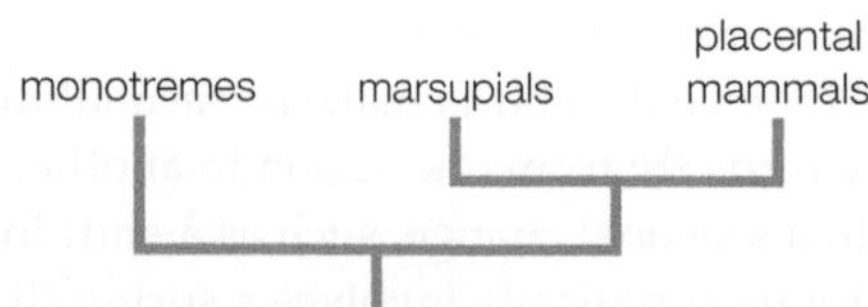
monotremes
marsupials
placental mammals

A Mammals in a Wyoming forest, about 63 million years ago.

B The giant placental mammal, *Indricotherium*, weighed as much as 20 tons and stood 5.5 meters (18 feet) tall at the shoulder. It lived in Asia between 20 and 35 million years ago.

FIGURE 25.23 After the dinosaurs; early results of the mammalian adaptive radiation.

Representatives of all three mammalian lineages lived alongside the dinosaurs. They were generally small bodied. About 65 million years ago many mammals perished in a mass extinction (the K-Pg extinction) that resulted in the dinosaurs' demise (Section 16.1). However, in the aftermath of this event, mammals underwent a great adaptive radiation. FIGURE 25.23 shows some results of this early diversification.

Monotremes—Egg-Laying Mammals

The egg-laying mammals, or **monotremes,** are the oldest surviving mammal lineage and they are thought to retain the most ancestral traits. Female monotremes lay and incubate eggs that have a leathery shell like that of lizard eggs. Offspring hatch in a relatively undeveloped state—tiny, hairless, and blind. Young cling to the mother or are held in a skin fold on her belly. Milk oozes from openings on the mother's skin; monotremes do not have nipples. The name monotreme comes from the Greek *mono trema,* which means "single hole" and refers to the cloaca that is present in all monotremes.

Montremes are the least diverse group of modern mammals, with only five species. Four are echidnas (FIGURE 25.24A) that live in Australia or New Guinea.

mammal Animal with hair or fur; females feed young with milk secreted by mammary glands.

monotreme (MON-oh-treem) Egg-laying mammal; for example, an echidna or platypus.

A Echidna (spiny anteater).

B Platypus with its young.

FIGURE 25.24 **Monotremes.**

Echidnas live on land, where they feed on insects and earthworms. The fifth species is the platypus, a cat-sized semiaquatic Australian animal that has a beaverlike tail, a ducklike bill, and webbed feet (FIGURE 25.24B). Platypuses burrow into riverbanks using claws that are exposed when they retract the web on their feet. Both males and females have spurs on their hind feet. The male's spurs produce venom that is used to fend off predators and rival males. Platypuses use their bill to search for freshwater shrimp and aquatic insects. The bill is highly sensitive both to touch and to electrical signals that are produced by the muscle contractions of their prey.

CREDITS: ; (23A-B) Painting © Ely Kish; (24A) D. & V. Blagden/ANT Photo Library; (24B) Jean-Philippe Varin/Science Source.

A Kangaroo with a single juvenile in its pouch.

B A North American oppossum with its young.

FIGURE 25.25 **Marsupials.**

Marsupials—Pouched Mammals

Marsupials are pouched mammals. Young marsupials develop for a brief period inside their mother's body, where they are nourished by egg yolk and by nutrients that diffuse from maternal tissues. They are born while still blind and tiny, when their limbs have just begun to develop. After birth, they must use these stubby limbs to crawl along their mother's body to a permanent pouch on her ventral surface. Once inside the pouch, they attach to a nipple, suckle, and grow. Like monotremes, marsupials have a cloaca.

About 300 species of marsupials live in Australia and on nearby islands. They include the plant-eating kangaroos (FIGURE 25.25A) and koalas, and the Tasmanian devil, a carnivore the size of a small dog. About 100 marsupial species live in South or Central America. Most are opossums. North America's only native marsupial is also an opposum (FIGURE 25.25B).

Placental Mammals

The young of **placental mammals** develop for an extended period inside their mother's body, where they are nourished by means of a placenta. The mammalian **placenta** is an organ that develops from membranes around the embryo during pregnancy. It attaches the embryo to the wall of the mother's uterus (womb) and allows materials to diffuse between the maternal and embryonic bloodstreams. Placental mammals are born at a much later developmental stage than monotremes. After birth, young placental mammals suckle milk from nipples on their mother's ventral surface. Placental mammals have separate openings for the urinary, reproductive, and excretory systems.

With more than 4,000 species, placental mammals are the dominant mammals in most land habitats and the only mammals that live in the seas. FIGURE 25.26 shows examples of major, currently accepted orders.

About 40 percent of placental mammals are rodents (order Rodentia). They include rats, mice, hamsters, squirrels, beavers, porcupines, and guinea pigs. Short generation times and a diverse diet probably contributed to the rodents' success. All rodents have teeth specialized for gnawing. A rodent's incisors grow continually, to prevent them from being worn down.

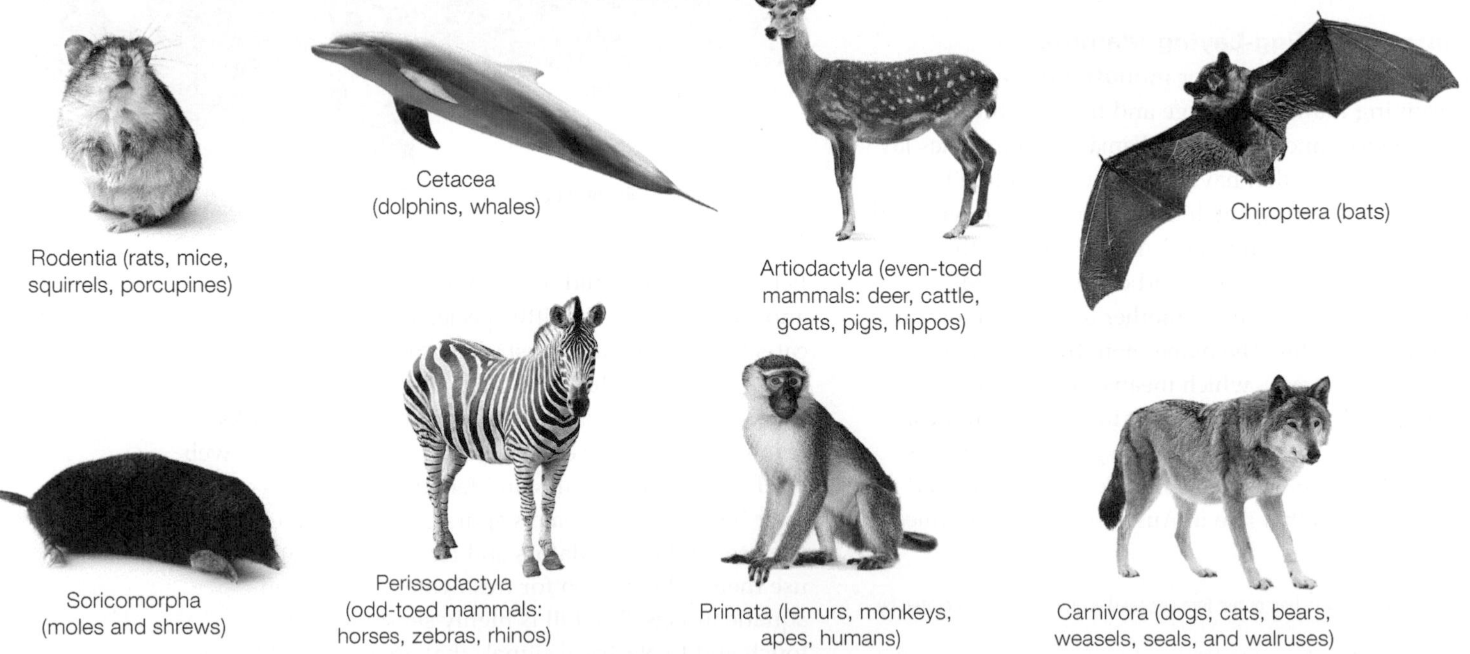

Rodentia (rats, mice, squirrels, porcupines)
Cetacea (dolphins, whales)
Artiodactyla (even-toed mammals: deer, cattle, goats, pigs, hippos)
Chiroptera (bats)
Soricomorpha (moles and shrews)
Perissodactyla (odd-toed mammals: horses, zebras, rhinos)
Primata (lemurs, monkeys, apes, humans)
Carnivora (dogs, cats, bears, weasels, seals, and walruses)

FIGURE 25.26 **Representatives of major orders of placental mammals.**

About 1,200 species of bats constitute the second most diverse order (Chiroptera). Most are nocturnal (active at night). Bats are the only mammals capable of sustained flight. Their wings consist of membranes of skin stretched between bones. Old World bats are larger and eat fruit, whereas New World bats are smaller and most are insect eaters.

Moles and shrews make up the next most diverse order (Soricomorpha). Most moles are adapted to a burrowing way of life, with strong forelimbs for tunneling, reduced eyes, and no external ear flaps. They eat earthworms and insects. Shrews resemble mice with spiky teeth. Moles and shrews were historically grouped with other small insect eaters such as hedgehogs in an order (Insectivora) that was dismantled when gene comparisons showed its members were not closely related.

"Carnivora" means meat eater, and members of this order have enlarged canine teeth suited to tearing at flesh. They include cats, dogs, wolves, bears, foxes, and weasels, as well as marine pinnipeds (fin-footed animals) such as seals, sea lions, and walruses.

Whales and dolphins (Cetacea) have adapted to life in the sea. Section 16.4 described some fossils demonstrating the cetacean transition from land dwellers to marine mammals. As far as we know, the blue whale, which can be 30 meters (100 feet) long, is the largest mammal that ever lived.

Cetaceans are close relatives of even-toed mammals (Artiodactyla). Most large mammalian grazers such as cattle, deer, and goats are members of this order, as are pigs and hippos. Horses, zebras, and rhinos are grazers too, but they belong to the odd-toed order (Perissodactyla). Both orders of grazing mammals subsist on a diet of plant material that they digest with the assistance of microbes that live in their gut.

Humans are members of the order Primates. Chapter 26 explores the unique adaptations of this group and the history of the human lineage.

marsupial mammal Mammal in which young are born at an early stage and complete development in a pouch on the mother's surface.
placenta (pluh-SENT-uh) Of placental mammals, organ that forms during pregnancy and allows diffusion of substances between the maternal and embryonic bloodstreams.
placental mammal Mammal in which a mother and her embryo exchange materials by means of an organ called the placenta.

TAKE-HOME MESSAGE 25.8

- ✔ Mammals produce milk and have hair or fur.
- ✔ The first mammals evolved while dinosaurs lived, but the group underwent an adaptive radiation after extinction of the dinosaurs.
- ✔ Five species of egg-laying mammals (monotremes) persist. Pouched mammals (marsupials) are more diverse. They live mainly in Australia and South America.
- ✔ Placental mammals are the most diverse mammal group. Their young develop to an advanced stage inside the mother.

25.1 Very Early Birds (revisited)

Archaeopteryx was the first early bird fossil discovered but many other fossil birds are now known. Consider *Confuciusornis sanctus*, a fossil species that lived about 120 million years ago. It had a beak and a short tailbone like that of a modern bird (**FIGURE 25.7A**). However, unlike wings of nearly all modern birds, those of *C. sanctus* had claws at their tips.

Another fossil takes us back even farther in time to about 130 million years ago and supports the hypothesis that birds descended from dinosaurs. In 1994, a farmer in China discovered a fossil of a small dinosaur with short forelimbs and a long tail (**FIGURE 25.7B**). Unlike most dinosaurs, this one was covered with tiny filaments that resemble downy feathers of modern birds. Researchers named the farmer's fuzzy find *Sinosauropteryx prima*, which means "first Chinese feathered dragon." Given its shape and lack of long feathers, *S. prima* was certainly unable to fly. Its fuzzy feathers probably functioned as insulation, as they do in modern birds.

Some fossil feathers such as those of the dinosaur *Sinosauropteryx* contain microscopic granules of the type that contain pigment in modern bird feathers. In birds, different shaped granules contain different shades of melanin, a pigment that comes in red to brown shades. By studying the distribution and shape of pigment granules in *Sinosauropteryx* fossils, scientists concluded this animal had a reddish color and a tail with alternating red and white bands. ●

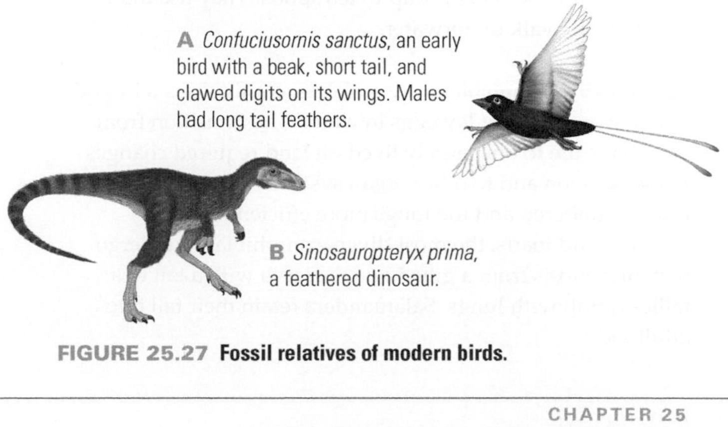

A *Confuciusornis sanctus*, an early bird with a beak, short tail, and clawed digits on its wings. Males had long tail feathers.

B *Sinosauropteryx prima*, a feathered dinosaur.

FIGURE 25.27 Fossil relatives of modern birds.

Section 25.1 Fossils provide evidence of evolutionary transitions between major animal groups. For example, fossils show that feathers, a defining trait of modern birds, evolved in a dinosaur ancestor of birds. Fossils also document how birds lost their long tail and how a jaw with teeth was replaced by a beak.

Section 25.2 Four embryonic features define the **chordates**: a **notochord**, a dorsal hollow nerve cord, a pharynx with gill slits, and a tail that extends past the anus. Depending on the group, these features may or may not persist in adults. Two groups of marine invertebrates, **tunicates** and **lancelets**, are chordates. However, most chordates are **vertebrates**, meaning they have a backbone as part of their **endoskeleton**. Early vertebrates that swam in the seas gave rise to **tetrapods** that walked on land. **Amniotes** are a tetrapod subgroup that adapted to a life spent entirely on land.

Section 25.3 **Fishes** are gilled aquatic vertebrates. The earliest fishes such as the ostracoderms were jawless. Lampreys and hagfishes are modern jawless fishes. Some lampreys are parasites of other fish as adults. Hagfishes are scavengers that produce slime when stressed.

Jaws evolved through a modification of gill-supporting bones in a jawless ancestor. Jawed fishes were the first vertebrates to have **scales** and paired fins. Placoderms, a now-extinct group of armored jawed fishes, dominated seas during the Devonian period.

Cartilaginous fishes such as sharks and rays are jawed fishes with a cartilage skeleton. As in hagfishes and lampreys, the reproductive, digestive, and urinary systems share a single opening called the **cloaca**.

Bony fishes have a skeleton that consists of bone as well as cartilage. A bony cover overlays their gills, and there are usually separate openings serving the reproductive, digestive, and urinary systems. **Ray-finned fishes** are bony fishes that have thin, flexible fins and a **swim bladder**. They include teleosts, the most diverse lineage of modern fishes. A whole-genome duplication may have contributed to teleost diversity. **Lobe-finned fishes** have thick fins with bony supports. Coelacanths and lungfishes are in this group. Lungfishes have lungs and are the sister group to tetrapods. They use their lobed fins to walk underwater.

Section 25.4 **Amphibians** are scaleless, carnivorous tetrapods that live on land, but lay eggs in water. The transition from an aquatic life to one largely lived on land required changes to the skeleton and to other organ systems. The heart became three-chambered and the lungs more efficient.

Frogs and toads, the most diverse amphibians, undergo metamorphosis from a gilled aquatic larva with a tail to a tailless adult with lungs. Salamanders retain their tail into adulthood.

Section 25.5 Amniotes are tetrapods adapted to life on land by waterproof skin, highly efficient kidneys, and special eggs that allow embryos to develop away from water. There are two lineages. Reptilia includes the birds and all animals commonly called reptiles, including the extinct **dinosaurs**. Synapsida includes the mammals. Some dinosaurs may have been **endotherms**, which regulate their temperature as birds and mammals do. Modern nonbird reptiles are **ectotherms**.

Section 25.6 Lizards and snakes collectively constitute the most diverse lineage of modern reptiles. Snakes evolved from lizards and some retain vestiges of legs. Most snakes and lizards are predators and some are venomous.

Turtles are reptiles with a keratin-covered shell and a toothless beak. Fossils show that their shell evolved through a widening of the ribs.

Crocodilians are predators that spend much of their time in water. Unlike other reptiles, crocodilians have a four-chambered heart that prevents oxygen-poor and oxygen-rich blood from mixing. They are the closest living relatives of birds and like birds they provide parental care to the young.

Section 25.7 **Birds** are the only living animals with feathers. **Feathers** are derived from ancestral scales. They provide insulation and waterproofing. They also function in courtship and flight.

The bird body has many adaptations to flight. Front limbs are wings that are moved by muscles attached to a keeled sternum. Bones are lightweight, and a system of air sacs keeps air flowing continually through the lungs. Birds have a heart with four chambers. Most male birds do not have a penis, so reproduction involves cloaca-to-cloaca contact. Bird eggs must be incubated to develop properly, and many birds also provide additional care to the young.

Perching birds are the most diverse lineage. Some birds migrate seasonally, typically moving between a breeding area and an area where they spend the winter. Penguins and ratites are two lineages of flightless birds.

Sections 25.8 Female **mammals** nourish their young with milk secreted by mammary glands. Mammals have fur or hair, and more than one kind of tooth. The lineage evolved while dinosaurs were present, but underwent an adaptive radiation after dinosaurs became extinct.

There are three modern lineages of egg-laying mammals (**monotremes**), pouched mammals (**marsupials**), and **placental mammals**. Placental mammals are the most diverse group. A **placenta** is an organ that facilitates exchange of substances between the embryonic and maternal blood.

Placental mammals are born more fully developed than other groups. They have become the dominant mammal group in most environments. The most diverse placental lineages are rodents, bats, and the moles and shrews.

SELF-QUIZ

Answers in Appendix VII

1. List the four distinguishing chordate traits.
2. Which of the traits listed above are retained by an adult lancelet?
3. Vertebrate jaw bones evolved from _______ .
   a. gill supports b. ribs c. scales d. ear bones
4. Both cartilaginous and bony fishes have _______ .
   a. jaws
   b. a bony skeleton
   c. lungs
   d. a swim bladder
   e. a four-chambered heart
   f. all of the above
5. Tetrapods evolved from _______ .
   a. sharks
   b. teleosts
   c. lobe-finned fishes
   d. placoderms
6. Turtles, lizards, and birds belong to one major lineage of amniotes, and _______ belong to another.
   a. sharks
   b. frogs and toads
   c. mammals
   d. salamanders
7. Amniotes are adapted to life on land by _______ .
   a. waterproof skin
   b. internal fertilization
   c. efficient kidneys
   d. amniote eggs
   e. both a and c
   f. all of the above
8. The closest living relatives of hagfishes are _______ .
   a. snakes
   b. sharks
   c. lampreys
   d. placoderms
9. Among living animals, only birds have _______ .
   a. a cloaca
   b. amniote eggs
   c. feathers
   d. a four-chambered heart
10. Feathers and hair consist mainly of _______ .
    a. chitin
    b. cartilage
    c. keratin
    d. collagen
11. Unlike *Archaeopteryx,* modern birds have _______ .
    a. a long bony tail
    b. a toothless beak
    c. a two-chambered heart
    d. feathers
12. _______ have scaleless skin and a heart with three chambers.
    a. Amphibians
    b. Mammals
    c. Ray-finned fishes
    d. Snakes
13. Match each structure with its description.

___	cloaca	a. adjusts buoyancy
___	swim bladder	b. multipurpose opening
___	fur or hair	c. supports body
___	placenta	d. insulates a mammal
___	endoskeleton	e. waterproofs amniote skin
___	keratin	f. connects mother and her developing offspring

14. Match the organisms with the appropriate description.

___	lancelets	a. pouched mammals
___	lampreys	b. most diverse mammal lineage
___	amphibians	c. feathered amniotes
___	lizards	d. egg-laying mammals
___	birds	e. extinct jawed fishes
___	sharks	f. ectothermic amniotes
___	monotremes	g. cartilaginous fishes
___	marsupials	h. first land tetrapods
___	placoderms	i. living jawless fishes
___	placental mammals	j. invertebrate chordates

15. Arrange the groups in order in which they evolved.

___	1 (earliest)	a. Jawless fishes
___	2	b. Birds
___	3	c. Dinosaurs
___	4	d. Amphibians
___	5 (most recent)	e. Jawed fishes

CRITICAL THINKING

1. In 1798, a stuffed platypus specimen was delivered to the British Museum. Reports that it laid eggs created much confusion. To modern biologists, a platypus is clearly a mammal. It has fur and the females produce milk. Young animals have typical mammalian teeth that are replaced by hardened pads of the "bill" as the animal matures. Why do you think modern biologists can more easily accept that a mammal can have some reptilelike traits than scientists who were considering this animal in the late 1700s?

2. Controlling for body size, would you expect a flightless bird to produce eggs that are larger than, smaller than, or the same size as eggs produced by a bird that flies? Explain your reasoning?

3. Why is it more difficult to determine the sex of a newly hatched canary than a newborn puppy?

4. The diversity of modern vertebrates was shaped in part by mass extinctions. The most severe extinction event we know of occured at the border between the Permian and Triassic periods about 250 million years ago. An estimated 96 percent of the marine species and 70 percent of land species disappeared in what is sometimes called "The Great Dying." The cause of this mass extinction remains under investigation. Suggested causes include an asteroid impact, extreme volcanic activity, or a sudden release of carbon dioxide from deposits on the sea floor. Which groups of vertebrates were present when this mass extinction event occurred? Which evolved afterward?

26 HUMAN EVOLUTION

CORE CONCEPTS

Evolution

Evolution underlies the unity and diversity of life. Shared features provide evidence that all primates are linked by lines of descent from a common ancestor. Among modern species, humans are most closely related to the chimpanzees and bonobo. Humans did not evolve from these apes, but rather share a common ancestor with them. Uniquely human traits evolved after our lineage branched off from this common ancestor.

Structure and Function

The three-dimensional form and arrangement of biological structures give rise to their function and interactions. The structure of an animal body is an adaptation to a particular habitat. The wide range of motion in primate limbs and forward-facing eyes are adaptations to life in the trees. The transition to walking upright on the ground took place over millions of years and reshaped the skeleton of humans and their now extinct relatives.

Process of Science

The field of biology consists of and relies upon experimentation and the collection and analysis of scientific evidence. Science addresses only testable ideas about observable events and processes. The fossil record indicates that humans evolved in Africa, and results of genetic comparisons among modern humans in different parts of the world confirms this view. Extraction of DNA from fossils has shown that interbreeding with closely related species has left genetic marks on the modern human gene pool.

Links to Earlier Concepts

This chapter focuses on primates, an order of placental mammals (Sections 25.8). It is based mainly on information derived from fossils (16.4) and from genetic comparisons (18.4). We discuss primate adaptive traits (16.3) and consider examples of both morphological convergence and divergence (18.3).

26.1 A Bit of a Neanderthal

Paleoanthropology, the scientific study of prehistoric humans and their relatives, was born in the mid-1800s, when scientists found humanlike fossils in Germany's Neander Valley. The scientists hypothesized that the fossils were of an extinct human relative, which they named *Homo neanderthalensis*. At the time the fossils were discovered, their age was unknown. We now know that they dated to about 40,000 years ago.

Since the discovery of those first Neanderthal bones, many other fossils with similar features have been unearthed, including a few nearly complete skeletons. Analysis of these many fossil finds confirmed that Neanderthals were a distinct ancient species, and evolutionary biologists now consider Neanderthals our closest extinct relatives.

Compared with modern humans (*Homo sapiens*), Neanderthals had a shorter, stockier build, with thicker bones and bulkier muscles. A reconstruction of a Neanderthal male, based on material from multiple fossils, stands about 164 centimeters (5 feet 4 inches) tall (**FIGURE 26.1**). Neanderthals had a braincase that

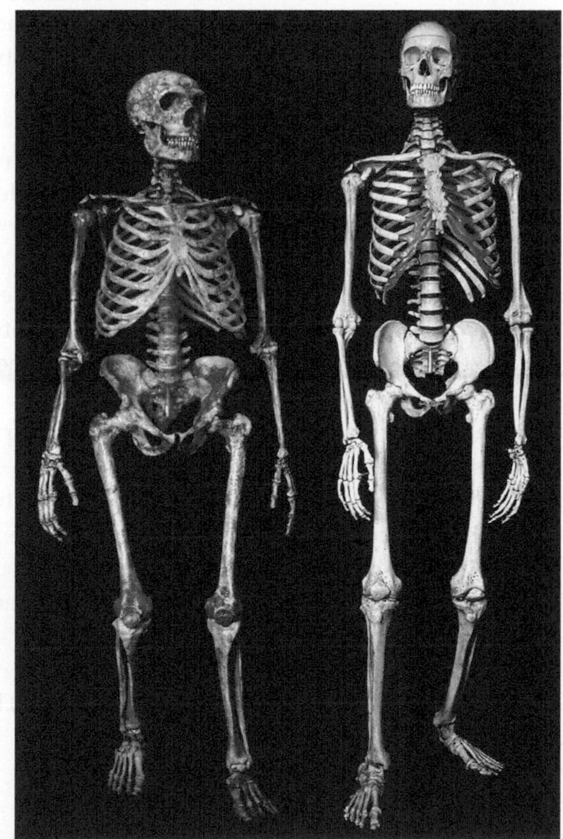

Homo neanderthalensis
Homo sapiens

FIGURE 26.1 Skeletal anatomy of Neanderthal male and a modern human male. The Neanderthal skeleton is a reconstruction based on multiple fossils. Each color denotes a different fossil.

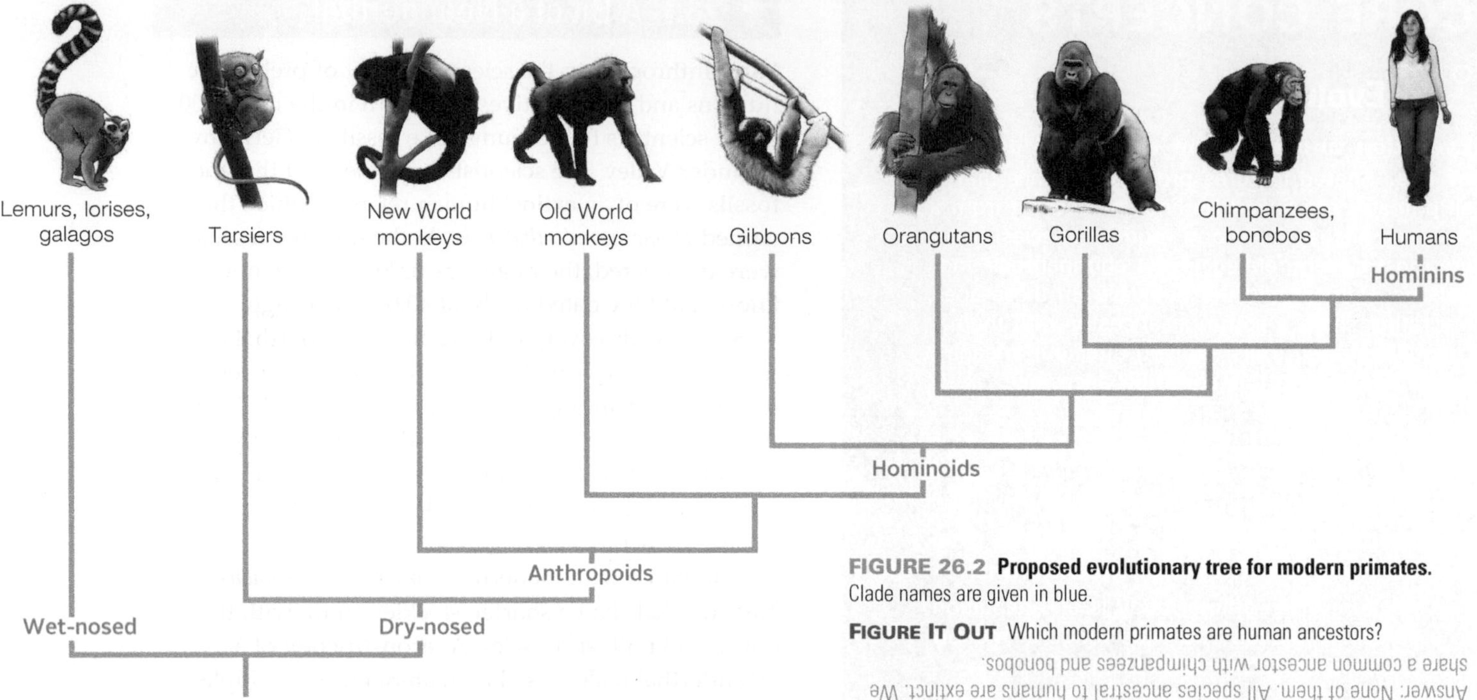
Lemurs, lorises, galagos
Tarsiers
New World monkeys
Old World monkeys
Gibbons
Orangutans
Gorillas
Chimpanzees, bonobos
Humans
Hominins
Hominoids
Anthropoids
Wet-nosed
Dry-nosed

FIGURE 26.2 Proposed evolutionary tree for modern primates. Clade names are given in blue.

FIGURE IT OUT Which modern primates are human ancestors?

Answer: None of them. All species ancestral to humans are extinct. We share a common ancestor with chimpanzees and bonobos.

was longer and lower than that of modern humans, but their brain was as big as ours or bigger. Their face had pronounced brow ridges and a large nose with widely spaced nostrils.

We know from fossils that Neanderthals lived in the Middle East, Europe, and in central Asia as far east as Siberia. The last known population lived in seaside caves in Gibraltar until perhaps as recently as 28,000 years ago.

In some regions, Neanderthals existed side by side with our own species (*Homo sapiens*) for thousands of years. People have long speculated about whether the two species met and mated. Scientists are less interested in the logistics of such matings than in whether the matings produced fertile offspring whose descendants survive to this day. In other words, they want to know whether Neanderthals contributed in any substantive way to the modern human gene pool.

This question was answered in 2010, when an international team of scientists headed by Svante Pääbo sequenced DNA extracted from Neanderthal fossils discovered in Croatia, Russia, Spain, and Germany. The sequence data they obtained represents much of the Neanderthal genome. Pääbo and his team compared this genome with homologous portions of genomes from five modern human populations and from chimpanzees. The results revealed that many of us do, in fact, have a Neanderthal in our family tree. Genomes of people from Europe, China, and Papua New Guinea share with the Neanderthal genome certain DNA sequence variants that are not present in people from Africa or in chimpanzees. This shows that the human–Neanderthal matings that left their mark on the human gene pool took place in the Middle East after *H. sapiens* began venturing out of Africa, but before they dispersed to Europe, Asia, and elsewhere. ●

26.2 Primates: Our Order

LEARNING OBJECTIVES

- Describe the traits that characterize primates.
- Compare the bodies of an Old World monkey, a New World monkey, and an ape, giving examples of each.

Primate Characteristics

Primates are the order of placental mammals that includes humans, apes, monkeys, and close relatives. FIGURE 26.2 shows relationships among living groups of primates. This lineage arose in warm forests, and many traits characteristic of the group are adaptations to life among the branches. Primate shoulders have an extensive range of motion that facilitates climbing. Unlike most mammals, a primate can extend its arms out to its sides, reach above its head, and rotate its forearm at the elbow. With the exception of humans, all living primates have both hands and feet capable of grasping. A typical mammal has claws, but in primates the tips of fingers and toes have touch-sensitive pads protected by flattened nails.

A Wet-nosed primate. A lemur, has a wet nose and cleft upper lip.

B Dry-nosed primate. A tarsier, has a dry nose and an intact upper lip.

FIGURE 26.3 Wet-nosed and dry-nosed primates.

A Old World monkey, a baboon with a long nose and a short nonprehensile tail.

B New World monkey, a squirrel monkey with a flat face and long, prehensile tail.

FIGURE 26.4 Monkeys.

Most mammals have eyes that are widely spaced and set toward the side of the skull, but primate eyes tend to be at the front of the head. As a result, both eyes view the same area, each from a somewhat different vantage point. The brain receives slightly differing signals from the two eyes and integrates them to produce a three-dimensional mental image. The result is excellent depth perception, which comes in handy when a primate is leaping or swinging from branch to branch.

Compared to other mammals, primates have a large brain for their body size. The regions of the brain devoted to vision and to information processing are expanded, and the area devoted to smell is reduced.

Primates have a varied diet, and their teeth reflect this lack of specialization; they have all four types of mammalian teeth (Section 25.8).

Most primates spend their life in a social group that includes adults of both sexes. Females usually give birth to only one or two young at a time and provide care for an extended period after birth.

Origins and Lineages

Primates probably arose before the demise of the dinosaurs, some time between 85 and 66 million years ago. An early divergence among primates created two suborders: the wet-nosed primates (Strepsirrhini) and the dry-nosed primates (Haplorhini).

Modern representatives of the wet-nosed suborder include lemurs of Madagascar, lorises of India and Asia, and galagos of Africa. Wet-nosed primates have a typical mammalian nose. Like a dog, they have nostrils set in an area of moist skin, and their cleft upper lip attaches tightly to the underlying gum (FIGURE 26.3A). Having a wet nose enhances the ability to detect odors. Wet-nosed primates eat mainly fruits and tree sap.

Humans and most other modern primates belong to the dry-nosed suborder. Their nostrils are set in an area of dry, hairy skin. The upper lip is not cleft, and its attachment to the underlying gum is greatly reduced. A movable upper lip allows a wider range of facial expressions and vocalizations than an anchored one.

Tarsiers, the oldest surviving dry-nosed lineage (FIGURE 26.3B), are small, nocturnal carnivores that live on southern Asian islands. They are eat insects, lizards, and smaller mammals. Tarsiers were traditionally grouped with lemurs as prosimians (meaning "before monkeys") because both tarsiers and lemurs have claws on some digits. However, DNA comparisons revealed that tarsiers are more similar to the other dry-nosed primates than they are to lemurs.

The dry-nosed lineage known as **anthropoids** includes monkeys, apes, and humans; *anthropoid* means humanlike. Nearly all anthropoids are diurnal (active during the day) and have good eyesight, including color vision. Most feed primarily on plant material.

There are two lineages of monkeys: Old World monkeys and New World monkeys (FIGURE 26.4). Old World monkeys live in Africa, the Middle East, and Asia. They have a long nose with close-set, downward facing nostrils. The tail, if present is relatively short. Some Old World monkeys are tree-climbing forest dwellers. Others, such as baboons, spend most of their time on the ground in grasslands and deserts.

anthropoids Primate lineage that includes monkeys, apes, and humans.
primates Mammalian order that includes lemurs, tarsiers, monkeys, apes, and humans.

A Gibbon B Orangutan C Gorilla D Chimpanzee

FIGURE 26.5 Modern ape diversity. A lesser ape (A) and three of the great apes (B–D).

New World monkeys tend to be smaller than Old World monkeys. They have a flatter face and a nose with widely separated nostrils that open to the side. New World monkeys live in the forests of Central and South America and feed primarily on fruits. A long tail helps them maintain their balance. In some species, the tail is prehensile, meaning it can grasp things.

New World monkeys arrived in South America from western Africa about 35 million years ago. Ancestors of South American rodents such as guinea pigs arrived from Africa at about the same time. Both groups of animals probably crossed the Atlantic on rafts of vegetation. This journey may not have been as daunting as it sounds, because at that time the continents were closer than they are today. Also, until 40 million years ago, sea level fluctuations periodically exposed islands. Crossing the ocean probably took many generations, with animals colonizing one island after another.

Hominoids are a family of tailless anthropoids with an upright posture and a relatively large brain. All modern apes, including humans, belong to this group.

TAKE-HOME MESSAGE 26.2

✔ Primates traits such as a flexible shoulder joint and grasping hands with nails are adaptations to a climbing lifestyle.

✔ Compared to other mammals, primates have a relatively large brain with more area devoted to vision and less to smell.

✔ An early divergence divided the wet-nosed suborder and the dry-nosed suborders. Lemurs are wet-nosed primates; tarsiers and anthropoid primates (monkeys, apes, humans) are dry-nosed primates.

26.3 Hominoids

LEARNING OBJECTIVES

- List the types of modern nonhuman apes, and explain which of them are our closest relatives.
- Describe some of the anatomical differences between modern humans and apes.

Hominoid Origins and Divergences

Hominoids are not descended from any existing group of monkeys, but they do share a common ancestor with the Old World monkeys. That common ancestor lived as estimated 30 million years ago.

Modern Apes

About 15 species of small apes called gibbons inhabit Southeast Asian forests. They use their elongated arms and permanently curved fingers to swing from limb to limb (FIGURE 26.5A). Gibbons live in family groups with similarly sized males and females. Depending on the species, adults weigh 7 to 14 kilograms (15 to 30 lb).

Gibbons are sometimes referred to as "lesser apes," in comparison with the larger apes, or "great apes." Unlike gibbons, all great apes show sexual dimorphism in size; males are substantially larger than females.

The forest-dwelling orangutan (*Pongo pygmaeus*) of Sumatra and Borneo is the only surviving Asian great ape. Like gibbons, orangutans are tree dwellers, but they live solitary lives and climb slowly using all four limbs (FIGURE 26.5B). Males are twice as big as females and can weigh 80 to 90 kg (176 to 198 lb).

Gorillas, chimpanzees, and bonobos are great apes that are native to central Africa, live in social groups, and spend most of their time on the ground. When walking, they lean forward and support their weight on their knuckles (**FIGURE 26.5C**). Gorillas (*Gorilla gorilla*), the largest living primates, live in forests and feed mainly on leaves. A male gorilla weighs 135 to 180 kg (300 to 400 lb), a female about half that.

Chimpanzees (*Pan troglodytes*) (**FIGURE 26.5D**) and the bonobos, or pygmy chimpanzees (*Pan paniscus*), are our closest living relatives. Our lineage diverged from theirs between 6 million and 10 million years ago. Chimpanzees and bonobos feed mainly on fruit, but also prey on insects and small mammals, including monkeys. The two species differ in their social behavior, with chimpanzees engaging in more intraspecific aggression and bonobos in more nonreproductive sexual acts. Chimpanzees and bonobos are a bit smaller than humans, with males weighing about 90 to 130 pounds (40 to 60 kilograms).

A Human–Great Ape Comparison

Humans differ from other apes in many ways. Compared to a chimpanzee, we have a flatter face, a smaller jaw with reduced canine teeth, and a larger brain relative to body size. On average, a human brain is about three times larger than a chimpanzee brain.

We also walk differently. When gorillas and chimpanzees walk, they lean forward and put their weight on their knuckles. By contrast, humans walk upright. Habitual upright walking is called **bipedalism**.

Evolution of bipedalism involved many skeletal changes (**FIGURE 26.6**). A knuckle-walking ape such as a gorilla has a C-shaped backbone, flat feet that are capable of grasping, and a spinal cord that enters near the back of the skull. By contrast, the human backbone has an S-shaped curve that keeps our head centered over our feet. Our feet have a pronounced arch and a nonopposable big toe. Our spinal cord enters at the skull's base, rather than at its rear.

Human hands also differ from those of nonhuman apes. A human thumb is longer, stronger, and more maneuverable than that of a chimpanzee or gorilla. Many monkeys and apes wrap their hand around objects in a power grip, but only humans routinely use a precision grip, pinching together the tips of the thumb and forefinger for fine manipulation of objects:

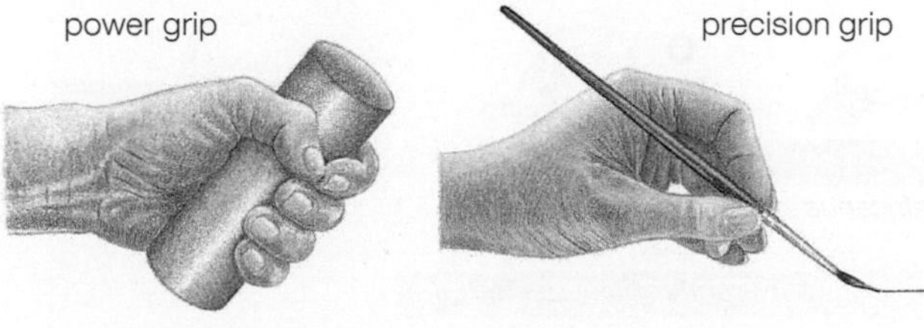
power grip
precision grip

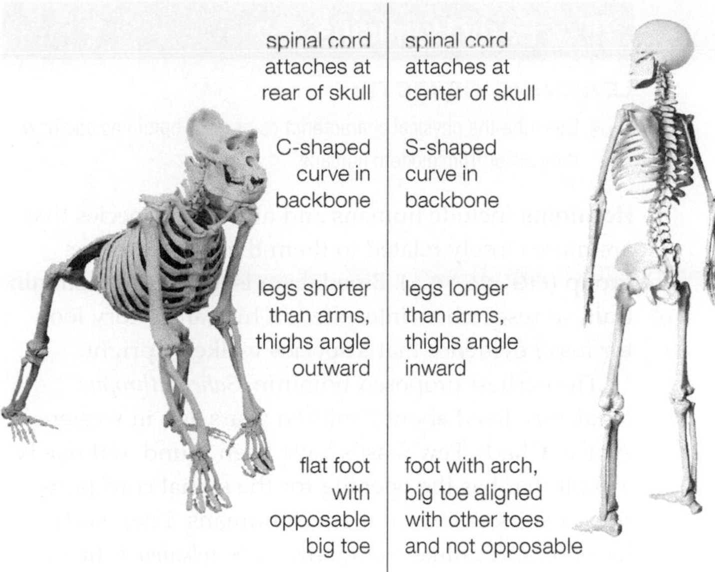
spinal cord attaches at rear of skull
spinal cord attaches at center of skull
C-shaped curve in backbone
S-shaped curve in backbone
legs shorter than arms, thighs angle outward
legs longer than arms, thighs angle inward
flat foot with opposable big toe
foot with arch, big toe aligned with other toes and not opposable

FIGURE 26.6 Gorilla–human skeletal comparison. Skeleton of a knuckle-walking gorilla (left) and a bipedal human (right).

Humans are the only living primates that do not have a thick coat of body hair. We have about the same number of body hairs as other primates, but ours are shorter and finer, giving our skin a more naked appearance. The reduction in body hair may have benefited our bipedal ancestors by making it easier for them to cool their body during daytime walks or runs. Sweat evaporates more quickly from bare skin than from under thick hair. Humans have a much greater density of sweat glands than other primates.

The many differences between humans and apes did not arise all at once. Rather, different traits changed at different rates. As you will see in the next sections, our ancestors walked upright for many millions of years before their brain size began to increase.

TAKE-HOME MESSAGE 26.3

✔ Hominoids share a common ancestor with Old World monkeys, but they are generally larger than monkeys, lack a tail, and have an upright stance.

✔ Modern hominoids include Asian apes, African apes, and humans. Our closest living relatives are the chimpanzees and bonobos.

✔ Upright walking (bipedalism) evolved in the lineage leading to humans. This lineage also shows trends toward increased brain size, increased maneuverability of the thumb, and a decrease in body hair.

bipedalism (bi-PEE-duh-lism) Habitual upright walking.
hominoids Tailless primate lineage; includes humans and other apes.

26.4 Early Hominins

LEARNING OBJECTIVE

- Describe the physical characteristics of early hominins and how they differ from modern humans.

Hominins include humans and all extinct species that are more closely related to them than to any other group (**FIGURE 26.7**). Bipedalism is a defining hominin trait, so researchers interested in human history look for fossil evidence that a species walked upright.

The earliest proposed hominin, *Sahelanthropus tchadensis*, lived about 7 million years ago in western Africa (Chad). Few fossils have been found, but one is a skull that has the opening for the spinal cord positioned at its base as in modern humans. Like modern humans and unlike chimpanzees, *S. tchadensis* had a flat face and small canine teeth.

Another proposed early hominin, *Orrorin tugenensis*, lived about 6 million years ago in East Africa. Two sturdy fossilized femurs (thighbones) are cited as evidence that this species stood upright.

Two species of chimpanzee-sized *Ardipithecus* that lived in East Africa are also likely hominins. *A. kadabba* is known from a few fossils that date from 5.8 to 5.2 million years ago. *A. ramidus* left many fossils, including a nearly complete female skeleton known as Ardi. *A. ramidus* had smaller teeth than a chimpanzee, and its pelvis was more humanlike than apelike. It walked upright on the ground, but long arms, curved fingers, and a splayed big toe indicate that it also walked on all fours along branches (**FIGURE 26.8A**).

Australopiths

The best known early hominins are **australopiths**, an informal group comprising two genera of hominins that lived in Africa from about 4 million to 1.2 million years ago. *Australopithecus* species were small boned. Their fossil history reveals trends toward smaller teeth and improvements in the ability to walk upright, but little increase in brain size.

Fossil footprints in Tanzania document the passage of a bipedal species 3.6 million years ago. We can tell from the prints that the walkers' feet had a pronounced arch and a big toe in line with the other toes—both adaptations to upright walking. The prints were probably made by *Australopithecus afarensis*, an australopith that lived in Tanzania and other parts of eastern Africa from 3.9 to 3 million years ago. A partial *A. afarensis* skeleton known as Lucy (**FIGURE 26.8B**) is the best known representative of this species.

A. afarensis males stood about 1.5 to 1.8 meters (5 to 5.5 feet) tall, and females 1 meter (3 feet) tall. Members of this species had a pelvis and legs suited to upright walking, although their gait may not have been as fluid as that of modern humans. Long arms and curved fingers indicate that *A. afarensis* also spent a lot of time climbing among tree branches.

Australopithecus sediba, a species recently discovered in South Africa, may be the most humanlike australopith discovered to date (**FIGURE 26.8C**). *A. sediba*, which lived about 2 million years ago had remarkably humanlike hands; they were probably capable of a precision grip. In addition, the *A. sediba* brain, although chimpanzee-sized and largely apelike, has a frontal region that resembles that of modern humans.

A. afarensis and *A. sediba* are considered possible ancestors of modern humans, but the hominin family tree is bushy, with many branches that certainly did not lead to modern species. The species known as robust australopiths were on one of these dead-end branches. These heavy-bodied species such as *Paranthropus boisei* had a skull with a pronounced crest, to which large jaw

FIGURE 26.7 Timeline for hominin evolution. Different colors denote different genera.

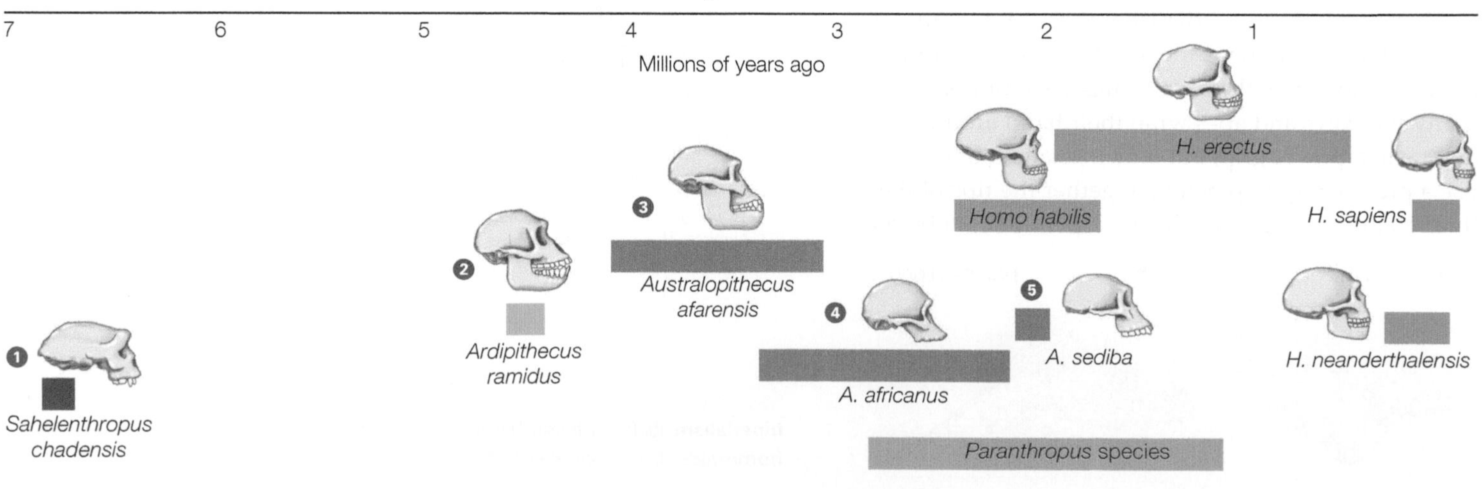

7
6
5
4
3
2
1
Millions of years ago
1 Sahelenthropus chadensis
2 Ardipithecus ramidus
3 Australopithecus afarensis
4 A. africanus
5 A. sediba
Paranthropus species
Homo habilis
H. erectus
H. sapiens
H. neanderthalensis

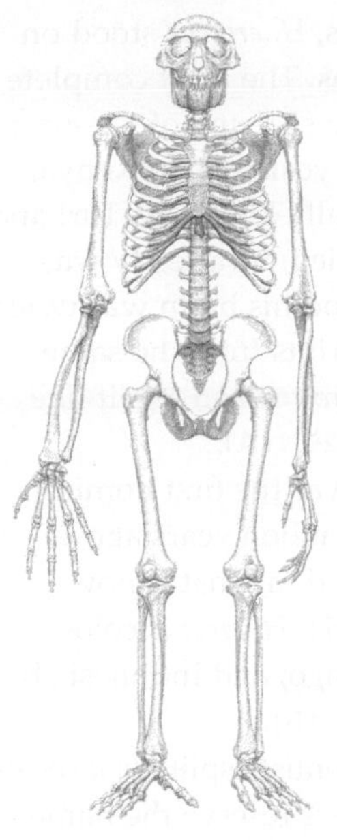

A Sketch of *Ardipithecus ramidus* skeleton

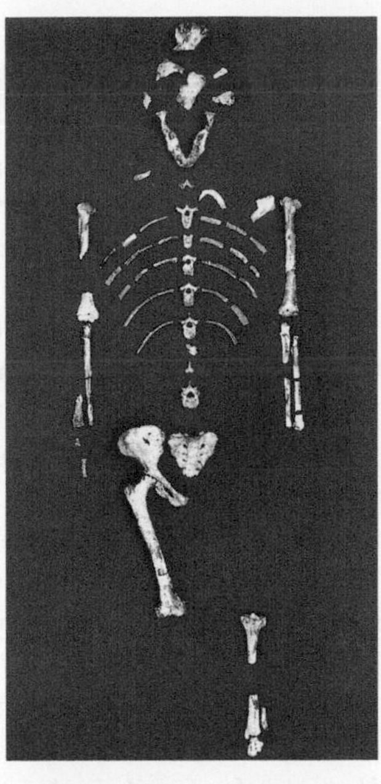

B *Australopithecus afarensis* fossil skeleton (Lucy)

C Reconstruction of *Australopithecus sediba*

FIGURE 26.8 Three early hominins with extensive fossil records. All are considered possible ancestors of modern humans.

muscles attached (**FIGURE 26.9**). This suggests their diet was rich in difficult-to-chew nuts and seeds.

Factors Favoring Bipedalism

What advantages could have driven the hominin shift from walking on all fours to bipedalism? A trend toward a drier and hotter climate was probably a factor. Hominins originated at a time when Africa's lush rain forests were giving way to woodlands interspersed with grassy plains. In this altered habitat, an ability to move efficiently across open ground between trees would have been favored, and upright walking can be energy efficient. Human walkers use less energy than chimpanzees who walk on all fours. Bipedalism also keeps the body cooler. A bipedal animal gains less heat from the ground than a four-legged walker and intercepts less warming sunlight. In addition, an upright stance makes it easier to scan the horizon for predators, and hands freed from locomotion can be used to gather and carry food and other items.

TAKE-HOME MESSAGE 26.4

✔ Hominins include modern humans and extinct members of their lineage. Fossils that may be hominins date back as long as 7 million years ago.

✔ *Ardipithecus* is the oldest hominin genus for which there is extensive fossil evidence. Australopiths are a diverse group of African hominins that probably include human ancestors.

✔ Early hominins were chimpanzee-like in their size and brain capacity, but they walked upright.

FIGURE 26.9 Reconstructed skull of a robust australopith, *Paranthropus boisei*. Robust australopiths lived alongside likely ancestors of modern humans, but are not themselves our ancestors.

26.5 Early Humans

LEARNING OBJECTIVES

- Describe the anatomy of *Homo erectus* and *Homo habilis*.
- Give examples of cultural traits exhibited by early humans.

Classifying Fossils—Lumpers and Splitters

When it comes to classifying fossil species, evolutionary biologists fall into two general camps. The "lumpers" look for similarities among fossils and tend to assign fossils to relatively few species. Lumpers support this approach by pointing out that members of a single species can vary widely in their traits. For example, modern humans share many traits, but vary greatly in their height. By contrast, "splitters" focus on

australopith Informal name for chimpanzee-sized hominins that lived in Africa from 4 million to 1.2 million years ago.

hominins Modern humans and their closest extinct relatives.

CREDITS: (8A) Splash News/Newscom/Splash News/London/UNITED KINGDOM; (8B) Dr. Donald Johanson, Institute of Human Origins; (8C) Courtesy of John Gurche; (9) Pascal Goetgheluck/Science Source.

differences between fossils and often name new species on the basis of subtle differences. In support of their approach, splitters point out that many modern species differ only slightly in their morphology. For example, chimpanzees and bonobos appear very similar, although bonobos are a bit smaller.

From a scientific standpoint, both approaches are equally valid. Both lumpers and splitters formulate hypotheses about how a particular fossil relates to other known fossils. They then seek evidence that will support or refute their hypotheses. Additional fossil discoveries often provide this evidence.

In this section, we discuss early fossils of our own genus: *Homo*. Lumpers assign these fossils to two species; splitters divide them into four or more.

Homo habilis

At this time, the oldest named member of the genus *Homo* is ***Homo habilis*** (FIGURE 26.10). Fossils of *H. habilis* date from 2.3 million years to 1.4 million years before the present. The species was named based on a fossil discovered in Kenya in 1964. At that time, tool production was considered a diagnostic trait of the genus *Homo*. The name *Homo habilis* means "handy man" and is a reference to stone tools that were found near the fossil.

Classification of *H. habilis* continues to inspire debate. *H. habilis* resembles some australopiths in body proportions and brain size, and some scientists think it should be classified in *Australopithecus*. Other scientists argue for the species' inclusion in the genus *Homo*, pointing out that the hands and arms are similar to those of modern humans. Anatomical variation among the fossils has also resulted in another dispute. Some splitters contend that the fossils currently classified as *H. habilis* actually include remains from two different species. They place larger-brained, longer-faced individuals in the species *Homo rudolfensis*.

Homo habilis/rudolfensis persisted in eastern Africa until at least 1.4 million years ago. Thus, it overlapped in time and range with some australopiths and with *Homo erectus*, which we turn to next.

Homo erectus

The earliest known human with body proportions similar to our own, was ***Homo erectus***, which arose in Africa by about 2 million years ago. The species name means "upright man," and like us, *H. erectus* stood on legs that were longer than its arms. The most complete *H. erectus* fossil found thus far is a skeleton of a young male who lived about 1.5 million years ago in Kenya. Although this individual, informally known as Turkana boy, was under age 14 when he died, he already was 1.60 meters (5 feet 2 inches) tall and his brain was twice the size of a chimpanzee's. Footprints from the same region and time suggest that *H. erectus* had a gait like that of modern humans (FIGURE 26.11A).

As far as we know, *H. erectus* was the first hominin to venture out of Africa. By 1.75 million years ago, a population had become established in what is now Dmanisi in the Republic of Georgia. *H. erectus* colonized China by 1.7 million years ago, and Indonesia by 1.6 million years ago (FIGURE 26.11B).

As with *Homo habilis*, some scientists split *Homo erectus* fossils into two species. Splitters reserve the name *H. erectus* for fossils in Asia and refer to the similar African fossils as *H. ergaster*, meaning "working man."

Early Culture

Culture is a set of learned behaviors that are passed from one individual to another, and from one generation to the next. The earliest hominin cultural trait for which we have fossil evidence is toolmaking. By 3.3 million years ago, hominins were chipping sharp flakes from larger rocks to create cutting tools and shaping rocks to serve as hammers and anvils. Given that the oldest fossils of *Homo* date to about 2.8 million years ago, some scientists now think it is possible that australopiths were the first to produce stone tools.

The earliest stone tools were sharp flakes chipped from larger rocks. Toolmaking improved about 1.4 million years ago, when African *H. erectus* began to use a succession of carefully placed strikes to sculpt tools in a variety of shapes.

H. erectus used stone tools to cut up scavenged animal carcasses, to scrape meat from bones, and to extract marrow. Some researchers have suggested that increased meat consumption may have been a necessary prerequisite for the evolution of a larger brain. Meat is a more concentrated source of energy and

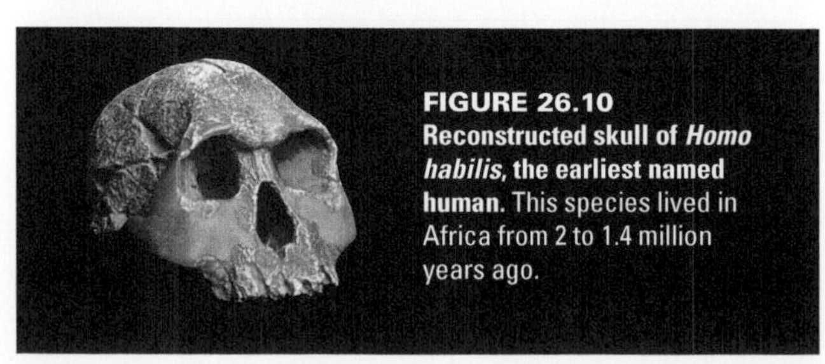

FIGURE 26.10
Reconstructed skull of *Homo habilis*, the earliest named human. This species lived in Africa from 2 to 1.4 million years ago.

culture Learned behaviors transmitted between individuals and down through generations.

Homo erectus Early human species that arose in Africa by about 2 million years ago; some populations migrated to other regions.

Homo habilis Earliest named human species; lived in Africa from about 2.3 to 1.4 million years ago.

CREDIT: (10) Pascal Goetgheluck/Science Source.

protein than plant foods, so it can better fuel development and maintenance of a large brain.

Chimpanzees use gestures and calls to communicate, and early humans probably did the same. However, it is unlikely that *H. habilis* or *H. erectus* had a spoken language. Evolution of a capacity for speech involved changes in the brain, remodeling of the larynx (voice box), and an improved ability to control air flow through the vocal tract. To date, scientists have not found fossil evidence of these necessary anatomical modifications in either early species of *Homo*.

TAKE-HOME MESSAGE 26.5

✔ *Homo habilis*, the most ancient named human species, lived in Africa. It was similar in many respects to australopiths, but its hands and arms more closely resemble those of modern humans.

✔ *Homo erectus* is known from fossils found in Africa, Europe, and Asia. It had a significantly larger brain than earlier hominins and a gait similar to modern humans.

✔ Early humans made stone tools, but probably did not have a spoken language.

26.6 Recent Human Lineages

LEARNING OBJECTIVES

- Describe what is known about Neanderthals and about the other recently discovered members of our genus .
- Explain the evidence that our species originated in Africa.

Homo erectus is considered the likely ancestor of our own species, Neanderthals, and some other more recently discovered hominins.

Neanderthals

Homo neanderthalensis is an extinct hominin that is known from many fossils in the Middle East, Europe, and Asia. Neanderthals are our closest known relatives, and gene comparisons between Neanderthals and modern humans inicate that the two lineages diverged more than 500,000 years ago. The oldest known Neanderthal fossils date to about 400,000 years ago.

The common conception of Neanderthals as brutish cavemen with poor posture arose from an early reconstruction based on a fossil of an individual deformed by arthritis. More recent reconstructions reveal Neanderthals were shorter than modern humans, but stood upright. They lived in regions where winters are cold, and a short, stocky body minimized the surface area available for heat loss. Modern Arctic peoples have a similar body shape. Maintaining body temperature in a cool environment burns a lot of energy and requires a lot of calories.

A 1.5-million-year-old footprints from Kenya indicate a smooth gait.

B Reconstruction based on a 700,000-year-old skull found in China.

FIGURE 26.11 *Homo erectus*, the first hominin for which we have evidence outside of Africa.

The most recent evidence of Neanderthals dates to 28,000 years ago. Neanderthals may have been outcompeted by newly arrived *H. sapiens* or killed by diseases these migrants brought with them. Climate changes and volcanic eruptions may have contributed to their demise by altering the abundance of the animals they hunted. Meat made up the bulk of the Neanderthal diet, so they would have been more affected by such declines than *H. sapiens*, who had a more varied diet.

Denisovans

Denisovans are a recently discovered lineage closely related to Neanderthals. Their existence came to light in 2010, when researchers found a 50,000-year-old fossil pinky finger in Denisova cave in Siberia. Analysis of DNA from this fossil revealed that it was from a female who was more similar to Neanderthals than to modern humans, but was genetically distinct from typical Neanderthals. A few additional Denisovan fossils have since been found in Sibera, including a 110,000-year-old tooth. However, no major bones have been discovered, so we know little about what Denisovans looked like. At this writing, Denisovans have not been assigned to a new species, although they are sometimes referred to by the temporary designation *Homo* sp. *altai*.

Flores Hominins

In 2003, scientists discovered 50,000-year-old hominin fossils on the Indonesian island of Flores. The fossil individuals were only a meter tall, with a heavy brow and a small brain. Scientists who found the fossils assigned them to a new species, *Homo floresiensis*, which they believe descended from *Homo erectus*.

CREDITS: (11A) Professor Matthew Bennett, Bournemouth University/Science Source; (11B) Philippe Plailly & Atelier Daynes/Science Source.

Data Analysis Activities

Neanderthal Hair Color The *MC1R* gene regulates pigmentation in humans (Sections 13.5 and 14.2), so loss-of-function mutations in this gene affect hair and skin color. A person with two mutated alleles for this gene makes more of the reddish melanin than the brownish melanin, resulting in red hair and pale skin. DNA extracted from two Neanderthal fossils contains a mutated *MC1R* allele that has not yet been found in humans. To see how the Neanderthal mutation affects the function of the *MC1R* gene, Carles Lalueza-Fox and her team introduced the allele into cultured monkey cells (**FIGURE 26.12**).

1. How did *MCR1* activity in monkey cells with the mutant allele differ from that in cells with the normal allele?
2. What does this imply about the mutation's effect on Neanderthal hair color?
3. What purpose do the cells with the gene for green fluorescent protein serve in this experiment?

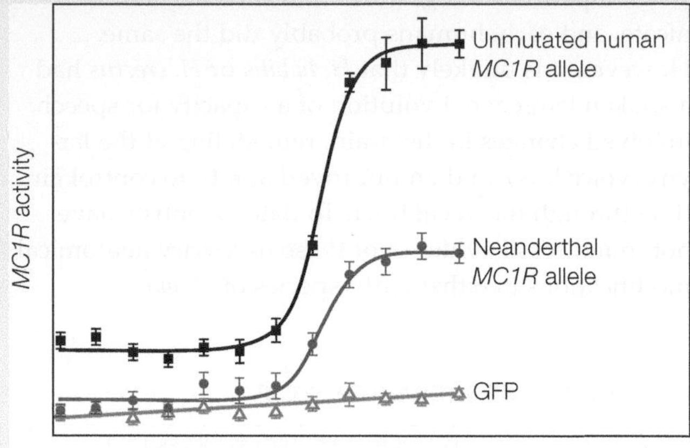
Unmutated human
MC1R allele
Neanderthal
MC1R allele
GFP
MC1R activity
Concentration of MC1R-stimulating hormone

FIGURE 26.12 ***MC1R* activity.** Activity is shown in monkey cells transgenic for an unmutated *MC1R* gene, the Neanderthal *MC1R* allele, or the gene for green fluorescent protein (GFP). GFP is not related to *MC1R*.

The media nicknamed these small-bodied hominins "hobbits," a reference to the similarly small beings in the *Lord of the Rings* books. Some scientists initially suggested that the fossils were of modern humans that had been dwarfed by some sort of disease or disorder. However, recently discovered 700,000-year-old Flores fossils indicate that *H. floresiensis* most likely is a distinct species and that it evolved from Asian *H. erectus*. Species that colonize islands often become smaller than their mainland ancestors.

Homo naledi

Fossils of more than 15 individuals discovered in an underground chamber in South Africa in 2013 may be members of another *Homo* species. The first researchers to analyze the fossils assigned them to a new species, *Homo nadeli*, and suggested that the discovery site was a burial chamber. Other scientists are skeptical about the burial hypothesis and think the fossils may be *H. erectus*. The fossil individuals stood about 1.5 meters (5 feet) tall and had feet like modern humans, curved fingers that would have been useful for climbing in trees, and a brain smaller than that of most *H. erectus*. The fossils are an estimated 236,000 to 335,000 years old.

Homo Sapiens

Anatomically modern humans belong to the species ***Homo sapiens***. Compared to *H. erectus*, *H. sapiens* have a higher, rounder skull, a larger brain, and a flatter face with smaller jawbones and teeth. Unlike other hominins, they have a chin, which is a protruding area of thickened bone in the middle of our lower jawbone.

Much evidence suggests our species arose in Africa some time before 200,000 years ago. The oldest *H. sapiens* fossils known are two partial male skulls from Ethiopia, in East Africa. Known as Omo I and Omo II, they date to 195,000 years ago. Fossil remains of two adult males and a child who lived 160,000 years ago were also unearthed in this region.

Genetic evidence also points to an Africa origin. Modern Africans are more genetically diverse than people of any other region. This diversity indicates that our species has existed in Africa for a very long time—long enough to accumulate a greater number of random mutations than populations of other regions. Furthermore, most genetic variations seen in people native to regions outside of Africa are subsets of the variation found in Africa. This is evidence that founder effects (Section 17.6) occurred after some people left Africa to colonize the rest of the world.

We know from fossils that some modern humans ventured from Africa into the Middle East about 100,000 years ago. However, genetic evidence indicates that modern non-African populations are descendants of individuals that began leaving Africa about 60,000 years ago. Our species expanded its range gradually, as

Denisovans (deh-NEE-so-vans) Recently discovered hominins closely related to Neanderthals.
Homo sapiens Anatomically modern humans; only surviving *Homo* species.
Homo neanderthalensis Neanderthals. Extinct hominins that lived in the Middle East, Europe, and Asia; the closest relatives of modern humans.

small groups ventured away from their homelands. By mapping the frequency of maternal and paternal genetic markers in modern peoples, geneticists have created a picture of when and where people moved.

The distribution of genetic markers indicates that people crossed from the Horn of Africa onto the Arabian Peninsula, then moved along the coast to India, reaching Southeast Asia and Australia by 50,000 years ago. Later, another group traveled through the Middle East into south central Asia. Offshoots from this lineage peopled North Asia and Europe. The ancestors of modern Native Americans were East Asians who crossed from Siberia into North America perhaps as early as 23,000 years ago. By about 14,500 years ago, some of them had reached Chile in South America.

The Leaky Replacement Model

Until relatively recently, it was thought that as modern humans spread across the world, they simply replaced earlier hominins. However, new genetic evidence of interbreeding has led to what is called the leaky replacement model (**FIGURE 26.13**). This model states that as modern humans spread across the globe, some genes from other hominins they encountered "leaked" into the human population as a result of occasional interbreeding. Early in the human dispersal out of Africa, some people encountered and successfully interbred with Neanderthals, most likely in the Middle East ❶. As a result of this interbreeding, modern humans of non-African heritage have a bit of Neanderthal DNA in their genome, whereas Africans do not. Another hybridization took place about 40,000 years ago, when humans moving through Asia on their way to Melanesia (Papua New Guinea and adjacent islands) interbred with Denisovans ❷. As a result of this later event, people of Melanesian descent carry genetic markers of both Neanderthals and Denisovans.

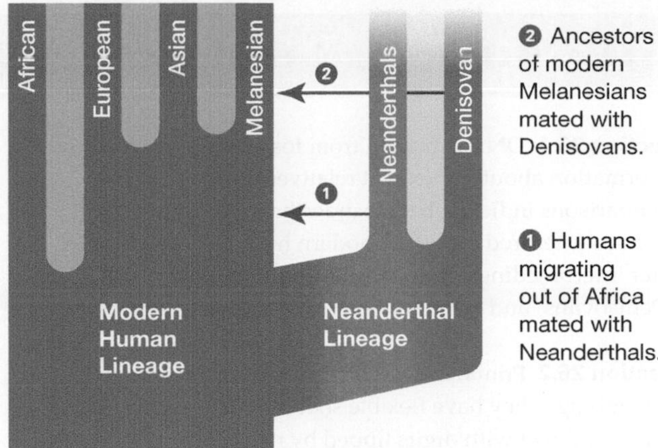
African
European
Asian
Melanesian
Neanderthals
Denisovan
Modern Human Lineage
Neanderthal Lineage
❷ Ancestors of modern Melanesians mated with Denisovans.
❶ Humans migrating out of Africa mated with Neanderthals.

FIGURE 26.13 The leaky replacement model for human evolution. Arrows indicate two instances in which interbreeding introduced genes that persist in some modern human populations.

TAKE-HOME MESSAGE 26.6

- ✔ Neanderthals are the closest relatives of modern humans. They arose more than 400,000 years ago and became extinct about 28,000 years ago. They have a robust fossil record.
- ✔ Denisovans are a poorly understood, genetically distinct group related to Neanderthals. Their existence was discovered when DNA from a fossil was sequenced..
- ✔ *Homo floresiensis* from Indonesia and *Homo naledi* from South Africa are recently named members of the genus *Homo.* Fossils indicate that both were small in stature and had relatively small brains.
- ✔ Fossils and genetic comparisons indicate that our species, *Homo sapiens,* arose in Africa more than 200,000 years ago.
- ✔ As our species expanded its range out of Africa, some individuals mated with Neanderthals and others with Denisovans. Thus some modern humans carry genes inherited from these now-extinct groups.

26.1 A Bit of a Neanderthal (revisited)

Analyzing the genomes of modern humans provides information about interspecies breeding events that resulted in descendants who survived to the present. However, genetic analysis of fossil DNA has uncovered evidence of some interspecies liaisons that were genetic dead ends.

Consider what researchers discovered when they analyzed DNA from a fossilized *Homo sapiens* male who lived in Romania about 40,000 years ago. This individual's genome included an unusually high amount of Neanderthal DNA, leading researchers to conclude that he must have had a Neanderthal ancestor only four to six generations back. Although the human–Neanderthal mating presumably took place in Europe, the Neanderthal genes seen in the fossil individual are not found in modern Europeans. Apparently none of his descendants survived to the present.

Analysis of fossil genomes has also showed that genes flowed in both directions. The genome of a fossil Neanderthal from Siberia was recently found to contain some human genes, indicating that this individual had a *H. sapiens* ancestor. ●

STUDY GUIDE

Section 26.1 DNA extracted from fossils can provide information about our extinct relatives. DNA sequence comparisons indicate that Neanderthals and modern humans interbred just after modern humans left Africa. A later interbreeding between a subgroup of Neanderthals (Denisovans) and humans took place in Asia.

Section 26.2 **Primates** are a mammalian order adapted to climbing. They have flexible shoulder joints, grasping hands and feet with digits tipped by nails, and good depth perception. They rely on vision more than smell.

Lemurs and their relatives are wet-nosed primates that have a typical mammalian nose and a fixed, cleft upper lip. Tarsiers, monkeys, apes, and humans belong to the dry-nosed lineage; they have a dry-nose and a movable upper lip.

Anthropoids include monkeys and hominoids (which include humans and other apes). The **hominoids** do not have a tail. Hominoids are not descended from Old World monkeys, but rather share and ancestor with them.

Section 26.3 Modern nonhuman apes include the gibbons and orangutans of Asia, and the gorillas, chimpanzees, and bonobos of Africa. Chimpanzees and bonobos are our closest living relatives. Both live in social groups and are sexually dimorphic. Our lineage and the chimpanzee/bonobo lineage diverged 6–10 million years ago.

Gorillas, chimpanzees, and bonobos are knuckle walkers. By contrast, humans exhibit **bipedalism**, meaning they habitually walk and stand upright. The human skeleton has numerous adaptations related to bipedalism. The position of the hole that attaches our spinal cord to our brain is at the base of our skull, we have a foot with an arch and a nonopposable big toe, and our backbone has an S-shaped curve. Humans also have a much larger brain and more flexible hands than any living ape. The traits that characterize humans did not evolve simultaneously; upright walking evolved long before larger brain size.

Section 26.4 **Hominins** include modern humans and their extinct bipedal relatives. The earliest proposed hominins date to 6 or 7 million years ago and are known from fossil fragments found in Africa. *Ardipithecus* lived in Africa 5.8 to 5.2 million years ago and left many fossils, including an almost complete skeleton of a female. Her skeletal anatomy suggests that she walked upright when on the ground, but also walked on all fours along branches.

Footprints dating to 3.6 million years ago are evidence of the passage of a fully bipedal species. It was most likely one of the **australopiths**, a group of hominins that lived in Africa from 4 to 1.2 million years ago and left many fossils. Some australopiths such as *Australopithecus afarensis* and *A. sediba* are considered probable ancestors of humans. Australopiths were upright walkers with chimpanzee-sized brains.

Bipedalism evolved at a time when the African climate was warming and grasslands were replacing forests. Walking upright increases the efficiency of movement on the ground, keeps a body cooler than four-legged walking, and makes it easier to carry food or other material.

Section 26.5 Humans are members of the genus *Homo*. The earliest human species, ***Homo habilis***, appeared in Africa by 2.3 million years ago and survived until 1.4 million years ago. *H. habilis* had an australopith-like build and a small brain.

Homo erectus evolved by 1.9 million years ago. Compared to *H. habilis*, this species had a significantly larger brain. *H. erectus* also had longer legs than arms, as in modern humans. Some *H. erectus* dispersed out of Africa, and fossils of this species have been found in Europe, China, and Indonesia.

Stone tool production is an aspect of **culture**, a set of learned behaviors passed down from one generation to the next. Evidence of stone tool use dates back to 3.3 million years ago, suggesting it probably began among the australopiths. Hominins used tools to scrape meat from animal bones. By one hypothesis, increased meat consumption provided energy necessary to develop larger bodies and brains. *H. habilis* and *H. erectus* probably did not have a spoken language.

Section 26.6 *Homo neanderthalensis* and *H. sapiens* are descendants of *Homo erectus*. They diverged from a common ancestor more than 500,000 years ago. ***Homo neanderthalensis*** lived in the Middle East, Europe, and Asia. Compared to modern humans, Neanderthals had a bulkier body and larger brain. They may have had a language. Sequencing of DNA from a Siberian pinkie bone led to the discovery of Denisovans, a relative of Neanderthals.

Fossils of short, small-brained hominins that lived on an Indonesian island between 700,000 years and 50,000 years ago have been named as *Homo floresiensis*. Another short, small-brained species, *Homo nadeli*, is known from fossils in South Africa that date to 236,000 to 335,000 years ago.

The earliest fossils of our own species, ***Homo sapiens***, are from central Africa and date to 195,000 years ago. *H. sapiens* is distinguished by a larger brain, flatter face, smaller jaw, and a bony protrusion on the lower jaw (a chin). By 115,000 years ago, *H. sapiens* had expanded its range south into South Africa, and by 110,000 years ago there was a population in the Middle East. However, it was migrations that began about 60,000 years ago that dispersed our species across the globe.

The greatest diversity of humans occurs in Africa, where the species has lived the longest. The genomes of people native to other regions contain a subset of this diversity.

The leaky replacement model states that as modern humans spread across the globe, some genes from other hominins they encountered "leaked" into the human population as a result of occasional interbreeding.

SELF-QUIZ

Answers in Appendix VII

1. New World monkeys ______ .
   a. lack a tail
   b. are bipedal
   c. are dry-nosed primates
   d. include human ancestors

2. The closest relatives of bonobos are ______ .
   a. chimpanzees
   b. humans
   c. tarsiers
   d. Old World monkeys

3. An S-shaped backbone is an adaptation to ______ .
   a. tool use
   b. climbing trees
   c. bipedalism
   d. eating meat

4. Compared to the other living apes, modern humans have ______ .
   a. a smaller brain
   b. shorter, finer body hairs
   c. greater sexual dimorphism in body size
   d. shorter, weaker thumbs

5. The 3.6-million-year-old footprints left by bipedal walkers in Tanzania were probably made by ______ .
   a. australopiths
   b. Neanderthals
   c. modern humans
   d. *Homo erectus*

6. The position where a spinal cord enters the skull provides evidence about whether a fossil species ______ .
   a. was nocturnal
   b. was carnivorous
   c. walked upright
   d. could speak

7. Australopiths are ______ .
   a. a type of monkey
   b. wet-nosed primates
   c. extinct hominins
   d. descendants of *H. erectus*

8. The oldest known named species of *Homo* is ______ .
   a. *H. sapiens*
   b. *H. habilis*
   c. *H. erectus*
   d. *H. floresiensis*

9. A prominent chin is typical of ______ .
   a. *Homo sapiens*
   b. *Homo habilis*
   c. *Homo erectus*
   d. *Homo floresiensis*

10. *Homo floresiensis* fossils were discovered in ______ .
    a. Australia
    b. China
    c. Indonesia
    d. South Africa

11. DNA sequencing of a fossil finger revealed the existence of ______ .
    a. Neanderthals
    b. *Homo habilis*
    c. *Australopiticus sediba*
    d. Dennisovans

12. ______ lived in Europe, Asia, and the Middle East.
    a. *Ardipithecus ramidus*
    b. *Homo habilis*
    c. *Australopithecus sediba*
    d. *Homo neanderthalensis*

13. The greatest genetic diversity of modern humans is found among the people of ______ .
    a. Europe
    b. Asia
    c. Africa
    d. North America

14. Match each group with its description.

___ hominins	a. first to be found outside Africa
___ tarsiers	b. modern humans
___ *Homo erectus*	c. defined by bipedalism
___ *Homo sapiens*	d. primates, but not anthropoids
___ *Homo habilis*	e. short-statured "hobbits"
___ *Homo floresiensis*	f. name means "handy man"

15. Place the events in order.

___ 1 (earliest)	a. Neanderthals become extinct
___ 2	b. Old World monkeys evolve
___ 3	c. first apes evolve
___ 4	d. some *Homo erectus* leave Africa
___ 5	e. divergence of lineages leading to humans and to chimpanzees
___ 6	f. *Homo sapiens* evolve in Africa
___ 7 (most recent)	g. divergence of wet-nosed and dry-nosed primates

CRITICAL THINKING

1. Male aggression is rare in bonobo society and common in chimpanzee society. Various authors have argued that either one species or the other should be considered a model for "natural" human behavior. Explain why, from the standpoint of divergence time, there is no reason to think that one of these species is a better model for human behavior than the other.

2. Think about the pattern of human dispersal. Given what you know about the founder effect, would you expect populations native to South America to be more or less genetically diverse than those native to North America? Explain your reasoning.

3. Many modern people have some Neanderthal DNA in their genome, but the Neanderthal alleles are not uniformly distributed across their genome. A person of European or Asian ancestry typically has 1.5 to 2 percent Neanderthal DNA in their autosomes. However, researchers have never found any Neanderthal alleles on the Y chromosome of a modern human. By one hypothesis, Neanderthal Y alleles disappeared because they were incompatible with *H. sapiens* genes. Explain how reduced fitness in hybrids arising from genetic incompatibility could have, over time, led to the elimination of the Neanderthal Y alleles from the *H. sapiens* gene pool.

27 PLANT TISSUES

CORE CONCEPTS

Pathways of Transformation

Organisms exchange matter and energy with the environment in order to grow, maintain themselves, and reproduce.

Living things have various strategies to eliminate wastes and to obtain nutrients and energy for use in biological processes. Plant leaves are specialized to intercept sunlight and to exchange gases for photosynthesis and respiration. Roots are specialized to take up water and nutrients from soil.

Evolution

Evolution underlies the unity and diversity of life.

All plant parts consist of the same tissues, but the arrangement of the tissues differs between monocots and eudicots. Plants specialized for different environments have different adaptations for acquiring and retaining water and nutrients, supporting growth and reproduction, and optimizing photosynthetic performance.

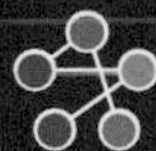

Systems

Complex properties arise from interactions among components of a biological system.

Stacks of cells in vascular tissue form conducting tubes that deliver water and nutrients to all of a plant's living cells. All plant parts arise from continually dividing undifferentiated cells in the tips of roots and shoots. The activity of these cells lengthens plant parts. Thickening arises from divisions of undifferentiated cells in cylindrical tissues that run lengthwise through older structures.

Links to Earlier Concepts

This chapter builds on the discussion of flowering plant anatomy (Section 22.3) in the context of life's organization (1.2). We also revisit carbohydrates (3.3), amyloplasts (4.7), plant cell walls and cuticle (4.9), photosynthesis (6.1), stomata (6.5), differentiation (8.7 and 10.2), and stem cells (11.5).

27.1 Sequestering Carbon in Forests

A 2,000-year-old giant sequoia tree is just cranking out wood. Growth does not slow with age in long-lived trees; rather, more and more wood is produced until the tree dies. That may be because a tree's leaf area increases as its crown expands over a long life span.

Consider how plants are specialized to absorb carbon dioxide (CO_2) from the air via photosynthesis. In an actively growing plant, photosynthetically produced sugars are continually remodeled into other carbon-containing compounds that end up in the plant's tissues. Biologists recently studied a 3,200-year-old giant sequoia tree in California's Sequoia National Park (**FIGURE 27.1**). By climbing and measuring this tree, they calculated that it holds more than 54,000 cubic feet (1,500 cubic meters) of wood and bark. A lot of carbon is locked up in that one old tree!

After a tree dies, its tissues decompose and the carbon in them is released to the atmosphere. However, compounds such as lignin and cellulose that waterproof and reinforce plant parts are relatively stable, so plant matter decomposes more slowly than other organic materials. Thus, carbon in sturdy plant tissues such as wood can stay out of the atmosphere for many thousands of years.

This finding that very old trees continue making wood throughout their lifetime is very relevant for carbon offsets, which are financial instruments designed to reduce global emissions of carbon dioxide and other carbon-based gases. You learned in Chapter 6 that humans release a lot of carbon dioxide by burning fossil fuels and other plant-derived materials. As a result of these activities, the amount of CO_2 in the atmosphere is increasing exponentially, with unintended and potentially catastrophic effects on Earth's climate.

FIGURE 27.1 Researchers study a giant sequoia (*Sequoiadendron giganteum*) **in California's Sequoia National Park.** Old trees like this one continue producing wood—a lot of it—until they die.

CREDITS: (opposite) Dan Legere; (1) MICHAEL NICHOLS/National Geographic Creative.

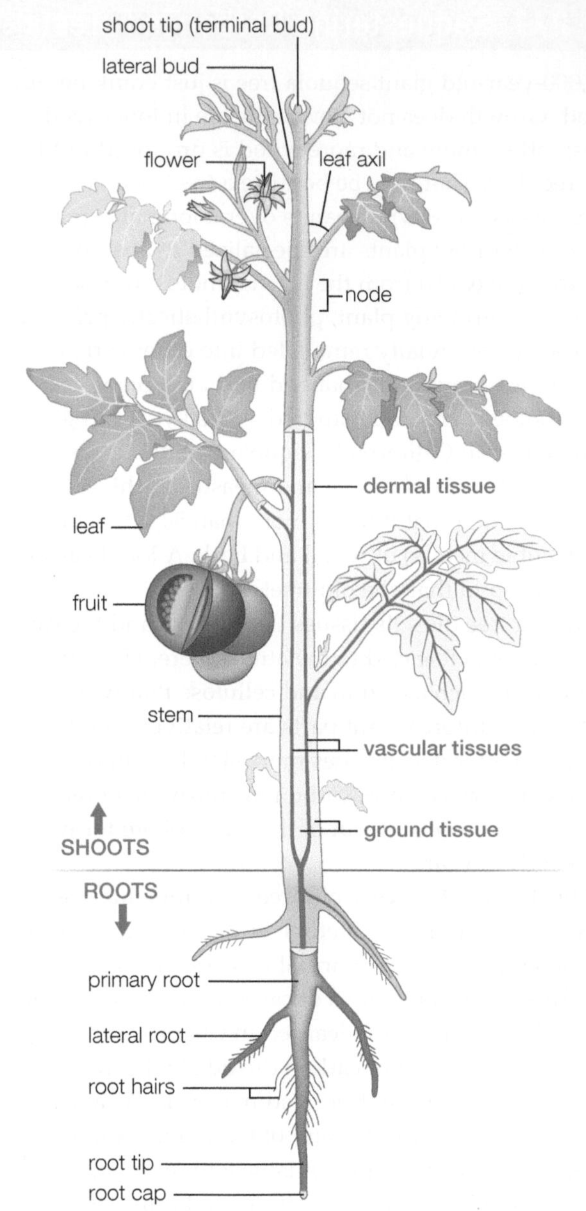
shoot tip (terminal bud)
lateral bud
flower
leaf axil
node
dermal tissue
leaf
fruit
stem
vascular tissues
ground tissue
SHOOTS
ROOTS
primary root
lateral root
root hairs
root tip
root cap

FIGURE 27.2 General body plan of a tomato plant. Vascular tissues (purple) conduct water and solutes. They thread through ground tissues that make up most of the plant body. Dermal tissue covers the plant's surfaces.

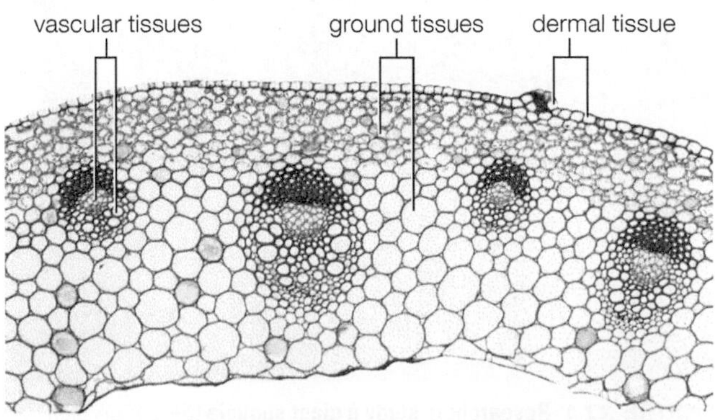
vascular tissues
ground tissues
dermal tissue

FIGURE 27.3 Roots and shoots consist of dermal, ground, and vascular tissues. This is a stem of buttercup (*Ranunculus*).

Companies and individuals buy carbon offsets to "offset" activities that release CO_2 and other carbon-containing greenhouse gases into the atmosphere. The funds are then used to support projects aimed at reducing emissions of the gases, or activities that remove them from the atmosphere. ●

27.2 The Plant Body

LEARNING OBJECTIVES

- Describe the two organ systems of a flowering plant.
- Explain the functions of the three types of plant tissue systems.
- List some differences between eudicots and monocots.

With more than 260,000 species, angiosperms—flowering plants—dominate the plant kingdom. Magnoliids, eudicots (true dicots), and monocots (Section 22.9) are the major angiosperm groups. In this chapter, we focus mainly on the structure of eudicots and monocots. Eudicots include flowering shrubs and trees, vines, and many nonwoody plants such as tomatoes and dandelions. Lilies, orchids, grasses, and palms are examples of monocots.

Like cells of most other multicelled organisms, those in plants are organized as tissues, organs, and organ systems (Section 1.2). The body of a vascular plant consists of two organ systems: shoots and roots (**FIGURE 27.2**). In a typical plant, shoots are above the ground and roots are below it.

Roots and shoots are composed of three tissue systems: dermal, vascular, and ground (**FIGURE 27.3**). The **dermal tissue system** consists of tissues that cover and protect the plant's exposed surfaces. The **vascular tissue system** includes the tissues that carry water and nutrients from one part of the plant body to another. The **ground tissue system** is essentially everything that is not part of the dermal or vascular tissue systems. Most cells that carry out basic processes necessary for survival—growth, photosynthesis, and storage—are part of the ground tissue system.

Monocots and eudicots have the same types of cells and tissues, but the two lineages differ in many aspects of their organization and hence in their structure (**FIGURE 27.4**). The names of the groups refer to one difference, the number of seed leaves, or **cotyledons**, in their embryos. Monocot seeds have one cotyledon;

cotyledon (cot-uh-LEE-dun) Seed leaf of a flowering plant embryo.
dermal tissue system Tissues that cover and protect a plant's surfaces.
ground tissue system All tissues that are not dermal or vascular tissue; makes up the bulk of the plant body. Ground tissues carry out basic processes of survival such as photosynthesis, storage, and growth.
vascular tissue system All tissues that carry water and nutrients through the plant body.

FIGURE 27.4 Some of the structural features that differ between eudicots and monocots.

Eudicots	Monocots

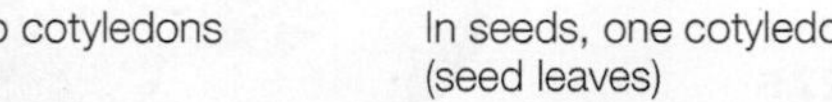

In seeds, two cotyledons (seed leaves)

In seeds, one cotyledon (seed leaves)

Flower parts in fours or fives (or multiples of four or five)

Flower parts in threes (or multiples of three)

Leaf veins usually forming a netlike array

Leaf veins usually running parallel with one another

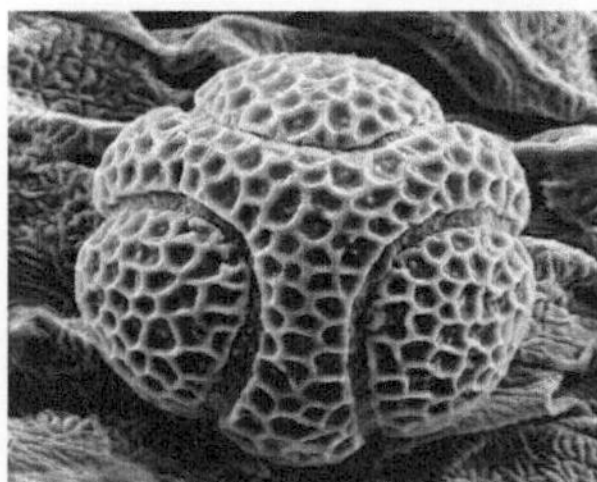

Pollen grains with three pores or furrows

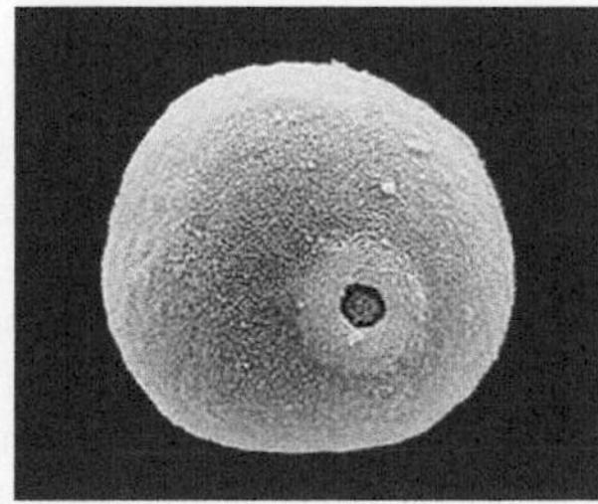

Pollen grains with one pore or furrow

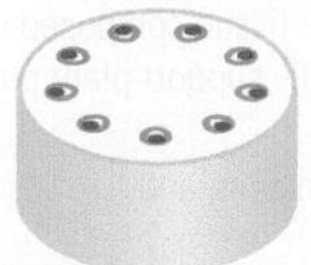

Vascular bundles in a ring in ground tissue of stem

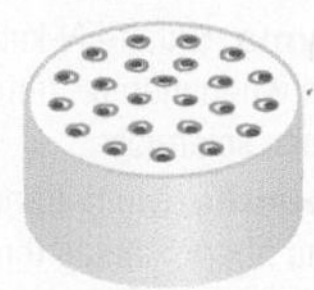

Vascular bundles all through ground tissue of stem

eudicot seeds have two (Chapter 29 returns to embryonic development in plants). As you will see, the tissue organization inside shoots and roots also differs between eudicots and monocots.

TAKE-HOME MESSAGE 27.2

✔ Roots and shoots compose the plant body. Both structures consist of ground, vascular, and dermal tissue systems.

✔ Ground tissues make up most of a plant. Vascular tissues that thread through ground tissue distribute water and nutrients. Dermal tissues cover and protect plant surfaces.

✔ Eudicots and monocots have the same types of tissues, but differ somewhat in their pattern of tissue organization.

27.3 Plant Tissues

LEARNING OBJECTIVES

- Explain the function of epidermal tissue.
- Describe components and functions of plant vascular tissues.

Plant tissues that consist of one cell type are called simple tissues (**TABLE 27.1**). Dermal and vascular tissues are complex tissues, which means they consist of two or more cell types. Some plant tissues include the remains of dead cells, the rigid walls of which lend

TABLE 27.1

Components of Simple and Complex Plant Tissues

Tissue Type	Main Components	Main Functions
Simple Tissues		
Ground tissues		
Parenchyma	Parenchyma cells	Photosynthesis; storage; secretion; tissue repair
Collenchyma	Collenchyma cells	Pliable structural support
Sclerenchyma	Sclerenchyma cells (fibers or sclereids)	Structural support
Complex Tissues		
Dermal tissues		
Epidermis	Epidermal cells (including specialized types) and their secretions	Secretion of cuticle; protection; control of gas exchange and water loss
Periderm	Cork cambium; cork	Forms protective cover on older stems and roots
Vascular tissues		
Xylem	Tracheids; vessel elements; parenchyma cells; sclerenchyma cells	Water-conducting tubes; structural support
Phloem	Sieve elements, parenchyma cells; sclerenchyma cells	Sugar-conducting tubes and their supporting cells

support to the mature structure. **FIGURE 27.5** shows some locations of simple and complex tissues in a stem.

Simple Tissues

Parenchyma **Parenchyma** is a simple tissue that consists of parenchyma cells. The shape of these cells varies, but all typically have a thin, flexible cell wall. Parenchyma has specialized roles that vary with its location in the plant. In stems and roots, for example, parenchyma cells store proteins, starch, oils, and water. The photo above left shows parenchyma in the ground tissue of a buttercup root; starch-packed amyloplasts are visible inside these cells. Parenchyma cells remain alive at maturity and retain the ability to divide. As you will see in Section 27.7, all plant tissues arise from a special type of parenchyma. Other functions include structural support, nectar secretion, storage, and photosynthesis. A photosynthetic tissue called **mesophyll** consists of parenchyma cells that are filled with chloroplasts (left).

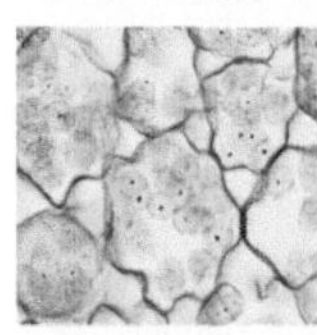

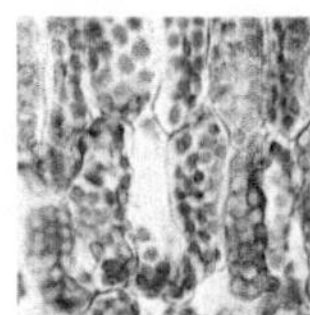

Collenchyma **Collenchyma** is a simple tissue that provides pliable structural support to rapidly growing plant parts such as young stems. Collenchyma cells remain alive at maturity. A complex polysaccharide called pectin imparts flexibility to their walls, which is thickened unevenly where three or more of the cells abut (left).

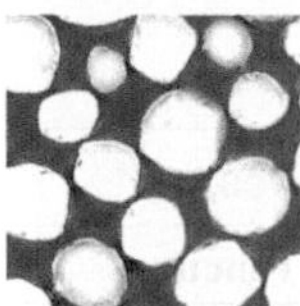

Sclerenchyma Variably shaped cells of **sclerenchyma** die after they mature. Their thick cell walls, which contain a high proportion of durable polymers such as cellulose and lignin, lend sturdiness to plant parts and help them resist stretching and compression. Fibers are long, tapered sclerenchyma cells; they occur in bundles (above left) that support and protect vascular tissues. Fibers are used to make cloth, rope, paper, and other commercial products. Sclereids (below left) are sclerenchyma cells that prevent the collapse of soft tissues; these cells make pear flesh gritty. Sclereids in nut shells and other hard seed coats make these structures difficult to breach.

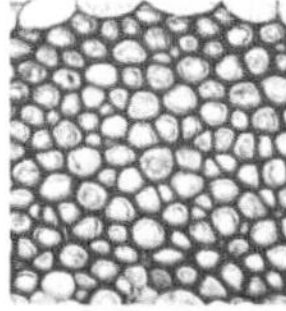

Complex Tissues

Dermal Tissue The first dermal tissue to form on a plant is **epidermis**, which in most species consists of a single layer of epidermal cells on the plant's outer surface. These cells deposit a waterproof, protective cuticle onto their outward-facing cell walls. The cuticle helps the plant conserve water and repel pathogens. Hairs and other outgrowths of epidermal cells are common (the photo above left shows some of these structures on the surface of a coleus leaf). Epidermis of leaves and young stems includes specialized cells such as those that form stomata (Section 6.5). Plants close these tiny gaps to limit water loss from internal tissues. In older woody stems and roots, a dermal tissue called periderm replaces epidermis. The photo below left shows multiple cell types in periderm of an elder stem.

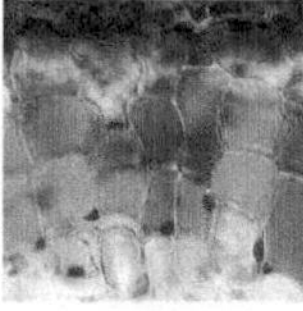

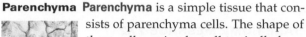

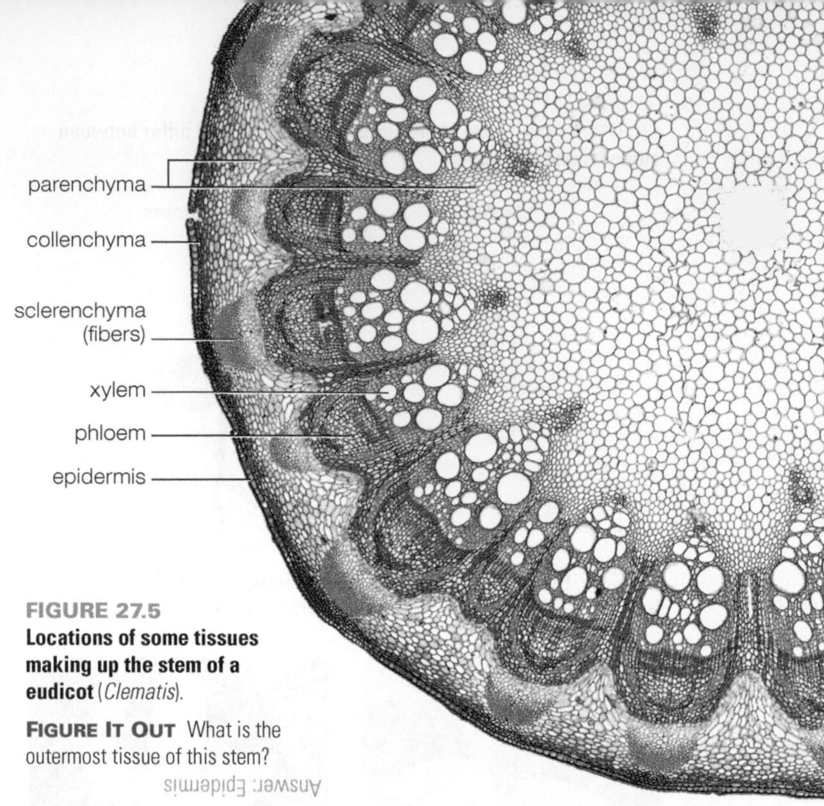
parenchyma
collenchyma
sclerenchyma
(fibers)
xylem
phloem
epidermis

FIGURE 27.5
Locations of some tissues making up the stem of a eudicot (*Clematis*).

FIGURE IT OUT What is the outermost tissue of this stem?

Answer: Epidermis

collenchyma (coal-EN-kuh-muh) In plants, simple tissue composed of living cells with unevenly thickened walls; provides flexible support.

epidermis (epp-ih-DUR-miss) Dermal tissue; outermost layer of a young plant.

mesophyll (MEZ-uh-fill) Photosynthetic parenchyma.

node A region of stem where leaves attach and new shoots form.

parenchyma (puh-REN-kuh-muh) In plants, simple tissue composed of living cells with functions that depend on location.

phloem (FLOW-um) Complex vascular tissue of plants; its living sieve elements compose sieve tubes that distribute sugars and other organic solutes.

sclerenchyma (sklair-EN-kuh-muh) In plants, simple tissue composed of cells that die when mature; their tough walls structurally support plant parts. Includes fibers, sclereids.

vascular bundle Multistranded bundle of xylem, phloem, and sclerenchyma fibers running through a stem or leaf.

xylem (ZY-lum) Complex vascular tissue of plants; cell walls of its dead tracheids and vessel elements form tubes that distribute water and mineral ions.

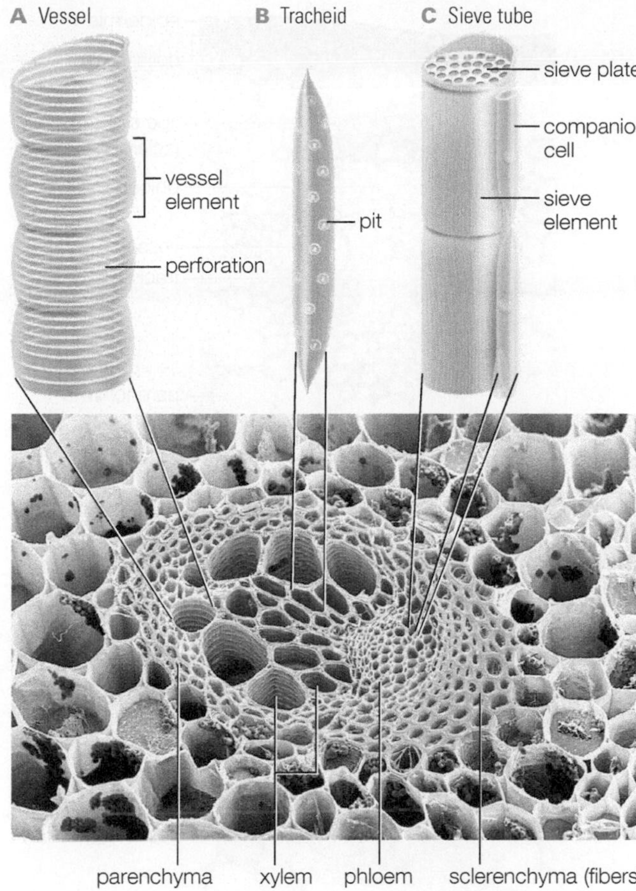
A Vessel
B Tracheid
C Sieve tube
sieve plate
companion cell
vessel element
sieve element
pit
perforation
parenchyma
xylem
phloem
sclerenchyma (fibers)

FIGURE 27.6 Plant vascular tissues. Tracheids and vessel elements make up the conducting tubes of xylem; sieve elements and companion cells make up phloem. Vascular tissues are bundled with fibers and parenchyma.

Vascular Tissue Vascular tissues, which distribute water and nutrients through the plant body, are complex tissues. The two types of vascular tissues, xylem and phloem, are composed of elongated conducting tubes. These tubes are bundled with sclerenchyma fibers and parenchyma.

Xylem is the vascular tissue that conducts water and minerals dissolved in it. Xylem pipelines are stacks of two types of cells: vessel elements (FIGURE 27.6A) and tracheids (FIGURE 27.6B). Both types of cells are dead in mature tissue, but their stiff, waterproof walls form tubes that thread through the plant. In addition to conducting water, xylem tubes lend structural support to plant parts. Interconnecting perforations and pits in the walls of adjacent cells allow water to move laterally between the tubes as well as vertically through them (Section 28.4 returns to xylem).

Phloem is the vascular tissue that conducts sugars and other organic solutes. Phloem pipelines are called sieve tubes (FIGURE 27.6C). A sieve tube is a stack of living cells—sieve elements—connected end to end at perforated sieve plates. An associated parenchyma cell provides each sieve element with metabolic support, and also transfers sugars into it. The sugars move through the sieve tubes to all parts of the plant (Section 28.5 returns to phloem).

TAKE-HOME MESSAGE 27.3

✔ In plants, simple tissues (parenchyma, collenchyma, and sclerenchyma) consist of one type of cell.

✔ Parenchyma has diverse roles, including photosynthesis. Collenchyma and sclerenchyma support and strengthen plant parts.

✔ In plants, complex tissues (dermal tissue and vascular tissue) have two or more cell types.

✔ Dermal tissues (epidermis and periderm) cover and protect plant surfaces.

✔ Vascular tissues conduct water and solutes through ground tissue.

27.4 Stems

LEARNING OBJECTIVES

- Describe vascular bundles and their arrangement in the stems of monocots and eudicots.
- Using appropriate examples, list some types of modified stems and explain their defining features.

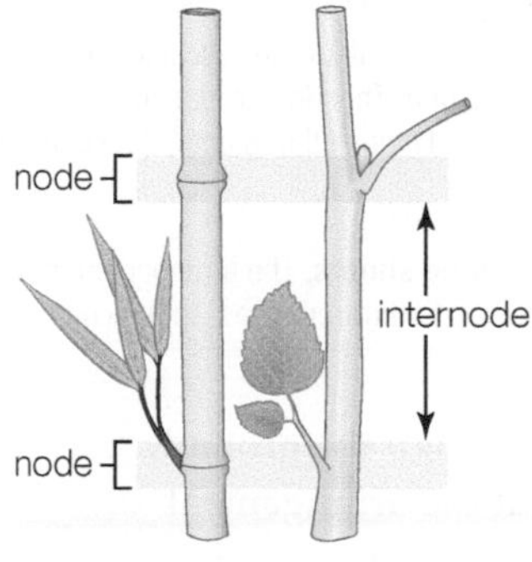
node
node
internode

FIGURE 27.7
Typical stems of monocots (left) and eudicots (right).

Stems form the basic framework of a flowering plant, providing support and keeping leaves positioned for photosynthesis. Depending on the species, they grow above or below the soil, and may be specialized for storage or asexual reproduction. Stems characteristically have **nodes**, which are regions of the stem where leaves attach (FIGURE 27.7). New shoots (and, in some cases, roots) can form at nodes. The region of stem between two nodes is called an internode.

Internal Structure

Xylem and phloem are organized as long, multistranded **vascular bundles** that run lengthwise through a stem. The main function of vascular bundles is to conduct water, ions, and nutrients between different parts of the plant. Some components of the bundles—fibers and the lignin-reinforced walls of tracheids—also play an important role in supporting upright stems of vascular plants.

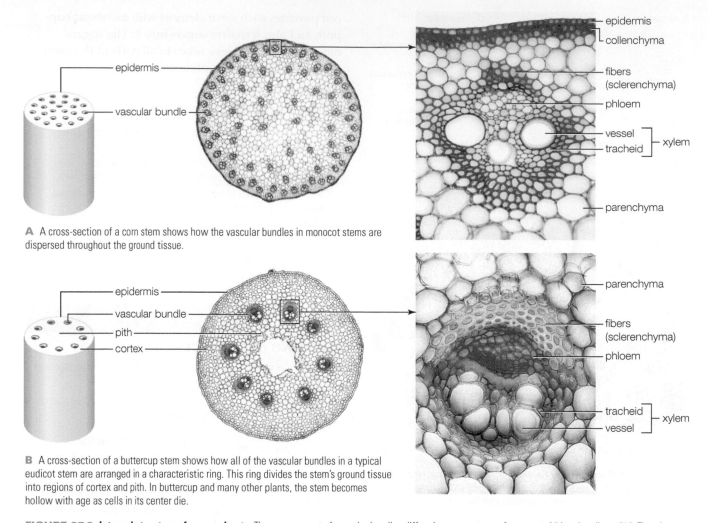

A A cross-section of a corn stem shows how the vascular bundles in monocot stems are dispersed throughout the ground tissue.

B A cross-section of a buttercup stem shows how all of the vascular bundles in a typical eudicot stem are arranged in a characteristic ring. This ring divides the stem's ground tissue into regions of cortex and pith. In buttercup and many other plants, the stem becomes hollow with age as cells in its center die.

FIGURE 27.8 Internal structure of young shoots. The arrangement of vascular bundles differs between stems of monocots (**A**) and eudicots (**B**). The photos show stem sections that have been stained with different dyes, so the colors of the cell types vary.

Vascular bundles extend through the ground tissue of stems and leaves, but the arrangement of the bundles differs between monocots and eudicots. In monocot stems, vascular bundles are typically distributed throughout the ground tissue (**FIGURE 27.8A**). By contrast, all of the vascular bundles in a typical eudicot stem has are arranged in a characteristic ring (**FIGURE 27.8B**). The ring divides the stem's ground tissue into distinct regions of pith (inside the ring) and cortex (outside the ring).

stolon

Stem Specializations

Stolons Stolons are stems that branch from the main stem of the plant and grow horizontally on the ground or just under it. They are commonly called runners because in many plants they "run" along the surface of the soil. Stolons may look like roots, but they have nodes (roots do not have nodes). Roots and shoots that sprout from the nodes develop into new plants.

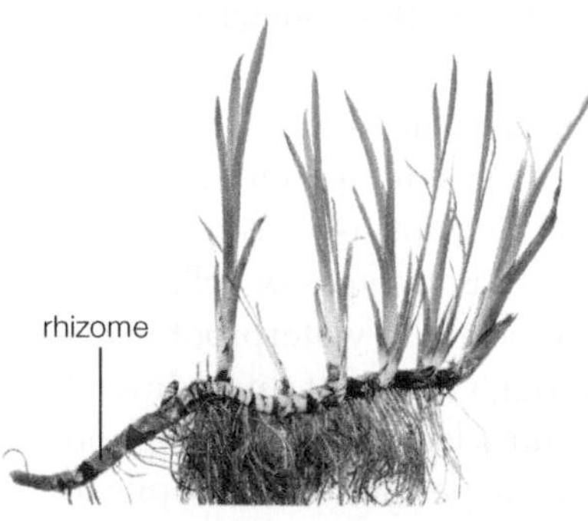
rhizome

Rhizomes Ginger, irises, and some grasses have rhizomes, which are fleshy stems that typically grow under the soil and parallel to its surface. In many plants, the main stem is a rhizome, and it often serves as the primary region for storing food. Shoots that sprout from nodes grow aboveground for photosynthesis and flowering.

Bulbs A bulb consists of a short, flattened stem (a basal plate) encased in overlapping layers of thickened modified leaves called scales. The lower portion of

CREDITS: (8A, middle) Garry DeLong/Science Source; (8A, right) Wim van Egmond/Visuals Unlimited; (8B, middle and right) ISM/Phototake; art (8A-B) © Cengage Learning; (in text) vilax/Shutterstock; dabjola/Shutterstock.

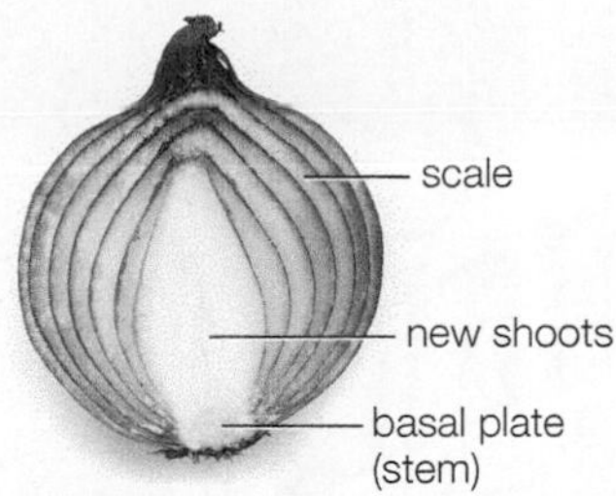
scale
new shoots
basal plate
(stem)

a bulb's basal plate gives rise to roots; the upper portion, to scales and new shoots. Bulb scales contain starch and other substances that a plant holds in reserve during times when conditions in the environment are unfavorable for growth. When favorable conditions return, the plant uses these stored substances to sustain rapid growth. The dry, paperlike outer scale of many bulbs is a protective covering.

Corms A corm is a swollen base of a stem with a papery or netlike covering. Corms are like bulbs in

that they store nutrients during times when conditions in the environment are unfavorable for growth, and they have a basal plate that gives rise to roots. Unlike a bulb, however, a corm is solid—it has no fleshy scales—and nodes form in rings on its outer surface. New shoots sprout from the nodes.

Tubers Stem tubers are thick, fleshy storage structures that form on the stolons or rhizomes of some plant species. Most are underground and temporary. Unlike

bulbs and corms, tubers have no basal plate; all new shoots and roots form at nodes on their surface. The photo on the left shows how potatoes, which are stem tubers, grow on stolons of *Solanum tuberosum* plants. The "eyes" of a potato are its nodes.

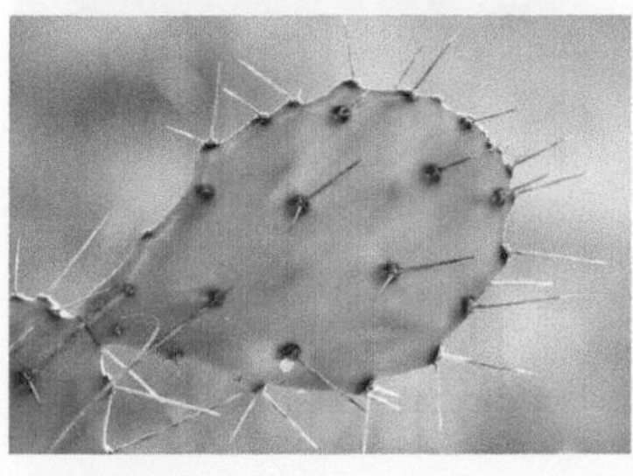

Cladodes Many types of cacti and other succulents have cladodes, which are flattened, photosynthetic stems specialized to store water. The cladodes of some plants appear leaflike, but most are unmistakably fleshy. Flowers, spikes, small leaves, new cladodes, or entire plants may form at nodes.

TAKE-HOME MESSAGE 27.4

✔ Multistranded vascular bundles, which consist of xylem, phloem, and sclerenchyma fibers, run through stems and leaves. The arrangement of vascular bundles differs between eudicots and monocots.

✔ Stolons, rhizomes, bulbs, corms, stem tubers, and cladodes are modified stems with various roles in storage and reproduction.

27.5 Leaves

LEARNING OBJECTIVES

- Draw the general arrangement of tissues in a typical eudicot leaf.
- List some of the variations in eudicot leaf structure.
- Describe the main structural differences between the leaves of typical monocots and eudicots.

Leaves are the main organs of photosynthesis in most flowering plant species. They also function in gas exchange, and they are the major site of evaporative water loss. Typical leaves are thin, with a high surface-to-volume ratio. Leaves of most monocots are long and narrow, and the base of the leaf wraps around the stem to form a sheath around it (**FIGURE 27.9A**). In most eudicots, a short stalk called a petiole attaches the leaf to a stem at a leaf axil, which is the upper angle between the stem and petiole. Modified leaves of many eudicots carry out special functions (**FIGURE 27.9B**).

FIGURE 27.9 External structure of leaves.

A Typical leaves of monocots (left) and eudicots (right).

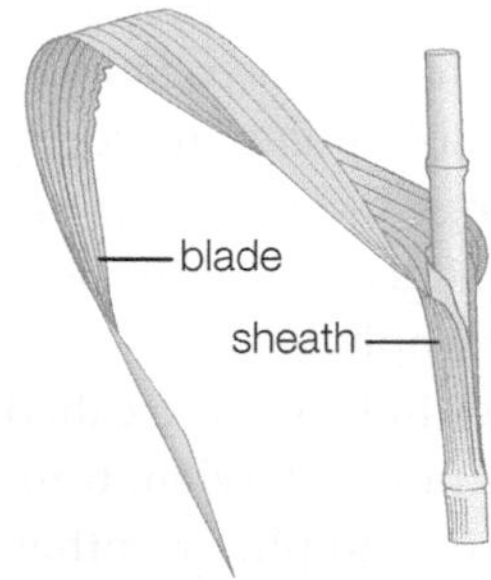
blade
sheath

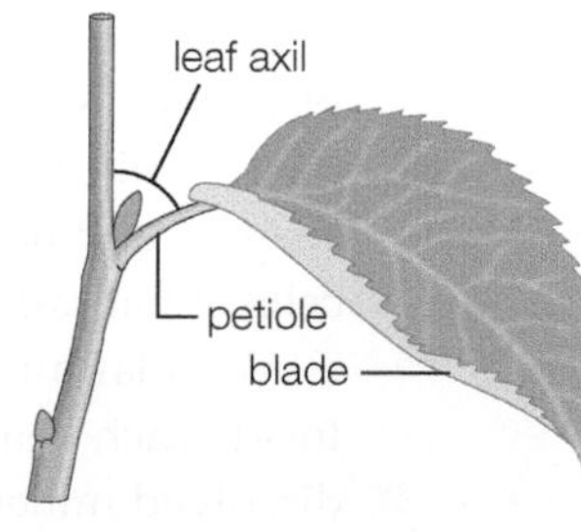
leaf axil
petiole
blade

B Example of specialized leaf form in a eudicot. Carnivorous plants of the genus *Nepenthes* grow in places where nitrogen is scarce. They secrete acids and protein-digesting enzymes into fluid in a cup-shaped modified leaf. The enzymes release nitrogen from small animals (such as insects, frogs, and mice) that are attracted to odors from the fluid and then drown in it. The leaves then take up the released nitrogen.

CREDITS: (in text) Aaron Amat/Shutterstock; © Ian Young, www.srgc.org.uk; Chase Studio/Science Source; Four Oaks/Shutterstock. (9A) © Cengage Learning; (9B) Perennou Nuridsany/Science Source.

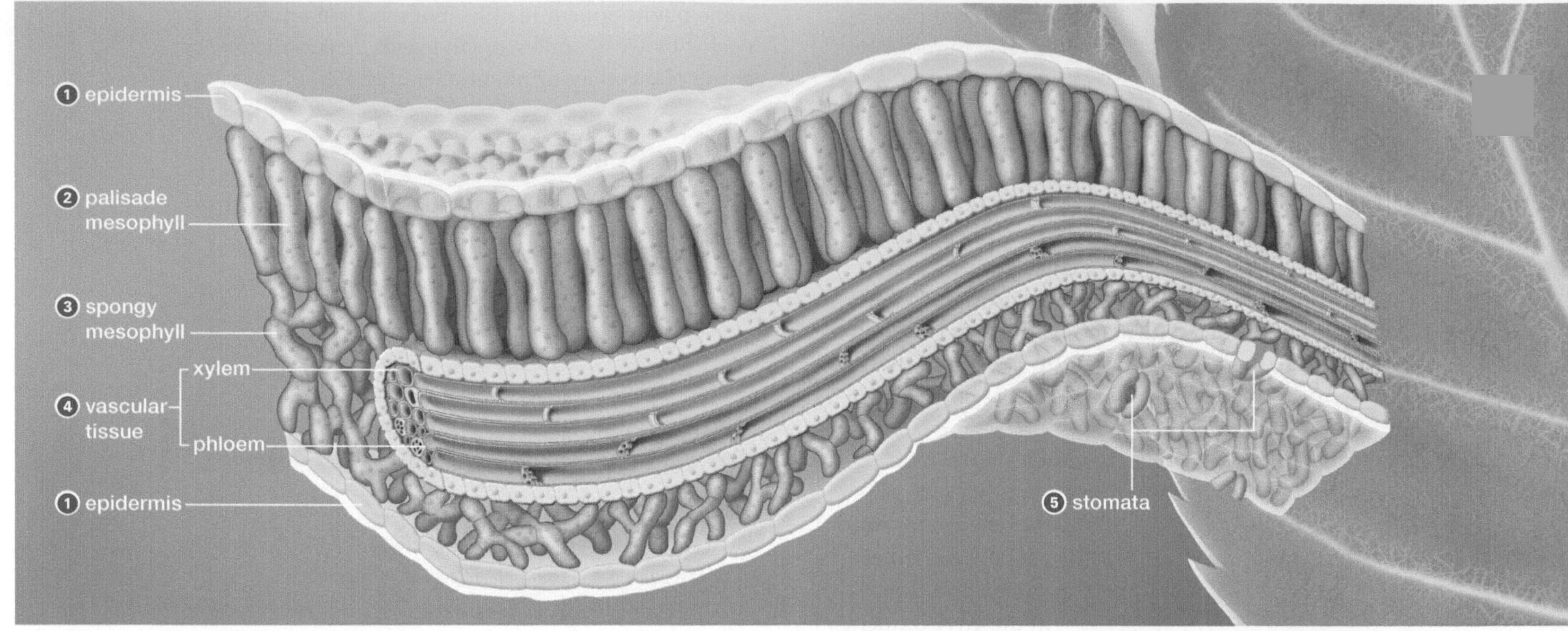
① epidermis
② palisade mesophyll
③ spongy mesophyll
xylem
④ vascular tissue
phloem
① epidermis
⑤ stomata

FIGURE 27.10 Anatomy of a eudicot leaf.

The leaf surface is epidermis with a secreted layer of cuticle ❶. The bulk of the leaf is mesophyll, a type of photosynthetic parenchyma. Eudicot leaves often have two distinct layers of this tissue: palisade mesophyll with elongated cells ❷, and spongy mesophyll with irregularly shaped cells ❸.

❹ Inside leaf veins, vascular bundles of xylem (blue) and phloem (pink) transport materials to and from photosynthetic cells.

❺ Gas exchanges between air inside and outside of the leaf occur at stomata.

FIGURE 27.10 illustrates the internal structure of a typical eudicot leaf. The bulk of the leaf consists of photosynthetic parenchyma—mesophyll—that lies between an upper and a lower layer of epidermis ❶. Eudicot leaves are typically oriented perpendicular to the sun's rays, and have two distinct layers of mesophyll. The uppermost layer is palisade mesophyll ❷. Cells in this layer are elongated, and have more chloroplasts than cells of the spongy mesophyll layer below them. Cells of spongy mesophyll ❸ are irregularly shaped, and have larger air spaces between them than cells of palisade mesophyll.

Vascular bundles in leaves are called **leaf veins.** Inside each vein, strands of xylem transport water and dissolved mineral ions to photosynthetic cells, while strands of phloem transport products of photosynthesis (sugars) away from them ❹. Sclerenchyma fibers stiffen a vein and provide support for the leaf's softer tissues. In most eudicot leaves, large veins branch into a network of minor veins (see Figure 27.4).

The upper and lower surfaces of a leaf are sheets of epidermis one cell thick. Either or both surfaces may be smooth, sticky, or slimy, and have epidermal cell outgrowths such as hairs, scales, spikes, hooks, and other structures (FIGURE 27.11). Epidermal cells secrete a translucent, waxy cuticle that slows water loss. A leaf's upper surface, which typically receives the most direct sunlight, may have a thicker cuticle than the lower surface, which tends to be shaded. The lower surface usually has more stomata ❺. These openings can be closed to prevent water loss, or opened to allow gases to cross the epidermis. Carbon dioxide needed for photosynthesis enters a leaf through open stomata, then diffuses through air spaces to mesophyll cells. Oxygen released by photosynthesis diffuses in the opposite direction.

The internal structure of leaves differs somewhat between eudicots and monocots (FIGURE 27.12). Blades of grass and other monocot leaves that grow vertically can intercept light from all directions. Unlike eudicot

FIGURE 27.11 Example of leaf epidermis specialization: trichomes, which are outgrowths of epidermal cells. Glandular types such as the ones on the surface of this marijuana leaf secrete substances that deter plant-eating animals. Marijuana trichomes produce a chemical (tetrahydrocannabinol, or THC) that has a psychoactive effect in humans.

fibrous root system Root system composed of an extensive mass of similar-sized adventitious roots; typical of monocots.
leaf vein A vascular bundle in a leaf.
taproot system In eudicots, an enlarged primary root together with all of the lateral roots that branch from it.

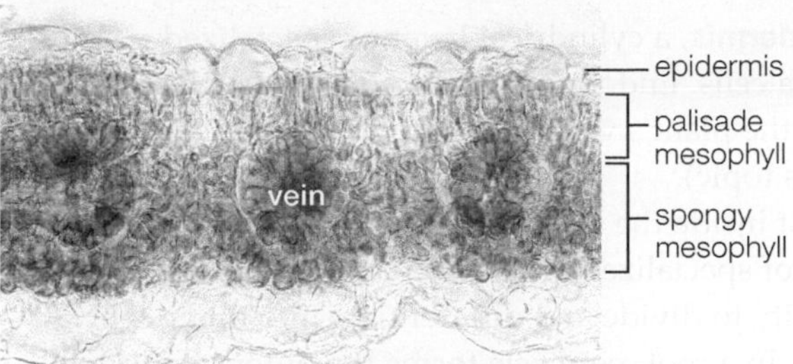
epidermis
palisade mesophyll
vein
spongy mesophyll

A Leaf section of a eudicot: coastal plain yellowtops (*Flaveria bidentis*).

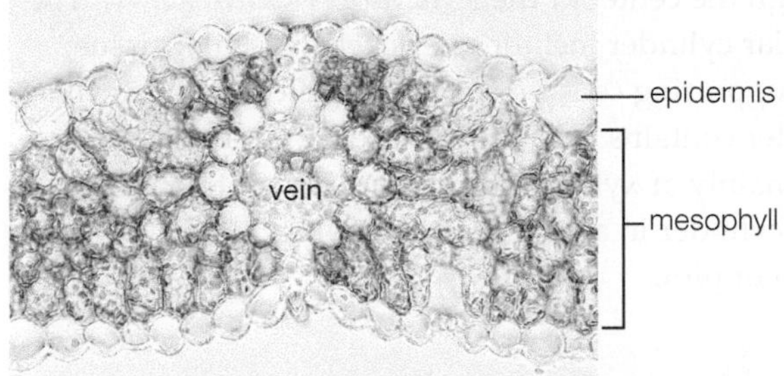
epidermis
vein
mesophyll

B Leaf section of a monocot: barley (*Hordeum vulgare*).

FIGURE 27.12 Comparing the arrangement of cells and tissues in monocot and eudicot leaves.

leaves, monocot leaves typically have a single layer of mesophyll. The arrangement of veins also differs. A eudicot leaf has branching veins; almost all of the veins in a monocot leaf are similar in length and run parallel with the leaf's long axis.

TAKE-HOME MESSAGE 27.5

✔ Leaves are structurally adapted to intercept sunlight and exchange gases for photosynthesis. Specialized forms allows some leaves to carry out special functions.

✔ Leaf components include veins (bundles of vascular tissue), mesophyll (photosynthetic cells), and cuticle-secreting epidermis.

✔ A typical eudicot leaf has two layers of mesophyll; a typical monocot leaf has one.

27.6 Roots

LEARNING OBJECTIVES

- Distinguish between a fibrous root system and a taproot system.
- Name the tissues that compose a typical root, and compare their arrangement in monocot and eudicot roots.

External Structure

The main function of a root is to take up water and mineral ions from soil (Chapter 28 returns to this topic). Roots also anchor a plant in soil, and in many species they store nutrients. The roots of a typical plant are as extensive as its shoots, and often more so. An extensive root system provides the plant with a very large surface area for absorbing soil water. Roots branch from existing roots. In some plants, they can also form on stems or leaves, in which case they are called adventitious roots. In a few species, adventitious roots form aboveground at nodes on trunks or lower branches. These roots are called prop roots.

When a seed germinates, the first structure to emerge from it is a root. In typical monocots, this primary root is quickly replaced with a mat of adventitious roots that arise from the developing stem. All of the roots in the resulting **fibrous root system** are roughly equal in diameter—there is no dominant, central root (FIGURE 27.13A). Roots in some fibrous systems are branched.

In typical eudicots, the primary root that emerges from a seed thickens and lengthens to become a taproot. A taproot can give rise to other, lateral-branching roots, but even so it usually remains the largest, dominant root. A eudicot taproot together with its lateral root branchings constitutes a **taproot system** (FIGURE 27.13B). In some eudicots, the taproot is eventually replaced by adventitious roots, so a taproot system develops into a fibrous root system. Taproots differ from root tubers, which are swollen, modified lateral roots used for storing nutrients in some monocots and eudicots.

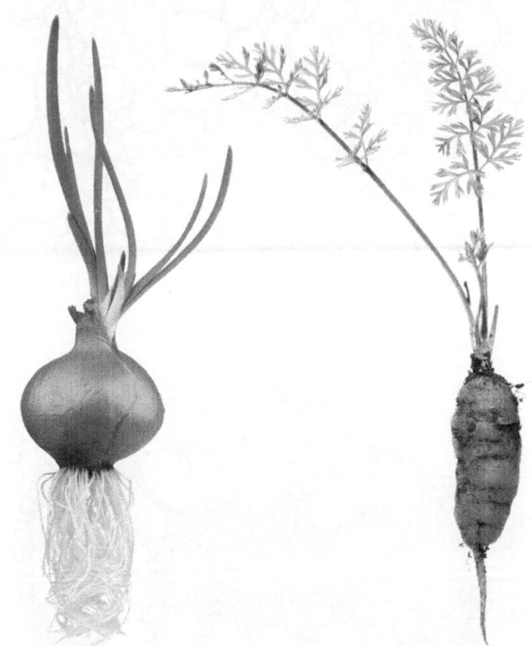

A In a fibrous root system, similar-sized adventitious roots branch from the stem. A fibrous root system is typical of onions and other monocots.

B Taproot systems consist of a large main root together with any lateral roots that branch from it. Carrots are the taproots of a variety of *Daucus carota*, a eudicot.

FIGURE 27.13 Comparing root systems.

Dermal tissue on the surface of roots is the plant's absorptive interface with soil. Epidermal cells on young roots often send out very thin extensions called **root hairs**. These structures are tiny, but a huge number of them form. Collectively, they greatly increase the root's surface area, thus maximizing its ability to absorb soil water.

Internal Structure

Water and substances dissolved in it enter a root by crossing epidermal cells and seeping into parenchyma that forms the root's region of cortex. The water moves between and through parenchyma cells until it reaches **endodermis**, a cylindrical layer of specialized parenchyma cells. Endodermal cells control which solutes enter the plant's vascular system (Section 28.3 returns to this topic).

Just inside the endodermis is **pericycle**, a thin layer of specialized parenchyma cells that retain the capacity to divide and differentiate into other cell types. In a root, pericycle forms the outer boundary of the **vascular cylinder**, or stele, that runs lengthwise through the center of the structure (**FIGURE 27.14**). The vascular cylinder includes pericycle, vascular tissue, and supporting cells. In a typical eudicot, the vascular cylinder contains very little pith or none at all; it consists mainly of xylem and phloem. By contrast, the vascular cylinder in a typical monocot has a large central region of pith.

FIGURE 27.14 Zooming in on the vascular cylinder. The arrangement of tissues in the vascular cylinder differs between monocot and eudicot roots. The photos show root sections that have been stained with different dyes, so the colors of the cell types vary.

A Eudicot (buttercup) root in cross-section. In this and other typical eudicots, the vascular cylinder is mainly vascular tissue, with little or no pith.

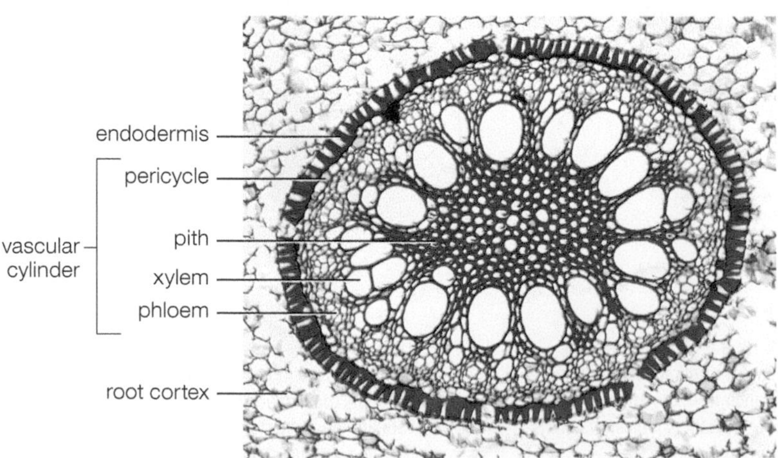
endodermis
pericycle
vascular cylinder
pith
xylem
phloem
root cortex

B Monocot (iris) root in cross-section. In this and other typical monocots, the vascular cylinder includes a large central region of pith.

TAKE-HOME MESSAGE 27.6

- ✔ Taproot systems consist of an enlarged primary root and lateral roots that branch from it. Fibrous root systems consist of adventitious roots that arise from stems.
- ✔ Roots provide a large surface area for absorbing water and dissolved mineral ions from soil.
- ✔ A vascular cylinder that runs lengthwise through each root contains pipelines of xylem and phloem.

27.7 Patterns of Growth

LEARNING OBJECTIVES

- Compare plant primary and secondary growth.
- Explain how roots and shoots lengthen.
- Describe secondary growth and its origin.
- Explain the formation of tree rings.

A young plant grows by mainly lengthening; an older woody plant grows by lengthening and thickening. During both processes, tissues arise from the activity of **meristems**, which are regions of undifferentiated parenchyma cells that can divide continually during a growing season. Meristem cells are analogous to stem cells in animals: When they divide, some of their descendants remain undifferentiated, and others differentiate (Section 11.5). Like stem cells, differentiating meristem cells give rise to specialized tissues as they divide and mature.

Primary Growth

During **primary growth**, a plant's roots and shoots lengthen, and it produces soft parts such as leaves. Primary growth originates at **apical meristems**, each a mass of meristem cells at the tip of a shoot or root.

Cells at the center of the mass are completely undifferentiated, and they divide continually by mitosis as the plant grows. The divisions push some of the meristem cells away from the tip of the shoot or root. These displaced cells begin to differentiate. Cells behind the mass of apical meristem elongate as they mature and take on their specialized function. The meristem remains at the tip of the lengthening structure, continually renewing itself and also producing more cells that divide and differentiate behind it.

Primary Growth in Shoots In a shoot, primary growth originates at its tip—its **terminal bud** (FIGURE 27.15A, B). A terminal bud contains apical meristem ❶. In an actively growing shoot, cells in the apical meristem divide continuously. The divisions push some of the dividing cells toward the base of the shoot, and these cells begin to differentiate. Depending on a cell's location in relationship to the shoot's surface and to other cells, it will become a component of one of three primary meristem tissues: Cells in the shoot's interior form ground meristem ❷ and procambium ❸; those that reach the outer surface form protoderm ❹. Further divisions and differentiation of cells in these primary meristems give rise to mature tissues: ground meristem, to ground tissue ❺; procambium, to vascular tissue ❻; and protoderm, to dermal tissue ❼.

As a typical shoot lengthens, masses of tissue near the terminal bud bulge out and then develop into leaves or new shoots. These structures form in orderly tiers along the lengthening stem. As they do, **lateral buds** (also called axillary buds) form in the leaf axils. Lateral buds also contain apical meristem cells, but this tissue may stay dormant until hormonal signals trigger the cells to divide. Depending on these signals, apical meristem in a lateral bud can give rise to leaves, a flower, or a stem that branches from the main stem (Chapter 30 returns to this topic).

apical meristem (A-pih-cull) Meristem in the tip of a shoot or root. Associated with primary growth.
endodermis In a plant root, a sheet of cells just outside the vascular cylinder. Controls uptake of solutes into the plant's vascular system.
lateral bud Axillary bud; forms in a leaf axil.
meristem (MARE-ih-stem) Zone of undifferentiated parenchyma cells, the divisions of which give rise to new tissues during plant growth.
pericycle (PAIR-ih-sigh-cull) Thin layer of undifferentiated parenchyma cells. In roots, forms the outer layer of the vascular cylinder.
primary growth Lengthening of young shoots and roots; originates at apical meristems.
root hairs Hairlike, absorptive extensions of a root epidermis cell; many of them form on young roots.
terminal bud Tip of an actively growing shoot; contains apical meristem.
vascular cylinder Stele. Of a root, central column that consists of vascular tissue, pericycle, and supporting cells.

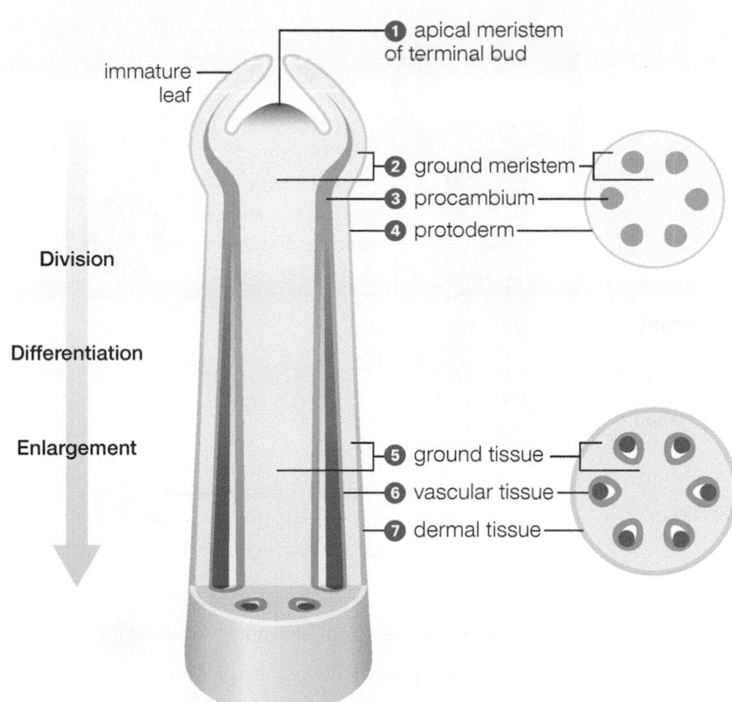
❶ apical meristem of terminal bud
immature leaf
❷ ground meristem
❸ procambium
❹ protoderm
Division
Differentiation
Enlargement
❺ ground tissue
❻ vascular tissue
❼ dermal tissue

A New tissues arise by divisions of apical meristem cells in the shoot's terminal bud. Dividing meristem cells that are displaced away from the shoot's tip begin to differentiate as protoderm, procambium, and ground meristem. Cells in these primary meristem tissues continue to divide until they fully differentiate into dermal, vascular, and ground tissues, respectively. The stem elongates as the maturing cells enlarge.

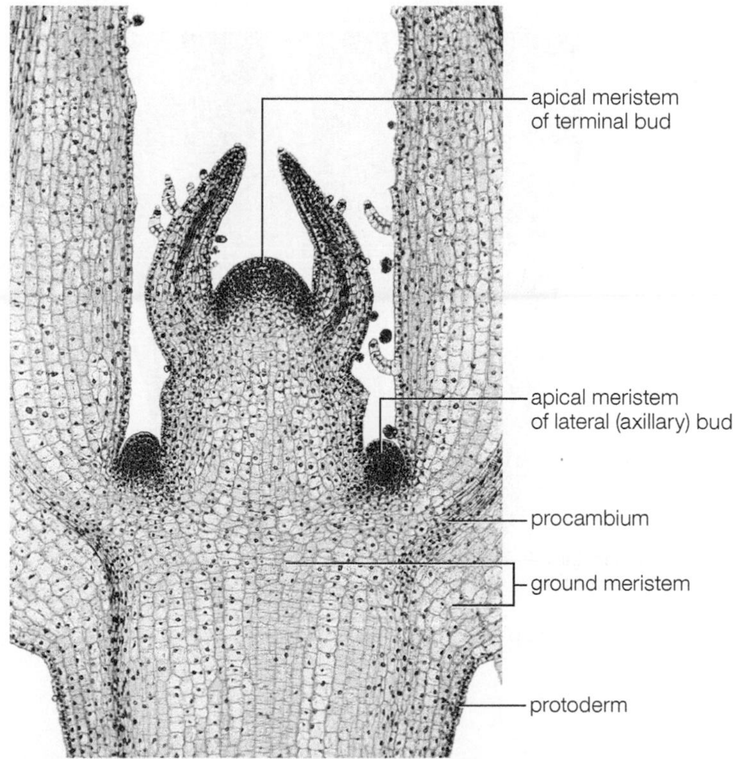
apical meristem of terminal bud
apical meristem of lateral (axillary) bud
procambium
ground meristem
protoderm

B Locations of tissues making up a growing shoot tip of *Coleus*, a eudicot. Cell nuclei are stained red (nuclei take up most of the volume of the smaller cells in the darker regions).

FIGURE 27.15 Primary growth in shoots.

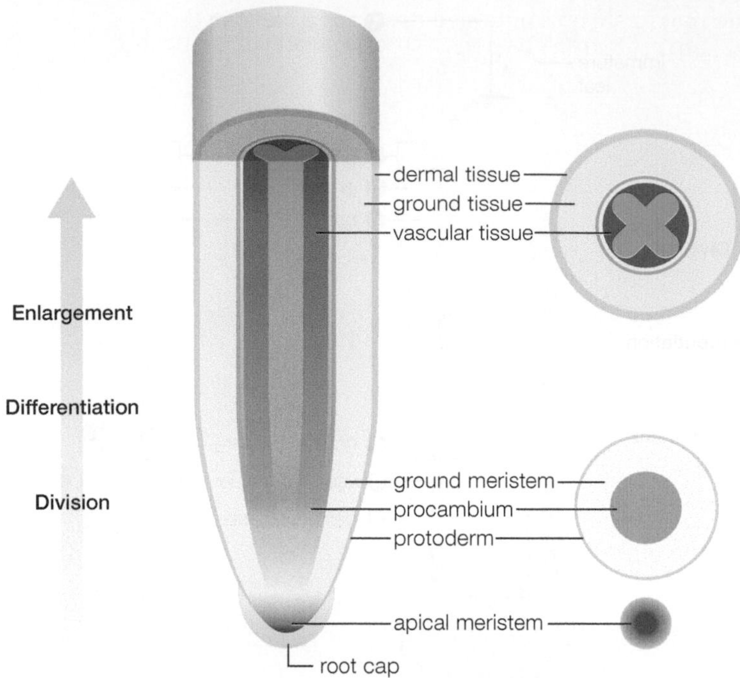

dermal tissue
ground tissue
vascular tissue
Enlargement
Differentiation
Division
ground meristem
procambium
protoderm
apical meristem
root cap

A New tissues arise by divisions of apical meristem cells in the root's terminal bud. Dividing meristem cells that are displaced away from the root's tip begin to differentiate as protoderm, procambium, and ground meristem. Cells in these primary meristem tissues continue to divide until they fully differentiate into dermal, vascular, and ground tissues, respectively. The root elongates as the maturing cells enlarge.

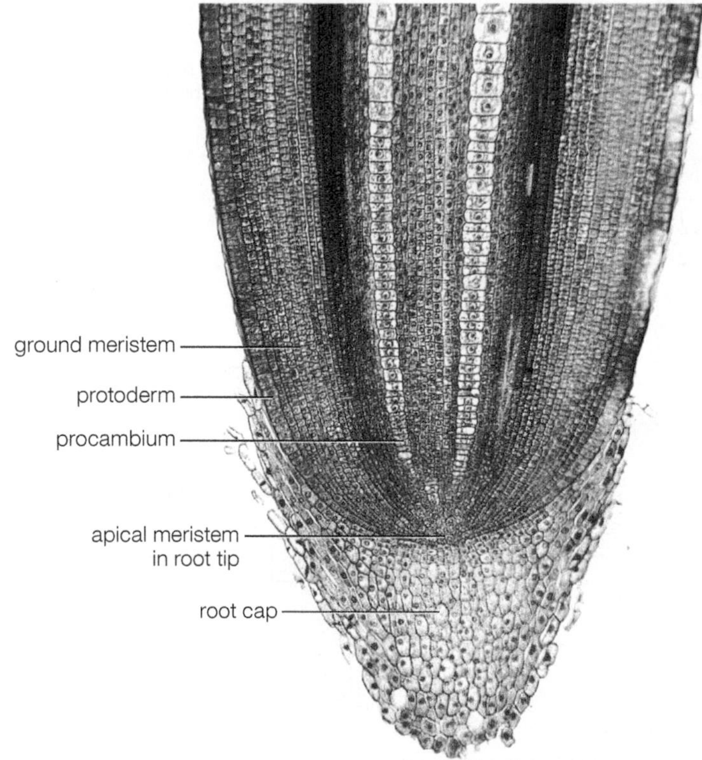

ground meristem
protoderm
procambium
apical meristem in root tip
root cap

B Locations of apical meristem and primary meristem tissues in a root tip of onion (*Allium*), a monocot.

FIGURE 27.16 Primary growth in roots.

Primary Growth in Roots Root primary growth begins with dividing apical meristem cells in root tips (FIGURE 27.16). Dividing cells behind the meristem begin the process of differentiation. Depending on its location in relationship to the bud's surface and to other cells, a cell will become a component of protoderm, ground meristem, or procambium; which, in turn, give rise to dermal, ground, and vascular tissues. Some protoderm cells give rise to a root cap, which is a dome-shaped mass of cells that protects the tip of the root as it grows through soil. Cells of the root cap are shed continually as the root lengthens.

Unlike shoots, roots have no nodes. Lateral roots that form on an existing root originate with pericycle cells in the root's vascular cylinder. These cells have meristematic activity, and they can divide in a direction perpendicular to the long axis of the root. A lateral root is arising from pericycle in the corn root shown on the left.

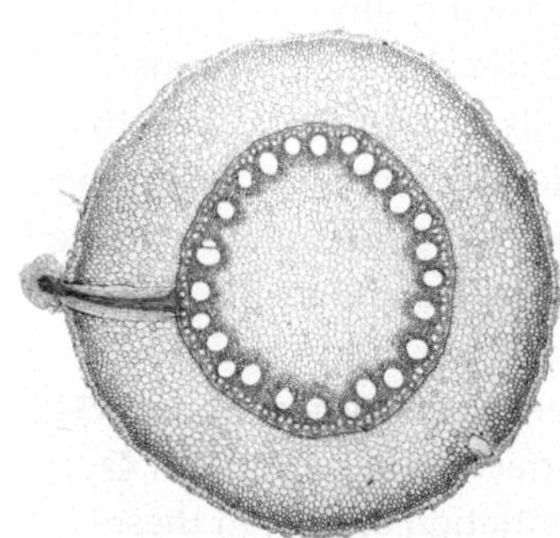

Secondary Growth

Lateral Meristems Many species undergo **secondary growth**, during which their shoots and roots thicken and become woody. In eudicots and gymnosperms, the thickening originates at **lateral meristems**, which are cylindrical layers of meristem that run lengthwise through shoots and roots (FIGURE 27.17). Two lateral meristems form as stems and roots age: vascular cambium and cork cambium (*cambium* is the Latin word for change). Both arise from pericycle. **Vascular cambium** produces secondary vascular tissue between cortex and pith. When cells of vascular cambium divide, some of them differentiate. Those that differentiate on the outer (cortex) side of this lateral meristem become part of a narrow zone of secondary phloem. Those that

bark Informal term for all living and dead tissues outside the ring of vascular cambium in woody plants.
cork Tissue that waterproofs, insulates, and protects the surfaces of woody stems and roots.
cork cambium (CAM-bee-um) Lateral meristem that produces cork.
lateral meristem Cylindrical sheet of meristem that runs lengthwise through shoots and roots; associated with secondary growth (thickening). Vascular cambium or cork cambium.
periderm Dermal tissue that replaces epidermis during secondary growth.
secondary growth Thickening of older stems and roots; originates at lateral meristems.
vascular cambium (CAM-bee-um) Lateral meristem that produces secondary xylem and phloem.
wood Secondary xylem.

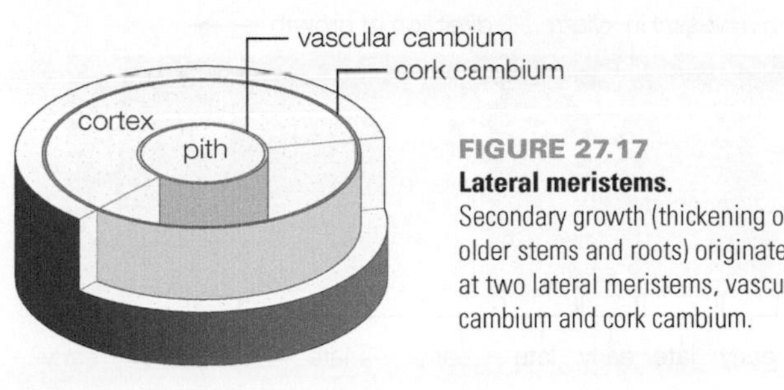
vascular cambium
cork cambium
cortex
pith

FIGURE 27.17 Lateral meristems.
Secondary growth (thickening of older stems and roots) originates at two lateral meristems, vascular cambium and cork cambium.

differentiate on the inner (pith) side become part of secondary xylem (**FIGURE 27.18**).

Secondary xylem is called **wood**, and it accumulates during secondary growth. The accumulation displaces the vascular cambium toward the outer surface of the root or shoot. Continued divisions of cells in the vascular cambium maintain its cylindrical shape.

Another lateral meristem eventually forms in the cortex, just under the dermal tissue of the structure. This lateral meristem is **cork cambium**, so named because it produces a tissue called cork on its outer surface. **Cork** consists of densely packed dead cells with waxy walls; it protects, insulates, and waterproofs the surface of the stem or root. Cork is part of **periderm**, the dermal tissue that replaces epidermis on the surfaces of older stems and roots. Periderm includes cork cambium as well as the cork it produces. What we call **bark** is an informal term for periderm and all other living and dead tissues that lie outside the cylinder of vascular cambium (**FIGURE 27.19**).

FIGURE 27.18 Secondary growth at the vascular cambium in a eudicot stem.

In older stems and roots, secondary growth (thickening) occurs simultaneously with primary growth (lengthening) at terminal and lateral buds.

Cells of vascular cambium divide during secondary growth. Some of these cells remain undifferentiated; others differentiate as divisions continue. Differentiating cells on the outer surface of the vascular cambium produce secondary phloem. Differentiating cells on the inner surface of the vascular cambium (orange) produce secondary xylem. As the inner core of xylem expands, it displaces the cylinder of vascular cambium toward the surface of the shoot or root.

Cells that differentiate on the pith (inner) side of the vascular cambium become part of xylem (X_1, X_2, X_3).

Cells that differentiate on the cortex (outer) side of the vascular cambium become part of phloem (P_1, P_2, P_3).

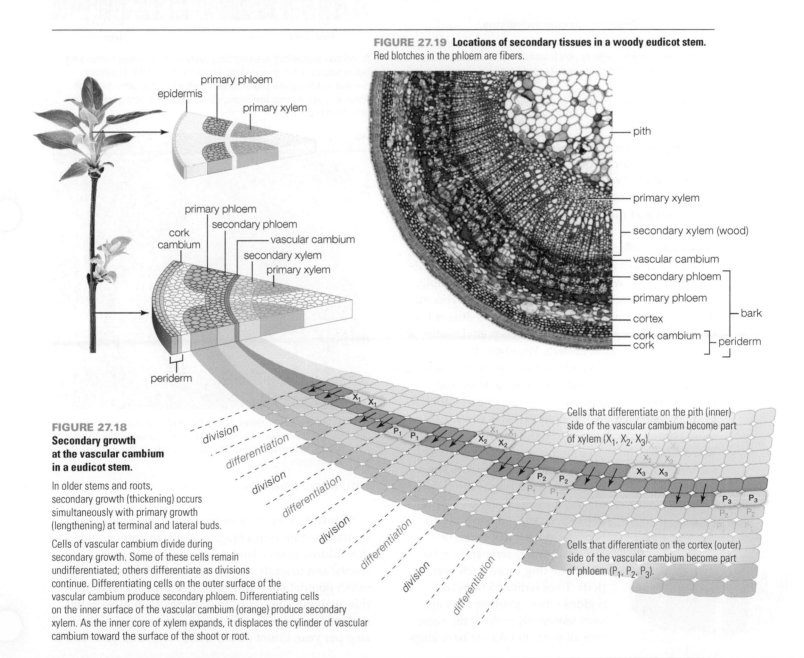

FIGURE 27.19 Locations of secondary tissues in a woody eudicot stem. Red blotches in the phloem are fibers.

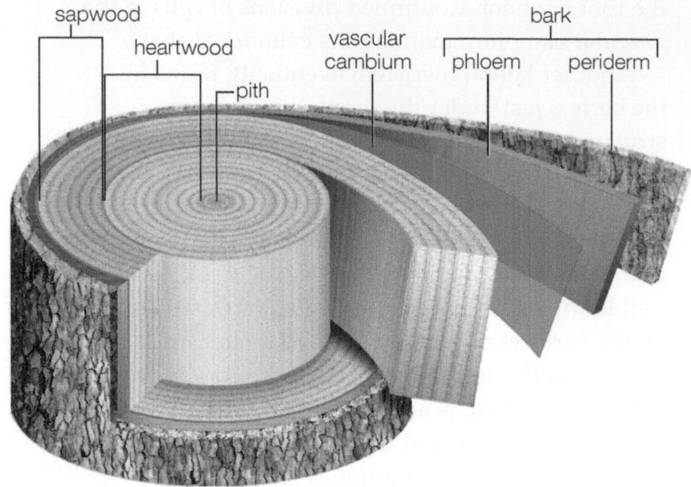
sapwood
heartwood
pith
vascular cambium
bark
phloem
periderm

FIGURE 27.20 Structure of an older eudicot stem.

Secondary xylem (wood) continues to accumulate over time. As a tree ages, the oldest xylem may no longer be able to transport fluid because its tubes have become clogged. This older xylem is called heartwood. Sapwood is still-functional secondary xylem.

FIGURE IT OUT Why is sapwood on the external side of heartwood?

Answer: Because it formed more recently.

Over years of growth, the conducting tubes of secondary xylem can become plugged so they no longer transport fluid. Wood like this is called heartwood. Sapwood is the region of still-functional xylem between heartwood and the vascular cambium (FIGURE 27.20).

Tree Rings A tree adjusts growth in response to seasonal changes in its environment, and these adjustments affect the way it produces wood. For example, many eudicot trees native to temperate zones become dormant during winter. No growth occurs during this period. Dormancy ends in spring, when environmental conditions favor rapid growth. Wood produced during rapid growth—early wood—consists of large-diameter, thin-walled xylem. Growth slows in summer, and wood produced during this time—late wood—consists of small-diameter, thick-walled xylem. Early wood is more porous, less dense, and lighter in color than late wood. These differences are often visible in a cross-section of an older tree, as bands called growth rings or tree rings. Each ring is composed of a layer of early and late wood (FIGURE 27.21A); the transitions between rings mark cycles of growth. In most temperate-zone trees, one ring forms each year (left). Trees native to tropical regions where weather does not vary seasonally grow at the same rate all year, and do not have rings.

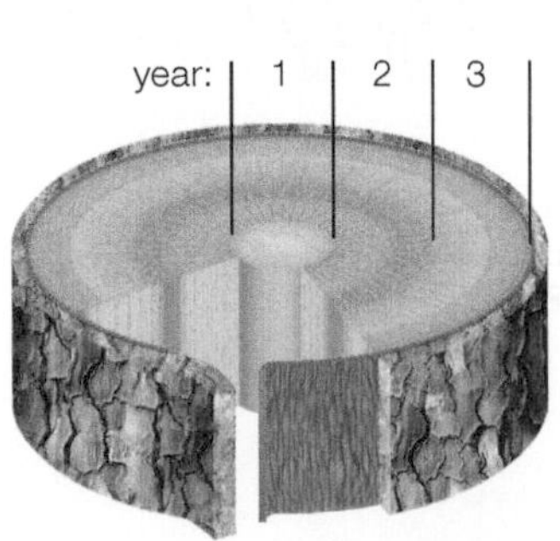
year: 1 2 3

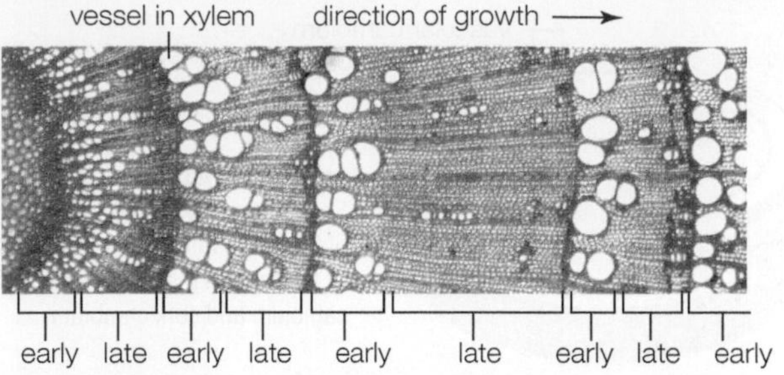
vessel in xylem
direction of growth
early late early late early late early late early

A Each tree ring is a band of early wood and late wood.

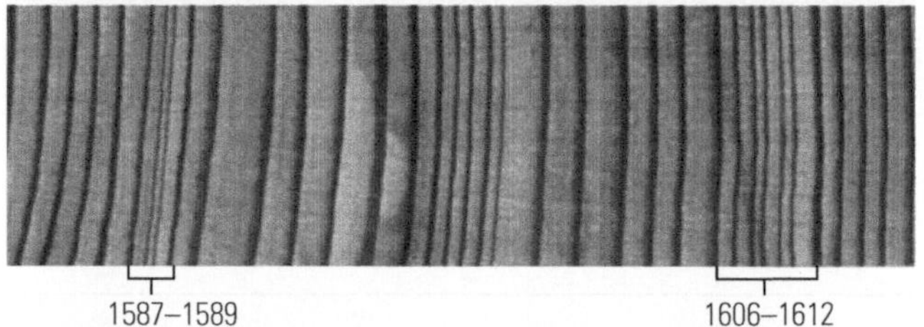
1587–1589
1606–1612

B Relative thickness of tree rings are used to estimate annual rainfall long before records of climate were kept. This is a section of a bald cypress tree that was living near English colonists when they first settled in North America. Narrower rings that formed during abnormally slow growth mark years of severe drought.

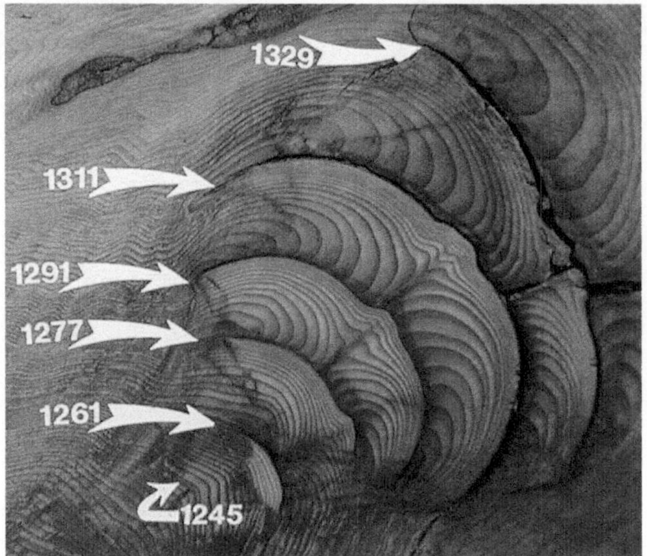
1329
1311
1291
1277
1261
1245

C Scars that disrupt the normal, circular pattern of tree rings are evidence of past fires. This photo shows fire scars in the wood of an ancient giant sequoia from California's Sequoia National Park; numbers indicate the year a particular ring was laid down.

FIGURE 27.21 Tree rings.

Tree rings may be used to estimate average annual rainfall; to date archaeological ruins; to gather evidence of wildfires, floods, landslides, and glacier movements; and to study the ecology and effects of parasitic insect populations. How can a tree provide all of this information? Long-lived tree species such as sequoias and bristlecone pines add wood over centuries, one ring per year. Count an old tree's rings, and you have

Data Analysis Activities

Tree Rings Reveal Droughts El Malpais National Monument, in west central New Mexico, has pockets of vegetation that have been surrounded by lava fields for about 3,000 years, so they have escaped wildfires, grazing animals, agricultural activity, and logging. Henri Grissino-Mayer generated a 2,129-year annual precipitation record using tree ring data from living and dead trees in this park (**FIGURE 27.22**).

1. Around A.D. 770, the Mayan civilization began to suffer a massive population loss, particularly in the southern lowlands of Mesoamerica. The El Malpais tree ring data show a drought during that time. Was it more or less severe than the Dust Bowl drought in 1933–1939?
2. One of the worst population catastrophes ever recorded occurred in Mesoamerica between 1519 and A.D. 1600, when around 22 million people native to the region died. Which period between 137 B.C. and 1992 had the most severe drought? How long did that drought last?

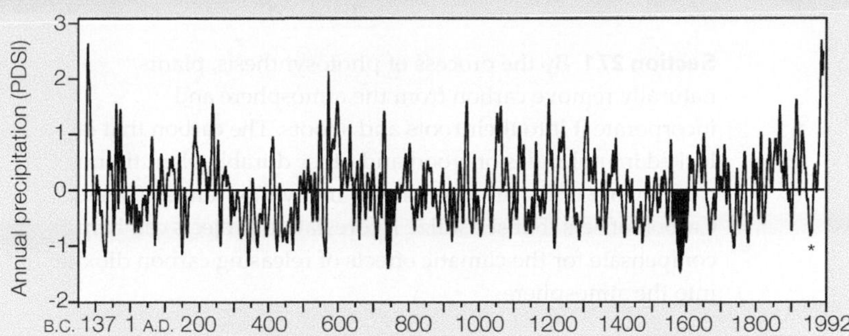
Annual precipitation (PDSI)
3
2
1
0
-1
-2
*
B.C. 137
1 A.D.
200
400
600
800
1000
1200
1400
1600
1800
1992

FIGURE 27.22 Annual precipitation record for 2,129 years, inferred from compiled tree ring data in El Malpais National Monument, New Mexico. Data were averaged over 10-year intervals; the graph correlates with other indicators of rainfall collected in all parts of North America. PDSI, Palmer Drought Severity Index: 0, normal rainfall; increasing numbers mean increasing excess of rainfall; decreasing numbers mean increasing severity of drought.

* A severe drought contributed to a series of catastrophic dust storms that turned the midwestern United States into a "dust bowl" between 1933 and 1939.

an idea of its age. If you know the year in which the tree was cut, you can determine when a particular ring formed by counting the rings backward from the outer edge. Thickness, isotope content, and other features of a ring offer clues about the environmental conditions that prevailed during the year it formed (**FIGURE 27.21B**).

Consider how more fires tend to occur in warmer, drier conditions, so an increased frequency of fires can be evidence of a period of drought. Very old trees often bear the scars of many forest fires in their rings (**FIGURE 27.21C**). In 2010, researchers used the rings and fire scars of ancient trees in California's Sequoia National Park to reconstruct a 3,000-year history of the region's climate. They discovered a dramatic increase in the number of forest fires between A.D. 800 and A.D. 1300. This 500-year period coincided with the Medieval Warm Period, an anomaly in climate that had previously been documented in some other parts of the world. Taken together, the evidence suggests that this warming period was a global climate pattern that caused worldwide drought.

TAKE-HOME MESSAGE 27.7

- ✔ All plant growth arises at meristems, which are regions of undifferentiated cells in roots and shoots. Divisions of these cells renew meristem; they also produce cells that differentiate and give rise to all other tissues in the plant.
- ✔ Primary growth (lengthening) originates at apical meristems in the tips of roots and shoots.
- ✔ In eudicots and gymnosperms, secondary growth (thickening) originates at vascular cambium and cork cambium, each a cylinder of lateral meristem that runs lengthwise through older roots and shoots.
- ✔ Wood is accumulated secondary xylem. Bark comprises all living and dead tissue outside of the vascular cambium.
- ✔ Tree rings are bands of early and late wood laid down one per year by temperate zone trees. Each band reflects conditions in the environment during the time it formed.

27.1 Sequestering Carbon in Forests (revisited)

Some carbon offsets finance endeavors such as increasing plant density in existing forests, and replanting deforested areas. Efforts like these typically cost less than other methods of removing carbon from the atmosphere, in part because they take advantage of the ability of plants to do this naturally. Forests worldwide absorb about 3 billion tons of CO_2 from the atmosphere every year, about one-third of the amount released by human activities.

Critics point out that carbon offsets supporting reforestation must be purchased well in advance of their benefit. Most trees grow very slowly, so many decades can pass before a tree sequesters a substantial amount of carbon in its wood. Carbon sequestration in forests may also be impermanent: There is no guarantee that reforested trees will survive to maturity, or that the carbon they have sequestered will not be released again, for example by fire or by logging. ●

CREDIT: (22) Grissino-Mayer, H. A 2129-year reconstruction of precipitation for northwestern New Mexico, USA. (1996) *Tree Rings, Environment and Humanity*: 191-204.

STUDY GUIDE

Section 27.1 By the process of photosynthesis, plants naturally remove carbon from the atmosphere and incorporate it into their roots and shoots. The carbon that is locked in molecules of wood and other durable plant tissues can stay out of the atmosphere for many thousands of years. Carbon offsets that subsidize reforestation projects can help compensate for the climatic effects of releasing carbon dioxide into the atmosphere.

Section 27.2 A plant body consists of roots and shoots. Roots and shoots, in turn, consist of ground, dermal, and vascular tissue systems. Tissues that protect plant surfaces make up the **dermal tissue system**. The **vascular tissue system** conducts water and nutrients through the plant body. Tissues that are neither dermal nor vascular are part of the **ground tissue system**. Ground tissue includes most of the cells that carry out essential metabolic processes such as photosynthesis.

Monocots and eudicots have the same tissues organized in different ways. For example, embryos of monocots have one **cotyledon** (seed leaf); those of eudicots have two.

Section 27.3 **Parenchyma**, **collenchyma**, and **sclerenchyma** are simple tissues, which means each consists of one type of cell. All are ground tissues. Parenchyma stays alive at maturity; **mesophyll** is photosynthetic parenchyma. Collenchyma remains alive in mature tissue; its cells have sturdy, flexible walls that support fast-growing plant parts. Fibers and sclereids are sclerenchyma cells. These cells die at maturity, and their lignin-reinforced walls remain and support the plant.

Dermal tissues and vascular tissues are complex tissues, which means they consist of multiple cell types. Stomata open across **epidermis**, a dermal tissue that covers soft plant parts. Periderm replaces epidermis in woody plants. Xylem and phloem are vascular tissues. Tracheids and vessel elements that make up **xylem** are dead at maturity; the stiff, waterproof walls that remain after these cells die form pipelines for water and dissolved minerals. **Phloem** conducts sugars through sieve tubes that consist of stacks of living sieve elements.

Section 27.4 **Vascular bundles** extending through stems conduct water and nutrients between different parts of the plant, and they also provide structural support. In most eudicot stems, vascular bundles form a ring that divides ground tissue into cortex and pith. In monocot stems, the vascular bundles are distributed throughout the ground tissue. **Nodes**, which are regions where new shoots and roots form, are characteristic of stems. Stem specializations such as rhizomes, corms, tubers, bulbs, cladodes, and stolons are adaptations for storage and/or reproduction.

Section 27.5 Leaves, which are specialized for photosynthesis, contain mesophyll and vascular bundles (**leaf veins**) between upper and lower epidermis. Eudicots typically have two layers of mesophyll; monocots do not. Water vapor and gases cross cuticle-covered epidermis at stomata.

Section 27.6 Roots are specialized for absorbing water and mineral ions from soil. Tiny **root hairs** greatly increase root surface area. Typical monocots have a **fibrous root system** that consists of similar-sized adventitious roots. Typical eudicots have a **taproot system**, which is an enlarged primary root together with its lateral root branchings.

A **vascular cylinder**, or stele, extends lengthwise through each root. A layer of **pericycle** forms its outer boundary. The vascular cylinder is enclosed by **endodermis** that helps control which solutes enter the vascular system.

Section 27.7 All plant tissues originate at **meristems**, which are regions of undifferentiated cells that can divide continuously during growth. The divisions renew the meristem and also produce other cells that divide and differentiate behind it.

Primary growth (lengthening) arises at **apical meristems**. Dividing and differentiating cells behind apical meristem form three primary meristem tissues. Cells in these tissues fully differentiate into ground, vascular, and dermal tissues. Apical meristem in **terminal buds** of shoots and **lateral** (axillary) **buds** in leaf axils gives rise to primary growth in stems. Apical meristem in root tips gives rise to primary growth in roots. Lateral roots arise from divisions of meristematic pericycle cells in a vascular cylinder.

Secondary growth (thickening) arises at cylinders of **lateral meristem** (**vascular cambium** and **cork cambium**) that run lengthwise through older stems and roots. Vascular cambium produces secondary xylem (**wood**) on its inner surface, and secondary phloem on its outer surface. Cork cambium gives rise to **cork**, which is part of **periderm**. **Bark** is an informal term for all tissues outside the cylinder of vascular cambium.

A tree ring is a layer of early and late wood. In trees native to temperate zones, one ring forms per year. Tree rings indicate the age of such trees, and also hold information about prevailing environmental conditions while the rings were forming, such as the availability of water.

SELF-QUIZ

Answers in Appendix VII

1. In plants, fibers are a type of _______ cell.
   a. parenchyma
   b. sclerenchyma
   c. collenchyma
   d. mesophyll
2. Parenchyma is a type of _______ .
   a. ground tissue
   b. vascular tissue
   c. stem
   d. apical meristem
3. Which of the following cell types remain alive in mature plant tissue?
   a. sclerenchyma
   b. sieve elements
   c. tracheids
   d. vessel elements

4. All of the vascular bundles inside a typical ______ are arranged in a ring.
   a. monocot stem c. monocot root
   b. eudicot stem d. eudicot root
5. Which of the following is false?
   a. Stems have vascular bundles, and roots have vascular cylinders.
   b. Lateral roots can arise from nodes on a root.
   c. Primary growth (lengthening) can occur at an axillary bud.
   d. Apical meristem gives rise to epidermal tissue, ground tissue, and vascular tissue.
   e. Wood is xylem that forms during secondary growth.
6. Epidermis and periderm are ______ tissues.
   a. ground b. vascular c. dermal
7. A vascular bundle in a leaf is called ______ .
   a. a vascular cylinder c. a vein
   b. mesophyll d. vascular cambium
8. Typically, vascular tissue is organized as ______ in stems and as ______ in roots.
   a. multiple vascular bundles; one vascular cylinder
   b. one vascular bundle; multiple vascular cylinders
   c. one vascular cylinder; multiple vascular bundles
   d. multiple vascular cylinders; one vascular bundle
9. Is an onion a root or a stem?
10. In a(n) ______ , the primary root is typically the largest.
    a. lateral meristem c. fibrous root system
    b. adventitious root system d. taproot system
11. Root hairs ______ .
    a. conduct water from cortex to aboveground shoots
    b. increase the root's surface area for absorption
    c. anchor the plant in soil
12. Roots and shoots lengthen through activity at ______ .
    a. apical meristems c. vascular cambium
    b. lateral meristems d. cork cambium
13. The activity of lateral meristems ______ older roots and stems.
    a. lengthens b. thickens c. both a and b
14. Tree rings occur ______ .
    a. when there are droughts during the time the rings form
    b. where environmental conditions influence xylem cell size
    c. if heartwood alternates with sapwood
    d. as epidermis replaces periderm

15. Match the plant parts with the best description.

___ vascular cambium	a.	ground tissue in roots and stems
___ mesophyll	b.	stem structure
___ wood	c.	a lateral meristem
___ cortex	d.	photosynthetic parenchyma
___ stoma	e.	secondary xylem
___ potato	f.	between epidermal cells
___ cotyledon	g.	only one in monocot seeds

CRITICAL THINKING

1. Is the plant with the yellow flower above a eudicot or a monocot? What about the plant with the purple flower?
2. Oscar and Lucinda meet in a tropical rain forest and fall in love, and he carves their initials into the bark of a tree. They never do get together, though. Ten years later, still heartbroken, Oscar searches for the tree. Given what you know about primary and secondary growth, will he find the carved initials higher relative to ground level? If he goes berserk and chops down the tree, what kinds of growth rings will he see?
3. Aboveground plant surfaces are typically covered with a waxy cuticle. Why do roots lack this protective coating?
4. Why do eudicot trees tend to be wider at the base than at the top?
5. Was the section in the micrograph shown at right taken from a stem or a root? Monocot or eudicot? How can you tell?

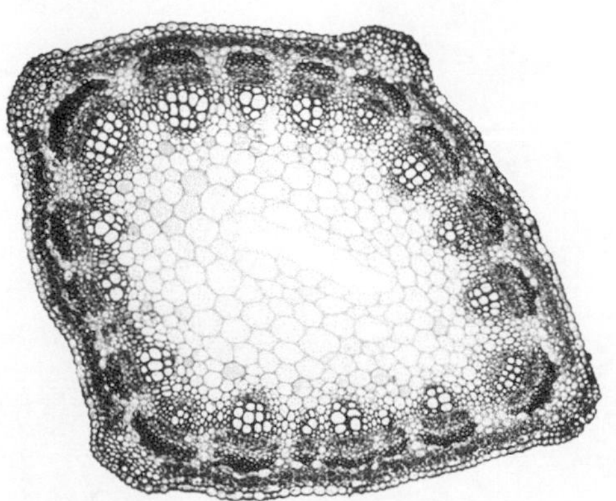

CREDITS: (in text) Edward S. Ross; Mike Clayton/ University of Wisconsin Department of Botany.

28 PLANT NUTRITION AND TRANSPORT

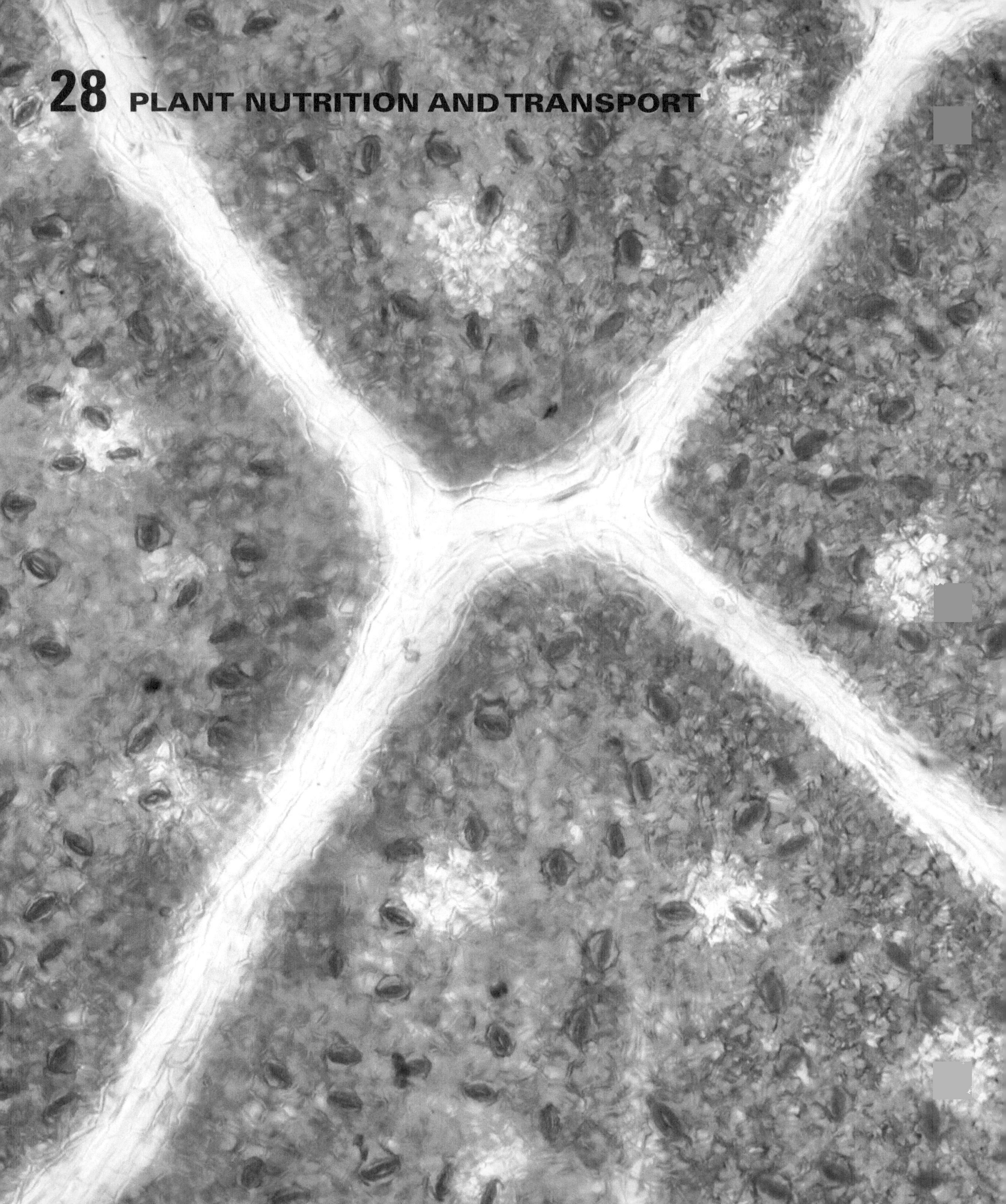

CORE CONCEPTS

 Pathways of Transformation

Organisms exchange matter and energy with the environment in order to grow, maintain themselves, and reproduce.

Nutrients required for plant growth are available in water, air, and soil. Open stomata allow plants to exchange gases with air, but they also allow the loss of water by evaporation. Water must be continually taken up from soil to replace this loss.

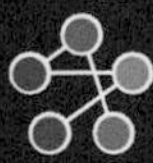 Systems

Complex properties arise from interactions among components of a biological system.

Passive and active mechanisms drive the movement of fluids through conducting tubes of vascular tissue in plants. Water's special properties allow it to be drawn upward through a plant, from roots to shoots, through narrow conduits formed by dead xylem cells. Osmosis and active transport drive the movement of sugars through stacks of living phloem cells, from regions of the plant where sugars are being produced to regions where they are being used.

Evolution

Evolution underlies the unity and diversity of life.

Evolutionary adaptations of plants include external specializations that facilitate a root's ability to take up water and nutrients from soil. Coevolutionary interactions with soil microorganisms enhance the efficiency of this uptake. Internal specializations allow selectivity over substances that enter vascular tissue for distribution to the rest of the plant. Vessel elements are evolutionarily derived from tracheids: Both structures reflect the properties of water that must be conducted through them from soil to the tips of shoots.

Links to Earlier Concepts

This chapter expands on your knowledge of plant vascular tissues (Section 27.3). Before beginning, be sure you understand concepts of membrane permeability, osmosis, osmotic pressure, and turgor (Section 5.8). You will revisit properties of water (2.5), plant cell walls and plasmodesmata (4.9), cofactors (5.6), membrane transport (5.9), light-dependent reactions of photosynthesis (6.4), stomata (6.5), nitrogen fixation (20.7), mycorrhizae (23.7), and plant endodermis (27.6).

28.1 Leafy Cleanup

From 1940 until the 1970s, the United States Army tested and disposed of weapons at J-Field, Aberdeen Proving Ground in Maryland. Obsolete chemical weapons and explosives were burned in open pits, together with plastics, solvents, and other wastes. Lead, arsenic, mercury, and other metals heavily polluted the soil and groundwater. So did hazardous organic compounds, including one called TCE (trichloroethylene). TCE is toxic to humans and other animals: It damages the nervous system, lungs, and liver, and exposure to large amounts can be fatal.

To clean up the soil at J-Field and protect nearby Chesapeake Bay, the Army and the Environmental Protection Agency turned to phytoremediation: the use of plants to remove or degrade environmental contaminants. At J-Field, the Army planted poplar trees (*Populus trichocarpa*) to cleanse groundwater. Like other vascular plants, poplar trees take up water through their roots; along with the water come substances dissolved in it—including TCE and other chemical pollutants. TCE does not harm poplars. The trees break down some of the toxin, but release most into the atmosphere. Airborne TCE is the lesser of two evils: It breaks down more quickly in air than it does in water. By 2001, the trees had removed 60 pounds of TCE from the soil at J-Field, and they had helped foster communities of soil microorganisms that were also breaking it down. Researchers estimate that J-Field will be mostly cleansed of TCE by 2030.

With metal pollutants, the best phytoremediation strategies use plants that take up toxins and store them in aboveground tissues. The toxin-laden plant parts can then be harvested for safe disposal. This strategy is being used in the Fukushima region of Japan, where a nuclear accident in 2011 released huge amounts of radioactive material that still heavily contaminates the area. Sunflowers (*Helianthus*), which are quite effective at taking up toxic metals, have been removing radioactive cesium from contaminated soil in a 30-mile radius around the damaged Fukushima nuclear reactors. Sunflowers were also used in phytoremediation of radioactive soil around Hiroshima and Chernobyl, so it is perhaps not surprising that the sunflower is an international symbol of nuclear disarmament.

CREDITS: (opposite) toeytoey/Shutterstock; (in text) lan 2010/Shutterstock.

Data Analysis Activities

TCE Uptake by Transgenic Plants Plants used for phytoremediation take up pollutants, then transport the chemicals to cells that store, detoxify, or release them to the air. Researchers are now engineering transgenic plants that have an enhanced ability to carry out these tasks.

Sharon Doty and her colleagues published the results of their efforts to genetically engineer plants for phytoremediation of soil and air contaminated with organic solvents. The researchers used *Agrobacterium tumefaciens* (Section 15.7) to deliver a mammalian gene into poplar plants. The gene encodes cytochrome P450, an enzyme involved in the breakdown of a range of organic molecules, including solvents such as TCE. **FIGURE 28.1** shows data from one test on the resulting transgenic plants.

1. How many transgenic plants did the researchers test?
2. In which group did the researchers see the slowest rate of TCE uptake? The fastest?
3. On day 6, what was the difference between the TCE content of air around planted transgenic plants and that around vector control plants?
4. Assuming no other experiments were done, what two explanations are there for the results of this experiment? What other control might the researchers have used?

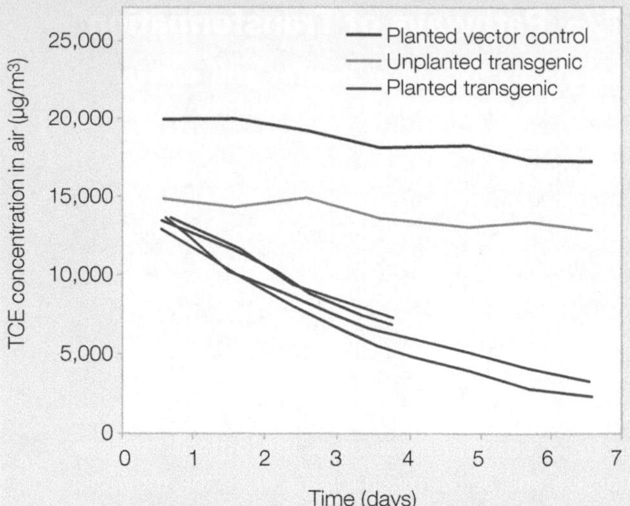

Planted vector control
Unplanted transgenic
Planted transgenic
TCE concentration in air (μg/m³)
25,000
20,000
15,000
10,000
5,000
0
0
1
2
3
4
5
6
7
Time (days)

FIGURE 28.1 TCE uptake from air by transgenic poplar plants. Individual potted plants were kept in separate sealed containers with an initial level of TCE (trichloroethylene) around 15,000 micrograms per cubic meter of air. Samples of the air in the containers were taken daily and measured for TCE content. Controls included a tree transgenic for a T_i plasmid with no cytochrome P450 in it (vector control), and a bare-root transgenic tree (one that was not planted in soil).

28.2 Plant Nutrients

LEARNING OBJECTIVES

- Give some examples of plant macronutrients and their sources.
- Describe the components of soil and their importance for plant growth.
- Explain two ways that soil can lose nutrients.

Sixteen elements are required for growth and maintenance of a plant body (**TABLE 28.1**). Nine are macronutrients, meaning they are required in amounts above 0.5 percent of a plant's dry weight. Carbon, oxygen, and hydrogen are macronutrients, as are nitrogen, phosphorus, and sulfur (components of proteins and nucleic acids), potassium and calcium (which affect processes such as cell signaling), and magnesium (a component of chlorophyll). Chlorine, iron, boron, manganese, zinc, copper, and molybdenum are micronutrients, meaning they make up traces of the plant's dry weight. Some micronutrients serve as cofactors (Section 5.6).

Properties of Soil

Three nutrients required by plants are abundantly available in air and water: Carbon dioxide (CO_2) in air provides carbon and oxygen, and water (H_2O) provides hydrogen and more oxygen. The remaining elements occur in soil. Plants take them up in the form of mineral ions dissolved in soil water.

Soils vary in their nutrient content and other properties, so some support plant growth better than others. Soil consists mainly of mineral particles—sand, silt, and clay—that form by the weathering of rocks. Sand grains are about one millimeter in diameter. Silt particles are hundreds or thousands of times smaller than sand grains, and clay particles are even smaller. Soils rich in clay can retain dissolved nutrients that might otherwise trickle past roots too quickly to be absorbed. This is because tiny clay particles carry a negative charge that attracts and traps positively charged mineral ions in soil water. However, clay particles can pack so tightly that they exclude air—and the oxygen in it. Cells in a vascular plant's roots, like cells in its aboveground parts, require oxygen for aerobic respiration.

Sand and silt intervene between clay particles, so they loosen soil. **Loam** is soil with roughly equal proportions of sand, silt, and clay. Most plants do

humus (HUE-muss) Partially decayed organic material in soil.
leaching Process by which water moving through soil removes nutrients from it.
loam Soil with roughly equal amounts of sand, silt, and clay.
soil erosion Loss of soil under the force of wind and water.
topsoil Uppermost soil layer; contains the most organic matter and nutrients for plant growth.

TABLE 28.1

Plant Nutrients and Their Functions

Macronutrient	Example of Function
Carbon, Hydrogen, Oxygen	Main components of all biological molecules
Nitrogen	Component of proteins, nucleic acids, cofactors, photosynthetic pigments
Potassium	Enzyme activation, pH, stomata opening
Calcium	Cell signaling
Magnesium	Component of chlorophyll, many enzymes and cofactors
Phosphorus	Component of phospholipids, nucleic acids (particularly ATP and other coenzymes)
Sulfur	Component of some proteins, cofactors
Micronutrient	**Example of Function**
Chlorine	Photosystem II cofactor; signaling
Iron	Component of chlorophyll and many enzymes of electron transfer chains
Boron	Cross-links polysaccharides in cell walls
Manganese	Component of photosystem II and many coenzymes
Zinc	Cofactor for many enzymes; component of many proteins
Copper	Cofactor in many proteins and enzymes of electron transfer chains
Molybdenum	Cofactor required for nitrogen fixation

Note: All mineral elements contribute to water–solute balance.

O horizon Fallen leaves and other organic litter on the soil surface.
A horizon Topsoil. The uppermost layer of soil, topsoil is typically darker than other soil layers because it is rich in organic matter.
B horizon Subsoil. Mainly accumulated clay, minerals, and organic matter that have leached down from topsoil.
C horizon Parent material. Weathered, broken rock.
R horizon Bedrock (solid rock).

FIGURE 28.2 Examples of soil horizons.

best in loam that contains between 10 and 20 percent **humus**, which is partially decomposed organic material—fallen leaves, feces, and so on. Humus supports plant growth because it releases nutrients as it breaks down, and its negatively charged organic acids can trap positively charged mineral ions in soil water. Humus also swells and shrinks as it absorbs and releases water, and these changes in size aerate soil by opening spaces for air to penetrate.

Humus tends to accumulate in swamps, bogs, and other perpetually wet areas because waterlogged soil holds little air. Humus can form in anaerobic conditions, but its breakdown is especially slow. Soils in these areas often contain more than 90 percent humus. Very few types of plants can grow in these soils.

How Soils Change

Soils develop over thousands of years, and they exist in different stages of development in different regions. Most form in layers called horizons that are distinct in color and other properties (**FIGURE 28.2**). The A horizon (the uppermost) is **topsoil**, and it contains the greatest amount of organic matter. The roots of most plants grow most densely in this layer. Topsoil and other horizons can vary in depth and composition among the regions. Grasslands typically have a deep layer of topsoil; tropical forests do not.

Minerals, salts, and other molecules dissolve in water as it filters through soil. **Leaching** is the process by which water flow removes nutrients from soil and carries them away. Leaching is fastest in sandy soils, which do not trap nutrients as well as clay soils.

The movement of wind and water can remove soil from an area, a process called **soil erosion**. Poor farming practices result in erosion. In 1800, for example, European settlers who arrived in what is now Providence Canyon, Georgia, plowed the land straight up and down the hills. The furrows made excellent conduits for rainwater, which carved deep crevices that made even better rainwater conduits. The area became useless for farming by 1850. It now consists of about 445 hectares (1,100 acres) of deep canyons that continue to expand at the rate of about 2 meters (6 feet) per year.

TAKE-HOME MESSAGE 28.2

- ✔ Plants require sixteen elements. All are available in water, air, and soil.
- ✔ Soil consists mainly of mineral particles: sand, silt, and clay.
- ✔ Clay retains mineral ions in soil water. Sand and silt that intervene between clay particles allow air to penetrate soil.
- ✔ Humus aerates soil and retains mineral ions.
- ✔ Most plants grow best in loams (soils with equal proportions of sand, silt, and clay) that contain 10 to 20 percent humus.
- ✔ Leaching and erosion remove nutrients from soil.

28.3 Root Adaptations for Nutrient Uptake

LEARNING OBJECTIVES

- Trace the paths that water and mineral ions travel as they move from soil into a root's vascular cylinder.
- Explain how endodermis helps control the amounts and types of substances that enter xylem.
- Describe the formation of mycorrhizae and root nodules, and how they benefit the organisms involved.

The Function of Endodermis

Water moves from soil, through a root's epidermis and cortex, to the vascular cylinder (**FIGURE 28.3**). Osmosis (Section 5.8) drives this movement, because fluid in the plant typically contains more solutes than soil water. After soil water enters the vascular cylinder, xylem distributes it to the rest of the plant.

Most of the soil water that enters a root moves through cell walls (Section 4.9) of adjacent plant cells ❶. Cell walls form a more or less continuous conduit through the cortex of a root, so the water can seep through them from epidermis to vascular cylinder without ever entering a cell.

Although soil water can reach the vascular cylinder by diffusing through cell walls, it cannot enter the cylinder the same way. Why not? A vascular cylinder is surrounded by endodermis, a tissue that consists of a single layer of tightly packed endodermal cells (Section 27.6). Adjacent endodermal cells secrete lignin into their walls. The lignin spans the cell walls in a band that connects tightly to each cell's plasma membrane. That band is called a **Casparian strip**, and it forms a waterproof barrier between endodermal cells ❷. A Casparian strip prevents water from crossing epidermis by diffusion through endodermal cell walls (**FIGURE 28.4**). It ensures that soil water can enter the plant's vascular system only by passing through the cytoplasm of an endodermal cell.

Soil water can enter an endodermal cell's cytoplasm by diffusing through its cell wall and plasma membrane, or via plasmodesmata. Once inside the cell, the water diffuses through pericycle cells and into a conducting tube of xylem.

Unlike water molecules, ions cannot cross lipid bilayers on their own (Section 5.8). Mineral ions in soil water enter or exit a root cell only through transport proteins in its plasma membrane. Thus, transport proteins in root cell membranes control the types and amounts of ions that the plant takes up from soil. This mechanism offers protection from some toxic substances. For example, iron ions are toxic to plants in high quantities, so control over uptake of this nutrient is critical in soils with a high iron content. Other

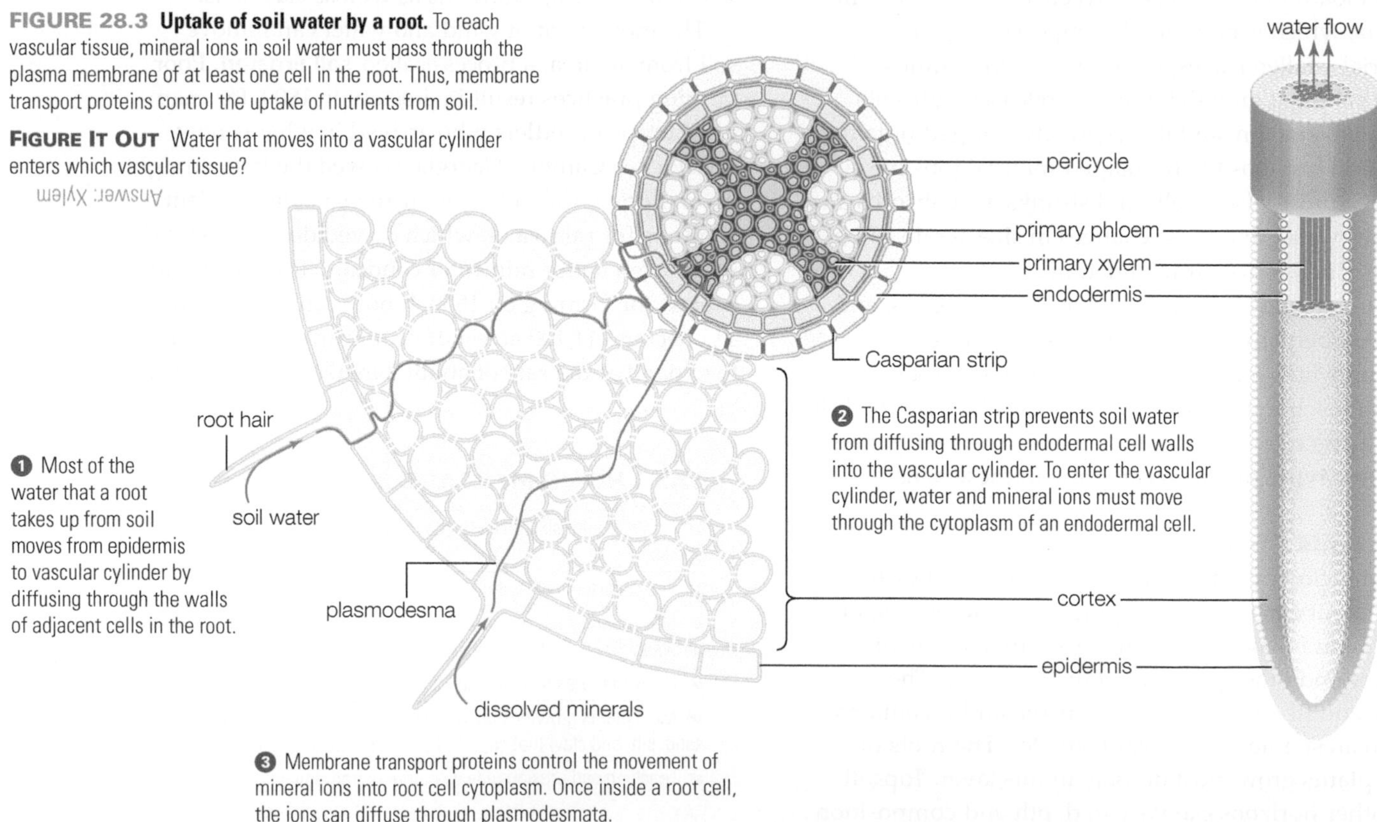
water flow
pericycle
primary phloem
primary xylem
endodermis
Casparian strip
root hair
soil water
plasmodesma
dissolved minerals
cortex
epidermis
❶ Most of the water that a root takes up from soil moves from epidermis to vascular cylinder by diffusing through the walls of adjacent cells in the root.
❷ The Casparian strip prevents soil water from diffusing through endodermal cell walls into the vascular cylinder. To enter the vascular cylinder, water and mineral ions must move through the cytoplasm of an endodermal cell.
❸ Membrane transport proteins control the movement of mineral ions into root cell cytoplasm. Once inside a root cell, the ions can diffuse through plasmodesmata.

FIGURE 28.3 Uptake of soil water by a root. To reach vascular tissue, mineral ions in soil water must pass through the plasma membrane of at least one cell in the root. Thus, membrane transport proteins control the uptake of nutrients from soil.

FIGURE IT OUT Water that moves into a vascular cylinder enters which vascular tissue?

Answer: Xylem

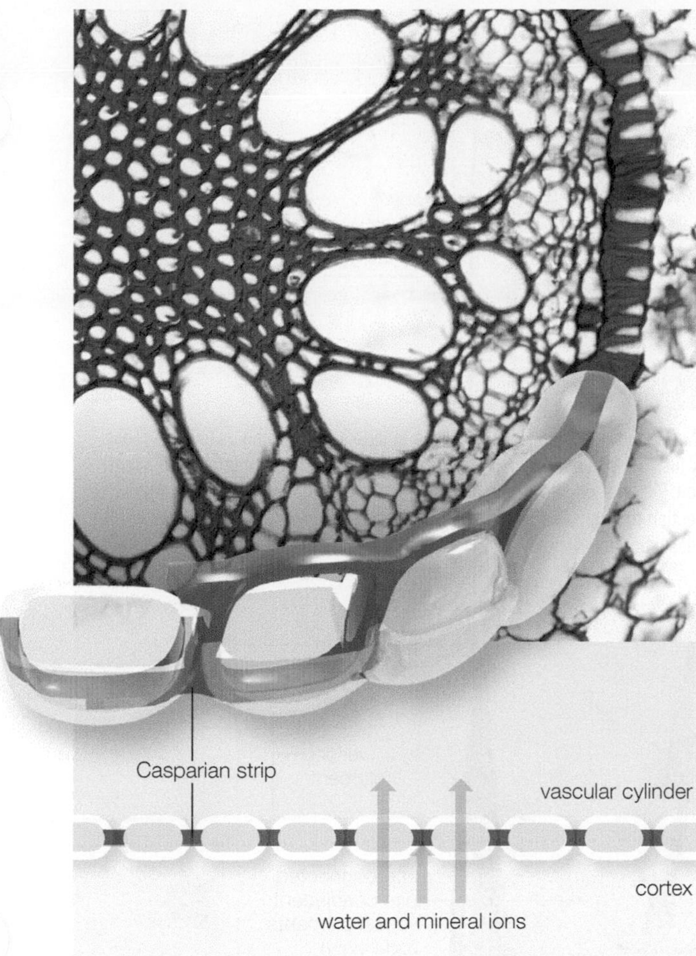
Casparian strip
vascular cylinder
cortex
water and mineral ions

FIGURE 28.4 The Casparian strip.

The micrograph shows a vascular cylinder in the root of an iris plant. Parenchyma cells that make up endodermis are specialized to secrete a waxy substance into their walls wherever they touch. The secretions, which form a Casparian strip, prevent soil water from diffusing through endodermal cell walls to enter the vascular cylinder.

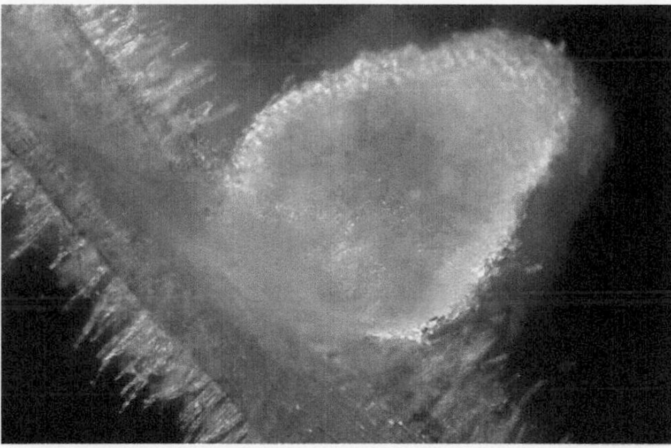

A A root nodule. The red color comes from leghemoglobin, a molecule produced by infected plant cells that is similar to hemoglobin. Leghemoglobin binds to oxygen molecules, thus maintaining a low level of free oxygen inside a root nodule. Free oxygen destroys the activity of the bacterial enzyme that fixes nitrogen; leghemoglobin releases just enough of it to sustain aerobic respiration of bacterial in the nodule. Note the root hairs.

B Soybean plants in nitrogen-poor soil show how root nodules affect growth. The darker green plants growing in the rows on the right were infected with nitrogen-fixing *Rhizobium* bacteria.

FIGURE 28.5 Root nodules. Root nodules form by a beneficial interaction between *Rhizobium* bacteria and plant roots.

mineral ions are more concentrated in the root than in soil, so they will not spontaneously diffuse into the root. These nutrients must be actively transported into root cells. In young roots, most control over ion uptake occurs at the membrane of root hairs ❸. Mineral ions that have entered the cytoplasm of any root cell can diffuse through plasmodesmata until they enter pericycle cells. From there, the ions can be loaded into xylem.

Beneficial Microorganisms

Most flowering plants maximize their ability to take up nutrients by taking part in relationships such as mycorrhizae. As Section 23.7 explained, a mycorrhiza is a mutually beneficial interaction between a root and a fungus that grows on or in it. Filaments of the fungus (hyphae) form a velvety cloak around roots or penetrate their cells. Collectively, the hyphae have a large surface area, so they absorb mineral ions from a larger volume of soil than roots alone. The fungus takes up some sugars and nitrogen-rich compounds from root cells. In return, the root cells get some scarce minerals that the fungus is better able to absorb.

Legumes and some other plants can have mutually beneficial relationships with nitrogen-fixing bacteria (Section 20.7). The roots of these plants release certain compounds into the soil that are recognized by compatible bacteria. The bacteria respond by releasing signaling molecules that, in turn, trigger the roots to grow around and encapsulate them inside swellings called **root nodules** (**FIGURE 28.5**). The association benefits both parties. The plants require a lot of nitrogen, but

Casparian strip Waterproof band that seals abutting cell walls of root endodermal cells. Helps the plant control the amounts and types of substances that it takes up from soil water.
root nodules Of some plant roots, swellings that contain and support beneficial nitrogen-fixing bacteria.

cannot use nitrogen gas (N≡N, or N_2) that is abundant in air. The bacteria in root nodules convert this gas to ammonia (NH_3), which is a form of nitrogen that the plant can use. In return for this valuable nutrient, the plant provides a low-oxygen environment for the anaerobic bacteria, and shares its photosynthetically produced sugars with them.

TAKE-HOME MESSAGE 28.3

- ✔ Roots absorb water and nutrients from soil.
- ✔ A Casparian strip forms a waterproof barrier between adjacent endodermal cells. The barrier prevents water and ions from entering the vascular cylinder by diffusion through cell walls.
- ✔ Mineral ions must pass through the plasma membrane of at least one root cell in order to enter the vascular cylinder. Thus, membrane transport proteins control the amounts and types of ions that enter the vascular system of the plant.
- ✔ Root function is enhanced by specializations such as root hairs and beneficial relationships with soil microorganisms.
- ✔ A root nodule is a structure that forms by mutually beneficial interactions between a plant and nitrogen-fixing bacteria.

28.4 Movement of Water In Plants

LEARNING OBJECTIVES

- Describe the way xylem's function arises from its structure.
- Explain how water can move from the roots of a vascular plant to leaves that may be hundreds of feet above them.
- Describe some plant adaptations that reduce water loss.
- Explain why and how stomata open.

Water that enters a root travels to the rest of the plant inside conducting tubes of xylem, which consist of the stacked, interconnected walls of dead cells (Section 27.3). Rings or spirals of lignin in the walls of these cells imparts strength to the tubes, which in turn impart strength to the structure they service. Water flows lengthwise through xylem tubes, and also laterally, from one tube to another.

As xylem tissue is maturing, its still-living cells deposit secondary wall material on the inner surface of their primary wall (Section 4.9). The material is deposited unevenly, leaving the primary wall uncovered in patterns such as rows of spots or bands that match up between adjacent cells. When the cells die, the uncovered areas form cavities called pits (**FIGURE 28.6A**). The primary wall exposed at the pits has pores where plasmodesmata once connected the adjacent cells. In the mature tissue, water moves through these pores laterally from one xylem tube to another. Pits are often bordered, which means that a circular flap of secondary wall extends over the edge of the exposed primary

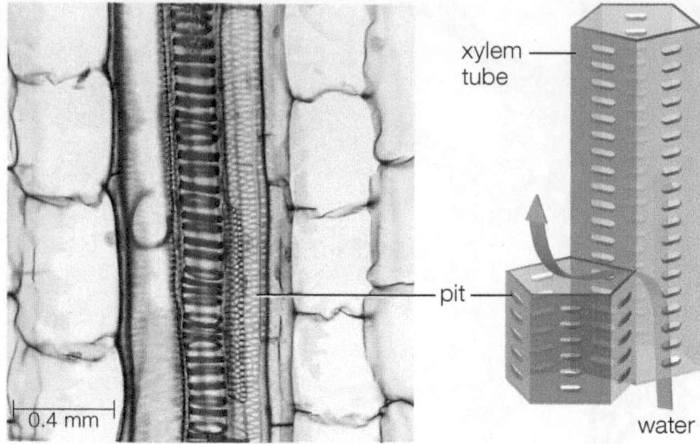
xylem tube
pit
0.4 mm
water

A Cells in tubes of xylem meet side to side at rows of holes (pits) in their walls. The pits match up, so water can flow through them into an adjacent cell. Pits in a xylem vessel are visible in the micrograph (left) of a lengthwise section through a corn stem. Rings of lignin deposited in the walls of an older vessel are also visible.

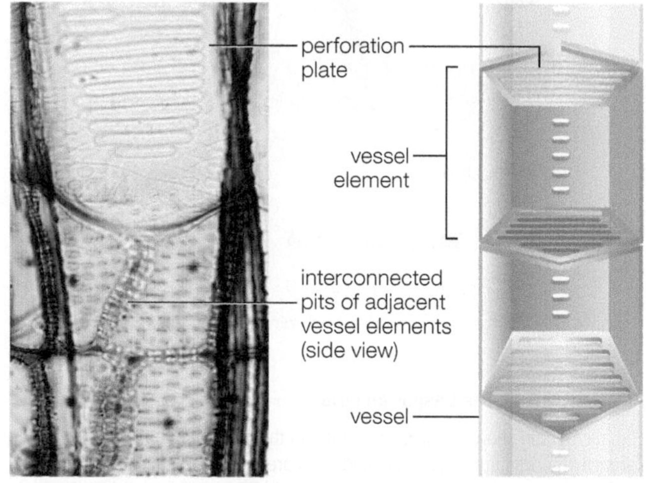
perforation plate
vessel element
interconnected pits of adjacent vessel elements (side view)
vessel

B Xylem vessels consist of stacked vessel elements connected end to end at perforation plates. Water flows vertically between cells through these plates. Left, the perforation plate in this micrograph of a section of Japanese holly has ladderlike bars. The bars break up large air bubbles in fluid moving lengthwise through a vessel.

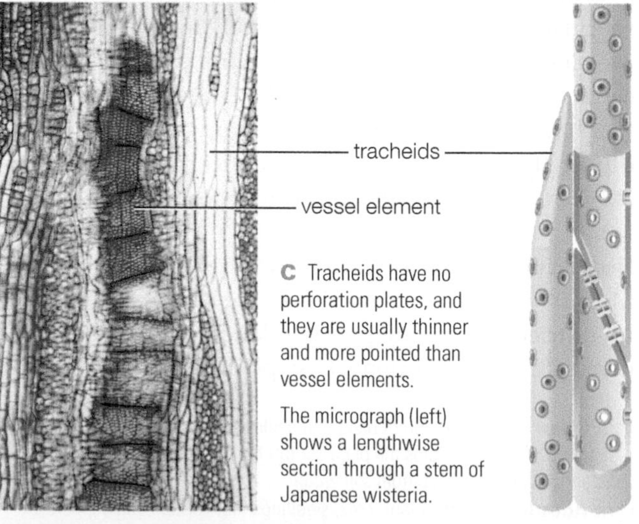
tracheids
vessel element

C Tracheids have no perforation plates, and they are usually thinner and more pointed than vessel elements.

The micrograph (left) shows a lengthwise section through a stem of Japanese wisteria.

FIGURE 28.6 Pipelines of xylem.

wall. The types and concentrations of ions in water moving through a bordered pit determines the rate of flow through it.

Angiosperms have two types of xylem tubes, each composed of one cell type: vessel elements or tracheids. Vessels, the main water-conducting tubes in angiosperms, consist of **vessel elements** stacked end to end (**FIGURE 28.6B**). In maturing xylem tissue, enzymes digest very large holes in the end walls of stacked vessel elements (where they meet) just before they die. The resulting perforation plates allow water to flow freely through vessel elements in a stack. Some perforation plates have ladderlike bars that break up air bubbles in water flowing through larger tubes.

Vessel elements are evolutionarily derived from tracheids, and their improved flow efficiency probably contributed to the success of angiosperms (most gymnosperms have only tracheids). **Tracheids**, like vessel elements, die in stacks to form water-conducting tubes. Tracheids are narrower than vessel elements, so they are more resistant to vertical compression. Thus, in addition to conducting water, tracheids have a substantial role in structural support. Unlike vessel elements, tracheids have no perforation plates; their ends are typically pointy and closed (**FIGURE 28.6C**). Pits in the narrowed ends of stacked tracheids match up. Water flows vertically through the tube by moving through these matched pits.

Cohesion–Tension Theory

How does water in xylem move upward, from roots to leaves? Tracheids and vessel elements that compose xylem tubes are dead, so these cells cannot expend any energy to pump water upward against gravity. Rather, the upward movement of water in vascular plants occurs as a consequence of evaporation and cohesion (Section 2.5). By the **cohesion–tension theory**, water is pulled upward through a plant by air's drying power, which creates a continuous negative pressure called tension. **FIGURE 28.7** illustrates this mechanism, which begins with water evaporating from leaves and stems ❶. The evaporation of water from aboveground plant parts is called **transpiration**. Transpiration's effect on water inside a plant is a bit like what happens when you suck a drink through a straw. Transpiration exerts negative pressure (it pulls) on water. Because the water molecules are connected by hydrogen bonds, a pull on one tugs all of them (cohesion). Thus, the negative pressure (tension) created by transpiration pulls on entire columns of water that fill xylem tubes ❷. The tension extends all the way from leaves that may be hundreds of feet in the air, down through stems, into young roots where water is being taken up from soil ❸.

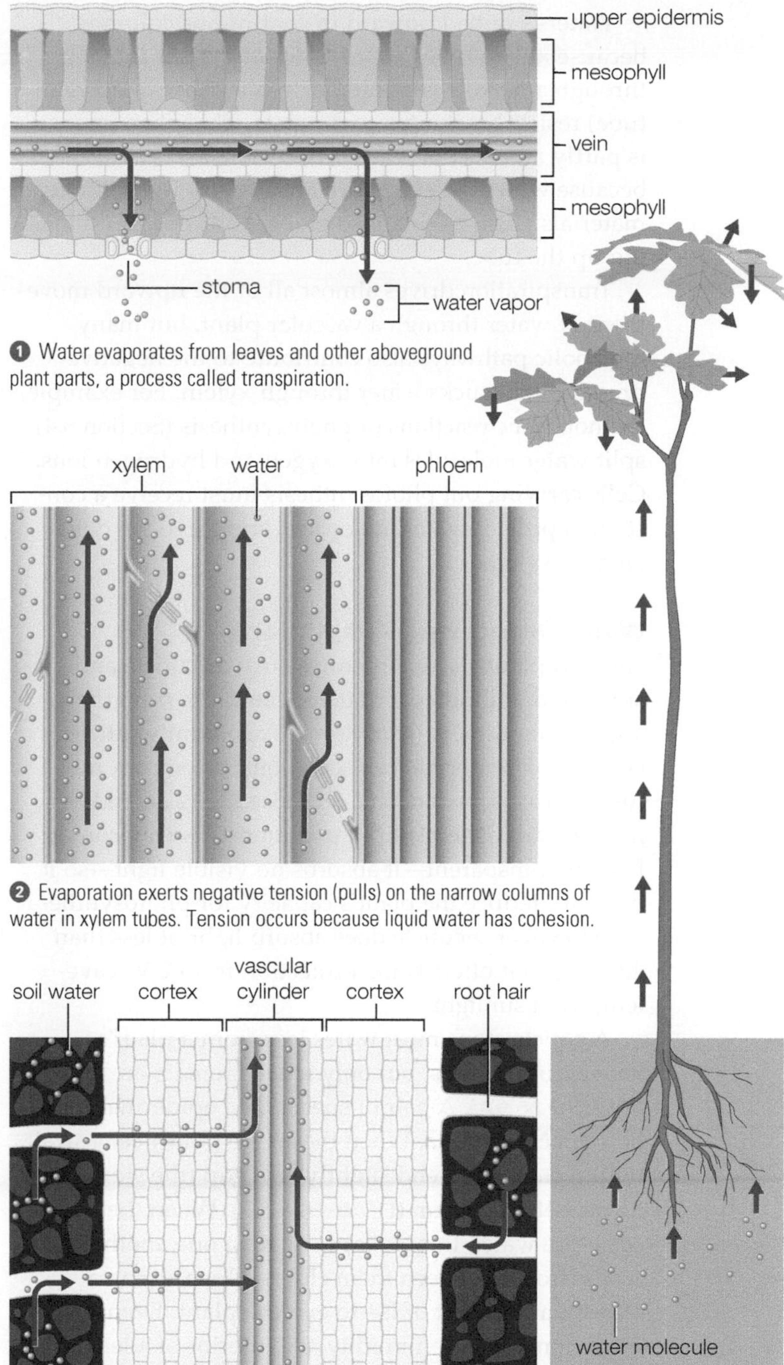
upper epidermis
mesophyll
vein
mesophyll
stoma
water vapor
xylem
water
phloem
soil water
cortex
vascular cylinder
cortex
root hair
water molecule

❶ Water evaporates from leaves and other aboveground plant parts, a process called transpiration.

❷ Evaporation exerts negative tension (pulls) on the narrow columns of water in xylem tubes. Tension occurs because liquid water has cohesion.

❸ The tension inside xylem tubes extends from leaves to roots, where water molecules are being taken up from the soil.

FIGURE 28.7 Cohesion–tension theory.

cohesion–tension theory Explanation of how transpiration creates a tension that pulls a cohesive column of water upward through xylem.

tracheids (TRAY-key-idz) Tapered cells of xylem that die when mature; their interconnected, pitted walls remain and form water-conducting tubes.

transpiration (trans-purr-A-shun) Evaporation of water from aboveground plant parts.

vessel elements Of xylem, cells that form in stacks and die when mature; their pitted walls remain to form water-conducting tubes. Each tube consists of a stack of vessel elements that meet end to end at perforation plates.

Water is pulled upward in continuous columns because xylem tubes are narrow, and water moving through a narrow conduit (such as a straw or a xylem tube) resists breaking into droplets. This phenomenon is partly an effect of water's cohesion, and partly because water molecules are attracted to hydrophilic materials (such as cellulose in the walls of xylem) making up the tube.

Transpiration drives almost all of the upward movement of water through a vascular plant, but many metabolic pathways also contribute to the negative pressure that sucks water through xylem. For example, the noncyclic reactions of photosynthesis (Section 6.4) split water molecules into oxygen and hydrogen ions. Cells carrying out photosynthesis must receive a constant supply of water molecules, and these are delivered by xylem.

Water-Conserving Adaptations

All land plants have structures such as cuticle and stomata that limit evaporative water loss from their aboveground parts (**FIGURE 28.8**). A plant's cuticle consists of epidermal cell secretions: a complex mixture of waxes, pectin, and cellulose fibers embedded in cutin, an insoluble lipid polymer. This waterproof layer is transparent—it absorbs no visible light—so it does not reduce the plant's capacity for photosynthesis. However, a cuticle does absorb light of less than 400 nm, so it offers some protection from UV wavelengths in sunlight.

A cuticle stops most water loss from a plant's aboveground parts, but only when stomata are closed (**FIGURE 28.9A**). A pair of specialized epidermal cells borders each stoma. When these two **guard cells** swell with water, they bend slightly so a gap (the stoma) forms between them (**FIGURE 28.9B**). When the guard cells lose water, they collapse against one another, so the stoma between them closes. Open stomata allow water—a lot of it—to exit the plant. Even under conditions of high humidity, the interior of a leaf or stem contains more water vapor than air. Thus, water vapor diffuses out of a stoma whenever it is open. Indentations, ledges, or other structures that reduce air movement next to a stoma decrease the rate of evaporation from it. Even so, around 95 percent of the water taken up from soil is lost by transpiration from open stomata.

Water is a necessity for land plants, so why do they let almost all of it evaporate away? A cuticle is not only waterproof, it is also gasproof. When a plant's stomata are closed, it cannot exchange enough carbon dioxide and oxygen with the air to support critical metabolic processes. Opening and closing stomata allow a plant to balance its requirement for water with its requirement to exchange gases. A lot more water is lost through open stomata than gases are exchanged, however, and this is the reason for the fantastic amount of water loss by transpiration. During the day, each open stoma loses about 400 molecules of water for every molecule of carbon dioxide it takes in.

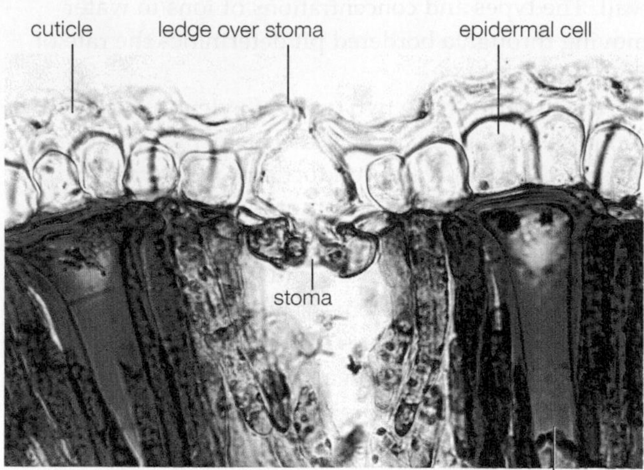
cuticle
ledge over stoma
epidermal cell
stoma
mesophyll cell

FIGURE 28.8 Water-conserving structures in a leaf of the Pincushion Hakea tree (*Hakea laurina*).

In this plant, leaf stomata are recessed beneath a ledge of cuticle. The ledge creates a small cup that traps moist air above the stoma.

How Stomata Work Stomata open or close based on an integration of cues that include humidity, light intensity, the level of carbon dioxide inside the leaf, and hormonal signals from other parts of the plant. These cues trigger changes in osmotic pressure (Section 5.8) that affect turgor in guard cells.

Consider how, like all other cells of all multicelled eukaryotes, guard cells maintain voltage—a charge difference—across their plasma membrane. Membrane voltage arises from active transport proteins such as sodium–potassium pumps that maintain ion gradients across the plasma membrane (Section 5.9). Each ion gradient contributes to the overall charge of cytoplasm and extracellular fluid. In most cases, the overall charge of cytoplasm is lower than the overall charge of extracellular fluid, and any change in this membrane voltage triggers a response inside the cell.

In C3 and C4 plants, a voltage change across guard cell membranes causes stomata to open at sunrise (**FIGURE 28.9C**). The voltage change occurs in response to light absorbed by receptor proteins called phototropins. Light-activated phototropins initiate cellular events that phosphorylate ATP synthases in guard cell membranes. Remember from Section 6.5 that hydrogen ion flow through these transport proteins drives the formation of ATP during the light reactions of photosynthesis. In this case, phosphorylation drives the reverse reaction, in which ATP synthases pump hydrogen ions (H^+) out of guard cell cytoplasm ❶. The outflow of positively charged hydrogen ions decreases the overall charge of the cytoplasm, and increases the overall charge of extracellular fluid. The resulting voltage change across guard cell plasma membranes causes gated transport proteins in them to open. Potassium ions follow the charge gradient and flow into the cell through the opened gates ❷. Water follows the ions by osmosis. Turgor in the guard cells increases as they plump up with water, and the gap between them opens ❸. Carbon dioxide in air diffuses into the plant's tissues, and photosynthesis begins.

TAKE-HOME MESSAGE 28.4

- ✔ Water moves through a plant, from roots to leaves, inside continuous tubes of xylem. Each tube consists of interconnected, pitted secondary walls of dead tracheids or vessel elements.
- ✔ Transpiration, the evaporation of water from aboveground plant parts, pulls on cohesive columns of water inside xylem tubes. This tension pulls water upward through the plant, from roots to shoots.
- ✔ Water-conserving structures are key to the survival of all plants that have aboveground parts.
- ✔ When stomata are open, gas exchange required for metabolism can occur, but the plant also loses a lot of water.
- ✔ A cuticle minimizes water loss when stomata are open.
- ✔ Environmental cues and conditions inside the plant trigger stomata to open or close. A stoma opens when the guard cells that define it become plump with water. It closes when the cells lose water and collapse against each other.

28.5 Movement of Organic Compounds in Plants

LEARNING OBJECTIVES

- Describe the function of phloem.
- Describe the structure of sieve tubes and the role of companion cells in their functioning.
- Explain the steps involved in translocation of sucrose.

Phloem is the vascular tissue that conducts the products of photosynthesis—sugars—through the plant (Section 27.3). Conducting tubes of phloem are called **sieve tubes**, and each is composed of a stack of living

guard cell One of a pair of cells that define a stoma across the epidermis of a leaf or other plant part.
sieve tube Sugar-conducting tube of phloem; consists of stacked sieve elements.

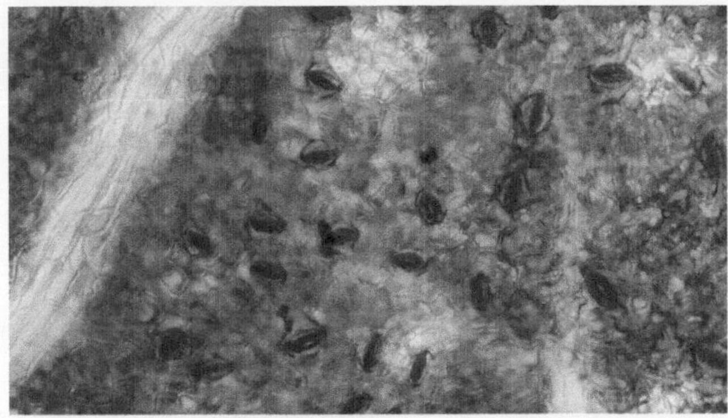

A Stomata on the surface of a leaf. When these tiny pores are closed, they prevent gas exchange between the plant's internal tissues and air. White areas are vascular tissue in leaf veins.

20 µm
guard cells

This stoma is open. When guard cells swell with water, they bend so that a gap (the stoma) opens up between them. The gap allows the plant to exchange gases with air.

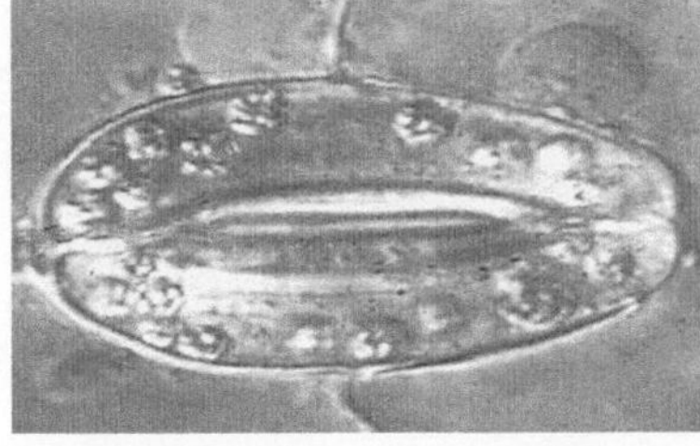

This stoma is closed. When the guard cells lose water, they collapse against each other so the gap between them closes. A closed stoma limits water loss. It also limits gas exchange.

B Whether a stoma is open or closed depends on how much water is plumping up guard cells. The round structures inside the cells are chloroplasts. Guard cells are the only type of plant epidermal cell with these organelles.

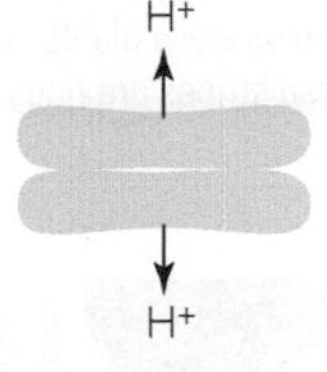
H+
H+

❶ Light-triggered phosphorylation of ATP synthases causes these transport proteins to pump hydrogen ions (H^+) out of guard cells.

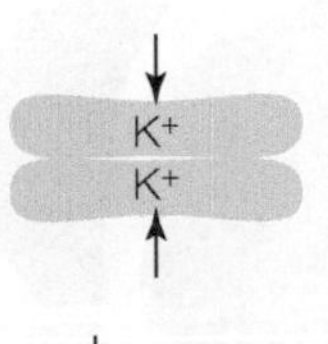
K+
K+

❷ The resulting change in voltage across guard cell plasma membranes opens gated transport proteins in them. Potassium ions enter the cell through the open gates.

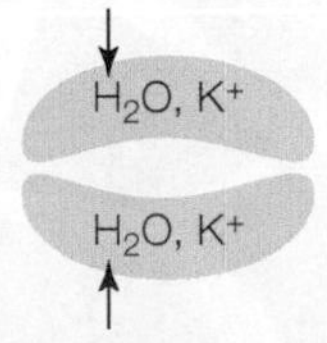
H2O, K+
H2O, K+

❸ Water follows the potassium ions by osmosis. Turgor increases inside the guard cells, and the stoma opens.

C Light triggers an increase in osmotic pressure that opens stomata.

FIGURE 28.9 Stomata function.

cells called **sieve elements** (FIGURE 28.10). Unlike the cells that make up xylem tubes, sieve elements have no pits in their sides; fluid flows through their ends only.

As phloem tissue matures, plasmodesmata connecting the sieve elements in a stack enlarge up to 100 times their original size, leaving very large holes in the cell walls. The resulting perforated end walls, which are called sieve plates after their appearance, separate stacked sieve elements (FIGURE 28.11). The stacked cells' plasma membranes then merge through matching holes in their sieve plates, so their cytoplasm becomes continuous throughout the sieve tube. The tube becomes an open conduit for fluid when the sieve elements lose most of their organelles (including the nucleus, Golgi bodies, and almost all of its ribosomes).

How does a sieve element stay alive without organelles? Each sieve element has an associated **companion cell** that arises by division of the same parent cell. The companion cell retains its nucleus and other components at maturity. Many plasmodesmata connect the cytoplasm of the two cells, so the companion cell can provide all metabolic functions necessary to sustain its paired sieve element.

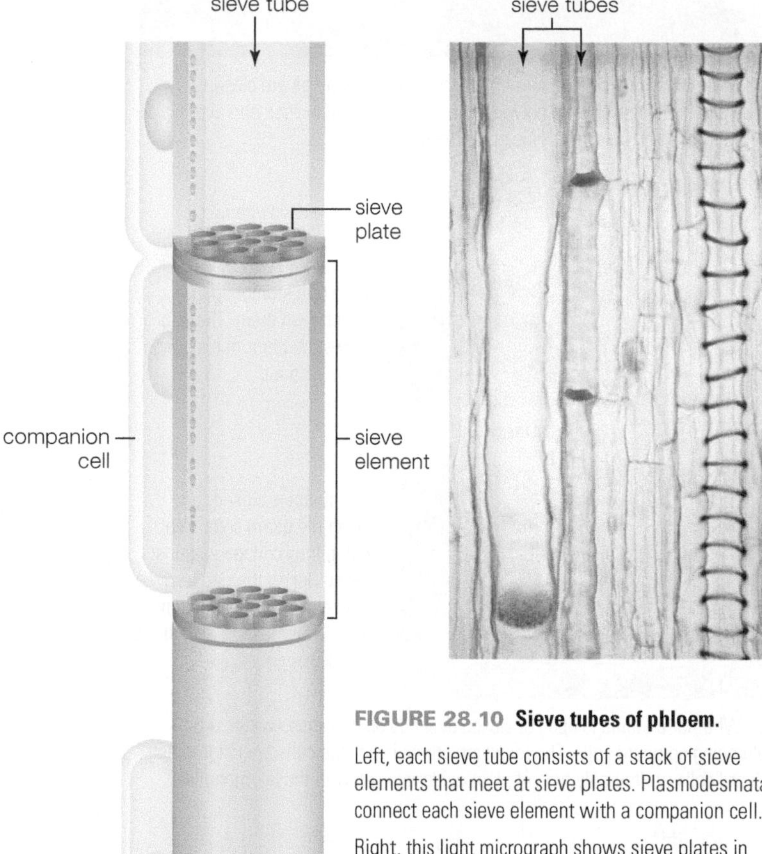
sieve tube
sieve tubes
sieve plate
companion cell
sieve element

FIGURE 28.10 Sieve tubes of phloem.

Left, each sieve tube consists of a stack of sieve elements that meet at sieve plates. Plasmodesmata connect each sieve element with a companion cell.

Right, this light micrograph shows sieve plates in two columns of phloem. The red stripes are rings of lignin in a xylem vessel.

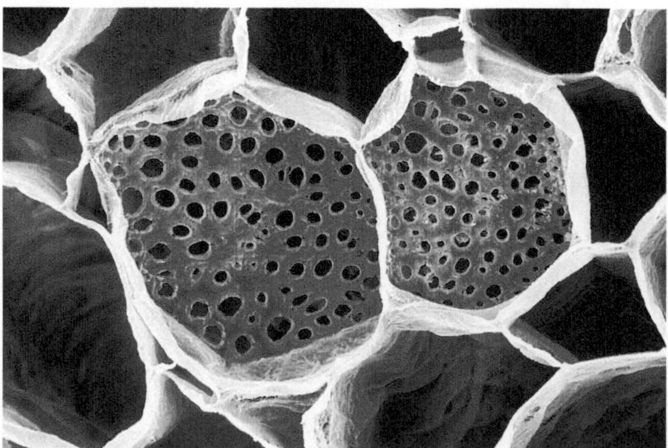

FIGURE 28.11 Sieve plates. This scanning electron micrograph shows sieve plates on the ends of two side-by-side sieve-tube elements. Sieve plates are the perforated end walls of stacked sieve elements.

Pressure Flow Theory

A plant produces sugar molecules during photosynthesis. Some of these molecules are used by or stored in the cells that make them; the rest are conducted to other parts of the plant through sieve tubes. The movement of sugars and other organic molecules through phloem is called **translocation**.

Inside sieve tubes, fluid rich in sucrose and other sugars flows from a **source** (a region of plant tissue where sugars are being produced or released from storage) to a **sink** (a region where sugars are being broken down or put into storage for later use). Photosynthetic tissues are typical source regions; tissues of developing shoots and fruits are typical sink regions.

Why does sugar-rich fluid flow from source to sink? A pressure gradient between the two regions drives the movement (FIGURE 28.12). Consider mesophyll cells in a leaf, which produce far more sugar than they use for their own metabolism. Photosynthesis in these cells sets up the leaf as a source region. The excess sugar molecules they produce diffuse into adjacent companion cells through plasmodesmata. Depending on the species and the source region, sugar molecules may also be pumped into a companion cell through active transport proteins.

companion cell In phloem, a specialized cell that provides metabolic support to its partnered sieve element.

pressure flow theory Explanation of how a difference in turgor between source and sink regions pushes sugar-rich fluid through a sieve tube.

sieve elements Living cells that compose sugar-conducting sieve tubes of phloem. Each sieve tube consists of a stack of sieve elements that meet end to end at sieve plates.

sink Region of plant tissue where sugars are being used or put into storage.

source Region of plant tissue where sugars are being produced or released from storage.

translocation Movement of organic compounds through phloem.

CREDITS: (10) M. I. Walker/Science Source; (11) © J.C. Revy/ ISM/ Phototake.

Once inside a companion cell, sugar molecules diffuse through plasmodesmata into a sieve element ❶. The sugar loaded into sieve tubes at a source region increases the solute concentration of a sieve tube's cytoplasm, so it becomes hypertonic with respect to the surrounding cells. Water follows the sugar by osmosis ❷, moving into the sieve element from the surrounding cells.

The rigid cell walls of a mature sieve element cannot expand very much, so the influx of water raises the pressure inside the cell. Pressure inside a sieve tube can be very high—up to five times higher than that of an automobile tire. Pressure that a fluid exerts against a structure that contains it is called turgor (Section 5.8). High fluid pressure (turgor) inside a sieve tube pushes sugar-rich cytoplasm from one sieve element to the next, toward a sink region where turgor is lower ❸. Pressure inside a sieve tube decreases at a sink because sugars leave the tube in this region. Sugar molecules diffuse into companion cells, and then move into adjacent sink cells (either through plasmodesmata or by active transport). Water follows, again by osmosis ❹, so turgor inside the sieve element decreases. This explanation of how turgor pushes sugar-rich fluid inside a sieve tube from source to sink is called the **pressure flow theory**.

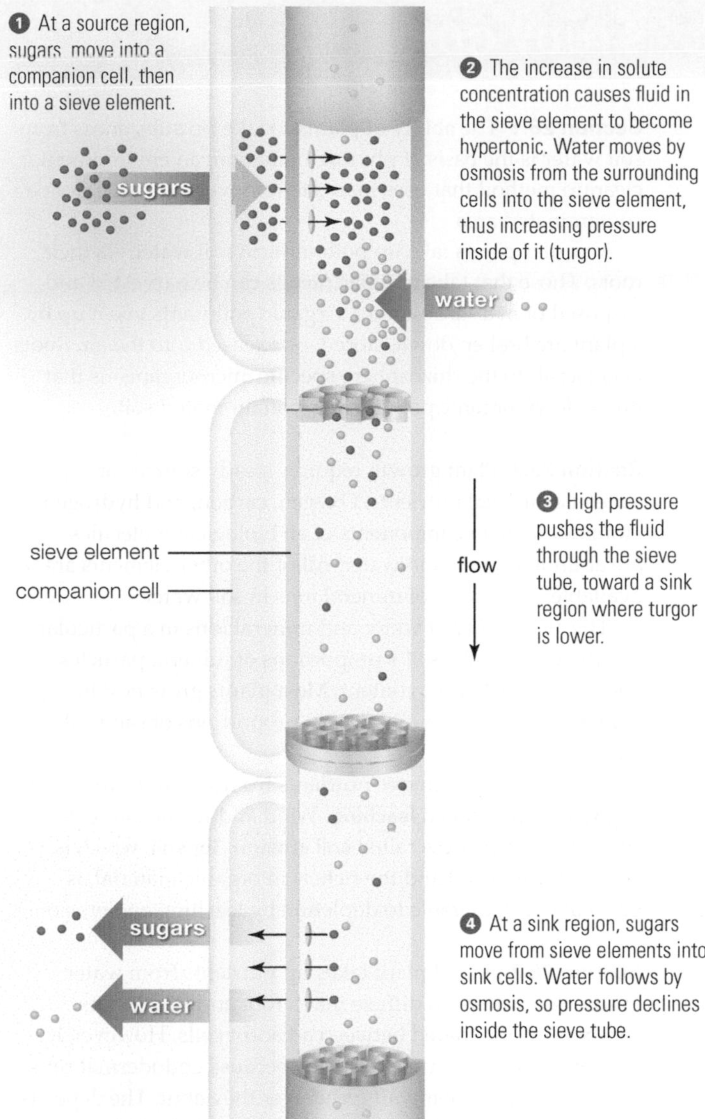
❶ At a source region, sugars move into a companion cell, then into a sieve element.
sugars
❷ The increase in solute concentration causes fluid in the sieve element to become hypertonic. Water moves by osmosis from the surrounding cells into the sieve element, thus increasing pressure inside of it (turgor).
water
sieve element
companion cell
flow
❸ High pressure pushes the fluid through the sieve tube, toward a sink region where turgor is lower.
sugars
water
❹ At a sink region, sugars move from sieve elements into sink cells. Water follows by osmosis, so pressure declines inside the sieve tube.

FIGURE 28.12 Translocation of organic compounds in phloem from source to sink. Water molecules are represented by blue balls; sugar molecules, by red balls. Sugars move into and out of a sieve element via its associated companion cell.

TAKE-HOME MESSAGE 28.5

✔ Sugars produced by photosynthesis are distributed through the plant inside sieve tubes of phloem.

✔ A sieve tube consists of stacked sieve elements separated by porous end walls (sieve plates). Each sieve tube has a companion cell that provides metabolic support and moves sugars into and out of it.

✔ The movement of sugars into and out of a sieve tube at source and sink regions sets up a pressure gradient that pushes sugar-rich fluid through a sieve tube.

28.1 Leafy Cleanup (revisited)

Compared to other methods of cleaning up toxic waste sites, phytoremediation is usually less expensive—and it is more appealing to neighbors. Consider Ford Motor Company's Rouge Center in Dearborn, Michigan, where decades of steelmaking left topsoil contaminated with highly carcinogenic organic compounds known as polycyclic aromatic hydrocarbons, or PAHs. With funding from the automaker, researchers developed a phytoremediation system based on native plants, which also boosted a sitewide initiative to restore the facility's native wildlife habitat. The site's plantings are attracting insects, birds, and other wildlife while aggressively accelerating the natural degradation of toxins. Phytoremedial breakdown of PAH occurs at rhizospheres, each a millimeter-wide region of soil adjacent to a root's surface. The rhizosphere is a zone of intense interactions between the root and soil microorganisms. Roots dynamically alter physical and chemical characteristics of the soil in a rhizosphere, and these alterations attract and support microorganisms that are of particular benefit to the plant at a particular time. Plants growing in PAH-contaminated soil can use these tactics to recruit specific bacterial species that degrade the toxin. ●

STUDY GUIDE

Section 28.1 The ability of plants to take up substances from soil water is the basis of phytoremediation, an environmental cleanup method that uses plants to remove pollutants from a contaminated area.

Vascular plants take up pollutants in soil water via their roots. Those that take up toxic metals can be harvested and disposed of in a safe manner. Organic pollutants taken up by a plant are broken down, stored, or released into the air. Roots also recruit to the rhizosphere specific microorganisms that break down organic pollutants in contaminated soil.

Section 28.2 Plant growth requires steady sources of sixteen elemental nutrients. Oxygen, carbon, and hydrogen atoms—the main components of all biological molecules—are abundant in air and water. All of the other elements are available in the form of mineral ions in soil water.

The availability of water and mineral ions in a particular soil depends on the soil's proportions of mineral particles, and also on its **humus** content. Most plants grow best in **loams**, which have roughly equal proportions of sand, silt, and clay.

Water flow can remove nutrients from soil and carry them away, a process called **leaching**. Wind and water can remove soil entirely, a process called **soil erosion**. **Topsoil**, which is the top layer of soil and the richest in organic material, is particularly vulnerable to depletion by leaching and erosion.

Section 28.3 A land plant takes up nutrients from water in soil. Soil water can diffuse into a root and through its cortex via walls shared between adjacent cells. However, it cannot cross endodermis this way, because endodermal cells deposit lignin in their walls wherever they abut. The deposits form a **Casparian strip** that makes a watertight seal between the plasma membrane of adjacent endodermal cells. The Casparian strip prevents soil water from entering the vascular cylinder by diffusing through endodermal cell walls. To enter the vascular center, water and ions in it must first pass through the cytoplasm of an endodermal cell.

Water can enter endodermal cell cytoplasm by diffusing directly across plasma membrane of any cell in the root, and then through plasmodesmata into endodermal cell cytoplasm. Ions cannot. Ions enter an endodermal cell via transport proteins in its plasma membrane. They can also enter an endodermal cell through plasmodesmata, from adjacent cells in root cortex that have taken up the ions via transport proteins in their own plasma membranes. Thus, transport proteins in root cell plasma membranes control the plant's uptake of ionic substances in soil water.

Many plants form mutually beneficial relationships with microorganisms in soil. Fungi associate with young roots in mycorrhizae, which enhance the plant's ability to absorb mineral ions from soil. Nitrogen-fixing bacteria that a plant encapsulates in **root nodules** convert nitrogen in air to ammonia, which is a form of nitrogen that the plant can use.

Section 28.4 Water and solutes flow through xylem tubes from roots to shoot tips. Each tube consists of the pitted, lignin-reinforced cell walls of **tracheids** or **vessel elements** that formed in stacks and then died. Perforation plates separate the vessel elements in a stack.

The **cohesion–tension** theory explains how water moves through a plant: **Transpiration** (the evaporation from aboveground plant parts, mainly at stomata) pulls water from roots to shoots through xylem tubes. Cohesion helps water in these tubes to resist breaking into droplets under the tension exerted by the pull. Cohesion also keeps the water from breaking into droplets, so it moves in continuous columns.

A cuticle helps a plant conserve water; stomata help it balance water conservation with gas exchange required for metabolism. A stoma, which is a gap between two **guard cells**, may be surrounded by an indentation, protrusions, or other specializations that reduce airflow around it.

Environmental and internal signals cause stomata to open or close. The signals trigger cellular events that cause guard cells to pump ions into or out of their cytoplasm. Water then follows the ions by osmosis. Water moving into guard cells plumps them, which opens the stoma between them. Water diffusing out of the cells causes them to collapse against each other, so the stoma closes.

Section 28.5 Sugars move through a plant by **translocation** in phloem's **sieve tubes**, which consist of stacked **sieve elements** separated by perforated sieve plates. Each sieve tube has an associated **companion cell** that loads sugars into it at a **source** (a region where sugars are being produced or released from storage). By the **pressure flow theory**, the movement of sugar-rich fluid through a sieve tube is driven by a pressure gradient between the source and a **sink** (a region where sugars are being broken down or put into storage).

SELF-QUIZ

Answers in Appendix VII

1. The main source(s) of hydrogen and oxygen for plants is (are) ________.
   a. soil and air
   b. water and soil
   c. water and fertilizer
   d. water and air

2. Decomposing organic matter in soil is called ________.
   a. clay
   b. humus
   c. topsoil
   d. silt
   e. sand
   f. leaching

3. Most water moves from soil to vascular cylinder ______.
   a. through root hairs
   b. between root cells
   c. through root cell walls
   d. in xylem

4. The conducting tubes in a vascular cylinder are ______.
   a. endodermis
   b. pericycle
   c. root cortex
   d. xylem only
   e. xylem and phloem
   f. companion cells

5. A _______ strip between abutting endodermal cell walls forces water and solutes to move through these cells rather than around them.
   a. cutin c. cohesion
   b. Casparian d. cellulose

6. The nutrition of some plants is enhanced by a mutually beneficial interaction between a root and soil bacteria. The interaction triggers formation of a _______ .
   a. root nodule c. root hair
   b. mycorrhiza d. hypha

7. Water evaporation from plant parts is called _______ .
   a. translocation c. transpiration
   b. respiration d. tension

8. Water transport from roots to leaves occurs by _______ .
   a. a pressure gradient inside sieve tubes
   b. different solutes at source and sink regions
   c. the pumping force of xylem vessels
   d. evaporation, tension, and cohesion

9. Tracheids are part of _______ .
   a. cortex c. phloem
   b. mesophyll d. xylem

10. Sieve tubes are part of _______ .
   a. cortex c. phloem
   b. mesophyll d. xylem

11. With stomata closed, a waterproof cuticle _______ .
   a. minimizes water loss through plant surfaces
   b. permits gas exchange between the plant and the air
   c. both a and b

12. When guard cells swell, _______ .
   a. transpiration ceases c. stomata open
   b. sugars enter phloem d. root cells die

13. In phloem, organic compounds flow through _______ .
   a. collenchyma cells c. vessels
   b. sieve elements d. tracheids

14. Match each term with the appropriate concept.

___ stomata	a. separates cells in xylem tubes
___ nutrient	b. takes up soil water and nutrients
___ hydrogen bonds	c. required element
___ sink	d. cohesion in xylem tubes
___ root	e. sugars unloaded from sieve tubes
___ sieve plate	f. distributes sugars
___ perforation plate	g. separates cells in phloem tubes
___ xylem	h. distributes water
___ phloem	i. important for gas exchange

15. Sugar transport from leaves to roots occurs by _______ .
   a. a pressure gradient inside sieve tubes
   b. different solutes at source and sink regions
   c. the pumping force of xylem vessels
   d. evaporation, tension, and cohesion

CRITICAL THINKING

1. Nitrogen deficiency stunts plant growth and causes leaves to turn yellow and then die. Why does nitrogen deficiency cause these symptoms? *Hint:* Think about which biological molecules incorporate nitrogen atoms.

2. You just returned home from a three-day vacation. Your severely wilted plants tell you they were not watered before you left. Use the cohesion–tension theory of water transport to explain what happened to them.

3. When you dig up a plant to move it from one spot to another, the plant is more likely to survive if some of the soil around the roots is transferred to the new location along with the plant. Make a hypothesis that explains this observation, and then design an experiment to test your hypothesis.

4. If a plant's stomata are made to stay open at all times, or closed at all times, it will die. Why?

5. What are the structures shown in the micrographs in **FIGURE 28.13**? In which plant tissue(s) can these structures be found?

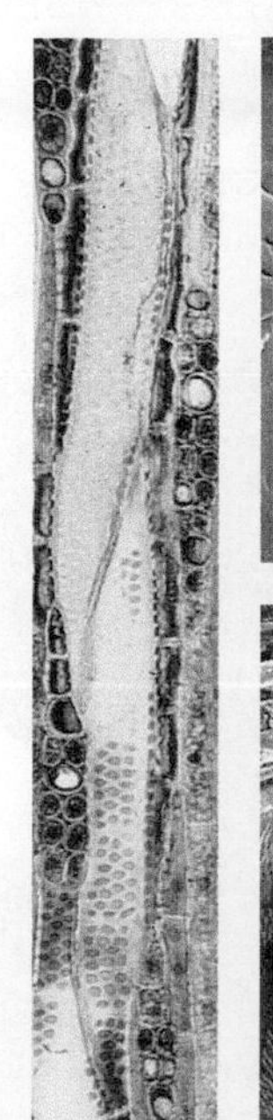
A

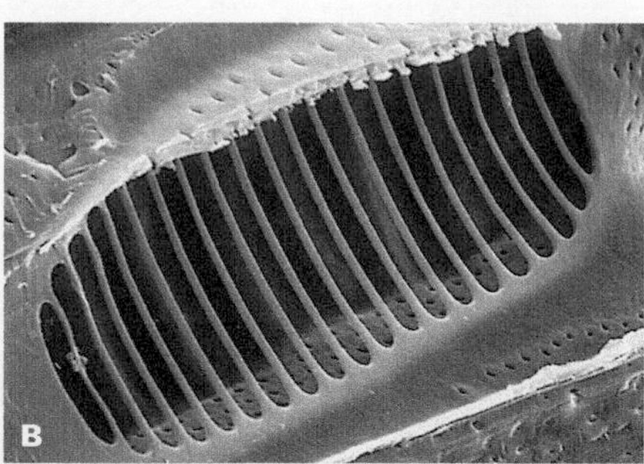
B

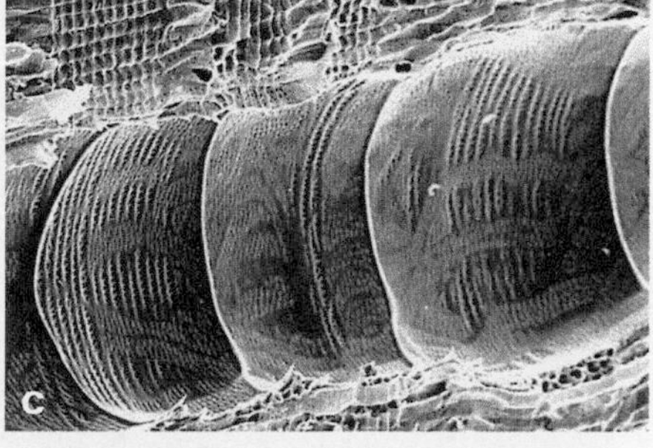
C

FIGURE 28.13 Name the mystery structures.

CREDITS: (13A) H. A. Core, W. A. Cote, and A. C. Day, *Wood Structure and Identification*, 2nd Ed., Syracuse University Press, 1979; (13B) Alison W. Roberts, University of Rhode Island; (13C) H. A. Core, W. A. Cote, and A. C. Day, *Wood Structure and Identification*, 2nd Ed., Syracuse University Press, 1979.

29 LIFE CYCLES OF FLOWERING PLANTS

CORE CONCEPTS

Systems

Complex properties arise from interactions among components of a biological system.

Sexual reproduction in angiosperms involves the transfer of pollen among flowers, which, in many species, occurs via ecological interactions with animals. Animal pollinators that visit a flower typically receive some type of reward that reinforces the pollinator's memory of the flower. In plants, as in animals, processes of sexual reproduction depend on the exchange of chemical signals.

Evolution

Evolution underlies the unity and diversity of life.

Sexual reproduction in flowering plants reflects an evolutionary dependence on environmental agents of pollination and seed dispersal. Floral traits reflect the sensory abilities of coevolved pollinators; angiosperm species that depend on the same pollinator often share the same sets of floral traits (an example of convergent evolution). Developmental programs triggered by environmental cues adapt plant species for life in certain habitats.

Structure and Function

The three-dimensional form and arrangement of biological structures give rise to their function and interactions.

Gamete production and fertilization in flowering plants occurs in structures set aside for this purpose. Some floral structures allow pollination to occur only by coevolved animal pollinators. Structural specializations of fruits allow seed dispersal by particular vectors such as wind, water, or animals.

Links to Earlier Concepts

You may want to review what you have learned about meiosis in sexual reproduction (Sections 12.2 and 12.3), and the life cycles of angiosperms (22.8). We revisit membrane proteins (5.7), clones (8.1), asexual reproduction (11.2), floral identity genes (10.3), polyploidy (14.6), gametophytes (22.2), pollen (22.6), monocots and eudicots (22.9, 27.2), stem nodes (27.4), and pericycle (27.7). Evolutionary themes of this chapter include factors that reduce genetic diversity (17.6), mechanical isolation (17.7), coevolution (17.9), and convergent evolution (18.3).

29.1 Plight of the Honeybee

In the fall of 2006, commercial beekeepers worldwide began to notice something was amiss in their honeybee hives. All of the adult worker bees were missing, but there were no bee corpses to be found. The bees had suddenly (and quite unusually) abandoned their queen and many living larvae. The few young workers that remained were reluctant to feed on abundant pollen and honey stored in their own hive or even in other abandoned hives.

A third of the colonies did not survive the following winter. By spring, the phenomenon had a name—colony collapse disorder (CCD)—and to this day it continues to affect honeybee populations. Sixty percent of the beehives in the United States collapsed in 2008; intensive management practices have since reduced losses to around thirty percent per year.

CCD is a problem that affects more than just bees and honey lovers. Nearly all of our food crops are the fruits of angiosperms (Section 22.8). Like other seed-bearing plants, angiosperms produce pollen (Section 22.3), and their reproductive success requires the transfer of pollen from one flower to another. Fruits do not form from unpollinated flowers, and in some species they only form when flowers receive pollen from a different plant of the same species. Even plants with flowers that can self-pollinate tend to make bigger fruits and more of them when they are cross-pollinated.

Many plant species depend on insect pollinators, but honeybees are especially important in this regard. Honeybees efficiently pollinate a wide variety of plants, and they are the only ones that tolerate living in artificial hives transported by truck to wherever our crops require pollination. The potential loss of their portable pollination service is a huge threat to our agricultural economy. Currently, about half of the world's leading crops—hundreds of billions of dollars worth—depend on pollination by honeybees.

CCD is important because it directly affects our food supply, but populations of wild bees and other insect pollinators are also dwindling. Habitat loss may be a factor, but human activities that spread diseases, pests, and pesticides implicated in CCD are also harming other invertebrates.

Flowering plants rose to dominance in part because they coevolved with insects and other animals that pollinate them. Most flowers are specialized to attract and be pollinated by a specific type (or even a specific species) of animal. Those adaptations become handicaps if coevolved pollinator populations decline.

29.2 Floral Structure and Function

LEARNING OBJECTIVES

- List the components of a complete flower and their function.
- Describe some structural variations in flower form.
- Using suitable examples, explain the relationship between floral traits and coevolved pollination vectors.

A **flower** is a reproductive structure of an angiosperm (FIGURE 29.1). The parts of a flower are modified leaves that develop from a thickened region of stem called a receptacle ❶. A typical flower consists of four rings (whorls) of modified leaves. The outermost whorl is a **calyx**, which is a ring of leaflike **sepals** ❷. Sepals are photosynthetic and inconspicuous; they enclose and protect internal tissues before the flower opens. Just inside the calyx is the **corolla** (from the Latin *corona*, or crown), which is a ring of petals ❸. **Petals** are typically the largest and most brightly colored parts of a flower.

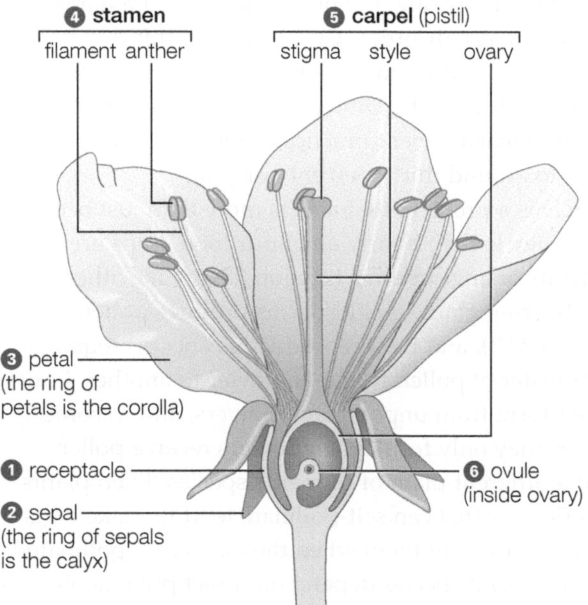

❹ stamen
filament anther
❺ carpel (pistil)
stigma style ovary
❸ petal
(the ring of petals is the corolla)
❶ receptacle
❷ sepal
(the ring of sepals is the calyx)
❻ ovule
(inside ovary)

FIGURE 29.1 Anatomy of a typical flower: cherry (*Prunus avium*). Male gametophytes are produced by stamens; female gametophytes, by carpels. A stamen consists of a pollen-bearing anther on a slender filament. A carpel consists of ovary, stigma, and style.

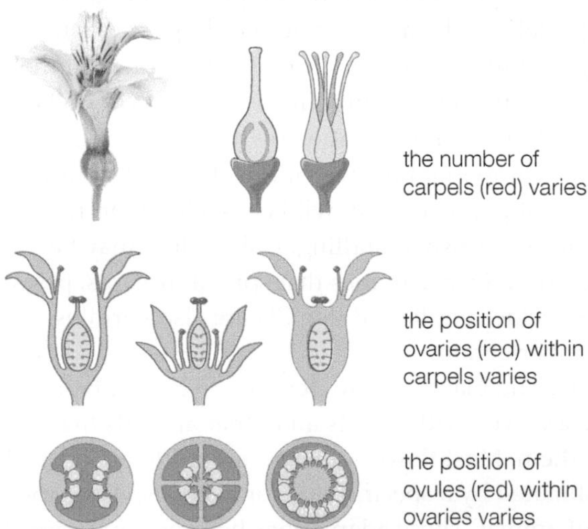

the number of carpels (red) varies
the position of ovaries (red) within carpels varies
the position of ovules (red) within ovaries varies

FIGURE 29.2 Variation in reproductive structures. In flowers of many eudicots, one ovule forms inside one ovary inside one carpel. Other species have different patterns.

Inside the corolla is a whorl of **stamens**, the reproductive organs that produce the plant's male gametophytes (gamete-making structures, Section 22.2). A typical stamen is a thin filament with an **anther** at the tip ❹. A typical anther consists of four pouches called pollen sacs. **Pollen grains**, which are immature male gametophytes, form inside pollen sacs.

The modified leaves making up a flower's innermost whorl fold and fuse into one or more **carpels**: reproductive organs that produce female gametophytes ❺. The flowers of some species have one carpel; those of other species have several (FIGURE 29.2). A carpel (or a compound structure that consists of multiple fused carpels) is commonly called a pistil.

The upper region of a carpel is a sticky or hairy **stigma** that is specialized to receive pollen grains. Typically, the stigma sits on top of a slender stalk called a style. The lower, swollen region of a carpel is the **ovary**, which contains one or more ovules. In seed plants, an **ovule** ❻ is the part of the carpel where the female gametophyte forms. As you will see in Section 29.6, seeds are mature ovules.

Expression of floral identity genes (Section 10.3) determines the fate of a flower's four whorls. Alleles

anther (AN-thur) Part of the stamen that produces pollen grains.
calyx A flower's outer, protective whorl of sepals.
carpel (CAR-pull) Floral reproductive organ that produces the female gametophyte; consists of an ovary, stigma, and usually a style.
corolla A flower's whorl of petals; forms within sepals and encloses reproductive organs.
flower Specialized reproductive structure of an angiosperm.
ovary In flowering plants, the enlarged base of a carpel, inside which one or more ovules form.
ovule (AH-vule) Of a seed plant, the part of the carpel where the female gametophyte forms.
petal Unit of a flower's corolla; often showy and conspicuous.
pollen grain Immature male gametophyte of a seed plant.
pollination The arrival of pollen on a pollen-receiving reproductive part of a seed plant.
pollination vector Animal or environmental agent that moves pollen grains from one plant to another.
pollinator An animal pollination vector.
sepal Unit of a flower's calyx; typically photosynthetic and inconspicuous.
stamen (STAY-men) Floral reproductive organ that consists of an anther and, in most species, a filament.
stigma Upper part of a carpel; adapted to receive pollen.

A The solitary flower of the lady's slipper orchid (*Paphiopedilum*) is irregular.

B A flower of hyacinth (*Hyacinthus orientalis*) is an elongated inflorescence.

C A daisy (*Gerbera jamesonii*) is a composite flower with many individual florets.

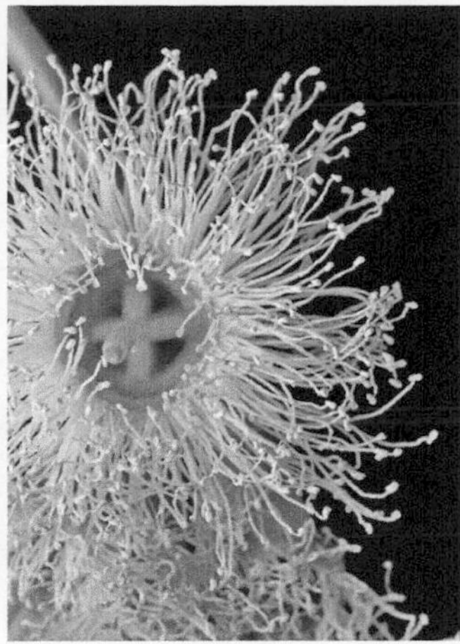

D A flower of eucalyptus (*Eucalyptus robusta*) is incomplete because it has no petals.

FIGURE 29.3 **Examples of structural variation in flowers.**

of these and other master regulators give rise to a tremendous variety of floral forms. Some flowers are single blossoms (FIGURE 29.3A); others are clusters called inflorescences (FIGURE 29.3B). A composite flower is an inflorescence of many flowers ("florets") clustered as a single head (FIGURE 29.3C). A "regular" flower is radially symmetric, which means if it were cut like a pie, the pieces would be roughly identical. "Irregular" flowers are not radially symmetric. A flower with all four sets of modified leaves (sepals, petals, stamens, and carpels) is a "complete" flower. An "incomplete" flower lacks one or more of these structures (FIGURE 29.3D).

Pollination

Pollination is the arrival of pollen on a stigma or other pollen-receiving reproductive part of a seed plant. "Perfect" flowers have both stamens and carpels, so they can potentially pollinate themselves. "Imperfect" flowers, which lack stamens or carpels (FIGURE 29.4), cannot. Self-pollination can be adaptive because it does not require a partner, but it can lower a population's genetic diversity (Section 17.6) and reduce the quality of fruit and offspring (FIGURE 29.5). Accordingly, many plants have adaptations that encourage or even require cross-pollination by another individual. For example, some perfect flowers release pollen only after their stigma is no longer receptive to being fertilized. In other species, stamen-bearing and carpel-bearing flowers form on different plants.

Pollination is essential to sexual reproduction in flowering plants. The diversity of flower form reflects a dependence on **pollination vectors**: animals or environmental agents that transfer pollen from anther to stigma. Animal pollination vectors are called **pollinators**. Fragrance and other flower traits not directly related to reproduction are evolutionary adaptations that attract coevolved pollinators (Section 17.9). An animal attracted to a particular flower often picks up pollen on a visit, then inadvertently transfers it

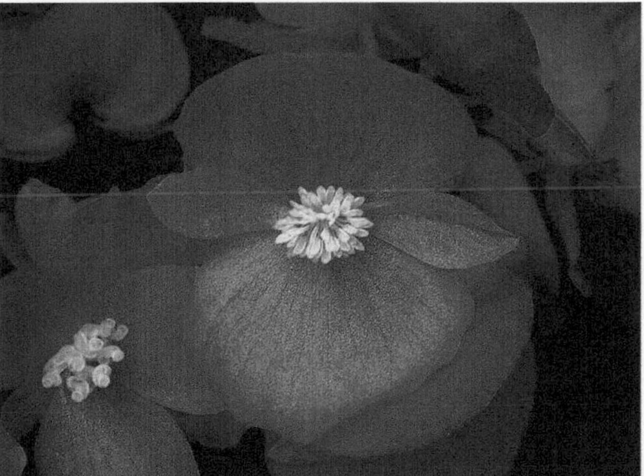

FIGURE 29.4 **Imperfect flowers.**
In this species of begonia, female blossoms (with no stamens, left) form on the same plants as male blossoms (with no carpels, right).

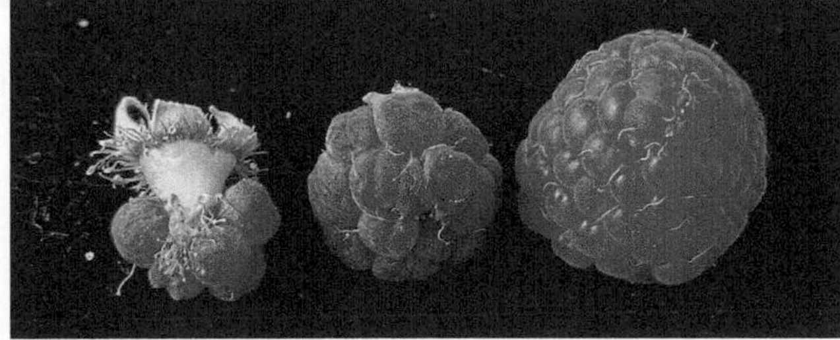

FIGURE 29.5 **Self-pollination versus cross-pollination.**
The two raspberries on the left formed from self-pollinated flowers; the one on the right, from an insect-pollinated flower. Cross-pollination produces larger fruit with more seeds.

CREDITS: (3A) Salawin Chanthapan/Shutterstock; (3B–D) Brian Johnston; (4) Lauren Doyle; (5) Courtesy of James H. Cane, USDA-ARS Bee Biology and Systematics Lab, Utah State University, Logan, UT.

TABLE 29.1

Pollination Syndromes

Flower Trait:	Animal Pollination Vector:						
	Bats	**Bees**	**Beetles**	**Birds**	**Butterflies**	**Flies**	**Moths**
Color:	Dull white, green, purple	Bright white, yellow, blue, UV	Dull white or green	Scarlet, orange, red, white	Bright, such as red, purple	Pale, dull, dark brown or purple	Pale/dull red, pink, purple, white
Odor:	Strong, musty, emitted at night	Fresh, mild, pleasant	None to strong	None	Faint, fresh	Putrid	Strong, sweet, emitted at night
Nectar:	Abundant, hidden	Usually	Sometimes, not hidden	Ample, deeply hidden	Ample, deeply hidden	Usually absent	Ample, deeply hidden
Pollen:	Ample	Limited, often sticky, scented	Ample	Modest	Limited	Modest	Limited
Shape:	Regular, bowl-shaped, closed during the day	Shallow with landing pad; tubular	Large, bowl-shaped	Large funnel-shaped cups, strong perch	Narrow tube with spur; wide landing pad	Shallow, funnel-shaped or trap-like and complex	Regular; tube-shaped with no lip
Examples:	Banana, agave	Larkspur, violet	Magnolia, dogwood	Fuschia, hibiscus	Phlox	Skunk cabbage, philodendron	Tobacco, lily, some cacti

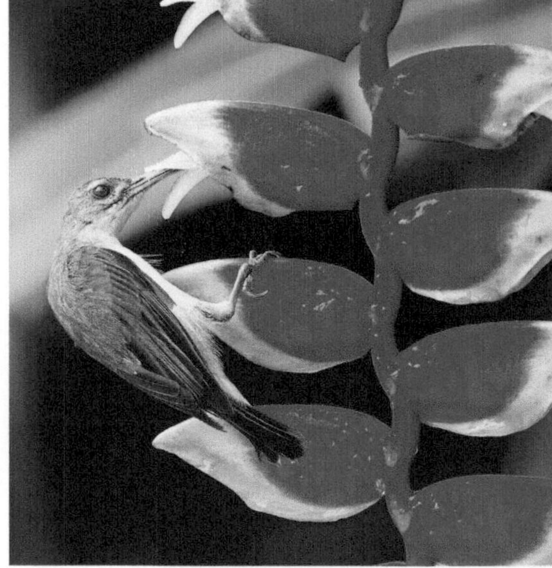

A Birds.
A shower of pollen from a *Heliconia* flower sprays the face of a spiderhunter (*Arachnothera*) sipping nectar from it.

B Mammals.
A lesser long-tongued fruit bat sips nectar from (and pollinates) flowers of wild banana.

C Insects.
Female burnet moths perch on purple blossoms—preferably pincushion flowers—when they are ready to mate. The visual combination attracts male moths; the activity of the moths pollinates the flowers.

FIGURE 29.6 Examples of animals pollinating flowers.

to another flower on a later visit. The more specific the attraction, the more efficient the transfer of pollen among plants of the same species. Thus, a flower's traits typically reflect the sensory abilities and preferences of its coevolved pollinator. In an example of convergent evolution (Section 18.3), the flowers of angiosperm species that depend on the same pollinator (bees, for example, or birds) often share a certain set of traits called a "pollination syndrome" (**TABLE 29.1**).

Consider that bees have a keen sense of smell, and they can see ultraviolet (UV) light; bee-pollinated flowers tend to be fragrant, with a bull's-eye pattern of UV-reflecting pigments in their petals. Bees can also sense distinct patterns of electric charge on the surfaces of bee-pollinated flowers. Among other purposes, these electric fields help the bees discriminate between flowers of different species.

Birds have excellent color vision and a relatively poor sense of smell, so many bird-pollinated flowers are bright and unscented (**FIGURE 29.6A**). Flowers

nectar Sweet fluid exuded by some flowers; rewards pollinators.

pollinated by night-flying bats, which have a good sense of smell but poor vision, tend to be large and light-colored, with a strong nighttime fragrance (**FIGURE 29.6B**). Not all flowers smell sweet; odors like dung or rotting flesh beckon beetles and flies.

Rewards offered by a flower reinforce a pollinator's memory of a visit, and encourage the animal to seek out other flowers of the same species. Many flowers exude a sweet fluid called **nectar** that is prized by many pollinators. Nectar is the only food for most adult butterflies, and it is a favorite food of hummingbirds. Honeybees convert nectar to honey, which helps feed the bees and their larvae through the winter. Bees also collect protein-rich pollen for food, and many beetles feed primarily on pollen. Pollen is the only reward for a pollinator of roses, poppies, and other flowers that produce no nectar. Some flowers offer other, less typical rewards (**FIGURE 29.6C**).

A specific relationship between plant and pollinator benefits both. For example, a flower that captivates the attention of an animal has a pollinator that spends its time seeking (and pollinating) only those flowers. Pollinators benefit when they receive an exclusive supply of a reward. Consider how floral specializations can prevent pollination by all but one type of pollinator. Nectar or pollen at the bottom of a long floral tube may be accessible only to an insect with a matching feeding device, for example. In some plants, only a specific pollinator can trip a floral mechanism that brings stamens on stigma into contact with the animal's body (the white sage plants in Section 17.7 offer an example).

Grasses and other plants pollinated by wind do not benefit by attracting animals, so they do not waste resources making scented, brilliantly colored, or nectar-laden flowers. Wind-pollinated flowers are typically small, nonfragrant, and green, with insignificant petals. These flowers typically have large stamens that release pollen grains by the billions, insurance in numbers that some of the grains will reach a receptive stigma.

TAKE-HOME MESSAGE 29.2

- ✔ Flowers are the reproductive structures of angiosperms. The different parts of a flower (sepals, petals, stamens, and carpels) are modified leaves.
- ✔ Stamens produce male gametophytes. They typically consist of a filament with an anther at the tip. Pollen forms inside the anthers.
- ✔ Carpels typically consist of stigma, style, and ovary. Female gametophytes form in an ovule inside the ovary.
- ✔ Flowers vary in structure. Their shape, pattern, color, and fragrance attract coevolved pollinators.
- ✔ A floral reward such as pollen or nectar that attracts specific pollinators increases the efficiency of pollination.

Data Analysis Activities

Who's the Pollinator? *Massonia depressa* is a low-growing succulent plant native to the desert of South Africa. The dull-colored flowers of this monocot develop at ground level, have tiny petals, emit a yeasty aroma, and produce a thick, jellylike nectar. These traits led researchers to suspect that desert rodents such as gerbils pollinate this plant. The researchers trapped rodents in areas where *M. depressa* grows and checked them for pollen (**FIGURE 29.7A,B**). They also put some plants in wire cages that excluded mammals, but not insects, to see whether fruits and seeds would form in the absence of rodents (**FIGURE 29.7C**).

1. How many rodents were captured? Of these, how many showed some evidence of ingesting *M. depressa* pollen?
2. Would this evidence alone be sufficient to conclude that rodents are the main pollinators of this plant?
3. How did the average number of seeds produced by caged plants compare with that of control plants?
4. Do these data support the hypothesis that rodents are required for pollination of *M. depressa*? Why or why not?

20 mm

A The dull, petalless, ground-level flowers of *Massonia depressa* are accessible to rodents, who push their heads through the stamens to reach the nectar at the bottom of floral cups. Note the pollen on the gerbil's snout.

Type of rodent	Number caught	Number with pollen on snout	Number with pollen in feces
Namaqua rock rat	4	3	2
Cape spiny mouse	3	2	2
Hairy-footed gerbil	4	2	4
Cape short-eared gerbil	1	0	1
African pygmy mouse	1	0	0

B Evidence of visits to *M. depressa* by rodents.

	Mammals allowed access to plants	Mammals excluded from plants
Percent of plants that set fruit	30.4	4.3
Average number of fruits per plant	1.39	0.47
Average number of seeds per plant	20.0	1.95

C Fruit and seed production of *M. depressa* with and without visits by mammals. Mammals were excluded from plants by wire cages with openings large enough for insects to pass through. Twenty-three plants were tested in each group.

FIGURE 29.7 Testing pollination of *M. depressa* by rodents.

CREDIT: (7) Steven D. Johnson.

29.3 A New Generation Begins

LEARNING OBJECTIVE

- Draw the life cycle of a typical flowering plant, explain the roles of meiosis and mitosis in it.

Flowers are produced by the angiosperm sporophyte, a diploid, spore-producing plant body that grows by mitosis from a fertilized egg (Section 22.2). Spores that form by meiosis in flowers develop into haploid gametophytes, which produce gametes. FIGURE 29.8 and FIGURE 29.9 show the life cycle of a typical angiosperm.

The production of female gametes begins when an ovule starts growing on the inner wall of an ovary ❶. In a common pattern, a cell in the ovule enlarges, then undergoes meiosis and cytoplasmic division to form four haploid **megaspores** ❷. Three of the four megaspores disintegrate, and the remaining one undergoes three rounds of mitosis without cytoplasmic division. One cell results from these divisions, and it has eight haploid nuclei ❸.The cytoplasm of this cell divides unevenly, forming a seven-celled embryo sac—the female gametophyte ❹. One of the cells is called the central cell. This cell is much larger than the others and has two nuclei ($n + n$). Another cell in the embryo sac is the egg. In most species, the female gametophyte is surrounded by one or two protective outer layers called integuments.

The production of male gametes begins as masses of diploid, spore-producing cells form by mitosis inside anthers. Walls typically develop around the masses, so four pollen sacs form ❺. Each cell inside the pollen sacs undergoes meiosis and cytoplasmic division to form four haploid **microspores** ❻. Mitosis and differentiation of a microspore produce a pollen grain ❼. A pollen grain consists of two cells, one inside the cytoplasm of the other, enclosed by a durable, protective coat. After the pollen grains form, they enter a period of suspended metabolism—**dormancy**—before being released from the anther when the pollen sacs split open ❽.

A pollen grain that lands on a receptive stigma **germinates**, which means it resumes metabolic activity after dormancy. When a pollen grain germinates,

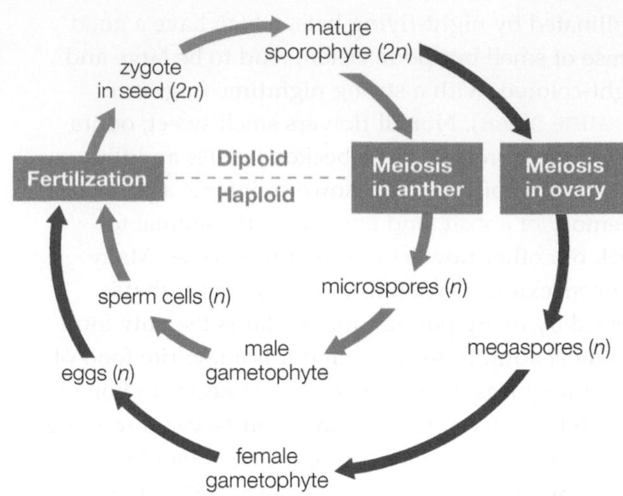
mature
sporophyte (2n)
zygote
in seed (2n)
Fertilization
Diploid
Haploid
Meiosis
in anther
Meiosis
in ovary
sperm cells (n)
microspores (n)
male
gametophyte
megaspores (n)
eggs (n)
female
gametophyte

FIGURE 29.8 **Overview of the life cycle of a typical flowering plant.** Compare Figure 29.9.

its outer cell develops into a tubular outgrowth called a pollen tube ❾. The inner cell undergoes mitosis to produce two sperm cells (the male gametes) within the pollen tube. The mature male gametophyte is the pollen tube together with its contents of male gametes ❿.

The pollen tube elongates at its tip, burrowing into the tissues of the style and carrying the two sperm cells toward the ovule. A pollen tube that reaches the ovule penetrates the female gametophyte and releases the two sperm cells into it ⓫. (Many pollen tubes may grow down into the carpel, but usually only one penetrates the female gametophyte.)

Flowering plants undergo **double fertilization**, in which one of the sperm cells delivered by the pollen tube fuses with (fertilizes) the egg and forms a diploid zygote. The other sperm cell fuses with the central cell to form a triploid ($3n$) cell. This cell gives rise to triploid **endosperm**, a nutritious tissue that forms only in seeds of flowering plants. When the seed sprouts, endosperm will sustain rapid growth of the seedling until its new leaves begin photosynthesis.

TAKE-HOME MESSAGE 29.3

✔ In most flowering plants, a cell in an ovule gives rise to the female gametophyte, which is a seven-celled embryo sac that contains a central cell with two nuclei and a (haploid) egg.

✔ A pollen grain that germinates on a stigma develops into the male gametophyte, which consists of a pollen tube and two (haploid) sperm cells. The pollen tube grows into the carpel, enters an ovule, and releases the sperm cells into the female gametophyte.

✔ Double fertilization occurs when one of the sperm cells delivered into the gametophyte fuses with the egg, producing a diploid zygote. The other sperm fuses with the central cell, producing a triploid cell that will give rise to endosperm.

dormancy Period of temporarily suspended metabolism.
double fertilization Fertilization as it occurs in flowering plants. Two cells are fertilized: One sperm cell fuses with the egg to form the zygote; the second sperm cell fuses with the central cell to form a triploid ($3n$) cell that gives rise to endosperm.
endosperm Nutritive tissue in the seeds of flowering plants.
germinate To resume metabolic activity after dormancy.
megaspore Of seed plants, haploid spore that forms in an ovule and gives rise to an egg-producing gametophyte.
microspore Of seed plants, haploid spore that gives rise to a pollen grain.

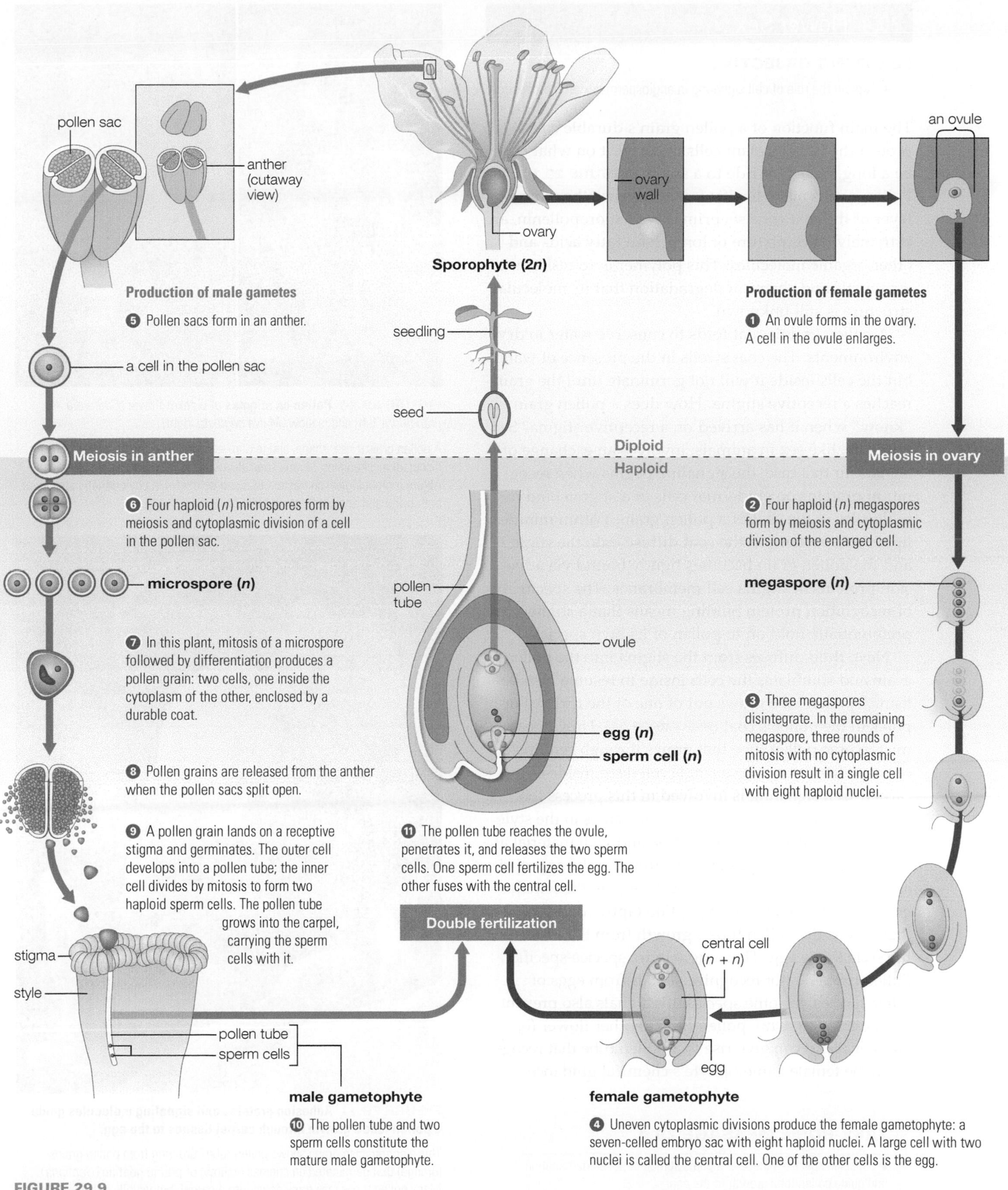
pollen sac
anther
(cutaway
view)
Sporophyte (2n)
ovary
wall
ovary
an ovule
Production of male gametes
5 Pollen sacs form in an anther.
seedling
Production of female gametes
1 An ovule forms in the ovary. A cell in the ovule enlarges.
a cell in the pollen sac
seed
Meiosis in anther
Diploid
Haploid
Meiosis in ovary
6 Four haploid (n) microspores form by meiosis and cytoplasmic division of a cell in the pollen sac.
2 Four haploid (n) megaspores form by meiosis and cytoplasmic division of the enlarged cell.
microspore (n)
pollen
tube
megaspore (n)
ovule
7 In this plant, mitosis of a microspore followed by differentiation produces a pollen grain: two cells, one inside the cytoplasm of the other, enclosed by a durable coat.
3 Three megaspores disintegrate. In the remaining megaspore, three rounds of mitosis with no cytoplasmic division result in a single cell with eight haploid nuclei.
egg (n)
sperm cell (n)
8 Pollen grains are released from the anther when the pollen sacs split open.
9 A pollen grain lands on a receptive stigma and germinates. The outer cell develops into a pollen tube; the inner cell divides by mitosis to form two haploid sperm cells. The pollen tube grows into the carpel, carrying the sperm cells with it.
11 The pollen tube reaches the ovule, penetrates it, and releases the two sperm cells. One sperm cell fertilizes the egg. The other fuses with the central cell.
Double fertilization
central cell
(n + n)
stigma
style
pollen tube
sperm cells
egg
male gametophyte
female gametophyte
10 The pollen tube and two sperm cells constitute the mature male gametophyte.
4 Uneven cytoplasmic divisions produce the female gametophyte: a seven-celled embryo sac with eight haploid nuclei. A large cell with two nuclei is called the central cell. One of the other cells is the egg.

FIGURE 29.9
Life cycle of a typical eudicot. Compare Figure 29.8.

29.4 Flower Sex

LEARNING OBJECTIVE

- Explain the role of cell signaling in angiosperm sexual reproduction.

The main function of a pollen grain's durable coat is to protect the two dormant cells inside of it on what may be a long, turbulent ride to a stigma (FIGURE 29.10). Pollen grains make terrific fossils because the outer layer of the coat consists primarily of sporopollenin, an extremely hard mixture of long-chain fatty acids and other organic molecules. This polymer is so resistant to enzymatic and chemical degradation that its molecular structure is still unknown.

A pollen grain's coat folds to conserve water in dry environments. The coat swells in the presence of water, but the cells inside it will not germinate until the grain reaches a receptive stigma. How does a pollen grain "know" when it has arrived on a receptive stigma? Sex in plants, like sex in animals, involves an exchange of signals. In this case, the signaling begins when recognition proteins on epidermal cells of a stigma bind to molecules in the coat of a pollen grain. Within minutes, lipids and proteins in the coat diffuse onto the stigma, and the pollen grain becomes tightly bound via adhesion proteins in stigma cell membranes. The specificity of recognition protein binding means that a stigma can preferentially hold on to pollen of its own species.

Next, fluid diffuses from the stigma into the pollen grain and stimulates the cells inside to resume metabolism. A pollen tube grows out of one of the furrows or pores in the pollen's coat (FIGURE 29.11). How does a microscopic pollen tube that grows through centimeters of tissue find its way to a single cell deep inside of the carpel? Cell signaling is involved in this process too. Adhesion proteins and signaling molecules in the style direct the deposition of membrane and wall material at the tube's tip, so the pollen tube grows in the appropriate direction. Two cells flanking the egg in the female gametophyte secrete a polypeptide (aptly named LURE) that guides the pollen tube's growth from the bottom of the style to the egg. These signals are species-specific: Pollen tubes do not recognize signals from eggs of other species. In some species, the signals also prevent self-pollination: Only pollen from another flower (or another plant) can give rise to a pollen tube that recognizes the female gametophyte's chemical guidance.

TAKE-HOME MESSAGE 29.4

✔ Species-specific molecular signals stimulate pollen germination and guide pollen tube growth to the egg.

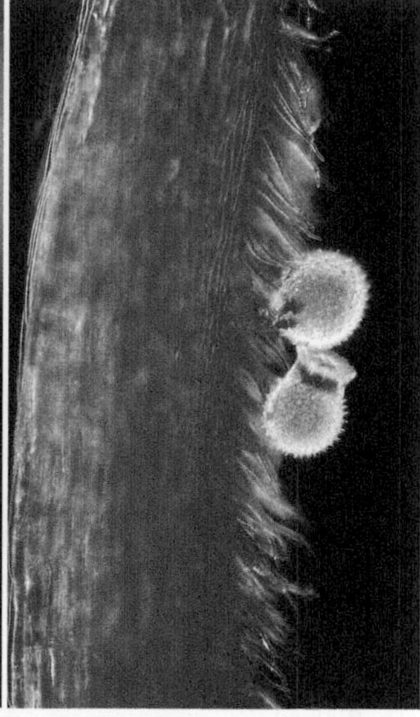

FIGURE 29.10 **Pollen on stigmas** of blanket flower (*Gaillardia grandiflora*, left) and mallow (*Malva neglecta*, right).

A pollen grain's size, shape, and texture are species-specific adaptations for dispersal mechanisms, stigma specializations, and environmental conditions. All are morphological characters that can be useful in phylogenetic studies of both living and ancient flowering plants.

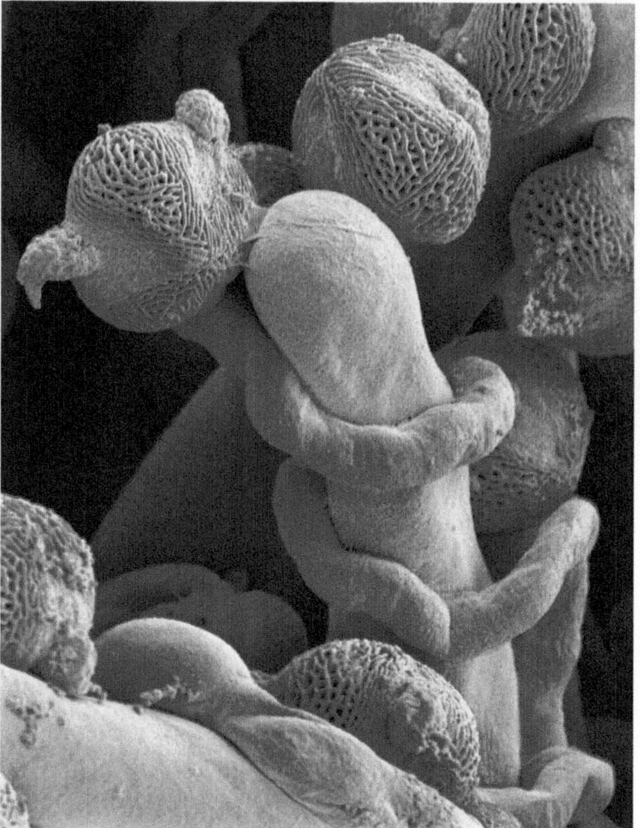

FIGURE 29.11 **Adhesion proteins and signaling molecules guide a pollen tube's growth through carpel tissues to the egg.**

This electron micrograph shows pollen tubes growing from pollen grains (orange) that germinated on stigmas (yellow) of prairie gentian (*Gentiana*). Many pollen tubes may grow down into a carpel, but usually only one penetrates the female gametophyte.

29.5 Seed Formation

LEARNING OBJECTIVES

- Compare the development of typical eudicot and monocot embryos.
- Explain the function of seeds and how they form.

In most flowering plants, double fertilization produces a zygote and a triploid ($3n$) cell, and both immediately begin mitotic divisions. The zygote develops into an embryo sporophyte, and the triploid cell develops into endosperm (**FIGURE 29.12A,B**). As the embryo matures, the ovule's integuments develop into a tough seed coat. A mature ovule is a **seed**, a self-contained package that consists of an embryo sporophyte, its reserves of food, and the seed coat. The seed may undergo dormancy until it receives signals that environmental conditions are appropriate for germination.

As a seed is forming inside an ovule, the parent plant transfers nutrients into it. These nutrients accumulate in endosperm mainly as starch with some lipids and proteins. In typical monocots, nutrients stay in endosperm as the seed matures. By contrast, a maturing eudicot embryo transfers nutrients from the endosperm to its cotyledons (seed leaves). In a mature eudicot seed, most nutrients are stored in the two enlarged cotyledons (**FIGURE 29.12C,D**). When a seed sprouts, nutrients stored in its endosperm or cotyledons will sustain rapid growth of the seedling until new leaves form and begin photosynthesis.

Nutrients in endosperm and cotyledons also nourish humans and other animals. For example, we cultivate many cereals—monocot grasses such as corn, wheat, rye, oats, and barley—for their nutritious seeds. A cereal seed's embryo (the "germ") contains most of its protein and vitamins, and the seed coat (the "bran") contains most of its minerals and fiber. Milling removes the bran and germ, leaving only the starch-packed endosperm. Popcorn is an intact corn seed; its moist endosperm steams when heated, causing pressure to build inside the seed coat until it bursts (pops).

We also cultivate many eudicots for their seeds. For example, beans, lentils, and peas are eudicot seeds valued for their high protein and starch content; seeds of coffee and cacao yield tasty treats.

TAKE-HOME MESSAGE 29.5

✔ After double fertilization in a flowering plant, a zygote develops into an embryo sporophyte, endosperm becomes enriched with nutrients, and a tough seed coat forms.

✔ A seed is a mature ovule, and it consists of an embryo sporophyte, its food reserves, and a seed coat.

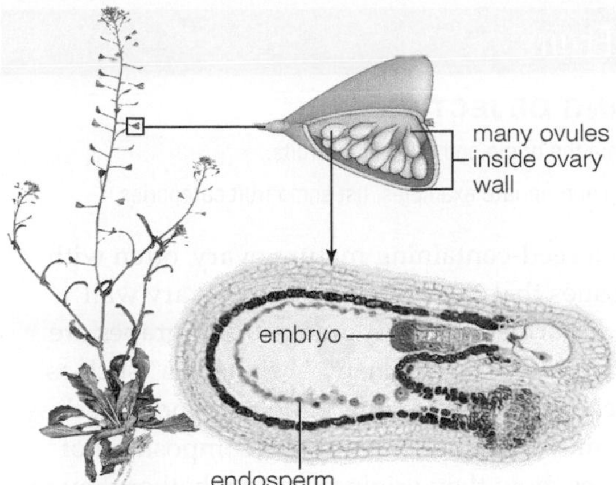
many ovules inside ovary wall
embryo
endosperm

A An embryo and endosperm begin to develop after double fertilization in an ovule of *Capsella*.

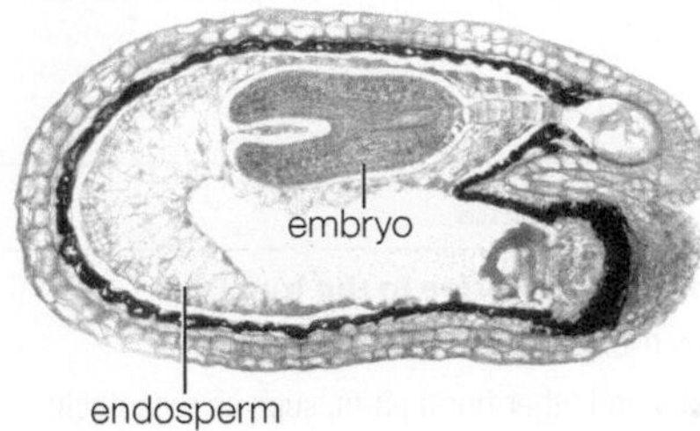
embryo
endosperm

B The embryo is heart-shaped when its two cotyledons start forming. Endosperm tissue expands as the parent plant transfers nutrients into it.

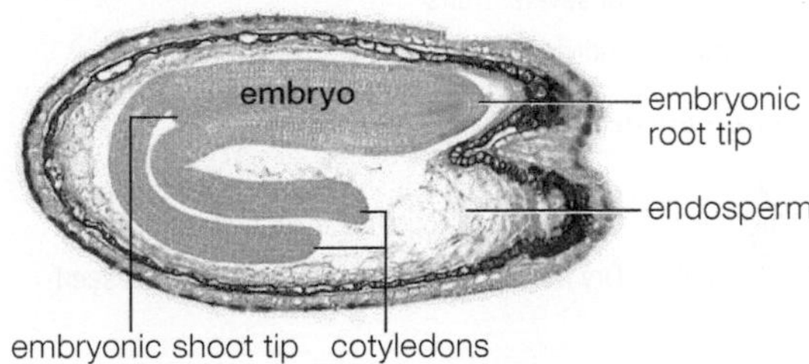
embryo
embryonic root tip
endosperm
embryonic shoot tip
cotyledons

C In eudicots like *Capsella*, nutrients are transferred from endosperm into two cotyledons as the embryo matures. The developing embryo becomes shaped like a torpedo when the enlarging cotyledons bend.

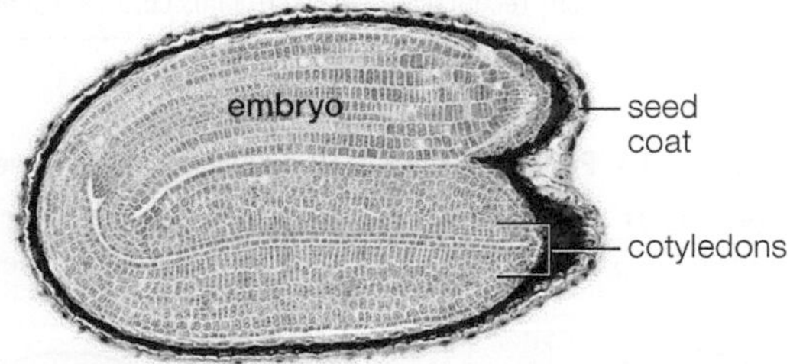
embryo
seed coat
cotyledons

D A tough seed coat forms around the mature embryo and the enlarged cotyledons.

FIGURE 29.12 Embryonic development of a eudicot: shepherd's purse (*Capsella*).

seed Mature ovule of an angiosperm. Consists of an embryo sporophyte, nutritive tissue, and a protective seed coat.

CREDITS: (12A, C, D) Michael Clayton, University of Wisconsin; (12B) David T. Webb.

29.6 Fruits

LEARNING OBJECTIVES

- Explain the origin and function of fruits.
- Using appropriate examples, list some fruit categories.

A **fruit** is a seed-containing mature ovary, often with fleshy tissues that developed from the ovary wall as the seed formed. Apples, oranges, and grapes are familiar fruits, but so are many "vegetables" such as beans, peas, tomatoes, grains, eggplant, and squash.

Fruits may be categorized by the composition of their tissues, how they originate, and whether they are fleshy (juicy) or dry (TABLE 29.2). "True" fruits such as oranges develop only from the ovary wall and its contents (FIGURE 29.13A). "Accessory" fruits have tissues derived from other floral parts (petals, sepals, stamens, or receptacle) that expand along with the developing ovary. An apple is an example of an accessory fruit; most of its flesh is an enlarged receptacle (FIGURE 29.13B).

"Simple" fruits are derived from one ovary or a few fused ovaries of one flower. Cherries, beans, and acorns are examples. By contrast, "aggregate" fruits such as blackberries are derived from multiple unfused ovaries of one flower that mature as a cluster. A "multiple" fruit such as a pineapple or a fig develops as a unit from several individually pollinated flowers.

Cherries, almonds, and olives are fleshy fruits, as are individual fruits of blackberries and other related species. To a botanist, a "berry" is a fleshy fruit produced from one ovary; grapes, tomatoes, citrus fruits, pumpkins, watermelons, and cucumbers are examples. Apples and pears are pomes: fruits in which fleshy tissues derived from the receptacle enclose a core derived from the ovary wall.

A dry fruit has an ovary wall that dries out as it matures. These fruits can be dehiscent or indehiscent. The wall of a dehiscent fruit splits along a seam to release the seeds inside when they are mature; pea pods are examples. By contrast, a mature indehiscent fruit may have a seam, but its wall does not split open: Its (typically single) seed is dispersed inside the intact fruit. Acorns and grains are indehiscent fruits, as are the fruits of sunflowers, maples, and strawberries. Strawberries are not berries, and their fruits are not fleshy. The red, juicy part of a strawberry is an enlarged receptacle with individual indehiscent dry fruits on its surface.

TABLE 29.2

Three Ways to Classify Fruits

What parts of the flower gave rise to the fruit?	
True fruit	Only ovary wall and its contents
Accessory fruit	Ovary and other floral parts, such as receptacle
How did the fruit arise?	
Simple fruit	One flower, single or fused carpels
Aggregate fruit	One flower, several unfused carpels; becomes cluster of several fruits
Multiple fruit	Individually pollinated flowers fuse
Is the fruit dry or fleshy?	
Dry	
Dehiscent	Dry fruit wall splits on seam to release seeds
Indehiscent	Dry fruit wall does not split; usually one seed
Fleshy	
Drupe	Fleshy fruit, one seed enclosed by a hard pit
Berry	Fleshy fruit, often many seeds, no pit
Pepo: Berry with tough, thick outer rind	
Hesperidium: Berry with leathery rind, partitioned sections	
Pome	Fleshy receptacle tissue surrounds core with seeds

FIGURE 29.13 Parts of a fruit develop from parts of a flower.

FIGURE IT OUT The flower that gave rise to the orange in **A** had how many carpels?

Answer: Eight

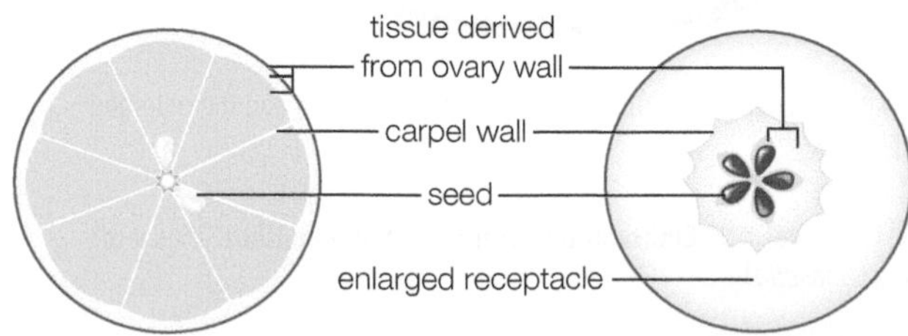

tissue derived from ovary wall
carpel wall
seed
enlarged receptacle

A The tissues of an orange develop from the ovary wall.

B The flesh of an apple is an enlarged receptacle.

A cherry is a true fruit, a simple fruit, and a fleshy drupe.

A blackberry is an accessory fruit and an aggregate of many small, fleshy drupes.

A pineapple is an accessory fruit and a multiple fruit formed by the fusion of many small, fleshy berries.

A pea pod is a true fruit, a simple fruit and a dehiscent dry fruit.

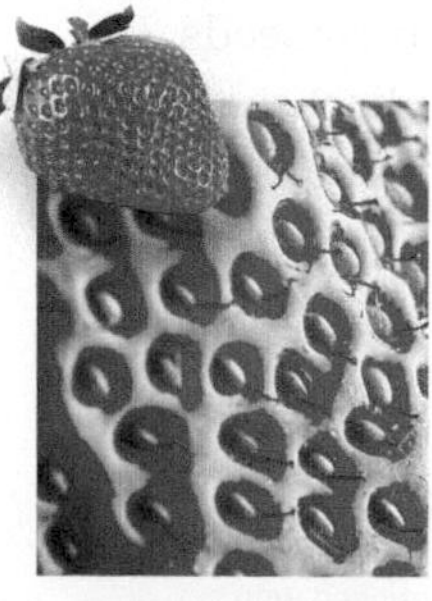

A strawberry is an accessory fruit and an aggregate of many individual indehiscent dry fruits.

A Wind-dispersed fruits. Left, dry outgrowths of the ovary wall of a maple fruit form "wings" that catch the wind and spin the seeds away from the parent tree. Right, wind that lifts the hairy modified sepals of a dandelion fruit may carry the attached seed miles away from the parent plant.

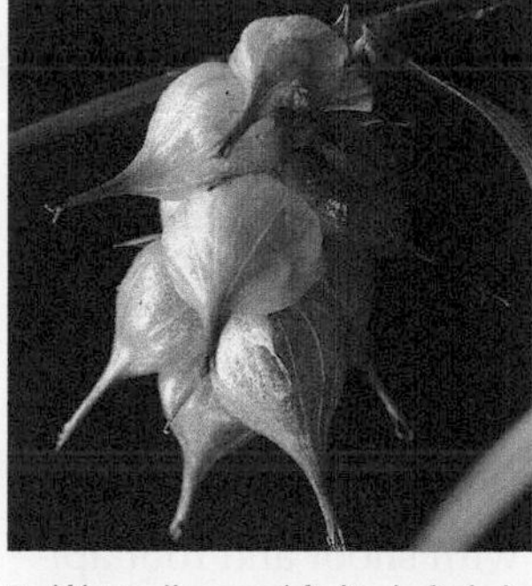

B Water-dispersed fruits. Left, fruits of sedges native to American marshlands have seeds encased in a bladderlike envelope that floats. Right, buoyant fruits of the coconut palm have tough, waterproof husks. Coconuts can float for thousands of miles in seawater.

C Animal-dispersed fruits. Left, curved spines attach cocklebur fruits to the fur of animals (and clothing of humans) that brush past it. Right, the red, fleshy fruits of hawthorn plants are an important food source for cedar waxwings, which disperse the fruits' seeds in feces.

D The dry, dehiscent fruits of California poppy (*Eschscholzia californica*) and some other species spread their own seeds when they split open suddenly along their center seam. The movement propels the seeds through the air, away from the parent plant.

FIGURE 29.14 Some fruit adaptations that aid seed dispersal.

The function of a fruit is to disperse seeds. Dispersal increases reproductive success by minimizing competition for resources among parent and offspring. Just as flower structure is adapted to certain pollination vectors, so is fruit structure adapted to certain dispersal vectors: environmental agents of movement such as wind or water, or mobile organisms such as birds or insects. For example, fruits dispersed by wind are lightweight with breeze-catching specializations (**FIGURE 29.14A**). Fruits dispersed by water have water-repellent outer layers, and they float (**FIGURE 29.14B**). Many fruits have specializations that facilitate dispersal by animals (**FIGURE 29.14C**). Some have hooks or spines that stick to the feathers, feet, fur, or clothing of more mobile species. Colorful, fleshy, or fragrant fruits attract birds and mammals that can scatter seeds. The animal may eat the fruit and discard the seeds, or eat the seeds along with the fruit. Seeds of apricots, plums, and many other fruits contain toxins that are released if the seed is crushed, thus discouraging animals from chewing them up completely and killing the embryo. However, abrasion of the seed coat by teeth or by digestive enzymes in an animal's gut can help the seed germinate after it departs in feces. A few types of fruit have mechanical specializations that spread their seeds even without use of a dispersal vector (**FIGURE 29.14D**).

TAKE-HOME MESSAGE 29.6

- ✔ As embryos develop into seeds, tissues around them develop into a fruit.
- ✔ A fruit is a mature ovary, with or without accessory tissues that develop from other parts of the flower.
- ✔ We can categorize a fruit in terms of how it originated, its composition, and whether it is dry or fleshy.
- ✔ A fruit's function is seed dispersal. Specializations are adaptations for particular dispersal vectors such as water, wind, or animals.

fruit Mature ovary of a flowering plant, often with accessory parts; encloses a seed or seeds.

CREDITS: (14A) left, R. Carr; right, JupiterImages Corporation; (14B) left, T. M. Jones; right, Alex James Bramwell/Shutterstock; (14C) left, © Robert H. Mohlenbrock © USDA-NRCS PLANTS Database; right, Vishnevskiy Vasily/Shutterstock; (14D) Trudi Davidoff, www.WinterSown.org.

29.7 Early Development

LEARNING OBJECTIVES

- Explain why some seeds undergo dormancy, and give some examples of triggers for germination.
- Describe the process of seed germination.
- Compare the pattern of early growth that occurs in eudicots and monocots after germination.

An embryonic plant complete with shoot and root apical meristems forms as part of a seed (**FIGURE 29.15**). A tiny section of embryonic stem called the hypocotyl separates the embryonic shoot (the plumule) from the embryonic root (radicle). As the seed matures, the embryo may dry out and enter a period of dormancy. A dormant embryo can rest in its protective seed coat for many years before it resumes metabolic activity and germinates.

Breaking Dormancy

Seed dormancy is a climate-specific adaptation that allows germination to occur when conditions in the environment are most likely to support the growth of a tender seedling. For example, seeds of many annual plants native to cold winter regions are dispersed in autumn. If the seeds germinated immediately, the seedlings would not survive the upcoming winter. Instead, the seeds remain dormant until spring, when milder temperatures and longer day length favor growth. By contrast, seeds of many plants in regions near the equator germinate as soon as they mature. In these regions, conditions favorable for growth prevail year round, so dormancy is not adaptive.

Other than the presence of water, the triggers for germination differ by species. Some seed coats are so dense that they must be abraded or broken (by being chewed, for example) before germination can occur. Seeds of many cool-climate plants require exposure to freezing temperatures; those of some lettuce species, to bright light. In seeds of some species native to regions that have periodic wildfires, germination is inhibited by light and enhanced by smoke; in others, germination does not occur unless the seeds have been previously burned. Such requirements are evolutionary adaptations for life in a particular environment.

Germination typically begins with water seeping into the seed. The seed's internal tissues absorb water and swell, causing the seed coat to rupture. After the seed coat breaks, air diffuses into the embryo's cells. The water also provokes a series of events that activate hydrolysis enzymes, which begin to break down stored starches into sugar subunits (we return to this topic in Section 30.4). Together with oxygen in the air, the released glucose allows the embryo's cells to begin aerobic respiration. Energy released from the sugar breakdown fuels rapid divisions of apical meristem cells, and the embryonic plant begins to grow. Germination ends when the first part of the embryo—the embryonic root, or radicle—emerges from the seed coat to become a primary (first) root.

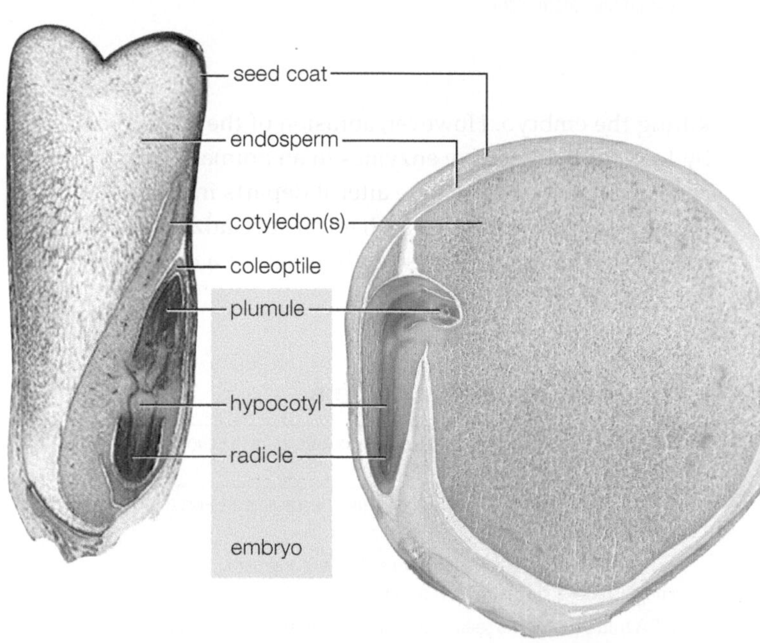
seed coat
endosperm
cotyledon(s)
coleoptile
plumule
hypocotyl
radicle
embryo

FIGURE 29.15 Anatomy of a seed.
Left, corn (*Zea mays*), a monocot; right, pea (*Pisum sativum*), a eudicot.

As dormancy ends, cell divisions resume mainly at apical meristems of the plumule (the embryonic shoot) and radicle (the embryonic root). These two structures are separated by a section of embryonic stem called the hypocotyl. In monocots, the plumule is protected by a sheathlike coleoptile.

After Germination

Remember that monocots and eudicots differ in some details of their structure (Section 27.2). For example, monocot embryos have one cotyledon; eudicot embryos have two. The pattern of early growth after germination also differs between the two groups. In corn and other monocots, a rigid sheath called a **coleoptile** surrounds and protects the plumule. In a typical monocot pattern of development (**FIGURE 29.16**), germination is followed by the emergence of the radicle and coleoptile from the seed coat ❶. The radicle, which is now a primary root, grows down into the soil ❷ as the coleoptile grows upward ❸. When the coleoptile reaches the surface of the soil, it stops growing. The plumule then emerges from the coleoptile as it develops into the plant's primary (first) shoot ❹. The single

coleoptile (coal-ee-OPP-till) In monocots, a rigid sheath that protects the plumule (embryonic shoot).

CREDIT: (15) left, Michael Clayton, University of Wisconsin; right, Jubal Harshaw/Shutterstock.

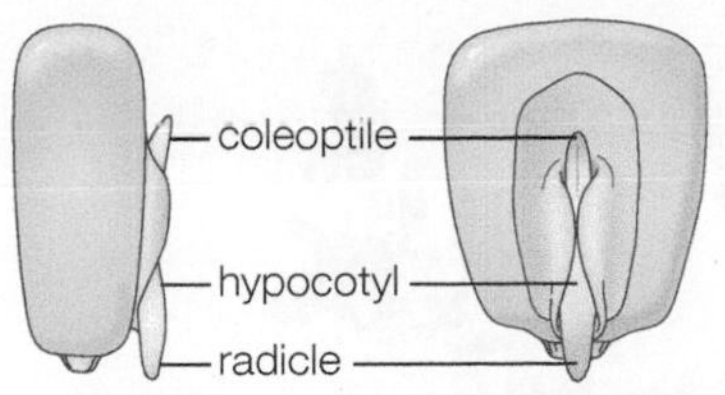
coleoptile
hypocotyl
radicle
primary shoot
coleoptile

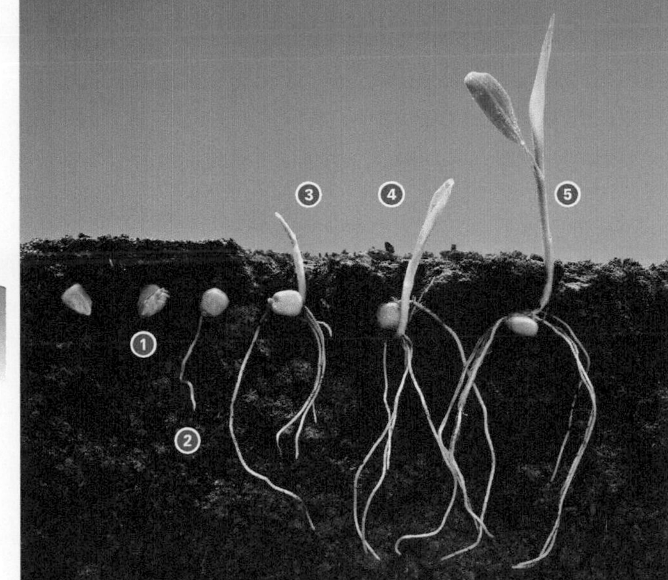
1
2
3
4
5

❶ At the end of germination of a corn grain (seed), the radicle and coleoptile emerge from the seed coat.

❷ The radicle develops into a primary root that grows down into the soil.

❸ The coleoptile grows up and opens a channel through the soil to the surface, where it stops growing. Note the adventitious roots; these will replace the primary root in a fibrous root system typical of monocots.

❹ The plumule develops into a primary shoot that emerges from the coleoptile.

❺ The single cotyledon remains under the soil, transferring nutrients to the seedling until its new leaves can produce enough sugars by photosynthesis to sustain growth.

FIGURE 29.16 Early growth of corn (*Zea mays*), a typical monocot.

1
2
3
4
5

❶ A bean's seed coat splits and the radicle emerges.

❷ The radicle develops into a primary root that grows down into the soil as the hypocotyl emerges from the seed and bends in the shape of a hook.

❸ The bent hypocotyl lengthens, dragging the two cotyledons upward.

❹ When the hypocotyl reaches the surface of the soil, exposure to sunlight causes it to straighten. Primary leaves emerge from between the cotyledons and begin photosynthesis.

❺ Cotyledons typically undergo a period of photosynthesis, then wither and fall off the lengthening stem.

FIGURE 29.17 Early growth of the common bean, a typical eudicot.

cotyledon, which typically stays beneath the soil, functions mainly to transfer endosperm breakdown products to the seedling until leaves form on the shoot and begin photosynthesis ❺.

In eudicot seedlings, the primary shoot is not protected by a coleoptile. In a common eudicot pattern of development (FIGURE 29.17), the radicle emerges from the seed ❶, then begins to develop into a primary root that grows down into the soil. As this occurs, the hypocotyl emerges from the seed and bends into the shape of a hook ❷. The bent hypocotyl lengthens and pulls the cotyledons upward ❸. When the hypocotyl reaches the soil surface, exposure to light causes it to straighten ❹. Primary leaves emerge from between the cotyledons and begin photosynthesis. The cotyledons typically undergo a period of photosynthesis before shriveling ❺. Eventually, the cotyledons fall off the lengthening stem, and the young plant's new leaves produce all of its food.

TAKE-HOME MESSAGE 29.7

- ✔ Seed dormancy and germination requirements are evolutionary adaptations for life in a particular climate.
- ✔ Specific environmental cues trigger germination in many species.
- ✔ After a seed germinates, nutrients stored in endosperm (in monocots) or cotyledons (in eudicots) support the seedling's growth until new leaves begin photosynthesis.
- ✔ The pattern of early development differs between monocots and eudicots. For example, a coleoptile protects monocot plumules.

CREDITS: (16) Scott Sinklier/Getty Images; (17) Bogdan Wankowicz/Shutterstock.

29.8 Asexual Reproduction of Flowering Plants

LEARNING OBJECTIVES

- Describe some agricultural applications of vegetative reproduction.
- Explain why some fruits are seedless.

Many flowering plants can reproduce asexually by a natural process called **vegetative reproduction,** in which new roots and shoots grow from extensions or pieces of a parent plant. Each new plant is a clone, a genetic replica of its parent. New plants can form via roots and shoots that sprout from nodes on stems, or in some cases from pericycle in roots (**FIGURE 29.18**). Entire forests of quaking aspen trees (*Populus tremuloides*) are actually stands of clones that arose from root suckers—shoots that sprout from shallow lateral roots. One quaking aspen forest in Utah consists of about 47,000 shoots and stretches for 107 acres. This stand of clones, which has been named Pando, is estimated to be around 80,000 years old!

A Potatoes are modified stems; new roots and shoots sprout from their nodes, which we call "eyes" (Section 27.4). The new plants are clones, genetically identical to the parent.

B Root suckers—shoots that sprout from roots—arise from pericycle in the vascular cylinder of some plants (the photo shows suckers forming on an aspen root). Suckers can give rise to new plants that are clones of the parent.

FIGURE 29.18 Examples of asexual reproduction in plants.

Agricultural Applications

For thousands of years, we humans have been taking advantage of the natural capacity of many plants to reproduce asexually. For example, almost all houseplants, woody ornamentals, and orchard trees are clones grown from stem fragments (cuttings) of a parent plant. Propagating a plant from cuttings may be as simple as jamming a broken stem into soil, a method that harnesses the root- and shoot-forming ability of nodes on a stem. Plants grown from cuttings may mature more quickly when they are grafted (induced to fuse with and become supported by another plant).

Propagating a plant from cuttings ensures that offspring will have the same desirable traits as the parent. Consider familiar orchard apples, which are descendants of a wild species native to central Asia. The chromosome number of wild apples doubled between 50 and 60 million years ago, and subsequent chromosomal recombinations and rearrangements made a genetic mashup of duplicated genes. Today, each apple tree is highly heterozygous, and this is the reason why apples do not breed true for fruit traits. Every seed—even ones from the same apple—will produce a tree that bears unique fruit (**FIGURE 29.19**).

In fact, most apple trees grown from seed produce uniquely unpalatable fruit. In the early 1800s, John Chapman (known as Johnny Appleseed) grew millions of apple trees from seed in the midwestern United States. He sold the trees to homesteading settlers, who made hard cider from the otherwise inedible apples. About one of every hundred trees produced tasty fruit. Its lucky owner would patent the tree, then propagate it from cuttings grafted onto other apple trees. An estimated 16,000 varieties of edible apples were discovered during this time. Few of them remain; only 15 varieties now account for 90 percent of the apples sold in U.S. grocery stores. All are clones of Chapman's original trees, and all are still grafted.

Grafting is also used to increase the hardiness of a desirable plant. In 1862, the plant louse *Phylloxera* was accidentally introduced into France via imported American grapevines. European grapevines had little resistance to this tiny insect, which attacks and kills the root systems of the vines. By 1900, *Phylloxera* had destroyed two-thirds of the vineyards in Europe, thus devastating the wine-making industry for decades. Today, French vintners routinely graft their prized grapevines onto *Phylloxera*-resistant American vines.

tissue culture propagation Laboratory method in which embryonic plants are grown from individual meristem cells.
vegetative reproduction Growth of new roots and shoots from extensions or fragments of a parent plant; a form of asexual reproduction.

CREDITS: (18A) HUAJI/Shutterstock; (18B) B. Campbell/USDA;

Grafting allows us to propagate plants that produce "seedless" fruit, which means the seeds are underdeveloped or missing. Fruit can form on some plants even in the absence of fertilization, and fruits that develop from unfertilized flowers are seedless. Plants with mutations that cause ovules or embryos to abort during development also make seedless fruit; seedless grapevines and navel orange trees are like this. All commercially produced bananas are seedless because the plants that bear them are triploid ($3n$). During meiosis I, the three chromosomes cannot be divided equally between the two spindle poles, so viable gametes do not form and neither do seeds. Grafting bananas and other monocots is notoriously difficult, so seedless banana plants are propagated from shoots that sprout from nodes on their corms (Section 27.4), or by tissue culture. With **tissue culture propagation**, individual cells (typically from meristem) are coaxed to divide and form embryos in a laboratory. This technique can yield millions of genetically identical offspring from a single parent plant. It is frequently used in research aimed at improving food crops, and also to propagate rare or hybrid ornamental plants.

Plant breeders can artificially increase the frequency of polyploidy by treating plants with colchicine, a microtubule poison that disrupts spindle function during meiosis. Tetraploid ($4n$) plants produced by colchicine treatment are then crossed with diploid ($2n$) individuals. The resulting offspring are triploid and sterile: They make seedless fruit on their own or after pollination (but not fertilization) by a diploid plant. Seedless watermelons are produced this way.

FIGURE 29.19 Wild apples. Color, flavor, size, sweetness, and texture vary in the fruits of wild apple trees (21 are shown in this photo). Apple trees are grafted because they do not breed true for these valuable traits.

TAKE-HOME MESSAGE 29.8

✔ Many plants propagate themselves asexually when new shoots grow from a parent plant or pieces of it. The offspring are clones.

✔ Humans propagate plants asexually for agricultural or research purposes, for example by grafting or tissue culture.

29.1 Plight of the Honeybee (revisited)

A single honeybee visits hundreds, sometimes thousands, of flowers a day to find nectar and pollen. The insect uses "scent memory" to remember the location of these rewards. Then it finds its way back to the hive, navigating distances up to 8 kilometers (5 miles) to communicate the location of the flowers to other bees. Bees' ability to rapidly learn, remember, and communicate with each other makes them efficient pollinators.

Long-term exposure to low levels of pesticides might impair a bee's ability to carry out its pollen mission. Many researchers suspect that synthetic pesticides called neonicotinoids are an important factor in colony collapse disorder. Neonicotinoids are now the most widely used insecticides in the United States, and they are also used in other countries—about 2.5 million acres of croplands in total. The chemicals are systemic, which means they are taken up by all tissues of a plant treated with them, including the nectar and pollen that honeybees collect from flowers as they forage.

Neonicotinoids mimic the neurotoxic (nerve-killing) effect of nicotine, a natural insecticide made by plants such as tobacco, and they impair a bee's learning and memory. "Honeybees learn to associate floral colors and scents with the quality of food rewards," says Geraldine Wright, a neuroscientist at Newcastle University in England. Neonicotinoids affect the neurons involved in these behaviors, so bees exposed to these pesticides are likely to have difficulty communicating with other members of the colony. Neonicotinoids also increase bees' vulnerability to infection in amounts too tiny to detect in the insects themselves. Even dewdrops that form on neonicotinoid-treated plants apparently contain enough of the insecticide to affect bees.

CREDIT: (19) Photo by Peggy Greb, USDA, ARS.

STUDY GUIDE

Section 29.1 Reproduction in seed plants depends on the transfer of pollen from one flower to another. Angiosperms in particular depend on animal pollinators, and honeybees are critical for production of food crops worldwide. Colony collapse disorder (CCD) is an ongoing, unexplained syndrome that is wiping out a huge number of honeybee hives. Declines in populations of bees and other pollinators negatively affect plant populations as well as other animal species that depend on the plants, including humans. Widely used neonicotinoid pesticides may contribute to CCD.

Section 29.2 **Flowers** are the reproductive structures of angiosperms. A typical flower consists of four whorls (rings) of modified leaves that form from a region of stem called a receptacle. Expression of master regulators during development of the flower gives rise to the identity of each whorl. The outer whorl is a **calyx** (a ring of **sepals**). The calyx surrounds a **corolla** (a ring of **petals**), which in turn surrounds a ring of **stamens**. The innermost whorl consists of one or more **carpels**.

A typical carpel consists of a **stigma**, a style, and an **ovary** inside which one or more **ovules** develop. The female gametophyte forms inside an ovule. A typical stamen consists of an **anther** on a thin filament. Anthers produce **pollen grains**, which are immature male gametophytes.

In angiosperms, **pollination** is the arrival of pollen on a receptive stigma. A flower's shape, pattern, color, and fragrance reflect an evolutionary relationship with a particular **pollination vector**, often a coevolved animal **pollinator**. Pollinators may receive **nectar**, pollen, or another reward for visiting a flower.

Sections 29.3, 29.4 Meiosis of diploid cells inside pollen sacs of anthers produces haploid **microspores**. Each microspore develops into a pollen grain that is released from a pollen sac after a period of **dormancy**.

Meiosis and cytoplasmic division of a cell in an ovule produce four haploid **megaspores**, one of which gives rise to the female gametophyte. One of the seven cells of the gametophyte is the egg; another is the diploid central cell.

An interplay of species-specific molecular signals trigger a pollen grain to **germinate** on a receptive stigma and develop into a pollen tube that contains two sperm cells. Other molecular signals guide the pollen tube's growth through the carpel to the ovule. The pollen tube then penetrates the ovule and releases the sperm cells into it. In **double fertilization**, one sperm cell fertilizes the egg, forming a zygote; the other fuses with the central cell and gives rise to triploid **endosperm**.

Section 29.5 As a zygote develops into an embryo, the ovule's protective outer layers (integuments) develop into a seed coat, and endosperm collects nutrients from the parent plant. In eudicots, nutrients are transferred from endosperm into two cotyledons. A **seed** is a mature ovule: an embryo sporophyte and endosperm enclosed within a seed coat. Nutrients in endosperm or cotyledons nourish the forthcoming seedling. They also make seeds a nutritious food source for animals.

Section 29.6 As an embryo sporophyte develops, the ovary wall and sometimes other tissues mature into a **fruit** that encloses the seed. Fruit specializations are adaptations for seed dispersal by specific vectors such as wind, water, or animals.

Section 29.7 Seeds often undergo a period of dormancy that does not end until species-specific environmental cues trigger germination. The radicle emerges from the seed coat at the end of germination; other patterns of early development vary. For example, monocot plumules are sheathed by a **coleoptile**; eudicot hypocotyls form a hook that pulls cotyledons up through soil.

Section 29.8 Many angiosperms can reproduce asexually by **vegetative reproduction**. These plants (and seedless plants) can be propagated by grafting. We propagate some valuable ornamentals by **tissue culture propagation**.

SELF-QUIZ

Answers in Appendix VII

1. The arrival of pollen grains on a receptive stigma is called ________ .
   a. germination   c. pollination
   b. fertilization   d. propagation
2. An animal pollinator may be rewarded by ________ when it visits a flower of a coevolved plant (choose all that apply).
   a. pollen   c. hormones
   b. nectar   d. fruit
3. The ________ of a flower contains one or more ovaries in which eggs develop, fertilization occurs, and seeds mature.
   a. pollen sac   c. receptacle
   b. carpel   d. sepal
4. In flowers, the structures that produce male gametophytes are called ________ ; the structures that produce female gametophytes are called ________ .
   a. carpels; stamens   c. pollen grains; flowers
   b. anthers; carpels   d. megaspores; microspores
5. Meiosis of cells in pollen sacs forms haploid ________ .
   a. megaspores   c. stamens
   b. microspores   d. sporophytes
6. The ________ are parts of a mature seed.
   a. stamen, anther, and pollen grain
   b. embryo, food reserves, and seed coat
   c. stigma, style, and ovary
   d. male and female gametophytes

7. Choose the statement that is true.
   a. All flowers are pollinated by bees.
   b. Apple trees are propagated by grafting because they are triploid.
   c. Carpels function to attract pollinators.
   d. A coleoptile protects the plumule of eudicot seedlings.
   e. The arrival of pollen on a receptive seed is called pollination.
   f. In plants, sperm cells are male gametes.
   g. Nutrients in fruits nourish new seedlings.

8. The seed coat forms from the _______ .
   a. integuments c. endosperm
   b. coleoptile d. sepals

9. Seeds are mature _______ ; fruits are mature _______ .
   a. ovaries; ovules c. ovules; ovaries
   b. ovules; stamens d. stamens; ovaries

10. Jessica is preparing a plate of fruits for a party and cuts open a cantaloupe (*Cucumis melo*). A tough outer rind and soft, juicy tissue enclose many seeds (left). Knowing her friends will ask her to categorize this fruit, she panics, runs to her biology book, and opens it to Table 29.2. What does she find out?

11. Cotyledons develop as part of _______ .
    a. carpels c. embryo sporophytes
    b. accessory fruits d. flowers

12. Exposure to _______ can trigger seed germination.
    a. light c. smoke
    b. cold d. all can be triggers

13. A new plant forms from a stem that broke off of the parent plant. This is an example of _______ .
    a. nodal cloning c. asexual reproduction
    b. exocytosis d. tissue culture propagation

14. Match the terms with the most suitable description.

___ ovule	a. pollen tube together with its contents
___ receptacle	b. consists of seven cells, one with two nuclei
___ double fertilization	c. after fertilization, develops into a seed
___ anther	d. embryonic shoot
___ plumule	e. pollen sacs inside
___ mature female gametophyte	f. swollen stem; base of flower
___ mature male gametophyte	g. formation of zygote and first cell of endosperm

CENGAGE brain .com To access course materials, please visit www.cengagebrain.com.

15. Domesticated banana plants produce seedless fruit because they are _______ .
    a. triploid d. propagated by grafting
    b. monocots e. treated with colchicine
    c. endangered f. infected with *Phylloxera*

CRITICAL THINKING

1. Label the parts of the flower shown above.

2. All but one species of large birds native to New Zealand's tropical forests are now extinct. Numbers of the one surviving species, the kereru (*Hemiphaga novaeseelandiae*), are declining rapidly due to habitat loss, poaching, predation, and interspecies competition that wiped out the other native birds. The keruru is the only remaining dispersal agent for several native trees that produce big seeds and fruits, mainly because it is the only remaining species that can swallow big fruits (left) and expel big seeds whole. One of these trees, the puriri (*Vitex lucens*), is New Zealand's most valued hardwood. Explain, in terms of natural selection, what would happen to puriri trees in New Zealand if the kereru becomes extinct.

3. Is the seedling shown on the right a monocot or eudicot? How can you tell?

CREDITS: (in text) Yellowj/Shutterstock; Bell/Alamy; Adam Gryko/Shutterstock

30 COMMUNICATION STRATEGIES IN PLANTS

CORE CONCEPTS

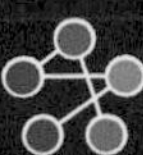

Systems

Complex properties arise from interactions among components of a biological system.
The timing and coordination of specific molecular and cellular events are regulated by mechanisms that govern gene expression. Many types of environmental cues can trigger internal molecular signals in plants. In turn, these signals result in adjustments to an interplay of hormones and other molecules that affect gene expression. Plant hormones can have synergistic or opposing effects that differ in different tissues. Specialized physiological defenses help individual plants survive environmental stresses and attacks by herbivores and pathogens.

Evolution

Evolution underlies the unity and diversity of life.
Plant hormones that operate the same way in evolutionarily distant groups are evidence of shared ancestry. Secondary metabolites that defend a plant are adaptations to coevolved pathogens and herbivores. Many of these compounds have beneficial effects in humans.

Information Flow

Living systems store, retrieve, transmit, and respond to information essential for life.
Plants produce and respond to hormones, which operate by binding to receptors that effect a response inside the cell. As in animals, hormones orchestrate development, and feedback mechanisms involving hormones are part of developmental and homeostatic control. Unlike animals, plants continue to develop after maturity, in response to environmental cues. Most developmental and metabolic processes in plants are influenced by networks of signaling involving multiple hormones with overlapping effects.

Links to Earlier Concepts

This chapter explores plant growth (Section 27.7) in terms of cell signaling (13.6). You may want to review plant cell structure and function (4.7, 4.9, 27.6, 28.3, 28.4), membrane properties (5.7, 5.8–5.10), and control of gene expression (10.2, 10.3, 10.5). You will also revisit starch (3.3), pigments (6.3), bacterial flagella (4.3), redox reactions (5.5), circadian rhythms (10.4), senescence (11.5), pathogens (20.8, 23.7), and symbionts (23.7, 28.3).

30.1 Prescription: Chocolate

For several centuries, the small islands of the San Blas Archipelago just off the coast of Panama have been inhabited by the Kuna, a tribe of indigenous South Americans. Despite their lack of access to medicine, the Kuna have almost no incidence of the hypertension (high blood pressure and its associated cardiovascular problems) that plagues about one-quarter of people in mainland populations.

Scientists studying the Kuna hypothesized that the resistance to hypertension had a genetic basis, so in 1990 they began to search for an allele unique to the tribe that affects blood pressure. At the time, the human genome was being sequenced (Section 15.4), and genomic comparisons were becoming a routine avenue of research. However, a decade later, researchers still had not found the allele, and began to look for other correlations. They discovered that genetics really had nothing to do with the low incidence of hypertension in the Kuna. All else being equal, the most striking difference between island-dwelling populations and their mainland relatives is that the islanders consume an unusually large amount of cocoa. The Kuna tend to drink cocoa instead of water, and they live in a hot tropical climate so they drink a lot of it—a minimum of five cups per day.

Cocoa is made from cacao beans, which are the seeds of the *Theobroma cacao* tree. The seeds and young leaves of this tree have a particularly high content of flavonoids. Flavonoids are secondary metabolites, so named because they are not required for the immediate survival of the organism that makes them (Section 22.9). A flavonoid called epicatechin is probably responsible for the unusual absence of heart disease among the Kuna. Cocoa is rich in epicatechin, as are other forms of minimally processed dark chocolate. Like caffeine, resveratrol, curcumin (in coffee, grapes, and turmeric, respectively) and many other secondary metabolites, epicatechin has medicinal effects in humans. It influences a wide range of biological activities in the body, and it has a demonstrably protective effect against oxidative tissue damage that typically occurs after a stroke or a heart attack. A dose of epicatechin can reduce the heart muscle damage that a heart attack causes by 52 percent, and reduce the brain damage that a stroke causes by 32 percent. As an added benefit, epicatechin enhances memory and kills cancer cells.

epicatechin

Plants do not make this seemingly miraculous compound for our benefit. Epicatechin serves a role in plant immunity. Pigments, scents, and other chemicals that

CREDITS: (opposite) Stephen Rees/Shutterstock; (in text) top, © Cengage Learning; bottom, Dmitry Melnikov/Shutterstock.

attract pollinators are also secondary metabolites, as are compounds that deter herbivores and pathogens, attract symbiotic organisms, or inhibit growth in competing individuals. These chemicals are all part of the means by which plants interact with their environments. ●

30.2 Chemical Signaling in Plants

LEARNING OBJECTIVES

- Describe the role of hormones in plants.
- Explain why the same hormone can elicit different responses in different cells.

In animals, most development occurs before adulthood. By contrast, development in plants continues throughout the individual's lifetime. Plants, unlike animals, cannot move away from unfavorable conditions. Each plant adapts to its environment by altering growth patterns. Adaptive changes in form and physiology are triggered by environmental cues such as temperature, gravity, night length, availability of water and nutrients, and the presence of pathogens or herbivores. This developmental flexibility depends on extensive coordination among individual cells. Cells in different tissues and even in different parts of a plant coordinate their activities by communicating with one another. As an example, a leaf being chewed by a caterpillar can signal other parts of the plant to produce appropriate caterpillar-deterring chemicals. Cell-to-cell communication in plants involves **plant hormones**: extracellular signaling molecules that exert effects at very low concentrations.

Hormones operate in both plants and animals, but the origin and effects of these molecules differ between the two groups. For example, animals have organs specialized for hormone secretion; plants do not. All cells in a plant have the ability to make and release hormones. Animal hormones are defined by their ability to elicit an effect in a distant tissue. Similarly, a plant hormone may be released far from the tissue it affects. For example, cells in an actively growing root release a hormone that keeps shoot tips actively growing too. However, plant hormones also act locally, as when a cell in a ripening fruit releases a hormone that affects itself as well as its neighbors. Unlike animals, plants do not have a circulatory system that moves hormones through the body. The movement of hormones through a plant occurs by diffusion (for example through plasmodesmata between adjacent cells), and by fluid transport through xylem or phloem to more distant areas of the plant body. A few plant hormones move from cell to cell via active transport proteins.

All hormones work by binding to receptor proteins on target cells: When a receptor binds to a hormone, it triggers a change in cellular activities (Section 5.7). A receptor can trigger different responses in different cells, so the same hormone can have different effects. Consider how some plant hormones act as hormones in animals, and vice versa. Both plants and mammals make estrogen, progesterone, testosterone, and other similar steroid hormones, for example, but the effects of these molecules differ between the two groups. In mammals, they function as sex hormones; in plants, they are part of stress responses. The difference does not stem from differences in the structure of the hormones, but rather in the way the receptors for these compounds work in cells that bear them.

Plant hormones typically have different effects in different parts of the plant body. For example, one type of hormone alters expression of genes governing anther and pollen development in cells of a developing flower. In leaves, the same hormones are part of signaling pathways that cause cells to die in response to pathogen infection.

As you will see in the following sections, a plant cell's response to a hormone often varies with the hormone's concentration. This response may involve development of a plant part; the direction and rate of growth; defense responses; circadian rhythms; the timing and duration of flowering; fruit and seed formation; aging; and initiating and breaking dormancy, to give some examples. However, one process is rarely controlled by one hormone. Rather, each process is influenced by multiple hormones in a complex network of signaling. A plant cell's activities are driven by an integration of hormonal signals: One hormone can enhance or oppose another's effects on the same cell. Many plant hormones inhibit their own expression, a homeostatic mechanism of negative feedback that maintains ongoing activities and circadian rhythms (Section 10.4). Positive feedback loops in which a plant hormone promotes its own transcription are part of intermittent processes.

TAKE-HOME MESSAGE 30.2

✔ Plant development and tissue functioning depend on cell-to-cell communication. This communication depends on hormones: signaling molecules that coordinate activities among cells in different parts of the plant body.

✔ Hormones are involved in all aspects of growth, development, and function in plants. A plant hormone may affect different cells in different ways, and different hormones may have synergistic or opposing effects on the same cell.

✔ Cells that bear receptors for a hormone—and thus can respond to it—may be in the same tissue as the hormone-releasing cell, or in another region of the plant body.

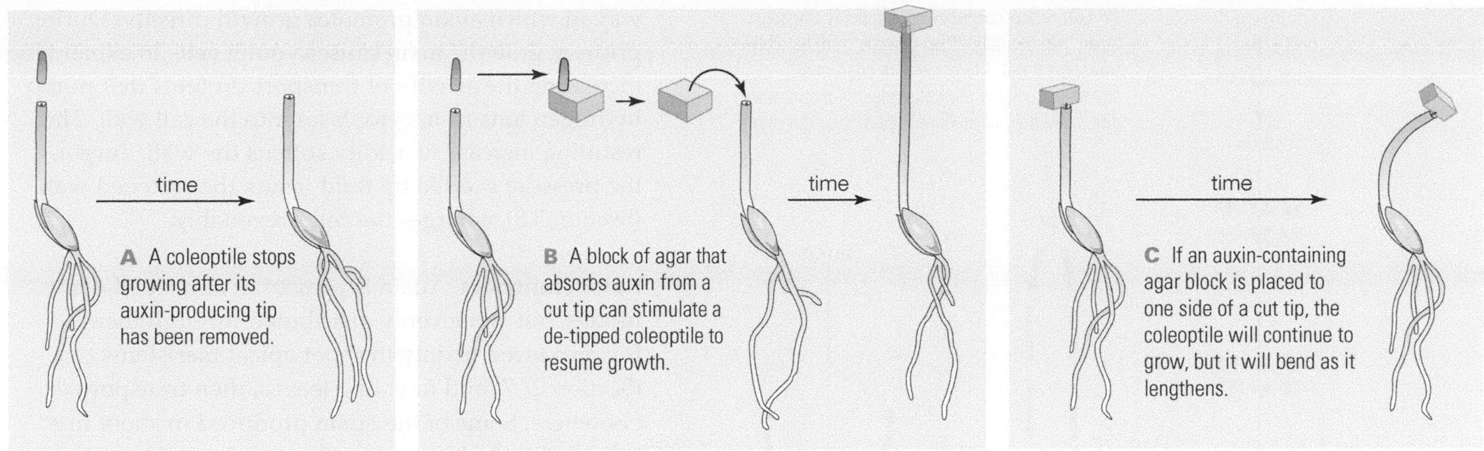
time
A A coleoptile stops growing after its auxin-producing tip has been removed.
B A block of agar that absorbs auxin from a cut tip can stimulate a de-tipped coleoptile to resume growth.
time
C If an auxin-containing agar block is placed to one side of a cut tip, the coleoptile will continue to grow, but it will bend as it lengthens.
time

FIGURE 30.1 Experiments showing that a coleoptile lengthens in response to auxin produced in its tip.

30.3 Auxin and Cytokinin

LEARNING OBJECTIVES

- Explain auxin's effect on growth and how it causes plant cells to enlarge.
- Describe the unique way that auxin can move through a plant.
- Summarize the interaction of auxin and cytokinin in apical dominance.

Auxin

There are a few naturally occurring auxins, but the one that occurs most frequently in plants is IAA (indole-3-acetic acid), a small molecule derived from the amino acid tryptophan. In most cases, the term **auxin** refers to IAA. Auxin was first discovered for its ability to promote growth in plants (**FIGURE 30.1**), and its name is derived from the Greek word for growth. It was later shown to have a critical role in all aspects of plant development (**TABLE 30.1**). Auxin is involved in polarity and tissue patterning in the embryo, formation of plant parts, differentiation of vascular tissues, formation of lateral roots, and, as you will see in later sections, shaping the plant body in response to environmental stimuli.

Auxin exerts many of its effects by influencing the levels of other plant hormones. For example, auxin produced in a shoot's tip supports growth in the shoot's stem and leaves. In the stem, it induces synthesis of a plant hormone (gibberellin) that stimulates growth; in the leaves, it causes breakdown of another plant hormone (cytokinin) that inhibits cell division. Auxin has a different effect in roots, where it interacts with a third hormone (ethylene) to inhibit lengthening and to stimulate the formation of lateral roots and root hairs.

TABLE 30.1

Some Effects of Auxin and Cytokinin

Auxin
Coordinates effects of other plant hormones during development of shoots and roots
Promotes growth by cell enlargement
Stimulates formation of roots on a shoot
Stimulates division of shoot apical meristem cells
Promotes apical dominance
Cytokinin
Stimulates development of lateral buds
Inhibits formation of lateral roots
Stimulates differentiation of cells in root apical meristem

Humans take advantage of auxin's effects on plant growth for many purposes. Application of auxin to a cut stem causes roots to form on it, so this hormone is often sold as a rooting compound. Auxin is also used as an herbicide. Applied to leaves, it causes uncontrolled cell division in the plant's apical meristems. The resulting growth is unsustainable: Stems curl as they become too long to be supported properly, leaves wither as the plant's resources are diverted to inappropriate growth, and the plant dies. Agent Orange, a defoliant used extensively during the Vietnam War, consisted of a combination of two synthetic auxins. One of those auxins, called 2,4-D, is now widely used as an herbicide on lawns and in cornfields, where it kills eudicots but not monocots.

Researchers are still working out the mechanisms of many of auxin's effects, but they do understand one

auxin (OX-in) Plant hormone that causes cell enlargement; also has a central role in growth by coordinating the effects of other hormones. Indole-3-acetic acid (IAA) is the most common auxin.

plant hormone Extracellular signaling molecule of plants; exerts an effect on target cells at very low concentration.

auxin

auxin

A Left, influx carriers (blue) actively transport auxin into the cell; efflux carriers (red) actively transport it out. Efflux carriers positioned asymmetrically in the cells' plasma membranes direct the flow of auxin in a particular direction.

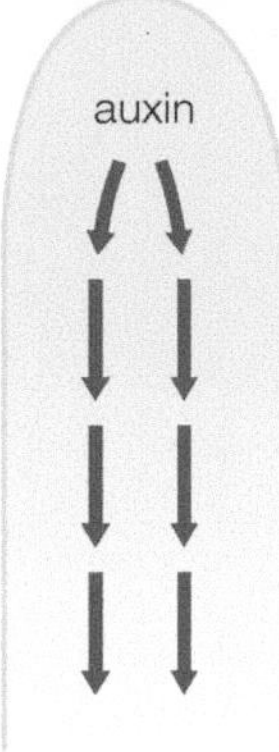

auxin

B Auxin efflux carriers in cells of a lengthening shoot tip direct auxin downward through the stem.

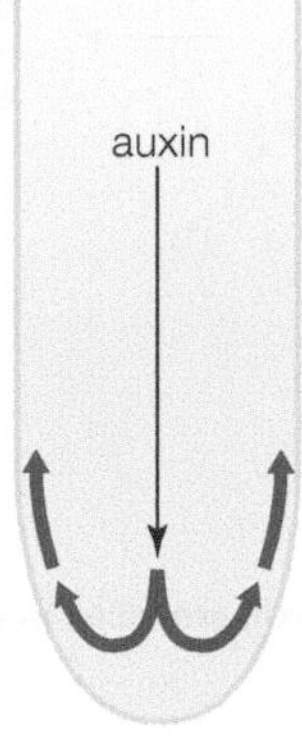

auxin

C Auxin enters a root tip in vascular tissue. Efflux carriers direct its flow through epidermal cells in the root tip.

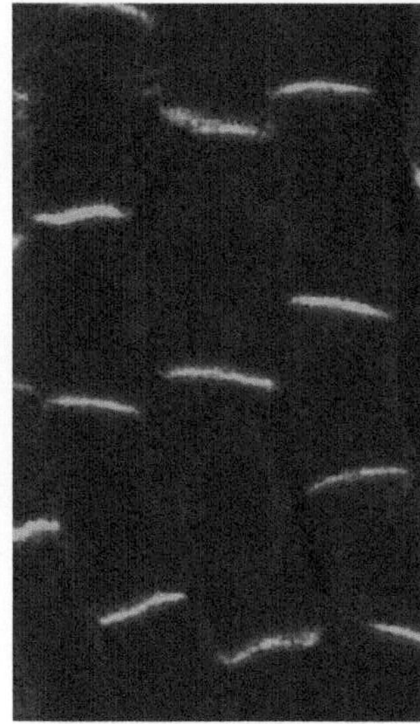

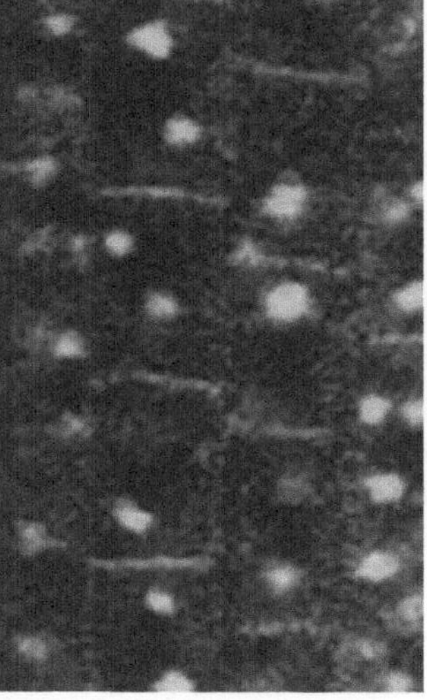

D Green fluorescence marks the location of auxin efflux carriers in these micrographs of *Arabidopsis* roots. Efflux carriers that cluster at one end of a cell membrane (left) direct the flow of auxin through a tissue. The polar positioning of these carriers in plant cell membranes is dynamic: In cells treated with a chemical that interrupts membrane trafficking between ER and Golgi bodies, efflux carriers accumulate quickly in cytoplasmic vesicles (right).

FIGURE 30.2 **Polar transport of auxin.**
The directional transport of auxin is driven by asymmetrical positioning of plasma membrane active transport proteins called efflux carriers.

apical dominance Effect in which a lengthening shoot tip inhibits the growth of lateral buds.
cytokinin (site-oh-KINE-in) Plant hormone that promotes development of lateral buds, and inhibits the formation of lateral roots, among other effects. Often opposes auxin's effects.

way in which auxin promotes growth directly: During primary growth, auxin causes young cells to expand by increasing the activity of transport proteins that pump hydrogen ions from cytoplasm into the cell wall. The resulting increase in acidity softens the wall. Turgor, the pressure exerted by fluid inside the softened wall (Section 5.8), enlarges the cell irreversibly.

Polar Transport Auxin is present in almost all plant tissues, but is unevenly distributed through them. It is produced mainly in shoot apical meristems (Section 27.7) and in young leaves, then transported elsewhere. Some of the auxin produced in shoot tips is loaded into phloem, travels to roots, and then is unloaded into root cells. A unique cellular mechanism also distributes the hormone directionally through adjacent cells. Auxin is pumped into and out of cells by active transport proteins called influx carriers and efflux carriers, respectively. Unlike most other membrane proteins, efflux carriers are not always distributed evenly around a cell's plasma membrane. When efflux carriers in adjacent cells "point" in the same direction, they direct the flow of auxin through a tissue (FIGURE 30.2A). In an actively lengthening shoot, for example, most of the efflux carriers are positioned on the bottom of the cells (toward the base of the stem). Thus, auxin flows from the shoot's apical meristem down into the stem (FIGURE 30.2B). In young root tips, efflux carriers direct the flow of auxin in the opposite direction, from the root's tip toward the root–shoot interface (FIGURE 30.2C).

Polar transport is unique among plant hormones, and it establishes auxin gradients that span tissues making up the plant's parts. Auxin affects expression of other hormones, so these gradients set up localized patterns of hormone production that vary even from one cell layer to the next. The patterns change over time because efflux carriers are continually recycled by membrane trafficking (Section 5.10 and FIGURE 30.2D). This active process allows auxin-directed development to be flexible and responsive.

Cytokinin

The plant hormone **cytokinin** affects division and differentiation of meristem cells. It also stimulates development of lateral buds, and inhibits development of lateral roots. Cytokinin is synthesized locally in stems, but most of it is produced in roots. The hormone is transported from roots to shoots in xylem.

Cytokinin and auxin influence one another's expression, a homeostatic mechanism that dynamically regulates their relative concentrations. They also oppose one another's effects in many processes.

CREDIT: (2D) Geldner N, Friml J, Stierhof YD, Jürgens G, and Palme K., "Auxin transport inhibitors block PIN1 cycling and vesicle trafficking," *Nature*, 2001 Sep 27;413(6854):425–8.

Consider how auxin and cytokinin help balance the growth of apical and lateral buds on a shoot. When a shoot is lengthening, its lateral buds are typically dormant, an effect called **apical dominance** (**FIGURE 30.3**). If the shoot's tip breaks off, its lateral buds begin to grow, an effect exploited by gardeners who remove shoot tips to make a plant bushier. Auxin maintains apical dominance mainly by regulating the production and transport of cytokinin in a stem, and by promoting expression of an enzyme that breaks it down. Auxin, remember, is produced by a shoot's apical meristem and transported from cell to cell down through the stem ➊. The flow of auxin through these cells causes the stem to lengthen, and it also keeps the cytokinin level low. When the shoot's tip is removed, its source of auxin disappears. The auxin level declines in the stem, allowing the cytokinin level to rise ➋. The cytokinin moves into lateral buds and stimulates cell divisions of apical meristem inside them ➌. The newly active meristem cells produce auxin, which is then transported away from the bud tips and down the stem. The auxin flow causes the lateral buds to lengthen ➍.

Cytokinin also opposes auxin's effect on lateral root formation, which requires an auxin gradient. Lateral roots grow from pericycle cells (Section 27.6). Only a few pericycle cells give rise to lateral roots, however. This is because, in roots, the level of cytokinin is normally high compared with auxin. (Most cytokinin is produced in roots; most auxin is produced in shoots.) Cytokinin inhibits production of auxin efflux carriers and their insertion into the membrane. Thus, in most pericycle cells (where the cytokinin to auxin ratio is high), an auxin gradient does not form, and neither does a lateral root. Lateral root formation is initiated by pericycle cells in which the auxin to cytokinin ratio increases. The increase results in efflux carriers being inserted into the cells' plasma membrane. The carriers set up an auxin gradient, and a lateral root forms.

TAKE-HOME MESSAGE 30.3

✔ Auxin directly promotes cell enlargement during primary growth. It also coordinates the effects of other plant hormones involved in growth and development.

✔ A dynamically regulated polar distribution system sets up auxin concentration gradients across a plant's tissues.

✔ Auxin gradients affect growth and development in localized patterns that can change over time.

✔ Cytokinin stimulates development of lateral buds and inhibits formation of lateral roots, among other effects.

✔ Cytokinin and auxin have antagonistic effects. The balance of auxin and cytokinin governs processes such as apical dominance and lateral root formation.

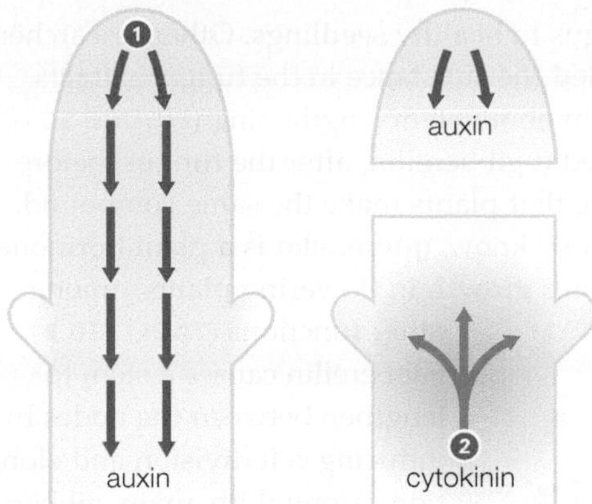
1
auxin
auxin
2
cytokinin

➊ Auxin being transported away from shoot apical meristem supports stem lengthening. Its presence in the stem keeps the level of cytokinin low.

➋ Removing the tip also removes the source of auxin, so auxin flow ends in the stem. As the auxin level declines, the cytokinin level rises.

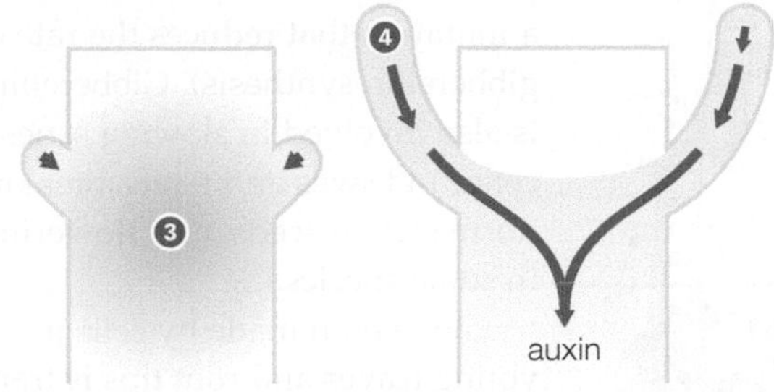
4
3
auxin

➌ The rise in the cytokinin level in the stem stimulates cell division in apical meristem of lateral buds. The cells begin to produce auxin.

➍ Auxin is transported away from the lateral buds and down through the stem. This movement causes the lateral buds to lengthen.

FIGURE 30.3 Interaction of auxin and cytokinin in the release of apical dominance.

The loss of a shoot's tip ends its main supply of auxin—a signal that breaks dormancy in lateral buds.

30.4 Gibberellin

LEARNING OBJECTIVES

- Describe the way gibberellin causes plant growth.
- Explain gibberellin's role in seed germination.

In 1926, researcher E. Kurosawa was studying what Japanese call *bakane*, the "foolish seedling" effect. The stems of rice plants infected with a fungus, *Gibberella fujikuroi*, grew twice the length of uninfected seedlings. Kurosawa discovered that he could make this lengthening happen experimentally by applying extracts of the fungus to healthy seedlings. Other researchers later purified the substance in the fungal extracts that brought about stem lengthening (**FIGURE 30.4**). They named it gibberellin, after the fungus, before discovering that plants make the same compound.

As we now know, **gibberellin** is a plant hormone that promotes growth in flowering plants, among other functions (**TABLE 30.2**). Gibberellin causes a stem to lengthen between the nodes by inducing cell division and elongation. Along with auxin, gibberellin stimulates expansion along the long axis of a plant organ. Thus, it increases the length of the organ, and of the plant itself (the short stature of Mendel's dwarf pea plants (Section 13.3) is the result of a mutation that reduces the rate of gibberellin synthesis). Gibberellin is also involved in slowing senescence in leaves and fruits, breaking dormancy in seeds, and flowering in some species.

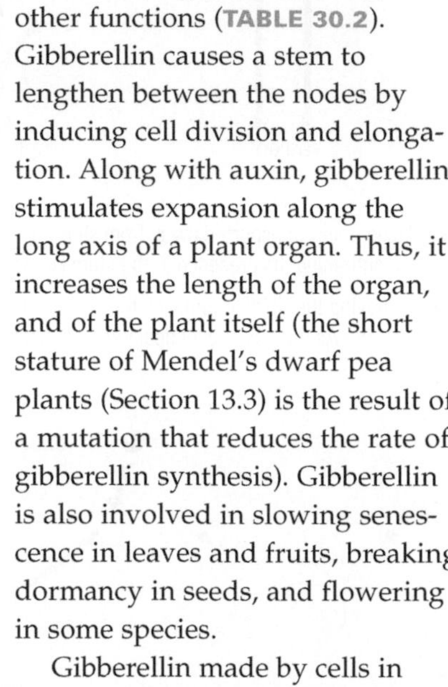

FIGURE 30.4
Stem-lengthening effect of gibberellin. The three tall cabbage plants to the right of the ladder were treated with gibberellin. The two short plants were not treated.

Gibberellin made by cells in young leaves and root tips is transported through phloem to the rest of the plant. Seeds also make this hormone. Seedless grapes tend to be smaller than seeded varieties because they have tiny, underdeveloped seeds that do not produce normal amounts of gibberellin. Farmers spray their seedless grape plants with synthetic gibberellin to increase fruit size.

Gibberellin works by inhibiting inhibitors, thus removing the brakes on some cellular processes. When giberrellin receptors bind to the hormone, they trigger cellular events that destroy a set of repressors in the nucleus (Section 10.2). The genes that these proteins repress are still being studied, but their products have overlapping functions involving cell proliferation and expansion, fertility, and germination.

Consider how gibberellin works during germination of a barley seed. Water absorbed by the seed causes cells of the embryo to release the hormone, which then diffuses into the aleurone, a single layer of living cells that surrounds endosperm (**FIGURE 30.5**). Gibberellin causes these cells to destroy a repressor that inhibits expression of the gene for amylase (the enzyme that breaks the bonds between glucose monomers in starch, Section 3.3). The cells start producing amylase and releasing it into the endosperm's interior, where the enzyme breaks down stored starch molecules. The embryo then uses the released glucose for aerobic respiration, which fuels rapid growth of the embryonic shoot and root.

TABLE 30.2

Some Effects of Gibberellin
Stimulates cell division, elongation in stems
Mobilizes food reserves in germinating seeds
Stimulates flowering in some plants
Delays senescence

A Absorbed water causes cells of a barley embryo to release gibberellin, which diffuses through the seed into the aleurone layer of the endosperm.

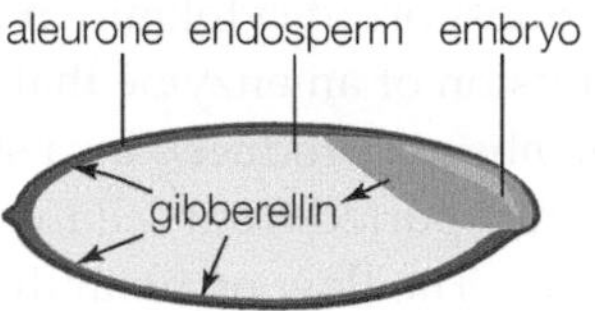
aleurone
endosperm
embryo
gibberellin

B Gibberellin causes cells of the aleurone layer to produce amylase. This enzyme diffuses into the starch-packed endosperm.

C Amylase that enters endosperm breaks down starch into its glucose subunits. The glucose diffuses into the embryo and is used for aerobic respiration. Energy released by the reactions of aerobic respiration fuels meristem cell divisions in the embryo.

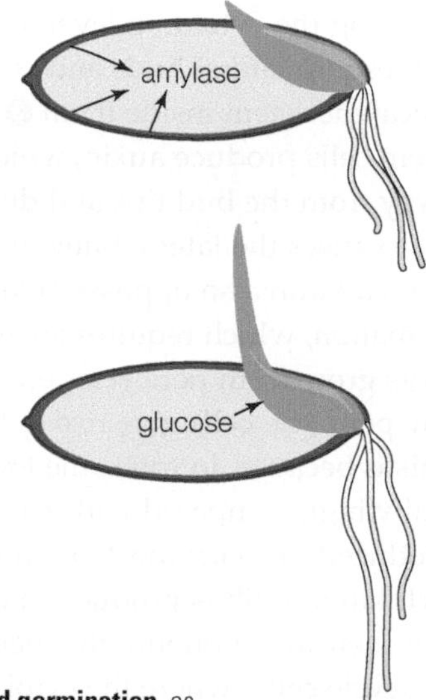
amylase
glucose

FIGURE 30.5 Gibberellin and germination, as illustrated in a seed of barley (a monocot).

TAKE-HOME MESSAGE 30.4

✔ Gibberellin causes lengthening in plants by stimulating cell division and elongation along the long axis of stems and other organs.

✔ Gibberellin is part of a mechanism that mobilizes nutrients in a seed during germination.

abscisic acid (ABA) (ab-SIH-sick) Plant hormone that has a major role in stress responses; also inhibits germination, among other effects.
abscission (ab-SIH-zhun) Process by which plant parts are shed.
gibberellin (jib-er-ELL-in) Plant hormone that induces stem elongation; also helps seeds break dormancy, among other effects.

30.5 Abscisic Acid and Ethylene

LEARNING OBJECTIVES

- Describe in general terms the mechanism by which abscisic acid helps minimize water loss.
- Explain feedback loops involved in ethylene synthesis.
- Describe the role of ethylene in fruit ripening.

Abscisic Acid

The plant hormone **abscisic acid (ABA)** was named because its discoverers thought it mainly mediated **abscission**, the process by which a plant sheds leaves or other parts. It was later discovered to have a much greater role in plant stress responses (**TABLE 30.3**). Temperature extremes, water shortages, and other environmental stresses trigger an increase in ABA synthesis, release, and transport, so the level of this hormone rises in the plant's tissues. In turn, the increased level of ABA induces expression of genes that help the plant survive the adverse conditions.

When ABA binds to its receptors, it triggers a cascade of reactions that activate hundreds of transcription factors. In turn, these transcription factors alter the expression of thousands of genes—around 10 percent of the total number in plants and a much larger proportion than any other plant hormone. In general, ABA inhibits the expression of genes involved in growth, and enhances expression of genes involved in metabolism, stress responses, and embryonic development.

Consider how ABA encourages cells to produce NADPH oxidase. This plasma membrane enzyme transfers electrons from NADPH inside the cell to oxygen molecules outside the cell—a redox reaction (Section 5.5) that produces free radicals and hydrogen peroxide. These reactive substances very quickly activate another plasma membrane enzyme, which, in turn, produces a burst of nitric oxide (NO). Nitric oxide is a gas that functions as a hormone in plants (and also in animals), and its release in a burst is a critical part of plant stress responses.

TABLE 30.3

Some Effects of Abscisic Acid (ABA) and Ethylene

Abscisic acid (ABA)	Ethylene
Mediates stress responses	Stimulates fruit ripening
Closes stomata	Stimulates abscission
Inhibits shoot growth	Involved in breaking dormancy
Inhibits seed germination	Involved in stress responses

You learned in Section 28.4 about a mechanism that opens stomata at sunrise. An ABA-mediated NO burst is part of a related mechanism that closes stomata when a plant is stressed by drought (a lack of water). When soil dries out, root cells release ABA. The hormone travels through the plant's vascular system to leaves and stems, where it binds to receptors on guard cells (**FIGURE 30.6**). ABA binding triggers these cells to release NO. The NO burst activates transport proteins that allow positively charged calcium ions (Ca^{2+}) to flow into guard cell cytoplasm ❶. The calcium ion concentration in cytoplasm rises, and this has two effects. First, it activates another set of transport proteins that pump chloride (Cl^-) and other negatively charged ions out of the cells ❷. Second, the inflow of positively charged calcium ions (and the exit of negatively charged ions) increases the overall charge of cytoplasm, and decreases the overall charge of extracellular fluid. The resulting voltage change across guard cell membranes causes gated transport proteins in them to open. Potassium ions follow the charge gradient and flow out of the cell through the opened gates ❸. Water follows the ions by osmosis, and stomata close as the guard cells lose turgor and collapse against one another ❹. Note that the release of nitric oxide also triggers production of an enzyme that breaks down ABA—a feedback loop that dynamically controls ABA signaling.

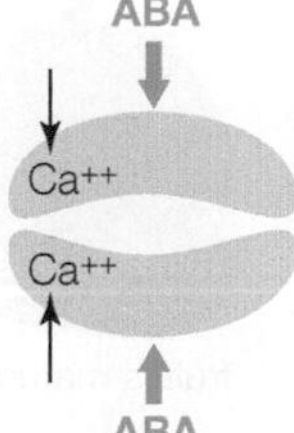
ABA
Ca++
Ca++
ABA

❶ ABA binds to its receptor on guard cell membranes. The binding triggers an NO burst that activates calcium transport proteins, so these ions enter cytoplasm.

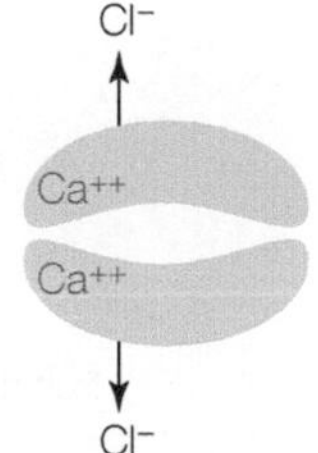
Cl−
Ca++
Ca++
Cl−

❷ The influx of calcium ions activates transport proteins that pump negatively charged ions such as chlorine (Cl^-) out of the cells.

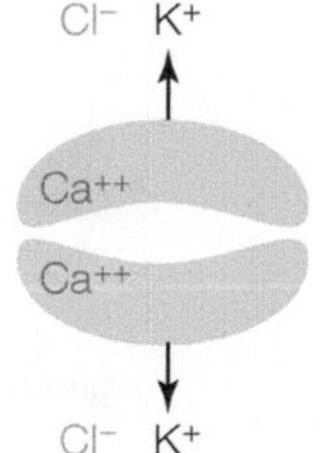
Cl− K+
Ca++
Ca++
Cl− K+

❸ The shift in charge across the guard cell plasma membranes opens gated transport proteins that allow potassium ions to exit the cells.

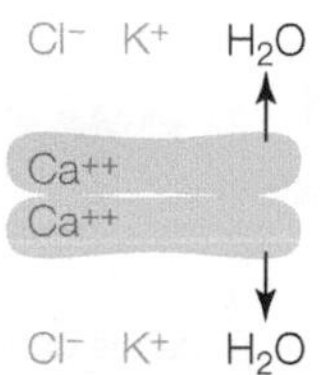
Cl− K+ H2O
Ca++
Ca++
Cl− K+ H2O

❹ Water follows the solutes by osmosis. The guard cells lose turgor and collapse against one another, so the stoma closes.

FIGURE 30.6 ABA causes stomata to close by triggering a series of events that decreases turgor in guard cells.

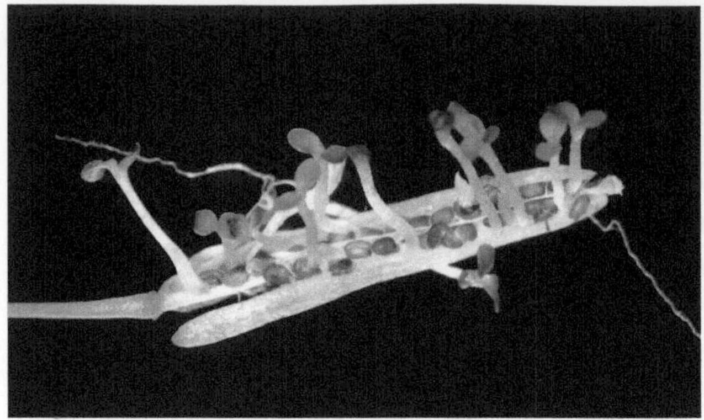

FIGURE 30.7 ABA prevents seed germination. Part of the ABA signaling pathway in this *Arabidopsis* plant has been knocked out. Its seeds germinated before they had a chance to disperse.

In addition to mediating stress responses, ABA has an important role in embryo maturation, fruit ripening, and germination. ABA accumulates in a seed as it forms and matures. The hormone prevents premature germination (**FIGURE 30.7**), in part by inhibiting expression of genes involved in cell wall loosening and expansion—both critical processes for growth of an embryonic plant. ABA also inhibits expression of genes involved in gibberellin synthesis. Thus, a seed cannot germinate until its ABA level declines. When the ABA level falls in a seed, cells of the embryo start expressing genes that result in cell wall softening and enlargement. Transcriptional control over gibberellin synthesis genes is also lifted. Hydrogen peroxide (produced by the activity of NADPH oxidase) enhances expression of these genes, so germination begins as enlarging cells of the embryo start using carbohydrates in endosperm.

Ethylene

The plant hormone **ethylene** is a gas, and it is part of regulatory pathways that govern a wide range of metabolic and developmental processes. Cells in all parts of a plant can produce it.

Two enzymes are involved in ethylene synthesis, and each occurs in multiple versions encoded by slightly different genes. Expression of some of these genes is inhibited by ethylene (a negative feedback loop); expression of the others is enhanced by ethylene (a positive feedback loop). Negative feedback loops maintain a low level of ethylene that helps fine-tune ongoing metabolic and developmental processes such as growth and cell expansion. For example, in roots, ethylene produced in a negative feedback loop enhances the transcription of genes involved in producing auxin and its efflux carriers (remember, auxin and ethylene inhibit lengthening in roots). Positive feedback loops produce surges of ethylene required for intermittent processes such as germination, defense responses, and abscission.

One ethylene feedback loop involves fruit ripening. ABA accumulates in a fruit as it forms and enlarges, eventually triggering production of ethylene in a positive feedback loop. The resulting ethylene surge enhances transcription of genes whose products have several effects associated with ripening (**FIGURE 30.8**). The fruit changes color as chloroplasts are converted to chromoplasts (Section 4.7); inside the organelles, red, orange, yellow, or purple accessory pigments are produced as green chlorophylls break down. Firm cell walls and middle lamellae (Section 4.9) break down. Starch and organic acids are converted to sugars, and aromatic molecules are produced. The resulting color change, softening, and increased palatability attract animals that can disperse the fruit's seeds.

Because ethylene is a gas, it can diffuse from one fruit to initiate ripening in another. Humans use this property to artificially ripen several types of fruit.

FIGURE 30.8 Ethylene production during strawberry formation and ripening. Oscillations are normal circadian cycles.

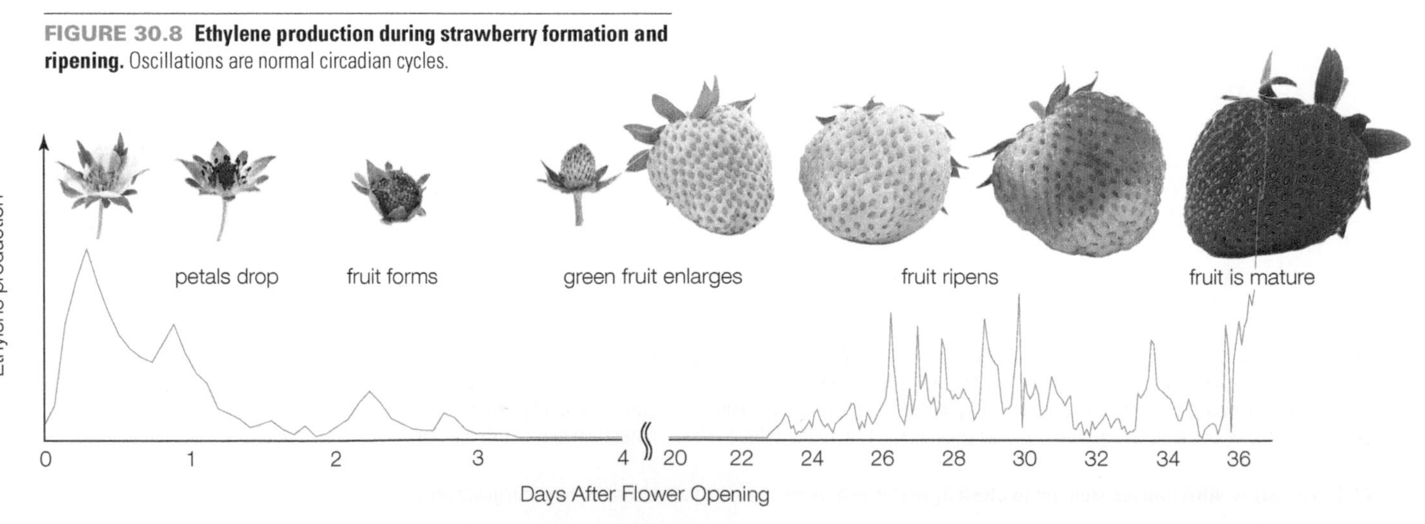
Ethylene production
petals drop
fruit forms
green fruit enlarges
fruit ripens
fruit is mature
0
1
2
3
4
20
22
24
26
28
30
32
34
36
Days After Flower Opening

CREDITS: (7) Image courtesy Kazuo NAKASHIMA, Ph.D. Taishi Umezawa, Kazuo Nakashima, Takuya Miyakawa, Takashi Kuromori, Masaru Tanokura, Kazuo Shinozaki, and Kazuko Yamaguchi-Shinozaki. Molecular Basis of the Core Regulatory Network in ABA Responses: Sensing, Signaling and Transport. *Plant Cell Physiol* (2010) 51(11): 18211839 first published online October 26, 2010. doi:10.1093/pcp/pcq156; (8) from left, (1) Madlen/Shutterstock; (2) Westend61/Westend61/Superstock; (3) Alena Brozova/Shutterstock; (4) Anest/Shutterstock; (5-8) Alena Brozova/Shutterstock.

Hard, unripe fruit can be transported long distances with less damage than soft, ripe fruit. Upon arrival at its final destination, the unripe fruit is exposed to synthetic ethylene, which jump-starts the positive feedback ripening loop. If you purchase unripe fruit, you can encourage ripening by storing it in a closed container with ripe, ethyelene-emitting apples or bananas.

TAKE-HOME MESSAGE 30.5

✔ Abscisic acid plays a major role in stress responses, including stomata closure.

✔ ABA also inhibits germination and growth, and has a role in fruit ripening and embryonic development.

✔ Ethylene produced in negative feedback loops participates in ongoing metabolic and developmental processes.

✔ Ethylene produced in positive feedback loops is involved in intermittent processes such as fruit ripening.

30.6 Movement

LEARNING OBJECTIVE

- Use examples to explain how plant parts move in response to environmental stimuli.

Plants cannot move themselves from place to place, but movement of plant parts is common. Reversible changes in turgor can move a structure, as can a change in a growth pattern (such as when cells on one side of a stem start to elongate more than cells on the other side). Plant movement occurs in response to environmental stimuli such as light or contact, but only some of that movement occurs in a direction that depends on the stimulus (as when a stem bends toward a light source). A directional movement toward or away from a stimulus that triggers it is a **tropism**.

Environmental Triggers

Following are examples of environmental cues that can provoke movements in plants.

Gravity Even if a seedling is turned upside down just after germination, its primary root and shoot will curve so the root grows down and the shoot grows up (FIGURE 30.9). This and any other growth response to gravity is called **gravitropism**. A root or shoot bends because of shifts in the direction of local auxin transport. In shoots, auxin enhances cell elongation. When auxin flow is directed toward one side of a shoot, the shoot bends away from that side. Auxin has the opposite effect in roots: When it is directed toward one side of a growing root, the root bends toward that side.

In plants and many other organisms, the ability to sense gravity is based on organelles called **statoliths**. Plant statoliths are amyloplasts (plastids filled with dense grains of starch, Section 4.7). Amyloplasts occur in root cap cells, and also in endodermal cells that border vascular tissues in shoots. Starch grains are heavier than cytoplasm, so statoliths tend to sink to the lowest region of the cell, wherever that is (FIGURE 30.10). A shift in statolith position causes auxin efflux carriers to be redistributed to the down-facing side of the root or

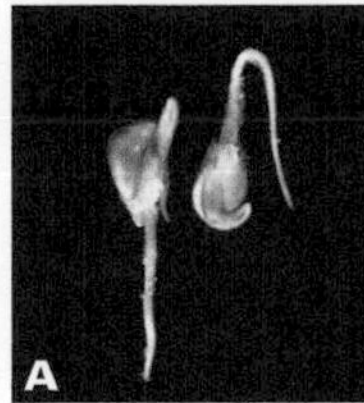
A

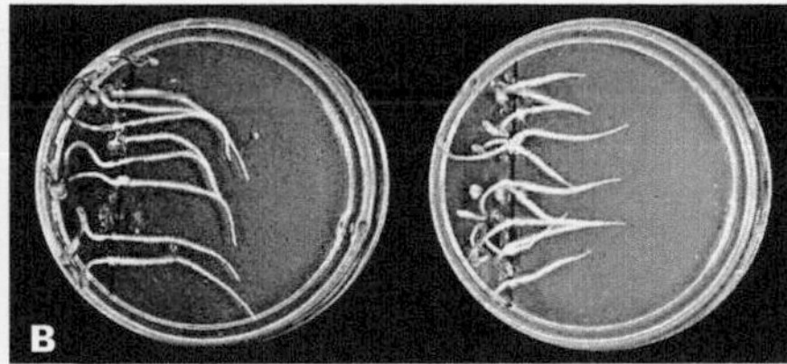
B

FIGURE 30.9 Gravitropism.

A Regardless of how a corn seed is oriented in soil, the seedling's primary root always grows down, and its primary shoot always grows up.

B Left, these seedlings were rotated 90° counterclockwise after they germinated. The plants adjusted to the change by redistributing auxin, and the direction of growth shifted as a result. Right, auxin transport was inhibited in these seedlings. They were also rotated 90° after germination, but the direction of growth did not shift. Mutations in genes that encode auxin efflux carriers have the same effect.

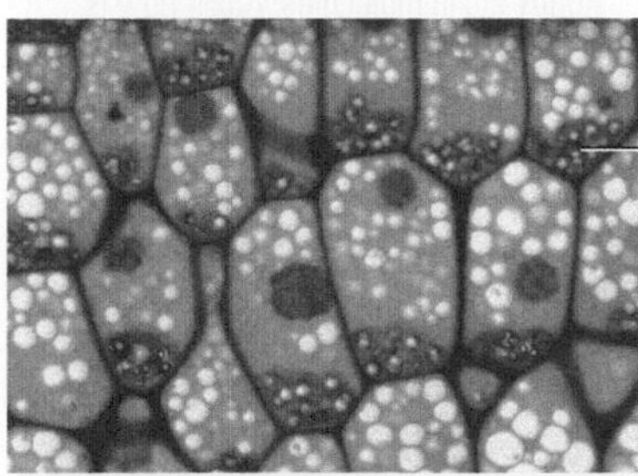
statoliths

A This micrograph shows heavy, starch-packed statoliths settled on the bottom of gravity-sensing cells in a corn root cap.

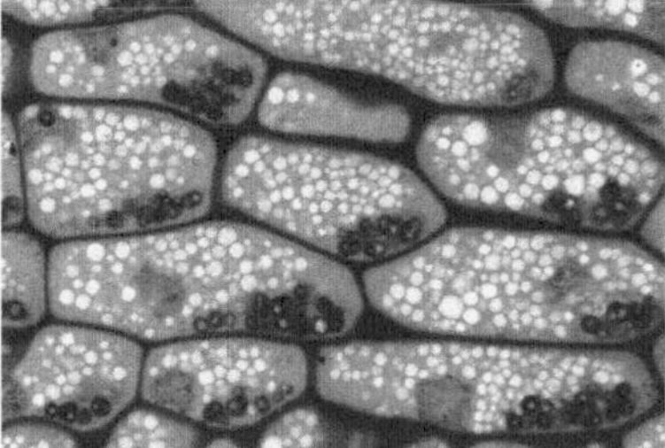

B This micrograph was taken ten minutes after the root in **A** was rotated 90°. The statoliths are already settling to the new "bottom" of the cells.

FIGURE 30.10 Statoliths.

FIGURE IT OUT In which direction was this root rotated?

Answer: Counterclockwise

ethylene (ETH-ill-een) Gaseous plant hormone involved in regulating growth and cell expansion. Participates in germination, abscission, ripening, and stress responses.
gravitropism (grah-vih-TRO-pizm) Directional response to gravity.
statolith Amyloplast involved in sensing gravity.
tropism (TROPE-izm) A directional movement toward or away from a stimulus that triggers it.

CREDITS: (9A) Michael Clayton, University of Wisconsin; (9B) Muday, GK and P. Haworth (1994) Tomato root growth, gravitropism, and lateral development: Correlations with auxin transport. *Plant Physiology and Biochemistry, 32,* 193–203, with permission from Elsevier Science; (10A, B) Micrographs courtesy of Randy Moore from "How Roots Respond to Gravity," M. L. Evans, R. Moore, and K. Hasenstein, *Scientific American*, December 1986.

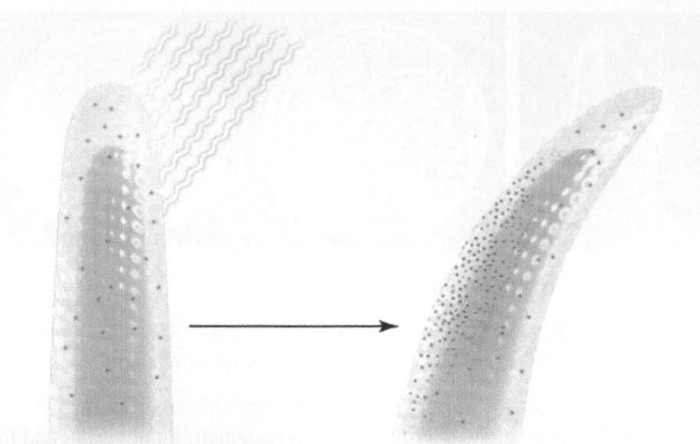

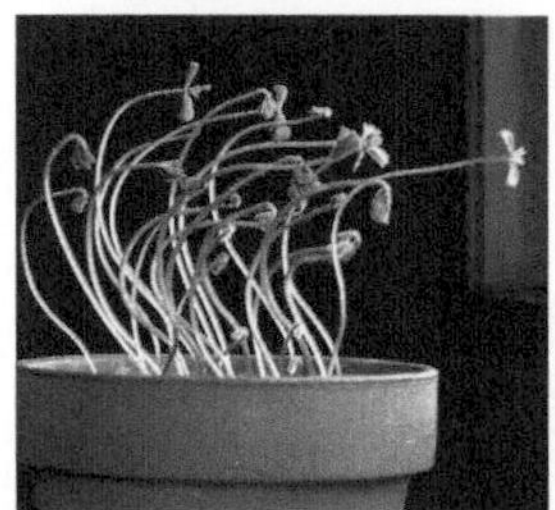

FIGURE 30.11 Phototropism. Above, auxin-mediated differences in cell elongation between two sides of a shoot induce bending toward light. Red dots signify auxin molecules. Auxin flow is directed toward a shaded side, so cells on that side lengthen more. The photo on the left shows what happens when a plant is exposed to a directional light source.

stem. As the direction of auxin flow through the cells changes, the hormone is directed toward the structure's down-facing side. The adjustment in auxin flow results in a directional response.

Light A **phototropism** is a directional response to light. Many plants can change the orientation of leaves or other parts in a direction that depends on a light source. For example, an elongating stem curves toward a directional light source, thus optimizing the amount of light captured by leaves for photosynthesis in low-light environments. In shoots, phototropism is mediated by nonphotosynthetic pigments called phototropins, which absorb blue light (these pigments are also part of a mechanism by which stomata open at sunrise). Light-energized phototropins cause auxin to be directed away from the illuminated side of a stem. Cells on the shaded side elongate more than cells on the illuminated side, and the difference causes the entire structure to bend toward the light as it lengthens (**FIGURE 30.11**).

In another example of phototropism, chloroplasts are dragged from one position to another along actin filaments of the cytoskeleton. In low-intensity light, the chloroplasts are directed toward leaf surfaces and oriented perpendicular to the light source, thus maximizing their exposure to light for photosynthesis. In high-intensity light, chloroplasts move in the opposite direction—toward internal tissues of the leaf where they are more shielded, and oriented parallel to the light. Moving away from high-intensity light minimizes damage that can occur as a result of excess electrons accumulating in electron transfer chains of

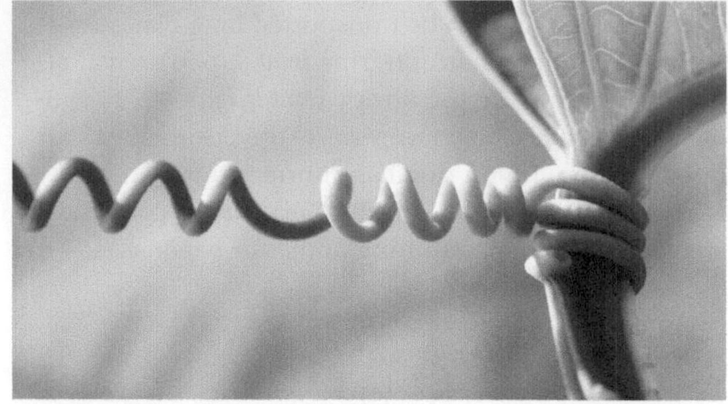

FIGURE 30.12 Thigmotropism. Mechanical stimulation causes a vine's tendrils to grow in a coil around an object.

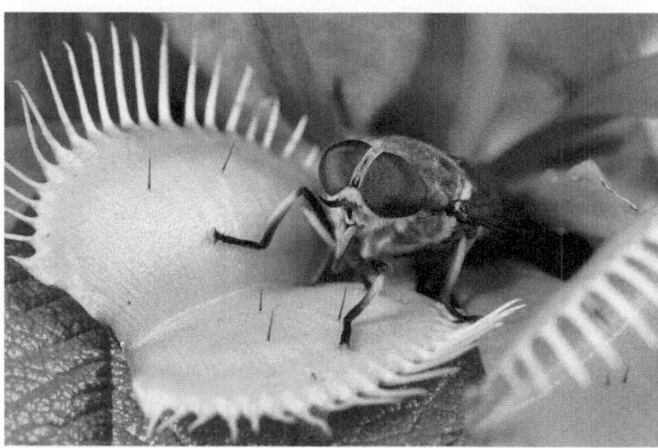

A The "trap" of a Venus flytrap plant is a highly modified leaf that is creased in the middle. Three sensitive hairs protrude from the upper surface of each half of the leaf.

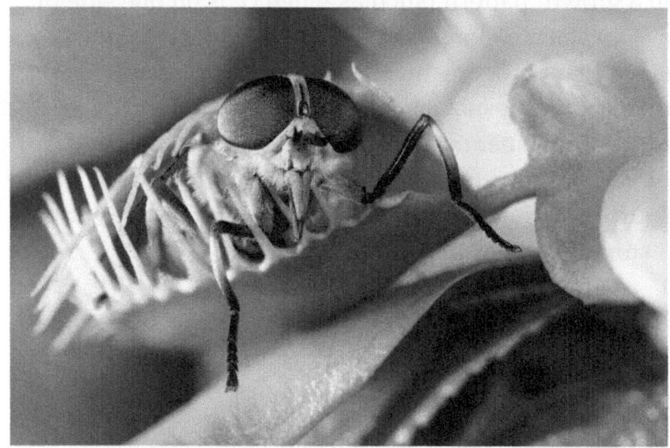

B An insect that brushes a hair two times in succession triggers a mechanism that causes the leaf to snap shut very quickly—even faster than a fly. Secreted enzymes slowly digest trapped prey, releasing nitrogen that the plant requires for growth. The trap will not spring shut unless a hair is brushed twice in quick succession, so raindrops or debris do not trigger it.

FIGURE 30.13 Movement of the Venus flytrap.

thigmotropism (thig-MA-truh-pizm) In plants, growth in a direction influenced by contact.
phototropism (foe-toe-TRO-pizm) Directional response to light.
phytochrome (FIGHT-uh-krome) A light-sensitive pigment that helps set plant circadian rhythms based on length of night.

CREDITS: (11) Cathlyn Melloan/Getty Images; (12) ZenShui/Michele Constantini/Getty Images; (13A, B) Linas T/Shutterstock.

photosynthesis light reactions (Section 6.5). Blue light drives the movement in both cases.

Leaves or flowers of some plants change position in response to the changing angle of the sun throughout the day, a response called heliotropism (from *helios*, the Greek word for sun). The mechanism that drives heliotropism is not understood, but it may be similar to phototropism because it has been observed to coincide with differential elongation of cells in stems, and it also occurs in response to blue light.

Contact A directional response to contact with an object is called **thigmotropism** (*thigma* means touch). We see the effects of thigmotropism in plants when a vine's tendrils coil around an object (FIGURE 30.12). The mechanism that gives rise to the response is not well understood, but it probably involves the immediate increase in cytoplasmic calcium concentration that typically accompanies contact. Plants have plasma membrane transport proteins that flood cytoplasm with calcium ions upon mechanical disturbance of the membrane. Several gene products that can sense calcium ions are involved in the response to the flood of ions, as well as nitric oxide and another plant hormone called jasmonic acid. These molecules trigger unequal growth rates of cells on opposite sides of the shoot tip, and this causes a growing shoot to coil around an object as it lengthens. A similar mechanism causes roots to grow away from contact, so they "feel" their way around rocks and other impassable objects in soil.

The fastest movements of plant parts occur in response to contact. For example, modified leaves of Venus flytraps and some other carnivorous plants spring shut in a fraction of a second upon being triggered by insect prey (FIGURE 30.13). The mechanism involves electrical signaling and the redistribution of auxin, but it is not well understood.

TAKE-HOME MESSAGE 30.6

✔ Plants adjust the direction and rate of growth in response to environmental stimuli such as gravity, light, and contact. Hormones mediate many of these responses.

30.7 Responses to Recurring Environmental Change

LEARNING OBJECTIVES

- Give an example of a circadian rhythm in plants.
- Explain how plants sense seasonal change, and give some examples of their responses.

Daily Change

Plants respond to cycles of environmental change with cycles of biological activity such as circadian rhythms (Section 10.4). Consider how a bean plant holds its leaves horizontally during the day but folds them close to its stem at night. Bean plants exposed to constant light or darkness for a few days will continue to move their leaves in and out of the "sleep" position at the time of sunrise and sunset (FIGURE 30.14). Similar mechanisms cause flowers of some plants to open only at certain times of day. For example, the flowers of plants pollinated mainly by night-flying bats open, secrete nectar, and release fragrance only at night. Periodically closing flowers protects delicate reproductive parts when the likelihood of pollination is lowest.

Phytochromes, cryptochromes, and other nonphotosynthetic pigments provide light input into internal circadian "clocks." **Phytochromes** absorb red light (660 nanometers) and far-red light (730 nanometers). Absorbing red light causes a phytochrome's structure to change from an inactive to an active form. Absorbing far-red light, which predominates in shade, changes the structure back to the inactive form. Blue light activates cryptochromes, which occur widely in organisms of all kingdoms. Plants use cryptochromes to control a variety of activities, including circadian ones. Human circadian clocks also use them. Some cryptochromes associate with other molecules and become magnetic when activated; they are part of a light-dependent magnetic compass in birds (Section 33.2 returns to this topic).

Seasonal Change

Except at the equator, the length of day varies with the season. Days are longer in summer than in winter, and the difference increases with latitude. These seasonal

1 A.M.

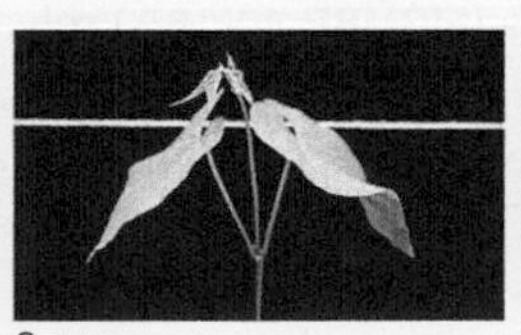
6 A.M.

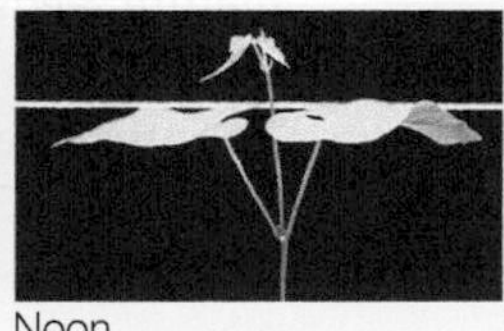
Noon

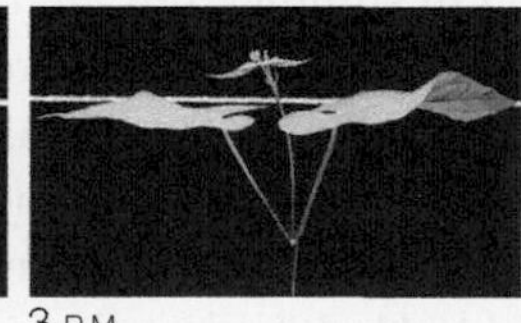
3 P.M.

10 P.M.

Midnight

FIGURE 30.14 **A circadian rhythm: leaf movements by a young bean plant.**

Physiologist Frank Salisbury kept this plant in darkness for twenty-four hours. Despite the lack of light cues, the leaves kept on folding and unfolding at sunrise (6 A.M.) and sunset (6 P.M.).

CREDIT: (14) Frank B. Salisbury.

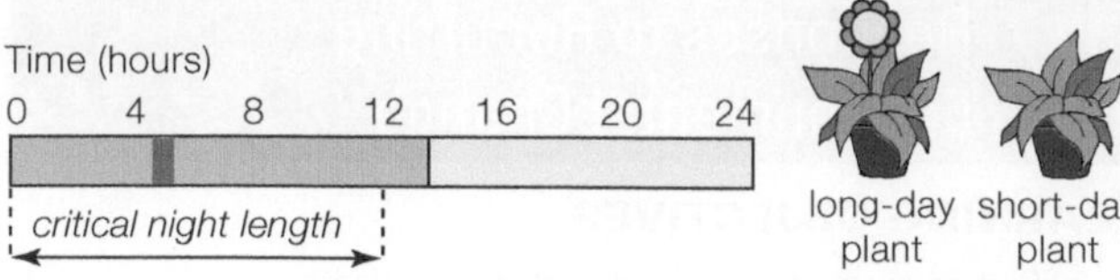

Time (hours)
0 4 8 12 16 20 24
critical night length
long-day plant
short-day plant

A A flash of red light at 660 nm interrupting a long night causes plants to respond as if the night were short. Long-day plants flower; short-day plants do not.

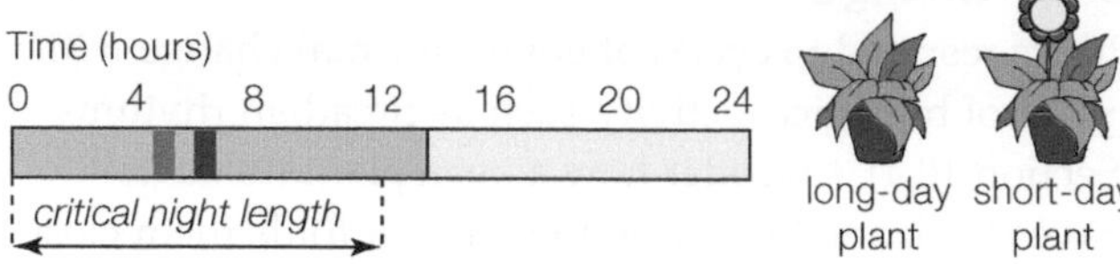

Time (hours)
0 4 8 12 16 20 24
critical night length
long-day plant
short-day plant

B A flash of far-red light at 730 nm cancels the effect of a red light flash at night. Short-day plants flower; long-day plants do not.

FIGURE 30.15 Experiments showing that long- or short-day plants flower in response to night length. Each horizontal bar represents 24 hours. Blue indicates night length; yellow, day length.

FIGURE IT OUT Which type of pigment detected the light flashes in these two experiments?

Answer: A phytochrome

changes in light availability trigger seasonally appropriate responses in plants. **Photoperiodism** refers to an organism's response to changes in the length of day relative to night.

Flowering is a photoperiodic response in many species. Such species are termed long-day or short-day plants depending on the season in which they flower. These names are somewhat misleading, as the main trigger for flowering is the length of night, not the length of day (**FIGURE 30.15**).

Photoperiodic flowering involves phytochromes and cryptochromes, as well as circadian cycles of regulatory gene expression. Inputs from both converge on a gene called *CO*, which encodes a transcription factor. All plants have this gene. In long-day plants, the transcription factor induces expression of the *FT* gene, which in turn activates expression of floral identity genes (Section 10.3). Irises and other long-day plants flower only when the hours of darkness fall below a critical value. In these plants, the expression of the *CO* gene and the activity of its product peak 8 to 10 hours after dawn—which is late afternoon in summer, but after dusk in other seasons. The transcription factor is broken down at night, so it does not accumulate to a high enough level to promote flowering except in summer.

Cabbage, spinach, and other edible long-day plants grown for their leaves are said to "bolt" when they switch from producing leaves (desirable) to producing flowers (undesirable). Farmers minimize bolting in these plants by growing them in short-day seasons.

Chrysanthemums and other short-day plants flower only when the hours of darkness are greater than a critical value. In these plants, the same *CO* gene product inhibits *FT* gene expression, and flowering is stimulated in a separate pathway involving gibberellin.

Some plants native to temperate regions flower only after exposure to a long period of cold in the preceding winter, a response called **vernalization**. Researchers suspect that plant cells perceive temperature via their plasma membrane, because the lipid composition and calcium ion permeability of plasma membranes vary with temperature. Expression of a gene called *VRN* increases with the duration of exposure to cold temperatures. The gene's transcription factor product promotes rapid flowering when days start lengthening in spring. In other plants, exposure to cold does not affect *VRN* expression.

Yearly cycles of abscission and dormancy are photoperiodic responses too. Plants that drop their leaves before dormancy are typically native to regions too dry or too cold for optimal growth during part of the year. For example, deciduous trees of the northeastern United States lose their leaves around September. The trees remain dormant during the months of harsh winter weather that would otherwise damage leaves and buds. Growth resumes when milder conditions return, around March. On the opposite side of the world, many tree species native to tropical monsoon forests of south Asia lose their leaves during the summer dry season, which is between November and May. Although the region receives a lot of rain annually, almost none of it falls during this period of the year. Dormancy offers the trees a way to survive the extended droughtlike conditions. New growth appears at the beginning of June, just in time to be supported by the ample water of monsoon rains.

Seasonal abscission of leaves and fruits is mediated by ethylene. Let's use a deciduous tree called a horse chestnut as an example. In this species, most root and shoot growth occurs between March and June. By July, the tree is producing fruits and seeds. In August, the growing season is coming to a close, and nutrients are being routed to stems and roots for storage during the forthcoming period of dormancy. Ripe fruits and mature seeds (**FIGURE 30.16A**) release ethylene that diffuses into nearby twigs, petioles, and fruit stalks. The ethylene triggers formation of a dense layer of small cells. Cells above this layer begin to produce

photoperiodism (foe-toe-PEER-ee-ud-is-um) Biological response to seasonal changes in the relative lengths of day and night.
vernalization (vurr-null-iz-A-shun) Stimulation of flowering in spring by long exposure to low temperature in winter.

enzymes that digest their own walls. The cells bulge as their walls soften, and separate from one another as the extracellular matrix that cements them together dissolves. Tissue in this abscission zone weakens, and the structure above it drops. A scar often remains where the structure had been attached (**FIGURE 30.16B**).

TAKE-HOME MESSAGE 30.7

✔ Recurring environmental cues trigger daily and seasonal cycles of activity in plants.

✔ In many species, flowering is a photoperiodic response to the length of night relative to the length of day, which varies seasonally in most places.

✔ Seasonal abscission and dormancy are evolutionary adaptations to climates with seasonal periods of harsh environmental conditions.

30.8 Responses to Stress

LEARNING OBJECTIVES

- Explain what happens when a pathogen enters a plant's tissues.
- Distinguish between a hypersensitive response and systemic acquired resistance in plants.
- Describe one way that plants defend themselves from herbivory.

A plant cannot run away from stressful situations: It either adjusts to adverse conditions or dies. Plant stressors can be abiotic (caused by nonliving environmental conditions such as temperature extremes or lack of water) or biotic (imposed by pathogens and herbivores). Hormones mediate plant defenses against both types of stressors.

Defenses Against Disease

You learned in Section 30.5 how ABA and nitric oxide are part of a response that causes a plant's stomata to close during a water shortage, thus minimizing water loss. ABA and nitric oxide also participate in plant responses to pathogens. Consider that receptors on plant cells recognize molecules specific to microbial agents of disease. Some of these receptors recognize flagellin, a protein component of bacterial flagella (Section 4.3). When flagellin binds to these receptors, it triggers a burst of ethylene synthesis. When ethylene is *not* present, ethylene receptors block transcription of a set of genes involved in defense. When ethylene *is* present, it binds to its own receptors and locks them in an inactive form that marks them for destruction. Thus, the ethylene burst lifts the brakes on production of molecules used in defense. The cell starts making more flagellin receptors—a positive feedback loop that makes the plant more sensitive to the presence of bacteria. If any additional flagellin binds to the

A Ripening fruits and maturing seeds of the horse chestnut emit ethylene, which causes abscission of plant parts (note the browning fruit stalks).

B Horse chestnut trees take their name from abscission zones that leave horseshoe-shaped scars in this species.

FIGURE 30.16 Abscission as part of the normal life cycle of deciduous plants such as the horse chestnut.

accumulated receptors, an ABA-mediated nitric oxide burst immediately closes stomata. Bacteria cannot penetrate plant epidermis; they can enter plant tissues only through wounds or open stomata. Thus, closing stomata defends the plant from bacterial invasion.

Mutualistic bacteria and fungi avoid triggering a plant's defense responses by exchanging chemical signals with it. Remember that *Rhizobium* bacteria in soil are attracted to compounds released from root cells (Section 28.3). Those compounds are flavonoids, a type of secondary metabolite you learned about in Section 30.1. *Rhizobium* bacteria respond to the flavonoids by producing molecules required for living in symbiosis with the plant. One of these molecules is secreted into the soil, and it is recognized by root cells as a signal to form a nodule.

When a pathogen does enter plant tissues, it triggers a large surge of hydrogen peroxide and nitric oxide that causes cells in the infected region to commit

FIGURE 30.17 Plant hypersensitive response to a pathogen. Brown spots are dead tissue where germinating fungi penetrated the leaf's epidermis, triggering a release of hydrogen peroxide and nitric oxide that resulted in plant cell suicide. This hypersensitive response can prevent the spread of a pathogen into healthy tissue.

suicide. The details of this "hypersensitive" response are unknown, but its effects are clear: It can prevent a pathogen from spreading to other parts of the plant because it often kills the pathogen along with the infected tissue (**FIGURE 30.17**). Grapevines native to America are more resistant than European cultivated grapevines to *Phylloxera* (Section 29.8) because their hypersensitive response to this insect is much stronger. American grapevines make more resveratrol, a phytoestrogen hormone that mediates the hypersensitive response in these plants.

A hypersensitive response is often ineffective against pathogens that gain nutrients by killing cells outright. Some sac fungi, for example, use toxins to kill their plant hosts, then absorb nutrients released from decomposing tissues. Localized exposure to one of these pathogens induces a type of immunity called systemic acquired resistance. **Systemic acquired resistance** is an enhanced resistance of the entire plant body to a wide range of fungi and other pathogens. A plant that has activated its systemic acquired resistance responds more quickly and more effectively to pathogen attack, and it also has an increased tolerance for abiotic stresses. Systemic acquired resistance persists for the plant's lifetime and for a few generations thereafter (an example of epigenetic inheritance, Section 10.5).

Systemic acquired resistance begins when an infected tissue releases small molecules that travel to other parts of the plant, where they trigger cells to produce a plant hormone called salicylic acid. Salicylic acid increases transcription of hundreds of genes involved in defense. One of the molecules produced in response to salicylic acid is epicatechin; this flavonoid prevents hardening of "pegs" that develop on germinating spores of many fungi. Fungi use these pegs to punch through plant epidermal tissue. Flavonoids related to epicatechin inhibit fatty acid synthesis in bacteria. Other gene products include lignin, which increases the strength of plant cell walls; enzymes that break down chitin in fungal cell walls; phytoestrogens that inhibit fungal metabolic enzymes and attract symbiotic soil bacteria; antioxidants that detoxify molecules produced in a hypersensitive response; and so on.

Defenses Against Herbivory

Plants have the ability to release into the air thousands of volatile (airborne) chemicals that can be used for communication with other plants and other organisms. The chemicals released depend on a plant's internal conditions and environmental cues. Consider what happens when an insect chews on a leaf. The resulting

systemic acquired resistance In plants, inducible whole-body resistance to a wide range of pathogens and abiotic stressors.

A Saliva of a tobacco budworm (*Heliothis virescens*) chewing on a leaf of a tobacco plant (*Nicotiana*) triggers the plant to emit a particular combination of 11 volatile secondary metabolites.

B Red-tailed wasps (*Cardiochiles nigriceps*), which parasitize tobacco budworms, recognize the unique chemical signature emitted by the plant. They follow the trail of chemicals back to the source.

C A wasp that finds a budworm deposits an egg inside of it. When the egg hatches, a larva emerges and begins to eat the budworm, which eventually dies.

FIGURE 30.18 A hormone-mediated defense against herbivory. Plant cells release a particular combination of volatile secondary metabolites in response to being chewed by a particular insect species. The chemical signature attracts wasps that parasitize the insect.

Data Analysis Activities

Volatile Secondary Metabolites in Plant Stress Responses In 2007, researchers Casey Delphia, Mark Mescher, and Consuelo De Moraes (pictured at left) published a study on the production of different volatile chemicals by tobacco plants in response to predation by two types of insects: western flower thrips and tobacco budworms. Their results are shown in **FIGURE 30.19**.

1. Which treatment elicited the greatest production of volatiles?
2. Which chemical was produced in the greatest amount? What was the stimulus?
3. Which one of the chemicals tested is most likely produced by tobacco plants in a nonspecific response to predation?
4. Are any chemicals produced in response to predation by budworms, but not in response to predation by thrips?

Volatile Compound Produced	Treatment					
	C	T	W	WT	HV	HVT
Myrcene	0	0	0	0	17	22
β-Ocimene	0	433	15	121	4,299	5,315
Linalool	0	0	0	0	125	178
Indole	0	0	0	0	74	142
Nicotine	0	0	233	160	390	538
β-Elemene	0	0	0	0	90	102
β-Caryophyllene	0	100	40	124	3,704	6,166
α-Humulene	0	0	0	0	123	209
Sesquiterpene	0	7	0	0	219	268
α-Farnesene	0	15	0	0	293	457
Caryophyllene oxide	0	0	0	0	89	166
Total	0	555	288	405	9,423	13,563

FIGURE 30.19 Volatile (airborne) compounds produced by tobacco plants in response to predation by different insects. Plants were untreated (C), attacked by thrips (T), mechanically wounded (W), mechanically wounded and attacked by thrips (WT), attacked by budworms (HV), or attacked by budworms and thrips (HVT). Values are in nanograms/day.

wounding of the leaf triggers production of several molecules, including the hormone jasmonic acid. Jasmonic acid increases transcription of several genes whose products include volatile secondary metabolites. The particular cocktail of chemicals produced depends on the species of plant—and the species chewing on it. In this case, the chemical signature attracts wasps that parasitize insect herbivores (**FIGURE 30.18**). Volatile secondary metabolites released in response to the herbivory are also detected by neighboring plants, which respond by increasing production of chemicals that make them less palatable.

TAKE-HOME MESSAGE 30.8

- ✔ Biotic and abiotic stressors provoke short-term and long-term defense responses in plants. These responses are mediated by hormones.
- ✔ Detection of pathogens on a plant's surface can trigger an ABA-mediated closure of stomata. Pathogens that enter internal tissues trigger a hypersensitive response that causes cell death.
- ✔ Systemic acquired resistance triggered by pathogen attacks increases a plant's ability to withstand biotic and abiotic stresses.
- ✔ In response to wounding, plants release a mixture of volatile chemicals that (for example) attracts parasites of herbivores.

30.1 Prescription: Chocolate (revisited)

Epicatechin has now been shown to have a beneficial effect on a variety of illnesses, including hypertension and other cardiovascular disorders, Parkinson's disease, obesity, and even acne. Clinical trials involving chocolate consumption as a medical intervention for these conditions are now under way.

Epicatechin probably works its magic in humans because it activates an enzyme, nitric oxide synthase, in endothelial cells making up the walls of blood vessels. This enzyme makes nitric oxide, which in turn widens the blood vessels by relaxing muscles in their walls. In mammals, nitric oxide is part of hormonal systems that regulate blood vessel width, and it is also necessary for proper function of the endothelial cells that produce it. Dysfunction of these cells is linked to cardiovascular disease, a primary cause of illness and death in the United States and in many other countries. Thus, increasing nitric oxide production in blood vessel cells is a major area of medical research.

Humans have no stomata to close, and plants have no blood vessels, but nitric oxide operates as a hormone on both structures. In fact, NADPH oxidase, the plasma membrane enzyme that produces nitric oxide in plants, also occurs in human white blood cells, where it produces a nitric oxide burst in response to detection of bacterial flagellin inside the body. This response is mediated by abscisic acid, just as it is in plants—and all animals.

STUDY GUIDE

Section 30.1 Some plant secondary metabolites attract pollinators or symbionts, or function in defense. A few of these compounds, including a number of flavonoids, are beneficial to human health. Signaling pathways common to plants and humans are evidence of life's common origin.

Section 30.2 Unlike animals, plants continue to develop through their lifetime in response to environmental cues. Developmental flexibility depends on cell-to-cell communication mediated by **plant hormones**. These hormones promote or arrest development in regions of a plant by stimulating or inhibiting cell division, differentiation, or enlargement (reviewed in **TABLE 30.4**). Some have roles in occasional responses such as pathogen defense. Hormones work together or in opposition, and many have different effects in different regions of the plant.

Section 30.3 **Auxin** promotes plant growth directly, and coordinates the effects of other hormones involved in growth. A unique transport system distributes auxin directionally. The polar distribution sets up auxin gradients that affect the growth and development of plant parts, such as when auxin gradients maintain **apical dominance** in a growing shoot tip. **Cytokinin** promotes development of lateral buds and inhibits formation of lateral roots. This hormone often opposes the effects of auxin, for example in apical dominance.

TABLE 30.4

Some Major Plant Hormones

Hormone	Examples of Effects
Abscisic acid (ABA)	Mediates environmental stress responses Inhibits shoot growth and seed germination
Auxin	Coordinates effects of other plant hormones during development of shoots and roots Promotes apical dominance Promotes growth by cell enlargement Mediates some tropisms
Cytokinin	Stimulates differentiation of cells in root apical meristem Stimulates development of lateral buds Inhibits formation of lateral roots
Ethylene	Involved in breaking dormancy Stimulates fruit ripening, abscission Involved in stress responses
Gibberellin	Stimulates cell division, elongation in stems Mobilizes food reserves in germinating seeds Stimulates flowering in some plants
Jasmonic acid	Induces expression of genes used in defense Involved in thigmotropisms
Nitric oxide	Abiotic and biotic stress defense responses Enhances germination Involved in thigmotropisms
Salicylic acid	Activates systemic acquired resistance

Section 30.4 **Gibberellin** lengthens stems between nodes by inducing cell division and elongation, and it is required for seed germination. It also stimulates production of enzymes that break down endosperm during seed germination.

Section 30.5 **Abscisic acid** (**ABA**) plays a part in **abscission**, but it has a much greater role in stress responses (including stomata closure in response to lack of water). ABA influences expression of thousands of genes, with effects that include the suppression of seed germination and shoot growth.

Ethylene is produced in negative and positive feedback loops. The negative feedback loops help regulate ongoing metabolism and development; bursts of ethylene produced in positive feedback loops function in special processes such as abscission, ripening, and defense responses.

Section 30.6 Plant movements can be directional or not. A **tropism** is a directional movement toward or away from a stimulus that provokes it. Tropisms occur in response to environmental stimuli such as gravity (**gravitropism**), light (**phototropism**), or contact (**thigmotropism**). **Statoliths** play a part in gravitropism.

Section 30.7 Input from light-activated **phytochromes** and other nonphotosynthetic pigments sets circadian rhythms. A **photoperiodism** is a response to seasonal change in the length of day relative to night. Recurring cycles of abscission and dormancy are adaptations to seasonal changes in weather. Some plants require prolonged exposure to cold before they can flower (**vernalization**).

Section 30.8 Hormones are part of plant responses to abiotic and biotic stresses. A nitric oxide burst that closes stomata or kills tissue is part of a hypersensitive defense response to pathogen detection. Pathogens also trigger **systemic acquired resistance**, which increases the plant's general resilience. Volatiles released by a plant are defenses against herbivory.

SELF-QUIZ

Answers in Appendix VII

1. An important difference between plant development and animal development is that ________ .
   a. only plant development depends on hormones
   b. plants continue to develop throughout their lifetime
   c. animals, but not plants, have a circulatory system
2. Plant hormones ________ .
   a. trigger the same responses in animals and plants
   b. are cues for tropisms
   c. do not participate in stress responses
   d. may have different effects in different tissues

3. A plant cell's response to a hormone depends on _____ .
   a. which hormone receptors it has
   b. the way its hormone receptors work inside the cell
   c. the presence or absence of other hormones
   d. all of the above

4. ________ is the hormone in most rooting compounds.
   a. Gibberellin c. Cytokinin
   b. Auxin d. ABA

5. A tropism is a _______ response.
   a. defensive c. dilational
   b. directional d. detrimental

6. Which of the following statements is false?
   a. Auxin and gibberellin promote stem elongation.
   b. Cytokinin triggers growth of lateral buds.
   c. Abscisic acid promotes water loss and dormancy.
   d. Ethylene promotes fruit ripening and abscission.

7. Which combination of hormones is best for promoting the formation of lateral roots from pericycle?
   a. A high auxin level and low cytokinin level
   b. A low auxin level and high cytokinin level
   c. A high level of auxin and cytokinin

8. Ethylene differs from other hormones in that it _______ .
   a. is a gas c. operates during ripening
   b. participates in defense d. none of the above

9. ________ are organelles involved in gravitropism.

10. In some plants, flowering is a _______ response.
    a. phototropic c. photoperiodic
    b. gravitropic d. thigmotropic

11. Stomata close in response to _______ .
    a. ABA c. bacterial flagellin
    b. nitric oxide d. all of the above

12. Match the response with its main trigger.
    ___ phototropism a. contact with an object
    ___ gravitropism b. blue light
    ___ thigmotropism c. long period of cold
    ___ photoperiodism d. gravity
    ___ vernalization e. sun position
    ___ heliotropism f. night length

13. Match the hormone with the description.
    ___ ethylene a. efflux carriers set up gradients
    ___ cytokinin b. produced in positive and negative feedback loops
    ___ auxin c. affects expression of more genes than any other hormone
    ___ gibberellin d. works antagonistically to auxin in apical dominance
    ___ ABA e. big in stems
    ___ nitric oxide f. active in bursts

14. Systemic acquired resistance ________ .
    a. occurs in response to pathogen detection
    b. causes the release of volatile chemicals
    c. is a hypersensitive response to pathogens
    d. defends the plant against herbivory
    e. does not involve hormones

15. Match the hormone with the observation.
    ___ ethylene a. Your cabbage plants form elongated flowering stalks.
    ___ cytokinin b. The potted plant in your room is leaning toward the window.
    ___ auxin c. The last of your apples is getting really mushy.
    ___ gibberellin d. The seeds of your roommate's marijuana plant do not germinate no matter what he does to them.
    ___ ABA e. Lateral buds are sprouting.
    ___ nitric oxide f. Your lettuce plants develop brown spots on their leaves.

CRITICAL THINKING

1. Professional gardeners often soak seeds in hydrogen peroxide before planting them. Why?

2. Photosynthesis sustains plant growth, and inputs of sunlight sustain photosynthesis. Why, then, do seedlings that germinate in a fully darkened room grow taller than seedlings of the same species that germinate in full sun?

3. Certain mutations in common wall cress (*Arabidopsis thaliana*) cause excess auxin production. Predict the impact of these mutations on a plant's phenotype.

4. Cattle in industrial dairy farms are typically given rBST, or recombinant bovine somatotropin. This animal hormone increases a cow's milk production, but there is concern that such hormones may have unforeseen effects on milk-drinking humans. There is no similar concern about plant hormones in the plant foods we eat. Why not?

5. The oat coleoptiles on the left have been modified: either cut or placed in a light-blocking tube. Which ones will still bend toward a directional light source?

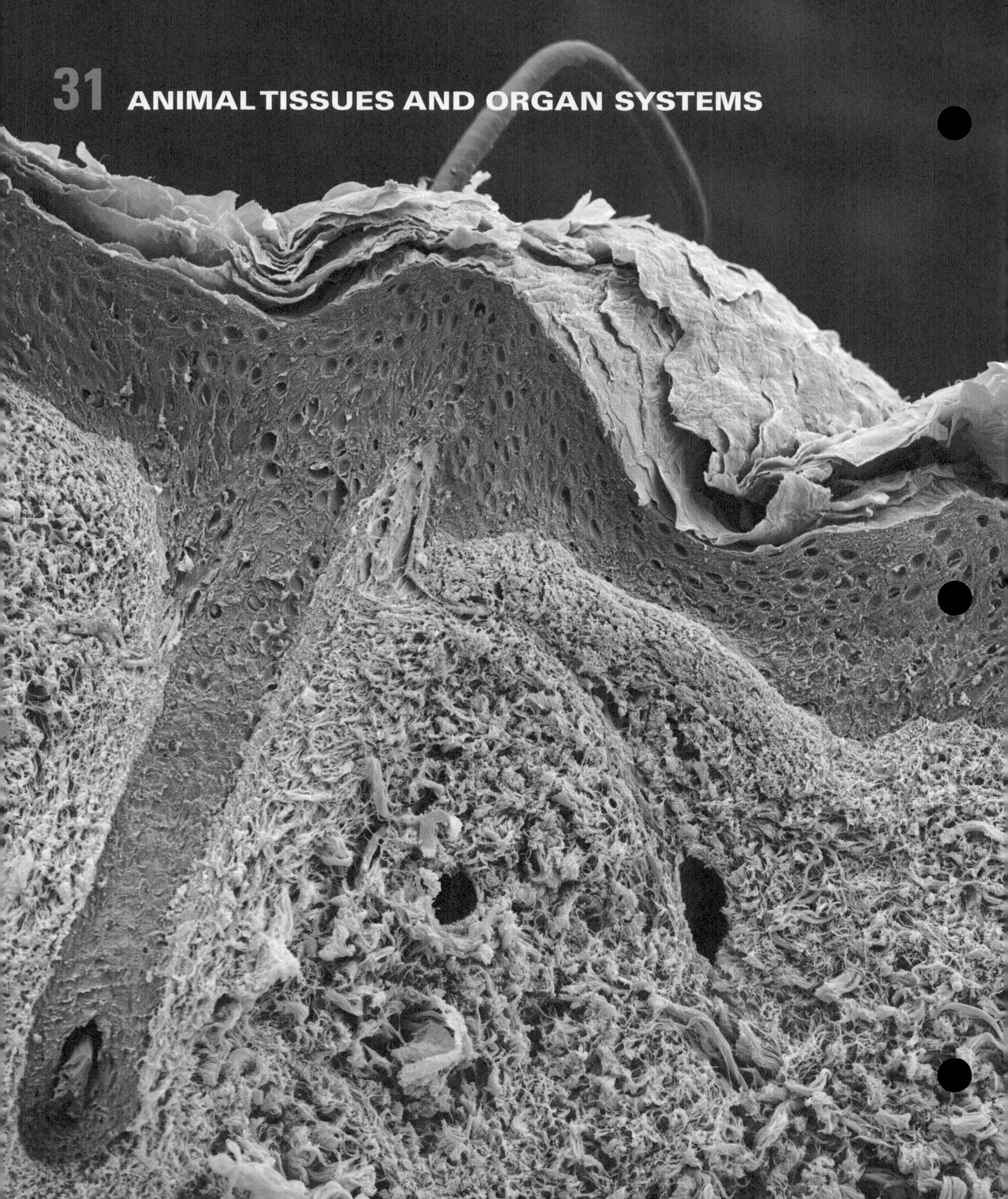

31 ANIMAL TISSUES AND ORGAN SYSTEMS

CORE CONCEPTS

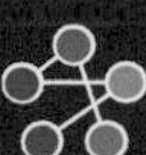

Systems

Complex properties arise from interactions among components of a biological system.

In nearly all animals, cells interact as tissues. Each animal tissue consists of specific cell types that collectively carry out a task or tasks. Tissues interact in organs, which in turn interact as components of organ systems. At all levels of organization, animal structure is shaped by natural selection, and constrained by physical and developmental factors.

Evolution

Evolution underlies the unity and diversity of life.

Shared core processes and features provide evidence that living things are related. Four types of tissues occur in all vertebrates. Epithelial tissues cover the body's surfaces and line its cavities. Connective tissues provide support. Muscle tissues bring about movement. Nervous tissue receives, integrates, and communicates information.

Structure and Function

The three-dimensional form and arrangement of biological structures give rise to their function and interactions.

Body cells vary in their size and shape. In some tissues, special junctions attach cells to one another. In other tissues, each cell is surrounded by a secreted matrix. Some cells are ciliated, and others have projections from their surface. The form of a cell suits its unique function, and each tissue is composed of specific cell types. Similarly, the function of each organ arises from the types of tissues within it and the manner in which they are arranged.

Links to Earlier Concepts

With this chapter, we begin our survey of the tissues and organ systems (Section 1.2) in animals. The chapter expands on the nature of animal body plans (24.2) and trends in vertebrate evolution (25.2). We see examples of cell features such as cilia (4.8), cell junctions, and a secreted extracellular matrix (4.11). We revisit the topic of cancer (8.6 and 11.6) and look again at the adaptive significance of human skin color (14.1).

31.1 Making Replacement Cells

All animals can replace some tissues lost to injury. Invertebrates have the greatest capacity for regeneration. For example, some sea stars can regrow their entire body from a single arm and a bit of central disk (**FIGURE 31.1**). No vertebrate can grow a body from a limb, but some salamanders can regrow a limb, and many salamanders and lizards can replace a lost tail. Mammals do not replace limbs or tails, although like other animals they replace lost skin and blood cells.

Stem cells are the key to producing new or replacement tissues. A **stem cell** is an unspecialized cell that can either divide to produce more stem cells or differentiate into one of the specialized cells that characterize specific body parts. All cells in an animal body "stem" from stem cells.

Stem cells vary in their potential to form a new individual or new tissues. A fertilized human egg and the cells produced by its first four divisions are said to be "totipotent," meaning they have the potential to develop into a new individual. Later divisions produce "pluripotent "embryonic stem cells. A pluripotent cell cannot develop into a new individual, but it can give rise to any of the cell types in a body.

Skin cells, blood cells, and other cells that your body replaces continually arise from adult stem cells. However, adults have relatively few stem cells capable of replacing cells in your heart or spinal cord. This is why death of cardiac muscle cells during a heart attack permanently weakens the heart, and why spinal cord injuries result in permanent paralysis.

In recent years, researchers have begun investigating ways to grow stem cells in culture and to direct their differentiation into cells that can provide replacement tissues for people who need them. The first clinical studies involving tissue derived from human stem cells are now under way.

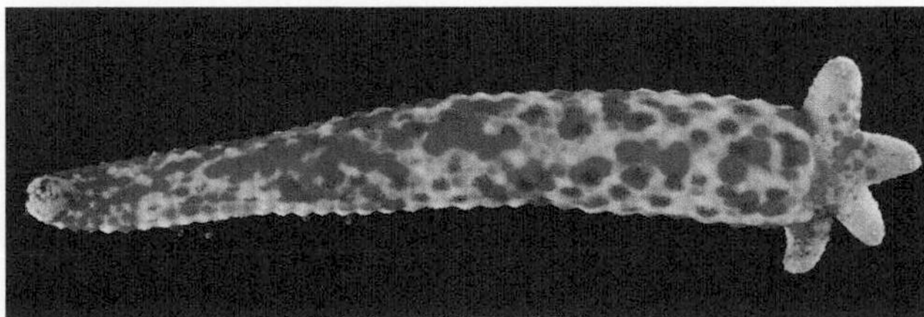

FIGURE 31.1 Regeneration of a sea star from a severed arm. Stem cells in the remaining bit of central disk are giving rise to replacement parts, including the four new arms visible on the right.

stem cell Cell that can divide to produce two stem cells or differentiate into some or all specific cell types.

31.2 Animal Body Plans

LEARNING OBJECTIVES

- List the four types of tissue in vertebrate bodies.
- Describe the types and locations of human body fluids.
- Give an example of a factor that constrains the evolution of animal body plans.

Levels of Organization

In all animals, development produces a body with multiple types of cells (FIGURE 31.2A). For example, a human body has about 200 different kinds of cells. In all animals more complex than sponges, the cells are organized as tissues (FIGURE 31.2B). A **tissue** is an array of cells of a specific type or types that interact in a collective task. Cells of a tissue are often anchored by a secreted extracellular matrix (Section 4.11).

Four types of tissue occur in all vertebrates (TABLE 31.1). Epithelial tissue covers body surfaces and lines internal cavities such as the digestive tract. Connective tissue holds body parts together and provides structural support. Muscle tissue moves a body or its parts. Nervous tissue detects stimuli and relays signals.

Each tissue contains unique cell types not present in other tissues. For example, contractile cells occur in muscle tissue, but not in nervous tissue or epithelial tissue. We discuss the four tissue types in detail in the sections that follow.

In most animals, tissues are organized into organs. An **organ** is a structural unit of two or more tissues organized in a specific way and capable of carrying out specific tasks. Consider a human heart (FIGURE 31.2C). The heart wall consists mostly of cardiac muscle tissue. Epithelial tissue covers the heart's internal and external surfaces. Heart valves consist mainly of connective tissue, and nervous tissue delivers signals to and from the heart. The coordinated activity of these four types of tissue allows the heart to undergo rhythmic contractions that move blood through the body at an appropriate pace.

In organ systems, two or more organs and other components interact physically, chemically, or both in a common task. For example, in the vertebrate circulatory system, the force generated by a beating heart (an organ) moves blood (a tissue) through blood vessels (organs), thereby transporting gases and solutes to and from all body cells (FIGURE 31.2D). Multiple organ systems sustain the body (FIGURE 31.2E).

Body Fluids

Fluids account for about 60 percent of your body weight (FIGURE 31.3). Most human body fluid is intracellular fluid, the fluid inside cells. **Extracellular fluid** is the environment in which body cells live. It bathes cells and provides them with the substances they require to stay alive. It also carries away cellular wastes. In vertebrates, extracellular fluid consists mainly of **interstitial fluid** (the fluid in spaces between cells) and plasma, the fluid portion of the blood.

Cells survive only if the fluid surrounding them remains within a narrow range of solute concentration and temperature. Maintaining conditions of the cell's environment within this range is an important aspect of homeostasis (Section 1.3).

Evolution of Animal Body Plans

An animal's structural traits (its anatomy) evolve in concert with its functional traits (its physiology). Both types of traits are genetically determined and vary among individuals. Genes for those traits that best help individuals survive and reproduce in their environment are passed on preferentially to the next generation. Over many generations, anatomical and

FIGURE 31.2 Levels of organization in a human body.

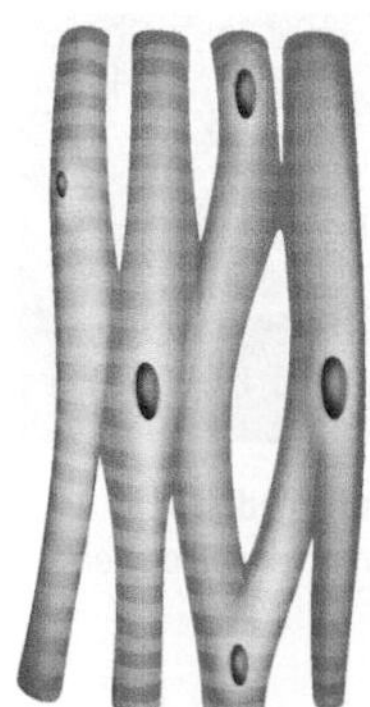

A Cell (cardiac muscle cells)

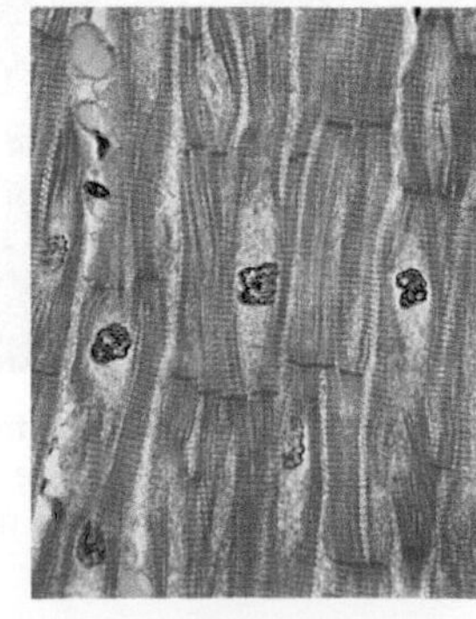

B Tissue (cardiac muscle)

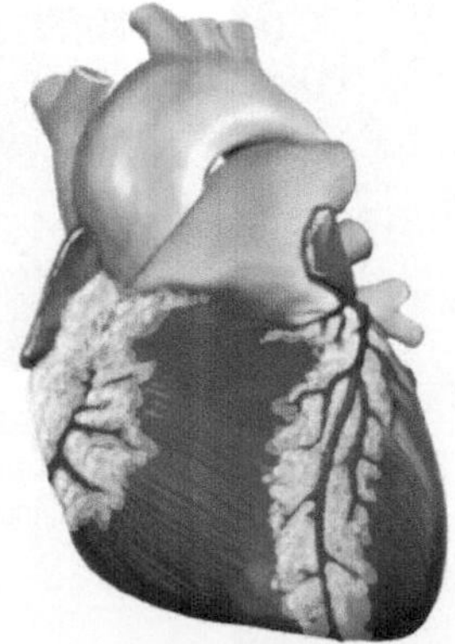

C Organ (heart)

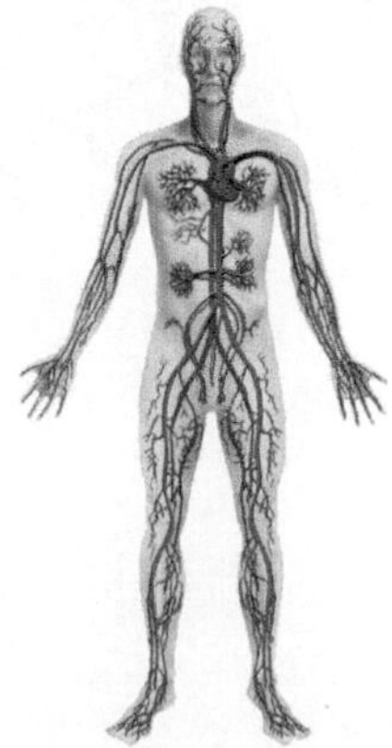

D Organ system (circulatory system)

E Body (human body)

TABLE 31.1

Summary of Human Tissue Types

Tissue Type	Functions	Examples
Epithelial tissue	Protection; secretion and/or absorption	Epidermis (outer layer of skin), endothelium (lining of blood vessels), lining of gut, lining of lung
Connective tissue	Structural support and connection	Loose connective tissue, dense connective tissue, adipose tissue (fat), bone, cartilage, blood
Muscle tissue	Contracts to bring about movement of the body or a body component	Skeletal muscle, cardiac muscle, smooth muscle
Nervous tissue	Detects stimuli, processes information, coordinates responses to stimuli	White matter and gray matter of the brain and spinal cord; peripheral nerves

structural traits evolve in ways that reflect their function in a specific environment.

Physical constraints can affect the evolution of body structure. Consider the challenges that vertebrates faced as they shifted from aquatic life to life on land (Section 25.4). Gases can only enter or leave an animal's body by diffusing across a moist surface. In an aquatic organism, the surrounding water both delivers oxygen and moistens the respiratory surface. By contrast, a land animal must extract oxygen from air, which can dry a respiratory surface. Evolution of lungs allowed vertebrates to maintain a moist respiratory surface inside their body. Cells inside the lung secrete the fluid that keeps this surface moist.

Evolutionary history also constrains body plans. Evolution by natural selection often involves incremental modifications of existing tissues or organs. Lungs first evolved from an outpouching of the digestive tract in a lineage of bony lobe-finned fish (Section 25.3). These fishes had gills, but evolution of a simple lung improved their ability to survive even in poorly oxygenated water. The fishes swallowed air to fill their lung, then oxygen diffused from the lung into the blood. Fishes with lungs were the ancestors of tetrapods, which were the first animals to colonize land (Section 25.4). A land-based life spurred further evolutionary modifications to lungs, including an increase in size, and internal partitions that increased the surface area for gas exchange. Although the details of lung structure changed when animals moved onto land, the anatomical connection to the gut did not. Thus, the human throat connects to both the digestive tract and respiratory tract. This dual connection sometimes allows food to go where air should, causing a person to choke. It would be safer if food and air entered the body through separate passageways. However, because evolution modifies existing structures, it sometimes does not produce the optimal body plan.

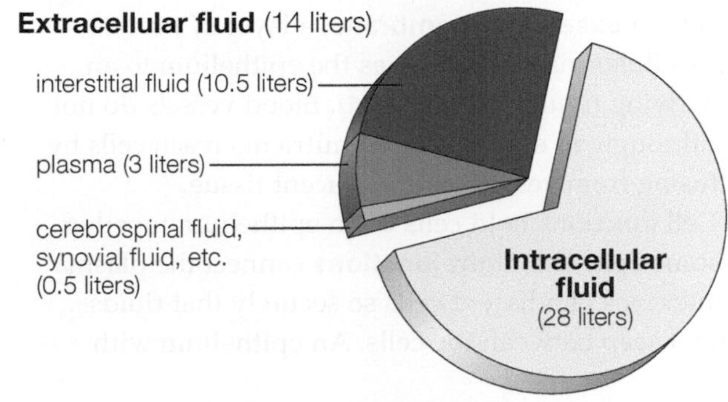

Extracellular fluid (14 liters)
interstitial fluid (10.5 liters)
plasma (3 liters)
cerebrospinal fluid, synovial fluid, etc. (0.5 liters)
Intracellular fluid
(28 liters)

FIGURE 31.3 Distribution of fluids in an adult man.

Similarly, a body framework can impose architectural constraints. For example, the ancestors of all modern land vertebrates had a body plan with four limbs. When wings evolved in birds and bats, they did so by modification of existing forelimbs, not by new limbs sprouting. Although it might be advantageous to have both wings and arms, no living or fossil vertebrate with this body plan has ever been discovered.

TAKE-HOME MESSAGE 31.2

✔ In most animals, cells are organized as tissues. Each tissue consists of cells of a specific type that cooperate in carrying out a particular task. Tissues are organized into organs, which in turn are components of organ systems.

✔ The animal body consists mainly of fluid. Maintaining the solute concentration and temperature of extracellular fluid is an important facet of homeostasis.

✔ Many anatomical traits evolved as solutions to physical challenges. However, these solutions are sometimes imperfect because evolution modifies existing structures.

extracellular fluid Of a multicelled organism, the body fluid outside of cells; serves as the body's internal environment.

interstitial fluid (in-ter-STIH-shul) Portion of the extracellular fluid that fills spaces between body cells.

organ In multicelled organisms, a structure that consists of tissues engaged in a collective task.

tissue In multicelled organisms, specialized cells organized in a pattern that allows them to perform a collective function.

31.3 Epithelial Tissue

LEARNING OBJECTIVES

- Describe the structure and function of epithelial tissues.
- Compare endocrine and exocrine glands.
- Explain why epithelial tissues are especially prone to cancer.

Epithelial tissue, or an epithelium (plural, epithelia), is a sheetlike layer of cells with no extracellular matrix between them. The surface of the epithelium that faces the outside world (or lines the interior of a body cavity or tube) is referred to as the apical surface. The opposite side of the epithelium (the basal surface) secretes a **basement membrane**, a layer of fibrous extracellular matrix that glues the epithelium to an underlying tissue (FIGURE 31.4). Blood vessels do not run through an epithelium, so nutrients reach cells by diffusing from vessels in an adjacent tissue.

Cell junctions hold cells of an epithelium together. In some epithelia, **tight junctions** connect the plasma membranes of adjacent cells so securely that fluids cannot seep between the cells. An epithelium with cells connected by tight junctions keeps fluid within a particular body compartment from seeping into underlying tissue. For example, tight junctions join the epithelial cells in the lining of the intestinal tract. The junctions allow this epithelium to function as a selective barrier. Substances in the intestines can enter the body's internal environment only by controlled movement across cells of the intestinal epithelium.

Epithelial tissues subject to mechanical stress such as skin epithelium have many **adhering junctions**. These particular junctions function like buttons that hold a shirt closed. They connect the plasma membranes of cells at distinct points but do not form a seal between the cells.

Simple squamous epithelium
- Lines blood vessels, the heart, and air sacs of lungs
- Allows substances to cross by diffusion

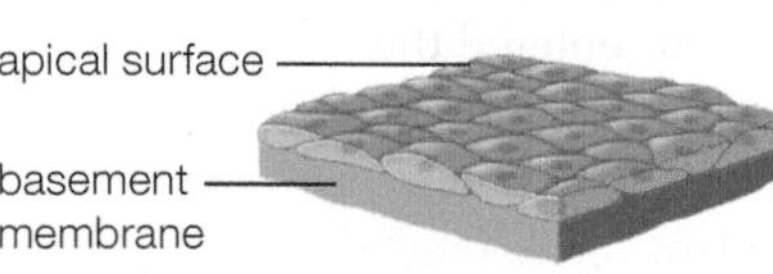
apical surface
basement membrane

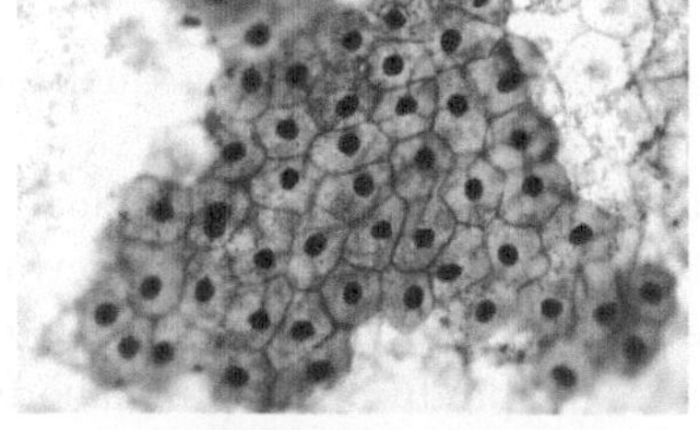

Simple cuboidal epithelium
- Lines kidney tubules, ducts of some glands, reproductive tract
- Functions in absorption and secretion, movement of materials

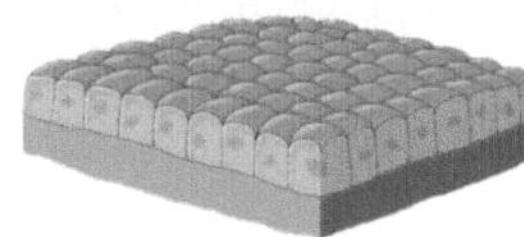

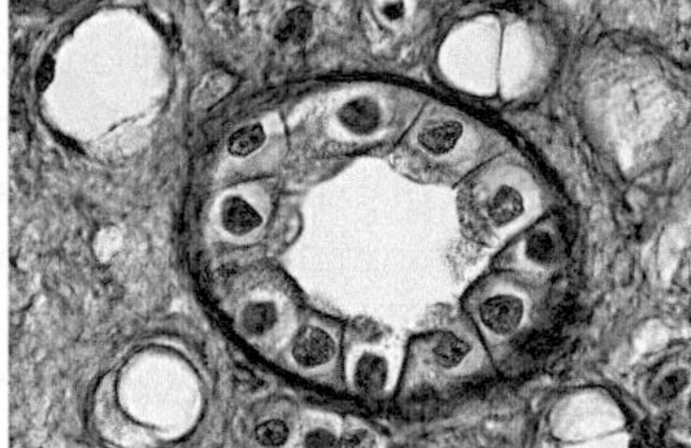

Simple columnar epithelium
- Lines some airways, parts of the gut
- Functions in absorption and secretion, protection

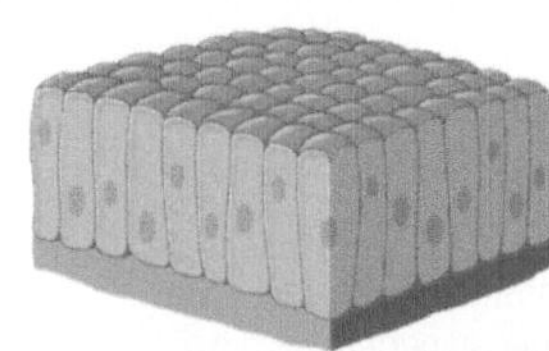

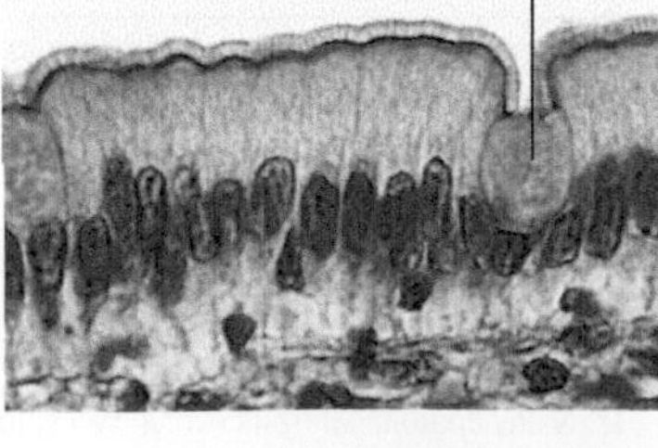
mucus-secreting gland cell

FIGURE 31.4 **Three types of simple epithelium.** In all, the basal surface secretes a noncellular basement membrane (blue) that attaches the epithelium to the underlying tissue.

Types of Epithelia

Epithelia differ in their thickness and in the shapes of their constituent cells. *Simple* epithelium is one cell thick, whereas *stratified* epithelium has multiple layers. Three types of simple epithelium are defined on the bases of cell shape. Cells in squamous epithelium are flattened or scalelike (FIGURE 31.4A). *Squama* is the Latin word for scale. Cells of cuboidal epithelium are short cylinders that look like cubes when viewed in cross section (FIGURE 31.4B). Cells in columnar epithelium are taller than they are wide (FIGURE 31.4C).

The structure of each type of epithelium reflects its function. Simple squamous epithelium is the thinnest type of epithelium, so gases and nutrients diffuse across it easily. This type of epithelium lines blood vessels and the inner surface of the lungs. By contrast, stratified squamous epithelium is thick and serves a protective function. For example, it makes up the outermost layer of human skin. Other stratified squamous epithelia line the mouth and the vagina.

Cuboidal and columnar epithelia function in absorption, secretion, and movement of materials. These epithelia often include cells that have microvilli. A **microvillus** is a nonmotile projection of the plasma membrane. It increases the area of a cell's free surface. For example, microvilli at the surface of the simple columnar epithelium that lines your intestines increase the area available for absorption of nutrients.

Cuboidal and columnar epithelia can also include cells that have cilia (Section 4.8) at their free surface. In the lining of your upper airways, ciliated columnar epithelium moves mucus away from lungs. In the lining of a woman's oviducts, ciliated columnar epithelium propels an egg toward the uterus (womb).

Some cuboidal and columnar epithelial cells secrete substances that function outside the cell. In most animals, such secretory cells cluster in glands. A gland is a multicelled structure that releases substances onto a body surface or into a body cavity or fluid. There

CREDIT: (4) top, Ray Simons/Science Source; middle, Ed Reschke/Photolibrary/Getty Images; bottom, Dr. Donald Fawcett/Visuals Unlimited, Inc.

Exocrine gland
parotid gland (secretes saliva)
parotid duct (delivers saliva to mouth)
Endocrine gland
capillary
cell that secretes hormone
thyroid gland (secretes hormones into blood)

FIGURE 31.5 Examples of an exocrine and an endocrine gland.

FIGURE IT OUT Which type of gland is ductless?

Answer: Endocrine glands are ductless and secrete hormones into the blood.

are two main types of glands (**FIGURE 31.5**) **Exocrine glands** have ducts or tubes that deliver their secretions onto an internal or external surface. Exocrine secretions include mucus, saliva, tears, digestive enzymes, earwax, and breast milk. **Endocrine glands** release signaling molecules called hormones into interstitial fluid. Hormones in interstitial fluid usually enter the blood, which distributes them throughout the body.

Specialized epithelial cells that produce large amounts of the protein keratin are the source of hair, nails, hooves, and feathers. Keratin is a fibrous structural protein, which means that molecules of this protein aggregate as long, insoluble strands. Large deposits of keratin harden a cell. The visible part of a hoof, hair, or feather consists primarily of the remains of keratin-rich cells. Other specialized cells produce pigments that impart color to these body parts, as well as to the skin.

Carcinomas—Epithelial Cell Cancers

Animals constantly renew their epithelial tissue. You shed and replace about 0.7 kilogram (1.5 pounds) of skin epithelia each year. Similarly, the lining of your intestine is replaced every four to six days. The continual cell divisions required to maintain these epithelia provide many opportunities for DNA replication errors that can lead to cancer-causing mutations. As a result, epithelium is the animal tissue most likely to become cancerous. An epithelial cell cancer is called a carcinoma. Nearly all skin cancers are carcinomas. Most breast cancers are carcinomas of epithelial cells that line the milk ducts or of the breast's glandular epithelium. Similarly, most cancers of the lungs, stomach, and colon arise in the epithelial lining of these organs.

TAKE-HOME MESSAGE 31.3

- ✔ Epithelia are sheetlike tissues that line a body's surfaces, including its cavities, ducts, and tubes. They function in protection, absorption, and secretion. Specialized epithelial cells that produce large amounts of keratin are the source of hair, nails, hooves, and feathers.
- ✔ Glands are secretory organs derived from epithelium.
- ✔ Epithelial tissues undergo continual turnover and are the tissues most prone to cancers.

31.4 Connective Tissues

LEARNING OBJECTIVES

- Describe the structure and function of connective tissues.
- Explain the role of collagen in connective tissue.

Connective tissue has cells scattered in an extracellular matrix of their own secretions. In most connective tissues, fibroblasts or closely related cells secrete an

adhering junctions Cell junction that fastens an animal cell to another cell, or to basement membrane.
basement membrane Secreted extracellular matrix that attaches an epithelium to an underlying tissue.
connective tissue Animal tissue with an extensive extracellular matrix; provides structural and functional support.
endocrine gland Ductless gland that secretes hormones into a body fluid.
epithelial tissue Sheetlike animal tissue that covers outer body surfaces and lines internal tubes and cavities.
exocrine gland Gland that secretes milk, sweat, saliva, or some other substance through a duct.
microvillus (plural, microvilli) Nonmotile projection from the plasma membrane of an epithelial cell; increase the cell's surface area.
tight junctions Cell junctions that hold the plasma membranes of adjacent animals cells together; collectively prevent fluids from leaking between cells.

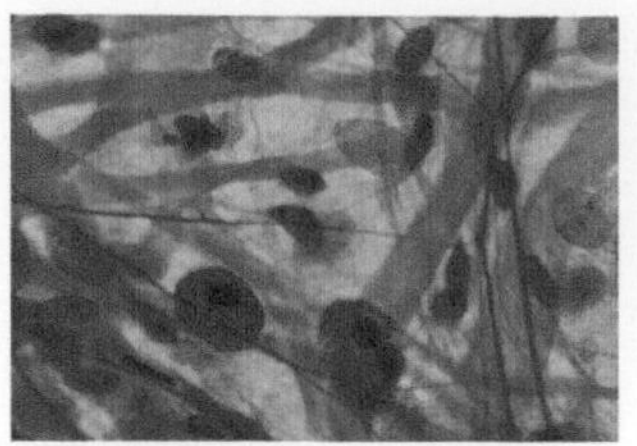
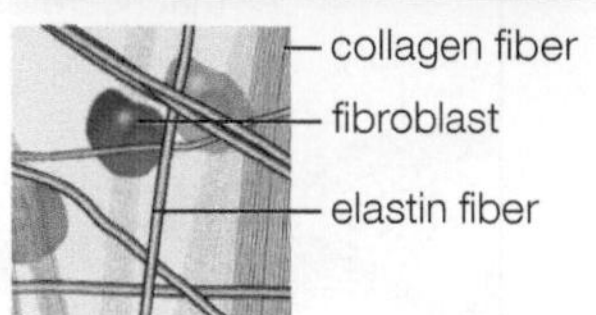
collagen fiber
fibroblast
elastin fiber

A Loose connective tissue
- Underlies most epithelia
- Provides elastic support; stores fluid

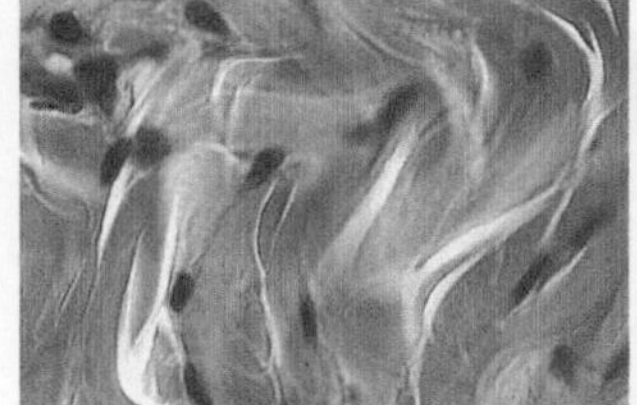
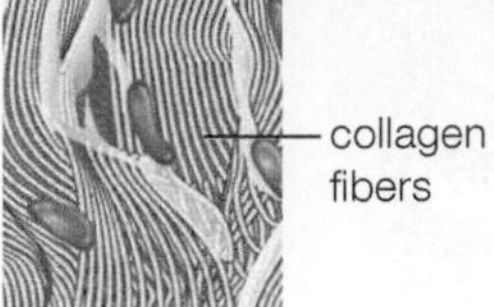
collagen fibers

B Dense, irregular connective tissue
- Deep layer of skin (dermis) and wall of gut, in kidney capsule
- Binds parts together; stretches in all directions

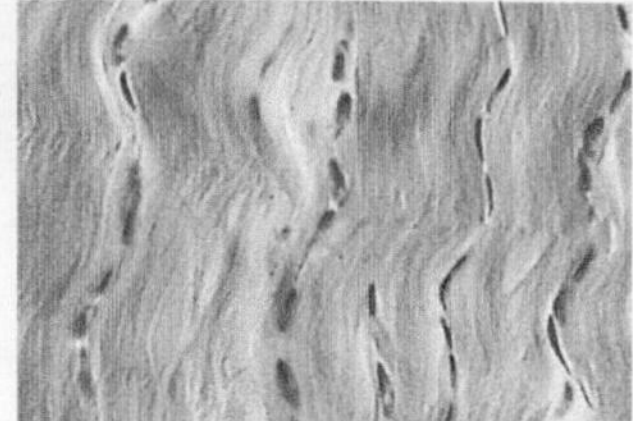
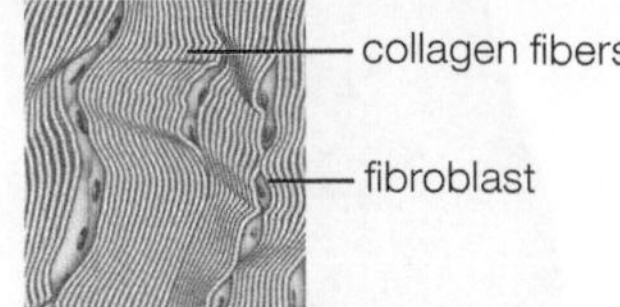
collagen fibers
fibroblast

C Dense, regular connective tissue
- Tendons and ligaments
- Fibrous attachment between parts; allows stretching, but in one direction only

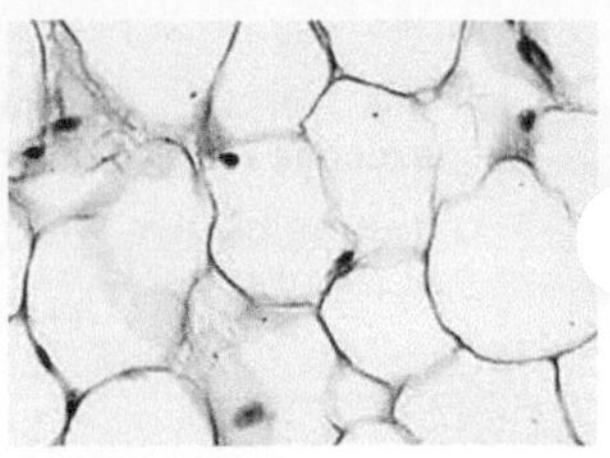
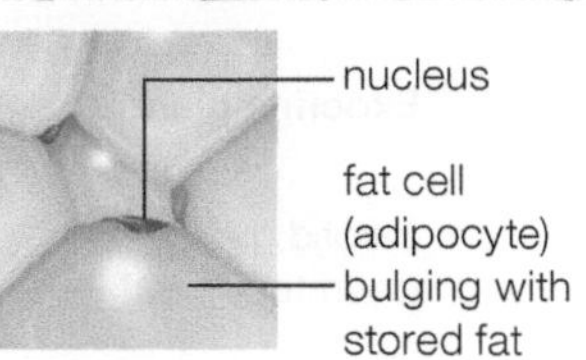
nucleus
fat cell (adipocyte) bulging with stored fat

D Adipose tissue
- Under skin and around the heart and kidneys
- Stores energy-rich lipids; insulates the body and cushions its parts

FIGURE 31.6 **Connective tissue structure and function.**

extracellular matrix containing polysaccharides and strands of collagen and elastin. Like keratin, collagen and elastin are fibrous proteins. Collagen is the most abundant protein in a vertebrate body. Elastin is springier than collagen. It abounds in tissues that stretch, and then spring back to their original shape.

Synthesizing collagen requires vitamin C, so a deficiency in this vitamin causes the nutritional disorder scurvy. Early in this disorder, connective tissue surrounding tiny blood vessels breaks down, so gums bleed and small bruiselike spots appear. If scurvy continues untreated, bones weaken. Collagen is the main component of scar tissue, so a person with scurvy heals poorly. Eventually, the disorder can be fatal.

Loose and Dense Connective Tissues

Loose and dense connective tissues have the same components (fibroblasts and a matrix with elastin and collagen fibers) but in different proportions. **Loose connective tissue** is the most abundant tissue in vertebrates. Its fibroblasts, collagen fibers, and elastin fibers are widely scattered in a gel-like matrix (FIGURE 31.6A). Loose connective tissue underlies most epithelia, and it surrounds nerves and blood vessels. It also forms the bulk of the mesentery, the sheet of tissue that holds the intestines in place within the abdomen.

In **dense connective tissue**, a matrix with a dense array of fibers surrounds the fibroblasts. In dense irregular connective tissue, fibers are arrayed randomly (FIGURE 31.6B). This randomness allows the tissue to withstand stretching in any direction. Dense, irregular connective tissue makes up the deepest layer of the skin (the dermis) and the walls of the digestive tract. It forms a protective capsule around the kidneys and the testes. Dense, regular connective tissue consists of fibroblasts in orderly rows between parallel, tightly packed bundles of collagen fibers (FIGURE 31.6C). This parallel organization maximizes the tensile strength of the tissue. Tendons and ligaments are made of dense, regular connective tissue. Tendons connect skeletal muscle to bones. Ligaments attach one bone to another.

Specialized Connective Tissues

Adipose tissue, cartilage, bone tissue, and blood are specialized connective tissues.

Adipose tissue consists mainly of cells (adipocytes) that store fatty acids as triglycerides. It is the body's main energy reservoir, acts as insulation, and provides padding. Many small blood vessels run through adipose tissue.

White fat cells are the most abundant cells in most adipose tissue. A vacuole with triglycerides fills their interior, restricting the nucleus and cytoplasm to a thin layer around the edge (FIGURE 31.6D). Some mammals, including humans, also have adipose tissue rich in brown fat cells. Brown fat cells have a centrally located nucleus, many small lipid droplets in the cytoplasm, and an abundance of specialized mitochondria that burn lipids to produce heat, rather than ATP.

Walruses and other marine mammals have a special type of adipose tissue called blubber (FIGURE 31.7). A thick coat of blubber insulates the animal's body. Even when the external temperature dips far below freezing, a walrus's core body temperature is about the same as yours. Fat is less dense than water, so being encased in blubber helps the animal float. Blubber includes

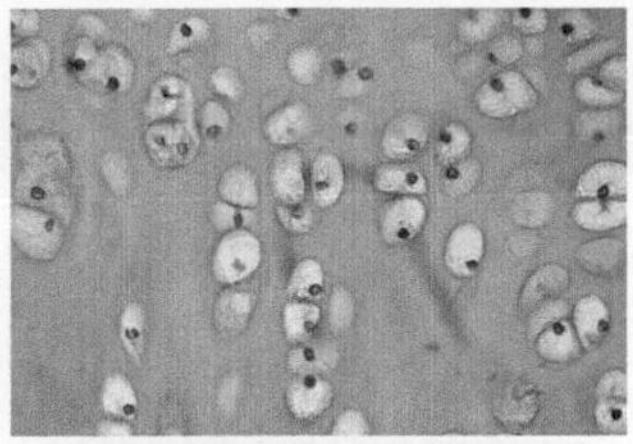
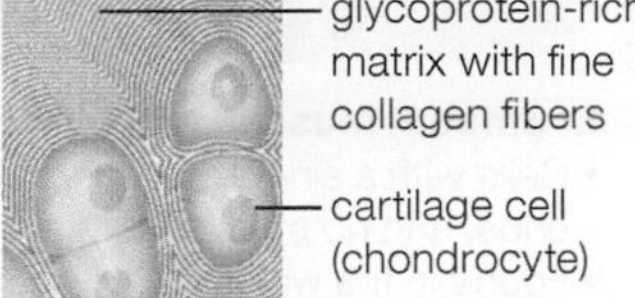
glycoprotein-rich matrix with fine collagen fibers
cartilage cell (chondrocyte)

E Cartilage
- Internal framework of ears, nose, airways; covers ends of bones
- Supports soft tissues; cushions, reduces friction at joints

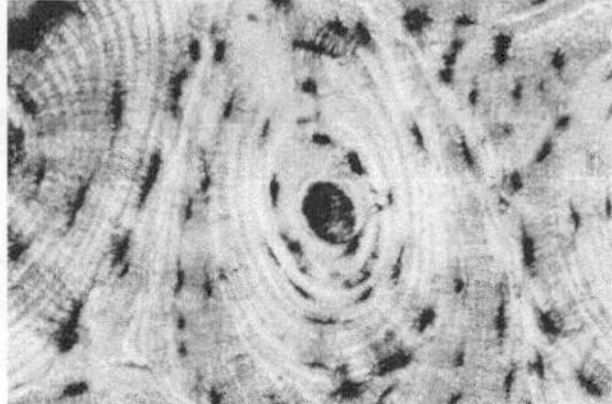
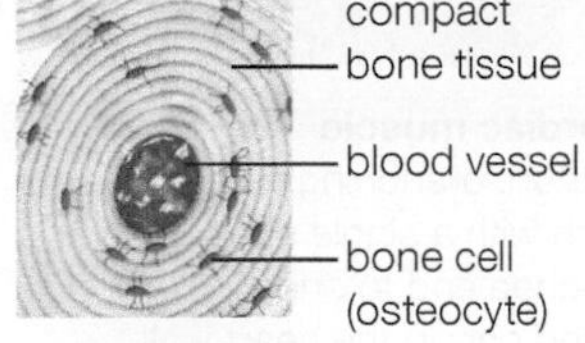
compact bone tissue
blood vessel
bone cell (osteocyte)

F Bone tissue
- Bulk of most vertebrate skeletons
- Protects soft tissues; functions in movement; stores minerals; produces blood cells

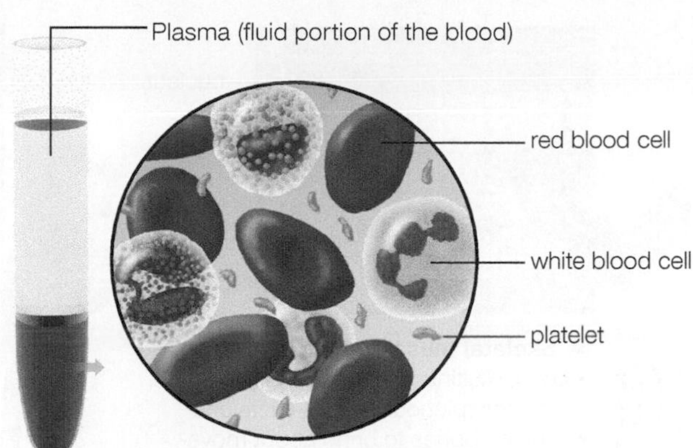
Plasma (fluid portion of the blood)
red blood cell
white blood cell
platelet

G Blood
- Flows through blood vessels and the heart
- Distributes essential gases and nutrients to cells; removes wastes from them

collagen and elastin fibers, so it spring backs when compressed. This springiness reduces the amount of energy required to swim.

Cartilage and bone are the main tissues of vertebrate skeletal systems. **Cartilage** has a matrix of collagen fibers and rubbery, compression-resistant proteoglycans. Living cartilage cells (chondrocytes) secrete the matrix (**FIGURE 31.6E**). Some fish such as sharks have a cartilage skeleton. In bony fishes and tetrapods, cartilage forms during development, then bone replaces most of it. After birth, cartilage still supports your nose, throat, and outer ears. It covers the ends of bones and reduces friction where they meet. Cartilage also acts as a shock absorber between segments of the backbone. Blood vessels do not extend through cartilage, as they do through other connective tissues, so cartilage is slow to heal if injured.

Bone tissue is a connective tissue in which living cells are surrounded by a collagen-rich matrix hardened by calcium and phosphorus (**FIGURE 31.6F**). It is the main component of bones. Bones support the body, function in movement, and protect internal organs. Some bones also produce blood cells.

FIGURE 31.7 Walruses with a thick layer of blubber. This special adipose tissue makes a walrus buoyant, helps it swim, and keeps it warm.

Blood (**FIGURE 31.6G**), the fluid in a vertebrate circulatory system, is considered a connective tissue because its cellular components (red blood cells, white blood cells, and platelets) descend from stem cells in bone. Red blood cells transport oxygen. White blood cells defend the body against pathogens. Platelets help blood clot. Blood cells and platelets drift along in plasma, a fluid extracellular matrix.

adipose tissue Animal tissue that specializes in fat storage.
blood Circulatory fluid; in vertebrates it is a fluid connective tissue consisting of plasma, red blood cells, white blood cells, and platelets.
bone tissue Animal tissue in which a mineral-hardened secreted matrix surrounds living cells.
cartilage Animal tissue in which a rubbery secreted matrix surrounds living cells.
dense connective tissue Animal tissue in which cells are surrounded by a matrix with densely packed fibers.
loose connective tissue Animal tissue in which cells and fibers are scattered through a gel-like matrix.

TAKE-HOME MESSAGE 31.4

- ✔ Loose and dense connective tissues underlie epithelia, form capsules, and connect muscle to bones or bones to one another.
- ✔ Adipose tissue is a specialized connective tissue that stores lipids.
- ✔ A vertebrate skeleton consists of two specialized connective tissues: rubbery cartilage and mineral-hardened bone.
- ✔ Blood is a connective tissue because blood cells form in bone. Blood cells are carried by plasma, the fluid portion of the blood.

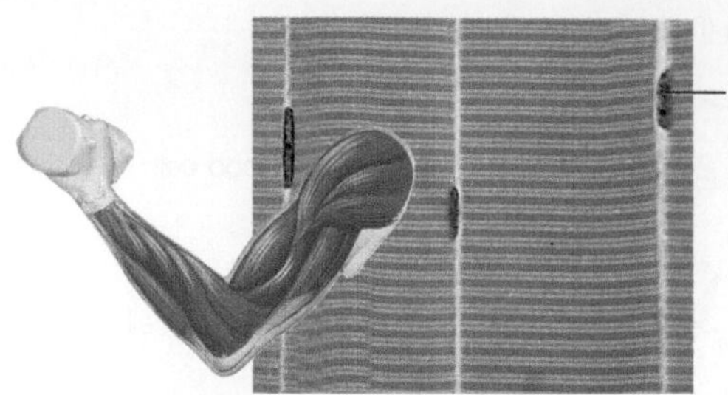
nucleus

nucleus
adjoining ends of cells

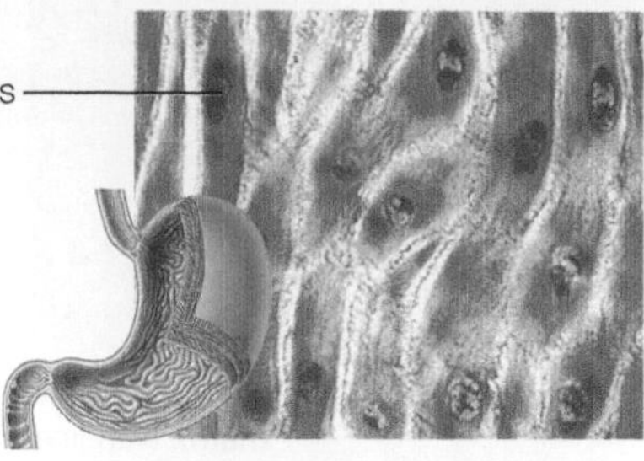
nucleus

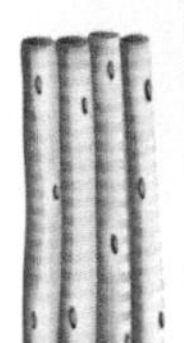

A Skeletal muscle
- Long, multinucleated, cylindrical cells with conspicuous stripes (striations)
- Pulls on bones to bring about movement, maintain posture
- Under voluntary control, but some activated in reflexes

B Cardiac muscle
- Striated, branching cells (each with a single nucleus) attached end to end
- Found only in the heart wall
- Contraction is not voluntary

C Smooth muscle
- Cells with a single nucleus, tapered ends, and no striations
- Found in the walls of arteries, the digestive tract, the reproductive tract, the bladder, and other organs
- Contraction is not voluntary

FIGURE 31.8 **Three types of muscle tissue.**

31.5 Muscle Tissue

LEARNING OBJECTIVES

- Explain the trait that defines muscle tissues.
- List the three types of muscle tissues and their functions.
- Explain what is meant by voluntary muscle.

Muscle tissue contains cells that contract (shorten) in response to stimulation, then passively lengthen when they relax. ATP provides the energy for contractions.

Skeletal muscle tissue makes up the muscles that pull on bones. It consists of parallel arrays of long, cylindrical cells called muscle fibers, which have a striated, or striped, appearance (FIGURE 31.8A). Muscle fibers are multinucleated and form by fusion of immature muscle cells during embryonic development. Skeletal muscle contracts reflexively, as when you pull your hand away after touching a hot object. More often, its contraction is deliberate, as when you reach for something. Thus, skeletal muscle is commonly described as "voluntary" muscle. By contrast, cardiac muscle and smooth muscle are said to be "involuntary" muscle because people cannot deliberately make these tissues contract.

Along with the liver, skeletal muscle is a major site for glycogen storage. Metabolic activity in skeletal muscles is the major source of body heat.

Cardiac muscle tissue occurs only in the heart wall (FIGURE 31.8B). Like skeletal muscle tissue, it has a striated appearance. Cardiac muscle consists of branching cells, each with a single nucleus. The cells are attached end to end by adhering junctions that hold them together during forceful contractions. Cardiac muscle cells are also connected by many **gap junctions**, cell junctions that form a channel across their plasma membranes. Signals that encourage cell contraction pass swiftly from cell to cell at gap junctions. The rapid flow of signals ensures that all cells in cardiac muscle tissue contract as a unit. Compared to other muscle tissues, cardiac muscle has far more mitochondria. These organelles provide a steady supply of ATP.

Smooth muscle tissue constitutes the wall of hollow internal organs of the digestive tract, urinary tract, and reproductive tract, as well as the wall of some blood vessels and the iris of the eye. Smooth muscle cells are unbranched, have a single elongated nucleus, and taper at both ends (FIGURE 31.8C). Contractile units are not arranged in an orderly repeating fashion, so smooth muscle does not appear striated. As in cardiac muscle, cells are connected by gap junctions. Smooth muscle contracts more slowly than skeletal muscle, but its contractions can be sustained longer.

TAKE-HOME MESSAGE 31.5

✔ Muscle tissue consists of cells that contract in response to nervous signals. This contraction requires ATP.

✔ Skeletal muscle, which is striated and under voluntary control, interacts with the skeleton to move a body or body part.

✔ Cardiac muscle makes up the bulk of the heart.

✔ Smooth muscle makes up the wall of hollow organs and some blood vessels.

cardiac muscle tissue Muscle tissue of the heart wall.
gap junction Cell junction that forms a closable channel across the plasma membranes of adjoining animal cells; allows ions that serve as signals to move between cells.
muscle tissue Tissue that consists mainly of contractile cells.
skeletal muscle tissue Muscle tissue that pulls on bones and moves body parts; under voluntary control.
smooth muscle tissue Muscle tissue that lines blood vessels and forms the wall of hollow organs.

CREDITS: (8A–B top) Ed Reschke; (8C top) Biophoto Associates/ Science Source; (8A–C bottom) BlueRingMedia/Shutterstock.

31.6 Nervous Tissue

LEARNING OBJECTIVES

- Explain the role of nervous tissue.
- Compare the functions of neurons and neuroglia.

Nervous tissue detects changes in the internal or external environment, integrates information, and controls the activity of muscle and glands. It is found in the brain, spinal cord, and nerves.

Nervous tissue consists of neurons and the cells that support them. A **neuron** is an electrically excitable cells that uses chemical signals to communicate with other neurons, with a muscle, or with a gland. A neuron has a central cell body that contains its nucleus and other organelles. Cytoplasmic extensions called dendrites and axons project from the cell body. Dendrites receive chemical signals, and the axon sends chemical signals (**FIGURE 31.9**).

Your nervous system includes more than 100 billion neurons. There are three types. Sensory neurons are excited by specific stimuli, such as light or pressure. Interneurons receive and integrate sensory information; these neurons allow the brain to store information and coordinate responses to stimuli. In vertebrates, interneurons occur mainly in the brain and spinal cord. (Some also occur in the nervous tissue that controls the digestive tract.) Motor neurons relay commands from the brain and spinal cord to glands and muscle cells.

Neurons are surrounded by **glial cells** (also called neuroglia or simply glia). At least half of the cells in the human brain are glial cells. Glia cells keep neurons positioned properly and provide them with nutrients. One type of glial cell (called the Schwann cell) wraps around the signal-sending axons of most motor neurons. Schwann cells act like electrical insulation, speeding the rate at which an electrical signal travels along the plasma membrane of the neuron.

The brain and spinal cord include two visually distinct types of nervous tissue: white matter and gray matter (**FIGURE 31.10**). White matter consists primarily of the signal-sending axons of neurons, together with the glial cells wrapped around them. Gray matter includes other types of glia, along with most neuron cell bodies and their signal-receiving dendrites. The brain's outermost layer is gray matter and its interior is primarily white matter.

glial cell (GLEE-uhl) In nervous tissue, one of the cell types that support and assist neurons.

nervous tissue Animal tissue composed of neurons and supporting cells; detects stimuli and controls responses to them.

neuron (NUHR-on) One of the cells that make up communication lines of a nervous system; transmits electrical signals along its plasma membrane and communicates with other cells through chemical messages.

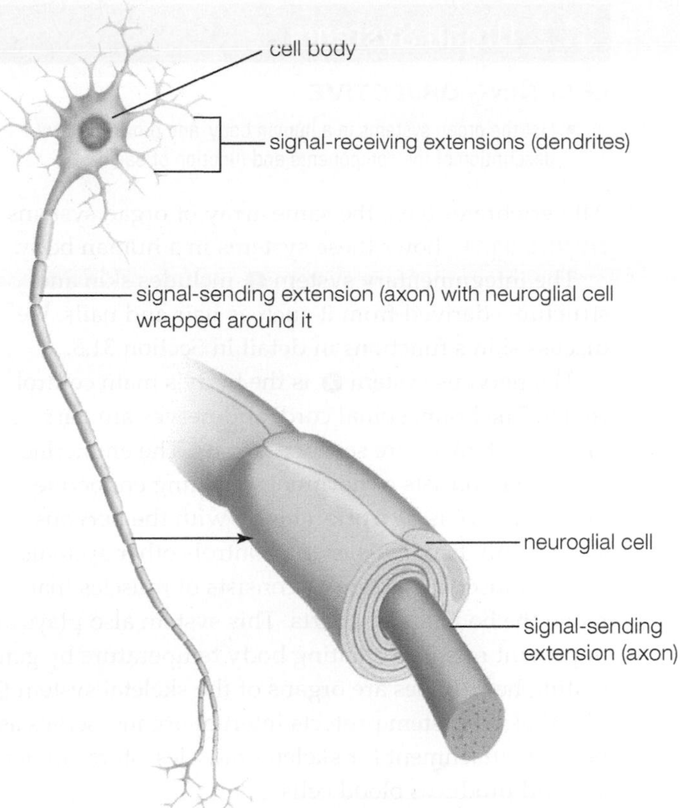
cell body
signal-receiving extensions (dendrites)
signal-sending extension (axon) with neuroglial cell wrapped around it
neuroglial cell
signal-sending extension (axon)

FIGURE 31.9 Structure of a motor neuron. The neuron has a cell body with a nucleus (visible as a dark spot) and cytoplasmic extensions. Many neuroglial cells wraps around and insulate the signal-sending axon.

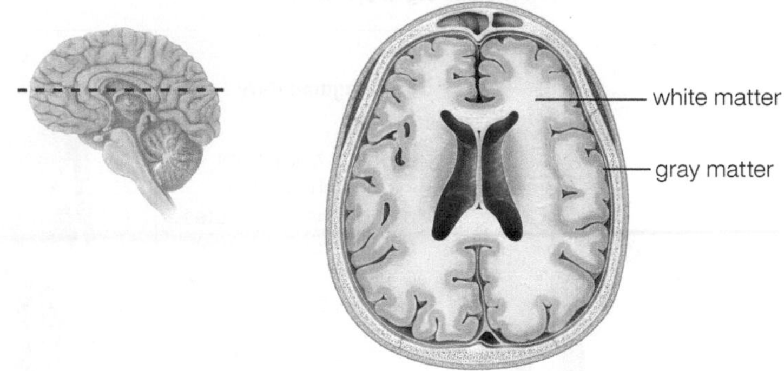
white matter
gray matter

FIGURE 31.10 Two types of nervous tissues in a human brain. A section through the brain shows the outer layer of gray matter and the inner layer of primarily white matter.

TAKE-HOME MESSAGE 31.6

✔ Nervous tissue consists of neurons and the glial cells that support them physically and metabolically.

✔ A neuron is an electrically excitable cell that communicates with other cells by means of chemical signals.

✔ Sensory neurons detect stimuli; interneurons integrate information; motor neurons relay commands to glands and muscle cells.

31.7 Organ Systems

LEARNING OBJECTIVE

- List the organ systems in a human body, and provide a brief description of the components and function of each.

All vertebrates have the same array of organ systems. **FIGURE 31.11** shows these systems in a human body.

The integumentary system ❶ includes skin and structures derived from it such as hair and nails. We discuss skin's functions in detail in Section 31.8.

The nervous system ❷ is the body's main control center. The brain, spinal cord, and nerves are part of this system, as are sensory organs. The endocrine system ❸ consists of hormone-secreting endocrine glands and cells. It works closely with the nervous system and, like that system, controls other systems.

The muscular system ❹ consists of muscles that move the body and its parts. This system also plays an important role in regulating body temperature by generating heat. Bones are organs of the skeletal system ❺. The skeletal system protects internal organs, serves as a point of attachment for skeletal muscles, stores minerals, and produces blood cells.

The circulatory system ❻ consists of the heart, blood vessels, and blood. It cooperates with the respiratory, digestive, and urinary systems in delivering oxygen and nutrients to cells and clearing away wastes (**FIGURE 31.12**). It also helps regulate body temperature by conveying heat from the body's warm core to skin. The lymphatic system ❼ has vessels that move fluid (lymph) from tissues to blood. Organs that help protect the body from pathogens are part of this system.

The respiratory system ❽ includes two lungs and the airways leading to them. It facilitates the exchange of gases between the air and the blood.

food, water intake
oxygen inhaled
Digestive System
Respiratory System
carbon dioxide exhaled
nutrients, water, solutes
oxygen
carbon dioxide
Circulatory System
Urinary System
water, solutes
excretion of food residues
transport of materials to and from cells
elimination of soluble wastes, excess water, and salts

FIGURE 31.12 Organ system interactions. These interactions keep the body supplied with essential substances and eliminate unwanted wastes.

FIGURE 31.11 Human organ systems.

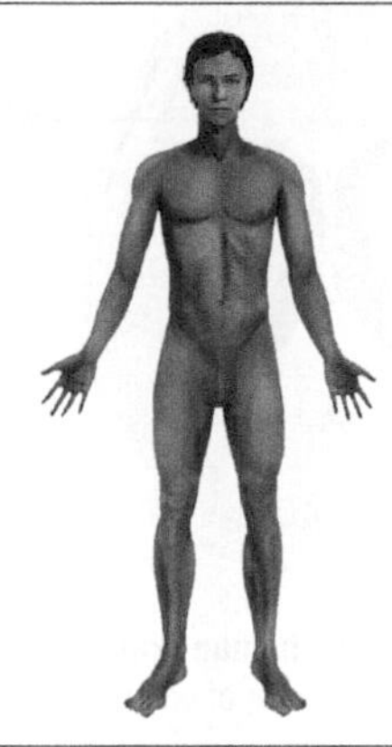

❶ Integumentary System

Protects body from injury, dehydration, pathogens; moderates temperature; excretes some wastes; detects external stimuli.

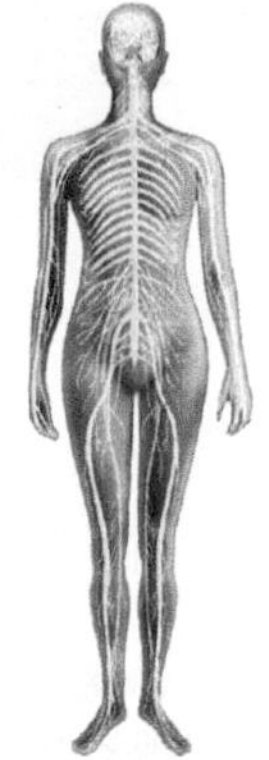

❷ Nervous System

Detects external and internal stimuli; coordinates responses to stimuli; integrates organ system activities.

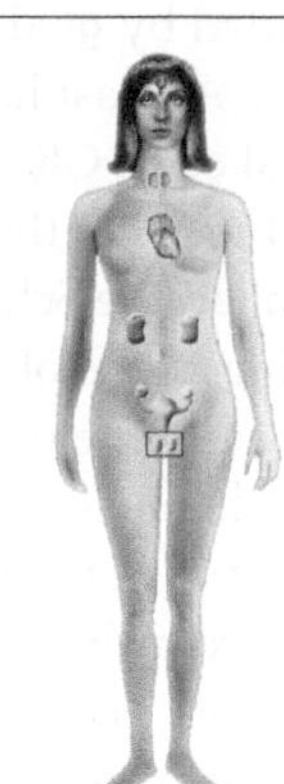

❸ Endocrine System

Secretes hormones that control activity of other organ systems. (Male testes added.)

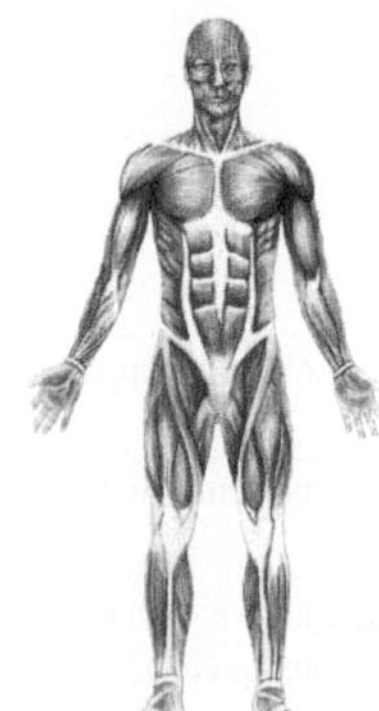

❹ Muscular System

Moves the body and its parts; maintains posture; produces heat to maintain body temperature.

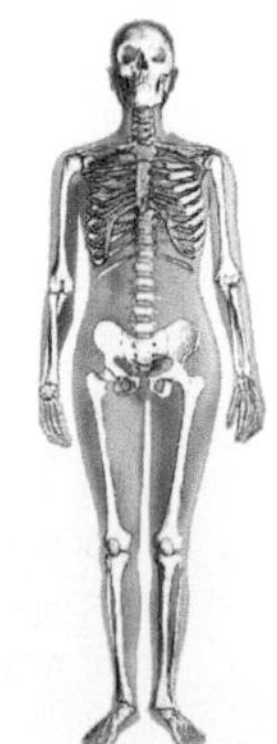

❺ Skeletal System

Supports and protects body parts; site of muscle attachment; produces red blood cells; stores minerals.

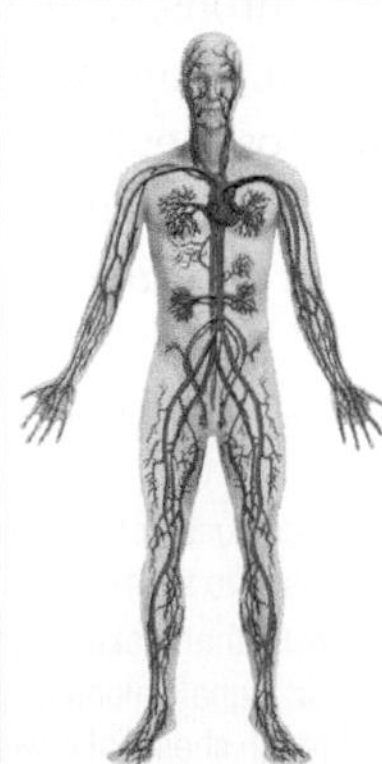

❻ Circulatory System

Distributes materials and heat throughout the body; helps maintain pH.

The digestive system takes in food, breaks it down, delivers nutrients to the blood, and eliminates undigested wastes ❾. It includes organs of the gut (esophagus, stomach, intestine), as well as glandular organs such as the liver and pancreas, which supply substances that function in digestion.

The urinary system consists of kidneys (organs that filter blood and create urine), the bladder, and tubes that deliver urine to the body surface for excretion ❿. It removes wastes from blood and adjusts blood volume and solute composition. Reproductive systems of both sexes include gamete-making organs (ovaries or testes) and ducts through which gametes travel ⓫.

Some organs are components of more than one organ system. For example, the pancreas is an endocrine gland and a digestive organ.

Many organs reside within a body cavity (**FIGURE 31.13**). The cranial cavity, which holds the brain, is continuous with the spinal cavity, which houses the spinal cord. Like other vertebrates, humans have a coelom. The diaphragm, a sheet of skeletal muscle, physically divides the human coelom into two cavities. The upper (thoracic) cavity holds the heart and lungs. The lower cavity (the abdominopelvic cavity) has two functional regions. Digestive organs, including the stomach, intestines, and liver, reside in the abdominal region (the abdominal cavity). The pelvic region (the pelvic cavity) holds the bladder and reproductive organs.

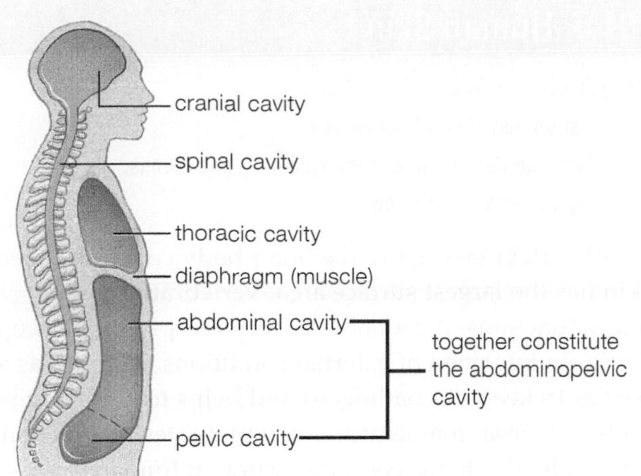
cranial cavity
spinal cavity
thoracic cavity
diaphragm (muscle)
abdominal cavity
together constitute the abdominopelvic cavity
pelvic cavity

FIGURE 31.13 Body cavities that hold organs.
FIGURE IT OUT Which body cavity or cavities are not part of the coelom?
Answer: The cranial cavity and spinal cavity.

TAKE-HOME MESSAGE 31.7

✔ Organs consist of multiple tissues and are themselves components of organ systems. Cooperative action of organ systems sustains the body as a whole.

✔ Many organs reside in body cavities. The thoracic cavity, abdominal cavity, and pelvic cavity are portions of the coelom.

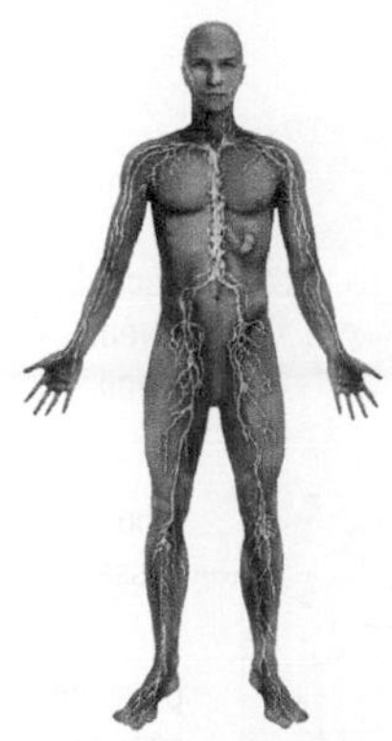

❼ Lymphatic System
Collects and returns tissue fluid to the blood; defends the body against infection, cancers.

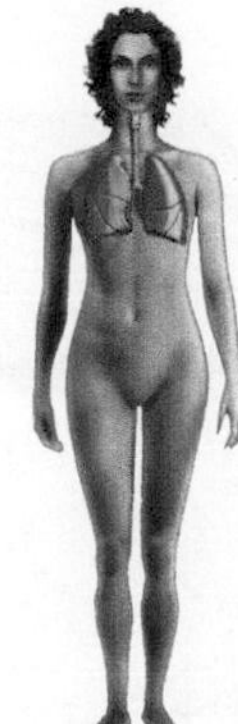

❽ Respiratory System
Takes in the oxygen for aerobic respiration; expels carbon dioxide released by this pathway.

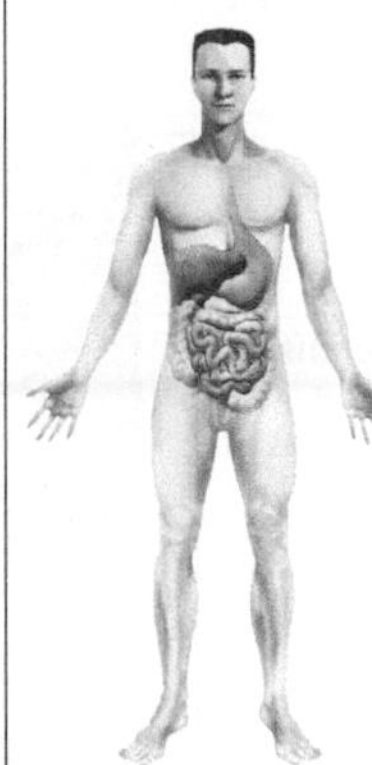

❾ Digestive System
Takes in food and water; breaks food down and absorbs needed nutrients, then eliminates food residues.

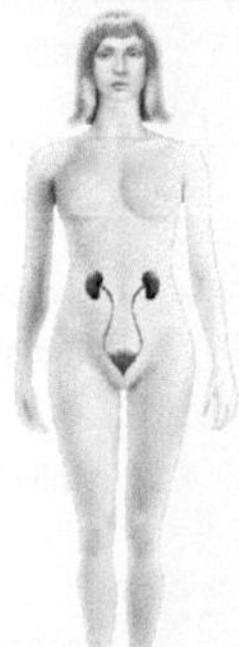

❿ Urinary System
Maintains volume and composition of blood; excretes excess fluid and wastes.

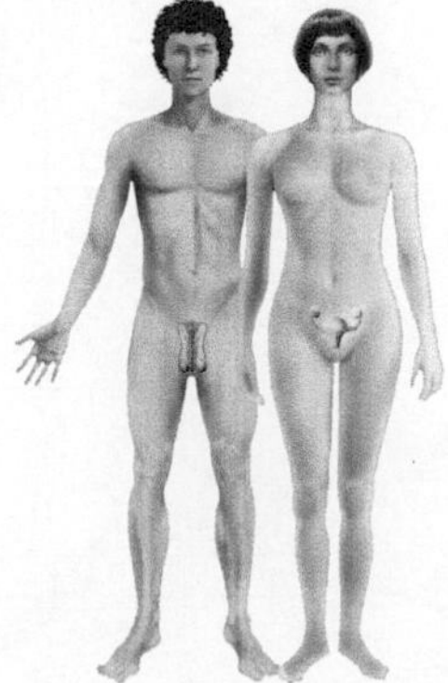

⓫ Reproductive System
Female: Produces eggs; nourishes and protects developing offspring.
Male: Produces sperm and transfers them to a female.

31.8 Human Skin

LEARNING OBJECTIVES

- List the functions of human skin.
- Describe the structure of the epidermis and dermis.
- Explain how a hair forms.

Of all vertebrate organs, the outer body covering called skin has the largest surface area. Vertebrate skin has many functions. It contains sensory receptors that keep the brain informed of external conditions. It serves as a barrier to keep out pathogens and helps to regulate the body's internal temperature. In vertebrates that live on land, skin also helps conserve water. In humans, reactions that produce vitamin D occur in the skin.

Structure of Skin

Human skin consists of two layers, a thin upper epidermis and the dermis beneath it (FIGURE 31.14). The hypodermis, a layer of loose connective and adipose tissue, underlies the skin.

Epidermis is a stratified squamous epithelium with an abundance of adhering junctions. Human epidermis consists mainly of keratinocytes, cells that synthesize the protein keratin. Keratinocytes in deep epidermal layers continually divide by mitosis, so new cells displace the older ones toward the skin's surface. As cells move upward, they travel farther and farther away from underlying connective tissues where blood vessels deliver nutrients and oxygen. Eventually, the cells die and become flattened. Dead keratinocytes accumulate at the surface of the skin, forming a tough protective layer that resists penetration by pathogens and helps the body conserve water.

The deepest epidermal layer contains melanocytes, which make pigments called melanins and donate them to keratinocytes. Variations in skin color arise from differences in the distribution and activity of melanocytes, and in the type of melanin they produce (Section 14.1). The importance of melanocyte activity can be seen with vitiligo, a skin disorder in which the destruction of these cells results in light patches of skin (FIGURE 31.16).

The **dermis** consists primarily of dense connective tissue with stretchy elastin fibers and supportive collagen fibers. Blood vessels, lymph vessels, and sensory receptors weave through the dermis. The dermis is much thicker than the epidermis, and more resistant to tearing. Leather is an animal dermis that has been treated with chemical preservatives.

Sweat glands, sebaceous glands, and hair follicles are pockets of epidermal cells that migrated into the dermis during early development. Sweat glands and sebaceous glands are exocrine glands. Sweat is mostly water. As it evaporates, it cools the skin's surface and helps keep the body from overheating. Sebaceous glands produce sebum, an oily mix of triglycerides, fatty acids, and other lipids. Sebum helps keep skin and hair soft. It also has antimicrobial properties.

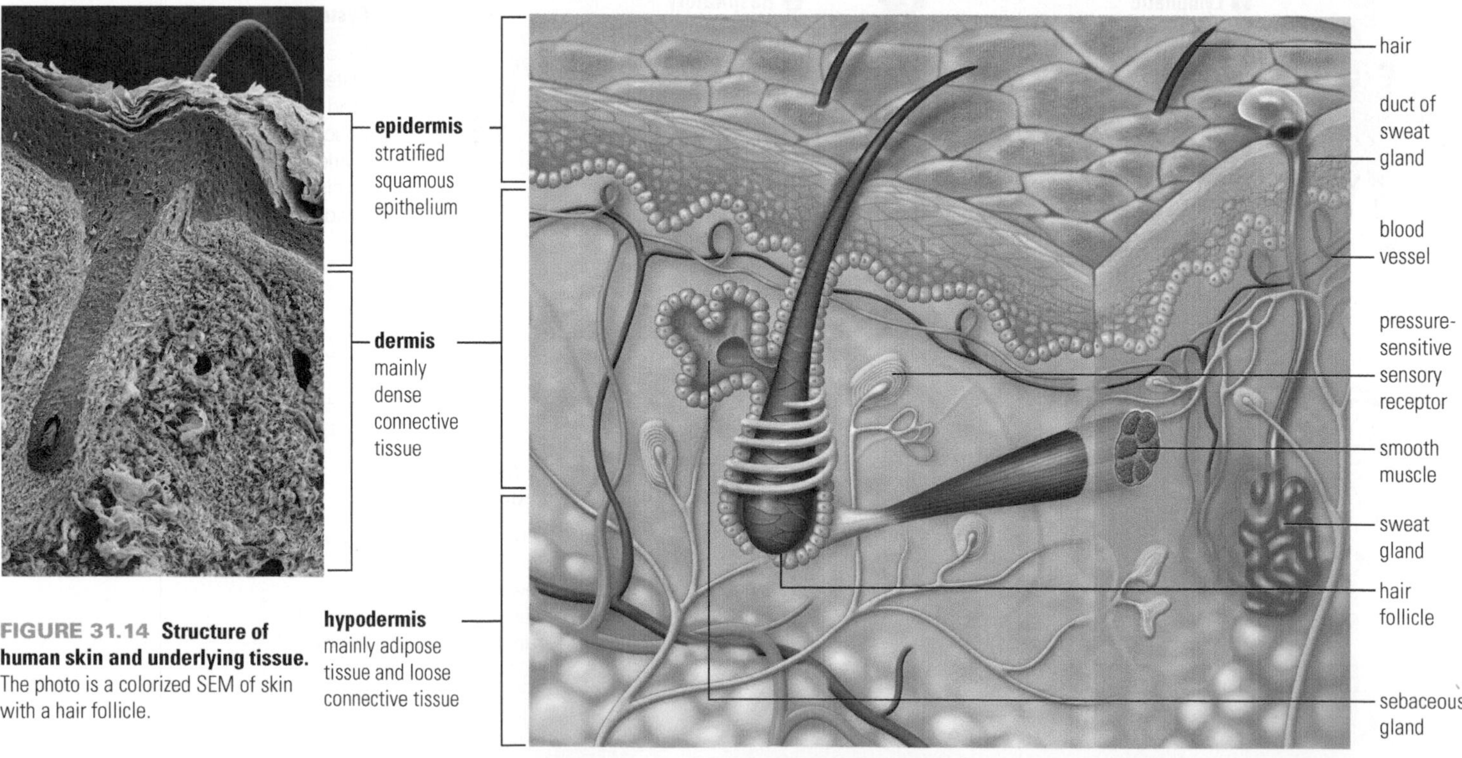

epidermis
stratified squamous epithelium
dermis
mainly dense connective tissue
hypodermis
mainly adipose tissue and loose connective tissue
hair
duct of sweat gland
blood vessel
pressure-sensitive sensory receptor
smooth muscle
sweat gland
hair follicle
sebaceous gland

FIGURE 31.14 **Structure of human skin and underlying tissue.** The photo is a colorized SEM of skin with a hair follicle.

Data Analysis Activities

Cultured Skin for Healing Wounds Diabetes is a disorder in which the blood sugar level is not properly controlled. Among other effects, it reduces blood flow to the lower legs and feet. As a result, about 3 million diabetes patients have ulcers (open wounds that do not heal) on their feet. Each year, about 80,000 require amputations.

Several companies provide cultured cell products designed to promote the healing of diabetic foot ulcers. **FIGURE 31.15** shows the results of a clinical experiment that tested the effect of one such cultured skin product versus standard treatment for diabetic foot wounds. Patients were randomly assigned to either the experimental treatment group or the control group, and their progress was monitored for 12 weeks.

1. What percentage of wounds had healed at 8 weeks when treated the standard way? When treated with cultured skin?
2. What percentage of wounds had healed at 12 weeks when treated the standard way? When treated with cultured skin?
3. How early was the healing difference between the control and treatment groups obvious?

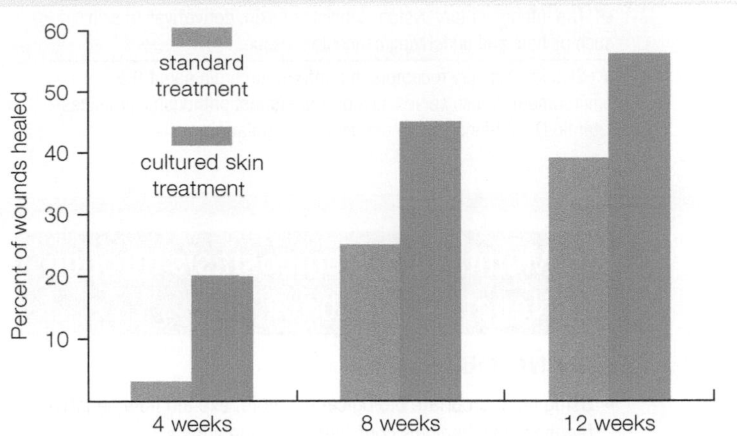

Percent of wounds healed
60
50
40
30
20
10
standard treatment
cultured skin treatment
4 weeks
8 weeks
12 weeks

FIGURE 31.15 **Treatment of diabetic food ulcers.** Results of a multicenter study of the effects of standard treatment versus use of a cultured cell product for diabetic foot ulcers. Bars show the percentage of foot ulcers that had completely healed.

FIGURE 31.16 **Vitiligo.** Lee Thomas, an African American television reporter, has vitiligo. The death of melanocytes has turned his hands white and produced white blotches on his face and arms.

The portion of a hair that you can see consists of the remains of cells that began life deep in the dermis, at the base of a hair follicle. Keratinocytes at a hair follicle's base divide every 24 to 72 hours, making them among the fastest-dividing cells in the body. As a result of these divisions, newly formed keratinocytes are continually forced away from the follicle's base. These cells are pushed through a sheath of cells that extends to the skin's surface. As keratinocytes make this journey, they take up pigment from melanocytes and they make and store keratin. Once filled with keratin, they die before ever reaching the skin's surface. The threadlike stack of dead, keratin-rich cells that eventually protrudes above the skin surface is what we call a shaft of hair.

A hair can "stand on end" because smooth muscle attaches to a sheath that surrounds the base of each hair shaft. When this muscle reflexively contracts in response to cold or fright, the hair is pulled upright.

Sunlight and the Skin

Melanin in the skin functions as a sunscreen, absorbing ultraviolet (UV) radiation that could damage DNA and other biological molecules. When skin is exposed to sunlight, melanocytes produce more melanin, resulting in a protective "tan." The dark skin color characteristic of people native to tropical regions provides protection from the intense sun in these regions.

Skin only produces the active form of vitamin D when it is exposed to specific UV wavelengths in sunlight. Thus, skin with fewer melanocytes became advantageous in some populations that moved to higher latitude regions, where sunlight is less intense, winter days are shorter, and more time is spent indoors or bundled in clothing. During cold, dark winters, having lighter skin makes it easier to get enough UV exposure to maintain an adequate level of vitamin D.

dermis Deep layer of skin that consists of connective tissue with nerves and blood vessels running through it.

epidermis Outermost tissue layer; in animals, the epithelial layer of skin.

TAKE-HOME MESSAGE 31.8

✔ The integumentary system consists of skin, derivatives of skin such as hair, and underlying connective tissue.

✔ Skin has sensory receptors that inform the brain about the environment. It also serves as a barrier against pathogens, produces vitamin D, and functions in temperature regulation.

31.9 Maintaining Homeostasis through Negative Feedback

LEARNING OBJECTIVE

- Using an appropriate biological example, explain how negative feedback mechanisms contribute to homeostasis.

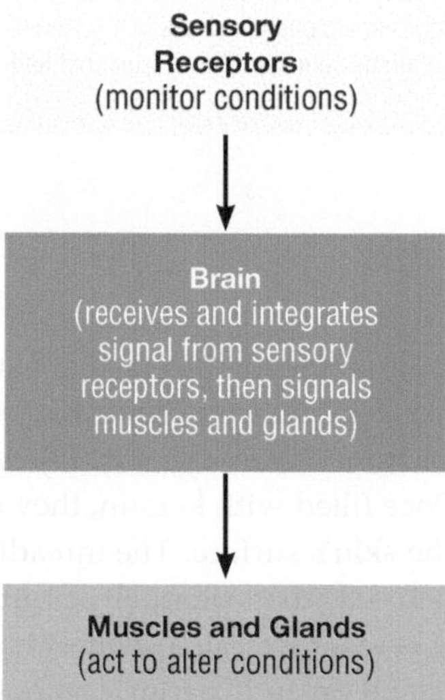

Sensory Receptors (monitor conditions)
Brain (receives and integrates signal from sensory receptors, then signals muscles and glands)
Muscles and Glands (act to alter conditions)

FIGURE 31.17 Components involved in vertebrate homeostasis.

Homeostasis, again, is the process of keeping conditions in a body within the range that the body's cells can tolerate. In vertebrates, it involves interactions among sensory receptors, the brain, and muscles and glands (**FIGURE 31.17**). A **sensory receptor** is a cell or cell component that detects a specific stimulus. Sensory receptors involved in homeostasis function like internal watchmen that monitor the body for change. Information from sensory receptors throughout the body flows to the brain. The brain evaluates this information, then signals specific muscles and glands to take the actions necessary to adjust conditions in the internal environment.

Homeostasis often involves **negative feedback**, a process in which a change causes a response that reverses the change. An air conditioner with a thermostat is a familiar nonbiological example of a negative feedback system. A person sets the air conditioner to a desired temperature. When the temperature rises above this preset point, a sensor in the air conditioner detects the change and turns the unit on. When the temperature declines to the desired level, the thermostat detects this change and turns off the air conditioner.

A negative feedback mechanism also keeps your internal temperature near 37°C (98.6°F). Consider what happens when you exercise on a hot day (**FIGURE 31.18**). Muscle activity generates heat, and your internal temperature rises. Sensory receptors in

negative feedback A change causes a response that reverses the change; important mechanism of homeostasis.

sensory receptor Cell or cell component that detects a specific type of stimulus.

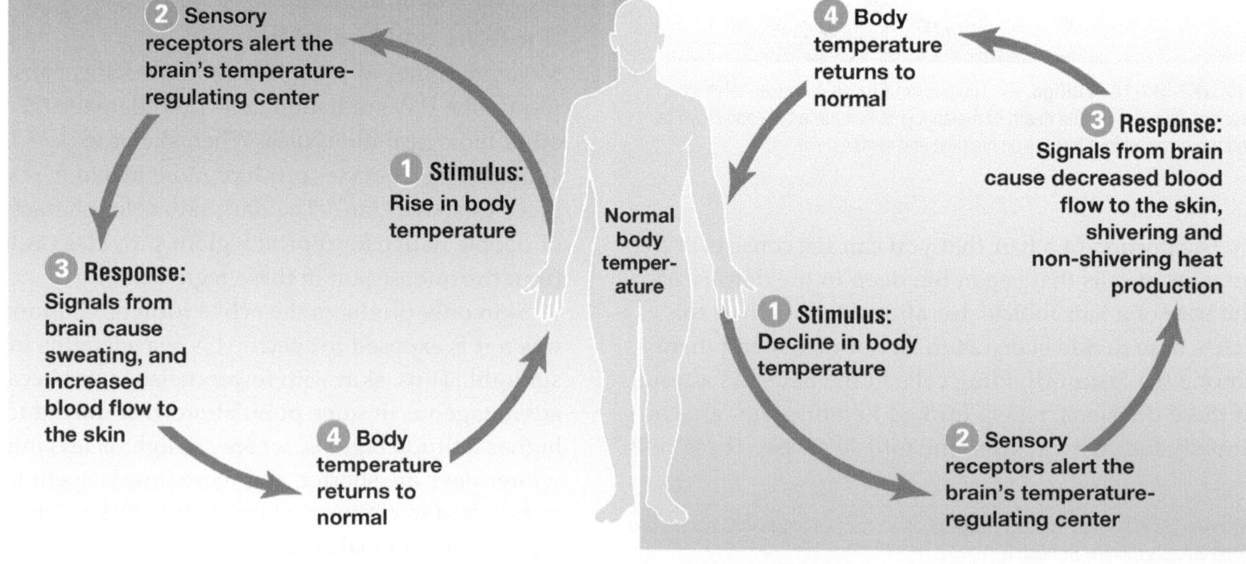

2 Sensory receptors alert the brain's temperature-regulating center
1 Stimulus: Rise in body temperature
3 Response: Signals from brain cause sweating, and increased blood flow to the skin
4 Body temperature returns to normal
Normal body temper-ature
4 Body temperature returns to normal
3 Response: Signals from brain cause decreased blood flow to the skin, shivering and non-shivering heat production
1 Stimulus: Decline in body temperature
2 Sensory receptors alert the brain's temperature-regulating center

A Response to an increase in body temperature.

B Response to a decrease in body temperature.

FIGURE 31.18 Negative feedback control of human body temperature.

the skin detect the increase and signal the brain, which sends signals that bring about a response. Blood flow shifts, so more blood from the body's hot interior flows to the skin. The shift maximizes the amount of heat given off to the surrounding air. At the same time, sweat glands increase their output. Evaporation of sweat helps cool the body surface. Breathing quickens and deepens, speeding the transfer of heat from the blood in your lungs to the air. As your rate of heat loss increases, temperature falls.

Sensory receptors also notify the brain when body temperature declines. The brain responds by sending signals that divert blood flow away from the skin and cause body hair to stand up. With prolonged cold, the brain commands skeletal muscles to contract 10 to 20 times a second. This shivering increases heat production by muscles. When you warm up, blood returns to your skin, you stop shivering, and your goose bumps subside.

Through the process of negative feedback, the body can prevent large variations in external temperature from causing similarly large changes inside the body.

TAKE-HOME MESSAGE 31.9

✔ With negative feedback an initial change causes a response that reverses that change. Negative feedback mechanisms contribute to homeostasis by preventing dramatic changes in internal conditions.

✔ Negative feedback involves sensory receptors that detect changes and signal the brain, and the muscles and glands that respond to signals from the brain.

31.1 Making Replacement Cells (revisited)

Stem cell researchers seek to find ways to replace dead or malfunctioning cells that a body cannot replace on its own. One way to do this is to produce the desired cell types from human embryonic stem cells, also known as hESCs (**FIGURE 31.19**). Human embryonic stem cells are derived either from human embryos that were produced for fertility treatments but never used or from nonviable embryos produced specifically for the purpose of stem cell harvest.

Several clinical trials are now testing the safety and effectiveness of replacement cells derived from hESCs. One clinical trial is looking into whether glial cells derived from hESCs can restore function in patients paralyzed by a recent spinal cord injury. Another is testing whether retina cells derived from hESCs can reverse a common type of adult-onset blindness called macular degeneration. Still another is investigating whether insulin-producing pancreatic cells derived from hESCs can be used to cure diabetes.

An alternative approach is to start with adult cells and cause them to reverse the process by which they differentiated. This dedifferentiation restores the cells to a pluripotent state. Adult cells that have been induced to behave like embryonic cells are called induced pluripotent stem cells (IPSCs). For therapeutic purposes, IPSCs like hESCs must be forced to differentiate into a specific cell type.

The first clinical experiment using specialized cells derived from IPSCs began only recently. The cells are being used to treat macular degeneration, the same type of blindness for which hESC-derived cells are being tested.

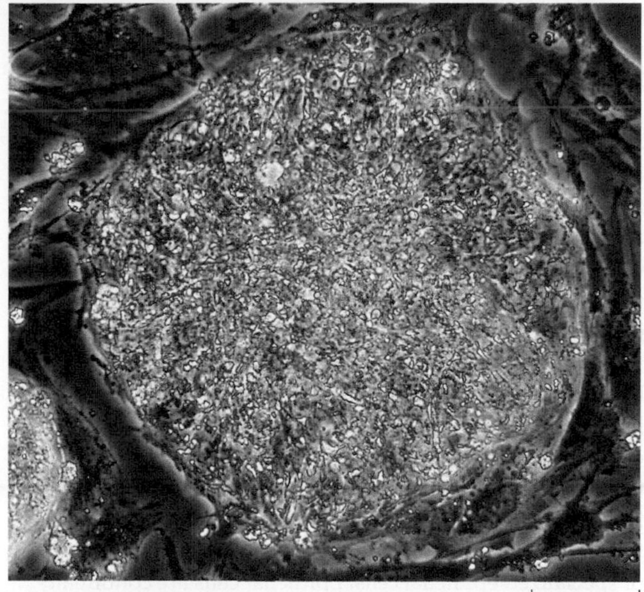
50 µm

FIGURE 31.19 Stem cells. A cluster of human embryonic stem cells (hESCs) growing in culture at the University of Pittsburgh.

Transdifferentiation is another method for creating replacement cells. This method converts one type of differentiated cell directly into another without going through a pluripotent stage. For example, cultured fibroblasts have been transdifferentiated into neurons and cardiac muscle cells. Similarly, cultured liver cells have been transdifferentiated into insulin-secreting pancreatic beta cells. Clinical trials of transdifferentiated cells are likely to begin within the next few years. Both induction of pluripotency and transdifferentiation can be carried out using a patient's own cells. ●

STUDY GUIDE

Section 31.1 All cells in an animal body are derived from **stem cells**, which can either divide or differentiate into a specialized cell type. The first divisions of a fertilized egg yield totipotent cells that can form any tissue or develop into a new individual. Later embryos have pluripotent cells that can still form any tissue. After birth, cells are less versatile. Researchers hope to use embryonic stem cells to produce new cells of types that are not normally replaced in adults. Induced pluripotent stem cells or transdifferentiated cells derived from adult cells may also be used for this purpose.

Section 31.2 In all animals more complex than sponges, cells are organized as **tissues**. Vertebrates have four types of tissues organized as **organs**. An organ system consists of two or more organs that interact in tasks that keep the whole body functioning smoothly.

An animal body consists largely of fluids. Most fluid is intracellular. **Extracellular fluid** serves as the body's internal environment. In humans, extracellular fluid consists mainly of **interstitial fluid** and plasma.

Physical factors, developmental factors, and evolutionary history constrain animal body plans.

Section 31.3 **Epithelial tissue** covers the body surface and lines its internal tubes and cavities. An epithelium has a free apical surface. Its basal surface secretes a **basement membrane** that attaches it to underlying connective tissue. Cells in an epithelium have little extracellular matrix between them. They connect to one another by way of **adhering junctions** or **tight junctions**. Simple epithelia are one cell thick; stratified epithelia have multiple layers. Some epithelial cells have cilia that move materials across their surface. Others have **microvilli** that increase their surface area for absorption or secretion. Gland cells produce secretions that act outside the cell. Ductless **endocrine glands** secrete hormones into the blood. **Exocrine glands** secrete products such as milk or saliva through ducts. Hair, fur, and nails are keratin-rich remains of specialized epithelial cells. Epithelial cells are continually shed and replaced, so epithelia are prone to cancer.

Section 31.4 **Connective tissues** "connect" tissues to one another, both functionally and structurally. They bind, organize, support, strengthen, protect, and insulate other tissues. All consist of cells in a secreted matrix. Loose and dense connective tissues have the same components (fibroblasts and a matrix with elastin and collagen fibers) but in different proportions. **Loose connective tissue** holds internal organs in place. Ligaments and tendons consist of **dense connective tissue**. Fat stored in **adipose tissue** is the body's main energy reservoir. Rubbery **cartilage** and mineral-hardened **bone tissue** are components of the vertebrate skeleton. **Blood** consists of fluid plasma, cells, and platelets. It is considered a connective tissue because blood cells and platelets arise from stem cells in bone.

Section 31.5 Contraction of muscle tissue moves a body or its parts. Muscle contraction is a response to signals from the nervous system and is fueled by ATP. **Skeletal muscle tissue** consists of long fibers with multiple nuclei and has a striated (striped) appearance. Skeletal muscles, which pull on bones, are under voluntary control. They have large stores of glycogen. The metabolic reactions carried out by skeletal muscle are the main source of body heat. **Cardiac muscle tissue**, found only in the heart wall, has a striated appearance. Cardiac muscle cells are branched and have a single nucleus. **Smooth muscle tissue** is found in the walls of tubular organs and some blood vessels. Its unbranched cells have a single nucleus, taper at both ends, and are not striated. Cardiac muscle and smooth muscle are not under voluntary control. In both, **gap junctions** allow signals to travel between adjacent cells, so many cells to contract as a unit.

Section 31.6 **Nervous tissue** makes up the communication lines of the body and is found in the brain, spinal cord, and nerves. It consists of **neurons** that send and receive communication signals and **glial cells** that support the neurons. A neuron has a central cell body and long cytoplasmic extensions that send and receive signals. Sensory neurons detect information, interneurons integrate and assess information about internal and external conditions, and motor neurons command muscles and glands. White matter and gray matter are two types of nervous tissue.

Section 31.7 An organ system consists of two or more organs that interact chemically, physically, or both in tasks that help keep individual cells as well as the whole body functioning. All vertebrates have the same set of organ systems. Many internal organs reside inside body cavities. The human thoracic, abdominal, and pelvic cavities are regions of our coelom. Organ systems interact to provide cells with the materials that they need and to rid the body of wastes.

Section 31.8 Skin is the human body's largest organ. It functions in temperature control, detection of shifts in external conditions, vitamin production, and defense against pathogens. The outermost layer of skin, the **epidermis**, is a stratified squamous epithelium consisting mainly of keratinocytes. Melanocytes produce the melanin that gives skin its color. The deeper **dermis** consists mainly of dense connective tissue and contains blood vessels, nerves, and muscles. Underlying the skin is the hypodermis, a layer of connective tissue and adipose cells.

Sweat glands, sebaceous glands, and hair follicles are collections of epidermal cells that descended into the dermis during development. The main component of sweat is water, which cools the body when it evaporates. Sebaceous glands secrete an oily sebum. The visible portion of a hair consists of the keratin-rich remains of cells that began their life deep inside the hair follicle.

Skin color varies with latitude. People native to tropical regions have darker skin than those farther from the equator. Lighter skin allows people who live in regions where exposure to sunlight is seasonally limited to produce sufficient vitamin D. Tanning is a protective response to the UV wavelengths in sunlight.

Section 31.9 Homeostasis requires **sensory receptors** that detect changes, an integrating center (the brain), and effectors (muscles and glands) that bring about responses. **Negative feedback** often plays a role in homeostasis: A change causes the body to respond in a way that reverses the change.

SELF-QUIZ

Answers in Appendix VII

1. _______ tissues are sheetlike with one free surface.
   a. Epithelial c. Nervous
   b. Muscle d. Connective

2. _______ keep fluid from leaking between cells.
   a. Tight junctions c. Gap junctions
   b. Adhering junctions d. Fibroblasts

3. Exocrine glands are specialized _______ tissue.
   a. epithelial c. nervous
   b. muscle d. connective

4. A rubbery secreted matrix of proteoglycans and collagen surrounds living cells in _______ .
   a. bone c. adipose tissue
   b. cartilage d. blood

5. Blood cells develop from stem cells in _______ .
   a. epidermis c. cartilage
   b. dermis d. bone

6. Your body's main energy reservoir is _______ .
   a. glycogen stored in cardiac muscle
   b. lipids stored in adipose tissue
   c. starch stored in skeletal muscle
   d. phosphorus stored in bone

7. Cytoplasmic extensions of _______ send and receive chemical messages.
   a. glial cells c. fibroblasts
   b. neurons d. melanocytes

8. _______ muscle pulls on bones and _______ muscle regulates the diameter of blood vessels.
   a. Skeletal/cardiac c. Skeletal/smooth
   b. Smooth/cardiac d. Smooth/skeletal

9. Straps of dense, regular connective tissue _______ .
   a. connect muscles to bones c. underlie the skin
   b. produce blood cells d. lack fibroblasts

10. _______ increase the surface area of some epithelial cells.
    a. Microfilaments c. Gap junctions
    b. Microvilli d. Adhering junctions

11. Tears are an _______ secretion released by specialized _______ tissue cells.
    a. endocrine; epithelial c. exocrine; epithelial
    b. endocrine; connective d. exocrine; connective

12. Cancers most commonly arise in _______ tissue.
    a. epithelial c. nervous
    b. muscle d. connective

13. The most abundant protein in your body is _______ .
    a. melanin c. collagen
    b. elastin d. keratin

14. Match each term with the most suitable description.
    ___ exocrine gland a. cell in nervous tissue
    ___ endocrine gland b. secretes a hormone
    ___ fibroblast c. secretes through a duct
    ___ melanocyte d. collagen-producing cell
    ___ neuron e. contracts involuntarily
    ___ smooth muscle f. pigment-producing cell
    ___ skeletal muscle g. main source of body heat
    ___ blood h. main cells in epidermis
    ___ keratinocyte i. fluid connective tissue
    ___ extracellular fluid j. includes interstitial fluid, lymph

15. With negative feedback, detection of a change brings about a response that _______ the change.
    a. reverses c. has no effect on
    b. accelerates d. mimics

CRITICAL THINKING

1. IPSCs are nearly identical to human embryonic stem cells in terms of gene expression, but there may be other ways in which they are not equivalent. For example, the telomeres of IPSCs often vary in length, with many IPSCs cells having telomeres shorter than those of embryonic. How might shortened telomeres affect the life-span of IPSCs or of differentiated cells derived from them?

2. Radiation and chemotherapy drugs preferentially kill cells that divide frequently, most notably cancer cells. These cancer treatments also cause hair to fall out. Why?

3. Each level of biological organization has emergent properties that arise from the interaction of its component parts. For example, cells have a capacity for inheritance that molecules making up the cell do not. What are some emergent properties of specific types of tissues?

32 NEURAL CONTROL

CORE CONCEPTS

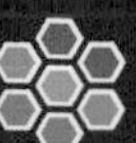

Structure and Function

The three-dimensional form and arrangement of biological structures give rise to their function and interactions. Transport proteins embedded in plasma membranes allow cells to establish electrical and concentration gradients across their plasma membrane. In neurons, controlled opening and closing of specialized tranport proteins allows transmission of electrical signals within a cell. Release of chemical signals and their reception by specialized proteins allows communication between cells.

Evolution

Evolution underlies the unity and diversity of life. Shared features provide evidence of descent from a common ancestor. In animals, the nervous system detects changes in the internal and external environment and issues the signals that bring about appropriate responses to those changes. The structure of animal nervous systems varies among groups. There has been an evolutionary trend toward more complex, centralized systems.

Systems

Complex properties arise from interactions among components of a biological system. The neurons and supporting cells of the vertebrate nervous system are organized into a brain, a spinal cord, and the many peripheral nerves that extend throughout the body. In humans, interactions among these components not only maintain homeostasis but also give rise to emotion, memory, a capacity for language, and consciousness.

Links to Earlier Concepts

The nervous signals discussed in this chapter involve receptor proteins (Section 5.7) and transport mechanisms (5.9, 5.10). They depend on ion gradients, which are a type of potential energy (5.2). We also reconsider health-related topics such as cancer (11.6) and alcohol abuse (5.1).

32.1 Impacts of Concussions

The human brain is surprisingly delicate. It is only about as firm as Jell-O or warm butter. Head impacts or rapid stops can cause this soft, squishy organ to slam against the inside of the bony skull, causing a mild traumatic brain injury called a concussion. Symptoms of a concussion can include confusion, dizziness, blurred vision, increased sensitivity to light, headache, impaired short-term memory, difficulty concentrating, irritability, nausea, altered sleep patterns, and a temporary loss of consciousness.

The Centers for Disease Control estimates that somewhere between 1.4 and 3.8 million concussions occur each year in the United States. Participating in contact sports raises the risk, but concussions also occur as a result of car accidents, workplace accidents, and simple falls. Military personnel deployed to combat zones often suffer concussions as a result of exposure to explosions.

Usually, the brain resumes normal function within 7 to 10 days after a concussion. However, repeated head injuries can cause a neurodegenerative disorder called chronic traumatic encephalopathy (CTE). Symptoms of CTE can include memory loss, emotional problems, depression, suicidal impulses, and dementia.

Concussions are an occupational hazard for professional football players (**FIGURE 32.1**). Physicians for the National Football League routinely diagnose between 100 and 150 concussions per season, and these injuries can have long-term consequences. In 2017, researchers investigating the possible connection between football and CTE reported that they had examined the brains of 111 deceased professional football players and found signs of CTE in 110 of them ●

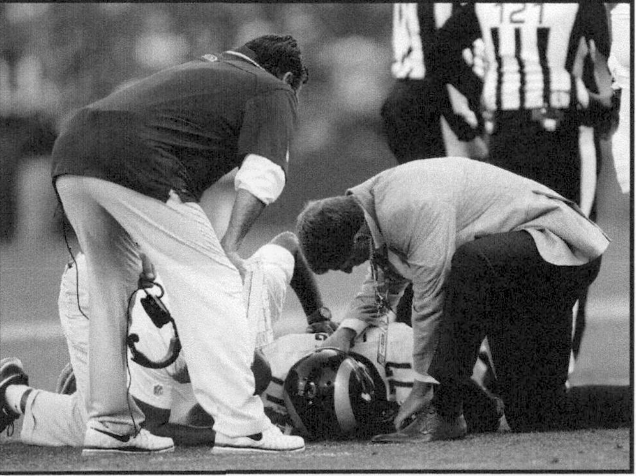

FIGURE 32.1 Football and concussions. Jake Long of the St. Louis Rams football team is examined after suffering a concussion during a game.

CREDITS: (opposite) Courtesy of Rekin Brain Science Institute; (1) Michael Zagaris/Getty Images.

32.2 Animal Nervous Systems

LEARNING OBJECTIVES

- List the three steps involved in intercellular communication.
- Explain how a sea anemone's nerve net differs from the nervous system of a flatworm in its structure and function.
- Name the two functional divisions of the vertebrate nervous system, and describe how they interact.

For an animal body to work as a whole, cells must constantly signal one another. In some tissues, gap junctions connect adjacent cells and allow direct transfer of signals between their cytoplasms. In other cases, intercellular communication involves secretion of signaling molecules into interstitial fluid (the fluid between cells). These molecules exert their effects when they bind to a receptor on or inside another cell. Any cell with receptors that bind and respond to a specific signaling molecule is a "target" of that molecule.

Secreted signaling molecules exert their effect by a three-step process. The molecules bind to a receptor protein on or in the target cell, the receptor transduces the signal (changes it into a form that affects target cell behavior), and the target cell responds:

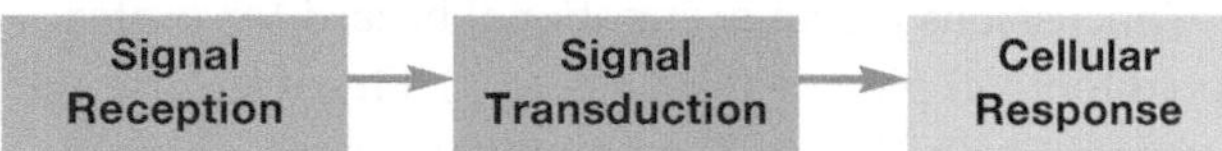
Signal Reception
Signal Transduction
Cellular Response

Two organ systems—the nervous system and the endocrine system—facilitate long-distance communication among cells of an animal body. The endocrine system coordinates activities by way of molecules called hormones that travel in the blood. We discuss the endocrine system in detail in Chapter 34. Here, we focus on the nervous system.

As Section 31.6 explained, neurons make up the communication lines of nervous systems. Neurons transmit electrical signals along their plasma membrane and also send chemical messages to other cells. In most animals, glial cells support the neurons.

Invertebrate Nervous Systems

Sponges do not have neurons or other cells organized as a nervous system.

Invertebrates with radial symmetry have a mesh of interconnected neurons called a **nerve net**. Information flows in all directions among cells of the nerve net, and there is no centralized, controlling organ that functions like a brain. Sea anemones and other cnidarians (Section 24.5) are among the simplest animals with a nerve net (**FIGURE 32.2A**). By causing cells in the body wall to contract, a cnidarian nerve net can alter the size of the mouth, change the body's shape, or shift the position of the tentacles.

Echinoderms (Section 24.11) are also radial, and they too have a nerve net. However, a echinoderm nerve net is a bit more complex than a cnidarian's because it includes nerves. A **nerve** is a bundle of neuron fibers (cytoplasmic extensions) wrapped in connective tissue. In sea stars, a ring of nerves surrounds the mouth and a nerve extends from this ring into each arm.

Most animals have bilateral symmetry, with paired organs arranged on either side of the main body axis (Section 24.2). Evolution of bilateral body plans was accompanied by a trend toward **cephalization**: the formation of a distinct head that has a large concentration of neurons.

Planarian flatworms are bilateral animals with a simple nervous system. A pair of ganglia in the head serves as an integrating center (**FIGURE 32.2B**). Each **ganglion** (plural, ganglia) is a cluster of neuron cell bodies. (A cell body is the part of the neuron that holds

increasing cephalization

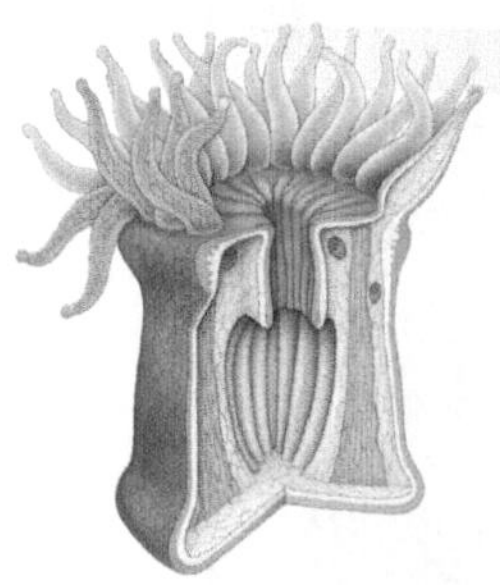

A Nerve net (purple) of sea anemone. No central organ integrates signals.

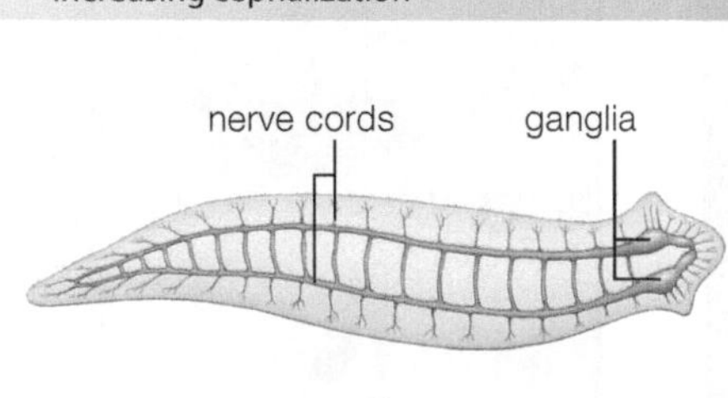
nerve cords
ganglia

B Planarian nervous system. Two ganglia in the head integrate information. Nerve cords extend the length of the body along the ventral (lower) surface.

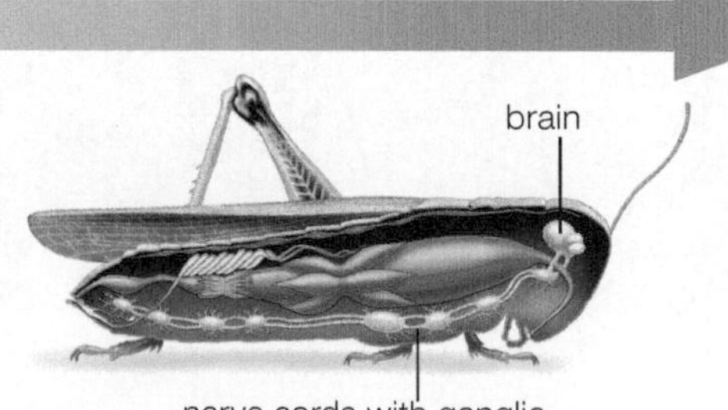
brain
nerve cords with ganglia

C Insect nervous system. A brain with hundreds of thousands of neurons connects to a ventral nerve cord that has a ganglion in each segment. The ganglia serve as local control centers. The head has a variety of sensory receptors.

FIGURE 32.2 Examples of invertebrate nervous systems.

FIGURE IT OUT Which of these organisms has/have a cephalized nervous system?

Answer: The planarian and grasshopper

the cell's nucleus and other organelles.) A planarian's ganglia receive signals from eyespots and from the chemical-detecting cells on the head. The ganglia connect to paired nerve cords that run along the animal's ventral (lower) surface. A **nerve cord** is a chain of ganglia that extends the length of a body. Nerve fibers extend out from the ganglia.

In annelids and arthropods, paired nerve cords run along the ventral surface and connect to a simple brain (**FIGURE 32.2C**). A **brain** is a central control organ of a nervous system. It receives and integrates sensory information, regulates internal processes, and sends out signals that bring about movements.

Organization of the nervous system varies among mollusk groups. Bivalves are brainless, with widely separated ganglia that serve as local control centers. On the other hand, cephalopods such as octopuses have the largest and most complex nervous system of any invertebrate group. An octopus has a proportionately larger brain than many fishes or reptiles.

The Vertebrate Nervous System

A dorsal nerve cord (a nerve cord that runs along the back) is one of the defining features of chordate embryos (Section 25.2). In vertebrates, the dorsal nerve cord evolved into a brain and spinal cord, which together constitute the **central nervous system** (**FIGURE 32.3**). Nerves that extend from the central nervous system through the rest of the body are the **peripheral nervous system**.

The human peripheral nervous system consists of 12 cranial nerves that connect to the brain and 31 spinal nerves that connect to the spinal cord (**FIGURE 32.4**). Most cranial nerves, and all spinal nerves, carry signals both to and from the central nervous system. Consider the sciatic nerve, which runs from the spinal cord, through the buttock, and down the leg. When someone touches your thigh, this nerve carries signals from receptors in the skin to the spinal cord. When you bend your knee, the same nerve carries commands for movement from the spinal cord to muscles in the leg.

TAKE-HOME MESSAGE 32.2

- ✔ Cnidarians and echinoderms have a simple nervous system—a nerve net with no central integrating organ.
- ✔ Flatworms have paired ganglia that serve as an integrating center. Some other invertebrates have brains.
- ✔ Bilateral invertebrates usually have a pair of ventral nerve cords. In contrast, chordates have one dorsal nerve cord.
- ✔ The vertebrate nervous system includes a well-developed brain, a spinal cord, and peripheral nerves.

CENTRAL NERVOUS SYSTEM
Brain, Spinal Cord

PERIPHERAL NERVOUS SYSTEM
Cranial and Spinal Nerves

Somatic Nerves
Nerves that carry signals to and from skeletal muscle, tendons, and the skin

Autonomic Nerves
Nerves that carry signals to smooth muscle, cardiac muscle, and glands

Sympathetic Division | Parasympathetic Division
Two sets of nerves that have opposing effects and often signal the same effectors

FIGURE 32.3 Functional divisions of vertebrate nervous systems.

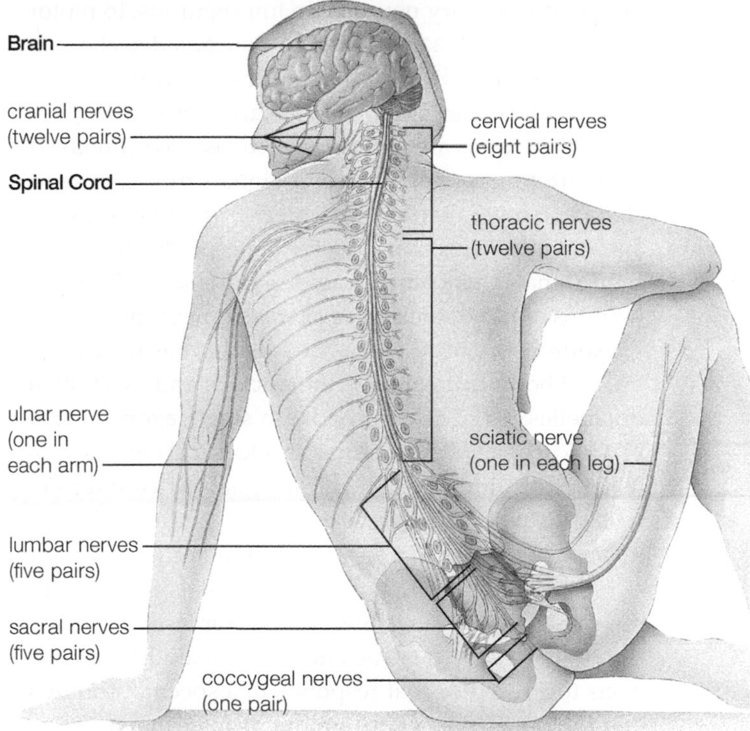
Brain
cranial nerves (twelve pairs)
Spinal Cord
cervical nerves (eight pairs)
thoracic nerves (twelve pairs)
ulnar nerve (one in each arm)
sciatic nerve (one in each leg)
lumbar nerves (five pairs)
sacral nerves (five pairs)
coccygeal nerves (one pair)

FIGURE 32.4 Examples of major human nerves.

brain Central control organ in a nervous system.
central nervous system Brain and spinal cord.
cephalization (SEF-uh-lih-ZAY-shun) Neurons become concentrated in the head of an animal.
ganglion (GANG-lee-on) Plural, ganglia. Cluster of neuron cell bodies.
nerve Neuron fibers bundled inside a sheath of connective tissue.
nerve cord Chain of ganglia that extends the length of a body.
nerve net Mesh of interacting neurons with no central control organ.
peripheral nervous system Nerves that extend through the body and carry signals to and from the central nervous system.

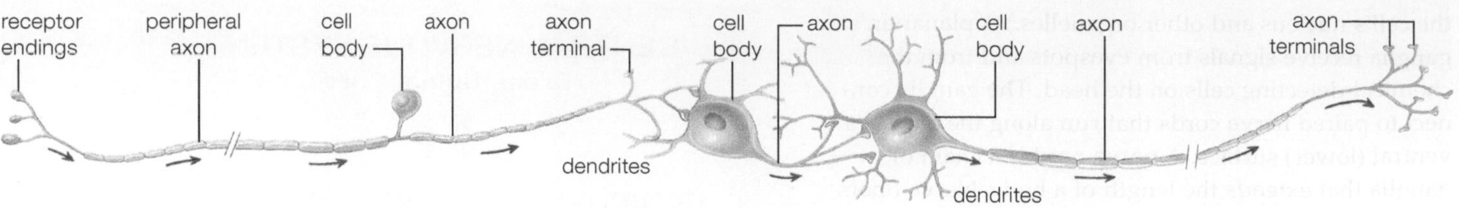
receptor endings
peripheral axon
cell body
axon
axon terminal
dendrites
cell body
axon
cell body
axon
axon terminals
dendrites

A Sensory neurons have a long axon with receptor endings at one end and axon terminals at the other.

B Interneurons have many dendrites and a short axon.

C Motor neurons have many dendrites and a long axon.

FIGURE 32.5 The three types of neurons. Arrows indicate the direction of information flow.

32.3 Cells of the Nervous System

LEARNING OBJECTIVE

- Compare the three types of neurons, and explain how differences in their structure reflect their functions.

Three Types of Neurons

In vertebrate nervous systems, information typically flows from sensory neurons, to interneurons, to motor neurons (**FIGURE 32.5**). **Sensory neurons** detect a stimulus such as light or touch. When activated by this stimulus, they usually signal an interneuron. Sensory neurons are also referred to as *afferent* neurons. **Interneurons** both receive signals from and send signals to other neurons. An interneuron may signal another interneuron or a motor neuron. **Motor neurons** send signals that control muscles and glands. Motor neurons are also referred to as *efferent* neurons.

Neuron structure varies, but all neurons have a central cell body that contains the nucleus and most other organelles. A typical neuron has a single **axon**, a cytoplasmic extension that transmits electrical signals along its length and releases chemical signals at axon terminals. The axon terminals are knoblike endings that lie in close proximity to a signal-receiving cell.

Sensory neurons do not receive signals from other neurons, so the axon is their only cytoplasmic extension (**FIGURE 32.5A**). One end of this axon has or connects to a receptor that responds to a specific stimulus and the other has signal-sending axon terminals.

Interneurons and motor neurons both receive and send signals, so they have both dendrites and an axon. **Dendrites** are cytoplasmic extensions that receive chemical signals from other neurons and convert them to electrical signals. Nearly all vertebrate interneurons are located in the brain or spinal cord. (Some interneurons are located in the wall of the digestive tract.) Interneurons have a short axon (**FIGURE 32.5B**). Motor neurons have a cell body in the brain or spinal cord, and an axon that extends out to the muscle or gland they control (**FIGURE 32.5C**).

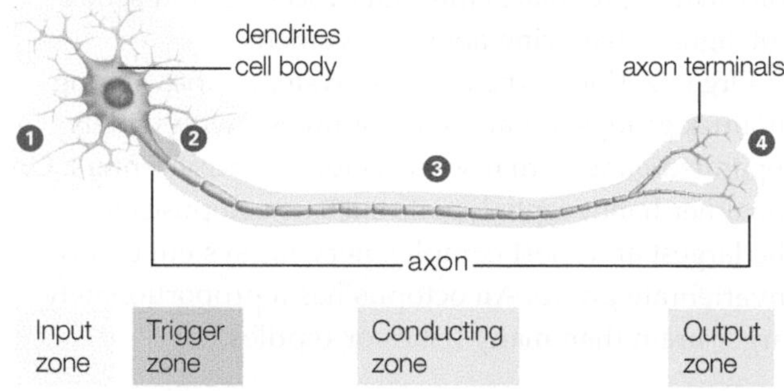
dendrites
cell body
axon terminals
1
2
3
4
axon
Input zone
Trigger zone
Conducting zone
Output zone

FIGURE 32.6 Functional zones of a motor neuron.

FIGURE 32.6 shows the functional zones of a motor neuron. Dendrites and the cell body are signal input zones, where chemical signals from other neurons arrive ❶. The region of axon nearest the cell body is a trigger zone ❷. Excitation of this region triggers flow of electrical signals along the conducting zone of the axon ❸. Axon terminals are the output zone: They release signaling molecules that influence other cells ❹.

Glial Cells

Glial cells, or neuroglia, act as a framework that holds neurons in place; *glia* means "glue" in Latin. Some neuroglia wrap around axons and speed transmission of nerve impulses (more about this in Section 32.4). Ciliated glial cells that line the fluid-filled cavities inside the brain and spinal cord keep nutrient-rich fluid moving around these organs. Other glia wrap around blood vessels of the brain and help to prevent bloodborne toxins from reaching this organ. We discuss additional roles of glial cells in Section 32.8.

TAKE-HOME MESSAGE 32.3

- ✔ All neurons have a cell body and a single axon that sends chemical signals to other cells.
- ✔ Interneurons and motor neurons have signal-receiving dendrites. A sensory neuron has no dendrites. One end of its axon has receptor endings that detect a specific stimulus.
- ✔ Glial cells provide structural and functional support to neurons.

32.4 Electrical Signaling in Neurons

LEARNING OBJECTIVES

- Describe the ion concentration gradients across the plasma membrane of a neuron at resting potential.
- Explain how and why membrane potential changes during an action potential.

All animal cells have an electrical gradient across their plasma membrane. The cytoplasmic fluid near this membrane contains more negatively charged ions and proteins than the interstitial fluid does. As in a battery, the separation of charge constitutes potential energy (Section 5.2). We measure potential energy as voltage, and the voltage across a cell's membrane is called **membrane potential**. Researchers determine the membrane potential across a neuron's plasma membrane by inserting one electrode into a neuron and another into the interstitial fluid just outside of it.

Resting Potential

The membrane potential of a neuron that is not being stimulated is the **resting potential**. This potential is usually about −70 millivolts. [A millivolt (mV) is one-thousandth of a volt.] The negative sign indicates the charge of the neuron's cytoplasm is more negative than that of the adjacent interstitial fluid.

What causes resting potential? The cytoplasm contains negatively charged proteins and amino acids that are not present in the interstitial fluid. The proteins cannot cross the neuron's plasma membrane and so remain trapped within the cell. The distribution of positively charged potassium ions (K^+) and positively charged sodium ions (Na^+) also influences resting potential. These ions move in and out of the neuron with the assistance of transport proteins. Sodium–potassium pumps actively move two potassium ions into the cell for every three sodium ions they move out (**FIGURE 32.7A**). By moving more positively charged ions out of the cell than into it, these pumps increase the electrical gradient across the membrane. The pumps also create concentration gradients for sodium and potassium ions across the membrane.

action potential Brief reversal of the voltage difference across the plasma membrane of a neuron or muscle cell.
axon Of a neuron, a cytoplasmic extension that transmits electrical signals along its length and secretes chemical signals at its endings.
dendrite Of a neuron, a cytoplasmic extension that receives chemical signals sent by other neurons and converts them to electrical signals.
interneuron Neuron that both receives signals from and sends signals to other neurons. Located mainly in the brain and spinal cord.
membrane potential Voltage difference across a cell membrane.
motor neuron Neuron that controls a muscle or gland.
resting potential Membrane potential of a neuron at rest.
sensory neuron Neuron that is activated when its receptor endings detect a specific stimulus, such as light or pressure.

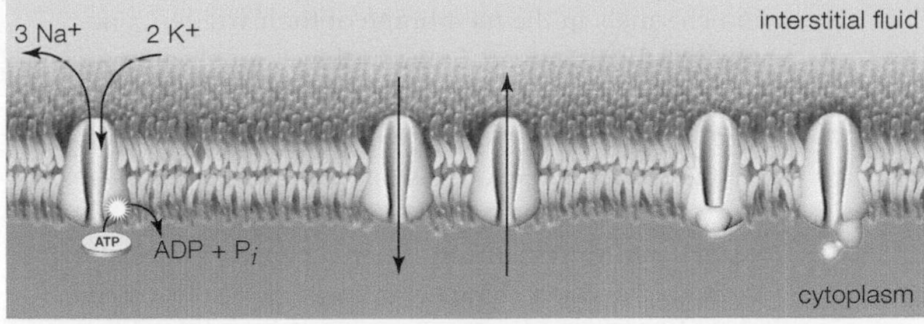

3 Na+
2 K+
interstitial fluid
ATP
ADP + Pi
cytoplasm

A Sodium–potassium cotransporters actively transport three Na^+ out of a neuron for every two K^+ they pump in. See Figure 5.28.

B Passive transporters allow K^+ ions to leak across the plasma membrane, following its concentration gradient.

C Voltage-gated channels for Na^+ or K^+ are closed in a neuron at rest (left), but open if the voltage changes (right).

FIGURE 32.7 Transport proteins in a neuron plasma membrane. Voltage-gated channels are only in the membrane of the axon.

Sodium ions pumped out of the cell cannot easily cross the plasma membrane of a resting neuron. However, some of the potassium ions pumped into the neuron exit through passive transport proteins in the membrane (**FIGURE 32.7B**). This movement of positively charged potassium down its concentration gradient and out of the cell also contributes to the charge difference across the membrane.

In summary, the cytoplasm of a resting neuron contains more negatively charged proteins and amino acids, fewer sodium ions (Na^+), and more potassium ions (K^+) than interstitial fluid (**FIGURE 32.8**).

The Action Potential

Neurons and muscle cells are said to be "excitable" because they can undergo an **action potential**—a brief reversal in the electrical gradient across the plasma membrane. This reversal occurs when sodium and potassium ions flow through voltage-gated ion channels. These channels are passive transport proteins whose shape varies depending on the membrane potential (**FIGURE 32.7C**). Neurons have voltage-gated

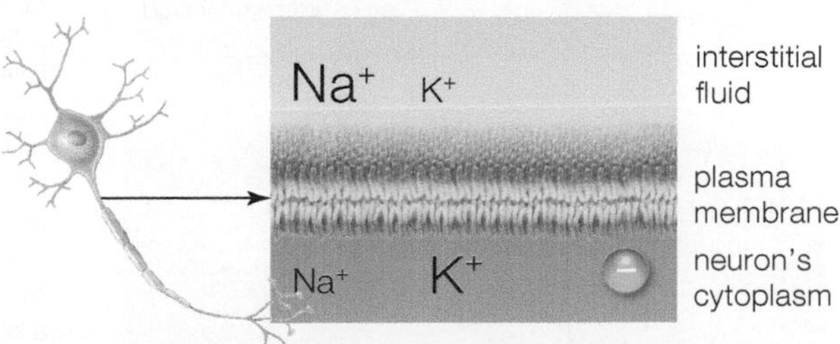

Na+
K+
interstitial fluid
plasma membrane
Na+
K+
neuron's cytoplasm

FIGURE 32.8 Distribution of ions in a neuron at resting potential. Larger text represents higher concentrations; the green ball represents negatively charged proteins that cannot cross the membrane.

ion channels in the membrane of their trigger zone and conducting zone. Some of the channels allow the passage of sodium ions, and others allow the passage of potassium ions. As you will see, these two types of voltage-gated ion channels open at different voltages.

Approaching Threshold When a neuron is at rest, all voltage-gated ion channels in its trigger zone remain closed (**FIGURE 32.9** ❶). For an action potential to occur, the cytoplasm in the neuron's trigger zone must reach **threshold potential**—the voltage at which the voltage-gated sodium channels will open.

The threshold potential of a neuron is less negative than its resting potential, so an influx of positive ions into a neuron's cytoplasm will push the neuron toward threshold. When enough positive ions enter a neuron's trigger zone, the neuron reaches threshold potential, its voltage-gated sodium channels open, and the action potential gets under way.

An All-or-Nothing Spike When voltage-gated sodium channels in the membrane open, they allow sodium ions (Na^+) to flow down their concentration and electrical gradients, to move from the interstitial fluid into the neuron ❷. The influx of sodium results in **positive feedback**, a response that intensifies the conditions that caused it to occur. In this case, the influx of sodium ions makes the membrane potential even more positive, which causes even more sodium channels to open, and so on:

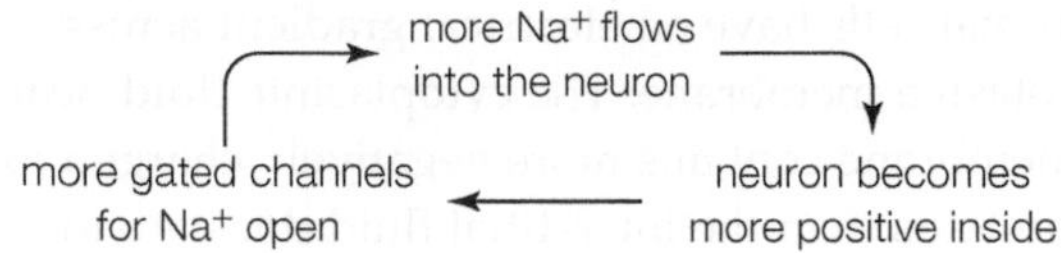
more Na^+ flows into the neuron
neuron becomes more positive inside
more gated channels for Na^+ open

An action potential is said to be an all-or-nothing event, because once threshold potential is reached, an action potential always occurs and all action potentials are the same size. After threshold potential is attained, the membrane potential quickly shoots up to about +30 mV. The positive membrane potential means that the cytoplasm contains more positively charged ions than the interstitial fluid does. Membrane potential cannot exceed +30 mV, because when it reaches this level, the voltage-gated sodium ion channels close, and the voltage-gated potassium ion channels open ❸.

Outward flow of potassium ions reduces membrane potential, thus causing the potassium channels to close. However, by the time these channels close, so many potassium ions have exited the cell that the membrane potential is a bit below resting potential. Rapid diffusion of potassium ions from the adjacent cytoplasm restores the potassium ion concentration, and membrane potential returns to its resting value.

Propagation Along the Axon An action potential is self-propagating. It affects one region of the axon after another as it moves from the trigger zone toward the axon terminals. After voltage-gated sodium channels in

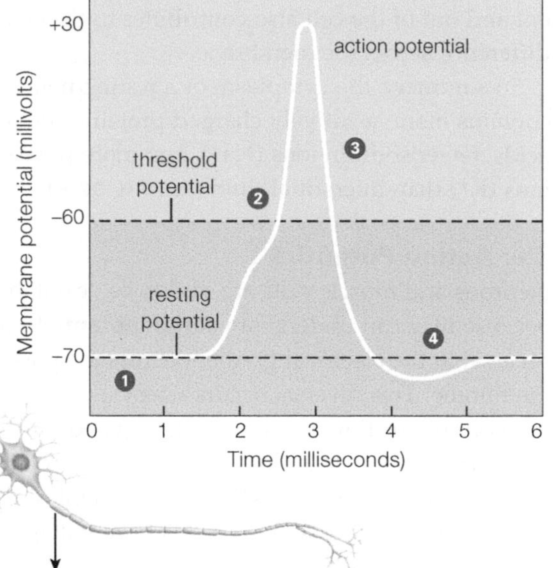
+30
-60
-70
Membrane potential (millivolts)
action potential
threshold potential
resting potential
❶
❷
❸
❹
0
1
2
3
4
5
6
Time (milliseconds)

FIGURE 32.9 Action potential. Right, plot of a neuron's membrane potential over time. The numbers correlate with those in the graphics below, which illustrate the events at one region of the axon membrane.

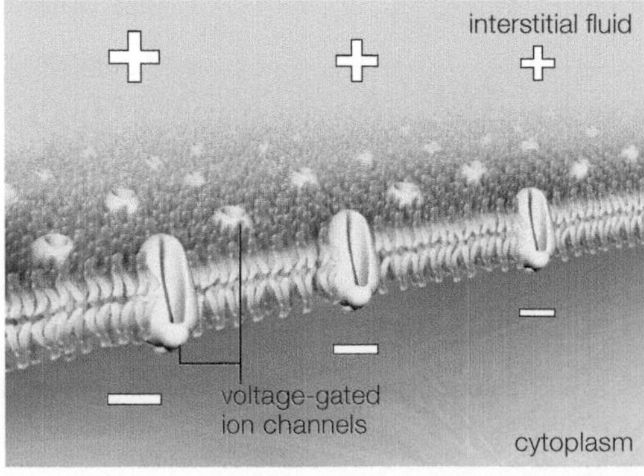
interstitial fluid
voltage-gated ion channels
cytoplasm

❶ At resting potential, all voltage-gated channels at a trigger zone are closed. White pluses and minuses indicate that the cytoplasm's charge is negative relative to that of the interstitial fluid.

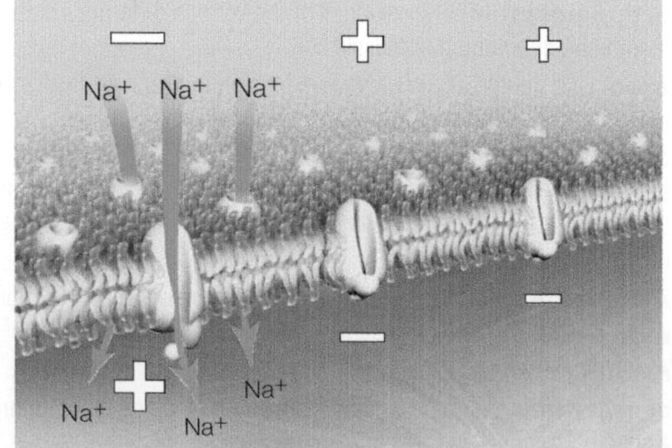
Na+ Na+ Na+
Na+
Na+
Na+

❷ At threshold potential, voltage-gated sodium (Na^+) channels in the trigger zone open, and Na^+ flows inward (blue arrows). This inward flow of positively charged ions causes a local reversal of the charge across the neuron membrane.

one region of an axon open, some of the sodium ions that rush inward diffuse into adjoining regions. There they cause the membrane potential to reach threshold level, thus setting in motion a new action potential ❹.

The action potential can only travel in one direction because after a voltage-gated sodium channel closes, there is a brief period during which it cannot reopen. However, voltage-gated sodium channels in regions farther along the axon's membrane can and do swing open as these regions reach threshold.

The Insulating Myelin Sheath

The speed at which an action potential travels along an axon depends in part on whether the axon is insulated. Most vertebrate axons have a **myelin sheath**, a discontinuous covering composed of myelin-producing glial cells that wrap around the axon one after the another. In the peripheral nervous system, myelin sheaths are formed by glial cells called Schwann cells (**FIGURE 32.10**). Myelin is a fatty material that functions like insulation around an electrical wire. In an unmyelinated axon, some positively charged ions that enter during an action potential leak out of the axon rather than moving along the axon to trigger a nearby action potential. A myelin sheath prevents this leakage.

A myelinated axon has voltage-gated channels for sodium only at nodes, which are gaps between the glial cells. When an action potential occurs at one node, the sodium ions flow into the neuron, then diffuse quickly through the axon until they reach the next node. Here, their arrival triggers the next action potential. By "jumping" from node to node, an action potential can travel along a myelinated axon as quickly as 150 meters per second. In unmyelinated axons, the maximum speed is about 10 meters per second.

Symptoms of multiple sclerosis provide some insight into the importance of myelin. Multiple sclerosis is a nervous system disorder that arises when a person's immune system mistakenly attacks and destroys myelin in the brain and spinal cord. As this myelin is replaced by scar tissue, the speed at which action potentials travel declines. The result is weakness, fatigue, impaired coordination, numbness, and vision problems.

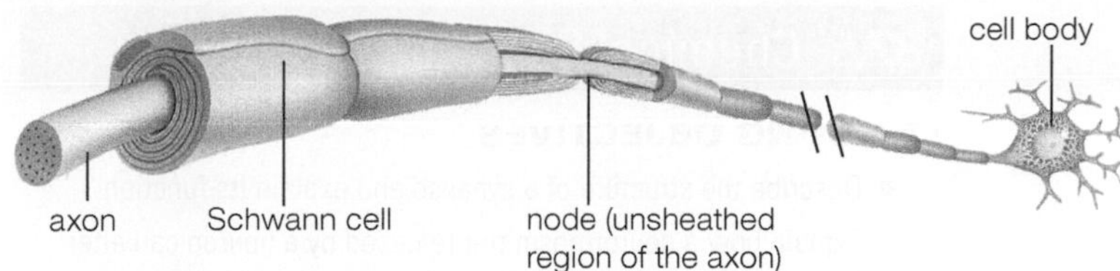
cell body
axon
Schwann cell
node (unsheathed region of the axon)

FIGURE 32.10 Myelinated axon of a peripheral nerve. Glial cells called Schwann cells form a discontinuous wrapping of myelin around such axons.

TAKE-HOME MESSAGE 32.4

✔ Membrane potential is the charge difference (voltage) across a neuron's plasma membrane.

✔ At resting potential, neuron cytoplasm is more negatively charged than the interstitial fluid. It also contains more potassium ions and fewer sodium ions than the interstitial fluid.

✔ An action potential begins when a stimulus causes gated sodium channels to open, allowing sodium to flow into the neuron. The reversal in the polarity across the membrane causes gated potassium channels to open, allowing potassium ions to flow out of the neuron.

✔ Action potentials move in one direction only because after gated sodium channels close, they cannot reopen immediately.

✔ The presence of an insulating myelin sheath speeds the conduction of action potentials along most vertebrate axons.

myelin sheath (MY-uh-lin) Of a vertebrate axon, a discontinuous covering composed of neuroglial cells; acts as insulation and thus speeds conduction along the axon.

positive feedback A response intensifies the conditions that caused its occurrence.

threshold potential Neuron membrane potential at which gated sodium channels open, causing an action potential to occur.

❸ When membrane potential reaches +30 mV, the voltage-gated Na^+ channels close and voltage-gated potassium (K^+) channels open. K^+ flows out of the axon (red arrows). Na^+ that diffuses through cytoplasm drives adjacent membrane to threshold potential, so Na^+ channels open there.

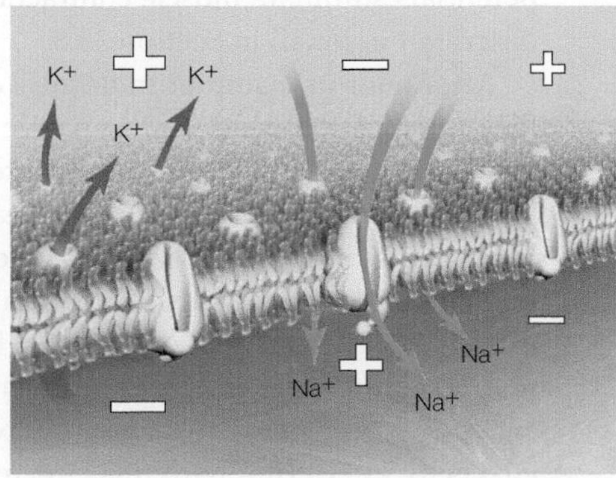
K+
K+
K+
Na+
Na+
Na+

❹ Voltage-gated K^+ channels in the trigger zone close. The action potential propagates along the axon as Na^+ diffusing through cytoplasm triggers voltage-gated Na^+ channels to open farther and farther away from the trigger zone.

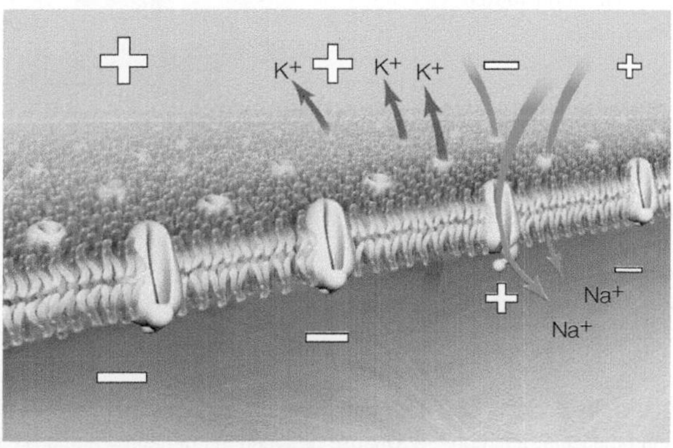
K+
K+
K+
Na+
Na+

CREDIT: (10) © Cengage Learning.

32.5 Chemical Signaling by Neurons

LEARNING OBJECTIVES

- Describe the structure of a synapse and explain its function.
- Explain how a neurotransmitter released by a neuron can alter the behavior of a postsynaptic cell.
- Describe the process of synaptic integration.

The Synapse

The region where an axon terminal signals another cell is called a **synapse**. Chemicals relay the message from one cell to the next. The signal-sending neuron at a synapse is referred to as the presynaptic cell. A fluid-filled synaptic cleft separates this neuron's axon terminal from the signal-receiving zone of the postsynaptic cell. FIGURE 32.11 shows a synapse between a motor neuron and a skeletal muscle fiber, a type of synapse called a **neuromuscular junction**.

An action potential arrives at a neuromuscular junction by traveling along the axon of a motor neuron to axon terminals ❶. Vesicles inside an axon terminal contain a **neurotransmitter** ❷, a type of signaling molecule that conveys messages between presynaptic and postsynaptic cells. A motor neuron releases the neurotransmitter acetylcholine (ACh). Other neurotransmitters are released by other types of neurons.

The plasma membrane of an axon terminal has voltage-gated calcium channels. When the neuron is resting, these channels are closed, and calcium pumps actively transport calcium ions (Ca^{+2}) out of the cell. Thus, the cytoplasm of a resting neuron has fewer calcium ions than the interstitial fluid. When an action potential arrives at an axon terminal, the voltage-gated calcium channels open. Calcium ions can then follow their gradient and diffuse from the interstitial fluid into the axon terminal ❸. The resulting increase in the calcium ion concentration of the neuron's cytoplasm causes neurotransmitter-filled vesicles to release their contents into the synaptic cleft by exocytosis ❹.

The plasma membrane of a postsynaptic cell has receptors that can bind neurotransmitter ❺. The ACh receptors at a neuromuscular junction are transport proteins that change shape when they bind ACh. Binding of ACh results in the opening of a sodium channel in the receptor protein. Sodium ions flow through that opening into the muscle cell ❻. Most neurotransmitters exert their effect by altering the movement of ions into or out of the postsynaptic cell.

Like a neuron, a muscle cell can undergo an action potential. An increase in a muscle cell's cytoplasmic sodium ion concentration drives the membrane potential toward threshold. Once threshold is reached, action potentials stimulate muscle contraction by a process described in detail in Section 35.8.

After neurotransmitter molecules do their work, they must be removed from synaptic clefts to make way for new signals. Neurotransmitter can be pumped back into presynaptic cells or into nearby glial cells. Some postsynaptic cells have enzymes in their plasma membrane that break down neurotransmitter. For example, the membrane of cells that bind and respond to ACh contains an enzyme that breaks down this enzyme. Nerve gases such as sarin exert their deadly effects by interfering with this enzyme. The resulting

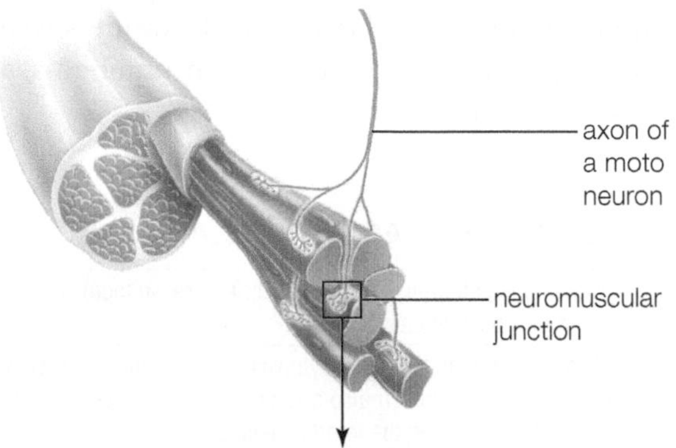

❶ Action potentials flow along the axon of a motor neuron to a neuromuscular junction, where an axon terminal forms a synapse with a muscle fiber.
axon of a moto neuron
neuromuscular junction

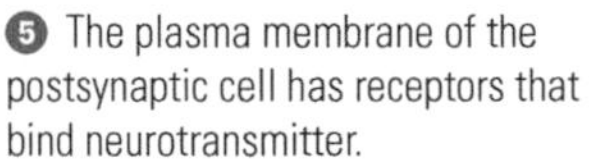

❷ The axon terminal stores neurotransmitter (green) inside synaptic vesicles.
❸ Arrival of an action potential at an axon terminal causes calcium ions (Ca^{+2}) to enter the neuron.
❹ Influx of Ca^{+2} causes exocytosis of synaptic vesicles, so neurotransmitter is released into the synaptic cleft.
axon terminal of motor neuron
plasma membrar of muscle fibe
synaptic vesicle
Ca^{+2}
synaptic cleft

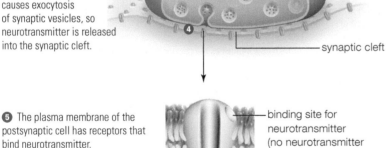

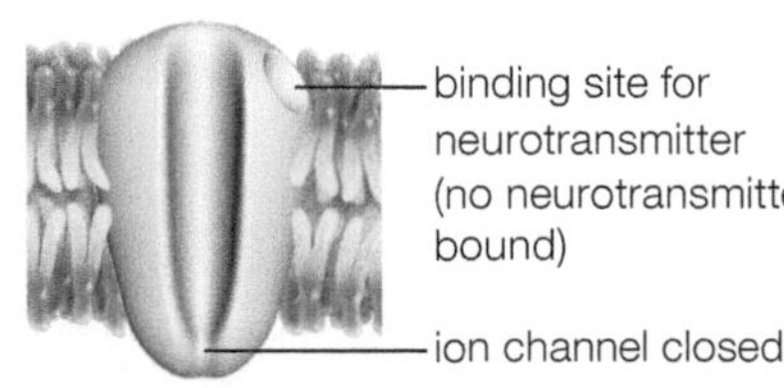

❺ The plasma membrane of the postsynaptic cell has receptors that bind neurotransmitter.
binding site for neurotransmitter (no neurotransmitter bound)
ion channel closed

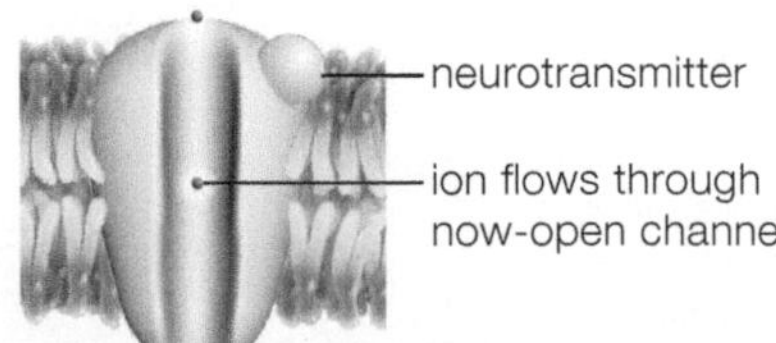

❻ Binding of neurotransmitter opens a channel through the receptor, allowing ions to flow into the postsynaptic cell.
neurotransmitter
ion flows through now-open channel

FIGURE 32.11 **Cell–cell communication at a synapse.** This example depicts a neuromuscular junction—a synapse between motor neuron and a skeletal muscle fiber.

accumulation of ACh in synaptic clefts causes muscle paralysis. Because ACh is used at synapses in the brain, as well as at neuromuscular junctions, sarin also causes confusion, headaches, and in some cases death.

Synaptic Integration

A postsynaptic cell receives messages from many neurons at the same time (**FIGURE 32.12**). Depending on the neurotransmitter released at a synapse and the type of receptor on the postsynaptic cell, an incoming signal may be excitatory or inhibitory. A signal that opens sodium gates has an excitatory effect because it pushes membrane potential closer to threshold. A signal that opens potassium gates has an inhibitory effect because it moves membrane potential away from threshold.

Incoming synaptic signals can amplify, dampen, or cancel one another's effects. All signals arriving at a neuron's input zone at the same time are summed, an effect called **synaptic integration** (**FIGURE 32.13**). An action potential will not occur in a postsynaptic neuron unless the summed signal is sufficient to drive the membrane potential to threshold level.

Competing signals cause membrane potential at the postsynaptic cell's input zone to rise and fall. When excitatory signals outweigh inhibitory ones, ions diffuse from the input zone into the trigger zone and drive the postsynaptic cell to threshold and an action potential occurs.

Neurons also integrate signals that arrive in quick succession from a single presynaptic cell. An ongoing stimulus can trigger a series of action potentials in a presynaptic cell, which will bombard a postsynaptic cell with waves of neurotransmitter.

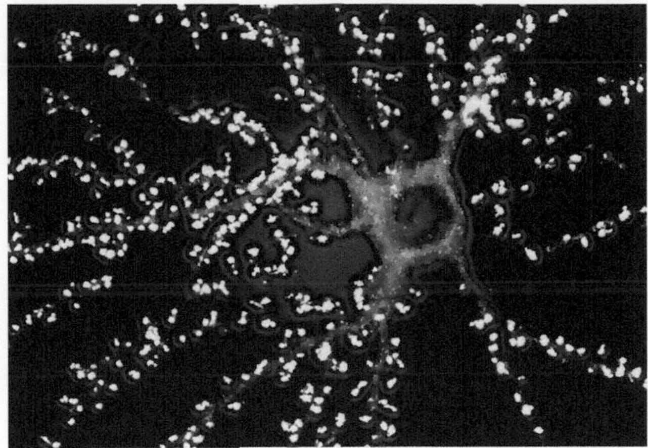

FIGURE 32.12 Synaptic density. This interneuron is stained with a yellow fluorescent dye that indicates the many locations where other neurons synapse on this cell.

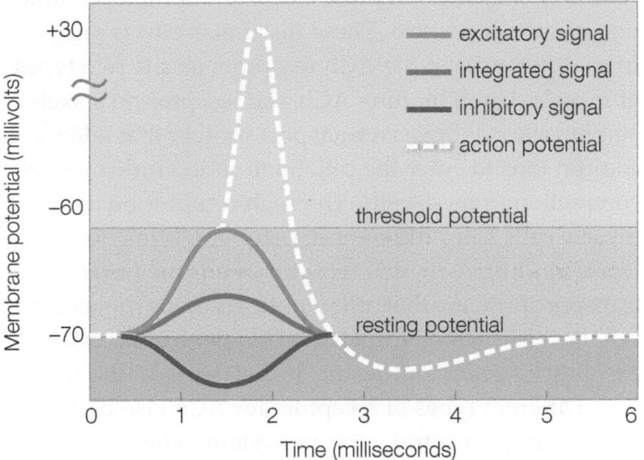
+30
-60
-70
Membrane potential (millivolts)
excitatory signal
integrated signal
inhibitory signal
action potential
threshold potential
resting potential
0
1
2
3
4
5
6
Time (milliseconds)

FIGURE 32.13 Synaptic integration. All excitatory and inhibitory signals arriving at a postsynaptic neuron's input zone at the same time are summed. The colored lines on the graph show the postsynaptic cell's response to an excitatory signal (green), to an inhibitory signal (red) and to both at once (blue). In this example, the excitatory signal would have been sufficient to trigger an action potential (white), but the inhibitory signal cancelled some of its effects.

FIGURE IT OUT Which colored line would you expect to see if a neurotransmitter caused an influx of potassium ions into the postsynaptic cell?

Answer: The red line. Influx of potassium drives the neuron away from threshold.

TAKE-HOME MESSAGE 32.5

- ✔ An action potential travels to the axon's terminals, where it stimulates exocytosis of vesicles that contain neurotransmitter.
- ✔ Neurotransmitters are chemical signals that can have excitatory or inhibitory effects on a postsynaptic cell. Synaptic integration is the summation of all excitatory and inhibitory signals arriving at a postsynaptic cell's input zone at the same time.
- ✔ For a synapse to function properly, neurotransmitter in the synaptic cleft must be removed after it has served its purpose.

neuromuscular junction Synapse between a motor neuron and the muscle it controls.
neurotransmitter Chemical signal released by axon terminals of a neuron; it binds to receptors on a postsynaptic cell.
synapse (SIN-aps) Region where a neuron's axon terminals transmit chemical signals to another cell.
synaptic integration The summation of excitatory and inhibitory signals by a postsynaptic cell.

32.6 Neurotransmitter Function

LEARNING OBJECTIVES

- Explain why a single neurotransmitter can have different effects on different types of cells.
- Give examples of the ways that drugs can alter synaptic function.

Discovery of Neurotransmitters

In the early 1920s, Austrian scientist Otto Loewi was investigating what controls heart rate. He surgically removed a frog heart while leaving the nerve that controls its rate of beating still attached. He put the heart in saline solution where it continued to beat. When Loewi stimulated the nerve, the heartbeat slowed a bit.

Loewi suspected that stimulation of the nerve caused release of a chemical signal that affected the heart. To test this hypothesis, he put two frog hearts into a saline-filled dish. When he stimulated the nerve connected to one heart, both hearts beat more slowly. As Loewi had expected, the nerve had released a chemical that not only affected the attached heart, but also diffused through the liquid and slowed the second heart. Later, another scientist identified the signaling chemical as acetylcholine (ACh).

Receptor Diversity

Different types of cells may respond differently to the same neurotransmitter. Consider that ACh stimulates contraction of skeletal muscle cells but inhibits contraction of cardiac muscle. These dissimilar effects arise from differences in the ACh receptors on the two types of muscle. Upon binding ACh, the receptor on a skeletal muscle cell opens a transport protein that allows sodium ions to enter the cell. Sodium ion influx causes contraction in these cells. The ACh receptor on cardiac muscle cells has a different structure and function. Upon binding ACh, this receptor opens membrane transport proteins that allow potassium ions to exit the cell. Potassium ion efflux has a dampening effect on excitatory signals and results in a slowing of the heart rate. Different types of receptors for ACh also occur on neurons of the central nervous system, where this neurotransmitter affects mood and memory.

Neurotransmitter Diversity

ACh is one of many human neurotransmitters (**TABLE 32.1**). In the brain, glutamate is the main excitatory neurotransmitter and GABA (gamma aminobutyric acid) is the main inhibitory neurotransmitter. Tranquilizers such as Xanax and Valium alleviate anxiety by enhancing effects of GABA.

Dopamine-secreting neurons in the brain govern body movements. Damage to dopamine-secreting neurons in the area governing motor control results in Parkinson's disease. Tremors are an early symptom. Later, the sense of balance becomes impaired, and voluntary movement, including speech, becomes difficult. Dopamine also plays a role in attention and reward-based learning. A lower than normal dopamine level occurs with attention deficit hyperactivity disorder (ADHD). Affected people can have trouble concentrating and controlling impulses. Drugs used to treat ADHD increase dopamine availability in the brain. For example, Ritalin (methylphenidate) acts by preventing the reuptake of dopamine at a synapse.

neuromodulator Chemical signal that is released by one neuron and affects multiple other neurons.

TABLE 32.1

Examples of Neurotransmitters and Their Effects

Neurotransmitter	Examples of Effects
Acetylcholine (ACh)	Induces skeletal muscle contraction; slows cardiac muscle contraction; affects mood and memory
Epinephrine and norepinephrine	Increase heart rate; dilate the pupils and airways; slow gut contractions; increase anxiety
Glutamate	Activates neurons of the CNS; has roles in memory and learning
GABA	Inhibits neurons of the CNS; influences motor control and anxiety
Dopamine	Involved in reward-based motivation and learning, motor control
Serotonin	Elevates mood; role in memory

Serotonin also acts in the brain, where it influences mood and memory. Low levels of serotonin result in depression—a persistent feeling of sadness and inability to experience pleasure. Antidepressants such as Zoloft, Prozac, and Paxil, increase the level of serotonin in the brain by blocking transport proteins that return it to axon terminals.

Some chemicals that neurons release function as **neuromodulators**, meaning they diffuse to and affect multiple neurons. The neuromodulator substance P, enhances pain perception. Neuromodulators called endorphins are natural painkillers. They are secreted by neurons in response to strenuous activity or injury. Endorphins also are released when people laugh, reach orgasm, or get a comforting hug or a relaxing massage.

Effects of Psychoactive Drugs

Psychoactive drugs are chemicals that enter the brain and affect signal transmission at synapses. Some, including antidepressants, are taken to restore normal function. Others are taken to alleviate pain, relieve stress, or simply for pleasure (**FIGURE 32.14**).

Psychoactive drugs operate by a variety of mechanisms. Some have a structure similar to a neurotransmitter or neuromodulator so they bind to and activate receptors for these molecules. Nicotine binds to and activates brain receptors for ACh. Narcotic analgesics, such as morphine, codeine, heroin, and oxycodone, bind to and activate receptors for natural painkillers. THC, the active ingredient in marijuana, binds to and activates receptors for neuromodulators called endocannabinoids. By doing so, it indirectly alters levels of multiple neurotransmitters including dopamine, serotonin, norepinephrine, and GABA.

Binding of a drug to a receptor does not always activate the receptor. Caffeine makes us alert by binding to and preventing normal action of receptors for adenosine, a neurotransmitter that causes drowsiness.

Other psychoactive drugs encourage or inhibit the release of neurotransmitter from a presynaptic cell. Ethanol does both. It slows motor response by inhibiting ACh release and makes us feel drowsy by encouraging adenosine release. It also stimulates the release of endorphins and GABA, resulting in a brief euphoria followed by depression. Still other drugs interfere with reuptake of neurotransmitter from the synaptic cleft. For example, cocaine interferes with reuptake of several neurotransmitters, including dopamine.

Drug addiction has many causes, but dopamine plays an important role in creating dependency. A surge of dopamine is released in the brain during behaviors such as eating, which enhances survival, or engaging in sex, which enhances reproduction. This release helps individuals learn to repeat beneficial behaviors. Drugs that trigger dopamine release or prevent its reuptake hijack this ancient learning pathway. Drug users inadvertently teach themselves that the drug is essential to their well-being.

Continual use of a drug often results in tolerance, meaning the effectiveness of a given dose of the drug decreases over time. Mechanisms of tolerance vary. Tolerance to a drug that causes the release of a particular neurotransmitter, for example, may arise from decreased synthesis of the neurotransmitter or its receptors. Tolerance can also involve a change in how neurotransmitter is cleared from a synaptic cleft.

FIGURE 32.14 Commonly used psychoactive drugs. Ethanol in beer, caffeine in coffee, and nicotine in tobacco.

TAKE-HOME MESSAGE 32.6

✔ Neurotransmitters vary widely in their effects. There are a variety of different neurotransmitters, and the same one may elicit different responses in different cells.

✔ Psychoactive drugs act by influencing communication at synapses. Many are addictive, and continued use of them can alter the function of the user's synapses.

Data Analysis Activities

Effect of MDMA on Neural Development Animal studies are often used to assess effects of prenatal exposure to illicit drugs. For example, Jack Lipton used rats to study the behavioral effect of prenatal exposure to MDMA, the active ingredient in the drug commonly called Ecstasy. He injected female rats with either MDMA or saline solution when they were 14 to 20 days pregnant. This is the period when their offsprings' brains were forming. When those offspring were 21 days old, Lipton tested their ability to adjust to a new environment. He placed each young rat in a new cage and used a photobeam system to record how much each rat moved around before settling down. FIGURE 32.15 shows his results.

1. Which rats moved around most (caused the most photobeam breaks) during the first 5 minutes in a new cage, those prenatally exposed to MDMA or the controls?
2. How many photobeam breaks did the MDMA-exposed rats make during their second 5 minutes in the new cage?
3. Which rats moved around most during the last 5 minutes?
4. Does this study support the hypothesis that MDMA affects a developing rat's brain?

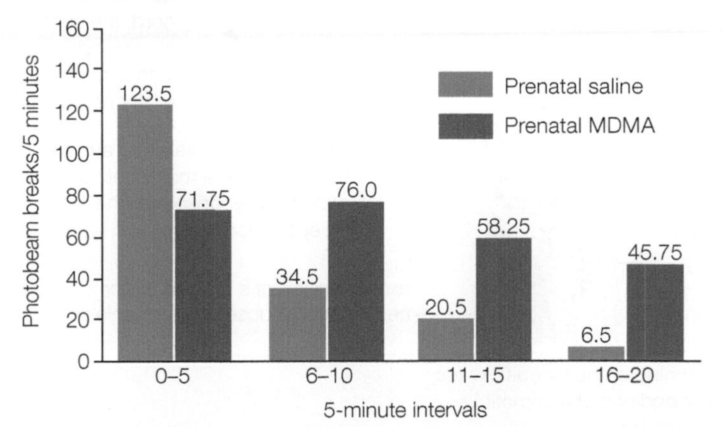

Photobeam breaks/5 minutes
160
140
120
100
80
60
40
20
0
123.5
71.75
76.0
34.5
58.25
20.5
45.75
6.5
Prenatal saline
Prenatal MDMA
0–5
6–10
11–15
16–20
5-minute intervals

FIGURE 32.15 Effect of prenatal exposure to MDMA on rats. Young (21-day-old) rats were place in an unfamiliar cage. Movements were detected when the rat interrupted a photobeam. Rats were monitored at 5-minute intervals for a total of 20 minutes. Blue bars are average numbers of photobeam breaks for rats whose mothers received saline; red bars are for rats whose mothers received MDMA.

32.7 The Peripheral Nervous System

LEARNING OBJECTIVES

- Describe the structure and location of peripheral nerves.
- Compare functions of somatic and autonomic nerves.
- Using several organs as examples, compare the effects of sympathetic and parasympathetic stimulation.

The human peripheral nervous system includes 31 pairs of spinal nerves that connect to the spinal cord and 12 pairs of cranial nerves that connect to the brain. Each nerve consists of axons of many neurons bundled together in a connective tissue sheath (**FIGURE 32.16**).

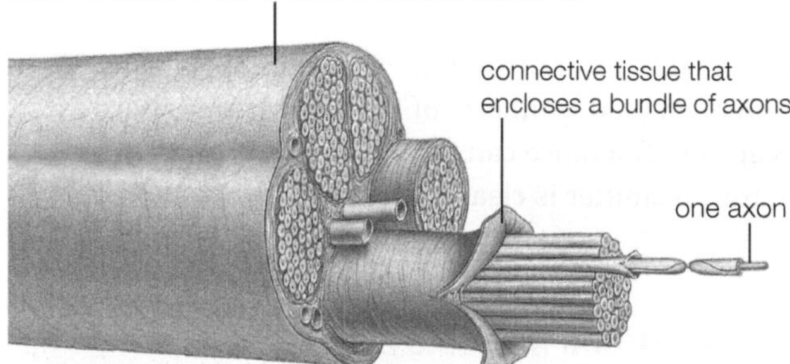
outer connective tissue sheath around the nerve
connective tissue that encloses a bundle of axons
one axon

FIGURE 32.16 Structure of a peripheral nerve.

FIGURE 32.17 The stretch reflex, a somatic reflex. This reflex involves sensory and motor neurons of the somatic division of the peripheral nervous system.

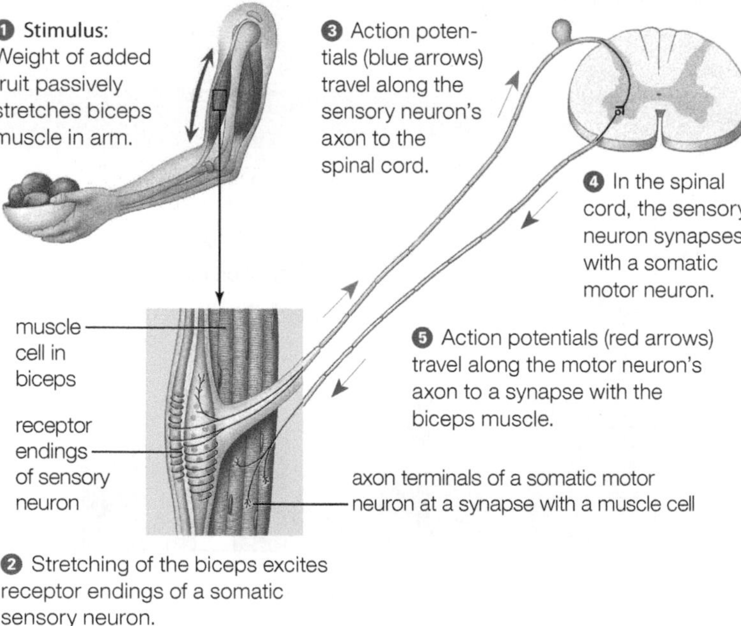
1 Stimulus: Weight of added fruit passively stretches biceps muscle in arm.
3 Action potentials (blue arrows) travel along the sensory neuron's axon to the spinal cord.
4 In the spinal cord, the sensory neuron synapses with a somatic motor neuron.
muscle cell in biceps
receptor endings of sensory neuron
5 Action potentials (red arrows) travel along the motor neuron's axon to a synapse with the biceps muscle.
axon terminals of a somatic motor neuron at a synapse with a muscle cell
2 Stretching of the biceps excites receptor endings of a somatic sensory neuron.
6 Response: Stimulation of the biceps muscle causes this muscle to contract, steadying the arm.

As previously noted, most peripheral nervous system axons associate with myelin-producing Schwann cells.

There are two functional divisions of the peripheral nervous system: the somatic nervous system and the autonomic nervous system (**TABLE 32.2**).

Somatic Nervous System

Somatic nerves of the peripheral system control skeletal muscles and inform the central nervous system about external conditions and the body's position. They allow you to make voluntary movements and to feel conscious sensations such as cold or touch.

The somatic nervous system also plays a role in some reflexes. A **reflex** is an automatic response to a stimulus, a movement or other action that does not require conscious thought. The stretch reflex is an example of a somatic reflex (**FIGURE 32.17**). Suppose you hold a bowl as someone drops fruit into it ❶. The added weight stretches the biceps muscle in your upper arm. Receptor endings of a somatic sensory neuron wrap around some cells of the biceps, and lengthening of the muscle excites this neuron ❷. As a result, an action potential travels along the sensory neuron's axon to the spinal cord ❸. In the spinal cord, the axon of the sensory neuron synapses with one of the somatic motor neurons that controls the biceps ❹. Excited by neurotransmitter from the sensory neuron, the motor neuron undergoes an action potential that travels along its axon to the biceps ❺. At a synapse with the biceps, the motor neuron's axon terminals release ACh. The biceps responds to the ACh by contracting, so your arm steadies itself ❻.

As this example illustrates, the somatic nervous system requires only one neuron to carry a signal to the CNS and one to carry a signal from it. In the somatic nervous system, there are no synapses outside the CNS.

Autonomic Nervous System

Autonomic nerves control glands and involuntary muscle (smooth muscle and cardiac muscle). They inform the CNS about conditions inside the body. The autonomic nervous system tells your brain when your

autonomic nerves Of the peripheral nervous system, nerves that relay signals to and from internal organs and to glands.
parasympathetic neurons Neurons of the autonomic system that encourage "housekeeping" tasks in the body.
reflex An automatic response to a stimulus, a movement or other action that does not require conscious thought.
somatic nerves Of the peripheral nervous system, nerves that control skeletal muscle and relay signals from sensory neurons in joints and skin to the central nervous system.
sympathetic neurons Neurons of the autonomic system that prepare the body for danger or excitement.

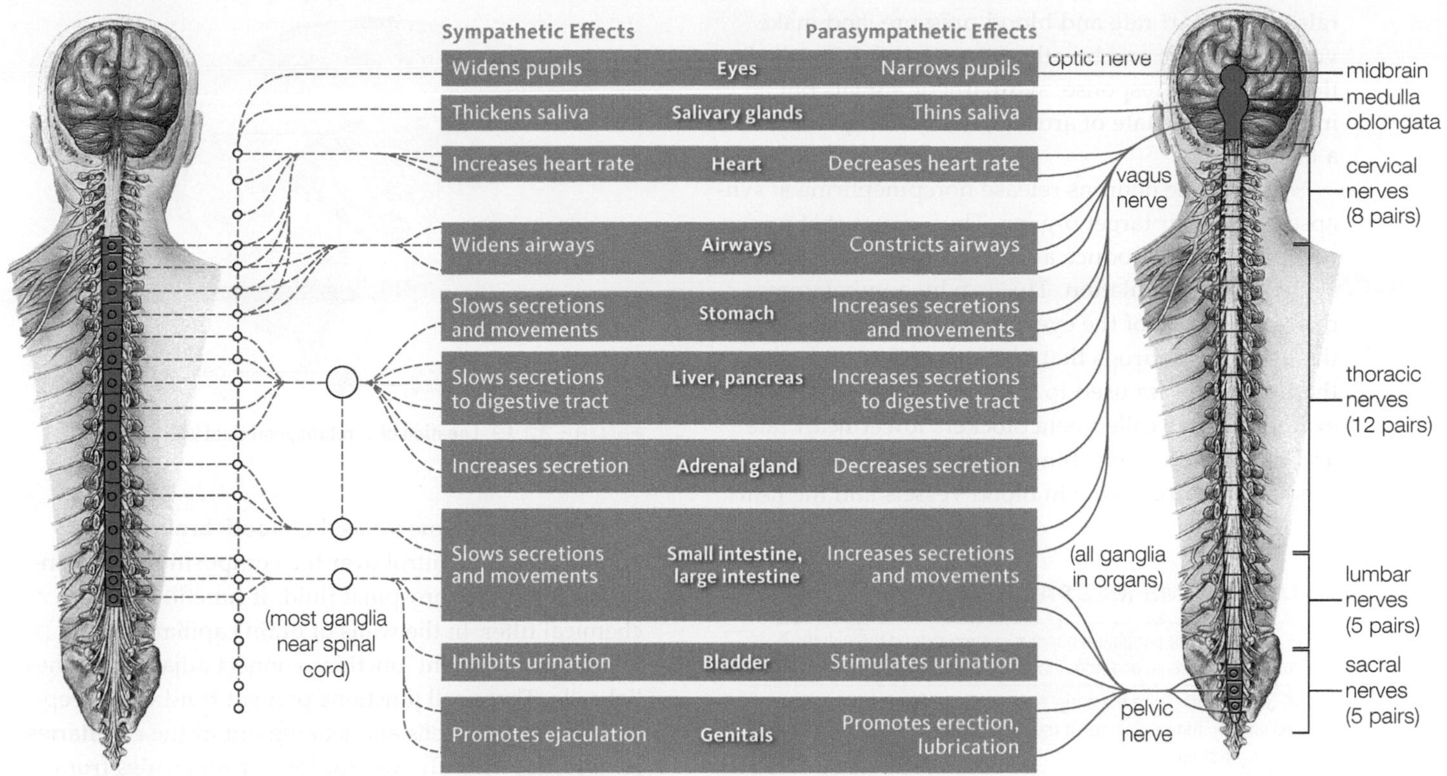
Sympathetic Effects
Parasympathetic Effects
Widens pupils
Eyes
Narrows pupils
Thickens saliva
Salivary glands
Thins saliva
Increases heart rate
Heart
Decreases heart rate
Widens airways
Airways
Constricts airways
Slows secretions and movements
Stomach
Increases secretions and movements
Slows secretions to digestive tract
Liver, pancreas
Increases secretions to digestive tract
Increases secretion
Adrenal gland
Decreases secretion
Slows secretions and movements
Small intestine, large intestine
Increases secretions and movements
Inhibits urination
Bladder
Stimulates urination
Promotes ejaculation
Genitals
Promotes erection, lubrication
(most ganglia near spinal cord)
optic nerve
vagus nerve
(all ganglia in organs)
pelvic nerve
midbrain
medulla oblongata
cervical nerves (8 pairs)
thoracic nerves (12 pairs)
lumbar nerves (5 pairs)
sacral nerves (5 pairs)

FIGURE 32.18 Effects of sympathetic and parasympathetic stimulation on organ and gland function.

FIGURE IT OUT What effect does stimulation of the vagus nerve (a parasympathetic nerve) have on the heart?

Answer: It decreases heart rate.

blood pressure drops too low and reflexively adjusts your heart rate to restore the correct blood pressure.

The autonomic system requires two neurons to carry a signal to the CNS or from the CNS. Consider how a message to increase the heart rate reaches the heart. The signal travels first along the axon of an autonomic neuron. The cell body of an autonomic neuron is in the spinal cord, and its axon extends out of the spinal cord to a ganglion some distance away. In the ganglion, the axon synapses with a second neuron. The axon of the second neuron carries that signal to the heart.

There are two categories of autonomic neurons: sympathetic and parasympathetic. Both service most organs, and they work antagonistically, meaning the signals from one type oppose signals from the other (FIGURE 32.18). Synaptic integration determines the outcome of the conflicting commands.

Signals from **parasympathetic neurons** dominate when you are calm. These neurons promote housekeeping tasks, such as digestion and urine formation, bringing about what is called a rest and digest response. Parasympathetic neurons release ACh at their synapses with target organs.

Signals from **sympathetic neurons** increase in times of stress, excitement, and danger. When you are startled, frightened, or angry, signals from these neurons

TABLE 32.2

Comparison of the Components of the Peripheral Nervous System

	Somatic Nervous System	Autonomic Nervous System	
Motor functions	Carries signals for voluntary or reflexive contraction from the CNS to skeletal muscle	Carries signals for involuntary contraction from the CNS to cardiac muscle and smooth muscle and carries CNS signals for gland activity to glands	
Sensory functions	Monitors the external environment and the position of the body; relays signals from receptors in skin and near joints to the CNS	Monitors the internal environment; relays signals from receptors in the walls of internal organs to the CNS	
Signaling pathway	Single-neuron pathway	Two-neuron pathway, with a synapse between neurons	
Subdivisions	None	**Parasympathetic division** Promotes maintenance tasks (resting and digesting)	**Sympathetic division** Prepares body for intensive activity (fight–flight)

CREDIT: (18) © Cengage Learning.

raise your heart rate and blood pressure, and make you sweat more and breathe faster. In what is called the fight–flight response, sympathetic signals put an individual in a state of arousal, ready to fight or make a fast getaway.

Sympathetic neurons release norepinephrine at synapses with their target organs. Thus, drugs that mimic norepinephrine produce an effect similar to that of sympathetic stimulation. This is why amphetamines dilate the pupils of the eye and increase heart rate. On the other hand, drugs that block the effect of sympathetic neurons are used to treat some disorders. For example, drugs called beta blockers lower heart rate and blood pressure by binding to and blocking receptors for norepinephrine in blood vessels and the heart.

TAKE-HOME MESSAGE 32.7

- ✔ Nerves of the peripheral nervous system extend through the body and relay signals to and from the central nervous system.
- ✔ Somatic nerves of the peripheral system control skeletal muscle and convey information about the external environment to the central nervous system.
- ✔ Autonomic nerves carry information to and from smooth muscle and cardiac muscle, and to glands.
- ✔ The two autonomic divisions (sympathetic and parasympathetic) have opposing effects.

32.8 Cells and Tissues of the Central Nervous System

LEARNING OBJECTIVES

- Explain how cerebrospinal fluid forms and the function of the blood–brain barrier.
- Compare the components of white matter and gray matter.

The vertebrate central nervous system (CNS) consists of the spinal cord and brain.

Meninges and the Cerebrospinal Fluid

Three layers of connective tissue called the **meninges** enclose and protect the organs of the central nervous system. **Cerebrospinal fluid** fills the space between the two innermost meninges, the spinal cord's central canal, and cavities called ventricles within the brain (FIGURE 32.19). This fluid serves as a liquid cushion for tissues of the brain and spinal cord. It consists of water with ions and nutrient molecules that have been selectively transported out of blood. It is rich in oxygen and in glucose, which neurons use for aerobic respiration. Movement of cilia on glial cells in the lining of brain ventricles keeps cerebrospinal fluid circulating.

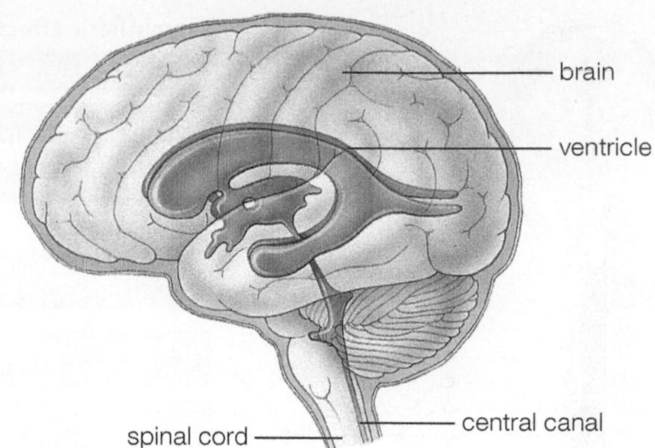
brain
ventricle
spinal cord
central canal

FIGURE 32.19 **Location of cerebrospinal fluid** (shown in blue).

A mechanism known as the **blood–brain barrier** maintains tight control over the composition and concentration of cerebrospinal fluid. It functions like a chemical filter. In the walls of brain capillaries (tiny blood vessels), tight junctions connect adjacent epithelial cells. These cell junctions prevent fluid from seeping between the cells and leaking out of the capillaries. As a result, the only way for ions or molecules from blood to enter the cerebrospinal fluid is by passing through cells of the capillary wall. Water and oxygen diffuse through these cells. Glucose and other essential nutrients and ions move across these cells with the assistance of transport proteins.

The blood–brain barrier prevents most drugs from entering the brain. Psychoactive drugs such as alcohol (ethanol) and cocaine are notable exceptions. These substances are small and lipid soluble, so they diffuse easily across the plasma membranes of both epithelial cells and glial cells. Once in the cerebrospinal fluid, they disrupt brain function by affecting synapse function, as described in Section 32.5.

Gray Matter and White Matter

The bulk of the brain and spinal cord consists of two visibly different tissues: gray matter and white matter. **Gray matter** is a pinkish gray tissue that includes neuron cell bodies, dendrites, axon terminals, and glial cells. All synapses of the central nervous system are located within the gray matter. **White matter** is mainly myelin-sheathed axons with their associated glial cells. In the central nervous system, a bundle of such axons is referred to as a nerve tract, rather than a nerve.

Glia of the Central Nervous System

The central nervous system contains four main types of glial cells: oligodendrocytes, ependymal cells, microglia, and astrocytes. Oligodendrocytes are the functional equivalent of the peripheral system's Schwann cells.

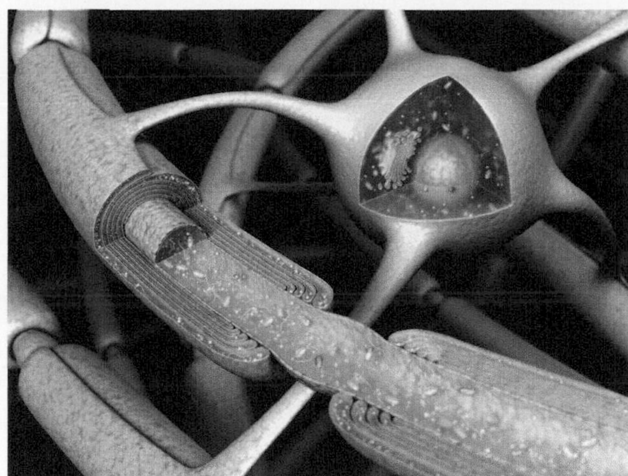

FIGURE 32.20 Oligodendrocyte. These neuroglia cells provide the myelin wrapping for axons in the brain and spinal cord.

They make the myelin that insulates axons in the central nervous system. However, the two types of myelin-making glial cells differ in some respects. An oligodendrocyte myelinates multiple axons, whereas a Schwann cell myellinates only one. An oligodendrocyte has a structure somewhat like an octopus, with a central cell body from which long, tentacle-like cytoplasmic processes extend out to wrap around axons (FIGURE 32.20). If the axon of a peripheral nerve is cut, Schwann cells help it regrow. Oligodendrocytes, on the other hand, do not promote regrowth of damaged axons. This is one reason why axons of a severed finger can grow back, but axons cut by a spinal cord injury cannot. Oligodendrocytes are the cells that are destroyed by the immune system in multiple sclerosis.

Ependymal cells are glial cells that line the brain's fluid-filled ventricles and the spinal cord's central canal. Waving cilia on their surface keep cerebrospinal fluid flowing in a consistent direction.

Microglia are, as the name implies, the smallest neuroglial cells. They continually survey the brain. If brain tissue is injured or infected, microglia become active; they move about, engulfing dead cells and cellular debris. They also produce chemical signals that alert the immune system to threats.

Star-shaped astrocytes are the brains most abundant glial cells. Astrocytes that wrap around blood vessels of the brain stimulate formation of the blood–brain barrier. Other astrocytes play a role at synapses, where they take up neurotransmitters that were released by neurons. In addition, astrocytes assist in immune defense, release lactate that fuels the activity of neurons, and synthesize nerve growth factor. Neurons are stopped in G1 of the cell cycle (Section 11.2) and cannot divide, but nerve growth factor encourages a neuron to form new synapses.

Being mature, differentiated cells that do not divide, neurons do not give rise to tumors. However, some types of glial cells are continually renewed and uncontrolled division of these cells sometimes causes a tumor called a glioma. Brain tumors can also arise from epithelial cells in the meninges or endocrine glands of the brain, such as the pituitary. In addition, arrival of metastatic cancer cells from elsewhere in the body can cause a tumor.

Most tumors that originate in the brain are not cancer. However, even a benign (noncancerous) tumor can pose a serious threat. Benign tumors do not metastasize (Section 11.6), but growth of a tumor within the confined space of the skull can put pressure on surrounding nervous tissue and damage neurons.

TAKE-HOME MESSAGE 32.8

✔ The brain and spinal cord are enclosed by the meninges and bathed by cerebrospinal fluid. The blood–brain barrier regulates the composition of the cerebrospinal fluid.

✔ The gray matter of the brain and spinal cord contains neuron cell bodies and dendrites. All synapses of CNS occur in the gray matter. White matter consists mainly of myelinated axons.

✔ A variety of glial cells assist neurons of the CNS. Uncontrolled division of glial cells can cause a brain tumor.

32.9 The Spinal Cord

LEARNING OBJECTIVE

- Describe the structure and function of the spinal cord.

Structure of the Spinal Cord

The **spinal cord** is the portion of the central nervous system that runs through the vertebral column (the backbone). It conveys signals between peripheral nerves and the brain. It also plays a role in reflexes that do not involve the brain. The stretch reflex, described in Section 32.7, is an example of a spinal reflex.

blood–brain barrier Protective mechanism that prevents unwanted substances from entering cerebrospinal fluid.
cerebrospinal fluid Clear fluid that surrounds the brain and spinal cord and fills cavities within the brain.
gray matter Central nervous system tissue that includes neuron cell bodies, dendrites, and axon terminals, as well as glial cells.
meninges (meh-NIN-jeez) Three layers of connective tissue that enclose the brain and spinal cord.
spinal cord Portion of central nervous system that connects peripheral nerves with the brain.
white matter Central nervous system tissue that consists primarily of myelinated axons with their associated glial cells.

The human spinal cord is a cylinder about as thick as a thumb (**FIGURE 32.21**). Its outermost portion is white matter. Gray matter fills a butterfly-shaped region in the cord's center. The vertebral column that encloses the spinal cord is a segmented structure consisting of individual bones called vertebrae.

Axons of the peripheral nerves connect to the spinal cord through small openings between the vertebrae. Regions where axons of peripheral nerves enter or exit the spinal cord are called nerve roots. Dorsal roots (roots toward the rear of the backbone) contain the axons of sensory neurons. Ventral roots (roots near the front of the backbone) contain the axons of motor neurons. Within the cord itself, axons of motor and sensory neurons run through different parts of the cord as separate tracts.

Interrupted Spinal Signaling

Physicians sometimes halt transmission of signals through the spinal cord temporarily to provide pain relief to a specific region of the body. Consider epidural anesthesia, the most common way of lessening pain during childbirth. During this procedure, a physician injects a painkiller into the fat-filled area between the spinal cord and the vertebra that encloses it. The drug is administered in a region where it will partially numb the body from the waist down, without otherwise impairing perception or motor function.

An injury to the spinal cord can result in the loss of sensation or motor function (paralysis). Severed peripheral nerves can regrow a bit, but tracts within the spinal cord do not. As a result, spinal cord injuries usually cause permanent disability. Symptoms depend on what portion of the spinal cord is damaged. Nerves carrying signals to and from the upper body are higher in the cord than nerves that service the lower body. An injury to the cord in the lower back often paralyzes the legs. An injury to higher spinal cord regions can paralyze all limbs, as well as muscles involved in breathing. In the United States, more than a million people are now paralyzed as a result of a spinal cord injury.

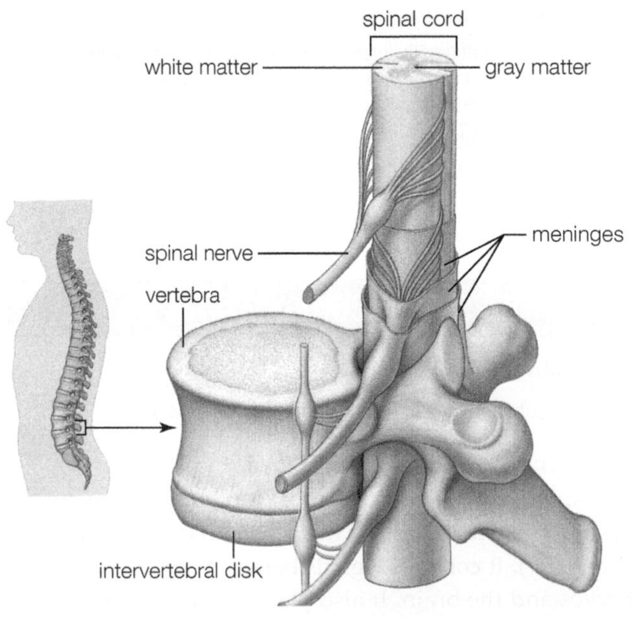

spinal cord
white matter
gray matter
meninges
spinal nerve
vertebra
intervertebral disk

FIGURE 32.21 **Location and organization of the spinal cord.**

TAKE-HOME MESSAGE 32.9

✔ The spinal cord connects peripheral nerves to the brain.

✔ Signals from sensory neurons enter the cord through the dorsal root of spinal nerves. Commands for responses go out along the ventral root of these nerves.

32.10 The Vertebrate Brain

LEARNING OBJECTIVES

- Explain the embryonic origin of the spinal cord and brain.
- Describe the location, function, and components of the hindbrain.
- List the components of the forebrain and describe their functions.

The Vertebrate Brain

In all vertebrates, the embryonic neural tube develops into a spinal cord and brain. During development, the brain becomes organized as three functional regions: forebrain, midbrain, and hindbrain (**FIGURE 32.22**).

The hindbrain is continuous with the spinal cord and is responsible for many reflexes and coordination. Fishes and amphibians have the most pronounced midbrain (**FIGURE 32.23**). Their forebrain sorts out sensory input and initiates motor responses. The midbrain is reduced in birds and mammals; their expanded forebrain took over what were midbrain functions.

Functional Anatomy of the Human Brain

An adult human brain weighs about 1,400 grams, or 3 pounds. It contains about 100 billion interneurons, and glial cells constitute about half of its volume. **FIGURE 32.24** shows the anatomy of a human brain.

Three structures make up the hindbrain. The spinal cord connects to the hindbrain's **medulla oblongata**, a region that influences heartbeat and breathing and also controls reflexes such as swallowing, coughing, vomiting, and sneezing. Just above the medulla oblongata is the **pons**, which also affects breathing. Pons means "bridge," a reference to the tracts that extend through the pons to the midbrain and functionally bridge these regions. The plum-sized **cerebellum** at the rear of the hindbrain controls posture, coordinates voluntary movements, and plays a role in learning new motor skills. The cerebellum is densely packed with neurons. It has as many as all other brain regions combined. A drunk person becomes uncoordinated because alcohol impairs the cerebellum. Police officers evaluate the extent of this impairment by asking a person suspected of driving while drunk to walk a straight line. Over the long term, excessive alcohol consumption kills cells in the cerebellum, so alcoholics often have a shuffling gait. On the other hand, routinely practicing an athletic skill improves function of the cerebellum.

The pons, medulla, and midbrain are referred to collectively as the **brain stem**. This is the most evolutionarily ancient part of the brain. The human midbrain is small but essential. It relays sensory information to the forebrain and facilitates reward-based learning. It also functions in voluntary movement. Death of dopamine-producing neurons in one region of the midbrain results in Parkinson's disease. Affected people first develop tremors (unintentional shaky movements). Later, all voluntary movement, including speech, may become difficult.

The forebrain is the largest area of the human brain. The **cerebrum** makes up the bulk of the forebrain. It receives sensory signals, integrates information, and initiates voluntary actions. A fissure divides the cerebrum into right and left hemispheres. The corpus callosum, a thick bundle of axons that crosses this fissure, relays signals between the two hemispheres.

Each hemisphere has a thin outer layer of gray matter called the cerebral cortex. Our large cerebral cortex is the part of the brain responsible for human capacities such as language and abstract thought. Section 32.11 discusses its functions in more detail.

The **thalamus** is a two-lobed forebrain structure that sorts sensory signals and sends them to the proper region of the cerebral cortex. It also influences sleep and wakefulness. Fatal familial insomnia is an extremely rare, fatal genetic disorder that damages the thalamus. Symptoms usually arise in middle age, when

brain stem The most evolutionarily ancient region of the vertebrate brain; includes the pons, medulla, and midbrain.
cerebellum (ser-uh-BELL-um) Hindbrain region responsible for posture and for coordinating voluntary movements.
cerebrum (suh-REE-brum) Forebrain region that, in humans, is essential to language, planning, and abstract thought.
medulla oblongata (mih-DOO-luh ob-long-GAH-tuh) Hindbrain region that controls breathing rhythm and reflexes such as coughing and vomiting.
pons (ponz) Hindbrain region that influences breathing and serves as a functional bridge to the adjacent midbrain.
thalamus Forebrain region that relays signals to the cerebral cortex.

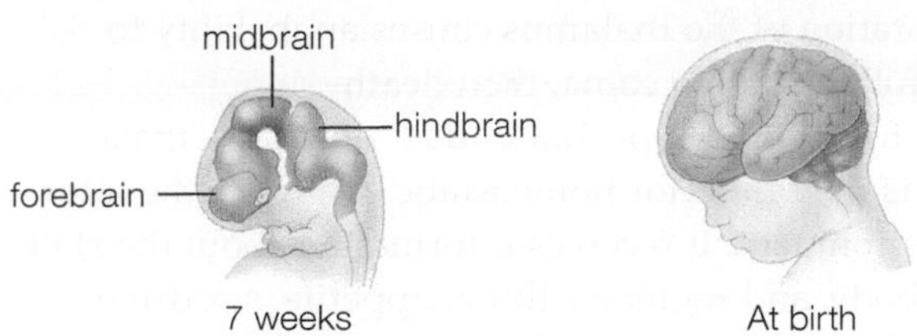
midbrain
hindbrain
forebrain
7 weeks
At birth

FIGURE 32.22 Human brain development. At 7 weeks, there is a hollow neural tube with regions that will develop into the forebrain, midbrain, and hindbrain.

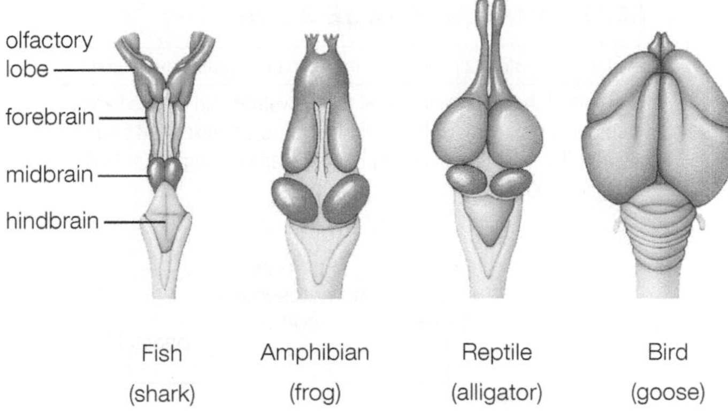
olfactory lobe
forebrain
midbrain
hindbrain
Fish (shark)
Amphibian (frog)
Reptile (alligator)
Bird (goose)

FIGURE 32.23 Vertebrate brains, dorsal views. Sketches are not to scale.

FIGURE 32.24
Anatomy of the human brain.

A Right, view of a normal brain from above, showing the two cerebral hemispheres divided by a deep fissure. The meninges have been removed from the right hemisphere.

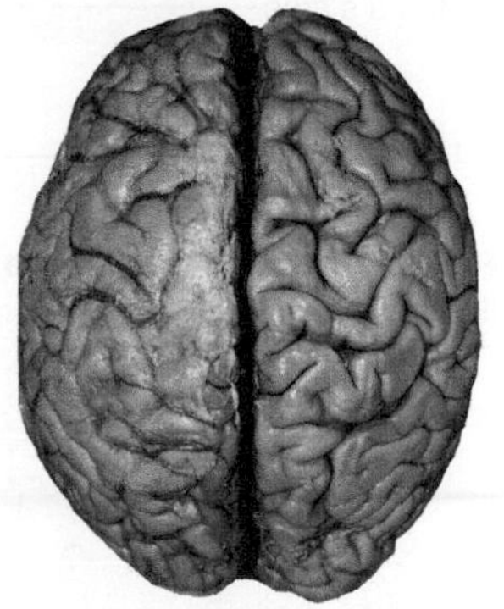

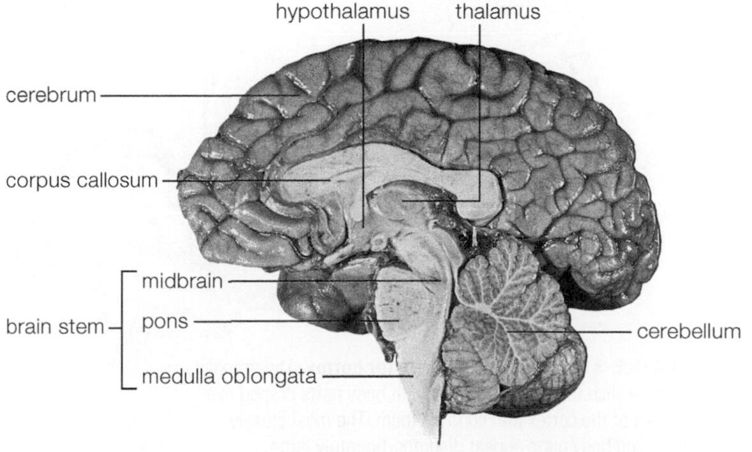
hypothalamus
thalamus
cerebrum
corpus callosum
brain stem
midbrain
pons
medulla oblongata
cerebellum

B The right half of a brain that has been sectioned along the longitudinal fissure.

deterioration of the thalamus causes an inability to sleep, followed by a coma, then death.

The forebrain's **hypothalamus** ("under the thalamus") is the center for homeostatic control of the internal environment. It receives information about the state of the body, and regulates thirst, appetite, sex drive, and body temperature. It also interacts with the nearby pituitary gland as a central control center for the endocrine system. We will consider the endocrine functions of the hypothalamus in detail in Chapter 34.

TAKE-HOME MESSAGE 32.10

✔ The vertebrate brain develops from the embryonic neural tube.

✔ Tissue of the embryonic neural tube develops into the hindbrain, forebrain, and midbrain. The hindbrain controls reflexes and coordination. The unique capacities of humans arise in regions of their enlarged forebrain.

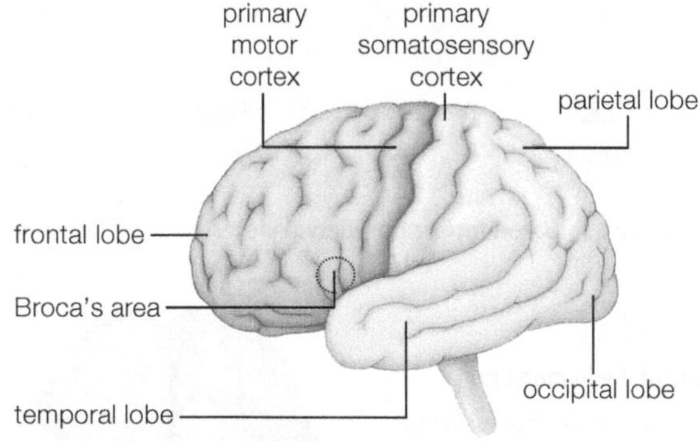
primary motor cortex
primary somatosensory cortex
parietal lobe
frontal lobe
Broca's area
temporal lobe
occipital lobe

FIGURE 32.25 Regions of the cerebral cortex. Prominent folds define the borders of the four cortical lobes, each indicated here by a different color.

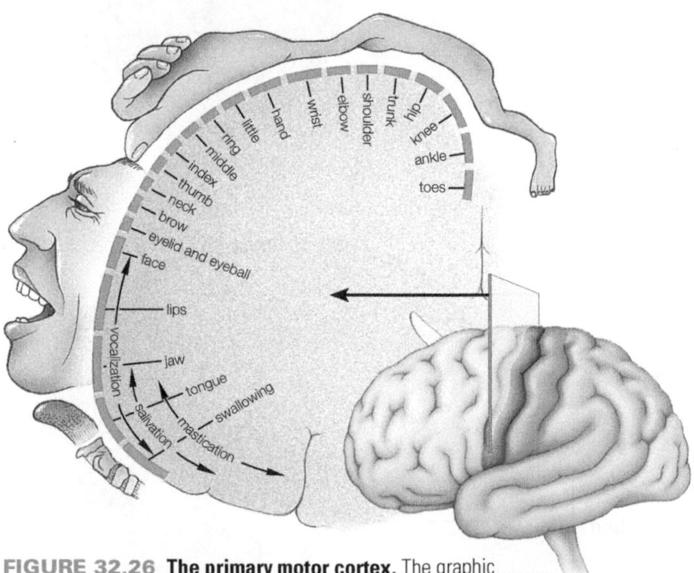
hip
trunk
shoulder
elbow
wrist
hand
little
ring
middle
index
thumb
neck
brow
eyelid and eyeball
face
lips
jaw
tongue
swallowing
mastication
salivation
vocalization
knee
ankle
toes

FIGURE 32.26 The primary motor cortex. The graphic shows a slice through this region with body parts draped over the part of the cortex that controls them. The most closely controlled body parts appear disproportionately large.

32.11 The Human Cerebral Cortex

LEARNING OBJECTIVE

- Describe the functions of the human cerebral cortex.

The **cerebral cortex**, the outermost portion of the cerebrum, is a 2-millimeter-thick layer of gray matter. Over the course of human evolution, this layer has become increasingly folded. Folds allowed expansion of our gray matter with a minimal increase in brain volume. (Keeping brain volume in check is important because a larger brain requires a larger skull, which can pose problems during childbirth.)

Prominent folds in the cortex are used as landmarks to define the four lobes of each cerebral hemisphere (FIGURE 32.25). The cortex of the frontal lobe allows us to make reasoned choices, concentrate on tasks, plan for the future, and behave appropriately in social situations. During the 1950s, more than 20,000 people had their frontal lobes deliberately damaged in a surgical procedure called frontal lobotomy. Lobotomy was used to treat mental illness, personality disorders, and even headaches. The procedure made patients calmer, but also blunted their emotions and impaired their ability to plan, concentrate, and behave appropriately.

The two cerebral hemispheres differ somewhat in their function. In most people, the cortex of the left hemisphere plays the leading role in movement and in language, whereas mathematical tasks, such as adding two numbers, typically activate cortical areas of the right hemisphere. Broca's area, the region that translates thoughts into speech, is in the cortex of the left frontal lobe. Damage to Broca's area usually prevents a person from speaking, but does not affect understanding of language. Note that, despite their differences, both hemispheres are flexible. If a stroke or injury damages one side of the brain, the other hemisphere can take on new tasks. People can even function with a single hemisphere.

Most functions involve many brain regions, but some parts of the cortex have primary responsibility for certain tasks. The **primary motor cortex** at the rear of each frontal lobe initiates voluntary movements. Neurons of this region are laid out like a map of the body, with body parts capable of fine movements taking up the greatest area (FIGURE 32.26).

The **primary somatosensory cortex** at the front of the parietal lobe also has neurons arranged like a map of the body. It processes sensory input from the skin and joints.

A primary visual cortex at the rear of each occipital lobe receives and processes signals from the eyes. A primary auditory cortex in the temporal lobes processes signals from the ears.

The brain is "cross wired," meaning each cerebral hemisphere receives input from and sends commands to the opposite side of the body. Commands to move your left arm originate in the motor cortex of your right hemisphere, and sensory signals from your left ear end up in the right hemisphere's temporal lobe.

TAKE-HOME MESSAGE 32.11

✔ The human cerebral cortex is a thin, highly folded layer of gray matter. It functions in social behavior, motor activity, sensory perception, reasoning, and language and speech.

32.12 Emotion and Memory

LEARNING OBJECTIVES

- Give examples of how components of the limbic system affect emotions.
- Compare the different types of memories.
- Describe some of the evidence that the hippocampus plays an important role in formation of long-term memories.

Most of the brain's gray matter is in the cerebral cortex, but a few structures deep in the brain are also composed of gray matter. Among other functions, these structures play a role in emotion and in memory.

The Emotional Brain

The **limbic system** is a collection of structures that encircle the upper part of the brain stem (FIGURE 32.27). This system governs emotions, assists in memory, and correlates organ activities with self-gratifying behavior such as eating and sex. The limbic system is sometimes described as our emotional-visceral brain, to contrast it with the cerebral cortex, which can often override the limbic system's "gut reactions."

Exactly how the structures of the limbic system give rise to different emotions is poorly understood, but we do know a bit about how its components function. For example, the hypothalamus summons up the physiological changes that accompany emotions. Signals from the hypothalamus make our heart pound and our palms sweat when we are fearful.

The cingulate gyrus is an arch-shaped region of tissue above the corpus callosum. It helps us to control our emotions and to learn from negative experiences. Defects in or damage to this region can cause symptoms of obsessive–compulsive disorder. A person with this disorder has recurrent unwanted thoughts and is driven by fear or anxiety to repeat behaviors.

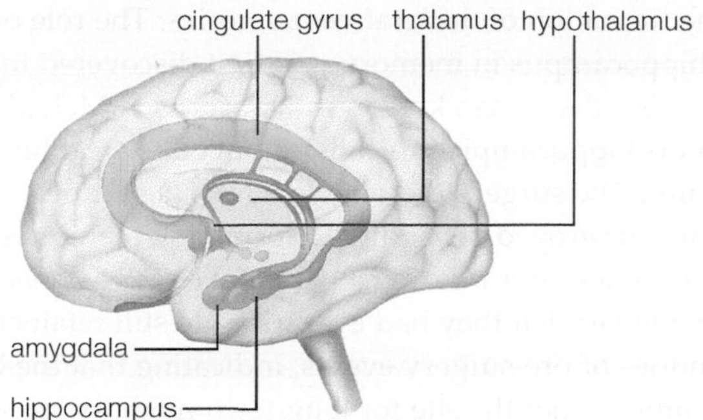
cingulate gyrus
thalamus
hypothalamus
amygdala
hippocampus

FIGURE 32.27 **Components of the limbic system.**

The adjacent, almond-shaped amygdala becomes active when we hear or see something that could be a threat. Its activity gives rise to our emotion of fear. The amygdala is often overactive in people who suffer from panic disorders.

Making Memories

At a cellular level, memory formation involves altering the number and placement of synaptic connections among existing neurons. The cerebral cortex receives information continually, but only a fraction of it is stored as a memory.

Memory forms in stages. Short-term memory lasts seconds to hours. This type of memory holds a few bits of information: a set of numbers, words of a sentence, and so forth. In long-term memory, larger chunks of information can be stored more or less permanently.

Different types of memories are stored and brought to mind by different mechanisms. Repetition of motor tasks creates skill memories that can be highly persistent. Learn to ride a bicycle, dribble a basketball, or play the piano, and you are unlikely to lose that skill. Forming skill memories involves neurons in the cerebellum, which coordinates motor activity. By contrast, declarative memory stores facts and impressions. It helps you remember how a lemon smells, that a quarter is worth more than a dime, where your classes are held, and how to find your way home.

The **hippocampus** (plural, hippocampi), a structure adjacent to the amygdala, plays an essential role

cerebral cortex Outer gray matter layer of the cerebrum.
hippocampus Brain region essential to formation of declarative memories.
hypothalamus (HI-poe-THAL-uh-mus) Forebrain region that controls processes related to homeostasis; control center for endocrine functions.
limbic system Group of structures deep in the brain that function in expression of emotion.
primary motor cortex Region of brain's frontal lobes that governs voluntary movements.
primary somatosensory cortex Region of the brain's parietal lobes that receives sensory information from skin and joints.

in the formation of declarative memories. The role of the hippocampus in memory was first discovered in the 1950s after a man known in reports as H.M. had both his hippocampi surgically removed to treat his seizures. The surgery did alleviate H.M.'s seizures, but also destroyed his ability to form new memories. Five minutes after meeting a person, H.M. was unable to remember that they had ever met. He still retained memories of pre-surgery events, indicating that the hippocampus is not the site for long-term memory storage.

Additional evidence of the hippocampus's role in memory comes from people with Alzheimer's disease. The hippocampus is one of the first brain regions destroyed by this disorder. Like H.M., people in early-stage Alzheimer's usually have impaired short-term memory, but retain their memories of long-ago events.

The hippocampus is one of the very few brain regions where new neurons continue to arise during adulthood. It is not yet clear how producing new neurons assists in memory formation. By one hypothesis, continued neuron turnover helps make new memories by disrupting connections among existing cells.

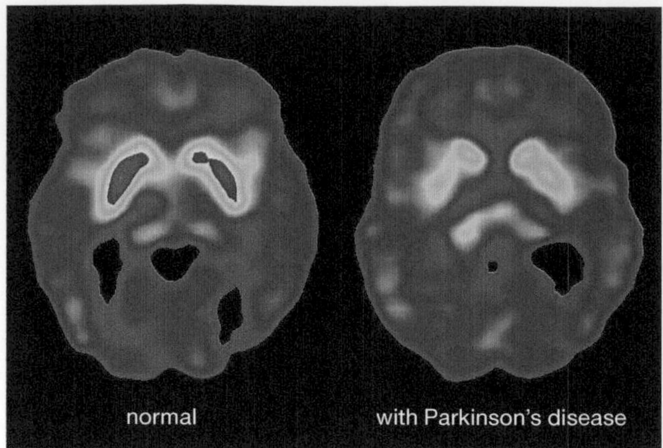
normal
with Parkinson's disease

FIGURE 32.28 PET scan images from a normal brain and from the brain of a person affected by Parkinson's disease. Red and yellow indicate areas of high glucose uptake by dopamine-secreting neurons. Death of these neurons in structures called basal nuclei gives rise to the symptoms of Parkinson's disease.

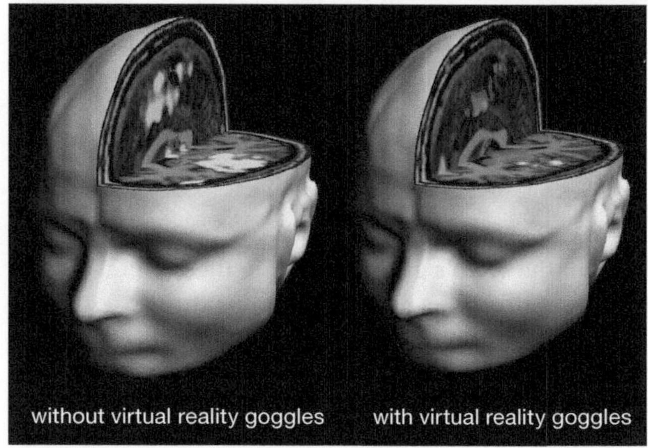
without virtual reality goggles
with virtual reality goggles

FIGURE 32.29 Pain-related brain activity with or without virtual reality goggles as revealed by fMRI. The goggles allow patients to enter a virtual world, distracting them as they undergo a painful procedure. Note that use of the goggles reduced the extent of high-activity areas (areas colored red, orange, and yellow).

TAKE-HOME MESSAGE 32.12

✔ The brain structures collectively described as the limbic system affect emotions and contribute to memory.

✔ Forming memories involves alterations to synaptic connections. Skill memory and declarative memory involve different regions of the brain. The hippocampus is essential to formation of new declarative memories.

32.13 Studying Brain Function

LEARNING OBJECTIVE

- Describe methods that scientists use to study brain function.

Scientists can learn about brain function by studying the behavior of individuals such as H.M. who have suffered brain damage in a specific region. Additional information comes from techniques that allow observations of brain activity in living people and from direct examination of brain tissue in the deceased.

Observing Electrical Activity

Electroencephalography detects voltage fluctuations in neurons of the cerebral cortex by way of electrodes attached to the scalp. It records these fluctuations as an electroencephalogram (EEG): a graph of the brain's electrical activity over time. EEGs provide information about a person's level of wakefulness, so they are frequently used to monitor people during sleep studies. (There are different stages of sleep, each characterized by a specific EEG pattern.) EEGs provide information about the level of neural activity across the entire cerebrum over intervals as short as a millisecond. However, they do not provide much information about the exact location of the active cells.

Monitoring Metabolism

Other methods allow scientists to study events in specific brain regions. Positron-emission tomography (PET) reveals areas of high activity in the brain by tracing uptake of radioactively labeled glucose. Glucose is the main energy source for brain cells, and active cells take up more than resting cells. **FIGURE 32.28** shows results of a PET study of Parkinson's disease.

Functional magnetic resonance imaging (fMRI) reveals details of brain activity by detecting changes in blood flow. Active brain cells use more oxygen than resting cells, and increased blood flow provides the necessary oxygen. **FIGURE 32.29** shows results of an fMRI study in which virtual reality goggles were used to reduce the patient's perception of pain.

Examining Brain Tissue

To study the structure, genetics, and biochemistry of human brain cells, scientists need brain tissue. "Brain banks" are repositories of such tissue. The largest one, the Harvard Brain Tissue Resource Center, has thousands of donated brains, most from people who had degenerative brain disorders (such as Parkinson's disease) or psychiatric illnesses (such as schizophrenia). This federally funded center supplies tissue samples to researchers who investigate the underlying causes of these conditions. Recently, microscopic analysis of brain samples provided by the center allowed researchers to determine why schizophrenics tend to have an unusually small prefrontal cortex. The study showed that schizophrenics have the normal number of brain cells in their prefrontal cortex, but these cells are unusually tightly packed, with fewer synapses among them. Examination of donated brain tissue is also integral to the ongoing study of how physical trauma affects the brain, the topic we return to below.

TAKE-HOME MESSAGE 32.13

- ✔ Observing people with damage to specific brain regions can provide information about the function of these regions.
- ✔ An EEG is a record of the electrical activity in a brain over time.
- ✔ PET scans and functional MRIs show how activity of different brain regions varies.
- ✔ Access to donated brain tissue in brain banks allows researchers to investigate structural, genetic, and biochemical differences between healthy and disordered brains.

32.1 Impacts of Concussions (revisited)

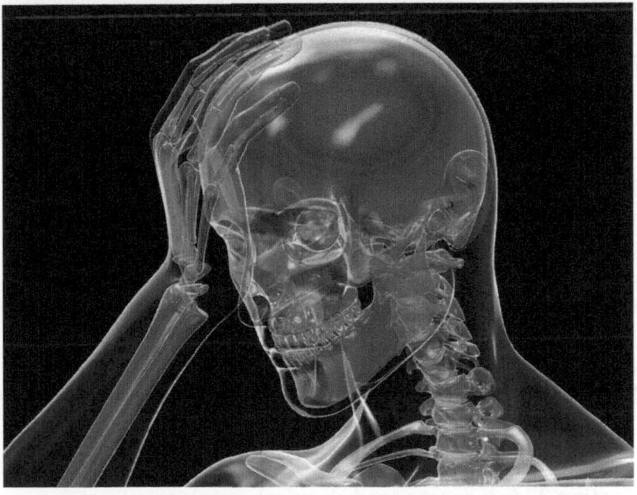

The brain fits relatively tightly into the skull, but there is a small cerebrospinal fluid-filled gap between the two. As a result, the brain can move a bit within the skull. An impact or sudden stop can cause enough movement to tear, bruise, and stretch brain tissue, causing a concussion. Axons become twisted and torn, impairing their ability to signal. Tiny blood vessels are torn too, compromising the blood–brain barrier and reducing blood flow to the brain.

The mechanical stress also results in spontaneous action potentials, causing neurotransmitters from the brain's neurons to flood the cerebrospinal fluid. Glutamate, the most abundant neurotransmitter in the brain, has an excitatory effect on neurons. When it pours into the cerebrospinal fluid, the brain's neurons undergo additional spontaneous action potentials. Collectively, these events disrupt normal ion concentration gradients across neuron plasma membranes.

Energy metabolism is also disrupted. In response to alteration of the normal concentration gradients, sodium–potassium pumps and other active transport pumps go to work overtime, and fueling their activity uses up a neuron's ATP. Even under ordinary circumstances, a neuron's mitochondria operate near their peak capacity, so they cannot increase their ATP output to help out. Thus, to meet the increased demand for ATP, the neuron must step up its rate of glycolysis and lactate fermentation. The resulting accumulation of lactate (an acid) further impairs neuron function.

With all of these problems, it is no surprise that the brain does not function normally for a period after a concussion. There is no treatment for concussion, other than physical and mental rest. In most cases, the brain heals itself within about 10 days. During recovery, it is especially important to avoid additional head trauma. Even a seemingly slight injury to the healing brain can result in increased swelling that can lead to permanent paralysis and, in some cases, death. The threat of this "second impact syndrome" makes it all the more important that an initial concussion be properly diagnosed. If a concussion is suspected, a person should halt physical activities and see a doctor as soon as possible. ●

CREDIT: (in text) BioMedical/Shutterstock.

STUDY GUIDE

Section 32.1 A concussion is a type of traumatic brain injury that typically disrupts brain function for about 10 days. Repeated blows to the head can cause irreversible damage and result in chronic traumatic encephalopathy (CTE).

Section 32.2 Neurons are electrically excitable cells that signal other cells by means of chemicals. Cnidarians have a **nerve net**. Most other animals have a bilateral nervous system with **cephalization**, which means they have paired **ganglia** or a **brain** at the head end. Chordates have a dorsal **nerve cord**. A vertebrate **central nervous system** (CNS) consists of the brain and spinal cord. **Nerves** that run through the body and connect to the CNS constitute the **peripheral nervous system**.

Sections 32.3 Vertebrates have three types of neurons. **Sensory neurons** detect stimuli. **Interneurons** relay signals between neurons. **Motor neurons** signal muscles and glands. A neuron's **dendrites** receive signals and its **axon** transmits signals. Glial cells assist and support neurons.

Section 32.4 Separation of charges across the plasma membrane causes the **membrane potential**, which is measured as voltage. At **resting potential**, the interior of the neuron is more negative than interstitial fluid. If the trigger zone of an axon reaches **threshold potential**, voltage-gated sodium ion channels open and an **action potential** begins. Flow of sodium ions into the axon results in **positive feedback**. The increase in positive charge causes more and more sodium ion channels to open. The resulting change in potential causes voltage-gated potassium channels to open, allowing potassium to exit the axon. The outward flow of potassium and diffusion of ions within the axon restores resting potential.

All action potentials are the same size and travel away from the cell body and toward the axon terminals. Most vertebrate axons have a **myelin sheath** that insulates them and speeds the conduction of axon potentials.

Section 32.5 Neurons send chemical signals to cells at **synapses**. A motor neuron communicates with a muscle fiber at a type of synapse called a **neuromuscular junction**. Arrival of an action potential at a presynaptic cell's axon terminal triggers release of a **neurotransmitter**. Neurotransmitter diffuses to receptors on a postsynaptic cell and binds to them. A postsynaptic cell often receives signals from many presynaptic cells; its response is determined by **synaptic integration** of all of these signals.

Section 32.6 Different neurons produce different neurotransmitters. Receptors for a particular neurotransmitter can vary between cells, so the same neurotransmitter can elicit different effects in different targets. Some neurons produce **neuromodulators** that affect multiple neurons at the same time. Production of too little or too much neurotransmitter can result in a disorder such as depression. Psychoactive drugs interfere with signaling at synapses.

Section 32.7 Nerves of the peripheral nervous system are bundles of myelinated axons wrapped in connective tissue. **Somatic nerves** control skeletal muscles. They play a role in **reflexes** that involve these muscles. **Autonomic nerves** control internal organs and glands.

There are two types of neurons in the autonomic system. In times of stress or danger, **sympathetic neurons** elicit a fight–flight response. During less stressful times, **parasympathetic neurons** encourage resting and digesting. Organs receive signals from both types of autonomic neurons.

Section 32.8 The spinal cord and brain are organs of the central nervous system. They are enclosed by **meninges** and cushioned by **cerebrospinal fluid**. The **blood–brain barrier** controls the composition of cerebrospinal fluid.

There are two visibly different types of tissue in the central nervous system. **White matter** contains myelinated axons. **Gray matter** contains neuron cell bodies, dendrites, axon terminals, and glial cells. Synapses are in the gray matter.

Sections 32.9, 32.10 A vertebrate embryo's neural tube develops into the **spinal cord** and brain. The spinal cord runs through the vertebral column. It connects some peripheral nerves to the brain. It also acts in spinal reflexes.

Evolutionarily, the **brain stem** is the oldest brain tissue. It includes the **pons** and **medulla oblongata**, which control reflexes involved in breathing and other essential tasks. The **cerebellum** in the hindbrain coordinates motor activities.

In the forebrain, the cerebrum consists of two hemispheres connected by the corpus callosum. The forebrain also includes the **thalamus** and **hypothalamus**. The thalamus regulates sleep and waking. The hypothalamus is a major control center for activities that maintain homeostasis.

Sections 32.11, 32.12 The **cerebral cortex** is a layer of gray matter. The frontal lobe is essential to planning and normal social behavior. Broca's area in the left frontal lobe is required for speech. The **primary motor cortex** at the rear of the frontal lobe controls voluntary movement. The **primary somatosensory cortex** in the parietal lobe receives sensory input from the skin and joints.

The cerebral cortex interacts with the **limbic system**, which governs emotion, and with the **hippocampus**, which is essential to memory. These regions of gray matter lie deep within the brain.

Section 32.13 EEGs record the electrical activity of neurons in the brain. PET scans and fMRI studies pinpoint areas of high neuron activity. Brain banks are respositories of donated brain tissue that can be used for research.

SELF-QUIZ

Answers in Appendix VII

1. _______ relay messages from the brain and spinal cord to muscles and glands.
   a. Motor neurons c. Interneurons
   b. Sensory neurons d. Neuroglia

2. When a neuron is at rest, _______ .
   a. it is at threshold potential
   b. voltage-gated sodium channels are open
   c. sodium–potassium pumps are operating
   d. it contains more sodium ions than the surrounding interstitial fluid

3. An action potential begins when _______ .
   a. a neuron reaches threshold potential
   b. voltage-gated potassium gates open
   c. voltage-gated sodium gates close
   d. sodium–potassium pumps stop operating

4. Most human axons have a myelin sheath that _______ .
   a. prevents action potentials from moving backward
   b. increases the speed of action potential conduction
   c. consists of specialized interneurons
   d. increases the size of action potentials

5. Neurotransmitters are released by _______ .
   a. axon terminals c. dendrites
   b. a neuron cell body d. glial cells

6. What chemical is released by the axon terminals of a motor neuron at a neuromuscular junction?
   a. ACh c. dopamine
   b. serotonin d. epinephrine

7. Which neurotransmitter is important in reward-based learning and drug addiction?
   a. ACh c. dopamine
   b. serotonin d. epinephrine

8. Skeletal muscles are controlled by _______ .
   a. sympathetic neurons
   b. parasympathetic neurons
   c. somatic motor neurons
   d. somatic sensory neurons

9. When you sit quietly on the couch and read, output from _______ neurons prevails.
   a. sympathetic b. parasympathetic

10. In the central nervous system, all synapses are in the _______ .
   a. cerebellum c. gray matter
   b. brain stem d. spinal cord

11. The brain and spinal cord develop from the embryonic _______ .
   a. notocord c. neural tube
   b. gill slits d. gut

12. An injury _______ in the spinal cord can lead to paralysis of all limbs.
   a. high c. low

13. _______ deep in the brain plays a role in emotion.
   a. The cerebellum c. The limbic system
   b. Broca's area d. The somatosensory cortex

14. Commands to move your right arm start in the _______.
   a. left frontal lobe c. right temporal lobe
   b. right occipital lobe d. left parietal lobe

15. Match each item with its description.
   ___ gray matter a. fatty insulating material
   ___ neurotransmitter b. connects the hemispheres
   ___ pons c. protects brain and spinal cord from some toxins
   ___ corpus callosum d. type of signaling molecule
   ___ cerebral cortex e. allows spoken language
   ___ myelin f. brain stem structure
   ___ Broca's area g. controls language, reasoning
   ___ blood–brain barrier h. cell bodies and dendrites

CENGAGE brain .com To access course materials, please visit www.cengagebrain.com.

CRITICAL THINKING

1. Some survivors of disastrous events have post-traumatic stress disorder (PTSD). Symptoms include nightmares about the experience and suddenly feeling as if the event is recurring. Brain-imaging studies of people with PTSD showed that their hippocampus was shrunken and their amygdala unusually active. Given these changes, what other brain functions might be disrupted in PTSD?

2. Most tumors that originate in the brain are not cancer, but growth of any tumor within the confined space of the skull can put pressure on surrounding nervous tissue and thus harm neurons. Treating brain tumors with drugs is difficult because of the blood–brain barrier. Lipid-soluble drugs can penetrate this barrier, but they get into and affect all other cells too. Explain why.

3. Hansen's disease, also known as leprosy, arises when bacteria infect the Schwann cells that wrap around axons of peripheral nerves to form a myelin shealth. Attacks by the immune system on the bacteria-infected glial cells causes a loss of myelin in peripheral nerves. How is Hansen's disease similar to multiple sclerosis? Why do the symptoms differ? Why are people with Hansen's disease unusually prone to injuring themselves?

33 SENSORY PERCEPTION

CORE CONCEPTS

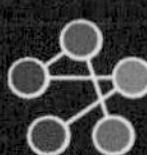 Systems

Complex properties arise from interactions among components of a biological system.
Sensory organs such as eyes and ears are complex structures. In addition to sensory receptors, they contain a variety of tissues that collectively ensure that the proper stimulus reaches these receptors, and that nerve signals from these receptors are sent on their way toward the brain. Input from sense organs throughout our body contributes to our perception of our world.

Evolution

Evolution underlies the unity and diversity of life.
The types of sensory receptors an animal has, the range of stimuli that those receptors respond to, and the location of those receptors vary among animal groups. These variations reflect sensory adaptations to different environments. Most animals can sense temperature, touch, tissue injury, light, and the presence of some chemicals. Some also have a capacity to sense other stimuli such as magnetic or electric fields.

 Structure and Function

The three-dimensional form and arrangement of biological structures give rise to their function and interactions.
The three-dimensional form of a sensory receptor allows it to respond to a specific stimulus, such as mechanical energy, light energy, or the binding of a particular chemical. Variations in the form of a sensory receptor make it sensitive to a specific type of stimulus, such as a certain wavelength of light or a particular odorant molecule.

Links to Earlier Concepts

This chapter explains how the cone snail venom discussed in Section 24.1 prevents pain. It draws on your knowledge of action potentials (32.4), neuromodulators (32.6), and the human brain (32.10–32.12). We also refer back to morphological convergence (18.3) and other phenomena of vertebrate evolution (25.6).

33.1 Neuroprostheses

All sensory input travels toward the brain in the form of action potentials, which are electrical signals. As a result, it is possible to elicit sensory sensations using devices that electrically stimulate sensory nerves. Devices that assist users by manipulating their nervous system are known as neuroprostheses. (A prosthesis is a medical device that replaces or assists a missing or damaged body part.)

The most widely used neuroprostheses are cochlear implants, devices that replace the function of the inner ear in people who are deaf. Cochlear implants first became available about 30 years ago. Today, more than 350,000 people around the world use them.

A cochlear implant has a microphone that sits just outside each ear (**FIGURE 33.1**). The microphone picks up sounds and sends them to a processor that converts the sounds into electrical signals. Electrodes in the internal portion of the device pick up these signals and generate action potentials in the appropriate regions of the inner ear's sound-detecting cochlea. These action potentials travel along the auditory nerves to the brain.

Sound perception with an implant is not the same as normal hearing, and people's ability to comprehend spoken language using the implant varies. Congenitally deaf individuals who receive a cochlear implant later in life typically have a less successful result than those who received implants as young children. This is because a brain that does not receive auditory input reorganizes itself. Brain regions that would normally function in hearing do not simply sit idle, but rather are recruited to function in vision. Thus, deaf people often have more acute vision than those with normal hearing. After this reorganization has occurred early in life, it is largely irreversible. The brain of a deaf adult who receives a cochlear implant cannot once again reorganize itself to begin processing auditory information in an area now devoted to other senses.

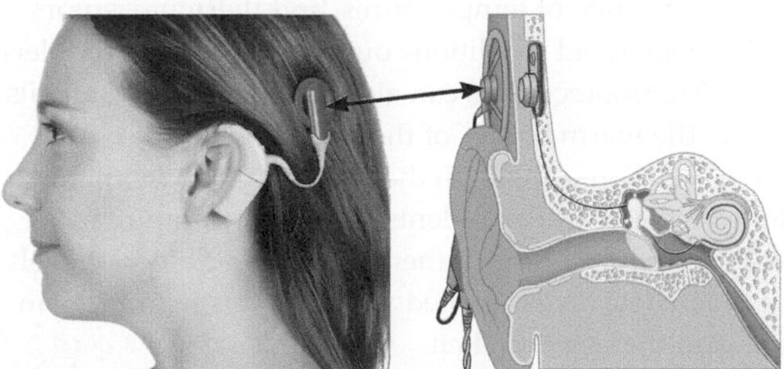

FIGURE 33.1 Cochlear implant. This device transduces sounds into electrical signals that flow along the auditory nerve to the brain. The external transmitter (indicated by the arrows) sends information about sounds to the internal portion of the device.

CREDITS: (opposite) Dieter Hawlan/Shutterstock; (1) left, iStock.com/ELizabethHoffmann; right, Dorling Kindersley/Getty Images.

33.2 Overview of Sensory Pathways

LEARNING OBJECTIVES

- Using appropriate examples, describe the different types of sensory receptors.
- Explain how the brain determines the location and intensity of a stimulus.
- Explain what occurs during sensory adaptation.

The sensory portion of a vertebrate nervous system includes sensory neurons that become excited in the presence of specific stimuli, nerves that carry information about the stimulus to the brain, and brain regions that process this information. In the context of sensory systems, "stimulus" refers to some aspect of the internal or external environment that has the capacity to excite a sensory neuron.

Sensory Diversity

Detection of stimuli begins with excitation of sensory receptors. A sensory receptor can be either a sensory neuron or a specialized epithelial cell that responds to a stimulus by exciting a sensory neuron. In either case, detection of a stimulus by a sensory receptor results in excitation of a sensory neuron.

We classify sensory receptors by the types of stimuli they respond to. Five types occur in most animals:

Mechanoreceptors respond to mechanical energy. Some detect shifts in a body's position or acceleration, others respond to touch or to stretching of a muscle, and still others respond to vibrations caused by pressure waves. Sound is a type of pressure wave, so auditory receptors are a type of mechanoreceptor. Auditory receptors of different animals vary in the frequency that they detect. Whales and elephants produce and detect ultra-low frequency sounds that humans cannot hear. Bats emit and respond to sounds too high for us to perceive (**FIGURE 33.2A**).

Thermoreceptors respond to a specific temperature or to a temperature change. All animals tolerate only a limited range of temperatures, and thermoreceptors help them avoid conditions outside their range of tolerance. Thermoreceptors can also help predatory animals locate the warm bodies of their prey. Some snakes have thermoreceptors on their head that allow them to detect any nearby rodents. Similarly, vampire bats use thermoreceptors on their nose to find blood vessels of their prey; a vessel filled with blood is warmer than the skin that surrounds it.

Chemoreceptors detect specific molecules in an environment. They function in the senses of taste and smell. Nearly all animals have chemoreceptors that help them locate nutrients and avoid ingesting

A Mechanoreceptors in a vampire bat's ear allow the bat to detect high-pitched sound waves (ultrasound). Thermoreceptors in the skin of the bat's nose help it locate veins filled with warm blood.

B Chemoreceptors embedded in the skin of a snail's two pairs of tentacles allow the snail to detect chemicals associated with food. Eyes at the tips of the two long tentacles contain photoreceptors that detect light.

C Magnetoreceptors allow pigeons to detect variations in Earth's magnetic field. They use this information to navigate.

FIGURE 33.2 Sensory diversity.

poisons (FIGURE 33.2B). Other types of chemoreceptors monitor the internal environment. For example, chemoreceptors in some of your arteries monitor the level of CO_2 in your blood.

Photoreceptors detect light energy. Humans detect only visible light, but insects and some other animals, including rodents, also respond to ultraviolet (UV) light. Some flowers have UV-absorbing pigments arranged in patterns that are invisible to us, but draw the attention of the insects that pollinate them.

Pain receptors, also called nociceptors, detect tissue damage. They have a protective function and are often involved in reflexes that minimize further harm.

Other types of sensory receptors are less widespread among animals. For example, some animals have electroreceptors that can detect electrical signals. Such receptors help some fish, amphibians, and even platypuses detect electrical signals produced by the nerves and hearts of their prey. As another example, animals such as pigeons, sea turtles, and honeybees have magnetoreceptors that aid their navigation by allowing them to detect Earth's magnetic field (FIGURE 33.2C).

From Sensing to Sensation to Perception

In animals that have a brain, processing of sensory signals gives rise to **sensation**, which is the awareness of some stimulus.

When a sensory receptor is stimulated sufficiently, it causes an action potential in a sensory neuron. Action potentials, remember, are always the same size (Section 32.4). The brain gathers additional information about a stimulus from three types of data: (1) the nerve pathway that was triggered, (2) the number of axons in the pathway that are undergoing action potentials, and (3) the number of axons recruited by the stimulus.

An animal's brain interprets action potentials on the basis of where they originate. This is why you may "see stars" if you press on your eyes in a dark room. The pressure accidentally causes action potentials to travel along an optic nerve to the brain, which interprets all signals from this nerve as "light."

A strong stimulus causes a receptor to generate action potentials more often and longer than a weak signal does. The same receptors are stimulated by a whisper and a whoop. Your brain interprets the difference by variations in the frequency of the incoming signals (FIGURE 33.3). In addition, a strong stimulus recruits more sensory receptors, compared with a weak stimulus. A gentle tap on the arm activates fewer receptors than a slap.

Stimulus duration also affects how the stimulus is interpreted. In **sensory adaptation**, sensory neurons stop generating action potentials (or make fewer of them) despite continued stimulation. Walk into a house where an apple pie is in the oven, and you will notice the sweet scent of baking apples immediately. Then, within a few minutes, the scent seems to lessen. The odor does not actually change in intensity, but chemoreceptors in your nose adapt to it.

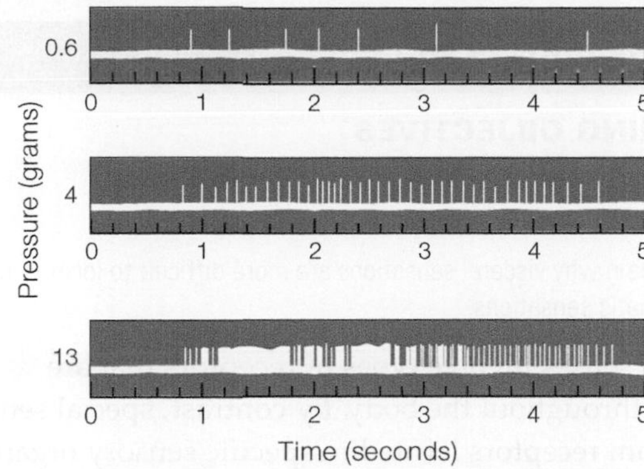

Pressure (grams)
0.6
4
13
0
1
2
3
4
5
Time (seconds)

FIGURE 33.3 Variations in action potential frequency. Vertical bars above each thick horizontal line represent action potentials in a pressure-sensitive mechanoreceptor in the skin. A rod was pressed against skin with varying amounts of pressure. The stronger the pressure, the more action potentials (white bars) occur per second.

Sensory perception arises when the brain assigns meaning to sensory signals. Consider what happens when you watch a person walking away from you. As the distance between the two of you increases, the image of the person on your eye becomes smaller and smaller. You perceive this change in sensation as evidence of increasing distance between you and the person, rather than evidence that the person is shrinking.

TAKE-HOME MESSAGE 33.2

✔ Sensory neurons undergo action potentials in response to specific stimuli. Different kinds of sensory receptors respond to different types of stimuli.

✔ Action potentials are all the same size, but which axons are responding, how many are responding, and the frequency of action potentials provide the brain with information about stimulus location and strength.

chemoreceptor (KEEM-oh-ree-SEP-tor) Sensory receptor that responds to the presence of a specific chemical.
mechanoreceptor Sensory receptor that responds to pressure, position, or acceleration.
pain receptor Sensory receptor that responds to tissue damage.
photoreceptor Sensory receptor that responds to light.
sensation Awareness of a stimulus.
sensory adaptation Slowing or halting of a sensory response to an ongoing stimulus.
sensory perception The meaning a brain derives from a sensation.
thermoreceptor Temperature-sensitive sensory receptor.

33.3 General Senses

LEARNING OBJECTIVES

- Compare general senses and special senses.
- Describe somatic sensations and how they arise.
- Explain why visceral sensations are more difficult to localize than somatic sensations.

General senses involve types of receptors that are located throughout the body. By contrast, special senses arise from receptors located in specific sensory organs, such as eyes or ears. Smell, taste, vision, hearing, and the sense of equilibrium (balance) are special senses. We consider them in later sections of this chapter. Here we focus on the general senses.

Sensory neurons responsible for **somatic sensations** are located in skin, muscle, tendons, and joints. These sensations are easily localized to a specific part of the body. In contrast, **visceral sensations**, which arise from excitation of neurons in the walls of soft internal organs, are often difficult to pinpoint. It is easy to determine exactly where someone is touching you, but difficult to say exactly where you feel nausea.

The Somatosensory Cortex

Signals from the sensory neurons involved in somatic sensation travel along axons to the spinal cord, then along tracts in the spinal cord to the brain. The signals end up in the somatosensory cortex, a part of the cerebral cortex. Like the motor cortex (Section 32.11), the somatosensory cortex has neurons arrayed like a map of the body (**FIGURE 33.4**).

As an example of the types of receptors that signal the somatosensory cortex, consider those in the human skin (**FIGURE 33.5**). Receptor endings of sensory neurons are either surrounded by some sort of capsule or free (unenclosed) in the tissue. Some of the free nerve endings in the derm is coil around the roots of hair follicles and can detect even the slightest movement of the hair. Other free nerve endings detect temperature changes or tissue damage. Free nerve endings in skeletal muscles, tendons, joints, and walls of internal organs give rise to sensations that range from itching, to a dull ache, to sharp pain.

Concentric layers of epithelial tissue wrap around the receptor endings of many sensory neurons in the skin. These encapsulated receptors are named for the scientists who first described them. Meissner's corpuscles and Pacinian corpuscles detect touch and pressure in hairless skin regions such as fingertips, palms, and the soles of the feet. Meissner's corpuscles in the upper dermis are small and respond to light touches. Pacinian corpuscles respond to more pressure. They are larger, deeper in the dermis, and also occur near joints and in the wall of some organs. Ruffini endings are encapsulated receptors that are found in skin throughout the body and detect its stretching.

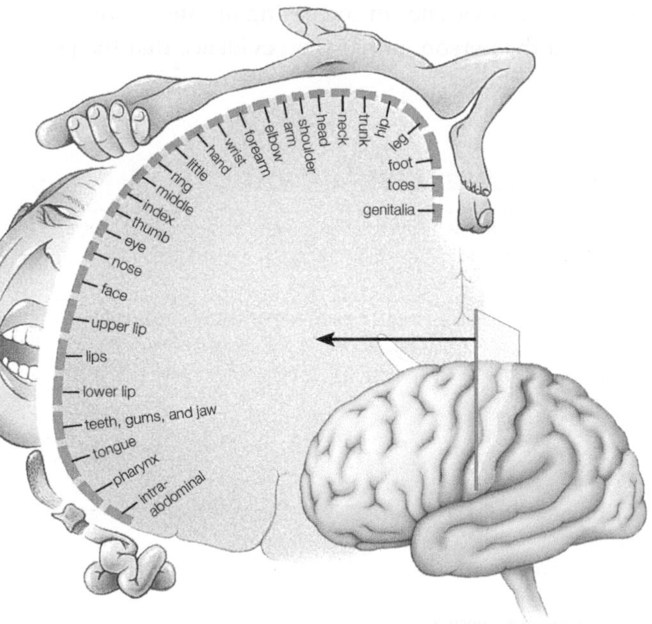

FIGURE 33.4 Representation of body regions in the human primary somatosensory cortex. This brain region is a narrow strip of the cerebral cortex (shown here in yellow) that runs from the top of the head to just above each ear. Body parts that are disproportionately large in the "body" mapped onto this brain correspond to body regions with the most sensory receptors, such as the fingertips, face, and lips. Body parts that have relatively fewer sensory neurons, such as thighs, are disproportionately small. Compare with Figure 32.26.

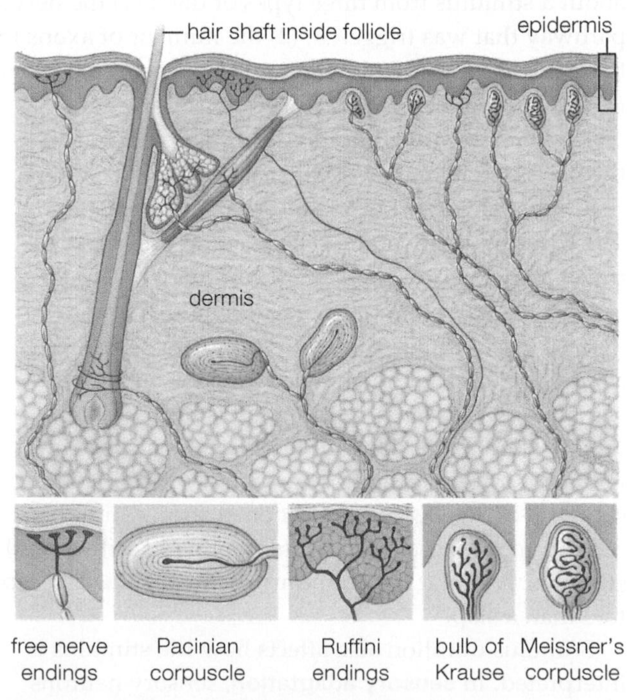
hair shaft inside follicle
epidermis
dermis
free nerve endings
Pacinian corpuscle
Ruffini endings
bulb of Krause
Meissner's corpuscle

FIGURE 33.5 Sensory receptors in human skin.

Compared to Meissner's and Pacinian corpuscles, Ruffini endings adapt to continued stimulation more slowly. If you hold a stone in your hand, Ruffini endings continue to inform your brain the stone is still there even after other receptors have stopped responding. Still another encapsulated receptor, called the bulb of Krause, responds to touch and to cold. Bulbs of Krause occur in skin throughout the body but are especially abundant in skin of the lips, tongue, nipples, and genitals.

The somatosensory cortex also receives signals from mechanoreceptors that respond to the stretching of tendons and muscles. This information helps your brain keep track of where your body parts are located in space. The muscle spindle fibers that take part in the stretch reflex (Section 32.7) are one such type of receptor. The more a muscle stretches, the more frequently stretch receptors fire. Nearly all skeletal muscles and smooth muscles contain muscle spindles.

Pain

Pain is the perception of a tissue injury. Somatic pain begins with signals from pain receptors in skin, skeletal muscles, joints, and tendons. Visceral pain is associated with organs inside body cavities. It occurs as a response to a smooth muscle spasm, inadequate blood flow to an organ, overstretching of a hollow organ such as the stomach, and other abnormal conditions.

Injured or distressed body cells release local signaling molecules such as histamine and prostaglandins. These molecules stimulate neighboring pain receptors, increasing the likelihood that the receptors will send signals along their axons to the spinal cord. In the spinal cord, the axons synapse with interneurons that relay signals about pain to the somatosensory cortex.

The neuromodulator substance P enhances pain perception by making spinal interneurons more likely to send signals about pain to the sensory cortex. People with fibromyalgia, a disorder characterized by chronic pain in muscles and joints, tend to have an elevated level of substance P in their cerebrospinal fluid. The pain-reducing effect of endorphins arises from their ability to slow release of substance P.

Pain-relieving drugs (analgesics) interfere with pain perception. Aspirin reduces pain by slowing prostaglandin production. Synthetic opiates such as morphine mimic the activity of endorphins. The drug ziconotide, a chemical first discovered in the venom of a cone snail (Section 24.1), blocks calcium channels in axon terminals of pain receptor neurons. Blocking these channels prevents the neurons from releasing neurotransmitter and inhibits transmission of pain signals.

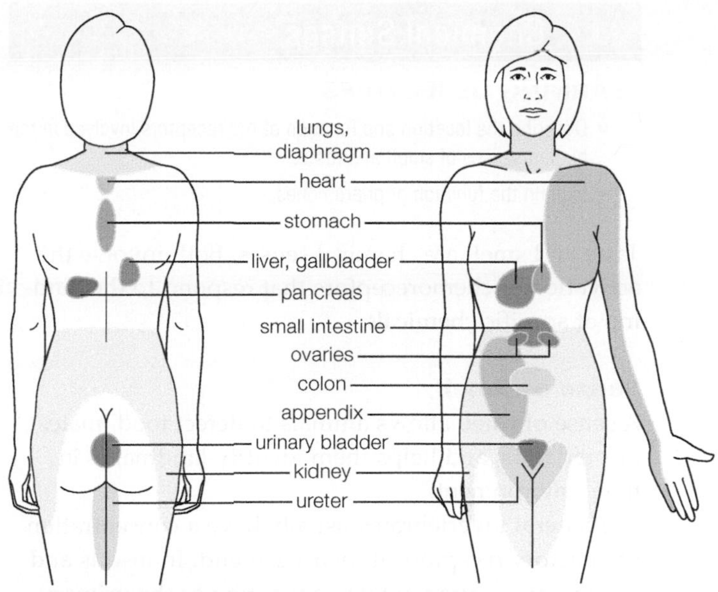
lungs,
diaphragm
heart
stomach
liver, gallbladder
pancreas
small intestine
ovaries
colon
appendix
urinary bladder
kidney
ureter

FIGURE 33.6 Sites of referred pain. Colored regions indicate the area that the brain interprets as affected when specific internal organs are actually distressed.

The brain sometimes mistakenly interprets signals from receptors in internal organs as if they were from receptors in the skin or joints. The result is referred pain. The classic example is a pain that radiates from the chest across the shoulder and down the left arm during a heart attack (**FIGURE 33.6**). The arm is not affected, so why does it hurt? Referred pain occurs because each part of the spinal cord receives sensory input from both skin and internal organs. Skin encounters more painful stimuli than internal organs, so pain signals from skin flow more frequently along the neural pathway to the brain. The brain tends to attribute signals that arrive along a particular pathway to their most common source, even if they originate elsewhere.

TAKE-HOME MESSAGE 33.3

- ✔ Somatic sensations originate at sensory receptors in skin, skeletal muscle, and joints. They travel along sensory neuron axons to the spinal cord, then to the somatosensory cortex.
- ✔ Visceral sensations originate with the stimulation of sensory neurons in the walls of organs. These signals are relayed to the spinal cord, and then to the brain.
- ✔ Pain is the sensation associated with tissue damage. Referred pain occurs because the brain sometimes misinterprets visceral pain as if it were caused by a problem in the skin or a joint.

pain Perception of tissue injury.
somatic sensations Sensations such as touch and pain that arise when sensory neurons in skin, muscle, or joints are activated.
visceral sensations Sensations that arise when sensory neurons associated with organs inside body cavities are activated.

33.4 Chemical Senses

LEARNING OBJECTIVES

- Describe the location and function of the receptors involved in the human senses of smell and taste.
- Explain the function of pheromones.

Taste and smell are chemical senses. Both involve the activation of chemoreceptors that respond to the binding of specific chemicals.

Sense of Smell

A sense of smell allows animals to detect food, mates, or predators, and helps them identify landmarks in their environment.

Bilateral invertebrates usually have a concentration of olfactory receptors at their head end. In insects and crustaceans, paired antennae function as the primary olfactory organs. In snails and slugs, a pair of sensory tentacles serve this purpose.

In humans and other vertebrates, receptor endings of sensory receptors extend into the lining of a nasal cavity, where receptor proteins in their plasma membrane bind dissolved odorant molecules (**FIGURE 33.7** ❶). Humans have hundreds of types of chemoreceptive sensory neurons, each with endings that respond to one kind of odorant molecule. Binding of that molecule to the receptor protein alters the protein's shape and sets in motion reactions that culminate in an action potential in the sensory neuron.

Axons of the olfactory sensory neurons extend through the base of the skull and into the brain's olfactory bulb, where they synapse on interneurons ❷. Each interneuron receives signals from many sensory neurons that all detect the same odorant molecule ❸. In response to excitatory signals from these cells, the interneurons send signals to other brain regions such as the limbic system and the cerebral cortex ❹.

An average person can discriminate between and recall about 10,000 different odors. When you recognize an odor, such as the scent of a rose, you are responding to a distinctive mix of odorant molecules. These molecules excite a unique subset of your nose's sensory neurons, thus triggering a unique pattern of excitation in the cerebral cortex. Through past experience, your brain has learned to associate this pattern of excitatory signals with its source.

Vertebrates vary in their ability to detect odors. For example, dogs are known for their acute sense of smell. They detect some odorant molecules at concentrations 10,000 to 100,000 times lower than humans can. A variety of factors contribute to dogs' greater olfactory capacity. Adjusting for body size, a dog has a far larger area of nasal epithelium. This epithelium has both a

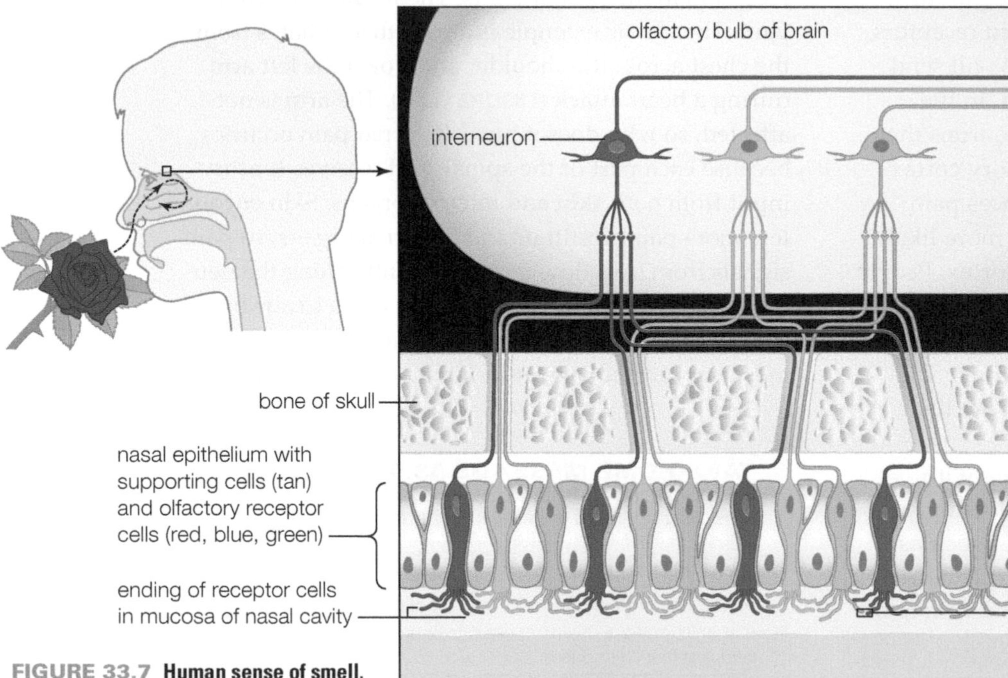

olfactory bulb of brain
interneuron
bone of skull
nasal epithelium with supporting cells (tan) and olfactory receptor cells (red, blue, green)
ending of receptor cells in mucosa of nasal cavity

❹ Interneurons in the olfactory lobe relay signals from sensory neurons to other regions of the brain, including the limbic system and cerebral cortex.

❸ In the olfactory bulb, axons of sensory neurons synapse on interneurons. Each interneuron receives signals only from cells with the same type of receptor.

❷ An odorant molecule binds to a receptor protein on a sensory neuron, initiating an action potential that travels along the neuron's axon to the olfactory bulb in the brain.

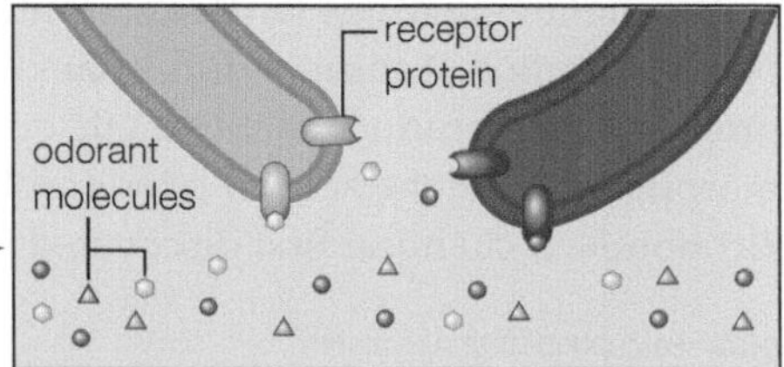

receptor protein
odorant molecules

FIGURE 33.7 Human sense of smell. Odorant molecules excite a sensory neuron when they bind to chemoreceptors in its plasma membrane.

❶ Inhaled odorant molecules bind to receptor proteins on chemoreceptive sensory neurons in the nasal cavity. A receptor protein binds only one type of odorant molecule, and each cell has only one type of receptor. This art shows three types of receptor cells (coded red, green, and blue), but there are hundreds.

denser array of sensory neurons and a greater variety of olfactory receptor proteins. Differences in the speed of air flow through the nose also play a role. A dog's nose has a recessed area where air flow slows, making it easier for olfactory receptors to bind inhaled odorant molecules. The human nasal cavity, like that of other primates, has no equivalent recessed region.

Sense of Taste

Like olfactory receptors, taste receptors are chemoreceptors that detect chemicals dissolved in fluid, but they differ in their function, location, and structure. An animal uses its taste receptors to decide whether to ingest or reject a potential food item.

An octopus tastes with receptors in suckers on its tentacles; a fly tastes with receptors in its antennae and feet. Some catfishes have taste receptors in skin all over their body, with the greatest number in their barbs (fleshy projections near the mouth).

In humans and other mammals, most taste receptors are located in taste buds in the lining of the mouth and on the upper surface of the tongue. Taste buds are located in specialized epithelial structures, or papillae, that are visible as raised bumps (**FIGURE 33.8**). Each taste bud contains taste receptor cells (specialized epithelial cells) and neurons. Receptor-covered microvilli of the taste receptor cells extend out through a pore, and thus come in contact with food molecules in saliva.

The sensation of taste begins with the binding of a molecule to receptor proteins on the microvilli of a taste receptor cell. Binding of the molecule excites this cell, which in turn excites a sensory neuron within the taste bud, which relays an action potential along its axon to the brain.

In humans, perceived taste arises from a combination of signals produced by different types of taste receptors. We have receptors that respond to *sweetness* (elicited by glucose and the other simple sugars), *sourness* (acids), *saltiness* (sodium chloride or other salts), *bitterness* (plant toxins, including alkaloids), *umami* (elicited by amino acids such as glutamate, which is found in cheese and aged meat), and—as recently discovered—*fattiness* (fatty acids). All mammals generally have the same types of taste receptors, although cats, dolphins, and some other carnivores have lost the ability to detect sweetness.

Each taste receptor cell is most sensitive to one of the six primary tastes, but taste buds in all regions of the tongue include all six types of cells. Thus, contrary to popular belief, the ability to detect specific tastes does not map to specific regions of the tongue.

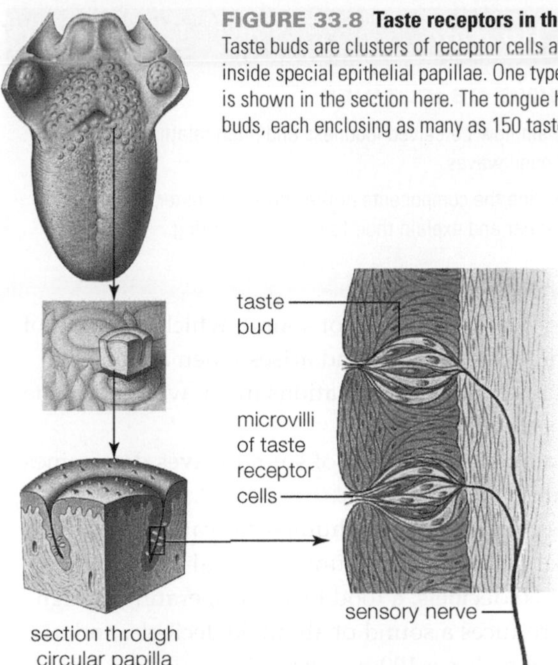
taste bud
microvilli of taste receptor cells
sensory nerve
section through circular papilla

FIGURE 33.8 Taste receptors in the human tongue. Taste buds are clusters of receptor cells and supporting cells inside special epithelial papillae. One type, a circular papilla, is shown in the section here. The tongue has about 5,000 taste buds, each enclosing as many as 150 taste receptor cells.

Pheromones—Chemical Messages

Many animals release and detect chemicals that function in communication. A **pheromone** is a signaling molecule that is secreted by one individual and affects the behavior of other members of its species. For example, male silk moths have antennae that detect a sex pheromone secreted by female moths. Male moths will fly upwind toward the pheromone's source.

Reptiles and most mammals make pheromones. Their **vomeronasal organ** is a collection of sensory neurons in the nasal cavity that bind and respond to pheromones. Humans and our closest primate relatives have a reduced version of this organ. The extent to which humans detect and respond to pheromones remains a matter of investigation. We discuss the role of pheromones in animal communication in more detail in Chapter 43.

TAKE-HOME MESSAGE 33.4

✔ Smell and taste involve stimulation of chemoreceptors by the binding of specific molecules.

✔ Humans have six types of taste receptors and hundreds of types of olfactory receptors.

✔ Many kinds of animals make and detect pheromones, chemicals that function in intraspecific communication.

pheromone (FER-uh-moan) Chemical that serves as a communication signal among members of an animal species.

vomeronasal organ Pheromone-detecting organ of some vertebrates.

33.5 Hearing

LEARNING OBJECTIVES

- Explain how perceived loudness and pitch relate to properties of sound waves.
- Describe the components of the vertebrate outer, middle, and inner ear and explain their functions in hearing.

Properties of Sound

Hearing is the perception of sound, which is a form of mechanical energy. A sound arises when a vibrating object causes pressure variations in air, water, or some other medium.

The amplitude (height) of sound waves determines the volume of a sound (FIGURE 33.9A). We measure loudness in decibels. The human ear can detect a 1-decibel difference in volume. Normal conversation is about 60 decibels, a food blender operating at high speed produces a sound of about 90 decibels, and a chain saw makes a 100-decibel noise.

A sound's frequency determines its pitch, which is measured as the number of wave cycles per second, or hertz (Hz). The more waves per second, the higher the pitch (FIGURE 33.9B).

The Vertebrate Ear

Water readily transfers vibrations to body tissues, so fishes do not require elaborate ears to detect sounds.

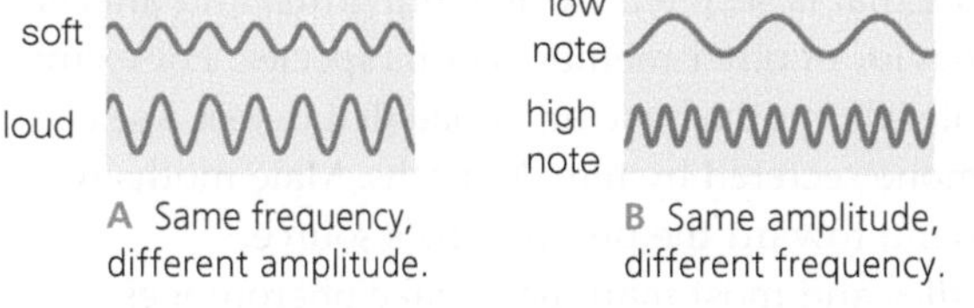
soft
loud
A Same frequency, different amplitude.
low note
high note
B Same amplitude, different frequency.

FIGURE 33.9 **Properties of sound.**

When vertebrates left water for land, their capacity to collect and amplify vibrations evolved in response to a new environmental challenge: The transfer of sound waves to body tissues is less efficient in air than in water. Certain features of mammalian ears are adaptations that increase the efficiency of sound transfer (FIGURE 33.10). For example, unlike amphibians and reptiles, most mammals have an **outer ear** that funnels sound inward ❶. A skin-covered flap of cartilage, called the pinna is the visible part of your outer ear. It collects sound waves and directs them into the auditory canal, an air-filled passage that connects to the middle ear.

The **middle ear** amplifies sound waves and transmits them to the inner ear. An **eardrum**, or tympanic membrane, first evolved in amphibians. It is visible as a round area on each side of their head. Sound waves cause an eardrum to vibrate. In mammals, an air-filled cavity behind the eardrum holds three small bones called the malleus, incus, and stapes ❷. The names refer to their shapes. In Latin *malleus* means hammer, *incus* means anvil, and *stapes* means stirrup. The middle ear bones transmit the force of sound waves from the eardrum onto the surface of the oval window, an elastic membrane that is the boundary between the middle and inner ear.

The **inner ear** includes structures involved in hearing and in balance. We discuss the sense of balance in Section 33.6. Here, we focus on the role of the sound-detecting **cochlea**, a pea-sized, fluid-filled structure that resembles a coiled snail shell. (The Greek *koklias* means snail.) Membranes divide the interior of the cochlea into three fluid-filled ducts ❸. When sound waves make the three bones of the middle ear vibrate, the stirrup pushes against the oval window. With

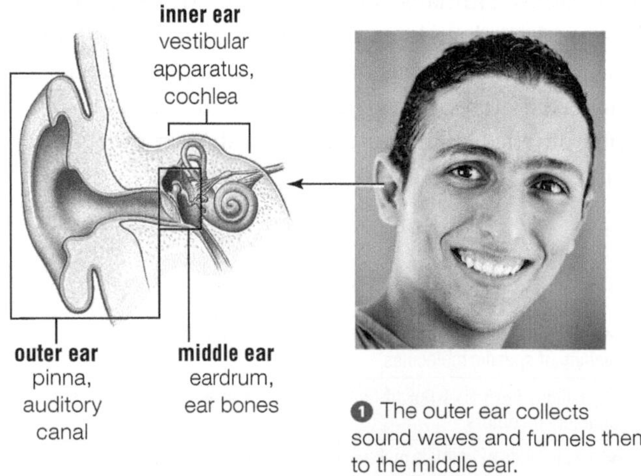
inner ear
vestibular apparatus, cochlea
outer ear
pinna, auditory canal
middle ear
eardrum, ear bones

❶ The outer ear collects sound waves and funnels them to the middle ear.

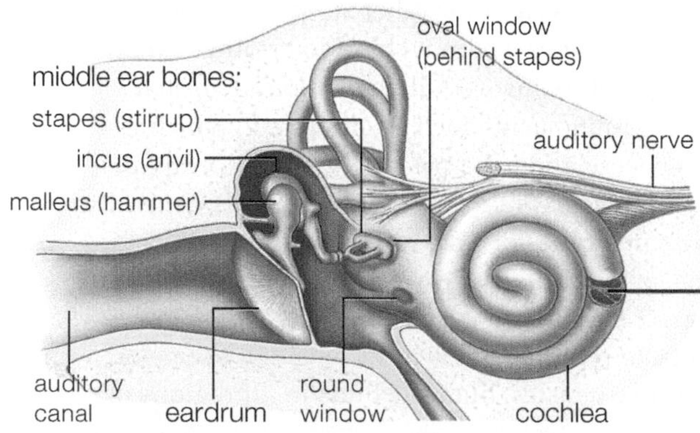
oval window (behind stapes)
middle ear bones:
stapes (stirrup)
incus (anvil)
malleus (hammer)
auditory nerve
auditory canal
eardrum
round window
cochlea

❷ The middle ear amplifies sound. Arrival of sound waves via the auditory canal causes the eardrum to vibrate. Middle ear bones transmit these vibrations to the inner ear's fluid-filled cochlea.

FIGURE 33.10 **Structure and function of the human ear.**

CREDITS: (9) © Cengage Learning; (10) 1. Zurijeta/Shutterstock; 4. Medtronic Xomed; 5. University of Miami Ear Institute.

each pulse of the sound wave, the oval window bows inward into the cochlea. This movement creates a pressure wave in the fluid of cochlea. As pressure waves travel through the fluid, they make the lower wall of the cochlear duct (the basilar membrane) vibrate.

The **organ of Corti**, an acoustical organ with arrays of hair cells, sits on top of the basilar membrane ❹. Each auditory hair cell is a mechanoreceptor that has a tuft of stereocilia at one end. Stereocilia are nonmotile projections similar in structure to microvilli. A hair cell's stereocilia project into a tectorial membrane above it.

When fluid movement in the cochlea causes vibration of the basilar and tectorial membranes, the stereocilia of the hair cells sandwiched between these membranes are forced to bend ❺. Bending of a hair cell's stereocilia opens mechanically sensitive gated ion channels. The resulting influx of ions causes the cell to undergo an action potential. The auditory nerve conveys signals from the hair cells to the brain.

The number of hair cells that fire and the frequency of their signals inform the brain about the volume of a sound. The louder the sound, the more action potentials flow along the auditory nerve to the brain. The brain determines the pitch of a sound by assessing which part of the basilar membrane is vibrating the most. The basilar membrane is not uniform along its length. It is stiff and narrow near the oval window and broader and more flexible deeper into the coil. High-pitched sounds make the stiff, narrow, closer-in part of the basilar membrane vibrate most. Low-pitched sounds cause vibrations mainly in the wide, flexible part close to the membrane's tip. The more vibrations there are, the more hair cells in that region fire.

cochlea (COCK-lee-uh) Coiled, fluid-filled structure in the inner ear that holds the mechanoreceptors involved in hearing.

eardrum Membrane that vibrates in response to pressure waves and functions in hearing; tympanic membrane.

inner ear Of the mammalian ear, region that contains the fluid-filled vestibular apparatus and cochlea.

middle ear Of the mammalian ear, region that contains the eardrum and the tiny bones that transfer sound to the inner ear.

organ of Corti Portion of the cochlea that contains the sound-detecting mechanoreceptors (sensory hair cells).

outer ear Of the mammalian ear, external pinna and the air-filled auditory canal.

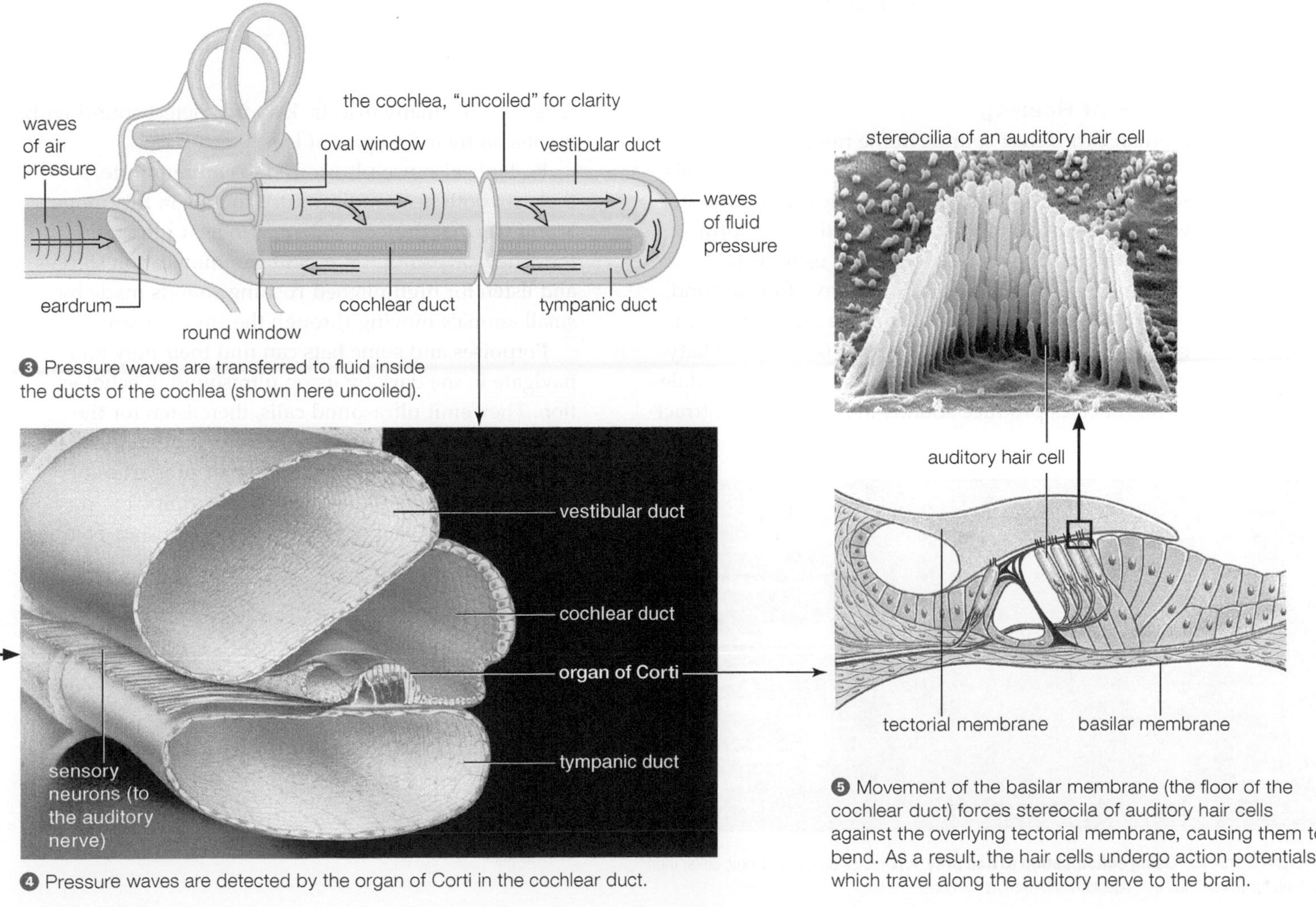
the cochlea, "uncoiled" for clarity
waves of air pressure
oval window
vestibular duct
waves of fluid pressure
eardrum
cochlear duct
tympanic duct
round window
stereocilia of an auditory hair cell
auditory hair cell
vestibular duct
cochlear duct
organ of Corti
tympanic duct
sensory neurons (to the auditory nerve)
tectorial membrane
basilar membrane

❸ Pressure waves are transferred to fluid inside the ducts of the cochlea (shown here uncoiled).

❹ Pressure waves are detected by the organ of Corti in the cochlear duct.

❺ Movement of the basilar membrane (the floor of the cochlear duct) forces stereocila of auditory hair cells against the overlying tectorial membrane, causing them to bend. As a result, the hair cells undergo action potentials, which travel along the auditory nerve to the brain.

Data Analysis Activities

Occupational Hearing Loss Frequent exposure to loud noise of a particular pitch can cause loss of hair cells in the part of the cochlea that responds to that pitch. People who work with or around noisy machinery are at risk for such frequency-specific hearing loss. Taking precautions such as using ear plugs to reduce sound exposure is important. Noise-induced hearing loss can be prevented, but once it occurs it is irreversible because dead or damaged hair cells are not replaced.

FIGURE 33.11 shows the threshold decibel levels at which sounds of different frequencies can be detected by an average 25-year-old carpenter, a 50-year-old carpenter, and a 50-year-old who has not been exposed to on-the-job noise. Sound frequencies are given in hertz (cycles per second). The more cycles per second, the higher the pitch.

1. Which sound frequency was most easily detected by all three people?
2. How loud did a 1,000-hertz sound have to be for the 50-year-old carpenter to detect it?
3. Which of the three people had the best hearing in the range of 4,000 to 6,000 hertz? Which had the worst?
4. Based on these data, would you conclude that the hearing decline in the 50-year-old carpenter was caused by age or by job-related noise exposure?

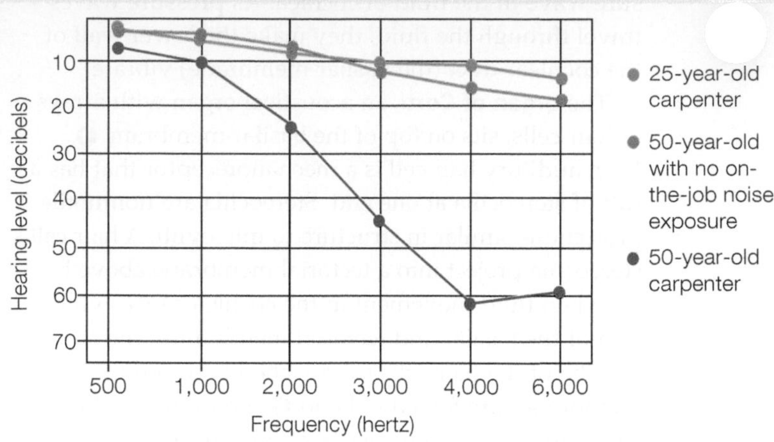
Hearing level (decibels)
10
20
30
40
50
60
70
500
1,000
2,000
3,000
4,000
6,000
Frequency (hertz)
25-year-old carpenter
50-year-old with no on-the-job noise exposure
50-year-old carpenter

FIGURE 33.11 **Effects of age and occupational noise exposure on hearing.** The graph shows the threshold hearing capacities (in decibels) for sounds of different frequencies (given in hertz) in a 25-year-old carpenter (blue), a 50-year-old carpenter (red), and a 50-year-old who did not have any on-the-job noise exposure (brown).

Range of Hearing

Human ears detect sounds in the range of 20 to 20,000 hertz. Many animals hear sounds outside this range. Elephants, giraffes, and whales communicate with members of their own species through infrasound, which is pitched too low for us to hear.

Some insects communicate by way of ultrasound, which is too high for us to hear. The animal with the most impressive ultrasonic hearing is a tropical katydid. It can detect sounds of up to 150,000 hertz. Males of this species produce sounds in this range to attract females. Like many insects, katydids detect sound with organs on their first pair of legs.

Rodents also use ultrasound to communicate, so they can both hear and produce sounds in this range. Mammals that prey on rodents can hear ultrasound too. They eavesdrop on rodent communication signals and listen for high-pitched rustling sounds made by small animals moving through the environment.

Porpoises and some bats can find their prey and navigate in the dark by using ultrasound in echolocation. They emit ultrasound calls, then listen for the echoes that bounce back from objects around them (FIGURE 33.12). The timing and other properties of these echoes allow the animals to determine the position, size, and speed of motion of these unseen objects.

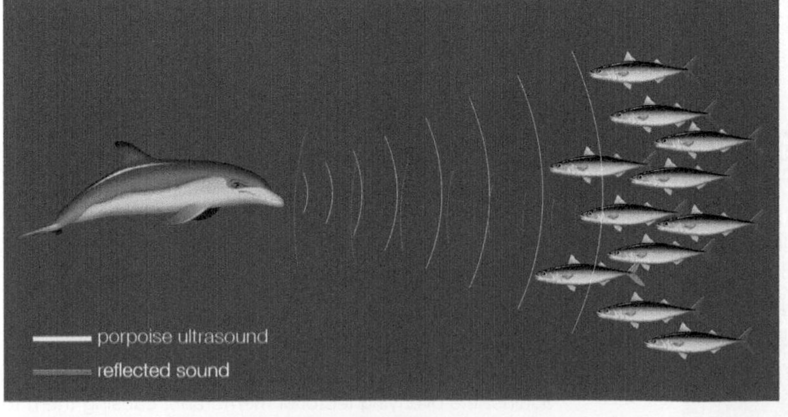
porpoise ultrasound
reflected sound

FIGURE 33.12 **Using echolocation to find prey.** A porpoise emits ultrasonic calls, then listens for the echoes that bounce back from its fish prey.

TAKE-HOME MESSAGE 33.5

✔ The human outer ear collects sound waves and directs them to the middle ear. In the middle ear, vibrations of the eardrum are transmitted via movement of small bones. These bones set up pressure waves in fluid inside the inner ear's cochlea.

✔ Inside the cochlea, hair cells in the organ of Corti convert pressure waves into action potentials that travel along the auditory nerve to the brain.

✔ Some animals can hear sounds outside the range of human hearing.

CREDITS: (11) left, Chris Newbert; right, After M. Gardiner, *The Biology of Vertebrates*, McGraw Hill, 1972; (12) © Cengage Learning.

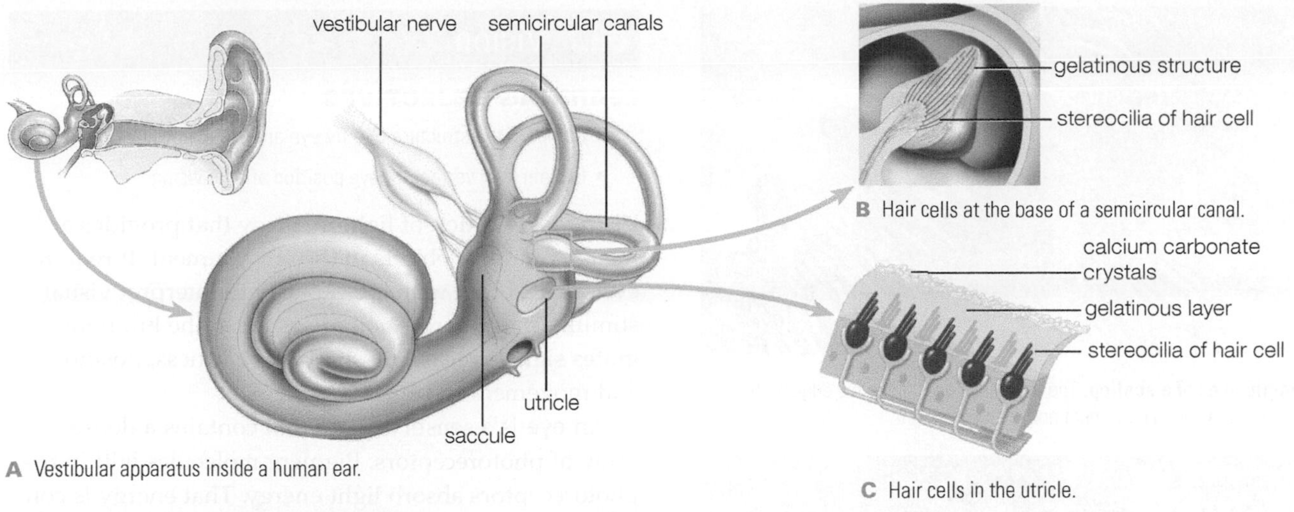
vestibular nerve
semicircular canals
utricle
saccule
gelatinous structure
stereocilia of hair cell
calcium carbonate crystals
gelatinous layer
stereocilia of hair cell

A Vestibular apparatus inside a human ear.

B Hair cells at the base of a semicircular canal.

C Hair cells in the utricle.

FIGURE 33.13 Human organs of equilibrium.

33.6 Balance and Equilibrium

LEARNING OBJECTIVES

- Describe the vertebrate organs of equilibrium.
- Explain the difference between dynamic and static equilibrium.

Organs of equilibrium monitor the body's position and motion. They are the basis for what we commonly call our sense of balance. Many invertebrates have one or more statocysts that provide information about position by responding to gravity. A statocyst is a hollow, fluid-filled sphere containing one or more dense particles that sink to the lowest point in the sphere. Mechanoreceptors in the walls of the sphere become excited when the particle or particles press on them.

In vertebrates, the **vestibular apparatus** in the inner ear houses the organs of equilibrium (**FIGURE 33.13A**). The fluid-filled vestibular apparatus consists of three semicircular canals and two sacs, called the saccule and utricle. Like the organ of Corti, organs of the vestibular apparatus contain hair cells. Fluid movement inside the various canals and sacs causes action potentials in the hair cells. A vestibular nerve then relays the signals from these cells to the brain.

Dynamic Equilibrium

The three semicircular canals of the vestibular apparatus function in dynamic equilibrium, the special sense related to angular movement and rotation of the head. The canals are oriented at right angles to one another, so rotation of the head in any combination of directions (front/back, up/down, or left/right) causes the fluid inside them to move. At the base of each semicircular canal is a bulging region called the ampulla. The ampulla contains hair cells whose stereocilia are embedded in a jellylike mass (**FIGURE 33.13B**). Movement of fluid inside the canal causes this mass to shift, thus bending the stereocilia and triggering action potentials.

The brain receives and integrates signals from the semicircular canals on both sides of the head. By comparing the number and frequency of action potentials coming from both sides, the brain gives rise to a sense of dynamic equilibrium. Among other things, your sense of dynamic equilibrium allows you to keep your eyes locked on an object even when you swivel your head or nod.

Static Equilibrium

Organs in the saccule and utricle act in the sense of static equilibrium. These organs help the brain keep track of the head's position and how fast it is moving in a straight line. They also help you keep your head upright and maintain normal posture. Inside the saccule and utricle, a jellylike layer weighted with calcium carbonate overlies hair cells (**FIGURE 33.13C**). When you tilt your head, or start or stop moving, the weighted mass shifts, bending the stereocilia of the hair cells and altering the cells' rate of action potentials.

organs of equilibrium Sensory organs that respond to body position and motion.

vestibular apparatus System of fluid-filled sacs and canals in the inner ear; contains the organs of equilibrium.

TAKE-HOME MESSAGE 33.6

✔ Mechanoreceptors in the fluid-filled vestibular apparatus of the human inner ear detect the body's position in space and when we start or stop moving.

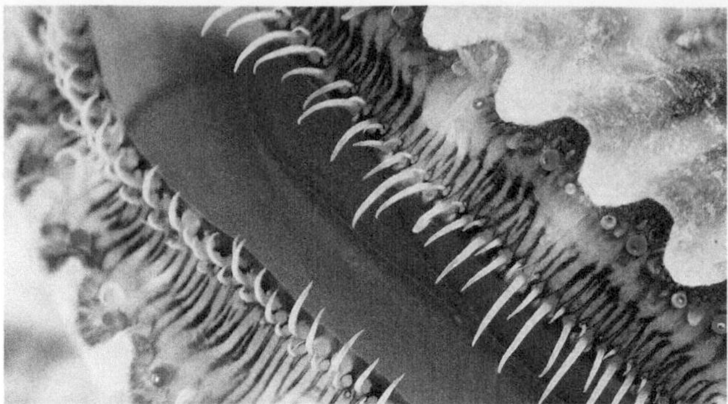
A Simple eyes of a scallop. The many blue eyes along the edge of the mantle are sensitive to movement and changes in light.

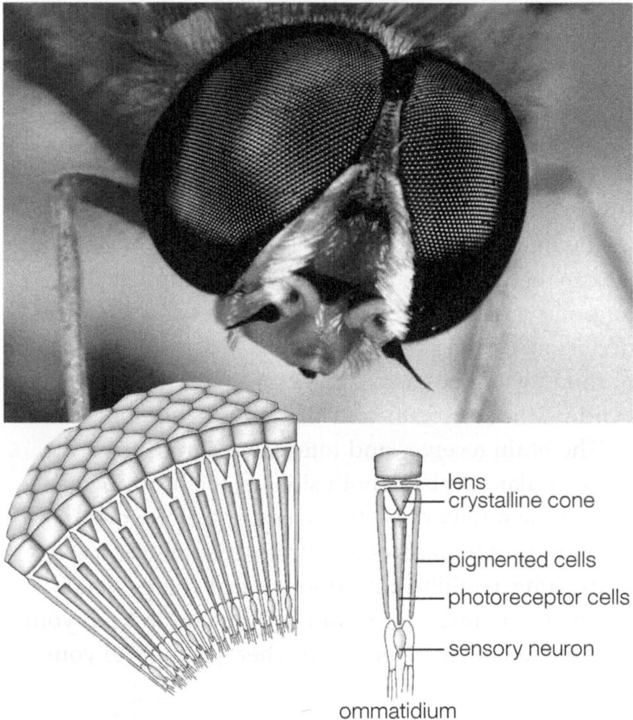
lens
crystalline cone
pigmented cells
photoreceptor cells
sensory neuron
ommatidium
B Compound eye of a fly. There are many densely packed units called ommatidia, each with its own lens.

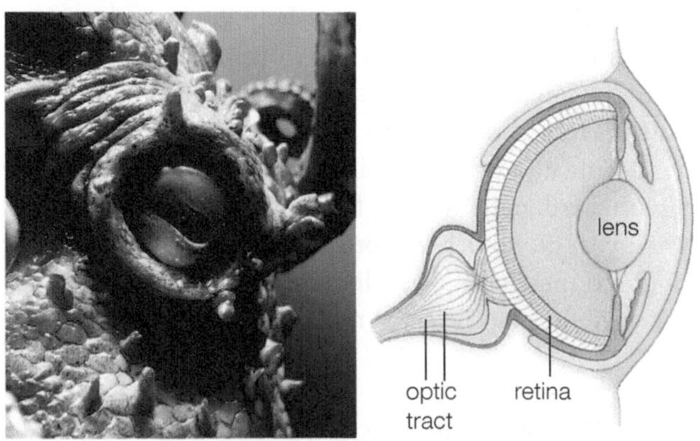
lens
optic tract
retina
C Camera eye of an octopus. A single lens that focuses light on a retina. Axons of sensory neurons in the retina form an optic tract that relays information to the brain.

FIGURE 33.14 Invertebrate eyes.

33.7 Vision

LEARNING OBJECTIVES

- Compare the structure of a fly eye and an octopus eye.
- Explain how vertebrate eye position affects vision.

Vision is detection of light in a way that provides a mental image of objects in the environment. It requires eyes and a brain with the capacity to interpret visual stimuli. Image perception arises when the brain integrates signals regarding shapes, brightness, positions, and movement of visual stimuli.

An **eye** is a sensory organ that contains a dense array of photoreceptors. Pigment molecules within the photoreceptors absorb light energy. That energy is converted to signals in the form of action potentials that travel to the brain.

Invertebrate Eyes

Some invertebrates, including earthworms, do not have eyes, but they do have photoreceptors dispersed under the epidermis or clustered in parts of it.

Scallops have as many as 60 eyes along the border of their mantle (**FIGURE 33.14A**). Each eye has a **lens,** a transparent disk-shaped structure that bends light rays from any point in the visual field so they converge on photoreceptors. Despite its many eyes, a scallop cannot form a mental image of its surroundings because it has no brain to integrate visual information.

Insects have **compound eyes** consisting of many separate units called ommatidia, each with its own lens (**FIGURE 33.14B**). The insect brain constructs an image based on the intensity of light detected by the different units. Compound eyes do not provide the clearest vision, but they are highly sensitive to movement.

Cephalopod mollusks such as squids and octopuses have the most complex eyes of any invertebrate (**FIGURE 33.14C**). Their **camera eyes** have an adjustable opening that allows light to enter a dark chamber. Each eye has a single lens that focuses incoming light onto a **retina**, a tissue densely packed with photoreceptors. The retina of a camera eye is analogous to the light-sensitive film used in a traditional film

camera eye Eye with an adjustable opening and a single lens that focuses light on a retina.
compound eye Eye with many units, each having its own lens.
conjunctiva (con-junk-TIE-va) Mucous membrane that lines the inner surface of the eyelids and folds back to cover the eye's sclera.
eye Sensory organ that incorporates a dense array of photoreceptors.
lens Disk-shaped structure that bends light rays so they fall on an eye's photoreceptors.
retina In an eye, a layer densely packed with photoreceptors.
vision Perception of visual stimuli based on light focused on a retina and image formation in the brain.

camera. Signals from the photoreceptors in each eye travel along one of the two optic tracts to the brain. Compared to compound eyes, camera eyes can produce a more sharply defined and detailed image.

Vertebrate Eyes

Like cephalopods, vertebrates have camera eyes. Vertebrates are only distantly related to cephalopod mollusks, so camera eyes are presumed to have evolved independently in the two lineages. This is an example of convergent evolution (Section 18.3).

All vertebrates have a pair of eyes, but how these eyes are positioned varies. Most non-predatory birds and mammals have eyes on the sides of their head (**FIGURE 33.15A**). These animals have monocular vision, meaning each eye views a different area. Monocular vision provides a wide field of view, which is useful for surveying your surroundings for predators. By contrast, predators such as eagles and wolves tend to have eyes positioned at the front of their skull (**FIGURE 33.15B**). This arrangement allows for binocular vision, meaning the same visual field is viewed simultaneously from two slightly different positions. The brain compares the overlapping information it receives to determine the distance between objects in the field of view. Binocular vision provides enhanced depth perception, which is useful when pursuing prey.

Animals that are active primarily at night often have large eyes relative to their body size. A large eye captures more available light than a smaller one. Light-reflecting layers evolved in many animals that are active under low-light conditions. When light shines on the eyes of these animals, the reflection from this layer makes their eyes appear to glow, a phenomenon known as "eyeshine" (**FIGURE 33.16**). Sharks, alligators, canines, felines, dolphins, deer, and rodents are among the many animals that have reflective eyes. The location and structure of the reflective material varies among groups, indicating that this adaptation arose independently in the different lineages.

> **TAKE-HOME MESSAGE 33.7**
>
> ✔ Some animals such as earthworms have photoreceptors that detect light but do not form any sort of image.
>
> ✔ The compound eye of insects has many individual units, each with its own lens. This eye produces a fuzzy, mosaic image, but it is highly sensitive to movement.
>
> ✔ A camera eye with an adjustable opening and a lens that focuses light on a photoreceptor-rich retina provides a highly detailed image. Camera eyes evolved independently in cephalopod mollusks and vertebrates.
>
> ✔ Among vertebrates, the position of the eyes determines the dimensions of the field of view and the extent of depth perception.

A Impala with eyes at the sides of its head.

B Wolf with eyes at the front of its head.

FIGURE 33.15 Variations in eye position. Eyes positioned at the side of the head provide the widest field of view. Eyes at the front of the head provide a narrower field of view but enhance depth perception.

FIGURE 33.16 Eyeshine of a black-footed ferret. The reflective layer in the eyes of this nocturnal predator enhances its night vision.

33.8 Human Vision

LEARNING OBJECTIVES

- List the structures light travels through to reach the photoreceptors of a human eye.
- Describe how a human eye adjusts its focus for different distances.
- Compare the functions of the two types of photoreceptors.

Anatomy of the Human Eye

Each human eyeball sits inside a protective, cuplike, bony cavity called the orbit. Skeletal muscles that run from the rear of the eye to the bones of the orbit move the eyeball up and down or side to side.

Eyelids, eyelashes, and tears all help protect the eye's delicate tissues. Reflexive blinking spreads a film of tears over the eyeball's exposed surface. Tears are secreted by exocrine glands in the eyelids. A protective mucous membrane called the **conjunctiva** lines the inner surface of the eyelids and folds back to cover most of the eye's outer surface. Conjunctivitis, commonly called pinkeye, is an inflammation of the conjunctiva. It is caused by a viral or bacterial infection of this membrane.

The eyeball itself is spherical, with a three-layered structure (**FIGURE 33.17**). A dense, white, fibrous **sclera** covers most of the eye's outer surface. However, the very front of the eye is covered by a **cornea** composed of transparent proteins called crystallins.

The eye's middle layer includes the choroid, iris, and ciliary body. The blood vessel–rich **choroid** is darkened by the brownish pigment melanin. Its dark color prevents light reflection within the eyeball. Attached to the choroid, and suspended behind the cornea, is a muscular, doughnut-shaped **iris**. It too contains melanin. Whether your eyes are blue, brown, or green depends on the amount of melanin in your iris.

Light enters the eye's interior through the **pupil**, an opening at the center of the iris. The smooth muscle of the iris reflexively adjusts pupil diameter in response to light conditions. The pupil shrinks in bright light and widens in low light, thus regulating the amount of light that enters the eye.

Behind the pupil is the lens, a flexible, transparent disk about 1 centimeter (1/2 inch) across. The lens is biconvex, meaning it curves outward on both sides. The lens is ringed by the ciliary body, a structure that consists of secretory cells, smooth muscle, and fibers that attach the muscle to the lens.

The eye has two internal chambers. The ciliary body produces a fluid called aqueous humor that fills the anterior chamber and bathes the iris and lens. A jelly-like vitreous body fills the larger chamber behind the lens. The innermost layer of the eye, the retina, is at the rear of this chamber. The retina contains the photoreceptors that detect light.

The cornea and lens both bend incoming light. When light reflected from objects in the environment passes through these structures, it converges on the retina and forms an image of the objects. This image is upside down and a mirror image of the real world (**FIGURE 33.18**). However, because the brain makes the necessary adjustments, you perceive objects in their correct orientation.

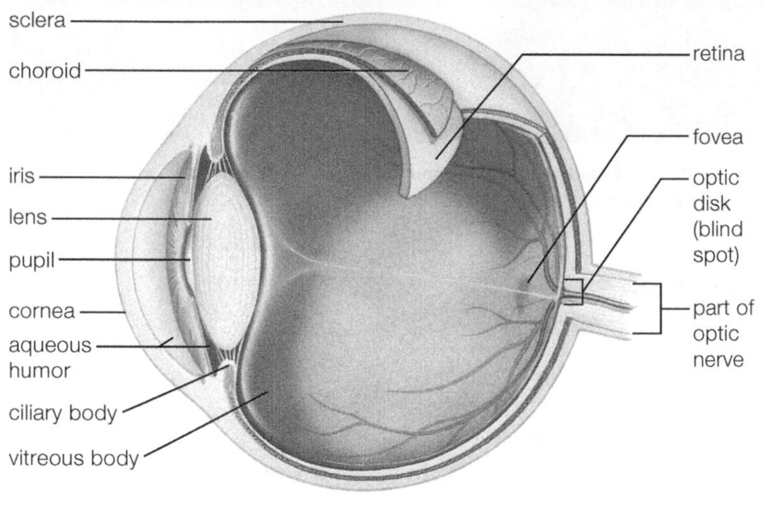

sclera
choroid
retina
fovea
iris
lens
pupil
optic disk (blind spot)
cornea
part of optic nerve
aqueous humor
ciliary body
vitreous body

FIGURE 33.17 Components and structure of the human eye.

Wall of eyeball (three layers)	
Outer layer	*Sclera:* Protects eyeball
	Cornea: Helps the lens focus light
Middle layer	*Choroid:* Its blood vessels nutritionally support wall cells; its pigments lessen light scattering
	Iris: Smooth muscle that adjusts pupil size
	Pupil: Opening that allows light into eye
	Ciliary body: Its muscles control the lens shape; its fine fibers hold lens in place
Inner layer	*Retina:* Absorbs, transduces light energy
Interior of eyeball	
Lens	Focuses light on photoreceptors
Aqueous humor	Transmits light, maintains fluid pressure
Vitreous body	Transmits light, supports retina, maintains the shape of the eyeball

FIGURE 33.18 Pattern of retinal stimulation in the human eye.

The curved, transparent cornea changes the trajectory of light rays that enter the eye. As a result, light rays that fall on the retina produce a pattern that is upside down and inverted left to right.

K

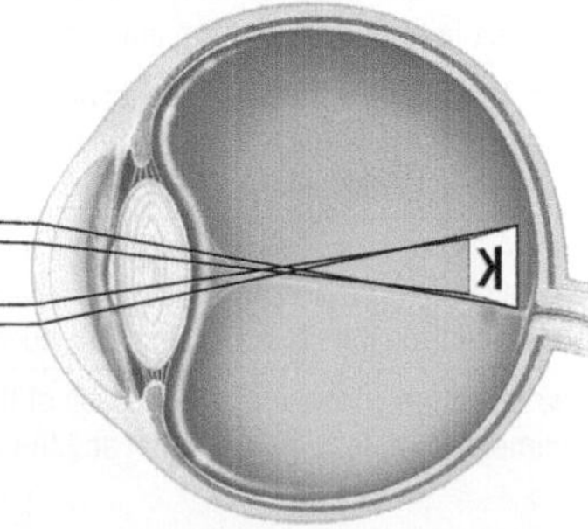

Focusing Mechanisms

When you see an object, you are perceiving light rays reflected from that object. Light rays reflected from near and distant objects hit the eye at different angles. By the process of **visual accommodation**, the shape or position of a lens is adjusted so that incoming light from any object forms an image on the retina, rather than in front of or behind it. Without such adjustments, only objects at a fixed distance would be in focus. Objects closer or farther away would appear fuzzy.

In fishes and amphibians, an eye's lens can be shifted forward or back, but its shape does not change. Extending or decreasing the distance between the lens and retina keeps light focused on the retina.

In humans and other amniotes, a ring-shaped **ciliary muscle** (a part of the ciliary body) adjusts the shape of the lens. Consider that the curvature of the lens determines the extent to which light rays bend. A flat lens focuses light from a distant object on the retina. However, the lens must be rounder to focus light from close objects. Gaze into the distance and the ciliary muscle encircling the lens relaxes, the ciliary fibers connecting this muscle to the lens are drawn taut, and the resulting tension flattens the lens (**FIGURE 33.19A**). Focus on a close object such as a computer screen and the ciliary muscle contracts, fibers connecting the muscle to the lens slacken, and the lens becomes rounder (**FIGURE 33.19B**).

A variety of defects can prevent light rays from converging properly on the retina. With astigmatism, vision is blurred at all distances by an unevenly curved cornea or lens. With nearsightedness (myopia), distant objects are out of focus. In young people, nearsightedness usually occurs because the distance between the retina and the cornea or lens is too long (**FIGURE 33.20A**) or the ciliary muscles contract too much. With farsightedness (hyperopia), close objects are out of focus. This occurs if the distance between the retina and the cornea or lens is too short (**FIGURE 33.20B**) or the ciliary muscles are weak. Either way, light rays reflected from nearby objects get focused behind the retina. Many people who have normal vision early in life become farsighted after age 40 because their lens becomes less flexible.

Changes to the lens can cloud it, producing a condition called a cataract. Smoking, steroid use, and diabetes promote cataract formation, as does excessive exposure to UV radiation. Typically, both eyes are affected. When a cataract first appears, it scatters light and blurs vision. Eventually, the lens may become fully opaque, causing blindness. Cataract surgery can restore normal vision by replacing a clouded lens with a clear synthetic lens.

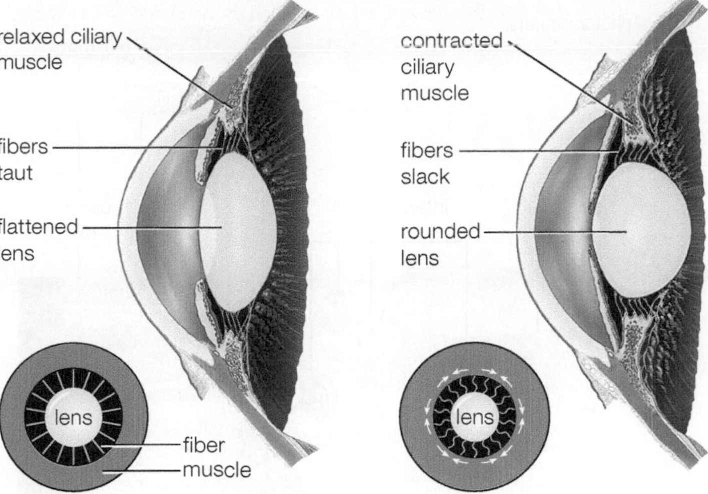
relaxed ciliary muscle
fibers taut
flattened lens
lens
fiber
muscle
contracted ciliary muscle
fibers slack
rounded lens
lens

A Relaxed ciliary muscle pulls fibers taut; the lens is stretched into a flatter shape that focuses light from a distant object on the retina.

B Contracted ciliary muscle allows fibers to slacken; the lens rounds up and focuses light from a close object on the retina.

FIGURE 33.19 Visual accommodation. Ciliary muscle encircles the lens and attaches to it by means of elastic fibers. Contracting or relaxing the ciliary muscle alters the tension on the elastic fibers and thus changes the shape of the lens.

FIGURE IT OUT The thicker a lens, the more it bends light. Does the lens bend light more with distance vision or close vision?

Answer: Close vision

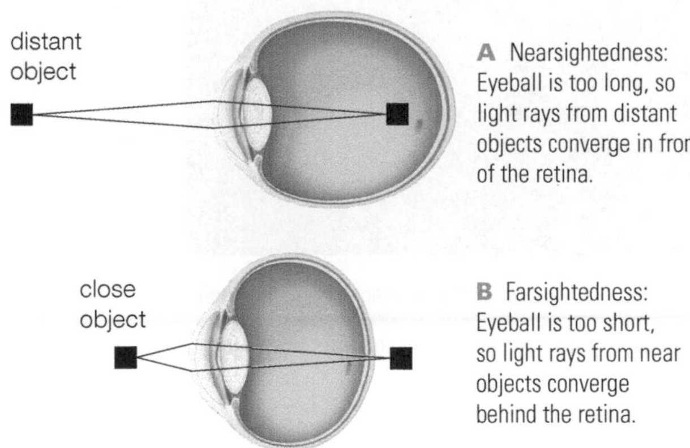
distant object
close object

A Nearsightedness: Eyeball is too long, so light rays from distant objects converge in front of the retina.

B Farsightedness: Eyeball is too short, so light rays from near objects converge behind the retina.

FIGURE 33.20 Focusing problems caused by abnormal eye shape.

choroid (KOR-oyd) Blood vessel–rich layer of the middle eye. It is darkened by the brownish pigment melanin to prevent light scattering.
ciliary muscle Ring-shaped muscle of the eye that encircles the lens and attaches to it by short fibers.
cornea (KOR-nee-uh) Clear, protective covering at the front of the vertebrate eye.
iris (EYE-ris) Circular muscle that adjusts the shape of the pupil to regulate how much light enters the eye.
pupil Adjustable opening that allows light into a camera eye.
sclera Fibrous white layer that covers most of the eyeball.
visual accommodation Process of adjusting lens shape or position so that light from an object falls on the retina.

A Two types of photoreceptors.

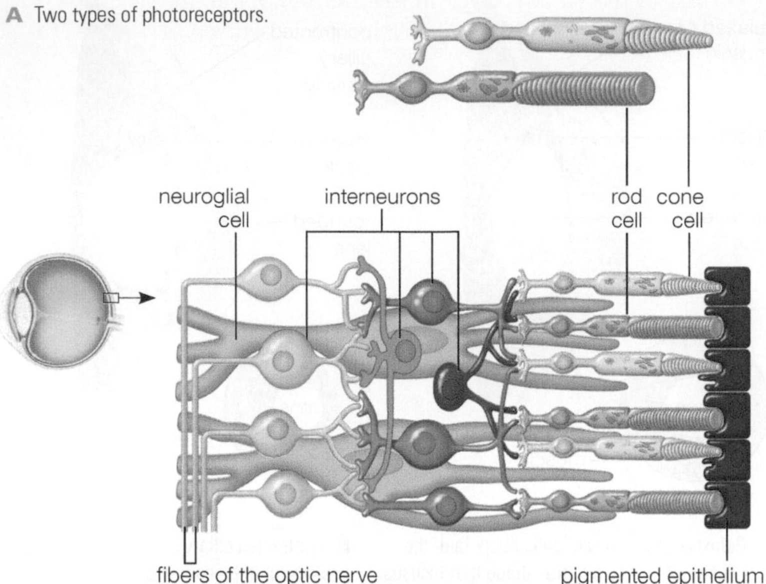
neuroglial cell
interneurons
rod cell
cone cell
fibers of the optic nerve
pigmented epithelium

B Multilayered organization of the retina.

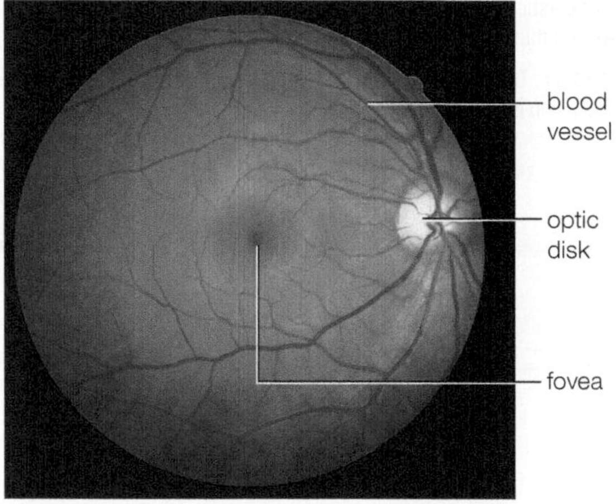
blood vessel
optic disk
fovea

C Surface view of the retina (as seen through the pupil).

FIGURE 33.21 **Structure and components of the human retina.**

The Photoreceptors

As previously noted, the cornea and lens bend light so that it falls on the retina where it can excite photoreceptors. There are two differently shaped types of photoreceptors, rod cells and cone cells (**FIGURE 33.21A**). In both, one end of the cell consists of stacks of membranous disks with many molecules of visual pigment embedded in the membrane.

cone cell Photoreceptor that provides sharp vision and allows detection of color.

fovea (FOE-vee-uh) Retinal region where cone cells are most concentrated.

rod cell Photoreceptor that is active in dim light; provides coarse perception of image and detects motion.

A vertebrate visual pigment consists of a protein (an opsin) bound to a cofactor called retinal. Retinal is derived from vitamin A, so a diet deficient in vitamin A can result in impaired vision. The different properties of rods and cones arise from differences in their opsin component.

Rod cells, the most abundant photoreceptors, detect dim light and are responsible for the coarse image we associate with night vision. They are also highly sensitive to motion. In the human eye, the rods tend to be concentrated at the edges of the retina. All rods have the same pigment (rhodopsin), which is excited by exposure to green-blue light.

Cone cells provide acute daytime vision and allow us to detect colors. There are three types of cone cells, each with a slightly different form of the cone pigment iodopsin. One cone pigment absorbs mainly red light, another absorbs mainly blue, and a third absorbs green. Normal human color vision requires all three types of cones. When one or more types of cones are missing or impaired, the result is color blindness (Section 14.4).

The greatest density of cone cells is in the **fovea**, a pit in the central region of the retina. With normal vision, images fall mainly on the fovea. The fovea resides in a pigmented region of the retina called the macula. Light from objects right in front of you is focused on the macula.

Age-related macular degeneration (AMD) is a disorder in which destruction of cells in the macula impairs vision. In developed countries, it is the leading cause of blindness among older adults. Destruction of photoreceptors in the macula clouds the center of the visual field more than the periphery.

Signal Transduction to Visual Processing

The retina has a multilayered organization (**FIGURE 33.21B**), with rod cells and cone cells situated beneath several layers of interneurons that function in visual processing. Glia cells (called Müller glia) span the retina. They structurally support the photoreceptors and function like pipes for light. At the surface of the retina, the funnel-shaped end of the glial cell captures light. The light travels through the cell to illuminate rods and cones deeper in the retina.

Absorption of light by the visual pigment in a photoreceptor causes the pigment to change shape. This shape change triggers a chain of reactions within the cell that culminates in a change in the photoreceptor's membrane potential. Thus the photoreceptor transduces light energy into electric energy.

Action potentials flow from a rod cell or cone cell to interneurons in the layers above. These interneurons integrate signals from the photoreceptors and send signals to interneurons called ganglion cells. Axons of ganglion cells constitute the fibers of the optic nerve.

Glaucoma is a disorder in which excess aqueous humor in the eyeball puts pressure on ganglion cells and damages them. Although glaucoma is most often diagnosed in old age, conditions that give rise to the disorder start to develop long before symptoms arise. Screening for glaucoma allows doctors to detect increased fluid pressure inside the eye before the damage to ganglion cells becomes severe. They can then manage the disorder with medication, surgery, or both.

The optic nerve exits the retina through a region called the optic disk (**FIGURE 33.21C**). This region lacks photoreceptors, so it cannot respond to light and its presence results in a blind spot. We all have a blind spot in each eye, but it does not impair our vision because information missed by one eye is provided to the brain by the other.

Each optic nerve ends in a part of the thalamus that integrates incoming signals. From here, signals are conveyed to the visual cortex in the occipital lobe. The occipital lobe is where the final integration of signals produces visual sensations.

Signals from the right visual field of both eyes are conveyed to the left hemisphere's visual cortex. Signals from the left visual field of both eyes end up in the right hemisphere's visual cortex.

TAKE-HOME MESSAGE 33.8

- ✔ The cornea at the front of the eye bends light rays entering the eye's interior through the pupil. The diameter of the pupil changes depending on the amount of available light.
- ✔ The lens focuses light onto the retina. Ciliary muscle adjusts the shape of the lens to accommodate either near or distance vision.
- ✔ When stimulated by light, photoreceptors (rod cells and cone cells) send signals to neurons in the layer of retina above them. These neurons process signals, then send messages to the brain.
- ✔ The visual cortex integrates signals from the eyes and gives rise to visual sensations.

33.1 Neuroprostheses (revisited)

Cochlear implants allow people to hear despite a damaged or defective cochlea. Similarly, retinal implants assist people whose photoreceptors have been damaged. These implants, sometimes described as artificial retinas, are a relatively recent invention. The first such device to be approved, the Argus II retinal prosthesis system, became available for sale in Europe in 2011 and in the United States in 2013. It is currently approved for use only in adults with retinosis pigmentosa, a rare, inherited disease in which degeneration of the retina causes blindness. However, the Argus II or similar devices may eventually prove helpful to patients blinded by macular degeneration or other disorders.

The Argus II system consists of a pair of glasses with a video camera that captures a scene and sends information to a processing unit mounted on the glasses. The processor translates the visual information to digital instructions that are sent wirelessly to an electrode-bearing implant on the surface of the retina. The electrodes in the implant respond to these signals by electrically exciting the optic nerve, triggering action potentials that travel to the brain. Other systems that work in a similar manner are also being tested.

As with cochlear implants, the Argus II retinal implant cannot completely restore normal function. A previously blind person who uses this device can perceive differences between dark and light areas and even recognize letters, but only if they are quite large (**FIGURE 33.22**). All vision is in black and white.

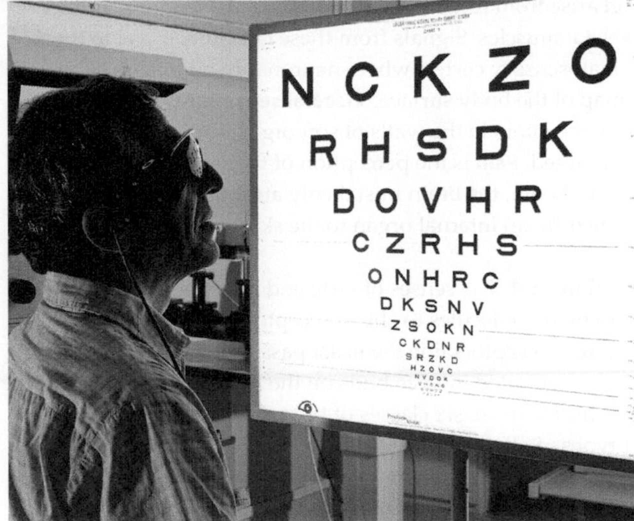
N C K Z O
R H S D K
D O V H R
C Z R H S
O N H R C
D K S N V
Z S O K N
C K D N R

FIGURE 33.22 Vision testing in a patient with an Argus II retinal implant. This previously blind patient can now read the largest letters on this chart.

However, researchers expect updates to the device, development of competing devices, and improved processing mechanisms to further enhance function.

Researchers are also developing prosthetic hands that will give amputees not only an ability to grip and hold materials, but also restore some sensation of touch. The prostheses convert information about pressure to electrical signals, which cause action potentials to travel along nerves to the somatosensory cortex. ●

CREDIT: (22) Philippe Psaila/Science Source.

Section 33.1 Devices that electrically stimulate the nerves that carry sensory information to the brain can be used to restore some degree of hearing, sight, or touch to people in whom these senses are impaired.

Section 33.2 The types of sensory receptors that an animal has determine the types of stimuli it can detect and respond to. Stimulation of a sensory receptor causes a sensory neuron to undergo an action potential. **Mechanoreceptors** respond to mechanical energy such as touch. **Pain receptors** respond to tissue damage. **Thermoreceptors** are sensitive to temperature. **Chemoreceptors** respond when a dissolved chemical binds. **Photoreceptors** respond to light.

The brain evaluates action potentials from sensory receptors based on which nerves deliver them, their frequency, and the number of axons firing in any given interval. Continued stimulation of a receptor may lead to **sensory adaptation** (a diminished response). **Sensation** is detection of a stimulus, whereas **sensory perception** involves assigning meaning to some sensation.

Section 33.3 **Somatic sensations** are easy to pinpoint and arise from receptors in the skin, joints, tendons, and skeletal muscles. Signals from these receptors travel to the somatosensory cortex, where neurons are organized like a map of the body surface. **Visceral sensations** originate from receptors in the walls of soft organs and are less easily pinpointed. **Pain** is the perception of tissue damage. With referred pain, the brain mistakenly attributes signals that come from an internal organ to the skin or muscles.

Section 33.4 The senses of taste and olfaction (smell) involve binding of molecules to chemoreceptors. In vertebrates, olfactory receptors line the nasal passages. Taste receptors are concentrated in taste buds on the tongue and lining of the mouth. There are six classes of taste receptors, but hundreds of types of olfactory receptors.

Pheromones are chemical signals that act as social cues among many animals. A **vomeronasal organ** functions in detection of pheromones in many vertebrates.

Section 33.5 Hearing is perception of sound. Sound waves are pressure waves. Variations in the amplitude of these waves are perceived as differences in loudness, and variations in wave frequency are perceived as differences in pitch.

Human ears have three functional regions. The pinna of the **outer ear** collects sound waves. The **middle ear** contains the **eardrum** and a set of tiny bones that amplify sound waves and transmit them to the inner ear. In the **inner ear**, pressure waves elicit action potentials inside the **cochlea**. The cochlea's **organ of Corti** contains hair cells, the mechanoreceptors responsible for hearing. Hair cells undergo an action potential when pressure waves traveling through fluid inside the cochlea cause their stereocilia to bend. The brain gauges the loudness of a sound by the number of action potentials the sound elicits. It determines a sound's pitch by which part of the cochlea the signals arrive from.

Section 33.6 **Organs of equilibrium** detect effects of gravity and acceleration. Statocysts play this role in some invertebrates. Human organs of equilibrium are in the **vestibular apparatus**, a system of fluid-filled sacs and canals in the inner ear.

Our sense of dynamic equilibrium arises when body movements cause shifts in the fluid inside the three semicircular canals. Static equilibrium depends on signals from hair cells that lie beneath a weighted, jellylike mass in the vestibular apparatus. A shift in head position or a sudden stop or start shifts this mass.

Section 33.7 An **eye** is a sensory organ that contains a dense array of photoreceptors. **Vision** requires eyes and a brain capable of processing the visual information that arrives from the eyes. Insects have a **compound eye**, with many individual units. Each unit has a **lens**, a structure that bends light so that it falls on photoreceptors. Like squids and octopuses, humans have **camera eyes**, with an adjustable opening that lets in light and a single lens that focuses images on a **retina** with a dense array of photoreceptors.

Depth perception arises when two forward-facing eyes send the brain information about the same viewed area. Large eyes with an internal light-reflecting layer adapt some animals to low-light conditions.

Section 33.8 A human eye sits in a bony orbit and is protected by eyelids lined by the **conjunctiva**. Most of the eye is covered by the **sclera**, or white of the eye. The clear, curved **cornea** at the front of the eye bends incoming light. Light enters the eye's interior through the **pupil** in the center of the **iris**. It then passes through the lens. The lens focuses an image on the retina. The retina sits on a pigmented **choroid** that minimizes reflections inside the eye.

With **visual accommodation**, the **ciliary muscle** adjusts the shape of the lens so that light reflected from a near or distant object focuses on the retina.

Humans have two types of photoreceptors. **Rod cells** detect dim light and are important in coarse vision and peripheral vision. **Cone cells** detect bright light and provide a sharp image. There are three types of cone cells, each containing a visual pigment that responds to a different color. Thus cones are responsible for color vision. The greatest concentration of cone cells is in the portion of the retina called the **fovea**. There are no photoreceptors in the eye's blind spot, the area where the optic nerve begins.

Signals from rod cells and cone cells are sent to other cells in the retina. These cells process visual information before sending it to the brain. Visual signals travel to the cerebral cortex along two optic nerves.

SELF-QUIZ

Answers in Appendix VII

1. The pain of heartburn is an example of a _______ .
   a. somatic sensation c. sensory adaptation
   b. visceral sensation d. spinal reflex
2. _______ is defined as a decrease in the response to an ongoing stimulus.
   a. Perception c. Sensory adaptation
   b. Visual accommodation d. Somatic sensation
3. Which is a somatic sensation?
   a. taste c. touch e. a through c
   b. smell d. hearing f. all of the above
4. Chemoreceptors play a role in the sense of _______ .
   a. taste c. touch e. both a and b
   b. smell d. hearing f. all of the above
5. In the _______ , neurons are arranged like maps that correspond to different parts of the body surface.
   a. retina c. basilar membrane
   b. somatosensory cortex d. occipital lobe
6. Mechanoreceptors in the _______ send signals to the brain about the body's position relative to gravity.
   a. eye b. ear c. tongue d. nose
7. The middle ear functions in _______ .
   a. detecting shifts in body position
   b. transmitting sound waves
   c. sorting sound waves out by frequency
8. Substance P _______ .
   a. increases pain-related signals
   b. is a natural painkiller
   c. is the active ingredient in aspirin
9. The organ of Corti contains receptors that signal in response to _______ .
   a. heat b. sound c. light d. pheromones
10. Night vision begins with stimulation of _______ .
    a. hair cells c. cone cells
    b. rod cells d. neuroglia
11. Visual accommodation involves adjustment to the shape or position of the _______ .
    a. conjunctiva c. orbit
    b. retina d. lens
12. When you view a close object, your lens gets _______ .
    a. flatter c. darker
    b. rounder d. cloudier
13. Defective or missing _______ cause color blindness.
    a. hair cells c. cone cells
    b. rod cells d. neuroglia
14. _______ causes the pupil to widen.
    a. Low light b. Bright light
15. Match each structure with its description.

___ cataract	a. protects eyeball
___ cochlea	b. transmits vibration to bone
___ eardrum	c. functions in balance
___ lens	d. detects pheromones
___ sclera	e. interferes with vision
___ fovea	f. contains chemoreceptors
___ taste bud	g. focuses rays of light
___ Pacinian corpuscle	h. has the most cones
___ pinna	i. collects sound waves
___ vestibular apparatus	j. sorts out sound waves
___ vomeronasal organ	k. detects touch

CRITICAL THINKING

1. Like other nocturnal carnivores, the ferret shown in Figure 33.16 has light-reflecting material in its choroid. Explain why the presence of reflective material in this layer of the eye maximizes the degree to which light excites photoreceptors. Explain also why having reflective material in this layer causes the perceived image to be somewhat blurry.

2. A compound extracted from the leaves of the shrub *Stevia rebaudiana* is a natural sugar substitute. The compound is 300 times sweeter than sugar, but has a slight bitter aftertaste. Given what you know about taste receptors, explain how a compound can be perceived as both sweet and bitter.

3. Most bats eat insects or fruit. Vampire bats, however, suck blood from birds or mammals. Like some snakes, and unlike any other mammals, vampire bats have thermoreceptors that can detect body heat given off by prey. Is the heat-detecting ability of vampire bats and snakes a homologous trait or an analogous one? Explain your reasoning.

4. If a mammal injures its leg, the resulting pain discourages the animal from putting too much weight on the leg while it is healing. An injured insect shows no such shielding response when its leg is injured. Some have cited the lack of such a shielding response as evidence that insects do not feel pain. On the other hand, insects do produce substances similar to one of our natural painkillers. Is the presence of these compounds in insects sufficient evidence to conclude that they do perceive pain as we do?

34 ENDOCRINE CONTROL

CORE CONCEPTS

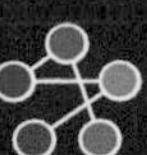

Systems

Complex properties arise from interactions among components of a biological system.

Cells interact with one other through communication signals. Hormones, the chemical signaling molecules of the endocrine system, allow long-distance communication. By coordinating the activities of widely separated cells they support the function of the body as a whole.

Evolution

Evolution underlies the unity and diversity of life.

Shared core processes and features provide evidence of shared ancestry. Common mechanisms of cell-to-cell communication offer examples. All vertebrates have the same types of endocrine glands. Some vertebrate hormones also have counterparts among the invertebrates, indicating that hormonal signaling evolved early in animal evolution.

Structure and Function

The three-dimensional form and arrangement of biological structures give rise to their function and interactions.

Hormone function arises from interactions between hormones and specific target cells. These interactions begin with the binding of a hormone to a specific receptor. Chemical properties of hormones determine whether they bind to a receptor at the target cell's surface or enter the cell and bind to a receptor inside it. This interaction between the hormone and its receptor influences other cellular components and gives rise to the hormone's observable effects.

Links to Earlier Concepts

This chapter focuses on endocrine glands, which were introduced in Section 31.3. Knowing the properties of steroids (3.4), proteins (3.6), and cell membranes (5.7) will help you understand hormone action. Among other topics, you will learn how hormones affect human glucose metabolism (7.6) and arthropod molting (24.10).

34.1 Endocrine Disruptors

We live in a world awash in synthetic chemicals. We drink from plastic bottles, wear clothing made of synthetic fabrics, slather ourselves with synthetic skin care products, and eat food treated with synthetic pesticides. Huge numbers of human-made compounds are used to make computers and other electronic gadgets. Synthetic chemicals enter our bodies when we ingest them, inhale them, or absorb them across our skin.

We have learned by sad experience that some synthetic chemicals threaten animal and human health. For example, years after we started using them, we realized that DDT (a pesticide) and PCBs (used in electronic products, caulking, and solvents) are endocrine disruptors. **Endocrine disruptors** are chemical or mixtures of chemicals that interfere with the function of the endocrine system. In the United States, use of DDT was banned in 1972 and PCBs in 1979. However, both chemicals were in wide use for years and are highly stable, so they still persist in the environment.

Other endocrine disruptors remain in use. Phthalates (pronounced THAL-ates) are synthetic chemicals used to make plastics more flexible and to stabilize fragrances in scented products. Experimental exposure of male rats to phthalates lowers their level of testosterone. (Testosterone is the main sex hormone in males.) Similarly, men who have a high concentration of phthalates in their blood are more likely to have a low testosterone level and decreased sperm quality. In women, a high phthalate level during pregnancy is correlated with an increased likelihood of giving birth to a son who has an undescended testicle.

In the United States, concerns about the health effects of phthalates have led to restrictions on their use in plastic pacifiers, teething toys, and other plastic products intended for children under age 12. However, phthalates remain in many artificially scented products such as air fresheners, laundry detergents, fabric softeners, dryer sheets. They are also in personal care products such as shampoos, cosmetics, nail polishes, and bubble baths. Phthalates from any of these products can cross the skin and end up in the bloodstream.

Concern about the endocrine-disrupting effects of phthalates led the American Academy of Pediatrics to recommend that parents use only unscented products on infants and young children. Similarly, many obstetricians recommend that women who are pregnant, trying to become pregnant, or breast-feeding use personal care products specifically labeled as phthalate-free.

endocrine disruptor Chemical or mixture of chemicals that interferes with the function of the endocrine system.

34.2 The Vertebrate Endocrine System

LEARNING OBJECTIVES

- Compare hormones and neurotransmitters.
- Describe how hormones were discovered.

Signals That Travel in the Blood

Two organ systems facilitate long-distance communication among cells of an animal body—the nervous system and the endocrine system. This chapter focuses on the **endocrine system**, which consists of all of the body's hormone-secreting glands and cells. **FIGURE 34.1** and **TABLE 34.1** provide an overview of the human endocrine system.

Unlike a neurotransmitter (a chemical signal of the nervous system), an **animal hormone** is secreted by an endocrine cell and enters the blood. Compared to neurotransmitters, hormones travel farther, exert their effects on more cells, and persist longer in the body.

Like neurotransmitters, hormones are chemical signals that affect target cells by a three-step process. The molecules bind to a target cell's receptor, the receptor transduces the signal (changes it into a form that affects target cell behavior), and the target cell responds:

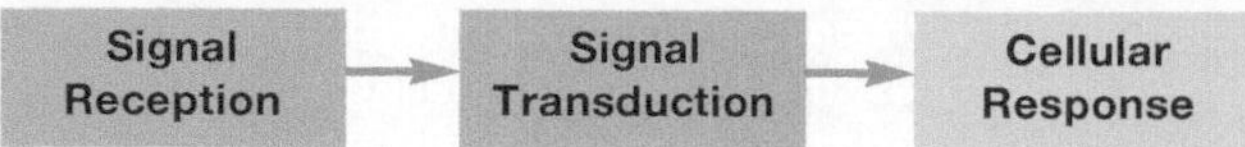
Signal Reception
Signal Transduction
Cellular Response

Discovery of Hormones

Hormones were first discovered in the early 1900s by physiologists William Bayliss and Ernest Starling. They were studying how the secretion of pancreatic juices is

animal hormone Intercellular signaling molecule secreted by an endocrine gland or cell; travels in the blood to target cells.

endocrine system System of hormone-producing glands and secretory cells of a vertebrate body.

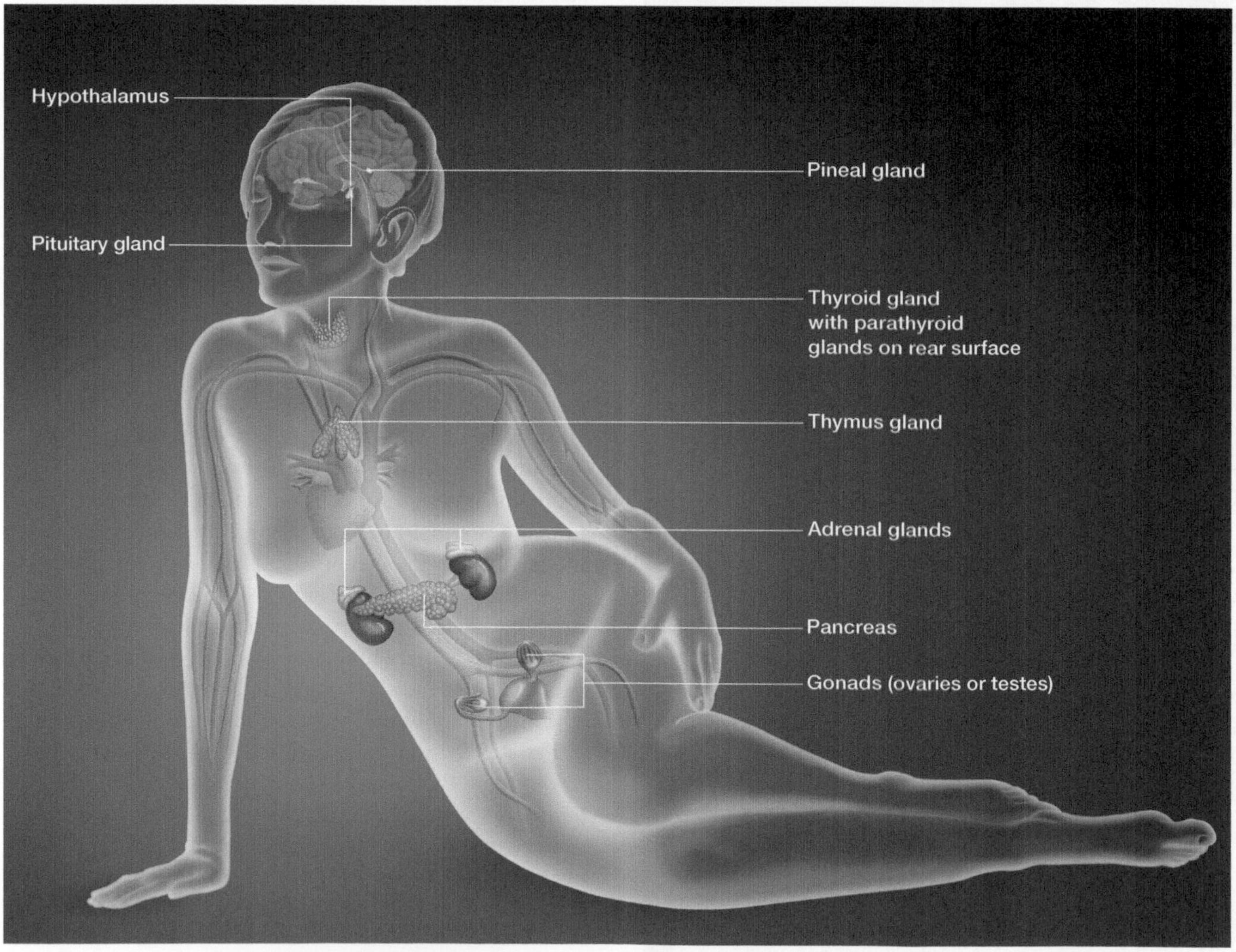

FIGURE 34.1 **Major glands of the human endocrine system.**

TABLE 34.1

Examples of Human Endocrine Glands, Their Hormones, and Hormone Actions

Gland	Hormone(s)	Main Target(s)	Primary Actions
Hypothalamus	Releasing and inhibiting hormones	Anterior pituitary	Encourage or discourage release of other hormones
Posterior pituitary gland	Antidiuretic hormone (ADH; vasopressin)	Kidneys	Induces water conservation as required to control extracellular fluid volume and solute concentrations
	Oxytocin (OT)	Mammary glands	Induces milk movement into secretory ducts
		Reproductive tract	Induces muscle contractions during childbirth and orgasm
Anterior pituitary gland	Adrenocorticotropic hormone (ACTH)	Adrenal cortex	Stimulates release of cortisol
	Thyroid-stimulating hormone (TSH)	Thyroid gland	Stimulates release of thyroid hormone
	Follicle-stimulating hormone (FSH)	Ovaries, testes	In females, stimulates estrogen secretion, egg maturation; in males, helps stimulate sperm formation
	Luteinizing hormone (LH)	Ovaries, testes	In females, stimulates progesterone secretion, ovulation, corpus luteum formation; in males, stimulates testosterone secretion, sperm release
	Prolactin (PRL)	Mammary glands	Stimulates and sustains milk production
	Growth hormone (GH)	Most cells	Promotes growth in young; induces protein synthesis, cell division in adults
Pineal gland	Melatonin	Brain	Affects sleep–wake cycles, seasonal changes
Thyroid gland	Thyroid hormone	Most cells	Regulates metabolism; has roles in growth, development
	Calcitonin	Bone	Lowers calcium level in blood
Parathyroid glands	Parathyroid hormone (PTH)	Bone, kidney	Elevates calcium level in blood
Pancreas	Insulin	Liver, muscle, adipose tissue	Promotes cell uptake of glucose and glycogen synthesis; lowers glucose level in blood
	Glucagon	Liver	Promotes glycogen breakdown; raises blood glucose
Adrenal glands			
Adrenal cortex	Cortisol*	Most cells	Promotes breakdown of glycogen, fats, and proteins as energy sources; raises blood level of glucose
	Aldosterone*	Kidney	Promotes sodium reabsorption (sodium conservation); helps control the body's salt–water balance
Adrenal medulla	Epinephrine (adrenaline) and norepinephrine	Most cells	Promote fight–flight response
Gonads			
Testes (in males)	Testosterone*	Sex organs	Required in sperm formation, development of genitals, maintenance of sexual traits, growth, and development
Ovaries (in females)	Estrogens*	Uterus, breasts	Required for egg maturation and release; preparation of uterine lining for pregnancy and its maintenance in pregnancy; genital development
	Progesterone*	Uterus	Prepares and maintains uterine lining for pregnancy; stimulates development of breast tissues

* Denotes steroid hormones; all other hormones are derived from amino acids.

regulated. Bayliss and Starling knew that the presence of food mixed with stomach acid in the small intestine stimulates the pancreas to secrete a bicarbonate buffer. To find out how the small intestine communicated with the pancreas, the scientists did an experiment. They surgically altered a laboratory animal, cutting all the nerves that carry signals to and from its small intestine. Even after these nerves had been cut, the presence of food and acid in the small intestine caused the pancreas to secrete bicarbonate. This result indicated that communication between the small intestine and the pancreas does not involve nerves.

Next, Starling and Bayliss tested whether the small intestine produces a signal that travels in the blood. They exposed small intestinal cells to acid, then made an extract of those cells. When this extract was injected into the blood of another animal, that animal's pancreas secreted bicarbonate. This result confirmed that exposure to acid causes the small intestine to release a chemical signal. This blood-borne signal triggers the pancreas to secrete bicarbonate into the gut.

The signaling substance discovered by Starling and Bayliss is now called secretin. Identifying its mode of action supported a hypothesis that dated back centuries: Blood carries internal secretions that influence the activities of the body's organs.

Starling coined the term "hormone," to describe blood-borne glandular secretions such as secretin. (The Greek *hormon* means "to set in motion.") Later, researchers identified additional hormones and their sources. In addition to major endocrine glands, many internal organs such as the small intestine and heart have cells that produce and release hormones.

Conversely, some of the major endocrine glands also have functions unrelated to hormone secretion. For example, the pancreas secretes hormones into the blood, but also secretes digestive enzymes into the small intestine. The gonads (ovaries and testes) produce sex hormones and also make gametes.

TAKE-HOME MESSAGE 34.2

✔ Hormones are intercellular signaling molecules that travel in the bloodstream.

✔ Collectively, hormone-secreting glands and cells make up an endocrine system.

34.3 The Nature of Hormone Action

LEARNING OBJECTIVES

- Compare the most common mechanisms of action of protein hormones and steroid hormones.
- Describe the role of a second messenger.
- Explain why mutations that alter hormone receptors can influence the response to a hormone.

There are two major types of hormones, those derived from amino acids and those derived from cholesterol.

Hormones Derived from Amino Acids

Amino acid–derived hormones include amine hormones (modified amino acids), peptide hormones (short chains of amino acids), and protein hormones (longer chains of amino acids).

Peptide hormones and protein hormones are polar, so they dissolve easily in the blood. Like other polar molecules, protein and peptide hormones cannot diffuse across a lipid bilayer. They act by binding to receptors at the plasma membrane of a target cell.

When a peptide or protein hormone binds to a receptor in the plasma membrane, a second messenger transduces the signal into events within the cell (FIGURE 34.2A). A **second messenger** is a molecule that forms inside a cell in response to an external signal, and it triggers a change in cellular activities. Formation of the second messenger sets in motion a cascade of reactions that bring about the target cell's response. Reactions triggered by formation of the second messenger usually result in activation of an enzyme that is already present in the cytoplasm in its inactive form. Effects of second messenger formation sometimes extend into the nucleus, bringing about changes in gene expression. However, peptide and protein hormones cannot enter a nucleus and interact directly with molecules that affect gene expression.

Steroid Hormones

A **steroid hormone** is derived from cholesterol and is thus nonpolar and lipid soluble. Being lipid soluble, it can diffuse across a target cell's plasma membrane. A steroid hormone typically binds to a receptor inside the cell, forming a hormone–receptor complex (FIGURE 34.2B). This complex functions in the nucleus, where it directly affects gene expression by binding to a promoter in the target cell's DNA. Recall that a promoter is a region where RNA polymerase binds (Section 9.2). Depending on the hormone, the receptor, and the cell, binding of the complex to a promoter can increase or decrease the rate of transcription of a nearby gene or set of genes.

Hormone Receptors

Many hormones target more than one type of cell, and they elicit a different response in each cell type. For example, ADH (antidiuretic hormone) affects urine formation when it binds to receptors on vertebrate kidney cells. It also triggers smooth muscle contraction when it binds to receptors on smooth muscle cells in blood vessel walls. These differing responses to ADH result from differences in the structure of ADH receptors.

A cell can only respond to a hormone for which it has appropriate and functional receptors. All hormone

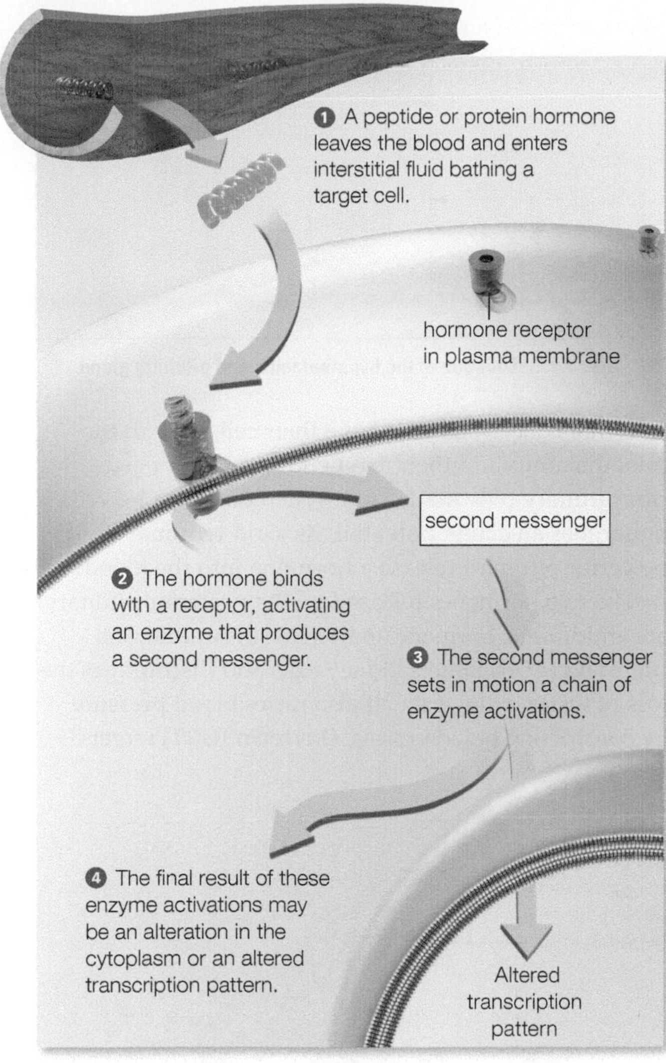
1 A peptide or protein hormone leaves the blood and enters interstitial fluid bathing a target cell.
hormone receptor in plasma membrane
second messenger
2 The hormone binds with a receptor, activating an enzyme that produces a second messenger.
3 The second messenger sets in motion a chain of enzyme activations.
4 The final result of these enzyme activations may be an alteration in the cytoplasm or an altered transcription pattern.
Altered transcription pattern

A Protein and peptide hormones act by way of a second messenger.

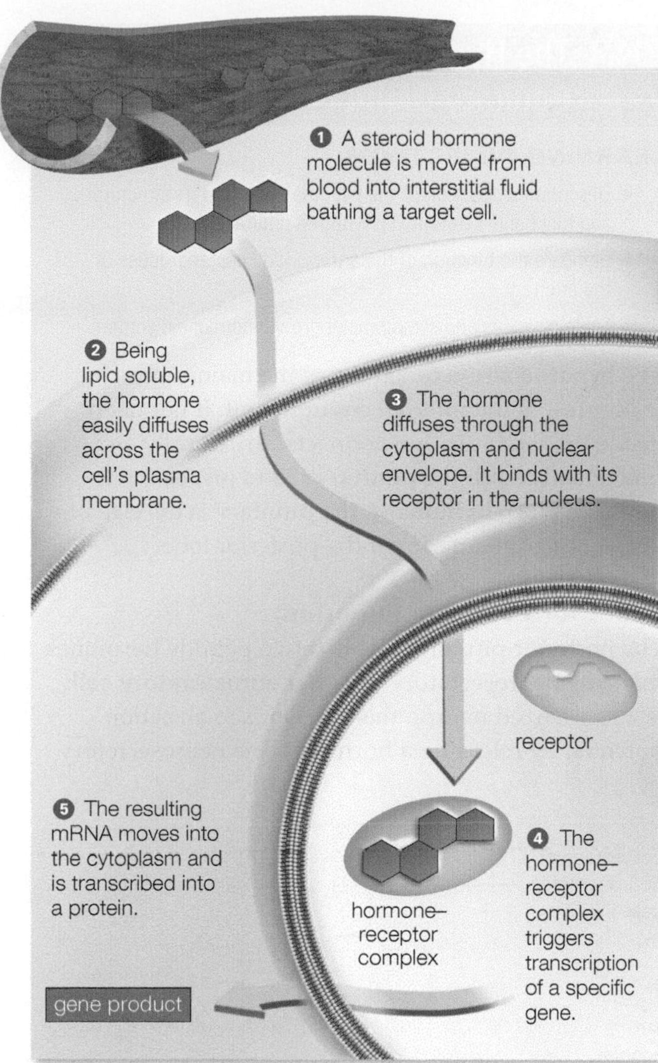
1 A steroid hormone molecule is moved from blood into interstitial fluid bathing a target cell.
2 Being lipid soluble, the hormone easily diffuses across the cell's plasma membrane.
3 The hormone diffuses through the cytoplasm and nuclear envelope. It binds with its receptor in the nucleus.
receptor
5 The resulting mRNA moves into the cytoplasm and is transcribed into a protein.
hormone–receptor complex
4 The hormone–receptor complex triggers transcription of a specific gene.
gene product

B Steroid hormones can enter and act inside a target cell.

FIGURE 34.2 Mechanisms of hormone action.

FIGURE IT OUT Why does protein hormone action require a second messenger?

Answer: Protein hormones cannot enter target cells.

receptors are proteins. Mutations that result in missing or defective receptors, or alter their rate of production, also affect the response to a hormone, even if the hormone is present in normal levels. Consider the effects of total androgen insensitivity syndrome. Affected people are genetically male (XY), so they make and secrete the male sex hormone testosterone. However, a mutation makes their testosterone receptors nonfunctional. As a result, their body behaves as if testosterone is not present. Testes form during embryonic development, but do not descend into the scrotum. At birth, the genitals appear female, so these individuals are typically identified as girls. Often their condition is not discovered until puberty. Being genetically male, they do not have ovaries or a uterus. Like other males, they make only a tiny amount of the female sex hormones, so they do not develop breasts or menstruate.

TAKE-HOME MESSAGE 34.3

- ✔ Hormones exert their effects by binding to protein receptors, either inside a cell or at the plasma membrane.
- ✔ Peptide and protein hormones cannot enter cells; they bind to a receptor at the plasma membrane. Often they trigger formation of a second messenger, a molecule that relays a signal into the cell.
- ✔ Steroid hormones can enter a cell, and they often act in the nucleus, where they alter the expression of specific genes.

amino acid–derived hormone An amine (modified amino acid), peptide, or protein that functions as a hormone.

second messenger Molecule that forms inside a cell when a hormone binds to a receptor in the plasma membrane; sets in motion reactions that alter activity inside the cell.

steroid hormone Lipid-soluble hormone derived from cholesterol.

34.4 The Hypothalamus and Pituitary Gland

LEARNING OBJECTIVES

- Describe the location of the hypothalamus and its structural and functional relationship with the pituitary gland.
- Compare the function of the anterior and posterior lobes of the pituitary.
- Explain causes and symptoms of growth hormone disorders.

The **hypothalamus** functions as the main center for control of the internal environment. It lies deep inside the forebrain and connects, structurally and functionally, with the **pituitary gland** just below it (FIGURE 34.3). In humans, the pituitary is the size of a pea and has an anterior and a posterior lobe.

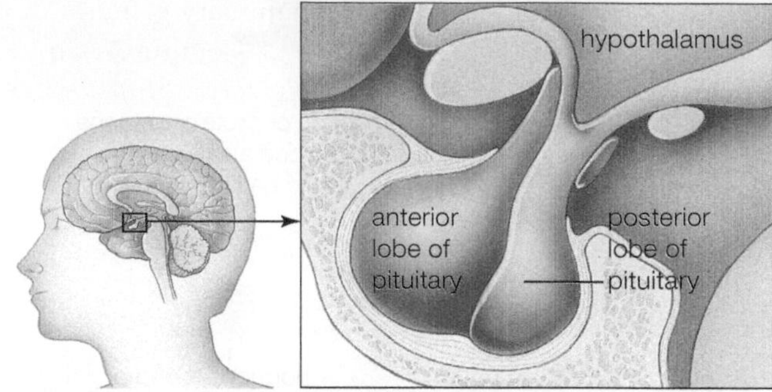

hypothalamus
anterior lobe of pituitary
posterior lobe of pituitary

FIGURE 34.3 **Location of the hypothalamus and pituitary gland.**

Posterior Pituitary Function

The posterior pituitary releases two peptide hormones made by neurosecretory cells. A **neurosecretory cell** is a specialized neuron that responds to an action potential by releasing a hormone. The neurosecretory cells that concern us here have their cell body in the hypothalamus and their axon endings in the posterior pituitary (FIGURE 34.4A). When one of these cells undergoes an action potential, its axon terminals in the posterior pituitary release a hormone into the blood.

The two hormones released by the posterior pituitary are antidiuretic hormone and oxytocin. **Antidiuretic hormone (ADH)** targets kidney cells and discourages the loss of water in the urine. It also raises blood pressure by constricting blood vessels. **Oxytocin (OCT)** targets

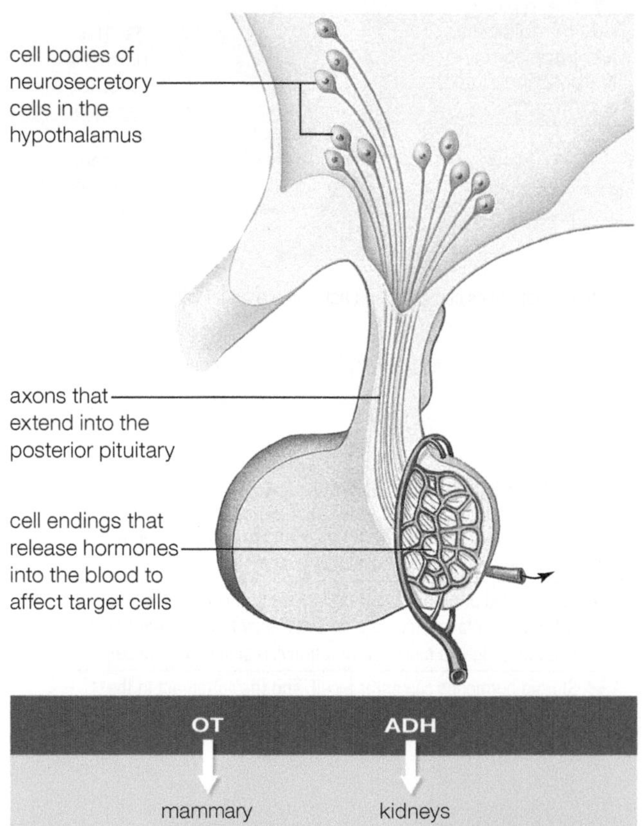

cell bodies of neurosecretory cells in the hypothalamus
axons that extend into the posterior pituitary
cell endings that release hormones into the blood to affect target cells
OT
ADH
mammary glands, uterus
kidneys

A **Posterior pituitary function.** Axon endings in the posterior lobe of the pituitary secrete two hormones made by cell bodies in the hypothalamus.

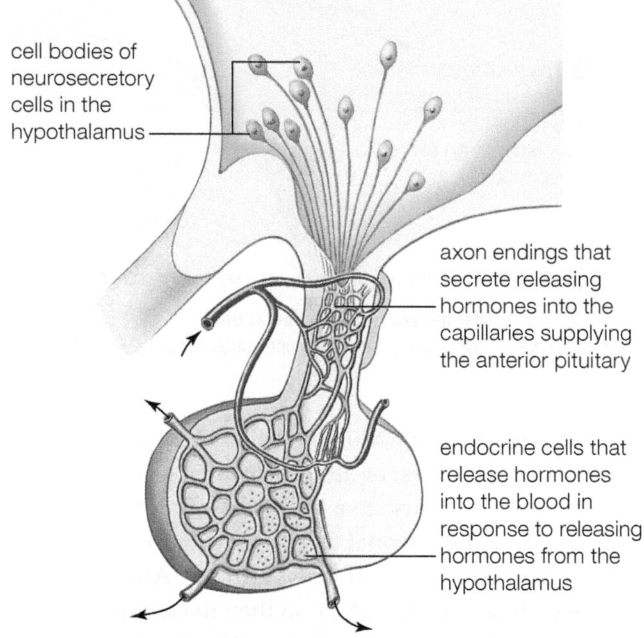

cell bodies of neurosecretory cells in the hypothalamus
axon endings that secrete releasing hormones into the capillaries supplying the anterior pituitary
endocrine cells that release hormones into the blood in response to releasing hormones from the hypothalamus

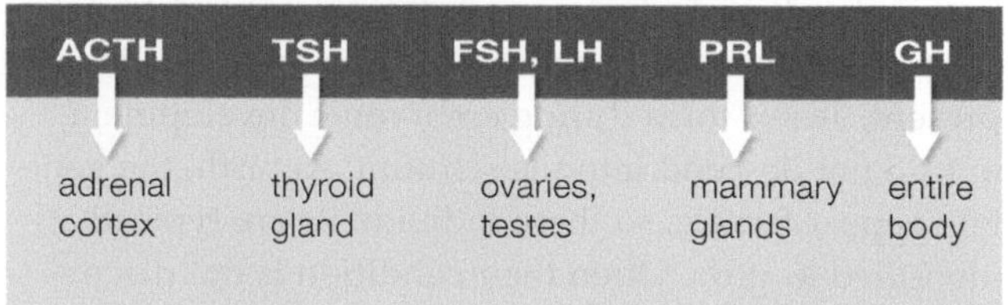

ACTH
TSH
FSH, LH
PRL
GH
adrenal cortex
thyroid gland
ovaries, testes
mammary glands
entire body

B **Anterior pituitary function.** Endocrine cells in the anterior lobe of the pituitary produce six hormones and secrete them in response to hypothalamic releasing hormones.

FIGURE 34.4 **Pituitary function.** Both lobes of the pituitary interact with the adjacent hypothalamus.

cells in the reproductive tract and mammary glands (breasts). It causes rhythmic contractions of muscles in the reproductive tract during orgasm (sexual climax) and contractions of the uterus (womb) during childbirth. When a woman is breastfeeding, contractions caused by oxytocin force milk out of the breast through milk ducts. Chapters 41 and 42 discuss oxytocin's reproduction-related functions in more detail. Both antidiuretic hormone and oxytocin also act in the brain, where they affect social bonds, a topic we consider in Chapter 43.

Anterior Pituitary Function

Anterior pituitary hormones are peptides that are produced within the anterior pituitary. Their release is regulated by hormones from the hypothalamus (**FIGURE 34.4B**). Most hypothalamic hormones that regulate the anterior pituitary are **releasing hormones**, meaning they encourage secretion of another hormone. The hypothalamus also produces inhibiting hormones that affect the anterior pituitary. An **inhibiting hormone** discourages the secretion of another hormone.

The anterior pituitary itself produces four releasing hormones that affect other endocrine glands. **Adrenocorticotropic hormone (ACTH)** stimulates the adrenal glands to reduce cortisol. (*Tropic* comes from the Greek for "stimulating.") **Thyroid-stimulating hormone (TSH)** causes the thyroid gland to secrete thyroid hormone. **Follicle-stimulating hormone (FSH)** and **luteinizing hormone (LH)** affect sex hormone secretion and production of gametes by gonads (a male's testes or a female's ovaries).

The anterior pituitary also produces two other hormones. The first, **prolactin (PRL)**, targets cells in the breast's mammary glands. Prolactin contributes to breast development at puberty and governs milk production after a woman gives birth. The second, **growth hormone (GH)**, targets cells throughout the body. It affects metabolism, causing a decrease in fat storage and an increase in synthesis of muscle proteins. It also encourages production of new bone and cartilage.

Hormonal Growth Disorders

Disturbances of growth hormone production or function alter normal growth patterns. Normally, GH production surges during teenage years, causing a growth spurt, then declines with age. Excessive secretion of growth hormone during childhood leads to pituitary gigantism. An affected person has a normal body form but is unusually tall.

When excessive growth hormone secretion continues into or begins during adulthood, the result is acromegaly. With this disorder, continued deposition of new bone and cartilage enlarges and eventually deforms the hands, feet, and face. Skin thickens, and the lips and tongue increase in size. Internal organs are also affected; the heart may become enlarged. The most common cause of gigantism and acromegaly is a pituitary tumor (**FIGURE 34.5**).

VIRGINIA

FIGURE 34.5 Excessive growth hormone. Sultan Kosen, the tallest living man, with his medical team at the University of Virginia. Kosen had a pituitary tumor that caused excessive growth hormone secretion. Treatment halted his growth at 8 feet, 3 inches (2.5 meters).

Insufficient GH or a lack of functional GH receptors during childhood can result in a type of dwarfism. Affected individuals are unusually small but normally proportioned. Note that the more common types of dwarfism involve mutations in genes that affect bone

adrenocorticotropic hormone (ACTH) (ah-DREEN-o-KOR-tah-ko-TROP-ik) Anterior pituitary hormone that stimulates cortisol secretion by the adrenal glands.

antidiuretic hormone (ADH) (AN-tee-die-ur-ET-ik) Posterior pituitary hormone that affects urine formation and blood pressure.

follicle-stimulating hormone (FSH) Anterior pituitary hormone that acts on the gonads to stimulate secretion of sex hormones.

growth hormone (GH) Anterior pituitary hormone that regulates growth and metabolism.

hypothalamus Forebrain region that controls processes related to homeostasis; produces and controls release of some hormones.

inhibiting hormone Hormone that deters release of a different hormone by its target endocrine cells.

luteinizing hormone (LH) (LOO-tee-in-iz-ing) Anterior pituitary hormone that acts on gonads to stimulate secretion of sex hormones.

neurosecretory cell Specialized neuron that secretes a hormone into the blood in response to an action potential.

oxytocin (ox-ee-TOE-sin) Posterior pituitary hormone that acts on smooth muscle of the reproductive tract and contractile cells of milk ducts.

pituitary gland (pih-TOO-eh-TAIR-ee) Pea-sized endocrine gland in the forebrain; interacts closely with the adjacent hypothalamus.

prolactin (pro-LAC-tin) Anterior pituitary hormone that stimulates milk production by mammary glands.

releasing hormone Hormone that stimulates release of a different hormone by its target endocrine cells.

thyroid-stimulating hormone (TSH) Anterior pituitary hormone; stimulates secretion of thyroid hormone by the thyroid.

formation rather than GH function. Individuals with these genes have disproportionately short limbs.

Human growth hormone is now produced through genetic engineering (Section 15.6). Injections of recombinant human growth hormone (rhGH) can increase the growth rate of children who have a naturally low GH level. However, such treatment is expensive and it remains controversial. Some people object to treating short stature as a defect to be cured.

TAKE-HOME MESSAGE 34.4

✔ Some secretory neurons of the hypothalamus make hormones (ADH, OT) that move through axons into the posterior pituitary, which releases them.

✔ Other hypothalamic neurons produce releasing or inhibiting hormones that are carried by the blood into the anterior pituitary. These hormones regulate the secretion of anterior pituitary hormones (ACTH, TSH, LH, FSH, PRL, and GH).

34.5 The Pineal Gland

LEARNING OBJECTIVES

- Describe how light influences the pineal gland.
- Give examples of melatonin's effects.

Like the hypothalamus and pituitary, the **pineal gland** lies deep inside the brain (**FIGURE 34.6**). It is shaped like a pine cone and is about as big as a grain of rice. The pineal gland secretes the hormone **melatonin**, but only during conditions of low light or darkness. When the retina is responding to light, nervous signals suppress melatonin secretion.

The amount of light varies with the time of day, so melatonin secretion follows a **circadian rhythm**, meaning it varies cyclically over an approximately 24-hour interval. In turn, melatonin secretion influences other circadian rhythms. At night, increased melatonin secretion causes a decline in body temperature and we become sleepy. Just after sunrise, melatonin secretion decreases, body temperature rises, and we awaken. Melatonin's sleep-inducing effects stem from its effects on target cells in the hypothalamus.

Night shift work and poor sleep habits can disrupt human melatonin secretion and increase the risk of cancer. Melatonin normally protects against cancer in two ways. First, it directly inhibits growth of some cancers. Second, it regulates secretion of sex hormones that can encourage growth of some cancers.

In temperate regions, day length varies seasonally, so melatonin secretion changes with the seasons. In many animals, changes in melatonin level trigger seasonal changes in appearance or behavior. For example, melatonin level influences the timing of seasonal coat color changes in hares and some other mammals that live where winters are cold and snowy (**FIGURE 34.7**). These animals have a dark summer coat, and a white winter one.

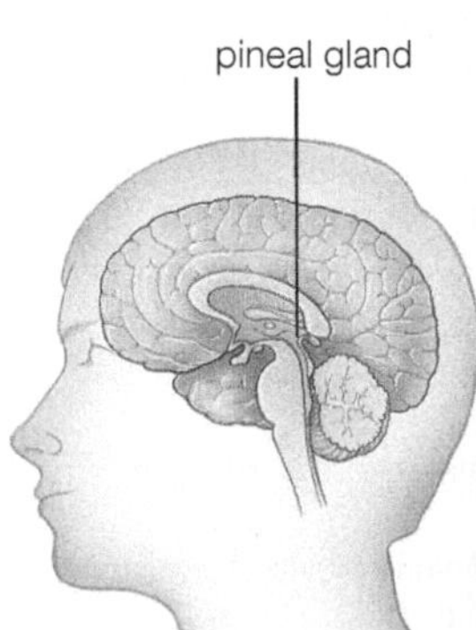
pineal gland

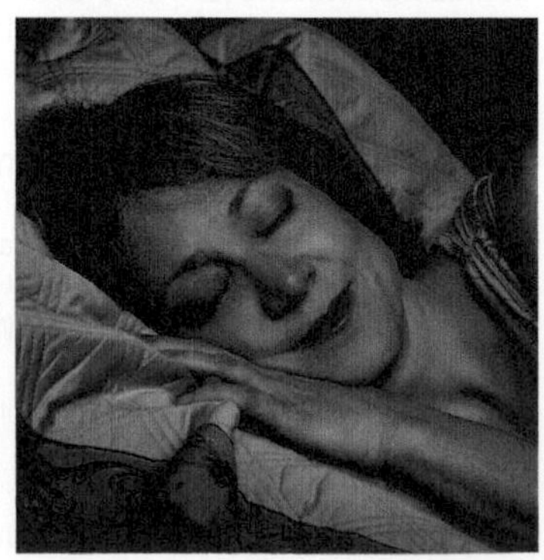

FIGURE 34.6 Human pineal gland. This gland releases melatonin when the retina is not being stimulated by light. Typically, melatonin secretion peaks at about 2 A.M., during sleep.

FIGURE 34.7 Melatonin-induced coat color change. This hare's white winter coat is growing in. A white coat will camouflage the animal against the snow.

TAKE-HOME MESSAGE 34.5

✔ Melatonin secreted by the pineal gland during periods of darkness encourages sleepiness and helps set the body's internal clock.

34.6 Thyroid and Parathyroid Glands

LEARNING OBJECTIVES

- Describe the feedback loop that controls release of thyroid hormone.
- Explain how diet can affect thyroid function.
- Describe how the parathyroid gland regulates blood calcium.

Feedback Control of Thyroid Hormone

The human **thyroid gland** lies at the base of the neck, attached to the trachea, or windpipe (**FIGURE 34.8**).

It secretes two iodine-containing amines (triiodothyronine and thyroxine) that we refer to collectively as **thyroid hormone**. Thyroid hormone increases the metabolic activity of cells throughout the body.

The anterior pituitary gland and hypothalamus regulate thyroid hormone secretion by a negative feedback loop (FIGURE 34.9). A low blood concentration of thyroid hormone causes the hypothalamus to secrete thyroid-releasing hormone (TRH) ❶. This releasing hormone causes the anterior pituitary to secrete thyroid-stimulating hormone (TSH) ❷. TSH in turn stimulates the secretion of thyroid hormone ❸. When the blood level of thyroid hormone rises, secretion of TRH and TSH declines ❹.

Thyroid Disorders and Disrupters

A deficiency in thyroid hormone, a condition called hypothyroidism, results in a reduced metabolic rate. Symptoms of hypothyroidism include fatigue, depression, increased sensitivity to cold, and weight gain. Hypothyroidism can arise from a TSH deficiency, a dietary deficiency in the iodine required to produce thyroid hormone, or an immune disorder in which the body mistakenly attacks the thyroid.

By contrast, an excess of thyroid hormone (hyperthyroidism) causes nervousness and irritability, chronic fever, and weight loss. In some people, it also induces tissues behind the eyeball to swell, making the eyes bulge. Hyperthyroidism usually arises as a result of an immune disorder or a thyroid tumor. Both hyperthyroidism and hypothyroidism can cause goiter, an abnormal enlargement of the thyroid gland.

Thyroid hormone has important developmental effects. In humans, maternal hypothyroidism during pregnancy can cause cretinism, a syndrome of stunted growth and impaired mental capacity. Cretinism can also result from hypothyroidism during infancy or early childhood.

In frogs, a surge in thyroid hormone triggers metamorphosis from a tadpole (the larval form) to an adult. If a tadpole's thyroid gland is removed, it will keep growing, but will not undergo metamorphosis. Tadpoles exposed to thyroid-disrupting chemicals develop abnormally, so they are used to test substances suspected of interfering with thyroid function.

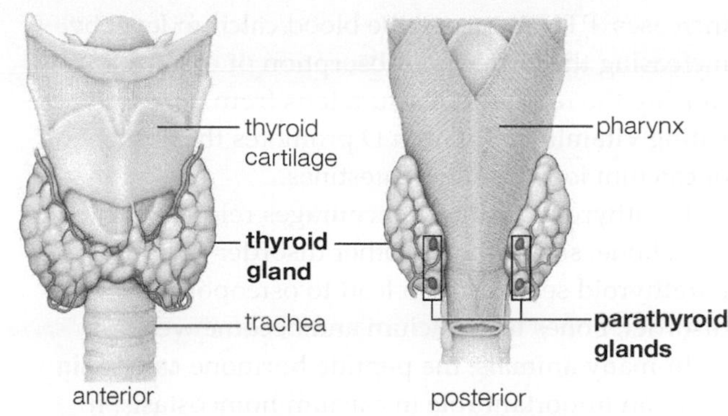

thyroid cartilage
thyroid gland
trachea
anterior
pharynx
parathyroid glands
posterior

FIGURE 34.8 Human thyroid and parathyroid glands.

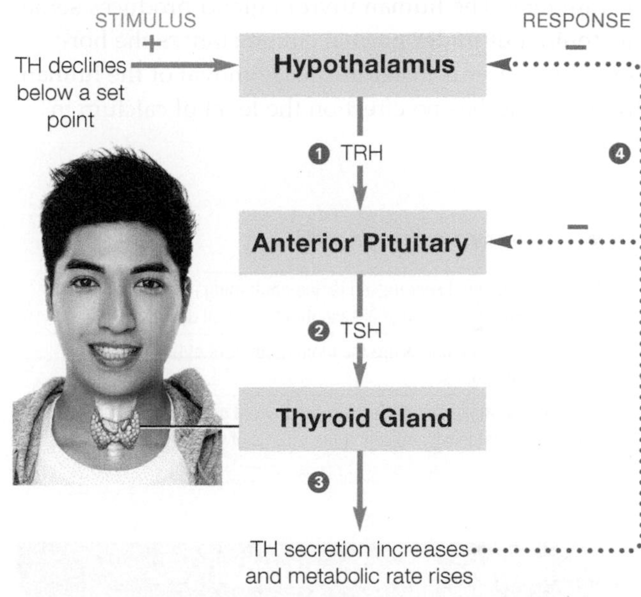

STIMULUS
+
TH declines below a set point
Hypothalamus
❶ TRH
Anterior Pituitary
❷ TSH
Thyroid Gland
❸
TH secretion increases and metabolic rate rises
RESPONSE
−
❹
−

FIGURE 34.9 Negative feedback control of thyroid hormone secretion. This feedback loop involves the hypothalamus and the pituitary's anterior lobe.

FIGURE IT OUT What effect does a high thyroid hormone level have on hormone secretion by the hypothalamus?

Answer: It inhibits secretion of TRH.

Hormonal Control of Blood Calcium

Humans have four **parathyroid glands**, each about the size of a grain of rice, on the thyroid's posterior surface. These glands produce **parathyroid hormone (PTH)**, a peptide hormone that is the main regulator of calcium ion concentration in the blood.

A high concentration of calcium ions in the blood discourages the secretion of PTH. When the blood's calcium ion concentration declines, PTH secretion

circadian rhythm Physiological change that repeats itself on an approximately 24-hour cycle.
melatonin (mell-uh-TOE-nin) Pineal gland hormone that regulates sleep–wake cycles and seasonal changes.
parathyroid glands Four small endocrine glands whose hormone product increases the level of calcium in blood.
parathyroid hormone (PTH) Hormone that regulates the concentration of calcium ions in the blood.
pineal gland (pie-KNEE-ul) Endocrine gland in the brain that secretes melatonin under low-light or dark conditions.
thyroid gland Endocrine gland at the base of the neck; produces thyroid hormone, which increases metabolism.
thyroid hormone (TH) Iodine-containing hormones that collectively increase metabolic rate and play a role in development.

CREDITS: (8) © Cengage Learning; (9) leungchopan/Shutterstock.

increases. PTH increases the blood calcium level by increasing the kidneys' reabsorption of calcium, causing the release of calcium ions from bone and activating vitamin D. Vitamin D promotes the absorption of calcium ions from the intestines.

Parathyroid hormone encourages release of calcium from bone, so a tumor or other disorder that increases parathyroid secretion can lead to osteoporosis. In this disorder, bones lose calcium and become weak.

In many animals, the peptide hormone **calcitonin** plays an important role in calcium homeostasis. It lowers the level of blood calcium by discouraging the breakdown of bone and increasing urinary excretion of calcium ions. The human thyroid gland produces some calcitonin, but under normal circumstances the hormone's role seems insignificant. Removal of the human thyroid gland has no effect on the level of calcium in the blood.

TAKE-HOME MESSAGE 34.6

✔ The thyroid gland secretes an iodine-containing hormone that regulates metabolic rate and is essential to normal development.

✔ The parathyroid glands are the main regulators of blood calcium level in human adults.

✔ In many animals, calcitonin from the thyroid gland contributes to calcium homeostasis by opposing the effects of parathyroid hormone.

34.7 Pancreatic Hormones

LEARNING OBJECTIVES

- Describe how insulin and glucagon collectively regulate the level of glucose in the blood.
- Compare the two types of diabetes.

Regulation of Blood Sugar

The **pancreas** is an organ with both exocrine and endocrine functions. It sits just behind the stomach in the abdominal cavity (FIGURE 34.10). The bulk of the pancreas consists of exocrine cells that secrete digestive enzymes into the small intestine. Scattered among these exocrine cells are clusters of endocrine cells known as the pancreatic islets.

The most abundant cells in pancreatic islets are beta cells. Beta cells secrete **insulin**, a peptide hormone that causes its target cells to take up and store glucose. After a meal, a rise in the blood's concentration of glucose stimulates beta cells to release insulin. Insulin's main targets are cells of the liver, skeletal muscle, and adipose tissue. In the liver and skeletal muscle, insulin triggers glucose uptake and glycogen synthesis. In all target cells, it encourages synthesis of fats and proteins and inhibits their breakdown. These cellular responses to insulin collectively lower the level of glucose in the blood (FIGURE 34.10 ❶–❺).

Pancreatic islets also contain alpha cells. These endocrine cells secrete the peptide hormone **glucagon** when the blood concentration of glucose falls. The glucagon binds to its receptors on liver cells, which respond by activating enzymes that break glycogen into glucose subunits. This response to glucagon raises the level of glucose in blood (FIGURE 34.10 ❻–❿). By working in opposition, glucagon and insulin maintain the blood's glucose level within a range that keeps cells throughout the body functioning properly.

Diabetes

Diabetes mellitus is a common metabolic disorder. Its name can be loosely translated as "passing honey-sweet water." Diabetics have sweet urine because their liver, fat, and muscle cells do not take up and store glucose as they should. The resulting high blood sugar, or hyperglycemia, disrupts normal metabolism throughout the body. Vision and hearing loss, skin infections, poor circulation, and nerve damage are among the many effects of diabetes. There are two main types of diabetes mellitus.

Type 1 Diabetes Type 1 diabetes accounts for only 5 to 10 percent of all reported cases, but it is the most dangerous. Type 1 diabetes develops after white blood cells mistakenly identify insulin-secreting beta cells as foreign and destroy them. Commonly, symptoms first appear during childhood or adolescence. Thus, this metabolic disorder is known as juvenile-onset diabetes.

All affected individuals must monitor their blood sugar concentration carefully and inject themselves with insulin. Prior to the discovery of insulin, most people with Type 1 diabetes died within a year of diagnosis. Purified animal insulin became available in the 1920s, greatly increasing survival. Today, Type 1 diabetics can use recombinant human insulin produced by genetically engineered bacteria or yeast.

Type 2 Diabetes Type 2 diabetes is the more common form of the disorder. Insulin levels are normal or even high. However, because target cells have an impaired response to the hormone the level of blood sugar

calcitonin (cal-sih-TOE-nin) Thyroid hormone that encourages bones to take up and incorporate calcium.

glucagon (GLUE-kah-gon) Pancreatic hormone that causes cells to break down glycogen and release glucose.

insulin Pancreatic hormone that causes cells to take up glucose and store it as glycogen.

pancreas Organ that secretes digestive enzymes into the small intestine and hormones into the blood.

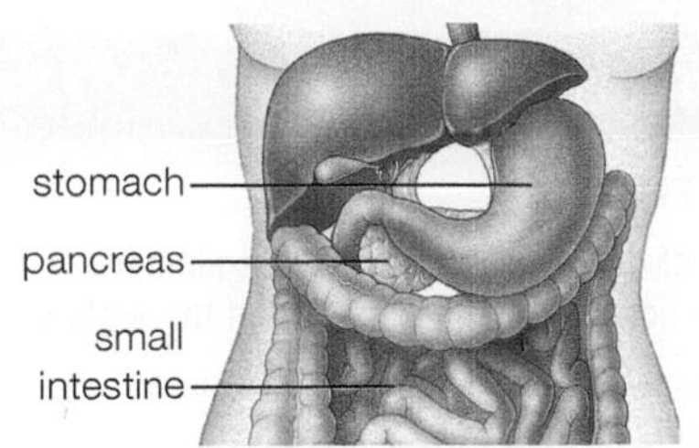
stomach
pancreas
small intestine

FIGURE 34.10 The pancreas.

Above, the location of the pancreas. Right, how cells that secrete insulin and glucagon work antagonistically to adjust the level of glucose in the blood.

1. After a meal, glucose enters blood faster than cells can take it up, so blood glucose increases.
2. The increase stops pancreatic alpha cells from secreting glucagon, and 3. stimulates pancreatic beta cells to secrete insulin.
4. Insulin encourages glucose uptake by adipose cells and muscle cells. It also causes liver and muscle cells to produce more glycogen.
5. As a result, the blood level of glucose declines to its normal level.
6. Between meals, blood glucose declines as cells take it up and use it for metabolism.
7. The decrease encourages glucagon secretion, and 8. slows insulin secretion.
9. In the liver, glucagon causes cells to break glycogen down into glucose, which enters the blood.
10. As a result, blood glucose increases to the normal level.

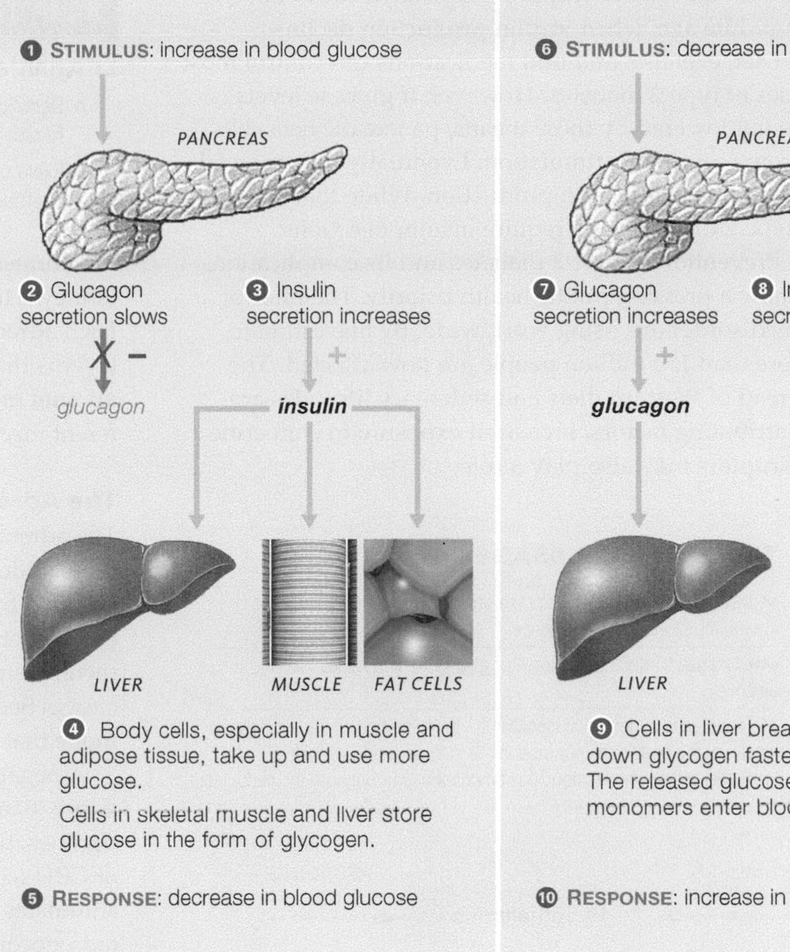
1 STIMULUS: increase in blood glucose
PANCREAS
2 Glucagon secretion slows
3 Insulin secretion increases
glucagon
insulin
LIVER
MUSCLE
FAT CELLS
4 Body cells, especially in muscle and adipose tissue, take up and use more glucose.
Cells in skeletal muscle and liver store glucose in the form of glycogen.
5 RESPONSE: decrease in blood glucose
6 STIMULUS: decrease in blood glucose
PANCREAS
7 Glucagon secretion increases
8 Insulin secretion slows
glucagon
insulin
LIVER
9 Cells in liver break down glycogen faster. The released glucose monomers enter blood.
10 RESPONSE: increase in blood glucose

Data Analysis Activities

Effects of BPA on Insulin Secretion Bisphenol A (BPA) is an endocrine disruptor that may increase the risk of type 2 diabetes. Angel Nadal suspected that BPA disrupts insulin metabolism by activating an estrogen receptor on pancreatic islet cells. FIGURE 34.11 shows the results of one experiments. Cultured cells from human pancreatic islets were exposed either to BPA or to DPN, a chemical that binds to the estrogen receptor and activates it. Cells were then exposed to glucose, and their insulin secretion was monitored.

1. Consider the two groups of cells exposed to glucose alone. How did the concentration of glucose affect their insulin secretion?
2. How did treating cells with DPN or with BTA alter the response to glucose concentration?
3. How were the effects of DPN and BPA similar? How did they differ?
4. Is this data consistent with the hypothesis that BPA alters human insulin secretion by binding to and activating the estrogen receptor?

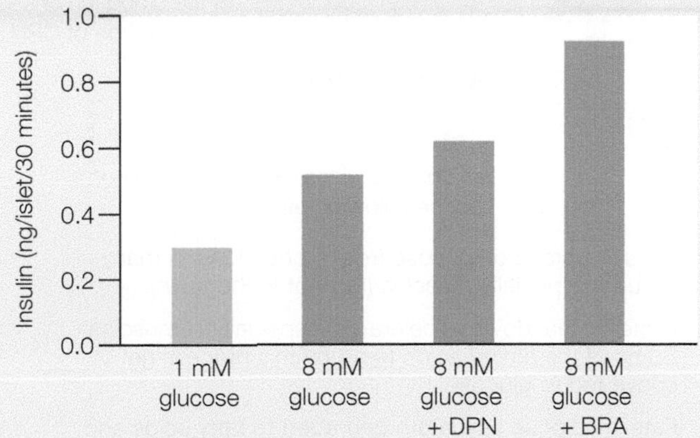
Insulin (ng/islet/30 minutes)
1.0
0.8
0.6
0.4
0.2
0.0
1 mM glucose
8 mM glucose
8 mM glucose + DPN
8 mM glucose + BPA

FIGURE 34.11 Effects of BPA and DPN on glucose-stimulated insulin secretion. DPN is a chemical known to bind and activate estrogen receptors on pancreatic beta cells. A glucose concentration of 8 millimolar (mM) is equivalent to that of the blood after a meal.

remains elevated. Symptoms typically start to develop in middle age, when insulin production declines.

Diet, exercise, and oral medications can control most cases of type 2 diabetes. However, if glucose levels are not lowered by these means, pancreatic beta cells receive continual stimulation. Eventually they may falter, and so will insulin production. When that happens, a type 2 diabetic may require insulin injections.

Prevention of type 2 diabetes and its complications is now a pressing public health priority. The rates of this disorder are rising worldwide. By one estimate, more than 150 million people are now affected. The spread of Western diets and sedentary lifestyles are contributing factors. Increased exposure to endocrine disruptors may also play a role.

TAKE-HOME MESSAGE 34.7

✔ Insulin encourages cells to take up and store more glucose, so it lowers the blood level of glucose.

✔ Glucagon triggers breakdown of glycogen, so it raises blood glucose.

✔ Diabetes is a metabolic disorder in which the body does not make insulin or the body does not respond to it. As a result, cells do not take up sugar as they should, causing complications throughout the body.

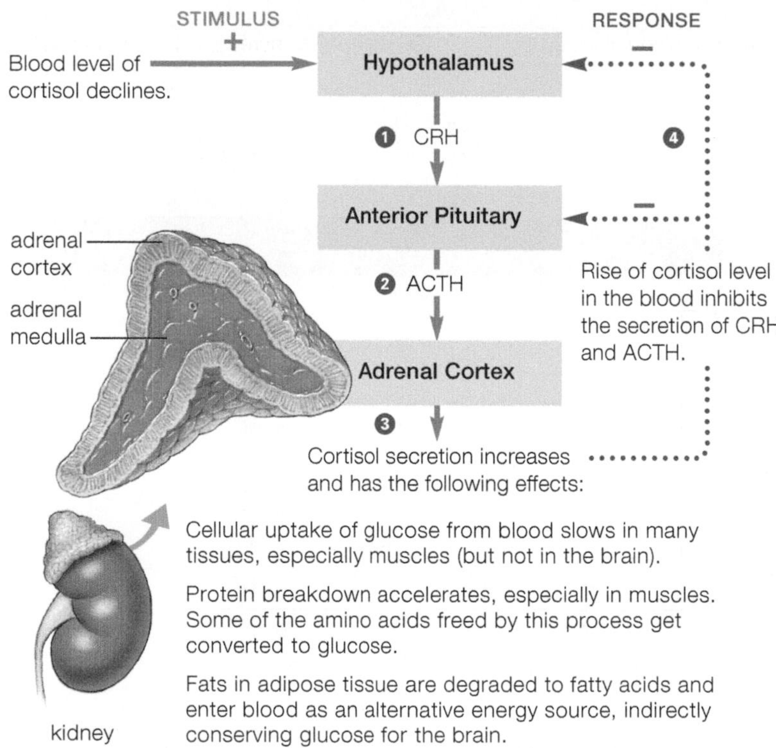

STIMULUS
+
Blood level of cortisol declines.
RESPONSE
Hypothalamus
−
1 CRH
4
Anterior Pituitary
−
adrenal cortex
adrenal medulla
2 ACTH
Rise of cortisol level in the blood inhibits the secretion of CRH and ACTH.
Adrenal Cortex
3
Cortisol secretion increases and has the following effects:
Cellular uptake of glucose from blood slows in many tissues, especially muscles (but not in the brain).
Protein breakdown accelerates, especially in muscles. Some of the amino acids freed by this process get converted to glucose.
Fats in adipose tissue are degraded to fatty acids and enter blood as an alternative energy source, indirectly conserving glucose for the brain.
kidney

FIGURE 34.12 **Structure and function of the human adrenal gland.** The diagram shows a negative feedback loop that governs cortisol secretion.

FIGURE IT OUT What effect would a decrease in ACTH have on the rate of fat breakdown in adipose tissue?

Answer: A decrease in ACTH would cause a decrease in cortisol secretion and less fat breakdown.

34.8 The Adrenal Glands

LEARNING OBJECTIVES

- Describe the location of the adrenal glands, and explain the functions of the hormones produced in each part of these glands.
- Explain how the hormonal effects of long-term stress can contribute to poor health.

Vertebrates have two **adrenal glands**, one above each kidney. (*Ad–* means near, and *renal* refers to the kidney.) Each adrenal gland is the size of a big grape. Its outer layer is the **adrenal cortex**, and its inner portion is the **adrenal medulla**. The two regions are controlled by different mechanisms and secrete different substances.

The Adrenal Cortex

The adrenal cortex releases steroid hormones. One of these, aldosterone, controls sodium and water reabsorption by kidneys. (Section 40.5 explains this process in detail.) The adrenal cortex also produces and secretes small amounts of sex hormones, which we discuss in Section 34.9. Here we focus on **cortisol**, a hormone that affects metabolism and immune responses.

A negative feedback loop regulates cortisol secretion (FIGURE 34.12). When the cortisol concentration declines, the hypothalamus increases its secretion of CRH (corticotropin-releasing hormone) ❶. CRH stimulates the anterior pituitary to secrete ACTH (adrenocorticotropic hormone) ❷, which in turn causes the adrenal cortex to release cortisol ❸. When the cortisol level becomes sufficiently high, the hypothalamus stops secreting CRH, the anterior pituitary stops secreting ACTH, and cortisol secretion slows ❹.

Cortisol helps ensure that adequate glucose is available to the brain by inducing liver cells to break down their store of glycogen and suppressing uptake of glucose by most cells. Cortisol also induces adipose cells to break down fats, and skeletal muscles to break down proteins. The products of these breakdown reactions (fatty acids and amino acids) function as energy sources (Section 7.6).

With injury, illness, or anxiety, the nervous system overrides the negative feedback loop that normally regulates cortisol, and the cortisol level soars. In the short term, this response helps supply enough glucose to the brain when food intake is likely to be low. A heightened cortisol level also suppresses inflammatory responses, thus lessening inflammation-related pain.

The Adrenal Medulla

The adrenal medulla contains specialized neurons of the sympathetic division (Section 32.7). Like other sympathetic neurons, those in the adrenal medulla release norepinephrine and epinephrine. However, in this case, the norepinephrine and epinephrine enter blood and function as hormones, rather than acting as neurotransmitters at a synapse. Epinephrine and norepinephrine released into the blood have the same effect on a target organ as direct stimulation by a sympathetic nerve.

Remember that sympathetic stimulation plays a role in the fight–flight response. Epinephrine and norepinephrine dilate the pupils, increase breathing rate, and make the heart beat faster. They prepare the body to deal with an exciting or dangerous situation.

Stress, Elevated Cortisol, and Health

When an animal is frightened or under physical stress, the nervous system triggers increased secretion of cortisol, epinephrine, and norepinephrine. As these hormones find their targets, they divert resources from longer-term tasks and help the body deal with the immediate threat. This stress response is highly adaptive for short periods of time, as when an animal is fleeing from a predator.

Ongoing stress can cause problems. Consider what Robert Sapolsky and his Kenyan colleagues found when they looked into how the social position of olive baboons (*Papio anubis*) influences hormone levels. Olive baboons live in large troops with a clearly defined dominance hierarchy. Individuals on top of the hierarchy get first access to food, grooming, and sexual partners. Those at the bottom of the hierarchy must continually relinquish resources to a higher-ranking baboon or face attack (**FIGURE 34.13**). Not surprisingly, the low-ranking baboons tend to have chronically high cortisol levels.

Physiological effects of high cortisol impair growth as well as immune system function, sexual function, and cardiovascular function. Chronically high cortisol levels also harm cells in the hippocampus, a brain region central to memory and learning (Section 32.12).

We see the impact of a long-term elevated cortisol level in humans affected by Cushing's syndrome, or hypercortisolism. This rare metabolic disorder can be triggered by an adrenal gland tumor, oversecretion of ACTH by the anterior pituitary, or chronic use of the drug cortisone. Doctors often prescribe cortisone to relieve chronic pain, inflammation, or other health problems. The body converts it to cortisol.

adrenal cortex Outer portion of adrenal gland; secretes aldosterone and cortisol.

adrenal gland Endocrine gland located atop the kidney; secretes aldosterone, cortisol, epinephrine, and norepinephrine.

adrenal medulla Inner portion of adrenal gland; secretes epinephrine and norepinephrine.

cortisol Adrenal cortex hormone that influences metabolism and immunity; secretions rise with stress.

gonads Primary reproductive organs (ovaries or testes); produce gametes and sex hormones.

FIGURE 34.13 Cortisol and stress. A dominant baboon (right) raising the stress level—and cortisol level—of a less dominant member of its troop.

Symptoms of hypercortisolism include a puffy, rounded "moon face"and deposition of fat in the torso. Blood pressure and blood glucose rise. White blood cell counts decline, so affected people are more prone to infections. Thin skin, decreased bone density, and muscle loss are common. Patients with the highest cortisol level have the greatest reduction in the size of the hippocampus and the most impaired memory.

Cortisol Deficiency

Tuberculosis and other diseases can damage adrenal glands, and slow or halt cortisol secretion. The result is Addison's disease, or hypocortisolism. In developed countries, this disorder more often arises after autoimmune attacks on the adrenal glands. President John F. Kennedy had this form of the disorder. Symptoms include fatigue, depression, weight loss, and darkening of the skin. If cortisol declines too much, blood sugar and blood pressure fall to life-threatening levels. Addison's disease is treated with synthetic cortisone.

TAKE-HOME MESSAGE 34.8

✔ The adrenal medulla releases epinephrine and norepinephrine, which prepare the body for excitement or danger.

✔ The adrenal cortex's main secretions are aldosterone, which affects urine concentration, and cortisol, which affects metabolism and stress responses. Cortisol secretion is governed by a feedback loop to the hypothalamus and pituitary, but stress breaks that loop and allows the level of cortisol to rise.

✔ Long-term cortisol elevation harms health. Insufficient cortisol can be fatal.

34.9 The Gonads

LEARNING OBJECTIVES

- List the main sex hormones made by males and by females and the organs that produce those hormones.
- Give examples of secondary sexual characteristics.

Gonads are primary reproductive organs, meaning they produce gametes (eggs or sperm). Vertebrate gonads also produce **sex hormones**, steroid hormones

CREDIT: (13) Mitch Reardon/Science Source.

that control sexual development and reproduction. FIGURE 34.14 shows the location of the human gonads. A male's gonads, his testes (singular, testis), secrete mainly testosterone. A female's gonads, her ovaries, secrete mainly estrogens and progesterone.

The hypothalamus and anterior pituitary control sex hormone secretion (FIGURE 34.15). In both sexes, the hypothalamus produces gonadotropin-releasing hormone (GnRH). This releasing hormone causes the anterior pituitary to secrete follicle-stimulating hormone (FSH) and luteinizing hormone (LH). FSH and LH target the gonads and cause them to produce and secrete sex hormones. FSH and LH are sometimes referred to as "gonadotropins." The suffix *-tropin* refers to a substance that has a stimulating effect. In this case, the effect is on the gonads.

As Section 10.3 explained, the presence of an *SRY* gene causes an early human embryo to produce testosterone and develop a male reproductive tract. Embryos that do not make testosterone or do not respond to it develop a female reproductive tract. During childhood, levels of sex hormones are low and, with the exception of their reproductive organs, males and females have similar bodies.

Sex hormone production increases during **puberty**, the period of development when reproductive organs mature and secondary sexual traits appear. **Secondary sexual traits** are traits that differ between the sexes but do not have a direct role in reproduction. When a girl undergoes puberty, her ovaries increase their estrogen production, causing fat deposition in breasts and on hips. Estrogens and progesterone also regulate egg formation and ready the uterus for pregnancy.

In males, a rise in testosterone output at puberty triggers the onset of sperm production and the development of secondary sexual traits. In humans, these traits include facial hair and a deep voice.

Testes secrete mostly testosterone, but they also make a little bit of estrogen and progesterone. The estrogen is necessary for sperm formation. Similarly, a female's ovaries make mostly estrogen and progesterone, but also a little testosterone. The presence of testosterone contributes to libido (the desire for sex).

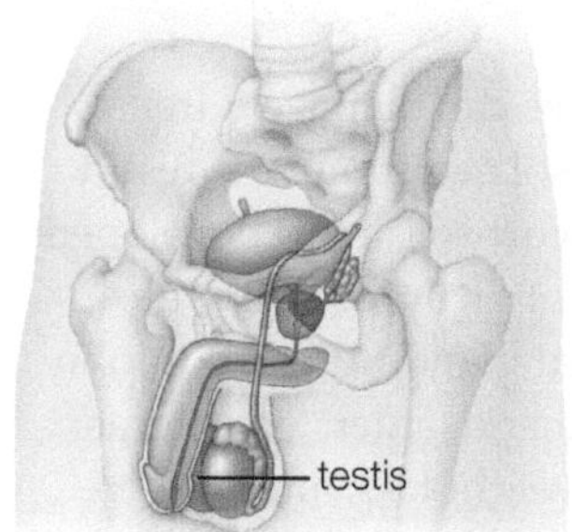
testis

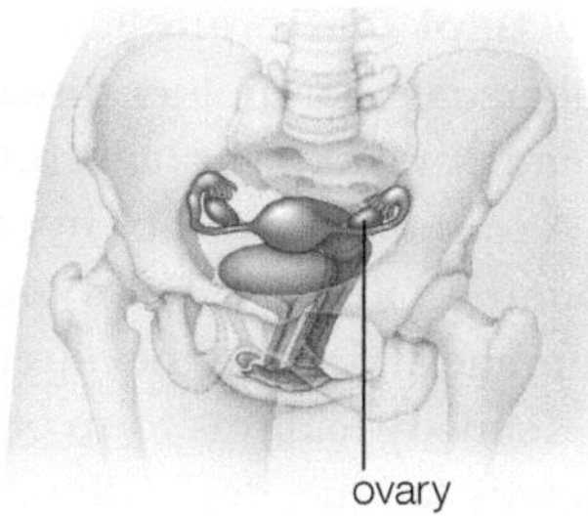
ovary

FIGURE 34.14 **Human gonads.** Testes (left) make sperm and the hormone testosterone. Ovaries (right) make eggs and the hormones estrogen and progesterone.

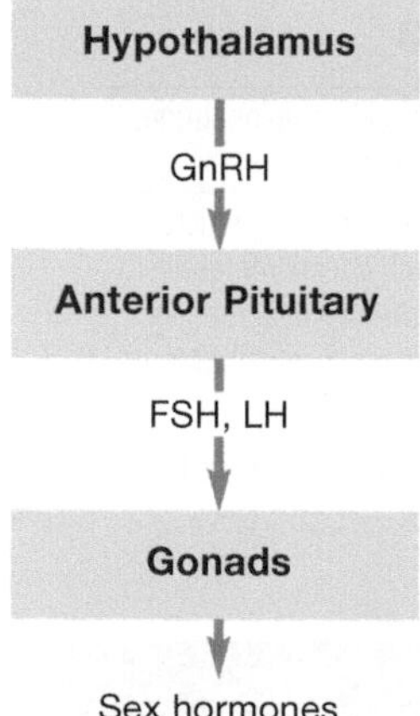
Hypothalamus
GnRH
Anterior Pituitary
FSH, LH
Gonads
Sex hormones

The hypothalamus produces gonadotropin-releasing hormone that acts on the anterior pituitary.

The anterior pituitary produces follicle-stimulating hormone and luteinizing hormone, both of which target cells in a male's testes or a female's ovaries.

The testes or ovaries produce sex hormones.

FIGURE 34.15 **Control of sex hormone secretion.**

TAKE-HOME MESSAGE 34.9

- ✔ Sex hormones are steroid hormones that influence the development of sexual traits and reproduction. Production of sex hormones increases at puberty.
- ✔ In both sexes, gonads secrete sex hormones in response to anterior pituitary hormones (FSH and LH), which are in turn released in response to a hypothalamic releasing hormone.
- ✔ A male's testes secrete mainly testosterone, and a female's ovaries secrete mainly estrogens and progesterone.

34.10 Invertebrate Hormones

LEARNING OBJECTIVE

- Give an example of a hormone unique to invertebrates, and describe its function.

We can trace the evolutionary roots of some vertebrate hormones and hormone receptors back to homologous molecules in invertebrates. Invertebrates such as sea stars, roundworms, annelids, and mollusks have steroid hormones and hormone receptors homologous to those of vertebrates. This homology suggests that steroid-based intercellular signaling evolved long ago, in the common ancestor of all bilateral animals. Invertebrates do not have the same glands as vertebrates do. They produce homologous hormones in

puberty Developmental stage during which human reproductive organs mature and begin to function.
secondary sexual traits Traits, such as the distribution of body fat, that differ between the sexes but do not have a direct role in reproduction.
sex hormone A steroid hormone that controls sexual development and reproduction; testosterone in males, estrogens or progesterone in females.

different glands. For example, female octopuses produce estrogens in a gland near their eye. In octopuses, as in vertebrates, estrogens and testosterone are sex hormones that function in reproduction.

On the other hand, some hormones are unique to invertebrate groups. Consider the hormones that control molting in arthropods. All arthropods have a hardened external cuticle that must be periodically shed as they grow (**FIGURE 34.16**). Although details vary among arthropod groups, molting is generally under the control of the steroid hormone **ecdysone**. When conditions favor molting, ecdysone causes the existing cuticle to separate from the epidermis and the muscles. At the same time, cells of the epidermis secrete a new cuticle underneath.

Chemicals that interfere with ecdysone function are sometimes used as insecticides. They kill insects by preventing normal molts. When such insecticides run off from fields and get into water, they can affect ecdysone signaling pathways in noninsect arthropods.

TAKE-HOME MESSAGE 34.10

✔ Some invertebrate hormones are homologous to those in vertebrates.

✔ Invertebrates also make hormones with no vertebrate counterpart. Ecdysone, which affects molting in arthropods, is one example.

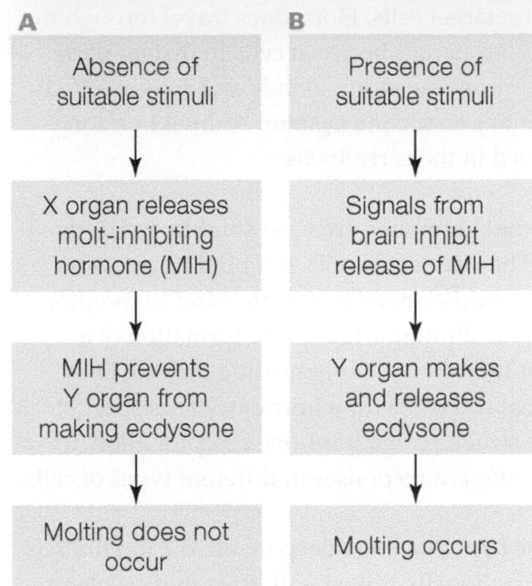
A
Absence of suitable stimuli
X organ releases molt-inhibiting hormone (MIH)
MIH prevents Y organ from making ecdysone
Molting does not occur
B
Presence of suitable stimuli
Signals from brain inhibit release of MIH
Y organ makes and releases ecdysone
Molting occurs

FIGURE 34.16 Hormonal control of molting in crustaceans.

Two hormone-secreting organs play a role, the X organ in the eye stalk and the Y organ at the base of the antennae.

A In the absence of environmental cues for molting, hormone from the X organ prevents molting.

B When stimulated by proper environmental cues such as day length or temperature, the brain sends nervous signals that inhibit X organ activity. With the X organ suppressed, the Y organ releases the ecdysone that stimulates molting.

The photo shows a newly molted prawn beside its old exoskeleton.

ecdysone (eck-DIE-sohn) Steroid hormone that controls molting in arthropods.

34.1 Endocrine Disruptors (revisited)

The Endocrine Society now lists more than 1,000 synthetic chemicals as known endocrine disruptors. The society is an international professional organization for medical physicians and scientists who study endocrine function. Their list will undoubtably grow as we learn more about the endocrine effects of additional chemicals. The concept of endocrine disruption is relatively new, so most synthetic chemicals now in use were approved as "safe" without ever being tested for possible endocrine effects.

Consider bisphenol-A (BPA), a synthetic chemical that has been in use since the 1950s. Like phthalates, it is a component of plastics. The endocrine-disrupting effects of BPA were discovered by accident in the 1990s by researchers studying yeast. The researchers detected an estrogen-like chemical in their yeast cultures. At first, they thought they had discovered that yeast produce an estrogen. However, additional studies showed the estrogen-like chemical was BPA. It was leaching out of the plastic bottles in which the yeast was grown.

BPA's estrogen-like effects and its ability to leach into a container's contents raised concerns. BPA-containing plastics are widely used in food and beverage containers. Does BPA from these sources enter the human food chain? A 2003–2004 study showed it does. A sampling of people in the United States revealed that more than 90 percent of them had BPA in their body.

The extent to which low-dose BPA exposure threatens human health remains unclear. Many studies have shown that BPA disrupts endocrine function in cultured human cells and in model animals such as rats and zebrafish. In addition, some epidemiological studies have found a correlation between human BPA level and an increased risk of type 2 diabetes. However, other studies have not found such a correlation.

Given the uncertainty, the U.S. Food and Drug Administration still allows BPA in food packaging. However, the U.S. Office of Disease Prevention and Health Promotion recommends reducing exposure to BPA as a way to promote public health.

CREDIT: (16) iStock.com/Bangkeaw.

STUDY GUIDE

Section 34.1 **Endocrine disruptors** are synthetic chemicals or chemical mixtures that interfere with the endocrine system. Phthalates and bisphenol-A (BPA) are examples.

Section 34.2 **Animal hormones** are signaling molecules that bind to receptors on target cells. Hormones travel through the blood and can convey signals between cells in distant parts of the body. All hormone-secreting glands and cells in a body constitute the animal's **endocrine system**. Animal hormones were first discovered in the early 1900s.

Section 34.3 **Steroid hormones** are lipid soluble and derived from cholesterol. They can enter cells and bind to receptors inside them. **Amino acid–derived hormones** bind to receptors in the cell membrane. Binding triggers the formation of a **second messenger** that elicits changes inside the cell. Only cells with functional receptors for a hormone can respond to it. Variation in the structure of a hormone receptor allow a hormone to elicit different responses in different types of cells.

Sections 34.4 The **hypothalamus** deep inside the forebrain is structurally and functionally linked with the **pituitary gland**. The pituitary gland has two lobes.

Axons of **neurosecretory cells** in the hypothalamus extend into the posterior pituitary, where they release antidiuretic hormone and oxytocin. **Antidiuretic hormone** concentrates the urine by acting in the kidney. **Oxytocin** targets contractile cells in mammary glands and the reproductive tract.

Releasing hormones and **inhibiting hormones** secreted by the hypothalamus control the secretion of hormones made by the anterior lobe of the pituitary. The anterior pituitary produces and releases four hormones that act on endocrine glands: **adrenocorticotropic hormone, thyroid-stimulating hormone, follicle-stimulating hormone**, and **luteinizing hormone**. It also produces **prolactin**, which encourages milk production, and **growth hormon**e, which has effects throughout the body. An excess of growth hormone can cause gigantism and acromegaly. A growth hormone deficiency can cause one type of dwarfism.

Sections 34.5 Exposure to light suppresses secretion of **melatonin** by the **pineal gland** deep inside the brain. Melatonin secretion, which occurs in a **circadian rhythm**, causes drowsiness and changes in body temperature. Melatonin also has a protective effect against cancer. In some temperate-zone animals, seasonal variations in appearance or behavior are regulated by changes in melatonin secretion.

Section 34.6 A feedback loop involving the anterior pituitary and hypothalamus governs secretion of **thyroid hormone** by the **thyroid gland** at the base of the neck. This iodine-containing hormone increases metabolic rate and is required for normal development. In frogs, thyroid hormone is essential for metamorphosis.

The **parathyroid glands** release **parathyroid hormone**, which acts on bone and kidney cells and raises the blood calcium level. **Calcitonin** secreted by the thyroid has the opposite effect in many animals. In humans, calcitonin does not play an important role in calcium homeostasis. It can, however, be used to treat osteoporosis.

Section 34.7 The **pancreas**, located inside the abdominal cavity, has both exocrine and endocrine functions. Beta cells secrete the hormone **insulin** when the blood glucose level is high. Insulin stimulates uptake of glucose by muscle and liver cells. When the blood glucose level is low, alpha cells secrete **glucagon**, a hormone that causes liver cells to break down glycogen and release glucose. The two hormones work in opposition to keep the blood glucose concentration within an optimal range.

Diabetes occurs when the body does not make insulin (type 1 diabetes) or its cells do not respond to it (type 2 diabetes). The resulting disruption of glucose metabolism harms cells throughout the body.

Section 34.8 Vertebrates have an **adrenal gland** atop each kidney. The gland's outer layer is the **adrenal cortex**. It secretes aldosterone, which acts in the kidney, and **cortisol**, the stress hormone. The cortisol level in blood is stabilized by a negative feedback loop involving the anterior pituitary and hypothalamus. In times of stress, the nervous system overrides this control and the blood cortisol level soars.

Long-term elevation of blood cortisol level, as a result of stress or a disorder, is harmful. A total lack of cortisol is fatal.

The inner part of the adrenal gland is the **adrenal medulla**. Norepinephrine and epinephrine released by neurons of the adrenal medulla influence organs as sympathetic stimulation does; they cause a fight–flight response.

Section 34.9 **Gonads** (ovaries and testes) make gametes and secrete the **sex hormones**. Ovaries secrete mostly estrogens and progesterone. Testes secrete mostly testosterone. Sex hormone output rises at **puberty**, and the increase in the concentration of these hormones encourages development of **secondary sexual traits** such as facial hair in males or rounded breasts in females.

In both sexes, follicle-stimulating hormone (FSH) and luteinizing hormone (LH) from the anterior pituitary govern sex hormone production. LH and FSH secretion are in turn controlled by gonadotropin-releasing hormone from the hypothalamus.

Section 34.10 Some invertebrate hormones are homologous to hormones in vertebrates, although they are made in different glands. Other invertebrate hormones such as **ecdysone**, a steroid hormone that regulates molting in arthropods, have no vertebrate counterpart.

SELF-QUIZ

Answers in Appendix VII

1. _______ are signaling molecules that travel through the blood and affect distant cells in the same individual.
   a. Hormones c. Pheromones
   b. Neurotransmitters d. Neuromodulators

2. A _______ is synthesized from cholesterol and can diffuse across the plasma membrane.
   a. steroid hormone c. peptide hormone
   b. protein hormone d. hormone receptor

3. Match each pituitary hormone with its target.
   ___ antidiuretic hormone a. gonads
   ___ oxytocin b. mammary glands, uterus
   ___ luteinizing hormone c. kidneys
   ___ growth hormone d. adrenal glands
   ___ adrenocorticotropic hormone e. most body cells

4. Releasing hormones secreted by the hypothalamus cause secretion of hormones by the _______ lobe of the pituitary gland.
   a. anterior b. posterior

5. In adults, an excess of _______ can cause acromegaly.
   a. melatonin c. insulin
   b. cortisol d. growth hormone

6. Low blood calcium triggers secretion by _______ .
   a. adrenal glands c. parathyroid glands
   b. ovaries d. the thyroid gland

7. _______ lowers blood sugar levels; _______ raises it.
   a. Glucagon; insulin b. Insulin; glucagon

8. With Type 2 diabetes, the most common type of diabetes, a person's body _______ .
   a. makes too much glucagon
   b. cannot make insulin
   c. does not respond normally to insulin
   d. has an autoimmune response to glucose

9. The _______ has both endocrine and exocrine functions.
   a. hypothalamus c. pineal gland
   b. parathyroid gland d. pancreas

10. Exposure to bright light discourages the secretion of _______ .
    a. glucagon c. thyroid hormone
    b. melatonin d. parathyroid hormone

11. Secretion of _______ suppresses immune responses.
    a. melatonin c. thyroid hormone
    b. antidiuretic hormone d. cortisol

12. In frogs, a lack of _______ prevents metamorphosis. In humans, a lack of the same hormone during development results in _______ .
    a. glucagon/diabetes
    b. thyroid hormone/cretinism
    c. growth hormone/acromegaly
    b. testosterone/androgen insensitivity syndorome

13. Which of the following is not a steroid hormone?
    a. testosterone d. edysone
    b. cortisol e. insulin

14. True or false? All hormones secreted by arthropods are also secreted by vertebrates.

15. Match the term listed at left with the most suitable description at right.
    ___ adrenal medulla a. makes sex hormones
    ___ thyroid gland b. source of melatonin
    ___ posterior lobe of the pituitary c. secretes hormones made in the hypothalamus
    ___ pancreatic islet d. source of epinephrine
    ___ pineal gland e. secretes insulin, glucagon
    ___ gonad f. function requires iodine

CRITICAL THINKING

1. Women who have been blind since birth almost never have breast cancer. A high level of one specific hormone is though to contribute to their low cancer rate. Which hormone is it and why are levels of this hormone unusually high in the blind?

2. A diabetic who injects too much insulin can lose consciousness. Explain how injecting excess insulin can impair brain function. Explain also why an injection of glucagon can restore normal function.

3. Endocrine disruptors can interfere with biological processes by mimicking hormones, activating or blocking the body's hormone receptors, disrupting the synthesis of hormones, or altering their breakdown. How are these mechanisms of action similar to those of the psychoactive drugs that act at synapses?

4. An arctic hare's coat color changes from white to brown each spring in response to a change in its melatonin level. This change is triggered by increasing daylength. As a result of ongoing global climate change, the arctic snow melt is occurring earlier and earlier. Explain why this change in the annual temperature pattern has increased the number of hares that are killed by predators.

35 STRUCTURAL SUPPORT AND MOVEMENT

CORE CONCEPTS

Systems

Complex properties arise from interactions among components of a biological system.
Vertebrate skeletal muscles move body parts by pulling on bones. These muscles often work in pairs, with the action of one opposing the action of the other. Signals from the nervous system initiate muscle contraction. Interactions among organized protein filaments in muscle cells bring about the contraction.

Evolution

Evolution underlies the unity and diversity of life.
All vertebrates have an internal skeleton. The structural and functional similarities among the components of this skeleton reflect a common ancestry. Natural selection leads to the evolution of structures that increase fitness in a specific environment, and evolved variations in body form allow efficent locomotion in different environments.

Pathways of Transformation

Organisms exchange matter and energy with the environment in order to grow, maintain themselves, and reproduce.
All animals can move their body parts, and most move about in search of food and mates. Moving from place to place requires overcoming the forces of friction and gravity. ATP-fueled muscle contractions exert force against a skeleton to bring about movements.

Links to Earlier Concepts

This chapter builds on your knowledge of connective (Section 31.4) and muscle (31.5) tissues. You will revisit muscular dystrophy (14.4), and bacterial toxins (20.8) that affect muscles, as well as active transport (5.9) and motor proteins (4.8). Nervous control of muscle (32.6) and the effects of hormones on bone (34.6) are also discussed. We also look again at adaptive changes to the vertebrate skeleton (25.2, 25.3).

35.1 Bulking Up

The more you use your muscles, the bulkier and more powerful they become. A mature skeletal muscle fiber cannot divide, but it can make more of the proteins involved in muscle contraction. Protein filaments inside a muscle fiber are continually built and broken down. Exercise tilts this process in favor of synthesis, so muscle cells get bigger and the muscle gets stronger. In addition, exercise encourages stem cells in skeletal muscle to divide and differentiate into muscle fibers.

Hormones affect muscle mass. One effect of the sex hormone testosterone is to encourage muscle cells to build more proteins. Men make much more testosterone than women, which is why men tend to be more muscular. Human growth hormone also stimulates synthesis of muscle proteins. Taking synthetic versions of these muscle-building (anabolic) hormones will bulk up muscles. Some athletes use synthetic hormones to increase their muscle size. However, most sport organizations prohibit the use of these drugs and penalize athletes who cheat by using them.

Genetic variations among individuals also influence muscle bulk. Myostatin is a regulatory protein that discourages growth of skeletal muscle. If a mutation prevents production of normal myostatin, the result is muscle hypertrophy (excessive growth of muscles).

Bully whippets have muscle hypertrophy. These dogs are homozygous for a mutation that prevents myostatin production. Whippets are bred for racing and most have a sleek body. By contrast, bully whippets are heavily muscled (**FIGURE 35.1**). Bully whippets are not allowed to race. However, whippets that are heterozygous for the mutant myostatin allele do race. Although they do not appear heavily muscled, they tend to win races more often than normal whippets ●.

FIGURE 35.1 Genetic variation in muscle mass. A bully whippet (right) is homozygous for a mutant myostatin allele, so she does not make functional myostatin and has bulky muscles. A whippet homozygous for the wild-type myostatin gene has a comparatively light musculature (left).

35.2 Animal Movement

LEARNING OBJECTIVES

- Give examples of adaptations that reduce the energy animals must expend to overcome friction and gravity when moving.
- Explain how squid locomotion differs from fish locomotion.
- List some structural adaptations that contribute to a cheetah's speed.

All animals move. During part or all of their life, they are capable of **locomotion**, which is self-propelled movement from place to place. As adults, even sessile animals (those that live fixed in place) usually move some body parts. Consider barnacles, which begin life as free-swimming larvae, then settle and secrete a protective shell. Although adults remain in one place, they still wave their feathery legs to capture food from the water around them. More typically animals shift their location on a daily basis as they look for food and escape from predators.

Animals use diverse mechanisms of locomotion. They swim through water; fly through air; run, walk, or crawl along surfaces; or burrow through soil and sediments. Despite these differences, the same physical law governs all movement: For every action, there is an equal and opposite reaction. For an animal to move in one direction, its muscles must exert a force in the opposite direction.

squid moves this way
water shoots this way

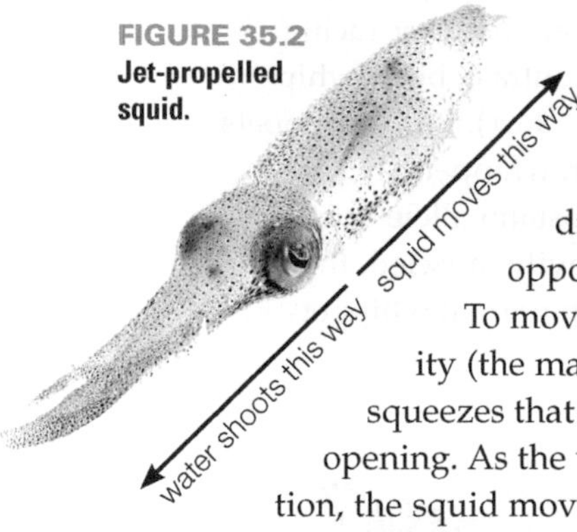

FIGURE 35.2 **Jet-propelled squid.**

Locomotion in Water

Jet propulsion by squids provides a simple example of how a force exerted in one direction moves an animal in the opposite direction (FIGURE 35.2). To move, a squid fills an internal cavity (the mantle cavity) with water, then squeezes that water out through a small opening. As the water shoots out in one direction, the squid moves in the opposite direction. By analogy, think of what happens if you inflate a balloon, then release it without tying it shut. As the rubber of the balloon contracts, force exerted by air rushing out makes the balloon shoot the opposite way.

More typically, aquatic animals that swim do so by pushing at water through body movements and movements of fins or flippers. Animals also move along surfaces, both underwater and on land, by pushing against those surfaces.

The effects of friction and gravity increase the amount of energy an animal must expend to move. Water is denser than air, so moving through it produces more friction. Think of how much harder it is to walk through waist-deep water than to walk on land. Aquatic animals that benefit by swimming fast typically have a streamlined body that reduces the effects of friction (FIGURE 35.3A). Aquatic animals are generally somewhat buoyant (prone to float), so gravity is less of a constraint for them than it is for land animals.

Locomotion on Land

Snails, snakes, and other animals that creep or slither along surfaces must expend energy to overcome friction between the body and that surface. Snails reduce friction by secreting a lubricating mucus from their large foot. A snake's smooth scales can have a similar effect. By varying the angle of the scales in different parts of its ventral (belly) surface, a snake can minimize friction everywhere except where it pushes off against the ground. In this region, the scales are oriented so they act like the tread on a tire. The treadlike scales grip the ground firmly so the snake can propel itself forward.

Animals that walk or run minimize friction by reducing the surface area in contact with the ground. Limbs partially or fully elevate their body and only some feet touch the ground at any given time.

The cheetah is the fastest land animal (FIGURE 35.3B). It can sprint as fast as 65 miles per hour (about 100 kilometers per hour), although it cannot sustain that speed for long. A variety of traits contribute to the cheetah's speed. Compared to other big cats, a cheetah has longer legs relative to its body size, a wider range of motion at its hips and shoulders, and a longer, more flexible backbone. Together, these features maximize the length of the cheetah's stride. The cheetah also has a unique sprinting stride: When its back legs land, its backbone becomes bent into a tight C shape. When the animal pushes off again, its spine recoils and helps to force the front of the body forward.

Elastic tissue plays a locomotive role in many animals. Consider kangaroos, which usually move by hopping. Each time a kangaroo lands, elastic material in its hind legs becomes compressed. The compression stores potential energy that is released when this material expands during the next hop.

Flight

The ability to fly evolved independently in insects, birds, mammals, and an extinct group of reptiles called pterosaurs. To fly, an animal must overcome gravity to become airborne, and also generate a force to propel itself forward. Wing shape is key to flight. In cross section, a bird wing has the same shape as an airplane wing. Its upper surface is convex (bulges outward), while the lower is flat or slightly concave (sunken).

CREDIT: (2) eye-blink/Shutterstock.

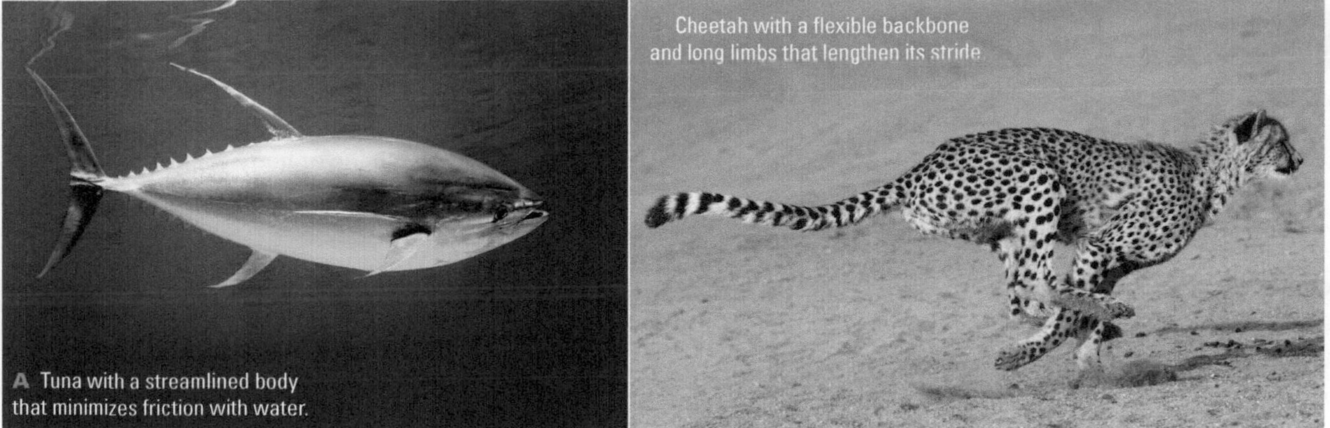
Cheetah with a flexible backbone
and long limbs that lengthen its stride
A Tuna with a streamlined body
that minimizes friction with water.

FIGURE 35.3 Predators with adaptations for speed.

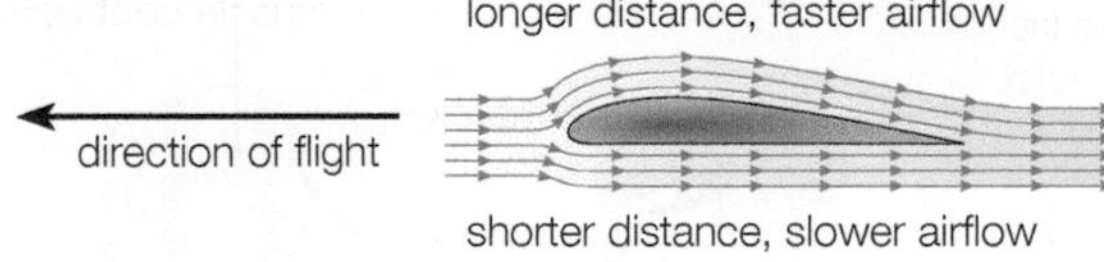
longer distance, faster airflow
direction of flight
shorter distance, slower airflow

FIGURE 35.4 Gull gliding. Air flows faster over the rounded upper surface of its wings than the flatter lower surface (right). This difference in air velocity between the two surfaces results in an upward force called lift.

Consider what happens when the bird is gliding (flying without flapping its wings). As the bird moves forward, air flowing over the upper surface of the wings moves faster than air flowing over the wings' lower surface. This difference in air velocity across the two surfaces creates lift, an upward force, that opposes gravity and keeps the bird aloft (**FIGURE 35.4**).

For its initial liftoff, the bird has to flap its wings. It spreads its feathers to maximize the size of the wings, then pulls the wings downward. As the air beneath the wings is forced downward, the bird moves in the opposite direction. The bird then folds its wings to minimize their size as it pulls them upward in preparation for another lift-producing downstroke.

locomotion Self-propelled movement from place to place.

Flying, jumping, swimming, and all other forms of locomotion involve interactions between muscles and a skeleton. These interactions, as well as the additional functions of the muscular and skeletal systems, are the focus of this chapter.

TAKE-HOME MESSAGE 35.2

- ✔ To move, an animal must expend energy to oppose the effects of gravity and friction. Structural adaptations minimize this energy cost.
- ✔ Streamlined bodies help aquatic animals reduce friction and move quickly through water. Secreted mucus and smooth scales reduce friction in animals that move along the ground.
- ✔ Elastic tissues that compress and recoil help animals minimize the energy they use to move.
- ✔ The shape of a wing creates the lift that allows flight.

35.3 Types of Skeletons

LEARNING OBJECTIVES

- Using appropriate examples, describe the three types of skeletons.
- Describe how interactions between muscles and a skeleton allow a fly to flap its wings and an earthworm to burrow through soil.
- Explain the function of the vertebrate pelvic girdle and pectoral girdle.

Muscles bring about movement by interacting with a skeleton. Your bony internal framework is one type of skeleton. In many other animals, muscles exert force against fluid-filled chambers or external hard parts.

Invertebrate Skeletons

A **hydrostatic skeleton** consists of fluid-filled internal chambers that receive the force of muscle contraction. Soft-bodied invertebrates such as sea anemones and worms have this type of skeleton. For example, an earthworm's coelom is divided into many fluid-filled chambers, one per segment (Section 24.7). Movement occurs when muscles exert force against these chambers (FIGURE 35.5). Two sets of muscles squeeze the chambers to alter the shape of body segments, much as squeezing a water-filled balloon changes the balloon's shape. Each earthworm segment is enclosed by a ring of circular muscle. When this muscle contracts, the segment gets longer and narrower. Longitudinal muscle runs the length of each segment; when this muscle contracts, the segment gets shorter and wider. Coordinated changes in the shape of the body segments move the earthworm through soil.

FIGURE 35.5 **Hydrostatic skeleton of an earthworm.** The worm moves when muscles put pressure on coelomic fluid in individual body segments, causing the segments to change shape.

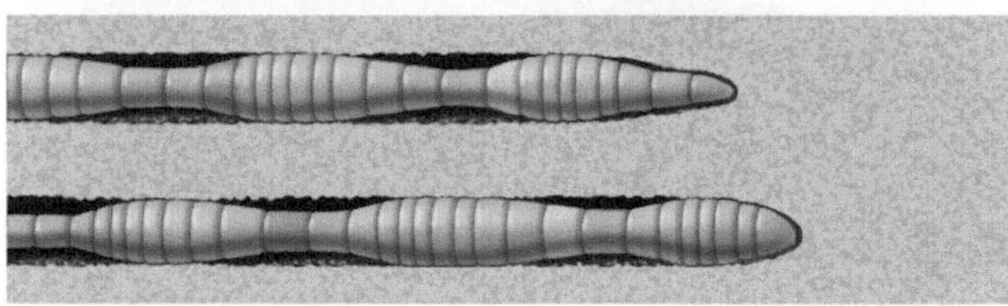

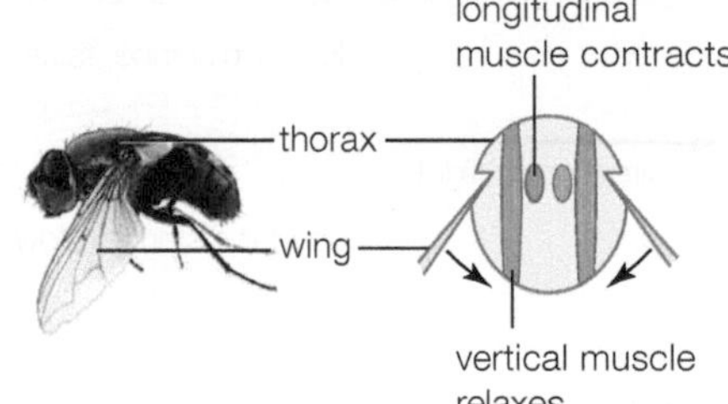
longitudinal muscle contracts
thorax
wing
vertical muscle relaxes

A The fly's wings pivot down when the vertical muscles relax and the longitudinal muscles contract, pulling the sides of the thorax inward.

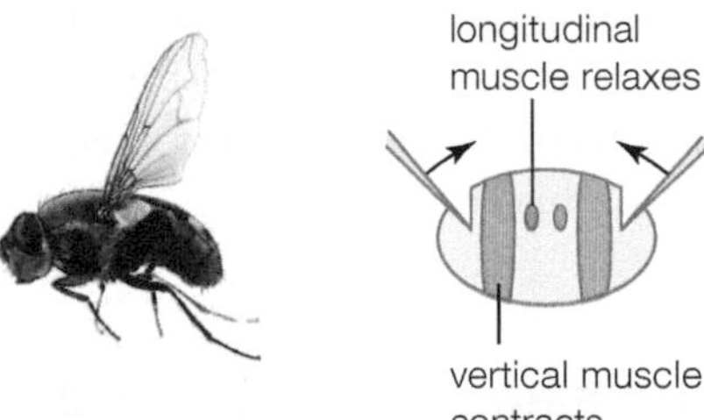
longitudinal muscle relaxes
vertical muscle contracts

B The wings pivot up when the longitudinal muscles relax and the vertical muscles contract, flattening the thorax.

FIGURE 35.6 **How a fly flaps its wings.** Wings move when muscles attached to the exoskeleton of the thorax alter its shape.

An **exoskeleton** is a cuticle, shell, or other hard external body part that receives the force of a muscle contraction. For example, muscles attached to the cuticle of a fly's thorax alter the shape of the thorax, causing the wings attached to the thorax to flap up and down (FIGURE 35.6). One advantage of an exoskeleton is its ability to protect the soft body tissues inside it. However, external skeletons have some drawbacks. An exoskeleton consists of noncellular secreted material, so it cannot grow with the animal. As an arthropod grows, it must periodically molt its old skeleton and replace it with a larger one. Repeatedly producing new exoskeleton uses resources that could otherwise be used for growth or reproduction.

An **endoskeleton** is an internal framework of hard parts. For example, echinoderms such as sea stars have an endoskeleton made of hard, calcium-rich plates.

The Vertebrate Endoskeleton

All vertebrates have an endoskeleton. In sharks and other cartilaginous fishes, it consists of cartilage, a rubbery connective tissue. Other vertebrate skeletons include some cartilage, but consist mostly of bone tissue (Section 31.4).

The term "vertebrate" refers to a **vertebral column**, or backbone, a feature common to all members of this group (FIGURE 35.7). The backbone supports the body, serves as an attachment point for muscles, and protects the spinal cord that runs through a canal inside it. Bony segments called **vertebrae** (singular, vertebra) make up

FIGURE 35.7 **Skeletal elements typical of early reptiles.**

axial skeleton

appendicular skeleton

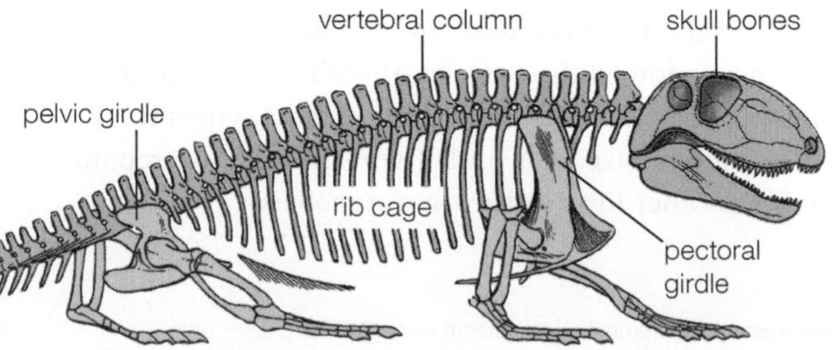
vertebral column
skull bones
pelvic girdle
rib cage
pectoral girdle

the backbone. **Intervertebral disks** of cartilage between vertebrae act as shock absorbers. The vertebral column and bones of the head and rib cage constitute the **axial skeleton**, the bones that make up the main axis of the body. The **appendicular skeleton** includes bones of the appendages (limbs or bony fins) and bones that connect the appendages to the axial skeleton (the pectoral girdle at the shoulders and the pelvic girdle at the hip).

You learned earlier how vertebrate skeletons have evolved over time. For example, jaws are derived from the gill supports of ancient jawless fishes (Section 25.3). As another example, bones in the limbs of land vertebrates are homologous to those in the pelvic and pectoral fins of lobe-finned fishes (Section 25.3).

For a closer look at vertebrate skeletal features, consider a human skeleton (**FIGURE 35.8**). The skull's flattened cranial bones fit together as a braincase, or cranium. Facial bones include cheekbones and other bones around the eyes, the bone that forms the bridge of the nose, and the bones of the jaw.

Ribs and the breastbone, or sternum, form a protective cage around the heart and lungs. Both males and females have 12 pairs of ribs.

The vertebral column extends from the base of the skull to the pelvic girdle. Viewed from the side, our backbone has an S shape that keeps our head and torso centered over our feet when we stand upright.

The scapula (shoulder blade) and clavicle (collarbone) are bones of the pectoral girdle. The clavicle transfers force from the arms to the axial skeleton. The upper arm has a single bone, the humerus. The forearm has two bones, the radius and ulna. Carpals are bones of the wrist, metacarpals are bones of the palm, and phalanges (singular, phalanx) are finger bones.

The pelvic girdle is a basin-shaped ring that encloses the pelvic cavity and supports the weight of the upper body when you stand. It includes bones of the two hips and (at the back) the sacrum and coccyx (tailbone).

Bones of the leg include the femur (thighbone), patella (kneecap), and the tibia and fibula (bones of the lower leg). Tarsals are ankle bones, and metatarsals are bones of the sole of the foot. Like the bones of the fingers, those of the toes are called phalanges.

TAKE-HOME MESSAGE 35.3

✔ In animals with a hydrostatic skeleton, muscle contractions alter the shape of fluid-filled chambers. In animals that have an exoskeleton, muscles pull on hard external parts. In animals that have an endoskeleton, muscles pull on hard internal parts.

✔ Vertebrates have an endoskeleton made of cartilage and (in most groups) bone.

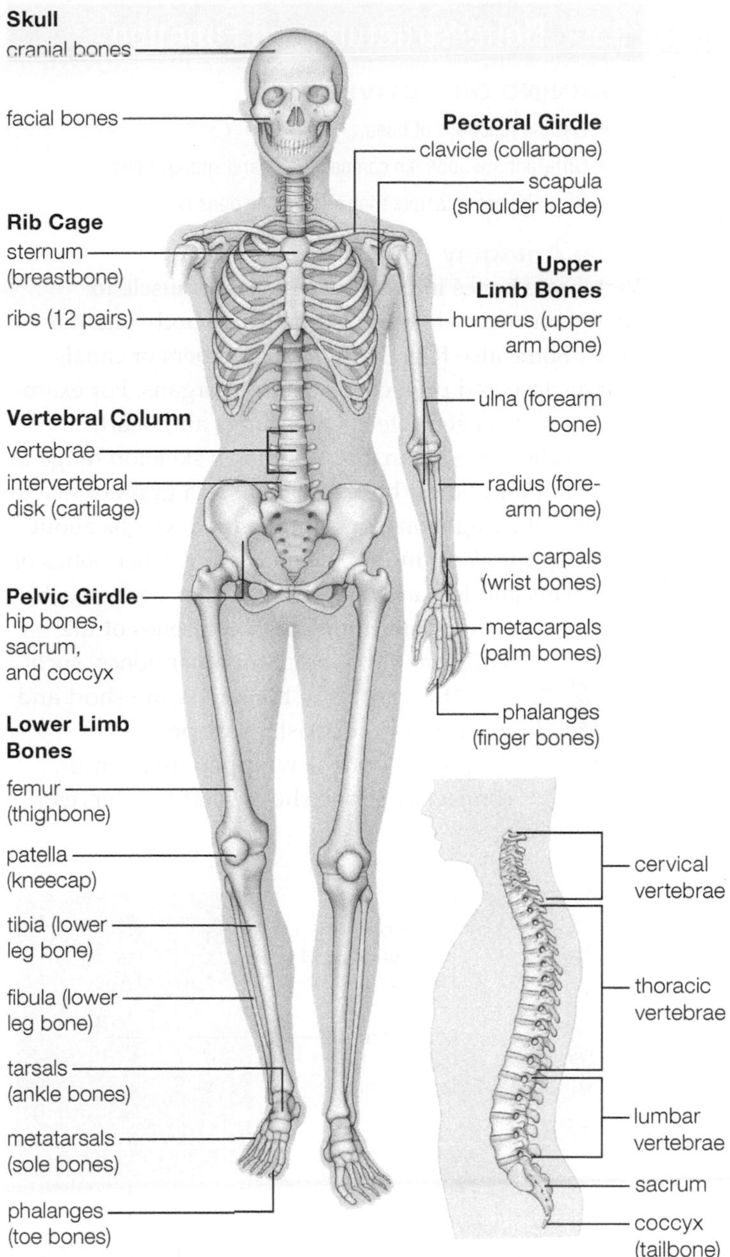
Skull
cranial bones
facial bones
Pectoral Girdle
clavicle (collarbone)
scapula (shoulder blade)
Rib Cage
sternum (breastbone)
ribs (12 pairs)
Upper Limb Bones
humerus (upper arm bone)
ulna (forearm bone)
Vertebral Column
vertebrae
intervertebral disk (cartilage)
radius (forearm bone)
carpals (wrist bones)
Pelvic Girdle
hip bones, sacrum, and coccyx
metacarpals (palm bones)
phalanges (finger bones)
Lower Limb Bones
femur (thighbone)
patella (kneecap)
tibia (lower leg bone)
fibula (lower leg bone)
tarsals (ankle bones)
metatarsals (sole bones)
phalanges (toe bones)
cervical vertebrae
thoracic vertebrae
lumbar vertebrae
sacrum
coccyx (tailbone)

FIGURE 35.8 **Human skeleton.** It consists of bone (tan) and cartilage (light blue). Lower right, side view of the vertebral column showing its S-shaped curve.

appendicular skeleton Of vertebrates, bones of the limbs or fins and the bones that connect them to the axial skeleton.

axial skeleton Of vertebrates, bones of the main body axis; the skull and backbone.

endoskeleton Internal skeleton consisting of hard parts.

exoskeleton Of some invertebrates, hard external parts that muscles attach to and move.

hydrostatic skeleton Of soft-bodied invertebrates, a fluid-filled chamber that muscles exert force against, redistributing the fluid.

intervertebral disk Cartilage disk between two vertebrae.

vertebrae (VER-tah-bray) Singular, vertebra. Bones of the backbone, or vertebral column.

vertebral column (ver-TEE-bral) Backbone.

35.4 Bone Structure and Function

LEARNING OBJECTIVES

- List the functions of bone.
- Differentiate between compact bone and spongy bone.
- Describe some factors that affect bone density.

Bone Anatomy

Vertebrate bones interact with skeletal muscle to change or maintain the position of the body and its parts. Some also form hardened chambers or canals that enclose and protect soft internal organs. For example, your rib cage protects your lungs and heart.

The 206 bones of an adult human's skeleton range in size from middle ear bones as small as a grain of rice to the massive thighbone, or femur, which weighs about a kilogram (2 pounds). The femur and other bones of arms and legs are long bones. Other bones, such as the ribs, sternum, and most bones of the skull, are flat bones. Still other bones, such as the carpals in the wrists, are short and roughly squarish in shape.

Each bone is wrapped in a dense connective tissue sheath that has nerves and blood vessels running through it. Bone tissue consists of bone cells in an extracellular matrix (Section 4.9). The matrix consists mainly of the protein collagen, and is hardened by calcium and phosphorus ions. Bones serve as reservoirs for these mineral ions. Storing these ions in bone and withdrawing them as needed helps the body maintain the ion concentrations in body fluids.

There are three types of bone cells. Their names all begin with *osteo-*, which is Greek for bone. **Osteoblasts** are bone-building cells. They secrete the components of the matrix. An adult bone has osteoblasts at its surface and in its internal cavities. **Osteocytes** are former osteoblasts that have become surrounded by the matrix they secreted. Osteocytes are the most abundant cells in adult bones. **Osteoclasts** are cells that break down the matrix by secreting acids and enzymes. They play a role in bone remodeling and repair.

A long bone such as a femur includes two types of bone tissue, compact bone and spongy bone (**FIGURE 35.9**). Compact bone forms the outer layer and shaft of the femur. It consists of many functional units called osteons, each having concentric rings of bone tissue, with bone cells in spaces between the rings. Nerves and blood vessels run through the osteon's central canal. Spongy bone lines the shaft and fills the knobby ends of long bones. It is strong yet lightweight; many open spaces riddle its hardened matrix.

The cavities inside a bone contain bone marrow. **Red marrow** is the major site of blood cell formation. In adults, red marrow fills the spaces in spongy bone. In children, it also fills that shaft of long bones. In adults, the shaft of long bones is filled by **yellow marrow**, which consists mainly of fat.

A The femur contains both compact and spongy bone.

nutrient canal

location of yellow marrow

compact bone

spongy bone (contains red marrow)

living bone cell between rings of secreted matrix

central canal (contains blood vessels, nerve)

one osteon

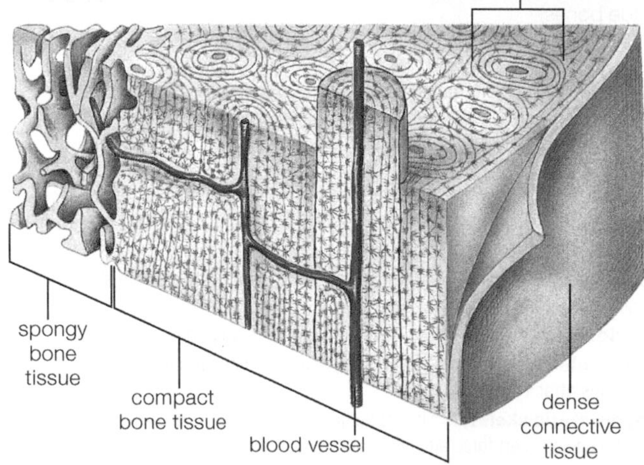
spongy bone tissue
compact bone tissue
blood vessel
dense connective tissue

B Cross section through a femur showing osteons. These cylindrical structures consist of concentric layers of compact bone.

FIGURE 35.9 Structure of a human adult femur, or thighbone.

Bone Development, Remodeling, and Repair

A cartilage skeleton forms in all vertebrate embryos. In sharks and other cartilaginous fishes, the cartilage skeleton persists into adulthood. In other vertebrates, embryonic cartilage serves as a model for a mostly bony skeleton. Before birth, mineralization of the cartilage model transforms most of it to bone.

Many bones continue to grow in size until early adulthood. Even in adults, bone remains a dynamic tissue that the body continually remodels. Microscopic fractures caused by normal body movements are repaired. In response to hormonal signals, osteoclasts dissolve portions of the matrix, releasing mineral ions into the blood. Osteoblasts secrete new matrix to replace that broken down by osteoclasts.

Bones and teeth contain most of the body's calcium. Parathyroid hormone, the main regulator of blood calcium, raises the concentration of calcium ions in the

Data Analysis Activities

Building Better Bones Tiffany, shown in **FIGURE 35.10**, was born with multiple fractures in her arms and legs. By age six, she had undergone surgery to correct more than 200 bone fractures. Her fragile, easily broken bones are symptoms of osteogenesis imperfecta (OI), a genetic disorder caused by a mutation in a gene for collagen. As bones develop, collagen forms a scaffold for deposition of mineralized bone tissue. The scaffold forms improperly in children with OI. **FIGURE 35.10** also shows the results of a test of a new drug. Treated children, all less than two years old, were compared to similarly affected children of the same age who were not treated with the drug.

1. An increase in vertebral area during the 12-month period of the study indicates bone growth. How many of the treated children showed such an increase?
2. How many of the untreated children showed an increase in vertebral area?
3. How did the rate of fractures in the two groups compare?
4. Do these results shown support the hypothesis that this drug, which slows bone breakdown, can increase bone growth and reduce fractures in young children with OI?

Treated child	Vertebral area in cm^2 (Initial)	Vertebral area in cm^2 (Final)	Fractures per year
1	14.7	16.7	1
2	15.5	16.9	1
3	6.7	16.5	6
4	7.3	11.8	0
5	13.6	14.6	6
6	9.3	15.6	1
7	15.3	15.9	0
8	9.9	13.0	4
9	10.5	13.4	4
Mean	11.4	14.9	2.6

Control child	Vertebral area in cm^2 (Initial)	Vertebral area in cm^2 (Final)	Fractures per year
1	18.2	13.7	4
2	16.5	12.9	7
3	16.4	11.3	8
4	13.5	7.7	5
5	16.2	16.1	8
6	18.9	17.0	6
Mean	16.6	13.1	6.3

FIGURE 35.10 A clinical trial of a drug treatment for osteogenesis imperfecta (OI). OI affects the child shown at right. Nine children with OI received the drug. Six others were untreated controls. Surface area of certain vertebrae was measured before and after treatment. Fractures occurring during the 12 months of the trial were also recorded.

blood by encouraging the breakdown of calcium-rich bone matrix. Other hormones also affect bone turnover. Growth hormone and sex hormones encourage bone deposition, whereas cortisol discourages it.

Until people are in their mid-twenties, matrix production by osteoblasts outpaces matrix breakdown by osteoclasts, so bone mass increases. Bones become denser and stronger. Later in life, bone mass declines as osteoblasts become less active.

Osteoporosis is a disorder in which bone loss exceeds bone formation. As a result, bones become more porous and more easily broken (**FIGURE 35.11**). Osteoporosis most often occurs in postmenopausal women; they produce little of the sex hormones that encourage bone deposition. However, it can occur in younger women and in men. To reduce your risk of osteoporosis, ensure that your diet provides adequate levels of calcium and of vitamin D, which facilitates calcium absorption from the gut. Avoid smoking and excessive alcohol intake (both slow bone deposition), and exercise regularly to encourage bone renewal.

Even healthy bones can break. When they do, white blood cells move in to clean up any debris. Additional tiny blood vessels grow into the damaged area, then bone repair begins. As with bone formation, repair starts with cartilage deposition. The cartilage fills the gap created by the break, then osteoblasts move in and replace the cartilage with new bone.

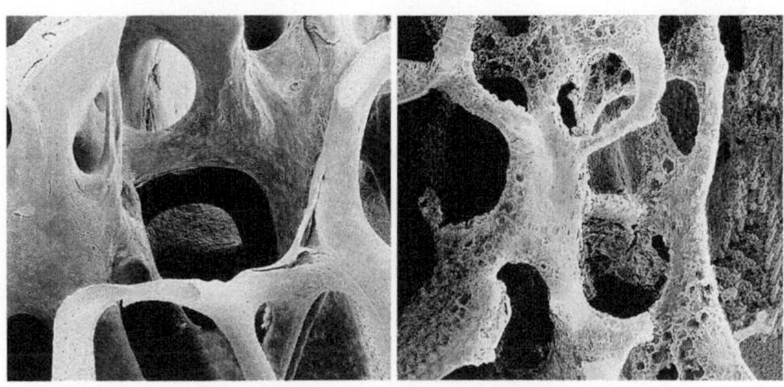

FIGURE 35.11 Bone health. Normal bone (left) and bone affected by osteoporosis (right). Osteoporosis means "porous bones."

osteoblast Bone-forming cell; it secretes a bone's matrix.
osteoclast Bone-digesting cell; it breaks down bone matrix.
osteocyte A mature bone cell; a former osteoblast.
red marrow Bone marrow that makes blood cells.
yellow marrow Bone marrow that is mostly fat; fills most long bones.

TAKE-HOME MESSAGE 35.4

- ✔ A bone is enclosed by a sheath of connective tissues and has an inner cavity containing marrow. Red marrow produces blood cells.
- ✔ Bone tissue consist of bone cells in a secreted extracellular matrix. The matrix contains the protein collagen and ions of calcium and phosphorus.
- ✔ A bone is continually remodeled by a hormone-regulated process.

35.5 Joint Structure and Function

LEARNING OBJECTIVE

- Using appropriate examples, describe the three types of joints.

A **joint** is an area of contact or near contact between bones. There are three types (FIGURE 35.12). At fibrous joints, dense, fibrous connective tissue holds bones firmly in place. Fibrous joints connect bones of the skull and hold teeth in their sockets in the jaw. At cartilaginous joints, bones are connected by cartilage. This flexible connection allows just a bit of movement. Cartilaginous joints connect vertebrae to one another and connect some ribs to the sternum.

Most joints, including the knees, hips, shoulders, wrists, and ankles, are synovial joints. At these joints, the cartilage-covered ends of bones are separated by a small space. Cords of dense, regular connective tissue called **ligaments** hold bones of a synovial joint in place and some form a capsule around the joint. The capsule's lining secretes lubricating synovial fluid.

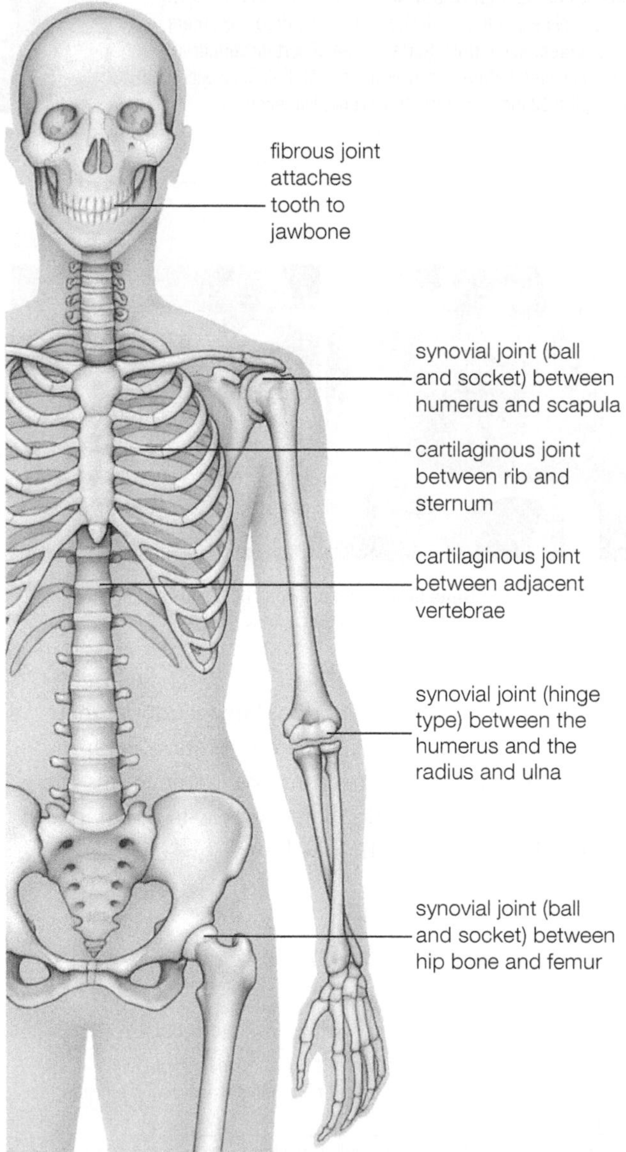
fibrous joint attaches tooth to jawbone
synovial joint (ball and socket) between humerus and scapula
cartilaginous joint between rib and sternum
cartilaginous joint between adjacent vertebrae
synovial joint (hinge type) between the humerus and the radius and ulna
synovial joint (ball and socket) between hip bone and femur

FIGURE 35.12 **Examples of the three types of joints.**

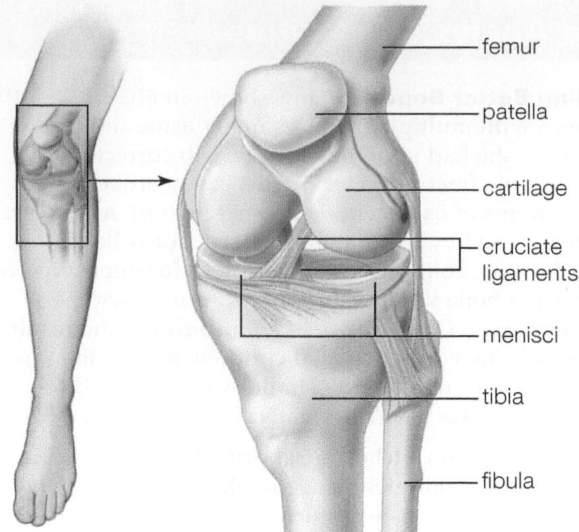
femur
patella
cartilage
cruciate ligaments
menisci
tibia
fibula

FIGURE 35.13 **The knee, a hinge-type synovial joint.** It is stabilized by ligaments and menisci (wedges of cartilage).

Synovial joints allow a variety of movements. Ball-and-socket joints at the shoulders and hips provide a wide range of rotational motion. At other synovial joints, including those in the wrist and ankle, bones glide past one another. Joints at the elbows and knees function like a hinged door; they allow the bones to move back and forth in one plane.

FIGURE 35.13 shows the knee joint. Cruciate ligaments cross one another in the center of the joint. (*Cruciate* means cross.) These ligaments stabilize the knee. If they are torn completely, bones may shift so the knee gives out when a person tries to stand. In addition to ligaments, the knee is stabilized by C-shaped wedges of cartilage called menisci (singular, meniscus). Each knee has two menisci.

Arthritis is the inflammation of a joint. The most common type of arthritis is osteoarthritis. It usually appears in older adults, whose cartilage has thinned at a frequently jarred joint or joints. Rheumatoid arthritis is an autoimmune disorder in which the immune system mistakenly attacks the fluid-secreting lining of synovial joints throughout the body.

TAKE-HOME MESSAGE 35.5

✔ Joints are areas where bones meet. In the most common type, synovial joints, the bones are separated by a small fluid-filled space and are held together by ligaments of fibrous connective tissue.

✔ Arthritis is chronic inflammation of a joint.

35.6 Skeletal Muscle Function

LEARNING OBJECTIVES

- Describe how muscles attach to and move bones.
- Using a suitable example, explain how muscles work in opposition.
- Define a sphincter and provide an example.

Vertebrate skeletal muscles are sometimes referred to as voluntary muscles because they function mainly in intentional movement. However, skeletal muscles also participate in reflexes such as the stretch reflex (Section 32.9). In addition, skeletal muscle plays a role in thermoregulation because muscle activity releases heat.

A sheath of dense connective tissue encloses each vertebrate skeletal muscle and extends beyond it to form a cordlike or straplike **tendon**. Most often, tendons attach each end of a muscle to different bone.

Consider the biceps brachii muscle in the upper arm (**FIGURE 35.14A**). *Biceps* means "two-headed" in Latin and refers to the fact that the upper portion of the biceps attaches to the scapula (shoulder blade) by way of two tendons. At the opposite end of the muscle, a single tendon attaches the biceps to the radius, a bone in the forearm.

Muscles can pull on bones, but they cannot push them. Thus, to achieve the greatest range of motion, muscles work in opposition: Motion generated by contraction of one muscle is reversed by contraction of another. The triceps in the upper arm opposes the biceps. When the biceps contracts, the triceps relaxes and the forearm is pulled toward the shoulder (**FIGURE 35.14B**). Contraction of the triceps coupled with relaxation of the biceps reverses this movement, extending the arm (**FIGURE 35.14C**).

You can feel the biceps contract if you extend one arm out, palm up, then place your other hand over the muscle and slowly bend your arm at the elbow. Although the biceps shortens only about a centimeter when it contracts, the forearm moves through a much greater distance. The elbow and many other joints function like a lever, a mechanism in which a rigid structure pivots about a fixed point. Use of a lever allows a small force, such as that exerted by a contracting biceps, to overcome a larger one, such as the gravitational force acting on the forearm.

Most skeletal muscles pull on bones, but a few tug on other tissues. Some skeletal muscles pull facial skin to bring about changes in expression. Others attach to and move the eyeballs. Skeletal muscle also composes some sphincters. A **sphincter** is a ring of muscle in a tubular organ or at a body opening. The sphincter of skeletal muscle that encircles the urethra (the tube through which urine exits the body) allows voluntary control of urination. Another at the anus allows control over defecation.

Some animals have boneless muscular organs capable of making complex movements. The mammalian tongue is one example. An elephant's trunk is another.

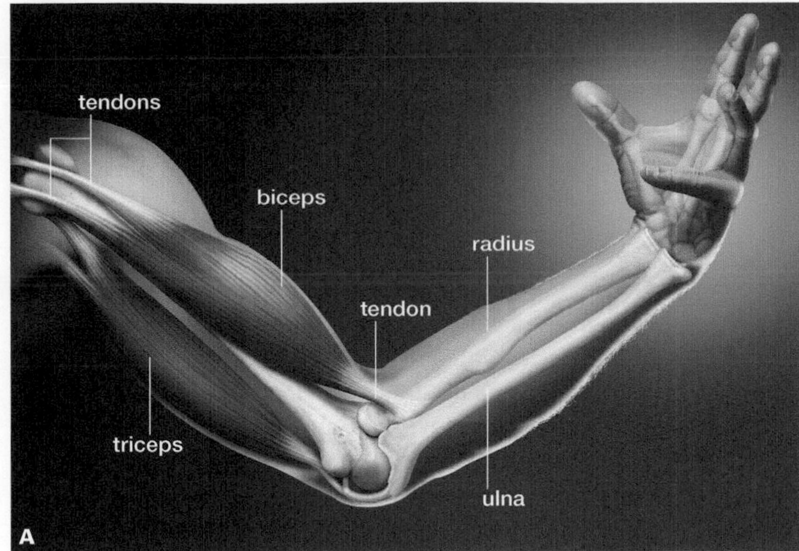
tendons
biceps
radius
tendon
triceps
ulna
A

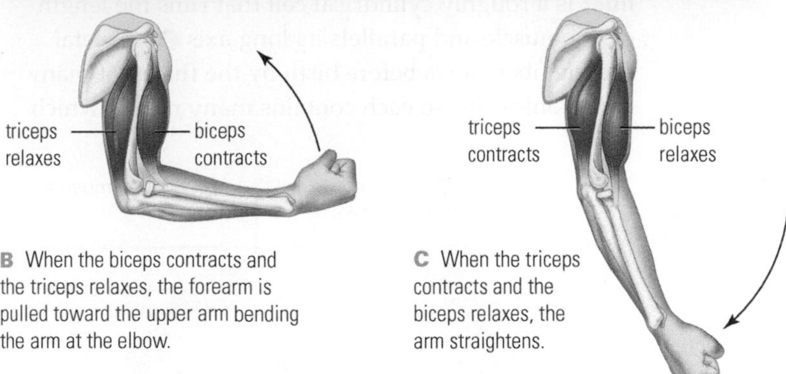
triceps relaxes
biceps contracts
triceps contracts
biceps relaxes

B When the biceps contracts and the triceps relaxes, the forearm is pulled toward the upper arm bending the arm at the elbow.

C When the triceps contracts and the biceps relaxes, the arm straightens.

FIGURE 35.14 Opposing muscles of the upper arm.
FIGURE IT OUT What bone does the biceps pull on? Answer: The radius

TAKE-HOME MESSAGE 35.6

✔ Most skeletal muscles pull on and move bones. Others pull on skin to alter facial expressions.

✔ Rings of skeletal muscle form sphincters that are under voluntary control.

joint Region where bones meet.
ligament Strap of dense connective tissue that holds bones together at a synovial joint.
sphincter Ring of muscle that controls passage of material through a tubular organ or a body opening.
tendon Strap of dense connective tissue that connects a skeletal muscle to bone.

35.7 How Muscle Contracts

LEARNING OBJECTIVES

- Describe the structure of a skeletal muscle.
- Explain how interactions among filaments cause a sarcomere to shorten.
- Draw a sarcomere in its relaxed and its contracted states.

Structure of Skeletal Muscle

Properties of skeletal muscle arise from the structure and arrangement of its component cells, which are known as fibers (FIGURE 35.15). As previously noted, a muscle is enclosed within a sheath of connective tissue ❶. This tissue extends beyond the muscle as a tendon. Additional connective tissue encloses each bundle of muscle fibers within the muscle ❷. A **skeletal muscle fiber** is a roughly cylindrical cell that runs the length of the muscle and parallels its long axis ❸. Skeletal muscle fibers form before birth by the fusion of many embryonic cells, so each contains many nuclei, which are positioned at the outer edges. A muscle fiber also contains multiple mitochondria that supply the ATP necessary for muscle contraction.

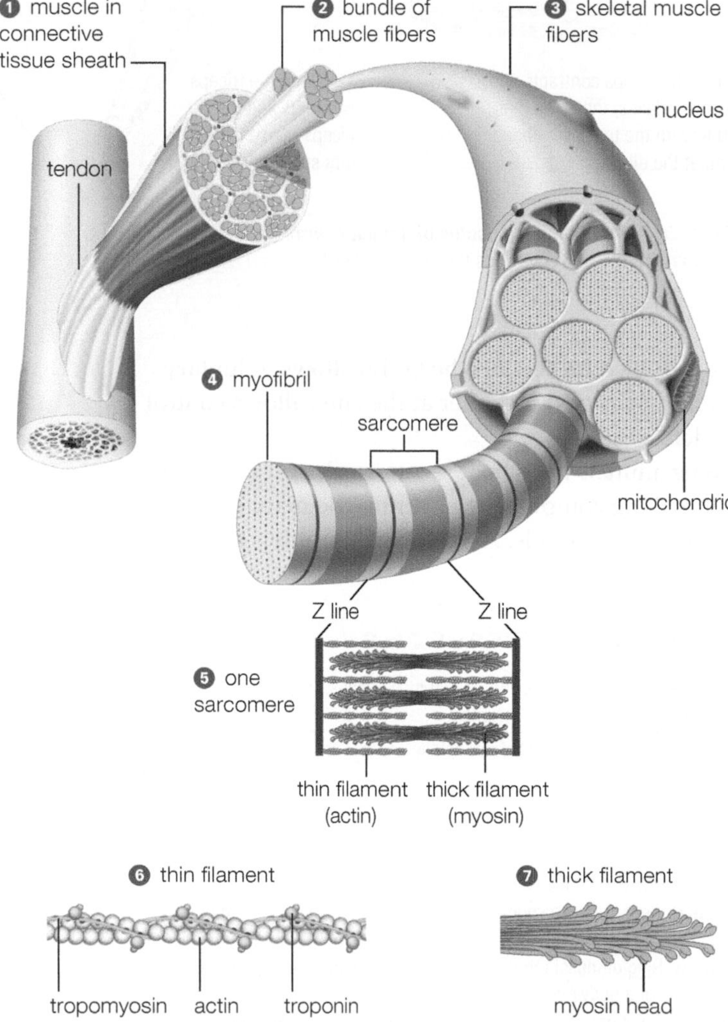
❶ muscle in connective tissue sheath
❷ bundle of muscle fibers
❸ skeletal muscle fibers
nucleus
tendon
❹ myofibril
sarcomere
mitochondrion
Z line
Z line
❺ one sarcomere
thin filament (actin)
thick filament (myosin)
❻ thin filament
tropomyosin
actin
troponin
❼ thick filament
myosin head

FIGURE 35.15 **Structure of a skeletal muscle.**

Thousands of threadlike structures called **myofibrils** fill the bulk of a muscle fiber's interior ❹. Each myofibril consists of many identical contractile units, called **sarcomeres**, attached end to end. Each end of a sarcomere is delineated by a Z line, a mesh of cytoskeletal elements that attach the sarcomere to its neighbors. The Z stands for *zwischen*, the German word for "between," and refers to the fact that a sarcomere is the region between two Z lines ❺.

Between a sarcomere's Z lines and perpendicular to them is an array of alternating thin and thick filaments. Thin filaments attach to the Z line and extend inward toward the center of the sarcomere. The main component of a thin filament is a string of beadlike actin molecules ❻. **Actin** is a globular protein, meaning it has a spherical form. Thin filaments also include two additional proteins (troponin and tropomyosin) that we will ignore for the moment. We discuss their function in Section 35.8.

Thick filaments reside at the center of the sarcomere, where they are flanked by the free ends of the thin filaments. Thick filaments consist of **myosin**, a motor protein that has a clublike head and a long tail ❼. The myosin head has enzymatic activity. The head can bind ATP and break it into ADP and phosphate. The head also has a site that can bind the actin of thin filaments.

Muscle fibers, myofibrils, thin filaments, and thick filaments all run parallel with a muscle's long axis. As a result, all sarcomeres in all fibers of a skeletal muscle pull in the same direction—they work together. Skeletal muscle and cardiac muscle appear striated because Z lines and other sarcomere components in all their fibers are aligned. Smooth muscle fibers have sarcomeres, but because the sarcomeres are not aligned, smooth muscle does not have a striped appearance.

The Sliding-Filament Model

The **sliding-filament model** explains how interactions between thick and thin filaments bring about muscle contraction. Neither actin nor myosin filaments change length, and the myosin filaments do not change position. Instead, myosin heads bind to actin filaments and slide them toward the center of a sarcomere. As the actin filaments are pulled inward, the ends of the sarcomere are drawn closer together, and the sarcomere shortens (FIGURE 35.16A).

FIGURE 35.16B provides a step-by-step look at muscle contraction, starting with the sarcomere in a resting muscle fiber. In a relaxed sarcomere, sites where myosin could bind to actin are blocked. The myosin heads

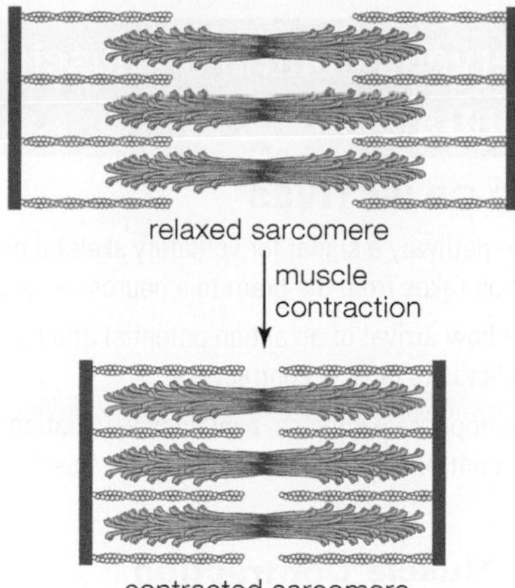
relaxed sarcomere
muscle contraction
contracted sarcomere

A (Above) During muscle contraction, thin filaments slide inward past thick filaments, reducing the width of the sarcomere.

B Right, molecular mechanism of contraction. For clarity, we show only two myosin heads. A head binds repeatedly to an actin filament and slides it toward the center of the sarcomere.

FIGURE 35.16 The sliding-filament model.

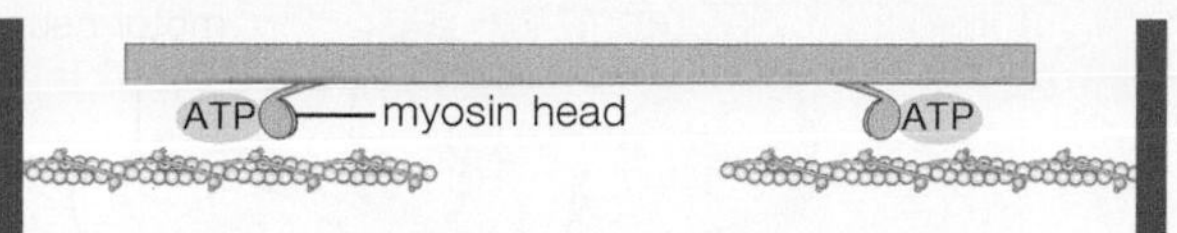
ATP
myosin head
ATP

❶ In a relaxed sarcomere, myosin-binding sites on actin are blocked. The myosin heads have bound ATP, and are in their low-energy conformation.

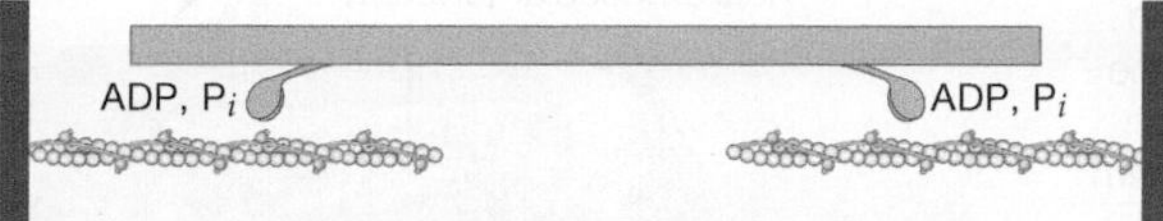
ADP, P_i
ADP, P_i

❷ The myosin heads remove a phosphate group from the ATP. Absorbing energy released by breaking the phosphate bond boosts the myosin heads to a high-energy conformation. ADP and phosphate remain bound to each head.

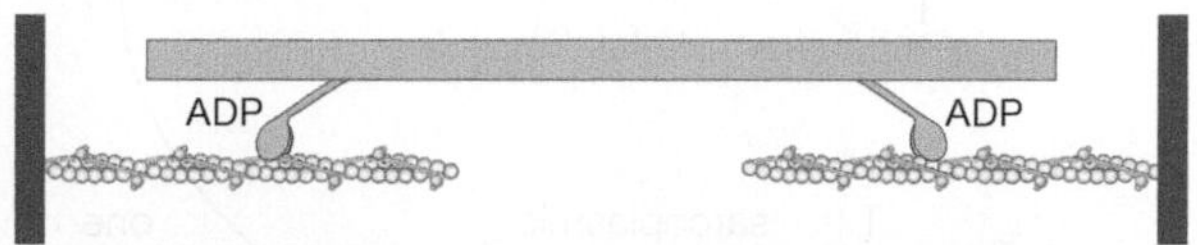
ADP
ADP

❸ A signal from the nervous system opens up the myosin-binding sites on actin. Each myosin head attaches to actin and releases its bound phosphate group, thus forming a cross-bridge between thick and thin filaments.

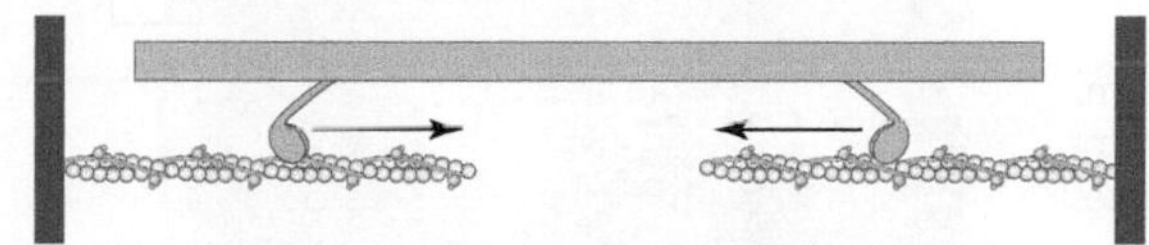

❹ The release of the phosphate group triggers a power stroke, in which the myosin heads snap inward and release ADP. As the heads contract, they pull the attached thin filaments inward.

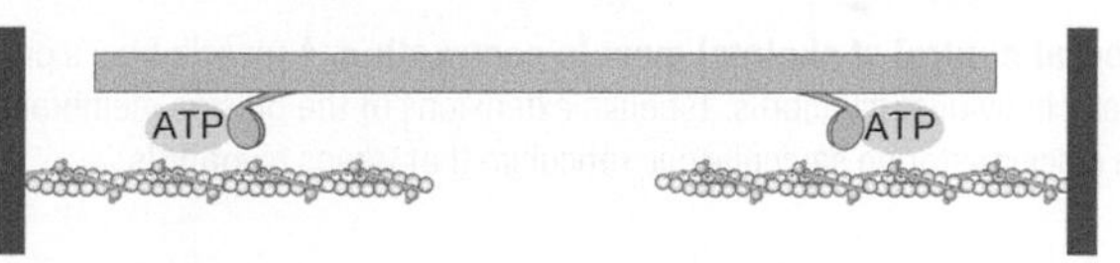
ATP
ATP

❺ Another ATP binds to each myosin head, causing it to release actin and return to its low-energy conformation.

of thick filaments have bound ATP molecules and are in a low-energy conformation ❶. Removing a phosphate group (P_i) from a bound ATP energizes a myosin head in a manner analogous to stretching a spring ❷. The myosin head remains in this high-energy conformation, with bound ADP and phosphate, until a signal from the nervous system excites the muscle (a process we consider in the Section 35.8). When this signal arrives, the myosin head releases its bound phosphate group and attaches to actin. This attachment constitutes a cross-bridge between the thin and thick filaments ❸. Attachment is followed by the power stroke ❹. Like a stretched spring returning to its original shape, the myosin head snaps back toward the sarcomere center. As the myosin head moves inward, it pulls the attached thin filament along with it.

During the power stroke, a myosin head releases bound ADP. Afterward, the head can bind to a new molecule of ATP and return to its low-energy conformation ❺. In the process, the myosin head releases its grip on actin, and the cross-bridge no longer exists. Loss of one cross-bridge does not allow a thin filament to slip backward because other cross-bridges hold it in place. During a contraction, many myosin heads repeatedly bind, move, and release an adjacent thin filament.

actin Spherical protein that plays a role in cell movements; the main component of thin filaments in a sarcomere.
myofibrils (my-oh-FIE-bril) Of a muscle fiber, threadlike protein structures consisting of contractile units (sarcomeres) arranged end to end.
myosin (MY-oh-sin) ATP-dependent motor protein; makes up the thick filaments in a sarcomere.
sarcomere Contractile unit of muscle.
skeletal muscle fiber Multinucleated contractile cell that contains mainly myofibrils and runs the length of a skeletal muscle.
sliding-filament model Explanation of how interactions among actin filaments and myosin filaments bring about muscle contraction.

TAKE-HOME MESSAGE 35.7

✔ Sarcomeres of skeletal muscle line up end to end in myofibrils that parallel a muscle fiber. These fibers, in turn, run parallel with the whole muscle. The parallel orientation of skeletal muscle components focuses a muscle's contractile force in a particular direction.

✔ ATP-driven interactions between thick (myosin) filaments and thin (actin) filaments shorten sarcomeres of a muscle cell.

✔ During muscle contraction, sarcomeres shorten because myosin filaments pull neighboring actin filaments inward toward the center of the sarcomere.

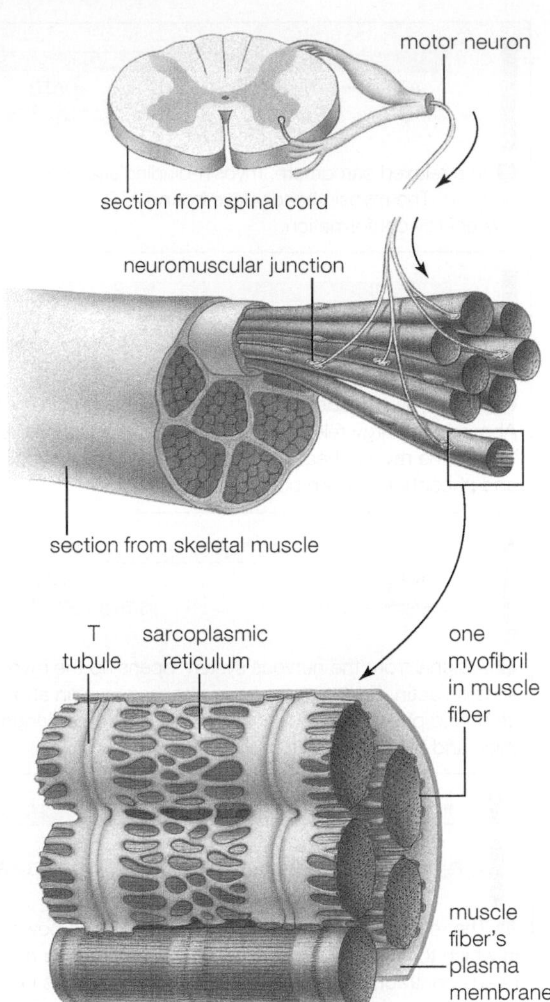

❶ A signal travels along the axon of a motor neuron, from the spinal cord to a skeletal muscle.
motor neuron
section from spinal cord
❷ At a neuromuscular junction, arrival of an action potential at the motor neuron's axon terminals causes them to release ACh. The ACh binds to receptors in a muscle fiber's plasma membrane, causing an action potential in that muscle fiber.
neuromuscular junction
section from skeletal muscle
T tubule
sarcoplasmic reticulum
one myofibril in muscle fiber
muscle fiber's plasma membrane

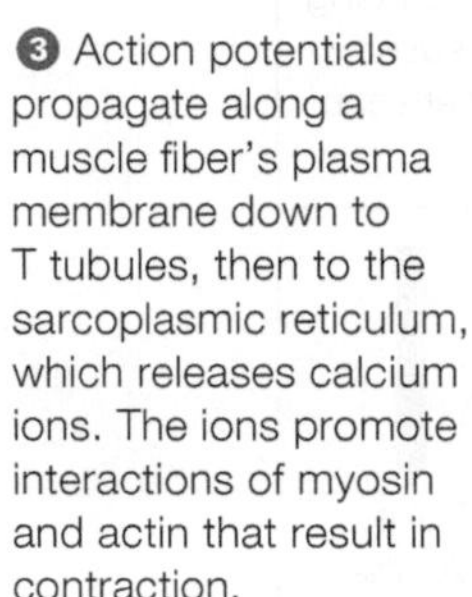

❸ Action potentials propagate along a muscle fiber's plasma membrane down to T tubules, then to the sarcoplasmic reticulum, which releases calcium ions. The ions promote interactions of myosin and actin that result in contraction.

FIGURE 35.17 Neural control of skeletal muscle contraction. A muscle fiber's plasma membrane encloses many individual myofibrils. Tubelike extensions of the plasma membrane (T tubules) connect to the calcium-storing sarcoplasmic reticulum that wraps myofibrils.

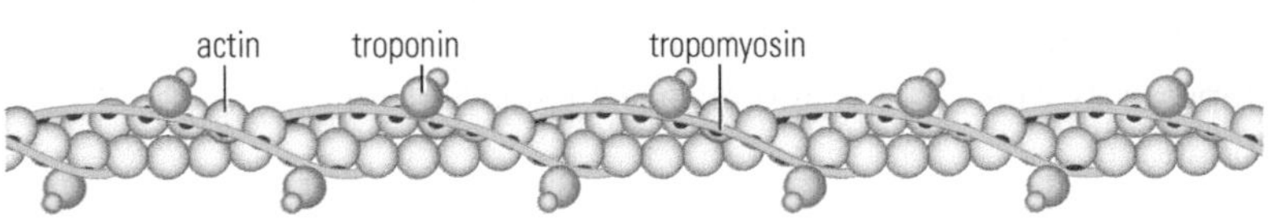

actin
troponin
tropomyosin

A Resting muscle. Calcium ion (Ca^{+2}) concentration is low and tropomyosin covers the myosin-binding sites on actin.

Ca^{+2}
exposed myosin-binding sites on actin

B Excited muscle. Ca^{+2} released from the sarcoplasmic reticulum binds to troponin. As a result troponin shifts and moves tropomyosin, thus exposing myosin-binding sites on actin.

FIGURE 35.18 Role of calcium in muscle contraction. The configuration of the thin filament proteins varies with the concentration of Ca^{+2} ions.

35.8 Nervous Control of Muscle Contraction

LEARNING OBJECTIVES

- Trace the pathway a signal for voluntary skeletal muscle contraction takes from the brain to a neuromuscular junction.
- Describe how arrival of an action potential at a neuromuscular junction leads to muscle contraction.
- Using appropriate examples, explain how variation in the number of fibers controlled by a motor unit affects muscle tension.

Initiating Muscle Contraction

Most commands for voluntary movement originate in the portion of the cerebral cortex known as the primary motor cortex (Section 32.11). Signals from this brain region travel to and excite a motor neuron, which has its cell body in the spinal cord (FIGURE 35.17 ❶). The axon of a motor neuron extends to a skeletal muscle and synapses with it at a neuromuscular junction ❷. Arrival of an action potential at this junction triggers the release of the neurotransmitter acetylcholine (ACh).

Like a neuron, a muscle fiber is excitable, meaning it is capable of undergoing an action potential. ACh released at a neuromuscular junction binds to receptors in the muscle fiber's plasma membrane. The binding triggers an action potential that travels along the plasma membrane, then down membranous extensions called T tubules. T tubules convey the action potential to the **sarcoplasmic reticulum**, a specialized smooth endoplasmic reticulum that wraps around myofibrils and stores calcium ions ❸. Arrival of an action potential opens voltage-gated channels in the sarcoplasmic reticulum, allowing calcium ions to follow their gradient and flood out of the organelle. The flow of ions raises the calcium concentration around the myofibrils.

The increase in calcium ion concentration allows actin and myosin filaments to interact by altering the configuration of two proteins: tropomyosin and troponin. These proteins are, along with actin, components of thin filaments. Tropomyosin is a fibrous protein; it forms long polymers that wrap around the actin of a thin filament. Globular molecules of troponin bind to each tropomyosin polymer at intervals along its length.

When a muscle is at rest, tropomyosin polymers block the myosin-binding sites on actin, so actin and myosin cannot interact (FIGURE 35.18A). With muscle excitation, calcium ions bind to troponin causing this protein to change its shape and pull the attached tropomyosin away from actin's myosin-binding sites (FIGURE 35.18B). With these binding sites cleared, actin can bind myosin, and the sliding action described in Section 35.7 takes place. After contraction, active transport proteins pump the calcium ions back into the sarcoplasmic reticulum.

Motor Units and Muscle Tension

A motor neuron has many axon endings that synapse on different fibers in a muscle. One motor neuron and all of the muscle fibers it synapses with constitute a **motor unit**. Stimulate a motor neuron, and all the muscle fibers of its motor unit will contract simultaneously. The motor neuron cannot make only some of the fibers it controls contract.

The mechanical force generated by a contracting muscle—the **muscle tension**—depends on the number of muscle fibers contracting. Some functions require more muscle tension than others, so the number of muscle fibers controlled by a single motor neuron varies. In motor units that bring about small, fine movements, one motor neuron synapses with only five or so muscle fibers. Motor units controlling eye muscles are like this. By contrast, muscles that must frequently exert a large amount of force have many fibers per motor unit. For example, the biceps of the arm has about 700 muscle fibers per motor unit. Having many fibers all pulling the same way at once increases the force a motor unit can generate.

Disrupted Control of Skeletal Muscle

Disruption of motor neuron signaling impairs muscle function. Consider what happens when *Clostridium tetani* bacteria colonize a wound. Toxin released by the bacteria prevents the release of GABA, a neurotransmitter that normally inhibits motor neuron signaling. Without GABA present, nothing dampens signals calling for muscle contraction, and symptoms of the disorder known as tetanus appear. The fists and jaw clench, which is why tetanus is sometimes called lockjaw. Similar clenching of back muscles locks the vertebral column in an abnormal arch (**FIGURE 35.19**). Untreated tetanus can be fatal. Vaccines have essentially eliminated the disease in the United States.

Another *Clostridium* species, *C. botulinum*, makes botulinum toxin. This toxin prevents motor neurons from releasing ACh, so it inhibits muscle contraction. Controlled doses of botulinum toxin, sold under the name "Botox" are used in some medical procedures. For example, Botox is injected into specific facial muscles to inhibit contractions that result in wrinkles.

Polio is an infectious disease in which a virus infects and destroys motor neurons. Effects range from temporary paralysis, to permanent paralysis, to death. Polio survivors are at risk for postpolio syndrome, a disorder characterized by muscle fatigue and progressive muscle weakness. Thanks to vaccines, no new cases of polio have arisen in the United States since 1979, but sporadic outbreaks continue in developing countries.

Motor neurons are also destroyed in amyotrophic lateral sclerosis (ALS). ALS is sometimes called Lou Gehrig's disease because this famous baseball player died of ALS in the 1930s. Affected people usually die of respiratory failure within a few years of diagnosis.

motor unit One motor neuron and the muscle fibers it controls.
muscle tension Force exerted by a contracting muscle.
sarcoplasmic reticulum Specialized endoplasmic reticulum in muscle cells; stores and releases calcium ions.

FIGURE 35.19 Disrupted motor neuron signaling. This 1809 painting depicts a wounded soldier dying of tetanus.

TAKE-HOME MESSAGE 35.8

✔ A skeletal muscle contracts in response to a signal from a motor neuron. Release of ACh at a neuromuscular junction causes an action potential in the muscle cell.

✔ An action potential results in release of calcium ions, which affect proteins attached to actin. Resulting changes in the shape and location of these proteins open the myosin-binding sites on actin, allowing cross-bridge formation.

✔ Each motor neuron signals multiple muscle fibers and causes them all to contract at once.

35.9 Muscle Metabolism

LEARNING OBJECTIVES

- Describe the three pathways by which muscles can produce the ATP they need to contract.
- Compare the three different types of muscle fibers
- Explain how prolonged inactivity harms health.

Energy-Releasing Pathways

Muscle contraction requires ATP, but muscle fibers store only a limited amount of this molecule. The fibers have a larger store of creatine phosphate, a molecule that can transfer a phosphate to ADP and form ATP

CREDITS: (19) Painting by Sir Charles Bell, 1809, courtesy of Royal College of Surgeons, Edinburgh.

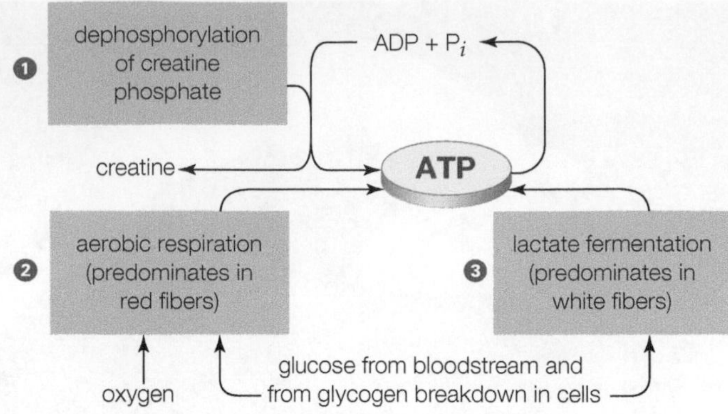
dephosphorylation of creatine phosphate
ADP + P_i
creatine
ATP
aerobic respiration (predominates in red fibers)
lactate fermentation (predominates in white fibers)
oxygen
glucose from bloodstream and from glycogen breakdown in cells

FIGURE 35.20 How muscles make ATP.

(**FIGURE 35.20** ❶). Phosphate transfers from creatine phosphate provide a quick source of energy. They can fuel muscle contraction until other pathways increase ATP output.

Some athletes take creatine supplements to increase the amount of creatine phosphate stored in muscle. Taking creatine supplements can enhance performance of tasks that require a quick burst of energy. However, such supplementation has no effect on endurance, and it may have harmful effects on the kidneys.

Aerobic respiration ❷ produces most of the ATP used by skeletal muscle during prolonged, moderate activity. Glucose derived from stored glycogen fuels five to ten minutes of activity, then the muscle fibers rely on glucose and fatty acids delivered by the blood. Fatty acids are the main fuel for activities that last more than half an hour.

Lactate fermentation is a muscle's third source of energy ❸. Some pyruvate is converted to lactate by the fermentation pathway even in resting muscle, but lactate fermentation steps up during exercise. This pathway produces less ATP than aerobic respiration, but has the advantage of operating even when the oxygen level in a muscle is low, as during strenuous exercise.

Types of Muscle Fibers

Muscle fibers can be classified as red fibers or white fibers depending on their primary mechanism of ATP production. In red fibers (also called oxidative fibers), aerobic respiration predominates. Red fibers have many mitochondria. An abundance of **myoglobin**, a protein that reversibly binds oxygen, gives them their red color. When the blood's oxygen level is high, oxygen diffuses into red muscle fibers and binds to myoglobin. When the blood oxygen level falls during periods of muscle activity, myoglobin releases oxygen for use in aerobic respiration. White muscle fibers have no myoglobin and few mitochondria. These fibers make ATP mainly by lactate fermentation.

Muscle fibers can also be subdivided into fast fibers or slow fibers depending on the speed with which their myosin converts ATP to ADP. All white fibers are fast fibers. They contract rapidly but do not produce sustained contractions. The muscles that move your eye consist mainly of white fibers.

Red fibers can be either fast or slow. Fast red fibers predominate in the human triceps muscle, which must often react quickly. Muscles involved in maintaining an upright posture, such as those in the back, consist mainly of slow red fibers.

For any given muscle, the relative proportions of fiber types vary among species and reflect the pattern of muscle usage. Limb muscles of cheetahs, which are renowned for their sprinting ability, contain a large proportion of white fibers. By contrast, the limb muscles of a loris (**FIGURE 35.21**) contain mostly slow red fibers. Lorises are tree-dwelling primates that avoid detection by creeping stealthily along branches. In the presence of a predator, a loris freezes in place. Its slow red fibers allow it to maintain its posture for long intervals without muscle fatigue.

Similarly, among human athletes, successful sprinters tend to have a higher-than-average percentage of fast, white fibers in their leg muscles whereas marathoners have a high percentage of slow, red fibers.

Effects of Exercise and Inactivity

Engaging in aerobic exercise—low intensity, but long duration—makes skeletal muscles more resistant to fatigue. It increases a muscle's blood supply by boosting growth of new capillaries, increases the number of mitochondria and amount of myoglobin in existing red muscle fibers, and encourages conversion of white fibers to red ones. Engaging in resistance exercise, such as weight lifting, encourages synthesis of additional actin and myosin filaments. The resulting increase in muscle mass allows for stronger contractions.

On the other hand, prolonged inactivity is hazardous to your health. There is increasing evidence that sitting for long periods results in unhealthy changes in metabolism. When you are seated, your leg muscles can relax completely. In their relaxed state, these muscles turn down their production of lipoprotein lipase (LPL), an enzyme that facilitates uptake of fatty acids and triglycerides from the blood. Decreased LPL activity raises the amount of lipids in the blood—thus increasing risk of cardiovascular disease and diabetes.

myoglobin (MY-oh-glow-bin) Muscle protein that reversibly binds oxygen.

Some studies suggest that the health risks associated with prolonged muscle inactivity persist even if a person also gets regular exercise. In other words, exercising for an hour each morning, although it improves your health in many respects, does not cancel out the negative metabolic effects of sitting in place for hours later in the day.

The best way to prevent health problems associated with a low LPL concentration is to avoid sitting for long intervals. When you must sit to carry out a task, get up every 20 minutes or so to walk around or otherwise exercise your legs. Any activity that requires your leg muscles to support your weight will lead to increased production of LPL.

TAKE-HOME MESSAGE 35.9

- ✔ Muscle contraction requires ATP. The ATP can be formed by a transfer of phosphate from creatine phosphate, aerobic respiration, or lactate fermentation.
- ✔ Aerobic respiration predominates in red fibers, which have oxygen-storing myoglobin. Lactate fermentation predominates in white fibers.
- ✔ Exercise increases blood flow to muscles, the number of mitochondria, production of actin and myosin, and the muscle's ability to take up lipids from the blood for use as an energy source.
- ✔ Sustained periods of muscle inactivity increase the risk of chronic disorders such as diabetes and cardiovascular disease.

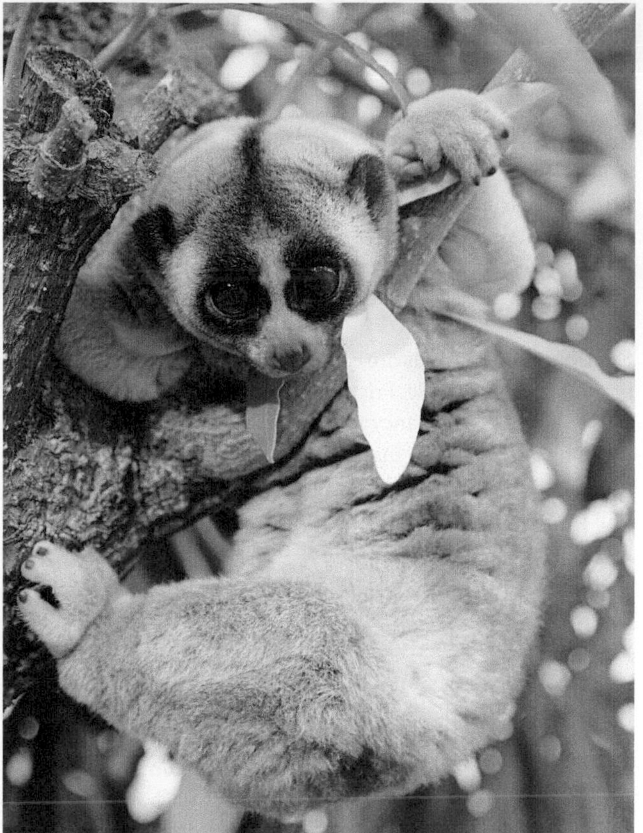

FIGURE 35.21 Loris, a slow-moving primate. Loris limbs have a large proportion of slow red fibers. Creeping cautiously along branches helps a loris escape the attention of predators.

35.1 Bulking Up (revisited)

To date, researchers have only identified one person who is totally deficient for myostatin, the protein that inhibits increases in muscle mass. This individual, a boy born in 1999, had highly muscular limbs even as an infant. He also has unusual strength. When he was only five years old, he could stand with his arms outstretched, holding a 3-kilogram (more than 5-pound) weight in each hand.

Knowing myostatin's inhibitory effect on human muscle growth, researchers are now trying to develop drugs that interfere with myostatin production or function. Their goal is to devise treatments that can prevent or reverse the effects of age-related muscle loss and muscular dystrophy.

Age-related muscle loss is a common problem. Muscle mass typically increases until age 40, then declines. Loss of muscle results in weakness and increases the risk of dangerous falls. Exercise can slow age-related muscle loss, but many people do not get adequate exercise.

Muscular dystrophies are a class of genetic disorders in which skeletal muscles progressively weaken. With Duchenne muscular dystrophy, symptoms begin to appear in childhood. A mutation of a gene on the X chromosome causes this disorder. This gene encodes dystrophin, a plasma membrane protein of muscle fibers. The altered dystrophin specified by the mutated allele allows foreign material to enter a muscle fiber, causing it to break down.

Affected boys usually begin to show signs of weakness by the time they are three years old and require a wheelchair when they are in their teens. Most die in their twenties of respiratory failure after the skeletal muscles involved in breathing become affected.

Tests of myostatin inhibitors in animals have shown these drugs increase the differentiation of muscle stem cells into muscle fibers. Recently begun clinical trials will determine whether the drugs can prevent or reverse muscle loss in muscular dystrophy and what, if any, negative side effects they produce.

STUDY GUIDE

Section 35.1 Muscle fibers do not divide, but they can enlarge by adding proteins. The regulatory protein myostatin discourages increases in muscle bulk, so mutations that alter myostatin's effect can increase muscle mass and muscle strength. Drugs that interfere with myostatin may be beneficial to people with muscle-wasting disorders.

Section 35.2 Animals are capable of **locomotion** during part or all of their life cycle. Friction and gravity oppose efforts to move. In water, buoyancy reduces the effects of gravity, but friction is greater than that on land. Streamlined bodies help swimming animals minimize drag in water. Many animals that creep along the ground reduce friction by having smooth body parts or an ability to secrete mucus. Air is less dense than water, so most land animals that walk, run, hop, or fly expend energy mainly to overcome gravity. The wing shape of flying animals helps lift them off the ground.

Section 35.3 Animals move when their muscles exert force against skeletal elements. In soft-bodied invertebrates, chambers full of fluid are a **hydrostatic skeleton**. Muscle contractions redistribute the fluid among the chambers. Arthropods have an **exoskeleton** of secreted hard parts at the body surface. An **endoskeleton** consists of hardened parts inside the body. Echinoderms and vertebrates have an endoskeleton.

All vertebrates have similar skeletal components. The vertebrate skull, vertebral column, and ribs constitute the **axial skeleton**. The **vertebral column** consists of **vertebrae** with **intervertebral disks** between them. Bony fins or limbs and the pectoral and pelvic girdles that attach them to the backbone constitute the **appendicular skeleton**.

Sections 35.4 Bones consist of living cells in a secreted matrix of collagen hardened with calcium and phosphorus. In addition to having a role in movement, bones store minerals and protect organs. The shaft of a long bone such as a femur consists of compact bone that contains **yellow marrow**. Lightweight spongy bone, as in the ends of long bones, contains **red marrow** that makes blood cells.

In a human embryo, bones develop from a cartilage model. Even in adults, bones are continually remodeled. **Osteoblasts** are cells that synthesize bone, whereas **osteoclasts** break bone down. **Osteocytes** are former osteoblasts enclosed in a matrix of their secretions.

Sections 35.5 A **joint** is an area of close contact between bones. Fibrous joints hold components tightly in place. Cartilaginous joints allow a small amount of movement. Synovial joints, the most common type of joint, allow the most movement. At a synovial joint, cartilage-covered ends of bones are separated by a tiny fluid-filled gap and held in place by **ligaments**.

Section 35.6 Skeletal muscles bring about voluntary movements of body parts, act in reflexes, function in breathing, and generate heat that warms the body. A sheath of connective tissue surrounds each skeletal muscle and extends beyond it as a **tendon**. Most often, the tendon connects the muscle to a bone. A muscle can only exert force in one direction; it can pull but not push. Some skeletal muscles work in pairs to oppose one another's actions. Skeletal muscles also function as **sphincters**.

Section 35.7 The internal organization of a skeletal muscle promotes a strong, directional contraction. Many **myofibrils** fill the interior of a **skeletal muscle fiber**. A myofibril consists of **sarcomeres**, units of muscle contraction, lined up along its length. Each sarcomere has a Z line at either end, with parallel arrays of thick and thin filaments between the Z lines.

The **sliding-filament model** describes how ATP-driven sliding of thin filaments past thick filaments shortens the sarcomere and brings about muscle contraction. Thin filaments consist mainly of the globular protein **actin**. Thick filaments are composed of the motor protein **myosin**. In an excited muscle, myosin binds to actin and uses energy released by the hydrolysis of ATP to move the bound thin filament toward the sarcomere center.

Section 35.8 A motor neuron and all the muscle fibers it controls constitute a **motor unit**. Signals from motor neurons result in action potentials in muscle fibers, which in turn cause the **sarcoplasmic reticulum** to release stored calcium ions. Flow of calcium into the cytoplasm causes two proteins associated with the thin filaments to shift in such a way that actin and myosin heads can interact and bring about a contraction that exerts **muscle tension**.

Diseases such as polio and toxins such as botulinum toxin (Botox) interfere with motor neuron function and cause skeletal muscle weakness and paralysis.

Section 35.9 Muscle fibers produce the ATP they require for contraction by way of three pathways: dephosphorylation of creatine phosphate, aerobic respiration, and lactate fermentation. Muscle fibers differ in which pathway predominates and how fast they hydrolyze the resulting ATP. Red fibers depend mainly on aerobic respiration. They have many mitochondria and contain **myoglobin**, a protein that can bind and release oxygen. Some react fast and others more slowly. White fibers react fast and produce most of their ATP by lactate fermentation.

Exercise can increase circulation to muscles, increase the number of myofibrils in muscle fibers, and enhance the ability of fibers to burn lipids as fuel. Long periods of uninterrupted sitting have a detrimental effect on the metabolic activity of fibers in leg muscles.

SELF-QUIZ

Answers in Appendix VII

1. An endoskeleton consists of _______ .
   a. a fluid in an internal space
   b. hardened plates at the surface of a body
   c. internal hard parts
   d. a fluid that surrounds the body

2. The main protein in bones is _______ .
   a. actin c. collagen
   b. myosin d. calcium

3. _______ joints are the most common joints in the body.
   a. Fibrous c. Smooth
   b. Cartilaginous d. Synovial

4. A ligament connects _______ .
   a. bones at a joint c. a muscle to a tendon
   b. a muscle to a bone d. a tendon to bone

5. Parathyroid hormone stimulates _______ .
   a. bone breakdown c. red blood cell formation
   b. bone deposition d. muscle contraction

6. The _______ attaches to the pelvic girdle.
   a. radius c. femur
   b. sternum d. tibia

7. The _______ is the basic unit of contraction.
   a. osteoblast c. myofibril
   b. sarcomere d. myosin filament

8. In sarcomeres, phosphate-group transfers from ATP activate _______ .
   a. actin c. troponin
   b. myosin d. Z bands

9. A sarcomere shortens when _______ .
   a. thick filaments shorten
   b. thin filaments shorten
   c. both thick and thin filaments shorten
   d. none of the above

10. White muscle fibers produce ATP mainly by _______ .
    a. aerobic respiration
    b. lactate fermentation
    c. phosphate transfer from creatine phosphate to ADP

11. Injection of *C. botulinum* toxin (Botox) _______ .
    a. causes irreversible muscle contraction
    b. stimulates formation of ATP from ADP
    c. prevents release of ACh by motor neurons
    d. blocks the actin binding site of myosin

12. A motor unit is _______ .
    a. a muscle and the bone it moves
    b. two muscles that work in opposition
    c. the amount a muscle shortens during contraction
    d. a motor neuron and the muscle fibers it controls

13. _______ from a motor neuron excites a muscle fiber.
    a. ACh c. GABA
    b. Calcium d. Phosphate

14. The sarcoplasmic reticulum stores and releases _______ .
    a. ACh c. ATP
    b. Calcium ions d. Phosphate ions

15. Match the words with their defining feature.

___ osteoblast	a. conveys excitatory signal
___ myofibril	b. in the hands
___ myoglobin	c. bone-forming cell
___ knee	d. main thin filament protein
___ actin	e. produces blood cells
___ red marrow	f. donates phosphate to ADP
___ metacarpal	g. binds and releases oxygen
___ T tubule	h. a synovial joint
___ creatine phosphate	i. composed of sarcomeres

CRITICAL THINKING

1. Continued strenuous activity can cause lactate to accumulate in muscles. After the activity stops, the lactate is converted into pyruvate and used as an energy source. Explain how pyruvate can be used to produce ATP.

2. After death, calcium pumps no longer function and the calcium ion concentration of muscle fiber cytoplasm increases. The result is rigor mortis—a state of postmortem muscle contraction. Explain why this contraction occurs and why it ends only when myosin heads begin to break down.

3. Bully whippets are homozygous for a deletion of two base pairs in the myostatin gene. This deletion changes an mRNA codon in the middle of the myostatin mRNA from UGU to UGA. Use your knowledge of the genetic code (Section 7.4) to determine the effect of this mutation on the structure of the resulting protein.

4. Athletes tend to have stronger bones than nonathletes, but different sports strengthen different bones. Volleyball, basketball, gymnastics, and soccer all thicken the femur, whereas swimming, skating, and bicycling have little effect on this bone. What does this tell you about the mechanism by which exercise strengthens bone?

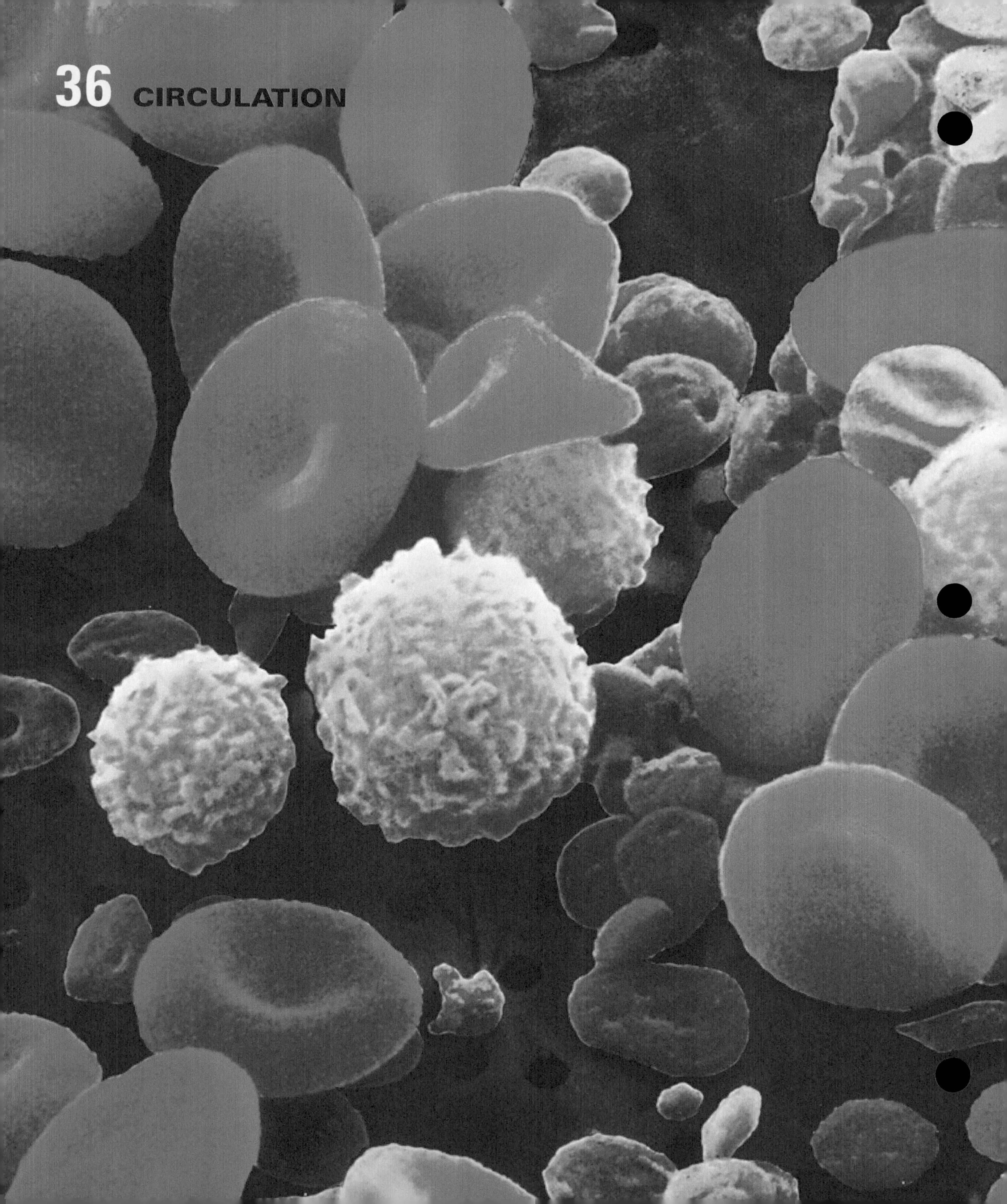

36 CIRCULATION

CORE CONCEPTS

Pathways of Transformation

Organisms exchange matter and energy with the environment in order to grow, maintain themselves, and reproduce.

All animals take up nutrients and oxygen from their environment, and all release carbon dioxide and other waste materials into their environment. In the simplest animals, materials move to or from a body surface to internal cells by means of diffusion alone. In most animals a circulatory system speeds the movement of materials within the body.

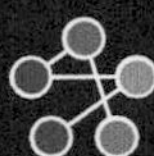

Systems

Complex properties arise from interactions among components of a biological system.

The cellular components of vertebrate blood arise in bone and are suspended in a liquid plasma rich in proteins. Circulation occurs when coordinated contractions of cardiac muscle cells force blood through a system of blood vessels. Adjustment to the width of some of these vessels alters the distribution of blood based on the body's metabolic requirements.

Evolution

Evolution underlies the unity and diversity of life.

Shared core processes and features provide evidence of shared ancestry. All vertebrates have a closed circulatory system that includes a single heart. In fishes, that heart has two chambers and pumps blood through a single circuit. Evolutionary modification of these chambers produced hearts that pump blood through two circuits.

Links to Earlier Concepts

This chapter expands on the discussion of circulatory systems in Section 24.7. You have already been introduced to blood (31.4) and cardiac muscle (31.5). You will draw on your knowledge of hemoglobin (3.2), diffusion and osmosis (5.8), endocytosis (5.10), autonomic nerves (32.7), sickle-cell anemia (9.6, 17.7), and malaria (21.5).

36.1 A Shocking Save

The heart is the body's most durable muscle. It begins to beat during the first month of human development and keeps beating for a lifetime. Each heartbeat is set in motion by an electrical signal generated by a natural pacemaker in the heart wall. In some people, this pacemaker malfunctions, causing what is called sudden cardiac arrest. Electrical signaling becomes disrupted, the heart stops beating, and blood flow halts. In the United States, sudden cardiac arrest occurs in more than 300,000 people per year. An inborn heart defect causes most cardiac arrests in people under age 36. In older people, heart disease usually causes the heart to stop functioning.

The chance of surviving sudden cardiac arrest rises by 50 percent when cardiopulmonary resuscitation (CPR) is started within four to six minutes of the arrest. With CPR, a person alternates mouth-to-mouth respiration that substitutes for breathing with chest compressions that keep the victim's blood moving. However, CPR cannot restart the heart. That requires a defibrillator, a device that delivers an electric shock to the chest and resets the natural pacemaker. You have probably seen this procedure depicted in hospital dramas.

Automated external defibrillators (AEDs) are increasingly available in public places. An AED is a device about the size of a laptop computer (**FIGURE 36.1**). It provides simple voice commands that describe how to attach electrodes to a person in distress. Once electrodes are in place, the device checks for a heartbeat and, if required, shocks the heart. A 2010 study found that use of an AED before arrival of emergency medical services increased the odds of survival by about 75 percent.

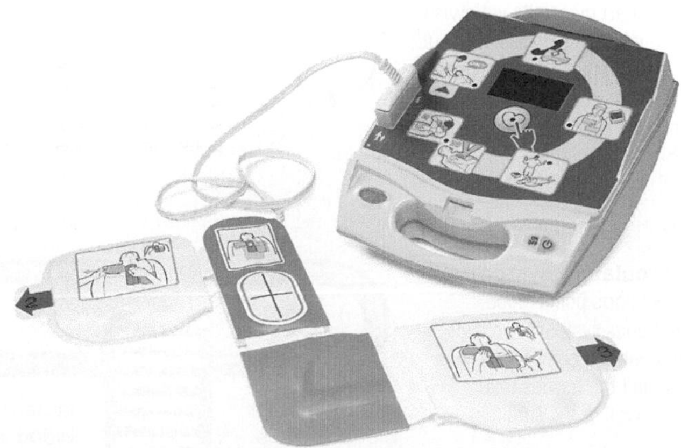

FIGURE 36.1 One type of automated external defibrillator (AED). Such devices are designed to be simple enough to be used by a trained layperson. AEDs are increasingly available in public places, but they only make a difference if someone uses them.

CREDITS: (opposite) National Cancer Institute/Science Source; (1) Courtesy of ZOLL Medical Corporation.

36.2 Circulatory Systems

LEARNING OBJECTIVES

- Using appropriate examples, explain how some animals survive without a circulatory system.
- Describe an open circulatory system and a closed circulatory system.
- Compare the path of blood flow in a fish and a mammal.
- Discuss the advantages of a four-chambered heart.

All animals must keep their cells supplied with nutrients and oxygen, and all must dispose of cellular wastes. Some invertebrates, including cnidarians and flatworms (Sections 24.5 and 24.6), rely on diffusion alone to accomplish these tasks. In such animals, nutrients and gases reach cells by diffusing across a body surface and then diffusing through the interstitial fluid (the fluid between cells). Wastes diffuse in the opposite direction. Diffusion occurs too slowly to efficiently move materials a long distance, so animals that rely on diffusion to distribute materials have a body plan in which all cells are close to a body surface. Evolution of circulatory systems allowed more complex body plans.

Open and Closed Circulatory Systems

A **circulatory system** is an organ system that speeds the distribution of materials within an animal body. It includes one or more **hearts** (muscular pumps) that propel fluid through a system of vessels that extends through the body.

Different types of circulatory systems evolved in different animal lineages. Arthropods and most mollusks have an **open circulatory system**. In such systems, a heart or hearts pump fluid called **hemolymph** into open-ended vessels (**FIGURE 36.2A**). Hemolymph leaves the vessels and mixes with the interstitial fluid, where it makes direct exchanges with cells before being drawn back into the heart through pores.

Annelids, cephalopod mollusks, and all vertebrates have a **closed circulatory system**, in which a heart or hearts pump fluid through a continuous network of vessels (**FIGURE 36.2B**). The fluid pumped through such a system is called **blood**. A closed circulatory system moves substances faster than an open one. It is "closed" in that blood does not leave blood vessels to bathe tissues. Instead, exchanges between blood and other tissues take place across the thin walls of small-diameter blood vessels called **capillaries**.

Evolution of Vertebrate Circulation

All vertebrates have a closed circulatory system, with a single heart. However, the structure of the heart and the circuits through which blood flows vary among vertebrate groups.

In most fishes, the heart has two chambers, and blood flows in a single circuit (**FIGURE 36.3A**). One chamber, an **atrium** (plural, atria), receives blood. From there, blood enters a **ventricle**, a chamber that pumps blood out of the heart. Pressure exerted by ventricular contractions drives blood through a series of vessels, into capillaries in each gill, then into capillaries in body tissues and organs, and finally back to the heart. Pressure imparted to blood by ventricular contraction dissipates as blood travels through capillaries, so blood is not under much pressure when it leaves the gill capillaries, and even less as it travels toward the heart.

Adapting to life on land involved coordinated modifications of the respiratory and circulatory systems. Amphibians and most reptiles have a three-chambered

A Open circulatory system. A grasshopper's heart pumps its yellowish hemolymph through a large vessel and out into tissue spaces. Hemolymph mingles with interstitial fluid, exchanges materials, and then reenters the heart through openings in the heart wall.

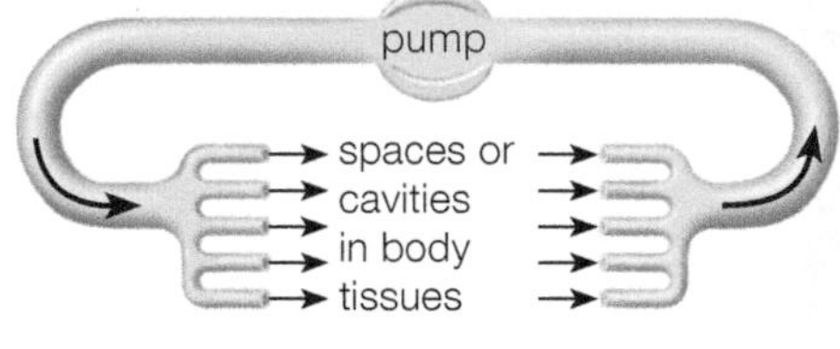
pump
spaces or
cavities
in body
tissues

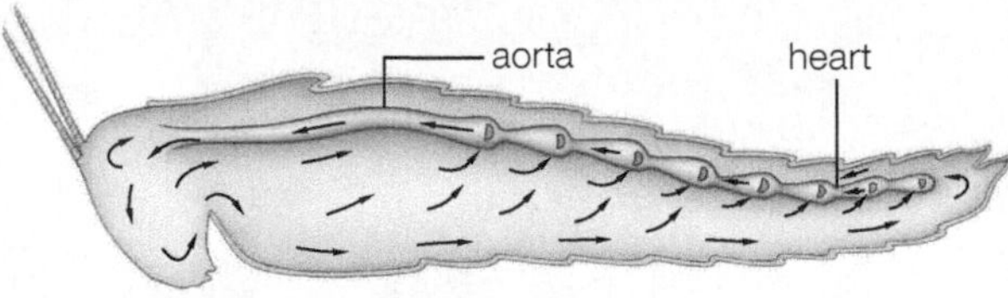
aorta
heart

B Closed circulatory system. An earthworm's hearts pump blood through a continuous system of vessels that extend through the body. Exchanges between blood and the tissues take place across the wall of the smallest vessels.

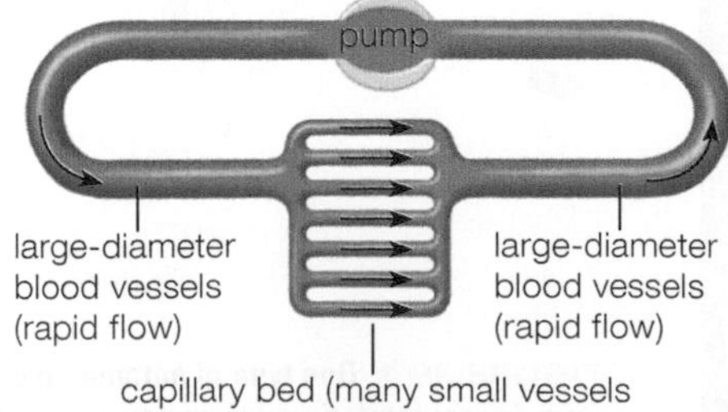
pump
large-diameter
blood vessels
(rapid flow)
large-diameter
blood vessels
(rapid flow)
capillary bed (many small vessels
that serve as a diffusion zone)

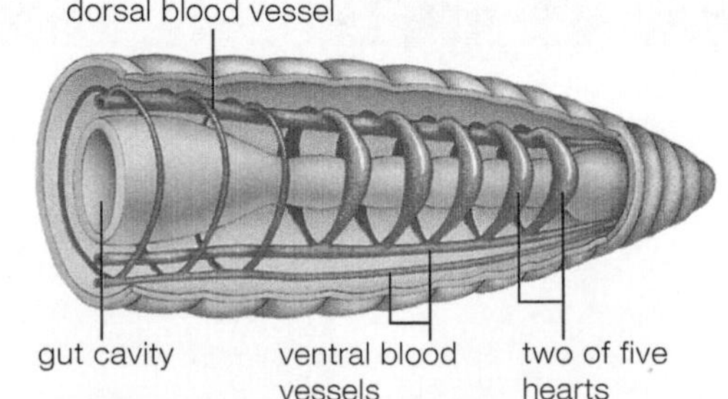
dorsal blood vessel
gut cavity
ventral blood
vessels
two of five
hearts

FIGURE 36.2 Comparison of open and closed circulatory systems.

CREDITS: (2A, B) left, © Cengage Learning; (2A, B) right, After M. Labarbera and S. Vogel, American Scientist, 1982, 70:54–60.

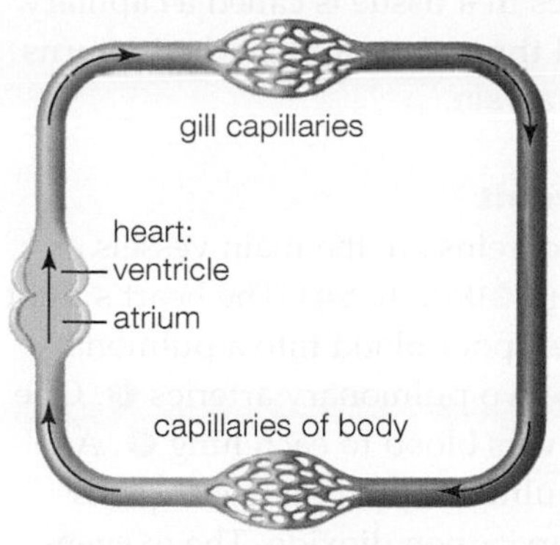
gill capillaries
heart:
ventricle
atrium
capillaries of body

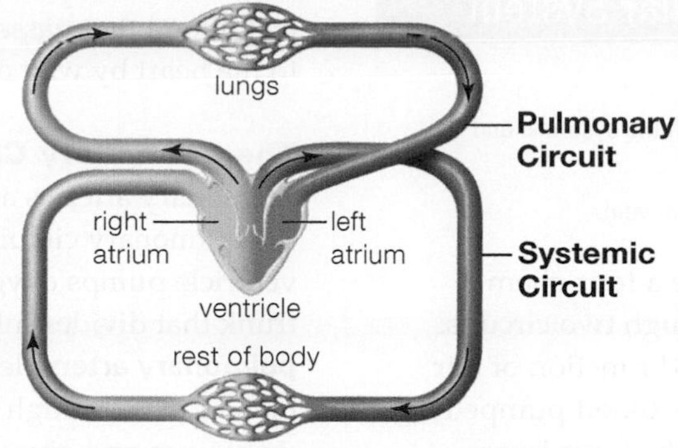
lungs
right atrium
left atrium
ventricle
rest of body
Pulmonary Circuit
Systemic Circuit

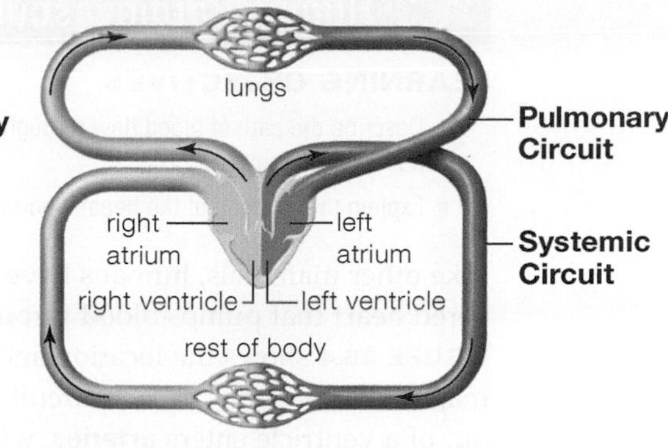
lungs
right atrium
left atrium
right ventricle
left ventricle
rest of body
Pulmonary Circuit
Systemic Circuit

A The fish heart has one atrium and one ventricle. The force imparted by the ventricle's contraction propels blood through a single circuit.

B In amphibians and most reptiles, the heart has three chambers: two atria and one ventricle. Blood flows in two partially separated circuits. Oxygenated blood and oxygen-poor blood mix a bit in the ventricle.

C In birds and mammals, the heart has four chambers: two atria and two ventricles. Oxygenated blood and oxygen-poor blood do not mix.

FIGURE 36.3 Variation in vertebrate circulatory systems. Vessels carrying oxygenated blood are red, and those carrying deoxygenated blood are blue.

heart, with two atria emptying into one ventricle (**FIGURE 36.3B**). A three-chambered heart speeds blood flow by moving blood through two partially separated circuits. The force of one contraction propels blood through the **pulmonary circuit**—to the lungs, and then back to the heart. (*Pulmo* is Latin for lung.) A second contraction sends now-oxygenated blood through the longer **systemic circuit**. This circuit extends through capillaries in body tissues and returns to the heart.

In birds and mammals, the single ventricle has been divided into two. The four-chambered heart has two atria and two ventricles (**FIGURE 36.3C**). With two fully separate circuits, only oxygen-rich blood flows to body tissues. As an additional advantage, blood pressure can be regulated independently in each circuit. Strong contraction of the heart's left ventricle moves blood quickly through the long systemic circuit. At the same time, the right ventricle can contract more gently, protecting the delicate lung tissue that would be blown apart by higher pressure.

The four-chambered heart of mammals and birds is an example of morphological convergence (Section 18.3). Birds and mammals do not share an ancestor with such a heart. Rather, this trait evolved independently in the two groups. The enhanced blood flow associated with a four-chambered heart supports the high metabolism of these endothermic animals (animals that regulate their body by varying their production of metabolic heat). Endotherms have higher energy needs than comparably sized ectotherms because they must produce heat to maintain their body temperature. Rapid blood flow in an endotherm's body delivers the large amount of oxygen required to sustain the aerobic reactions that generate heat.

TAKE-HOME MESSAGE 36.2

✔ Most animals have a circulatory system that speeds the distribution of substances through the body.

✔ Some invertebrates have an open circulatory system; other invertebrates and all vertebrates have a closed circulatory system, in which blood always remains enclosed within the heart or blood vessels.

✔ Fish have a one-circuit circulatory system. All other vertebrates have a short pulmonary circuit that carries blood to and from the lungs, and a longer systemic circuit that moves blood to and from the body's other tissues.

✔ A four-chambered heart evolved independently in birds and mammals. Such a heart allows strong contraction of one ventricle to speed blood through the systemic circuit, while a weaker contraction of the other ventricle protects lung tissue.

atrium (AY-tree-um) Heart chamber that receives blood.
blood Circulatory fluid of a closed circulatory system.
capillary A narrow, thin-walled blood vessel in a closed circulatory system; exchanges with interstitial fluid take place across its walls.
circulatory system Organ system consisting of a heart or hearts and fluid-filled vessels that distribute substances through a body.
closed circulatory system Circulatory system in which blood flows through a continuous system of vessels.
heart Muscular organ that pumps fluid through a circulatory system.
hemolymph (HEEM-oh-limf) Fluid that circulates in an open circulatory system.
open circulatory system Circulatory system in which hemolymph leaves vessels and flows through spaces in body tissues.
pulmonary circuit (PUL-mon-erry) Circuit through which blood flows from the heart to the lungs and back.
systemic circuit (sis-TEM-ick) Circuit through which blood flows from the heart to the body tissues and back.
ventricle (VEN-trick-uhl) Heart chamber that receives blood from an atrium and pumps it out of the heart

36.3 Human Cardiovascular System

LEARNING OBJECTIVES

- Describe the path of blood flow through the systemic and pulmonary circuits.
- Explain the function of the hepatic portal vein.

Like other mammals, humans have a four-chambered heart that pumps blood through two circuits. FIGURE 36.4 shows the location and function of our major blood vessels. In each circuit, blood pumped out of a ventricle enters **arteries**, which are large-diameter blood vessels that carry blood away from the heart. Arteries branch into smaller vessels, which in turn branch into capillaries inside specific tissues. A network of capillaries in a tissue is called a capillary bed. Blood that passed through a capillary bed returns to the heart by way of a **vein**.

FIGURE 36.4 **Major vessels of the human cardiovascular system.** Vessels carrying oxygenated blood are shown in red; those carrying oxygen-poor blood in blue. With the exception of the aortas and vena cavae,. all labeled vessels occur on both sides of the body. For example, you have a left and right jugular vein, and a left and right carotid artery.

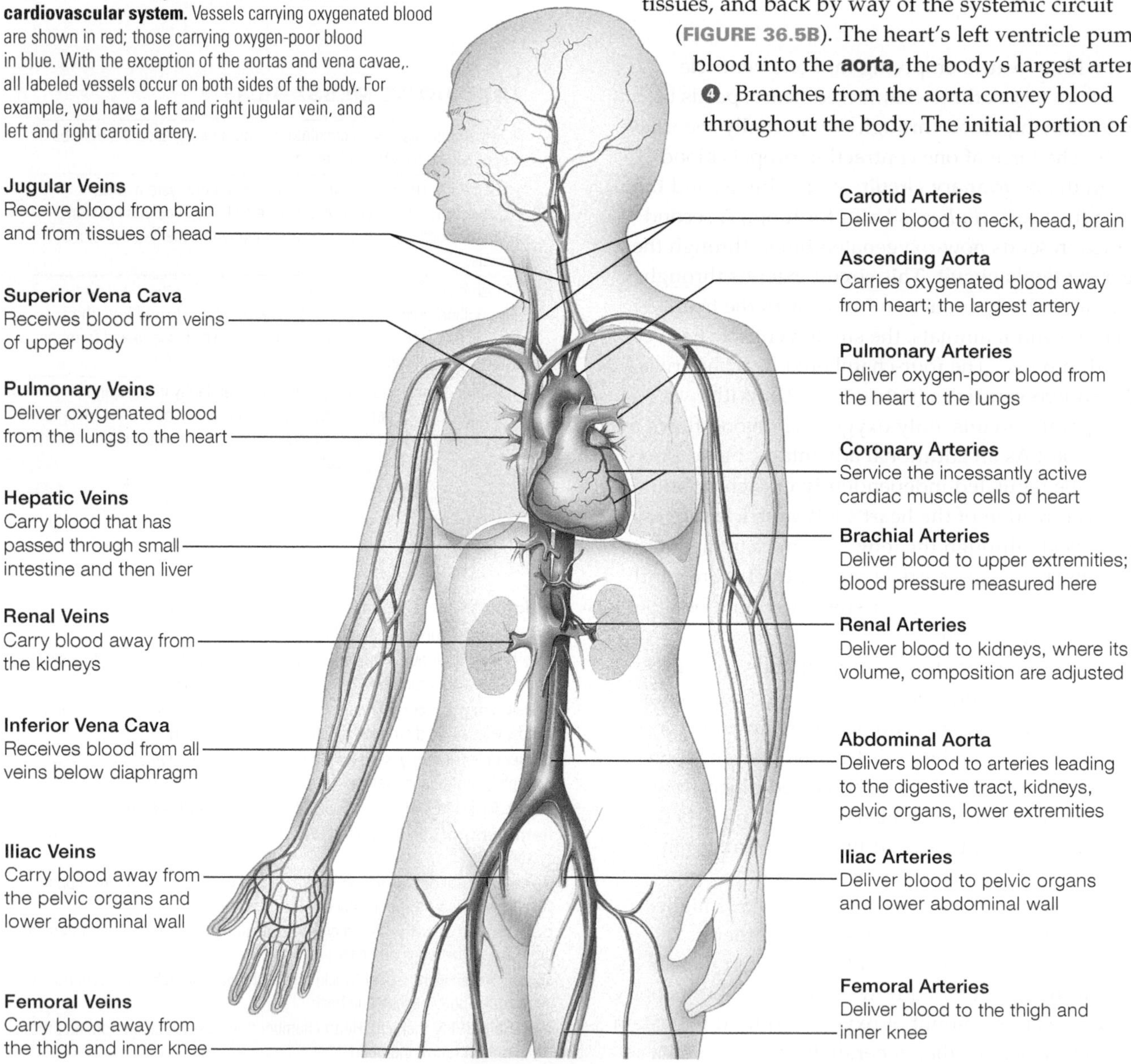

Jugular Veins
Receive blood from brain and from tissues of head

Superior Vena Cava
Receives blood from veins of upper body

Pulmonary Veins
Deliver oxygenated blood from the lungs to the heart

Hepatic Veins
Carry blood that has passed through small intestine and then liver

Renal Veins
Carry blood away from the kidneys

Inferior Vena Cava
Receives blood from all veins below diaphragm

Iliac Veins
Carry blood away from the pelvic organs and lower abdominal wall

Femoral Veins
Carry blood away from the thigh and inner knee

Carotid Arteries
Deliver blood to neck, head, brain

Ascending Aorta
Carries oxygenated blood away from heart; the largest artery

Pulmonary Arteries
Deliver oxygen-poor blood from the heart to the lungs

Coronary Arteries
Service the incessantly active cardiac muscle cells of heart

Brachial Arteries
Deliver blood to upper extremities; blood pressure measured here

Renal Arteries
Deliver blood to kidneys, where its volume, composition are adjusted

Abdominal Aorta
Delivers blood to arteries leading to the digestive tract, kidneys, pelvic organs, lower extremities

Iliac Arteries
Deliver blood to pelvic organs and lower abdominal wall

Femoral Arteries
Deliver blood to the thigh and inner knee

The Pulmonary Circuit

Pulmonary arteries and veins are the main vessels of the pulmonary circuit (FIGURE 36.5A). The heart's right ventricle pumps oxygen-poor blood into a pulmonary trunk that divides into two pulmonary arteries ❶. One **pulmonary artery** delivers blood to each lung ❷. As blood flows through pulmonary capillaries, it picks up oxygen and gives up carbon dioxide. The oxygen-enriched blood then returns to the heart in **pulmonary veins** ❸, which empty into the heart's left atrium.

The Systemic Circuit

Oxygenated blood travels from the heart, to body tissues, and back by way of the systemic circuit (FIGURE 36.5B). The heart's left ventricle pumps blood into the **aorta**, the body's largest artery ❹. Branches from the aorta convey blood throughout the body. The initial portion of the aorta (the ascending aorta) carries blood toward the head. Carotid arteries, which service the brain, and coronary arteries, which service heart muscle, receive blood from the ascending aorta. The aorta then turns and descends through the thorax, continuing into the abdomen. Branches from the descending portion of the aorta supply most internal organs and the lower limbs. For example, the descending aorta supplies blood to renal arteries that deliver blood to the kidneys, and femoral arteries that carry blood into each leg.

In the systemic circuit, blood gives up oxygen and picks up carbon dioxide as it flows through capillaries. It then returns to the heart via two large veins ❺. The **superior vena cava** returns blood from the head, neck, upper trunk, and arms. Smaller veins such as the carotid veins and coronary veins drain into the superior vena cava. The longer **inferior vena cava** returns blood from the lower trunk and legs. Renal veins and femoral veins are among the vessels that drain into the inferior vena cava.

In most cases, blood flows through only one capillary bed before returning to the heart. However, blood that passes through the capillaries in the small intestine enters the hepatic portal vein, a vein that delivers it to a capillary bed in the liver ❻. Journeying through these two capillary beds allows blood to pick up glucose and other substances absorbed from the small intestine, and deliver them to the liver. The liver stores some of the absorbed glucose as glycogen (Section 3.3). It also breaks down some ingested toxins, including alcohol (Section 5.1).

TAKE-HOME MESSAGE 36.3

✔ The pulmonary circuit carries oxygen-poor blood from the heart through the pulmonary arteries and to capillaries in the lungs. Pulmonary veins return oxygenated blood to the heart.

✔ The systemic circuit carries oxygenated blood from the heart out the aorta, through branching arteries and to capillaries throughout the body. It returns oxygen-poor blood to the heart by way of veins.

✔ Most blood traveling through the systemic circuit passes through one capillary bed, but blood that flows through capillaries in the intestines also flows through capillaries in the liver.

aorta (ay-OR-tuh) Largest artery; carries oxygenated blood away from the heart.
artery Large-diameter blood vessel that carries blood away from the heart.
inferior vena cava (VEE-nah CAY-vuh) Vein that delivers blood from the lower body to the heart.
pulmonary artery Vessel that carries blood from the heart to a lung.
pulmonary vein Vessel that carries blood from a lung to the heart.
superior vena cava (VEE-nah CAY-vuh) Vein that delivers blood from the upper body to the heart.
vein Large-diameter vessel that returns blood to the heart.

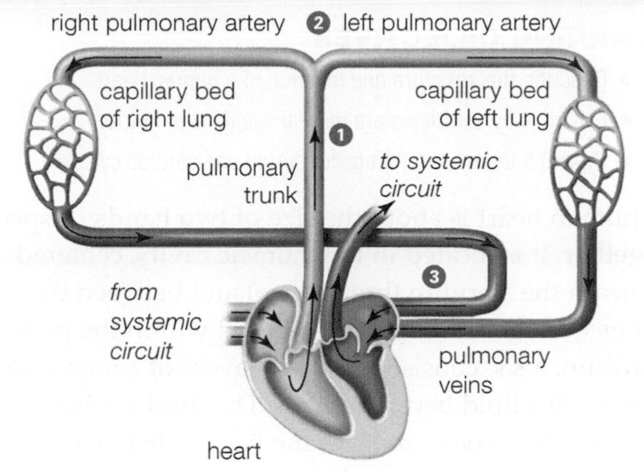
A Pulmonary Circuit
right pulmonary artery
❷ left pulmonary artery
capillary bed of right lung
capillary bed of left lung
❶
pulmonary trunk
to systemic circuit
❸
from systemic circuit
pulmonary veins
heart

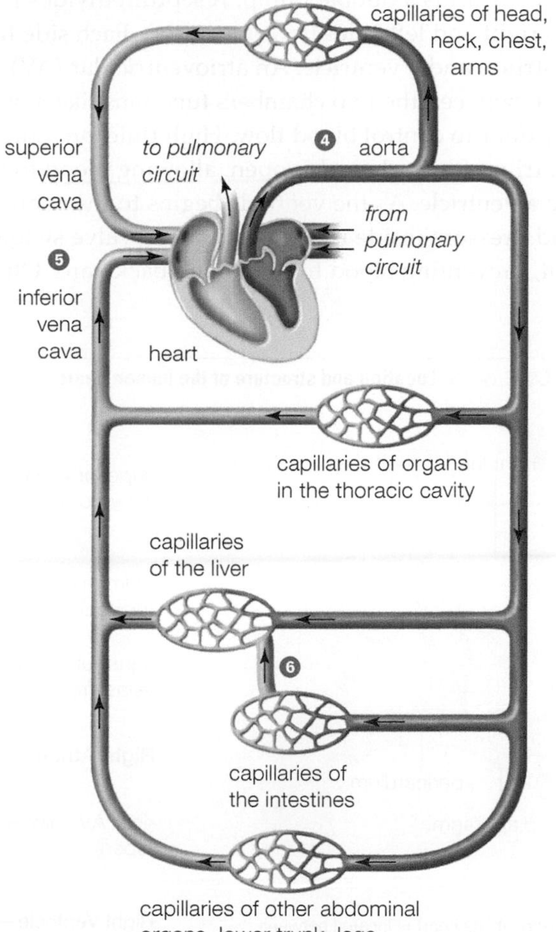
B Systemic Circuit
capillaries of head, neck, chest, arms
superior vena cava
to pulmonary circuit
❹ aorta
from pulmonary circuit
❺
inferior vena cava
heart
capillaries of organs in the thoracic cavity
capillaries of the liver
❻
capillaries of the intestines
capillaries of other abdominal organs, lower trunk, legs

FIGURE 36.5 Circuits of the human cardiovascular system. Vessels carrying oxygenated blood are shown in red and those carrying deoxygenated blood in blue.

FIGURE IT OUT Which circuit includes the hepatic portal vein, the vein that carries blood from the intestine to the liver?

Answer: The systemic circuit

36.4 The Human Heart

LEARNING OBJECTIVES

- Describe the structure and location of a human heart.
- Explain why ventricles are more muscular than atria.
- Describe the events that occur during one cardiac cycle.

A human heart is about the size of two hands clasped together. It is located in the thoracic cavity, centered beneath the sternum (breastbone) and between the lungs (**FIGURE 36.6A**). It is enclosed within the pericardium, a sac consisting of two layers of connective tissue with fluid between them. The fluid between the sac's two layers reduces the friction between them when the heart changes shape during contractions.

The heart wall consists mostly of cardiac muscle cells. Endothelium, a simple squamous epithelium (Section 31.3), lines the heart. It also lines blood vessels.

The heart is a double pump. A septum divides it into right and left sides (**FIGURE 36.6B**). Each side has an atrium and a ventricle. An atrioventricular (AV) valve between the two chambers functions like a one-way door to control blood flow. High fluid pressure in an atrium forces the valve open, allowing blood to flow into a ventricle. As the ventricle begins to the contract, fluid pressure inside it rises and the AV valve swings shut, preventing blood from moving backward. Other one-way valves control the flow of blood from ventricles into the pulmonary trunk and the aorta.

Oxygen-poor blood delivered to the right atrium by the superior and inferior venae cavae flows through the right AV valve into the right ventricle. The right ventricle pumps this blood through the pulmonary valve into the pulmonary arteries, and through the pulmonary circuit.

Oxygenated blood returns to the left atrium via pulmonary veins. This blood flows through the left AV valve into the left ventricle. The left ventricle pumps the blood through the aortic valve into the aorta. From here, the oxygenated blood flows to tissues of the body.

The Cardiac Cycle

The series of events that occur from the onset of one heartbeat to another are collectively called the **cardiac cycle** (**FIGURE 36.7**). During this cycle, the heart's chambers alternate through **diastole** (relaxation) and **systole** (contraction). First, the relaxed atria expand with blood ❶. Fluid pressure forces AV valves to open and blood to flow into the relaxed ventricles, which expand as the atria contract ❷. Once the ventricles fill, the AV valves close. The ventricles then contract causing the aortic and pulmonary valves to open. Blood flows through these valves and out of the ventricles ❸. Now emptied, the ventricles relax while the atria fill ❹.

FIGURE 36.6 Location and structure of the human heart.

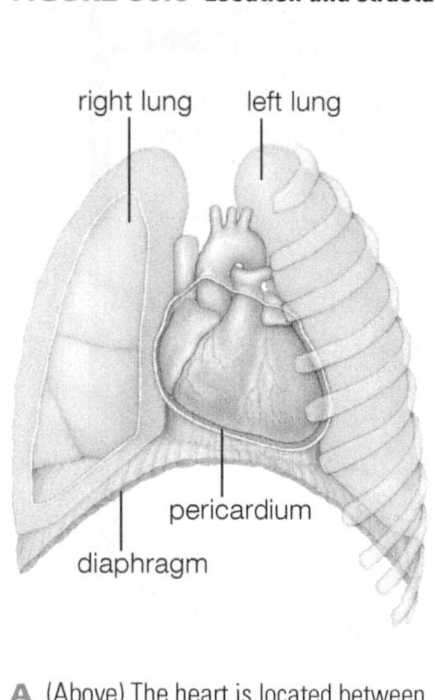
right lung
left lung
pericardium
diaphragm

A (Above) The heart is located between the lungs and is protected by the sternum and the ribs.

B (Right) Cutaway view of a human heart. Red arrows indicate the flow of oxygenated blood; blue arrows, oxygen-poor blood.

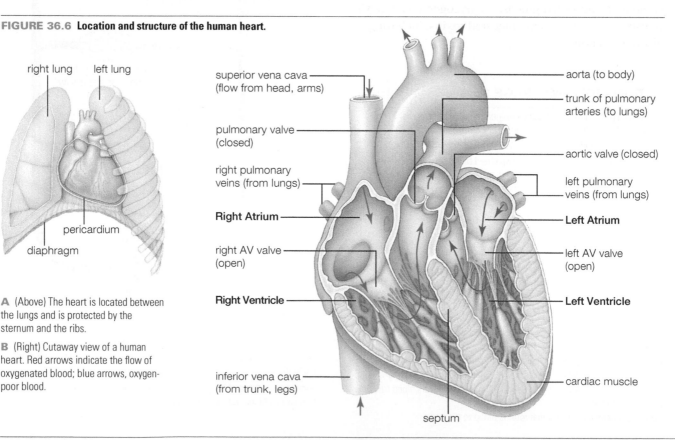

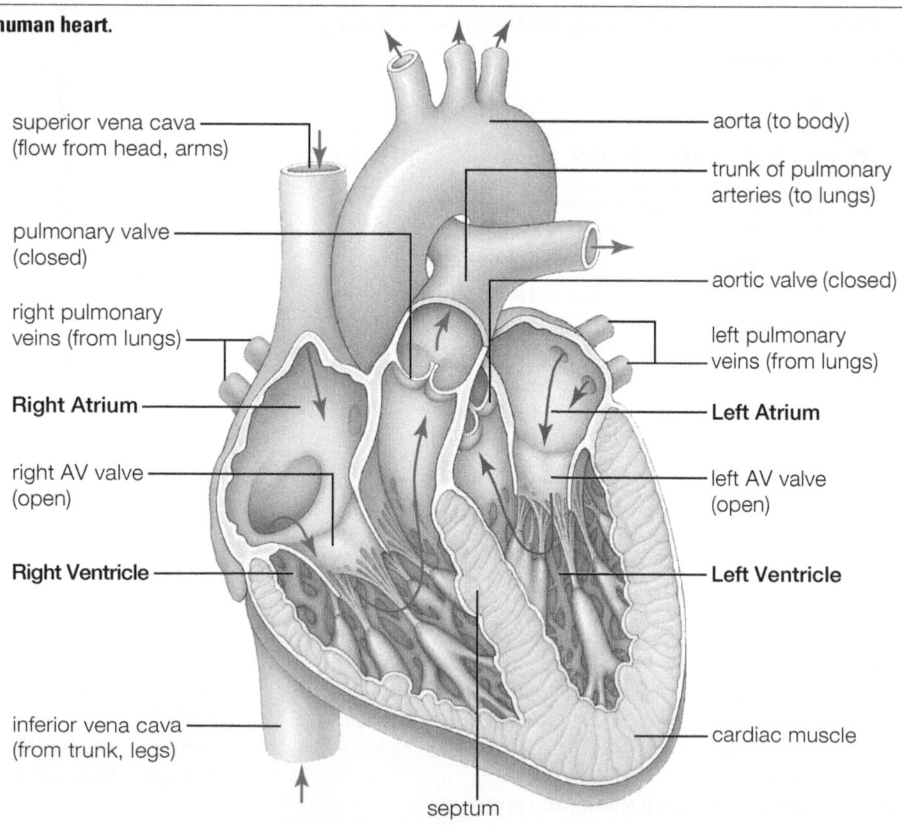
superior vena cava
(flow from head, arms)
pulmonary valve
(closed)
right pulmonary
veins (from lungs)
Right Atrium
right AV valve
(open)
Right Ventricle
inferior vena cava
(from trunk, legs)
septum
aorta (to body)
trunk of pulmonary
arteries (to lungs)
aortic valve (closed)
left pulmonary
veins (from lungs)
Left Atrium
left AV valve
(open)
Left Ventricle
cardiac muscle

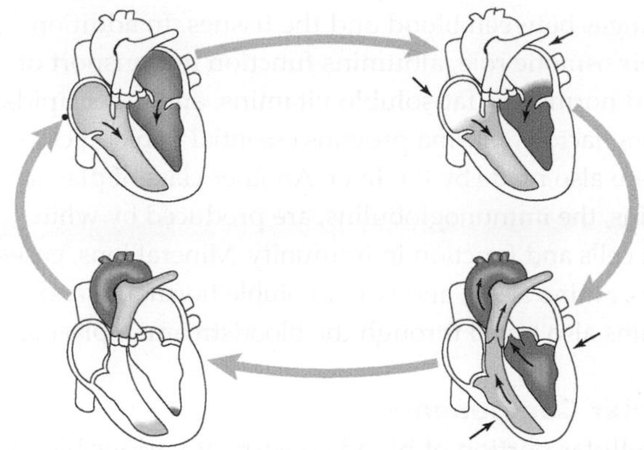
1 Relaxed atria fill. Fluid pressure opens AV valves, so blood flows into relaxed ventricles.
2 Atria contract, squeezing more blood into the relaxed ventricles.
4 Blood flows into arteries, decreasing pressure in ventricles. The pressure decline causes aortic and pulmonary valves to close.
3 Ventricles start to contract. Rising pressure pushes AV valves shut. As the pressure continues to rise, aortic and pulmonary valves open.

FIGURE 36.7 The cardiac cycle.

During the cardiac cycle a "lub-dup" sound can be heard through the chest wall. The "lub" is the sound of the heart's AV valves closing. The "dup" is the sound of the heart's aortic and pulmonary valves closing.

Blood circulation is driven entirely by contracting ventricles. Atrial contraction only helps fill the ventricles with blood. The structure of the cardiac chambers reflects their different functions. Atria need only generate enough force to squeeze blood into the ventricles, so they have relatively thin walls. By contrast, contraction of a ventricle has to produce enough pressure to propel blood through an entire cardiovascular circuit. Thus, ventricle walls are more thickly muscled. The left ventricle, which pumps blood throughout the long systemic circuit, has thicker walls than the right ventricle, which pumps blood only to the lungs and back.

Setting the Pace for Contraction

Cardiac muscle consists of muscle cells with arrays of sarcomeres that contract by the same sliding-filament mechanism described earlier for skeletal muscle (Section 35.7).

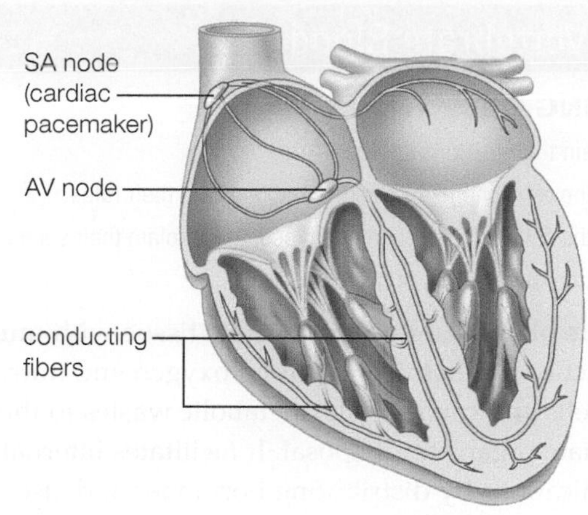
SA node (cardiac pacemaker)
AV node
conducting fibers

FIGURE 36.8 Cardiac conduction system.

The heart rate is set by the **sinoatrial (SA) node**, a clump of specialized cells in the wall of the right atrium (**FIGURE 36.8**). The SA node is the cardiac pacemaker. About 70 times a minute, it generates an action potential. The action potential travels from one cardiac muscle cell to the next by way of gap junctions. A gap junction is a tunnel-like structure that connects the cytoplasm of two adjacent cells (Section 4.9). The presence of gap junctions allows ions, and thus action potentials, to spread swiftly across the heart.

Spread of an action potential across the atria causes them to contract. Simultaneously, the action potential travels through specialized noncontractile muscle fibers to a clump of cells called the atrioventricular node. The **atrioventricular (AV) node** is the only place where action potentials can cross to ventricles. The time it takes for an action potential to cross the AV node allows blood from atria to fill ventricles before they contract.

From the AV node, the action potential travels along conducting fibers in the heart's septum. The fibers extend to the heart's lowest point and up the ventricle walls. As an action potential spreads from the base of the ventricles upward, both ventricles contract from the bottom up. The resulting wringing motion ejects blood into the aorta and pulmonary arteries.

TAKE-HOME MESSAGE 36.4

- ✔ The four-chambered heart is a muscular pump partitioned into two halves, each with an atrium and a ventricle.
- ✔ Forceful contraction of the ventricles provides the driving force for blood circulation. One-way valves control the flow of blood from atria to ventricles and from ventricles into arteries.
- ✔ The SA node is the cardiac pacemaker. Its spontaneous, rhythmic signals make cardiac muscle cells of the heart wall contract in a coordinated fashion.

---

atrioventricular (AV) node (AY-tree-oh-ven-TRICK-you-lur) Clump of cells that serves as the electrical bridge between the atria and ventricles.
cardiac cycle Sequence of contraction and relaxation of heart chambers that occurs with each heartbeat.
diastole (die-ASS-toll-ee) Relaxation phase of the cardiac cycle.
sinoatrial (SA) node Cardiac pacemaker; cluster of specialized cells whose spontaneous rhythmic signals trigger contractions.
systole (SIS-toll-ee) Contractile phase of the cardiac cycle.

36.5 Vertebrate Blood

LEARNING OBJECTIVES

- Explain the functions of blood.
- List the components of plasma, and describe their roles.
- List the cellular components of blood, and explain their sources.
- Describe the process of hemostasis.

Vertebrate blood is a fluid connective tissue with multiple functions. It delivers essential oxygen and nutrients to cells and carries cells' metabolic wastes to the appropriate organs for disposal. It facilitates internal communications by distributing hormones and also serves as a highway for cells and proteins that protect and repair tissues. In birds and mammals, blood helps maintain a stable internal temperature by distributing heat generated by muscle activity to the skin, where it can be lost to the surroundings.

A human adult has about 5 liters of blood (a bit more than 5 quarts). Blood is, as the saying goes, thicker than water. Dissolved substances and suspended cells contribute to its viscosity. **FIGURE 36.9** shows the components of human blood.

Plasma

The fluid portion of blood, the **plasma**, constitutes about 50 to 60 percent of the blood volume. Plasma is mostly water with dissolved plasma proteins. The majority of these proteins are albumins, which are water-soluble proteins made by the liver. The high albumin content of blood helps create a solute concentration gradient that draws water into capillaries. As you will see in Section 36.8, this osmosis plays an important role in exchanges between blood and the tissues. In addition to their osmotic role, albumins function in transport of steroid hormones, fat-soluble vitamins, and other lipids. Clotting factors, plasma proteins essential to blood clotting, are also made by the liver. Another class of plasma proteins, the immunoglobulins, are produced by white blood cells and function in immunity. Mineral ions, gases, sugars, amino acids, and water-soluble hormones and vitamins also travel through the bloodstream in plasma.

FIGURE 36.9 Components of human blood.

Plasma (fluid portion)
Water (92% of plasma volume)
Plasma proteins (albumins, clotting factors, immunoglobulins)
Ions, sugars, lipids, amino acids, hormones, vitamins, dissolved gases
Cellular components
Red blood cells (transport oxygen and some carbon dioxide)
White blood cells (act in housekeeping and defense)
Platelets (act in blood clotting)
red blood cell
white blood cell
platelet

Cellular Components

The cellular portion of blood consists of various blood cells and platelets. All cellular components of blood descend from stem cells in red bone marrow.

Red Blood Cells **Red blood cells**, or erythrocytes, transport oxygen from lungs to aerobically respiring cells and help move waste carbon dioxide away from them. They are the most abundant cells in blood, accounting for 40 to 50 percent of the blood volume.

In mammals, red blood cells lose their nucleus, mitochondria, and other organelles as they mature. A mature human red blood cell is a flexible disk with a depression at its center. The cell's flexibility allows it to slip easily through narrow blood vessels, and its thinness facilitates gas exchange.

The interior of a mature red blood cell is filled with hemoglobin, a protein whose structure you learned about in Section 3.5. Most oxygen that enters the blood travels to tissues while bound to the heme group of hemoglobin. In addition to hemoglobin, a mature red blood cell contains sugars, RNAs, and other molecules that sustain it for about 120 days. Ongoing replacements keep the red blood cell count at a fairly stable level. However, during reproductive years, women typically have a lower red blood cell count than men, because women lose blood during menstruation.

The kidney hormone **erythropoietin** stimulates red blood cell production. When the number of red blood cells decreases, oxygen delivery to the kidneys and other tissues declines. In the kidneys, this decline encourages secretion of erythropoietin, which in turn increases red blood cell production by bones.

Inherited variations in molecules at the surface of red blood cells are the basis for blood typing. Section 13.5 described the genetic basis of ABO blood types. Variation in another surface protein, called Rh factor, is the basis for Rh blood types. Most people are Rh positive, meaning they have the Rh factor. People who are Rh negative do not make the Rh protein.

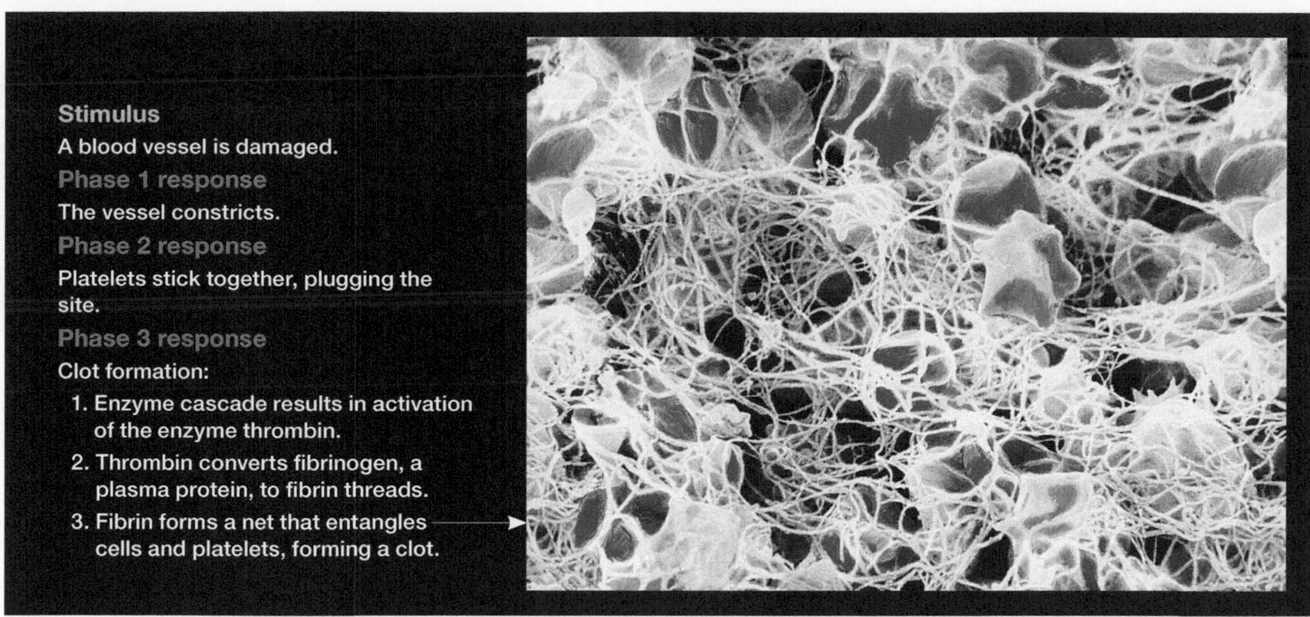
Stimulus
A blood vessel is damaged.
Phase 1 response
The vessel constricts.
Phase 2 response
Platelets stick together, plugging the site.
Phase 3 response
Clot formation:
1. Enzyme cascade results in activation of the enzyme thrombin.
2. Thrombin converts fibrinogen, a plasma protein, to fibrin threads.
3. Fibrin forms a net that entangles cells and platelets, forming a clot.

FIGURE 36.10 How blood clots. The micrograph shows the final phase—blood cells and platelets in a fibrin net.

White Blood Cells White blood cells, or leukocytes, carry out ongoing housekeeping tasks and function in defense. There are several kinds, and they differ in their size, nuclear shape, and staining traits, as well as function. We discuss the types and roles of white blood cells in detail in Chapter 37, but here is a brief preview. Neutrophils, the most abundant white blood cells, are phagocytes that engulf bacteria and cellular debris. Eosinophils attack larger parasites, such as worms. Basophils and mast cells secrete chemicals that have a role in inflammation. Monocytes circulate in the blood for a few days, then move into tissues, where they develop into phagocytic cells known as macrophages. Macrophages interact with lymphocytes to bring about immune responses. The two types of lymphocytes, B lymphocytes and T lymphocytes, protect the body against specific pathogens. B lymphocytes secrete the immunoglobulins that are a component of plasma.

Platelets A **platelet** is a membrane-wrapped fragment of cytoplasm that arises when a large cell (a megakaryocyte) breaks up. Hundreds of thousands of platelets circulate in the blood, ready to take part in **hemostasis**. This process stops blood loss from an injured vessel and provides a framework for repairs.

Hemostasis begins when an injured vessel constricts (narrows), reducing blood loss (**FIGURE 36.10**). Platelets adhere to the injured site and release substances that attract more platelets. Plasma proteins convert blood to a gel and form a clot. Clot formation involves a cascade of enzyme reactions. Fibrinogen (a clotting factor) is converted to fibrin by the enzyme thrombin, which circulates in blood as the inactive precursor prothrombin. Prothrombin (also a clotting factor) is activated by an enzyme that is activated by another enzyme, and so on.

Hemophilia is a genetic disorder in which a mutation affects the function or production of a clotting factor. The result is impaired clotting. A diet deficient in vitamin K can impair clotting too, because vitamin K is a cofactor for activation of some clotting factors.

erythropoietin (eh-rith-row-POY-eh-tin) Kidney hormone that stimulates production of red blood cells.
hemostasis (he-moh-STAY-sis) Process by which blood clots in response to injury.
plasma Fluid portion of blood.
platelet Cell fragment that helps blood clot.
red blood cell Hemoglobin-filled blood cell that transports oxygen; an erythrocyte.
white blood cell Vertebrate blood cell with a role in housekeeping tasks and defense; a leukocyte.

TAKE-HOME MESSAGE 36.5

- ✔ Blood consists mainly of plasma, a protein-rich fluid that transports wastes, gases, and nutrients.
- ✔ Blood cells and platelets form in bone marrow and are transported in plasma.
- ✔ Red blood cells contain hemoglobin that carries oxygen from lungs to tissues.
- ✔ White blood cells help defend the body from pathogens.
- ✔ Platelets are cell fragments that, together with clotting factors in plasma, play a role in clotting.

36.6 Arteries and Arterioles

LEARNING OBJECTIVES

- Describe how an artery's structure relates to its function.
- Explain how arterioles adjust blood flow to body regions, and describe some factors that trigger these adjustments.
- Explain what you feel when you check a person's pulse.

Rapid Transport in Arteries

Blood pumped out of ventricles flows into arteries. These large-diameter vessels are ringed by smooth muscle and have an outer covering of highly elastic connective tissue (FIGURE 36.11A). Like all other blood vessels and the heart, they are lined by endothelium.

Elastic properties of an artery help keep blood flowing, even when ventricles relax. When a ventricle contracts, the pressure exerted by the blood forced into an artery causes the wall of that artery to bulge outward. Then, as the ventricle relaxes, the artery wall springs back like a rubber band that has been stretched. When the artery wall recoils, the inward movement of the wall pushes blood inside the artery a bit farther away from the heart.

The brief expansion of an artery with each ventricular contraction is referred to as a **pulse**. You can feel a pulse by placing your finger on a pulse point, a body region where an artery runs close to the body surface. For example, to feel the pulse in your radial artery, put your fingers on your inner wrist near the base of your thumb (FIGURE 36.12).

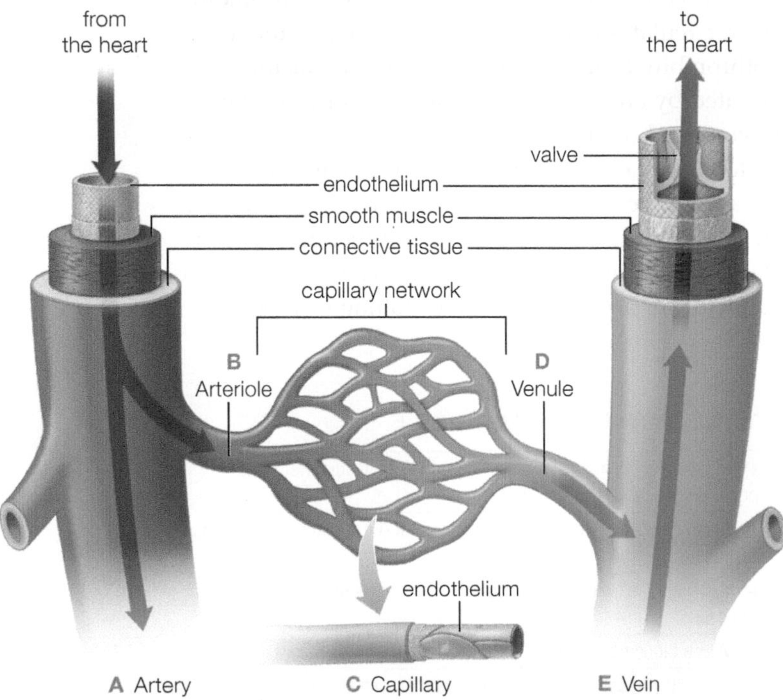
from the heart
to the heart
valve
endothelium
smooth muscle
connective tissue
capillary network
B Arteriole
D Venule
endothelium
A Artery
C Capillary
E Vein

FIGURE 36.11 **Arrangement and structure of human blood vessels.**

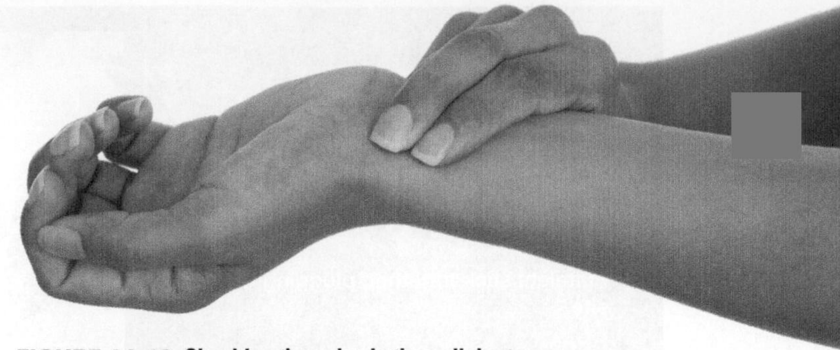

FIGURE 36.12 **Checking the pulse in the radial artery.** This artery delivers blood to the hand.

FIGURE IT OUT Which chamber of the heart contracts to produce the increase in the pressure that is detected as a pulse at this pulse point?

Answer: The left ventricle

Adjusting Flow at Arterioles

All blood pumped out of the right ventricle flows through pulmonary arteries to your lungs. In the systemic circuit, about 20 percent of the blood flows through the carotid arteries to the brain. The distribution of the remaining 80 percent of the systemic blood flow varies depending on the body's needs. For example, blood flow to the intestines increases when you are digesting a meal.

Distribution of blood to body regions is adjusted by altering the diameter of **arterioles**, the blood vessels that branch from an artery and deliver blood to capillaries. Each arteriole (FIGURE 36.11B) is ringed by smooth muscle that responds to commands from the central nervous system. Activation of the sympathetic nervous system triggers a fight–flight response (Section 32.7). This response includes **vasodilation** (widening) of arterioles in limbs, so more blood flows to skeletal muscles. At the same time, **vasoconstriction** (narrowing) of arterioles that deliver blood to the digestive tract decreases their share of the blood supply. By contrast, parasympathetic stimulation directs blood in a way that facilitates resting and digesting.

Arterioles also adjust blood flow in response to metabolic activity in nearby tissue. During exercise, skeletal muscle uses up oxygen and releases carbon dioxide and lactic acid. Arterioles delivering oxygenated blood to the muscle widen in response to these changes.

TAKE-HOME MESSAGE 36.6

✔ Arteries are thick-walled, large-diameter vessels that transport large volumes of blood away from the heart. Arterioles carry blood from an artery to capillaries.

✔ The body adjusts the distribution of blood flow in the systemic circuit by altering the diameter of arterioles. Both signals from the nervous system and local changes in metabolism can trigger changes in arteriole diameter.

36.7 Blood Pressure

LEARNING OBJECTIVES

- Describe how blood pressure varies across the systemic circuit.
- Explain how blood pressure is measured and the difference between systolic and diastolic pressure.

Blood pressure is pressure exerted by blood against the wall of the vessel that encloses it. The right ventricle contracts less forcefully than the left ventricle, so blood entering the pulmonary circuit is under less pressure than blood entering the systemic circuit. In both circuits, blood pressure is highest in arteries and declines over the course of the circuit (FIGURE 36.13).

Blood pressure is usually measured in the brachial artery of the upper arm (FIGURE 36.14). Two pressures are recorded. **Systolic pressure**, the highest pressure of a cardiac cycle, occurs as contracting ventricles force blood into the arteries. **Diastolic pressure**, the lowest blood pressure of a cardiac cycle, occurs when ventricles are relaxed. We measure blood pressure in millimeters of mercury (mm Hg), a standard unit for describing pressure, and record it as systolic value/diastolic value. Normal blood pressure is about 120/80 mm Hg, or "120 over 80."

Blood pressure depends on the total blood volume, how much blood ventricles pump out (cardiac output), and how much the arterioles are dilated. Vasodilation of arterioles lowers blood pressure, and vasoconstriction of arterioles raises it.

When blood pressure rises or falls, sensory receptors in the aorta and in carotid arteries signal the medulla oblongata in the hindbrain. In a reflexive response, this brain region calls for changes in cardiac output and arteriole diameter that restore normal pressure. This reflex is a short-term response to abnormal blood pressure. Over the longer term, kidneys adjust the blood pressure by regulating the amount of fluid lost in urine and thus determining the total blood volume.

Inability to regulate blood pressure can result in hypertension, in which resting blood pressure remains above 140/90 mmHg. Chronic high blood pressure makes the heart work harder and damages kidney capillaries, increasing the risk of heart disease or kidney failure.

TAKE-HOME MESSAGE 36.7

✔ Blood pressure is the fluid pressure exerted against a vessel wall. It is recorded as systolic/diastolic pressure.

✔ Adjustments to arteriole diameter, cardiac output, and blood volume regulate blood pressure.

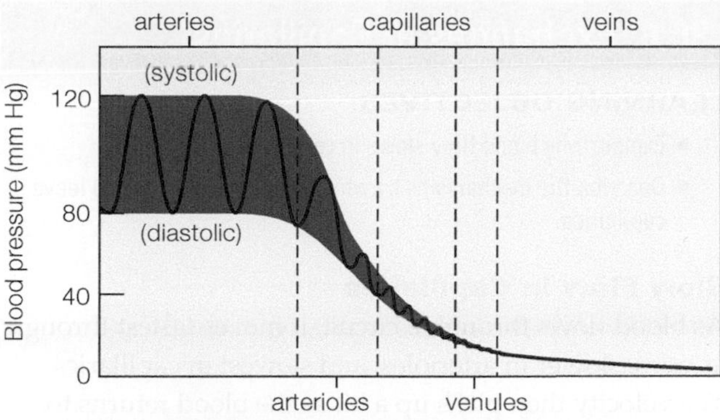
arteries
capillaries
veins
(systolic)
(diastolic)
Blood pressure (mm Hg)
120
80
40
0
arterioles
venules

FIGURE 36.13 Fluid pressure in the systemic circuit. Systolic pressure occurs when ventricles contract; diastolic, when ventricles relax.

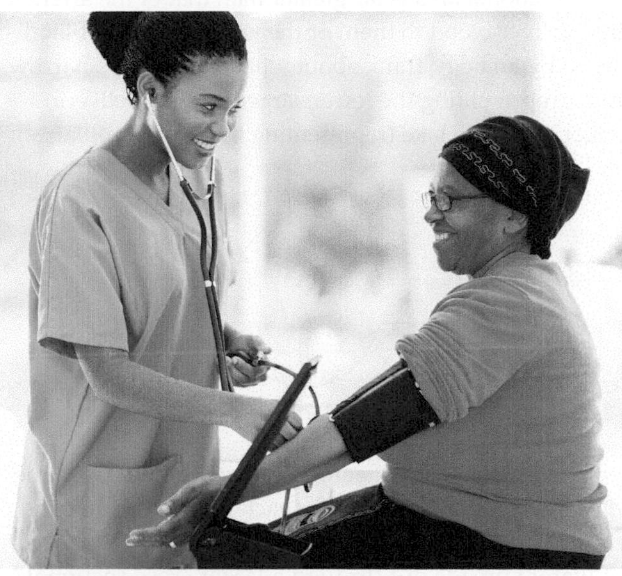

FIGURE 36.14 Measuring blood pressure. Typically, a hollow inflatable cuff attached to a pressure gauge is wrapped around the upper arm. The cuff is inflated with air, putting pressure on blood vessels of the arm. Eventually, this pressure becomes so high that it cuts off blood flow through the brachial artery, the main artery of the upper arm.

As air in the cuff is slowly released, blood begins to spurt through the artery when the left ventricle contracts and blood pressure is highest. Spurts of blood flow can be heard through a stethoscope as soft tapping sounds. The blood pressure at this point is the systolic pressure. It is typically about 120 mm Hg. Millimeters of mercury (mm Hg) is a standardized unit of pressure.

More air is released from the cuff. Eventually the sounds stop because blood is flowing continuously, even when the left ventricle is the most relaxed. The pressure when the sounds stop is the diastolic pressure, the lowest pressure during a cardiac cycle—usually about 80 mm Hg.

arteriole Blood vessel that conveys blood from an artery to capillaries.
blood pressure Pressure exerted by blood against a vessel wall.
diastolic pressure (die-ah-STAHL-ic) Blood pressure when ventricles are relaxed.
pulse Brief stretching of artery walls that occurs when ventricles contract.
systolic pressure (sis-STAHL-ic) Blood pressure when ventricles are contracting.
vasoconstriction Narrowing of a blood vessel when smooth muscle that rings it contracts.
vasodilation Widening of a blood vessel when smooth muscle that rings it relaxes.

36.8 Exchanges at Capillaries

LEARNING OBJECTIVES

- Explain why blood flow slows in capillaries.
- Describe the mechanisms by which substances enter and leave capillaries.

Slow Flow in Capillaries

As blood flows through a circuit, it moves fastest through arteries, slower in arterioles, and slowest in capillaries. The velocity then picks up a bit as the blood returns to the heart. The slowdown in capillaries occurs because the body has tens of billions of capillaries, and their collective cross-sectional area is far greater than that of the arterioles that deliver blood to them or the veins that carry blood away. By analogy, think about what happens if a narrow river (representing the fewer, larger vessels) delivers water to a wide lake (representing the many capillaries):

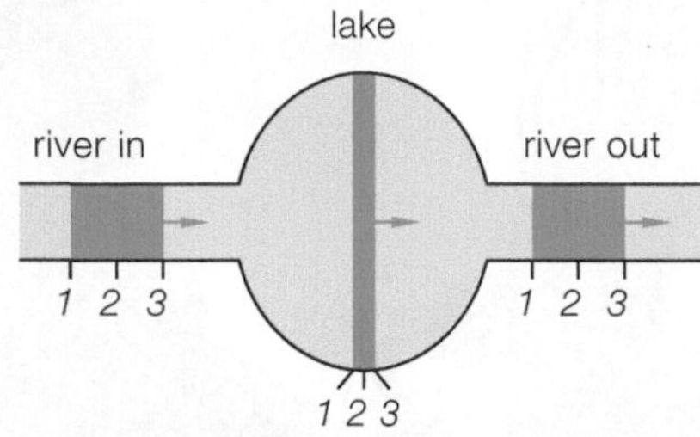
lake
river in
river out
1 2 3
1 2 3
1 2 3

The flow rate is constant, with an identical volume moving from points 1 to 3 in each interval. However, flow velocity decreases in the lake. Here, the volume of water spreads out through a larger cross-sectional area, so it flows forward a shorter distance during any specified interval. Slow flow through small capillaries enhances the rate of exchanges between blood and interstitial fluid. The more time blood spends in a capillary, the greater the opportunity for those exchanges to take place.

Mechanisms of Capillary Exchange

A capillary wall consists of a single layer of endothelial cells. At the end of the capillary nearest the arteriole there are spaces between cells so the capillary wall is a bit leaky. Here, pressure exerted by the beating heart forces plasma fluid out between cells and into the surrounding interstitial fluid (**FIGURE 36.15** ❶). Blood cells, platelets, and plasma proteins, including albumin, are too large to slip through the spaces between cells, so they remain in the capillary.

All along the length of the capillary, oxygen released by red blood cells diffuses from blood into interstitial fluid. Membrane proteins transport nutrients such as glucose in the same direction ❷. Carbon dioxide (CO_2) diffuses from interstitial fluid into the capillary, and other metabolic wastes are transported into it ❸.

Near the venous end of the capillary bed, blood pressure is lower. Here, the main factor influencing fluid movement is the high protein content of the plasma. Plasma is hypertonic relative to interstitial fluid, so water moves by osmosis into the capillary ❹.

As a result of all the leaking and osmotic movement of fluid, there is a small net outward flow from a capillary bed into interstitial fluid. Fluid lost from the blood by this process is returned by the lymphatic system. Interstitial fluid enters lymph capillaries ❺, which drain into lymphatic ducts that return fluid to veins near the heart.

TAKE-HOME MESSAGE 36.8

✔ Blood flow slows in capillaries, where exchanges between the blood and interstitial fluid occur.

✔ Fluid rich in oxygen and nutrients leaks out between cells of the capillary wall. Gases diffuse across capillary walls, and other substances cross with the aid of transport proteins.

FIGURE 36.15 Forces affecting capillary exchange. A single capillary is shown at left, a capillary bed at the right.

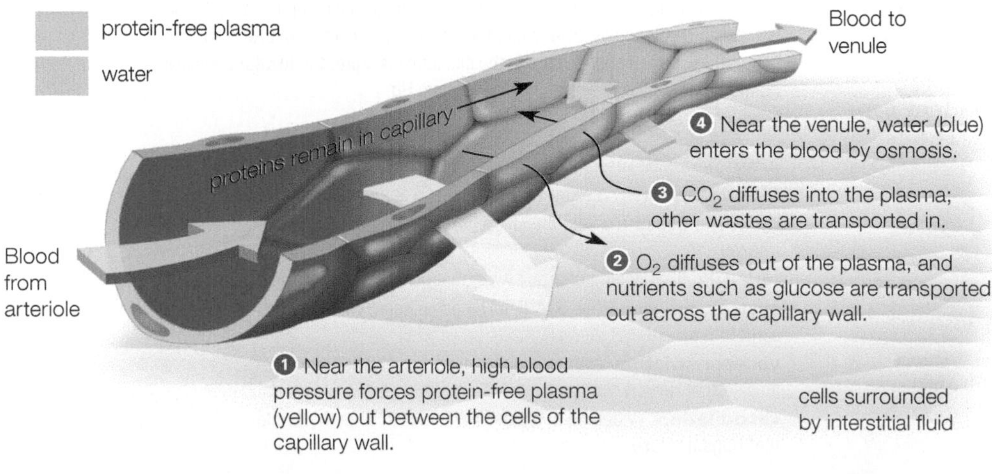
protein-free plasma
water
Blood to venule
proteins remain in capillary
❹ Near the venule, water (blue) enters the blood by osmosis.
❸ CO_2 diffuses into the plasma; other wastes are transported in.
❷ O_2 diffuses out of the plasma, and nutrients such as glucose are transported out across the capillary wall.
Blood from arteriole
❶ Near the arteriole, high blood pressure forces protein-free plasma (yellow) out between the cells of the capillary wall.
cells surrounded by interstitial fluid

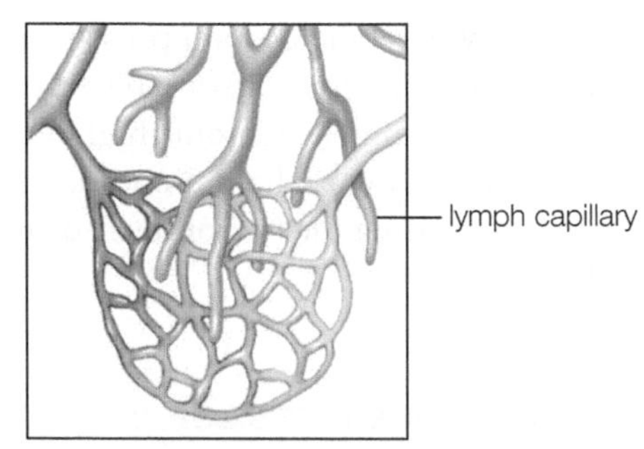
lymph capillary

❺ Some fluid leaked by blood capillaries enters neighboring lymph capillaries. This fluid, now called lymph, returns to the bloodstream when large lymph vessels drain into veins at the base of the neck near the heart.

36.9 Back to the Heart

LEARNING OBJECTIVES

- Describe the functions of venules and veins.
- Explain the mechanisms that keep blood in the veins moving toward the heart.

Blood from multiple capillaries flows into a **venule**, a small vessel that carries blood to a vein. The tiniest venules connect directly to capillaries and have no smooth muscle in their walls. White blood cells that leave the circulation to act in the tissues usually exit the blood by squeezing out between cells in the wall of such a venule. Larger venules and the veins have both elastic tissue and smooth muscle in their wall.

Of all blood vessels, veins hold the greatest volume of blood. When you are at rest, your veins contain about 60 percent of your total blood volume. Veins also have the lowest blood pressure. By the time blood reaches veins, most of the pressure imparted by ventricular contractions has dissipated.

Several mechanisms help blood at low pressure move back toward the heart. First, veins have one-way valves that prevent backflow. These valves shut when blood starts to reverse direction. For example, valves in the large veins of your leg prevent blood from moving downward in response to gravity when you stand. All vertebrates typically have the same types of arteries and veins, but the number and location of valves in those veins varies (**FIGURE 36.16**).

Smooth muscle in a vein's wall facilitates blood flow too. When this muscle contracts, the vein stiffens so it cannot hold as much blood, pressure on blood in the vein rises, and the blood is forced toward the heart.

Skeletal muscles used in limb movements also help move blood through veins. When these muscles contract, they bulge and press on neighboring veins, squeezing the blood inside them toward the heart (**FIGURE 36.17**). Exercise-induced deep breathing also raises pressure inside veins. The lungs and thoracic cavity expand during inhalation, forcing adjacent organs against veins. The resulting increase in pressure forces blood in a vein forward through a valve.

TAKE-HOME MESSAGE 36.9

✔ Venules connect capillaries to veins.

✔ Veins are the body's main blood reservoir. The amount of blood in the veins changes depending on activity level.

✔ Blood pressure in veins is low. One-way valves, contractions of skeletal muscle, and respiratory muscle action all help move the blood toward the heart.

FIGURE 36.16 Moving blood against the force of gravity. When a giraffe lowers its head to drink, gravity pulls oxygen-poor blood in its veins toward the brain. Seven valves in each jugular vein prevent backflow through the neck. Humans, who seldom lower their head below their heart, have only one valve in their jugular vein.

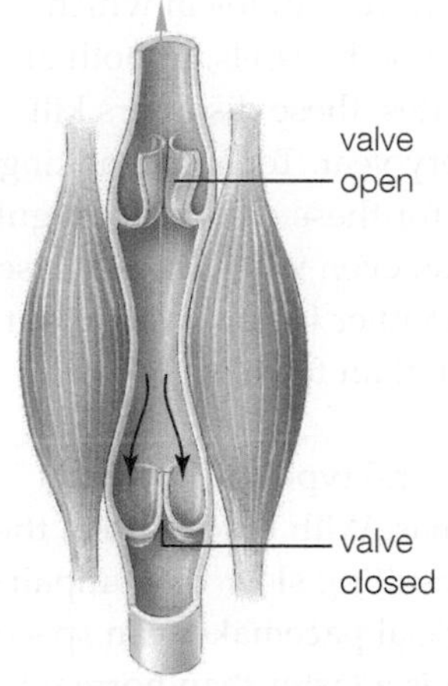

blood flow to heart
valve open
valve closed

When skeletal muscles contract, they bulge and press on neighboring veins. This puts pressure on the blood in the vein, forcing it forward through the pressure-sensitive valves.

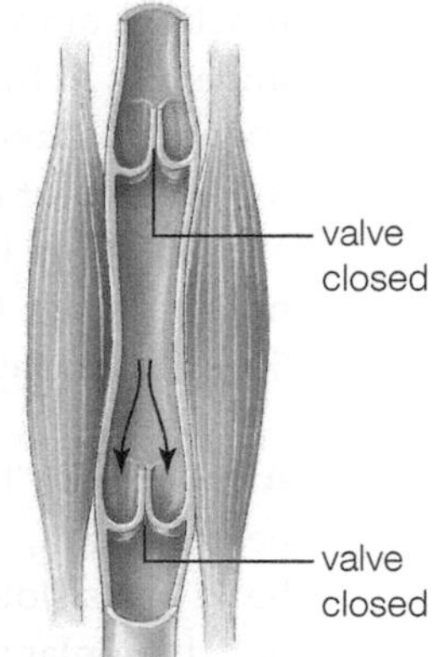

valve closed
valve closed

When skeletal muscles relax, the pressure in neighboring veins declines and pressure-sensitive valves shut, preventing blood from moving backward.

FIGURE 36.17 How skeletal muscle activity encourages blood flow through veins.

36.10 Blood and Cardiovascular Disorders

LEARNING OBJECTIVE

- Describe the causes and effects of blood and cardiovascular disorders.

Altered Blood Cell Count

In anemias, red blood cells are few or somehow impaired. Shortness of breath, fatigue, and chills are common symptoms. Anemia has many causes. It can arise as a result of blood loss from a wound or an infection by a pathogen that kills red blood cells. For example, the protist that causes malaria enters red blood cells, divides inside them, and then causes the cells to break apart. A diet with too little

venule Blood vessel that conveys blood from a capillary to a vein.

iron can cause anemia by preventing synthesis of heme. Sickle-cell anemia arises from a mutation that causes hemoglobin to change shape at a low oxygen concentration (Section 3.6). Thalassemias occur when mutations disrupt synthesis of a globin chain of hemoglobin (Section 9.6).

If stem cells in bone become cancerous, the result can be overproduction of red blood cells (polycythemia) or nonfunctional white blood cells (leukemia). Lymphomas are cancers that originate in B or T lymphocytes. Excessive divisions of the cancerous lymphocytes can produce tumors in lymph nodes and other parts of the lymphatic system.

Cardiovascular Disorders

Cardiovascular disorders are conditions in which functions of the heart, the blood vessels, or both are impaired. In the United States, these disorders kill about a million people every year. Tobacco smoking tops the list of risk factors for these disorders. Regular exercise helps lower the risk even when the exercise is not strenuous. Gender is another factor. Until about age 50, males are at greater risk than females.

Arrhythmias There are several types of abnormal heart rhythm, or arrhythmias. With bradycardia, the heart beats slowly. If the resulting slow flow impairs health, implanting an artificial pacemaker can speed the heart rate. Tachycardia is a faster than normal heart rate. Many people experience palpitations, which are occasional episodes of tachycardia. Palpitations can be brought on by stress, drugs such as caffeine, an overactive thyroid, excercise, or an underlying heart problem.

Atrial fibrillation is an arrhythmia in which the atria do not contract normally, but instead quiver. This slows blood flow and increases the risk of clot formation.

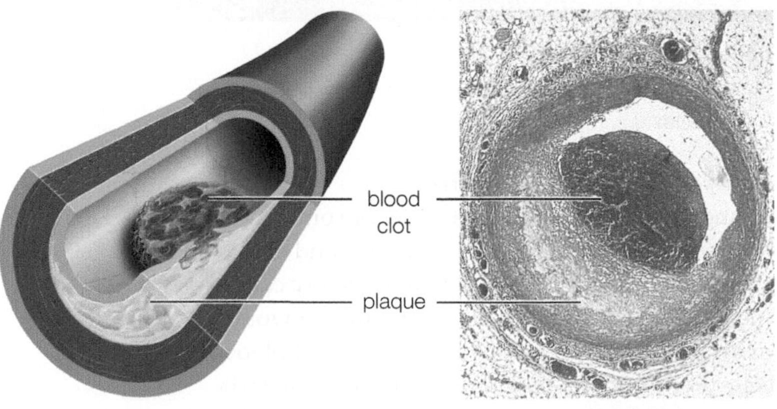
blood
clot
plaque

FIGURE 36.18 Atherosclerosis. Plaque accumulates inside an artery making it more likely to rupture. If a rupture does occur, a clot forms at the rupture site and further narrows the vessel.

Ventricular contraction is the driving force for blood circulation, so ventricular fibrillation is the most dangerous arrhythmia. Ventricles quiver, and pumping falters or stops, causing loss of consciousness and—if a normal rhythm is not restored—death. A defibrillator can often restore the heart's normal rhythm by resetting the SA node.

Hypertension High blood pressure (above 140/90) is called hypertension. Often its cause is unknown. Heredity is a factor, and African Americans have an elevated risk. Diet also plays a role; in some people high salt intake causes water retention that raises blood pressure. Hypertension is sometimes described as a silent killer, because people often are unaware they have it. Hypertension makes the heart work harder than normal, which can cause it to enlarge and to function less efficiently. High blood pressure also increases risk of other cardiovascular disorders.

Atherosclerosis Atherosclerosis is a cardiovascular disorder in which the arterial wall thickens, narrowing the vessel's interior diameter. A buildup of lipids in an artery's endothelial lining attracts white blood cells, which invade the vessel wall. Smooth muscle cells divide repeatedly and, together with the white blood cells, form a fibrous cap. Eventually, a mass called an atherosclerotic plaque bulges into the vessel's interior, narrowing the vessel's interior diameter and slowing blood flow. A plaque also makes an artery wall brittle and encourages clot formation (**FIGURE 36.18**).

Cholesterol plays a role in atherosclerosis. Most cholesterol in blood is bound to protein carriers forming complexes called low-density lipoproteins, or LDLs. LDLs deliver cholesterol to cells throughout the body. A lesser amount of cholesterol is bound in high-density lipoproteins, or HDLs. Cells in the liver take up HDLs and use them to form bile, which the liver secretes into the small intestine. Eventually, bile leaves the body in feces. Thus, HDLs help lower the body's cholesterol level. Having a high LDL level, a low HDL level, or a combination of the two is associated with an elevated risk of atherosclerosis.

Venous disorders Sometimes one or more valves in a vein become damaged, causing blood to accumulate in that vein. Damaged valves cause varicose veins, which are bulging and twisted. Varicose veins can also arise because of an inherited weakness of the vein wall. Chronic high blood pressure and an occupation that requires prolonged standing also raise the risk of varicose veins.

CREDIT: (18) Biophoto Associates/Science Source.

Data Analysis Activities

Risks of Hypertension Eleni Rapsomaniki and her colleagues analyzed medical records of 1.25 million people in the United Kingdom to see how hypertension affects the risk of cardiovascular disorders. They looked at records of people who at 30 years old had no history of heart disease or stroke, although some had high blood pressure. By looking at which of these people later had a heart attack or stroke, the researchers were able to estimate the lifetime risks of these disorders in people with and without high blood pressure. FIGURE 36.19 shows the researchers' estimates of the lifetime risk for heart attack and for transient ischemic attack (TIA), a type of stroke.

1. What is the risk that someone without hypertension at age 30 will have a heart attack by age 85? What is the comparable risk for someone with hypertension at age 30?
2. What is the risk that someone without hypertension at age 30 will have a heart attack by age 75? What is the comparable risk for someone with hypertension at age 30?
3. Which is a person who has hypertension at age 30 more likely to have by age 60, a heart attack or a TIA?

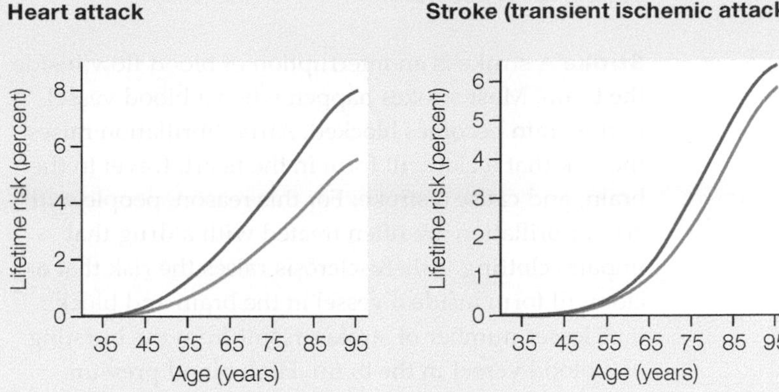

Heart attack
Stroke (transient ischemic attack)
Lifetime risk (percent)
0
2
4
6
8
35 45 55 65 75 85 95
Age (years)
Lifetime risk (percent)
0
1
2
3
4
5
6
35 45 55 65 75 85 95
Age (years)
Normal blood pressure
Hypertension

FIGURE 36.19 **Lifetime risks of heart attack and TIA (a type of stroke), with and without hypertension at age 30.** A transient ischaemic attack (TIA), sometimes called a mini-stroke, is caused by a temporary blockage of a blood vessel that supplies blood to the brain.

When blood pools inside a vein, it may clot. A clot that forms in a vessel and remains in place is called a thrombus. A clot or part of a clot that breaks loose and travels through blood vessels to a new location is an embolus. For example, a pulmonary embolus (an embolus in an artery in the lung) usually arises after a blood clot that formed in a vein in the thigh breaks off and travels through the heart into the lung. It is dangerous because it can block blood flow to lung tissue.

Heart Disease With heart disease, atherosclerosis affects one or more of the coronary arteries (the arteries that supply blood to heart muscle). A heart attack occurs when a coronary artery is completely blocked, most commonly by a clot. If the blockage is not removed, cardiac muscle cells die, permanently weakening the heart. Clot-dissolving drugs can restore blood flow, but only if they are given immediately after the onset of an attack.

Clogged coronary arteries can be treated with a bypass or angioplasty. With coronary bypass surgery, doctors open a person's chest and use a blood vessel from elsewhere in the body to divert blood around the clogged coronary artery (FIGURE 36.20A). In laser angioplasty, laser beams vaporize plaques. In balloon angioplasty, doctors inflate a small balloon in

FIGURE 36.20 **Treatments for blocked coronary arteries.**

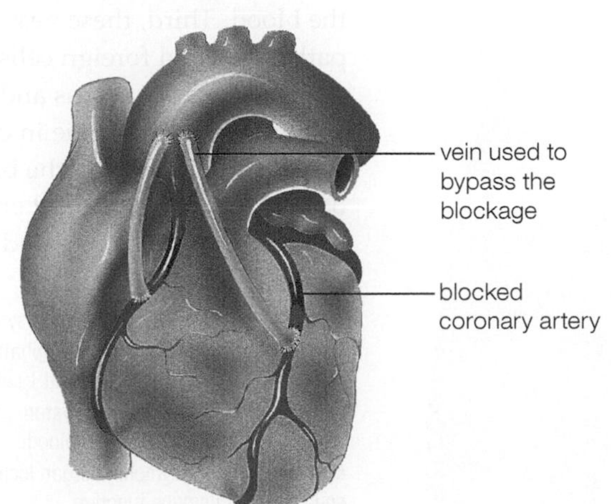

vein used to bypass the blockage
blocked coronary artery

A Coronary bypass surgery. Veins from another part of the body are used to divert blood past the blockages. This illustration shows a "double bypass," in which veins are placed to divert blood around two blocked coronary arteries.

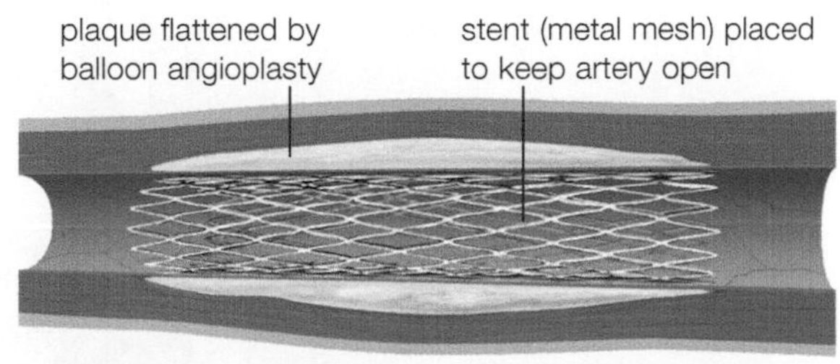

plaque flattened by balloon angioplasty
stent (metal mesh) placed to keep artery open

B Balloon angioplasty and the placement of a stent. After a balloonlike device is inflated in an artery to open it and flatten the plaque, a tube of metal (the stent) is inserted and left in place to keep the artery open.

a blocked artery to flatten the plaques. A wire mesh tube called a stent is then inserted to keep the vessel open (**FIGURE 36.20B**).

Stroke A stroke is an interruption of blood flow inside the brain. Most strokes happen when a blood vessel in the brain becomes blocked. Atrial fibrillation raises the risk that a clot will form in the heart, travel to the brain, and cause a stroke. For this reason, people with atrial fibrillation are often treated with a drug that impairs clotting. Atherosclerosis raises the risk that a clot will form inside a vessel in the brain and block it. A lesser number of strokes result from the bursting of a blood vessel in the brain. High blood pressure increases the risk of this type of stroke.

As with a heart attack, a suspected stroke requires immediate medical attention. The acronym FAST can help you remember how to evaluate and assist someone who may be having a stroke.

F stands for face; a stroke causes one side of the face to droop, especially when the person smiles. *A* stands for arms; a stroke makes it difficult to hold one's arms out at the same level. *S* stands for speech; a stroke often causes slurred speech. *T* stands for time; if any signs of stroke are detected, minimizing the time to treatment increases the odds of survival and, in survivors, of avoiding brain damage. Note that stroke symptoms vary. Someone undergoing a stroke may show one, some, or all of the symptoms listed abobe.

TAKE-HOME MESSAGE 36.10

✔ Changes in the number or quality of blood cells can alter blood's ability to carry out its functions.

✔ Problems with the cardiac pacemaker cause arrhythmias.

✔ Atherosclerosis and hypertension raise the risk of heart attack and stroke.

FIGURE 36.21 **Components of the lymphatic system.**

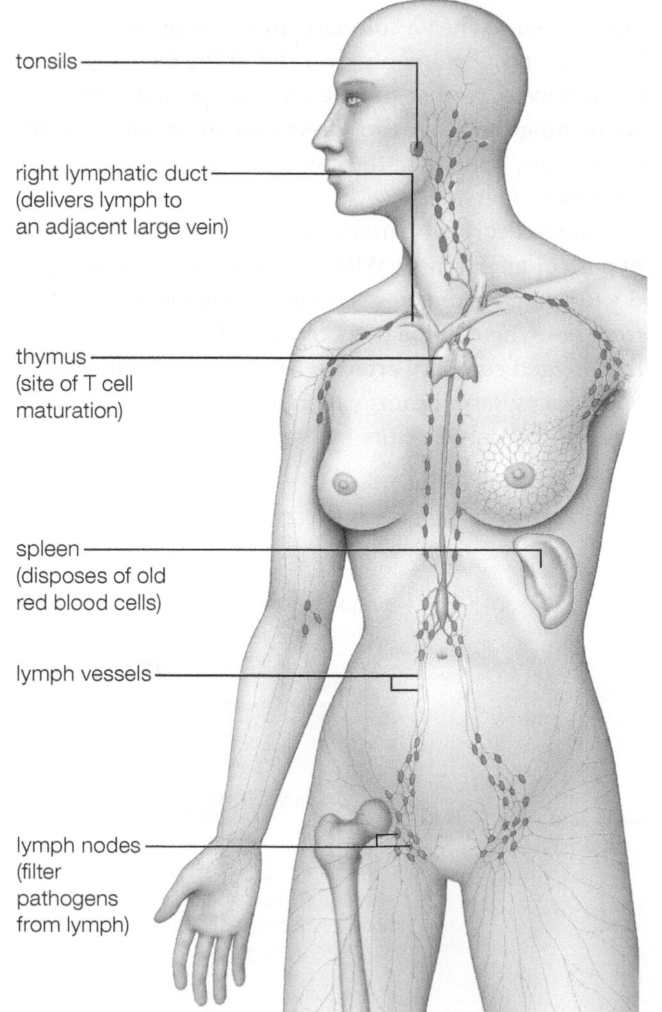
tonsils
right lymphatic duct
(delivers lymph to
an adjacent large vein)
thymus
(site of T cell
maturation)
spleen
(disposes of old
red blood cells)
lymph vessels
lymph nodes
(filter
pathogens
from lymph)

36.11 Interactions with the Lymphatic System

LEARNING OBJECTIVES

- Describe the components of the lymph vascular system and their functions.
- Explain how lymph is propelled through the lymph vessels.
- Explain the roles of lymph nodes, spleen, and thymus.

Lymph Vascular System

The **lymph vascular system** collects water and solutes from the interstitial fluid, then delivers them to the circulatory system. The system includes lymph capillaries and larger vessels (**FIGURE 36.21**). Fluid that moves through this system is referred to as **lymph**.

The lymph vascular system serves three functions. First, its vessels return fluid that leaked out of capillaries to the circulatory system. Second, lymph vessels deliver fats absorbed from food in the small intestine to the blood. Third, these vessels transport cellular debris, pathogens, and foreign cells to lymph nodes, where white blood cells assess and respond to this material.

Lymph capillaries lie in close proximity to blood capillaries throughout the body. A lymph capillary begins as a finger-shaped ending in a tissue. Gaps between the cells in the ending open sporadically as a

lymph Fluid in the lymph vascular system.
lymph node Small mass of lymphatic tissue through which lymph filters; contains many lymphocytes (B and T cells).
lymph vascular system System of vessels that takes up interstitial fluid and carries it (as lymph) to the blood.
spleen Fist-sized lymphoid organ located in the upper abdomen; filters blood and enhances immune function.

result of normal body movements, allowing interstitial fluid to enter the lymph capillary.

Lymph capillaries merge to form larger-diameter lymph vessels. Two mechanisms move lymph through these vessels. First, slow wavelike contractions of smooth muscle in the walls of large lymph vessels propel lymph forward. Second, as with veins, the bulging of adjacent skeletal muscles helps move fluid along. Like veins, the lymph vessels have one-way valves that prevent backflow.

The largest lymph vessels converge on collecting ducts that empty into veins near the heart. Each day these ducts deliver about 3 liters of fluid to the blood.

Lymphoid Organs and Tissues

Lymphoid organs and tissues associate with the lymph vascular system. They contain many lymphocytes (B cells and T cells) that help the body respond to injury and infection.

Lymph nodes are oval-shaped structures located at intervals along lymph vessels. They filter lymph before it returns to the blood (**FIGURE 36.21C**). Each node has a capsule of connective tissue and contains large numbers of lymphocytes and other white blood cells that survey the passing lymph for potential threats.

The white blood cells called T lymphocytes become capable of recognizing and responding to particular pathogens while in the thymus gland. The gland makes hormones necessary for T cell maturation.

The fist-sized **spleen**, which resides in the upper-left portion of the abdomen, is the largest lymphoid organ. During prenatal development it produces red blood cells. After birth, it filters worn-out and damaged red blood cells from the many blood vessels that branch through it. White blood cells in the spleen detect and respond to pathogens in the blood and lymph that filter through it. People can survive without a spleen, but they become more vulnerable to infections.

Tonsils are patches of lymphoid tissue. There are two tonsils the back of the throat (lingual tonsils) and others (called adenoids) at the rear of the nasal cavity. Tonsils help the body respond quickly to inhaled pathogens. The lining of the intestine and the appendix also include lymphoid tissue. Lymphocytes in these tissues help the body respond to food-borne pathogens.

Mechanisms by which lymphocytes in the lymphoid organs detect and respond to threats are discussed in Chapter 37.

TAKE-HOME MESSAGE 36.11

✔ The lymph vascular system consists of vessels that collect and deliver excess water and solutes from interstitial fluid to blood. It also carries absorbed fats to the blood, and delivers pathogens to lymph nodes.

✔ The system's lymphoid organs, including lymph nodes, have specific roles in body defenses.

36.1 A Shocking Save (revisited)

Most cardiac arrests do not occur in a hospital, so the presence of a bystander willing to carry out CPR or to use an AED often means the difference between life and death. Sadly, although most cardiac arrests are witnessed, only about 15 percent of victims get CPR before trained personnel arrive.

Most people have an understandable reluctance to engage in mouth-to-mouth contact with a stranger. This can be a problem with traditional CPR, which calls for a rescuer to alternate between exhaling into a victim's mouth to inflate the lungs and providing chest compressions. A new procedure called CCR (cardiocerebral resuscitation) relies on chest compressions alone to move air into and out of the lungs. Proponents of CCR argue that as long as a person's airway is clear, chest compressions will move enough air into and out of a victim's lungs to oxygenate the blood flowing to the person's heart and brain. Early studies indicate that CCR may be as good as or even better than traditional CPR at saving most people who experience a sudden cardiac arrest.

Another promising development is the rising availability of AEDs. In 2010, approximately 200,000 AEDs were purchased for installation in public places. The use of such devices is already saving lives. A recent study that looked at the outcome of nearly 14,000 out-of-hospital cardiac arrests concluded that, compared to no treatment, bystander use of an AED increased the likelihood of surviving to discharge from the hospital by about 75 percent. That's the good news. The bad news is that most people still do not know what an AED is, where they are available, or how simple it is to use one. The photo at the upper right shows the symbol for an AED in a public place.

AED
Automated
External
Defibrillator
Sign in an airport indicating where an AED is stored.

STUDY GUIDE

Section 36.1 When the heart stops pumping, blood flow halts and brain cells begin to die from lack of oxygen. CPR can keep some oxygenated blood moving to cells, but it cannot restart a heart. Reestablishing the normal rhythm requires a shock from a defibrillator.

Section 36.2 A **circulatory system** moves substances through a body by way of a fluid transport medium. Some invertebrates have an **open circulatory system**, in which **hemolymph** pumped out of vessels mixes with interstitial fluid. In other invertebrates and all vertebrates, a **closed circulatory system** confines **blood** inside a **heart** and blood vessels. Exchanges between blood and interstitial fluid take place across walls of **capillaries**.

In fish, a heart with one **atrium** and one **ventricle** pumps blood through a single circuit. A two-circuit system evolved in concert with tetrapod lungs. In tetrapods, blood travels to the lungs and back by way of a **pulmonary circuit**. The longer **systemic circuit** conveys blood to other body tissues and back. In birds and mammals, these two circuits are completely separate.

Section 36.3 **Arteries** receive blood from the heart. The **aorta** is the largest artery of the systemic circuit. **Pulmonary arteries** carry oxygen-poor blood to lungs. Exchanges take place across the walls of capillaries, then **veins** carry blood back to the heart. The **superior vena cava** and **inferior vena cava** return blood from the systemic circuit. **Pulmonary veins** return oxygenated blood from the lungs. Most blood flows through one capillary system, but blood in intestinal capillaries also flows through liver capillaries.

Section 36.4 A human heart has two halves, each with an atrium above a ventricle. Oxygen-poor blood from the body enters the right atrium and moves through a valve into the right ventricle. The right ventricle pumps blood through a valve into the pulmonary circuit. After picking up oxygen in the lungs, blood enters the heart's left atrium and moves through a valve into the left ventricle. The left ventricle pumps the blood through a valve into the systemic circuit.

During one **cardiac cycle**, all heart chambers undergo rhythmic relaxation (**diastole**) and contraction (**systole**). Contraction of the ventricles provides the force that propels blood through blood vessels.

The **sinoatrial (SA) node** in the right atrium wall is the cardiac pacemaker. The **atrioventricular (AV) node** serves as an electrical bridge to the ventricles. The delay between atrial contraction in response to SA signaling and ventricular contraction in response to signals arriving via the AV node allows the ventricles to fill fully before they contract.

Section 36.5 Blood consists of **plasma**, blood cells, and platelets. Plasma is mostly water. Plasma proteins include albumin and clotting factors, both made by the liver, and immunoglobulins made by white blood cells. Plasma also contains ions, hormones, nutrients, and dissolved gases. Cellular components of the blood are made in red bone marrow. **Red blood cells** contain hemoglobin, a protein that functions in oxygen transport. The kidney hormone **erythropoietin** encourages their synthesis. **White blood cells** help defend the body against damaged cells and pathogens. **Platelets** function in **hemostasis** (clotting).

Sections 36.6, 36.7 **Blood pressure** results from the force exerted on blood by ventricular contraction, and it declines as blood proceeds through a circuit. It is usually recorded as **systolic pressure** over **diastolic pressure**. A **pulse** is a brief expansion of an artery caused by ventricular contraction. Stretching and recoil of elastic tissue in arteries helps keep blood moving between contractions.

Arterioles are the main site for regulation of blood flow through the systemic circuit. **Vasodilation** of arterioles supplies more blood to a region. **Vasconstriction** decreases inward flow. Adjustments in arteriole diameter alter blood pressure in the short term. Adjustments in urine output affect it over the longer term.

Section 36.8 Blood flow slows in capillaries because of their collectively large cross-sectional area. Substances leave a capillary by diffusion, through transport proteins, or in fluid that seeps out between cells. Loss of fluid at the arterial end of a capillary is more or less balanced by return of water by osmosis at the capillary's venuous end. The slight excess of fluid that enters the interstitial fluid is returned to the circulation by way of the lymphatic circulation.

Section 36.9 Blood from capillaries enters **venules**, which then empty into veins. Veins are the body's main blood reservoir. Blood pressure in the veins is low, but one-way valves and the pressure exerted on veins by skeletal muscle contractions and respiratory movements keep blood moving toward the heart.

Section 36.10 Some disorders affect the number or function of blood cells. Anemia is a disorder in which red blood cells are few in number or impaired. Abnormal heart rhythms can slow or halt blood flow. Ventricular fibrillation is especially dangerous. Flow is also impaired when atherosclerosis narrows the interior of a blood vessel. A heart attack occurs when the blood supply to heart muscle is interrupted. A stroke occurs when blood flow to the brain is impaired.

Section 36.11 Some fluid that leaves capillaries enters the **lymph vascular system**. The fluid, now called **lymph**, is filtered by **lymph nodes**. The **spleen** filters the blood and removes any old red blood cells. Tonsils and adenoids are lymphoid organs that help fend off pathogens. The thymus is the site of T cell (T lymphocyte) maturation.

SELF-QUIZ

Answers in Appendix VII

1. The circulatory system of an insect ________.
   a. is a closed system
   b. includes a four-chambered heart
   c. circulates hemolymph
   d. moves blood through four separate circuits.

2. In ________, blood flows through a single circuit.
   a. fishes
   b. amphibians
   c. birds
   d. mammals

3. The systemic circuit of mammals ________.
   a. carries blood to the lungs and back
   b. is shorter than the pulmonary circuit
   c. carries blood to body tissues and back
   d. allows oxygenated and oxygen-poor blood to mix

4. The plasma protein albumin is made by ________.
   a. white blood cells   c. the heart
   b. red blood cells   d. the liver

5. Platelets function in ________.
   a. oxygen transport   c. thermal regulation
   b. blood clotting   d. movement of lymph

6. Most oxygen in blood is transported ________.
   a. in red blood cells   c. in platelets
   b. in white blood cells   d. bound to LDLs

7. Blood flows directly from the left atrium to ________.
   a. the aorta   c. the right atrium
   b. the left ventricle   d. the pulmonary arteries

8. Contraction of the ________ drives blood out into the aorta and through the systemic circuit.
   a. left atrium   c. right atrium
   b. left ventricle   d. right ventricle

9. Blood pressure is highest in the ________ and lowest in the ________.
   a. arteries; veins   c. veins; arteries
   b. arterioles; venules   d. capillaries; arterioles

10. In a person at rest, the largest volume of blood is in ________.
    a. arteries   c. veins
    b. capillaries   d. arterioles

11. Which of the following has the thickest wall?
    a. left atrium   c. right atrium
    b. left ventricle   d. right ventricle

12. Lymph nodes filter ________.
    a. blood   c. plasma
    b. lymph   d. urine

13. Which artery carries oxygen-poor blood?

14. Match the terms with their definition.

___ anemia		a. heart quivers
___ hypertension		b. impaired clotting
___ thrombus		c. cancer of bone marrow
___ hemophilia		d. too few red cells
___ stroke		e. high blood pressure
___ atrial fibrillation		f. clot that stays in place
___ leukemia		g. interrupted blood flow in brain

15. Match the terms with their definition.

___	bone marrow	a. largest artery
___	spleen	b. cardiac pacemaker
___	vitamin K	c. source of all blood cells
___	HDL	d. inside red blood cells
___	SA node	e. carries cholesterol to liver
___	erythropoietin	f. hormone made by kidney
___	aorta	g. lymphoid organ
___	hemoglobin	h. brings blood to left atrium
___	vena cava	i. cofactor for clotting factor

CRITICAL THINKING

1. Suppose a blood clot forms in the leg, then breaks free as an embolus. Is it more likely to become stuck in a lung or in the brain? Explain your reasoning.

2. An atrial septal defect is a birth defect in which there is an opening in the septum between the left and right atria. People with such a defect typically have more than a normal amount of blood in the right atrium. Why does blood flow from left to right across the septum rather than in the opposite direction? The excess of blood can cause hypertension in one circuit. Which one?

3. Chronic alcoholism damages the liver. In people who have alcohol-induced liver damage, the blood concentration of albumin can be unusually low. Explain why a low concentration of blood albumin can cause edema, a condition in which excess fluid accumulates in body tissues.

4. Recombinent human erythropoietin (Epo) was developed for for medical use. For example, it can be beneficial to people who have kidney damage and do not produce enough erythropoietin on their own. However, Epo is also used by some athletes as an illicit drug that boosts performance. Explain why Epo can enhance performance in some athletic events.

CREDIT: (22) National Vital Statistics System—Mortality (NVSS-M), NCHS, CDC.

37 IMMUNITY

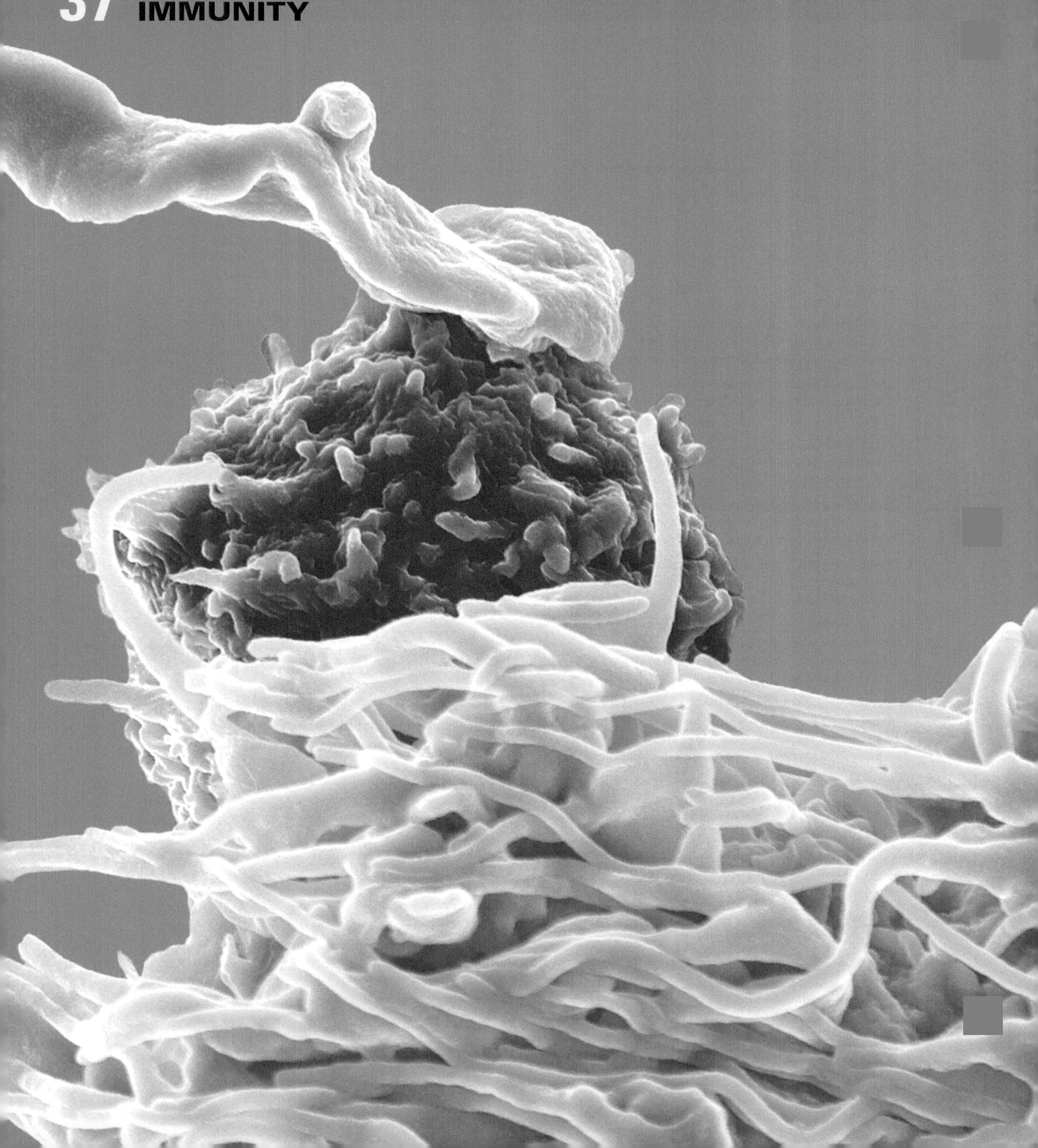

CORE CONCEPTS

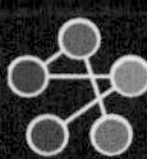

Systems

Complex properties arise from interactions among components of a biological system.
Vertebrate immunity comprises dynamic, multilevel responses to threats. Infections, toxins, and allergens that disrupt the body at the cellular level can affect the overall health of the organism. Feedback mechanisms inherent in immune responses maintain the fine line between eliminating a threat and damaging the body.

Evolution

Evolution underlies the unity and diversity of life.
Shared core processes and features provide evidence that all living things are descended from a common ancestor. All plants and animals have a set of general mechanisms that defend the body from invasion by pathogens. Vertebrates have an extra level of protection in which defense responses are tailored to combat specific pathogens.

Information Flow

Living systems store, retrieve, transmit, and respond to information essential for life.
Vertebrate adaptive immunity involves interactions among a network of specialized cells. The interactions involve communication via chemical signaling molecules. The activation of innate immunity triggers an adaptive immune response via this type of communication.

Links to Earlier Concepts

This chapter discusses bacteria and viruses (Sections 4.3, 20.1–20.4, 20.7) in the context of immunity. Before starting, be sure you understand cell signaling (34.2), and the components and functioning of the circulatory (36.5, 36.6, 36.8, 36.10) and lymphatic systems (36.11). The chapter also revisits chitin (3.3), cell structure and function (4.3, 4.5, 4.9, 30.8), membrane proteins (5.7) and membrane-crossing mechanisms (5.8–5.10), ABO blood typing (13.5), dideoxynucleotides (15.4), SCID and gene therapy (15.9), symbiosis (19.7), normal microbiota (20.8), animal tissues and organs (31.2–31.4, 31.8, 32.10), the vertebrate nervous system (31.6, 32.6, 32.8), and thyroid disorders (34.6).

37.1 Community Immunity

A preparation used to elicit immunity to a specific agent of disease is called a **vaccine**. A typical vaccine is either injected or given orally, and it contains a weakened, killed, or fragmented pathogen.

The first vaccine was developed as a result of desperate attempts to survive epidemics of smallpox, a severe disease that kills up to one-third of the people it infects. Before 1880, no one knew what caused infectious diseases or how to protect anyone from getting them, but there were clues. In the case of smallpox, survivors seldom contracted the disease a second time. They were said to be immune—protected from infection.

At the time, the idea of acquiring immunity to smallpox was extremely appealing. People had been risking their lives on it by poking into their skin bits of smallpox scabs or threads soaked in pus from smallpox sores. Some survived the crude practices, but many others did not.

By 1774, it had long been known that dairymaids usually did not get smallpox after they had contracted cowpox, a mild disease that affects humans as well as cattle. An English farmer collected pus from a cowpox sore on a cow's udder, and poked it into the arm of his pregnant wife and two small children. All survived the smallpox epidemic, but were from then on subject to rock peltings by neighbors convinced they would turn into cows.

Twenty years later, the English physician Edward Jenner injected liquid from a cowpox sore into the arm of a healthy boy. Six weeks later, Jenner injected the boy with liquid from a smallpox sore. The boy did not get smallpox. Jenner's experiment showed directly that the agent of cowpox elicits immunity to smallpox. Jenner named his procedure "vaccination," after the Latin word for cowpox (vaccinia). Though still controversial, Jenner's vaccine spread quickly through Europe, then to the rest of the world. The last known case of naturally occurring smallpox was in 1977. Use of the vaccine eradicated the disease.

Today, we know that vaccination with cowpox is effective against smallpox because the two diseases are caused by closely related viruses. Vaccines for many other infectious diseases have since reduced suffering and deaths overwhelmingly, but only where they are available—and used. Public confidence is a necessary part of their success. When enough individuals refuse vaccination, outbreaks of preventable and sometimes fatal diseases occur.

vaccine (vax-SEEN) A preparation introduced into the body in order to elicit immunity to a specific pathogen.

CREDIT: (opposite) Dr. Olivier Schwartz, Institut Pasteur/Science Source.

37.2 Integrated Responses to Threats

LEARNING OBJECTIVES

- Explain immunity in terms of antigens and receptors.
- Describe the three lines of defense in vertebrate immunity.
- Use examples to describe the main functions of white blood cells.

You continually cross paths with huge numbers of viruses, bacteria, fungi, parasitic worms, and other agents of disease, but humans coevolved with these pathogens so your body has defenses against them. The evolution of **immunity**, an organism's capacity to resist and combat infection, began even before multicelled eukaryotes evolved from free-living cells. Mutations introduced variations in membrane proteins of different cell lineages; as multicellularity evolved, so did mechanisms of identifying such molecules as self, or belonging to one's own body. Central to these mechanisms are receptor proteins: membrane proteins that trigger a change in cell activity in response to a specific stimulus (Section 5.7).

By about 1 billion years ago, the ability to recognize nonself molecules (that occur on other organisms) had also evolved. Cells of modern animals have a set of receptors that collectively can recognize around 1,000 nonself molecules called pathogen-associated molecular patterns, or **PAMPs**. (Plants and some algae have PAMP receptors too.) As their name suggests, PAMPs occur mainly on or in microorganisms that can be pathogenic. PAMPs include peptidoglycan of bacterial cell walls and flagellin of bacterial flagella, chitin of fungal cell walls, double-stranded RNA of some viruses, and so on. A PAMP is an example of **antigen**, any molecule or particle that is recognized by the body as nonself. Not all antigens are PAMPs: Molecules on body cells of other individuals can be antigenic, for example.

Three Lines of Defense

Anatomical and physiological barriers that prevent most microorganisms from entering the internal environment are the body's first line of defense against pathogens (FIGURE 37.1). A microorganism that breaches these barriers activates mechanisms of innate immunity, the second line of defense. **Innate immunity** comprises a set of immediate, general defenses that help protect all multicelled organisms from infection. These defenses are initiated by tissue injury and by PAMP receptors that bind to antigen-bearing particles.

In vertebrates, the activation of innate immunity triggers adaptive immunity, the third line of defense. **Adaptive immunity** comprises a set of immune defenses that can be tailored to fight billions of specific

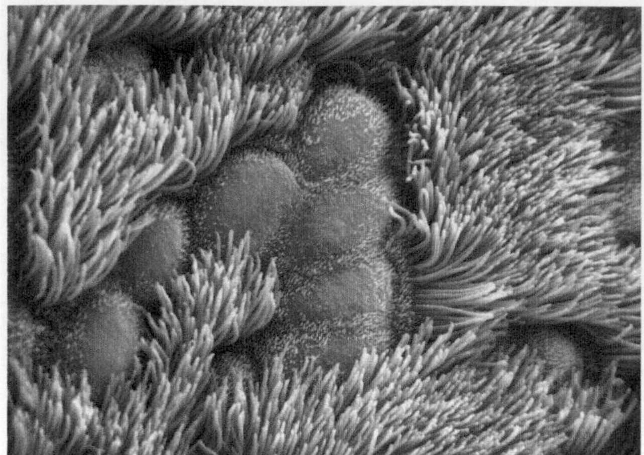

FIGURE 37.1 A physical barrier to infection.
In the lining of the airways leading to human lungs, goblet cells (gold) secrete mucus that traps bacteria and particles. Cilia (pink) on other cells sweep the mucus toward the throat for disposal.

TABLE 37.1

Comparing Features of Innate and Adaptive Immunity

	Innate Immunity	Adaptive Immunity
Response time	Immediate	About a week
Antigen recognition	About 1,000 PAMPs	Billions of potential antigens
Specificity	General response	Response specific to pathogen that triggers it
Persistence	None	Long-term

adaptive immunity In vertebrates, system of immune defenses that can be tailored to fight specific pathogens as an organism encounters them during its lifetime.
antigen (AN-tih-jenn) A molecule or particle that the immune system recognizes as nonself.
B cell B lymphocyte. Leukocyte that can make antibodies in an adaptive immune response.
basophil (BASE-uh-fill) Granular, circulating white blood cell.
cytokines (SITE-uh-kynes) Cell-to-cell signaling molecules secreted by leukocytes to coordinate activities during immune responses.
dendritic cell (den-DRIT-ick) Phagocytic leukocyte that specializes in antigen presentation.
eosinophil (ee-uh-SIN-uh-fill) Granular, circulating leukocyte that targets large, multicelled parasites.
immunity The body's ability to resist and fight infections.
innate immunity In all multicelled organisms, set of immediate, general defenses against infection.
macrophage (MACK-roe-faje) Leukocyte that specializes in phagocytosis.
mast cell Granular leukocyte that stays anchored in tissues.
neutrophil (NEW-tra-fill) Granular, circulating leukocyte; most abundant leukocyte in blood.
NK cell Natural killer cell. Leukocyte that can kill cancer cells undetectable by cytotoxic T cells.
PAMPs Pathogen-associated molecular patterns. Molecules found on or in microorganisms that can be pathogenic.
T cell T lymphocyte. Leukocyte that plays a central role in adaptive immunity; some kinds target infected or cancerous body cells.

pathogens that an individual may encounter during its lifetime. Two types of adaptive immune responses (antibody-mediated and cell-mediated) target different types of threats. Innate and adaptive immunity have different features (TABLE 37.1), but they work together.

The Defenders

Vertebrate white blood cells (leukocytes, Section 36.5) participate in both innate and adaptive immune responses. Almost all of these cells spend at least part of their life circulating in blood and lymph, and many populate the lymph nodes, spleen, and other solid tissues. White blood cells communicate by secreting and responding to chemical signaling molecules. These molecules, which include proteins and polypeptides called **cytokines**, allow cells throughout the body to coordinate activities during an immune response. Interleukins, interferons, and tumor necrosis factors are examples of cytokines.

All white blood cells are phagocytic—they carry out phagocytosis (Section 5.10) to some degree. A phagocytic leukocyte (a phagocyte) engulfs and digests pathogens, dead cells, or other particles. All white blood cells also have the ability to process and present antigen, a mechanism that alerts the adaptive immune system to threats (Section 37.6 returns to this topic). Aside from these similarities, different types of white blood cells specialize in different tasks (FIGURE 37.2). **Neutrophils** are the most abundant leukocytes in blood, so they are often the first responders at a site of infection or tissue damage. Monocytes also circulate in blood, but during an infection they migrate into solid tissue and differentiate into macrophages and dendritic cells. **Macrophages** are "professional" phagocytes that engulf essentially everything except undamaged body cells. **Dendritic cells** specialize in antigen presentation.

Some white blood cells have secretory vesicles called granules, and these can contain cytokines, local signaling molecules, destructive enzymes, and toxins such as hydrogen peroxide. A cell degranulates (releases the contents of its granules) in response to a trigger such as antigen binding. Neutrophils have granules, as do **eosinophils**, which are leukocytes that can target parasites too big for phagocytosis. **Basophils** and **mast cells** degranulate in response to injury as well as antigen. Mast cells, unlike other white blood cells, are permanently anchored in tissues. These cells also degranulate in response to signaling molecules secreted by cells of the endocrine and nervous systems.

Lymphocytes are a special category of white blood cells that carry out adaptive immune responses. **B cells** (B lymphocytes) make antibodies in one type of adaptive response. **T cells** (T lymphocytes) play a central

Neutrophil Most abundant white blood cell. Leukocyte that circulates in blood and migrates into damaged or infected tissues, where it participates in inflammatory response by degranulating.

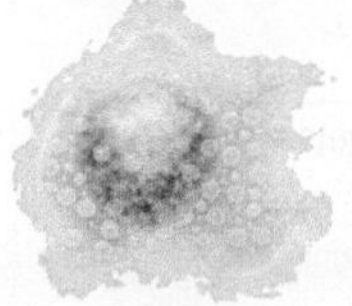
Macrophage Starts out as a circulating monocyte that enters damaged tissue and differentiates. Major phagocytic cell that also presents antigen.

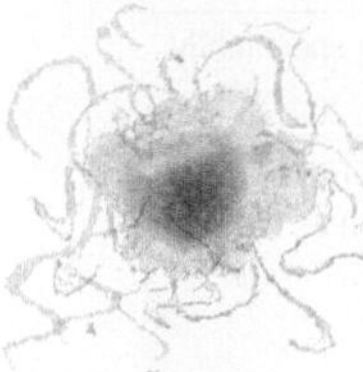
Dendritic cell Starts out as a circulating monocyte that enters damaged tissue and differentiates. Major antigen-presenting cell that is also phagocytic.

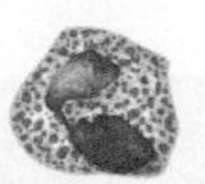
Eosinophil Circulates in blood; migrates into damaged or infected tissues, where it participates in inflammatory response. Granules contain enzymes effective against multicellular parasites.

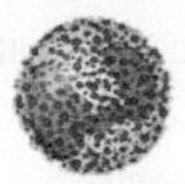
Basophil Circulates in blood; migrates into damaged or infected tissues, where it participates in inflammatory response by degranulating.

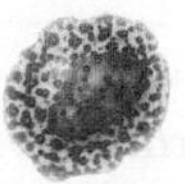
Mast cell Anchored in tissues. Participates in inflammation and allergies by degranulating.

Lymphocytes (below) carry out adaptive immune responses.

B cell Only type of cell that produces antibodies; participates in antibody-mediated adaptive responses.

T cell Several kinds play a central role in all adaptive immune responses.

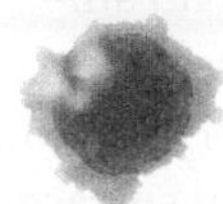
Natural killer (NK) cell Kills cancer cells undetectable by other white blood cells.

FIGURE 37.2 **Preview of leukocytes** (white blood cells). Details such as lobed nuclei and cytoplasmic granules are revealed by staining.

role in all adaptive immune responses. Cytotoxic T cells kill infected or cancerous body cells. **NK cells** (natural killer cells) can kill cancerous body cells that cytotoxic T cells cannot detect.

Invertebrates have no white blood cells and no adaptive immunity, but they do have other specialized immune cells and robust innate immunity. In these organisms, PAMP receptor binding triggers immune cell phagocytosis and the release of antimicrobial compounds. Some invertebrate immune cells also surround and encapsulate parasites that are too big for phagocytosis. Immune mechanisms are complex and they vary widely, as this example illustrates. For clarity, the remainder of the chapter focuses on immunity in humans and other mammals.

TAKE-HOME MESSAGE 37.2

- ✔ Vertebrate innate and adaptive immunity work together to combat infection and injury.
- ✔ Innate immune mechanisms are fast, general responses to tissue damage and invading microorganisms.
- ✔ Adaptive immunity is a system of defenses that can specifically target billions of different antigens.
- ✔ Different types of white blood cells (leukocytes) operate in both innate and adaptive immunity; signaling molecules such as cytokines integrate their activities.

37.3 Surface Barriers

LEARNING OBJECTIVES

- Use examples to explain some benefits and risks associated with microbiota.
- Describe the body's natural barriers and how they prevent microorganisms from entering the internal environment.

Biological Barriers

Your skin is in constant contact with the external environment, so it picks up many microorganisms. It normally teems with more than 4,000 species of bacteria, fungi, viruses, and archaea: up to 25 million microorganisms on every square inch of your external surfaces. They tend to flourish in warm, moist areas such as the armpit. Huge populations inhabit structures that open on the body's surface, including the eyes, nose, mouth, and anal and genital openings.

Microorganisms that typically live on human surfaces, including the interior tubes and cavities of the digestive and respiratory tracts, are called normal flora or normal microbiota (Section 20.8). Our body surfaces provide microorganisms with a stable environment and nutrients. Competition for attachment sites can deter more dangerous species from colonizing—and potentially penetrating—our epithelial tissues.

The human body has at least as many microorganisms as human cells. Most of them inhabit the digestive tract, and normally they make up a large and diverse community there. Gut microbiota are not just random passengers; they live in symbiosis with us (Section 19.7). These microorganisms help us digest food, and they make essential nutrients such as vitamins K and B_{12}. They also play a significant role in human health that goes far beyond nutrition. Mounting evidence suggests that species composition and diversity of gut microbiota have a major influence on our physiology, metabolism, mood, and immune function. Imbalances in this community have been associated with conditions that include diabetes, obesity, inflammatory bowel disease, and cancer.

Microbiota are helpful only on body surfaces.

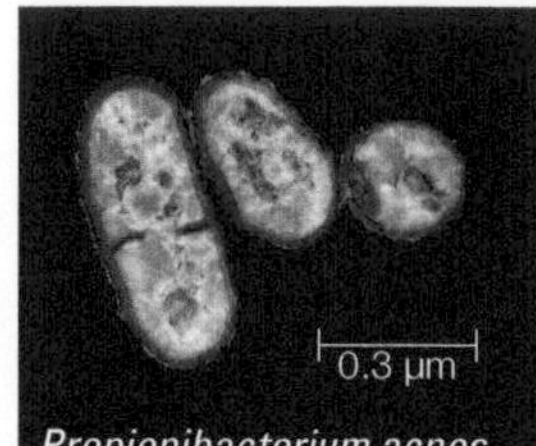
0.3 µm
Propionibacterium acnes

Consider a major constituent of microbiota, *Propionibacterium acnes* (left). This bacterium feeds on sebum, a greasy mixture of fats, waxes, and glycerides that lubricates hair and skin. Glands in the skin secrete sebum into hair follicles. During puberty, an increase in sex hormone production triggers an increase in sebum production. Excess sebum combines with dead, shed skin cells and blocks the openings of hair follicles. *P. acnes* can survive on the surface of the skin, but far prefers anaerobic habitats such as the interior of blocked hair follicles. There, the cells multiply to tremendous numbers. Secretions of their flourishing populations leak into internal tissues of the follicles and initiate inflammation (we return to inflammation in the next section). The resulting pustules are called acne.

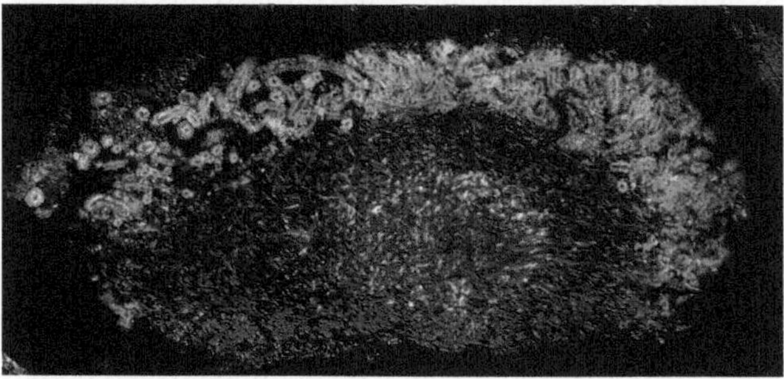

A Biofilm community in a chunk of human dental plaque. Six different genera of oral bacteria are indicated in different colors; each tiny dot or rod is a single bacterial cell.

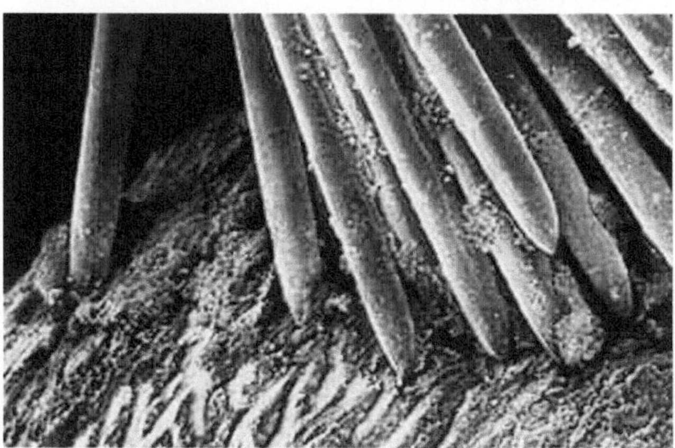

B Toothbrush bristles scrubbing plaque on the surface of a tooth.

FIGURE 37.3 Dental plaque.

CREDITS: (in text) Kwangshin Kim/Science Source; (3A) Blair Rossetti; (3B) www.zahnarzt-stuttgart.com.

About 60 species of microorganisms that inhabit the mouth are part of **dental plaque**, a thick biofilm of various bacteria and occasional archaea, their extracellular products, and saliva glycoproteins. Plaque sticks tenaciously to teeth (**FIGURE 37.3**). Some of the bacteria in plaque carry out lactate fermentation. The lactate they produce is acidic enough to dissolve minerals that make up the tooth, causing holes called cavities.

Tight junctions (Section 4.9) normally fasten gum epithelium to teeth. These junctions make a tight seal that excludes oral microorganisms. As we age, the connective tissue beneath the epithelium thins, so the seal weakens. Deep pockets can form between gums and teeth, and a nasty collection of anaerobic bacteria and archaea tends to accumulate in them. The microorganisms secrete destructive enzymes and acids that cause inflammation of the surrounding gum, a condition called periodontitis. Periodontal wounds are an open door to the circulatory system and its arteries. All genera of oral bacteria have been found in atherosclerotic plaque (Section 36.10). What role oral microorganisms play in atherosclerosis is not yet clear, but one thing is certain: They contribute to the inflammation that fuels coronary artery disease.

Microbiota can cause or worsen disease when they invade tissues. Serious illnesses associated with microbiota include pneumonia, ulcers, colitis, whooping cough, meningitis, abscesses of the lung and brain, and cancers of the colon, stomach, and intestine. The bacterial agent of tetanus, *Clostridium tetani*, is considered a normal inhabitant of human intestines. The bacteria responsible for diphtheria, *Corynebacterium diphtheriae*, was normal skin flora before widespread use of the vaccine eradicated the disease. *Staphylococcus aureus*, a resident of human skin and linings of the mouth, nose, throat, and intestines, is also a leading cause of human bacterial disease.

Physiological and Anatomical Barriers

In contrast to body surfaces, the blood and tissue fluids of healthy people are typically free of microorganisms (sterile). Surface barriers (**TABLE 37.2**) usually prevent microorganisms from entering the body's internal environment. The tough outer layer of skin (Section 31.8) is an example. Microorganisms flourish on skin's waterproof, oily surface, but they rarely penetrate its thick layer of dead cells (**FIGURE 37.4**). The thinner epithelial tissues that line the body's interior tubes and cavities also have surface barriers. For example, sticky mucus secreted by cells of these linings can trap microorganisms. The mucus contains **lysozyme**, an enzyme that kills bacteria. In the sinuses and respiratory tract, the coordinated beating of cilia sweeps away trapped microorganisms before they have a chance to breach the delicate walls of these structures (ciliated epithelium of the respiratory tract is shown in Figure 37.1).

Your mouth is a particularly inviting habitat for microorganisms because it offers plenty of nutrients and warm, moist pockets suitable for colonization. Accordingly, it harbors huge populations of microbiota that can resist lysozyme in saliva. Most microorganisms that get swallowed are killed in the stomach by gastric fluid, a potent brew of protein-digesting enzymes and acid. Those that survive passage to the small intestine are usually killed by bile salts secreted into the fluid there. The hardy ones that reach the large intestine must compete with about 500 resident species that are specialized to live there and have already established large populations. Any that disrupt this resident microbial community are typically flushed out by diarrhea.

Lactate produced by *Lactobacillus* bacteria (Section 20.7) helps keep the vaginal pH below the range of tolerance of other bacteria and most fungi.

dental plaque On teeth, a thick biofilm composed of various microorganisms, their extracellular products, and saliva glycoproteins.
lysozyme (LICE-uh-zime) Antibacterial enzyme in body secretions such as saliva and mucus.

TABLE 37.2

Examples of Surface Barriers

Biological	Established populations of microbiota
Physiological	Secretions (sebum, other waxy coatings, mucus); low pH of gastric juices, urinary and vaginal tracts; lysozyme
Anatomical	Epithelia that line tubes and cavities such as the gut and eye sockets; skin; broomlike action of cilia; flushing action of tears, saliva, urination, diarrhea

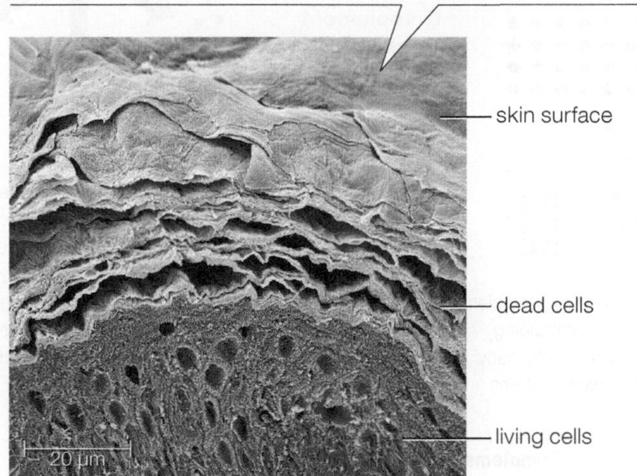
skin surface
dead cells
living cells
20 µm

FIGURE 37.4 A surface barrier to infection: the epidermis of human skin. Its thick, waterproof layer of dead cells usually keeps normal skin flora from penetrating internal tissues.

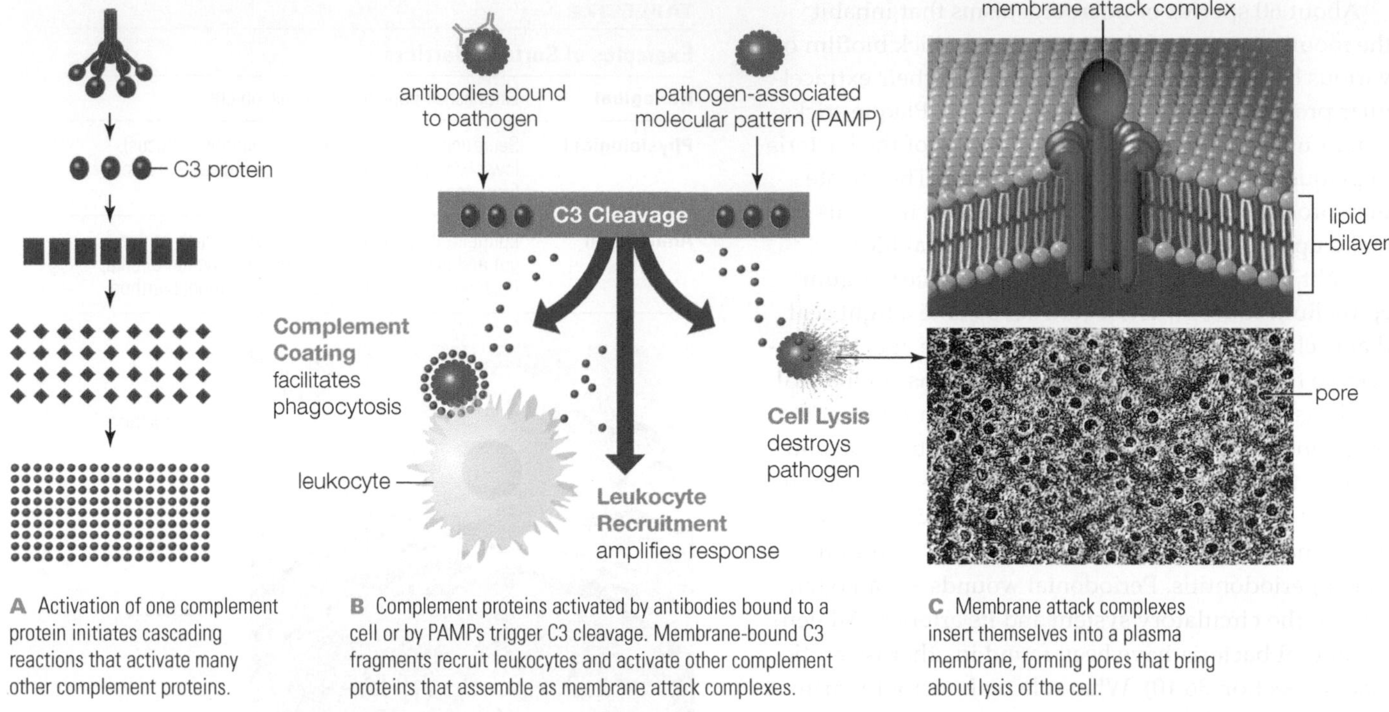
C3 protein
antibodies bound to pathogen
pathogen-associated molecular pattern (PAMP)
C3 Cleavage
Complement Coating facilitates phagocytosis
leukocyte
Leukocyte Recruitment amplifies response
Cell Lysis destroys pathogen
membrane attack complex
lipid bilayer
pore

A Activation of one complement protein initiates cascading reactions that activate many other complement proteins.

B Complement proteins activated by antibodies bound to a cell or by PAMPs trigger C3 cleavage. Membrane-bound C3 fragments recruit leukocytes and activate other complement proteins that assemble as membrane attack complexes.

C Membrane attack complexes insert themselves into a plasma membrane, forming pores that bring about lysis of the cell.

FIGURE 37.5 Complement activation.

Urination's flushing action usually prevents pathogens from colonizing the urinary tract.

TAKE-HOME MESSAGE 37.3

✔ A microorganism can cause infection if it enters the internal environment by penetrating skin or other protective barriers at the body's surfaces.

✔ Skin's tough epidermis is one physical barrier to infection. Resident populations of normal microbiota deter more dangerous microorganisms from colonizing skin and the linings of internal tubes and cavities.

✔ Lysozyme-containing mucus and other secretions offer more protection, as do the waving action of cilia and the flushing action of urine and diarrhea.

37.4 Mechanisms of Innate Immunity

LEARNING OBJECTIVES

- Describe the role of complement in innate immunity.
- Describe the process of inflammation.
- Explain how fever occurs and its advantages.

What happens if a pathogen slips by surface defenses and enters the body's internal environment? A second line of defense—the fast-acting, general mechanisms of innate immunity—can stop or slow a pathogen's proliferation for a few days until adaptive immunity peaks.

Complement

Complement is a set of proteins that circulate in inactive form in blood and tissue fluid; when activated, complement is part of responses against extracellular pathogens. Activated complement proteins mark antigenic particles for uptake by phagocytic leukocytes, and they can also destroy foreign cells.

Complement was named because the proteins were identified by their ability to "complement" the action of antibodies in an adaptive immune response (Section 37.7 returns to these responses). Mammals have about 30 types of complement proteins, and some of them recognize and bind to antibodies clustered on the surface of a cell. The binding initiates a series of reactions that ultimately results in cleavage of a complement protein called C3.

C3 is extremely abundant, so its cleavage products accumulate very quickly. Some of the C3 fragments attach directly to cell membranes of invading pathogens as well as body cells nearby. These fragments become enzymatic when they attach to a membrane, and they begin to activate other complement proteins, which activate other complement proteins, and so on (**FIGURE 37.5**). The cascading reactions quickly produce a huge amount of activated complement in the vicinity of the membrane's surface. These proteins diffuse into surrounding tissues to form a concentration gradient. Leukocytes can follow this

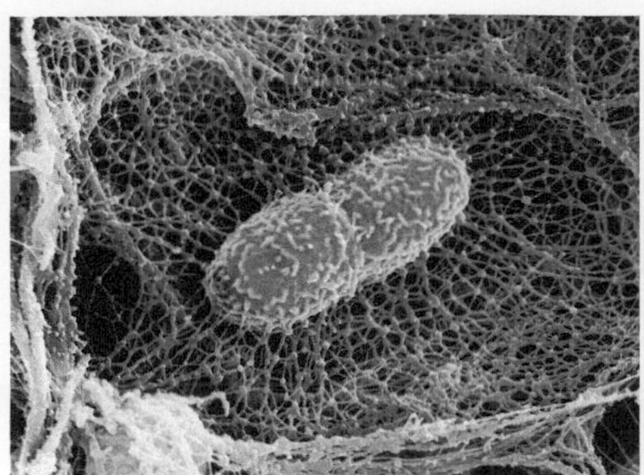

A These *Klebsiella* bacteria (pink) in lung tissue have been ensnared in a neutrophil net.

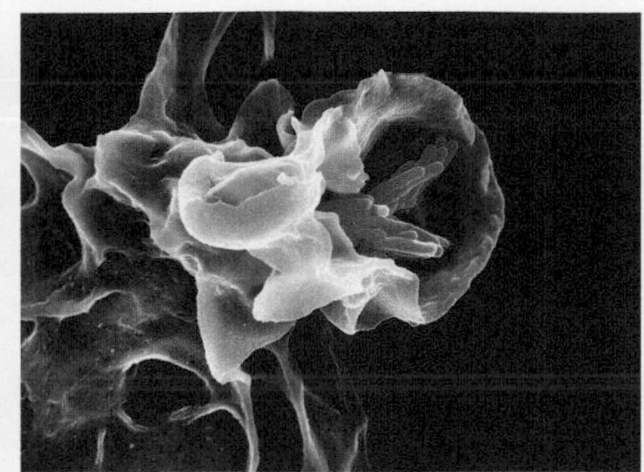

B Macrophage caught in the process of engulfing tuberculosis bacteria (red).

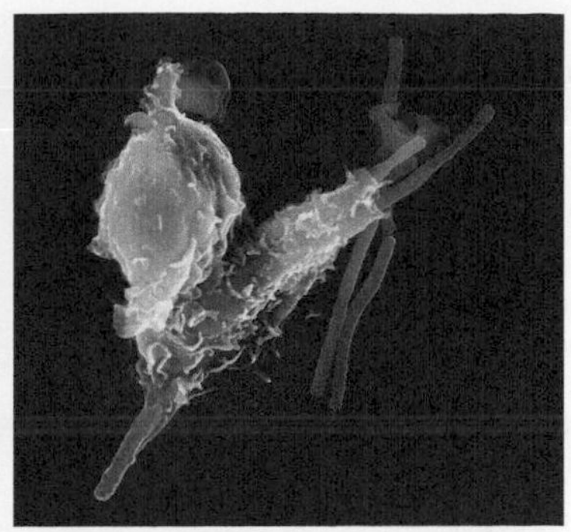

C Dendritic cell engulfing anthrax bacteria (*Bacillus anthracis*, red).

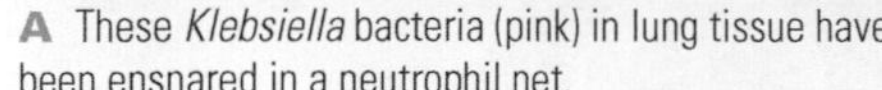

FIGURE 37.6 Some phagocytic leukocytes and their role in innate immunity.

gradient back to its origin: the antibody-coated cell that triggered the complement cascade.

Some activated complement proteins bind to microorganisms, damaged cells, and debris, forming a complement coating that enhances uptake of these particles by phagocytes. Other complement proteins assemble into structures called membrane attack complexes that puncture plasma membranes. Ions in extracellular fluid that flow through these channels trigger cell lysis (Section 20.3).

Years after complement was named, researchers discovered that some complement proteins function independently of antibodies. For example, the complement protein C1Q (shown at left) has six binding sites for simple carbohydrates found on the surface of a variety of pathogens. C1Q can bind directly to *Candida albicans* (a yeast); HIV and influenza virus; *Salmonella* and *Streptococcus* bacteria; and *Leishmania*, a protozoan parasite, for example. When all six of the protein's binding sites are occupied, it activates an enzyme that cleaves C3, thus initiating the complement cascade. Cytoplasmic and mitochondrial proteins that leak out of damaged body cells also trigger a complement cascade.

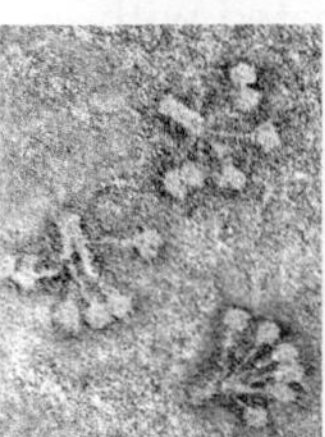

Normal body cells continuously produce proteins that inactivate complement, thus preventing a complement cascade from spreading too far into healthy tissue. Microorganisms do not make these inhibitory proteins, so they are singled out for destruction.

Phagocytosis

Neutrophils, macrophages, and dendritic cells have receptors for cytokines and activated complement. These receptors allow the phagocytes to follow a chemical trail—a type of movement called **chemotaxis**—leading to an affected tissue.

Neutrophils Neutrophils are the most abundant leukocyte, but they are also short-lived. Their turnover rate is phenomenal: The bone marrow of an adult human normally produces about 50 billion new neutrophils each day. This abundance is one reason that neutrophils are among the first responders to injury or infection, collecting within minutes at a site of tissue damage. After a neutrophil engulfs a microorganism, it releases the contents of its granules into the vesicle containing the engulfed pathogen (Section 4.5), thus destroying the microorganism. Neutrophils also degranulate when their receptors bind to antigens or signaling molecules. Enzymes and toxins released from the granules into extracellular fluid destroy all cells in the vicinity (including healthy body cells). Neutrophils literally explode in response to a certain combination of signaling molecules and complement, in the process ejecting their nuclear DNA and associated proteins along with the contents of their granules. The mixture solidifies into a net, trapping pathogens near antimicrobial compounds (**FIGURE 37.6A**). Neutrophil nets are very effective at killing bacteria.

Macrophages and Dendritic Cells Macrophages in interstitial fluid engulf microorganisms (**FIGURE 37.6B**), cell debris, and other particles, but they are more than just scavengers. Upon engulfing antigen, a macrophage secretes interleukins and other cytokines that alert the immune system to the presence of invading pathogens.

Dendritic cells (**FIGURE 37.6C**) patrol tissues that contact the external environment, such as the lining of respiratory airways. Phagocytosis by these cells protects the lungs from pathogens and harmful particles. However, the main function of dendritic cells

chemotaxis (key-mo-TAX-iss) Cellular movement in response to a chemical stimulus.

complement A set of proteins that circulate in inactive form, and when activated play a role in immune responses against extracellular pathogens.

CREDITS: (6A) SPL/Science Source; (6B) SPL/Science Source; (6C) Volker Brinkmann; (Intext) Courtesy K.B.M. Reid.

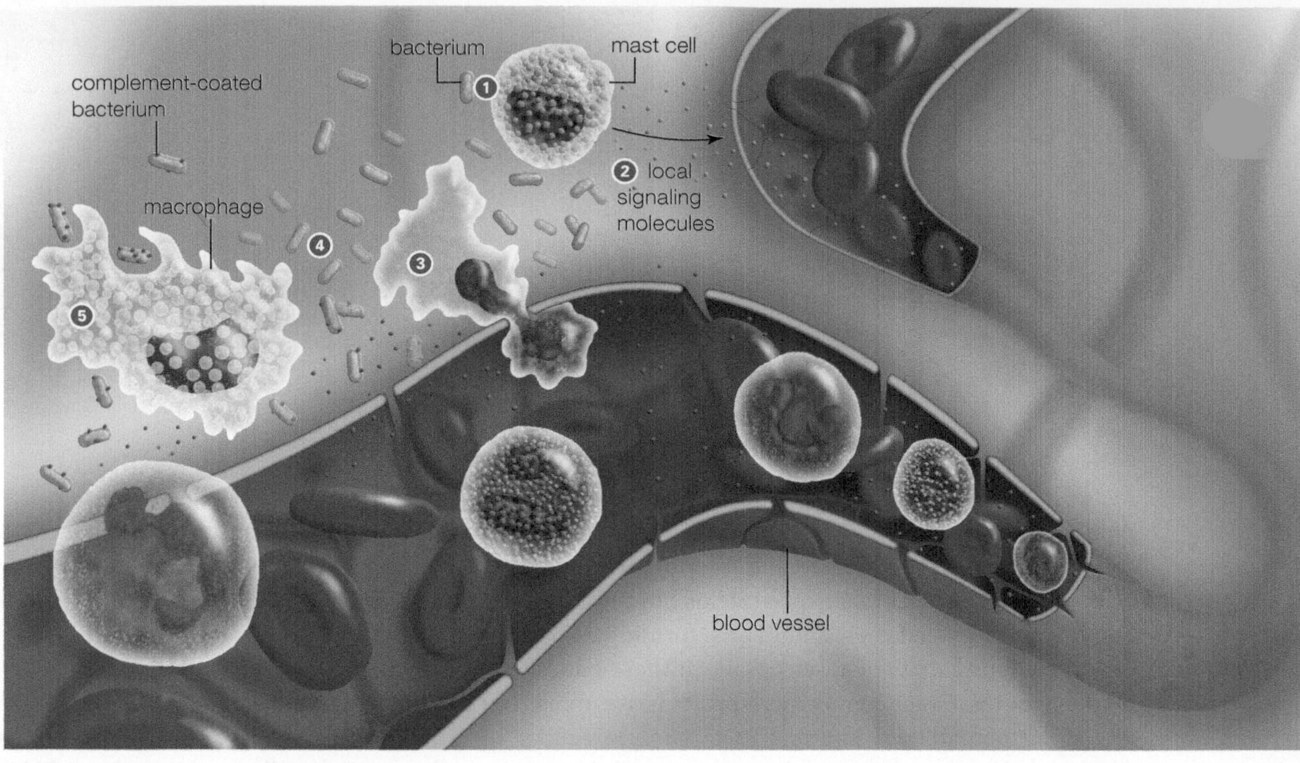
bacterium
mast cell
complement-coated bacterium
local signaling molecules
macrophage
blood vessel

FIGURE 37.7 Example of inflammation as a response to bacterial infection.

❶ PAMP receptors on mast cells in the tissue recognize and bind to bacteria.

❷ The mast cells release prostaglandins and histamines (blue dots) that cause arterioles to widen. The resulting increase in blood flow, which reddens and warms the tissue, hastens the arrival of phagocytes.

❸ The signaling molecules also increase capillary permeability, which allows phagocytes arriving in the bloodstream to move quickly through the vessel walls into the tissue. The tissue swells with fluid as plasma proteins seep out of the leaky capillaries.

❹ Meanwhile, bacterial antigens have triggered complement cascades, and invading bacteria have become coated with complement (purple dots).

❺ Phagocytic leukocytes in the tissue recognize and engulf the complement-coated bacteria.

is to present antigen to T cells in an adaptive immune response (more about how this works in Section 37.7).

Inflammation

Complement activation and cytokines released by leukocytes trigger two hallmarks of infection: inflammation and fever. **Inflammation** is a fast, local response that simultaneously destroys infected or damaged tissue and jump-starts the healing process (FIGURE 37.7). Inflammation begins when basophils, mast cells, or neutrophils degranulate, releasing the contents of their granules into an affected tissue. Degranulation occurs in response to a number of stimuli, such as PAMP receptor binding to a bacterial antigen ❶, or complement receptor binding to C3 fragments. Mast cells also degranulate in response to neuromodulators (Section 32.6) released at a site of tissue damage (FIGURE 37.8A).

A degranulating leukocyte releases cytokines. It also releases prostaglandins and histamines ❷, local signaling molecules that have two effects. First, they cause nearby arterioles to widen, so blood flow to the area increases. The increased flow speeds the arrival of circulating leukocytes attracted to the cytokines. Second, prostaglandins and histamines cause spaces to open up between the cells making up the walls of nearby capillaries. Leukocytes arriving in the bloodstream can move quickly into tissues by squeezing through the spaces ❸. Lymphocytes also exit capillaries by moving directly through cells of the vessels' walls (FIGURE 37.8B).

By the time circulating leukocytes enter an affected tissue, any invading cells have become coated with activated complement ❹. A complement coating makes the invaders easy targets for phagocytosis ❺.

As the capillary walls become more permeable, plasma proteins begin to escape into interstitial fluid. The fluid becomes hypertonic with respect to blood, and water follows by osmosis. The tissue swells with excess fluid, putting pressure on nerves. This pressure results in pain, as do prostaglandins released by the degranulating leukocytes. Other outward symptoms of inflammation include redness and warmth, which are indications of the area's increased blood flow.

Inflammation continues as long as its triggers do. When these stimuli subside, for example after invading bacteria have been cleared from an infected tissue, macrophages produce compounds that suppress inflammation and promote tissue repair. If the stimulus persists, inflammation becomes chronic. Chronic inflammation is not a normal condition. It does not benefit the body; rather, it causes or contributes to diseases such as asthma, Crohn's disease, rheumatoid arthritis, atherosclerosis, diabetes, and cancer.

Fever

Fever is a temporary, internally induced rise in body temperature above the normal 37°C (98.6°F) that often occurs in response to infection or serious injury. Some cytokines stimulate brain cells to make and release prostaglandins, which act on the hypothalamus (Section 32.10) to raise the body's internal (core) temperature set point. As long as core temperature remains below the new set point, the hypothalamus sends out signals that cause blood vessels in the skin to constrict, thus reducing heat loss from the skin. The signals also increase the metabolic rate, quickening heartbeat and respiration, and causing reflexive movements called shivering, or "chills," that boost the metabolic heat output of muscles. Together, these responses raise the body's internal temperature. If core temperature rises too much, sweating and flushing quickly lower it, thus maintaining the body's new temperature set point.

A fever is a sign that the body is fighting something, so it should never be ignored. However, a fever of 40.6°C (105°F) or less does not necessarily require treatment in an otherwise healthy adult. Body temperature usually will not rise above that value, but if it does, immediate hospitalization is recommended. Brain damage or death can occur if the core temperature reaches 42°C (107.6°F).

Fever enhances almost every step in innate and adaptive immunity. For example, fever speeds the production of white blood cells while greatly reducing the rate of viral replication and bacterial division. It also increases the phagocytic activity of white blood cells, boosts their production of antigen receptors, and enhances their recruitment to an affected area. Fever even increases the effectiveness of chemicals released by degranulating white blood cells.

TAKE-HOME MESSAGE 37.4

- ✔ Complement can be activated by the presence of antigen inside the body and by tissue damage.
- ✔ Activated complement recruits phagocytic leukocytes, coats all cells in an affected area, and triggers lysis of foreign cells.
- ✔ Phagocytic leukocytes engulf antigen-bearing particles, then release cytokines and other molecules.
- ✔ Complement activation, antigen detection, or tissue damage can trigger inflammation and fever.
- ✔ Inflammation occurs when granular leukocytes release prostaglandins and histamines at a site of tissue damage or infection. These local signaling molecules increase blood flow and enhance recruitment of additional leukocytes to the site.
- ✔ Fever enhances immune mechanisms while simultaneously slowing pathogen growth.

fever A temporary, internally induced rise in core body temperature above the normal set point.

inflammation A local response to tissue damage or infection; characterized by redness, warmth, swelling, and pain.

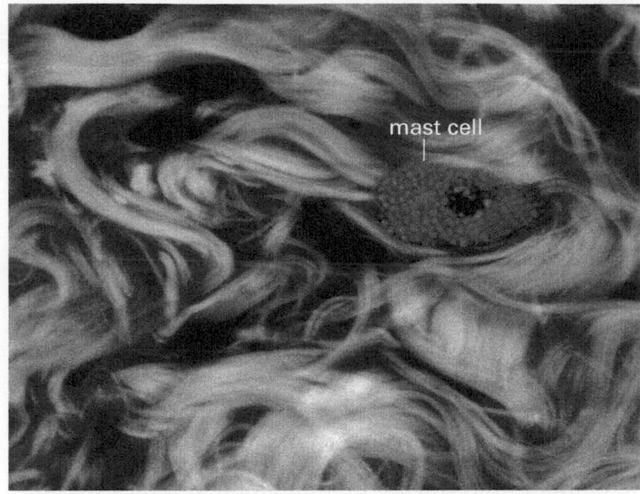
mast cell

A This human mast cell is embedded in connective tissue of the conjunctiva (Section 33.8). If it degranulates, for example in response to neuromodulators released at a site of tissue damage, the molecules it expels trigger inflammation. These molecules also loosen the tightly wound collagen fibers (pink, white) so phagocytes can pass through the tissue to reach the affected site.

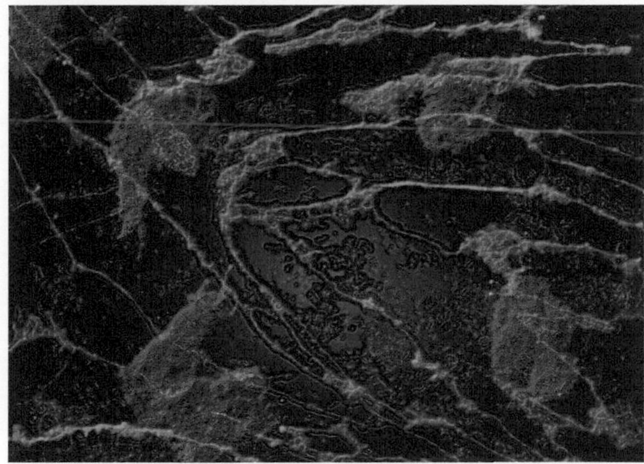

B T cells and NK cells (red) migrating through membranes and cytoplasm of endothelial cells (green). The ability to move directly through other cells allows lymphocytes to exit blood vessels quickly.

FIGURE 37.8 Leukocytes operating in innate immunity.

37.5 Antigen Receptors

LEARNING OBJECTIVES

- Explain how T cell receptors recognize antigen.
- Describe how an antibody's structure gives rise to its function.
- Explain the way antigen receptor diversity arises.

If innate immunity mechanisms do not quickly rid the body of an invading pathogen, an infection may become established in internal tissues. By that time, long-lasting mechanisms of adaptive immunity have already begun to target the invaders specifically. These mechanisms are triggered by cells that detect antigen via their antigen receptors. All body cells have one type

CREDITS: (8A) Mr. Donald Pottle,The Schepens Eye Research Institute; (8B) David Becker/ Science Source.

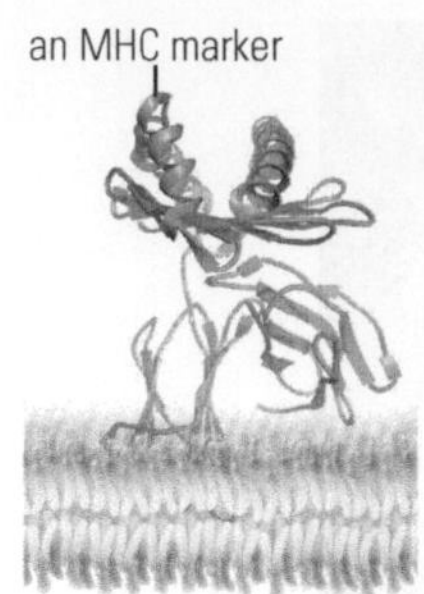
an MHC marker

of antigen receptor: plasma membrane proteins that recognize PAMPs. T cells also have antigen receptors called **T cell receptors**, or TCRs. Part of a TCR recognizes antigen as nonself. Another part recognizes certain proteins in the plasma membrane of body cells as self. These self-proteins are called **MHC markers** (left) after the major histocompatibility complex genes that encode them. MHC genes have thousands of alleles, so the cells of even closely related individuals rarely bear the same MHC markers.

Antibodies are another type of antigen receptor. **Antibodies** are Y-shaped proteins made and secreted by B cells. Antibodies circulate in blood and can enter interstitial fluid during inflammation, but they do not kill pathogens directly. Instead, they activate complement and facilitate phagocytosis. Antibody binding to pathogens and toxins can prevent these threats from attaching to healthy body cells. Bound to ailing body cells, antibodies recruit cell-killing lymphocytes (more about this in Section 37.7).

An antibody consists of four polypeptides: two identical "light" (small) chains and two identical "heavy" (large) chains (FIGURE 37.9A). Each chain has a region that can vary (the variable region) and one that does not (the constant region). When the four chains fold up together as an antibody, the variable regions form two antigen-binding sites at the tips of the Y. These sites have a specific distribution of bumps, grooves, and charge, and they bind only to antigen that has a complementary distribution of bumps, grooves, and charge (FIGURE 37.9B). Antigen-binding sites vary greatly among antibodies, so they are called hypervariable regions.

An antibody's constant region determines its structural identity, or class: IgG, IgA, IgE, IgM, or IgD (Ig stands for immunoglobulin, another name for antibody). The different classes serve different functions (TABLE 37.3). Most of the antibodies circulating in the bloodstream and tissue fluids are IgG, which binds pathogens, neutralizes toxins, and activates complement. IgG is the only antibody that can cross the placenta to protect a fetus before its own immune system becomes active.

IgA is the main antibody in mucus and other exocrine gland secretions (Section 31.3). IgA is secreted as a dimer (two antibodies bound together), which makes the molecule stable enough to function in harsh environments such as the interior of the digestive tract. There, IgA encounters pathogens before they have a chance to contact body cells. Leukocytes recognize IgA bound to antigen as a signal to initiate inflammation.

TABLE 37.3

Structural Classes of Antibodies

Class		Description
Secreted antibodies		
IgG		Main antibody in blood; binds pathogens, neutralizes toxins, activates complement. Can cross placenta to protect fetus.
IgA		Abundant in exocrine gland secretions (e.g., tears, saliva, milk, mucus), where it occurs in dimeric form (shown). Interferes with binding of pathogens to body cells; signals leukocytes to initiate inflammation.
Membrane-bound antibodies		
IgE		Anchored to surface of basophils, mast cells, eosinophils, and some dendritic cells. IgE binding to antigen induces anchoring cell to release histamines and cytokines. Factor in allergies and asthma.
IgD		B cell receptor.
IgM		B cell receptor, as a monomer. Also secreted as polymers (left) that are especially effective at binding antigen and activating complement.

FIGURE 37.9 **Antibody structure.**

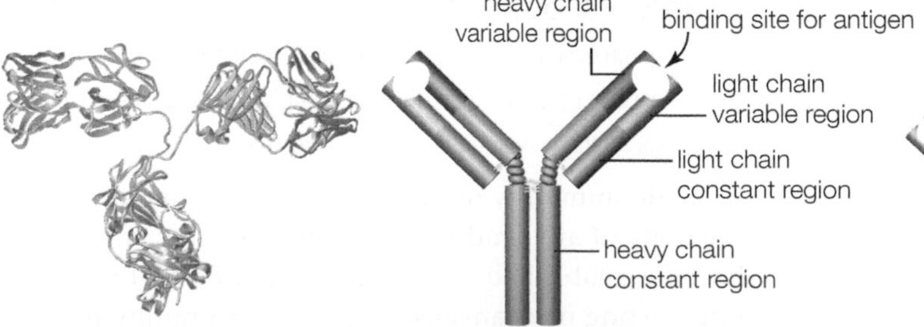
heavy chain
variable region
binding site for antigen
light chain
variable region
light chain
constant region
heavy chain
constant region

A An antibody molecule consists of four polypeptide chains joined in a Y-shaped configuration. The chains fold up so two antigen-binding sites form at the tips of the Y.

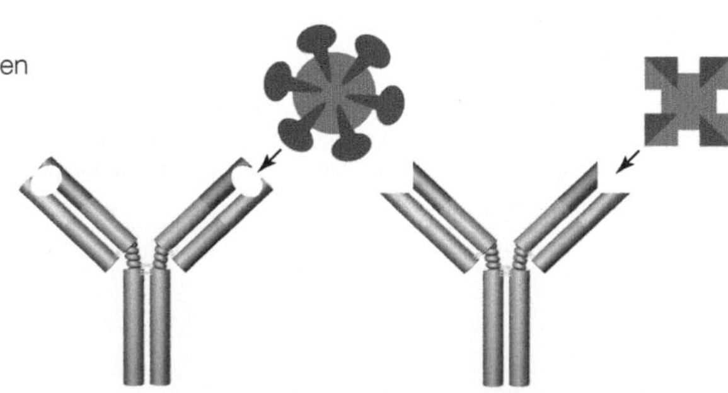

B The antigen-binding sites of each antibody bind only to an antigen that has a complementary distribution of bumps, grooves, and charge.

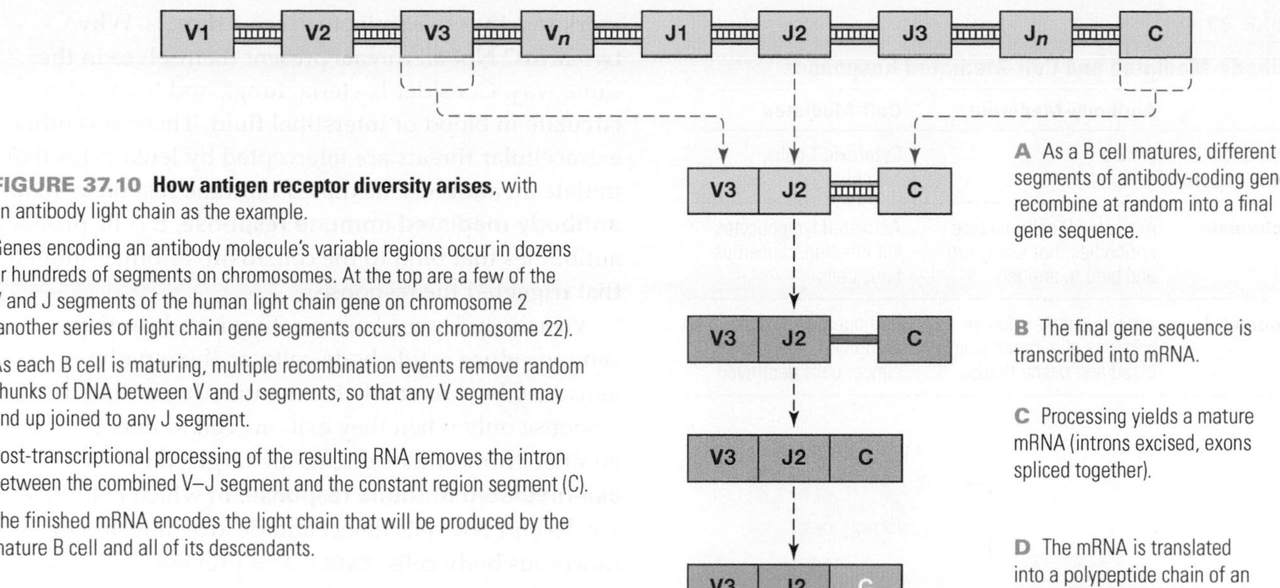
V1
V2
V3
Vn
J1
J2
J3
Jn
C
V3
J2
C
A As a B cell matures, different segments of antibody-coding genes recombine at random into a final gene sequence.
B The final gene sequence is transcribed into mRNA.
C Processing yields a mature mRNA (introns excised, exons spliced together).
D The mRNA is translated into a polypeptide chain of an antibody molecule.

FIGURE 37.10 How antigen receptor diversity arises, with an antibody light chain as the example.

Genes encoding an antibody molecule's variable regions occur in dozens or hundreds of segments on chromosomes. At the top are a few of the V and J segments of the human light chain gene on chromosome 2 (another series of light chain gene segments occurs on chromosome 22).

As each B cell is maturing, multiple recombination events remove random chunks of DNA between V and J segments, so that any V segment may end up joined to any J segment.

Post-transcriptional processing of the resulting RNA removes the intron between the combined V–J segment and the constant region segment (C).

The finished mRNA encodes the light chain that will be produced by the mature B cell and all of its descendants.

IgE made and secreted by B cells is taken up by mast cells and basophils, and then incorporated into their plasma membranes. These cells degranulate when their membrane-bound IgE binds to antigen.

IgM or IgD antibodies produced by a new B cell stay attached to the cell's plasma membrane as **B cell receptors**. The base of each receptor is embedded in the lipid bilayer of the cell's plasma membrane, and the two arms of the Y project into the extracellular environment. IgM is also secreted as polymers of five or six antibodies. The polymers are especially efficient at binding antigen and activating complement.

Antigen Receptor Diversity

Humans can make billions of unique B and T cell receptors. The diversity arises because the genes that encode the variable regions of these receptors occur in segments, and there are dozens or even hundreds of different versions of each segment (**FIGURE 37.10**). The gene segments become spliced together during B and T cell differentiation, but which version of each segment makes it into the finished antigen receptor gene of a particular cell is random. As a B or T cell differentiates, it ends up with one out of about 2.5 billion potential combinations of gene segments. When the resulting gene is expressed, the cell produces thousands of identical receptors, all of which can recognize the same specific antigen.

Like all other blood cells, new lymphocytes form in bone marrow. B cells and NK cells mature before leaving the marrow. Immature T cells leave the marrow and migrate to the thymus gland (a lymphatic organ, Section 36.11), where they encounter hormones that stimulate them to mature.

TAKE-HOME MESSAGE 37.5

✔ Random recombination events during B cell or T cell differentiation can collectively give rise to billions of different antigen receptors.

✔ Each lymphocyte makes antigen receptors that recognize one specific antigen.

✔ T cell receptors can discriminate between self (MHC markers) and nonself (antigen).

✔ Antibodies released into the circulatory system can activate complement and facilitate phagocytosis.

✔ B cell receptors are antibodies that are not secreted; they remain embedded in the B cell's plasma membrane.

37.6 Overview of Adaptive Immunity

LEARNING OBJECTIVES

- Describe the two types of adaptive immune responses and explain the advantages of having both.
- Distinguish between effector cells and memory cells.
- Explain the four defining characteristics of adaptive immunity.
- Describe antigen processing.

Two Arms of Adaptive Immunity

Like a boxer's one-two punch, adaptive immunity has two separate arms: two types of responses that

antibody (AN-tee-baa-dee) Y-shaped antigen receptor protein made only by B cells.
B cell receptor Antigen receptor on the surface of a B cell; IgM or IgD that stays anchored in the B cell's plasma membrane.
MHC markers Cell surface self-proteins of vertebrates.
T cell receptor (TCR) Special antigen receptor that only T cells have; recognizes antigen as nonself, and MHC markers as self.

TABLE 37.4

Antibody-Mediated and Cell-Mediated Responses

	Antibody-Mediated	Cell-Mediated
Primary Effector Cells	B cells	Cytotoxic T cells, NK cells
Mechanism	Activated B cells secrete antibodies that recognize and bind to antigen	Activated lymphocytes kill infected/cancerous body cells
Outcome(s)	Antigenic molecules or particles eliminated from blood and tissue fluids	Pathogen eliminated along with ailing cells; cancer cells destroyed

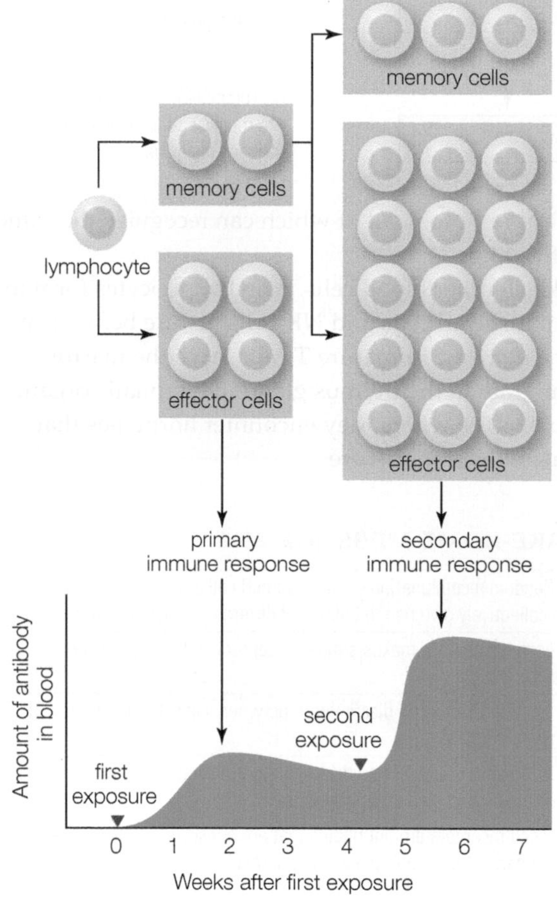

memory cells
memory cells
lymphocyte
effector cells
effector cells
primary immune response
secondary immune response
Amount of antibody in blood
first exposure
second exposure
0
1
2
3
4
5
6
7
Weeks after first exposure

FIGURE 37.11 Primary and secondary immune responses.

A first exposure to an antigen triggers a primary immune response in which effector cells fight the infection. Memory cells that also form initiate a faster, stronger secondary response if the antigen returns at a later time.

antibody-mediated immune response Adaptive immune response in which B cells produce antibodies that bind to an antigen.

cell-mediated immune response Adaptive immune response in which cytotoxic T cells and NK cells kill infected or cancerous body cells.

effector cell Antigen-sensitized lymphocyte that forms in an immune response and acts immediately.

memory cell Long-lived, antigen-sensitized lymphocyte that can carry out a secondary immune response.

work together to eliminate diverse threats. Why two arms? Not all threats present themselves in the same way. Consider bacteria, fungi, and toxins that circulate in blood or interstitial fluid. These and other extracellular threats are intercepted by leukocytes that initiate an antibody-mediated immune response. In an **antibody-mediated immune response,** B cells produce antibodies that bind to the cell, toxin, or other antigen that triggered the response.

Viruses and some bacteria, fungi, and protists can reproduce inside body cells, so these pathogens may be vulnerable to an antibody-mediated response only when they exit one cell to infect another. Intracellular threats are targeted by the **cell-mediated immune response,** in which cytotoxic T cells and NK cells detect and destroy infected or cancerous body cells. TABLE 37.4 previews and compares features of the two types of adaptive responses; Sections 37.7 and 37.8 detail their mechanisms.

Effector cells and memory cells form during all adaptive immune responses. **Effector cells** are lymphocytes that act at once in a primary immune response. **Memory cells** are long-lived lymphocytes reserved for possible future encounters with the same threat. If the same antigen is detected in the body later, memory cells carry out a faster, stronger secondary response (FIGURE 37.11).

Lymphocytes interact with other leukocytes to bring about the four defining characteristics of adaptive immunity:

1. *Self/nonself recognition,* based on the ability of T cell receptors to recognize self (in the form of MHC markers), and that of all antigen receptors to recognize nonself (in the form of antigen).
2. *Specificity,* which means that adaptive immune responses are tailored to combat specific antigens.
3. *Diversity,* which refers to the diversity of antigen receptors that an individual can make. The ability to produce billions of different antigen receptors confers the ability to counter billions of different threats.
4. *Memory,* the capacity of the adaptive immune system to "remember" an antigen via memory cells. It takes about a week for B and T cells to respond in force the first time an antigen is detected. If the same antigen shows up later, the secondary response is faster and stronger.

Antigen Processing

A new lymphocyte is "naive," which means that its receptors have never bound to an antigen. B cell receptors can bind directly to antigen, but T cell receptors cannot. T cell receptors recognize and bind

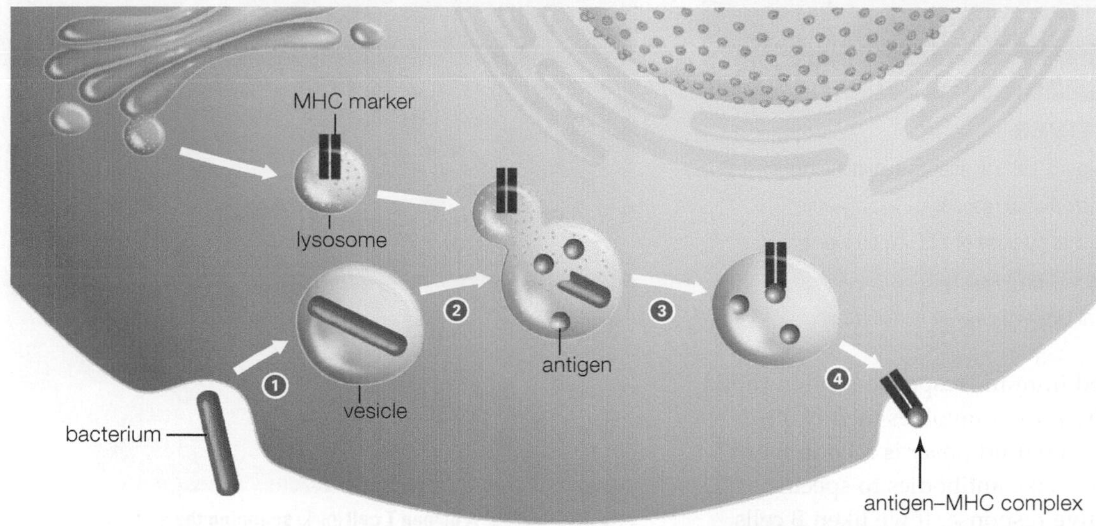
MHC marker
lysosome
bacterium
vesicle
antigen
antigen–MHC complex
1
2
3
4

FIGURE 37.12 Antigen processing.

This drawing shows what happens in a dendritic cell or other phagocyte after it engulfs a bacterium.

1 An endocytic vesicle forms around the bacterium as it is engulfed by a phagocyte.

2 The vesicle fuses with a lysosome, which contains enzymes and MHC markers.

3 Lysosomal enzymes digest the bacterium into molecular bits. Fragments of bacterial proteins and polysaccharides (antigens) bind to the MHC markers.

4 The vesicle fuses with the cell's plasma membrane by exocytosis. When it does, the antigen–MHC complexes become displayed on the cell's surface.

only to antigen that is displayed on the surface of another cell. All body cells can display antigen. White blood cells that display antigen are called antigen-presenting cells, and these have a special role in adaptive immune responses.

The mechanism by which a white blood cell prepares antigen for display is called antigen processing. It begins when a leukocyte engulfs a pathogen, ailing cell, or other antigen-bearing particle (**FIGURE 37.12**). A vesicle that contains the particle forms in the phagocyte's cytoplasm 1 and fuses with a lysosome 2. Remember from Section 4.5 that lysosomes are vesicles filled with powerful enzymes. These enzymes now proceed to break down the ingested particle into molecular bits. Fragments of proteins and polysaccharides bind to MHC markers to form antigen–MHC complexes. These complexes are anchored in the vesicle's membrane 3. The vesicle moves to and fuses with the plasma membrane 4, so the antigen–MHC complexes become displayed (presented) at the cell's surface.

Naive T cells often encounter antigen-presenting cells in the spleen or lymph nodes. Phagocytic leukocytes engulf antigen-bearing particles, and then migrate to these lymphatic organs (Section 36.11) as they start to present antigen. Each day, billions of T cells filter through the lymph nodes and spleen. As they do, they come into close contact with arrays of antigen-presenting cells that have taken up residence in these organs (**FIGURE 37.13**). As you will see shortly, T cells with receptors that bind to an antigen–MHC complex stimulate other white blood cells to divide and differentiate in an adaptive immune response.

During an infection, the lymph nodes swell because leukocytes accumulate inside them. When you are ill, you may notice your swollen lymph nodes as tender lumps under the jaw or elsewhere in your body.

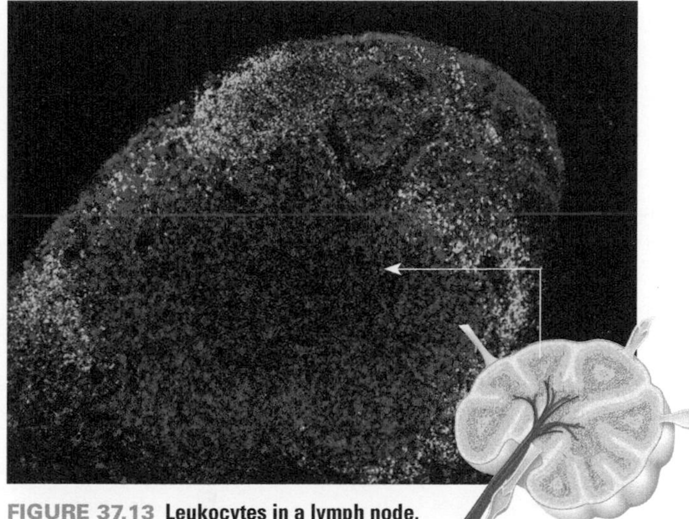

FIGURE 37.13 Leukocytes in a lymph node.

The fluorescence micrograph above shows naive T cells (blue) that are passing through a lymph node and interacting with populations of resident dendritic cells (red) and B cells (green).

TAKE-HOME MESSAGE 37.6

- ✔ Adaptive immune responses involve interactions between lymphocytes and phagocytic leukocytes.
- ✔ Vertebrate adaptive immunity is defined by self/nonself recognition, specificity, diversity, and memory.
- ✔ The two arms of adaptive immunity work together. Antibody-mediated responses target antigen in blood or interstitial fluid; cell-mediated responses target altered body cells.
- ✔ Effector and memory cells form during adaptive responses. Effector cells act immediately. Memory cells are set aside; if the antigen enters the body again at a later time, these cells initiate a faster, stronger secondary response.
- ✔ T cell receptors can recognize their specific antigen only in conjunction with MHC markers displayed on the surface of an antigen-presenting cell. The recognition process occurs in lymph nodes and the spleen.

37.7 Adaptive Immunity I: An Antibody-Mediated Response

LEARNING OBJECTIVES

- Explain the role of helper T cells in adaptive immune responses.
- Describe an antibody-mediated response.
- Explain the technique and purpose of ABO blood typing.
- Explain why a vaccine elicits immunity.
- Distinguish between active and passive immunization.

An antibody-mediated immune response is often called a humoral response because it involves mainly elements of blood and tissue fluid (*umor* is a Latin word for body fluid). B cells make antibodies to specific antigens in this adaptive response. If we liken B cells to assassins, then each one has a genetic assignment to liquidate a particular target: an antigen-bearing extracellular pathogen or toxin. Antibodies are a B cell's molecular bullets, as the following example illustrates.

Suppose that you accidentally nick your finger. Being opportunists, some *Staphylococcus aureus* cells on your skin immediately enter the cut and invade your internal environment. Complement in interstitial fluid quickly binds to peptidoglycan (a PAMP) in the cell walls of the bacteria, and complement activation cascades begin. Within an hour, complement-coated bacteria tumbling along in lymph vessels have become trapped by a lymph node in your neck. There, they encounter millions of naive B cells maturing in the node. One of the B cells makes antigen receptors that recognize an antigen in *S. aureus* cell walls: a protein called ClfA ("clumping factor A"). The B cell's receptors bind to one of the bacteria, and the bacterium's complement coating stimulates the B cell to engulf it (**FIGURE 37.14** ❶). The B cell is now activated (it is no longer naive).

Meanwhile, some of the invading *S. aureus* cells have been secreting metabolic products into the tissues around your cut. The secretions attract phagocytes,

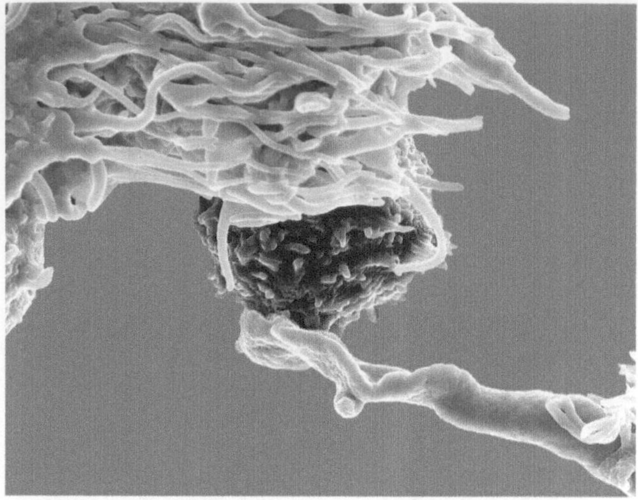

FIGURE 37.15 A human T cell (pink) **scanning the surface of a dendritic cell** (blue). The T cell will initiate an adaptive immune response if it recognizes antigen–MHC complexes presented by the dendritic cell.

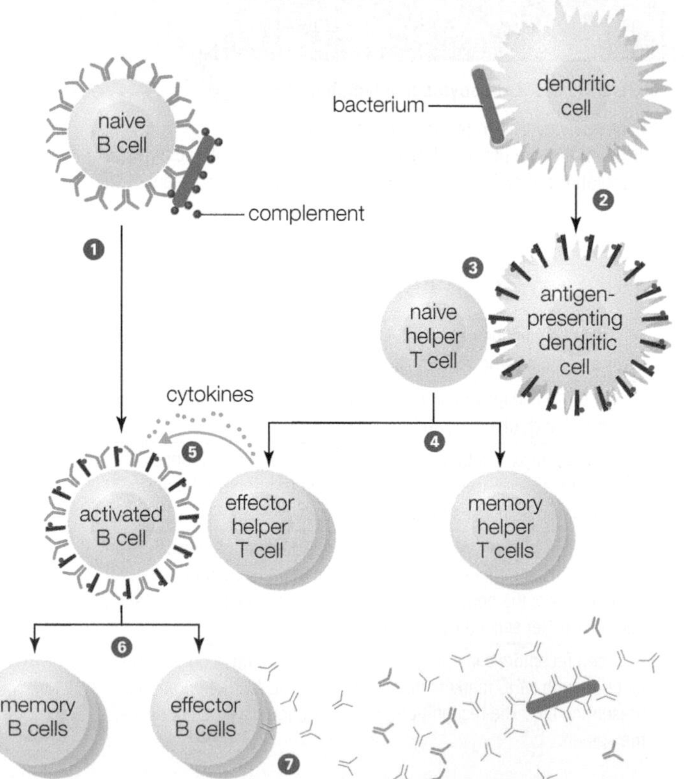
naive B cell
bacterium
dendritic cell
complement
1
2
3
naive helper T cell
antigen-presenting dendritic cell
cytokines
4
5
activated B cell
effector helper T cell
memory helper T cells
6
memory B cells
effector B cells
7

FIGURE 37.14 An example of an antibody-mediated immune response.

❶ The B cell receptors (green) on a naive B cell bind to an antigen on the surface of a bacterium (red). The bacterium's complement coating (purple dots) encourages the B cell to engulf and digest it. Bacterial fragments (red dots) bound to MHC markers (blue sticks) become displayed on the surface of the B cell.

❷ A dendritic cell engulfs and digests the same kind of bacterium that the B cell encountered. Bacterial fragments bound to MHC markers become displayed on the surface of the dendritic cell.

❸ The receptors of a naive helper T cell recognize bacterial antigen displayed by the dendritic cell. The two cells interact and disengage.

❹ The T cell begins to divide. Its descendants differentiate into effector helper T cells and memory helper T cells.

❺ The receptors of one of the effector helper T cells recognize and bind to antigen displayed by the B cell. Binding causes the T cell to secrete cytokines (blue dots).

❻ The cytokines induce the B cell to undergo repeated mitotic divisions. Its many descendants differentiate into effector B cells and memory B cells.

❼ The effector B cells begin making and secreting huge numbers of IgA, IgG, or IgE antibodies, all of which recognize the same antigen as the original B cell receptor. The new antibodies circulate throughout the body and bind to any remaining bacteria. The binding facilitates phagocytosis and can prevent the bacteria from attaching to body cells.

including a dendritic cell that engulfs several bacteria. The dendritic cell migrates to the lymph node in your neck. By the time it reaches its destination, the cell has digested the bacteria and is displaying their fragments as antigens bound to MHC markers ❷.

T cells are constantly filtering through the lymph node. One of them recognizes and binds to a fragment of ClfA displayed by the dendritic cell (FIGURE 37.15). This T cell is called a **helper T cell** because it activates other lymphocytes in an adaptive immune response. Within minutes of contacting the dendritic cell, the helper T cell stops its migratory behavior. A tight connection called an immunological synapse forms between the two cells (FIGURE 37.16).

The helper T cell and the dendritic cell interact for about 24 hours and then disengage ❸. The T cell is activated by this interaction, and it returns to the circulatory system and divides repeatedly. A huge population of identical helper T cells forms. These clones mature into effector and memory cells ❹, each with receptors that recognize ClfA. The army of T cells now starts filtering through your lymphatic organs.

Let's now go back to that lymph node in your neck. By now, the B cell has digested the engulfed bacterium, and it is displaying bits of *S. aureus* bound to MHC molecules on its plasma membrane. Receptors on the helper T cell clones recognize ClfA antigen displayed by the B cell. One of the T cells binds to the B cell, and like long-lost friends, the two cells stay together for a while and communicate ❺. A message that is communicated consists of cytokines secreted by the helper T cell. The cytokines stimulate the B cell to undergo repeated divisions after the two cells disengage. A huge clonal population of descendant cells forms, and these B cells differentiate into effector and memory cells ❻. As they mature, the B cells switch antibody classes, which means they start producing antibodies with a different constant region. Instead of making membrane-bound B cell receptors, they now make and release IgA, IgG, or IgE. The variable regions of the antibodies are unchanged, so their antigen-binding specificity remains the same: All can recognize and bind to ClfA ❼.

By a theory called clonal selection, the lymphocytes that give rise to clonal populations are "selected" during an immune response because antigen binds most tightly to their receptors. B cells or T cells with receptors that do not bind antigen do not divide; those

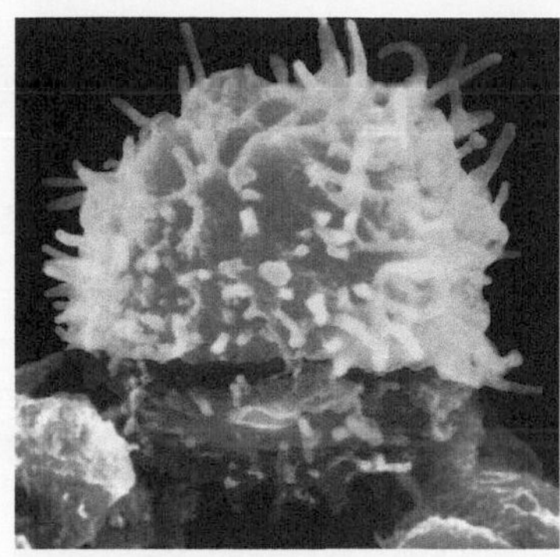

FIGURE 37.16 **An immunological synapse**: zooming in on the point of contact between a T cell and an antigen-presenting cell.

Left, a model shows a T cell receptor (yellow) bound to an antigen–MHC complex on an antigen-presenting cell. The antigen is shown in red; the MHC marker, in blue. The composite micrograph on the right shows the point of contact between the two cells. Adhesion proteins (red ring) around the T cell receptors (green spot) fasten the cells together.

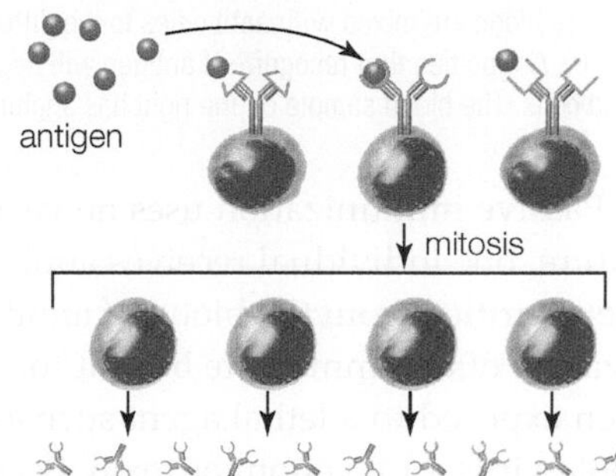

Antigen binds only to a matching B cell receptor.
antigen
B cell that binds antigen undergoes mitosis.
mitosis
Many effector B cells form and secrete many antibodies.

FIGURE 37.17 **Example of clonal selection**, with B cells. Only lymphocytes with receptors that bind to antigen divide and mature. Lymphocytes with receptors that best fit an antigen are favored in this process.

that bind antigen loosely do not form huge populations (FIGURE 37.17).

Tremendous numbers of antibodies now circulate throughout the body and attach themselves to any *S. aureus* cells. An antibody coating prevents the bacteria from attaching to body cells, and it facilitates phagocytosis to speed disposal. Antibodies also glue the foreign cells together into clumps, a process called **agglutination**. The clumps are quickly removed from the circulatory system by the liver and spleen.

Immunization

Immunization refers to procedures designed to induce an antibody-mediated response against particular pathogens. In active immunization, a preparation that contains antigen—a vaccine—elicits a primary immune response just as an infection would. A second vaccination, which is called a booster, elicits a secondary immune response for enhanced protection from the pathogen. Additional boosters may maintain immunity for the individual's lifetime.

agglutination (uh-glue-tih-NAY-shun) The clumping together of antigen-bearing cells or particles by antibodies.

helper T cell Type of T cell that activates other lymphocytes; required for all adaptive immune responses.

immunization Any procedure designed to induce immunity to a specific disease.

TABLE 37.5

Antibodies to H Antigen

ABO Blood Type	H Antigen on Red Blood Cells	Antibodies Present
A	A	anti-B
B	B	anti-A
AB	both A and B	none
O	neither A nor B	anti-A, anti-B

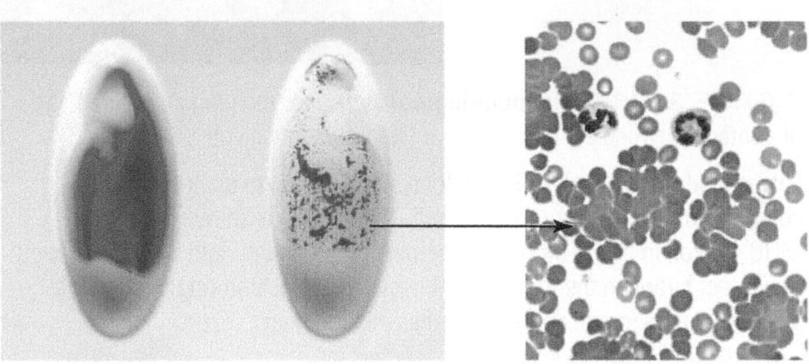

FIGURE 37.18 **ABO blood typing test.** In such tests, samples of a patient's blood are mixed with antibodies to the different types of H antigens (A or B). Antibodies that recognize H antigen will agglutinate (clump) the red blood cells. The blood sample on the right has agglutinated.

Passive immunization uses no vaccines. In this procedure, one individual receives antibodies that have been purified from the blood of another. Passive immunization offers immediate benefit for someone who has been exposed to a lethal agent such as tetanus, rabies, Ebola virus, or a venom or toxin. Because the antibodies were not made by the recipient's lymphocytes, effector and memory cells do not form, so benefits last only as long as the injected antibodies do.

Antibodies in ABO Blood Typing

As you learned in Section 13.5, a carbohydrate on red blood cell membranes determines your blood type. This carbohydrate is called "H antigen," and it occurs in two forms. People with one form of the antigen have type A blood; people with the other form have type B blood. People with both forms have type AB blood; those with neither have type O.

Early in life, each individual starts making antibodies that recognize molecules foreign to the body, including any nonself form of the H antigen (TABLE 37.5). If a blood transfusion becomes necessary, it is especially important to know which H antigens your blood cells carry. A transfusion of incompatible red blood cells can cause a potentially fatal transfusion reaction in which the recipient's antibodies recognize and bind to H antigens on the transfused cells. The binding activates complement, which punctures the membranes of the foreign cells, thus releasing a massive amount of hemoglobin that can very quickly cause the kidneys to fail.

Identifying red blood cell surface antigens helps prevent pairing of incompatible transfusion donors and recipients, and also alerts physicians to blood incompatibility problems that may arise during pregnancy. A typical blood typing test involves mixing drops of a patient's blood with antibodies to the different forms of red blood cell antigens. Agglutination occurs when the cells bear H antigens recognized by the antibodies (FIGURE 37.18).

TAKE-HOME MESSAGE 37.7

- ✔ During an antibody-mediated response, antigen-presenting phagocytes interact with and activate helper T cells.
- ✔ Interaction with a compatible helper T cell triggers a B cell to divide and differentiate. A huge clonal population of antibody-secreting B cells forms.
- ✔ Active immunization elicits an antibody-mediated immune response that protects against a specific disease.

37.8 Adaptive Immunity II: The Cell-Mediated Response

LEARNING OBJECTIVES

- Describe a cell-mediated response and how it differs from an antibody-mediated response.
- Name the lymphocytes that recognize ailing body cells and how the recognition occurs in each case.

A cell-mediated response targets infected or cancerous body cells. This type of adaptive response does not involve antibodies; it is carried out by lymphocytes that can recognize ailing or unfamiliar body cells. These cells display molecules not found on normal body cells. For example, foreign body cells display nonself proteins; cancer cells display abnormal proteins; and body cells infected with viruses or other intracellular pathogens display polypeptides of the infecting agent. All of these cells are killed in a cell-mediated response, and the killers are antigen-sensitized **cytotoxic T cells**. If B cells are like assassins, then cytotoxic T cells specialize in cell-to-cell combat.

Cytotoxic T Cells: Activation and Action

A typical cell-mediated immune response starts in interstitial fluid during inflammation, when dendritic cells engulf sick body cells or their remains (FIGURE 37.19), then migrate to a lymph node or the spleen. By the time the dendritic cells reach their destination, they are displaying antigen (abnormal or foreign protein fragments from the ingested cell) together with MHC markers ❶. These antigen-MHC complexes are inspected by millions of naive cytotoxic T cells ❷

CREDIT: (18) left, Ikordela/Shutterstock; right, Kristine Krafts, M.D.,PathologyStudent.com.

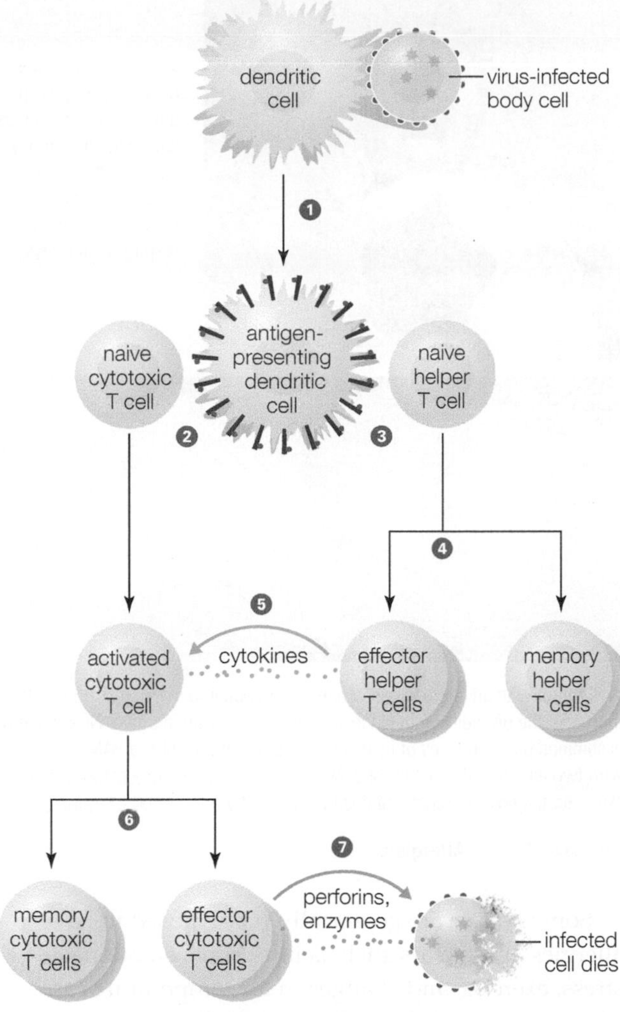
dendritic cell
virus-infected body cell
1
naive cytotoxic T cell
antigen-presenting dendritic cell
naive helper T cell
2
3
4
5
activated cytotoxic T cell
cytokines
effector helper T cells
memory helper T cells
6
7
memory cytotoxic T cells
effector cytotoxic T cells
perforins, enzymes
infected cell dies

FIGURE 37.19 Example of a cell-mediated immune response targeting virus-infected body cells.

❶ A dendritic cell engulfs a virus-infected cell and processes antigens. Protein fragments of the virus (red) bind to MHC markers (blue), and the complexes become displayed at the dendritic cell's surface. The dendritic cell, now an antigen-presenting cell, migrates to a lymph node.

❷ Receptors on a naive cytotoxic T cell bind to viral antigen displayed by the dendritic cell. The interaction activates the cytotoxic T cell.

❸ Receptors on a naive helper T cell bind to viral antigen displayed by the dendritic cell. The interaction activates the helper T cell.

❹ The activated helper T cell divides again and again. Its many descendants mature as effector and memory cells, each with T cell receptors that recognize the virus.

❺ The effector helper T cells secrete cytokines.

❻ The cytokines induce the activated cytotoxic T cell to divide again and again. Its many descendants differentiate into effector and memory cells. Each cell bears T cell receptors that recognize the virus.

❼ The new effector cytotoxic T cells circulate throughout the body. They kill any body cell that displays viral antigen on its surface.

FIGURE IT OUT What do the red spots represent?

Answer: Virus particles

cytotoxic T cell Type of T cell that kills ailing or cancerous body cells during a cell-mediated immune response.

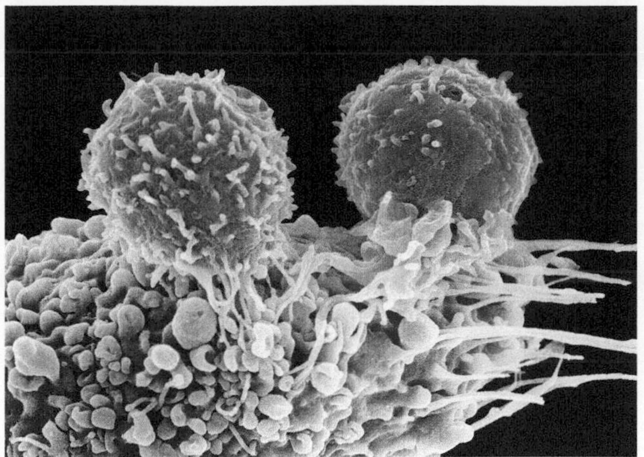

FIGURE 37.20 Cytotoxic T cells (pink) **killing a cancer cell.**

and naive helper T cells ❸ that are migrating through the lymphatic organ. Some of the T cells have receptors that recognize antigen presented by the dendritic cells. These T cells interact with the dendritic cells for awhile and then disengage. The T cells are now activated.

Activated helper T cells return to the circulatory system and undergo repeated mitotic divisions. Their many descendants mature as effector and memory cells ❹. The new effector cells recognize and interact with antigen-presenting macrophages. Any macrophage that interacts with an activated helper T cell increases its production of cytokines and pathogen-busting enzymes and toxins.

The new effector helper T cells also secrete cytokines ❺. Any cytotoxic T cells that have been activated by interacting with an antigen–presenting dendritic cell (❷) recognize these cytokines as a signal to divide repeatedly, and their many descendants mature as effector and memory cells ❻.

The new effector cytotoxic T cells have receptors that recognize abnormal or foreign proteins displayed by that first ailing body cell. These cytotoxic T cells now circulate throughout blood and interstitial fluid, and bind to any other body cell displaying the same proteins (FIGURE 37.20). Binding causes a cytotoxic T cell to release protein-digesting enzymes and small molecules called perforins. Perforins, like complement proteins, can assemble into structures that penetrate a cell's plasma membrane and form large channels through it. The channels allow the T cell's enzymes to enter the body cell and cause it to lyse (burst) or commit suicide ❼.

As occurs in an antibody-mediated response, memory cells form in a primary cell-mediated response. If the antigen appears in the body again at a later time, these memory cells will mount a faster, stronger secondary response.

The Role of Natural Killer (NK) Cells

Remember that part of a T cell receptor recognizes antigen, and the other recognizes the MHC markers on body cells as "self." Only cells that have intact MHC markers can be recognized and killed by cytotoxic T cells. However, some infections and cancers alter body cells so much that part or all of their MHC markers are missing. NK cells (natural killer cells) are crucial for eliminating such cells from the body. Unlike cytotoxic T cells, NK cells can kill body cells that lack MHC markers. Cytokines secreted by helper T cells stimulate NK cell division, and the resulting populations of NK cells recognize and attack any body cells that have antibodies bound to them. NK cells also recognize certain proteins displayed by body cells that are under stress. Stressed body cells with altered or missing MHC markers are killed; those with normal MHC markers are not.

NK cells are often considered to be part of innate immunity because their antigen receptors do not arise from genetic recombination. However, they also have features associated with adaptive immunity: activation by cytokines, for example, and memory.

TAKE-HOME MESSAGE 37.8

✔ In a cell-mediated immune response, activated cytotoxic T cells kill infected or ailing body cells with intact MHC markers.

✔ NK cells can kill infected or ailing body cells lacking MHC markers.

37.9 When Immunity Goes Wrong

LEARNING OBJECTIVES

- Describe the symptoms of allergies and how they arise.
- Using an appropriate example, explain autoimmunity.
- Give some examples of the ways in which pathogens can evade mechanisms of innate and adaptive immunity.
- Use examples to explain immune deficiency.

Despite built-in quality controls and redundancies in function, immunity does not always work as well as it should. Complexity is part of the problem, because there are simply more opportunities for failure to occur in systems with many components. Even a small failure in immune function can have a major effect on health.

Overly Vigorous Responses

Allergies An **allergen** is a normally harmless substance that triggers an immune response in some people. Common allergens include drugs, foods, pollen, dust mite feces, fungal spores, and insect venom. Sensitivity to an allergen is called an **allergy**.

A Hay fever is caused by allergy to grass pollen. Symptoms such as sneezing and a runny nose are a result of inflammation of the mucous membranes in respiratory airways.

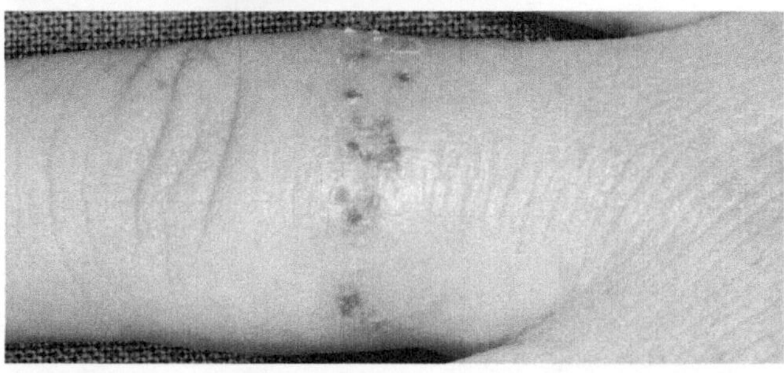

B Contact with an allergen can cause skin to become itchy and irritated. In this case, the offending substance was nickel (a metal) in a ring. Nickel allergy is common because many of us have an allele that specifies a PAMP receptor with two amino acid substitutions. Nickel normally binds to a carbohydrate in skin, and the resulting compound is recognized by the altered receptor.

FIGURE 37.21 Allergies.

Some people are genetically predisposed to have allergies, but factors such as infections, emotional stress, exercise, and changes in air temperature can trigger or worsen them. Typically, a first exposure to an allergen stimulates an antibody-mediated response targeting the allergen. Some of the IgE produced during this response becomes anchored to mast cells and basophils. Upon a later exposure, the anchored IgE binds to the allergen. Binding triggers the anchoring cell to degranulate and release histamines that initiate inflammation. If the allergen is detected by mast cells in the lining of the respiratory tract, the resulting inflammation constricts the airways and increases mucus secretion; sneezing, stuffed-up sinuses, and a drippy nose result (**FIGURE 37.21A**). Antihistamines relieve these symptoms by dampening the effects of histamines. Other drugs can inhibit mast cell degranulation, thus preventing histamine release.

Physical contact with some allergens can provoke an inflammatory response characterized by an intensely

allergen (AL-er-jen) A normally harmless substance that provokes an immune response in some people.

allergy Sensitivity to an allergen.

autoimmune response Immune response that targets one's own body tissues.

itchy, irritated patch of skin (**FIGURE 37.21B**). Oils produced by poison oak and similar plants commonly trigger contact allergies. Metal ions or small compounds such as these oils penetrate skin and covalently bond to molecules normally found in epidermal tissue. In some people, the resulting hybrid compounds bind to PAMP receptors on mast cells and keratinocytes. Keratinocytes are the main cellular constituents of epidermis (Section 31.8), and they also function as immune cells: When antigen binds to their PAMP receptors, these cells release cytokines that trigger inflammation and a cell-mediated response. Effector and memory T cells that form in the response migrate into dermis all over the body, a process that takes about two weeks. Upon later contact with the allergen, these cells release cytokines that trigger a widespread inflammatory reaction in skin, peaking a day or two after exposure. Each subsequent exposure triggers a faster and more severe response.

Acute Illnesses Immune defenses that eliminate a threat can also damage body tissues, so mechanisms that limit these defenses are always in play. Consider how some complement proteins become activated spontaneously, even in the absence of infection or tissue damage. Without inhibitory molecules that inactivate these proteins, complement cascades would occur constantly, with disastrous effects on body tissues.

Acute illnesses can arise when mechanisms that limit immune responses fail. For example, exposure to an allergen sometimes causes a rapid and severe allergic reaction called anaphylaxis. Huge amounts of inflammatory molecules, including histamines and prostaglandins, are released all at once in all parts of the body. Too much fluid leaks from blood into tissues, causing a sudden and dramatic drop in blood pressure (a reaction called shock). Rapidly swelling tissues (above) constrict the airways and may block them. Anaphylaxis is rare but life-threatening and requires immediate treatment. It may occur at any time upon exposure to even a tiny amount of allergen. Risks include any prior allergic reaction.

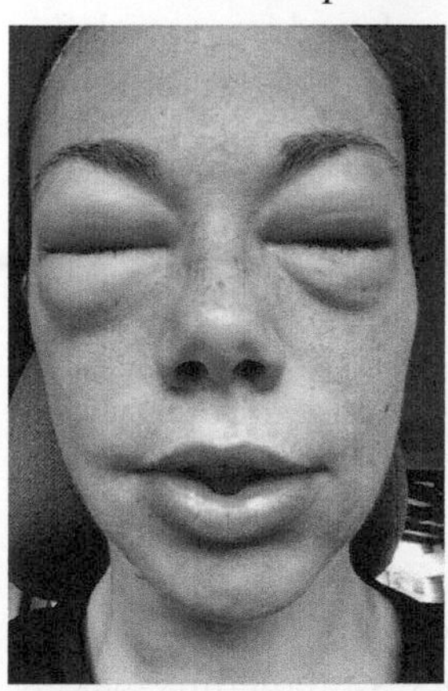

Severe episodes of asthma or septic shock occur when too many neutrophils degranulate at once. Some infections cause leukocytes to flood the body with cytokines, triggering an exaggerated immune response called "cytokine storm." The cytokine overdose activates more leukocytes, which release more cytokines, and so on—a positive feedback loop that can result in massive inflammation, organ failure, and death. Viruses such as H5N1 and Ebola (Section 20.4) have an unusually high mortality rate because they trigger a cytokine storm in their hosts.

TABLE 37.6

Examples of Autoimmune Disorders

Disorder	Autoantibody Target	Affected Area
Addison disease	Cytochrome P450 enzyme	Adrenal glands
Celiac disease	Collagen	Small Intestine
Dermatomyositis	tRNA synthesis enzyme	Muscles, skin
Diabetes mellitus type 1	Islet proteins or insulin	Pancreas
Goodpasture's syndrome	Type IV collagen	Kidney, lung
Graves' disease	TSH receptor	Thyroid
Guillain-Barré syndrome	Lipids of ganglia	Peripheral nervous system
Hashimoto's disease	TH synthesis proteins	Thyroid
Idiopathic thrombocytopenic purpura	Platelet glycoproteins	Blood (platelets)
Lupus erythematosus	DNA, nuclear proteins	Connective tissue
Multiple sclerosis	Myelin proteins	Central nervous system
Myasthenia gravis	Acetylcholine receptors	Neuromuscular junctions
Pemphigus vulgaris	Cadherin	Skin
Pernicious anemia	Parietal cell glycoprotein	Gut epithelium
Polymyositis	tRNA synthesis enzyme	Muscles
Primary biliary cirrhosis	Nuclear pore proteins, mitochondria	Liver
Rheumatoid arthritis	Constant region of IgG	Joints
Scleroderma	Topoisomerase	Arteriole endothelium
Wegener's granulomatosis	Enzyme in neutrophil granules	Blood vessels

Autoimmunity People usually do not make antibodies to molecules that occur on their own healthy body cells. Quality control mechanisms built into the thymus and other lymphoid organs weed out T cells and B cells with receptors that recognize one's own body proteins. Thymus cells snip small polypeptides from a variety of body proteins and attach them to MHC markers. Maturing T cells that bind too strongly to one of these peptide–MHC complexes have TCRs that recognize a self-protein; those that bind too weakly to the complexes do not recognize MHC markers. Both types of cells die. If this mechanism fails, lymphocytes that do not discriminate between self and nonself may be produced. Such lymphocytes can mount an **autoimmune response**, which is an immune response that targets one's own body tissues. Autoimmunity is beneficial when a cell-mediated response targets cancer cells, but in most other cases it is not (**TABLE 37.6**).

The neurological disorder called multiple sclerosis occurs when self-reactive T cells mount a cell-mediated response targeting myelin in the brain and spinal cord (Section 32.8). Symptoms include weakness, loss of balance, paralysis, and blindness. Some alleles of MHC genes increase susceptibility, but a bacterial or viral infection may trigger the disorder.

Autoantibodies (antibodies that recognize self-proteins) produced by self-reactive B cells can damage the body or disrupt its functioning. In Graves' disease, autoantibodies that bind stimulatory receptors on the thyroid gland cause it to produce excess thyroid hormone, which quickens the body's overall metabolic rate (Section 34.6). Antibodies are not part of the feedback loops that normally regulate thyroid hormone production, so antibody binding continues unchecked, the thyroid continues to release too much hormone, and the metabolic rate spins out of control. Symptoms of Graves' disease include uncontrollable weight loss; rapid, irregular heartbeat; sleeplessness; pronounced mood swings; and bulging eyes.

Immune Evasion

Sometimes, the immune system fails to eliminate a pathogen from the body. In most cases, the failure occurs because evolved defenses allow a persistent pathogen to evade its host's immune system.

Consider how antigen recognition is a key factor in adaptive immunity. Fast replication coupled with a high mutation rate allow pathogenic bacteria and viruses to evolve quickly, and a normal adaptive immune response provides plenty of selection pressure to drive the change. Recombination among different strains of viruses (or bacteria) adds even more evolutionary flexibility.

Antigen receptors, as you know, are quite specific, so even a small structural change in a viral envelope protein or bacterial carbohydrate can render a whole army of lymphocytes useless. In the few days it takes to make a new army of lymphocytes that recognize the changed molecule, a pathogen may be free to multiply unchecked.

Interfering with host adaptive immune responses is a common pathogen strategy. Some bacteria produce molecules that disrupt complement function or prevent antibody binding; extra-slimy capsules and biofilms can discourage phagocytosis by white blood cells. A toxin produced by *Staphyloccocus aureus* inhibits phagocyte movement. Herpes simplex viruses, which cause cold sores and genital herpes, are persistent because they inhibit antigen presentation. The bacterial species that causes tuberculosis disrupts T cell receptor signaling, as does the one that causes leprosy. HIV (the virus that causes AIDS) and Epstein–Barr virus (which causes mononucleosis) suppress the expression of MHC genes, so cytotoxic T cells cannot recognize infected cells.

AIDS Acquired immune deficiency syndrome. A secondary immune deficiency that develops as the result of infection by HIV.

TABLE 37.7

Global HIV Cases

Region	People Living with HIV	New HIV Infections
Sub-Saharan Africa	25,800,000	1,400,000
Asia and the Pacific	5,000,000	340,000
Western/Central Europe and North America	2,400,000	85,000
Latin America	1,700,000	87,000
Central Asia/Eastern Europe	1,500,000	140,000
Caribbean Islands	280,000	13,000
Middle East/North Africa	240,000	22,000
Approx. worldwide total	36,920,000	2,087,000

Source: Joint United Nations Programme HIV/AIDS, 2015 report

Immune Deficiency

Insufficient immune function (immune deficiency, also called immunodeficiency) renders an individual vulnerable to infections by opportunistic agents that are typically harmless to those in good health. Primary immune deficiencies, which are present at birth, are the outcome of mutations. Severe combined immunodeficiency (SCID) is an example (Section 15.9). Secondary immune deficiency is the loss of immune function after exposure to a virus or other outside agent.

AIDS

Acquired immunodeficiency syndrome, or **AIDS**, is the most common secondary immune deficiency. AIDS occurs as a result of infection with HIV (Section 20.1). Worldwide, almost 37 million individuals are infected with this virus (**TABLE 37.7** and **FIGURE 37.22**).

AIDS is a syndrome of disorders that occurs because HIV cripples the immune system, thus making the body susceptible to infections by other pathogens and to rare forms of cancer.

A person newly infected with HIV appears to be in good health, perhaps fighting a cold or "the flu." If the infection is untreated, symptoms eventually emerge that foreshadow AIDS: fever, many enlarged lymph nodes, chronic fatigue and weight loss, and drenching night sweats. Then, infections caused by normally harmless microorganisms strike. Yeast infections of the mouth, esophagus, and vagina often occur, as

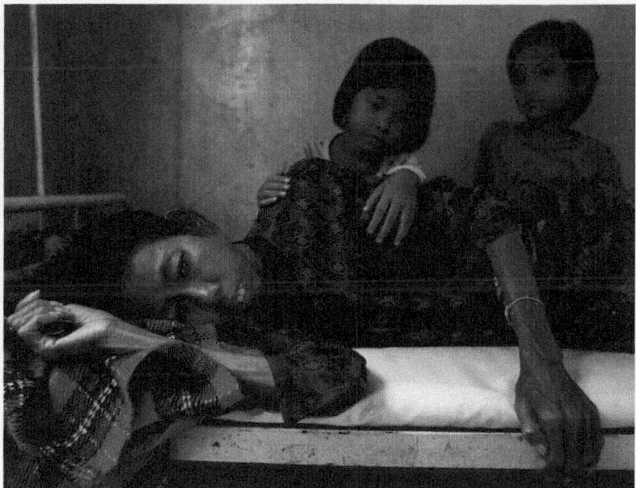

FIGURE 37.22 In Cambodia, a mother lies dying of AIDS in front of her children. This photo was taken in 2002. Today, the country has the highest incidence of AIDS in Southeast Asia.

well as a form of pneumonia caused by the fungus *Pneumocystis jirovecii*. Gastrointestinal inflammation due to infection by a yeast or virus causes diarrhea. Colored lesions that erupt are evidence of Kaposi's sarcoma, a type of cancer that is common among AIDS patients but rare among the general population. Other cancers are also common, as are infections by cancer-causing viruses such as Epstein–Barr virus. These medical problems are relatively uncommon in people with healthy immune systems.

HIV Revisited When HIV particles enter the body, dendritic cells engulf them. The dendritic cells then migrate to lymph nodes, where they present HIV antigen to naive T cells. Armies of HIV-neutralizing antibodies and HIV-specific cytotoxic T cells form.

We have just described a typical adaptive immune response. It rids the body of most—but not all—of the virus. HIV persists in a few helper T cells in a few lymph nodes. For years or even decades, antibodies keep the level of HIV in the blood low, and cytotoxic T cells kill most of the HIV-infected cells. During this stage, infected people often have no symptoms of AIDS, but they can pass the virus to others.

Eventually, the level of virus-neutralizing antibodies plummets, and the production of T cells slows. Why this occurs is still a major topic of research, but its effect is certain: The immune system becomes progressively less effective at fighting the virus. The number of virus particles rises, and more and more helper T cells become infected. Lymph nodes begin to swell with infected T cells, and the battle tilts as the body makes fewer replacement helper T cells and immunity fails. Other types of viruses replicate more quickly than HIV, but the immune system eventually demolishes them. HIV demolishes the immune system. Secondary infections and tumors kill the patient.

Transmission HIV is not transmitted by casual contact; most infections are the result of having unprotected sex with an infected partner. The virus occurs in semen and vaginal secretions, and it can enter a sexual partner through epithelial linings of the penis, vagina, rectum, and mouth. The risk of transmission increases by the type of sexual act; for example, anal sex carries 50 times the risk of oral sex. Infected mothers can transmit HIV to a child during pregnancy, labor, delivery, or breast-feeding. HIV also travels in tiny amounts of infected blood in syringes shared by intravenous drug abusers, or by hospital patients in less developed countries. Many people have become infected via blood transfusions, but this transmission route is becoming rarer because most blood is now tested prior to use for transfusions.

Testing Most AIDS tests check blood, saliva, or urine for antibodies that bind to HIV antigens. These antibodies are detectable in 99 percent of infected people within three months of exposure to the virus. One test that cross checks antibodies with viral antigen can detect infection three weeks after exposure. Currently, the only reliable tests are performed in clinical settings; home test kits are generally less accurate than clinical tests. A false negative result may cause an infected person to unknowingly transmit the virus.

Treatment There is no way to rid the body of HIV, no cure for those already infected. Of the twenty or so drugs approved by the FDA to treat the symptoms of AIDS, most target processes unique to retroviral replication. HIV injects its RNA genome and viral enzymes into a host cell (Section 20.3). One of the enzymes, reverse transcriptase, uses the viral RNA as a template to assemble DNA. RNA nucleotide analogs such as AZT dideoxynucleotides stop the DNA synthesis reaction when they substitute for normal nucleotides (Section 15.4), so these molecules are used as drugs to slow the progression of AIDS.

Protease inhibitors used as HIV drugs prevent viral enzymes from carrying out post-translational cleavage of viral proteins; the uncut proteins cannot assemble into new viruses. A three-drug "cocktail" of one protease inhibitor plus two reverse transcriptase inhibitors is currently the most successful AIDS therapy. The drugs are somewhat toxic, but they have changed the typical course of the disease from a short-term death sentence to a long-term, often manageable illness. Unfortunately,

CREDIT: (22) Masaru Goto/The World Bank.

Data Analysis Activities

Cervical Cancer Incidence in HPV-Positive Women A persistent infection with one of about 10 strains of genital HPV (human papillomavirus) is the main risk factor for cervical cancer. The virus spreads easily by sexual contact, but vaccines that prevent infection have been available since 2006. The vaccines consist of viral proteins that self-assemble into virus-like particles. The particles are not infectious (they contain no viral DNA), but their component proteins trigger an immune response that can prevent HPV infection and the cervical cancer it causes.

In 2003, Michelle Khan and her coworkers published results of their 10-year study correlating HPV status with cervical cancer incidence in women (**FIGURE 37.23**). All 20,514 participants were free of cervical cancer when the study began.

1. At 110 months into the study, what percentage of women who were not infected with any type of cancer-causing HPV had cervical cancer? What percentage of women who were infected with HPV16 also had cancer?
2. In which group would women infected with both HPV16 and HPV18 fall?
3. Is it possible to estimate from this graph the overall risk of cervical cancer that is associated with infection of cancer-causing HPV of any type?

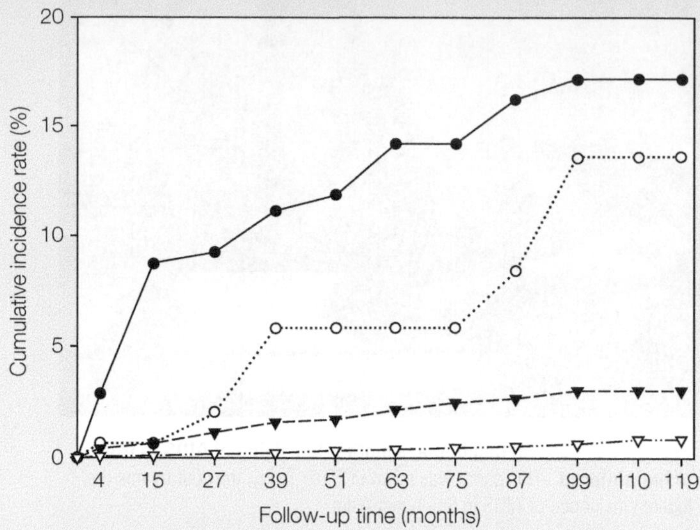
Cumulative incidence rate (%)
20
15
10
5
0
4
15
27
39
51
63
75
87
99
110
119
Follow-up time (months)

FIGURE 37.23 Cumulative incidence rate of cervical cancer correlated with HPV status.

- ● HPV16 positive
- ○ HPV16 negative and HPV18 positive
- ▼ All other cancer-causing HPV types combined
- ▽ No cancer-causing HPV type was detected.

strains of the virus resistant to the most common drugs are becoming common. HIV has a very high mutation rate—the highest known for any biological entity.

Vaccines We still have no vaccine effective against HIV infection. The virus acquires mutations quickly, and antibodies produced during a normal adaptive response exert selection pressure on it. Thus, a very large number of HIV variants exist. A successful vaccine must efficiently recognize and neutralize all variants, including those that have not yet arisen. Immunization with live, weakened virus is an effective vaccine in chimpanzees, but the risk of infection from the vaccination itself far outweighs its potential benefits in humans. Other types of vaccines have been notoriously ineffective against HIV.

One promising strategy involves reverse engineering HIV antibodies isolated from people with AIDS. These antibodies are being collected and studied in painstaking detail in order to discover the exact parts of the virus they recognize. Those parts of the virus are being used to create new vaccines. Genes encoding the three antibodies are also being inserted into viral vectors for use in gene therapy (Section 15.9), the idea being that the vector will deliver the genes into body cells, which will then start producing antibodies.

Prevention Preventive use of a drug that contains two reverse transcriptase inhibitors has recently been shown to greatly reduce the HIV infection rate in high-risk populations. However, education is still our best option for halting the global spread of AIDS. In most circumstances, HIV infection is the consequence of unprotected sex or use of a shared needle for intravenous drugs. Programs that teach people how to avoid these unsafe behaviors are having an effect on the spread of the virus, but overall, our global battle against AIDS is not being won.

TAKE-HOME MESSAGE 37.9

- ✔ An allergy is an immune response to an allergen—something that is ordinarily harmless to most people.
- ✔ When mechanisms that normally limit immune responses fail, severe allergic reactions and other overly vigorous immune responses can be the outcome. These responses can be life-threatening.
- ✔ Autoimmune disorders occur when an immune response targets a person's own healthy body cells.
- ✔ Immune deficiency, which can be inherited or triggered by environmental factors, causes an individual to be especially vulnerable to infections.
- ✔ AIDS is a secondary immune deficiency caused by HIV infection. HIV infects lymphocytes and so cripples the human immune system.

CREDIT: (23) Michelle Khan et al., "The Elevated 10-year Risk of Cervical Precancer and Cancer in Women with Human Papillomavirus (HPV) Type 16 or 18 and the Possible Utility of Type Specific HPV Testing in Clinical Practice; *Journal of the National Cancer Institute*, Vol. 97, No. 14, July 20, 2005.

37.1 Community Immunity (revisited)

Government health departments in all developed countries recommend a series of vaccinations for healthy children. TABLE 37.8 shows the CDC's immunization schedule for children in the United States; other countries have similar schedules for the same vaccines. These worldwide vaccination programs have greatly reduced suffering and deaths from many preventable, dangerous diseases.

However, even with easy access to vaccines, more and more parents are choosing not to vaccinate their children. Most cite fears about vaccine safety. Is this concern justified? Like all medical treatments, vaccines are not risk-free. Consider the MMR vaccine, which prevents infection from viruses that cause measles, mumps, and rubella. A rash or mild fever are relatively common side effects. However, one frequently cited concern, an increased risk of the neurological disorder autism, has no scientific basis. The 1998 publication that first proposed this risk was later shown to contain falsified data, and the lead author of the paper lost his medical license as a result of this fraud. Rigorous studies of thousands of individuals have repeatedly failed to show a link between vaccinations and autism.

Compare the risk of mild side effects from a vaccination to the risk posed by contracting and spreading measles. Measles is a very serious and sometimes fatal disease. It is caused by a virus that spreads easily from person to person in air, and stays infectious for hours in air or on solid surfaces. Measles is also highly contagious: An unvaccinated person exposed to measles has a 90 percent chance of contracting the disease; vaccination lowers the chance to 5 percent. An infected person is contagious for four days before the telltale rash appears, so there is no way to avoid infection by being vigilant.

Consider Lola-May, who was three years old when she attended a birthday party for a child who later became ill with measles. Her mother Rachel said, "I'd chosen not to vaccinate Lola-May or perhaps not put as much thought into it as I should've done." Lola became ill a few days later. Cold-like symptoms gave way to a barking cough, difficulty breathing, agonizing ear pain, and uncontrollable fever. Then the rash appeared. Worsening symptoms required hospitalization, intravenous antibiotics, and other treatments to keep her lungs working. "When you're looking at a child who's sick and knowing... that the decision you made inevitably has caused this to happen... the guilt... is awful," said Rachel. Lola survived, but measles left her with permanent hearing loss and a perforated eardrum that prevents the use of hearing aids. "She just has to struggle. It's affected her speech and language, which has affected her confidence... I kind of thought that if everyone else had done it, I perhaps didn't need to. And that complacency cost us so much as a family."

TABLE 37.8

Recommended Immunization Schedule for Children

Vaccine	Age of Vaccination
HepB (Hepatitis B)	Birth, 1 and 6 months
Rotavirus (RV)	2, 4, and 6 months
DTaP (diphtheria, tetanus, pertussis)	2, 4, 6, and 9–18 months; 4–6 years
HiB (*Haemophilus influenzae*)	2, 4, 6, and 18 months
PCV13 (Pneumococcus)	2, 4, 6, and 12 months
IPV (Inactivated poliovirus)	2, 4, and 6 months; 4–6 years
Influenza	Yearly, 6 months and older
MMR (measles, mumps, rubella)	12 months; 4–6 years
Varicella (chicken pox)	12 months; 4–6 years
HepA (Hepatitis A, 2 doses)	12 and 18 months
HPV (Human papillomavirus, 3 doses)	11 and 12 years
Meningococcal	11, 16 years

Source: Centers for Disease Control and Prevention (CDC), 2017

Other children who contract measles are not nearly as lucky as Lola-May. The virus outright kills one or two of every thousand children it infects. Optic nerve infection results in blindness; encephalitis (dangerous swelling of the brain) can cause permanent intellectual disability or death. Rarely, a fatal central nervous system disease appears decades after a measles infection has apparently resolved.

Those of us who live in developed countries are fortunate to have access to vaccines that can prevent terrible diseases such as measles. Decades of widespread vaccination have all but removed the immediate threat of infection, so it is perhaps easy to take our regional immunity for granted. Outbreaks are common in some parts of the world, and global travel means pathogens travel easily between countries.

A vaccination program requires comprehensive participation to be effective; measles, for example, spreads quickly through communities in which fewer than 92 percent of the people are vaccinated. Even people who have been vaccinated are not completely exempt from risk during an outbreak because no vaccine is 100 percent effective. Some have no choice but to rely on community immunity: children who are too young to be vaccinated, for example, or people with immune deficiencies. Refusing a vaccine attaches a serious risk to a child, and also to others. Parents who choose not to vaccinate their children are also choosing to risk the lives of everyone else in their community.

Section 37.1 **Vaccines** that elicit immunity to specific diseases have overwhelmingly reduced suffering and deaths from severe diseases. Vaccination programs are an important part of worldwide health initiatives, but they are only effective with widespread participation.

Section 37.2 **Immunity** is the body's ability to resist and fight infections. **Antigens** such as **PAMPs** trigger **innate immunity**, a set of general defenses that can prevent pathogens from becoming established inside the body. The activation of innate immunity triggers **adaptive immunity**, a system of defenses that can specifically target billions of different antigens. Signaling molecules called **cytokines** help coordinate the activities of white blood cells (leukocytes) such as **dendritic cells**, **mast cells**, **macrophages**, **neutrophils**, **basophils**, and **eosinophils**. Lymphocytes (**B cells**, **T cells**, and **NK cells**) are leukocytes with special roles in immune responses.

Section 37.3 Microbiota (microorganisms that normally colonize body surfaces, including the linings of tubes and cavities) offer a biological barrier to infection. Microbiota rarely cause disease unless they penetrate inner tissues. Anatomical and physiological barriers (such as **lysozyme**) also help keep microorganisms such as those in **dental plaque** on the outside of the body.

Section 37.4 Microorganisms that breach surface barriers trigger innate immunity. The presence of antigen in the body activates **complement** in cascading reactions. Phagocytic leukocytes (phagocytes) can follow gradients of activated complement back to an infected tissue, a movement called **chemotaxis**. Activated complement proteins coat cells and particles to enhance uptake by phagocytes, and also kill cells by puncturing their plasma membrane. Phagocytes that engulf a microorganism or encounter damaged tissue release cytokines and local signaling molecules that initiate **inflammation**. Increased blood flow and capillary permeability speed delivery of more phagocytes to the affected area. Prolonged exposure to an inflammation-provoking stimulus can cause chronic inflammation. Cytokines trigger **fever** that enhances innate immune responses.

Section 37.5 Vertebrate antigen receptors collectively have the ability to recognize billions of specific antigens, a diversity that arises from random splicing of antigen receptor genes. **T cell receptors** are the basis of self/nonself discrimination; these receptors recognize antigen only when it is displayed together with **MHC markers**. **B cell receptors** are **antibodies** that have not been released from a B cell. Antibody binding facilitates phagocytosis of antigen-bearing particles, and activates complement.

Section 37.6 B cells and T cells carry out adaptive immune responses. The four main characteristics of these responses are self/nonself recognition, specificity (the ability to be customized for specific antigens), diversity (the potential to intercept a tremendous variety of pathogens), and memory. **Antibody-mediated** and **cell-mediated immune responses** work together to rid the body of a specific pathogen. **Effector cells** act in an immune response as soon as they form. **Memory cells** that also form are reserved for a later encounter with the same antigen, in which case they trigger a faster, stronger secondary response. In all adaptive responses, phagocytes engulf, process, and present antigen to T cells in the spleen or lymph nodes.

Section 37.7 **Helper T cells**, which activate other lymphocytes, are central participants in all adaptive immune responses. Antibodies that recognize a specific antigen are secreted by B cells during an antibody-mediated immune response. An **immunization** induces an antibody-mediated response to a pathogen. ABO blood typing, an **agglutination** test, reveals the form of H antigen on a person's red blood cells.

Section 37.8 In a cell-mediated immune response, **cytotoxic T cells** kill body cells that have been altered by infection or cancer. **NK cells** kill ailing body cells that cannot be detected by cytotoxic T cells.

Section 37.9 **Allergens** are normally harmless substances that induce immune responses; sensitivity to an allergen is called **allergy**. A malfunction in the immune system or in its checks and balances can cause dangerous acute illnesses, or chronic and sometimes deadly disorders. In an **autoimmune response**, a body's own cells are inappropriately recognized as foreign and attacked. Immune deficiency is a reduced capacity to mount an immune response. **AIDS** is caused by the human immunodeficiency virus (HIV). An initial adaptive immune response to infection rids the body of most—but not all—of the virus. The virus infects T cells and other leukocytes, so it eventually cripples adaptive immunity.

SELF-QUIZ

Answers in Appendix VII

1. _______ trigger(s) immune responses.
   a. Cytokines
   b. Lysozyme
   c. Antibodies
   d. Antigens
   e. Histamines
   f. MHC markers

2. Which of the following is *not* among the first line of defenses against infection?
   a. skin
   b. lysozyme in saliva
   c. resident populations of bacteria
   d. acidic gastric fluid
   e. complement activation
   f. flushing action of diarrhea

3. Activated complement proteins _______ .
   a. puncture cells
   b. promote inflammation
   c. attract macrophages
   d. all of the above

4. Which is/are *not* part of innate immunity?
   a. phagocytic cells
   b. fever
   c. histamines
   d. complement activation
   e. inflammation
   f. cytokines
   g. presenting antigen
   h. all take part

5. Which is/are *not* part of adaptive immunity?
   a. phagocytic cells
   b. antigen-presenting cells
   c. MHC markers
   d. complement activation
   e. antigen receptors
   f. cytokines
   g. antibodies
   h. all take part

6. Choose the characteristics of adaptive immunity.
   a. self/nonself recognition
   b. immediate response
   c. set of general defenses
   d. antigen memory
   e. diverse antigen receptors
   f. fixed number of PAMPs
   g. tailored for specific antigens

7. Antibodies are ________ .
   a. antigen receptors
   b. made only by B cells
   c. proteins
   d. all of the above

8. A dendritic cell engulfs a bacterium, then presents bacterial bits on its surface along with a(n) ________ .
   a. MHC marker
   b. antibody
   c. T cell receptor
   d. antigen

9. Antibody-mediated responses are most effective against ________ .
   a. intracellular pathogens
   b. extracellular pathogens
   c. cancerous cells
   d. both a and c

10. Cell-mediated responses are most effective against ________ .
    a. intracellular pathogens
    b. extracellular pathogens
    c. cancerous cells
    d. both a and c

11. A vaccination works by ________ .
    a. curing a disease
    b. triggering an allergy
    c. revealing the type of H antigen on red blood cells
    d. eliciting an antibody-mediated response

12. ________ are targets of cytotoxic T cells.
    a. Extracellular virus particles in blood
    b. Virus-infected body cells or tumor cells
    c. Parasitic worms in the liver
    d. Bacterial cells in pus
    e. Pollen grains in nasal mucus

13. Match the immune cell with the function.

___ dendritic cell	a. professional phagocyte
___ B cell	b. antigen-presenter
___ helper T cell	c. activates other lymphocytes
___ NK cell	d. makes antibodies
___ macrophage	e. kills cells lacking MHC markers

14. Which combination of the following types of antibodies and immune cells is central to hay fever?
    a. IgE and mast cells
    b. IgG and basophils
    c. IgA and lymphocytes

15. Match the immunity concept with the best description.

___ anaphylactic shock	a. recognizes antigen
___ immune memory	b. insufficient immune response
___ immune deficiency	c. general defense mechanism
___ autoimmunity	d. immune response against one's own body
___ antigen receptor	e. secondary response
___ inflammation	f. acute allergic reaction

CRITICAL THINKING

1. A flu shot is a vaccine for several strains of influenza virus. This year, you get the shot and "the flu." What happened? (There are at least three explanations.)

2. Drugs with names that end in "–mab" consist of antibodies, and the research that yields these medications typically begins with mice. A mouse is injected with a molecule of interest: a human cytokine, for example, or a receptor implicated in cancer. The resulting antibody-mediated response produces a set of antibodies that recognize different parts of the molecule. B cells are harvested from the mouse's spleen, fused with cancerous B cells, and then cloned—isolated and cultivated separately. Each of the resulting cell lines produces identical antibodies that bind to one part of the injected molecule. These antibodies are called monoclonal antibodies, and they can be purified and studied for use as drugs.

   Monoclonal antibodies can be effective drugs, but only in the immediate term. Antibodies produced by one's own immune system can last up to about six months in the bloodstream, but monoclonals delivered in passive immunization often last for less than a week. Why the difference?

3. The immune system immediately attacks a transplanted tissue or organ, so transplant recipients require immunosuppressive drugs. What lymphocyte is responsible for this response, which is called transplant rejection?

4. Use Figure 37.2 to identify the leukocytes in the micrograph on the left.

   ___ neutrophil ___ lymphocyte
   ___ eosinophil ___ basophil

38 RESPIRATION

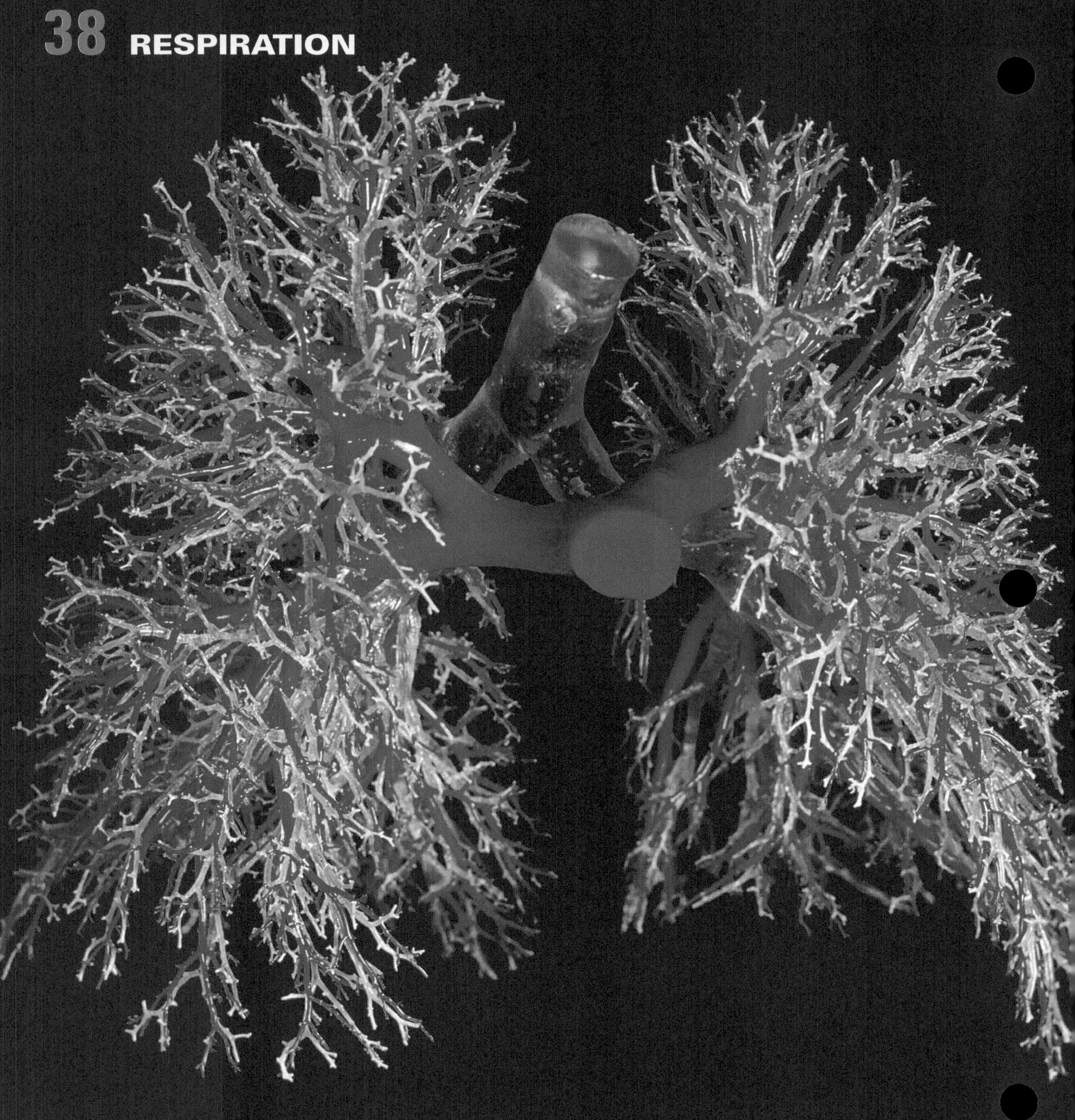

CORE CONCEPTS

Pathways of Transformation

Organisms exchange matter and energy with the environment in order to grow, maintain themselves, and reproduce.

Respiration is a physiological process that moves oxygen and carbon dioxide between body tissues and the external environment. Aquatic animals exchange gases with water, as by way of gills. Air-breathing invertebrates have a system of tubes that deliver air into the body. Air-breathing vertebrates exchange gases inside their lungs.

Evolution

Evolution underlies the unity and diversity of life.

Shared core processes and features provide evidence that living things are related. Vertebrate lungs evolved from outpouchings of the gut wall in fishes. Evolutionary modifications increased the efficiency of lungs, especially in mammals and birds. Birds have the most efficient lungs, which they require to support their high metabolism.

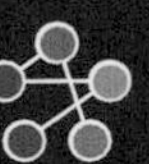

Systems

Complex properties arise from interactions among components of a biological system.

Interactions between the respiratory, circulatory, and nervous systems involve feedback control mechanisms. These homeostatic mechanisms allow a human body to respond dynamically to internal and external conditions. The brain monitors the gas content of the blood and adjusts the rate and depth of breathing accordingly, thus matching oxygen delivery to cells with the cellular activity level.

Links to Earlier Concepts

Understanding diffusion (Section 5.8) and aerobic respiration (7.2) will help you understand gas exchange. This chapter revisits the role of red blood cells (36.5) and hemoglobin (3.2) and how evolutionary changes that accompanied the move of vertebrates onto land (25.6) allow respiration in specific environments.

38.1 Carbon Monoxide—A Stealthy Poison

Carbon monoxide poisoning is the most common type of fatal inhalation poisoning both in the United States and worldwide. Carbon monoxide (CO) is a colorless, odorless gas released when organic material burns. Being lighter than air, it rises from its point of release.

The incidence of accidental CO poisonings typically increases during the winter, when people close up their homes and start using wood stoves and heaters that burn oil, gas, or kerosene. The number of CO poisoning also spikes after a power outage, when people turn to gasoline-powered generators.

Inhaled carbon monoxide exerts its adverse effect by interfering with gas exchange. CO binds to hemoglobin to form carboxyhemoglobin (COHb). Hemoglobin has a very high affinity for CO, binding it more than 200 times more tightly than oxygen. In addition to blocking many oxygen-binding sites, CO causes hemoglobin to hold on to any oxygen that it does bind. Thus, when CO is present, the blood holds less oxygen and releases little of what it does hold to tissues.

The organs most affected by CO poisoning are those with the highest oxygen needs—the brain and heart. At a low level, CO causes headache, fatigue, irritability, mild nausea, and a shortness of breath. The symptoms are often mistakenly attributed to an infectious disease such as a flu. However, unlike a flu, CO poisoning does not cause a fever. A person with no underlying heart or lung problems will begin to experience symptoms of CO poisoning when the level of COHb in the blood reaches about 15 percent. As this level increases, a person becomes dizzy, light-headed, and confused. Chest pain often occurs. The pain is a signal that heart cells are starved of oxygen. Continued exposure to CO results in seizures and loss of consciousness, followed by coma, cardiac arrest, and death.

A nonsmoker who lives in an area with little air pollution typically has a blood COHb level of 1 to 2 percent. People exposed to air pollution from heavy traffic have heightened blood levels of CO, as do smokers. Smoking cigarettes increases blood COHb by about 5 percent for each pack smoked per day.

Over the short term, a blood COHb level as low as 4 to 6 percent can decrease a healthy, young person's capacity to exercise. Over the longer term, an increased COHb level, such as that seen in smokers, damages the heart, which must work overtime to make up for CO's suffocating effects on respiratory function.

CREDITS: (opposite) Martin Dohrn/Royal College of Surgeons/Science Source.

38.2 The Nature of Respiration

LEARNING OBJECTIVES

- Explain why animal cells need oxygen and how they produce waste carbon dioxide.
- Describe the properties common to all respiratory surfaces.
- Give examples of adaptations that facilitate gas exchange.
- Compare the advantages and disadvantages of air and water as respiratory media.

Sites of Gas Exchange

In Chapter 7 you learned about aerobic respiration, an energy-releasing pathway that requires oxygen (O_2) and produces carbon dioxide (CO_2) as summarized in the equation below:

$$\underset{\text{glucose}}{C_6H_{12}O_6} + \underset{\text{oxygen}}{6O_2} \longrightarrow \underset{\text{carbon dioxide}}{6CO_2} + \underset{\text{water}}{6H_2O}$$

This chapter focuses on **respiration**, the physiological processes that supply body cells with oxygen from the environment and deliver waste carbon dioxide from cells to the environment. Respiration depends upon the tendency of gaseous oxygen and carbon dioxide to follow their concentration gradients and diffuse between external and internal environments.

Gases enter and leave the animal body by diffusing across a thin, moist layer of cells called the **respiratory surface** (FIGURE 38.1A). A typical respiratory surface is only one or two cell layers thick. The respiratory surface is thin because gases diffuse quickly only over very short distances. It must be moist because a gas can only diffuse across the lipid bilayer of a plasma membrane when it is in solution. Bubbles of gas cannot cross a plasma membrane.

A second exchange of gases occurs internally, at the plasma membrane of body cells (FIGURE 38.1B). Oxygen diffuses from the interstitial fluid into a cell, and carbon dioxide diffuses in the opposite direction. In invertebrates without a circulatory system, oxygen that crosses the respiratory surface reaches body cells by diffusion. In many invertebrates and all vertebrates, a circulatory system speeds movement of gases between the respiratory surface and body cells.

Factors Affecting Gas Exchange

Several factors affect the rate at which gases cross an animal's respiratory surface.

Size matters: The greater the area of a respiratory surface, the more molecules can cross the surface at once. This is why the area of a respiratory surface is often surprisingly large relative to the size of the animal's body. Extensive branching and folding commonly allow an extensive respiratory surface to fit into a small volume.

Gas concentration gradients across the respiratory surface affect the rate of gas exchange: The steeper the gradient across the membrane, the faster diffusion proceeds. As a result, many animals use muscle movements to keep oxygen-rich air or water from their environment flowing over their respiratory surface. In animals with a circulatory system, this system continuously delivers circulatory fluid (blood or hemolymph) to the respiratory surface and moves it away from that surface after gas exchange has taken place.

Respiratory proteins steepen the oxygen concentration gradient at the respiratory surface of many animals. A **respiratory protein** is a protein that transports oxygen. It has a metal ion or ions that bind oxygen where the oxygen concentration is high and release it in regions where oxygen is scarce. When an oxygen molecule is bound to a respiratory protein, it no longer contributes to the oxygen concentration of the blood or hemolymph. Thus, respiratory proteins lower the effective oxygen concentration in circulatory fluid and encourage diffusion of oxygen into that fluid.

Hemoglobins are iron-containing respiratory proteins that transport oxygen in the blood of vertebrates and annelids such as earthworms. They give the blood of these animals its red color. Some other invertebrates, including octopuses and crabs, have a copper-containing respiratory protein called hemocyanin. Hemocyanin is colorless when deoxygenated, and blue when carrying oxygen.

Respiratory Medium—Air or Water?

The respiratory medium, the environmental substance with which an animal exchanges gases, can be either water or air (FIGURE 38.2). Water is 50 times more viscous (thicker) than air, so moving it over a respiratory surface requires more effort than moving air. Oxygen

FIGURE 38.1 Two sites of gas exchange. In some invertebrates, gases diffuse between the two sites. In others, and all vertebrates, a circulatory system facilitates movement of gases.

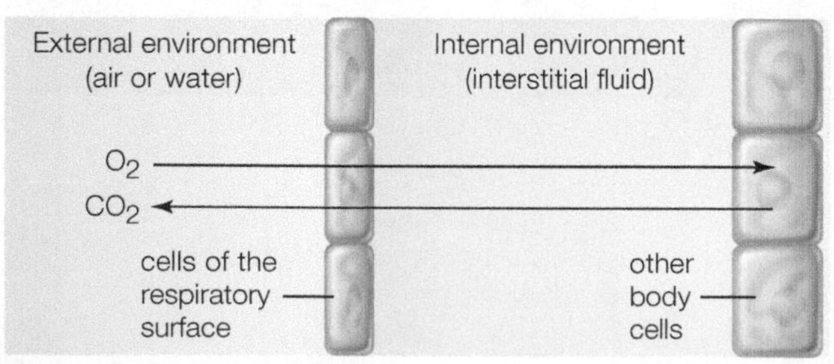
External environment
(air or water)
Internal environment
(interstitial fluid)
O_2
CO_2
cells of the
respiratory
surface
other
body
cells

A Cells of the respiratory surface exchange gases with the external and internal environment.

B Other body cells exchange gases with the internal environment.

Respiratory Medium: Water

- High viscosity, more difficult to move
- Slower diffusion rate for gases
- Lower O_2 concentration
- O_2 concentration varies with flow rate and temperature

Respiratory Medium: Air

- Low viscosity, easier to move
- Higher diffusion rate for gases
- Higher O_2 concentration
- O_2 concentration constant at a given altitude

FIGURE 38.2 **Properties of water and air as respiratory media.**

also diffuses more slowly through water than through air. In addition, at any given temperature, a volume of water holds much less oxygen than an equivalent volume of air. As a result of these factors, obtaining sufficient oxygen from water requires a greater expenditure of energy than obtaining it from air.

Air is also a more reliable source of oxygen. At present, the atmosphere consists of 21 percent oxygen everywhere on Earth. The atmosphere is less dense at high altitudes than low ones, but at any given altitude, the proportion of gases is constant. By contrast, the proportion of oxygen in aquatic environments varies widely. Warm water holds less oxygen than cold water, and more oxygen dissolves in fast-moving water than still water. Thus, changes in water temperature or flow rate affect oxygen availability. When water temperature rises or moving water becomes stagnant, aquatic species that have high oxygen needs may suffocate.

hemoglobin (HEEM-o-glow-bin) Iron-containing respiratory protein.
respiration Physiological process by which an animal body supplies cells with oxygen and disposes of their waste carbon dioxide.
respiratory protein Protein that reversibly binds oxygen when the oxygen concentration is high and releases it when oxygen concentration is low. Hemoglobin is an example.
respiratory surface Moist surface across which gases are exchanged between animal cells and the external environment.

TAKE-HOME MESSAGE 38.2

✔ Respiration supplies cells with oxygen for aerobic respiration and removes waste carbon dioxide.

✔ Gases are exchanged by diffusion across a respiratory surface, which is a thin, moist membrane.

✔ The area of a respiratory surface and the steepness of the gas concentration gradients across it influence the rate of exchange.

✔ Air and water have different properties as respiratory media. Aquatic animals have to expend more energy to obtain oxygen, and the amount of oxygen available can be a limiting factor.

38.3 Invertebrate Respiration

LEARNING OBJECTIVE

- Using appropriate examples, describe the ways that invertebrates exchange gases with their environment.

Animals that lack both respiratory and circulatory organs are either small and flat or, when larger, have cells arranged in thin layers. These animals live in aquatic or continually damp land environments where their gas exchange needs are met entirely by diffusion across a body surface. For example, cnidarians and flatworms exchange gases across their external surface and across the surface of their gastrovascular cavity.

Gilled Invertebrates

Gills evolved independently in several groups of aquatic invertebrates. **Gills** are filamentous or platelike respiratory organs that increase the surface area available for gas exchange with water.

Gills may be internal or external. Most aquatic mollusks have a gill inside their mantle cavity (**FIGURE 38.3A**). Sea slugs, which have gills on their body surface (**FIGURE 38.3B**), are an exception.

Aquatic arthropods such as lobsters and crabs have feathery gills beneath their exoskeleton, where the delicate tissues are protected from damage. These gills evolved as modifications of branches on the limbs.

Air-Breathing Invertebrates

Some land-dwelling invertebrates such as earthworms and nematodes rely solely on gas exchange across their outer body surface. The need to keep their body surface damp restricts these animals to moist locations.

A **lung** is an air-filled internal organ that functions in gas exchange. Snails and slugs that spend time on land have a lung in their mantle cavity instead of, or in addition to, a gill. A pore at the side of the body opens to allow air into the mantle cavity and shuts to conserve water.

Insects exchange gases with the environment by means of a **tracheal system**, a set of branching, chitin-reinforced tubes that convey air directly to tissues deep in the body (**FIGURE 38.3C**). Air enters the tubes through spiracles, which are small openings across the integument. Spiracles can be opened or closed to prevent water loss. The tips of the finest tracheal branches deliver gases to hemolymph in which gases can dissolve. Oxygen and carbon dioxide diffuse between hemolymph and adjacent cells. Tracheal tubes end next to cells, so most insects have no need for a respiratory protein such as hemoglobin to carry gases.

Spiders and scorpions typically have one or more book lungs in addition to, or instead of, tracheal tubes. In a book lung, air and hemolymph exchange gases across thin sheets of tissue (**FIGURE 38.3D**). The air enters the body through a spiracle.

FIGURE 38.3 Invertebrate respiratory organs.

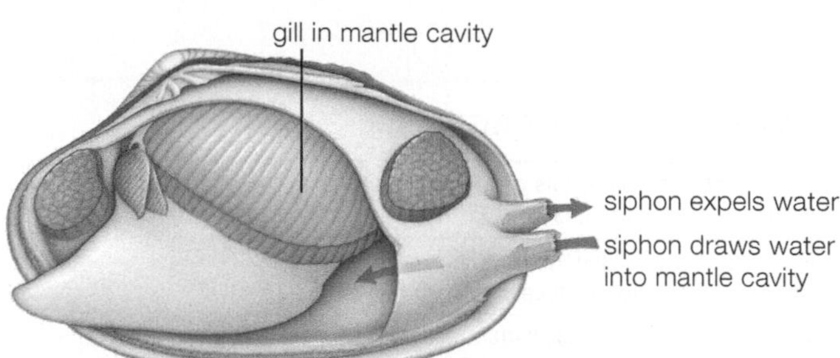
gill in mantle cavity
siphon expels water
siphon draws water into mantle cavity

A Internal gills of a clam, an aquatic mollusk.

B External gills of a sea slug, an aquatic mollusk.

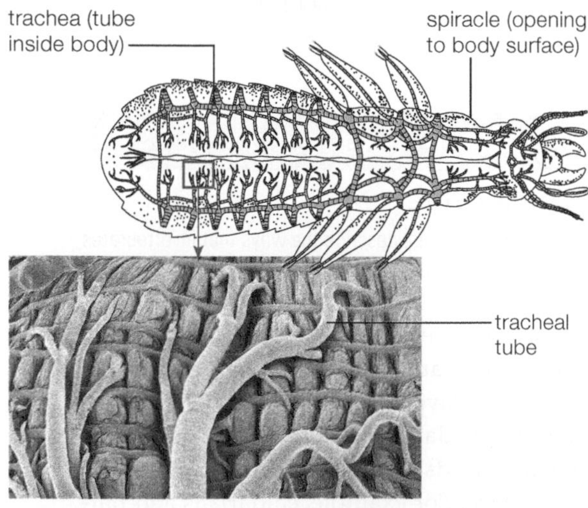
trachea (tube inside body)
spiracle (opening to body surface)
tracheal tube

C Tracheal system of an insect. Air-filled tracheal tubes reinforced by rings of chitin carry air deep inside the body.

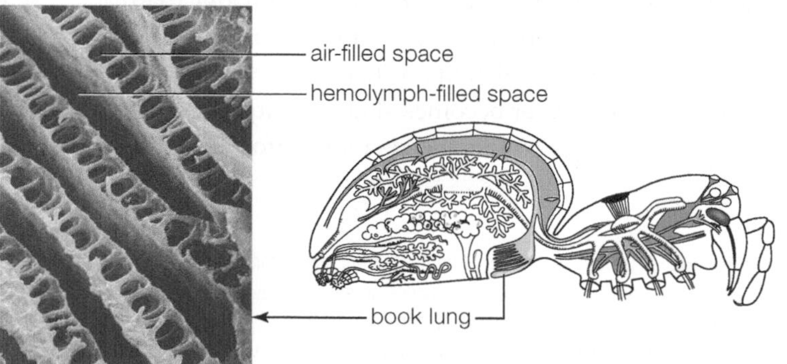
air-filled space
hemolymph-filled space
book lung

D Book lung of a spider. The lung contains many thin sheets of tissue, somewhat like the pages of a book. As hemolymph moves through spaces between the "pages," it exchanges gases with air in adjacent spaces.

CREDITS: (3A) Douglas Faulkner/Sally Faulkner Collection; (3B) iStockphoto.com/cbpix; (3C) Susumu Nishinaga/Science Source; (3D) © D. E. Hill.

TAKE-HOME MESSAGE 38.3

✔ Some invertebrates do not have respiratory organs. All gas exchange takes place across the body wall.

✔ Gills are filamentous organs that increase the surface area available for gas exchange in an aquatic habitat. They may be inside the body or extensions of the body surface.

✔ Some land snails and slugs have an air-filled lung inside their mantle cavity.

✔ Land arthropods have tracheal tubes or book lungs, respiratory organs that bring air deep inside their body.

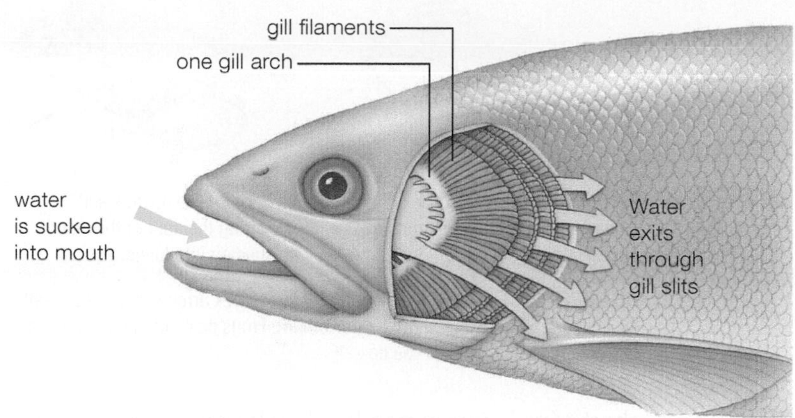

gill filaments
one gill arch
water
is sucked
into mouth
Water
exits
through
gill slits

A Bony fish with its gill cover removed. Water flows in through the mouth, over the gills, then out through gill slits. Each gill has bony gill arches with many thin gill filaments attached.

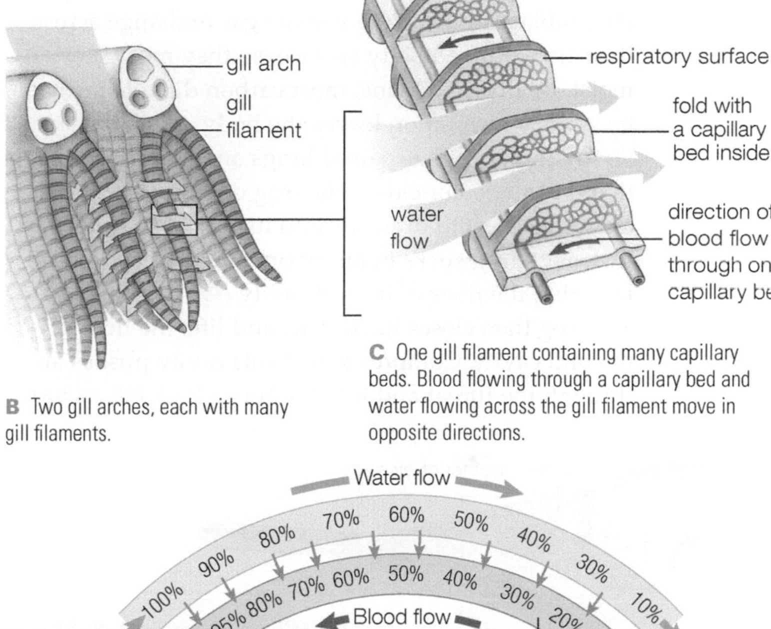

gill arch
gill
filament
respiratory surface
fold with
a capillary
bed inside
water
flow
direction of
blood flow
through one
capillary bed
Water flow
100%
90%
80%
70%
60%
50%
40%
30%
10%
100%
95%
80%
70%
60%
50%
40%
30%
20%
10%
Blood flow
Blood
capillary

B Two gill arches, each with many gill filaments.

C One gill filament containing many capillary beds. Blood flowing through a capillary bed and water flowing across the gill filament move in opposite directions.

D Countercurrent exchange enhances oxygen flow from water into a capillary. Percentages indicate the degree of oxygenation of water (blue) and blood (red). Along the entire length of the capillary, oxygen flows down its concentration gradient from water into blood.

FIGURE 38.4 **Gills of a bony fish.**

38.4 Vertebrate Respiration

LEARNING OBJECTIVES

- Describe the countercurrent flow of blood and water across fish gills, and explain how it enhances the efficiency of gas exchange.
- Compare the mechanisms by which frogs and mammals fill their lungs with air.
- Explain why birds can extract more oxygen from air than mammals can.

Respiration in Fishes

All fishes have gill slits that open across the pharynx (the throat region). In jawless fishes such as lampreys and cartilaginous fishes such as sharks, the gills are uncovered and visible at the body surface. However, most bony fishes have a moveable gill cover.

Gas exchange occurs when water flows outward over the gills (**FIGURE 38.4A**). A bony fish sucks water inward by opening its mouth, closing the cover over each gill, and contracting muscles that enlarge the oral cavity. To force that water out over its gills, the fish closes its mouth, opens its gill covers, and contracts muscles that reduce the size of the oral cavity.

A fish gill consists of bony gill arches, each with multiple gill filaments attached (**FIGURE 38.4B**). A gill filament contains many capillaries where gases in water are exchanged with gases in blood.

Countercurrent exchange maximizes the efficiency of gas exchange across fish gills. It is a mechanism by which a substance is transferred between two fluids that are flowing in opposite directions on either side of a semipermeable membrane. Water flowing over gills and blood flowing through gill capillaries travel in opposite directions (**FIGURE 38.4C**). As a result, the water adjacent to a gill capillary always has more oxygen than the blood flowing inside that capillary (**FIGURE 38.4D**). This persistent concentration gradient causes oxygen to diffuse from water to blood along the entire length of the capillary.

Paired Lungs of Tetrapods

Most tetrapods have paired lungs. Vertebrate lungs evolved from outpouchings of the gut wall in some fishes. Lungs may have helped these fishes survive

countercurrent exchange Exchange of substances between two fluids that are moving in opposite directions.
gill Filamentous or branching respiratory organ of some aquatic animals; may be internal or external.
lung Internal gas-exchange organ in some air-breathing animals.
tracheal system (TRAY-key-ul) Of insects and some other land arthropods, tubes that convey gases between the body surface and internal tissues.

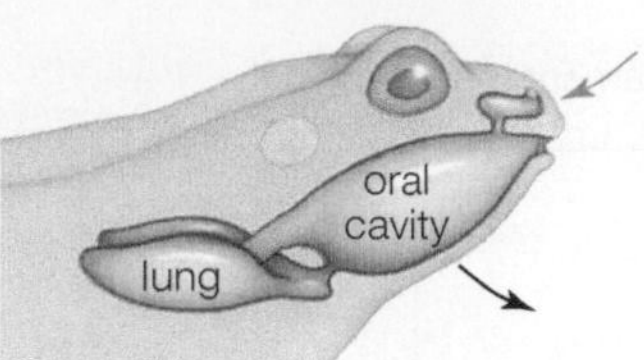
oral
cavity
lung

A The frog lowers the floor of its mouth, pulling air into the oral cavity through its nostrils.

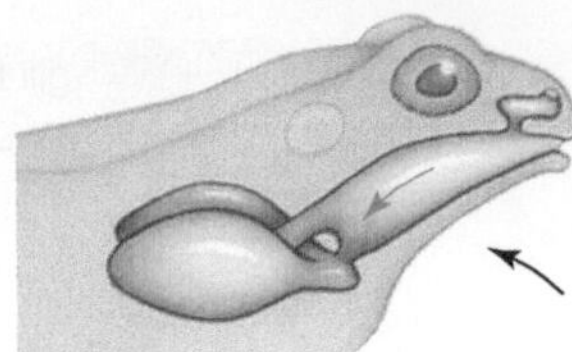

B Closing the nostrils and elevating the floor of the mouth pushes air into lungs.

FIGURE 38.5 How a frog fills its lungs. Black arrows show body wall movements. Blue arrows show air movement. Frogs push air into their lungs, rather than sucking it in as you do.

short trips between ponds. They became increasingly important as tetrapods moved onto land (Section 25.4).

Amphibian larvae typically have external gills that are replaced by paired lungs as the animal develops. Amphibians also carry out some gas exchange across their thin-skinned body surface, so they must stay moist. In all amphibians, most carbon dioxide formed by aerobic respiration leaves the body across the skin.

An adult frog has paired lungs and a pair of nostrils that it can open or close. The frog does not use muscles of its chest to draw air into its lungs, as you do. Instead, a frog sucks in air by opening its nostrils and lowering the floor of its oral cavity (**FIGURE 38.5A**). The frog then closes its nostrils and lifts the floor of the oral cavity. Compression of this cavity pushes air through the throat and into the lungs (**FIGURE 38.5B**).

lungs posterior air sacs

anterior air sacs

A (Above) Large inflatable air sacs attach to two small, inelastic lungs. The lining of tiny air-filled tubes in the lungs is the respiratory surface.

anterior air sacs lungs posterior air sacs

Inhalation 1

Exhalation 1

Inhalation 2

Exhalation 2

B (Right) Schematic diagram of showing the one-way movement of a volume of air (blue) through a bird's respiratory system. It takes two breaths for air to pass through the entire system.

FIGURE 38.6 Bird respiration. One-way flow of air through a bird's lungs increases the efficiency of gas exchange.

Reptiles, birds, and mammals are amniotes with a waterproof skin (Section 25.5). Their only respiratory surface is inside their two lungs.

Birds have the most efficient respiratory system of all vertebrates, meaning they are able to take up the most oxygen from any given volume of air. Their small, inelastic lungs do not expand and contract. Instead, large expandable air sacs attached to the lungs inflate and deflate (**FIGURE 38.6A**). No gas exchange occurs across the air sacs. Their function is to provide a continual one-way flow of fresh, oxygen-rich air through a bird's lungs.

It takes two breaths for an inhaled volume of air to travel through a bird's respiratory system (**FIGURE 38.6B**). The first inhalation draws fresh air into posterior air sacs. During the first exhalation, this fresh air moves from posterior air sacs through tiny tubes in the lungs. The lining of these tubes is the bird's respiratory surface. During the second inhalation, the air that was in the lungs moves into anterior air sacs. It exits these air sacs during the second exhalation.

In mammals, lungs inflate when muscles increase the size of the thoracic cavity. As the cavity expands, pressure in the lungs declines and air is sucked inward. The inhaled air flows through increasingly smaller airways until it reaches tiny sacs called **alveoli**, where gases are exchanged. During exhalation, air retraces its path, flowing out the same way it came in. Mammalian lungs never not deflate completely, so a bit of stale air remains behind even after exhalation.

TAKE-HOME MESSAGE 38.4

✔ Fishes exchange gases with water flowing over their gills. Movement of water and blood in opposite directions enhances this exchange.

✔ Amphibians exchange gases across their skin and push air from their mouth into their lungs. Amniotes suck air into their lungs by expanding the size of their thoracic cavity.

✔ Air flows into and out of mammalian lungs, but it flows continually through bird lungs.

38.5 Human Respiratory System

LEARNING OBJECTIVES

- List the structures that inhaled air passes through as it flows from the human nasal cavity to an alveolus.
- Describe the location of the epiglottis, and explain its function.
- Describe the location and structure of the human lungs.

The respiratory system functions in gas exchange, but it has additional roles. We speak or sing as air moves past our vocal cords. We have a sense of smell because inhaled molecules stimulate olfactory receptors in the nose. Cells in nasal passages and other airways intercept

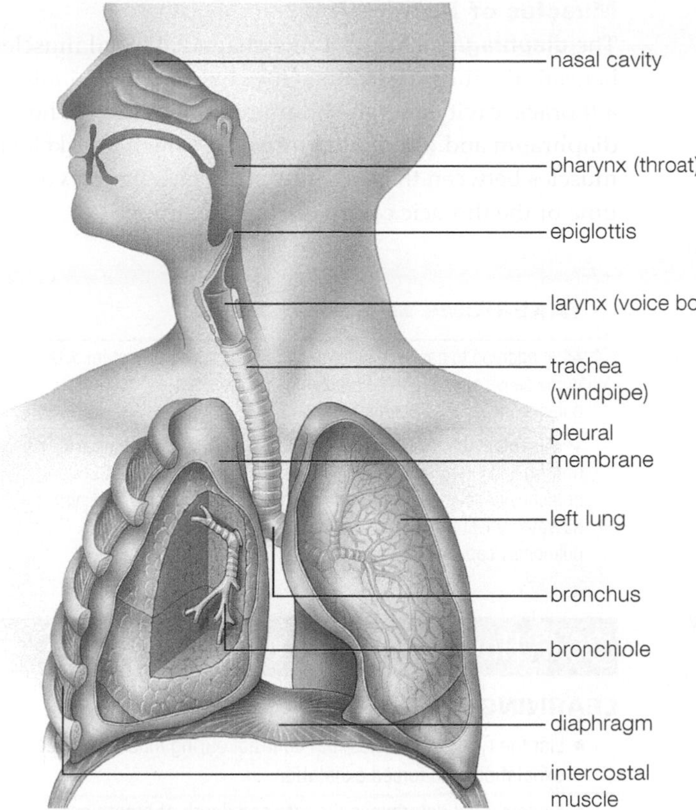
nasal cavity
pharynx (throat)
epiglottis
larynx (voice box)
trachea
(windpipe)
pleural
membrane
left lung
bronchus
bronchiole
diaphragm
intercostal
muscle
A Airways and respiratory muscles.

C Cast of airways (white) and blood vessels (red) in human lungs.

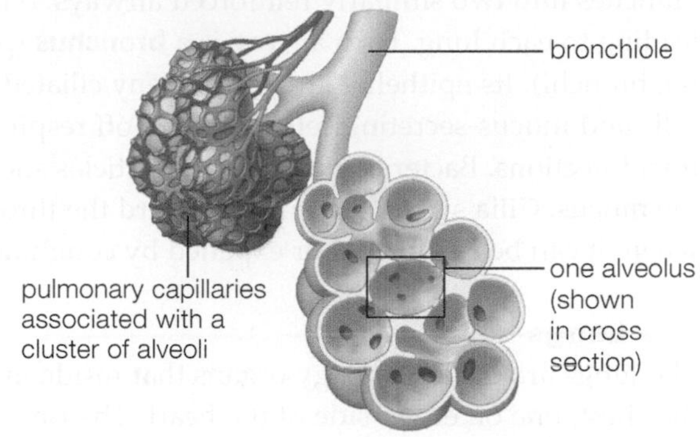
bronchiole
pulmonary capillaries
associated with a
cluster of alveoli
one alveolus
(shown
in cross
section)
B Close-up view of alveoli and pulmonary capillaries.

FIGURE 38.7 Human respiratory system and associated structures.

and neutralize airborne pathogens. The respiratory system contributes to acid–base balance by expelling waste CO_2 that can make the blood more acid. Controls over breathing also help maintain body temperature; water evaporating from airways has a cooling effect.

The Respiratory Tract

Take a deep breath. Now look at **FIGURE 38.7A** to get an idea of where the air went. If you are healthy and sitting quietly, air probably entered through your nose. As air moves through your nostrils, tiny hairs filter out large particles. Mucus secreted by some cells of the nasal lining captures most fine particles and airborne chemicals. Waving cilia on other cells sweep away captured contaminants.

Air from the nostrils enters the nasal cavity, where it is warmed and moistened. From here, it flows next into the **pharynx**, or throat. It continues to the **larynx**, a short airway commonly known as the voice box because it contains a pair of vocal cords. A vocal cord is skeletal muscle with a cover of mucus-secreting epithelium. Contraction of vocal cords narrows the gap between them, which is the **glottis** (**FIGURE 38.8**).

When the glottis is wide open, air flows through it silently. When muscle contraction narrows the glottis, flow of air outward through the tighter gap makes vocal cords vibrate so they produce sounds. The

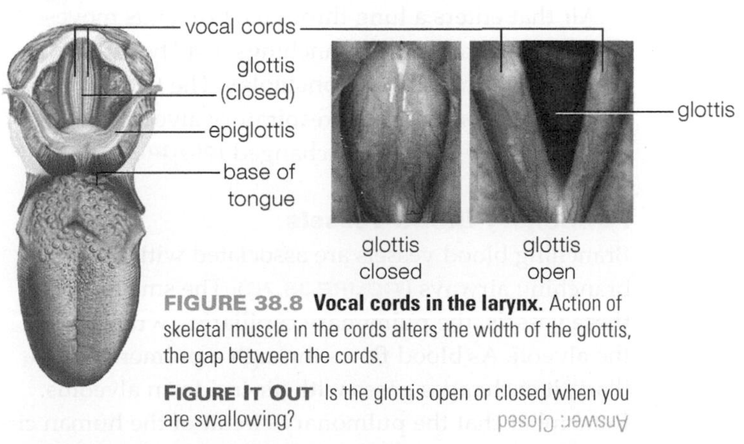
vocal cords
glottis
(closed)
epiglottis
base of
tongue
glottis
glottis
closed
glottis
open
FIGURE 38.8 Vocal cords in the larynx. Action of skeletal muscle in the cords alters the width of the glottis, the gap between the cords.

FIGURE IT OUT Is the glottis open or closed when you are swallowing?

Answer: Closed

alveolus (al-VEE-oh-luss) Plural, alveoli. In a lung, a tiny sac at the tip of a bronchiole; site of gas exchange.

glottis Opening formed when vocal cords in the larynx relax.

larynx (LAIR-inks) Short airway containing the vocal cords (voice box).

pharynx (FAIR-inks) Passage between the mouth and digestive tract; throat.

CREDITS: (7A, B) © Cengage Learning; (7C) Martin Dohrn/Royal College of Surgeons/Science Source; (8) Photographs, Courtesy of Kay Elemetrics Corporation.

tension on the cords and the position of the larynx determine the sound's pitch. To get a feel for how this works, place one finger on your "Adam's apple," the thyroid cartilage that sticks out at the front of your neck. Hum a low note, then a high one. You will feel the vibration of your vocal cords and how laryngeal muscles shift the position of your larynx.

At the entrance to the larynx is a flap of tissue called the **epiglottis**. When the epiglottis points up, air can move into or out of the **trachea**, or windpipe. When you swallow, the larynx elevates, the glottis closes, and the epiglottis bends backward to cover the larynx entrance, so food and fluids enter the esophagus. The esophagus connects the pharynx to the stomach.

The trachea is reinforced by rings of cartilage. It branches into two similarly reinforced airways, one leading to each lung. Each airway is a **bronchus** (plural, bronchi). Its epithelial lining has many ciliated cells and mucus-secreting cells that fend off respiratory tract infections. Bacteria and airborne particles stick to the mucus. Cilia sweep the mucus toward the throat, where it can be swallowed or expelled by coughing.

The Lungs

The lungs are conical spongy organs that reside in the chest, one on either side of the heart. The rib cage encloses and protects the lungs, and a two-layer-thick pleural membrane covers them. The pleural membrane's outer layer attaches to the wall of the chest cavity and upper surface of the diaphragm. Its inner layer attaches to the outer surface of the lungs. A secreted pleural fluid fills the small space between the pleural membrane's inner and outer layers. The fluid reduces friction between the membrane's two layers when the lungs expand and contract during breathing.

Air that enters a lung through a bronchus moves through finer and finer branchings of a "bronchial tree." The branches are **bronchioles**. The tips of the finest bronchioles end in the respiratory alveoli, the tiny air sacs where gases are exchanged (**FIGURE 38.7B**).

Pulmonary Blood Vessels

Branching blood vessels are associated with the branching airways (**FIGURE 38.7C**). The smallest of these vessels, the pulmonary capillaries, wrap around the alveoli. As blood flows through a pulmonary capillary, it exchanges gases with air inside an alveolus. Remember that the pulmonary circuit of the human circulatory system pumps oxygen-poor blood to the lungs and returns oxygen-rich blood to the heart. Systemic circulation then transports oxygen-rich blood to body tissues and returns oxygen-poor, carbon dioxide–rich blood to the heart.

Muscles of Respiration

The **diaphragm**, a broad dome-shaped skeletal muscle beneath the lungs, partitions the coelomic cavity into a thoracic cavity and an abdominopelvic cavity. The diaphragm and **intercostal muscles**, which are skeletal muscles between the ribs, interact to change the volume of the thoracic cavity during breathing.

TAKE-HOME MESSAGE 38.5

✔ In addition to gas exchange, the human respiratory system acts in the sense of smell, voice production, body defenses, acid–base balance, and temperature regulation.

✔ Air enters through the nose or mouth. It flows through the pharynx (throat) and larynx (voice box) to a trachea that branches into two bronchi, one to each lung. Inside each lung, additional branching airways deliver air to alveoli, where gases are exchanged with pulmonary capillaries.

38.6 How We Breathe

LEARNING OBJECTIVES

- List the types of muscles that contract during inhalation, passive exhalation, and forced exhalation.
- Explain what determines the rate and depth of breathing.

The Respiratory Cycle

A **respiratory cycle** is one breath in (inhalation) and one breath out (exhalation). Inhalation is always active, meaning muscle contractions drive it. During the respiratory cycle, changes in the volume of the lungs and thoracic cavity alter air pressure inside the lungs.

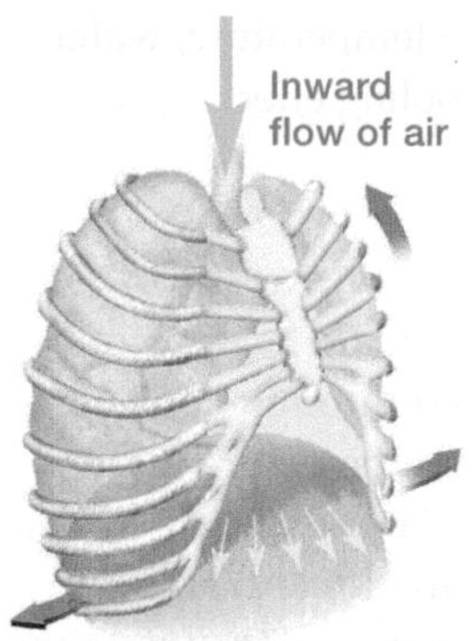
Inward
flow of air

A Inhalation.

The thoracic cavity and lungs expand as the diaphragm contracts and moves downward, External intercostal muscles contract, lifting the rib cage.

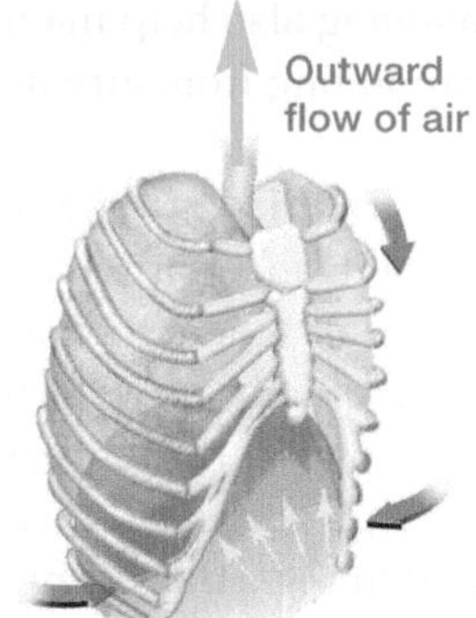
Outward
flow of air

B Exhalation.

The thoracic cavity and lungs shrink as the diaphragm relaxes and moves upward. External intercostal muscles relax lowering the rib cage.

FIGURE 38.9 Changes in the size of the thoracic cavity during the respiratory cycle.

FIGURE IT OUT What effect does contraction of the diaphragm have on the volume of the thoracic cavity? Answer: It increases the volume.

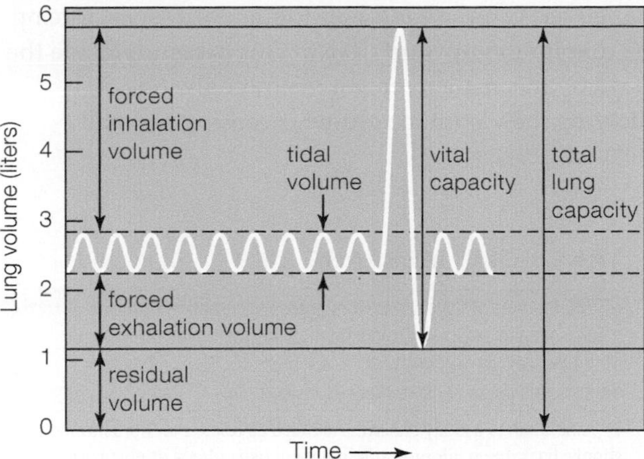
6
5
4
3
2
1
0
Lung volume (liters)
forced inhalation volume
tidal volume
vital capacity
total lung capacity
forced exhalation volume
residual volume
Time

FIGURE 38.10 Respiratory volumes. In normal breathing, the tidal volume of air entering and leaving the lungs is only 0.5 liter. Lungs never deflate completely. Even with a forced exhalation, a residual volume of air remains in them.

When you inhale, the diaphragm flattens and moves downward, and external intercostal muscles contract and lift the rib cage up and outward (**FIGURE 38.9A**). Together, these actions cause the thoracic cavity to expand; as it does, so do the lungs. Pressure in the alveoli falls below atmospheric pressure, and air follows the pressure gradient, flowing to the alveoli.

Exhalation is usually passive. When muscles that cause inhalation relax, the lungs passively recoil and lung volume decreases. This decrease in volume compresses alveoli, so air pressure inside them increases. Air follows this pressure gradient and flows out of the lungs (**FIGURE 38.9B**).

Exhalation becomes active only during vigorous exercise, or when you consciously blow air outward. During active exhalation, internal intercostal muscles contract, pulling the thoracic wall inward and downward. At the same time, muscles of the abdominal wall contract, causing intra-abdominal pressure to increase and exerting an upward-directed force on the diaphragm. These actions cause the volume of the thoracic cavity to decrease, so air is forced out of the lungs.

Respiratory Volumes

Total lung volume, the maximum amount of air that the lungs can hold, averages 5.7 liters in men and 4.2 liters in women. Usually lungs are less than half full. **Vital capacity**, the maximum volume that can move in and out in one cycle, is one measure of lung health. **Tidal volume**—the volume that moves in and out in a normal respiratory cycle—is about 0.5 liter (**FIGURE 38.10**). Your lungs never fully deflate, so the air inside them always is a mix of freshly inhaled air and stale air left behind during the previous exhalation. Even so, there is plenty of oxygen for gas exchange.

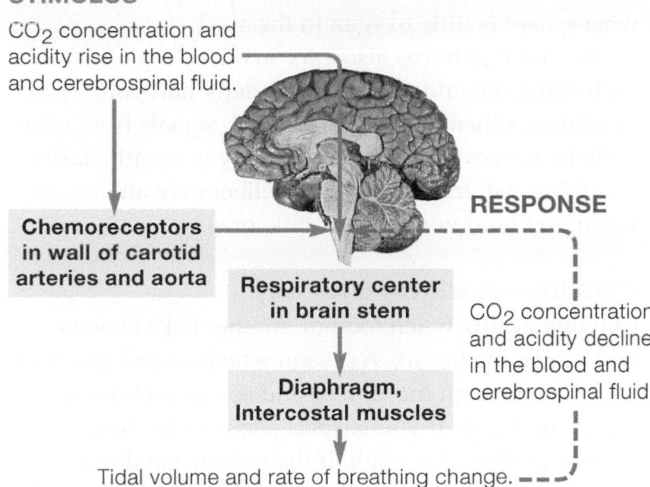
STIMULUS
CO_2 concentration and acidity rise in the blood and cerebrospinal fluid.
Chemoreceptors in wall of carotid arteries and aorta
Respiratory center in brain stem
Diaphragm, Intercostal muscles
Tidal volume and rate of breathing change.
RESPONSE
CO_2 concentration and acidity decline in the blood and cerebrospinal fluid.

FIGURE 38.11 Respiratory response to an increase in CO_2. When increased CO_2 excites chemoreceptors in arteries and the brain, the brain's respiratory center calls for deeper and more frequent breaths. As a result, more CO_2 is expelled.

Control of Breathing

Neurons in the medulla oblongata of the brain stem act as the pacemaker for inhalation, initiating an action potential 10–20 times per minute. Nerves deliver the action potentials to the diaphragm and intercostal muscles, these muscles contract, and you inhale. Between signals, the muscles relax and you exhale.

Breathing patterns change with activity level. Active muscle cells produce additional CO_2 that enters blood, where it combines with water to form carbonic acid (Section 2.6). Chemoreceptors in the walls of carotid arteries and the aorta detect the rise in acidity and signal the brain (**FIGURE 38.11**). In response, you breathe faster and deeper, so more carbon dioxide is expelled.

Chemoreceptors in the artery walls also signal the medulla oblongata when the O_2 concentration in the blood falls to a life-threatening level. This control mechanism usually comes into play only in people

bronchiole (BRONG-key-ohl) Airway that leads from a bronchus to the alveoli.
bronchus (BRON-cuss) Plural, bronchi. Airway that connects the trachea to a lung.
diaphragm (DIE-uh-fram) Skeletal muscle between the thoracic and abdominal cavities; contracts during inhalation.
epiglottis (ep-uh-GLOT-iss) Tissue flap that covers airway during swallowing to prevent food from entering airways.
intercostal muscles Skeletal muscles between the ribs; help change the volume of the thoracic cavity during breathing.
respiratory cycle One inhalation and one exhalation.
tidal volume Volume of air that flows into and out of the lungs during a normal inhalation and exhalation.
trachea (TRAY-key-uh) Airway between the pharynx and bronchi; windpipe.
vital capacity Maximum volume of air that can be moved in and out of the lungs with forced inhalation and exhalation.

with severe lung diseases and at very high altitudes, where there is little oxygen in the air.

Breathing patterns also vary in other ways. Reflexes such coughing and swallowing briefly halt your breathing. When you are frightened, signals from sympathetic nerves (Section 32.7) make you breathe faster. Breathing patterns can also be deliberately altered, as when you hold your breath, talk, or sing.

Choking—A Blocked Airway

Choking occurs when food or another object enters and blocks the larynx. A person who is choking cannot move air through the larynx, and so cannot breathe, cough, or speak. If you suspect someone is choking, encourage them to cough. If the person can do so, their airway is not fully obstructed and coughing will probably clear it.

If the person is choking, a rescuer can help clear the airway and keep the person from dying from lack of air. First aid for choking involves two types of maneuvers. Blows to the back can help dislodge the foreign material with little risk of injury to internal organs (FIGURE 38.12A). If back blows are ineffective, thrusts to the choker's abdomen (the Heimlich maneuver) can raise intra-abdominal pressure, forcing the diaphragm upward. Raising air pressure inside the lungs by this maneuver can dislodge the object, allowing the victim to resume breathing normally (FIGURE 38.12B).

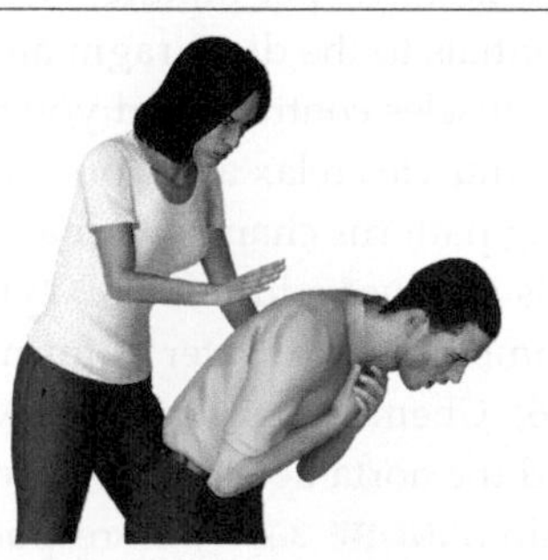

A Back blows. Have the person lean forward. Use the heel of your hand to strike between the shoulder blades.

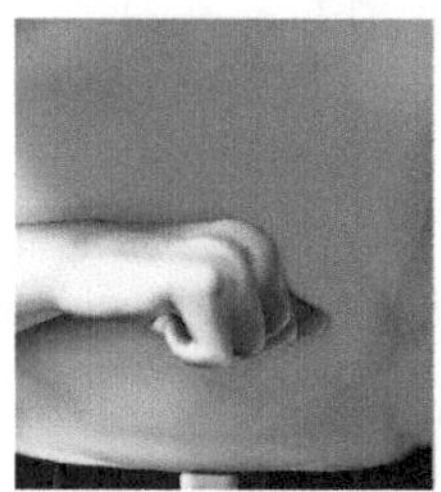

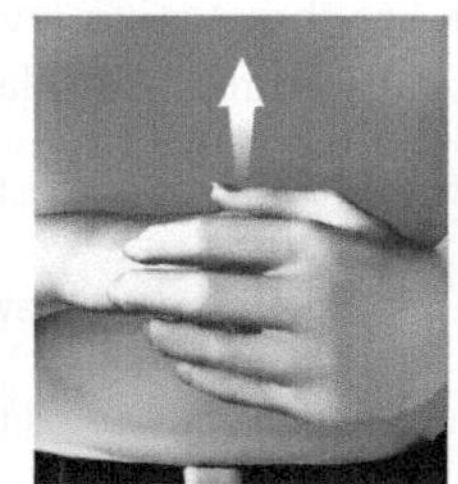

B Abdominal thrusts. Stand behind the person and place one fist below the rib cage, just above the navel, with your thumb facing inward. Cover the fist with your other hand and thrust inward and upward with both fists.

FIGURE 38.12 **Maneuvers to assist a conscious adult who is choking.** The Red Cross recommends that rescuers ask if a victim is choking and wants help. If the person nods, the rescuer should alternate between a series of five back blows and five abdominal thrusts.

TAKE-HOME MESSAGE 38.6

✔ Inhalation is always an active process. Contraction of the diaphragm and external intercostal muscles increases the volume of the thoracic cavity. This reduces air pressure in alveoli below atmospheric pressure, so air moves inward.

✔ Exhalation is usually passive. As muscles relax, the thoracic cavity shrinks back down, air pressure in alveoli rises above atmospheric pressure, and air moves out.

✔ Only some of the air in the lungs is replaced with each breath. The lungs are never fully emptied of air.

✔ The brain controls the rate and depth of breathing.

38.7 Gas Exchange and Transport

LEARNING OBJECTIVES

- Describe the structure of the respiratory membrane.
- Compare transport of oxygen and carbon dioxide in the blood.
- Describe the structure of hemoglobin and the conditions that encourage it to bind or to release oxygen.

The Respiratory Membrane

Gases diffuse between an alveolus and a pulmonary capillary at the lung's **respiratory membrane**. This thin membrane consists of alveolar epithelium, capillary endothelium, and the fused basement membranes of both epithelia (FIGURE 38.13). Alveolar epithelium contains squamous cells and secretory cells. The secretory cells release a substance (called a surfactant) that lubricates alveolar walls, preventing them from sticking together when the alveolus inflates.

Oxygen and carbon dioxide diffuse passively across squamous cells of the respiratory membrane. Which way these gases move depends on their concentration gradients across the membrane, or as we say for gases, partial pressure gradients. The partial pressure of a gas is its contribution to the pressure exerted by a mix of gases. It is measured in millimeters of mercury (mm Hg). Just as a solute tends to diffuse in response to its concentration gradient, so does a gas. If the partial pressure of a gas differs between two regions, the gas will diffuse from the region of higher partial pressure to the region of lower partial pressure.

Oxygen Transport

Inhaled air that reaches alveoli has a higher partial pressure of O_2 than does blood in pulmonary

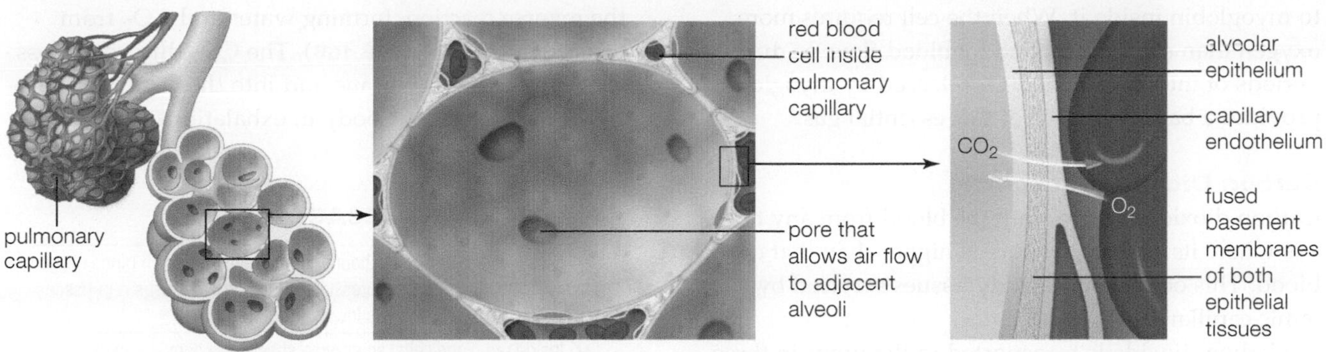
pulmonary
capillary
red blood
cell inside
pulmonary
capillary
pore that
allows air flow
to adjacent
alveoli
CO_2
O_2
alveolar
epithelium
capillary
endothelium
fused
basement
membranes
of both
epithelial
tissues

A Surface view of alveoli and associated pulmonary capillaries.

B Cutaway view of one alveolus and adjacent pulmonary capillaries.

C Three components of the respiratory membrane.

FIGURE 38.13 The respiratory membrane in human lungs.

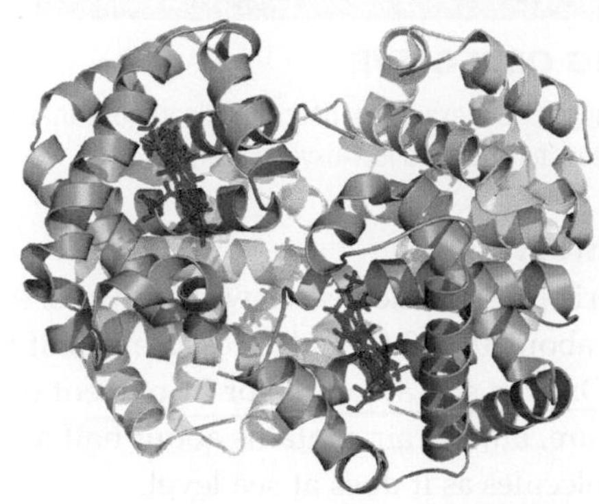

FIGURE 38.14 Structure of hemoglobin. The oxygen-transporting protein of red blood cells consists of four globin chains (green and blue), each associated with an iron-containing heme group (red).

capillaries. As a result, O_2 tends to diffuse from the air into the blood of these capillaries. Most O_2 that enters the blood diffuses into red blood cells, where it binds to hemoglobin. Hemoglobin has four globin subunits, each with an iron-containing heme group that can bind one molecule of O_2 (**FIGURE 38.14**). When O_2 is bound to one or more of hemoglobin's heme groups, we refer to the molecule as **oxyhemoglobin**.

Heme binds oxygen reversibly and releases it where the partial pressure of O_2 is lower than that in the alveoli. This occurs in the body tissues serviced by systemic capillaries: In **FIGURE 38.15**, compare the O_2 partial pressures inside alveoli and at the start of systemic capillaries. Metabolically active tissues also have other characteristics that encourage oxygen release: high temperature, low pH, and a high partial pressure of carbon dioxide.

Myoglobin, also an iron-containing respiratory protein, helps cardiac muscle and some skeletal muscles take up oxygen. Structurally, myoglobin resembles a globin chain, but its heme binds oxygen more tightly than globin does. The O_2 that hemoglobin releases near a cardiac muscle cell diffuses into the cell and binds

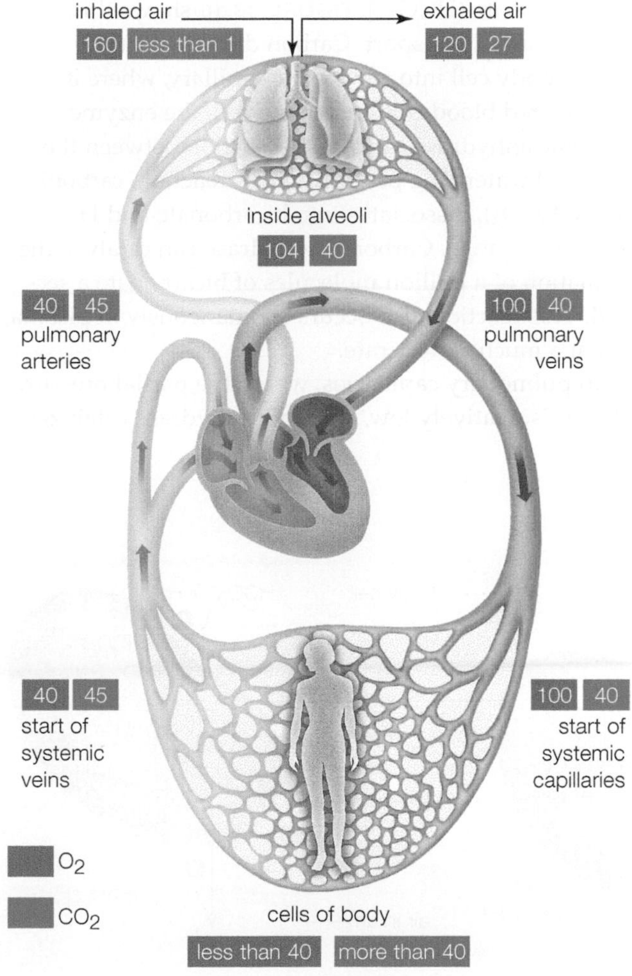
inhaled air
160
less than 1
exhaled air
120
27
inside alveoli
104
40
40
45
pulmonary
arteries
100
40
pulmonary
veins
40
45
start of
systemic
veins
100
40
start of
systemic
capillaries
O_2
CO_2
cells of body
less than 40
more than 40

FIGURE 38.15 Partial pressures (in mm Hg) of oxygen and carbon dioxide in the atmosphere, blood, and tissues.

oxyhemoglobin Hemoglobin with oxygen bound to it.
respiratory membrane Membrane consisting of alveolar epithelium, capillary endothelium, and their fused basement membranes; site of gas exchange in lungs.

to myoglobin inside it. When the cell requires more oxygen than can be supplied by blood flow, as during periods of intense exercise, O_2 released by myoglobin provides a backup source of this essential gas.

Carbon Dioxide Transport

Carbon dioxide diffuses into the blood from any tissue where its partial pressure is higher than that of the blood. This occurs in the body tissues serviced by systemic capillaries.

Carbon dioxide is transported to the lungs in three forms. About 10 percent remains dissolved in plasma. Another 30 percent reversibly binds with hemoglobin and forms carbaminohemoglobin ($HbCO_2$). However, most CO_2 that diffuses into the plasma is transported as bicarbonate (HCO_3^-). **FIGURE 38.16** shows this mechanism of transport. Carbon dioxide diffuses from a body cell into a systemic capillary, where it enters a red blood cell. Inside the cell, the enzyme carbonic anhydrase catalyzes a reaction between the CO_2 and water. The product of this reaction, carbonic acid (H_2CO_3), dissociates into bicarbonate and H^+ (**FIGURE 38.16A**). Carbonic anhydrase can catalyze the formation of a million molecules of bicarbonate a second. This reaction also occurs spontaneously in plasma, but at a much slower rate.

In pulmonary capillaries, where the partial pressure of CO_2 is relatively low, carbonic anhydrase catalyzes the reverse reaction, forming water and CO_2 from bicarbonate (**FIGURE 38.16B**). The CO_2 diffuses across the respiratory membrane and into the air in an alveolus. It then leaves the body in exhalations.

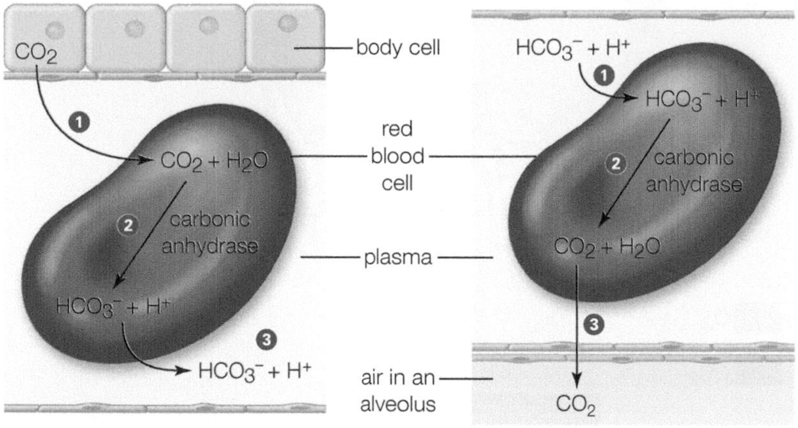

A At a systemic capillary bed:

1. CO_2 diffuses from a body cell into plasma, then into a red blood cell.
2. Carbonic anhydrase catalyzes formation of bicarbonate (HCO_3^-).
3. Bicarbonate diffuses into plasma.

B At a pulmonary capillary bed:

1. Bicarbonate diffuses from plasma into a red blood cell.
2. Carbonic anhydrase catalyzes formation of water and CO_2.
3. CO_2 diffuses across respiratory membrane into air in an alveolus.

FIGURE 38.16 Main mechanism of carbon dioxide transport and exchange. A lesser amount of CO_2 travels to the lungs bound to hemoglobin.

TAKE-HOME MESSAGE 38.7

✔ Most oxygen in blood is bound to hemoglobin, which binds oxygen in alveoli where O_2 partial pressure is high, and releases it in tissues where O_2 partial pressure is lower.

✔ Inside red blood cells, an enzyme speeds the conversion of carbon dioxide to bicarbonate. Most carbon dioxide is transported in blood in the form of bicarbonate.

38.8 Respiratory Adaptations

LEARNING OBJECTIVE

- Using appropriate examples, describe adaptations that facilitate respiratory function in a low-oxygen environment.

High Climbers

Atmospheric pressure decreases with altitude. At 5,500 meters, or about 18,000 feet, air pressure is half that at sea level. Oxygen still accounts for 21 percent of the total pressure, but the air contains about half as many oxygen molecules as it does at sea level.

When people who live at a low altitude ascend too fast to a high altitude, delivery of oxygen to their cells plummets. Hypoxia, or cellular oxygen deficiency, is the result. In an acute compensatory response to hypoxia, the brain signals the heart and respiratory muscles to work harder. People breathe faster and more deeply than usual (they hyperventilate). As a result, CO_2 is exhaled faster than it forms, and ion balances in the cerebrospinal fluid get skewed. Shortness of breath, a pounding heart, dizziness, nausea, and vomiting are symptoms of the resulting altitude sickness.

A healthy person who normally lives at a low altitude can become physiologically adjusted to a high one, but it takes time. Through **acclimatization**, the body makes long-term adjustments in cardiac output and the rate and magnitude of breathing. Hypoxia also stimulates kidney cells to secrete more erythropoietin. This hormone induces stem cells in the bone marrow to divide repeatedly and induces descendant cells to differentiate as red blood cells.

The increased number of circulating red blood cells improves the oxygen-delivery capacity of blood, but it also puts a strain on the heart. The more blood cells there are, the thicker the blood, and the harder the heart has to work to propel blood through the circulatory system. The heart enlarges, and its stronger-than-normal contractions increase blood pressure, putting a

llama hemoglobin
typical range for hemoglobin in most mammals
human hemoglobin
Heme binding site saturation (%)
0
20
40
60
80
100
O_2 partial pressure (mm Hg)

FIGURE 38.17 Saturation curve for hemoglobin of humans, llamas, and other mammals. FIGURE IT OUT At what partial pressure of oxygen do half the heme groups in human blood have oxygen bound?

Answer: 30 mm Hg

person at risk for the health problems associated with chronic hypertension (Section 36.7).

Long-established, high-altitude peoples have adaptations that allow them to function in a low-oxygen environment. For example, people of the Andes mountains have an unusually high concentration of hemoglobin in their blood. This trait allows them to maximize the amount of oxygen that they acquire from each breath. People who live on the Tibetan plateau do not have an unusually high level of hemoglobin, but they breathe unusually rapidly. Tibetans also make more nitric oxide. This gas encourages blood vessels to widen, increasing blood flow through lung capillaries. Although Andean and Tibetan populations face the same selective pressure (low oxygen), each evolved a different mechanism of countering that pressure. In Andeans, the blood composition was altered; in Tibetans, respiratory rate and blood flow were affected.

Many animals that live at a high altitude have hemoglobin that is unusually good at binding oxygen when the oxygen concentration is low. Llamas, which live high in the Andes, are an example (FIGURE 38.17).

Deep Divers

Water pressure increases with depth. Human divers using tanks of compressed air risk nitrogen narcosis, sometimes called "raptures of the deep." As divers descend, gaseous nitrogen (N_2) dissolves in their interstitial fluid. In neurons, this dissolved nitrogen can disrupt signaling, causing euphoria and drowsiness. Effects increase with increasing depth.

Returning to the surface from a deep dive also has risks. As a diver ascends, the decrease in pressure causes N_2 to move from interstitial fluid into the blood, from which it is eliminated in exhalations. If a diver rises too fast, N_2 bubbles form in the diver's body fluids. The resulting decompression sickness, also known as "the bends," usually begins with joint pain. Bubbles of N_2 slow the flow of blood to organs. If such bubbles form in the brain, heart, or lungs, the result can be fatal.

Humans who train to dive without oxygen tanks can remain submerged for about three minutes. So far, the human free-diving record is about 200 meters (700 feet). Other animals dive much deeper (FIGURE 38.18). As an air-breathing animal dives deeper and deeper, the weight of more and more water presses on its body. Lungs fully inflated with air would collapse inward under this pressure, so most diving animals move air out of their lungs and into cartilage-reinforced airways before a deep dive.

The longest dive recorded for a leatherback turtle lasted a little more than an hour. Sperm whales stay submerged for two hours. How does a diving animal, whose lungs are emptied of air and who has no access to the surface, supply its cells with the oxygen they need to survive? First, before the animal dives, it inhales and exhales deeply. A sperm whale blows out about 80–90 percent of the air in its lungs with each exhalation; you exhale only about 15 percent. Deep breaths keep oxygen pressure in alveoli high, so more oxygen diffuses into the blood.

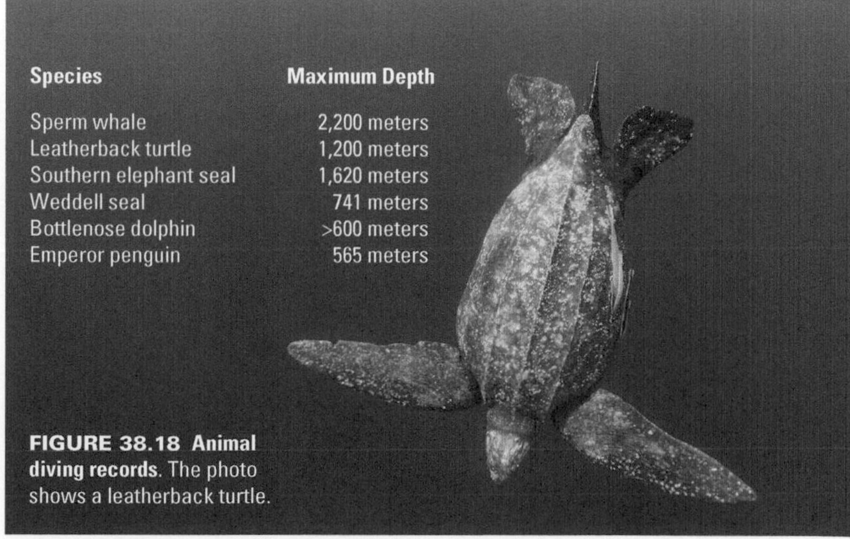

Species	Maximum Depth
Sperm whale	2,200 meters
Leatherback turtle	1,200 meters
Southern elephant seal	1,620 meters
Weddell seal	741 meters
Bottlenose dolphin	>600 meters
Emperor penguin	565 meters

FIGURE 38.18 Animal diving records. The photo shows a leatherback turtle.

acclimatization Physiological adjustment of a body to a different environment; e.g., after moving from sea level to a high-altitude habitat.

Second, diving animals can store large amounts of oxygen inside their blood and muscles. They tend to have a large blood volume relative to their body size, a high red blood cell count, and a considerable amount of myoglobin in their muscles. A skeletal muscle of a sperm whale has seven times as much as the comparable muscle of a dog.

Third, during a dive, oxygen is preferentially distributed to the heart, brain, and other organs that require an uninterrupted supply of ATP. This preferential distribution is made possible by a system of valves that control flow to blood vessels in specific tissues. To make the best use of the limited oxygen supply, metabolic rate and heart rate decrease, as do cellular oxygen uptake and carbon dioxide formation.

Finally, whenever possible, a diving animal makes the most of its oxygen stores by sinking and gliding instead of actively swimming. It conserves energy by avoiding unnecessary movements.

TAKE-HOME MESSAGE 38.8

✔ Air at high altitudes provides less oxygen than air at low altitudes. Animals and humans that live at a high altitude have adaptations that facilitate the uptake of oxygen.

✔ Animals that make long dives have adaptations that allow them to store oxygen and to shift the distribution of blood flow during a dive.

38.9 Respiratory Diseases and Disorders

LEARNING OBJECTIVES

- Explain the possible causes of apnea, and why it is dangerous.
- Describe common respiratory disorders and their causes.
- Describe the effects of smoking on lung health.

Interrupted Breathing

A tumor or damage to the brain stem's medulla oblongata can affect respiratory controls and cause apnea. In this disorder, breathing repeatedly stops and restarts spontaneously, especially during sleep. Sleep apnea also occurs when the tongue, tonsils, or other soft tissue obstructs the upper airways. Breathing may stop for up to several seconds many times each night, preventing deep sleep and causing daytime fatigue. Sleep apnea raises the risk of heart attacks and strokes because blood pressure soars when breathing stops. Obstructive sleep apnea can be reduced by wearing a mask that delivers pressurized air during sleep. Severe cases require removal of tissue that blocks airways.

Sudden infant death syndrome (SIDS) occurs when an infant does not awaken from an episode of apnea. A defect in the medulla oblongata, the brain's center for control of respiration, is associated with SIDS. Maternal smoking during pregnancy heightens the risk, and infants who sleep on their stomach or in a sitting position (as in a car seat or stroller) are at higher risk than those who sleep on their back.

Lung Diseases and Disorders

Worldwide, about one in three people is currently infected by bacteria that can cause tuberculosis (TB). Most of these people have no symptoms, but about 10 percent of them develop "active TB." They cough up bloody mucus, have chest pain, and find breathing difficult. If untreated, active TB can be fatal. Antibiotics can cure most infections, if taken diligently. Unfortunately, multi-drug-resistant strains of *Mycobacterium tuberculosis* are increasing in frequency.

Pneumonia is a general term for lung inflammation caused by an infection. Bacteria, viruses, and fungi can cause pneumonia. Typical symptoms include a cough, an aching chest, shortness of breath, and fever. An x-ray reveals lungs filled with fluid and white blood cells instead of air. Treatment and outcome depend on the type of pathogen.

Asthma occurs when an inhaled allergen or irritant triggers inflammation that constricts the airways so breathing becomes difficult. A tendency to have asthma is inherited, but avoiding potential irritants such as cigarette smoke and air pollutants can reduce the frequency of asthma attacks. An acute asthma attack is treated with inhaled drugs that cause dilation of smooth muscle around the airways.

Bronchitis is an inflammation of the ciliated, mucus-producing epithelium of the bronchi. The inflamed cells secrete extra mucus that triggers coughing. Bacteria colonize the mucus, leading to more inflammation, more mucus, and more coughing. Infectious bronchitis most often arises as a result of a viral infection of the upper respiratory tract. Chronic bronchitis most often occurs in tobacco smokers.

Emphysema is a disorder in which the thin, elastic alveolar walls disintegrate. As these walls disappear, the area of the respiratory surface declines. Over time, the lungs become distended and inelastic, causing a constant feeling of being short of breath.

Smoking and Vaping

Smoking tobacco has a wide range of negative health effects. It increases the risk of lung infections and many types of cancer. Tobacco smoke contains more than 40 carcinogens (cancer-causing chemicals), and more

Data Analysis Activities

Smoking and Lung Function Data from the Coronary Artery Risk Development in Young Adults (CARDIA) study was used to assess the effects of smoking on lung function. In 1985-86, the study enrolled about 5,000 people aged 18 to 30 years old in a long term study. Over the next 20 years, study subjects were periodically asked about their habits and their health, including their lung function, was assessed. **FIGURE 38.19** is a model based on data from this study, of how lifetime tobacco and marijuana use affects forced expiration.

1. How did the effects of low-level tobacco smoking on forced expiration volume differ from those of low level marijuana smoking?
2. Based on this model, how would smoking two packs of cigarettes a day for 20 years (40 pack-years of exposure) affect FEV_1? What level of marijuana smoking would produce an equivalent effect?
3. Relatively few people had the highest level of lifetime marijuana exposure. How might additional data from such users alter the model?

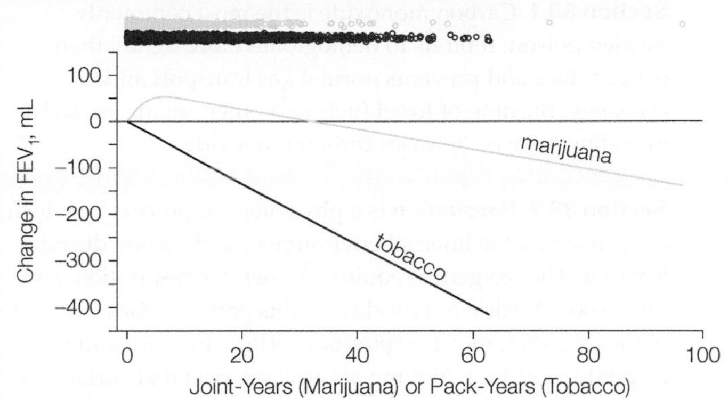

Change in FEV_1, mL
100
0
−100
−200
−300
−400
marijuana
tobacco
0
20
40
60
80
100
Joint-Years (Marijuana) or Pack-Years (Tobacco)

FIGURE 38.19 Associations between marijuana or tobacco smoking and the change in forced expiration volume (FEV). One joint-year means an average of one joint smoked daily for one year. One pack-year is an average of one pack smoked per day for one year. Dots at the top of the graph show the distribution of observations. Lines below show the modeled association. Blue is marijuana and black is tobacco. The change in FEV_1 is given as milliliters of air.

than 80 percent of lung cancers occur in smokers. Once diagnosed with lung cancer, the majority of smokers die within a year. Tobacco smokers also have a higher risk of cardiovascular disease.

Less is known about the long-term health effects of smoking marijuana. Marijuana smoke contains carbon monoxide and an assortment of carcinogens, including arsenic and ammonia. However, the few studies that have been done have not found any increased risk for lung cancer in people who smoke only marijuana. On the other hand, people who smoke both marijuana and tobacco do seem to have more respiratory problems than those who smoke only tobacco.

Effects of vaping—using an electronic device to produce an inhalable aerosol—are largely unknown. Vaping delivers far lower doses of carcinogens than smoking standard cigarettes, and no carbon monoxide. However, users still inhale toxins such as formaldehyde, as well as tiny lung-damaging particles.

TAKE-HOME MESSAGE 38.9

- ✔ Apnea, or interrupted breathing, is caused by tissue obstructing airways or a defective respiratory control center.
- ✔ Tuberculosis is a potentially fatal bacterial disease. Pneumonia can be caused by many different pathogens.
- ✔ In asthma and bronchitis, airways become inflamed and constricted. In emphysema, alveolar sacs become distended and inelastic.
- ✔ Smoking tobacco increases the risk of lung cancer and disorders.

38.1 Carbon Monoxide—A Stealthy Poison (revisited)

Carbon monoxide binds tightly to hemoglobin, but it does not bind irreversibly. If a person with carbon monoxide poisoning is moved into fresh air, about half of the CO bound to hemoglobin in his or her blood will be replaced by oxygen after four to five hours. The speed at which oxygen replaces carbon monoxide at binding sites can be accelerated by exposing the person to 100 percent oxygen.

To minimize your carbon monoxide exposure, do not smoke and avoid being in a closed area with people who do. Have wood-burning stoves and fireplaces and fossil fuel–burning heaters checked annually. Remember that any fossil fuel–powered engine releases CO and should not be allowed to run indoors. If you can smell exhaust from a car, boat, heater, stove, or other device, you are also inhaling CO. ●

STUDY GUIDE

Section 38.1 Carbon monoxide is the most commonly inhaled poison. It binds to hemoglobin more tightly than oxygen does and prevents normal gas transport and exchange. Burning of fossil fuels and other organic material, including tobacco, releases carbon monoxide.

Section 38.2 **Respiration** is a physiological process by which oxygen enters the internal environment and carbon dioxide leaves it. The oxygen is required for aerobic respiration, and the carbon dioxide is a product of this pathway. Gas exchange occurs at a thin, moist **respiratory surface**. Enlarging the respiratory surface, moving air or water past that surface, or having **hemoglobin** or another **respiratory protein** speeds the rate of gas exchange.

Air-breathing animals have a more constant supply of oxygen than aquatic animals and do not have to expend as much energy in respiratory functions.

Section 38.3 Flatworms and cnidarians do not have special respiratory or circulatory organs. They rely on diffusion of gases across the body surface and through their body. In earthworms, the circulatory system moves gases to and from the body surface where gas exchange occurs. **Gills** enhance respiration in aquatic invertebrates such as clams and crabs. Some land snails and slugs have a **lung**. The insect **tracheal system** consists of chitin-reinforced tubes that carry air from a spiracle deep inside the body. Some spiders have a book lung, in which gas exchange occurs across thin sheets of tissue.

Section 38.4 Water flowing over fish gills exchanges gases with blood flowing in the opposite direction inside gill capillaries. This **countercurrent exchange** is highly efficient.

Most adult amphibians have lungs and also exchange gases across the skin. Frogs push air into their lungs by compressing the air-filled oral cavity.

Reptiles, birds, and mammals rely on lungs for gas exchange. In mammals, gas exchange occurs in tiny sacs called **alveoli**. Birds have a more efficient system. Inflation and deflation of air sacs connected to their inelastic lungs keep air flowing continually through tubes in the lungs. The lining of these tubes is the respiratory surface.

Section 38.5 In humans, air flows through the nose and mouth into the **pharynx**, then the **larynx**, and the **trachea** (windpipe). The larynx contains the vocal cords, movements of which alter the size of the opening (the **glottis**) between them. When you swallow, the position of the **epiglottis** at the entrance to the larynx shifts, keeping food out of the trachea.

The trachea branches into two **bronchi** that enter the lungs. These two airways branch into **bronchioles**. At the ends of the finest bronchioles are thin-walled alveoli, where gases are exchanged with the blood in pulmonary capillaries.

The **diaphragm** at the base of the thoracic cavity and the **intercostal muscles** between ribs are involved in breathing.

Section 38.6 A **respiratory cycle** is one inhalation and one exhalation. Inhalation is always active. As muscle contractions expand the thoracic cavity, pressure in lungs decreases below atmospheric pressure, causing air to flow into the lungs. Exhalation is usually passive. As muscles relax, the thoracic cavity shrinks and air flows out of the lungs.

Tidal volume, the amount of air that normally flows in and out during one respiratory cycle is less than **vital capacity**. The lungs never fully deflate.

The medulla oblongata in the brain stem adjusts the rate and magnitude of breathing. If a person is choking, a blow to the back or pushing on the abdomen can raise pressure in the lungs and expel the blocking object.

Section 38.7 In human lungs, gases diffuse across a thin **respiratory membrane** that separates the air in alveoli from the blood in pulmonary capillaries. As red blood cells pass through these vessels, hemoglobin binds O_2 to form **oxyhemoglobin**. In systemic capillaries, hemoglobin releases O_2, which diffuses into cells.

Also in systemic capillaries, CO_2 diffuses from cells into the blood. Most CO_2 reacts with water inside red blood cells to form bicarbonate. The enzyme carbonic anhydrase catalyzes this reaction, which is reversed in the lungs. There, CO_2 forms and is expelled from the body in exhalations.

Section 38.8 The amount of available oxygen declines with altitude. Physiological changes that occur in response to high altitude are called **acclimatization**. They include altered breathing patterns and an increase in erythropoietin, a hormone that stimulates red blood cell formation. Over the longer term, genetic changes can adapt a population to life at a high altitude. Llamas, which live at a high altitude, have hemoglobin that is unusually good at binding oxygen when the oxygen partial pressure is low. Tibetans and people living in the Andes have both adapted to a high altitude, but in different ways.

A variety of adaptive mechanisms allow some turtles and marine mammals to hold their breath for long periods while making deep dives.

Section 38.9 Problems with signals from the medulla oblongata can cause apnea and sudden infant death syndrome (SIDS). Apnea in adults can also occur if soft tissues block an airway during sleep.

Tuberculosis and pneumonia arise when pathogens infect the lungs. Bronchitis and asthma arise when lungs become inflamed. Emphysema occurs when walls between alveoli break down, decreasing the surface area available for gas exchange. Smoking tobacco is the leading cause of emphysema. It also increases the risk of other respiratory disorders and of lung cancer.

SELF-QUIZ

Answers in Appendix VII

1. The respiratory protein hemocyanin _______ .
   a. contains iron
   b. occurs only in vertebrates
   c. transports oxygen in some invertebrates
   d. colors hemolymph red

2. In a _______ , air flows continually across the respiratory surface.
   a. fish c. frog
   b. bird d. mammal

3. The tracheal tubes of insects carry _______ to tissues deep inside the body.
   a. hemolymph c. air
   b. blood d. water

4. In human lungs, gas exchange occurs at the _______ .
   a. bronchi c. alveoli
   b. bronchioles d. pleural sacs

5. Which holds the most dissolved oxygen?
   a. warm, still water c. cold, running water
   b. warm, running water d. cold, still water

6. When you breathe quietly, inhalation is _______ and exhalation is _______ .
   a. passive; passive c. passive; active
   b. active; active d. active; passive

7. During inhalation _______ .
   a. the thoracic cavity expands
   b. the diaphragm relaxes
   c. atmospheric pressure declines
   d. the glottis is closed

8. _______ binds to hemoglobin even more strongly than oxygen does.
   a. Carbon dioxide c. Oxyhemoglobin
   b. Carbon monoxide d. Nitrogen

9. Most oxygen transported in human blood _______ .
   a. is bound to hemoglobin
   b. combines with carbon to form carbon dioxide
   c. is in the form of bicarbonate
   d. is dissolved in the plasma

10. At high altitudes, _______ .
    a. nitrogen bubbles out of the blood
    b. hemoglobin has fewer oxygen-binding sites
    c. there are fewer O_2 molecules per unit volume of air than at low altitudes
    d. there is more carbon monoxide than at sea level

11. Countercurrent flow allows _______ to maximize their oxygen uptake from water.
    a. fishes c. frogs
    b. flatworms d. whales

12. _______ in arteries sense changes in the acidity of the blood.
    a. Mechanoreceptors c. Photoreceptors
    b. Neurotransmitters d. Chemoreceptors

13. Vital capacity is the _______ .
    a. amount of hemoglobin in a red blood cell
    b. amount of time a person can hold his or her breath
    c. maximum amount of air a person can move in and out with one breath
    d. amount of oxygen in an exhalation

14. Match the disorders with their descriptions.

___	tuberculosis	a. breathing halts
___	emphysema	b. alveoli walls break down
___	asthma	c. airways constrict
___	apnea	d. bacterial disease

15. Match the words with their descriptions.

___	trachea	a. muscle of respiration
___	pharynx	b. gap between vocal cords
___	epiglottis	c. throat
___	larynx	d. windpipe
___	bronchi	e. voice box
___	bronchiole	f. keeps food out of airway
___	glottis	g. each leads to a lung
___	diaphragm	h. delivers air to alveoli

CRITICAL THINKING

1. The red blood cell enzyme carbonic anhydrase contains a zinc cofactor. A diet deficient in zinc does not reduce the number of red blood cells, but it impairs respiratory function by reducing CO_2 output. Explain why.

2. A developing fetus obtains oxygen from its mother's blood. In an organ called the placenta, fetal capillaries run through and exchange substances with maternal blood. The hemoglobin made by a fetus has different properties than that made after birth. At low oxygen levels, fetal hemoglobin binds oxygen more strongly than normal hemoglobin does. How does fetal hemoglobin's somewhat higher affinity for oxygen benefit a fetus?

3. Icefish live in the extremely cold ocean near Antarctica. They are the only vertebrates that do not make hemoglobin. They have a large heart and take up oxygen across their skin, as well as their gills. Why are icefish thought to be especially threatened by ongoing ocean warming?

39 DIGESTION AND NUTRITION

CORE CONCEPTS

 Pathways of Transformation

Organisms exchange matter and energy with the environment in order to grow, maintain themselves, and reproduce.

All animals are heterotrophs that meet their energy needs and nutritional requirements by taking in food from their environment. In most animals, mechanical and chemical digestion of food by digestive organs breaks food down into subunits that can be absorbed into the body's internal environment.

 Evolution

Evolution underlies the unity and diversity of life.

Shared core processes and features provide evidence that all living things are descended from a common ancestor. In all vertebrates, digestion takes place inside a tubular digestive tract that extends through the body. Specialized anatomical and physiological traits, including variations in structure, size, and shape of organs along the digestive tract, are adaptations to different types of diet.

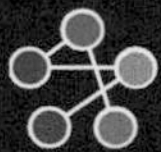 Systems

Complex properties arise from interactions among components of a biological system.

Specialized components of the vertebrate digestive system interact to take in food, break food down into absorbable nutrients, absorb materials, and eliminate any unabsorbed waste. Homones and nervous signals coordinate interactions among digestive organs and between these organs and the rest of the body.

Links to Earlier Concepts

This chapter explains how your body digests organic polymers (Section 3.2) and obtains vitamins and minerals required to make coenzymes (5.6), hemoglobin (38.7), and some hormones (34.7). You will see how low pH (2.6) and enzymes (5.4) break down food, and how active transport (5.9, 5.10) moves products of digestion across cell membranes.

 39.1 Breaking It Down

Food provides both the energy and the raw materials you need to build and maintain your body. Your digestive system breaks the large, complex molecules in food into smaller organic molecules. Enzymes that carry out hydrolysis reactions are essential to this process. As Section 3.2 explained, a hydrolysis reaction breaks an organic polymer (a molecule composed of many subunits) into monomers (the individual subunits). The name of each hydrolytic enzyme begins with a prefix that refers to the enzyme's substrate and ends with *-ase*. For example, lactase breaks down lactose (the main sugar in milk) into the two simple sugars glucose and galactose.

Nearly everyone makes lactase as an infant, but most people make little or none as adults. The result is lactose intolerance. Affected people experience cramping, diarrhea, and flatulence (gas) after drinking milk or eating milk products.

Most people who do make lactose as adults are of Northern European descent. The mutation that confers their lactase persistence arose about 10,000 years ago. It was probably adaptive in Northern European populations because these people had milk-providing livestock, periodic crop failures encouraged alternative sources of nutrition, and the cool climate prevented milk from spoiling too fast.

Sucrase-isomaltase is another digestive enzyme. It breaks down sucrose (table sugar) and maltose (the sugar in grains). People with congenital sucrase-isomaltase deficiency (CSID) are unable to make this enzyme, so they suffer from cramping and diarrhea after eating sugar or grains.

Among people of European descent about 1 in 5,000 individuals has CSID. By contrast, in some Inuit populations up to 10 percent are affected. The Inuit (previously referred to as Eskimos) live in the Arctic, where their traditional diet consists almost entirely of fish and meat. These foods contain little sucrose or maltose. As a result, selection favoring an ability to make sucrase-isomaltase has historically been lower in the Inuit than in people whose diet includes a greater proportion of plant foods.

Scientists have known since the 1960s that the Inuit are more likely than the general population to be intolerant of sucrose. In 2014, they discovered the genetics behind this intolerance. Inuit have a higher frequency of a sucrase-isomaltase allele with a frameshift mutation that prevents synthesis of this protein. A physician can now use a genetic test to determine whether an Inuit's digestive problems stem from possession of this allele or from some other source.

39.2 Animal Digestive Systems

LEARNING OBJECTIVES

- Explain how a sponge digests food and distributes nutrients.
- Compare the structure of a gastrovascular cavity and a complete digestive tract, and explain the advantages of the latter.
- Provide examples of regional specializations in a digestive tract.

All animals are heterotrophs that take food into their body and digest it. In this context, digestion means breaking food down into chemical components that cells of an animal body can use as sources of energy and as building blocks. Essential molecules and elements that an animal must obtain from its food are referred to as nutrients.

Intracellular Digestion in Sponges

Animals evolved from a colonial protist (Section 21.8). In such protists, individual cells take in food by phagocytosis, break it down inside food vacuoles, then expel wastes by exocytosis. Sponges, which do not have tissues or organs, feed in a similar manner (**FIGURE 39.1**). Flagellated collar cells at the surface of the sponge capture food by waving their flagellum. The movement draws water through a sievelike collar of microvilli that rings the flagellum. Food particles that collect on the collar are taken into the cell, where they are either digested or passed along to motile amoeboid cells. The amoeboid cells digest food and distribute nutrients to cells throughout the sponge body.

FIGURE 39.1 Food-processing cells of a sponge. Digestion is intracellular.

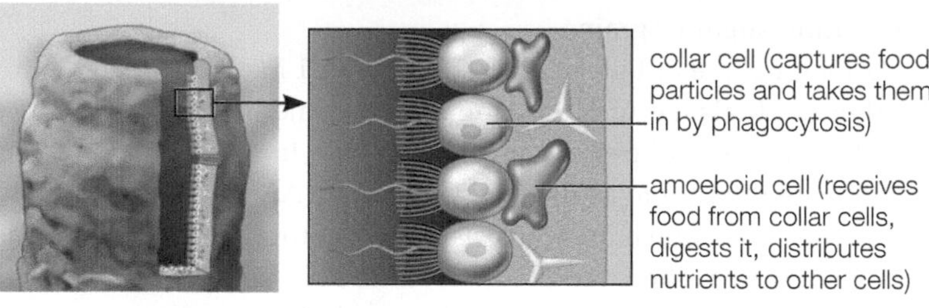
collar cell (captures food particles and takes them in by phagocytosis)
amoeboid cell (receives food from collar cells, digests it, distributes nutrients to other cells)

Extracellular Digestion

In most animals, digestion is extracellular. Food is broken down within the body, but inside a hollow sac or tube that opens to the external environment. In animals with tissues, nutrition involves four tasks:

1. *Ingestion.* Taking food into a digestive organ.
2. *Mechanical and chemical digestion.* Breaking food down into components that can be absorbed into the body. Mechanical digestion breaks food into fragments, thus increasing the surface area exposed to digestive enzymes. Chemical digestion breaks large polymers into component subunits.
3. *Absorption.* Moving nutrient molecules across the wall of a digestive organ and into the body's internal environment.
4. *Elimination.* Expelling any leftover material that was not digested and absorbed.

Sac or Tube?

Flatworms such as planarians digest their food inside a saclike **gastrovascular cavity** that also functions in gas exchange. A gastrovascular cavity has a single opening through which food enters and wastes leave. In flatworms, the opening is at the tip of a muscular pharynx that conveys food to and from the digestive cavity (**FIGURE 39.2A**). The two-way traffic in a gastrovascular cavity makes digestion somewhat inefficient. Food must be broken down, its nutrients absorbed, and any wastes eliminated before new food can enter.

A gastrovascular cavity is sometimes referred to as an incomplete digestive tract. By contrast, a **complete digestive tract** is a tubular digestive tract that has two openings. Many invertebrates, including earthworms, have a complete digestive tract (**FIGURE 39.2B**). A coelomate animal with a complete digestive tract has a tube-within-a-tube body plan (**FIGURE 39.2C**).

Evolution of a complete digestive tract was an important innovation. As food travels the length of the digestive tract, it passes through regions specialized

FIGURE 39.2 Complete versus incomplete digestive tracts. In both, digestion occurs within the body, but outside of cells.

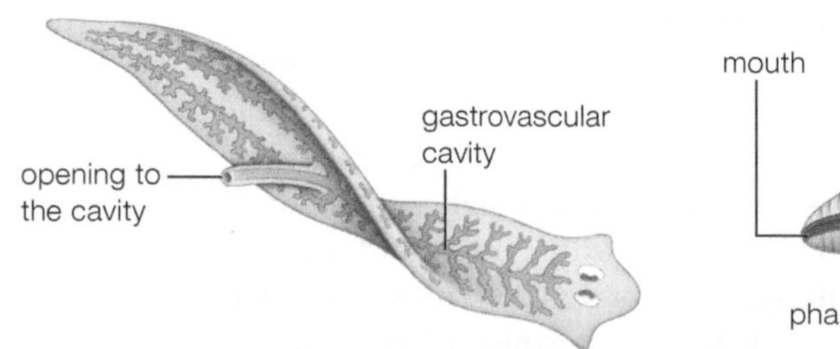
gastrovascular cavity
opening to the cavity

A Incomplete digestive tract of a flatworm.The branching gastrovascular cavity has a single opening.

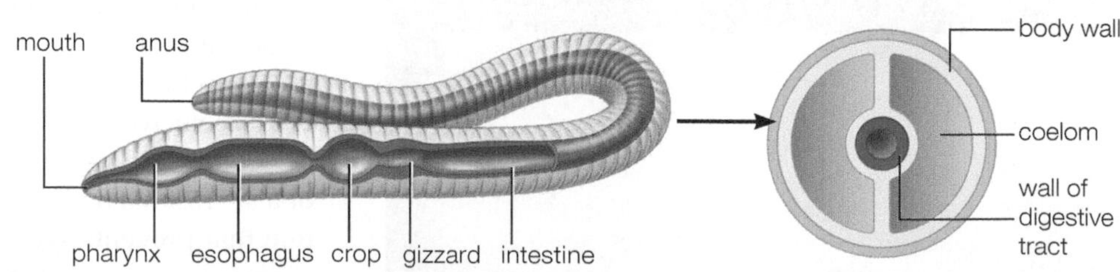
mouth
anus
pharynx
esophagus
crop
gizzard
intestine
body wall
coelom
wall of digestive tract

B Complete digestive tract of an earthworm. There are two openings (mouth and anus) and specialized regions in between.

C Cross section illustrating the tube-within-a-tube body plan.

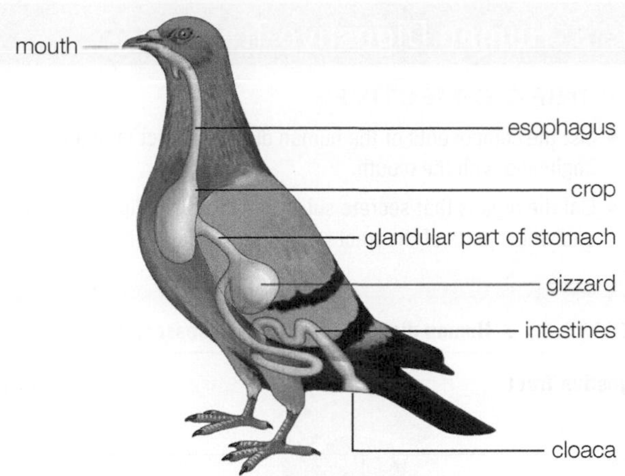
mouth
esophagus
crop
glandular part of stomach
gizzard
intestines
cloaca

FIGURE 39.3 Complete digestive tract of a pigeon. Food enters through the mouth and wastes exit through the cloaca.

for food storage, food breakdown, nutrient absorption, and waste elimination. Unlike a gastrovascular cavity, a tubular gut can carry out all of these tasks simultaneously. Like you, an earthworm can take in new food even when wastes from the food it ingested earlier has not yet exited the digestive tract.

Regional Specializations

In a complete digestive tract, a tubular organ called the **esophagus** receives food from the pharynx (throat). In some animals, including earthworms and many birds, part of the esophagus has become enlarged to serve as a food-storing pouch called the **crop** (FIGURE 39.3).

The esophagus delivers food to the **stomach**, a muscular digestive organ that chemically and mechanically breaks down food. In earthworms, birds, and other animals that lack teeth, a portion of the stomach may be specialized as a **gizzard**, an organ that grinds up food and begins the process of mechanical digestion. In the earthworm gizzard, hard bits of dirt ingested by the worm break up food. In birds, the gizzard is a muscular chamber with a hardened lining of glycoprotein. Seeds are difficult to grind up, so seed-eating birds have larger gizzards than those with other diets.

Ruminants such as cattle, goats, sheep, antelope, and deer are hoofed grazers that feed on plant leaves and stems. Like other animals, ruminants do not make enzymes that can break down cellulose. However, their four-chambered stomach allows them to survive on a cellulose-rich diet (FIGURE 39.4). The first two chambers of their stomach contain microbes that break down cellulose. As a ruminant feeds, solids accumulate in the second stomach chamber, forming a "cud" that is regurgitated—moved back into the mouth for a second round of chewing. Nutrient-rich fluid moves from the second chamber to the third and fourth chambers, and finally to the intestine.

All animals with a complete digestive tract have an **intestine,** an organ in which most chemical digestion is carried out and from which nutrients are absorbed. The

complete digestive tract Tubelike digestive system; food enters through one opening, and wastes leave through another.

crop Of birds and some invertebrates, an enlarged portion of the esophagus that stores food.

esophagus (eh-SOFF-ah-gus) Tubular organ that connects the pharynx to the stomach.

gastrovascular cavity Of some invertebrates, a saclike digestive cavity that also functions in gas exchange.

gizzard Of birds and some other animals, a digestive chamber lined with a hard material that grinds up food.

intestine Tubular organ in which chemical digestion is completed and from which most nutrients are absorbed.

stomach Tubular, muscular organ that functions in chemical and mechanical digestion.

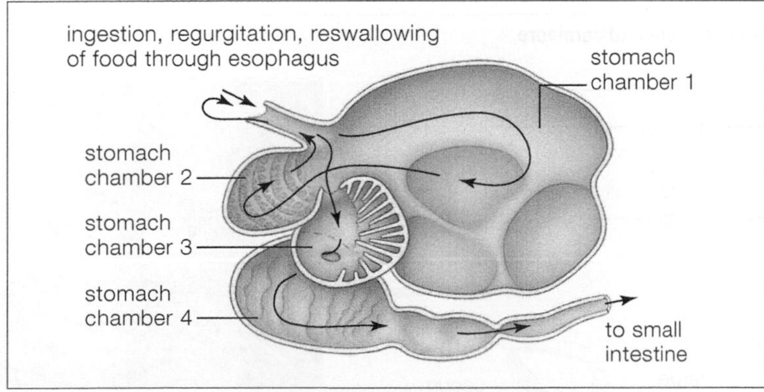
ingestion, regurgitation, reswallowing of food through esophagus
stomach chamber 1
stomach chamber 2
stomach chamber 3
stomach chamber 4
to small intestine

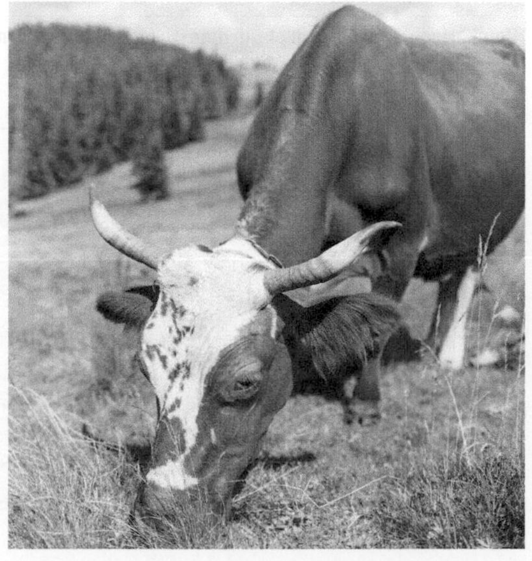

FIGURE 39.4 Multiple-chambered stomach of a ruminant. Microbes in the first two stomach chambers break down cellulose in plant cell walls. Solids accumulate in the second chamber, forming "cud" that is regurgitated—moved back into the mouth for a second round of chewing. Nutrient-rich fluid moves from the second chamber to the third and fourth chambers, and finally to the intestine.

CREDITS: (3) © Cengage Learning; (4) left, Based on A. Romer and T. Parsons, *The Vertebrate Body*, Sixth Edition, Saunders Publishing Company, 1986; right, Skumer/Shutterstock.

mammalian intestine has two distinct regions, the large intestine and the small intestine. Most nutrient absorption occurs in the small intestine.

The length of the intestine reflects an animal's diet. As previously noted, animals do not make enzymes that can break down cellulose. Thus, herbivores enlist the assistance of microbes that live in their digestive tract to help them digest their food. As a result, the intestine of an herbivore such as a rabbit is proportionately much longer than that of a carnivore such as a wolf (FIGURE 39.5). Rabbits also have a large cecum. The cecum is a pouch at the start of the large intestine. In rabbits, it is filled with bacteria that aid in digestion of cellulose-rich plant material. The cecum of a wolf is relatively small.

In rabbits, wolves, and other placental mammals, digestive wastes exit through an **anus**, an opening that serves this purpose alone. By contrast, amphibians, reptiles, and birds have a **cloaca**, a multipurpose opening that also releases urinary waste and functions in reproduction.

TAKE-HOME MESSAGE 39.2

✔ Digestive systems mechanically and chemically degrade food into small molecules that can be absorbed into the internal environment. These systems also expel the undigested residues from the body.

✔ Some invertebrates have a gastrovascular cavity, or incomplete digestive system. It is a sac with a single opening.

✔ Some invertebrates and all vertebrates have a complete digestive system. It is a tube with a mouth at one end, an anus or cloaca at the other, and regional specializations in between.

✔ Some structural differences among the digestive tracts of different animal groups are adaptations that allow the animals to exploit different types of foods.

39.3 Human Digestive Tract

LEARNING OBJECTIVES

- List the components of the human digestive tract in order beginning with the mouth.
- List the organs that secrete substances into the digestive tract and which parts of the tract receive their secretions.

FIGURE 39.6 Human digestive tract and accessory organs.

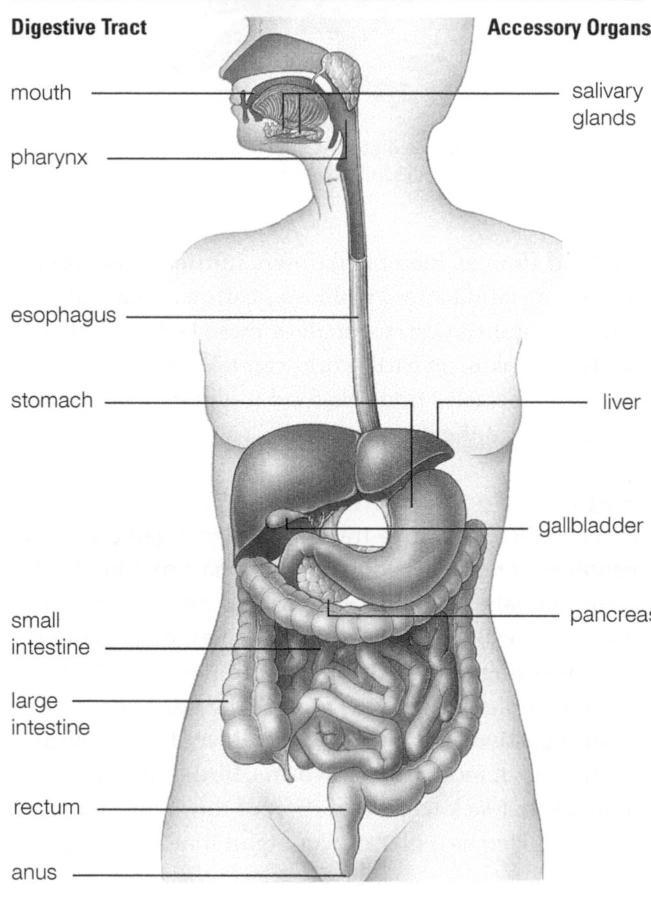
Digestive Tract
Accessory Organs
mouth
salivary glands
pharynx
esophagus
stomach
liver
gallbladder
pancreas
small intestine
large intestine
rectum
anus

FIGURE 39.5 Comparison of the length of the digestive tract in a mammalian herbivore and carnivore.

A Digestive tract of a rabbit, an herbivore.

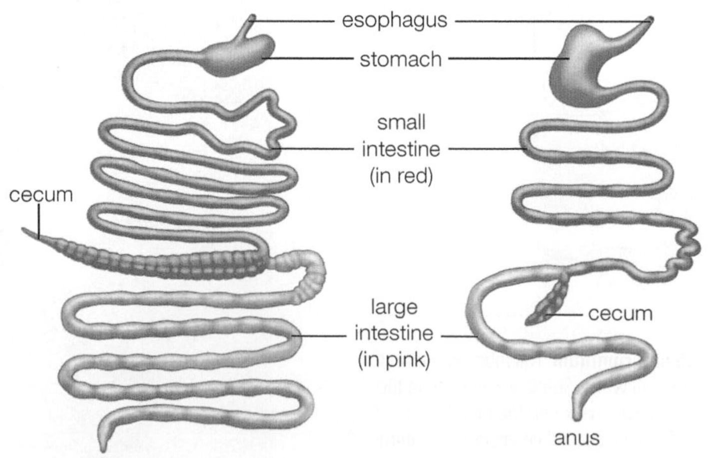
esophagus
stomach
small intestine (in red)
cecum
large intestine (in pink)
cecum
anus
anus

B Digestive tract of a wolf, a carnivore.

The human digestive tract, also referred to as the alimentary canal, extends from the mouth to the anus (FIGURE 39.6). If this tube were to be laid out in a straight line, it would extend 6.5 to 9 meters (21 to 30 feet). Extensive coiling and folding allow this lengthy tube to fit inside the body. Mucus-secreting epithelium (mucosa) lines the digestive tract, and accessory digestive organs such as the salivary glands, liver, and pancreas secrete substances into its interior, or **lumen**. These secretions assist in chemical digestion.

Food processing begins inside the mouth, or oral cavity. Swallowing moves food into the pharynx (throat), from which it proceeds into the esophagus. The esophagus extends down through the thoracic cavity, running behind the trachea. It then passes through an opening in the diaphragm to extend into the abdominal cavity, where it connects to the stomach.

The stomach empties into the **small intestine**, which completes the process of chemical digestion and absorbs most nutrients. Secretions from the liver and pancreas assist the small intestine in these tasks. The second region of the intestine, the **large intestine**, absorbs fluid and concentrates waste. The terminal portion of the large intestine, the **rectum**, stores waste until it is expelled from the body through the anus.

Material moves through the digestive tract by a process called **peristalsis**. Peristalsis occurs as smooth muscle in the wall of the digestive tract contracts and relaxes in rhythmic waves, propelling the material inside it farther along the tube (FIGURE 39.7).

Sphincters at various points along the digestive tract control passage of material through the tract. A sphincter is a ring of muscle that can relax and open to allow substances through a passageway, or contract to close off the passage.

TAKE-HOME MESSAGE 39.3

✔ Humans have a complete digestive system with a tubular, mucosa-lined digestive tract.

✔ Accessory organs positioned adjacent to the digestive tract secrete substances into its interior. These substances aid in digestion of food or absorption of nutrients.

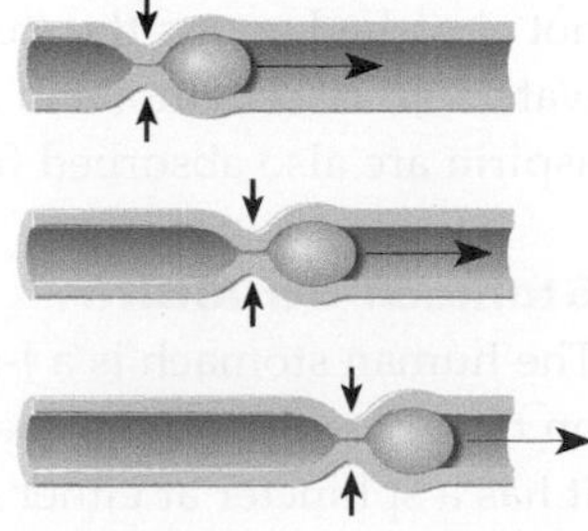

FIGURE 39.7 **Peristalsis.** Rings of smooth muscle contraction (indicated by the inward pointing arrows) travel in the mouth-to-anus direction along the digestive tract. The contraction narrows the interior of the tract, pushing the material inside it (green) through the tube.

39.4 Chewing and Swallowing

LEARNING OBJECTIVES

- Compare the teeth of mammalian herbivores and carnivores.
- List the components of saliva, and describe their functions.

Mechanical digestion, the smashing of food into smaller pieces, begins in the mouth. So does chemical digestion, the enzymatic breakdown of food into molecular subunits.

As a species, humans are omnivores, meaning we eat both plant and animal material. Our teeth reflect our omnivorous heritage. We have all four types of mammalian teeth, and all are equally well developed (FIGURE 39.8A). By contrast, mammals that are carnivores (meat eaters) have enlarged canine teeth for piercing and tearing flesh (FIGURE 39.8B). Mammals that are herbivores (plant eaters) have reduced canine teeth (FIGURE 39.8C). Premolars of carnivores have a narrow, bladelike surface that helps them cut through meat. Those of herbivores and omnivores like us are broad and flat, the better for grinding plant material.

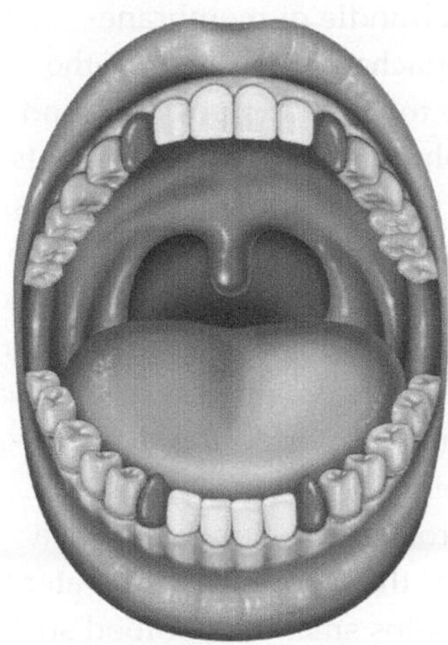

A Adult human. Humans have all four tooth types and all are equally large.

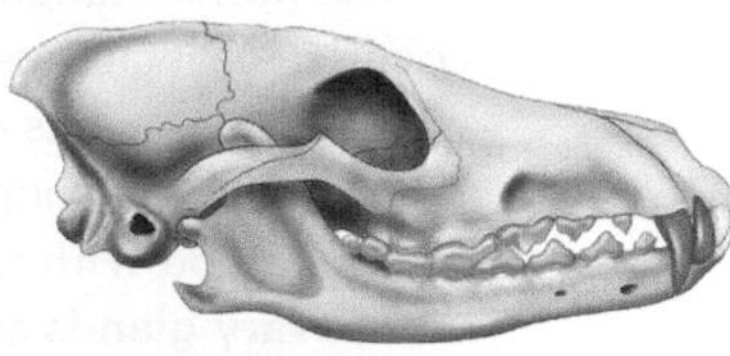

B Carnivore with enlarged canine teeth and sharp premolars.

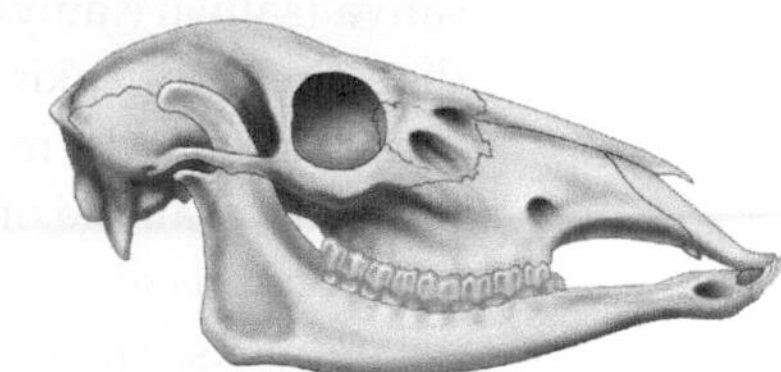

C Herbivore with reduced canines, and large, broad molars and premolars.

molar | premolar | canine | incisor

FIGURE 39.8 **Variations in mammalian tooth patterns.**

anus Body opening that serves solely as the exit for wastes from a complete digestive tract.
cloaca (cluh-AY-kuh) Body opening that serves as the exit for digestive and urinary waste; also functions in reproduction.
large intestine Region of the intestine that receives material from the small intestine, absorbs water, and concentrates and stores waste.
lumen (LOO-men) The space within a tubular or hollow organ.
peristalsis (pehr-ih-STAL-sis) Wavelike smooth muscle contractions that propel food through the digestive tract.
rectum Terminal part of the large intestine; stores digestive waste.
small intestine Region of the intestine that receives material from the stomach; site of most digestion and absorption.

Data Analysis Activities

Human Adaptation to a Starchy Diet The human *AMY-1* gene encodes salivary amylase, an enzyme that breaks down starch. The number of copies of this gene varies, and people who have more copies generally make more enzyme. In addition, the average number of *AMY-1* copies differs among cultural groups.

George Perry and his colleagues hypothesized that duplications of the *AMY-1* gene would be selectively advantageous in cultures in which starch is a large part of the diet. To test this hypothesis, the scientists compared the number of copies of the *AMY-1* gene among members of seven cultural groups that differed in their traditional diets. **FIGURE 39.9** shows their results.

1. Starchy tubers are a mainstay of Hadza hunter–gatherers in Africa, whereas fishing sustains Siberia's Yakut. Almost 60 percent of Yakut had fewer than 5 copies of the *AMY1* gene. What percentage of the Hadza had fewer than 5 copies?
2. None of the Mbuti (rain-forest hunter–gathers) had more than 10 copies of *AMY-1*. Did any European Americans?
3. Do these data support the hypothesis that a starchy diet favors duplications of the *AMY-1* gene?

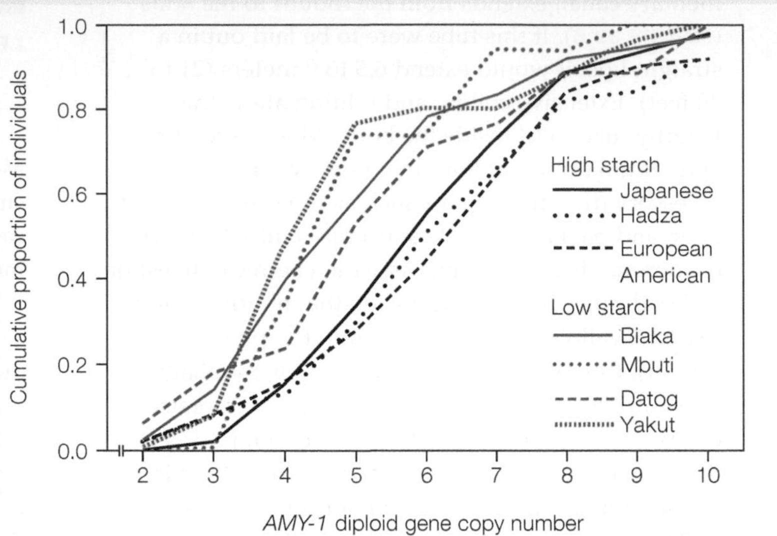

Cumulative proportion of individuals
1.0
0.8
0.6
0.4
0.2
0.0
2
3
4
5
6
7
8
9
10
AMY-1 diploid gene copy number
High starch
Japanese
Hadza
European American
Low starch
Biaka
Mbuti
Datog
Yakut

FIGURE 39.9 Number of copies of the *AMY-1* gene among members of cultures with traditional high-starch or low-starch diets. The Hadza, Biaka, Mbuti, and Datog are tribes in Africa. The Yakut live in Siberia.

The human tongue is a bundle of membrane-covered skeletal muscle attached to the floor of the mouth. Movements of the tongue help position food where teeth can chop or shred it. Tongue movements also mix food with saliva secreted by salivary glands.

Salivary glands are exocrine glands that open into the mouth beneath the tongue and on the inner surface of the cheeks next to the upper molars. An enzyme in saliva (salivary amylase) begins the process of chemical digestion by breaking starch into disaccharides (two-sugar units). Like many other reactions of digestion, this is an example of hydrolysis (Section 3.3). Saliva also contains glycoproteins that combine with water to form mucus. The mucus helps small bits of food stick together in easy-to-swallow clumps.

The presence of food at the back of the throat triggers a swallowing reflex. The throat, remember, opens onto both the digestive and respiratory tracts (Section 38.5). When you swallow, the flap of cartilage called the epiglottis flops down to cover the opening to the airway and the vocal cords constrict, so the route between the pharynx and larynx is blocked. This reflex keeps food from accidentally entering the trachea and choking you.

During swallowing, a sphincter in the upper esophagus opens to allow food into the esophagus, then closes behind it. The sound made during a burp, or belch, arises when previously swallowed air flows outward through this sphincter, causing it to vibrate.

TAKE-HOME MESSAGE 39.4

- ✔ Teeth mechanically break food into smaller particles.
- ✔ The tongue positions food and mixes it with saliva. Enzymes in the saliva begin the chemical digestion of carbohydrates.
- ✔ Swallowing is a reflex. During swallowing, the epiglottis covers the trachea to prevent food from entering this airway.

39.5 The Stomach

LEARNING OBJECTIVES

- Describe the structure of the stomach.
- List the components of gastric fluid, and describe their roles.
- Explain what triggers the secretion of gastric hormones, and describe their effects .

The vertebrate stomach has three functions. First, it stores food and controls the rate of passage to the small intestine. Second, it mechanically breaks food into smaller pieces. Third, it secretes substances that begin the chemical digestion of proteins. Nutrients are not absorbed across the stomach wall, although some water can be. Ethanol and some other drugs such as aspirin are also absorbed from the stomach.

Stomach Structure

The human stomach is a J-shaped muscular sac located on the left side of the upper abdomen (**FIGURE 39.10**). It has a sphincter at either end. When the stomach is

CREDIT: (9) George Perry, et al., Diet and evolution of human amylase gene copy number variation, Nature Genetics 39, 1256–1260 (2007).

empty, its inner surface is highly folded. As it fills with food, these folds smooth out, increasing the stomach's capacity. An average adult's stomach can expand enough to hold about 1 liter of fluid.

Like other portions of the digestive tract, the stomach has a multilayered structure. Its outermost layer (the serosa) consists primarily of connective tissue. Beneath the serosa are layers of smooth muscle. Another layer of connective tissue lies beneath the muscle. It supports the innermost layer, the mucosa.

Components and Function of Gastric Fluid

The stomach mucosa is a glandular epithelium dotted with tiny indentations called gastric pits. At the base of these pits are gastric glands that include chief cells and parietal cells. Chief cells secrete pepsinogen, an inactive form of the protein-digesting enzyme pepsin. Parietal cells secrete H^+ and Cl^- ions. These ions combine in the stomach lumen to form hydrochloric acid. Between the gastric pits are regions of mucosa that contain many mucus-secreting cells. Alkaline mucus secreted by these cells coats the mucosa and protects it from the damaging effects of acid and pepsin.

The acid, pepsin, and mucus released into the stomach by cells of the mucosa collectively constitute the **gastric fluid**. Your stomach produces about 2 liters of gastric fluid every day. Contractions of smooth muscle in the stomach wall mix gastric fluid with food to form a semiliquid mass called **chyme**.

Chemical digestion of proteins begins in the stomach when the high acidity of chyme denatures proteins (Section 3.5), exposing their peptide bonds. The high acidity also converts the protein pepsinogen into its active form—pepsin. Pepsin is an enzyme that cleaves the peptide bonds between amino acids, thus chopping proteins into polypeptides.

Gastric Hormones

The stomach steps up or slows down its secretion of acid depending on when and what you eat. The arrival of food in the stomach, especially protein, triggers endocrine cells in the uppermost region of the stomach to secrete the hormone **gastrin** into the blood. Gastrin acts on acid-secreting cells of the stomach lining, encouraging them to increase their output. When the stomach is empty, gastrin secretion and acid secretion decline. This prevents excess acidity from damaging the stomach wall.

Conversely, an empty stomach increases its secretion of the hormone ghrelin. **Ghrelin** is a peptide hormone that stimulates the appetite. When people attempt to reduce their weight by eating less, their ghrelin output rises. This is one reason it is difficult to stick to a diet.

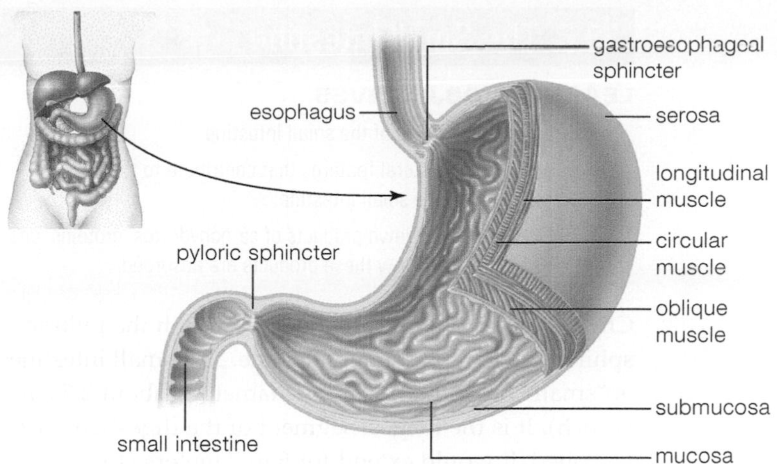

gastroesophageal sphincter
esophagus
serosa
longitudinal muscle
circular muscle
pyloric sphincter
oblique muscle
submucosa
small intestine
mucosa

FIGURE 39.10 **Location and structure of the human stomach.**

Stomach Ulcers and Reflux

A stomach ulcer is an open, inflamed wound in the mucosa. Most stomach ulcers occur after acid-tolerant *Helicobacter pylori* bacteria infect cells of the mucosa. The bacteria degrade the protective mucus and release chemicals that cause the stomach to secrete extra acid that damages the mucosa. Infection by *H. pylori* also increases the risk of stomach cancer sixfold. In 1994, the World Health Organization classified *H. pylori* as the first known bacterial carcinogen.

On the other hand, people infected by *H. pylori* are less likely than the uninfected to have gastroesophageal reflux disease (GERD). The main symptom of GERD is a pain in the upper abdomen, commonly referred to as heartburn. The pain occurs when acid from the stomach splashes back into the esophagus. Repeated exposure to acid can damage the tissue of the esophagus and raise the risk of esophageal cancer.

TAKE-HOME MESSAGE 39.5

✔ The stomach receives food from the esophagus and stretches to store it. Contractions of stomach muscles break up food and mix it with acidic gastric fluid. They also move the resulting mixture (the chyme) into the small intestine.

✔ Chemical digestion of proteins begins in the stomach.

chyme (kime) Mix of food and gastric fluid.

gastric fluid Fluid secreted by the stomach lining; contains acid, mucus, and the protein-digesting enzyme pepsin.

gastrin Hormone secreted by the stomach in response to the presence of food; encourages secretion of acid into the stomach.

ghrelin (GRELL-in) Hormone secreted by the stomach when empty; stimulates appetite.

salivary gland Exocrine gland that secretes saliva into the mouth.

39.6 The Small Intestine

LEARNING OBJECTIVES

- Explain the functions of the small intestine.
- Describe the structural features that contribute to the large surface area of the small intestine.
- List the final breakdown products of carbohydrates, proteins, and lipids, and explain how these products are absorbed.

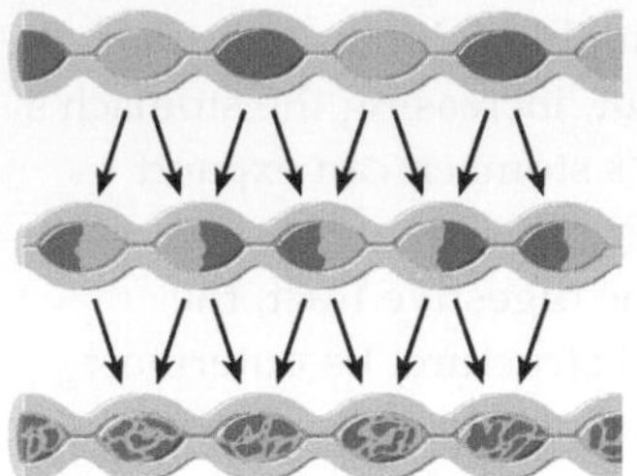

FIGURE 39.12 Segmentation contraction in the small intestine. Contractions of rings of muscle in the wall of the small intestine cause chyme to slosh back and forth, mixing it with digestive enzymes and forcing it against the wall of the small intestine.

Chyme forced out of the stomach through the pyloric sphincter enters the small intestine. The small intestine is "small" only in terms of its diameter—about 2.5 cm (1 inch). It is the longest segment of the digestive tract. Uncoiled, it would extend for 5 to 7 meters (16 to 23 feet). There are three regions of the small intestine. The first region, the duodenum, receives food from the stomach. The next region is the jejunum. The terminal region, the ileum, empties into the large intestine.

An Enormous Surface Area

Most digestion and nutrient absorption take place at the surface of the small intestine, which is highly folded (**FIGURE 39.11A**). Unlike the folds of the empty stomach, those of the small intestine are permanent.

The surface of each intestinal fold is covered by **villi** (singular, villus), hairlike multicelled projections about 1 millimeter long (**FIGURE 39.11B,C**). The millions of villi that project from the intestinal lining give it a furry or velvety appearance. Blood vessels and lymph vessels run through the interior of each villus.

Most of the epithelial cells at the surface of a villus have even tinier projections called **microvilli** (singular, microvillus). The 1,700 or so microvilli at the surface of a cell make its outer edge look like a brush. Thus, these cells are sometimes called **brush border cells** (**FIGURE 39.11D, E**). Brush border cells function in both digestion and absorption. Digestive enzymes at the surface of a microvillus break down sugars, polypeptides, and nucleotides. Transport proteins at this surface facilitate the movement of nutrients into the microvillus.

As a result of its many folds and projections the lining of the small intestine is surprising large—about the size of a tennis court. This large surface area maximizes the number of membrane-embedded digestive enzymes and transport proteins that come into contact with the chyme.

Movements of the small intestine also facilitate digestion and absorption. Like the stomach wall, the wall of the small intestine has multiple layers of smooth muscle. Action of these muscles pushes chyme back and forth in a pattern called segmentation (**FIGURE 39.12**). Segmentation mixes chyme with digestive enzymes and sloshes the mixture against the wall of the small intestine, where nutrients can be absorbed. Peristalsis propels chyme through the small intestine.

Interactions with Accessory Organs

The lumen of the small intestine receives acidic chyme from the stomach, enzymes and bicarbonate from the

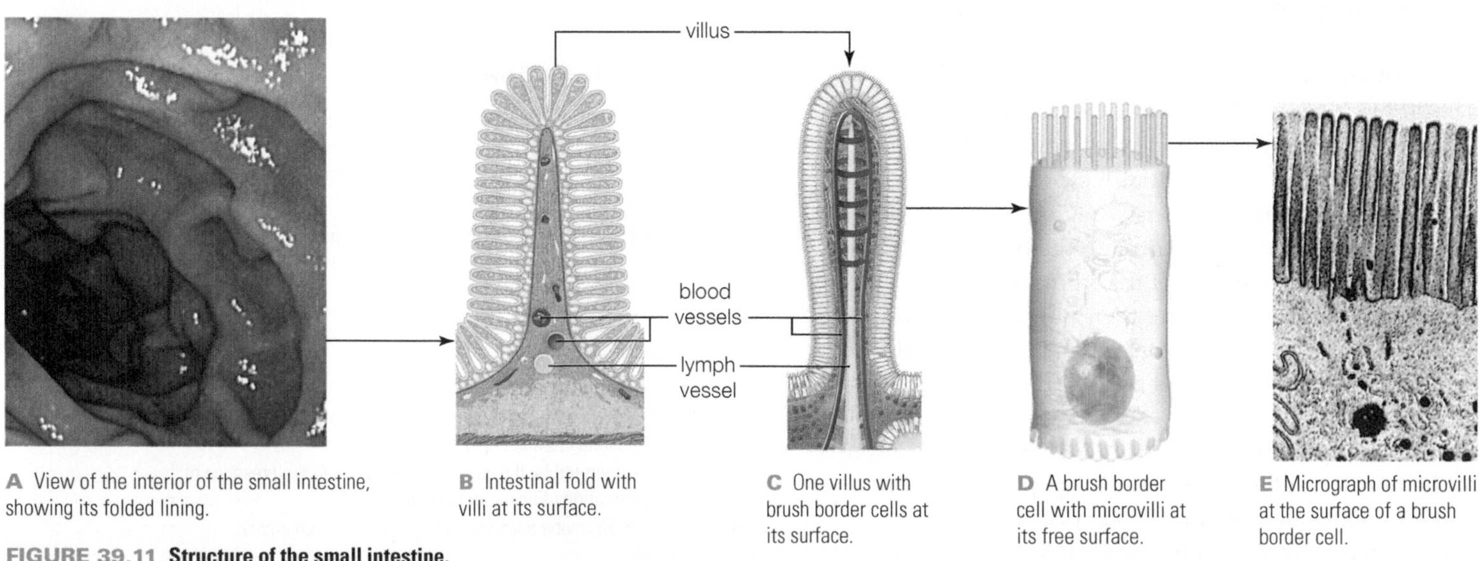

villus
blood vessels
lymph vessel

A View of the interior of the small intestine, showing its folded lining.

B Intestinal fold with villi at its surface.

C One villus with brush border cells at its surface.

D A brush border cell with microvilli at its free surface.

E Micrograph of microvilli at the surface of a brush border cell.

FIGURE 39.11 Structure of the small intestine.
FIGURE IT OUT Are microvilli multicelled or smaller than a cell?

Answer: Microvilli are smaller than a cell. Villi are multicellular.

TABLE 39.1

Summary of Chemical Digestion

	Enzymes Present	Enzyme Source	Enzyme Substrate	Main Breakdown Products
Carbohydrate Digestion				
Mouth, stomach	Salivary amylase	Salivary glands	Polysaccharides	Disaccharides
Small intestine	Pancreatic amylase	Pancreas	Polysaccharides	Disaccharides
	Disaccharidases	Intestinal lining	Disaccharides	**Monosaccharides*** (such as glucose)
Protein Digestion				
Stomach	Pepsins	Stomach lining	Proteins	Protein fragments
Small intestine	Trypsin, chymotrypsin	Pancreas	Proteins	Protein fragments
	Carboxypeptidase	Pancreas	Protein fragments	**Amino acids***
	Aminopeptidase	Intestinal lining	**Amino acids***	
Lipid Digestion				
Small intestine	Lipase	Pancreas	Triglycerides	**Free fatty acids, monoglycerides***

*** Breakdown products that can be absorbed into the internal environment.**

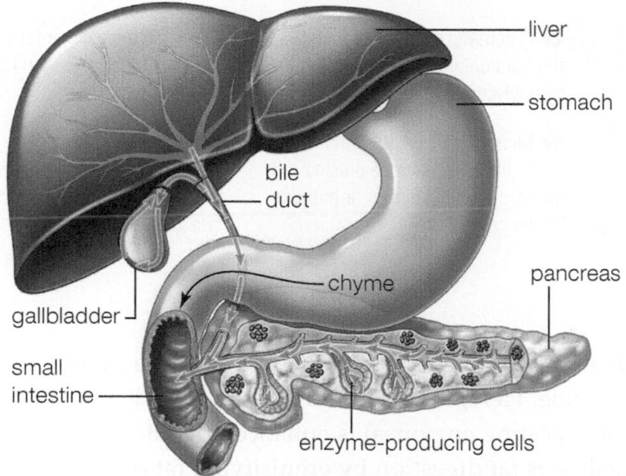
liver
stomach
bile
duct
pancreas
chyme
gallbladder
small
intestine
enzyme-producing cells

FIGURE 39.13 Organs that contribute to the contents of the small intestine.

pancreas, and bile from the gallbladder (**FIGURE 39.13**). Pancreatic enzymes work in concert with enzymes at the surface of brush border cells to complete the breakdown of organic compounds into absorbable subunits (**TABLE 39.1**). Bicarbonate secreted by the pancreas raises the pH of chyme high enough that digestive enzymes can function properly.

The **liver** is the body's largest internal organ, and it has a variety of functions. Here, we focus on its role in digestion. The liver produces **bile**, a fluid mixture that facilitates fat digestion. Bile contains cholesterol, salts, and bilirubin, a yellow pigment derived from the breakdown of hemoglobin.

Bile produced by the liver moves through a duct into the **gallbladder**, where it is concentrated and stored. After you eat a fatty meal, contraction of the gall bladder forces bile through a duct and into the small intestine. Here, it aids fat digestion by emulsifying fat droplets. (Emulsification is a process by which a fatty substance is dispersed in a liquid).

The bile salts used in fat digestion are recycled. They are absorbed from the final portion of the small intestine (the ileum) and returned to the liver by the blood.

In some people, components of bile accumulate to form pebble-like gallstones in the gallbladder. Most gallstones are harmless, but some block or become lodged in the bile duct. In this case, the gallbladder or gallstones are usually removed surgically. After removal of a gallbladder, all bile from the liver drains directly into the small intestine.

Hormones of the Small Intestine

The small intestine secretes two hormones with a role in digestion. **Secretin** is secreted in response to entry of acidic chyme. It encourages the pancreas to secrete bicarbonate into the small intestine. **Cholecystokinin** is secreted when fats and proteins enter the small intestine. This hormone encourages the delivery of

bile Liquid mixture of salts, pigments, and cholesterol that aids fat emulsification in the small intestine. Produced in the liver, stored and concentrated in the gallbladder, and secreted into the small intestine.
brush border cell In the lining of the small intestine, an epithelial cell with microvilli at its surface.
cholecystokinin (KOH -leh-sis-toe- KIE -nin) Small intestine hormone that encourages release of pancreatic enzymes and bile into the small intestine; also suppresses appetite.
gallbladder Organ that stores and concentrates bile.
liver Large organ that makes bile, stores glycogen, and detoxifies blood.
microvilli Singular, microvillus. Projections from the plasma membrane of some epithelial cells; increase the cell's surface area.
secretin (seh-KREE-tin) Small intestine hormone that encourages release of bicarbonate by the pancreas.
villi Singular, villus. Multicelled projections at the surface of each fold in the small intestine.

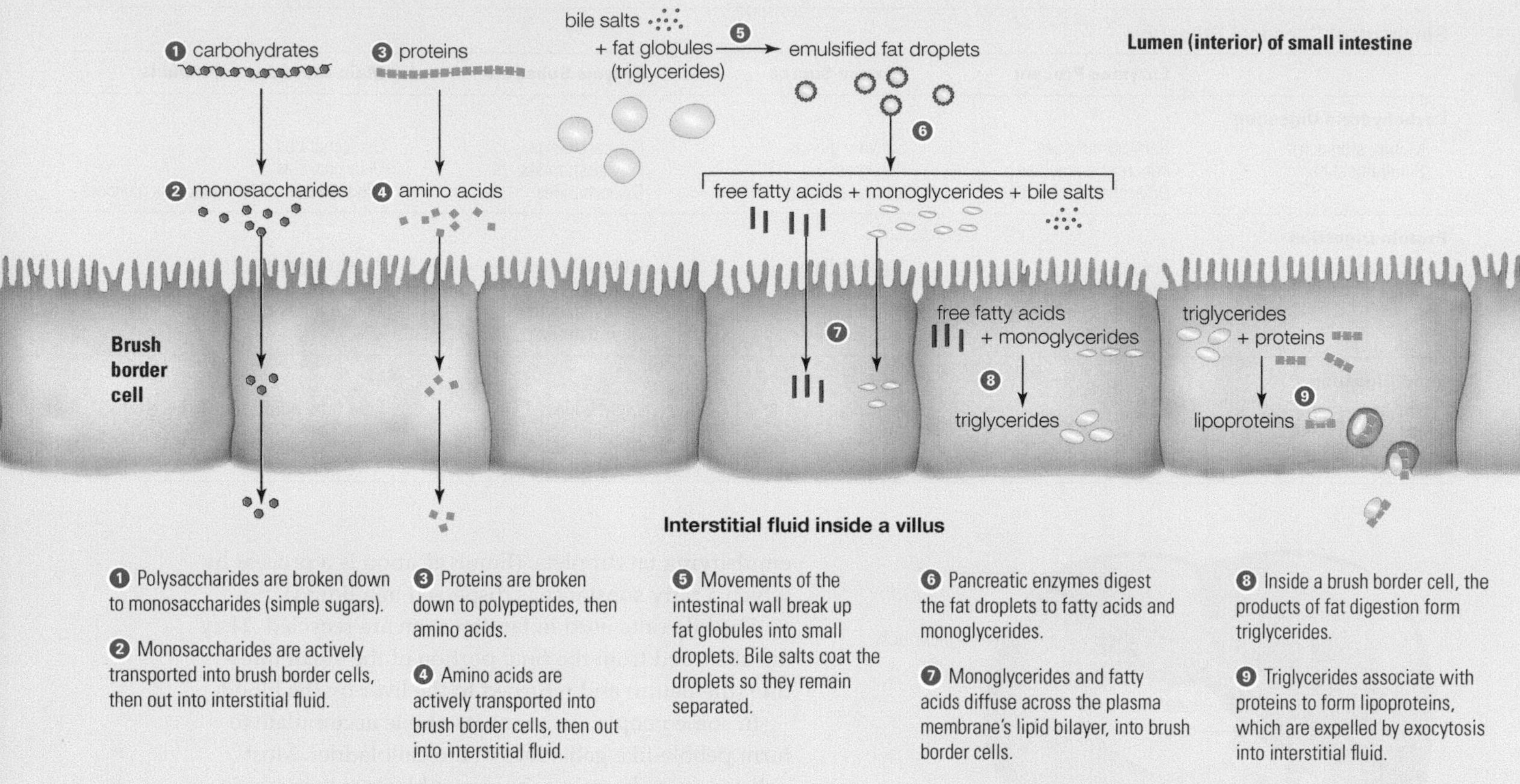

Lumen (interior) of small intestine
1 carbohydrates
3 proteins
bile salts
+ fat globules
(triglycerides)
5
emulsified fat droplets
6
2 monosaccharides
4 amino acids
free fatty acids + monoglycerides + bile salts
Brush border cell
7
free fatty acids + monoglycerides
8
triglycerides
triglycerides + proteins
9
lipoproteins
Interstitial fluid inside a villus
1 Polysaccharides are broken down to monosaccharides (simple sugars).
2 Monosaccharides are actively transported into brush border cells, then out into interstitial fluid.
3 Proteins are broken down to polypeptides, then amino acids.
4 Amino acids are actively transported into brush border cells, then out into interstitial fluid.
5 Movements of the intestinal wall break up fat globules into small droplets. Bile salts coat the droplets so they remain separated.
6 Pancreatic enzymes digest the fat droplets to fatty acids and monoglycerides.
7 Monoglycerides and fatty acids diffuse across the plasma membrane's lipid bilayer, into brush border cells.
8 Inside a brush border cell, the products of fat digestion form triglycerides.
9 Triglycerides associate with proteins to form lipoproteins, which are expelled by exocytosis into interstitial fluid.

FIGURE 39.14 Summary of digestion and absorption in the small intestine.

pancreatic enzymes and bile into the small intestine. Cholecystokinin also suppresses appetite.

Digestion and Absorption

Of Carbohydrates Recall that salivary amylase breaks polysaccharides into disaccharides. In the small intestine, pancreatic amylase and enzymes at the surface of brush border cells split disaccharides into monosaccharides such as glucose (**FIGURE 39.14** ❶). Active transport proteins move the monosaccharides from the lumen into a brush border cell, then out of that cell and into the interstitial fluid inside the villus ❷. From here, these simple sugars enter the blood.

Of Proteins Protein digestion began in the stomach, where pepsin broke proteins into polypeptides. In the small intestine, pancreatic proteases such as trypsin and chymotrypsin cut these polypeptides into smaller fragments. Enzymes at the surface of the brush border cells then break these fragments first into smaller peptides, then into amino acids ❸. Like monosaccharides, amino acids are transported into brush border cells, then into interstitial fluid ❹. From here, the amino acids enter the blood.

Of Fats Fat digestion occurs entirely in the small intestine. Here, the effectiveness of pancreatic lipases (fat-digesting enzymes) is enhanced by bile. Bile facilitates fat digestion by emulsifying fat droplets. Triglycerides from food tend to clump together as fat globules. Muscle contractions break the globs into smaller droplets, then bile salts coat the droplets so they remain separated ❺. Compared to big globules, many small droplets present a much greater surface area to lipases that break triglycerides into fatty acids and monoglycerides ❻.

Being lipid soluble, fatty acids and monoglycerides produced by fat digestion can enter a villus by diffusing through the lipid bilayer of brush border cells ❼. Inside these cells, the compounds become recombined into triglycerides ❽ that are packaged together with proteins into lipoprotein particles ❾. Lipoproteins enter lymph vessels, which carry them to the blood.

Of Water Each day, eating and drinking adds 1 to 2 liters of fluid into your small intestine. Secretions from your stomach, the liver and pancreas, and the intestinal lining add another 6 to 7 liters. The small intestine absorbs about 80 percent of this fluid. Transport of salts, sugars, and amino acids across brush border cells creates an osmotic gradient, and water moves down that gradient from chyme into the interstitial fluid.

TAKE-HOME MESSAGE 39.6

✔ Extensive folding and many projections (villi and microvilli) greatly increase the surface area of the small intestine.

✔ Entry of chyme into the small intestine causes this organ to release the hormones secretin and cholecystokinin.

✔ Chemical digestion is completed in the small intestine. Enzymes from the pancreas and enzymes embedded in the membrane of brush border cells break large molecules into smaller, absorbable subunits.

✔ Small subunits (monosaccharides, amino acids, fatty acids, and monoglycerides) enter the internal environment when they are absorbed into the interstitial fluid inside a villus.

✔ Most water that enters the digestive tract is absorbed across the wall of the small intestine.

39.7 The Large Intestine

LEARNING OBJECTIVES

- Describe the structure and function of the large intestine.
- Describe defecation and what triggers it.
- Explain why the large intestine contains many microbes and the beneficial effects of these microbes.

Concentrating and Expelling Waste

Not everything that enters the small intestine can or should be absorbed. Peristalsis propels indigestible material, dead bacteria and mucosal cells, inorganic substances, and some water from the small intestine into the large intestine. The large intestine concentrates wastes by transporting sodium ions from its lumen, across its lining, and into the interstitial fluid. Water follows by osmosis. The resulting compacted digestive waste is **feces**.

The human large intestine is wider than the small intestine, but shorter—only about 1.5 meters (5 feet) long. Its first segment is the cecum, which in humans is a cup-shaped pouch (**FIGURE 39.15**). A short, tubular **appendix** projects from the cecum. Appendicitis occurs when pathogenic bacteria infect the appendix. This disorder requires prompt surgical treatment to prevent the inflamed appendix from bursting.

The cecum connects to the **colon**, the longest portion of the large intestine. Material that enters the colon from the cecum moves up the right side of the abdomen inside the ascending colon, crosses the abdomen inside the transverse colon, and moves down the left side of the abdomen inside the descending colon. It then enters the sigmoid colon, an S-shaped segment that delivers feces to the rectum where it is stored.

appendix (ah-PEN-diks) Short, tubular projection from the first part of the large intestine; reservoir for beneficial bacteria.

colon Longest portion of the large intestine. Region between the cecum and the rectum.

feces (FEE-seez) Unabsorbed food material and cellular waste that is expelled from the digestive tract.

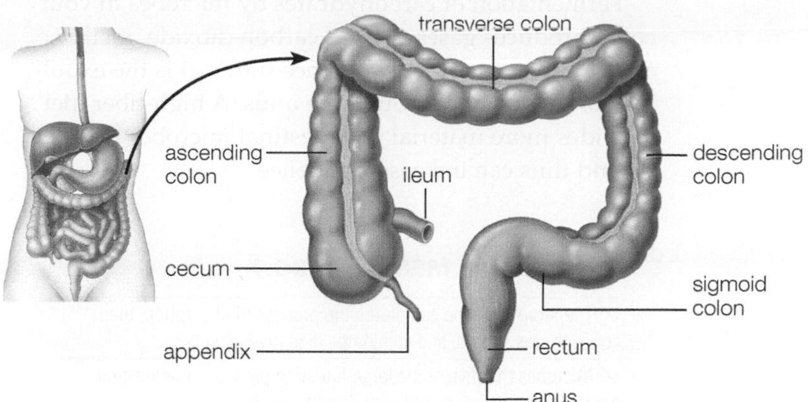

transverse colon
ascending colon
ileum
descending colon
cecum
sigmoid colon
appendix
rectum
anus

FIGURE 39.15 **Location and regions of the human large intestine.**

After a meal, signals from autonomic nerves cause the colon to contract, propelling feces into the rectum. The resulting stretching of the rectum activates a defecation reflex that opens a sphincter of smooth muscle at the base of the rectum. Voluntary contraction of a sphincter of skeletal muscle at the anus provides control over the timing of defecation (expulsion of feces).

Healthy adults typically defecate once a day, on average. Emotional stress, a diet low in fiber, minimal exercise, dehydration, and some medications can lead to constipation. This means defecation occurs fewer than three times a week, is difficult, and yields small, hardened, dry feces. Chronic constipation should be discussed with a physician.

Beneficial Microbes

Compared with other regions of the digestive tract, material moves more slowly through the large intestine, which also has a moderate pH. These conditions favor growth of bacteria and archaea that are part of our normal intestinal microbiota.

Bacteria in our large intestine are essential to our health. They synthesize some essential vitamins that we absorb across the intestinal lining. They ferment plant carbohydrates that we lack the enzymes to digest. Fermentation by bacteria converts these carbohydrates into short-chain fatty acids (SCFAs). These molecules are essential to our health. They are the main fuel for ATP production by cells of the intestinal lining. SCFAs also discourage growth of colon cancers, promote immune function, and have beneficial effects on metabolism throughout the body.

Fermentation of carbohydrates by microbes in your colon produces gases such as carbon dioxide, methane, and hydrogen sulfide. Flatulence (farting) is the expulsion of these gases through the anus. A high-fiber diet provides more material for intestinal microbes to dine on and thus can increase flatulence.

TAKE-HOME MESSAGE 39.7

✔ The large intestine completes the process of absorption, then concentrates, stores, and eliminates digestive wastes.

✔ Microbes that live in the large intestine produce vitamins and short-chain fatty acids that contribute to our health.

39.8 Nutritional Requirements

LEARNING OBJECTIVES

- Explain how the body uses the breakdown products of carbohydrates, proteins, and fats.
- Provide an example of a water-soluble vitamin and a fat-soluble vitamin, and explain why each is essential.
- Give an example of an essential mineral, and its role in the body.

Macronutrients

Dietary carbohydrates, fats, and proteins are called macronutrients because we require these substances in large amounts. Macronutrients function both as sources of energy and raw materials. As Section 7.7 explained, breakdown products of these molecules serve as reactants in the energy-releasing reactions of aerobic respiration. FIGURE 39.16 rounds out this picture by illustrating the routes by which the body uses organic compounds obtained from food.

Carbohydrates Fruits, vegetables, and grains should make up the largest proportion of your diet. These foods provide sugars and starches that are hydrolyzed quickly to release glucose, your cells' primary source of energy. When the amount of glucose absorbed exceeds the body's immediate needs, the excess is stored. Blood that flows through capillaries in the small intestine carries glucose-rich blood to the liver, which stores glucose as glycogen. Liver and adipose cells also use glucose to build fats.

Carbohydrates you cannot digest provide essential fiber. There are two types of fiber. Soluble fiber consists of polysaccharides that form a gel when mixed with water. Eating foods high in soluble fiber helps lower one's cholesterol level and reduces risk of heart disease. Insoluble fiber such as cellulose does not dissolve. It enters the colon more or less intact and helps prevent constipation. Dietary fiber also provides food for the microbes in your large intestine.

Many modern diets include an inadequate amount of fiber and an excess of added sugars. The American Heart Association recommends that women consume no more than 24 grams of sugar per day and men no more than 36 grams, from all sources. A typical can of soda has more than 30 grams of sugar.

Fats Per gram, fats provide more energy than any another nutrient. However, they require more steps to break down, so they do not provide the same quick energy boost that carbohydrates do. Fats also provide components required for the synthesis of cell membranes and steroid hormones. They also help you take up fat-soluble vitamins. Your body can remodel carbohydrates to make most of the fatty acids it needs,

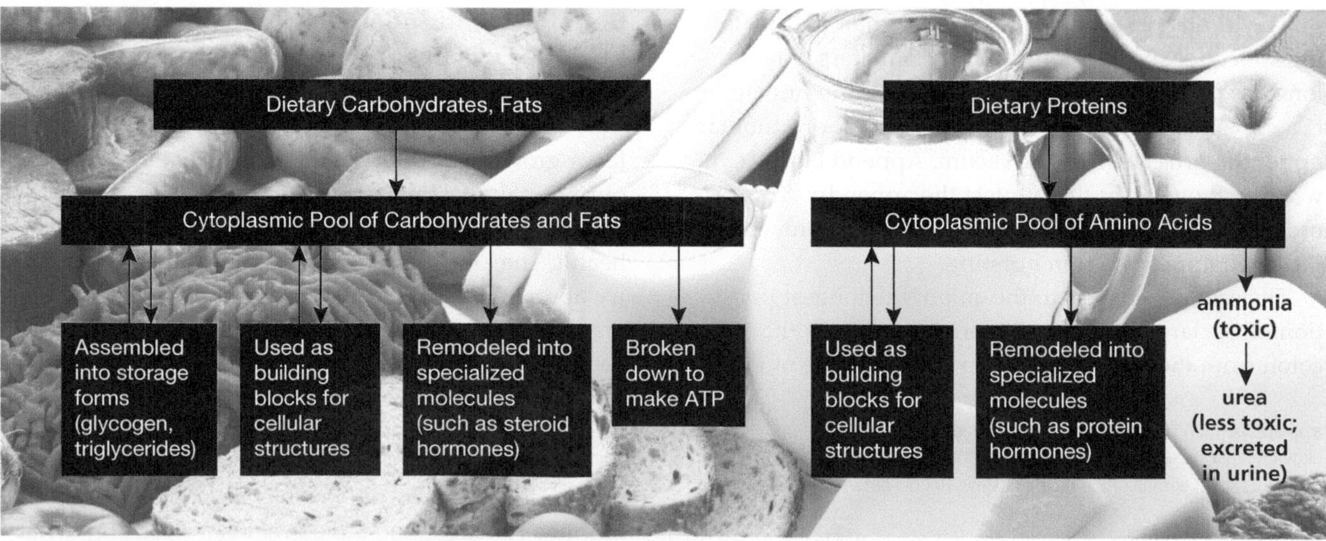

Dietary Carbohydrates, Fats
Dietary Proteins
Cytoplasmic Pool of Carbohydrates and Fats
Cytoplasmic Pool of Amino Acids
Assembled into storage forms (glycogen, triglycerides)
Used as building blocks for cellular structures
Remodeled into specialized molecules (such as steroid hormones)
Broken down to make ATP
Used as building blocks for cellular structures
Remodeled into specialized molecules (such as protein hormones)
ammonia (toxic)
urea (less toxic; excreted in urine)

FIGURE 39.16 **How cells use dietary carbohydrates, fats, and proteins.**

TABLE 39.2

Main Types of Dietary Lipids

Polyunsaturated Fatty Acids: Liquid at room temperature; essential for health.
- Omega-3 fatty acids
  - Alpha-linolenic acid and its derivatives
  - Sources: Nut oils, vegetable oils, oily fish
- Omega-6 fatty acids
  - Linoleic acid and its derivatives
  - Sources: Nut oils, vegetable oils, meat

Monounsaturated Fatty Acids: Liquid at room temperature. Main dietary source is olive oil. Beneficial in moderation.

Saturated Fatty Acids: Solid at room temperature. Main sources are meat and dairy products, palm and coconut oils. Excessive intake may raise risk of heart disease.

***Trans* Fatty Acids (Hydrogenated Fats):** Solid at room temperature. Manufactured from vegetable oils and used in many processed foods. Excessive intake may raise risk of heart disease.

but some must come from foods. **Essential fatty acids** cannot be synthesized by the body, so these fat components must be obtained from a dietary source such as nuts, seeds, and vegetable oils.

Essential fatty acids are polyunsaturated, meaning their tails have two or more double bonds (Section 3.3). There are two types of essential fatty acids: omega-3 fatty acids and omega-6 fatty acids (**TABLE 39.2**). Omega-3 fatty acids seem to have special health benefits. Studies suggest that a diet high in omega-3 fatty acids can reduce the risk of cardiovascular disease, lessen the inflammation associated with rheumatoid arthritis, and help diabetics control their blood glucose.

Trans fats should be avoided. A diet rich in these artificially modified fats increases the risk of heart disease. Most nutritionists also recommend moderating one's intake of foods rich in saturated fat in order to minimize the risk of cardiovascular disease. Such foods include eggs, red meat, and whole-fat dairy products. Packaged products have food labels designed to inform consumers about the food's fat content, as well as other nutritional values (**FIGURE 39.17**)

Nutrition Facts

Serving Size 1 cup (228g)
Servings Per Container 2

Amount Per Serving	
Calories 250	Calories from Fat 110
	% Daily Value*
Total Fat 12g	18%
Saturated Fat 3g	15%
Trans Fat 1.5g	
Cholesterol 30mg	10%
Sodium 470mg	20%
Total Carbohydrate 31g	10%
Dietary Fiber 0g	0%
Sugars 5g	
Protein 5g	
Vitamin A	4%
Vitamin C	2%
Calcium	20%
Iron	4%

* Percent Daily Values are based on a 2,000 calorie diet. Your Daily Values may be higher or lower depending on your calorie needs:

	Calories:	2,000	2,500
Total Fat	Less than	65g	80g
Sat Fat	Less than	20g	25g
Cholesterol	Less than	300mg	300mg
Sodium	Less than	2,400mg	2,400mg
Total Carbohydrate		300g	375g
Dietary Fiber		25g	30g

Check serving size. A package often holds more than one serving, but nutritional information is given per serving.

Avoid foods in which a large proportion of calories comes from fat.

Choose foods that provide a low percentage of the recommended maximum of saturated fat, *trans* fat, cholesterol, and sodium. 20% or more is high.

Choose foods that are high in dietary fiber and low in sugar.

If you eat meat, you probably get more than enough protein.

Choose foods that provide a high percentage of your daily vitamin and mineral requirements.

This part of the label shows recommended intake of nutrients for two levels of calorie intake. Keeping fat and salt intake below recommended levels and dietary fiber above recommended levels decreases risk of some chronic health problems.

FIGURE 39.17 How to read a food label. Information on a food label can be used to ensure that you get the nutrients you need without exceeding recommended limits on less healthy substances such as salt and *trans* fats. This hypothetical label is for a ready-to-eat macaroni and cheese product.

Proteins Your body uses amino acids from dietary proteins to build peptides and proteins and to make nucleotides. Most organs do not routinely break down amino acids for a source of energy, but the liver does. Here, the amino group is separated from the amino acid's carbon skeleton, which is then used to fuel the Krebs cycle. The toxic ammonia (NH_3) produced by amino acid breakdown is converted to urea, a somewhat less toxic compound that is excreted in urine.

Essential amino acids are those your body cannot make and must obtain from food. Animal proteins are usually complete, meaning they contain all essential amino acids in the ratio that a human needs. Most plant proteins are incomplete, meaning they lack or contain little of one or more essential amino acids. You can meet your amino acid requirements with a plant-based diet, but doing so requires combining foods. Amino acids missing from one food must be provided by others. As an example, rice and beans together provide all necessary amino acids, but rice alone or beans alone does not. Complementary sources of amino acids do not need to be eaten simultaneously, but can instead be "combined" over the course of a day or two.

essential amino acid Amino acid that the body cannot make and must obtain from food.

essential fatty acid Fatty acid that the body cannot make and must obtain from food.

CREDITS: (17) USDA.

TABLE 39.3

Major Vitamins: Sources, Functions, and Effects of Deficiencies

Vitamin	Main Dietary Sources	Main Functions	Symptoms of Deficiency
Fat-Soluble Vitamins			
A	Orange fruits and vegetables, leafy greens, egg yolk	Component of visual pigments; maintains epithelia	Night blindness, skin problems
D	Fatty fish, egg yolk	Aids uptake and use of calcium	Weak/soft bones, rickets in children
E	Nuts, seeds, vegetable oils	Antioxidant, aids fat absorption; helps maintain cell membranes	Muscle weakness, nerve damage
K	Green vegetables	Needed for blood clotting	Impaired blood clotting
Water-Soluble Vitamins			
B_1 (thiamin)	Meats, nuts, legumes	Coenzyme in carbohydrate metabolism	Beri beri (neurological and heart problems)
B_2 (riboflavin)	Meats, eggs, nuts, milk, green vegetables	Component of the coenzyme FAD	Anemia, sores in mouth, sore throat
B_3 (niacin)	Meats, fish, dairy products, nuts, legumes	Component of the coenzyme NAD	Skin and mucous membrane sores, gut pain, diarrhea, psychosis, dementia
B_6	Meats, fish, starchy vegetables, noncitrus fruits	Coenzyme in protein metabolism	Anemia, sores on lips, depression, impaired immune function
B_7 (biotin)	Meats, fish, nuts, legumes, whole grains	Coenzyme in many reactions	Hair loss, dry skin, dry eyes, fatigue, insomnia, depression
B_9 (folic acid)	Meats, fruits, green vegetables, whole grains	Coenzyme in nucleic acid synthesis and amino acid metabolism	Anemia, sores in mouth; deficiency during pregnancy causes neurological birth defects
B_{12}	Meats, seafood, dairy products	Coenzyme in amino acid synthesis	Anemia, fatigue, neurological problems
C (ascorbic acid)	Fruits (especially citrus) and vegetables	Required for collagen synthesis, antioxidant	Scurvy (anemia, bleeding gums, impaired wound healing, swollen joints)

Vitamins

Vitamins are organic substances that are required in the diet in very small amounts. TABLE 39.3 lists the major vitamins we need. Vitamins A, D, E, and K are fat soluble. Heat has little effect on these vitamins, which are abundant in both cooked and fresh foods. Fat-soluble vitamins can be stored in the body's own fat, so they do not need to be eaten daily. By contrast, water-soluble vitamins such as vitamins B and C are generally not stored, so they must be eaten more frequently. Water-soluble vitamins are also more sensitive to heat, so they are more easily destroyed by cooking.

The body remodels vitamin A into the visual pigment made by rod cells (Section 33.8). In the United States, milk and soy milk products are typically fortified with vitamin A, so a deficiency in this vitamin is rare. In less developed countries, vitamin A deficiency is the most common cause of childhood blindness.

Vitamin D increases calcium uptake from the gut and encourages retention of calcium in bone. It can be obtained from dietary sources or made in the skin. Vitamin D production by skin requires exposure to sunlight. People with dark skin make less vitamin D than lighter-skinned people, and so are more likely to be deficient in this vitamin. A deficiency of vitamin D while bones are growing can result in rickets. Affected individuals have unusually soft, weak bones, short stature, and, in extreme cases, misshapen bones.

Vitamin E is an antioxidant (Section 5.6). Being lipid soluble, it can enter lipid bilayers and it plays an important role in protecting the components of cell membranes from oxidation.

Vitamin K is a coenzyme that assists enzymes involved in blood clotting. (K denotes the German word *koagulation*.) Absorption of vitamin K produced by bacteria in the large intestine supplements vitamin K derived from food.

B vitamins are used to build the coenzymes that function in a variety of essential pathways. For example, vitamin B_3 (niacin) is used to make NAD, and vitamin B_2 (riboflavin) is used to make FAD. Both of these coenzymes play vital roles in aerobic respiration, as summarized in Section 7.5.

Vitamin C is an antioxidant, and it is needed to make collagen, the body's most abundant protein. A deficiency causes scurvy, a disorder in which skin and bones deteriorate and wounds are slow to heal.

Note that most people can obtain all the vitamins they need from a well-balanced diet. Some studies have even found an increased death rate among people who take supplemental antioxidants (beta-carotene,

TABLE 39.4

Major Minerals: Sources, Functions, and Effects of Deficiencies

Mineral	Main Dietary Sources	Main Functions	Symptoms of Deficiency
Calcium	Dairy products, dark green vegetables	Component of bones and teeth; needed for neuron and muscle action	Stunted growth, fragile bones, nerve impairment
Phosphorus	Whole grains, poultry, red meat	Component of bones, teeth; part of nucleic acids, ATP, phospholipids	Fragile bones, muscle weakness
Sodium	Table salt	Water balance, nerve function	Muscle cramps
Potassium	Meats, fruits, vegetables, grains	Water balance, nerve function	Muscle weakness
Sulfur	Dietary proteins	Component of proteins	Stunted growth
Chlorine	Table salt, vegetables	Acts in water balance, nerve function, formation of stomach acid (HCl)	Muscle cramps, stunted growth, poor appetite
Magnesium	Grains, legumes, nuts dairy products	Coenzyme with role in ATP–ADP cycle; roles in muscle, nerve function	Impaired nerve function
Iodine	Seafood, iodized salt	Component of thyroid hormone	Thyroid and metabolic disorders
Iron	Meats, fish, grains, legumes, green leafy vegetables	Cofactor, component of heme and electron carriers	Iron-deficiency anemia, impaired immune function
Fluorine	Fluoridated water, seafood	Bone, tooth maintenance	Tooth decay

vitamin A, and vitamin E). If you take vitamin supplements, be sure the quantities are not excessive.

Minerals

Minerals are elements that are essential in the diet in small amounts (TABLE 39.4). The seven minerals required in the largest quantities are calcium, phosphorus, potassium, sulfur, sodium, chlorine, and magnesium. Calcium and phosphorus, the body's most abundant minerals, are components of teeth and bones. Calcium also plays a role in intercellular signaling, and phosphorus is also used to make nucleic acids. Sodium and potassium are important in nerve function, sulfur is a component of some proteins, chlorine is a component of hydrochloric acid (HCl) in gastric fluid, and magnesium is a cofactor. Deficiencies in any of these seven elements are rare.

Note that salt contains two essential minerals, sodium and chloride, but both are easily obtained in sufficient quantity from unsalted foods. Eating foods with added salt elevates the body's sodium level, which raises the risk of high blood pressure in some people. Food labels show sodium content as a percentage of the recommended daily maximum for a person who has normal blood pressure.

Iodine, another essential mineral, is necessary to produce thyroid hormone, which has roles in development and metabolism (Section 34.6). Iodine is abundant in fish, shellfish, and seaweeds. A deficiency in this mineral causes goiter, which is an enlarged thyroid gland, and disrupts metabolism. In the United States, iodized salt makes iodine deficiency rare.

Iron is an essential component of heme, the chemical group that binds to oxygen in hemoglobin (Section 38.7) and myoglobin. Heme is also a cofactor for many enzymes, including critical components of electron transfer chains. Worldwide, iron is the mineral most commonly deficient in the diet.

A small amount of fluoride is required to keep bones and teeth strong. In the United States, most public drinking water supplies are fluoridated to help prevent tooth decay.

TAKE-HOME MESSAGE 39.8

- ✔ Carbohydrates are the body's main sources of energy.
- ✔ Fats are broken down for energy and used to make membrane components and steroid hormones.
- ✔ Proteins provide amino acids for building peptides, proteins, and nucleotides.
- ✔ Vitamins are organic molecules with an essential role in metabolism.
- ✔ Minerals are inorganic substances with an essential role.

mineral With regard to nutrition, an element required in small amounts for normal metabolism.

vitamin Organic substance required in small amounts for normal metabolism.

39.9 Maintaining a Healthy Weight

LEARNING OBJECTIVES

- Explain how BMI is measured, what it measures, and why it is only a crude measure of health risks.
- Explain what resting metabolic rate is, factors that affect it, and how it influences the risk of obesity.

What Is a Healthy Weight?

Body mass index (BMI) is an indirect measurement of body fat that is based on an individual's height and weight. You can calculate your body mass index with this formula:

$$\text{BMI} = \frac{\text{weight (pounds)} \times 703}{\text{height (inches)}^2}$$

Generally, individuals with a BMI of 25 to 29.9 are said to be overweight. A score of 30 or more indicates **obesity**: an overabundance of fat in adipose tissue that may lead to severe health problems. Conversely, a BMI of 18.5 or lower is considered dangerously underweight. A low BMI can result from anorexia nervosa, an eating disorder in which people restrict their food intake well below a healthy level because of an irrational fear of weight gain.

To maintain your weight, you must balance your caloric intake and your energy output. Energy stored in food is expressed as kilocalories, or Calories (with a capital C). One kilocalorie equals 1,000 calories, which are units of heat energy.

Here is a way to calculate roughly how many kilocalories you should take in daily to maintain a preferred weight. First, multiply the weight (in pounds) by 10 if you are not active physically, by 15 if you are moderately active, and by 20 if you are highly active. Next, subtract one of the following amounts from the multiplication result:

Age:	*Subtract:*
25–34	0
35–44	100
45–54	200
55–64	300
Over 65	400

For example, if you are 25 years old, highly active, and weigh 120 pounds, you will require 120 × 20 = 2,400 kilo calories daily to maintain weight. Take in more and you gain weight; take in less and you lose it.

Even at rest, you use energy in essential processes such as breathing, generating body heat, fighting off pathogens, moving material though your digestive tract, and circulating your blood. The rate at which you burn energy in these activities is your **basal metabolic rate**. This rate varies with temperature. The colder your environment, the more energy is required to maintain a normal body temperature. Basal metabolic rate also varies among individuals. Big, muscular bodies use more energy than smaller, less muscular ones. Men tend to be more muscular than women of the same weight, so their metabolic rate is higher. In both sexes, metabolic rate slows with age. Thyroid hormone level influences basal metabolic rate too. People with a high level of thyroid hormone have a higher metabolic rate than those with a lower level.

TABLE 39.5

Energy Expenditures

Activity	Energy expended (Calories/kilogram of body weight/hour)
Sitting quietly	1
Standing	2
Walking slowly	3
Bicycling	4–7
Swimming	6–10
Basketball game	8
Running (5 mph)	8

Dieting can influence basal metabolic rate. Often, when people eat less, resting metabolic rate declines. This mechanism evolved to promote survival when food became scarce. However, it is a source of great frustration to dieters who have chosen to limit their food intake in the hope of losing weight.

Exercise can help you lose weight because it requires energy. **TABLE 39.5** shows energy expenditure during various activities. Exercise that increases the body's muscle mass also helps increase resting metabolic rate.

Why Is Obesity Unhealthy?

Obesity increases the risk for a long list of diseases including high blood pressure, type 2 diabetes, heart disease, stroke, gallbladder disease, osteoarthritis, sleep apnea, and cancers of the uterus, breast, prostate gland, kidney, and colon. An obese person's internal organs can be squashed by fat (**FIGURE 39.18**), impairing the organs' function. For example, excessive fat in the abdomen impairs the diaphragm's ability to descend downward during inhalation, so breathing becomes more difficult.

Obesity also impairs function at the cellular level. Adipose cells of people who are at a healthy weight

basal metabolic rate Rate at which the body uses energy when you are at rest.

body mass index (BMI) Measure that indicates an individual's degree of body fat based on height and weight.

obesity Overabundance of body fat, which increases health risks.

hold a moderate amount of triglycerides. In obese people, adipose cells become overstuffed. Like cells stressed in other ways, the overstuffed adipose cells respond by sending out chemical signals that summon up an inflammatory response (Section 37.4). The resulting chronic inflammation harms organs throughout the body and increases the risk of cancer. Overstuffed adipose cells also increase their secretion of chemical messages that interfere with insulin's effects increasing the risk of type 2 diabetes.

Causes of Obesity

Geneticists have pinpointed some genes that predispose people to obesity. For example, about 16 percent of people of European ancestry are homozygous for an allele of the *fto* gene that predisposes them to obesity. Compared to people with two low-risk alleles, those homozygous for the high-risk allele are twice as likely to be obese. The function of the protein encoded by the *fto* gene is unknown. We do, however, know that the gene is expressed most strongly in the brain.

Genetic differences can explain why one person is more likely than another to be overweight, but cannot explain a national trend toward weight gain. In 1960, only about 15 percent of people in the United States were obese; since then, the obesity rate has more than doubled. Many factors contributed to this increase. The proportion of meals consumed outside the home increased, as did the average portion size of those meals. Soda consumption increased, and physical activity decreased. We now spend more time in front of televisions and computers, and fewer of us have jobs that require physical exertion.

Preventing obesity is important. Once a person becomes obese, dieting alone can seldom restore a normal body weight over the long term. A variety of drugs can reduce weight, but all have negative side effects and must be taken continually to prevent rebound weight gain. Surgical procedures that reduce stomach volume produce a dramatic sustainable weight loss, but these drastic interventions are expensive and can result in serious complications.

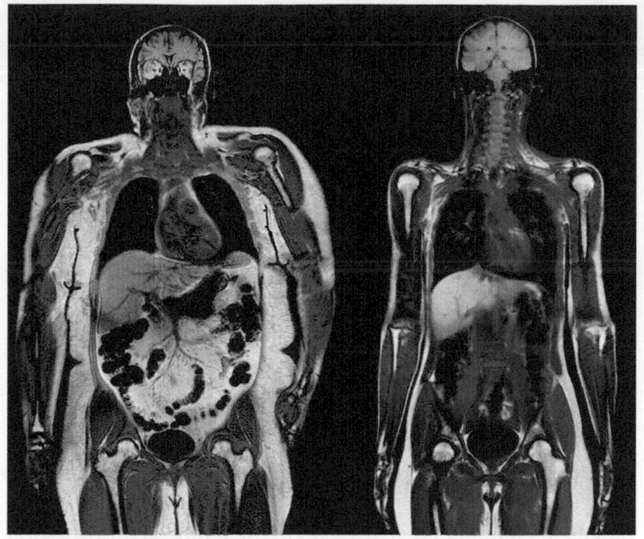

FIGURE 39.18 Obesity. MRIs (magnetic resonance images) of an obese woman (left) and a woman of normal body weight (right). With obesity, the abdomen fills with fat that squashes internal organs.

TAKE-HOME MESSAGE 39.9

- ✔ A person who balances caloric intake with energy expenditures will maintain current weight.
- ✔ We burn calories even at rest. Basal metabolic rate varies with age, sex, and hormone level.
- ✔ Obesity raises the risk of many health problems.
- ✔ Genetic differences affect the tendency toward obesity.

39.1 Breaking It Down (revisited)

Exogenous enzymes (enzymes made outside the body) can improve the ability to digest specific substances. For example, lactase supplements such as Lactaid allow lactose-intolerant people to drink milk and eat milk products without any ill effects. Such supplements contain lactase from a fungus (*Aspergillus*).

Similarly, supplemental alpha-galactosidase can break down certain oligosaccharides found in beans and green vegetables such as cabbage and broccoli. The enzyme alpha-galactosidase is not normally present in the human small intestine; it is not a mammalian digestive enzyme. Thus, in the absence of supplemental alpha-galactosidase the oligosaccharides proceed undigested to the large intestine. Here, their fermentation releases gases that cause flatulence. When supplemental alpha-galactosidase is present in the small intestine, the oligosaccharides are broken into simpler sugars and absorbed. Beano, a supplement marketed as a flatulence preventer, contains an alpha-galactosidase. As with Lactaid, the enzyme is derived from a species of *Aspergillus* fungus. ●

STUDY GUIDE

Section 39.1 Breaking down food to release nutrients requires the action of digestive enzymes. The ability to produce some digestive enzymes such as lactase varies among individuals. Use of enzyme supplements allows digestion of materials that would otherwise be undigestable.

Section 39.2 A digestive system breaks food down into molecules small enough to be absorbed into the internal environment. It also stores and eliminates unabsorbed materials. Some invertebrates have a **gastrovascular cavity**: a saclike gut with a single opening.

Most animals and all vertebrates have a **complete digestive tract** with a mouth at one end and an **anus** or **cloaca** at the other. An **esophagus** carries food to the **stomach** for digestion, then absorption occurs in the **intestine**. Variations in the structure of vertebrate digestive systems are adaptations to particular diets. For example, a bird's **crop** stores food and its **gizzard** grinds food up. The multiple stomachs of ruminants allow them to digest cellulose-rich plant parts. A large cecum filled with bacteria does the same for rabbits.

Section 39.3 The human digestive system includes the digestive tract and accessory organs that secrete material into its **lumen**. Food is moved along the tract by **peristalsis**, and sphincters regulate this passage. The human intestine has two functional regions: The **small intestine** carries out digestion and most nutrient absorption. The **large intestine** concentrates undigested wastes, which are stored in the **rectum** until they are expelled.

Section 39.4 Digestion starts in the mouth, where teeth break food into bits and movements of the tongue mix it with saliva from **salivary glands**. The number and arrangement of teeth vary among mammals and adapt different species to different diets. Saliva contains the enzyme salivary amylase, which begins the process of starch digestion. The presence of food in the back of the throat triggers a swallowing reflex that moves food into the esophagus. During swallowing, the epiglottis folds over to cover the trachea.

Section 39.5 The stomach is a J-shaped muscular sac with a sphincter at either end. Protein digestion begins in the stomach. The stomach mucosa secretes **gastric fluid** that contains hydrochloric acid, pepsin (a protein-digesting enzyme), and mucus. Gastric fluid mixes with food to form **chyme**. Production of gastric fluid is triggered by **gastrin**, a stomach hormone released in response to the arrival of food. When empty, the stomach produces **ghrelin**, a hormone that increases appetite.

Section 39.6 The small intestine is the longest part of the digestive tract and has the largest surface area. Its highly folded lining has **villi** at its surface. Each multicelled villus has a covering of **brush border cells**. These cells have **microvilli** that increase their surface area for digestion and absorption.

Chemical digestion is completed in the small intestine through the action of enzymes from the pancreas, bile from the gallbladder, and enzymes embedded in the plasma membrane of brush border cells.

Carbohydrates are broken into monosaccharides, which are actively transported across brush border cells and enter the blood. Similarly, proteins are broken into amino acids, which are transported across these cells and enter blood.

Bile made in the **liver** and stored and concentrated in the **gallbladder** aids in the emulsification of fats. Monoglycerides and fatty acids diffuse into the brush border cells. Here, they recombine as triglycerides, which get a protein coat and are moved by exocytosis into interstitial fluid. The lipoproteins then enter lymph vessels that deliver them to blood.

The small intestine is also the site of most water absorption. Water moves out of the gut by osmosis.

The hormone **secretin** is secreted by the small intestine in response to the arrival of acidic chyme. It stimulates the release of bicarbonate from the pancreas. The hormone **cholecystokinin** is secreted in response to proteins and fats. It stimulates secretion of pancreatic enzymes and and bile into the small intestine.

Section 39.7 Absoption of water from the large intestine concentrates digestive waste. The **appendix** is a short extension from the first part of the large intestine (the cecum). The longest portion of the large intestine, the **colon**, compacts undigested solid wastes as **feces**. Feces are stored in the rectum, a stretchable region just before the anus. Bacteria that live in the small intestine benefit our health by producing essential vitamins and short-chain fatty acids (SCFAs).

Section 39.8 Organic macromolecules (carbohydrates, proteins, and fats) serve as sources of energy and raw materials. Excess sugar is either stored as glycogen in the liver or used to make fat. The body builds most fatty acids and amino acids, but **essential fatty acids** and **essential amino acids** must be obtained from food. The diet must also include **vitamins**, which are small organic molecules, and **minerals**, which are inorganic. Plant-based diets can meet all human nutritional needs if foods are combined properly.

Excessive amounts of sugar, salt, and saturated fat are associated with increased health risks, whereas a diet high in fiber and polyunsaturated fats has health benefits.

Section 39.9 Maintaining body weight requires balancing energy intake and output. **Body mass index** indicates whether a given height and weight is healthy. **Basal metabolic rate**, the energy expended at rest, varies with age and other factors. With **obesity**, fat deposits press on internal organs and overstuffing of adipose cells leads to chronic inflammation. Obesity has a genetic component.

SELF-QUIZ

Answers in Appendix VII

1. A digestive system functions in ________.
   a. secreting enzymes
   b. absorbing compounds
   c. eliminating wastes
   d. all of the above

2. The enzyme pepsin is activated in the ________.
   a. mouth
   b. stomach
   c. small intestine
   d. large intestine

3. Most nutrients are absorbed in the ________.
   a. mouth
   b. stomach
   c. small intestine
   d. large intestine

4. A bird's keratin-lined ________ helps it grind up food.
   a. pharynx
   b. crop
   c. gizzard
   d. gastrovascular cavity

5. Monosaccharides and amino acids absorbed from the small intestine enter ________.
   a. blood vessels
   b. lymph vessels
   c. fat droplets
   d. bile

6. Bacteria in the ________ make essential vitamins.
   a. stomach
   b. small intestine
   c. colon
   d. esophagus

7. The low pH of the ________ aids in protein digestion.
   a. stomach
   b. small intestine
   c. large intestine
   d. esophagus

8. Most water that enters the digestive tract is absorbed across the lining of the ________.
   a. stomach
   b. small intestine
   c. large intestine
   d. esophagus

9. Iron is the ________ most often deficient in the diet.
   a. macronutrient
   b. vitamin
   c. mineral
   d. enzyme

10. Eating a healthy diet entirely of plant foods requires combining foods to obtain all essential ________.
    a. saturated fats
    b. monosaccharides
    c. enzymes
    d. amino acids

11. Tiny projections called ________ increase the surface area of a brush border cell.
    a. villi
    b. microvilli
    c. cilia
    d. flagella

12. ________ are considered healthy fats that may reduce the risk of heart disease.
    a. Saturated fats
    b. *Trans* fats
    c. Glycolipids
    d. Omega-3 fatty acids

13. A very low BMI indicates a person is ________.
    a. obese
    b. vitamin C deficient
    c. underweight
    d. lactose intolerant

14. Match each substance with its description.

___ gastrin	a. enzyme that acts in the stomach
___ secretin	b. produced by colon bacteria
___ bicarbonate	c. breaks down polysaccharides
___ bile	d. raises the pH of chyme
___ salivary amylase	e. stimulates bicarbonate secretion by the pancreas
___ pepsin	f. hormone made by stomach
___ short chain fatty acids (SCFAs)	g. emulsifies fats

15. Match each organ with its digestive function in humans.

___ gallbladder	a. final stop for digestive waste
___ salivary gland	b. makes bile
___ colon	c. compacts undigested residues
___ liver	d. adds enzymes to small intestine
___ esophagus	e. delivers food to the stomach
___ rectum	f. stores, secretes bile
___ stomach	g. secretes gastric fluid
___ pancreas	h. secretes enzyme that begins starch digestion

CRITICAL THINKING

1. Some people who have gallstones experience pain after they eat, with fatty meals causing the greatest discomfort. Why are fatty meals most likely to trigger gallbladder pain?

2. A python can survive by eating a large meal once or twice a year. When it does eat, microvilli in its small intestine lengthen fourfold and its stomach pH drops from 7 to 1. Explain the benefits of these changes.

3. Although rabbits cannot break cellulose down into simple sugars, the bacteria that live in their cecum can. However, for a rabbit to make use of the nutrients released by bacterial action, it must eat its feces. Why is this second round of ingestion necessary?

4. Cholera is a disease caused by bacteria that make cholera toxin. In the presence of this toxin, chloride ion channels in cells of the small intestine lining are stuck in their open position. As a result, an excess of chloride ions (Cl^-) flows from these cells into the lumen of the small intestine. Movement of Cl^- creates a charge gradient, so sodium ions (Na^+) also flow into the lumen. Explain why the movement of these ions into the lumen results in the life-threatening diarrhea that characterizes cholera.

40 MAINTAINING THE INTERNAL ENVIRONMENT

CORE CONCEPTS

Pathways of Transformation

Organisms exchange matter and energy with the environment in order to grow, maintain themselves, and reproduce.

Maintaining an internal enviroment that differs from an external environment requires constant movement of substances across cell membranes via osmosis, diffusion, and membrane transport proteins. These movements are central to processes that rid the animal body of metabolic wastes.

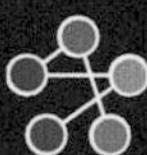

Systems

Complex properties arise from interactions among components of a biological system.

Interactions and coordination among specialized organs and organ systems contribute to the overall functioning of the animal body. These interactions include various strategies for regulating solute balance and body temperature. Internal and external factors trigger homeostatic responses in organs. These responses in turn affect the entire body and contribute to the survival and normal function of the individual.

Evolution

Evolution underlies the unity and diversity of life.

Animals respond to some types of changes in the environment through evolved physiological mechanisms. Negative feedback mechanisms that are triggered by change (in temperature and solute balance, for example) return conditions in the internal environment back to target set points. These mechanisms maintain the internal environment's conditions within optimal ranges.

Links to Earlier Concepts

This chapter's discussion of fluid balance draws on your knowledge of pH (2.6), osmosis (5.8), aerobic respiration (7.2), protein metabolism (39.9), body fluids (31.2), chemoreceptors (33.2) and the adrenal glands (34.8). Its discussion of temperature regulation refers to properties of water (2.5), forms of energy (5.2), and negative feedback controls (31.9).

40.1 Urine Tests

Clear or cloudy? A lot or a little? Asking about and examining urine is an ancient art. About 3,000 years ago in India, the pioneering healer Susruta reported that some patients produced an excessive amount of sweet-tasting urine. In time, the disorder that causes these symptoms was named diabetes mellitus, which means "passing honey-sweet water." Doctors still diagnose diabetes by testing the sugar level in urine, although they have now replaced the taste test with a chemical analysis.

Modern physicians routinely check the pH and solute concentrations of urine to monitor their patients' health. Acidic urine suggests metabolic problems. Alkaline urine can indicate an infection. Damaged kidneys produce urine with an unusually high concentration of the plasma protein albumin. An excess of certain salts can result from dehydration or trouble with the hormones that control kidney function. Urine tests can also detect chemicals produced by cancer of the prostate gland or urinary tract.

A variety of over-the-counter urine tests are available to determine hormone levels. If a woman wants to become pregnant, she can use one test to keep track of the amount of luteinizing hormone, or LH, in her urine. About midway through a menstrual cycle, LH triggers the release of an egg from an ovary. Another over-the-counter urine test can reveal whether a woman has become pregnant.

If you use tobacco, alcohol, nicotine, marijuana, cocaine, or some other psychoactive drug, your urine tells the tale. The liver filters the active ingredients of these drugs out of the blood and converts them to other compounds, which we call metabolites. Drug metabolites that form in the liver enter the blood and travel to the kidney.

Each day, your kidneys filter all of the blood in an your body many times. As kidneys filter the blood, they move excess solutes, hormones, and toxins such as drug metabolites into the forming urine. Depending on the type of drug and the dosage and frequency of its use, eliminating all drug metabolites from the body can take a few days to nearly a month. Age also has an effect. A healthy 35-year-old eliminates drug metabolites from the body about twice as fast as a healthy 85-year-old.

So far in this unit, we have considered the organ systems that keep cells supplied with oxygen, nutrients, water, and other substances. We turn now to the mechanisms that maintain the composition, volume, and temperature of the internal environment.

CREDITS: (opposite) Chappell, Mark / Animals Animals; (in text) Lawrence Lawry/Science Source.

40.2 Fluid Volume and Composition

LEARNING OBJECTIVES

- Explain the ways that animals gain and lose water.
- Compare the amounts of water and energy used in the excretion of urea, uric acid, and ammonia.

By weight, an animal consists mostly of water, with dissolved salts and other solutes. Fluid outside cells—the extracellular fluid—is the cells' environment. In vertebrates, interstitial fluid (fluid between cells) and plasma (the fluid portion of the blood) constitute the bulk of the extracellular fluid (Section 31.2).

To keep the solute composition and volume of extracellular fluid within the ranges cells can tolerate, water and solute gains must equal water and solute losses.

Water Gains and Losses

For all animals, some water enters the body in food and some is lost in elimination of digestive waste. In aquatic animals, water also moves into or out of the body by osmosis across the body surface (Section 5.8). In land animals, evaporation of water from the body surface, especially the respiratory surface, is a major mechanism of water loss.

Metabolic wastes

All animals gain solutes from food, and their metabolic reactions produce waste solutes, particularly carbon dioxide and ammonia. Carbon dioxide diffuses out across the body surface or leaves with the help of respiratory organs.

Ammonia (NH_3) is a toxic metabolic waste product formed primarily by the breakdown of proteins (FIGURE 40.1). Mechanisms that rid the body of ammonia vary among animal groups, and these adaptations reflect the availability of water in different environments. Flushing toxic ammonia from a body requires a large amount of water, so most animals that excrete ammonia without modifying it are aquatic.

FIGURE 40.1 Nitrogenous waste products.

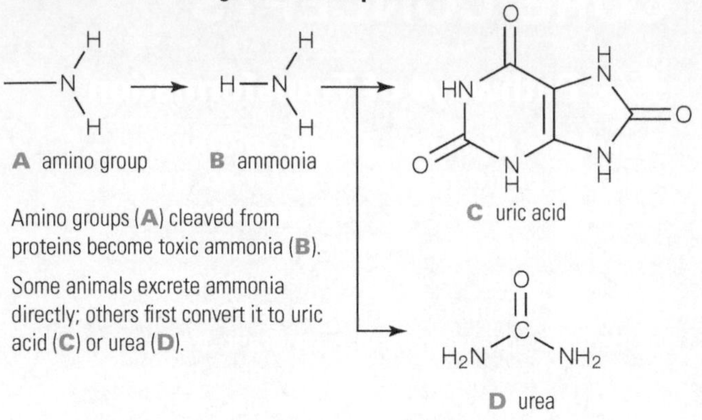

A amino group

B ammonia

C uric acid

D urea

Amino groups (A) cleaved from proteins become toxic ammonia (B).

Some animals excrete ammonia directly; others first convert it to uric acid (C) or urea (D).

By contrast, most land animals carry out energy-requiring reactions that convert ammonia to urea or uric acid. Compared to ammonia, **urea** is a less toxic compound that can be excreted in a smaller volume of water. Urea is the main nitrogenous waste excreted by mammals and by adult amphibians.

Uric acid is a nontoxic compound that is excreted as crystals mixed with just enough water to produce a thick, white paste. Producing uric acid from ammonia requires the greatest investment of energy and yields the greatest water savings. Insects excrete uric acid, as do lizards, snakes, and birds.

TAKE-HOME MESSAGE 40.2

✔ Animals must maintain the volume and composition of their interstitial fluid within a limited range.

✔ Protein breakdown produces toxic ammonia. Some animals convert this waste solute to urea or uric acid before excreting it.

✔ Uric acid is excreted as crystals, so its excretion requires less water than ammonia or urea.

40.3 Excretory Organs

LEARNING OBJECTIVE

- Compare the types of excretory organs, and explain the similarity in how they function.

Many animals have specialized organs that regulate the concentration of solutes and water in their body fluids. The functional units of these organs are tubules. Body fluid enters one end of the tubule and its volume and composition are adjusted as it proceeds through the tubule. Essential solutes and water leave the tubule and return to the body fluid, whereas waste solutes remain in the tubule or are transported into it. The terminal end of the tubule delivers fluid with a high concentration of waste solutes to the body surface or to an organ that conveys it to the body surface.

ammonia Nitrogen-containing waste compound (NH_3) that is produced when protein is broken down.

kidney Organ of the vertebrate urinary system that filters blood, adjusts its composition, and forms urine.

Malpighian tubules (mal–PIG–ee–an) Excretory organs of insects; deliver waste solutes to the digestive tract.

nephridia (nef–RID–ee–ah) Singular, nephridium. Excretory organs of earthworms.

protonephridia (PRO–toh–nef–RID–ee–ah) Singular, protonephridium. Excretory organs of planarian flatworms.

urea (yuh-REE-uh) Nitrogen-containing compound derived from ammonia; excreted by mammals.

uric acid (YUR-ik) Nitrogen-containing compound derived from ammonia; excreted by insects, birds, lizards, and snakes.

urine Mixture of water and soluble wastes formed and excreted by the vertebrate urinary system.

Planarian Protonephridia

In planarian flatworms, branching excretory organs called **protonephridia** rid the body of excess water and soluble wastes (FIGURE 40.2). Planarians have a higher solute concentration than the water around them, so water continually enters their body by osmosis. Movement of cilia on specialized cells (flame cells) draws fluid into a tubular branch of a protonephridium. As fluid flows through the tubule, essential solutes are returned to the body fluid. Excess water and unwanted solutes are expelled from the tubule where it ends at a pore in the body wall.

Earthworm Nephridia

An earthworm is a segmented annelid with a fluid-filled body cavity (a coelom) and a closed circulatory system. Most body segments have a pair of tubular excretory organs called **nephridia** that collect coelomic fluid (FIGURE 40.3). As fluid flows through a nephridium, essential solutes and some water exit the nephridium and enter adjacent blood vessels. Ammonia remains in the tube and exits the body through a pore.

Arthropod Malpighian Tubules

Land arthropods such as insects conserve water by converting ammonia to uric acid. The uric acid is taken up from hemolymph (the circulatory fluid) by Malpighian tubules (FIGURE 40.4). A **Malpighian tubule** is a tubular excretory organ that delivers uric acid and other solutes to the waste-processing portion of the digestive tract in the animal's abdomen. The waste solutes are then expelled from the body along with digestive wastes.

Vertebrate Kidneys

In vertebrates, the tubules that function in excretion are inside a pair of organs called kidneys. The **kidneys** filter the blood and produce urine. **Urine** is a mixture of water and soluble wastes. We describe the human kidney, explain how it produces urine forms, and explore the other components of the human urinary system in Sections 40.4 and 40.5.

TAKE-HOME MESSAGE 40.3

- ✔ Excretory organs contain tubules that take up body fluid, return essential solutes to that fluid, then excrete waste solutes and water.
- ✔ Planarian protonephridia and earthworm nephridia are tubules that excrete waste through a pore in the body wall.
- ✔ An insect's Malpighian tubules deliver uric acid to the digestive tract for excretion.
- ✔ A vertebrate's paired kidneys filter the blood and produce urine.

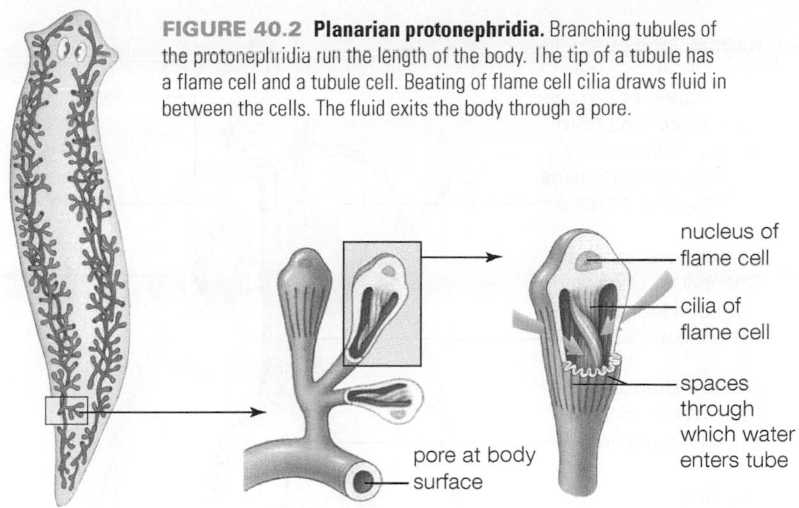
nucleus of
flame cell
cilia of
flame cell
spaces
through
which water
enters tube
pore at body
surface

FIGURE 40.2 Planarian protonephridia. Branching tubules of the protonephridia run the length of the body. The tip of a tubule has a flame cell and a tubule cell. Beating of flame cell cilia draws fluid in between the cells. The fluid exits the body through a pore.

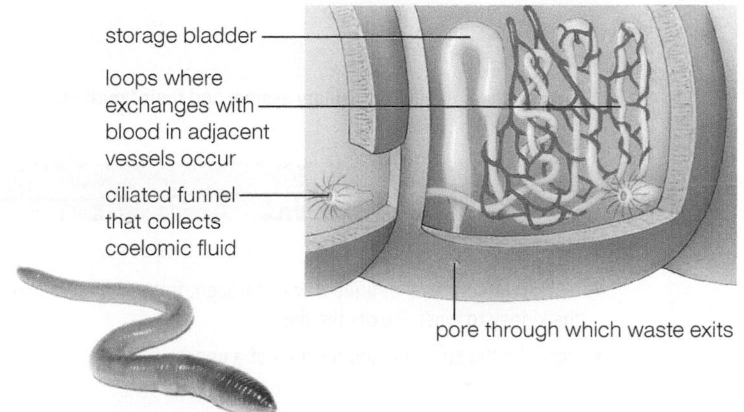
storage bladder
loops where
exchanges with
blood in adjacent
vessels occur
ciliated funnel
that collects
coelomic fluid
pore through which waste exits

FIGURE 40.3 Earthworm nephridia. Coelomic fluid enters a nephridium (green). As fluid travels through the tubule, essential solutes leave this tube and enter adjacent blood vessels (red). Ammonia-rich waste exits the body through a pore.

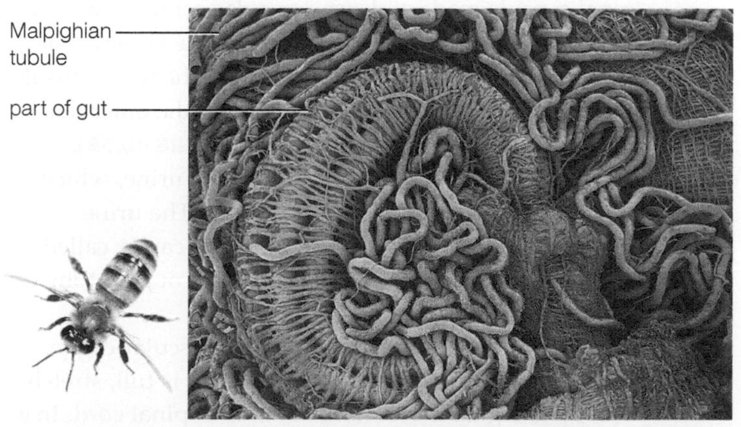
Malpighian
tubule
part of gut

FIGURE 40.4 Insect Malpighian tubules. A honeybee's Malpighian tubules (gold) are outpouchings of the gut (pink). Uric acid and other waste solutes move from the hemolymph into the tubule interior. The tubules deliver the wastes to the gut for elimination through the anus.

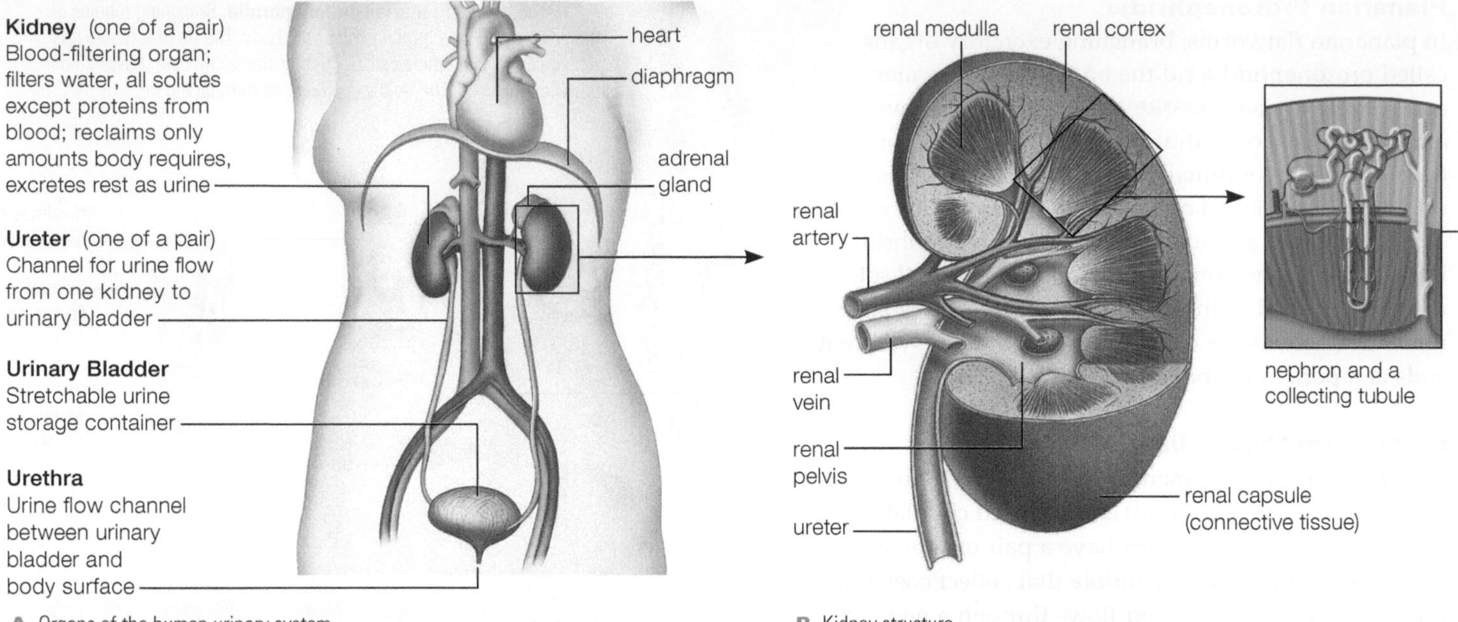
Kidney (one of a pair)
Blood-filtering organ; filters water, all solutes except proteins from blood; reclaims only amounts body requires, excretes rest as urine
Ureter (one of a pair)
Channel for urine flow from one kidney to urinary bladder
Urinary Bladder
Stretchable urine storage container
Urethra
Urine flow channel between urinary bladder and body surface
heart
diaphragm
adrenal gland
renal medulla
renal cortex
renal artery
renal vein
renal pelvis
ureter
renal capsule (connective tissue)
nephron and a collecting tubule

A Organs of the human urinary system.

B Kidney structure.

FIGURE 40.5 Components of the human urinary system and their functions.

40.4 The Human Urinary System

LEARNING OBJECTIVES

- List the structures that urine travels through, from when it leaves the kidney to when it exits the body.
- Describe the structure and function of a nephron.

Organs of the Urinary System

FIGURE 40.5A shows the human urinary system. Like all vertebrates, humans have a pair of kidneys. The bean-shaped, fist-sized human kidneys are located just beneath the peritoneum (the membrane that lines the abdominal cavity) and to the left and right of the backbone. A renal capsule made of dense, irregular connective tissue (Section 31.4) is the outermost layer of the kidney. *Renal* means "relating to the kidneys." Inside the renal capsule, kidney tissue is divided into two zones, the outer renal cortex and the inner renal medulla (FIGURE 40.5B).

A kidney filters blood and produces urine, which contains urea and other waste solutes. The urine formed in a kidney collects in a central cavity called the renal pelvis. From there, it flows through a tubular **ureter** to the bladder.

The **urinary bladder** is a hollow, muscular organ that stores the urine. When the bladder is full, stretch receptors signal motor neurons in the spinal cord. In a reflex action, these neurons cause smooth muscle in the bladder wall to contract. At the same time, sphincters encircling the urethra relax. The **urethra** conveys urine to the body surface for excretion from the body. After age two or three, the brain can override the spinal reflex and prevent urine from flowing through the urethra at inconvenient times.

A male's urethra runs the length of his penis, and it conveys urine and semen at different times. A sphincter cuts off urine flow during erections. In females, the urethra opens onto the body surface near the vagina.

Tubular Structure of the Kidneys

A kidney has more than 1 million **nephrons**, each a microscopically small tubule that consists of simple epithelium. The wall of a nephron is just one cell thick, so substances cross it easily by diffusion or active transport. The nephron is the functional unit of the kidney—each nephron filters blood and produces urine.

A nephron begins in the kidney cortex, where the tubule wall balloons out and folds back on itself to form the cup-shaped structure called **Bowman's capsule** (FIGURE 40.5C). Beyond Bowman's capsule, the tubule twists a bit, then straightens as a **proximal tubule** (the part of the kidney tubule in closest proximity to the start of the nephron). The next portion of the tubule is called the **loop of Henle**. It makes a hairpin turn, descending into the renal medulla, then ascending back into the renal cortex. In the cortex, the tubule continues as the **distal tubule** (the part of the tubule most distant from the start of the nephron).

Distal tubules of several nephrons drain into a **collecting tubule**. Collecting tubules extend through the kidney medulla and open into the renal pelvis.

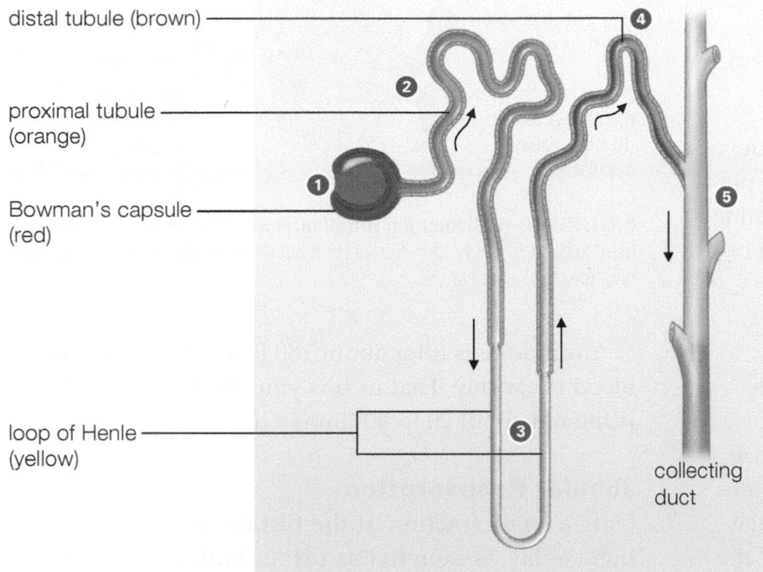
distal tubule (brown)
proximal tubule (orange)
Bowman's capsule (red)
loop of Henle (yellow)
collecting duct
1
2
3
4
5

C Tubular components of the nephron. Arrows indicate direction of fluid flow.

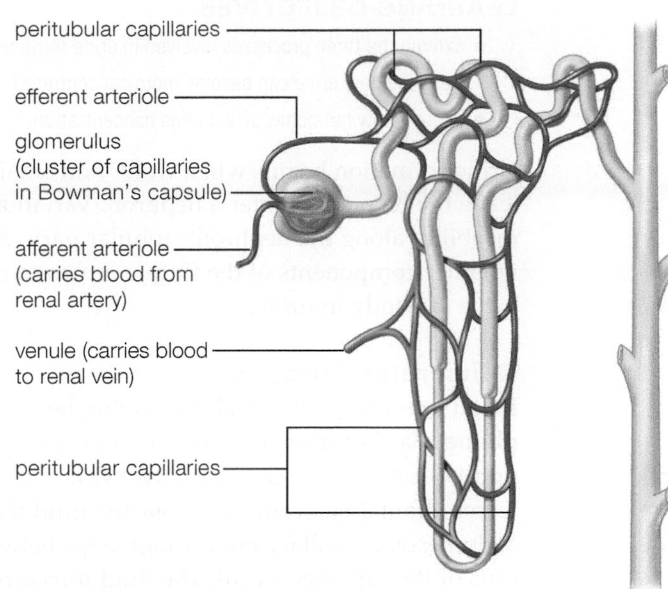
peritubular capillaries
efferent arteriole
glomerulus (cluster of capillaries in Bowman's capsule)
afferent arteriole (carries blood from renal artery)
venule (carries blood to renal vein)
peritubular capillaries

D Blood vessels associated with the nephron.

Blood Vessels of the Kidneys

A renal artery supplies blood to each kidney. Inside a kidney, the renal artery branches into afferent arterioles, each leading to a glomerulus. A **glomerulus** (plural, glomeruli) is a dense cluster of capillaries that sits within the cuplike Bowman's capsule of a nephron (**FIGURE 40.5D**).

Capillaries of a glomerulus are unusually leaky. Gaps between the cells in the wall of glomerular capillaries make these capillaries about 100 times more permeable than a typical capillary. As blood flows through a glomerulus, blood pressure forces some fluid out through the gaps and into the Bowman's capsule. This filtering action gives the glomerulus its name. *Glomerulus* is the Greek word for filter.

The portion of the blood that is not filtered into a Bowman's capsule continues on into an efferent arteriole. This arteriole branches into the **peritubular capillaries** that thread around the nephron (*peri*– means around). Peritubular capillaries are the site for exchanges between fluid flowing through kidney tubules and the blood. From the peritubular capillaries, blood continues into venules, then to the renal vein.

Bowman's capsule In the kidney, the cup-shaped portion of a nephron that encloses the glomerulus and receives filtrate from it.
collecting tubule Kidney tubule that receives filtrate from several nephrons and delivers it to the renal pelvis.
distal tubule Portion of kidney tubule that delivers filtrate to a collecting tubule.
glomerulus (gloh-MER-yuh-lus) Plural, glomeruli. In the kidney, a cluster of capillaries enclosed by Bowman's capsule.
loop of Henle U-shaped portion of a kidney tubule; it extends deep into the renal medulla.
nephron Functional unit of the kidney; filters blood and forms urine.
peritubular capillaries Network of capillaries that surrounds and exchanges substances with a kidney tubule.
proximal tubule Portion of kidney tubule that receives filtrate from Bowman's capsule.
ureter (YOOR-et-er) Tube that carries urine from a kidney to the bladder.
urethra (you-REE-thrah) Tube through which urine expelled from the bladder flows out of the body.
urinary bladder Hollow, muscular organ that stores urine.

TAKE-HOME MESSAGE 40.4

- ✔ Vertebrates have two kidneys that filter the blood and form urine.
- ✔ Urine flows out of the kidney through two ureters and into a hollow, muscular urinary bladder. When the bladder contracts, urine flows out of the body through the urethra.
- ✔ Each kidney contains microscopic tubules called nephrons. A nephron begins at the kidney cortex with a cuplike structure called Bowman's capsule, extends as a proximal tubule, continues as a loop of Henle, and finally a distal tubule. The distal tubule connects to a collecting duct that empties into the renal pelvis.
- ✔ A renal artery delivers blood to each kidney. Inside the kidney, arterioles that branch from this artery supply the clusters of capillaries called glomeruli.
- ✔ A glomerulus acts as a filter. Fluid that filters out of a glomerulus enters the first portion of a kidney tubule (Bowman's capsule). As this fluid continues through the tubule and to a collecting duct, it exchanges substances with blood that is flowing through adjacent peritubular capillaries.
- ✔ Peritubular capillaries return blood to the renal vein.

40.5 How Urine Forms

LEARNING OBJECTIVES

- Explain the three processes involved in urine formation.
- Describe how urine can become more concentrated than blood.
- Explain how hormones affect urine concentration.

Urine formation begins when water and small solutes leave the blood and enter a nephron. Variations in permeability along the nephron's tubular parts determine whether components of the filtrate return to blood or leave the body in urine.

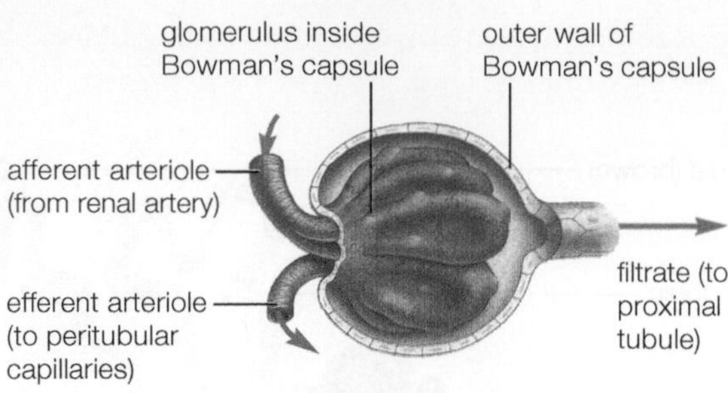
glomerulus inside Bowman's capsule
outer wall of Bowman's capsule
afferent arteriole (from renal artery)
efferent arteriole (to peritubular capillaries)
filtrate (to proximal tubule)

FIGURE 40.6 Glomerular filtration. Pressure exerted on blood by the beating heart forces protein-free plasma out of the glomerular capillaries and into Bowman's capsule.

Glomerular Filtration

Blood pressure generated by a beating heart drives **glomerular filtration**, the first step in urine formation (FIGURE 40.6 and FIGURE 40.7 ❶). The pressure forces about 20 percent of the plasma fluid that enters a glomerular capillary out through gaps between the cells of the capillary's wall. The fluid forced out of the capillary and into the interior of Bowman's capsule is referred to as the filtrate. It contains small molecules and ions. However, blood cells, platelets, and most plasma proteins are too large to be filtered out of the capillary. These components of the blood flow on into the efferent arteriole, along with the 80 percent of plasma fluid that did not leave the glomerulus.

Your kidneys filter about 180 liters (45 gallons) of blood every day. That means your entire blood volume is filtered about 20 to 40 times a day.

Tubular Reabsorption

Only a small fraction of the filtrate, about 1.5 liters a day, is excreted as urine. **Tubular reabsorption** returns the vast majority of the filtered water and solutes to the blood. The cell-free filtrate that enters Bowman's capsule drains into the proximal tubule. Reabsorption begins in the proximal tubule ❷. Active transport proteins move sodium ions (Na^+), chloride ions (Cl^-), potassium ions (K^+), and nutrients such as glucose across the tubule

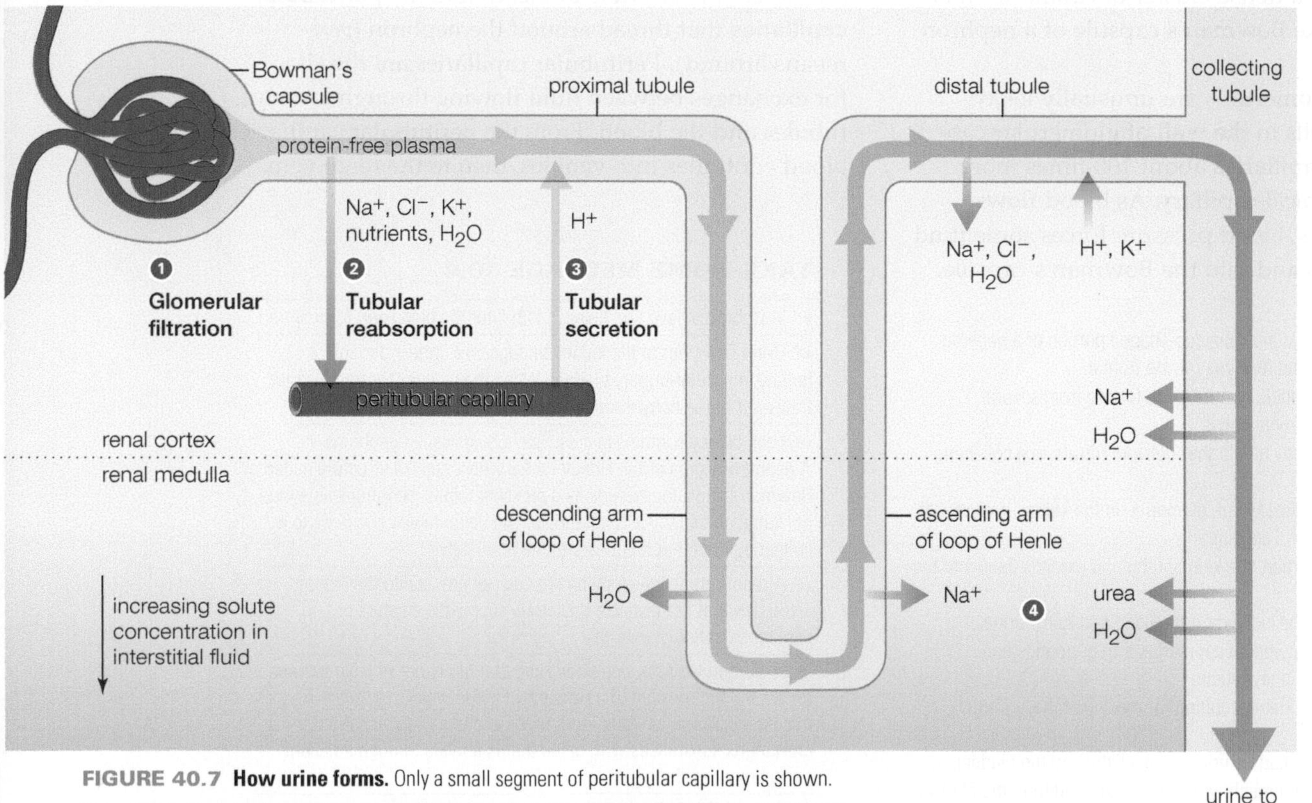
Bowman's capsule
proximal tubule
distal tubule
collecting tubule
protein-free plasma
Na^+, Cl^-, K^+, nutrients, H_2O
H^+
Na^+, Cl^-, H_2O
H^+, K^+
❶ Glomerular filtration
❷ Tubular reabsorption
❸ Tubular secretion
peritubular capillary
Na^+
H_2O
renal cortex
renal medulla
descending arm of loop of Henle
ascending arm of loop of Henle
increasing solute concentration in interstitial fluid
H_2O
Na^+
❹
urea
H_2O
urine to renal pelvis

❶ **Glomerular filtration** Cell-free plasma forced out of glomerular capillaries enters Bowman's capsule.

❷ **Tubular reabsorption** Essential ions, nutrients, water, and some urea in the filtrate return to the blood. Green arrows indicate reabsorption.

❸ **Tubular secretion** Wastes and excess ions are moved from the blood into the filtrate. Blue arrows indicate secretion.

❹ A high concentration of sodium and urea in the interstitial fluid deep in the medulla draws water out of collecting ducts by osmosis, thus concentrating filtrate as urine.

FIGURE 40.7 How urine forms. Only a small segment of peritubular capillary is shown.

FIGURE IT OUT What process moves H^+ from peritubular capillaries into the distal tubule?

Answer: Tubular secretion

wall and into peritubular capillaries. Water follows the solutes by osmosis, so it moves in the same direction.

Most water and nutrients are reabsorbed from the proximal tubule, but reabsorption occurs all along the tubule. About 99 percent of the water that filters into Bowman's capsule returns to blood via tubular reabsorption. All glucose and amino acids that enter Bowman's capsule return to blood the same way, as do most sodium ions, chloride ions, and bicarbonate.

Tubular Secretion

Tubular secretion is the movement of substances from the blood inside peritubular capillaries into the filtrate in a tubule ❸. Membrane proteins in the walls of peritubular capillaries actively transport substances into the interstitial fluid, from which they are similarly transported across the epithelium of the kidney tubule and into the filtrate. Substances that undergo tubular secretion include hydrogen ions (H^+) and breakdown products of foreign organic molecules such as drugs, food additives, and pesticides.

Concentrating the Filtrate

Urine concentration varies. However, even the most dilute urine has more solutes than plasma or typical interstitial fluid. In the kidney's cortex, the concentration of solutes in the interstitial fluid is similar to that of plasma, but as a tubule descends into the medulla, the interstitial fluid around it becomes increasingly hypertonic. This regional difference in tonicity is essential to the kidney's ability to concentrate the filtrate as urine. For water to leave a kidney tubule by osmosis, the interstitial fluid around that region of tubule must be hypertonic relative to the filtrate inside the tubule (the interstitial fluid must have a higher solute concentration than the filtrate).

The osmotic gradient within the kidney medulla arises because cells in different regions of the nephron wall differ in the types of transport proteins embedded in their plasma membranes. Consider the differences between the two arms of the loop of Henle. The descending arm of the loop of Henle is permeable to water, but impermeable to sodium ions. The loop of Henle's ascending arm is largely impermeable to water, but its cells actively transport sodium ions from the filtrate into the interstitial fluid. The two arms of the loop of Henle are in close proximity to one another, so as filtrate flows through the loop, the pumping of sodium into the interstitial fluid by the ascending loop causes water to leave the adjacent descending loop by osmosis. As a result, filtrate that flows through the loop of Henle becomes increasing concentrated as it travels through the descending arm, then increasingly dilute as it flows through the ascending arm.

Filtrate loses water once again as it passes through the collecting tubule. Like the loop of Henle, the collecting tubule extends deep into the medulla. In the deepest part of the medulla, urea pumped out of the collecting tubule and sodium pumped out of the ascending loop of Henle make the interstitial fluid hypertonic relative to filtrate in the collecting tubule. Thus, as filtrate descends through this tubule, the increasing solute concentration of the interstitial fluid around the tubule draws water outward by osmosis ❹.

The body can adjust how much water is reabsorbed by adjusting the permeability of kidney tubules, a process we discuss in Section 40.6.

glomerular filtration (gloh-MER-yuh-lur) First step in urine formation: protein-free plasma forced out of glomerular capillaries by blood pressure enters Bowman's capsule.

tubular reabsorption Substances move from the filtrate inside a kidney tubule into the peritubular capillaries.

tubular secretion Substances move out of peritubular capillaries and into the filtrate in kidney tubules.

TAKE-HOME MESSAGE 40.5

✔ During glomerular filtration, pressure generated by the beating heart drives some plasma out of the glomerular capillaries and into kidney tubules. By the process of tubular reabsorption, most water, some ions, glucose, and other solutes are moved out of the filtrate and return to the blood in peritubular capillaries.

✔ By the process of tubular secretion, active transport proteins move waste products and solutes such as H^+ from peritubular capillaries into the filtrate inside a nephron.

✔ Differential permeability of the two arms of the loop of Henle sets up a concentration gradient in the interstitial fluid that draws water out of the collecting tubule, concentrating the filtrate as urine.

40.6 Regulating Solute Levels

LEARNING OBJECTIVES

- Describe the effects of aldosterone and ADH on the kidney.
- Explain the role of the kidney in acid–base balance.

Extracellular fluids serve as the environment for body cells, so maintenance of their volume and composition is an important aspect of homeostasis. The kidneys are integral to this essential task.

Fluid Volume and Tonicity

When you take in a large amount of sodium, or do not drink enough fluid to offset your body's loss of water, the concentration of the extracellular fluid rises. Osmoreceptors (a type of chemoreceptor) in the brain detect this change and alert a region of the hypothalamus referred to as the thirst center. Exactly how activation of the hypothalamus gives rise to the perception of thirst is not understood.

By contrast, the role of the hypothalamus in the hormonal response to thirst is well documented. A negative feedback loop involving the hypothalamus and pituitary governs secretion of **antidiuretic hormone (ADH)**, a hormone that acts on kidneys to encourage water reabsorption (**FIGURE 40.8**). Recall that axons of some hormone-producing hypothalamic neurons extend into the posterior pituitary (Section 34.4). When osmoreceptors activated by a rise in sodium signal the hypothalamus ❶, an action potential travels along these axons to the pituitary, where it causes the release of ADH ❷. This hormone binds to receptors in the kidney's distal tubules and collecting tubules, making these tubule regions more permeable to water ❸. When ADH is present, more water is reabsorbed and less ends up in urine ❹. Reabsorption of extra water dilutes the extracellular fluid, bringing its sodium concentration back to the optimal level. Osmoreceptors detect this change and stop signaling the hypothalamus, so ADH secretion ceases ❺.

Triggers other than a rise in sodium concentration also stimulate the hypothalamus and result in ADH secretion. In the event of heavy blood loss, for example, receptors in the heart's atria sense a decline in blood pressure and signal the hypothalamus. Stress, heavy exercise, or vomiting also cause internal changes that can, in turn, trigger a rise in ADH output.

ADH increases water reabsorption by stimulating the insertion of proteins called aquaporins into the plasma membrane of cells in the distal and collecting tubules. An aquaporin is a passive transport protein that forms a pore through which water (but not solutes) can flow freely across the membrane. When ADH binds to a tubule cell, vesicles that hold aquaporin subunits move toward the cells' plasma membrane. As these vesicles fuse with the membrane, the subunits assemble themselves into functional aquaporins. Once in place, aquaporins increase the ability of water to leave the filtrate.

A decrease in the volume of the extracellular fluid results in lower blood pressure. This decline in pressure causes cells in the kidney's arterioles to release renin, an enzyme that sets in motion a complex chain of reactions (**FIGURE 40.9**). Renin converts angiotensinogen, a protein secreted by the liver into the blood, into angiotensin I ❶. Another enzyme converts angiotensin I to angiotensin II ❷, which acts in the adrenal glands atop the kidneys. The adrenal cortex responds to angiotensin II by secreting the hormone **aldosterone** into the blood ❸. Aldosterone acts on the collecting tubules, where it encourages sodium reabsorption ❹. Because water follows the sodium by osmosis, aldosterone increases water reabsorption. Retention of additional water raises blood pressure.

Note that both ADH and aldosterone cause urine to become more concentrated, although they do so by different mechanisms.

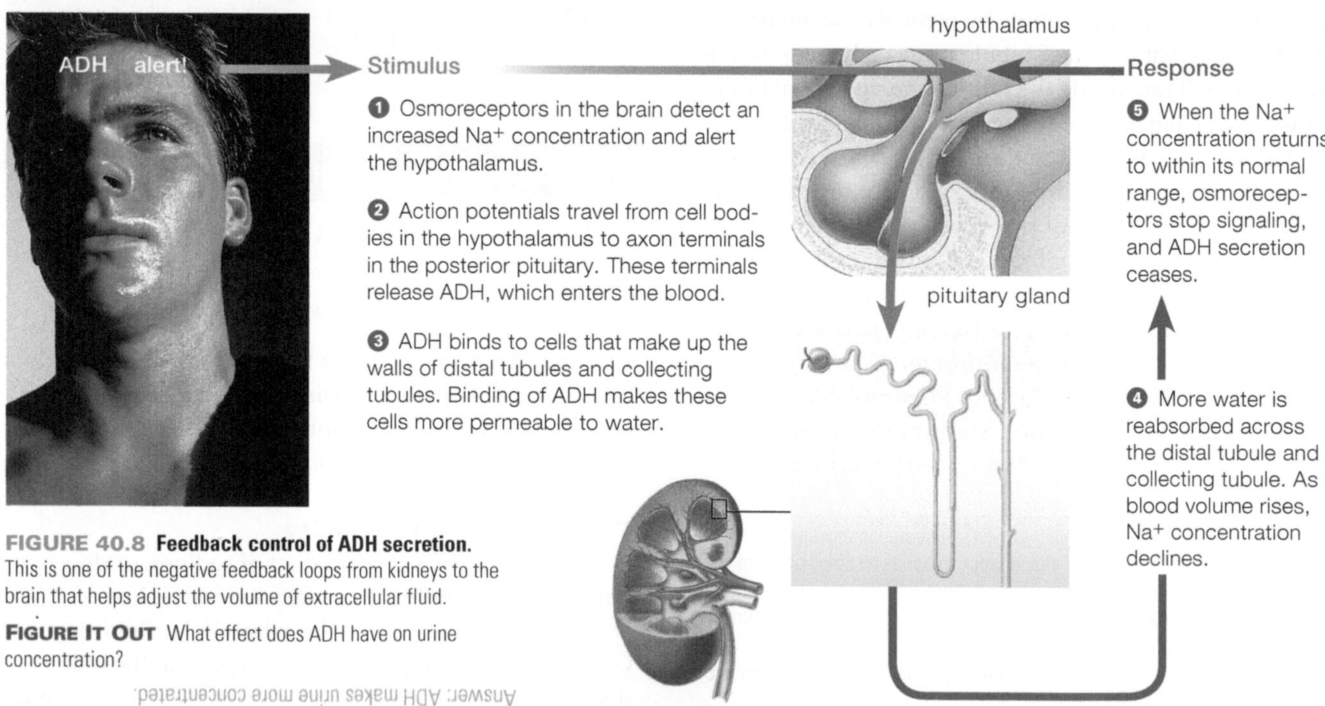
ADH alert!
Stimulus
❶ Osmoreceptors in the brain detect an increased Na^+ concentration and alert the hypothalamus.
❷ Action potentials travel from cell bodies in the hypothalamus to axon terminals in the posterior pituitary. These terminals release ADH, which enters the blood.
❸ ADH binds to cells that make up the walls of distal tubules and collecting tubules. Binding of ADH makes these cells more permeable to water.
hypothalamus
pituitary gland
Response
❺ When the Na^+ concentration returns to within its normal range, osmoreceptors stop signaling, and ADH secretion ceases.
❹ More water is reabsorbed across the distal tubule and collecting tubule. As blood volume rises, Na^+ concentration declines.

FIGURE 40.8 Feedback control of ADH secretion. This is one of the negative feedback loops from kidneys to the brain that helps adjust the volume of extracellular fluid.

FIGURE IT OUT What effect does ADH have on urine concentration?

Answer: ADH makes urine more concentrated.

CREDIT: (8) Evan Cerasoli.

❶ Angiotensinogen made by the liver circulates in the blood. It is converted to angiotensin I by renin, an enzyme released by kidney arterioles when the blood pressure declines.

❷ Another enzyme converts angiotensin I to angiotensin II.

❸ Among its actions, angiotensin II encourages aldosterone secretion by the adrenal cortex.

❹ Aldosterone acts on kidneys to increase Na^+ reabsorption; water follows by osmosis.

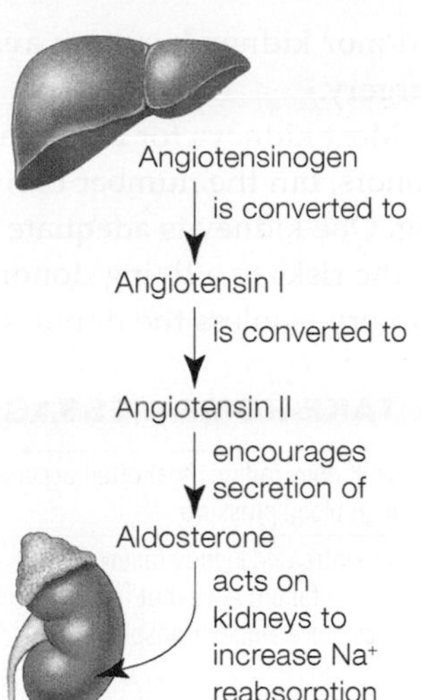
Angiotensinogen
is converted to
Angiotensin I
is converted to
Angiotensin II
encourages secretion of
Aldosterone
acts on kidneys to increase Na+ reabsorption

FIGURE 40.9 The renin–angiotensin–aldosterone system. Angiotensin II also encourages secretion of ADH and causes an increase in thirst, thus elevating blood volume and pressure.

Atrial natriuretic peptide (ANP) is a hormone that makes urine more dilute by inhibiting aldosterone secretion. Muscle cells in the heart's right atrium release ANP when a high blood volume causes the atrial walls to stretch.

Acid–Base Balance

Metabolic reactions such as protein breakdown and lactate fermentation add hydrogen ions (H^+) to the extracellular fluid. Despite these additions, a healthy body can maintain its H^+ concentration within a tight range, a state known as acid–base balance. Buffer systems and adjustments to the activity of respiratory and urinary systems are essential to this balance.

A buffer system involves substances that reversibly bind and release H^+ or OH^- (Section 2.6). In the body, buffer systems minimize pH changes when acidic or basic molecules enter or leave the extracellular fluid.

The pH of human extracellular fluid usually stays between 7.35 and 7.45. In the absence of a buffer, adding acids to this fluid would decrease its pH. In the presence of bicarbonate, some hydrogen ions combine with bicarbonate to form carbonic acid, which dissociates into carbon dioxide (CO^2) and water:

$$\underset{\text{bicarbonate}}{H^+ + HCO_3^-} \rightleftharpoons \underset{\text{carbonic acid}}{H_2CO_3} \rightleftharpoons CO_2 + H_2O$$

Thus bonded, H^+ does not contribute to the pH of the extracellular fluid.

The kidneys contribute to acid–base balance by adjusting the tubular reabsorption of bicarbonate and the tubular secretion of H^+. A decline in pH causes an increase in bicarbonate reabsorption. The bicarbonate moves into peritubular capillaries, where it buffers excess acid. At the same time, tubular secretion of H^+ increases. H^+ leaves the blood and enters the filtrate, where it combines with phosphate or ammonia. The resulting compounds are excreted in the urine.

TAKE-HOME MESSAGE 40.6

✔ Hormones that act on the kidney alter the amount of water reabsorbed. Antidiuretic hormone concentrates the urine by increasing water reabsorption. Aldosterone increases salt reabsorption, and water follows, so it concentrates the urine.

✔ The kidney helps to maintain the pH of the extracellular fluid by adjusting the reabsorption of bicarbonate and secretion of H^+.

40.7 Impaired Kidney Function

LEARNING OBJECTIVES

- Describe where and how kidney stones form.
- Explain the causes of kidney failure and how it is treated.

Kidney stones are a common urinary problem. A kidney stone is a hard deposit that forms when uric acid, calcium, and other wastes settle out of urine and collect in the renal pelvis. Most kidney stones wash away in urine, but sometimes one lodges in a ureter or the urethra and causes severe pain. A stone that slows or blocks urine flow raises the risk of infections and kidney damage.

The vast majority of serious kidney problems arise as complications of diabetes mellitus or high blood pressure. Both disorders damage all capillaries, including those of the nephrons. Some genetic disorders raise the risk of major kidney damage, as do high doses of some drugs and exposure to toxins such as lead.

Kidney function is measured in terms of the rate of filtration through glomerular capillaries. Kidney failure occurs when the filtration rate falls by half. Failure of both kidneys can be fatal because wastes build up in the blood and interstitial fluid. The resulting changes in pH and changes in the concentrations of other ions, most notably Na^+ and K^+, interfere with metabolism.

Kidney dialysis can restore proper solute concentrations in a person who has kidney failure. "Dialysis" refers to exchanges of solutes across a semipermeable membrane between two solutions

aldosterone Adrenal hormone that makes kidney tubules more permeable to sodium; encourages sodium reabsorption, thus increasing water reabsorption and concentrating the urine.

antidiuretic hormone (ADH) Hormone released in the posterior pituitary; makes kidney tubules more permeable to water; encourages water reabsorption, thus concentrating the urine.

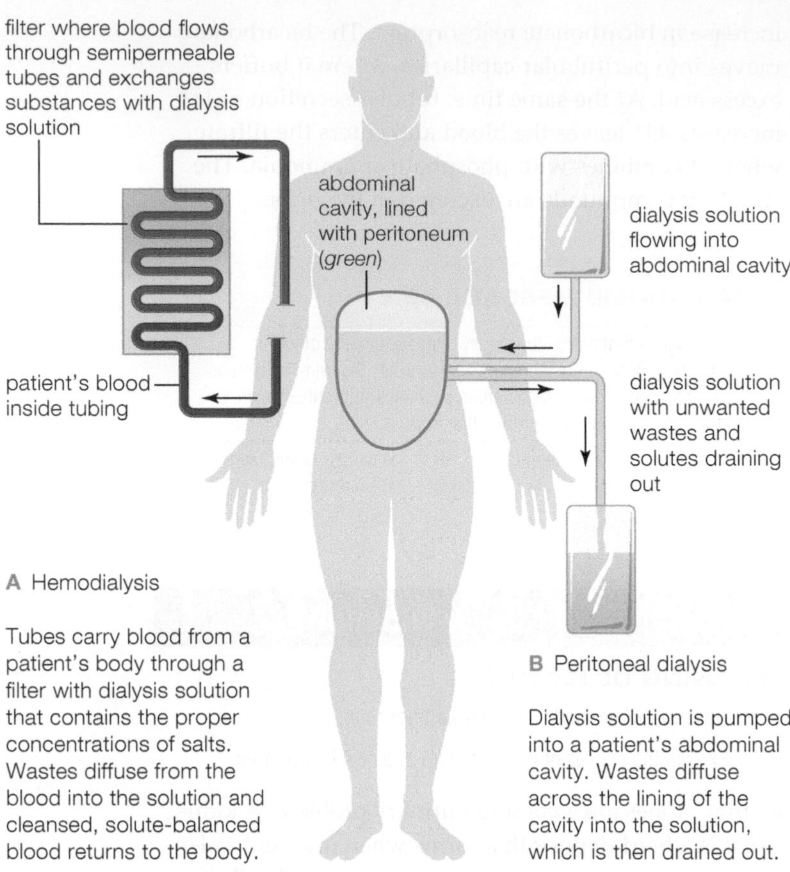
filter where blood flows through semipermeable tubes and exchanges substances with dialysis solution
abdominal cavity, lined with peritoneum (green)
dialysis solution flowing into abdominal cavity
patient's blood inside tubing
dialysis solution with unwanted wastes and solutes draining out
A Hemodialysis
Tubes carry blood from a patient's body through a filter with dialysis solution that contains the proper concentrations of salts. Wastes diffuse from the blood into the solution and cleansed, solute-balanced blood returns to the body.
B Peritoneal dialysis
Dialysis solution is pumped into a patient's abdominal cavity. Wastes diffuse across the lining of the cavity into the solution, which is then drained out.

FIGURE 40.10 Two types of kidney dialysis.

(FIGURE 40.10). Kidney dialysis can keep a person alive through an episode of temporary kidney failure. When kidney damage is permanent, dialysis must be continued for the rest of a person's life, or until a donor kidney becomes available for transplant surgery.

Most kidneys for transplants come from deceased donors, but the number of living donors is increasing. One kidney is adequate to maintain good health, so the risks to a living donor are mainly related to the surgery—unless the donor's remaining kidney fails.

TAKE-HOME MESSAGE 40.7

✔ Kidney failure most often occurs as a complication of diabetes or high blood pressure.

✔ Untreated kidney failure is fatal. Dialysis can keep a person with kidney failure alive, but it must be continued until a person dies or receives a kidney transplant.

40.8 Excretory Adaptations

LEARNING OBJECTIVES

- Compare the challenges that freshwater and marine bony fishes face with regard to maintaining their solute concentration.
- Describe the adaptations that allow desert kangaroo rats to survive without drinking water.

All vertebrates have kidneys, but excretory physiology varies among vertebrate groups. The differences reflect adaptations that suit animals to specific environments.

Fluid Regulation in Bony Fishes

Body fluids of bony fishes are less salty than seawater, but saltier than freshwater (FIGURE 40.12). A marine bony fish continually loses water by osmosis across its

Data Analysis Activities

Pesticide Residues in Urine To carry the USDA's organic label (right), food must be produced without synthetic pesticides that farmers often use on conventionally grown fruits, vegetables, and many grains.

USDA ORGANIC

Chensheng Lu of Emory University used urine testing to see whether eating organic food has a significant effect on the levels of pesticides in children's bodies (FIGURE 40.11). Over the course of 15 days, Lu and his colleagues collected the urine of 23 children (aged 3 to 11) and tested it for breakdown products of two synthetic pesticides. The children ate their normal diet of conventionally grown foods for three days, switched to organic versions of the same foods and drinks for five days, then returned to their conventional diet for a week.

1. During which phase of the experiment did the children's urine contain the lowest level of the malathion metabolite?

Study Phase	No. of Samples	Malathion Metabolite Mean (μg/liter)	Malathion Metabolite Maximum (μg/liter)	Chlorpyrifos Metabolite Mean (μg/liter)	Chlorpyrifos Metabolite Maximum (μg/liter)
1. Conventional	87	2.9	96.5	7.2	31.1
2. Organic	116	0.3	7.4	1.7	17.1
3. Conventional	156	4.4	263.1	5.8	25.3

FIGURE 40.11 Concentrations of metabolites of two pesticides (malathion and chlorpyrifos) in children's urine during the three different phases of the study. The difference in the mean level of metabolites between the organic and conventional phases of the study was statistically significant.

2. During which phase of the experiment was the maximum level of the chlorpyrifos metabolite detected?
3. Did switching to an organic diet lower the amount of pesticide residues excreted by the children?

body surfaces. It replaces this lost water by gulping seawater, then pumping salt out across its gills. Like other vertebrates, bony fishes have a pair of kidneys that produce urine. A freshwater bony fish produces a large volume of dilute urine because water continually enters its body by osmosis, while a marine bony fish produces a smaller volume of concentrated urine as water continually exits its body by osmosis. Solutes lost in urine are offset by solutes absorbed from the gut, and actively transported in across the gills.

Kangaroo Rats and Water Scarcity

Some vertebrates have adaptations that allow them to live where freshwater is in short supply. Consider the desert kangaroo rat, a small rodent that lives in deserts of southwestern North America (**FIGURE 40.13**). These rats obtain all the water they need from their diet of seeds. All the moisture in the seeds is absorbed by the digestive tract, so the feces are dry. Food also serves as a source of metabolic water (the water released by aerobic respiration).

A desert kangaroo rat has unusually efficient kidneys that reabsorb nearly all water from the filtrate, thus producing urine with a syruplike consistency. Kangaroo rat nephrons have a loop of Henle that is proportionately much longer than that of other mammals. The longer this loop, the greater the concentration gradient in the kidney and the more water can be reabsorbed.

TAKE-HOME MESSAGE 40.8

✔ Bony fishes can gain or lose water by osmosis across their skin. Pumping solutes in or out across the gills helps them maintain the solute concentration of their body fluids.

✔ Desert kangaroo rats are adapted to their desert environment by highly efficient kidneys and other mechanisms of water conservation.

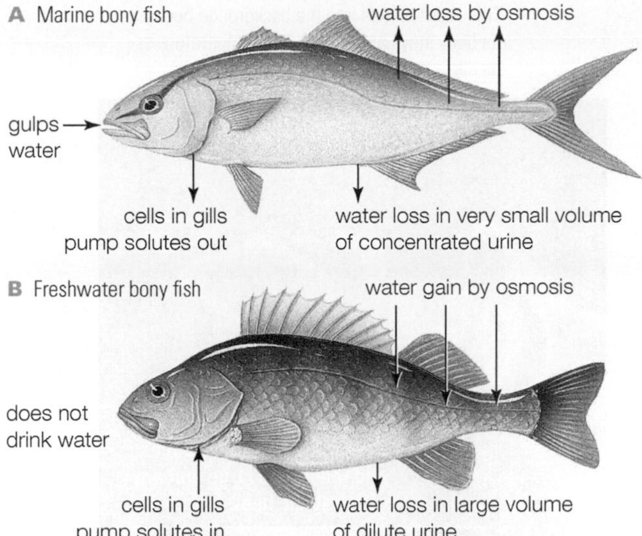
A Marine bony fish
water loss by osmosis
gulps water
cells in gills pump solutes out
water loss in very small volume of concentrated urine
B Freshwater bony fish
water gain by osmosis
does not drink water
cells in gills pump solutes in
water loss in large volume of dilute urine

FIGURE 40.12 Fluid–solute balance in bony fishes.

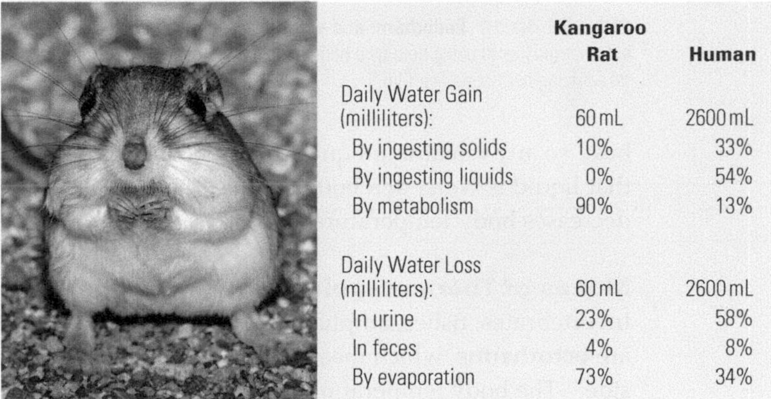

	Kangaroo Rat	Human
Daily Water Gain (milliliters):	60 mL	2600 mL
By ingesting solids	10%	33%
By ingesting liquids	0%	54%
By metabolism	90%	13%
Daily Water Loss (milliliters):	60 mL	2600 mL
In urine	23%	58%
In feces	4%	8%
By evaporation	73%	34%

FIGURE 40.13 Water gains and losses for a desert kangaroo rat and a human. In both species, fluid gains must balance fluid losses.

40.9 Heat Gains and Losses

LEARNING OBJECTIVES

- List the ways in which a body gains and loses heat.
- Compare the thermoregulatory mechanism of endotherms and ectotherms, and give examples of each.

We turn now to another aspect of homeostasis, maintaining the body's core temperature. Enzymes that carry out the metabolic reactions vital to life function best within a limited range of temperature (Section 5.3). Thus, thermoregulation, the maintenance of the body's core temperature within a limited range, is just as essential as maintaining solute concentrations.

How the Core Temperature Can Change

Metabolic reactions release heat, and this heat affects an animal's body temperature. Animals also gain heat from, and lose heat to, their surroundings. An animal's internal temperature is stable only when the metabolic heat it produces and the heat it gains from the environment balance heat losses to the environment.

Four processes affect heat gain and loss. **Thermal radiation** is emission of heat from a warm object into the space around it. At rest, a human adult radiates as much heat as a 100-watt incandescent lightbulb. **Conduction** is the transfer of heat within an object or among objects that contact one another. An animal loses heat when it contacts a cooler object, and gains heat when it contacts a warmer one. In **convection**, heated air or water moves away from a hot object, thus carrying heat away from the source. In **evaporation**, a liquid becomes a gas. Energy that powers this process typically comes from the liquid itself in the form of

conduction Of heat: the transfer of heat within an object or between two objects in contact with one another.
convection Transfer of heat by moving molecules of air or water.
evaporation Transition of a liquid to a gas.
thermal radiation Emission of heat from an object.

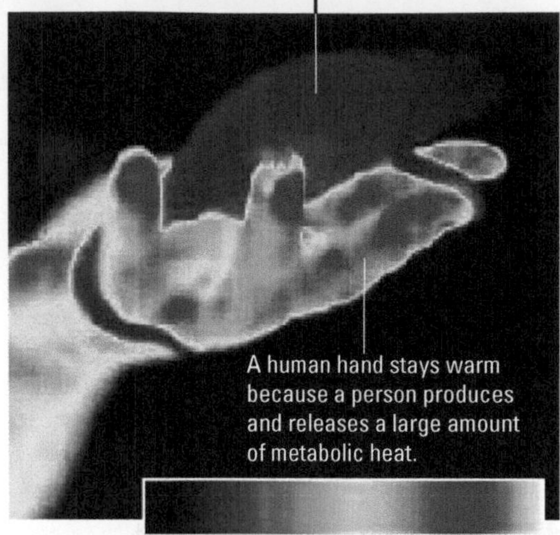
This lizard blends into the background because its body stays about the same temperature as its surroundings.
A human hand stays warm because a person produces and releases a large amount of metabolic heat.
cooler -------------------------> warmer

FIGURE 40.14 Endotherm and ectotherm. Photo of a chameleon (an ectothermic lizard) being held by a human (an endotherm). The image was taken using heat-sensitive film.

heat, so any remaining liquid cools (Section 2.5). When that liquid is water at a body surface, the cooling decreases body temperature.

Modes of Thermoregulation

Invertebrates, fishes, amphibians, and nonbird reptiles are **ectotherms**, which means "heated from the outside." The body temperature of ectotherms fluctuates with the temperature of the external environment. That is why the chameleon in **FIGURE 40.14** is about the same temperature as its surroundings. Ectotherms typically have a low metabolic rate, and they lack insulating fur, hair, or feathers. They regulate internal temperature by altering their position, rather than their metabolism. A lizard, for example, warms itself by basking on a rock in a sunny spot or cools itself by retreating into a shady burrow.

Most birds and mammals are **endotherms**, which means "heated from within." Compared to ectotherms, endotherms maintain their body temperature within a narrower range and have a higher metabolic rate. For example, a mouse uses 30 times more energy than a lizard of the same body weight. Metabolic heat production allows endotherms to be active in a wider range of temperatures than ectotherms. Fur, hair, or feathers insulate endotherms by minimizing heat transfers.

Some birds and mammals are **heterotherms**, animals that keep their core temperature constant some of the time, but allow it to fluctuate at other times. For example, hummingbirds have a very high metabolic rate when foraging for nectar during the day. At night, the birds' metabolic activity decreases so much that their body may become almost as cool as the surroundings.

TAKE-HOME MESSAGE 40.9

- ✔ Animals can gain heat from the environment, or lose heat to it. They can also generate heat by metabolic reactions.
- ✔ Fishes, amphibians, and reptiles are ectotherms; their body temperature varies with that of their environment.
- ✔ Most birds and mammals are endotherms that maintain their body temperature largely by adjusting their production of metabolic heat.

40.10 Responses to Cold and Heat

LEARNING OBJECTIVES

- Describe the mechanisms that allow animals to maintain their core body temperature when the temperature of their environment fluctuates.

Responses to Cold

Most animals that remain active when it is very cold are endothermic vertebrates. When outside temperature declines, thermoreceptors in their skin send action potentials to the hypothalamus, the brain region responsible for thermoregulation. A signal from the hypothalamus causes vessels that deliver blood to the skin to contract, so less metabolic heat reaches the body surface (**TABLE 40.1**). At the same time, muscle contractions make hair, fur, or feathers "stand up." This response creates a layer of still air next to skin, reducing the amount of heat lost by convection.

Skeletal muscles are a vertebrate's main source of metabolic heat. With prolonged exposure to cold, the hypothalamus commands these muscles to contract 10 to 20 times each second. This **shivering response** increases heat production, but it has a high energy cost.

brown adipose tissue Of some mammals, a specialized adipose tissue with mitochondria that produce heat, rather than ATP.
ectotherm Animal whose body temperature varies with that of its environment; it adjusts its internal temperature by altering its behavior.
endotherm Animal that maintains its internal temperature mainly by adjusting its metabolism; for example, a bird or mammal.
estivation An animal becomes dormant during a hot, dry season.
heterotherm Animal that sometimes maintains its temperature by producing metabolic heat, and at other times allows its temperature to fluctuate with the environment.
hibernation An animal becomes dormant during a cold season.
nonshivering heat production Increase in metabolic heat production that results when brown adipose tissue releases energy as heat, rather than storing it in ATP.
shivering response Rhythmic muscle contractions that generate metabolic heat in response to prolonged exposure to cold.

Brown adipose tissue helps heat the body of many mammals, including humans. This specialized adipose tissue has mitochondria that use the energy released by fatty acid oxidation to produce heat, rather than ATP. The resulting increase in metabolic heat output is called **nonshivering heat production**. Exposure to cold increases secretion of thyroid hormone (Section 34.6), which encourages nonshivering heat production. People who make too little thyroid hormone cannot increase their nonshivering heat production, so they often feel cold.

Some animals survive a seasonal cold period through **hibernation**, a state of prolonged inactivity during which the animal's metabolic rate and body temperature decline dramatically. Depending on the species, the animal may reawaken to feed periodically on stored food or subsist entirely on fat and glycogen accumulated during the prior season. Mammalian hibernators include some bats, bears, and rodents. Many reptiles and amphibians also hibernate during winter, as do some insects.

Responses to Heat

Some desert reptiles and amphibians survive a hot, dry season through **estivation**, a state of prolonged inactivity and reduced metabolism similar to hibernation.

Among vertebrates that remain active in the heat, the hypothalamus triggers responses that decrease core temperature (TABLE 40.1). The animal becomes less active, so its skeletal muscles produce less heat. Blood delivery to the skin increases, so more metabolic heat escapes into the surroundings. Many vertebrates also reduce their temperature by panting, which increases evaporative cooling from their respiratory tract.

Sweating is the main mechanism of cooling in large hoofed mammals and in primates (including humans). Sweat glands are exocrine glands that release water and solutes through a pore at the skin's surface. Evaporation of the water in sweat cools the skin.

TABLE 40.1

Vertebrate Responses to Heat and Cold

Stimulus	Main Responses	Outcome
Heat stress	Widening of blood vessels in skin; behavioral adjustments; in some species, sweating, panting	Dissipation of heat from body
	Decreased muscle action	Heat production decreases
Cold stress	Narrowing of blood vessels in skin; behavioral adjustments (e.g., minimizing surface parts exposed)	Conservation of body heat
	Increased muscle action; shivering; nonshivering heat production	Heat production increases

Humans have a denser array of sweat glands than other primates. For example, we have about ten times the number of sweat glands of a chimpanzee. Our impressive sweating capacity is thought to have arisen when our early ancestors spent much of their time walking upright across the hot, sunny savannas of equatorial Africa. Under these conditions, an increased capacity to cool the body by sweating is selectively advantageous. Our sparse body hair relative to other primates is also a heat-related adaptation. A thick coat of hair slows evaporative cooling and insulates the body, thus reducing the heat lost by convection.

TAKE-HOME MESSAGE 40.10

✔ Some endotherms can remain active when it is cold. Feathers or fur slow their heat loss. Shivering and (in mammals) nonshivering heat production warm the body.

✔ Panting and sweating are vertebrate mechanisms of increasing heat loss through evaporative cooling.

✔ Some animals become dormant during a cold season (hibernators) or a hot, dry season (estivators).

40.1 Urine Testing (revisited)

Urine specific gravity is a measure of the total number of solutes in the urine. It rises if water lost in urine, exhalations, and sweat is not replenished by drinking.

As urine specific gravity increases, so does the risk of a heat-related disorder. In healthy people, such disorders usually occur when overexertion is coupled with dehydration. As a result, body temperature rises. A rise in core body temperature above 104°F (40°C) results in heat stroke, a potentially fatal condition. Symptoms of heat stroke include nausea, headache, a racing heart, confusion, and a decrease in the capacity to sweat despite ongoing heat.

To stay safe outside on a hot day, drink plenty of water and avoid excessive exercise. Avoid direct sunlight, wear a hat and light-colored clothing, and use sunscreen. Sunburn impairs the skin's ability to transfer heat to the air. Also keep in mind that the sweat glands of a person who is not normally exposed to heat produce less sweat than those of a person who has routine heat exposure.

STUDY GUIDE

Section 40.1 The composition of urine provides information about health and about chemicals taken into the body. Hormones and drugs that enter the blood are filtered out by kidneys and excreted in the urine.

Section 40.2 Plasma and interstitial fluid are the main components of extracellular fluid. Maintaining the volume and composition of this fluid is an essential aspect of homeostasis. Animals must balance solute and fluid gains with solute and fluid losses. Those living in water lose or gain water by osmosis. On land, the main challenge is avoiding dehydration. All animals must eliminate metabolic wastes such as the **ammonia** produced by the breakdown of proteins and nucleic acids. Some animals excrete ammonia, but doing so involves loss of a great deal of water. Most animals that excrete ammonia are aquatic. Converting ammonia to **urea** requires a bit of extra energy and saves some water. Placental mammals such as humans excrete urea. Converting ammonia to **uric acid** saves even more water, but requires the most energy. Insects, birds, lizards, and snakes excrete uric acid.

Section 40.3 Most animals have excretory organs with a tubular structure. Body fluids enter one end of the tube. As this fluid moves through the tube, water and essential solutes leave the tube and return to the body fluid. Wastes continue to the end of the tube and are eliminated from the body.

Planarian flatworms have a system of branching tubules called **protonephridia**. Movement of cilia draws body fluid into the tubule and wastes leave the body through a pore in the body wall. Earthworms have a pair of **nephridia** in each body segment. A nephridium takes up coelomic fluid, adjusts its content, and releases waste through a pore in the body wall. Insects have **Malpighian tubules** that take up hemolymph and deliver waste solute to the digestive tract, from which they are excreted.

Vertebrates have a pair of **kidneys** that filter the blood and produce **urine**.

Section 40.4 A human urinary system consists of two kidneys, a pair of **ureters**, a **urinary bladder**, and the **urethra**. Kidney **nephrons** are the tiny tubules that produce urine. Each nephron starts at **Bowman's capsule** in a kidney's outer region, or renal cortex. Bowman's capsule and capillaries of the **glomerulus** that it cups around serve as a blood-filtering unit. Fluid filtered out of the glomerulus drains from Bowman's capsule into the nephron's **proximal tubule**. It next travels through the **loop of Henle**, the portion of the nephron that descends into and ascends from the renal medulla. The final portion of the nephron, the **distal tubule**, delivers filtrate into a **collecting tubule**. From here it enters the renal pelvis.

Most filtrate that enters Bowman's capsule is reabsorbed into the **peritubular capillaries** that are adjacent to the kidney tubule. The fluid not returned to blood is excreted as urine.

Section 40.5 Urine formation begins when **glomerular filtration** delivers protein-free plasma into a kidney tubule. Most water and solutes are returned to the blood by **tubular reabsorption**. Substances that are not reabsorbed, and substances added to the filtrate by **tubular secretion**, end up in the urine. The amount of water reabsorbed across the distal tubule and collecting tubule can be altered and affects the concentration of the urine.

Section 40.6 Hormones regulate the urine's concentration and composition. A region of the hypothalamus regulates thirst. In response to an increase in the solute concentration, the hypothalamus signals the pituitary gland to release **antidiuretic hormone (ADH)**, a hormone that increases the reabsorption of water through passive transport proteins called aquaporins. **Aldosterone**, a hormone secreted by the adrenal cortex, increases sodium reabsorption. Water follows the sodium by osmosis. Thus, both antidiuretic hormone and aldosterone concentrate the urine.

The urinary system helps regulate acid–base balance by eliminating H^+ in urine and reabsorbing bicarbonate.

Section 40.7 Kidney stones are hard deposits of material that form in the renal pelvis. Diabetes and hypertension are the most common causes of kidney failure. When both kidneys fail, dialysis or a kidney transplant is required to sustain life.

Section 40.8 Freshwater fishes take in water by osmosis, so they produce a large volume of urine. By contrast, marine fishes lose water by osmosis and produce little urine. Fish can take in or excrete solutes across their gills.

Kangaroo rats are mammals that do not need to drink water. Their kidney has a very long loop of Henle that allows them to produce highly concentrated urine.

Section 40.9 Animals produce metabolic heat. They also gain or lose heat by **thermal radiation**, **conduction**, and **convection** and lose it by **evaporation**. **Ectotherms** such as reptiles control core temperature mainly by behavior; their temperature varies with that of the external environment. **Endotherms** (most mammals and birds) regulate temperature by controlling production and loss of metabolic heat. **Heterotherms** control core temperature only part of the time.

Section 40.10 The hypothalamus is the main center for vertebrate temperature control. Increased delivery of blood to the skin, sweating, and panting are responses to heat. Exposure to cold makes blood vessels in the skin constrict, causes hair to stand upright, and elicits a **shivering response**. In mammals, **nonshivering heat production** by **brown adipose tissue** also provides heat. Some animals enter **hibernation** to survive a cold season or **estivation** to survive a hot, dry season.

SELF-QUIZ

Answers in Appendix VII

1. An insect's _______ deliver nitrogen-rich waste to its digestive tract.
   a. nephridia
   b. nephrons
   c. Malpighian tubules
   d. contractile vacuoles

2. Bowman's capsule, the start of the tubular part of a nephron, is located in the _______ .
   a. renal cortex
   b. renal medulla
   c. renal pelvis
   d. renal artery

3. Fluid that enters Bowman's capsule flows directly into the _______ .
   a. renal artery
   b. proximal tubule
   c. distal tubule
   d. loop of Henle

4. Blood pressure forces water and small solutes into Bowman's capsule during _______ .
   a. glomerular filtration
   b. tubular reabsorption
   c. tubular secretion
   d. both a and c

5. Kidneys return most of the water and small solutes back to blood by way of _______ .
   a. glomerular filtration
   b. tubular reabsorption
   c. tubular secretion
   d. both a and b

6. ADH binds to receptors on distal tubules and collecting tubules, making them _______ permeable to _______ .
   a. more; water
   b. less; water
   c. more; sodium
   d. less; sodium

7. Increased sodium reabsorption _______ .
   a. will make urine more concentrated
   b. will make urine more dilute
   c. is stimulated by aldosterone
   d. both a and c

8. Match each structure with a function.

___	ureter	a. start of nephron
___	Bowman's capsule	b. delivers urine to body surface
___	urethra	c. carries urine from kidney to bladder
___	collecting tubule	d. secretes ADH
___	glomerulus	e. target of aldosterone
___	pituitary gland	f. cluster of leaky capillaries

9. Secretion of H^+ into kidney tubules _______ .
   a. makes extracellular fluid less acidic
   b. makes urine less acidic
   c. can cause acidosis
   d. is regulated by aldosterone

10. Body fluids of a marine bony fish have a solute concentration that is _______ its surroundings.
    a. higher than
    b. equal to
    c. lower than

11. Which has a higher metabolic rate?
    a. an endotherm
    b. an ectotherm

12. Compared to other primates, humans have a(n) _______ capacity for sweating and _______ body hair.
    a. increased/increased
    b. increased/decreased
    c. decreased/increased
    d. decreased/decreased

13. The main control center for maintaining mammalian body temperature is in the _______ .
    a. anterior pituitary
    b. renal cortex
    c. adrenal gland
    d. hypothalamus

14. How does the human body respond to an increase in core body temperature?
    a. Decreased blood flow to the skin
    b. Increased sweating
    c. Increased activity of brown adipose tissue
    d. Secretion of aldosterone

15. Which organelle in brown adipose tissue gives this tissue its heightened capacity to produce heat?
    a. mitochondria
    b. endoplasmic reticulum
    c. Golgi bodies
    d. ribosomes

CRITICAL THINKING

1. Like desert rodents such as the kangaroo rat, marine mammals such as dolphins have highly efficient kidneys that produce only a tiny amount of very concentrated urine. What common selective pressure shaped this trait in these seemingly dissimilar animals?

2. Compared to closely related species that live in warmer areas, cold dwellers tend to have larger body size and smaller appendages. Think about heat transfers between animals and the habitat, then explain why smaller appendages and larger body size are advantageous in very cold climates.

3. When marathoners or other endurance athletes sweat heavily and drink a large amount of pure water, their sodium level drops. The resulting condition, called "water intoxication," can be fatal. Why is maintaining the sodium level so important?

4. Drinking alcohol inhibits ADH secretion. What effect will drinking a beer have on the permeability of kidney tubules to sodium? To water?

41 ANIMAL REPRODUCTION

CORE CONCEPTS

Evolution

Evolution underlies the unity and diversity of life.

All animals reproduce. The optimal reproductive method varies with the environment. Some animals can reproduce asexually, but most reproduce sexually. In most species each individual is either male or female, although in some an individual can produce both eggs and sperm. Fertilization may be internal or external, and young may develop in the environment or in a parent's body.

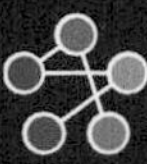

Systems

Complex properties arise from interactions among components of a biological system.

Sexual reproduction is the process by which a new individual forms through the union of two gametes. Signals from the endocrine system regulate formation of gametes, and nervous signals allow two individuals to interact in such a way that their gametes can unite. At a cellular level, interactions between sperm and the egg precede fertilization.

Process of Science

The field of biology consists of and relies upon experimentation and the collection and analysis of scientific evidence.

Scientific investigations into mechanisms of reproduction opened the way to applied technologies. Such studies allowed the development of birth control methods and treatments that can assist infertile couples. Similarly, studies of sexually transmitted diseases have uncovered the organisms responsible for these disorders and assisted in the development of treatments.

Links to Earlier Concepts

This chapter expands on mechanisms of reproduction (Sections 11.2,12.1) and gamete formation (12.2). It touches on human sex determination (10.3) and the function of the hypothalamus and pituitary (34.4), sex hormones (34.9), and autonomic nerves (32.7). We conclude with a discussion of HPV (11.6), HIV (20.3), and other sexually transmitted pathogens.

41.1 Assisted Reproduction

***In vitro* fertilization** (IVF) is an assisted reproduction method in which an egg and sperm are combined outside the body. *In vitro* means "in glass," and refers to the laboratory glassware in which fertilization takes place. Prior to an IVF procedure, a woman is given hormones to encourage maturation of multiple eggs and to prevent her from releasing them naturally. Then, mature eggs are removed from her body using a hollow needle. The eggs are combined with a partner's or donor's sperm to allow fertilization. After fertilization, each zygote undergoes mitotic divisions, forming a ball of cells (a blastocyst) that can be placed in a woman's uterus (womb) to develop to term.

The first child conceived by IVF was born in 1978. At that time, much of the public and many scientists were appalled by the idea of what newspapers referred to as "test tube babies." Scientists worried that this new, unnatural procedure would produce children with psychological and genetic defects. Ethicists warned about the societal implications of manipulating human embryos.

Despite these initial reservations, IVF is now in wide use worldwide. More than 3 million people have been born as a result of this procedure. The first test tube babies have become adults and are now having children of their own.

Research into IVF opened the way to other reproduction technologies. If a man makes sperm, but does not ejaculate properly or does not ejaculate enough sperm to allow fertilization by normal means, intracytoplasmic sperm injection can put his sperm inside his partner's egg. If a woman cannot produce viable eggs, she can obtain donated eggs and use IVF to produce an embryo that she can carry to term. A woman who has normal eggs but cannot sustain a pregnancy herself can have an embryo conceived by IVF implanted in a surrogate mother.

Another recent development is the ability to "bank" eggs, that is, to freeze them and store them for later use. Human sperm have been stored this way for decades, but the Food and Drug Administration did not approve an egg-freezing procedure until 2012. The ability to store eggs after their extraction from a donor has already cut the cost of using donated eggs for IVF. It also has given women who face the loss of fertility as a result of a medical condition or aging a way to retain the possibility of reproducing later in life.

***in vitro* fertilization** Assisted reproductive technology in which eggs and sperm are united outside the body.

CREDIT: (opposite) iStockphoto.com/Brett_Hondow.

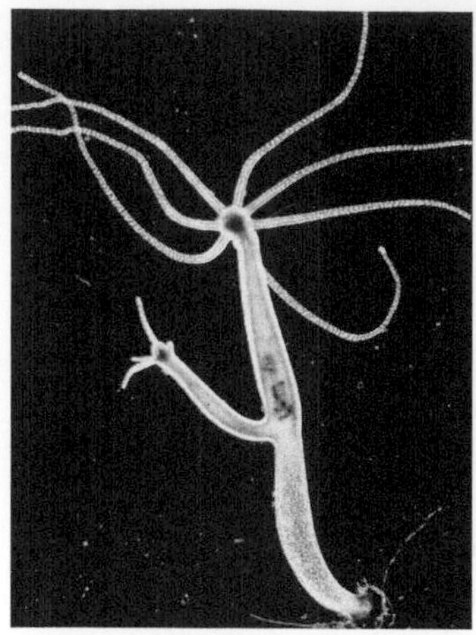

A Asexual reproduction in a hydra. A new individual (left) is budding from its parent.

B Sexual reproduction in banana slugs. Each slug donates sperm to and receives sperm from its partner.

C Sexual reproduction in elephants. The male is inserting his penis into the female. Eggs will be fertilized, and the offspring will develop inside the mother's body, nourished by nutrients delivered by her bloodstream.

FIGURE 41.1 **Examples of animal reproductive mechanisms.**

41.2 Modes of Animal Reproduction

LEARNING OBJECTIVES

- Using appropriate examples, describe some of the ways that animals reproduce asexually.
- Describe the different types of hermaphrodites.
- Compare internal and external fertilization.
- Using appropriate examples, describe the ways in which developing animal offspring are nourished.

Asexual Versus Sexual Reproduction

With **asexual reproduction**, a single individual produces offspring (Section 11.2). Mutations aside, all offspring are genetic replicas (clones) of the parent and thus identical to one another. Asexual reproduction is advantageous in a stable environment where the gene combination that makes a parent successful is likely to do the same for its offspring.

Invertebrates reproduce by a variety of asexual mechanisms. In some species, a new individual grows on the body of its parent, a process called budding. For example, new hydras bud from existing ones (FIGURE 41.1A). Many corals reproduce by fragmentation, in which a piece of the parent breaks off and develops into a new animal. Some flatworms reproduce by transverse fission: The worm divides in two, leaving one piece headless and the other tailless. Each piece then grows the missing body parts.

Parthenogenesis is a mechanism of asexual reproduction in which female offspring develop from unfertilized eggs. Some invertebrates, fishes, amphibians, lizards, and one bird (the turkey) can produce offspring by parthenogenesis. No mammal is known to reproduce asexually by natural means.

Most animals can reproduce sexually. With **sexual reproduction**, two parents produce haploid gametes that combine at fertilization. An **egg** is a female gamete, and a **sperm** is a male gamete. As a result of crossing over and random assortment of chromosomes during meiosis, each gamete receives a different combination of maternal or paternal alleles (Section 12.4). When egg and sperm combine, the result is a genetically unique individual with alleles from both parents.

Producing offspring that differ from both parents and from one another can be advantageous in a changing environment (Section 12.1). By reproducing sexually, a parent increases the likelihood that some of its offspring will have a combination of alleles that suits them to conditions in a changed environment.

Some animals reproduce asexually when conditions are stable and favorable, but switch to sexual reproduction when conditions begin to change. Consider aphids, a type of plant-sucking insect. In early spring, when tender plant parts are plentiful and aphids are not, a female aphid can give birth to several smaller

asexual reproduction Reproductive mode by which offspring arise from a single parent only.
egg Female gamete.
external fertilization Sperm and eggs are released into the external environment and meet there.
hermaphrodite (herm-AFF-roh-dyte) Animal that produces both eggs and sperm, either simultaneously or at different times in its life.
internal fertilization Fertilization of eggs inside a female's body.
placenta (pluh-SEN-tah) Organ that favilitates exchange between the bloodstreams of a developing embryo and its mother.
sexual reproduction Reproductive mode by which offspring arise from two parents and inherit genes from both.
sperm Male gamete.

CREDITS: (1A) Biophoto Associates/Science Source; (1B) Courtesy of Christine Evers; (1C) IndustryAndTravel/Shutterstock.

A Beetle depositing fertilized eggs on a leaf. Yolk in the eggs provides the nutrients that sustain development of the young.

B Snake (adder) giving birth to young that recently hatched in her body. The young were nourished by egg yolk, not by their mother.

C An elk examines her newborn calf. The placenta, the organ that allowed nutrients from her blood to diffuse into the calf's blood, is visible at the left. It is expelled at birth.

FIGURE 41.2 Mechanisms of nourishing the developing young.

clones of herself every day by parthenogenesis. In autumn, when plants prepare for dormancy, food for aphids becomes scarce and competition among them increases. At this time, the aphids switch reproductive modes to produce offspring by sexual means.

Variations on Sexual Reproduction

Sexually reproducing individuals that can make both eggs and sperm are called **hermaphrodites**. Tapeworms and some roundworms are simultaneous hermaphrodites. These worms produce eggs and sperm at the same time, and they can fertilize themselves. Earthworms, land snails, and slugs are simultaneous hermaphrodites too, but they require a partner (FIGURE 41.1B). Some fishes are sequential hermaphrodites, meaning individuals switch from one sex to another over the course of a lifetime. However, the overwhelming majority of vertebrate species have separate sexes that are fixed for life; each individual remains either male or female.

Most aquatic invertebrates, bony fishes, and amphibians have **external fertilization**; they release their gametes into water, and fertilization occurs there. With **internal fertilization**, sperm fertilize an egg inside the female's body (FIGURE 41.1C). Internal fertilization occurs in all cartilaginous fishes and in most land animals, including insects and the amniotes (reptiles, birds, and mammals).

After internal fertilization, a female may lay eggs or retain them inside her body while they develop. In all birds and most insects, most development occurs outside the mother's body (FIGURE 41.2A). In many sharks, snakes, and lizards, embryos develop while enclosed by an egg sac in the mother's body. The eggs hatch inside the mother shortly before she gives birth (FIGURE 41.2B).

A developing animal requires nutrients. Embryos of most animals are nourished solely by yolk, a thick fluid rich in proteins and lipids that is deposited in an egg as it forms. Eggs that take longer to hatch have larger amounts of yolk than eggs that hatch quickly. Kiwi birds have the longest incubation period of any bird—11 weeks. Their eggs are two-thirds yolk by volume.

By contrast, placental mammals produce almost yolkless eggs and nourish their embryos by means of a placenta (FIGURE 41.2C). A **placenta** is an organ that forms during pregnancy and facilitates the exchange of substances between the maternal and embryonic bloodstreams. Placenta-like organs also evolved in several groups of live-bearing fishes and in skinks, which are a type of live-bearing lizard. This independent evolution of similar body parts in different lineages is an example of morphological convergence (Section 18.3).

TAKE-HOME MESSAGE 41.2

✔ Most animals reproduce sexually, but some reproduce only asexually or can switch between asexual and sexual reproduction.

✔ External fertilization is typical in aquatic animals. Land-dwelling animals typically fertilize eggs in the female's body.

✔ Yolk deposited in an egg as it forms sustains development in most sexually reproducing animal species. In placental mammals, nutrients diffuse from maternal blood into the blood of the developing young.

CREDITS: (2A) iStockphoto.com/Brett_Hondow; (2B) Tony Phelps/Nature Picture Library; (2C) NPS Yellowstone/Becky Wyman.

41.3 Organs of Sexual Reproduction

LEARNING OBJECTIVES

- Explain the function of gonads.
- Compare the processes of spermatogenesis and oogenesis.

Gonads, Ducts, and Glands

In most animals, gamete production occurs inside special reproductive organs called **gonads**. Eggs are produced in gonads called **ovaries**, and sperm in gonads called **testes** (singular, testis).

The number, structure, and location of gonads varies among animal groups. Jellies typically have gonads on the lining of their gastrovascular cavity (FIGURE 41.3A). A sea star has a pair of ovaries or testes in each arm. An octopus has a single testis or ovary near the rear of its head, inside the mantle cavity. A fruit fly has a pair of ovaries or testes inside its abdomen (FIGURE 41.3B).

Most vertebrates have a pair of either ovaries or testes. Ovaries, when present, are always in the abdominal cavity. In fish, reptiles, and birds the testes are also internal. By contrast, most placental mammals have external testes. During embryonic development, their testes descend into a pouch that suspends them outside the abdominal cavity, just beneath the pelvic girdle.

Many animals reproduce only during a specific breeding period. In such species, gonads often shrink or even disappear during the time they are not in use.

For fertilization to occur, gametes must exit from the gonads. In cnidarians such as jellies, the gonads open onto the gastrovascular cavity, from which gametes exit the body. However, in most animals and all vertebrates, newly formed gametes enter into a system of ducts that conveys them to the surface of the body. In live-bearing animals, young develop inside specialized organs derived from such ducts.

In addition to gonads and the ducts that convey gametes, animal reproductive systems often include glands that support gamete development or facilitate fertilization. We delve into the structure and function of the various vertebrate reproductive glands in the sections of this chapter that describe the human reproductive organs.

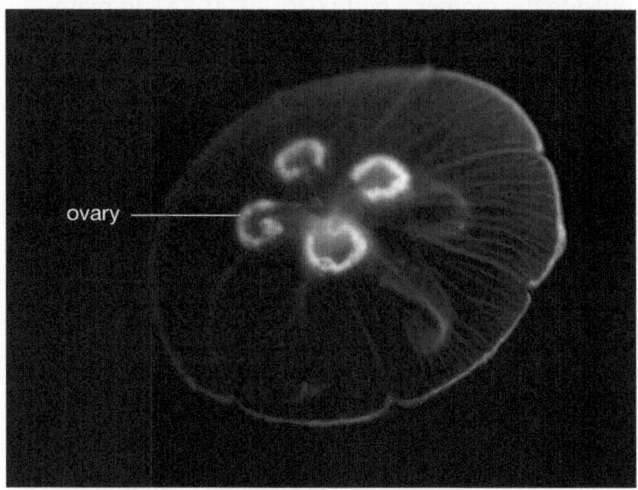
ovary

A Moon jelly (*Aurelia*), with four ringlike ovaries visible through its translucent body.

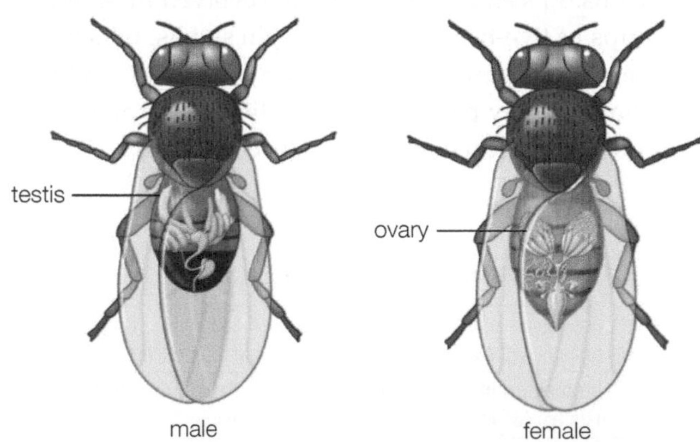
testis
ovary
male
female

B Fruit flies with paired gonads. A series of ducts convey sperm (produced in testes) and eggs (produced in ovaries) out of the body.

FIGURE 41.3 **Examples of animal gonads.**

How Gametes Form

All gonads contain germ cells. These diploid cells give rise to gametes by division. Because germ cells are diploid and eggs and sperm are haploid, the chromosome number must be halved by meiosis (Section 12.2).

Animal sperm production is called **spermatogenesis** (FIGURE 41.4A). It begins with mitosis of a spermatogonium, a diploid germ cell inside the testes. Mitosis and cytoplasmic division of spermatogonia produces primary spermatocytes, which are also diploid cells ❶. Each primary spermatocyte undergoes meiosis I and cytoplasmic division to yield two haploid secondary spermatocytes ❷. A secondary spermatocyte completes meiosis II and undergoes cytoplasmic division to produce two haploid spermatids ❸. The resulting four haploid spermatids then mature into four sperm ❹.

Animal egg production is called **oogenesis** (FIGURE 41.4B). It begins with mitosis and cytoplasmic division of an oogonium, which is a diploid germ cell in an ovary ❶. The resulting diploid cells are the primary oocytes. An **oocyte** is an immature animal egg.

A primary oocyte undergoes meiosis I to produce two haploid nuclei, then undergoes unequal cytoplasmic division to yield two differently sized cells ❷. One cell, the secondary oocyte, has a haploid nucleus and the bulk of the cytoplasm from the primary oocyte. The other cell, a **polar body**, has an identical haploid nucleus, but only a tiny bit of cytoplasm. It has no further role in reproduction and will degenerate.

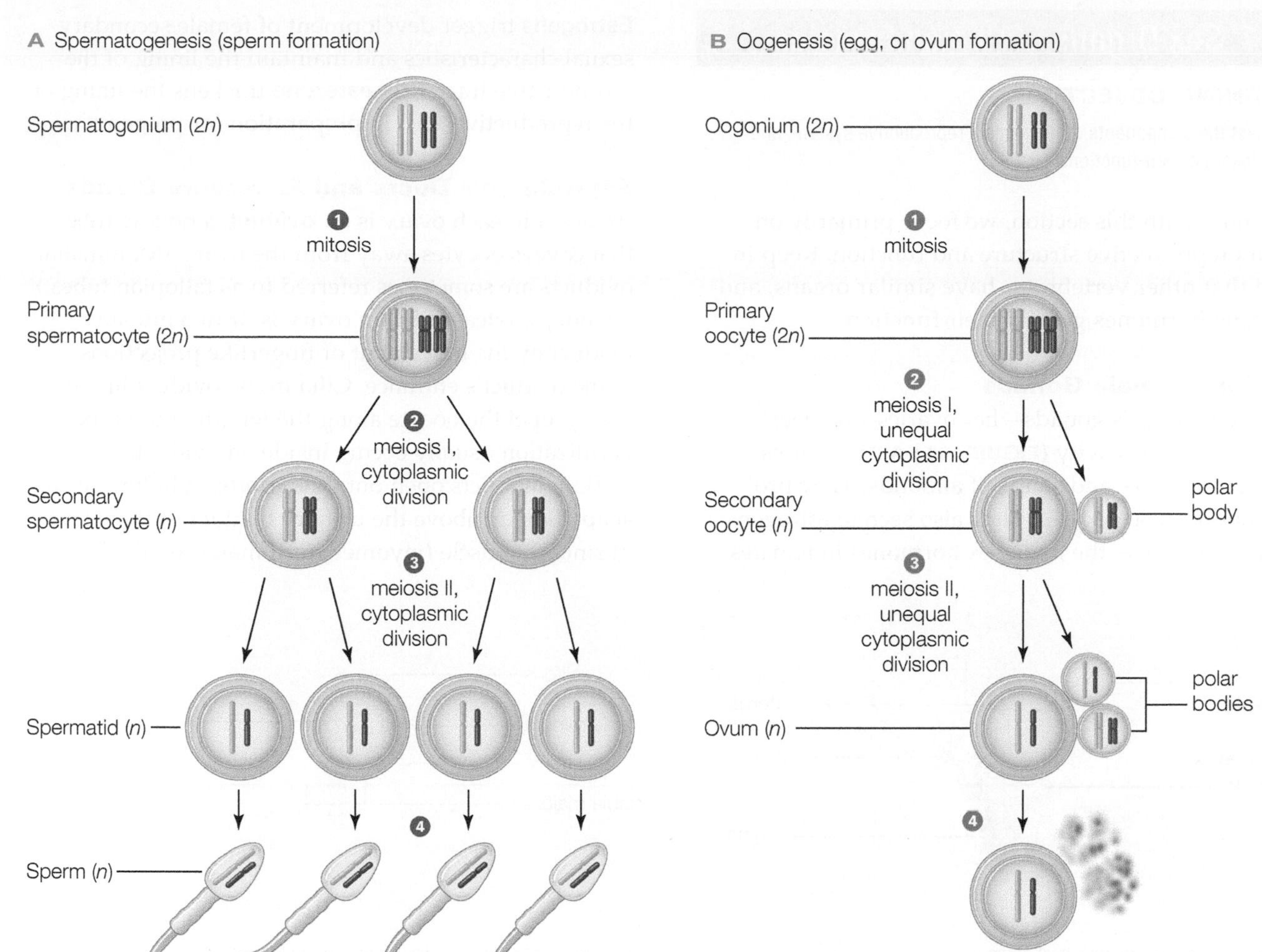

A Spermatogenesis (sperm formation)
Spermatogonium (2n)
1 mitosis
Primary spermatocyte (2n)
2 meiosis I, cytoplasmic division
Secondary spermatocyte (n)
3 meiosis II, cytoplasmic division
Spermatid (n)
4
Sperm (n)
B Oogenesis (egg, or ovum formation)
Oogonium (2n)
1 mitosis
Primary oocyte (2n)
2 meiosis I, unequal cytoplasmic division
Secondary oocyte (n)
polar body
3 meiosis II, unequal cytoplasmic division
Ovum (n)
polar bodies
4

FIGURE 41.4 Gamete formation in animals.

FIGURE IT OUT How does cytoplasmic division in spermatogenesis differ from that during oogenesis?

Answer: Spermatogenesis involves equal divisions of cytoplasm, and oogenesis involves unequal cytoplasmic divisions.

The secondary oocyte undergoes meiosis II, followed by unequal cytoplasmic division ❸. One cell receives a haploid nucleus with an unduplicated set of chromosomes, along with the bulk of the cytoplasm. This cell is the mature egg, or **ovum** (plural, ova). The other cell receives an identical nucleus, but little cytoplasm. Like the first polar body, this second polar body has no further role and will degenerate ❹. Thus, oogenesis produces only a single ovum from each primary oocyte. An ovum is always much larger than a sperm of the same species.

gonad (GO-nad) Animal reproductive organ that produces gametes.
oocyte (oh-oh-SITE) Immature animal egg.
oogenesis (oh-oh-JEN-eh-sis) Egg production in an animal.
ovary Egg-producing reproductive organ.
ovum (OH-vum) Plural, ova. Mature animal egg.
polar body Tiny cell produced by unequal cytoplasmic division during egg production in animals.
spermatogenesis (sper-mah-toh-JEN-eh-sis) Sperm production in an animal.
testis Plural, testes. Sperm-producing reproductive organ.

TAKE-HOME MESSAGE 41.3

✔ In animals, sexual reproduction begins with production of gametes. Testes (male gonads) produce sperm. Ovaries (female gonads) produce eggs.

✔ Meiosis of germ cells in the gonads gives rise to gametes. During spermatogenesis, meiosis of a male germ cell produces four sperm. Oogenesis involves unequal cytoplasmic divisions, so meiosis of a female germ cell yields only one large ovum (egg).

✔ In addition to gonads, most animals have a system of ducts that convey gametes to the body surface, as well as glands that nourish or otherwise support the gametes.

41.4 Sex Organs of Human Females

LEARNING OBJECTIVE

- List the components of the female reproductive system, and describe their functions.

Beginning with this section, we focus primarily on human reproductive structure and function. Keep in mind that other vertebrates have similar organs, and the same hormones govern their function.

Ovaries—Female Gonads

A human female's gonads—her ovaries—lie deep inside her pelvic cavity (**FIGURE 41.5A,B**). Ovaries are about the size and shape of almonds. They produce and release oocytes. They also secrete estrogens and progesterone, the main sex hormones in females.

Estrogens trigger development of female secondary sexual characteristics and maintain the lining of the reproductive tract. **Progesterone** thickens the lining of the reproductive tract in preparation for pregnancy.

Reproductive Ducts and Accessory Glands

Adjacent to each ovary is an **oviduct**, a hollow tube that coveys oocytes away from the ovary. (Mammalian oviducts are sometimes referred to as fallopian tubes.) An oocyte released by an ovary is drawn into an oviduct by the movement of fingerlike projections at the oviduct's entrance. Cilia in the oviduct lining then propel the oocyte along the length of the tube. Fertilization usually occurs inside an oviduct.

Both oviducts open onto the **uterus**, a hollow, pear-shaped organ above the urinary bladder. A thick layer of smooth muscle (myometrium) makes up the bulk of

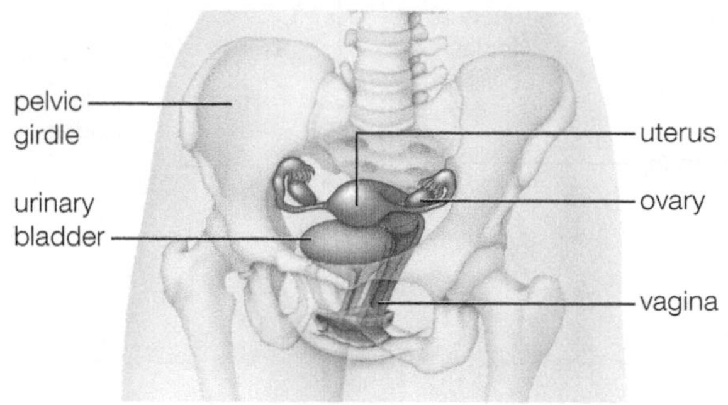

pelvic girdle
urinary bladder
uterus
ovary
vagina

A Location of female reproductive organs.

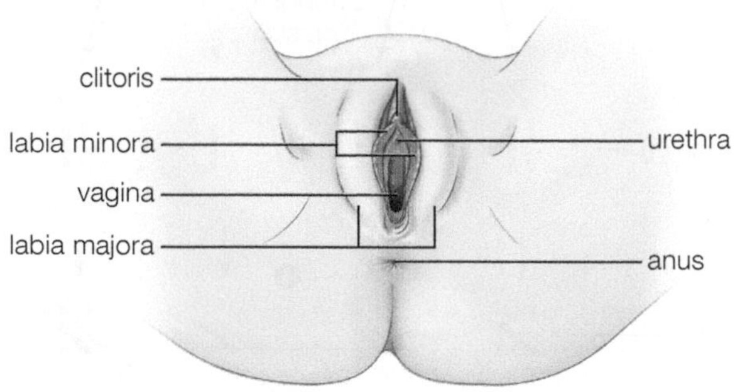

clitoris
labia minora
vagina
labia majora
urethra
anus

C External sex organs, collectively referred to as the vulva.

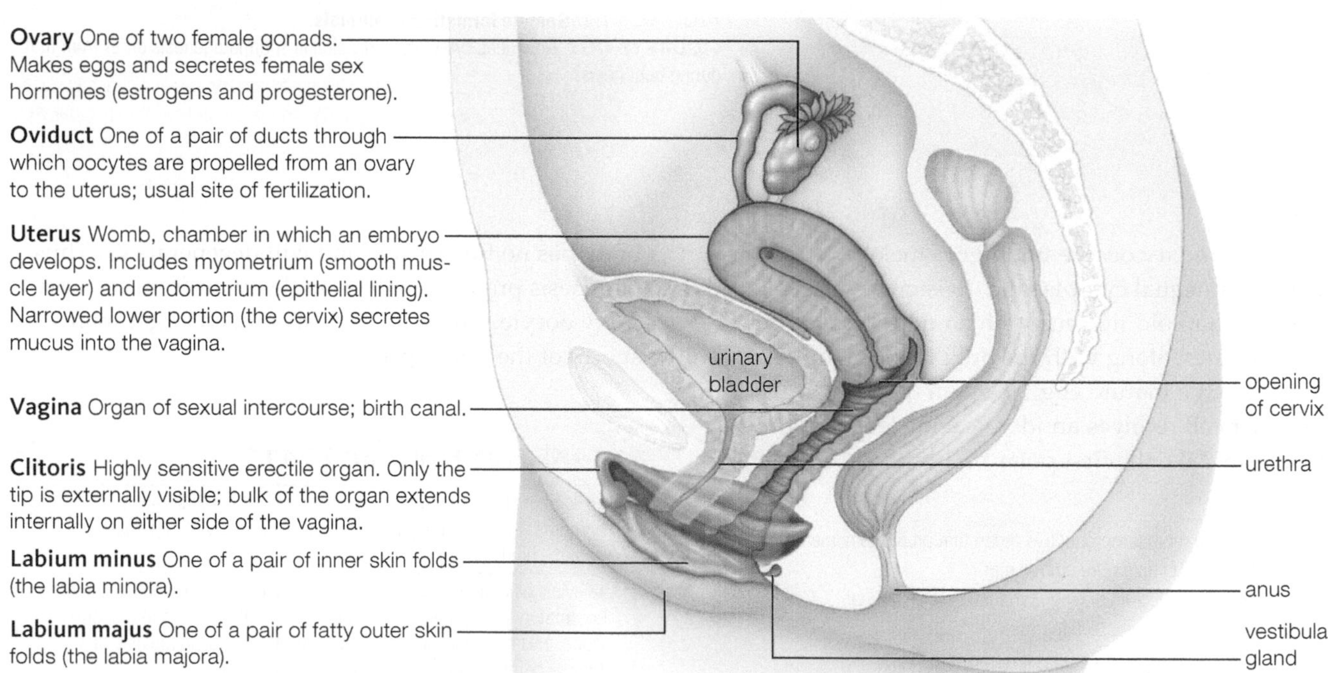

Ovary One of two female gonads. Makes eggs and secretes female sex hormones (estrogens and progesterone).
Oviduct One of a pair of ducts through which oocytes are propelled from an ovary to the uterus; usual site of fertilization.
Uterus Womb, chamber in which an embryo develops. Includes myometrium (smooth muscle layer) and endometrium (epithelial lining). Narrowed lower portion (the cervix) secretes mucus into the vagina.
Vagina Organ of sexual intercourse; birth canal.
Clitoris Highly sensitive erectile organ. Only the tip is externally visible; bulk of the organ extends internally on either side of the vagina.
Labium minus One of a pair of inner skin folds (the labia minora).
Labium majus One of a pair of fatty outer skin folds (the labia majora).
urinary bladder
opening of cervix
urethra
anus
vestibular gland

B Human female reproductive organs, shown in longitudinal section.

FIGURE 41.5 Components of the reproductive system in a human female.

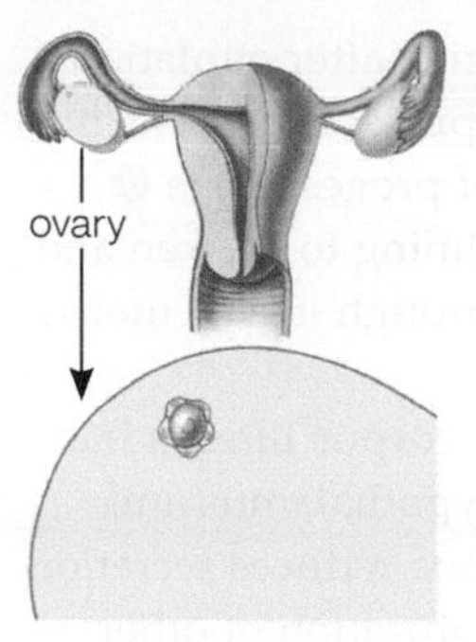
ovary

① An ovary has many immature follicles, each consisting of a primary oocyte and surrounding follicle cells.

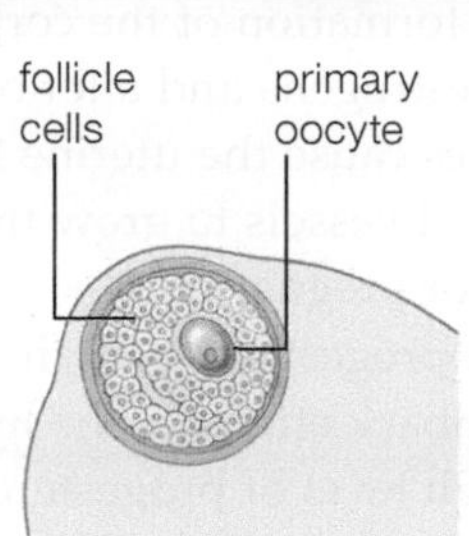
follicle cells
primary oocyte

② A fluid-filled cavity begins to form in the follicle's cell layer.

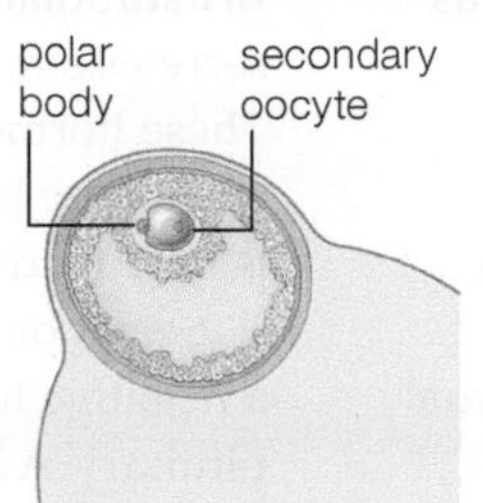
polar body
secondary oocyte

③ The primary oocyte completes meiosis I and divides unequally, forming a secondary oocyte and a polar body.

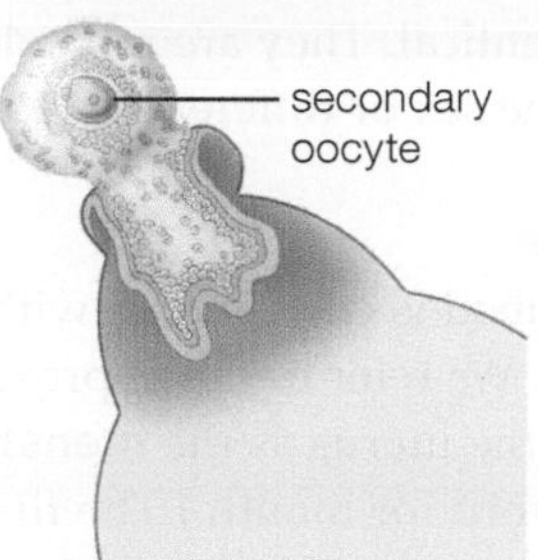
secondary oocyte

④ Ovulation. The mature follicle ruptures, releasing a secondary oocyte coated with secreted proteins and follicle cells.

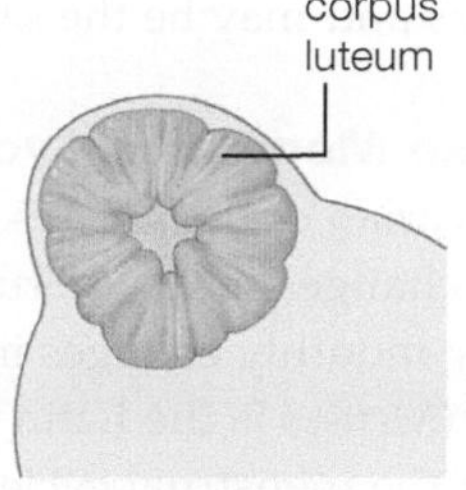
corpus luteum

⑤ A corpus luteum develops from follicle cells left behind after ovulation.

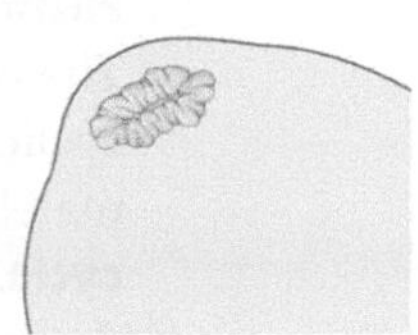

⑥ If pregnancy does not occur, the corpus luteum degenerates.

FIGURE 41.6 Ovarian cycle. The mature follicle is about the size of a pea and projects from the ovary's surface.

the uterine wall. The uterine lining (endometrium) consists of glandular epithelium, connective tissues, and blood vessels. The lowest portion of the uterus, a narrowed region called the **cervix**, opens into the vagina. The **vagina**, which extends from the cervix to the body's surface, serves both as the organ of intercourse and the birth canal.

Externally visible organs of the reproductive tract are called genitals. Female genitals include two pairs of liplike skin folds that enclose the openings of the vagina and urethra (FIGURE 41.5C). Adipose tissue fills the thick outer folds, the labia majora. Thin inner folds are the labia minora. The clitoris lies near the anterior junction of the labia minora. It contains erectile tissue and is highly sensitive to tactile stimulation.

TAKE-HOME MESSAGE 41.4

✔ A woman's paired ovaries produce oocytes (eggs). They also make and secrete the hormones estrogen and progesterone.

✔ An oviduct connects each ovary to the uterus. The cervix at the lower end of the uterus opens onto the vagina. The vagina serves as the female organ of intercourse and as the birth canal.

41.5 Female Reproductive Cycles

LEARNING OBJECTIVES

- Describe the human ovarian cycle and menstrual cycle.
- Explain how hormones regulate female reproductive cycles.
- Explain what an estrous cycle is.

Human Ovarian Cycle

In humans, all production of oocytes from germ cells occurs before birth. A girl is born with about 2 million **primary oocytes**—oocytes that entered meiosis but stopped in prophase I, rather than completing meiosis. At puberty (the stage of development when reproductive organs mature) hormonal changes prompt oocytes to mature, one at a time, in an approximately 28-day ovarian cycle. FIGURE 41.6 shows this cycle. An oocyte and cells around it are an **ovarian follicle** ①. As the cycle begins, the follicle enlarges and a fluid-filled cavity forms around it ②. The primary oocyte completes meiosis I and undergoes unequal cytoplasmic division to produce a secondary oocyte and a polar body ③. The **secondary oocyte** begins meiosis II, but halts in metaphase II. It will not complete meiosis until fertilization.

About two weeks after the follicle begins to mature, its wall ruptures and **ovulation** occurs: The secondary oocyte, polar body, and surrounding follicle cells are ejected into the adjacent oviduct ④.

After ovulation, cells of the ruptured follicle develop into a hormone-secreting **corpus luteum** ⑤. (*Corpus luteum* means "yellow body" in Latin.) If pregnancy does not occur, the corpus luteum breaks down ⑥, and a new follicle will begin to mature.

Occasionally, two oocytes mature and are released at the same time. If each egg is later fertilized by a

cervix Narrow part of uterus that connects to the vagina.
corpus luteum (COR-pus LOO-tee-um) Hormone-secreting structure that forms from follicle cells left behind after ovulation.
estrogens Hormones secreted by ovaries; cause development of secondary sexual traits and maintain the reproductive tract's lining.
ovarian follicle Immature animal egg and the surrounding cells.
oviduct Duct that conveys eggs away from an animal ovary.
ovulation Release of a secondary oocyte from an ovary.
primary oocyte Oocyte that has halted in prophase I of meiosis.
progesterone Hormone secreted by ovaries; prepares the reproductive tract for pregnancy.
secondary oocyte Oocyte that has halted in metaphase II of meiosis; released from an ovary at ovulation.
uterus Muscular chamber where offspring develop; womb.
vagina Female organ of intercourse and birth canal.

different sperm, the result is fraternal twins. **Fraternal twins** are not genetically identical. They are related as siblings and may be the same sex or different sexes.

Human Menstrual Cycle

The ovarian cycle just described is coordinated with cyclic changes in the uterus. We refer to the approximately monthly changes in the uterus as the **menstrual cycle**. (*Menses* is the Latin word for month.) The first day of the menstrual cycle is marked by the onset of **menstruation**, which is the flow of bits of uterine lining and some blood from the uterus, through the cervix, and out of the vagina.

Hormonal Control of Monthly Cycles

Hormones control the ovarian and menstrual cycles (TABLE 41.1 and FIGURE 41.7). As the cycles begin, secretion of **gonadotropin-releasing hormone (GnRH)** by the hypothalamus stimulates release of FSH and LH from the pituitary ❶. **Follicle-stimulating hormone (FSH)** stimulates maturation of an ovarian follicle ❷. The interval of follicle maturation before ovulation is the follicular phase of the cycle. During this time, cells around the oocyte secrete estrogens ❸ that stimulate the endometrium to thicken ❹.

As the level of estrogens increases, a positive feedback loop occurs. Increased estrogen causes the hypothalamus to release more GnRH, so the pituitary releases more FSH, and the follicle grows and produces more estrogens. The rise in estrogens encourages the anterior pituitary to release more **luteinizing hormone (LH)**, which in females has roles in ovulation and formation of the corpus luteum.

At about the midpoint in the cycle, the high level of estrogens causes a surge in LH secretion ❺. The surge of LH stimulates the primary oocyte to complete meiosis I and undergo cytoplasmic division. It also causes the follicle to swell and burst. Thus, the midcycle surge of LH is the trigger for ovulation ❻.

estrous cycle In most female mammals, a recurring cyclic variation in sexual receptivity.

follicle-stimulating hormone (FSH) Anterior pituitary hormone with roles in ovarian follicle maturation and sperm production.

fraternal twins Twins that arise when two eggs mature and are fertilized at the same time; related as siblings.

gonadotropin-releasing hormone (GnRH) Hypothalamic hormone that induces the pituitary to release hormones (LH and FSH) that regulate hormone secretion by the gonads.

luteinizing hormone (LH) Anterior pituitary hormone with roles in ovulation, corpus luteum formation, and sperm production.

menopause Permanent cessation of menstrual cycles.

menstrual cycle (MEN-struhl) Reproductive cycle in which the uterus lining thickens and then, if pregnancy does not occur, is shed.

menstruation (MEN-strew-ay-shun) Flow of shed uterine tissue out of the vagina.

The luteal phase of the cycle begins after ovulation. LH stimulates formation of the corpus luteum, which secretes some estrogens and a lot of progesterone ❼. These hormones cause the uterine lining to thicken and encourage blood vessels to grow through it. The uterus is now ready for pregnancy ❽.

Secretion of progesterone by the corpus luteum has a negative feedback effect on the hypothalamus and pituitary. A high level of progesterone reduces secretion of FSH and LH, thus preventing maturation of other follicles and inhibiting additional ovulations.

If fertilization does not occur, the level of LH continues to decline, causing the corpus luteum to degenerate. Breakdown of the corpus luteum causes estrogen and progesterone levels to plummet ❾. In the uterus, the decline in estrogens and progesterone causes degeneration of the uterine lining, and menstruation begins. Blood and endometrial tissue flow out of the vagina for three to six days. At the same time, the pituitary increases secretion of FSH and LH once again.

During their reproductive years, many women regularly experience discomfort before they menstruate. Breasts become tender because estrogens and progesterone cause milk ducts to widen. Body tissues swell because premenstrual changes influence aldosterone secretion. (As Section 40.6 explained, aldosterone stimulates reabsorption of sodium and, indirectly, water.) Cycle-associated hormonal changes can also cause depression, irritability, anxiety, headaches, and insomnia. Regular recurrence of these symptoms is known as premenstrual syndrome (PMS). The use of oral contraceptives, which minimize hormone swings, can prevent PMS.

During menstruation, cells in the uterine lining secrete prostaglandins, and these local signaling molecules stimulate contractions of smooth muscle in the uterine wall. Many women do not feel the muscle contractions, but others experience a dull ache or sharp pains commonly known as menstrual cramps. Severe pain and heavy bleeding during menstruation are not normal and may be caused by benign tumors in the uterus called fibroids.

A woman enters **menopause** when all the follicles in her ovaries have either been released during menstrual cycles or have disintegrated as a result of aging. With no follicles left to mature, production of estrogen and progesterone is diminished and menstrual cycles cease. Interestingly, menopause is rare among animals. It is known only in humans and two species of whales.

Animal Estrous Cycles

All female placental mammals need to replace their uterine lining on a cyclic basis. However, most do not

TABLE 41.1

Events of a 28-Day Ovarian/Menstrual Cycle

Phase	Events	Day of Cycle
Follicular phase	Menstruation; endometrium breaks down	1–5
	Follicle matures in ovary; endometrium rebuilds	6–13
Ovulation	Oocyte released from ovary	14
Luteal phase	Corpus luteum forms, secretes progesterone; the endometrium thickens and develops	15–28

menstruate. They reabsorb their endometrial lining rather than shedding it.

Most female mammals also have an **estrous cycle**, in which sexual receptivity occurs only when the female can conceive. The interval of sexual receptivity is known as estrus and is commonly referred to as "heat." The length of estrous cycles varies among animal species. In dogs, it is about six months; in mice, it is six days.

In many mammals, changes in a female's body alert males of her species to the fact she is in estrus. Often a female's labia swell and she produces a discharge containing chemical signals that help attract potential mates. For example, when a female dog is in heat, her vagina produces a thin discharge containing chemicals that attract male dogs.

TAKE-HOME MESSAGE 41.5

- ✔ In humans, egg maturation and preparation of the uterus for pregnancy occur in an approximately monthly cycle.
- ✔ The onset of the cycle is marked by menstruation, during which the lining of the uterus is shed.
- ✔ Following menstruation, FSH secreted by the pituitary encourages maturation of an ovarian follicle. The maturing follicle produces estrogen.
- ✔ At the midpoint in the cycle, the high estrogen level triggers a surge in LH secretion, which in turn triggers ovulation.
- ✔ After ovulation, the remains of the ovarian follicle becomes the corpus luteum. Progesterone secreted by the corpus luteum stimulates proliferation of the uterine lining.
- ✔ If pregnancy does not occur, the corpus luteum degenerates and menstruation occurs.
- ✔ Most female mammals do not menstruate and are sexually receptive only when they can conceive (during estrus).

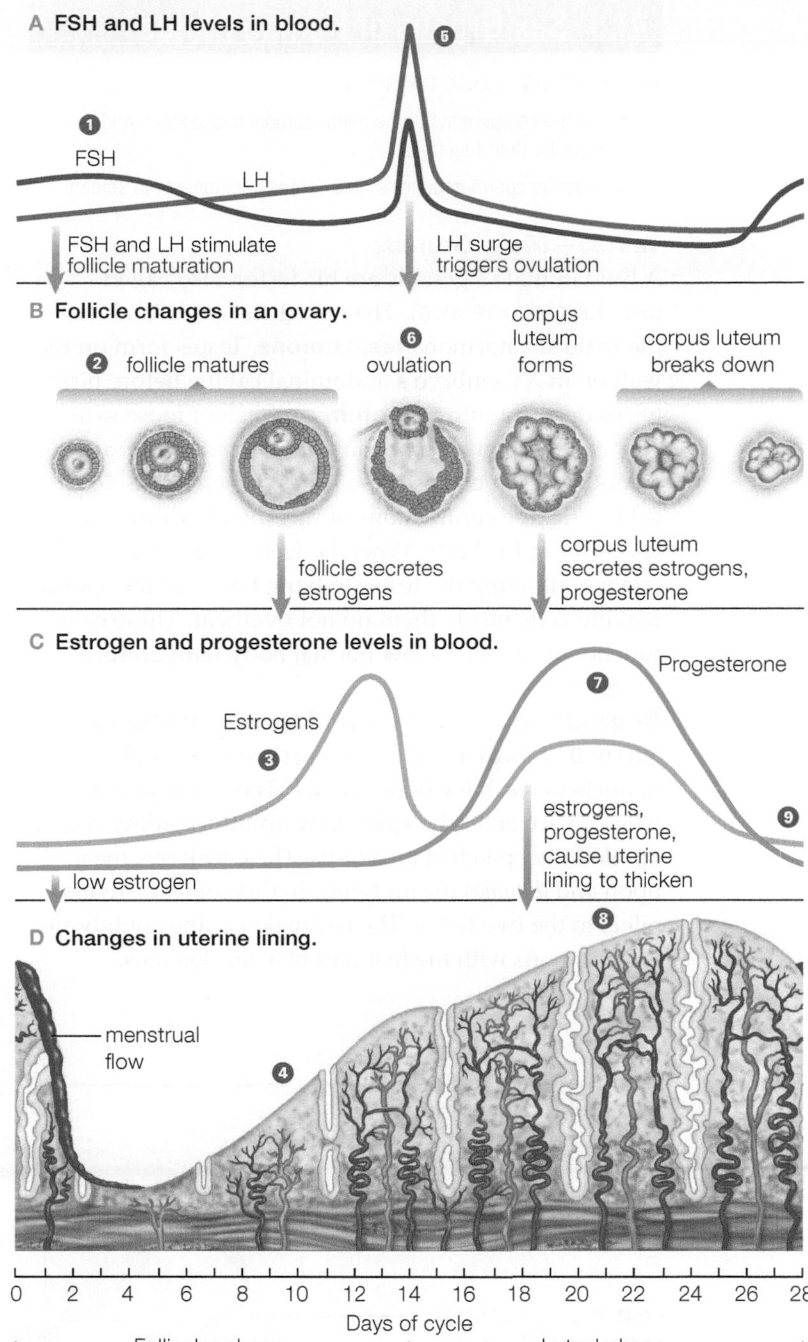

A FSH and LH levels in blood.
1
FSH
LH
5
FSH and LH stimulate follicle maturation
LH surge triggers ovulation
B Follicle changes in an ovary.
2 follicle matures
6 ovulation
corpus luteum forms
corpus luteum breaks down
follicle secretes estrogens
corpus luteum secretes estrogens, progesterone
C Estrogen and progesterone levels in blood.
Estrogens
3
Progesterone
7
estrogens, progesterone, cause uterine lining to thicken
9
low estrogen
8
D Changes in uterine lining.
menstrual flow
4
0 2 4 6 8 10 12 14 16 18 20 22 24 26 28
Days of cycle
Follicular phase
Luteal phase

FIGURE 41.7 Changes in the ovary and uterus correlated with changing hormone levels. We start with the onset of menstrual flow on day 1 of a 28-day menstrual cycle.

(**A,B**) Prompted by GnRH from the hypothalamus, the anterior pituitary secretes FSH and LH, which stimulate a follicle to grow and an oocyte to mature in an ovary. A midcycle surge of LH triggers ovulation and the formation of a corpus luteum. A decline in FSH after ovulation stops more follicles from maturing.

(**C,D**) Early in the cycle, estrogen from a maturing follicle stimulates repair and rebuilding of the endometrium. After ovulation, the corpus luteum secretes some estrogen and more progesterone that primes the uterus for pregnancy.

41.6 Sex Organs of Human Males

LEARNING OBJECTIVES

- List the components of the male reproductive system, and describe their functions.
- Describe spermatogenesis, and list the components of semen.

Testes—Male Gonads

A human male's gonads are his testes, also called testicles (FIGURE 41.8). They produce sperm and make the male sex hormone **testosterone**. Testes form on the wall of an XY embryo's abdominal cavity. Before birth, testes descend into a **scrotum**, a pouch of loose skin suspended below the pelvic girdle. Inside this pouch, smooth muscle encloses the testes. When a man feels cold, reflexive contractions of this muscle draw his testes closer to his body. When he feels warm, the muscle relaxes, allowing the testes to hang lower so the sperm-making cells inside them do not overheat. These cells function best a bit below normal body temperature.

Reproductive Ducts and Accessory Glands

Sperm that form in the testes must travel through a series of ducts to reach the body surface. The journey begins when sperm enter the **epididymis** (plural, epididymides), a coiled duct perched on a testis. The Greek *epi–* means upon and *didymos* means twins. In this context, "twins" refers to the two testes. The last region of the epididymis is continuous with the first part of a vas deferens. In Latin, *vas* means vessel, and *deferens*, means to carry away. Each **vas deferens** (plural, vasa deferentia) is a duct that carries sperm away from an epididymis and to a short ejaculatory duct. Ejaculatory ducts deliver sperm to the urethra, the duct that extends through a male's penis to open at the body surface.

The **penis** is the male organ of intercourse. It has a rounded head (the glans) at the end of a narrower shaft. Nerve endings in the glans make it highly sensitive to touch. Normally, when a man is not sexually excited, a retractable tube of skin called the foreskin covers the glans of his penis. In some cultures, male infants undergo circumcision, an elective surgical procedure that removes the foreskin.

Three elongated cylinders of spongy tissue fill the interior of the penis. When a male is sexually excited, blood flows into the spongy tissue faster than it flows out. Fluid pressure rises and the normally limp penis becomes stiff and erect.

Sperm stored in the epididymides and first part of the vasa deferentia continue their journey toward the body surface only when a male reaches the peak of sexual excitement and ejaculates. During ejaculation, rhythmic smooth muscle contractions propel sperm and accessory gland secretions out of the body as a thick, white fluid called semen.

Semen is a mix of sperm, proteins, nutrients, ions, and signaling molecules. Sperm constitute less than 5 percent of its volume; the bulk of it consists of

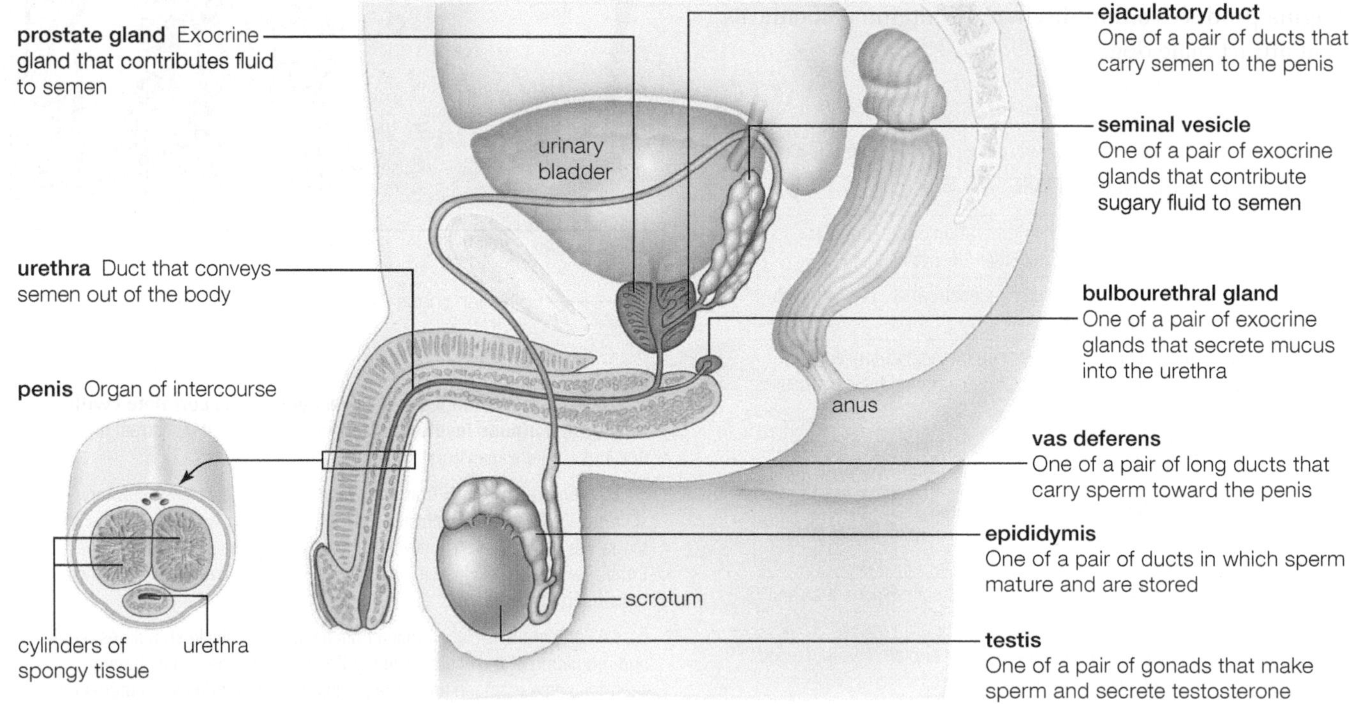

prostate gland Exocrine gland that contributes fluid to semen
urinary bladder
urethra Duct that conveys semen out of the body
penis Organ of intercourse
cylinders of spongy tissue
urethra
scrotum
anus
ejaculatory duct One of a pair of ducts that carry semen to the penis
seminal vesicle One of a pair of exocrine glands that contribute sugary fluid to semen
bulbourethral gland One of a pair of exocrine glands that secrete mucus into the urethra
vas deferens One of a pair of long ducts that carry sperm toward the penis
epididymis One of a pair of ducts in which sperm mature and are stored
testis One of a pair of gonads that make sperm and secrete testosterone

FIGURE 41.8 **Components of the reproductive system in a human male.**

secretions from accessory glands. Seminal vesicles, which are exocrine glands near the base of the bladder, secrete fructose-rich fluid into the vasa deferentia. Sperm use the fructose (a sugar) as their energy source. The prostate gland, which encircles the urethra, is the other major contributor to semen volume. Its slightly alkaline secretions help raise the pH of the female reproductive tract, making this passage more hospitable to sperm.

Prior to ejaculation, two pea-sized bulbourethral glands secrete mucus into the urethra. This mucus helps clear the urethra of residual urine.

Germ Cells to Sperm Cells

A testis is smaller than a golf ball, yet it contains more than 100 meters of **seminiferous tubules** (**FIGURE 41.9**). Diploid male germ cells line the inner wall of each tubule. At puberty, these cells begin to divide by mitosis to produce primary spermatocytes. The spermatocytes undergo meiosis to produce round, haploid cells called spermatids, which then differentiate into specialized cells with a long flagellum—sperm. A seminiferous tubule also contains big, elongated cells called nurse cells (or Sertoli cells) that support developing sperm. The testes' testosterone-secreting cells (Leydig cells) reside between the seminiferous tubules.

Like oogenesis, spermatogenesis is governed by LH and FSH. LH binds to Leydig cells stimulating them to secrete testosterone. FSH targets nurse cells. In combination with testosterone, it causes nurse cells to produce chemicals essential to sperm development.

Newly formed sperm cannot swim and are moved into the epididymis by smooth muscle contractions. Cilia of the epididymis lining push the immature sperm farther along this duct. The sperm mature and become mobile as they move through the epididymis. Sperm formation takes about 65 days.

FIGURE 41.10 shows the structure of a mature sperm. It is a haploid cell with a "head" that is packed full of DNA and tipped by an enzyme-containing cap called the **acrosome**. The enzymes of the acrosome will help the sperm penetrate an oocyte. At its other end, the sperm has a flagellum that it uses to swim toward an egg. The sperm does not have ribosomes, endoplasmic reticulum, or Golgi bodies. However, its midsection contains many mitochondria that supply the ATP required for flagellar movement.

A sexually mature male makes sperm continually, so on any given day he has millions of them in various stages of development. A sperm stored in the epididymis remains viable for about a month. After that, the sperm breaks down and its components are reabsorbed and reused.

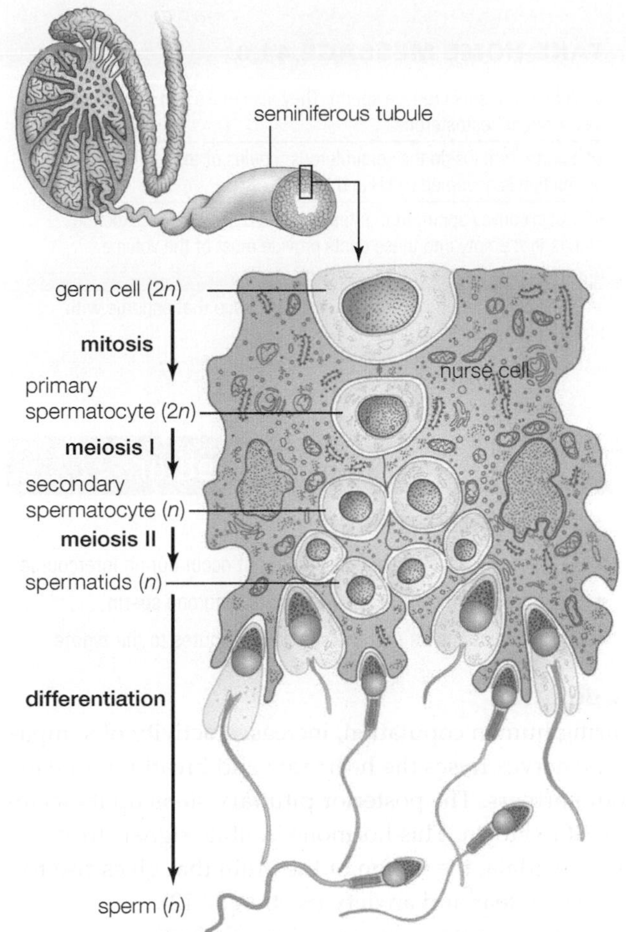
seminiferous tubule
germ cell (2n)
mitosis
nurse cell
primary
spermatocyte (2n)
meiosis I
secondary
spermatocyte (n)
meiosis II
spermatids (n)
differentiation
sperm (n)

FIGURE 41.9 **Human spermatogenesis.**

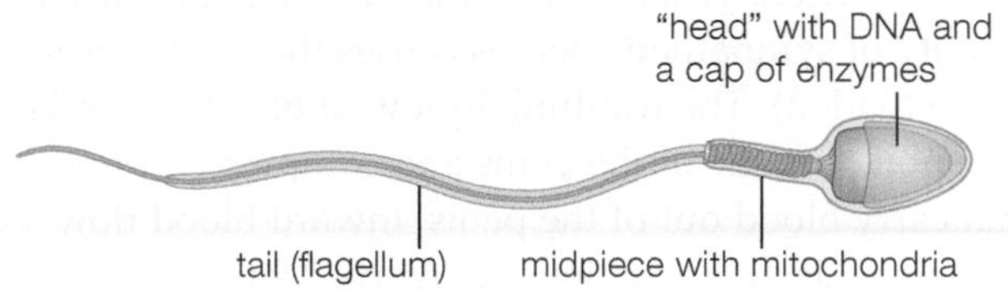
"head" with DNA and
a cap of enzymes
tail (flagellum)
midpiece with mitochondria

FIGURE 41.10 **Structure of a mature human sperm.**

acrosome Enzyme-containing cap on the "head" of a mammalian sperm.
epididymis (epp-ih-DID-ih-muss) Plural, epididymides. One of two ducts in which human sperm mature; each empties into a vas deferens.
penis Male organ of intercourse.
scrotum Pouch of skin that encloses a human male's testes.
semen Sperm mixed with secretions from seminal vesicles and the prostate gland.
seminiferous tubules (sehm-in-IF-er-us) Inside a testis, coiled tubules that contain male germ cells and produce sperm.
testosterone Main hormone produced by testes; required for sperm production and development of male secondary sexual traits.
vas deferens (vas DEF-er-ens) Plural, vasa deferentia. One of a pair of long ducts that convey mature sperm toward the body surface.

TAKE-HOME MESSAGE 41.6

✔ A pair of testes produce sperm. They also make and secrete the sex hormone testosterone.

✔ Sperm form inside the seminiferous tubules of testes. Their production is governed by LH and FSH.

✔ Ducts convey sperm from a testis to the body surface. Exocrine glands that empty into these ducts provide most of the volume of semen.

✔ The penis contains cylinders of spongy tissue that engorge with blood during sexual excitement.

41.7 Bringing Gametes Together

LEARNING OBJECTIVES

- Describe the physiological changes that occur during intercourse.
- Explain why fertilization requires more than one sperm.
- List the components that the sperm contributes to the zygote.

Copulation

During human copulation, increased activity of sympathetic nerves raises the heart rate and breathing rate in both partners. The posterior pituitary steps up its secretion of oxytocin. This hormone inhibits signals from the amygdala, the region of the brain that gives rise to feelings of fear and anxiety (Section 32.12).

For males, intercourse requires an erection. When a male is not sexually aroused, his penis is limp, because the arteries that transport blood to its spongy tissue are constricted. When he becomes aroused, increased activity of sympathetic nerves causes these arteries to dilate (widen). The resulting inflow of blood expands the spongy tissue of the penis and compresses veins that carry blood out of the penis. Inward blood flow now exceeds outward flow, so internal fluid pressure increases. This pressure enlarges and stiffens the penis so it can be inserted into a female's vagina.

The ability to obtain and sustain an erection peaks during the late teens. As a male ages, he may have episodes of erectile dysfunction. With this disorder, the penis does not stiffen enough for intercourse. Drugs prescribed for erectile dysfunction act by relaxing smooth muscle in the walls of arterioles that supply the penis. However, they also dilate blood vessels elsewhere in the body and so should not be taken without a prescription.

When a woman becomes sexually excited, blood flow to the vaginal wall, labia, and clitoris increases. Glands in the cervix secrete mucus, and glands on the labia produce a lubricating fluid. The labial glands are the equivalent of a male's bulbourethral glands. The vagina itself does not have any glandular tissue. It is moistened by mucus from the cervix and by plasma fluid that seeps out between epithelial cells of the vaginal lining.

Continued mechanical stimulation of the penis or clitoris can lead to orgasm. During orgasm, endorphins flood the brain and evoke feelings of pleasure. At the same time, a surge of oxytocin causes rhythmic contractions of smooth muscle in both the male and female reproductive tract. In males, orgasm is usually accompanied by ejaculation, in which contracting muscles force the semen out of the penis. Ejaculation can put 300 million sperm into the vagina.

The Sperm's Journey

While still inside a male's body, sperm make ATP by breaking down the fructose in seminal vesicle secretions. Once inside a female reproductive tract, they must take up carbohydrates from their environment to fuel their movement.

To travel from the vagina into the uterus, a sperm must swim through the cervical canal, a tunnel through the center of the cervix. During parts of the reproductive cycle when a woman's estrogen level is low, a thick plug of acidic mucus bars the passage through this canal. As the woman's estrogen level rises in the prelude to ovulation, the cervix becomes more sperm-friendly. It begins to secrete a thinner, more alkaline mucus that contains glucose, a sugar that sperm can use as fuel. Even so, swimming through cervical mucus is challenging, so only the strongest sperm pass through the cervix and enter the uterus.

Once sperm are in the uterus, contractions of smooth muscle in the uterine wall help them move toward the oviducts. Sperm enter both oviducts, but because ovulation occurs in only one ovary at a time, half of the sperm will continue onward without any possibility of ever encountering an egg. An egg does produce chemicals that attract sperm, but this attraction is effective only at close range.

Fertilization

Fertilization usually happens in the upper part of an oviduct (**FIGURE 41.11** ❶). Of the millions of sperm ejaculated into the vagina, only a few hundred make it this far. Sperm live for about three days after ejaculation, so fertilization can occur even if intercourse takes place a few days before ovulation.

A secondary oocyte released at ovulation retains a wrapping of follicle cells. Beneath those cells is the zona pellucida, a coat composed of glycoproteins secreted by the oocyte ❷. Sperm make their way between the follicle cells to the zona pellucida. The plasma membrane of the sperm's head has receptors that bind species-specific proteins in this coat.

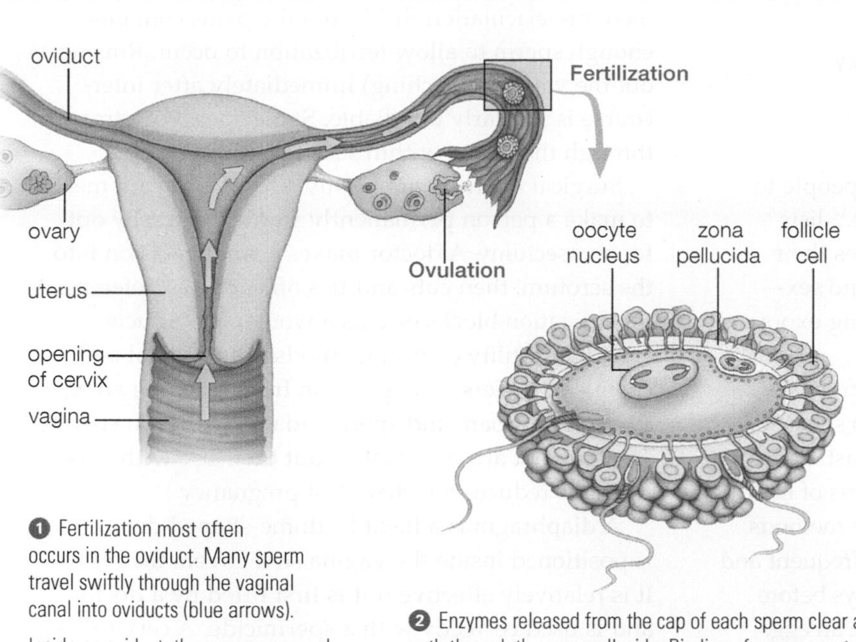
oviduct
Fertilization
ovary
uterus
opening of cervix
vagina
Ovulation
oocyte nucleus
zona pellucida
follicle cell

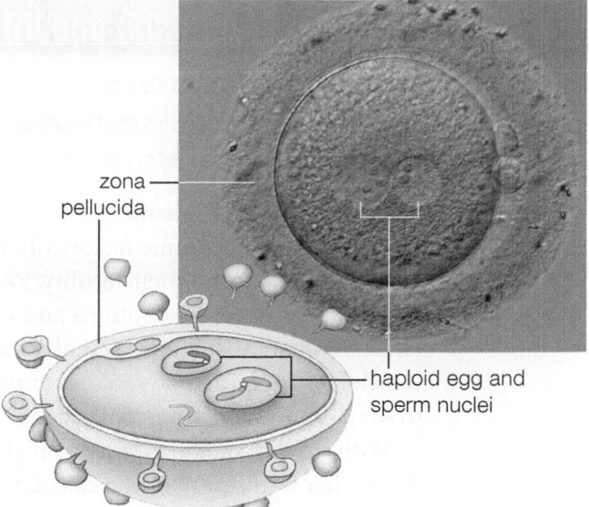
zona pellucida
haploid egg and sperm nuclei

❶ Fertilization most often occurs in the oviduct. Many sperm travel swiftly through the vaginal canal into oviducts (blue arrows).

Inside an oviduct, the sperm surround a secondary oocyte that was released by ovulation.

❷ Enzymes released from the cap of each sperm clear a path through the zona pellucida. Binding of one sperm to the membrane of the secondary oocyte causes the oocyte to release substances that alter the zona pellucida and prevent other sperm from binding.

❸ The sperm is drawn into the oocyte, and the oocyte nucleus completes meiosis II. Later, both paternal and maternal nuclear membranes break down and chromosomes become arranged on a bipolar spindle in preparation for the first mitotic division.

FIGURE 41.11 Events in human fertilization.
The light micrograph on the right shows a fertilized human oocyte.

FIGURE IT OUT In the micrograph, what are the small cells on the right, just beneath the zona pellucida?

Answer: The polar bodies.

Binding of a sperm to the zona pellucida triggers release of protein-digesting enzymes from the acrosome on the sperm's head. The collective effect of the enzymes released by many sperm clears a passage through the glycoprotein layer, thus opening the way to the oocyte's plasma membrane. When receptors in this membrane bind a sperm's plasma membrane, the entire sperm is drawn head first into the oocyte.

Typically only one sperm enters the oocyte. Its entry causes the oocyte to secrete proteins that alter the consistency of the zona pellucida making it difficult for other sperm to bind. In the rare case in which multiple sperm penetrate a human oocyte, the resulting zygote cannot survive.

Although it only takes one sperm to fertilize an egg, it takes the presence of many to release enough enzymes to clear a way to the egg plasma membrane. This is why a man who has healthy sperm but a low sperm count can be functionally infertile.

Remember that the secondary oocyte released at ovulation was halted in metaphase of meiosis II. Binding of the sperm membrane to the oocyte membrane causes the oocyte to complete meiosis. Unequal cytoplasmic division follows, producing a single mature egg—an ovum—and the second polar body ❸.

Chromosomes in the egg and sperm nuclei become the genetic material of the zygote (the first cell of the new individual). The sperm also supplies a single centriole. Recall from Section 4.8 that centrioles help microtubules organize themselves. The sperm's centriole replicates to form the pair of centrioles that draw maternal and paternal nuclei together and organize the spindle for the zygote's first mitotic division.

Sperm mitochondria and the flagellum enter the oocyte too, but they are typically broken down. Thus, mitochondria are inherited only from the mother.

TAKE-HOME MESSAGE 41.7

- ✔ Sexual arousal involves signals from the endocrine and nervous systems.
- ✔ Ejaculation releases millions of sperm into the vagina. Sperm travel through the uterus toward the oviducts, where fertilization most often occurs.
- ✔ Penetration of a secondary oocyte by a single sperm causes the oocyte to complete meiosis II and prevents additional sperm from penetrating the oocyte.
- ✔ The DNA of the sperm, along with that of the oocyte, become the genetic material of the zygote.

CREDIT: (11) Courtesy of Elizabeth Sanders, Women's Specialty Center, Jackson, MS.

41.8 Contraception and Infertility

LEARNING OBJECTIVES

- Describe the mechanisms used to prevent pregnancy.
- List the possible causes of infertility.

Birth Control Options

Emotional and economic factors often lead people to seek ways to control their fertility. **TABLE 41.2** lists common contraceptive options and compares their effectiveness. Most effective is abstinence—no sex—which has the added advantage of preventing exposure to sexually transmissible pathogens.

With rhythm methods, a woman abstains from sex during her fertile period. She calculates when she is fertile by recording how long menstrual cycles last, checking her temperature daily, monitoring the thickness of her cervical mucus, or some combination of these methods. However, cycles vary, so miscalculations are frequent and sperm deposited in the vagina up to three days before ovulation can survive long enough to fertilize an egg.

TABLE 41.2

Common Methods of Contraception

Method	Mechanism of Action	Pregnancy Rate*
Abstinence	Avoid intercourse entirely	0% per year
Rhythm method	Avoid intercourse when female is fertile	25% per year
Withdrawal	End intercourse before male ejaculates	27% per year
Vasectomy	Cut or close off male's vasa deferentia	<1% per year
Tubal ligation	Cut or close off female's oviducts	<1% per year
Condom	Enclose penis, block sperm entry to vagina	15% per year
Diaphragm, cervical cap	Cover cervix, block sperm entry to uterus	16% per year
Spermicides	Kill sperm	29% per year
Intrauterine device	Prevent sperm entry to uterus or prevent implantation	<1% per year
Oral contraceptives	Prevent ovulation	<1% per year
Hormone patches, implants, or injections	Prevent ovulation	<1% per year
Emergency contraception pill	Prevent ovulation	15–25% per use**

** Percentage of users who get pregnant despite consistent, correct use*

*** Not meant for regular use*

Removing the penis from the vagina before ejaculation (withdrawal) requires great willpower. In some men, pre-ejaculation fluid from the penis contains enough sperm to allow fertilization to occur. Rinsing out the vagina (douching) immediately after intercourse is similarly unreliable. Some sperm can travel through the cervix within seconds of ejaculation.

Surgical methods are highly effective but are meant to make a person permanently sterile. Men may opt for a vasectomy. A doctor makes a small incision into the scrotum, then cuts and ties off each vas deferens. A tubal ligation blocks or cuts a woman's oviducts.

Other fertility control methods use physical and chemical barriers to stop sperm from reaching an egg. Spermicidal foam and spermicidal jelly poison sperm. They are not always reliable, but their use with barrier methods reduces the chance of pregnancy.

A diaphragm is a flexible, dome-shaped device that is positioned inside the vagina so it covers the cervix. It is relatively effective if it is first fitted by a doctor and is used correctly with a spermicide. A cervical cap is a similar but smaller device.

Condoms are thin sheaths that are worn over the penis or used to line the vagina during intercourse. Reliable brands can be 95 percent effective when used correctly with a spermicide. Condoms made of latex also offer some protection against sexually transmitted diseases. However, even the best ones can tear or leak, reducing effectiveness.

An intrauterine device, or IUD, is inserted into the uterus by a physician. Some IUDs cause cervical mucus to thicken so sperm cannot swim through it. Others shed copper that interferes with implantation.

Oral contraceptives—birth control pills—are the most common fertility control method in developed countries. "The Pill" is a mixture of synthetic estrogens and progesterone-like hormones that prevents both maturation of oocytes and ovulation. When taken diligently, oral contraception is highly effective. It also reduces menstrual cramps and lowers the risk of ovarian and uterine cancer. Side effects include nausea, headaches, weight gain, and increased risk of cancer of the breast, cervix, and liver.

Hormones can also be delivered by a birth control patch (an adhesive patch applied to skin), by injections, or by an implant. Injections are effective for several months, whereas an implant such as Implanon is effect for years. Both methods are quite effective, but may cause sporadic, heavy bleeding.

If a woman has unprotected sex or a condom breaks, she can turn to emergency contraception. Some so-called morning-after pills are available without a prescription to women over age 17. The most widely

Data Analysis Activities

Stuck Sperm Cells lining the epididymis secrete a glycoprotein (beta-defensin) that coats sperm and facilitates their passage through cervical mucus. There are two common alleles for human beta-defensin: a wild-type allele (*wt*) and an allele with a deletion (*del*). To find out if this genetic variation could affect fertility, Gary Cherr and his colleagues tested the ability of sperm from men with different genotypes to pass through a gel barrier. The barrier was designed to simulate cervical mucus. The researchers placed sperm on one side of the gel, then used a video camera to record how many passed into the gel over time. **FIGURE 41.12** illustrates their results.

1. After four minutes, how many sperm from men with a wild-type allele were in the camera's view on average?
2. How long did it take for the same number of sperm from men homozygous for the deletion to make it into the view of the camera?
3. Given these results, which genotype or genotypes would you expect to be associated with infertility?
4. The beta-defensin gene is on chromosome 20. What percentage of sperm from a heterozygous man would you expect to carry the *del* allele?

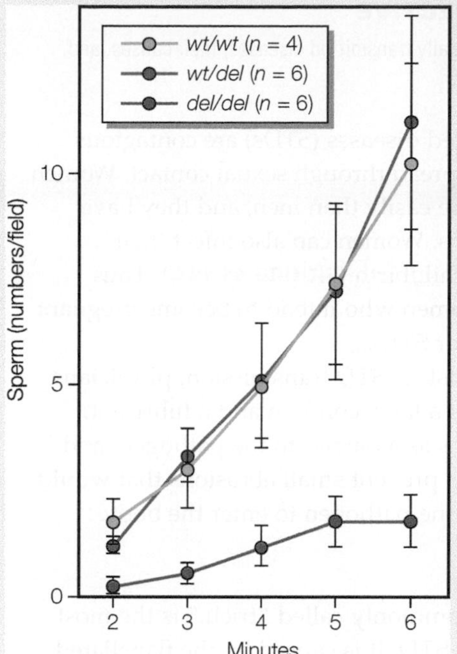
wt/wt (n = 4)
wt/del (n = 6)
del/del (n = 6)
Sperm (numbers/field)
0
5
10
2
3
4
5
6
Minutes

FIGURE 41.12 Effect of beta-defensin genotype on the ability of sperm to penetrate a gel that simulates cervical mucus. Sperm from 16 men was tested: 4 *wt*/*wt*, 6 *wt*/*del*, and 6 *del*/*del*.

Fresh sperm were placed on one side of the gel, and a video camera was focused on a region 2.75 mm into the gel. The number of sperm in this region was recorded at one-minute intervals for six minutes. Sperm numbers shown are averages. Bars indicate range.

used emergency contraceptive pill delivers a large dose of synthetic progesterone that prevents ovulation and interferes with fertilization. It is most effective when taken immediately after intercourse but has some effect up to three days later. Morning-after pills are not meant for regular use. They cause nausea, vomiting, abdominal pain, headache, and dizziness.

Infertility

About 10 percent of couples in the United States are infertile, which means they do not have a successful pregnancy despite a year of trying. Infertility increases with age. In about half of cases infertility arises in the female partner. The most common cause of female infertility is a hormonal disorder called polycystic ovarian syndrome. With this disorder, a follicle begins to mature, but ovulation does not occur; instead the follicle turns into a cyst (a fluid-filled sac).

Scarring of the oviducts can also cause female infertility. Such scarring can result from a sexually transmitted disease or from endometriosis. Endometriosis is a disorder in which endometrial tissue (tissue that normally lines the uterus) grows in the regions of the pelvis outside the uterus. Like normal endometrium, the misplaced tissue proliferates and is shed in response to hormones. The result is pain and scarring that can distort oviducts. Endometriosis can also make intercourse painful, thus lessening its frequency and the opportunities for conception.

If an embryo does form, it may not implant properly in the uterus. A tubal pregnancy occurs when the embryo mistakenly implants in an oviduct. Such a pregnancy threatens the life of the mother. Endometriosis and other uterine problems can interfere with the ability of the embryo to implant.

Male infertility arises when a man makes abnormal sperm or too few sperm, or when sperm are not ejaculated properly. Disorders that lower testosterone can prevent sperm maturation. Some genetic disorders can interfere with sperm motility. In males, as in females, sexually transmitted disease can scar ducts of the reproductive system, causing blockages that keep sperm from leaving the body. Male infertility can also arise if the valve between the bladder and urethra does not function properly, so sperm end up in the bladder rather than leaving the body through the urethra.

TAKE-HOME MESSAGE 41.8

- ✔ Couples can prevent pregnancy by avoiding sex entirely or abstaining when the woman is fertile.
- ✔ Hormones delivered by pills, patches, or injections can prevent a woman from ovulating.
- ✔ Barriers such as condoms or a diaphragm keep sperm and egg apart, as do surgical methods such as vasectomy or tubal ligation.
- ✔ Infertility can arise from hormonal disorders, blockages of reproductive ducts, and uterine problems that prevent implantation.

41.9 Sexually Transmitted Diseases

LEARNING OBJECTIVE

- List common sexually transmitted diseases, their causes, and their effects.

Sexually transmitted diseases (STDs) are contagious diseases that are spread through sexual contact. Women contract STDs more easily than men, and they have more complications. Women can also infect their offspring during childbirth (**FIGURE 41.13A**). Thus sexually active women who intend to become pregnant should be tested for STDs.

To reduce the risk of STD transmission, physicians recommend use of a latex condom and a lubricant. The condom serves as a barrier to the pathogen, and the lubricant helps prevent small abrasions that would make it easier for the pathogen to enter the body.

Trichomoniasis

Trichomoniasis, commonly called "trich," is the most common nonviral STD. It is caused by the flagellated protozoan *Trichomonas vaginalis*, which is shown in Figure 21.8. Unlike some other sexually transmitted pathogens, *T. vaginalis* does not colonize the mouth or the rectum.

In the United States, an estimated 3.7 million people currently have trichomoniasis. Most infections do not cause obvious symptoms, but some women have a yellowish discharge and a painful, itchy vagina. In pregnant women, an infection increases the risk of a premature birth. In both sexes, an untreated infection can damage the reproductive tract and cause infertility. A single dose of an antiprotozoal drug can quickly cure the infection.

Bacterial STDs

In the United States, the Centers for Disease Control (CDC) monitors the number of cases for three well-known bacterial STDs: chlamydia, gonorrhea, and syphilis. In 2015, this agency received reports of more than 1.5 million cases of chlamydia, about 395,000 reports of gonorrhea, and 24,000 reports of syphilis.

Infection by *Mycoplasma genitalium* is a recently recognized bacterial STD. There is no routine screening for *M. genitalium* infection and cases are not reported to the CDC, so its prevalence is not fully known. However, recent studies suggest that such infections are more common than gonorrhea.

All bacterial STDs can be spread by sexual contact with the penis, vagina, mouth, or anus of an infected partner. In men, these infections usually cause a discharge from the penis and painful urination. Women typically have no initial symptoms, but they can develop pelvic inflammatory disease (PID). This secondary outcome of bacterial STDs scars the female reproductive tract and can cause infertility and chronic pain. PID also raises the risk of a tubal pregnancy.

Syphilis can also have devastating effects throughout the body. The earliest symptom of syphilis is flattened, painless ulcers called chancres (pronounced "shankers") at the site of initial infection. As the bacteria spread, more widely distributed chancres appear (**FIGURE 41.13B**). A long-term infection can destroy organs causing joint pain, mental illness, blindness, and eventually death.

A bacterial STD can be cured with an appropriate antibiotic. However, use of an antibiotic selects for bacteria resistant to that medicine. Thus, the frequency of sexually transmitted bacterial pathogens that resist most commonly used antibiotics is on the rise.

Viral STDs

Unlike bacterial STDs, viral infections cannot be cured with medication. However, antiviral drugs can decrease the negative effects of infection.

Infection by human papillomaviruses (HPV) is widespread. Many HPV strains cause warts (bumplike growths), including those common on hands or feet. Some sexually transmitted strains of the virus cause genital warts (**FIGURE 41.13C**) and others can cause cancer. The overwhelming majority of cervical cancers in women and anal cancers in homosexual men are caused by HPV infection. Similarly, HPV introduced into the mouth by oral sex raises the risk of mouth and throat cancer.

Vaccinations and screenings are key to the fight against HPV-associated cancers. A vaccine can prevent HPV infection in both males and females, if given before viral exposure. A blood test can determine whether a person has been infected by one of the two strains that most commonly cause cancer, and a Pap test can detect early signs of cervical cancer.

About 45 million Americans have genital herpes caused by herpes simplex virus 2. An initial infection commonly causes small sores at the site where the virus entered the body. The sores heal, but the virus can be reactivated, causing tingling or itching with or without visible sores. Antiviral drugs cannot cure the infection, but they help sores heal faster and also lessen the likelihood of viral reactivation.

Transmission of the herpes virus from mother to child during birth is rare—the mother must have an active infection at the time of delivery—but it can have serious consequences. If untreated, an infant infected in this manner has a 40 percent chance of death. Those who survive remain infected and have a heightened

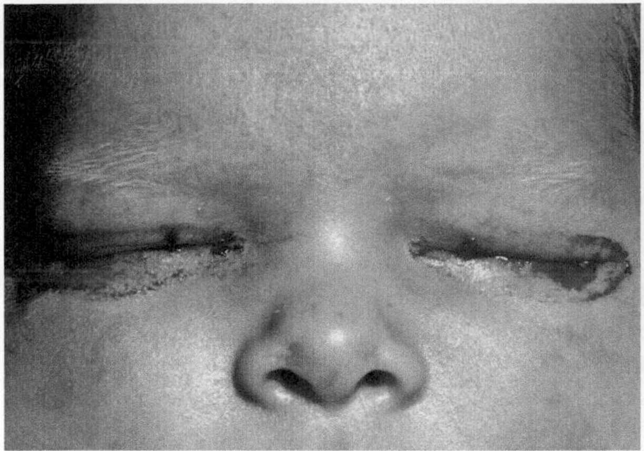

A Passage of a pathogen to offspring. This infant's eyes are infected by *Chlamydia* acquired from its mother during birth.

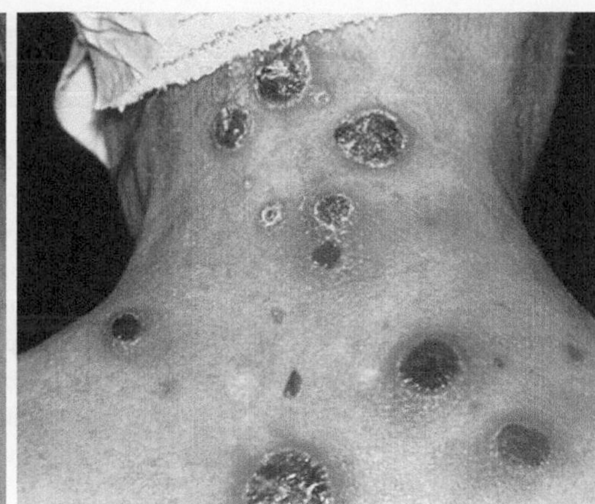

B Spread of a pathogen throughout the body. The chancres (open sores) are a sign of syphilis.

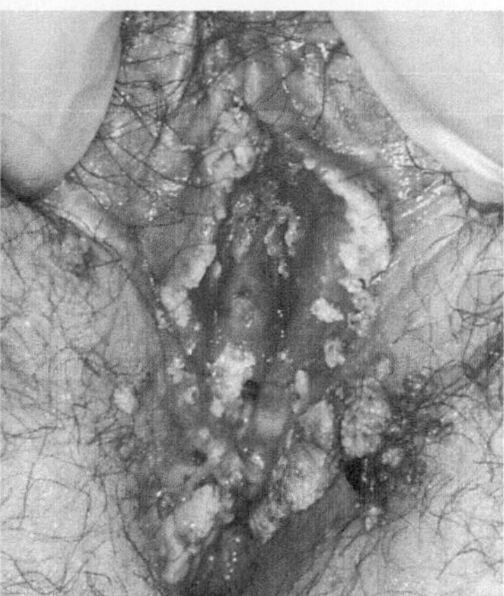

C Warty genitals and increased risk of cancer. The raised white warts on this woman's labia are a sign of HPV.

FIGURE 41.13 Negative effects of sexually transmitted diseases.

risk of health problems including blindness and impaired brain development. Women who become infected for the first time while pregnant are especially likely to have an active infection at delivery.

Infection by HIV (human immunodeficiency virus) can cause AIDS (acquired immunodeficiency syndrome). Sections 20.1 and 20.3 described the HIV virus, and Section 37.9 explained how its effects on the immune system can result in AIDS. With regard to sexual transmission, oral sex is least likely to pass on an infection. Unprotected anal sex is 5 times more dangerous than unprotected vaginal sex and 50 times more dangerous than oral sex.

If you think you may have been exposed to HIV, get tested as soon as possible. Treatment with antiviral drugs can prevent an HIV-infected person from developing AIDS and reduce his or her likelihood of passing the infection to others.

TAKE-HOME MESSAGE 41.9

- ✔ Viruses, bacteria, and protozoans cause common STDs.
- ✔ Effects of infection range from discomfort, to infertility, to systemic problems or cancer.
- ✔ Infected women can pass an STD to their offspring.
- ✔ Bacterial and protozoan pathogens can be killed by medications. Viral infections cannot be cured in this manner, but antiviral drugs reduce the effects of herpes and HIV infection.
- ✔ An HPV vaccine can prevent infection by this virus.

41.1 Assisted Reproduction (revisited)

The most common form of assisted reproduction is the use of donor sperm. In the United States, the Food and Drug Administration regulates sperm banks, requiring them to screen donors for good health, test them for STDs, take an extensive family history, and test semen for the presence of certain pathogens. It does not, however, mandate any type of genetic screening.

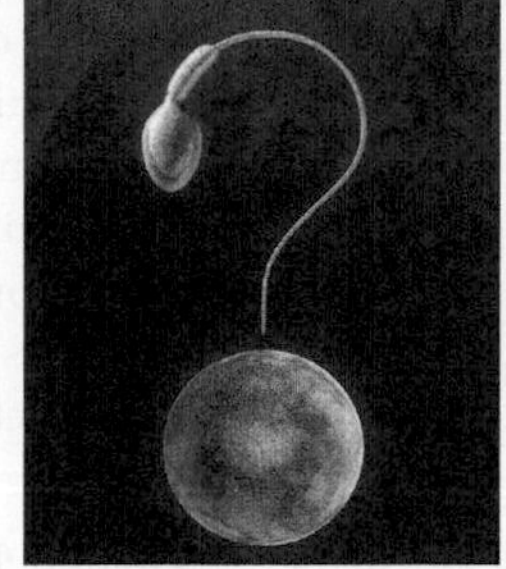

As a result, it is possible for young donors, who are healthy and capable of passing a physical exam, to unknowingly pass on recessive alleles for genetic disorders or alleles for dominant late-onset disorders. In one case, a man who donated sperm as a 23-year-old was later diagnosed with a genetic disorder that weakens the heart and raises the risk of early death. Of the man's 24 children, 22 fathered as a donor and 2 with his wife, 9 have the allele for this disorder. One died at age 2, and two had heart problems by age 15.

To minimize genetic risks, sperm banks now voluntarily screen donors for alleles most commonly associated with genetic disorders. However, even with expensive and extensive testing, they cannot eliminate the possibility of some less common harmful alleles from slipping by.

It's also important to remember that unique allele combinations affecting health can arise whenever people mate, whether with a spouse or an anonymous donor. A shuffling of the genetic cards is integral to sexual reproduction. ●

CREDITS: (13A) Western Ophthalmic Hospital/Science Source; (13B) CNRI/Science Source; (13C) Courtesy of CDC/Joe Millar; (in text) Lightspring/Shutterstock.

STUDY GUIDE

Section 41.1 Millions of people have been born as a result of assisted reproductive methods such as ***in vitro* fertilization** and sperm banking. The variety of reproductive options continues to expand.

Section 41.2 **Asexual reproduction** yields genetically identical copies (clones) of the parent. **Sexual reproduction** produces genetic variety in offspring, which can be advantageous in a changing environment.

Most animals that reproduce sexually have separate sexes, but some are **hermaphrodites** that produce both **eggs** and **sperm**. With **external fertilization**, gametes are released into water. Most animals on land have **internal fertilization**; gametes meet in a female's body.

Offspring may develop inside or outside the maternal body. Yolk helps nourish developing young of most animals. In placental mammals, young are sustained by nutrients delivered across the **placenta**.

Section 41.3 Gametes form in **gonads**— sperm in **testes** and eggs in **ovaries**. With **spermatogenesis**, meiosis and equal cytoplasmic divisions produce four haploid sperm. With **oogenesis**, meiosis I followed by unequal division produces an **oocyte** and a **polar body**. Meiosis II of this oocyte is followed by another unequal division to yield a second polar body and an **ovum**. In most animals, ducts convey mature gametes to the body surface and glands assist in reproductive function.

Sections 41.4, 41.5 Human ovaries produce eggs and sex hormones (**estrogens** and **progesterone**). Cilia move an oocyte through an **oviduct** to the **uterus** (womb). The **cervix** of the uterus opens into the **vagina**, which serves as the organ of intercourse and the birth canal.

From puberty until **menopause**, a woman has an approximately monthly **menstrual cycle**. **Gonadotropin-releasing hormone (GnRH)** secreted by the hypothalamus causes the pituitary to release **follicle-stimulating hormone (FSH)** and **luteinizing hormone (LH)**.

FSH causes an **ovarian follicle** to begin maturing. Follicle cells around a **primary oocyte**, which formed before birth, proliferate and secrete estrogens and progesterone. A midcycle surge of luteinizing hormone (LH) triggers **ovulation** of the now **secondary oocyte**. After ovulation, the **corpus luteum** secretes progesterone that primes the uterus for pregnancy. When the corpus luteum breaks down, **menstruation** occurs.

Fraternal twins arise when two oocytes mature during the same cycle and each is fertilized.

Most female mammals do not menstruate. They instead reabsorb the lining of their uterus. Most also have an **estrous cycle**, so they are sexually receptive only during intervals when they can conceive.

Section 41.6 A **scrotum** encloses the human testes, which are suspended beneath the pelvic girdle. The testes produce sperm and the hormone **testosterone**. Beginning with puberty, sperm form continually from germ cells inside the testes' **seminiferous tubules**. The sperm mature in an **epididymis** that opens into a **vas deferens**. Secretions from the seminal vesicles and prostate gland join with sperm to form **semen**. Semen is expelled from the body through the urethra that runs through the **penis**. Like oogenesis, spermatogenesis is governed by LH and FSH.

A mature sperm is a flagellated cell. The "head" of the sperm is capped by an enzyme-containing **acrosome**.

Section 41.7 Hormones and autonomic nerves govern physiological changes during arousal and intercourse. Ejaculation delivers millions of sperm into the vagina. To continue onward, sperm must swim through the mucus secreted by the cervix and enter the uterus. From the uterus, sperm enter a oviduct randomly. Fertilization normally occurs in the oviduct.

A secondary oocyte released at ovulation is surrounded by the zona pellucida, a layer of glycoproteins that bind sperm. Enzymes from the acrosomes of many sperm open a path through this layer to the oocyte's plasma membrane. Usually only one sperm penetrates a secondary oocyte. Binding of this sperm to the oocyte plasma membrane causes changes that prevent other sperm from binding and entering.

Fertilization causes the secondary oocyte to complete meiosis II. The nucleus of the resulting ovum and that of the sperm supply genetic material of the zygote.

Section 41.8 Humans can choose either to prevent pregnancy by abstaining from sex entirely or to reduce the chance of pregnancy by abstaining during a woman's fertile period, surgically severing reproductive ducts, placing a physical or chemical barrier at the entrance to the uterus, or administering synthetic female sex hormones to prevent ovulation.

Infertility can arise as a result of disorders that impair gamete production. Blocked reproductive ducts and uterine problems that interfere with implantation or development also affect fertility.

Section 41.9 Sexual interactions can pass viral, bacterial, and protozoan pathogens between partners. Infected mothers can transmit these pathogens to their offspring. The consequences of a sexually transmitted disease range from mild discomfort to sterility and systemic disease that can be fatal. Protozoan and bacterial STDs can be cured with medications, but there are no such treatments to cure viral STDs.

SELF-QUIZ

Answers in Appendix VII

1. Sexual reproduction ________ .
   a. requires formation of gametes by meiosis
   b. produces offspring identical in their traits
   c. occurs only in vertebrates
   d. all of the above

2. The ________ is a genetic dead end.
   a. polar body c. ovum
   b. oocyte d. sperm

3. The cervix is the entrance to the ________ .
   a. oviducts c. uterus
   b. vagina d. scrotum

4. Ovulation releases a(n) ________ into an oviduct.
   a. primary oocyte c. ovum
   b. secondary oocyte d. follicle cell

5. During a menstrual cycle, a midcycle surge of ________ from the ________ triggers ovulation.
   a. FSH; pituitary c. LH; pituitary
   b. progesterone; egg d. estrogen; follicle cells

6. The corpus luteum develops from ________ and secretes progesterone that causes the lining of the uterus to thicken.
   a. follicle cells c. a primary oocyte
   b. the polar body d. a secondary oocyte

7. Semen contains secretions from the ________ .
   a. adrenal gland c. prostate gland
   b. pituitary gland d. corpus luteum

8. Male germ cells divide by meiosis inside the ________ .
   a. urethra c. prostate gland
   b. seminiferous tubules d. vasa deferentia

9. A male attains an erection when ________ .
   a. muscles running the length of the penis contract
   b. Leydig cells release a surge of testosterone
   c. the posterior pituitary releases oxytocin
   d. spongy tissue inside the penis fills with blood

10. Binding of a sperm to an oocyte plasma membrane causes the oocyte to ________ .
   a. enter an oviduct c. secrete estrogen
   b. complete meiosis II d. expel its nucleus

11. Birth control pills deliver synthetic ________ .
   a. estrogens and progesterone
   b. LH and FSH
   c. testosterone
   d. oxytocin and prostaglandins

12. A vasectomy blocks or cut the ________ .
   a. oviducts
   b. urethra
   c. seminiferous tubules
   d. vas deferens

13. Match each disease with the type of organism that causes it. The choices can be used more than once.

___ chlamydial infection	a. bacteria
___ AIDS	b. protist
___ syphilis	c. virus
___ genital warts	
___ gonorrhea	
___ genital herpes	
___ trichomoniasis	

14. Match each structure with its description.

___ testis	a. conveys sperm out of body
___ epididymis	b. secretes semen components
___ labia majora	c. stores sperm
___ urethra	d. enclosed by scrotum
___ vagina	e. releases oocyte during ovulation
___ ovary	
___ oviduct	f. usual site of fertilization
___ prostate gland	g. lining of uterus
___ endometrium	h. fat-padded skin folds
	i. birth canal

15. Match each hormone with its source.

___ FSH and LH	a. pituitary gland
___ GnRH	b. ovaries
___ estrogens	c. hypothalamus
___ testosterone	d. testes

CRITICAL THINKING

1. Drugs that interfere with sympathetic nerve signals are often prescribed for men who have high blood pressure. How might such drugs impair sexual performance?

2. Fraternal twins are nonidentical siblings that form when two eggs mature, are released, and are fertilized at the same time. Explain why an unusually high level of FSH raises the likelihood of fraternal twins.

3. Some sperm mitochondria enter an egg during fertilization, but as sperm mature these mitochondria are tagged with a protein (ubiquitin) that marks them for destruction. What organelle carries out this destruction process?

4. Occasionally more than one sperm penetrates and enters a human secondary oocyte. The result is formation of a multiple spindles and disruption of the first mitotic division. Explain why the presence of multiple sperm has this lethal disruptive effect.

42 ANIMAL DEVELOPMENT

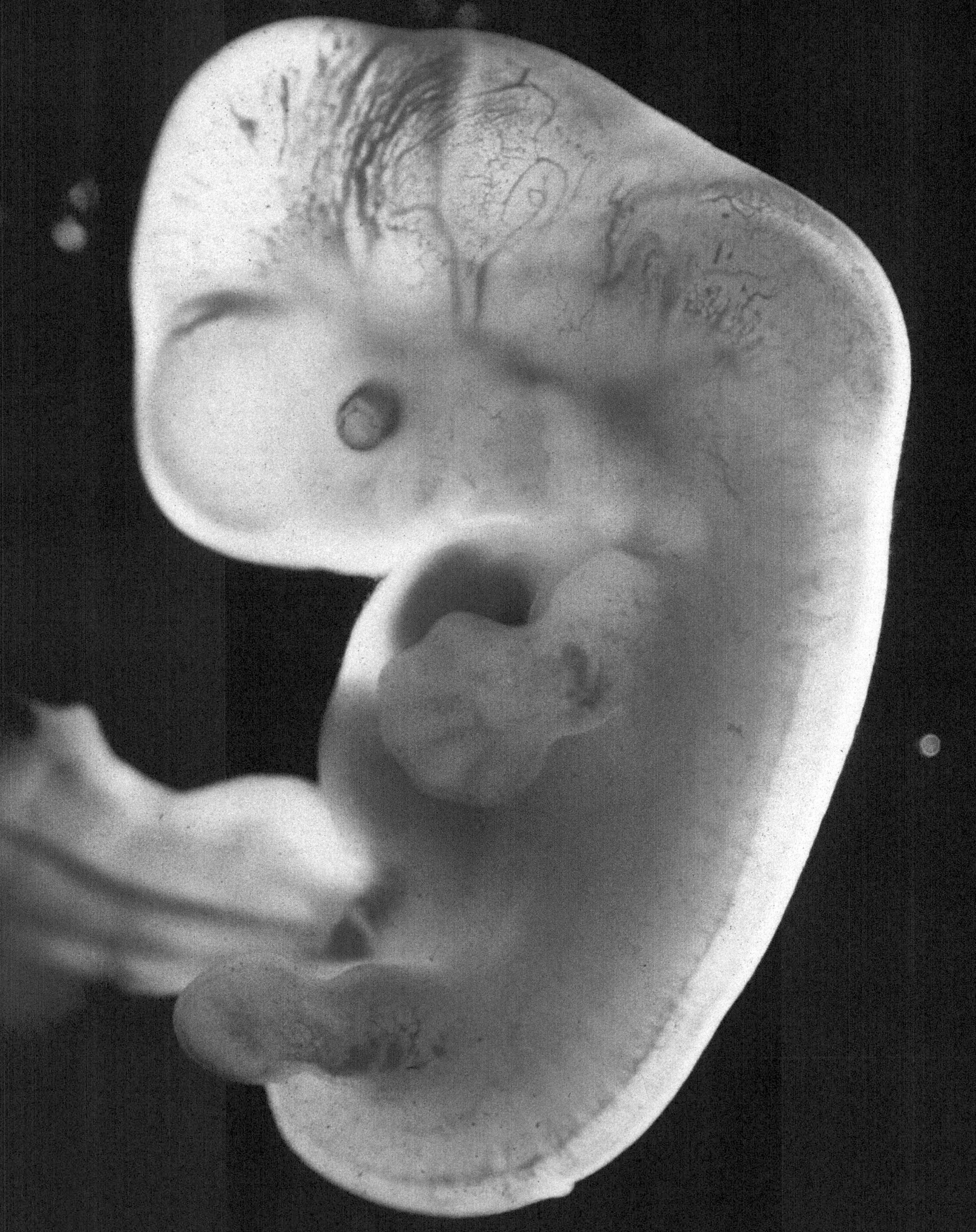

CORE CONCEPTS

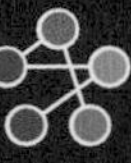

Systems

Complex properties arise from interactions among components of a biological system.

During development, cascades of master regulator expression shape a complex, multicelled body from a single-celled zygote. Localized patterns of gene expression trigger responses in embryonic cells. These responses, which include cell differentiation, cell movement, and controlled patterns of cell death, ultimately result in the formation of tissues and organs in specific regions at specific times.

Evolution

Evolution underlies the unity and diversity of life.

Similarities in embryonic development among all sexually reproducing animals are evidence of shared ancestry. Differences in body form among animal groups arise through modifications in the timing and regulation of developmental pathways.

Process of Science

The field of biology consists of and relies upon experimentation and the collection and analysis of scientific evidence.

Studies of animal development have shaped our understanding of the processes that underly human development. Experimental manipulation of animals such as fruit flies and amphibians has revealed the regulatory mechanisms that underlie development of all sexually reproducing animals.

Links to Earlier Concepts

This chapter continues the discussion of reproductive systems begun in Chapter 41. We revisit cell differentiation (10.2), master regulators (10.3), and embryonic tissue layers (24.2). You will be reminded how vertebrate body plans evolved (25.2–25.7), and will explore the development of humans, which are placental mammals (25.11).

42.1 Prenatal Problems

Development of a human infant from a single-celled zygote is a complicated process, so a lot can go wrong along the way. When it does, a pregnancy can end in miscarriage or stillbirth, or a child can be born with a birth defect. With miscarriage, death occurs before 20 weeks. Miscarriages that occur in the first month or two of development often go undetected, so their frequency is not known. The risk of miscarriage for recognized pregnancies is 12 to 15 percent. Most of these occur early in pregnancy. Death of a fetus after 20 weeks is less common and is called a stillbirth.

Disordered development that is not fatal can result in a birth defect, a structural malformation that is present at birth. In the United States, such defects occur in about 3 percent of births. A malformed heart, neural tube defects, and cleft lip or palate are common birth defects (**FIGURE 42.1**).

About half of miscarriages and between 5 and 10 percent of stillbirths result from chromosomal abnormalities (Section 14.6). These abnormalities, and thus the risk of an unsuccessful pregnancy, increase with both paternal and maternal age. Risk of birth defects also rises with parental age.

Some birth defects result from an environmental factor such as exposure to a teratogen. A **teratogen** is a toxin or infectious agent that interferes with development. Nicotine and alcohol are teratogens; prenatal (before birth) exposure to either raises the risk of cleft lip, as well as other defects. Prenatal exposure to the rubella virus can cause deafness and heart defects. Exposure to the Zika virus can cause microcephaly—an abnormally small brain and head.

Dietary deficiencies cause birth defects too. If a pregnant woman's diet includes too little iodine, her newborn may be affected by cretinism, a disorder that affects brain function and motor skills (Section 34.6). A maternal deficiency in folate (folic acid) can impair

teratogen A toxin or infectious agent that interferes with development and thus raises the risk of birth defects.

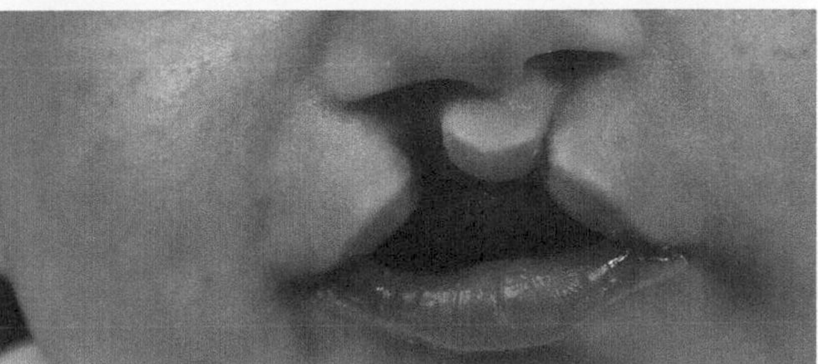

FIGURE 42.1 Cleft lip of an infant. This birth defect arises when tissues fail to fuse properly during early development.

CREDITS: (opposite) Lennart Nilsson/Bonnierforlagen AB; (1) Dr M.A. Ansary/Science Source.

neural tube formation. The neural tube develops into the brain and spinal cord.

Although scientists have identified many risk factors for birth defects, most such defects arise from unknown causes. The federally funded National Birth Defects Prevention Study (NBDPS) investigates the causes of birth defects. Since 1997, participating researchers have been interviewing and collecting genetic material from women who gave birth to a child with a birth defect, as well as from women who gave birth to normal children and can serve as controls. So far, over 35,000 women have participated, and the research is ongoing. ●

42.2 Stages of Animal Development

LEARNING OBJECTIVES

- Describe what occurs during cleavage and the product of this developmental process.
- Name the three embryonic tissue layers of vertebrate embryos, and describe the process by which they arise.
- Give examples of the structures derived from each of the three germ layers.

A General Model for Animal Development

Through studies of animals such as roundworms, fruit flies, fishes, and mice, researchers have come up with a general model for development. In all sexually reproducing animals, a new individual begins life as a zygote, the diploid cell that forms at fertilization (Section 12.2). Development from a zygote to an adult typically proceeds through a series of stages.

As an example of animal development, consider **FIGURE 42.2**, which shows the reproduction and development of a leopard frog. A female frog releases eggs into the water, and a male releases sperm onto the eggs. External fertilization produces the zygote. New cells arise when **cleavage** carves up a zygote by repeated mitotic cell divisions ❶. During cleavage, the number of cells increases, but the zygote's original volume remains unchanged. As a result, cells become more numerous but smaller.

Cells that form during cleavage are called blastomeres. They typically become arranged as a **blastula**: a ball of cells that encloses a cavity (a blastocoel) filled with cellular secretions ❷. Tight junctions hold the cells of the blastula together.

During **gastrulation**, cells of the blastula move about and become organized in layers. In cnidarians such as jellies, the gastrula has two primary tissue layers or **germ layers**. The outermost layer is **ectoderm**, and the inner layer is **endoderm.** However, in most animals (and all vertebrates) gastrulation produces a three-layered **gastrula** ❸. The gastrula has a layer

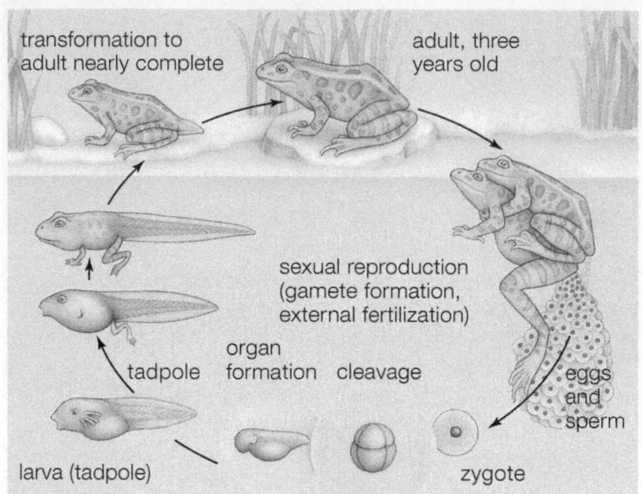
transformation to adult nearly complete
adult, three years old
sexual reproduction (gamete formation, external fertilization)
organ formation
cleavage
tadpole
eggs and sperm
larva (tadpole)
zygote

FIGURE 42.2 An example of stages in animal reproduction and development.

Above, overview of reproduction and development in the leopard frog. Opposite page, a closer look at each stage.

of **mesoderm** in between the outer ectoderm and the inner endoderm.

The three germ layers give rise to the same types of tissues and organs in all vertebrates (**TABLE 42.1**). Ectoderm, the outer germ layer, is the first to form. It develops into nervous tissue and the outer layer of skin or another body covering. Endoderm, the inner layer, is the source of the respiratory tract and gut lining. Mesoderm, the middle layer, is the source of muscles, connective tissues, and the circulatory system.

Specialized tissue and organs begin to form after the completion of gastrulation. Recall that a neural tube and a hollow dorsal notochord characterize chordate embryos (Section 25.2). These structures form very early in development ❹.

Many organs incorporate tissues derived from more than one germ layer. For example, the stomach's epithelial lining is derived from endoderm, and the smooth muscle that makes up the stomach wall develops from mesoderm. In frogs, as in some other animals, a larva (in this case a tadpole) undergoes metamorphosis, a remodeling of tissues into the adult form ❺.

TABLE 42.1

Derivatives of Vertebrate Germ Layers

Ectoderm (outer layer)	Outer layer (epidermis) of skin; nervous tissue
Mesoderm (middle layer)	Connective tissue of skin; skeletal, cardiac, smooth muscle; bone; cartilage; blood vessels; urinary system; peritoneum (coelom lining); gut organs; reproductive tract
Endoderm (inner layer)	Lining of gut and respiratory tract, and organs derived from these linings

❶ Here we show the first three divisions of cleavage, a process that carves up a zygote's cytoplasm. In this species, cleavage results in a blastula, a ball of cells with a fluid-filled cavity.

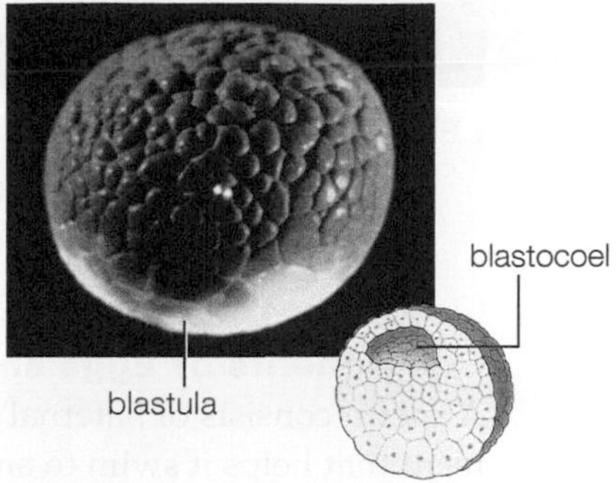

blastocoel
blastula

❷ Cleavage is over when the blastula forms.

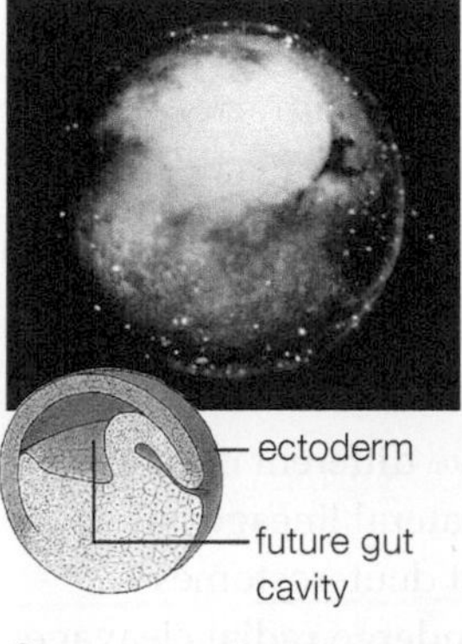

ectoderm
future gut cavity

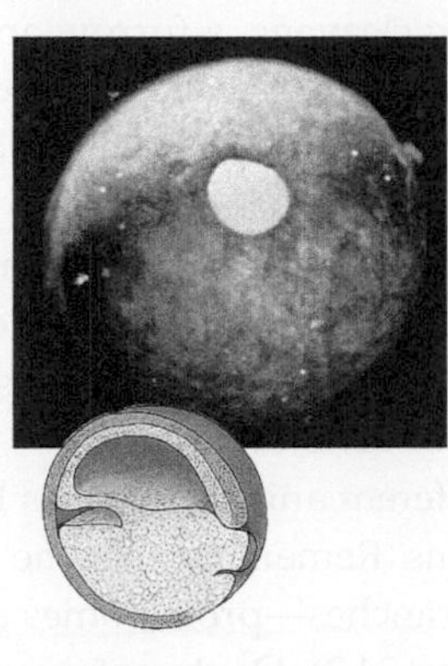

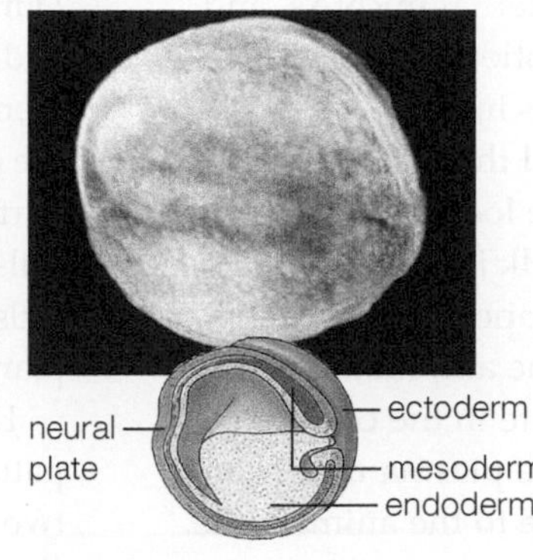

neural plate
ectoderm
mesoderm
endoderm

❸ The blastula becomes a three-layered gastrula—a process called gastrulation. At the dorsal lip (a fold of ectoderm above the first opening that appears in the blastula), cells migrate inward and start rearranging themselves.

neural tube
notochord
gut cavity

❹ Organs begin to form as a primitive gut cavity opens up. A neural tube, then a notochord and other organs, form from the primary tissue layers.

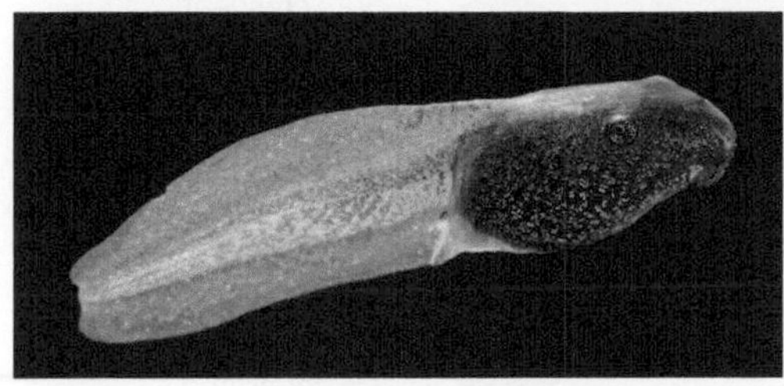

Tadpole, a swimming larva with segmented muscles and a notochord extending into a tail.

Limbs develop and the tail is absorbed during metamorphosis from the larval to the adult form.

❺ The frog's body form changes as it grows and its tissues specialize. The embryo becomes a tadpole, which metamorphoses into an adult.

TAKE-HOME MESSAGE 42.2

✔ A zygote undergoes cleavage, which increases the number of cells without increasing the size of the embryo. Cleavage ends with formation of a blastula.

✔ During gastrulation, cells of the blastula cells are rearranged to form a gastrula. In most animals, the gastrula has three layers: an outer ectoderm, a middle layer of mesoderm, and an inner layer of endoderm.

✔ After gastrulation, specialized tissues and organs begin to form.

✔ Continued growth and tissue specialization produce the adult body.

blastula (BLAST-yoo-luh) In animal development, a hollow ball of cells that forms as a result of cleavage.

cleavage In animal development,the process by which multiple mitotic divisions produce a blastula from a zygote.

ectoderm (ECK-toh-derm) Outermost primary tissue layer (germ layer) of an animal gastrula.

endoderm (EN-doh-derm) Innermost primary tissue layer (germ layer) of an animal gastrula.

gastrula (GAS-troo-luh) Three-layered developmental stage formed by gastrulation in an animal.

gastrulation (gas-troo-LAY-shun) In animal development, the process by which cell movements produce a three-layered gastrula.

germ layer Primary tissue layer that forms when a developing animal undergoes gastrulation.

mesoderm (MEZ-oh-derm) Middle tissue layer (germ layer) in a three-layered animal gastrula.

CREDITS: (2 #1–4) photos, Carolina Biological Supply Company; (#5) Ardea/Bevan, Brian/Animals Animals.

42.3 From Zygote to Gastrula

LEARNING OBJECTIVES

- Explain why cells of a blastula differ in their contents.
- Describe how experimental manipulation of amphibian embryos revealed the importance of cytoplasmic localization.

Components of Eggs and Sperm

A sperm consists of paternal DNA and a bit of equipment that helps it swim to and penetrate an egg. The egg has far more cytoplasm. It also has yolk proteins that will nourish the embryo, maternal mRNAs that encode proteins essential to early development, tRNAs and ribosomes to translate the maternal mRNAs, and the proteins required to build mitotic spindles.

Cytoplasmic localization occurs in all oocytes: Instead of being evenly distributed throughout egg, many cytoplasmic components are localized in one region. In a yolk-rich egg, most yolk is at one pole (called the vegetal pole) and the opposite pole (the animal pole) has little yolk. In some amphibian eggs, dark pigment molecules accumulate in the cell cortex, a cytoplasmic region just below the plasma membrane. Pigment is most concentrated close to the animal pole. After fertilization, the cortex rotates to reveal a gray crescent, a lightly pigmented region (**FIGURE 42.3A**).

Early in the 1900s, experiments by Hans Spemann showed that substances essential to development are localized in the gray crescent. In one experiment, Spemann separated the first two blastomeres formed at cleavage. Each cell had half of the gray crescent and developed normally (**FIGURE 42.3B**). In the next experiment, Spemann altered the cleavage plane (**FIGURE 42.3C**). One blastomere received all the gray crescent, and it developed normally. The other lacked gray crescent and halted development at the blastula stage.

Cleavage—Onset of Multicellularity

During cleavage, a furrow appears on the cell surface and defines the plane of the cut. Beneath the plasma membrane, a ring of microfilaments contracts, dividing the cell in two (Section 11.4). Cleavage puts different parts of the egg cytoplasm into different descendant cells. The types and proportions of substances that cells inherit, as well as their size, depend on where the plane of division occurs.

Different animal lineages have different cleavage patterns. Remember that the bilateral lineage has two branches—protostomes and deuterostomes (Section 24.2). Deuterostomes undergo radial cleavage: The plane of each cell division either parallels the cell's

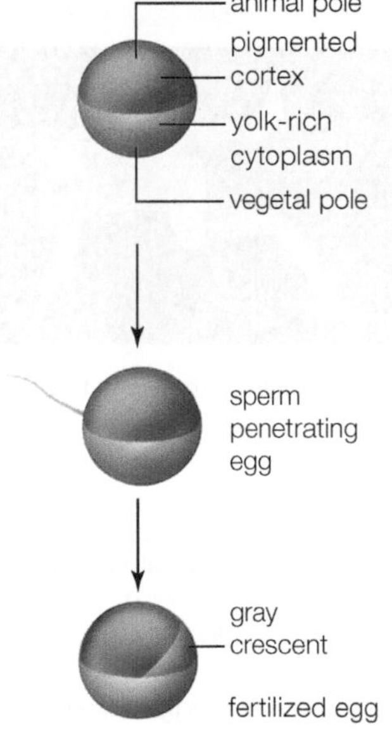

animal pole
pigmented cortex
yolk-rich cytoplasm
vegetal pole
sperm penetrating egg
gray crescent
fertilized egg

A Many amphibian eggs have a dark pigment concentrated in cytoplasm near the animal pole. At fertilization, the cytoplasm shifts and exposes a gray crescent-shaped region just opposite the sperm's entry point. The first cleavage normally distributes half of the gray crescent to each descendant cell.

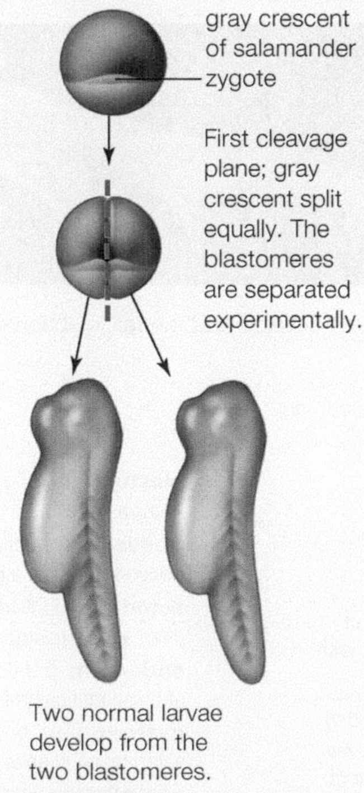

gray crescent of salamander zygote
First cleavage plane; gray crescent split equally. The blastomeres are separated experimentally.
Two normal larvae develop from the two blastomeres.

B In one experiment, the first two cells formed by normal cleavage were physically separated from each other. Each cell developed into a normal larva.

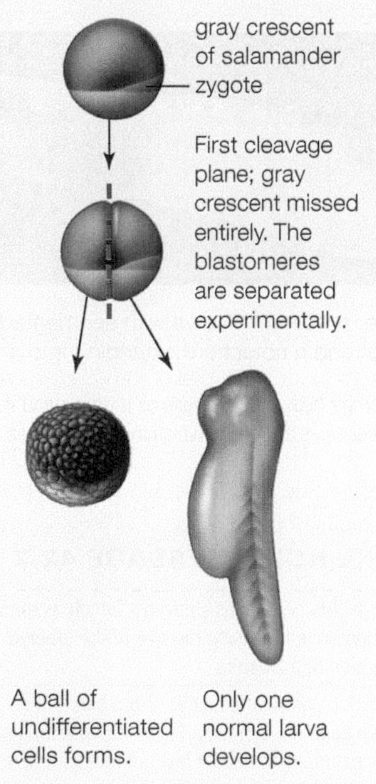

gray crescent of salamander zygote
First cleavage plane; gray crescent missed entirely. The blastomeres are separated experimentally.
A ball of undifferentiated cells forms.
Only one normal larva develops.

C In another experiment, a zygote was manipulated so one descendant cell received all the gray crescent. This cell developed normally. The other gave rise to an undifferentiated ball of cells.

FIGURE 42.3 Experimental demonstration of cytoplasmic localization in an amphibian oocyte.

FIGURE IT OUT Is the gray crescent essential to amphibian development?

Answer: Yes

FIGURE 42.4 Gastrulation in a fruit fly. In fruit flies (*Drosophila*), cleavage is restricted to the outermost region of cytoplasm; the interior is filled with yolk. The series of photographs, all cross sections, shows 16 cells (stained gold) migrating inward. The opening through which cells move in will become the fly's mouth. Descendants of the stained cells will form mesoderm. Movements shown in these photos occur during a period of less than 20 minutes.

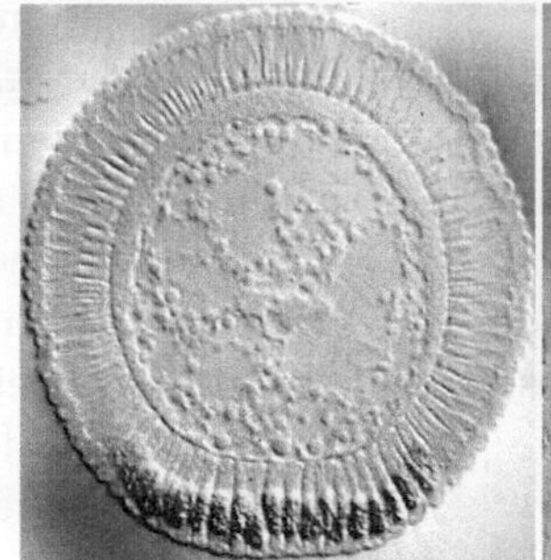

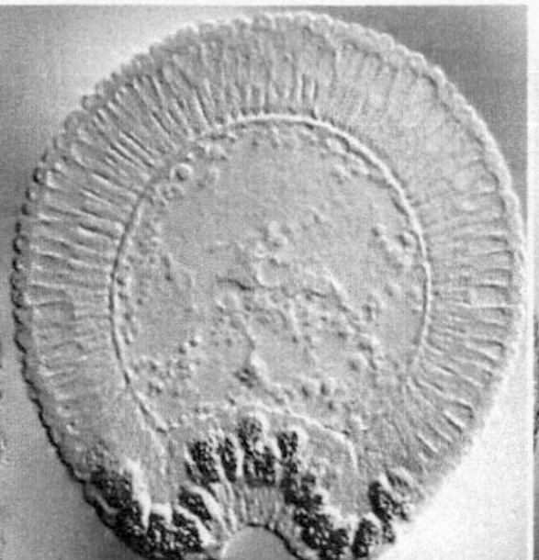

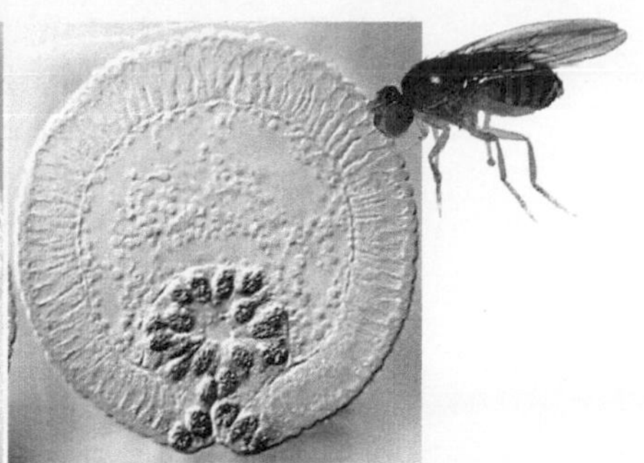

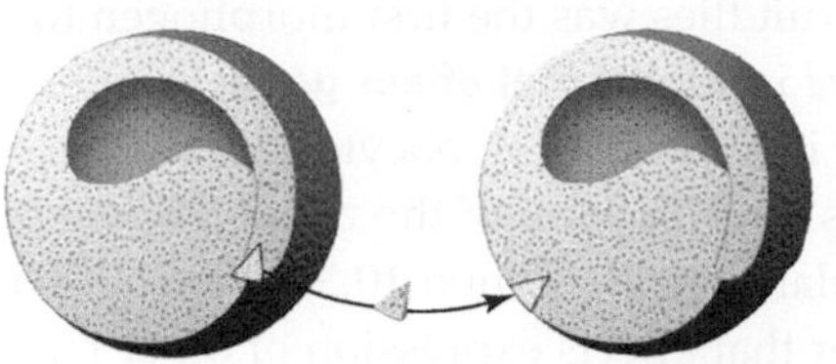

A Dorsal lip excised from donor embryo, grafted to novel site in another embryo.

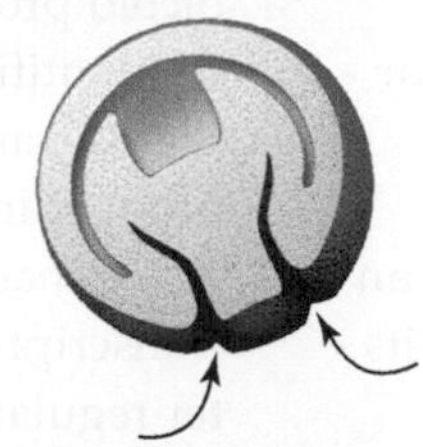

B Graft induces a second site of inward migration.

C The embryo develops into a "double" larva, with two heads, two tails, and two bodies joined at the belly.

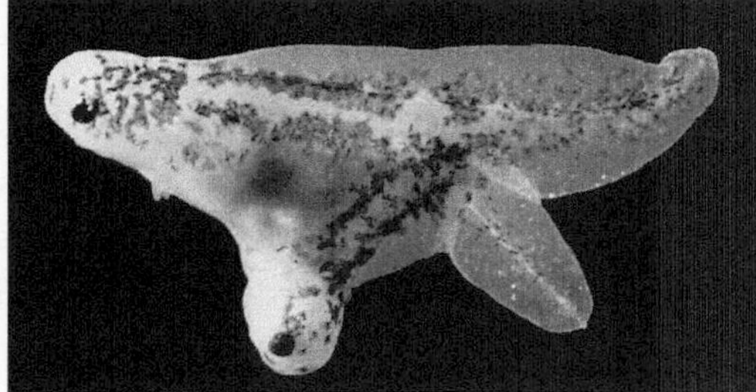

FIGURE 42.5 Experimental demonstration of the role of the dorsal lip in amphibian gastrulation. A dorsal lip region of a salamander embryo was transplanted to a different site in another embryo. A second set of body parts started to form.

polar axis or is perpendicular to it. In deuterostomes, cells undergo spiral cleavage, in which divisions occur at an oblique angle relative to the polar axis.

The amount of yolk also influences the pattern of division. Insects, frogs, fishes, and birds have yolk-rich eggs. In such eggs, a large volume of yolk slows or blocks some cuts. The result is fewer divisions of the yolky part of the egg than the less yolky part. By comparison, cleavage divides the nearly yolkless eggs of sea stars and mammals evenly.

Gastrulation

The blastula formed by cleavage contains hundreds to thousands of cells, depending on the species. During gastrulation, these cells move about and rearrange themselves to form multiple layers (FIGURE 42.4).

Hilde Mangold, one of Spemann's students, discovered how gastrulation is regulated in amphibians. Mangold knew that during gastrulation, cells of a salamander blastula move inward through an opening on the blastula's surface. Cells in the dorsal (upper) lip of the opening are descended from a zygote's gray crescent, and Mangold suspected that these cells made signals that caused gastrulation. To test her hypothesis, she transplanted dorsal lip material from one embryo to another (FIGURE 42.5A). As expected, cells migrated inward at the transplant site, as well as at the usual location (FIGURE 42.5B). As a result, a salamander larva with two joined sets of body parts developed (FIGURE 42.5C). Apparently, the transplanted cells signaled their new neighbors to develop in a novel way. This experiment also explains the results shown in FIGURE 42.3C. Without any gray crescent, an embryo does not have dorsal lip cells to signal the start of gastrulation, so development stops short.

TAKE-HOME MESSAGE 42.3

✔ Enzymes, mRNAs, yolk, and other materials are localized in specific parts of the cytoplasm of unfertilized eggs.

✔ Cleavage divides a fertilized egg into a number of small cells. The cells—blastomeres—inherit different parcels of cytoplasm that make them behave differently, starting at gastrulation.

42.4 Tissue and Organ Formation

LEARNING OBJECTIVES

- Explain the role of selective gene expression in development.
- Describe the process of embryonic induction, and provide an example.
- Give an example of how cell shape changes contribute to organ formation.
- Explain the process of apoptosis and its role in development.
- Describe how cells can shift their position during development.

cytoplasmic localization Accumulation of different materials in different regions of the egg cytoplasm.

Cell Differentiation

All cells in an embryo are descended from the same zygote, so all have the same genes. How then, do the specialized tissue and organs form? From gastrulation onward, different cell lineages express different subsets of genes. This selective gene expression is the key to **differentiation**, the process by which cell lineages become specialized in composition, structure, and function (Section 10.2).

An adult human body has about 200 differentiated cell types. As your eyes developed, cells of one lineage turned on genes for crystallin, a transparent protein that forms the lens in your eyes. No other cells in your body make crystallin.

Keep in mind that a differentiated cell still retains the entire genome. That is why it is possible to clone an adult animal—to create a genetic copy—from one of its differentiated cells (Section 8.7).

Embryonic Induction

Intercellular signals encourage selective gene expression. By the process known as **embryonic induction**, one subset of cells affects the developmental pathway of other cells. Induction often involves **morphogens**, which are substances (usually secreted proteins) that are produced in a specific region and affect gene expression in a concentration-dependent manner. Cells closest to the morphogen-producing region are exposed to a high concentration of morphogen and express one set of genes. Cells farther from the morphogen source are exposed to less morphogen, and as a result they express a different set of genes.

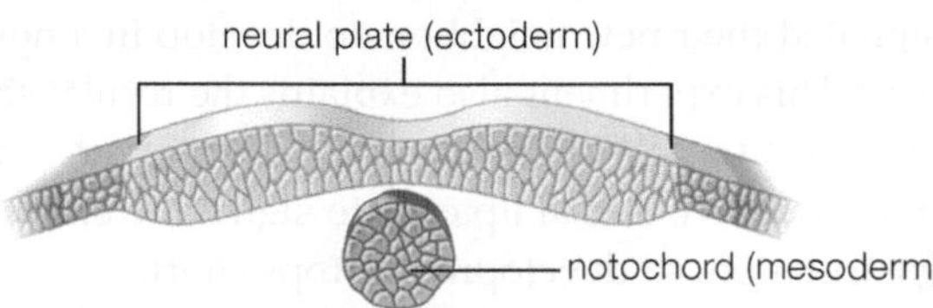
neural plate (ectoderm)
notochord (mesoderm)

❶ Chemical signals produced by notochord mesoderm induce the ectoderm above it to thicken and form a neural plate.

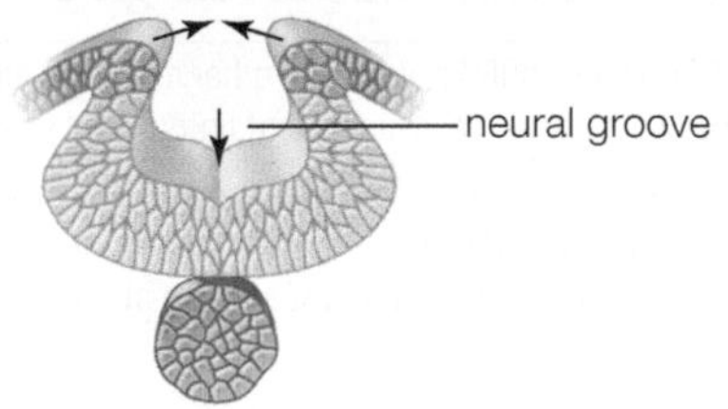
neural groove

❷ Changes in cell shape cause edges of the neural plate to fold in toward the plate center, forming a neural groove.

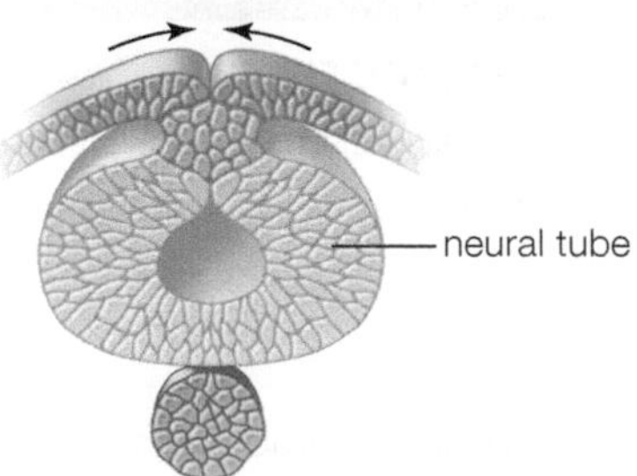
neural tube

❸ Further folding causes the edges of the neural plate to meet, forming the neural tube.

FIGURE 42.6 Neural tube formation in a vertebrate embryo.

Establishment of the Fruit Fly Body Axis The bicoid protein of fruit flies was the first morphogen to be identified. *Bicoid* is a **maternal effect gene**, a gene that is transcribed in a developing oocyte and whose product influences development of the zygote. *Bicoid* is also a master regulator gene (Section 10.3). It encodes a transcription factor that affects expression of other master regulators. The resulting cascade of gene expression culminates in the expression of homeotic genes in the anterior of the developing fly embryo. Recall that a homeotic gene is a master regulator whose expression results in the formation of a specific body part.

During egg production, bicoid mRNA accumulates at one end of an egg. Immediately after fertilization, this mRNA is translated into bicoid protein. The protein diffuses away from the site of translation, thus forming a gradient within the zygote. The orientation of this gradient determines the front-to-back axis of the developing individual: The head forms at the end where bicoid protein concentration is highest, and the abdomen at the end where the bicoid protein concentration is lowest.

Formation of the Vertebrate Neural Tube Development of the vertebrate neural tube also involves embryonic induction. The neural tube is the embryonic forerunner of the nervous system. Its development is induced by morphogens from the notochord, which formed earlier from mesoderm (FIGURE 42.6). The process begins when ectodermal cells overlying the notochord elongate, forming a thick neural plate ❶. The cell elongation results from the assembly of microtubules inside the cells. Next, actin filaments in cells at the edges of the neural plate contract, making these cells wedge shaped. This change in cell shape causes the edges of the plate to fold inward, forming a U-shaped structure called the neural groove ❷. Eventually the outer edges of the former neural plate meet at the midline, forming the neural tube ❸.

Apoptosis

Programmed cell death, or **apoptosis**, is a process by which a cell self-destructs in a controlled, predictable

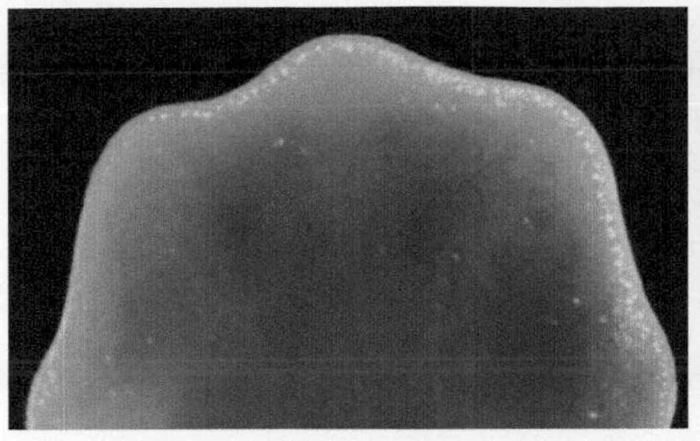
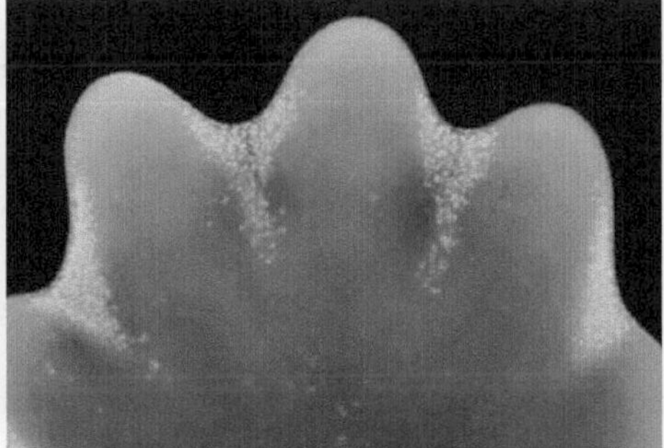

FIGURE 42.7 Apoptosis in the formation of a mouse paw. The yellow stain indicates regions where cells are undergoing apoptosis.

manner. During apoptosis, a cell activates enzymes that destroy its component parts. Some enzymes chop up cytoskeletal proteins and the histones that organize DNA. Others snip apart nucleic acids. As a result of this self-inflicted damage, the cell shrivels up and dies without spewing out its contents. The cell remains can then be engulfed and disposed of by other body cells. In a mature body, apoptosis plays a role in normal cell turnover and in defense against cancer.

During development, apoptosis shapes structures by eliminating excess cells, removing those cells that do not contribute to the expected final form. Consider how apoptosis shapes a mouse paw. At two weeks of development, the mouse paw is a paddle-like structure with the bones hidden inside (FIGURE 42.7). By day 16, apoptosis has removed many cells, leaving only thin webs of tissue between individual digits. By day 18, this webbing is completely gone and the individual digits are free. Apoptosis also makes the tail of a tadpole disappear during metamorphosis and eliminates the tail that forms early in human development.

Cell Migrations

Cell migrations are an essential part of development, For example, ectodermal cells that form at the top of the neural tube (a region called the neural crest) migrate outward to positions throughout the body. Descendants of neural crest cells include the neurons and glial cells of the peripheral nervous system and the melanocytes in skin.

apoptosis (ap-op-TOE-sis or ay-poe-TOE-sis) Mechanism of programmed cell death.
differentiation Process by which cells become specialized during development; occurs as different cells in an embryo begin to use different subsets of their DNA. .
embryonic induction Embryonic cells produce signals that alter the behavior of neighboring cells.
maternal effect gene Gene that is expressed during egg formation; its product later influences development of the zygote.
morphogen (MOR-foe-jen) Chemical signal that diffuses from its source and affects development in a concentration-dependent manner.

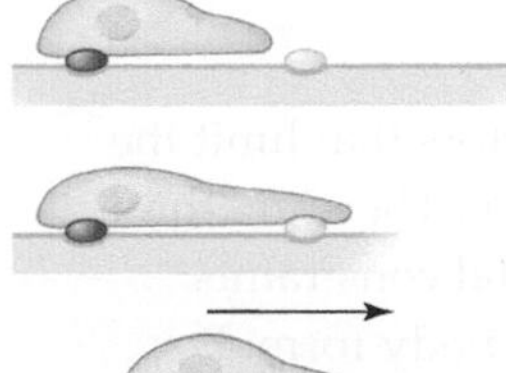
A cell adheres to one point and protrudes forward.
The protruding portion of the cell attaches to a new point.
The trailing portion of the cell releases its grip.

FIGURE 42.8 Cell migration in development.

Cells travel by inching along in an amoeba-like fashion (FIGURE 42.8). Assembly of actin microfilaments at one edge of a cell causes that portion of the cell to protrude. Adhesion proteins in the plasma membrane then anchor the advancing portion of the cell to a protein in the membrane of another cell or in the extracelluar matrix. Once the front of the cell is anchored, adhesion proteins in the trailing part of the cell release their grip, and the trailing region is drawn forward.

How does the cell know where to go? The cell may move in response to a concentration gradient of some chemical signal, or it may follow a "trail" of molecules that its adhesion proteins bind to. It stops migrating when it reaches a region where its adhesion proteins hold it tightly in place.

TAKE-HOME MESSAGE 42.4

✔ All cells in an embryo have the same genome, but they express different subsets of genes. This selective gene expression is the basis of cell differentiation. It results in cell lineages with characteristic structures and functions.

✔ Cytoplasmic localization results in concentration gradients of signaling proteins called morphogens. Morphogens activate sets of master regulators, whose products cause embryonic cells to form tissues and organs in specific places.

✔ Cell migrations, shape changes, and apoptosis help sculpt the form of the body and its parts.

CREDITS: (7) Courtesy of © Paul Martin; (8) From Russell/Wolfe/Hertz/Starr. *Biology*, 1e. © 2008 Cengage.

42.5 Evolutionary Developmental Biology

LEARNING OBJECTIVES

- Give examples of factors that constrain animal body plans.
- Using appropriate examples, explain how mutations that affect development can alter a body plan.

Evolutionary developmental biology is a field of biology concerned with the genetic basis for animal developmental pathways, how evolved variations in these pathways give rise to diverse body plans, and how developmental factors constrain this variation.

Constraints on Body Plans

Developmental constraints are factors that limit the types of body plans that can evolve. There are three general categories of developmental constraints.

First, physical constraints limit body form. For example, large body size cannot evolve in an animal that does not have circulatory and respiratory mechanisms sufficient to service cells far from the body surface. Also, an open circulatory system, in which blood is not constrained within vessels, cannot move fluid against gravity to the same extent that a closed system can. Thus, insects the size of hippos cannot evolve.

Second, an existing body framework imposes architectural constraints. For example, the common ancestor of all living land vertebrates had two pairs of limbs. Some vertebrates lost one or both pairs of limbs, but none have gained an extra pair. The wings of birds and bats evolved through modification of existing forelimbs. Although it would seem advantageous to have both wings and arms, no living or fossil vertebrate with such a body plan has ever been discovered.

Finally, there are phyletic constraints on body plans. These are constraints imposed by interactions among the genes that regulate development. Once master regulators evolved, their interactions determined the basic body form. Mutations that dramatically alter these interactions are usually lethal. Consider that early in development, the mesoderm on either side of the embryo's neural tube becomes divided into blocks of cells called **somites** that later develop into bones and skeletal muscles (FIGURE 42.9). A complex pathway involving many genes governs somite formation. Any mutation that disrupts this pathway and prevents somites formation is lethal to the embryo.

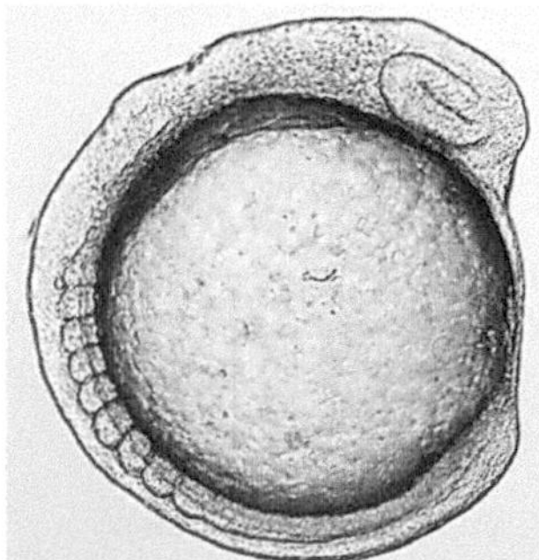

A Normal zebrafish embryo with somites—bumps of mesoderm that give rise to bone and muscle.

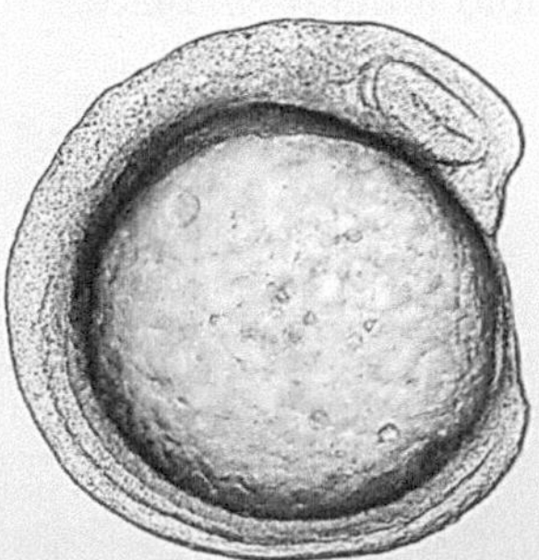

B Embryo with a mutation that prevents the formation of somites. It will die early in development.

FIGURE 42.9 **A lethal developmental mutation in zebrafish.**

Developmental Mutations

Evolutionary changes in body plans arise by modifications to existing developmental pathways rather than blazing entirely new genetic trails.

Modifications in the rate and/or the timing of development can alter body form. Consider how the length of a vertebrate body varies among species. All vertebrates develop somites, but the number of somites, and hence the number of vertebrae, varies. A typical lizard embryo has about 61 somites, whereas a cornsnake embryo has more than 300. Differences in body length between lizards and snakes arise from differences in the rate at which somites form. In snakes, somites form much more rapidly than they do in other vertebrates. As a result, the mesoderm is divided into more somites, which give rise to more vertebrae.

Changes in gene regulation also affect development. Consider the variation in beak sizes and shapes among the finches of the Galápagos Islands. All 14 species of these birds share a common ancestor, but different types of beaks adapt them to different food sources. Whether a species has a blunt beak or a pointed one is determined in large part by the genotype at the *ALX1* locus. The *ALX1* gene encodes a transcription factor (Section 10.2) that affects facial development. Among Galápagos finches, there are two alleles for *ALX1*; one produces a blunt beak and the other a pointy beak. Heterozygotes have a beak of intermediate shape. In most Galápagos finch species, either one or the other of these alleles has become fixed (all individuals of the species are homozygous for the same allele).

TAKE-HOME MESSAGE 42.5

✔ Animal body plans are constrained by physical and developmental factors. Once a developmental pathway evolves, drastic changes to genes that govern this pathway are generally lethal.

✔ Genetic changes that affect the rate or timing of developmental events can alter a body plan, as can other changes affecting the regulation of developmental processes.

somites Paired blocks of embryonic mesoderm that give rise to a vertebrate's muscle and bone.

42.6 Overview of Human Development

LEARNING OBJECTIVES

- Define how the terms embryo and fetus are used in discussing human prenatal development.
- Explain what is meant by a premature birth, and describe the associated risks.
- Summarize the main events of postnatal development.

Chapter 41 introduced the structure and function of human reproductive organs and explained how an egg and sperm meet at fertilization to form a zygote. The remaining sections of this chapter continue this story with an in-depth look at human development. In this section, we provide an overview of the process and define the stages that we will discuss. Prenatal (before birth) and postnatal (after birth) stages are listed in TABLE 42.2.

It takes about 5 trillion mitotic divisions to go from the single cell of a zygote to the 10 trillion or so cells of an adult human. The process gets under way during a pregnancy that lasts an average of 38 weeks from the time of fertilization.

The first cleavage occurs about 12 to 24 hours after fertilization. It takes about one week for a blastula (called a blastocyst in mammals) to form.

In discussing animals, the term **embryo** generally refers to an individual from the time of first cleavage to the time when it hatches or is born. With regard to humans, the term embryo is usually reserved for the period from 2 to 8 weeks after fertilization. From week nine until birth, the individual is referred to as a **fetus**. During embryonic development, the human body plan is set in place and organs form. During fetal development, the organs mature and begin to function.

We divide the prenatal period into three trimesters. The first trimester includes months one through three; the second trimester, months four through six; the third trimester, months seven through nine.

Births before 36 weeks after fertilization are considered premature. A fetus born earlier than 22 weeks rarely survives because its lungs are not yet fully mature. About half of births that occur before 26 weeks result in some sort of long-term disability.

After birth, the human body continues to grow and its body parts continue to change in proportion. FIGURE 42.10 shows body proportion changes during development. Sexual maturation occurs at puberty, and bones stop growing shortly thereafter. The brain is the last organ to become fully mature: Portions of the brain continue to develop until the individual is about 19 to 22 years old.

embryo In animals, a developing individual from first cleavage until hatching or birth; in humans, usually refers to an individual during weeks 2 to 8 of development.

fetus (FEE-tus) Developing human from about 9 weeks until birth.

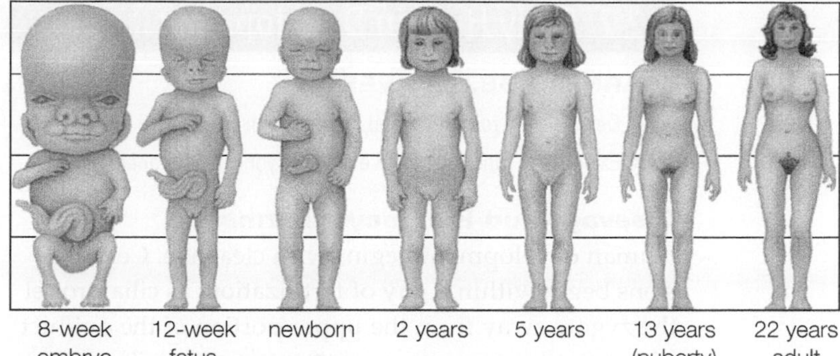
8-week embryo
12-week fetus
newborn
2 years
5 years
13 years (puberty)
22 years adult

FIGURE 42.10 **Observable, proportional changes in prenatal and postnatal periods of human development.** Changes in overall physical appearance are slow but noticeable until the teens.

TABLE 42.2

Stages of Human Development

Prenatal period	
Zygote	Single cell resulting from fusion of sperm nucleus and egg nucleus at fertilization.
Blastocyst (blastula)	Ball of cells with surface layer, fluid-filled cavity, and inner cell mass.
Embryo	All developmental stages from two weeks after fertilization until end of eighth week.
Fetus	All developmental stages from ninth week to birth (about 38 weeks after fertilization).
Postnatal period	
Newborn	Individual during the first two weeks after birth.
Infant	Individual from two weeks to 15 months.
Child	Individual from infancy to about 10 or 12 years.
Pubescent	Individual at puberty, when secondary sexual traits develop. For girls, occurs between 10 and 15 years; for boys, between 11 and 16 years.
Adolescent	Individual from puberty until about 3 or 4 years later; physical, mental, emotional maturation.
Adult	Early adulthood (between 18 and 25 years); bone formation and growth finished. Changes proceed slowly after this.
Old age	Aging processes result in expected tissue deterioration.

TAKE-HOME MESSAGE 42.6

✔ Humans undergo cleavage to form a blastula, which in mammals is called a blastocyst.

✔ Organs form during embryonic development. The fetal period, during which organs enlarge and begin to function, is the longest stage of prenatal development.

✔ Growth and development continues after an individual is born, ending shortly after puberty.

42.7 Early Human Development

LEARNING OBJECTIVES

- Describe the formation and implantation of a human blastocyst.
- Explain the function of the extraembryonic membranes.

Cleavage and Blastocyst Formation

Human development begins with cleavage. Cell divisions begin within a day of fertilization, as cilia propel the zygote away from the upper portion of the oviduct where fertilization occurred (FIGURE 42.11). The zygote divides by mitosis to form two cells, which become four, which become eight, and so on ❶.

Sometimes a cluster of four or eight cells splits in two and each cluster develops independently, resulting in **identical twins**, two individuals that have the same genome. More typically, all cells adhere tightly to one another as divisions continue.

By about five days after fertilization, the dividing cluster of cells has reached the uterus and formed a **blastocyst**, the mammalian blastula ❷. The blastocyst consists of an outer layer of cells, a cavity (the blastocoel) filled with the cells' fluid secretions, and an inner cell mass. Only the cells of the inner cell mass will give rise to the developing individual. Other cells of the blastocyst give rise to extraembryonic membranes that enclose, protect, and support the embryo.

Implantation and Formation of the Extra-embryonic Membranes

Up to this point, the developing individual has been nourished by nutrients it absorbed from maternal secretions in the oviduct and uterus. For development to continue, the blastocyst must implant itself in the wall of the uterus, where it can obtain nutrients from the maternal bloodstream. Implantation requires direct contact between the surface of the blastocyst and cells in the lining of the uterus. However, the blastocyst is encased by remnants of the zona pellucida; this protein coat must be shed before implantation can occur.

Once the blastocyst has escaped from its protein coat, it attaches to the lining of the uterus and begins burrowing into it ❸. During implantation, the inner cell mass develops into two flattened layers of cells that are collectively called an embryonic disk ❹. At the same time, extraembryonic membranes typical of amniotes begin to form. One of these membranes, called the **amnion**, will enclose a fluid-filled amniotic cavity between the embryonic disk and the blastocyst surface. Fluid in this cavity (amniotic fluid) acts as a buoyant cradle in which an embryo grows and moves freely, protected from temperature changes and mechanical impacts. As the amnion forms, other cells move around the inner wall of the blastocyst, forming the lining of a

FIGURE 42.11 Early embryonic development in humans. Graphics 3–5 depict a cross section through the uterus.

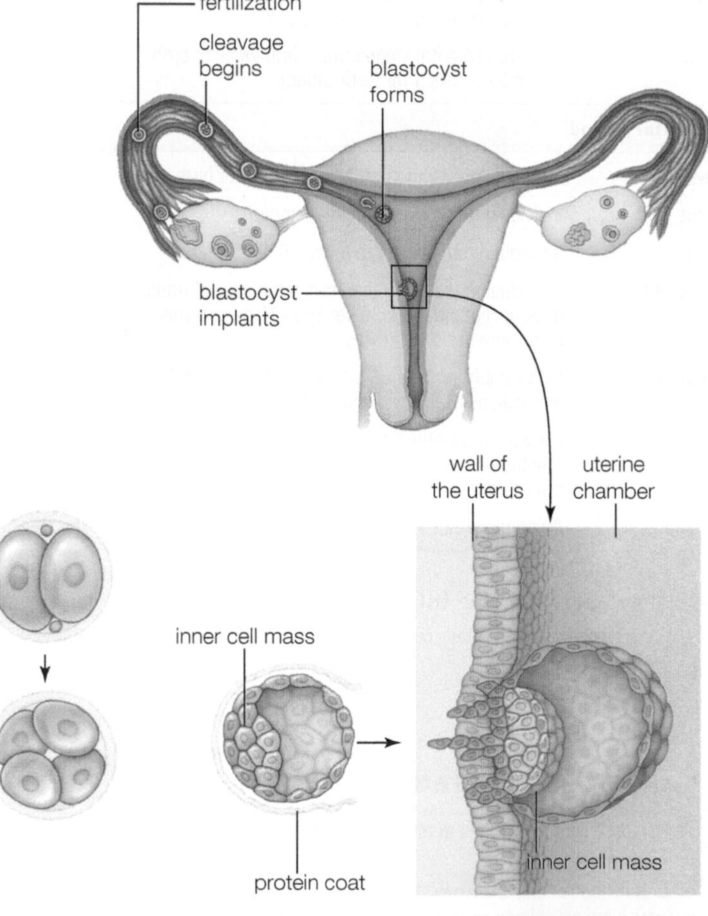
fertilization
cleavage begins
blastocyst forms
blastocyst implants
wall of the uterus
uterine chamber
inner cell mass
protein coat
inner cell mass

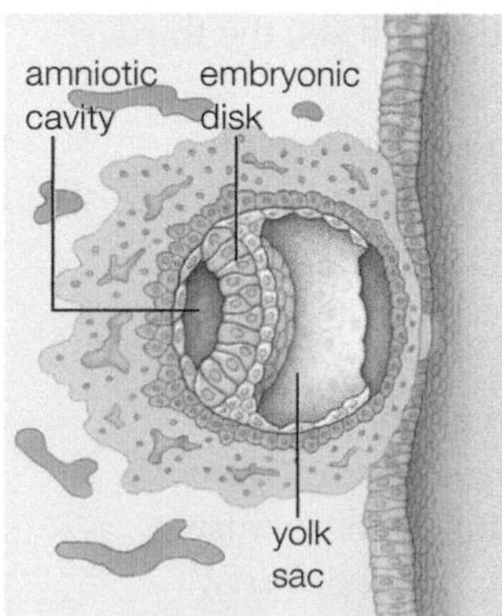
amniotic cavity
embryonic disk
yolk sac

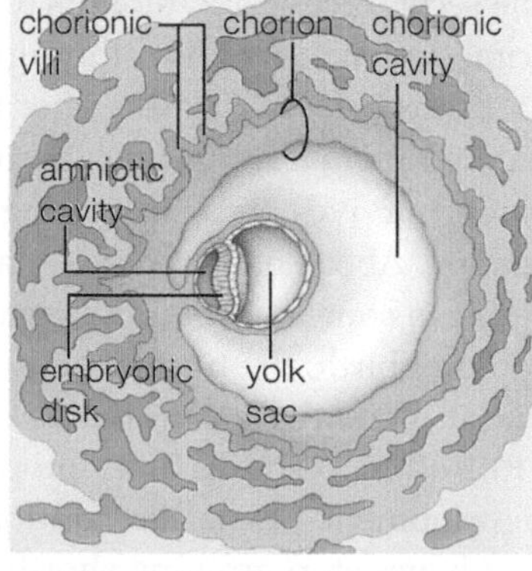
chorionic villi
chorion
chorionic cavity
amniotic cavity
embryonic disk
yolk sac

❶ **Days 1–4**
Cleavage divides the zygote into an increasing number of smaller cells.

❷ **Days 5–7**
The blastocyst forms, expands, then sheds its protein coat.

❸ **Days 8–9**
Implantation begins. The blastocyst attaches to the wall of the uterus and begins to burrow into it.

❹ **Days 10–11**
The inner cell mass develops into a two-layered embryonic disk. Extraembryonic membranes form.

❺ **Day 14**
Chorionic villi (fingerlike extensions of the chorion) begin to grow into blood-filled spaces in the lining of the uterus.

yolk sac. In reptiles and birds, this sac holds yolk that nourishes the developing embryo. In humans, it has no nutritional role. Some cells that form in a human yolk sac give rise to the embryo's first blood cells.

As implantation continues, spaces in the uterine tissue around the blastocyst become filled with blood that seeps from ruptured maternal capillaries. The outermost extraembryonic membrane—the **chorion**—develops many tiny fingerlike projections (chorionic villi) that extend into blood-filled maternal tissues ❺. Later, the chorion will become a part of the placenta.

Human chorionic gonadotropin (HCG), a hormone secreted by the chorion prevents degeneration of the corpus luteum and so maintains the uterine lining and prevents menstruation. By the beginning of the third week, HCG can be detected in a mother's blood or urine. At-home pregnancy tests have a region that changes color when exposed to urine that contains HCG. Later in a pregnancy, when the placenta has formed, it will also produce HCG.

An outpouching of the yolk sac will become the fourth extraembryonic membrane, the **allantois**. In humans, the allantois forms blood vessels of the umbilical cord, which is the structure that connects the fetus to the placenta.

Gastrulation and Onset of Organ Formation

Gastrulation occurs on about day 15. Cells migrate inward along the primitive streak, a groove that forms on the surface of the embryonic disk ❻. Shortly thereafter, the embryonic disk develops two folds that will merge to form a neural tube, the precursor of the spinal cord and brain ❼. Beneath the neural tube lies the notochord. As development continues, this flexible supporting rod will be replaced by the backbone.

Somites form by the end of the third week ❽. These paired segments of mesoderm will develop into the bones, skeletal muscles of the head and trunk, and overlying dermis of the skin.

Bands of tissue called pharyngeal arches form at the onset of the fourth week ❾. This tissue will later contribute to the pharynx, larynx, and the face, neck, mouth, and nose. In fishes, pharyngeal arches develop into gills, but a human embryo never has gills; it obtains oxygen by way of the placenta.

TAKE-HOME MESSAGE 42.7

- ✔ Cleavage produces a blastocyst that slips out of its protein coat and implants itself in the mother's uterus.
- ✔ During implantation, projections from the blastocyst extend into maternal tissues. These and other connections will support the developing embryo.
- ✔ The blastocyt's inner cell mass will become the embryo. Other blastocyst cells give rise to external membranes.
- ✔ After the embryonic disk undergoes gastrulation, the neural tube and notochord form. Then, somites and pharyngeal arches develop.

allantois (al-LAN-toh-is) Extraembryonic membrane that, in mammals, becomes part of the umbilical cord.
amnion (AM-nee-on) Extraembryonic membrane that encloses an amniote embryo and amniotic fluid.
blastocyst Mammalian blastula.
chorion (KOR-ee-on) Outermost extraembryonic membrane of amniotes; major component of the placenta in placental mammals.
identical twins Genetically identical individuals that develop after an early embryo splits in two.
human chorionic gonadotropin (HCG) (kor-ee-ON-ik go-NAD-oh-troe-pin) Hormone first secreted by the blastocyst, and later by the placenta; helps maintain the uterine lining during pregnancy.

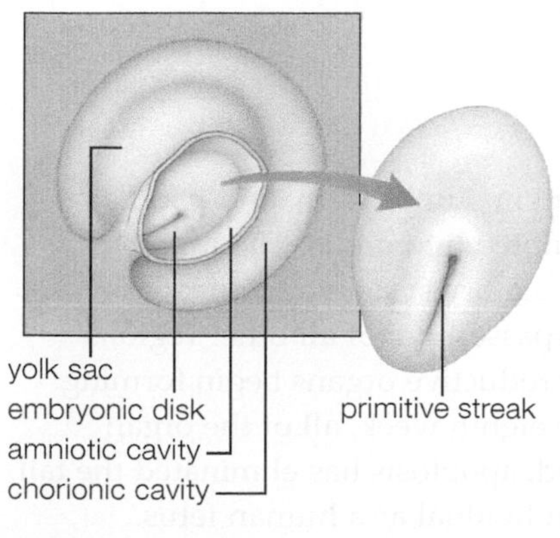
yolk sac
embryonic disk
amniotic cavity
chorionic cavity
primitive streak

❻ **Day 15**

Gastrulation begins at depression called the primitive streak.

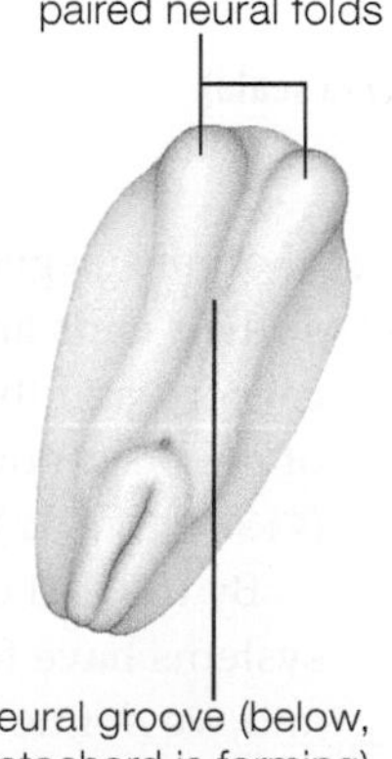
paired neural folds
neural groove (below, notochord is forming)

❼ **Day 16**

Neural tube begins to form.

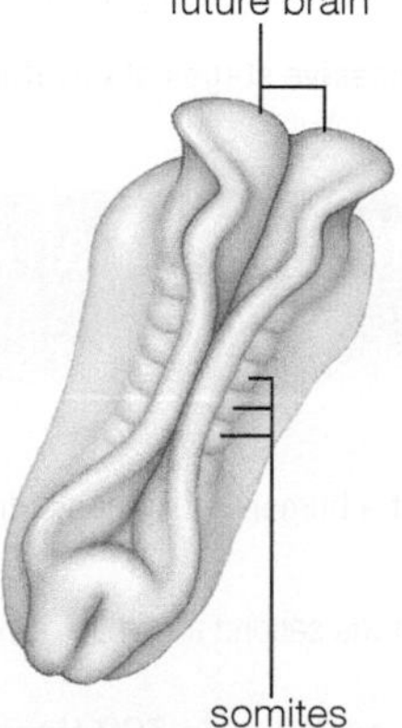
future brain
somites

❽ **Days 18–23**

Somites develop.

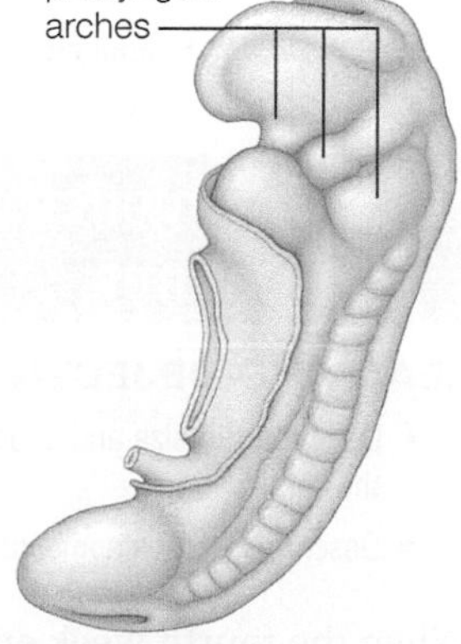
pharyngeal arches

❾ **Days 24–25**

Pharyngeal arches appear.

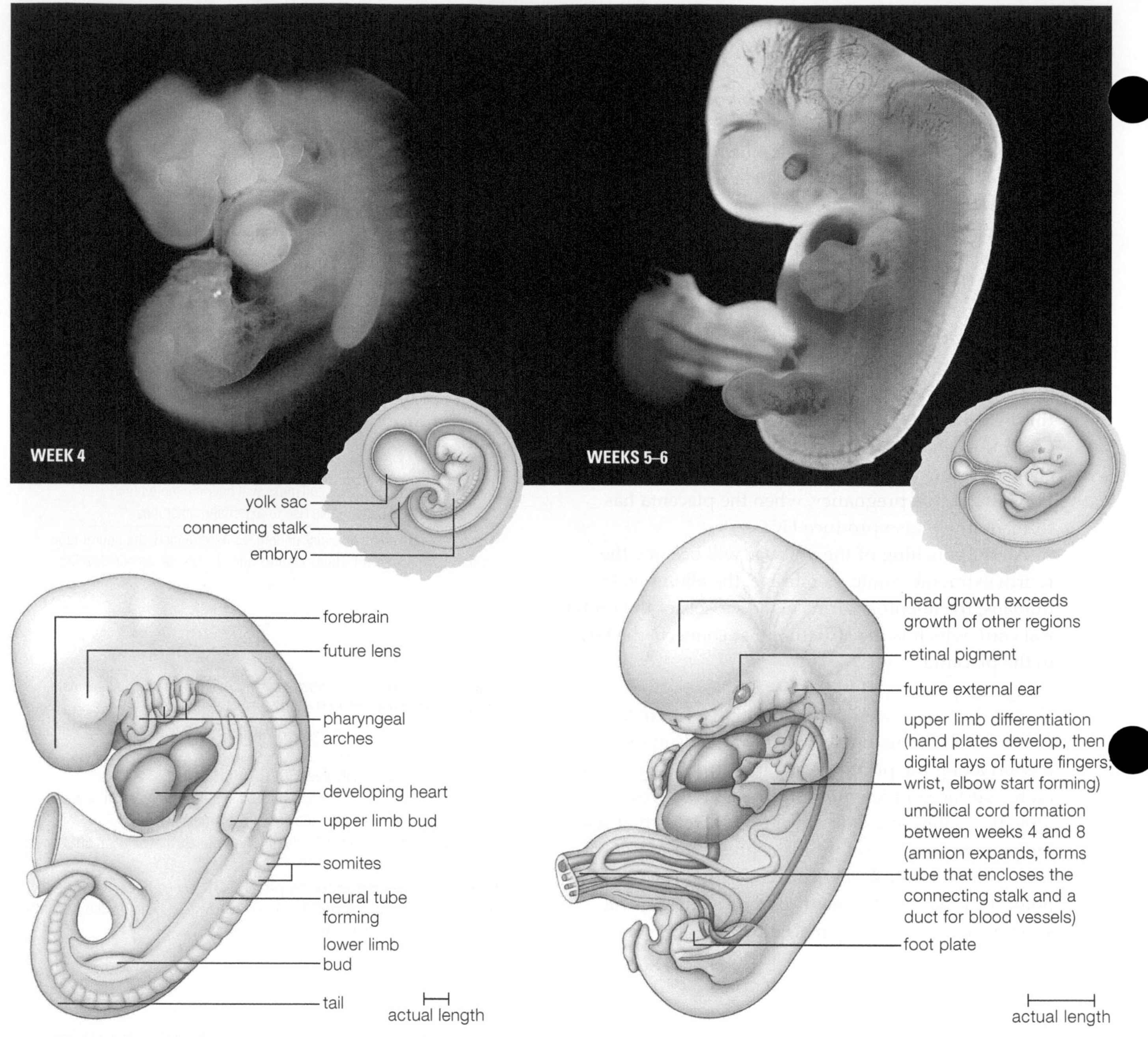

WEEK 4
yolk sac
connecting stalk
embryo
forebrain
future lens
pharyngeal arches
developing heart
upper limb bud
somites
neural tube forming
lower limb bud
tail
actual length
WEEKS 5–6
head growth exceeds growth of other regions
retinal pigment
future external ear
upper limb differentiation (hand plates develop, then digital rays of future fingers; wrist, elbow start forming)
umbilical cord formation between weeks 4 and 8 (amnion expands, forms tube that encloses the connecting stalk and a duct for blood vessels)
foot plate
actual length

FIGURE 42.12 Human embryo at successive stages of development (not to scale).

42.8 Emergence of Distinctly Human Features

LEARNING OBJECTIVES

- Describe the size and appearance of a human at the beginning of the fetal period.
- Describe the developmental events of the second and third trimester.

When the fourth week ends, the embryo is 500 times the size of a zygote, but still less than 1 centimeter long. Like other chordate embryos, it has a short tail that extends beyond its anus. Growth slows as details of organs begin to fill in. Limbs form, and paddles at their ends are sculpted into digits. The umbilical cord and the circulatory system develop. Growth of the head now surpasses that of all other regions (FIGURE 42.12). Reproductive organs begin forming.

By the end of the eighth week, all of the organ systems have formed, apoptosis has eliminated the tail, and we define the individual as a human fetus.

In the second trimester (months 3–6), obstetric sonography (ultrasound) can reveal the sex of a fetus. By four months, most neurons have formed. Reflexive movements begin as developing nerves and muscles connect.

CREDIT: (12) top photos, © Lennart Nilsson/Bonnierforlagen AB; art, © Cengage Learning.

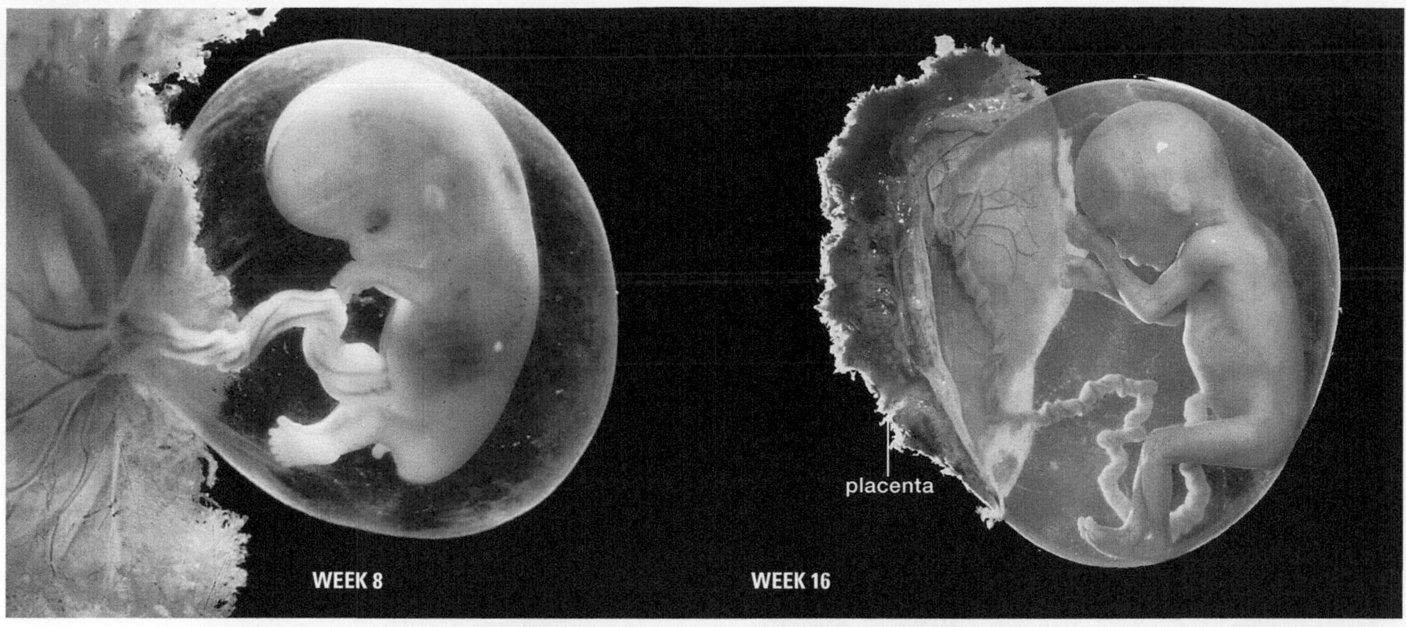
placenta
WEEK 8
WEEK 16

final week of embryonic period; embryo looks distinctly human compared to other vertebrate embryos

upper and lower limbs well formed; fingers and then toes have separated

primordial tissues of all internal, external structures now developed

tail has become stubby

actual length

Length: 16 centimeters (6.4 inches)
Weight: 200 grams (7 ounces)

WEEK 29
Length: 27.5 centimeters (11 inches)
Weight: 1,300 grams (46 ounces)

WEEK 38 (full term)
Length: 50 centimeters (20 inches)
Weight: 3,400 grams (7.5 pounds)

During fetal period, length measurement extends from crown to heel (for embryos, it is the longest measurable dimension, as from crown to rump).

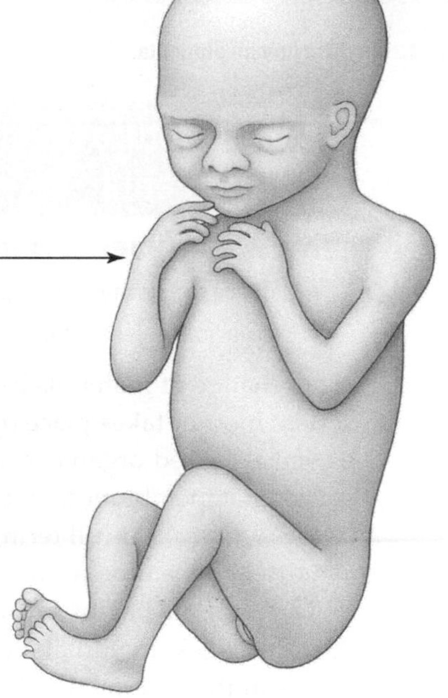

Legs kick, arms wave about, and fingers grasp. The fetus frowns, squints, puckers its lips, sucks, and hiccups. Soft fetal hair (lanugo) and a thick, cheesy coating (vernix) cover the skin. Most of the hair will be shed before birth.

Beginning at about five months, the fetal heartbeat can be heard clearly through a stethoscope positioned on the mother's abdomen.

By the seventh month, the fetus has begun not only to detect sounds, but also to remember some of them. Exposure to human singing and speech during this period may enhance language acquisition after birth. In this final trimester, the fetus opens its eyes, loses most of its fetal hair, and puts on a layer of fat. In preparation for birth, it practices muscle movements involved in breathing, inhaling amniotic fluid in place of air. Its lungs will later inflate with its first breath after birth.

TAKE-HOME MESSAGE 42.8

✔ The embryo takes on its human appearance by week 8 but remains tiny.

✔ In the fetal period, organs begin functioning and size increases dramatically.

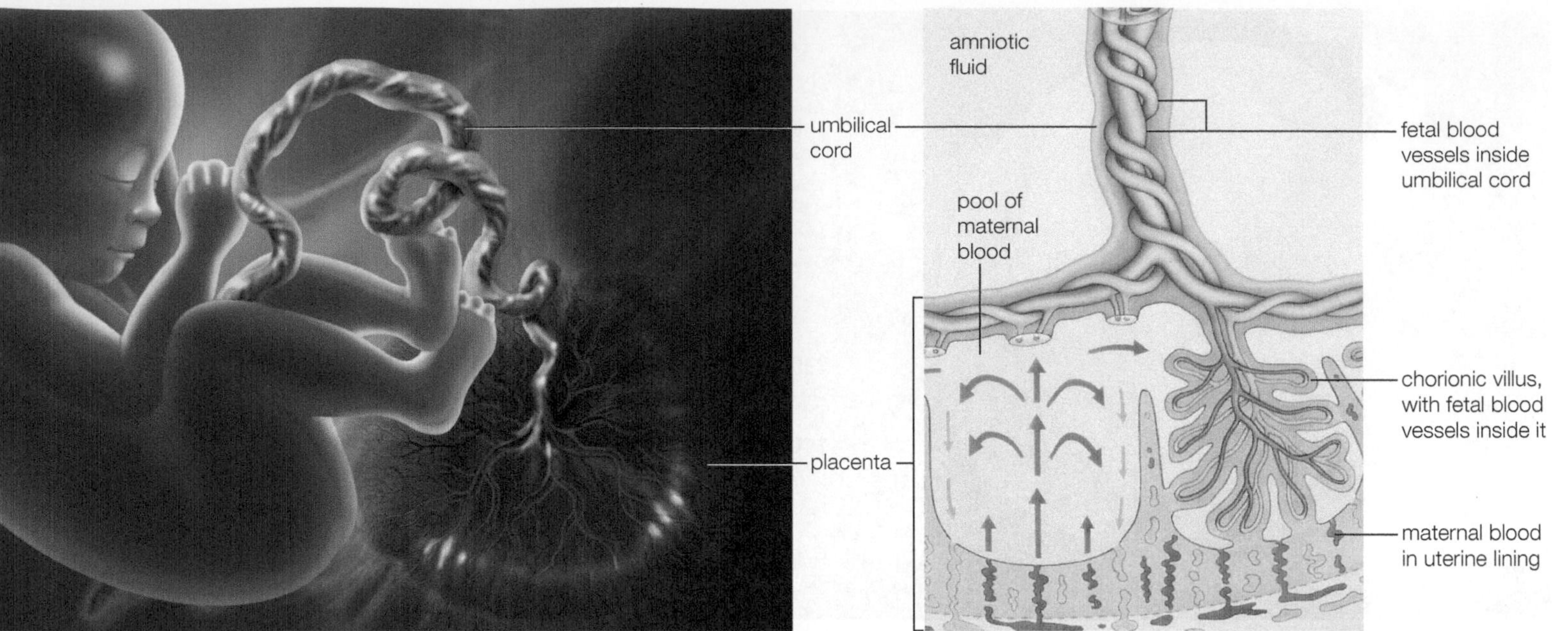
umbilical cord
placenta
amniotic fluid
fetal blood vessels inside umbilical cord
pool of maternal blood
chorionic villus, with fetal blood vessels inside it
maternal blood in uterine lining

Artist's depiction of the view inside the uterus, showing a fetus connected by an umbilical cord to the pancake-shaped placenta.

The placenta consists of maternal and fetal tissue. Fetal blood flowing in vessels of chorionic villi exchanges substances by diffusion with maternal blood around the villi. The bloodstreams do not mix.

FIGURE 42.13 The human placenta.

42.9 Structure and Function of the Placenta

LEARNING OBJECTIVES

- Describe the structure of the placenta.
- Explain how materials are exchanged across the placenta.

All exchange of materials between an embryo or fetus and its mother takes place by way of the placenta. This pancake-shaped organ consists of uterine lining, extraembryonic membranes, and embryonic blood vessels (**FIGURE 42.13**). At full term, a placenta covers about a quarter of the uterus's inner surface.

The placenta begins forming early in pregnancy. By the third week, maternal blood has begun to pool in spaces in the endometrial tissue. Chorionic villi—tiny fingerlike projections from the chorion—extend into the pools of maternal blood. Embryonic blood vessels extend through a coiled umbilical cord to the placenta and into the villi.

The maternal and embryonic bloodstreams never mix. Instead, substances move between them by diffusing across the walls of the embryonic vessels in the chorionic villi. Oxygen diffuses from maternal to embryonic blood, and carbon dioxide diffuses the opposite way. Transport proteins assist movement of essential nutrients from the maternal blood into embryonic blood vessels inside the villi.

The placenta also has a hormonal role. From the third month on, it produces HCG, progesterone, and estrogens. These hormones encourage the ongoing maintenance of the uterine lining. Once the placenta takes on this task, the corpus luteum degenerates.

One type of maternal antibody (IgG) is also transported from maternal blood into the fetal bloodstream. This sometimes causes problems if the mother and her child differ in their Rh blood type. An Rh negative woman lacks the Rh antigen on her red blood cells. When she gives birth to her first Rh positive child, exposure to the child's blood during childbirth can cause her to develop antibodies against the Rh antigen. During a subsequent pregnancy, these antibodies can cross the placenta and cause the mother's immune system to attack an Rh positive fetus.

Problems also arise when teratogens travel from mother to child by way of the placenta. The rubella virus and the Zika virus can cross the placenta. Toxins such as alcohol, nicotine, and mercury also are transmitted from mother to developing child in this way.

TAKE-HOME MESSAGE 42.9

✔ Blood vessels of the embryo's circulatory system extend through the umbilical cord to the placenta, where they run through pools of maternal blood.

✔ Maternal and embryonic blood do not mix; substances diffuse between the maternal and embryonic bloodstreams by crossing vessel walls.

42.10 Labor, Birth, and Lactation

LEARNING OBJECTIVES

- Describe normal childbirth, and compare it with a surgical delivery.
- List the components of milk, and describe their functions.

A mother's body changes as her fetus nears full term, at about 38 weeks after fertilization. Until the last few weeks, her firm cervix helped prevent the fetus from slipping out of her uterus prematurely. Now the connective tissue of the cervix softens and thins. These changes make the cervix more flexible in preparation for **labor**, the process by which a woman's body expels a fetus from her uterus.

Vaginal Birth

Typically, the amnion ruptures right before birth, so amniotic fluid drains out from the vagina. The cervical canal dilates, then contractions of uterine smooth muscle propel the fetus out of the body through the vagina (FIGURE 42.14).

A positive feedback mechanism operates during labor. When the fetus nears full term, it typically shifts position so that its head puts pressure on the mother's cervix. Receptors inside the cervix sense the pressure and signal the hypothalamus, which in turn signals the posterior lobe of the pituitary to secrete the hormone **oxytocin**. In a positive feedback loop, oxytocin binds to smooth muscle of the uterus, causing uterine contractions that push the fetus against the cervix. The added pressure triggers more oxytocin secretion, which causes more contractions and more cervical stretching:

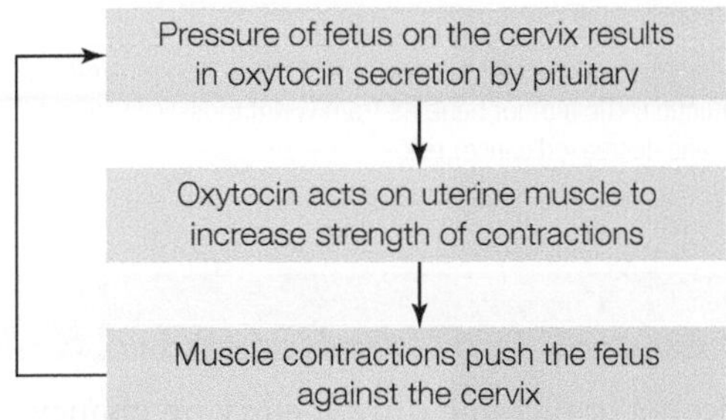
Pressure of fetus on the cervix results in oxytocin secretion by pituitary
Oxytocin acts on uterine muscle to increase strength of contractions
Muscle contractions push the fetus against the cervix

Uterine contractions continue to intensify until they push the fetus out of the mother's body. Synthetic oxytocin is sometimes given to induce labor or to increase the strength of contractions.

After the newborn emerges, continued muscle contractions expel the placenta from the uterus as the "afterbirth." The umbilical cord that connects the

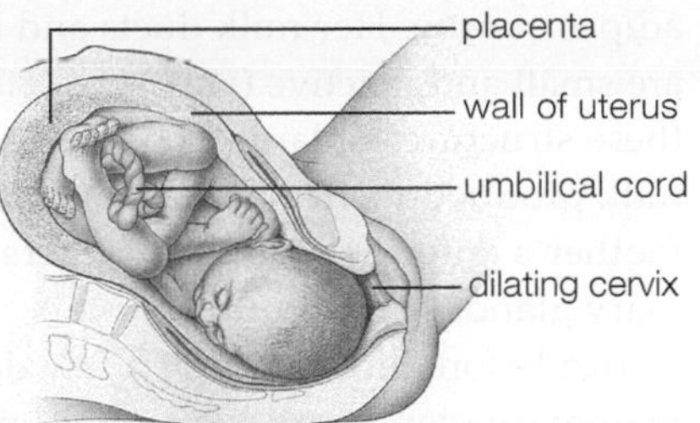
placenta
wall of uterus
umbilical cord
dilating cervix

A Fetus positioned for childbirth; its head is against the mother's cervix, which is dilating (widening).

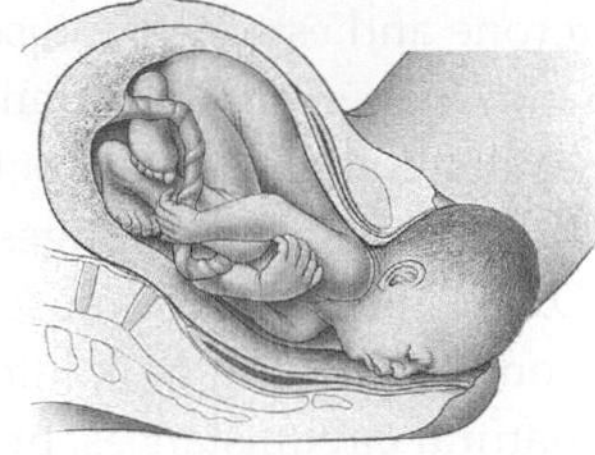

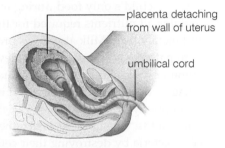

B Muscle contractions stimulated by oxytocin force the fetus out through the vagina.

placenta detaching from wall of uterus

umbilical cord

C The placenta detaches from the wall of the uterus and is expelled.

FIGURE 42.14 Normal birth and expulsion of the afterbirth. Afterbirth consists of the placenta, tissue fluid, and blood.

newborn to this mass of expelled tissue is clamped, cut short, and tied. The short stump of cord left in place withers and falls off. The navel (belly button) marks its former attachment site.

Surgical Delivery

In the United States, about 30 percent of births involve a cesarean section, sometimes called a C-section. During this surgical procedure, a physician cuts the mother's abdominal wall, then opens the uterus and removes the fetus. A variety of conditions can necessitate a surgical delivery. For example, a placenta that grows in the lower portion of the uterus and covers the cervix can make it impossible for the fetus to exit through the vagina. If the placenta dislodges too early or the umbilical cord kinks, blood flow to the fetus is compromised and surgery may be required to ensure its safety. A cesarean section is also used to prevent transmission of a sexually transmitted disease to an infant (Section 41.9).

Milk Production and Components

Like other mammals, humans nourish their newborns with milk produced by mammary glands. When a woman is not pregnant, her breasts consist mainly of

---

labor Expulsion of a placental mammal from its mother's uterus by muscle contractions.

oxytocin Posterior pituitary hormone that encourages contraction of smooth muscle of the reproductive tract.

adipose tissue. Her milk ducts and mammary glands are small and inactive (FIGURE 42.15). In pregnancy, these structures enlarge in preparation for **lactation**, or milk production. **Prolactin**, a hormone secreted by the mother's anterior pituitary, triggers growth of mammary glands during pregnancy.

Just before birth, and for a few days thereafter, mammary glands produce a clear, nutrient-rich fluid called colostrum. After a woman gives birth, a decline in progesterone and estrogens causes milk production to go into high gear. The stimulus of a newborn's suckling results in the release of both prolactin and oxytocin. Prolactin now encourages production of milk proteins. Oxytocin stimulates muscles around the milk glands to contract, forcing milk into the milk ducts.

Under natural circumstances, human breast milk will be a child's only food during infancy. Thus, it contains all nutrients required for the child's growth and metabolism. Milk also helps protect an infant from pathogens and encourages development of the immune system. In addition to maternal macrophages (phagocytic white blood cells) and immunoglobulin A (an antibody), milk contains a variety of antimicrobial compounds. One of these, lysozyme, is an enzyme that kills bacteria by destroying their cell walls. Another milk protein, lactoferrin, helps to prevent bacterial, viral, and fungal pathogens from attaching to and infecting human cells. The American Academy of Pediatricians recommends that women breast-feed exclusively for the first six months after birth and continue breast-feeding for the first year.

Breast-feeding benefits a mother as well as her child. The oxytocin released in response to suckling causes uterine contractions that help restore the uterus to its pre-pregnancy state. The hormonal effects of breast-feeding also slow the return of menstrual cycles, thus reducing the risk of a new pregnancy. Producing and secreting milk requires a lot of energy, so breast-feeding helps a woman lose excess weight that she may have gained during her pregnancy. Breast-feeding also lowers a woman's risk of cancers of the reproductive tract and protects against inherited forms of breast cancer that arise in premenopausal women.

lactation Milk production by a female mammal.
prolactin Anterior pituitary hormone that promotes milk production.

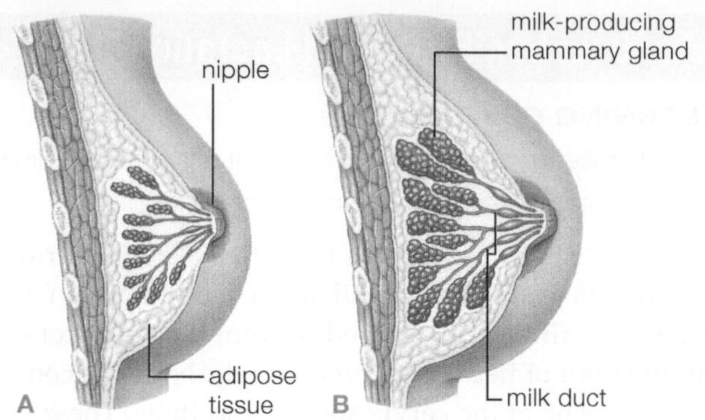
nipple
adipose tissue
A
milk-producing mammary gland
milk duct
B

FIGURE 42.15 **Cutaway views of (A) the breast of a woman who is not pregnant and (B) the breast of a lactating woman.**

TAKE-HOME MESSAGE 42.10

- ✔ During normal childbirth, the hormone oxytocin stimulates increasingly strong contractions of uterine smooth muscle. These contractions propel the child, and then the afterbirth, through the cervix and out of the vagina.
- ✔ Prolactin from the pituitary promotes mammary gland enlargement and milk production. Oxytocin released in response to suckling causes milk secretion.
- ✔ In addition to nourishing the newborn, breast-feeding protects it from infection. The mother benefits from weight loss, contraceptive effects, and decreased cancer risk.

42.1 Prenatal Problems (revisited)

As you now know, human development unfolds in a predictable manner, with specific events occurring at specific times. As a result, we can predict when specific birth defects are likely to arise (FIGURE 42.17). Exposure to teratogens during the first trimester, when organs are still forming, usually has the most dramatic negative effects.

With normal development, fusion of facial tissues during week 7 closes clefts in the developing upper lip and the palate (roof of the mouth). If this fusion does not proceed properly, the child is born with a cleft lip, cleft palate, or both. Each year, about 7,500 children born in the United States have such defects. Maternal use of alcohol or nicotine during early pregnancy increases the risk. Some genetic disorders such as van der Woude syndrome also increase the risk of such birth defects. This syndrome arises from a mutation in a gene for a transcription factor (*IRF6*) that is expressed in the face during early development.

Lip and palate clefts are not only a cosmetic concern, they can also interfere with feeding by making it impossible for the infant to suck milk from a nipple. Later, as the child matures, they can cause difficulties with speech. Surgical repair of facial clefts typically begins during a child's first year. Additional surgeries may be required as the child grows and develops.

Data Analysis Activities

Birth Defects and Multiple Births Carrying multiple offspring at the same time increases the risk of some birth defects. **FIGURE 42.16** shows the results of Yiwei Tang's study of birth defects reported in Florida from 1996 to 2000. Tang compared the incidence of various defects among single and multiple births. She calculated the relative risk for each type of defect based on type of birth, and corrected for other differences that might increase risk such as maternal age, income, race, and medical care during pregnancy. A relative risk of less than 1 means that multiple births pose less risk of that defect occurring. A relative risk greater than 1 means multiple births are more likely to have a defect.

1. What was the most common type of birth defect in the single-birth group?
2. Was that defect more or less common in the multiple-birth group?
3. Tang found that multiples have more than twice the risk of single newborns for one type of defect. Which type?
4. Does a multiple pregnancy increase the relative risk of chromosomal defects in offspring?

	Prevalence of Defect		Relative
	Multiples	Singles	Risk
Total birth defects	358.50	250.54	1.46
Central nervous system defects	40.75	18.89	2.23
Chromosomal defects	15.51	14.20	0.93
Gastrointestinal defects	28.13	23.44	1.27
Genital/urinary defects	72.85	58.16	1.31
Heart defects	189.71	113.89	1.65
Musculoskeletal defects	20.92	25.87	0.92
Fetal alcohol syndrome	4.33	3.63	1.03
Oral defects	19.84	15.48	1.29

FIGURE 42.16 Prevalence, per 10,000 live births, of various types of birth defects among multiple and single births. Relative risk for each defect is given after researchers adjusted for the mother's age, race, previous adverse pregnancy experience, education, Medicaid participation during pregnancy, as well as the infant's sex and number of siblings.

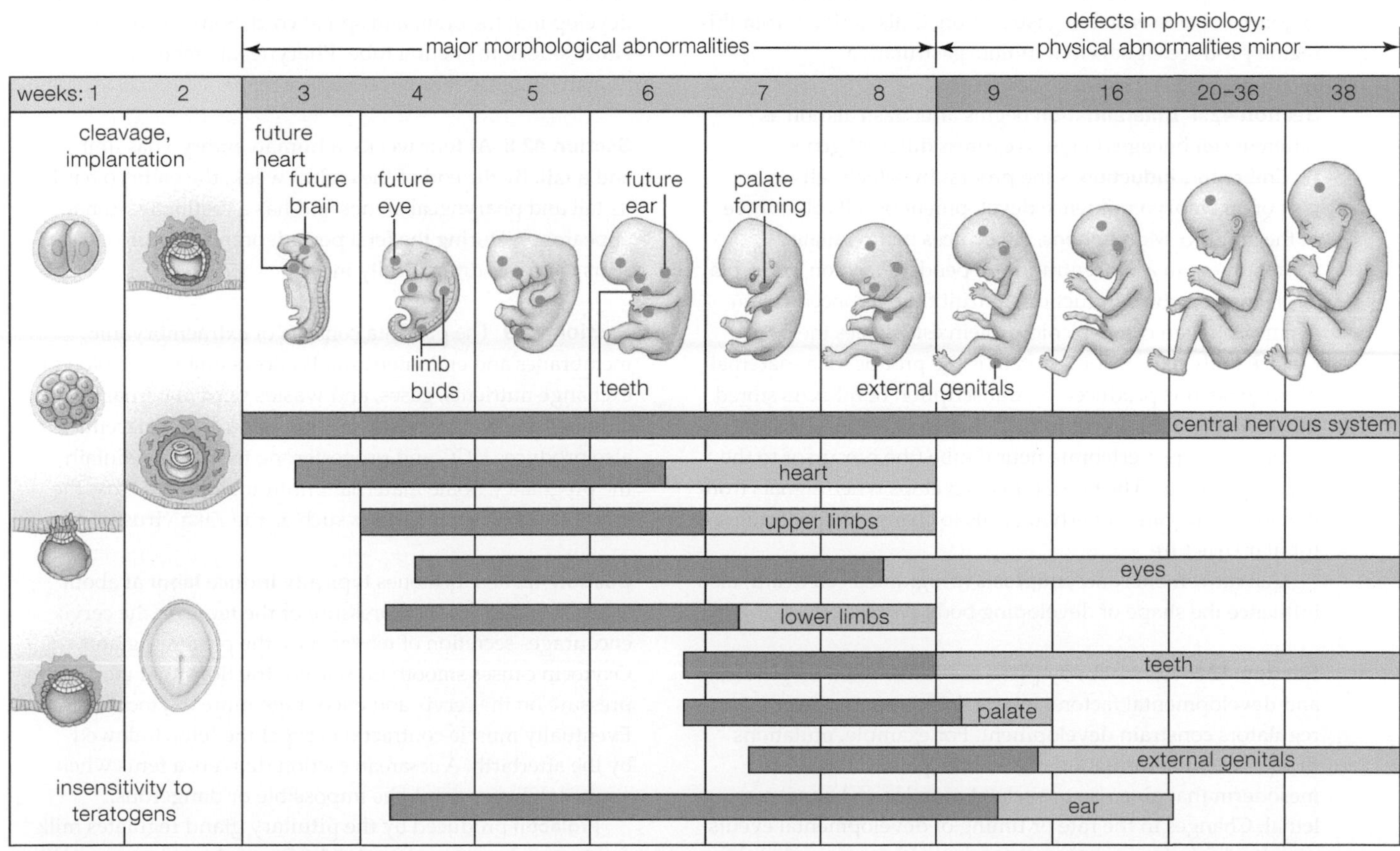

major morphological abnormalities
defects in physiology; physical abnormalities minor
weeks: 1
2
3
4
5
6
7
8
9
16
20–36
38
cleavage, implantation
future heart
future brain
future eye
future ear
palate forming
limb buds
teeth
external genitals
central nervous system
heart
upper limbs
eyes
lower limbs
teeth
palate
external genitals
ear
insensitivity to teratogens

FIGURE 42.17 Teratogen sensitivity. Teratogens are drugs, infectious agents, and environmental factors that cause birth defects. Dark blue signifies the highly sensitive period for an organ or body part; light blue signifies periods of less severe sensitivity. For example, the upper limbs are most sensitive to damage during weeks 4 through 6, and somewhat sensitive during weeks 7 and 8.

FIGURE IT OUT Is teratogen exposure in the 16th week more likely to affect the heart or the genitals?

Answer: Genitals

STUDY GUIDE

Section 42.1 Problems that arise during development can result in miscarriage, stillbirth, or birth defects. Such problems can have a genetic cause or arise as a result of prenatal exposure to a **teratogen**. The effect of a teratogen can vary depending on the time of exposure.

Section 42.2 All sexually reproducing animals have similar stages of development. After fertilization, **cleavage** produces a **blastula**. The blastula undergoes **gastrulation** to produce a **gastrula** with primary tissue layers, or **germ layers**. In vertebrates, three germ layers form: **ectoderm** (outer layer), **mesoderm** (middle layer), and **endoderm** (inner layer).

Section 42.3 **Cytoplasmic localization** occurs in all oocytes. Maternal RNAs and other components such as yolk are not randomly distributed. Cleavage distributes different portions of the egg cytoplasm, and those portions of these components, into different cells as a blastula forms. The cleavage pattern is genetically determined and differs somewhat between protostomes and deuterostomes. The amount of yolk in an egg also influences the pattern of cleavage.

Experimental manipulation of amphibian development demonstrated that material localized in one region of the zygote can be essential to gastrulation. Cells derived from this region produce signals that initiate gastrulation.

Section 42.4 **Differentiation** begins after gastrulation as different cell lineages begin to express different genes.

Embryonic induction is the process by which cells in one part of an embryo influence development of cells elsewhere in the embryo. **Morphogens**, substances that regulate development in a concentration-dependent fashion, often play a role in embryonic induction. In fruit flies, a concentration of a morphogen called bicoid protein establishes the front-to-back body axis. Bicoid protein is the product of a **maternal effect gene**. It is produced by transcription of mRNAs stored in the oocyte. Embryonic induction also plays a role in formation of the vertebrate neural tube (the precursor to the nervous system). The neural tube develops when signals from the notochord cause overlying cells to change shape forming a tubular structure.

Cell movements and **apoptosis** (programed cell death) also influence the shape of developing body parts.

Section 42.5 Animal body plans are constrained by physical and developmental factors. Interactions among master regulators constrain development. For example, mutations that prevent the formation of **somites** (paired blocks of mesoderm that give rise to skeletal muscles and bone) are lethal. Changes in the rate or timing of developmental events can alter body form.

Section 42.6 Human prenatal development occurs over a period of nine months. Organs take shape while the individual is an **embryo**, an interval that concludes at the end of the eighth week. For the remainder of the pregnancy, the **fetus** grows larger and organs mature and begin to function. Growth and development continue after birth (during the postnatal period).

Section 42.7 Human fertilization normally occurs in an oviduct. Cleavage begins as the fertilized egg is moved toward the uterus. If a cluster of cells splits early in cleavage, the result can be **identical twins**. More typically, cleavage produces a single **blastocyst** that implants in the uterine wall.

The blastocyst's inner cell mass of the blastocyst gives rise to the embryo. Cells in the blastocyst's outer layer give rise to extraembryonic membranes. The **amnion** is an extraembryonic membrane that encloses and protects the embryo in a fluid-filled sac. The **chorion** and **allantois** become part of the placenta, a structure that allows exchange of substances between maternal and fetal bloodstreams. After implantation, the chorion produces the hormone **human chorionic gonadotropin (HCG)**.

After implantation, gastrulation occurs. Cells from the blastocyst's surface migrate inward at a groove called the primitive streak. Next a neural tube forms. It will later develop into the brain and spinal cord. Somites form on either side of the neural tube. Pharyngeal arches form, but in humans, these arches never support gills.

Section 42.8 At four weeks, a human embryo has limbs and a tail. By the end of the eighth week, the embryo has lost its tail and pharyngeal arches and has a distinctly human appearance. During the fetal period, organs mature and the fetus increases dramatically in size.

Section 42.9 The placenta consists of extraembryonic membranes and endometrium. It allows embryonic blood to exchange nutrients, gases, and wastes with maternal blood, although the two bloodstreams never mix. The placenta also produces HCG and progesterone that help maintain the pregnancy. Some maternal immunoglobulins cross the placenta, as do some viruses such as the Zika virus.

Section 42.10 Hormones typically induce **labor** at about 38 weeks. During labor, pressure of the fetus on the cervix encourages secretion of **oxytocin** by the pituitary gland. Oxytocin causes smooth muscle contractions that increase the pressure on the cervix and encourage more oxytocin secretion. Eventually muscle contractions expel the fetus followed by the afterbirth. A cesarean section delivers a fetus when vaginal delivery would be impossible or dangerous.

Prolactin produced by the pituitary gland regulates milk production by mammary glands. Oxytocin encourages milk secretion during **lactation**. Milk contains not only essential nutrients, but also substances that enhance development of the immune system and help kill pathogens.

SELF-QUIZ

Answers in Appendix VII

1. The end product of cleavage is a ________ .
   a. gamete   c. gastrula
   b. blastula   d. zygote

2. The outermost germ layer in a vertebrate gastrula is the ________ .
   a. endoderm   c. mesoderm
   b. ectoderm   d. dermis

3. A morphogen ________ .
   a. is a master regulator gene
   b. has different effects at different concentrations
   c. maintains the placenta
   d. kills bacteria

4. Match each term with the most suitable description.
   ___ apoptosis   a. blastomeres form
   ___ embryonic induction   b. rearrangement of cells forms primary tissues
   ___ cleavage   c. cells die on cue
   ___ gastrulation   d. cells influence neighbors
   ___ implantation   e. blastocyst burrows into the uterus

5. In humans, fertilization typically occurs in the ________ .
   a. vagina   c. cervix
   b. uterus   d. oviduct

6. The ________ , a fluid-filled sac, surrounds and protects a human embryo and keeps it from drying out.
   a. amnion   c. yolk sac
   b. allantois   d. chorion

7. The placenta consists of ________ .
   a. embryonic tissue   c. paternal tissue
   b. maternal tissue   d. a combination of a and b

8. During the fetal period, ________ .
   a. gastrulation ends   c. a tail forms
   b. somites form   d. limb movements begin

9. Human milk contains ________ .
   a. sugars   c. antibodies
   b. lysozyme   d. all of the above

10. Match each hormone with its action(s).
    ___ prolactin   a. milk protein production
    ___ oxytocin   b. sustains corpus luteum
    ___ human chorionic gonadotropin   c. causes contraction of uterus, milk ducts

11. Pharyngeal arches of a human embryo will later develop into ________ .
    a. gills   c. structures of the head and neck
    b. lungs

12. ________ removes webs between developing digits.
    a. Gastrulation   c. Apoptosis
    b. Cleavage   d. Implantation

13. Most miscarriages ________ .
    a. occur in the first trimester
    b. are caused by exposure to a pathogen
    c. can be prevented by a cesarean section
    d. threaten the life of the mother

14. A pregnancy test detects ________ in the urine.
    a. lysozyme   c. oxytocin
    b. prolactin   d. human chorionic gonadotropin

15. Place these events in human development in the correct order.
    ___ 1   a. gastrulation is completed
    ___ 2   b. blastocyst forms
    ___ 3   c. tail is reabsorbed
    ___ 4   d. implantation occurs
    ___ 5   e. zygote forms
    ___ 6   f. neural tube forms
    ___ 7   g. maternal cervix softens

CRITICAL THINKING

1. The rubella virus causes German measles. If a pregnant woman is is infected by the rubella virus during the first trimester of her pregnancy, her child may be born deaf or blind. Infections later in pregnancy do not increase the risk of these effects. Explain why.

2. A nursing mother who has an alcoholic drink secretes alcohol into her milk for two to three hours afterward. Thus, feeding her child during this time will expose the child to alcohol. Alcohol also will decrease her oxytocin production. How will a decrease in this hormone affect the woman's ability to breast-feed?

3. To determine the effect of the bicoid gene, scientists injected bicoid mRNA into the posterior end of freshly laid fruit fly eggs. What effect would you expect such an injection to have on the flies' development?

4. If an embryo splits at the two-cell stage, each of the resulting identical twins will have its own placenta. If such a split occurs near the time of implantation, the identical twins may share a placenta. By contrast, fraternal twins never share placenta. Explain this difference between the two types of twins.

43 ANIMAL BEHAVIOR

CORE CONCEPTS

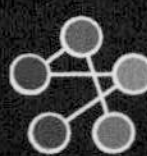

Systems

Complex properties arise from interactions among components of a biological system.

Genes and epigenetic mechanisms are the basis of innate (inherited) behaviors and the capacity for learning through experience. Natural selection acts on these traits, so it shapes behaviors as well as their timing and coordination. A population's overall survival depends on communication and cooperation among its members and is influenced by individual behavior.

Evolution

Evolution underlies the unity and diversity of life.

Many animal behaviors occur as responses to environmental cues and interactions with other organisms. Different species communicate information via visual, audible, tactile, electrical, or chemical signals that affect the behavior of other organisms. Genetic variation in these traits allows them to evolve under selection.

Pathways of Transformation

Organisms exchange matter and energy with the environment in order to grow, maintain themselves, and reproduce.

Most animals move about in search of food. Some animals make seasonal long-distance journeys, traveling between areas where the food they require is locally abundant. Evolved navigation systems facilitate these journeys. Animals also move in search of mates. Evolved communication systems ensure that individuals of the same species can meet and reproduce.

Links to Earlier Concepts

This chapter revisits environmental effects on phenotype (Section 13.6) and epigenetics (10.5). It draws on your knowledge of sensory and endocrine systems (Chapters 33 and 34) and provides examples of adaptation (16.3) and sexual selection (17.5).

43.1 Can You Hear Me Now?

Right whales are an endangered species. Hunting once drove the North Atlantic population of right whales to a low of about 100 animals; today it includes about 450. Unfortunately, this population lives near the northeastern coast of North America, where commercial ship traffic is heavy (**FIGURE 43.1**).

Marine biologist Susan Parks studies how North Atlantic right whales communicate, the effects of human-made noise on these whales, and ways to minimize any negative effects. By attaching recorders to right whales, Parks learned that when a ship is present, whales call more loudly and their calls become more shrill. They behave like humans at a noisy restaurant, shouting at one another in high-pitched voices to be heard above the din.

For humans, chronic exposure to noise is stressful, and the same may be true for whales. It would be impossible to ask tankers to stop their engines for days to see whether noise affects whales, but such a quiet period did occur immediately after the terrorist attacks of September 11, 2001. During this time, another scientist, Rosalind Rollins, was collecting whale feces to test the whales' level of cortisol, a hormone whose secretion increases with stress (Section 34.8). Parks and Rollins pooled data regarding ship-related noise and cortisol secretion and found that the brief decline in shipping noise was accompanied by a significant drop in the whales' cortisol level. In other years, there was no such drop in cortisol after September 11. These data suggest that the normally high noise level related to shipping causes a chronic stress response in right whales. Thus, noise-induced stress may be an important factor in slowing the recovery of this highly endangered species.

As this example shows, we can unknowingly interfere with animal signaling. Communication systems of whales and other animals evolved over countless generations in an environment free of human-generated noise, light, and other sensory distractions. Finding ways to meet human needs without unnecessarily disrupting the communication mechanisms of Earth's other species is an ongoing challenge.

M S C

FIGURE 43.1 Whales in a shipping channel. Ship noise can impair whales' ability to communicate by sound.

CREDITS: (opposite) Robert L. Kothenbeutel/Shutterstock; (1) iStockphoto.com/Jan-Otto.

43.2 Factors Affecting Behavior

LEARNING OBJECTIVES

- Explain the difference between proximate and ultimate causes of behavior.
- Using appropriate examples, explain how genetic and epigenetic mechanisms can give rise to differences in behavior.

Why do animals behave the way they do? In thinking about what drives animal behavior, biologists distinguish between "proximate" and "ultimate" causes. *Proximate* means "nearby or immediate;" proximate causes of a behavior include genes and epigenetic factors that specify the behavior, the physiological mechanisms that bring about the behavior, and environmental factors that influence and elicit the behavior. The ultimate causes of a behavior are the evolutionary factors that shaped it. Variations in behavior affect fitness, so heritable behavioral traits will evolve under natural selection.

Studying variations in behavior, either among members of a species or among closely related species, can help uncover the underlying causes of the behavior. A few examples will illustrate this approach.

Genetic Variation within a Species

Most behavior has a polygenic basis. However, there are some cases in which behavioral differences can be traced to allele differences at a single gene. The *foraging* gene, which determines the feeding behavior of fruit fly larvae, is one such gene. Fruit fly larvae are wingless, with a wormlike body, and they feed on yeast cells that grow on decaying fruit. In wild fruit fly populations, the larvae have two distinct types of feeding behavior. Larvae with the dominant allele at the *foraging* locus (larvae who are *FF* or *Ff*) have the "rover" phenotype. Rovers move around a lot as they feed, and they often leave one patch of food to seek another (**FIGURE 43.2A**). Larvae homozygous for the recessive allele at the foraging locus (*ff*) have the "sitter" phenotype; they tend to move little once they find a patch of yeast (**FIGURE 43.2B**).

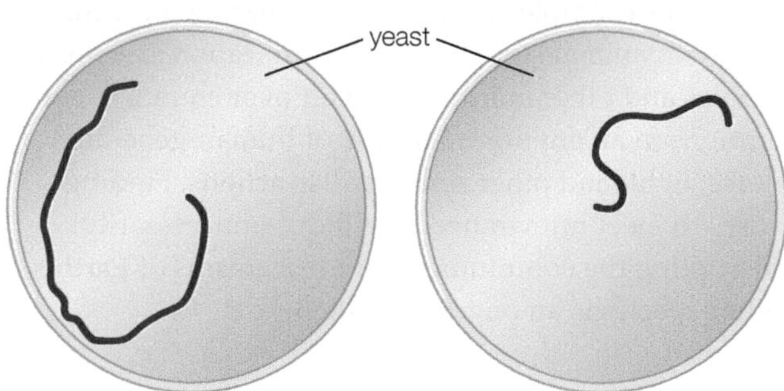
yeast

A Rovers (genotype *FF* or *Ff*) move often as they feed. When a rover's movements on a petri dish filled with yeast are traced for 5 minutes, the trail is relatively long.

B Sitters (genotype *ff*) move little as they feed. When a sitter's movements on a petri dish filled with yeast are traced for 5 minutes, the trail is relatively short.

FIGURE 43.2 Genetic polymorphism for foraging behavior in fruit fly larvae. When a larva is placed in the center of a yeast-filled plate, its genotype at the *foraging* locus influences whether it moves a little or a lot while it feeds. Black lines show a representative larva's path.

The ultimate cause of this behavioral variation in larval foraging behavior is selection related to competition for food. In experimental populations with limited food, both rovers and sitters are most likely to survive to adulthood when their foraging type is rare. A rover does best when surrounded by sitters, and vice versa. Presumably, when there are lots of larvae of one type, they all compete for food in the same way. Under these circumstances, a fly larva that behaves differently than the majority is at an advantage. Thus, frequency-dependent selection (Section 17.5) maintains both alleles of the *foraging* gene.

Genetic Differences between Species

Researchers sometimes investigate the genetic basis of a behavior by comparing related species. For example, studies of rodents called voles (**FIGURE 43.3A**) revealed that inherited differences in the number and distribution of certain hormone receptors influence mating and bonding behavior. Most voles are promiscuous, meaning both males and females have multiple mates and show no special affinity for individuals with whom they have mated. By contrast, prairie voles form a pair bond—a long-term, largely monogamous partnership. This species-specific difference in behavior arises from differences in how the voles respond to two hormones.

The hormone oxytocin is released during sexual intercourse, labor, and lactation in all mammals. In prairie voles, its action is essential to formation and maintenance of the pair bond. Inject the female member of an established prairie vole pair with a chemical that impairs oxytocin function, and she will no longer associate mainly with her partner. All vole species have similar numbers of oxytocin-producing cells. However, compared to members of promiscuous vole species, prairie voles have many more oxytocin receptors in the parts of the brain associated with social learning (**FIGURE 43.3B,C**).

Variations in the distribution of receptors for the hormone arginine vasopressin (AVP) also contribute to differences in the tendency to pair bond. Compared to members of promiscuous vole species, prairie voles have more AVP receptors in their brain. Additional evidence for the importance of AVP in pair-bonding behavior comes from an experiment in which the prairie vole AVP receptor gene was transferred into the

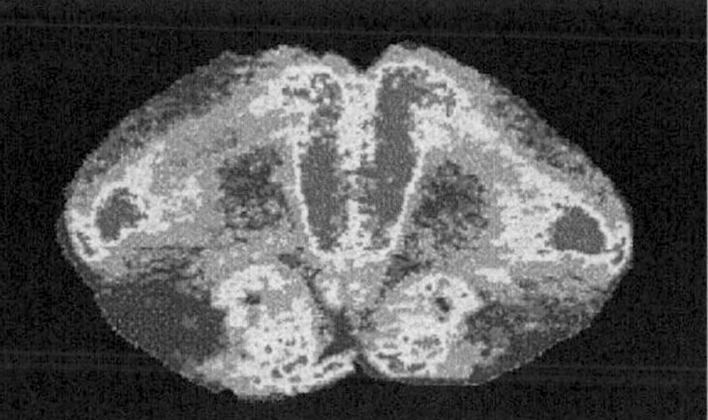

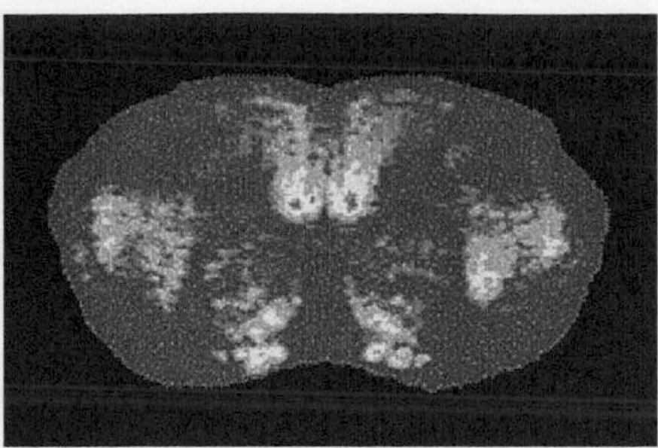

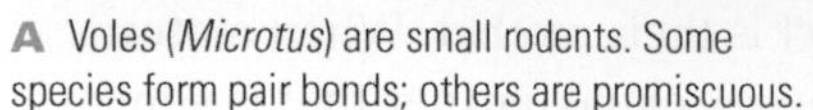

A Voles (*Microtus*) are small rodents. Some species form pair bonds; others are promiscuous.

B PET scan of a monogamous prairie vole's brain with many receptors for the hormone oxytocin (red).

C PET scan of a promiscuous prairie vole's brain with few hormone receptors for oxytocin.

FIGURE 43.3 Distribution of oxytocin receptors in two species of vole.

forebrain of male mice. Mice are naturally promiscuous. However, male mice genetically modified by the prairie vole gene gave up their playboy ways. After the procedure, they preferred a female with whom they had already mated over an unfamiliar female. These results confirmed the role of AVP in fostering monogamy among male rodents.

Environmental Effects

Phenotypic plasticity refers to the ability of an individual with a specific genotype to express different phenotypes in different environments. Many animals show **behavioral plasticity**, meaning their behavioral traits are affected by environmental factors.

In cichlids (a type of fish) a male's social environment affects his behavior and appearance. During the course of a male cichlid's life, he can switch back and forth between two social roles. When in the dominant role, he courts and breeds with females and engages in confrontations with other males. Dominant males have bright coloration and a dark eye bar (**FIGURE 43.4**). When in a subordinate role, a male cichlid flees from dominant males, does not court or breed, and has a drab, pale gray coloration.

Whether a male cichlid exhibits the dominant or subordinate phenotype depends on a variety of factors, most importantly the presence of a larger dominant male. Remove all larger dominant males from an environment, and a subordinate male will switch to the dominant phenotype. Within minutes, he becomes more colorful and begins chasing other males. Within hours, his previously shrunken testes enlarge and produce viable sperm.

Researchers suspected that the dramatic transformation between subordinate and dominant phenotypes involves epigenetic changes. Recall that epigenetic mechanisms cause changes in gene expression without altering the DNA sequence (Section 10.5). In eukaryotes, epigenetic changes often involve methylation of specific genes. To test the hypothesis that DNA methylation plays a role in the transition between subordinate and dominant roles, researchers injected some male cichlids who had been lifelong subordinates with methionine, which provides the methyl groups used for methylating DNA. They injected other lifelong subordinates with a chemical that inhibits DNA methylation. Pairs of similarly sized, former subordinates—one injected with methionine, the other with the methylation inhibitor—were then placed in a tank without other males. Within a day, one of the males had become dominant, and in the vast majority of cases it was the methionine-injected individual. An epigenetic effect, specifically methylation, seems to facilitate this change.

FIGURE 43.4 Social dominance. Dominant male cichlids with distinctive eye bars face off.

TAKE-HOME MESSAGE 43.2

- ✔ Biologists consider both the proximate and ultimate causes of animal behavior.
- ✔ Some differences in behavior among individuals or closely related species arise from genetic effects. Most behavior has a polygenic basis.
- ✔ Environmentally induced epigenetic effects can affect behavior.

behavioral plasticity The behavioral phenotype associated with a genotype depends on conditions in the animal's environment.

43.3 Instinct and Learning

LEARNING OBJECTIVES

- Compare instinctive and learned behavior.
- Explain the circumstances that favor evolution of instinctive behavior.
- Describe the types of learned behavior and give examples.

Instinctive Behavior

All animals are born with the capacity for **instinctive behavior**—an innate response to a specific and usually simple stimulus. The life cycle of cuckoo birds provides several examples of instinct at work. Cuckoo birds are "brood parasites," meaning they lays their eggs in the nest of other birds. A newly hatched cuckoo is blind, but contact with an object beside it stimulates an instinctive response. The hatchling maneuvers the object, which is usually one of its foster parents' eggs, onto its back, then shoves the object out of the nest (**FIGURE 43.5A**). If the cuckoo's foster siblings hatch before it does, the cuckoo will shove them out after it hatches. Instinctively shoving objects out of the nest ensures the cuckoo has its foster parents' undivided attention.

Instinctive responses are advantageous only if the triggering stimulus almost always signals the same situation. Doing away with an egg benefits a cuckoo chick because that egg always houses a future competitor for food. However, instinctive responses can open the way to exploitation. Cuckoos often look nothing like the chicks they displace, yet their foster parents feed them all the same (**FIGURE 43.5B**) The cuckoo chick exploits its foster parents' instinctive urge to fill a gaping mouth. Under most circumstances, that mouth belongs to one of their chicks.

Time-Sensitive Learning

Learned behavior is behavior that is altered by experience. Some instinctive behavior can be modified with learning. A garter snake's initial strikes at prey are instinctive, but the snake learns to avoid dangerous or unpalatable prey. Learning may occur throughout an animal's life, or it may be restricted to a critical period.

Imprinting is a form of instinctive learning that can occur only during a short, genetically determined interval in an animal's life. For example, baby geese learn to follow the large object that bends over them in response to their first peep (**FIGURE 43.6A**). With rare exceptions, this object is their mother. When mature, the geese will seek out a sexual partner that is similar to the imprinted object.

A genetic capacity to learn, combined with actual experiences in the environment, shapes most forms of behavior. For example, a male sparrow has an inborn capacity to recognize his species' song when he hears older males singing it. The young male uses these overheard songs as a guide to fill in details of his own song. Males reared alone sing an abnormal, simplified version of their species' song. So do males exposed only to the songs of other species.

The sparrow can only learn his species-specific song during a limited period early in life. To learn to sing normally, he must hear a male "tutor" of his own species during his first 50 or so days of life. Hearing a same-species tutor later will not improve his singing.

Most birds must also practice their song to perfect it. In one experiment, researchers temporarily paralyzed throat muscles of zebra finches who were beginning to sing. After being temporarily unable to practice, these birds never mastered their song. In contrast, temporary throat muscle paralysis in very young birds or adults did not impair later song production. Thus, in zebra finches, there is a critical period for song practice, as well as for song learning.

Conditioned Responses

Nearly all animals are lifelong learners. Most learn to associate certain stimuli with rewards and others with negative consequences.

With **classical conditioning**, an animal's involuntary response to a stimulus becomes associated with another stimulus that is presented at the same time. In the most famous example, Ivan Pavlov rang a bell whenever he fed a dog. Eventually, the dog's reflexive response to food—increased salivation—was elicited by the sound of the bell alone.

With **operant conditioning**, an animal modifies its voluntary behavior in response to consequences of the behavior. This type of learning was first described for behaviors acquired in a laboratory setting. For example, a rat that presses a lever in a laboratory cage and is rewarded with a food pellet becomes more likely to

A Young European cuckoo shoving its foster parents' egg from the nest. It pushes out any object beside it.

B Foster parent (left) feeds a huge cuckoo chick (right) in response to a simple cue: a gaping mouth.

FIGURE 43.5 Instinctive behavior.

press the lever again. A rat that receives a shock when it enters a particular area of a cage will quickly learn to avoid that area. Similarly, an animal in the wild will learn to repeat behaviors that provide food and to avoid those that cause it discomfort.

Other Types of Learned Behavior

With **habituation**, an animal learns by experience to ignore a repeated stimulus that has neither positive nor negative effects. For example, pigeons in cities learn not to flee from people who walk past them. Habituation is considered the simplest type of learning. Animals as simple as jellies can learn to ignore a repeated harmless touch.

Many animals are capable of **spatial learning**. They learn about the landmarks in their environment and form a sort of mental map. This map may be put to use when the animal needs to return home. For example, a fiddler crab foraging up to 10 meters (30 feet) away from its burrow is able to scurry straight home when it perceives a threat.

Animals also learn the details of their social landscape. Social learning allows them to recognize mates, offspring, or competitors by appearance, calls, odor, or some combination of cues. For example, two male lobsters usually fight when they meet for the first time (**FIGURE 43.6B**). After the fight, they will recognize one another by scent and behave accordingly, with the loser actively avoiding the winner.

With **observational learning**, an animal imitates the behavior of another individual. Consider how a chimpanzee strips leaves from branches to make a "fishing stick" for use in capturing termites as food. The chimpanzee inserts a fishing stick into a termite mound, then withdraws the stick and eats the termites that cling to it. Different groups of chimpanzees use different methods of tool-shaping and insect-fishing. In each group, young individuals learn by observing and imitating behavior of others (**FIGURE 43.6C**).

classical conditioning Animal's involuntary response to a stimulus becomes associated with another stimulus that is presented at the same time.
habituation Learning not to respond to a repeated stimulus.
imprinting Learning that can occur only during a specific interval in an animal's life.
instinctive behavior Innate response to a simple stimulus.
learned behavior Behavior that is modified by experience.
observational learning One animal acquires a new behavior by observing and imitating behavior of another.
operant conditioning Learning in which an animal's voluntary behavior is modified by the consequences of that behavior.
social learning Learning to recognize specific individuals of one's species.
spatial learning Learning landmarks allows formation of a mental map of the environment.

A **Imprinting.** Nobel laureate and animal behaviorist Konrad Lorenz with geese that imprinted on him. The inset photos shows a more common result of imprinting.

B **Social learning.** Captive male lobsters fight at their first meeting. The loser remembers the winner's scent and avoids him. Without another meeting, memory of the defeat lasts up to two weeks.

C **Observational learning.** Chimpanzees use sticks as tools for extracting tasty termites from a nest. The method of shaping the stick and catching termites is learned by observation and imitation.

FIGURE 43.6 Examples of learned behavior.

TAKE-HOME MESSAGE 43.3

- ✔ Instinctive behavior can initially be performed without any prior experience, and it is often elicited by a simple stimulus. Even instinctive behavior may be modified by experience.
- ✔ Certain types of learning can occur only at particular times during an individual's life.
- ✔ Learning alters both voluntary and involuntary behaviors.

CREDITS: (6A) © Nina Leen/TimePix/Getty Images; inset, © Exactostock/SuperStock; (6B) © Professor Jelle Atema, Boston University; (6C) Shah, Anup/Animals Animals.

43.4 Movements and Navigation

LEARNING OBJECTIVES

- Explain the difference between a taxis and a kinesis.
- Describe some of the ways that animals find their way when they migrate.

Taxis and Kinesis

All animals are motile (can move from place to place) during some part of their life cycle. Even simple animals such as planarian flatworms instinctively move toward some stimuli and away from others. An innate directional response to a stimulus is called a **taxis** (plural, taxes). For example, planarians are negatively phototactic (they move away from light) and positively geotactic (they move toward the pull of gravity). A stimulus may also cause an animal to increase or decrease its movements without regard for direction, a response called **kinesis**. For example, planarians are photokinetic, meaning they move more in light than in the dark. The planarian's innate responses to stimuli direct the worm toward a favorable habitat and help it to remain there.

Migration

Most animals move about daily to find food and avoid predators. Some also migrate. During a **migration**, an animal interrupts its daily pattern of activity to travel in a persistent manner toward a new habitat.

Many animal migrations involve seasonal movement to and from a breeding site. For example, birds that nest in the Arctic during the summer typically migrate to a temperate or tropical region to spend the winter. Over the course of a lifetime, a migratory bird may make these trips many times, traveling each year to breed in the region where it hatched. Marine turtles and some whales also make repeated long-distance journeys to breed at the site where their own life began.

For other animals, the migration to a breeding ground is a one-way trip. Consider the Atlantic eel, which spends most of its life in a European or North American river. Then, one fall, it migrates hundreds to thousands of kilometers to the Sargasso Sea, the region of the Atlantic where it hatched 10 to 30 years before. Here, it meets other eels, breeds, and dies. Currents distribute eel larvae throughout the North Atlantic. After the larvae develop into young eels, they swim to a coast and then make their way up a river.

Researchers have only begun to decipher how migrating animals find their way. Some animals have an innate magnetic compass, meaning they use latitudinal variations in Earth's magnetic field to determine direction. In the spring, a caged European robin will hop repeatedly toward the north, even if it has no view of the outdoors (FIGURE 43.7A). Obscuring the bird's vision disrupts this behavior, indicating that this bird detects directional differences in Earth's magnetic field with its eyes (FIGURE 43.7B). Magnetism-sensitive proteins called cryptochromes are thought to play a role in the European robin's compass. These proteins are present in cells of the bird's retina. In other animals, the mineral magnetite (iron oxide) is thought to be involved. For example, salmon have magnetite-containing cells in their nose.

Animals can also navigate by the sun and stars. The position of celestial bodies shifts over time, so the use of a sun- or star-based compass requires an innate sense of time. A bird that needs to head north must fly to the left of the sun in the morning, and to the right of the sun in the evening.

To find a specific site, an animal must know not only compass directions, but the location of its goal relative to its current location. It does no good to know where north is if you do not also know whether you are north or south of your destination. Localized variations in the Earth's magnetic field provide this information to some species such as European robins. An animal may also gauge its progress relative to visual or olfactory landmarks. For example, odor cues help guide eels as they travel from the mouth of the river where they spent their adulthood toward the open ocean.

FIGURE 43.7 Vision-based magnetic compass in European robins.

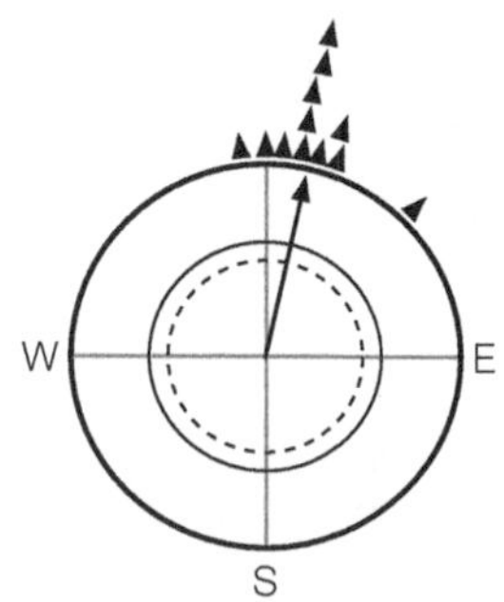
W
E
S

A Triangles indicate predominant direction of movements made by 12 robins tested indoors in the spring. Arrow indicates average result.

B A robin with a frosted lens in front of its right eye. Use of such lenses impairs the bird's ability to detect variations in Earth's magnetic field.

TAKE-HOME MESSAGE 43.4

✔ Innate responses to specific stimuli can influence the rate or direction of animal movements.

✔ Some animals migrate between sites seasonally. Navigating to a specific site requires a mental map and an ability to determine compass direction.

43.5 Communication Signals

LEARNING OBJECTIVES

- List the different types of communication signals, and give examples of each.
- Describe some ways that predators can take advantage of prey communication systems.

Evolution of Animal Communication

Communication signals are evolved cues that transmit information from one member of a species to another. Use of a communication signal becomes established and persists only if signaling benefits both the signal sender and the signal receiver. If signaling is disadvantageous for either party, then natural selection will tend to favor individuals that do not send or respond to it.

Types of Signals

Chemical signals called **pheromones** convey information among members of a species. There are two categories of pheromone. Signal pheromones cause a rapid shift in the receiver's behavior, whereas priming pheromones cause longer-term responses. Sex attractants that help males and females of many species find each other are signal pheromones. The alarm pheromone a honeybee emits when she perceives a threat to her hive is also a signal. Alarm pheromone causes other honeybees to rush out of the hive and attack a potential intruder. A chemical in the urine of male mice is a priming pheromone; it triggers ovulation in females.

Producing a pheromone requires less energy than calling or gesturing, but it also conveys less information. The chemical is either released or not. By contrast, properties of acoustical, visual, and tactile signals vary continuously, and so can convey more information.

Vocal signals help many male vertebrates, including songbirds, whales, frogs, and some fishes, attract prospective mates. Similarly, male crickets attract females by rubbing their legs together to chirp. Male cicadas attract females by using organs on their abdomen to make clicking sounds that can exceed 90 decibels.

Some birds and mammals give alarm calls that inform others of potential threats (**FIGURE 43.8A**). In many cases, the calls convey more than simply "Danger!" A prairie dog emits one type of bark when it detects an eagle and another when it sees a coyote.

Evolution shapes the properties of acoustical signals depending on whether or not the sender benefits by revealing its position. Sounds that lure mates are typically easily localized, whereas alarm calls are not.

Visual communication is most widespread in animals that have good eyesight and are active during the day. Bird courtship often involves both visual and acoustical signals. For example, courting grebes strike coordinated poses while making distinctive calls (**FIGURE 43.8B**). The display assures a prospective mate that the displaying individual is a member of the correct species and is in good health.

Threat displays advertise the displayer's good health too, but they serve a different purpose (**FIGURE 43.8C**). Males of many species compete for access to females. When two potential rivals meet, they often engage in a display that demonstrates their strength and how well armed they are. Most often, the males are not evenly matched, and the weaker individual retreats. Both males benefit by avoiding a fight that could lead to injury.

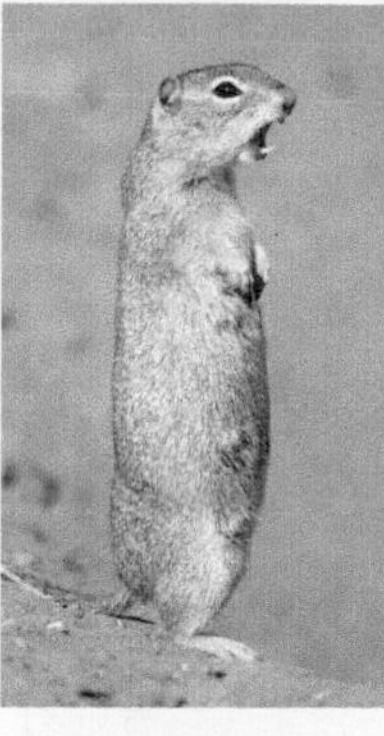
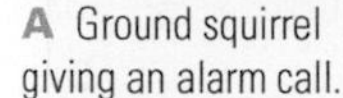

A Ground squirrel giving an alarm call.

B Courtship display in grebes; the birds move in unison.

C Threat display of a male collared lizard. This display allows rival males to assess one another's strengths without engaging in a potentially damaging fight.

FIGURE 43.8 Examples of vertebrate communication.

communication signal Chemical, acoustical, visual, or tactile cue that is produced by one member of a species and detected and responded to by other members of the same species.

kinesis (kin-EE-sis) Innate response in which an animal speeds up or slows its movement in reaction to a stimulus.

migration An animal ceases an ongoing pattern of daily activity to move in a persistent manner toward a new habitat.

pheromone (FER-uh-moan) Chemical that serves as a communication signal between members of an animal species.

taxis (TAX-iss) Innate response in which an animal moves toward or away from a stimulus.

Data Analysis Activities

Alarming Eyespots Section 1.6 described how a peacock butterfly will, when threatened, open its wings to reveal two large eyespots that are hidden when the butterfly is at rest. By one hypothesis, eyespots frighten a predatory bird by mimicking the eyes of the bird's predators. Alternatively, the sudden appearance of the spots may act by simply startling the bird. To differentiate between these two possibilities, Martin Olofsson presented peacock butterflies with or without eyespots painted over to domestic chickens. He then recorded whether the chickens gave an alarm call that is normally given upon sighting a ground predator. **FIGURE 43.9** shows the results.

1. When eyespots were visible, how many birds gave the alarm call? How many did not?
2. How did the number of alarm calls given differ when the butterflies' eyespots were hidden?
3. Does this data support the hypothesis that butterfly eyespots frighten birds by mimicking their predators?

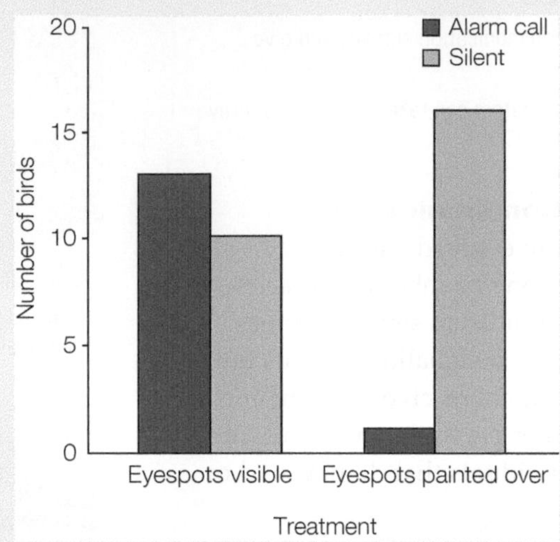

Alarm call
Silent
Number of birds
20
15
10
5
0
Eyespots visible
Eyespots painted over
Treatment

FIGURE 43.9 **Response of domestic chickens to the defense display of a peacock butterfly** (shown above). Butterflies were with or without eyespots painted over. All chickens were previously unfamiliar with these butterflies.

sun
food
hive
sun
120 °

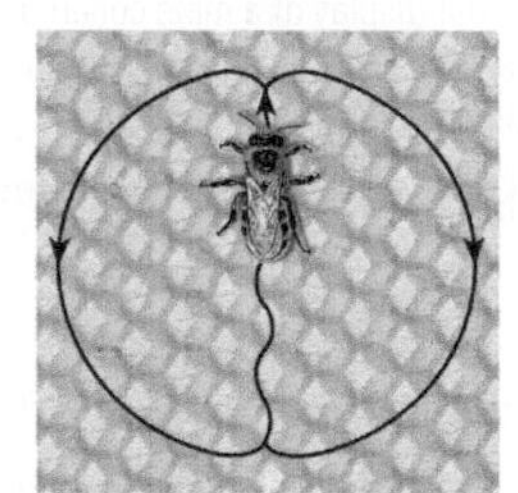

A When the waggle portion of the dance (wiggly portion) runs up the surface of the vertically oriented comb, workers fly in the direction of the sun to locate the food.

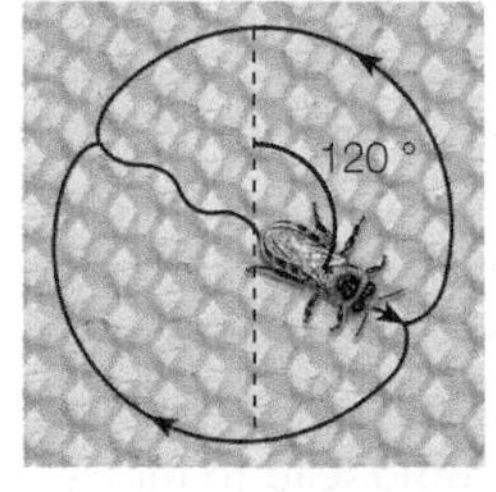

120 °

B When the waggle portion of the dance runs at an angle (in this case 120°) relative to vertical, the workers fly at the same angle relative to the position of the sun to locate the food.

FIGURE 43.10 **Honeybee waggle dance.** The angle of the waggle run, in which the bee wiggles back and forth, gives the direction of the food. The faster the dance is performed, the closer the food.

FIGURE IT OUT How will a waggle run be oriented if food is located in the opposite direction from the sun?

Answer: Straight down the comb.

With tactile displays, information is transmitted by touch. For example, after discovering food, a foraging honeybee worker returns to the hive and dances in the dark, surrounded by a crowd of other workers. If the food is more than 100 meters from the hive, it performs a waggle dance on a vertically orientated honeycomb, moving in a figure eight (**FIGURE 43.10**). During the waggle run portion of this dance, the dancer's orientation provides information about the direction of the food relative to the direction of the sun. For example, if the straight run of the dance proceeds straight up the honeycomb, food is in the same direction as the sun. How fast the dancer moves informs other bees about the distance to the food: the faster the movements, the closer the food.

Eavesdroppers and Counterfeiters

Predators can benefit by intercepting signals sent by their prey. For example, frog-eating bats locate male tungara frogs by listening for the frogs' mating calls. This poses a dilemma for the male frogs. Female tungara frogs prefer males who make complex calls, but complex calls are easier for bats to localize. Thus, when bats are near, male frogs call less, and with less flair. The subdued signal is a trade-off between competing pressures to attract a mate and to avoid being eaten.

Other predators lure prey with counterfeit signals. Fireflies are nocturnal beetles that attract mates with flashes of light. When a predatory female firefly sees the flash from a male of the prey species, she flashes back as if she were a female of his own species. If she lures the male close enough, she captures and eats him.

TAKE-HOME MESSAGE 43.5

✔ A communication signal transfers information from one individual to another individual of the same species. Such signals benefit both the signaler and the receiver.

✔ Signals have a potential cost. Some individuals of a different species benefit by intercepting signals or by mimicking them.

43.6 Mating and Parental Behavior

LEARNING OBJECTIVES

- Explain why genetic monogamy is rare among animals.
- Give examples of female choice and male-male competition.
- Using appropriate examples, explain some factors that determine whether parental care will evolve in a species and which parent will provide it.

Mating Systems

Animal mating systems have traditionally been categorized as promiscuous (multiple mates for both sexes), polygynous (multiple mates for males only), polyandrous (multiple mates for females only), or monogamous (a single mate for both sexes). In recent years, studies that integrate paternity analysis with behavioral observations have shown that many species do not always fit cleanly into these categories. Members of a species often vary in their behavior.

Some animals such as prairie voles form a pair bond, which means two individuals mate, preferentially spend time together, and cooperate in rearing offspring. However, participants in a pair bond may also seize opportunities to mate with other individuals. Thus, prairie voles are now described as socially monogamous, but genetically promiscuous. As a species, they form an exclusive social relationship with a single partner. However, the genetic composition of their offspring reflects an occasional tendency to mate with multiple partners.

Even social monogamy is rare in most animals, with the exception of birds. About 90 percent of birds are socially monogamous. Among this group, the vast majority of species that have been examined for paternity patterns are genetically promiscuous. This is not surprising as promiscuity benefits both males and females by increasing the genetic diversity among the offspring they produce.

Multiple matings can also provide another benefit: more offspring. This benefit usually applies most

lek Area where male animals perform courtship displays.

FIGURE 43.11 Male sage grouse displaying at a lek. To entice a female to choose him, the male dances, inflates the sacs on his chest, and makes gurgling calls.

strongly to males. Sperm are energetically inexpensive to produce, so a male's reproductive success is usually limited by access to mates. By contrast, the main limit on a female's reproductive success is her capacity to produce large, yolk-rich eggs or, in mammals, to carry developing young.

The selective benefits of promiscuity are offset somewhat when two parents must cooperate to rear the young. Promiscuity among females reduces males' probability of paternity, which in turn selects for males who contribute less parental care.

Female Choice and Male-Male Competition

When one sex exerts selection pressure on the other, we expect sexual selection to occur (Section 17.5).

In many insects and spiders, females choose among males based on the gifts of food that they offer. For example, a female nursery web spider is more likely to mate with a male if he approaches her with a gift of silk-wrapped prey. By choosing a male who can provide such a gift, a female receives a nutritional boost that helps her produce eggs.

In some birds such as sage grouse, males converge at a communal display ground called a **lek**. Each male at the lek performs a courtship display for females (**FIGURE 43.11**). The few males who are most attractive to the females mate many times, while their less attractive counterparts do not mate at all. It may seem strange for females to base their choice on a male's ability to sing and dance, but the healthiest males perform the most stunning displays. The quality of the display may thus indicate other less visible traits that make a male a desirable mate. In addition, males whose courtship behavior attracts many females will pass female-pleasing traits to their sons.

FIGURE 43.12 Male fiddler crab displaying. A male waves his one enlarged claw to attract a female to his burrow.

In other species, a male establishes a mating territory, an area that includes resources that females need for reproduction. A **territory** is any area from which an animal or group of animals actively exclude others. A male fiddler crab's territory is a stretch of shoreline in which he excavates a burrow. A male has one oversized claw (**FIGURE 43.12**), and during mating season, he stands outside his burrow waving it. The claw is used both in territorial disputes with other males and in a courtship display. Females attracted by a male's display check out the location and size of his burrow before mating with him. Burrow location and size are important because these factors affect the development of crab larvae.

In mammals such as elephant seals and elk, the strongest males hold large territories and mate with all females inside the area. Other males may never get to mate at all unless they sneak into an area under another male's control. Such mating systems result in a sexual dimorphism in size. Males, who must overpower a rival to mate, are much larger than females.

Parental Care

Parental care requires time and energy that an individual could otherwise invest in reproducing again. It arises only if the genetic benefit of providing care (increased offspring survival now) offsets the cost (reduced opportunity to produce other offspring). If a single parent can successfully rear offspring, individuals who spare themselves this cost by leaving parental duties to their mate will be at a selective advantage.

Most fishes provide no parental care, but in those that do, this duty usually falls to the male. In many species, males guard eggs until they hatch.

In some cases, male fishes retain eggs or young on or in their body. For example, male seahorses have a special brood pouch. Mouth brooding (housing developing eggs in the mouth) has also evolved in a variety of fish species. The prevalence of male parental care in fishes may be related to sex differences in how fertility changes with age. A female fish's fertility increases with age, whereas a male's does not. Thus, a female fish who invests energy in care forgoes more future reproduction than a male fish does.

In birds, two-parent care is most common. Chicks of birds that cooperate in care of their young tend to hatch while in a relatively helpless state. Chicks of sage grouse and other birds in which females alone provide care hatch when more fully developed.

In about 90 percent of mammals, the female rears young. In the remaining 10 percent, both sexes participate. No mammal relies solely on male parental care. Female mammals sustain developing young in their body, so they have a greater investment in newborns than males. Female mammals also nurse their young; males typically do not. Two fruit bat species are the exception: Both males and females of these species nurse offspring.

TAKE-HOME MESSAGE 43.6

✔ The positive effects of genetic diversity among offspring make monogamy rare.

✔ In many species, a few males monopolize mating opportunities either by enticing females to choose them or by defending a territory with many females.

✔ Whether parental care occurs and who delivers it depends on the benefits and costs of care to each parent.

43.7 Group Living

LEARNING OBJECTIVES

- Give examples of temporary and long-term grouping behavior.
- List the benefits and costs of living in a group.

Benefits of Grouping

Many animals benefit by forming short-term, unstable groups. One benefit is safety in numbers. Whenever animals cluster together, individuals at the margins of a group inadvertently shield others from predators. The **selfish herd effect** describes a temporary aggregation that arises when individual animals move near others to minimize their own risk of predation. Small fishes form a selfish herd when under attack (**FIGURE 43.13A**). Flocks of birds and herds of ungulates such as zebras show similar behavior.

In a group, multiple individuals can be on the alert for predators. In some cases, an animal that spots a threat will warn others of its approach. Birds, monkeys, meerkats, and prairie dogs are among the animals that make alarm calls in response to a predator.

CREDIT: (12) Ivan Kuzmin/Shutterstock.com.

A Sardines attempting to hide behind one another during an attack by predatory sailfish.

B Tent caterpillars on the surface of their communally produced tent. If threatened, the group wriggles and produces cyanide-laced vomit.

FIGURE 43.13 Safety in numbers. Clustering with others provides defense against predators.

Even in species that do not give alarm calls, individuals benefit from the vigilance of others. Often, when an animal notices another group member beginning to flee, it will do likewise. In what is called the "confusion effect," a predator has a more difficult time choosing one individual to chase when prey are scattering.

Not all prey groups flee from predators. Some act together to present a united defense. For example, when threatened by wolves, musk oxen stand back to back, presenting an imposing display of sharp horns.

Eastern tent caterpillars (**FIGURE 43.13B**) also engage in a cooperative defense. These moth larvae collaborate to spin a protective tent into which the group retreats when not feeding. When a predator or parasite approaches, the caterpillars thrash about as a group and regurgitate fluid laced with cyanide. (The cyanide is from their diet of leaves.)

Grouping also benefits the caterpillars by allowing them to share information about food. A caterpillar who finds a good feeding site leaves a pheromone trail as it returns to the communal tent. Other caterpillars follow this trail, feed, then add more pheromone to the trail. The more pheromone a trail has, the more likely hungry caterpillars are to follow it.

The eastern tent caterpillar's group defense and information sharing capacities have led some scientists to label them "social caterpillars." However, biologists have traditionally defined **social animals** as species in which individuals benefit by cooperating in a permanent multigenerational group. Tent caterpillars do not fit this definition because there is no generational overlap; all members of the group are larvae that hatched at about the same time. As this example shows, there is a range of what is considered social behavior. The most extreme social behavior occurs among ants, bees, and termites (we return to this topic in Section 43.8).

Wolves and lions are social animals who cooperate in catching prey. Cooperative hunting allows a group of predators to capture larger or faster prey. Even so, cooperative hunting is often no more efficient than hunting alone. In one study, researchers observed that a solitary lion catches prey about 15 percent of the time. Two lions hunting together catch prey twice as often as a solitary lion, but having to share the spoils of the hunt means the amount of food per lion is the same. When more lions join a hunt, the success rate per lion falls. Wolves have a similar pattern. Among carnivores that hunt cooperatively, hunting success is usually not the most important driver of group living. Individuals hunt together, but also cooperate in fending off scavengers, caring for one another's young, and defending their territory.

Another benefit of living in a social group is enhanced opportunities for observational learning, as when young chimpanzees learn how to make "fishing sticks" that they use to capture termites as food.

Costs of Grouping

In order for a grouping behavior to evolve and persist, the benefits of spending time in a group must offset the costs. The most obvious cost of group living is increased competition. Individuals that live in groups continually compete with one another for resources,

selfish herd effect A temporary group forms when individuals hide behind others to minimize their individual risk of predation.

social animal Animal that lives in a multigenerational group in which members, who are usually relatives, cooperate in some tasks.

territory Region that an animal or group of animals defends.

CREDITS: (13A) iStockphoto.com/NaluPhoto; (13B) StevenRussellSmithPhotos/Shutterstock.com.

ranging from food to mates. Group living can also increase the risk of predation and parasitism. Large groups are more conspicuous to predators than lone individuals, and parasites spread more easily when many hosts are in close proximity.

Often, costs and benefits of group living are not distributed equally among group members. Many animals that live in groups form a **dominance hierarchy**, a social system in which dominant animals receive a greater share of resources and breeding opportunities than subordinates. Why would a animal stay in a group where it is subordinate? Belonging to a group enhances the subordinate's chance of survival, and some subordinates do get a chance to reproduce in periods when food is abundant. A subordinate may also one day ascend to the dominant reproductive role.

TAKE-HOME MESSAGE 43.7

✔ Living in a social group can provide benefits, as through improved defenses, shared care of offspring, and greater access to food.

✔ Costs of group living include increased competition and increased vulnerability to infections.

43.8 Altruism and Eusociality

LEARNING OBJECTIVES

- Explain the ways that altruistic behavior can evolve.
- Describe the traits that define eusocial animals.
- Explain what determines whether a honeybee develops as a worker or a queen.
- Describe the social systems of honeybees, termites, and naked mole rats.

Evolution of Altruism

In many animal groups, individuals display altruism. **Altruism** is behavior that decreases the altruist's fitness but increases the fitness of other group members. Giving an alarm call is an example of an altruistic act; it puts the caller at an increased risk of predation but reduces the risk to others.

One way that altruistic behavior can evolve is through **kin selection**, a type of selection that favors individuals who have the highest inclusive fitness. An individual's **inclusive fitness** is the number of offspring that it could produce with no help from others, plus a fraction of the number of extra offspring produced by other individuals because of the altruist's help. The fraction is determined by the individual's relatedness to those whom it helps.

Field studies of social animals confirm that many altruistic behaviors are preferentially directed at relatives. For example, ground squirrels and prairie dogs are more likely to give an alarm call if they have close relatives living nearby. In hyenas, which often have intragroup fights, dominant females are more likely to risk injury by joining a fight to assist a relative than to assist a nonrelative.

Altruistic behavior can also arise through reciprocity, in which an individual assists group members who have previously assisted it or may assist it in the future. Consider vampire bats, which roost during the day in large groups. Hungry bats who have not successfully fed the night before are offered food (regurgitated blood) by other bats. Close kinship increases the likelihood of food sharing, but sharing also occurs among unrelated bats. Bats in need who previously offered food to many different recipients receive more offers of food than bats who were less generous.

Eusocial Animals

From a genetic standpoint, the greatest cost an individual can pay for living in a group is the failure to breed. Yet, **eusocial animals** live in a multigenerational family group in which sterile workers carry out tasks essential to the group's welfare, while other members of the group reproduce.

Many eusocial species are members of the order Hymenoptera, which includes ants, bees, and wasps. All ants are eusocial, as are some bees and wasps. In eusocial Hymenoptera all workers are females. Workers forage for food, maintain the hive, care for young, and defend the colony. Males are produced seasonally and function only in mating.

In all Hymenoptera, diploid females develop from fertilized eggs and haploid males from unfertilized eggs. As a result, female hymenopterans share 75 percent of their genes with their sisters. By one hypothesis, the unusually high relatedness among sisters may help explain why eusociality arose many times in this order.

Honeybees are eusocial Hymenoptera. The only egg-laying female in a honeybee colony is the queen, who is larger than the workers and anatomically distinct (**FIGURE 43.14A**). A female's diet as a larva determines whether she becomes a queen or a worker. A

altruism Behaving in a way that benefits others, despite a some risk or cost to oneself.

dominance hierarchy Social system in which resources and mating opportunities are unequally distributed within a group

eusocial animal Animal that lives in a multigenerational family group with a reproductive division of labor.

inclusive fitness Genetic contribution an individual makes by reproducing, plus a fraction of the contribution that it makes by facilitating the reproduction of its relatives.

kin selection Type of natural selection that favors individuals with the highest inclusive fitness.

larva destined to be a queen eats only a special glandular secretion called royal jelly. Larva that will become workers get a bit of royal jelly, plus pollen and honey. Components of the pollen and honey trigger epigenetic changes that result in the worker phenotype.

All of the more than 2,000 species of termites (order Isoptera) are eusocial. The termite queen is enormous (**FIGURE 43.14B**), far larger than her mate, the king, who also resides in the colony. Termite workers include both sterile males and females.

Only two species of eusocial vertebrates are known. Both are African mole-rats: mouse-sized rodents that live underground. The best studied is *Heterocephalus glaber*, the naked mole-rat (**FIGURE 43.14C**). Clans of this nearly hairless rodent live in dry parts of East Africa. A reproducing female (the queen) dominates the clan and mates with one to three males (the king or kings). Nonbreeding males and females protect and care for the queen, the king(s), and their offspring. Diggers excavate tunnels and chambers. When a digger finds an edible root, it hauls a bit back to the main chamber and chirps to alert other workers about the food source. Other sterile workers guard the colony.

Inbreeding, which increases genetic similarity among relatives, may have played a role in the evolution of naked mole-rat eusociality. Some colonies are highly inbred, so workers share an unusually high number of genes with the king and queen. However, ecological factors probably also play a role. According to one hypothesis, the mole-rat's arid habitat and patchy food supply favor a genetic predisposition to cooperate in digging burrows, searching for food, and fending off competitors. Individuals who strike out on their own are unlikely to be successful.

A Queen honeybee with her sterile daughters.

B A queen termite dwarfs her offspring and mate. Ovaries fill her enormous abdomen.

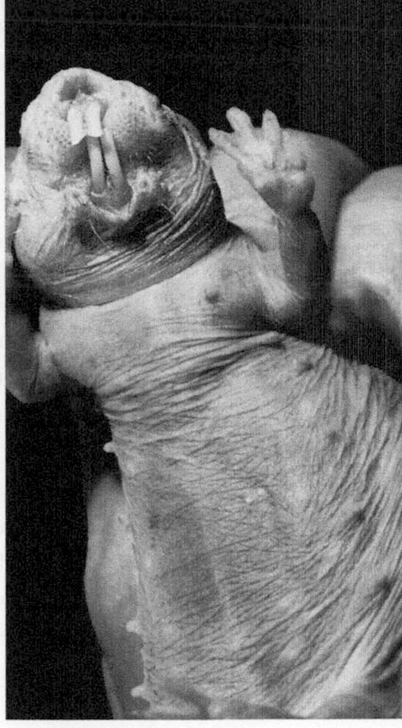

C Naked mole-rat queen.

FIGURE 43.14 **Three queens.** All are fertile females in eusocial species.

TAKE-HOME MESSAGE 43.8

- ✔ Altruistic behavior may be favored when individuals reciprocate or when they pass on genes indirectly, by helping relatives survive and reproduce.
- ✔ Mechanisms that increase relatedness among siblings may encourage the evolution of eusocial behavior.
- ✔ A harsh environment that favors group living can also favor the evolution of eusociality.

43.1 Can You Hear Me Now? (revisited)

The noise produced by human activities affects the behavior of many animals that communicate by sound. Like right whales, blue whales that feed off the coast of Southern California "yell" to be heard above shipping noise. However, these whales have an additional problem—sonar signals that military vessels use to locate submarines. The sonar signals inhibit blue whale calling, perhaps because they are similar in frequency to the calls of killer whales that prey on blue whales.

Land-dwelling animals alter their calls in response to noise too. Birds that live in cities produce louder, higher-pitched songs than their counterparts in rural environments and may begin singing earlier in the day, when it is less noisy. Frogs and grasshoppers living near heavily used roads also sing more loudly and at a higher pitch than members of the same species that live in quieter environments.

Even when animals are capable of altering their calls, human-induced noise can still drown out their signals. Thus, it is not surprising that some animals appear to suffer a decline in fitness as a result of the noise in their environment. For example, house sparrows who nest in noisy areas produce fewer young than those who nest in a tranquil setting.

CREDITS: (14A) Kenneth Lorenzen; (14B) wonderisland/Shutterstock.com; (14C) © Nicola Kountoupes/Cornell University; (in text revisited) Vishnevskiy Vasily/Shutterstock.

STUDY GUIDE

Section 43.1 Noise generated by human activities can interfere with animal communication systems that evolved in the absences of such noise. Exposure to noise can stress animals and in some cases lower their reproductive output.

Section 43.2 Studies of behavioral differences within a species or among closely related species can shed light on the proximate causes of a behavior (what triggers the behavior) and the ultimate causes (the selective forces that favored the behavior). Some behavioral differences such as variations in fruit fly foraging behavior arise from alleles of a single gene.

Behavioral traits can be influenced by an animal's environment. **Behavioral plasticity** allows an animal to respond appropriately to environmental pressures that can vary. Some environmentally induced variations in behavior arise as a result of epigenetic effects.

Section 43.3 **Instinctive behavior** occurs without a prior experience. It is often triggered as a response to a simple stimulus. Instinctive behavior is favored when the stimulus that triggers the behavior is consistently tied to a specific situation. Instinctive behavior is more easily misdirected than learned behavior.

Learned behavior arises in response to experience. **Imprinting** is time-sensitive learning. **Classical conditioning** forms an association between two stimuli, and **operant conditioning** forms one between a behavior and its outcome. With **habituation**, the simplest type of learning, an animal learns to disregard certain repeated stimuli. **Spatial learning** involves learning features of one's physical environment. **Social learning** involves learning features of of other individuals. **Observational learning** involves seeing and imitating a behavior.

Section 43.4 All animals move. **Kinesis** (random movement) and **taxis** (directional movement) are instinctual responses to stimuli. Many animals make a seasonal **migration**; the animal interrupts its usual activities to make a journey. Navigation requires the ability to form a mental map and to sense direction. Some animals have a magnetic compass that allows them to perceive Earth's magnetic field. Information about direction can also be determined from the position of the sun or stars, from visual landmarks, or from olfactory landmarks.

Section 43.5 **Communication signals** convey information between individuals of the same species. **Pheromones** are chemical communication signals. Pheromones are either secreted or not, so they convey less information than visual, acoustic, and tactile signals, which can vary in their properties. Signals have been shaped by evolution; for example, courtship calls are easily localized, whereas alarm calls are not. Predators may take advantage of the communication system of their prey either by intercepting signals or posing as a signaller.

Section 43.6 Genetic monogamy, in which a male and female mate only with each other, is extremely rare in animals. Social monogamy is rare in most groups, with the exception of birds. Both males and females benefit by mating with multiple partners. Depending on the species, females may choose males on the basis of resources they offer or may assess mate quality by their courtship performances, as in displays at a **lek**. When large numbers of females cluster in a defensible area, males may compete with one another to control a mating **territory**.

Parental care has reproductive costs in terms of reduced frequency of reproduction. It is adaptive when benefits to a present set of offspring offset this cost. In fish with parental care, the male is usually the caregiver. In birds, biparental care is most common. In mammals, females are usually the sole caregivers.

Section 43.7 Some animal species form temporary or long-term groups. The **selfish herd effect** describes how animals hide among others for protection. Animals that live in groups can also benefit by cooperating in tasks such as defense, obtaining food, or rearing the young. **Social animals** are multigenerational groups of cooperating individuals. Costs of group living include increased disease, increased, parasitism, and more competition for mates and resources. In groups that have a **dominance hierarchy**, costs and benefits are not distributed equally among group members. Dominant individuals receive a higher share of benefits than subordinates.

Section 43.8 **Altruism** is a behavior that benefits another at the altruist's expense. **Kin selection** can favor the evolution of altruism among relatives. By helping relatives an altruist can raise its **inclusive fitness**. Altruists help perpetuate the genes that lead to their altruism by promoting reproductive success of close relatives who share the same genes. Reciprocity can also encourage altruism. In such cases, individuals who engage in altruistic behavior can expect to later benefit from the altruism of others in the group.

Eusocial animals live in colonies that include overlapping generations and have a reproductive division of labor. Most colony members do not reproduce. Rather, they devote their life to maintaining and protecting the colony.

Eusocial Hymenoptera include all ants, as well as some bees and wasps. In all eusocial Hymenoptera, the workers that maintain the colony, protect it, and rear the young are female. The only function of males is mating. An unusual genetic system in Hymenoptera makes sisters unusually highly related to one another and may predispose species in this order to eusocial behavior.

All termites are eusocial as are two species of mole-rats. In these groups, there are sterile workers of both sexes. A harsh environment may encourage mole-rat social behavior; lone individuals are unlikely to survive.

SELF-QUIZ

Answers in Appendix VII

1. Whether a fruit fly is a sitter or a rover depends upon its _______ .
   a. diet as a larva
   b. age
   c. genotype at one specific locus
   d. oxytocin level

2. In voles, species with a higher concentration of receptors for the hormone oxytocin are more likely to _______ .
   a. form a pair bond
   b. engage in mating with multiple partners
   c. be eusocial
   d. form a dominance hierarchy

3. An animal that navigates by the stars needs _______ .
   a. an ability to sense Earth's magnetic field
   b. pheromone receptors
   c. an internal clock
   d. an acute sense of hearing

4. The honeybee dance language transmits information about distance to food by way of _______ signals.
   a. tactile   c. acoustical
   b. chemical   d. visual

5. A _______ is a chemical that conveys information between individuals of the same species.
   a. pheromone   c. hormone
   b. neurotransmitter   d. all of the above

6. In most _______ , males and females cooperate in care of the young.
   a. mammals   c. fishes
   b. birds   d. all of the above

7. Generally, living in a social group costs the individual in terms of _______ .
   a. competition for food, other resources
   b. vulnerability to contagious diseases
   c. competition for mates
   d. all of the above

8. Which of the following is an example of taxis?
   a. An earthworm moves away from light.
   b. Each spring, a bird travels hundreds of miles to its breeding grounds.
   c. A honeybee emits a pheromone that induces other bees to defend the hive.
   d. An increase in temperature increases the speed at which a butterfly flies.

9. Eusocial insects all _______ .
   a. have sterile male and female workers
   b. are members of the Hymenoptera
   c. have a reproductive division of labor
   d. have haploid males and diploid females

10. Helping other individuals at a reproductive cost to oneself can be adaptive if those helped are _______ .
   a. members of another species
   b. competitors for mates
   c. close relatives
   d. counterfeit signalers

11. A honeybee worker is a _______ .
   a. fertile female   c. fertile male
   b. sterile female   d. sterile male

12. A _______ is a change in the rate of random movement in response to a specific stimulus.
   a. epigenetic trait   c. kinesis
   b. taxis   d. migration

13. With _______ , the consequences of a voluntary behavior cause an animal to repeat or avoid that behavior.
   a. instinct   c. classical conditioning
   b. imprinting   d. operant conditioning

14. With behavioral plasticity, behavior varies depending on _______ .
   a. the time of day   c. environmental factors
   b. an individual's age   d. an individual's genotype

15. Match the terms with their most suitable description.

___ imprinting	a. time-limited learning
___ tactile display	b. requires staying in touch
___ selfish herd	c. communal display ground
___ habituation	d. learning not to respond
___ lek	e. hiding behind others
___ kinesis	f. random movement in response to a stimulus

CRITICAL THINKING

1. For billions of years, the only bright objects in the night sky were stars or the moon. Night-flying moths used these light sources to navigate in a straight line. Today, the instinct to fly toward bright objects causes moths to exhaust themselves fluttering around streetlights and banging against brightly lit windowpanes. This behavior is not adaptive, so why does it persist?

2. Many animals preferentially assist individuals with whom they are most familiar. Explain how such a preference could evolve through kin selection.

3. European cuckoos and North American brown-headed cowbirds are not close relatives, but both lay their eggs in the nest of other birds. Both also have an unusually simple song. Why is having a simple species-specific song an advantage for such birds?

4. When researchers gave pair-bonded voles methamphetamine in their drinking water, the voles' brains made less oxytocin. What effect do you think this hormonal change had on partner preference in these animals?

44 POPULATION ECOLOGY

CORE CONCEPTS

Systems

Complex properties arise from interactions among components of a biological system.

Properties of a population emerge from the characteristics of its members, change over time, and are affected by resource availability. Competition, cooperation, and other interactions among members of a population affect individuals and the population as a whole, as do interactions with the living and nonliving environment.

Evolution

Evolution underlies the unity and diversity of life.

A population's interactions with its living and nonliving environment involve the exchange of matter and energy. The finite availability of natural resources limits a population's size and influences its other characteristics. Competition for a limited resource can result in natural selection that changes the genetic makeup of a population. Such change is evolution.

Process of Science

The field of biology consists of and relies upon experimentation and the collection and analysis of scientific evidence.

Carefully designed experiments help researchers unravel cause-and-effect relationships in complex natural systems. Observations of natural populations illustrate the interplay of predictable and random events in population growth. Field experiments involving natural populations clarify the role of predation in shaping population characteristics.

Links to Earlier Concepts

This chapter considers the structure and growth of populations (Section 1.2). We refer back to sampling error (1.7), directional selection (17.4), gene flow (17.6), evolution of modern humans (26.6), smallpox vaccination (37.1), and animal social behavior (43.8).

44.1 Managing Canada Geese

Canada geese (*Branta canadensis*) were hunted to near extinction in the late 1800s. In the early 1900s, federal laws and international treaties were put in place to protect them and other migratory birds. In recent decades, the number of geese in the United States has soared. For example, Michigan had about 9,000 of these birds in 1970 and today has more than 300,000.

Canada geese are plant-eating birds, and they often congregate on the grassy lawns of golf courses and parks (**FIGURE 44.1**). They are considered pests because they damage lawns and produce slimy, green feces that soil shoes, stain clothing, and cloud ponds and lakes. Some goose feces also contain bacteria that can sicken people if ingested.

Controlling Canada goose numbers is a challenge, because several different populations spend time in the United States. A **population** is a group of organisms of the same species that live in the same area and interbreed. In the past, nearly all Canada geese seen in the United States were migratory. The geese nested in northern Canada, flew to the United States to spend the winter, then returned to Canada. The common name of the species reflects this tie to Canada.

Most Canada geese still migrate, but some populations have lost this trait. Canada geese breed where they were raised, and nonmigratory geese are generally descendants of birds introduced to a park or hunting preserve. In the winter, migratory birds often mingle with nonmigratory ones. For example, a bird that breeds in Canada and flies to Virginia for the winter finds itself among geese that have never left Virginia.

FIGURE 44.1 Canada geese in a park in Oakland, California.

population Group of organisms of the same species that live in the same area and interbreed.

CREDITS: (opposite) Sergey Krasnoshchokov/Shutterstock; (1) Courtesy of @ Joel Peter.

Migration is a strenuous and difficult process. Compared to a migratory goose, one that stays put in a resource-rich area can devote more energy to producing young. If the nonmigrant lives in a suburban or urban area, it also benefits from an unnatural abundance of food (grass) and an equally unnatural lack of predators. Not surprisingly, the biggest increases in Canada geese have been among nonmigratory birds that live where humans are plentiful.

In 2006, increasing complaints about Canada geese led the U.S. Fish and Wildlife Service to encourage wildlife managers to look for ways to reduce nonmigratory Canada goose populations, without unduly harming migratory birds. To do so, these biologists are studying which traits characterize each goose population, as well as how populations interact with one another, with other species, and with their physical environment. This sort of information is the focus of the field of population ecology. ●

44.2 Population Demographics

LEARNING OBJECTIVES

- Give some examples of population demographics.
- Describe two methods of estimating population demographics.

Ecology is the branch of biology that deals with interactions among organisms and their environment. Ecology is not the same as environmentalism, which is advocacy for protection of the environment. However, environmentalists often cite ecological studies when drawing attention to environmental concerns. Studying populations often involves the use of **demographics**—statistics that describe a population's traits. A population's demographics include its size, density, distribution, and age structure.

Population Size

Population size is the number of individuals in a population. It is often impractical to count all individuals, so biologists frequently use sampling techniques to estimate population size.

Plot sampling estimates the total number of individuals in an area on the basis of direct counts in a small part of the area. For example, ecologists might investigate the number of grass plants in a grassland or clams in a mudflat by measuring the number of individuals in several 1 meter by 1 meter square plots. The average number of individuals per plot is calculated, then multiplied by the size (in square meters) of the total area. The result is an estimate of the total population size. Plot sampling is most accurate for organisms that are not very mobile, in areas with uniform conditions.

Scientists use **mark–recapture sampling** to estimate the population size of mobile animals. With this technique, some number of animals in a population are captured, marked, then released. After allowing a sufficient time to pass for the marked individuals to meld back into the population, the scientists capture animals again. The proportion of marked animals in the second sample is taken to be representative of the proportion marked in the whole population. For example, suppose scientists capture, mark, and release 100 deer in an area. Later, the scientists return and again capture 100 deer. They find 50 of these deer were previously marked, implying that marked deer constitute half of the population. They then infer that the total population is 200 individuals.

Characteristics of individuals in a sample plot or captured group can also be used to infer characteristics of the population as a whole. For example, if

A Clumped distribution of hippopotamuses.

B Near-uniform distribution of nesting seabirds.

C Random distribution of dandelions.

FIGURE 44.2 Population distribution patterns.

Data Analysis Activities

Iguana Decline In 1987, Martin Wikelski began a long-term study of marine iguanas in the Galápagos Islands. He marked iguanas on two islands—Genovesa and Santa Fe—and collected data on how their body size, survival, and reproductive rates varied over time. He found that because iguanas eat algae and have no predators, deaths usually result from food shortages, disease, or old age. In January 2001, an oil tanker ran aground and leaked a small amount of oil into the waters near Santa Fe. FIGURE 44.3 shows the number of marked iguanas that Wikelski and his team counted in their study populations just before the spill and about a year later.

1. Which island had more marked iguanas at the time of the first census?
2. How much did the population size on each island change between the first and second census?
3. Wikelski concluded that changes on Santa Fe were the result of the oil spill, rather than sea temperature or other climate factors common to both islands. How would the census numbers be different from those he observed if an adverse event had affected both islands?

Number of marked iguanas
180
150
120
90
60
30
0
Jan Dec
Genovesa Island
Jan Dec
Santa Fe Island

FIGURE 44.3 Shifting numbers of marked marine iguanas on two Galápagos islands. An oil spill occurred near Santa Fe just after the January 2001 census (orange bars). A second census was carried out in December 2001 (green bars).

half the recaptured deer are of reproductive age, half of the population is assumed to share this trait. This extrapolation is based on the assumption that the individuals in the sample are representative of the general population.

Population Density and Distribution

Population density is the number of individuals per unit area or volume. Examples of population density include the number of dandelions per square meter of lawn or the number of amoebas per milliliter of pond water. **Population distribution** describes the location of individuals relative to one another. Members of a population may be clumped together, be an equal distance apart, or be distributed randomly.

Clumped Distribution Most populations have a clumped distribution, meaning members of the population are closer to one another than would be predicted by chance alone. A patchy distribution of resources encourages clumping, as when hippopotamuses gather in muddy river shallows (FIGURE 44.2A). Similarly, a cool, damp, north-facing slope may be covered with ferns, whereas an adjacent drier south-facing slope has none. Limited dispersal ability increases the likelihood of a clumped distribution: As the saying goes, the nut does not fall far from the tree. Asexual reproduction also results in clumping. It produces colonies of coral and vast stands of aspen trees. Finally, as Section 43.7 explained, some animals live in groups.

Near-Uniform Distribution Competition for limited resources can produce a near-uniform distribution, with individuals more evenly spaced than would be expected by chance. Creosote bushes in deserts of the American Southwest grow in this pattern. Competition for water among the root systems keeps the plants from growing in close proximity. Similarly, seabirds in breeding colonies often show a near-uniform distribution. Each bird aggressively repels others that get within reach of its beak (FIGURE 44.2B).

Random Distribution Members of a population are distributed randomly when resources are uniformly available, and proximity to others neither benefits nor harms individuals. For example, when wind-dispersed dandelion seeds land on the uniform environment of a suburban lawn, dandelion plants grow in a random pattern (FIGURE 44.2C).

demographics Statistics that describe a population.
ecology Study of how populations interact with one another and with their nonliving environment.
mark–recapture sampling Method of estimating population size of mobile animals by marking individuals, releasing them, then checking the proportion of marks among individuals later recaptured.
plot sampling Method of estimating population size of organisms that do not move much by making counts in small plots and extrapolating from this to the number in the larger area.
population density Number of individuals per unit area.
population distribution Describes whether individuals are clumped, uniformly dispersed, or randomly dispersed in an area.
population size Total number of individuals in a population.

Age Structure

The **age structure** of a population refers to the number of individuals in various age categories. Individuals are often categorized as pre-reproductive, reproductive, or post-reproductive. Members of the pre-reproductive category have a capacity to produce offspring when mature. Pre-reproductive and reproductive individuals constitute a population's **reproductive base**. The greater the size of a population's reproductive base, the greater its capacity for growth. We return to this concept when we discuss human population growth in Section 44.7.

Effects of Scale and Timing

The scale of the area sampled and the timing of a study can influence the observed demographics. For example, seabirds are spaced almost uniformly at a nesting site, but nesting sites are clumped along a shoreline. The birds have a clumped distribution during the breeding season, but are more widely and randomly dispersed when breeding is over. Similarly, the proportion of young birds is largest just after the breeding season.

Using Demographic Data

Wildlife managers can use demographic information to decide how to manage populations. For example, in considering how to manage Canada geese, wildlife managers began by evaluating the size, density, and distribution of nonmigratory populations. Based on this information, the U.S. Fish and Wildlife Service decided to allow destruction of some eggs and nests and increased hunting opportunities at times when migratory Canada geese are least likely to be present.

TAKE-HOME MESSAGE 44.2

- ✔ Each population has characteristic demographics, such as its size, density, distribution pattern, and age structure.
- ✔ Characteristics of the population as a whole are often inferred on the basis of a study of a smaller subsample.
- ✔ Environmental conditions and interactions among individuals can influence a population's demographics, which often change over time.

44.3 Modeling Population Growth

LEARNING OBJECTIVES

- Give the equation that describes exponential growth and explain what the terms in the equation mean.
- Define biotic potential and give examples of organisms that differ in their biotic potential.

In nature, populations continually change in size. Individuals are added to a population by births and by **immigration**, which is the arrival of new residents that previously belonged to another population. Individuals are removed from a population by deaths and by **emigration**, the departure of individuals who take up permanent residence elsewhere.

In many animal species, young of one or both sexes leave the area where they were born to breed elsewhere. The tendency of individuals to emigrate to a new breeding site is usually related to resource availability and crowding. As resources decline and crowding increases, the likelihood of emigration rises.

Zero to Exponential Growth

If we set aside the effects of immigration and emigration, we can define **zero population growth** as an interval during which the number of births is balanced by an equal number of deaths. As a result, population size remains unchanged, with no net increase or decrease in the number of individuals.

We can measure births and deaths in terms of rates per individual, or per capita. *Capita* means "head," as in a head count. Subtract a population's per capita death rate (d) from its per capita birth rate (b), and you have the **per capita growth rate**, or r:

b	$-$	d	$=$	r
(per capita birth rate)		(per capita death rate)		(per capita growth rate)

Imagine 2,000 mice living in the same field. If 1,000 mice are born each month, then the birth rate is 0.5 births per mouse per month (1,000 births/2,000 mice). If 200 mice die one way or another each month, then the death rate is 200/2,000 or 0.1 deaths per mouse per month. Thus, r is 0.5 – 0.1, or 0.4 per mouse per month.

The **exponential growth model** predicts growth of an idealized population that has no limits on population growth. In such a population, r remains constant and greater than zero. In any given interval, population's size will increase by the same proportion of its total as it did in the previous interval. Thus, we can calculate population growth (G) for each interval based on the number of individuals (N) and the per capita growth rate:

N	$\times$	r	$=$	G
(number of individuals)		(per capita growth rate)		(population growth per unit time)

Applying this equation to our hypothetical population of field mice, shows that after one month, our initial population of 2,000 mice will have increased to 2,800 mice (**FIGURE 44.4A**). In the next month, population size will expand by 1,120 individuals (2,800 × 0.4), bringing the total population size to 3,920. Graphing the increases against time results in a J-shaped curve, which is characteristic of exponential population growth (**FIGURE 44.4B**).

With exponential growth, the number of individuals born into the population increases each generation, although the per capita growth rate stays the same. Exponential population growth is analogous to compound interest in a bank account. The annual interest *rate* stays fixed, yet every year the *amount* of interest paid increases. The annual interest paid into the bank account adds to the balance, so the subsequent interest payment will be based on an increased balance.

Biotic Potential

The growth rate for a population under ideal conditions is its **biotic potential**. This is a theoretical rate at which the population would grow if shelter, food, and other essential resources were unlimited and there were no predators or pathogens. Factors that affect biotic potential include the age at which reproduction typically begins, how long individuals remain reproductive, and the number of offspring that are produced each time an individual reproduces. These factors are influenced by the environment, and vary by species. Microbes such as bacteria have some of the highest biotic potentials, whereas large-bodied mammals have some of the lowest. In the real world, populations seldom reach their biotic potential because of limiting factors. These factors are the topic of Section 44.4.

TAKE-HOME MESSAGE 44.3

✔ The size of a population depends on its rates of births, deaths, immigration, and emigration.

✔ A population will grow exponentially as long as the per capita birth rate (r) is constant and greater than zero. With exponential growth, population size increases at an ever accelerating rate.

✔ The biotic potential of a species is its maximum possible population growth rate under optimal conditions.

age structure Of a population, the number of individuals in each of several age categories.

biotic potential (by-AH-tick) Maximum possible population growth rate under optimal conditions.

emigration Movement of individuals out of a population.

exponential growth model Model of unlimited population growth. The population grows by a fixed percentage in each successive time interval; the size of each increase is determined by the current population size.

immigration Movement of individuals into a population.

per capita growth rate For some interval, the added number of individuals divided by the initial population size.

reproductive base Of a population, all individuals who are of reproductive age or younger.

zero population growth Interval in which births equal deaths.

Starting Size of Population		Net Monthly Increase	New Size of Population
2,000	× r =	800	2,800
2,800	× r =	1,120	3,920
3,920	× r =	1,568	5,488
5,488	× r =	2,195	7,683
7,683	× r =	3,073	10,756
10,756	× r =	4,302	15,058
15,058	× r =	6,023	21,081
21,081	× r =	8,432	29,513
29,513	× r =	11,805	41,318
41,318	× r =	16,527	57,845
57,845	× r =	23,138	80,983
80,983	× r =	32,393	113,376
113,376	× r =	45,350	158,726
158,726	× r =	63,490	222,216
222,216	× r =	88,887	311,103
311,103	× r =	124,441	435,544
435,544	× r =	174,218	609,762
609,762	× r =	243,905	853,667
853,667	× r =	341,467	1,195,134

A Increases in size over time. Note that the net increase becomes larger with each generation.

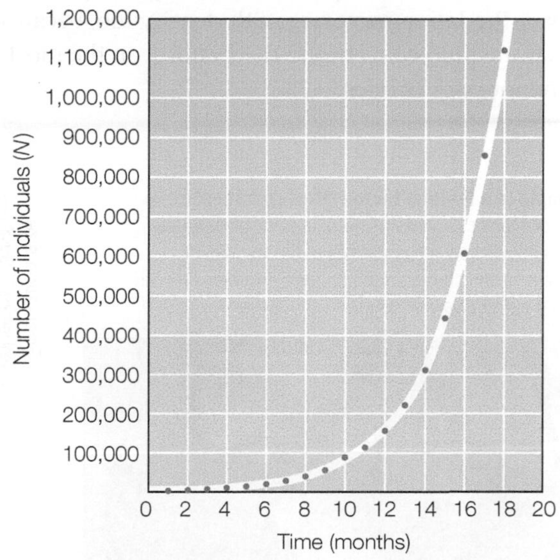
1,200,000
1,100,000
1,000,000
900,000
800,000
700,000
600,000
500,000
400,000
300,000
200,000
100,000
Number of individuals (N)
0 2 4 6 8 10 12 14 16 18 20
Time (months)

B Graphing numbers over time produces a J-shaped curve.

FIGURE 44.4 Exponential growth. Growth of a hypothetical population of mice with a per capita rate of growth (r) of 0.4 per mouse per month and an initial population size of 2,000.

CREDIT: (4) photo, Neil Burton/Shutterstock.

44.4 Limits on Population Growth

LEARNING OBJECTIVES

- Using appropriate examples, compare density-dependent limiting factors with density-independent limiting factors.
- Explain how limiting factors result in logistic growth.
- List some factors that affect carrying capacity.
- Explain why carrying capacity can vary over time.

Density-Dependent Limiting Factors

No population can grow exponentially forever. As the degree of crowding increases, **density-dependent limiting factors** cause birth rates to slow and/or death rates to rise, so the rate of population growth decreases. Density-dependent limiting factors include competition, predation, parasitism, and disease.

Any natural area has limited resources. Thus, as the number of individuals in an area increases, so does **intraspecific competition**: competition among members of the same species. As a result of increased competition, some individuals fail to secure what they need to survive and reproduce. Competition has a detrimental effect even on winners, because energy they use in competition for resources is not available for reproduction. Essential resources for which animals might compete include food, water, hiding places, and nesting sites (**FIGURE 44.5**). Plants compete for nutrients, water, and access to sunlight.

Parasitism and contagious disease increase with crowding because the closer individuals are to one another, the more easily parasites and pathogens can spread. Predation increases with density too, because predators often concentrate their efforts on the most abundant prey species.

Logistic Growth

Logistic growth occurs when density-dependent factors affect population size over time. Graphing logistic growth results in an S-shaped curve (**FIGURE 44.6**). When the population is small, density-dependent limiting factors have little effect and the population grows exponentially ❶. Then, as population size and the degree of crowding rise, limiting factors begin to slow growth ❷. Eventually, the population size levels off at the environment's carrying capacity ❸. **Carrying capacity (*K*)** is the maximum number of individuals that a population's environment can support indefinitely.

The equation that describes logistic growth is:

G	=	r_{max}	×	N	×	$(K - N)/K$
(population growth per unit time)		(maximum per capita population growth rate)		(number of individuals)		(proportion of resources not yet used)

In this equation, G is growth per unit time, r_{max} is the maximum possible per capita growth rate, and N is current population size. K is carrying capacity. The $(K - N) / K$ part of the equation represents the proportion of carrying capacity not yet used. As a population grows, this proportion decreases, so G becomes smaller and smaller. At carrying capacity $N = K$, so $(K - N) / K$ equals zero. Thus G is r_{max} multiplied by zero, which is zero.

Carrying capacity is species-specific, environment-specific, and can change over time ❹. For example, the carrying capacity for a plant species decreases when nutrients in the soil become depleted. Human activities also affect carrying capacity. For example, human

FIGURE 44.5 Example of a limiting factor. Wood ducks build nests only inside tree hollows of specific dimensions (left). In some places, lack of access to appropriate hollows now limits the size of the wood duck population. Adding artificial nesting boxes (right) to these environments can help increase the size of the duck populations.

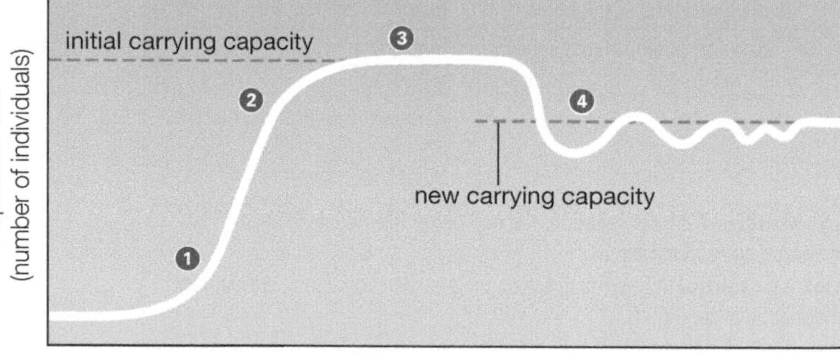
Population size (number of individuals)
initial carrying capacity
❶
❷
❸
❹
new carrying capacity

FIGURE 44.6 Logistic growth. Note the initial S-shaped curve.

❶ At low density, the population grows exponentially.

❷ As crowding increases, density-dependent limiting factors slow the rate of growth.

❸ Population size levels off at the carrying capacity for that species.

❹ Any change in the carrying capacity will result in a corresponding shift in population size.

FIGURE IT OUT How does adding nest boxes affect the carrying capacity for wood duck populations?

Answer: It increases carrying capacity.

CREDIT: (5) photo; left, G. K. Peck; right, Rick Leche, www.flickr.com/photos/rick_leche.

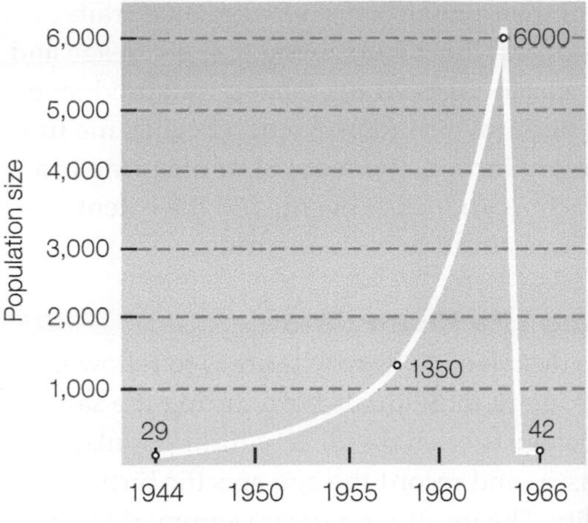

6,000
5,000
4,000
3,000
2,000
1,000
Population size
6000
1350
29
42
1944
1950
1955
1960
1966

FIGURE 44.7 Overshoot and crash. A reindeer herd introduced to a small island in 1944 increased in size exponentially, then crashed when depletion of the food supply was coupled with an especially cold and snowy winter in 1963–64. The estimated carrying capacity of this island for reindeer is about 1,000 animals.

harvest of horseshoe crabs has decreased the carrying capacity for red knot sandpipers, a type of migratory bird. Horseshoe crab eggs are the sandpipers' main food during their long-distance migration.

Density-Independent Factors

Sometimes, natural disasters or weather-related events affect population size. A volcanic eruption, hurricane, or flood can decrease population size. So can human-caused events such as an oil spill. These events are called **density-independent limiting factors**, because crowding does not influence the likelihood of their occurrence or the magnitude of their effect.

In nature, a population's size is often affected by a combination of density-dependent and density-independent factors. Consider what happened after the 1944 introduction of 29 reindeer to St. Matthew Island, an uninhabited island off the coast of Alaska. When biologist David Klein visited the island in 1957, he found 1,350 well-fed reindeer (**FIGURE 44.7**). Klein returned in 1963 and counted 6,000 reindeer. The population had soared far above the island's carrying capacity. A population can temporarily overshoot an environment's carrying capacity, but the high density cannot be sustained. Klein observed that some effects of density-dependent limiting factors were already apparent. For example, the average body size of the reindeer had decreased.

When Klein returned in 1966, only 42 reindeer survived. The single male had abnormal antlers and was thus unlikely to breed. There were no fawns. Klein figured out that thousands of reindeer had starved to death during the winter of 1963–1964. That winter was unusually harsh, with low temperatures, high winds, and 140 inches of snow. Most reindeer were already in poor condition as a result of increased competition, and they starved when deep snow covered their food. A population decline had been expected—a population that exceeds its carrying capacity usually falls back below that capacity—but bad weather magnified the extent of the crash. By the 1980s, there were no reindeer left on the island.

TAKE-HOME MESSAGE 44.4

- ✔ Carrying capacity is the maximum number of individuals of a population that can be sustained indefinitely by the resources in a given environment.
- ✔ With logistic growth, population growth is fastest during times of low density, then it slows as the population approaches carrying capacity.
- ✔ The effects of density-dependent factors such as disease result in a logistic growth pattern. Density-independent factors such as natural disasters also affect population size.

carrying capacity (*K*) Maximum number of individuals of a species that an environment can sustain.
density-dependent limiting factor Factor that increasingly limits population growth as population density increases.
density-independent limiting factor Factor that limits population growth to the same degree regardless of population density.
intraspecific competition Competition among members of a species.
logistic growth Density-dependent limiting factors cause population growth to slow as population size increases.

CREDIT: (7) photo, Sergey Krasnoshchokov/Shutterstock.

44.5 Life History Patterns

LEARNING OBJECTIVES

- Give some examples of life history traits.
- Distinguish between opportunistic and equilibrial species.
- Describe the three types of survivorship curves.

A population's growth rate is affected by its members' **life history**—the manner in which individuals allocate resources to growth, survival, and reproduction over the course of their lifetimes. Survival-related traits include the probability of surviving to a given age and of dying at specific ages. Traits related to reproduction include the age at which reproduction begins, the frequency of reproduction, the number of offspring produced by each reproductive event, and the extent of parental investment in each offspring.

TABLE 44.1

Life Table for an Annual Plant*

Age Interval	Survivorship (number surviving at start of interval)	Number dying during Interval	Death Rate (number dying/ number surviving)	"Birth" Rate in Interval (number of seeds per plant)
0–63	996	328	0.329	0
63–124	668	373	0.558	0
124–184	295	105	0.356	0
184–215	190	14	0.074	0
215–264	176	4	0.023	0
264–278	172	5	0.029	0
278–292	167	8	0.048	0
292–306	159	5	0.031	0.33
306–320	154	7	0.045	3.13
320–334	147	42	0.286	5.42
334–348	105	83	0.790	9.26
348–362	22	22	1.000	4.31
362–	0	0	0	0
		996		

* *Phlox drummondii;* data from W. J. Leverich and D. A. Levin, 1979.

Quantifying Life History Traits

One way to describe life history traits is to follow a **cohort**, a group of individuals born during the same interval. Ecologists often divide a natural population into age classes and record the age-specific birth rates and mortality. The resulting data are summarized in a life table, which can be used to project changes in the size of the population over time (**TABLE 44.1**). Data in life tables can inform decisions about how changes, such as harvesting a species or altering its environment, will affect a population's numbers.

Information about age-specific death rates can also be illustrated by a **survivorship curve**, a plot that shows how many members of a cohort remain alive over time. Ecologists have described three generalized types of curves. A type I curve is convex, indicating survivorship is high until late in life (**FIGURE 44.8A**). Humans and other large mammals that produce one or two young and care for them show this pattern. A diagonal type II curve indicates that the death rate of the population does not vary much with age (**FIGURE 44.8B**). In lizards, small mammals, and large birds, old individuals are about as likely to die of disease or predation as young ones. A type III curve is concave, indicating that the death rate for a population peaks early in life (**FIGURE 44.8C**). Marine animals that release eggs

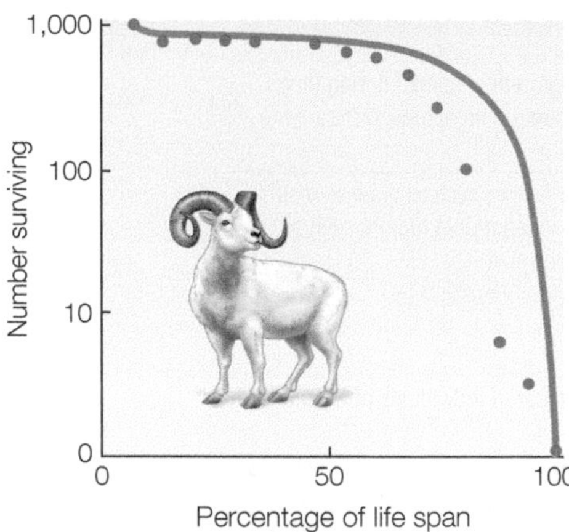
1,000
100
10
0
Number surviving
0
50
100
Percentage of life span

A Type I curve. Mortality is highest very late in life. Data for Dall sheep (*Ovis dalli*).

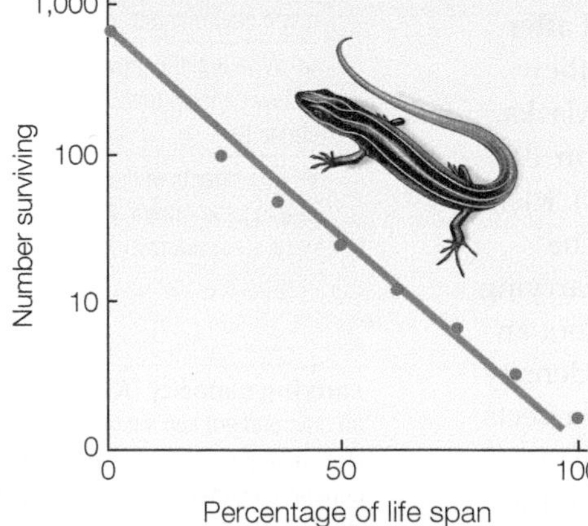
1,000
100
10
0
Number surviving
0
50
100
Percentage of life span

B Type II curve. Mortality does not vary with age. Data for five-lined skink (*Eumeces fasciatus*).

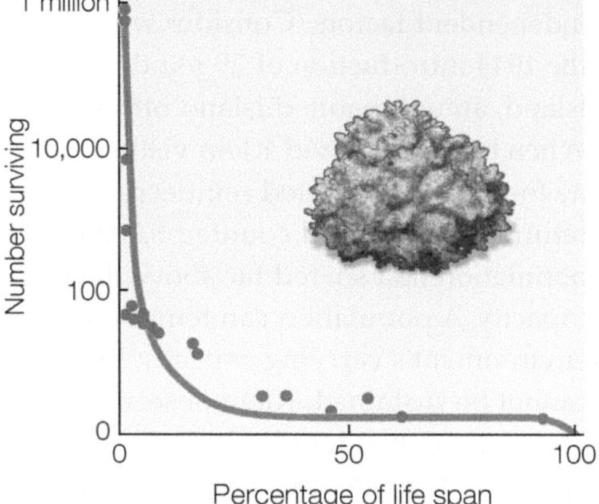
1 million
10,000
100
0
Number surviving
0
50
100
Percentage of life span

C Type III curve. Mortality is highest early in life. Data for a desert shrub (*Cleome droserifolia*).

FIGURE 44.8 Survivorship curves. Gray lines are theoretical curves. Red dots are data from field studies.

A Fly laying many eggs in a rotting tomato.

Opportunistic life history	Equilibrial life history
shorter development	longer development
early reproduction	later reproduction
fewer breeding episodes, many young per episode	more breeding episodes, few young per episode
less parental investment per young	more parental investment per young
higher early mortality, shorter life span	low early mortality, longer life span
result of *r*-selection	result of *K*-selection

B Elephant with its single calf.

FIGURE 44.9 Two types of life history. Most species have a mix of opportunistic and equilibrial life history traits.

into water have this type of curve, as do plants that release enormous numbers of tiny seeds.

Environmental Effects on Life History

To produce offspring, an individual must invest resources that it could otherwise use to grow and maintain itself. Species differ in the manner in which they distribute parental investment among offspring and over the course of their lifetime. Life history patterns vary continuously, but ecologists have described two theoretical extremes at either end of this continuum (FIGURE 44.9). Both maximize the number of offspring that are produced and survive to adulthood, but they do so under very different environmental conditions.

When a species lives where conditions vary in an unpredictable manner, its populations seldom reach carrying capacity. As a result, there is little competition for resources and most deaths occur as a result of density-independent factors. Such conditions favor an opportunistic life history, in which individuals produce as many offspring as possible, as fast as possible.

Opportunistic species are said to be subject to ***r*-selection**, because they maximize r_{max}, the per capita growth rate under optimal conditions. They have a short generation time and small body size. Opportunistic species usually have a type III survivorship curve. Weeds such as dandelions are examples of opportunistic plants. Flies are opportunistic animals. A female fly can lay hundreds of small eggs in a temporary food source (FIGURE 44.9A).

cohort Group of individuals born during the same time interval.
***K*-selection** Selection favoring traits that provide a competitive advantage when population size is near carrying capacity.
life history Schedule of how resources are allocated to growth, survival, and reproduction over a lifetime.
***r*-selection** Selection that favors traits that allow their bearers to produce the most offspring the most quickly.
survivorship curve Graph showing the decline in numbers of a cohort over time.

When a species lives in a stable environment, its populations often approach carrying capacity. Under these circumstances, the ability to successfully compete for resources affects reproductive success. Thus, an equilibrial life history, in which parents produce a few, high-quality offspring, is adaptive. Equilibrial species are shaped by ***K*-selection**, in which adaptive traits provide a competitive advantage when population size is near carrying capacity (*K*). Such species tend to have a large body and a long generation time. For example, a female elephant reaches maturity at the age of 10 to 14 years. She then produces only one large calf at a time and invests in the calf by nursing it after its birth (FIGURE 44.9B). Similarly, a coconut palm grows for years before beginning to produce a few coconuts at a time. In both elephants and coconut palms, a mature individual produces young for many years.

Some species have mixes of traits that cannot be explained by *r*-selection or *K*-selection alone. For example, century plants (a type of agave) and bamboo are large and long-lived, but reproduce only once. Such a strategy evolves when costs of successful reproduction exceeds a certain fraction of a species' resources, thus making future reproductive value small. In such circumstances, fitness is maximized by one reproductive episode. In century plants and bamboo, climate conditions that allow reproduction are rare, so selection favors expending all available energy when these conditions do arise.

TAKE-HOME MESSAGE 44.5

✔ Tracking a cohort reveals patterns of reproduction and mortality that can be summarized in a life table.

✔ Survivorship curves reveal differences in age-specific survival among species or among populations of the same species.

✔ Species that maximize offspring quantity are said to be *r*-selected, whereas those that maximize offspring quality are *K*-selected.

CREDITS: (9A) Richard Baker; (9B) ©iStockphoto.com/Serge_Vero.

44.6 Predation Effects on Life History

LEARNING OBJECTIVES

- Explain how predators act as selective agents that affect the life history patterns of prey species.
- Explain how the collection and analysis of life history data can help in conservation.

Many predators prefer prey of a specific size, and individuals of most prey species change in size over their lifetime. Thus, predation can affect life history traits of prey. When predators prefer large prey, prey individuals who reproduce when still small and young are at a selective advantage. When predators focus on small prey, fast-growing individuals have the selective advantage.

An Experimental Study

A long-term study by evolutionary biologists John Endler and David Reznick illustrates the effect of predation on life history traits. Endler and Reznick studied populations of guppies, small fishes native to shallow freshwater streams in the mountains of Trinidad (**FIGURE 44.10A**). The scientists focused their attention on a region where many small waterfalls stop guppies in one part of a stream from moving to another. As a result of these natural barriers, each stream holds several populations of guppies that have very little gene flow between them (Section 17.6).

Waterfalls also keep guppy predators from moving from one part of the stream to another. The main guppy predators, killifishes and cichlids, differ in size and prey preferences. The relatively small killifishes prey mostly on immature guppies, and ignore the larger adults (**FIGURE 44.10B**). Cichlids are bigger fish. They tend to pursue mature guppies and ignore small ones (**FIGURE 44.10C**).

Some parts of the streams hold one type of predator but not the other. Thus, different guppy populations face different predation pressures. Reznick and Endler discovered that guppies in regions with the large predator (the cichlid) grow faster and are smaller at maturity than guppies in regions with the small predator (the killifish). Guppies in populations hunted

FIGURE 44.10 Effects of predation on life history traits in guppies.

Guppy

A Biologist David Reznick at his study site, a freshwater stream in Trinidad that is home to guppies and their predators.

B Killifish prefer to prey on small guppies, thus selecting for individuals that grow quickly to large size before reproducing.

C Pike cichlids prefer to prey on large guppies, thus selecting for individuals that reproduce early, while still young and small.

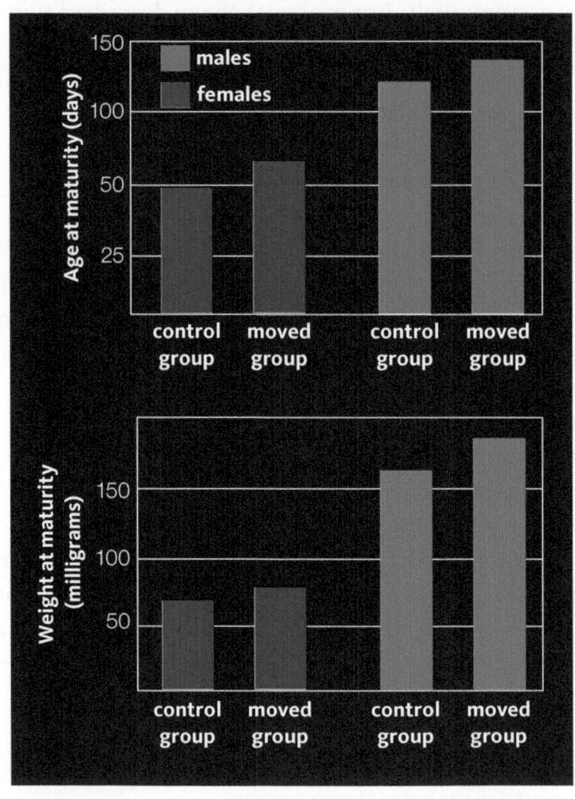
males
females
Age at maturity (days)
150
100
50
25
control group
moved group
control group
moved group
Weight at maturity (milligrams)
150
100
50
control group
moved group
control group
moved group

D Results of experimentally altering predator pressure on guppies. Some guppies (the control group) were native to a pool with pike cichlids and were never moved. Others guppies (the moved group) were transferred from the pool containing pike cichlids to a pool containing killifish. Data show the average age and weight at maturity for descendants of both groups 11 years after the study began.

CREDITS: (10A) Helen Rodd; (10A, insert) David Reznick/ University of California - Riverside; computer enhanced by Lisa Starr; (10B-C) Hippocampus Bildarchiv.

by the larger predator also reproduce earlier, have more offspring at a time, and breed more frequently.

Were these differences in life history traits genetic or the result of some environmental factor? To find out, the biologists collected guppies from both cichlid- and killifish-dominated streams. They reared the guppies in separate aquariums under identical predator-free conditions. Two generations later, the groups continued to show the differences observed in natural populations. Thus differences between guppies preyed on by different predators must have a genetic basis.

Reznick and Endler then made a prediction: If life history traits evolve in response to predation, then these traits will change when a population is exposed to a new predator that favors different prey traits. To test their prediction, they found a stream region with killifish but no guppies or cichlids. Here, they introduced guppies from a site where there were cichlids but no killifish. Thus, at the experimental site, guppies that had previously been preyed upon only by cichlids (which eat big guppies) were now exposed only to killifish (which eat small guppies). The control site was a downstream region where relatives of the transplanted guppies still coexisted with cichlids.

Reznick and Endler revisited the stream over the course of 11 years and 36 generations of guppies. Their data showed that guppies at the experimental site evolved (**FIGURE 44.10D**). Exposure to a previously unfamiliar predator altered the guppies' rate of growth, age at first reproduction, and other life history traits. By contrast, guppies at the control site showed no such changes. Reznick and Endler concluded that life history traits in guppies can evolve rapidly in response to the selective pressure exerted by predation.

Effects of Humans as Predators

Crash of the Atlantic Codfish Population Just as guppies evolved in response to predators, Atlantic codfish (*Gadus morhua*) evolved in response to human fishing. From the mid-1980s to early 1990s, the number of fishing boats targeting the North Atlantic population of codfish increased. As the yearly catch rose fishes that reproduce while young and small became increasingly common. The early-reproducing individuals were at an advantage because fishermen preferentially caught and kept larger fish (**FIGURE 44.11**).

Fishing pressure continued to rise until 1992, when declining cod numbers caused the Canadian government to ban cod fishing in some areas. That ban, and later restrictions, came too late to stop the Atlantic cod population from crashing. In some areas, the number of Atlantic cod declined by 97 percent and the population still shows no signs of recovery. Looking back, scientists determined that the shift toward earlier reproduction was an early sign that human fishing was exerting a strong selective pressure on the North Atlantic cod population. As a result, the fish were evolving toward a more *r*-selected strategy. Had biologists recognized what was happening, they might have been able to save the fishery and protect the livelihood of fishers and associated workers. Ongoing monitoring of the life history data for other economically important fishes may help prevent similar crashes in the future.

FIGURE 44.11 Humans as predators of Atlantic cod. Fishermen showing off a prized catch, a large Atlantic codfish. Both sport fishermen and commercial fishermen preferentially harvested the largest codfish.

Extinction of Wooly Mammoths

Wooly mammoths are extinct relatives of modern elephants. They disappeared from Siberia and North America about 10,000 years ago. The exact cause of their extinction remains a matter of debate, but evidence of life history changes suggests that human hunting pressure was a contributing factor.

The evidence of life history changes is preserved mammoth tusks. Chemical analysis of these tusks provides information about the age at which weaning occurred (the age at which a young mammoth stopped drinking its mother's milk). In the 30,000 years prior to the mammoths' extinction, weaning age declined by about 3 years. Poor nutrition usually increases the age of weaning. By contrast, hunting pressure—like fishing pressure—selects for earlier maturation and weaning.

TAKE-HOME MESSAGE 44.6

✔ When predators prefer large prey, prey individuals who reproduce when still small and young are at a selective advantage. When predators focus on small prey, fast-growing prey individuals have the selective advantage.

✔ Humans hunting can alter life history traits of the hunted species.

CREDIT: (11) Bruce Bornstein, www.captbluefin.com.

FIGURE 44.12 Human population growth. To check the current world population estimate, visit the U.S. Census website (www.census.gov/popclock).

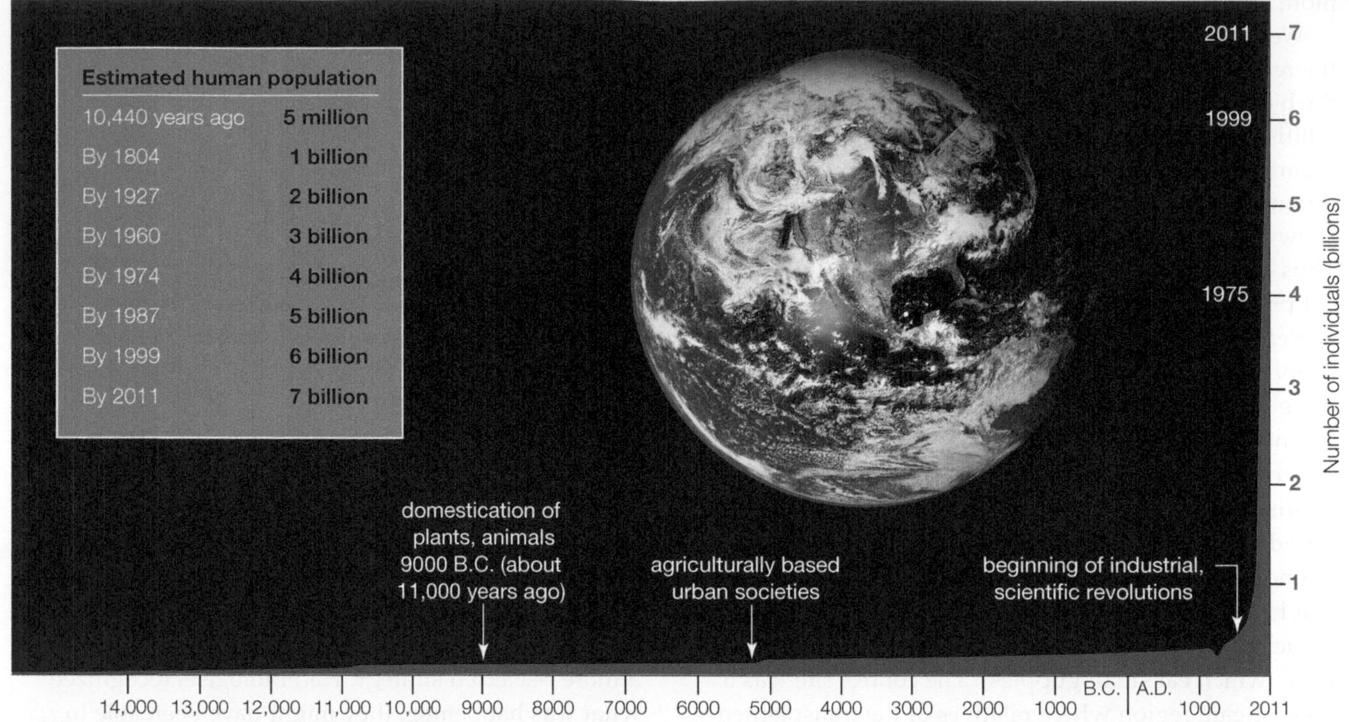

Estimated human population
10,440 years ago 5 million
By 1804 1 billion
By 1927 2 billion
By 1960 3 billion
By 1974 4 billion
By 1987 5 billion
By 1999 6 billion
By 2011 7 billion
2011
1999
1975
Number of individuals (billions)
7
6
5
4
3
2
1
domestication of plants, animals 9000 B.C. (about 11,000 years ago)
agriculturally based urban societies
beginning of industrial, scientific revolutions
B.C. A.D.
14,000 13,000 12,000 11,000 10,000 9000 8000 7000 6000 5000 4000 3000 2000 1000 1000 2011

44.7 Human Population Growth

LEARNING OBJECTIVES

- Describe the factors that allowed a rapid increase in the human population during the past 200 years.
- Explain why replacement fertility rate varies among regions.
- Explain how population growth is influenced by age structure.

For most of our history, the human population grew slowly. The growth rate began to increase about 10,000 years ago, and during the past two centuries, it soared (FIGURE 44.12). Three trends promoted the increases. First, humans migrated into new habitats and expanded into new climate zones. Second, they developed technologies that increased the carrying capacity of existing habitats. Third, they sidestepped some limiting factors that typically restrain population growth.

Expansions and Innovations

Modern humans evolved in Africa by 200,000 years ago, and by 43,000 years ago, their descendants were established in much of the world (Section 26.6). The invention of agriculture about 11,000 years ago provided a more dependable food supply than traditional hunting and gathering.

In the middle of the eighteenth century, people learned to harness energy in fossil fuels to operate machinery. This innovation opened the way to high-yielding mechanized agriculture and improved food distribution systems. Food production was further enhanced in the early 1900s, when the invention of synthetic nitrogen fertilizers increased crop yields. The invention of synthetic pesticides in the mid-1900s also contributed to increased food production.

Disease has historically dampened human population growth. During the mid-1300s, one-third of Europe's population was lost in a pandemic known as the Black Death. Beginning in the mid-1800s, an increased understanding of the link between microorganisms and illness led to improvements in food safety, sanitation, and medicine. People began to pasteurize foods and drinks, heating them to kill harmful bacteria. They also began to protect their drinking water. Advances in sanitation also lowered the death rate associated with medical treatment.

FIGURE 44.13 shows the result of improving nutrition, sanitation, and medicine in the United States: a shift in the survivorship curve. Infant mortality has plummeted and far more people survive to old age.

A worldwide decline in death rates without an equivalent drop in birth rates is responsible for the ongoing explosion in human population size. It took more than 100,000 years for the human population to reach 1 billion in number. The population is currently about 7 billion, and the United Nations estimates that it will reach 9 billion by 2050.

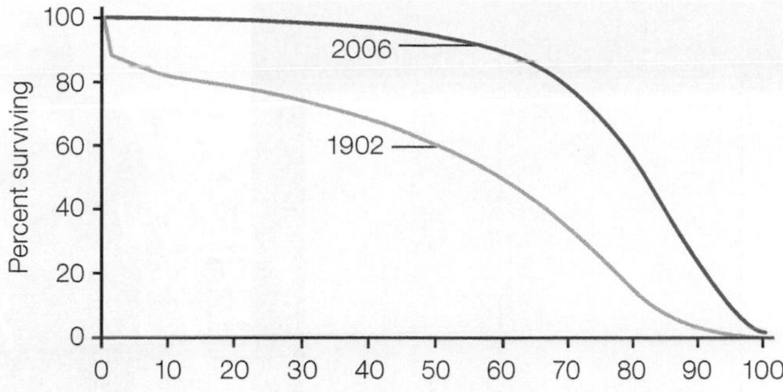
100
80
60
40
20
0
Percent surviving
2006
1902
0 10 20 30 40 50 60 70 80 90 100

FIGURE 44.13 Survivorship curves for the United States in 1902 and 2006. Since 1900, infant mortality has declined by more than 90 percent, and mortality during childbirth has declined 99 percent.

(CDC/NCHS, National Vital Statistics System and state death registration data)

Fertility and Age Structure

The **total fertility rate** of a human population is the average number of children born to a woman during her reproductive years. In 1950, the worldwide total fertility rate averaged 6.5. By 2015, this rate had declined to 2.5.

The total fertility rate is still above **replacement fertility rate**, which is the average number of children a woman must bear to replace herself with one daughter who reaches reproductive age. At present, the replacement rate is about 2.1 for developed countries and as high as 3 in some developing countries. It is higher in developing countries because more children die before reaching the age of reproduction. A population grows as long as its total fertility rate exceeds the replacement fertility rate.

Age structure affects a population's growth rate. FIGURE 44.14 shows the age structure for the world's three most populous countries. China and India have more than 1 billion people each. Next in line is the United States, with more than 320 million. These three countries differ in age structure, with India having the greatest proportion of young people. The broader the base of an age structure diagram, the greater the anticipated population growth.

Worldwide, about 1.9 billion people are about to enter their reproductive years. Even if every couple from this time forward has no more than two children, population growth will not slow for approximately 60 years.

replacement fertility rate Fertility rate at which each woman has, on average, one daughter who survives to reproductive age.

total fertility rate Average number of children the women of a population bear over the course of a lifetime.

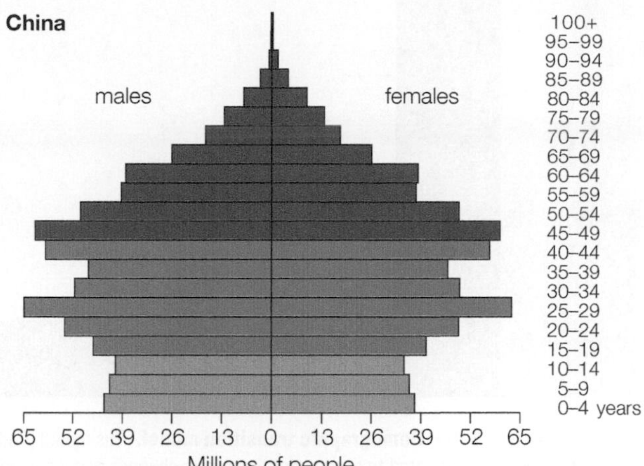
China
males
females
100+
95–99
90–94
85–89
80–84
75–79
70–74
65–69
60–64
55–59
50–54
45–49
40–44
35–39
30–34
25–29
20–24
15–19
10–14
5–9
0–4 years
65 52 39 26 13 0 13 26 39 52 65
Millions of people

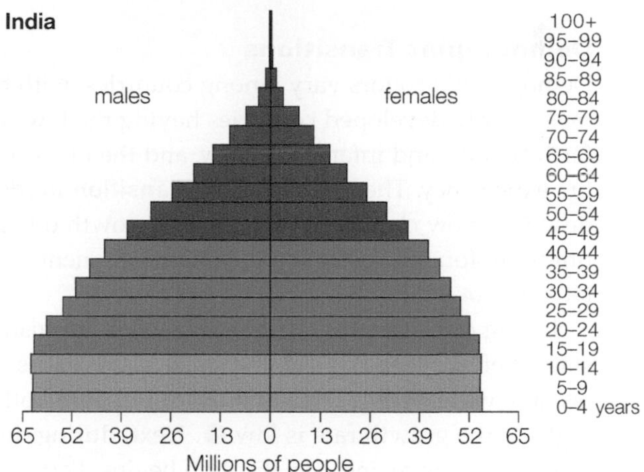
India
males
females
100+
95–99
90–94
85–89
80–84
75–79
70–74
65–69
60–64
55–59
50–54
45–49
40–44
35–39
30–34
25–29
20–24
15–19
10–14
5–9
0–4 years
65 52 39 26 13 0 13 26 39 52 65
Millions of people

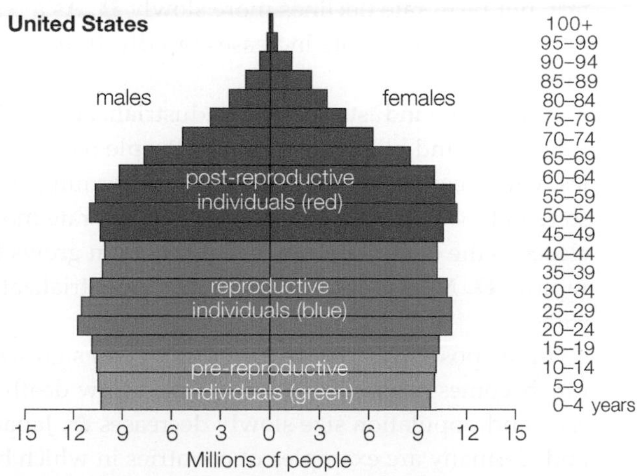
United States
males
females
post-reproductive individuals (red)
reproductive individuals (blue)
pre-reproductive individuals (green)
100+
95–99
90–94
85–89
80–84
75–79
70–74
65–69
60–64
55–59
50–54
45–49
40–44
35–39
30–34
25–29
20–24
15–19
10–14
5–9
0–4 years
15 12 9 6 3 0 3 6 9 12 15
Millions of people

FIGURE 44.14 Age structure for three countries. Green bars represent pre-reproductive individuals. The left side of each chart indicates males; the right side, females.

FIGURE IT OUT Which country has the largest number of women in the 45 to 49 age group?

Answer: China

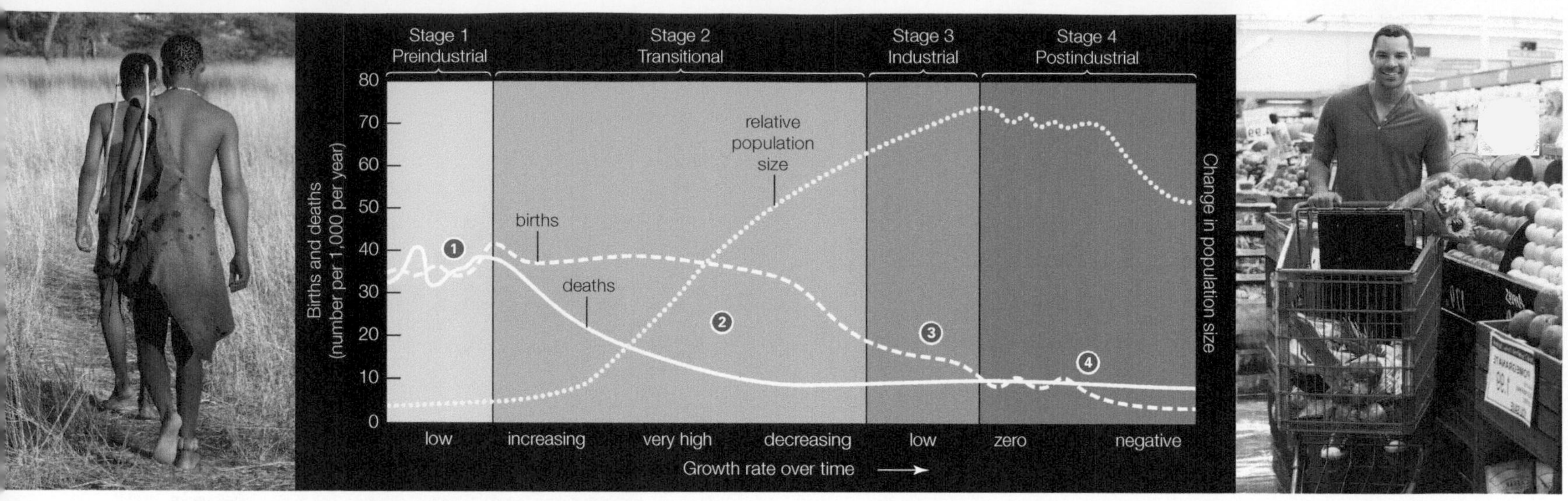
Stage 1
Preindustrial
Stage 2
Transitional
Stage 3
Industrial
Stage 4
Postindustrial
Births and deaths (number per 1,000 per year)
80
70
60
50
40
30
20
10
0
relative population size
births
deaths
Change in population size
low
increasing
very high
decreasing
low
zero
negative
Growth rate over time

FIGURE 44.15 Demographic transition model. This model predicts changes in population growth rates and sizes expected to accompany long-term changes economic development.

Demographic Transitions

Demographic factors vary among countries, with the most highly developed countries having the lowest fertility rates and infant mortality, and the highest life expectancy. The **demographic transition model** describes how changes in population growth often unfold in four stages of economic development (**FIGURE 44.15**).

Living conditions are harshest in the preindustrial stage, before technological and medical advances become widespread. Birth and death rates are both high, so the growth rate is low ❶. Next, during the transitional stage, industrialization begins. Food production and health care improve. Death rate drops fast, but birth rate declines more slowly ❷. As a result, population growth rate increases rapidly. India is in this stage.

During the industrial stage, industrialization is in full swing, and birth rate declines. People move to cities, where couples no longer need help farming and so tend to want smaller families. The birth rate moves closer to the death rate, and the population grows less rapidly ❸. Mexico is currently in the industrialization stage.

In the postindustrial stage, a population's growth rate becomes negative. Birth rate falls below death rate, and population size slowly decreases ❹. Japan and Germany are examples of countries in which birth rate has declined below the level required to maintain population size. In such countries, immigration is required to keep the population size stable.

The demographic transition model is based on the social changes that occurred when western Europe and North America industrialized. Whether it can accurately predict changes in modern developing countries remains to be seen. These countries receive aid from the fully industrialized nations, but also must compete with them in a global market.

Also, the decline in fertility rates that accompanies industrialization can pose problems.

demographic transition model Model describing changes in birth and death rates that occur as a region becomes industrialized.

ecological footprint Area of Earth's surface required to sustainably support a particular level of development and consumption.

Resource Consumption

What is Earth's carrying capacity for humans? There is no simple answer to this question. For one thing, we cannot predict what new technologies may arise or the effects they will have. For another, different types of societies require different amounts of resources to sustain them. On a per capita (per-individual) basis, people in highly developed countries use far more resources than those in less developed countries, and they also generate more waste and pollution.

Ecological footprint analysis is one widely used method of measuring and comparing human resource use. An **ecological footprint** is the amount of Earth's surface required to support a particular level of development and consumption in a sustainable fashion. It includes the amount of area required to grow crops, graze animals, produce forest products, catch fish, hold buildings, and take up any carbon emitted by burning fossil fuels.

In 2010, the per capita global footprint for the human population was 2.7 hectares, or about 6.5 acres.

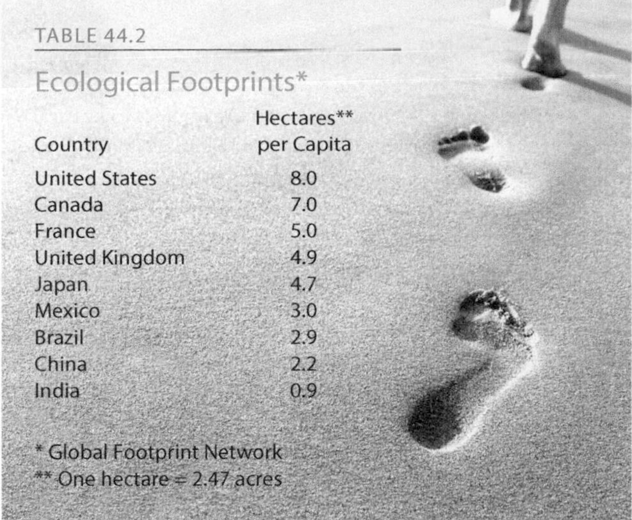
TABLE 44.2

Ecological Footprints*

Country	Hectares** per Capita
United States	8.0
Canada	7.0
France	5.0
United Kingdom	4.9
Japan	4.7
Mexico	3.0
Brazil	2.9
China	2.2
India	0.9

* Global Footprint Network
** One hectare = 2.47 acres

The world's two most populous countries, China and India, were below that average; the per capita footprint of the United States was about three times the average (**TABLE 44.2**). In other words, the lifestyle of an average person in the United States requires about three times as much of Earth's sustainable resources as the lifestyle of an average world citizen. It requires more than eight times the resources of a person in India. Consider that the United States accounts for about 5 percent of the world's population, yet it uses about 25 percent of the world's minerals and energy supply.

People of the United States are unlikely to lower their resource consumption to match that of India. In fact, billions of people in India, China, and other less developed nations dream that one day they or their offspring will enjoy the same type of lifestyle as the average American.

Ecological footprint analysis tells us that, with current technology, Earth may not have enough resources to make those dreams come true. The Global Footprint Network estimates that Earth does not have enough resources for everyone now alive to live like an average person in the United States. Ecological footprint analysis suggests that the human population is currently living well beyond its ecological means and is in the process of racking up a deficit that will take its toll on future generations.

TAKE-HOME MESSAGE 44.7

✔ Innovations such as the use of fire and clothing allowed early humans to expand into what would otherwise have been inhospitable habitats. Later, innovations in agriculture, sanitation, and medicine allowed a dramatic increase in the size of the human population.

✔ Fertility rates have been declining but remain above replacement level worldwide. Many young people are about to begin reproducing.

✔ Industrialization of less developed countries is predicted to decrease their birth rates over time, as it did for currently industrialized nations.

44.1 Managing Canada Goose (revisited)

Surging populations of Canada geese can pose a serious problem for air traffic. They are one of the species most commonly involved in collisions with aircraft. Consider what happened to a US Airways flight in 2009: Shortly after the plane took off from New York's LaGuardia airport, both engines failed. Fortunately, the pilot was able to land the plane in the nearby Hudson River, where boats safely unloaded all 155 people aboard (**FIGURE 44.16**).

After the emergency landing, investigators from the Federal Aviation Administration were asked to determine why both engines of a passenger plane failed, forcing the plane to land in the Hudson River. The pilot had reported a bird strike, and the investigators found bits of feather, bone, and muscle in the plane's wing flaps and engines. Samples of this tissue were sent to the Smithsonian Institute, which analyzed the DNA. Unique sequences in the DNA identified the tissue in both engines as Canada goose. One engine had female goose DNA. The other had male and female DNA.

FIGURE 44.16 **US Airways Flight 1549 floats in New York's Hudson River after collisions with Canada geese incapacitated both its engines.**

Researchers were even able to tell which population of geese these unlucky birds belonged to. The mix of hydrogen isotopes varies with latitude, so the isotope mix in a feather provides information about where that feather developed. The geese were migratory; their feathers had developed in Canada, not in New York.

STUDY GUIDE

Section 44.1 A **population** is a group of individuals of the same species that live in the same area and interbreed. Some Canada geese observed in the United States belong to a migratory population, and others are permanent residents.

Section 44.2 **Ecology** is the study of interactions among populations and between populations and their environment. **Demographics** are statistics used to describe populations. **Plot sampling** or **mark–recapture sampling** are methods of estimating **population size**. Other demographics include **population density** and **population distribution**, which is most often clumped. The **age structure** of a population is the proportion of individuals in each of several age categories. Individuals who are currently reproducing or will eventually reproduce constitute a population's **reproductive base**.

Section 44.3 **Immigration** and **emigration** affect population size. Ignoring migration, the per capita birth rate minus the per capita death rate is a population's **per capita growth rate** (r). **Zero population growth** occurs when birth rate equals death rate.

With **exponential growth**, the population size increases at a fixed rate. In each interval, the number of individuals added is some fixed proportion of the current population size and is described by the equation $G = r \times N$, where G is population growth and N is the number of individuals. With exponential growth, a graph of population size against time produces a J-shaped growth curve.

The maximum possible rate of increase under optimal conditions is a species' **biotic potential**. Conditions in nature are seldom optimal, so the actual rate of increase is generally lower than the biotic potential.

Section 44.4 Limiting factors constrain the growth of natural populations. **Logistic growth** occurs as a result of the constraints imposed by **density-dependent limiting factors** such as **intraspecific competition** and disease. With logistic growth, a population size rises exponentially, then levels off as the population nears **carrying capacity (*K*)**. Thus, a graph of population size over time yields an S-shaped curve. Carrying capacity is species specific, environment specific, and can change over time.

Density-independent limiting factors such as extreme weather events can also influence the growth rate of a population, but their effect does not vary with crowding.

Section 44.5 The time to maturity, number of reproductive events, number of offspring per event, and life span are aspects of an animal's **life history**. Life histories are often studied by following a **cohort**, a group of individuals that were born at the same time. **Survivorship curves** show the risk of death at various intervals. Three types of survivorship patterns are common: a high death rate late in life, a constant rate at all ages, or a high rate early in life.

Life histories have a genetic basis and are subject to natural selection. At low population density, ***r*-selection** favors quickly producing as many offspring as possible. At a higher population density, ***K*-selection** favors investing more time and energy in fewer, higher-quality offspring. Most species have traits that are intermediate between *r*-selected and *K*-selected traits.

Section 44.6 Predation affects life histories. Predation on large individuals selects for maturation and reproduction earlier in life. Human hunting or fishing pressure alters the life history of target species.

Section 44.7 The human population is currently more than 7 billion. Expansion into new habitats and the invention of agriculture opened the way for early increases in population size. Later, medical and technological innovations lowered death rates, while birth rates continued to increase.

The global **total fertility rate** is now declining, but it remains above the **replacement fertility rate** and is higher in developing nations than more developed ones. The age structure of countries varies, with the least developed countries having a broad reproductive base that will increase population size for some time.

The **demographic transition model** predicts that economic development may slow population growth. Some of the world's least developed countries remain in the earliest stages of their demographic transition.

World resource consumption will probably continue to rise because nations continue to develop, and a highly developed nation has a much larger **ecological footprint** than a developing one. However, with current technology, Earth does not have enough resources to support the existing population in the style of developed nations.

SELF-QUIZ

Answers in Appendix VII

1. Most commonly, individuals of a population show a ______ distribution within their habitat.
   a. clumped
   b. random
   c. nearly uniform
   d. none of the above

2. The rate at which population size grows or declines depends on the rate of ______ .
   a. births
   b. deaths
   c. immigration
   d. emigration
   e. available resources
   f. all of the above

3. Suppose 200 fish are marked and released in a pond. The following week, 200 fish are caught and 100 of them have marks. There are about ______ fish in this pond.
   a. 200
   b. 300
   c. 400
   d. 2,000

4. A population of worms is growing exponentially in a compost heap. Thirty days ago there were 300 worms and now there are 600. How many worms will there be 30 days from now, assuming conditions remain constant and resources are unlimited?
   a. 1,200 b. 1,600 c. 3,200 d. 6,400

5. For a given species, the maximum rate of increase per individual under ideal conditions is its ______ .
   a. biotic potential
   b. carrying capacity
   c. life history pattern
   d. age structure

6. ______ is a density-independent factor that influences population growth.
   a. Predation
   b. Infectious disease
   c. Resource competition
   d. Harsh weather

7. Species that live in unpredictable habitats are more likely to show traits that are favored by ______ .
   a. *r*-selection b. *K*-selection

8. When human or animal predators target the largest members of a prey population, prey individuals who ______ are at a selective advantage
   a. mature sexually while small
   b. grow rapidly and mature later
   c. reproduce a single time late in life
   d. do not reproduce

9. The human population is now about 7 billion. It reached 6 billion in ______ .
   a. 2007 b. 1999 c. 1802 d. 1350

10. Compared to the less developed countries, the highly developed ones have a higher ______ .
    a. death rate
    b. birth rate
    c. total fertility rate
    d. resource consumption rate

11. All members of a cohort are the same ______ .
    a. sex b. size c. age d. weight

12. The ecological footprint of a person in the United States is about ______ that of a person in India.
    a. half b. twice c. one-ninth d. nine times

13. The demographic transition model predicts a decline in population during the ______ stage.
    a. preindustrial
    b. transitional
    c. industrial
    d. postindustrial

14. A population in which the total fertility rate exceeds the replacement fertility rate is ______ .
    a. increasing in size
    b. decreasing in size
    c. not changing in size

15. Match each term with its most suitable description.

___ carrying capacity	a. maximum rate of increase per individual under ideal conditions
___ exponential growth	b. population growth plots out as an S-shaped curve
___ biotic potential	c. maximum number of individuals sustainable by the resources in a given environment
___ limiting factor	d. population growth plots out as a J-shaped curve
___ logistic growth	e. essential resource that restricts population growth when scarce

CRITICAL THINKING

1. Think back to Section 44.6. When researchers moved guppies from populations preyed on by cichlids to a habitat with killifish, the life histories of the transplanted guppies evolved. They came to resemble those of guppy populations preyed on by killifish. Males became gaudier; some scales formed larger, more colorful spots. How might a decrease in predation pressure on sexually mature fish allow this change?

2. Mountain gorillas are a highly endangered primate species, with 800 or so individuals surviving in the two remaining populations. A recent sampling of mountain gorilla genomes revealed a very low level of genetic diversity. Explain why a decline in genetic diversity often accompanies a decline in population size.

3. Bluebirds are songbirds that, like wood ducks, nest in cavities in trees. European house sparrows, which have been introduced to North America, also nest in such cavities. Explain how the arrival of the house sparrows affected the carrying capacity for bluebirds. What affect would you expect this change have on the number of bluebirds?

4. The age structure diagrams for two hypothetical populations are shown below. Describe the growth rate of each population and discuss the current and future social and economic problems that each is likely to face.

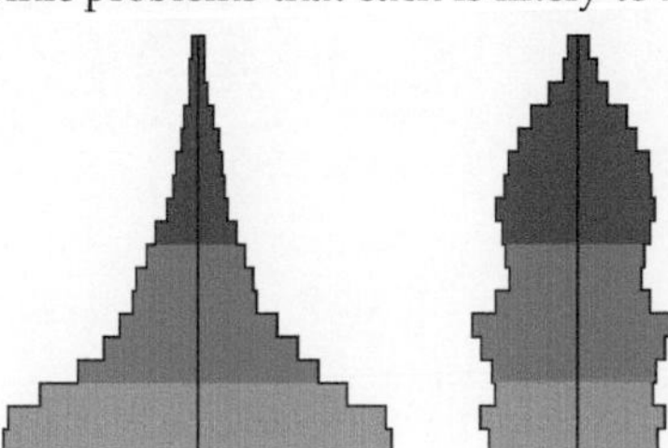

45 COMMUNITY ECOLOGY

CORE CONCEPTS

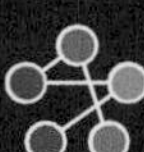

Systems

Complex properties arise from interactions among components of a biological system.
Interactions among organisms in a community involve the exchange of energy and matter. These interactions are complex: A change in any one component can affect the entire community. Community structure, including species variety, abundance, and patterns of distribution, changes over time as a result of disturbances, environmental change, and random factors.

Evolution

Evolution underlies the unity and diversity of life.
Each species is adapted to a specific habitat, and other species in its biological community are one component of that habitat. Cooperative behavior and symbiotic relationships contribute to the survival of participating populations. Competition between species is a selective pressure that may lead to adaptive diffences in resource use.

Process of Science

The field of biology consists of and relies upon experimentation and the collection and analysis of scientific evidence.
Carefully designing experiments helps researchers unravel cause-and-effect relationships in complex natural systems. The interactions among populations of a community, and the effects of disturbances or environmental change, can be modeled mathematically. However, random factors and disturbances that affect community structure make future change difficult to predict.

Links to Earlier Concepts

In this chapter, you will see how natural selection (Section 16.3) and coevolution (17.9) shape communities. You will revisit plant–pollinator interactions (22.8), the role of fungi as partners (23.7), and root nodules (28.3). Knowledge of biogeography (16.2) will help you understand how communities in different regions differ.

45.1 Fighting Foreign Fire Ants

Like most ants, red imported fire ants (*Solenopsis invicta*) nest in the ground (**FIGURE 45.1**). Accidentally step on one of their nests, and you will quickly realize your mistake. Fire ants defend their nest by stinging, and their venom causes a burning sensation that gives the ants their common name.

S. invicta is native to South America. The species first arrived in the southeastern United States in the 1930s, when the ants traveled as stowaways on a cargo ship. Since then, the ants have gradually expanded their range across the south and were accidentally transported to California and New Mexico.

Spread of *S. invicta* concerns ecologists because this species typically has a negative impact on a region's native species. For example, where *S. invicta* becomes prevalent, native ant species decline. In Texas, the *S. invicta*–induced decline of native ant species threatens the Texas horned lizard. Ants are a staple of this lizard's diet, but it cannot eat the red imported fire ants that have largely replaced its normal prey.

S. invicta has been implicated in population declines of some birds, including bobwhite quails and vireos (a type of songbird). The ants decrease the abundance of insects that the birds would normally feed to their young. They also feed on birds' eggs and on nestlings (**FIGURE 45.1**). Ground-nesting birds are at special risk.

Native plants feel the impact of an *S. invicta* invasion too. Species whose seeds are dispersed by native ants may decline when *S. invicta* replaces these natives. *S. invicta* interferes with pollination by displacing or preying on pollinators such as ground-nesting bees.

Species interactions such as competition between ant species or predation of ants on bird nestlings are one focus of community ecology. A **community** is all the species that live in some region. As you will see, species interactions and disturbances can shift community structure in both small and large ways, some predictable, and others unexpected.

FIGURE 45.1 An imported pest species. Red imported fire ants feeding on a quail egg. These South American ants are a threat to species in their adopted North American habitat.

community All species that live in a particular region.

CREDITS: (opposite) Ed Cesar/Science Source; (1) James Mueller.

45.2 What Factors Shape Community Structure?

LEARNING OBJECTIVES

- Using appropriate examples, describe some factors that influence community structure.
- Distinguish between the two components of species diversity.
- Explain symbiosis and give an example.

Each species lives in a specific type of place, which we call its **habitat**, and all species in a particular habitat constitute a community. The scale of a community is defined by a human observer, so one community can contain smaller communities or be a component of a larger one. Consider the community of microbial organisms that live inside the digestive tract of a termite. That termite is part of a larger community of organisms that live in and on a fallen log. The many log-dwellers are part of a still larger forest community.

Even communities that are similar in scale sometimes differ in their species diversity. There are two components to species diversity. The first, **species richness**, refers to the number of species. The second component, **species evenness**, describes the relative abundance of each species. For example, a pond that has five fish species in nearly equal numbers has a higher species diversity than a pond with one abundant fish species and four rare ones.

Community structure is dynamic, meaning the array of species and their relative abundances in the community tend to change over time. Communities change over a long time span as they form and then age. They also change suddenly as a result of natural or human-induced disturbances.

Abiotic (nonbiological) factors such as rainfall, sunlight intensity, and temperature affect community structure. Such factors vary along gradients in latitude, elevation, and—for aquatic habitats—depth.

Species interactions also influence community structure. In some cases, the effect is indirect. For example, when songbirds eat caterpillars, the birds directly reduce the abundance of caterpillars, but they also indirectly benefit the trees that the caterpillars feed on.

We define direct interactions by their effects on both participants (TABLE 45.1). Consider **commensalism**, in which one species benefits and the second is unaffected by the interaction. Epiphytes, which are plants that use the branches or trunk of another plant for structural support, have a commensal relationship with their host (FIGURE 45.2). Having an elevated perch benefits the epiphyte, while the host plant is unaffected.

Species interactions may be fleeting or a long-term relationship. **Symbiosis** means "living together," and in biology this term refers to a relationship in which two species have a prolonged close association that benefits at least one of them. Commensalism, mutualism, and parasitism can be symbiotic relationships, as when commensal, mutualistic, or parasitic microorganisms live inside an animal's gut.

Regardless of whether one species helps or harms another, two species that interact closely for generations can coevolve. Coevolution is an evolutionary process in which each species acts as a selective agent that shifts the range of variation in the other (Section 17.9).

commensalism Species interaction that benefits one species and neither helps nor harms the other.
habitat Type of environment in which a species typically lives.
species evenness Of a community, the relative abundance of species.
species richness Of a community, the number of species.
symbiosis (sim-by-OH-sis) One species lives in or on another in a commensal, mutualistic, or parasitic relationship.

TABLE 45.1
Direct Two-Species Interactions

Type of Interaction	Effect on Species 1	Effect on Species 2
Commensalism	Beneficial	None
Mutualism	Beneficial	Beneficial
Interspecific competition	Harmful	Harmful
Predation, herbivory, parasitism, parasitoidism	Beneficial	Harmful

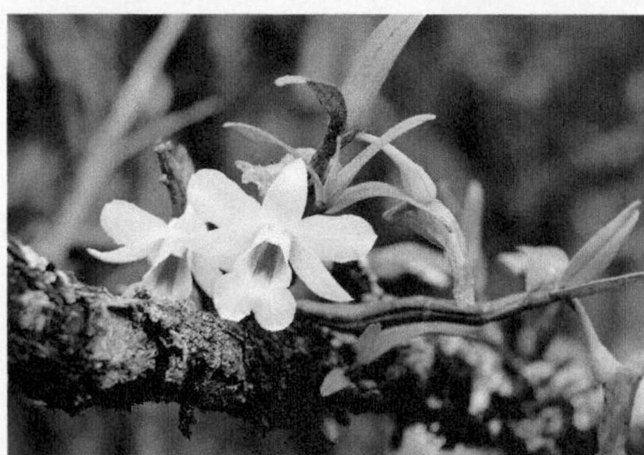

FIGURE 45.2 **Commensalism.** Epiphytic orchid growing on a tree branch. The tree provides orchids with an elevated perch from which they can capture sunlight, and the tree is neither helped nor harmed by the orchid's presence.

TAKE-HOME MESSAGE 45.2

✔ The types and relative abundances of species in a community are affected by physical factors and by species interactions.

✔ A species can be benefited, harmed, or unaffected by its interaction with another species.

45.3 Mutualism

LEARNING OBJECTIVES

- Using appropriate examples, describe the types of benefits that can be exchanged by partners in a mutualism.
- Explain why mutualism is considered reciprocal exploitation.

Mutualism is an interspecific interaction that benefits both species. Flowering plants and their pollinators are a familiar example. In some cases, coevolution of two species results in a mutual dependence. For example, there are several species of yucca plant and each is pollinated by a single species of yucca moth, whose larvae develop on that plant species alone (**FIGURE 45.3**). More often, mutualistic relationships are less exclusive. Most flowering plants have more than one pollinator, and most pollinators provide their service to more than one species of plant.

FIGURE 45.3 An obligate mutualism. Each species of yucca plant (left) has a relationship with a single species of yucca moth (right). After a female moth mates, she collects pollen from a yucca flower and places it on the stigma of another flower, then lays her eggs in that flower's ovary. Moth larvae develop in the fruit that develops from the floral ovary. When mature, the larvae gnaw their way out and disperse. Seeds that larvae did not eat give rise to new yucca plants.

Photosynthetic organisms often supply sugars to their nonphotosynthetic partners, as when plants lure pollinators with nectar. In addition, many plants make sugary fruits that attract seed-dispersing animals. Plants also provide sugars to mycorrhizal fungi and nitrogen-fixing bacteria (Section 28.3). The plants' fungal or bacterial symbionts return the favor by supplying their host with other essential nutrients. Similarly, photosynthetic dinoflagellates provide sugars to corals (Section 21.5), and photosynthetic bacteria and algae feed their fungal partner in a lichen (Section 23.7).

Animals often share ingested nutrients with mutualistic microorganisms that live in their digestive tract. For example, *Escherichia coli* bacteria living in your colon provide you with essential vitamin K. In return, they obtain a steady supply of food and a warm, moist place to live.

Other mutualisms involve protection. For example, an anemonefish and a sea anemone each fend off the other's predators (**FIGURE 45.4**). Ants protect bull acacia trees from leaf-eating insects, and in return the tree houses the ant in special hollow thorns and provides them with sugar-rich foods.

FIGURE 45.4 Mutual protection. Stinging tentacles of this sea anemone (*Heteractis magnifica*) protect its partner, a pink anemonefish (*Amphiprion perideraion*) from fish-eating predators. In return, the anemonefish chases away fish that eat sea anemone tentacles. The anemonefish secretes a special mucus that prevents the anemone from stinging it.

From an evolutionary standpoint, mutualism is best considered as a case of reciprocal exploitation. Each participant increases its own fitness by extracting a resource, such as protection or food, from its partner. If taking part in the mutualism has a cost, selection will favor individuals who minimize that cost. Consider nectar production, which is energetically costly for a flower, but serves as a necessary payoff to pollinators. A flower that produces the minimum amount of nectar necessary to keep pollinators coming will be at a selective advantage over one that expends additional energy to produce a more generous serving of nectar.

mutualism Species interaction that benefits both participants.

TAKE-HOME MESSAGE 45.3

✔ A mutualism benefits both participants.

✔ In some cases, two species form an exclusive partnership. In others, a species provides benefits to, and receives benefits from, multiple species.

✔ Participating in a mutualism has costs as well as benefits. Thus, selection favors individuals who maximize their benefits, while minimizing their costs.

CREDITS: (3) left, Harlo H. Hadow; right, Bob and Miriam Francis/Tom Stack & Associates; (4) © Thomas W. Doeppner.

45.4 Competitive Interactions

LEARNING OBJECTIVES

- Using appropriate examples, describe the two types of interspecific competition.
- Describe the process of competitive exclusion.
- Explain how character displacement allows resource partitioning.

Interspecific competition, the competition between species, is not usually as intense as competition within a species (intraspecific competition). The requirements of two species might be similar, but they will nearly always differ more than the requirements of two members of the same species.

A Interference competition between scavengers. A golden eagle attacks a fox with its talons to drive the fox away from a moose carcass.

B Exploitative competition between insect eaters. Sundew plants (left) and wolf spiders (right) both feed on insects. The presence of one species reduces the food available to the other.

FIGURE 45.5 Two types of interspecific competition.

The types of resources and environmental conditions an organism requires, and the manner in which it interacts with its living and nonliving environment, define its **ecological niche**. Aspects of an animal's niche include the temperature range it can tolerate, the species it eats and is eaten by, and the places it can breed. A description of a flowering plant's niche would include its soil, water, light, and pollinator requirements. The more similar the niches of two species are, the more intensely the species will compete.

Competition takes two forms, interference competition and exploitative competition. With interference competition, one species actively prevents another from accessing some resource. For example, one species of scavenger will often chase another away from a carcass (FIGURE 45.5A). As another example, some plants use chemical weapons against potential competition. Leaves of sagebrush plants, black walnut trees, and eucalyptus ooze aromatic chemicals that seep into the soil around the plant. These chemicals prevent other plants from germinating or growing.

With exploitative competition, competing species do not interact directly. However, by using the same resource, each reduces the amount of that resource available to the other. For example, wolf spiders and carnivorous plants called sundews both feed on insects in Florida swamps (FIGURE 45.5B). These two very different types of organisms do not come into contact, but they do compete for food. By catching insects, each reduces the nutrients available to the other.

Effects of Competition

Species compete most intensely when the supply of a shared resource is the main limiting factor for both (Section 44.4). In the early 1930s, G. F. Gause carried out a series of experiments that led him to describe **competitive exclusion**: Whenever two species require the same limited resource to survive or reproduce, the better competitor will drive the less competitive species to extinction in that habitat. Gause studied interactions between two species of ciliated protists (*Paramecium*) that compete for bacterial prey. He cultured the species separately and together (FIGURE 45.6). When the

FIGURE 45.6
Competitive exclusion. Growth curves for two *Paramecium* species when grown separately and together.

FIGURE IT OUT Which species is the better competitor?

Answer: P. aurelia

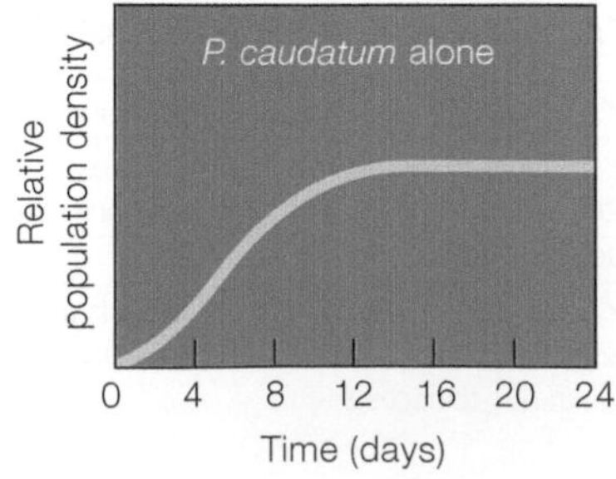
P. caudatum alone
Relative population density
0 4 8 12 16 20 24
Time (days)

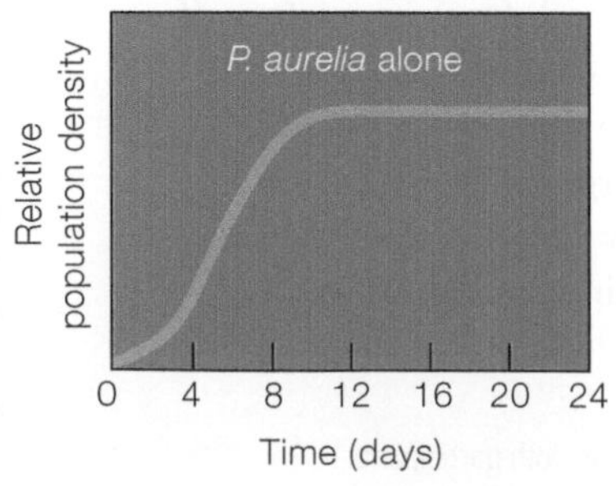
P. aurelia alone
Relative population density
0 4 8 12 16 20 24
Time (days)

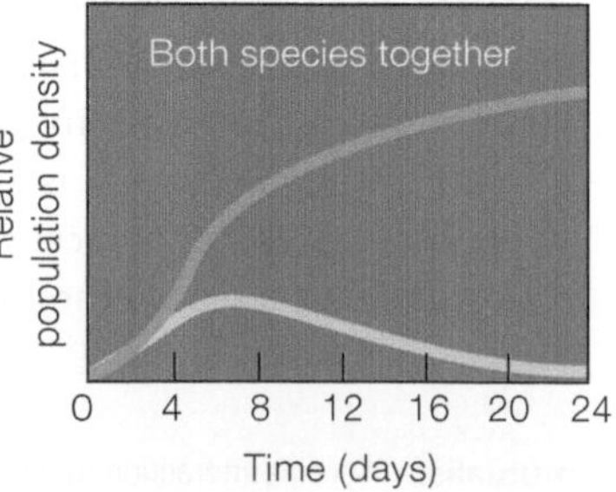
Both species together
Relative population density
0 4 8 12 16 20 24
Time (days)

species were grown together, population growth of one always outpaced the other, which died out.

When resource needs of competitors are not exactly the same, competing species can coexist, but the presence of each reduces the carrying capacity (Section 44.4) of the habitat for the other. For example, the reproductive success of a flowering plant is decreased by competition from other species that flower at the same time and rely on the same pollinators (**FIGURE 45.7**). Similarly, sundews grown in the presence of wolf spiders make fewer flowers than sundews in spider-free enclosures. Presumably, competition for insect prey reduces the energy the plants can devote to flowering.

Resource Partitioning

Resource partitioning is an evolutionary process by which species become adapted to use a shared limiting resource in a way that minimizes competition. For example, eight species of woodpecker coexist in Oregon forests. All feed on insects and nest in hollow trees, but the details of their foraging behavior and nesting preferences vary. Differences in nesting time also help reduce competitive interactions.

Resource partitioning arises as a result of directional selection on species who share a habitat and compete for a limiting resource. In each species, individuals who differ most from the competing species with regard to resource use will have the least competition and thus leave the most offspring. Over generations, directional selection leads to **character displacement**: The range of variation for one or more traits is shifted in a direction that lessens the intensity of competition for a limiting resource.

Consider the variation in body size among populations of two species of salamander in the genus *Plethodon* (**FIGURE 45.8**). In most parts of their ranges, the two species do not overlap and their body is quite similar. Where the species coexist and compete, their difference in body size becomes more pronounced. One species, *P. cinereus*, is smaller, and the other, *P. hoffmani*, is larger. The increased difference in body size may be related to a partitioning of food resources. In regions where the species overlap, *P. cinereus* tends to specialize on smaller insect prey and *P. hoffmani* on larger insects.

character displacement Outcome of competition between two species; similar traits that result in competition become dissimilar.
competitive exclusion Process whereby two species compete for a limiting resource, and one drives the other to local extinction.
ecological niche The resources and environmental conditions that a species requires.
interspecific competition Competition between two species.
resource partitioning Species adapt to access different portions of a limited resource; allows species with similar needs to coexist.

Mimulus

Lobelia

FIGURE 45.7 Competing for pollinators. *Mimulus* and *Lobelia* are purple-flowered plants that grow together in damp meadows. To test for competition, researchers grew *Mimulus* plants alone or with *Lobelia*. In mixed plots, pollinator visits to *Lobelia* plants often intervened between visits to *Mimulus*. As a result, *Mimulus* in mixed plots produced fewer seeds than those in *Mimulus*-only plots.

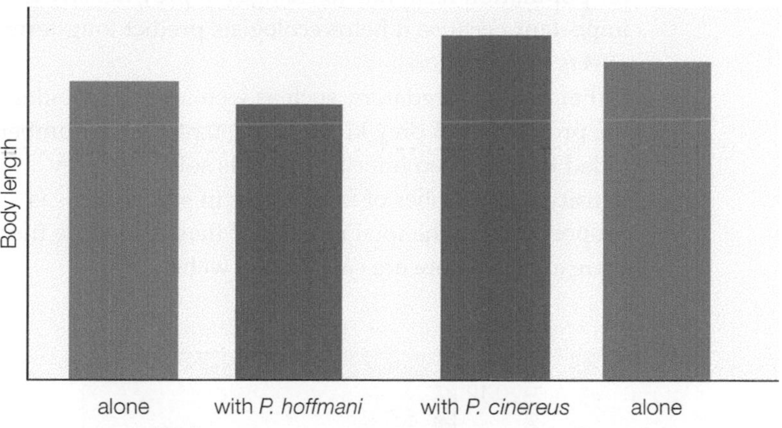
Body length
alone
with P. hoffmani
with P. cinereus
alone
P. cinereus
P. hoffmani

FIGURE 45.8 Character displacement in salamanders. Where two *Plethodon* species coexist, their average body lengths (purple bars) differ more than they do in habitats where each species lives alone (orange bars).

TAKE-HOME MESSAGE 45.4

✔ In some interspecific competitions, one species actively blocks another's access to a resource. In other interactions, one species is simply better than another at exploiting a shared resource.

✔ Interspecific competition reduces the reproductive success of both competitors. Thus, when two species compete, selection favors individuals whose needs are least like those of the competing species.

45.5 Predator–Prey Interactions

LEARNING OBJECTIVE

- Explain why and how predator abundance and prey abundance affect one another.

Predation is an interspecific interaction in which one species (the predator) captures, kills, and eats another (the prey). Predation removes a prey individual from the population immediately. Understanding predator–prey dynamics is important because such information allows scientists to make valid decisions about how to best manage threatened predator and prey species.

Functional Response to Prey Abundance

In any community, predator abundance and prey abundance are interconnected. Generally, an increase in the size of a predator population results in a decrease in the abundance of its prey. Understanding how a predator population responds to changes in prey density is important because it helps ecologists predict long-term effects of predation.

For passive predators, such as web-spinning spiders, the proportion of prey killed is constant, so the number killed in any given interval depends solely on prey density. The number of flies caught in spider webs is proportional to the total number of flies: The more flies there are, the more are captured in webs.

More typically, the rate at which prey are killed depends in part on the time it takes predators to process prey. For example, a wolf that just killed a caribou will not hunt another until it has eaten and digested the first one. As prey density increases, the rate of kills rises steeply at first because there are more prey to catch. The rate slows when predators are exposed to more prey than they can handle at once (**FIGURE 45.9**).

Cyclic Changes in Abundance

Predator and prey populations sometimes rise and fall in a cyclical fashion. **FIGURE 45.10** shows historical data for the numbers of lynx and their main prey, the snowshoe hare. Both populations rise and fall over an approximately ten-year cycle, with predator abundance lagging behind prey abundance. Field studies indicate that lynx numbers fluctuate mainly in response to hare numbers. However, experiments by Charles Krebs demonstrated that the size of the hare population is affected by the abundance of the hare's food as well as the number of lynx. Hare populations continue to rise and fall even when predators are experimentally excluded from areas.

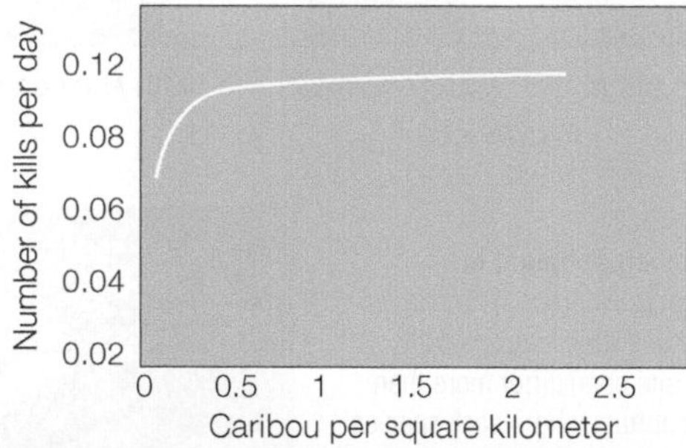
Number of kills per day
0.12
0.08
0.06
0.04
0.02
0
0.5
1
1.5
2
2.5
Caribou per square kilometer

FIGURE 45.9 Functional response of wolves to changes in caribou density. Wolf kills initially increase with prey density, but level off once the wolves' capacity to catch and eat prey is reached. The graph shows data that B. W. Dale and his coworkers compiled by observing wolf packs.

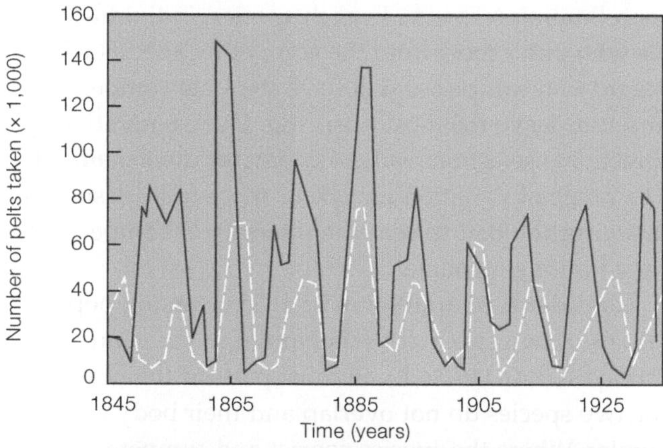
Number of pelts taken (× 1,000)
160
140
120
100
80
60
40
20
0
1845
1865
1885
1905
1925
Time (years)

FIGURE 45.10 Predator–prey cycles. The graph shows the abundance of the Canadian lynx (dashed line) and snowshoe hares (solid line), based on counts of pelts sold by trappers to Hudson's Bay Company from 1845 to 1925.

TAKE-HOME MESSAGE 45.5

✔ Predation removes individuals from the prey population.

✔ The numbers in predator and prey populations vary in complex ways that reflect the multiple levels of interaction in a community.

45.6 Evolutionary Arms Races

LEARNING OBJECTIVES

- Give examples of evolved defenses against predation and herbivory.
- Distinguish between the two types of mimicry.

Predator–Prey Evolutionary Arms Race

Predator and prey species exert selection pressure on one another. Suppose a mutation gives members of a prey species a more effective defense. Over generations, directional selection will cause this mutation to spread through the prey population. If some members of a predator population have a trait that makes them better at thwarting the improved defense, these predators and their descendants will be at an advantage. Thus, predators exert selection pressure that favors improved prey defense, which in turn exerts selection pressure on predators, and so the process continues over many generations.

You have already learned about some defensive adaptations. Many prey species have hard or sharp parts that make them difficult to eat. Think of a snail's shell or a sea urchin's spines. Cnidarians such as sea anemones and corals have stinging cells on their tentacles. Other prey contain chemicals that taste bad to predators or sicken them.

Most defensive toxins in animals are from the plants that they eat. For example, a monarch butterfly caterpillar takes up chemicals from the milkweed plant that it feeds on. A bird that later eats the butterfly will be sickened by these chemicals.

Well-defended prey often have **warning coloration**, a conspicuous color pattern that predators learn to avoid. Monarch butterflies are bright orange, and many stinging wasps and bees share a pattern of black and yellow stripes (**FIGURE 45.11A**). The similar appearance of bees and wasps is an example of **mimicry**, an evolutionary pattern in which one species comes to resemble another. The tendency of well-defended species to have similar coloration is called Müllerian mimicry, after the German naturalist who first described the phenomenon. Fritz Müller recognized that well-defended species who share the same predator would benefit by having a similar appearance. The more often a predator is stung by a black and yellow striped insect, the less likely it is to attack similar looking insects in the future.

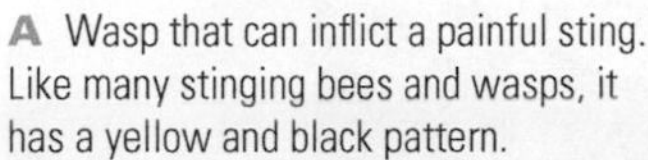

A Wasp that can inflict a painful sting. Like many stinging bees and wasps, it has a yellow and black pattern.

B Fly, which lacks a stinger, mimics the color pattern of stinging insects.

FIGURE 45.11 Warning coloration and mimicry.

FIGURE 45.12 Defense and counter defense. (**A**) *Eleodes* beetles defend themselves by spraying irritating chemicals at predators. (**B**) This defense is ineffective against grasshopper mice, who plunge the chemical-spraying end of the beetle into the ground, then devour the insect head first.

In Batesian mimicry, also named for the scientist who first described it, a species that lacks a defense mimics the appearance of a well-defended species. For example, some stingless insects resemble stinging bees or wasps (**FIGURE 45.11B**). These impostors benefit when predators avoid them after a painful encounter with the better-defended species that they mimic.

Stinging is an example of a defensive behavior that can repel a potential predator. Section 1.6 described how eyespots and a hissing sound protect some butterflies from predatory birds. Similarly, a lizard's tail may detach from the body and wiggle a bit as a distraction, allowing the rest of the lizard to escape.

Skunks squirt a foul-smelling, irritating repellent, as do some darkling beetles (**FIGURE 45.12**). Both skunks and beetles announce their intention to spray by assuming a distinctive posture, with their posterior end pointed toward the threat. Some beetles that cannot spray mimic this posture.

mimicry (MIM-ik-cree) A species evolves traits that make it similar in appearance to another species.

predation One species captures, kills, and eats another.

warning coloration In many well-defended or unpalatable species, bright colors, patterns, and other signals that predators learn to recognize and avoid.

FIGURE 45.13 Camouflage. Fleshy protrusions give a predatory scorpionfish the appearance of an algae-covered rock. When algae-eating fish come close for a nibble, they end up as prey.

Camouflage is a body shape, color pattern, or behavior that allows an individual to blend into its surroundings and avoid detection. Prey benefit when camouflage hides them from predators, and predators benefit when it hides them from prey (**FIGURE 45.13**).

Other predator adaptations include sharp teeth and claws that can pierce protective hard parts. Speedy prey select for faster predators. For example, the cheetah, the fastest land animal, can run 114 kilometers per hour (70 mph). Its preferred prey, Thomson's gazelles, run 80 kilometers per hour (50 mph).

Coevolution of Herbivores and Plants

With **herbivory**, an animal feeds on plants. The number and type of plants in a community can influence the number and type of herbivores present.

Two types of defenses have evolved in response to herbivory. Some plants have adapted to withstand and recover quickly from the loss of their parts. For example, prairie grasses are seldom killed by native grazers such as bison. The grasses have a fast growth rate and store enough resources in their roots to replace the shoots lost to grazers.

Other plants have traits that deter herbivory. Physical deterrents include spines, thorns, and tough leaves that are difficult to chew. Many plants produce secondary metabolites (Section 22.9) that are unpalatable to herbivores or sicken them. Ricin, the toxin made by castor bean plants (Section 9.1), makes all eukaryotic herbivores ill. Caffeine in coffee beans and nicotine in tobacco leaves are defenses against insects.

Capsaicin, the compound that makes some peppers "hot," is an evolved defense against seed-eating mammals. Rodents avoid pepper fruits, leaving them to be eaten by birds. The pepper benefits by deterring rodent seed eaters because rodents chew up and kill seeds, whereas birds excrete the seeds intact and alive.

When a well-defended plant is abundant, herbivores capable of overcoming the plant's defenses will be at a selective advantage. Such a selection process has given koalas the ability to feed on eucalyptus leaves, which are tough and contain toxic oils. Koalas have evolved specialized teeth that can shred the leaves and special liver enzymes that allow them to detoxify the oils.

TAKE-HOME MESSAGE 45.6

✔ In any community, predators and prey coevolve, as do plants and the herbivores that feed on them.

✔ Defensive adaptations in plants and prey can limit the ability of predators or herbivores to exploit some species in their community.

45.7 Parasites and Parasitoids

LEARNING OBJECTIVES

- Distinguish between parasites and parasitoids.
- Describe how parasites can reduce the population size of their hosts.
- Using an appropriate example, explain brood parasitism.
- Explain the benefits and risks of biological pest control.

Parasitism

With **parasitism**, one species (the parasite) benefits by feeding on another (the host), without killing it immediately. Endoparasites such as parasitic roundworms live and feed inside their host. An ectoparasite such as a tick feeds while attached to a host's external surface.

Parasitism has evolved in a diverse variety of groups. Bacterial, fungal, protistan, and invertebrate parasites feed on vertebrates. Lampreys (Section 25.3) attach to and feed on other fish. There are even a few parasitic plants that withdraw nutrients from other plants (**FIGURE 45.14**).

Although most parasites do not kill their hosts, parasitism can still decrease the size of a host population. The presence of parasites can weaken a host, making it more vulnerable to predation or less attractive to potential mates. Some parasites can make a host sterile or shift the sex ratio among its offspring.

Adaptations to a parasitic lifestyle include traits that allow the parasite to locate hosts and to feed undetected. For example, ticks that feed on mammals or birds move toward a source of heat and carbon dioxide, which may be a potential host. A chemical in tick saliva acts as a local anesthetic, preventing the host from noticing the feeding tick. Endoparasites often have adaptations that help them evade a host's immune defenses.

CREDIT: (13) David Burdick, NOAA.

Among hosts, traits that minimize the negative effects of parasites confer a selective advantage. For example, the allele that causes sickle-cell anemia persists at high levels in some human populations because having one copy of the allele increases the odds of surviving malaria (Section 17.5). Grooming and preening behavior are adaptations that minimize the impact of ectoparasites. Some animals produce chemicals that repel or interfere with parasite activity. For example, one type of seabird (the crested auklet) produces a citrus-scented secretion that repels mosquitoes.

Brood Parasites—Strangers in the Nest

With **brood parasitism**, one species benefits by having another raise its offspring. The European cuckoos described in Section 43.3 are brood parasites, as are North American cowbirds. Not having to invest in parental care allows a female cowbird to produce a large number of eggs, in some cases as many as 30 in a single reproductive season.

When the presence of brood parasites decreases the reproductive rate of the host species, selection favors host individuals that detect and eject foreign young. Some brood parasites counter this host defense by producing eggs that closely resemble those of their host.

Many insects are brood parasites. For example, the Alcon blue butterfly outsources care of its young to ants. Caterpillars of these butterflies smell and sound like ants. Worker ants, fooled by these false cues, carry the caterpillars into their nest, where ants care for them as if they were members of the colony, feeding them and protecting them from predators.

Parasitoids

Parasitoids are insects that lay their eggs in the bodies of other insects. Larvae that hatch from these eggs develop in the host's body, devour its tissue, and eventually kill it. The presence of parasitoids reduces the size of a host population in two ways. First, as the parasitoid larvae grow inside their host, they withdraw nutrients and prevent the host from developing normally and reproducing. Second, the presence of these larvae eventually leads to the death of the host.

Biological Pest Controls

Biological pest control is the practice of using a pest's natural enemies to reduce its numbers. Commercially raised parasites and parasitoids are often used as agents of biological pest control (**FIGURE 45.15**). The method of pest control has some advantages over pesticides. Most chemical insecticides kill a wide variety of insects, including helpful species. Insecticides also have negative effects on human health. By contrast, the parasites and parasitoids usually used as biological control agents target only a limited number of species. For a species to be an effective biological control, it must be adapted to take advantage of a specific host species and to survive in that species' habitat. The ideal biological control agent excels at finding the target host species, has a population growth rate comparable to the host's, and has offspring that disperse widely.

Introducing a species into a community as a biological control agent always entails some risks. The introduced control agent sometimes attacks nontargeted species in addition to, or instead of, those that they were expected to control. For example, some parasitoid wasps were introduced to Hawaii to control stinkbugs that feed on some Hawaiian crops. Instead, the

FIGURE 45.14 **A parasitic plant.** The orange strands are dodder (*Cuscuta*), a parasitic flowering plant, that has almost no chlorophyll. Dodder's modified roots penetrate a host plant's vascular tissues and absorb water and nutrients from them.

FIGURE 45.15 **Biological control agent.** A commercially raised parasitoid wasp about to deposit a fertilized egg in an aphid.

biological pest control Use of a pest's natural enemies to reduce its numbers.
brood parasitism One egg-laying species benefits by having another raise its offspring.
camouflage Body coloration, patterning, form, or behavior that helps predators or prey blend with the surroundings and possibly escape detection.
herbivory An animal feeds on plant parts.
parasitism Relationship in which one species withdraws nutrients from another species, without immediately killing it.
parasitoid An insect that lays eggs in another insect, and whose young devour their host from the inside.

Data Analysis Activities

Testing Biological Control Biological control agents are used to battle red imported fire ants. Researchers have enlisted the help of *Thelohania solenopsae*, a natural enemy of the ants. This microsporidian (Section 23.4) is a parasite that infects ants and shrinks the ovaries of the colony's egg-producing female (the queen). As a result, a colony dwindles in numbers.

Are these biological controls useful against imported fire ants? To find out, USDA scientists treated infested areas with either traditional pesticides or pesticides plus biological controls (both flies and the parasite). The scientists left some plots untreated as controls. FIGURE 45.16 shows the results.

1. How did population size in the control plots change during the first four months of the study?
2. How did population size in the two types of treated plots change during this same interval?
3. If this study had ended after the first year, would you conclude that biological controls had a major effect?
4. Would your conclusion differ at the end of the time period shown?

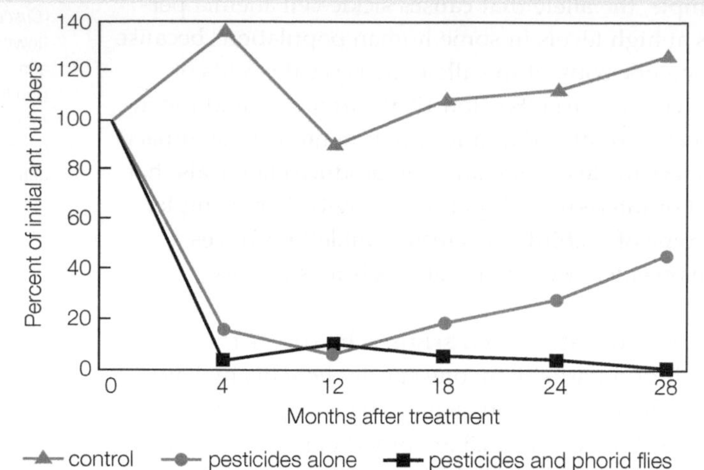
Percent of initial ant numbers
140
120
100
80
60
40
20
0
0
4
12
18
24
28
Months after treatment
control
pesticides alone
pesticides and phorid flies

FIGURE 45.16 A comparison of two methods of controlling red imported fire ants. The graph shows the numbers of red imported fire ants over a 28-month period. Orange triangles represent untreated control plots. Green circles are plots treated with pesticides alone. Black squares are plots treated with pesticide and biological control agents (parasitoid flies and a microsporidian parasite).

parasitoids decimated the population of koa bugs, Hawaii's largest native bug. Introduced parasitoids have also been implicated in ongoing declines of many native Hawaiian butterfly and moth populations.

TAKE-HOME MESSAGE 45.7

✔ Parasites reduce the reproductive rate of host individuals by withdrawing nutrients from them.

✔ Brood parasites reduce the reproductive rate of hosts by tricking them into caring for young that are not their own.

✔ Parasitoids reduce the number of host organisms by preventing reproduction and eventually killing the host.

45.8 How Communities Change

LEARNING OBJECTIVES

- Distinguish between primary and secondary succession.
- Explain why the exact course of succession can be unpredictable.
- Describe how the frequency and magnitude of disturbance influences the species richness of a community.
- Explain how indicator species can be used to monitor the environment.

Ecological Succession

The array of species in a community can change over time. Species often alter the habitat in ways that allow other species to come in and replace them. We call this type of change ecological succession. There are two types of ecological succession: primary succession and secondary succession.

Primary succession is a process that begins when pioneer species colonize a barren habitat with no soil, such as a new volcanic island or land exposed by the retreat of a glacier. **Pioneer species**, which often include mosses and lichens, are small, have a brief life cycle, and can tolerate intense sunlight, extreme temperature changes, and little or no soil. Some hardy annual flowering plants with wind-dispersed seeds are also frequent pioneers.

Many pioneers help build and improve the soil. In doing so, they often set the stage for their own replacement. For example, some pioneer species partner with nitrogen-fixing bacteria, so these species can grow in nitrogen-poor habitats. Seeds of later arrivals find shelter in mats of low-growing pioneer vegetation. Over time, the organic remains that accumulate add volume and nutrients to soil, thereby helping other species to become established. Later successional species often shade and eventually displace earlier ones.

In **secondary succession**, a disturbed area within a community recovers. For example, secondary succession occurs in abandoned agricultural fields and burned forests. Because improved soil is present from the start, secondary succession usually occurs more quickly than primary succession.

A The 1980 Mount Saint Helens eruption obliterated the community at the base of this Cascade volcano.

B The first pioneer species arrived less than a decade after the eruption.

FIGURE 45.17 A natural laboratory for studies of succession.

When the concept of ecological succession was first developed in the late 1800s, it was thought to be a predictable and directional process. Which species are present at each stage in succession was thought to be determined primarily by physical factors such as climate, altitude, and soil type. In this view, succession culminates in a "climax community," an array of species that persists over time and will be reconstituted in the event of a disturbance.

Ecologists now know that the species composition of a community changes in unpredictable ways. Communities do not journey along a well-worn path to a predetermined climax state. Random events determine the order in which species arrive in a habitat, and thus affect the course of succession.

Ecologists had an opportunity to investigate how random factors influence succession after the 1980 eruption of Mount Saint Helens leveled about 600 square kilometers (235 square miles) of forest in Washington State (FIGURE 45.17). They recorded the pattern of colonization and carried out experiments in plots inside the blast zone. The results of these and other studies showed that the types of pioneers that arrive first (a random event) influences which species are most likely to follow. Some pioneer species make a habitat more hospitable to some later successional species and less welcoming for others.

Effects of Disturbance

Physical and biological disturbances can alter the species composition of a community. The **intermediate disturbance hypothesis** states that species richness is greatest when physical and biological disturbances are moderate in their intensity or frequency (FIGURE 45.18). When disturbance is infrequent and of low intensity (does not remove many individuals), the most competitive species will exclude others, so diversity is low. By contrast, when disturbance occurs often or is of high intensity, most species present will be colonizers. With a moderate level of disturbance, the community will contain a mix of early and late successional species, and so will be most diverse.

In communities that repeatedly experience a particular type of disturbance, individuals that withstand or benefit from that disturbance have a selective advantage. For example, some plants in areas subject to periodic fires produce seeds that germinate only after a fire. Seedlings of these plants benefit from the lack

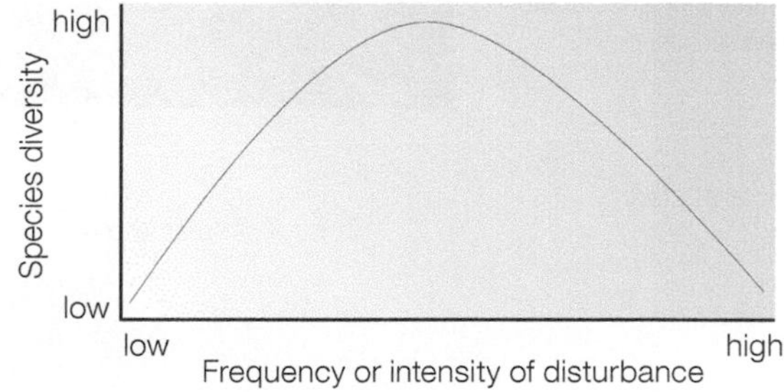

high
Species diversity
low
low
high
Frequency or intensity of disturbance

FIGURE 45.18 Intermediate disturbance hypothesis.

intermediate disturbance hypothesis Species richness is greatest in communities where disturbances are moderate in their intensity or frequency.
pioneer species Species that can colonize a new habitat.
primary succession A new community colonizes an area where there is no soil.
secondary succession A new community develops in a disturbed site where the soil that supported a previous community remains.

CREDITS: (17A) R. Barrick/USGS; (17B) USGS; (18) © Cengage Learning.

of competition for resources in newly burned areas. Other plants have an ability to resprout fast after a fire (**FIGURE 45.19**). Because different species respond differently to fire, the frequency of this disturbance affects competitive interactions. Human suppression of naturally occurring fires can alter the composition of a biological community. In the absence of fires, species whose numbers would otherwise be suppressed by fire can become dominant.

FIGURE 45.19 Adapted to disturbance. Some woody shrubs, such as this toyon, resprout from their roots after a fire. In the absence of occasional fire, toyons are outcompeted and displaced by species that grow faster but are less fire resistant.

FIGURE 45.20 The effect of sea star removal on species richness in tide pools. Experimental removal of the sea star *Pisaster* (right) from plots resulted in a decline in species richness (brown dots and line). Control plots from which *Pisaster* was not removed showed no comparable decline in species richness (green dots). These data indicate that the sea star is a keystone species.

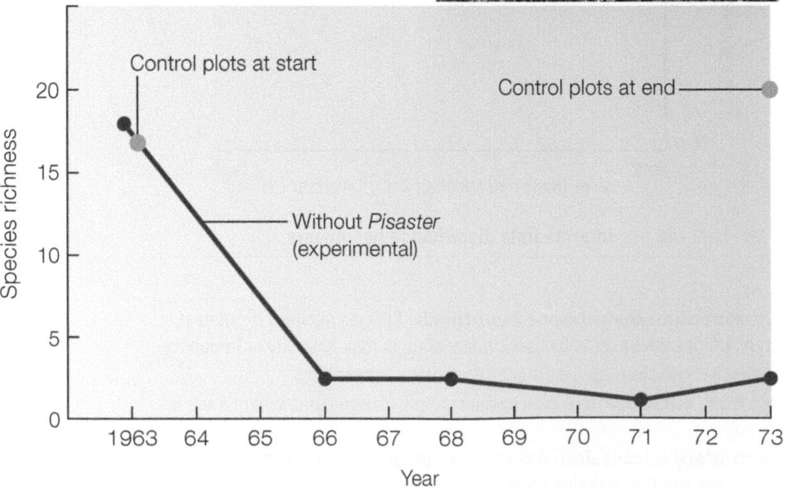

Control plots at start
Control plots at end
Without *Pisaster* (experimental)
Species richness
0
5
10
15
20
1963
64
65
66
67
68
69
70
71
72
73
Year

Indicator species are species that are especially intolerant of physical disturbance to their environment. They are the first to decline or disappear when conditions change, so they can provide an early warning of environmental degradation. For example, a decline in a trout population can be an early sign of problems in a stream, because trout are highly sensitive to pollutants and they cannot tolerate low oxygen levels. Similarly, some lichens that are intolerant of air pollution serve as indicators of air quality.

The Role of Keystone Species

A **keystone species** is a species that has a disproportionately large effect on a community relative to its abundance. Robert Paine coined this term after observing what happened after he removed one species (a sea star) from the rocky intertidal community on the California coast. The sea star, *Pisaster ochraceus*, preys on chitons, limpets, barnacles, and mussels. When Paine experimentally removed *Pisaster* from some plots, mussels took over, crowding out seven other species of invertebrates and reducing the diversity of algae living in the plots (**FIGURE 45.20**). In control plots, where *Pisaster* remained, there was no similar decline in diversity. Paine concluded that this sea star is a keystone species. It normally keeps species diversity in the intertidal zone high by preventing competitive exclusion of other invertebrates and of algae by mussels.

Keystone species that exert their effect by physically modifying a habitat, rather than through competition or food web effects, are usually referred to as "ecosystem engineers." Beavers are an example. A beaver is a large herbivorous rodent that cuts down trees by gnawing through their trunks. It uses the trees it fells to build a dam. Construction of a beaver dam creates a deep pool where a shallow stream would otherwise exist. By altering the physical conditions in a section of the stream, the beaver affects the types of fish and aquatic invertebrates that can live there.

Species Introductions

The arrival of a new species in a community can also cause dramatic changes. When you hear someone speaking enthusiastically about exotic species, you can safely bet the speaker is not an ecologist. An **exotic species** is a resident of an established community that dispersed from its home range and became established elsewhere. Many species do not do well outside their home range, but an exotic species becomes a permanent member of its new community. By contrast, an

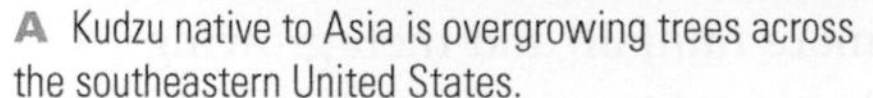

A Kudzu native to Asia is overgrowing trees across the southeastern United States.

B Gypsy moths native to Europe and Asia feed on oaks through much of the United States.

C Nutrias from South America are now in the Gulf States and in the Pacific Northwest.

FIGURE 45.21 Exotic species that are altering natural communities in the United States. To learn more visit the National Invasive Species Information Center online at www.invasivespeciesinfo.gov.

endemic species is one that evolved in a particular community and exists nowhere else.

The United States is now home to more than 4,500 exotic species. An estimated 25 percent of Florida's plant and animal species are exotics. In Hawaii, 45 percent are exotic. Some species were brought in for use as food crops, to brighten gardens, or to provide textiles. Others, including red imported fire ants, arrived with cargo from distant regions.

One of the most notorious exotic species is a vine called kudzu (*Pueraria lobata*). Native to Asia, it was introduced to the American Southeast as a food for grazers and to control erosion. However, it quickly became an invasive weed. Kudzu overgrows trees, telephone poles, houses, and almost everything else in its path (**FIGURE 45.21A**). Kudzu also poses a threat to agriculture; it serves as the main winter host for soybean rust, a fungal pathogen that can drastically reduce soybean yields.

Gypsy moths (*Lymantria dispar*) are an exotic species native to Europe and Asia. They entered the northeastern United States in the mid-1700s and now range into the Southeast, Midwest, and Canada. Gypsy moth caterpillars (**FIGURE 45.21B**) preferentially feed on oaks. Loss of leaves to gypsy moths can weaken trees, making them more susceptible to parasites and disease.

Large semiaquatic rodents called nutrias (*Myocastor coypus*) were imported from South America to be grown for their fur. They were released into the wild in many states. Along the Gulf of Mexico, nutrias now thrive in freshwater marshes (**FIGURE 45.21C**), and their voracious appetite threatens the native vegetation. In addition, their burrowing contributes to marsh erosion and damages levees, increasing risk of flooding.

TAKE-HOME MESSAGE 45.8

- ✔ Succession is a process in which one array of species replaces another over time. It can occur in a barren habitat (primary succession) or a region where a community previously existed (secondary succession). Physical factors affect succession, but so do species interactions and disturbances.
- ✔ Species diversity is highest in communities with an intermediate frequency and magnitude of disturbances. A change in the typical frequency of a disturbance can favor some species over others.
- ✔ Removing a single keystone species can have a disproportionately large effect on community structure.
- ✔ Introducing an exotic species can threaten native species.

endemic species (en-DEM-ick) A species found in the region where it evolved and nowhere else.

exotic species A species that evolved in one community and later became established in a different one.

indicator species A species that is especially sensitive to disturbance and can be monitored to assess the health of a habitat.

keystone species A species that has a disproportionately large effect on community structure.

CREDITS: (21A) Angelina Lax/Science Source; (21B) Photo by Scott Bauer, USDA/ARS; (21C) © Greg Lasley Nature Photography, www.greglasley.net.

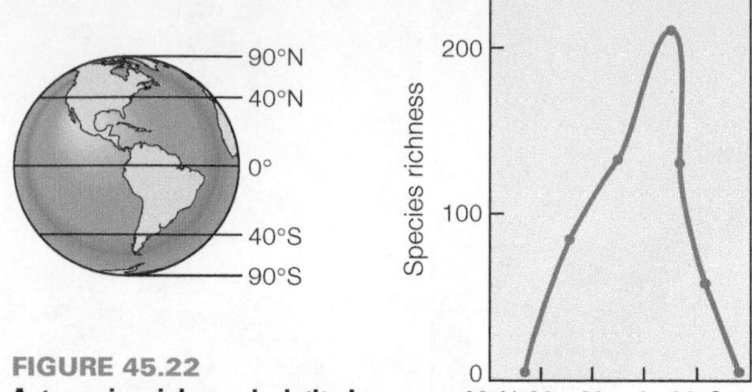
90°N
40°N
0°
40°S
90°S
Species richness
200
100
0
90°N 60 30 0 30°S 60

FIGURE 45.22
Ant species richness by latitude.

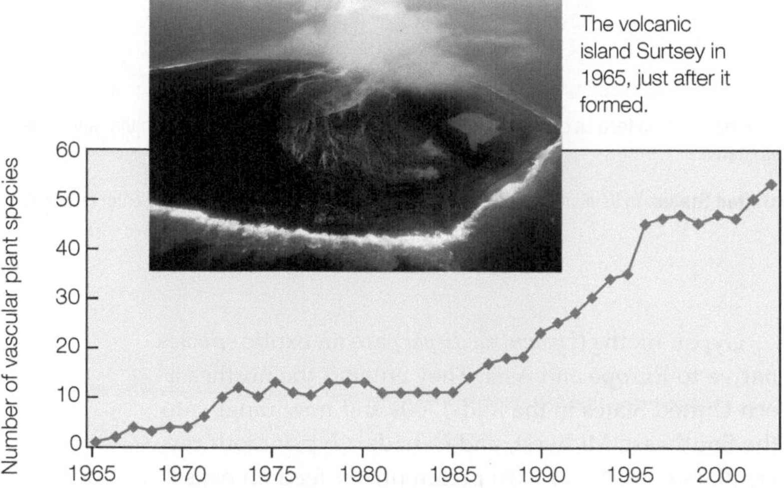
The volcanic island Surtsey in 1965, just after it formed.
Number of vascular plant species
60
50
40
30
20
10
0
1965 1970 1975 1980 1985 1990 1995 2000

FIGURE 45.23 Vascular plant colonization of Surtsey. Seagulls began nesting on the island in 1986.

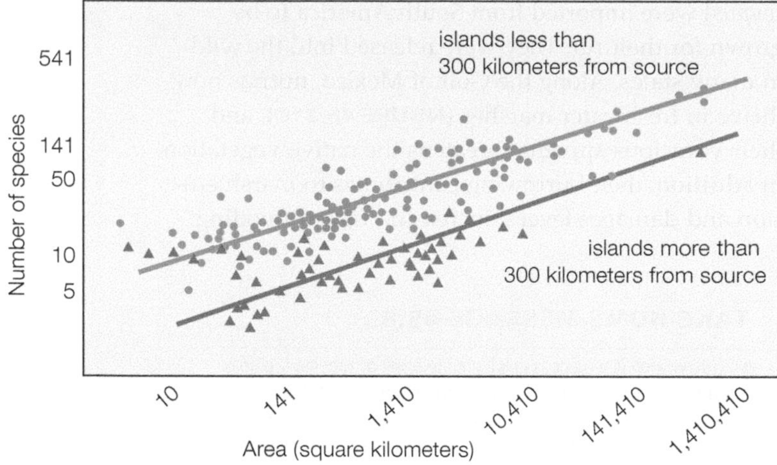
islands less than 300 kilometers from source
islands more than 300 kilometers from source
Number of species
541
141
50
10
5
10
141
1,410
10,410
141,410
1,410,410
Area (square kilometers)

FIGURE 45.24 Island biodiversity patterns.

Distance effect: Species richness on islands of a given size declines as distance from a source of colonists rises. Green circles are values for islands less than 300 kilometers from the colonizing source. Orange triangles are values for islands more than 300 kilometers (190 miles) from a source of colonists.

Area effect: Among islands the same distance from a source of colonists, larger islands tend to support more species than smaller ones.

45.9 Biogeographic Patterns in Community Structure

LEARNING OBJECTIVES

- Describe the formation and colonization of the island of Surtsey.
- Explain how island size and area affect the number of species on the island.

Latitudinal Patterns

Biogeography is the scientific study of the geographic distribution of species (Section 16.2). Perhaps the most striking pattern of species richness corresponds with distance from the equator. For most major plants and animal groups, the number of species is greatest in the tropics and declines from the equator to the poles (**FIGURE 45.22**). Consider just a few factors that help bring about and maintain this pattern.

First, tropical latitudes intercept more intense sunlight and receive more rainfall, and their growing season is longer. As one outcome, resource availability tends to be greatest and most reliable in the tropics, providing more opportunities for resource sharing and character displacement. Thus, the tropics support a degree of specialized interrelationships not possible where species are active for shorter periods.

Second, tropical communities have been established for a long time. Some temperate communities did not start forming until the end of the last ice age. The longer a community has been established, the more time there has been for speciation to occur within it.

Third, species richness may be self-reinforcing. The number of species of trees in tropical forests is much greater than in comparable forests at higher latitudes. Where more plant species compete and coexist, more species of herbivores also coexist, partly because no single herbivore species can overcome all the defenses of all plants. In addition, more predators and parasites can evolve in response to more kinds of prey and hosts. The same principles apply to tropical reefs.

Island Patterns

In the mid-1960s, volcanic eruptions formed a new island 33 kilometers (21 miles) from the coast of Iceland. The island was named Surtsey (**FIGURE 45.23**). Bacteria and fungi were early colonists. The first vascular plant became established on the island in 1965. Mosses appeared two years later and thrived. The

area effect Larger islands have more species than small ones.
distance effect Islands close to a mainland have more species than those farther away.
equilibrium model of island biogeography Model that predicts the number of species on an island based on the island's area and distance from the mainland.

first lichens were found five years after that. The rate of arrivals of new vascular plants picked up after a seagull colony became established in 1986.

The number of species on Surtsey will not continue increasing forever. How many species will there be when the number levels off? The **equilibrium model of island biogeography** addresses this question. According to this model, the number of species living on any island reflects a balance between immigration rates for new species and extinction rates for established ones. The distance between an island and a mainland source of colonists affects immigration rates. An island's size affects both immigration rates and extinction rates.

Consider first the **distance effect**: Islands far from a source of colonists receive fewer immigrants than those closer to a source. Most species cannot disperse very far, so they will not turn up far from a mainland.

Species richness is also shaped by the **area effect**: Big islands tend to support more species than small ones for several reasons. First, more colonists will happen upon a larger island simply by virtue of its size. Second, larger islands are more likely to offer a variety of habitats, such as high and low elevations. This variety makes it more probable that a new arrival will find a suitable habitat. Finally, big islands can support larger populations of species than small islands. The larger a population, the less likely it is to become locally extinct as the result of some random event.

FIGURE 45.24 illustrates how interactions between the distance effect and the area effect can influence the equilibrium number of species on islands.

Robert H. MacArthur and Edward O. Wilson first developed the equilibrium model of island biogeography in the late 1960s. Since then the model has been modified and its use has been expanded to help scientists understand what happens on habitat islands—areas of natural habitat surrounded by a "sea" of habitat that has been disturbed by humans. Many parks and wildlife preserves fit this description. Island-based models can help estimate the size of an area that must be set aside as a protected reserve to ensure survival of a species or a community.

TAKE-HOME MESSAGE 45.9

✔ Species richness of communities is highest in the tropics and lowest at the poles. Tropical habitats have the highest primary production, and tropical communities are often older than temperate ones.

✔ When a new island forms, species richness rises over time and then levels off. The size of an island and its distance from a colonizing source influence the number of species it supports when it reaches equilibrium.

45.1 Fighting Foreign Fire Ants (revisited)

Red imported fire ants first arrived in the southeast United States in the 1930s when some of the ants traveled as stowaways on a cargo ship. Since then, the species has gradually expanded its range across the south and has been accidentally introduced to California and New Mexico. More recently, it became established in the Caribbean, Australia, New Zealand, and several countries in Asia. Genetic comparisons among *S. invicta* populations revealed that the ants involved in these recent introductions originated in the southeastern United States, rather than South America.

Given the problems that the imported ants are causing in the United States, you might wonder what things are like in their native South America. The ants are not considered much of a concern there, in part because they are far less common. In that community coevolved parasites, predators, and diseases keep the ants' numbers in check.

Scientists have been testing some phorid flies from Brazil for use as a biological control agent (FIGURE 45.25A). The phorid flies are parasitoids, and *S. invicta* is their host. The female fly lays an egg in an ant's body. The egg hatches into a larva that grows in the ant and feeds on its soft tissues. Eventually the larva enters the ant's head and causes it to fall off (FIGURE 45.25B). The larva then undergoes metamorphosis to its adult form inside this somewhat gruesome changing room.

A Phorid fly attempting to lay its egg in a fire ant.

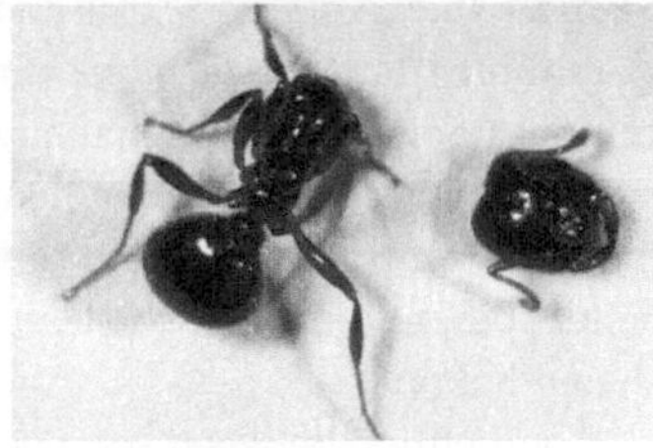

B Ant that lost its head after a phorid fly larva matured inside it.

FIGURE 45.25 **Phorid flies as biological control agents.**

Phorid flies are not expected to kill off all *S. invicta* in affected areas. Rather, the hope is that the flies will reduce the density of the invader's colonies. Ecologists are also exploring other options, such as introducing pathogens that infect *S. invicta* but not native ants.

CREDITS: (25A) Photo by Sanford Porter/USDA; (25B) USDA.

STUDY GUIDE

Section 45.1 All the species that live in an area constitute a **community**. Introduction of a new species to a community can have negative effects on the species that evolved in that community. Global trade has distributed species such as red imported fire ants from Brazil to new habitats where they are pests. Scientists have now imported some of the ants' natural enemies in attempts to control it.

Section 45.2 Each species occupies a certain **habitat** that is characterized by physical features and by the array of species living in it. The species diversity of a community has two components: **species richness** and **species evenness**.

Species interactions affect community structure. Direct species interactions are described in terms of their effect on the two participants. For example, **commensalism** is a direct interspecific interaction in which one species benefits and the other is unaffected.

A **symbiosis** is an interaction in which one species lives in or on another. The symbiotic interaction can be commensalism, mutualism, or parasitism.

Section 45.3 A **mutualism** is a direct interspecific interaction that benefits both participants. Some mutualists cannot complete their life cycle without the interaction. In others, the mutualism is beneficial but not essential.

Mutualism involves mutual exploitation. Mutualists who maximize their own benefits while limiting the cost of cooperating are at a selective advantage.

Section 45.4 A species' roles and requirements in its community define its **ecological niche**. With **interspecific competition**, two species that have similar niches are harmed by one another's presence. When two species both require the same limiting resource, **competitive exclusion** occurs. The better competitor drives the lesser one to extinction in their shared habitat.

Competition between species that have similar requirements results in directional selection. Over time, the competing species become less similar, an effect called **character displacement**. This divergence in traits facilitates **resource partitioning**.

Sections 45.5, 45.6 **Predation** benefits a predator at the expense of the prey it captures, kills, and eats. The functional response of predators to a rise in prey density varies with the type of predator.

Predators and their prey exert selective pressure on one another. Both predators and prey can benefit from **camouflage**. Some well-defended prey have **warning coloration**. With **mimicry**, well-defended species evolve a similar appearance or vulnerable species come to resemble better-defended ones.

With **herbivory**, an animal eats a plant or plant parts. Some plants have traits such as thorns that discourage herbivory. Others withstand herbivory by making a quick recovery from the loss of shoots.

Section 45.7 **Parasitism** involves feeding on a living host. It benefits the parasite at the expense of the host. Even when a parasite does not immediately kill a host, its effects can decrease the host population.

Parasitoids are insects that lay eggs in or on a host insect. Parasites and parasitoids are often used in **biological pest control**.

Animals that steal parental care from another species are referred to as **brood parasites**.

Section 45.8 Ecological succession is the sequential replacement of one array of species by another over time. **Primary succession** happens in newly formed, vacant habitats. **Secondary succession** occurs in disturbed ones. In both types of succession, the first species to become established are **pioneer species**. The pioneers may help, hinder, or have no effect on later colonists.

The idea that all communities eventually reach a predictable climax state has been replaced by models that emphasize the role of chance and disturbances. According to the **intermediate disturbance hypothesis**, disturbances of moderate intensity and frequency maximize species diversity. Some species are adapted to a particular type of periodic disturbance such as wildfires. Such species are at a competitive disadvantage if the disturbance stops occurring. An **indicator species** is especially sensitive to environmental change, so its presence or absence can provide early information about an environment's degradation.

Keystone species are especially important in maintaining the composition of a community. The removal of a keystone species or introduction of an **exotic species**—one that evolved in a different community—can alter community structure permanently. An **endemic species** is one that exists only in a limited geographic region, the region where it arose.

Section 45.9 Species richness, the number of species in a given area, varies with latitude. Species diversity tends to be greatest at low latitudes (near the equator).

The **equilibrium model of island biogeography** predicts the number of species that an island will sustain based on the **area effect** and the **distance effect**. This model predicts that larger islands will hold more species than smaller ones, and that islands closer to a source of colonists will have more species than islands more distant from such a source. Scientists can use this model to predict the number of species that islands of habitat such as parks can sustain.

SELF-QUIZ

Answers in Appendix VII

1. The type of physical environment in which a species typically lives is its ________.
   a. niche c. community
   b. habitat d. population

2. Which cannot be a symbiosis?
   a. mutualism c. commensalism
   b. parasitism d. interspecific competition

3. A tick is a(n) ________.
   a. parasitoid b. ectoparasite c. endoparasite

4. ________ can lead to resource partitioning.
   a. Mutualism c. Commensalism
   b. Parasitism d. Interspecific competition

5. Match the terms with the most suitable descriptions.
   ___ mutualism
   ___ parasitism
   ___ commensalism
   ___ predation
   ___ interspecific competition

   a. one free-living species feeds on another and usually kills it
   b. two species interact and both benefit by the interaction
   c. two species interact and one benefits while the other is neither helped nor harmed
   d. one species feeds on another that it lives on or in
   e. species both need a resource

6. Lizards that eat flies they catch on the ground and birds that catch and eat flies in the air are engaged in ________ competition.
   a. exploitative d. interspecific
   b. interference e. both a and d
   c. intraspecific f. both b and c

7. By a currently favored hypothesis, species richness of a community is greatest when physical disturbances are of ________ intensity and frequency.
   a. low c. high
   b. intermediate d. variable

8. With ________, one species evolves to look like another.

9. Growth of a forest in an abandoned corn field is an example of ________.
   a. primary succession c. secondary succession
   b. resource partitioning d. competitive exclusion

10. Species richness is greatest in communities ________.
    a. near the equator c. near the poles
    b. in temperate regions d. that formed recently

11. If you remove a species from a community, the population size of its main ________ is likely to increase.
    a. parasite b. competitor c. predator

12. ________ steal parental care.
    a. Mutualists c. Brood parasites
    b. Commensalists d. Predators

13. The oldest established land communities are ________.
    a. in the Arctic c. in the tropics
    b. in temperate zones d. on volcanic islands

14. Biological control of pest species ________.
    a. has no side effects c. uses natural enemies
    b. involves mutualists d. requires use of chemicals

15. Match the terms with the most suitable descriptions.
    ___ area effect
    ___ pioneer species
    ___ indicator species
    ___ keystone species
    ___ exotic species
    ___ resource partitioning

    a. native species with large effect
    b. first species established in a new habitat
    c. more species on large islands than small ones at same distance from the source of colonists
    d. species that is especially sensitive to changes in the environment
    e. allows competitors to coexist
    f. often outcompete, displace native species of established community

CRITICAL THINKING

1. With antibiotic resistance rising, researchers are looking for ways to reduce use of these drugs. Some cattle once fed antibiotic-laced food now get probiotic feed that can bolster populations of helpful bacteria in the animal's gut. The idea is that if a large population of beneficial bacteria is in place, then harmful bacteria cannot become established or thrive. Which ecological principle is guiding this practice?

2. Flightless birds on islands often have relatives on the mainland that can fly. The island species presumably evolved from fliers that, in the absences of predators, lost their ability to fly. Many island populations of flightless birds are in decline because rats have been introduced to their previously isolated island habitats. Despite current selection pressure in favor of flight, no flightless island bird species has regained the ability to fly. Why is this unlikely to happen?

3. Some attempts to use biological controls prove ineffective. Although the introduced control agent targets only the expected pest species, the control agent does not take hold and spread in its new habitat. Describe some community interactions that could slow or prevent an introduced biological control agent from becoming established in a new habitat.

46 ECOSYSTEMS

CORE CONCEPTS

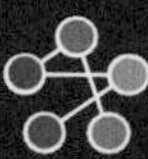

Systems

Complex properties arise from interactions among components of a biological system.
The interactions among organisms and between organisms and nonliving components of their environment give rise to the properties of an ecosystem. Ecosystems and their distribution can be affected by environmental catastrophes and geologic events. Human activities accelerate change in ecosystems at the local and global level.

Pathways of Transformation

Organisms exchange matter and energy with the environment in order to grow, maintain themselves, and reproduce.
In an ecosystem, materials and energy are transferred through food chains that connect as food webs. With each transfer or conversion, some energy disperses, so energy flows through an ecosystem in one direction. By contrast, substances such as water and nutrients continually cycle between the living and nonliving components of an ecosystem. Human activities can alter these cycles.

Process of Science

The field of biology consists of and relies upon experimentation and the collection and analysis of scientific evidence.
Nutrient cycles involve complex interactions between the living and nonliving components of ecosystems. Determining the effects of human activities on these cycles requires both measurements that detect ongoing changes and use of methods that provide insight into the state of the Earth in the distant past. Analysis of data from multiple sources helps researchers unravel cause-and-effect relationships in these complex biological systems.

Links to Earlier Concepts

This chapter builds on prior discussions of energy flow (Sections 1.3 and 5.2), carbon imbalances (6.1), algal blooms (21.5), plant nutrition (28.2), nitrogen fixation (20.7 and 28.3), and mycorrhizal fungi (23.7). Discussions of nutrient cycles and their disruption draw on your knowledge of isotopes (2.2), atomic bonds (2.4), and plate tectonics (16.5).

46.1 Too Much of a Good Thing

Carbon, phosphorus, and nitrogen all occur naturally in the environment and all are essential to life. Plants and other photosynthetic organisms obtain the carbon they need by taking up carbon dioxide from the air or water. Animals obtain carbon by eating plants or other animals. Plants meet their nitrogen and phosphorus requirements by taking up ionized forms of these elements from soil water. Animals then obtain phosphorus and nitrogen by eating plants or by eating other animals. Thus, carbon, phosphorus, and nitrogen taken up from the environment continually passes from one organism to another.

Although carbon dioxide, phosphorus, and nitrogen are present naturally, they can also be considered pollutants. A pollutant is a substance that disrupts natural processes when it is released into the environment by human activities.

Consider what happens when human activities increase the phosphate content of an aquatic habitat. In many areas, the soil contains little phosphorus, so it is the main factor limiting producer growth. This is why fertilizers typically include phosphorus. When the phosphorus from fertilized lawns or fields runs off, it pollutes rivers and lakes. Phosphate-containing detergents can be a source of phosphate pollution too. When these phosphate-rich compounds get into aquatic habitats, they have the same effect as fertilizer.

Addition of phosphate or some other nutrient to an aquatic ecosystem is called **eutrophication**. Eutrophication can occur slowly by natural processes, or fast as a result of human actions. In most lakes, the main limiting factor for aquatic algae and cyanobacteria is access to phosphorus. Thus, adding phosphorus to such a lake allows a population explosion of these organisms. The result is an algal bloom (Section 21.5), an explosion of growth that clouds water and can threaten other aquatic species.

We began using phosphate-rich products when we had no idea of their effects beyond greener lawns and cleaner clothes. Similarly, we began producing nitrogen fertilizers and releasing carbon dioxide from fossil fuels without understanding the potential effects of these activities. Today, we know that daily actions of millions of people can disrupt nutrient cycles that have been in operation since long before humans existed. Our species has a unique capacity to shape the environment to our liking. In doing so, we have become major players in global flows of energy and nutrients even before we fully comprehend how these systems work. ●

eutrophication Nutrient enrichment of an aquatic ecosystem.

CREDIT: (opposite) D. A. Rintoul.

46.2 The Nature of Ecosystems

LEARNING OBJECTIVES

- Distinguish between the producers and consumers in an ecosystem.
- Explain how the movement of energy in an ecosystem differs from the movement of nutrients.
- List some factors that influence primary productivity.
- Give an example of a food chain with four trophic levels.
- Explain what limits the length of food chains.

Overview of the Participants

An **ecosystem** is an array of organisms and their physical environment, all interacting through a one-way flow of energy and a cycling of nutrients. It is an open system, because it requires ongoing inputs of energy and nutrients to endure (**FIGURE 46.1**).

Earth's ecosystems are remarkably diverse in their characteristics. In climate, soil type, array of species, and other features, prairies differ from forests, which differ from tundra and deserts. Reefs differ from the open ocean, which differs from streams and lakes. Yet, despite these many differences, all ecosystems are alike in many aspects of their structure and function.

All ecosystems run on energy captured by **primary producers**. These autotrophs, or "self-feeders," obtain energy from a nonliving source—generally sunlight—and use it to build organic compounds from inorganic starting materials. Plants and phytoplankton are the main producers. Chapter 6 explained how they capture energy from the sun and use it in photosynthesis to build sugars from carbon dioxide and water.

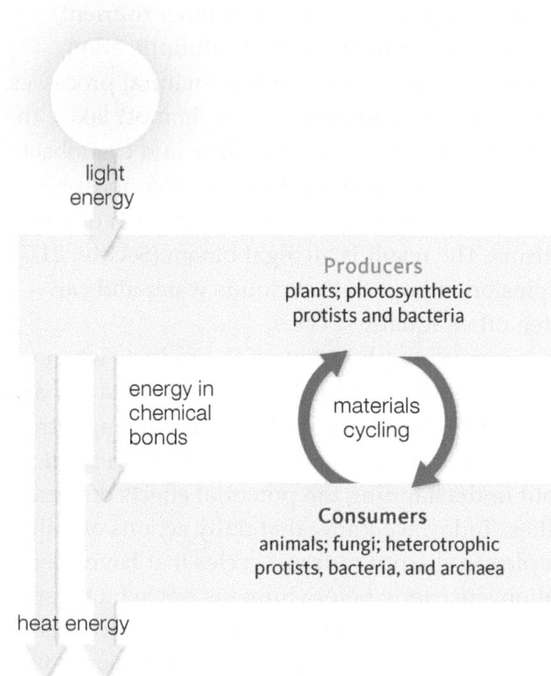
light
energy
Producers
plants; photosynthetic
protists and bacteria
energy in
chemical
bonds
materials
cycling
Consumers
animals; fungi; heterotrophic
protists, bacteria, and archaea
heat energy

FIGURE 46.1 Energy flow and materials cycling. This model is for ecosystems on land, in which energy flow starts with autotrophs that capture energy from the sun. Energy flows one way, into and out of the ecosystem. Nutrients get cycled among producers and consumers.

Consumers are heterotrophs that obtain energy and carbon by feeding on tissues, wastes, and remains of producers and one another. We can describe consumers by their diets. Herbivores eat plants. Carnivores eat the flesh of animals. Omnivores devour both animal and plant materials. Parasites live inside or on a living host and feed on its tissues. **Detritivores**, such as earthworms and crabs, dine on small particles of organic matter, or detritus. **Decomposers** feed on organic wastes and remains and break them down into inorganic building blocks. The main decomposers are bacteria and fungi.

Energy flows one way—into an ecosystem, through its many living components, then back to the physical environment (Section 5.2). Light energy captured by producers is converted to chemical bond energy in organic molecules, which is then released by metabolic reactions that give off heat. This is a one-way process because heat energy cannot be recycled; producers cannot convert heat into energy stored in chemical bonds.

In contrast, nutrients can be cycled continuously within an ecosystem. The cycle begins when producers take up hydrogen, oxygen, and carbon from inorganic sources, such as the air and water. Producers also take up dissolved nitrogen, phosphorus, and other minerals needed to make organic compounds. Nutrients move from producers into the consumers who eat them. After an organism dies, the action of decomposers releases nutrients from its remains into the environment, where they are once again taken up by producers.

Not all nutrients remain in an ecosystem; typically there are gains and losses. Mineral ions are added to an ecosystem when weathering processes break down rocks, and when winds blow in mineral-rich dust from elsewhere. Leaching and soil erosion can remove minerals (Section 28.2). Gains and losses of each mineral tend to balance out over time in most ecosystems.

Trophic Structure of Ecosystems

All organisms in an ecosystem take part in a hierarchy of feeding relationships called **trophic levels** (*troph* means "nourishment"). When one organism eats another, energy stored in chemical bonds is transferred from the eaten to the eater. All organisms at the same trophic level in an ecosystem are the same number of transfers away from the energy input into that system.

A **food chain** is a sequence of steps by which some energy captured by primary producers is transferred to

hawk

Fourth Trophic Level
Third-level consumer

sparrow

Third Trophic Level
Second-level consumer

grasshopper

Second Trophic Level
Primary consumer

big bluestem grass

First Trophic Level
Primary producer

FIGURE 46.2 Example of a food chain and corresponding trophic levels in tallgrass prairie, Kansas.

organisms at successively higher trophic levels. For example, big bluestem grass and other plants are the major primary producers in a tallgrass prairie (**FIGURE 46.2**). They are at this ecosystem's first trophic level. In one food chain, energy flows from bluestem grass to grasshoppers, to sparrows, and finally to hawks. Grasshoppers are primary consumers; they are at the second trophic level. Sparrows that eat grasshoppers are second-level consumers and at the third trophic level. Hawks are third-level consumers, and they are at the fourth trophic level.

At each trophic level, organisms interact with the same sets of predators, prey, or both. Omnivores feed at several levels, so we would partition them among different levels or assign them to a level of their own.

TAKE-HOME MESSAGE 46.2

- ✔ An ecosystem includes a community of organisms that interact with their physical environment by a one-way energy flow and a cycling of materials.
- ✔ Autotrophs tap into an environmental energy source and make their own organic compounds from inorganic raw materials. They are the ecosystem's primary producers.
- ✔ Autotrophs are at the first trophic level of a food chain, a linear sequence of feeding relationships that proceeds through one or more levels of heterotrophs, or consumers.

consumer Organism that gets energy and nutrients by feeding on tissues, wastes, or remains of other organisms; a heterotroph.
decomposer Organism that feeds on biological remains and breaks organic material down into its inorganic subunits.
detritivore Consumer that feeds on small bits of organic material.
ecosystem A community interacting with its environment.
food chain Description of who eats whom in one path of energy flow in an ecosystem.
primary producer In an ecosystem, an organism that captures energy from an inorganic source and stores it as biomass; first trophic level.
trophic level (TROE-fick) Position of an organism in a food chain.

46.3 The Nature of Food Webs

LEARNING OBJECTIVES

- Distinguish between the two types of food webs, and give examples of the habitats in which each predominates.
- Explain the factors that limit the length of food chains.

An organism that participates in one food chain usually takes part in others as well. Food chains of an ecosystem cross-connect as a **food web**. FIGURE 46.3 shows some participants in an arctic food web.

Ecosystems typically support two types of food webs. In **grazing food webs**, most primary producers are eaten by primary consumers. In **detrital food webs**, most energy in producers flows directly to detritivores.

Detrital food webs tend to predominate in most land ecosystems. For example, in an arctic ecosystem, primary consumers such as voles, lemmings, and hares eat some living plant parts. However, far more plant matter becomes detritus. Bits of dead plant material sustain detritivores such as roundworms and decomposers that include soil bacteria and fungi.

human (Inuk)
arctic wolf
arctic fox
Higher Trophic Levels
Carnivores feed on herbivores and, in some cases, on other carnivores.
snowy owl
gyrfalcon
ermine
Second Trophic Level
Herbivores are the primary consumers.
arctic hare
lemming
vole
mosquito
flea
Parasitic consumers feed at more than one trophic level.
First Trophic Level
Low-growing plants are the main producers
grasses, sedges
purple saxifrage
arctic willow
Detritivores and decomposers (nematodes, annelids, saprobic insects, protists, fungi, bacteria)

FIGURE 46.3 **A very small sampling of organisms in an arctic food web on land.**

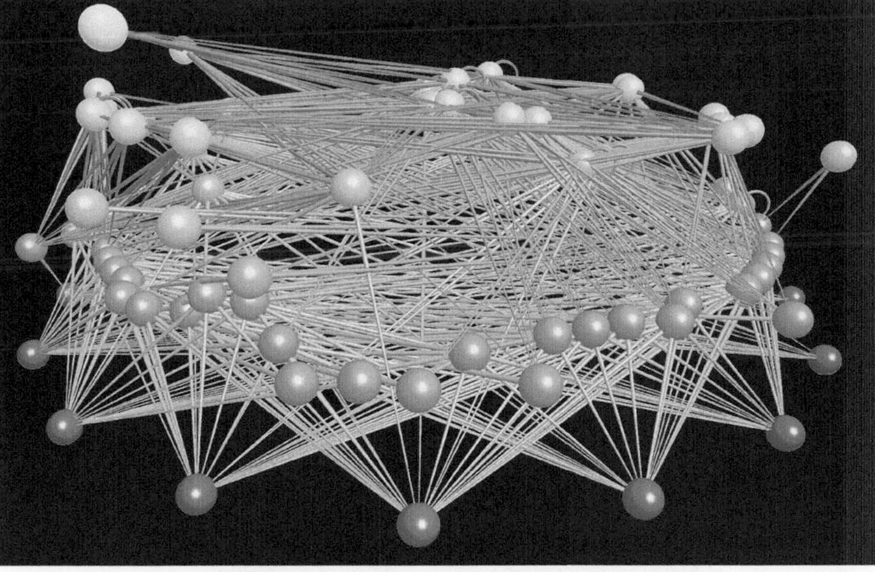

FIGURE 46.4 **Computer model (right) for a food web in East River Valley, Colorado (above).** Balls signify species. Their colors identify trophic levels, with producers (coded red) at the bottom and predators (yellow) at top. The connecting lines thicken, starting from an eaten species to the eater.

In marine ecosystems, most grazing food chains start with the consumption of phytoplankton (photosynthetic protists, archaea, and bacteria) by zooplankton (heterotrophic protists and tiny animals that drift or swim).

Detrital food chains and grazing food chains interconnect. For example, many animals at higher trophic levels eat both primary consumers and detritivores. Also, after consumers die, their tissues become food for detritivores and decomposers. Decomposers and detritivores also feed on the wastes from consumers.

How Many Transfers?

When ecologists looked at food webs for a variety of ecosystems, they discovered some common patterns. For example, the energy captured by producers usually passes through no more than four or five trophic levels. Even in ecosystems with many species, the number of transfers is limited. Remember that energy transfers are not very efficient (Section 5.2), so energy losses limit the length of a food chain.

Food chains tend to be shortest in habitats where conditions vary widely over time. Chains tend to be longer in stable habitats, such as the ocean depths where weather has no effect. The most complex webs tend to have a large variety of herbivores, as in grasslands. By comparison, food webs with fewer connections tend to have more carnivores.

Diagrams of food webs help ecologists predict how ecosystems will respond to change. Neo Martinez and his colleagues constructed the one shown in FIGURE 46.4. By comparing many food webs, they discovered that trophic interactions connect species more closely than previously thought. As Martinez concluded in a paper discussing his findings, "Everything is linked to everything else." He cautioned that extinction of any species in a food web may affect many other species.

TAKE-HOME MESSAGE 46.3

✔ Nearly all ecosystems include both grazing food webs and detrital food webs that interconnect as the system's food web. Tissues of living plants and other producers are the base for grazing food webs. By contrast, remains of producers are the base for detrital food webs.

✔ The cumulative energy losses from energy transfers between trophic levels limit the length of food chains.

46.4 Measuring Ecosystem Properties

LEARNING OBJECTIVES

- Distinguish between gross primary production and net primary production.
- Explain why an energy pyramid is always broadest at the bottom.

Primary Production

Flow of energy through an ecosystem begins with **primary production**: the rate at which producers capture energy (usually light energy) and convert it into chemical energy. The amount of energy captured by all the producers in an ecosystem is the system's gross

detrital food web (dee-TRITE-uhl) Food web in which most energy is transferred directly from producers to detritivores.
food web Set of cross-connecting food chains.
grazing food web Food web in which most energy is transferred from producers to grazers (herbivores).
primary production Rate at which an ecosystem's producers capture and store energy.

CREDIT: (4) left, Courtesy of Dr. Chris Floyd; right, Graphic created by FoodWeb3D program written by Rich Williams courtesy of the Webs on the Web project (www.foodwebs.org).

primary production. The portion of that energy used for growth and reproduction (rather than for maintenance) is the net primary production of the ecosystem. This is the energy available to the first-level consumers.

Factors that influence the rate of photosynthesis affect primary productivity. As a result, primary production differs among habitats and often varies seasonally (FIGURE 46.5). Per unit area, the net primary production on land tends to be higher than that in the oceans. However, because oceans cover about 70 percent of Earth's surface, marine producers account for nearly half of the global net primary production.

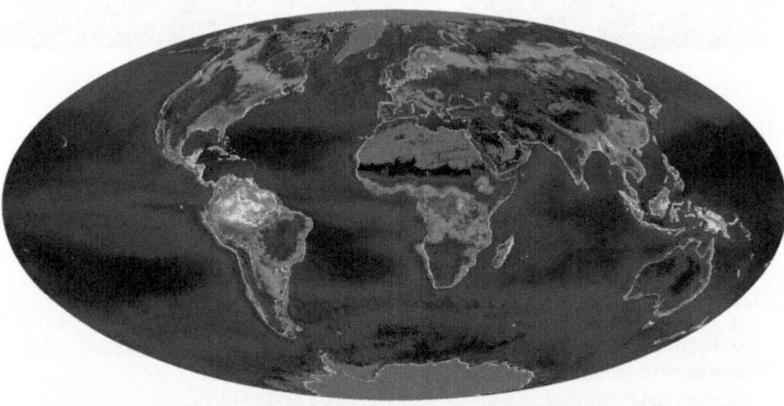

A Summary of annual net primary productivity on land and in the oceans.

North America
Atlantic Ocean in Winter
Africa
B
North America
Atlantic Ocean in Spring
Africa
C

B,C Seasonal changes in net primary productivity of the North Atlantic Ocean.

FIGURE 46.5 **Satellite data showing net primary production.** Productivity is coded as red (highest) down through orange, yellow, green, blue, and purple (lowest).

Ecological Pyramids

A food web diagram is one way of depicting the trophic relationships of species in a particular ecosystem. Ecological pyramid diagrams are another. In such diagrams, primary producers collectively form a base for successive tiers of consumers above them.

A **biomass pyramid** illustrates the dry weight of all organisms at each trophic level. FIGURE 46.6 shows the biomass pyramid for Silver Springs, an aquatic ecosystem in Florida. In this ecosystem, as in most others, primary producers constitute most of the biomass in the pyramid, and top carnivores account for very little. If you visited Silver Springs, you would see a lot of aquatic plants but very few gars (the fish that is the main top predator in this ecosystem). Similarly, if you walked through a prairie, you would see more grams of grass than of hawks.

There are some aquatic ecosystems in which the lowest tier of the biomass pyramid is also the narrowest. In these ecosystems, the main producers are bacteria and single-celled protists—organisms that reproduce fast, rather than investing in building a large body. Because of their quick turnover, a smaller biomass of phytoplankton can support a greater biomass of zooplankton.

An **energy pyramid** illustrates how the rate of energy flow diminishes as energy is transferred through an ecosystem. Sunlight energy is captured at the base (the primary producers) and declines with successive levels to the pyramid's tip (the top carnivores). Energy pyramids depict energy flow per unit of water (or land) per unit of time. They are always right side up, meaning the pyramid's base is its largest tier. FIGURE 46.7 shows both the energy pyramid for the Silver Springs ecosystem and the annual energy flow that this pyramid represents.

Ecological Efficiency

Only 5 to 30 percent of the energy in organisms at one trophic level ends up in the tissues of those at the next level up. Several factors reduce the efficiency of transfers. First, not all energy harvested by consumers is used to build biomass; some is lost as heat. Second, some components of a body may be unavailable to a consumer. The lignin and cellulose that reinforce bodies of most land plants pass largely undigested through

biomass pyramid Diagram that depicts the biomass (dry weight) in each of an ecosystem's trophic levels.

energy pyramid Diagram that depicts the energy that enters each of an ecosystem's trophic levels. Lowest tier of the pyramid, representing primary producers, is always the largest.

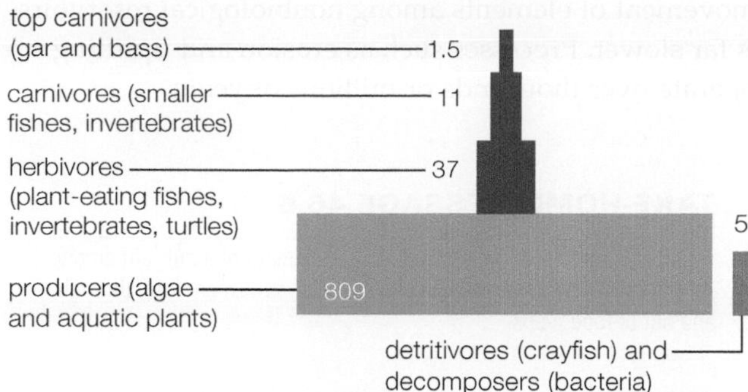
top carnivores (gar and bass)
1.5
carnivores (smaller fishes, invertebrates)
11
herbivores (plant-eating fishes, invertebrates, turtles)
37
5
producers (algae and aquatic plants)
809
detritivores (crayfish) and decomposers (bacteria)

FIGURE 46.6 Biomass pyramid. Biomass (in grams per square meter) for Silver Springs, a freshwater aquatic ecosystem in Florida. In this system, primary producers make up the bulk of the biomass.

some consumers, including humans. Similarly, many animals have some biomass tied up in an internal or external skeleton and hair, feathers, or fur—all of which are difficult for carnivores to digest.

Ecological efficiency is usually higher in aquatic ecosystems than on land, in part because the main aquatic producers—photosynthetic protists—do not make difficult-to-digest lignin as most land plants do. In addition, aquatic ecosystems usually have a higher proportion of ectotherms (such as fishes). This improves ecological efficiency because ectotherms lose less energy as heat than endotherms do.

People sometimes advocate for a vegetarian or vegan diet by touting the ecological benefits of "eating lower on the food chain." Such a diet minimizes the energy losses that occur during transfers between plants, livestock, and humans. When people eat plant materials, they obtain a larger proportion of the energy that the plant captured than they would if that same plant was used as food for livestock. When plants are used to feed livestock, only a very small percentage of the energy that was stored in the plant body ends up in the meat or dairy products that a person can eat. Thus, feeding a population of meat-eaters requires far greater crop production than sustaining a population of plant-eaters.

TAKE-HOME MESSAGE 46.4

- ✔ Primary producers capture energy and convert it into biomass. We measure this process as primary production.
- ✔ The primary production of Earth's land and ocean are approximately equal.
- ✔ A biomass pyramid depicts dry weight of organisms at each trophic level in an ecosystem. Its largest tier is usually producers, but the pyramid for some aquatic systems is inverted.
- ✔ An energy pyramid depicts the amount of energy that enters each level. Its largest tier is always at the bottom (producers).

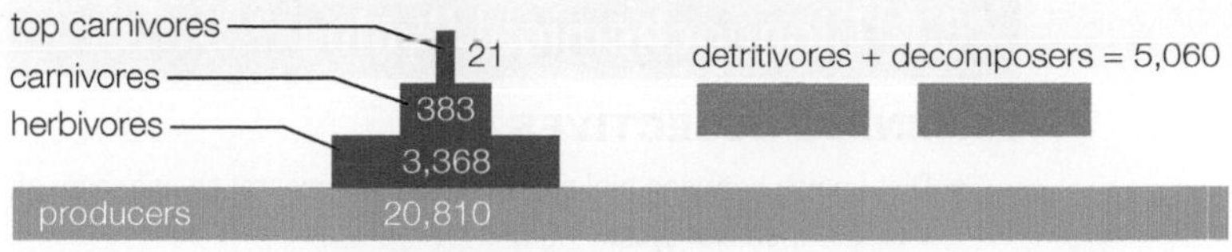
top carnivores
21
carnivores
383
herbivores
3,368
producers
20,810
detritivores + decomposers = 5,060

A Energy pyramid for the Silver Springs ecosystem. The width of each tier in the pyramid represents the amount of energy that enters each trophic level annually, as shown in detail below.

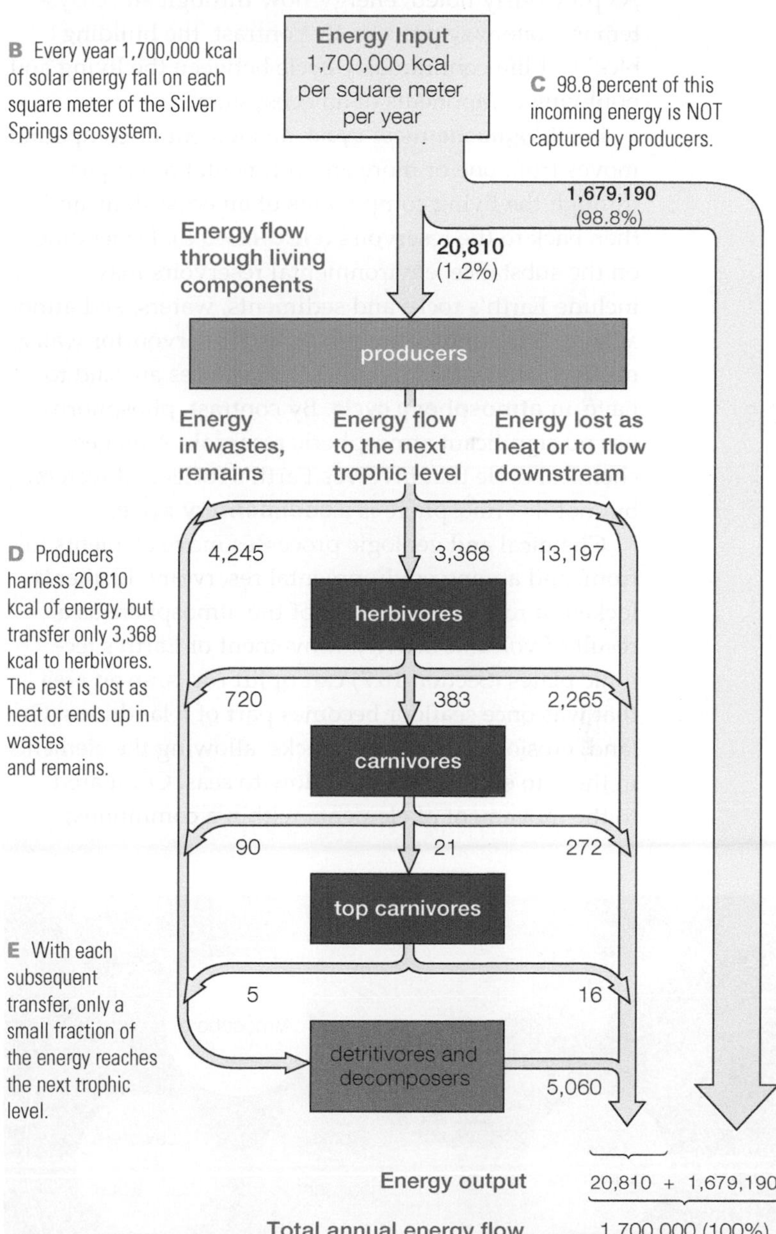
Energy Input 1,700,000 kcal per square meter per year
1,679,190 (98.8%)
Energy flow through living components
20,810 (1.2%)
producers
Energy in wastes, remains
Energy flow to the next trophic level
Energy lost as heat or to flow downstream
4,245
3,368
13,197
herbivores
720
383
2,265
carnivores
90
21
272
top carnivores
5
16
detritivores and decomposers
5,060
Energy output
20,810 + 1,679,190
Total annual energy flow
1,700,000 (100%)

B Every year 1,700,000 kcal of solar energy fall on each square meter of the Silver Springs ecosystem.

C 98.8 percent of this incoming energy is NOT captured by producers.

D Producers harness 20,810 kcal of energy, but transfer only 3,368 kcal to herbivores. The rest is lost as heat or ends up in wastes and remains.

E With each subsequent transfer, only a small fraction of the energy reaches the next trophic level.

FIGURE 46.7 Annual energy flow in Silver Springs measured in kilocalories (kcal) per square meter per year.

FIGURE IT OUT What percentage of the energy carnivores received directly from herbivores was later passed on to top carnivores?

Answer: 21/383 × 100 = 5.5 percent

46.5 Biogeochemical Cycles

LEARNING OBJECTIVES

- Distinguish between biological and environmental components of a biogeochemical cycle.
- Describe some processes by which an element is moved among environmental reservoirs.

A Biogeochemical Cycle

As previously noted, energy flow through an ecosystem is a one-way process. By contrast, the building blocks of life continuously cycle between the living and nonliving components of an ecosystem.

In a **biogeochemical cycle**, an element or compound moves from one or more environmental reservoirs, through the living components of an ecosystem, and then back to the reservoirs (**FIGURE 46.8**). Depending on the substance, environmental reservoirs may include Earth's rocks and sediments, waters, and atmosphere. The atmosphere serves as a reservoir for water, nitrogen, and carbon, so these substances are said to have an **atmospheric cycle**. By contrast, phosphorus has no significant atmospheric reservoir. A biogeochemical cycle that involves Earth's rocks and waters, but not its atmosphere is a **sedimentary cycle**.

Chemical and geologic processes move elements to, from, and among environmental reservoirs. Elements locked in rocks become part of the atmosphere as a result of volcanic activity. Movement of Earth's tectonic plates (Section 16.7) can uplift rocks, so an area that was once seafloor becomes part of a landmass. On land, erosion breaks down rocks, allowing the elements in them to enter rivers, and flow to seas. Compared to the movement of elements within a community, movement of elements among nonbiological reservoirs is far slower. Processes such as erosion and uplifting operate over thousands or millions of years.

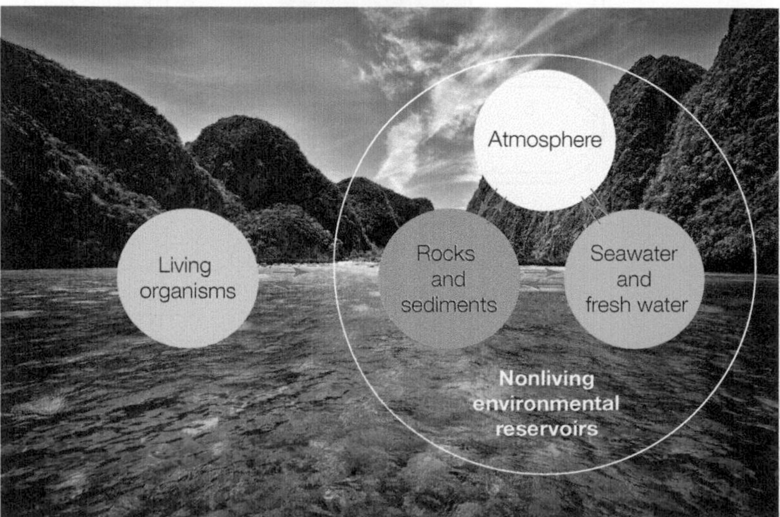
Atmosphere
Living organisms
Rocks and sediments
Seawater and fresh water
Nonliving environmental reservoirs

FIGURE 46.8 Generalized biogeochemical cycle. For all nutrients, the cumulative amount in all environmental reservoirs far exceeds the amount in living organisms.

TAKE-HOME MESSAGE 46.5

✔ A biogeochemical cycle is the slow movement of a nutrient among its environmental reservoirs and its speedier movement into, through, and out of food webs.

46.6 The Water Cycle

LEARNING OBJECTIVES

- List the three environmental reservoirs that hold the most water.
- Describe how water enters land food webs.
- Explain how water movement affects the flow of nutrients.
- Describe the ecological effects of overdrawing water from aquifers and rivers.

Reservoirs and Transfers

By weight, most organisms are water. The **water cycle** moves water from the ocean to the atmosphere, onto land, and back to the oceans (**FIGURE 46.9**). It is an atmospheric cycle. Sunlight energy drives the water cycle by causing evaporation, the conversion of liquid water to water vapor. Water vapor that enters the cool upper layers of the atmosphere condenses into droplets, forming clouds. When droplets get large and heavy enough, they fall as precipitation (rain, snow, and hail).

Oceans cover about 70 percent of Earth's surface, so most rainfall returns water directly to the oceans. On land, we define a **watershed** as a region in which all precipitation drains into a specific waterway. A watershed may be as small as a valley that feeds a stream, or as large as the 5.9 million square kilometers (2.3 million square miles) of the Amazon River Basin. The Mississippi River Basin watershed includes 41 percent of the continental United States.

Most precipitation that enters a watershed seeps into the ground to become groundwater. **Groundwater** consists of soil water and the water in aquifers. **Soil water** is the water that remains between soil particles. Plant roots tap into soil water as their water source. Soils differ in their water-holding capacity, with clay-rich soils holding the most water and sandy soils the least. Water that drains through soil often collects in **aquifers**. These natural underground reservoirs consist of layers of sand or porous rock that can hold water.

Precipitation that falls on impermeable rock or on saturated soil becomes **runoff**: It flows over the ground into streams. The flow of groundwater and surface water returns water to the oceans.

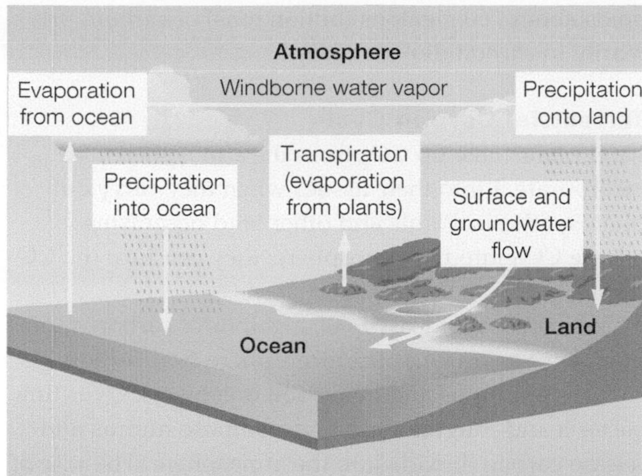
Atmosphere
Evaporation from ocean
Windborne water vapor
Precipitation onto land
Precipitation into ocean
Transpiration (evaporation from plants)
Surface and groundwater flow
Land
Ocean

FIGURE 46.9 The water cycle. Water moves from the ocean to the atmosphere, land, and back. The arrows identify processes that move water.

Movement of water results in the movement of soluble nutrients. Carbon, nitrogen, and phosphorus all have soluble forms that can be moved from place to place by flowing water. As water trickles through soil, it carries nutrient particles from topsoil into deeper soil layers. As a stream flows over limestone, water slowly dissolves the rock and carries carbonates back to the seas where the limestone formed. Flowing water can transport pollutants too; runoff from heavily fertilized lawns and agricultural fields carries dissolved phosphates and nitrates into streams and lakes.

Limited Freshwater

Earth has an enormous quantity of water, but the amount of freshwater available to meet human needs and sustain land ecosystems is severely limited. The vast majority of Earth's water (97 percent) is seawater, and most fresh water is frozen as ice (**TABLE 46.1**).

Aquifers supply about half of the drinking water in the United States, and many of these are in trouble. Overdrafts—removal of water from aquifers faster than natural processes replenish it—are common. Overdrawing water from an aquifer lowers the water table (the topmost level at which rock is saturated with water). When the water table falls in an inland area, wells that tap that aquifer can run dry. In coastal areas, overdrawing an aquifer allows saltwater to move in and contaminate the aquifer.

TABLE 46.1

Environmental Water Reservoirs

Reservoir	Volume (10^3 km^3)
Ocean	1,370,000
Polar ice, glaciers	29,000
Groundwater	4,000
Lakes, rivers	230
Atmosphere (water vapor)	14

The largest aquifer in the United States, the Ogallala aquifer, stretches from South Dakota to Texas and supplies irrigation water for 27 percent of the nation's crops. For the past 30 years, withdrawals from the Ogallala aquifer have exceeded replenishment by a factor of ten. As a result, the water table has dropped as much as 50 meters (150 feet) in some regions.

Rivers are another source of freshwater. However, water in many rivers is currently over allocated.The amount of water promised to various stakeholders such as cities and farmers exceeds the amount that currently flows through the river. In such rivers, diversion of water for human use results in lowered or nonexistent flow in some portion of the river.

Consider that for millions of years, the Colorado River flowed from its source high in the Rocky Mountains to the Gulf of Mexico. Today, the river often does not reach the sea; its waters instead irrigate cropland and flow into pipes that supply cities such as Los Angeles, Las Vegas, Phoenix, and Denver.

Diversion of river water for human uses can have a devastating effect on downstream biological communities that depend on the river. Rivers convey sediment and nutrients as well as water, so decreased flow alters ecosystems by slowing delivery of these materials to the river's delta, the region where the river approaches the sea.

aquifer (AH-qwih-fur) Porous rock layer that holds some groundwater.
biogeochemical cycle Movement of an element or compound amon environmental reservoirs and into and out of food webs.
atmospheric cycle Biogeochemical cycle that includes an atmospheric reservoir.
groundwater Soil water and water in aquifers.
runoff Water that flows over soil into streams.
sedimentary cycle Biogeochemical cycle that involves rocks and waters, but not the atmosphere.
soil water Water between soil particles.
water cycle Biogeochemical cycle in which water among the atmosphere, land, and waters and into an out of food webs.
watershed Land area that drains into a particular stream or river.

TAKE-HOME MESSAGE 46.6

- ✔ Water moves slowly from the world ocean—the main reservoir—through the atmosphere, onto land, then back to the ocean.
- ✔ Oceans contain most of Earth's water.
- ✔ Most freshwater is frozen as ice, so the availability of freshwater is limited. Excessive water withdrawals threaten many sources of drinking water.

46.7 The Carbon Cycle

LEARNING OBJECTIVES

- Compare the amount of carbon in Earth's sediments, air, and oceans.
- Describe how carbon enters marine and aquatic food webs.

After water, carbon is the most abundant substance in living things. In the **carbon cycle**, carbon moves among Earth's atmosphere, oceans, rocks, and soils, and into and out of food webs (**TABLE 46.2** and **FIGURE 46.10**). It is an atmospheric cycle because the atmosphere holds about 760 gigatons (billion tons) of carbon, primarily in carbon dioxide (CO_2).

TABLE 46.2

Annual Carbon Movement in Gigatons (Billions of Metric Tons)

From atmosphere to plants (carbon fixation)	120
From atmosphere to ocean	107
From ocean to atmosphere	105
From plants to atmosphere	60
From soil to atmosphere	60
From land to ocean in runoff	0.4
Burial in ocean sediments	0.1

Terrestrial Carbon Cycle

Land plants take up CO_2 from the atmosphere and incorporate it into their tissues when they carry out photosynthesis. Plants and other land organisms release CO_2 into the atmosphere; they produce the CO_2 during aerobic respiration.

Soil contains at least 1,600 gigatons of carbon, more than twice as much as the atmosphere. Soil carbon consists of humus and living soil organisms. Over time, bacteria and fungi in the soil decompose humus and release carbon dioxide into the atmosphere. The rate of decomposition increases with temperature. In the tropics, decomposition proceeds rapidly, so most carbon is stored in living plants, rather than in soil. By contrast, in temperate zone forests and grasslands, soil holds more carbon than living plants. Soil is most carbon-rich in the arctic, where low temperature hampers decomposition, and in peat bogs, where acidic, anaerobic conditions do the same.

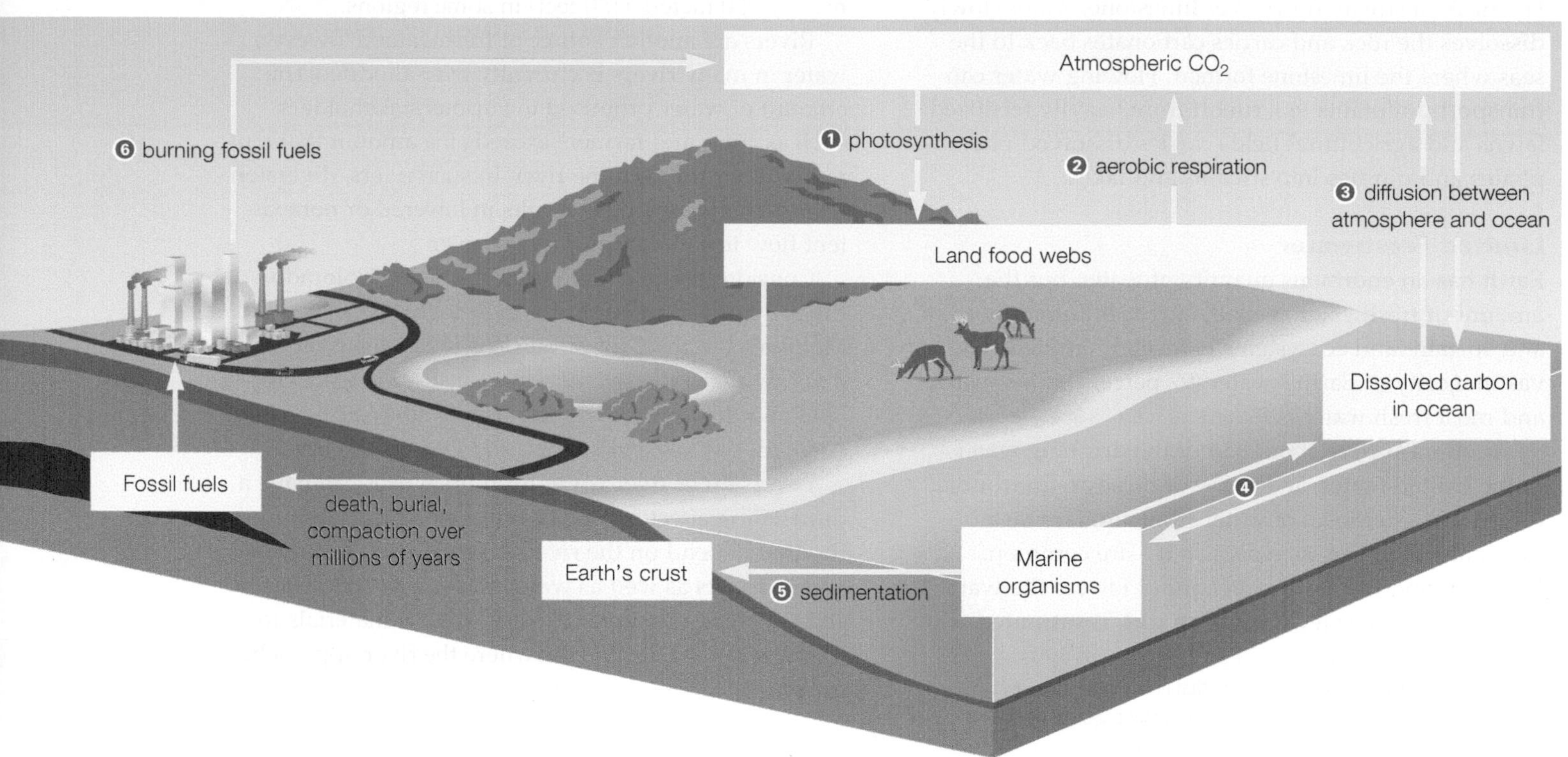
Atmospheric CO_2
❻ burning fossil fuels
❶ photosynthesis
❷ aerobic respiration
❸ diffusion between atmosphere and ocean
Land food webs
Dissolved carbon in ocean
❹
Fossil fuels
death, burial, compaction over millions of years
Earth's crust
❺ sedimentation
Marine organisms

FIGURE 46.10 The carbon cycle. Most carbon is in Earth's rocks, where it is largely unavailable to living organisms.

❶ Carbon enters land food webs when plants take up carbon dioxide from the air for use in photosynthesis.

❷ Carbon returns to the atmosphere as carbon dioxide when plants and other land organisms carry out aerobic respiration.

❸ Carbon diffuses between the atmosphere and the ocean. Bicarbonate forms when carbon dioxide dissolves in seawater.

❹ Marine producers take up bicarbonate for use in photosynthesis, and marine organisms release carbon dioxide from aerobic respiration.

❺ Many marine organisms incorporate carbon into their shells. After they die, these shells become part of the sediments. Over time, the sediments become carbon-rich rocks such as limestone and chalk in Earth's crust.

❻ Burning of fossil fuels derived from the ancient remains of plants and photosynthetic protists puts additional carbon dioxide into the atmosphere.

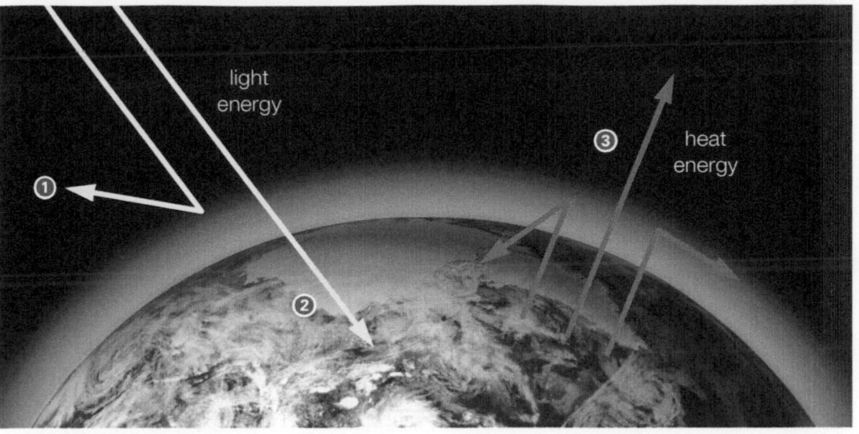
light energy
①
②
③
heat energy

FIGURE 46.11 Greenhouse effect.

❶ Earth's atmosphere reflects some sunlight energy back into space.

❷ More light energy reaches and warms Earth's surface.

❸ Earth's warmed surface emits heat energy. Some of this energy escapes through the atmosphere into space. The rest is absorbed and then emitted in all directions by greenhouse gases. Some emitted heat warms Earth's surface and lower atmosphere.

FIGURE IT OUT Do greenhouse gases reflect heat energy toward Earth?

Answer: No. The gases absorb heat energy, then reemit it in all directions.

Marine Carbon Cycle

Seawater holds about 40,000 gigatons of carbon, about 50 times as much as the atmosphere. The main form of carbon in seawater is bicarbonate (HCO_3^-), an ion that forms when CO_2 dissolves in water. Marine producers take up bicarbonate and convert it to CO_2 for use in photosynthesis. Marine organisms release CO_2 produced by aerobic respiration into the water. Some marine organisms such as foraminifera, shelled mollusks, and reef-building corals also store carbon in their calcium carbonate–hardened parts.

Marine sediments and sedimentary rocks are Earth's largest carbon reservoir by far, holding more than 65 million gigatons. Limestone and other sedimentary rocks form when the calcium carbonate–rich shells of marine organisms become compacted over millions of years. Such rocks become part of land ecosystems when movements of tectonic plates uplift portions of the seafloor. Carbon in rocks is not available to producers, so this vast store of carbon has little effect on ecosystems. The geologic part of the cycle is completed as erosion breaks down rocks and rivers carry dissolved carbon to the sea.

Carbon in Fossil Fuels

Fossil fuels such as coal, oil, and natural gas hold an estimated 5,000 gigatons of carbon. Deposits of fossil fuels formed over hundreds of millions of years from carbon-rich remains. High pressure and temperature transformed the remains of land plants to coal (Section 22.6). A similar process transformed the remains of plankton to oil and natural gas. Until recently, carbon in fossil fuels, like the carbon in rocks, had little impact on ecosystems. Currently, our burning of this fuel adds about 9 gigatons of CO_2 to the atmosphere every year.

carbon cycle Movement of carbon, mainly between the oceans, atmosphere, and living organisms.
greenhouse effect Warming of Earth that occurs when greenhouse gases absorb heat energy then reemit some of back toward the Earth.
greenhouse gas Atmospheric gas, such as carbon dioxide, that absorbs heat emitted by Earth's surface, then reradiates it.

TAKE-HOME MESSAGE 46.7

✔ The largest reservoir is sedimentary rock. Carbon moves into and out of this reservoir over very long time spans and is not available to organisms.

✔ Seawater is the largest reservoir of biologically available carbon. Marine producers take up bicarbonate and convert it to CO_2 for use in photosynthesis.

✔ On land, large amounts of carbon are stored in soil, especially in arctic regions and in peat bogs.

✔ The atmosphere holds less carbon than rocks, seawater, or soil. It serves as the source of CO_2 for land producers. Burning of fossil fuels adds carbon to this reservoir.

46.8 Greenhouse Gases and Climate Change

LEARNING OBJECTIVES

- Describe how atmospheric greenhouse gases affect the temperature of Earth's surface.
- Explain the causes of the ongoing increase in greenhouse gases and the effect of this increase on climate.

The Greenhouse Effect

Carbon dioxide is a **greenhouse gas**, an atmospheric gas whose ability to absorb and reradiate heat energy helps keep Earth warm enough to sustain life. The mechanism by which this occurs is referred to as the **greenhouse effect** (FIGURE 46.11). When radiant energy from the sun reaches Earth's atmosphere, some energy is reflected back into space ❶. However, more energy passes through the atmosphere and warms Earth's surface ❷. When the warmed surface radiates heat energy, greenhouse gases absorb some of that heat, then emit a portion of it back toward Earth ❸.

CREDIT: (11) NASA.

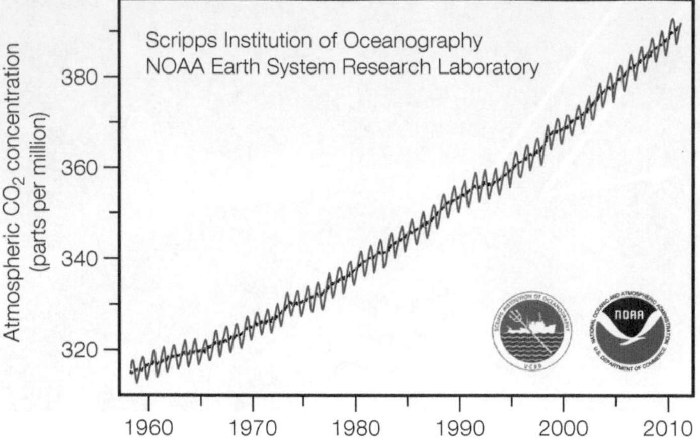

Scripps Institution of Oceanography
NOAA Earth System Research Laboratory
Atmospheric CO_2 concentration (parts per million)
380
360
340
320
1960
1970
1980
1990
2000
2010
NOAA

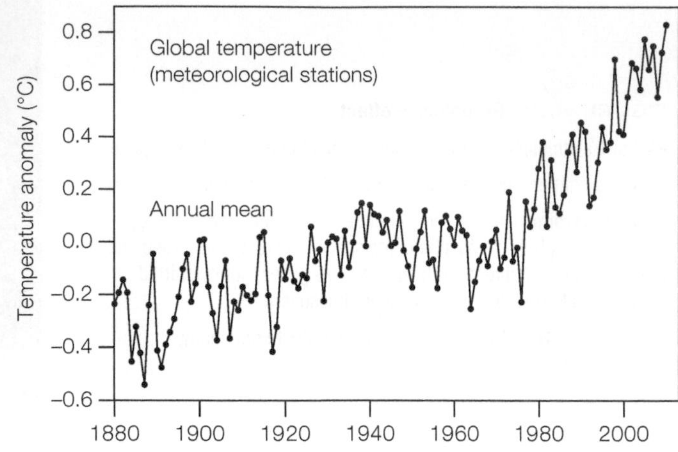

Global temperature (meteorological stations)
Annual mean
Temperature anomaly (°C)
0.8
0.6
0.4
0.2
0.0
−0.2
−0.4
−0.6
1880
1900
1920
1940
1960
1980
2000

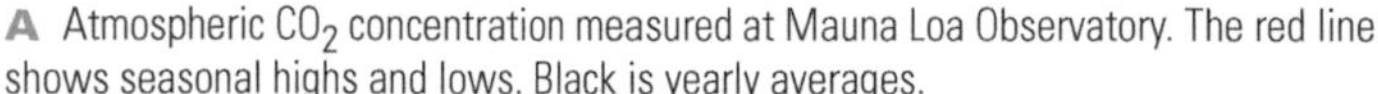

A Atmospheric CO_2 concentration measured at Mauna Loa Observatory. The red line shows seasonal highs and lows. Black is yearly averages.

B Global annual mean air temperature based on measurements at meteorological stations. The temperature anomaly (vertical axis) refers to the deviation from the mean temperature of 1951–1980.

FIGURE 46.12 Directly measured changes in atmospheric carbon dioxide and global temperature.

If greenhouse gases did not exist, heat energy emitted by Earth's surface would escape into space, leaving the planet cold and entirely lifeless.

Increasing Atmospheric Carbon Dioxide

In 1959, researchers began to measure the atmospheric concentration of CO_2 at an observatory near the top of Mauna Loa, Hawaii's highest volcano. This remote site 3,500 meters (11,000 feet) above sea level was chosen because it is largely free of local airborne contamination and is representative of atmospheric conditions for the Northern Hemisphere. For the first time, researchers could observe atmospheric carbon dioxide fluctuations for the entire hemisphere.

The researchers immediately noticed that there are seasonal lows and highs in atmospheric CO_2. The level of CO_2 is lowest in the summer, when primary production in the Northern Hemisphere is at its peak. It is highest in the winter, when photosynthesis declines but aerobic respiration and fermentation continue.

Over time, the researchers also noticed a long-term trend: The average annual level of CO_2 is increasing (**FIGURE 46.12A**). As additional sites around the world began to monitor atmospheric CO_2, they too detected an ongoing rise in this gas.

To put the current increase in perspective, scientists look at historical changes in atmospheric CO_2. Glacial ice provides one window into the past. Such ice forms when snow falls, then is compressed by new snowfalls above it. In some regions, layers of ice have been laid down one atop the next for hundreds of thousands of years. The result is sheets of ice more than a kilometer thick. To determine what conditions were like long ago, scientists use a hollow drill to remove a long ice core. Air bubbles trapped at different depths within the ice provide snapshots of atmospheric conditions at the time that ice formed. So far, analysis of ice cores has provided a history of atmospheric changes that extends back about 800,000 years.

Fossil foraminiferan shells provide information about CO_2 levels in the even more distant past. Foraminifera are single-celled protists that have lived in Earth's oceans for millions of years (Section 21.6). Throughout this time, they have taken up carbon and other elements from seawater and incorporated them into their shells. By studying the composition of fossilized shells, scientists can trace how atmospheric CO_2 changed over many millions of years.

Data from glacial ice, foraminiferan fossils, and other sources consistently show that atmospheric CO_2 has risen and fallen many times. These data also show that the current CO_2 concentration is the highest in at least 15 million years.

Causes of the Atmospheric Increase in CO_2

Some natural processes can cause a rise in atmospheric CO_2, but these processes do not seem to be responsible for the current increase. Volcanoes emit CO_2, but there has not been any increase in volcanic eruptions during the past century. Changes in the chemistry of the oceans can cause more CO_2 from the ocean to enter the atmosphere, but that does not seem to be happening. If the oceans were giving up additional CO_2 to the air, seawater would be becoming less acidic, because

global climate change Long-term change in Earth's climate.

CREDIT: (12) NASA Goddard Institute for Space Studies.

Data Analysis Activities

Changes in the Air To assess the impact of human activity on the carbon dioxide level in Earth's atmosphere, it helps to take a long view. One useful data set comes from deep core samples of Antarctic ice. The oldest ice core that has been fully analyzed dates back a bit more than 400,000 years. Air bubbles trapped in the ice provide information about the gas content in Earth's atmosphere at the time the ice formed. Combining ice core data with more recent direct measurements of atmospheric carbon dioxide—as in **FIGURE 46.13**—can help scientists put current changes in the atmospheric carbon dioxide into historical perspective.

1. What was the highest carbon dioxide level between 400,000 B.C. and 0 A.D.?
2. During this period, how many times did carbon dioxide reach a level comparable to that measured in 1980?
3. The Industrial Revolution occurred around 1800. What was the trend in carbon dioxide levels in the 800 years prior to this event? What about in the 175 years after it?
4. Was the rise in the carbon dioxide level between 1800 and 1975 larger or smaller than the rise between 1980 and 2013?

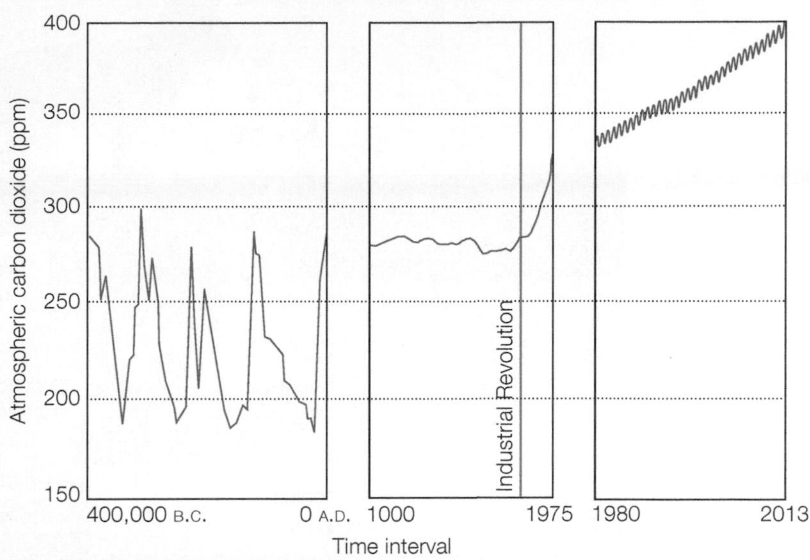

Atmospheric carbon dioxide (ppm)
400
350
300
250
200
150
Industrial Revolution
400,000 B.C.
0 A.D.
1000
1975
1980
2013
Time interval

FIGURE 46.13 Changes in atmospheric carbon dioxide levels (in parts per million). Direct measurements began in 1980. Earlier data are based on ice cores.

the water would contain less carbonic acid. (Carbonic acid forms when CO_2 dissolves in water.) However, the acidity of the oceans is increasing. This increase indicates that CO_2 from the air is entering the oceans, rather than the other way around. The ocean is acidifying most quickly at the surface, which also indicates that a rise in atmospheric CO_2 is driving this process.

A variety of evidence points to human activities, most notably burning fossil fuels, as the cause of the current increase in atmospheric CO_2. Consider that when these fuels burn, the carbon in them binds with oxygen from the atmosphere to form CO_2. Thus, if fossil fuel use was driving the increase in atmospheric CO_2, we would expect to see a corresponding decrease in atmospheric oxygen. Such a decrease in oxygen has in fact been documented.

Isotope data also indicates that the source of the additional atmospheric carbon dioxide is fossil fuels. As Section 2.2 explained, isotopes of an element have different numbers of protons and so have different mass numbers. There are three isotopes of carbon: C^{12}, C^{13}, and C^{14}. Fossil fuels contain carbon taken up in CO_2 by photosynthetic organisms, but because the carbon was taken up so long ago it now contains little C^{13} and no C^{14}. If the source of the rise in CO_2 over the past century was the burning of fossil fuel, we would expect a corresponding increase in the proportion of C^{12} relative to the heavier isotopes in the air. Scientists began looking at the carbon isotope ratios of the air in the 1990s and, as expected the data shows a rise in the proportion of C^{12} relative to C^{13} and C^{14}. Tree ring data, which can be used to determine historic carbon isotope ratios, indicates that the relative proportion of C^{12} in the air began to increase in the mid-1800s with the onset of the industrial revolution.

Changing Climate

Given the greenhouse effect, we would predict that an increase in the atmospheric concentration of carbon dioxide and other greenhouse gases would raise the temperature of Earth's surface. Evidence from a variety of sources indicate that such a rise in temperature is under way. We are in the midst of **global climate change**, a long-term alteration of Earth's climate. Global warming, an increase in Earth's average surface temperature, is one aspect of this change (**FIGURE 46.12B**). In the past 100 years, Earth's average temperature has risen by about 0.74°C (1.3°F), and the rate of warming is accelerating. The rate at which the temperature is rising doubled in the past 50 years.

Earth's climate has varied greatly over its long history. During ice ages, much of the planet was covered by glaciers. Other periods were warmer than the present, and tropical plants and coral reefs thrived at what are now cool latitudes. Scientists can correlate some historical large-scale temperature changes with shifts

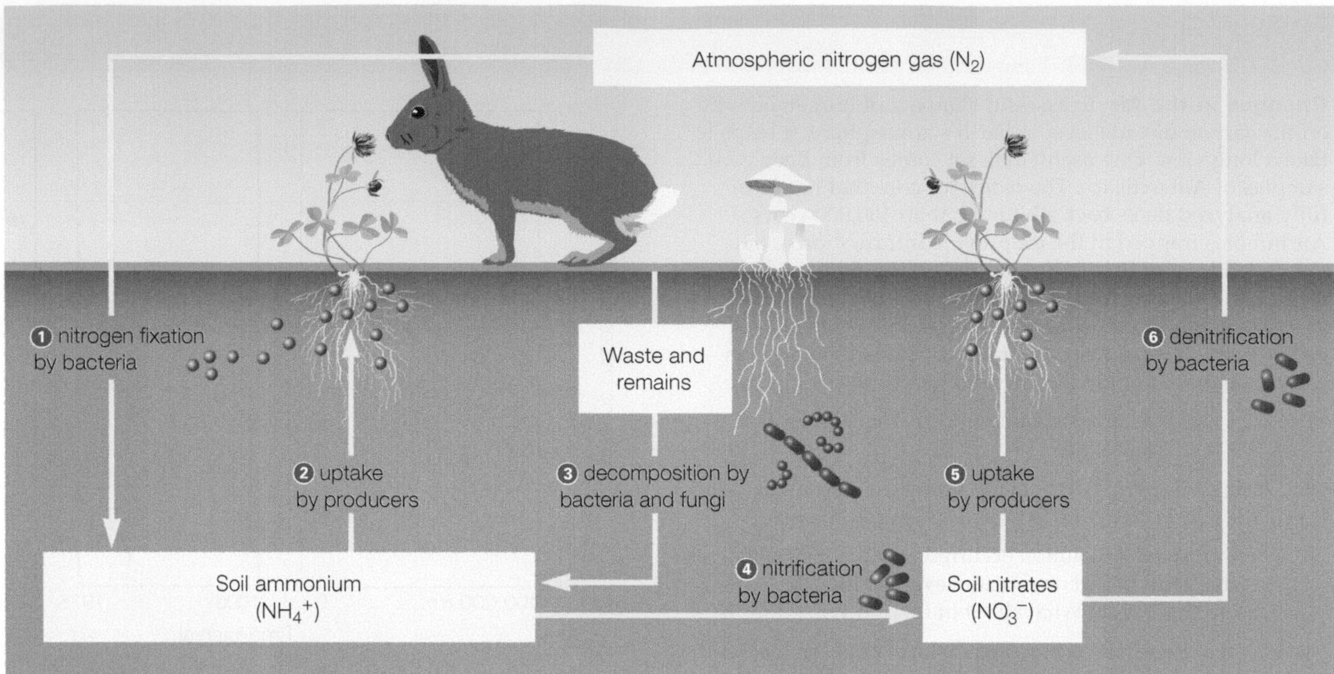

Atmospheric nitrogen gas (N_2)
1 nitrogen fixation by bacteria
2 uptake by producers
Waste and remains
3 decomposition by bacteria and fungi
4 nitrification by bacteria
5 uptake by producers
6 denitrification by bacteria
Soil ammonium (NH_4^+)
Soil nitrates (NO_3^-)

FIGURE 46.14 Nitrogen cycle on land. Nitrogen becomes available to plants through the activities of nitrogen-fixing bacteria. Other bacterial species cycle nitrogen to plants. They break down organic wastes to ammonium and nitrates.

in Earth's orbit, which varies in a regular fashion over 100,000 years, and Earth's tilt, which varies over 40,000 years. Changes in solar output, the arrangement of land masses and the frequency of volcanic eruptions also influence Earth's temperature. However, the overwhelming majority of scientists agree that these factors cannot explain the current climate change. There is broad scientific consensus that this climate change is the result of the ongoing rise in greenhouse gases.

A rise of a degree or two in average temperature may not seem like a big deal, but it is enough to increase the rate of glacial melting and raise sea level. Sea level is currently rising at a rate of about one-eighth of an inch (3.2 mm) per year, increasing the risk of catastrophic flooding in low-lying areas. Scientists expect the global mean sea level to rise at least 8 inches (0.2 meter) and perhaps as much as 6.6 feet (2.0 meters) by 2100.

Temperature of the land and seas affects evaporation, winds, and currents, so many weather patterns are changing as the temperature rises. For example, warmer temperatures are correlated with extremes in rainfall patterns: periods of drought interrupted by unusually heavy rains. An increase in sea temperature also makes strong hurricanes stronger.

Looking forward, atmospheric CO_2 is expected to continue to rise as large nations such as China and India become increasingly industrialized. However, efforts are under way to reduce the damage by increasing the efficiency of processes that require fossil fuels, shifting to alternative energy sources that do not release carbon, such as solar and wind power, and developing innovative ways to store carbon dioxide.

TAKE-HOME MESSAGE 46.8

✔ Carbon dioxide is one of the atmospheric gases that absorb heat and emit it toward Earth's surface, thus keeping the planet warm enough for life.

✔ The CO_2 level of the atmosphere is currently rising as a result of human activity, and global mean temperature is rising with it. Changes in temperature affect other climate factors such as patterns of rainfall.

46.9 Nitrogen Cycle

LEARNING OBJECTIVES

- Explain why eukaryotes cannot fix nitrogen.
- Distinguish between nitrogen fixation, ammonification, nitrification, and denitrification.
- Describe how bacteria are used in sewage treatment.
- Describe how human activities disrupt the nitrogen cycle.

Nitrogen moves in an atmospheric cycle known as the **nitrogen cycle**. The main nitrogen reservoir is the air, which is about 80 percent nitrogen gas. Nitrogen gas consists of two atoms of nitrogen held together

FIGURE 46.15 Spraying synthetic nitrogen fertilizer on a corn field to boost crop yield. The inset photo shows corn grown with adequate nitrogen (left) and in nitrogen-deficient soil (right).

by a triple covalent bond as N_2, or N≡N. Recall from Section 2.4 that a triple bond holds atoms together more strongly than a single or double bond.

Reactions That Drive the Nitrogen Cycle

All organisms use nitrogen to build ATP, nucleic acids, and proteins. Photosynthetic organisms also use it to build chlorophyll. Despite the universal need for nitrogen and the abundance of atmospheric nitrogen, no eukaryote can make use of nitrogen gas. Eukaryotes do not have an enzyme that can break the strong bond between the two nitrogen atoms.

Only certain prokaryotes can carry out **nitrogen fixation** (FIGURE 46.14). These nitrogen-fixing organisms break the bonds in N_2 and use the nitrogen atoms to form ammonia, which dissolves to form ammonium (NH_4^+) ❶.

You have already learned about some organisms that fix nitrogen. Nitrogen-fixing cyanobacteria live in aquatic habitats, soil, and as components of lichens (Sections 20.7 and 23.7). Another group of nitrogen-fixing bacteria forms nodules on the roots of legumes, a plant group that includes peas, beans, and their many undomesticated relatives (Section 28.3). Still other nitrogen-fixing bacteria live on their own in soil. Some deep-sea archaea can also fix nitrogen.

In addition to biological nitrogen fixation by prokaryotes, a small amount of ammonium forms as a result of lightning-fueled reactions in the atmosphere. Energy from the lightning causes nitrogen gas to react with atmospheric water vapor.

Plants take up ammonium from soil water ❷. Animals obtain nitrogen by eating plants or one another. Bacterial and fungal decomposers return the nitrogen in wastes and remains to the soil by a process called **ammonification** ❸.

Nitrification is a two-step process that converts ammonium to nitrates ❹. First, ammonia-oxidizing bacteria and archaea convert ammonium to nitrite (NO_2^-). Other bacteria then take up nitrite and use it in reactions that form nitrates (NO_3^-). Nitrates, like ammonium, are taken up and used by producers ❺. Nitrification is essential to ecosystem health because it prevents ammonium from accumulating to toxic concentrations. Humans make use of bacteria that carry out this process in sewage treatment plants. Sewage contains large amounts of ammonium formed from urea excreted in urine (Section 40.2).

Denitrification, an anaerobic reaction carried out mainly by bacteria, converts nitrates to nitrogen gas ❻. In ecosystems, denitrification can have important effects on productivity because it results in a decline in the amount of soluble nitrogen available to producers. In sewage treatment plants, denitrifying bacteria are used to remove nitrates from wastewater before the water is released into the environment.

ammonification (am-on-if-ih-CAY-shun) Breakdown of nitrogen-containing organic material resulting in the release of ammonia and ammonium ions.
denitrification (dee-nigh-triff-ih-CAY-shun) Conversion of nitrates or nitrites to gaseous forms of nitrogen.
nitrification (nigh-triff-ih-CAY-shun) Conversion of ammonium to nitrates.
nitrogen cycle Movement of nitrogen among the atmosphere, soil, and water, and into and out of food webs.
nitrogen fixation Incorporation of nitrogen from nitrogen gas into ammonia.

Human Effects on the Nitrogen Cycle

In the early 1900s, German scientists discovered a method of fixing atmospheric nitrogen and producing ammonium on an industrial scale. This process allowed the manufacture of synthetic nitrogen fertilizers that have boosted crop yields (**FIGURE 46.15**). However, use of these fertilizers, along with other human activities, has added large amounts of nitrogen-containing compounds to our air and water. Here we consider two such types of compounds and their effects.

Nitrous Oxide We know from air bubbles in ice cores that the atmospheric concentration of nitrous oxide (N_2O) remained about 270 parts per billion (ppb) for at least a thousand years before the industrial revolution. It is now about 325 ppb and rising. The ongoing increase is the result of burning of fossil fuels, use of synthetic nitrogen fertilizers, and industrial livestock production. Burning fossil fuel releases N_2O directly into the air. Chemical fertilizers and manure from livestock increase atmospheric N_2O by encouraging growth of bacteria that release this gas.

An increased concentration of atmospheric N_2O is a matter of concern for two reasons. First, N_2O is a greenhouse gas, and a highly persistent and effective one. It can remain in the atmosphere for more than 100 years, and it has a warming potential 300 times that of CO_2. Second, N_2O contributes to destruction of the ozone layer. As Section 19.6 explained, ozone high in the atmosphere protects life at Earth's surface from the damaging effects of ultraviolet radiation.

Nitrates Nitrate from fertilizers and animal manure runs off into streams and lakes or leaches (travels down through the soil) into groundwater. Ingested nitrate inhibits iodine uptake by the thyroid gland and it may increase the risk of thyroid cancer. It is also correlated with an increased risk for respiratory infections, diabetes, and some cancers.

TAKE-HOME MESSAGE 46.9

- ✔ The atmosphere is the main reservoir for nitrogen, but only nitrogen-fixing prokaryotes can access it.
- ✔ Plants take up ammonium and nitrates from soil. Ammonium is released by nitrogen-fixing bacteria and by fungal and bacterial decomposers. Bacteria and archaea produce nitrates.
- ✔ Nitrogen returns to the atmosphere when denitrifying bacteria convert soluble forms of nitrogen to nitrogen gas.
- ✔ Use of synthetic nitrogen fertilizers and fossil fuels has increased the amount of nitrous oxide (N_2O) in the air and added nitrates to the water.

46.10 The Phosphorus Cycle

LEARNING OBJECTIVES

- Describe the geochemical portion of the phosphorus cycle.
- Describe how phosphorus enters food webs.

Atoms of phosphorus are highly reactive, so phosphorus does not occur naturally in its elemental form. Most of Earth's phosphorus is bonded to oxygen as phosphate (PO_4^{3-}), an ion that abounds in rocks and sediments. In the **phosphorus cycle**, phosphorus passes quickly through food webs as it moves from land to ocean sediments, then slowly back to land (**FIGURE 46.16**). Because little phosphorus exists in a gaseous form and its major reservoir is sedimentary rock, the phosphorus cycle is described as a sedimentary cycle.

Weathering and erosion move phosphates from rocks into soil, lakes, and rivers ❶. Leaching and runoff carry dissolved phosphates to the ocean ❷. Here, most phosphorus comes out of solution and settles as deposits along continental margins ❸. Slow movements of Earth's crust can uplift these deposits onto land ❹, where weathering releases phosphates from rocks once again. Phosphate-rich rocks are mined for use in the industrial production of fertilizer.

All organisms require phosphorus. It is a component of ATP and of all nucleic acids and phospholipids. The biological portion of the phosphorus cycle begins when producers take up phosphate. Land plants take up dissolved phosphate from the soil water ❺. Land animals get phosphates by eating the plants or one another. Phosphorus returns to the soil in the wastes and remains of organisms ❻. Phosphate-rich droppings from seabird or bat colonies are collected and used as a natural fertilizer. Manure from livestock is also rich in phosphorus.

In the seas, phosphorus enters food webs when producers take up phosphate dissolved in seawater ❼. As on land, wastes and remains continually replenish the phosphorus supply ❽.

TAKE-HOME MESSAGE 46.10

- ✔ Rocks are the main phosphorus reservoir. Weathering puts phosphates into water. Producers take up dissolved phosphates.
- ✔ Phosphate-rich wastes are a natural fertilizer, and phosphate from rocks can be used to produce fertilizer on an industrial scale.

phosphorus cycle Movement of phosphorus among Earth's rocks and waters and into and out of food webs.

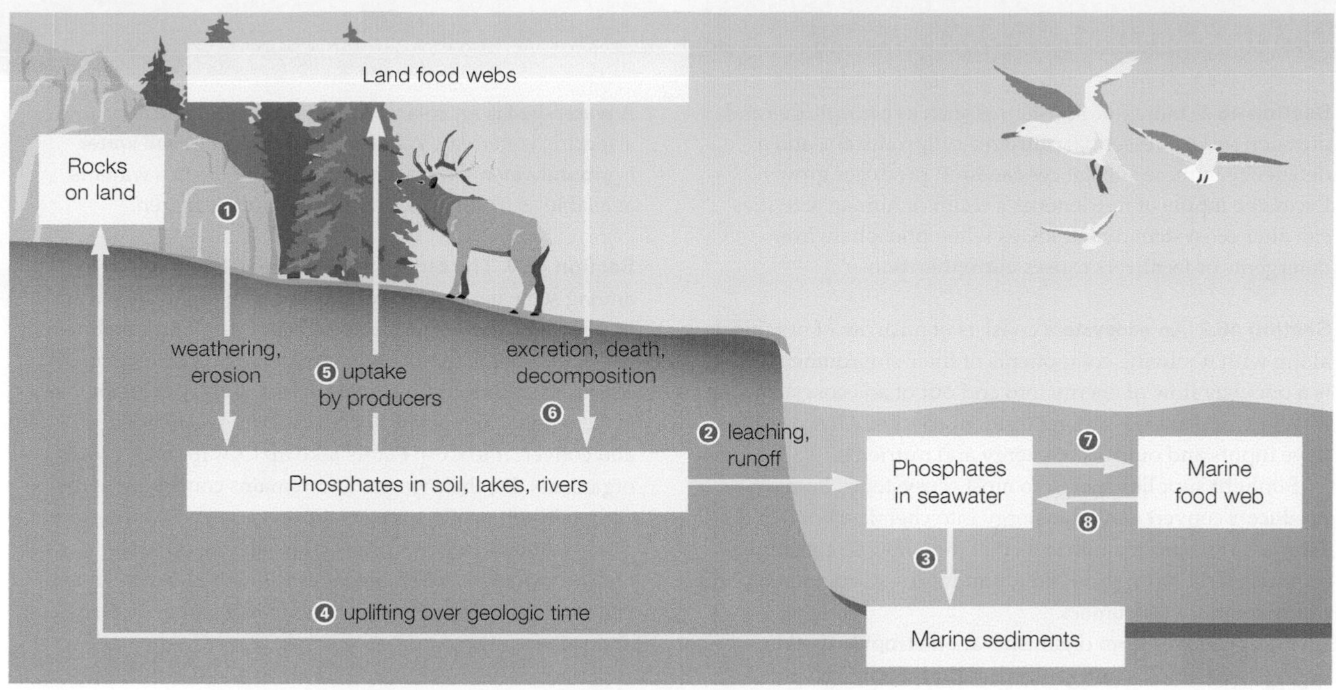
Land food webs
Rocks on land
1
weathering, erosion
5 uptake by producers
excretion, death, decomposition
6
2 leaching, runoff
Phosphates in soil, lakes, rivers
Phosphates in seawater
7
Marine food web
8
3
4 uplifting over geologic time
Marine sediments

FIGURE 46.16 The phosphorus cycle. In this sedimentary cycle, phosphorus moves mainly in the form of phosphate ions (PO_4^{3-}).

46.1 Too Much of a Good Thing (revisited)

Farmers use industrially produced fertilizers because such products are a relatively inexpensive way to enhance crop yield. However, nitrates and phosphates from these fertilizers can pollute waters, both locally and in distant regions. Fertilizer applied to fields in the Midwest not only pollutes local drinking water, it also flows into streams that feed the Mississippi River. The river then delivers its heavy load of phosphates and nitrates to the Gulf of Mexico.

Most excess phosphates and nitrogen entering the Mississippi come from agriculture, but suburbanites and city dwellers also contribute. In some communities, the sewer system receives both sewage and water from storm drains. Heavy rains can overwhelm such a system, causing an overflow that delivers raw sewage into streams. Rain-related overflows can be prevented by a system that keeps sewage and water from storm drains separate. However, such systems typically do not treat water from storm drains before discharging it. Thus, when fertilizer from lawns enters such a system, it is delivered untreated into the river.

Each summer, the excessive nutrient inputs from the Mississippi into the Gulf of Mexico result in formation of a "dead zone," an area of deep water where the oxygen level is too low to support most marine organisms. In recent years, the dead zone has encompassed an average area about the size of the state of Connecticut (about 14,000 km^2).

The dead zone forms after eutrophication causes an algal bloom (Section 21.5). The high nutrient level encourages a population explosion of marine algae and photosynthetic bacteria. When these organisms die, their remains drift down to the seafloor. Decomposition of these remains by bacteria then depletes the water of oxygen, making it impossible for most animals to survive. Fishes can swim away from the low-oxygen zone, but less mobile animals suffocate. There are also indirect effects, as when an increase in jellies leads to a decrease in the fish larvae on which the jellies prey. Jellies are among the few animals that thrive in anoxic waters. ●

Section 46.1 Inorganic substances such as phosphorus and nitrogen serve as essential nutrients for producers, and a deficiency in these substances can limit producer growth. Excessive inputs of nutrients as a result of human activities can alter ecosystem dynamics, as when phosphate from detergents or fertilizers causes **eutrophication**.

Section 46.2 An **ecosystem** consists of an array of organisms along with nonliving components of their environment. There is a one-way flow of energy into and out of an ecosystem, and a cycling of materials among resident species. All ecosystems have inputs and outputs of energy and nutrients.

Sunlight supplies energy to most ecosystems. **Primary producers** convert sunlight energy into chemical bond energy. They also take up the nutrients that they, and all consumers, require. Herbivores, carnivores, omnivores, **decomposers**, and **detritivores** are **consumers**.

Energy moves from organisms at one **trophic level** to organisms at another. Organisms are at the same trophic level if they are an equal number of steps away from the energy input into the ecosystem. A **food chain** shows one path of energy and nutrient flow among organisms. It depicts who eats whom.

Section 46.3 Food chains interconnect as **food webs**. In a **grazing food web**, most energy captured by producers flows to herbivores. In a **detrital food web**, most energy flows from producers directly to detritivores and decomposers. Both types of food webs interconnect in nearly all ecosystems. The efficiency of energy transfers is low, so most food chains have no more than four or five trophic levels.

Section 46.4 The **primary production** of an ecosystem is the rate at which producers capture and store energy in their tissues. It varies with climate, seasonal changes, and nutrient availability.

Energy pyramids and **biomass pyramids** depict how energy and organic compounds are distributed among the organisms of an ecosystem. All energy pyramids are largest at their base. If producers get eaten as fast as they reproduce, the biomass of consumers can exceed that of producers, so the biomass pyramid is upside down.

Section 46.5 In a **biogeochemical cycle**, an element or compound moves from an environmental reservoir, through organisms, then back to the environment. Chemical and geologic processes move the nutrients between their environmental reservoirs. In an **atmospheric cycl**e, the air is a reservoir. In a **sedimentary cycle**, it is not.

Section 46.6 In the **water cycle**, evaporation, condensation, and precipitation move water from its main reservoir—oceans—into the atmosphere, onto land, then back to oceans. **Runoff** is water that flows over ground into streams. A **watershed** is an area where all precipitation drains into a specific waterway. Water in **aquifers** and in **soil water** is **groundwater**. Only a small fraction of Earth's water is available as freshwater, and most of that is frozen.

Section 46.7 The **carbon cycle** moves carbon mainly among seawater, the air, soils, and living organisms in an **atmospheric cycle**. The largest carbon reservoir is sedimentary rocks, but living organisms cannot take up carbon from this reservoir. The largest reservoir of biologically available carbon is the ocean. Aquatic producers take up dissolved bicarbonate and convert it to CO_2. Plants take up CO_2 from air. When land organisms die, their wastes and remains contribute to the carbon in soil, which exceeds the amount in the atmosphere. The amount of carbon in soil differs among ecosystems, being highest in regions where decomposition is slowest. Human use of fossil fuels adds previously stored carbon to Earth's air and water.

Section 46.8 The **greenhouse effect** refers to the ability of **greenhouse gases** to absorb heat and reemit it, thus warming Earth's surface. CO_2 is a greenhouse gas and humans are putting increasing amounts of it into the atmosphere, mainly by burning fossil fuels. Direct measurements of the atmosphere, studies of air bubbles in ice cores, and analysis of the composition of fossil foraminiferan shells indicate that atmospheric CO_2 is currently at its highest level in at least 15 million years. A decline in atmospheric oxygen, the increasing acidity of the ocean, and a change in the carbon isotope ratios in the atmosphere indicate that the additional CO_2 is derived from the burning of fossil fuels, rather than some natural process that releases this gas. As one would expect, the ongoing rise in CO_2 is causing global warming, which is one aspect of **global climate change**.

Section 46.9 The **nitrogen cycle** is an atmospheric cycle. Air is the main reservoir for N_2, a gaseous form of nitrogen accessible only to nitrogen-fixing prokaryotes. By the process of **nitrogen fixation**, some bacteria and archaea take up N_2 and incorporate it into ammonia that producers can take up and use. Ammonia is also formed by the **ammonification** of organic wastes and remains by bacteria and fungi.

Bacteria also carry out **nitrification**, converting ammonium to nitrite and then nitrate, which plants can also take up. Nitrification prevents toxic ammonia excreted in wastes from accumulating. **Denitrification** of nitrate by bacteria converts nitrate in soil or water to a gaseous form.

Human activities add nitrogen-containing compounds to the air and water. Nitrous oxide is a greenhouse gas that also contributes to destruction of the protective ozone layer. It is formed by burning fossil fuels and by the activity of bacteria in habitats enriched by nitrogen fertilizer. Nitrate is a soluble compound that sometimes pollutes drinking water. Ingestion of excess nitrate is correlated with health problems.

Section 46.10 The **phosphorus cycle** is a sedimentary cycle. Earth's crust is the largest reservoir of phosphorus, a element that does not occur as a gas in any significant quantity. Producers cannot access the phosphate that is tied up in rocks; they obtain the phosphorus they need by taking up dissolved phosphates. Lack of phosphate often limits plant growth. To overcome this limitation, farmers apply fertilizer. Deposits of bat and bird wastes are mined as a natural phosphate-rich fertilizer, and phosphate-rich rocks are used to produce fertilizer on an industrial scale.

SELF-QUIZ

Answers in Appendix VII

1. In most ecosystems, the primary producers use energy from _______ to build organic compounds.
   a. sunlight
   b. heat
   c. breakdown of wastes and remains
   d. breakdown of inorganic substances in the habitat

2. Organisms at the lowest trophic level in a tallgrass prairie are all _______.
   a. two steps away from the original energy input
   b. autotrophs
   c. heterotrophs
   d. both a and b
   e. both a and c

3. All organisms at the top trophic level _______.
   a. capture energy from a nonliving source
   b. obtain carbon from a nonliving source
   c. would be at the top of an energy pyramid
   d. all of the above

4. Primary productivity is affected by _______.
   a. nutrient availability
   b. amount of sunlight
   c. temperature
   d. all of the above

5. Efficiency of energy transfers in aquatic ecosystems is typically higher than in land ecosystems because _____.
   a. aquatic food webs include more endotherms
   b. algae do not make lignin
   c. primary production cannot occur in water
   d. all of the above

6. Most of Earth's freshwater is _______.
   a. in lakes and streams
   b. in aquifers and soil
   c. frozen as ice
   d. in bodies of organisms

7. Earth's largest carbon reservoir is _______.
   a. the atmosphere
   b. sediments and rocks
   c. seawater
   d. living organisms

8. Carbon is released into the atmosphere by _______.
   a. photosynthesis
   b. the greenhouse effect
   c. burning fossil fuels
   d. fertilizer use

9. Greenhouse gases _______.
   a. slow the escape of heat energy from Earth into space
   b. are produced by natural and human activities
   c. are at higher levels than they were 100 years ago
   d. all of the above

10. The _______ cycle is a sedimentary cycle.
    a. phosphorus
    b. carbon
    c. nitrogen
    d. water

11. Earth's largest phosphorus reservoir is _______.
    a. the atmosphere
    b. the ocean
    c. sedimentary rock
    d. living organisms

12. Plant growth requires uptake of _______ from the soil.
    a. nitrogen
    b. carbon
    c. phosphorus
    d. both a and c
    e. all of the above

13. Nitrogen fixation converts _______ to _______.
    a. nitrogen gas; ammonia
    b. nitrates; nitrites
    c. ammonia; nitrogen gas
    d. ammonia; nitrates
    e. nitrogen gas; nitrogen oxides

14. Burning fossil fuel releases _______ into the air.
    a. carbon dioxide
    b. nitrous oxide
    c. phosphates
    d. a and b

15. Match each term with its most suitable description.

___ carbon dioxide	a. contains triple bond
___ nitrate	b. mined from sedimentary rock
___ phosphate	c. marine carbon source
___ nitrogen gas	d. soluble form of nitrogen
___ bicarbonate	e. greenhouse gas

CRITICAL THINKING

1. Where does your drinking water come from? An aquifer or an aboveground reservoir? What area is included within your watershed? Visit the Science in Your Watershed site at http://water.usgs.gov/wsc to find out.

2. Scientists study bubbles trapped in ancient glacial ice to determine how concentrations of nitrogen and carbon dioxide gas have changed over time. However, bubbles in glacial ice cannot provide information about changes in phosphorus. Explain why air samples are not useful for this purpose, and propose an alternative method to study how the amount of phosphorus in a region has changed over time.

3. Nitrogen-fixing bacteria live throughout the ocean, from its sunlit upper waters to 200 meters (650 feet) beneath its surface. Recall that nitrogen is a limiting factor in many habitats. What effect would an increase in populations of marine nitrogen-fixers have on carbon uptake and primary productivity in those waters?

47 THE BIOSPHERE

CORE CONCEPTS

 Pathways of Transformation

Organisms exchange matter and energy with the environment in order to grow, maintain themselves, and reproduce.
The main flow of energy through the biosphere starts with the capture of solar energy by way of photosynthesis. Regional differences in the amount of solar energy that reaches the Earth's surface, and the extent to which that energy varies with the seasons, give rise to regional differences in primary production.

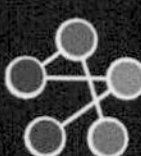 Systems

Complex properties arise from interactions among components of a biological system.
A biome consists of geographically separated regions that have a similar climate and soils. These regions support producers that are similarly adapted to the conditions within them. The distribution of different types of producers, in combination with differences in climate, determines the types of consumers that each biome includes.

Evolution

Evolution underlies the unity and diversity of life.
Evolved differences in the ability to withstand cold, heat, drought, fire, and salinity allow different species to thrive in different environments. Organisms found in the same biome in different parts of the world often have independently evolved adaptations to the environmental stresses that occur in that biome.

Links to Earlier Concepts

This chapter's topic is the biosphere, the highest level of organization in nature (Section 1.2). You will draw on your knowledge of soils (28.2), primary production (46.4), and food webs (46.3), as well as properties of water (2.5), eutrophication (46.1), coral reefs (24.5), and deep-sea hydrothermal vents (19.3).

47.1 Going with the Flow

Earth's air and seas are in constant circulation, distributing materials on a global scale. Consider what happened after a powerful earthquake that occurred off the northeast coast of Japan triggered a huge tsunami (tidal wave) in March of 2011. Together, the earthquake and tsunami killed more than 15,000 people and decimated coastal cities. Amid this destruction, a nuclear power plant on the shore of the city of Fukushima became damaged and released some radioactive material into the air and sea.

Prevailing winds carried radioisotopes accidentally released into the air at Fukushima eastward. Rain deposited the vast majority of this radioactive material in the Pacific Ocean, but some remained aloft and continued farther east. Radioisotopes from Fukushima were first detected on the west coast of North America about 60 hours after their release, and in Europe about a week later. Within 18 days, some radioisotopes released at Fukushima had circled the globe.

Materials move more slowly in the ocean. Many millions of tons of debris was dragged into the ocean by the tsunami. Most of this material sank, but some remained afloat and was carried along by surface currents, which also flow predominately east from Japan.

Scientists have been monitoring the movement of the floating material and recording when objects that are clearly tsunami debris turn up along the west coast of North America. Several Japanese boats lost during the tsunami have come ashore. One arrived in Washington state in early 2013 carrying five live fishes and a variety of invertebrates native to the tropical Pacific. California received its first verified tsunami debris the same year—a boat belonging to a high school that was destroyed by the tsunami. The boat was cleaned up by American high school students and returned to Japan.

Movement of radioactive water that entered the sea at the Fukushima nuclear plant is also under close scrutiny. This water contains cesium 137 (^{137}Cs), a radioisotope with a half-life of 30 years. Ocean currents carried ^{137}Cs-enriched water eastward from Japan but currents move much more slowly than winds. ^{137}Cs transported by currents was first detected on the shore of North America's west coast in 2015, and it is expected to continue to arrive through about 2020.

Although this water contains more ^{137}Cs than normal seawater, scientists think it is unlikely to pose a threat to human health. The water's ^{137}Cs concentration is expected to remain below the level that the Environmental Protection Agency currently allows in drinking water.

47.2 Global Air Circulation Patterns

LEARNING OBJECTIVES

- Explain why the amount of sunlight that reaches the ground varies with latitude (distance from the equator).
- Describe how latitudinal differences in sunlight energy lead to global air circulation patterns that affect climate.
- Explain why winds trace a curved path relative to Earth's surface.

The **biosphere** includes all places on Earth where life exists. The geographical distribution of species within the biosphere depends largely on climate. **Climate** refers to average weather conditions, such as cloud cover, temperature, humidity, and wind speed, over time. Regional climates differ because factors that influence winds and ocean currents vary from place to place. Such factors include the intensity of sunlight, the distribution of landmasses and seas, and elevation.

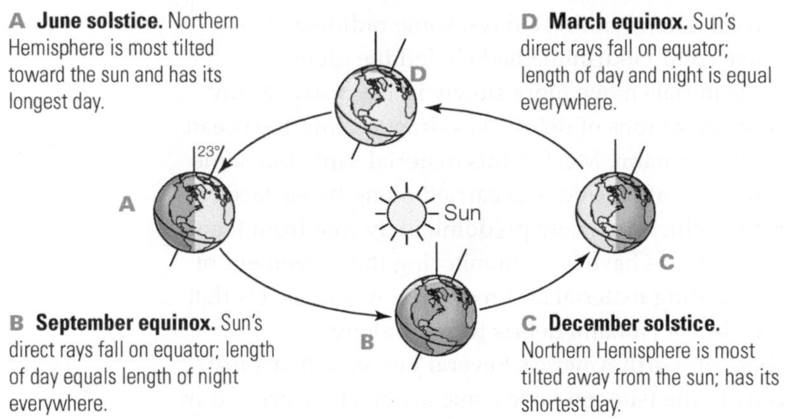
A **June solstice.** Northern Hemisphere is most tilted toward the sun and has its longest day.
D **March equinox.** Sun's direct rays fall on equator; length of day and night is equal everywhere.
23°
Sun
B **September equinox.** Sun's direct rays fall on equator; length of day equals length of night everywhere.
C **December solstice.** Northern Hemisphere is most tilted away from the sun; has its shortest day.

FIGURE 47.1 Effects of Earth's tilt and yearly rotation around the sun. The 23 degree tilt of Earth's axis causes the Northern Hemisphere to receive more intense sunlight and have longer days in summer than in winter.

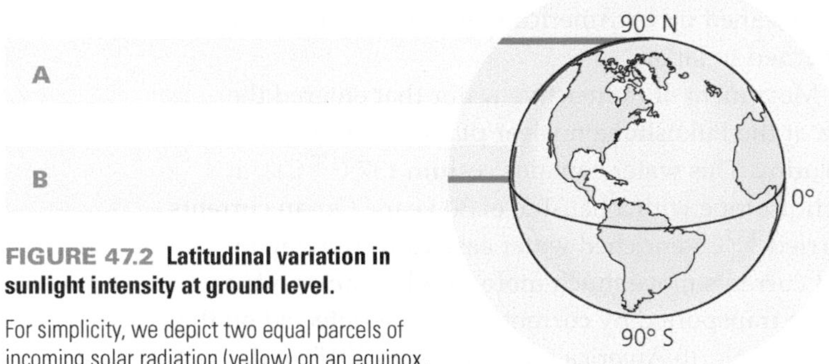
90° N
A
B
0°
90° S

FIGURE 47.2 Latitudinal variation in sunlight intensity at ground level.

For simplicity, we depict two equal parcels of incoming solar radiation (yellow) on an equinox, when incoming rays are perpendicular to Earth's axis. Rays that fall on high latitudes (**A**) have passed through more atmosphere (blue) than those that fall near the equator (**B**). Compare the length of the green lines. (Width of atmosphere is not to scale.)

In addition, energy in the rays that fall at the high latitude is spread over a greater area than energy that falls on the equator. Compare the length of the red lines.

As a result of these two factors, the amount of solar energy that reaches the Earth's surface decreases with increasing latitude.

Seasonal Effects

Each year, Earth rotates around the sun in an elliptical path (**FIGURE 47.1**). Seasonal changes in day length and temperature arise because Earth's axis is not perpendicular to the plane of this ellipse, but rather tilts about 23 degrees. In June, when the Northern Hemisphere is tipped toward the sun, this hemisphere receives more intense sunlight and has longer days than the Southern Hemisphere, which is tipped away from the sun. The June solstice is the day when the Northern Hemisphere receives the most sunlight (**FIGURE 47.1A**), and the Southern Hemisphere receives the least. On the December solstice, the opposite occurs (**FIGURE 47.1C**). Twice a year—on spring and autumn equinoxes—Earth's axis is perpendicular to incoming sunlight. On these days, every place on Earth has 12 hours of daylight and 12 hours of darkness (**FIGURES 47.1B** and **D**).

In each hemisphere, the extent of seasonal change in daylight increases with latitude. Latitude, the distance from the equator, is measured in degrees (°), with the equator being a latitude of 0° and the poles being 90°. In Honolulu, Hawaii (21° north of the equator) the longest day is 13 hours and the shortest is 10 hours. In Anchorage, Alaska (61° north of the equator) the longest day is 19 hours and the shortest 5.5 hours.

Air Circulation and Rainfall

On any given day, equatorial regions receive more sunlight energy than higher latitudes, so Earth's surface warms more at the equator than at the poles. There are two reasons for this difference in solar energy input. First, the atmosphere contains particles of dust, water vapor, and greenhouse gases that absorb some solar radiation or reflect it back into space. Sunlight traveling to high latitudes passes through more atmosphere to reach Earth's surface than light traveling to the equator, so less energy reaches the ground (**FIGURE 47.2A**). Second, energy in an incoming parcel of sunlight is dispersed over a smaller surface area at the equator than at the higher latitudes (**FIGURE 47.2B**).

Knowing about two properties of air can help you understand how regional differences in surface warming give rise to global air circulation and rainfall patterns. First, as air warms, it becomes less dense and rises. Hot air balloonists take advantage of this effect when they ascend by heating the air inside their balloon. Second, warm air can hold more water vapor than cooler air. This is why you can "see your breath" in cold weather. When you exhale, warm air with

biosphere All regions of Earth that can support life.
climate Average weather conditions in a region over a long period.

Initial Pattern of Air Circulation

D At the poles, cold air sinks and moves toward lower latitudes.

C Air rises again at 60° north and south, where air flowing poleward meets air coming from the poles.

B As the air flows toward higher latitudes, it cools and loses moisture as rain. At around 30° north and south latitude, the air sinks and flows north and south along Earth's surface.

A Warmed by energy from the sun, air at the equator picks up moisture and rises. It reaches a high altitude, and spreads north and south.

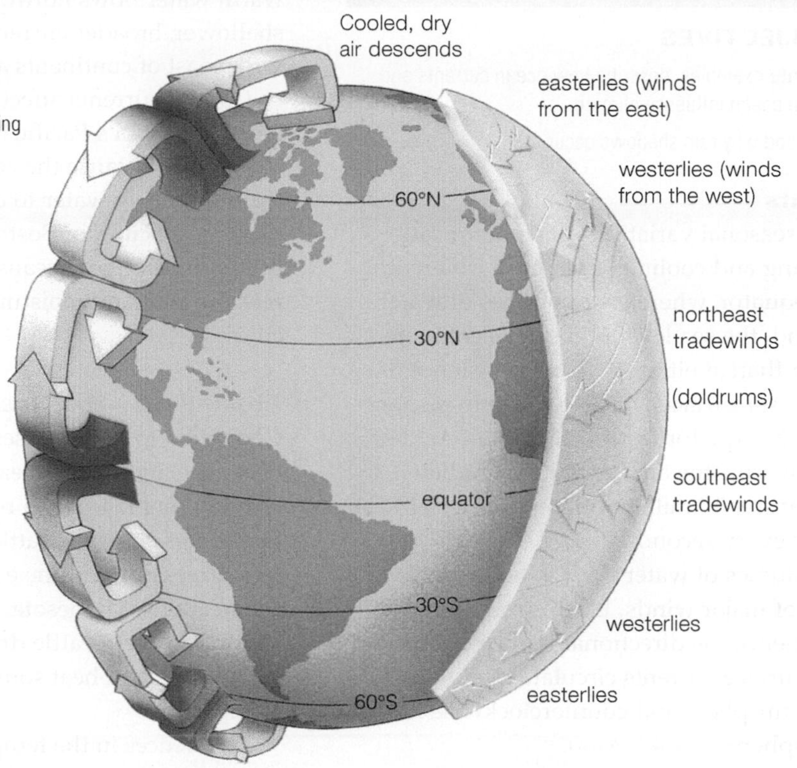
Cooled, dry air descends
easterlies (winds from the east)
westerlies (winds from the west)
60°N
30°N
northeast tradewinds
(doldrums)
southeast tradewinds
equator
30°S
westerlies
60°S
easterlies

Prevailing Wind Patterns

E Major winds near Earth's surface do not blow directly north and south because of Earth's rotation. Winds deflect to the right of their original direction in the Northern Hemisphere and to the left in the Southern Hemisphere.

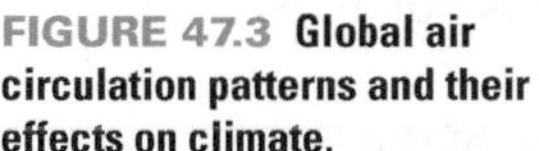

FIGURE 47.3 Global air circulation patterns and their effects on climate.

FIGURE IT OUT What is the direction of prevailing winds in the central United States?

Answer: From west to east.

moisture from your lungs cools, causing the water in that air to condense as tiny droplets.

The global air circulation pattern begins at the equator, where intense sunlight warms air and causes evaporation from the ocean. As a result, warm, moist air rises (FIGURE 47.3A). As this air ascends to higher altitudes, it moves north and south and cools, releasing moisture as rain that supports tropical rain forests.

By the time the air has reached 30° north or south of the equator, it has given up most moisture and cooled off, so it sinks back toward Earth's surface (FIGURE 47.3B). Many of the world's great deserts, including the Sahara, are about 30° from the equator.

As air continues along Earth's surface toward the poles, it again picks up heat and moisture. At a latitude of about 60°, it rises (FIGURE 47.3C). The resulting rains support temperate zone forests.

Cold, dry air descends near the poles (FIGURE 47.3D). Precipitation is sparse, and polar deserts form.

Surface Wind Patterns

The rising and falling of air masses set in motion the pattern of prevailing winds (FIGURE 47.3E). Air flows as wind across Earth's surface, moving from areas where air is sinking downward (at the poles and at 30° north and south latitude) toward areas where it is rising (at the equator and at 60° north and south latitude). Thus, winds in the tropics tend to blow toward the equator, whereas those in temperate regions tend to blow toward the poles.

Winds do not blow directly north and south because of Earth's rotation. As Earth rotates on its axis from west to east, points on the equator (where Earth is widest) travel faster than points nearer the poles (where Earth is narrower). Regional differences in the speed of rotation deflect winds east or west. In the tropics, prevailing winds (the trade winds) blow from the east toward the equator. In temperate zones, prevailing winds blow from west to east toward the poles. Winds are named for the direction from which they blow, so prevailing winds in the United States are westerlies.

TAKE-HOME MESSAGE 47.2

✔ Equatorial regions receive more sunlight than higher latitudes.

✔ Sunlight drives the rise of warm, moist air at the equator. Air cools as it moves north and south, releasing rains that support tropical forests. Deserts form where cool, dry air descends. Sunlight energy also drives moisture-laden air aloft at 60° north and south latitude. This air gives up moisture as it flows toward the equator or a pole.

✔ Major surface winds arise as air in the lower atmosphere flows from latitudes where air is sinking toward latitudes where air is rising. These winds trace a curved path relative to Earth's surface because of Earth's rotation.

47.3 Oceans, Landforms, and Climate

LEARNING OBJECTIVES

- Using appropriate examples, describe how ocean currents and proximity to the ocean influence climate.
- Explain where and why rain shadows occur.

Ocean Currents

Latitudinal and seasonal variations in sunlight cause differential heating and cooling of water in different regions. At the equator, where vast volumes of water warm and expand, the sea level is about 8 centimeters (3 inches) higher than at either pole. The existence of this slope causes sea surface water to move in response to gravity, from the equator toward the poles. As the water moves, it warms the air above it. At midlatitudes, oceans transfer 10 million billion calories of heat energy to the air every second!

Enormous volumes of water flow as ocean currents. The force of major winds, Earth's rotation, and topography influence the directional movement of these currents. Surface currents circulate clockwise in the Northern Hemisphere and counterclockwise in the Southern Hemisphere (**FIGURE 47.4**).

Swift, deep, and narrow currents of nutrient-poor water flow away from the equator along the east coast of continents. Along the east coast of North America, warm water flows north, as the Gulf Stream. Slower, shallower, broader currents of cold water parallel the west coast of continents and flow toward the equator.

Coastal currents affect coastal climates. Coasts of North America's Pacific Northwest are cool and foggy in summer because the cold California current chills the air, causing water to condense out as droplets. The coastal cities of Boston and Baltimore are hot and humid in summer because the warm Gulf Stream releases heat and moisture into the air over these cities.

Proximity to the Ocean

Compared to land masses, the oceans have a higher heat capacity, so they heat and cool more slowly than any adjacent land. As a result, proximity to an ocean moderates climate. Seattle, Washington, which is about 48 degrees north of the equator has milder winters than Minneapolis, Minnesota, which is 2 degrees farther south. Air over Seattle draws heat from the adjacent Pacific Ocean, a heat source that is not available to Minneapolis.

Differences in the temperature of the oceans and the shore give rise to coastal breezes. During the day, land warms faster than water. As air over the land warms

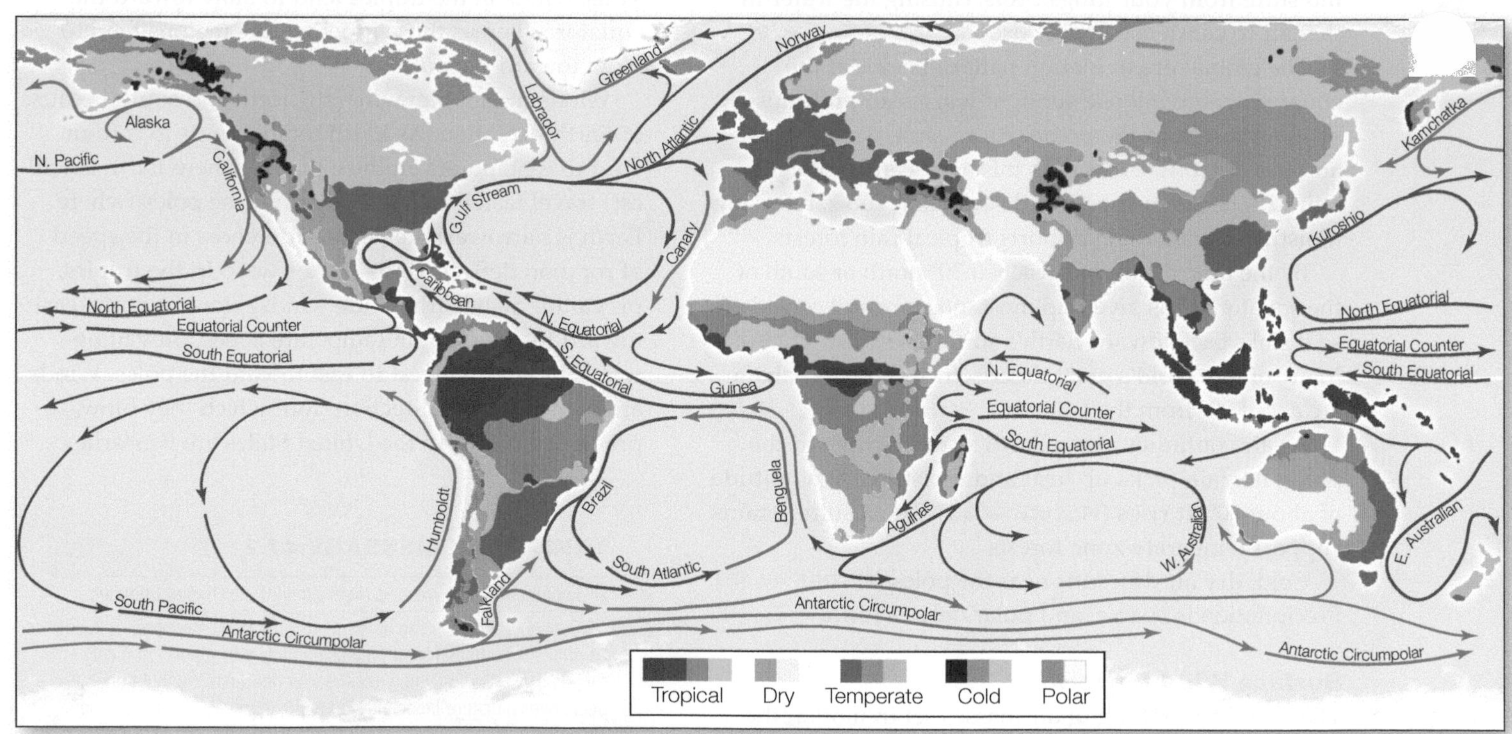

Norway
Greenland
Labrador
Alaska
N. Pacific
California
North Atlantic
Gulf Stream
Canary
Caribbean
Kamchatka
Kuroshio
North Equatorial
Equatorial Counter
South Equatorial
N. Equatorial
S. Equatorial
Guinea
N. Equatorial
Equatorial Counter
South Equatorial
North Equatorial
Equatorial Counter
South Equatorial
Humboldt
Brazil
Benguela
Agulhas
W. Australian
E. Australian
South Atlantic
South Pacific
Falkland
Antarctic Circumpolar
Antarctic Circumpolar
Antarctic Circumpolar
Tropical
Dry
Temperate
Cold
Polar

FIGURE 47.4 Major climate zones correlated with surface currents of the world ocean. Warm surface currents start moving from the equator toward the poles, but prevailing winds, Earth's rotation, gravity, the shape of ocean basins, and landforms influence the direction of flow. Water temperatures, which differ with latitude and depth, contribute to the regional differences in air temperature and rainfall.

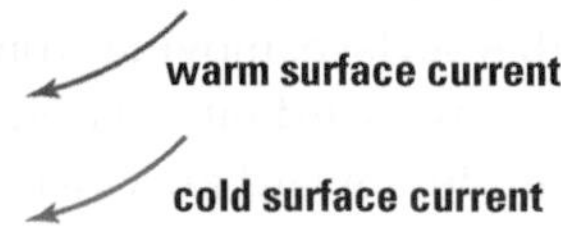

warm surface current
cold surface current

CREDIT: (4) NASA.

and rises, cooler air from offshore moves in to replace it (FIGURE 47.5A). After sundown, land cools more quickly than the water, so the breezes reverse direction (FIGURE 47.5B).

Differential heating of water and land also causes **monsoons**, which are winds that change direction seasonally. In the summer, the continental interior of Asia heats up, causing air to warm and rise above it. Moist air from over the warm Indian Ocean to the south moves in to replace the rising air, and this north-blowing wind delivers heavy rains. In the winter, the continental interior of Asia is cooler than the ocean. As a result, a cool, dry wind blows from the north toward southern coasts causing a seasonal drought.

Effects of Land Features

Mountains, valleys, and other surface features of the land affect climate too. Suppose you track a warm air mass after it picks up moisture off California's coast. This air moves inland as wind from the west and piles up against the Sierra Nevada, a high mountain range that parallels the coast. As the air rises in altitude, it cools and loses moisture as rain (FIGURE 47.6). The result is a **rain shadow**, a semiarid or arid region of sparse rainfall on the leeward side of high mountains. *Leeward* is the side facing away from the wind. The Himalayas, Andes, Rockies, and other great mountain ranges cast similar rain shadows.

monsoon (mon-SOON) Wind that reverses direction seasonally.
rain shadow Dry region downwind of a coastal mountain range.

A In afternoons, land is warmer than the sea, so a breeze blows onto shore.
cool air
warm air

B In evenings, the sea is warmer than land, so the breeze blows out to sea.

FIGURE 47.5 Coastal breezes.

TAKE-HOME MESSAGE 47.3

- ✔ Surface ocean currents set in motion by latitudinal differences in solar radiation distribute heat. These currents are affected by winds and by Earth's rotation.
- ✔ Collective effects of air masses, oceans, and landforms determine regional temperature and annual precipitation.

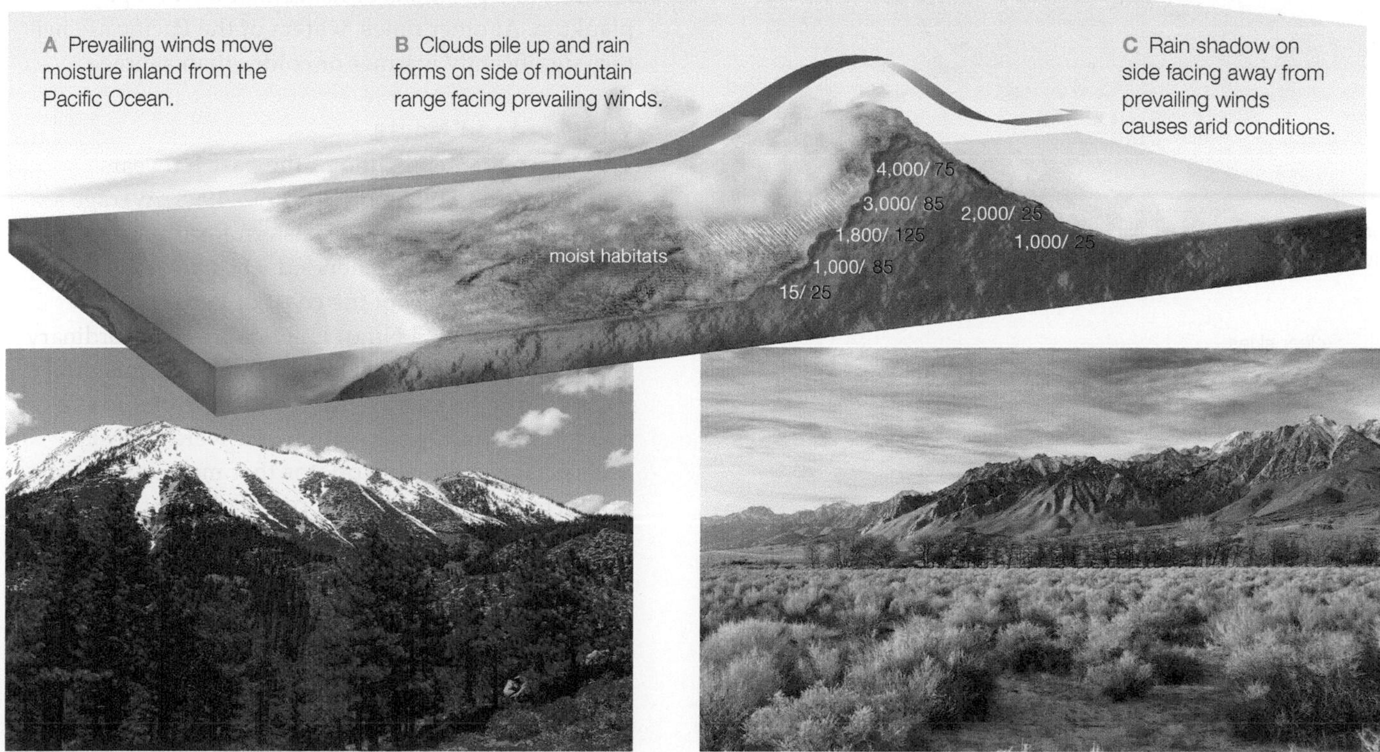
A Prevailing winds move moisture inland from the Pacific Ocean.
B Clouds pile up and rain forms on side of mountain range facing prevailing winds.
C Rain shadow on side facing away from prevailing winds causes arid conditions.
moist habitats
4,000/ 75
3,000/ 85
1,800/ 125
1,000/ 85
15/ 25
2,000/ 25
1,000/ 25

FIGURE 47.6 Rain shadow effect. White numbers signify elevations, in meters. Black numbers signify annual precipitation, in centimeters, averaged on both sides of the Sierra Nevada, a mountain range. Photos show the vegetation on the two sides of this mountain range.

47.4 The El Niño Southern Oscillation

LEARNING OBJECTIVES

- Compare weather conditions during an El Niño and a La Niña.
- Give examples of the biological effects of an El Niño.

The **El Niño Southern Oscillation**, or ENSO, is a naturally occurring, irregularly timed fluctuation in sea surface temperature and wind patterns in the equatorial Pacific. The two extremes of this oscillation are referred to as El Niño and La Niña. Their influence is felt most strongly during the winter and along the western coast of South America, but they affect weather patterns year-round and worldwide (TABLE 47.1).

The term El Niño means "baby boy" and refers to Jesus Christ; it was first used by Peruvian fishermen to describe local weather changes and a shortage of fishes that occurred in some winters and began around Christmas. During an El Niño, unusually warm water flows toward the west coast of South America, displacing the Humboldt Current that would otherwise bring up cooler, nutrient-rich water from the deep (FIGURE 47.7). Without this input of nutrients, marine primary producers decline in numbers. The dwindling producer populations and warming water cause a decrease in numbers of small, cold-water fishes, as well as the larger fishes that eat them. This is why Peruvian fishermen catch fewer fishes during an El Niño.

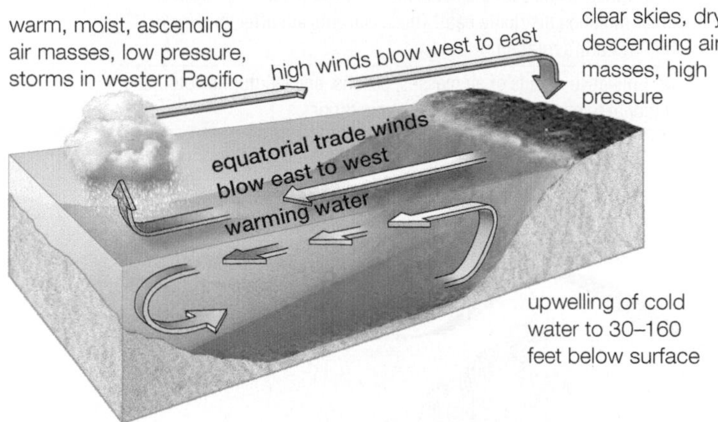
warm, moist, ascending air masses, low pressure, storms in western Pacific
high winds blow west to east
clear skies, dry descending air masses, high pressure
equatorial trade winds blow east to west
warming water
upwelling of cold water to 30–160 feet below surface

A Conditions during an average year (neither El Niño nor La Niña). Note the upward movement (upwelling) of cool, nutrient-rich water near the coast.

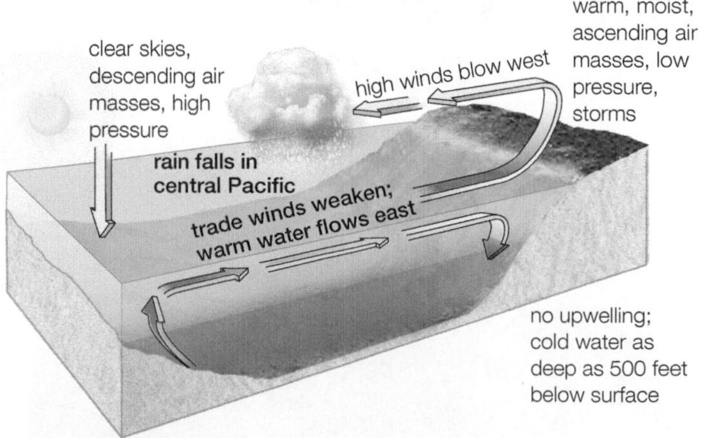
clear skies, descending air masses, high pressure
high winds blow west
warm, moist, ascending air masses, low pressure, storms
rain falls in central Pacific
trade winds weaken; warm water flows east
no upwelling; cold water as deep as 500 feet below surface

B Conditions during an El Niño. Note the lack of upwelling.

FIGURE 47.7 **Effects of El Niño on winds and currents in the equatorial Pacific.**

TABLE 47.1

Effects of El Niño and La Niña

El Niño	La Niña
Western Pacific waters warm	Western Pacific waters cool
Easterly trade winds weaken or reverse	Easterly trade winds strengthen
Less nutrient-rich cold water wells up along South America's west coast	More nutrient-rich cold water wells up along South America's west coast
More rain in western South America	Less rain in western South America
Less rain in Australia	More rain in Australia
Fewer North Atlantic hurricanes	More North Atlantic hurricanes

El Niño episodes persist for 6 to 18 months. Often they are followed by a La Niña, in which Pacific waters become cooler than usual. During a La Niña, cold nutrient-rich water flows toward the western coast of South America, phytoplankton populations rebound, and so do populations of fishes that feed on phytoplankton. At other times, waters of the Pacific are neither significantly warmer or colder than average.

Widespread Effects

The most extreme El Niño of the past 100 years occurred during the winter of 1997–1998. Average sea surface temperatures in the eastern Pacific soared and a plume of warm water extended 9,660 kilometers (6,000 miles) west from the coast of Peru.

The 1997–1998 El Niño/La Niña had extraordinary effects on the primary productivity in the equatorial Pacific. With the massive eastward flow of nutrient-poor warm water, photoautotrophs were almost undetectable in satellite photos that measure primary productivity. The dwindling populations of producers and the warming water caused decreases in populations of cold-water fishes, as well as the fish-eating animals that feed on them. During the 1997–1998 El Niño, about half of the sea lions on the Galápagos Islands starved to death. California's population of northern

El Niño Southern Oscillation Naturally occurring, irregularly timed fluctuation in sea surface temperature and wind patterns in the equatorial Pacific; affects weather worldwide.

Data Analysis Activities

Sea Temperatures To predict the effect of El Niño or La Niña events in the future, the National Oceanographic and Atmospheric Administration collects information about sea surface temperature (SST) and atmospheric conditions. They compare monthly temperature averages in the eastern equatorial Pacific Ocean to historical data and calculate the difference (the degree of anomaly) to determine if El Niño conditions, La Niña conditions, or neutral conditions are developing. El Niño is a rise in the average SST above 0.5°C. A decline of the same amount is La Niña. **FIGURE 47.8** shows data for 42 years.

1. When did the greatest positive temperature deviation occur during this time period?
2. What type of event, if any, occurred during the winter of 1982–1983? What about the winter of 2001–2002?
3. During a La Niña event, less rain than normal falls in the American West and Southwest. In the time interval shown, what was the longest interval without a La Niña event?
4. What type of conditions were in effect in the fall of 2007 when California suffered severe wildfires?

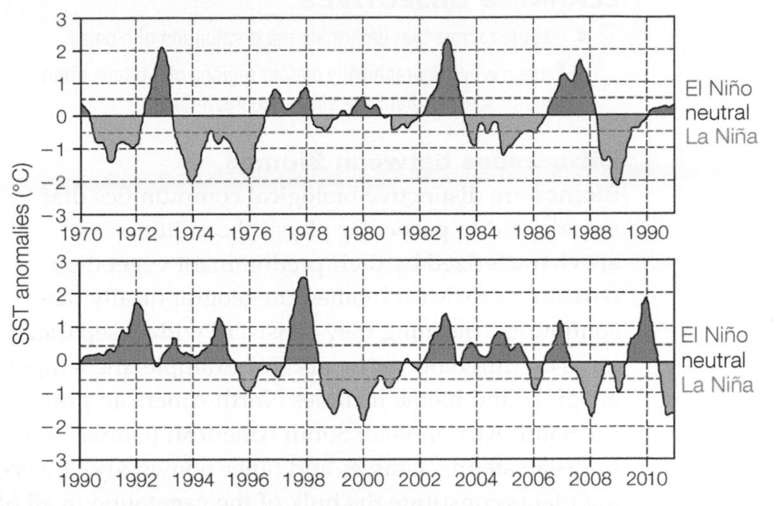

SST anomalies (°C)
3
2
1
0
−1
−2
−3
1970 1972 1974 1976 1978 1980 1982 1984 1986 1988 1990
El Niño
neutral
La Niña
1990 1992 1994 1996 1998 2000 2002 2004 2006 2008 2010
El Niño
neutral
La Niña

FIGURE 47.8 Sea surface temperature anomalies (differences from the historical mean) in the eastern equatorial Pacific Ocean. A rise above the dashed red line is an El Niño event, a decline below the blue line is La Niña.

fur seals also suffered a sharp decline. Primary productivity rebounded during the La Niña that followed.

Rainfall patterns shift during an El Niño. During the winter of 1997–1998, torrential rains caused flooding and landslides along the west coast of the Americas, while Australia and Indonesia suffered from drought-driven crop failures and raging wildfires. An El Niño typically brings cooler, wetter weather to the American Gulf states and reduces the likelihood of hurricanes.

Effects on Human Health

Outbreaks of some human diseases are more likely to occur during an El Niño. For example, the increased ocean temperature in the Pacific sometimes leads to an increased incidence of cholera, a disease that causes potentially deadly diarrhea. Copepods, a type of small crustacean, serve as a reservoir for cholera-causing bacteria between disease outbreaks. During an El Niño, the rise in the temperature of the ocean's surface results in a rise in the number of cholera-carrying copepods. An El Niño also increases the incidence of malaria in coastal South Asia and Latin America, because increased rain creates more standing water in which mosquitoes can breed.

Monitoring and Predicting

Among other duties, the United States National Oceanographic and Atmospheric Administration (NOAA) is charged with studying and predicting El Niño events. This agency collects and analyzes sea temperature data from a system of buoys moored in the tropical Pacific Ocean. The goal of the research is to determine how the El Niño Southern Oscillation affects global weather patterns and the extent of its effects. Such an understanding could help the scientists to develop a method of predicting when an El Niño or La Niña event is likely to occur and which regions are at a heightened risk for flooding, drought, hurricanes, or fires as a result. Predicting and planning for such occurrences could help prevent or minimize their harmful effects. Current data on sea surface temperature, as well as information about the monitoring program and its goals, are available on NOAA's website at www.elnino.noaa.gov.

TAKE-HOME MESSAGE 47.4

✔ The El Niño Southern Oscillation is an irregular fluctuation in conditions in the equatorial Pacific. The two extremes of this fluctuation are referred to as El Niño and La Niña.

✔ Changes in primary productivity that result from the El Niño Southern Oscillation have cascading effects on biological communities.

✔ Rainfall patterns associated with an El Niño raise the risk of some diseases such as cholera.

47.5 Biomes

LEARNING OBJECTIVES

- List the factors that determine the distribution of biomes.
- Explain why geographically distant regions of a biome often contain species that have similar characteristics.

Differences between Biomes

Biomes are distinctive biological communities that are adapted to particular physical conditions and are characterized by their predominant vegetation (**FIGURE 47.9**). Most biomes are geographically discontinuous, meaning they consist of widely separated areas on different continents. For example, the temperate grassland biome includes North American prairie, South African veld, South American pampa, and Eurasian steppe. Grasses and other nonwoody flowering plants constitute the bulk of the vegetation in all of these regions.

The type of biome characteristic of an area depends largely on rainfall and temperature. Desert biomes receive the least annual rainfall, grasslands and shrublands receive more, and forests receive the most. Warm deserts occur where temperatures soar highest and tundra where they drop the lowest.

Soils also influence biome distribution. Soils consist of a mixture of mineral particles and varying amounts of humus. Water and air fill spaces between soil particles. Properties of soils vary depending on the types, proportions, and compaction of particles. Deserts have sandy or gravelly, fast-draining soil with little topsoil. (Recall from Section 28.2 that topsoil is an uppermost soil layer that is rich in organic material and nutrients.) Topsoil is deepest in natural grasslands, where it can be more than one meter thick. This is why grasslands are often converted to agricultural uses.

Climate and soils influence primary production, so primary production varies greatly among biomes (**FIGURE 47.10**).

Similarities within a Biome

Unrelated species living in widely separated parts of a biome often have similar body structures that arose by the process of morphological convergence

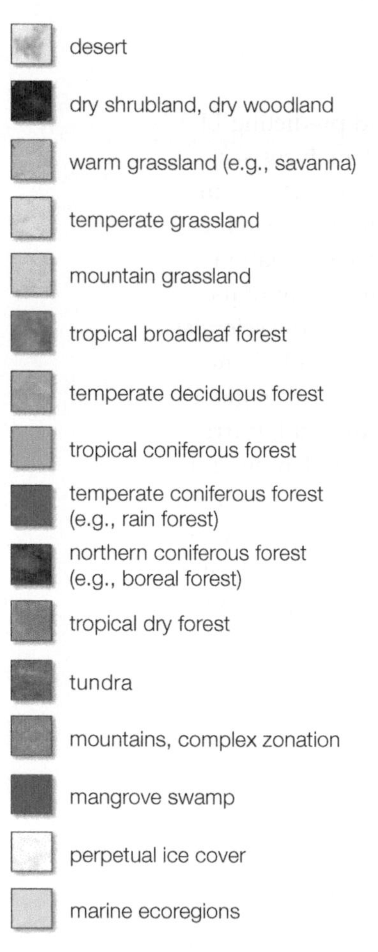
desert
dry shrubland, dry woodland
warm grassland (e.g., savanna)
temperate grassland
mountain grassland
tropical broadleaf forest
temperate deciduous forest
tropical coniferous forest
temperate coniferous forest (e.g., rain forest)
northern coniferous forest (e.g., boreal forest)
tropical dry forest
tundra
mountains, complex zonation
mangrove swamp
perpetual ice cover
marine ecoregions

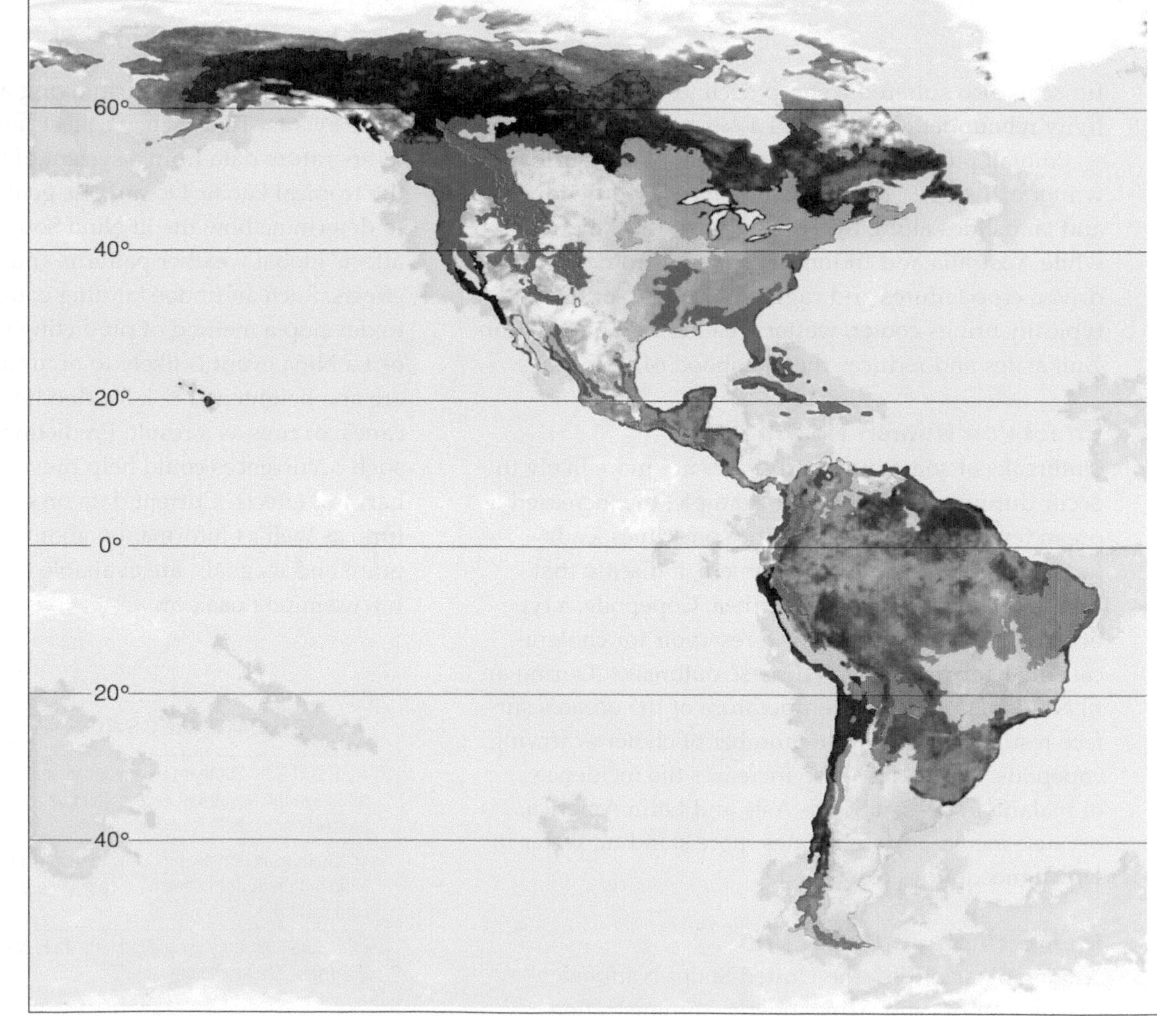
60°
40°
20°
0°
20°
40°

FIGURE 47.9 Global distribution of major categories of biomes and marine ecoregions.

CREDIT: (9) NASA.

(Section 18.3). For example, cacti with water-storing stems live in North American deserts and euphorbs with water-storing stems live in African deserts. Cacti and euphorbs do not share an ancestor with a water-storing stem. Rather, this feature evolved independently in the two groups as a result of similar selection pressures. Similarly, an ability to carry out C4 photosynthesis evolved independently in grasses growing in warm grasslands on different continents. C4 photosynthesis is more efficient than the more common C3 pathway under hot, dry conditions (Section 6.5).

TAKE-HOME MESSAGE 47.5

✔ Biomes are vast expanses of land dominated by distinct kinds of plants that support characteristic communities.

✔ The global distribution of biomes is a result of topography, climate, and evolutionary history.

biome (BY-ohm) Group of regions that may be widely separated but share a characteristic climate, soil composition, and dominant vegetation.

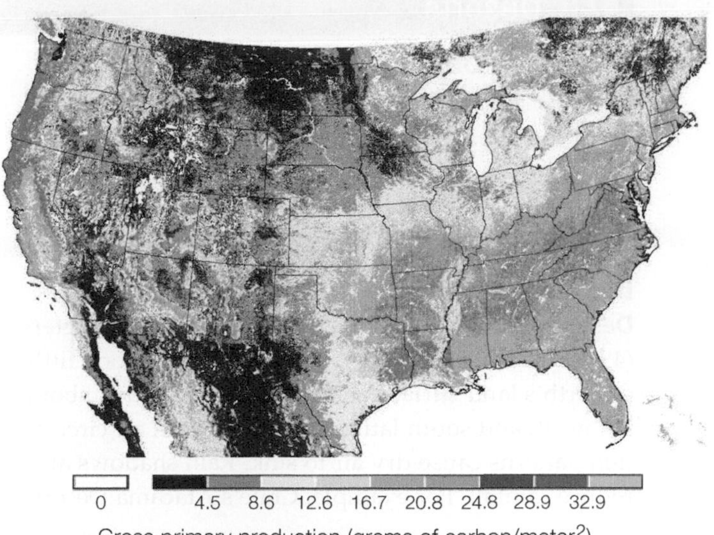
0
4.5 8.6 12.6 16.7 20.8 24.8 28.9 32.9
Gross primary production (grams of carbon/meter^2)

FIGURE 47.10 Remote satellite monitoring of gross primary productivity across the United States. The differences roughly correspond with variations in soil composition and moisture.

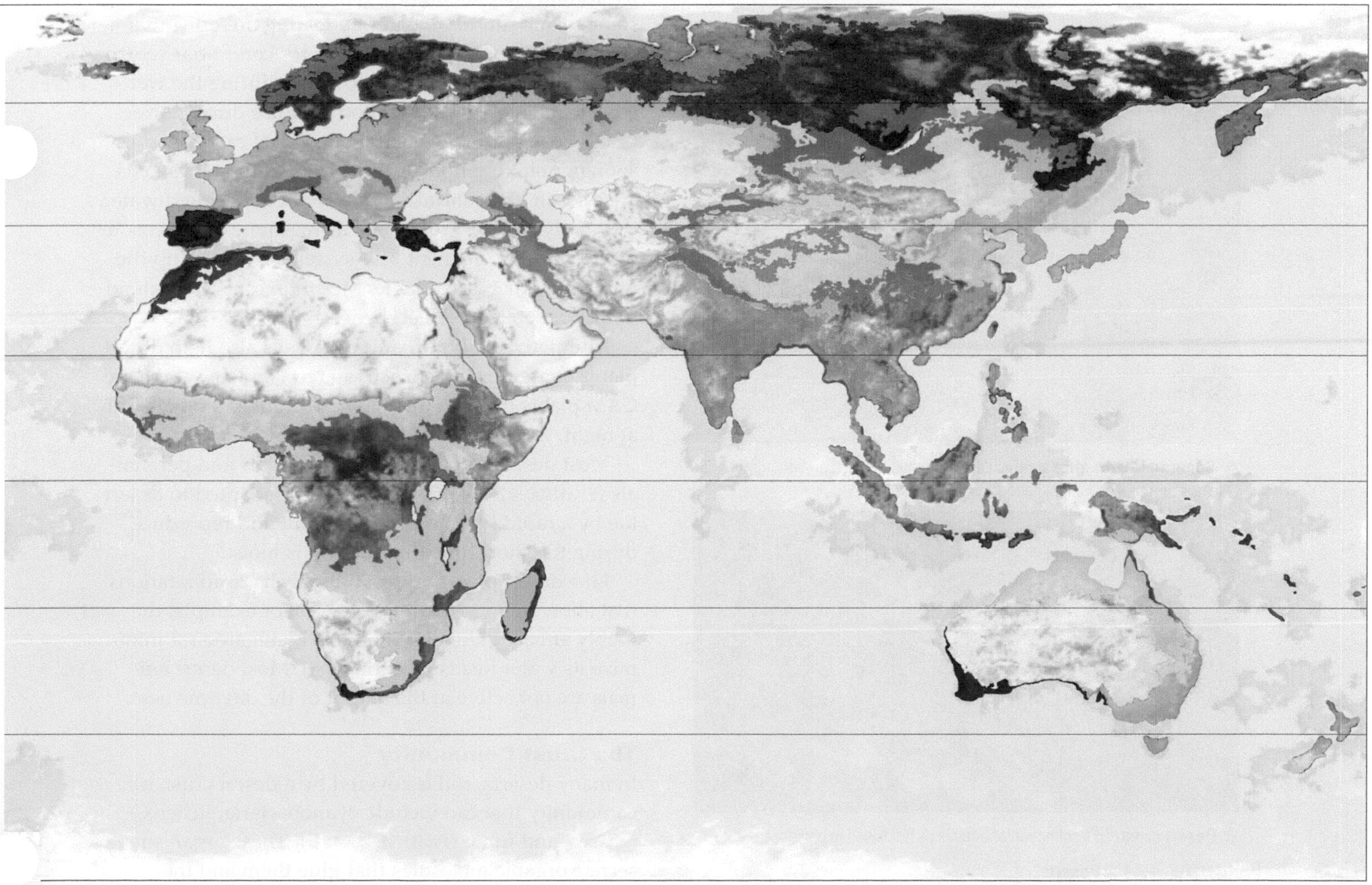

CREDIT: (10) NASA.

47.6 Deserts

LEARNING OBJECTIVES

- Describe the climate of desert biomes.
- List some traits that adapt plants to life in the desert.
- Describe the composition of the desert crust and explain its ecological importance.

Desert Locations and Conditions

Deserts receive an average of less than 10 centimeters (4 inches) of rain per year. They cover about one-fifth of Earth's land surface and many are located at about 30° north and south latitude, where global air circulation patterns cause dry air to sink. Rain shadows also reduce rainfall. For example, Chile's Atacama Desert is on the leeward side of the Andes, and the Himalayas prevent rain from falling in China's Gobi desert.

Lack of rainfall keeps the humidity in deserts low. With little water vapor to block the sun's rays, intense sunlight reaches and heats the ground. At night, the lack of insulating water vapor in the air allows the temperature to fall fast. As a result, deserts tend to have larger daily temperature shifts than other biomes.

Desert soils have little topsoil, the layer most important for plant growth. These soils often become somewhat salty, because rain that falls evaporates before seeping into the ground. Rapid evaporation allows any salt in rainwater to accumulate at the soil surface.

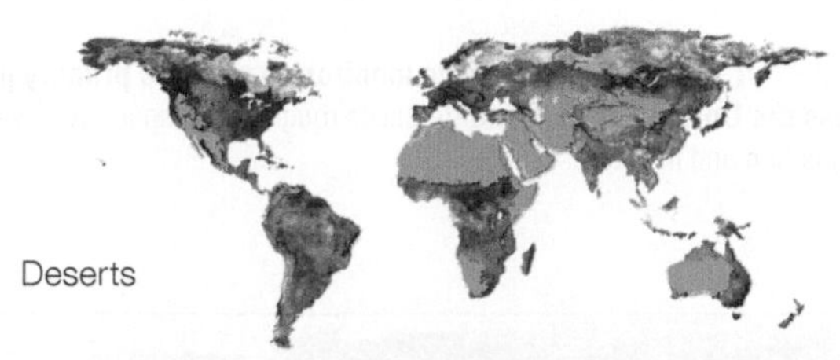
Deserts

A Sonoran Desert after a rain. Perennial saguaro cacti and drought-resistant shrubs grow beside annual wildflowers.

B Desert crust. This collection of organisms holds soil in place.

FIGURE 47.11 Deserts.

Adaptations to Desert Life

Despite their harsh conditions, most deserts support some plant life. Diversity is highest in regions where soil moisture is available in more than one season.

Many desert plants have adaptations that reduce water loss. Some have tiny leaves or no leaves. Spines or hairs serve both to deter herbivores and to reduce evaporation. By trapping water, they keep the humidity around stomata high. Where rains fall seasonally, some plants conserve water by leafing out only after a rain, then dropping leaves when dry conditions return.

Other desert plants take up water during the wet season and store it in their body for use during drier times. For example, the stem of a barrel cactus has a spongy pulp that holds water. The cactus stem swells after a rain, then shrinks as the plant uses stored water.

Woody desert shrubs such as mesquite and creosote have extensive, efficient root systems that take up the little water that is available. Mesquite roots can extend up to 60 meters (197 feet) beneath the soil surface.

Alternative carbon-fixing pathways also help desert plants conserve water. Cacti, agaves, and euphorbs are CAM plants (Section 6.5). They open their stomata only at night, when the temperature declines.

Most deserts contain a mix of annuals and perennials (**FIGURE 47.11A**). The annuals are adapted to desert life by a rapid life cycle. They sprout and reproduce during the short time that the soil is moist.

Like desert plants, desert animals have adaptations that allow them to conserve water. For example, the highly efficient kidneys of a desert kangaroo rat minimize its water needs (Section 40.8). Most desert animals are not active at the height of the daytime heat.

The Crust Community

In many deserts, soil is covered by a desert crust, a community that can include cyanobacteria, lichens, mosses, and fungi (**FIGURE 47.11B**). These organisms secrete organic molecules that glue them and the surrounding soil particles together. The crust benefits members of the larger desert community in important ways. Its cyanobacteria fix nitrogen and make ammonia available to plants. The crust also holds soil particles in place. When the fragile connections within the desert crust are broken, soil can blow away. Negative effects of such disturbance are heightened when windblown soil buries healthy crust in an undisturbed area, killing additional crust organisms and allowing more soil to take flight.

TAKE-HOME MESSAGE 47.6

✔ A desert gets little rain and has low humidity. There is plenty of sunlight, but lack of water prevents most plants from surviving here.

✔ The predominant plants in deserts have adaptations that allow them to reduce water lost by transpiration, store water, or access water deep below the soil surface.

✔ Desert soils are held in place by a community of organisms that form a desert crust.

47.7 Grasslands and Dry Shrublands

LEARNING OBJECTIVES

- Describe the dominant vegetation in grasslands and chaparral.
- Explain the role of fire and grazers in grasslands and chaparral.

Plants adapted to periodic lightning-ignited fires dominate grasslands, savanna, and chaparral.

Grasslands

Grasslands form in the interior of continents between deserts and temperate forests. Their soils are highly fertile, with a deep layer of topsoil that has been enriched by the decay of countless grass roots (**FIGURE 47.12A**). Annual rainfall is sufficient to prevent desert from forming, but not enough to support woodlands. The dominant plants are low-growing grasses and other nonwoody plants that tolerate strong winds, sparse and infrequent rain, and intervals of drought. Constant trimming by grazers, along with periodic fires, prevents trees and most shrubs from taking hold.

Temperate grasslands are warm in summer, but cold in winter. Annual rainfall is 25 to 100 centimeters (10 to 40 inches), with rains throughout the year. Grass roots extend profusely through the thick topsoil and help hold it in place, preventing erosion by the constant winds. North America's grasslands are shortgrass and tallgrass prairies (**FIGURE 47.12B**).

Tallgrass prairie has somewhat richer topsoil and slightly more frequent rainfall than shortgrass prairie. Before the arrival of Europeans, it covered about 140 million acres, mostly in Kansas. Nearly all tallgrass prairie has now been converted to cropland. North America's prairies once supported enormous herds of elk, pronghorn antelope, and bison that were prey to wolves. Today, these predators and prey are absent from most of their former range.

desert Biome with little rain and low humidity; plants that have water-storing and water-conserving adaptations predominate.
savanna (suh-VAN-uh) Tropical biome dominated by grasses with a few scattered shrubs and trees.
temperate grassland Temperate biome in the interior of continents; perennial grasses and other nonwoody plants adapted to grazing and fire predominate.

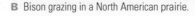

Temperate grasslands
and tropical savannas
Topsoil:
Alkaline, deep,
rich in humus
Percolating
water enriches
layer with
calcium
carbonates

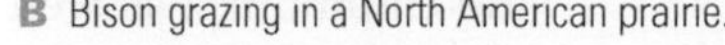

A Prairie soil profile. **B** Bison grazing in a North American prairie. **C** Wildebeest grazing on an African savanna.

FIGURE 47.12 Grasslands.

CREDITS: (in text, 12A) © Cengage Learning; (12B) U.S. Fish & Wildlife Service; (12C) Jonathan Scott/Planet Earth Pictures.

FIGURE 47.13 Dry shrubland. Chaparral in California.

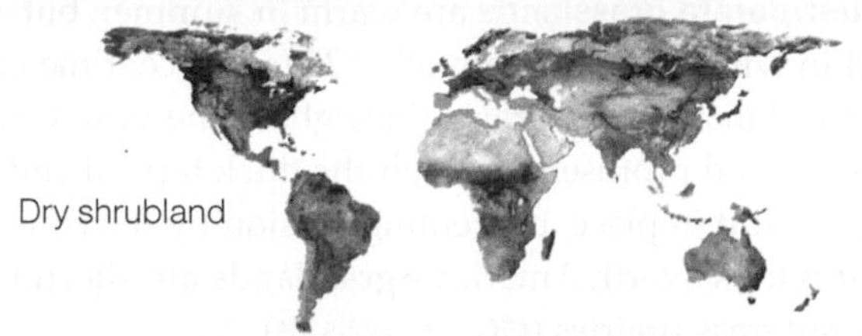
Dry shrubland

Tropical Savannas

Savannas are broad belts of grasslands with a few scattered shrubs and trees (**FIGURE 47.12C**). Savannas lie between the tropical forests and hot deserts of Africa, India, and Australia. Temperatures are warm year-round. During the rainy season, 90 to 150 centimeters (35 to 60 inches) of rain falls. Africa's savannas are famous for their abundant wildlife. Herbivores include giraffes, zebras, elephants, a variety of antelopes, and immense herds of wildebeests. Lions and hyenas are carnivores that eat the grazers.

Dry Shrublands

Fire-adapted shrubs dominate **dry shrublands**. This biome typically occurs along the western coast of continents, between 30° and 40° north or south latitude. Winters are mild and wet, with 25 to 60 centimeters (10 to 24 inches) of rain. Summers are hot and dry. California's dry shrublands, called chaparral, are the state's most extensive biological community (**FIGURE 47.13**). Dry shrubland also occurs in regions bordering the Mediterranean, as well as Chile, Australia, and South Africa.

Plants in dry shrublands tend to have small, leathery leaves that help them withstand the summer drought. Many make aromatic oils that help fend off insects, but also make them highly flammable.

TAKE-HOME MESSAGE 47.7

- ✔ Plants in grasslands and dry shrublands are adapted to fire.
- ✔ Grasslands form in the interior of continents. They have a deep topsoil and are often converted to cropland.
- ✔ Dry shrublands form in coastal areas with a mild, rainy winter and a hot, dry summer.

FIGURE 47.14 North American temperate deciduous forest.

Temperate deciduous forest

47.8 Broadleaf Forests

LEARNING OBJECTIVES

- Describe the types of broad-leaf forest biomes.
- Explain why tropical rainforests have high biodiversity.

Temperate Deciduous Forests

Deciduous broadleaf trees are angiosperms that lose their leaves seasonally. They dominate the **temperate deciduous forests** of eastern North America, western and central Europe, and parts of Asia, including Japan. In these regions, 50 to 150 centimeters (about 20 to 60 inches) of precipitation falls throughout the year. Winters are cool and summers are warm.

Growth of temperate deciduous forests is seasonal. Leaves often turn color before dropping in autumn (**FIGURE 47.14**). Winters are cold, and trees remain dormant while water is locked in snow and ice. In the spring, when conditions again favor growth, deciduous trees flower and put out new leaves. Also during the spring, leaves that were shed the prior autumn decay to form a rich humus. Rich soil and a somewhat open canopy that allows sunlight through creates conditions in which shorter understory plants can flourish.

The temperate deciduous forests of North America are the most species-rich examples of this biome. Different tree species characterize different regions of these forests. For example, Appalachian forests include mainly oaks, whereas beeches and maples dominate Ohio's forests. Animals in North American deciduous forests include grazing deer and seed-eating squirrels and chipmunks, as well as omnivores such as raccoons, opossums, and black bears. Native predators such as

CREDITS: (13) Jack Wilburn/Animals Animals; (14) Dean Pennala/Shutterstock; (in text) © Cengage Learning.

O horizon: Sparse litter

A–B horizons: Continually leached; iron, aluminum left behind impart red color to acidic soil

C horizon: Clays with silicates, other residues of weathering

FIGURE 47.15 Tropical rain forest. Forest soil profile at right.

wolves and mountain lions have been largely eliminated from the forests.

Tropical Rain Forests

Tropical rain forests trees occur between latitudes 10° north and south in equatorial Africa, the East Indies, Southeast Asia, South America, and Central America (**FIGURE 47.15**). Rain that falls throughout the year sums to an annual total of 130 to 200 centimeters (50 to 80 inches). Regular rains, combined with an average temperature of 25°C (77°F) and little variation in day length, allow photosynthesis to continue year-round. The broadleaf trees that dominate these forests lose leaves a few at a time, rather than shedding all leaves seasonally.

Of all land biomes, tropical forests have the greatest primary production. Per unit area, they remove more carbon dioxide from the atmosphere than any other forest or grassland. The high primary production supports an enormous variety of species. Tropical rain forest is the most species-rich biome, as well as the oldest. The biome's age may contribute to its diversity. Some rain forests have existed for more than 50 million years, a long interval in which many opportunities for speciation could have arisen. Compared to other land biomes, tropical rain forest has the greatest variety of plants, insects, birds, and primates.

The forest has a multilayer structure. Its broadleaf trees can stand 30 meters (100 feet) tall. The trees often form a closed canopy that prevents most sunlight from reaching the forest floor. Vines and epiphytes (plants that grow on another plant, but do not withdraw nutrients from it) thrive in shade beneath the canopy.

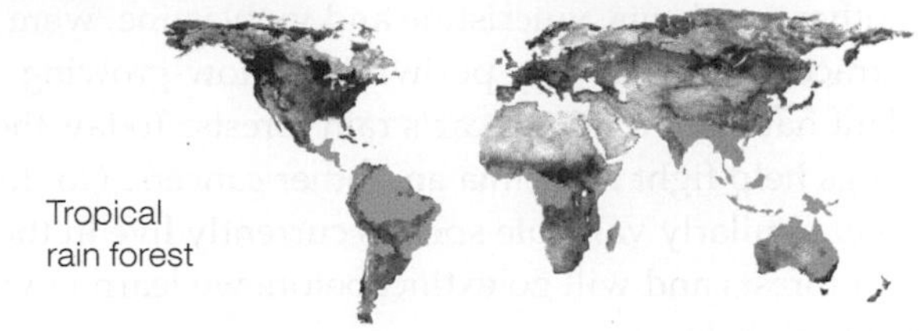
Tropical
rain forest

Although trees of tropical rain forests shed leaves continually, leaf litter does not accumulate because decomposition and mineral cycling happen fast in this warm, moist environment. Rapid decay of leaf litter provides the nutrients that sustain the forests' high primary production. The soil itself is highly weathered and heavily leached. Being a poor nutrient reservoir, this soil is not well suited to agriculture if the forest is removed. Nevertheless, deforestation is an ongoing threat to tropical rain forests. Most are located in developing countries with a rapidly growing human population that looks to the forest as a source of lumber, fuel, and potential cropland.

dry shrubland Biome dominated by a diverse array of fire-adapted shrubs; occurs in regions with cool, wet winters and a dry summer.

temperate deciduous forest Northern Hemisphere biome in which the main plants are broadleaf trees that lose their leaves in autumn and become dormant during cold winters.

tropical rain forest Highly productive and species-rich biome in which year-round rains and warmth support continuous growth of evergreen broadleaf trees.

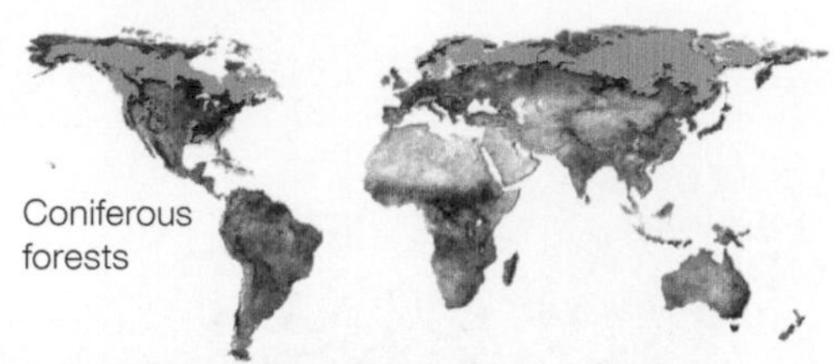
Coniferous
forests

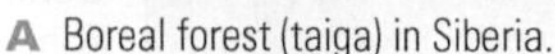

A Boreal forest (taiga) in Siberia.

B Montane coniferous forest near Mount Rainier, Washington.

FIGURE 47.16 Coniferous forests.

The great diversity of tropical rain forests means that loss of any portion of these forests affects many species. Among the potential losses are species with chemicals that could save human lives. Two chemotherapy drugs, vincristine and vinblastine, were extracted from the rosy periwinkle, a low-growing plant native to Madagascar's rain forests. Today, these drugs help fight leukemia and other cancers. No doubt other similarly valuable species currently live in the rain forests and will go extinct before we learn how they can help us.

TAKE-HOME MESSAGE 47.8

✔ Temperate broadleaf forests grow in the Northern Hemisphere where cold winters prevent year-round growth. Trees lose their leaves in autumn, then remain dormant during the winter.

✔ Year-round warmth and rains support tropical rain forests, the most productive, structurally complex, and species-rich land biome.

47.9 Coniferous Forests

LEARNING OBJECTIVES

- Describe the location and composition of the boreal forest biome.
- List some conifer adaptations to cold.

Conifers (gymnosperms with seed-bearing cones) are the main plants in coniferous forests. Conifers shed and replace their leaves, but they do so continually, not seasonally as deciduous trees do. As a group, conifers tolerate drought, cold, and nutrient-poor soils better than broadleaf trees. Conifer leaves are typically needle-shaped, with a thick, waxy cuticle and stomata that are sunk below the leaf surface. These adaptations help conifers conserve water during drought and at times when the ground is frozen. The shape of conifer leaves also helps the tree shed snow, so heavy snowfall does not break limbs. Conifer-dominated forests occur mainly in the Northern Hemisphere.

The most extensive land biome is the coniferous forest that sweeps across northern Asia, Europe, and North America (FIGURE 47.16A). It is referred to as **boreal forest**, or taiga, which means "swamp forest" in Russian. The conifers are mainly pine, fir, and spruce. Most rain falls in the summer, and little evaporates into the cool summer air. Winters are long, cold, and dry. Moose are dominant grazers in this biome.

Also in the Northern Hemisphere, montane coniferous forests extend southward through the great mountain ranges (FIGURE 47.16B). Spruce and fir dominate at the highest elevations. At lower elevations, the mix becomes firs and pines.

Conifers also dominate temperate lowlands along the Pacific coast from Alaska into northern California. These coniferous forests hold some of the world's tallest trees—Sitka spruce and coast redwoods. Other conifers dominate in the eastern United States. About a quarter of New Jersey is pine barrens, a mixed forest of pitch pines and scrub oaks that grow in sandy, acidic soil. Pine forest covers about one-third of the Southeast. Fast-growing loblolly pines dominate these forests and are a major source of lumber and wood pulp. The pines can survive periodic fires that kill most hardwood species. When fires are suppressed, hardwoods outcompete the pines.

TAKE-HOME MESSAGE 47.9

✔ Conifers prevail across the Northern Hemisphere's high-latitude forests, at high elevations, and in temperate regions with nutrient-poor soils.

47.10 Tundra

LEARNING OBJECTIVES

- Describe the climate and location of arctic and alpine tundra.
- Explain why the soil of arctic tundra contains a large amount of carbon.

Arctic Tundra

Arctic tundra forms between the polar ice cap and the belts of boreal forests in the Northern Hemisphere. Most is in northern Russia and Canada. Arctic tundra is Earth's youngest existing biome. Modern arctic tundra communities first became established about 10,000 years ago, when glaciers retreated at the end of the last ice age.

Conditions in arctic tundra are harsh. Snow blankets the ground for as long as nine months of the year. Annual precipitation is usually less than 25 centimeters (10 inches), and cold temperature keeps the snow that does fall from melting for much of the year. During a brief summer, plants grow quickly under the nearly continuous sunlight (**FIGURE 47.17**). Lichens and shallow-rooted, low-growing plants are the producers for food webs that include voles, arctic hares, caribou, arctic foxes, wolves, and brown bears. In the summer, these year-round residents are joined by enormous numbers of migratory birds that nest here.

Even at midsummer, only the surface layer of tundra soil thaws. Below the thawed soil lies **permafrost**, a layer of frozen soil that is 500 meters (1,600 feet) thick in some places. Permafrost acts as a barrier that prevents drainage, so the soil above it remains perpetually waterlogged. The cool, anaerobic conditions in this soil slow decay, so organic remains can build up. Organic matter in permafrost makes the arctic tundra one of Earth's greatest stores of carbon.

As global temperatures rise, the amount of frozen soil that melts each summer is increasing. With warmer temperatures, much of the snow and ice that would otherwise reflect sunlight is disappearing. As a result, newly exposed dark soil absorbs heat from the sun's rays, which encourages more melting.

Alpine Tundra

Alpine tundra occurs at high altitudes throughout the world (**FIGURE 47.18**). Even in the summer, some patches of snow persist in shaded areas, but there is no permafrost. The alpine soil is well drained, but thin, rocky and nutrient-poor. As a result, primary productivity is low.

Grasses and small-leafed, low shrubs grow in patches where soil has accumulated to a greater depth. These low-growing plants can withstand the strong winds that discourage the growth of trees.

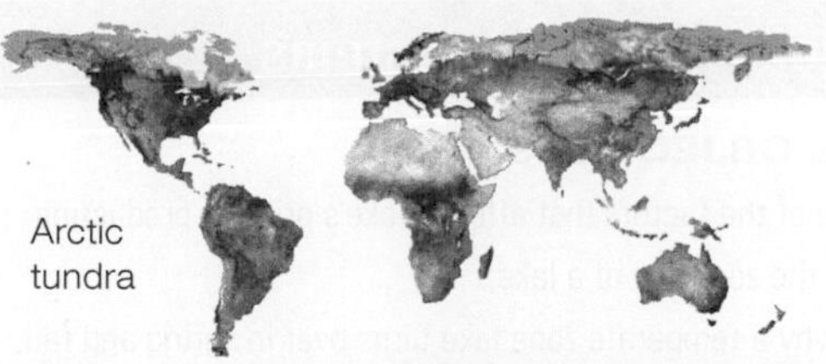
Arctic
tundra

FIGURE 47.17 Arctic tundra. Permafrost underlies the soil.

FIGURE 47.18 Alpine tundra. Low-growing plants at high altitude.

TAKE-HOME MESSAGE 47.10

✔ Arctic tundra prevails at high latitudes, where short, cold summers alternate with long, cold winters.

✔ Alpine tundra prevails in high, cold mountains across all latitudes.

alpine tundra Biome of low-growing, wind-tolerant plants adapted to high-altitude conditions.

arctic tundra Highest-latitude Northern Hemisphere biome, where low, cold-tolerant plants survive with only a brief growing season.

boreal forest (BOHR-ee-uhl) Extensive high-latitude forest of the Northern Hemisphere; conifers are the predominant vegetation.

permafrost Continually frozen soil layer that lies beneath arctic tundra and prevents water from draining.

47.11 Freshwater Ecosystems

LEARNING OBJECTIVES

- List some of the factors that affect a lake's primary production.
- Describe the zonation of a lake.
- Explain why a temperate zone lake turns over in spring and fall, and the effect that this turnover has on primary production.
- Explain the factors that affect the oxygen content of a river.

With this section, we turn our attention to Earth's waters. Freshwater and saltwater ecosystems cover more of Earth's surface than all land biomes combined. We begin here with freshwater (non-saline) ecosystems, continue to coasts in Section 47.12, then dive into the ocean. Note that the term "biome" was originally coined to apply only to a community on land, but many people now refer to the types of aquatic ecosystems as "aquatic biomes."

Littoral zone
Limnetic zone
Profundal zone
limit of effective light penetration

FIGURE 47.19 Lake zonation. A lake's littoral zone extends around the shore to a depth where rooted aquatic plants stop growing. Its limnetic zone is the open waters where light penetrates and photosynthesis occurs. Below that lies the cool, dark profundal zone, where detrital food chains predominate.

FIGURE 47.20 An oligotrophic lake. Crater Lake in Oregon is a collapsed volcano that filled with snow melt. It began filling about 7,700 years ago. From a geologic standpoint, it is a young lake.

Lakes

A lake is a body of standing freshwater. If it is sufficiently deep, it can be divided into zones that differ in their physical characteristics and species composition (**FIGURE 47.19**). Near shore is the littoral zone, from the Latin *litus* for "shore." Here, sunlight penetrates all the way to the lake bottom; aquatic plants and algae that attach to the bottom are the primary producers. The lake's open waters include an upper, well-lit limnetic zone, and—if the lake is deep—a dark profundal zone where light does not penetrate. Primary producers in the limnetic zone can include aquatic plants, green algae, diatoms, and cyanobacteria. These organisms serve as food for rotifers, copepods, and other types of zooplankton. In the profundal zone, where there is not enough light for photosynthesis, consumers feed on organic debris that drifts down from above.

Nutrient Content and Succession A lake undergoes succession; it changes over time (Section 45.8). A newly formed lake is oligotrophic: deep, clear, and nutrient-poor, with low primary productivity (**FIGURE 47.20**). Later, as sediments accumulate and plants take root, the lake becomes eutrophic. Eutrophication refers to processes, either natural or artificial, that enrich a body of water with nutrients (Section 46.1).

Seasonal Changes Temperate zone lakes undergo seasonal changes (**FIGURE 47.21**). During winter, a layer of ice forms at the surface of all but the largest and deepest lakes. Unlike most substances, water is denser as a liquid than as a solid (ice). As water cools, its density increases, until it reaches 4°C (39°F). Below this temperature, additional cooling decreases water's density—which is why ice floats on water (Section 2.5). In an ice-covered lake, water just under the ice is near its freezing point and at its lowest density. The densest (4°C) water resides at the bottom of the lake ❶.

In spring, the air warms and ice melts. When the resulting meltwater warms to 4°C, it sinks. Wind blowing across the surface of the lake assists in bringing about a **spring overturn**, during which oxygen-rich water at the surface of the lake moves downward while nutrient-rich water from the lake's depths moves up ❷.

fall overturn During the fall, waters of a temperate zone mix. Upper, oxygenated water cools, gets dense, and sinks; nutrient-rich water from the bottom moves up.

spring overturn In temperate zone lakes, a downward movement of oxygenated surface water and an upward movement of nutrient-rich water in spring.

thermocline Thermal stratification in a large body of water; a cool midlayer stops vertical mixing between warm surface water above it and cold water below it.

In the summer, a lake's waters form three layers that differ in their temperature and oxygen content ❸. The top layer is warm and oxygen-rich. It overlies the **thermocline,** a thin layer where temperature falls rapidly. Beneath the thermocline is the coolest water. The thermocline acts as a barrier that keeps the upper and lower layers from combining. As a result of the thermocline, oxygen from the lake's surface cannot reach the lake's depths, where decomposition is using up oxygen. At the same time, nutrients from those depths cannot escape into surface waters.

In autumn, the upper layer cools and sinks, and the thermocline disappears. During the **fall overturn**, oxygen-rich water moves down while nutrient-rich water moves up ❹.

Overturns influence primary productivity. After a spring overturn, longer day length and an abundance of nutrients support the greatest primary productivity. During the summer, vertical mixing ceases. Nutrients do not move up, and photosynthesis slows. By late summer, nutrient shortages limit growth. Fall overturn brings nutrients to the surface and favors a brief burst of photosynthesis. The burst ends as winter brings shorter days and the amount of sunlight declines.

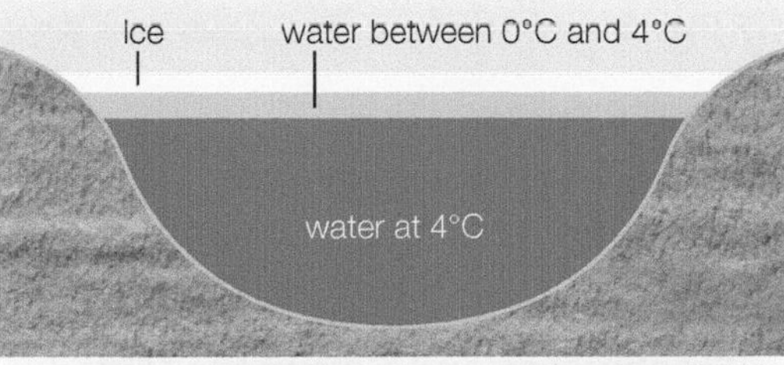
ice
water between 0°C and 4°C
water at 4°C

❶ Winter. Ice covers a thin layer of slightly warmer water just below it. Densest (4°C) water is at the bottom. Winds do not affect water under the ice, so there is little circulation.

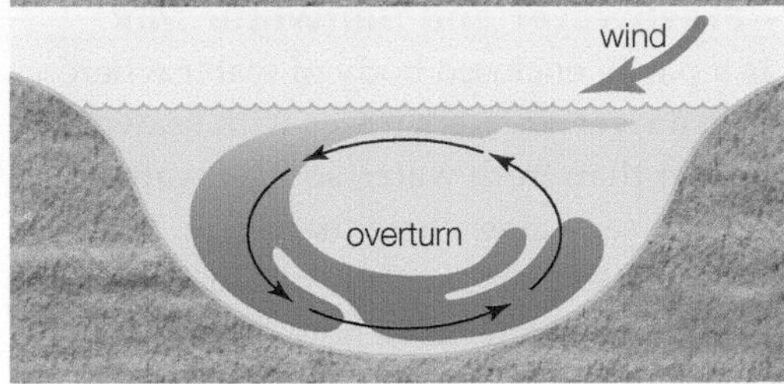
wind
overturn

❷ Spring. Ice thaws. Upper water warms to 4°C and sinks. Winds blowing across water create currents that help overturn water, bringing nutrients up from the bottom.

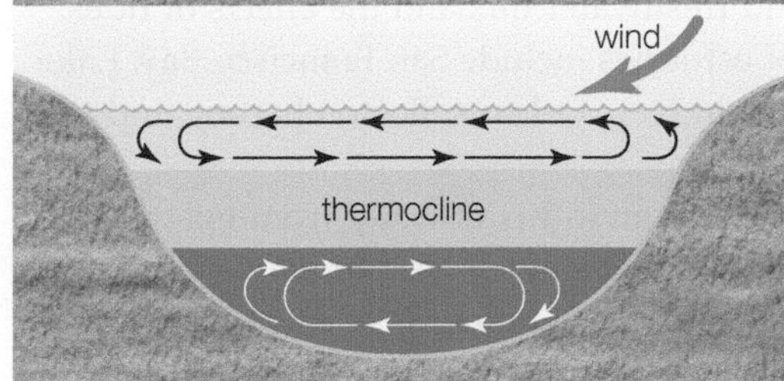
wind
thermocline

❸ Summer. Sun-warmed water floats on a thermocline, a layer across which temperature declines abruptly with depth. Upper and lower water do not mix because of this thermal boundary.

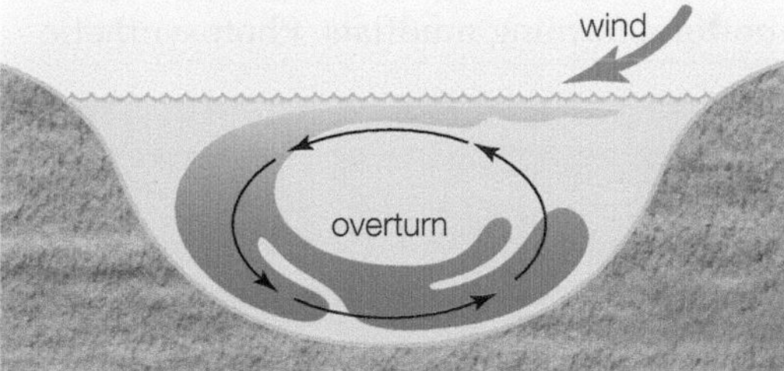
wind
overturn

❹ Fall. Upper water cools and sinks downward, eliminating the thermocline. Vertical currents mix water that was separated during the summer.

FIGURE 47.21 Seasonal changes in a temperate zone lake.

Streams and Rivers

Flowing-water ecosystems start as freshwater springs or seeps. Streams grow and merge as they flow downhill. Rainfall, snowmelt, geography, altitude, and shade affect flow volume and temperature. Minerals in submerged rocks dissolve in the water and affect its solute concentrations. Water in different parts of a river moves at different speeds, contains different solutes, and differs in temperature, so the species composition of a river varies along its length.

Dissolved Oxygen Content

The amount of oxygen dissolved in water is one of the most important factors affecting aquatic organisms. More oxygen dissolves in cooler, fast-flowing, turbulent water than in warmer, smooth-flowing or still water. Thus, an increase in water temperature or decrease in its flow rate can cause aquatic species with high oxygen needs to suffocate.

In freshwater habitats, aquatic larvae of mayflies and stoneflies are the first invertebrates to disappear when the oxygen content of the water decreases. These insect larvae are active predators that demand considerable oxygen, so they serve as indicator species (Section 45.8). Aquatic snails disappear, too. Declines in these invertebrates can have cascading effects on the fishes that feed on them. Fishes can also be more directly affected. Trout and salmon are especially intolerant of low oxygen. Carp (including goldfish) are among the most tolerant; they survive tepid, stagnant water in ponds.

No fishes survive when the oxygen content of water falls below 4 parts per million. Leeches persist, but most other invertebrates disappear. At the lowest oxygen concentration, annelids called sludge worms (*Tubifex*) often are the only animals. These worms are colored red by their large amount of hemoglobin. This abundance of hemoglobin adapts them to their low-oxygen habitat, where predators and competition for food are scarce.

TAKE-HOME MESSAGE 47.11

✔ Lakes have gradients in light, dissolved oxygen, and nutrients.

✔ Primary productivity varies with a lake's age and, in temperate zones, with the season.

✔ Different conditions along the length of a stream or river create habitat for different organisms.

47.12 Coastal Ecosystems

LEARNING OBJECTIVES

- Explain why estuaries are especially nutrient-rich.
- Compare the main types of food chains on rocky and sandy shores.

Estuaries—Freshwater and Saltwater Mix

An **estuary** is a partly enclosed body of water where freshwater from a river or rivers mixes with seawater. Seawater is denser than fresh water, so freshwater floats on top of the seawater where they meet. The size and shape of the estuary, and the rate at which freshwater flows into it, determine how quickly the saltwater and freshwater mix and the effects of tides. Examples of estuaries include San Francisco Bay, Lake Pontchartrain in New Orleans, Boston harbor, and the Chesapeake Bay.

In all estuaries, an influx of water from upstream continually replenishes nutrients and allows a high level of productivity. Incoming freshwater also carries silt. Where the velocity of water flow slows, the silt falls to the bottom, forming mudflats. Photosynthetic bacteria and protists in biofilms on mudflats often account for a large portion of an estuary's primary production. Plants adapted to withstand changes in water level and salinity also serve as producers.

A Salt marsh dominated by cordgrass (*Spartina*). Salt taken up in water by roots is excreted by glands on the leaves.

B Mangrove forest along a tropical coast. Specialized cells at the surface of exposed prop roots allow gas exchange with the air.

FIGURE 47.22 Two types of coastal wetlands.

Spartina is the dominant plant in the salt marshes of many estuaries along the Atlantic coast (**FIGURE 47.22A**). It is adapted to its habitat by an ability to withstand immersion during high tides and to tolerate salty, waterlogged, anaerobic soil. Modified parenchyma (Section 27.3) with hollow air spaces allows oxygen taken in by shoots to diffuse to roots. Salt taken up in water by roots is excreted by specialized glands on the leaves. As a result, the leaves are typically covered with salt crystals. The high salt content of *Spartina* and other estuary plants makes them unpalatable to most herbivores, so detrital food webs predominate.

Estuaries and tidal flats of tropical and subtropical latitudes often support nutrient-rich mangrove wetlands (**FIGURE 47.22B**). "Mangrove" is the common term for certain salt-tolerant woody plants that live in sheltered areas along tropical coasts. The plants have prop roots that extend out from their trunk and help the plant stay upright in the soft sediments. Specialized cells at the surface of some exposed roots allow gas exchange with air.

Estuaries provide important ecological services. Plants in estuaries help slow river flow, thus reducing the risk of flooding. Similarly, they help protect coasts from storm surges. Estuaries' waters serve as nurseries for many species of marine fishes and invertebrates. Migratory birds often stop over in estuaries as they travel from one region to another.

Human activities threaten estuary ecosystems. Pollution from farms and cities flows down rivers and into estuary waters. In the tropics, people have traditionally cut mangroves for firewood. A more recent threat is conversion of mangrove wetlands to shrimp farms. The shrimp mainly end up on dinner plates in the United States, Japan, and western Europe.

Rocky and Sandy Coasts

Rocky and sandy shores support ecosystems of the intertidal zone. As with lakes, an ocean's shoreline is described as its littoral zone. The littoral zone can be divided into three vertical regions that differ in their physical characteristics and diversity. The upper littoral zone, or splash zone, regularly receives ocean spray but is submerged only during the highest of high tides. It receives the most sun, but holds the fewest species. The midlittoral zone is typically covered by water during an average high tide and dry during a low tide. The lower littoral zone, exposed only during the lowest tide of the lunar cycle, has the most diversity.

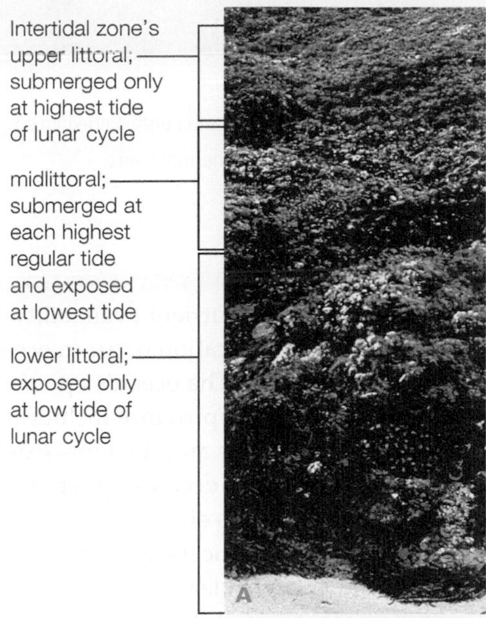
Intertidal zone's upper littoral; submerged only at highest tide of lunar cycle
midlittoral; submerged at each highest regular tide and exposed at lowest tide
lower littoral; exposed only at low tide of lunar cycle
A

FIGURE 47.23 Vertical zonation of a rocky shore.

You can easily see the zonation along a rocky shore (FIGURE 47.23). Barnacles live in the midlittoral zone. Algae clinging to rocks are primary producers for the prevailing grazing food web. The primary consumers include mussels and a variety of snails.

Zonation is less obvious on sandy shores where detrital food chains start with material washed ashore. Some crustaceans eat detritus in the upper littoral zone. Nearer to the water, other invertebrates feed as they burrow through the sand.

TAKE-HOME MESSAGE 47.12

✔ We find estuaries where rivers empty into seas. The rivers deliver nutrients that foster high productivity.

✔ Mangrove wetlands are common along shorelines in tropical latitudes.

✔ Rocky and sandy shores show zonation, with different zones exposed during different phases of the tidal cycle. Diversity is highest in the zone that is submerged most of the time.

47.13 Coral Reefs

LEARNING OBJECTIVES

- Describe where coral reefs occur.
- Explain what causes coral bleaching.

Coral reefs are wave-resistant formations that consist primarily of calcium carbonate secreted by generations of coral polyps. Reef-forming corals live mainly in shallow, clear, warm waters between latitudes 25° north and 25° south. About 75 percent of all coral reefs are in the Indian and Pacific Oceans. A healthy reef is home to living corals and a huge number of other species (FIGURE 47.24). Biologists estimate that about a quarter of all marine fish species associate with coral reefs.

FIGURE 47.24 Coral reef. A coral gets its color from the pigments of symbiotic dinoflagellates that live in its tissues.

The largest existing reef, Australia's Great Barrier Reef, parallels Queensland for 2,500 kilometers (1,550 miles), and is the largest example of biological architecture. Scientists estimate it began to form about 600,000 years ago. Today it is a string of reefs, some of them 150 kilometers (95 miles) across. The Great Barrier Reef supports about 500 coral species, 3,000 fish species, 1,000 kinds of mollusks, and 40 kinds of sea snakes.

Photosynthetic dinoflagellates live as mutualistic symbionts inside the tissues of all reef-building corals (Section 24.5). The dinoflagellates are protected within the coral's tissues and provide the coral polyp with oxygen and sugars that it depends on. When stressed, coral polyps expel the dinoflagellates. Dinoflagellates give a coral its color, so expelling these protists turns the coral white, an event called coral bleaching. When a coral is stressed for more than a short time, the dinoflagellate population in a coral's tissues cannot rebound and the coral dies, leaving only its bleached hard parts behind. The incidence of coral bleaching events has been increasing, most likely as a result of warming sea temperature.

Coral reefs are also under threat from many other factors such as discharge of sewage and other

estuary Highly productive ecosystem where nutrient-rich water from a river mixes with seawater.

coral reef Highly diverse marine ecosystem centered around reefs built by living corals that secrete calcium carbonate.

CREDITS: (23) Courtesy of J. L. Sumich, *Biology of Marine Life,* 7th ed., W. C. Brown, 1999; (24) John Easley, www.johneasley.com.

pollutants into coastal waters, human-induced erosion that clouds water with sediments, destructive fishing practices, and introduced invasive species.

Another threat to coral is ocean acidification, a decrease in seawater pH caused by the rise in atmospheric carbon dioxide (CO_2). Since the onset of the industrial revolution, acidity of the ocean has risen by about 30 percent. Increased acidity makes it more difficult for corals to take up the calcium carbonate they need to build their skeleton. It also makes it more difficult for other marine organisms to make calcium carbonate parts such as shells.

As a result of these various threats, 25 species of coral are now considered threatened or endangered. This means their population has declined so much that they are now at risk of extinction.

TAKE-HOME MESSAGE 47.13

✔ Coral reefs form by the action of living corals that lay down a calcium carbonate skeleton. Photosynthetic dinoflagellates in the coral's tissues are necessary for the coral's survival.

✔ Rising water temperature, pollutants, fishing, and exotic species contribute to loss of reefs.

✔ Declines in coral reefs will affect the enormous number of fishes and invertebrate species that make their home on or near the reefs.

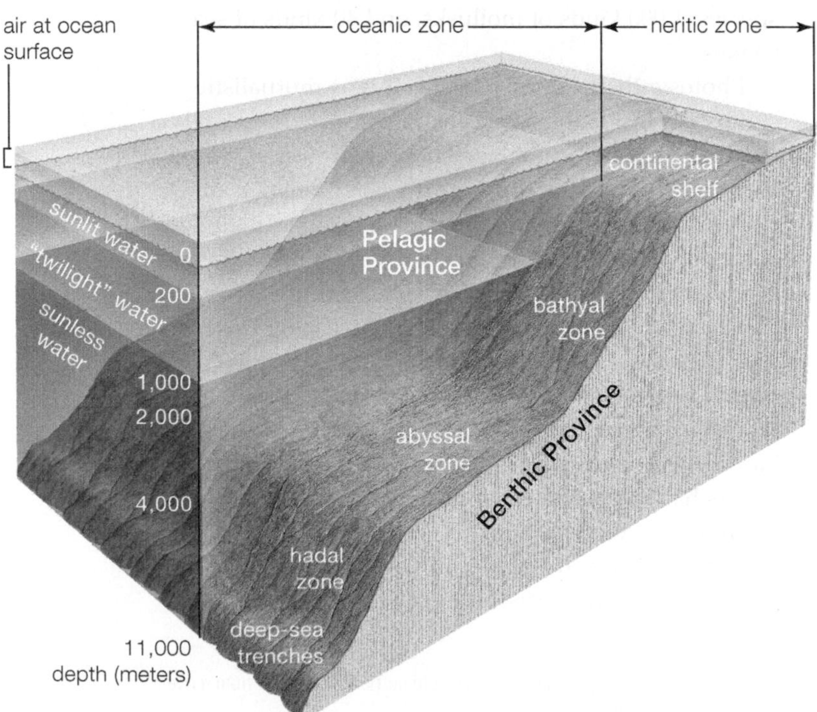
air at ocean surface
oceanic zone
neritic zone
continental shelf
sunlit water
"twilight" water
sunless water
0
200
1,000
2,000
4,000
11,000
depth (meters)
Pelagic Province
bathyal zone
abyssal zone
Benthic Province
hadal zone
deep-sea trenches

FIGURE 47.25 Oceanic zones. Zone dimensions are not to scale.

47.14 The Open Ocean

LEARNING OBJECTIVES

- Distinguish between the pelagic and benthic provinces.
- Describe the producers at hydrothermal vents.
- Describe a seamount.

Pelagic Ecosystems

Regional variations in light, nutrient availability, temperature, and oxygen concentration occur in oceans as well as lakes (**FIGURE 47.25**). The ocean's open waters are the **pelagic province.** This province includes the water over continental shelves and the more extensive waters farther offshore. In the ocean's upper, bright waters, phytoplankton such as single-celled algae and bacteria are the primary producers, and grazing food chains predominate. Even in clear water, light sufficient for photosynthesis cannot penetrate more than 200 meters beneath the sea surface. However, a bit of light may penetrate as much as 1,000 meters (3,000 feet) beneath the sea surface. In deeper waters, organisms live in continual darkness. Here, organic material that drifts down from above serves as the basis of detrital food chains.

Our knowledge of the deep pelagic zone is limited because human divers cannot survive at great depths, where the pressure exerted by the weight of the water above them is very high. Thus, exploration of deeper regions requires submersible vessels. Use of such technology has revealed some extraordinary life forms.

The Seafloor

The **benthic province** is the ocean bottom—its rocks and sediments. Benthic biodiversity is greatest on the margins of continents, or the continental shelves. The benthic province also includes some largely unexplored concentrations of biodiversity on seamounts and at hydrothermal vents.

Seamounts are undersea mountains that stand 1,000 meters or more tall, but are still below the sea surface (**FIGURE 47.26**). They attract large numbers of fishes and are home to many marine invertebrates. Like islands, seamounts often are home to species that evolved there and are found nowhere else. There are more than 30,000 seamounts, and scientists have just begun to document species that live on them.

benthic province The ocean's sediments and rocks.
hydrothermal vent Rocky, underwater opening where mineral-rich water heated by geothermal energy streams out.
pelagic province The ocean's waters.
seamount An undersea mountain.

The abundance of life at seamounts makes them attractive to commercial fishing vessels. Fishes and other organisms are often harvested by trawling, a fishing technique in which a large net is dragged along the bottom, capturing everything in its path. The process is ecologically devastating; trawled areas are stripped bare of life, and silt stirred up by the giant, weighted nets suffocates filter-feeders in adjacent areas.

At **hydrothermal vents**, hot water rich in dissolved minerals spews out from an opening on the ocean floor. The water is seawater that seeped into cracks in the ocean floor at the margins of tectonic plates and was heated by heat energy from within Earth. Minerals in this water settle out when it mixes with the cold deep-sea water. The community here does not run on sunlight energy. Rather, the main producers are chemoautotrophic bacteria and archaea that obtain energy by removing electrons from minerals. Food webs include diverse invertebrates, including large annelid tube worms (**FIGURE 47.27**).

Life exists even in the deepest sea. A remote-controlled submersible that sampled sediments in the deepest part of the ocean (the Mariana Trench) brought up foraminifera that live 11 kilometers (7 miles) below the surface.

FIGURE 47.26 Computer model of seamounts on the seafloor off the coast of Alaska. Patton Seamount, at the rear, stands 3.6 kilometers (about 2 miles) tall, with its peak about 240 meters (800 feet) below the sea surface.

FIGURE 47.27 Life at a hydrothermal vent on the seafloor. The giant tube worms can be longer than your arm. They do not eat. Rather, sulfur absorbed by the worm serves as the energy source for chemoautotrophic bacteria that live inside the worm and provide it with sugars.

TAKE-HOME MESSAGE 47.15

- ✔ Oceans have gradients in light, dissolved oxygen, and nutrients. Nearshore and well-lit zones are the most productive and species-rich.
- ✔ On the seafloor, pockets of diversity occur on seamounts and around hydrothermal vents.
- ✔ Animal life exists even in the deepest ocean.

47.1 Going With the Flow (revisited)

After the release of radioactive material from Fukishima, scientists monitored the movement of this material toward North America's Pacific coast. Canadian scientists enlisted citizen volunteers to collect seawater samples from along the coast of British Columbia. They also partnered with native American communities to assess the level of radiation in salmon. These efforts showed that ocean currents began delivering low levels of ^{137}Cs in 2015.

Another monitoring program involved documenting the level of radiation in kelp forests along the western coast of the United States. Participants collected kelp samples several times a year, then sent them to a government laboratory where the samples were tested for radioisotopes from Fukushima. Kelp was chosen because of its important role in the coastal ecosystem and because it tends to take up and concentrate radioactive material, making it easier to detect.

Bits of kelp tested only weeks after the accident at Fukushima showed the presence of iodine 131, a radioisotope produced in nuclear reactors, but rare in nature. Researchers assume the ^{131}I traveled east on the winds before falling into the sea and being taken up by the kelp. ^{131}I has a half-life of eight days, so it is no longer a concern and there is no evidence that exposure to it did the kelp any long-term harm.

Given the diluting effects of trans-ocean movement, the concentration of radioactive material in coastal water and on beaches is not expected to ever climb to a point where it poses a threat to human health. ●

CREDITS: (26) NOAA; (27) NOAA/Photo courtesy of Cindy Van Dover, Duke University Marine Lab.

Section 47.1 Earth's air and water circulate constantly and distribute energy and materials across the globe. In the aftermath of the 2011 earthquake and tsunami in Japan, these processes distributed radioactive material and debris released by the disaster. Because prevailing winds and currents flow west from Japan, this material was transported to the west coast of North America.

Section 47.2 Global air circulation patterns influence **climate** and the distribution of communities throughout the **biosphere**. The patterns are set into motion by latitudinal variations in incoming solar radiation. Tropical latitudes receive more sun than higher latitudes. Tropical latitudes also have more uniform daylength.

At the equator, warm air rises and loses moisture as rain. At 30° north and south of the equator, the now cooler and drier air descends. Air rises again at 60 ° north and south latitude, then descends again at the poles.

Although the main movement of air masses is from the equator toward the poles, Earth's rotation alters the direction of prevailing winds.

Section 47.3 Latitudinal and seasonal variations in sunlight warm up seawater and create surface currents that are affected by prevailing winds. The currents distribute heat energy worldwide and influence the weather patterns. Oceans lose and gain heat more slowly than land, so proximity to an ocean has a moderating effect on temperature and influences daily wind patterns.

Ocean currents, air currents, and landforms interact in shaping global temperature zones, as in regions where the presence of coastal mountains causes a **rain shadow** or where **monsoon** rains fall seasonally.

Section 47.4 The **El Niño Southern Oscillation** is an irregular fluctuation in the temperature of the equatorial Pacific Ocean. It triggers changes in rainfall and other weather patterns around the world. These changes can affect food chains and influence human health, as when warming waters increase the likelihood of cholera epidemics.

Section 47.5 **Biomes** are areas characterized by environmental conditions that support a particular type of vegetation. Many biomes include multiple discontinuous areas. Similarities among species in geographically separated portions of a biome can arise as a result of morphological convergence.

Section 47.6 **Deserts** form around latitudes 30° north and south, where rainfall is sparse and falls seasonally. Desert plants include annuals that grow fast after seasonal rains and perennials that are adapted to withstand a seasonal drought. Desert crust, a community of organisms in the upper soil layer, helps hold soil in place.

Sections 47.7 **Temperate grasslands** form in the interior of midlatitude continents. North America's grasslands are prairies. Prairies have a highly fertile soil, and most have now been converted to agricultural use. Africa has **savannas**, which include widely dispersed trees. Both grasslands and savannas support herds of grazing animals.

Southwest coasts of continents have cool, rainy winters and a hot, dry summer. They support **dry shrublands** such as California's chaparral. Grassland and shrubland plants are adapted to periodic fire.

Section 47.8 The broadleaf trees that dominate **temperate deciduous forests** shed their leaves all at once just before a cold winter prevents growth. By contrast, broadleaf trees in **tropical rain forests** can grow year-round. Tropical rain forest is the oldest existing biome. It is also the most productive and the most diverse. Rapid decomposition prevents the accumulation of leaf litter in these forests, so the underlying soil is poor.

Section 47.9 Conifers withstand cold and drought better than broadleaf trees. They dominate Northern Hemisphere high-latitude **boreal forests**, where winters are long and cold. These forests are Earth's most extensive land biome. Conifers also dominate some mountainous areas and areas where soils are poor.

Section 47.10 The northmost biome in the Northern Hemispherre is **arctic tundra**. The low-growing plants of arctic tundra grow only during a brief summer. The soil of arctic tundra overlies a layer of **permafrost** that stores a large amount of carbon.

Alpine tundra is a high-elevation biome that is dominated by similar low-growing plants.

Section 47.11 Most lakes, streams, and other aquatic ecosystems have gradients in the penetration of sunlight, water temperature, and in dissolved gases and nutrients. These characteristics vary over time and affect primary productivity.

In temperate zone lakes, a **spring overturn** and a **fall overturn** cause vertical mixing of waters and trigger a burst of productivity. In summer, a **thermocline** prevents upper and lower waters from mixing.

Sections 47.12–47.14 Coastal zones support diverse ecosystems. Among these, the coastal wetlands, **estuaries**, and **coral reefs** are especially productive.

Life persists throughout the ocean. Diversity is highest in sunlit waters at the top of the **pelagic province**, which is the ocean's waters. On the seafloor (in the **benthic province**), diversity is high near deep-sea **hydrothermal vents** and on **seamounts**. Even the deepest ocean sediments support life.

SELF-QUIZ

Answers in Appendix VII

1. The Northern Hemisphere is most tilted toward the sun on the _______ .
   a. March equinox
   b. September equinox
   c. June solstice
   d. December solstice

2. Which latitude will have the most hours of daylight on the summer solstice?
   a. 0° (the equator)
   b. 30° north
   c. 45° north
   d. 60° north

3. Warm air _______ and it holds _______ water than cold air.
   a. sinks; less
   b. rises; less
   c. sinks; more
   d. rises; more

4. A rain shadow is a reduction in rainfall _______ .
   a. on the inland side of a coastal mountain range
   b. during an El Niño event
   c. that results from global warming

5. The Gulf Stream is a current that flows _______ along the _______ coast of the United States.
   a. north to south; east
   b. north to south; west
   c. south to north; east
   d. south to north; west

6. _______ have a deep layer of nutrient-rich topsoil.
   a. Deserts
   b. Grasslands
   c. Rain forests
   d. Seamounts

7. Biome distribution depends on _______ .
   a. climate
   b. elevation
   c. soils
   d. all of the above

8. Grasslands most often are found _______ .
   a. at 30° north and south
   b. at high altitudes
   c. in interior of continents
   d. where fire is rare

9. Permafrost underlies _______ .
   a. arctic tundra
   b. alpine tundra
   c. boreal forest
   d. tallgrass prairie

10. Warm, still water holds _______ oxygen than cold, fast-flowing water.
    a. more
    b. less

11. Chemoautotrophic bacteria and archaea are the main primary producers for food webs _______ .
    a. in mangrove wetlands
    b. at seamounts
    c. on coral reefs
    d. at hydrothermal vents

12. Corals rely on symbiotic _______ for sugars.
    a. fungi
    b. amoebas
    c. dinoflagellates
    d. green algae

13. What biome borders boreal forest to the north?
    a. savanna
    b. taiga
    c. tundra
    d. chaparral

14. Unrelated species in geographically separated parts of a biome may resemble one another as a result of _______ .
    a. morphological divergence
    b. morphological convergence
    c. resource partitioning
    d. coevolution

15. Match the terms with the most suitable description.

___ tundra	a. broadleaf forest near equator
___ chaparral	b. partly enclosed by land; where fresh water and seawater mix
___ desert	c. African grassland with trees
___ savanna	d. low-growing plants at high latitudes or elevations
___ estuary	e. dry shrubland
___ boreal forest	f. at latitudes 30° north and south
___ prairie	g. mineral-rich, superheated water supports communities
___ tropical rain forest	h. conifers dominate
___ hydrothermal vents	i. North American grassland

CENGAGE brain .com To access course materials, please visit www.cengagebrain.com.

CRITICAL THINKING

1. London, England, is at the same latitude as Calgary in Canada's province of Alberta. However, the mean January temperature in London is 5.5°C (42°F), whereas in Calgary it is minus 10°C (14°F). Compare the locations of these two cities, and suggest a reason for this temperature difference.

2. Increased industrialization in China has environmentalists worried about air quality elsewhere. Are air pollutants emitted in Beijing more likely to end up in eastern Europe or the western United States? Why?

3. The use of off-road recreational vehicles may double in the next 20 years. Enthusiasts would like increased access to government-owned deserts. Some argue that it's the perfect place for off-roaders because "There's nothing there." Explain whether you agree, and why.

4. As global temperature rises as a result of increased atmospheric carbon dioxide, temperate zone lakes are freezing later in the winter and defrosting earlier in spring. How will these changes affect the timing of nutrient availability in these lakes? What effect would you expect this change in timing have on life in the lakes?

48 HUMAN IMPACTS ON THE BIOSPHERE

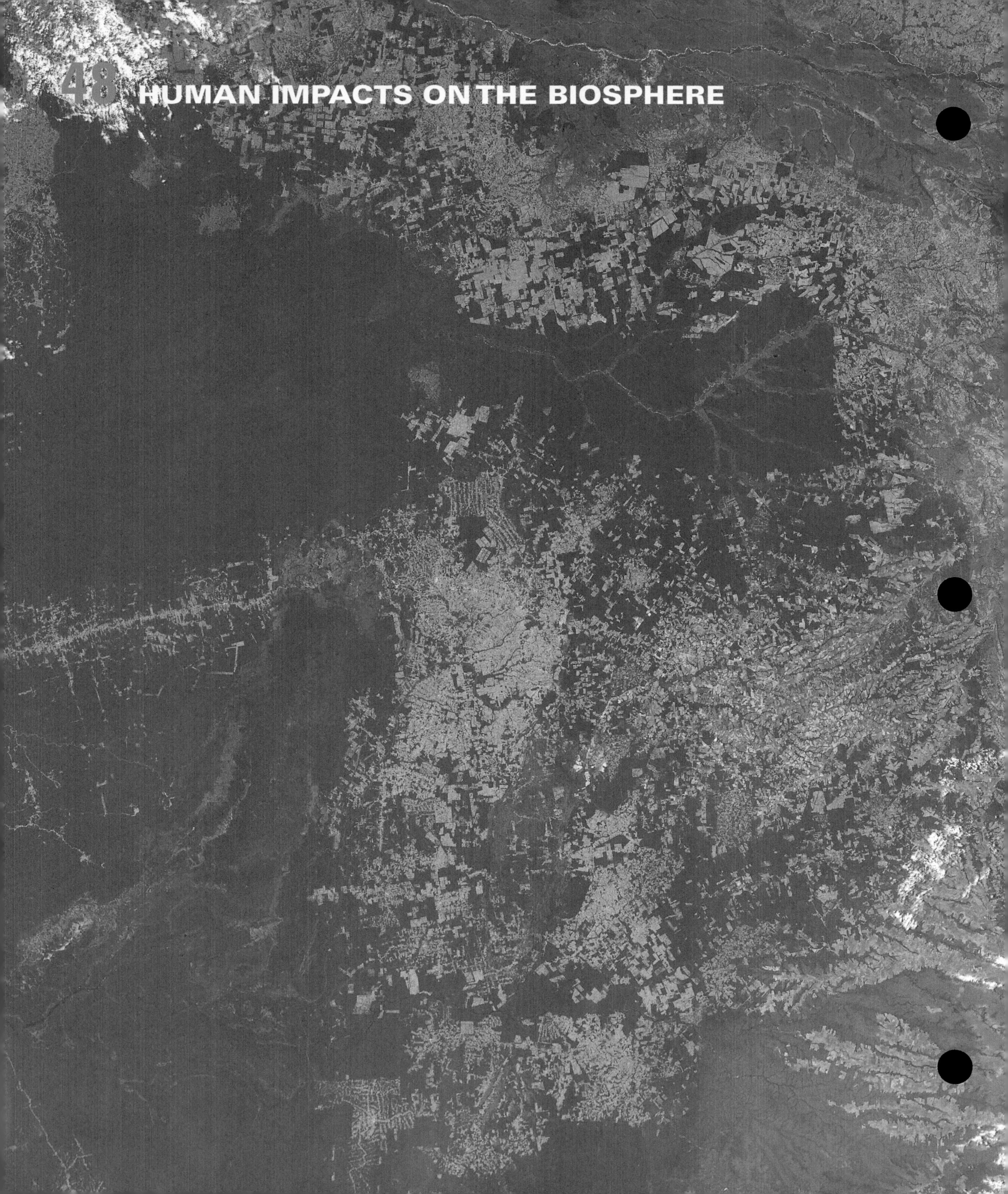

CORE CONCEPTS

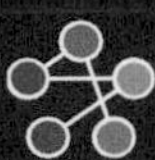

Systems

Complex properties arise from interactions among components of a biological system.

Human activities that alter one component of a habitat can have long-range and long-term effects. Plowing grasslands and cutting forests allow soil erosion that affects rainfall patterns, thus altering the habitat in a way that is difficult to reverse. Raising the temperature of the land and seas causes cascading effects by altering feeding, migration, and breeding patterns.

Evolution

Evolution underlies the unity and diversity of life.

All living species are products of an ongoing evolutionary process that stretches back billions of years, and each species has a unique combination of evolved traits that suit it to its habitat. Extinction removes that collection of traits from the world forever. Humans have increased the frequency of extinctions, and the extent and effects of these losses are not fully known or understood.

Process of Science

The field of biology consists of and relies upon experimentation and the collection and analysis of scientific evidence.

Carefully designing experiments helps researchers unravel cause-and-effect relationships in complex natural systems. Long-term observations can reveal patterns such as a decline in the ozone level or a rise in global temperature. Documenting biodiversity and how it changes over time is an enormous and ongoing process.

Links to Earlier Concepts

In this chapter, we consider how human activities are causing a mass extinction (Section 17.9). In doing so, we reconsider the ecological effects of acid rain (2.6), soil erosion (28.2), aquifer depletion (46.6), and global climate change (46.8).

48.1 Life in the Anthropocene

We live in a time when human activities are radically altering conditions on our planet. In light of the profound changes now taking place, geologists are considering designating a new interval in the geologic time scale. The proposed name for this interval, which includes the current day, is the **Anthropocene**. *Anthropo-* means "human," and *-cene* is the standard suffix for geologic epochs. The name denotes a period during which human activities alter the global environment so dramatically that our presence makes a distinctive imprint in the geologic record.

This imprint takes many forms. Our mining and use of metals such as aluminum and copper has distributed these materials across the globe in a pattern never seen before. At the same time, we are creating huge amounts of plastics and concrete (**FIGURE 48.1**), two materials that did not exist in prior geological periods. Sediments now being laid down in the seas and lakes contain synthetic pesticides and industrial chemicals, along with ashy products of fossil fuel combustion. These sediments will also contain an unprecedented abundance of nitrogen (a result of industrial nitrogen fixation) and an array of unusual radioisotopes (a result of testing nuclear weapons). Increasing atmospheric carbon dioxide, and the associated increase in temperature and in sea level, will also leave behind physical evidence. In addition, the fossil record will record a precipitous decline in biodiversity across the globe.

The many human-induced changes that are now occurring, the methods scientists use to document these changes, and the ways that we can address them are the focus of this chapter.

FIGURE 48.1 Concrete evidence of the Anthropocene. Cities throughout the world contain vast amounts of cement-based concrete that will endure long into the future.

Anthropocene (AN-thruh-puh-seen) The current geologic interval, in which human activities are leaving a global imprint on the Earth.

CREDITS: (opposite) NASA courtesy of the MODIS Rapid Response Team at Goddard Space Flight Center; (1) iStockphoto.com/CharlieTong.

48.2 The Extinction Crisis

LEARNING OBJECTIVES

- Describe what occurs during a mass extinction.
- Explain how human activities can lead to extinctions.

Mass Extinction

Extinction, like speciation, is a natural process. Species arise and become extinct on an ongoing basis. Based on several lines of evidence, scientists estimate that 99 percent of all species that have ever lived are now extinct. The rate of extinction picks up dramatically during a **mass extinction**, when a large proportion of Earth's organisms become extinct in a relatively short period of geologic time.

Five great mass extinctions mark the boundaries for the geologic time periods. As you learned in Section 16.1, the mass extinction at the end of the Cretaceous period was most likely caused by an asteroid impact. In terms of species losses, the greatest mass extinction occurred at the end of the Permian, when about 90 percent of the then-existing species perished. The cause of this extinction was most likely increased volcanic activity in what is now Siberia. Carbon dioxide and sulfur oxides released into the atmosphere by eruptions caused global acidification and warming. As the oceans warmed, methane (natural gas) that had been frozen in their depth thawed. The ocean became anoxic, and methane that entered the air caused explosive conflagrations on land.

TABLE 48.1

Global List of Threatened Species*

	Species	% Evaluated for Threats	Number Threatened
Vertebrates			
Mammals	5,560	100	1,194
Birds	11,121	100	1,460
Reptiles	10,450	52	1,090
Amphibians	7,635	86	2,067
Fishes	33,500	48	2,359
Invertebrates			
Insects	1,000,000	<1	623
Arachnids	102,248	<1	170
Crustaceans	47,000	7	460
Mollusks	85,000	9	978
Corals	2,175	40	5
Others	68,827	<1	132
Land Plants			
Mosses	16,236	<1	76
Ferns and allies	12,000	3	217
Gymnosperms	1,052	96	400
Angiosperms	268,000	8	10,972
Protists			
Green algae	6,050	>1	0
Red algae	7,104	>1	9
Brown algae	3,784	>1	6
Fungi			
Lichens	17,000	>1	7
Mushrooms	31,496	>1	21

* *IUCN–WCU Red List, available online at www.iucnredlist.org*

The Sixth Great Mass Extinction

We are presently in the midst of a sixth great mass extinction. Scientists estimate that the current extinction rate is 100 to 1,000 times above the typical background rate, putting it on a level with the five major extinction events of the past. Unlike previous extinction events, this one is not the inevitable result of a physical catastrophe such as an eruption or asteroid impact. Humans are the driving force behind the current rise in extinction rate, and our actions will determine the extent of the losses.

Estimates of the number of currently living species range from 5 million to 50 million. Of these, fewer than 2 million have been named. Simply put, we don't know what we have to lose. Scientists are racing to document Earth's diversity in the face of current threats. They are also attempting to determine which species are likely to become extinct if biodiversity is not protected.

For the purposes of conservation, a species is considered extinct if repeated, extensive surveys of its known range repeatedly fail to turn up signs of any individuals. It is "extinct in the wild" if the only known members of the species are in captivity. Some species are difficult to find, so occasionally a population of species previously thought to be extinct or extinct in the wild turns up. However, this is a rare occurrence.

An **endangered species** is one currently at a high risk of extinction in the wild. A **threatened species** is one that is likely to become endangered in the near future. Keep in mind that not all rare species are threatened or endangered. Some species have always been present only in low numbers.

In the United States, species are listed as endangered or threatened by the Fish and Wildlife Service (USFWS). The International Union for Conservation of Nature and Natural Resources (IUCN) monitors threats to species worldwide (**TABLE 48.1**).

---

endangered species Species that faces extinction in all or a part of its range.

mass extinction Event in which a large proportion of Earth's organisms become extinct in a relatively short period of geologic time.

threatened species Species likely to become endangered in the near future.

A White abalone. **B** Black rhinoceros.

FIGURE 48.2 Two species endangered by overharvesting.

A Texas blind salamander.

B Buffalo clover.

FIGURE 48.3 Two species endangered by habitat loss.

Causes of Declining Biodiversity

A variety of human activities are contributing to the increased extinction rate.

Hunting and Overharvesting Hunting directly reduces a species' numbers. For example, evidence implicates the arrival of humans in the prehistoric decline of many large animals, collectively referred to as megafauna, in Australia and North America. In Australia, the arrival of humans about 40,000 years ago correlates with the onset of extinctions of the continent's largest marsupials, birds, and lizards. In North America, arrival of humans by about 15,000 years ago was followed by the extinction of large herbivores such as camels, giant ground sloths, and mammoths and mastodons (relatives of elephants).

Some more recent extinctions were certainly caused by humans. The World Conservation Union has compiled a list of more than 800 documented extinctions that occurred since 1500. The passenger pigeon is one of them. When European settlers first arrived in North America, they found between 3 and 5 billion passenger pigeons. In the 1800s, commercial hunting of these birds for food caused a steep decline in their number. The last wild passenger pigeon was shot in 1900, and the last captive member of the species died in 1914.

We continue to overharvest species. The overfishing of the Atlantic codfish population, described in Section 44.6, is one recent example. Consider also the plight of the white abalone, a gastropod mollusk native to kelp forests off the coast of California (**FIGURE 48.2A**). Heavy harvesting of this species during the 1970s reduced the population to about 1 percent of its original size. In 2001, the white abalone became the first invertebrate listed as endangered by the USFWS.

Although some white abalone remain in the wild, their population density is too low for effective reproduction. Abalones release their gametes into the water, a strategy that succeeds only when many individuals live within a few meters of one another. At this time, the species' only hope for survival is an ongoing program of captive breeding. If this program succeeds, individuals will be reintroduced to the wild.

Species are overharvested not only as food, but also for use in traditional medicine, for the pet trade, and for ornamentation. Some orchids prized by collectors have become nearly extinct in the wild. Elephants are killed for their ivory tusks, and black rhinos (**FIGURE 48.2B**) for their horns. The majority of elephant tusks end up in China, in the form of decorative carved objects. Rhino horn is used as dagger handles and as a traditional (but ineffective) treatment for impotence.

Habitat Degradation and Fragmentation Each species requires a specific habitat, so degradation, fragmentation, or destruction of that habitat also reduces population numbers. An endemic species, a species that lives only in the region where it evolved, is more likely to go extinct as a result of habitat loss than a species with a more widespread distribution. Also, a species with highly specific needs is more likely to go extinct than one that is a generalist. For example, giant pandas are endemic to China's bamboo forests and feed almost entirely on bamboo. As bamboo forests have disappeared, so have pandas. Their population, which may once have been as high as 100,000 animals, is now reduced to about 1,600 animals in the wild.

In Texas, habitat degradation threatens Texas blind salamanders (**FIGURE 48.3A**). The salamander is endemic to the Edwards Aquifer, a series of water-filled, underground limestone formations. Like other aquifers, the salamander's home is threatened by excess withdrawal of water and by water pollution.

CREDITS: (2A) John Butler, NOAA; (2B) © George Sanker/naturepl.com; (3A) Joe Fries, U.S. Fish & Wildlife Service; (3B) USDA Forest Service/Photo by Sarena Selbo.

Species introductions (Section 45.8) can also degrade essential habitat. For example, many ground-nesting birds that live on islands are endangered by predators such as rats or snakes that arrived on the island as stowaways. Exotic species can also harm natives by competing with them, as when red imported fire ants eat food that would otherwise sustain native ants. Introduced pathogens can decimate species that did not coevolve with them and hence lack defenses.

Decline or loss of one species sometimes endangers others. Consider buffalo clover, a plant that was abundant when many buffalo grazed in the American Midwest (**FIGURE 48.3B**). The clover thrived at the edges of woodlands, where buffalo enriched the soil with their droppings and helped disperse the clover's seeds. Like most endangered species, buffalo clover faces a number of simultaneous threats. In addition to the loss of buffalo, the clover is threatened by competition from introduced plants, attacks by introduced insects, and development of its habitat for housing.

Such development often fragments habitats. It creates islands of suitable habitat scattered across what was once a continuous range. Buildings, roads, and fences divide up a habitat, thus breaking a large population into smaller populations in which inbreeding is more likely. Recall that inbreeding can have deleterious genetic effects (Section 17.6). Fragmentation also prevents individuals from accessing essential resources. For example, buildings and roads now prevent spotted turtles endemic to the eastern United States from moving among the different wetlands they use to feed and hibernate. The wetlands still exist, but turtles can no longer move between them as needed.

TAKE-HOME MESSAGE 48.2

✔ Previous mass extinctions occurred when an event such as an asteroid impact or a series of volcanic eruptions caused a global catastrophe. Current extinctions are the result of human actions.

✔ Humans directly decrease species numbers by overharvesting. They indirectly harm species by destroying, degrading, or fragmenting essential habitat.

48.3 Harmful Land Use Practices

LEARNING OBJECTIVES

- Distinguish between desertification and deforestation.
- Explain why desertification and deforestation alter the local climate.

Human activities have the potential not only to harm individual species, but also to transform entire biomes.

A Giant dust cloud about to descend on a farm in Kansas during the 1930s, when a large part of the southern Great Plains was known as the Dust Bowl.

Africa

B Modern-day dust cloud blowing from Africa's Sahara desert out into the Atlantic Ocean. The Sahara's area is increasing as a result of a long-term drought, overgrazing, and stripping of woodlands for firewood.

FIGURE 48.4 Dust storms, a result of desertification.

Desertification

Deserts naturally expand and contract over long periods as a result of fluctuations in climate. However, poor agricultural practices that encourage soil erosion sometimes lead to rapid shifts from grassland or woodland to desert. As the human population increases, more and more people are forced to farm in areas ill-suited to agriculture. Others allow their livestock to overgraze grasslands, killing grass plants. Such practices can result in **desertification**, the conversion of grassland or woodlands to desertlike conditions.

Consider what happened during the mid-1930s, when large portions of prairie on the southern Great Plains of the United States were plowed under to plant crops. Plowing exposed the deep prairie topsoil to the force of the region's constant winds. Coupled with a drought, the result was an economic and ecological

desertification Conversion of a grassland or woodland to desert.

CREDITS: (4A) NOAA; (4B) Satellite image courtesy of GeoEye.

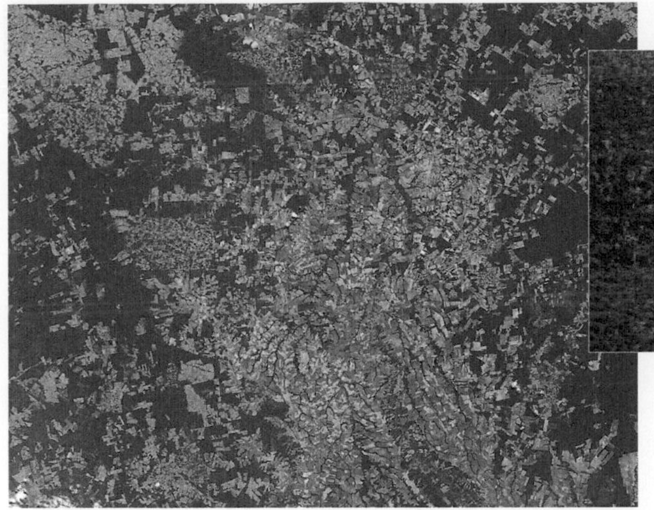

FIGURE 48.5 **Tropical deforestation.** An satellite image (left) and a closer aerial veiw (above) of deforestation in Brazil, a country where tropical woodlands and forest have been cleared at an alarming rate, to create soybean farms and pastures for cattle.

disaster. Winds carried more than a billion tons of topsoil aloft as sky-darkening dust clouds, causing the region be known as the Dust Bowl (FIGURE 48.4A). Tons of displaced soil from the Dust Bowl fell to earth as far away as New York City and Washington, D.C.

Desertification remains a threat in developing countries. In Africa, the Sahara desert is expanding south into the Sahel region. Ongoing overgrazing of the Sahel strips grasslands of their vegetation and allows winds to erode the soil. Winds carry soil particles aloft and westward, carrying them as far away as the southern United States and the Caribbean (FIGURE 48.4B).

In China's northwestern regions, overplowing and overgrazing have expanded the Gobi desert so that dust clouds periodically darken skies above Beijing. Winds carry some of this dust across the Pacific to the United States. In an attempt to hold back the desert, China is planting billions of trees as a "Green Wall."

In a positive feedback cycle, drought encourages desertification, which results in more drought. Plants cannot thrive in a region where the topsoil has blown away. With less transpiration by plants (Section 28.4), less water enters the air, so local rainfall decreases.

The best way to prevent desertification is to avoid farming in areas subject to high winds and periodic drought. If these areas must be used, farming methods that do not repeatedly disturb the soil can minimize the risk of desertification.

Deforestation

The amount of forested land is currently stable or increasing in North America, Europe, and China, but tropical forests continue to disappear at an alarming rate. In Brazil, increases in the export of soybeans and free-range beef have helped make the country the world's seventh-largest economy. However, this

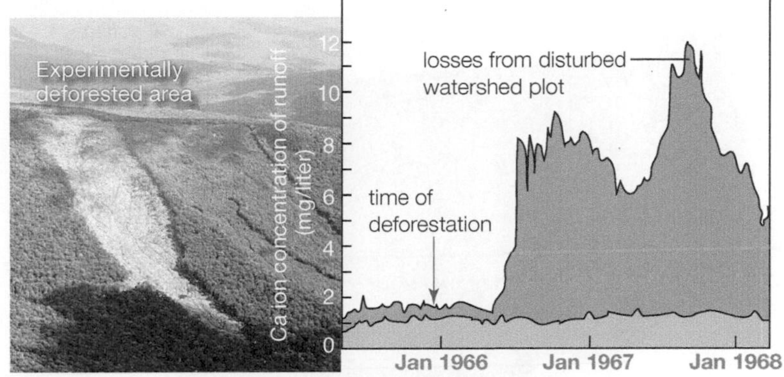

Experimentally deforested area
Ca ion concentration of runoff (mg/liter)
0
2
4
6
8
10
12
losses from disturbed watershed plot
time of deforestation
Jan 1966
Jan 1967
Jan 1968

FIGURE 48.6 **Deforestation and nutrient losses from soil.** After experimental deforestation of a portion of forest in the Hubbard Brook watershed, calcium (Ca) ion levels in runoff increased sixfold (gray). An undisturbed plot in the same forest showed no corresponding increase during this time (green).

economic expansion has come at the expense of the country's woodlands and forests (FIGURE 48.5).

Deforestation has detrimental effects beyond the immediate destruction of forest organisms. Deforestation encourages flooding because rather than being taken up by tree roots, water runs off into streams. Deforestation of hilly areas raises the risk of landslides. Tree roots tend to hold soil in place, so their removal makes waterlogged soil more likely to slide. Deforestation also reduces the fertility of soils by increasing runoff. FIGURE 48.6 shows results of an experiment in which scientists deforested a region in New Hampshire and monitored the nutrients in runoff. Deforestation increased the loss of nutrients such as calcium, which are essential for healthy plant growth.

Like desertification, deforestation affects local weather. In hot weather, forests remain cooler than adjacent nonforested areas because trees shade the ground and transpiration causes evaporative cooling. When a forest is cut down, daytime temperatures rise

CREDITS: (5) left, NASA courtesy of the MODIS Rapid Response Team at Goddard Space Flight Center; right, Frontpage/Shutterstock; (6) left, USDA Forest Service, Northeastern Research Station; right, © Cengage Learning.

and reduced transpiration results in less rainfall. Once a tropical forest is logged, the resulting nutrient losses and drier, hotter conditions can make it impossible for tree seeds to germinate or for seedlings to survive. Thus, deforestation can be difficult to reverse.

Deforestation also contributes to global climate change in two ways. First, trees that are cut to clear land for other uses are often burned, releasing carbon into the atmosphere. Second, conversion of forest to cropland or pasture decreases the rate at which carbon is taken up by plants.

TAKE-HOME MESSAGE 48.3

✔ Desertification and deforestation not only destroy habitat, they allow increased erosion and can cause changes in local weather patterns.

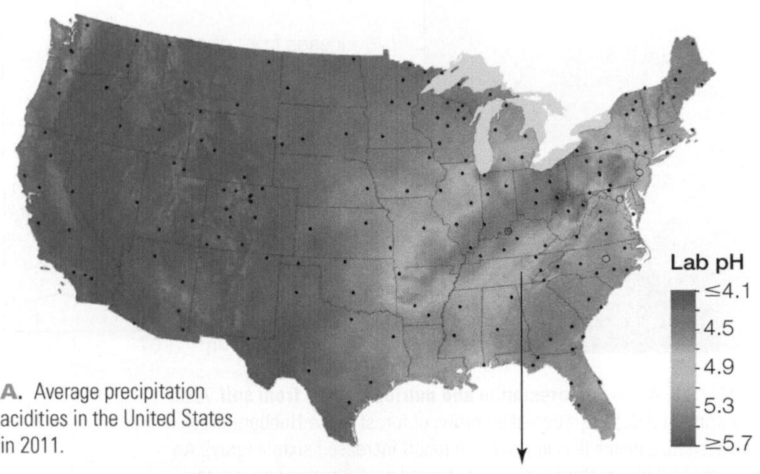
Lab pH
≤4.1
4.5
4.9
5.3
≥5.7

A. Average precipitation acidities in the United States in 2011.

B. Dying trees in Great Smoky Mountains National Park, where acid rain harms leaves and causes loss of nutrients from soil.

FIGURE 48.7 Acid rain. FIGURE IT OUT Is rain more acidic on the East Coast or the West Coast?

Answer: The East Coast

48.4 Effects of Pollutants

LEARNING OBJECTIVES

- Describe how plastic in the sea harms marine organisms.
- Compare the negative effects of acid, ammonium, and mercury deposition.

Pollutants are natural or synthetic substances released into soil, air, or water in greater than natural amounts. Some come from a few easily identifiable sites, or point sources. A factory that discharges pollutants into the air or water is a point source. Pollutants that come from point sources are the easiest to control: Identify the sources, and you can take action there. Dealing with pollution from nonpoint sources is more challenging. Such pollution stems from the widespread release of a pollutant. For example, oil that runs off from roads and driveways is a nonpoint source of water pollution.

Atmospheric Deposition

Pollutants that enter the atmosphere can harm organisms when they fall to Earth as dust or in precipitation.

Sulfur dioxides and nitrogen oxides are common air pollutants. Most sulfur dioxide pollution comes from coal-burning power plants. Combustion of gasoline and oil releases most nitrogen oxides. In dry weather, airborne sulfur and nitrogen oxides coat dust particles. Dry acid deposition occurs when this coated dust falls to the ground. Wet acid deposition, or **acid rain**, occurs when pollutants combine with water and fall as acidic precipitation. The pH of unpolluted rainwater is 5 or higher (Section 2.6). Acid rain can be ten times more acidic (**FIGURE 48.7A**).

Acid rain that falls on or drains into waterways, ponds, and lakes can harm aquatic organisms. A low pH prevents fish eggs from developing and kills adult fishes. When acid rain falls on a forest, it burns tree leaves and alters the forest's soil. As acidic water drains through the soil, positively charged hydrogen ions displace positively charged nutrient ions such as calcium, causing nutrient loss. The acidity also causes soil particles to release aluminum and other metals that can harm plants. The combination of poor nutrition and exposure to toxic metals weakens trees, making them more susceptible to insects and pathogens, and thus more likely to die. Effects are most pronounced at

acid rain Low-pH rain formed when sulfur dioxide and nitrogen oxides mix with water vapor in the atmosphere.

bioaccumulation An organism accumulates increasing amounts of a chemical pollutant in its tissues over the course of its lifetime.

biological magnification A chemical pollutant becomes increasingly concentrated as it moves up through food chains.

CREDITS: (7A) Courtesy of National Atmospheric Deposition; (7B) Frederica Georgia/Science Source.

higher elevations where trees are frequently exposed to clouds of acidic droplets (**FIGURE 48.7B**).

Sulfur dioxides and nitrogen oxides that enter the air can also combine with gaseous ammonia released by fertilizers and animal wastes. When the resulting ammonium salts fall to Earth, they act like fertilizer. Addition of ammonium salts to soil allows plants with high nitrogen needs to thrive where they would otherwise be excluded, thus altering the array of species in an ecosystem.

Emissions from coal-burning power plants and factories are the main source of mercury in the atmosphere. When mercury falls to Earth, microorganisms combine it with carbon to form methylmercury, which is toxic to animals. Mercury does not harm plants.

Biological Accumulation and Magnification

Some chemical pollutants can build up inside an organism's body. By the process of **bioaccumulation**, an organism's tissues store a pollutant taken up from the environment, causing the amount in the body to increase over time. The ability of some plants to bioaccumulate toxic substances makes them useful in phytoremediation of polluted soils (Section 28.1).

In animals, hydrophobic chemical pollutants ingested or absorbed across the skin accumulate in fatty tissues. The amount of pollutant in an animal's body increases over time, so long-lived species tend to have a higher amount of fat-soluble pollutants than shorter-lived ones. Within a species, old individuals have a higher pollutant load than younger ones.

Trophic level also affects the pollutant concentration in an organism's tissues. By the process of **biological magnification**, concentration of a chemical increases as the pollutant moves up a food chain. **FIGURE 48.8**

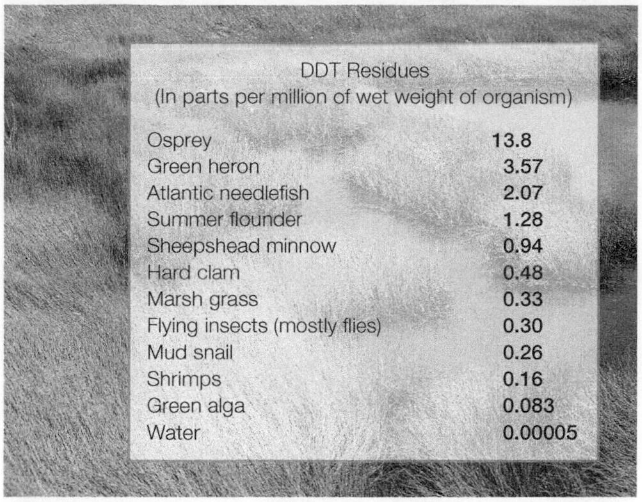

DDT Residues
(In parts per million of wet weight of organism)

Osprey	13.8
Green heron	3.57
Atlantic needlefish	2.07
Summer flounder	1.28
Sheepshead minnow	0.94
Hard clam	0.48
Marsh grass	0.33
Flying insects (mostly flies)	0.30
Mud snail	0.26
Shrimps	0.16
Green alga	0.083
Water	0.00005

FIGURE 48.8 Biological magnification of DDT in an estuary.
Reported in 1967 by George Woodwell, Charles Wurster, and Peter Isaacson.

Data Analysis Activities

Bioaccumlation in tuna Bluefin tuna are among the fishes that were exposed to radioactive material released into the sea by the damaged Fukushima nuclear power plant (Section 47.1). These tuna breed in the western Pacific, and some migrate to the Eastern Pacific when they are 1 to 2 years old. Daniel Madigan suspected that migrating tuna would transport radioactive material to the waters off the coast of California. To test this hypothesis, he and his collaborators analyzed the concentration of cesium radioisotopes (^{137}Cs and ^{134}Cs) in bluefin tuna and yellowfin tuna caught near San Diego in August 2011. Yellowfin tuna do not cross the Pacific; they spend their lives in California's coastal waters. The scientists compared these data with the radioisotope content of bluefin tuna that had been caught near San Diego in 2008. **FIGURE 48.9** shows their results.

1. How did the radioisotope content of the bluefin tuna caught in August of 2011 compare to that of bluefins caught in 2008 (before the Fukushima accident)?
2. In August of 2011, how did the isotope content of blue fin tuna compare to that of yellowfin tuna?
3. Bluefin tuna and yellowfin tuna are similarly sized top predators. Why did Madigan compare species that were similar in this way?
4. Does the data support the hypothesis that bluefin tuna transported radioisotopes from Fukushima in their body?

			Mean radioisotope concentration (Bq per kilogram)	
Tuna type, date sampled	**Mean body mass (kg dry)**	**Mean age (years)**	**^{137}Cs**	**^{134}Cs**
Bluefin, 2011 ($n = 15$)	1.5	1.5	6.3	4
Bluefin, 2008 ($n = 5$)	1.5	1.4	1.4	0
Yellowfin, 2011 ($n = 5$)	1.9	1.2	1.1	0

FIGURE 48.9 Radioactive cesium (Cs) concentration in two species of tuna caught along the coast of California.

^{137}Cs has a half-life of 30 years and some persists in the Pacific from weapons testing. ^{134}Cs has a half-life of 2 years and was undetectable in the Pacific prior to the accident at Fukushima. Becquerels (Bq) is a measure of radioactivity.

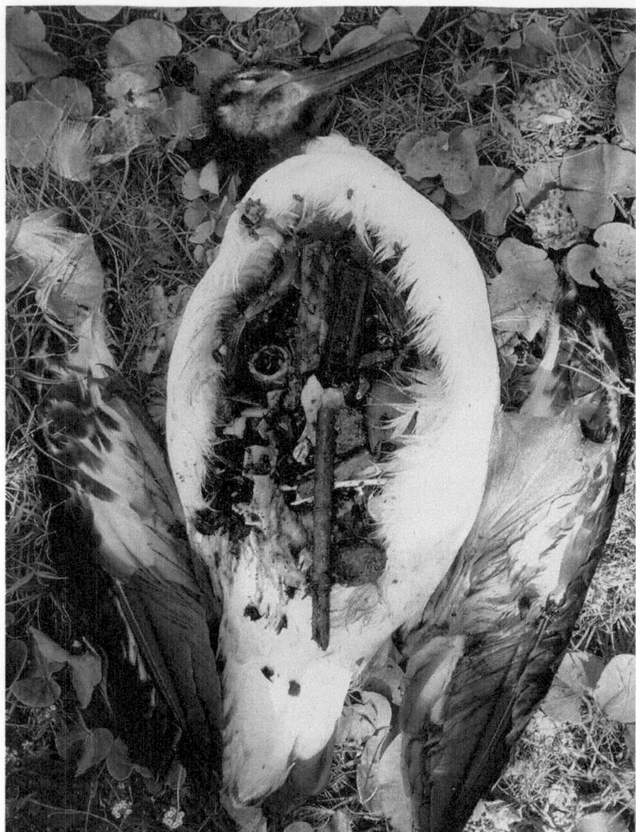

FIGURE 48.10 Death by plastic. Scientists found more than 300 pieces of plastic in this albatross chick. One piece had punctured the bird's gut, causing its death. The chick was fed plastic by its parents, who gathered the material from the ocean surface, mistaking it for food.

provides data documenting biological magnification of DDT (an insecticide) in a salt marsh ecosystem during the 1960s. Notice that the concentration of DDT in ospreys, which are fish-eating birds, was 276,000 times higher than the concentration in the water. As a result of bioaccumulation and biological magnification, even very low environmental concentrations of pollutants can have detrimental effects on a species.

Bioaccumulation and magnification of other pollutants such as methylmercury continue to pose a threat to wildlife and human health (Section 2.1).

Talking Trash

Seven billion people use and discard a lot of stuff. Historically, unwanted material was buried in the ground or dumped out at sea. Trash was out of sight, and also out of mind. We now know that burying garbage can contaminate groundwater, as when lead from discarded batteries seeps into aquifers. We also know that solid waste dumped into oceans harms marine life (**FIGURE 48.10**). In the United States, solid municipal waste can no longer legally be dumped at sea. Nevertheless, plastic constantly enters our coastal waters. Foam cups and containers from fast-food outlets, plastic shopping bags, plastic water bottles, and other litter ends up in storm drains. From there it enters streams and rivers that convey it to the sea.

Ocean currents can carry bits of plastic for thousands of miles. Currents also cause accumulations of plastic in certain areas of the ocean. Consider the Great Pacific Garbage Patch. The media often describes this region of the north central Pacific as an "island of trash." In fact, the plastic here is not easily visible. The garbage patch is a region where vast amounts of confetti-like plastic particles swirl slowly around an area as large as the state of Texas. The small bits of plastic absorb and concentrate toxic compounds such as pesticides and industrial chemicals from the seawater around them, making the plastic even more harmful to marine organisms that mistakenly eat it.

TAKE-HOME MESSAGE 48.4

- ✔ Acid rain can make a lake or soil too acidic for life to thrive.
- ✔ Even small amounts of fat-soluble chemicals can accumulate in tissues and be magnified from one trophic level to the next.
- ✔ Animals can be harmed when they mistake indigestible trash that enters waterways and oceans for food.

48.5 Ozone Depletion and Pollution

LEARNING OBJECTIVES

- Explain the importance of the ozone layer.
- List the pollutants that affect the ozone layer.
- Describe what has been done to slow ozone layer thinning.

Depletion of the Ozone Layer

In the upper atmosphere, between 17 and 27 kilometers (10.5 and 17 miles) above sea level, is a region of high ozone (O_3) concentration referred to as the ozone layer. The ozone layer benefits living organisms by absorbing most ultraviolet (UV) radiation from incoming sunlight. UV radiation causes mutations (Section 8.6).

In the mid-1970s, scientists noticed that Earth's ozone layer was thinning. Its thickness had always varied seasonally, but now the average level was declining steadily from year to year. By the mid-1980s, spring ozone thinning over Antarctica was so pronounced that people were referring to the lowest-ozone region as an "ozone hole" (**FIGURE 48.11A**).

Declining ozone quickly became an international concern. With a thinner ozone layer, people would be exposed to more UV radiation, the main cause of skin cancers. Higher UV levels also harm wildlife, which do not have the option of avoiding sunlight. In addition,

CREDIT: (10) Claire Fackler/NOAA.

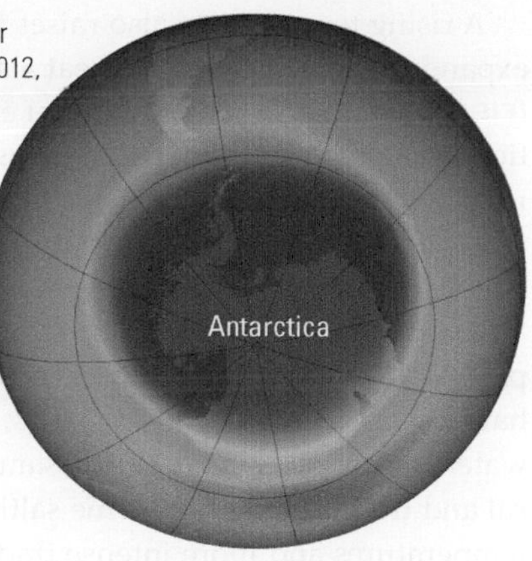
Antarctica

A Ozone levels in the upper atmosphere in September 2012, the Antarctic spring.

Purple indicates the least ozone, with blue, green, and yellow indicating increasingly higher levels.

Check the current status of the ozone hole at NASA's website (http://ozonewatch.gsfc.nasa.gov/).

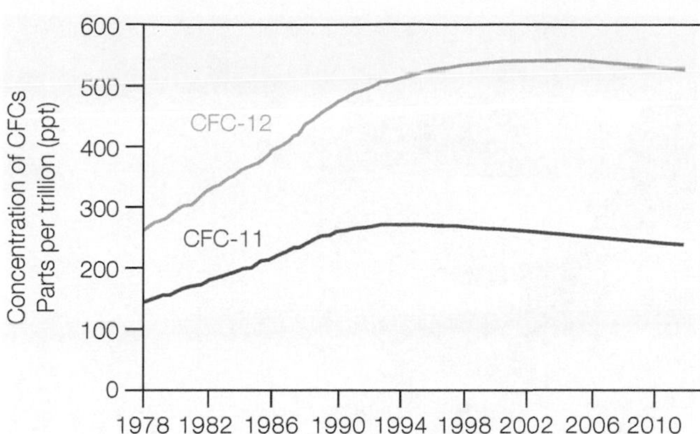
Concentration of CFCs
Parts per trillion (ppt)
600
500
400
300
200
100
0
CFC-12
CFC-11
1978 1982 1986 1990 1994 1998 2002 2006 2010

B Concentration of two CFCs in the upper atmosphere. These pollutants destroy ozone. A worldwide ban on CFCs has halted the rise in atmospheric CFC concentration.

FIGURE 48.11 Ozone and CFCs.

exposure to higher than normal UV levels affects plants and other producers, slowing the rate of photosynthesis and the release of oxygen into the air.

Chlorofluorocarbons, or CFCs, are major ozone destroyers. These odorless gases were once widely used as propellants in aerosol cans, as coolants, and in solvents and plastic foam. In response to the threat posed by ozone thinning, countries worldwide agreed in 1987 to phase out production of CFCs and other ozone-destroying chemicals. As a result of that agreement (the Montreal Protocol), concentrations of CFCs in the atmosphere are no longer rising dramatically (**FIGURE 48.11B**). However, because CFCs break down slowly, scientists expect them to continue to significantly impair the ozone layer for several decades.

Today, the main ozone-depleting pollutant entering the atmosphere is nitrous oxide (N_2O). This gas is released when people burn fossil fuels and when bacteria feed on fertilizers and animal waste (Section 46.9).

Near-Ground Ozone Pollution

Near the ground, there is naturally little ozone, so it is considered a pollutant. Ozone irritates the eyes and respiratory tracts of humans and wildlife. It also interferes with plant growth.

Ground-level ozone forms when nitrogen oxides and volatile organic compounds released by burning or evaporating fossil fuels are exposed to sunlight. Warm temperatures speed the reaction. Thus, ground-level ozone tends to vary daily (being higher in the daytime) and seasonally (being higher in the summer).

To help reduce ozone pollution, avoid putting fossil fuels or their combustion products into the air at times that favor ozone production. On hot, sunny, still days, postpone filling a gas tank or using gas-powered appliances until evening, when there is less sunlight to power conversion of pollutants to ozone.

TAKE-HOME MESSAGE 48.5

✔ Certain synthetic chemicals destroy ozone in the upper atmosphere's protective ozone layer. This layer serves as a protective shield against UV radiation.

✔ Evaporation and burning of fossil fuels increases the amount of ozone in the air near the ground, where ozone is considered a harmful pollutant.

48.6 Effects of Global Climate Change

LEARNING OBJECTIVES

- Explain why a rise in global temperature raises the sea level.
- Describe some ways in which global climate change can alter biological communities.

Ongoing climate change affects ecosystems throughout the world. Most notably, average temperatures are increasing, especially at temperate and polar latitudes.

Temperature changes have wide-reaching biological impacts. Temperature is an important cue for many temperate zone species. Abnormally warm spring temperatures are causing deciduous trees to put leaves out earlier, and spring-blooming plants to flower earlier. Animal migration times and breeding seasons are also shifting. Species arrays in biological communities are changing as warmer temperatures allow some species to expand their range to higher latitudes or elevations that were previously too cold for them.

Not all species can move or disperse quickly to a new location, so warmer temperatures are expected to drive some of these species to extinction. In tropical seas, warming water is stressing reef-building corals

CREDITS: (11A) NASA Ozone Watch; (11B) From www.esrl.noaa.gov.

a. 1941 photo of Muir Glacier in Alaska.

b. 2004 photo of the same region.

FIGURE 48.12 Melting glaciers, evidence of a warming world. Water from melting glaciers contributes to a rising sea level.

and increasing the frequency of coral bleaching events (Section 47.13). In the arctic, seasonal sea ice now forms later and melts earlier, stressing polar bears. During late spring and early fall, regions where these bears previously would have walked across sea ice to gain access to their seal prey are now open water.

The warming trend is also expected to affect human health. Deaths from heat stroke are likely to increase, and some infectious diseases will become more widespread. One way of predicting how warmer global temperatures will affect the incidence of disease is to look at what happens during an El Niño event, when sea temperature rises. As Section 47.4 explained, the incidence of cholera and malaria rises during an El Niño. In addition, elevated temperature and CO_2 concentration are expected to increase production of allergy-inducing plant pollen and fungal spores.

A rising temperature also raises the sea level. Water expands as it is heated, and heat also melts glaciers (**FIGURE 48.12**). Together, thermal expansion and addition of meltwater from glaciers cause the sea level to rise. In the past century, the sea level has risen about 20 centimeters (8 inches). As a result, some coastal wetlands are disappearing under water.

The composition of seawater is changing too. In the past 50 years, the upper waters of the Atlantic Ocean have become less salty at the poles, where more meltwater is flowing in. During the same period, subtropical and tropical waters became saltier as warmer ocean temperatures and more intense trade winds encouraged evaporation. Altered rainfall patterns have also promoted salinity changes; annual precipitation has risen at high latitudes and fallen at low ones.

Seawater is becoming more acidic too, because when CO_2 dissolves in a solution, the solution's pH declines. Increased acidity can harm marine life by making less calcium carbonate available. Groups ranging from foraminifera, to corals, to bivalve mollusks require calcium carbonate to build their hard parts.

TAKE-HOME MESSAGE 48.6

- ✔ Warming water, a rising sea level, and changes in the composition of seawater pose threats to aquatic species.
- ✔ On land, warming temperatures are altering the timing of flowering and migrations, and the distribution of some species.
- ✔ Increased warmth will also encourage the spread of some human pathogens.

48.7 Conservation Biology

LEARNING OBJECTIVES

- List the three levels of biodiversity.
- Explain the benefits of protecting biodiversity.
- Explain what a biodiversity hot spot is and give an example.
- Using an appropriate example, explain the goal of ecological restoration.

The Value of Biodiversity

Every nation has several forms of wealth: material wealth, cultural wealth, and biological wealth. Biological wealth is called **biodiversity**. We measure a region's biodiversity at three levels: the genetic diversity within species, species diversity, and ecosystem diversity. Biodiversity is currently declining at all three levels, in all regions.

Conservation biology is the field that addresses these declines by surveying the range of biodiversity, and finding ways to maintain and use biodiversity to

CREDITS: (12) National Snow and Ice Data Center;

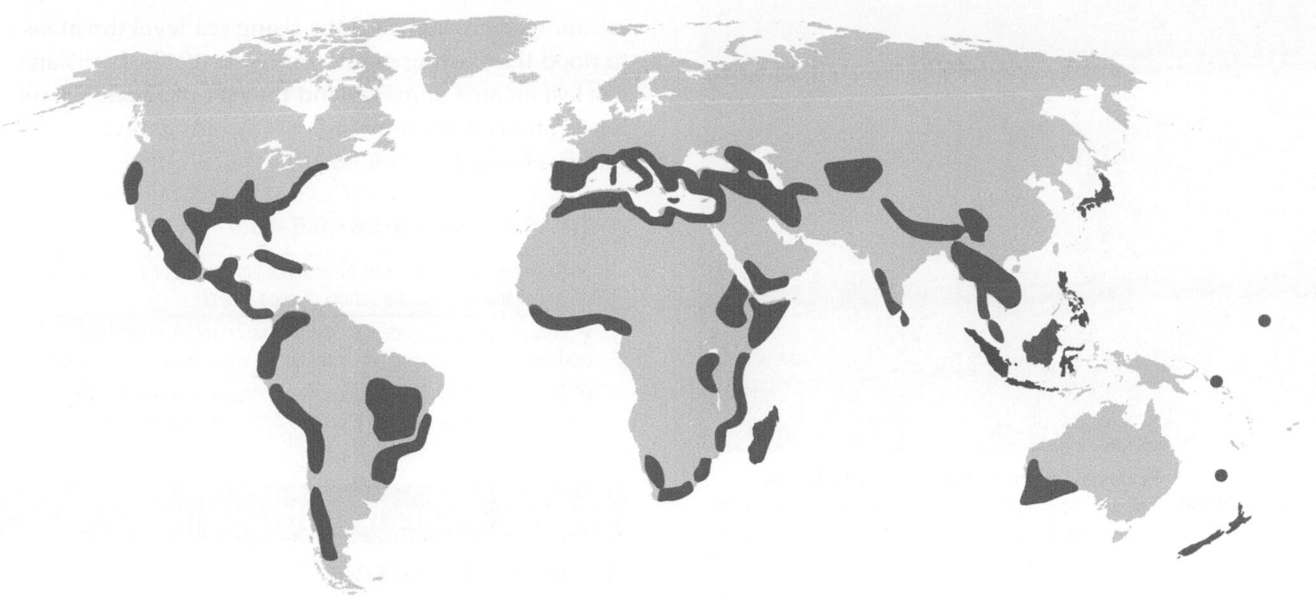

FIGURE 48.13 **Biodiversity hotspots.** These areas are home to large numbers of endemic plants and have lost much of their native vegetation.

benefit human populations. The aim is to conserve biodiversity by encouraging people to value it and use it in ways that do not destroy it.

Why should we protect biodiversity? From a selfish standpoint, doing so is an investment in our future. Healthy ecosystems are essential to our survival. Other organisms produce the oxygen we breathe and the food we eat. They remove waste carbon dioxide from the air and decompose and detoxify wastes. Plants take up rain and hold soil in place, preventing erosion and reducing the risk of flooding. Compounds in wild species can serve as medicines. Wild relatives of crop plants are reservoirs of genetic diversity that plant breeders draw on to protect and improve crops.

FIGURE 48.14 **Life in a biodiversity hot spot.** The California Floristic Province in North America is home to many endemic species such as giant sequoias (left) and the northern spotted owl (right).

Setting Priorities

Protecting biological diversity can pose a challenge. People often oppose environmental protections because they fear that such measures will have adverse economic consequences. Conservation biologists help us make difficult choices about how to most efficiently protect biodiversity. They identify **biodiversity hot spots**, areas of high biodiversity that are under the greatest threat. As of 2017, the Critical Ecosystems Partnership Fund has designated 36 global biodiversity hot spots (FIGURE 48.13). Each of these areas has more than 1,500 species of endemic vascular plants (plants found only in that region) and has lost more than 70 percent of its vegetation as a result of human activities. Although the combined area of these biodiversity hot spots constitutes about 2 percent of the world's land mass, they collectively contain more than half of the world's endemic plant species and 40 percent of its endemic vertebrates. The biodiversity hot spots include areas in 140 nations.

As you might expect, many hot spots are in tropical regions of developing countries. However, there are two biodiversity hot spots in the United States. The California Floristic province on the west coast is home to a diverse array of endemic conifers, including the giant sequoia (FIGURE 48.14). Endangered northern spotted owls nest in the region's old-growth forests. The Coastal Plain hot spot in the southeast extends

biodiversity Of a region, the genetic variation within its species, variety of species, and variety of ecosystems.
biodiversity hot spot Threatened region with great biodiversity that is considered a high priority for conservation efforts.
conservation biology Field of applied biology that surveys and documents biodiversity and seeks ways to maintain and use it.

FIGURE 48.15 Ecological restoration in Louisiana's Sabine National Wildlife Refuge. The squares contain sediment that was brought in and planted with marsh grass. Restoring marsh in areas that had become open water increases animal abundance and species richness of the area.

along the Gulf and Atlantic coasts. Venus flytrap, an insect-eating plant, is one of the plants endemic to this region. Endangered vertebrates in this region include the red-cockcaded woodpecker, which nests only in old-growth pines, and the Florida bonneted bat.

Preservation and Restoration

Worldwide, many ecologically important regions have been protected in ways that benefit local people. Consider the Monteverde Cloud Forest of Costa Rica, which is within the Mesoamerica biodiversity hot spot. During the 1970s, George Powell was studying birds in this forest, which was rapidly being cleared. Powell decided to buy part of the forest as a nature sanctuary. His efforts inspired individuals and conservation groups to donate funds, and much of the forest is now protected as a private nature reserve. The reserve's animals include more than 100 mammal species, 400 bird species, and 120 species of amphibians and reptiles. It is one of the few habitats left for jaguars and ocelots. A tourism industry centered on the reserve provides economic benefits to local people.

Sometimes, an ecosystem is so damaged, or there is so little of it left, that conservation alone is not enough to sustain biodiversity. **Ecological restoration** is work designed to renew or recreate a natural ecosystem that has been degraded or destroyed, fully or in part. Restoration work in Louisiana's coastal wetlands is an example. More than 40 percent of the coastal wetlands in the United States are in Louisiana. The marshes are an ecological and economic treasure, but they are troubled. Dams and levees built upstream of the marshes keep back sediments that would normally replenish sediments lost to the sea. Channels cut through the marshes for oil exploration and production have encouraged erosion, and the rising sea level threatens to flood the existing plants. Since the 1940s, Louisiana has lost an area of marshland the size of Rhode Island. Restoration efforts now under way aim to reverse some of those losses (FIGURE 48.15).

TAKE-HOME MESSAGE 48.7

- ✔ Biodiversity is the genetic diversity of individuals of a species, the variety of species, and the variety of ecosystems.
- ✔ Conservation biologists identify threatened regions with high biodiversity and prioritize which should be first to receive protection.
- ✔ Through ecological restoration, we re-create or renew a biologically diverse ecosystem that has been destroyed or degraded.

48.8 Reducing Negative Impacts

LEARNING OBJECTIVE

- Describe some ways that individuals can help sustain biodiversity.

Ultimately, the health of life on Earth depends on our ability to live within our limits. The goal is to live sustainably, which means meeting the needs of the present generation without reducing the ability of future generations to meet their own needs.

Promoting sustainability begins with recognizing the environmental consequences of one's lifestyle. People in industrial nations use huge quantities of resources, and the extraction of these resources has negative effects on biodiversity. In the United States, the size of the average family has declined since the 1950s, while the size of the average home has doubled. All of the materials used to build and furnish those larger homes must be extracted from the environment.

For example, an average new home contains about 500 pounds of copper in its wiring and plumbing. Where does that copper come from? Like most other mineral elements used in manufacturing, most copper is mined from the ground. Surface mining strips an area of vegetation and soil, creating an ecological dead zone (FIGURE 48.16). It also emits dust into the air, creates mountains of rocky waste, and can contaminate nearby waterways.

Nonrenewable mineral resources are also used in electronic devices such as phones, computers, televisions, and gaming consoles. Reducing consumption by fixing existing products is a sustainable resource use, as is recycling. Obtaining nonrenewable materials by recycling reduces the need for extraction of those resources, and it also keeps materials out of landfills.

Reducing your energy use is another way to promote sustainability. Fossil fuels such as petroleum, natural gas, and coal supply most of the energy used by developed countries. You already know that burning

CREDIT: (15) Diane Borden-Bilot, U.S. Fish and Wildlife Service.

FIGURE 48.16 Resource extraction. Bingham copper mine near Salt Lake City, Utah. This open pit mine is 4 kilometers (2.5 miles) wide and 1,200 meters (0.75 miles) deep. It is the largest human-made excavation on Earth.

FIGURE 48.17 Volunteers restore the Little Salmon River in Idaho. Their work allows salmon to migrate upstream to their breeding site.

these nonrenewable fuels contributes to global warming and acid rain. In addition, extracting and transporting these fuels have negative impacts. Oil harms species when it leaks from pipelines, ships, or wells. Similarly, nuclear power plants can have accidents that release dangerous radioactive substances. Renewable energy sources do not produce greenhouse gases, but they have drawbacks too. Wind turbines harm birds and bats. Panels used to collect solar energy are made of nonrenewable mineral resources, and manufacturing them generates pollutants.

To take a more active role in preserving biodiversity, learn about the threats to ecosystems in your own area. Support efforts to preserve and restore local biodiversity. Many ecological restoration projects are supervised by trained biologists but carried out primarily through the efforts of volunteers (**FIGURE 48.17**). Keep in mind that unthinking actions of billions of individuals are the greatest threat to biodiversity. Each of us may have little impact, but our collective behavior, for good or for bad, will determine the future of the planet.

ecological restoration Work aimed at to renew or recreate a natural ecosystem that has been degraded or destroyed, fully or in part.

TAKE-HOME MESSAGE 48.8

- ✔ Extraction of natural resources for human uses has side effects that threaten biodiversity.
- ✔ You can save energy and other resources by reducing energy consumption and by recycling and reusing materials.
- ✔ Rehabilitating degraded or destroyed natural areas can help sustain or renew biodiversity.

48.1 Life in the Anthropocene (revisited)

To document the rapid changes occurring during the Anthropocene, scientists need to know the current population size and distribution for species. Given the vast number of existing species, gathering such information is a task too great to be accomplished by scientists alone. Fortunately, modern technology makes it easy to crowdsource data—to gather and organize information collected independently by a large number of individuals. Anyone with a camera can photograph organisms that they encounter, then upload information about their sighting to an online database. Researchers then access this information.

iNaturalist

Consider the iNaturalist project, an online social network for sharing information about biodiversity. The goal of iNaturalist is to foster a public appreciation of biodiversity, while gathering scientifically valuable biodiversity data. The iNaturalist app (available for iOS and Android) allows users to upload photos of organisms they observe, along with the time and location of the observation. Users do not need to know exactly what they saw. The iNaturalist community includes experts who assist with species identifications. To learn more, visit iNaturalist.org, check out the iNaturalist page on Facebook, or download the iNaturalist app.

CREDITS: (16) © Lee Prince/Shutterstock.com; (17) Mountain Visions/NOAA; (in text) iNaturalist.org.

Section 48.1 Human activities are altering the Earth in ways that will be evident in the geologic record. The current interval in which these alterations are occurring is referred to as the **Anthropocene**. Scientists are working to document the current number and distribution of species so as to record the changes that occur during the Anthropocene.

Section 48.2 The geologic record shows that there have been five great **mass extinctions**. The current rate of species loss is high enough to suggest that a sixth mass extinction is under way. Unlike previous mass extinctions, which are attributed to physical causes such as an asteroid impact or a volcanic eruption, the current rise in extinctions is the result of human actions and thus may be preventable.

Arrival of humans in Australia and North America may have had a role in the extinction of the megafauna on these continents. Many more recent extinctions certainly resulted from human activity.

Endangered species currently face a high risk of extinction. **Threatened species** are likely to become endangered in the near future. Endemic species, which evolved in one place and are present only in that habitat, are highly vulnerable to extinction. Species with specialized needs are more vulnerable to extinction than generalists.

Human activities reduce the population size of other species and thus increase the risk of extinctions. Hunting and overharvesting directly decreases the number of individuals. In some such cases, it causes the population to decline below the level required for reproduction to occur.

Human activities can also cause habitat loss, degradation, and fragmentation. Release of pollutants, species introductions, and extinction of species can all degrade a habitat for other species. Construction of homes and roads can fragment a habitat. In many cases, a species becomes endangered because of multiple factors.

Section 48.3 A combination of drought and poor agricultural practices can lead to **desertification**. Desertification occurred in the American Great Plains during the 1930s and is currently a threat in parts of China and Africa. Once plant cover is removed, topsoil can blow away making it difficult for plants to become reestablished.

Deforestation is a problem in some regions, including parts of South America. In addition to harming forest species, it increases the risk of flooding, encourages the loss of nutrients from the soil, and changes the local weather. Deforestation also contributes to large-scale climate change.

Section 48.4 Pollutants released from point sources are easier to control than those released from nonpoint sources. Burning fossil fuels, especially coal, releases sulfur dioxide and nitrogen oxides into the atmosphere. These pollutants mix with water vapor in the air, then fall to earth as **acid rain**. The resulting increase in the acidity of soils and waters can weaken or kill organisms. Ammonium salts and mercury also enter ecosystems by way of atmospheric deposition. Ammonium salts act like fertilizer, altering the array of plants in an ecosystem. Mercury is toxic to animals but does not affect plants.

Some chemical pollutants undergo **bioaccumulation**. The pollutant becomes stored in an organism's tissues, so older animals have a higher pollutant concentration than younger ones. With **biological magnification**, a pollutant increases in concentration as it is passed up a food chain. As a result of bioaccumulation and biological magnification, the concentration of a pollutant in the tissues of an organism can be far greater than the concentration of that pollutant in the environment.

Solid waste is another form of pollution. Plastic that enters the oceans travels on ocean currents, breaks into small fragments, and absorbs toxic chemicals. Ingestion of plastic harms marine organisms.

Section 48.5 The ozone layer in the upper atmosphere protects against incoming UV radiation. Chemicals called CFCs were banned when they were found to cause thinning of the ozone layer. Near the ground, where the ozone concentration is naturally low, ozone emitted as a result of fossil fuel use is considered a pollutant. It irritates animal respiratory tracts and interferes with photosynthesis.

Section 48.6 Global climate change is raising the sea level, altering the salinity of upper ocean waters, and making the sea more acidic. It is also affecting the distribution and behavior of terrestrial species, allowing some to move into higher elevations or latitudes and altering the timing of migrations and flowering. Global climate change is also expected to have negative effects on human health by increasing heat-related deaths and encouraging outbreaks of some infectious diseases.

Section 48.7 We recognize three levels of **biodiversity**: genetic diversity, species diversity, and ecosystem diversity. All are threatened. The field of **conservation biology** surveys the range of biodiversity, investigates its origins, and identifies ways to maintain and use it in ways that benefit human populations.

Given that resources are limited, biologists attempt to identify **biodiversity hot spots**, regions rich in endemic species and under a high level of threat. These regions are then considered a priority for preservation efforts. When an ecosystem has been totally or partially degraded, **ecological restoration** can help restore biodiversity.

Section 48.8 Individuals can help sustain biodiversity by limiting their energy use and material consumption. Reuse and recycling can help protect species by reducing the use of destructive practices that are required to extract resources.

SELF-QUIZ

Answers in Appendix VII

1. True or false? Most species that evolved have already become extinct.

2. Passenger pigeons were driven to extinction _______ .
   a. by introduced pathogens
   b. by overharvesting
   c. as a result of global warming
   d. by competition with introduced birds

3. An _______ species has population levels so low it is at great risk of extinction in the near future.
   a. endemic c. indicator
   b. endangered d. exotic

4. Sulfur dioxide released by coal-burning power plants contributes to _______ .
   a. ozone destruction c. acid rain
   b. sea level rise d. desertification

5. As a result of _______ , an old animal usually has more pollutants in its body than a young one.
   a. bioaccumulation b. biological magnification

6. With biological magnification, a _______ will have the highest pollutant load.
   a. producer c. secondary consumer
   b. primary consumer d. top-level consumer

7. An increase in the size of the ozone hole would be expected to _______ .
   a. increase skin cancers
   b. lower sea level
   c. increase respiratory disorders
   d. increase primary production in the arctic

8. True or false? All pollutants are synthetic chemicals.

9. Global climate change is causing _______ .
   a. a decrease in sea level c. acid rain
   b. glacial melting d. all of the above

10. Use of CFCs was banned in order to _______ .
    a. lower sea level c. discourage desertification
    b. slow glacial melting d. prevent ozone destruction

11. A highly threatened region that is home to many unique species is a(n) _______ .
    a. ecoregion c. biodiversity hot spot
    b. biome d. community

12. Biodiversity refers to _______ .
    a. genetic diversity c. ecosystem diversity
    b. species diversity d. all of the above

13. Restoring a marsh that has been damaged by human activities is an example of _______ .
    a. globalization c. ecological restoration
    b. bioaccumulation d. biological magnification

14. Individuals help sustain biodiversity by _______ .
    a. reducing consumption c. recycling materials
    b. reusing materials d. all of the above

15. Match the terms with the most suitable description.

___ ozone	a. tree loss alters rainfall pattern and is difficult to reverse
___ biodiversity	b. can increase dust storms
___ acid rain	c. evolved in one region and remains there
___ endemic species	d. coal-burning is a major cause
___ biological magnification	e. results in highest pollution level at top trophic level
___ global climate change	f. good up high, bad nearby
___ deforestation	g. cause of rising sea level
___ desertification	h. genetic, species, and ecosystem diversity

CRITICAL THINKING

1. Two arctic marine mammals that live in the same waters differ in the level of pollutants in their bodies. Bowhead whales have a lower pollutant load than ringed seals. What are some factors that might explain this difference?

2. In one seaside community in New Jersey, the U.S. Fish and Wildlife Service suggested trapping and removing feral cats (domestic cats that live in the wild). The goal was to protect some endangered wild birds (plovers) that nested on the town's beaches. Many residents were angered by the proposal, arguing that the cats have as much right to be there as the birds. Do you agree? Why or why not?

3. In some areas, the barrier built to prevent uncontrolled movement of people across the border between the United States and Mexico consists of miles of 6-meter-tall steel beams placed four inches apart. Discuss the ecological effects of such a barrier. What types of organisms would be most adversely affected by this habitat fragmentation?

4. The magnitude of acid rain's effects can be influenced by the properties of the rock that the rain runs over. Acid rain is least likely to significantly acidify lakes in regions where there is a large amount of calcium carbonate–rich rock. How does the presence of these rocks mitigate the effects of acid rain?

Appendix I Periodic Table of the Elements

Atomic number → 11
Symbol → Na
Atomic weight → 22.99

(Atomic weight is a ratio: the mass of an atom relative to 1/12 the mass of a carbon 12 atom. Atomic weights given in parentheses are those of the most stable or best known isotopes of radioactive elements.)

Period	1	2	3	4	5	6	7	8	9	10	11	12	13	14	15	16	17	18
1	1 H 1.008																	2 He 4.0026
2	3 Li 6.94	4 Be 9.0122											5 B 10.81	6 C 12.011	7 N 14.007	8 O 15.999	9 F 18.998	10 Ne 20.180
3	11 Na 22.990	12 Mg 24.305											13 Al 26.982	14 Si 28.085	15 P 30.974	16 S 32.06	17 Cl 35.45	18 Ar 39.948
4	19 K 39.098	20 Ca 40.078	21 Sc 44.956	22 Ti 47.867	23 V 50.942	24 Cr 51.996	25 Mn 54.938	26 Fe 55.845	27 Co 58.933	28 Ni 58.693	29 Cu 63.546	30 Zn 65.38	31 Ga 69.723	32 Ge 72.63	33 As 74.922	34 Se 78.96	35 Br 79.904	36 Kr 83.798
5	37 Rb 85.468	38 Sr 87.62	39 Y 88.906	40 Zr 91.224	41 Nb 92.906	42 Mo 95.96	43 Tc (97.91)	44 Ru 101.07	45 Rh 102.91	46 Pd 106.42	47 Ag 107.87	48 Cd 112.41	49 In 114.82	50 Sn 118.71	51 Sb 121.76	52 Te 127.60	53 I 126.90	54 Xe 131.29
6	55 Cs 132.91	56 Ba 137.33	* 71 Lu 174.97	72 Hf 178.49	73 Ta 180.95	74 W 183.84	75 Re 186.21	76 Os 190.23	77 Ir 192.22	78 Pt 195.08	79 Au 196.97	80 Hg 200.59	81 Tl 204.38	82 Pb 207.2	83 B 208.98	84 Po (208.98)	85 At (209.99)	86 Rn (222.02)
7	87 Fr (223.02)	88 Ra (226.03)	** 103 Lr (262.11)	104 Rf (265.12)	105 Db (268.13)	106 Sg (271.13)	107 Bh (270)	108 Hs (277.15)	109 Mt (276.15)	110 Ds (281.16)	111 Rg (280.16)	112 Cn (285.17)	113 Nh (284.18)	114 Fl (289.19)	115 Mc (288.19)	116 Lv (293)	117 Ts (294)	118 Og (294)

***Lanthanoids**	57 La 138.91	58 Ce 140.12	59 Pr 140.91	60 Nd 144.24	61 Pm (144.91)	62 Sm 150.36	63 Eu 151.96	64 Gd 157.25	65 Tb 158.93	66 Dy 162.50	67 Ho 164.93	68 Er 167.26	69 Tm 168.93	70 Yb 173.05
****Actinoids**	89 Ac (227.03)	90 Th 232.04	91 Pa 231.04	92 U 238.03	93 Np (237.05)	94 Pu (244.06)	95 Am (243.06)	96 Cm (247.07)	97 Bk (247.07)	98 Cf (251.08)	99 Es (252.08)	100 Fm (257.10)	101 Md (258.10)	102 No (259.10)

1	Hydrogen	H	25	Manganese	Mn	49	Indium	In	73	Tantalum	Ta	97	Berkelium	Bk
2	Helium	He	26	Iron	Fe	50	Tin	Sn	74	Tungsten	W	98	Californium	Cf
3	Lithium	Li	27	Cobalt	Co	51	Antimony	Sb	75	Rhenium	Re	99	Einsteinium	Es
4	Beryllium	Be	28	Nickel	Ni	52	Tellurium	Te	76	Osmium	Os	100	Fermium	Fm
5	Boron	B	29	Copper	Cu	53	Iodine	I	77	Iridium	Ir	101	Mendelevium	Md
6	Carbon	C	30	Zinc	Zn	54	Xenon	Xe	78	Platinum	Pt	102	Nobelium	No
7	Nitrogen	N	31	Gallium	Ga	55	Cesium	Cs	79	Gold	Au	103	Lawrencium	Lr
8	Oxygen	O	32	Germanium	Ge	56	Barium	Ba	80	Mercury	Hg	104	Rutherfordium	Rf
9	Fluorine	F	33	Arsenic	As	57	Lanthanum	La	81	Thallium	Tl	105	Dubnium	Db
10	Neon	Ne	34	Selenium	Se	58	Cerium	Ce	82	Lead	Pb	106	Seaborgium	Sg
11	Sodium	Na	35	Bromine	Br	59	Praseodymium	Pr	83	Bismuth	Bi	107	Bohrium	Bh
12	Magnesium	Mg	36	Krypton	Kr	60	Neodymium	Nd	84	Polonium	Po	108	Hassium	Hs
13	Aluminum	Al	37	Rubidium	Rb	61	Promethium	Pm	85	Astatine	At	109	Meitnerium	Mt
14	Silicon	Si	38	Strontium	Sr	62	Samarium	Sm	86	Radon	Rn	110	Darmstadtium	Ds
15	Phosphorus	P	39	Yttrium	Y	63	Europium	Eu	87	Francium	Fr	111	Roentgenium	Rg
16	Sulfur	S	40	Zirconium	Zr	64	Gadolinium	Gd	88	Radium	Ra	112	Copernicium	Cn
17	Chlorine	Cl	41	Niobium	Nb	65	Terbium	Tb	89	Actinium	Ac	113	Nihonium	Nh
18	Argon	Ar	42	Molybdenum	Mo	66	Dysprosium	Dy	90	Thorium	Th	114	Flerovium	Fl
19	Potassium	K	43	Technetium	Tc	67	Holmium	Ho	91	Protactinium	Pa	115	Moscovium	Mc
20	Calcium	Ca	44	Ruthenium	Ru	68	Erbium	Er	92	Uranium	U	116	Livermorium	Lv
21	Scandium	Sc	45	Rhodium	Rh	69	Thulium	Tm	93	Neptunium	Np	117	Tennessine	Ts
22	Titanium	Ti	46	Palladium	Pd	70	Ytterbium	Yb	94	Plutonium	Pu	118	Oganesson	Og
23	Vanadium	V	47	Silver	Ag	71	Lutetium	Lu	95	Americium	Am			
24	Chromium	Cr	48	Cadmium	Cd	72	Hafnium	Hf	96	Curium	Cm			

Appendix II The Amino Acids

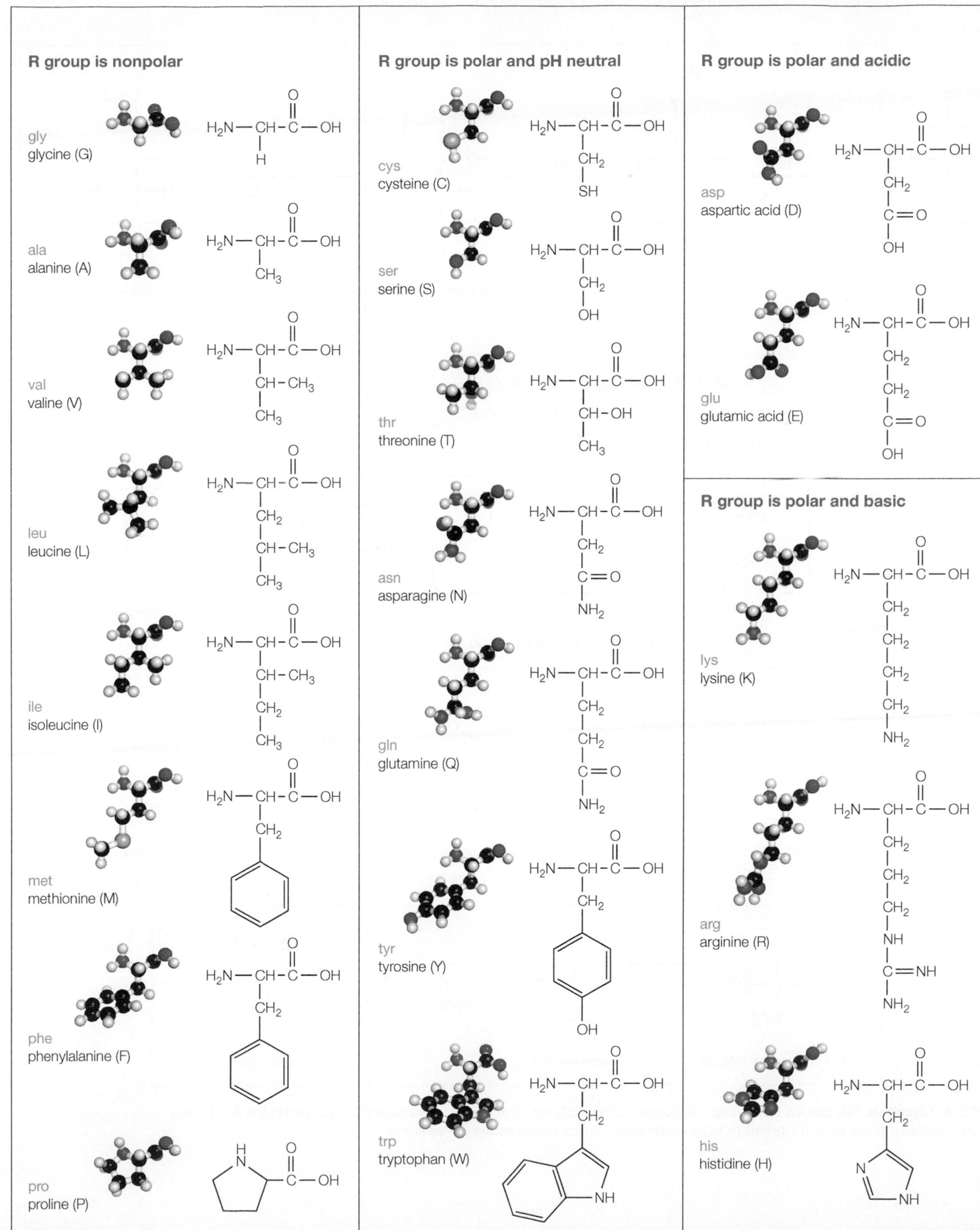

R group is nonpolar
gly
glycine (G)
ala
alanine (A)
val
valine (V)
leu
leucine (L)
ile
isoleucine (I)
met
methionine (M)
phe
phenylalanine (F)
pro
proline (P)
R group is polar and pH neutral
cys
cysteine (C)
ser
serine (S)
thr
threonine (T)
asn
asparagine (N)
gln
glutamine (Q)
tyr
tyrosine (Y)
trp
tryptophan (W)
R group is polar and acidic
asp
aspartic acid (D)
glu
glutamic acid (E)
R group is polar and basic
lys
lysine (K)
arg
arginine (R)
his
histidine (H)

GLYCOLYSIS

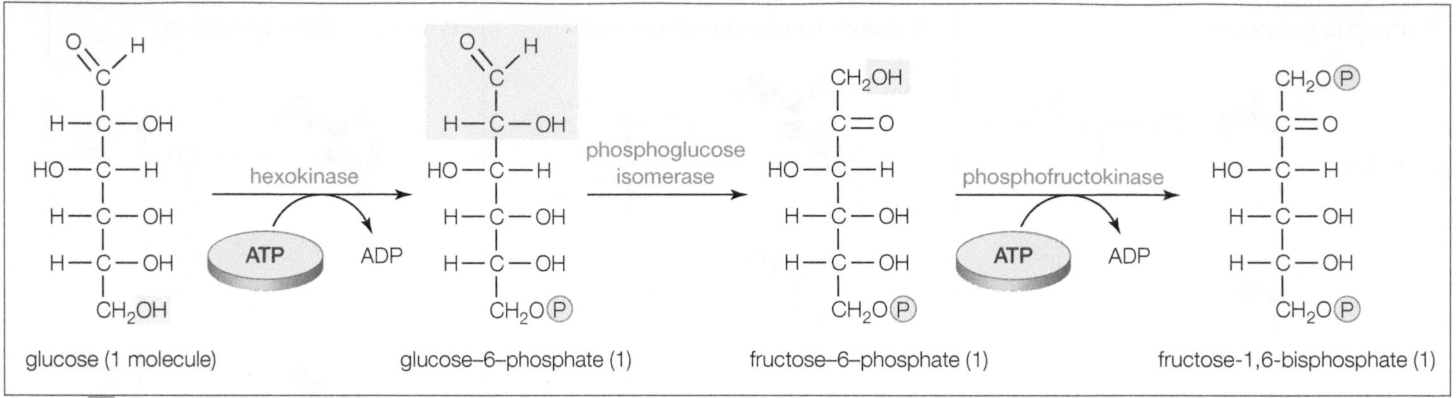
hexokinase
ATP
ADP
phosphoglucose isomerase
phosphofructokinase
ATP
ADP
glucose (1 molecule)
glucose–6–phosphate (1)
fructose–6–phosphate (1)
fructose-1,6-bisphosphate (1)

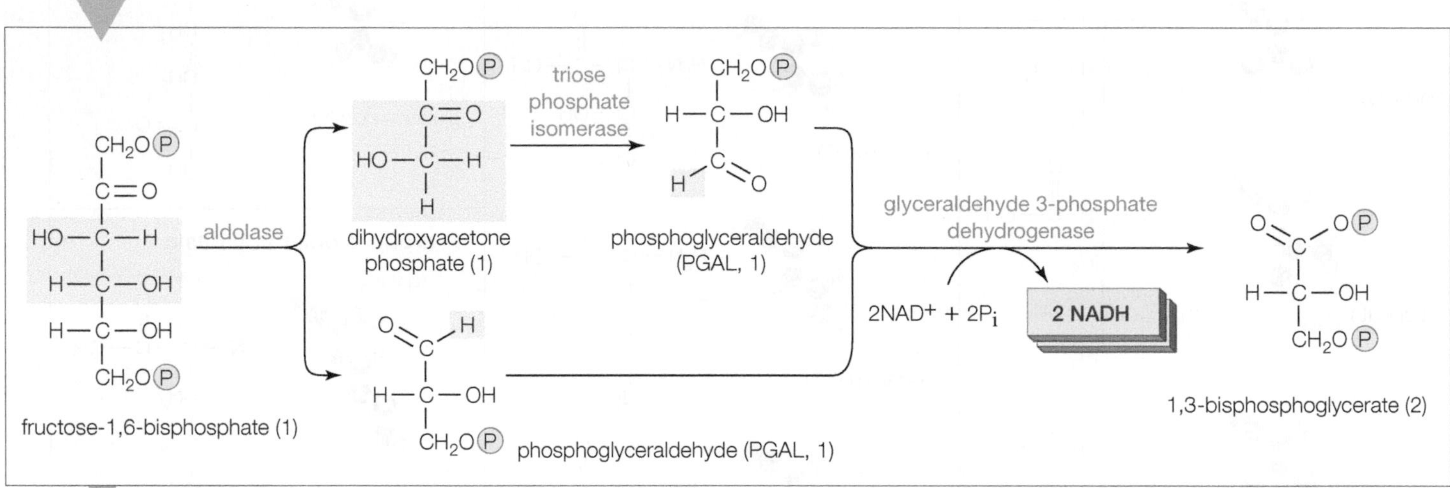
aldolase
triose phosphate isomerase
fructose-1,6-bisphosphate (1)
dihydroxyacetone phosphate (1)
phosphoglyceraldehyde (PGAL, 1)
glyceraldehyde 3-phosphate dehydrogenase
2NAD+ + 2Pi
2 NADH
1,3-bisphosphoglycerate (2)
phosphoglyceraldehyde (PGAL, 1)

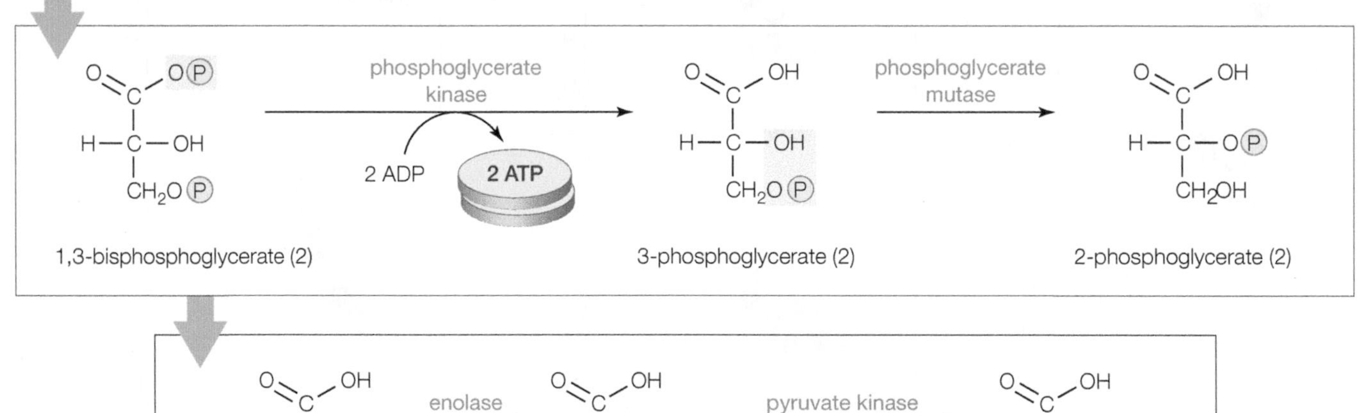
phosphoglycerate kinase
2 ADP
2 ATP
phosphoglycerate mutase
1,3-bisphosphoglycerate (2)
3-phosphoglycerate (2)
2-phosphoglycerate (2)
enolase
pyruvate kinase
2 ADP
2 ATP
2-phosphoglycerate (2)
phosphoenolpyruvate (2)
pyruvate (2)

FIGURE A Glycolysis. This pathway breaks down one glucose molecule into two 3-carbon pyruvate molecules for a net yield of two ATP. Enzyme names are indicated in green; parts of substrate molecules undergoing chemical change are highlighted in blue.

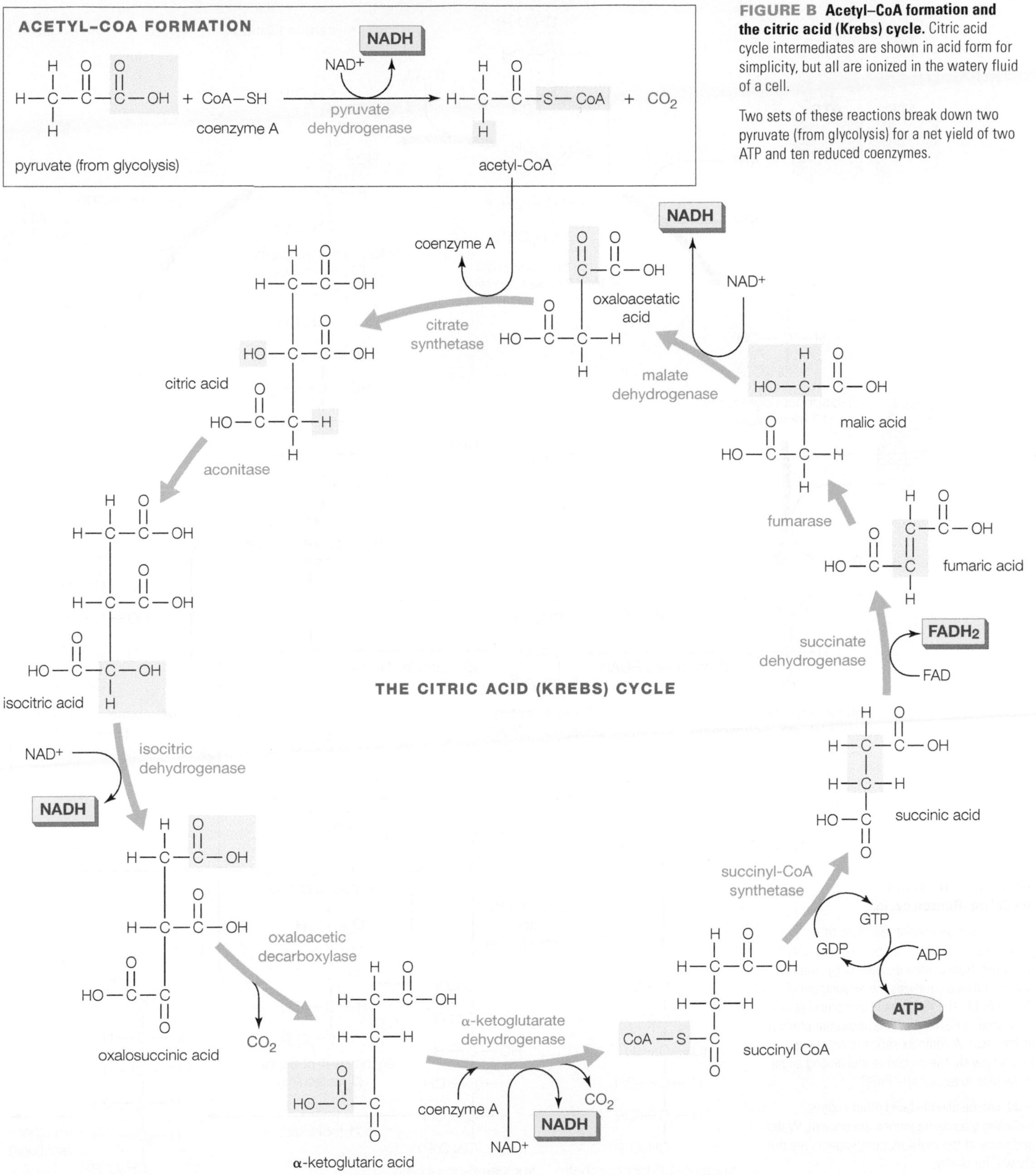

FIGURE B Acetyl–CoA formation and the citric acid (Krebs) cycle. Citric acid cycle intermediates are shown in acid form for simplicity, but all are ionized in the watery fluid of a cell.

Two sets of these reactions break down two pyruvate (from glycolysis) for a net yield of two ATP and ten reduced coenzymes.

ACETYL–COA FORMATION

pyruvate (from glycolysis) + CoA–SH (coenzyme A) → (pyruvate dehydrogenase; NAD+ → NADH) → acetyl-CoA + CO_2

THE CITRIC ACID (KREBS) CYCLE

oxaloacetatic acid + acetyl-CoA → (citrate synthetase; coenzyme A released) → citric acid

citric acid → (aconitase) → isocitric acid

isocitric acid → (isocitric dehydrogenase; NAD+ → NADH) → oxalosuccinic acid

oxalosuccinic acid → (oxaloacetic decarboxylase; CO_2) → α-ketoglutaric acid

α-ketoglutaric acid → (α-ketoglutarate dehydrogenase; coenzyme A; NAD+ → NADH; CO_2) → succinyl CoA

succinyl CoA → (succinyl-CoA synthetase; GDP → GTP; ADP → ATP) → succinic acid

succinic acid → (succinate dehydrogenase; FAD → $FADH_2$) → fumaric acid

fumaric acid → (fumarase) → malic acid

malic acid → (malate dehydrogenase; NAD+ → NADH) → oxaloacetatic acid

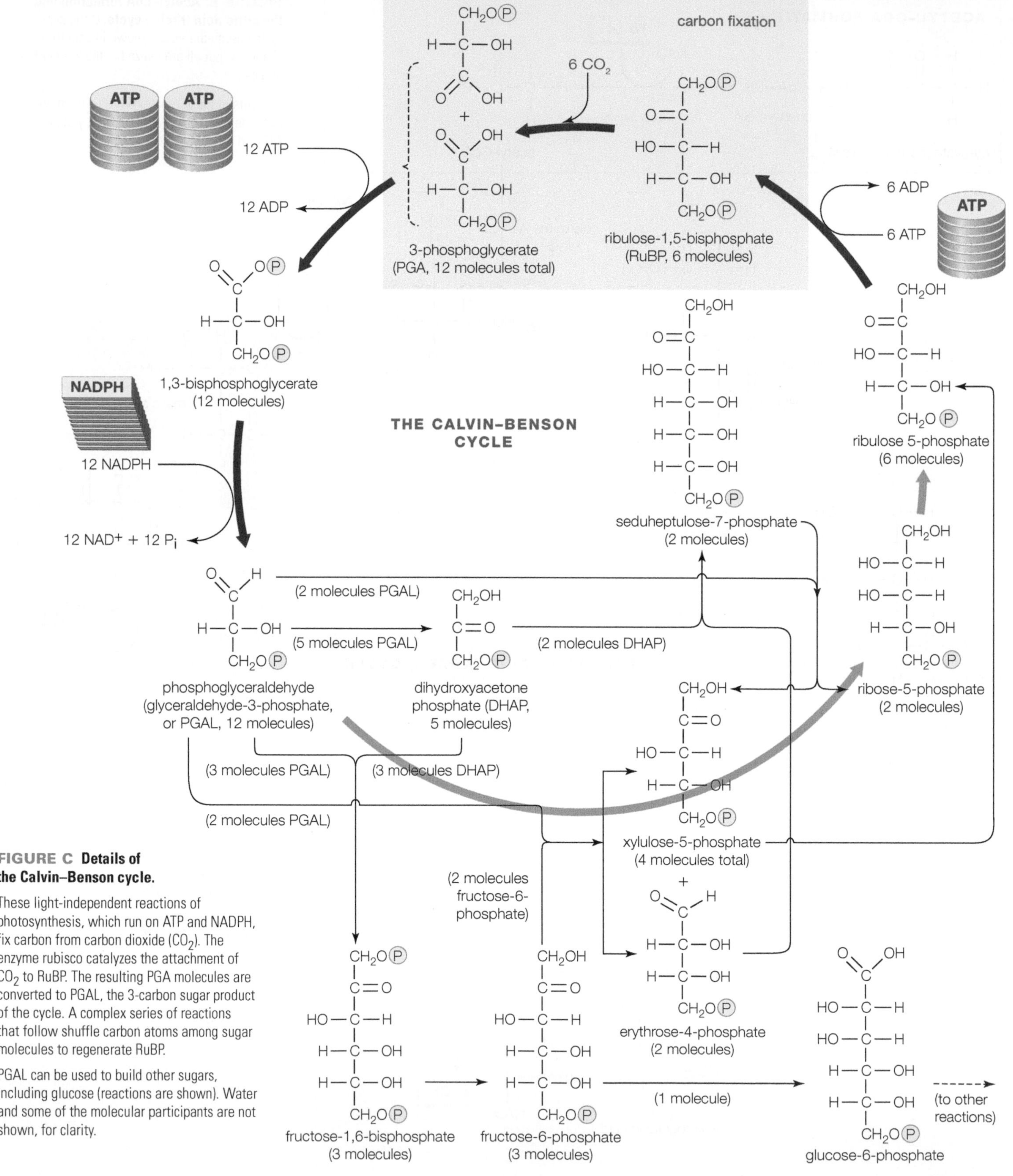
carbon fixation
6 CO_2
ATP
ATP
12 ATP
12 ADP
3-phosphoglycerate
(PGA, 12 molecules total)
ribulose-1,5-bisphosphate
(RuBP, 6 molecules)
6 ADP
6 ATP
ATP
1,3-bisphosphoglycerate
(12 molecules)
NADPH
12 NADPH
12 NAD+ + 12 P_i
THE CALVIN–BENSON CYCLE
seduheptulose-7-phosphate
(2 molecules)
ribulose 5-phosphate
(6 molecules)
(2 molecules PGAL)
(5 molecules PGAL)
(2 molecules DHAP)
phosphoglyceraldehyde
(glyceraldehyde-3-phosphate,
or PGAL, 12 molecules)
dihydroxyacetone
phosphate (DHAP,
5 molecules)
ribose-5-phosphate
(2 molecules)
(3 molecules PGAL)
(3 molecules DHAP)
(2 molecules PGAL)
xylulose-5-phosphate
(4 molecules total)
(2 molecules
fructose-6-
phosphate)
erythrose-4-phosphate
(2 molecules)
(1 molecule)
(to other
reactions)
fructose-1,6-bisphosphate
(3 molecules)
fructose-6-phosphate
(3 molecules)
glucose-6-phosphate

FIGURE C Details of the Calvin–Benson cycle.

These light-independent reactions of photosynthesis, which run on ATP and NADPH, fix carbon from carbon dioxide (CO_2). The enzyme rubisco catalyzes the attachment of CO_2 to RuBP. The resulting PGA molecules are converted to PGAL, the 3-carbon sugar product of the cycle. A complex series of reactions that follow shuffle carbon atoms among sugar molecules to regenerate RuBP.

PGAL can be used to build other sugars, including glucose (reactions are shown). Water and some of the molecular participants are not shown, for clarity.

Appendix IV A Plain English Map of the Human Chromosomes

chymotrypsin
Rh blood group antigen
cannabinoid receptor (peripheral)
tRNAs (gly, glu)
endorphin receptor δ
T cell leukemia virus receptor
glucose transporter
leptin receptor: obesity
amylase
thyroid stimulating hormone (TSH) beta chain
receptor for IgG
tRNA (asn)
receptor for IgG
peptidoglycan (PAMP) receptor
interleukin-6 receptor
lamin A: Hutchinson-Gilford progeria
receptors for IgE, IgG
receptors for IgA, IgM
feline leukemia virus receptor
flagellin (PAMP) receptor
presenilin 2: Alzheimer's disease
actin (cytoplasmic)
1
follicle-stimulating hormone (FSH) receptor
luteinizing hormone (LH) receptor
anthrax toxin receptor
antibody light chain gene clusters
interleukin-1
lactase
Cowden syndrome, macrocephaly, cancer
synesthesia
glucagon
fibronectin
2
ghrelin
oxytocin receptor
TRH receptor
tRNA (arg)
bacterial DNA (PAMP) receptor
camptodactyly
rhodopsin
TSH releasing hormone
male pattern baldness
tRNA (val)
somatostatin
3
Huntington disease
achondroplasia
Ellis-van Creveld syndrome
lipoprotein (PAMP) receptor
casein
anthrax toxin receptor 2
alcohol dehydrogenase
albinism
interleukin-2: SCID
aldosterone receptor
fibrinogen
double-stranded RNA (PAMP) receptor
4
dopamine transporter
telomerase
growth hormone receptor: dwarfism
prolactin receptor
interleukins 3, 4, 5, 9, and 13
tRNAs: val, pro, thr, and lys
5
prolactin
histone proteins
hemochromatosis
tRNAs: val, arg
tRNAs: val, ser, thr, met
tumor necrosis factor
MHC markers
alpha chain of FSH, LH, TSH, and HCG
cannabinoid receptor, brain
estrogen receptor
endorphin receptor, μ
6
interleukin-6
cytochrome c
EGF receptor
elastin
calcitonin receptor
collagen I alpha-2: osteogenesis imperfecta
CFTR: cystic fibrosis
leptin: obesity
blue cone opsin: blue-deficient colorblindness
trypsin: pancreatitis
7
gonadotropin-releasing hormone (GnRH)
Werner syndrome
tissue plasminogen activator
endorphin receptor, κ
corticotropin-releasing hormone (CRH)
hypertrichosis
Burkitt lymphoma
thyroid hormones
8
TYRP1
interferons
galactosemia
alcohol dehydrogenase, cytoplasmic: alcohol flushing reaction
Friedreich ataxia
fructose intolerance
ABO blood group
9
vitamin B12 receptor
mannose (PAMP) receptor
perforin
10
insulin: diabetes
beta globin
parathyroid hormone
calcitonin
follicle stimulating hormone (FSH) beta chain
PAX6: aniridia
thrombin
catalase
pepsinogen
folate receptor
tyrosinase: albinism
Tourette syndrome
APOA5
11
GABA transporter
CD4 helper T cell antigen
KRAS: leukemia, other cancers
keratins
collagen II
vitamin D receptor: rickets
Hoxc6
lysozyme
gamma interferon
phenylketonuria
cryptochrome I
alcohol dehydrogenase, mitochondria: alcohol flushing reaction
12
rRNA gene repeats 1
BRCA2: breast cancer
severe gastroesophageal reflux
13
rRNA gene repeats 2
estrogen receptor 2
presenilin 1: Alzheimer's
TSH receptor: hypo- and hyperthyroidism
antibody heavy chain gene cluster
14
rRNA gene repeats 3
Prader-Willi/Angelman syndrome
antibody heavy chain variable region gene cluster
fibrillin: Marfan syndrome
Tay-Sachs disease
15
hemoglobin alpha chain gene cluster: thalassemias
DNAse I
VKORC1: Warfarin resistance
MC1R
16
Canavan disease
capsaicin receptor
p53: cancer
tRNA: gly, arg, leu, gln
aurora B kinase
neurofibromatosis
serotonin transporter: OCD, anxiety
thyroid hormone receptor
keratins
BRCA1: cancer
collagen I alpha-1: osteogenesis imperfecta
growth hormone (GH)
glucagon receptor
17
ACTH receptor
laminin alpha-1
carpel tunnel syndrome, familial
obesity, autosomal dominant
cytochrome b5: methemoglobinemia
18
leutinizing hormone (LH)
tRNAs: gly, val
LDL receptor
insulin receptor: diabetes
receptor for IgE
Cushing syndrome
interferons
APOE
receptor for IgG
human chorionic gonadotropin (HCG) beta chain
receptor for IgA
19
antidiuretic hormone
oxytocin
prion protein: Creutzfeldt-Jakob disease
GH releasing hormone: acromegaly
20
rRNA gene repeats 4
amyloid beta precursor protein: Alzheimer's disease
interferon receptors
superoxide dismutase: amyotrophic lateral sclerosis (ALS)
21
rRNA gene repeats 5
antibody light chain variable region gene cluster
RAD53: cancer
myoglobin
22
SRY
Y
dystrophin: Duchenne muscular dystrophy
MAO-B
testosterone receptor: androgen insensitivity
neuroligin 3: autism
IL2RG: SCID, X-linked
XIST
factor IX: hemophilia B
fragile X syndrome
red cone opsin: red-deficient colorblindness
green cone opsin: green-deficient colorblindness
incontinentia pigmenti
factor VIII: hemophilia A
X
© 2002 Susan Offner/SK45176-02

FIGURE D Genes on human chromosomes. Characteristic bands appear after treatment with a stain called Giemsa.

More than 45,000 genes code for RNA or protein products; some of the genes and gene products discussed in this book are indicated. Genetic disorders and other phenotypes associated with single-gene mutations are in red. Annotations per NCBI *Homo sapiens* release 107, genome assemblies GRCh38.p2 and CHM1_1.1.

Appendix V Restless Earth—Life's Changing Geologic Stage

This NASA map summarizes the tectonic and volcanic activity of Earth during the past 1 million years.

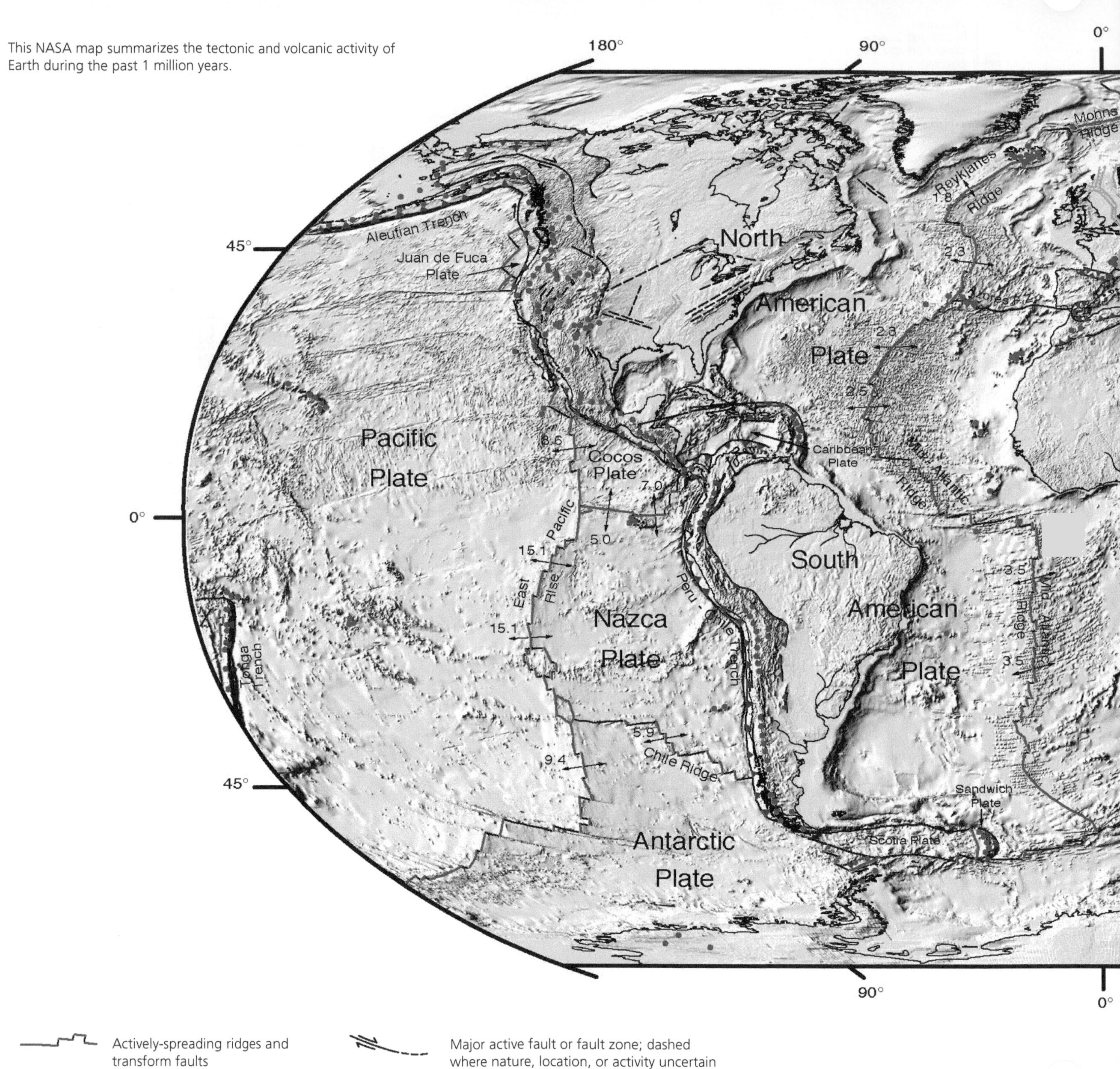

180°
90°
0°
45°
0°
45°
90°
0°
Mohns Ridge
Reykjanes Ridge
7.8
Aleutian Trench
Juan de Fuca Plate
North American Plate
2.3
Azores
2.8
2.5
Pacific Plate
8.6
Cocos Plate
Caribbean Plate
Mid-Atlantic Ridge
7.0
5.0
15.1
East Pacific Rise
South American Plate
Peru-Chile Trench
Nazca Plate
3.5
Mid-Atlantic Ridge
15.1
Tonga Trench
3.5
5.9
Chile Ridge
9.4
Sandwich Plate
Scotia Plate
Antarctic Plate

Actively-spreading ridges and transform faults

1.4 Total spreading rate, cm/year

Major active fault or fault zone; dashed where nature, location, or activity uncertain

Normal fault or rift; hachures on down-thrown side

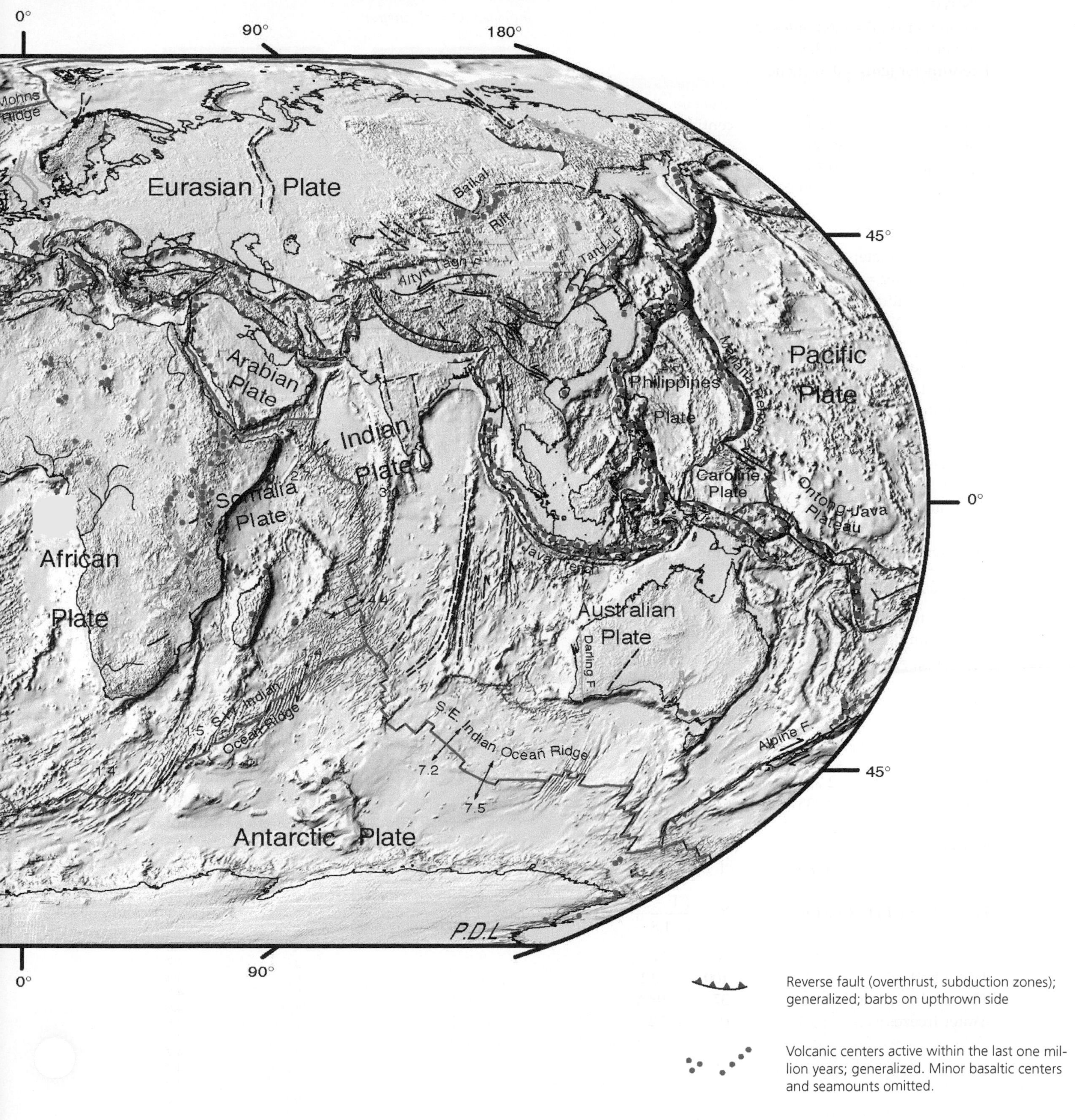

0°
90°
180°
45°
0°
45°
0°
90°
Mohns Ridge
Eurasian Plate
Baikal Rift
Altyn Tagh F.
Tan-Lu
Arabian Plate
Indian Plate
Philippines Plate
Mariana Trench
Pacific Plate
Caroline Plate
Ontong-Java Plateau
Somalia Plate
African Plate
Java Trench
Australian Plate
Darling F.
S.W. Indian Ocean Ridge
S.E. Indian Ocean Ridge
Alpine F.
Antarctic Plate
P.D.L.
2.7
3.0
4.4
1.4
1.5
1.4
7.2
7.5
Reverse fault (overthrust, subduction zones); generalized; barbs on upthrown side
Volcanic centers active within the last one million years; generalized. Minor basaltic centers and seamounts omitted.

Appendix VI Units of Measure

LENGTH

1 kilometer (km) = 0.62 miles (mi)
1 meter (m) = 39.37 inches (in)
1 centimeter (cm) = 0.39 inches

To convert	*multiply by*	*to obtain*
inches	2.25	centimeters
feet	30.48	centimeters
centimeters	0.39	inches
millimeters	0.039	inches

AREA

1 square kilometer = 0.386 square miles
1 square meter = 1.196 square yards
1 square centimeter = 0.155 square inches

VOLUME

1 cubic meter = 35.31 cubic feet
1 liter = 1.06 quarts
1 milliliter = 0.034 fluid ounces = 1/5 teaspoon

To convert	*multiply by*	*to obtain*
quarts	0.95	liters
fluid ounces	28.41	milliliters
liters	1.06	quarts
milliliters	0.03	fluid ounces

WEIGHT

1 metric ton (mt) = 2,205 pounds (lb) = 1.1 tons (t)
1 kilogram (kg) = 2.205 pounds (lb)
1 gram (g) = 0.035 ounces (oz)

To convert	*multiply by*	*to obtain*
pounds	0.454	kilograms
pounds	454	grams
ounces	28.35	grams
kilograms	2.205	pounds
grams	0.035	ounces

TEMPERATURE

Celsius (°C) to Fahrenheit (°F): $°F = 1.8\,(°C) + 32$

Fahrenheit (°F) to Celsius: $°C = \dfrac{(°F - 32)}{1.8}$

	°C	°F
Water boils	100	212
Human body temperature	37	98.6
Water freezes	0	32

Relevant section numbers provided for each answer

CHAPTER 1

1.	a	1.2
2.	c	1.2
3.	c	1.3
4.	c	1.3
5.	d	1.3
6.	a	1.3
7.	a,d,e	1.2–1.5
8.	b	1.4
9.	a,b	1.4
10.	domains	1.5
11.	b	1.6
12.	b	1.7
13.	b	1.8
14.	a	1.6
15.	c	1.6
	e	1.7
	b	1.1, 1.5
	d	1.6
	a	1.6
	f	1.3

CHAPTER 2

1.	a	2.2
2.	b	2.2
3.	d	2.2
4.	d	2.2
5.	a	2.3, 2.6
6.	a,c,b	2.3
7.	a	2.3
8.	a	2.4
9.	c	2.4
10.	c	2.5
11.	b	2.5
12.	d	2.6
13.	a	2.6
14.	c	2.6
15.	c	2.5
16.	c	2.4
	b	2.1
	d	2.4
	a	2.1, 2.2
	e	2.3
	f	2.5
	i	2.6
	h	2.4
	g	2.5

CHAPTER 3

1.	c	3.2
2.	d	3.2
3.	b	3.2
4.	a	3.3
5.	starch, cellulose, glycogen	3.3
6.	double bonds	3.4
7.	False. Unhealthy trans fats are unsaturated. They have double bonds that do not kink fatty acid tails, so these molecules can pack tightly.	3.1, 3.4
8.	b	3.4
9.	e	3.4
10.	d	3.5, 3.6
11.	d	3.5
12.	d	3.5
13.	a. polypeptide	3.5
	b. carbohydrate	3.3
	c. amino acid	3.5
	d. fatty acid	3.4
14.	c	3.4
	e	3.3
	d	3.4
	g	3.6
	a	3.5
	b	3.6
	f	3.3
15.	g	3.5
	a	3.4
	b	3.5
	c	3.4
	d	3.6
	e	3.3
	f	3.6
	i	3.5
	h	3.3

CHAPTER 4

1.	c	4.2
2.	c	4.2
3.	b	4.2
4.	a	4.2
5.	False	5.3
6.	a	4.3
7.	b	4.4
8.	a	4.5
9.	a	4.5
10.	c,b,d,a	4.5
11.	a	4.9
12.	a	4.9
13.	b	4.4, 4.5–4.7
14.	d	4.9
15.	h	4.6
	g	4.7
	k	4.8
	f	4.5
	e	4.5
	a	4.5
	d	4.5
	c	4.7
	j	4.8
	i	4.9
	b	4.5

Appendix VII Answers to Self-Quizzes and Genetics Problems

CHAPTER 5

1. b 5.2
2. c 5.2
3. d 5.2
4. a 5.3
5. c 5.3
6. c 5.6
7. temperature, pH, or salt concentration 5.4
8. d 5.5
9. c 5.5
10. a 5.6
11. a 5.8
12. c 5.8
13. b 5.9
14. a 5.8
15. c 5.3
e 5.10
h 5.7
i 5.5
b 5.3
a 5.6
g 5.9
d 5.4
f 5.5
k 5.2
j 5.8

CHAPTER 6

1. Autotroph: weed; heterotrophs: cat, bird, caterpillar 6.2
2. a 6.2
3. b 6.2
4. c 6.3
5 a 6.2
6. d 6.4
7. b 6.2, 6.4
8. b 6.4
9. b 6.4
10. c 6.5
11. b 6.5
12. e 6.5
13. a 6.5
14. c 6.5
15. f 6.5
h 6.5
g 6.4
d 6.4
e 6.5
b 6.2
a 6.3
c 6.2

CHAPTER 7

1. False 7.2
2. d 7.2
3. b 7.4
4. c 7.2
5. b 7.3,7.4
6. b 7.3
7. c 7.4
8. c 7.4
9. d,b,c,a 7.2, 7.4
10. f 7.5
11. c 7.5
12. c 7.5
13. f 7.6
14. b 7.2
c 7.5
a 7.3
d 7.5
15. b 7.3
d 7.2
a 7.2, 7.4
c 7.3, 7.4
e 7.2, 7.3
f 7.1

CHAPTER 8

1. b 8.3
2. b 8.3
3. b 8.3
4. b 8.3
5. a 8.4
6. b 8.4
7. b 8.4
8. a 8.5
9. f 8.5
10. a 8.5
11. b 8.5
12. d 8.4
13. d 8.2
b 8.1
a 8.3
i 8.4
e 8.5
g 8.6
c 8.7
f 8.3, 8.6
h 8.5
14. d 8.6
15. a 8.7

CHAPTER 9

1. c 9.2
2. b 9.2, 9.3
3. a 9.2
4. b 9.2
a 9.3
c 9.2
5. a 9.2
6. b 9.2, 9.4
7. d 9.3
8. c 9.4
9. b 9.3
10. a 9.4
11. a 9.4
12. a 9.3
13. c 9.3
14. c 9.5
15. c 9.4
b 9.2, 9.3
e 9.5
a 9.3
f 9.4
d 9.3
j 9.4
g 9.1
h 9.6
i 9.6

CHAPTER 10

1. d 10.2
2. b 10.2
3. b 10.2
4. b 10.2
5. d 10.2, 10.3
6. c 10.2, 10.3
7. c 10.3
8. b 10.3
9. b 10.3
10. b 10.3
11. b 10.3
12. c 10.3
13. b 10.4
14. c 10.3
15. e 10.4
i 10.4
d 10.3
j 10.2, 10.3
f 10.2
c 10.5
h 10.3
g 10.2
a 10.3
b 10.4

CHAPTER 11

1. b 11.2
2. Two 11.2
3. interphase, prophase, metaphase, anaphase, telophase 11.3
4. a 11.2
5. e 11.2
6. c 11.3
7. c 11.2
8. a 11.2
9. c 11.2
10. a 11.2
11. d 11.4
12. c 11.4
f 11.3
a 11.6
g 11.4
b 11.4
e 11.6
d 11.3
h 11.5
13. d 11.3
b 11.3
c 11.3
e 11.2
a 11.3
f 11.4
14. d 11.6
15. a 11.6

CHAPTER 12

1. b 12.1
2. b 12.2
3. c 12.2
4. d 12.2
5. b 12.2
6. a 12.2
7. Sister chromatids are still attached. 12.3
8. c 12.2
9. c 12.3
10. b 12.4
11. a 12.4
12. c 12.3
13. b 12.5
14. c 12.3
h 12.3
a 12.2
g 12.4
f 12.2
e 12.2
b 12.1
d 12.5
15. c 12.4

CHAPTER 13

1. b 13.2
2. a 13.2
3. b 13.4
4. c 13.3
5. c 13.3
6. a 13.3
7. b 13.3
8. d 13.4
9. c 13.4
10. F 13.5–13.7
11. c 13.5
12. b 13.5
13. d 13.7
14. b 13.4
d 13.3
a 13.2
c 13.2

CHAPTER 14

1. b 14.2
2. b 14.2
3. a 14.2
4. b 14.3
A person who is homozygous for a Tay–Sachs allele does not live long enough to reach reproductive age.
5. d 14.3, 14.4
The trait could be an outcome of both parents carrying a recessive allele on an autosome, or the mom carrying a recessive allele on an X chromosome.
6. a 14.4
7. d 14.4
8. b 14.6
9. c 14.6
10. c 14.6
e 14.5
f 14.6
b 14.5
a 14.2
d 14.5

CHAPTER 15

1. c 15.2
2. a 15.2
3. b 15.2
4. b 15.3
5. c 15.3
6. b 15.4
7. b 15.4
8. b 15.5
9. e,a,d,b,c 15.2-15.4
10. d 15.5
11. c 15.6
12. True 15.6, 15.7
13. a 15.3
14. b 15.9
15. c 15.5
f 15.7
d 15.3
b 15.1
a 15.6
e 15.5
g 15.9
16. a 15.3
d 15.2
c 15.2
e 15.4
b 15.2

CHAPTER 16

1. b 16.2
2. a 16.2
3. c 16.3
4. 66 16.1, 16.5
5. c 16.2
6. b 16.3
7. a 16.3
8. b 16.4
9. a 16.4
10. c 16.5
11. d 16.5
12. Gondwana 16.5
13. a 16.5
14. a, c, d, e, f 16.1, 16.2, 16.4, 16.5
15. g 16.2
f 16.5
d 16.4
e 16.3
a 16.3
b 16.2
c 16.2

CHAPTER 17

1. a 17.2
2. b 17.2
3. d 17.2
4. 0.25 17.3
5. a 17.7
6. d 17.5
7. d 17.5
8. b 17.6
9. d 17.6
10. d 17.5, 17.7
11. allopatric speciation 17.8
12. False. 17.6
Inbreeding can maintain a harmful allele in a gene pool, but not increase its frequency.
13. c 17.8
14. f 17.2, 17.4
15. c 17.6
d 17.5
e 17.9
b 17.6
a 17.9
f 17.9

CHAPTER 18

1. a 18.2
2. d 18.2
3. c 18.2
4. d 18.2
5. c 18.2
6. a 18.2
7. b 18.2
8. d 18.3
9. d 18.3
10. c 18.2
11. b 18.4
12. b 18.4
13. e 18.3, 18.4
14. c 18.4
15. b 18.5
16. b 18.2
g 18.2
a 18.5
h 18.3
d 18.2
e 18.4
f 18.3
i 18.3
c 18.3

CHAPTER 19

1.	b	19.2
2.	d	19.2
3.	b	19.3
4.	a	19.4
5.	b	19.4
6.	d	19.4
7.	b	19.6
8.	d	19.6
9.	d	19.7
10.	d	19.7
11.	a	19.7
12.	c	19.5
13.	c	19.7
14.	d	19.7
15.	f	19.2
	b	19.2
	c	19.4
	a	19.6
	d	19.7
	e	19.7

CHAPTER 20

1.	a	20.2
2.	c	20.2
3.	b	20.3
4.	b	20.3
5.	a	20.5
6.	c	20.5
7.	c	20.7
8.	d	20.6
9.	c	20.4
10.	c	20.6
11.	b	20.9
12.	b	20.5
13.	c	20.8
14.	b	20.9
15.	d	20.9
	e	20.5
	b	20.3
	f	20.3
	g	20.9
	a	20.7
	c	20.8

CHAPTER 21

1.	c	21.2
2.	b	21.3
3.	b	21.4, 21.6
4.	a	21.5
5.	b	21.4
6.	a	21.2, 21.7
7.	c	21.4
8.	d	21.5
9.	c	21.3
10.	c	21.8
11.	c	21.7
12.	c	21.8
13.	b	21.5
14.	e	21.6
	c	21.6
	a	21.4
	b	21.7
	d	21.7
15.	d	21.3
	c	21.4
	a	21.5
	b	21.4
	h	21.8
	g	21.7
	f	21.8
	e	21.7

CHAPTER 22

1.	c	22.2
2.	c	22.3
3.	a	22.3
4.	False	22.5
5.	a	22.4
6.	b	22.2, 22.3
7.	c	22.5
8.	a	22.6
9.	a	22.5
10.	b	22.6
11.	a	22.5
12.	b	22.9
13.	d	22.3
	c	22.5
	a	22.7
	b	22.8
14.	c	22.6
	h	22.3
	a	22.2
	b	22.2
	e	22.8
	f	22.8
	d	22.5
	g	22.5
15.	d, a, b, c	22.2

CHAPTER 23

1.	c	23.2, 23.3
2.	a	23.7
3.	a	23.2
4.	c	23.5
5.	d	23.6
6.	b	23.2
7.	c	23.6
8.	d	23.6
9.	a	23.3
10.	b	23.7
11.	c	23.4
12.	a	23.7
13.	a	23.4
14.	b	23.5
15.	d	23.6
	b	23.2
	a	23.7
	f	23.5
	g	23.4
	c	23.7
	e	23.4, 23.7
	i	23.2
	h	23.7

CHAPTER 24

1.	False	24.2
2.	a	24.2
3.	c	24.2
4.	d	24.10
5.	a	24.5
6.	d	24.7
7.	b	24.8
8.	c	24.10
9.	d	24.4
10.	b	24.8
11.	d	24.10
12.	c	24.10
13.	d	24.9
14.	e	24.8
	b	24.10
	a	24.11
	c	24.5
	g	24.10
	h	24.11
	f	24.7
	d	24.10
	i	24.6
15.	g	24.11
	c	24.4
	a	24.5
	b	24.6
	h	24.9
	f	24.7
	d	24.10
	e	24.8

CHAPTER 25

1.	notochord, hollow dorsal nerve cord, pharyngeal gill slits, tail extending past anus	25.2
2.	All of them	25.2
3.	a	25.3
4.	a	25.3
5.	c	25.3
6.	c	25.5
7.	f	25.5
8.	c	25.3
9.	c	25.7
10.	c	25.8
11.	b	25.7
12.	a	25.4
13.	b	25.3
	a	25.3
	d	25.8
	f	25.8
	c	25.2
	e	25.5
14.	j	25.2
	i	25.3
	h	25.4
	f	25.6
	c	25.7
	g	25.3
	d	25.9
	a	25.8
	e	25.3
	b	25.8
15.	a	25.3
	e	25.3
	d	25.4
	c	25.5
	b	25.8

CHAPTER 26

1.	c	26.2
2.	a	26.2
3.	c	26.3
4.	b	26.3
5.	a	26.4
6.	c	26.3
7.	c	26.4
8.	b	26.5
9.	a	26.6
10.	c	26.6
11.	d	26.6
12.	d	26.6
13.	c	26.6
14.	c	26.4
	d	26.2
	a	26.5
	b	26.6
	f	26.5
	e	26.6
15	g	26.2
	b	26.2
	c	26.2
	e	26.3
	d	26.5
	f	26.6
	a	26.6

CHAPTER 27

1.	b	27.3
2.	a	27.3
3.	b	27.3
4.	a	27.4
5.	b	27.4, 27.7
6.	c	27.3
7.	c	27.5
8.	a	27.4. 27.6
9.	A stem	27.4
10.	d	27.6
11.	b	27.6
12.	a	27.7
13.	b	37.7
14.	b	27.7
15.	c	27.7
	d	27.2, 27.5
	e	27.7
	a	27.6
	f	27.3
	b	27.4
	g	27.2

CHAPTER 28

1.	d	28.2
2.	b	28.2
3.	c	28.3
4.	e	28.3,28.5
5.	b	28.3
6.	a	28.3
7.	c	28.4
8.	d	28.4
9.	d	28.4
10.	c	28.5
11.	a	28.4
12.	c	28.4
13.	b	28.5
14.	i	28.4
	c	28.2
	d	28.4
	e	28.5
	b	28.3
	g	28.5
	a	28.3
	h	28.4
	f	28.5
15.	a	28.5

CHAPTER 29

1.	c	29.2
2.	a,b	29.2
3.	b	29.2
4.	b	29.2*
5.	b	29.3
6.	b	29.5*
7.	f	29.3
8.	a	29.5
9.	c	29.5, 29.6
10.	A cantaloupe is a type of berry called a pepo.	29.6
11.	c	29.5
12.	d	29.7
13.	c	29.8
14.	c	29.5
	f	29.2
	g	29.3
	e	29.3
	d	29.6
	b	29.3
	a	29.3
15.	a	29.8

CHAPTER 30

1.	b	30.2
2.	d	30.2
3.	d	30.2
4.	b	30.3
5.	b	30.6
6.	c	30.5
7.	a	30.5
8.	d	30.5
9.	statoliths	30.6
10.	c	30.7
11.	d	30.5, 30.7
12.	b	30.6
	d	30.6
	a	30.6
	f	30.7
	c	30.7
	e	30.6
13.	b	30.5
	d	30.3
	a	30.3
	e	30.4
	c	30.5
	f	30.5
14.	a	30.8
15.	c	30.5
	e	30.3
	b	30.6
	a	30.4
	d	30.5
	f	30.8

CHAPTER 31

1.	a	31.3
2.	a	31.3
3.	a	31.3
4.	b	31.4
5.	d	31.4
6.	b	31.4
7.	b	31.6
8.	c	31.5
9.	a	31.4
10.	b	31.3
11.	c	31.3
12.	a	31.3
13.	c	31.4
14.	c	31.3
	b	31.3
	d	31.4
	f	31.8
	a	31.6
	e	31.5
	g	31.5
	i	31.4
	h	31.8
	j	31.2
15.	a	31.9

CHAPTER 32

1.	a	32.3
2.	c	32.4
3.	a	32.4
4.	b	32.4
5.	a	32.5
6.	a	32.5
7.	c	32.6
8.	c	32.7
9.	b	32.7
10.	c	32.8
11.	c	32.10
12.	a	32.9
13.	c	32.12
14.	a	32.11
15.	h	32.8
	d	32.10
	f	32.10
	b	32.10
	g	32.11
	a	32.4
	e	32.11
	c	32.8

CHAPTER 33

1.	b	33.3
2.	c	33.2
3.	c	33.3
4.	e	33.4
5.	b	33.3
6.	b	33.6
7.	b	33.5
8.	a	33.3
9.	b	33.5
10.	b	33.8
11.	d	33.8
12.	b	33.8
13.	c	33.8
14.	a	33.8
15.	e	33.8
	j	33.5
	b	33.5
	g	33.7
	a	33.8
	h	33.8
	f	33.4
	k	33.3
	i	33.5
	c	33.6
	d	33.4

CHAPTER 34

1.	a	34.2
2.	a	34.3
3.	c, b,a,e,d	34.4
4.	a	34.4
5.	d	34.4
6.	c	34.6
7.	b	34.7
8.	c	34.7
9.	d	34.7
10.	b	34.5
11.	d	34.8
12.	b	34.6
13.	e	34.3
14.	False	34.10
15.	d	34.8
	f	34.6
	c	34.4
	e	34.7,
	b	34.5, a 34.9

CHAPTER 35

1.	c	35.3
2.	c	35.4
3.	d	35.5
4.	a	35.5
5.	a	35.4
6.	c	35.3
7.	b	35.7
8.	b	35.7
9.	d	35.7
10.	b	35.9
11.	c	35.8
12.	d	35.8
13.	a	35.8
14.	b	35.8
15.	c	35.4
	i	35.7
	g	35.9
	h	35.5
	d	35.7
	e	35.4
	b	35.3
	a	35.8
	f	35.9.

CHAPTER 36

1.	c	36.2
2.	a	36.2
3.	c	36.2
4.	d	36.5
5.	b	36.5
6.	a	36.5
7.	b	36.4
8.	b	36.4
9.	a	36.7
10.	c	36.9
11.	b	36.4
12.	b	36.11
13.	pulmonary artery	36.3
14.	d	36.10
	e	36.10
	f	36.10
	b	36.5
	g	36.10
	a	36.10
	c	36.10
15.	c	36.5
	g	36.11
	i	36.5
	e	36.10
	b	36.4
	f	36.5
	a	36.3
	d	36.5
	h	36.4

CHAPTER 37

1.	d	37.2
2.	e	37.3
3.	d	37.4
4.	g	37.4
5.	h	37.7, 37.8
6.	a,d,e,g	37.6
7.	d	37.5
8.	a	37.5, 37.7, 37.8
9.	b	37.6
10.	d	37.6
11.	d	37.7
12.	b	37.8
13.	b	37.2, 37.6
	d	37.5
	c	37.7
	e	37.8
	a	37.2
14.	a	37.9
15.	f	37.9
	e	37.6
	b	37.9
	d	37.9
	a	37.5
	c	37.4

CHAPTER 38

1.	c	38.2
2.	b	32.4
3.	c	32.3
4.	c	38.7
5.	c	38.2
6.	d	38.6
7.	a	38.6
8.	b	38.1
9.	a	38.7
10.	c	38.8
11.	a	38.4
12.	d	38.6
13.	c	38.6
14.	d	38.9
	b	38.9
	c	38.9
	d	38.9
15.	d	38.5
	c	38.5
	f	38.5
	e	38.5
	g	38.5
	h	38.5
	b	38.5
	a	38.5

CHAPTER 39

1.	d	39.2
2.	b	39.5
3.	c	39.6
4.	c	39.2
5.	a	39.6
6.	c	39.7
7.	a	39.5
8.	b	39.6
9.	c	39.8
10.	d	39.8
11.	b	39.6
12.	d	39.8
13.	c	39.9
14.	f	39.5
	e	39.6
	d	39.6
	g	39.6
	c	39.4
	a	39.5
	b	39.7
15.	f	39.6
	h	39.4
	c	39.7
	b	39.6
	e	39.2
	a	39.7
	g	39.5
	d	39.6

CHAPTER 40

1.	c	40.3
2.	a	40.4
3.	b	40.5
4.	a	40.5
5.	b	40.5
6.	a	40.5
7.	d	40.6
8.	c	40.4
	a	40.4
	b	40.4
	e	40.6
	f	40.4
	d	40.6
9.	a	40.6
10.	c	40.8
11.	a	40.9
12.	b	40.10
13.	d	40.10
14.	b	40.10
15.	a	40.10

CHAPTER 41

1.	a	41.2
2.	a	41.3
3.	c	41.4
4.	b	41.5
5.	c	41.5
6.	a	41.5
7.	c	41.6
8.	b	41.6
9.	d	41.7
10.	b	41.7
11.	a	41.8
12.	d	41.8
13.	a,c,a,c,a,c,b	41.9
14.	d	41.6
	c	41.6
	h	41.4
	a	41.6
	i	41.4
	e	41.4
	f	41.7
	b	41.6
	g	41.4
15.	a	41.5
	c	41.5
	b	41.5
	d	41.6

CHAPTER 42

1.	b	42.2
2.	b	42.2
3.	b	42.4
4.	c	42.4
	d	42.4
	a	42.4
	b	42.2
	e	42.7
5.	d	42.7
6.	a	42.7,
7.	d	42.9
8.	d	42.8
9.	d	42.10
10.	a	42.10
	c	42.10
	b	42.7
11.	c	42.7
12.	c	42.4
13.	a	42.1
14.	d	42.7
15.	e	42.7
	b	42.7
	d	42.7
	a	42.7
	f	42.7
	c	42.8
	g	42.10

CHAPTER 43

1.	c	43.2
2.	a	43.2
3.	c	43.4
4.	a	43.5
5.	a	43.5
6.	b	43.6
7.	d	43.7
8.	a	43.4
9.	c	43.8
10.	c	43.8
11.	b	43.8
12.	c	43.4
13.	d	43.3
14.	c	43.2
15.	a	43.3
	b	43.5
	e	43.7
	d	43.3
	c	43.6
	f	43.4

CHAPTER 44

1.	a	44.2
2.	f	44.4, 44.5
3.	c	44.2
4.	a	44.3
5.	a	44.3
6.	d	44.4
7.	a	44.5
8.	a	44.6
9.	b	44.7
10.	d	44.7
11.	c	44.5
12.	d	44.7
13.	d	44.7
14.	a	44.7
15.	c	44.4
	d	44.3
	a	44.3
	e	44.4
	b	44.4

CHAPTER 45

1.	b	45.2
2.	d	45.2
3.	b	45.7
4.	d	45.4
5.	b	45.3
	d	45.7
	c	45.2
	a	45.5
	e	45.4
6.	e	45.4
7.	b	45.8
8.	mimicry	45.6
9.	c	45.8
10.	a	45.9
11.	b	45.4
12.	c	45.7
13.	c	45.9
14.	c	45.7
15.	c	45.9
	b	45. 8
	d	45.8
	a	45.8
	f	45.8
	e	45.4

CHAPTER 46

1. a	46.2
2. b	46.2
3. c	46.2
4. d	46.4
5. b	46.3
6. c	46.6
7. b	46.7
8. c	46.7
9. d	46.8
10. a	46.10
11. c	46.10
12. d	46.9, 46.10
13. a	46.9
14. d	46.7, 46.9
15. e	46.8
d	46.9
b	46.10
a	46.9
c	46.9

CHAPTER 47

1. c	47.2
2. d	47.2
3. d	47.2
4. a	47.3
5. c	47.3
6. b	47.7
7. d	47.5
8. c	47.7
9. a	47.10
10. b	47.11
11. d	47.14
12. c	47.12
13. c	47.10
14. b	47.5
15 d	47.10
e	47.7
f	47.6
c	47.7
b	47.12
h	47.9
i	47.7
a	47.8
g	47.14

CHAPTER 48

1. True.	48.2
2. b	48.2
3. b	48.2
4. c	48.4
5. a	48.4
6. d	48.4
7. a	48.5
8. False	48.4
9. b	48.6
10. d	48.5
11. c	48.7
12. d	48.7
13. c	48.7
14. d	48.8
15. f	48.5
h	48.7
d	48.4
c	48.2
e	48.4
g	48.6
a	48.3
b	48.3

CHAPTER 13: GENETICS

1. Yellow is recessive. Because F1 plants have a green phenotype and must be heterozygous, green must be dominant over the recessive yellow.

2. **a.** *AB*
b. *AB*, *aB*
c. *Ab*, *ab*
d. *AB*, *Ab*, *aB*, *ab*

3. **a.** All offspring will be *AaBB*.
b. 1/4 *AABB* (25% each genotype)
1/4 *AABb*
1/4 *AaBB*
1/4 *AaBb*
c. 1/4 *AaBb* (25% each genotype)
1/4 *Aabb*
1/4 *aaBb*
1/4 *aabb*
d. 1/16 *AABB* (6.25% of genotype)
1/8 *AaBB* (12.5%)
1/16 *aaBB* (6.25%)
1/8 *AABb* (12.5%)
1/4 *AaBb* (25%)
1/8 *aaBb* (12.5%)
1/16 *AAbb* (6.25%)
1/8 *Aabb* (12.5%)
1/16 *aabb* (6.25%)

4. **a.** *ABC*
b. *ABC*, *aBC*
c. *ABC*, *aBC*, *ABc*, *aBc*
d. *ABC*, *aBC*, *AbC*, *abC*, *ABc*, *aBc*, *Abc*, *abc*

5. A mating of two *ML* cats yields 1/4 *MM*, 1/2 *MLM*, and 1/4 *MLML*. Because *MLML* is lethal, the probability that any one kitten among the survivors will be heterozygous is 2/3.

6. A mating between a mouse from a truebreeding, white-furred strain and a mouse from a truebreeding, brown-furred strain would provide you with the most direct evidence. Because truebreeding strains typically are homozygous for a trait being studied, all F1 offspring from this mating should be heterozygous. Record the phenotype of each F1 mouse, then let them mate with one another. Assuming only one gene locus is involved, these are possible outcomes for the F1 offspring:

a. All F1 mice are brown, and their F2 offspring segregate: 3 brown : 1 white. *Conclusion*: Brown is dominant to white.
b. All F1 mice are white, and their F2 offspring segregate: 3 white : 1 brown. *Conclusion*: White is dominant to brown.
c. All F1 mice are tan, and the F2 offspring segregate: 1 brown : 2 tan : 1 white. *Conclusion*: The alleles at this locus show incomplete dominance.

7. **a.** Both parents are heterozygous (*Aa*). Their children may be albino (*aa*) or unaffected (*AA* or *Aa*).
b. All are homozygous recessive (*aa*).
c. Homozygous recessive (*aa*) father, and heterozygous (*Aa*) mother. The albino child is *aa*, the unaffected children *Aa*.

8. The data reveal that these genes do not assort independently because the observed ratio is very far from the 9:3:3:1 ratio expected with independent assortment. Instead, the results can be explained if the genes are located close to each other on the same chromosome.

9. **a.** 1/2 red 1/2 pink 0 white
b. 0 red All pink 0 white
c. 1/4 red 1/2 pink 1/4 white
d. 0 red 1/2 pink 1/2 white

CHAPTER 14: GENETICS

1. Autosomal dominant. This pattern fits the pedigree: Both parents would necessarily be heterozygous for a dominant allele associated with the trait because one of their offspring does not have the trait. The second generation female who reproduces must also be heterozygous for the allele because her mate does not have the trait, and two offspring of the union also do not have the trait.

   No other pattern fits this pedigree. The allele associated with the trait could not be inherited in an autosomal recessive pattern. If it were, then both individuals in the first generation would be homozygous for it, so all of their offspring would also be homozygous for it and have the trait. One of their daughters does not have the trait, however.

   The allele associated with the trait could not be inherited in an X-linked pattern, regardless of whether it was dominant or recessive. If the allele were on the X chromosome, the male in the first generation would necessarily have the allele because he has the trait. If the allele was recessive, then the female in the first generation would necessarily be homozygous for it because she has the trait. All of the individuals in the second generation would have the trait. However, one of them does not, so the allele could not be recessive. If the allele was dominant, then all of the female offspring in the second generation would have the trait because all would have inherited their father's X chromosome with the dominant allele. However, one of the females does not have the trait, so the allele could not be on the X chromosome and dominant.

2. A son has a 100% chance of inheriting the trait. A daughter has a zero percent chance of having the trait, but a 50 percent chance of being a carrier.

3. 50 percent.

4. A daughter could develop DMD only if she inherited two recessive alleles associated with it one from each parent. Males who carry the allele are unlikely to father children because they develop the disorder and die early in life.

5. a. A male produces two kinds of gametes: one with an X chromosome and the allele, and the other with a Y chromosome.
   b. A female homozygous for an X chromosome allele produces one type of gamete, which carries the allele.
   c. A female heterozygous for an X chromosome allele produces two types of gametes: one that carries the allele, and another that carries the partnered allele on the homologous chromosome.

6. This disorder is most likely to be inherited in an autosomal dominant pattern. Genes on both chromosomes of a homologous autosome pair are expressed, so excess protein will be produced even in heterozygous people.

7. Y-linked dominant.

8. a. anaphase I or anaphase II.
   b. Translocation Down syndrome occurs when a chromosome breaks and reattaches to another chromosome. In most cases, a chromosome 21 breaks and becomes attached to a chromosome 14. Even if all the genes on chromosome 21 function normally, gene dosage is the issue, and Down syndrome results. 14.6

9. The most likely explanation for this outcome is that a crossover between her two chromosomes during meiosis generated an X chromosome that carries neither allele.

Glossary of Biological Terms

abscisic acid (ABA) (ab-SIH-sick) Plant hormone with a major role in stress responses such as stomata closure. **504**

abscission (ab-SIH-zhun) Process by which plant parts are shed. **504**

acclimatization Physiological adjustment of a body to a different environment; e.g., after moving from sea level to a high-altitude habitat. **673**

acid rain Low-pH rain formed when sulfur dioxide and nitrogen oxides mix with water vapor in the atmosphere. **856**

acid Substance that releases hydrogen ions in water. **33**

acrosome Enzyme-containing cap on the "head" of a mammalian sperm. **725**

actin Spherical protein that plays a role in cell movements; the main component of thin filaments in a sarcomere. **607**

action potential Brief reversal of the voltage difference across the plasma membrane of a neuron or muscle cell. **539**

activation energy Minimum amount of energy required to start a reaction. **81**

activator Transcription factor that increases the rate of transcription when it binds to a promoter or enhancer. **163**

active site Pocket in an enzyme where substrates react and are converted to products. **83**

active transport Energy-requiring mechanism in which a transport protein pumps a solute across a cell membrane against its concentration gradient. **92**

adaptation *See* adaptive trait. **253**

adaptive immunity In vertebrates, system of immune defenses that can be tailored to fight specific pathogens. **636**

adaptive radiation Macroevolutionary pattern in which a lineage undergoes a rapid burst of genetic divergences that gives rise to many species. **286**

adaptive trait A form of a heritable trait that enhances an individual's fitness. An evolutionary adaptation. **253**

adenosine triphosphate *See* ATP.

adhering junction Cell junction that fastens an animal cell to another cell, or to basement membrane. **69, 521**

adhesion protein Plasma membrane protein that helps cells stick together in animal tissues. Some types form adhering junctions and tight junctions. **88**

adipose tissue Specialized animal connective tissue that stores fat. **523**

adrenal cortex Outer portion of an adrenal gland; secretes aldosterone and cortisol. **591**

adrenal gland Endocrine gland located atop the kidney; secretes aldosterone, cortisol, epinephrine, and norepinephrine. **591**

adrenal medulla Inner portion of adrenal gland; secretes epinephrine and norepinephrine. **591**

adrenocorticotropic hormone (ACTH) (ah-DREEN-o-KOR-tah-ko-TROP-ik) Anterior pituitary hormone that stimulates cortisol secretion by the adrenal glands. **585**

aerobic (air-OH-bick) Involving or occurring in the presence of oxygen. **115**

aerobic respiration Oxygen-requiring cellular respiration; breaks down glucose and produces ATP, carbon dioxide, and water. **116**

age structure Of a population, the number of individuals in each of several age categories. **775**

agglutination (uh-glue-tih-NAY-shun) The clumping together of antigen-bearing cells or particles by antibodies. Basis of blood typing tests. **649**

AIDS Acquired immune deficiency syndrome. A secondary immune deficiency that develops as the result of infection by HIV. **654**

alcoholic fermentation Anaerobic sugar breakdown pathway that produces ATP, carbon dioxide, and ethanol. **124**

aldosterone Adrenal hormone that makes kidney tubules more permeable to sodium; encourages sodium reabsorption, thus increasing water reabsorption and concentrating the urine. **707**

algal bloom Population explosion of photosynthetic cells in an aquatic habitat. **345**

allantois (al-LAN-toh-is) Extraembryonic membrane that, in mammals, becomes part of the umbilical cord. **745**

allele frequency Abundance of a particular allele among all copies of the gene in a gene pool; proportion of chromosomes that carry the allele. **270**

alleles (uh-LEELZ) Forms of a gene with slightly different DNA sequences; may encode different versions of the gene's product. **188**

allergen (AL-er-jen) A normally harmless substance that provokes an immune response in some people. **652**

allergy Sensitivity to an allergen. **652**

allopatric speciation Speciation pattern in which a physical barrier that interrupts gene flow between populations fosters genetic divergences. **282**

allosteric regulation (al-oh-STARE-ick) Regulation of enzyme activity by the binding of a specific ion or molecule outside the active site. **85**

alpine tundra Biome of low-growing, wind-tolerant plants that are adapted to high-altitude conditions. **841**

alternation of generations Life cycle in which both haploid and diploid multicelled bodies form; occurs in some algae and all land plants. **341**

alternative splicing Post-transcriptional RNA modification process in which exons are rearranged and/or joined in different combinations. **151**

altruism Behaving in a way that benefits others, despite some risk or cost to oneself. **766**

alveolate (al-VEE-o-late) Member of a protist lineage having small sacs beneath the plasma membrane; dinoflagellate, ciliate, or apicomplexan. **345**

alveolus (al-VEE-oh-luss) Plural, alveoli. In a lung, a tiny sac at the tip of a bronchiole; site of gas exchange. **667**

amino acid (uh-ME-no) Small organic compound that is a monomer of proteins. Consists of a carboxyl group, an amine group, and one of 20 side groups (R), all typically bonded to the same carbon atom. **46**

amino acid–derived hormone An amine (modified amino acid), peptide, or protein that functions as a hormone. **583**

ammonia Inorganic compound consisting of nitrogen and hydrogen (NH_3); a metabolic product of protein breakdown. **700**

ammonification (am-on-if-ih-CAY-shun) Process by which bacteria and archaea break down nitrogen-containing organic material, thus releasing ammonia and ammonium ions. **821**

amnion (AM-nee-on) Extraembryonic membrane that encloses an amniote embryo and amniotic fluid. **745**

amniote (AM-nee-oat) Vertebrate with eggs that enclose the embryo within waterproof membranes; a reptile, bird, or mammal. **417**

amoeba (uh-ME-buh) Single-celled, unwalled protist that extends pseudopods to move and to capture prey. **351**

amoebozoans (uh-me-buh-ZOE-ans) Members of the eukaryotic supergroup Amoeboza; includes amoebas and slime molds. **351**

amphibian Tetrapod with a three-chambered heart and scaleless skin; typically develops in water, then lives on land as an air-breathing carnivore. **423**

anaerobic (an-air-OH-bick) Occurring in or requiring the absence of oxygen. **115**

analogous structures Similar body structures that evolved separately in different lineages (by convergent evolution). **295**

anaphase (ANN-uh-faze) Stage of mitosis during which sister chromatids separate and move toward opposite spindle poles. **179**

aneuploidy (AN-you-ploy-dee) Condition of having too many or too few copies of a particular chromosome. **225**

angiosperms (AN-gee-o-sperms) Seed plants that produce flowers and fruits. **369**

animal A multicelled heterotroph that has unwalled cells, develops through a series of stages, and moves about during part or all of its life. **8, 391**

animal hormone Intercellular signaling molecule secreted by an endocrine gland or cell; travels in the blood to target cells. **580**

annelid Segmented worm with a coelom, complete digestive system, and closed circulatory system. A polychaete, leech, or oligochaete. **399**

antenna Plural, antennae. Of some arthropods, a sensory structure on the head that detects touch and odors. **403**

anther (AN-thur) Of a flower, the part of the stamen that produces pollen grains. **482**

Anthropocene (AN-thruh-puh-seen) The current geologic interval, in which human activities are leaving a global imprint on the Earth. **851**

anthropoids Primate lineage that includes monkeys, nonhuman apes, and humans. **437**

antibody (AN-tee-baa-dee) Y-shaped antigen receptor protein made only by B cells; immunoglobulin. **645**

antibody-mediated immune response Adaptive immune response in which B cells produce antibodies that bind to an antigen. **646**

anticodon In a tRNA, set of three nucleotides that base-pairs with an mRNA codon. **152**

antidiuretic hormone (ADH) (AN-tee-die-ur-ET-ik) Hormone released in the posterior pituitary; makes kidney tubules more permeable to water; encourages water reabsorption, thus concentrating the urine. **585, 707**

antigen (AN-tih-jenn) A molecule or particle that the immune system recognizes as nonself. **636**

antioxidant Substance that prevents oxidation of other molecules. **86**

anus Body opening that serves solely as the exit for wastes from a complete digestive tract. **683**

aorta (ay-OR-tuh) Largest artery; carries oxygenated blood away from the heart. **619**

apical dominance Effect in which a lengthening shoot tip inhibits the growth of lateral buds. **502**

apical meristem (A-pih-cull) Meristem in the tip of a shoot or root. Associated with primary growth. **459**

apicomplexan (ay-PEE-com-PLEKS-an) Single-celled alveolate protist that lives as a parasite inside animal cells; some cause malaria or toxoplasmosis. **345**

apoptosis (ap-op-TOE-sis or ay-poe-TOE-sis) Programmed cell death. **741**

appendicular skeleton Of vertebrates, bones of the limbs or fins and the bones that connect these appendages to the axial skeleton. **601**

appendix (ah-PEN-diks) Short, tubular projection from the first part of the large intestine (the cecum); reservoir for beneficial bacteria. **689**

aquifer (AH-qwih-fur) Porous rock layer that holds some groundwater. **815**

arachnids (uh-RAK-nidz) Land-dwelling chelicerate arthropods with four pairs of walking legs; spiders, scorpions, mites, and ticks. **405**

archaea (are-KEY-uh) Singular, archaeon. Group of prokaryotes that are more closely related to eukaryotes than to bacteria. **8, 324**

archaeplastids (ark-ee-PLAST-ids) Members of the eukaryotic supergroup Archaeplastida; includes the red algae, two lineages of green algae, and the land plants. **349**

Archean eon (ahr-KEY-an) Precambrian interval extending from 4 billion years ago to 2.5 billion years ago. First eon for which there is evidence of life. **308**

arctic tundra Highest-latitude Northern Hemisphere biome, where low, cold-tolerant plants survive with only a brief growing season. **841**

area effect Big islands have more species than small islands because big islands present a larger target for colonists and include more habitats. **802**

arteriole Blood vessel that conveys blood from an artery to capillaries. **625**

artery Large-diameter blood vessel that carries blood away from the heart. **619**

arthropod Invertebrate with jointed legs and a hard exoskeleton that is periodically molted; for example, an insect or crustacean. **403**

ascus Plural, asci. Of sac fungi, a saclike cell in which spores form by meiosis. **381**

asexual reproduction Reproductive mode by which offspring arise from a single parent only. **177, 716**

astrobiology Scientific study of life's origin and distribution in the universe. **303**

atmospheric cycle Biogeochemical cycle that includes an atmospheric reservoir. **815**

atom Fundamental building block of matter; consists of varying numbers of protons, neutrons, and electrons. The smallest unit of a substance. **4**

atomic number Number of protons in the atomic nucleus; determines the element. **24**

ATP Adenosine triphosphate. (uh-DEN-uh-seen) Nucleotide monomer of RNA; consists of an adenine base, a ribose sugar, and three phosphate groups. **49**

ATP/ADP cycle Process by which cells use and regenerate ATP. ADP forms when ATP loses a phosphate group. ATP forms when ADP is phosphorylated. **86**

atrioventricular (AV) node (AY-tree-oh-ven-TRICK-you-lur) Clump of cells that serves as the electrical bridge between the atria and ventricles. **621**

atrium (AY-tree-um) Heart chamber that receives blood from veins and pumps it into a ventricle. **617**

australopith Informal name for chimpanzee-sized hominins that lived in Africa from 4 million to 1.2 million years ago. **441**

autoimmune response Immune response that targets one's own body tissues. **652**

autonomic nerves Of the peripheral nervous system, nerves that relay signals between the central nervous system and internal organs and glands. **546**

autosome Chromosome of a pair that is the same in males and females; a chromosome that is not a sex chromosome. **138**

autotroph (AH-toe-trof) Producer. Organism that makes its own food using energy from the environment and carbon from inorganic molecules. **101**

auxin (OX-in) Plant hormone that causes cell enlargement; also has a central role in growth by coordinating the effects of other hormones. **501**

axial skeleton Of vertebrates, bones of the main body axis; the skull and backbone. **601**

axon Of a neuron, a cytoplasmic extension that transmits electrical signals along its length and secretes chemical signals at its endings. **539**

B cell B lymphocyte. Leukocyte that makes antibodies in an antibody-mediated immune response. **636**

B cell receptor Antigen receptor on the surface of a B cell; IgM or IgD that stays anchored in the B cell's plasma membrane. **645**

bacteria Singular, bacterium. The most diverse and well-studied domain of prokaryotes (single-celled organisms that lack a nucleus). **8, 324**

bacteriophage (bac-TEER-ee-oh-fayj) Virus that infects bacteria. **132, 318**

balanced polymorphism Maintenance of two or more alleles of a gene at high frequency in a population. **276**

bark Informal term for all living and dead tissues outside the ring of vascular cambium in woody plants. **460**

Barr body Condensed and inactivated X chromosome in a cell of a female mammal. The other X chromosome is active. **167**

basal body (BASE-ull) Organelle that develops from a centriole. **67**

basal metabolic rate Rate of energy use in a body at rest. **694**

base Substance that accepts hydrogen ions in water. **33**

base-pair substitution Mutation in which a single base pair changes. **156**

basement membrane Extracellular matrix that attaches epithelium to underlying tissue. **69, 521**

basophil (BASE-uh-fill) Granular, circulating white blood cell that migrates into tissues and degranulates in response to injury or antigen. **636**

behavioral plasticity The behavioral phenotype associated with a genotype depends on conditions in the animal's environment. **757**

bell curve Bell-shaped curve; typically results from graphing frequency versus distribution for a trait that varies continuously. **210**

benthic province The ocean's sediments and rocks. **846**

big bang theory Well-supported hypothesis that the universe originated by a nearly instant distribution of matter through space. **304**

bilateral symmetry Having paired structures so the right and left halves are mirror images. **391**

bile Liquid mixture of salts, pigments, and cholesterol that aids fat emulsification in the small intestine. Produced in the liver, stored and concentrated in the gallbladder, and secreted into the small intestine. **687**

binary fission Cell reproduction process of bacteria and archaea; a type of asexual reproduction. **324**

bioaccumulation An organism accumulates increasing amounts of a chemical pollutant in its tissues over the course of its lifetime. **856**

biodiversity Of a region, the genetic variation within its species, variety of species, and variety of ecosystems. **861**

biodiversity hot spot Threatened region with great biodiversity; such regions are considered a high priority for conservation efforts. **861**

biofilm Community of microorganisms living within a shared mass of secreted slime. **58**

biogeochemical cycle Movement of an element or compound among environmental reservoirs and into and out of food webs. **815**

biogeography Study of patterns in the geographic distribution of species and communities. **251**

biological magnification A chemical pollutant becomes increasingly concentrated as it moves up through food chains. **856**

biological pest control Use of a pest's natural enemies to reduce its numbers. **797**

biology The scientific study of life. **3**

bioluminescence (BY-oh-loom-ih-NESS-sense) Light produced by a living organism. **345**

biomass pyramid Diagram that depicts the biomass (dry weight) in each trophic level of an ecosystem. **812**

biome (BY-ohm) Group of regions that may be widely separated but share a characteristic climate, soil composition, and dominant vegetation. **835**

biosphere (BY-oh-sfeer) All regions of Earth that can support life. **4, 828**

biotic potential (by-AH-tick) Maximum possible population growth rate under optimal conditions. **775**

bipedalism (bi-PEE-duh-lism) Habitual upright walking. **439**

bird Feathered, endothermic amniote with a beak and wings. **426**

blastocyst Mammalian blastula. **745**

blastula (BLAST-yoo-luh) In animal development, a hollow ball of cells that forms as a result of cleavage. **737**

blood Circulatory fluid of a closed circulatory system. **523, 617**

blood–brain barrier Protective mechanism that prevents unwanted substances from entering cerebrospinal fluid. **549**

blood pressure Pressure exerted by blood against a vessel wall. **625**

BMI *See* body mass index.

body mass index (BMI) Measure that estimates an individual's degree of body fat based on height and weight. **694**

bone tissue Animal connective tissue in which a mineral-hardened secreted matrix surrounds living cells. **523**

bony fish Fish that has a skeleton consisting primarily of bone; a ray-finned fish or lobe-finned fish. **421**

boreal forest (BOHR-ee-uhl) Conifer-dominated forest that covers extensive high latitude regions of the Northern Hemisphere. **841**

bottleneck Reduction in population size so severe that it reduces genetic diversity. **278**

Bowman's capsule In the kidney, the cup-shaped portion of a nephron that encloses the glomerulus and receives filtrate from it. **703**

brain Central control organ in a nervous system. **537**

brain stem The most evolutionarily ancient region of the vertebrate brain; includes the pons, medulla, and midbrain. **551**

bronchiole (BRONG-key-ohl) In a lung, one of the airways that leads from a bronchus to the alveoli. **669**

bronchus (BRON-cuss) Plural, bronchi. Airway that connects the trachea to a lung. **669**

brood parasitism One egg-laying species benefits by having another raise its offspring. **797**

brown adipose tissue Of some mammals, a specialized adipose tissue with mitochondria that primarily produce heat, rather than ATP. **710**

brown algae Multicelled, marine, stramenopile protists with chloroplasts that contain a brown accessory pigment (fucoxanthin). **343**

brush border cell In the lining of the small intestine, an epithelial cell with microvilli at its surface. **687**

bryophyte (BRY-oh-fite) Member of one of the early-evolving plant lineages that has a gametophyte-dominant life cycle; a moss, liverwort, or hornwort. **357**

budding In yeasts, a mechanism of asexual reproduction by which a small cell forms on a parent, then is released. **381**

buffer Set of chemicals that can keep the pH of a solution stable by alternately donating and accepting ions that contribute to pH. **33**

C3 plant Type of plant that uses only the Calvin–Benson cycle to fix carbon. **109**

C4 plant Type of plant that minimizes photorespiration by fixing carbon twice, in two cell types. **109**

calcitonin (cal-sih-TOE-nin) Thyroid hormone that encourages bones to take up and incorporate calcium. **588**

Calvin–Benson cycle Cyclic carbon-fixing pathway that builds sugars from carbon dioxide; light-independent reactions of photosynthesis. **109**

calyx A flower's outer, protective whorl of sepals. **482**

CAM plant Type of C4 plant that minimizes photorespiration by fixing carbon twice, at different times of day. **110**

camera eye Eye with an adjustable opening and a single lens that focuses light on a retina. **571**

camouflage Body coloration, patterning, form, or behavior that helps predators or prey blend with the surroundings and possibly escape detection. **797**

cancer Disease that occurs when a malignant neoplasm physically and metabolically disrupts body tissues. **182**

capillary A narrow, thin-walled blood vessel in a closed circulatory system; exchanges with interstitial fluid take place across its walls. **617**

carbohydrate (car-bow-HI-drait) Molecule that consists primarily of carbon, hydrogen, and oxygen atoms in a 1:2:1 ratio. A sugar, or a polymer of sugars. **42**

carbon cycle Biogeochemical cycle in which carbon moves between the oceans, atmosphere, and living organisms. **817**

carbon fixation Process in which carbon from an inorganic source such as carbon dioxide gets incorporated (fixed) into an organic molecule. **109**

cardiac cycle Sequence of contraction and relaxation of heart chambers that occurs with each heartbeat. **621**

cardiac muscle tissue Muscle tissue of the heart wall. **524**

carpel In flowering plants, a reproductive structure that produces female gametophytes; consists of a stigma, a style, and an ovary. **369, 482**

carrying capacity (***K***) Maximum number of individuals of a species that an environment can sustain. **777**

cartilage Animal connective tissue in which a rubbery secreted matrix surrounds living cells. **523**

cartilaginous fish Jawed fish with a skeleton that consists mainly or entirely of cartilage; for example, a shark or ray. **421**

Casparian strip Waterproof band that seals abutting cell walls of root endodermal cells, thus preventing water from seeping through cell walls into the vascular cylinder. Helps the plant regulate its uptake of mineral ions. **471**

catalysis (cut-AL-ih-sis) The acceleration of a reaction rate by a molecule that is unchanged by participating in the reaction. **83**

catastrophism Idea that catastrophic geologic events have periodically shaped Earth's surface. **251**

cDNA Complementary strand of DNA synthesized from an RNA template by the enzyme reverse transcriptase. **232**

cell Smallest unit of life. All have a plasma membrane, cytoplasm, and DNA. **4**

cell cortex Region of cytoplasm just inside the plasma membrane; often contains a mesh of microfilaments and associated proteins. **67**

cell cycle The collective series of intervals and events of a cell's life, from the time it forms until it divides. **177**

cell junction Structure that connects a cell to another cell or to extracellular matrix. **69**

cell plate Disk-shaped structure that forms and partitions descendant cells during cytokinesis in a plant cell. **180**

cell theory Set of principles that constitute the foundation of modern biology: every living organism consists of one or more cells; the cell is the basic structural and functional unit of life; all cells come from division of preexisting cells; and all cells pass hereditary material (DNA) to offspring. **57**

cell wall Rigid but permeable structure that surrounds the plasma membrane of some cells. **58**

cell-mediated immune response Adaptive immune response in which cytotoxic T cells and NK cells kill infected or cancerous body cells. **646**

cellular respiration Pathway that breaks down an organic molecule to form ATP and includes an electron transfer chain. **115**

cellular slime mold Soil-dwelling protist that feeds as solitary cells, but congregates under adverse conditions to form a cohesive unit that develops into a fruiting body. **351**

cellulose (SELL-you-low-ss) Crosslinked polysaccharide of glucose monomers. Tough and insoluble, it is the major structural material in plants. **42**

central nervous system Brain and spinal cord. **537**

central vacuole Very large fluid-filled vesicle of plant cells. **62**

centriole (SEN-tree-ole) Barrel-shaped organelle from which microtubules lengthen. **67**

centromere (SEN-truh-meer) Of a duplicated eukaryotic chromosome, constricted region where sister chromatids attach to each other. **136**

cephalization (SEF-uh-lih-ZAY-shun) Evolutionary trend whereby nerve and sensory cells become concentrated in an animal's front end. **391, 537**

cerebellum (ser-uh-BELL-um) Hindbrain region responsible for posture and for coordinating voluntary movements. **551**

cerebral cortex Outer gray matter layer of the cerebrum. **553**

cerebrospinal fluid Clear fluid that surrounds the brain and spinal cord and fills cavities within the brain. **549**

cerebrum (suh-REE-brum) Forebrain region that, in humans, is essential to language, planning, and abstract thought. **551**

cervix Narrow part of the uterus that connects to the vagina. **721**

character displacement Outcome of competition between two species; similar traits that result in competition become dissimilar. **793**

character Quantifiable, heritable trait such as the number of segments in a backbone or the nucleotide sequence of ribosomal RNA. **293**

charge Electrical property in which opposite charges attract, and like charges repel. **24**

charophyte algae (kar-UH-fight) Green algal lineage that includes the ancestors of land plants. **349**

chelicerates (kell-ISS-er-ates) Arthropod group that has specialized feeding structures (chelicerae) and no antennae. Includes the horseshoe crabs and arachnids. **405**

chemical bond Attractive force that joins atoms in molecules. **27**

chemoautotroph (keem-oh-AWE-toe-trof) Organism that uses carbon dioxide as its carbon source and obtains energy by oxidizing inorganic molecules. **326**

chemoheterotroph (keem-oh-HET-ur-o-trof) Organism that obtains both energy and carbon by breaking down organic compounds. **326**

chemoreceptor KEEM-oh-ree-SEP-tor) Sensory receptor that responds to the presence of a specific chemical. **561**

chemotaxis (key-mo-TAX-iss) Cellular movement in response to a chemical stimulus. **641**

chitin (KIE-tin) Nitrogen-containing polysaccharide that composes fungal cell walls and arthropod exoskeletons. **42**

chlamydias (klah-MID-ee-ahs) Tiny bacteria that live as intracellular parasites of vertebrates. **329**

chlorophyll *a* (KLOR-uh-fil) Main photosynthetic pigment in plants. **103**

chlorophyte algae (KLOR-uh-fight) Most diverse lineage of green algae. **349**

chloroplast (KLOR-uh-plast) Organelle of photosynthesis in the cells of plants and photosynthetic protists. **65**

choanoflagellates (ko-an-oh-FLAJ-el-ets) Heterotrophic protists closely related to animals; collared flagellates that strain food from water. **351**

cholecystokinin (KOH -leh-sis-toe- KIE -nin) Small intestine hormone that encourages release of pancreatic enzymes and bile into the small intestine; also suppresses appetite. **687**

chordate (CORE-date) Animal with an embryo that has a notochord, dorsal nerve cord, pharyngeal gill slits, and a tail that extends beyond the anus. For example, a lancelet or a vertebrate. **417**

chorion (KOR-ee-on) Outermost extraembryonic membrane of amniotes; major component of the placenta in placental mammals. **745**

choroid (KOR-oyd) Blood vessel–rich layer of the middle eye. The brownish pigment melanin darkens it and prevents light scattering. **573**

chromatin Collective term for all of the DNA and associated proteins in a cell nucleus. **61**

chromosome A structure that consists of DNA together with associated proteins; carries part or all of a cell's genetic information. **136**

chromosome number The total number of chromosomes in a cell of a given species. **138**

chyme (kime) Mix of food and gastric fluid. **685**

chytrids (KIH-trids) Fungi that produce flagellated spores. **379**

ciliary muscle Ring-shaped muscle of the eye that encircles the lens and attaches to it by short fibers. **573**

ciliate (SILL-ee-et) Single-celled or colonial protist with multiple cilia. **345**

cilium (SILL-ee-uh) Plural, cilia. Short, hairlike structures on the plasma membrane of some eukaryotic cells. Often occur in clumps that beat in unison. **67**

circadian rhythm (sir-KAY-dee-un) A biological activity that is repeated about every 24 hours. **169, 587**

circulatory system Organ system consisting of a heart or hearts and the fluid-filled vessels that distribute substances through a body. **617**

citric acid cycle Cyclic pathway that reduces many coenzymes by breaking down acetyl–CoA; part of aerobic respiration. Also called the Krebs cycle. **120**

clade (CLAYD) A group whose members share a defining derived character. **293**

cladistics (cluh-DISS-ticks) Making hypotheses about evolutionary relationships among clades. **293**

cladogram (CLAD-oh-gram) Evolutionary tree diagram that shows how a group of clades are related. **293**

classical conditioning An animal's involuntary response to a stimulus becomes associated with another stimulus that is presented at the same time. **759**

cleavage furrow In a dividing animal cell, the indentation where cytoplasmic division will occur. **180**

cleavage In animal development, the process by which multiple mitotic divisions produce a blastula from a zygote. **737**

climate Average weather conditions in a region over a long period. **828**

cloaca (klo-AY-kuh) In some vertebrates, a body opening that functions in elimination of digestive and urinary waste, as well as in reproduction. **419, 683**

clone Genetic copy of an organism. **131**

cloning vector *See* vector.

closed circulatory system System in which blood flows to and from a heart or hearts through a continuous series of vessels. **399, 617**

club fungi Fungi that have septate hyphae and reproduce sexually by producing spores in club-shaped cells. **381**

cnidarian (nigh-DARE-ee-en) Radially symmetrical invertebrate that has tentacles with stinging cells (cnidocytes). **395**

cnidocyte (NIGH-duh-site) Stinging cell unique to cnidarians. **395**

coal Fossil fuel formed over millions of years by compaction and heating of plant remains. **365**

cochlea (COCK-lee-uh) In the inner ear, the coiled, fluid-filled structure that holds the mechanoreceptors involved in hearing. **567**

codominance The full and separate phenotypic effects of two alleles are apparent in heterozygous individuals. **206**

codon (CO-don) In an mRNA, a nucleotide base triplet that codes for an amino acid or stop signal during translation. **152**

coelom (SEE-lum) In many animals, a body cavity that surrounds the gut and is lined with tissue derived from mesoderm. **391**

coenzyme An organic cofactor, e.g., NAD. **86**

coevolution The joint evolution of two closely interacting species; macroevolutionary pattern in which each species is a selective agent for traits of the other. **286, 369**

cofactor (KO-fack-ter) Coenzyme or metal ion that associates with an enzyme and is necessary for its function. **86**

cohesion (ko-HE-zhun) Property of a substance that arises from the tendency of its molecules to resist separating from one another. **31**

cohesion–tension theory Explanation of how transpiration creates a tension that pulls a cohesive column of water upward through xylem. **473**

cohort Group of individuals born during the same time interval. **779**

coleoptile (coal-ee-OPP-till) In monocots, a rigid sheath that protects the plumule (embryonic shoot) as it grows up through soil. **492**

collecting tubule Kidney tubule that receives filtrate from several nephrons and delivers it to the renal pelvis. **703**

collenchyma (coal-EN-kuh-muh) In plants, simple tissue composed of living cells with unevenly thickened walls. **452**

colon Longest portion of the large intestine. Region between the cecum and the rectum. **689**

colonial organism Organism composed of many integrated cells, each capable of living and reproducing on its own. **338**

colonial theory of animal origins Well-supported hypothesis that the first animals evolved from a colonial protist. **392**

commensalism Species interaction that benefits one species and neither helps nor harms the other. **790**

communication signal Chemical, acoustical, visual, or tactile cue that is produced by one member of a species and detected and responded to by other members of the same species. **761**

community All populations of all species that live in a defined region. **4, 789**

companion cell In phloem, a specialized cell that provides metabolic support to its partnered sieve element. **476**

comparative morphology (more-FALL-uh-jee) Scientific study of similarities and differences in body plans. **251**

competitive exclusion Process whereby two species compete for a limiting resource, and one drives the other to local extinction. **793**

complement In vertebrates, a set of proteins that circulate in inactive form in blood, and when activated, participate in immunity. **641**

complete digestive tract Tubelike digestive system; food enters through one opening, and wastes leave through another. **391, 681**

compound Molecule that has atoms of more than one element. **29**

compound eye Of some arthropods, an eye that consists of many individual units, each with a lens; excels at detecting movement. **403, 571**

concentration Amount of solute per unit volume of solution. **31**

condensation Chemical reaction in which an enzyme builds a large molecule from smaller subunits; water also forms. **40**

conduction Of heat: the transfer of heat within an object or between two objects in contact with one another. **709**

cone cell Photoreceptor that provides sharp vision and allows detection of color. **574**

conifer Gymnosperm with nonmotile sperm and woody cones; for example, a pine. **366**

conjugation (con-juh-GAY-shun) Mechanism of horizontal gene transfer in which one bacterial or archaeal cell passes a plasmid to another. **324**

conjunctiva (con-junk-TIE-va) Mucous membrane that lines the inner surface of the eyelids and folds back to cover the eye's sclera. **571**

connective tissue Animal tissue with an extensive extracellular matrix; provides structural and functional support. **521**

conservation biology Field of applied biology that surveys and documents biodiversity and seeks ways to maintain and use it. **861**

consumer (kun-SUE-murr)) Organism that gets energy and nutrients by feeding on tissues, wastes, or remains of other organisms; a heterotroph. **7, 809**

continuous variation Range of small differences in a trait. **210**

contractile vacuole In freshwater protists, an organelle that collects and expels excess water. **338**

control group Of an experiment, group of individuals identical to an experimental group except for the independent variable under investigation. **13**

convection Transfer of heat by moving molecules of air or water. **709**

convergent evolution Morphological convergence. Macroevolutionary pattern in which similar body parts evolve separately in different lineages. **295**

coral reef Highly diverse marine ecosystem centered around reefs built by living corals that secrete calcium carbonate. **845**

cork cambium (CAM-bee-um) Lateral meristem that produces cork. **460**

cork Tissue that waterproofs, insulates, and protects the surfaces of woody stems and roots. **460**

cornea (KOR-nee-uh) Clear, protective covering at the front of the vertebrate eye. **573**

corolla A flower's whorl of petals; forms within sepals and encloses reproductive organs. **482**

corpus luteum (COR-pus LOO-tee-um) Hormone-secreting structure that forms from the ovarian follicle cells left behind after ovulation. **721**

cortisol Adrenal cortex hormone that influences metabolism and immunity; secretions rise with stress. **591**

cotyledon (cot-uh-LEE-dun) Seed leaf; structure that stores nutrients in a flowering plant embryo. Monocots have one; eudicots have two. **450**

countercurrent exchange Exchange of substances between two fluids that are moving in opposite directions. **665**

covalent bond (ko-VAY-lent) Type of chemical bond in which two atoms share a pair of electrons. **29**

critical thinking Act of evaluating information before accepting it. **11**

crop Of birds and some invertebrates, an enlarged portion of the esophagus that stores food. **681**

crossing over Process by which homologous chromosomes exchange corresponding segments of DNA during prophase I of meiosis. **192**

crustaceans (krus-TAY-shuns) Mostly marine arthropod group that has two pairs of antennae and a calcium-stiffened exoskeleton. **407**

culture Learned behaviors transmitted between individuals and down through generations. **442**

cuticle Secreted covering at a body surface. **69, 359**

cyanobacteria (sigh-AN-no-BAK-tee-ree-ah) Oxygen-producing photosynthetic bacteria. **329**

cycad (SIGH-cad) Tropical or subtropical gymnosperm with flagellated sperm, palmlike leaves, and fleshy seeds. **366**

cytokines (SITE-uh-kynes) Cell-to-cell signaling molecules secreted by leukocytes to coordinate activities during immune responses. **636**

cytokinesis (site-oh-kye-KNEE-sis) Cytoplasmic division of eukaryotic cells. **180**

cytokinin (site-oh-KINE-in) Plant hormone that promotes development of lateral buds, and inhibits the formation of lateral roots, among other effects. **502**

cytoplasm (SITE-uh-plaz-um) In a prokaryote, collective term for everything enclosed by the plasma membrane. In a eukaryote, everything between the plasma membrane and the nucleus. **55**

cytoplasmic localization As an egg forms, different materials accumulate in different regions of its cytoplasm. **739**

cytoskeleton (sigh-toe-SKEL-uh-ton) Network of protein filaments that support, organize, and move eukaryotic cells and their internal structures. **67**

cytosol (SITE-uh-sall) Jellylike mixture of water and solutes that is enclosed by a cell's plasma membrane. **55**

cytotoxic T cell Type of T cell that kills ailing or cancerous body cells during a cell-mediated immune response. **651**

data (DAY-tuh) Experimental result(s). **13**

decomposer Organism that breaks down organic wastes and remains into their inorganic subunits. **326, 809**

deductive reasoning Using a general idea to make a conclusion about a specific case. **13**

deletion Mutation in which one or more nucleotides are lost. **156**

demographic transition model Model describing changes in birth and death rates that occur as a region becomes industrialized. **784**

demographics Statistics that describe a population. **773**

denaturation (dee-nay-turr-AY-shun) Loss of a protein's three-dimensional shape, as by a shift in temperature. **49**

dendrite Of a neuron, a cytoplasmic extension that receives chemical signals sent by other neurons and converts them to electrical signals. **539**

dendritic cell (den-DRIT-ick) Phagocytic leukocyte that specializes in antigen presentation during adaptive immune responses. **636**

Denisovans (deh-NEE-so-vans) Recently discovered hominins closely related to Neanderthals. **444**

denitrification (dee-nigh-triff-ih-CAY-shun) Conversion of nitrates or nitrites to gaseous forms of nitrogen. **821**

dense connective tissue Animal tissue in which cells are surrounded by a matrix with densely packed collagen and elastin fibers. **523**

density-dependent limiting factor Factor that increasingly limits population growth as population density increases. **777**

density-independent limiting factor Factor that limits population growth to the same degree regardless of population density. **777**

dental plaque On teeth, a thick biofilm composed of various microorganisms, their extracellular products, and saliva glycoproteins. **639**

deoxyribonucleic acid *See* DNA.

dependent variable In an experiment, a variable that is presumably affected by an independent variable being tested. **13**

derived character A trait shared by all members of a clade but not in any of the clade's ancestors. **293**

dermal tissue system Tissues that cover and protect a plant's surfaces. **450**

dermis Deep layer of skin that consists of connective tissue with nerves and blood vessels running through it. **529**

desert Biome with little rain and low humidity; plants with water-storing and water-conserving adaptations predominate. **837**

desertification Conversion of a grassland or woodland to desert. **854**

detrital food web (dee-TRITE-uhl) Food web in which most energy is transferred directly from producers to detritivores. **811**

detritivore Consumer that feeds on small bits of organic material. **809**

deuterostomes (DUE-ter-oh-stomes) Lineage of bilateral animals in which the second opening that forms on the embryo surface develops into the mouth. **391**

development (dih-VELL-up-ment) Processes by which the first cell of a multicelled organism gives rise to an adult. **7**

diaphragm (DIE-uh-fram) Dome-shaped skeletal muscle that separates the thoracic and abdominal cavities; contracts during inhalation. **669**

diastole (die-ASS-toll-ee) Relaxation phase of the cardiac cycle. **621**

diastolic pressure (die-ah-STAHL-ic) Blood pressure when ventricles are relaxed. **625**

diatoms (DIE-ah-tom) Single-celled photosynthetic stramenopile protists that have two-part silica shell. **343**

differentiation Process by which cells become specialized during development as they begin to use different subsets of their DNA. **142, 741**

diffusion (dif-YOU-zhun) Spontaneous spreading of molecules or atoms through a fluid or gas. **90**

dihybrid cross Cross between two individuals identically heterozygous for alleles of two genes; for example $AaBb \times AaBb$. **204**

dikaryotic (die-kary-OTT-ik) Having two different haploid nuclei (n+n). **377**

dinoflagellate (di-no-FLAJ-o-let) Single-celled, aquatic protist with cellulose plates and two flagella; moves with a whirling motion; may be heterotrophic or photosynthetic. **345**

dinosaurs Group of reptiles that became extinct at the end of the Cretaceous; included the ancestors of birds. **425**

diploid (DIP-loyd) Having two of each type of chromosome characteristic of the species ($2n$). **138**

directional selection Mode of natural selection in which a form of a trait at one end of a range of variation is adaptive. **272**

disaccharide (die-SACK-uh-ride) Carbohydrate that consists of two monosaccharide monomers. E.g., sucrose. **42**

disease vector Animal that transmits a pathogen beween hosts. **321**

disruptive selection Mode of natural selection in which extreme forms of a trait are adaptive, and intermediate forms are not. **275**

distal tubule Portion of kidney tubule that delivers filtrate from the Loop of Henle to a collecting tubule. **703**

distance effect Islands close to a mainland have more species than islands farther away. **802**

divergent evolution Morphological divergence. Evolutionary pattern in which a body part of an ancestor changes in its descendants. **295**

DNA Deoxyribonucleic (dee-ox-ee-rye-bo-new-CLAY-ick) acid. Nucleic acid that carries hereditary information. Double helix structure. **7, 49**

DNA cloning Set of laboratory methods that uses living cells to mass-produce targeted DNA fragments. **232**

DNA library Collection of cells that host different fragments of foreign DNA, often representing an organism's entire genome. **235**

DNA ligase (LIE-gayce) Enzyme that seals gaps or breaks in double-stranded DNA. **138**

DNA polymerase Enzyme that carries out DNA replication. Uses a DNA template to assemble a complementary strand of DNA. **138**

DNA profiling Identifying an individual by analyzing the unique parts of his or her DNA. **238**

DNA replication Process by which a cell duplicates its DNA. **138**

DNA sequence Order of nucleotide bases in a strand of DNA. **136**

DNA sequencing *See* sequencing.

dominance hierarchy Social system in which resources and mating opportunities are unequally distributed within a group **766**

dominant Refers to an allele that masks the effect of a recessive allele paired with it in heterozygous individuals. **200**

dormancy Period of temporarily suspended metabolism. **486**

dosage compensation Any mechanism of equalizing gene expression between males and females. **167**

double fertilization Fertilization as it occurs in flowering plants. One sperm cell fertilizes the egg, producing a zygote; the second sperm cell fuses with the central cell, producing a triploid cell that gives rise to endosperm. **369, 486**

dry shrubland Biome dominated by a diverse array of fire-adapted shrubs; occurs in regions with cool, wet winters and a dry summer. **839**

duplication Repeated section of a chromosome. **222**

eardrum Membrane that vibrates in response to pressure waves and functions in hearing; the tympanic membrane. **567**

ecdysone (eck-DIE-sohn) Steroid hormone that controls molting in arthropods. **593**

echinoderms (eh-KYE-nuh-derms) Deuterostome invertebrates that, as adults, have radial symmetry, a water–vascular system, and an endoskeleton of embedded spines and plates. Larvae are bilateral. **411**

ECM *See* extracellular matrix. **69**

ecological footprint Area of Earth's surface required to sustainably support a particular level of development and consumption. **784**

ecological niche The resources and environmental conditions that a species requires. **793**

ecological restoration Work aimed at renewing or recreating a natural ecosystem that has been degraded or destroyed, fully or in part. **863**

ecology Study of how populations interact with one another and with their nonliving environment. **773**

ecosystem A community interacting with its environment. **4, 809**

ectoderm (ECK-toh-derm) Outermost primary tissue layer (germ layer) of an animal gastrula. **737**

ectotherm Animal whose body temperature varies with that of its environment; it adjusts its internal temperature by altering its behavior. **425, 710**

effector cell Antigen-sensitized lymphocyte that forms in an immune response and acts immediately. **646**

egg Female gamete. **716**

El Niño Southern Oscillation Naturally occurring, irregularly timed fluctuation in sea surface temperature and wind patterns in the equatorial Pacific; affects weather worldwide. **832**

electron Negatively charged subatomic particle. **24**

electron transfer chain In a cell membrane, a series of enzymes and other molecules that accept and give up electrons in turn, thus releasing the energy of the electrons in small, usable steps. **85**

electron transfer phosphorylation Process in which electron flow through electron transfer chains drives ATP formation. **104**

electronegativity Measure of the ability of an atom to pull electrons away from other atoms. **29**

electrophoresis (eh-lek-troh-fur-EE-sis) Laboratory technique that separates DNA fragments by size. **236**

element A pure substance that consists only of atoms with the same number of protons. **24**

embryo In animals, a developing individual from first cleavage until hatching or birth; in humans, usually refers to an individual during weeks 2 to 8 of development. **743**

embryonic induction Embryonic cells produce signals that alter the behavior of neighboring cells. **741**

embryophytes (EM-bree-oh-fite) Land plants; clade of multicelled, photosynthetic species that protect and nourish the embryo on the parental body. **357**

emergent property (ee-MERGE-ent) A characteristic of a system that does not appear in any of the system's components. **4**

emigration Movement of individuals out of a population. **775**

endangered species Species that faces extinction in all or a part of its range. **852**

endemic species (en-DEM-ick) A species found in the region where it evolved and nowhere else. **801**

endergonic (end-er-GON-ick) Describes a reaction that requires a net input of free energy to proceed. **81**

endocrine disruptor Chemical or mixture of chemicals that interferes with the function of the endocrine system. **579**

endocrine gland Ductless gland that secretes hormones into a body fluid. **521**

endocrine system System of hormone-producing glands and secretory cells of a vertebrate body. **580**

endocytosis (en-doe-sigh-TOE-sis) Process by which a cell takes in a small amount of extracellular fluid (and its contents) by the ballooning inward of the plasma membrane. **94**

endoderm (EN-doh-derm) Innermost primary tissue layer (germ layer) of an animal gastrula. **737**

endodermis In a plant root, a sheet of cells just outside the vascular cylinder. A Casparian strip prevents water from seeping through their walls. **459**

endomembrane system Multifunctional network of membrane-enclosed organelles (endoplasmic reticulum, Golgi bodies, and vesicles). **62**

endoplasmic reticulum (ER) (en-doh-PLAZ-mick ruh-TICK-you-lum) System of sacs and tubes that is a continuous extension of the nuclear envelope. Comprises smooth ER, rough ER. **62**

endoskeleton Internal skeleton consisting of hard parts. **411, 417, 601**

endosperm Triploid (3*n*) nutritive tissue in an angiosperm seed. **369, 486**

endospore Spore (resting structure) formed by some soil bacteria; contains a dormant cell and is highly resistant to adverse conditions. **326**

endosymbiont hypothesis (en-doh-SIM-by-ont) Hypothesis that mitochondria and chloroplasts evolved from bacteria. **311**

endotherm Animal that maintains its temperature by adjusting its production of metabolic heat; for example, a bird or mammal. **425, 710**

endotoxin Substance that is integral to the membrane of a Gram-negative bacterial pathogen and causes symptoms of disease in the bacteria's host. **330**

energy pyramid Diagram that depicts the energy that enters each of an ecosystem's trophic levels. The bottom tier of the pyramid, representing primary producers, is always the largest. **812**

energy The capacity to do work. **78**

enhancer Binding site for an activator. **163**

entropy (EN-truh-pee) Measure of how much the energy of a system is dispersed. **78**

enzyme Organic molecule that speeds a reaction without being changed by it. **40**

eosinophil (ee-uh-SIN-uh-fill) Granular, circulating leukocyte that targets multicelled parasites. **636**

epidemic Disease outbreak that occurs in a limited region. **323**

epidermis Outermost tissue layer; in animals, the epithelial layer of skin. In plants, dermal tissue. **452, 529**

epididymis (epp-ih-DID-ih-muss) Plural, epididymides. One of two ducts in which human sperm mature; each empties into a vas deferens. **725**

epigenetic Refers to potentially heritable modifications to DNA that affect gene expression without changing the DNA sequence. **170**

epiglottis (ep-uh-GLOT-iss) Tissue flap that covers the trachea during swallowing and thus prevents food from entering the airway. **669**

epistasis (epp-ih-STAY-sis) Form of polygenic inheritance in which the effect of an allele on a trait masks the effect of a different gene. **206**

epithelial tissue Sheetlike animal tissue that covers outer body surfaces and lines internal tubes and cavities. **521**

equilibrium model of island biogeography Model that predicts the number of species on an island based on the island's area and distance from the mainland. **802**

ER *See* endoplasmic reticulum.

erythropoietin (eh-rith-row-POY-eh-tin) Kidney hormone that stimulates production of red blood cells. **623**

esophagus (eh-SOFF-ah-gus) Tubular organ that connects the pharynx to the stomach. **681**

essential amino acid Amino acid that the body cannot make and must obtain from food. **691**

essential fatty acid Fatty acid that the body cannot make and must obtain from food. **691**

estivation An animal becomes dormant during a hot, dry season. **710**

estrogens Hormones secreted by ovaries; cause development of secondary sexual traits and maintain the reproductive tract's lining. **721**

estrous cycle In most female mammals, a recurring cyclic variation in sexual receptivity. **722**

estuary Highly productive ecosystem where nutrient-rich water from a river mixes with seawater. **845**

ethylene (ETH-ill-een) Gaseous plant hormone involved in regulating growth and cell expansion, among other functions. **507**

eudicots (you DIE cots) Most diverse lineage of angiosperms; members have two seed leaves, branching leaf veins. **371**

euglenid (you-GLEEN-id) Free-living, flagellated, excavate protist with mitochondria; most live in fresh water; e.g., *Euglena*. **341**

eukaryote (you-CARE-ee-oat) An organism whose cells characteristically have a nucleus; a protist, fungus, plant, or animal. **8**

eusocial animal Animal that lives in a multigenerational family group with a reproductive division of labor. **766**

eutrophication Nutrient enrichment of an aquatic ecosystem. **807**

evaporation Transition of a liquid to a vapor. **31, 709**

evolution Change in a line of descent. **251**

evolutionary tree Branching diagram of ancestral connections. **293**

exaptation (eggs-app-TAY-shun) A trait that has been evolutionarily repurposed for a new use. **286**

excavates Protist in a supergroup defined by their feeding groove; all are single, unwalled flagellated cells; includes diplomonads, parabasalids, trypanosomes, and euglenids. **341**

exergonic (ex-er-GON-ick) Describes a reaction that ends with a net release of free energy. **81**

exocrine gland Gland that secretes milk, sweat, saliva, or some other substance through a duct. **521**

exocytosis (ex-oh-sigh-TOE-sis) Process by which a cell expels a vesicle's contents to extracellular fluid. **94**

exon Gene segment that remains in an RNA after post-transcriptional modification. **151**

exoskeleton Of some invertebrates, hard external parts that muscles attach to and move. **403, 601**

exotic species A species that evolved in one community and later became established in another. **801**

exotoxin Substance that is secreted by a bacterial pathogen and causes symptoms of disease in the bacteria's host. **330**

experiment A test designed to support or falsify a prediction. **13**

experimental group Group of individuals who have a certain characteristic or receive a certain treatment (as compared with a control group). **13**

exponential growth model Model of unlimited population growth. The population grows by a fixed percentage each successive time interval; the size of each increase is determined by the current population size. **775**

external fertilization Sperm and eggs are released into the external environment and combine there. **716**

extinct Refers to a species that no longer has living members. **286**

extracellular fluid Of a multicelled organism, the body fluid outside of cells; serves as the body's internal environment. **519**

extracellular matrix (ECM) Complex mixture of substances secreted by a cell onto its surface; composition and function vary by cell type. **69**

extreme halophile Organism adapted to life in a highly salty environment. **332**

extreme thermophile Organism adapted to life in a high-temperature environment. **332**

eye Sensory organ that incorporates a dense array of photoreceptors. **571**

eyespot In some protists, a pigmented organelle that detects light. **338**

facilitated diffusion Passive transport mechanism in which a solute moves across a membrane through a transport protein. **92**

fall overturn During the fall, waters of a temperate zone lake mix. Upper, oxygenated water cools, becomes dense, and sinks; nutrient-rich water from the bottom moves up. **842**

fat A triglyceride. **44**

fatty acid Lipid that consists of a carboxyl group "head" and a hydrocarbon "tail." Saturated types have only single bonds linking the carbons in their tails. **42**

feather Filamentous keratin stucture derived from ancestral scales. **426**

feces (FEE-seez) Unabsorbed food material and cellular waste that is expelled from the digestive tract. **689**

feedback inhibition Regulatory mechanism in which a change that results from some activity decreases or stops the activity. **85**

fermentation Anaerobic glucose-breakdown pathway that produces ATP without the use of an electron transfer chain. Includes glycolysis. **116**

fertilization Fusion of two gametes to form a zygote. **188**

fetus (FEE-tus) Developing human from about 9 weeks until birth. **743**

fever A temporary, internally induced rise in core body temperature above the normal set point. **643**

fibrous root system Root system composed of an extensive mass of similar-sized adventitious roots; typical of monocots. **456**

first law of thermodynamics Energy cannot be created or destroyed. **78**

fishes Aquatic vertebrates that have gills and lack limbs; jawless fishes, cartilaginous fishes, and bony fishes. **419**

fitness Degree of adaptation to an environment, as measured by the individual's relative genetic contribution to future generations. **253**

fixed Refers to an allele for which all members of a population are homozygous. **278**

flagellum (fluh-JEL-um) Plural, flagella. Long, slender cellular structure used for motility. **58**

flatworm Acoelomate, unsegmented worm; a planarian, fluke, or tapeworm. **397**

flower Specialized reproductive structure of an angiosperm. **369, 482**

fluid mosaic Model of a cell membrane as a two-dimensional fluid of mixed composition. **88**

follicle-stimulating hormone (FSH) Anterior pituitary hormone with roles in ovarian follicle maturation and sperm production. **585, 722**

food chain Description of who eats whom in one path of energy flow in an ecosystem. **809**

food web Set of cross-connecting food chains. **811**

foraminifera (for-am-in-IF-er-ah) Heterotrophic single-celled protists with a porous calcium carbonate shell and long cytoplasmic extensions. **347**

fossil Physical evidence of an organism that lived in the ancient past. **255**

founder effect After a small group of individuals founds a new population, allele frequencies in the new population differ from the original population. **278**

fovea (FOE-vee-uh) Retinal region where cone cells are most concentrated. **574**

fraternal twins Twins that arise when two eggs mature and are fertilized at the same time; related as siblings. **722**

free radical Atom with an unpaired electron. Most are highly reactive and can damage biological molecules. **27**

frequency-dependent selection Natural selection in which a trait's adaptive value depends on its frequency in a population. **276**

fruit Mature ovary of a flowering plant, often with accessory parts; encloses a seed or seeds. **369, 491**

functional group An atom (other than hydrogen) or small molecular group bonded to a carbon of an organic compound; imparts a chemical property. **38**

fungus Plural, fungi. Eukaryotic heterotroph with chitin-containing cell walls; obtains nutrients by extracellular digestion and absorption. **8, 377**

gallbladder Organ that stores and concentrates bile. **687**

gametangium (gam-et-TAN-gee-um) Plural, gametangia. Of a plant gametophyte, the organ that produces gametes. **361**

gamete (GAM-eat) Mature, haploid reproductive cell; e.g., a sperm. **188**

gametophyte (gam-EET-o-fite) Haploid, multicelled body that produces gametes; forms during the life cycle of all land plants and some algae. **357**

ganglion (GANG-lee-on) Plural, ganglia. Cluster of neuron cell bodies. **397, 537**

gap junction Cell junction that forms a closable channel across the plasma membranes of adjoining animal cells; allows ions that serve as signals to move between cells. **69, 524**

gastric fluid Fluid secreted by the stomach lining; contains acid, mucus, and the protein-digesting enzyme pepsin. **685**

gastrin Hormone secreted by the stomach in response to the presence of food; encourages secretion of acid into the stomach. **685**

gastrovascular cavity In some invertebrates, a saclike cavity that functions in digestion and in gas exchange. **391, 681**

gastrula (GAS-troo-luh) Three-layered developmental stage formed by gastrulation in an animal. **737**

gastrulation (gas-troo-LAY-shun) In animal development, the process by which cell movements produce a three-layered gastrula. **737**

gene (JEEN) Unit of information encoded in the sequence of nucleotide bases in DNA. Encodes an RNA or protein product. **149**

gene expression Multistep process of converting information in a gene to an RNA or protein product. Includes transcription and translation. **149**

gene flow The movement of alleles into and out of a population. **278**

gene pool All alleles of all genes in a population; pool of genetic resources. **268**

gene therapy Treating a disease or genetic defect by repairing or replacing a gene in an affected individual. **243**

genetic code Complete set of 64 mRNA codons. **152**

genetic drift Change in allele frequency due to chance alone. **278**

genetic engineering Process by which an individual's genome is changed deliberately in order to change phenotype. **240**

genetic equilibrium Theoretical state in which an allele's frequency never changes in a population's gene pool. **270**

genome An organism's complete set of genetic material. **235**

genomics The study of whole-genome structure and function. **238**

genotype (JEEN-oh-type) The particular set of alleles that is carried by an individual's chromosomes. **200**

genus (JEE-nuss) Plural, genera. A group of species that share a unique set of traits. **8**

geologic time scale Chronology of Earth's history; correlates geologic and evolutionary events. **260**

germ cell Immature reproductive cell that gives rise to haploid gametes when it divides. **188**

germ layer Primary tissue layer that forms when a developing animal undergoes gastrulation. **737**

germinate To resume metabolic activity after dormancy. **486**

ghrelin (GRELL-in) Hormone secreted by the stomach when empty; stimulates appetite. **685**

gibberellin (jib-er-ELL-in) Plant hormone that induces stem elongation and helps seeds break dormancy, among other effects. **504**

gill Filamentous or branching respiratory organ of some aquatic animals; may be internal or external. **400, 665**

ginkgo Deciduous gymnosperm with flagellated sperm, fan-shaped leaves, and fleshy seeds. **366**

gizzard Of birds and some other animals, a digestive chamber lined with a hard material that grinds up food. **681**

glial cell (GLEE-uhl) In nervous tissue, one of the cell types that support and assist neurons. **525**

global climate change Long-term change in Earth's climate. **818**

glomeromycete (GLOW-mer-o-MY-seat) Fungus with hyphae that grow inside the wall of a plant root cell. **379**

glomerular filtration (gloh-MER-yuh-lur) First step in urine formation: protein-free plasma forced out of glomerular capillaries by blood pressure enters Bowman's capsule. **705**

glomerulus (gloh-MER-yuh-lus) Plural, glomeruli. In the kidney, a cluster of capillaries enclosed by Bowman's capsule. **703**

glottis Opening formed when vocal cords in the larynx relax. **667**

glucagon (GLUE-kah-gon) Pancreatic hormone that causes cells to break down glycogen and release glucose. **588**

glycogen (GLIE-ko-jen) Highly branched polysaccharide of glucose monomers. Principal form of stored sugars in animals. **42**

glycolysis (gly-COLL-ih-sis) Set of reactions that convert glucose to two pyruvate. First stage of aerobic respiration and fermentation. **119**

GMO (genetically modified organism) Organism whose genome has been modified by genetic engineering. **240**

gnetophyte (NEAT-o-fite) Shrubby or vinelike gymnosperm, with nonmotile sperm; for example, *Ephedra*. **366**

Golgi body (GOAL-jee) Membrane-enclosed organelle that modifies proteins and lipids, then packages the finished products into vesicles. **62**

gonad (GO-nad) Animal reproductive organ that produces gametes and sex hormones. **591, 719**

gonadotropin-releasing hormone (GnRH) Hypothalamic hormone that induces the pituitary to release hormones (LH and FSH) that regulate hormone secretion by the gonads. **722**

Gondwana (gond-WAN-uh) Supercontinent that existed before Pangea, more than 500 million years ago. **259**

Gram-positive bacteria Lineage of bacteria with a thick cell wall that is colored purple by Gram staining. **329**

gravitropism (grah-vih-TRO-pizm) Directional response to gravity. **507**

gray matter Central nervous system tissue that includes neuron cell bodies, dendrites, and axon terminals, as well as glial cells. **549**

grazing food web Food web in which most energy is transferred from producers to grazers (herbivores). **811**

greenhouse effect Warming of Earth that occurs when greenhouse gases absorb heat energy then reemit some of it back toward the Earth. **817**

greenhouse gas Atmospheric gas, such as carbon dioxide, that absorbs heat emitted by Earth's surface, then reradiates it. **817**

ground tissue system All tissues that are not dermal or vascular tissue; makes up the bulk of the plant body. **450**

groundwater Soil water and water in aquifers. **815**

growth In multicelled eukaryotes, an increase in the number, size, and volume of cells. **7**

growth factor Molecule that stimulates mitosis and differentiation. **182**

growth hormone (GH) Anterior pituitary hormone that regulates growth and metabolism. **585**

guard cell One of a pair of cells that define a stoma. **475**

gymnosperm (JIM-no-sperm) Seed plant that does not make flowers or fruits; for example, a conifer. **366**

habitat Type of environment in which a species typically lives. **790**

habituation Learning not to respond to a repeated stimulus. **759**

Hadean eon (HAY-dee-un) Precambrian interval that extends from Earth's formation to 4 billion years ago. **304**

half-life Characteristic time it takes for half of a quantity of a radioisotope to decay. **262**

haploid (HAP-loyd) Having one copy of each type of chromosome (n). **188**

heart Muscular organ that pumps fluid through a circulatory system. **617**

helper T cell Type of T cell that activates other lymphocytes. **649**

hemoglobin (HEEM-o-glow-bin) Iron-containing respiratory protein. **663**

hemolymph (HEEM-oh-limf) Fluid that circulates in an open circulatory system. **617**

hemostasis (he-moh-STAY-sis) Process by which blood clots in response to injury. **623**

herbivory An animal feeds on plant parts. **797**

hermaphrodite (herm-AFF-roh-dyte) Animal that produces both eggs and sperm, either simultaneously or at different times in its life. **393, 716**

heterotherm Animal that sometimes maintains its temperature by producing metabolic heat, and at other times allows its temperature to fluctuate with the environment. **710**

heterotroph (HET-er-oh-trof) Consumer. Organism that obtains carbon from organic compounds assembled by other organisms. **101**

heterozygous (het-er-uh-ZYE-guss) Genotype in which homologous chromosomes have different alleles at the same locus. **200**

hibernation An animal becomes dormant during a cold season. **710**

hippocampus Brain region essential to formation of declarative memories. **553**

histone (HISS-tone) Type of protein that associates with DNA and structurally organizes eukaryotic chromosomes. **136**

homeostasis (home-ee-oh-STAY-sis) Process in which organisms keep their internal conditions within tolerable ranges by sensing and responding appropriately to change. **7**

homeotic gene (home-ee-OTT-ic) Type of master regulator; its expression results in the formation of a specific body part during development. **165**

hominins Modern humans and their closest extinct relatives. **441**

hominoids Tailless primate lineage; includes humans and other apes. **439**

Homo erectus Early human species that arose in Africa by about 2 million years ago; some populations migrated to other regions. **442**

Homo habilis Earliest named human species; lived in Africa from about 2.3 to 1.4 million years ago. **442**

Homo neanderthalensis Neanderthals. Extinct hominins that lived in the Middle East, Europe, and Asia; the closest relatives of modern humans. **444**

Homo sapiens Anatomically modern humans; only surviving *Homo* species. **444**

homologous chromosomes In the nucleus of a eukaryotic cell, chromosomes with the same length, shape, and genes. **177**

homologous structures Body structures that are similar in different lineages because they evolved in a common ancestor. **293**

homozygous (ho-mo-ZYE-guss) Genotype in which homologous chromosomes have the same allele at the same locus. **200**

horizontal gene transfer Transfer of genetic material by a mechanism other than inheritance from a parent or parents. **324**

human chorionic gonadotropin (HCG) (kor-ee-ON-ik go-NAD-oh-troe-pin) Hormone first secreted by the blastocyst, and later by the placenta; helps maintain the uterine lining during pregnancy. **745**

human microbiota (my-crow-by-O-tuh) All the microbes (cellular organisms and viruses) that can live on or in humans. **317**

humus (HUE-muss) Partially decayed organic material in soil. **468**

hybrid (HI-brid) A heterozygous individual. **200**

hybridization (hi-brih-die-ZAY-shun) Spontaneous establishment of base-pairing between two nucleic acid strands. **138**

hydrocarbon Compound that consists only of carbon and hydrogen atoms. **38**

hydrogen bond Attraction between a covalently bonded hydrogen atom and another atom taking part in a separate covalent bond. **31**

hydrolysis (hy-DRAWL-uh-sis) Water-requiring chemical reaction in which an enzyme breaks a molecule into smaller subunits. **40**

hydrophilic Describes a substance that dissolves easily in water. **31**

hydrophobic Describes a substance that resists dissolving in water. **31**

hydrostatic skeleton Of soft-bodied invertebrates, a fluid-filled chamber that muscles exert force against, redistributing the fluid. **395, 601**

hydrothermal vent Rocky, underwater opening where mineral-rich water heated by geothermal energy streams out. **306, 846**

hypertonic Describes a fluid that has a high overall solute concentration relative to another fluid. **90**

hypha (HI-fah) Component of a fungal mycelium; a filament made up of cells arranged end to end. **377**

hypothalamus (HI-poe-THAL-uh-mus) Forebrain region that controls processes related to homeostasis; control center for endocrine functions. **553, 585**

hypothesis (high-POTH-uh-sis) Testable explanation of a natural phenomenon. **13**

hypotonic Describes a fluid that has a low overall solute concentration relative to another fluid. **90**

identical twins Genetically identical individuals that develop after an early embryo splits in two. **745**

immigration Movement of individuals into a population. **775**

immunity The body's ability to resist and fight infections. **636**

immunization Any procedure designed to induce immunity to a disease. **649**

imprinting Learning that can occur only during a specific interval in an animal's life. **759**

***in vitro* fertilization** Assisted reproductive technology in which eggs and sperm are united outside the body. **715**

inbreeding Mating among close relatives. **278**

inclusive fitness Genetic contribution an individual makes by reproducing, plus a fraction of the contribution that it makes by facilitating the reproduction of its relatives. **766**

incomplete dominance One allele is not fully dominant over another, so the heterozygous phenotype is an intermediate blend between the two homozygous phenotypes. **206**

independent variable A variable that is controlled by an experimenter in order to explore its relationship to a dependent variable. **13**

indicator species A species that is especially sensitive to disturbance and can be monitored to assess the health of a habitat. **801**

induced-fit model Substrate binding to an active site improves the fit between the two; this fit is necessary for catalysis. **83**

inductive reasoning Drawing a conclusion based on observation. **13**

inferior vena cava (VEE-nah CAY-vuh) Vein that delivers blood from the lower body to the heart. **619**

inflammation A local response to tissue damage or infection; characterized by redness, warmth, swelling, and pain. Part of innate immunity. **643**

inheritance (in-HAIR-ih-tunce) Transmission of DNA to offspring. **7**

inhibiting hormone Hormone that discourages release of a different hormone by its target endocrine cells. **585**

innate immunity In all multicelled organisms, set of immediate, general defenses against infection. **636**

inner ear Of the mammalian ear, region that contains the fluid-filled vestibular apparatus and cochlea. **567**

insects Six-legged arthropods with two antennae and two compound eyes. The most diverse class of animals. **407**

insertion Mutation in which one or more nucleotides are inserted. **156**

instinctive behavior Innate response to a simple stimulus. **759**

insulin Pancreatic hormone that causes cells to take up glucose and store it as glycogen. **588**

intercostal muscles Skeletal muscles between the ribs; help change the volume of the thoracic cavity during breathing. **669**

intermediate disturbance hypothesis Species richness is greatest in communities where disturbances are moderate in intensity or frequency. **799**

intermediate filament Stable cytoskeletal element of animals and some protists; different types are assembled from different fibrous proteins. **67**

internal fertilization Fertilization of eggs inside a female's body. **716**

interneuron Neuron that both receives signals from and sends signals to other neurons. Located mainly in the brain and spinal cord. **539**

interphase In a eukaryotic cell cycle, the interval during which a cell grows, roughly doubles the number of its cytoplasmic components, and replicates its DNA in preparation for division. **177**

interspecific competition Competition between two species. **793**

interstitial fluid (in-ter-STIH-shul) Portion of the extracellular fluid that fills spaces between body cells. **519**

intervertebral disk Cartilage disk between two vertebrae. **601**

intestine Tubular organ in which chemical digestion is completed and from which most nutrients are absorbed. **681**

intraspecific competition Competition among members of a species. **777**

intron Gene segment that intervenes between exons and is removed during post-transcriptional modification. **151**

inversion Structural rearrangement of a chromosome in which part of the DNA has become oriented in the reverse direction. **222**

invertebrate Animal that does not have a backbone. **389**

ion (EYE-on) An atom or molecule that carries a net charge. **27**

ionic bond (eye-ON-ick) Type of chemical bond in which a strong mutual attraction links ions of opposite charge. **29**

iris (EYE-ris) Circular muscle that adjusts the shape of the pupil to regulate how much light enters the eye. **573**

iron–sulfur world hypothesis Hypothesis that the metabolic reactions that led to the first cells took place on the porous surface of iron–sulfide-rich rocks at hydrothermal vents. **306**

isotonic Describes two fluids with identical solute concentrations. **90**

isotopes (ICE-oh-topes) Forms of an element that differ in the number of neutrons their atoms carry. **24**

joint Region where bones meet. **605**

***K*-selection** Selection favoring traits that provide a competitive advantage when population size is near carrying capacity. **779**

karyotype (CARE-ee-uh-type) Image of a cell's chromosomes arranged by size, length, shape, and centromere location. **138**

key innovation An evolutionary adaptation that gives its bearer the opportunity to exploit a particular environment much more efficiently or in a new way. **286**

keystone species A species that has a disproportionately large effect on community structure. **801**

kidney Organ of the vertebrate urinary system that filters blood, adjusts its composition, and forms urine. **700**

kin selection Type of natural selection that favors individuals with the highest inclusive fitness. **766**

kinesis (kin-EE-sis) Innate response in which an animal speeds up or slows its movement in reaction to a stimulus. **761**

kinetic energy (kih-NEH-tick) The energy of motion. **78**

knockout Technique of inactivating an organism's gene. **165**

Krebs cycle *See* citric acid cycle.

labor Expulsion of a placental mammal from its mother's uterus by muscle contractions. **749**

lactate fermentation Anaerobic sugar breakdown pathway that produces ATP and lactate. **124**

lactation Milk production by a female mammal. **750**

lancelet (LANCE-uh-lit) Invertebrate chordate with a fishlike shape; retains all the defining embryonic chordate traits into adulthood. **417**

large intestine Region of the intestine that receives material from the small intestine, absorbs water, and concentrates and stores waste. **683**

larva Plural, larvae. In some animal life cycles, a sexually immature form that differs from the adult in its body plan. **393**

larynx (LAIR-inks) Short airway containing the vocal cords (voice box). **667**

lateral bud Axillary bud; forms in a leaf axil. **459**

lateral meristem Cylindrical sheet of meristem that runs lengthwise through shoots and roots; associated with secondary growth (thickening). **460**

law of independent assortment During meiosis, alleles at one gene locus on homologous chromosomes tend to be distributed into gametes independently of alleles at other loci. **204**

law of nature A generalization that describes a consistent natural phenomenon that has an incomplete scientific explanation. **18**

law of segregation During meiosis, two alleles at any gene locus are distributed into separate gametes. **203**

leaching Process by which water moving through soil removes nutrients from it. **468**

leaf vein A vascular bundle in a leaf. **456**

learned behavior Behavior that is modified by experience. **759**

lek Area where male animals perform courtship displays. **763**

lens Disk-shaped structure that bends light rays so they fall on an eye's photoreceptors. **571**

lethal mutation Mutation that alters phenotype so drastically that it causes death. **268**

lichen (LIE-kin) Composite organism consisting of a fungus and a single-celled alga or a cyanobacterium. **383**

life history Schedule of how resources are allocated to growth, survival, and reproduction over a lifetime. **779**

ligament Strap of dense connective tissue that holds bones together at a synovial joint. **605**

light-dependent reactions First stage of photosynthesis; collectively convert light energy to chemical energy of ATP. **101**

light-independent reactions Carbon-fixing second stage of photosynthesis; collectively assemble sugars from carbon dioxide and water. **101**

lignin (LIG-nin) Material that strengthens cell walls of vascular plants. **69, 359**

limbic system Group of structures deep in the brain that function in expression of emotion. **553**

lineage (LINN-edge) Line of descent. **251**

linkage group A set of genes whose alleles do not assort independently into gametes. **204**

lipid bilayer Double layer of lipids arranged tail-to-tail; structural foundation of all cell membranes. **44**

lipid A fatty acid derivative (such as a triglyceride), steroid, or wax. **42**

liver Large organ that makes bile, stores glycogen, and detoxifies blood. **687**

loam Soil with roughly equal amounts of sand, silt, and clay. **468**

lobe-finned fish Bony fish that has fleshy fins supported by bones. **421**

locomotion Self-propelled movement from place to place. **599**

locus (LOW-cuss) A gene's location on a chromosome. **200**

logistic growth Density-dependent limiting factors cause population growth to slow as population size increases. **777**

loop of Henle U-shaped portion of a kidney tubule; it extends deep into the renal medulla. **703**

loose connective tissue Animal connective tissue in which cells and fibers of collagen and elastin are scattered through a gel-like matrix. **523**

lumen (LOO-men) The space within a tubular or hollow organ. **683**

lung Respiratory organ in which blood exchanges gases with air. **400, 665**

luteinizing hormone (LH) (LOO-tin-aye-zing) Anterior pituitary hormone with roles in ovulation, corpus luteum formation, and sperm production. **585, 722**

lymph Fluid in the lymph vascular system. **630**

lymph node Small mass of lymphatic tissue that filters lymph; contains many lymphocytes (B and T cells). **630**

lymph vascular system System of vessels that takes up interstitial fluid and carries it to the blood as lymph. **630**

lysogenic pathway (lice-oh-JEN-ik) Bacteriophage replication mechanism in which viral DNA becomes integrated into the host's chromosome and is passed to the host's descendants. **321**

lysosome (LICE-uh-sohm) Enzyme-filled vesicle that breaks down cellular wastes and debris. **62**

lysozyme (LICE-uh-zime) Antibacterial enzyme in body secretions such as saliva and mucus. **639**

lytic pathway (LIH-tik) Bacteriophage replication mechanism in which a virus replicates in its host and kills it quickly. **321**

macroevolution Evolutionary change in taxa above the species level; large-scale evolutionary patterns. **284**

macrophage (MACK-roe-faje) Leukocyte that specializes in phagocytosis. **636**

Malpighian tubules (mal–PIG–ee–an) Water-conserving excretory organs of insects; deliver waste solutes to the digestive tract. **407**, **700**

mammal Animal with hair or fur; females feed young with milk secreted by mammary glands. **429**

mantle In mollusks, extension of the body wall. **400**

mark–recapture sampling Method of estimating population size of mobile animals by marking individuals, releasing them, then checking the proportion of marks among individuals later recaptured. **773**

marsupial mammal Mammal in which young are born at an early stage and complete development in a pouch on the mother's surface. **431**

mass extinction Event in which a large proportion of Earth's organisms become extinct in a relatively short period of geologic time. **852**

mass number Total number of protons and neutrons in the atomic nucleus. **24**

mast cell Granular leukocyte that stays anchored in tissues. **636**

master regulator Gene whose expression ultimately changes cells in a lineage from one type to other, more differentiated types. **165**

maternal effect gene Gene that is expressed during egg formation; its product later influences development of the zygote. **741**

mechanoreceptor Sensory receptor that responds to pressure, position, or acceleration. **561**

medulla oblongata (mih-DOO-luh ob-long-GAH-tuh) Hindbrain region that controls breathing rhythm and reflexes such as coughing and vomiting. **551**

medusa (meh-DUE-suh) Plural, medusae. Dome-shaped cnidarian body; fringed by tentacles. **395**

megaspore Haploid spore formed in ovule of seed plants; gives rise to an egg-producing gametophyte. **365**, **486**

meiosis (my-OH-sis) Nuclear division process required for sexual reproduction. Halves the chromosome number for forthcoming gametes. **188**

melatonin (mell-uh-TOE-nin) Pineal gland hormone secreted during darkness; regulates sleep–wake cycles and seasonal changes. **587**

membrane potential Voltage difference across a cell membrane. **539**

memory cell Long-lived, antigen-sensitized lymphocyte that can carry out a secondary immune response. **646**

meninges (meh-NIN-jeez) Three layers of connective tissue that enclose the brain and spinal cord. **549**

menopause Permanent cessation of menstrual cycles. **722**

menstrual cycle (MEN-struhl) Reproductive cycle in which the uterus lining thickens and then, if pregnancy does not occur, is shed. **722**

menstruation (MEN-strew-ay-shun) Flow of shed uterine tissue out of the vagina. **722**

meristem (MARE-ih-stem) Zone of undifferentiated parenchyma cells, the divisions of which give rise to new tissues during plant growth. **459**

mesoderm (MEZ-oh-derm) Middle tissue layer (germ layer) in a three-layered animal gastrula. **737**

mesophyll (MEZ-uh-fill) Photosynthetic parenchyma of plants. **452**

metabolic pathway A series of enzyme-mediated reactions by which cells build, remodel, or break down an organic molecule. **85**

metabolism All of the enzyme-mediated reactions in a cell. **40**

metamonads Anaerobic excavate protists, including diplomonads and parabasalids; have multiple flagella and no mitochondria. **341**

metamorphosis Remodeling of body form during the transition from larva to adult. **405**

metaphase (MEH-tuh-faze) Stage of mitosis at which all chromosomes are aligned midway between spindle poles. **179**

metastasis (meh-TASS-tuh-sis) The process in which malignant cells spread from one part of the body to another. *See* cancer. **182**

methanogen Organism that produces methane gas as a metabolic by-product. **332**

MHC markers Cell surface self-proteins of vertebrates. **645**

microevolution Change in allele frequency. **270**

microfilament Reinforcing cytoskeletal element involved in cell movement; fiber of actin subunits. **67**

microspore Haploid spore formed in pollen sacs of seed plants; gives rise to a sperm-producing gametophyte (a pollen grain). **365**, **486**

microsporidia Single-celled, spore-forming fungi that are intracellular animal parasites. **379**

microtubule (my-crow-TUBE-yule) Cytoskeletal element that participates in cellular movement. Hollow filament of tubulin subunits. **67**

microvillus Plural, microvilli. Nonmotile projection from the plasma membrane of an epithelial cell; increases the cell's surface area. **521**, **687**

middle ear Of the mammalian ear, region that contains the eardrum and the tiny bones that transfer sound to the inner ear. **567**

migration An animal ceases an ongoing pattern of daily activity to move in a persistent manner toward a new habitat. **761**

mimicry (MIM-ik-cree) A species evolves traits that make it similar in appearance to another species. **795**

mineral With regard to nutrition, an element required in small amounts for normal metabolism. **693**

mitochondrion (my-tuh-CON-dree-un) Double-membraned organelle that produces ATP by aerobic respiration in eukaryotes. **65**

mitosis (my-TOE-sis) Mechanism of nuclear division that maintains the chromosome number. **177**

model An analogous system used for testing hypotheses. **13**

molecular clock Method of comparing the number of neutral mutations in the genomes of two lineages to estimate how long ago they diverged. **297**

molecule (MAUL-ick-yule) Two or more atoms bonded together. **4**

mollusk Invertebrate with a reduced coelom and a mantle, includes chitons, bivalves, gastropods, and cephalopods. **400**

molting Shedding and replacement of an animal body part. **402**

monocots Highly diverse angiosperm lineage; includes plants such as grasses that have one seed leaf and parallel veins. **371**

monohybrid cross Cross between two individuals identically heterozygous for alleles of one gene; for example $Aa \times Aa$. **203**

monomer Molecule that is a subunit of a polymer. **40**

monophyletic group (ma-no-fill-EH-tick) A group including an ancestor in which a derived character evolved, together with all of its descendants. **293**

monosaccharide (mon-oh-SACK-uh-ride) Simple sugar; carbohydrate that consists of one sugar unit. **42**

monotreme (MON-oh-treem) Egg-laying mammal; for example, an echidna or platypus. **429**

monsoon (mon-SOON) Wind that reverses direction seasonally. **831**

morphogen (MOR-foe-jen) Chemical signal that diffuses from its source and affects development in a concentration-dependent manner. **741**

motor neuron Neuron that controls a muscle or gland. **539**

motor protein Type of energy-using protein that interacts with cytoskeletal elements to move a cell's parts or the whole cell. **67**

motor unit One motor neuron and the muscle fibers it controls. **609**

mRNA (messenger RNA) RNA that carries a protein-building message. **151**

multicellular organism Organism composed of interdependent cells that vary in their structure and function. **338**

multiple allele system Gene for which three or more alleles persist in a population at relatively high frequency. **206**

muscle tension Force exerted by a contracting muscle. **609**

muscle tissue Tissue that consists mainly of contractile cells. **524**

mutation Permanent change in the DNA sequence of a chromosome. **140**

mutualism Species interaction that benefits both participants. **383, 791**

mycelium (my-SEAL-ee-um) Mass of threadlike filaments (hyphae) that make up the body of a multicelled fungus. **377**

mycorrhiza (my-kuh-RYE-zah) Mutually beneficial partnership between a fungus and a plant root. **381**

myelin sheath (MY-uh-lin) Of a vertebrate axon, a discontinuous covering composed of neuroglial cells; acts as insulation and thus speeds conduction along the axon. **541**

myofibrils (my-oh-FIE-bril) Of a muscle fiber, threadlike protein structures consisting of contractile units (sarcomeres) arranged end to end. **607**

myoglobin (MY-oh-glow-bin) Muscle protein that reversibly binds oxygen. **610**

myosin (MY-oh-sin) ATP-dependent motor protein; makes up the thick filaments in a sarcomere. **607**

myriapods (MEER-ee-uh-pods) Land-dwelling arthropod group with two antennae and an elongated body with many segments; the millipedes and centipedes. **405**

natural killer cell *See* NK cell.

natural selection A major mechanism of evolution: the differential survival and reproduction of individuals of a population based on differences in shared, heritable traits. Outcome of environmental pressures. **253**

nectar Sweet fluid exuded by some flowers; rewards pollinators. **484**

negative feedback A change causes a response that reverses the change; important mechanism of homeostasis. **530**

neoplasm (KNEE-oh-plaz-um) An accumulation of abnormally dividing cells. **182**

nephridia (nef–RID–ee–ah) Singular, nephridium. Excretory organs of earthworms. **700**

nephron Functional unit of the kidney; filters blood and forms urine. **703**

nerve Neuron fibers bundled inside a sheath of connective tissue. **537**

nerve cord Chain of ganglia that extends the length of a body. **397, 537**

nerve net Mesh of interacting neurons with no central control organ. **395, 537**

nervous tissue Animal tissue composed of neurons and supporting cells; detects stimuli and controls responses to them. **525**

neuromodulator Chemical signal that is released by one neuron and affects multiple other neurons. **544**

neuromuscular junction Synapse between a motor neuron and the muscle it controls. **543**

neuron (NUHR-on) One of the cells that make up communication lines of a nervous system; transmits electrical signals along its plasma membrane and communicates with other cells through chemical messages. **525**

neurosecretory cell Specialized neuron that secretes a hormone into the blood in response to an action potential. **585**

neurotransmitter Chemical signal released by axon terminals of a neuron; it binds to receptors on a postsynaptic cell. **543**

neutral mutation A mutation that has no effect on survival or reproduction. **268**

neutron (NOO-tron) Uncharged subatomic particle that occurs in the atomic nucleus. **24**

neutrophil (NEW-tra-fill) Granular, circulating leukocyte; most abundant leukocyte in blood. **636**

nitrification (nigh-triff-ih-CAY-shun) Conversion of ammonium to nitrates. **821**

nitrogen cycle Movement of nitrogen among the atmosphere, soil, and water, and into and out of food webs. **821**

nitrogen fixation Incorporation of nitrogen from nitrogen gas into ammonia (NH_3). **326, 821**

NK cell Natural killer cell. Leukocyte that can kill cancer cells undetectable by cytotoxic T cells. **636**

node A region of stem where leaves attach and new shoots form. **452**

nondisjunction Failure of chromosomes to separate properly during mitosis or meiosis. **225**

nonshivering heat production Increase in metabolic heat production that results when brown adipose tissue releases energy as heat, rather than storing it in ATP. **710**

notochord (NO-toe-cord) Stiff rod of connective tissue that runs the length of the body in chordate larvae or embryos. **417**

nuclear envelope Double membrane that constitutes the outer boundary of the nucleus. Pores in the membrane control which molecules can cross it. **61**

nucleic acid (new-CLAY-ick) Polymer of nucleotides; DNA or RNA. **49**

nucleoid (NEW-klee-oyd) Of a bacterium or archaeon, region of cytoplasm where the DNA is concentrated. **58, 324**

nucleolus (new-KLEE-oh-luss) In a cell nucleus, a dense, irregularly shaped region where ribosome subunits are assembled. **61**

nucleoplasm (NEW-klee-oh-plaz-um) Viscous fluid inside the nucleus. **61**

nucleosome (NEW-klee-uh-sohm) Unit of chromosomal organization: a length of DNA wound twice around histone proteins. **136**

nucleotide (NEW-klee-uh-tide) Small molecule with a five-carbon sugar, a nitrogen-containing base, and phosphate groups. Monomer of nucleic acids. **49**

nucleus (NEW-klee-us) Plural, nuclei. Of an atom, core area occupied by protons and neutrons. Of a eukaryotic cell, double-membraned organelle that holds the DNA. **24, 55**

nutrient (NEW-tree-unt) A substance that an organism acquires from the environment to support growth and survival. **7**

obesity Overabundance of body fat. Increases health risks. **694**

observational learning One animal acquires a new behavior by observing and imitating behavior of another. **759**

oncogene (ON-ko-jean) Gene with a mutation that results in the overabundance or overactivity of a molecule that stimulates cell division. **182**

oocyte (oh-oh-SITE) Immature animal egg. **719**

oogenesis (oh-oh-JEN-eh-sis) Egg production in an animal. **719**

open circulatory system Circulatory system in which fluid (hemolymph) leaves vessels and flows through spaces in body tissues before returning to a heart. **400, 617**

operant conditioning Learning in which an animal's voluntary behavior is modified by the consequences of that behavior. **759**

operator Part of an operon; a DNA binding site for a repressor. **169**

operon (OPP-er-on) Group of genes together with a promoter–operator DNA sequence that controls their transcription. **169**

opisthokonts (oh-PIS-thuh-konts) Members of the eukaryotic super group Opisthokonta: fungi, animals, and the protistan relatives of these groups. **351**

organ In multicelled organisms, a structure that consists of tissues engaged in a collective task. **4**

organ of Corti Portion of the cochlea that contains the sound-detecting mechanoreceptors (sensory hair cells). **567**

organ of equilibrium Sensory organ that responds to body position and motion. **569**

organ system In multicelled organisms, a set of interacting organs that carry out a particular body function. **4**

organelle (or-guh-NEL) Structure that carries out a specialized metabolic function inside a cell. **55**

organic Describes a compound that consists mainly of carbon and hydrogen. **38**

organism (ORG-uh-niz-um) An individual that consists of one or more cells. **4**

osmosis (oz-MOE-sis) Diffusion of water across a selectively permeable membrane; occurs in response to a difference in solute concentration. **90**

osmotic pressure (oz-MAH-tick) Amount of turgor that prevents osmosis into a hypertonic fluid. **90**

osteoblast Bone-forming cell; it secretes a bone's matrix. **603**

osteoclast Bone-digesting cell; it breaks down bone matrix. **603**

osteocyte A mature bone cell; a former osteoblast. **603**

outer ear Of the mammalian ear, external pinna and the air-filled auditory canal. **567**

ovarian follicle Immature animal egg and the surrounding cells. **721**

ovary In flowering plants, the enlarged base of a carpel, inside which one or more ovules form and eggs are fertilized. In animals, an egg-producing reproductive organ. **369, 482, 719**

oviduct Duct that conveys eggs away from an animal ovary. **721**

ovulation Release of a secondary oocyte from an ovary. **721**

ovule (AH-vule) Of seed plants, the part of the carpel in which the female gametophyte forms and develops. Matures into a seed after fertilization. **482, 365**

ovum (OH-vum) Plural, ova. Mature animal egg. **719**

oxyhemoglobin Hemoglobin with oxygen bound to it. **671**

oxytocin (ox-ee-TOE-sin) Posterior pituitary hormone that acts on smooth muscle of the reproductive tract and contractile cells of milk ducts. **585, 749**

ozone layer (OH-zone) Upper atmospheric region rich in ozone (O_3) that screens out incoming ultraviolet (UV) radiation. **310**

pain Perception of tissue injury. **563**

pain receptor Sensory receptor that responds to tissue damage. **561**

PAMPs Pathogen-associated molecular patterns. **636**

pancreas Organ that secretes digestive enzymes into the small intestine and hormones (insulin and glycogen) into the blood. **588**

pandemic Disease outbreak with cases worldwide. **323**

Pangea (pan-JEE-uh) Supercontinent that began to form about 300 million years ago; broke up 100 million years later. **259**

parapatric speciation Speciation pattern in which two populations become different species while in contact along a common border. **284**

parasitism Relationship in which one species withdraws nutrients from another species, without immediately killing it. **797**

parasitoid An insect that lays eggs in another insect, and whose young devour their host from the inside. **797**

parasympathetic neurons Neurons of the autonomic system that encourage "housekeeping" tasks in the body. **546**

parathyroid glands Four small endocrine glands whose hormone product increases the level of calcium in blood. **587**

parathyroid hormone (PTH) Hormone that regulates the concentration of calcium ions in the blood. **587**

parenchyma (puh-REN-kuh-muh) In plants, simple tissue composed of thin-walled living cells; has various specialized functions depending on location. **452**

passive transport Membrane-crossing mechanism that requires no energy input. E.g., osmosis, facilitated diffusion. **92**

pathogen Cellular organism or virus that causes disease in its host. **317**

PCR *See* polymerase chain reaction.

peat bog High-latitude community dominated by *Sphagnum* moss. **361**

pedigree Chart that marks the appearance of a phenotype through generations of a family tree. **217**

pelagic province The ocean's waters. **846**

pellicle Outer layer of plasma membrane and elastic proteins that protects and gives shape to many unwalled, single-celled protists. **338**

penis Male organ of intercourse. **725**

peptide bond A bond between the amine group of one amino acid and the carboxyl group of another. Joins amino acids in proteins. **46**

peptide Short chain of amino acids linked by peptide bonds. **46**

per capita growth rate For some interval, the added number of individuals divided by the initial population size. **775**

pericycle (PAIR-ih-sigh-cull) Thin layer of undifferentiated parenchyma cells. In roots, forms the outer layer of the vascular cylinder and can give rise to lateral roots. **459**

periderm Dermal tissue that replaces epidermis during secondary growth of plants. **460**

periodic table Tabular arrangement of elements by their atomic number. **24**

peripheral nervous system Nerves that extend through the body and carry signals to and from the central nervous system. **537**

peristalsis (pehr-ih-STAL-sis) Wavelike smooth muscle contractions that propel food through the digestive tract. **683**

peritubular capillaries Network of capillaries that surrounds and exchanges substances with a kidney tubule. **703**

permafrost Continually frozen soil layer that lies beneath arctic tundra and prevents water from draining. **841**

peroxisome (purr-OX-uh-sohm) Enzyme-filled vesicle that breaks down molecules and has other specialized functions in animal organs. **62**

petal Unit of a flower's corolla; often showy and conspicuous. **482**

pH Measure of the number of hydrogen ions in a fluid. **33**

phagocytosis (fag-oh-sigh-TOE-sis) "Cell eating"; type of receptor-mediated endocytosis in which a cell engulfs a large solid particle such as another cell. **94**

pharynx (FAHR-inks) Tube connecting the mouth to the digestive tract. **397, 667**

phenotype (FEEN-oh-type) An individual's observable traits. **200**

pheromone (FER-uh-moan) Chemical that serves as a communication signal among members of an animal species. **565, 761**

phloem (FLOW-um) Complex vascular tissue of plants; distributes photosynthetically produced sugars and other organic solutes through the plant body. **359, 452**

phospholipid (foss-foe-LIP-id) A lipid with two (hydrophobic) fatty acid tails and a (hydrophilic) phosphate group in its head. **44**

phosphorus cycle Movement of phosphorus among Earth's rocks and waters and into and out of food webs. **822**

phosphorylation Addition of a phosphate group to an organic molecule. **86**

photoautotroph (foe-toe-AWE-toe-trof) Organism that obtains carbon from inorganic sources such as carbon dioxide and energy from light. **326**

photoheterotroph (foe-toe-HET-ur-o-trof) Organism that obtains carbon from organic compounds and energy from light. **326**

photolysis (foe-TALL-ih-sis) Process by which light energy breaks down a molecule. **107**

photoperiodism (foe-toe-PEER-ee-ud-is-um) Biological response to seasonal changes in the relative lengths of day and night. **510**

photoreceptor Sensory receptor that responds to light. **561**

photorespiration Pathway of light-independent reactions that occurs when rubisco attaches oxygen instead of carbon dioxide to ribulose bisphosphate. **109**

photosynthesis (foe-toe-SIN-thuh-sis) Process by which most autotrophs use light energy to make sugars from carbon dioxide and water. **7, 101**

photosystem Large protein complex in the thylakoid membrane; converts light energy to chemical energy in photosynthesis. **104**

phototropism (foe-toe-TRO-pizm) Directional response to light. **508**

phylogeny (fie-LA-juh-knee) Evolutionary history of a species or group of species. **293**

phytochrome (FIGHT-uh-krome) A light-sensitive pigment that helps set plant circadian rhythms based on length of night. **508**

phytoplankton (FIGHT-o-plank-tun) Photosynthetic members of the plankton; tiny, aquatic autotrophs. **341**

pigment An organic molecule that can absorb light of certain wavelengths. Reflected light imparts a characteristic color. **103**

pilus (PIE-luss) Plural, pili. Protein filament that projects from the surface of some prokaryotic cells. **58, 324**

pineal gland (pie-KNEE-ul) Endocrine gland in the brain that secretes melatonin under low-light or dark conditions. **587**

pinocytosis (pin-oh-sigh-TOE-sis) Endocytic pathway by which fluid and materials in bulk are brought into the cell. **94**

pioneer species Species that can colonize a new habitat. **799**

pituitary gland (pih-TOO-eh-TAIR-ee) Pea-sized endocrine gland in the forebrain; interacts closely with the adjacent hypothalamus. **585**

placenta (pluh-SEN-tuh) Of placental mammals, organ that forms during pregnancy and allows diffusion of substances between the maternal and embryonic bloodstreams. **431, 716**

placental mammal Mammal in which a mother and her developing embryo exchange materials by means of an organ called the placenta. **431**

plankton Collection of tiny organisms that drift and swim in lakes and seas. **341**

plant A multicelled, typically photosynthetic eukaryotic producer that forms and nurtures embryos on the parental body. **8, 357**

plant hormone Extracellular signaling molecule of plants; exerts an effect on target cells at very low concentration. **501**

plasma Fluid portion of blood. **623**

plasma membrane (PLAZ-muh) Membrane that encloses a cell and separates it from the external environment. **55**

plasmid (PLAZ-mid) In some prokaryotes, small extrachomosomal circle of DNA that carries a few genes. **58, 324**

plasmodesmata (plaz-mow-dez-MA-tah) Singular, plasmodesma. Cell junction that forms an open channel between the cytoplasm of adjacent plant cells. **70**

plasmodial slime mold (plaz-MOH-dee-ul) Soil-dwelling protist that feeds as a multinucleated mass and develops into a fruiting body under adverse conditions. **351**

plastid One of several types of double-membraned organelles that functions in photosynthesis, pigmentation, or storage in plants and algal cells; for example, a chloroplast or amyloplast. **65**

plate tectonics theory Theory that Earth's outermost layer of rock is cracked into plates, the slow movement of which conveys continents to new locations over geologic time. **259**

platelet Cell fragment that helps blood clot. **623**

pleiotropic (ply-uh-TROH-pick) Refers to a gene that affects multiple traits. **206**

plot sampling Method of estimating population size of organisms that do not move much by making counts in small plots and extrapolating from this to the number in the larger area. **773**

polar body Tiny cell produced by unequal cytoplasmic division during egg production in animals. **719**

polarity Separation of charge into positive and negative regions. **29**

pollen grain Walled, immature male gametophyte of a seed plant. **359, 482**

pollen sac Of seed plants, sporangium where microspores form and develop into pollen grains. **365**

pollination The arrival of pollen on a stigma or other pollen-receiving reproductive part of a seed plant. **365, 482**

pollination vector Animal or environmental agent that moves pollen grains from one plant to another. **482**

pollinator An animal pollination vector. **369, 482**

polygenic inheritance (pall-ee-JENN-ick) Pattern of inheritance in which the form of a single trait is collectively determined by alleles of several genes. **206**

polymer Molecule that consists of repeated monomers. **40**

polymerase *See* DNA polymerase, RNA polymerase.

polymerase chain reaction (**PCR**) Laboratory method that rapidly generates many copies of a specific section of DNA. **235**

polyp (PAH-lup) A pillarlike cnidarian body topped with tentacles. **395**

polypeptide Long chain of amino acids linked by peptide bonds. **46**

polyploidy (PALL-ee-ploy-dee) Condition of having three or more of each type of chromosome characteristic of the species. **225**

polysaccharide (paul-ee-SACK-uh-ride) Carbohydrate that consists of hundreds or thousands of monosaccharide monomers. **42**

pons (ponz) Hindbrain region that influences breathing and serves as a functional bridge to the adjacent midbrain. **551**

population A group of interbreeding individuals of the same species who live in a defined area and breed with one another more often than they breed with members of other populations. **4, 268, 771**

population density Number of individuals per unit area. **773**

population distribution Description of how individuals are dispersed in an area. **773**

population size Total number of individuals in a population. **773**

positive feedback A response intensifies the conditions that caused its occurrence. **541**

potential energy Energy stored in the position or arrangement of a system's components. **78**

Precambrian era Interval from Earth's formation to 541 million years ago. **304**

predation One species captures, kills, and eats another. **795**

prediction A statement, based on a hypothesis, about a condition that should occur if the hypothesis is correct. **13**

pressure flow theory Explanation of how a difference in turgor between source and sink regions pushes sugar-rich fluid through a sieve tube. **476**

primary endosymbiosis Evolution of an organelle from bacteria that entered a host cell and lived inside it. **338**

primary growth Of plants, lengthening of young shoots and roots; originates at apical meristems. **459**

primary motor cortex Region of brain's frontal lobes that governs voluntary movements. **553**

primary oocyte Oocyte that has halted in prophase I of meiosis. **721**

primary producer In an ecosystem, an organism that captures energy from an inorganic source and stores it as biomass; first trophic level. **809**

primary production Rate at which an ecosystem's producers capture and store energy. **811**

primary somatosensory cortex Region of the brain's parietal lobes that receives sensory information from skin and joints. **553**

primary succession A new community colonizes an area where there is no soil. **799**

primary wall The first cell wall of young plant cells. **69**

primates Mammalian order that includes lemurs, tarsiers, monkeys, apes, and humans. **437**

primer Short, single strand of DNA or RNA that base-pairs with a specific DNA sequence and serves as an attachment point for DNA polymerase. **138**

prion (PREE-on) Infectious protein. **49**

probability The chance that a particular outcome of an event will occur; depends on the total number of outcomes possible. **17**

probe Short fragment of DNA labeled with a tracer; designed to hybridize with a base sequence of interest. **235**

producer Autotroph. An organism that makes its own food using energy and nonbiological raw materials from the environment. **7**

product A molecule that is produced by a reaction. **81**

progesterone Hormone secreted by ovaries; prepares the reproductive tract for pregnancy. **721**

prokaryote (pro-CARE-ee-oat) Informal term for a single-celled organism without a nucleus; a member of the domain Bacteria or Archaea. **8, 324**

prolactin (pro-LAC-tin) Anterior pituitary hormone that stimulates milk production by mammary glands. **585, 750**

promoter Short sequence of bases that functions as a binding site in DNA for molecules that carry out transcription. **149**

prophase (PRO-faze) Stage of mitosis during which chromosomes condense and become attached to a newly forming spindle. **179**

protein (PRO-teen) Organic molecule that consists of one or more polypeptides folded into a specific shape. **46**

proteobacteria (PRO-tee-o-bak-TEAR-ee-ah) Most diverse bacterial lineage; includes species that carry out photosynthesis and that fix nitrogen, as well as some human pathogens. **329**

Proterozoic eon (pro-ter-oh-ZOE-ik) Final eon of the Precambrian era; 2.5 billion to 541 million years ago. **310**

protist Common term for a eukaryote that is not a plant, animal, or fungus. **8, 337**

proto-oncogene Gene that, by mutation, can become an oncogene. **182**

protocell Membranous sac that contains interacting organic molecules; hypothesized to have formed prior to the earliest life-forms. **307**

proton (PRO-tawn) Positively charged subatomic particle that occurs in the nucleus of all atoms. **24**

protonephridia (PRO–toh–nef–RID–ee–ah) Singular, protonephridium. Excretory organs of planarian flatworms. **700**

protostomes (PRO-toe-stomes) Lineage of bilateral animals in which the first opening on the embryo surface develops into the mouth. **391**

proximal tubule Portion of kidney tubule that receives filtrate from Bowman's capsule and delivers it to the loop of Henle. **703**

pseudocoelom (SUE-doe-see-lum) Animal body cavity that is only partially lined with tissue derived from mesoderm. **391**

pseudopod (SUE-doh-pod) A temporary protrusion that helps some eukaryotic cells move and engulf prey. **67**

pseudoscience A claim, argument, or method that is presented as if it were scientific, but is not. **18**

puberty Developmental stage during which human reproductive organs mature and begin to function. **592**

pulmonary artery Vessel that carries blood from the heart to a lung. **619**

pulmonary circuit (PUL-mon-erry) Circuit through which blood flows from the heart to the lungs and back. **617**

pulmonary vein Vessel that carries blood from a lung to the heart. **619**

pulse Brief stretching of artery walls that occurs when ventricles contract. **625**

Punnett square (PUN-it) Diagram used to predict the genotypic and phenotypic outcomes of a cross. **203**

pupil Adjustable opening that allows light into a camera eye. **573**

pyruvate (pie-ROO-vait) Three-carbon product of glycolysis. **119**

***r*-selection** Selection that favors traits that allow their bearers to produce the most offspring the most quickly. **779**

radial symmetry Having parts arranged around a central axis, like spokes around a wheel. **391**

radioactive decay Process by which atoms of a radioisotope emit energy and/or subatomic particles when their nucleus spontaneously breaks up. **24**

radioisotope An isotope with an unstable nucleus. **24**

radiolaria (ray-dee-oh-LAIR-ee-ah) Heterotrophic single-celled protists with a porous silica shell and long cytoplasmic extensions. **347**

radiometric dating Method of estimating the age of a rock or a fossil by measuring the content and proportions of a radioisotope and its daughter elements. **262**

radula In many mollusks, a tonguelike organ hardened with chitin. **400**

rain shadow Dry region downwind of a coastal mountain range. **831**

ray-finned fish Bony fish that has fin supports derived from skin. **421**

reactant (ree-ACK-tunt) A molecule that enters a reaction and is changed by participating in it. **81**

reaction Process of molecular change. **40**

receptor protein Membrane protein that triggers a change in cell activity in response to a specific stimulus such as binding a hormone. **88**

recessive Refers to an allele with an effect that is masked by a dominant allele on the homologous chromosome. **200**

recombinant DNA (ree-COM-bih-nent) A DNA molecule that contains genetic material from more than one organism. **232**

rectum Terminal region of the large intestine; stores digestive waste. **683**

red algae Photosynthetic protists; typically multicelled, with chloroplasts containing red accessory pigments (phycobilins). **349**

red blood cell Hemoglobin-filled blood cell that transports oxygen; an erythrocyte. **623**

red marrow Bone marrow that makes blood cells. **603**

redox reaction (REE-docks) Oxidation–reduction reaction; an electron transfer. The electron donor is oxidized, and the electron acceptor is reduced. **85**

reflex An automatic response to a stimulus, a movement or other action that does not require conscious thought. **546**

releasing hormone Hormone that stimulates release of a different hormone by its target endocrine cells. **585**

replacement fertility rate Fertility rate at which each woman has, on average, one daughter who survives to reproductive age. **783**

repressor Transcription factor that slows or stops transcription. **163**

reproduction (ree-pruh-DUCK-shun) Processes by which organisms produce offspring. **7**

reproductive base Of a population, all individuals who are of reproductive age or younger. **775**

reproductive cloning Laboratory procedure that produces animal clones from a single cell. **142**

reproductive isolation The end of gene flow between populations. **280**

resource partitioning Species adapt to access different portions of a limited resource; allows species with similar needs to coexist. **793**

respiration Physiological process by which an animal body supplies cells with oxygen and disposes of their waste carbon dioxide. **663**

respiratory cycle One inhalation and one exhalation. **669**

respiratory membrane Membrane consisting of alveolar epithelium, capillary endothelium, and their fused basement membranes; site of gas exchange in lungs. **671**

respiratory protein Protein that reversibly binds oxygen when the oxygen concentration is high and releases it when oxygen concentration is low. **663**

respiratory surface Moist surface across which gases are exchanged between animal cells and the external environment. **663**

resting potential Membrane potential of a neuron at rest. **539**

restriction enzyme Type of enzyme that cuts DNA at a specific base sequence. **232**

retina In an eye, a layer densely packed with photoreceptors. **571**

retrovirus Virus having an RNA genome that is used as a template to produce double-stranded viral DNA within a host cell. **321**

reverse transcriptase (trance-CRYPT-ace) An enzyme that uses mRNA as a template to make a strand of cDNA. **232**

rhizarians (riz-ahr-EE-ans) Members of the eukaryote supergroup Rhizaria; single-celled protists with cytoplasmic extensions protruding through a secreted sievelike shell; e.g., foraminifera and radiolaria. **347**

rhizoid Threadlike structure that anchors a bryophyte. **361**

rhizome Horizontal stem that grows along or under the ground. **363**

ribonucleic acid *See* RNA.

ribosome (RYE-buh-sohm) Organelle of protein synthesis. An intact ribosome consists of two subunits, each composed of rRNA and proteins. **58**

ribozyme (RYE-boh-zime) RNA that functions as an enzyme. **307**

RNA interference Mechanism in which translation of a particular mRNA is suppressed by other RNA molecules. **163**

RNA polymerase Enzyme that carries out transcription. **151**

RNA Ribonucleic acid. (rye-bo-new-CLAY-ick) Nucleic acid that carries out protein synthesis, among other roles. Most are single-stranded. **49**

RNA world hypothesis Hypothesis that RNA served as the genetic information of early life. **307**

rod cell Photoreceptor that is active in dim light; provides coarse perception of image and detects motion. **574**

root hairs Hairlike, absorptive extensions of a root epidermis cell; many of them form on young roots. **459**

root nodules Of some plant roots, swellings that contain and support beneficial nitrogen-fixing bacteria. **471**

roundworm Nematode. An unsegmented pseudocoelomate worm with a cuticle that is molted periodically as the animal grows. **402**

rRNA (ribosomal RNA) RNA component of ribosomes. **149**

rubisco Ribulose bisphosphate carboxylase. Carbon-fixing enzyme of the Calvin–Benson cycle. **109**

runoff Water that flows over soil into streams. **815**

sac fungi Most diverse group of fungi; hyphae are septate and sexual reproduction produces spores inside a saclike cell (an ascus). **381**

salivary gland Exocrine gland that secretes saliva into the mouth. **685**

salt Compound that releases ions other than H^+ and OH^- when it dissolves in water. **31**

sampling error A difference between results derived from testing an entire group, and results derived from testing a subset of the group. **17**

SAR supergroup Major lineage of eukaryotes; includes three protist lineages, the stramenopiles, alveolates, and rhizarians. **343**

sarcomere Contractile unit of muscle. **607**

sarcoplasmic reticulum Specialized endoplasmic reticulum in muscle cells; stores and releases calcium ions. **609**

saturated fat Triglyceride with three saturated fatty acid tails. **44**

savanna (suh-VAN-uh) Tropical biome dominated by grasses with a few scattered shrubs and trees. **837**

scales In some vertebrates, flattened structures that grow from and sometimes cover the skin. **419**

science Systematic study of the physical universe. **11**

scientific method Making, testing, and evaluating hypotheses about the natural world. **13**

scientific theory A hypothesis that has not been falsified after many years of rigorous testing, and is useful for making predictions about a wide range of phenomena. **17**

sclera Fibrous white layer that covers most of the eyeball. **573**

sclerenchyma (sklair-EN-kuh-muh) In plants, simple tissue composed of cells that die when mature; their tough walls structurally support plant parts. Includes fibers, sclereids. **452**

SCNT *See* somatic cell nuclear transfer.

scrotum Pouch of skin that encloses a human male's testes. **725**

seamount An undersea mountain. **846**

second law of thermodynamics Energy tends to disperse spontaneously. **78**

second messenger Molecule that forms inside a cell when a hormone binds to a receptor in the plasma membrane; its formation sets in motion reactions that alter activity inside the cell. **583**

secondary endosymbiosis Evolution of a chloroplast from a protist that itself contains chloroplasts that arose by primary endosymbiosis. **338**

secondary growth Thickening of older stems and roots; originates at lateral meristems. **460**

secondary metabolite Molecule that is produced by an organism but does not play any known role in its metabolism. Some serve as defense against predation. **371**

secondary oocyte Oocyte that has halted in metaphase II of meiosis; released from an ovary at ovulation. **721**

secondary sexual traits Traits, such as the distribution of body fat, that differ between the sexes but do not have a direct role in reproduction. **592**

secondary succession A new community develops in a disturbed site where the soil that supported a previous community remains. **799**

secondary wall Lignin-reinforced wall that forms inside the primary wall of a mature plant cell. **69**

secretin (seh-KREE-tin) Small intestine hormone that encourages release of bicarbonate by the pancreas. **687**

sedimentary cycle Biogeochemical cycle that involves rocks and waters, but not the atmosphere. **815**

seed Mature ovule of an angiosperm. Consists of an embryo sporophyte, nutritive tissue, and a protective seed coat. **359, 489**

seed plant Plant that produces seeds and pollen; an angiosperm or gymnosperm. **357**

selfish herd effect A temporary group forms when individuals hide behind others to minimize their individual risk of predation. **765**

semen Sperm mixed with secretions from seminal vesicles and the prostate gland. **725**

semiconservative replication Describes the process of DNA replication in which one strand of each copy of a DNA molecule is new, and the other is a strand of the original DNA. **138**

seminiferous tubules (sehm-in-IF-er-us) Inside a testis, coiled tubules that contain male germ cells and produce sperm. **725**

senescent (suh-NESS-ent) Refers to a metabolically active but malfunctioning cell stuck in an irreversibly arrested cell cycle. **180**

sensation Awareness of a stimulus. **561**

sensory adaptation Slowing or halting of a sensory response to an ongoing stimulus. **561**

sensory neuron Neuron that is activated when its receptor endings detect a specific stimulus, such as light or pressure. **539**

sensory perception The meaning a brain derives from a sensation. **561**

sensory receptor Cell or cell component that detects a specific type of stimulus. **530**

sepal Unit of a flower's calyx; typically photosynthetic and inconspicuous. **482**

sequencing Method of determining DNA sequence. **236**

sessile animal Animal that lives fixed in place on some surface. **393**

sex chromosome Chromosome involved in determining anatomical sex; member of a chromosome pair that differs between males and females. **138**

sex hormone A steroid hormone that controls sexual development and reproduction; testosterone in males, estrogens or progesterone in females. **592**

sexual dimorphism A trait that differs between males and females of a species. **276**

sexual reproduction Reproductive mode by which offspring arise from two parents and inherit genes from both. **187, 716**

sexual selection Mode of natural selection in which some individuals outreproduce others of a population because they are better at securing mates. **276**

shell model Model of electron distribution in an atom. **27**

shivering response Rhythmic muscle contractions that generate metabolic heat in response to prolonged exposure to cold. **710**

short tandem repeat In chromosomal DNA, sequences of a few nucleotides repeated multiple times in a row. Used in DNA profiling. **210**

sieve elements Living cells that compose sieve tubes of phloem. **476**

sieve tube Sugar-conducting tube of phloem; consists of stacked sieve elements. **475**

sink Region of plant tissue where sugars are being used or put into storage. **476**

sinoatrial (SA) node Cardiac pacemaker; cluster of specialized cells whose spontaneous rhythmic signals trigger contractions. **621**

sister chromatids (KROME-uh-tid) The two attached DNA molecules of a duplicated eukaryotic chromosome, attached at the centromere. **136**

sister groups The two lineages that emerge from a node on a cladogram. **293**

skeletal muscle fiber Multinucleated contractile cell that contains mainly myofibrils and runs the length of a skeletal muscle. **607**

skeletal muscle tissue Muscle tissue that pulls on bones and moves body parts; under voluntary control. **524**

sliding-filament model Explanation of how interactions among actin filaments and myosin filaments bring about muscle contraction. **607**

small intestine Region of the intestine that receives material from the stomach; site of most digestion and absorption. **683**

smooth muscle tissue Muscle tissue that lines blood vessels and forms the wall of hollow organs. **524**

SNP (single-nucleotide polymorphism) One-nucleotide DNA sequence variation carried by a measurable percentage of a population. **231**

social animal Animal that lives in a multigenerational group in which members, who are usually relatives, cooperate in some tasks. **765**

social learning An animal learns to recognize specific individuals of its species. **759**

soil erosion Loss of soil under the force of wind and water. **468**

soil water Water between soil particles. **815**

solute (SAWL-yute) A dissolved substance. **31**

solution Uniform mixture of solute completely dissolved in solvent. **31**

solvent Liquid in which other substances dissolve. **31**

somatic cell nuclear transfer (SCNT) Reproductive cloning method in which the DNA of a donor's body cell is transferred into an enucleated egg. **142**

somatic nerves Of the peripheral nervous system, nerves that control skeletal muscle and relay signals from sensory neurons in joints and skin to the central nervous system. **546**

somatic sensations Sensations such as touch and pain that arise when sensory neurons in skin, muscle, or joints are activated. **563**

somites Paired blocks of embryonic mesoderm that give rise to a vertebrate's muscle and bone. **742**

sorus (SORE-us) Plural, sori. Cluster of spore-producing capsules of a fern. **363**

source Region of plant tissue where sugars are being produced or released from storage. **476**

spatial learning Learning about landmarks allows formation of a mental map of the environment. **759**

speciation (spee-see-A-shun) Emergence of a new species; the splitting of one lineage into two. **280**

species (SPEE-sheez) Unique type of organism designated by genus name and specific epithet. Of sexual reproducers, a group of individuals that produce fertile offspring when they breed, and do not interbreed with other groups. **3**

species evenness Of a community, the relative abundance of species. **790**

species richness Of a community, the number of species. **790**

specific epithet Second part of a species name. **8**

sperm Male gamete. **716**

spermatogenesis (sper-mah-toh-JEN-eh-sis) Sperm production in an animal. **719**

sphincter Ring of muscle that controls passage of material through a tubular organ or a body opening. **605**

spinal cord Portion of central nervous system that connects peripheral nerves with the brain. **549**

spindle (SPIN-dull) Temporary structure that moves chromosomes during nuclear division; consists of microtubules. **179**

spirochetes (spy-row-KEY-teez) Lineage of bacteria shaped like a stretched-out spring. **329**

spleen Fist-sized lymphoid organ located in the upper abdomen; filters blood and enhances immune function. **630**

sponge Aquatic invertebrate that has no body symmetry, tissues, or organs; feeds by filtering food from the water. **393**

sporangium (spor-AN-gee-um) Plural, sporangia. A structure in which haploid spores form by meiosis. **361**

sporophyte (SPOR-o-fite) Diploid, multicelled body that produces spores; forms during the life cycle of all land plants and some algae. **357**

spring overturn In temperate zone lakes, a downward movement of oxygenated surface water and an upward movement of nutrient-rich water in spring. **842**

stabilizing selection Mode of natural selection in which an intermediate form of a trait is adaptive, and extreme forms are not. **275**

stamen (STAY-men) Floral reproductive organ that consists of an anther and, in most species, a filament. **369, 482**

starch Coiled polysaccharide of glucose monomers. Principal form of stored sugars in plants. **42**

stasis (STAY-sis) Macroevolutionary pattern in which a lineage persists with little or no change over evolutionary time. **286**

statistically significant Refers to a result that is statistically unlikely to have occurred by chance alone. **17**

statolith Amyloplast involved in sensing gravity. **507**

stem cell Cell that can divide to produce two stem cells or differentiate into some or all specific cell types. **517**

steroid (STARE-oyd) Lipid with four carbon rings, no fatty acid tails. **44**

steroid hormone Lipid-soluble hormone derived from cholesterol. **583**

stigma Of a flower, upper part of a carpel. Adapted to receive pollen. **482**

stomach Tubular, muscular organ that functions in chemical and mechanical digestion. **681**

stomata (STOH-ma-tuh) Singular, stoma. Closable gap across a plant's cuticle and epidermis; opens for gas exchange, closes to prevent water loss. **109, 359**

stramenopiles (struh-men-O-piles) Protist lineage that includes the photosynthetic diatoms and brown algae, as well as the heterotrophic water molds. **343**

strobilus (STROH-bill-us) Plural, strobili. Of some nonflowering plants such as horsetails and cycads, a cluster of spore-producing structures. **363**

stroma (STROH-muh) Cytosol-like fluid between the thylakoid membrane and the two outer membranes of a chloroplast. **101**

stromatolite (stroh-MAT-oh-lite) Rocky, layered structure that forms over time as a mat of aquatic prokaryotic cells traps sediments. **309**

substrate Molecule that an enzyme acts upon and converts to a product; reactant in an enzyme-catalyzed reaction. **83**

substrate-level phosphorylation ATP formation by the transfer of a phosphate group from a phosphorylated molecule to ADP. **119**

superior vena cava (VEE-nah CAY-vuh) Vein that delivers blood from the upper body to the heart. **619**

surface-to-volume ratio Relationship in which the volume of an object increases with the cube of the diameter, and the surface area increases with the square. Limits cell size. **57**

survivorship curve Graph showing the decline in numbers of a cohort over time. **779**

suspension feeder Animal that filters food from water around it. **393**

swim bladder In ray-finned fishes, a gas-filled sac that allows the fish to regulate its buoyancy. **421**

symbiosis (sim-by-OH-sis) One species lives in or on another in a commensal, mutualistic, or parasitic relationship. **790**

sympathetic neurons Neurons of the autonomic system that prepare the body for danger or excitement. **546**

sympatric speciation Speciation pattern in which genetic changes within a population lead to reproductive isolation. Occurs in the absence of a physical barrier to gene flow. **282**

synapse (SIN-aps) Region where a neuron's axon terminals transmit chemical signals to another cell. **543**

synaptic integration The summation of excitatory and inhibitory signals by a postsynaptic cell. **543**

systemic acquired resistance In plants, inducible whole-body resistance to a wide range of pathogens and abiotic stressors. **512**

systemic circuit (sis-TEM-ick) Circuit through which blood flows from the heart to the body tissues and back. **617**

systole (SIS-toll-ee) Contractile phase of the cardiac cycle. **621**

systolic pressure (sis-STAHL-ic) Blood pressure when ventricles are contracting. **625**

T cell T lymphocyte. Leukocyte with a central role in adaptive immunity. **636**

T cell receptor (TCR) Special antigen receptor that only T cells have; recognizes antigen as nonself, and MHC markers as self. **645**

taproot system In eudicots, an enlarged primary root together with all of the lateral roots that branch from it. **456**

taxis (TAX-iss) Innate response in which an animal moves toward or away from a stimulus. **761**

taxon (TAX-on) Plural, taxa. A rank of organisms that share a unique set of traits; a species, genus, family, order, phylum, kingdom, or domain. **11**

taxonomy (tax-ON-oh-me) The practice of naming and classifying species. **8**

TCR *See* T cell receptor.

telomere (TELL-a-meer) Region of noncoding DNA at the end of a chromosome. Proteins that bind to it protect the chromosome from end damage. **180**

telophase (TELL-uh-faze) Stage of mitosis during which chromosomes arrive at opposite spindle poles, decondense, and become enclosed by a new nuclear envelope. **179**

temperate deciduous forest Northern Hemisphere biome in which the main plants are broadleaf trees that lose their leaves in autumn and become dormant during cold winters. **839**

temperate grassland Temperate biome in the interior of continents; perennial grasses and other nonwoody plants adapted to grazing and fire predominate. **837**

temperature Measure of molecular motion. **31**

tendon Strap of dense connective tissue that connects a skeletal muscle to bone. **605**

teratogen A toxin or infectious agent that interferes with development and thus raises the risk of birth defects. **735**

terminal bud Tip of an actively growing shoot; contains apical meristem. **459**

territory Region that an animal or group of animals defends. **765**

testcross Method of determining the genotype of an individual with a dominant phenotype: a cross between the individual and another individual known to be homozygous recessive. **203**

testis Plural, testes. Sperm-producing reproductive organ. **719**

testosterone Main hormone produced by testes; required for sperm production and development of male secondary sexual traits. **725**

tetrapod (TEH-trah-pod) Vertebrate that has four bony limbs or is a descendant of a four-legged ancestor. **417**

thalamus Forebrain region that relays signals to the cerebral cortex. **551**

thermal radiation Emission of heat from an object. **709**

thermocline Thermal stratification in a large body of water; a cool midlayer prevents vertical mixing between warm surface water above it and cold water below it. **842**

thermoreceptor Temperature-sensitive sensory receptor. **561**

thigmotropism (thig-MA-truh-pizm) In plants, growth in a direction influenced by contact. **508**

threatened species Species likely to become endangered in the near future. **852**

threshold potential Neuron membrane potential at which gated sodium channels open, causing an action potential to occur. **541**

thylakoid membrane (THIGH-la-koyd) Inner membrane system that carries out light-dependent reactions in chloroplasts and cyanobacteria. **101**

thyroid gland Endocrine gland at the base of the neck; produces thyroid hormone, which increases metabolism. **587**

thyroid hormone (TH) Iodine-containing hormones that collectively increase metabolic rate and play a role in development. **587**

thyroid-stimulating hormone (TSH) Anterior pituitary hormone; stimulates secretion of thyroid hormone by the thyroid. **585**

tidal volume Volume of air that flows into and out of the lungs during a normal inhalation and exhalation. **669**

tight junction Cell junction that fastens together the plasma membrane of adjacent animal cells. Collectively prevent fluids from leaking between the cells. **69, 521**

tissue culture propagation Laboratory method in which embryonic plants are grown from individual meristem cells. **494**

tissue In multicelled organisms, specialized cells organized in a pattern that allows them to perform a collective function. **4**

topsoil Uppermost soil horizon; contains the most organic matter and nutrients for plant growth. **468**

total fertility rate Average number of children the women of a population bear over the course of a lifetime. **783**

tracer A substance that can be traced via its detectable component. **24**

trachea (TRAY-key-uh) Airway between the pharynx and bronchi; windpipe. **669**

tracheal system (TRAY-key-ul) Of insects and some other land arthropods, tubes that convey gases between the body surface and internal tissues. **665**

tracheids (TRAY-key-idz) Tapered cells that form xylem tubes. **473**

trait An inherited characteristic of an organism or species. **8**

transcription factor Regulatory protein that binds to DNA and influences transcription; e.g, an activator or repressor. **163**

transcription Process in which a gene is copied (transcribed) into RNA form. **149**

transduction In bacteria and archaea, a mechanism of horizontal gene transfer in which a bacteriophage transfers DNA between cells. **324**

transformation In bacteria and archaea, a type of horizontal gene transfer in which DNA is taken up from the environment. **324**

transgenic (trans-JEN-ick) Refers to a genetically modified organism that carries a gene from a different species. **240**

transition state During a reaction, the moment at which substrate bonds are at the breaking point and the reaction will run spontaneously. **83**

translation Process in which a polypeptide is assembled from amino acids in the order specified by codons in an mRNA. **149**

translocation In genetics, structural rearrangement of a chromosome in which a broken piece has become reattached in the wrong location. In plants, movement of organic compounds through phloem. **222, 476**

transpiration (trans-purr-A-shun) Evaporation of water from plant parts. **473**

transport protein Protein that helps specific ions or molecules to cross a membrane. **88**

transposable element (transposon) Segment of DNA that can move spontaneously within or between chromosomes. **222**

transposon *See* transposable element.

triglyceride (tri-GLISS-a-ride) A molecule with three fatty acids bonded to a glycerol; a fat. **44**

tRNA (transfer RNA) RNA that delivers amino acids to a ribosome during translation. **149**

trophic level (TROE-fick) Position of an organism in a food chain. **809**

tropical rain forest Highly productive and species-rich biome in which year-round rains and warmth support continuous growth of evergreen broadleaf trees. **839**

tropism (TROPE-izm) A directional movement toward or away from a stimulus that triggers it. **507**

trypanosome (trih-PAN-o-SOHM) Parasitic excavate protist with a single mitochondrion and an undulating membrane. **341**

tubular reabsorption Process by which substances move from the filtrate inside a kidney tubule into the peritubular capillaries. **705**

tubular secretion Process by which substances move out of peritubular capillaries and into the filtrate in kidney tubules. **705**

tumor (TOO-murr) A neoplasm that forms a lump. **182**

tunicate (TOON-ih-kit) Marine invertebrate chordate with a fish-shaped, swimming larva that undergoes metamorphosis to become a barrel-shaped, sessile adult. **417**

turgor (TR-gr) Pressure that a fluid exerts against a structure that contains it. **90**

uniformitarianism Idea that gradual repetitive processes occurring over long time spans shaped Earth's surface. **253**

unsaturated fat Triglyceride with one or more unsaturated fatty acid tails. **44**

urea (yuh-REE-uh) Nitrogen-containing compound derived from ammonia; the main nitrogen-containing waste excreted by mammals. **700**

ureter (YOOR-et-er) Tube that carries urine from a kidney to the bladder. **703**

urethra (you-REE-thrah) Tube through which urine expelled from the bladder flows out of the body. **703**

uric acid (YUR-ik) Nitrogen-containing compound derived from ammonia. **700**

urinary bladder Hollow, muscular organ that stores urine. **703**

urine Mixture of water and soluble wastes that is formed and excreted by the vertebrate urinary system. **700**

uterus Muscular chamber where offspring develop; womb. **721**

vaccine (vax-SEEN) A preparation introduced into the body in order to elicit immunity to a specific pathogen. **635**

vacuole (VAK-you-ole) Large, fluid-filled vesicle that isolates or breaks down waste, debris, toxins, or food. **62**

vagina Female organ of intercourse and birth canal. **721**

variable (VAIR-ee-uh-bull) In an experiment, a characteristic or event that differs among individuals or over time. **13**

vas deferens (vas DEF-er-ens) Plural, vasa deferentia. One of a pair of long ducts that convey mature sperm toward the body surface. **725**

vascular bundle Multistranded bundle of xylem, phloem, and sclerenchyma fibers running through a stem or leaf. **452**

vascular cambium (CAM-bee-um) Of plants, lateral meristem that produces secondary xylem and phloem. **460**

vascular cylinder Stele. Of a root, central column that consists of vascular tissue, pericycle, and supporting cells. **459**

vascular plant Plant having specialized tissues (xylem and phloem) that transport water and sugar within the plant body. **357**

vascular tissue In vascular plants, tissue (xylem and phloem) that distributes water and nutrients through the plant body. **359**

vascular tissue system All tissues that carry water and nutrients through the plant body. *See* phloem, xylem. **450**

vasoconstriction Narrowing of a blood vessel when smooth muscle that rings it contracts. **625**

vasodilation Widening of a blood vessel when smooth muscle that rings it relaxes. **625**

vector In biotechnology, a DNA molecule that can accept foreign DNA and be replicated inside a host cell. *See also* disease vector. **232**

vegetative reproduction Growth of new roots and shoots from extensions or fragments of a parent plant; a form of asexual reproduction. **494**

vein Large-diameter vessel that returns blood to the heart. **619**

ventricle (VEN-trick-uhl) Heart chamber that receives blood from an atrium and pumps it out of the heart **617**

venule Blood vessel that conveys blood from a capillary to a vein. **627**

vernalization (vurr-null-iz-A-shun) Stimulation of flowering in spring by long exposure to low temperature in winter. **510**

vertebrae (VER-tah-bray) Singular, vertebra. Bones of the backbone, or vertebral column. **601**

vertebral column (ver-TEE-bral) Backbone. **601**

vertebrate Animal with a backbone. **417**

vesicle (VESS-ih-cull) Saclike, membrane-enclosed organelle; different kinds store, transport, or break down their contents. **62**

vessel elements Cells that form water-conducting vessels of xylem. **473**

vestibular apparatus System of fluid-filled sacs and canals in the inner ear; contains the organs of equilibrium. **569**

villi Singular, villus. Multicelled projections at the surface of each fold in the small intestine. **687**

viral reassortment Two related viruses infect the same individual simultaneously and swap genes. **323**

virion (VEER-ee-on) Free viral particle. **318**

virus Noncellular, infectious particle of protein and nucleic acid; replicates only in a host cell. **318**

visceral sensations Sensations that arise when sensory neurons associated with organs inside body cavities are activated. **563**

vision Perception of visual stimuli based on light focused on a retina and image formation in the brain. **571**

visual accommodation Process of adjusting lens shape or position so that light from an object falls on the retina. **573**

vital capacity Maximum volume of air that can be moved in and out of the lungs with forced inhalation and exhalation. **669**

vitamin Organic substance required in small amounts for normal metabolism. **693**

vomeronasal organ Pheromone-detecting organ of some vertebrates. **565**

warning coloration In many well-defended or unpalatable species, bright colors, patterns, and other signals that predators learn to recognize and avoid. **795**

water cycle Biogeochemical cycle in which water moves among the atmosphere, land, and waters and into an out of food webs. **815**

water molds Heterotrophic stramenopiles that grow as a mesh of nutrient-absorbing filaments. **343**

water–vascular system In echinoderms, a system of fluid-filled tubes and tube feet that function in locomotion. **411**

watershed Land area that drains into a particular stream or river. **815**

wavelength Distance between the crests of two successive waves. **103**

wax Water-repellent substance that is a complex, varying mixture of lipids with long fatty acid tails bonded to long-chain alcohols. **44**

white blood cell Vertebrate blood cell with a role in housekeeping tasks and defense; a leukocyte. **623**

white matter Central nervous system tissue that consists primarily of myelinated axons with their associated glial cells. **549**

wood Secondary xylem. Forms inside the cylinder of vascular cambium in an older stem or root. **460**

X chromosome inactivation Developmental shutdown of one of the two X chromosomes in the cells of female mammals. *See also* Barr body. **167**

xylem (ZY-lum) Complex vascular tissue of plants; distributes water and mineral ions through the plant body. **359, 452**

yeast Single-celled fungus. **377**

yellow marrow Bone marrow that is mostly fat; fills most long bones. **603**

zero population growth Interval in which births equal deaths. **775**

zooplankton (zoh-ah-PLANK-ton) Heterotrophic members of the plankton; tiny, aquatic heterotrophs. **341**

zygote (ZYE-goat) Diploid cell that forms when two gametes fuse; the first cell of a new individual **188**

zygote fungus Fungus that forms a zygospore during sexual reproduction; hyphae do not have septa. **379**

Index

Page numbers followed by an f or t indicate figures and tables. ■ indicates human health topics; ■ indicates environmental topics. Bold terms indicate major topics.

B

C

E

M

N

O

S

T

W

ISBN-13: 978-1337408417
ISBN-10: 1337408417
90000
9 781337 408417